## Some Fundamental Constants[a]

| Quantity | Symbol | Value[b] |
|---|---|---|
| Atomic mass unit | u | $1.660\ 540\ 2(10) \times 10^{-27}$ kg |
| | | $931.494\ 32(2\ 8)$ MeV/$c^2$ |
| Avogadro's number | $N_A$ | $6.022\ 136\ 7(36) \times 10^{23}$ particles/mol |
| Bohr magneton | $\mu_B = \dfrac{e\hbar}{2m_e}$ | $9.274\ 015\ 4(31) \times 10^{-24}$ J/T |
| Bohr radius | $a_0 = \dfrac{\hbar^2}{m_e e^2 k_e}$ | $5.291\ 772\ 49\ (24) \times 10^{-11}$ m |
| Boltzmann's constant | $k_B = R/N_A$ | $1.380\ 658\ (12) \times 10^{-23}$ J/K |
| Compton wavelength | $\lambda_C = \dfrac{h}{m_e c}$ | $2.426\ 310\ 58(2\ 2) \times 10^{-12}$ m |
| Coulomb constant | $k_e = \dfrac{1}{4\pi\epsilon_0}$ | $8.987\ 551\ 787 \times 10^9$ N·m$^2$/C$^2$ (exact) |
| Deuteron mass | $m_d$ | $3.343\ 586\ 0(20) \times 10^{-27}$ kg |
| | | $2.013\ 553\ 214\ (24)$ u |
| Electron mass | $m_e$ | $9.109\ 389\ 7(54) \times 10^{-31}$ kg |
| | | $5.485\ 799\ 03(1\ 3) \times 10^{-4}$ u |
| | | $0.510\ 999\ 06(1\ 5)$ MeV/$c^2$ |
| Electron-volt | eV | $1.602\ 177\ 33(4\ 9) \times 10^{-19}$ J |
| Elementary change | e | $1.602\ 177\ 33(4\ 9) \times 10^{-19}$ C |
| Gas constant | R | $8.314\ 510\ (70)$ J/K·mol |
| Gravitational constant | G | $6.672\ 59(8\ 5) \times 10^{-11}$ N·m$^2$/kg$^2$ |
| Hydrogen ground state energy | $E_1 = -\dfrac{e^2 k_e}{2a_0}$ | $-13.605\ 698\ (40)$ eV |
| Josephson frequency–voltage ratio | $2e/h$ | $4.835\ 976\ 7(14) \times 10^{14}$ Hz/V |
| Magnetic flux quantum | $\Phi_0 = \dfrac{h}{2e}$ | $2.067\ 834\ 61(6\ 1) \times 10^{-15}$ T·m$^2$ |
| Neutron mass | $m_n$ | $1.674\ 928\ 6(10) \times 10^{-27}$ kg |
| | | $1.008\ 664\ 904\ (14)$ u |
| | | $939.565\ 63(2\ 8)$ MeV/$c^2$ |
| Nuclear magneton | $\mu_n = \dfrac{e\hbar}{2m_p}$ | $5.050\ 786\ 6(17) \times 10^{-27}$ J/T |
| Permeability of free space | $\mu_0$ | $4\pi \times 10^{-7}$ T·m/A (exact) |
| Permittivity of free space | $\epsilon_0 = 1/\mu_0 c^2$ | $8.854\ 187\ 817 \times 10^{-12}$ C$^2$/N·m$^2$ (exact) |
| Planck's constant | $h$ | $6.626\ 075\ (40) \times 10^{-34}$ J·s |
| | $\hbar = h/2\pi$ | $1.054\ 572\ 66(6\ 3) \times 10^{-34}$ J·s |
| Proton mass | $m_p$ | $1.672\ 623\ (10) \times 10^{-27}$ kg |
| | | $1.007\ 276\ 470\ (12)$ u |
| | | $938.272\ 3(28)$ MeV/$c^2$ |
| Rydberg constant | $R_H$ | $1.097\ 373\ 153\ 4(13) \times 10^7$ m$^{-1}$ |
| Speed of light in vacuum | c | $2.997\ 924\ 58 \times 10^8$ m/s (exact) |

[a] These constants are the values recommended in 1986 by CODATA, based on a least-squares adjustment of data from different measurements. For a more complete list, see E. R. Cohen and B. N. Taylor, *Rev. Mod. Phys.* 59:1121, 1987.

[b] The numbers in parentheses for the values above represent the uncertainties of the last two digits.

## Solar System Data

| Body | Mass (kg) | Mean Radius (m) | Period (s) | Distance from the Sun (m) |
|------|-----------|-----------------|------------|---------------------------|
| Mercury | $3.18 \times 10^{23}$ | $2.43 \times 10^6$ | $7.60 \times 10^6$ | $5.79 \times 10^{10}$ |
| Venus | $4.88 \times 10^{24}$ | $6.06 \times 10^6$ | $1.94 \times 10^7$ | $1.08 \times 10^{11}$ |
| Earth | $5.98 \times 10^{24}$ | $6.37 \times 10^6$ | $3.156 \times 10^7$ | $1.496 \times 10^{11}$ |
| Mars | $6.42 \times 10^{23}$ | $3.37 \times 10^6$ | $5.94 \times 10^7$ | $2.28 \times 10^{11}$ |
| Jupiter | $1.90 \times 10^{27}$ | $6.99 \times 10^7$ | $3.74 \times 10^8$ | $7.78 \times 10^{11}$ |
| Saturn | $5.68 \times 10^{26}$ | $5.85 \times 10^7$ | $9.35 \times 10^8$ | $1.43 \times 10^{12}$ |
| Uranus | $8.68 \times 10^{25}$ | $2.33 \times 10^7$ | $2.64 \times 10^9$ | $2.87 \times 10^{12}$ |
| Neptune | $1.03 \times 10^{26}$ | $2.21 \times 10^7$ | $5.22 \times 10^9$ | $4.50 \times 10^{12}$ |
| Pluto | $\approx 1.4 \times 10^{22}$ | $\approx 1.5 \times 10^6$ | $7.82 \times 10^9$ | $5.91 \times 10^{12}$ |
| Moon | $7.36 \times 10^{22}$ | $1.74 \times 10^6$ | — | — |
| Sun | $1.991 \times 10^{30}$ | $6.96 \times 10^8$ | — | — |

## Physical Data Often Used[a]

| | |
|---|---|
| Average Earth–Moon distance | $3.84 \times 10^8$ m |
| Average Earth–Sun distance | $1.496 \times 10^{11}$ m |
| Average radius of the Earth | $6.37 \times 10^6$ m |
| Density of air (20°C and 1 atm) | $1.20$ kg/m$^3$ |
| Density of water (20°C and 1 atm) | $1.00 \times 10^3$ kg/m$^3$ |
| Free-fall acceleration | $9.80$ m/s$^2$ |
| Mass of the Earth | $5.98 \times 10^{24}$ kg |
| Mass of the Moon | $7.36 \times 10^{22}$ kg |
| Mass of the Sun | $1.99 \times 10^{30}$ kg |
| Standard atmospheric pressure | $1.013 \times 10^5$ Pa |

[a] These are the values of the constants as used in the text.

## Some Prefixes for Powers of Ten

| Power | Prefix | Abbreviation | Power | Prefix | Abbreviation |
|-------|--------|--------------|-------|--------|--------------|
| $10^{-24}$ | yocto | y | $10^1$ | deka | da |
| $10^{-21}$ | zepto | z | $10^2$ | hecto | h |
| $10^{-18}$ | atto | a | $10^3$ | kilo | k |
| $10^{-15}$ | femto | f | $10^6$ | mega | M |
| $10^{-12}$ | pico | p | $10^9$ | giga | G |
| $10^{-9}$ | nano | n | $10^{12}$ | tera | T |
| $10^{-6}$ | micro | $\mu$ | $10^{15}$ | peta | P |
| $10^{-3}$ | milli | m | $10^{18}$ | exa | E |
| $10^{-2}$ | centi | c | $10^{21}$ | zetta | Z |
| $10^{-1}$ | deci | d | $10^{24}$ | yotta | Y |

# Principles *of* Physics

*A Calculus-Based Text*

third edition

# Principles *of* Physics

## *A Calculus-Based Text*

### THIRD EDITION

**Raymond A. Serway**

*James Madison University*

**John W. Jewett, Jr.**

*California State Polytechnic University—Pomona*

HARCOURT COLLEGE PUBLISHERS

Fort Worth   Philadelphia   San Diego   New York   Orlando   Austin
San Antonio   Toronto   Montreal   London   Sydney   Tokyo

Publisher: Emily Barrosse
Publisher: John Vondeling
Marketing Strategist: Kathleen S. McLellan
Developmental Editor: Ed Dodd
Project Editor: Bonnie Boehme
Production Manager: Charlene Catlett Squibb
Art Director and Text Designer: Carol Bleistine

**Cover Image and Credit:** Competitor on luge course./© *Tony Stone Images*

**Frontmatter Images and Credits:** Title page: Downhill skier *(Jean Y. Ruszniewski/Stone);*
p. vii: Drag racing *(George Lepp/Stone);* p. viii: Rolling cylinder *(Henry Leap and Jim Lehman);*
p. ix: Plane breaking the sound barrier *(Courtesy of U.S. Navy. Photo by Ensign John Gay);*
p. x: Power lines *(Telegraph Colour Library/FPG);* p. xi: Rainbow *(Mark D. Phillips/*
*PhotoResearchers, Inc.);* p.xii: Whole-body bone scan *(Alexander Tsiaras/Science Source/Photo*
*Researchers, Inc.);* p. xiii: Twin Falls on the Island of Kauai, Hawaii *(Bruce Byers/ FPG);*
p. xiv: Rock climbers *(Scott Markewitz/FPG);* p. xvi: Edwin Aldrin *(Courtesy of NASA);*
p. xvii: Roller coaster *(Frank Cezus/FPG International);* p. xix: Windmills *(Billy Hustace/Stone);*
p. xxi: Glass shattering *(© 1992 Ben Rose/The Image Bank);* p. xxiv: Sky surfer *(Jump Run*
*Productions/Image Bank);* p. xxv: Stonehenge *(© John Serafin, Peter Arnold, Inc.);*
p. xxvi: Wine bottle defying gravity *(Charles D. Winters).*

PRINCIPLES OF PHYSICS, Third Edition
ISBN: 0-03-027157-6
Library of Congress Catalog Card Number: 00-110798

*Address for domestic orders:*
Harcourt College Publishers, 6277 Sea Harbor Drive, Orlando, FL 32887-6777
1-800-782-4479
e-mail  collegesales@harcourt.com

*Address for international orders:*
International Customer Service, Harcourt, Inc.
6277 Sea Harbor Drive, Orlando FL 32887-6777
(407) 345-3800
Fax  (407) 345-4060
e-mail  hbintl@harcourt.com

*Address for editorial correspondence:*
Harcourt College Publishers,  Public Ledger Building, Suite 1250,
150 S. Independence Mall West, Philadelphia, PA 19106-3412

*Web Site Address*
http://www.harcourtcollege.com

Printed in the United States of America

1234567890  048   10 987654321

# Dedication

IN MEMORY OF

## John Vondeling

*a dear friend and companion,*
*for his great wisdom and enthusiasm as a publisher,*
*and for his thoughtful guidance through the years*

AND

## Sally Kusch

*a first-class project editor,*
*who worked so diligently on early versions of this*
*and other physics textbooks.*

*We shall miss them.*

# Contents Overview

# Contents

# Preface

**P**rinciples of Physics is designed for a one-year introductory calculus-based physics course for engineering and science students and for premed students taking a rigorous physics course. This third edition contains many new pedagogical features—most notably, a contextual approach to enhance motivation, an increased emphasis on avoiding misconceptions, and a problem-solving strategy that uses a modeling approach. Based on comments from users of the second edition and reviewers' suggestions, a major effort was made to improve organization, clarity of presentation, precision of language, and accuracy throughout.

This project was conceived because of well-known problems in teaching the introductory calculus-based physics course. The course content (and hence the size of textbooks) continues to grow, while the number of contact hours with students has either dropped or remained unchanged. Furthermore, traditional one-year courses cover little if any 20th-century physics.

In preparing this textbook, we were motivated by the spreading interest in reforming this course, primarily through the efforts of the Introductory University Physics Project (IUPP) sponsored by the American Association of Physics Teachers and the American Institute of Physics. The primary goals and guidelines of this project are to

- Reduce course content following the "less may be more" theme;
- Incorporate contemporary physics naturally into the course;
- Organize the course in the context of one or more "story lines";
- Treat all students equitably.

Recognizing a need for a textbook that could meet these guidelines several years ago, we studied the various proposed IUPP models and the many reports from IUPP committees. Eventually, one of us (RAS) became actively involved in the review and planning of one specific model, initially developed at the U.S. Air Force Academy, entitled "A Particles Approach to Introductory Physics." Part of the summer of 1990 was spent at the Academy working with Colonel James Head and Lt. Col. Rolf Enger, the primary authors of the Particles model, and other members of that department. This most useful collaboration was the starting point of this project.

The coauthor (JWJ) became involved with the IUPP model called "Physics in Context," developed by John Rigden (American Institute of Physics), David Griffiths (Oregon State University), and Lawrence Coleman (University of Arkansas at Little Rock). This involvement led to the contextual overlay that is used in this book and described in detail later in the Preface.

The combined IUPP approach in this book has the following features:

- It is an evolutionary approach (rather than a revolutionary approach), which should meet the current demands of the physics community.

- It deletes many topics in classical physics (such as alternating current circuits and optical instruments) and places less emphasis on rigid body motion, optics, and thermodynamics.
- Some topics in 20th-century physics, such as special relativity, energy quantization, and the Bohr model of the hydrogen atom, are introduced early in the textbook.
- A deliberate attempt is made to show the unity of physics.
- As a motivational tool, the textbook connects physics principles to interesting social issues, natural phenomena, and technological advances.

## OBJECTIVES

This introductory physics textbook has two main objectives: to provide the student with a clear and logical presentation of the basic concepts and principles of physics, and to strengthen an understanding of the concepts and principles through a broad range of interesting applications to the real world. To meet these objectives, we have emphasized sound physical arguments and problem-solving methodology. At the same time, we have attempted to motivate the student through practical examples that demonstrate the role of physics in other disciplines, including engineering, chemistry, and medicine.

## CHANGES IN THE THIRD EDITION

A number of changes and improvements have been made in the third edition of this text. Many of these are in response to current trends in science education and to comments and suggestions provided by the reviewers of the manuscript and instructors using the first two editions. The following represent the major changes in the third edition:

**Content**   While the overall content of the textbook is similar to that of the second edition, several changes were implemented. A global approach to energy and energy transfer is introduced in Chapter 6 and has been incorporated throughout the book. A discussion of molar specific heats of gases has been added to Chapter 17. Also in Chapter 17 the first law of thermodynamics is written in the form $\Delta E_{\text{int}} = Q + W$, rather than the common expression that appears in many physics textbooks, $\Delta E_{\text{int}} = Q - W$. This form follows naturally from the global approach to energy introduced in Chapter 6 and is consistent with the form of the law that most chemistry books use. The use of this form of the first law follows a recommendation made by a committee appointed by the American Physical Society. Finally, many sections have been streamlined, deleted, or combined with other sections to allow for a more balanced presentation.

**Organization**   We have incorporated a "context overlay" scheme into the textbook, in response to the "Physics in Context" approach in the IUPP. This new feature adds interesting applications of the material covered in the third edition to real issues. We have developed this feature to be flexible, so that the instructor who does not wish to follow the contextual approach can simply ignore the additional contextual features without sacrificing complete coverage of the existing material. We feel, though, that the benefits students will gain from this approach will be many.

The context overlay organization divides the text into eight sections, or "Contexts," after Chapter 1, as follows:

| Context Number | Context | Physics Topics | Chapters |
|---|---|---|---|
| 1 | Mission to Mars | Classical mechanics | 2–11 |
| 2 | Earthquakes | Vibrations and waves | 12–14 |
| 3 | Search for the *Titanic* | Fluids | 15 |
| 4 | Global Warming | Thermodynamics | 16–18 |
| 5 | Lightning | Electricity | 19–21 |
| 6 | Magnetic Levitation Vehicles | Magnetism | 22–23 |
| 7 | Lasers | Optics | 24–27 |
| 8 | The Cosmic Connection | Modern physics | 28–31 |

Each Context begins with an introduction, leading to a "central question" that motivates study within the Context. The final section of each chapter is a "Context Connection," which discusses how the material in the chapter relates to the Context and the central question. The final chapter in each Context is followed by a "Context Conclusion." Each conclusion uses the principles learned in the context to respond fully to the central question. Each chapter, as well as the Context Conclusions, includes problems related to the context material.

**Pitfall Prevention** These new features are placed in the margins of the text and address common student misconceptions and situations in which students often follow unproductive paths. Over 200 Pitfall Preventions are provided to help students avoid common mistakes and misunderstandings.

**Quick Quizzes** Several Quick Quiz questions are included in each chapter to provide students opportunities to test their understanding of the physical concepts presented. The questions require students to make decisions on the basis of sound reasoning. Some of them help students overcome common misconceptions. Answers to all Quick Quiz questions are found at the end of each chapter.

**Modeling** A modeling approach, based on four types of models commonly used by physicists, is introduced to help students understand they are solving problems that approximate reality. They must then learn how to test the validity of the model. This approach also helps students see the unity in physics, as a large fraction of problems can be solved with a small number of models. A general problem-solving strategy using the modeling approach is introduced in Chapter 1.

**Alternative Representations** Emphasis is placed on alternative representations of the information, including mental, pictorial, graphical, tabular, and mathematical representations. Many problems are easier to solve if the information is presented in alternative ways, to reach the many different methods students use to learn.

**Line-by-Line Revision** The text has been carefully edited to improve clarity of presentation and precision of language. We hope that the result is a book both accurate and enjoyable to read.

**Problems** In an effort to improve clarity and quality, the end-of-chapter problems were substantially revised. Approximately 40% of the problems (about 575) are new to this edition, and most of these new problems are at the intermediate level (as identified by blue problem numbers). Many problems require students to make order-of-magnitude calculations. All problems have been carefully edited and reworded where necessary. Solutions to approximately 20% of the end-of-chapter problems are included in the *Student Solutions Manual and Study Guide*. Boxed numbers identify these problems. A smaller subset of solutions will be posted on the World Wide Web (**http://www.harcourtcollege.com/physics**) and will be accessible to students and instructors using *Principles of Physics*. The web icon identifies these problems. See the next section for a complete description of other features of the problem set.

**Web Notes**   Useful World Wide Web addresses are provided as marginal notes (indicated by a $\boxed{\textbf{WEB}}$ icon) to encourage students to explore extensions of the material beyond what is covered in the text. In particular, the Contexts allow for rich opportunities for further explorations on the Web.

**Biomedical Applications**   For biology and premed students,  icons point the way to various practical and interesting applications of physical principles to biology and medicine.

## TEXT FEATURES

Most instructors would agree that the textbook selected for a course should be the student's primary guide for understanding and learning the subject matter. Furthermore, the textbook should be easily accessible as well as styled and written to facilitate instruction and learning. With these points in mind, we have included many pedagogical features that are intended to enhance the textbook's usefulness to both students and instructors. These features are as follows:

**Style**   To facilitate rapid comprehension, we have attempted to write the book in a clear, logical, and engaging style. The somewhat informal and relaxed writing style is intended to increase reading enjoyment. New terms are carefully defined, and we have tried to avoid the use of jargon.

**Previews**   Most chapters begin with a brief preview that includes a discussion of the particular chapter's objectives and content.

**Important Statements and Equations**   Most important statements and definitions are set in boldface type or are highlighted with a background screen for added emphasis and ease of review. Similarly, important equations are highlighted with a tan background screen to facilitate location.

**Problem-Solving Hints**   We have included general strategies for solving the types of problems featured both in the examples and in the end-of-chapter problems. This feature helps students identify necessary steps in solving problems and eliminate any uncertainty they might have. Problem-solving strategies are highlighted with a light blue-gray screen for emphasis and ease of location.

**Marginal Notes**   Comments and notes appearing in the margin can be used to locate important statements, equations, and concepts in the text.

**Illustrations and Tables**   The readability and effectiveness of the text material and worked examples are enhanced by the large number of figures, diagrams, photographs, and tables. Full color adds clarity to the artwork and makes illustrations as realistic as possible. For example, vectors are color coded, and curves in graphs are drawn in color. The three-dimensional appearance of many illustrations has been improved in this third edition. The color photographs have been carefully selected, and their accompanying captions have been written to serve as an added instructional tool.

**Mathematical Level**   We have introduced calculus gradually, keeping in mind that students often take introductory courses in calculus and physics concurrently. Most steps are shown when basic equations are developed, and reference is often made to mathematical appendices at the end of the textbook. Vector products are discussed in detail later in the text, where they are needed in physical applications.

The dot product is introduced in Chapter 6, which addresses work and energy; the cross product is introduced in Chapter 10, which deals with rotational dynamics.

**Worked Examples**  A large number of worked examples of varying difficulty are presented to promote students' understanding of concepts. In many cases, the examples serve as models for solving the end-of-chapter problems. Because of the increased emphasis on understanding physical concepts, many examples are conceptual in nature. The examples are set off in boxes, and the answers to examples with numerical solutions are highlighted with a tan screen.

**Worked Example Exercises**  Many of the worked examples are followed immediately by exercises with answers. These exercises are intended to promote interactivity between the student and the textbook and to immediately reinforce the student's understanding of concepts and problem-solving techniques. The exercises represent extensions of the worked examples.

**Questions**  Questions requiring verbal responses are provided at the end of each chapter. Over 500 questions are included in this edition. Some questions provide the student with a means of self-testing the concepts presented in the chapter. Others could serve as a basis for initiating classroom discussions. Answers to selected questions are included in the *Student Solutions Manual and Study Guide*.

**Significant Figures**  Significant figures in both worked examples and end-of-chapter problems have been handled with care. Most numerical examples and problems are worked out to either two or three significant figures, depending on the accuracy of the data provided.

**Problems**  The end-of-chapter problems are more numerous in this edition and more varied (in all, over 1 800 problems are given throughout the text). For the convenience of both the student and the instructor, about two thirds of the problems are keyed to specific sections of the chapter, including Context Connection sections. The remaining problems, labeled "Additional Problems," are not keyed to specific sections. An  icon identifies problems dealing with applications to the life sciences and medicine. One or more problems in each chapter ask students to make an order-of-magnitude calculation based on their own estimated data. Other types of problems are described in more detail below. Answers to odd-numbered problems are provided at the end of the book.

Usually, the problems within a given section are presented so that the straight-forward problems (those with black problem numbers) appear first; these straightforward problems are followed by those of increasing difficulty. For ease of identification, the numbers of intermediate-level problems are printed in blue, and those of challenging problems are printed in magenta.

Solutions to approximately 20% of the problems in each chapter are in the *Student Solutions Manual and Study Guide*. Among these, selected problems are identified with **web** icons and have their solutions posted on the World Wide Web at **http://www.harcourtcollege.com/physics.**

**Review Problems**  Many chapters include review problems requiring the student to relate concepts covered in the chapter to those discussed in previous chapters. These problems could be used by students in preparing for tests and by instructors for special assignments and classroom discussions.

**Paired Problems**  As an aid for students learning to solve problems symbolically, paired numerical and symbolic problems are included in Chapters 1 through 4 and 16 through 21. Paired problems are identified by a common tan background screen.

**Computer- and Calculator-Based Problems**   Most chapters include one or more problems whose solution requires the use of a computer or graphing calculator. Modeling of physical phenomena enables students to obtain graphical representations of variables and to perform numerical analyses.

**Units**   The international system of units (SI) is used throughout the text. The British engineering system of units (conventional system) is used only to a limited extent in the chapters on mechanics and thermodynamics.

**Summaries**   Each chapter contains a summary that reviews the important concepts and equations discussed in that chapter.

**Appendices and Endpapers**   Several appendices are provided at the end of the textbook. Most of the appendix material represents a review of mathematical concepts and techniques used in the text, including scientific notation, algebra, geometry, trigonometry, differential calculus, and integral calculus. Reference to these appendices is made throughout the text. Most mathematical review sections in the appendices include worked examples and exercises with answers. In addition to the mathematical reviews, the appendices contain tables of physical data, conversion factors, atomic masses, and the SI units of physical quantities, as well as a periodic table of the elements and a list of Nobel prize recipients. Other useful information, including fundamental constants and physical data, planetary data, a list of standard prefixes, mathematical symbols, the Greek alphabet, and standard abbreviations of units of measure, appears on the endpapers.

## ANCILLARIES

The ancillary package has been updated substantially and streamlined in response to suggestions from users of the second edition. The most essential changes in the student package are a *Student Solutions Manual and Study Guide* with a tighter focus on problem-solving, the *Student Tools CD-ROM*, and the *Saunders Core Concepts in Physics CD-ROM* developed by Archipelago Productions. Instructors will find increased support for their teaching efforts with new electronic materials.

### Student Ancillaries

***Student Solutions Manual and Study Guide***   by John R. Gordon, Ralph McGrew, and Raymond A. Serway. This two-volume manual features detailed solutions to approximately 20% of the end-of-chapter problems from the textbook. Boxed numbers identify those problems in the textbook whose complete solutions are found in the manual. The manual also features a list of important equations and concepts, as well as answers to selected end-of-chapter questions.

***Student Tools CD-ROM***   This CD-ROM contains tools designed to enhance the learning of physical concepts and train students to become better problem-solvers. It includes a textbook version of the highly acclaimed Interactive Physics™ software by MSC Working Knowledge, more than 100 Interactive Physics™ simulations keyed to appropriate worked examples and selected end-of-chapter problems (as identified by the 🖳 icon), and support for working those end-of-chapter problems that require the use of computers.

***Saunders Core Concepts in Physics CD-ROM***   This CD-ROM package developed by Archipelago Productions applies the power of multimedia to the introductory physics course, offering full-motion animation and video, engaging interactive graphics, clear and concise text, and guiding narration. *Saunders Core Concepts in Physics CD-ROM* focuses on those concepts students usually find most difficult in the course, drawing from topics in mechanics, thermodynamics, electric fields, magnetic fields, and optics. The animations and graphics are presented to aid the student in developing accurate conceptual models of difficult topics—topics often too complex to be explained in words or chalkboard illustrations. The CD-ROM also presents step-by-step explorations of problem-solving strategies and provides animations of problems in order to promote conceptual understanding and sharpen problem-solving skills. Textbook topics further explored on the CD-ROM are identified by marginal icons that give the appropriate module and screen number(s). Students should look to the CD-ROM for help in understanding these topics.

*Student Web Site* Students will have access to an abundance of material at **http://www.harcourtcollege.com/physics.** The Web Site features special topic essays by guest authors, practice problems with answers, and optional topics that accompany selected chapters of the textbook. Also included are selected solutions from the *Student Solutions Manual and Study Guide* and a glossary that includes more than 300 physics terms. Students also can take practice quizzes in our Practice Exercises and Testing area.

*Physics Laboratory Manual, Second Edition* by David Loyd. Updated and redesigned, this manual supplements the learning of basic physical principles while introducing laboratory procedures and equipment. Each chapter includes a prelaboratory assignment, objectives, an equipment list, the theory behind the experiment, step-by-step experimental procedures, and questions. A laboratory report form is provided for each experiment so that students can record data and make calculations. Students are encouraged to apply statistical analysis to their data so they can develop the ability to judge the validity of their results.

*So You Want to Learn Physics: A Preparatory Course with Calculus* by Rodney Cole. This introductory-level book is useful to those students who need additional mathematics preparation before or during a calculus-based course in physics. The friendly, straightforward style makes it easier to understand how mathematics is used in the context of physics.

*Life Science Applications for Physics* by Jerry Faughn. This supplement provides examples, readings, and problems from the biological sciences as they relate to physics. Topics include "Friction in Human Joints," "Physics of the Human Circulatory System," "Physics of the Nervous System," and "Ultrasound and Its Applications." This supplement is useful in those courses taken by a significant number of premed students.

## Instructor's Ancillaries

*Instructor's Manual with Solutions* by Ralph McGrew, Jeffery Saul, and Charles Teague. This manual consists of complete, worked solutions to all the problems in the textbook. The solutions to problems new to the third edition are marked for easy identification by the instructor. New to this edition of the manual are suggestions on how to teach difficult topics and help students overcome common misconceptions. These suggestions are based on recent research in physics education.

*Instructor's Web Site* The instructor's area at **http://www.harcourtcollege.com/physics** includes a listing of overhead transparencies; a guide to relevant experiments in David Loyd's *Physics Laboratory Manual, Second Edition;* a correlation guide between sections in *Principles of Physics* and modules in the *Saunders Core Concepts in Physics CD-ROM;* supplemental problems with answers; optional topics to accompany selected chapters of the textbook; and a syllabus generator.

*Instructor's Resource CD-ROM*   This CD-ROM accompanying the third edition of *Principles of Physics* has been created to provide instructors with an exciting new tool for classroom presentation. The CD-ROM contains a collection of graphics files of line art from the textbook. These files can be opened directly or can be imported into a variety of presentation packages. The labels for each piece of art have been enlarged and boldfaced to facilitate classroom viewing. The CD-ROM also contains electronic files of the *Instructor's Manual* and *Test Bank.*

## Instructor Options for Online Homework

**WebAssign: A Web-Based Homework System**   WebAssign is a Web-based homework delivery, collection, grading, and recording service developed at North Carolina State University. Instructors who sign up for WebAssign can assign frequent homework to their students, using questions and problems taken directly from *Principles of Physics.* WebAssign gives students immediate feedback on their homework and helps them to master information and skills, leading to greater competence and better grades. WebAssign can free instructors from the drudgery of grading homework and recording scores, allowing them to devote more time to meeting with students and preparing classroom presentations.

WebAssign is being used in scores of educational institutions by tens of thousands of students and hundreds of instructors. Most of the numerical problems that can be assigned have different values, so that each student has a unique problem to solve. This feature motivates independent thinking within the context of collaborative learning.

Details about and a demonstration of WebAssign are available at **http:// wasnet01ws.physics.ncsu.edu/info/.** For more information about ordering this service, contact WebAssign at **webassign@ncsu.edu**

**CAPA: A Computer-Assisted Personalized Approach**   CAPA is a network system for learning, teaching, assessment, and administration. It provides students with personalized problem sets, quizzes, and examinations consisting of qualitative conceptual problems and quantitative problems. CAPA was developed through a collaborative effort of the Physics–Astronomy, Computer Science, and Chemistry Departments at Michigan State University. Students are given instant feedback and relevant hints via the Internet and may correct errors without penalty before an assignment's due date. The system records each student's participation and performance on assignments, quizzes, and examinations; records are available online to both the individual student and his or her instructor. For more information, visit the CAPA Web site at: **http://capa4.lite.msu.edu/homepage/**

**Homework Service**   With this service, instructors can reduce their grading workload by assigning thought-provoking homework problems using the World Wide Web. Instructors browse problem banks, select those they wish to assign to their students, and then let the Homework Service take over the delivery and grading. This system was developed and is maintained by Fred Moore at the University of Texas **(moore@physics.utexas.edu).** Students download their unique problems, submit their answers, and obtain immediate feedback; if students' answers are incorrect, they can resubmit them. This rapid grading feature facilitates effective learning. After the due date of their assignments, students can obtain the solutions to their problems. Minimal online connect time is required. The Homework Service uses algorithm-based problems: This means that each student solves sets of problems different from those given to other students. Details about and a demonstration of this service are available at **http://hw.ph.utexas.edu/hw.html**

*Printed Test Bank*　by Edward Adelson. Contains approximately 2,000 multiple-choice questions. It is provided for the instructor who does not have access to a computer.

*Computerized Test Bank*　Available in Windows™ and Macintosh® formats, the *Computerized Test Bank* contains more than 2,000 multiple-choice questions, representing every chapter of the text. The *Test Bank* enables the instructor to create many unique tests by allowing the editing of questions and the addition of new questions. The software program solves all problems and prints each answer on a separate grading key. All questions have been reviewed for accuracy.

*Overhead Transparency Acetates*　This collection of transparencies consists of 200 full-color figures from the text and features large print for easy viewing in the classroom.

*Instructor's Manual for Physics Laboratory Manual*　by David Loyd. Each chapter contains a discussion of the experiment, teaching hints, answers to selected questions, and a post-laboratory quiz with short-answer and essay questions. It also includes a list of the suppliers of scientific equipment and a summary of the equipment needed for each of the laboratory experiments in the manual.

---

Harcourt College Publishers may provide complementary instructional aids and supplements or supplement packages to those adopters qualified under our adoption policy. Please contact your sales representative for more information. If as an adopter or potential user you receive supplements you do not need, please return them to your sales representative or send them to

Attn: Returns Department
Troy Warehouse
465 South Lincoln Drive
Troy, MO 63379

---

## TEACHING OPTIONS

Although some topics found in traditional textbooks have been omitted from this textbook, instructors may find that the current text still contains more material than can be covered in a two-semester sequence. For this reason, we would like to offer the following suggestions. If you wish to place more emphasis on contemporary topics in physics, you should consider omitting parts or all of Chapters 15, 16, 17, 18, 24, 25, and 26. On the other hand, if you wish to follow a more traditional approach that places more emphasis on classical physics, you could omit Chapters 9, 11, 28, 29, 30, and 31. Either approach can be used without any loss in continuity. Other teaching options would fall somewhere between these two extremes by choosing to omit some or all of the following sections, which can be considered optional:

| | | | |
|---|---|---|---|
| 3.6 | Relative Velocity | 12.6 | Damped Oscillations |
| 7.7 | Energy Diagrams and Stability of Equilibrium | 12.7 | Forced Oscillations |
| | | 14.7 | Nonsinusoidal Wave Patterns |
| 9.9 | General Relativity | 15.8 | Other Applications of Fluid Dynamics |
| 10.11 | Rolling of Rigid Bodies | | |

## ACKNOWLEDGMENTS

The third edition of this textbook was prepared with the guidance and assistance of many professors who reviewed part or all of the manuscript, the pre-revision text, or both. We wish to acknowledge the following scholars and express our sincere appreciation for their suggestions, criticisms, and encouragement:

Yildirim M. Aktas, *University of North Carolina — Charlotte*
Alfonso M. Albano, *Bryn Mawr College*
Michael Bass, *University of Central Florida*
James Carolan, *University of British Columbia*
Kapila Clara Castoldi, *Oakland University*
Michael Dennin, *University of California, Irvine*
Madi Dogariu, *University of Central Florida*
William Fairbank, *Colorado State University*
Marco Fatuzzo, *University of Arizona*
Patrick Gleeson, *Delaware State University*
Christopher M. Gould, *University of Southern California*
James D. Gruber, *Harrisburg Area Community College*
John B. Gruber, *San Jose State University*
Gail Hanson, *Indiana University*
Dieter H. Hartmann, *Clemson University*
Michael J. Hones, *Villanova University*
Roger M. Mabe, *United States Naval Academy*

Thomas P. Marvin, *Southern Oregon University*
Martin S. Mason, *College of the Desert*
Wesley N. Mathews, Jr., *Georgetown University*
Ken Mendelson, *Marquette University*
Allen Miller, *Syracuse University*
John W. Norbury, *University of Wisconsin — Milwaukee*
Romulo Ochoa, *The College of New Jersey*
Melvyn Oremland, *Pace University*
Steven J. Pollock, *University of Colorado — Boulder*
Rex D. Ramsier, *The University of Akron*
Charles R. Rhyner, *University of Wisconsin — Green Bay*
Dennis Rioux, *University of Wisconsin — Oshkosh*
Gregory D. Severn, *University of San Diego*
Shirvel Stanislaus, *Valparaiso University*
Randall Tagg, *University of Colorado at Denver*
Robert Watkins, *University of Virginia*

This book was carefully checked for accuracy by Edward Gibson (California State University, Sacramento), Chris Vuille (Embry–Riddle Aeronautical University), and Ronald Jodoin (Rochester Institute of Technology).

We thank the following people for their suggestions and assistance during the preparation of earlier editions of this textbook:

Edward Adelson, *Ohio State University*
Subash Antani, *Edgewood College*
Harry Bingham, *University of California, Berkeley*
Anthony Buffa, *California Polytechnic State University, San Luis Obispo*
Ralph V. Chamberlin, *Arizona State University*
Gary G. DeLeo, *Lehigh University*
Alan J. DeWeerd, *Creighton University*
Gordon Emslie, *University of Alabama at Huntsville*
Donald Erbsloe, *United States Air Force Academy*
Philip Fraundorf, *University of Missouri — St. Louis*
Todd Hann, *United States Military Academy*
Gerald Hart, *Moorhead State University*
Richard W. Henry, *Bucknell University*
Laurent Hodges, *Iowa State University*

Joey Huston, *Michigan State University*
Herb Jaeger, *Miami University*
David Judd, *Broward Community College*
Thomas H. Keil, *Worcester Polytechnic Institute*
V. Gordon Lind, *Utah State University*
David Markowitz, *University of Connecticut*
John W. McClory, *United States Military Academy*
L. C. McIntyre, Jr., *University of Arizona*
Alan S. Meltzer, *Rensselaer Polytechnic Institute*
Roy Middleton, *University of Pennsylvania*
Clement J. Moses, *Utica College of Syracuse University*
Anthony Novaco, *Lafayette College*
Desmond Penny, *Southern Utah University*
Prabha Ramakrishnan, *North Carolina State University*
Rogers Redding, *University of North Texas*

Perry Rice, *Miami University*
Janet E. Seger, *Creighton University*
Antony Simpson, *Dalhousie University*
Harold Slusher, *University of Texas at El Paso*

J. Clinton Sprott, *University of Wisconsin at Madison*
Cecil Thompson, *University of Texas at Arlington*
Chris Vuille, *Embry–Riddle Aeronautical University*
James Whitmore, *Pennsylvania State University*

We are indebted to the developers of the IUPP models, "A Particles Approach to Introductory Physics" and "Physics in Context," upon which much of the pedagogical approach in this textbook is based.

Ralph McGrew coordinated the end-of-chapter problems. Problems new to this edition were written by Michael Browne, Michael Hones, Robert Forsythe, John Jewett, Ralph McGrew, Laurent Hodges, Boris Korsunsky, Richard Cohen, John DiNardo, Ronald Bieniek, and Raymond Serway. Robert Beichner and John Gerty contributed ideas for problems. Students Eric Peterman, Karl Payne, and Alexander Coto made corrections in problems from the previous edition, as did instructors Vasili Haralambous, Frank Hayes, Eugene Mosca, David Aspnes, and Erika Hermon.

We are grateful to John R. Gordon and Ralph McGrew for writing the *Student Solutions Manual and Study Guide,* and we thank Michael Rudmin for its attractive layout. Ralph McGrew, Jeffery Saul, and Charles Teague have prepared an excellent *Instructor's Manual,* and we thank them (and Michael Rudmin, again, for his excellent work on its layout). During the development of this text, the authors benefited from many useful discussions with colleagues and other physics instructors, including Robert Bauman, William Beston, Don Chodrow, Jerry Faughn, John R. Gordon, Kevin Giovanetti, Dick Jacobs, Harvey Leff, Clem Moses, Dorn Peterson, Joseph Rudmin, and Gerald Taylor. Special thanks and recognition go to the professional staff at Harcourt College Publishers—in particular, Ed Dodd, Frank Messina, Bonnie Boehme, Carol Bleistine, and Kathleen S. McLellan. We are most appreciative of the proofreading by Margaret Mary Anderson, the final copy editing by Linda Davoli, the excellent artwork produced by Rolin Graphics, and the dedicated photo research efforts of Dena Digilio-Betz. We shall miss our good friend, the late John Vondeling, who was legendary as a publisher of high-quality instructional products for science education.

Finally, we are deeply indebted to our wives and children for their love, support, and long-term sacrifices.

**Raymond A. Serway**
Leesburg, Virginia

**John W. Jewett, Jr.**
Pomona, California

It is appropriate to offer some words of advice that should benefit you, the student. Before doing so, we assume you have read the Preface, which describes the various features of the text that will help you through the course.

## HOW TO STUDY

Very often instructors are asked, "How should I study physics and prepare for examinations?" There is no simple answer to this question, but we would like to offer some suggestions based on our own experiences in learning and teaching over the years.

First and foremost, maintain a positive attitude toward the subject matter, keeping in mind that physics is the most fundamental of all natural sciences. Other science courses that follow will use the same physical principles, so it is important that you understand and are able to apply the various concepts and theories discussed in the text.

The Contexts in the text will help you understand how the physical principles relate to real issues, phenomena, and applications. Be sure to read the Context Introductions, Context Connection sections in each chapter, and Context Conclusions. These will be most helpful in motivating your study of physics.

## CONCEPTS AND PRINCIPLES

It is essential that you understand the basic concepts and principles before attempting to solve assigned problems. You can best accomplish this goal by carefully reading the textbook before you attend your lecture on the covered material. When reading the text, you should jot down those points that are not clear to you. We've purposely left wide margins in the text to give you space for doing this. Also be sure to make a diligent attempt at answering the questions in the Quick Quizzes as you come to them in your reading. We have worked hard to prepare questions that help you judge for yourself how well you understand the material. Pay careful attention to the many Pitfall Preventions throughout the text. These will help you avoid misconceptions, mistakes, and misunderstandings as well as maximize the efficiency of your time by minimizing adventures along fruitless paths. During class, take careful notes and ask questions about those ideas that are unclear to you. Keep in mind that few people are able to absorb the full meaning of scientific material after only one reading. Several readings of the text and your notes may be necessary. Your lectures and laboratory work supplement reading of the textbook and should clarify some of the more difficult material. You should minimize your memorization of material. Successful memorization of passages from the text, equations, and derivations does not necessarily indicate that you understand the material. Your understanding of the material will be enhanced through a combination of efficient study habits, discussions with other students and with instructors, and your ability to solve the problems presented in the textbook. Ask questions whenever you feel clarification of a concept is necessary.

## STUDY SCHEDULE

It is important for you to set up a regular study schedule, preferably a daily one. Make sure you read the syllabus for the course and adhere to the schedule set by your instructor. The lectures will be much more meaningful if you read the corresponding textual material before attending them. As a general rule, you should devote about two hours of study time for every hour you are in class. If you are having trouble with the course, seek the advice of the instructor or other students who have taken the course. You may find it necessary to seek further instruction from experienced students. Very often, instructors offer review sessions in addition to regular class periods. It is important that you avoid the practice of delaying study until a day or two before an exam. More often than not, this approach has disastrous results. Rather than undertake an all-night study session, briefly review the basic concepts and equations and get a good night's rest. If you feel you need additional help in understanding the concepts, in preparing for exams, or in problem-solving, we suggest that you acquire a copy of the *Student Solutions Manual and Study Guide* that accompanies this textbook; this manual should be available at your college bookstore.

## USE THE FEATURES

You should make full use of the various features of the text discussed in the preface. For example, marginal notes are useful for locating and describing important equations and concepts, and **boldfaced** type indicates important statements and definitions. Many useful tables are contained in the Appendices, but most are incorporated in the text where they are most often referenced. Appendix B is a convenient review of mathematical techniques.

Answers to odd-numbered problems are given at the end of the textbook, answers to Quick Quizzes are located at the end of each chapter, and answers to selected end-of-chapter questions are provided in the *Student Solutions Manual and Study Guide.* The exercises (with answers) that follow some worked examples represent extensions of those examples; in most of these exercises, you are expected to perform a simple calculation. Their purpose is to test your problem-solving skills as you read through the text. Problem-Solving Strategies and Hints are included in selected chapters throughout the text and give you additional information about how you should solve problems. The Table of Contents provides an overview of the entire text, while the Index enables you to locate specific material quickly. Footnotes sometimes are used to supplement the text or to cite other references on the subject discussed.

After reading a chapter, you should be able to define any new quantities introduced in that chapter and to discuss the principles and assumptions used to arrive at certain key relations. The chapter summaries and the review sections of the *Student Solutions Manual and Study Guide* should help you in this regard. In some cases, it may be necessary for you to refer to the index of the text to locate certain topics. You should be able to correctly associate with each physical quantity the symbol used to represent that quantity and the unit in which the quantity is specified. Furthermore, you should be able to express each important relation in a concise and accurate prose statement.

## PROBLEM-SOLVING

R. P. Feynman, Nobel laureate in physics, once said, "You do not know anything until you have practiced." In keeping with this statement, we strongly advise that you develop the skills necessary to solve a wide range of problems. Your ability to solve problems will be one of the main tests of your knowledge of physics; therefore, you should try to solve as many problems as possible. It is essential that you understand basic concepts and principles before attempting to solve problems. It is good practice to try to find alternative solutions to the same problem. For example, you can solve problems in mechanics using Newton's laws, but very often an alternative method that draws on energy considerations is more direct. You should not deceive yourself into thinking you understand a problem merely because you have seen it solved in class. You must be able to solve the problem and similar problems on your own.

The approach to solving problems should be carefully planned. A systematic plan is especially important when a problem involves several concepts. First, read the problem several times until you are confident you understand what is being asked. Look for any key words that will help you interpret the problem and perhaps allow you to make certain assumptions. Your ability to interpret a question properly is an integral part of problem-solving. Second, you should acquire the habit of writing down the information given in a problem and those quantities that need to be found; for example, you might construct a table listing both the quantities given and the quantities to be found. This procedure is sometimes used in the worked examples of the textbook. Finally, after you have decided on the method you feel is appropriate for a given problem, proceed with your solution. General problem-solving strategies of this type are included in the text and are highlighted with a light blue-gray screen. We have also developed a general problem-solving strategy, making use of models, to help guide you through complex problems. This strategy is located at the end of Chapter 1. If you follow the steps of this procedure, you will find it easier to come up with a solution and also gain more from your efforts.

Often, students fail to recognize the limitations of certain equations or physical laws in a particular situation. It is very important that you understand and remember the assumptions underlying a particular theory or formalism. For example, certain equations in kinematics apply only to a particle moving with constant acceleration. These equations are not valid for describing motion whose acceleration is not constant, such as the motion of an object connected to a spring or the motion of an object through a fluid.

## EXPERIMENTS

Physics is a science based on experimental observations. In view of this fact, we recommend that you try to supplement the text by performing various types of "hands-on" experiments, either at home or in the laboratory. For example, the common Slinky™ toy is excellent for studying traveling waves; a ball swinging on the end of a long string can be used to investigate pendulum motion; various masses attached to the end of a vertical spring or rubber band can be used to determine their elastic nature; an old pair of Polaroid sunglasses and some discarded lenses and a magnifying glass are the components of various experiments in optics; and the approximate measure of the acceleration due to gravity can be determined simply by measuring with a stopwatch the time it takes for a ball to drop from a known height. The list of such experiments is endless. When physical models are not available, be imaginative and try to develop models of your own.

## NEW MEDIA

We strongly encourage you to use one or more of the following multimedia products that accompany this textbook. It is far easier to understand physics if you see it in action, and these new materials will enable you to become a part of that action.

***Student Tools CD-ROM*** You may purchase the dual-platform (Windows™- and Macintosh®-compatible) *Student Tools CD-ROM*. In addition to support for working those end-of-chapter problems that require use of a computer, this CD-ROM contains a textbook version of the Interactive Physics™ program by MSC Working Knowledge. Interactive Physics™ simulations are keyed to the following figures, worked examples, Quick Quizzes, and end-of-chapter problems (identified in the text with the 🖼 icon).

***Saunders Core Concepts in Physics CD-ROM***    Alternatively or in addition, you may purchase the *Saunders Core Concepts in Physics CD-ROM* developed by Archipelago Productions. This CD-ROM provides a complete multimedia presentation of selected topics in mechanics, thermodynamics, electromagnetism, and optics. It contains more than 350 movies—both animated and live video—that bring to life laboratory demonstrations, "real-world" examples, graphic models, and step-by-step explanations of essential mathematics. Those CD-ROM modules that supplement the material in *Principles of Physics* are identified in the margin of the text by the  icon.

It is our sincere hope that you too will find physics an exciting and enjoyable experience and that you will profit from this experience, regardless of your chosen profession. Welcome to the exciting world of physics!

> *The scientist does not study nature because it is useful; he studies it because he delights in it, and he delights in it because it is beautiful. If nature were not beautiful, it would not be worth knowing, and if nature were not worth knowing, life would not be worth living.*
>
> **Henri Poincaré**

# List of Life Science Applications/Problems

# About the Authors

**RAYMOND A. SERWAY** received his doctorate at Illinois Institute of Technology and is Professor Emeritus at James Madison University. In 1990, he received the Madison Scholar Award at James Madison University, where he taught for 17 years. Dr. Serway began his teaching career at Clarkson University, where he conducted research and taught from 1967 to 1980. He was the recipient of the Distinguished Teaching Award at Clarkson University in 1977 and of the Alumni Achievement Award from Utica College in 1985. As Guest Scientist at the IBM Research Laboratory in Zurich, Switzerland, he worked with K. Alex Müller, 1987 Nobel Prize recipient. Dr. Serway also was a visiting scientist at Argonne National Laboratory, where he collaborated with his mentor and friend, Sam Marshall. In addition to earlier editions of this textbook, Dr. Serway is the co-author of *Physics for Scientists and Engineers, Fifth Edition; College Physics, Fifth Edition;* and *Modern Physics, Second Edition.* He also is the author of the high-school textbook *Physics,* published by Holt, Rinehart, & Winston. In addition, Dr. Serway has published more than 40 research papers in the field of condensed matter physics and has given more than 70 presentations at professional meetings. Dr. Serway and his wife Elizabeth enjoy traveling, golfing, and spending quality time with their four children and five grandchildren.

**JOHN W. JEWETT, JR.,** earned his doctorate at Ohio State University, specializing in optical and magnetic properties of condensed matter. He is currently Professor of Physics at California State Polytechnic University—Pomona. Throughout his teaching career, Dr. Jewett has been active in promoting science education. In addition to receiving four National Science Foundation grants, he helped found and direct the Southern California Area Modern Physics Institute (SCAMPI). He also is the director of Science IMPACT (Institute for Modern Pedagogy and Creative Teaching), which works with teachers and schools to develop effective science curricula. Dr. Jewett's honors include four Meritorious Performance and Professional Promise awards, selection as Outstanding Professor at California State Polytechnic University for 1991–1992, and the Excellence in Undergraduate Physics Teaching Award from the American Association of Physics Teachers (AAPT) in 1998. He has given many presentations both domestically and abroad, including multiple presentations at national meetings of the AAPT. Dr. Jewett is the author of *The World of Physics—Mysteries, Magic, and Myth,* which provides many connections between physics and everyday experiences. In addition to his work as the co-author for the third edition of *Principles of Physics,* he is also contributing author for *Physics for Scientists and Engineers, Fifth Edition,* and co-author of *Global Issues,* a four-volume set of instruction manuals in integrated science for high school. Dr. Jewett enjoys playing piano, traveling, and collecting antique quack medical devices, as well as spending time with his wife Lisa and their children.

# Pedagogical Color Chart

## Mechanics

Displacement and
position vectors

Linear (**v**) and angular (**ω**)
velocity vectors

Velocity component vectors

Force vectors (**F**)

Force component vectors

Acceleration vectors (**a**)

Acceleration component vectors

Linear (**p**) and
angular (**L**)
momentum vectors

Torque vectors (**τ**)

Linear or rotational
motion directions

Springs

Pulleys

## Electricity and Magnetism

Electric fields

Magnetic fields

Positive charges

Negative charges

Resistors

Batteries and other
dc power supplies

Switches

Capacitors

Inductors (coils)

Voltmeters

Ammeters

Galvanometers

ac Generators

Ground symbol

## Light and Optics

Light rays

Lenses and prisms

Mirrors

Objects

Images

# An Invitation to Physics

P hysics, the most fundamental physical science, is concerned with the basic principles of the universe. It is the foundation on which engineering, technology, and the other sciences—astronomy, biology, chemistry, and geology—are based. The beauty of physics lies in the simplicity of its fundamental theories and in the manner in which just a small number of basic concepts, equations, and assumptions can alter and expand our view of the world around us.

*Classical physics,* developed prior to 1900, includes the theories, concepts, laws, and experiments in classical mechanics, thermodynamics, and electromagnetism. For example, Galileo Galilei (1564–1642) made significant contributions to classical mechanics through his work on the laws of motion with constant acceleration. In the same era, Johannes Kepler (1571–1630) used astronomical observations to develop empirical laws for the motions of planetary bodies.

The most important contributions to classical mechanics, however, were provided by Isaac Newton (1642–1727), who developed classical mechanics as a systematic theory and was one of the originators of the calculus as a mathematical tool. Although major developments in classical physics continued in the 18th century, thermodynamics and electromagnetism were not developed until the latter part of the 19th century, principally because the apparatus for controlled experiments was either too crude or unavailable until then. Although many electric and magnetic phenomena had been studied earlier, the work of James Clerk Maxwell (1831–1879) provided a unified theory of electromagnetism. In this text we shall treat the various disciplines of classical physics in separate sections; however, we will see that the disciplines of mechanics and electromagnetism are basic to all the branches of physics.

A major revolution in physics, usually referred to as *modern physics*, began near the end of the 19th century. Modern physics developed mainly because many physical phenomena could not be explained by classical physics. The two most important developments in this modern era were the theories of relativity and quantum mechanics. Einstein's theory of relativity completely revolutionized the traditional concepts of space, time, and energy. Einstein's theory correctly describes the motion of objects moving at speeds comparable to the speed of light. The theory of relativity also shows that the speed of light is the upper limit of the speed of an object and that mass and energy are related. Quantum mechanics was formulated by a number of distinguished scientists to provide descriptions of physical phenomena at the atomic level.

Scientists continually work at improving our understanding of fundamental laws, and new discoveries are made every day. In many research areas a great deal of overlap exists among physics, chemistry, and biology. Evidence for this overlap is seen in the names of some subspecialties in science — biophysics, biochemistry, chemical physics, biotechnology, and so on. Numerous technological advances in recent times are the result of the efforts of many scientists, engineers, and technicians. Some of the most notable developments in the latter half of the 20th century are (1) space missions to the moon and other planets, (2) microcircuitry and high-speed computers, (3) sophisticated imaging techniques used in scientific research and medicine, and (4) several remarkable accomplishments in genetic engineering. The impact of such developments and discoveries on our society has indeed been great, and future discoveries and developments will very likely be exciting, challenging, and of great benefit to humanity.

In order to investigate the impact of physics on developments in our society, we will use a *contextual* approach to the study of the content in this textbook. The book is divided into eight *Contexts*, which relate the physics to social issues, natural phenomena, or technological applications, as outlined here:

| Chapters | Context |
|---|---|
| 2–11 | Mission to Mars |
| 12–14 | Earthquakes |
| 15 | Search for the *Titanic* |
| 16–18 | Global Warming |
| 19–21 | Lightning |
| 22–23 | Magnetic Levitation Vehicles |
| 24–27 | Lasers |
| 28–31 | The Cosmic Connection |

The Contexts provide a storyline for each section of the text, which will help provide relevance and motivation for studying the material.

The Hubble Space Telescope in the final stages of construction before launch. (*Lockheed Missiles and Space Co., Inc.*)

Each Context begins with a *central question,* which forms the focus for the study of the physics in the Context. The final section of each chapter is a "Context Connection," in which the material in the chapter is explored with the central question in mind. At the end of each Context, a Context Conclusion brings together all of the principles necessary to respond as fully as possible to the central question.

In the first chapter, we investigate some of the mathematical fundamentals and problem-solving strategies that we will use in our study of physics. The first Context, *Mission to Mars,* is introduced in Chapter 2, where the principles of classical mechanics are applied to the problem of transferring a spacecraft from Earth to Mars.

# chapter 1

These controls in the cockpit of a commercial aircraft assist the pilot in maintaining control over the *velocity* of the aircraft — how fast it is traveling and in what direction it is traveling — allowing it to land safely. Quantities that are defined by both a magnitude and a direction, such as velocity, are called ***vectors.***

*(Mark Wagner/Stone)*

# Introduction and Vectors

The goal of physics is to provide a quantitative understanding of certain basic phenomena that occur in our Universe. Physics is a science based on experimental observations and mathematical analyses. The main objective behind such experiments and analyses is to develop theories that explain the phenomenon being studied and to relate those theories to other established theories. Fortunately, it is possible to explain the behavior of various physical systems using relatively few fundamental laws. Analytical procedures require the expression of those laws in the language of mathematics, the tool that provides a bridge between theory and experiment. In this chapter we shall discuss a few mathematical concepts and techniques that will be used throughout the text.

## 1.1 • STANDARDS OF LENGTH, MASS, AND TIME

If we measure a certain quantity and wish to describe it to someone, a unit for the quantity must be specified and defined. For example, it would be meaningless for a visitor from another planet to talk to us about a length of 8 "glitches" if we did not

- *Definition of the meter*

- *Definition of the kilogram*

**WEB**

Visit the International Bureau of Weights and Measures at **www.bipm.fr**

**WEB**

For information on the SI system of units, visit **www.bipm.fr/enus/3_SI/**

- *Definition of the second*

know the meaning of the unit glitch. On the other hand, if someone familiar with our system of measurement reports that a wall is 2.0 meters high and our unit of length is defined to be 1.0 meter, we then know that the height of the wall is twice our fundamental unit of length. Likewise, if we are told that a person has a mass of 75 kilograms and our unit of mass is defined as 1.0 kilogram, then that person has a mass 75 times larger than our fundamental unit of mass. An international committee has agreed on a system of definitions and standards to describe fundamental physical quantities. It is called the **SI system** (Système International) of units. Its units of length, mass, and time are the meter, kilogram, and second, respectively.

### Length

In A.D. 1120, King Henry I of England decreed that the standard of length in his country would be the yard, and that the yard would be precisely equal to the distance from the tip of his nose to the end of his outstretched arm. Similarly, the original standard for the foot adopted by the French was the length of the royal foot of King Louis XIV. This standard prevailed until 1799, when the legal standard of length in France became the **meter,** defined as one ten-millionth of the distance from the equator to the North Pole.

Many other systems have been developed in addition to those just discussed, but the advantages of the French system have caused it to prevail in most countries and in scientific circles everywhere. Until 1960, the length of the meter was defined as the distance between two lines on a specific bar of platinum–iridium alloy stored under controlled conditions. This standard was abandoned for several reasons, a principal one being that the limited accuracy with which the separation between the lines can be determined does not meet the current requirements of science and technology. The definition of the meter was modified to be equal to 1 650 763.73 wavelengths of orange–red light emitted from a krypton-86 lamp. In October 1983, the meter was redefined to be the **distance traveled by light in a vacuum during a time of 1/299 792 458 second.** This arises from the establishment of the speed of light in a vacuum as exactly 299 792 458 meters per second.

### Mass

Mass represents a measure of the resistance of an object to changes in its motion. The SI unit of mass, the **kilogram,** is defined as the **mass of a specific platinum–iridium alloy cylinder kept at the International Bureau of Weights and Measures at Sèvres, France.** At this point, we should add a word of caution. Many beginning students of physics tend to confuse the physical quantities called *weight* and *mass*. For the present we shall not discuss the distinction between them; they will be clearly defined in later chapters. For now you should note that they are distinctly different quantities.

### Time

Before 1960, the standard of time was defined in terms of the average length of a solar day in the year 1900. (A solar day is the time interval between successive appearances of the Sun at the highest point it reaches in the sky each day.) The basic unit of time, the **second,** was defined to be $(1/60)(1/60)(1/24) = 1/86\ 400$ of the average solar day. In 1967 the second was redefined to take advantage of the great precision obtainable with a device known as an atomic clock, which uses the

NIST-7

(*Left*) Stonehenge, a popular tourist attraction in England, is an early timepiece. For example, at the summer solstice, an observer at the center of the stone circle sees the Sun rising directly over a marker stone. (*©John Serafin, Peter Arnold, Inc.*) (*Right*) A modern timepiece, an atomic clock at the National Institute of Standards and Technology. This device keeps time with an accuracy of about 3 millionths of a second per year. (*Courtesy of National Institute of Standards and Technology, U.S. Department of Commerce*)

characteristic frequency of the cesium-133 atom as the "reference clock." The second is now defined as **9 192 631 770 times the period of oscillation of radiation from the cesium atom.** It is possible today to purchase clocks and watches that receive radio signals from the atomic clock in Colorado, which the clock or watch uses to continuously reset itself to the correct time.

### Approximate Values for Length, Mass, and Time

Approximate values of various lengths, masses, and time intervals are presented in Tables 1.1, 1.2, and 1.3, respectively. Note the wide range of values for these quan-

**WEB**

For more information on Stonehenge, visit **www.stonehenge.org.uk**

**WEB**

Visit the National Institute of Standards and Technology at **www.NIST.gov**

| TABLE 1.1 | Approximate Values of Some Measured Lengths | |
|---|---|---|
| | | Length (m) |
| Distance from Earth to most remote quasar known | | $1.4 \times 10^{26}$ |
| Distance from Earth to most remote normal galaxies known | | $4 \times 10^{25}$ |
| Distance from Earth to nearest large galaxy (M 31, the Andromeda galaxy) | | $2 \times 10^{22}$ |
| Distance from Sun to nearest star (Proxima Centauri) | | $4 \times 10^{16}$ |
| One lightyear | | $9.46 \times 10^{15}$ |
| Mean orbit radius of Earth | | $1.5 \times 10^{11}$ |
| Mean distance from Earth to Moon | | $3.8 \times 10^{8}$ |
| Distance from equator to North Pole | | $1 \times 10^{7}$ |
| Mean radius of Earth | | $6.4 \times 10^{6}$ |
| Typical altitude of orbiting Earth satellite | | $2 \times 10^{5}$ |
| Length of a football field | | $9.1 \times 10^{1}$ |
| Length of a housefly | | $5 \times 10^{-3}$ |
| Size of smallest dust particles | | $1 \times 10^{-4}$ |
| Size of cells of most living organisms | | $1 \times 10^{-5}$ |
| Diameter of a hydrogen atom | | $1 \times 10^{-10}$ |
| Diameter of a uranium nucleus | | $1.4 \times 10^{-14}$ |
| Diameter of a proton | | $1 \times 10^{-15}$ |

**PITFALL PREVENTION 1.3**
**Reasonable values**

The generation of intuition about typical values of quantities suggested here is critical. An important step in solving problems is to think about your result at the end of a problem and determine if it seems reasonable. If you are calculating the mass of a housefly and arrive at a value of 100 kg, this is *unreasonable*—there is an error somewhere. If you are calculating the length of a spacecraft on a launch pad and end up with a value of 10 cm, this is *unreasonable*—look for a mistake.

### TABLE 1.2

**Masses of Various Objects (Approximate Values)**

| | Mass (kg) |
|---|---|
| Visible universe | $10^{52}$ |
| Milky Way galaxy | $10^{42}$ |
| Sun | $2 \times 10^{30}$ |
| Earth | $6 \times 10^{24}$ |
| Moon | $7 \times 10^{22}$ |
| Shark | $3 \times 10^2$ |
| Human | $7 \times 10^1$ |
| Frog | $1 \times 10^{-1}$ |
| Mosquito | $1 \times 10^{-5}$ |
| Bacterium | $1 \times 10^{-15}$ |
| Hydrogen atom | $1.67 \times 10^{-27}$ |
| Electron | $9.11 \times 10^{-31}$ |

### TABLE 1.4

**Some Prefixes for Powers of Ten**

| Power | Prefix | Abbreviation |
|---|---|---|
| $10^{-24}$ | yocto | y |
| $10^{-21}$ | zepto | z |
| $10^{-18}$ | atto | a |
| $10^{-15}$ | femto | f |
| $10^{-12}$ | pico | p |
| $10^{-9}$ | nano | n |
| $10^{-6}$ | micro | $\mu$ |
| $10^{-3}$ | milli | m |
| $10^{-2}$ | centi | c |
| $10^{-1}$ | deci | d |
| $10^{3}$ | kilo | k |
| $10^{6}$ | mega | M |
| $10^{9}$ | giga | G |
| $10^{12}$ | tera | T |
| $10^{15}$ | peta | P |
| $10^{18}$ | exa | E |
| $10^{21}$ | zetta | Z |
| $10^{24}$ | yotta | Y |

### TABLE 1.3  Approximate Values of Some Time Intervals

| | Interval (s) |
|---|---|
| Age of the Universe | $5 \times 10^{17}$ |
| Age of the Earth | $1.3 \times 10^{17}$ |
| Time since the fall of the Roman Empire | $5 \times 10^{12}$ |
| Average age of a college student | $6.3 \times 10^8$ |
| One year | $3.2 \times 10^7$ |
| One day (time for one revolution of Earth about its axis) | $8.6 \times 10^4$ |
| Time between normal heartbeats | $8 \times 10^{-1}$ |
| Period of audible sound waves | $1 \times 10^{-3}$ |
| Period of typical radio waves | $1 \times 10^{-6}$ |
| Period of vibration of an atom in a solid | $1 \times 10^{-13}$ |
| Period of visible light waves | $2 \times 10^{-15}$ |
| Duration of a nuclear collision | $1 \times 10^{-22}$ |
| Time for light to cross a proton | $3.3 \times 10^{-24}$ |

tities.* You should study the tables and begin to generate an intuition for what is meant by a mass of 100 kilograms, for example, or by a time interval of $3.2 \times 10^7$ seconds.

Systems of units commonly used in science, commerce, manufacturing, and everyday life are the (1) SI system, in which the units of length, mass, and time are the meter (m), kilogram (kg), and second (s), respectively, and (2) the British engineering system (sometimes called the conventional system), in which the units of length, mass, and time are the foot (ft), slug, and second, respectively. Throughout most of this text we shall use SI units because they are almost universally accepted in science and industry. We will make limited use of conventional units in the study of classical mechanics.

Some of the most frequently used prefixes for the powers of ten and their abbreviations are listed in Table 1.4. For example, $10^{-3}$ m is equivalent to 1 millimeter (mm), and $10^3$ m is 1 kilometer (km). Likewise, 1 kg is $10^3$ grams (g), and 1 megavolt (MV) is $10^6$ volts (V).

## 1.2 • DENSITY AND ATOMIC MASS

The variables length, time, and mass are examples of *fundamental quantities*. A much larger list of variables contains *derived quantities*. These are quantities that can be expressed as a combination of fundamental quantities. Common examples are *area*, which is a combination of two lengths, and *speed*, which is a combination of a length and a time interval.

In this section, we will investigate another derived quantity, **density.** The density $\rho$ (Greek letter rho; a table of the letters in the Greek alphabet is provided at the back of the book) of any substance is defined as its *mass per unit volume:*

* If you are unfamiliar with the use of powers of ten (scientific notation), you should review Appendix B.1.

$$\rho \equiv \frac{m}{V} \qquad [1.1]$$

which is a combination of mass and three lengths. For example, aluminum has a density of $2.70 \times 10^3$ kg/m$^3$, and lead has a density of $11.3 \times 10^3$ kg/m$^3$. An extreme difference in density can be imagined by thinking about holding a 10-centimeter (cm) cube of Styrofoam in one hand and a 10-cm cube of lead in the other. A list of densities for various substances is given in Table 1.5.

All ordinary matter consists of atoms, and each atom is made up of electrons and a nucleus. Practically all of the mass of an atom is contained in the nucleus, which consists of protons and neutrons. Thus, we can understand one reason why the densities of the elements differ—different elements have different numbers of protons and neutrons.

The numbers of protons and neutrons are related to the **atomic mass** of an element, which is defined as the mass of a single atom of the element measured in **atomic mass units** u, where 1 u $= 1.660\ 540\ 2 \times 10^{-27}$ kg. The atomic mass of lead is 207 u and that of aluminum is 27.0 u. However, the ratio of atomic masses, 207 u/27.0 u $= 7.67$, does not correspond to the ratio of densities, $(11.3 \times 10^3$ kg/m$^3)/(2.70 \times 10^3$ kg/m$^3) = 4.19$. The discrepancy is due to the difference in atomic spacings and atomic arrangements in the crystal structures of the two elements.

| TABLE 1.5 | |
|---|---|
| **Densities of Various Substances** | |
| Substance | Density $\rho$ (kg/m$^3$) |
| Platinum | $21.45 \times 10^3$ |
| Gold | $19.3 \times 10^3$ |
| Uranium | $18.7 \times 10^3$ |
| Lead | $11.3 \times 10^3$ |
| Copper | $8.93 \times 10^3$ |
| Iron | $7.86 \times 10^3$ |
| Aluminum | $2.70 \times 10^3$ |
| Magnesium | $1.75 \times 10^3$ |
| Water | $1.00 \times 10^3$ |
| Air at atmospheric pressure | $0.001\ 2 \times 10^3$ |

### Example 1.1 How Many Atoms in the Cube?

A solid cube of aluminum (density 2.70 g/cm$^3$) has a volume of 0.200 cm$^3$. It is known that 27.0 g of aluminum contains $6.02 \times 10^{23}$ atoms. How many aluminum atoms are contained in the cube?

**Solution** Because density equals mass per unit volume, the mass of the cube is

$$m = \rho V = (2.70 \text{ g/cm}^3)(0.200 \text{ cm}^3) = 0.540 \text{ g}$$

To solve this problem, we will set up a ratio based on the fact that the mass of a sample of material is proportional to the number of atoms contained in the sample. This technique of solving by ratios is very powerful and should be studied and understood so that it can be applied in future problem solving. Let us express our proportionality as $m = kN$, where $m$ is the mass of the sample, $N$ is the number of atoms in the sam-

ple, and $k$ is an unknown proportionality constant. We write this relationship twice, once for the actual sample of aluminum in the problem and once for a 27.0-g sample, and then we divide the first equation by the second:

$$m_{\text{sample}} = kN_{\text{sample}} \longrightarrow \frac{m_{\text{sample}}}{m_{27.0\,\text{g}}} = \frac{kN_{\text{sample}}}{kN_{27.0\,\text{g}}}$$
$$m_{27.0\,\text{g}} = kN_{27.0\,\text{g}}$$

Notice that the unknown proportionality constant $k$ cancels, so we do not need to know its value. We now substitute the values:

$$\frac{0.540 \text{ g}}{27.0 \text{ g}} = \frac{N_{\text{sample}}}{6.02 \times 10^{23} \text{ atoms}}$$

$$N_{\text{sample}} = \frac{(0.540 \text{ g})(6.02 \times 10^{23} \text{ atoms})}{27.0 \text{ g}}$$

$$= 1.20 \times 10^{22} \text{ atoms}$$

## 1.3 • DIMENSIONAL ANALYSIS

The word *dimension* has a special meaning in physics. It denotes the physical nature of a quantity. Whether a distance is measured in units of feet or meters or furlongs, it is a distance. We say its dimension is *length*.

When using ratios to solve a problem, keep in mind that *ratios come from equations.* Students often just say, "the ratio of this to this is equal to the ratio of that to that" without thinking about the actual mathematical relationships. For example, a student may say, "the ratio of the areas of two circles is equal to the ratio of their radii." This ignores the fact that the area depends on the *square* of the radius. By using the technique of dividing equations in Example 1.1, these errors can be avoided.

The symbols used in this book to specify the dimensions* of length, mass, and time are L, M, and T, respectively. We shall often use square brackets [ ] to denote the dimensions of a physical quantity. For example, in this notation the dimensions of velocity $v$ are written $[v] = L/T$, and the dimensions of area $A$ are $[A] = L^2$. The dimensions of area, volume, velocity, and acceleration are listed in Table 1.6, along with their units in the two common systems. The dimensions of other quantities, such as force and energy, will be described as they are introduced in the text.

In many situations, you may be faced with having to derive or check a specific equation. Although you may have forgotten the details of the derivation, a useful and powerful procedure called *dimensional analysis* can be used as a consistency check, to assist in the derivation, or to check your final expression. Dimensional analysis makes use of the fact that **dimensions can be treated as algebraic quantities.** For example, quantities can be added or subtracted only if they have the same dimensions. Furthermore, the terms on both sides of an equation must have the same dimensions. By following these simple rules, you can use dimensional analysis to help determine whether an expression has the correct form, because the relationship can be correct only if the dimensions on the two sides of the equation are the same.

To illustrate this procedure, suppose you wish to derive an expression for the position $x$ of a car at a time $t$ if the car starts from rest at $t = 0$ and moves with constant acceleration $a$. In Chapter 2 we shall find that the correct expression for this special case is $x = \frac{1}{2} at^2$. Let us check the validity of this expression from a dimensional analysis approach.

The quantity $x$ on the left side has the dimension of length. For the equation to be dimensionally correct, the quantity on the right side must also have the dimension of length. We can perform a dimensional check by substituting the basic dimensions for acceleration, $L/T^2$, and time, T, into the equation $x = \frac{1}{2}at^2$. That is, the dimensional form of the equation $x = \frac{1}{2}at^2$ can be written as

$$L = \frac{L}{\mathcal{T}^2} \cdot \mathcal{T}^2 = L$$

The units of time cancel as shown, leaving the unit of length. Notice that the number $\frac{1}{2}$ in the equation has no units, so it does not enter into the dimensional analysis.

| TABLE 1.6 | Units of Area, Volume, Velocity, and Acceleration | | | |
|---|---|---|---|---|
| **System** | **Area** $(L^2)$ | **Volume** $(L^3)$ | **Velocity** $(L/T)$ | **Acceleration** $(L/T^2)$ |
| SI | $m^2$ | $m^3$ | m/s | $m/s^2$ |
| British engineering | $ft^2$ | $ft^3$ | ft/s | $ft/s^2$ |

* The *dimensions* of a variable will be symbolized by a capitalized, nonitalic letter, such as, in the case of length, L. The *symbol* for the variable itself will be italicized, such as $L$ for the length of an object, or $t$ for time.

### Example 1.2 Analysis of an Equation

Show that the expression $v_f = v_i + at$ is dimensionally correct, where $v_f$ and $v_i$ represent speeds at two instants of time, $a$ is acceleration, and $t$ is an instant of time.

**Solution** The dimensions of the speeds are

$$[v_f] = [v_i] = \frac{L}{T}$$

and the dimensions of acceleration are $L/T^2$. Thus, the dimensions of $at$ are

$$[at] = \frac{L}{T^2} \cdot T = \frac{L}{T}$$

and the expression is dimensionally correct. On the other hand, if the expression were given as $v_f = v_i + at^2$, it would be dimensionally *incorrect*. Try it and see!

## 1.4 • CONVERSION OF UNITS

Sometimes it is necessary to convert units from one system to another, or to convert within a system, for example, from kilometers to meters. Equalities between SI and conventional units of length are as follows:

$$1 \text{ mile} = 1\,609 \text{ m} = 1.609 \text{ km} \qquad 1 \text{ ft} = 0.304\,8 \text{ m} = 30.48 \text{ cm}$$

$$1 \text{ m} = 39.37 \text{ in.} = 3.281 \text{ ft} \qquad 1 \text{ in.} = 0.025\,4 \text{ m} = 2.54 \text{ cm}$$

A more complete list of equalities can be found in Appendix A.

Units can be treated as algebraic quantities that can cancel each other. To perform a conversion, a quantity can be multiplied by a **conversion factor,** which is a fraction equal to 1, with numerator and denominator having different units, to provide the desired units in the final result. For example, suppose we wish to convert 15.0 in. to centimeters. Because 1 in. = 2.54 cm, we multiply by a conversion factor that is the appropriate ratio of these equal quantities, and find that

$$15.0 \text{ in.} = (15.0 \text{ in.}) \left( \frac{2.54 \text{ cm}}{1 \text{ in.}} \right) = 38.1 \text{ cm}$$

**PITFALL PREVENTION 1.5**
**Always include units**

When performing calculations, make it a habit to include the units on every quantity and carry the units through the entire calculation. Avoid the temptation to drop the units during the calculation steps and then apply the appropriate unit to the number that results for an answer. By including the units in every step, you can detect errors if the units for the answer are incorrect.

(*Left*) Conversion of miles to kilometers. (*Right*) This automobile speedometer gives speed readings in both miles per hour and kilometers per hour. You should confirm the equality for a few readings on the dial. (*Paul Silverman, Fundamental Photographs*)

where the ratio in parentheses is equal to 1. Notice that we choose to put the unit of an inch in the denominator and it cancels with the unit in the original quantity. The remaining unit is the centimeter, which is our desired result.

---

### Example 1.3   The Density of a Cube

The mass of a solid cube is 856 g, and each edge has a length of 5.35 cm. Determine the density $\rho$ of the cube in SI units.

**Solution**   We will convert the mass and length to SI units before calculating the density. Because 1 kg = 1 000 g and 1 m = 100 cm, the mass $m$ and volume $V$ in SI units are

$$m = (856 \text{ g}) \left( \frac{1.00 \text{ kg}}{1.00 \times 10^3 \text{ g}} \right) = 0.856 \text{ kg}$$

$$L = (5.35 \text{ cm}) \left( \frac{1.00 \text{ m}}{1.00 \times 10^2 \text{ cm}} \right) = 5.35 \times 10^{-2} \text{ m}$$

Now, the volume of the cube is

$$V = L^3 = (5.35 \times 10^{-2} \text{ m})^3 = (5.35)^3 \times 10^{-6} \text{ m}^3$$
$$= 1.53 \times 10^{-4} \text{ m}^3$$

Therefore the density of the cube is

$$\rho = \frac{m}{V} = \frac{0.856 \text{ kg}}{1.53 \times 10^{-4} \text{ m}^3} = \boxed{5.59 \times 10^3 \text{ kg/m}^3}$$

---

## 1.5 • ORDER-OF-MAGNITUDE CALCULATIONS

It is often useful to compute an approximate answer to a given physical problem even when little information is available. This answer can then be used to determine whether a more precise calculation is necessary. Such an approximation is usually based on certain assumptions, which must be modified if greater precision is needed. Thus, we will sometimes refer to an *order of magnitude* of a certain quantity as the power of ten of the number that describes that quantity. Usually, when an order-of-magnitude calculation is made, the results are reliable to within about a factor of 10. If a quantity increases in value by three orders of magnitude, this means that its value increases by a factor of $10^3 = 1\ 000$. We use the symbol $\sim$ for "is on the order of." Thus,

$$0.008\ 6 \sim 10^{-2} \qquad 0.002\ 1 \sim 10^{-3} \qquad 700 \sim 10^3$$

The spirit of attempting order-of-magnitude calculations, sometimes referred to as "guesstimates" or "ball-park figures," is captured by the following quotation: "Make an estimate before every calculation, try a simple physical argument . . . before every derivation, guess the answer to every puzzle." *

---

### Example 1.4   The Number of Atoms in a Solid

Estimate the number of atoms in 1 cm³ of a solid.

**Solution**   From Table 1.1 we note that the diameter $d$ of an atom is about $10^{-10}$ m. Let us assume that the atoms in the solid are spheres of this diameter. Then the volume of each sphere is about $10^{-30}$ m³ (more precisely, volume = $4\pi r^3/3 = \pi d^3/6$, where $r = d/2$). Therefore, because

1 cm³ = $10^{-6}$ m³, the number of atoms in the solid is on the order of $10^{-6}/10^{-30} = 10^{24}$ atoms.

A more precise calculation would require knowledge of the density of the solid and the mass of each atom. However, our estimate agrees with the more precise calculation to within a factor of 10.

---

* E. Taylor and J. A. Wheeler, *Spacetime Physics*, San Francisco, W. H. Freeman, 1966, p. 60.

**Example 1.5** How Much Gas Do We Use?

Estimate the number of gallons of gasoline used by all U.S. cars each year.

**Solution** Because there are about 280 million people in the United States, an estimate of the number of cars in the country is $7 \times 10^7$ (assuming one car and four people per family). We can also estimate that the average distance traveled per year is $1 \times 10^4$ miles. If we assume gasoline consumption of 0.05 gal/mi (equivalent to 20 miles per gallon), each car uses about $5 \times 10^2$ gal/year. Multiplying this by the total number of cars in the United States gives an estimated total consumption of about $10^{11}$ gal, which corresponds to a yearly consumer expenditure on the order of $10^2$ billion dollars. This is probably a low estimate because we haven't accounted for commercial consumption.

## 1.6 • SIGNIFICANT FIGURES

When certain quantities are measured, the measured values are known only to within the limits of the experimental uncertainty. The value of the uncertainty can depend on various factors, such as the quality of the apparatus, the skill of the experimenter, and the number of measurements performed. The number of **significant figures** in a measurement can be used to express something about the uncertainty.

As an example of significant figures, consider the population of New York state, as reported in a published road atlas: 18 044 505. Notice that this number reports the population *to the level of one individual*. We would describe this number as having eight significant figures. Can the population really be this accurate? First of all, is the census process accurate enough to measure the population to one individual? By the time this number was actually published, had the number of births and immigrations into the state balanced the number of deaths and emigrations out of the state, so that the change in the population is exactly zero? This same number for the population is reported in the 1994 edition of the road atlas *and* the 1998 edition—could it be a reasonable expectation that the population did not change by even one person in four years?

The claim that the population is measured and known to the level of one individual is unjustified. We would describe this by saying that *there are too many significant figures in the measurement.* To account for the inherent uncertainty in the census-taking process and the inevitable changes in population by the time the number is read in the road atlas, it might be better to report the population as something like 18.0 million. This number has three significant figures, rather than the eight significant figures in the published population. The fact that the identical populations are reported in the 1994 and 1998 editions suggests that these measurements might come from the census performed every ten years. Thus, for this number to be somewhat valid for a ten-year cycle of editions, it might be better to report the population as $2 \times 10^7$, which has only one significant figure.

Let us look at a more scientific example. Suppose we are asked in a laboratory experiment to measure the area of a rectangular plate using a meter stick as a measuring instrument. Let us assume that the accuracy to which we can measure a particular dimension of the plate is ±0.1 cm. If the length of the plate is measured to be 16.3 cm, we can claim only that its length lies somewhere between 16.2 cm and 16.4 cm. In this case, we say that the measured value has three significant figures. Likewise, if its width is measured to be 4.5 cm, the actual value lies between 4.4 cm and 4.6 cm. This measured value has only two significant figures. Note that the significant figures include the first estimated digit. Thus, we could write the measured values as 16.3 ± 0.1 cm and 4.5 ± 0.1 cm.

Suppose now that we would like to find the area of the plate by multiplying the two measured values. If we were to claim that the area is $(16.3 \text{ cm})(4.5 \text{ cm}) = 73.35 \text{ cm}^2$, our answer would be unjustifiable because it contains four significant figures, which is greater than the number of significant figures in either of the measured lengths. The following is a good rule of thumb to use in determining the number of significant figures that can be claimed:

When multiplying several quantities, the number of significant figures in the final answer is the same as the number of significant figures in the quantity having the lowest number of significant figures. The same rule applies to division.

Applying this rule to the previous multiplication example, we see that the answer for the area can have only two significant figures because the length of 4.5 cm has only two significant figures. Thus, all we can claim is that the area is $73 \text{ cm}^2$, realizing that the value can range between $(16.2 \text{ cm})(4.4 \text{ cm}) = 71 \text{ cm}^2$ and $(16.4 \text{ cm})(4.6 \text{ cm}) = 75 \text{ cm}^2$.

Zeros may or may not be significant figures. Those used to position the decimal point in such numbers as 0.03 and 0.0075 are not significant. Thus there are one and two significant figures, respectively, in these two values. When the positioning of zeros comes after other digits, however, there is the possibility of misinterpretation. For example, suppose the mass of an object is given as 1 500 g. This value is ambiguous because we do not know whether the last two zeros are being used to locate the decimal point or whether they represent significant figures in the measurement. To remove this ambiguity, it is common to use scientific notation to indicate the number of significant figures. In this case, we would express the mass as $1.5 \times 10^3$ g if the measured value has two significant figures, $1.50 \times 10^3$ g if it has three significant figures, and $1.500 \times 10^3$ g if it has four. Likewise, 0.000 150 should be expressed in scientific notation as $1.5 \times 10^{-4}$ if it has two significant figures or as $1.50 \times 10^{-4}$ if it has three significant figures. The three zeros between the decimal point and the digit 1 in the number 0.000 150 are not counted as significant figures because they are present only to locate the decimal point. In general, a **significant figure** in a measurement is a reliably known digit (other than a zero used to locate the decimal point) or the first estimated digit.

For addition and subtraction, the number of decimal places must be considered when you are determining how many significant figures to report.

When numbers are added or subtracted, the number of decimal places in the result should equal the smallest number of decimal places of any term in the sum.

For example, if we wish to compute $123 + 5.35$, the answer is 128 and not 128.35. If we compute the sum $1.000 \ 1 + 0.000 \ 3 = 1.000 \ 4$, the result has the correct number of decimal places; consequently it has five significant figures even though one of the terms in the sum, 0.000 3, has only one significant figure. Likewise, if we perform the subtraction $1.002 - 0.998 = 0.004$, the result has only one significant figure even though one term has four significant figures and the other has three. In this book, **most of the numerical examples and end-of-chapter problems will yield answers having three significant figures.**

If the number of significant figures in the result of an addition or subtraction must be reduced, a general rule for rounding off numbers states that the last digit retained is to be increased by 1 if the last digit dropped is greater than 5. If the last digit dropped is less than 5, the last digit retained remains as it is. If the last digit dropped is equal to 5, the remaining digit should be rounded to the nearest even number. (This helps avoid accumulation of errors in long arithmetic processes.)

---

### Example 1.6  Installing a Carpet

A carpet is to be installed in a room whose length is measured to be 12.71 m (four significant figures) and whose width is measured to be 3.46 m (three significant figures). Find the area of the room.

**Solution**  If you multiply 12.71 m by 3.46 m on your calculator, you will obtain an answer of 43.976 6 m². How many of

these numbers should you claim? Our rule of thumb for multiplication tells us that you can claim only the number of significant figures in the quantity with the smallest number of significant figures. In this example, that number is three (in the width 3.46 m), so we should express our final answer as 44.0 m².

---

## 1.7 • COORDINATE SYSTEMS

Many aspects of physics deal in some way or another with locations in space. For example, the mathematical description of the motion of an object requires a method for specifying the position of the object. Thus, we first discuss how to describe the position of a point in space. This is done by means of coordinates in a graphical representation. A point on a line can be located with one coordinate; a point in a plane is located with two coordinates; and three coordinates are required to locate a point in space.

A coordinate system used to specify locations in space consists of:

- A fixed reference point $O$, called the origin
- A set of specified axes or directions with an appropriate scale and labels on the axes
- Instructions that tell us how to label a point in space relative to the origin and axes

One convenient coordinate system that we will use frequently is the *cartesian coordinate system*, sometimes called the *rectangular coordinate system.* Such a system in two dimensions is illustrated in Figure 1.1. An arbitrary point in this system is labeled with the coordinates $(x, y)$. Positive $x$ is taken to the right of the origin, and positive $y$ is upward from the origin. Negative $x$ is to the left of the origin, and negative $y$ is downward from the origin. For example, the point $P$, which has coordinates $(5, 3)$, may be reached by going first 5 m to the right of the origin and then 3 m above the origin (*or* by going 3 m above the origin and then 5 m to the right). Similarly, the point $Q$ has coordinates $(-3, 4)$, which correspond to going 3 m to the left of the origin and 4 m above the origin.

Sometimes it is more convenient to represent a point in a plane by its *plane polar coordinates* $(r, \theta)$, as in Figure 1.2a. In this coordinate system, $r$ is the length of the line from the origin to the point, and $\theta$ is the angle between that line and a fixed axis, usually the positive $x$ axis, with $\theta$ measured counterclockwise. From the right triangle in Figure 1.2b, we find $\sin \theta = y/r$ and $\cos \theta = x/r$. (A review of trigonometric functions is given in Appendix B.4.) Therefore, starting with plane polar coordinates, one can obtain the cartesian coordinates through the equations

$$x = r \cos \theta \qquad \text{[1.2]}$$

$$y = r \sin \theta \qquad \text{[1.3]}$$

See the *Core Concepts in Physics CD-ROM,* Screen 2.2

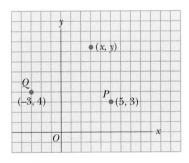

**Figure 1.1**

Designation of points in a cartesian coordinate system. Each square in the $xy$ plane is 1 m on a side. Every point is labeled with coordinates $(x, y)$.

(a)

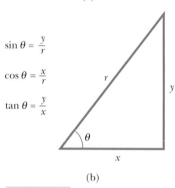

$\sin \theta = \dfrac{y}{r}$

$\cos \theta = \dfrac{x}{r}$

$\tan \theta = \dfrac{y}{x}$

(b)

**Figure 1.2**

(a) The plane polar coordinates of a point are represented by the distance $r$ and the angle $\theta$, where $\theta$ is measured in a clockwise direction from the positive $x$ axis. (b) The right triangle used to relate $(x, y)$ to $(r, \theta)$.

 See Screen 2.3

• *Definition of displacement along a line*

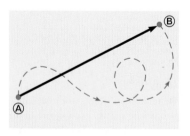

**Figure 1.3**

As a particle moves from Ⓐ to Ⓑ along an arbitrary path represented by the broken line, its displacement is a vector quantity shown by the arrow drawn from Ⓐ to Ⓑ.

Furthermore, it follows that

$$\tan \theta = \frac{y}{x} \tag{1.4}$$

and

$$r = \sqrt{x^2 + y^2} \tag{1.5}$$

You should note that these expressions relating the coordinates $(x, y)$ to the coordinates $(r, \theta)$ apply only when $\theta$ is defined as in Figure 1.2a, where positive $\theta$ is an angle measured *counterclockwise* from the positive $x$ axis. Other choices are made in navigation and astronomy. If the reference axis for the polar angle $\theta$ is chosen to be other than the positive $x$ axis, or the sense of increasing $\theta$ is chosen differently, then the corresponding expressions relating the two sets of coordinates will change.

## 1.8 • VECTORS AND SCALARS

Each of the physical quantities that we shall encounter in this text can be placed in one of two categories: It is either a scalar or a vector. A **scalar** is a quantity that is completely specified by a positive or negative number with appropriate units. On the other hand, a **vector** is a physical quantity that must be specified by both magnitude and direction.

The number of apples in a basket is an example of a scalar quantity. If you are told there are 38 apples in the basket, this completely specifies the information; no specification of direction is required. Other examples of scalars are temperature, volume, mass, and time intervals. The rules of ordinary arithmetic are used to manipulate scalar quantities—they can be freely added and subtracted (assuming that they have the same units!), multiplied and divided.

Force is an example of a vector quantity. To completely describe the force on an object, we must specify both the direction of the applied force and the magnitude of the force.

Another simple example of a vector quantity is the **displacement** of a particle, defined as its *change in position.* Suppose the particle moves from some point Ⓐ to a point Ⓑ along a straight path, as in Figure 1.3. This displacement can be represented by drawing an arrow from Ⓐ to Ⓑ, where the arrowhead represents the direction of the displacement and the length of the arrow represents the magnitude of the displacement. If the particle travels along some other path from Ⓐ to Ⓑ, such as the broken line in Figure 1.3, its displacement is still the vector from Ⓐ to Ⓑ. The vector displacement along any indirect path from Ⓐ to Ⓑ is defined as being equivalent to the displacement represented by the direct path from Ⓐ to Ⓑ. The magnitude of the displacement is the shortest distance between the end points. Thus, the **displacement of a particle is completely known if its initial and final coordinates are known.** The path need not be specified. In other words, the **displacement is independent of the path,** if the end points of the path are fixed.

It is important to note that the **distance** traveled by a particle is distinctly different from its displacement. The distance traveled (a scalar quantity) is the length of the path, which in general can be much greater than the magnitude of the displacement (see Fig. 1.3).

If the particle moves along the $x$ axis from position $x_i$ to position $x_f$, as in Figure 1.4, its displacement is given by $x_f - x_i$. (The indices $i$ and $f$ refer to the initial

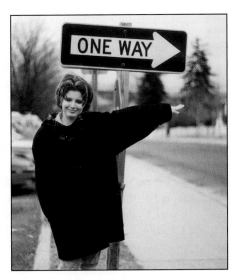

(a) The number of apples in the basket is one example of a scalar quantity. Can you think of other examples? *(Superstock)* (b) Jennifer pointing in the right direction. *(Photo by Raymond A. Serway)*

and final values.) We use the Greek letter delta ($\Delta$) to denote the *change* in a quantity. Therefore, we define the change in the position of the particle (the displacement) as

$$\Delta x \equiv x_f - x_i \qquad \text{[1.6]}$$

From this definition we see that $\Delta x$ is positive if $x_f$ is greater than $x_i$ and negative if $x_f$ is less than $x_i$. For example, if a particle changes its position from $x_i = -5$ m to $x_f = 3$ m, its displacement is $\Delta x = 8$ m.

Many physical quantities in addition to displacement are vectors. They include velocity, acceleration, force, and momentum, all of which will be defined in later chapters. In this text we will use boldface letters, such as **A**, to represent vectors. Another common notation for vectors with which you should be familiar is the use of an arrow over the letter: $\vec{A}$. This notation is useful in handwriting on paper or a chalkboard, where using bold letters is inconvenient.

The magnitude of the vector **A** is written $A$ or, alternatively, $|\mathbf{A}|$. The magnitude of a vector is always positive and carries the units of the quantity that the vector represents, such as meters for displacement or meters per second for velocity. Vectors combine according to special rules, which will be discussed in Sections 1.9 and 1.10.

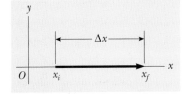

**Figure 1.4**

A particle moving along the $x$ axis from $x_i$ to $x_f$ undergoes a displacement $\Delta x = x_f - x_i$.

### THINKING PHYSICS 1.1

Consider your commute to work or school in the morning. Which is larger, the distance you travel or the magnitude of the displacement vector?

**Reasoning** Unless you have a very unusual commute, the distance traveled *must* be larger than the magnitude of the displacement vector. The distance includes all of the twists and turns you make in following the roads from home to work or school. On the other hand, the magnitude of the displacement vector is the length of a

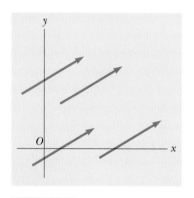

**Figure 1.5**

These four representations of vectors are equal because all four vectors have the same magnitude and point in the same direction.

 See Screen 2.4

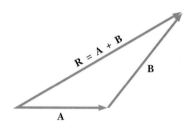

**Figure 1.6**

When vector **B** is added to vector **A**, the resultant **R** is the vector that runs from the tail of **A** to the tip of **B**.

**PITFALL PREVENTION 1.6**

**Vector addition versus scalar addition**

 Keep in mind that **A** + **B** = **C** is very different from $A + B = C$. The first is a vector sum, which must be handled carefully, such as with the graphical method described here. The second is a simple algebraic addition of numbers that is handled with the normal rules of arithmetic.

straight line from your home to work or school. This is often described informally as "the distance as the crow flies." The only way that the distance could be the same as the magnitude of the vector is if your commute is a perfect straight line, which is highly unlikely! The distance could *never* be less than the magnitude of the displacement vector, because the shortest distance between two points is a straight line.

## 1.9 • SOME PROPERTIES OF VECTORS

**Equality of Two Vectors** Two vectors **A** and **B** are defined to be equal if they have the same units, the same magnitude, and the same direction. That is, **A** = **B** only if $A = B$ *and* **A** and **B** point in the same direction. For example, all the vectors in Figure 1.5 are equal even though they have different starting points. This property allows us to translate a vector parallel to itself in a diagram without affecting the vector.

**Addition** When two or more vectors are added together, they must *all* have the same units. For example, it would be meaningless to add a velocity vector to a displacement vector because they are different physical quantities. Scalars obey the same rule. For example, it would be meaningless to add time intervals and temperatures.

The rules for vector sums are conveniently described using geometry. To add vector **B** to vector **A**, first draw a diagram of vector **A** on graph paper, with its magnitude represented by a convenient scale, and then draw vector **B** to the same scale with its tail starting from the tip of **A**, as in Figure 1.6. The *resultant vector* **R** = **A** + **B** is the vector drawn from the tail of **A** to the tip of **B**. If these vectors are displacements, **R** is the single displacement that has the same effect as the displacements **A** and **B** performed one after the other.

This process is known as the *triangle method of addition*, because the three vectors can be geometrically modeled as the sides of a triangle. An alternative graphical procedure for adding two vectors, known as the *parallelogram rule of addition*, is shown in Figure 1.7a. In this construction, the tails of the two vectors **A** and **B** are together, and the resultant vector **R** is the diagonal of a parallelogram formed with **A** and **B** as its sides.

When vectors are added, the sum is independent of the order of the addition. This can be seen for two vectors from the geometric construction in Figure 1.7b

(a)

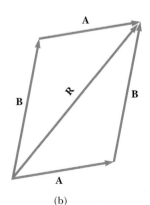

(b)

**Figure 1.7**

(a) In this construction, the resultant **R** is the diagonal of a parallelogram having sides **A** and **B**. (b) This construction shows that **A** + **B** = **B** + **A**—vector addition is commutative.

and is known as the **commutative law of addition:**

$$\mathbf{A} + \mathbf{B} = \mathbf{B} + \mathbf{A} \qquad [1.7]$$

If three or more vectors are added, their sum is independent of the way in which they are grouped. A geometric demonstration of this for three vectors is given in Figure 1.8. This is called the **associative law of addition:**

$$\mathbf{A} + (\mathbf{B} + \mathbf{C}) = (\mathbf{A} + \mathbf{B}) + \mathbf{C} \qquad [1.8]$$

Geometric constructions can also be used to add more than three vectors. This is shown in Figure 1.9 for the case of four vectors. The resultant vector sum **R** = **A** + **B** + **C** + **D** is the *vector that closes the polygon formed by the vectors being added*. In other words, **R** is the *vector drawn from the tail of the first vector to the tip of the last vector*. Again, the order of the summation is unimportant.

Thus we conclude that a vector is a quantity that has both magnitude and direction and also obeys the laws of vector addition described in Figures 1.6 to 1.9.

**Negative of a Vector** The negative of the vector **A** is defined as the vector that, when added to **A**, gives zero for the vector sum. That is, **A** + (−**A**) = 0. The vectors **A** and −**A** have the same magnitude but opposite directions.

**Subtraction of Vectors** The operation of vector subtraction makes use of the definition of the negative of a vector. We define the operation **A** − **B** as vector −**B** added to vector **A**:

$$\mathbf{A} - \mathbf{B} = \mathbf{A} + (-\mathbf{B}) \qquad [1.9]$$

A diagram for subtracting two vectors is shown in Figure 1.10.

**Multiplication of a Vector by a Scalar** If a vector **A** is multiplied by a positive scalar quantity $s$, the product $s\mathbf{A}$ is a vector that has the same direction as **A** and magnitude $sA$. If $s$ is a negative scalar quantity, the vector $s\mathbf{A}$ is directed opposite to **A**. For example, the vector $5\mathbf{A}$ is five times greater in magnitude than **A** and has the same direction as **A**. On the other hand, the vector $-\frac{1}{3}\mathbf{A}$ has one third the magnitude of **A** and points in the direction opposite **A** (because of the negative sign).

- *Commutative law*

- *Associative law*

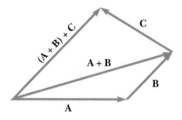

**Figure 1.8**

Geometric constructions for verifying the associative law of addition.

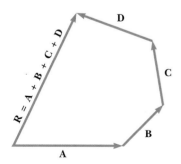

**Figure 1.9**

Geometric construction for summing four vectors. The resultant vector **R** closes the polygon and points from the tail of the first vector to the tip of the final vector.

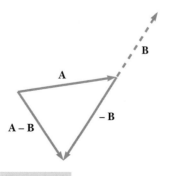

**Figure 1.10**

This construction shows how to subtract vector **B** from vector **A**: Add the vector −**B** to vector **A**. The vector −**B** is equal in magnitude and opposite to the vector **B**.

**Multiplication of Two Vectors**  Two vectors **A** and **B** can be multiplied in two different ways to produce either a scalar or a vector quantity. The **scalar product** (or dot product) **A** · **B** is a scalar quantity equal to $AB \cos \theta$, where $\theta$ is the angle between **A** and **B**. The **vector product** (or cross product) **A** × **B** is a vector quantity whose magnitude is equal to $AB \sin \theta$. We shall discuss these products more fully in Chapters 6 and 10, where they are first used.

### Quick Quiz 1.1

The magnitudes of two vectors **A** and **B** are $A = 8$ units and $B = 3$ units. What are the largest and smallest possible values for the magnitude of the resultant vector **R** = **A** + **B**?

### Quick Quiz 1.2

If vector **B** is added to vector **A**, under what circumstances does the resultant vector **A** + **B** have magnitude $A + B$? Under what circumstances is the resultant vector equal to zero?

See Screen 2.5

---

**PITFALL PREVENTION 1.7**

**Component vectors versus components**

The vectors $\mathbf{A}_x$ and $\mathbf{A}_y$ are the *component vectors* of **A**. These should not be confused with the scalars $A_x$ and $A_y$, which we shall always refer to as the *components* of **A**.

---

**PITFALL PREVENTION 1.8**

**x Components**

Equation 1.10 for the $x$ and $y$ components of a vector associates the cosine of the angle with the $x$ component and the sine of the angle with the $y$ component. This occurs *solely* because we chose to measure the angle with respect to the $x$ axis, so don't memorize these equations. Invariably, you will face a problem in the future in which the angle is measured with respect to the $y$ axis, and the equations will be incorrect. It is much better to always think about which side of the triangle containing the components is adjacent to the angle and which side is opposite, and then to assign the sine and cosine accordingly.

---

## 1.10 • COMPONENTS OF A VECTOR AND UNIT VECTORS

The geometric method of adding vectors is not the recommended procedure in situations where great precision is required, or in three-dimensional problems, because we are forced to draw them on two-dimensional paper. In this section we describe a method of adding vectors that makes use of the *projections* of a vector along the axes of a rectangular coordinate system.

Consider a vector **A** lying in the $xy$ plane and making an arbitrary angle $\theta$ with the positive $x$ axis, as in Figure 1.11a. The vector **A** can be represented by its rectangular **components,** $A_x$ and $A_y$. The component $A_x$ represents the projection of **A** along the $x$ axis, and $A_y$ represents the projection of **A** along the $y$ axis. The components of a vector, which are scalar quantities, can be positive or negative. For example, in Figure 1.11a, $A_x$ and $A_y$ are both positive. The absolute values of the components are the magnitudes of the associated **component vectors** $\mathbf{A}_x$ and $\mathbf{A}_y$.

Figure 1.11b shows the component vectors again, but with the $y$ component vector shifted so that it is added vectorially to the $x$ component vector. This diagram shows us two important features. First, a vector is equal to the sum of its component vectors. Thus, the combination of the component vectors is a valid substitute for the actual vector. The second feature is that the vector and its component vectors form a right triangle. Thus, we can let the triangle be a model for the vector and can use right triangle trigonometry to analyze the vector. The legs of the triangle are of lengths proportional to the components (depending on what scale factor you have chosen), and the hypotenuse is of a length proportional to the magnitude of the vector.

From Figure 1.11b and the definition of the sine and cosine of an angle, we see that $\cos \theta = A_x/A$ and $\sin \theta = A_y/A$. Hence, the components of **A** are given by

$$A_x = A \cos \theta \quad \text{and} \quad A_y = A \sin \theta \qquad \textbf{[1.10]}$$

It is important to note that when using these component equations, $\theta$ must be measured counterclockwise from the positive $x$ axis. From our triangle, it follows that the magnitude of **A** and its direction are related to its components through the Pythagorean theorem and the definition of the tangent function:

$$A = \sqrt{A_x^2 + A_y^2} \qquad \text{[1.11]}$$

$$\tan \theta = \frac{A_y}{A_x} \qquad \text{[1.12]}$$

To solve for $\theta$, we can write $\theta = \tan^{-1}(A_y/A_x)$, which is read "$\theta$ equals the angle the tangent of which is the ratio $A_y/A_x$." *Note that the signs of the components $A_x$ and $A_y$ depend on the angle $\theta$.* For example, if $\theta = 120°$, $A_x$ is negative and $A_y$ is positive. On the other hand, if $\theta = 225°$, both $A_x$ and $A_y$ are negative. Figure 1.12 summarizes the signs of the components when **A** lies in the various quadrants.

If you choose reference axes or an angle other than those shown in Figure 1.11, the components of the vector must be modified accordingly. In many applications it is more convenient to express the components of a vector in a coordinate system having axes that are not horizontal and vertical but are still perpendicular to each other. Suppose a vector **B** makes an angle $\theta'$ with the $x'$ axis defined in Figure 1.13. The components of **B** along these axes are given by $B_{x'} = B \cos \theta'$ and $B_{y'} = B \sin \theta'$, as in Equation 1.10. The magnitude and direction of **B** are obtained from expressions equivalent to Equations 1.11 and 1.12. Thus, we can express the components of a vector in *any* coordinate system that is convenient for a particular situation.

Vector quantities are often expressed in terms of unit vectors. **A unit vector is a dimensionless vector with a magnitude of one and is used to specify a given direction.** Unit vectors have no other physical significance. They are used simply as a bookkeeping convenience in describing a direction in space. We will use the symbols **i**, **j**, and **k** to represent unit vectors pointing in the $x$, $y$, and $z$ directions, respectively. Thus, the unit vectors **i**, **j**, and **k** form a set of mutually perpendicular vectors as shown in Figure 1.14a, where the magnitude of each unit vector equals one; that is, $|\mathbf{i}| = |\mathbf{j}| = |\mathbf{k}| = 1$.

Consider a vector **A** lying in the $xy$ plane, as in Figure 1.14b. The product of the component $A_x$ and the unit vector **i** is the component vector $A_x\mathbf{i}$ parallel to the $x$ axis with magnitude $A_x$. Likewise, $A_y\mathbf{j}$ is a component vector of magnitude $A_y$ parallel to the $y$ axis. When using the unit form of a vector, we are simply multiplying a vector (the unit vector) by a scalar (the component). Thus, the unit-vector notation for the vector **A** is written

$$\mathbf{A} = A_x\mathbf{i} + A_y\mathbf{j} \qquad \text{[1.13]}$$

(a)

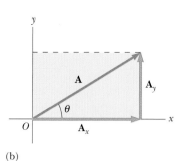

(b)

**Figure 1.11**

(a) A vector **A** lying in the $xy$ plane can be represented by its component vectors $\mathbf{A}_x$ and $\mathbf{A}_y$. (b) The $y$ component vector $\mathbf{A}_y$ can be moved to the right so that it adds to $\mathbf{A}_x$. The vector sum of the component vectors is **A**. These three vectors form a right triangle.

**PITFALL PREVENTION 1.9**

**Tangents on calculators**

 Generally, the inverse tangent function on calculators provides an angle between $-90°$ and $+90°$. As a consequence, if the vector you are studying lies in the second or third quadrant, the angle measured from the positive $x$ axis will be the angle your calculator returns plus $180°$.

| | $y$ | |
|---|---|---|
| $A_x$ negative | | $A_x$ positive |
| $A_y$ positive | | $A_y$ positive |
| $A_x$ negative | | $A_x$ positive |
| $A_y$ negative | | $A_y$ negative |

**Figure 1.12**

The signs of the components of a vector **A** depend on the quadrant in which the vector is located.

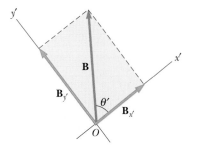

**Figure 1.13**

The components of vector **B** in a coordinate system that is tilted.

(a)

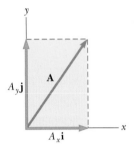

**Figure 1.14**

(a) The unit vectors, **i**, **j**, and **k** are directed along the *x*, *y*, and *z* axes, respectively. (b) A vector **A** lying in the *xy* plane has component vectors $A_x\mathbf{i}$ and $A_y\mathbf{j}$ where $A_x$ and $A_y$ are the components of **A**.

Now suppose we wish to add vector **B** to vector **A**, where **B** has components $B_x$ and $B_y$. The procedure for performing this sum is to simply add the *x* and *y* components separately. The resultant vector **R** = **A** + **B** is therefore

$$\mathbf{R} = (A_x + B_x)\mathbf{i} + (A_y + B_y)\mathbf{j} \qquad [1.14]$$

Thus, the components of the resultant vector are given by

$$R_x = A_x + B_x$$
$$R_y = A_y + B_y \qquad [1.15]$$

The magnitude of **R** and the angle it makes with the *x* axis can then be obtained from its components using the relationships

$$R = \sqrt{R_x^2 + R_y^2} = \sqrt{(A_x + B_x)^2 + (A_y + B_y)^2} \qquad [1.16]$$

$$\tan\theta = \frac{R_y}{R_x} = \frac{A_y + B_y}{A_x + B_x} \qquad [1.17]$$

The procedure just described for adding two vectors **A** and **B** using the component method can be checked using a diagram like Figure 1.15.

The extension of these methods to three-dimensional vectors is straightforward. If **A** and **B** both have *x*, *y*, and *z* components, we express them in the form

$$\mathbf{A} = A_x\mathbf{i} + A_y\mathbf{j} + A_z\mathbf{k}$$
$$\mathbf{B} = B_x\mathbf{i} + B_y\mathbf{j} + B_z\mathbf{k}$$

The sum of **A** and **B** is

$$\mathbf{R} = \mathbf{A} + \mathbf{B} = (A_x + B_x)\mathbf{i} + (A_y + B_y)\mathbf{j} + (A_z + B_z)\mathbf{k} \qquad [1.18]$$

The same procedure can be used to add three or more vectors.

If a vector **R** has *x*, *y*, and *z* components, the magnitude of the vector is

$$R = \sqrt{R_x^2 + R_y^2 + R_z^2}$$

The angle $\theta_x$ that **R** makes with the *x* axis is given by

$$\cos\theta_x = \frac{R_x}{R}$$

with similar expressions for the angles with respect to the *y* and *z* axes.

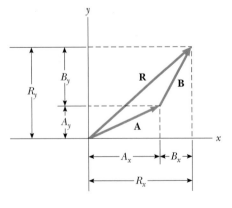

**Figure 1.15**

A geometric construction showing the relation between the components of the resultant **R** of two vectors and the individual components.

### THINKING PHYSICS 1.2

You may have asked someone directions to a destination in a city and been told something like, "walk 3 blocks east and then 5 blocks south." If so, are you experienced with vector components?

**Reasoning** Yes, you are! Although you may not have thought of vector component language when you heard these directions, this is exactly what the directions represent. The perpendicular streets of the city reflect an *xy* coordinate system—we can assign the *x* axis to the east-west streets, and the *y* axis to the north-south streets. Thus, the comment of the person giving you directions can be translated as, "Undergo a displacement vector that has an *x* component of $+3$ blocks and a *y* component of $-5$ blocks." You would arrive at the same destination by undergoing the *y* component first, followed by the *x* component, demonstrating the commutative law of addition.

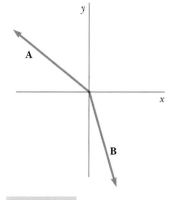

**Figure 1.16**

(Quick Quiz 1.3)

### Quick Quiz 1.3

Figure 1.16 shows two vectors lying in the *xy* plane. Determine the signs of (a) the *x* components of **A** and **B**, (b) the *y* components of **A** and **B**, and (c) the *x* and *y* components of **A** + **B**.

### Quick Quiz 1.4

If one component of a vector is not zero, can the magnitude of the vector be zero? Explain.

### Quick Quiz 1.5

If **A** + **B** = 0, what can you say about the components of the two vectors?

### Quick Quiz 1.6

Under what circumstances can the component of a vector be equal to the magnitude of the vector?

### Example 1.7 The Sum of Two Vectors

Find the sum of two vectors **A** and **B** lying in the *xy* plane and given by

$$\mathbf{A} = 2.00\mathbf{i} + 3.00\mathbf{j} \quad \text{and} \quad \mathbf{B} = 5.00\mathbf{i} - 4.00\mathbf{j}$$

**Solution** It might be helpful for you to draw a diagram of the vectors to clarify what they look like on the *xy* plane. Using the rule given by Equation 1.14, we solve this problem mathematically as follows. Note that $A_x = 2.00$, $A_y = 3.00$, $B_x = 5.00$, and $B_y = -4.00$. Therefore, the resultant vector **R** is

$$\mathbf{R} = \mathbf{A} + \mathbf{B} = (2.00 + 5.00)\mathbf{i} + (3.00 - 4.00)\mathbf{j}$$
$$= 7.00\mathbf{i} - 1.00\mathbf{j}$$

or

$$R_x = 7.00, \ R_y = -1.00$$

The magnitude of **R** is

$$R = \sqrt{R_x^2 + R_y^2} = \sqrt{(7.00)^2 + (-1.00)^2} = \sqrt{50.0} = 7.07$$

**EXERCISE** Find the angle $\theta$ that the resultant vector **R** makes with the positive *x* axis.

*Answer* $352°$

## Example 1.8   The Resultant Displacement

A particle undergoes three consecutive displacements: $\Delta\mathbf{r}_1 =$ (1.50**i** + 3.00**j** − 1.20**k**) cm, $\Delta\mathbf{r}_2 =$ (2.30**i** − 1.40**j** − 3.60**k**) cm, and $\Delta\mathbf{r}_3 =$ (−1.30**i** + 1.50**j**) cm. Find the components of the resultant displacement and its magnitude.

**Solution**   We use Equation 1.18 for three vectors:

$\mathbf{R} = \Delta\mathbf{r}_1 + \Delta\mathbf{r}_2 + \Delta\mathbf{r}_3 = (1.50 + 2.30 - 1.30)\mathbf{i}$ cm

$+\ (3.00 - 1.40 + 1.50)\mathbf{j}$ cm $+\ (-1.20 - 3.60 + 0)\mathbf{k}$ cm

$=\ (2.50\mathbf{i} + 3.10\mathbf{j} - 4.80\mathbf{k})$ cm

That is, the resultant displacement has components $R_x =$ 2.50 cm, $R_y =$ 3.10 cm, and $R_z =$ − 4.80 cm. Its magnitude is

$$R = \sqrt{R_x^2 + R_y^2 + R_z^2}$$
$$= \sqrt{(2.50 \text{ cm})^2 + (3.10 \text{ cm})^2 + (-4.80 \text{ cm})^2} = 6.24 \text{ cm}$$

## Example 1.9   Taking a Hike

A hiker begins a two-day trip by first walking 25.0 km due southeast from her car. She stops and sets up her tent for the night. On the second day she walks 40.0 km in a direction 60.0° north of east, at which point she discovers a forest ranger's tower.

   (a) Determine the components of the hiker's displacements in the first and second days.

**Solution**   If we denote the displacement vectors on the first and second days by **A** and **B**, respectively, and use the car as the origin of coordinates, we obtain the vectors shown in the diagram in Figure 1.17. Notice that the resultant vector **R** can be drawn in the diagram to provide you with an approximation of the final result of the two hikes.

   Displacement **A** has a magnitude of 25.0 km and is 45.0° southeast. Its components are

$A_x = A\cos(-45.0°) = (25.0 \text{ km})(0.707) = 17.7 \text{ km}$

$A_y = A\sin(-45.0°) = (25.0 \text{ km})(-0.707) = -17.7 \text{ km}$

The positive value of $A_x$ indicates that the $x$ coordinate increased in this displacement. The negative value of $A_y$ indicates that the $y$ coordinate decreased in this displacement. Notice in the diagram of Figure 1.17 that vector **A** lies in the fourth quadrant, consistent with the signs of the components we calculated.

   The second displacement **B** has a magnitude of 40.0 km and is 60.0° north of east. Its components are

$B_x = B\cos 60.0° = (40.0 \text{ km})(0.500) = 20.0 \text{ km}$

$B_y = B\sin 60.0° = (40.0 \text{ km})(0.866) = 34.6 \text{ km}$

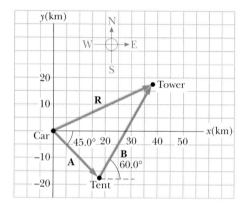

**Figure 1.17**

(Example 1.9) The total displacement of the hiker is the vector **R** = **A** + **B**.

   (b) Determine the components of the hiker's total displacement for the trip.

**Solution**   The resultant displacement vector for the trip, **R** = **A** + **B**, has components given by

$R_x = A_x + B_x = 17.7 \text{ km} + 20.0 \text{ km} = 37.7 \text{ km}$

$R_y = A_y + B_y = -17.7 \text{ km} + 34.6 \text{ km} = 16.9 \text{ km}$

In unit-vector form, we can write the total displacement as

$$\mathbf{R} = (37.7\mathbf{i} + 16.9\mathbf{j}) \text{ km}$$

**EXERCISE**   Determine the magnitude and direction of the total displacement.

*Answer*   41.3 km, 24.1° north of east from the car.

## 1.11 • MODELING, ALTERNATIVE REPRESENTATIONS, AND PROBLEM-SOLVING STRATEGY

Most courses in general physics require the student to learn the skills of problem solving, and examinations usually include problems that test such skills. This section describes some useful ideas that will enable you to enhance your understanding of physical concepts, increase your accuracy in solving problems, eliminate initial panic or lack of direction in approaching a problem, and organize your work.

One of the primary problem-solving methods in physics is to form an appropriate **model** of the problem. **A model is a simplified substitute for the real problem that allows us to solve the problem in a relatively simple way.** As long as the predictions of the model agree to our satisfaction with the actual behavior of the real system, the model is valid. If the predictions do not agree, then the model must be refined or replaced with another model. The power of modeling is in its ability to reduce a wide variety of very complex problems to a limited number of classes of problems that can be approached in similar ways.

In science, a model is very different from, for example, an architect's scale model of a proposed building, which appears as a smaller version of what it represents. A scientific model is a theoretical construct and may have no visual similarity to the physical problem. A simple application of modeling is presented in Example 1.10, and we shall encounter many more examples of models as the text progresses.

Models are needed because the actual operation of the Universe is extremely complicated. Suppose, for example, we are asked to solve a problem about the Earth's motion around the Sun. The Earth is very complicated, with many processes occurring simultaneously. These include weather processes, seismic activity, and ocean movements, as well as the multitude of processes involving human activity. Trying to maintain knowledge and understanding of all these processes is an impossible task.

The modeling approach recognizes that none of these processes affects the motion of the Earth around the Sun to a measurable degree. Thus, all of these details are ignored. In addition, as we shall find in a later chapter, the size of the Earth does not affect the gravitational force between the Earth and the Sun—only the masses of the Earth and Sun and the distance between them determine this force. In a simplified model, the Earth is imagined to be a particle, an object with mass but zero size. This replacement of an extended object by a particle is called the **particle model,** which is used extensively in physics. By analyzing the motion of a particle with the mass of the Earth in orbit around the Sun, we find that the predictions of the motion of the particle are in excellent agreement with the actual motion of the Earth.

The two primary conditions for using the particle model are as follows:

- The size of the actual object is of no consequence in the analysis of its motion
- Any internal processes occurring in the object are of no consequence in the analysis of its motion

Both of these conditions are in action in modeling the Earth as a particle—its radius is not a factor in determining its motion, and internal processes such as thunderstorms, earthquakes, and manufacturing processes can be ignored.

Four categories of models used in this book will help us to understand and solve physics problems. The first category is the **geometric model.** In this model,

- *Models*

- *Geometric model*

we form a geometric construction that represents the real situation. We then set aside the real problem and perform an analysis of the geometric construction. Consider a popular problem in elementary trigonometry, as in the following example.

## Example 1.10   Finding the Height of a Tree

You wish to find the height of a tree, but cannot measure it directly. You stand 50.0 m from the tree, and determine that a line of sight from the ground to the top of the tree makes an angle of 25.0° with the ground. How tall is the tree?

**Solution**   Figure 1.18 shows the tree and a right triangle corresponding to the information in the problem superimposed over it. (We assume that the tree is exactly perpendicular to a perfectly flat ground.) In the triangle, we know the length of the horizontal leg and the angle between the hypotenuse and the horizontal leg. We can find the height of the tree by calculating the length of the vertical leg. We do this with the tangent function:

$$\tan \theta = \frac{opposite\ side}{adjacent\ side} = \frac{h}{50.0\ \text{m}}$$

$$h = (50.0\ \text{m})\tan \theta = (50.0\ \text{m})\tan 25.0° = \boxed{23.3\ \text{m}}$$

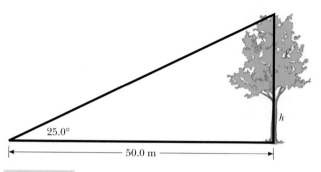

**Figure 1.18**

(Example 1.10) The height of a tree can be found by measuring the distance from the tree and the angle of sight to the top above the ground. This problem is a simple example of geometrically *modeling* the actual problem.

You may have performed a problem very similar to Example 1.10, but never thought about the notion of modeling. From the modeling approach, however, once we draw the triangle in Figure 1.18, the triangle is a geometric model of the real problem—it is a *substitute*. Until we reach the end of the problem, we no longer imagine the problem to be about a tree, but to be about a triangle. We use trigonometry to find the vertical leg of the triangle, leading to a value of 23.3 m. Because this leg *represents* the height of the tree, we can now return to the original problem and claim that the height of the tree is 23.3 m.

Other examples of geometric models include modeling the Earth as a perfect sphere, a pizza as a perfect disk, a meter stick as a long rod with no thickness, and an electric wire as a long straight cylinder.

Example 1.10 is an example of the modeling approach in which the modeling is so subtle that you may not even notice it. In fact, you should be able to solve this problem with no knowledge of modeling. More complicated physics problems that you will encounter, however, may appear to be very difficult using a "traditional" approach to problem solving, but can be simplified profoundly with a modeling approach.

- *Simplification model*

The particle model is an example of the second category of models, which we will call **simplification models.** In a simplification model, details that are not significant in determining the outcome of the problem are ignored. When we study rotation in Chapter 10, objects will be modeled as *rigid bodies*. All of the molecules in a rigid body maintain their exact positions with respect to one another. This is a simplification model, and we adopt it because a spinning rock is much easier to analyze than a spinning block of gelatin, which is *not* a rigid body. Other simplification models will assume that quantities such as friction forces either remain constant or are proportional to some power of the object's speed. If the predictions of the model agree with experimental measurements on the real situation, the model is valid. If there are discrepancies between the predictions of the model and the experimental results, we may have to modify the model and analyze the problem again.

The third category is that of **analysis models,** which are problems we have solved before. An important technique in problem solving is to cast a new problem into a form similar to one we have already solved. We will see our first analysis models in Chapter 2.

The fourth category of models is **structural models.** These are generally used to understand the behavior of a system that is far different in scale from our macroscopic world—either much smaller or much larger—so that we cannot interact with it directly. As an example, the notion of a hydrogen atom as an electron in a circular orbit around a proton is a structural model of the atom. We will discuss this model and structural models in general in Chapter 11.

Intimately related to the notion of modeling is that of forming alternative **representations** of the problem. **A representation is a method of viewing or presenting the information related to the problem.** Scientists must be able to communicate complex ideas to individuals without scientific backgrounds. The best representation to use in conveying the information successfully will vary from one individual to the next. Some will be convinced by a well-drawn graph, and others will require a picture. Physicists are often persuaded to agree with a point of view by examining an equation, but nonphysicists may not be convinced by this mathematical representation of the information.

• *Representations*

A word problem, such as those at the ends of the chapters in this book, is one representation of a problem. In the "real world" that you will enter after graduation, the initial representation of a problem may be just an existing situation, such as the effects of global warming, or a patient in danger of dying. You may have to identify the important data and information, and then cast the situation into an equivalent word problem!

As a familiar example of alternative representations, consider how you might represent a song. As you listen to a song, you form a *mental representation* of the song. This representation includes any emotional responses you have to the song, any memories the song evokes, any analysis of the chord structure you might be performing, and so on. The song also has an *audio representation*. This is related to the actual physical variations in pressure in the air as the song is played. Your ear–brain combination is sensitive to this representation and converts the audio representation into a mental representation. Another representation is the *digital representation,* referring to the pattern of pits in the compact disc for the song. This representation requires a compact disc player to interpret the representation and convert it to an audio representation. Yet another representation of the song is the sheet music for the song—the *written representation*. This representation can be interpreted only by a trained musician and is subject to variation among different musicians. By using the voice or a musical instrument, the written representation can be converted to an audio representation.

All four representations convey the information about the song, but in very different ways. This is what we try to do in solving physics problems—think about the information in the problem in several different ways to help us understand and solve it. Several types of representations can be of help in this endeavor:

• **Mental representation**—From the description of the problem, imagine a scene that describes what is happening in the word problem, and let time

A pictorial representation of a pop foul being hit by a baseball player.

progress so that you understand the situation and can predict what changes will occur in the situation. This step is critical in approaching *every* problem.

- **Pictorial representation**—Drawing a picture of the situation described in the word problem can be of great assistance in understanding the problem. In Example 1.10, the pictorial representation in Figure 1.18 allows us to identify the triangle as a geometric model of the problem. In architecture, a blueprint is a pictorial representation of a proposed building.

    Generally, a pictorial representation describes *what you would see* if you were observing the problem. For example, Figure 1.19 shows a pictorial representation of a baseball player hitting a short pop foul. Any coordinate axes included in your pictorial representation will be in two dimensions: $x$ and $y$ axes.

- **Simplified pictorial representation**—It is often useful to redraw the pictorial representation without complicating details by applying a simplification model. This is similar to the discussion of the particle model described earlier. In a pictorial representation of the Earth in orbit around the Sun, you might draw the Earth and the Sun as spheres, with possibly some attempt to draw continents to identify which sphere is the Earth. In the simplified pictorial representation, the Earth and the Sun would be drawn simply as dots, representing particles. Figure 1.20 shows a simplified pictorial representation corresponding to the pictorial representation of the baseball trajectory in Figure 1.19. The notations $v_x$ and $v_y$ refer to the components of the velocity vector for the baseball. We shall use such simplified pictorial representations throughout the book.

- **Graphical representation**—In some problems, drawing a graph that describes the situation can be very helpful. In mechanics, for example, position–time graphs can be of great assistance. Similarly, in thermodynamics, pressure–volume graphs are essential to understanding. Figure 1.21 shows a graphical representation of the position as a function of time of a block on the end of a vertical spring as it oscillates up and down. Such a graph is helpful for understanding simple harmonic motion, which we study in Chapter 12.

    This is different from a pictorial representation, which is also a two-dimensional display of information, but whose axes, if any, represent *length* coordinates. In a graphical representation, the axes may represent any two related variables. For example, a graphical representation may have axes that are tem-

A simplified pictorial representation for the situation shown in Figure 1.19.

A graphical representation of the position as a function of time of a block hanging from a spring and oscillating.

perature and time. Thus, in comparison to a pictorial representation, a graphical representation is generally *not* something you would see when observing the problem.

- **Tabular representation**—It is sometimes helpful to organize the information in tabular form to help make it clearer. For example, some students find that making tables of known quantities and unknown quantities is helpful. The periodic table is an extremely useful tabular representation of information in chemistry and physics.
- **Mathematical representation**—This is often the ultimate goal in solving a problem. You want to move from the information contained in the word problem, through various representations of the problem that allow you to understand what is happening, to one or more equations that represent the situation in the problem and can be solved mathematically for the desired result.

As we study physics in this textbook, we shall use modeling and alternative representations in numerical examples to help you learn these powerful techniques.

An important way to become a skilled problem solver is to adopt a problem-solving strategy. Many chapters in this text include a section labeled "Problem-Solving Strategies" that should help you through the rough spots. The following steps are commonly used to develop a general problem-solving strategy for numerical problems:

1. Read the problem carefully at least twice. Be sure you understand the nature of the problem before proceeding further. Imagine a movie, running in your mind, of what happens in the problem. This step allows you to set up the mental representation of the problem.
2. Draw a suitable diagram with appropriate labels and coordinate axes if needed. This provides the pictorial representation. Simplify the problem by drawing a simplified pictorial representation, which eliminates the unnecessary details. Use a simplification model to remove additional unnecessary details if the conditions for the model are satisfied. If appropriate, generate a graphical representation. If you find it helpful, generate a tabular representation. If it helps you solve the problem, identify a useful geometric model from the diagrams.
3. If the problem can be cast into a form similar to one you have previously analyzed, identify an analysis model for the problem. From the model, identify the basic physical principle or principles that are involved, listing the knowns and unknowns. Select a basic relationship or derive an equation that can be used to find the unknown, and symbolically solve the equation for the unknown. This step allows you to establish a mathematical representation of the problem.
4. Substitute the given values in the problem, along with the appropriate units, into the equation(s). Obtain a numerical value with units for the unknown.
5. Think about your final result. You can have confidence in your result if questions such as the following can be properly answered: Do the units match? Is the answer of a reasonable order of magnitude? Is the positive or negative sign proper and meaningful? Is the result consistent with your initial mental representation of the problem?

Although the problem-solving strategy we just described may look quite complicated, it may not be necessary to perform all of these steps for a given problem. Examples in the early chapters of this text focus on how to apply these steps explicitly to help you become an effective problem solver. Once you have developed an organized system for examining problems and extracting relevant information, you will become a more confident problem solver in physics as well as in other areas.

## SUMMARY

Mechanical quantities can be expressed in terms of three fundamental quantities—**length, mass,** and **time**—which in the SI system have the units **meters** (m), **kilograms** (kg), and **seconds** (s), respectively. It is often useful to use the method of **dimensional analysis** to check equations and to assist in deriving expressions.

The **density** of a substance is defined as its mass per unit volume.

**Vectors** are quantities that have both magnitude and direction and obey the vector law of addition. **Scalars** are quantities that add algebraically.

Two vectors **A** and **B** can be added using either the triangle method or the parallelogram rule. In the triangle method (see Fig. 1.6), the vector **R** = **A** + **B** runs from the tail of **A** to the tip of **B**. In the parallelogram method (see Fig. 1.7a), **R** is the diagonal of a parallelogram having **A** and **B** as its sides.

The $x$ component $A_x$ of the vector **A** is equal to its projection along the $x$ axis of a coordinate system, where $A_x = A \cos \theta$, and $\theta$ is the angle **A** makes with the $x$ axis. Likewise, the $y$ component $A_y$ of **A** is its projection along the $y$ axis, where $A_y = A \sin \theta$.

If a vector **A** has an $x$ component equal to $A_x$ and a $y$ component equal to $A_y$, the vector can be expressed in unit-vector form as **A** = $A_x$**i** + $A_y$**j**. In this notation, **i** is a unit vector in the positive $x$ direction and **j** is a unit vector in the positive $y$ direction. Because **i** and **j** are unit vectors, $|\mathbf{i}| = |\mathbf{j}| = 1$. In three dimensions, a vector can be expressed as **A** = $A_x$**i** + $A_y$**j** + $A_z$**k**, where **k** is a unit vector in the $z$ direction.

The resultant of two or more vectors can be found by resolving all vectors into their $x$, $y$, and $z$ components and adding their components:

$$\mathbf{R} = \mathbf{A} + \mathbf{B} = (A_x + B_x)\mathbf{i} + (A_y + B_y)\mathbf{j} + (A_z + B_z)\mathbf{k} \quad [1.18]$$

Problem-solving skills and physical understanding can be improved by **modeling** the problem and by constructing different **representations** of the problem. Models helpful in solving problems include **geometric, simplification,** and **analysis models.** Scientists use **structural models** to understand systems larger or smaller in scale than those with which we normally have direct experience. Helpful representations include the **mental, pictorial, simplified pictorial, graphical, tabular,** and **mathematical representations.**

## QUESTIONS

1. Suppose that the three fundamental standards of the metric system were length, *density,* and time rather than length, *mass,* and time. The standard of density in this system is to be defined as that of water. What considerations about water would need to be addressed to make sure that the standard unit of density is as accurate as possible?

2. What types of natural phenomena could serve as alternative time standards?

3. The height of a horse is sometimes given in units of "hands." Why is this a poor standard of length?

4. Express the following quantities using the prefixes given in Table 1.4: (a) $3 \times 10^{-4}$ m, (b) $5 \times 10^{-5}$ s, (c) $72 \times 10^2$ g.

5. As one moves upward in atomic number in the periodic table, the atomic masses of the elements increase. Because the atoms are becoming more and more massive, why

doesn't the *density* of elemental materials increase in the same way?

6. Suppose that two quantities $A$ and $B$ have different dimensions. Determine which of the following arithmetic operations *could* be physically meaningful: (a) $A + B$, (b) $A/B$, (c) $B - A$, (d) $AB$.

7. What accuracy is implied in an order-of-magnitude calculation?

8. Apply an order-of-magnitude calculation to an everyday situation you might encounter. For example, how far do you walk or drive each day?

9. Estimate your age in seconds.

10. Which of the following are vectors and which are not: force, temperature, the volume of water in a can, the ratings of a TV show, the height of a building, the velocity of a sports car, the age of the Universe?

**11.** A vector **A** lies in the *xy* plane. For what orientations of **A** will both of its components be negative? For what orientations will its components have opposite signs?

**12.** A book is moved once around the perimeter of a tabletop with the dimensions 1.0 m × 2.0 m. If the book ends up at its initial position, what is its displacement? What is the distance traveled?

**13.** While traveling along a straight interstate highway you notice that the mile marker reads 260. You travel until you reach the 150-mile marker and then retrace your path to the 175-mile marker. What is the magnitude of your resultant displacement from the 260-mile marker?

**14.** If the component of vector **A** along the direction of vector **B** is zero, what can you conclude about the two vectors?

**15.** If **A** = **B**, what can you conclude about the components of **A** and **B**?

**16.** Is it possible to add a vector quantity to a scalar quantity? Explain.

**17.** A roller coaster travels 135 ft at an angle of 40.0° above the horizontal. How far does it move horizontally and vertically?

**18.** The resolution of vectors into components is equivalent to replacing the original vector with the sum of two vectors, whose sum is the same as the original vector. An infinite number of pairs of vectors will satisfy this condition; we choose a pair with one vector parallel to the *x* axis and the second parallel to the *y* axis. What difficulties would be introduced by defining components relative to axes that are not perpendicular—for example, the *x* axis and a *y* axis oriented at 45° to the *x* axis?

**19.** Identify the type of model (geometric, simplification, structural) represented by each of the following: (a) In its orbit around the Sun, the Earth is treated as a particle. (b) The distance the Earth travels around the Sun is calculated as $2\pi$ times the Earth–Sun distance. (c) The atomic structure of a solid material is imagined to consist of small masses (atoms) connected to neighboring masses by springs. (d) As an object is dropped, air resistance is ignored. (e) The volume of water in a bottle is estimated by calculating the volume of a cylinder. (f) A baseball is hit by a bat. In studying the motion of the baseball, any distortion of the ball while it is in contact with the bat is not considered. (g) In the early 20th century, the atom was proposed to consist of electrons in orbit around a very small but massive nucleus.

# PROBLEMS

1, 2, 3 = straightforward, intermediate, challenging     □ = full solution available in the *Student Solutions Manual and Study Guide*

**web** = solution posted at **http://www.harcourtcollege.com/physics/**      = computer useful in solving problem

 = Interactive Physics     ▪ = paired numerical/symbolic problems      = life science application

*Note:* Consult the endpapers, appendices, and tables in the text whenever necessary in solving problems. For this chapter Appendix B.3 may be particularly useful. Answers to odd-numbered problems are in the back of the book.

## Section 1.2 Density and Atomic Mass

**1.** The standard kilogram is a platinum–iridium cylinder 39.0 mm in height and 39.0 mm in diameter. What is the density of the material?

**2.** The mass of the planet Saturn is $5.64 \times 10^{26}$ kg, and its radius is $6.00 \times 10^{7}$ m. Calculate its average density.

**3.** A major motor company displays a die-cast model of its first automobile, made from 9.35 kg of iron. To celebrate its 100th year in business, a worker will recast the model in gold from the original dies. What mass of gold is needed to make the new model?

**4.** What mass of a material with density $\rho$ is required to make a hollow spherical shell having inner radius $r_1$ and outer radius $r_2$?

**5.** On your wedding day your lover gives you a gold ring of mass 3.80 g. Fifty years later its mass is 3.35 g. On the average, how many atoms were abraded from the ring during each second of your marriage? The atomic mass of gold is 197 u.

**6.** A structural I-beam is made of steel. A view of its cross-section and its dimensions are shown in Figure P1.6. The

**Figure P1.6**

density of steel is $7.56 \times 10^3$ kg/m$^3$. (a) What is the mass of a section 1.50 m long? (b) Assume that the atoms are predominantly iron, with atomic mass 55.9 u. How many atoms are in this section?

7. For many electronic applications, such as in computer chips, it is desirable to make components as small as possible to keep the temperature of the components low and increase the speed of the device. Thin metallic coatings (films) can be used instead of wires to make electrical connections. Gold is especially useful because it does not oxidize readily. Its atomic mass is 197 u. A gold film can be no thinner than the size of a gold atom. Calculate the minimum coating thickness, assuming that a gold atom occupies a cubical volume in the film that is equal to the volume it occupies in a large piece of metal. This geometric model yields a result of the correct order of magnitude.

## Section 1.3 Dimensional Analysis

8. Which of the following equations are dimensionally correct? (a) $v_f = v_i + ax$; (b) $y = (2 \text{ m}) \cos(kx)$, where $k = 2 \text{ m}^{-1}$.

9. The position of a particle when moving under uniform acceleration is some function of the elapsed time and the acceleration. Suppose we write this position as $x = ka^m t^n$, where $k$ is a dimensionless constant. Show by dimensional analysis that this expression is satisfied if $m = 1$ and $n = 2$. Can this analysis give the value of $k$?

10. Figure P1.10 shows a *frustrum of a cone.* Of the following mensuration (geometrical) equations, which describes (a) the total circumference of the flat circular faces, (b) the volume, and (c) the area of the curved surface? (i) $\pi(r_1 + r_2)[h^2 + (r_1 - r_2)^2]^{1/2}$, (ii) $2\pi(r_1 + r_2)$, (iii) $\pi h(r_1^2 + r_1 r_2 + r_2^2)$

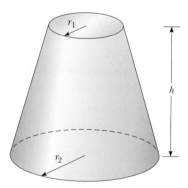

**Figure P1.10**

11. The consumption of natural gas by a company satisfies the empirical equation $V = 1.50 \, t + 0.008 \, 00 \, t^2$, where $V$ is the volume in millions of cubic feet and $t$ is the time in months. Express this equation in units of cubic feet and seconds. Put the proper units on the coefficients. Assume a month is 30.0 days.

## Section 1.4 Conversion of Units

12. Suppose your hair grows at the rate of 1/32 in. per day. Find the rate at which it grows in nanometers per second. Because the distance between atoms in a molecule is on the order of 0.1 nm, your answer suggests how rapidly layers of atoms are assembled in this protein synthesis.

13. Assume that it takes 7.00 min to fill a 30.0-gal gasoline tank. (a) Calculate the rate at which the tank is filled in gallons per second. (b) Calculate the rate at which the tank is filled in cubic meters per second. (c) Determine the time, in hours, required to fill a 1-m$^3$ volume at the same rate. (1 U.S. gal = 231 in.$^3$)

14. An ore loader moves 1 200 tons/h from a mine to the surface. Convert this to lb/s, using 1 ton = 2 000 lb.

15. A *section* of land has an area of 1 square mile and contains 640 acres. Determine the number of square meters in 1 acre.

16. At the time of this book's printing, the U. S. national debt is about \$6 trillion. (a) If payments were made at the rate of \$1 000 per second, how many years would it take to pay off the debt, assuming no interest were charged? (b) A dollar bill is about 15.5 cm long. If six trillion dollar bills were laid end to end around the Earth's equator, how many times would they encircle the Earth? Take the radius of the Earth at the equator to be 6 378 km. (*Note:* Before doing any of these calculations, try to guess at the answers. You may be very surprised.)

17. One gallon of paint (volume = $3.78 \times 10^{-3}$ m$^3$) covers an area of 25.0 m$^2$. What is the thickness of the paint on the wall?

18. A hydrogen atom has a diameter of approximately $1.06 \times 10^{-10}$ m, as defined by the diameter of the spherical electron cloud around the nucleus. The hydrogen nucleus has a diameter of approximately $2.40 \times 10^{-15}$ m. (a) For a scale model, represent the diameter of the hydrogen atom by the length of an American football field (100 yards = 300 feet), and determine the diameter of the nucleus in millimeters. (b) The atom is how many times larger in volume than its nucleus?

19. One cubic meter (1.00 m$^3$) of aluminum has a mass of $2.70 \times 10^3$ kg, and 1.00 m$^3$ of iron has a mass of $7.86 \times 10^3$ kg. Find the radius of a solid aluminum sphere that will balance a solid iron sphere of radius 2.00 cm on an equal-arm balance.

20. Let $\rho_{Al}$ represent the density of aluminum and $\rho_{Fe}$ that of iron. Find the radius of a solid aluminum sphere that balances a solid iron sphere of radius $r_{Fe}$ on an equal-arm balance.

### Section 1.5 Order-of-Magnitude Calculations

**21.** Find the order of magnitude of the number of Ping-Pong balls that would fit into a typical-size room (without being crushed). In your solution, state the quantities you measure or estimate and the values you take for them.

**web**

**22.** An automobile tire is rated to last for 50 000 miles. To an order of magnitude, through how many revolutions will it turn? In your solution, state the quantities you measure or estimate and the values you take for them.

**23.** Compute the order of magnitude of the mass of a bathtub half full of water. Compute the order of magnitude of the mass of a bathtub half full of pennies. In your solution, list the quantities you take as data and the value you measure or estimate for each.

**24.** Soft drinks are commonly sold in aluminum cans. To an order of magnitude, how many such containers are thrown away or recycled each year by U.S. consumers? How many tons of aluminum does this represent? In your solution, state the quantities you measure or estimate and the values you take for them.

### Section 1.6 Significant Figures

**25.** Carry out the following arithmetic operations: (a) the sum of the measured values 756, 37.2, 0.83, and 2.5; (b) the product $0.003\ 2 \times 356.3$; (c) the product $5.620 \times \pi$.

**26.** The *tropical year*, the time from vernal equinox to vernal equinox, is the basis for our calendar. It contains 365.242 199 days. Find the number of seconds in a tropical year.

**27.** How many significant figures are in the following numbers: (a) $78.9 \pm 0.2$, (b) $3.788 \times 10^9$, (c) $2.46 \times 10^{-6}$, (d) 0.005 3?

**28.** The radius of a solid sphere is measured to be $(6.50 \pm 0.20)$ cm, and its mass is measured to be $(1.85 \pm 0.02)$ kg. Determine the density of the sphere in kilograms per cubic meter and the uncertainty in the density.

**29.** A sidewalk is to be constructed around a swimming pool that measures $(10.0 \pm 0.1)$ m by $(17.0 \pm 0.1)$ m. If the sidewalk is to measure $(1.00 \pm 0.01)$ m wide by $(9.0 \pm 0.1)$ cm thick, what volume of concrete is needed, and what is the approximate uncertainty of this volume?

### Section 1.7 Coordinate Systems

**30.** The polar coordinates of a point are $r = 5.50$ m and $\theta = 240°$. What are the cartesian coordinates of this point?

**31.** A fly lands on one wall of a room. The lower left-hand corner of the wall is selected as the origin of a two-dimensional cartesian coordinate system. If the fly is located at the point having coordinates $(2.00, 1.00)$ m, (a) how far is it from the lower left-hand corner of the room? (b) what is its location in polar coordinates?

**32.** Two points in the $xy$ plane have cartesian coordinates $(2.00, -4.00)$ m and $(-3.00, 3.00)$ m. Determine (a) the distance between these points and (b) their polar coordinates.

**33.** If the polar coordinates of the point $(x, y)$ are $(r, \theta)$, determine the polar coordinates for the points: (a) $(-x, y)$, (b) $(-2x, -2y)$, and (c) $(3x, -3y)$.

### Section 1.8 Vectors and Scalars

### Section 1.9 Some Properties of Vectors

**34.** Each of the displacement vectors **A** and **B** shown in Figure P1.34 has a magnitude of 3.00 m. Find graphically (a) **A** + **B**, (b) **A** − **B**, (c) **B** − **A**, (d) **A** − 2**B**. Report all angles counterclockwise from the positive $x$ axis.

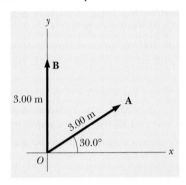

**Figure P1.34** Problems 34 and 47.

**35.** A skater glides along a circular path of radius 5.00 m. If he coasts around one half of the circle, find (a) the magnitude of the displacement vector and (b) how far the person skated. (c) What is the magnitude of the displacement if he skates all the way around the circle?

**web**

**36.** A plane flies from base camp to lake A, a distance of 280 km at a direction of 20.0° north of east. After dropping off supplies, it flies to lake B, which is 190 km at 30.0° west of north from lake A. Graphically determine the distance and direction from lake B to the base camp.

**37.** A roller coaster car moves 200 ft horizontally and then rises 135 ft at an angle of 30.0° above the horizontal. It then travels 135 ft at an angle of 40.0° downward. What is its displacement from its starting point? Use graphical techniques.

### Section 1.10 Components of a Vector and Unit Vectors

**38.** Find the horizontal and vertical components of the 100-m displacement of a superhero who flies from the top of a tall building following the path shown in Fig. P1.38.

**39.** A vector has an $x$ component of $-25.0$ units and a $y$ component of 40.0 units. Find the magnitude and direction of this vector.

**Figure P1.38**

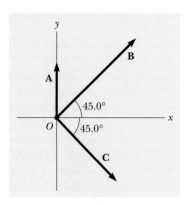

**Figure P1.43**

**40.** Given the vectors $\mathbf{A} = 2.00\mathbf{i} + 6.00\mathbf{j}$ and $\mathbf{B} = 3.00\mathbf{i} - 2.00\mathbf{j}$, (a) draw the vector sum $\mathbf{C} = \mathbf{A} + \mathbf{B}$ and the vector difference $\mathbf{D} = \mathbf{A} - \mathbf{B}$. (b) Calculate $\mathbf{C}$ and $\mathbf{D}$, first in terms of unit vectors and then in terms of polar coordinates, with angles measured with respect to the $+x$ axis.

**41.** A person going for a walk follows the path shown in Figure P1.41. The total trip consists of four straight-line paths. At the end of the walk, what is the person's resultant displacement measured from the starting point?

**Figure P1.41**

**42.** Vector $\mathbf{A}$ has $x$ and $y$ components of $-8.70$ cm and $15.0$ cm, respectively; vector $\mathbf{B}$ has $x$ and $y$ components of $13.2$ cm and $-6.60$ cm, respectively. If $\mathbf{A} - \mathbf{B} + 3\mathbf{C} = 0$, what are the components of $\mathbf{C}$?

**43.** Three displacement vectors of a croquet ball are shown in Figure P1.43, where $|\mathbf{A}| = 20.0$ units, $|\mathbf{B}| = 40.0$ units, and $|\mathbf{C}| = 30.0$ units. Find (a) the resultant in unit-vector notation and (b) the magnitude and direction of the resultant displacement.

**44.** Consider the displacement vectors $\mathbf{A} = (3\mathbf{i} + 3\mathbf{j})$ m, $\mathbf{B} = (\mathbf{i} - 4\mathbf{j})$ m, and $\mathbf{C} = (-2\mathbf{i} + 5\mathbf{j})$ m. Use the component method to determine (a) the magnitude and direction of the vector $\mathbf{D} = \mathbf{A} + \mathbf{B} + \mathbf{C}$, and (b) the magnitude and direction of $\mathbf{E} = -\mathbf{A} - \mathbf{B} + \mathbf{C}$.

**45.** Consider two vectors $\mathbf{A} = 3\mathbf{i} - 2\mathbf{j}$ and $\mathbf{B} = -\mathbf{i} - 4\mathbf{j}$. Calculate (a) $\mathbf{A} + \mathbf{B}$, (b) $\mathbf{A} - \mathbf{B}$, (c) $|\mathbf{A} + \mathbf{B}|$, (d) $|\mathbf{A} - \mathbf{B}|$, (e) the directions of $\mathbf{A} + \mathbf{B}$ and $\mathbf{A} - \mathbf{B}$.

**46.** The helicopter view in Figure P1.46 shows two people pulling on a stubborn mule. Find (a) the single force that is equivalent to the two forces shown, and (b) the force that a third person would have to exert on the mule to make the resultant force equal to zero. The forces are measured in units of newtons.

**Figure P1.46**

**47.** Use the component method to add the vectors $\mathbf{A}$ and $\mathbf{B}$ shown in Figure P1.34. Express the resultant $\mathbf{A} + \mathbf{B}$ in unit-vector notation.

**48.** A billiard ball undergoes two displacements. The first has a magnitude of 150 cm and makes an angle of 120° with the positive *x* axis. The *resultant* displacement has a magnitude of 140 cm and is directed at an angle of 35.0° to the positive *x* axis. Find the magnitude and direction of the second displacement.

**49.** The vector **A** has *x*, *y*, and *z* components of 8.00, 12.0, and − 4.00 units, respectively. (a) Write a vector expression for **A** in unit-vector notation. (b) Obtain a unit-vector expression for a vector **B** one-fourth the length of **A** pointing in the same direction as **A**. (c) Obtain a unit-vector expression for a vector **C** three times the length of **A** pointing in the direction opposite the direction of **A**.

**50.** In an assembly operation illustrated in Figure P1.50, a robot moves an object first straight upward and then also to the east, around an arc forming one quarter of a circle of radius 4.80 cm that lies in an east–west vertical plane. The robot then moves the object upward and to the north, through a quarter of a circle of radius 3.70 cm that lies in a north–south vertical plane. Find (a) the magnitude of the total displacement of the object, and (b) the angle the total displacement makes with the vertical.

**Figure P1.50**

**51.** Vector **B** has *x*, *y*, and *z* components of 4.00, 6.00, and 3.00 units, respectively. Calculate the magnitude of **B** and the angles that **B** makes with the coordinate axes.

**52.** If **A** = (6.00**i** − 8.00**j**) units, **B** = (− 8.00**i** + 3.00**j**) units, and **C** = (26.0**i** + 19.0**j**) units, determine *a* and *b* such that *a***A** + *b***B** + **C** = 0.

**53.** Vectors **A** and **B** have equal magnitudes of 5.00. If the sum of **A** and **B** is the vector 6.00**j**, determine the angle between **A** and **B**.

## Section 1.11 Modeling, Alternative Representations, and Problem-Solving Strategy

**54.** A surveyor measures the distance across a straight river by the following method: Starting directly across from a tree on the opposite bank, she walks 100 m along the riverbank to establish a baseline. Then she sights across to the tree. The angle from her baseline to the tree is 35.0°. How wide is the river?

**55.** A crystalline solid consists of atoms stacked up in a repeating lattice structure. Consider a crystal as shown in Figure P1.55a. The atoms reside at the corners of cubes of side *L* = 0.200 nm. One piece of evidence for the regular arrangement of atoms comes from the flat surfaces along which a crystal separates, or cleaves, when it is broken. Suppose this crystal cleaves along a face diagonal, as shown in Figure P1.55b. Calculate the spacing *d* between two adjacent atomic planes that separate when the crystal cleaves.

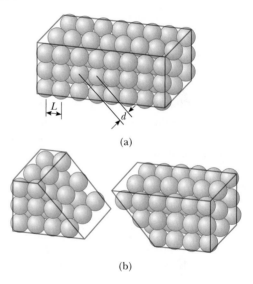

(a)

(b)

**Figure P1.55**

### Additional Problems

**56.** When data are known to three significant figures, we write 6.379 m = 6.38 m and 6.374 m = 6.37 m. When a number ends in 5, we write 6.375 m = 6.38 m. We could equally well write 6.375 m = 6.37 m, "rounding down" instead of "rounding up," because we would change the number 6.375 by equal increments in both cases, but we choose the convention of rounding to an even number. Now consider an order-of-magnitude estimate, in which factors rather than increments are important. We write 500 m ~ $10^3$ m because 500 differs from 100 by a factor of 5 but it differs from 1000 only by a factor of 2. We write 437 m ~ $10^3$ m and 305 m ~ $10^2$ m. What length differs from 100 m and from 1000 m by equal factors, so that we could equally well choose to represent its order of magnitude either as ~ $10^2$ m or as ~ $10^3$ m?

**57.** The basic function of the carburetor of an automobile is to "atomize" the gasoline and mix it with air to promote rapid combustion. As an example, assume that 30.0 cm³ of gasoline is atomized into *N* spherical droplets, each with a radius of $2.00 \times 10^{-5}$ m. What is the total surface area of these *N* spherical droplets?

58. In physics it is important to use mathematical approximations. Demonstrate for yourself that for small angles ($< 20°$)

$$\tan \alpha \approx \sin \alpha \approx \alpha = \pi \alpha'/180°$$

where $\alpha$ is in radians and $\alpha'$ is in degrees. Use a calculator to find the largest angle for which $\tan \alpha$ may be approximated by $\sin \alpha$ if the error is to be less than 10.0%.

59. There are nearly $\pi \times 10^7$ s in one year. Find the percentage error in this approximation, where "percentage error" is defined as

$$\frac{|\text{ assumed value } - \text{ true value}|}{\text{true value}} \times 100\%$$

60. A child loves to watch as you refill a transparent plastic bottle with shampoo. Every horizontal cross-section of the bottle is a circle, but the diameters of the circles have different values, so that the bottle is much wider in some places than in others. You pour in bright green shampoo with a constant volume flow rate of 16.5 cm³/s. At what rate is the fluid level in the bottle rising (a) at a point where the diameter of the bottle is 6.30 cm and (b) at a point where the diameter is 1.35 cm?

61. The distance from the Sun to the nearest star is $4 \times 10^{16}$ m. The Milky Way galaxy is roughly a disk of diameter $\sim 10^{21}$ m and thickness $\sim 10^{19}$ m. Find the order of magnitude of the number of stars in the Milky Way. Assume the distance between the Sun and our nearest neighbor is typical.

62. An air-traffic controller notices two aircraft on his radar screen. The first is at altitude 800 m, horizontal distance 19.2 km, and 25.0° south of west. The second aircraft is at altitude 1 100 m, horizontal distance 17.6 km, and 20.0° south of west. What is the distance between the two aircraft? (Place the *x* axis west, the *y* axis south, and the *z* axis vertical.)

63. Two vectors **A** and **B** have precisely equal magnitudes. For the magnitude of **A** + **B** to be larger than the magnitude of **A** − **B** by the factor *n*, what must be the angle between them?

64. A pirate has buried his treasure on an island with five trees, located at the following points: (30.0 m, − 20.0 m), (60.0 m, 80.0 m), (− 10.0 m, − 10.0 m), (40.0 m, − 30.0 m), and (− 70.0 m, 60.0 m), all measured relative to some origin, as in Figure P1.64. His ship's log instructs you to start at tree A and move toward tree B, but to cover only one-half the distance between A and B. Then move toward tree C, covering one-third the distance between your cur-

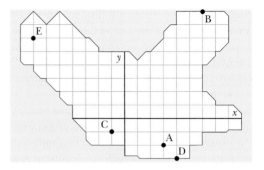

**Figure P1.64**

rent location and C. Next move toward D, covering one-fourth the distance between where you are and D. Finally move toward E, covering one-fifth the distance between you and E, stop, and dig. (a) Assume you have correctly determined the order in which the pirate labeled the trees as A, B, C, D, and E, as shown in the figure. What are the coordinates of the point where his treasure is buried? (b) Now rearrange the order of the trees [e.g., B(30 m, − 20 m), A(60 m, 80 m), E(− 10 m, − 10 m), C(40 m, − 30 m), and D(− 70 m, 60 m)], and repeat the calculation to show that the answer does not depend on the order in which the trees are labeled.

65. Consider a game in which *N* children position themselves at equal distances around the circumference of a circle. At the center of the circle is a rubber tire. Each child holds a rope attached to the tire and, at a signal, pulls on his rope. All children exert forces of the same magnitude *F*. In the case $N = 2$, it is easy to see that the net force on the tire will be zero, because the two oppositely directed force vectors add to zero. Similarly, if $N = 4$, 6, or any even integer, the resultant force on the tire must be zero, because the forces exerted by each pair of oppositely positioned children will cancel. When an odd number of children are around the circle, it is not so obvious that the total force on the central tire will be zero. (a) Calculate the net force on the tire in the case $N = 3$, by adding the components of the three force vectors. Choose the *x* axis to lie along one of the ropes. (b) Determine the net force for the general case where *N* is any integer, odd or even, greater than one. Proceed as follows: Assume that the total force is not zero. Then it must point in some particular direction. Let every child move one position clockwise. Give a reason that the total force must then have a direction turned clockwise by $360°/N$. Argue that the total force must nevertheless be the same as before. Explain that the contradiction proves that the magnitude of the force is zero. This problem illustrates a widely use-

ful technique of proving a result "by symmetry"—by using a bit of the mathematics of *group theory*. The particular situation is actually encountered in physics and chemistry when an array of electric charges (ions) exerts electric forces on an atom at a central position in a molecule or in a crystal.

66. A rectangular parallelepiped has dimensions $a$, $b$, and $c$, as in Figure P1.66. (a) Obtain a vector expression for the face diagonal vector $\mathbf{R}_1$. What is the magnitude of this vector? (b) Obtain a vector expression for the body diagonal vector $\mathbf{R}_2$. Note that $\mathbf{R}_1$, $c\mathbf{k}$, and $\mathbf{R}_2$ form a right triangle, and prove that the magnitude of $\mathbf{R}_2$ is $\sqrt{a^2 + b^2 + c^2}$.

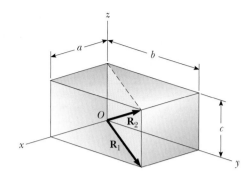

**Figure P1.66**

## ANSWERS TO QUICK QUIZZES

1.1 The largest possible magnitude of the resultant occurs when the two vectors are in the same direction. In this case, the magnitude of the resultant is the sum of the magnitudes of $\mathbf{A}$ and $\mathbf{B}$: $R = A + B = 11$ units. The smallest possible magnitude of the resultant occurs when the two vectors are in opposite directions, and the magnitude is the absolute value of the difference of the magnitudes of $\mathbf{A}$ and $\mathbf{B}$: $R = |A - B| = 5$ units.

1.2 The resultant has magnitude $A + B$ when $\mathbf{A}$ is oriented in the same direction as $\mathbf{B}$. The resultant vector is zero only when the two vectors $\mathbf{A}$ and $\mathbf{B}$ have the same magnitude and are oriented in opposite directions.

1.3 (a) Because $\mathbf{A}$ is oriented toward the left, its $x$ component is negative. Because $\mathbf{B}$ is oriented toward the right, its $x$ component is positive. (b) Because $\mathbf{A}$ is oriented upward, its $y$ component is positive. Because $\mathbf{B}$ is oriented down-

ward, its $y$ component is negative. (c) The resultant vector $\mathbf{R} = \mathbf{A} + \mathbf{B}$ lies in the third quadrant, so both $x$ and $y$ components are negative.

1.4 No. The magnitude of a vector $\mathbf{A}$ is equal to $\sqrt{A_x^2 + A_y^2 + A_z^2}$. Therefore, if any component is nonzero, $A$ cannot be zero.

1.5 The fact that $\mathbf{A} + \mathbf{B} = 0$ tells you that $\mathbf{A} = -\mathbf{B}$. Therefore, the components of the two vectors must have equal magnitudes and opposite signs: $A_x = -B_x$, $A_y = -B_y$, and $A_z = -B_z$.

1.6 The only circumstances under which this can happen is if the vector is oriented along the positive direction for one of the three axes: $x$, $y$, or $z$. Then the magnitude of the vector is the same as the component associated with its direction and the other components are zero.

By permission of John Hart and Field Enterprises, Inc.

# Mission to Mars

I n our opening Context, we shall investigate the physics necessary to send a spacecraft from Earth to Mars. If the two planets were sitting still in space, millions of kilometers apart, this would be a difficult enough proposition, but keep in mind that we are launching the spacecraft from a moving object, the Earth, and are aiming at a moving target, Mars. Despite these apparent difficulties, we can use the principles of physics to plan a successful mission.

Travel in space began in the early 1960s, with the launch of human-occupied spacecraft in both the United States and the Soviet Union. The first human to ride into space was Yuri Gagarin, who made a one-orbit trip in 1961 in the

The roving robot in the Mars Pathfinder mission explores the area of the Martian surface near the landing spot. *(NASA)*

Space shuttles have been operating since the 1980s, delivering satellites to orbit, carrying astronauts on repair missions for the Hubble Space Telescope, and providing appropriate environmental conditions for scientific experimentation. *(NASA)*

Soviet spacecraft Vostok. Competition between the two countries resulted in a "space race," which led to the successful landing of American astronauts on the Moon in 1969.

In the 1970s, the Viking Project landed spacecraft on Mars to analyze the soil for signs of life. These tests were inconclusive.

American efforts in the 1980s focused on the development and implementation of the space shuttle system, a reusable space transportation system.

The United States returned to Mars in the 1990s, with Mars Global Surveyor, designed to perform careful mapping of the Martian surface, and Mars Pathfinder, which landed on Mars and deployed a roving robot to analyze rocks and soil. In 1999, Mars Polar Lander was launched to land near the polar ice cap and search for water. Mars Surveyor 2001 is scheduled for launch in early 2001. Not all trips have been successful—Mars Climate Orbiter was lost in 1999 due to communication errors between the builder of the spacecraft and the mission control team.

Many individuals dream of one day establishing colonies on Mars. This is far in the future—we are still learning much about Mars today and have yet taken only a handful of trips to the planet. Travel to

Mars is still not an everyday occurrence, although we learn more from each mission. In this Context, we address the central question,

*How can we undertake a successful transfer of a spacecraft from Earth to Mars?*

## chapter 2

At the Stanford Linear Accelerator Center, electrons are accelerated over a 2-mile distance to nearly the speed of light.

*(Courtesy of Stanford Linear Accelerator Center and the U.S. Department of Energy)*

# Motion in One Dimension

**D**ynamics is the study of the motion of an object and the relationship of this motion to such physical concepts as force and mass. Before beginning our study of dynamics, however, it is important to be able to *describe* motion using the concepts of space and time without regard to the causes of the motion. This portion of mechanics is called *kinematics*. In this chapter we shall consider motion along a straight line, that is, one-dimensional motion. Chapter 3 extends our discussion to two-dimensional motion.

From everyday experience, we recognize that motion represents continuous change in the position of an object. If you are driving from New York to Los Angeles, your position on the Earth's surface is changing. If you travel at an average speed of 50 miles per hour (mi/h), how long will it take to make this trip? This is the kind of question we can answer with the material in this chapter.

The movement of an object through space (translation) may be accompanied by the rotation or vibration of the object. Such motions can be quite complex. It is often possible to simplify matters, however, by temporarily ignoring rotation and internal motions of the moving object. This is the simplification model that we call the particle model, discussed in Chapter 1. In many situations, an *object* can be treated as a *particle* if the only motion being considered is translation through space. We will use the particle model extensively in this and the next few chapters.

# 2.1 • AVERAGE VELOCITY

 See the *Core Concepts in Physics CD-ROM,* Screen 3.2

We begin our study of kinematics with the notion of average velocity. You may be familiar with a similar notion, average speed, from experiences with driving. If you drive your car 100 miles according to your odometer, and it takes 2.0 hours to do this, your average speed is $(100 \text{ mi})/(2.0 \text{ h}) = 50 \text{ mi/h}$. For a particle moving through a distance $d$ in a time interval $\Delta t$, the average speed $\bar{v}$ is mathematically defined as

$$\bar{v} = \frac{d}{\Delta t} \qquad \text{[2.1]}$$

Speed is not a vector, so there is no direction associated with average speed.

Average velocity may be a little less familiar to you, due to the vector nature of the velocity. Let us start by imagining the motion of a particle, which, through the particle model, can represent the motion of many types of objects. We shall restrict our study at this point to one-dimensional motion along the *x* axis.

The motion of a particle is completely specified if the position of the particle in space is known at all times. Consider a car moving back and forth along the *x* axis, and imagine that we take data on the position of the car every 10 seconds. Figure 2.1a is a *pictorial* representation of this one-dimensional motion that shows the positions of the car at 10-second intervals. The six data points we have recorded are represented by the letters Ⓐ through Ⓕ. Table 2.1 is a *tabular* representation of the motion. It lists the data as entries for position at each time. The black dots in Figure 2.1b show a *graphical* representation of the motion. Such a plot is often called a **position–time graph.** The curved line in Figure 2.1b cannot be unambiguously drawn through our six data points because we have no information about what happened between the data points. The curved line is, however, a *possible* graphical representation of the position of the car at all instants of time during the 50 seconds.

If a particle is moving during a time interval $\Delta t = t_f - t_i$, the displacement of the particle is described as $\Delta \mathbf{x} = \mathbf{x}_f - \mathbf{x}_i = (x_f - x_i)\mathbf{i}$. (Recall that displacement is defined as the change in the position of the particle, which is equal to its final position value minus its initial position value.) Because we are considering only one-dimensional motion in this chapter, we shall drop the vector notation at this point and pick it up again in Chapter 3. The direction of a vector in this chapter will be indicated by means of a positive or negative sign.

The **average velocity** $\bar{v}_x$ of the particle is defined as the ratio of its displacement $\Delta x$ to the time interval $\Delta t$:

$$\bar{v}_x = \frac{\Delta x}{\Delta t} = \frac{x_f - x_i}{t_f - t_i} \qquad \text{[2.2]}$$

where the subscript *x* indicates motion along the *x* axis. From this definition we see that average velocity has the dimensions of length divided by time—m/s in SI units and ft/s in conventional units. The average velocity is *independent* of the path taken between the initial and final points. This independence is a major difference from the average speed discussed at the beginning of this section. The average velocity is independent of path because it is proportional to the displacement $\Delta x$, which depends only on the initial and final coordinates of the particle. Average speed (a scalar) is found by dividing the *distance* traveled by the time interval, whereas average velocity (a vector) is the *displacement* divided by the time interval. Thus, average velocity gives us no details of the motion—it only gives us the result

**TABLE 2.1**

**Positions of the Car at Various Times**

| Position | $t$ (s) | $x$ (m) |
|---|---|---|
| Ⓐ | 0 | 30 |
| Ⓑ | 10 | 52 |
| Ⓒ | 20 | 38 |
| Ⓓ | 30 | 0 |
| Ⓔ | 40 | −37 |
| Ⓕ | 50 | −53 |

**PITFALL PREVENTION 2.1**

**Average speed and average velocity**

 The magnitude of the average velocity is *not* the average speed. Consider a particle moving from the origin to $x = 10$ m and then back to the origin in a time interval of 4.0 s. The magnitude of the average velocity is zero, because the particle ends the time interval at the same position at which it started—the displacement is zero. The average speed, however, is the total distance divided by the time interval: $20 \text{ m}/4.0 \text{ s} = 5.0 \text{ m/s}$.

On this basketball court, players run back and forth for the entire game. The average speed of the players over the duration of the game is nonzero. The average velocity of the players over the duration of the game is approximately zero because they keep returning to the same point over and over again. *(Ken White/Allsport)*

(b)

(c)

**Figure 2.1**

(a) A pictorial representation of the motion of a car. The positions of the car at six instants of time are shown and labeled. (b) A graphical representation, known as a position–time graph, of the motion of the car in part (a). The average velocity $\bar{v}_x$ in the interval $t = 0$ to $t = 10$ s is obtained from the slope of the straight line connecting the points Ⓐ and Ⓑ. (c) A velocity–time graph of the motion of the car in part (a).

of the motion. Finally, note that the average velocity in one dimension can be positive or negative, depending on the sign of the displacement. (The time interval $\Delta t$ is always positive.) If the $x$ coordinate of the particle increases during the time interval (i.e., if $x_f > x_i$), then $\Delta x$ is positive and $\bar{v}_x$ is positive. This corresponds to an average velocity in the positive $x$ direction. On the other hand, if the coordinate decreases over time ($x_f < x_i$), $\Delta x$ is negative; hence, $\bar{v}_x$ is negative. This corresponds to an average velocity in the negative $x$ direction.

The average velocity can also be interpreted geometrically, as seen in the graphical representation in Figure 2.1b. A straight line can be drawn between any two points on the curve. Figure 2.1b shows such a line drawn between points Ⓐ and Ⓑ. Using a geometric model, this line forms the hypotenuse of a right triangle of height $\Delta x$ and base $\Delta t$. The slope of the hypotenuse is the ratio $\Delta x/\Delta t$. Therefore, we see that the **average velocity of the particle during the time interval $t_i$ to $t_f$ is equal to the slope of the straight line joining the initial and final points on the position–time graph.** For example, the average velocity of the car between points Ⓐ and Ⓑ is $\bar{v}_x = (52\text{ m} - 30\text{ m})/(10\text{ s} - 0) = 2.2\text{ m/s}$.

We can also identify a geometric interpretation for the total displacement during the time interval. Figure 2.1c shows the velocity–time graphical representation of the motion in Figures 2.1a and 2.1b. The time interval has been divided into small increments of duration $\Delta t_n$. During each of these increments, if we model the velocity as constant during the short increment, the displacement of the particle is given by $\Delta x_n = v_n \Delta t_n$. Geometrically, the product on the right side of this expression represents the area of a thin rectangle associated with each time increment in Figure 2.1c—the height of the rectangle (from the time axis) is $v_n$ and the width is $\Delta t_n$. The total displacement of the particle will be the sum of the displacements during each of the increments:

$$\Delta x \approx \sum_n \Delta x_n = \sum_n v_n \Delta t_n$$

This is an approximation because we have modeled the velocity as constant in each increment, which is not the case. The term on the right represents the total area of all the thin rectangles. Now let us take the limit of this expression as the time increments shrink to zero, in which case the approximation becomes exact:

$$\Delta x = \lim_{\Delta t_n \to 0} \sum_n \Delta x_n = \lim_{\Delta t_n \to 0} \sum_n v_n \Delta t_n$$

In this limit, the sum of the areas of all of the very thin rectangles becomes equal to the area under the curve. Thus, **the displacement of a particle during the time interval $t_i$ to $t_f$ is equal to the area under the curve between the initial and final points on the velocity–time graph.**

**PITFALL PREVENTION 2.3**
**Modeling with triangles**

Notice that we are modeling with a space-based triangle even though one of the axes in the graph in Figure 2.1b is a time axis. Thus, we say that the base of the triangle is "of length" $\Delta t$, even though this is not a physical length. This is a more dramatic example of geometric modeling than that in Example 1.10, in which both legs of the triangle could be identified with actual physical lengths. The scale factor for a graph allows for a conversion between the length of the triangle leg in time units and an actual physical length.

**PITFALL PREVENTION 2.4**
**Slopes of graphs**

The word *slope* is often used in reference to the graphs of physical data. Regardless of what data are plotted, the word *slope* will represent the ratio of the change in the quantity represented on the vertical axis to the change in the quantity represented on the horizontal axis. Remember that *a slope has units* (unless both axes have the same units). Thus, the units of the slope in Figure 2.1b are m/s, the units of velocity. Furthermore, avoid any temptation to generally define the slope as the tangent of the angle between the line and the horizontal axis. This is *only* true in the limited case in which both axes have the same units and the same scale factor.

## Example 2.1  Calculate the Average Velocity

A particle moving along the $x$ axis is located at $x_i = 12$ m at $t_i = 1$ s and at $x_f = 4$ m at $t_f = 3$ s. Find its displacement and average velocity during this time interval.

**Solution**  First establish the mental representation—imagine the particle moving along the axis. Based on the information in the problem, which way is it moving? You may find it useful to draw a pictorial representation, but for this simple example, we will go straight to the mathematical representation. The displacement is

$$\Delta x = x_f - x_i = 4\text{ m} - 12\text{ m} = \boxed{-8\text{ m}}$$

The average velocity is, according to Equation 2.2,

$$\bar{v}_x = \frac{\Delta x}{\Delta t} = \frac{4\text{ m} - 12\text{ m}}{3\text{ s} - 1\text{ s}} = \boxed{-4\text{ m/s}}$$

Because the displacement is negative for this time interval, we conclude that the particle has moved to the left, toward decreasing values of $x$. Is this consistent with your mental representation? Keep in mind that it may not have *always* been moving to the left. We only have information about its location at two points in time. After $t_i = 1$ s, it could have moved to the right, turned around, and ended up farther to the left than its original position by the time $t_f = 3$ s. To be completely confident that we know the motion of the particle, we would need to have information about its location at *every* instant of time.

**Example 2.2**    A jogger runs in a straight line, with a magnitude of average velocity of 5.00 m/s for 4.00 min, and then with a magnitude of average velocity of 4.00 m/s for 3.00 min. (a) What is the magnitude of the final displacement from her initial position? (b) What is the magnitude of her average velocity during this entire time interval of 7.00 min?

**Solution**    (a) The fact that this problem involves a jogger is not important—we model the jogger as a particle. We have data for two separate portions of the motion, so we use these data to find the displacement for each portion, using Equation 2.2:

$$\bar{v}_x = \frac{\Delta x}{\Delta t} \quad \longrightarrow \quad \Delta x = \bar{v}_x \Delta t$$

$$\Delta x_{\text{portion 1}} = (5.00 \text{ m/s})(4.00 \text{ min})\left(\frac{60 \text{ s}}{1 \text{ min}}\right)$$

$$= 1.20 \times 10^3 \text{ m}$$

$$\Delta x_{\text{portion 2}} = (4.00 \text{ m/s})(3.00 \text{ min})\left(\frac{60 \text{ s}}{1 \text{ min}}\right)$$

$$= 7.20 \times 10^2 \text{ m}$$

We add these two displacement to find the total displacement of $1.92 \times 10^3$ m.

(b) We now have the data we need to find the average velocity for the entire time interval using Equation 2.2:

$$\bar{v}_x = \frac{\Delta x}{\Delta t} = \frac{1.92 \times 10^3 \text{ m}}{7.00 \text{ min}}\left(\frac{1 \text{ min}}{60 \text{ s}}\right) = 4.57 \text{ m/s}$$

Notice, as indicated in Pitfall Prevention 2.2, that the average velocity is *not* calculated as the simple arithmetic mean of the two velocities given in the problem.

## 2.2  •  INSTANTANEOUS VELOCITY

See Screen 3.3

We would like to be able to define the velocity of a particle at a particular instant in time, rather than during a finite interval of time. For example, you may drive your car through a displacement of magnitude 40 miles and it may take exactly 1 hour to do this, from 1:00:00 PM to 2:00:00 PM. Then your average velocity is of magnitude 40 mi/h for the 1-h interval. But how fast were you going at the particular *instant* of time 1:20:00 PM? It is likely that your velocity varied during the trip, due to hills, traffic lights, slow drivers, and the like, so that there was not a single velocity maintained during the entire hour of travel. The velocity of a particle at any instant of time is called the **instantaneous velocity.**

Consider again the motion of the car shown in Figure 2.1a. Figure 2.2a is the graphical representation again, with two blue lines representing average velocities over very different time intervals. The dark blue line represents the average velocity we calculated earlier over the interval from Ⓐ to Ⓑ. The light blue line represents the average velocity over the much longer interval Ⓐ to Ⓕ. How well does either of these represent the instantaneous velocity at point Ⓐ? In Figure 2.1a, the car begins to move to the right, which we identify as a positive velocity. The average velocity from Ⓐ to Ⓕ is *negative* (because the slope of the line from Ⓐ to Ⓕ is negative), so this clearly is not an accurate representation of the instantaneous velocity at Ⓐ. The average velocity from interval Ⓐ to Ⓑ is *positive*, so this velocity at least has the right sign.

In Figure 2.2b, we show the result of drawing the lines representing the average velocity of the car as point Ⓑ is brought closer and closer to point Ⓐ. As this occurs, the slope of the blue line approaches that of the green line, which is the line drawn tangent to the curve at point Ⓐ. As Ⓑ approaches Ⓐ, the time interval that includes point Ⓐ becomes infinitesimally small. Thus, the average velocity over this interval as the interval shrinks to zero can be interpreted as the instantaneous velocity at point Ⓐ. Furthermore, the slope of the line tangent to the curve

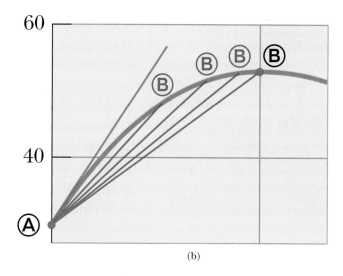

(a)                                    (b)

**Figure 2.2**

(a) Position–time graph for the motion of the car in Figure 2.1. (b) An enlargement of the upper left-hand corner of the graph in part (a) shows how the blue line between positions Ⓐ and Ⓑ approaches the green tangent line as point Ⓑ is moved closer to point Ⓐ.

at Ⓐ is the instantaneous velocity at the time $t_A$. In other words, **the instantaneous velocity $v_x$ equals the limiting value of the ratio $\Delta x/\Delta t$ as $\Delta t$ approaches zero:*

$$v_x \equiv \lim_{\Delta t \to 0} \frac{\Delta x}{\Delta t}$$

In the calculus notation, this limit is called the *derivative* of $x$ with respect to $t$, written $dx/dt$:

$$v_x \equiv \lim_{\Delta t \to 0} \frac{\Delta x}{\Delta t} = \frac{dx}{dt} \qquad [2.3]$$

The instantaneous velocity can be positive, negative, or zero. When the slope of the position–time graph is positive, such as at point Ⓐ in Figure 2.3, $v_x$ is positive. At point Ⓒ, $v_x$ is negative because the slope is negative. Finally, the instantaneous velocity is zero at the peak Ⓑ (the turning point), where the slope is zero. From here on, we shall usually use the word *velocity* to designate instantaneous velocity.

The instantaneous speed of a particle is defined as the magnitude of the instantaneous velocity vector. Hence, by definition, *speed* can never be negative.

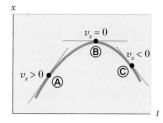

**Figure 2.3**

In the position–time graph shown, the velocity is positive at Ⓐ, where the slope of the tangent line is positive; the velocity is zero at Ⓑ, where the slope of the tangent line is zero; and the velocity is negative at Ⓒ, where the slope of the tangent line is negative.

**Quick Quiz 2.1**

Are members of the highway patrol more interested in your average speed or your instantaneous speed as you drive?

* Note that the displacement $\Delta x$ also approaches zero as $\Delta t$ approaches zero. As $\Delta x$ and $\Delta t$ become smaller and smaller, however, the ratio $\Delta x/\Delta t$ approaches a value equal to the *true* slope of the line tangent to the $x$ versus $t$ curve.

## Quick Quiz 2.2

Under what conditions is the magnitude of the average velocity of a particle moving in one dimension smaller than the average speed over some time interval?

If you are familiar with the calculus, you should recognize that specific rules exist for taking the derivatives of functions. These rules, which are listed in Appendix B.6, enable us to evaluate derivatives quickly.

Suppose $x$ is proportional to some power of $t$, such as

$$x = At^n$$

where $A$ and $n$ are constants. (This is a very common functional form.) The derivative of $x$ with respect to $t$ is

$$\frac{dx}{dt} = nAt^{n-1}$$

For example, if $x = 5t^3$ we see that $dx/dt = 3(5)t^{3-1} = 15t^2$.

### THINKING PHYSICS 2.1

Consider the following motions of an object in one dimension: (a) A ball is thrown directly upward, rises to its highest point, and falls back into the thrower's hand. (b) A race car starts from rest and speeds up to 100 m/s along a straight line. (c) A *Voyager* spacecraft drifts through empty space at constant velocity. Are there any instants of time in the motion of these objects at which the instantaneous velocity at the instant and the average velocity over the entire interval are the same? If so, identify the point(s).

**Reasoning** (a) The average velocity over the entire interval for the thrown ball is zero—the ball returns to the starting point at the end of the time interval. There is one point at which the instantaneous velocity is zero—at the top of the motion. (b) The average velocity for the motion of the race car cannot be evaluated unambiguously with the information given, but its magnitude must be some value between 0 and 100 m/s. Because the magnitude of the instantaneous velocity of the car will have every value between 0 and 100 m/s at some time during the interval, there must be some instant at which the instantaneous velocity is equal to the average velocity over the entire interval. (c) Because the instantaneous velocity of the spacecraft is constant, its instantaneous velocity at *any* time and its average velocity over *any* time interval are the same.

### THINKING PHYSICS 2.2

Does an automobile speedometer measure average or instantaneous speed?

**Reasoning** Based on the fact that changes in the speed of the automobile are reflected in changes in the speedometer reading, we might be tempted to say that a speedometer measures instantaneous speed. This would not be quite correct, however. The reading on a speedometer is related to the rotation of the wheels. Often, a rotating magnet, whose rotation is related to that of the wheels, is used in a speedometer. The reading is based on the *time interval* between rotations of the magnet. Thus, the reading is an average speed over this interval. This interval is generally very short, so that the average speed over the interval is a good approximation to the instantaneous speed at any instant within the interval. As long as the changes in speed are gradual, this approximation is very good. If the automobile brakes to a quick stop, however, the speedometer may not be able to keep up with the rapid changes in speed, and the measured average speed over the time interval may be different from the instantaneous speed at the end of the interval.

## Example 2.3   The Limiting Process

The position of a particle moving along the $x$ axis varies in time according to the expression* $x = 3t^2$, where $x$ is in meters and $t$ is in seconds. Find the velocity in terms of $t$ at any time.

**Solution**   The position–time graphical representation for this motion is shown in Figure 2.4. We can compute the velocity at any time $t$ by using the definition of the instantaneous velocity. If the initial coordinate of the particle at time $t$ is $x_i = 3t^2$, the coordinate at a later time $t + \Delta t$ is

$$x_f = 3(t + \Delta t)^2 = 3[t^2 + 2t\,\Delta t + (\Delta t)^2]$$
$$= 3t^2 + 6t\,\Delta t + 3(\Delta t)^2$$

Therefore, the displacement in the time interval $\Delta t$ is

$$\Delta x = x_f - x_i = (3t^2 + 6t\,\Delta t + 3(\Delta t)^2) - (3t^2)$$
$$= 6t\,\Delta t + 3(\Delta t)^2$$

The average velocity in this time interval is

$$\overline{v}_x = \frac{\Delta x}{\Delta t} = \frac{6t\,\Delta t + 3(\Delta t)^2}{\Delta t} = 6t + 3\,\Delta t$$

To find the instantaneous velocity, we take the limit of this expression as $\Delta t$ approaches zero. In doing so, we see that the term $3\,\Delta t$ goes to zero; therefore,

$$v_x = \lim_{\Delta t \to 0} \frac{\Delta x}{\Delta t} = \boxed{6t}$$

Notice that this expression gives us the velocity at *any* general time $t$. It tells us that $v_x$ is increasing linearly in time. It is then a straightforward matter to find the velocity at some

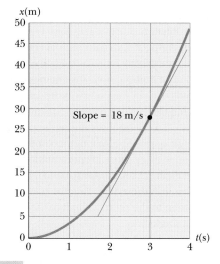

**Figure 2.4**

(Example 2.3) Position–time graph for a particle having an $x$ coordinate that varies in time according to $x = 3t^2$. Note that the instantaneous velocity at $t = 3.0$ s is obtained from the slope of the blue line tangent to the curve at this point.

specific time from the expression $v_x = 6t$ by substituting the value of the time. For example, at $t = 3.0$ s, the velocity is $v_x = 6(3) = 18$ m/s. Again, this can be checked from the slope at $t = 3.0$ s (the blue line in Figure 2.4).

We can also find $v_x$ by taking the first derivative of $x$ with respect to time, as in Equation 2.3. In this example, $x = 3t^2$, and we see that $v_x = dx/dt = 6t$, in agreement with our result from taking the limit explicitly.

---

* Simply to make it easier to read, we write the equation as $x = 3t^2$ rather than as $x = (3.00\ \text{m/s}^2)\,t^{2.00}$. When an equation summarizes measurements, consider its coefficients to have as many significant digits as other data quoted in a problem. Also consider its coefficients to have the units required for dimensional consistency. When we start our clocks at $t = 0$, we usually do not mean to limit precision to a single digit. Consider any zero value in this book to have as many significant figures as you need.

## Example 2.4   Average and Instantaneous Velocity

A particle moves along the *x* axis. Its *x* coordinate varies with time according to the expression $x = -4t + 2t^2$, where *x* is in meters and *t* is in seconds. The position–time graph for this motion is shown in Figure 2.5.

(a) Determine the displacement of the particle in the time intervals *t* = 0 to *t* = 1 s and *t* = 1 s to *t* = 3 s.

**Reasoning**   The Thinking Physics examples in this book have a response called "Reasoning." Some numerical examples, such as this one, will also have a Reasoning section to emphasize the fact that you should make a habit of *thinking* about a problem and its solution before diving into the equations. This problem provides a graphical representation of the motion in Figure 2.5. In your mental representation, note that the particle moves in the negative *x* direction for the first second of motion, stops instantaneously at *t* = 1 s, and then heads back in the positive *x* direction for *t* > 1 s. Remember that this is a one-dimensional problem, so the curve in Figure 2.5 does not represent the path the particle follows through space—be sure not to confuse a graphical representation with a pictorial representation of the motion in space (see Figure 2.1 for a comparison). In your mental representation, you should imagine the particle moving to the left and then to the right, with all of the motion taking place along a single line.

**Solution**   In the first time interval (Ⓐ to Ⓑ) we set $t_i = 0$ and $t_f = 1$ s. Because $x = -4t + 2t^2$, the displacement during the first time interval is

$$\Delta x_{AB} = x_f - x_i = -4(1) + 2(1)^2 - [-4(0) + 2(0)^2]$$
$$= \boxed{-2 \text{ m}}$$

Likewise, in the second time interval (Ⓑ to Ⓓ) we can set $t_i = 1$ s and $t_f = 3$ s. Therefore, the displacement in this interval is

$$\Delta x_{BD} = x_f - x_i = -4(3) + 2(3)^2 - [-4(1) + 2(1)^2]$$
$$= \boxed{8 \text{ m}}$$

These displacements can also be read directly from the position–time graph (see Fig. 2.5).

(b) Calculate the average velocity in the time intervals *t* = 0 to *t* = 1 s and *t* = 1 s to *t* = 3 s.

**Solution**   In the first time interval, $\Delta t = t_f - t_i = 1$ s. Therefore, using Equation 2.2 and the result from (a) gives

$$\bar{v}_x = \frac{\Delta x_{AB}}{\Delta t} = \frac{-2 \text{ m}}{1 \text{ s}} = \boxed{-2 \text{ m/s}}$$

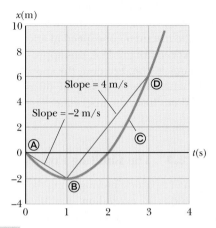

**Figure 2.5**

(Example 2.4) Position–time graph for a particle having an *x* coordinate that varies in time according to $x = -4t + 2t^2$.

Likewise, in the second time interval, $\Delta t = 2$ s; therefore,

$$\bar{v}_x = \frac{\Delta x_{BD}}{\Delta t} = \frac{8 \text{ m}}{2 \text{ s}} = \boxed{4 \text{ m/s}}$$

These values agree with the slopes of the lines joining these points in Figure 2.5.

(c) Find the instantaneous velocity of the particle at *t* = 2.5 s (point Ⓒ).

**Solution**   We can find the instantaneous velocity at any time *t* by taking the first derivative of *x* with respect to *t*:

$$v_x = \frac{dx}{dt} = \frac{d}{dt}(-4t + 2t^2) = -4 + 4t$$

Thus, at *t* = 2.5 s, we find that

$$v_x = -4 + 4(2.5) = \boxed{6 \text{ m/s}}$$

We can also obtain this result by measuring the slope of the position–time graph at *t* = 2.5 s. Do you see any symmetry in the motion? For example, are there points at which the speed is the same? Is the velocity the same at these points?

## 2.3 • ANALYSIS MODELS—THE PARTICLE UNDER CONSTANT VELOCITY

As mentioned in Section 1.11, the third category of models used in this book is that of **analysis models.** Such models help us analyze the situation in a physics problem and guide us toward the solution. **An analysis model is a problem we have solved before—it is a description of either (1) the behavior of some physical entity or (2) the interaction between that entity and the environment.** When you encounter a new problem, you should identify the fundamental details of the problem and attempt to recognize which of the types of problems you have already solved might be used as a model for the new problem. For example, suppose an automobile is moving along a straight freeway at a constant speed. Is it important that it is an automobile? Is it important that it is a freeway? If the answers to both questions are "no," then we model the situation as a particle under constant velocity, which we will discuss in this section.

This method is somewhat similar to the common practice in the legal profession of finding "legal precedents." If a previously resolved case can be found that is very similar legally to the present one, it is offered as a model, and an argument is made in court to link them logically. The finding in the previous case can then be used to sway the finding in the present case. We will do something similar in physics. For a given problem, we search for a "physics precedent"—a model with which we are already familiar and that can be applied to the present problem.

We shall generate analysis models based on four fundamental simplification models. The first simplification model is the particle model discussed in Chapter 1. We will look at a particle under various behaviors and environmental interactions. Further analysis models are introduced in later chapters with simplification models of a *system*, a *rigid body*, and a *wave*. Once we have introduced these analysis models, we shall see that they appear again and again later in the book in different situations.

Let us use Equation 2.2 to build our first analysis model for solving problems. We imagine a particle moving with a constant velocity. This model can be applied in *any* situation in which an entity that can be modeled as a particle is moving with constant velocity. This situation occurs frequently, so this is an important model.

If the velocity of a particle is constant, its instantaneous velocity at any instant during a time interval is the same as the average velocity over the interval, $v_x = \bar{v}_x$. Thus, we start with Equation 2.2 to generate an equation to be used in the mathematical representation of this situation:

$$v_x = \bar{v}_x = \frac{\Delta x}{\Delta t} \quad \longrightarrow \quad \Delta x = x_f - x_i = v_x \Delta t$$

$$\longrightarrow \quad x_f = x_i + v_x \Delta t$$

This equation tells us that the position of the particle is given by the sum of its original position $x_i$ at time $t = 0$ plus the displacement $v_x \Delta t$ that occurs during the time interval $\Delta t$. In practice, we usually choose the time at the beginning of the interval to be $t_i = 0$ and the time at the end of the interval to be $t_f = t$, so our equation becomes

$$x_f = x_i + v_x t \qquad \text{(for constant } v_x\text{)} \qquad \qquad \textbf{[2.4]}$$

• *A particle under constant velocity*

This is the primary equation used in the model of a particle under constant velocity—it can be applied to particles or objects if they can be modeled as particles.

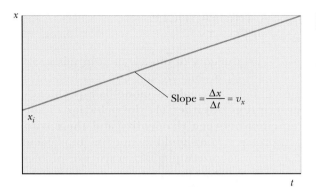

**Figure 2.6**

**Figure 2.6**

Position–time graph for a particle under constant velocity. The value of the constant velocity is the slope of the line.

Figure 2.6 is a graphical representation of the particle under constant velocity. On this position–time graph, the slope of the line representing the motion is constant and equal to the magnitude of the velocity. This is consistent with the mathematical representation, Equation 2.4, which is the equation of a straight line. The slope of the straight line is $v_x$ and the $y$ intercept is $x_i$ in both representations.

### Example 2.5   Modeling a Runner as a Particle

A runner has her track coach determine her velocity while she runs at a constant rate. The coach starts the stopwatch at the moment the runner passes and stops it after the runner has passed another point 20 m away. The time interval indicated on the stopwatch is 4.4 s.

(a) What is the runner's velocity?

**Solution**   We model the runner as a particle, as we did in Example 2.2 because the size of the runner and the movement of arms and legs are unnecessary details. This, in combination with the fact that the velocity is stated as being constant, allows us to use Equation 2.4 to find the velocity:

$$v_x = \frac{x_f - x_i}{\Delta t} = \frac{20 \text{ m} - 0}{4.4 \text{ s}} = \boxed{4.5 \text{ m/s}}$$

(b) What is the position of the runner after 10 s has passed?

**Solution**   In this part of the problem, we use Equation 2.4 to find the position of the particle at the time $t = 10$ s. Using the velocity found in part (a),

$$x_f = x_i + v_x t = 0 + (4.5 \text{ m/s})(10 \text{ s}) = \boxed{45 \text{ m}}$$

The mathematical manipulations for the particle under constant velocity stem from Equation 2.2 and its descendent, Equation 2.4. These equations can be used to solve for any variable in the equations that happens to be unknown if the other variables are known. For example, in part (b) of Example 2.5, we find the position when the velocity and the time are known. Similarly, if we know the velocity and the final position, we could use Equation 2.4 to find the time at which the runner is at this position. We shall present more examples of a particle under constant velocity in Chapter 3.

A particle under constant velocity moves with a constant speed along a straight line. Now consider a particle moving with a constant speed along a curved path. The primary equation for this situation is Equation 2.1, with the average speed replaced by the constant speed. As an example, imagine a particle moving at a constant speed in a circular path. If the speed is 5.00 m/s, and the radius of the path is 10.0 m, we can calculate the time required to complete one trip around the circle:

$$v = \bar{v} = \frac{d}{\Delta t} \quad \longrightarrow \quad \Delta t = \frac{d}{v} = \frac{2\pi r}{v} = \frac{2\pi(10.0\ \text{m})}{5.00\ \text{m/s}} = 12.6\ \text{s}$$

## 2.4 • ACCELERATION

When the velocity of a particle changes with time, the particle is said to be *accelerating*. For example, the speed of a car increases when you "step on the gas"; the car slows down when you apply the brakes; it changes direction when you turn the wheel—all of these are accelerations. We will need a precise definition of acceleration for our studies of motion.

Suppose a particle moving along the $x$ axis has a velocity $v_{xi}$ at time $t_i$ and a velocity $v_{xf}$ at time $t_f$. The **average acceleration** $\bar{a}_x$ of a particle in the time interval $\Delta t = t_f - t_i$ is defined as the ratio $\Delta v_x/\Delta t$, where $\Delta v_x = v_{xf} - v_{xi}$ is the *change* in velocity in this time interval:

$$\bar{a}_x \equiv \frac{v_{xf} - v_{xi}}{t_f - t_i} = \frac{\Delta v_x}{\Delta t} \tag{2.5}$$

• *Definition of average acceleration*

In drag racing, acceleration is a highly desired quantity. In a distance of 1/4 mi, speeds of over 320 mi/h are reached, with the entire distance being covered in under 5 s. *(George Lepp/Stone)*

Thus, **acceleration is a measure of how rapidly the velocity is changing.** Acceleration is a vector quantity having dimensions of length divided by (time)$^2$, or L/T$^2$. Some of the common units of acceleration are meters per second per second (m/s$^2$) and feet per second per second (ft/s$^2$). For example, an acceleration of 2 m/s$^2$ means that the velocity changes by 2 m/s during each second of time that passes.

In some situations, the value of the average acceleration may be different for different time intervals. It is therefore useful to define the **instantaneous acceleration** as the limit of the average acceleration as $\Delta t$ approaches zero, analogous to the definition of instantaneous velocity discussed in Section 2.2:

$$a_x \equiv \lim_{\Delta t \to 0} \frac{\Delta v_x}{\Delta t} = \frac{dv_x}{dt} \tag{2.6}$$

• *The instantaneous acceleration is the derivative of the velocity with respect to time*

That is, the **instantaneous acceleration** equals the derivative of the velocity with respect to time, which by definition is the slope of the velocity–time graph. Note that if $a_x$ is positive, the acceleration is in the positive $x$ direction, whereas negative $a_x$ implies acceleration in the negative $x$ direction. This does not necessarily mean that the particle is *moving* in the negative $x$ direction, a point which we shall address in more detail shortly. From now on, we use the term *acceleration* to mean instantaneous acceleration.

Because $v_x = dx/dt$, the acceleration can also be written

$$a_x = \frac{dv_x}{dt} = \frac{d}{dt}\left(\frac{dx}{dt}\right) = \frac{d^2x}{dt^2} \tag{2.7}$$

that is, the acceleration equals the *second derivative* of the position with respect to time.

(a)

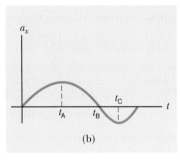

(b)

### Figure 2.7

The instantaneous acceleration can be obtained from the velocity–time graph (a). At each instant, the acceleration in the $a_x$ versus $t$ graph (b) equals the slope of the line tangent to the $v_x$ versus $t$ curve.

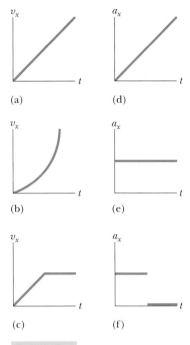

### Figure 2.8

(Quick Quiz 2.3) Parts (a), (b), and (c) are velocity–time graphs of objects in one-dimensional motion. The possible acceleration–time graphs of each object are shown in scrambled order in parts (d), (e), and (f).

Figure 2.7 shows how the acceleration–time curve can be derived from the velocity–time curve. In these diagrams, the acceleration of a particle at any time is simply the slope of the velocity–time graph at that time. Positive values of the acceleration correspond to those points (between $t_A$ and $t_B$) where the velocity in the positive $x$ direction is increasing in magnitude (the particle is speeding up), or those points (between $t = 0$ and $t_A$) where the velocity in the negative $x$ direction is decreasing in magnitude (the particle is slowing down). The acceleration reaches a maximum at time $t_A$, when the slope of the velocity–time graph is a maximum. The acceleration then goes to zero at time $t_B$, when the velocity is a maximum (i.e., when the velocity is momentarily not changing and the slope of the $v$ versus $t$ graph is zero). Finally, the acceleration is negative when the velocity in the positive $x$ direction is decreasing in magnitude (between $t_B$ and $t_C$) or when the velocity in the negative direction is increasing in magnitude (after $t_C$).

### Quick **Quiz 2.3**

Using Figure 2.8, match each of the velocity–time graphical representations on the left with the acceleration–time graphical representation on the right that best describes the motion.

As an example of the computation of acceleration, consider the pictorial representation of the motion of a car in Figure 2.9. In this case, the velocity of the car

### Figure 2.9

The velocity of the car decreases from 30 m/s to 15 m/s in a time interval of 2.0 s.

has changed from an initial value of 30 m/s to a final value of 15 m/s in a time interval of 2.0 s. The average acceleration during this time interval is

$$\overline{a}_x \equiv \frac{15 \text{ m/s} - 30 \text{ m/s}}{2.0 \text{ s}} = -7.5 \text{ m/s}^2$$

The negative sign in this example indicates that the acceleration vector is in the negative x direction (to the left in Figure 2.9). For the case of motion in a straight line, the direction of the velocity of an object and the direction of its acceleration are related as follows. **When the object's velocity and acceleration are in the same direction, the object is speeding up in that direction. On the other hand, when the object's velocity and acceleration are in opposite directions, the speed of the object decreases in time.**

To help with this discussion of the signs of velocity and acceleration, let us take a peek ahead to Chapter 4, where we shall relate the acceleration of an object to the *force* on the object. We will save the details until that later discussion, but for now, let us borrow the notion that **force is proportional to acceleration:**

$$\mathbf{F} \propto \mathbf{a}$$

This proportionality indicates that acceleration is caused by force. What's more, as indicated by the vector notation in the proportionality, force and acceleration are in the same direction. Thus, let us think about the signs of velocity and acceleration by forming a mental representation in which a force is applied to the object to cause the acceleration. Again consider the case in which the velocity and acceleration are in the same direction. This is equivalent to an object moving in a given direction, and experiencing a force that pulls on it in the same direction. It is clear in this case that the object speeds up! If the velocity and acceleration are in opposite directions, the object moves one way and a force pulls in the opposite direction. In this case, the object slows down! It is very useful to equate the direction of the acceleration in these situations to the direction of a force because it is easier from our everyday experience to think about what effect a force will have on an object than to think only in terms of the direction of the acceleration.

**PITFALL PREVENTION 2.6**
**Negative acceleration**

Keep in mind that *negative acceleration does not necessarily mean that an object is slowing down*. If the acceleration is negative, and the velocity is negative, the object is speeding up!

**PITFALL PREVENTION 2.7**
**Deceleration**

The word *deceleration* has a common popular connotation as *slowing down*. When combined with the misconception in Pitfall Prevention 2.6 that negative acceleration means slowing down, the situation can be further confused by the use of the word *deceleration*. We will not use this word in this text.

**Quick Quiz 2.4**

If a car is traveling eastward and slowing down, what is the direction of the force on the car? Explain.

**Quick Quiz 2.5**

Can a particle with constant acceleration ever stop and stay stopped?

**Example 2.6  Average and Instantaneous Acceleration**

The velocity of a particle moving along the x axis varies in time according to the expression $v_x = (40 - 5t^2)$, where t is in seconds.

(a) Find the average acceleration in the time interval $t = 0$ to $t = 2.0$ s.

**Solution**  Build your mental representation from the mathematical expression given for the velocity. For example, which way is the particle moving at $t = 0$? How does the velocity change in the first few seconds—does it move faster or slower? The velocity–time graphical representation for this function is given in Figure 2.10. The velocities at $t_i = t_A = 0$ and $t_f = t_B = 2.0$ s are found by substituting these values of $t$ into the expression given for the velocity:

$$v_{xA} = 40 - 5t_A^2 \text{ m/s} = 40 - 5(0)^2 \text{ m/s} = 40 \text{ m/s}$$

$$v_{xB} = 40 - 5t_B^2 \text{ m/s} = 40 - 5(2.0)^2 \text{ m/s} = 20 \text{ m/s}$$

Therefore, the average acceleration in the specified time interval $\Delta t = t_B - t_A$ is

$$\bar{a}_x \equiv \frac{20 \text{ m/s} - 40 \text{ m/s}}{2.0 \text{ s}} = \boxed{-10 \text{ m/s}^2}$$

The negative sign is consistent with the fact that the slope of the line joining the initial and final points on the velocity–time graph is negative.

(b) Determine the acceleration at $t = 2.0$ s.

**Solution** Because this question refers to a specific instant of time, it is asking for an instantaneous acceleration. The velocity at time $t$ is $v_{xi} = 40 - 5t^2$, and the velocity at time $t + \Delta t$ is

$$v_{xf} = 40 - 5(t + \Delta t)^2 = 40 - 5t^2 - 10t\,\Delta t - 5(\Delta t)^2$$

Therefore, the change in velocity over the time interval $\Delta t$ is

$$\Delta v_x = v_{xf} - v_{xi} = -10t\,\Delta t - 5(\Delta t)^2$$

Dividing this expression by $\Delta t$ and taking the limit of the result as $\Delta t$ approaches zero gives the acceleration at *any* time $t$:

$$a_x \equiv \lim_{\Delta t \to 0} \frac{\Delta v_x}{\Delta t} = \lim_{\Delta t \to 0} (-10t - 5\Delta t) = -10t$$

Therefore, at $t = 2.0$ s we find that

$$a_x = (-10)(2.0) \text{ m/s}^2 = \boxed{-20 \text{ m/s}^2}$$

This result can also be obtained by measuring the slope of the velocity–time graph at $t = 2.0$ s (see Fig. 2.10) or by taking the derivative of the velocity expression.

**EXERCISE** When struck by a club, a golf ball initially at rest acquires a speed of 31.0 m/s. If the ball is in contact with the club for 1.17 ms, what is the magnitude of the average acceleration of the ball?

*Answer* $2.65 \times 10^4$ m/s$^2$

**Figure 2.10**

(Example 2.6) The velocity–time graph for a particle moving along the $x$ axis according to the relation $v_x = (40 - 5t^2)$. The acceleration at $t = 2.0$ s is obtained from the slope of the blue tangent line at that time.

**EXERCISE** A particle moves with a velocity given by the expression $v = 4t^2$. (a) What is the acceleration of the particle at $t = 0.25$ s? (b) Is the particle speeding up or slowing down at this instant? Why?

*Answer* (a) 2.0 m/s$^2$ (b) Speeding up, because both the velocity and acceleration are in the positive direction.

**EXERCISE** A particle moves with a velocity $v = -4.0$ m/s at time $t = 2.0$ s. At $t = 3.0$ s, its velocity is $-1.0$ m/s. The acceleration during this time interval is constant. (a) What is the sign of the acceleration during this interval? (b) What is the magnitude of the acceleration during this interval? (c) Is the particle speeding up or slowing during this interval? Why?

*Answer* (a) Positive (b) 3.0 m/s$^2$ (c) Slowing down, because the acceleration is in a direction opposite to the velocity. This result is consistent with the fact that the magnitude of the velocity is smaller at the end of the time interval than at the beginning.

## 2.5 • MOTION DIAGRAMS

The concepts of velocity and acceleration are often confused with each other, but in fact they are quite different quantities. It is instructive to make use of the specialized pictorial representation called a **motion diagram** to describe the velocity and acceleration vectors while an object is in motion.

**Figure 2.11**

(a) Motion diagram for a car moving at constant velocity. (b) Motion diagram for a car whose constant acceleration is in the direction of its velocity. The velocity vector at each instant is indicated by a red arrow, and the constant acceleration vector by a violet arrow. (c) Motion diagram for a car whose constant acceleration is in the direction *opposite* the velocity at each instant.

A *stroboscopic photograph* of a moving object shows several images of the object, taken as the strobe light flashes at a constant rate. Figure 2.11 represents three sets of strobe photographs of cars moving along a straight roadway in a single direction, from left to right. The time intervals between flashes of the stroboscope are equal in each part of the diagram. To distinguish between the two vector quantities, we use red for velocity vectors and violet for acceleration vectors in Figure 2.11. The vectors are sketched at several instants during the motion of the object. Let us describe the motion of the car in each diagram.

In Figure 2.11a, the images of the car are equally spaced—the car moves through the same displacement in each time interval. Thus, the car moves with *constant positive velocity* and has *zero acceleration*. We could model the car as a particle and describe it as a particle under constant velocity.

In Figure 2.11b, the images of the car become farther apart as time progresses. In this case, the velocity vector increases in time because the car's displacement between adjacent positions increases as time progresses. Thus, the car is moving with a *positive velocity* and a *positive acceleration*. The velocity and acceleration are in the same direction. In terms of our earlier force discussion, imagine a force pulling on the car in the same direction it is moving—it speeds up.

In Figure 2.11c, we interpret the car as slowing down as it moves to the right because its displacement between adjacent positions decreases as time progresses. In this case, the car moves initially to the right with a *positive velocity* and a *negative acceleration*. The velocity vector decreases in time and eventually reaches zero. (This type of motion is exhibited by a car that skids to a stop after applying its brakes.) From this diagram we see that the acceleration and velocity vectors are *not* in the same direction. The velocity and acceleration are in opposite directions.

In terms of our earlier force discussion, imagine a force pulling on the car opposite to the direction it is moving—it slows down.

The violet acceleration vectors in Figures 2.11b and 2.11c are all of the same length. Thus, these diagrams represent a motion with constant acceleration. This important type of motion is discussed in the next section.

## 2.6 • THE PARTICLE UNDER CONSTANT ACCELERATION

See Screen 3.4

If the acceleration of a particle varies in time, the motion may be complex and difficult to analyze. A very common and simple type of one-dimensional motion occurs when the acceleration is constant, such as for the motion of the cars in Figures 2.11b and 2.11c. In this case, the average acceleration over any time interval equals the instantaneous acceleration at any instant of time within the interval. Consequently, the velocity increases or decreases at the same rate throughout the motion. The particle under constant acceleration forms the basis of a common analysis model that we can apply to appropriate problems and is often used to model situations such as falling objects and braking cars.

If we replace $\bar{a}_x$ with the constant $a_x$ in Equation 2.5, we find that

$$a_x = \frac{v_{xf} - v_{xi}}{t_f - t_i}$$

For convenience, let $t_i = 0$ and $t_f$ be any arbitrary time $t$. With this notation, we can express the acceleration as

$$a_x = \frac{v_{xf} - v_{xi}}{t}$$

or, solving for $v_{xf}$,

$$v_{xf} = v_{xi} + a_x t \qquad \text{(for constant } a_x\text{)} \qquad [2.8]$$

This expression enables us to predict the velocity at *any* time $t$ if the initial velocity and constant acceleration are known. It is the first of four equations that can be used to solve problems in which a particle is under constant acceleration. A graphical representation of position versus time for this motion is shown in Figure 2.12a. The velocity–time graph shown in Figure 2.12b is a straight line, the slope of which is the constant acceleration $a_x$. This is consistent with the fact that $a_x = dv_x/dt$ is a constant. From this graph and from Equation 2.8, we see that the velocity at any time $t$ is the sum of the initial velocity $v_{xi}$ and the change in velocity $a_x t$ due to the acceleration. The graph of acceleration versus time (Fig. 2.12c) is a straight line with a slope of zero, because the acceleration is constant. If the acceleration were negative, the slope of Figure 2.12b would be negative and the horizontal line in Figure 2.12c would be below the time axis.

We can generate another equation for constant acceleration by recalling a result from Section 2.1, that the displacement of a particle is the area under the curve on a velocity–time graph. Because the velocity varies linearly with time (Fig. 2.12b), the area under the curve is the sum of a rectangular area (under the horizontal dashed line in Fig. 2.12b) and a triangular area (from the horizontal dashed line upward to the curve). Thus,

$$\Delta x = v_{xi}\,\Delta t + \tfrac{1}{2}(v_{xf} - v_{xi})\Delta t$$

(a)

(b)

(c)

**Figure 2.12**

A particle moving along the $x$ axis with constant acceleration $a_x$; (a) the position–time graph, (b) the velocity–time graph, and (c) the acceleration–time graph.

This can be simplified as follows:

$$\Delta x = (v_{xi} + \tfrac{1}{2} v_{xf} - \tfrac{1}{2} v_{xi})\Delta t = \tfrac{1}{2}(v_{xi} + v_{xf})\Delta t$$

In general, from Equation 2.2, the displacement for a time interval is

$$\Delta x = \bar{v}_x \Delta t$$

Comparing these last two equations, we find that the average velocity in any time interval is the arithmetic mean of the initial velocity $v_{xi}$ and the final velocity $v_{xf}$:

$$\bar{v}_x = \tfrac{1}{2}(v_{xi} + v_{xf}) \qquad \text{(for constant } a_x) \qquad \textbf{[2.9]}$$

*(handwritten: avg velocity)*

- *Average velocity for a particle under constant acceleration*

Remember that this expression is valid only when the acceleration is constant, that is, when the velocity varies linearly with time.

We now use Equations 2.2 and 2.9 to obtain the position as a function of time. Again we choose $t_i = 0$, at which time the initial position is $x_i$. This gives

$$\Delta x = \bar{v}_x \Delta t = \tfrac{1}{2}(v_{xi} + v_{xf})t$$

$$x_f = x_i + \tfrac{1}{2}(v_{xi} + v_{xf})t \qquad \text{(for constant } a_x) \qquad \textbf{[2.10]}$$

*(handwritten: final position = 1st position + change in position; $x_i + \Delta x$ ... + change in position)*

- *Position as a function of velocity and time*

We can obtain another useful expression for the position by substituting Equation 2.8 for $v_{xf}$ in Equation 2.10:

*(handwritten: integrate accel)*

$$x_f = x_i + \tfrac{1}{2}[v_{xi} + (v_{xi} + a_x t)]t$$

$$x_f = x_i + v_{xi}t + \tfrac{1}{2} a_x t^2 \qquad \text{(for constant } a_x) \qquad \textbf{[2.11]}$$

*(handwritten: $\frac{a}{\downarrow} \, at + v_i$ ; $\frac{1}{2}a t^2 + v_i t + x_i$)*

- *Position as a function of time for a particle under constant acceleration*

Note that the position at any time $t$ is the sum of the initial position $x_i$, the displacement $v_{xi}t$ that would result if the velocity remained constant at the initial velocity, and the displacement $\tfrac{1}{2}a_x t^2$ due to the fact that the particle is accelerating. Consider again the position–time graph for motion under constant acceleration shown in Figure 2.12a. The curve representing Equation 2.11 is a parabola, as shown by the $t^2$ dependence in the equation. The slope of the tangent to this curve at $t = 0$ equals the initial velocity $v_{xi}$, and the slope of the tangent line at any time $t$ equals the velocity at that time.

Finally, we can obtain an expression that does not contain the time by substituting the value of $t$ from Equation 2.8 into Equation 2.10. This gives

$$x_f = x_i + \tfrac{1}{2}(v_{xi} + v_{xf})\left(\frac{v_{xf} - v_{xi}}{a_x}\right) = x_i + \frac{v_{xf}^2 - v_{xi}^2}{2a_x}$$

*(handwritten: $v_f = v_i + at$, $t = \frac{v_f - v_i}{a}$ ; $x_f = x_i + \frac{v_f^2 - v_i^2}{2a}$)*

$$v_{xf}^2 = v_{xi}^2 + 2a_x(x_f - x_i) \qquad \text{(for constant } a_x) \qquad \textbf{[2.12]}$$

- *Velocity as a function of position for a particle under constant acceleration*

This is *not* an independent equation, because it arises from combining Equations 2.8 and 2.10. It is useful, however, for those problems in which a value for the time is not involved.

If motion occurs in which the constant value of the acceleration is *zero*, then

$$\left.\begin{array}{l} v_{xf} = v_{xi} \\ x_f = x_i + v_{xi}t \end{array}\right\} \qquad \text{when } a_x = 0$$

That is, when the acceleration is zero, the velocity remains constant and the position changes linearly with time. In this case, the particle under constant *acceleration* becomes the particle under constant *velocity* (Equation 2.4).

*(handwritten top right:)*

$$\Delta x = v_i \Delta t + \tfrac{1}{2}(v_f - v_i)\Delta t$$
$$= v_i \Delta t + \tfrac{1}{2}(v_f - \tfrac{1}{2}v_i)\Delta t$$
$$= \Delta t(\tfrac{1}{2}v_f + \tfrac{1}{2}v_i)$$
$$= \tfrac{1}{2}\Delta t(v_f + v_i)$$

| TABLE 2.2 | Kinematic Equations for Motion in a Straight Line Under Constant Acceleration |
|---|---|
| **Equation** | **Information Given by Equation** |
| $v_{xf} = v_{xi} + a_x t$ | Velocity as a function of time |
| $x_f = x_i + \frac{1}{2}(v_{xf} + v_{xi})t$ | Position as a function of velocity and time |
| $x_f = x_i + v_{xi}t + \frac{1}{2}a_x t^2$ | Position as a function of time |
| $v_{xf}^2 = v_{xi}^2 + 2a_x(x_f - x_i)$ | Velocity as a function of position |

*Note:* Motion is along the *x* axis. At $t = 0$, the position of the particle is $x_i$ and its velocity is $v_{xi}$.

Equations 2.8, 2.10, 2.11, and 2.12 are four kinematic expressions that may be used to solve any problem in one-dimensional motion of a particle (or an object that can be modeled as a particle) with constant acceleration. Keep in mind that these relationships were derived from the definitions of velocity and acceleration, together with some simple algebraic manipulations and the requirement that the acceleration be constant. It is often convenient to choose the initial position of the particle as the origin of the motion, so that $x_i = 0$ at $t = 0$. We will see cases, however, in which we must choose the value of $x_i$ to be something other than zero.

The four kinematic equations for the particle under constant acceleration are listed in Table 2.2 for convenience. The choice of which kinematic equation or equations you should use in a given situation depends on what is known beforehand. Sometimes it is necessary to use two of these equations to solve for two unknowns, such as the position and velocity at some instant. You should recognize that the quantities that vary during the motion are velocity $v_{xf}$, position $x_f$, and time *t*. The other quantities—$x_i$, $v_{xi}$, and $a_x$—are *parameters* of the motion and remain constant.

Solving the exercises and problems will give you a great deal of practice in the use of these equations. You will often discover that more than one method can be used to obtain a solution.

---

**PROBLEM-SOLVING STRATEGY**  |  **Particle Under Constant Acceleration**

The following procedure is recommended for solving problems that involve an object undergoing a constant acceleration:

1. Think about what is going on physically in the problem. Establish the mental representation.
2. Simplify the problem as much as possible. Confirm that the problem involves either a particle or an object that can be modeled as a particle and that it is moving with a constant acceleration.
3. Construct an appropriate pictorial representation, such as a motion diagram, or a graphical representation.
4. Make sure all the units in the problem are consistent. That is, if positions are measured in meters, be sure that velocities have units of m/s and accelerations have units of m/s².
5. Choose a coordinate system to be used throughout the problem.

6. Set up the mathematical representation. Choose an instant to call the "initial" time $t = 0$ and another the "final" time $t$. Let your choice be guided by what you know about the particle and what you want to know about it. The initial instant need not be when the particle starts to move and the final instant will only rarely be when the particle stops moving. Identify all the quantities given in the problem and a separate list of those to be determined. A tabular representation of these quantities may be helpful to you. Select from the list of kinematic equations the one or ones that will enable you to determine the unknowns. Solve the equations.

7. Once you have determined your result, check to see if your answers are consistent with the mental and pictorial representations and that your results are realistic.

## Example 2.7   Accelerating an Electron

An electron in the cathode-ray tube of a television set enters a region in which it accelerates uniformly in a straight line from a speed of $3.00 \times 10^4$ m/s to a speed of $5.00 \times 10^6$ m/s in a distance of 2.00 cm. For what length of time is the electron accelerating?

**Reasoning**   For this example, we shall identify the individual steps in the Problem-Solving Strategy. In subsequent examples, you should identify the portions of the solution that correspond to each step. For step 1, think about the electron moving through space. Note that it is moving faster at the end of the interval than before, so imagine it speeding up as it covers the 2.00-cm displacement. In step 2, ignore the fact that this is an electron and that it is in a television. The electron is easily modeled as a particle and the phrase "accelerates uniformly" tells us that this is a particle under constant acceleration. In step 3, all of the parts of Figure 2.12 represent the motion of the particle as a function of time, although you may want to graph velocity versus position because no time is given in the problem. In step 4, note that all units are metric, although we must convert 2.00 cm to meters to put all units in SI. In step 5, we make the simple choice of the x axis lying along the straight line mentioned in the text of the problem.

**Solution**   We are now ready to move on to step 6, in which we develop the mathematical representation of the problem. Notice that no acceleration is given in the problem and that

the time interval is requested, which provides a hint that we should use an equation that does not involve acceleration. We can find the time at which the particle is at the end of the 2.00-cm distance from Equation 2.10:

$$x_f = x_i + \tfrac{1}{2}(v_{xi} + v_{xf})t \quad \longrightarrow \quad t = \frac{2(x_f - x_i)}{v_{xi} + v_{xf}}$$

$$t = \frac{2(0.0200 \text{ m})}{3.00 \times 10^4 \text{ m/s} + 5.00 \times 10^6 \text{ m/s}}$$

$$= 7.95 \times 10^{-9} \text{ s}$$

Finally, we check if the answer is reasonable (step 7). The final speed is on the order of $10^6$ m/s. Let us estimate the time required to move 1 cm at this speed:

$$\Delta t = \frac{\Delta x}{v} \approx \frac{0.01 \text{ m}}{10^6 \text{ m/s}} \approx 10^{-8} \text{ s} = 10 \times 10^{-9} \text{ s}$$

This is the same order of magnitude as our answer, so this provides confidence that our answer is reasonable.

**EXERCISE**   A hockey puck sliding on a frozen lake comes to rest after traveling 200 m. Its initial speed was 3.00 m/s. (a) What was its acceleration if it is assumed to have been constant? (b) How long was it in motion? (c) What was its speed after traveling 150 m?

*Answer*   (a) $-2.25 \times 10^{-2}$ m/s$^2$ (b) 133 s (c) 1.50 m/s

---

## Example 2.8   Watch Out for the Speed Limit!

A car traveling at a constant velocity of magnitude 45.0 m/s passes a trooper hidden behind a billboard. One second after the speeding car passes the billboard, the trooper sets out

from the billboard to catch it, accelerating at a constant rate of 3.00 m/s$^2$. How long does it take her to overtake the speeding car?

**Reasoning** We will point out again in this example steps in the Problem-Solving Strategy. A pictorial representation (step 3) of the situation is shown in Figure 2.13. Establish the mental representation (step 1) of this situation for yourself; in the following solution, we will go straight to the mathematical representation. As you become more proficient at solving physics problems, a quick thought about the mental representation may be enough to allow you to skip pictorial representations and go right to the mathematics. Let us model the speeding car as a particle under constant velocity and the trooper as a particle under constant acceleration (step 2). We shall ignore the fact that these are cars—we will imagine the speeder and the trooper as point particles undergoing the motion described in the problem.

**Solution** Note that all units are in the same system (step 4). To solve this problem algebraically, we will write an expression for the position of each vehicle as a function of time. It is convenient to choose the origin at the position of the billboard (step 5) and take $t_B = 0$ as the time the trooper begins moving (step 6). At that instant, the speeding car has already traveled a distance of 45.0 m, because it has traveled at a constant speed of $v_x = 45.0$ m/s for 1.00 s—it is at point Ⓑ in Figure 2.13. Thus, the initial position of the speeding car is $x_i = x_B = 45.0$ m. We do *not* choose $t = 0$ as the time at which the car passes the trooper (point Ⓐ in Figure 2.13), because then the acceleration of the trooper is not constant during the problem. Her acceleration is $a_x = 0$ for the first second and then 3.00 m/s$^2$ after that. Thus, we could not model the trooper as a particle under constant acceleration with this choice.

Because the car moves with constant velocity, its acceleration is zero, and applying Equation 2.4 gives us

$$x_f = x_B + v_x t \longrightarrow x_{car} = 45.0 \text{ m} + (45.0 \text{ m/s})t$$

Note that at $t = 0$, this expression gives the car's correct initial position, $x_{car} = 45.0$ m.

For the trooper, who starts from the origin at $t = 0$, we have $x_i = 0$, $v_{xi} = 0$, and $a_x = 3.00$ m/s$^2$. Hence, from Equation 2.11 for a particle under constant acceleration, the position of the trooper as a function of time is

$$v_{x\,car} = 45.0 \text{ m/s}$$
$$a_{x\,car} = 0$$
$$a_{x\,trooper} = 3.00 \text{ m/s}^2$$

$t_A = -1.00$ s   $t_B = 0$   $t_C = ?$
Ⓐ   Ⓑ   Ⓒ

**Figure 2.13**

(Example 2.8) A speeding car passes a hidden trooper. The trooper catches up to the car at point Ⓒ.

$$x_f = x_i + v_{xi}t + \tfrac{1}{2}a_x t^2 \longrightarrow x_{trooper} = \tfrac{1}{2}a_x t^2$$
$$= \tfrac{1}{2}(3.00 \text{ m/s}^2)t^2$$

The trooper overtakes the car at the instant that $x_{trooper} = x_{car}$, which is at position Ⓒ in Figure 2.13:

$$\tfrac{1}{2}(3.00 \text{ m/s}^2)t^2 = 45.0 \text{ m} + (45.0 \text{ m/s})t$$

This gives the quadratic equation (dropping the units)

$$1.50t^2 - 45.0t - 45.0 = 0$$

whose positive solution is $t = 31.0$ s.

From your everyday experience, is this value reasonable (step 7)? (For help in solving quadratic equations, see Appendix B.2.)

**EXERCISE** In the preceding solution, the problem was solved in the mathematical representation. This problem can also be solved using a graphical representation. On the *same* graph, plot the position-versus-time function for each vehicle and from the intersection of the two curves determine the time at which the trooper overtakes the speeding car.

## 2.7 • FREELY FALLING OBJECTS

It is well known that all objects, when dropped, fall toward the Earth with nearly constant acceleration. Legend has it that Galileo Galilei first discovered this fact by observing that two different weights dropped simultaneously from the Leaning Tower of Pisa hit the ground at approximately the same time. (Air resistance plays a role in the falling of an object, but, for now, we shall model falling objects as if they are falling through a vacuum—a simplification model.) Although there is some doubt that this particular experiment was actually carried out, it is well established that Galileo did perform many systematic experiments on objects moving on inclined planes. Through careful measurements of distances and time intervals, he was able to show that the displacement from an origin of an object starting from rest is proportional to the square of the time the object is in motion. This observation is consistent with one of the kinematic equations we derived for a particle under constant acceleration (Eq. 2.11, with $v_{xi} = 0$). Galileo's achievements in mechanics paved the way for Newton in his development of the laws of motion.

You might want to try the following experiment. Drop a coin and a crumpled-up piece of paper simultaneously from the same height. Do they hit the floor at the same time? In the idealized case of no air resistance, the two objects experience the same motion and hit the floor at the same time. In a real (nonideal) experiment, air resistance cannot be ignored, and there will be a small time difference between their arrivals at the floor. In the idealized case, however, where air resistance *is* ignored, such motion is referred to as *free-fall*. If this same experiment could be conducted in a good vacuum, where air friction is truly negligible, the paper and coin would fall with the same acceleration, regardless of the shape or weight of the paper, even if the paper were still flat. This point is illustrated very convincingly in Figure 2.14, which is a photograph of an apple and a feather falling in a vacuum. On August 2, 1971, such an experiment was conducted on the Moon by astronaut David Scott. He simultaneously released a geologist's hammer and a falcon's feather, and in unison they fell to the lunar surface. This demonstration surely would have pleased Galileo!

We shall denote the magnitude of the free-fall acceleration with the symbol $g$, representing a vector acceleration **g**. At the surface of the Earth, $g$ is approximately 9.80 m/s$^2$, or 980 cm/s$^2$, or 32 ft/s$^2$. Unless stated otherwise, we shall use the value 9.80 m/s$^2$ when doing calculations. Furthermore, we shall assume that the vector **g** is directed downward toward the center of the Earth.

**WEB**

The following site contains images of Galileo's experimental notes on motion: **www.imss.fi.it/ms72**

- *Free-fall acceleration*
  **g** = *9.80 m/s$^2$ down*

**Figure 2.14**

An apple and a feather, released from rest in a 4-ft vacuum chamber, fall at the same rate, regardless of their masses. Ignoring air resistance, all objects fall to the Earth with the same acceleration of magnitude 9.80 m/s$^2$, as indicated by the violet arrows in this multiflash photograph. The velocity of the two objects increases linearly with time, as indicated by the series of red arrows. (© 1993 James Sugar/Black Star)

When we use the expression *freely falling object,* we do not necessarily mean an object dropped from rest. A freely falling object is an object moving freely under the influence of gravity alone, regardless of its initial motion. Thus, objects thrown upward or downward and those released from rest are all freely falling objects once they are released!

It is important to emphasize that any freely falling object experiences an acceleration directed *downward,* as shown in the strobe photograph of a falling apple and a falling feather (see Fig. 2.14). If we ignore air resistance (the apple and feather are falling in a vacuum chamber in Figure 2.14), then a freely falling object (released from rest or projected with a vertical velocity) moves in one dimension with constant acceleration. Therefore, we can model the object as a particle under constant acceleration.

In previous examples in this chapter, the particles were truly undergoing constant acceleration, as stated in the problem. Thus, it may have been difficult to understand the need for modeling. Now, we can begin to see the need for modeling—we really are *modeling* a real falling object with an analysis model. Notice that we are (1) ignoring air resistance and (2) assuming that the free-fall acceleration is constant. Thus, the model of a particle under constant acceleration is a *replacement* for the real problem, which could be more complicated. If air resistance and any variation in *g* are small, however, the model should make predictions that agree closely with the real situation.

The equations developed in Section 2.6 for objects moving with constant acceleration can be applied to the falling object. The only necessary modification that we need to make in these equations for freely falling objects is to note that the motion is in the vertical direction, so we will use *y* instead of *x*, and that the acceleration is downward and of magnitude 9.80 m/s². Thus, for a freely falling object we commonly take $a_y = -g = -9.80$ m/s², where the negative sign indicates that the acceleration of the object is downward. The choice of negative for the downward direction is arbitrary, but common.

### Quick Quiz 2.6

A ball is thrown upward. While the ball is in free-fall, (a) what happens to its speed? (b) Does its acceleration increase, decrease, or remain constant?

### THINKING PHYSICS 2.3

A sky diver steps out of a stationary helicopter. A few seconds later, another sky diver steps out, so that they both fall along the same vertical line. Ignore air resistance, so that both sky divers fall with the same acceleration and let us model the sky divers as particles under constant acceleration. Does the vertical separation distance between them stay the same? Does the difference in their speeds stay the same?

**Reasoning**   At any given instant of time, the speeds of the jumpers are definitely different, because one had a head start over the other. In any time interval, however, each jumper increases his or her speed by the same amount, because they have the same acceleration. Thus, the difference in speeds remains the same. The first jumper will always be moving with a higher speed than the second. In a given time interval, then, the first jumper will have a larger displacement than the second. Thus, the separation distance between them increases.

## Example 2.9   Try to Catch the Dollar

Emily challenges David to catch a dollar bill as follows. She holds the bill vertically, as in Figure 2.15, with the center of the bill between David's index finger and thumb. David must catch the bill after Emily releases it without moving his hand downward. The reaction time of most people is at best about 0.2 s. Who would you bet on?

**Reasoning**   Place your bets on Emily. There is a time delay between the instant Emily releases the bill and the time David reacts and closes his fingers. We model the bill as a particle. When released, the bill will probably flutter downward to the floor due to the effects of the air, but for the very early part of its motion, we will assume that it can be modeled as a particle falling through a vacuum. Because the bill is in free-fall and undergoes a downward acceleration of magnitude 9.80 m/s², in 0.2 s it falls a distance of $y = \frac{1}{2} g t^2 \approx 0.2$ m = 20 cm. This distance is about twice the distance between the center of the bill and its top edge ($\approx 8$ cm). Thus, David will be unsuccessful. You might want to try this "trick" on one of your friends.

**Figure 2.15**

(Example 2.9)   *(George Semple)*

## Example 2.10   Not a Bad Throw for a Rookie!

A stone is thrown at point Ⓐ from the top of a building with an initial velocity of 20.0 m/s straight upward. The building is 50.0 m high, and the stone just misses the edge of the roof on its way down, as in the pictorial representation of Figure 2.16. Determine (a) the time at which the stone reaches its maximum height, (b) the maximum height above the rooftop, (c) the time at which the stone returns to the level of the thrower, (d) the velocity of the stone at this instant, (e) the velocity and position of the stone at $t = 5.00$ s, and (f) the position of the stone at $t = 6.00$ s. How does the model fail in this last part of the problem?

**Solution**   Think about the mental representation—the stone rises upward, slowing down. It stops momentarily (point Ⓑ) and then begins to fall downward again. During the entire motion, it is accelerating downward because the gravitational force is always pulling downward on it. Ignoring air resistance, we model the stone as a particle under constant acceleration.

   (a) To begin the mathematical representation, we find the time at which the stone reaches the maximum height, point Ⓑ. We use the vertical modification of Equation 2.8, noting

that $v_{yf} = 0$ at the maximum height:

$$v_{yf} = v_{yi} + a_y t \longrightarrow 0 = 20.0 \text{ m/s} + (-9.80 \text{ m/s}^2) t_\text{B}$$

$$t_\text{B} = \frac{20.0 \text{ m/s}}{9.80 \text{ m/s}^2} = \boxed{2.04 \text{ s}}$$

   (b) This value of time can be substituted into Equation 2.11, to give the maximum height measured from the position of the thrower:

$$\begin{aligned} y_{\max} = y_\text{B} &= y_\text{A} + v_{y\text{A}}t + \tfrac{1}{2}a_y t_\text{A}^2 \\ &= 0 + (20.0 \text{ m/s})(2.04 \text{ s}) + \tfrac{1}{2}(-9.80 \text{ m/s}^2)(2.04 \text{ s})^2 \\ &= \boxed{20.4 \text{ m}} \end{aligned}$$

   (c) When the stone is back at the height of the thrower, the $y$ coordinate is zero. From Equation 2.11, letting $y_f = y_\text{C} = 0$, we obtain the expression

$$y_\text{C} = y_\text{A} + v_{y\text{A}}t + \tfrac{1}{2}a_y t_\text{C}^2 \longrightarrow 0 = 20.0 t_\text{C} - 4.90 t_\text{C}^2$$

This is a quadratic equation and has two solutions for $t_\text{C}$. The equation can be factored to give

$$t_\text{C}(20.0 - 4.90 t_\text{C}) = 0$$

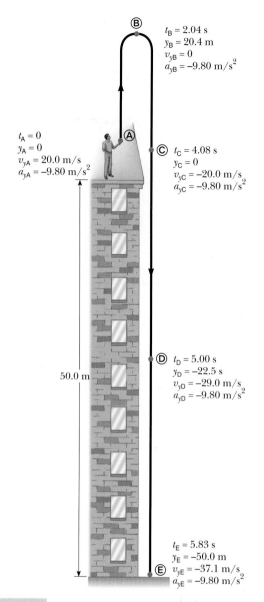

$t_A = 0$
$y_A = 0$
$v_{yA} = 20.0$ m/s
$a_{yA} = -9.80$ m/s$^2$

**B**   $t_B = 2.04$ s
$y_B = 20.4$ m
$v_{yB} = 0$
$a_{yB} = -9.80$ m/s$^2$

**A**

**C**  $t_C = 4.08$ s
$y_C = 0$
$v_{yC} = -20.0$ m/s
$a_{yC} = -9.80$ m/s$^2$

50.0 m

**D**  $t_D = 5.00$ s
$y_D = -22.5$ s
$v_{yD} = -29.0$ m/s
$a_{yD} = -9.80$ m/s$^2$

$t_E = 5.83$ s
$y_E = -50.0$ m
**E**  $v_{yE} = -37.1$ m/s
$a_{yE} = -9.80$ m/s$^2$

**Figure 2.16**

(Example 2.10) Position, velocity, and acceleration at various in-stants of time for a freely falling particle initially thrown upward with a velocity $v_y = 20.0$ m/s.

One solution is $t_C = 0$, corresponding to the time the stone starts its motion. The other solution—the one we are after—is $t_C = 4.08$ s. Note that this is twice the value for $t_B$. The fall from **B** to **C** is the reverse of the rise from **A** to **B** and the stone takes exactly the same time to undergo each part of the motion.

(d) The value for $t_C$ found in part (c) can be inserted into Equation 2.8 to give

$$v_{yC} = v_{yA} + a_y t_C$$
$$= 20.0 \text{ m/s} + (-9.80 \text{ m/s}^2)(4.08 \text{ s})$$
$$= -20.0 \text{ m/s}$$

Note that the velocity of the stone when it arrives back at its original height is equal in magnitude to its initial velocity but opposite in direction. This, along with the equal time inter-vals noted at the end of part (c), indicates that the motion to this point is symmetric.

(e) From Equation 2.8, the velocity at **D** after 5.00 s is

$$v_{yD} = v_{yA} + a_y t_D$$
$$= 20.0 \text{ m/s} + (-9.80 \text{ m/s}^2)(5.00 \text{ s})$$
$$= -29.0 \text{ m/s}$$

We can use Equation 2.11 to find the position of the stone at $t = 5.00$ s:

$$y_D = y_A + v_{yA}t + \tfrac{1}{2}a_y t_D^2$$
$$= 0 + (20.0 \text{ m/s})(5.00 \text{ s}) + \tfrac{1}{2}(-9.80 \text{ m/s}^2)(5.00 \text{ s})^2$$
$$= -22.5 \text{ m}$$

(f) We use Equation 2.11 again to find the position of the stone at $t = 6.00$ s:

$$y = 0 + (20.0 \text{ m/s})(6.00 \text{ s}) + \tfrac{1}{2}(-9.80 \text{ m/s}^2)(6.00 \text{ s})^2$$
$$= -56.4 \text{ m}$$

The failure of the model lies in the fact that the building is only 50.0 m high, so that the stone cannot be at a position 6.4 m below ground. Our model did not include the fact that the ground exists at $y = -50.0$ m, so the mathematical rep-resentation gives us an answer that is not consistent with our expectations in this case.

**EXERCISE**   Find (a) the velocity of the stone just before it hits the ground at **E** and (b) the total time the stone is in the air.

*Answer*   (a) $-37.1$ m/s (b) 5.83 s

context connection
## 2.8 • LIFTOFF ACCELERATION

We now have our first opportunity to address a Context in a closing section, as we will do in each remaining chapter. Our present Context is *Mission to Mars,* and our central question is, *How can we undertake a successful transfer of a spacecraft from Earth to Mars?*

The first step on our mission to Mars is to lift the spacecraft from the surface of the Earth and place it in orbit. Before the rocket engines are fired and the launch system begins to move, the velocity and acceleration of the spacecraft are both zero. When the engines fire, an upward force is applied to the launch system, resulting in an upward acceleration of the launch system and the attached spacecraft.

The acceleration of the launch system is not uniform. There are two primary reasons for this. The first is that the force from the rocket engines may not be constant, as adjustments are made to keep the launch system moving according to the preflight plan. The second reason is that the mass of the launch system decreases as the fuel is expended. As we shall learn in Chapter 4, the acceleration of an object is related to its mass. As the launch system loses mass, it becomes easier for the engines to accelerate it, and its acceleration increases.

Early launch systems, such as those for the Apollo, Gemini, and Mercury missions, were rather harsh on astronauts, with launch accelerations as high as 78 m/s². This can be expressed as 8g, or eight times the free-fall acceleration. As a result, astronauts felt pressed into their flight seats with a force many times their weight! The space shuttle launch system is much easier on astronauts, with a maximum acceleration of about 3g at the time of solid rocket booster separation.

If you watch a shuttle launch on television, you can measure the time interval between the beginning of motion of the shuttle and the passing of the bottom of the shuttle by the top of the launch tower. From this time, and knowledge of the height of the launch tower, you can calculate the acceleration and speed of the shuttle, as in the following example.

| WEB |

For information on Mars exploration, visit
**cmex=www.arc.nasa.gov**

### Example 2.11 Space Shuttle Acceleration

During a shuttle launch (Fig. 2.17), the bottom of the launch system clears the launch tower at a time of 6.0 s after liftoff. The height of the launch tower from the initial position of the bottom of the launch system is 75 m. Determine (a) the acceleration, assumed constant, of the launch system during these 6.0 s; and (b), the velocity of the launch system as the bottom clears the tower.

**Reasoning**   The shuttle is a large object, but we shall model it as a particle. The position of this particle is chosen at the bottom of the launch system because it is the bottom about which we have information. Let us assume that the acceleration is constant so that we can model the bottom of the launch system as a particle under constant acceleration and use the kinematic equations. If the acceleration is not truly

constant during this time interval, what we shall calculate is the *average* acceleration during the interval.

**Solution**   (a) Using Equation 2.11 and realizing that the initial velocity is zero (because the launch system begins from rest) and defining the initial position of the bottom of the launch system as the origin, we have

$$y_f = y_i + v_{yi}t + \tfrac{1}{2}a_y t^2 = 0 + 0 + \tfrac{1}{2}a_y t^2$$

$$\longrightarrow \quad a_y = \frac{2y_f}{t^2} = \frac{2(75 \text{ m})}{(6.0 \text{ s})^2} = \boxed{4.2 \text{ m/s}^2}$$

(b) We use the kinematic equation for velocity, Equation 2.8:

$$v_{yf} = v_{yi} + a_y t = 0 + (4.2 \text{ m/s}^2)(6.0 \text{ s}) = \boxed{25 \text{ m/s}}$$

This velocity, of course, is in the upward direction.

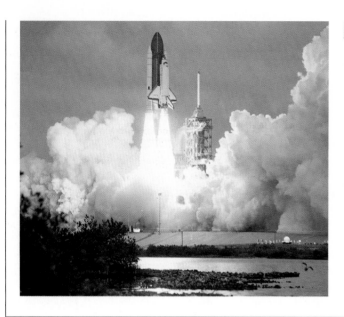

**Figure 2.17**

(Example 2.11) A space shuttle is launched from Cape Canaveral. *(Courtesy of NASA)*

Notice that the acceleration of the shuttle in Example 2.11 is only about 0.4*g*. As mentioned earlier, as the shuttle moves toward solid rocket booster separation, at about 120 seconds into the launch, the acceleration rises to about 3*g*.

As we investigate two-dimensional motion in the next chapter, we shall look at the next step in our mission after our spacecraft is launched from the surface of the Earth. We shall consider the details of its circular parking orbit around the Earth as it waits for the proper time to be launched to Mars.

## SUMMARY

The **average speed** of a particle during some time interval is equal to the ratio of the distance *d* traveled by the particle and the time interval $\Delta t$:

$$\bar{v} = \frac{d}{\Delta t} \qquad [2.1]$$

The **average velocity** of a particle moving in one dimension during some time interval is equal to the ratio of the displacement $\Delta x$ and the time interval $\Delta t$:

$$\bar{v}_x \equiv \frac{\Delta x}{\Delta t} \qquad [2.2]$$

The **instantaneous velocity** of a particle is defined as the limit of the ratio $\Delta x / \Delta t$ as $\Delta t$ approaches zero:

$$v_x \equiv \lim_{\Delta t \to 0} \frac{\Delta x}{\Delta t} = \frac{dx}{dt} \qquad [2.3]$$

The **speed** of a particle is defined as the magnitude of the instantaneous velocity vector.

If the velocity $v_x$ is constant, the preceding equations can be modified and used to solve problems describing the motion of a particle under constant velocity:

$$v_x = \frac{\Delta x}{\Delta t} \qquad [2.4]$$

$$x_f = x_i + v_x t$$

The **average acceleration** of a particle moving in one dimension during some time interval is defined as the ratio of the change in its velocity $\Delta v_x$ and the time interval $\Delta t$:

$$\bar{a}_x \equiv \frac{\Delta v_x}{\Delta t} \qquad [2.5]$$

The **instantaneous acceleration** is equal to the limit of the ratio $\Delta v_x / \Delta t$ as $\Delta t \to 0$. By definition, this equals the derivative of $v_x$ with respect to $t$, or the time rate of change of the velocity:

$$a_x \equiv \lim_{\Delta t \to 0} \frac{\Delta v_x}{\Delta t} = \frac{dv_x}{dt} \qquad [2.6]$$

The slope of the tangent to the *x*-versus-*t* curve at any instant gives the instantaneous velocity of the particle.

The slope of the tangent to the *v*-versus-*t* curve gives the instantaneous acceleration of the particle.

The **equations of kinematics** for a particle moving along the *x* axis with uniform acceleration $a_x$ (constant in magnitude and direction) are

$$v_{xf} = v_{xi} + a_x t \qquad [2.8]$$

$$x_f = x_i + \tfrac{1}{2}(v_{xi} + v_{xf})t \qquad [2.10]$$

$$x_f = x_i + v_{xi}t + \tfrac{1}{2}a_x t^2 \qquad [2.11]$$

$$v_{xf}^2 = v_{xi}^2 + 2a_x(x_f - x_i) \qquad [2.12]$$

These equations can be used to solve problems describing the motion of a particle under constant acceleration.

An object falling freely experiences an acceleration directed toward the center of the Earth. If air friction is ignored, and if the altitude of the motion is small compared with the Earth's radius, then one can assume that the magnitude of the free-fall acceleration $g$ is constant over the range of motion, where $g$ is equal to $9.80 \text{ m/s}^2$, or $32 \text{ ft/s}^2$. Assuming $y$ to be positive upward, the acceleration is given by $-g$, and the equations of kinematics for an object in free-fall are the same as those already given, with the substitutions $x \to y$ and $a_y \to -g$.

## QUESTIONS

**1.** The average velocity for a particle moving in one dimension has a positive value. Is it possible for the instantaneous velocity of the particle to have been negative at any time in the interval? Suppose the particle started at the origin, $x = 0$. If the average velocity is positive, could the particle ever have been in the $-x$ region?

**2.** If the average velocity of an object is zero in some time interval, what can you say about the displacement of the object for that interval?

**3.** Can the instantaneous velocity of an object during a time interval ever be greater in magnitude than the average velocity over the entire interval? Can it ever be less?

**4.** Consider the following combinations of signs and values for velocity and acceleration of a particle with respect to a one-dimensional *x* axis:

| Velocity | Acceleration |
|---|---|
| a. Positive | Positive |
| b. Positive | Negative |
| c. Positive | Zero |
| d. Negative | Positive |
| e. Negative | Negative |
| f. Negative | Zero |
| g. Zero | Positive |
| h. Zero | Negative |

Describe what a particle is doing in each case, and give a real-life example for an automobile on an east-west one-dimensional axis, with east considered the positive direction.

**5.** Can the equations of kinematics (Eqs. 2.8 through 2.12) be used in a situation where the acceleration varies in time? Can they be used when the acceleration is zero?

**6.** A student at the top of a building of height $h$ throws one ball upward with an initial speed $v_i$ and then throws a second ball downward with the same initial speed. How do the final velocities of the balls compare when they reach the ground?

**7.** Two cars are moving in the same direction in parallel lanes along a highway. At some instant, the velocity of car A exceeds the velocity of car B. Does this mean that the acceleration of A is greater than that of B? Explain.

**8.** Car A traveling south from New York to Miami has a speed of 25 m/s. Car B traveling west from New York to Chicago also has a speed of 25 m/s. Are their velocities equal? Explain.

**9.** You drop a ball from a window on an upper floor of a building. It strikes the ground with velocity $v$. You now repeat the drop, but you have a friend down on the street, who throws another ball upward at velocity $v$. Your friend throws the ball upward at exactly the same time that you drop yours from the window. At some location, the balls pass each other. Is this location at the halfway point between window and ground, above this point, or below this point?

**10.** Galileo experimented with balls rolling down inclined planes in order to reduce the acceleration along the plane and thus reduce the rate of descent of the balls. Suppose the angle that the incline makes with the horizontal is $\theta$. How would you expect the acceleration along the plane to decrease as $\theta$ decreases? What specific trigonometric dependence on $\theta$ would you expect for the acceleration?

**11.** A ball rolls in a straight line along the horizontal direction. Using motion diagrams (or multiflash photographs) as in Figure 2.11, describe the velocity and acceleration of

**Figure Q2.13** See Question 13.

the ball for each of the following situations: (a) The ball moves to the right at a constant speed. (b) The ball moves from right to left and continually slows down. (c) The ball moves from right to left and continually speeds up. (d) The ball moves to the right, first speeding up at a constant rate and then slowing down at a constant rate.

12. A plant growing rapidly next to a building doubles in height each week. At the end of the 25th day, the plant reaches the height of the building. At what time was the plant one fourth the height of the building?

13. A pebble is dropped into a water well, and the splash is heard 16 s later, as illustrated in the cartoon strip. What is the approximate distance from the rim of the well to the water's surface?

14. The motion of the Earth's crustal plates is described by a model referred to as *plate tectonic motion*. Measurements indicate that coastal portions of southern California have northward plate tectonic speeds of 2.5 cm per year. Estimate the time it would take for this motion to carry southern California to Alaska.

15. A heavy object falls from a height $h$ under the influence of gravity. It is released at $t = 0$ and strikes the ground at time $t$. Ignore air resistance. (a) When the object is at height $0.5h$, is the time earlier than $0.5t$, later than $0.5t$ or equal to $0.5t$? (b) When the time is $0.5t$, is the height of the object larger than $0.5h$, smaller than $0.5h$, or equal to $0.5h$?

# PROBLEMS

1, 2, 3 = straightforward, intermediate, challenging      □ = full solution available in the *Student Solutions Manual and Study Guide*

**web** = solution posted at **http://www.harcourtcollege.com/physics/**      🖥 = computer useful in solving problem

🌐 = Interactive Physics      ▪ = paired numerical/symbolic problems      ⛏ = life science application

### Section 2.1   Average Velocity

1. The position of a pinewood derby car is observed at various times; the results are summarized in the following table. Find the average velocity of the car for (a) the first second, (b) the last 3 s, and (c) the entire period of observation.

| $t$ (s) | 0 | 1.0 | 2.0 | 3.0 | 4.0 | 5.0 |
|---|---|---|---|---|---|---|
| $x$ (m) | 0 | 2.3 | 9.2 | 20.7 | 36.8 | 57.5 |

2. A motorist drives north for 35.0 min at 85.0 km/h and then stops for 15.0 min. He next continues north, traveling 130 km in 2.00 h. (a) What is his total displacement? (b) What is his average velocity?

3. The displacement versus time for a certain particle moving along the $x$ axis is shown in Figure P2.3. Find the average velocity in the time intervals (a) 0 to 2 s, (b) 0 to 4 s, (c) 2 s to 4 s, (d) 4 s to 7 s, (e) 0 to 8 s.

4. A particle moves according to the equation $x = 10t^2$, where $x$ is in meters and $t$ is in seconds. (a) Find the aver-

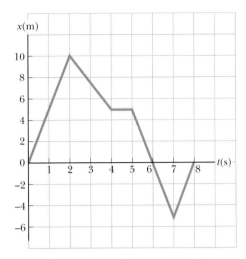

**Figure P2.3**  Problems 3 and 9.

age velocity for the time interval from 2.00 s to 3.00 s. (b) Find the average velocity for the time interval from 2.00 s to 2.10 s.

5. A person first walks at a constant speed $v_1$ along a straight line from $A$ to $B$ and then back along the line from $B$ to $A$ at a constant speed $v_2$. What are (a) her average speed over the entire trip and (b) her average velocity over the entire trip?

## Section 2.2   Instantaneous Velocity

6. The position of a particle moving along the $x$ axis varies in time according to the expression $x = 3t^2$, where $x$ is in meters and $t$ is in seconds. Evaluate its position (a) at $t = 3.00$ s and (b) at $3.00$ s $+ \Delta t$. (c) Evaluate the limit of $\Delta x/\Delta t$ as $\Delta t$ approaches zero, to find the velocity at $t = 3.00$ s.

7. A position–time graph for a particle moving along the $x$ web axis is shown in Figure P2.7. (a) Find the average velocity

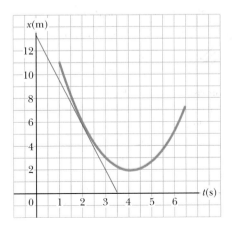

**Figure P2.7**

in the time interval $t = 1.5$ s to $t = 4.0$ s. (b) Determine the instantaneous velocity at $t = 2.0$ s by measuring the slope of the tangent line shown in the graph. (c) At what value of $t$ is the velocity zero?

8. (a) Use the data in Problem 1 to construct a smooth graph of position versus time. (b) By constructing tangents to the $x(t)$ curve, find the instantaneous velocity of the car at several instants. (c) Plot the instantaneous velocity versus time and, from this, determine the average acceleration of the car. (d) What was the initial velocity of the car?

9. Find the instantaneous velocity of the particle described in Figure P2.3 at the following times: (a) $t = 1.0$ s, (b) $t = 3.0$ s, (c) $t = 4.5$ s, and (d) $t = 7.5$ s.

## Section 2.3   Analysis Models — The Particle Under Constant Velocity

10. A truck tractor pulls two trailers at a constant speed of 100 km/h. It takes 0.600 s for the big rig to completely pass onto a bridge 400 m long. For what duration of time is all or part of the truck-trailer combination on the bridge?

11. A hare and a tortoise compete in a race over a course 1.00 km long. The tortoise crawls straight and steadily at its maximum speed of 0.200 m/s toward the finish line. The hare runs at its maximum speed of 8.00 m/s toward the goal for 0.800 km and then stops to tease the tortoise. How close to the goal can the hare let the tortoise approach before resuming the race, which the tortoise wins in a photo finish? Assume that, when moving, both animals move steadily at their respective maximum speeds.

## Section 2.4   Acceleration

12. A 50.0-g superball traveling at 25.0 m/s bounces off a brick wall and rebounds at 22.0 m/s. A high-speed camera records this event. If the ball is in contact with the wall for 3.50 ms, what is the magnitude of the average acceleration of the ball during this time interval? (*Note:* 1 ms $= 10^{-3}$ s.)

13. Figure P2.13 shows a graph of $v_x$ versus $t$ for the motion of a motorcyclist as he starts from rest and moves along the road in a straight line. (a) Find the average acceleration for the time interval $t_0 = 0$ to $t_1 = 6.0$ s. (b) Estimate the time at which the acceleration has its greatest positive value and the value of the acceleration at that instant. (c) When is the acceleration zero? (d) Estimate the maximum negative value of the acceleration and the time at which it occurs.

**Figure P2.13**

14. An object moves along the *x* axis according to the equation $x(t) = (3.00t^2 - 2.00t + 3.00)$ m. Determine (a) the average speed between $t = 2.00$ s and $t = 3.00$ s, (b) the instantaneous speed at $t = 2.00$ s and at $t = 3.00$ s, (c) the average acceleration between $t = 2.00$ s and $t = 3.00$ s, and (d) the instantaneous acceleration at $t = 2.00$ s and $t = 3.00$ s.

15. A particle moves along the *x* axis according to the equation $x = 2.00 + 3.00t - t^2$, where *x* is in meters and *t* is in seconds. At $t = 3.00$ s, find (a) the position of the particle, (b) its velocity, and (c) its acceleration.

16. A student drives a moped along a straight road as described by the velocity-versus-time graph in Figure P2.16. Sketch this graph in the middle of a sheet of graph paper. (a) Directly above your graph, sketch a graph of position versus time, aligning the time coordinates of the two graphs. (b) Sketch a graph of the acceleration versus time directly below the $v_x$-*t* graph, again aligning the time coordinates. On each graph, show the numerical values of *x* and $a_x$ for all points of inflection. (c) What is the acceleration at $t = 6$ s? (d) Find the position (relative to the starting point) at $t = 6$ s. (e) What is the moped's final position at $t = 9$ s?

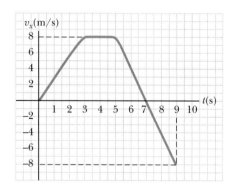

**Figure P2.16**

17. A particle starts from rest and accelerates as shown in Figure P2.17. Determine: (a) the particle's speed at $t = 10$ s and at $t = 20$ s and (b) the distance traveled in the first 20 s.

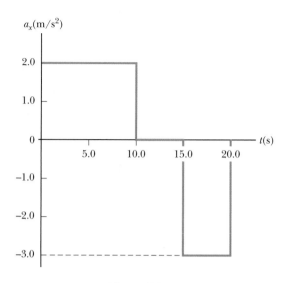

**Figure P2.17**

## Section 2.5 Motion Diagrams

18. Draw motion diagrams for (a) an object moving to the right at constant speed, (b) an object moving to the right and speeding up at a constant rate, (c) an object moving to the right and slowing down at a constant rate, (d) an object moving to the left and speeding up at a constant rate, and (e) an object moving to the left and slowing down at a constant rate. (f) How would your drawings change if the changes in speed were not uniform; that is, if the speed were not changing at a constant rate?

## Section 2.6 The Particle Under Constant Acceleration

19. A speedboat moving at 30.0 m/s approaches a no-wake buoy marker 100 m ahead. The pilot slows the boat with a constant acceleration of $-3.50$ m/s$^2$ by reducing the throttle. (a) How long does it take the boat to reach the buoy? (b) What is the velocity of the boat when it reaches the buoy?

20. A certain automobile manufacturer claims that its super-deluxe sports car will accelerate from rest to a speed of 42.0 m/s in 8.00 s. Under the (improbable) assumption that the acceleration is constant, (a) determine the acceleration of the car. (b) Find the distance the car travels in the first 8.00 s. (c) What is the speed of the car 10.0 s after it begins its motion if it continues to move with the same acceleration?

21. An object moving with uniform acceleration has a velocity of 12.0 cm/s in the positive *x* direction when its *x* coordinate is 3.00 cm. If its *x* coordinate 2.00 s later is $-5.00$ cm, what is its acceleration?

22. The minimum distance required to stop a car moving at 35.0 mi/h is 40.0 ft. What is the minimum stopping dis-

**Figure P2.29** (*Left*) Col. John Stapp on the rocket sled. (*Courtesy of the U.S. Air Force*)
(*Right*) Col. Stapp's face is contorted by the stress of rapid negative acceleration. (*Photri, Inc.*)

tance for the same car moving at 70.0 mi/h, assuming the same rate of acceleration?

**23.** The initial velocity of an object is 5.20 m/s. What is its velocity after 2.50 s (a) if it accelerates uniformly at 3.00 m/s$^2$ and (b) if it accelerates uniformly at $-3.00$ m/s$^2$?

**24.** A particle moves along the $x$-axis. Its position is given by the equation $x = 2.0 + 3.0t - 4.0t^2$ with $x$ in meters and $t$ in seconds. Determine (a) its position at the instant when it changes direction and (b) its velocity when it returns to the position it had at $t = 0$.

**25.** A jet plane comes in for a landing with a speed of 100 m/s and can accelerate at a maximum rate of $-5.00$ m/s$^2$ as it comes to rest. (a) From the instant the plane touches the runway, what is the minimum time it needs before it can come to rest? (b) Can this plane land at a small tropical island airport where the runway is 0.800 km long?

**26.** A car is approaching a hill at 30.0 m/s when its engine suddenly fails just at the bottom of the hill. The car moves with a constant acceleration of $-2.00$ m/s$^2$ while coasting up the hill. (a) Write equations for the position along the slope and for the velocity as functions of time, taking $x = 0$ at the bottom of the hill, where $v_i = 30.0$ m/s. (b) Determine the maximum distance the car travels up the hill.

**27.** The driver of a car slams on the brakes when he sees a tree blocking the road. The car slows uniformly with an acceleration of $-5.60$ m/s$^2$ for 4.20 s, making straight skid marks 62.4 m long, ending at the tree. With what speed does the car then strike the tree?

**28.** *Help! One of our equations is missing!* We describe constant-acceleration motion with the variables $v_{xf}$, $x_f$, and $t$, and parameters $v_{xi}$, $a_x$, and $x_i$. Of the equations in Table 2.2, the first does not involve $x_f$ or $x_i$. The second does not contain $a_x$, the third omits $v_{xf}$, and the last leaves out $t$. So to complete the set there should be an equation *not* involving $v_{xi}$. Derive it from the others. Use it to solve Problem 27 in one step.

**29.** For many years the world's land speed record was held by Colonel John P. Stapp, USAF. On March 19, 1954, he rode a rocket-propelled sled that moved down a track at 632 mi/h. He and the sled were safely brought to rest in 1.40 s. (Fig. P2.29) Determine (a) the negative acceleration he experienced and (b) the distance he traveled during this negative acceleration.

**30.** In a linear accelerator, an electron is accelerated to 1.00% of the speed of light in 40.0 m before it coasts 60.0 m to a target. (a) What is the electron's acceleration during the first 40.0 m? (b) How long does the total flight take?

**31.** A truck on a straight road starts from rest, accelerating at 2.00 m/s$^2$ until it reaches a speed of 20.0 m/s. Then the truck travels for 20.0 s at constant speed until the brakes are applied, stopping the truck in a uniform manner in an additional 5.00 s. (a) How long is the truck in motion? (b) What is the average velocity of the truck for the motion described?

**32.** A ball starts from rest and accelerates at 0.500 m/s$^2$ while moving down an inclined plane 9.00 m long. When it reaches the bottom, the ball rolls up another plane, where, after moving 15.0 m, it comes to rest. (a) What is the speed of the ball at the bottom of the first plane? (b) How long does it take to roll down the first plane? (c) What is the acceleration along the second plane? (d) What is the ball's speed 8.00 m along the second plane?

**33.** Speedy Sue, driving at 30.0 m/s, enters a one-lane tunnel. She then observes a slow-moving van 155 m ahead traveling at 5.00 m/s. Sue applies her brakes but can accelerate only at $-2.00$ m/s$^2$ because the road is wet. Will there be a collision? If so, determine how far into the tunnel and at what time the collision occurs. If not, determine the distance of closest approach between Sue's car and the van.

## Section 2.7 Freely Falling Objects

> *Note:* In all problems in this section, ignore the effects of air resistance.

**34.** A golf ball is released from rest from the top of a very tall building. Calculate (a) the position and (b) the velocity of the ball after 1.00 s, 2.00 s, and 3.00 s.

**35.** A student throws a set of keys vertically upward to her sorority sister, who is in a window 4.00 m above. The keys are caught 1.50 s later by the sister's outstretched hand. (a) With what initial velocity were the keys thrown? (b) What was the velocity of the keys just before they were caught?

**36.** A ball is thrown directly downward, with an initial speed of 8.00 m/s from a height of 30.0 m. After what time interval does the ball strike the ground?

**37.** A baseball is hit so that it travels straight upward after being struck by the bat. A fan observes that it takes 3.00 s for the ball to reach its maximum height. Find (a) its initial velocity and (b) the height reached by the ball.

**38.** *Every morning at seven o'clock*
*There's twenty terriers drilling on the rock.*
*The boss comes around and he says, "Keep still*
*And bear down heavy on the cast-iron drill*
*And drill, ye terriers, drill." And drill, ye terriers, drill.*
*It's work all day for sugar in your tea*
*Down beyond the railway. And drill, ye terriers, drill.*
*The foreman's name was John McAnn.*
*By God, he was a blamed mean man.*
*One day a premature blast went off*
*And a mile in the air went big Jim Goff. And drill  . . .*
*Then when next payday came around*
*Jim Goff a dollar short was found.*
*When he asked what for, came this reply:*
*"You were docked for the time you were up in the sky." And drill  . . .*
—American folksong
What was Goff's hourly wage? State the assumptions you make in computing it.

**39.** A daring ranch hand sitting on a tree limb wishes to drop vertically onto a horse galloping under the tree. The constant speed of the horse is 10.0 m/s, and the distance from the limb to the level of the saddle is 3.00 m. (a) What must be the horizontal distance between the saddle and limb when the ranch hand makes his move? (b) How long is he in the air?

**40.** A boy stands at the edge of a bridge 20.0 m above a river and throws a stone straight down with a speed of 12.0 m/s. He throws another pebble straight upward with the same speed so that it misses the edge of the bridge on the way back down and falls into the river. For each stone find (a) the velocity as it reaches the water and (b) the average velocity while it is in flight.

**41.** A ball is thrown vertically upward from the ground with an initial speed of 15.0 m/s. (a) How long does it take the ball to reach its maximum altitude? (b) What is its maximum altitude? (c) Determine the velocity and acceleration of the ball at $t = 2.00$ s.

**42.** The height of a helicopter above the ground is given by $h = 3.00t^3$, where $h$ is in meters and $t$ is in seconds. After 2.00 s, the helicopter pilot lets go of a small mailbag. How long after its release does the mailbag reach the ground?

## Section 2.8 Context Connection — Liftoff Acceleration

**43.** Jules Verne in 1865 suggested sending people to the Moon by firing a space capsule from a 220-m-long cannon with a final velocity of 10.97 km/s. What would have been the unrealistically large acceleration experienced by the space travelers during launch? Compare your answer with the free-fall acceleration 9.80 m/s$^2$.

## *Additional Problems*

**44.** Figure P2.44 represents part of the performance data of a car owned by a proud physics student. (a) Calculate from the graph the total distance traveled. (b) What distance does the car travel between the times $t = 10$ s and $t = 40$ s? (c) Draw a graph of its acceleration versus time between $t = 0$ and $t = 50$ s. (d) Write an equation for $x$ as a function of time for each phase of the motion, represented by (i) $0a$, (ii) $ab$, (iii) $bc$. (e) What is the average velocity of the car between $t = 0$ and $t = 50$ s?

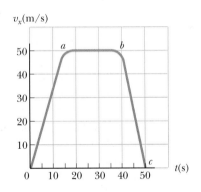

**Figure P2.44**

**45.** A test rocket is fired vertically upward from a well. A catapult gives it an initial velocity of 80.0 m/s at ground level. Its engines then fire and give it an upward acceleration of 4.00 m/s$^2$ until it reaches an altitude of 1000 m. At that point its engines fail and the rocket goes into free-fall, with an acceleration of $-9.80$ m/s$^2$. (a) How long is the rocket in motion above the ground? (b) What is its maximum altitude? (c) What is its velocity just before it collides with the Earth? (*Hint:* Consider the motion while the engine is operating separate from the free-fall motion.)

**46.** A motorist drives along a straight road at a constant speed of 15.0 m/s. Just as she passes a parked motorcycle police officer, the officer starts to accelerate at 2.00 m/s² to overtake her. Assuming the officer maintains this acceleration, (a) determine the time it takes the police officer to reach the motorist. Find (b) the speed and (c) the total displacement of the officer as he overtakes the motorist.

**47.** Setting new world records in a 100-m race, Maggie and Judy cross the finish line in a dead heat, both taking 10.2 s. Accelerating uniformly, Maggie took 2.00 s and Judy 3.00 s to attain maximum speed, which they maintained for the rest of the race. (a) What was the acceleration of each sprinter? (b) What were their respective maximum speeds? (c) Which sprinter was ahead at the 6.00-s mark and by how much?

**48.** A commuter train travels between two downtown stations. Because the stations are only 1.00 km apart, the train never reaches its maximum possible cruising speed. During rush hour the engineer minimizes the travel time $t$ between the two stations by accelerating at $a_1 = 0.100$ m/s² for a time $t_1$ and then immediately braking with acceleration $a_2 = -0.500$ m/s² for a time $t_2$. Find the minimum time of travel $t$ and the time $t_1$.

**49.** A hard rubber ball, released at chest height, falls to the pavement and bounces back to nearly the same height. When it is in contact with the pavement, the lower side of the ball is temporarily flattened. Suppose that the maximum depth of the dent is on the order of 1 cm. Compute an order-of-magnitude estimate for the maximum acceleration of the ball while it is in contact with the pavement. State your assumptions, the quantities you estimate, and the values you estimate for them.

**50.** At NASA's John H. Glenn research center in Cleveland, Ohio, free-fall research is performed by dropping experiment packages from the top of an evacuated shaft 145 m high. Free-fall imitates the so-called microgravity environment of a satellite in orbit. (a) What is the maximum time for free-fall if a package falls the entire 145 m? (b) Actual NASA specifications allow for a 5.18-s free-fall time. How far do the packages drop and (c) what is their speed at 5.18 s? (d) What constant acceleration is required to stop a package in the distance remaining in the shaft after its 5.18-s fall?

**51.** An inquisitive physics student and mountain climber climbs a 50.0-m cliff that overhangs a calm pool of water. He throws two stones vertically downward, 1.00 s apart, and observes that they cause a single splash. The first stone has an initial speed of 2.00 m/s. (a) How long after release of the first stone do the two stones hit the water? (b) What was the initial velocity of the second stone? (c) What is the velocity of each stone at the instant the two hit the water?

**52.** A rock is dropped from rest into a well. (a) If the sound of the splash is heard 2.40 s later, how far below the top of the well is the surface of the water? The speed of sound in air (at the ambient temperature) is 336 m/s. (b) If the travel time for the sound is ignored, what percentage error is introduced when the depth of the well is calculated?

**53.** To protect his food from hungry bears, a boy scout raises his food pack with a rope that is thrown over a tree limb at height $h$ above his hands. He walks away from the vertical rope with constant velocity $v_{boy}$, holding the free end of the rope in his hands (Fig. P2.53). (a) Show that the speed $v$ of the food pack is given by $x(x^2 + h^2)^{-1/2} v_{boy}$, where $x$ is the distance he has walked away from the vertical rope. (b) Show that the acceleration $a$ of the food pack is $h^2(x^2 + h^2)^{-3/2} v_{boy}^2$. (c) What values do the acceleration and velocity $v$ have shortly after he leaves the point under the pack ($x = 0$)? (d) What values do the pack's velocity and acceleration approach as the distance $x$ continues to increase?

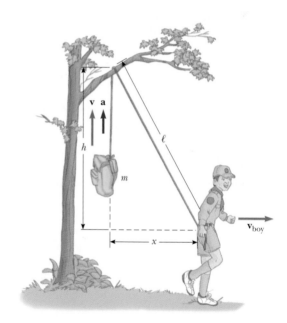

**Figure P2.53** Problems 53 and 54.

**54.** In Problem 53, let the height $h$ equal 6.00 m and the speed $v_{boy}$ equal 2.00 m/s. Assume that the food pack starts from rest. (a) Tabulate and graph the speed–time graph. (b) Tabulate and graph the acceleration–time graph. (Let the range of time be from 0 s to 5.00 s and the time intervals be 0.500 s.)

**55.** Liz rushes down onto a subway platform to find her train already departing. She stops and watches the cars go by. Each car is 8.60 m long. The first moves past her in 1.50 s and the second in 1.10 s. Find the constant acceleration of the train.

56. Astronauts on a distant planet toss a rock into the air. With the aid of a camera that takes pictures at a steady rate, they record the height of the rock as a function of time as given in the following table. (a) Find the average velocity of the rock in the time interval between each measurement and the next. (b) Using these average velocities to approximate instantaneous velocities at the midpoints of the time intervals, make a graph of velocity as a function of time. Does the rock move with constant acceleration? If so, plot a straight line of best fit on the graph and calculate its slope to find the acceleration.

| TABLE P2.56 | Height of a Rock versus Time | | |
|---|---|---|---|
| Time (s) | Height (m) | Time (s) | Height (m) |
| 0.00 | 5.00 | 2.75 | 7.62 |
| 0.25 | 5.75 | 3.00 | 7.25 |
| 0.50 | 6.40 | 3.25 | 6.77 |
| 0.75 | 6.94 | 3.50 | 6.20 |
| 1.00 | 7.38 | 3.75 | 5.52 |
| 1.25 | 7.72 | 4.00 | 4.73 |
| 1.50 | 7.96 | 4.25 | 3.85 |
| 1.75 | 8.10 | 4.50 | 2.86 |
| 2.00 | 8.13 | 4.75 | 1.77 |
| 2.25 | 8.07 | 5.00 | 0.58 |
| 2.50 | 7.90 | | |

57. Two objects, *A* and *B*, are connected by a rigid rod that has a length *L*. The objects slide along perpendicular guide rails, as shown in Figure P2.57. If *A* slides to the left with a constant speed *v*, find the speed of *B* when $\alpha = 60.0°$.

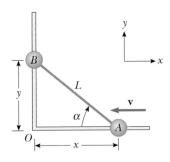

**Figure P2.57**

## ANSWERS TO QUICK QUIZZES

**2.1** A member of the highway patrol is interested in your instantaneous speed. Regardless of your speeds at all other times, if your instantaneous speed at the instant that it is measured is higher than the speed limit, you may receive a speeding ticket.

**2.2** If the particle moves along a line without changing direction, the displacement and distance over any time interval are the same. As a result, the magnitude of the average velocity and the average speed are the same. If the particle reverses direction, however, the displacement would be less than the distance. In turn, the magnitude of the average velocity would be smaller than the average speed.

**2.3** Graph (a) has a constant slope, indicating a constant acceleration; this is represented by graph (e). Graph (b) represents a speed that is increasing constantly but not at a uniform rate. Thus, the acceleration must be increasing, and the graph that best indicates this is (d). Graph (c) depicts a velocity that first increases at a constant rate, indicating constant acceleration. Then the velocity stops increasing and becomes constant, indicating zero acceleration. The best match to this is graph (f).

**2.4** If the car is slowing down, the direction of its acceleration is opposite to its direction of motion. Thus, the acceleration and, therefore, the force are directed to the west.

**2.5** No. For an accelerating particle to stop at all, the velocity and acceleration must have opposite signs, so that the speed is decreasing. Given that this is the case, the particle will eventually come to rest. If the acceleration remains constant, however, the particle must begin to move again, opposite to the direction of its original motion. If the particle comes to rest and then stays at rest, the acceleration has become zero at the moment the motion stops. This is the case for a braking car—the acceleration is negative and goes to zero as the car comes to rest.

**2.6** (a) As it travels upward, its speed decreases by 9.80 m/s during each second of its motion. When it reaches the peak of its motion, its speed becomes zero. As the ball moves downward, its speed increases by 9.80 m/s each second. (b) The acceleration of the ball remains constant while it is in the air. The magnitude of its acceleration is the free-fall acceleration, $g = 9.80$ m/s$^2$.

chapter

**3**

**A welder cuts holes through a heavy metal construction beam with a hot torch. The sparks generated in the process all follow a path that is described by a curve called a** *parabola.*

*(©The Telegraph Colour Library/FPG)*

# Motion in Two Dimensions

I n this chapter we shall study the kinematics of an object that can be modeled as a particle moving in a plane. This is two-dimensional motion. Some common examples of motion in a plane are the motions of satellites in orbit around the Earth, projectiles, such as a thrown baseball, and the motion of electrons in uniform electric fields. We shall also study a particle in uniform circular motion and discuss various aspects of particles moving in curved paths.

## 3.1 • THE POSITION, VELOCITY, AND ACCELERATION VECTORS

In Chapter 2 we found that the motion of a particle moving along a straight line is completely specified if its position is known as a function of time. Now let us extend this idea to motion in the $xy$ plane. We will find equations for position and velocity that are the same as those in Chapter 2, except for their vector nature.

We begin by describing the position of a particle with a **position vector r,** drawn from the origin of a reference frame to the location of the particle in the $xy$

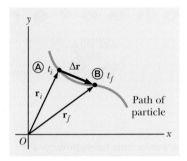

**Figure 3.1**

A particle moving in the *xy* plane is located with the position vector **r** drawn from the origin to the particle. The displacement of the particle as it moves from Ⓐ to Ⓑ in the time interval $\Delta t = t_f - t_i$ is equal to the vector $\Delta \mathbf{r} = \mathbf{r}_f - \mathbf{r}_i$.

plane, as in Figure 3.1. At time $t_i$, the particle is at the point Ⓐ, and at some later time $t_f$, the particle is at Ⓑ, where the subscripts *i* and *f* refer to initial and final values. As the particle moves from Ⓐ to Ⓑ in the time interval $\Delta t = t_f - t_i$, the position vector changes from $\mathbf{r}_i$ to $\mathbf{r}_f$. As we learned in Chapter 2, the displacement of a particle is the difference between its final position and initial position:

$$\Delta \mathbf{r} \equiv \mathbf{r}_f - \mathbf{r}_i \qquad [3.1]$$

The direction of $\Delta \mathbf{r}$ is indicated in Figure 3.1.

The **average velocity** $\bar{\mathbf{v}}$ of the particle during the time interval $\Delta t$ is defined as the ratio of the displacement to the time interval:

$$\bar{\mathbf{v}} \equiv \frac{\Delta \mathbf{r}}{\Delta t} \qquad [3.2]$$

Because displacement is a vector quantity and the time interval is a scalar quantity, we conclude that the average velocity is a *vector* quantity directed along $\Delta \mathbf{r}$. The average velocity between points Ⓐ and Ⓑ is *independent of the path* between the two points. This is because the average velocity is proportional to the displacement, which in turn depends only on the initial and final position vectors and not on the path taken between those two points. As with one-dimensional motion, if a particle starts its motion at some point and returns to this point via any path, its average velocity is zero for this trip because its displacement is zero.

Consider again the motion of a particle between two points in the *xy* plane, as shown in Figure 3.2. As the time intervals over which we observe the motion become smaller and smaller, the direction of the displacement approaches that of the line tangent to the path at the point Ⓐ.

The **instantaneous velocity** **v** is defined as the limit of the average velocity $\Delta \mathbf{r}/\Delta t$ as $\Delta t$ approaches zero:

$$\mathbf{v} \equiv \lim_{\Delta t \to 0} \frac{\Delta \mathbf{r}}{\Delta t} = \frac{d\mathbf{r}}{dt} \qquad [3.3]$$

- *Definition of instantaneous velocity*

That is, the instantaneous velocity equals the derivative of the position vector with respect to time. The direction of the instantaneous velocity vector at any point in a particle's path is along a line that is tangent to the path at that point and in the direction of motion. The magnitude of the instantaneous velocity is called the *speed*.

As a particle moves from point Ⓐ to point Ⓑ along some path as in Figure 3.3, its instantaneous velocity changes from $\mathbf{v}_i$ at time $t_i$ to $\mathbf{v}_f$ at time $t_f$.

The **average acceleration** $\bar{\mathbf{a}}$ of a particle over a time interval is defined as the ratio of the change in the instantaneous velocity $\Delta \mathbf{v}$ to the time interval $\Delta t$:

$$\bar{\mathbf{a}} \equiv \frac{\mathbf{v}_f - \mathbf{v}_i}{t_f - t_i} = \frac{\Delta \mathbf{v}}{\Delta t} \qquad [3.4]$$

**PITFALL PREVENTION 3.1**

**Vector addition**

The vector addition that was discussed in Chapter 1 involved displacement vectors. Because we are familiar with movements through space due to our everyday experience, the addition of displacement vectors can be understood easily. The notion of vector addition can be applied to *any* type of vector quantity. Figure 3.3, for example, shows the addition of velocity vectors using the tip-to-tail approach.

Because the average acceleration is the ratio of a vector quantity $\Delta \mathbf{v}$ and a scalar quantity $\Delta t$, we conclude that $\bar{\mathbf{a}}$ is a vector quantity directed along $\Delta \mathbf{v}$. As indicated in Figure 3.3, the direction of $\Delta \mathbf{v}$ is found by adding the vector $-\mathbf{v}_i$ (the negative of $\mathbf{v}_i$) to the vector $\mathbf{v}_f$, because by definition $\Delta \mathbf{v} \equiv \mathbf{v}_f - \mathbf{v}_i$.

The **instantaneous acceleration a** is defined as the limiting value of the ratio $\Delta\mathbf{v}/\Delta t$ as $\Delta t$ approaches zero:

$$\mathbf{a} \equiv \lim_{\Delta t \to 0} \frac{\Delta\mathbf{v}}{\Delta t} = \frac{d\mathbf{v}}{dt}$$    [3.5]

• *Definition of instantaneous acceleration*

That is, the instantaneous acceleration equals the derivative of the velocity vector with respect to time.

It is important to recognize that various changes can occur that represent a particle undergoing an acceleration. First, the magnitude of the velocity vector (the speed) may change with time as in straight-line (one-dimensional) motion. Second, the direction of the velocity vector may change with time as its magnitude remains constant. Finally, both the magnitude and the direction of the velocity vector may change.

## Quick Quiz 3.1

(a) Can an object accelerate if its speed is constant? (b) Can an object accelerate if its velocity is constant?

## Quick Quiz 3.2

(a) Can a particle have a constant velocity and a varying speed? (b) Can a particle have a constant speed and a varying velocity? Give examples of a positive answer and explain a negative answer.

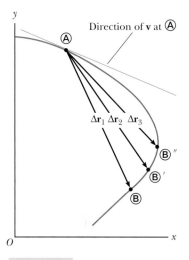

**Figure 3.2**

As a particle moves between two points, its average velocity is in the direction of the displacement vector $\Delta\mathbf{r}$. As the end point of the path is moved from Ⓑ to Ⓑ′ to Ⓑ″, the respective displacements and corresponding time intervals become smaller and smaller. In the limit that the end point approaches Ⓐ, $\Delta t$ approaches zero and the direction of $\Delta\mathbf{r}$ approaches that of the line tangent to the curve at Ⓐ. By definition, the instantaneous velocity at Ⓐ is in the direction of this tangent line.

## THINKING PHYSICS 3.1

The gas pedal in an automobile is commonly called the *accelerator*. Are there any other controls that could also be considered accelerators?

**Reasoning**    The gas pedal is called the accelerator because the common usage of the word *acceleration* refers to an *increase in speed*. The scientific definition, however, is that acceleration occurs *whenever the velocity changes in any way*. Thus, the *brake pedal* can also be considered an accelerator because it causes the car to slow down. The *steering wheel* is also an accelerator because it changes the direction of the velocity vector.

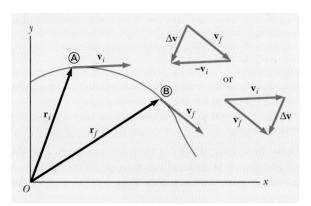

**Figure 3.3**

A particle moves from position Ⓐ to position Ⓑ. Its velocity vector changes from $\mathbf{v}_i$ at time $t_i$ to $\mathbf{v}_f$ at time $t_f$. The vector addition diagrams at the upper right show two ways of determining the vector $\Delta\mathbf{v}$ from the initial and final velocities.

## 3.2 • TWO-DIMENSIONAL MOTION WITH CONSTANT ACCELERATION

Let us consider two-dimensional motion during which the magnitude and direction of the acceleration remain unchanged. In this situation, we shall investigate motion as a two-dimensional version of the analysis in Section 2.6.

The motion of a particle can be determined if its position vector $\mathbf{r}$ is known at all times. The position vector for a particle moving in the $xy$ plane can be written

$$\mathbf{r} = x\mathbf{i} + y\mathbf{j} \qquad \text{[3.6]}$$

where $x$, $y$, and $\mathbf{r}$ change with time as the particle moves. If the position vector is known, the velocity of the particle can be obtained from Equations 3.3 and 3.6:

$$\mathbf{v} = \frac{d\mathbf{r}}{dt} = \frac{dx}{dt}\mathbf{i} + \frac{dy}{dt}\mathbf{j} = v_x\mathbf{i} + v_y\mathbf{j} \qquad \text{[3.7]}$$

Because we are assuming that $\mathbf{a}$ is constant in this discussion, its components $a_x$ and $a_y$ are also constants. Therefore, we can apply the equations of kinematics to the $x$ and $y$ components of the velocity vector separately. Substituting $v_x = v_{xf} = v_{xi} + a_x t$ and $v_y = v_{yf} = v_{yi} + a_y t$ into Equation 3.7 gives

$$\mathbf{v}_f = (v_{xi} + a_x t)\mathbf{i} + (v_{yi} + a_y t)\mathbf{j}$$
$$= (v_{xi}\mathbf{i} + v_{yi}\mathbf{j}) + (a_x\mathbf{i} + a_y\mathbf{j})t$$

$$\mathbf{v}_f = \mathbf{v}_i + \mathbf{a}t \qquad \text{[3.8]}$$

• *Velocity vector as a function of time for constant acceleration*

This result states that the velocity $\mathbf{v}_f$ of a particle at some time $t$ equals the vector sum of its initial velocity $\mathbf{v}_i$ and the additional velocity $\mathbf{a}t$ acquired at time $t$ as a result of its constant acceleration. This is the same as Equation 2.8, except for its vector nature.

Similarly, from Equation 2.11 we know that the $x$ and $y$ coordinates of a particle moving with constant acceleration are

$$x_f = x_i + v_{xi}t + \tfrac{1}{2}a_x t^2 \qquad \text{and} \qquad y_f = y_i + v_{yi}t + \tfrac{1}{2}a_y t^2$$

Substituting these expressions into Equation 3.6 gives

$$\mathbf{r}_f = (x_i + v_{xi}t + \tfrac{1}{2}a_x t^2)\mathbf{i} + (y_i + v_{yi}t + \tfrac{1}{2}a_y t^2)\mathbf{j}$$
$$= (x_i\mathbf{i} + y_i\mathbf{j}) + (v_{xi}\mathbf{i} + v_{yi}\mathbf{j})t + \tfrac{1}{2}(a_x\mathbf{i} + a_y\mathbf{j})t^2$$

• *Position vector as a function of time for constant acceleration*

$$\mathbf{r}_f = \mathbf{r}_i + \mathbf{v}_i t + \tfrac{1}{2}\mathbf{a}t^2 \qquad \text{[3.9]}$$

This equation implies that the final position vector $\mathbf{r}_f$ is the vector sum of the initial position vector $\mathbf{r}_i$ plus a displacement $\mathbf{v}_i t$, arising from the initial velocity of the particle, and a displacement $\tfrac{1}{2}\mathbf{a}t^2$, resulting from the uniform acceleration of the particle. It is the same as Equation 2.11, except for its vector nature.

Pictorial representations of Equations 3.8 and 3.9 are shown in Figures 3.4a and 3.4b. Note from Figure 3.4b that $\mathbf{r}_f$ is generally not along the direction of $\mathbf{v}_i$ or $\mathbf{a}$, because the relationship between these quantities is a vector expression. For the same reason, from Figure 3.4a we see that $\mathbf{v}_f$ is generally not along the direction of $\mathbf{v}_i$ or $\mathbf{a}$. Finally, if we compare the two figures, we see that $\mathbf{v}_f$ and $\mathbf{r}_f$ are not in the same direction.

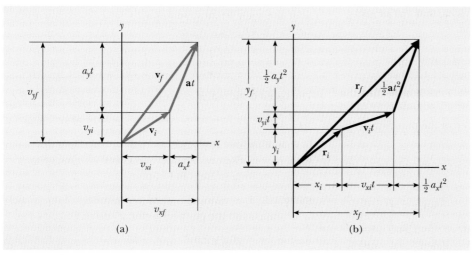

**Figure 3.4**

Vector representations and components of (a) the velocity and (b) the position of a particle under constant acceleration **a**.

Because Equations 3.8 and 3.9 are *vector* expressions, we may also write their $x$ and $y$ component equations:

$$\mathbf{v}_f = \mathbf{v}_i + \mathbf{a}t \quad \longrightarrow \quad \begin{cases} v_{xf} = v_{xi} + a_x t \\ v_{yf} = v_{yi} + a_y t \end{cases}$$

$$\mathbf{r}_f = \mathbf{r}_i + \mathbf{v}_i t + \tfrac{1}{2}\mathbf{a}t^2 \quad \longrightarrow \quad \begin{cases} x_f = x_i + v_{xi}t + \tfrac{1}{2}a_x t^2 \\ y_f = y_i + v_{yi}t + \tfrac{1}{2}a_y t^2 \end{cases}$$

These components are illustrated in Figure 3.4. In other words, **two-dimensional motion having constant acceleration is equivalent to two *independent* motions in the $x$ and $y$ directions having constant accelerations $a_x$ and $a_y$.** Motion in the $x$ direction does not affect motion in the $y$ direction, and vice-versa. Thus, there is no new model for a particle under two-dimensional constant acceleration—the appropriate model is just the one-dimensional particle under constant acceleration applied twice, in the $x$ and $y$ directions separately!

## Example 3.1 Motion in a Plane

A particle moves through the origin of an $xy$ coordinate system at $t = 0$ with initial velocity $\mathbf{v}_i = (20\mathbf{i} - 15\mathbf{j})$ m/s. The particle moves in the $xy$ plane with an acceleration $\mathbf{a} = 4.0\mathbf{i}$ m/s$^2$. (a) Determine the components of velocity as a function of time and the total velocity vector at any time.

**Reasoning** Establish the mental representation—think about what the particle is doing. From the given information, we see that the particle starts off moving to the right and downward and accelerates only toward the right—what will the particle do under these conditions? It may help if you draw a pictorial representation. Because the ac-

celeration is only in the $x$ direction, the moving particle can be modeled as one under constant acceleration in the $x$ direction, and one under constant velocity in the $y$ direction.

**Solution** (a) With $v_{xi} = 20$ m/s and $a_x = 4.0$ m/s$^2$, the equations of kinematics give us for the $x$ direction,

$$v_{xf} = v_{xi} + a_x t = (20 + 4.0t)$$

Also, with $v_{yi} = -15$ m/s and $a_y = 0$,

$$v_{yf} = v_{yi} = -15 \text{ m/s}$$

Therefore, using these results and noting that the velocity vector **v** has two components, we find

$$\boxed{\mathbf{v}_f = v_{xf}\mathbf{i} + v_{yf}\mathbf{j} = [(20 + 4.0t)\mathbf{i} - 15\mathbf{j}]}$$

Note that only the $x$ component varies in time, reflecting the fact that acceleration occurs only in the $x$ direction.

(b) Calculate the velocity and speed of the particle at $t = 5.0$ s.

**Solution**   Taking $t = 5.0$ s, the velocity expression from (a) gives

$$\mathbf{v}_f = \{[20 + 4(5.0)]\mathbf{i} - 15\mathbf{j}\} \text{ m/s} = \boxed{(40\mathbf{i} - 15\mathbf{j}) \text{ m/s}}$$

That is, at $t = 5.0$ s, $v_{xf} = 40$ m/s and $v_{yf} = -15$ m/s. Knowing these two components for this two-dimensional motion, we know the numerical value of the velocity vector. To deter-

mine the angle $\theta$ that $\mathbf{v}_f$ makes with the $x$ axis, we use the fact that $\tan \theta = v_{yf}/v_{xf}$, or

$$\theta = \tan^{-1}\left(\frac{v_{yf}}{v_{xf}}\right) = \tan^{-1}\left(\frac{-15 \text{ m/s}}{40 \text{ m/s}}\right) = -21°$$

The speed is the magnitude of $\mathbf{v}_f$:

$$v_f = |\mathbf{v}_f| = \sqrt{v_{xf}^2 + v_{yf}^2} = \sqrt{(40)^2 + (-15)^2} \text{ m/s} = \boxed{43 \text{ m/s}}$$

In examining our result, we find that $v_f > v_i$. Does this make sense to you? (Remember step 5—**Think**—in the general problem-solving strategy in Chapter 1.)

**EXERCISE**   Determine the $x$ and $y$ coordinates of the particle at any time $t$ and the position vector at this time.

*Answer*   $x_f = 20t + 2.0t^2$   $y_f = -15t$
$\mathbf{r}_f = (20t + 2.0t^2)\mathbf{i} - 15t\mathbf{j}$

---

## 3.3 • PROJECTILE MOTION

See the *Core Concepts in Physics CD-ROM*, Screen 3.5

• *Assumptions of projectile motion*

Anyone who has observed a baseball in motion (or, for that matter, any object thrown into the air) has observed projectile motion. The ball moves in a curved path when thrown at some angle with respect to the Earth's surface. This very common form of motion is surprisingly simple to analyze if the following two assumptions are made in building a model for these types of problems: (1) the free-fall acceleration $g$ is constant over the range of motion and is directed downward,* and (2) the effect of air resistance is negligible.† With these assumptions, the path of a projectile, called its *trajectory,* is *always* a parabola. **We shall use a simplification model based on these assumptions throughout this chapter.**

If we choose our reference frame such that the $y$ direction is vertical and positive upward, then $a_y = -g$ (as in one-dimensional free-fall) and $a_x = 0$ (because the only possible horizontal acceleration is due to air resistance, and it is ignored). Furthermore, let us assume that at $t = 0$, the projectile leaves the origin (point Ⓐ, $x_i = y_i = 0$) with speed $v_i$, as in Figure 3.5. If the vector $\mathbf{v}_i$ makes an angle $\theta_i$ with the horizontal, we can identify a right triangle in the diagram as a geometric model, and from the definitions of the cosine and sine functions we have

$$\cos \theta_i = \frac{v_{xi}}{v_i} \quad \text{and} \quad \sin \theta_i = \frac{v_{yi}}{v_i}$$

---

* This approximation is reasonable as long as the range of motion is small compared with the radius of the Earth ($6.4 \times 10^6$ m). In effect, this approximation is equivalent to assuming that the Earth is flat within the range of motion considered, and that the maximum height of the object is small compared to the radius of the Earth.

† This approximation is often *not* justified, especially at high velocities. In addition, the spin of a projectile, such as a baseball, can give rise to some very interesting effects associated with aerodynamic forces (for example, a curve ball thrown by a pitcher).

**Figure 3.5**

The parabolic path of a projectile that leaves the origin (point Ⓐ) with a velocity $\mathbf{v}_i$. The velocity vector $\mathbf{v}$ changes with time in both magnitude and direction. The change in the velocity vector is the result of acceleration in the negative $y$ direction. The $x$ component of velocity remains constant in time because no acceleration occurs in the horizontal direction. The $y$ component of velocity is zero at the peak of the path (point Ⓒ).

Therefore, the initial $x$ and $y$ components of velocity are

$$v_{xi} = v_i \cos \theta_i \quad \text{and} \quad v_{yi} = v_i \sin \theta_i$$

Substituting these expressions into Equations 3.8 and 3.9 with $a_x = 0$ and $a_y = -g$ gives the velocity components and position coordinates for the projectile at any time $t$:

$$v_{xf} = v_{xi} = v_i \cos \theta_i = \text{constant} \qquad \textbf{[3.10]} \qquad \bullet \quad \textit{Horizontal velocity component}$$

$$v_{yf} = v_{yi} - gt = v_i \sin \theta_i - gt \qquad \textbf{[3.11]} \qquad \bullet \quad \textit{Vertical velocity component}$$

$$x_f = x_i + v_{xi}t = (v_i \cos \theta_i)\, t \qquad \textbf{[3.12]} \qquad \bullet \quad \textit{Horizontal position component}$$

$$y_f = y_i + v_{yi}t - \tfrac{1}{2}gt^2 = (v_i \sin \theta_i)\, t - \tfrac{1}{2}gt^2 \qquad \textbf{[3.13]} \qquad \bullet \quad \textit{Vertical position component}$$

From Equation 3.10 we see that $v_{xf}$ remains constant in time and is equal to $v_{xi}$; there is no horizontal component of acceleration. Thus, we model the horizontal motion as a particle moving with constant velocity. For the $y$ motion, note that the equations for $v_{yf}$ and $y_f$ are similar to Equations 2.8 and 2.11 for freely falling objects. Thus, we can apply the model of a particle moving with constant acceleration to the $y$ component. In fact, *all* of the equations of kinematics developed in Chapter 2 are applicable to projectile motion.

If we solve for $t$ in Equation 3.12 and substitute this expression for $t$ into Equation 3.13, we find that

$$y_f = (\tan \theta_i) x_f - \left( \frac{g}{2v_i^2 \cos^2 \theta_i} \right) x_f^2 \qquad \textbf{[3.14]}$$

This strobe photograph of two balls released simultaneously illustrates both free fall (red ball) and projectile motion (yellow ball). The yellow ball was projected horizontally, while the red ball was released from rest. Can you explain why both balls reach the floor simultaneously? *(Richard Megna, Fundamental Photographs)*

which is valid for angles in the range $0 < \theta_i < \pi/2$. This expression is of the form $y = ax - bx^2$, which is the equation of a parabola that passes through the origin. Thus, we have proven that the trajectory of a projectile can be geometrically modeled as a parabola. The trajectory is *completely* specified if $v_i$ and $\theta_i$ are known.

The vector expression for the position of the projectile as a function of time follows directly from Equation 3.9, with $\mathbf{a} = \mathbf{g}$:

$$\mathbf{r}_f = \mathbf{r}_i + \mathbf{v}_i t + \tfrac{1}{2}\mathbf{g}t^2$$

- *Vector expression for the position of a projectile*

This equation gives the same information as the combination of Equations 3.12 and 3.13 and is plotted in Figure 3.6. Note that this expression for $\mathbf{r}_f$ is consistent with Equation 3.13, because the expression for $\mathbf{r}_f$ is a vector equation and $\mathbf{a} = \mathbf{g} = -g\mathbf{j}$ when the upward direction is taken to be positive.

The position of a particle can be considered the superposition of its original position $\mathbf{r}_i$, the term $\mathbf{v}_i t$, which would be the displacement if no acceleration were present, and the term $\tfrac{1}{2}\mathbf{g}t^2$, which arises from the acceleration due to gravity. In other words, if no gravitational acceleration occurred, the particle would continue to move along a straight path in the direction of $\mathbf{v}_i$.

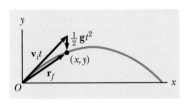

**Figure 3.6**

The position vector **r** of a projectile whose initial velocity at the origin is $\mathbf{v}_i$. The vector $\mathbf{v}_i t$ would be the position vector of the projectile if gravity were absent. To this is added the vector $\tfrac{1}{2}\mathbf{g}t^2$ due to the downward gravitational acceleration of the projectile.

### Horizontal Range and Maximum Height of a Projectile

Let us assume that a projectile is launched over flat ground from the origin at $t = 0$ with a positive $v_y$ component, as in Figure 3.7. There are two special points that are interesting to analyze: the peak point Ⓐ, which has cartesian coordinates $(R/2, h)$, and the landing point Ⓑ, having coordinates $(R, 0)$. The distance $R$ is called the *horizontal range* of the projectile, and $h$ is its *maximum height*. Because of the symmetry of the trajectory, the projectile is at the maximum height $h$ when its $x$ position is half of the range $R$. Let us find $h$ and $R$ in terms of $v_i$, $\theta_i$, and $g$.

We can determine $h$ by noting that at the peak, $v_{yA} = 0$. Therefore, Equation 3.11 can be used to determine the time $t_1$ it takes to reach the peak:

$$t_1 = \frac{v_i \sin \theta_i}{g}$$

Substituting this expression for $t_1$ into Equation 3.13 and replacing $y_f$ with $h$ gives $h$ in terms of $v_i$ and $\theta_i$:

$$h = (v_i \sin \theta_i) \frac{v_i \sin \theta_i}{g} - \frac{1}{2}g\left(\frac{v_i \sin \theta_i}{g}\right)^2$$

$$h = \frac{v_i^2 \sin^2 \theta_i}{2g} \qquad \text{[3.15]}$$

Notice from the mathematical representation how you could increase the peak height $h$: You could launch the projectile with a larger initial velocity, at a higher angle, or at a location with lower free-fall acceleration, such as on the Moon. Is this consistent with your mental representation of this situation?

The range $R$ is the horizontal distance traveled in twice the time it takes to reach the peak, that is, in a time $2t_1$. (This can be seen by setting $y_f = 0$ in Equation 3.13 and solving the quadratic for $t$. One solution of this quadratic is $t = 0$, and the second is $t = 2t_1$.) Using Equation 3.12 and noting that $x_f = R$ at $t = 2t_1$, we find that

$$R = (v_i \cos \theta_i)2t_1 = (v_i \cos \theta_i) \frac{2v_i \sin \theta_i}{g} = \frac{2v_i^2 \sin \theta_i \cos \theta_i}{g}$$

Because $\sin 2\theta = 2 \sin \theta \cos \theta$, $R$ can be written in the more compact form

$$R = \frac{v_i^2 \sin 2\theta_i}{g} \qquad \text{[3.16]}$$

Notice from the mathematical expression how you could increase the range $R$: You could launch the projectile with a larger initial velocity or at a location with lower free-fall acceleration, such as on the Moon. Is this consistent with your mental representation of this situation?

The range also depends on the angle of the initial velocity vector. The maximum possible value of $R$ from Equation 3.16 is given by $R_{max} = v_i^2/g$. This result follows from the fact that the maximum value of $\sin 2\theta_i$ is unity, which occurs when $2\theta_i = 90°$. Therefore, $R$ is a maximum when $\theta_i = 45°$.

Figure 3.8 illustrates various trajectories for a projectile of a given initial speed. As you can see, the range is a maximum for $\theta_i = 45°$. In addition, for any $\theta_i$ other than $45°$, a point with coordinates $(R, 0)$ can be reached by using either one of two complementary values of $\theta_i$, such as $75°$ and $15°$. Of course, the maximum height and the time of flight will be different for these two values of $\theta_i$.

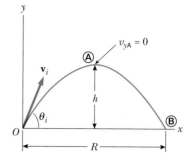

**Figure 3.7**

A projectile launched from the origin at $t = 0$ with an initial velocity $\mathbf{v}_i$. The maximum height of the projectile is $h$, and its horizontal range is $R$. At Ⓐ, the peak of the trajectory, the projectile has coordinates $\left(\dfrac{R}{2}, h\right)$.

**PITFALL PREVENTION 3.2**

**The height and range equations**

Keep in mind that Equations 3.15 and 3.16 are useful for calculating $h$ and $R$ only for a symmetric path, as shown in Figure 3.7. If the path is not symmetric, *do not use these equations.* The general expressions given by Equations 3.10 through 3.13 are the *more important* results because they give the coordinates and velocity components of the projectile at *any* time $t$ for *any* trajectory.

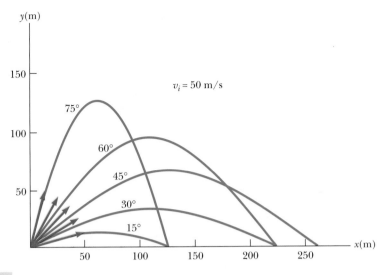

**Figure 3.8**

A projectile launched from the origin with an initial speed of 50 m/s at various angles of projection. Note that complementary values of $\theta_i$ will result in the same value of $R$.

**PITFALL PREVENTION 3.3**

**Acceleration at the highest point**

As discussed in Pitfall Prevention 2.10 in Chapter 2, many people have trouble with the very topmost point of the trajectory of a projectile, claiming that the acceleration of the ball at the highest point is zero. This interpretation arises from a confusion between zero vertical velocity and zero acceleration. Even though the ball has momentarily come to rest at the highest point (vertically—in two dimensions, it is still moving horizontally), it is still accelerating because the gravitational force is still acting. If the ball were to experience zero acceleration at the highest point, then the velocity would not change—the ball would continue to move horizontally once it reached the highest point! This doesn't happen because the acceleration is indeed not equal to zero.

**PROBLEM-SOLVING STRATEGY** | **Projectile Motion**

We suggest that you use the following approach when solving projectile motion problems:

1. Select a coordinate system.
2. Resolve the initial velocity vector into $x$ and $y$ components.
3. Treat the horizontal motion and the vertical motion independently.
4. Analyze the horizontal motion of the projectile as a particle under constant velocity.
5. Analyze the vertical motion of the projectile as a particle under constant acceleration.

**Quick Quiz 3.3**

Suppose you run at constant speed and wish to throw a ball so that you can catch it as it comes back down. In what direction should you throw the ball?

**Quick Quiz 3.4**

As a projectile moves in its parabolic path, is there any point along its path where the velocity and acceleration vectors are (a) perpendicular to each other? (b) parallel to each other?

### THINKING PHYSICS 3.2

A home run is hit in a baseball game. The ball is hit from home plate into the stands, along a parabolic path. What is the acceleration of the ball (a) while it is rising, (b) at the highest point of the trajectory, and (c) while it is descending after reaching the highest point? Ignore air resistance.

**Reasoning** The answers to all three parts are the same—the acceleration is that due to gravity, $a_y = -9.80$ m/s$^2$, because the gravitational force is pulling downward on the ball during the entire motion. During the rising part of the trajectory, the downward acceleration results in the decreasing positive values of the vertical component of the velocity of the ball. During the falling part of the trajectory, the downward acceleration results in the increasing negative values of the vertical component of the velocity.

## Example 3.2 That's Quite an Arm

A stone is thrown from the top of a building at an angle of 30.0° to the horizontal and with an initial speed of 20.0 m/s, as in Figure 3.9. If the height of the building is 45.0 m, (a) how long is the stone "in flight"?

**Reasoning** Looking at the pictorial representation in Figure 3.9, it is clear that this is not a symmetric trajectory. Thus, we cannot use Equations 3.15 and 3.16. We use the more general approach described by the Problem-Solving Strategy and represented by Equations 3.10 to 3.13.

**Solution** The initial $x$ and $y$ components of the velocity are

$$v_{xi} = v_i \cos \theta_i = (20.0 \text{ m/s})(\cos 30.0°) = 17.3 \text{ m/s}$$

$$v_{yi} = v_i \sin \theta_i = (20.0 \text{ m/s})(\sin 30.0°) = 10.0 \text{ m/s}$$

To find $t$, we use the vertical motion, in which we model the particle as one moving with constant acceleration. We use Equation 3.13 with $y_f = -45.0$ m and $v_{yi} = 10.0$ m/s (we have chosen the top of the building as the origin, as in Figure 3.9):

$$y_f = y_i + v_{yi}t - \tfrac{1}{2}gt^2$$

$$-45.0 \text{ m} = 0 + (10.0 \text{ m/s})t - \tfrac{1}{2}(9.80 \text{ m/s}^2)t^2$$

Solving the quadratic equation for $t$ gives, for the positive root, $t = 4.22$ s. Does the negative root have any physical meaning? (Can you think of another way of finding $t$ from the information given?)

(b) What is the speed of the stone just before it strikes the ground?

**Solution** The $y$ component of the velocity just before the stone strikes the ground can be obtained using Equation 3.11, with $t = 4.22$ s:

$$v_{yf} = v_{yi} - gt$$

$$= 10.0 \text{ m/s} - (9.80 \text{ m/s}^2)(4.22 \text{ s}) = -31.4 \text{ m/s}$$

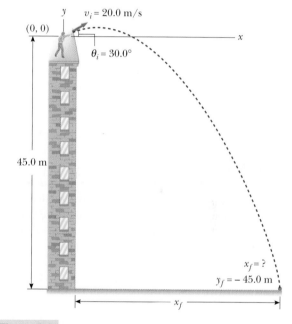

**Figure 3.9**

(Example 3.2)

In the horizontal direction, the appropriate model is the particle under constant velocity. Because $v_{xf} = v_{xi} = 17.3$ m/s, the speed as the stone strikes the ground is

$$v_f = \sqrt{v_{xf}^2 + v_{yf}^2} = \sqrt{(17.3)^2 + (-31.4)^2} \text{ m/s} = 35.9 \text{ m/s}$$

**EXERCISE** Where does the stone strike the ground?

*Answer* 73.0 m from the base of the building

## Example 3.3   The Stranded Explorers

An Alaskan rescue plane drops a package of emergency rations to a stranded party of explorers, as shown in the pictorial representation in Figure 3.10. If the plane is traveling horizontally at 40.0 m/s at a height of 100 m above the ground, where does the package strike the ground relative to the point at which it is released?

**Reasoning**   We ignore air resistance, so we model this as a particle in two-dimensional free-fall, which, as we have seen, is modeled by a combination of a particle moving with constant velocity in the *x* direction and a particle moving with constant acceleration in the *y* direction. The coordinate system for this problem is selected as shown in Figure 3.10, with the positive *x* direction to the right and the positive *y* direction upward.

**Solution**   Consider first the horizontal motion of the package. From Equation 3.12, the position is given by $x_f = x_i + v_{xi}t$. The initial *x* component of the package velocity is the same as that of the plane when the package is released, 40.0 m/s. We define the initial position $x_i = 0$ right under the plane at the instant the package is released. Thus,

$$x_f = x_i + v_{xi}t = 0 + (40.0 \text{ m/s})t$$

If we know *t*, the time at which the package strikes the ground, we can determine $x_f$, the distance traveled by the package in the horizontal direction. At present, however, we have no information about *t*. To find *t*, we turn to the equations for the vertical motion of the package, modeling the particle as being under constant acceleration. We know that at the instant the package hits the ground, its *y* coordinate is − 100 m. We also know that the initial component of velocity $v_{yi}$ of the package in the vertical direction is zero, because the package was released with only a horizontal component of velocity. From Equation 3.13, we have

$$y_f = y_i + v_{yi}t - \tfrac{1}{2}gt^2$$

$$- 100 \text{ m} = 0 + 0 - \tfrac{1}{2}(9.80 \text{ m/s}^2)t^2$$

$$t^2 = 20.4 \text{ s}^2$$

$$t = 4.52 \text{ s}$$

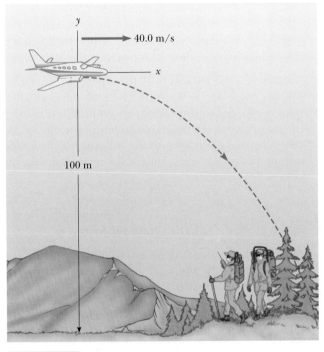

**Figure 3.10**

(Example 3.3)

This value for the time at which the package strikes the ground is substituted into the equation for the *x* coordinate to give us

$$x_f = (40.0 \text{ m/s})(4.52 \text{ s}) = \boxed{181 \text{ m}}$$

The package hits the ground 181 m to the right of the point at which it was dropped in Figure 3.10.

**EXERCISE**   What are the horizontal and vertical components of the velocity of the package just before it hits the ground?

*Answer*   $v_{xf} = 40.0$ m/s;   $v_{yf} = -44.3$ m/s

## Example 3.4   Javelin Throwing at the Olympics

An athlete throws a javelin a distance of 80.0 m at the Olympics held at the equator, where $g = 9.78$ m/s². Four years later, the Olympics are held at the North Pole, where $g = 9.83$ m/s². Assuming that the thrower provides the javelin with exactly the same initial velocity as he did at the equator, how far does the javelin travel at the North Pole?

**Reasoning**   In the absence of any information about how the javelin is affected by moving through the air, we adopt the free-fall model for the javelin. Track and field events are normally held on flat fields. Thus, we surmise that the javelin returns to the same vertical position from which it was thrown and, therefore, the trajectory is symmetric. This al-

A javelin can be thrown over a very long distance by a world class athlete. *(Tony Duffy/Allsport)*

lows us to use Equations 3.15 and 3.16 to analyze the motion. The difference in range is due to the difference in the free-fall acceleration at the two locations. This difference is due to the rotational motion of the Earth.

**Solution**   We use Equation 3.16 to express the range of the particle at each of the two locations:

$$R_{equator} = \frac{v_i^2 \sin 2\theta_i}{g_{equator}}$$

$$R_{North\ Pole} = \frac{v_i^2 \sin 2\theta_i}{g_{North\ Pole}}$$

We divide the first equation by the second to establish a relationship between the ratio of the ranges and the ratio of the free-fall accelerations. Because the problem states that the same initial velocity is provided to the javelin at both locations, $v_i$ and $\theta_i$ are the same in the numerator and denomina-

tor of the ratio, which gives us:

$$\frac{R_{equator}}{R_{North\ Pole}} = \frac{\left(\dfrac{v_i^2 \sin 2\theta_i}{g_{equator}}\right)}{\left(\dfrac{v_i^2 \sin 2\theta_i}{g_{North\ Pole}}\right)} = \frac{g_{North\ Pole}}{g_{equator}}$$

We can now solve this equation for the range at the North Pole and substitute the numerical values:

$$R_{North\ Pole} = \frac{g_{equator}}{g_{North\ Pole}} R_{equator} = \frac{9.78\ m/s^2}{9.83\ m/s^2}(80.0\ m)$$

$$= 79.6\ m$$

Notice one of the advantages of this powerful technique of setting up ratios—we do not need to know the magnitude $(v_i)$ nor the direction $(\theta_i)$ of the initial velocity. As long as they are the same at both locations, they cancel in the ratio.

**EXERCISE**   Suppose that the javelin thrower wished to match his 80.0-m throw at the equator in the Olympics held at the North Pole. By what percentage must he increase the magnitude of the initial velocity of the javelin to throw it 80.0 m at the North Pole?

*Answer*   0.26%

**EXERCISE**   It has been said that in his youth George Washington threw a silver dollar across a river. Assuming that the river was 75 m wide, (a) what *minimum initial speed* was necessary to throw the coin across the river, and (b) how long was the coin in flight?

*Answer*   (a) 27 m/s (b) 3.9 s

## 3.4 • THE PARTICLE IN UNIFORM CIRCULAR MOTION

See Screen 3.6

Figure 3.11a shows a car moving in a circular path with *constant speed v.* Such motion is called **uniform circular motion** and serves as the basis for a new group of problems that we can solve.

It is often surprising to students to find that **even though an object moves at a constant speed in a circular path, it still has an acceleration.** To see why, consider the defining equation for average acceleration, $\bar{\mathbf{a}} = \Delta\mathbf{v}/\Delta t$ (Eq. 3.4). The acceleration depends on *the change in the velocity vector.* Because velocity is a vector quantity, an acceleration can be produced in two ways, as mentioned in Section 3.1: by a change in the *magnitude* of the velocity or by a change in the *direction* of the velocity. The latter situation is occurring for an object moving with constant speed in a circular path. The velocity vector is always tangent to the path of the object and perpendicular to the radius of the circular path. We now show that the acceleration vector in uniform circular motion is always perpendicular to the path and always points toward the center of the circle. An acceleration of this nature is

**Acceleration of a particle in uniform circular motion**

Many students have trouble with the notion of a particle moving in a circular path at constant speed and yet having an acceleration. This is due to the everyday interpretation of acceleration as meaning *speeding up* or *slowing down*. But remember that acceleration is defined as a change in the *velocity*, not a change in the *speed*. In circular motion, the velocity vector is changing in direction, so there is indeed an acceleration.

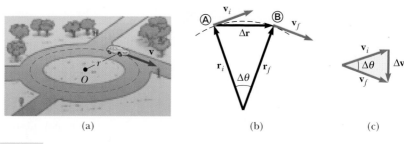

(a) (b) (c)

**Figure 3.11**

(a) A car moving along a circular path at constant speed is in uniform circular motion. (b) As the particle moves from Ⓐ to Ⓑ, its velocity vector changes from $\mathbf{v}_i$ to $\mathbf{v}_f$. (c) The construction for determining the direction of the change in velocity $\Delta\mathbf{v}$, which is toward the center of the circle for small $\Delta\theta$.

called a **centripetal acceleration** (*centripetal* means *center seeking*), and its magnitude is

- *Magnitude of centripetal acceleration*

$$a_c = \frac{v^2}{r} \qquad [3.17]$$

where $r$ is the radius of the circle. The subscript on the acceleration symbol reminds us that the acceleration is centripetal.

Let us first argue conceptually that the acceleration must be perpendicular to the path followed by the particle. If this were not true, there would be a component of the acceleration parallel to the path and, therefore, parallel to the velocity vector. Such an acceleration component would lead to a change in the speed of the object, which we model as a particle, along the path. But this is inconsistent with our setup of the problem—the particle moves with constant speed along the path. Thus, for *uniform* circular motion, the acceleration vector can only have a component perpendicular to the path, which is toward the center of the circle.

To derive Equation 3.17, consider the pictorial representation of the position and velocity vectors in Figure 3.11b. In addition, the figure shows the vector representing the change in position, $\Delta\mathbf{r}$. The particle follows a circular path, part of which is shown by the dotted curve. The particle is at Ⓐ at time $t_i$, and its velocity at that time is $\mathbf{v}_i$; it is at Ⓑ at some later time $t_f$, and its velocity at that time is $\mathbf{v}_f$. Let us also assume that $\mathbf{v}_i$ and $\mathbf{v}_f$ differ only in direction; their magnitudes are the same (i.e., $v_i = v_f = v$, because it is *uniform* circular motion). To calculate the acceleration of the particle, let us begin with the defining equation for average acceleration (Eq. 3.4):

$$\bar{\mathbf{a}} = \frac{\mathbf{v}_f - \mathbf{v}_i}{t_f - t_i} = \frac{\Delta\mathbf{v}}{\Delta t}$$

In Figure 3.11c, the velocity vectors in Figure 3.11b have been redrawn tail to tail. The vector $\Delta\mathbf{v}$ connects the tips of the vectors, representing the vector addition, $\mathbf{v}_f = \mathbf{v}_i + \Delta\mathbf{v}$. In Figures 3.11b and 3.11c, we can identify triangles that can serve as geometric models to help us analyze the motion. The angle $\Delta\theta$ between the two position vectors in Figure 3.11b is the same as the angle between the velocity vectors in Figure 3.11c, because the velocity vector $\mathbf{v}$ is always perpendicular to the position vector $\mathbf{r}$. Thus, the two triangles are *similar*. (Two triangles are similar if the angle between any two sides is the same for both triangles and if the ratio of the lengths of these sides is the same.) This enables us to write a relationship between the lengths of the sides for the two triangles:

$$\frac{|\Delta \mathbf{v}|}{v} = \frac{|\Delta \mathbf{r}|}{r}$$

where $v = v_i = v_f$ and $r = r_i = r_f$. This equation can be solved for $|\Delta \mathbf{v}|$ and the expression so obtained can be substituted into $\bar{\mathbf{a}} = \Delta \mathbf{v}/\Delta t$ (Eq. 3.4) to give the magnitude of the average acceleration over the time interval for the particle to move from Ⓐ to Ⓑ:

$$|\bar{\mathbf{a}}| = \frac{v}{r} \frac{|\Delta \mathbf{r}|}{\Delta t}$$

Now imagine that points Ⓐ and Ⓑ in Figure 3.11b become extremely close together. As Ⓐ and Ⓑ approach each other, $\Delta t$ approaches zero, and the ratio $|\Delta \mathbf{r}|/\Delta t$ approaches the speed $v$. In addition, the average acceleration becomes the instantaneous acceleration at point Ⓐ. Hence, in the limit $\Delta t \to 0$, the magnitude of the acceleration is

$$a_c = \frac{v^2}{r}$$

Thus, in uniform circular motion, the acceleration is directed inward toward the center of the circle and has magnitude $v^2/r$.

In many situations it is convenient to describe the motion of a particle moving with constant speed in a circle of radius $r$ in terms of the **period** $T$, which is defined as the time required for one complete revolution. In the time $T$ the particle moves a distance of $2\pi r$, which is equal to the circumference of the particle's circular path. Therefore, because its speed is equal to the circumference of the circular path divided by the period, or $v = 2\pi r/T$, it follows that

$$T \equiv \frac{2\pi r}{v} \qquad \text{[3.18]}$$

The particle in uniform circular motion is a very common physical situation and is useful as an analysis model for problem solving. If a particle, or an object that can be modeled as a particle, moves in a circular path at a uniform speed, we can use the results of the preceding discussion, with the important equations represented by Equation 3.17 for the centripetal acceleration and Equation 3.18 for the period.

**PITFALL PREVENTION 3.5**

**Centripetal acceleration is not constant**

We derived the magnitude of the centripetal acceleration vector and found it to be constant for uniform circular motion. But *the centripetal acceleration vector is not constant*. It always points toward the center of the circle, but continuously changes direction as the particle moves around the circular path.

- *Period of a particle in uniform circular motion*

---

**THINKING PHYSICS 3.3**

An airplane travels from Los Angeles to Sydney, Australia. After cruising altitude is reached, the instruments on the plane indicate that the ground speed holds rock-steady at 700 km/h and that the heading of the airplane does not change. Is the velocity of the airplane constant during the flight?

**Reasoning** The velocity is not constant, due to the curvature of the Earth. Even though the speed does not change, and the heading is always toward Sydney (is this actually true?), the airplane travels around a significant portion of the circumference of the Earth. Thus, the direction of the velocity vector does indeed change. We could extend this by imagining that the airplane passes over Sydney and continues (assuming it has enough fuel!) around the Earth until it arrives at Los Angeles again. It is impossible for an airplane to have a constant velocity (relative to the Universe, not to the Earth's surface) and return to its starting point.

### Example 3.5 The Centripetal Acceleration of the Earth

What is the centripetal acceleration of the Earth as it moves in its orbit around the Sun?

**Solution** We shall model the Earth as a particle and approximate the Earth's orbit as circular (it's actually slightly elliptical, as we discuss in Chapter 11). Although we don't know the orbital speed of the Earth, with the help of Equation 3.18, we can recast Equation 3.17 in terms of the period

of the Earth's orbit, which we know is one year:

$$a_c = \frac{v^2}{r} = \frac{\left(\dfrac{2\pi r}{T}\right)^2}{r} = \frac{4\pi^2 r}{T^2}$$

$$= \frac{4\pi^2 (1.5 \times 10^{11}\ \text{m})}{(1\ \text{yr})^2}\left(\frac{1\ \text{yr}}{3.15 \times 10^7\ \text{s}}\right)^2$$

$$= 6.0 \times 10^{-3}\ \text{m/s}^2$$

## 3.5 • TANGENTIAL AND RADIAL ACCELERATION

*See Screen 3.6*

Let us consider the motion of a particle along a curved path where the velocity changes both in direction and in magnitude, as described in Figure 3.12. In this situation, the velocity vector is always tangent to the path; however, the acceleration vector **a** is at some angle to the path. At each of three points Ⓐ, Ⓑ, and Ⓒ in Figure 3.12, we draw dotted circles that form geometric models of circular paths for a portion of the actual path at each point. The radius of the model circles is equal to the radius of curvature of the path at each point.

As the particle moves along the curved path in Figure 3.12, the direction of the total acceleration vector **a** changes from point to point. This vector can be resolved into two components, based on an origin at the center of the model circle: a radial component $a_r$ along the radius of the model circle, and a tangential component $a_t$ perpendicular to this radius. The *total* acceleration vector **a** can be written as the vector sum of the component vectors:

• *Total acceleration*

$$\mathbf{a} = \mathbf{a}_r + \mathbf{a}_t \qquad\qquad [3.19]$$

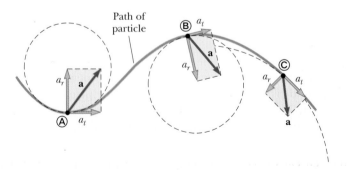

**Figure 3.12**

The motion of a particle along an arbitrary curved path lying in the *xy* plane. If the velocity vector **v** (always tangent to the path) changes in direction and magnitude, the acceleration vector **a** has a tangential component $a_t$ and a radial component $a_r$.

**The tangential acceleration arises from the change in the speed of the particle,** and has a magnitude given by

$$a_t = \frac{d|\mathbf{v}|}{dt}$$

[3.20]  •  *Tangential acceleration*

**The radial acceleration is due to the change in direction of the velocity vector** and is given by

$$a_r = -a_c = -\frac{v^2}{r}$$

•  *Centripetal acceleration*

where $r$ is the radius of curvature of the path at the point in question, which is the radius of the model circle. We recognize the radial component of the acceleration as the centripetal acceleration discussed in Section 3.4. The negative sign indicates that the direction of the centripetal acceleration is toward the center of the model circle, opposite the direction of the radial unit vector $\hat{\mathbf{r}}$, which always points away from the center of the circle.

Because $\mathbf{a}_r$ and $\mathbf{a}_t$ are perpendicular component vectors of $\mathbf{a}$, it follows that $a = \sqrt{a_r^2 + a_t^2}$. At a given speed, $a_r$ is large when the radius of curvature is small (as at points Ⓐ and Ⓑ in Fig. 3.12) and small when $r$ is large (such as at point Ⓒ). The direction of $\mathbf{a}_t$ is either in the same direction as $\mathbf{v}$ (if $v$ is increasing) or opposite $\mathbf{v}$ (if $v$ is decreasing).

In the case of uniform circular motion, where $v$ is constant, $a_t = 0$ and the acceleration is always radial, as described in Section 3.4. In other words, uniform circular motion is a special case of motion along a curved path. Furthermore, if the direction of $\mathbf{v}$ doesn't change, then no radial acceleration occurs and the motion is one-dimensional ($a_r = 0$, but $a_t$ may not be zero).

## 3.6 • RELATIVE VELOCITY

 See Screen 3.7

In Section 1.7, we discussed the need for a fixed reference point as the origin of a coordinate system used to locate the position of a point. We have made observations of position, velocity, and acceleration of a particle with respect to this reference point. Now imagine that we have two observers making measurements of a particle located in space, and one of the observers moves with respect to the other at constant velocity. Each observer can define a coordinate system with an origin fixed with respect to him or her. The origins of the two coordinate systems are in motion with respect to each other. In this section, we explore how we relate the measurements of one observer to that of the other.

As an example, consider two cars, a red one and a blue one, moving on a highway in the same direction, both with speeds of 60 mi/h, as in Figure 3.13. If we identify the red car as a particle to be observed, an observer on the side of the road measures a speed for this car of 60 mi/h. Now consider an observer riding in the blue car. This observer looks out the window and sees that the red car is always in the same position with respect to the blue car. Thus, this observer measures a speed for the red car of *zero*. This simple example demonstrates that speed measurements differ in different frames of reference. Both observers look at the same particle (the red car) and arrive at different values for its speed. Both are correct; the difference in their measurements is due to the relative velocity of their frames of reference.

**Figure 3.13**

Two observers measure the speed of the red car. Observer $O$ is standing on the ground beside the highway. Observer $O'$ is in the blue car.

Let us now generate a mathematical representation that will allow us to calculate one observer's measurements from the other's. Consider a particle located at point $P$ in an $xy$ plane, as shown in Figure 3.14. Imagine that the motion of this particle is being observed by two observers. Observer $O$ is in reference frame S. Observer $O'$ is in reference frame S', which moves with velocity $\mathbf{v}_{O'O}$ with respect to S, where the subscripts indicate the following: The first subscript describes what is being observed, and the second describes who is doing the observing. Thus, $\mathbf{v}_{O'O}$ is the velocity of observer $O'$ as measured by observer $O$. At $t = 0$, the origins of the reference frames coincide. Thus, modeling the origin of S' as a particle under constant velocity, the origins of the two reference frames are separated by a displacement $\mathbf{v}_{O'O}t$ at time $t$. This displacement is shown in Figure 3.14. Also shown in the figure are the position vectors $\mathbf{r}_{PO}$ and $\mathbf{r}_{PO'}$ for point $P$ from each of the two origins. These are the position vectors that the two observers would use to describe the location of point $P$, using the same subscript notation. From the diagram, we see that these three vectors form a vector addition triangle:

$$\mathbf{r}_{PO} = \mathbf{r}_{PO'} + \mathbf{v}_{O'O}t$$

Notice the order of subscripts in this expression. The subscripts on the left side are the same as the first and last subscripts on the right. The second and third subscripts on the right are both $O'$. These subscripts are helpful in analyzing these types of situations. On the left, we are looking at the position vector that points directly to $P$ from $O$, as described by the subscripts. On the right, the same point $P$ is

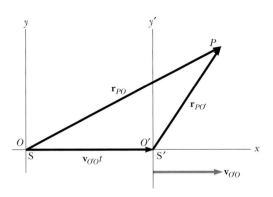

**Figure 3.14**

Position vectors for an event occurring at point $P$ for two observers. Observer $O'$ is moving to the right at speed $v_{O'O}$ with respect to observer $O$.

located by first going to *P* from *O′* and then describing where *O′* is relative to *O*, again as suggested by the subscripts.

Let us now differentiate this expression with respect to time to find an expression for the velocity of a particle located at point *P*:

$$\frac{d}{dt}(\mathbf{r}_{PO}) = \frac{d}{dt}(\mathbf{r}_{PO'} + \mathbf{v}_{O'O}t) \quad \longrightarrow \quad \mathbf{v}_{PO} = \mathbf{v}_{PO'} + \mathbf{v}_{O'O} \qquad \text{[3.21]}$$

This expression relates the velocity of the particle as measured by *O* to that measured by *O′* and the relative velocity of the two reference frames.

In the one-dimensional case, this equation reduces to

$$v_{PO} = v_{PO'} + v_{O'O}$$

Often, this equation is expressed in terms of the observer *O′*, as follows:

$$v_{PO'} = v_{PO} - v_{O'O} \qquad \text{[3.22]}$$

• *Relative velocity in one dimension*

and is called the **relative velocity**—the velocity of a particle as measured by a moving observer (moving with respect to another observer). In our car example, observer *O* is standing on the side of the road. Observer *O′* is in the blue car. Both observers are measuring the speed of the red car, which is located at point *P*. Thus,

$$v_{PO} = 60 \text{ mi/h}$$

$$v_{O'O} = 60 \text{ mi/h}$$

and, using Equation 3.22,

$$v_{PO'} = v_{PO} - v_{O'O} = 60 \text{ mi/h} - 60 \text{ mi/h} = 0$$

The result of our calculation agrees with our previous intuitive discussion.

This equation will be used in Chapter 9, when we discuss special relativity. We shall find that this simple expression is valid for low-speed particles but is no longer valid when the particle or observers are moving at speeds close to the speed of light.

## Example 3.6 A Boat Crossing a River

A boat heading due north crosses a wide river with a speed of 10.0 km/h relative to the water. The river has a current such that the water moves with uniform speed of 5.00 km/h due east relative to the ground. (a) What is the velocity of the boat relative to a stationary observer on the side of the river?

**Reasoning** It is often useful to use subscripts other than *O* and *P* that make it easy to identify the observers and the object being observed. Observer *O* is standing on the side of the river. Because he is at rest with respect to the Earth, we will use the subscript E for this observer. Let us identify an imaginary observer *O′* at rest in the water, floating with the current. Because he is at rest with respect to the water, we will use the subscript w for this observer. Both observers are looking at the boat, denoted by the subscript b. We can identify the velocity of the boat relative to the water as $\mathbf{v}_{bw} = 10.0\mathbf{j}$ km/h. The velocity of the water relative to the Earth is that of the current in the river, $\mathbf{v}_{wE} = 5.00\mathbf{i}$ km/h.

**Solution** We are looking for the velocity of the boat relative to the Earth, so, from Equation 3.21,

$$\mathbf{v}_{bE} = \mathbf{v}_{bw} + \mathbf{v}_{wE} = (10.0\mathbf{j} + 5.00\mathbf{i}) \text{ km/h}$$

This vector addition is shown in Figure 3.15a. The speed of the boat relative to the observer on shore is found from the

(a)

(b)

**Figure 3.15**

(Example 3.6)

Pythagorean theorem:

$$v_{bE} = \sqrt{v_{bw}^2 + v_{wE}^2} = \sqrt{(10.0)^2 + (5.00)^2} \text{ km/h}$$
$$= 11.2 \text{ km/h}$$

The direction of the velocity vector can be found with the inverse tangent function:

$$\theta = \tan^{-1}\left(\frac{v_{wE}}{v_{bw}}\right) = \tan^{-1}\left(\frac{5.00}{10.0}\right) = 26.6°$$

(b) At what angle should the boat be headed if it is to travel directly north across the river, and what is the speed of the boat relative to the Earth?

**Solution**   We now want $\mathbf{v}_{bE}$ to be pointed due north, as shown in Figure 3.15b. From the vector triangle,

$$\theta = \sin^{-1}\left(\frac{v_{wE}}{v_{bw}}\right) = \sin^{-1}\left(\frac{5.00 \text{ km/h}}{10.0 \text{ km/h}}\right) = 30.0°$$

The speed of the boat relative to the Earth is

$$v_{bE} = \sqrt{v_{bw}^2 - v_{wE}^2} = \sqrt{(10.0)^2 - (5.00)^2} \text{ km/h}$$
$$= 8.66 \text{ km/h}$$

### context connection
## 3.7 • CIRCULAR ORBITS

In Chapter 2, we considered some information about the liftoff phase of our spacecraft on its way to Mars. Now let us look at the next phase of its motion, in which the spacecraft is placed in a parking orbit around the Earth. We shall assume that the parking orbit is circular so that the spacecraft can be modeled as a particle in uniform circular motion.

The spacecraft is actually in free-fall, but it never moves closer to the Earth. This is a difficult concept to accept, but it might help to remember a discussion from Section 3.3 in which it was mentioned that if there were no gravitational acceleration, the particle would continue to move along a straight path in the direc-

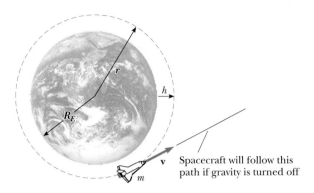

**Figure 3.16**

If the Earth's gravity were to be turned off, a satellite initially in orbit would follow the straight line path shown.

tion of its original velocity. The gravitational force causes it to "fall" below the straight path. In the case of the spacecraft, if gravity were to be turned off, the straight line path would take the spacecraft tangentially off its orbit into space, as in Figure 3.16. The gravitational force causes the spacecraft to "fall" below this straight line path. With a spacecraft in orbit, the speed of the spacecraft is designed so that this falling below the straight line path keeps the spacecraft at a fixed height $h$ above the Earth's surface.

We can use the model of a particle in uniform circular motion to analyze the orbit of a spacecraft around the Earth, as in the following example.

**WEB**

For a *New York Times* retrospective on *Sputnik,* the first Earth satellite, visit **http://www.nytimes.com/ partners/aol/special/sputnik/**

### Example 3.7    A Spacecraft in Orbit

A spacecraft is in orbit at a height of 200 km above the surface of the Earth. The period of its orbit is 88.2 min.

(a) What is the centripetal acceleration of the spacecraft?

**Solution**    It is not stated explicitly that the orbit is circular, but we make that assumption based on the fact that only one height above the Earth's surface is given. We shall use the equation derived in Example 3.5 relating the centripetal acceleration to the period. We first recognize that the radius of the circular orbit of the spacecraft is the height above the Earth's surface *plus* the radius of the Earth. Thus,

$$a_c = \frac{4\pi^2 r}{T^2}$$

$$= \frac{4\pi^2(2.00 \times 10^5 \text{ m} + 6.37 \times 10^6 \text{ m})}{(88.2 \text{ min})^2}\left(\frac{1 \text{ min}}{60 \text{ s}}\right)^2$$

$$= 9.26 \text{ m/s}^2$$

Notice that this is slightly less than the free-fall acceleration at the surface of the Earth. This should not be a surprise, because the spacecraft is simply *falling* as it orbits the Earth, so its centripetal acceleration *should* be equal to the free-fall acceleration. The value is less than that at the surface due to the decrease in the free-fall acceleration as one moves higher off the surface of the Earth.

(b) What is the speed of the spacecraft in its orbit?

**Solution**    We use Equation 3.18 to find the speed:

$$v = \frac{2\pi r}{T} = \frac{2\pi(2.00 \times 10^5 \text{ m} + 6.37 \times 10^6 \text{ m})}{(88.2 \text{ min})}\left(\frac{1 \text{ min}}{60 \text{ s}}\right)$$

$$= 7.80 \times 10^3 \text{ m/s}$$

This is equivalent to over 17,000 mi/h.

The motion of our spacecraft in orbit, with the rocket engine turned off, is completely governed by gravity, resulting in a circular orbit around the Earth. We must have control over the motion of our spacecraft beyond that provided by gravity. In the next chapter, we shall see how to use the rocket engine to change our velocity from an initial vector to a final desired vector.

## SUMMARY

If a particle moves with *constant* acceleration **a** and has velocity **v**$_i$ and position **r**$_i$ at $t = 0$, its velocity and position vectors at some later time $t$ are

$$\mathbf{v}_f = \mathbf{v}_i + \mathbf{a}t \qquad [3.8]$$

$$\mathbf{r}_f = \mathbf{r}_i + \mathbf{v}_i t + \tfrac{1}{2}\mathbf{a}t^2 \qquad [3.9]$$

For two-dimensional motion in the $xy$ plane under constant acceleration, these vector expressions are equivalent to two component expressions, one for the motion along $x$ and one for the motion along $y$.

**Projectile motion** is a special case of two-dimensional motion under constant acceleration, where $a_x = 0$ and $a_y = -g$. In this case, the horizontal components of Equations 3.8 and 3.9 reduce to those of a particle under constant velocity:

$$v_{xf} = v_{xi} = \text{constant} \qquad [3.10]$$

$$x_f = x_i + v_{xi}t \qquad [3.12]$$

and the vertical components of Equations 3.8 and 3.9 are those of a particle under constant acceleration:

$$v_{yf} = v_{yi} - gt \qquad [3.11]$$

$$y_f = y_i + v_{yi}t - \tfrac{1}{2}gt^2 \qquad [3.13]$$

where $v_{xi} = v_i \cos\theta_i$, $v_{yi} = v_i \sin\theta_i$, $v_i$ is the initial speed of the projectile, and $\theta_i$ is the angle **v**$_i$ makes with the positive $x$ axis.

A particle moving in a circle of radius $r$ with constant speed $v$ undergoes a **centripetal acceleration** because the direction of **v** changes in time. The magnitude of this acceleration is

$$a_c = \frac{v^2}{r} \qquad [3.17]$$

and its direction is always toward the center of the circle.

If a particle moves along a curved path in such a way that the magnitude and direction of **v** change in time, the particle has an acceleration vector that can be described by two components: (1) a radial component $a_r = -a_c$ arising from the change in direction of **v**, and (2) a tangential component $a_t$ arising from the change in magnitude of **v**.

If an observer $O'$ is moving with velocity **v**$_{O'O}$ with respect to observer $O$, their measurements of the velocity of a particle located at point $P$ are related according to

$$\mathbf{v}_{PO} = \mathbf{v}_{PO'} + \mathbf{v}_{O'O} \qquad [3.21]$$

The velocity **v**$_{PO'}$ is called the **relative velocity**—the velocity of a particle as measured by a moving observer (moving at constant velocity with respect to another observer).

## QUESTIONS

1. If you know the position vectors of a particle at two points along its path and also know the time it took to move from one point to the other, can you determine the particle's instantaneous velocity? Its average velocity? Explain.

2. Explain whether the following particles have an acceleration: (a) a particle moving in a straight line with constant speed and (b) a particle moving around a curve with constant speed.

3. Correct the following statement: "The racing car rounds the turn at a constant velocity of 90 miles per hour."

4. A spacecraft drifts through space at a constant velocity. Suddenly a gas leak in the side of the spacecraft causes a constant acceleration of the spacecraft in a direction perpendicular to the initial velocity. The orientation of the spacecraft does not change, so that the acceleration remains perpendicular to the original direction of the velocity. What is the shape of the path followed by the spacecraft in this situation?

5. A ball is projected horizontally from the top of a building. One second later another ball is projected horizontally from the same point with the same velocity. At what point in the motion will the balls be closest to each other? Will the first ball always be traveling faster than the second ball? What will be the time difference between when the balls hit the ground? Can the horizontal projection velocity of the second ball be changed so that the balls arrive at the ground at the same time?

6. At the end of a pendulum's arc, its velocity is zero. Is its acceleration also zero at that point?

7. If a rock is dropped from the top of a sailboat's mast, will it hit the deck at the same point whether the boat is at rest or in motion at constant velocity?

8. Two projectiles are thrown with the same magnitude of initial velocity, one at an angle $\theta$ with respect to the level ground and the other at angle $90° - \theta$. Both projectiles will strike the ground at the same distance from the projection point. Will both projectiles be in the air for the same time interval?

9. A projectile is launched at some angle to the horizontal with some initial speed $v_i$, and air resistance is ignored. Is

the projectile a freely falling body? What is its acceleration in the vertical direction? What is its acceleration in the horizontal direction?

10. State which of the following quantities, if any, remain constant as a projectile moves through its parabolic trajectory: (a) speed, (b) acceleration, (c) horizontal component of velocity, (d) vertical component of velocity.

11. The maximum range of a projectile occurs when it is launched at an angle of 45.0° with the horizontal, if air resistance is ignored. If air resistance is not ignored, will the optimum angle be greater or less than 45.0°? Explain.

12. An object moves in a circular path with constant speed $v$. (a) Is the velocity of the object constant? (b) Is its acceleration constant? Explain.

13. A projectile is launched on the Earth with some initial velocity. Another projectile is launched on the Moon with the *same* initial velocity. Ignoring air resistance, which projectile has the greater range? Which reaches the greater altitude? (Note that the free-fall acceleration on the Moon is about 1.6 m/s².)

14. A coin on a table is given an initial horizontal velocity such that it ultimately leaves the end of the table and hits the floor. At the instant the coin leaves the end of the table, a ball is released from the same height and falls to the floor. Explain why the two objects hit the floor simultaneously, even though the coin has an initial velocity.

15. Describe how a driver can steer a car traveling at constant speed so that (a) the acceleration is zero or (b) the magnitude of the acceleration remains constant.

16. An ice skater executes a figure eight, consisting of two equal, tangent circular paths. During the first loop she increases her speed uniformly, and during the second loop she moves at a constant speed. Make a sketch of her acceleration vector at several points along the path of motion.

17. Construct motion diagrams showing the velocity and acceleration of a projectile at several points along its path if (a) the projectile is launched horizontally and (b) the projectile is launched at an angle $\theta$ with the horizontal.

18. A baseball is thrown such that its initial $x$ and $y$ components of velocity are known. Ignoring air resistance, describe how you would calculate, at the instant the ball reaches the top of its trajectory, (a) its coordinates, (b) its velocity, and (c) its acceleration. How would these results change if air resistance were taken into account?

# PROBLEMS

## Section 3.1 The Position, Velocity, and Acceleration Vectors

1. A motorist drives south at 20.0 m/s for 3.00 min, then
**web** turns west and travels at 25.0 m/s for 2.00 min, and finally travels northwest at 30.0 m/s for 1.00 min. For this 6.00-min trip, find (a) the total vector displacement, (b) the average speed, and (c) the average velocity. Let the positive $x$ axis point east.

2. Suppose that the position vector for a particle is given as a function of time by $\mathbf{r}(t) = x(t)\mathbf{i} + y(t)\mathbf{j}$, with $x(t) = at + b$ and $y(t) = ct^2 + d$, where $a = 1.00$ m/s, $b = 1.00$ m, $c = 0.125$ m/s², and $d = 1.00$ m. (a) Calculate the average velocity during the time interval from $t = 2.00$ s to $t = 4.00$ s. (b) Determine the velocity and the speed at $t = 2.00$ s.

3. A golf ball is hit off a tee at the edge of a cliff. Its $x$ and $y$ coordinates as functions of time are given by the following expressions:

$$x = (18.0 \text{ m/s})\,t$$

and

$$y = (4.00 \text{ m/s})\,t - (4.90 \text{ m/s}^2)\,t^2$$

(a) Write a vector expression for the ball's position as a function of time, using the unit vectors $\mathbf{i}$ and $\mathbf{j}$. By taking derivatives, obtain expressions for (b) the velocity vector $\mathbf{v}$ as a function of time and (c) the acceleration vector as a function of time. Next use unit-vector notation to write expressions for (d) the position, (e) the velocity, and (f) the acceleration of the golf ball, all at $t = 3.00$ s.

## Section 3.2 Two-Dimensional Motion with Constant Acceleration

4. At $t = 0$, a particle moving in the $xy$ plane with constant acceleration has a velocity of $\mathbf{v}_i = (3.00\mathbf{i} - 2.00\mathbf{j})$ m/s and is at the origin. At $t = 3.00$ s, the particle's velocity is $\mathbf{v}_f = (9.00\mathbf{i} + 7.00\mathbf{j})$ m/s. Find (a) the acceleration of the particle and (b) its coordinates at any time $t$.

5. A fish swimming in a horizontal plane has velocity $\mathbf{v}_i = (4.00\mathbf{i} + 1.00\mathbf{j})$ m/s at a point in the ocean where the dis-

placement from a certain rock is $\mathbf{r}_i = (10.0\mathbf{i} - 4.00\mathbf{j})$ m. After the fish swims with constant acceleration for 20.0 s, its velocity is $\mathbf{v}_f = (20.0\mathbf{i} - 5.00\mathbf{j})$ m/s. (a) What are the components of the acceleration? (b) What is the direction of the acceleration with respect to unit vector $\mathbf{i}$? (c) If the fish maintains constant acceleration, where is it at $t = 25.0$ s, and in what direction is it moving?

6. A particle initially located at the origin has an acceleration of $\mathbf{a} = 3.00\mathbf{j}$ m/s$^2$ and an initial velocity of $\mathbf{v}_i = 5.00\mathbf{i}$ m/s. Find (a) the vector position and velocity at any time $t$ and (b) the coordinates and speed of the particle at $t = 2.00$ s.

7. It is not possible to see very small objects, such as viruses, using an ordinary light microscope. An electron microscope can view such objects using an electron beam instead of a light beam. Electron microscopy has proved invaluable for investigations of viruses, cell membranes and subcellular structures, bacterial surfaces, visual receptors, chloroplasts, and the contractile properties of muscles. The "lenses" of an electron microscope consist of electric and magnetic fields that control the electron beam. As an example of the manipulation of an electron beam, consider an electron traveling away from the origin along the $x$ axis in the $xy$ plane with initial velocity $\mathbf{v}_i = v_i\mathbf{i}$. As it passes through the region $x = 0$ to $x = d$, the electron experiences acceleration $\mathbf{a} = a_x\mathbf{i} + a_y\mathbf{j}$, where $a_x$ and $a_y$ are constants. Assume that $v_i = 1.80 \times 10^7$ m/s, $a_x = 8.00 \times 10^{14}$ m/s$^2$, and $a_y = 1.60 \times 10^{15}$ m/s$^2$. Determine at $x = d = 0.010\ 0$ m (a) the position of the electron, (b) the velocity of the electron, (c) the speed of the electron, and (d) the direction of travel of the electron (i.e., the angle between its velocity and the $x$ axis).

8. The determined coyote is out once more in pursuit of the elusive roadrunner. The coyote wears a pair of Acme jet-powered roller skates, which provide a constant horizontal acceleration of 15.0 m/s$^2$ (Fig. P3.8). The coyote starts at rest 70.0 m from the brink of a cliff at the instant the roadrunner zips past him in the direction of the cliff. (a) If the roadrunner moves with constant speed, determine the minimum speed he must have in order to reach the cliff

before the coyote. At the edge of the cliff, the roadrunner escapes by making a sudden turn, and the coyote continues straight ahead. His skates remain horizontal and continue to operate while he is in flight, so that the coyote's acceleration is $(15.0\ \mathbf{i} - 9.80\mathbf{j})$ m/s$^2$. (b) If the cliff is 100 m above the flat floor of a canyon, determine where the coyote lands in the canyon. (c) Determine the components of the coyote's impact velocity.

## Section 3.3   Projectile Motion

*Note*: Ignore air resistance in all problems and take $g = 9.80$ m/s$^2$ at the Earth's surface.

9. In a local bar, a customer slides an empty beer mug down the counter for a refill. The bartender is momentarily distracted and does not see the mug, which slides off the counter and strikes the floor 1.40 m from the base of the counter. If the height of the counter is 0.860 m, (a) with what velocity did the mug leave the counter, and (b) what was the direction of the mug's velocity just before it hit the floor?

10. In a local bar, a customer slides an empty beer mug down the counter for a refill. The bartender is momentarily distracted and does not see the mug, which slides off the counter and strikes the floor at distance $d$ from the base of the counter. If the height of the counter is $h$, (a) with what velocity did the mug leave the counter, and (b) what was the direction of the mug's velocity just before it hit the floor?

11. A tennis player standing 12.6 m from the net hits the ball at 3.00° above the horizontal. To clear the net, the ball must rise at least 0.330 m. If the ball just clears the net at the apex of its trajectory, how fast was the ball moving when it left the racquet?

12. An astronaut on a strange planet finds that she can jump a maximum horizontal distance of 15.0 m if her initial speed is 3.00 m/s. What is the free-fall acceleration on the planet?

13. A ball is tossed from an upper-story window of a building. The ball is given an initial velocity of 8.00 m/s at an angle of 20.0° below the horizontal. It strikes the ground 3.00 s later. (a) How far horizontally from the base of the building does the ball strike the ground? (b) Find the height from which the ball was thrown. (c) How long does it take the ball to reach a point 10.0 m below the level of launching?

14. A cannon with a muzzle speed of 1 000 m/s is used to start an avalanche on a mountain slope. The target is 2 000 m from the cannon horizontally and 800 m above the cannon. At what angle, above the horizontal, should the cannon be fired?

Coyoté    Roadrunner
Stupidus    Delightus

BEEP
BEEP

**Figure P3.8**

15. A place-kicker must kick a football from a point 36.0 m (about 40 yards) from the goal, and half the crowd hopes the ball will clear the crossbar, which is 3.05 m high. When kicked, the ball leaves the ground with a speed of 20.0 m/s at an angle of 53.0° to the horizontal. (a) By how much does the ball clear or fall short of clearing the crossbar? (b) Does the ball approach the crossbar while still rising or while falling?

16. One strategy in a snowball fight is to throw a snowball at a high angle over level ground. While your opponent is watching the first one, a second snowball is thrown at a low angle timed to arrive before or at the same time as the first one. Assume both snowballs are thrown with a speed of 25.0 m/s. The first one is thrown at an angle of 70.0° with respect to the horizontal. (a) At what angle should the second snowball be thrown to arrive at the same point as the first? (b) How many seconds later should the second snowball be thrown after the first to arrive at the same time?

17. The speed of a projectile when it reaches its maximum height is one half of its speed when it is at half its maximum height. What is the initial projection angle of the projectile?

18. The small archerfish (length 20–25 cm) lives in brackish waters of southeast Asia. This aptly named creature captures its prey by shooting a stream of water drops at an insect, either flying or at rest. After the bug falls into the water, the fish gobbles it up. The archerfish has high accuracy at distances of 1.2 m to 1.5 m and sometimes makes hits at distances up to 3.5 m. A groove in the roof of its mouth, along with a curled tongue, forms a tube that enables the fish to impart high velocity to the water in its mouth when it suddenly closes its gill flaps. Suppose the archerfish shoots at a target 2.00 m away, at an angle of 30.0° above the horizontal. With what speed must the water stream be launched if it is not to drop more than 3.00 cm vertically on its path to the target?

19. A firefighter at a distance $d$ from a burning building directs a stream of water from a fire hose at angle $\theta_i$ above the horizontal as in Figure P3.19. If the initial speed of the stream is $v_i$, at what height $h$ does the water strike the building?

20. A soccer player kicks a rock horizontally off a 40.0-m high cliff into a pool of water. If the player hears the sound of the splash 3.00 s later, what was the initial speed given to the rock? Assume the speed of sound in air to be 343 m/s.

21. A playground is on the flat roof of a city school, 6.00 m above the street below. The vertical wall of the building is 7.00 m high, forming a 1.00-m-high railing around the playground. A ball falls to the street below, and a passerby returns it by launching it at an angle of 53.0° above the horizontal at a point 24.0 m from the base of the building

**Figure P3.19** (*Frederick McKinney/FPG International*)

**Figure P3.22** *(Left, Jed Jacobsohn/Allsport; right, Bill Lee/Dembinsky Photo Associates)*

wall. The ball takes 2.20 s to reach a point vertically above the wall. Find (a) the speed at which the ball was launched, (b) the vertical distance by which the ball clears the wall, and (c) the distance from the wall to the point on the roof where the ball lands.

22. A basketball star covers 2.80 m horizontally in a jump to dunk the ball (Fig. P3.22). His motion through space can be modeled as that of a particle at his *center of mass,* which we will define in Chapter 8. His center of mass is at elevation 1.02 m when he leaves the floor. It reaches a maximum height of 1.85 m above the floor and is at elevation 0.900 m when he touches down again. Determine (a) his time of flight (his "hang time"), (b) his horizontal and (c) vertical velocity components at the instant of takeoff, and (d) his takeoff angle. (e) For comparison, determine the hang time of a white-tailed deer making a jump with center-of-mass elevations $y_i = 1.20$ m, $y_{max} = 2.50$ m, $y_f = 0.700$ m.

23. A fireworks rocket explodes at height $h$, the peak of its vertical trajectory. It throws out burning fragments in all directions, but all at the same speed $v$. The bits of ash fall to the ground without air resistance. Find the smallest angle that the final velocity of an impacting fragment makes with the horizontal.

## Section 3.4 The Particle in Uniform Circular Motion

24. The orbit of the Moon about the Earth is approximately circular, with a mean radius of $3.84 \times 10^8$ m. It takes 27.3 days for the Moon to complete one revolution about the Earth. Find (a) the mean orbital speed of the moon and (b) its centripetal acceleration.

**25.** The athlete shown in Figure P3.25 rotates a 1.00-kg discus along a circular path of radius 1.06 m. The maximum speed of the discus is 20.0 m/s. Determine the magnitude of the maximum radial acceleration of the discus.

**Figure P3.25** *(Allsport)*

26. From information on the endsheets of this book, compute the radial acceleration of a point on the surface of the Earth at the equator, due to the rotation of the Earth about its axis.

27. A tire 0.500 m in radius rotates at a constant rate of 200 rev/min. Find the speed and acceleration of a small

stone lodged in the tread of the tire (on its outer edge). (*Hint:* In one revolution, the stone travels a distance equal to the circumference of its path, $2\pi r$.)

28. As their booster rockets separate, space shuttle astronauts typically feel accelerations up to $3g$, where $g = 9.80$ m/s$^2$. In their training, astronauts ride in a device where they experience such an acceleration as a centripetal acceleration. Specifically, the astronaut is fastened securely at the end of a mechanical arm which then turns at constant speed in a horizontal circle. Determine the rotation rate, in revolutions per second, required to give an astronaut a centripetal acceleration of $3.00g$ while in circular motion with radius 9.45 m.

## Section 3.5 Tangential and Radial Acceleration

29. A train slows down as it rounds a sharp horizontal turn, slowing from 90.0 km/h to 50.0 km/h in the 15.0 s that it takes to round the bend. The radius of the curve is 150 m. Compute the acceleration at the moment the train speed reaches 50.0 km/h. Assume it continues to slow down at this time at the same rate.

30. A point on a rotating turntable 20.0 cm from the center accelerates from rest to a final speed of 0.700 m/s in 1.75 s. At $t = 1.25$ s, find the magnitude and direction of (a) the radial acceleration, (b) the tangential acceleration, and (c) the total acceleration of the point.

31. Figure P3.31 represents the total acceleration of a particle moving clockwise in a circle of radius 2.50 m at a certain time. At this instant, find (a) the radial acceleration, (b) the speed of the particle, and (c) its tangential acceleration.

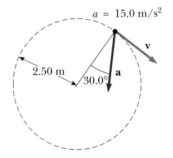

**Figure P3.31**

32. A ball swings in a vertical circle at the end of a rope 1.50 m long. When the ball is 36.9° past the lowest point on its way up, its total acceleration is $(-22.5\mathbf{i} + 20.2\mathbf{j})$ m/s$^2$. At that instant, (a) sketch a vector diagram showing the components of its acceleration, (b) determine the magnitude of its radial acceleration, and (c) determine the speed and velocity of the ball.

## Section 3.6 Relative Velocity

33. A river has a steady speed of 0.500 m/s. A student swims upstream a distance of 1.00 km and swims back to the starting point. If the student can swim at a speed of 1.20 m/s in still water, how long does the trip take? Compare this with the time the trip would take if the water were still.

34. How long does it take an automobile traveling in the left lane at 60.0 km/h to pull alongside a car traveling in the right lane at 40.0 km/h if the cars' front bumpers are initially 100 m apart?

35. The pilot of an airplane notes that the compass indicates a heading due west. The airplane's speed relative to the air is 150 km/h. If there is a wind of 30.0 km/h toward the north, find the velocity of the airplane relative to the ground.

36. A child in danger of drowning in a river is being carried downstream by a current that has a speed of 2.50 km/h. The child is 0.600 km from shore and 0.800 km upstream of a boat landing when a rescue boat sets out. (a) If the boat proceeds at its maximum speed of 20.0 km/h relative to the water, what heading relative to the shore should the pilot take? (b) What angle does the boat velocity make with the shore? (c) How long does it take the boat to reach the child?

37. A science student is riding on a flatcar of a train traveling along a straight horizontal track at a constant speed of 10.0 m/s. The student throws a ball into the air along a path that he judges to make an initial angle of 60.0° with the horizontal and to be in line with the track. The student's professor, who is standing on the ground nearby, observes the ball to rise vertically. How high does she see the ball rise?

38. Two swimmers, Alan and Beth, start together at the same point on the bank of a wide stream that flows with a speed $v$. Both move at the same speed $c$ ($c > v$), relative to the water. Alan swims downstream a distance $L$ and then upstream the same distance. Beth swims so that her motion relative to the Earth is perpendicular to the banks of the stream. She swims the distance $L$ and then back the same distance, so that both swimmers return to the starting point. Which swimmer returns first? (*Note:* First guess the answer.)

## Section 3.7 Context Connection — Circular Orbits

39. The astronaut orbiting the Earth in Figure P3.39 (see next page) is preparing to dock with a Westar VI satellite. The satellite is in a circular orbit 600 km above the Earth's surface, where the free-fall acceleration is 8.21 m/s$^2$. Take the radius of the Earth as 6400 km. Determine the speed of the satellite and the time required to complete one orbit around the Earth.

**Figure P3.39** *(Courtesy of NASA)*

40. An astronaut on the surface of the Moon fires a cannon to launch an experiment package, which leaves the barrel moving horizontally. (a) What must be the muzzle speed of the probe so that it travels completely around the Moon and returns to its original location? (b) How long does this trip around the Moon take? Assume that the free-fall acceleration on the Moon is one-sixth that on the Earth.

### Additional Problems

41. When baseball players throw the ball in from the outfield, they usually allow it to take one bounce before it reaches the infield, on the theory that the ball arrives sooner that way. Suppose that the angle at which a bounced ball leaves the ground is the same as the angle at which the outfielder threw it, as in Figure P3.41, but that the ball's speed after the bounce is one half of what it was before the bounce. (a) Assuming the ball is always thrown with the same initial speed, at what angle $\theta$ should the fielder throw the ball to make it go the same distance $D$ with one bounce (blue path) as a ball thrown upward at 45.0° with no bounce (green path)? (b) Determine the ratio of the times for the one-bounce and no-bounce throws.

42. A ball on the end of a string is whirled around in a horizontal circle of radius 0.300 m. The plane of the circle is 1.20 m above the ground. The string breaks and the ball lands 2.00 m (horizontally) away from the point on the ground directly beneath the ball's location when the string breaks. Find the radial acceleration of the ball during its circular motion.

43. A projectile is launched up an incline (incline angle $\phi$) with an initial speed $v_i$ at an angle $\theta_i$ with respect to the horizontal ($\theta_i > \phi$), as shown in Figure P3.43. (a) Show that the projectile travels a distance $d$ up the incline, where

$$d = \frac{2v_i^2 \cos\theta_i \sin(\theta_i - \phi)}{g\cos^2\phi}.$$

(b) For what value of $\theta_i$ is $d$ a maximum, and what is that maximum value?

**Figure P3.43**

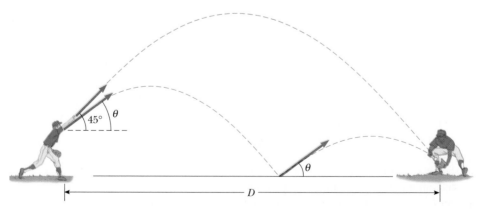

**Figure P3.41**

44. A fisherman sets out upstream from Metaline Falls on the Pend Oreille River in northwestern Washington State. His small boat, powered by an outboard motor, travels at a constant speed $v$ in still water. The water flows at a lower constant speed $v_w$. After traveling upstream for 2.00 km, his ice chest falls out of the boat. He notices that the chest is missing only after he has gone upstream for another 15.0 minutes. At that point he turns around and heads back downstream, all the time traveling at the same speed relative to the water. He catches up with the floating ice chest just as it is about to go over the falls at his starting point. How fast is the river flowing? Solve this problem in two ways. (a) First, use the Earth as a reference frame. (With respect to the Earth, the boat travels upstream at speed $v - v_w$ and downstream at $v + v_w$.) (b) Second, use the water as the reference frame. (This is a simpler and more elegant approach. It has important application in many more complicated problems, such as calculating the motion of rockets and Earth satellites and analyzing the scattering of subatomic particles from massive targets.)

45. A home run is hit in such a way that the baseball just clears the top row of bleachers, 21.0 m high, located 130 m from home plate. The ball is hit at an angle of 35.0° to the horizontal, and air resistance is negligible. Find (a) the initial speed of the ball, (b) the time it takes the ball to reach the cheap seats, and (c) the velocity components and the speed of the ball when it passes over the top row. (Assume the ball is hit at a height of 1.00 m above the ground.)

46. A small massive ball is suspended at the bottom end of a light cord 1.00 m long, with the top end of the cord fixed. The ball is set swinging around in a vertical circle. Let $\theta$ represent the instantaneous angle the cord makes with the vertical. When the ball is in the two positions where the cord is horizontal, $\theta = 90°$ and $\theta = 270°$, its speed is 5.00 m/s. (a) Find the magnitude of the ball's radial acceleration and tangential acceleration for these positions. (b) Draw vector diagrams to determine the direction of the total acceleration for these two positions. (c) Calculate the magnitude and direction of the total acceleration.

47. A basketball player who is 2.00 m tall is standing on the floor 10.0 m from the basket, as in Figure P3.47. If he shoots the ball at a 40.0° angle with the horizontal, at what initial speed must he throw so that it goes through the hoop without striking the backboard? The basket height is 3.05 m.

48. A ball is thrown with an initial speed $v_i$ at an angle $\theta_i$ with the horizontal. The horizontal range of the ball is $R$, and the ball reaches a maximum height $R/6$. In terms of $R$ and $g$, find (a) the time the ball is in motion, (b) the ball's speed at the peak of its path, (c) the initial vertical component of its velocity, (d) its initial speed, and (e) the angle

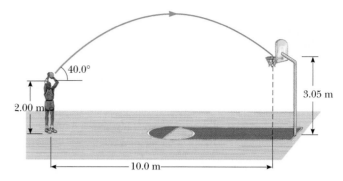

**Figure P3.47**

$\theta_i$. Suppose the ball is thrown at the same initial speed found in (d), but at the angle appropriate for reaching the greatest height that it can. (f) Find this height. Suppose the ball is thrown at the same initial speed but at the angle for greatest possible range. (g) Find this maximum horizontal range.

49. A person standing at the top of a hemispherical rock of radius $R$ kicks a ball (initially at rest on the top of the rock) to give it horizontal velocity $\mathbf{v}_i$ as in Figure P3.49. (a) What must be its minimum initial speed if the ball is never to hit the rock after it is kicked? (b) With this initial speed, how far from the base of the rock does the ball hit the ground?

**Figure P3.49**

50. Your grandfather is copilot of a bomber, flying horizontally over level terrain, with a speed of 275 m/s relative to the ground, at an altitude of 3 000 m. (a) The bombardier releases one bomb. How far will it travel horizontally between its release and its impact on the ground? Ignore the effects of air resistance. (b) The bombardier is inexperienced and does not call, "Bombs away!" Consequently, the pilot maintains the plane's original course, altitude, and speed. Where will the plane be when the bomb hits the ground? (c) At what angle from the vertical was the telescopic bomb sight set so that the bomb hits the target seen in the sight at the time of release?

51. A quarterback throws a football toward a receiver with an initial speed of 20.0 m/s, at an angle of 30.0° above the horizontal. At that instant, the receiver is 20.0 m from the quarterback. In what direction and with what constant speed should the receiver run in order to catch the football at the level at which it was thrown?

52. A truck loaded with cannonball watermelons stops suddenly to avoid running over the edge of a washed-out bridge (Fig. P3.52). The quick stop causes a number of melons to fly off the truck. One melon rolls over the edge with an initial speed $v_i = 10.0$ m/s in the horizontal direction. A cross-section of the bank has the shape of the bottom half of a parabola with its vertex at the edge of the road, and described by the equation $y^2 = 16x$, where $x$ and $y$ are measured in meters. What are the $x$ and $y$ coordinates of the melon when it splatters on the bank?

$v_i = 10.0$ m/s

**Figure P3.52**

53. A car is parked on a steep incline overlooking the ocean. The incline makes an angle of 37.0° below the horizontal. The negligent driver leaves the car in neutral, and the parking brakes are defective. The car rolls from rest down the incline with a constant acceleration of 4.00 m/s², traveling 50.0 m to the edge of a vertical cliff. The cliff is 30.0 m above the ocean. Find (a) the speed of the car when it reaches the edge of the cliff and the time it takes to get there, (b) the velocity of the car when it lands in the ocean, (c) the total time the car is in motion, and (d) the position of the car when it lands in the ocean, relative to the base of the cliff.

54. A catapult launches a rocket at an angle of 53.0° above the horizontal with an initial speed of 100 m/s. The rocket engine immediately starts a burn and for 3.00 s the rocket moves along its initial line of motion with an acceleration of 30.0 m/s². Then its engine fails, and the rocket proceeds to move in free fall. Find (a) the maximum altitude reached by the rocket, (b) its total time of flight, and (c) its horizontal range.

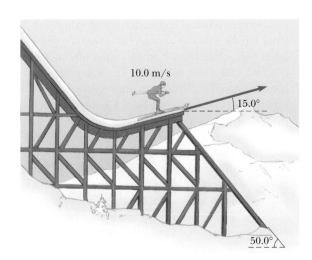

10.0 m/s

15.0°

50.0°

**Figure P3.55**

55. A skier leaves the ramp of a ski jump with a velocity of 10.0 m/s, 15.0° above the horizontal, as in Figure P3.55. The slope is inclined at 50.0°, and air resistance is negligible. Find (a) the distance from the ramp to where the jumper lands and (b) the velocity components just before the landing. (How do you think the results might be affected if air resistance were included? Note that jumpers lean forward in the shape of an airfoil, with their hands at their sides, to increase their distance. Why does this work?)

56. In a television picture tube (a cathode-ray tube) electrons are emitted with velocity $\mathbf{v}_i$ from a source at the origin of coordinates. The initial velocities of different electrons make different angles $\theta$ with the $x$ axis. As they move a distance $D$ along the $x$ axis, the electrons are acted on by a constant electric field, giving each a constant acceleration $\mathbf{a}$ in the $x$ direction. At $x = D$ the electrons pass through a circular aperture, oriented perpendicular to the $x$ axis. At the aperture, the velocity imparted to the electrons by the electric field is much larger than $\mathbf{v}_i$ in magnitude. Show that the velocity vectors of electrons going through the aperture radiate from a certain point on the $x$ axis, which is not the origin. Determine the location of this point. This point is called a *virtual source* and is important in determining where the electron beam hits the screen of the tube.

57. Do not hurt yourself; do not strike your hand against anything. Within these limitations, describe what you do to give your hand a large acceleration. Compute an order-of-magnitude estimate of this acceleration, stating the quantities you measure or estimate and their values.

**58.** An enemy ship is on the east side of a mountain island, as shown in Figure P3.58. The enemy ship has maneuvered to within 2 500 m of the 1 800-m-high mountain peak and only can shoot projectiles with an initial speed of 250 m/s. If the western shoreline is horizontally 300 m from the peak, what are the distances from the western shore at which a ship can be safe from the bombardment of the enemy ship?

**Figure P3.58**

## ANSWERS TO QUICK QUIZZES

**3.1** (a) Yes. An object moving along a curved path at constant speed is accelerating because the direction of the velocity vector is changing. (b) No. If the velocity is constant, there is no change in the velocity and the acceleration is zero.

**3.2** (a) No. If the velocity is constant, the magnitude of the instantaneous velocity vector, which is the speed, is constant. (b) Yes. An electron traveling around a curved path at a constant speed has a varying velocity because the direction of the velocity vector is changing.

**3.3** You should simply throw it straight up in the air. Because the ball is moving along with you, it will follow a parabolic trajectory with a horizontal component of velocity that is the same as yours.

**3.4** (a) At the top of its flight, **v** is horizontal and **a** is vertical. This is the only point at which the velocity and acceleration vectors are perpendicular. (b) If the object is thrown straight up or down, then **v** and **a** will be parallel throughout the downward motion. Otherwise, the velocity and acceleration vectors are never parallel.

# chapter 4

A locomotive pulls on the cars of a long train. The acceleration of a given car depends on forces between that car and the one in front and the one behind.

*(Bruce Hands/Stone)*

# The Laws of Motion

In the preceding two chapters on kinematics, we described the motion of particles based on the definitions of position, velocity, and acceleration. Aside from our discussion of gravity for objects in free-fall, we did not address what causes an object to move as it does. We would like to be able to answer general questions related to the causes of motion, such as "What mechanism causes changes in motion?" and "Why do some objects accelerate at higher rates than others?" In this first chapter on *dynamics,* we shall describe the change in motion of particles, using the concepts of force and mass. We then discuss the three fundamental laws of motion, which are based on experimental observations and were formulated about three centuries ago by Sir Isaac Newton.

## 4.1 • THE CONCEPT OF FORCE

Everyone has a basic understanding of the concept of force as a result of everyday experiences. When you push or pull an object, you exert a force on it. You exert a force when you throw or kick a ball. In these examples, the word *force* is associated with the result of muscular activity and with some change in the state of motion of an object. Forces do not always cause an object to move, however. For example, as you sit reading this book, the gravitational force acts on your body, and yet you remain stationary. You can push on a heavy block of stone and yet fail to move it.

**Figure 4.1**

Some examples of forces applied to various objects. In each case a force is exerted on the particle or object within the boxed area. The environment external to the boxed area provides this force.

This chapter is concerned with the relation between the force on an object and the change in motion of that object. If you pull on a spring, as in Figure 4.1a, the spring stretches. If the spring is calibrated, the distance it stretches can be used to measure the strength of the force. If a child pulls on a wagon, as in Figure 4.1b, the wagon moves. When a football is kicked, as in Figure 4.1c, it is both deformed and set in motion. These are all examples of a class of forces called *contact forces.* That is, they represent the result of physical contact between two objects.

• *Contact forces*

Another class of forces, which do not involve physical contact between two objects but act through empty space, are known as *field forces.* The gravitational force between two objects that causes the free-fall acceleration described in Chapters 2 and 3 is an example of this class of force, and is illustrated in Figure 4.1d. This gravitational force keeps objects bound to the Earth and gives rise to what we commonly call the *weight* of an object. The planets of our solar system are bound under the action of gravitational forces. Another common example of a field force is the electric force that one electric charge exerts on another electric charge, as in Figure 4.1e. These charges might be an electron and proton forming a hydrogen atom. A third example of a field force is the force that a bar magnet exerts on a piece of iron, as shown in Figure 4.1f.

• *Field forces*

The distinction between contact forces and field forces is not as sharp as you may have been led to believe by the preceding discussion. At the atomic level, all

(a)  (b)  (c)  (d)

**Figure 4.2**

The vector nature of a force is tested with a spring scale. (a) A downward vertical force $\mathbf{F}_1$ elongates the spring 1.00 cm. (b) A downward vertical force $\mathbf{F}_2$ elongates the spring 2.00 cm. (c) When $\mathbf{F}_1$ and $\mathbf{F}_2$ are applied simultaneously, the spring elongates by 3.00 cm. (d) When $\mathbf{F}_1$ is downward and $\mathbf{F}_2$ is horizontal, the combination of the two forces elongates the spring $\sqrt{(1.00)^2 + (2.00)^2} = \sqrt{5.00}$ cm.

the forces classified as contact forces turn out to be due to electric (field) forces of the type illustrated in Figure 4.1e. Nevertheless, in understanding macroscopic phenomena, it is convenient to use both classifications of forces.

It is sometimes convenient to use the linear deformation of a spring to measure force, as in the case of a common spring scale. Suppose a force is applied vertically to a spring that has a fixed upper end, as in Figure 4.2a. We can calibrate the spring by defining the unit force $\mathbf{F}_1$ as the force that produces an elongation of 1.00 cm. If a force $\mathbf{F}_2$, applied as in Figure 4.2b, produces an elongation of 2.00 cm, the magnitude of $\mathbf{F}_2$ is 2.00 units. If the two forces $\mathbf{F}_1$ and $\mathbf{F}_2$ are applied simultaneously, as in Figure 4.2c, the elongation of the spring is 3.00 cm because the forces are applied in the same direction and their magnitudes add. If the two forces $\mathbf{F}_1$ and $\mathbf{F}_2$ are applied in perpendicular directions, as in Figure 4.2d, the elongation is $\sqrt{(1.00)^2 + (2.00)^2}$ cm $= \sqrt{5.00}$ cm $= 2.24$ cm. The single force $\mathbf{F}$ that would produce this same elongation is the vector sum of $\mathbf{F}_1$ and $\mathbf{F}_2$, as described in Figure 4.2d. That is, $|\mathbf{F}| = \sqrt{F_1^2 + F_2^2} = 2.24$ units, and its direction is $\theta = \tan^{-1}(-0.500) = -26.6°$. **Because forces have been experimentally verified to behave as vectors, you must use the rules of vector addition to obtain the net force on an object.**

## 4.2 • NEWTON'S FIRST LAW

See the *Core Concepts in Physics CD-ROM,* Screen 4.2

Before about 1600, scientists felt that the natural state of matter was the state of rest, because moving objects on the Earth tended to come to rest if a force on them was not maintained. (Today, of course, we recognize that friction can cause a moving object to come to rest.) Galileo took a different approach to motion and the natural state of matter. He devised thought experiments, such as that of an object moving on a frictionless surface, and concluded that it is not the nature of an object to stop once set in motion; rather, it is an object's nature to resist changes in its motion. In his words, "Any velocity, once imparted to a moving body, will be rigidly maintained as long as the external causes of retardation are removed."

This new approach to motion was later formalized by Newton in a statement that has come to be known as **Newton's first law of motion:**

> In the absence of external forces, an object at rest remains at rest and an object in motion continues in motion with a constant velocity (that is, with a constant speed in a straight line).

In simpler terms, we can say that **when no force acts on a body, its acceleration is zero.** Thus, an object has a tendency to maintain its original state of motion in the absence of a force. This tendency is called **inertia,** and Newton's first law is sometimes called the **law of inertia.** It is not a common experience to observe Newton's first law, because it is difficult to completely eliminate all forces on an object, especially the force of friction.

An example of almost uniform motion on a nearly frictionless plane is the motion of a light disk on a layer of air, as in Figure 4.3. If the disk is given an initial velocity, it coasts a great distance before coming to rest. This idea is used in the game of air hockey, in which the disk makes many collisions with the walls of the air hockey table before coming to rest.

Consider a spacecraft traveling in space, far removed from any planets or other matter. The spacecraft requires some propulsion system to change its velocity. If the propulsion system is turned off when the spacecraft reaches a velocity **v**, however, the spacecraft "coasts" in space with that velocity, and the astronauts enjoy a "free ride" (i.e., no propulsion system is required to keep them moving at the velocity **v**).

Finally, recall our discussion in Chapter 2 about the proportionality between force and acceleration:

$$\mathbf{F} \propto \mathbf{a}$$

Newton's first law tells us that the velocity does not change if no force acts on an object—the object maintains its state of motion. The preceding proportionality tells us that if a force *does* act, then a change does occur in the motion, measured by the acceleration. This notion will form the basis of Newton's second law, and we shall provide more details on this concept shortly.

## Inertial Frames

Newton's first law defines a special set of reference frames called inertial frames. **An inertial frame of reference is one in which Newton's first law is valid.** For example, suppose you are on a perfectly smoothly moving train, observing a stationary air hockey puck on a perfectly level operating air hockey table. If the train were moving at a constant velocity, you would see the puck remain at rest, obeying Newton's law. If the train were to accelerate in the forward direction, however, you would see the puck accelerate toward the back of the train—the same thing happens when something on your dashboard slides off when you step on the gas pedal. The train moving at constant velocity represents an inertial frame. The puck at rest remains at rest, and Newton's first law is obeyed. The accelerating train is not an inertial frame. According to you as the observer on the train, there

**Isaac Newton** (1642–1727)

Newton, an English physicist and mathematician, was one of the most brilliant scientists in history. Before the age of 30, he formulated the basic concepts and laws of mechanics, discovered the law of universal gravitation, and invented the mathematical methods of calculus. As a consequence of his theories, Newton was able to explain the motions of the planets, the ebb and flow of the tides, and many special features of the motions of the Moon and the Earth. He also interpreted many fundamental observations concerning the nature of light. His contributions to physical theories dominated scientific thought for two centuries and remain important today. *(Giraudon/Art Resource)*

**v** = constant

Air flow

Electric blower

**Figure 4.3**

Air hockey takes advantage of Newton's first law to make the game more exciting. A disk moving on a column of air is an approximation of uniform motion, that is, motion in which the acceleration is zero.

**PITFALL PREVENTION 4.1**

**Newton's first law**

Newton's first law does *not* say what happens for an object with *zero net force*, that is, multiple forces that cancel; it says what happens *in the absence of a force*. This subtle but important difference allows us to define force as that which causes a change in the motion. The description of an object under the effect of forces that balance is covered by Newton's second law.

is no force on the puck, yet it accelerates from rest toward the back of the train, violating Newton's first law.

Any reference frame that moves with constant velocity with respect to an inertial frame is itself an inertial frame. The Earth is not an inertial frame because of its orbital motion about the Sun and rotational motion about its own axis, both of which represent centripetal accelerations. As the Earth travels in its nearly circular orbit about the Sun, it experiences a centripetal acceleration of about $6.0 \times 10^{-3}$ m/s$^2$ toward the Sun, as found in Example 3.5. In addition, because the Earth rotates about its own axis once every 24 h, a point on the equator experiences an additional centripetal acceleration of $3.4 \times 10^{-2}$ m/s$^2$ toward the center of the Earth. These accelerations are small compared with *g*, however, and can often be ignored—this is a simplification model. In most situations we shall model a reference frame on or near the Earth's surface as an inertial frame.

**Quick Quiz 4.1**

Is it possible to have motion in the absence of a force? Is it possible to have force in the absence of motion?

**Quick Quiz 4.2**

(a) If a single force acts on an object, does the object accelerate? (b) If an object experiences an acceleration, is a force acting on it? (c) If an object experiences no acceleration, is no force acting on it?

See Screen 4.3

## 4.3 • INERTIAL MASS

Consider playing catch with either a basketball or a bowling ball. If both are coming toward you with the same velocity, which would be easier to catch? Which one would be harder to throw? The bowling ball has more mass than the basketball, so it will be harder to catch and to throw. When you either catch or throw the ball, you are changing its motion. Newton's first law tells us that if we make no attempt to change the motion of an object (i.e., if we apply no force on it), then its motion remains uniform. What if we *do* attempt to change its motion by applying a force on the object, as we do with the basketball and bowling ball? We find that the object resists this change. We quantify this resistance as the *mass* of the object, and, in particular, we call this the inertial mass. **Inertial mass is the measure of an object's resistance to a change in motion in response to an external force.** The bowling ball has more inertial mass than the basketball, so it is harder to change the motion of the bowling ball by catching or throwing it. Inertial mass is different in definition from gravitational mass, which we discuss later in this chapter, but has the same value, so we call them both simply *mass*.

• *Inertial mass*

A quantitative measurement of mass can be made by comparing the accelerations that a given force produces on different objects. Suppose a force acting on an object of mass $m_1$ produces an acceleration $\mathbf{a}_1$, and the *same force* acting on an object of mass $m_2$ produces an acceleration $\mathbf{a}_2$. The ratio of the two masses is defined as the *inverse* ratio of the magnitudes of the accelerations produced by the same force:

$$\frac{m_1}{m_2} \equiv \frac{a_2}{a_1} \qquad \qquad [4.1]$$

If one mass is standard and known—say, 1 kg—the mass of an unknown object can be obtained from acceleration measurements. For example, if the standard 1-kg object undergoes an acceleration of 3 m/s$^2$ under the influence of some force, an object experiencing an acceleration of 1.5 m/s$^2$ under the action of the same force must have a mass of 2 kg.

Mass is an inherent property of an object, independent of the object's surroundings and of the method used to measure it. It is an experimental fact that mass is a scalar quantity. As a scalar quantity, mass obeys the rules of ordinary arithmetic. That is, several masses can be combined in simple numerical fashion—if you combine a 3-kg mass with a 5-kg mass, the total mass is 8 kg. This can be verified experimentally by comparing the acceleration of each object produced by a known force with the acceleration of the combined system using the same force.

Mass should not be confused with weight. **Mass and weight are two different quantities.** What we call the weight of an object is equal to the magnitude of the gravitational force exerted by the planet on which the object resides. For example, a person who weighs 180 lb on Earth weighs only about 30 lb on the Moon. On the other hand, the mass of an object is the same everywhere. A given object exhibits a fixed amount of resistance to changes in motion regardless of its location. An object of mass 2 kg on Earth also has a mass of 2 kg on the Moon.

• *Mass and weight are different quantities*

## 4.4 • NEWTON'S SECOND LAW—THE PARTICLE UNDER A NET FORCE

See Screen 4.4

Newton's first law explains what happens to an object when no force acts on it: It either remains at rest or moves in a straight line with constant speed. This allows us to define an inertial frame of reference. It also allows us to identify force as that which changes motion. Newton's second law answers the question of what happens to an object that has a nonzero net force acting on it, based on our discussion of mass in the preceding section.

Imagine you are pushing a block of ice across a frictionless horizontal surface. When you exert some horizontal force **F**, the block moves with some acceleration **a**. Experiments show that if you apply a force twice as large to the same object, the acceleration doubles. If you increase the applied force to 3**F**, the original acceleration is tripled, and so on. From such observations, we conclude that **the acceleration of an object is directly proportional to the net force acting on it.** We alluded to this proportionality in our discussion of acceleration in Chapter 2. We are now ready to extend that discussion.

These observations and those in Section 4.3 relating mass and acceleration are summarized in **Newton's second law:**

> The acceleration of an object is directly proportional to the net force acting on it and inversely proportional to its mass.

We write this as

$$\mathbf{a} \propto \frac{\Sigma \mathbf{F}}{m}$$

where $\Sigma\mathbf{F}$ is the **net force,** which is the vector sum of *all* forces acting on the object of mass *m*. If the object consists of a system of individual elements, the net force is the vector sum of all forces *external* to the system. Any *internal* forces—that

**PITFALL PREVENTION 4.2**

**Force is the cause of changes in motion**

Be sure that you are clear on the role of force. Many times, students make the mistake of thinking that force is the cause of motion. But we can have motion in the absence of forces, as described in Newton's first law. Be sure to understand that force is the cause of *changes* in motion.

is, forces between elements of the system—are not included because they do not affect the motion of the entire system. The net force is also sometimes called the *resultant force,* the *sum of the forces,* the *total force,* or the *unbalanced force.*

Newton's second law in mathematical form is a statement of this relationship that makes the preceding proportionality an equality:*

- *Newton's second law*

$$\sum \mathbf{F} = m\mathbf{a} \qquad\qquad [4.2]$$

Note that Equation 4.2 is a *vector* expression and hence is equivalent to the following three component equations:

$$\sum F_x = ma_x \qquad \sum F_y = ma_y \qquad \sum F_z = ma_z \qquad [4.3]$$

**PITFALL PREVENTION 4.3**

**ma is not a force**

Equation 4.2 does *not* say that the product *m***a** is a force. All forces on an object are added vectorially to generate the net force on the left side of the equation. This net force is then equated to the product of the mass of the object and the acceleration that results from the net force. Do *not* include an "*m***a** force" in your analysis.

### Quick Quiz 4.3

If an object experiences a net force in the *x* direction, does it *move* in the *x* direction?

Newton's second law introduces us to a new analysis model—the particle under a net force. If a particle, or an object that can be modeled as a particle, is under the influence of a net force, then Equation 4.2, the mathematical statement of Newton's second law, can be used to describe its motion. The acceleration is constant if the net force is constant. Thus, the particle under a net force may have its motion predicted as a particle under constant acceleration. The net force does not have to be constant, however. We shall investigate situations in this chapter and the next involving both constant and varying forces.

### Unit of Force

The SI unit of force is the **newton,** which is defined as the force that, when acting on a 1-kg mass, produces an acceleration of 1 m/s$^2$.

From this definition and Newton's second law, we see that the newton can be expressed in terms of the fundamental units of mass, length, and time:

$$1 \text{ N} \equiv 1 \text{ kg} \cdot \text{m/s}^2 \qquad\qquad [4.4]$$

The units of mass, acceleration, and force are summarized in Table 4.1. Most of the calculations we shall make in our study of mechanics will be in SI units. Equalities between units in the SI and British engineering systems are given in Appendix A.

**PITFALL PREVENTION 4.4**

**Newton's second law**

The protocol for using Newton's second law is to add *all* of the forces on the object vectorially and then find the resultant acceleration. Do *not* find individual accelerations for each force and then add the accelerations vectorially.

| TABLE 4.1 | Units of Mass, Acceleration, and Force | | |
|---|---|---|---|
| **System of Units** | **Mass (M)** | **Acceleration (L/T$^2$)** | **Force (ML/T$^2$)** |
| SI | kg | m/s$^2$ | N = kg·m/s$^2$ |
| British engineering | slug | ft/s$^2$ | lb = slug·ft/s$^2$ |

---

\* Equation 4.2 is valid only when the speed of the object is much less than the speed of light. We will treat the relativistic situation in Chapter 9.

## THINKING PHYSICS 4.1

In a train, such as that in the opening photograph of this chapter, the cars are connected by *couplers*. The couplers between the cars exert forces on the cars as the train is pulled by the locomotive in the front. Imagine that the train is speeding up in the forward direction. As you imagine moving from the locomotive to the caboose, does the force exerted by the couplers *increase, decrease,* or *stay the same?* What if the engineer applies the brakes? How does the force vary from locomotive to caboose in this case? (Assume that the only brakes applied are those on the engine.)

**Reasoning**   The force *decreases* from the front of the train to the back. The coupler between the locomotive and the first car must apply enough force to accelerate all of the remaining cars. As we move back along the train, each coupler is accelerating less mass behind it. The last coupler only has to accelerate the caboose, so it exerts the smallest force. If the brakes are applied, the force decreases from front to back of the train also. The first coupler, at the back of the locomotive, must apply a large force to slow down all of the remaining cars. The final coupler must only apply a force large enough to slow down the mass of the caboose.

## Example 4.1   An Accelerating Hockey Puck

A 0.30-kg hockey puck slides on the horizontal frictionless surface of an ice rink. It is struck simultaneously by two different hockey sticks. The two constant forces that act on the puck as a result of the hockey sticks are parallel to the ice surface and are shown in the pictorial representation in Figure 4.4. The force **F**₁ has a magnitude of 5.0 N, and **F**₂ has a magnitude of 8.0 N. Determine the acceleration of the puck while it is in contact with the two sticks.

**Solution**   The puck is modeled as a particle under a net force. We first find the components of the net force. The component of the net force in the *x* direction is

$$\sum F_x = F_{1x} + F_{2x} = F_1 \cos 20° + F_2 \cos 60°$$
$$= (5.0 \text{ N})(0.940) + (8.0 \text{ N})(0.500) = 8.7 \text{ N}$$

The component of the net force in the *y* direction is

$$\sum F_y = F_{1y} + F_{2y} = -F_1 \sin 20° + F_2 \sin 60°$$
$$= -(5.0 \text{ N})(0.342) + (8.0 \text{ N})(0.866) = 5.2 \text{ N}$$

Now we use Newton's second law in component form to find the *x* and *y* components of acceleration:

$$a_x = \frac{\sum F_x}{m} = \frac{8.7 \text{ N}}{0.30 \text{ kg}} = 29 \text{ m/s}^2$$

$$a_y = \frac{\sum F_y}{m} = \frac{5.2 \text{ N}}{0.30 \text{ kg}} = 17 \text{ m/s}^2$$

The acceleration has a magnitude of

$$a = \sqrt{(29)^2 + (17)^2} \text{ m/s}^2 = \boxed{34 \text{ m/s}^2}$$

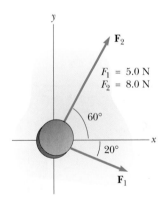

**Figure 4.4**

(Example 4.1) A hockey puck moving on a frictionless surface accelerates in the direction of the net force, $\sum \mathbf{F} = \mathbf{F}_1 + \mathbf{F}_2$.

and its direction is

$$\theta = \tan^{-1}\frac{a_y}{a_x} = \tan^{-1}\frac{17 \text{ m/s}^2}{29 \text{ m/s}^2} = \boxed{30°}$$

relative to the positive *x* axis.

**EXERCISE**   Determine the components of a third force that, when applied to the puck along with the two in Figure 4.4, causes the puck to have zero acceleration.

*Answer*   $F_x = -8.7 \text{ N}, F_y = -5.2 \text{ N}$

**EXERCISE**  A 6.0-kg object undergoes an acceleration of 2.0 m/s². (a) What is the magnitude of the net force acting on the object? (b) If this same force is applied to a 4.0-kg object, what acceleration does it produce?

*Answer*   (a) 12 N (b) 3.0 m/s²

**EXERCISE**  A $1.80 \times 10^3$-kg car is traveling in a straight line with a speed of 25.0 m/s. What is the magnitude of the constant horizontal force needed to bring the car to rest in a distance of 80.0 m?

*Answer*   $7.03 \times 10^3$ N

---

**PITFALL PREVENTION 4.5**

**Weight "of an object"**

We are familiar with the common phrase, the "weight of an object." Keep in mind that, despite this everyday phrase, weight is not a property of an object. It is a measure of the gravitational force between the object and the Earth. Thus, weight is a property of a *system* of items—the object and the Earth. As the system changes, for example, by lifting the object well off the surface, the weight changes, due to the decreasing value of *g*. We could also move the object into a new system, consisting of the object and Mars, for example. Then the weight will be very different from that on the Earth because of the different value of *g* on Mars.

## 4.5 • THE GRAVITATIONAL FORCE AND WEIGHT

We are well aware that all objects are attracted to the Earth. The force exerted by the Earth on an object is the **gravitational force $F_g$**. This force is directed toward the center of the Earth.* The magnitude of the gravitational force is called the **weight** $F_g$ of the object.

We have seen in Chapters 2 and 3 that a freely falling object experiences an acceleration **g** directed toward the center of the Earth. A freely falling object has only one force on it, the gravitational force, so the net force on the object is equal to the gravitational force:

$$\sum \mathbf{F} = \mathbf{F}_g$$

Because the acceleration of a freely falling object is equal to the free-fall acceleration **g**, it follows that

$$\sum \mathbf{F} = m\mathbf{a} \quad \longrightarrow \quad \mathbf{F}_g = m\mathbf{g}$$

or, in magnitude,

$$F_g = mg \qquad [4.5]$$

**WEB**

For information and many links to *Apollo 11* sites, visit **nssdc.gsfc.nasa.gov/planetary/lunar/apollo11info.html** For general information about NASA projects, visit **www.nasa.gov**

Astronaut Edwin E. Aldrin, Jr., walking on the Moon after the *Apollo 11* lunar landing. The life-support system strapped to his back weighed 300 lb on Earth. During his training, a 50-lb mock-up system was used to simulate the reduction of gravity on the Moon. Although this effectively simulated the reduced weight of the unit on the Moon, it did not correctly represent the unchanging mass. The system was easier to accelerate during the training on Earth by jumping or twisting than it was on the Moon. *(Courtesy of NASA)*

---

\* This statement represents a simplification model in that it ignores the fact that the mass distribution of the Earth is not perfectly spherical.

Because it depends on *g*, weight varies with location, as we mentioned in Section 4.3. Objects weigh less at higher altitudes than at sea level because *g* decreases with increasing distance from the center of the Earth. Hence, weight, unlike mass, is not an inherent property of an object. For example, if an object has a mass of 70 kg, then its weight in a location where $g = 9.80$ m/s$^2$ is $mg = 686$ N. At the top of a mountain where $g = 9.76$ m/s$^2$, the object's weight would be 683 N. Thus, if you want to lose weight without going on a diet, climb a mountain or weigh yourself at 30 000 ft during an airplane flight.

Because $F_g = mg$, we can compare the masses of two objects by measuring their weights with a spring scale. At a given location (so that *g* is fixed) the ratio of the weights of two objects equals the ratio of their masses.

Equation 4.5 quantifies the gravitational force on the object, but notice that this equation does not require the object to be moving. Even for a stationary object, or an object on which several forces act, Equation 4.5 can be used to calculate the magnitude of the gravitational force. This results in a subtle shift in the interpretation of *m* in the equation. The mass *m* in Equation 4.5 is playing the role of determining the strength of the gravitational attraction between the object and the Earth. This is a completely different role from that previously described for mass, that of measuring the resistance to changes in motion in response to an external force. Thus, we call *m* in this type of equation the **gravitational mass.** Despite this quantity being different from inertial mass, it is one of the experimental conclusions in Newtonian dynamics that gravitational mass and inertial mass have the same value.

### Quick Quiz 4.4

Suppose you are talking by interplanetary telephone to your friend, who lives on the Moon. He tells you that he has just won a piece of gold weighing one newton in a contest. Excitedly, you tell him that you entered the Earth version of the same contest and also won a newton of gold! Who's richer?

### Quick Quiz 4.5

A baseball of mass *m* is thrown upward with some initial speed. If air resistance is ignored, what is the magnitude of the force on the ball (a) when it reaches half its maximum height and (b) when it reaches its peak?

## THINKING PHYSICS 4.2

In the absence of air friction, it is claimed that all objects fall with the same acceleration. A heavier object is pulled to the Earth with more force than a light object. Why does the heavier object not fall faster?

**Reasoning**  It is indeed true that the heavier object is pulled with a larger force. The *strength* of the force is determined by the gravitational mass of the object. The *resistance* to the force and, therefore, to the change in motion of the object, is represented by the inertial mass. Thus, if an object has twice as much mass as another, it is pulled to the Earth with twice the force, but it also exhibits twice the resistance to having its motion changed. These factors cancel, so that the change in motion, the acceleration, is the same for all objects, regardless of mass.

See Screen 4.5

• *A statement of Newton's third law*

## 4.6 • NEWTON'S THIRD LAW

Newton's third law conveys the notion that forces are always interactions between two objects: **If two objects interact, the force $\mathbf{F}_{12}$ exerted by object 1 on object 2 is equal in magnitude but opposite in direction to the force $\mathbf{F}_{21}$ exerted by object 2 on object 1:**

$$\mathbf{F}_{12} = -\mathbf{F}_{21} \qquad\qquad [4.6]$$

When it is important to designate forces as interactions between two objects, we will use this subscript notation, where $\mathbf{F}_{ab}$ means "the force exerted *by* a *on* b." The third law, illustrated in Figure 4.5a, is equivalent to stating that **forces always occur in pairs,** or that **a single isolated force cannot exist.** The force that object 1 exerts on object 2 may be called the *action force* and the force of object 2 on object 1 the *reaction force.* In reality, either force can be labeled the action or reaction force. **The action force is equal in magnitude to the reaction force and opposite in direction. In all cases, the action and reaction forces act on different objects and must be of the same type.** For example, the force acting on a freely falling projectile is the gravitational force exerted by the Earth on the projectile $\mathbf{F}_g = \mathbf{F}_{Ep}$ (E = Earth, p = projectile), and the magnitude of this force is *mg*. The reaction to this force is the gravitational force exerted by the projectile on the Earth $\mathbf{F}_{pE} = -\mathbf{F}_{Ep}$. The reaction force $\mathbf{F}_{pE}$ must accelerate the Earth toward the projectile just as the action force $\mathbf{F}_{Ep}$ accelerates the projectile toward the Earth. Because the Earth has such a large mass, however, its acceleration due to this reaction force is negligibly small.

Another example of Newton's third law in action is shown in Figure 4.5b. The force $\mathbf{F}_{hn}$ exerted by the hammer on the nail (the action) is equal in magnitude and opposite the force $\mathbf{F}_{nh}$ exerted by the nail on the hammer (the reaction). This latter force stops the forward motion of the hammer when it strikes the nail.

You directly experience the third law if you slam your fist against a wall or kick a football with your bare foot. You can feel the force back on your fist or your foot. You should be able to identify the action and reaction forces in these cases.

**PITFALL PREVENTION 4.8**

**Newton's third law**

This is such an important and often misunderstood notion that it will be repeated here in a Pitfall Prevention. Newton's third law action and reaction forces act on *different* objects. Two forces acting on the same object, even if they are equal in magnitude and opposite in direction, *cannot* be an action–reaction pair.

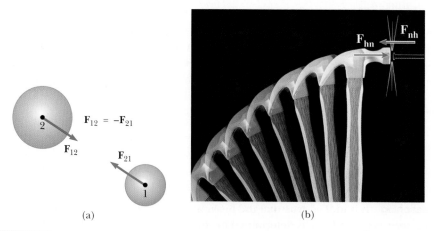

(a)

(b)

**Figure 4.5**

Newton's third law. (a) The force $\mathbf{F}_{12}$ exerted by object 1 on object 2 is equal in magnitude and opposite in direction to the force $\mathbf{F}_{21}$ exerted by object 2 on object 1. (b) The force $\mathbf{F}_{hn}$ exerted by the hammer on the nail is equal in magnitude and opposite in direction to the force $\mathbf{F}_{nh}$ exerted by the nail on the hammer. (*John Gillmoure, The Stock Market*)

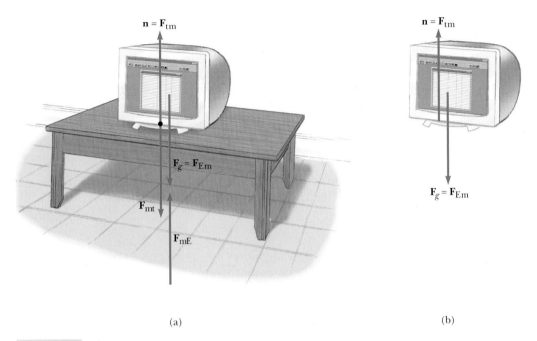

(a)                                                    (b)

### Figure 4.6

(a) When a computer monitor is sitting on a table, several forces are acting. (b) The free-body diagram for the monitor. The forces acting on the monitor are the normal force $\mathbf{n} = \mathbf{F}_{tm}$ and the gravitational force $\mathbf{F}_g = \mathbf{F}_{Em}$. These are the only forces acting on the monitor (notice "m" as the second subscript on these forces, but no other forces) and the only forces that should appear in a free-body diagram for the monitor. The other forces in part (a) are as follows. The reaction force to $\mathbf{n}$ is the force $\mathbf{F}_{mt}$ of the monitor on the table. The reaction to $\mathbf{F}_g$ is the force $\mathbf{F}_{mE}$ of the monitor on the Earth.

The Earth exerts a gravitational force $\mathbf{F}_g$ on any object. If the object is a computer monitor at rest on a table, as in the pictorial representation in Figure 4.6a, the reaction force to $\mathbf{F}_g = \mathbf{F}_{Em}$ is the force exerted by the monitor on the Earth $\mathbf{F}_{mE} = -\mathbf{F}_{Em}$. The monitor does not accelerate because it is held up by the table. The table exerts on the monitor an upward force $\mathbf{n} = \mathbf{F}_{tm}$, called the **normal force.**\* This force prevents the monitor from falling through the table; it can have any value needed, up to the point of breaking the table. From Newton's second law, we see that, because the monitor has zero acceleration, it follows that $\Sigma\mathbf{F} = \mathbf{n} - m\mathbf{g} = 0$, or $n = mg$. The normal force balances the gravitational force on the monitor, so that the net force on the monitor is zero. The reaction to $\mathbf{n}$ is the force exerted by the monitor downward on the table, $\mathbf{F}_{mt} = -\mathbf{F}_{tm}$.

Note that the forces acting on the monitor are $\mathbf{F}_g$ and $\mathbf{n}$, as shown in Figure 4.6b. The two reaction forces $\mathbf{F}_{mE}$ and $\mathbf{F}_{mt}$ are exerted on objects other than the monitor. Remember, the two forces in an action–reaction pair always act on two different objects.

Figure 4.6 illustrates an extremely important difference between a pictorial representation and a simplified pictorial representation for solving problems involving forces. Figure 4.6a shows many of the forces in the situation—those on the monitor, one on the table, and one on the Earth. Figure 4.6b, by contrast, shows only the forces on *one object*, the monitor. This is a critical pictorial representation

• *Normal force*

**PITFALL PREVENTION 4.9**

**n does not always equal mg**

In the situation shown in Figure 4.6, we find that $n = mg$. There are many situations in which the normal force has the same magnitude as the gravitational force, but *do not* adopt this as a general rule—this is a common student pitfall. If the problem involves an object on an incline, if there are applied forces with vertical components, or if there is a vertical acceleration of the system, then $n \neq mg$. *Always* apply Newton's second law to find the relationship between $n$ and $mg$.

\*The word *normal* is used because the direction of $\mathbf{n}$ is always *perpendicular* to the surface.

- *A free-body diagram is a critical step when applying Newton's laws*

called a **free-body diagram.** When analyzing a particle under a net force, we are interested in the net force on one object, an object of mass *m*, which we will model as a particle. Thus, a free-body diagram helps us to isolate only those forces on the object and eliminate the other forces from our analysis. The free-body diagram can be simplified further, if you wish, by representing the object, such as the monitor in this case, as a particle, by simply drawing a dot.

**PITFALL PREVENTION 4.10**

**Free-body diagrams**

The *most important* step in solving a problem using Newton's laws is to draw a proper simplified pictorial representation—the free-body diagram. Be sure to draw only those forces that act on the object you are isolating. Be sure to draw *all* forces acting on the object, including any field forces, such as the gravitational force.

### Quick Quiz 4.6

If a sports car collides head-on with a massive truck, which vehicle experiences the greater force? Which vehicle experiences the greater acceleration?

### Quick Quiz 4.7

What force causes an automobile to move? A propeller-driven airplane? A rocket? A rowboat?

### Quick Quiz 4.8

A large man and a small boy stand facing each other on frictionless ice. They put their hands together and push toward each other, so that they move away from each other. Who exerts the larger force? Who experiences the large acceleration? Who moves away with the higher velocity? Who moves over a longer distance while their hands are in contact?

### THINKING PHYSICS 4.3

A horse pulls on a sled with a horizontal force, causing it to accelerate as in Figure 4.7a. Newton's third law says that the sled exerts a force of equal magnitude and opposite direction on the horse. In view of this, how can the sled accelerate—don't these forces cancel?

**Reasoning**  When applying Newton's third law, it is important to remember that the forces involved act on different objects. Notice that the force exerted by the horse acts *on the sled,* whereas the force exerted by the sled acts *on the horse.* Because these forces act on different objects, they cannot cancel.

The horizontal forces exerted on the *sled* alone are the forward force **F** exerted by the horse and the backward force of friction **f**$_\text{sled}$ between sled and surface (Fig. 4.7b). When **F** exceeds **f**$_\text{sled}$, the sled accelerates to the right.

The horizontal forces exerted on the *horse* alone are the forward friction force **f**$_\text{horse}$ from the ground and the backward force **F** exerted by the sled (Fig. 4.7c). The resultant of these two forces causes the horse to accelerate. When **f**$_\text{horse}$ exceeds **F**, the horse accelerates to the right.

The two forces described in the text of the problem act on different objects, so they cannot cancel. If we consider the horse and sled as a system, the two forces described are *internal* to the system and, therefore, cannot affect the motion of the system.

(a)           (b)           (c)

**Figure 4.7**

(Thinking Physics 4.3)

## 4.7 • APPLICATIONS OF NEWTON'S LAWS

See Screen 4.6

In this section we present some simple applications of Newton's laws to objects that are either in equilibrium ($\mathbf{a} = 0$) or accelerating under the action of constant external forces. We shall assume that the objects behave as particles so that we need not worry about rotational motion or other complications. In this section we also apply some additional simplification models. We ignore the effects of friction for those problems involving motion. This is equivalent to stating that the surfaces are *frictionless*. We usually ignore the masses of any ropes or strings involved. In this approximation, the magnitude of the force exerted at any point along a string is the same at all points along the string. In problem statements, the terms *light* and *of negligible mass* are used to indicate that a mass is to be ignored when you work the problem. These two terms are synonymous in this context.

When we apply Newton's laws to an object, we shall be interested only in those external forces that act *on the object*. For example, in Figure 4.6 the only external forces acting on the monitor are $\mathbf{n}$ and $\mathbf{F}_g$. The reactions to these forces, $\mathbf{F}_{mt}$ and $\mathbf{F}_{mE}$, act on the table and on the Earth, respectively, and do not appear in Newton's second law as applied to the monitor. The importance of drawing a proper free-body diagram to ensure that you are considering the correct forces cannot be stressed enough.

When an object such as a block is being pulled by a rope or string attached to it, the rope exerts a force on the object. The magnitude of this force is called the **tension** in the rope. Its direction is along the rope, away from the object.

Consider a crate being pulled to the right on a frictionless, horizontal surface, as in Figure 4.8a. Suppose you are asked to find the acceleration of the crate and the force the floor exerts on it. First, note that the horizontal force being applied to the crate acts through the rope. The force the rope exerts on the crate is denoted by the symbol $\mathbf{T}$ and its magnitude is the tension in the rope.

Because we are interested only in the motion of the crate, we must be able to *identify any and all external forces acting on it*. These are illustrated in the free-body diagram in Figure 4.8b. In addition to the force $\mathbf{T}$, the free-body diagram for the crate includes the gravitational force $\mathbf{F}_g$ and the normal force $\mathbf{n}$ exerted by the floor on the crate. The *reactions* to the forces we have listed—namely, the force exerted by the crate on the rope, the force exerted by the crate on the Earth, and the force exerted by the crate on the floor—are not included in the free-body diagram because they act on *other* objects and not on the crate.

Now let us apply Newton's second law to the crate. First we must choose an appropriate coordinate system. In this case it is convenient to use the coordinate

(a)

(b)

**Figure 4.8**

(a) A crate being pulled to the right on a frictionless surface. (b) The free-body diagram that represents the external forces on the crate.

Rock climbers depend on the tension forces in ropes for their safety. *(Scott Markewitz/FPG)*

system shown in Figure 4.8b, with the *x* axis horizontal and the *y* axis vertical. We can apply Newton's second law in the *x* direction, *y* direction, or both, depending on what we are asked to find in the problem. In addition, we may be able to use the equations of motion for the particle under constant acceleration that we discussed in Chapter 2. You should use these equations only when the acceleration is constant, however, which is the case if the net force is constant. For example, if the force **T** in Figure 4.8 is constant, then the acceleration in the *x* direction is also constant, because **a** = **T**/*m*. Hence, if we need to find the position or the velocity of the crate at some instant of time, we can use the equations of motion with constant acceleration.

### The Particle in Equilibrium

Objects that are either at rest or moving with constant velocity are said to be in **equilibrium.** From Newton's second law with **a** = 0, this condition of equilibrium can be expressed as

$$\sum \mathbf{F} = 0 \qquad\qquad [4.7]$$

This statement signifies that the vector sum of all the forces (the net force) acting on an object in equilibrium is zero.* If a particle is subject to forces but exhibits an acceleration of zero, we use Equation 4.7 to analyze the situation, as we shall see in some following examples.

Usually, the problems we encounter in our study of equilibrium are easier to solve if we work with Equation 4.7 in terms of the components of the external forces acting on an object. By this we mean that, in a two-dimensional problem, the sum of all the external forces in the *x* and *y* directions must separately equal zero; that is,

$$\sum F_x = 0 \qquad\qquad \sum F_y = 0 \qquad\qquad [4.8]$$

The extension of Equations 4.8 to a three-dimensional situation can be made by adding a third component equation, $\sum F_z = 0$.

In a given situation, we may have balanced forces on an object in one direction, but unbalanced forces in the other. Thus, for a given problem, we may need to model the object as a particle in equilibrium for one component and a particle under a net force for the other.

### Quick Quiz 4.9

Consider the two situations shown in Figure 4.9, in which no acceleration occurs. In both cases, all individuals pull with a force of magnitude *F* on a rope attached to a spring scale. Is the reading on the spring scale in part (i) of the figure (a) greater than, (b) less than, or (c) equal to the reading in part (ii)?

### The Accelerating Particle

In a situation in which a nonzero net force is acting on an object, the object is accelerating, and we use Newton's second law in order to determine the features of the motion:

$$\sum \mathbf{F} = m\mathbf{a}$$

---

* This is only one condition of equilibrium for an object. A second condition of equilibrium is a statement of rotational equilibrium. This condition will be discussed in Chapter 10.

In practice, this equation is broken into components, so that two (or three) equations can be handled independently. The representative suggestions and problems that follow should help you to solve problems of this kind.

(i)

(ii)

**Figure 4.9**

(Quick Quiz 4.9) (i) An individual pulls with a force of magnitude *F* on a spring scale attached to a wall. (ii) Two individuals pull with forces of magnitude *F* in opposite directions on a spring scale attached between two ropes.

---

**PROBLEM-SOLVING STRATEGY** | **Particle Under a Net Force**

The following procedure is recommended when dealing with problems involving the application of Newton's second law for a particle under a net force:

1. **Think**—establish your mental representation of the situation. Draw a diagram of the system.
2. Isolate the object whose motion is being analyzed. Draw a free-body diagram for this object, showing all external forces acting on it. For systems containing more than one object, draw a *separate* diagram for each object.
3. Establish convenient coordinate axes for each object and find the components of the forces along those axes. Apply Newton's second law, $\Sigma \mathbf{F} = m\mathbf{a}$, in the *x* and *y* directions for each object. If the object is in equilibrium in either direction, set the right hand side of Newton's second law equal to zero.
4. Solve the component equations for the unknowns. Remember that, in order to obtain a complete solution, you must have as many independent equations as you have unknowns.
5. If appropriate to the situation (the net force is constant), and necessary to respond to the problem, use the equations of kinematics from Chapter 2 to find the unknowns.

---

**PITFALL PREVENTION 4.12**

**Again, *ma* is not a force.**

Because of the frequent occurrence of this misconception, it bears repeating that the product *m***a** is *not* a force. Avoid the temptation to include an "*m***a** force" in your free-body diagram.

We now embark on a series of examples that demonstrate how to solve problems involving a particle under a net force. You should read and study these examples very carefully.

**THINKING PHYSICS 4.4**

You have most likely had the experience of standing in an elevator that accelerates upward as it leaves to move toward a higher floor. In this case, you *feel* heavier. If you are standing on a bathroom scale at the time, it *measures* a force larger than your weight. Thus, you have tactile and measured evidence that may lead you to believe that you are heavier in this situation. *Are* you heavier?

**Reasoning** No, you are not—the gravitational force on you is unchanged. To provide the upward acceleration, the floor or the bathroom scale must apply an upward force larger than the gravitational force. It is this larger force that you feel, which you interpret as feeling heavier. A bathroom scale measures this force, so its reading increases.

## Example 4.2   A Traffic Light at Rest

A traffic light weighing 122 N hangs from a cable tied to two other cables fastened to a support, as in Figure 4.10a. The upper cables make angles of 37.0° and 53.0° with the horizontal. These upper cables are not as strong as the vertical cable, and will break if the tension in them exceeds 100 N. Does the traffic light remain in this situation, or will one of the cables break?

**Reasoning**   Let us assume that the cables do not break so that no acceleration of any sort occurs in any direction. Thus, we use the model of a particle in equilibrium for both $x$ and $y$ components. We shall construct two free-body diagrams. The first of these is for the traffic light, shown in Figure 4.10b; the second is for the knot that holds the three cables together, as in Figure 4.10c. The knot is a convenient point to choose because all the forces in which we are interested act through this point. Because the acceleration of the system is zero, we can use the equilibrium conditions that the net force on the light is zero, and the net force on the knot is zero.

**Solution**   Considering Figure 4.10b, we apply the equilibrium condition in the $y$ direction, $\Sigma F_y = 0 \rightarrow T_3 - F_g = 0$. This leads to $T_3 = F_g = 122$ N. Thus, the force $\mathbf{T}_3$ exerted by the vertical cable balances the weight of the light.

Considering the knot next, we choose the coordinate axes as shown in Figure 4.10c and resolve the forces into their $x$ and $y$ components, as shown in the following table:

| Force | $x$ component | $y$ component |
|-------|---------------|---------------|
| $\mathbf{T}_1$ | $-T_1 \cos 37.0°$ | $T_1 \sin 37.0°$ |
| $\mathbf{T}_2$ | $T_2 \cos 53.0°$ | $T_2 \sin 53.0°$ |
| $\mathbf{T}_3$ | $0$ | $-122$ N |

Equations 4.8 give us:

(1)   $\displaystyle\sum F_x = T_2 \cos 53.0° - T_1 \cos 37.0° = 0$

(2)   $\displaystyle\sum F_y = T_1 \sin 37.0° + T_2 \sin 53.0° - 122\text{ N} = 0$

We solve (1) for $T_2$ in terms of $T_1$ to give

$$T_2 = T_1 \left( \frac{\cos 37.0°}{\cos 53.0°} \right) = 1.33\, T_1$$

This value for $T_2$ is substituted into (2) to give

$$T_1 \sin 37.0° + (1.33\, T_1)(\sin 53.0°) - 122\text{ N} = 0$$

$$T_1 = \boxed{73.4\text{ N}}$$

Then, we calculate $T_2$:

$$T_2 = 1.33\, T_1 = \boxed{97.4\text{ N}}$$

Both of these values are less than 100 N (just barely for $T_2$!), so the cables do not break.

**EXERCISE**   If the angles of the cables can be adjusted, in what situation does $T_1 = T_2$?

*Answer*   When the angles the cables make with the horizontal are equal.

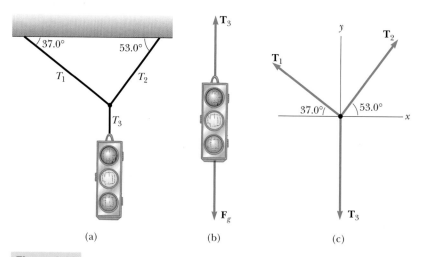

(a)                (b)                (c)

**Figure 4.10**

(Example 4.2) (a) A traffic light suspended by cables. (b) The free-body diagram for the traffic light. (c) The free-body diagram for the knot in the cable.

## Example 4.3 A Sled on Frictionless Snow

A child on a sled is released on a frictionless, inclined hill of angle $\theta$, as in Figure 4.11a. (a) Determine the acceleration of the sled after it is released.

**Reasoning** We identify the system of the sled and the child as our object of interest. We model the system as a particle of mass $m$. Newton's second law can be used to determine the acceleration of the particle. First, we construct the free-body diagram for the particle as in Figure 4.11b. The only forces on the particle are the normal force **n** acting perpendicularly to the incline and the gravitational force $m_{\mathbf{g}}$ acting vertically downward. For problems of this type involving inclines, it is convenient to choose the coordinate axes with $x$ *along* the incline and $y$ *perpendicular* to it. Then, we replace $m_{\mathbf{g}}$ by a combination of a component vector of magnitude $mg\sin\theta$ along the *positive x* axis (down the incline) and one of magnitude $mg\cos\theta$ in the *negative y* direction.

**Solution** Applying Newton's second law in component form to the particle and noting that $a_y = 0$ gives

$$(1) \quad \sum F_x = mg\sin\theta = ma_x$$

$$(2) \quad \sum F_y = n - mg\cos\theta = 0$$

From (1) we see that the acceleration along the incline is provided by the component of the gravitational force parallel to the incline, which gives us

$$(3) \quad \boxed{a_x = g\sin\theta}$$

Note that the acceleration given by (3) is *independent* of the mass of the particle—it depends only on the angle of inclination and on $g$. From (2) we conclude that the component of the gravitational force perpendicular to the incline is *balanced* by the normal force; that is, $n = mg\cos\theta$. (Notice, as

pointed out in Pitfall Prevention 4.9, that $n$ does not equal $mg$ in this case.)

***Special Cases*** When $\theta = 90°$, (3) gives us $a_x = g$ and (2) gives us $n = 0$. This case corresponds to the particle in free-fall. (For our choice of coordinate system, positive $x$ is in the downward direction when $\theta = 90°$, which is why the acceleration is $+ g$ rather than $- g$.) When $\theta = 0°$, $a_x = 0$ and $n = mg$ (its maximum value). This corresponds to the situation in which the particle is on a level surface and not accelerating.

This technique of looking at special cases of limiting situations is often useful in checking an answer. In this situation, if the angle $\theta$ goes to 90°, we know intuitively that the object should be falling parallel to the surface of the incline. The fact that (3) mathematically reduces to $a_x = g$ when $\theta = 90°$ gives us confidence in our answer. It doesn't prove that the answer is correct, but if the acceleration does not reduce to $g$, this would tell us that the answer is incorrect.

(b) Suppose the sled is released from rest at the top of the hill, and the distance from the front of the sled to the bottom of the hill is $d$. How long does it take the front of the sled to reach the bottom, and what is its speed just as it arrives at that point?

**Solution** In part (a), we found $a_x = g\sin\theta$, which is constant. Hence, we can model the system as a particle under constant acceleration for the motion parallel to the incline. We use Equation 2.11, $x_f = x_i + v_{xi}t + \frac{1}{2}a_xt^2$, to describe the position of the front edge of the sled. We define the initial position as $x_i = 0$ and the final position as $x_f = d$. Because the sled starts sliding from rest, $v_{xi} = 0$. With these values, Equation 2.11 becomes simply $d = \frac{1}{2}a_xt^2$, or

$$(4) \quad t = \sqrt{\frac{2d}{a_x}} = \sqrt{\frac{2d}{g\sin\theta}}$$

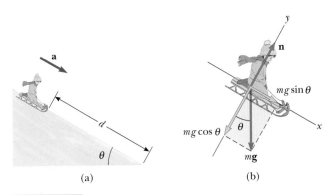

(a)

(b)

### Figure 4.11

(Example 4.3) (a) A child on a sled sliding down a frictionless incline. (b) The free-body diagram for the system.

### Figure 4.12

A child holding a sled at rest on a frictionless hill.

This answers the first question as to the length of time to reach the bottom. Now, to determine the speed when the sled arrives at the bottom, we use Equation 2.12, $v_{xf}^2 = v_{xi}^2 + 2a_x(x_f - x_i)$ with $v_{xi} = 0$, and we find that $v_{xf}^2 = 2a_x d$, or

$$(5) \quad v_{xf} = \sqrt{2a_x d} = \boxed{\sqrt{2gd \sin \theta}}$$

As with the acceleration parallel to the incline, $t$ and $v_{xf}$ are *independent* of the mass of the sled and child.

**EXERCISE**   A child holds a sled weighing 77.0 N at rest on a frictionless snow-covered incline at 30.0° as in Figure 4.12. Find (a) the magnitude of the force the child must exert on the rope, and (b) the magnitude of the force the incline exerts on the sled.

*Answer*   (a) 38.5 N (b) 66.7 N

## Example 4.4   The Atwood Machine

When two objects with unequal masses are hung vertically over a light, frictionless pulley as in Figure 4.13a, the arrangement is called an *Atwood machine*. The device is sometimes used in the laboratory to measure the free-fall acceleration. Calculate the magnitude of the acceleration of the two objects and the tension in the string.

**Reasoning**   Think about the mental representation suggested by Figure 4.13a—as one object moves upward, the other object moves downward. Because the objects are connected by an inextensible string, they must have the same magnitude of acceleration. The objects in the Atwood machine are subject to the gravitational force as well as to the forces exerted by the strings connected to them. We model the objects as particles under a net force. The free-body diagrams for the two objects are shown in Figure 4.13b. Two forces act on each object: the upward force **T** exerted by the string and the downward gravitational force. In a problem such as this in which the pulley is modeled as massless and frictionless, the tension in the string on both sides of the pulley is the same. If the pulley has mass or is subject to a friction force, the tensions on either side are not the same and the situation requires the techniques of Chapter 10.

In these types of problems, involving strings that pass over pulleys, we must be careful about the sign convention. Notice that if $m_1$ goes up, then $m_2$ goes down. Thus, $m_1$ going up and $m_2$ going down should be represented equivalently as far as a sign convention is concerned. We can do this by defining our sign convention with up as positive for $m_1$ and down as positive for $m_2$, as shown in Figure 4.13a.

With this sign convention, the magnitude of the net force exerted on $m_1$ is $T - m_1 g$, while the magnitude of the net force exerted on $m_2$ is $m_2 g - T$. We have chosen the signs of the forces to be consistent with the choices of the positive direction for each object.

**Figure 4.13**

(Example 4.4) The Atwood machine. (a) Two objects connected by a light string over a frictionless pulley. (b) The free-body diagrams for $m_1$ and $m_2$.

**Solution** When Newton's second law is applied to $m_1$, we find

(1) $\quad \sum F_y = T - m_1 g = m_1 a$

Similarly, for $m_2$ we find

(2) $\quad \sum F_y = m_2 g - T = m_2 a$

When (2) is added to (1), $T$ cancels and we have

$$-m_1 g + m_2 g = m_1 a + m_2 a$$

Solving this for the acceleration $a$,

(3) $\quad a = \left( \dfrac{m_2 - m_1}{m_1 + m_2} \right) g$

If $m_2 > m_1$, the acceleration given by (3) is positive — $m_1$ goes up and $m_2$ goes down. Is this consistent with your mental representation? If $m_1 > m_2$, the acceleration is negative and the masses move in the opposite direction.

If (3) is substituted into (1), we find

(4) $\quad T = \left( \dfrac{2 m_1 m_2}{m_1 + m_2} \right) g$

**Special Cases** When $m_1 = m_2$, (3) and (4) give us $a = 0$ and $T = m_1 g = m_2 g$, as we would intuitively expect for the balanced case. Also, if $m_2 \gg m_1$, $a \approx g$ (a freely falling object) and $T \approx 0$. For such a large mass $m_2$, we would expect $m_1$ to have little effect, so that $m_2$ is simply falling. Thus, our results are consistent with our intuitive predictions in both of these limiting situations.

**EXERCISE** Find the acceleration and tension of an Atwood machine in which $m_1 = 2.00$ kg and $m_2 = 4.00$ kg.

*Answer* $\quad a = 3.27$ m/s$^2$, $T = 26.1$ N

## Example 4.5 One Block Pushes Another

Two blocks of masses $m_1$ and $m_2$, with $m_1 > m_2$, are placed in contact with each other on a frictionless, horizontal surface, as in Figure 4.14a. A constant horizontal force **F** is applied to $m_1$ as shown. (a) Find the magnitude of the acceleration of the system of two blocks.

**Reasoning and Solution** Both blocks must experience the *same* acceleration because they are in contact with each other and remain in contact with each other. We model the system of both blocks as a particle under a net force. Because **F** is the only horizontal force exerted on the system, we have

$$\sum F_x (\text{system}) = F = (m_1 + m_2) a$$

(1) $\quad a = \dfrac{F}{m_1 + m_2}$

(b) Determine the magnitude of the contact force between the two blocks.

**Reasoning and Solution** The contact force is internal to the system of two blocks. Thus, we cannot find this force by modeling the whole system as a single particle. We now need to treat each of the two blocks individually as a particle under a net force. We first construct a free-body diagram for each block, as shown in Figures 4.14b and 4.14c, where the contact force is denoted by **P**. From Figure 4.14c we see that the only horizontal force acting on $m_2$ is the contact force

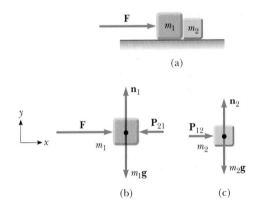

**Figure 4.14**

(Example 4.5)

$\mathbf{P}_{12}$ (the force exerted by $m_1$ on $m_2$), which is directed to the right. Applying Newton's second law to $m_2$ gives

(2) $\quad \sum F_x = P_{12} = m_2 a$

Substituting the value of the acceleration $a$ given by (1) into (2) gives

(3) $\quad P_{12} = m_2 a = \left( \dfrac{m_2}{m_1 + m_2} \right) F$

From this result, we see that the contact force $P_{12}$ is *less* than the applied force $F$. This is consistent with the fact that the force required to accelerate $m_2$ alone must be less than the force required to produce the same acceleration for the system of two blocks. Compare this with the forces in the couplers in the train of Thinking Physics 4.1.

It is instructive to check this expression for $P_{12}$ by considering the forces acting on $m_1$, shown in Figure 4.14b. The horizontal forces acting on $m_1$ are the applied force **F** to the right and the contact force $\mathbf{P}_{21}$ to the left (the force exerted by $m_2$ on $m_1$). From Newton's third law, $\mathbf{P}_{21}$ is the reaction to

$\mathbf{P}_{12}$, so $P_{21} = P_{12}$. Applying Newton's second law to $m_1$ gives

$$(4) \quad \sum F_x = F - P_{21} = F - P_{12} = m_1 a$$

Solving for $P_{12}$ and substituting the value of $a$ from (1) into (4) gives

$$P_{12} = F - m_1 a = F - m_1 \left( \frac{F}{m_1 + m_2} \right) = \left( \frac{m_2}{m_1 + m_2} \right) F$$

This agrees with (3), as it must.

---

**Quick Quiz 4.10**

Imagine that the force **F** in Example 4.5 is applied toward the left on the right-hand mass $m_2$. Is the magnitude of the force $\mathbf{P}_{12}$ now (a) greater than, (b) less than, or (c) equal to the magnitude of the force $\mathbf{P}_{12}$ in Example 4.5?

---

### Example 4.6   Weighing a Fish in an Elevator

A person weighs a fish on a spring scale attached to the ceiling of an elevator, as shown in Figure 4.15. Show that if the elevator accelerates, the spring scale reads a weight different from the true weight of the fish.

**Reasoning**   An observer on the accelerating elevator is not in an inertial frame. We need to analyze this situation in an inertial frame, so let us imagine observing it from the stationary ground. We model the fish as a particle under a net force. The external forces acting on the fish are the downward gravitational force $\mathbf{F}_g$ and the upward force **T** exerted on it by the hook hanging from the bottom of the scale. (It might be more fruitful in your mental representation to imagine that the hook is a string connecting the fish to the scale.) Because the tension is the same everywhere in the hook supporting the fish, the hook pulls downward with a force of magnitude $T$ on the spring scale. Thus, the tension $T$ in the hook is also the reading of the spring scale.

If the elevator is either at rest or moves at constant velocity, the fish is not accelerating and is a particle in equilibrium, which gives us $\sum F_y = T - mg = 0 \rightarrow T = mg$. If the elevator accelerates either up or down, however, the tension is no longer equal to the weight of the fish because $T - mg$ does not equal zero.

**Solution**   If the elevator accelerates with an acceleration **a** relative to an observer in an inertial frame outside the elevator, Newton's second law applied to the fish in the vertical di-

rection gives us:

$$\sum F_y = T - mg = ma_y$$

which leads to

$$(1) \quad T = mg + ma_y$$

Thus, we conclude from (1) that the scale reading $T$ is greater than the weight $mg$ if **a** is upward as in Figure 4.15a. Furthermore, we see that $T$ is less than $mg$ if **a** is downward as in Figure 4.15b. For example, if the weight of the fish is 40.0 N, and **a** is upward with $a_y = 2.00$ m/s$^2$, then the scale reading is

$$T = mg + ma_y = mg\left(1 + \frac{a_y}{g}\right)$$
$$= (40.0 \text{ N})\left(1 + \frac{2.00 \text{ m/s}^2}{9.80 \text{ m/s}^2}\right) = \boxed{48.2 \text{ N}}$$

If $a_y = -2.00$ m/s$^2$, so that **a** is downward, then

$$T = mg\left(1 + \frac{a_y}{g}\right) = (40.0 \text{ N})\left(1 + \frac{-2.00 \text{ m/s}^2}{9.80 \text{ m/s}^2}\right) = \boxed{31.8 \text{ N}}$$

Hence, if you buy a fish in an elevator, make sure the fish is weighed while the elevator is at rest or accelerating downward!

***Special Case***   If the cable breaks, the elevator falls freely so that $a_y = -g$, and from (1) we see that the tension $T$ is zero; that is, the fish appears to be weightless.

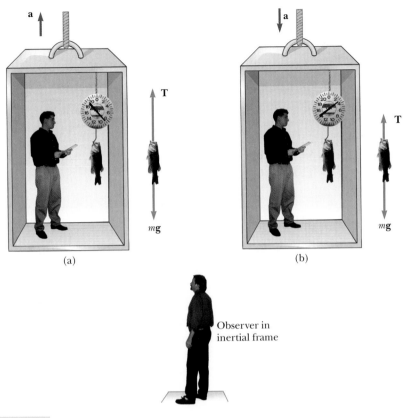

**Figure 4.15**

(Example 4.6) (a) When the elevator accelerates *upward,* the spring scale reads a value *greater* than the true weight of the fish. (b) When the elevator accelerates *downward,* the spring scale reads a value *less* than the true weight.

 **context connection**

## 4.8 • CONTROLLING THE SPACECRAFT IN EMPTY SPACE

Our entire trip from Earth to Mars will be governed by the laws of dynamics. We shall look at the details of the planned trip in the Context Conclusion, following Chapter 11. In this section, we present an example of a *midcourse correction.* As you read this discussion, note that all three of Newton's laws are involved.

Suppose we are in a situation such as that shown in Figure 4.16, such that our spacecraft is on the wrong trajectory toward our destination. Our actual trajectory

**Figure 4.16**

Our spacecraft is on a trajectory different from the desired trajectory. How shall we fire the engines to put the spacecraft on the correct trajectory?

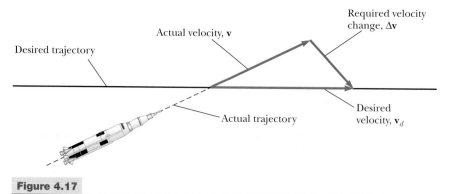

**Figure 4.17**

We set up the three velocity vectors required to solve the problem posed by Figure 4.16—the actual velocity, the desired velocity, and the change in velocity.

is different from the desired trajectory. Newton's *first* law tells us that we will continue on our incorrect trajectory in the absence of a force. How can we return our spacecraft to the correct trajectory? In what direction should we fire the rocket engines? We can answer these questions by first setting up a pictorial representation showing the velocity vectors, as in Figure 4.17.

The vector $\Delta\mathbf{v}$ for the change in velocity in Figure 4.17 indicates the direction of the acceleration that we must impart to the rocket, because $\mathbf{a} = \Delta\mathbf{v}/\Delta t$. Thus, we must fire the rockets in exactly the opposite direction at the time we cross the desired trajectory, as shown in Figure 4.18. (We imagine that there are thrusters on the side of the spacecraft that allow it to rotate in space to the position shown.) This is an application of Newton's *third* law—the rocket exerts on the exhaust gases a force $\mathbf{F}_{\text{rg}}$ (the force *by* the rocket *on* the gases) in one direction, and the gases exert a force $\mathbf{F}_{\text{gr}} = -\mathbf{F}_{\text{rg}}$ in the opposite direction on the rocket. We must also fire the rocket for just the right amount of time to give us the desired change in velocity. We can find the required engine burn time from Newton's *second* law, because we know the desired change in velocity and we assume that we know the

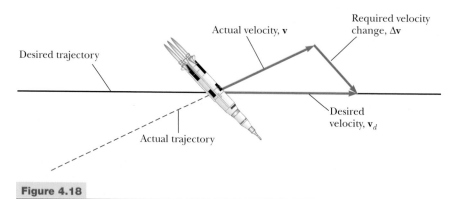

**Figure 4.18**

We must fire the engines in the direction opposite to the change in velocity vector so that, by Newton's third law, the rocket experiences a force in the direction of this vector.

mass $m$ of the rocket. We write Newton's second law for components along the direction of $\Delta \mathbf{v}$ in Figure 4.18:

$$F_{gr} = ma = m\frac{\Delta v}{\Delta t} \longrightarrow \Delta t = \frac{m\Delta v}{F_{gr}}$$

As long as the midcourse correction is fairly minor, the required burn time $\Delta t$ can be approximated with this equation because we know the mass $m$ of the rocket and the force $F_{gr}$ from the engine, and we found the required $\Delta v$ from Figure 4.18.

For larger corrections, two major complications arise, which we have left out of this simple discussion and which may require a more refined model:

- The mass of the rocket changes for longer burns, so that $\Sigma\mathbf{F} = m\mathbf{a}$ is not the appropriate form of Newton's second law. (The more appropriate form involves momentum, discussed in Chapter 8.) This may require the use of calculus.
- If the time $\Delta t$ for the engine burn is relatively long, by the time the desired velocity is reached, the rocket may be on a parallel trajectory above the desired trajectory in Figure 4.16. Thus, the timing of the initiation of the engine burn must be handled carefully.

We shall not concern ourselves with these complications because we are restricting ourselves to a relatively simple approach to our contextual problem.

In the next chapter, we investigate more details about the gravitational force. This is the only force on our spacecraft once the rocket engines have been shut down, so it is an important factor in our analysis of our trip to Mars.

## SUMMARY

**Newton's first law** states that, if no force is exerted on an object, an object at rest remains at rest, and an object in uniform motion in a straight line maintains that motion.

Newton's first law defines an **inertial frame of reference,** which is a frame in which Newton's first law is valid.

**Newton's second law** states that the acceleration of an object is directly proportional to the net force acting on the object and inversely proportional to the object's mass. Thus, the net force on an object equals the product of the mass of the object and its acceleration,

$$\Sigma\mathbf{F} = m\mathbf{a} \qquad [4.2]$$

The **weight** of an object is equal to the product of its mass (a scalar quantity) and the magnitude of the free-fall acceleration, or

$$F_g = mg \qquad [4.5]$$

If the acceleration of an object is zero, then the particle is in equilibrium, with the appropriate equations being

$$\Sigma F_x = 0 \qquad \Sigma F_y = 0 \qquad [4.8]$$

**Newton's third law** states that if two objects interact, the force exerted on object 1 by object 2 is equal in magnitude but opposite in direction to the force exerted on object 2 by object 1. Thus, an isolated force cannot exist in nature.

## QUESTIONS

1. Draw a free-body diagram for each of the following objects: (a) a projectile in motion in the presence of air resistance, (b) a rocket leaving the launch pad with its engines operating, (c) an athlete running along a horizontal track.

2. In the motion picture *It Happened One Night* (Columbia Pictures, 1934), Clark Gable is standing inside a stationary bus in front of Claudette Colbert, who is seated. The bus suddenly starts moving forward and Clark falls into Claudette's lap. Why did this happen?

3. A passenger sitting in the rear of a bus claims that he was injured when the driver slammed on the brakes, causing a suitcase to come flying toward the passenger from the front of the bus. If you were the judge in this case, what disposition would you make? Why?

**4.** As you sit in a chair, the chair pushes up on you with a normal force. This force is equal to your weight and in the opposite direction. Is this force the Newton's third law reaction to your weight?

**5.** A space explorer is in a spacecraft moving through space, far from any planet or star. She notices a large rock, taken as a specimen from an alien planet, floating around the cabin of the spacecraft. Should she push it gently toward a storage compartment or kick it toward the compartment? Why?

**6.** The observer in the accelerating elevator of Example 4.6 would claim that the "weight" of the fish is *T*, the scale reading. This is obviously wrong. Why does this observation differ from that of a person outside the elevator, at rest with respect to the Earth?

**7.** Identify the action–reaction pairs in the following situations: a man takes a step; a snowball hits a girl in the back; a baseball player catches a ball; a gust of wind strikes a window.

**8.** While a football is in flight, what forces act on it? What are the action–reaction pairs while the football is being kicked and while it is in flight?

**9.** A rubber ball is dropped onto the floor. What force causes the ball to bounce back into the air?

**10.** What is wrong with the statement "Because the car is at rest, no forces are acting on it"? How would you correct this sentence?

**11.** If gold were sold by weight, would you rather buy it in Denver or in Death Valley? If it were sold by mass, at which of the two locations would you prefer to buy it? Why?

**12.** A weight lifter stands on a bathroom scale. He pumps a barbell up and down. What happens to the reading on the bathroom scale as this is done? Suppose he is strong enough to actually *throw* the barbell upward. How does the reading on the scale vary now?

**13.** The mayor of a city decides to fire some city employees because they will not remove the sags from the cables that support the city traffic lights. If you were a lawyer, what defense would you give on behalf of the employees? Who do you think would win the case in court?

**14.** In a tug-of-war between two athletes, each athlete pulls on the rope with a force of 200 N. What is the tension in the rope? If the rope does not move, what force does each athlete exert against the ground?

**15.** Suppose a truck loaded with sand accelerates at 0.5 m/s$^2$ on a highway. If the driving force on the truck remains constant, what happens to the truck's acceleration if it leaks sand at a constant rate through a hole in its bottom?

**16.** Can an object exert a force on itself? Argue for your answer.

## PROBLEMS

1, 2, 3 = straightforward, intermediate, challenging    □ = full solution available in the *Student Solutions Manual and Study Guide*

**web** = solution posted at **http://www.harcourtcollege.com/physics/**    ⬛ = computer useful in solving problem

🖳 = Interactive Physics    ⬛ = paired numerical/symbolic problems    ⬛ = life science application

### Section 4.3 Inertial Mass

**1.** A force **F** applied to an object of mass $m_1$ produces an acceleration of 3.00 m/s$^2$ . The same force applied to a second object of mass $m_2$ produces an acceleration of 1.00 m/s$^2$. (a) What is the value of the ratio $m_1/m_2$? (b) If $m_1$ and $m_2$ are combined, find their acceleration under the action of the force **F**.

**2.** (a) A car with a mass of 850 kg is moving to the right with a constant speed of 1.44 m/s. What is the total force on the car? (b) What is the total force on the car if it is moving to the left?

### Section 4.4 Newton's Second Law — The Particle Under a Net Force

**3.** An object of mass 3.00 kg undergoes an acceleration given by **a** = (2.00**i** + 5.00**j**) m/s$^2$. Find the vector expression for the net force acting on it and the magnitude of the net force.

**4.** The largest-caliber antiaircraft gun operated by the German air force during World War II was the 12.8-cm Flak 40. This weapon fired a 25.8-kg shell with a muzzle speed of 880 m/s. What propulsive force was necessary to attain the muzzle speed within the 6.00-m barrel? (Assume the shell moves horizontally with constant acceleration and neglect friction.)

**5.** Two forces, $\mathbf{F}_1 = (-6\mathbf{i} - 4\mathbf{j})$ N and $\mathbf{F}_2 = (-3\mathbf{i} + 7\mathbf{j})$ N, act on a particle of mass 2.00 kg that is initially at rest at coordinates (-2.00 m, +4.00 m). (a) What are the components of the particle's velocity at *t* = 10.0 s? (b) In what direction is the particle moving at *t* = 10.0 s? (c) What displacement does the particle undergo during the first 10.0 s? (d) What are the coordinates of the particle at *t* = 10.0 s?

6. Two people pull as hard as they can on horizontal ropes attached to a boat that has a mass of 200 kg. If they pull in the same direction, the boat has an acceleration of 1.52 m/s² to the right. If they pull in opposite directions, the boat has an acceleration of 0.518 m/s² to the left. What is the force exerted by each person on the boat? (Disregard any other horizontal forces on the boat.)

7. Two forces $F_1$ and $F_2$ act on a 5.00-kg object. If $F_1 =$ 20.0 N and $F_2 = 15.0$ N, find the accelerations in (a) and (b) of Figure P4.7.

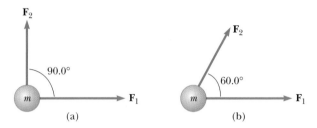

**Figure P4.7**

8. Three forces, given by $F_1 = (-2.00\mathbf{i} + 2.00\mathbf{j})$ N, $F_2 = (5.00\mathbf{i} - 3.00\mathbf{j})$ N, and $F_3 = (-45.0\mathbf{i})$ N, act on an object to give it an acceleration of magnitude 3.75 m/s². (a) What is the direction of the acceleration? (b) What is the mass of the object? (c) If the object is initially at rest, what is its speed after 10.0 s? (d) What are the velocity components of the object after 10.0 s?

9. A 3.00-kg object is moving in a plane, with its $x$ and $y$ coordinates given by $x = 5t^2 - 1$ and $y = 3t^3 + 2$, where $x$ and $y$ are in meters and $t$ is in seconds. Find the magnitude of the net force acting on this object at $t = 2.00$ s.

## Section 4.5 The Gravitational Force and Weight

10. A woman weighs 120 lb. Determine (a) her weight in newtons and (b) her mass in kilograms.

11. If a man weighs 900 N on Earth, what would he weigh on Jupiter, where the acceleration due to gravity is 25.9 m/s²?

12. The gravitational force on a baseball is $-F_g\mathbf{j}$. A pitcher throws the baseball with velocity $v\mathbf{i}$ by uniformly accelerating it straight forward horizontally for a time $t$. If the ball starts from rest, (a) through what distance does it accelerate before its release? (b) What force does the pitcher exert on the ball?

13. An electron of mass $9.11 \times 10^{-31}$ kg has an initial speed of $3.00 \times 10^5$ m/s. It travels in a straight line, and its speed increases to $7.00 \times 10^5$ m/s in a distance of 5.00 cm. Assuming its acceleration is constant, (a) determine the force exerted on the electron and (b) compare this force with the weight of the electron, which we ignored.

14. Define one pound as the weight of an object of mass 0.453 592 37 kg at a location where the acceleration of gravity is 32.174 0 ft/s². Express the pound as one quantity with one SI unit.

15. The distinction between mass and weight was discovered after Jean Richer transported pendulum clocks from Paris to French Guiana in 1671. He found that they ran slower there quite systematically. The effect was reversed when the clocks returned to Paris. How much weight would you personally lose in traveling from Paris, where $g = 9.809\ 5$ m/s², to Cayenne, where $g = 9.780\ 8$ m/s²? (We will consider how the free-fall acceleration influences the period of a pendulum in Section 12.4.)

16. Besides its weight, a 2.80-kg object is subjected to one other constant force. The object starts from rest and in 1.20 s undergoes a displacement of $(4.20$ m$)\mathbf{i} - (3.30$ m$)\mathbf{j}$, where the direction of $\mathbf{j}$ is the upward vertical direction. Determine the other force.

## Section 4.6 Newton's Third Law

17. You stand on the seat of a chair and hop off. (a) During the time you are in flight down to the floor, the Earth is lurching up toward you with an acceleration of what order of magnitude? In your solution explain your logic. Model the Earth as a perfectly solid object. (b) The Earth moves up through a distance of what order of magnitude?

18. The average speed of a nitrogen molecule in air is about $6.70 \times 10^2$ m/s, and its mass is $4.68 \times 10^{-26}$ kg. (a) If it takes $3.00 \times 10^{-13}$ s for a nitrogen molecule to hit a wall and rebound with the same speed but moving in the opposite direction, what is the average acceleration of the molecule during this time interval? (b) What average force does the molecule exert on the wall?

19. A 15.0-lb block rests on the floor. (a) What force does the floor exert on the block? (b) If a rope is tied to the block and run vertically over a pulley, and the other end is attached to a free-hanging 10.0-lb weight, what is the force of the floor on the 15.0-lb block? (c) If we replace the 10.0-lb weight in part (b) with a 20.0-lb weight, what is the force exerted by the floor on the 15.0-lb block?

## Section 4.7 Applications of Newton's Laws

20. A bag of cement of weight 325 N hangs from three wires as suggested in Figure P4.20. Two of the wires make angles $\theta_1 = 60.0°$ and $\theta_2 = 25.0°$ with the horizontal. If the system is in equilibrium, find the tensions $T_1$, $T_2$, and $T_3$ in the wires.

21. A bag of cement of weight $F_g$ hangs from three wires as shown in Figure P4.20. Two of the wires make angles $\theta_1$

and $\theta_2$ with the horizontal. If the system is in equilibrium, show that the tension in the left-hand wire is

$$T_1 = F_g \cos \theta_2 / \sin(\theta_1 + \theta_2)$$

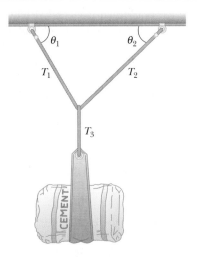

**Figure P4.20** Problems 20 and 21.

22. You are a judge in a children's kite-flying contest, and two children will win prizes for the kites that pull most strongly and least strongly on their strings. To measure string tensions, you borrow a weight hanger, some slotted weights, and a protractor from your physics teacher, and use the following protocol, illustrated in Figure P4.22: Wait for a child to get her kite well controlled, hook the hanger onto the kite string about 30 cm from her hand, pile on weight until that section of string is horizontal, record the mass required, and record the angle between the horizontal and the string running up to the kite. (a) Explain how this method works. As you construct your explanation, imagine that the children's parents ask you about your method, that they might make false assumptions about your ability without concrete evidence, and that your explanation is an

opportunity to give them confidence in your evaluation technique. (b) Find the string tension if the mass is 132 g and the angle of the kite string is 46.3°.

23. A 1.00-kg object is observed to accelerate at 10.0 m/s² in a
web direction 30.0° north of east (Fig. P4.23). The force $\mathbf{F}_2$ acting on the object has a magnitude of 5.00 N and is directed north. Determine the magnitude and direction of the force $\mathbf{F}_1$ acting on the object.

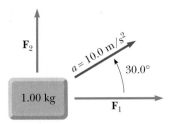

**Figure P4.23**

24. A simple accelerometer is constructed by suspending an object of mass $m$ from a string of length $L$ that is tied to the roof of a car. As the car accelerates the string makes a constant angle of $\theta$ with the vertical. (a) Assuming that the string mass is negligible compared to $m$, derive an expression for the car's acceleration in terms of $\theta$ and show that it is independent of the mass $m$ and the length $L$. (b) Determine the acceleration of the car when $\theta = 23.0°$.

25. The systems shown in Figure P4.25 are in equilibrium. If the spring scales are calibrated in newtons, what do they

(a)

(b)

(c)

**Figure P4.25**

**Figure P4.22**

read? (Ignore the masses of the pulleys and strings, and assume the incline is frictionless.)

26. A 5.00-kg object placed on a frictionless, horizontal table is connected to a cable that passes over a pulley and then is fastened to a hanging 10.0-kg object, as in Figure P4.26. Draw free-body diagrams of both objects. Find the acceleration of the two objects and the tension in the cable.

5.00 kg

10.0 kg

**Figure P4.26**

27. Two objects, of mass $m_1$ and $m_2$, situated on a frictionless, horizontal surface, are connected by a light string. A force **F** is exerted on one of the objects to the right (Fig. P4.27). Determine the acceleration of the system and the tension $T$ in the string.

$m_1$ $T$ $m_2$ **F**

**Figure P4.27**

28. Draw a free-body diagram of a block that slides down a frictionless plane having an inclination of $\theta = 15.0°$ (Fig. P4.28). If the block starts from rest at the top and the length of the incline is 2.00 m, find (a) the acceleration of the block and (b) its speed when it reaches the bottom of the incline.

$\theta$

**Figure P4.28** Problems 28 and 29

29. A block is given an initial speed of 5.00 m/s up a frictionless 20.0° incline (see Fig. P4.28). How far up the incline does the block slide before coming to rest?

30. One of the great dangers to mountain climbers is an avalanche, in which a large mass of snow and ice breaks loose and goes on an essentially frictionless ride down a mountainside on a cushion of compressed air. If you were on a 30.0° mountain slope and an avalanche started 400 m

up the slope, how much time would you have to get out of the way?

31. Two objects of mass 3.00 kg and 5.00 kg are connected by a light string that passes over a frictionless pulley, as in Figure 4.13a. Determine (a) the tension in the string, (b) the acceleration of each object, and (c) the distance each object will move in the first second of motion if they start from rest.

32. In the Atwood machine shown in Figure 4.13a, $m_1 = 2.00$ kg and $m_2 = 7.00$ kg. The masses of the pulley and string are negligible. The pulley turns without friction and the string does not stretch. The object of mass $m_1$ is released with a sharp push that sets it into motion at $v_i = 2.40$ m/s downward. (a) How far will $m_1$ descend below its initial level? (b) Find the velocity of $m_1$ after 1.80 s.

33. In the system shown in Figure P4.33, a horizontal force of magnitude $F_x$ acts on the 8.00-kg object. The horizontal surface is frictionless. (a) For what values of $F_x$ does the 2.00-kg object accelerate upward? (b) For what values of $F_x$ is the tension in the cord zero? (c) Plot the acceleration of the 8.00-kg object versus $F_x$. Include values of $F_x$ from $-100$ N to $+100$ N.

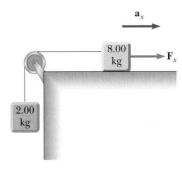

$\mathbf{a}_x$

8.00 kg $\mathbf{F}_x$

2.00 kg

**Figure P4.33**

34. Two objects are connected by a light string that passes over a frictionless pulley, as in Figure P4.34. If the incline

$m_1$ $m_2$

$\theta$

**Figure P4.34**

is frictionless and if $m_1 = 2.00$ kg, $m_2 = 6.00$ kg, and $\theta = 55.0°$, find (a) the accelerations of the objects, (b) the tension in the string, and (c) the speed of each object 2.00 s after being released from rest. Draw free-body diagrams for both objects.

35. A 72.0-kg man stands on a spring scale in an elevator. Starting from rest, the elevator ascends, attaining its maximum speed of 1.20 m/s in 0.800 s. It travels with this constant speed for the next 5.00 s. The elevator then undergoes a uniform acceleration in the negative y direction for 1.50 s and comes to rest. What does the spring scale register (a) before the elevator starts to move? (b) during the first 0.800 s? (c) while the elevator is traveling at constant speed? (d) during the time it is slowing down?

36. A frictionless incline is 10.0 m long and inclined at 35.0°. A particle starts at the bottom with an initial speed of 5.00 m/s up the incline. When it reaches the point at which it momentarily stops, a second particle is released from the top of this incline with an initial speed $v_i$ down the incline. Both particles reach the bottom of the incline at the same moment. (a) Determine the distance that the first particle traveled up the incline. (b) Determine the initial speed of the second particle.

## Section 4.8   Context Connection — Controlling the Spacecraft in Empty Space

37. To model a spacecraft, a toy rocket engine is securely fastened to a large puck, which can glide with negligible friction over a horizontal surface, taken as the *xy* plane. The 4.00-kg puck has a velocity of 3.00**i** m/s at one instant. Eight seconds later, its velocity is to be $(8.00\mathbf{i} + 10.0\mathbf{j})$ m/s. Assuming the rocket engine exerts a constant horizontal force, find (a) the components of the force and (b) its magnitude.

### *Additional Problems*

38. An object of mass $m_1$ on a frictionless horizontal table is connected to an object of mass $m_2$ through a very light pulley $P_1$ and a light fixed pulley $P_2$ as shown in Figure P4.38. (a) If $a_1$ and $a_2$ are the accelerations of $m_1$ and $m_2$,

respectively, what is the relation between these accelerations? Express (b) the tensions in the strings and (c) the accelerations $a_1$ and $a_2$ in terms of the masses $m_1$ and $m_2$, and *g*.

39. An inventive child named Pat wants to reach an apple in a tree without climbing the tree. Sitting in a chair connected to a rope that passes over a frictionless pulley (Fig. P4.39), Pat pulls on the loose end of the rope with such a force that the spring scale reads 250 N. Pat's true weight is 320 N, and the chair weighs 160 N. (a) Draw free-body diagrams for Pat and the chair considered as separate systems, and another diagram for Pat and the chair considered as one system. (b) Show that the acceleration of the system is *upward* and find its magnitude. (c) Find the force Pat exerts on the chair.

**Figure P4.39**

40. Three blocks are in contact with one another on a frictionless, horizontal surface, as in Figure P4.40. A horizontal force **F** is applied to $m_1$. If $m_1 = 2.00$ kg, $m_2 = 3.00$ kg, $m_3 = 4.00$ kg, and $F = 18.0$ N, draw a separate free-body diagram for each block and find (a) the acceleration of the blocks, (b) the *net* force on each block, and (c) the magnitudes of the contact forces between the blocks. (d) You are working on a construction project. A coworker is nailing up plasterboard on one side of a light partition, and you are on the opposite side, providing "backing" by leaning against the wall with your back push-

**Figure P4.38**

**Figure P4.40**

ing on it. Every blow makes your back sting. The supervisor helps you to place a heavy block of wood between the wall and your back. Using the situation analyzed in parts (a), (b), and (c) as a model, explain how this makes your job more comfortable.

41. A high diver of mass 70.0 kg jumps off a board 10.0 m above the water. If his downward motion is stopped 2.00 s after he enters the water, what average upward force did the water exert on him?

42. Any device that allows you to alter the force you exert is a kind of *machine*. Some machines, such as the prybar or the inclined plane, are very simple. Some machines do not even look like machines. An example is the following: Your car is stuck in the mud, and you can't pull hard enough to extract it. You have a long cable, however, which you connect between your front bumper and the trunk of a stout tree and pull taut as you tie it around the tree. You now pull sideways on the cable at its midpoint, exerting a force *f*. Each half of the cable is displaced through a small angle $\theta$ from the straight line between the ends of the cable. Deduce an expression for the magnitude of the force acting on the car. (b) Evaluate the cable tension for the case where $\theta = 7.00°$ and $f = 100$ N.

43. Two blocks of mass 3.50 kg and 8.00 kg are connected by a massless string passing over a frictionless pulley (Fig. P4.43). The inclines are frictionless. Find (a) the magnitude of acceleration of each block and (b) the tension in the string.

3.50 kg       8.00 kg

35.0°       35.0°

**Figure P4.43**

44. If you jump from a desk and land stiff-legged on a concrete floor, the chance is good that you will break a leg. To see how this happens, consider the average force stopping your body when you drop from rest from a height of 1.00 m and stop in a much shorter distance *d*. Your leg is likely to break where the cross-sectional area of the bone (the tibia) is smallest. This is at a point just above the ankle, where the cross-sectional area of one bone is about 1.60 cm². A bone fractures when the compressive stress on it exceeds about $1.60 \times 10^8$ N/m². If you land on both legs, the maximum force your ankles can safely exert on the rest of your body is then about

$$2\,(1.60 \times 10^8 \text{ N/m}^2)\,(1.60 \times 10^{-4} \text{ m}^2) = 5.12 \times 10^4 \text{ N}.$$

Calculate the minimum stopping distance *d* that does not result in a broken leg if your mass is 60.0 kg. Don't try it! Bend your knees!

45. An inquisitive physics student, wishing to combine pleasure with scientific inquiry, rides on a roller coaster sitting on a bathroom scale. (Do not try this yourself on a roller coaster that forbids loose heavy packages.) The bottom of the seat in the roller coaster car is in a plane parallel to the track. The seat has a perpendicular back and a seat belt that fits around the student's chest in a plane parallel to the bottom of the seat. The student lifts his feet from the floor, so that the scale reads his weight, 200 lb, when the car is horizontal. At one point during the ride, the car zooms with negligible friction down a straight slope inclined at 30.0° below the horizontal. What does the scale read there?

46. An 8.40-kg object slides down a fixed, frictionless inclined plane. Use a computer to determine and tabulate the normal force exerted on the object and its acceleration for a series of incline angles (measured from the horizontal) ranging from 0° to 90° in 5° increments. Plot a graph of the normal force and the acceleration as functions of the incline angle. In the limiting cases of 0° and 90°, are your results consistent with the known behavior?

47. An object of mass *M* is held in place by an applied force **F** and a pulley system as shown in Figure P4.47. The pulleys are massless and frictionless. Find (a) the tension in each section of rope, $T_1$, $T_2$, $T_3$, $T_4$, and $T_5$, and (b) the magnitude of **F**. (*Hint:* Draw a free-body diagram for each pulley.)

**Figure P4.47**

48. A mobile is formed by supporting four metal butterflies of equal mass $m$ from a string of length $L$. The points of support are evenly spaced a distance $\ell$ apart as shown in Figure P4.48. The string forms an angle $\theta_1$ with the ceiling at each end point. The center section of string is horizontal. (a) Find the tension in each section of string in terms of $\theta_1$, $m$, and $g$. (b) Find the angle $\theta_2$, in terms of $\theta_1$, that the sections of string between the outside butterflies and the inside butterflies form with the horizontal. (c) Show that the distance $D$ between the end points of the string is

$$D = \frac{L}{5}\left(2\cos\theta_1 + 2\cos\left[\tan^{-1}\left(\tfrac{1}{2}\tan\theta_1\right)\right] + 1\right)$$

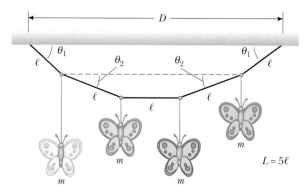

**Figure P4.48**

49. **Review Problem.** A block of mass $m = 2.00$ kg is released from rest at $h = 0.500$ m above the surface of a table, at the top of a $\theta = 30.0°$ incline as shown in Figure P4.49. The frictionless incline is fixed on a table of height $H = 2.00$ m. (a) Determine the acceleration of the block as it slides down the incline. (b) What is the speed of the block as it leaves the incline? (c) How far from the table will the block hit the floor? (d) How much time has elapsed between when the block is released and when it hits the

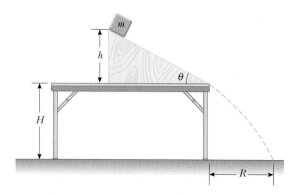

**Figure P4.49**   Problems 49 and 56.

floor? (e) Does the mass of the block affect any of the preceding calculations?

50. A student is asked to measure the acceleration of a cart on a "frictionless" inclined plane, using an air track, a stopwatch, and a meter stick. (Figure 4.11 shows a similar situation with a sled sliding on an incline on frictionless snow.) The height of the incline is measured to be 1.774 cm, and the total length of the incline is measured to be $d = 127.1$ cm. Hence, the angle of inclination $\theta$ is determined from the relation $\sin\theta = 1.774/127.1$. The cart is released from rest at the top of the incline, and its position $x$ along the incline is measured as a function of time, where $x = 0$ refers to the initial position of the cart. For $x$ values of 10.0 cm, 20.0 cm, 35.0 cm, 50.0 cm, 75.0 cm, and 100 cm, the measured times at these positions (averaged over five runs) are 1.02 s, 1.53 s, 2.01 s, 2.64 s, 3.30 s, and 3.75 s, respectively. Construct a graph of $x$ versus $t^2$, and perform a linear least-squares fit to the data. Determine the acceleration of the cart from the slope of this graph, and compare it with the value you obtain using $a = g\sin\theta$, where $g = 9.80$ m/s$^2$.

51. What horizontal force must be applied to the cart shown in Figure P4.51 in order that the blocks remain stationary relative to the cart? Assume all surfaces, wheels, and pulley are frictionless. (*Hint:* Note that the force exerted by the string accelerates $m_1$.)

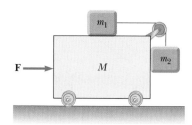

**Figure P4.51**   Problems 51 and 52.

52. Initially the system of objects shown in Figure P4.51 is held motionless. All surfaces, pulley, and wheels are frictionless. Let the force **F** be zero and assume that $m_2$ can move only vertically. At the instant after the system of objects is released, find (a) the tension $T$ in the string, (b) the acceleration of $m_2$, (c) the acceleration of $M$, and (d) the acceleration of $m_1$. (*Note:* The pulley accelerates along with the cart.)

53. A van accelerates down a hill (Fig. P4.53), going from rest to 30.0 m/s in 6.00 s. During the acceleration, a toy ($m = 0.100$ kg) hangs by a string from the van's ceiling. The acceleration is such that the string remains perpendicular to the ceiling. Determine (a) the angle $\theta$ and (b) the tension in the string.

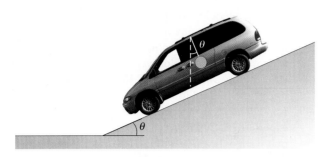

**Figure P4.53**

54. Cam mechanisms are used in many machines. For example, cams open and close the valves in your car engine to admit gasoline vapor to each cylinder and to allow the escape of exhaust. The principle of a cam is illustrated in Figure P4.54, showing a follower rod (also called a

**Figure P4.54**

pushrod) of mass $m$ resting on a wedge of mass $M$. The sliding wedge duplicates the function of a rotating eccentric disk on a camshaft in your car. Assume that no friction exists between the wedge and the base, between the pushrod and the wedge, or between the rod and the guide through which it slides. When the wedge is pushed to the left by the force **F**, the rod moves upward and does something such as opening a valve. By varying the shape of the wedge, the motion of the follower rod can be made quite complex, but assume that the wedge makes a constant angle of $\theta = 15.0°$. Suppose you want the wedge and the rod to start from rest and move with constant acceleration, with the rod moving upward 1.00 mm in 8.00 ms. Take $m = 0.250$ kg and $M = 0.500$ kg. What must be the magnitude $F$ of the force applied to the wedge?

55. An orbiting shuttle spacecraft is in a nearly constant state of free-fall and hence produces a near zero-gravity environment inside it. The spacecraft experiences only small accelerations due to atmospheric drag and gaseous efflux from equipment. These accelerations produce a microgravity environment within the shuttle cabin, where the free-fall acceleration is measured in units of $\mu g$, where $1 \, \mu g = 1 \times 10^{-6} \, g$. On Earth, physics students sometimes drop their calculators. Find the time of fall and the impact speed of a calculator dropped through 1.30 m (a) at the Earth's surface and (b) in the cabin of a spacecraft where the free-fall acceleration is 8.00 $\mu g$.

56. In Figure P4.49 the incline has mass $M$ and is fastened to the stationary horizontal tabletop. The block of mass $m$ is placed near the bottom of the incline and is released with a quick push that sets it sliding upward. It stops near the top of the incline, as shown in the figure, and then slides down again, always without friction. Find the force the tabletop exerts on the incline throughout this motion.

## ANSWERS TO QUICK QUIZZES

**4.1** Motion requires no force. Newton's first law says an object in motion continues to move by itself in the absence of external forces. It is possible for forces to act on an object with no resulting motion if the forces are balanced.

**4.2** (a) If a single force acts on an object, it must accelerate. (b) If an object accelerates, at least one force must act on it. (c) If an object has no acceleration, you cannot conclude that no forces act on it. In this case, you can only say that the *net* force on the object is zero.

**4.3** If the object begins at rest or is moving with a velocity with only an $x$ component, the net force in the $x$ direction will cause the object to continue to move in the $x$ direction. In any other case, however, the motion of the object will involve velocity components in directions other than $x$. Thus, the direction of the velocity vector will not generally be along the $x$ axis. What we can say with confidence is that a net force in the $x$ direction causes the object to *accelerate* in the $x$ direction.

**4.4** Because the value of $g$ is smaller on the Moon than on the Earth, more mass of gold would be required to represent 1 newton of weight on the Moon. Thus, your friend on the Moon is richer, by about a factor of 6!

**4.5** The only force acting on the baseball during its motion is the downward gravitational force. The magnitude of this force is $mg$ at all points in the motion, including the specific points mentioned in (a) and (b).

**4.6** The car and truck experience forces that are equal in magnitude but in opposite directions, according to Newton's third law. A calibrated spring scale placed between the colliding vehicles reads the same whichever way it faces. Because the car has the smaller mass, it experiences the greater acceleration.

**4.7** The force causing an automobile to move is the friction between the tires and the roadway as the automobile attempts to push the roadway backward. Notice that forces from the engine are not directly what accelerate the automobile. The force driving a propeller-driven airplane forward is the force of the air on the propeller as the rotating propeller pushes the air backward. The force propelling a rocket is the force of the exhausted gases from the back of the rocket, as the rocket pushes the gases backward. In a rowboat, the rower pushes the water backward with the oars. As a result, the water pushes forward on the oars, and the force is applied to the boat at the oarlocks. All of these situations involve the propulsion of an object due to Newton's third law.

**4.8** According to Newton's third law, the force of the man on the boy and the force of the boy on the man are third law reaction forces, so they must be equal in magnitude. Both individuals experience the same magnitude of force, so the boy, with the smaller mass, experiences the larger acceleration. Both individuals experience an acceleration over the same time interval. The larger acceleration of the boy results in his moving away from the interaction with a larger velocity. Again, because of the larger acceleration, the boy moves through a larger distance during the interval while the hands are in contact.

**4.9** (c). The scale is in equilibrium in both situations, so it experiences a net force of zero. Because each individual pulls with a force $F$ and there is no acceleration, each individual is in equilibrium. Therefore, the tension in the ropes must be equal to $F$. In case (i), the individual on the right pulls with force $F$ on a spring mounted rigidly to a brick wall. The resulting tension $F$ in the rope causes the scale to read a force $F$. In case (ii), the individual on the left can be modeled as simply holding the rope tightly while the individual on the right pulls. Thus, the individual on the left is doing the same thing that the wall does in case (i). The resulting scale reading is the same whether a wall or a person is holding the left side of the scale.

**4.10** (a). With the force applied toward the left on $m_2$, the contact force alone must accelerate $m_1$. (In the original example, the contact force alone accelerated $m_2$.) Because $m_1 > m_2$, this will require more force, so the magnitude of $\mathbf{P}_{12}$ is greater.

# Calvin and Hobbes                    by Bill Watterson

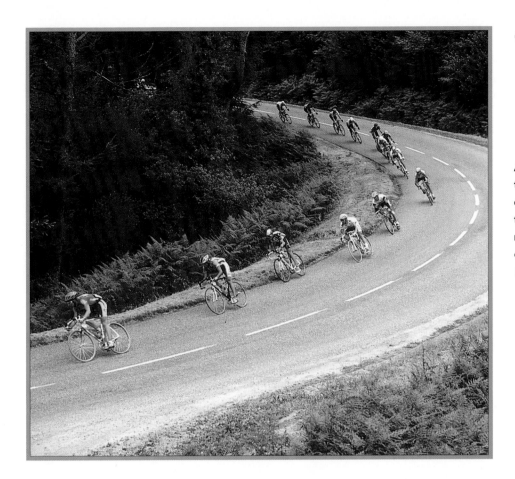

As these cyclists round a sharp turn in the Tour de France, they depend on the force of friction between the bicycle tires and the roadway to change the direction of their velocity.

*(Alex Livesey/Allsport)*

# More Applications of Newton's Laws

I n Chapter 4 we introduced Newton's laws of motion and applied them to situations in which we ignored friction. In this chapter, we shall expand our investigation to objects moving in the presence of friction, which will allow us to model situations more realistically. Such objects include those moving on rough surfaces and those moving through viscous media such as liquids and air. We also apply Newton's laws to the dynamics of circular motion so that we can understand more about various types of objects moving in circular paths.

## 5.1 • FORCES OF FRICTION

When an object moves either on a surface or through a viscous medium such as air or water, there is resistance to the motion because the object interacts with its surroundings. We call such resistance a **force of friction.** Forces of friction are very important in our everyday lives. They allow us to walk or run and are necessary for the motion of wheeled vehicles.

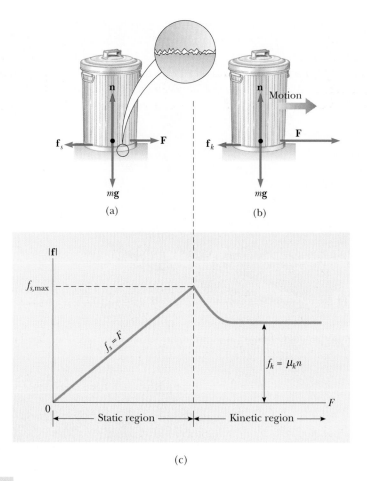

(a)    (b)

(c)

### Figure 5.1

(a) The force of static friction $\mathbf{f}_s$ between a trash can and a concrete patio is opposite the applied force $\mathbf{F}$. The magnitude of the force of static friction equals that of the applied force. (b) When the magnitude of the applied force exceeds the magnitude of the force of kinetic friction $\mathbf{f}_k$, the trash can accelerates to the right. (c) A graph of the magnitude of the friction force versus that of the applied force. In our model, the force of kinetic friction is independent of the applied force and the relative speed of the surfaces. Note that $f_{s,\text{max}} > f_k$.

Imagine you are working in your garden and have filled a trash can with yard clippings. You then try to drag the trash can across the surface of your concrete patio as in Figure 5.1a. This is a *real* surface, not an idealized, frictionless surface in a simplification model. If we apply an external horizontal force $\mathbf{F}$ to the trash can, acting to the right, the trash can remains stationary if $\mathbf{F}$ is small. The force that counteracts $\mathbf{F}$ and keeps the trash can from moving acts to the left and is called the **force of static friction $\mathbf{f}_s$**. As long as the trash can is not moving, $f_s = F$. Thus, if $\mathbf{F}$ is increased, $\mathbf{f}_s$ also increases. Likewise, if $\mathbf{F}$ decreases, $\mathbf{f}_s$ also decreases. Experiments show that the friction force arises from the nature of the two surfaces; because of their roughness, contact is made only at a few points, as shown in the magnified view of the surface in Figure 5.1a.

If we increase the magnitude of $\mathbf{F}$, as in Figure 5.1b, the trash can eventually slips. When the trash can is on the verge of slipping, $f_s$ is a maximum as shown in Figure 5.1c. When $F$ exceeds $f_{s,\text{max}}$, the trash can moves and accelerates to the right. When the trash can is in motion, the friction force is less than $f_{s,\text{max}}$ (Fig. 5.1c). We call the friction force for an object in motion the **force of kinetic friction** $\mathbf{f}_k$. The net force $F - f_k$ in the $x$ direction produces an acceleration to the right, according to Newton's second law. If $F = f_k$, the acceleration is zero, and the trash can moves to the right with constant speed. If the applied force is removed, the friction force acting to the left provides an acceleration of the trash can in the $-x$ direction and eventually brings it to rest.

Experimentally, one finds that, to a good approximation, both $f_{s,\text{max}}$ and $f_k$ for an object on a surface are proportional to the normal force exerted by the surface on the object—thus we adopt a simplification model in which this approximation is assumed to be exact. The assumptions in this simplification model can be summarized as follows:

- The magnitude of the force of static friction between any two surfaces in contact can have the values

$$f_s \leq \mu_s n \qquad [5.1]$$

where the dimensionless constant $\mu_s$ is called the **coefficient of static friction** and $n$ is the magnitude of the normal force. The equality in Equation 5.1 holds when the surfaces are on the verge of slipping, that is, when $f_s = f_{s,\text{max}} \equiv \mu_s n$. This situation is called *impending motion*. The inequality holds when the component of the applied force parallel to the surfaces is less than this value.

- The magnitude of the force of kinetic friction acting between two surfaces is

$$f_k = \mu_k n \qquad [5.2]$$

where $\mu_k$ is the **coefficient of kinetic friction.** In our simplification model, this coefficient is independent of the relative speed of the surfaces.

- The values of $\mu_k$ and $\mu_s$ depend on the nature of the surfaces, but $\mu_k$ is generally less than $\mu_s$. Typical values of $\mu$ range from around 0.05 to 1.5. Table 5.1 lists some reported values.

- The direction of the friction force on an object is opposite to the actual motion (kinetic friction) or the impending motion (static friction) of the object relative to the surface with which it is in contact.

**PITFALL PREVENTION 5.1**

**The equal sign is used in limited situations**

In Equation 5.1, the equal sign is used *only* when the surfaces are just about to break free and begin sliding. Do not fall into the common trap of using $f_s = \mu_s n$ in *any* static situation.

- *Force of static friction*

- *Force of kinetic friction*

**PITFALL PREVENTION 5.2**

**Friction equations**

Keep in mind that Equations 5.1 and 5.2 are *not* vector equations. They are relationships between the *magnitudes* of the vectors representing the friction and normal forces. Because the friction and normal forces are perpendicular to each other, the vectors cannot be related by a multiplicative constant.

| **TABLE 5.1** | **Coefficients of Friction**[a] | |
|---|---|---|
| | $\mu_s$ | $\mu_k$ |
| Steel on steel | 0.74 | 0.57 |
| Aluminum on steel | 0.61 | 0.47 |
| Copper on steel | 0.53 | 0.36 |
| Rubber on concrete | 1.0 | 0.8 |
| Wood on wood | 0.25–0.5 | 0.2 |
| Glass on glass | 0.94 | 0.4 |
| Waxed wood on wet snow | 0.14 | 0.1 |
| Waxed wood on dry snow | — | 0.04 |
| Metal on metal (lubricated) | 0.15 | 0.06 |
| Ice on ice | 0.1 | 0.03 |
| Teflon on Teflon | 0.04 | 0.04 |
| Synovial joints in humans | 0.01 | 0.003 |

[a] All values are approximate.

**PITFALL PREVENTION 5.3**

**The direction of the friction force**

Be careful with the direction of the friction force. Sometimes, an incorrect statement about the friction force between an object and a surface is made—"the friction force on an object is opposite to its motion or impending motion"—rather than the correct phrasing, "the friction force on an object is opposite to its motion or impending motion *relative to the surface.*" An example is a package placed on the bed of a pickup truck. As the truck accelerates forward, the natural inertia of the package tends to "leave it behind." Thus, the impending motion relative to the floor of the truck is toward the rear, and the friction force exerted on the package is forward, in the same direction in which it moves. If the truck accelerates rapidly enough, the package slides backward in the bed of the truck. In this case, we have real motion of the object rather than impending motion (relative to the surface), and the motion is backward. The friction force, then, is forward.

The approximate nature of Equations 5.1 and 5.2 is easily demonstrated by trying to arrange for an object to slide down an incline at constant speed. Especially at low speeds, the motion is likely to be characterized by alternate stick and slip episodes. These assumptions represent a simplification model, developed so that we can solve problems involving friction in a relatively straightforward way.

Now that we have identified the characteristics of the friction force, we can include the friction force in the net force on an object in the model of a particle under a net force.

### Quick Quiz 5.1

You press your physics textbook flat against a vertical wall with your hand. What is the direction of the friction force on the book exerted by the wall? (a) Downward. (b) Upward. (c) Out from the wall. (d) Into the wall.

### Quick Quiz 5.2

A crate is located at the center of a flatbed truck. The truck accelerates toward the east, and the crate moves with it, not sliding on the bed of the truck. What is the direction of the friction force exerted by the bed of the truck on the crate? (a) To the west. (b) To the east. (c) There is no friction force because the crate is not sliding.

### Quick Quiz 5.3

You are playing with your daughter in the snow. She is sitting on a sled and asking you to slide her across a flat, horizontal field. You have a choice of pushing her from behind, by applying a force downward on her shoulders at 30° below the horizontal (Fig. 5.2a), or attaching a rope to the front of the sled and pulling with a force at 30° above the horizontal (Fig 5.2b). Which would be easier for you and why?

(a)

(b)

**Figure 5.2**

(Quick Quiz 5.3) A father tries to push his daughter on a sled over snow by (a) pushing downward on her shoulders, or (b) pulling upward on a rope attached to the sled. Which is easier?

## THINKING PHYSICS 5.1

In the motion picture *The Abyss* (Twentieth Century Fox, 1990), an underwater oil exploration rig is located at the ocean bottom in very deep water. It is connected to a ship on the ocean surface by an "umbilical cord" as suggested in Figure 5.3a. On the ship, the umbilical cord is attached to a gantry. During a hurricane, the gantry structure breaks loose from the ship, falls into the water, and sinks to the bottom, passing over the edge of an extremely deep abyss. As a result, the rig is dragged by the umbilical cord along the ocean bottom as described in Figure 5.3b. As the rig approaches the edge of the abyss, however, it is not pulled over the edge, but stops just short of the edge as shown in Figure 5.3c. Is this purely a cinematic edge-of-the-seat situation, or do the principles of physics suggest why the moving rig does not topple over the edge?

**Reasoning**   Physics can explain this phenomenon. While the rig is being pulled across the ocean floor (Fig. 5.3b), it is pulled by the section of the umbilical cord that is almost horizontal—almost parallel to the ocean floor. Thus, the rig is subject to two horizontal forces: the tension in the umbilical cord pulling it forward and

**WEB**

For information about
*The Abyss*, visit
**www.sciflicks.com/the__abyss**

**Figure 5.3**

(Thinking Physics 5.1)

**Normal force and weight**

In Example 5.1, we find that the magnitudes of the normal force and the gravitational force are equal, $n = mg$. As pointed out in Pitfall Prevention 4.9 in Chapter 4, this is *not* a general result. This is such a common student pitfall that we point it out again here. *Always* use Newton's second law to find the relation between $n$ and $mg$; it will give you the correct relationship. Consult Examples 5.2 and 5.4 for cases in which $n$ is not equal to $mg$.

friction with the ocean floor pulling back. Let us assume that these forces are equal, so that the rig moves with constant speed. As the rig nears the edge of the abyss, the angle the umbilical cord makes with the horizontal increases. As a result, the component of the force from the cord parallel to the ocean floor decreases and the downward vertical component increases. As a result of the increased vertical force, the rig is pulled downward more strongly to the ocean floor, increasing the normal force on it and, in turn, increasing the friction force between the rig and the ocean floor. Thus, with less force pulling it forward (from the umbilical cord) and more force opposite to the motion (due to friction), the rig slows down. By the time the rig reaches the edge of the abyss, the force from the umbilical cord is almost *straight down* (Fig. 5.3c), resulting in little forward force. Furthermore, this large downward force pulls the rig into the ocean floor, resulting in a very large friction force that stops the rig.

If the coefficient of friction between the rig and ocean floor is relatively small, the forward velocity of the rig could be large enough so that the rig would be projected off the edge.

## Example 5.1   The Skidding Truck

The driver of an empty speeding truck slams on the brakes and skids to a stop through a distance $d$. (a) If the truck carries a heavy load such that its mass is doubled, what would be its skidding distance if it starts from the same initial speed?

(b) If the initial speed of the empty truck is halved, what would be the skidding distance?

**Solution**   (a) Figure 5.4 shows a free-body diagram for the skidding truck. The only force in the horizontal direction is the friction force, which is assumed to be constant in our simplification model for friction. Thus, from Newton's second law

$$\sum F_x = f_k = ma$$

where $m$ is the mass of the truck. In the vertical direction, there is no acceleration, so we model the truck as a particle in equilibrium:

$$\sum F_y = n - mg = 0 \longrightarrow n = mg$$

Finally, from the relation between the friction force and the normal force, we combine these two equations:

$$f_k = \mu_k n \longrightarrow ma = \mu_k(mg) \longrightarrow a = \mu_k g$$

Because both $\mu_k$ and $g$ are constant, the acceleration of the truck is constant. We therefore model the truck as a particle under constant acceleration. We use Equation 2.12 to find the position of the truck when the velocity is zero:

$$v_{xf}^2 = v_{xi}^2 + 2a_x(x_f - x_i) \longrightarrow 0 = v_{xi}^2 + 2(\mu_k g)(x_f - 0)$$

$$x_f = d = \frac{v_{xi}^2}{2\mu_k g}$$

We can argue from the mathematical representation as follows. The expression for the skidding distance $d$ does not in-

**Figure 5.4**

(Example 5.1) A truck skids to a stop.

clude the mass. Thus, the truck skids the same distance regardless of the mass. Conceptually, we can argue that the truck with twice the mass requires twice the friction force to exhibit the same acceleration and stop in the same distance. The normal force is equal to the doubled weight, and the friction force is proportional to the doubled normal force!

(b) This part of the problem is a comparison problem and can be solved by a ratio technique. We write the result from part (a) for the skidding distance $d$ twice, once for the original situation, and once for the halved initial velocity:

$$d_1 = \frac{v_{1xi}^2}{2\mu_k g}$$

$$d_2 = \frac{v_{2xi}^2}{2\mu_k g} = \frac{(\frac{1}{2}v_{1xi})^2}{2\mu_k g} = \frac{1}{4}\frac{v_{1xi}^2}{2\mu_k g}$$

Dividing the first equation by the second, we have

$$\frac{d_1}{d_2} = 4 \quad \longrightarrow \quad d_2 = \tfrac{1}{4}d_1$$

Notice that a reduction of half the velocity reduces the skidding distance by 75%! This is an important safety consideration associated with the possibility of an accident when driving at high speed.

---

## Example 5.2 Experimental Determination of $\mu_s$ and $\mu_k$

The following is a simple method of measuring coefficients of friction. Suppose a block is placed on a rough surface inclined relative to the horizontal, as shown in Figure 5.5. The incline angle $\theta$ is increased until the block starts to move. (a) How is the coefficient of static friction related to the critical angle $\theta_c$ at which the block begins to move?

(b) How could we find the coefficient of kinetic friction?

**Solution** (a) The forces on the block, as shown in Figure 5.5, are the gravitational force $m\mathbf{g}$, the normal force **n**, and the force of static friction $\mathbf{f}_s$. As long as the block is not moving, these forces are balanced and the block is in equilibrium. We choose a coordinate system with the positive $x$ axis parallel to the incline and downhill and the positive $y$ axis upward perpendicular to the incline. Applying Newton's second law in component form to the block gives

$$(1) \quad \sum F_x = mg \sin \theta - f_s = 0$$

$$(2) \quad \sum F_y = n - mg \cos \theta = 0$$

These equations are valid for any angle of inclination $\theta$. At the critical angle $\theta_c$ at which the block is on the verge of slipping, the friction force has its maximum magnitude $\mu_s n$, so we rewrite (1) and (2) as

$$(3) \quad mg \sin \theta_c = \mu_s n$$

$$(4) \quad mg \cos \theta_c = n$$

Dividing (3) by (4), we have

$$\tan \theta_c = \mu_s$$

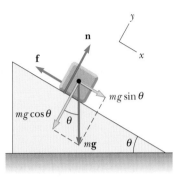

**Figure 5.5**

(Example 5.2) A block on an adjustable incline is used to determine the coefficients of friction.

Thus, the coefficient of static friction is equal to the tangent of the angle at which the block begins to slide.

(b) Once the block begins to move, the magnitude of the friction force is the kinetic value $\mu_k n$, which is smaller than that of the force of static friction. As a result, if the angle is maintained at the critical angle, the block accelerates down the incline. To restore the equilibrium situation in Equation (1), with $f_s$ replaced by $f_k$, the angle must be reduced to a value $\theta'_c$ such that the block slides down the incline at *constant speed*. In this situation, Equations (3) and (4), with $\theta_c$ replaced by $\theta'_c$ and $\mu_s$ by $\mu_k$, give us

$$\tan \theta'_c = \mu_k$$

---

## Example 5.3 Connected Objects

A ball and a cube are connected by a light string that passes over a frictionless light pulley, as in Figure 5.6a. The coefficient of kinetic friction between the cube and the surface is 0.30. Find the acceleration of the two objects and the tension in the string.

**Reasoning** In the mental representation, imagine the ball moving downward and the cube sliding to the right, both accelerating from rest. We set up a simplified pictorial representation by drawing the free-body diagrams for the two objects as in Figures 5.6b and 5.6c. For the ball, no forces are exerted in the horizontal direction, and we apply Newton's second law in the vertical direction. For the cube, the acceleration is only horizontal, so we know the cube is in equilibrium in the vertical direction. We shall make use of the fact that the magnitude of the force of kinetic friction acting on the cube is proportional to the normal force according to $f_k = \mu_k n$. Because the pulley is light (massless) and frictionless,

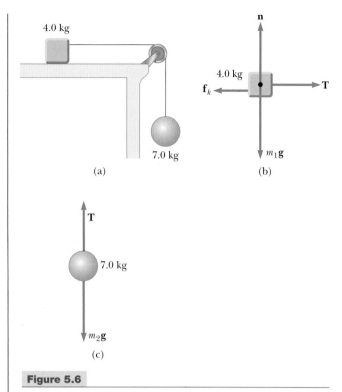

**Figure 5.6**

(Example 5.3) (a) Two objects connected by a light string that passes over a frictionless pulley. (b) Free-body diagram for the sliding cube. (c) Free-body diagram for the hanging ball.

the tension in the string is the same on both sides of the pulley. Because the tension acts on both objects, this is the common quantity that applies to both objects and allows us to combine separate equations for the two objects into one equation.

**Solution**    Let us address the cube of mass $m_1$ first. Newton's second law applied to the cube in component form, with the positive $x$ direction to the right, gives

$$\sum F_x = m_1 a \quad \longrightarrow \quad T - f_k = m_1 a$$
$$\sum F_y = 0 \quad \longrightarrow \quad n - m_1 g = 0$$

where $T$ is the tension in the string. Because $f_k = \mu_k n$ and $n = m_1 g$ from the equilibrium equation for the $y$ direction, we have $f_k = \mu_k m_1 g$. Therefore, from the equation for the $x$ direction,

$$(1) \quad \boxed{T = \mu_k m_1 g + m_1 a}$$

Now we apply Newton's second law to the ball moving in the vertical direction. Because the ball moves downward when the cube moves to the right, we choose the downward direction as positive for the ball:

$$(2) \quad \sum F_y = m_2 a \quad \longrightarrow \quad m_2 g - T = m_2 a$$

Substituting the expression for $T$ from (1) into (2) gives us

$$m_2 g - (\mu_k m_1 g + m_1 a) = m_2 a$$

$$a = \frac{m_2 - \mu_k m_1}{m_1 + m_2} g$$

Now, substituting the known values,

$$a = \frac{7.0 \text{ kg} - 0.30(4.0 \text{ kg})}{7.0 \text{ kg} + 4.0 \text{ kg}}(9.80 \text{ m/s}^2) = \boxed{5.2 \text{ m/s}^2}$$

This is the *magnitude* of the acceleration of each of the two objects. For the ball, the acceleration *vector* is downward, and the vector is toward the right for the cube. When the magnitude of the acceleration is substituted into (1), we find the tension:

$$T = 0.30(4.0 \text{ kg})(9.80 \text{ m/s}^2) + (4.0 \text{ kg})(5.2 \text{ m/s}^2) = \boxed{33 \text{ N}}$$

**EXERCISE**    If the coefficient of kinetic friction were very large, the ball and cube could move with constant velocity. What coefficient is required for this situation to exist?

*Answer*    1.75

---

## Example 5.4    The Sliding Crate

A warehouse worker places a crate on a sloped surface that is inclined at 30.0° with respect to the horizontal (Fig. 5.7a). If the crate slides down the incline with an acceleration of magnitude $g/3$, determine the coefficient of kinetic friction between the crate and the surface.

**Solution**    Figure 5.7b shows the forces acting on the crate. The $x$ axis is chosen parallel to the incline and the $y$ axis per-

pendicular. From Newton's second law,

$$(1) \quad \sum F_x = ma \quad \longrightarrow \quad mg \sin \theta - f_k = ma$$
$$(2) \quad \sum F_y = 0 \quad \longrightarrow \quad n - mg \cos \theta = 0$$

The friction force is $f_k = \mu_k n$ and, from (2), we find that $n = mg \cos \theta$. Thus, the friction force can be expressed as $f_k = \mu_k mg \cos \theta$. Substituting this into (1),

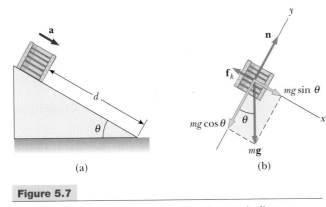

(a)

(b)

**Figure 5.7**

(Example 5.4) (a) A crate of mass $m$ slides down an incline.
(b) Free-body diagram for the sliding crate.

$$mg \sin \theta - \mu_k mg \cos \theta = ma \longrightarrow \mu_k = \frac{g \sin \theta - a}{g \cos \theta}$$

Substituting the known values, we have

$$\mu_k = \frac{\cancel{g} \sin 30.0° - \frac{1}{3}\cancel{g}}{\cancel{g} \cos 30.0°} = \frac{(0.500 - 0.333)}{0.867} = \boxed{0.192}$$

**EXERCISE** A block moves up a 45.0° incline with constant speed under the action of a force of 15.0 N applied *parallel* to the incline. If the coefficient of kinetic friction is 0.300, determine (a) the weight of the block and (b) the minimum force parallel to the incline required to allow it to move *down* the incline at constant speed.

*Answer* (a) 16.3 N (b) 8.07 N up the incline

## 5.2 • NEWTON'S SECOND LAW APPLIED TO A PARTICLE IN UNIFORM CIRCULAR MOTION

Solving problems involving friction is just one of many applications of Newton's second law. Let us now consider another common situation, associated with a particle in uniform circular motion. In Chapter 3, we found that a particle moving in a circular path of radius $r$ with uniform speed $v$ experiences a centripetal acceleration of magnitude

$$a_c = \frac{v^2}{r}$$

The acceleration vector with this magnitude is directed toward the center of the circle and is *always* perpendicular to $\mathbf{v}$.

According to Newton's second law, if an acceleration occurs, a net force must be causing it. Because the acceleration is toward the center of the circle, the net force must be toward the center of the circle. Thus, when a particle travels in a circular path, a force must be acting *inward* on the particle that causes the circular motion. We investigate the forces causing this type of acceleration in this section.

Consider an object of mass $m$ tied to a string of length $r$ and being whirled in a horizontal circular path on a frictionless table top as in Figure 5.8. Let us assume that the object moves with constant speed. The natural tendency of the object is to move in a straight-line path, according to Newton's first law; however, the string prevents this motion along a straight line by exerting a radial force $\mathbf{F}_r$ on the object to make it follow a circular path. This force, which is the tension in the string, is directed along the length of the string toward the center of the circle as shown in Figure 5.8.

In this discussion, the tension in the string causes the circular motion. There are many other examples of forces causing objects to move in circular paths. These forces can be any of our familiar forces causing an object to follow a circular path and, therefore, to have a centripetal acceleration.

Regardless of the nature of the force acting on the particle in circular motion, we can apply Newton's second law to the particle along the radial direction:

$$\sum F = ma_c = m\frac{v^2}{r} \qquad \text{[5.3]}$$

**See the *Core Concepts in Physics CD-ROM*, Screen 4.7**

• *Centripetal acceleration*

**PITFALL PREVENTION 5.5**

**Centrifugal force**

The phrase "centrifugal force" is often heard, and described as a force pulling *outward* on an object moving in a circular path. It is likely that you have felt this effect, perhaps on an amusement park ride, or during a sharp turn in your car. This is not a real force, however. To convince yourself of this, remember that forces are always interactions between objects. If you are feeling a "centrifugal force," with what other object are you interacting? You cannot identify another object because this is not a real force. This is a "fictitious force" that occurs as a result of your being in a noninertial reference frame. (See "Inertial Frames" in Section 4.2—the acceleration of the puck on the table in the accelerating train is caused by another "fictitious force.")

**PITFALL PREVENTION 5.6**

**Centripetal force**

The force causing centripetal acceleration is called *centripetal force* in some textbooks. This is a pitfall for many students. Giving the force causing circular motion a name—centripetal force—leads many students to consider this as a new *kind* of force, rather than a new *role* for force. A common mistake in force diagrams is to draw in all of the usual forces and then add another vector for the centripetal force. But it is not a separate force—it is simply one of our familiar forces *acting in the role of causing a circular motion*. For the motion of the Earth around the Sun, for example, the "centripetal force" is *gravity*. For a rock whirled on the end of a string, the "centripetal force" is the *tension* in the string. For an amusement park patron pressed against the inner wall of a rapidly rotating circular room, the "centripetal force" is the *normal force* from the wall. After this discussion, we shall no longer use the phrase *centripetal force*.

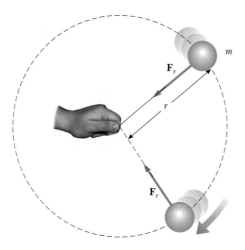

**Figure 5.8**

Overhead view of a ball moving in a circular path in a horizontal plane. A force $\mathbf{F}_r$ directed toward the center of the circle keeps the ball moving in its circular path.

In general, an object can move in a circular path under the influence of various types of forces, or a *combination* of forces, as we shall see in some of the examples that follow.

If the force acting on an object vanishes, the object no longer moves in its circular path; instead it moves along a straight-line path tangent to the circle. This idea is illustrated in Figure 5.9 for the case of the ball whirling in a circle at the end of a string. If the string breaks at some instant, the ball moves along the straight-line path tangent to the circle at the point where the string breaks.

**PITFALL PREVENTION 5.7**

**Direction of travel when the string is cut**

Study Figure 5.9 carefully. Many students have a misconception that the ball moves *radially* away from the center of the circle when the string is cut. The velocity of the ball is *tangent* to the circle. By Newton's first law, the ball simply continues to move in the direction that it is moving just as the force from the string disappears.

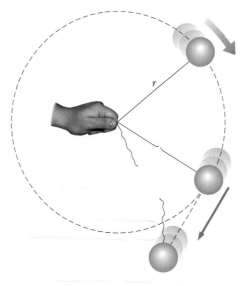

**Figure 5.9**

When the string breaks, the ball moves in the direction tangent to the circular path.

**Quick Quiz 5.4**

You are riding on a Ferris wheel that is rotating with constant speed. The car in which you ride always maintains its correct upward orientation—it does not invert. What is the direction of your centripetal acceleration when you are (a) at the top of the wheel? (b) At the bottom of the wheel? What is the direction of the normal

On a Ferris wheel, the force of the seat upward on a rider varies with the position of the rider. Where is this force the largest?  *(Color Box/FPG)*

force on you from the seat when you are (c) at the top of the wheel? (d) At the bottom of the wheel? (e) At which point, top or bottom, is the normal force the largest in magnitude?

### Quick Quiz 5.5

A car travels on a circular roadway of radius $r$. The roadway is flat, as in Figure 5.10a. The car travels at a high speed $v$, such that the friction force causing the centripetal acceleration is the maximum possible value. If this same car is now driven on another flat circular roadway of radius $2r$, and the coefficient of friction between the tires and the roadway is the same as on the first roadway, what is the maximum speed of the car such that it does not slide off the roadway?

#### Figure 5.10

(a) (Quick Quiz 5.5) Free-body diagram for a car traveling on a curved roadway that is not banked. The center of curvature of the roadway is to the left. (b) (Thinking Physics 5.2) Free-body diagram for a car traveling on a curved roadway that is banked.

The cars of a corkscrew roller coaster must travel in tight loops. The normal force from the track contributes to the centripetal acceleration. The gravitational force, because it remains constant in direction, is sometimes in the same direction as the normal force, but is sometimes in the opposite direction. (*Robin Smith/Tony Stone Images*)

**WEB**

For discussions of the physics behind several amusement park rides, visit **www.learner.org/exhibits/ parkphysics/**

### THINKING PHYSICS 5.2

High-speed curved roadways are *banked*—that is, the roadway is tilted toward the inside of tight curves. Why is this?

**Reasoning**   As an automobile rounds a curve, there must be a force on the car providing its centripetal acceleration. In the rolling motion of each tire, the bit of rubber meeting the road is instantaneously at rest relative to the road. It is prevented from skidding outward by a static friction force that acts radially inward, enabling the car to move in its circular path. This static friction force is parallel to the roadway surface and perpendicular to the centerline of the car, as shown in Figure 5.10a for a car on a flat roadway. This is the only horizontal force—the weight and normal force are both vertical. If the curve is tight or the speed is high, the required friction force may be larger than the maximum possible static friction force. In this case, Newton's first law takes over, and the car skids toward the side of the curved road. This situation is even more likely to occur if the road is icy, so that the friction coefficient is reduced. Banking the roadway "tilts" the normal force away from the vertical, so that it has a horizontal component toward the center of the circular roadway, as shown in Figure 5.10b. Thus, the combination of the normal force and the friction force works together to provide the needed centripetal acceleration. The larger the acceleration, the larger is the banking angle, to provide a larger component of the normal force in the horizontal direction.

### THINKING PHYSICS 5.3

The Copernican theory of the solar system is a model in which the planets are assumed to travel around the Sun in circular orbits. Historically, this was a break from the Ptolemaic theory, a model in which the Earth was at the center. When the Copernican theory was proposed, a natural question arose—what keeps the Earth and other planets moving in their paths around the

Sun? An interesting response to this question comes from Richard Feynman*—"In those days, one of the theories proposed was that the planets went around because behind them there were invisible angels, beating their wings and driving the planets forward. . . . It turns out that in order to keep the planets going around, the invisible angels must fly in a different direction. . . ." What did Feynman mean by this?

**Reasoning**  The question asked by those at the time of Copernicus indicates that they did not have a proper understanding of inertia as described by Newton's first law. At that time in history, before Galileo and Newton, the interpretation was that *motion* was caused by force. This is different from our current understanding that *changes in motion* are caused by force. Thus, it was natural for Copernicus's contemporaries to ask what force propelled a planet in its orbit. According to our current understanding, it is equally natural for us to realize that no force tangent to the orbit is necessary—the motion simply continues due to inertia.

Thus, in Feynman's imagery, we do not need the angels pushing the planet *from behind*. We need the angels to push *inward*, to provide the centripetal acceleration associated with the orbital motion of the planet. Of course, the angels are not real, from a scientific point of view—they are a metaphor for the *gravitational force*.

## Example 5.5  How Fast Can It Spin?

An object of mass 0.500 kg is attached to the end of a cord whose length is 1.50 m. The object is whirled in a horizontal circle as in Figure 5.8. If the cord can withstand a maximum tension of 50.0 N, what is the maximum speed the object can have before the cord breaks?

**Solution**  Because the magnitude of the force that provides the centripetal acceleration of the object in this case is the tension $T$ exerted by the cord on the object, Newton's second law gives us for the inward radial direction,

$$\sum F_r = ma_c \longrightarrow T = m\frac{v^2}{r}$$

Solving for the speed $v$, we have

$$v = \sqrt{\frac{Tr}{m}}$$

The maximum speed that the object can have corresponds to the maximum value of the tension. Hence, we find

$$v_{max} = \sqrt{\frac{T_{max}r}{m}} = \sqrt{\frac{(50.0 \text{ N})(1.50 \text{ m})}{0.500 \text{ kg}}} = \boxed{12.2 \text{ m/s}}$$

**EXERCISE**  Calculate the tension in the cord if the speed of the object is 5.00 m/s.

*Answer*  8.33 N

## Example 5.6  The Conical Pendulum

A small object of mass $m$ is suspended from a string of length $L$. The object revolves in a horizontal circle of radius $r$ with constant speed $v$, as in Figure 5.11a. (Because the string sweeps out the surface of a cone, the system is known as a *conical pendulum*.) Find (a) the speed of the object, and (b) the period of revolution, defined as the time needed to complete one revolution.

**Solution**  (a) The free-body diagram for the object of mass $m$ is shown in Figure 5.11b, where the force **T** exerted by the string has been resolved into a vertical component $T\cos\theta$ and a horizontal component $T\sin\theta$ acting toward the center of rotation. Because the object does not accelerate in the ver-

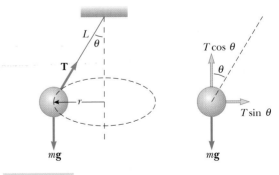

**Figure 5.11**

(Example 5.6) The conical pendulum and its free-body diagram.

* R. P. Feynman, R. B. Leighton, M. Sands, *The Feynman Lectures on Physics*, Vol. 1, Reading, Massachusetts, Addison-Wesley Publishing Co., 1963, p. 7-2.

tical direction, we model it as a particle in equilibrium in the vertical direction:

$$(1) \quad \sum F_y = 0 \quad \longrightarrow \quad T\cos\theta - mg = 0$$
$$\longrightarrow \quad T\cos\theta = mg$$

In the horizontal direction, we have a centripetal acceleration. Because the force that provides the centripetal acceleration in this example is the component $T\sin\theta$, from Newton's second law we have

$$(2) \quad \sum F_r = ma_c \quad \longrightarrow \quad T\sin\theta = m\frac{v^2}{r}$$

By dividing (2) by (1), we eliminate $T$ and find that

$$\tan\theta = \frac{v^2}{rg} \quad \longrightarrow \quad v = \sqrt{rg\tan\theta}$$

From a triangle we can construct in the pictorial representation in Figure 5.11a, we note that $r = L\sin\theta$; therefore,

$$v = \sqrt{Lg\sin\theta\tan\theta}$$

(b) The object is traveling at constant speed around its circular path. Because the object travels a distance of $2\pi r$ (the circumference of the circular path) in a time equal to the period of revolution $T$ (not to be confused with the magnitude $T$ of the tension force), we find

$$(3) \quad T = \frac{2\pi r}{v} = \frac{2\pi r}{\sqrt{rg\tan\theta}} = 2\pi\sqrt{\frac{L\cos\theta}{g}}$$

The intermediate algebraic steps used in obtaining (3) are left to the reader. Note that the period is independent of the mass $m$!

## Example 5.7    What Is the Maximum Speed of the Car?

A 1 500-kg car moving on a flat, horizontal road negotiates a curve whose radius is 35.0 m (Fig. 5.12a). If the coefficient of static friction between the tires and the dry pavement is 0.500, find the maximum speed the car can have in order to make the turn successfully.

**Solution**   The car is an extended object with four friction forces acting on it, one on each wheel, but we shall model it as a particle with only one net friction force. Figure 5.12b shows a free-body diagram for the car. From Newton's second law in the horizontal direction, we have

$$(1) \quad \sum F_x = ma \quad \longrightarrow \quad f_s = m\frac{v^2}{r}$$

The maximum speed that the car can have around the curve corresponds to the speed at which it is on the verge of skidding toward the side of the road. At this point, the friction force has its maximum value

$$f_{s,max} = \mu_s n$$

In the vertical direction, no acceleration occurs, so

$$\sum F_y = 0 \quad \longrightarrow \quad n - mg = 0$$

Thus, the magnitude of the normal force equals the weight in this case, and we find

$$f_{s,max} = \mu_s mg$$

(a)

(b)

**Figure 5.12**

(Example 5.7) (a) The force of static friction directed toward the center of the curve keeps the car moving in a circular path. (b) Free-body diagram for the car.

Substituting this expression into (1), we find the maximum speed:

$$\mu_s mg = m\frac{v_{max}^2}{r} \longrightarrow v_{max} = \sqrt{\mu_s gr}$$

Substituting the numerical values,

$$v_{max} = \sqrt{(0.500)(9.80 \text{ m/s}^2)(35.0 \text{ m})} = \boxed{13.1 \text{ m/s}}$$

This is equivalent to 29.3 mi/h, which is less than a typical nonfreeway speed of 35 mi/h. Thus, this roadway could benefit greatly from some banking!

**EXERCISE** On a wet day, the car described in this example begins to skid on the curve when its speed reaches 8.00 m/s. What is the coefficient of static friction in this case?

*Answer* 0.187

## Example 5.8 Let's Go Loop-the-Loop

A pilot of mass $m$ in a jet aircraft executes a "loop-the-loop" maneuver as illustrated in Figure 5.13a. The aircraft moves in a vertical circle of radius 2.70 km at a *constant speed* of 225 m/s. Determine the force exerted by the seat on the pilot at (a) the bottom of the loop and (b) the top of the loop. Express the answers in terms of the weight $mg$ of the pilot.

**Reasoning** This is the first numerical example in which the force causing the centripetal acceleration is a *combination* of forces rather than a single force. We shall model the pilot as a particle under a net force and analyze the situation at the bottom and top of the circular path.

**Solution** (a) The free-body diagram for the pilot at the bottom of the loop is shown in Figure 5.13b. The forces acting on the pilot are the downward gravitational force $m\mathbf{g}$ and the upward normal force $\mathbf{n}_{bot}$ exerted by the seat on the pilot. Because the net upward force at the bottom that provides the centripetal acceleration has a magnitude $n_{bot} - mg$,

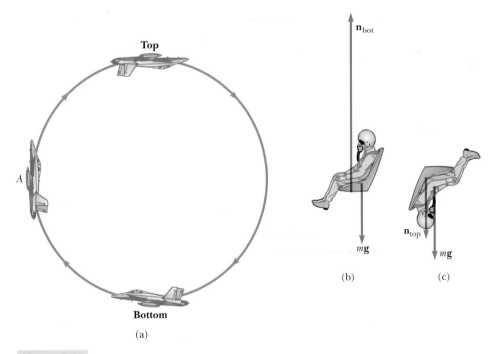

**Figure 5.13**

(Example 5.8) (a) An aircraft executes a loop-the-loop maneuver as it moves in a vertical circle at constant speed. (b) Free-body diagram for the pilot at the bottom of the loop. In this position, the pilot experiences a force from the seat that is larger than his weight. (c) Free-body diagram for the pilot at the top of the loop. Here the force from the seat is smaller than his weight.

Newton's second law for the radial (upward) direction gives

$$\Sigma F_y = ma \longrightarrow n_{bot} - mg = m\frac{v^2}{r}$$

$$n_{bot} = mg + m\frac{v^2}{r} = mg\left[1 + \frac{v^2}{rg}\right]$$

Substituting the values given for the speed and radius gives

$$n_{bot} = mg\left[1 + \frac{(225 \text{ m/s})^2}{(2.70 \times 10^3 \text{ m})(9.80 \text{ m/s}^2)}\right] = \boxed{2.91\, mg}$$

Hence, the force exerted by the seat on the pilot at the bottom of the loop is *greater* than his weight by a factor of 2.91.

(b) The free-body diagram for the pilot at the top of the loop is shown in Figure 5.13c. At this point, both the gravitational force and the force $\mathbf{n}_{top}$ exerted by the seat on the pilot act *downward*, so the net force downward that provides the centripetal acceleration has a magnitude $n_{top} + mg$. Applying Newton's second law gives

$$\Sigma F_y = ma \longrightarrow n_{top} + mg = m\frac{v^2}{r}$$

$$n_{top} = m\frac{v^2}{r} - mg = mg\left[\frac{v^2}{rg} - 1\right]$$

$$n_{top} = mg\left[\frac{(225 \text{ m/s})^2}{(2.70 \times 10^3 \text{ m})(9.80 \text{ m/s}^2)} - 1\right] = \boxed{0.911\, mg}$$

In this case, the force exerted by the seat on the pilot is *less* than the weight by a factor of 0.911. Hence, the pilot feels lighter at the top of the loop.

**EXERCISE**   Calculate the force on the pilot causing the centripetal acceleration if the aircraft is at point *A* in Figure 5.13a, midway up the loop.

*Answer*   $1.91\, mg$ directed toward the right.

**EXERCISE**   An automobile moves at constant speed over the crest of a hill. The driver moves in a vertical circle of radius 18.0 m. At the top of the hill, she notices that she barely remains in contact with the seat. Find the speed of the vehicle.

*Answer*   13.3 m/s

## 5.3 • NONUNIFORM CIRCULAR MOTION

In Chapter 3 we found that if a particle moves with varying speed in a circular path, there is, in addition to the radial component of acceleration, a tangential component of magnitude $dv/dt$. Therefore, the net force acting on the particle must also have a radial and a tangential component as shown in Figure 5.14. That is, because the total acceleration is $\mathbf{a} = \mathbf{a}_r + \mathbf{a}_t$, the total force exerted on the particle is $\Sigma\mathbf{F} = \Sigma\mathbf{F}_r + \Sigma\mathbf{F}_t$. The component vector $\Sigma\mathbf{F}_r$ is directed toward the center of the circle and is responsible for the centripetal acceleration. The component vector $\Sigma\mathbf{F}_t$ tangent to the circle is responsible for the tangential acceleration, which causes the speed of the particle to change with time.

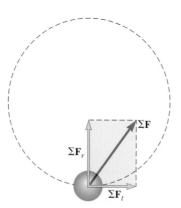

**Figure 5.14**

When the net force acting on a particle moving in a circular path has a tangential component vector $\Sigma\mathbf{F}_t$, its speed changes. The total force on the particle also has a component vector $\Sigma\mathbf{F}_r$ directed toward the center of the circular path. Thus, the total force is $\Sigma\mathbf{F} = \Sigma\mathbf{F}_t + \Sigma\mathbf{F}_r$.

## Example 5.9 Follow the Rotating Ball

A small sphere of mass $m$ is attached to the end of a cord of length $R$, which rotates under the influence of the gravitational force in a *vertical* circle about a fixed point $O$, as in Figure 5.15a. Let us determine the tension in the cord at any instant when the speed of the sphere is $v$ and the cord makes an angle $\theta$ with the vertical.

**Reasoning** First note that the speed is *not* uniform because a tangential component of acceleration arises from the gravitational force on the sphere. Thus, although this example is similar to Example 5.8, it is not identical. From the free-body diagram in Figure 5.15a, we see that the only forces acting on the sphere are the gravitational force $m\mathbf{g}$ and the force $\mathbf{T}$ exerted by the cord.

**Solution** We resolve $m\mathbf{g}$ into a tangential component $mg \sin \theta$ and a radial component $mg \cos \theta$. Applying Newton's second law for the tangential direction gives

$$\sum F_t = ma_t \longrightarrow mg \sin \theta = ma_t$$

$$a_t = g \sin \theta$$

This component causes $v$ to change in time, because $a_t = dv/dt$.

Applying Newton's second law to the forces in the radial direction (for which the outward direction is positive), we find

$$\sum F_r = ma_r \longrightarrow mg \cos \theta - T = -m\frac{v^2}{r}$$

$$T = m\left(\frac{v^2}{R} + g \cos \theta\right)$$

At the bottom of the path, where $\theta = 0$, we see that because $\cos 0 = 1$,

$$T_{\text{bot}} = m\left(\frac{v_{\text{bot}}^2}{r} + g\right)$$

This is the *maximum* value of $T$—as the sphere passes through the bottom point, the string is under the most tension. This is of interest to trapeze artists, because their support wires must withstand this largest tension at the bottom of the swing, as well as to Tarzan, in choosing a nice strong vine on which to swing to withstand this force.

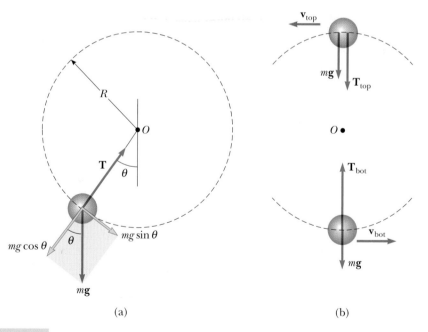

(a)                                    (b)

### Figure 5.15

(Example 5.9) (a) Forces acting on a sphere of mass $m$ connected to a cord of length $R$ and rotating in a vertical circle centered at $O$. (b) Forces acting on the sphere when it is at the top and bottom of the circle. The tension has its maximum value at the bottom and its minimum value at the top.

## 5.4 • MOTION IN THE PRESENCE OF VELOCITY-DEPENDENT RESISTIVE FORCES

See Screen 4.9

Earlier we described the friction force between a moving object and the surface along which it moves. So far, we have ignored any interaction between the object and the *medium* through which it moves. Now let us consider the effect of a medium such as a liquid or gas. The medium exerts a **resistive force R** on the object moving through it. You feel this force if you ride in a car at high speed with your hand out the window—the force you feel pushing your hand backward is the resistive force of the air rushing past the car. The magnitude of this force depends on the relative speed between the object and the medium, and the direction of **R** on the object is always opposite the direction of motion of the object relative to the medium. Some examples are the air resistance associated with moving vehicles (sometimes called air drag), the force of the wind on the sails of a sailboat, and the viscous forces that act on objects sinking through a liquid.

Generally, the magnitude of the resistive force increases with increasing speed. The resistive force can have a complicated speed dependence. In the following discussions, we consider two simplification models that allow us to analyze these situations. The first model assumes that the resistive force is proportional to the velocity; this is approximately the case for objects that fall through a liquid with low speed and for very small objects, such as dust particles, that move through air. The second model treats situations for which we assume that the magnitude of the resistive force is proportional to the square of the speed of the object. Large objects, such as a skydiver moving through air in free-fall, experience such a force.

### Model 1: Resistive Force Proportional to Object Velocity

At low speeds, the resistive force acting on an object that is moving through a viscous medium is effectively modeled as being proportional to the object's velocity. The mathematical representation of the resistive force can be expressed as

$$\mathbf{R} = -b\mathbf{v} \qquad [5.4]$$

where **v** is the velocity of the object and $b$ is a constant that depends on the properties of the medium and on the shape and dimensions of the object. If the object is a sphere of radius $r$, $b$ is proportional to $r$. The negative sign represents the fact that the resistive force is opposite the velocity.

Consider a sphere of mass $m$ released from rest in a liquid, as in Figure 5.16a. Assuming the only forces acting on the sphere are the resistive force **R** and the weight $m\mathbf{g}$ we can describe its motion using Newton's second law.* Considering the vertical motion and choosing the downward direction to be positive, we have

$$\sum F_y = ma_y \quad \longrightarrow \quad mg - bv = m\frac{dv}{dt}$$

Dividing this equation by the mass $m$ gives

$$\frac{dv}{dt} = g - \frac{b}{m}v \qquad [5.5]$$

(a)

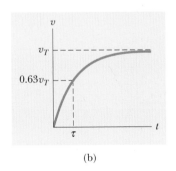

(b)

**Figure 5.16**

(a) A small sphere falling through a viscous fluid. (b) The speed–time graph for an object falling through a viscous medium. The object reaches a terminal speed $v_T$, and the time constant $\tau$ is the time it takes to reach $0.63v_T$.

* A *buoyant* force is also acting on the submerged object. This force is constant and equal to the weight of the displaced fluid, as will be discussed in Chapter 12. The effect of this force can be modeled by changing the apparent weight of the sphere by a constant factor, so we can ignore it here.

Equation 5.5 is called a *differential equation*—it includes both the speed $v$ and the derivative of the speed. The methods of solving such an equation may not be familiar to you as yet. However, note that if we define $t = 0$ when $v = 0$, the resistive force is zero at this time and the acceleration $dv/dt$ is simply $g$. As $t$ increases, the speed increases, the resistive force increases, and the acceleration decreases. Thus, this problem is one in which both the velocity and acceleration of the particle are not constant.

The acceleration becomes zero when the increasing resistive force eventually balances the weight. At this point, the object reaches its **terminal speed** $v_T$ and from then on it continues to move with zero acceleration. After this point, the motion is that of a particle under constant velocity. The terminal speed can be obtained from Equation 5.5 by setting $a = dv/dt = 0$. This gives

•  *Terminal speed*

$$mg - bv_T = 0 \longrightarrow \boxed{v_T = \frac{mg}{b}} \quad \text{terminal velocity}$$

The expression for $v$ that satisfies Equation 5.5 with $v = 0$ at $t = 0$ is

$$v = \frac{mg}{b}(1 - e^{-bt/m}) = v_T(1 - e^{-t/\tau}) \qquad [5.6]$$

where $e = 2.71828$ is the base of the natural logarithm. This expression for $v$ can be verified by substituting this solution back into Equation 5.5. (Try it!) This function is plotted in Figure 5.16b.

The mathematical representation of the motion (Equation 5.6) indicates that the terminal speed is never reached, because the exponential function never becomes exactly equal to zero. For all practical purposes, however, when the exponential function is very small, the speed of the particle can be approximated as being constant.

We cannot compare different objects by means of the time to reach terminal speed, because, as we have just discussed, this time is infinite for all objects! We need some means to compare these exponential behaviors for different objects. We do this with a parameter called the **time constant.** The time constant $\tau = m/b$ that appears in Equation 5.6 is the time it takes for the factor in parentheses in Equation 5.6 to become equal to $1 - e^{-1} = 0.632$. Thus, the time constant represents the time required for the object to reach 63.2% of its terminal speed (Fig. 5.16b).

## Example 5.10  A Sphere Falling in Oil

A small sphere of mass 2.00 g is released from rest in a large vessel filled with oil. The sphere approaches a terminal speed of 5.00 cm/s. Determine (a) the time constant $\tau$ and (b) the time it takes the sphere to reach 90.0% of its terminal speed.

**Solution**  (a) Because the terminal speed is given by $v_T = mg/b$, the coefficient $b$ is

$$b = \frac{mg}{v_T} = \frac{(2.00 \times 10^{-3} \text{ kg})(9.80 \text{ m/s}^2)}{5.00 \times 10^{-2} \text{ m/s}} = 0.392 \text{ N·s/m}$$

Therefore, the time constant $\tau$ is

$$\tau = \frac{m}{b} = \frac{2.00 \times 10^{-3} \text{ kg}}{0.392 \text{ N·s/m}} = 5.10 \times 10^{-3} \text{ s}$$

(b) The speed of the sphere as a function of time is given by Equation 5.6. To find the time $t$ at which the sphere is traveling at a speed of $0.900v_T$, we set $v = 0.900v_T$, substitute this into Equation 5.6, and solve for $t$:

$$0.900v_T = v_T(1 - e^{-t/\tau})$$

$$1 - e^{-t/\tau} = 0.900$$

$$e^{-t/\tau} = 0.100$$

$$-\frac{t}{\tau} = \ln 0.100 = -2.30$$

$$t = 2.30\tau = 2.30(5.10 \times 10^{-3}\ \mathrm{s})$$

$$= 11.7 \times 10^{-3}\ \mathrm{s} = \boxed{11.7\ \mathrm{ms}}$$

**EXERCISE**   What is the sphere's speed through the oil at $t = 11.7$ ms? Compare this value with the speed the sphere would have if it were falling in a vacuum and so influenced only by gravity.

*Answer*   4.50 cm/s in oil versus 11.5 cm/s in free-fall

## Model 2: Resistive Force Proportional to Object Speed Squared

For large objects moving at high speeds through air, such as airplanes, skydivers, and baseballs, the magnitude of the resistive force is modeled as being proportional to the square of the speed:

$$R = \tfrac{1}{2}D\rho Av^2 \qquad [5.7]$$

where $\rho$ is the density of air, $A$ is the cross-sectional area of the moving object measured in a plane perpendicular to its velocity, and $D$ is a dimensionless empirical quantity called the *drag coefficient*. The drag coefficient has a value of about 0.5 for spherical objects moving through air but can be as high as 2 for irregularly shaped objects.

Consider an airplane in flight that experiences such a resistive force. Equation 5.7 shows that the force is proportional to the density of air and hence decreases with decreasing air density. Because air density decreases with increasing altitude, the resistive force on a jet airplane flying at a given speed will decrease with increasing altitude. Thus, airplanes tend to fly at very high altitudes to take advantage of this reduced resistive force. This allows them to fly faster for a given engine thrust. Of course, this higher speed *increases* the resistive force, in proportion to the square of the speed, so a balance is struck between flying at higher altitude and higher speed.

Now let us analyze the motion of a falling object subject to an upward air resistive force whose magnitude is given by Equation 5.7. Suppose an object of mass $m$ is released from rest, as in Figure 5.17, from the position $y = 0$. The object experiences two external forces: the downward gravitational force $m\mathbf{g}$ and the upward resistive force $\mathbf{R}$. (There is also an upward buoyant force that we ignore, due to its small value.) Hence, using Newton's second law,

$$\sum F = ma \quad \longrightarrow \quad mg - \tfrac{1}{2}D\rho Av^2 = ma \qquad [5.8]$$

Solving for $a$, we find that the object has a downward acceleration of magnitude

$$a = g - \left(\frac{D\rho A}{2m}\right)v^2 \qquad [5.9]$$

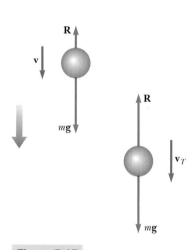

**Figure 5.17**

An object falling through air experiences a resistive drag force $\mathbf{R}$ and a gravitational force $\mathbf{F}_g = m\mathbf{g}$. The object reaches terminal speed (on the right) when the net force acting on it is zero, that is, when $\mathbf{R} = -\mathbf{F}_g$, or $R = mg$. Before this occurs, the acceleration varies with speed according to Equation 5.9.

Aerodynamic car. A streamlined body reduces air drag and increases fuel efficiency. *(Courtesy of Ford Motor Company)*

| TABLE 5.2 | Terminal Speeds for Various Objects Falling Through Air | | |
|---|---|---|---|
| Object | Mass (kg) | Cross-sectional Area (m$^2$) | $v_T$ (m/s)$^a$ |
| Skydiver | 75 | 0.70 | 60 |
| Baseball (radius 3.7 cm) | 0.145 | $4.2 \times 10^{-3}$ | 33 |
| Golf ball (radius 2.1 cm) | 0.046 | $1.4 \times 10^{-3}$ | 32 |
| Hailstone (radius 0.50 cm) | $4.8 \times 10^{-4}$ | $7.9 \times 10^{-5}$ | 14 |
| Raindrop (radius 0.20 cm) | $3.4 \times 10^{-5}$ | $1.3 \times 10^{-5}$ | 9.0 |

$^a$ The drag coefficient $D$ is assumed to be 0.5 in each case.

Because $a = dv/dt$, this is another differential equation that provides us with the speed as a function of time.

Again, we can calculate the terminal speed $v_T$ using the fact that when the gravitational force is balanced by the resistive force, the net force is zero and therefore the acceleration is zero. Setting $a = 0$ in Equation 5.9 gives

$$g - \left( \frac{D\rho A}{2m} \right) v_T^2 = 0$$

$$v_T = \sqrt{\frac{2mg}{D\rho A}} \qquad \qquad \textbf{[5.10]}$$

Table 5.2 lists the terminal speeds for several objects falling through air, all computed on the assumption that the drag coefficient is 0.5.

### Quick Quiz 5.6

A baseball and a basketball, having the same mass, are dropped through air from rest from the same height above the ground. Which ball strikes the ground first?

### Quick Quiz 5.7

Consider a sky surfer falling through air, as in Figure 5.18, before reaching her terminal speed. As the speed of the sky surfer increases, what happens to her acceleration? What is her acceleration once she reaches terminal speed?

## 5.5 • NUMERICAL REPRESENTATIONS OF PARTICLE DYNAMICS*

So far, our mathematical representations for physical situations have involved analytical functions that could be represented as continuous curves on graph paper. This is due to the fact that we have adopted simplification models that have allowed the mathematical equations to be solved relatively simply. In the "real world," however, complications often arise that make analytical solutions for many practical situations difficult and perhaps beyond the mathematical abilities of most students taking introductory physics. For example, the force may depend on the

* The authors are most grateful to Colonel James Head of the U.S. Air Force Academy for preparing this section.

**Figure 5.18**

(Quick Quiz 5.7) A sky surfer takes advantage of the upward force of the air on her board. (*Jump Run Productions/Image Bank*) position of the particle, as in cases where the variation in the gravitational force with height must be taken into account. The force may also vary with speed, as we have seen in cases of resistive forces caused by motion through a liquid or gas. The force may depend on both position and speed, as in the case of an object falling through air where the resistive force depends on speed and on height (air density). In rocket motion, the mass changes with time, so even if the force is constant, the acceleration is not.

Another complication arises because the equations relating acceleration, velocity, position, and time are differential, not algebraic, equations. Differential equations are usually solved using integral calculus and other special techniques that you may not have mastered.

So how does one proceed to describe the motion of a particle under these complicated conditions? One answer is to solve such problems on personal computers, using elementary numerical methods. We describe one such technique in this section, which introduces us to the notion of a **numerical representation** of a problem. In a numerical representation, the calculations reduce to simple mathematical steps involving only addition, subtraction, multiplication, and division. Using the numerical representation, the position of a particle can be estimated step by step at many closely spaced instants. Of course, the cost of using this approach is that the calculations must be performed a very large number of times, which can be done with computers.

As we have seen in this and the preceding chapter, the study of dynamics of a particle focuses on describing the position, velocity, and acceleration due to a net force as functions of time. Cause-and-effect relationships exist between these quantities: A net force causes an acceleration, an acceleration causes velocity to change, a velocity causes position to change. Therefore, the study of motion usually begins with an evaluation of the net force on the particle.

In this section, we confine our discussion to one-dimensional motion, so boldface notation is not used for vector quantities. If a particle of mass $m$ moves under the influence of a net force $\Sigma F$, Newton's second law tells us that the acceleration of the particle is given by $a = \Sigma F/m$. In general, we can then obtain the analytical solution to a dynamics problem (the position of the particle at any time) by using the following procedure:

- Sum all the forces on the particle to obtain the net force $\Sigma F$.
- Use this force to determine the acceleration, using $a = \Sigma F/m$.
- Use this acceleration to determine the velocity from $dv/dt = a$.
- Use this velocity to determine the position from $dx/dt = v$.

This procedure is straightforward for some physical situations, such as ones described by the model of a particle under constant acceleration. Let us now consider how we might solve a problem that is not so straightforward. Suppose, for example, that the acceleration is so complicated that we cannot easily integrate the acceleration function to find the velocity function. One approach to a numerical representation is the Euler method, named after the Swiss mathematician Leonhard Euler (1707–1783).

### The Euler Method

In the **Euler method** of solving differential equations, derivatives are approximated as ratios of finite differences. Let us assume that we have determined the net force $\Sigma F(x, v, t)$ on the particle. The acceleration of the particle is determined from Newton's second law,

$$a(x, v, t) = \frac{\Sigma F(x, v, t)}{m} \qquad \text{[5.11]}$$

Notice that the net force, and therefore the acceleration, may depend explicitly on the position, velocity, or time. This is what makes the problem difficult to solve analytically.

Considering a small increment of time $\Delta t$, the relationship between velocity and acceleration may be approximated as

$$a(t) \approx \frac{\Delta v}{\Delta t} = \frac{v(t + \Delta t) - v(t)}{\Delta t}$$

Rearranging this equation, we find that the velocity of the particle at the end of the period $\Delta t$ is approximately equal to the velocity at the beginning of the period plus the acceleration during the interval multiplied by $\Delta t$:

$$v(t + \Delta t) \approx v(t) + a(t)\,\Delta t \qquad \text{[5.12]}$$

Because the acceleration is a function of time, this estimate of $v(t + \Delta t)$ will be accurate only if the time interval $\Delta t$ is short enough that the change in acceleration during this interval is very small. Thus, because the time interval we are considering is very short, we use the model of a particle under constant acceleration for this calculation. We assume that, despite the fact that the acceleration is not constant, it does not change by very much in a very short time interval.

The position can be found in the same manner:

$$v(t) \approx \frac{\Delta x}{\Delta t} = \frac{x(t + \Delta t) - x(t)}{\Delta t}$$

$$x(t + \Delta t) \approx x(t) + v(t)\,\Delta t \qquad \text{[5.13]}$$

Here, we assume the velocity to be constant, because it does not change by very much in a very short time interval. It may be tempting to add the term $\frac{1}{2}a(\Delta t)^2$ to this result to make it look like the familiar kinematics equation, but this term is not included in the Euler method of integration because $\Delta t$ is assumed to be so small that $(\Delta t)^2$ is nearly zero.

If the acceleration at any instant $t$ is known, the particle's velocity and position at $(t + \Delta t)$ can be calculated from Equations 5.12 and 5.13. The calculation can then proceed in a series of finite steps to determine the velocity and position at any later time. At the end of each time interval, the new value of the force is determined from the functional expression for the force, Equation 5.11 is used to find the new acceleration, Equation 5.12 is used to find the new velocity, and Equation 5.13 is used to find the new position.

It is convenient to set up the numerical solution to this kind of problem by numbering the steps and entering the calculations in a tabular representation. Table 5.3 illustrates how to do this in an orderly way. Many small increments can be taken, and accurate results can usually be obtained with the help of a computer. The equations provided in the table can be entered into a spreadsheet and the calculations performed row by row to determine the velocity, position, and acceleration as functions of time. The calculations can also be carried out using a programming language or with commercially available mathematics packages for personal computers. Graphical representations of velocity versus time or position versus time can be displayed to help you visualize the motion.

| TABLE 5.3 | | The Euler Method for Solving Dynamics Problems | | |
|---|---|---|---|---|
| **Step** | **Time** | **Position** | **Velocity** | **Acceleration** |
| 0 | $t_0$ | $x_0$ | $v_0$ | $a_0 = F(x_0, v_0, t_0)/m$ |
| 1 | $t_1 = t_0 + \Delta t$ | $x_1 = x_0 + v_0\,\Delta t$ | $v_1 = v_0 + a_0\,\Delta t$ | $a_1 = F(x_1, v_1, t_1)/m$ |
| 2 | $t_2 = t_1 + \Delta t$ | $x_2 = x_1 + v_1\,\Delta t$ | $v_2 = v_1 + a_1\,\Delta t$ | $a_2 = F(x_2, v_2, t_2)/m$ |
| 3 | $t_3 = t_2 + \Delta t$ | $x_3 = x_2 + v_2\,\Delta t$ | $v_3 = v_2 + a_2\,\Delta t$ | $a_3 = F(x_3, v_3, t_3)/m$ |
| . | . | . | . | . |
| . | . | . | . | . |
| . | . | . | . | . |
| $n$ | $t_n = t_{n-1} + \Delta t$ | $x_n = x_{n-1} + v_{n-1}\,\Delta t$ | $v_n = v_{n-1} + a_{n-1}\,\Delta t$ | $a_n = F(x_n, v_n, t_n)/m$ |

The Euler method is completely reliable for infinitesimally small time increments, but for practical reasons a finite increment size must be chosen. For the finite difference approximation of Equation 5.11 to be valid, the time increment must be small enough that the acceleration can be approximated as being constant during the increment. We can determine an appropriate size for the time increment by examining the particular problem that is being investigated. The criterion for the size of the time increment may need to be changed during the course of the motion. In practice, however, we usually choose a time increment appropriate to the initial conditions and use the same value throughout the calculations.

The size of the time increment influences the accuracy of the result, but unfortunately it is not easy to determine the accuracy of a solution by the Euler method without a knowledge of the correct analytical solution. One method of determining the accuracy of the numerical solution is to repeat the calculations with a smaller time increment and compare results. If the two calculations agree to a certain number of significant figures, you can assume that the results are correct to that precision. The solution to Problem 27 at the end of the chapter, in the *Student Solutions Manual and Study Guide* or on the publisher's Web site, illustrates the use of the Euler method.

## 5.6 • THE FUNDAMENTAL FORCES OF NATURE

We have described a variety of forces experienced in our everyday activities, such as the gravitational force that acts on all objects at or near the Earth's surface and the force of friction as one surface slides over another. Newton's second law tells us how to relate the forces to the acceleration of the object or particle.

In addition to these familiar macroscopic forces in nature, forces also act in the atomic and subatomic world. For example, atomic forces within the atom are responsible for holding its constituents together, and nuclear forces act on different parts of the nucleus to keep its parts from separating.

Until recently, physicists believed that there were four fundamental forces in nature: the gravitational force, the electromagnetic force, the nuclear force, and the weak force. We shall discuss these individually and then consider the current view of fundamental forces.

### The Gravitational Force

The **gravitational force** is the mutual force of attraction between any two objects in the Universe. It is interesting and rather curious that, although the gravitational force can be very strong between macroscopic objects, it is inherently the weakest of all the fundamental forces. For example, the gravitational force between the electron and proton in the hydrogen atom has a magnitude on the order of $10^{-47}$ N, whereas the electromagnetic force between these same two particles is on the order of $10^{-7}$ N.

In addition to his contributions to the understanding of motion, Newton studied gravity extensively. **Newton's law of universal gravitation** states that every particle in the Universe attracts every other particle with a force that is directly proportional to the product of the masses of the particles and inversely proportional to the square of the distance between them. If the particles have masses $m_1$ and $m_2$ and are separated by a distance $r$, as in Figure 5.19, the magnitude of the gravitational force is

$$F_g = G \frac{m_1 m_2}{r^2} \qquad [5.14]$$

where $G = 6.67 \times 10^{-11}$ N·m$^2$/kg$^2$ is the **universal gravitational constant.** More detail on the gravitational force will be provided in Chapter 11.

## The Electromagnetic Force

The **electromagnetic force** is the force that binds atoms and molecules in compounds to form ordinary matter. It is much stronger than the gravitational force. The force that causes a rubbed comb to attract bits of paper and the force that a magnet exerts on an iron nail are electromagnetic forces. Essentially all forces at work in our macroscopic world, apart from the gravitational force, are manifestations of the electromagnetic force. For example, friction forces, contact forces, tension forces, and forces in elongated springs are consequences of electromagnetic forces between charged particles in proximity.

The electromagnetic force involves two types of particles: those with positive charge and those with negative charge. (More information on these two types of charge is provided in Chapter 19.) Unlike the gravitational force, which is always an attractive interaction, the electromagnetic force can be either attractive or repulsive, depending on the charges on the particles.

**Coulomb's law** expresses the magnitude of the electrostatic force* $F_e$ between two charged particles separated by a distance $r$:

$$F_e = k_e \frac{q_1 q_2}{r^2} \qquad [5.15]$$

where $q_1$ and $q_2$ are the charges on the two particles, measured in units called *coulombs* (C), and $k_e$ ($= 8.99 \times 10^9$ N·m$^2$/C$^2$) is the Coulomb constant. Note that the electrostatic force has the same mathematical form as Newton's law of universal gravitation (see Eq. 5.14), with charge playing the mathematical role of mass, and the Coulomb constant being used in place of the universal gravitational constant. The electrostatic force is attractive if the two charges have opposite signs, and repulsive if the two charges have the same sign, as indicated in Figure 5.20.

The smallest amount of isolated charge found in nature (so far) is the charge on an electron or proton. This fundamental unit of charge is given the symbol $e$

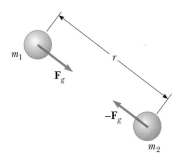

**Figure 5.19**

Two particles with masses $m_1$ and $m_2$ attract each other with a force of magnitude $Gm_1 m_2/r^2$.

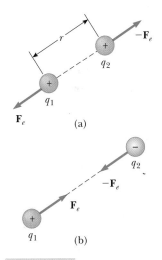

**Figure 5.20**

Two point charges separated by a distance $r$ exert an electrostatic force on each other given by Coulomb's law. (a) When the charges are of the same sign, the charges repel each other. (b) When the charges are of opposite sign, the charges attract each other.

**PITFALL PREVENTION 5.8**
**The symbol *e***

Take care not to confuse the symbol $e$ for the fundamental electric charge with that for the base of the natural logarithm $e$. Both are represented by exactly the same symbol.

---

\* The electrostatic force is the electromagnetic force between two electric charges that are at rest. If the charges are moving, magnetic forces are also present; these will be studied in Chapter 22.

and has the magnitude $e = 1.60 \times 10^{-19}$ C. Theories developed in the latter half of the 20th century propose that protons and neutrons are made up of smaller particles called **quarks,** which have charges of either $\frac{2}{3}e$ or $-\frac{1}{3}e$ (discussed further in Chapter 31). Although experimental evidence has been found for such particles inside nuclear matter, free quarks have never been detected.

### The Nuclear Force

An atom, as we currently model it, consists of an extremely dense positively charged nucleus surrounded by a cloud of negatively charged electrons, with the electrons attracted to the nucleus by the electric force. Because all nuclei except those of hydrogen are combinations of positively charged protons and neutral neutrons (collectively called nucleons), why does the repulsive electrostatic force between the protons not cause nuclei to break apart? Clearly, there must be an attractive force that counteracts the strong electrostatic repulsive force and is responsible for the stability of nuclei. This force that binds the nucleons to form a nucleus is called the **nuclear force.** Unlike the gravitational and electromagnetic forces, which depend on distance in an inverse-square fashion, the nuclear force is extremely short range; its strength decreases very rapidly outside the nucleus and is negligible for separations greater than approximately $10^{-14}$ m. For separations of approximately $10^{-15}$ m (a typical nuclear dimension), the nuclear force is about two orders of magnitude stronger than the electrostatic force.

- *The nuclear force*

### The Weak Force

- *The weak force*

The **weak force** is a short-range force that tends to produce instability in certain nuclei. It was first observed in naturally occurring radioactive substances and was later found to play a key role in most radioactive decay reactions. The weak force is about $10^{25}$ times stronger than the gravitational force and about $10^{12}$ times weaker than the electromagnetic force.

### The Current View of Fundamental Forces

For years physicists have searched for a simplification scheme that would reduce the number of fundamental forces in nature. In 1967 physicists predicted that the electromagnetic force and the weak force, originally thought to be independent of each other and both fundamental, are in fact manifestations of one force, now called the **electroweak** force. The prediction was confirmed experimentally in 1984. We shall discuss this more fully in Chapter 31.

We also now know that protons and neutrons are not fundamental particles; current models of protons and neutrons theorize that they are composed of simpler particles called quarks, as mentioned previously. The quark model has led to a modification of our understanding of the nuclear force. Scientists now define a **strong force** as that force which binds the quarks to one another in a nucleon (proton or neutron). This force is also referred to as a **color force,** in reference to a property of quarks called "color," which we shall investigate in Chapter 31. The previously defined nuclear force, that which acts between nucleons, is now interpreted as a secondary effect of the strong force between the quarks.

Scientists believe that the fundamental forces of nature are closely related to the origin of the Universe. The Big Bang theory states that the Universe began with a cataclysmic explosion 15 to 20 billion years ago. According to this theory,

**PITFALL PREVENTION 5.9**

**The strong force**

Historically, before the quark model was proposed, the force between nucleons was called the strong force. Now that the quark model is well accepted, the force between quarks is identified as the strong force. Thus, as you read other books, you may encounter the phrase "strong force" as meaning either the force between nucleons or the force between quarks. In this book, we use the phrase "nuclear force" to mean the force between nucleons and "strong force" to be the force between quarks.

the first moments after the Big Bang saw such extremes of energy that all the fundamental forces were unified into one force. Physicists are continuing their search for connections among the known fundamental forces—connections that could eventually prove that the forces are all merely different forms of a single superforce. This fascinating search continues to be at the forefront of physics.

## 5.7 • THE GRAVITATIONAL FIELD

When Newton first published his theory of gravitation, his contemporaries found it difficult to accept the concept of a force that one object could exert on another without anything happening in the space between them. How was it possible for two objects with mass to interact even though they were not in contact with each other? Although Newton himself could not answer this question, his theory was considered a success because it satisfactorily explained the motions of the planets.

An alternative mental representation of the gravitational force is to think of the gravitational interaction as a two-step process involving a *field* as discussed in Section 4.1. First, one object (a *source mass*) creates a **gravitational field g** throughout the space around it. Then a second object (a *test mass*) of mass $m$ residing in this field experiences a force $\mathbf{F}_g = m\mathbf{g}$. In other words, we model the *field* as exerting a force on the test mass rather than the source mass exerting the force directly. The gravitational field is defined by

$$\mathbf{g} \equiv \frac{\mathbf{F}_g}{m}$$

[5.16]  • *Definition of gravitational field*

That is, the gravitational field at a point in space equals the gravitational force that a test mass $m$ experiences at that point divided by the mass. Consequently, if $\mathbf{g}$ is known at some point in space, a particle of mass $m$ experiences a gravitational force $\mathbf{F}_g = m\mathbf{g}$ when placed at that point. We shall also use the model of a particle in a field for electricity and magnetism in later chapters, where it plays a much larger role than it does for gravity.

As an example, consider an object of mass $m$ near the Earth's surface. The gravitational force on the object is directed toward the center of the Earth and has a magnitude $mg$. Thus, the gravitational field experienced by the object at some point has a magnitude equal to the free-fall acceleration at that point. Because the gravitational force on the object has a magnitude $GM_E m/r^2$ (where $M_E$ is the mass of the Earth), the field $\mathbf{g}$ at a distance $r$ from the center of the Earth is

$$\mathbf{g} = \frac{\mathbf{F}_g}{m} = -\frac{GM_E}{r^2}\hat{\mathbf{r}}$$

[5.17]  • *Gravitational field of the Earth*

where $\hat{\mathbf{r}}$ is a unit vector pointing radially outward from the Earth, and the negative sign indicates that the field vector points toward the center of the Earth, as shown in Figure 5.21a. Note that the field vectors at different points surrounding the spherical mass vary in both direction and magnitude. In a small region near the Earth's surface, $\mathbf{g}$ is approximately constant and the downward field is uniform, as indicated in Figure 5.21b. Equation 5.17 is valid at all points outside the Earth's surface, assuming that the Earth is spherical and that rotation can be ignored. At the Earth's surface, where $r = R_E$, $\mathbf{g}$ has a magnitude of 9.80 m/s$^2$.

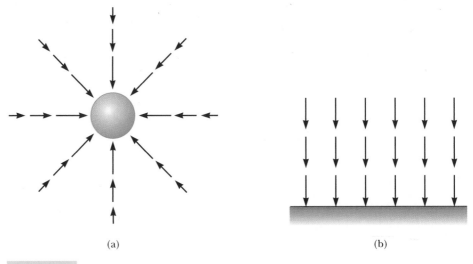

(a)                                    (b)

**Figure 5.21**

(a) The gravitational field vectors in the vicinity of a uniform spherical mass vary in both direction and magnitude. (b) The gravitational field vectors in a small region near the Earth's surface are uniform; that is, they all have the same direction and magnitude.

**context connection**

## 5.8 • THE EFFECT OF GRAVITY ON OUR SPACECRAFT

In the previous chapter, we considered the expenditure of fuel to cause a mid-course correction in the trajectory of our spacecraft on its way to Mars. Of course, we would like to use as little fuel as possible, because we don't want to have to depend on taking large amounts of heavy fuel with us. Most of our trip to Mars will be governed by the gravitational force between the spacecraft and the Sun. We will see the full details of this in the Context Conclusion after Chapter 11.

For now, let us just perform two calculations from Newton's law of universal gravitation that will be of interest to us. First, let's assume that we have fired ourselves out of the parking orbit around the Earth in Chapter 3 and are far enough away from the Earth so that the main gravitational interaction is with the Sun. How strong is this gravitational force? We can use data for the Sun and Newton's law of universal gravitation to find out. Let us assume that our spacecraft has a mass of $1.50 \times 10^3$ kg. Let us also assume that the position of the spacecraft is halfway between the orbits of Earth and Mars, which will put us at a distance of $1.89 \times 10^{11}$ m from the Sun:

$$F_g = G \frac{M_{Sun} m_{spacecraft}}{r^2}$$

$$= (6.67 \times 10^{-11} \text{ N} \cdot \text{m}^2/\text{kg}^2) \frac{(1.99 \times 10^{30} \text{ kg})(1.50 \times 10^3 \text{ kg})}{(1.89 \times 10^{11} \text{ m})^2} = 5.56 \text{ N}$$

This is a very small force; on our spacecraft, this results in an acceleration of only $3.7 \times 10^{-3}$ m/s². As the spacecraft moves farther from the Sun, this force becomes even smaller. But this force will act on the spacecraft *continuously* for a long time, so it can have a profound effect on the path of the spacecraft as we shall see later.

For our second calculation, let us imagine that we are on the Martian surface. Because Mars differs from Earth in radius and mass, the free-fall acceleration will be different. As discussed at the end of the previous section, the free-fall acceleration and the gravitational field are numerically equal, so let us calculate the magnitude of the gravitational field on the surface of Mars, using data for the planet:

$$g = \frac{GM_{\text{Mars}}}{r_{\text{Mars}}^2} = \frac{(6.67 \times 10^{-11}\,\text{N} \cdot \text{m}^2/\text{kg}^2)(6.42 \times 10^{23}\,\text{kg})}{(3.37 \times 10^6\,\text{m})^2} = 3.77\,\text{m/s}^2$$

Thus, objects on Mars will weigh significantly less than they do on the surface of the Earth, where $g = 9.80\,\text{m/s}^2$. We will need to plan for this in our activities after landing on Mars!

Earlier in this section, we mentioned the need for *fuel* for the rocket engines of the spacecraft. The fuel represents a source of *energy* for the spacecraft. In the next chapter, we will begin our study of the extremely important notions of energy and energy transfer.

## SUMMARY

Forces of friction are complicated, but we design a simplification model for friction that allows us to analyze motion that includes the effects of friction. The **maximum force of static friction** $f_{s,\text{max}}$ between two surfaces is proportional to the normal force between the surfaces. This maximum force occurs when the surfaces are on the verge of slipping. In general, $f_s \leq \mu_s n$, where $\mu_s$ is the **coefficient of static friction** and $\mathbf{n}$ is the normal force. When an object slides over a rough surface, the **force of kinetic friction** $\mathbf{f}_k$ is opposite the direction of the velocity of the object relative to the surface and is also proportional to the normal force on the object. The magnitude of this force is given by $f_k = \mu_k n$, where $\mu_k$ is the **coefficient of kinetic friction.** Usually, $\mu_k < \mu_s$.

Newton's second law, applied to a particle moving in uniform circular motion, states that the net force in the inward radial direction must equal the product of the mass and the centripetal acceleration:

$$\sum F = ma_c = \frac{mv^2}{r} \qquad \textbf{[5.3]}$$

An object moving through a liquid or gas experiences a **resistive force** that is velocity dependent. This resistive force, which opposes the motion, generally increases with speed. The force depends on the shape of the object and on the properties of the medium through which the object is moving. In the limiting case for a falling object, when the resistive force balances the weight ($a = 0$), the object reaches its **terminal speed.**

The fundamental forces existing in nature can be expressed as the following four: the gravitational force, the electromagnetic force, the nuclear force, and the weak force.

Rather than considering the gravitational force as a direct interaction between two objects, we can imagine that one object sets up a **gravitational field g** in space. A second object in this field experiences a force $\mathbf{F}_g = m\mathbf{g}$ when placed in this field.

## QUESTIONS

1. A book is pushed upward on a ramp (non-frictionless) so that it slides up to a point and then slides back down to the starting point. Does it take the same time to go up as it does to come down?

2. Pushing on a heavy box that is at rest requires some force **F** to start its motion. Once it is sliding, however, a smaller force is required to maintain that motion. Why?

3. Suppose you are driving a car along a highway at a high speed. Why should you avoid slamming on your brakes if you want to stop in the shortest distance? That is, why should you keep the wheels turning as you brake?

4. Twenty people are playing tug-of-war, with ten people on a team. The teams are so evenly matched that neither team wins. After they give up the game, they notice that one of the team members' cars is mired in mud. They attach the tug-of-war rope to the bumper of the car and the 20 people try to pull the car from the mud with the rope. The friction force between the car and the mud is huge, the car does not move, and the rope breaks. Why did the rope break in this situation when it did not break when 20 people pulled on it in the tug-of-war?

5. What causes a rotary lawn sprinkler to turn?

**6.** It has been suggested that rotating cylinders about 10 mi in length and 5 mi in diameter be placed in space and used as colonies. The purpose of the rotation is to simulate gravity for the inhabitants. Explain this concept for producing an effective gravity.

**7.** Consider a rotating space station, spinning with just the right speed such that the centripetal acceleration on the inner surface is *g*. Thus, astronauts standing on this inner surface would feel pressed to the surface as if they were pressed into the floor because of the Earth's gravitational force. Suppose an astronaut in this station holds a ball above her head and "drops" it to the floor. Will the ball fall just like it would on the Earth?

**8.** A pail of water can be whirled in a vertical path such that none is spilled. Why does the water stay in the pail, even when the pail is above your head?

**9.** How would you explain the force that pushes a rider toward the side of a car as the car rounds a corner?

**10.** Why does a pilot tend to black out when pulling out of a steep dive?

**11.** An object executes circular motion with a constant speed whenever a net force of constant magnitude acts perpendicular to the velocity. What happens to the speed if the force is not perpendicular to the velocity?

**12.** Suppose a race track is built on the Moon. For a given speed of a car around the track, would the roadway have to be banked at a steeper angle or a shallower angle than the same roadway on the Earth?

**13.** A skydiver in free-fall reaches terminal speed. After the parachute is opened, what parameters change to decrease this terminal speed?

**14.** Consider a small raindrop and a large raindrop falling through the atmosphere. Compare their terminal speeds. What are their accelerations when they reach terminal speed?

**15.** On long journeys, jet aircraft usually fly at high altitudes of about 30 000 ft. What is the main advantage of flying at these altitudes from an economic viewpoint?

**16.** If someone told you that astronauts are weightless in orbit because they are beyond the pull of gravity, would you accept the statement? Explain.

**17.** "If the current position and velocity of every particle in the Universe were known, together with the laws describing the forces that particles exert on one another, then the whole future of the Universe could be calculated. The future is determinate. Free will is an illusion." Do you agree with this thesis? Argue for or against it.

## PROBLEMS

1, 2, 3 = straightforward, intermediate, challenging      ☐ = full solution available in the *Student Solutions Manual and Study Guide*

**web** = solution posted at **http://www.harcourtcollege.com/physics/**      🖥 = computer useful in solving problem

🌐 = Interactive Physics      ▦ = paired numerical/symbolic problems      ◪ = life science application

### Section 5.1   Forces of Friction

**1.** A 25.0-kg block is initially at rest on a horizontal surface. A horizontal force of 75.0 N is required to set the block in motion. After it is in motion, a horizontal force of 60.0 N is required to keep the block moving with constant speed. Find the coefficients of static and kinetic friction from this information.

**2.** A car is traveling at 50.0 mi/h on a horizontal highway. (a) If the coefficient of static friction between road and tires on a rainy day is 0.100, what is the minimum distance in which the car will stop? (b) What is the stopping distance when the surface is dry and $\mu_s = 0.600$?

**3.** To meet a U.S. Postal Service requirement, footwear must have a coefficient of static friction of 0.5 or more when in contact with a specified tile surface. A typical athletic shoe has a coefficient of 0.800. In an emergency, what is the minimum time for a person starting from rest to move 3.00 m on a tile surface if she is wearing (a) footwear meeting the Postal Service minimum? (b) A typical athletic shoe?

**4.** Before 1960 it was believed that the maximum attainable coefficient of static friction for an automobile tire was less than 1. Then about 1962, three companies independently developed racing tires with coefficients of 1.6. Since then, tires have improved, as illustrated in this problem. According to the 1990 *Guinness Book of Records*, the shortest time in which a piston-engine car initially at rest has covered a distance of one-quarter mile is 4.96 s. This record was set by

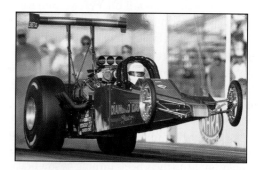

**Figure P5.4** *(Mike Powell/AllSport USA)*

Shirley Muldowney in September 1989. (a) Assume that, as in Figure P5.4, the rear wheels lifted the front wheels off the pavement. What minimum value of $\mu_s$ is necessary to achieve the record time? (b) Suppose Muldowney were able to double her engine power, keeping other things equal. How would this change affect the elapsed time?

5. Consider a large truck carrying a heavy load, such as steel beams. A significant hazard for the driver is that the load may slide forward, crushing the cab, if the truck stops suddenly in an accident or even in braking. Assume, for example, that a 10 000-kg load is located on the flat bed of a 20 000-kg truck moving at 12.0 m/s. Assume the load is not tied down to the truck and assume that the coefficient of friction between the load and the truck bed is 0.500. (a) Calculate the minimum stopping distance for the truck for which the load will not slide forward relative to the truck. (b) Is any piece of data unnecessary for the solution?

6. A woman at an airport is towing her 20.0-kg suitcase at constant speed by pulling on a strap at an angle of $\theta$ above the horizontal (Fig. P5.6). She pulls on the strap with a 35.0-N force, and the frictional force on the suitcase is 20.0 N. Draw a free-body diagram of the suitcase. (a) What angle does the strap make with the horizontal? (b) What normal force does the ground exert on the suitcase?

**Figure P5.6**

7. A 3.00-kg block starts from rest at the top of a 30.0° incline
**web** and slides a distance of 2.00 m down the incline in 1.50 s. Find (a) the magnitude of the acceleration of the block, (b) the coefficient of kinetic friction between block and plane, (c) the friction force acting on the block, and (d) the speed of the block after it has slid 2.00 m.

8. A 9.00-kg hanging weight is connected by a string over a light pulley to a 5.00-kg block that is sliding on a flat table (Fig. P5.8). If the coefficient of kinetic friction is 0.200, find the tension in the string.

5.00 kg

9.00 kg

**Figure P5.8**

9. Two blocks connected by a rope of negligible mass are being dragged by a horizontal force **F** (see Fig. P4.27). Suppose that $F = 68.0$ N, $m_1 = 12.0$ kg, $m_2 = 18.0$ kg, and the coefficient of kinetic friction between each block and the surface is 0.100. (a) Draw a free-body diagram for each block. (b) Determine the tension $T$ and the magnitude of the acceleration of the system.

10. To determine the coefficients of friction between rubber and various surfaces, a student uses a rubber eraser and an incline. In one experiment the eraser begins to slip down the incline when the angle of inclination is 36.0° and then moves down the incline with constant speed when the angle is reduced to 30.0°. From these data, determine the coefficients of static and kinetic friction for this experiment.

11. A block of mass 3.00 kg is pushed up against a wall by a force **P** that makes a 50.0° angle with the horizontal as shown in Figure P5.11. The coefficient of static friction between the block and the wall is 0.250. Determine the possible values for the magnitude of **P** that allow the block to remain stationary.

50.0°

P

**Figure P5.11**

## Section 5.2 Newton's Second Law Applied to a Particle in Uniform Circular Motion

12. In the Bohr model of the hydrogen atom, the speed of the electron is approximately $2.20 \times 10^6$ m/s. Find (a) the force acting on the electron as it revolves in a circular orbit of radius $0.530 \times 10^{-10}$ m and (b) the centripetal acceleration of the electron.

13. A light string can support a stationary hanging load of 25.0 kg before breaking. A 3.00-kg mass attached to the string rotates on a horizontal, frictionless table in a circle of radius 0.800 m. What range of speeds can the mass have before the string breaks?

14. A merry-go-round makes one complete revolution in 12.0 s. If a 45.0-kg child sits on the horizontal floor of the merry-go-round 3.00 m from the center, find (a) the child's acceleration and (b) the horizontal force of friction that acts on the child. (c) What minimum coefficient of static friction is necessary to keep the child from slipping?

15. A crate of eggs is located in the middle of the flat bed of a pickup truck as the truck negotiates an unbanked curve in the road. The curve may be regarded as an arc of a circle of radius 35.0 m. If the coefficient of static friction between crate and truck is 0.600, how fast can the truck be moving without the crate sliding?

16. Consider a conical pendulum with an 80.0-kg bob on a 10.0-m wire making an angle of 5.00° with the vertical (Fig. P5.16). Determine (a) the horizontal and vertical components of the force exerted by the wire on the pendulum and (b) the radial acceleration of the bob.

**Figure P5.16**

## Section 5.3   Nonuniform Circular Motion

17. Tarzan ($m = 85.0$ kg) tries to cross a river by swinging from a vine. The vine is 10.0 m long, and his speed at the bottom of the swing (as he just clears the water) is 8.00 m/s. Tarzan doesn't know that the vine has a breaking strength of 1 000 N. Does he make it safely across the river?

web

18. A car of mass $m$ passes over a bump in a road that follows the arc of a circle of radius $R$ as in Figure P5.18. (a) What

**Figure P5.18**

force does the road exert on the car as the car passes the highest point of the bump if the car travels at a speed $v$? (b) What is the maximum speed the car can have as it passes this highest point before losing contact with the road?

19. A pail of water is rotated in a vertical circle of radius 1.00 m. What is the minimum speed of the pail when it is upside down at the top of the circle if no water is to spill out?

20. A hawk flies in a horizontal arc of radius 12.0 m at a constant speed of 4.00 m/s. (a) Find its centripetal acceleration. (b) It continues to fly along the same horizontal arc but increases its speed at the rate of 1.20 m/s$^2$. Find the acceleration (magnitude and direction) under these conditions.

21. A roller coaster at the Six Flags Great America amusement park in Gurnee, Illinois, incorporates some of the latest design technology and some basic physics. Each vertical loop, instead of being circular, is shaped like a teardrop (Fig. P5.21). The cars ride on the inside of the loop at the top, and the speeds are high enough to ensure that the cars remain on the track. The biggest loop is 40.0 m high, with a maximum speed of 31.0 m/s (nearly 70 mph) at the bottom. Suppose the speed at the top is 13.0 m/s and the corresponding centripetal acceleration is $2g$. (a) What is the radius of the arc of the teardrop at the top? (b) If the total mass of the cars plus people is $M$, what force does the rail exert on this total mass at the top? (c) Suppose the roller coaster had a circular loop of radius 20.0 m. If the

**Figure P5.21** *(Frank Cezus/FPG International)*

cars have the same speed, 13.0 m/s at the top, what is the centripetal acceleration at the top? Comment on the normal force at the top in this situation.

## Section 5.4 Motion in the Presence of Velocity-Dependent Resistive Forces

22. A small piece of Styrofoam packing material is dropped from a height of 2.00 m above the ground. Until it reaches terminal speed, the magnitude of its acceleration is given by $a = g - bv$. After falling 0.500 m, the Styrofoam effectively reaches terminal speed, and then takes 5.00 s more to reach the ground. (a) What is the value of the constant $b$? (b) What is the acceleration at $t = 0$? (c) What is the acceleration when the speed is 0.150 m/s?

23. A small, spherical bead of mass 3.00 g is released from rest at $t = 0$ in a bottle of liquid shampoo. The terminal speed is observed to be $v_T = 2.00$ cm/s. Find (a) the value of the constant $b$ in Equation 5.5, (b) the time $\tau$ it takes to reach $0.632v_T$, and (c) the value of the resistive force when the bead reaches terminal speed.

24. (a) Estimate the terminal speed of a wooden sphere (density 0.830 g/cm$^3$) falling through air if its radius is 8.00 cm. (b) From what height would a freely falling object reach this speed in the absence of air resistance?

25. A motorboat cuts its engine when its speed is 10.0 m/s and coasts to rest. The equation describing the motion of the motorboat during this period is $v = v_i e^{-ct}$, where $v$ is the speed at time $t$, $v_i$ is the initial speed, and $c$ is a constant. At $t = 20.0$ s, the speed is 5.00 m/s. (a) Find the constant $c$. (b) What is the speed at $t = 40.0$ s? (c) Differentiate the expression for $v(t)$ and thus show that the acceleration of the boat is proportional to the speed at any time.

26. Assume that the resistive force on a speed-skater is given by $f = - kmv^2$, where $k$ is a constant and $m$ is the skater's mass. The skater crosses the finish line of a straight-line race with speed $v_f$ and then slows down by coasting on his skates. Show that the skater's speed at any time $t$ after crossing the finish line is $v(t) = v_f/(1 + ktv_f)$.

## Section 5.5 Numerical Representations of Particle Dynamics

27. A hailstone of mass $4.80 \times 10^{-4}$ kg falls through the air and experiences a net force given by

$$F = - mg + Cv^2$$

where $C = 2.50 \times 10^{-5}$ kg/m. (a) Calculate the terminal speed of the hailstone. (b) Use Euler's method of numerical analysis to find the speed and position of the hailstone at 0.2-s intervals, taking the initial speed to be zero. Continue the calculation until the hailstone reaches 99% of terminal speed.

28. A 3.00-g leaf is dropped from a height of 2.00 m above the ground. Assume that the net downward force exerted on the leaf is given by $F = mg - bv$, where the drag factor is $b = 0.030\ 0$ N·s/m. (a) Calculate the terminal speed of the leaf. (b) Use Euler's method of numerical analysis to find the speed and position of the leaf, as functions of time, from the instant it is released until 99% of terminal speed is reached. (*Hint:* Try $\Delta t = 0.005$ s.)

29. A 0.142-kg baseball is dropped from rest and has a terminal speed of 42.5 m/s (95 mph). (a) If a baseball experiences a drag force of magnitude $R = Cv^2$, what is the value of the constant $C$? (b) What is the magnitude of the drag force when the speed of the baseball is 36.0 m/s? (c) Use a computer to determine the motion of a baseball thrown vertically upward at an initial speed of 36 m/s. What maximum height does the ball reach? How long is it in the air? What is its speed just before it hits the ground?

30. Consider a 10.0-kg projectile launched with an initial speed of 100 m/s, at an elevation angle of 35.0°. The resistive force is given by $\mathbf{R} = - b\mathbf{v}$, where $b = 10.0$ N·s/m. (a) Use a numerical method to determine the horizontal and vertical coordinates of the projectile as functions of time. (b) What is the range of this projectile? (c) Determine the elevation angle that gives the maximum range for the projectile. (*Hint:* Adjust the elevation angle by trial and error to find the greatest range.)

## Section 5.6 The Fundamental Forces of Nature

31. Two identical isolated particles, each of mass 2.00 kg, are separated by a distance of 30.0 cm. What is the magnitude of the gravitational force exerted by one particle on the other?

32. In a thundercloud there may be electric charges of $+ 40.0$ C near the top of the cloud and $- 40.0$ C near the bottom of the cloud. These charges are separated by 2.00 km. What is the electric force on the top charge?

33. Find the order of magnitude of the gravitational force you exert on another person 2 m away. In your solution state the quantities you measure or estimate and their values.

## Section 5.7 The Gravitational Field

34. The Moon travels around the Earth in a nearly circular orbit of radius $3.84 \times 10^8$ m with a period of 27.3 days. (a) What is the speed of the Moon relative to the Earth? (b) Calculate the centripetal acceleration of the Moon from its orbital speed. (c) Use the law of universal gravitation to calculate the gravitational field created by the Earth at the location of the Moon. *Historical note:* Newton was the first person to solve this problem. He compared the results of parts (b) and (c) to check the validity of his law of gravitation, and found them to "answer pretty nearly."

**35.** When a falling meteor is at a distance above the Earth's surface of 3.00 times the Earth's radius, what is its free-fall acceleration due to the gravitational force exerted on it?

## Section 5.8   Context Connection—The Effect of Gravity on Our Spacecraft

**36.** A satellite of mass 300 kg is in a circular orbit around the Earth at an altitude equal to the Earth's mean radius. Find (a) the satellite's orbital speed, (b) the period of its revolution, and (c) the gravitational force acting on it.

**37.** Whenever two Apollo astronauts were on the surface of the Moon, a third astronaut orbited the Moon. Assume the orbit to be circular and 100 km above the surface of the Moon. If the mass of the Moon is $7.40 \times 10^{22}$ kg and its radius is $1.70 \times 10^6$ m, determine (a) the orbiting astronaut's acceleration, (b) his orbital speed, and (c) the period of the orbit.

## *Additional Problems*

**38.** Consider the three connected objects shown in Figure P5.38. If the inclined plane is frictionless, and the system is in equilibrium, find (in terms of $m$, $g$, and $\theta$) (a) the mass $M$ and (b) the tensions $T_1$ and $T_2$. If the value of $M$ is double the value found in part (a), find (c) the acceleration of each object, and (d) the tensions $T_1$ and $T_2$. If the coefficient of static friction between $m$ and $2m$ and the inclined plane is $\mu_s$, and the system is in equilibrium, find (e) the minimum value of $M$ and (f) the maximum value of $M$. (g) Compare the values of $T_2$ when $M$ has its minimum and maximum values.

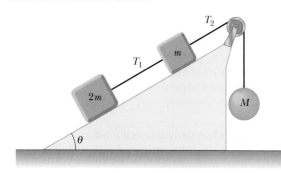

**Figure P5.38**

**39.** A crate of weight $F_g$ is pushed by a force **P** on a horizontal floor. (a) If the coefficient of static friction is $\mu_s$ and **P** is directed at angle $\theta$ below the horizontal, show that the minimum value of $P$ that will move the crate is given by

$$P = \frac{\mu_s F_g \sec \theta}{1 - \mu_s \tan \theta}$$

(b) Find the minimum value of $P$ that can produce motion when $\mu_s = 0.400$, $F_g = 100$ N, and $\theta = 0°$, 15.0°, 30.0°, 45.0°, and 60.0°.

**40.** A 1.30-kg toaster is not plugged in. The coefficient of static friction between the toaster and a horizontal countertop is 0.350. To make the toaster start moving, you carelessly pull on its electric cord. (a) For the cord tension to be as small as possible, you should pull at what angle above the horizontal? (b) With this angle, how large must the tension be?

**41.** A 2.00-kg aluminum block and a 6.00-kg copper block are connected by a light string over a frictionless pulley. They sit on a steel surface, as shown in Figure P5.41, where $\theta = 30.0°$. When they are released from rest, will they start to move? If so, determine (a) their acceleration and (b) the tension in the string. If not, determine the sum of the magnitudes of the forces of friction acting on the blocks.

**Figure P5.41**

**42.** The system shown in Figure P4.43 (Chapter 4) has an acceleration of magnitude 1.50 m/s². Assume the coefficients of kinetic friction between block and incline are the same for both inclines. Find (a) the coefficient of kinetic friction and (b) the tension in the string.

**43.** A block of mass $m = 2.00$ kg rests on the left edge of a block of mass $M = 8.00$ kg. The coefficient of kinetic friction between the two blocks is 0.300, and the surface on which the 8.00-kg block rests is frictionless. A constant horizontal force of magnitude $F = 10.0$ N is applied to the 2.00-kg block, setting it in motion as shown in Figure P5.43a. If the distance $L$ that the leading edge of the

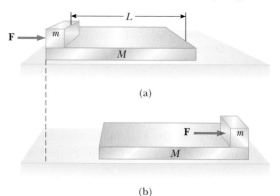

**Figure P5.43**

smaller block travels on the larger block is 3.00 m, (a) how long will it take before this block makes it to the right side of the 8.00-kg block, as shown in Figure P5.43b? (Note: Both blocks are set into motion when **F** is applied.) (b) How far does the 8.00-kg block move in the process?

44. A 5.00-kg block is placed on top of a 10.0-kg block (Fig. P5.44). A horizontal force of 45.0 N is applied to the 10-kg block, and the 5-kg block is tied to the wall. The coefficient of kinetic friction between all moving surfaces is 0.200. (a) Draw a free-body diagram for each block, and identify the action-reaction forces between the blocks. (b) Determine the tension in the string attached to the wall and the magnitude of the acceleration of the 10-kg block.

**Figure P5.44**

45. **Review Problem.** The roof of a building is sloped at an angle of 37.0° with the horizontal. A student throws a Frisbee onto the roof. It strikes with a speed of 15.0 m/s and does not bounce, but slides straight up the incline. The coefficient of kinetic friction between the Frisbee and the roof is 0.400. The Frisbee slides 10.0 m up the roof to its peak, where it goes into free fall, following a parabolic trajectory with negligible air resistance. Determine the maximum height the Frisbee reaches above the point where it struck the roof.

46. A student builds and calibrates an accelerometer, which she uses to determine the speed of her car around a certain unbanked highway curve. The accelerometer is a plumb bob with a protractor that she attaches to the roof of her car. A friend riding in the car with her observes that the plumb bob hangs at an angle of 15.0° from the vertical when the car has a speed of 23.0 m/s. (a) What is the centripetal acceleration of the car rounding the curve? (b) What is the radius of the curve? (c) What is the speed of the car if the plumb bob deflection is 9.00° while rounding the same curve?

47. An air puck of mass $m_1$ is tied to a string and allowed to revolve in a circle of radius $R$ on a frictionless horizontal table. The other end of the string passes through a hole in the center of the table, and an object of mass $m_2$ is tied to it (Fig. P5.47). The suspended object remains in equilibrium while the puck on the tabletop revolves. What are (a)

**Figure P5.47**

the tension in the string? (b) The net force acting on the puck? (c) The speed of the puck?

48. Materials such as rubber tires and shoe soles are tested for coefficients of static friction using an apparatus called a James tester. The pair of surfaces for which $\mu_s$ is to be measured are labeled B and C in Figure P5.48. Sample C is attached to a foot D at the lower end of a pivoting arm E, which makes an angle $\theta$ with the vertical. The upper end of the arm is hinged at F to a vertical rod G, which slides freely in a guide H fixed to the frame of the apparatus and supports a load I of 36.4 kg. The hinge pin at F is also the axle of a wheel that can roll vertically on the frame. All of the moving parts have weights negligible in comparison to the 36.4-kg load. The pivots are nearly frictionless. The test surface B is attached to a rolling platform A. An operator slowly moves the platform to the left in the picture until the sample C suddenly slips over surface B. At the critical point where sliding motion is ready to begin the operator notes the angle $\theta_s$ of the pivoting

**Figure P5.48**

arm. (a) Make a free-body diagram of the pin at F, which is in equilibrium under three forces. These forces are the weight of the load I, a horizontal normal force exerted by the frame, and a force of compression directed upward along the arm E. (b) Draw a free-body diagram of the foot D and sample C, considered as one system. (c) Determine the normal force that the test surface B exerts on the sample for any angle $\theta$. (d) Show that $\mu_s = \tan \theta_s$. (e) The protractor on the tester can record angles as large as 50.2°. What is the largest coefficient of friction it can measure?

49. Because the Earth rotates about its axis, a point on the equator experiences a centripetal acceleration of 0.033 7 m/s$^2$, but a point at the poles experiences no centripetal acceleration. (a) Show that at the equator the gravitational force on an object must exceed the normal force required to support the object. That is, show that the object's true weight exceeds its apparent weight. (b) What is the apparent weight at the equator and at the poles of a person having a mass of 75.0 kg? (Assume the Earth is a uniform sphere and take $g = 9.800$ N/kg.)

50. An engineer wishes to design a curved exit ramp for an interstate highway in such a way that a car will not have to rely on friction to round the curve without skidding (Fig. P5.50). Suppose that a typical car rounds the curve with a speed of 30 mi/h (13.4 m/s) and that the radius of the curve is 50.0 m. At what angle should the curve be banked?

51. A space station, in the form of a wheel 120 m in diameter, rotates to provide an "artificial gravity" of 3.00 m/s$^2$ for persons walking around on the inner wall of the outer rim. Find the rate of rotation of the wheel (in revolutions per minute) that will produce this effect.

52. A 9.00-kg object starting from rest falls through a viscous medium and experiences a resistive force $\mathbf{R} = -b\mathbf{v}$, where $\mathbf{v}$ is the velocity of the object. If the object's speed reaches one-half its terminal speed in 5.54 s, (a) determine the terminal speed. (b) At what time is the speed of the object three-fourths the terminal speed? (c) How far has the object traveled in the first 5.54 s of motion?

53. An amusement park ride consists of a large vertical cylinder that spins about its axis fast enough that any person inside is held up against the wall when the floor drops away (Fig. P5.53). The coefficient of static friction between a person and the wall is $\mu_s$, and the radius of the cylinder is $R$. (a) Show that the maximum period of revolution necessary to keep the person from falling is given by $T = (4\pi^2 R\mu_s/g)^{1/2}$. (b) Obtain a numerical value for $T$ if $R = 4.00$ m and $\mu_s = 0.400$. How many revolutions per minute does the cylinder make?

(a)

(b)

**Figure P5.50**

**Figure P5.53**

54. The expression $F = arv + br^2v^2$ gives the magnitude of the resistive force (in newtons) exerted on a sphere of radius $r$ (in meters) exerted by a stream of air moving at speed $v$ (in meters per second), where $a$ and $b$ are constants with appropriate SI units. Their numerical values are $a = 3.10 \times 10^{-4}$ and $b = 0.870$. Using this expression, find the terminal speed for water droplets falling under

their own weight in air, taking the following values for the drop radii: (a) 10.0 $\mu$m, (b) 100 $\mu$m, (c) 1.00 mm. Note that for (a) and (c) you can obtain accurate answers without solving a quadratic equation, by considering which of the two contributions to the air resistance is dominant and ignoring the lesser contribution.

55. A model airplane of mass 0.750 kg flies in a horizontal circle at the end of a 60.0-m control wire, with a speed of 35.0 m/s. Compute the tension in the wire if it makes a constant angle of 20.0° with the horizontal. The forces exerted on the airplane are the pull of the control wire, its own weight, and aerodynamic lift, which acts at 20.0° inward from the vertical as shown in Figure P5.55.

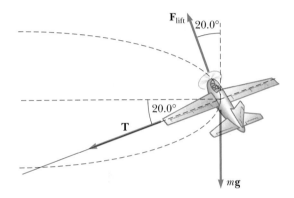

**Figure P5.55**

56. A single bead can slide with negligible friction on a wire that is bent into a circular loop of radius 15.0 cm, as in Figure P5.56. The circle is always in a vertical plane and rotates steadily about its vertical diameter with (a) a pe-

riod of 0.450 s. The position of the bead is described by the angle $\theta$ that the radial line, from the center of the loop to the bead, makes with the vertical. At what angle up from the bottom of the circle can the bead stay motionless relative to the turning circle? (b) Repeat the problem if the period of the circle's rotation is 0.850 s.

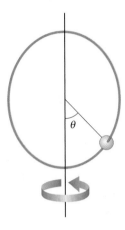

**Figure P5.56**

57. If a single constant force acts on an object of mass $m$ that moves on a straight line, the object's velocity is a linear function of time. The equation $v = v_i + at$ gives its velocity $v$ as a function of time, where $a$ is its constant acceleration. Now assume that as the object moves through a resistive medium, its speed decreases as described by the equation $v = v_i - kx$, where $x$ is the distance it has traveled since it had initial speed $v_i$ and $k$ is a constant coefficient. Find the law describing the total force acting on this object.

## ANSWERS TO QUICK QUIZZES

**5.1** (b). Friction forces are always parallel to the surfaces in contact, which, in this case, are the wall and the cover of the book. This tells us that the friction force is either upward or downward. Because the tendency of the book is to fall due to gravity, the friction force must be in the upward direction.

**5.2** (b). The static friction force between the bottom surface of the crate and the surface of the truck bed is the net horizontal force on the crate that causes it to accelerate. It is in the same direction as the acceleration of the crate, to the east.

**5.3** It is easier to attach the rope and pull. In this case, there is a component of your applied force that is upward. This reduces the normal force between the sled and the snow. In turn, this reduces the friction force between the sled and

the snow, making it easier to move. If you push from behind with a force having a downward component, the normal force is larger, the friction force is larger, and the sled is harder to move.

**5.4** The centripetal acceleration is always directed toward the center of the circle. Thus, at the top (a), the centripetal acceleration is downward and at the bottom (b), it is upward. The normal force is always perpendicularly outward from the surface, so the normal force from the seat on you is upward at both the top (c) and the bottom (d). (e) The normal force is largest at the bottom because it is in the same direction as the centripetal acceleration.

**5.5** On both roadways, the maximum friction force on the car is the same. From Newton's second law, this means that the centripetal acceleration is the same on both roadways.

Consequently, from the expression for centripetal acceleration, the maximum speed is proportional to the square root of the radius. Because we have doubled the radius, the maximum speed is $\sqrt{2}v$.

**5.6** The baseball, because it has a smaller cross-sectional area, experiences a smaller resistive force as it falls through the air. Thus, the baseball strikes the ground first.

**5.7** The forces exerted on the sky surfer are the downward gravitational force $m\mathbf{g}$ and an upward force of air resistance $\mathbf{R}$, the magnitude of which is less than her weight before she reaches terminal speed. As her downward speed increases, the force of air resistance increases. The vector sum of the gravitational force and the force of air resistance gives a total force that decreases with time, so her acceleration decreases. Once she reaches terminal speed, the two forces balance each other, the net force is zero, and her acceleration is zero.

**A large earth-moving machine does work on a pile of dirt by moving it to a new location.**

*(© 1995 Paul Steel/The Stock Market)*

# Energy and Energy Transfer

In the preceding chapters, we approached the motion of an object with quantities such as *position, velocity, acceleration,* and *force,* with which you are familiar from everyday life. We developed a number of models using these notions that allow us to solve a variety of problems. Some problems that, in theory, could be solved with Newton's laws, are very difficult to solve in practice, but can be made much simpler with a different approach. In this and the following two chapters, we shall investigate this new approach, which will introduce us to new analysis models for problem solving. This approach includes definitions of quantities that may not be familiar to you. You may be familiar with some quantities, but they may have more specific meanings in physics than in everyday life. We begin this discussion by exploring the notion of *energy.*

Energy is present in the Universe in various forms. *Every* physical process in the Universe involves energy and energy transfers or transformations. Thus, energy is an extremely important concept to understand. Unfortunately, despite its importance, it cannot be easily defined. The variables in previous chapters were relatively concrete—we have everyday experience with velocities and forces, for example. Although the *notion* of energy is more abstract, we do have *experiences* with energy, such as running out of gasoline or losing our electrical service if we forget to pay the bill.

The concept of energy can be applied to the dynamics of a mechanical system without resorting to Newton's laws. This "energy approach" to describing motion is especially useful when the force acting on a particle is not constant; in such a case, the acceleration is not constant, and we cannot apply the particle under constant acceleration model we developed in Chapter 2. Particles in nature are often subject to forces that vary with the particles' positions. These forces include gravitational forces and the force exerted on an object attached to a spring. We shall describe techniques for treating such situations with the help of an important concept called the *work–kinetic energy theorem*. The work–kinetic energy theorem is a special case of a global approach to problems involving energy and energy transfer. This approach extends well beyond physics and can be applied to biological organisms, technological systems, and engineering situations.

All of our analysis models in the earlier chapters were based on the motion of a particle or an object modeled as a particle. We begin our study of our new approach by identifying a new simplification model for a *system*.

## 6.1 • SYSTEMS AND ENVIRONMENTS

As mentioned in the introduction, an important new concept we will need in studying energy is that of a **system.** This is a simplification model, in that we focus our attention on a small region of the Universe and ignore details of the rest of the Universe outside the system. A critical skill in applying the energy approach to problems in the next three chapters is *correctly identifying the system*. A valid system may

- Be a single object or particle
- Be a collection of objects or particles
- Be a region of space (e.g., the interior of an automobile engine combustion cylinder)
- Vary in size and shape (e.g., a rubber ball that deforms upon striking a wall)

A **system boundary,** which is an imaginary surface (often but not necessarily coinciding with a physical surface), divides the Universe between the system and the **environment** of the system.

As an example, imagine a force applied to an object in empty space. We can define the object as the system, as in the first item in the previous list. The force applied to it influences the system from the environment, which acts across the system boundary. We will see how to analyze this situation from a system approach in a subsequent section of this chapter.

Another example occurs in Example 5.3. Here the system can be defined as the combination of the ball, the cube, and the string, consistent with the second item of the previous list. The influence from the environment includes the gravitational forces on the ball and the cube, the normal and friction forces on the cube, and the force of the pulley on the string. The forces exerted by the string on the ball and the cube are internal to the system and, therefore, are not included as an influence from the environment.

- *A system*

One of the most important steps to take in solving a problem using the energy approach is to correctly identify the system of interest. Be sure this is the *first* step you take in solving a problem.

- *The environment*

## 6.2 • WORK DONE BY A CONSTANT FORCE

See the *Core Concepts in Physics CD-ROM,* Screen 5.1

Let us begin our analysis of systems by introducing a term whose meaning in physics is distinctly different from its everyday meaning. This new term is **work.** Imagine that you are trying to push a heavy sofa across your living room floor. If you push on the sofa *and* it moves through a displacement, then you have done work on the sofa.

Consider a particle, which we identify as the system, that undergoes a displacement $\Delta \mathbf{r}$ along a straight line while acted on by a constant force $\mathbf{F}$ that makes an angle $\theta$ with $\Delta \mathbf{r}$, as in Figure 6.1. The force has accomplished something—it has moved the particle—so we say that work was done on the particle.

Notice that we only know the force and the displacement given in the description of the situation. We have no information about how long it took for this displacement to occur, nor any information about velocities or accelerations. This hints at the power of the energy approach, as well as how different it will be from our approach in previous chapters. We do not need these values to find the work done. Let us now define the work done on the system if the force is constant:

---

The work $W$ done by an agent exerting a constant force on a system is the product of the component $F \cos \theta$ of the force along the direction of the displacement of the point of application of the force and the magnitude $\Delta r$ of the displacement:

$$W \equiv F \Delta r \cos \theta \qquad \text{[6.1]}$$

---

Work is a scalar quantity—no direction is associated with it. Its units are those of force multiplied by length; therefore, the SI unit of work is the **newton·meter** (N·m). The newton·meter, when it refers to work or energy, is called the **joule** (J).

From the definition in Equation 6.1, we see that a force does no work on a system if the point of application of the force does not move. In the mathematical representation, if $\Delta r = 0$, Equation 6.1 gives $W = 0$. In the mental representation, imagine pushing on the sofa mentioned earlier—if it doesn't move, no work has been done on the sofa. Of course, the work is also zero if the applied force is zero—if you don't push, you don't do any work!

Also note from Equation 6.1 that the work done by a force is zero when the force is perpendicular to the displacement. That is, if $\theta = 90°$, then $\cos 90° = 0$ and $W = 0$. For example, consider the free-body diagram for a block moving across a frictionless surface in Figure 6.2. The work done by the normal force and the gravitational force on the block during its horizontal displacement are both zero for the same reason—they are both perpendicular to the displacement.

In general, we will consider work done on a system. For example, in Chapter 17, we will consider work done in compressing a gas, which is modeled as a system of particles. For now, however, let us restrict our attention to a system consisting of a single particle. In the case of a single particle, the displacement of the point of application of the force is necessarily the same as the displacement of the particle.

**Figure 6.1**

If an object undergoes a displacement $\Delta \mathbf{r}$, the work done by the constant force $\mathbf{F}$ on the object is $(F \cos \theta)\Delta r$.

- *Work done by a constant force*

- *The joule*

**PITFALL PREVENTION 6.2**

**What's the correct angle?**

Keep in mind that the angle $\theta$ in Equation 6.1 is the angle between the force vector and the displacement vector. A given problem may provide a geometric construction that includes an angle, but this may not be the angle you want for Equation 6.1. Be sure you have correctly identified the proper angle $\theta$ before calculating the work.

The weight lifter does no work on the weights as he holds them at rest.
*(Gerard Vandystadt/Photo Researchers, Inc.)*

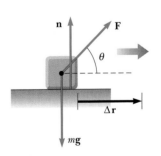

**Figure 6.2**

When an object is displaced horizontally on a flat table, the normal force $\mathbf{n}$ and the gravitational force $m\mathbf{g}$ do no work.

**Figure 6.3**

A force of **F** is applied to an object, which undergoes a displacement to the right. In each of the four cases, the magnitudes of the force and displacement are the same.

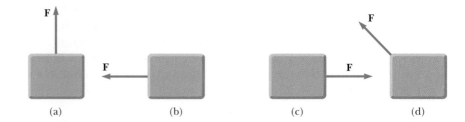

(a)                    (b)                    (c)                    (d)

**PITFALL PREVENTION 6.3**

**What is being displaced?**

Note that the displacement in Equation 6.1 is that of the *point of application of the force*. For a force applied to a particle, it is the same as the displacement of the particle. For extended objects, however, this may differ from the displacement of the object. For example, imagine climbing a ladder. The force causing your body to be displaced upward is the force from the rungs upward on your feet. But this force undergoes no displacement, even though your body does. This force does no work! For now, we will restrict our attention to work done on particles, but keep this distinction in mind. We will see how to address problems like ladder climbing shortly.

**PITFALL PREVENTION 6.4**

**Work is done by . . . on . . .**

In Pitfall Prevention 6.1, we noted the importance of identifying the system. It is also important to identify the interaction of the system with the environment. When discussing work, try always to use the phrase, "the work done by . . . on. . . ." After "by," insert the part of the environment that is interacting directly with the system. After "on," insert the system. For example, "the work done by the hammer on the nail" identifies the nail as the system and the force from the hammer represents the interaction with the environment. This is similar to our discussion in Chapter 4 of taking care to say "the force exerted by . . . on. . . ."

In general, a particle may be moving under the influence of several forces. In that case, because work is a scalar quantity, the total work done as the particle undergoes some displacement is the algebraic sum of the work done by each of the forces.

The sign of the work depends on the direction of **F** relative to **Δr**. The work done by the applied force is positive when the vector component of magnitude $F \cos \theta$ is in the *same direction* as the displacement. For example, when an object is lifted, the work done by the applied force on the object is positive because the lifting force is upward, that is, in the same direction as the displacement. When the vector component of magnitude $F \cos \theta$ is in the direction *opposite* the displacement, $W$ is *negative*. In the case of the object being lifted, for instance, the work done by the gravitational force on the object is negative.

If a constant applied force **F** acts parallel to the direction of the displacement, then $\theta = 0$ and $\cos 0 = 1$. In this case, Equation 6.1 gives

$$W = F \Delta r \qquad \textbf{[6.2]}$$

Both Equations 6.1 and 6.2 are special cases of a more generalized definition of work. Both equations assume a constant force, and Equation 6.2 assumes that the force is parallel to the displacement. In the next two sections, we shall consider the situation in which a force is not parallel to the displacement and the more general case of a varying force.

**Quick Quiz 6.1**

Consider a tug-of-war, in which the two teams pulling on the rope are evenly matched, so that no motion takes place. Is work done on the rope? On the pullers? On the ground? Is work done on any system?

**Quick Quiz 6.2**

Figure 6.3 shows four situations in which a force is applied to an object. In all four cases, the force has the same magnitude, and the displacement of the object is to the right and of the same magnitude. Rank the situations in order of the work done by the force on the object, from most positive to most negative.

**THINKING PHYSICS 6.1**

A person slowly lifts a heavy box of mass $m$ a vertical height $h$, and then walks horizontally at constant velocity a distance $d$ while holding the box, as in Figure 6.4. Determine the work done (a) by the person and (b) by the gravitational force on the box in this process.

**Figure 6.4**

(Thinking Physics 6.1) A person lifts a heavy box of mass $m$ a vertical distance $h$ and then walks horizontally at constant velocity a distance $d$.

**Reasoning**   (a) Assuming that the person lifts the box with a force of magnitude equal to the weight of the box $mg$, the work done by the person on the box during the vertical displacement is $W = F\,\Delta r = (mg)(h) = mgh$, because the force in this case is in the same direction as the displacement. For the horizontal displacement, we assume that the acceleration of the box is approximately zero. As a result, the work done by the person on the box during the horizontal displacement of the box is zero because the horizontal force is approximately zero, and the force to support the box's weight in this process is perpendicular to the displacement. Thus, the net work done by the person during the complete process is $mgh$.

(b) The work done by the gravitational force on the box during the vertical displacement of the box is $-mgh$, because this force is opposite the displacement. The work done by the gravitational force is zero during the horizontal displacement because this force is perpendicular to the displacement. Hence, the net work done by the gravitational force for the complete process is $-mgh$. The net work done by all forces on the box is zero, because $+mgh + (-mgh) = 0$.

### THINKING PHYSICS 6.2

Roads going up mountains are formed into *switchbacks*, with the road weaving back and forth along the face of the slope, so that any portion of the roadway has only a gentle rise. Does this require that an automobile climbing the mountain do any less work than if it were driving on a roadway that runs straight up the slope? Why are the switchbacks used?

**Reasoning**   If we ignore the effects of rolling friction on the tires of the car, the same amount of work would be done in driving up the switchbacks and driving straight up the mountain because the weight of the car is moved upward against the gravitational force by the same vertical distance in each case. So why do we use the switchbacks? The answer lies in the force required, not the work. The force needed from the engine to follow a gentle rise is much less than that required to drive straight up the hill. Roadways running straight uphill would require redesigning engines so as to enable them to apply much larger forces. This is similar to the ease with which a heavy object can be rolled up a ramp into a moving van truck, compared with lifting the object straight up from the ground.

**PITFALL PREVENTION 6.5**
**Cause of the displacement**

When we calculate the work done by a force on an object, we are *not* necessarily saying that the force is the cause of the displacement. In some cases, a causal relationship exists, as in our simple example of applying a horizontal force to an object sitting on a frictionless table. But in the case of the work done by the gravitational force in Thinking Physics 6.1, the gravitational force is clearly not causing upward motion of the box! Work done by a force can be calculated regardless of whether the force causes the displacement.

### Example 6.1   Mr. Clean

A man cleaning his apartment pulls a vacuum cleaner with a force of magnitude $F = 50.0$ N. The force makes an angle of 30.0° with the horizontal as shown in Figure 6.5. The vacuum cleaner is displaced 3.00 m to the right. Calculate the work done by the 50.0-N force on the vacuum cleaner.

50.0 N

**n**

30.0°

*m***g**

**Figure 6.5**

(Example 6.5) A vacuum cleaner being pulled at an angle of 30.0° with the horizontal.

**Solution**   Using the definition of work (Equation 6.1), we have

$$W = (F \cos \theta) \Delta r = (50.0 \text{ N})(\cos 30.0°)(3.00 \text{ m})$$
$$= 130 \text{ N} \cdot \text{m} = \boxed{130 \text{ J}}$$

Note that the normal force **n**, the gravitational force $m\mathbf{g}$, and the upward component of the applied force do *no* work because they are perpendicular to the displacement.

**EXERCISE**   Find the work done by the man on the vacuum cleaner if he pulls it 3.00 m with a horizontal force of 32.0 N.

*Answer*   96.0 J

**EXERCISE**   If a person lifts a 20.0-kg bucket very slowly from a well and does 6.00 kJ of work on the bucket, how deep is the well?

*Answer*   30.6 m

**EXERCISE**   A 65.0-kg woman climbs a flight of 20 stairs, each 23.0 cm high. How much work is done by the gravitational force on the woman in the process?

*Answer*   − 2.93 kJ

## 6.3 • THE SCALAR PRODUCT OF TWO VECTORS

See Screen 2.6

Based on Equation 6.1, it is convenient to express the definition of work in terms of a **scalar product** of the two vectors **F** and **Δr**. The scalar product was introduced briefly in Section 1.9. We repeat its definition here:

• *Scalar product of any two vectors* **A** *and* **B**

> The scalar product of any two vectors **A** and **B** is a scalar quantity equal to the product of the magnitudes of the two vectors and the cosine of the angle $\theta$ between them:
>
> $$\mathbf{A} \cdot \mathbf{B} \equiv AB \cos \theta \qquad \text{[6.3]}$$
>
> where $\theta$ is the smaller of the two angles between **A** and **B**, as in Figure 6.6.

Note that **A** and **B** need not have the same units. The units of the scalar product are simply the product of the units of the two vectors. Because of the dot symbol, the scalar product is often called the *dot product*.

Notice that the right-hand side of Equation 6.3 has the same structure as the right-hand side of Equation 6.1. This allows us to write the definition of work as the scalar product **F · Δr**. Thus, we can express Equation 6.1 as

• *Work expressed as a scalar product*

$$W = \mathbf{F} \cdot \Delta\mathbf{r} = F \Delta r \cos \theta \qquad \text{[6.4]}$$

Before continuing with our discussion of work, let us investigate some properties of the dot product, because we will need to use it later in the book as well. From Equation 6.3 we see that the scalar product is *commutative*. That is,

• *The order of the vectors in the scalar product can be reversed*

$$\mathbf{A} \cdot \mathbf{B} = \mathbf{B} \cdot \mathbf{A} \qquad \text{[6.5]}$$

In addition, the scalar product obeys the *distributive law of multiplication*, so that

$$\mathbf{A} \cdot (\mathbf{B} + \mathbf{C}) = \mathbf{A} \cdot \mathbf{B} + \mathbf{A} \cdot \mathbf{C} \qquad [6.6]$$

The dot product is simple to evaluate from Equation 6.3 when $\mathbf{A}$ is either perpendicular or parallel to $\mathbf{B}$. If $\mathbf{A}$ is perpendicular to $\mathbf{B}$ ($\theta = 90°$), then $\mathbf{A} \cdot \mathbf{B} = 0$. (The equality $\mathbf{A} \cdot \mathbf{B} = 0$ also holds in the more trivial case when either $\mathbf{A}$ or $\mathbf{B}$ is zero.) If $\mathbf{A}$ and $\mathbf{B}$ point in the same direction ($\theta = 0°$), then $\mathbf{A} \cdot \mathbf{B} = AB$. If $\mathbf{A}$ and $\mathbf{B}$ point in opposite directions ($\theta = 180°$), then $\mathbf{A} \cdot \mathbf{B} = -AB$. The scalar product is negative when $90° < \theta < 180°$.

The unit vectors $\mathbf{i}$, $\mathbf{j}$, and $\mathbf{k}$, which were defined in Chapter 1, lie in the positive $x$, $y$, and $z$ directions, respectively, of a right-handed coordinate system. Therefore, it follows from the definition of $\mathbf{A} \cdot \mathbf{B}$ that the scalar products of these unit vectors are given by

$$\mathbf{i} \cdot \mathbf{i} = \mathbf{j} \cdot \mathbf{j} = \mathbf{k} \cdot \mathbf{k} = 1 \qquad [6.7]$$

$$\mathbf{i} \cdot \mathbf{j} = \mathbf{i} \cdot \mathbf{k} = \mathbf{j} \cdot \mathbf{k} = 0 \qquad [6.8]$$

Two vectors $\mathbf{A}$ and $\mathbf{B}$ can be expressed in component form as

$$\mathbf{A} = A_x\mathbf{i} + A_y\mathbf{j} + A_z\mathbf{k} \qquad \mathbf{B} = B_x\mathbf{i} + B_y\mathbf{j} + B_z\mathbf{k}$$

Therefore, using these expressions, Equations 6.7 and 6.8 reduce the scalar product of $\mathbf{A}$ and $\mathbf{B}$ to

$$\mathbf{A} \cdot \mathbf{B} = A_xB_x + A_yB_y + A_zB_z \qquad [6.9]$$

This equation and Equation 6.3 are alternative but equivalent expressions for the scalar product. Equation 6.3 is useful if you know the magnitudes and directions of the vectors, while Equation 6.9 is useful if you know the components of the vectors. In the special case where $\mathbf{A} = \mathbf{B}$, we see that

$$\mathbf{A} \cdot \mathbf{A} = A_x^2 + A_y^2 + A_z^2 = A^2$$

### Quick Quiz 6.3

Which of the following statements is true about the relationship between the dot product of two vectors and the product of the magnitudes of the vectors? (a) $\mathbf{A} \cdot \mathbf{B}$ is larger than $AB$; (b) $\mathbf{A} \cdot \mathbf{B}$ is smaller than $AB$; (c) $\mathbf{A} \cdot \mathbf{B}$ could be larger or smaller than $AB$, depending on the angle between the vectors; (d) $\mathbf{A} \cdot \mathbf{B}$ could be equal to $AB$.

**PITFALL PREVENTION 6.6**
### Work is a scalar

In Equation 6.4, we are defining the work in terms of two vectors, but keep in mind that *work is a scalar*—no direction is associated with it. In fact, all types of energy and energy transfer are scalars. This is one of the nice features of the energy approach—we don't need to concern ourselves so much about vectors!

• *Scalar products of unit vectors*

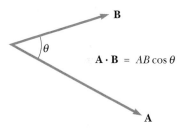

$$\mathbf{A} \cdot \mathbf{B} = AB\cos\theta$$

**Figure 6.6**

The scalar product $\mathbf{A} \cdot \mathbf{B}$ equals the magnitude of $\mathbf{A}$ multiplied by the magnitude of $\mathbf{B}$ and the cosine of the angle between $\mathbf{A}$ and $\mathbf{B}$.

### Example 6.2   The Scalar Product

The vectors $\mathbf{A}$ and $\mathbf{B}$ are given by $\mathbf{A} = 2\mathbf{i} + 3\mathbf{j}$ and $\mathbf{B} = -\mathbf{i} + 2\mathbf{j}$.
   (a) Determine the scalar product $\mathbf{A} \cdot \mathbf{B}$.

**Solution**   We will evaluate the scalar product directly using the unit vector notation:

$$\mathbf{A} \cdot \mathbf{B} = (2\mathbf{i} + 3\mathbf{j}) \cdot (-\mathbf{i} + 2\mathbf{j})$$
$$= -2\mathbf{i} \cdot \mathbf{i} + 2\mathbf{i} \cdot 2\mathbf{j} - 3\mathbf{j} \cdot \mathbf{i} + 3\mathbf{j} \cdot 2\mathbf{j}$$
$$= -2 + 6 = \boxed{4}$$

where we have used the facts that $\mathbf{i} \cdot \mathbf{i} = \mathbf{j} \cdot \mathbf{j} = 1$ and $\mathbf{i} \cdot \mathbf{j} =$

$\mathbf{j} \cdot \mathbf{i} = 0$. The same result is obtained using Equation 6.9 directly, where $A_x = 2$, $A_y = 3$, $B_x = -1$, and $B_y = 2$. Note that the result has no units because no units were specified on the original vectors $\mathbf{A}$ and $\mathbf{B}$.
   (b) Find the angle $\theta$ between $\mathbf{A}$ and $\mathbf{B}$.

**Solution**   The magnitudes of $\mathbf{A}$ and $\mathbf{B}$ are given by

$$A = \sqrt{A_x^2 + A_y^2} = \sqrt{(2)^2 + (3)^2} = \sqrt{13}$$
$$B = \sqrt{B_x^2 + B_y^2} = \sqrt{(1)^2 + (2)^2} = \sqrt{5}$$

Using Equation 6.3 and the result from (a) gives

$$\cos\theta = \frac{\mathbf{A}\cdot\mathbf{B}}{AB} = \frac{4}{\sqrt{13}\sqrt{5}} = \frac{4}{\sqrt{65}} = 0.496$$

$$\theta = \cos^{-1} 0.496 = \boxed{60.3°}$$

**EXERCISE**   For $\mathbf{A} = 4\mathbf{i} + 3\mathbf{j}$ and $\mathbf{B} = -\mathbf{i} + 3\mathbf{j}$, find (a) $\mathbf{A}\cdot\mathbf{B}$ and (b) the angle between $\mathbf{A}$ and $\mathbf{B}$. Assume the vector components are known to three significant figures.

*Answer*   (a) 5.00   (b) 71.6°

**EXERCISE**   As a particle moves from the origin to $(3.0\mathbf{i} - 4.0\mathbf{j})$ m, it is acted upon by a force given by $(4.0\mathbf{i} - 5.0\mathbf{j})$ N. Calculate the work done by this force on the particle as it moves through the given displacement.

*Answer*   32 J

## 6.4 • WORK DONE BY A VARYING FORCE

See Screen 5.2

Consider a particle being displaced along the $x$ axis under the action of a force of magnitude $F_x$ in the $x$ direction, which varies with position, as in the graphical representation in Figure 6.7. The particle is displaced in the direction of increasing $x$ from $x = x_i$ to $x = x_f$. In such a situation, we cannot use Equation 6.1 to calculate the work done by the force because this relationship applies only when $\mathbf{F}$ is constant in magnitude and direction. As seen in Figure 6.7, we do not have a *single* value of the force to substitute into Equation 6.1. If, however, we imagine that the particle undergoes a small displacement $\Delta r = \Delta x$, shown in Figure 6.7a, then the $x$ component $F_x$ of the force is approximately constant over this interval, and we can approximate the work done by the force for this small displacement as

$$W_1 \approx F_x \Delta x \qquad [6.10]$$

This quantity is just the area of the shaded rectangle in Figure 6.7a. If we imagine that the curve described by $F_x$ versus $x$ is divided into a large number of such intervals, then the total work done for the displacement from $x_i$ to $x_f$ is approximately equal to the sum of a large number of such terms:

$$W \approx \sum_{x_i}^{x_f} F_x \Delta x$$

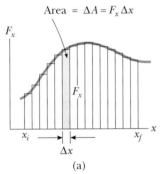

Area $= \Delta A = F_x \Delta x$

(a)

If the displacements $\Delta x$ are allowed to approach zero, then the number of terms in the sum increases without limit, but the value of the sum approaches a definite value equal to the area under the curve bounded by $F_x$ and the $x$ axis in Figure 6.7b. As you probably have learned in calculus, this limit of the sum is called an *integral* and is represented by

$$\lim_{\Delta x \to 0} \sum_{x_i}^{x_f} F_x \Delta x = \int_{x_i}^{x_f} F_x \, dx$$

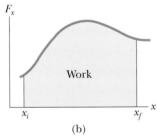

Work

(b)

**Figure 6.7**

(a) The work done by a force of magnitude $F_x$ for the small displacement $\Delta x$ is $F_x \Delta x$, which equals the area of the shaded rectangle. The total work done for the displacement from $x_i$ to $x_f$ is approximately equal to the sum of the areas of all the rectangles. (b) The work done by the variable force $F_x$ as the particle moves from $x_i$ to $x_f$ is *exactly* equal to the area under this curve.

The limits on the integral $x = x_i$ to $x = x_f$ define what is called a **definite integral.** (An *indefinite integral* is the limit of a sum over an unspecified interval. Appendix B.7 gives a brief description of integration.) This definite integral is numerically equal to the area under the curve of $F_x$ versus $x$ between $x_i$ and $x_f$. Therefore, we can express the work done by $F_x$ for the displacement of the particle from $x_i$ to $x_f$ as

$$W = \int_{x_i}^{x_f} F_x \, dx \qquad [6.11]$$

This equation reduces to Equation 6.1 when $F_x = F \cos\theta$ is constant and $x_f - x_i = \Delta x$.

If more than one force acts on a particle, the **total work done is just the scalar addition of the work done by each force.** Alternatively, the **total work is the work done by the net force.** If we express the net force in the $x$ direction as $\Sigma F_x$, then the *net work* done on the particle as it moves from $x_i$ to $x_f$ is

$$W_{\text{net}} = \int_{x_i}^{x_f} \left( \Sigma F_x \right) dx \qquad [6.12]$$

The most general case for calculating work occurs when several forces are acting, the net force is not constant, and the net force is not parallel to the displacement. In this case, the expression for the work is

$$W_{\text{net}} = \int_{\mathbf{r}_1}^{\mathbf{r}_2} \left( \Sigma \mathbf{F} \right) \cdot d\mathbf{r}$$

• *Work done by a variable net force*

---

• **Example 6.3    Calculating Total Work Done from a Graph**

A force acting on a particle varies with $x$ as shown in Figure 6.8. Calculate the work done by the force on the particle as it moves from $x = 0$ to $x = 6.0$ m.

**Solution**   The work done by the force is equal to the area under the curve from $x = 0$ to $x = 6.0$ m. We can model this geometrically as the area of the rectangular section from Ⓐ to Ⓑ combined with the area of the triangular section that extends from Ⓑ to Ⓒ. The area of the rectangle is $(5.0\text{ N})(4.0\text{ m}) = 20$ J, and the area of the triangle is $\frac{1}{2}(5.0\text{ N})(2.0\text{ m}) = 5.0$ J. Therefore, the total work done is  25 J.

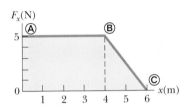

**Figure 6.8**

(Example 6.3) The force acting on a particle is constant for the first 4.0 m of motion and then decreases linearly with $x$ from $x_B = 4.0$ m to $x_C = 6.0$ m. The net work done by this force is the area under this curve.

---

## Work Done by a Spring

A common physical system for which the force varies with position is shown in Figure 6.9. A block on a horizontal, frictionless surface is connected to a spring. If the spring is stretched or compressed a small distance $x$ from its equilibrium position $x = 0$, the spring exerts a force on the block given by

$$F_s = -kx \qquad [6.13]$$

See Screen 5.3

• *Hooke's law*

where $k$ is a positive constant called the *force constant* (or *spring constant* or *stiffness constant*) of the spring. This force law for springs is known as **Hooke's law.** For many springs, Hooke's law can describe the behavior very accurately, provided that the displacement is not too large. The value of $k$ is a measure of the stiffness of the spring. Stiff springs have larger $k$ values, and weak springs have smaller $k$ values. We shall employ a simplification model in which all springs, unless specified otherwise, obey Hooke's law.

The negative sign in Equation 6.13 signifies that the force exerted by the spring is always directed *opposite* the displacement from the equilibrium position $x = 0$. For example, when $x > 0$, such that the block is pulled to the right and the spring is stretched, as in Figure 6.9a, the spring force is to the left, or negative.

**Figure 6.9**

The force exerted by a spring on a block varies with the block's displacement from the equilibrium position $x = 0$. (a) When $x$ is positive (stretched spring), the spring force is to the left. (b) When $x$ is zero (natural length of the spring), the spring force is zero. (c) When $x$ is negative (compressed spring), the spring force is to the right. (d) Graph of $F_s$ versus $x$ for the block–spring system. The work done by the spring force as the block moves from $-x_{max}$ to 0 is the area of the shaded triangle, $\frac{1}{2}kx_{max}^2$.

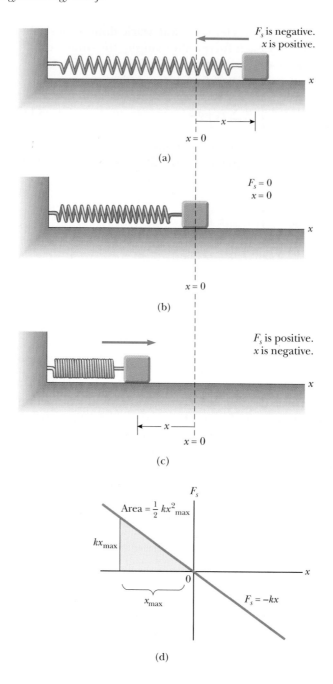

When $x < 0$, and the spring is compressed as in Figure 6.9c, the spring force is to the right, or positive. Of course, when $x = 0$, as in Figure 6.9b, the spring is unstretched and $F_s = 0$. Because the spring force always acts toward the equilibrium position, it is sometimes called a *restoring force*.

If the block is displaced to a position $-x_{max}$ and then released, it moves from $-x_{max}$ through zero to $+x_{max}$ (assuming a frictionless surface), and then turns around and returns to $-x_{max}$. The details of this oscillating motion will be given in Chapter 12. For our purposes here, let us calculate the work done by the spring force on the block as the block moves from $x_i = -x_{max}$ to $x_f = 0$. Applying the particle model to the block and using Equation 6.11, we have

$$W_s = \int_{x_i}^{x_f} F_s \, dx = \int_{-x_{max}}^{0} (-kx) \, dx = \tfrac{1}{2}kx_{max}^2 \qquad \text{[6.14]}$$

• *Work done by a spring*

The work done by the spring force on the block is positive because the spring force is in the same direction as the displacement (both are to the right). If we consider the work done by the spring force on the block as the block continues to move from $x_i = 0$ to $x_f = x_{max}$, we find that $W_s = -\tfrac{1}{2}kx_{max}^2$. This work is negative because for this part of the motion the displacement is to the right and the spring force is to the left. Therefore, the *net* work done by the spring force as the block moves from $x_i = -x_{max}$ to $x_f = x_{max}$ is *zero*.

If we plot $F_s$ versus $x$, as in Figure 6.9d, we arrive at the same results. The work calculated in Equation 6.14 is equal to the area of the shaded triangle in Figure 6.9d, with base $x_{max}$ and height $kx_{max}$. This area is $\tfrac{1}{2}kx_{max}^2$.

If the block undergoes an *arbitrary* displacement from $x = x_i$ to $x = x_f$, the work done by the spring force is

$$W_s = \int_{x_i}^{x_f} (-kx) \, dx = \tfrac{1}{2}kx_i^2 - \tfrac{1}{2}kx_f^2 \qquad \text{[6.15]}$$

From this equation we see that the work done by the spring force on the block is zero for any motion that ends where it began ($x_i = x_f$). We shall make use of this important result in Chapter 7, where we describe the motion of this system in more detail. Equation 6.15 also shows that the work done by the spring force is zero when the block moves between any two symmetric locations, $x_i = -x_f$. Consider the curve representing the spring force in Figure 6.9d; if the block moves from $x = -x_{max}$ to $x = +x_{max}$, the total work is zero because we are adding a positive area (for $-x_{max} < x < 0$) to a negative area (for $0 < x < +x_{max}$) of equal magnitude.

Equations 6.14 and 6.15 describe the work done by the spring force on the block. Now consider the work done by an *external agent* on the block as the spring is stretched *very slowly* from $x_i = 0$ to $x_f = x_{max}$, as in Figure 6.10. This work can be easily calculated by noting that the *applied force* $\mathbf{F}_{app}$ is of equal magnitude and opposite direction to the spring force $\mathbf{F}_s$ at any value of the displacement (because the block is not accelerating), so that $F_{app} = -(-kx) = +kx$. The work done by this applied force (the external agent) on the block is therefore

$$W_{F_{app}} = \int_{0}^{x_{max}} F_{app} \, dx = \int_{0}^{x_{max}} kx \, dx = \tfrac{1}{2}kx_{max}^2$$

Note that this work is equal to the negative of the work done by the spring force on the block for this displacement.

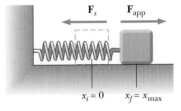

**Figure 6.10**

A block being pulled from $x_i = 0$ to $x_f = x_{max}$ on a frictionless surface by a force $\mathbf{F}_{app}$. If the process is carried out very slowly, the applied force is equal in magnitude and opposite in direction to the spring force at all times.

---

• ## Example 6.4    Work Required to Stretch a Spring

One end of a horizontal spring ($k = 80$ N/m) is held fixed while an external force is applied to the free end, stretching it slowly from $x_A = 0$ to $x_B = 4.0$ cm. (a) Find the work done by the external force on the spring.

**Solution**    Because we have not been told otherwise, we assume the spring obeys Hooke's law. We place the zero refer-

ence of the coordinate axis at the free end of the unstretched spring. The applied force is $F_{app} = kx = (80$ N/m$)(x)$. The work done by $F_{app}$ is the area of the triangle from 0 to 4.0 cm in Figure 6.11:

$$W = \tfrac{1}{2}kx_B^2 = \tfrac{1}{2}(80 \text{ N/m})(0.040 \text{ m})^2 = \boxed{0.064 \text{ J}}$$

(b) Find the additional work done in stretching the spring from $x_B = 4.0$ cm to $x_C = 7.0$ cm.

**Solution**    The work done in stretching the spring the additional amount is the darker shaded area between these limits in Figure 6.11. Geometrically, it is the difference in area between the large and small triangles:

$$W = \tfrac{1}{2}kx_C{}^2 - \tfrac{1}{2}kx_B{}^2$$
$$= \tfrac{1}{2}(80 \text{ N/m}) [(0.070 \text{ m})^2 - (0.040 \text{ m})^2] = \boxed{0.13 \text{ J}}$$

Using calculus, we find the same result:

$$W = \int_{x_B}^{x_C} F_{app}\, dx = \int_{0.04 \text{ m}}^{0.07 \text{ m}} (80 \text{ N/m})x\, dx$$

$$= \tfrac{1}{2}(80 \text{ N/m})(x^2)\Big|_{0.04 \text{ m}}^{0.07 \text{ m}}$$

$$W = \tfrac{1}{2}(80 \text{ N/m}) [(0.070 \text{ m})^2 - (0.040 \text{ m})^2] = 0.13 \text{ J}$$

**EXERCISE**    If an applied force varies with position according to $F_x = 3.00x^2 - 5.00$, where $x$ is in meters, how much work

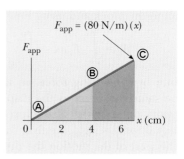

**Figure 6.11**

(Example 6.4) A graph of the applied force required to stretch a spring that obeys Hooke's law versus the elongation of the spring.

is done by this force on an object that moves from $x = 4.00$ m to $x = 7.00$ m?

*Answer*    264 J

## 6.5 • KINETIC ENERGY AND THE WORK–KINETIC ENERGY THEOREM

  See Screens 5.4 & 5.7

Now that we have explored various means of evaluating the work done by a force on a system, let us begin to explore the power of this approach. Solutions to problems using Newton's second law can be difficult if the forces in the problem are complicated. The alternative that enables us to understand and solve such problems relates the speed of the particle to its displacement under the influence of some net force from the environment. As we shall see in this section, if the work done by the net force on a particle can be calculated for a given displacement, the change in the particle's speed is easy to evaluate. Let's see how this is done.

Figure 6.12 shows an object modeled as a particle of mass $m$ moving to the right along the $x$ axis under the action of a net force $\Sigma \mathbf{F}$, also to the right. If the particle is moved through a displacement $\Delta x = x_f - x_i$, the work done by the force $\Sigma \mathbf{F}$ on the particle is

$$W_{net} = \int_{x_i}^{x_f} \Sigma F\, dx \qquad [6.16]$$

Using Newton's second law, we can substitute for the magnitude of the net force $\Sigma F = ma$, and then perform the following chain-rule manipulations on the integrand:

$$W_{net} = \int_{x_i}^{x_f} ma\, dx = \int_{x_i}^{x_f} m\frac{dv}{dt}\, dx = \int_{x_i}^{x_f} m\frac{dv}{dx}\frac{dx}{dt}\, dx = \int_{x_i}^{x_f} mv\, dv \qquad [6.17]$$

$$W_{net} = \tfrac{1}{2}mv_f{}^2 - \tfrac{1}{2}mv_i{}^2$$

This equation was generated for the specific situation of one-dimensional motion, but it is a general result. It tells us that the work done by the net force on a particle of mass $m$ is equal to the difference between the initial and final values of

**Figure 6.12**

An object modeled as a particle undergoes a displacement of magnitude $\Delta x$ and a change in speed under the action of a constant net force $\Sigma \mathbf{F}$.

a quantity $\frac{1}{2}mv^2$. This quantity $\frac{1}{2}mv^2$ is so important that we give it a special name—
the **kinetic energy** $K$ of a particle of mass $m$ moving with a speed $v$ is defined as

$$K \equiv \tfrac{1}{2}mv^2 \qquad [6.18]$$

- *Kinetic energy is energy associated with the motion of a particle*

Kinetic energy is a scalar quantity and has the same units as work. For example, an object of mass 2.0 kg moving with a speed of 4.0 m/s has a kinetic energy of 16 J. Table 6.1 shows some typical kinetic energies for various objects.

It is often convenient to write Equation 6.17 in the form

$$W_{\text{net}} = K_f - K_i = \Delta K \qquad [6.19]$$

- *The work–kinetic energy theorem*

Equation 6.19 is an important result known as the **work–kinetic energy theorem:**

> When work is done on a system and the only change in the system is in its speed, the work done by the net force equals the change in kinetic energy of the system.

The work–kinetic energy theorem indicates that the speed of a particle increases if the net work done on it is positive, because the final kinetic energy will be greater than the initial kinetic energy. The speed decreases if the net work is negative, because the final kinetic energy will be less than the initial kinetic energy.

The work–kinetic energy theorem will clarify some results we saw earlier in this chapter that may have seemed odd. In Thinking Physics 6.1, a person lifts a block and moves it horizontally. At the end of the Reasoning, it was mentioned that the net work done by all forces on the block is zero. This may seem strange, but it is correct. If we choose the block as the system, the net force on the system is zero because the upward lifting force is modeled as being equal to the gravitational force. Thus, the net force is zero, and zero net work is done, which is consistent with the fact that the kinetic energy of the block does not change. It may seem incorrect that no work was done, because something changed—the block was lifted—but this is correct *because we chose the block as the system.* If we had chosen the block and the Earth as the system, then we would have a different result—the work done on this system is not zero. We will explore this idea in the next chapter.

In Section 6.4, we also saw a result of zero work done, when we let a spring push a block from $x_i = -x_m$ to $x_f = x_m$. This is zero for a different reason from that for lifting the block. It is due to the combination of positive work and an equal amount of negative work done by the *same* force. It is also different from the

| TABLE 6.1 | Typical Kinetic Energies of Various Objects |
|---|---|
| **Object** | **Kinetic Energy (J)** |
| Earth | $3 \times 10^{33}$ |
| Moon | $4 \times 10^{28}$ |
| Space shuttle in orbit | $3 \times 10^{12}$ |
| Human sprinter | $4 \times 10^{3}$ |
| Flying bee | $6 \times 10^{-1}$ |
| Snail | $6 \times 10^{-8}$ |
| Electron in computer monitor | $4 \times 10^{-15}$ |

block example in that the speed of the block is continually changing. The work–kinetic energy theorem only refers to the initial and final points for the speeds—it does not depend on details of the path followed between these points. We shall use this concept often in the remainder of this and the next chapter.

## Example 6.5   A Block Pulled on a Frictionless Surface

A 6.00-kg block initially at rest is pulled to the right along a horizontal frictionless surface by a constant, horizontal force of 12.0 N, as in Figure 6.13. Find the speed of the block after it has moved 3.00 m.

**Reasoning**   The block is the system, and three external forces interact with it. Neither the gravitational force nor

the normal force does work on the block because these forces are vertical and the displacement is horizontal. In the horizontal direction, there is no friction. Thus, the only external force that we must consider in the calculation is the 12.0-N force.

**Solution**   The work done by the 12.0-N force is

$$W = F\Delta x = (12.0\text{ N})(3.00\text{ m}) = 36.0\text{ N}\cdot\text{m} = 36.0\text{ J}$$

Using the work–kinetic energy theorem and noting that the initial kinetic energy is zero, we find

$$W = K_f - K_i = \tfrac{1}{2}mv_f^2 - 0$$

$$v_f = \sqrt{\frac{2W}{m}} = \sqrt{\frac{2(36.0\text{ J})}{6.00\text{ kg}}} = \boxed{3.46\text{ m/s}}$$

Notice that an energy calculation such as this will give you only the speed of the particle, not the velocity. In many cases, this is all you need. If you need the direction of the velocity vector, you may need to analyze the pictorial representation or perform other calculations.

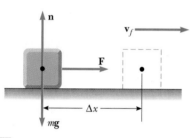

**Figure 6.13**

(Example 6.5) A block is pulled to the right by a constant horizontal force on a frictionless surface.

## Example 6.6   Speed of a Block on a Spring

A block of mass 1.6 kg is attached to a spring that has a force constant of $1.0 \times 10^3$ N/m as in Figure 6.14. The spring is compressed a distance of 2.0 cm, and the block is released from rest. Calculate the speed of the block as it passes

through the equilibrium position $x = 0$ if the surface is frictionless.

**Solution**   We identify the block as the system. Then, the force from the spring is the interaction with the environment that does work on the block. We use Equation 6.14 to find the work done by the spring on the block with $x_i = -x_{\text{max}} = -2.0\text{ cm} = -2.0 \times 10^{-2}$ m:

$$W_s = \tfrac{1}{2}kx_{\text{max}}^2 = \tfrac{1}{2}(1.0 \times 10^3\text{ N/m})(2.0 \times 10^{-2}\text{ m})^2 = 0.20\text{ J}$$

Using the work–kinetic energy theorem with $v_i - 0$ gives

$$W_s = \tfrac{1}{2}mv_f^2 - \tfrac{1}{2}mv_i^2 = \tfrac{1}{2}mv_f^2 = 0$$

$$v_f = \sqrt{\frac{2W_s}{m}} = \sqrt{\frac{2(0.20\text{ J})}{1.6\text{ kg}}} = \boxed{0.50\text{ m/s}}$$

**Figure 6.14**

(Example 6.6) A spring is compressed and pushes an object through 2.0 cm on a frictionless table.

## Example 6.7    Dropping a Block onto a Spring

A spring that has a force constant of $1.00 \times 10^3$ N/m is placed on a table in a vertical position as in Figure 6.15. A block of mass 1.60 kg is held 1.00 m above the free end of the spring. The block is dropped from rest so that it falls vertically onto the spring. By what distance does the spring compress?

**Solution**   We identify the block as the system. We identify the initial condition as the release of the block from the height $y_i = h = 1.00$ m above the free end of the spring. The final condition occurs when the block is momentarily at rest with the spring compressed its maximum distance. For this condition, the block is located at $y_f = -d$, where $d$ is the distance by which the spring is compressed. The net work on

**Figure 6.15**

(Example 6.7) A mass is dropped onto a vertical spring, causing it to compress.

the block during this displacement is done by gravity (positive work) and the spring force (negative work):

$$W_{net} = \mathbf{F}_g \cdot \Delta \mathbf{r} - \tfrac{1}{2}kd^2 = (-mg)\mathbf{j} \cdot (-d - h)\mathbf{j} - \tfrac{1}{2}kd^2$$
$$= mg(h + d) - \tfrac{1}{2}kd^2$$
$$= (1.60 \text{ kg})(9.80 \text{ m/s}^2)(1.00 \text{ m} + d)$$
$$\quad - \tfrac{1}{2}(1.00 \times 10^3 \text{ N/m})d^2$$
$$= -500d^2 + 15.7d + 15.7$$

The change in kinetic energy of the block is zero because it is at rest at both the initial and final conditions. Thus, from the work–kinetic energy theorem, the work done by the net force must be equal to zero:

$$-500d^2 + 15.7d + 15.7 = 0$$

This quadratic equation can be solved, and the solutions are $d = 0.19$ m and $d = -0.16$ m. Because we have chosen the value of $d$ as a positive number, by claiming that $y = -d$ is *below* the initial position of the end of the spring, we need to choose the positive root, $d = 0.19$ m. (The negative root gives the position for the final condition as $y = -d = -(-0.16 \text{ m}) = +0.16$ m, which is the position above the *initial* $y = 0$ at which the spring again comes to rest in its oscillation, assuming that the block remains attached to the spring. These two positions are symmetric around $y = -0.016$ m, which is where the block would rest in equilibrium on the spring, according to Hooke's law.)

## 6.6 • THE NONISOLATED SYSTEM

We have seen a number of examples in which an object, modeled as a particle, is acted on by various forces, resulting in a change in its kinetic energy. This very simple situation is the first example of the **nonisolated system,** which is an important new analysis model for us. Physical problems for which this model is appropriate involve systems that interact with or are influenced by their environment, causing some kind of change in the system.

The work–kinetic energy theorem is our first introduction to the nonisolated system. The interaction is the work done by the external force, and the quantity related to the system that changes is its kinetic energy. Because the energy of the system changes, we conceptualize the work as a means of **energy transfer—work has the effect of transferring energy between the system and the environment.** If positive work is done on the system, energy is transferred to the system, whereas negative work indicates that energy is transferred from the system to the environment.

So far, we have discussed kinetic energy as the only type of energy in a system. We now argue the existence of a second type of energy. Consider a situation in

• *Work is energy transfer*

**Figure 6.16**

A book sliding to the right on a horizontal surface slows down in the presence of a force of kinetic friction acting to the left. The initial velocity of the book is $\mathbf{v}_i$, and its final velocity is $\mathbf{v}_f$. The normal force and gravitational force are not included in the diagram because they are perpendicular to the direction of motion and therefore do not influence the speed of the book.

• *Internal energy*

• *Methods of energy transfer*

which an object slides along a surface with friction. Clearly, work is done by the friction force because there is a force and a displacement. Keep in mind, however, that our equations for work involve the displacement of the *point of application of the force*. The friction force is spread out over the entire contact area of an object sliding on a surface, so the force is not localized at a point. In addition, the magnitudes of the friction forces at various points are constantly changing as spot welds occur, the surface and the object deform locally, and so on. The points of application of the friction force on the object are jumping all over the face of the object in contact with the surface. In general, this means that the displacement of the point of application of the friction force (assuming we could calculate it!) is not the same as the displacement of the object.*

The work–kinetic energy theorem is valid for a particle or an object that can be modeled as a particle. When an object is extended and cannot be treated as a particle, however, things can become more complicated.

Let us imagine the book in Figure 6.16 sliding to the right on the surface of a heavy table and slowing down due to the friction force. Suppose the *surface* is the system. Then the friction force from the sliding book does work on the surface. The force on the surface is to the right and the displacement of the point of application of the force is to the right—the work is positive. But the surface is not moving after the book has stopped. Positive work has been done on the surface, yet the kinetic energy of the surface does not increase. Is this a violation of the work–kinetic energy theorem?

It is indeed a violation, because this situation does not fit the description of the conditions given for the work–kinetic energy theorem. Work is done on the system of the surface, but the result of that work is *not* an increase in kinetic energy. From your everyday experience with sliding over surfaces with friction, you can probably guess that the surface will be *warmer* after the book slides over it (rub your hands together briskly to experience this!). Thus, the work done has gone into warming the surface rather than causing it to increase in speed. We use the phrase **internal energy** $E_{\text{int}}$ for the energy associated with an object's temperature. (We will see a more general definition for internal energy in Chapter 17.) In this case, the work done on the surface does indeed represent energy transferred into the system, but it appears in the system as internal energy rather than kinetic energy.

We have now seen two methods of storing energy in a system: kinetic energy, related to motion of the system, and internal energy, related to its temperature. We have seen only one way to transfer energy into the system so far—work. Next, we introduce a few other ways to transfer energy into or out of a system, which will be studied in detail in other sections of the book. We will focus on the following six methods (Fig. 6.17) for transferring energy between the environment and the system.

**Work** (present chapter) is a method of transferring energy to a system by applying a force to the system and causing a displacement of the point of application of the force, as we have seen in the previous sections (Fig. 6.17a).

**Mechanical waves** (Chapter 13) are a means of transferring energy by allowing a disturbance to propagate through air or another medium. This is the method by which energy leaves a radio through the loudspeaker—sound—and by which energy enters your ears to stimulate the hearing process (Fig. 6.17b). Mechanical waves also include seismic waves and ocean waves.

---

* For more details on energy transfer situations involving forces of kinetic friction, see B. A. Sherwood and W. H. Bernard, *Am J Physics*, 52:1001, 1984; and R. P. Bauman, *Physics Teacher*, 30:264, 1992.

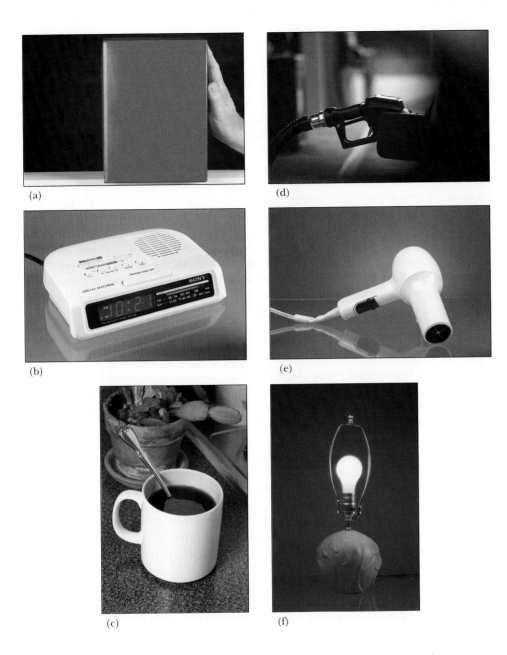

(a)

(d)

(b)

(e)

(c)

(f)

**Figure 6.17**

Energy transfer mechanisms. (a) Energy is transferred to the block by *work;* (b) energy leaves the radio by *mechanical waves;* (c) energy transfers up the handle of the spoon by *heat;* (d) energy enters the automobile gas tank by *matter transfer;* (e) energy enters the hair dryer by *electrical transmission;* and (f) energy leaves the light bulb by *electromagnetic radiation.* (*George Semple*)

**PITFALL PREVENTION 6.9**

**Heat is not a form of energy**

The word *heat* is one of the most misused words in our popular language. Note that heat is a method of *transferring* energy, it is *not* a form of storing energy. Thus, phrases such as "heat content," "the heat of the summer," and "the heat escaped" all represent *incorrect* uses of this word. We shall study heat in detail in Chapter 17.

**Heat** (Chapter 17) is a method of transferring energy by means of microscopic collisions; for example, the end of a metal spoon in a cup of coffee becomes hot because fast-moving electrons and atoms in the bowl of the spoon bump into slower ones in the nearby part of the handle (Fig. 6.17c). These particles move faster because of the collisions and bump into the next group of slow particles. Thus, the internal energy of the end of the spoon rises from energy transfer due to this bumping process. This process, also called *thermal conduction,* is caused by a temperature difference between two regions in space.*

* Many textbooks use the term *heat* to include *conduction, convection,* and *radiation.* Conduction is the only one of these three processes driven by a temperature difference alone, so we will restrict heat to this process in this book. Convection and radiation are included in other types of energy transfer in our list of six.

In **matter transfer** (Chapter 17), matter physically crosses the boundary of the system, carrying energy with it. Examples include filling the system of your automobile tank with gasoline (Fig. 6.17d) and carrying energy to the rooms of your home by means of circulating warm air from the furnace. Matter transfer occurs in several situations and is introduced in Chapter 17 by means of one example, *convection*.

**Electrical transmission** (Chapter 21) involves energy transfer by means of electric currents. This is how energy transfers into your stereo system or any other electrical device (Fig. 6.17e).

**Electromagnetic radiation** (Chapter 24) refers to electromagnetic waves such as light, microwaves, radio waves, and so on (Fig. 6.17f). Examples of this method of transfer include energy going into your baked potato in your microwave oven and light energy traveling from the Sun to the Earth through space.*

One of the central features of the energy approach is the notion that **we can neither create nor destroy energy—energy is *conserved*. Thus, if the amount of energy in a system changes, it can *only* be due to the fact that energy has crossed the boundary by a transfer mechanism such as one of the methods listed.** This is a general statement of the principle of **conservation of energy.** We can describe this idea mathematically as follows:

- *Conservation of energy—the continuity equation for energy*

$$\Delta E_{\text{system}} = \sum H \qquad \text{[6.20]}$$

where $E_{\text{system}}$ is the total energy of the system, including all methods of energy storage (kinetic, internal, and another to be discussed in Chapter 7) and $H$ is the amount of energy transferred across the system boundary by a transfer mechanism. Two of our transfer mechanisms have well-established symbolic notations. For work, $H_{\text{work}} = W$, as we have seen in the current chapter, and for heat, $H_{\text{heat}} = Q$, which we will see in detail in Chapter 17. The other four members of our list do not have established symbols, so we will call them $H_{\text{MW}}$ (mechanical waves), $H_{\text{MT}}$ (matter transfer), $H_{\text{ET}}$ (electrical transmission), and $H_{\text{ER}}$ (electromagnetic radiation).

We have seen how to calculate work in this chapter and the other types of transfers will be discussed in subsequent chapters. Equation 6.20 is called the **continuity equation for energy.** The full expansion of Equation 6.20, with kinetic and internal energy as the storage mechanisms, is

$$\Delta K + \Delta E_{\text{int}} = W + Q + H_{\text{MW}} + H_{\text{MT}} + H_{\text{ET}} + H_{\text{ER}}$$

This is the primary mathematical representation of the energy analysis of the non-isolated system. In most cases, this reduces to a much simpler equation because some of the terms are zero. A continuity equation arises in any situation where the change in a quantity in a system occurs solely because of transfers across the boundary (because the quantity is conserved), several examples of which occur in areas of physics, as we shall see.

This concept is no more complicated in theory than is that of balancing your checking account statement. If your account is the system, the change in the account balance for a given month is the sum of all the transfers—deposits, withdrawals, fees, interest, and checks written. It may be useful for you to think of energy as the *currency of nature!*

On a wind farm, the moving air does work on the blades of the windmills. The energy transferred into the system of the windmill appears as kinetic energy of the blades and the rotor of an electrical generator attached to the axis of the windmill. Energy is transferred out of the system of the windmill by electrical transmission. *(Billy Hustace/Stone)*

---

\* Electromagnetic radiation and work done by field forces are the only energy transfer mechanisms that do not require molecules of the environment to be available at the system boundary. Thus, systems surrounded by a vacuum (such as planets) can only exchange energy with the environment by means of these two possibilities.

Suppose a force is applied to a nonisolated system and the point of application of the force moves through a displacement. Suppose further that the only effect on the system is to increase its speed. Then the only transfer mechanism is work (so that $\Sigma\, H$ in Equation 6.20 reduces to just $W$) and the only kind of energy in the system that changes is the kinetic energy (so that $\Delta E_{\text{system}}$ reduces to just $\Delta K$). Equation 6.20 then becomes

$$\Delta K = W$$

which is the work–kinetic energy theorem, Equation 6.19. This theorem is a special case of the more general continuity equation for energy. We shall see several more examples of other special cases of the continuity equation for energy in future chapters.

## Quick Quiz 6.4

By what transfer mechanisms does energy enter and leave (a) your television set? (b) Your gasoline-powered lawn mower? (c) Your hand-cranked pencil sharpener?

### THINKING PHYSICS 6.3

An automobile strikes a tree and comes to rest. The total energy in the system (the car) is now less than before. How did the energy leave the system, and in what form is energy left in the system?

**Reasoning**    Energy transferred out of the system in a number of ways. The automobile did some *work* on the tree during the collision, causing the tree to deform. A large crash occurred during the collision, representing transfer of energy by *sound* (mechanical waves). If the gasoline tank leaks after the collision, the automobile is losing energy by means of *matter transfer*. After the car has come to rest, the system has no more kinetic energy. The automobile has more *internal energy* because its temperature is likely to be higher after the collision than before. We could express this process in terms of the continuity equation for energy as

$$\Delta E_{\text{system}} = \Sigma\, H \quad\longrightarrow\quad \Delta K + \Delta E_{\text{int}} = W + H_{\text{MW}} + H_{\text{MT}}$$

where $\Delta K$ is negative and $\Delta E_{\text{int}}$ is positive. All three of the energy transfer terms are negative because energy left the system by all three mechanisms.

### THINKING PHYSICS 6.4

A toaster is turned on. Discuss the forms of energy and energy transfer occurring in the coils of the toaster.

**Reasoning**    We identify the coils as the system. The energy that changes in the system is *internal energy* because the temperature of the coils rises. The energy transfer mechanism for energy coming into the coils is *electrical transmission* through the wire plugged into the wall. Energy is transferring out of the coils by *electromagnetic radiation* because the coils are hot and glowing. Some transfer of energy also occurs by *heat* from the hot surfaces of the coils into the air. We could express this process in terms of the continuity equation for energy as

$$\Delta E_{\text{system}} = \Sigma\, H \quad\longrightarrow\quad \Delta E_{\text{int}} = Q + H_{\text{ET}} + H_{\text{ER}}$$

After a short warm-up period, the coils will stabilize in temperature, and the internal energy will no longer change. In this situation, the energy input and output will be balanced:

$$0 = Q + H_{ET} + H_{ER} \quad \longrightarrow \quad -H_{ET} = Q + H_{ER}$$

Note that $Q$ and $H_{ER}$ are both negative because they represent energy leaving the system, and $H_{ET}$ is positive—energy continues to enter the system by electrical transmission.

### THINKING PHYSICS 6.5

Soft steel can be made red hot by continued hammering on it. This doesn't work as well for hard steel. Why is there this difference?

**Reasoning**   The rise in temperature of the soft steel is an example of transferring energy into a system by work and having it appear as an increase in the internal energy of the system. This works well for the soft steel because it is *soft*. This softness results in a deformation of the steel under the blow of the hammer. Thus, the point of application of the force is displaced by the hammer, and positive work is done on the steel. With the hard steel, less deformation occurs; thus, there is less displacement of the point of application of the force, and less work done on the steel. The soft steel is therefore better at absorbing energy from the hammer by means of work, and its temperature rises more rapidly.

## 6.7 • SITUATIONS INVOLVING KINETIC FRICTION

In the preceding section, we discussed the complicated nature of the friction force as an object slides across a surface. For these kinds of situations, Newton's second law is still valid for the system, even though the work–kinetic energy theorem is not. In the case of a nondeformable object like our block in Figure 6.12 sliding on the surface,* we can handle this in a relatively straightforward way, as follows.

Consider a situation in which a constant force is applied to the block and follow a similar procedure to that in going from Equation 6.16 to Equation 6.17. We start by multiplying Newton's second law (*x* component only) by a displacement of the block:

$$\left(\sum F_x\right)\Delta x = (ma_x)\,\Delta x \qquad \text{[6.21]}$$

For an object under constant acceleration, we know that the following relationships (Eqs. 2.8 and 2.10) are valid:

$$a_x = \frac{v_f - v_i}{t} \qquad \Delta x = \tfrac{1}{2}(v_i + v_f)t$$

where $v_i$ is the speed at $t = 0$ and $v_f$ is the speed at time $t$. Substituting these expressions into Equation 6.21 gives

$$\left(\sum F_x\right)\Delta x = m\left(\frac{v_f - v_i}{t}\right)\tfrac{1}{2}(v_i + v_f)t$$

$$\left(\sum F_x\right)\Delta x = \tfrac{1}{2}mv_f^2 - \tfrac{1}{2}mv_i^2$$

---

* The overall shape of the block remains the same, which is why we say it is nondeformable. On a microscopic level, however, the face of the block is deformed as it slides over the surface.

This *looks* like the work–kinetic energy theorem, but *the left-hand side has not been called work*. The quantity $\Delta x$ is the displacement of the block; it is *not* the displacement of the point of application of the friction force. The work–kinetic energy theorem can be applied only to a particle; the previous equation is true for an extended nondeformable object experiencing a constant force.

Let us now apply this equation to a block that has been projected across a surface. We imagine that the block has an initial velocity and slows down due to friction as the only force in the horizontal direction. The net force on the block is the kinetic friction force $\mathbf{f}_k$, which is directed opposite to the displacement $\Delta x$. Thus,

$$\left(\sum F_x\right)\Delta x = -f_k\,\Delta x = \tfrac{1}{2}mv_f^2 - \tfrac{1}{2}mv_i^2 = \Delta K \qquad \text{[6.22]}$$

which mathematically describes the decrease in kinetic energy due to the friction force.

If there are other forces besides friction acting on the object, the change in kinetic energy is the sum of that due to the other forces from the work–kinetic energy theorem and that due to friction:

$$\Delta K = -f_k\,\Delta x + \sum W_{\text{other forces}} \qquad \text{[6.23]}$$

Now, consider the larger system consisting of the block *and* the surface as the block slows down under the influence of a friction force alone. No work is done across the boundary of this system—the system does not interact with the environment. No other types of energy transfer occur across the boundary of the system, assuming we ignore the inevitable sound the sliding object makes—this is a simplification model! In this case, Equation 6.20 becomes

$$\Delta K + \Delta E_{\text{int}} = 0$$

The change in kinetic energy of this system is the same as the change in kinetic energy of the system of the block in Equation 6.22 because the block is the only part of the block–surface system that is moving. Thus,

$$\Delta K + \Delta E_{\text{int}} = 0 \quad \longrightarrow \quad -f_k\,\Delta x + \Delta E_{\text{int}} = 0$$

$$\Delta E_{\text{int}} = f_k\,\Delta x \qquad \text{[6.24]}$$

Therefore, the **increase in internal energy of the system is equal to the product of the friction force and the displacement of the block.**

The bottom line is that a friction force transforms kinetic energy to internal energy, and the increase in internal energy is equal to the decrease in kinetic energy.

### Quick Quiz 6.5

You are traveling along a freeway at 65 mi/h. Your car has kinetic energy. You now brake to a stop because of congestion in traffic. Where is the kinetic energy that your car once had? That is, what form does it now take, and where is this form of energy located?

### THINKING PHYSICS 6.6

In many situations, friction forces tend to reduce the kinetic energy of an object; however, friction forces can sometimes increase an object's kinetic energy. Describe a situation in which friction causes an increase in kinetic energy.

---

**PITFALL PREVENTION 6.10**
**It sure looks like work**

Although the left-hand side of Equation 6.22 does indeed look like work—force multiplied by displacement—we don't call it work, as discussed in the text. So despite the similarity, avoid saying that you are "calculating the work done by friction." The product $-f_k\,\Delta x$ represents "the change in kinetic energy of the system due to friction."

- *The change in kinetic energy of an object due to friction and other forces*

- *The increase in internal energy of a system due to friction*

**Reasoning**   If a crate is located on the bed of a truck, and the truck accelerates to the east, the static friction force exerted on the crate by the truck acts to the east to give the crate the same acceleration as the truck (assuming the crate doesn't slip). Another example is a car that accelerates because of the friction forces exerted on the car's tires by the road. These forces act in the direction of the car's motion, and the sum of these forces causes an increase in the car's kinetic energy.

### Example 6.8   A Block Pulled on a Rough Surface

A block of mass 6.00 kg initially at rest is pulled to the right by a constant horizontal force with magnitude $F = 12.0$ N (Fig. 6.18). The coefficient of kinetic friction between the block and the surface is 0.150. Find the speed of the block after it has moved 3.00 m. (This is Example 6.5, modified so that the surface is no longer frictionless.)

**Solution**   We define the system as the block and apply Equation 6.23:

$$\Delta K = -f_k \, \Delta x + \sum W_{\text{other forces}} = -\mu_k n \, \Delta x + F \, \Delta x$$

The block is modeled as a particle in equilibrium in the vertical direction, so that $n = mg$. Thus,

$$\Delta K = -\mu_k mg \, \Delta x + F \, \Delta x$$

Evaluating $\Delta K$ and solving for $v_f$, we have

$$\Delta K = \tfrac{1}{2}mv_f^2 - \tfrac{1}{2}mv_i^2 = -\mu_k mg \, \Delta x + F \, \Delta x$$

$$v_f = \sqrt{\left(\frac{2}{m}\right)\left(-\mu_k mg \, \Delta x + F \, \Delta x + \tfrac{1}{2}mv_i^2\right)}$$

Substituting the numerical values, we find,

$$v_f = \sqrt{\left(\frac{2}{6.00 \text{ kg}}\right)\left[-(0.150)(6.00 \text{ kg})(9.80 \text{ m/s}^2)(3.00 \text{ m}) + (12.0 \text{ N})(3.00 \text{ m}) + 0\right]}$$

$$= \boxed{1.78 \text{ m/s}}$$

Notice that this is less than the value calculated in Example 6.5, due to the effect of the friction force.

**Figure 6.18**

(Example 6.8) A block is pulled to the right by a constant horizontal force on a surface with friction.

## 6.8 • POWER

See Screen 5.8

We discussed transfers of energy across the boundary of a system by a number of methods. From a practical viewpoint, it is interesting to know not only the amount of energy transferred to a system but also the *rate* at which the energy is transferred. The time rate of energy transfer is called **power.**

We shall focus on work as our particular energy transfer method in this discussion, but keep in mind that the notion of power is valid for *any* means of energy transfer. If an external force is applied to an object (for which we will adopt the particle model), and if the work done by this force is $W$ in the time interval $\Delta t$, then the **average power** during this interval is defined as

• *Average power when work is done*

$$\overline{\mathcal{P}} \equiv \frac{W}{\Delta t} \qquad\qquad \text{[6.25]}$$

The **instantaneous power** $\mathcal{P}$ at a particular point in time is the limiting value of the average power as $\Delta t$ approaches zero:

$$\mathcal{P} \equiv \lim_{\Delta t \to 0} \frac{W}{\Delta t} = \frac{dW}{dt} \qquad \text{[6.26]}$$

where we represent the infinitesimal value of the work done by $dW$. We know from Equation 6.4 that we can write the infinitesimal amount of work done over a displacement $d\mathbf{r}$ as $dW = \mathbf{F} \cdot d\mathbf{r}$. Therefore, the instantaneous power can be written

$$\mathcal{P} = \frac{dW}{dt} = \mathbf{F} \cdot \frac{d\mathbf{r}}{dt} = \mathbf{F} \cdot \mathbf{v} \qquad \text{[6.27]}$$

where we have used the fact that $\mathbf{v} = d\mathbf{r}/dt$.

In general, power is defined for any type of energy transfer. The most general expression for power is, therefore

$$\mathcal{P} = \frac{dE}{dt} \qquad \text{[6.28]}$$

• *General expression for power*

where $dE/dt$ is the rate at which energy is crossing the boundary of the system by a given transfer mechanism.

The SI unit of power is joules per second (J/s), also called a **watt** (W) (after James Watt):

$$1 \text{ W} = 1 \text{ J/s} = 1 \text{ kg} \cdot \text{m}^2/\text{s}^3$$

• *The watt*

The unit of power in the British engineering system is the **horsepower** (hp):

$$1 \text{ hp} \equiv 550 \text{ ft} \cdot \text{lb/s} \equiv 746 \text{ W}$$

A new unit of energy can now be defined in terms of the unit of power. One **kilowatt hour** (kWh) is the energy transferred in 1 h at the constant rate of 1 kW. The numerical value of 1 kWh of energy is

$$1 \text{ kWh} = (10^3 \text{ W})(3600 \text{ s}) = 3.60 \times 10^6 \text{ J}$$

It is important to realize that a kilowatt hour is a unit of energy, not power. When you pay your electric bill, you are buying energy, and the amount of energy transferred by electrical transmission into a home during the period represented by the electric bill is usually expressed in kilowatt hours. For example, your bill may state that you used 900 kWh of energy during a month, and you are being charged at the rate of 10¢ per kWh. Your obligation is then $90 for this amount of energy. As another example, suppose an electric bulb is rated at 100 W. In 1.00 h of operation, it would have energy transferred to it by electrical transmission in the amount of $(0.100 \text{ kW})(1.00 \text{ h}) = 0.100 \text{ kWh} = 3.60 \times 10^5 \text{ J}$.

## Quick Quiz 6.6

An older model car accelerates from 0 to speed $v$ in 10 s. A newer, more powerful sports car accelerates from 0 to $2v$ in the same time. What is the ratio of powers for the two cars if they have the same mass?

**PITFALL PREVENTION 6.11**

**Be careful with power**

Students can be confused about power in two ways. First, do not confuse the symbol W for the watt with the italic symbol $W$ for work. Second, students sometimes forget that the watt represents a rate of energy transfer and try to talk about "watts per second." We have seen rates expressed in earlier chapters as "something per second," as in the case of velocity and acceleration. Here, we have another rate, but the action of defining the joule per second as a watt has "hidden" the "per second" phrase in the definition. Always remember that the watt is *the same as* a joule per second.

## Example 6.9   Power Delivered by an Elevator Motor

A 1 000-kg elevator carries a maximum load of 800 kg. A constant friction force of 4 000 N retards its motion upward, as in Figure 6.19. (a) What is the minimum power delivered by the motor to lift the elevator at a constant speed of 3.00 m/s?

**Solution**   We use two analysis models for the elevator. First, we model it as a particle in equilibrium because it moves at constant speed. The motor must supply the force **T** that results in the tension in the cable that pulls the elevator upward. From Newton's second law and from the fact that $a = 0$ because $v$ is constant, we have

$$T - f - Mg = 0$$

(a)                                   (b)

**Figure 6.19**

(Example 6.9) (a) A motor lifts an elevator car. (b) Free-body diagram for the elevator. The motor exerts an upward force **T** on the supporting cables. The magnitude of this force is $T$, the tension in the cables, which is applied in the upward direction on the elevator. The downward forces on the elevator are the friction force **f** and the gravitational force $\mathbf{F}_g = M\mathbf{g}$.

where $M$ is the *total* mass (elevator plus load), equal to 1 800 kg. Therefore,

$$T = f + Mg$$
$$= 4.00 \times 10^3\,\text{N} + (1.80 \times 10^3\,\text{kg})(9.80\,\text{m/s}^2)$$
$$= 2.16 \times 10^4\,\text{N}$$

Now, we model the elevator as a nonisolated system. Work is being done on it by the tension force (as well as other forces). We can use Equation 6.27 to evaluate the power delivered by the motor, which is the rate at which work is done on the elevator by the tension force. Because **T** is in the same direction as **v**, we have

$$\mathcal{P} = \mathbf{T} \cdot \mathbf{v} = Tv$$
$$= (2.16 \times 10^4\,\text{N})(3.00\,\text{m/s}) = 6.48 \times 10^4\,\text{W}$$
$$= \boxed{64.8\,\text{kW}}$$

Because **T** is the force the motor applies to the cable, the preceding result represents the rate at which energy is being transferred out of the motor by doing work on the cable.

(b) What power must the motor deliver at any instant if it is designed to provide an upward acceleration of 1.00 m/s²?

**Solution**   In this case, we expect the tension to be larger than in part (a) because the cable must now cause an upward acceleration of the elevator. Applying Newton's second law to the elevator gives

$$T - f - Mg = Ma$$
$$T = M(a + g) + f$$
$$= (1.80 \times 10^3\,\text{kg})(1.00\,\text{m/s}^2 + 9.80\,\text{m/s}^2)$$
$$+ 4.00 \times 10^3\,\text{N}$$
$$= 2.34 \times 10^4\,\text{N}$$

Therefore, using Equation 6.27, we have for the required power

$$\mathcal{P} = Tv = \boxed{(2.34 \times 10^4\,v)}$$

where $v$ is the instantaneous speed of the elevator in meters per second. Hence, the power required increases with increasing speed.

### context connection

## 6.9 • A PROBE TO THE SUN

Suppose that a secondary purpose of our mission to Mars is to release a probe as we leave the Earth that will fall into the Sun, sending solar data back to Earth as it rushes toward its demise. The design plan for the probe is that we will release it from our spacecraft in such a way that it has zero velocity relative to the Sun. The probe has a mass of 200 kg and no rocket engine, so it simply falls into the Sun from rest due to the gravitational force from the Sun. We will form a simplification model by ignoring all other gravitational forces on the probe from other planets. Our question is this: With what speed will the probe arrive at the surface of the Sun?

If we imagine trying to model the probe as a particle under a net force, we run into trouble. Newton's second law allows us to find the acceleration of the probe. But because the gravitational force is not constant, the acceleration is not constant. We haven't discussed a kinematics model for a particle under varying acceleration. Thus, the problem facing us is complicated. In theory, we could solve this problem of Newton's second law using calculus, but the energy approach gives us a different path to follow, which will be very fruitful for us in solving future problems.

We will model the probe as a system consisting of a single particle. The gravitational force from the Sun acts as the interaction with the environment, doing work on the probe. The result of this work is the increasing kinetic energy of the probe as it falls toward the Sun. (Internal energy also increases as the probe warms up on approach to the Sun, but this is due to electromagnetic radiation—this is a separate process and will not be addressed here.) Thus, this situation is a candidate for use of the work–kinetic energy theorem.

Let us first calculate the amount of work done by the gravitational force on the probe as it falls to the surface of the Sun. The force on the probe is given by Newton's law of universal gravitation. The initial separation of the probe and the Sun is the radius of the Earth's orbit, $1.50 \times 10^{11}$ m. The final separation is the radius of the Sun, $6.96 \times 10^{8}$ m. We use Equation 6.11 in the radial direction from the Sun to find the work done by the varying gravitational force:

$$W = \int_{r_i}^{r_f} F_r \, dr = \int_{r_i}^{r_f} - G \frac{M_{\mathrm{Sun}} m_{\mathrm{probe}}}{r^2} \, dr = - GM_{\mathrm{Sun}} m_{\mathrm{probe}} \int_{r_i}^{r_f} \frac{1}{r^2} \, dr$$

$$= - GM_{\mathrm{Sun}} m_{\mathrm{probe}} \left( -\frac{1}{r} \right) \bigg|_{r_i}^{r_f} = GM_{\mathrm{Sun}} m_{\mathrm{probe}} \left( \frac{1}{r_f} - \frac{1}{r_i} \right)$$

In this calculation, the radial component of the gravitational force has been substituted as a negative expression because the radial component vector $\mathbf{F}_r$ is toward the Sun, but the radial unit vector $\hat{\mathbf{r}}$ points away from the Sun.

Now, let us substitute the numerical values for the masses and the initial and final separation distances:

$$W = (6.67 \times 10^{-11} \text{ N} \cdot \text{m}^2/\text{kg}^2)(1.99 \times 10^{30} \text{ kg})(200 \text{ kg})$$

$$\times \left( \frac{1}{6.96 \times 10^8 \text{ m}} - \frac{1}{1.50 \times 10^{11} \text{ m}} \right)$$

$$= 3.80 \times 10^{13} \text{ J}$$

Finally, we set this work done on the probe equal to the change in the kinetic energy of the probe, according to the work–kinetic energy theorem:

$$W = \Delta K = \tfrac{1}{2} m_{\mathrm{probe}} \, v_f^2 - \tfrac{1}{2} m_{\mathrm{probe}} \, v_i^2 = \tfrac{1}{2} m_{\mathrm{probe}} \, v_f^2 - 0 \quad \longrightarrow \quad v_f = \sqrt{\frac{2W}{m_{\mathrm{probe}}}}$$

Substituting the numerical values, we have,

$$v_f = \sqrt{\frac{2(3.80 \times 10^{13} \text{ J})}{200 \text{ kg}}} = 6.16 \times 10^5 \text{ m/s}$$

This solution is about as much effort as that necessary in finding the acceleration from Newton's second law and using calculus to find the final velocity. In the next chapter, however, we will see some types of problems that are very tedious using calculus but become extremely simple when the energy approach is used. In particular, we shall use an energy approach to find an upper limit to the speed with which we need to leave the Earth to arrive at Mars.

## SUMMARY

A **system** can be a single particle, a collection of particles, or a region of space. A **system boundary** separates the system from the **environment.** Many physics problems can be solved by considering the interaction of a system with its environment.

The **work** done by a *constant* force $\mathbf{F}$ on a particle is defined as the product of the component of the force in the direction of the particle's displacement and the magnitude of the displacement. If $\mathbf{F}$ makes an angle $\theta$ with the displacement $\Delta\mathbf{r}$, the work done by $\mathbf{F}$ is

$$W \equiv F\,\Delta r\cos\theta \qquad [6.1]$$

The **scalar,** or **dot, product** of any two vectors $\mathbf{A}$ and $\mathbf{B}$ is defined by the relationship

$$\mathbf{A}\cdot\mathbf{B} \equiv AB\cos\theta \qquad [6.3]$$

where the result is a scalar quantity and $\theta$ is the angle between the directions of the two vectors. The scalar product obeys the commutative and distributive laws.

The scalar product allows us to write the work done by a constant force $\mathbf{F}$ on a particle as

$$W = \mathbf{F}\cdot\Delta\mathbf{r} \qquad [6.4]$$

The work done by a *varying* force acting on a particle moving along the *x* axis from $x_i$ to $x_f$ is

$$W \equiv \int_{x_i}^{x_f} F_x\,dx \qquad [6.11]$$

where $F_x$ is the component of force in the *x* direction. If several forces act on the particle, the net work done by all forces is the sum of the individual amounts of work done by each force.

The **kinetic energy** of a particle of mass *m* moving with a speed *v* is

$$K = \tfrac{1}{2}mv^2 \qquad [6.18]$$

The **work–kinetic energy theorem** states that when work is done on a system and the only change in the system is in its speed, the net work done on the system by external forces equals the change in kinetic energy of the system:

$$W_{\text{net}} = K_f - K_i = \tfrac{1}{2}mv_f^2 - \tfrac{1}{2}mv_i^2 \qquad [6.19]$$

For a nonisolated system, we can equate the change in the total energy stored in the system to the sum of all the transfers across the system boundary:

$$\Delta E_{\text{system}} = \sum H \qquad [6.20]$$

This is the **continuity equation for energy.** Methods of energy transfer ($H$) include **work** ($H = W$), **mechanical waves** ($H_{\text{MW}}$), **heat** ($H = Q$), **matter transfer** ($H_{\text{MT}}$), **electrical transmission** ($H_{\text{ET}}$), and **electromagnetic radiation** ($H_{\text{ER}}$). Storage mechanisms ($E_{\text{system}}$) seen in this chapter include **kinetic energy** $K$ and **internal energy** $E_{\text{int}}$. The continuity equation arises because **energy is conserved**—we can neither create nor destroy energy. The work–kinetic energy theorem is a special case of the continuity equation for energy in situations in which work is the only transfer mechanism and kinetic energy is the only type of energy storage in the system.

In the case of an object sliding through a displacement $\Delta x$ over a surface with friction, the change in kinetic energy of the system is

$$\Delta K = -f_k\,\Delta x + \sum W_{\text{other forces}} \qquad [6.23]$$

where $f_k$ is the force of kinetic friction and $\sum W_{\text{other forces}}$ is the work done by all forces other than friction.

**Average power** is the time rate of energy transfer. If we use work as the energy transfer mechanism,

$$\overline{\mathcal{P}} \equiv \frac{W}{\Delta t} \qquad [6.25]$$

If an agent applies a force $\mathbf{F}$ to an object moving with a velocity $\mathbf{v}$, the **instantaneous power** delivered by that agent is

$$\mathcal{P} \equiv \frac{dW}{dt} = \mathbf{F}\cdot\mathbf{v} \qquad [6.27]$$

Because power is defined for any type of energy transfer, the general expression for power is

$$\mathcal{P} = \frac{dE}{dt} \qquad [6.28]$$

# QUESTIONS

1. Explain why the work done by the force of sliding friction is negative when an object undergoes a displacement on a rough surface.

2. When a particle rotates in a circle, a force acts on it directed toward the center of rotation. Why is it that this force does no work on the particle?

3. When a punter kicks a football, is he doing any work on the ball while his toe is in contact with it? Is he doing any work on the ball after it loses contact with his toe? Are any forces doing work on the ball while it is in flight?

4. A team of furniture movers wishes to load a truck using a ramp from the ground to the rear of the truck. One of the movers claims that less work would be required to load the truck if the length of the ramp were increased, reducing the angle of the ramp with respect to the horizontal. Is his claim valid? Explain.

5. Cite two examples in which a force is exerted on an object without doing any work on the object.

6. As a simple pendulum swings back and forth, the forces acting on the suspended mass are the gravitational force, the tension in the supporting cord, and air resistance. (a) Which of these forces, if any, does no work on the pendulum? (b) Which of these forces does negative work at all times during its motion? (c) Describe the work done by the gravitational force while the pendulum is swinging.

7. If the dot product of two vectors is positive, does this imply that the vectors must have positive rectangular components?

8. A hockey player pushes a puck with his hockey stick over frictionless ice. If the puck starts from rest and ends up moving at speed $v$, has the hockey player performed any work on the puck in the reference frame of the ice? An ant is on the puck and hangs on tightly while the puck is accelerated. In the reference frame of the ant, has any work been done by the hockey stick on the puck? Is the work–kinetic energy theorem satisfied for each observer?

9. Can kinetic energy be negative? Explain.

10. One bullet has twice the mass of a second bullet. If both are fired so that they have the same speed, which has more kinetic energy? What is the ratio of the kinetic energies of the two bullets?

11. If the speed of a particle is doubled, what happens to its kinetic energy?

12. A spring is stretched by a distance $x_{max}$, resulting in work $W$ being done by the external agent. The spring is now cut in half and stretched by the same distance. How much work is done in this case?

13. You are working in a library, reshelving books. You lift a book from the floor to the top shelf. The kinetic energy of the book on the floor was zero, and the kinetic energy of the book on the top shelf is zero, so no change occurs in kinetic energy. Yet you did some work in lifting the book. Is the work–kinetic energy theorem violated?

14. What can be said about the speed of a particle if the net work done on it is zero?

15. Can the average power over a time interval ever equal the instantaneous power at an instant within the interval? Explain.

16. In Example 6.9, does the required power increase or decrease as the force of friction is reduced?

17. Sometimes, physics words are used in popular literature in interesting ways. For example, consider a description of a rock falling from the top of a cliff as "gathering force as it falls to the beach below." What does the phrase "gathering force" mean, and can you repair this phrase?

# PROBLEMS

1, 2, 3 = straightforward, intermediate, challenging    □ = full solution available in the *Student Solutions Manual and Study Guide*

**web** = solution posted at **http://www.harcourtcollege.com/physics/**    💻 = computer useful in solving problem

🖳 = Interactive Physics    ▦ = paired numerical/symbolic problems    ▨ = life science application

## Section 6.2   Work Done by a Constant Force

1. A block of mass 2.50 kg is pushed 2.20 m along a frictionless horizontal table by a constant 16.0-N force directed 25.0° below the horizontal. Determine the work done on the block by (a) the applied force, (b) the normal force exerted by the table, and (c) the gravitational force. (d) Determine the total work done on the block.

2. A shopper in a supermarket pushes a cart with a force of 35.0 N directed at an angle of 25.0° downward from the horizontal. Find the work done by the shopper on the cart as he moves down an aisle 50.0 m long.

3. Batman, whose mass is 80.0 kg, is dangling on the free end of a 12.0-m rope, the other end of which is fixed to a tree limb above. He is able to get the rope in motion as only Batman knows how, eventually getting it to swing enough that he can reach a ledge when the rope makes a 60.0° angle with the vertical. How much work is done by the gravitational force in this maneuver?

4. A raindrop of mass $3.35 \times 10^{-5}$ kg falls vertically at constant speed under the influence of gravity and air resistance. After the drop has fallen 100 m, what is the work done on the raindrop (a) by gravity and (b) by air resistance?

## Section 6.3    The Scalar Product of Two Vectorsz

5. Vector **A** has a magnitude of 5.00 units, and **B** has a magnitude of 9.00 units. The two vectors make an angle of 50.0° with each other. Find **A · B**.

6. For any two vectors **A** and **B**, show that $\mathbf{A} \cdot \mathbf{B} = A_x B_x + A_y B_y + A_z B_z$. (*Hint:* Write **A** and **B** in unit vector form and use Equations 6.7 and 6.8.)

> In problems 7 through 10, calculate numerical answers to three significant figures.

7. A force $\mathbf{F} = (6\mathbf{i} - 2\mathbf{j})$ N acts on a particle that undergoes a displacement $\Delta\mathbf{r} = (3\mathbf{i} + \mathbf{j})$ m. Find (a) the work done by the force on the particle and (b) the angle between **F** and $\Delta\mathbf{r}$.
   web

8. Find the scalar product of the vectors in Figure P6.8.

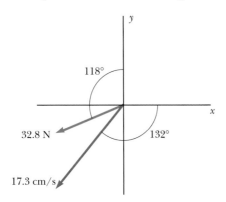

**Figure P6.8**

9. Using the definition of the scalar product, find the angles between (a) $\mathbf{A} = 3\mathbf{i} - 2\mathbf{j}$ and $\mathbf{B} = 4\mathbf{i} - 4\mathbf{j}$; (b) $\mathbf{A} = -2\mathbf{i} + 4\mathbf{j}$ and $\mathbf{B} = 3\mathbf{i} - 4\mathbf{j} + 2\mathbf{k}$; (c) $\mathbf{A} = \mathbf{i} - 2\mathbf{j} + 2\mathbf{k}$ and $\mathbf{B} = 3\mathbf{j} + 4\mathbf{k}$.

10. For $\mathbf{A} = 3\mathbf{i} + \mathbf{j} - \mathbf{k}$, $\mathbf{B} = -\mathbf{i} + 2\mathbf{j} + 5\mathbf{k}$, and $\mathbf{C} = 2\mathbf{j} - 3\mathbf{k}$, find $\mathbf{C} \cdot (\mathbf{A} - \mathbf{B})$.

## Section 6.4    Work Done by a Varying Force

11. A particle is subject to a force $F_x$ that varies with position as in Figure P6.11. Find the work done by the force on the object as it moves (a) from $x = 0$ to $x = 5.00$ m, (b) from $x = 5.00$ m to $x = 10.0$ m, and (c) from $x = 10.0$ m to $x = 15.0$ m. (d) What is the total work done by the force over the displacement $x = 0$ to $x = 15.0$ m?
    web

12. The force (in newtons) acting on a particle is $F_x = (8.00x - 16.0)$, where $x$ is in meters. (a) Make a plot of this force versus $x$ from $x = 0$ to $x = 3.00$ m. (b) From your graph, find the net work done by this force on the particle as it moves from $x = 0$ to $x = 3.00$ m.

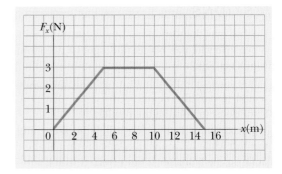

**Figure P6.11**    Problems 11 and 22.

13. When a 4.00-kg object is hung vertically on a certain light spring that obeys Hooke's law, the spring stretches 2.50 cm. If the 4.00-kg object is removed, (a) how far will the spring stretch if a 1.50-kg object is hung on it, and (b) how much work must an external agent do to stretch the same spring 4.00 cm from its unstretched position?

14. An archer pulls her bowstring back 0.400 m by exerting a force that increases uniformly from zero to 230 N. (a) What is the equivalent force constant of the bow? (b) How much work does the archer do in pulling the bow?

15. A 6 000-kg freight car rolls along rails with negligible friction. The car is brought to rest by a combination of two coiled springs, as illustrated in Figure P6.15. Both springs obey Hooke's law with $k_1 = 1\,600$ N/m and $k_2 =$

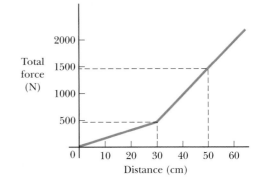

**Figure P6.15**

3 400 N/m. After the first spring compresses a distance of 30.0 cm, the second spring acts with the first to increase the force as additional compression occurs, as shown in the graph. If the car is brought to rest 50.0 cm after first contacting the two-spring system, find the car's initial speed.

16. A 100-g bullet is fired from a rifle having a barrel 0.600 m long. Assuming the origin is placed where the bullet begins to move, the force (in newtons) exerted on the bullet by the expanding gas is $15\,000 + 10\,000x - 25\,000x^2$, where $x$ is in meters. (a) Determine the work done by the gas on the bullet as the bullet travels the length of the barrel. (b) If the barrel is 1.00 m long, how much work is done, and how does this value compare to the work calculated in part (a)?

17. A force (in newtons) $\mathbf{F} = (4x\mathbf{i} + 3y\mathbf{j})$ acts on an object as the object moves in the $x$ direction from the origin to $x = 5.00$ m. Find the work done on the object by the force.

18. A small object of mass $m$ is pulled to the top of a frictionless half-cylinder (of radius $R$) by a cord that passes over the top of the cylinder, as illustrated in Figure P6.18. (a) If the object moves at a constant speed, show that $F = mg\cos\theta$. (*Hint:* If the object moves at constant speed, the component of its acceleration tangent to the cylinder must be zero at all times.) (b) By directly integrating $W = \int \mathbf{F} \cdot d\mathbf{r}$, find the work done in moving the object at constant speed from the bottom to the top of the half-cylinder.

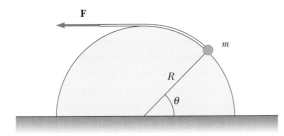

**Figure P6.18**

## Section 6.5 Kinetic Energy and the Work–Kinetic Energy Theorem

19. A 0.600-kg particle has a speed of 2.00 m/s at point $A$ and kinetic energy of 7.50 J at point $B$. What is (a) its kinetic energy at $A$? (b) Its speed at $B$? (c) The total work done on the particle as it moves from $A$ to $B$?

20. A 0.300-kg ball has a speed of 15.0 m/s. (a) What is its kinetic energy? (b) If its speed were doubled, what would be its kinetic energy?

21. A 3.00-kg object has an initial velocity $\mathbf{v}_i = (6.00\mathbf{i} - 2.00\mathbf{j})$ m/s. (a) What is its kinetic energy at this time? (b) Find the total work done on the object if its velocity changes to $(8.00\mathbf{i} + 4.00\mathbf{j})$ m/s. (*Hint:* Remember that $v^2 = \mathbf{v} \cdot \mathbf{v}$.)

22. A 4.00-kg particle is subject to a total force that varies with position as shown in Figure P6.11. The particle starts from rest at $x = 0$. What is its speed at (a) $x = 5.00$ m, (b) $x = 10.0$ m, (c) $x = 15.0$ m?

23. A 2 100-kg pile driver is used to drive a steel I-beam into the ground. The pile driver falls 5.00 m before coming into contact with the top of the beam, and it drives the beam 12.0 cm farther into the ground before coming to rest. Using energy considerations, calculate the average force the beam exerts on the pile driver while the pile driver is brought to rest.

24. You can think of the work–kinetic energy theorem as a second theory of motion, parallel to Newton's laws in describing how outside influences affect the motion of an object. In this problem, work out parts (a) and (b) separately from parts (c) and (d) to compare the predictions of the two theories. In a rifle barrel, a 15.0-g bullet is accelerated from rest to a speed of 780 m/s. (a) Find the work done on the bullet. (b) If the rifle barrel is 72.0 cm long, find the magnitude of the average total force acted on it, as $F = W/(\Delta r \cos\theta)$. (c) Find the constant acceleration of a bullet that starts from rest and gains a speed of 780 m/s over a distance of 72.0 cm. (d) If the bullet has mass 15.0 g, find the total force that acted on it as $\Sigma F = ma$.

25. A block of mass 12.0 kg slides from rest down a frictionless 35.0° incline and is stopped by a strong spring with $k = 3.00 \times 10^4$ N/m. The block slides 3.00 m from the point of release to the point where it comes to rest against the spring. When the block comes to rest, how far has the spring been compressed?

26. In the neck of the picture tube of a certain black-and-white television set, an electron gun contains two charged metallic plates 2.80 cm apart. An electric force accelerates each electron in the beam from rest to 9.60% of the speed of light over this distance. (a) Determine the kinetic energy of the electron as it leaves the electron gun. Electrons carry this energy to a phosphorescent material on the inner surface of the television screen, making it glow. For an electron passing between the plates in the electron gun, determine (b) the magnitude of the constant force acting on the electron, (c) the acceleration, and (d) the time of flight.

27. An Atwood machine (see Fig. 4.13a) supports objects with masses of 0.200 kg and 0.300 kg. The objects are held at rest beside each other and then released. Ignoring friction, what is the speed of each object at the instant it has moved 0.400 m?

## Section 6.7 Situations Involving Kinetic Friction

28. A sledge loaded with bricks has a total mass of 18.0 kg and is pulled at constant speed by a rope. The rope is inclined at 20.0° above the horizontal, and the cart moves a distance of 20.0 m on a horizontal surface. The coefficient of

kinetic friction between the sledge and surface is 0.500. (a) What is the tension in the rope? (b) How much work is done on the sledge by the rope? (c) What is the increase in internal energy in the sledge and the surface due to friction?

29. A 40.0-kg box initially at rest is pushed 5.00 m along a rough, horizontal floor with a constant applied horizontal force of 130 N. If the coefficient of friction between box and floor is 0.300, find (a) the work done by the applied force, (b) the increase in internal energy due to friction, (c) the work done by the normal force, (d) the work done by gravity, (e) the change in kinetic energy of the box, and (f) the final speed of the box.

30. A 15.0-kg block is dragged over a rough, horizontal surface by a 70.0-N force acting at 20.0° above the horizontal. The block is displaced 5.00 m, and the coefficient of kinetic friction is 0.300. Find the work done by (a) the 70-N force, (b) the normal force, and (c) the gravitational force. (d) What is the increase in internal energy due to friction? (e) Find the total change in the block's kinetic energy.

31. A sled of mass $m$ is given a kick on a frozen pond. The kick imparts to it an initial speed of 2.00 m/s. The coefficient of kinetic friction between sled and ice is 0.100. Use energy considerations to find the distance the sled moves before it stops.

32. A 2.00-kg block is attached to a spring of force constant 500 N/m as in Figure 6.9. The block is pulled 5.00 cm to the right of equilibrium and released from rest. Find the speed of the block as it passes through equilibrium if (a) the horizontal surface is frictionless and (b) the coefficient of friction between block and surface is 0.350.

33. A crate of mass 10.0 kg is pulled up a rough incline with an initial speed of 1.50 m/s. The pulling force is 100 N parallel to the incline, which makes an angle of 20.0° with the horizontal. The coefficient of kinetic friction is 0.400, and the crate is pulled 5.00 m. (a) How much work is done by gravity? (b) Determine the increase in internal energy due to friction. (c) How much work is done by the 100-N force? (d) What is the change in kinetic energy of the crate? (e) What is the speed of the crate after being pulled 5.00 m?

## Section 6.8 Power

34. Make an order-of-magnitude estimate of the power a car engine contributes to speeding the car up to highway speed. As a concrete example, consider your own car if you use one. In your solution state the physical quantities you take as data and the values you measure or estimate for them. The mass of the vehicle is given in the owner's manual. If you do not wish to estimate for a car, consider a bus or truck that you specify.

35. A 700-N Marine in basic training climbs a 10.0-m vertical rope at a constant speed in 8.00 s. What is his power output?

36. A certain automobile engine delivers $2.24 \times 10^4$ W (30.0 hp) to its wheels when moving at a constant speed of 27.0 m/s ($\approx 60$ mi/h). What is the resistive force acting on the automobile at that speed?

37. A skier of mass 70.0 kg is pulled up a slope by a motor-driven cable. (a) How much work is required to pull him a distance of 60.0 m up a 30.0° slope (assumed frictionless) at a constant speed of 2.00 m/s? (b) A motor of what power is required to perform this task?

38. A 650-kg elevator starts from rest. It moves upward for 3.00 s with constant acceleration until it reaches its cruising speed of 1.75 m/s. (a) What is the average power of the elevator motor during this period? (b) How does this power compare with the motor power when the elevator moves at its cruising speed?

39. Energy is conventionally measured in Calories as well as in joules. One Calorie in nutrition is equal to one kilocalorie, which we will define in Chapter 17 as 1 kcal = 4186 J. Metabolizing one gram of fat can release 9.00 kcal. A student decides to try to lose weight by exercising. She plans to run up and down the stairs in a football stadium as fast as she can and as many times as necessary. Is this in itself a practical way to lose weight? To evaluate the program, suppose she runs up a flight of 80 steps, each 0.150 m high, in 65.0 s. For simplicity, ignore the energy she uses in coming down (this is small). Assume that a typical efficiency for human muscles is 20.0%. This means that when your body converts 100 J from metabolizing fat, 20 J goes into doing mechanical work (climbing stairs). The remainder leaves the body by heat. Assume the student's mass is 50.0 kg. (a) How many times must she run the flight of stairs to lose 1 lb of fat? (b) What is her average power output, in watts and in horsepower, as she runs up the stairs?

40. For saving energy, bicycling and walking are far more efficient means of transportation than is travel by automobile. For example, when riding at 10.0 mi/h a cyclist uses food energy at a rate of about 400 kcal/h above what he would use if merely sitting still. (In exercise physiology, power is often measured in kilocalories per hour rather than in watts. Here 1 kcal = 1 dietary Calorie = 4186 J.) Walking at 3.00 mi/h requires about 220 kcal/h. It is interesting to compare these values with the energy consumption required for travel by car. Gasoline yields a value of about $1.30 \times 10^8$ J/gal. Find the fuel economy in equivalent miles per gallon for a person (a) walking and (b) bicycling.

## Section 6.9 Context Connection — A Probe to the Sun

41. A payload is fired from the Earth's surface with an initial speed of $2.00 \times 10^4$ m/s. What will its speed be when it is very far from the Earth? (Ignore friction.)

42. (a) Determine the amount of work that must be done on a 100-kg object to elevate it to a height of 1 000 km above the Earth's surface. (b) Determine the amount of additional work required to put the object into a circular orbit at this elevation.

**Additional Problems**

43. A baseball outfielder throws a 0.150-kg baseball at a speed of 40.0 m/s and an initial angle of 30.0°. What is the kinetic energy of the baseball at the highest point of its trajectory?

44. The direction of an arbitrary vector **A** can be completely specified with the angles $\alpha$, $\beta$, and $\gamma$ that the vector makes with the x, y, and z axes, respectively. If $\mathbf{A} = A_x\mathbf{i} + A_y\mathbf{j} + A_z\mathbf{k}$, (a) find expressions for $\cos\alpha$, $\cos\beta$, and $\cos\gamma$ (these are known as *direction cosines*), and (b) show that these angles satisfy the relation $\cos^2\alpha + \cos^2\beta + \cos^2\gamma = 1$. (*Hint:* Take the scalar product of **A** with **i**, **j**, and **k** separately.)

45. A 4.00-kg particle moves along the x axis. Its position varies with time according to $x = t + 2.0t^3$, where x is in meters and t is in seconds. Find (a) the kinetic energy at any time t, (b) the acceleration of the particle and the force acting on it at time t, (c) the power being delivered to the particle at time t, and (d) the work done on the particle in the interval $t = 0$ to $t = 2.00$ s.

46. A traveler at an airport takes an escalator up one floor, as in Figure P6.46. The moving staircase would itself carry him upward with vertical velocity component v between entry and exit points separated by height h. However, while the escalator is moving, the hurried traveler climbs the steps of the escalator at a rate of n steps/s. Assume that the height of each step is $h_s$. (a) Determine the amount of work done by the traveler during his escalator ride, given that his mass is m. (b) Determine the work the escalator motor does on this person.

47. When a certain spring is stretched beyond its proportional limit, the restoring force of the spring satisfies the equation $F = -kx + \beta x^3$. If $k = 10.0$ N/m and $\beta = 100$ N/m³, calculate the work done by this force when the spring is stretched 0.100 m.

48. In a control system, an accelerometer consists of a 4.70-g object sliding on a low-friction horizontal rail. A low-mass spring attaches the object to a flange at one end of the rail. When subject to a steady acceleration of 0.800g, the mass is to assume a location 0.500 cm away from its equilibrium position. Find the force constant required for the spring.

49. A single constant force **F** acts on a particle of mass m. The particle starts at rest at $t = 0$. (a) Show that the instantaneous power delivered by the force at any time t is $(F^2/m)t$. (b) If $F = 20.0$ N and $m = 5.00$ kg, what is the power delivered at $t = 3.00$ s?

50. A particle is attached between two identical springs on a horizontal frictionless table. Both springs have force constant k and are initially unstressed. (a) If the particle is pulled a distance x along a direction perpendicular to the initial configuration of the springs, as in Figure P6.50, show that the force exerted on the particle by the springs is

$$\mathbf{F} = -2kx\left(1 - \frac{L}{\sqrt{x^2 + L^2}}\right)\mathbf{i}$$

(b) Determine the amount of work done by this force in moving the particle from $x = A$ to $x = 0$.

**Figure P6.46** *(Ron Chapple/FPG)*

Top view

**Figure P6.50**

51. A 200-g block is pressed against a spring of force constant 1.40 kN/m until the block compresses the spring 10.0 cm. The spring rests at the bottom of a ramp inclined at 60.0° to the horizontal. Using energy considerations, determine

how far up the incline the block moves before it stops (a) if there is no friction between the block and the ramp and (b) if the coefficient of kinetic friction is 0.400.

52. When different weights are hung on a spring, the spring stretches to different lengths as shown in the following table. (a) Make a graph of the applied force versus the extension of the spring. By least-squares fitting, determine the straight line that best fits the data. (You may not want to use all the data points.) (b) From the slope of the best fit line, find the force constant $k$. (c) If the spring is extended to 105 mm, what force does it exert on the suspended weight?

| $F$ (N) | $L$ (mm) |
|---|---|
| 2.0 | 15 |
| 4.0 | 32 |
| 6.0 | 49 |
| 8.0 | 64 |
| 10 | 79 |
| 12 | 98 |
| 14 | 112 |
| 16 | 126 |
| 18 | 149 |
| 20 | 175 |
| 22 | 190 |

53. **Review Problem.** Two constant forces act on a 5.00-kg object moving in the $xy$ plane, as shown in Figure P6.53. Force $\mathbf{F}_1$ is 25.0 N at 35.0°, while $\mathbf{F}_2 = 42.0$ N at 150°. At time $t = 0$, the object is at the origin and has velocity $4.00\mathbf{i} + 2.50\mathbf{j}$ m/s. (a) Express the two forces in unit-vector notation. Use unit-vector notation for your other answers. (b) Find the total force on the object. (c) Find the object's acceleration. Now, considering the instant $t = 3.00$ s, (d) find the object's velocity, (e) its position, (f) its kinetic energy from $\frac{1}{2}mv_f^2$, and (g) its kinetic energy from

$$\frac{1}{2}mv_i^2 + \sum \mathbf{F} \cdot \Delta\mathbf{r}$$

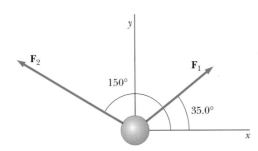

**Figure P6.53**

54. A 0.400-kg particle slides around a horizontal track. The track has a smooth vertical outer wall forming a circle with a radius of 1.50 m. The particle is given an initial speed of

8.00 m/s. After one revolution, its speed has dropped to 6.00 m/s because of friction with the rough floor of the track. (a) Find the energy transformed to internal energy due to friction in one revolution. (b) Calculate the coefficient of kinetic friction. (c) What is the total number of revolutions the particle makes before stopping?

55. The ball launcher in a pinball machine has a spring with a force constant of 1.20 N/cm (Fig. P6.55). The surface on which the ball moves is inclined 10.0° with respect to the horizontal. If the spring is initially compressed 5.00 cm, find the launching speed of a 100-g ball when the plunger is released. Friction and the mass of the plunger are negligible.

**Figure P6.55**

56. In diatomic molecules, the constituent atoms exert attractive forces on one another at large distances and repulsive forces at short distances. For many molecules, the Lennard–Jones law is a good approximation to the magnitude of these forces:

$$F = F_0\left[2\left(\frac{\sigma}{r}\right)^{13} - \left(\frac{\sigma}{r}\right)^{7}\right]$$

where $r$ is the center-to-center distance between the atoms in the molecule, $\sigma$ is a length parameter, and $F_0$ is the force when $r = \sigma$. For an oxygen molecule, the force is $F_0 = 9.60 \times 10^{-11}$ N and $\sigma = 3.50 \times 10^{-10}$ m. Determine the work done by this force if the atoms are pulled apart from $r = 4.00 \times 10^{-10}$ m to $r = 9.00 \times 10^{-10}$ m.

57. A horizontal string is attached to a 0.250-kg object lying on a rough, horizontal table. The string passes over a light frictionless pulley and a 0.400-kg object is then attached to its free end. The coefficient of sliding friction between the 0.250-kg object and table is 0.200. Using the work–kinetic energy theorem, determine (a) the speed of the objects after each has moved 20.0 m from rest and (b) the mass that must be added to the 0.250-kg object so that, given an initial velocity, the objects continue to move at a constant speed. (c) What mass must be removed from the 0.400-kg object so that the same constant-speed outcome as in part (b) is achieved?

58. As it plows a parking lot, a snowplow pushes an ever-growing pile of snow in front of it. Suppose a car moving through the air is similarly modeled as a cylinder pushing a growing plug of air in front of it. The originally stationary air is set into motion at the constant speed $v$ of the cylinder, as in Figure P6.58. In a time $\Delta t$, a new disk of air

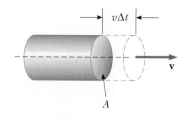

**Figure P6.58**

of mass $\Delta m$ must be moved a distance $v\Delta t$ and hence must be given a kinetic energy $\frac{1}{2}(\Delta m)v^2$. Using this model, show that the automobile's power loss due to air resistance is $\frac{1}{2}\rho Av^3$ and that the resistive force acting on the car is $\frac{1}{2}\rho Av^2$, where $\rho$ is the density of air. Compare this result with the empirical expression $\frac{1}{2}D\rho Av^2$ for the resistive force.

59. A particle moves along the $x$ axis from $x = 12.8$ m to $x = 23.7$ m under the influence of a force

$$F = \frac{375}{x^3 + 3.75x}$$

where $F$ is in newtons and $x$ is in meters. Using numerical integration, determine the total work done by this force during this displacement. Your result should be accurate to within 2%.

## ANSWERS TO QUICK QUIZZES

**6.1** Because no motion is taking place, the point of application of the forces from the pullers experiences no displacement. Thus, no work is done on the rope. For the same reason, no work is being done on the pullers or the ground. Work is being done only within the bodies of the pullers. For example, the heart of each puller is applying forces on the blood to move it through the body.

**6.2** (c), (a), (d), (b). The work in (c) is positive and of the largest possible value because the angle between the force and the displacement is zero. The work done in (a) is zero because the force is perpendicular to the displacement. In (d) and (b), negative work is done by the applied force because in both cases there is a component of the force opposite the direction of the displacement. Situation (b) is the most negative value because the angle between the force and the displacement is 180°.

**6.3** (d). Choice (a) represents something that cannot be true because the value of $\cos\theta$ cannot be larger than 1. For choice (b), although the dot product *can* be smaller than $AB$, it is not *necessarily* smaller. Part (c) cannot be true for the same reason as part (a). Choice (d) is true if the angle between **A** and **B** is zero.

**6.4** (a) Enter: electrical transmission, electromagnetic radiation (for an antenna-based set); Leave: electromagnetic radiation (from the screen), mechanical waves (sound from the speaker); heat (from warm surfaces into the air). (b) Enter: matter transfer (gasoline); Leave: mechanical waves (sound), work (on grass), heat (from warm surfaces into the air). (c) Enter: work (when you crank it); Leave: mechanical waves (sound), work (on pencil).

**6.5** The original kinetic energy of your car has been transformed to internal energy, located primarily in the brakes. Because your brakes are warmer than the surrounding air, the energy will leave the brakes by heat, going into the air and warming it. Eventually, this energy spreads throughout the atmosphere!

**6.6** Because the time periods are the same for both cars, we need only compare the energy transferred from the engine. Because the sports car is moving twice as fast as the older car at the end of the time interval, it has four times the kinetic energy. Thus, four times as much energy was transferred from the engine, and the engine must have expended four times the power.

A strobe photograph of a pole
vaulter. During this process,
several types of energy
transformations occur.
*(©Harold E. Edgerton/Courtesy of Palm
Press, Inc.)*

# Potential Energy

I n Chapter 6 we introduced the concepts of kinetic energy, which is associated
with the motion of an object or a particle, and internal energy, which is associated with the temperature of a system. In this chapter we introduce another
form of energy, called *potential energy,* which is associated with the configuration of a system of two or more interacting objects or particles. This new type of
energy will provide us with a powerful and universal fundamental principle.

## 7.1 • POTENTIAL ENERGY OF A SYSTEM

In Chapter 6, we defined a system in general but focused our attention on single
particles under the influence of an external force. In this chapter, we will consider
systems of two or more objects or particles interacting via a force that is *internal* to
the system. The kinetic energy of such a system is the algebraic sum of the kinetic
energies of all members of the system. In some systems, however, one object may be
so massive that it can be modeled as stationary and its kinetic energy can be ignored. For example, if we consider a ball–Earth system as the ball falls to the

ground, the kinetic energy of the system can be considered as only the kinetic energy of the ball. The Earth moves so slowly in this process that we can ignore its kinetic energy. (We will justify this in Chapter 8.) On the other hand, the kinetic energy of a system of two electrons must include the kinetic energies of both particles.

Let us imagine a system consisting of a book and the Earth, interacting via the gravitational force. We do some work on the system by lifting the book slowly through a height $\Delta y = y_b - y_a$, as in Figure 7.1. According to the continuity equation for energy (Eq. 6.20) introduced in Chapter 6, this work on the system must appear as an increase in energy of the system. The book is at rest before we perform the work and is at rest after we perform the work; thus, the kinetic energy of the system does not change. There is no reason the temperature of the book or the Earth should change, so the internal energy of the system experiences no change.

Because the energy change of the system is not in the form of kinetic energy or internal energy, it must appear as some other form of energy storage. After lifting the book, we could release it and let it fall to the ground. Notice that the book (and, therefore, the system) will now have kinetic energy, and its origin is in the work that was done in lifting the book. While the book was at the highest point, the energy of the system had the *potential* to become kinetic energy, but did not do so until the book was allowed to fall. Thus, we call the energy storage mechanism before we release the book **potential energy.** We will find that a potential energy can be associated with a number of types of forces. In this particular case, we are discussing **gravitational potential energy.**

Let us now derive an expression for the gravitational potential energy associated with an object at a given location above the surface of the Earth. To do this, consider an external agent lifting an object of mass $m$ from an initial height $y_a$ above the ground to a final height $y_b$ as in Figure 7.1. We assume that the lifting is done slowly, with no acceleration, so that the lifting force can be modeled as equal to the weight of the object—the object is in equilibrium and moving at constant velocity. The work done by the external agent on the system (object and Earth) as the object undergoes this upward displacement is given by the product of the upward force $\mathbf{F} = m\mathbf{g}$ and the displacement $\Delta\mathbf{r} = \Delta y\mathbf{j}$:

$$W = (m\mathbf{g}) \cdot \Delta\mathbf{r} = (mg\mathbf{j}) \cdot [(y_b - y_a)\mathbf{j}] = mgy_b - mgy_a \qquad [7.1]$$

Notice how similar this equation is to Equation 6.17 in the preceding chapter. In each equation, the work done on a system equals a difference between the final and initial values of a quantity. Of course, both equations are nothing more than special cases of the continuity equation for energy. In Equation 6.17, the work represents a transfer of energy into the system, and the increase in energy of the system is kinetic in form. In Equation 7.1, the work represents a transfer of energy into the system, and the system energy appears in a different form, which we have called gravitational potential energy.

Thus, we can represent the quantity $mgy$ to be the gravitational potential energy $U_g$:

$$U_g \equiv mgy \qquad [7.2]$$

The units of gravitational potential energy are joules, the same as those of work and kinetic energy. Potential energy, like work and kinetic energy, is a scalar quantity. Note that Equation 7.2 is valid only for objects near the surface of the Earth, where $g$ is approximately constant.

See the *Core Concepts in Physics CD-ROM*, Screen 5.3

**Figure 7.1**

The work done by an external agent on the system of the book and the Earth as the book is lifted from $y_a$ to $y_b$ is equal to $mgy_b - mgy_a$.

**PITFALL PREVENTION 7.1**

**Potential energy belongs to a system**

Keep in mind that potential energy is always associated with a *system* of two or more interacting objects. In the gravitational case, in which a small object moves near the surface of the Earth, we may sometimes refer to the potential energy "associated with the object" rather than the more proper "associated with the system," because the Earth does not move significantly. We will not, however, refer to the potential energy "of the object" because this clearly ignores the role of the Earth in the potential energy.

• *Gravitational potential energy*

Using our definition of gravitational potential energy, we can now rewrite Equation 7.1 as

$$W = \Delta U_g$$

which mathematically describes that fact that the work done on the system in this situation appears as a change in the gravitational potential energy of the system.

The gravitational potential energy depends only on the vertical height of the object above the surface of the Earth. Thus, the same amount of work is done on an object–Earth system whether the object is lifted vertically from the Earth or it starts at the same point and is pushed up a frictionless incline, ending up at the same height. This can be shown in a mathematical representation by reperforming the work calculation in Equation 7.1 with a displacement having both vertical and horizontal components:

$$W = (m\mathbf{g}) \cdot \Delta\mathbf{r} = (mg\mathbf{j}) \cdot [(x_b - x_a)\mathbf{i} + (y_b - y_a)\mathbf{j}] = mgy_b - mgy_a$$

where no term involving $x$ appears in the final result because $\mathbf{j} \cdot \mathbf{i} = 0$.

In solving problems, it is necessary to choose a reference location at which to set the gravitational potential energy equal to some reference value, which is normally zero. The choice of zero level is completely arbitrary because the important quantity is the *difference* in potential energy, and this difference is independent of the choice of zero level.

It is often convenient to choose the surface of the Earth as the reference position for zero potential energy, but this is not essential. Often, the statement of the problem suggests a convenient level to use.

### Quick Quiz 7.1

Can the gravitational potential energy of a system be negative?

Elevator

Counterweight

**Figure 7.2**

(Thinking Physics 7.1)

### THINKING PHYSICS 7.1

Elevators that are lifted by a cable have a counterweight, as shown in Figure 7.2. The elevator and the counterweight are connected by a cable that passes over a drive wheel. The counterweight moves in the direction opposite to that of the elevator. What is the purpose of the counterweight?

**Reasoning**   Without the counterweight, the elevator motor would need to lift the weight of the elevator and the people riding it. This would represent a large increase in gravitational potential energy of the elevator–Earth system, which means a large transfer of energy to the system by work from the motor. With the counterweight, as the elevator moves upward, the counterweight moves down. Thus, the net change in separation distance between the total mass of the elevator–counterweight combination and the Earth is small. This results in a much smaller change in the system's potential energy than without the counterweight and less work from the motor.

**EXERCISE**   What is the gravitational potential energy, relative to the ground, associated with a 0.15-kg baseball at the top of a 100-m-tall building?

*Answer*   147 J

## 7.2 • THE ISOLATED SYSTEM

See Screen 5.9

The introduction of potential energy allows us to generate a powerful and universally applicable principle for solving problems that are difficult to solve with Newton's laws. Let us develop this new principle by thinking about the book–Earth system in Figure 7.1 again. After we have lifted the book, there is gravitational potential energy stored in the system, which we can calculate from the work done by the external agent on the system, using $W = \Delta U_g$.

Let us now shift our focus to the work done on the book alone (Fig. 7.3) as it falls back to its original height under the influence of the gravitational force alone. As the book falls from $y_b$ to $y_a$, the work done by the gravitational force on the book is

$$W_{\text{on book}} = (m\mathbf{g}) \cdot \Delta\mathbf{r} = (-mg\mathbf{j}) \cdot (y_a - y_b)\mathbf{j} = mgy_b - mgy_a \qquad [7.3]$$

From the work–kinetic energy theorem of Chapter 6, the work done on the book is also

$$W_{\text{on book}} = \Delta K_{\text{book}}$$

Therefore, equating these two expressions for the work done on the book

$$\Delta K_{\text{book}} = mgy_b - mgy_a \qquad [7.4]$$

Now, let us relate each side of this equation to the *system* of the book and the Earth. For the right-hand side,

$$mgy_b - mgy_a = -(mgy_a - mgy_b) = -(U_f - U_i) = -\Delta U_g$$

where $U_g$ is the gravitational potential energy of the system. Because the book is the only part of the system that is moving, the left-hand side of Equation 7.4 becomes

$$\Delta K_{\text{book}} = \Delta K$$

where $K$ is the kinetic energy of the system. Thus, replacing each side of Equation 7.4 with its system equivalent, the equation becomes

$$\Delta K = -\Delta U_g \qquad [7.5]$$

This equation can be manipulated to provide a very important result for a new analysis model. First, we bring the change in potential energy to the left side of the equation:

$$\Delta K + \Delta U_g = 0$$

Notice that this is in the form of the continuity equation. On the left, we have a sum of changes of the energy stored in the system. The right-hand side in the continuity equation is the sum of the transfers across the boundary of the system. This is equal to zero in the present case because our book–Earth system is isolated from the environment.

Let us now write the changes in energy explicitly:

$$(K_f - K_i) + (U_f - U_i) = 0 \quad \longrightarrow \quad \boxed{K_f + U_f = K_i + U_i} \qquad [7.6]$$

We define the sum of kinetic and potential energies as **mechanical energy:**

$$E_{\text{mech}} = K + U_g$$

Thus, Equation 7.6 is a statement of **conservation of mechanical energy** for an **isolated system.** An isolated system is one for which no energy transfers occur

**Figure 7.3**

The work done by the gravitational force on the book as the book falls from $y_b$ to $y_a$ is equal to $mgy_b - mgy_a$.

**PITFALL PREVENTION 7.2**

**Mechanical energy in an isolated system**

Two comments are needed here. First, the isolated system model goes far beyond Equation 7.6. This is only the *mechanical energy* version of this model. We will see shortly how to include internal energy. In later chapters, we will see other isolated systems and generate new versions (and associated equations) related to such quantities as momentum, angular momentum, and electric charge. Second, Equation 7.6 is true for only one of two categories of forces, called *conservative forces,* which is discussed in the next section.

• *Conservation of mechanical energy for an isolated system*

Twin Falls on the Island of Kauai, Hawaii. The gravitational potential energy of the water–Earth system when the water is at the top of the falls is converted into kinetic energy as the water descends. *(Bruce Byers, FPG)*

across the boundary. Thus, the energy in the system is conserved—the sum of the kinetic and potential energies remains constant.

For the gravitational situation that we are describing, the conservation of mechanical energy equation can be written as

$$\tfrac{1}{2}mv_f^2 + mgy_f = \tfrac{1}{2}mv_i^2 + mgy_i$$

As the book falls to the Earth, the book–Earth system loses potential energy and gains kinetic energy, such that the total of the two types of energy always remains constant.

We will see other types of potential energy besides gravitational, so we can write the general form of the definition for mechanical energy without a subscript on $U$:

$$E_{\text{mech}} \equiv K + U \qquad\qquad [7.7]$$

### Quick Quiz 7.2

Three identical balls are thrown from the top of a building, all with the same initial speed. The first ball is thrown horizontally, the second at some angle above the horizontal, and the third at some angle below the horizontal as in Figure 7.4. Ignoring air resistance, rank the speeds of the balls as they reach the ground.

**Figure 7.4**

(Quick Quiz 7.2) Three identical balls are thrown with the same initial speed from the top of a building.

# THINKING PHYSICS 7.2

You have graduated from college and are designing roller coasters for a living. You design a roller coaster in which a car is pulled to the top of a hill of height $h$ and then, starting from a momentary rest, rolls freely down the hill and upward toward the peak of the next hill, which is at height $1.1h$. Will you have a long career in this business?

**Reasoning**   Your career will probably not be long, because this roller coaster will not work! At the top of the first hill, the roller coaster car has no kinetic energy, and the gravitational potential energy of the car–Earth system is that associated with a height $h$. If the car were to reach the top of the next hill, the system would have higher potential energy, that associated with height $1.1h$. This would violate the principle of conservation of mechanical energy. If this coaster were actually built, the car would move upward on the second hill to a height $h$ (ignoring the effects of friction), stop short of the peak, and then start rolling backward, becoming trapped between the two hills.

## Example 7.1   Ball in Free-Fall

A ball of mass $m$ is dropped from rest at a height $h$ above the ground, as in Figure 7.5. (a) Determine the speed of the ball when it is at a height $y$ above the ground.

**Reasoning**   We will consider the ball and the Earth as the system. We adopt the simplification model in which the ball encounters no air resistance. Thus, the ball and the Earth do not experience any forces from the environment, and we can use the principle of conservation of mechanical energy. Initially, the system has potential energy and no kinetic energy.

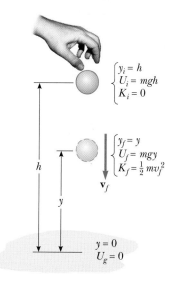

$$\begin{cases} y_i = h \\ U_i = mgh \\ K_i = 0 \end{cases}$$

$$\begin{cases} y_f = y \\ U_f = mgy \\ K_f = \tfrac{1}{2}mv_f^2 \end{cases}$$

$\mathbf{v}_f$

$y = 0$
$U_g = 0$

**Figure 7.5**

(Example 7.1) A ball is dropped from rest at a height $h$ above the ground. Initially, the total energy of the ball–Earth system is gravitational potential energy, equal to $mgh$ relative to the ground. At the elevation $y$, the total system energy is the sum of kinetic and potential energies.

As the ball falls, the total mechanical energy of the system (the sum of kinetic and potential energies) remains constant and equal to its initial potential energy. The potential energy of the system decreases, and the kinetic energy of the system (which is due only to the ball) increases.

**Solution**   Before the ball is released from rest at a height $h$ above the ground, the kinetic energy of the system is $K_i = 0$ and the potential energy is $U_i = mgh$, where the $y$ coordinate is measured from ground level. When the ball is at a distance $y$ above the ground, its kinetic energy is $K_f = \tfrac{1}{2}mv_f^2$ and the potential energy of the system is $U_f = mgy$. Applying Equation 7.6, we have

$$K_i + U_i = K_f + U_f$$

$$0 + mgh = \tfrac{1}{2}mv_f^2 + mgy$$

$$v_f^2 = 2g(h - y)$$

$$v_f = \sqrt{2g(h - y)}$$

(b) Determine the speed of the ball at $y$ if it is given an initial speed $v_i$ at the initial altitude $h$.

**Solution**   In this case, the initial energy includes kinetic energy of the ball equal to $\tfrac{1}{2}mv_i^2$ and Equation 7.6 gives

$$\tfrac{1}{2}mv_i^2 + mgh = \tfrac{1}{2}mv_f^2 + mgy$$

$$v_f^2 = v_i^2 + 2g(h - y)$$

$$v_f = \sqrt{v_i^2 + 2g(h - y)}$$

This result is consistent with Equation 2.12 (Chapter 2), $v_{yf}^2 = v_{yi}^2 - 2g(y_f - y_i)$, for a particle under constant acceleration, where $y_i = h$. Furthermore, this result is valid even if the initial velocity is at an angle to the horizontal (the projectile situation), as discussed in Quick Quiz 7.2.

## Example 7.2  One Way to Lift an Object

Two blocks are connected by a massless cord that passes over two frictionless pulleys, as in Figure 7.6. One end of the cord is attached to an object of mass $m_1 = 3.00$ kg that is a distance $R = 1.20$ m from the pulley on the left. The other end of the cord is connected to a block of mass $m_2 = 6.00$ kg resting on a table. From what angle $\theta$ (measured from the vertical) must the 3.00-kg mass be released in order to just lift the 6.00-kg block off the table?

**Reasoning**  Several models and concepts are needed to solve this problem. First, we consider the 3.00-kg object and the Earth as an isolated system. Even though an external force—that of the cord—is exerted on the 3.00-kg object, this force is always perpendicular to the displacement during the swing of the object, so no work is done across the boundary of the system by this force. Thus, we use conservation of mechanical energy to find the speed of the 3.00-kg object at the bottom of the circular path as a function of $\theta$ and the radius of the path. (In Example 5.9, we found that the tension in the cord is largest as the object passes through the lowest point.) Next, we use the particle under a net force model on the 3.00-kg object at the bottom of its path. At this position, Newton's second law will allow us to find the tension force exerted by the cord as a function of the given parameters. This force is also applied to the 6.00-kg block through the string. We apply the particle under a net force model again to the block. We note that the block lifts off the table when the force exerted on it by the cord exceeds the gravitational force acting on the block. This procedure enables us to find the required angle for the 3.00-kg object.

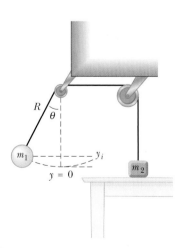

**Figure 7.6**

(Example 7.2)

**Solution**  Applying conservation of energy to the system of the 3.00-kg object and the Earth gives

$$K_i + U_i = K_f + U_f$$
$$(1) \qquad 0 + m_1 g y_i = \tfrac{1}{2} m_1 v^2 + 0$$

where $v$ is the speed of the object at the bottom of its path. (*Note:* $K_i = 0$ because the object starts from rest and $U_f = 0$ because we define the potential energy of the system as zero when the object is at the bottom of the circle.) From the geometry in Figure 7.6, we see that $y_i = R - R\cos\theta = R(1 - \cos\theta)$. Using this relation in (1) gives

$$(2) \qquad v^2 = 2gR(1 - \cos\theta)$$

Now we apply Newton's second law to the 3.00-kg object when it is at the bottom of the circular path:

$$T - m_1 g = m_1 \frac{v^2}{R}$$

$$(3) \qquad T = m_1 g + m_1 \frac{v^2}{R}$$

Notice that we have also used the centripetal acceleration equation for a particle in uniform circular motion here. The speed of the object is not uniform, but we can use the expression for the instantaneous acceleration at any speed.

This same tension force is transmitted to the 6.00-kg block, and if it is to be just lifted off the table, the normal force on it becomes zero. In this situation, Newton's second law gives us $T - m_2 g = 0$ or $T = m_2 g$. Using this condition, together with (2) and (3), gives

$$m_2 g = m_1 g + 2 m_1 g (1 - \cos\theta)$$

Solving for $\cos\theta$ and substituting in the given parameters, we find

$$\cos\theta = \frac{3 m_1 - m_2}{2 m_1} = \frac{3(3.00 \text{ kg}) - 6.00 \text{ kg}}{2(3.00 \text{ kg})} = 0.500$$

$$\theta = \boxed{60.0°}$$

**EXERCISE**  If the initial angle is $\theta = 40.0°$, find the speed of the 3.00-kg object and the tension in the cord when the 3.00-kg object is at the bottom of its circular path.

*Answer*  2.35 m/s; 43.2 N

## 7.3 • CONSERVATIVE AND NONCONSERVATIVE FORCES

In the preceding section, we showed that the mechanical energy of a system is conserved in a process in which the force between members of the system is the gravitational force. The gravitational force is one example of a category of forces for which the mechanical energy of a system is conserved. These are called **conservative forces.** The other possibility for energy storage in a system besides kinetic and potential is internal energy. Thus, a conservative force, for our purposes in mechanics, is a **force between members of a system that causes no transformation of mechanical energy to internal energy within the system.** If no energy is transformed to internal energy, the mechanical energy of the system is conserved, as described by Equation 7.6.

If a force is conservative, the work done by such a force has a special property as the members of the system move in response either to the force itself or to an external force: **The work done by a conservative force does not depend on the path followed by the members of the system, and depends only on the initial and final configurations of the system.**

• *Definition of a conservative force*

From this property, it follows that: **The work done by a conservative force when a member of the system is moved through a closed path is equal to zero.**

These statements can be mathematically demonstrated and serve as general mathematical definitions of conservative forces. Both of these statements can be seen for the gravitational force from Equation 7.3. The work done is expressed only in terms of the initial and final heights, with no indication of what path is followed. If the path is closed, the initial and final heights are the same in Equation 7.3, and the work is equal to zero.

Another example of a conservative force is the force of a spring on an object attached to the spring, where the spring force is given by Hooke's law, $F_s = -kx$. As we learned in Chapter 6 (Eq. 6.15), the work done by the spring force is

$$W_s = \tfrac{1}{2}kx_i^2 - \tfrac{1}{2}kx_f^2$$

where the initial and final positions of the object are measured from its equilibrium position $x = 0$, at which the spring is unstretched. Again we see that $W_s$ depends only on the initial and final coordinates of the object and is zero for any closed path. Hence, the spring force is conservative.

In Section 7.1, we discussed the notion of an external agent lifting a book and storing energy as potential energy in the book–Earth system. In Section 6.4, we discussed an external agent pulling a block attached to a spring from $x = 0$ to $x = x_{max}$ and calculated the work that was done on the system as $\tfrac{1}{2}kx_{max}^2$. This is another situation, like that of the book in Section 7.1, in which work is done on a system but there is no change in kinetic energy of the system. Thus, the energy must be stored in the block–spring system as potential energy. The **elastic potential energy** associated with the spring force is defined by

$$U_s \equiv \tfrac{1}{2}kx^2 \qquad\qquad [7.8]$$

• *Elastic potential energy*

The elastic potential energy can be considered as the energy stored in the deformed spring (one that is either compressed or stretched from its equilibrium position). To visualize this, consider Figure 7.7a, which shows an undeformed spring on a frictionless, horizontal surface. When the block is pushed against the spring (Fig. 7.7b), compressing the spring a distance $x$, the elastic potential energy stored in the spring is $\tfrac{1}{2}kx^2$. When the block is released, the spring returns to its original

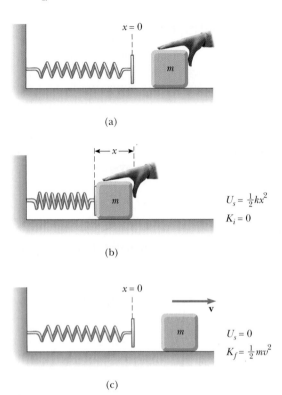

(a)

$$U_s = \frac{1}{2}kx^2$$
$$K_i = 0$$

(b)

$$U_s = 0$$
$$K_f = \frac{1}{2}mv^2$$

(c)

**Figure 7.7**

(a) An undeformed spring on a friction-less horizontal surface. (b) A block of mass *m* is pushed against a spring, compressing it through a distance *x*. (c) When the block is released from rest, the elastic potential energy stored in the system is transformed to kinetic energy of the block.

length, applying a force to the block. This force does work on the block, resulting in kinetic energy of the block (Fig. 7.7c). The elastic potential energy stored in the spring is zero whenever the spring is undeformed $(x = 0)$. Finally, because the elastic potential energy is proportional to $x^2$, we see that $U_s$ is always positive in a deformed spring.

In comparison to a conservative force, a **nonconservative force** in mechanics is **a force between members of a system that is not conservative,** that is, it causes transformation of mechanical energy to internal energy within the system. A common nonconservative force in mechanics is the friction force. If we consider a system consisting of a block and a surface and imagine an initially sliding block coming to a rest due to friction, we see the result of a nonconservative force. Initially, the system has kinetic energy (of the block). Afterward, nothing is moving. The friction force between the block and the surface transformed the mechanical energy into internal energy—the block and surface are both warmer than before.

Let us return to the notion of the work done over a path. The two statements claimed for conservative forces are not true for nonconservative forces. As an example, consider Figure 7.8. Suppose you displace a book between two points on a table. If the book is displaced in a straight line along the blue path between points Ⓐ and Ⓑ in Figure 7.8, you do a certain amount of work against the kinetic friction force to keep the book moving at a constant speed. Now, imagine that you push the book along the red semicircular path in Figure 7.8. You perform more work against friction along this longer path than along the straight path. The work done depends on the path, so the friction force cannot be conservative.

In Chapter 6, we discussed the fact that we cannot calculate the work done by friction because the actual friction force behaves in a very messy way. In this ex-

• *Definition of a nonconservative force*

**Figure 7.8**

The work done against the force of friction depends on the path taken as the book is moved from Ⓐ to Ⓑ and hence friction is a nonconservative force. The work required is greater along the red path than the blue path.

ample the particle model is not valid. Another example of a situation with non-particle objects is seen for deformable objects. Suppose, for example, that a rubber ball is flattened against a brick wall. The ball deforms during the pushing process. If the ball is suddenly released, it will jump away from the wall. The force causing the ball to accelerate is the normal force of the wall on the ball. But the point of application of this force does not move through space—it stays fixed at the point of contact between the ball and the wall. Thus, no work is done on the ball. Yet the ball has kinetic energy afterward. The work–kinetic energy theorem does not correctly describe this situation. It is more valuable to apply the isolated system model to this situation. With the ball as the system, there is no transfer of energy across the boundary as the ball springs off the wall. There is a *transformation* of energy, from elastic potential energy (stored in the ball when it was flattened) to kinetic energy. In the same way, if a skateboarder pushes off a wall to start rolling, no work is done by the force from the wall—the kinetic energy is transformed within the system from potential energy stored in the body from previous meals.

We also discussed in Chapter 6 the difficulties associated with the nonconservative force of friction in energy calculations. Recall that the result of a friction force is to transform kinetic energy in a system to internal energy, and the increase in internal energy is equal to the decrease in kinetic energy. If a potential energy is associated with the system, then the decrease in *mechanical* energy, kinetic plus potential, equals the increase in internal energy in the isolated system. Thus, we have

$$-f_k\,\Delta x = \Delta K + \Delta U = \Delta E_{\text{mech}} = -\Delta E_{\text{int}} \qquad \textbf{[7.9]}$$

We can recast this expression by putting all forms of energy storage on one side of the equation:

$$\Delta K + \Delta U + \Delta E_{\text{int}} = \Delta E_{\text{system}} = 0 \qquad \textbf{[7.10]}$$

This gives us the most general expression of the continuity equation for an isolated system. Note that $\Delta K$ may represent more than one term if two or more parts of the system are moving. Also, $\Delta U$ may represent more than one term if different types of potential energy (e.g., gravitational and elastic) are associated with the system. Equation 7.10 is equivalent to

$$\boxed{K + U + E_{\text{int}} = \text{constant}} \qquad \textbf{[7.11]}$$

• *General conservation of energy for an isolated system*

which claims that the total energy (kinetic, potential, and internal) of an isolated system is conserved. No violation of this critical conservation principle has ever been observed. If we consider the Universe as an isolated system, this statement claims that there is a fixed amount of energy in our Universe, and all processes within the Universe represent transformations of energy from one type to another.

### Quick Quiz 7.3

A block of mass $m$ is projected across a horizontal surface with a speed $v$. It slides until it stops due to the friction force between the block and the surface. The surface is now tilted at $30°$, and the block is projected up the surface with the same initial speed $v$. When the block stops, how does the decrease in mechanical energy of the block–Earth system compare with that when the block slid over the horizontal surface? (a) It is the same. (b) It is larger. (c) It is smaller. (d) The relationship cannot be determined.

---

**PROBLEM-SOLVING STRATEGY** | **Isolated Systems**

---

Many problems in physics can be solved using the principle of conservation of energy for an isolated system. The following procedure should be used when you apply this principle:

1. Define your system, which may consist of more than one object and may or may not include springs or other possibilities for storage of potential energy. Choose situations to represent the initial and final conditions of the system.

2. Determine if any energy transfers occur across the boundary of your system. If so, use the nonisolated system model, $\Delta E_{system} = \Sigma H$. If not, use the isolated system model, $\Delta E_{system} = 0$, and continue with the following steps.

3. For each object that changes elevation, select a reference position for the zero point of gravitational potential energy. For a spring, the zero point of elastic potential energy is the position of the end of the spring when the spring force is zero. If there is more than one conservative force, write an expression for the potential energy associated with each force.

4. Determine whether any nonconservative forces are present. Remember that if friction or air resistance is present, mechanical energy *is not conserved.*

5. If mechanical energy is *conserved,* you can write the total initial system energy $E_i$ at some point as the sum of the kinetic and potential energy at that point. Then write a similar expression for the total system energy at the final point that is of interest. Because mechanical energy is *conserved,* equate the two total energies and solve for the quantity that is unknown.

6. If nonconservative forces are present (and thus mechanical energy is not conserved), first write expressions for the total initial and total final mechanical energies. In this case, the difference between the total final mechanical energy and the total initial mechanical energy equals the energy transformed to or from internal energy by the nonconservative forces.

---

## Example 7.3   Crate Sliding Down a Ramp

A 3.00-kg crate slides down a ramp at a loading dock. The ramp is 1.00 m in length, and inclined at an angle of 30.0°, as in Figure 7.9. The crate starts from rest at the top and experiences a constant friction force of magnitude 5.00 N. Use energy methods to determine the speed of the crate when it reaches the bottom of the ramp.

**Reasoning**   We define the system as the crate, the Earth, and the ramp. This is an isolated system. If we had chosen the crate and the Earth as the system, we would need to use the nonisolated system model because the friction force between the crate and the ramp is an external influence. There would be work done across the boundary as well as flow of energy by heat between the crate and the ramp. This would

be a difficult problem to solve. In general, if a friction force acts, it is easiest to define the system so that the friction force is an internal force.

**Solution**   Because $v_i = 0$ for the crate, the initial kinetic energy of the system is zero. If the $y$ coordinate is measured from the bottom of the ramp, then $y_i = (1.00 \text{ m})\sin 30° = 0.500$ m for the crate. The total mechanical energy of the crate–Earth–ramp system when the crate is at the top is therefore the gravitational potential energy:

$$E_i = U_i = mgy_i$$

When the crate reaches the bottom, the gravitational potential energy of the system is *zero* because the elevation of

**Figure 7.9**

(Example 7.3) A crate slides down a ramp under the influence of gravity. The potential energy of the crate–Earth system decreases while the kinetic energy of the crate increases.

the crate is $y_f = 0$. The total mechanical energy when the crate is at the bottom is therefore kinetic energy:

$$E_f = K_f = \tfrac{1}{2}mv_f^2$$

We cannot say that $E_f = E_i$ in this case, however, because a nonconservative force—the force of friction—reduces the mechanical energy of the system. In this case, the change in mechanical energy for the system is $\Delta E = -f_k \Delta r$, where $\Delta r$ is the displacement of the crate along the ramp. With $f_k = 5.00$ N and $\Delta r = 1.00$ m, we have

$$\Delta E_{\text{mech}} = -f_k \Delta r = -(5.00\ \text{N})(1.00\ \text{m}) = -5.00\ \text{J}$$

This is the change in the mechanical energy of the system. Because $\Delta E_{\text{mech}} = \Delta K + \Delta U = \tfrac{1}{2}mv_f^2 - mgy_i$ in this situation, Equation 7.9 gives

$$-f_k\Delta r = \tfrac{1}{2}mv_f^2 - mgy_i \longrightarrow \tfrac{1}{2}mv_f^2 = mgy_i - f_k\Delta r$$

$$\tfrac{1}{2}mv_f^2 = (3.00\ \text{kg})(9.80\ \text{m/s}^2)(0.500\ \text{m}) - 5.00\ \text{J} = 9.70\ \text{J}$$

$$v_f = \sqrt{\frac{2(9.70\ \text{J})}{3.00\ \text{kg}}} = 2.54\ \text{m/s}$$

**EXERCISE**  Use Newton's second law to find the acceleration of the crate along the ramp, and the equations of kinematics to determine the final speed of the crate.

*Answer*  3.23 m/s²; 2.54 m/s

**EXERCISE**  If the ramp is assumed to be frictionless, find the final speed of the crate and its acceleration along the ramp.

*Answer*  3.13 m/s; 4.90 m/s²

---

### Example 7.4  Motion on a Curved Track

A child of mass $m$ takes a ride on an irregularly curved slide of height $h = 2.00$ m, as in Figure 7.10. The child starts from rest at the top. (a) Determine the speed of the child at the bottom, assuming no friction is present.

**Reasoning**  We will define the system as the child and the Earth, and model the child as a particle. The normal force **n** does no work on the system because this force is always perpendicular to each element of the displacement. Furthermore, because no friction is present, no work is done by friction across the boundary of the system. Thus, we will use the isolated system model with no friction forces, for which mechanical energy is conserved; that is, $K + U = $ constant.

**Solution**  If we measure the $y$ coordinate for the child from the bottom of the slide, then $y_i = h$, $y_f = 0$, and we have for the system,

$$K_i + U_i = K_f + U_f$$

$$0 + mgh = \tfrac{1}{2}mv_f^2 + 0$$

$$v_f = \sqrt{2gh}$$

**Figure 7.10**

(Example 7.4) If the slide is frictionless, the speed of the child at the bottom depends only on the height of the slide.

The result is the same as if the child simply fell vertically through a distance *h*! In this example, *h* = 2.00 m, giving

$$v_f = \sqrt{2gh} = \sqrt{2(9.80 \text{ m/s}^2)(2.00 \text{ m})} = \boxed{6.26 \text{ m/s}}$$

(b) If a friction force acts on the 20.0-kg child, and he arrives at the bottom of the slide with a speed *v_f* = 3.00 m/s, by how much does the mechanical energy of the system decrease due to this force?

**Solution** We define the system as the child, the Earth, and the slide. In this case, a nonconservative force acts within the system and mechanical energy is *not* conserved. We can find

the change in mechanical energy due to friction, given that the final speed at the bottom is known:

$$\Delta E_{mech} = K_f + U_f - K_i - U_i = \tfrac{1}{2}mv_f^2 + 0 - 0 - mgh$$

$$\Delta E_{mech} = \tfrac{1}{2}(20.0 \text{ kg})(3.00 \text{ m/s})^2$$
$$- (20.0 \text{ kg})(9.80 \text{ m/s}^2)(2.00 \text{ m})$$
$$= \boxed{-302 \text{ J}}$$

The change in mechanical energy $\Delta E_{mech}$ is negative because friction reduces the mechanical energy of the system. The change in internal energy in the system is + 302 J.

---

## Example 7.5 Block–Spring Collision

A block of mass 0.800 kg is given an initial velocity $v_A$ = 1.20 m/s to the right and collides with a light spring of force constant *k* = 50.0 N/m, as in Figure 7.11. (a) If the surface is frictionless, calculate the maximum compression of the spring after the collision.

**Reasoning** We define the system as the block and the spring. No transfers of energy occur across the boundary of this system, so we use the isolated system model. Before the collision, when the block is at Ⓐ, for example, the system has kinetic energy due to the moving block and the spring is uncompressed, so the potential energy stored in the system is zero. Thus, the total energy of the system before the collision is $\tfrac{1}{2}mv_A^2$. After the collision, and when the spring is fully compressed at point Ⓒ, the mass is momentarily at rest and has zero kinetic energy, whereas the potential energy stored in the spring has its maximum value $\tfrac{1}{2}kx_C^2$. The total mechanical energy of the system is conserved because no nonconservative forces act within the system.

**Solution** Because mechanical energy is conserved, the mechanical energy of the system before the collision must equal the mechanical energy of the system when the spring is fully compressed, or

$$\tfrac{1}{2}mv_A^2 = \tfrac{1}{2}kx_C^2$$

$$x_C = \sqrt{\frac{m}{k}}\, v_A = \sqrt{\frac{0.800 \text{ kg}}{50.0 \text{ N/m}}}(1.20 \text{ m/s}) = \boxed{0.152 \text{ m}}$$

(b) If a constant force of kinetic friction acts between the block and the surface with $\mu_k$ = 0.500, and if the speed of the block just as it collides with the spring is $v_A$ = 1.20 m/s, what is the maximum compression in the spring?

**Solution** We define the system as the block, the spring, and the surface. In this case, mechanical energy of the system is

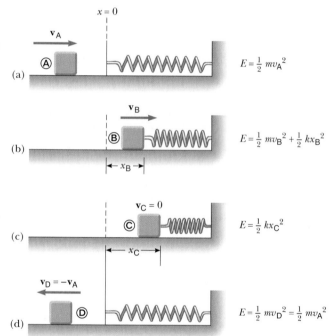

**Figure 7.11**

(Example 7.5) A block sliding on a smooth, horizontal surface collides with a light spring. (a) Initially the mechanical energy of the system is all kinetic energy. (b) The mechanical energy is the sum of the kinetic energy of the block and the elastic potential energy in the spring. (c) The mechanical energy is entirely potential energy. (d) The mechanical energy is transformed back to the kinetic energy of the block. The total energy of the block–spring system remains constant throughout the motion.

*not* conserved because of friction. The magnitude of the friction force is

$$f_k = \mu_k n = \mu_k mg = 0.500(0.800 \text{ kg})(9.80 \text{ m/s}^2) = 3.92 \text{ N}$$

where we have found $n = mg$ from Newton's second law in the vertical direction. Therefore, the decrease in mechanical energy due to friction as the block is displaced from $x_i = 0$ to the point $x_f = x_{max}$ at which the block stops is

$$\Delta E_{mech} = -f_k x_{max} = -3.92 x_{max}$$

Equation 7.9 gives

$$\Delta E_{mech} = E_f - E_i = (0 + \tfrac{1}{2}kx_{max}^2) - (\tfrac{1}{2}mv_A^2 + 0)$$

Substituting the numerical values and dropping the units, we have

$$-3.92 x_{max} = \frac{50.0}{2}x_{max}^2 - \tfrac{1}{2}(0.800)(1.20)^2$$

$$25.0 x_{max}^2 + 3.92 x_{max} - 0.576 = 0$$

Solving the quadratic equation for $x_{max}$ gives $x_{max} = 0.0924$ m and $x_{max} = -0.249$ m. We choose the positive root $x_{max} =$ 0.0924 m because the block must be to the right of the origin when it comes to rest. Note that 0.0924 m is less than the distance obtained in the frictionless case (a). This result is what we expect because friction retards the motion of the system.

**EXERCISE** A 3.00-kg block starts at a height $h = 60.0$ cm on a plane that has an inclination angle of 30.0°. Upon reaching the bottom, the block slides along a horizontal surface. If the coefficient of friction for both surfaces is $\mu_k = 0.200$, how far does the block slide on the horizontal surface before coming to rest?

*Answer* 1.96 m

**EXERCISE** A child starts from rest at the top of a slide of height 4.00 m. (a) What is her speed at the bottom if the incline is frictionless? (b) If she reaches the bottom with a speed of 6.00 m/s, what percentage of the total mechanical energy when she was at the top of the slide was transformed to internal energy as a result of friction?

*Answer* (a) 8.85 m/s (b) 54.1%

## 7.4 • CONSERVATIVE FORCES AND POTENTIAL ENERGY

Let us return to the falling book discussed in Section 7.2. We found that the work done on the book by the gravitational force can be expressed as the negative of the difference between two quantities that we called the initial and final potential energies of the system of the book and the Earth:

$$W_{on\ book} = mgy_b - mgy_a = -\Delta U \qquad [7.12]$$

This is the hallmark of a conservative force—we can identify a potential energy function, such that the work done by the force on a member of the system in which the force acts depends only on the difference in the initial and final values of the function. This cannot be done for a nonconservative force because the work done depends on the particular path followed between the initial and final points.

For a conservative force, this notion allows us to generate a mathematical relationship between a force and its potential energy function. From the definition of work, we can write Equation 7.12 for a general force in the $x$ direction as

$$W = \int_{x_i}^{x_f} F_x\,dx = -\Delta U = -(U_f - U_i) = -U_f + U_i \qquad [7.13]$$

• *Work done by a conservative force*

Thus, the potential energy function can be written as

$$U_f = -\int_{x_i}^{x_f} F_x\,dx + U_i \qquad [7.14]$$

• *Finding the potential energy from the force*

This expression allows us to calculate the potential energy function associated with a conservative force if we know the force function. The value of $U_i$ is often taken to be zero at some arbitrary reference point. It really doesn't matter what value we assign to $U_i$ because any value simply shifts $U_f$ by a constant, and it is the *change* in potential energy that is physically meaningful.

As an example, let us calculate the potential energy function for the spring force. We model the spring as obeying Hooke's law, so the force the spring exerts is $F_s = -kx$. The potential energy stored in a block–spring system is

$$U_f = -\int_{x_i}^{x_f}(-kx)\,dx + U_i = \tfrac{1}{2}kx_f^2 - \tfrac{1}{2}kx_i^2 + U_i$$

As mentioned earlier, we can choose the value of the zero of potential energy arbitrarily. Let us choose $U_i = 0$ when the block is at the position $x_i = 0$. Then,

$$U_f = \tfrac{1}{2}kx_f^2 - \tfrac{1}{2}kx_i^2 + U_i = \tfrac{1}{2}kx_f^2 - 0 + 0 \quad\longrightarrow\quad U_f = U_s = \tfrac{1}{2}kx^2$$

This is the potential energy function we have already recognized (see Eq. 7.8) for a spring that obeys Hooke's law.

In the previous discussion, we have seen how to find a potential energy function if we know the force function. Let us now turn this around. Suppose we know the potential energy function. Can we find the force function? We start from the basic definition of work done by a conservative force for an infinitesimal displacement $d\mathbf{r} = dx\mathbf{i}$ in the *x* direction:

$$dW = \mathbf{F}\cdot d\mathbf{r} = \mathbf{F}\cdot dx\mathbf{i} = F_x\,dx = -dU$$

This can be rewritten as

- *Finding the force from the potential energy*

$$F_x = -\frac{dU}{dx} \qquad\qquad [7.15]$$

Thus, the **conservative force acting between parts of a system equals the negative derivative of the potential energy associated with that system.**[*]

In the case of an object located a distance *y* above some reference point, the gravitational potential energy function is given by $U_g = mgy$, and it follows from Equation 7.15 that (considering the *y* direction rather than *x*)

$$F_y = -\frac{dU_g}{dy} = -\frac{d}{dy}(mgy) = -mg$$

which is the correct expression for the gravitational force.

## 7.5 • THE NONISOLATED SYSTEM IN STEADY STATE

We have seen two approaches related to systems so far. In a nonisolated system, the energy stored in the system changes due to transfers across the boundaries of the system. Thus, nonzero terms occur on both sides of the continuity equation for energy, $\Delta E_{system} = \Sigma\, H$. For an isolated system, no energy transfer takes place across the boundary, so the right-hand side of the continuity equation is zero; that is, $\Delta E_{system} = 0$.

---

[*] In three dimensions, the appropriate expression is

$$\mathbf{F} = -\mathbf{i}\frac{\partial U}{\partial x} - \mathbf{j}\frac{\partial U}{\partial y} - \mathbf{k}\frac{\partial U}{\partial z}$$

where $\partial U/\partial x$, and so on are *partial derivatives*. In the language of vector calculus, $\mathbf{F}$ equals the negative of the *gradient* of the scalar potential energy function $U(x, y, z)$.

Another possibility exists that we have not yet addressed. It is possible for no change to occur in the energy of the system even though nonzero terms are present on the right-hand side of the continuity equation, $0 = \Sigma\ H$. This can only occur if the rate at which energy is entering the system is equal to the rate at which it is leaving. In this case, the system is in steady state under the effects of two or more competing transfers, and we describe this as a **nonisolated system in steady state.** The system is nonisolated because it is interacting with the environment, but it is in steady state because the system energy remains constant.

We could identify a number of examples of this type of situation. First, consider your home as a nonisolated system. Ideally, you would like to keep the temperature of your home constant for the comfort of the occupants. Thus, your goal is to keep the internal energy in the home fixed.

The energy transfer mechanisms for the home are numerous, as we can see in Figure 7.12. Solar electromagnetic radiation is absorbed by the roof and walls of the home and enters the home through the windows. Energy enters by electrical transmission to operate electrical devices. Leaks in the walls, windows, and doors allow warm or cold air to enter and leave, carrying energy across the boundary of the system by matter transfer. Matter transfer also occurs if any devices in the home operate from natural gas—energy is carried in with the gas. Energy transfers through the walls, windows, floor, and roof by heat due to temperature differences between the inside and outside of the home. Thus, we have a variety of transfers, but the energy in the home remains constant in the idealized case. In reality, this is a system in *quasi-steady state* because some small temperature variations actually occur over a 24-h period, but we can imagine an idealized situation that conforms to the nonisolated system in steady-state model.

- *The nonisolated system in steady state*

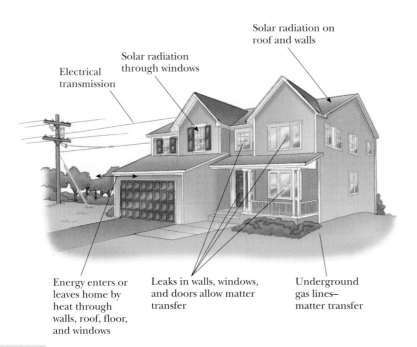

**Figure 7.12**

Energy enters and leaves a home by several mechanisms. The home can be modeled as a nonisolated system in steady state.

As a second example, consider the Earth as a system. Because the Earth is located in the vacuum of space, the only possible types of energy transfers involve no contact between the system and external molecules in the environment. As mentioned in Chapter 6, only two types of transfer do not depend on contact with molecules: work done by field forces and electromagnetic radiation. The Earth as a system exchanges energy with the rest of the Universe only by means of electromagnetic radiation (ignoring work done by field forces, and some small matter transfer due to cosmic ray particles and meteoroids entering the system and spacecraft leaving the system!). The primary input radiation is that from the Sun, and the output radiation is primarily infrared radiation emitted from the ground. These transfers are balanced so that the Earth maintains a constant temperature. The transfers are not *exactly* balanced, however, so the Earth is in quasi-steady state—the temperature does appear to be changing. The change in temperature is very gradual and currently appears to be in the positive direction. This is the essence of the social issue of global warming.

In this chapter and Chapter 6, we explored three analysis models related to systems. We described these models only in terms of energy and energy transfers. Other physical quantities can transfer across the boundary of a system, and we will return to these system models again throughout the book. Additional quantities for which we will discuss versions of system models include momentum (Chapter 8), angular momentum (Chapter 10), volume of fluid (Chapter 15), and electric charge (Chapter 19).

### Quick Quiz 7.4

Model your body as a nonisolated system in steady state. What types of energy transfer occur?

## 7.6 • POTENTIAL ENERGY FOR GRAVITATIONAL AND ELECTRIC FORCES

Earlier in this chapter we introduced the concept of gravitational potential energy, that is, the energy associated with a system of objects interacting via the gravitational force. We emphasized the fact that the gravitational potential energy function, Equation 7.2, is valid only when the object of mass $m$ is near the Earth's surface. We would like to find a more general expression for the gravitational potential energy that is valid for all separation distances. Because the free-fall acceleration $g$ varies as $1/r^2$, it follows that the general dependence of the potential energy function of the system on separation distance is more complicated than our simple expression, Equation 7.2.

Consider a particle of mass $m$ moving between two points Ⓐ and Ⓑ above the Earth's surface, as in Figure 7.13. The gravitational force on the particle due to the Earth, first introduced in Section 5.6, can be written in vector form as

$$\mathbf{F}_g = -\frac{GM_E m}{r^2}\,\hat{\mathbf{r}} \qquad [7.16]$$

where $\hat{\mathbf{r}}$ is a unit vector directed from the Earth toward the particle, and the negative sign indicates that the force is downward toward the Earth. This expression shows that the gravitational force depends on the polar coordinate $r$. Further-

In Section 5.6, we discussed the gravitational force between two *particles*. In Equation 7.16, we present the gravitational force between a particle and an extended object, the Earth. We could also express the gravitational force between two extended objects, such as the Earth and the Sun. In these kinds of situations, remember that $r$ is measured *between the centers* of the objects. Be sure *not* to measure $r$ from the surface of the Earth.

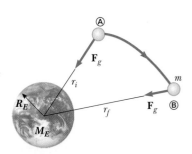

**Figure 7.13**

As a particle of mass $m$ moves from Ⓐ to Ⓑ above the Earth's surface, the potential energy of the particle–Earth system changes according to Equation 7.16.

more, the gravitational force is conservative. Equation 7.14 gives

$$U_f = -\int_{r_i}^{r_f} F(r)\,dr + U_i = GM_E m \int_{r_i}^{r_f} \frac{dr}{r^2} + U_i = GM_E m \left(-\frac{1}{r}\right)_{r_i}^{r_f} + U_i$$

or

$$U_f = -GM_E m \left(\frac{1}{r_f} - \frac{1}{r_i}\right) + U_i \qquad \textbf{[7.17]}$$

As always, the choice of a reference point for the potential energy is completely arbitrary. It is customary to locate the reference point where the force is zero. Letting $U_i \rightarrow 0$ as $r_i \rightarrow \infty$, we obtain the important result

$$U_g = -\frac{GM_E m}{r} \qquad \textbf{[7.18]}$$

Because of our choice of the reference point for zero potential energy, the function $U_g$ is always negative (Fig. 7.14).

Although Equation 7.18 was derived for the particle–Earth system, it can be applied to *any* two particles. For *any pair* of particles of masses $m_1$ and $m_2$ separated by a distance $r$, the gravitational force of attraction is given by Equation 5.14, and the gravitational potential energy of the system of two particles is

$$U_g = -\frac{Gm_1 m_2}{r} \qquad \textbf{[7.19]}$$

This expression also applies to larger objects *if they are spherically symmetric*, as first shown by Newton.

Equation 7.19 shows that the gravitational potential energy for any pair of particles varies as $1/r$ (whereas the force between them varies as $1/r^2$). Furthermore, the potential energy is *negative* because the force is attractive and we have chosen the potential energy to be zero when the particle separation is infinity. Because the force between the particles is attractive, we know that an external agent must do positive work to increase the separation between the two particles. The work done by the external agent produces an increase in the potential energy as the two particles are separated. That is, $U_g$ becomes less negative as $r$ increases.

When the two particles are separated by a distance $r$, an external agent would have to supply an energy *at least* equal to $+Gm_1 m_2/r$ in order to separate the particles by an infinite distance. (In practice, any distance larger than $r$ by at least three orders of magnitude can be modeled as an infinite distance.) It is convenient to think of the absolute value of the gravitational potential energy defined this way as the *binding energy* of the system. An external agent needs to provide this energy to separate the particles by an infinite distance. If the external agent supplies an energy *greater than* the binding energy $Gm_1 m_2/r$, the additional energy of the system is in the form of kinetic energy when the particles are at an infinite separation.

We can extend this concept to three or more particles. In this case, the total potential energy of the system is the sum over all *pairs* of particles. Each pair contributes a term of the form given by Equation 7.19. For example, if the system contains three particles, as in Figure 7.15, we find that

$$U_{\text{total}} = U_{12} + U_{13} + U_{23} = -G\left(\frac{m_1 m_2}{r_{12}} + \frac{m_1 m_3}{r_{13}} + \frac{m_2 m_3}{r_{23}}\right) \qquad \textbf{[7.20]}$$

The absolute value of $U_{\text{total}}$ represents the work needed to separate the particles by an infinite distance.

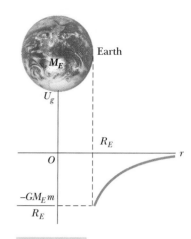

**Figure 7.14**

Graph of the gravitational potential energy $U_g$ versus $r$ for a particle above the Earth's surface. The potential energy of the system goes to zero as $r$ approaches infinity.

• *Gravitational potential energy*

**PITFALL PREVENTION 7.4**

**Gravitational potential energy**

Be careful! Equation 7.19 looks similar to Equation 7.15 for the gravitational force, but there are two major differences. The gravitational force is a vector, whereas the gravitational potential energy is a scalar. The gravitational force varies as the *inverse square* of the separation distance, whereas the gravitational potential energy varies as the simple *inverse* of the separation distance.

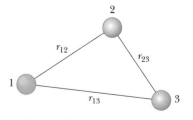

**Figure 7.15**

Three interacting particles.

## THINKING PHYSICS 7.3

Why is the Sun hot?

**Reasoning**   The Sun was formed when a cloud of gas and dust coalesced, due to gravitational attraction, into a massive astronomical object. Let us define this cloud as our system and model the gas and dust as particles. Initially, the particles of the system were widely scattered, representing a large amount of gravitational potential energy. In principle, we could calculate this potential energy by summing the gravitational potential energy for all pairs of particles in the system:

$$U = \sum_{\text{all pairs } i, j} - G \frac{m_i m_j}{r}$$

As the particles moved together to form the Sun, the gravitational potential energy of the system decreased. According to the isolated system model, this potential energy was transformed to kinetic energy as the particles fell toward the center. As the speeds of the particles increased, many collisions occurred between particles, randomizing their motion and transforming the kinetic energy to internal energy, representing an increase in temperature. If enough particles come together, the temperature can rise to a point at which nuclear reactions occur, and the object becomes a star. If not enough particles are present, the temperature rises, but not to a point at which nuclear reactions occur—the object is a planet if it is in orbit around a star. Jupiter is an example of a large planet that might have been a star if more particles had been available.

## Example 7.6   The Change in Potential Energy

A particle of mass $m$ is displaced through a small vertical distance $\Delta y$ near the Earth's surface. Show that the general expression for the change in gravitational potential energy reduces to the familiar relationship $\Delta U_g = mg \, \Delta y$.

**Solution**   We can express Equation 7.17 in the form

$$\Delta U_g = - GM_E m \left( \frac{1}{r_f} - \frac{1}{r_i} \right) = GM_E m \left( \frac{r_f - r_i}{r_i r_f} \right)$$

If both the initial and final positions of the particle are close to the Earth's surface, then $r_f - r_i = \Delta y$ and $r_i r_f \approx R_E^2$. (Recall that $r$ is measured from the center of the Earth.) The change in potential energy therefore, becomes

$$\Delta U_g \approx \frac{GM_E m}{R_E^2} \Delta y = mg \, \Delta y$$

where we have used the fact that $g = GM_E / R_E^2$.

**EXERCISE**   A satellite of the Earth has a mass of 100 kg and its altitude is $2.0 \times 10^6$ m. (a) What is the gravitational potential energy of the satellite–Earth system? (b) What is the magnitude of the force on the satellite?

*Answer*   (a) $- 4.8 \times 10^9$ J   (b) $5.7 \times 10^2$ N

In Chapter 5, we discussed the electrostatic force between two point particles, which is Coulomb's law,

$$F_e = k_e \frac{q_1 q_2}{r^2} \qquad \text{[7.21]}$$

Because this looks so similar to Newton's law of universal gravitation, we would expect that the generation of a potential energy function for this force would proceed in a similar way. This is indeed the case, and this procedure results in the **electric potential energy** function,

$$U_e = k_e \frac{q_1 q_2}{r}$$ [7.22]

As with the gravitational potential energy, the electric potential energy is defined as zero when the charges are infinitely far apart. Comparing this expression with that for the gravitational potential energy, we see the obvious differences in the constants and the use of charges instead of masses, but there is one more difference. The gravitational expression has a negative sign, but the electrical expression doesn't. For systems of objects that experience an attractive force, the potential energy decreases as the objects are brought closer together. Because we have defined zero potential energy at infinite separation, all real separations are finite and the energy must decrease from a value of zero. Thus, all potential energies for systems of objects that attract must be negative. In the gravitational case, attraction is the only possibility. The constant, the masses, and the separation distance are all positive, so the negative sign must be included explicitly, as it is in Equation 7.19.

The electric force can be either attractive or repulsive. Attraction occurs between *unlike* charges. Thus, for the two charges in Equation 7.22, one is positive and one is negative if the force is attractive. The product of the charges provides the negative sign for the potential energy mathematically, and we do not need an explicit negative sign in the potential energy expression. In the case of like charges, either a product of two negative charges or two positive charges will be positive, leading to a positive potential energy. This is reasonable, because to cause repelling particles to move together from infinite separation requires work to be done on the system, so the potential energy increases.

## 7.7 • ENERGY DIAGRAMS AND STABILITY OF EQUILIBRIUM

The motion of a system can often be understood qualitatively by analyzing a graphical representation of the system's potential energy curve. An **energy diagram** shows the potential energy of the system as a function of the position of one of the members of the system (or as a function of the separation distance between two members of the system). Consider the potential energy function for the block–spring system, given by $U_s = \frac{1}{2}kx^2$. This function is plotted versus $x$ in Figure 7.16a.

The spring force is related to $U_s$ through Equation 7.15:

$$F_s = -\frac{dU_s}{dx} = -kx$$

That is, the force is equal to the negative of the *slope* of the $U_s$ versus $x$ curve. When the block is placed at rest at the equilibrium position ($x = 0$), where $F_s = 0$, it will remain there unless some external force acts on it. If the spring in Figure 7.16b is stretched to the right from equilibrium, $x$ is positive and the slope $dU_s/dx$ is positive; therefore, $F_s$ is negative and the block accelerates back toward $x = 0$. If the spring is compressed, $x$ is negative and the slope is negative; therefore, $F_s$ is positive and again the block accelerates toward $x = 0$.

From this analysis, we conclude that the $x = 0$ position is one of **stable equilibrium.** That is, any movement away from this position results in a force directed

• *Stable equilibrium*

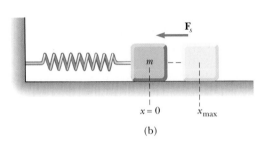

$U_s = \frac{1}{2}kx^2$

(a)

(b)

**Figure 7.16**

(a) Potential energy as a function of $x$ for the block–spring system shown in part (b). The block oscillates between the turning points, which have the coordinates $x = \pm x_{max}$. The restoring force exerted by the spring always acts toward $x = 0$, the position of stable equilibrium.

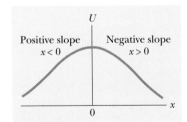

**Figure 7.17**

A plot of $U$ versus $x$ for a particle that has a position of unstable equilibrium, located at $x = 0$. For any finite displacement of the particle, the force on the particle is directed away from $x = 0$.

back toward $x = 0$. (We described this type of force in Chapter 6 as a *restoring force*.) In general, **positions of stable equilibrium correspond to those values of $x$ for which $U(x)$ has a relative minimum value on an energy diagram.**

From Figure 7.16 we see that if the block is given an initial displacement $x_{max}$ and released from rest, the total initial energy of the system is the potential energy stored in the spring, given by $\frac{1}{2}kx_{max}^2$. As motion commences, the system acquires kinetic energy and loses an equal amount of potential energy. From an energy viewpoint, the energy of the system cannot exceed $\frac{1}{2}kx_{max}^2$; therefore, the block must stop at the points $x = \pm x_{max}$ and, because of the spring force, accelerate toward $x = 0$. The block oscillates between the two points $x = \pm x_{max}$, called the *turning points*. The block cannot be farther from equilibrium than $\pm x_{max}$ because the potential energy of the system beyond these points would be larger than the total energy, an impossible situation in classical physics. Because there is no transformation of mechanical energy to internal energy (no friction), the block oscillates between $-x_{max}$ and $+x_{max}$ forever. (We shall discuss these oscillations further in Chapter 12).

Now consider an example in which the curve of $U$ versus $x$ is as shown in Figure 7.17. In this case, $F_x = 0$ at $x = 0$, and so the particle is in equilibrium at this point. However, this is a position of **unstable equilibrium** for the following reason. Suppose the particle is displaced to the *right* from the origin. Because the slope is negative for $x > 0$, $F_x = -dU/dx$ is positive and the particle accelerates away from $x = 0$. Now suppose the particle is displaced to the *left* from the origin. In this case the force is negative because the slope is positive for $x < 0$, and the particle again accelerates away from the equilibrium position. The $x = 0$ position in this situation is called a position of unstable equilibrium because, for any displacement from this point, the force pushes the particle farther away from equilibrium. In fact, the force pushes the particle toward a position of lower potential energy of the system. A ball placed on the top of an inverted spherical bowl is in a position of unstable equilibrium. If the ball is displaced slightly from the top and released, it will roll off the bowl. In general, **positions of unstable equilibrium correspond to those values of $x$ for which $U(x)$ has a relative maximum value on an energy diagram.**\*

Finally, a situation may arise in which $U$ is constant over some region, and hence $F = 0$. A point in this region is called a position of **neutral equilibrium.** Small displacements from this position produce neither restoring nor disrupting forces. A ball lying on a flat horizontal surface is an example of an object in neutral equilibrium.

• *Unstable equilibrium*

• *Neutral equilibrium*

### context connection
## 7.8 • ESCAPE SPEED FROM THE SUN

Our central question is based on the requirements for sending a spacecraft from Earth to Mars. We have discussed some initial steps toward this goal: launching the spacecraft from the surface of the Earth and placing it in a circular parking orbit around the Earth. The next step is to fire our rocket engines and begin our journey to Mars. But with what speed should we leave the parking orbit?

We cannot answer this question yet because it requires information not covered until later chapters. But we can establish an *upper limit* to the required speed. Let us ask this question: With what speed would we need to leave the orbit to be able to completely escape from the gravitational force of the Sun? Surely, the speed necessary to travel to Mars must be less than this speed.

Using our notions of energy, we can find the speed necessary to escape from the Sun. We will model the spacecraft and the Sun as an isolated system. As a simplification model, we will ignore the interaction between the spacecraft and all planets and stars in the universe, including Earth. In this isolated system, the total amount of energy must remain constant. Before the rocket engines are fired, the system has kinetic energy due to the motion of the spacecraft in its orbit around the Sun, gravitational potential energy related to the force between the spacecraft and the Sun, and energy stored in the fuel in the rocket. Let us simplify things further by imagining that the spacecraft does not move very far relative to the Sun while the rocket engine is burning. Thus, we will imagine the rocket engine fires for a very short time to give the rocket some speed and then is shut off, and the rocket drifts through space.

After the firing, the spacecraft–Sun system has kinetic energy and gravitational potential energy, and no nonconservative forces act. The energy in the fuel that was burned has been transformed into kinetic energy of the spacecraft. To completely escape from the Sun, the spacecraft must keep drifting away, approaching an infinite separation distance from the Sun. The minimum speed to produce this is found by imagining that the spacecraft keeps drifting away forever, with its speed asymptotically approaching zero. The mechanical energy of the system after the rocket engine is fired must equal the mechanical energy at infinite separation. For a spacecraft of mass $m$, this gives us

$$\tfrac{1}{2}mv^2_{\text{at Earth}} - G\frac{mm_{\text{Sun}}}{r_{\text{Earth–Sun}}} = \tfrac{1}{2}mv^2_{\text{at infinity}} - G\frac{mm_{\text{Sun}}}{r_{\text{infinity}}}$$

As the separation distance approaches infinity, the last term on the right approaches zero. If the velocity asymptotically approaches zero as the separation

---

\* You can mathematically test whether an extreme of $U$ is stable or unstable by examining the sign of $d^2U/dx^2$.

distance approaches infinity, the kinetic energy term on the right also becomes zero. With this condition, the speed at the position of the Earth becomes the escape speed $v_{esc}$. Thus, we have,

$$\frac{1}{2}mv_{esc}^2 - G\frac{mm_{Sun}}{r_{Earth-Sun}} = 0 \quad \longrightarrow \quad v_{esc} = \sqrt{2G\frac{m_{Sun}}{r_{Earth-Sun}}} \qquad \text{[7.23]}$$

When we substitute values for the mass of the Sun and the initial distance of the spacecraft from the Sun, we have

$$v_{esc} = \sqrt{2(6.67 \times 10^{-11}\,\text{N}\cdot\text{m}^2/\text{kg}^2)\frac{1.99 \times 10^{30}\,\text{kg}}{1.50 \times 10^{11}\,\text{m}}} = 4.21 \times 10^4\,\text{m/s}$$

Thus, we must move with at least this speed from Earth orbit to escape from the Sun. As mentioned previously, the speed necessary to send the spacecraft to Mars is less than this value. In fact, it is about one-tenth this value.

We have calculated an upper limit to a speed with which we need to leave the Earth and have indicated that we can determine the actual speed with material from upcoming chapters. Assuming that we know this required speed, how do we determine how much fuel to burn to attain this speed? We will do this in the next chapter by considering isolated systems again, but in terms of a new quantity called *momentum*. This quantity is critical to determining our fuel requirements for our trip to Mars.

## SUMMARY

If a particle of mass $m$ is elevated a distance $y$ from a reference point $y = 0$ near the Earth's surface, the **gravitational potential energy** of the particle–Earth system can be defined as

$$U_g \equiv mgy \qquad \text{[7.2]}$$

The **total mechanical energy of a system** is defined as the sum of the kinetic energy and potential energy:

$$E_{mech} \equiv K + U \qquad \text{[7.7]}$$

If no energy transfers occur across the boundary of the system, the system is modeled as an **isolated system.** In this model, the principle of **conservation of mechanical energy** states that the total mechanical energy of the system is constant if all of the forces in the system are conservative:

$$K_i + U_i = K_f + U_f \qquad \text{[7.6]}$$

A force is **conservative** if the work it does on a particle is independent of the path the particle takes between two given points. A conservative force in mechanics does not cause a transformation of mechanical energy to internal energy. A force that does not meet these criteria is said to be **nonconservative.**

The **elastic potential energy** stored in a spring of force constant $k$ is

$$U_s \equiv \tfrac{1}{2}kx^2 \qquad \text{[7.8]}$$

If some of the forces acting within a system are not conservative, the mechanical energy of the system does not remain constant. In the case of a common nonconservative force, friction, the change in mechanical energy is equal to the product of the kinetic friction force and the displacement:

$$-f_k\,\Delta x = \Delta K + \Delta U \qquad \text{[7.9]}$$

This decrease in mechanical energy in the system is equal to the increase in internal energy,

$$\Delta E_{int} = f_k\,\Delta x \qquad \text{[7.9]}$$

A potential energy function $U$ can be associated only with a conservative force. If a conservative force $\mathbf{F}$ acts on a particle that moves along the $x$ axis from $x_i$ to $x_f$, the **change in the potential energy equals the negative of the work done by that force:**

$$U_f - U_i = -\int_{x_i}^{x_f} F_x\,dx \qquad \text{[7.14]}$$

If we know the potential energy function, a conservative force is given by the negative of the derivative of the potential energy function:

$$F_x = -\frac{dU}{dx} \qquad \text{[7.15]}$$

In some situations, a system may have energy crossing the boundary with no change in the energy stored in the system. In this case, the energy input in any time interval equals the energy output, and we describe this system as a **nonisolated system in steady state.**

The **gravitational potential energy** associated with two particles or uniform spherical distributions of mass separated by a distance $r$ is

$$U_g = -\frac{Gm_1m_2}{r} \qquad [7.19]$$

where $U_g$ is taken to approach zero as $r \to \infty$.

The **electric potential energy** associated with two charged particles separated by a distance $r$ is

$$U_e = k_e\frac{q_1q_2}{r} \qquad [7.22]$$

where $U_e$ is taken to approach zero as $r \to \infty$.

In an energy diagram, a point of **stable equilibrium** is one at which the potential energy is a minimum. A point of **unstable equilibrium** is one at which the potential energy is a maximum. **Neutral equilibrium** exists if the potential energy function is constant.

## QUESTIONS

1. One person drops a ball from the top of a building while another person at the bottom observes its motion. Will these two people agree on the value of the gravitational potential energy of the ball–Earth system? On the change in potential energy? On the kinetic energy?

2. Discuss the changes in mechanical energy of an object–Earth system in (a) lifting the object, (b) holding the object at a fixed position, and (c) lowering the object slowly. Include the muscles in your discussion.

3. In Chapter 6, the work–kinetic energy theorem, $W = \Delta K$, was introduced. This equation states that work done on a system appears as a change in kinetic energy. This is a special case equation—it is valid if no changes occur in any other type of energy, such as potential or internal. Give some examples in which work is done but the change in energy is not that of kinetic energy.

4. A skier prepares to take off down a snow-covered hill. Is it correct to say that gravity provides the energy for the trip down the hill?

5. You walk leisurely up a flight of stairs. You then return to the bottom and run up the same set of stairs as fast as you can. Have you performed a different amount of work in these two cases?

6. A rock is thrown vertically in the air. Ignoring air friction, it takes the same time to go up as it does to come down. Is this still true in the presence of air friction?

7. A driver brings an automobile to a stop. If the brakes lock, so that the car skids, where is the original kinetic energy of the car, and in what form is it, after the car stops? Answer the same question for the case in which the brakes do not lock but the wheels continue to turn.

8. In an earthquake, a large amount of energy is "released" and spreads outward, potentially causing severe damage.

In what form does this energy exist before the earthquake, and by what energy transfer mechanism does it travel?

9. You ride a bicycle. In what sense is your bicycle solar powered?

10. A bowling ball is suspended from the ceiling of a lecture hall by a strong cord. The bowling ball is drawn away from its equilibrium position and released from rest at the tip of the demonstrator's nose. If the demonstrator remains stationary, explain why she will not be struck by the ball on its return swing. Would the demonstrator be safe if she pushed the ball as she released it?

11. A pile driver is a device used to drive objects into the Earth by repeatedly dropping a heavy weight on them. By how much does the energy of the pile driver–Earth system increase when the weight it drops is doubled? (Assume the weight is dropped from the same height each time.)

12. Our body muscles exert forces when we lift, push, run, jump, and so forth. Are these forces conservative?

13. A block is connected to a spring suspended from the ceiling. If the block is set in motion and air resistance is ignored, describe the energy transformations that occur within the system consisting of the block and spring.

14. What would the curve of $U$ versus $x$ look like if a particle were in a region of neutral equilibrium?

15. A ball rolls on a horizontal surface. Is the ball in stable, unstable, or neutral equilibrium?

16. Discuss the energy transformations that occur during the operation of an automobile.

17. A ball is thrown straight up into the air. At what position is its kinetic energy a maximum? At what position is the gravitational potential energy of the ball–Earth system a maximum?

# PROBLEMS

$1, 2, 3$ = straightforward, intermediate, challenging    □ = full solution available in the *Student Solutions Manual and Study Guide*

**web** = solution posted at **http://www.harcourtcollege.com/physics/**    💻 = computer useful in solving problem

📲 = Interactive Physics    ▓ = paired numerical/symbolic problems    📰 = life science application

## Section 7.1    Potential Energy of a System

**1.** A 1 000-kg roller coaster is initially at the top of a rise, at point Ⓐ. It then moves 135 ft, at an angle of 40.0° below the horizontal, to a lower point Ⓑ. (a) Choose point Ⓑ to be the zero level for gravitational potential energy. Find the potential energy of the roller coaster–Earth system at points Ⓐ and Ⓑ and the change in potential energy as the coaster moves. (b) Repeat part (a), setting the zero reference level at point Ⓐ.

**2.** A 400-N child is in a swing that is attached to ropes 2.00 m long. Find the gravitational potential energy of the child–Earth system relative to the child's lowest position when (a) the ropes are horizontal, (b) the ropes make a 30.0° angle with the vertical, and (c) the child is at the bottom of the circular arc.

**3.** A person with a remote mountain cabin plans to install her own hydroelectric plant. A nearby stream is 3.00 m wide and 0.500 m deep. Water flows at 1.20 m/s over the brink of a waterfall 5.00 m high. The manufacturer promises only 25.0% efficiency in converting the potential energy of the water–Earth system into electric energy. Find the power she can generate. (Large-scale hydroelectric plants, with a much larger drop, are more efficient.)

## Section 7.2    The Isolated System

**4.** A particle of mass 0.500 kg is shot from *P* as shown in Figure P7.4. The particle has an initial velocity $\mathbf{v}_i$ with a horizontal component of 30.0 m/s. The particle rises to a maximum height of 20.0 m above *P*. Using the law of conservation of energy, determine (a) the vertical component of $\mathbf{v}_i$, (b) the work done by the gravitational force on the particle during its motion from *P* to *B*, and (c) the horizontal and the vertical components of the velocity vector when the particle reaches *B*.

**Figure P7.4**

**5.** A bead slides without friction around a loop-the-loop (Fig. P7. 5). The bead is released from a height $h = 3.50R$. (a) What is its speed at point *A*? (b) How large is the normal force on it at point *A* if its mass is 5.00 g?

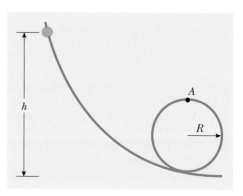

**Figure P7.5**

**6.** Dave Johnson, the bronze medalist at the 1992 Olympic decathlon in Barcelona, leaves the ground at the high jump with vertical velocity component 6.00 m/s. How far does his center of mass move up as he makes the jump?

**7.** A simple pendulum consists of a small object suspended by a string. The object is modeled as a particle. The string, with its top end fixed, has negligible mass and does not stretch. In the absence of air friction, the system oscillates by swinging back and forth in a vertical plane. If the string is 2.00 m long and makes an initial angle of 30.0° with the vertical, calculate the speed of the particle (a) at the lowest point in its trajectory and (b) when the angle is 15.0°.

**8.** A circus trapeze consists of a bar suspended by two parallel ropes, each of length $\ell$, allowing performers to swing in a vertical circular arc (Figure P7.8). Suppose a performer with mass $m$ holds the bar and steps off an elevated platform, starting from rest with the ropes at an angle of $\theta_i$ with respect to the vertical. Suppose the size of the performer's body is small compared with the length $\ell$, that she does not pump the trapeze to swing higher, and that air resistance is negligible. (a) Show that when the ropes make an angle $\theta$ with the vertical, the performer must exert a force of magnitude $mg(3 \cos \theta - 2 \cos \theta_i)$ in order to hang on. (b) Determine the angle $\theta_i$ for which the force needed to hang on at the bottom of the swing is twice the performer's weight.

**Figure P7.8**

9. Two objects are connected by a light string passing over a light, frictionless pulley as shown in Figure P7.9. The 5.00-kg object is released from rest. Using the law of conservation of energy, (a) determine the speed of the 3.00-kg object just as the 5.00-kg object hits the ground. (b) Find the maximum height to which the 3.00-kg object rises.

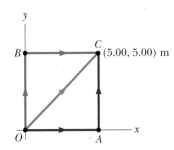

$m_1 = 5.00 \text{ kg}$

$m_2 = 3.00 \text{ kg}$   $h = 4.00 \text{ m}$

**Figure P7.9**

10. A light rigid rod is 77.0 cm long. Its top end is pivoted on a low-friction horizontal axle. The rod hangs straight down at rest, with a small massive ball attached to its bottom end. You strike the ball, suddenly giving it a horizontal velocity so that it swings around in a full circle. What minimum speed at the bottom is required to make the ball go over the top of the circle?

**Section 7.3** Conservative and Nonconservative Forces

11. A system contains one particle that is free to move. At time $t_i$, the kinetic energy of the particle is 30.0 J and the po-

tential energy of the system is 10.0 J. At some later time $t_f$, the kinetic energy of the particle is 18.0 J. (a) If only conservative forces act on the particle, what are the potential energy and the total energy of the system at time $t_f$? (b) If the potential energy of the system at time $t_f$ is 5.00 J, are any nonconservative forces acting on the particle? Explain.

12. (a) Suppose that a constant force acts on an object. The force does not vary with time, nor with the position or the velocity of the object. Start with the general definition for work done by a force

$$W = \int_i^f \mathbf{F} \cdot d\mathbf{r}$$

and show that the force is conservative. (b) As a special case, suppose that the force $\mathbf{F} = (3\mathbf{i} + 4\mathbf{j})$ N acts on a particle that moves from $O$ to $C$ in Figure P7.12. Calculate the work done by $\mathbf{F}$ if the particle moves along each one of the three paths $OAC$, $OBC$, and $OC$. (Your three answers should be identical.)

*y*

$B$ ——→ $C$ (5.00, 5.00) m

$O$ —→ $A$   *x*

**Figure P7.12**  Problems 12 and 13.

13. A force acting on a particle moving in the *xy* plane is given by $\mathbf{F} = (2y\mathbf{i} + x^2\mathbf{j})$ N, where *x* and *y* are in meters. The particle moves from the origin to a final position having coordinates $x = 5.00$ m and $y = 5.00$ m, as in Figure P7.12. Calculate the work done by $\mathbf{F}$ along (a) *OAC*, (b) *OBC*, (c) *OC*. (d) Is $\mathbf{F}$ conservative or nonconservative? Explain.

14. In her hand a softball pitcher swings a ball of mass 0.250 kg around a vertical circular path of radius 60.0 cm before releasing it. The pitcher maintains a component of force on the ball of constant magnitude 30.0 N in the direction of motion around the complete path. The speed of the ball at the top of the circle is 15.0 m/s. If she releases the ball at the bottom of the circle, what is its speed upon release?

15. The world's biggest locomotive is the MK5000C, a 160-metric-ton behemoth driven by the most powerful engine ever used for rail transportation, a Caterpillar diesel capable of 5 000 hp. Such a huge machine provides certain gains in efficiency, but its large inertia presents challenges as well. The engine driver finds that the locomotive handles

differently from conventional ones, notably in braking and climbing hills. To investigate the loss of speed that can occur on an uphill stretch of track, consider the locomotive pulling no train, but traveling at 27.0 m/s on a level track while operating with an output power 1 000 hp. It comes to a 5.00% grade (a slope that rises 5.00 m for every 100 m along the track). If the throttle is not advanced, so that the power level is held steady, to what value will the speed drop? Assume that friction forces do not depend on the speed.

16. A glider of mass 0.150 kg moves on a horizontal friction-less air track. It is permanently attached to one end of a massless horizontal spring, which has a force constant of 10.0 N/m. The other end of the spring is fixed. The glider is moved so as to compress the spring by 0.180 m and then released from rest. Calculate the speed of the glider at the point where it has moved (a) 0.180 m from its starting point, so that the spring is momentarily exerting no force, and (b) 0.250 m from its starting point.

17. An object of mass $m$ starts from rest and slides down a friction-less incline of angle $\theta$. After sliding a distance $d$, it contacts the end of an unstressed spring of negligible mass as shown in Figure P7.17. The object slides an additional distance $x$ as it is brought momentarily to rest by compression of the spring (of force constant $k$). Find the initial separation $d$ between the object and the end of the spring.

**Figure P7.17**

18. A 120-g object is attached to the bottom end of an un-stressed spring. The spring is hanging vertically and has a force constant 40.0 N/m. The object is dropped. (a) What is its maximum speed? (b) How far does it drop before coming to rest momentarily?

19. A block of mass 0.250 kg is placed on top of a light vertical spring of force constant 5 000 N/m and pushed downward, so that the spring is compressed by 0.100 m. After the block is released from rest, it travels upward and then leaves the spring. To what maximum height above the point of release does it rise?

20. A boy in a wheelchair (total mass 47.0 kg) wins a race with a skateboarder. He has speed 1.40 m/s at the crest of a slope 2.60 m high and 12.4 m long. At the bottom of the slope his speed is 6.20 m/s. If air resistance and rolling resistance can be modeled as a constant friction force of 41.0 N, find the work he did in pushing forward on his wheels during the downhill ride.

21. The coefficient of friction between the 3.00-kg block and the surface in Figure P7.21 is 0.400. The system starts from rest. What is the speed of the 5.00-kg ball when it has fallen 1.50 m?

3.00 kg

5.00 kg

**Figure P7.21**

22. An 80.0-kg skydiver jumps out of a balloon at an altitude of 1 000 m and opens the parachute at an altitude of 200 m. (a) Assuming that the total retarding force on the diver is constant at 50.0 N with the parachute closed and constant at 3 600 N with the parachute open, what is the speed of the diver when he lands on the ground? (b) Do you think the skydiver will be injured? Explain. (c) At what height should the parachute be opened so that the final speed of the skydiver when he hits the ground is 5.00 m/s? (d) How realistic is the assumption that the total retarding force is constant? Explain.

23. A 5.00-kg block is set into motion up an inclined plane with an initial speed of 8.00 m/s (Fig. P7.23). The block comes to rest after traveling 3.00 m along the plane, which is inclined at an angle of 30.0° to the horizontal. For this motion determine (a) the change in the block's kinetic energy, (b) the change in the potential energy of the block–Earth system, and (c) the friction force exerted on the block (assumed to be constant). (d) What is the coefficient of kinetic friction?

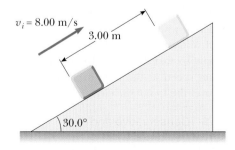

$v_i = 8.00$ m/s

3.00 m

30.0°

**Figure P7.23**

24. A toy cannon uses a spring to project a 5.30-g soft rubber ball. The spring is originally compressed by 5.00 cm and has a force constant of 8.00 N/m. When it is fired, the ball moves 15.0 cm through the horizontal barrel of the cannon, and a constant frictional force of 0.032 0 N exists between the barrel and the ball. (a) With what speed does the projectile leave the barrel of the cannon? (b) At what point does the ball have maximum speed? (c) What is this maximum speed?

25. A 1.50-kg object is held 1.20 m above a relaxed massless vertical spring with a force constant of 320 N/m. The object is dropped onto the spring. (a) How far does it compress the spring? (b) How far does it compress the spring if the same experiment is performed on the Moon, where $g = 1.63$ m/s²? (c) Repeat part (a), but this time assume a constant air resistance force of 0.700 N acts on the mass during its motion.

26. A 75.0-kg skysurfer is falling straight down with terminal speed 60.0 m/s. Determine the rate at which the skysurfer–Earth system is losing mechanical energy.

## Section 7.4 Conservative Forces and Potential Energy

27. A single conservative force acts on a 5.00-kg particle. The equation $F_x = (2x + 4)$ N describes the force, where $x$ is in meters. As the particle moves along the $x$ axis from $x = 1.00$ m to $x = 5.00$ m, calculate (a) the work done by this force, (b) the change in the potential energy of the system, and (c) the kinetic energy of the particle at $x = 5.00$ m if its speed is 3.00 m/s at $x = 1.00$ m.

**web**

28. A single conservative force acting on a particle varies as $\mathbf{F} = (-Ax + Bx^2)\mathbf{i}$ N, where $A$ and $B$ are constants and $x$ is in meters. (a) Calculate the potential energy function $U(x)$ associated with this force, taking $U(x) = 0$ at $x = 0$. (b) Find the change in potential energy of the system and the change in kinetic energy of the particle as it moves from $x = 2.00$ m to $x = 3.00$ m.

29. The potential energy of a system of two particles separated by a distance $r$ is given by $U(r) = A/r$, where $A$ is a constant. Find the radial force $\mathbf{F}_r$ that each particle exerts on the other.

**web**

30. A potential energy function for a two-dimensional force on a particle is of the form $U = 3x^3y - 7x$. Find the force that acts on the particle at the point $(x, y)$.

## Section 7.6 Potential Energy for Gravitational and Electric Forces

31. A satellite of the Earth has a mass of 100 kg and is at an altitude of $2.00 \times 10^6$ m. (a) What is the potential energy of the satellite–Earth system? (b) What is the magnitude of the gravitational force exerted by the Earth on the satellite? (c) What force does the satellite exert on the Earth?

32. How much energy is required to move a 1 000-kg mass from the Earth's surface to an altitude twice the Earth's radius?

33. At the Earth's surface a projectile is launched straight up at a speed of 10.0 km/s. To what height will it rise? Ignore air resistance.

34. A system consists of three particles, each of mass 5.00 g, located at the corners of an equilateral triangle with sides of 30.0 cm. (a) Calculate the gravitational potential energy of the system. (b) If the particles are released simultaneously, where will they collide?

## Section 7.7 Energy Diagrams and Stability of Equilibrium

35. For the potential energy curve shown in Figure P7.35, (a) determine whether the force $F_x$ is positive, negative, or zero at the five points indicated. (b) Indicate points of stable, unstable, and neutral equilibrium. (c) Sketch the curve for $F_x$ versus $x$ from $x = 0$ to $x = 9.5$ m.

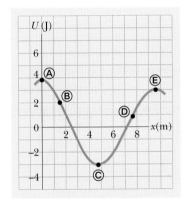

**Figure P7.35**

36. A particle moves along a line where the potential energy of the system depends on the position $r$ of the particle as graphed in Figure P7.36. In the limit as $r$ increases without bound, $U(r)$ approaches $+1$ J. (a) Identify each equilibrium position for this particle. Indicate whether each is a point of stable, unstable, or neutral equilibrium. (b) The particle will be bound if the total energy of the system is in what range? Now suppose that the system has energy $-3$ J. Determine (c) the range of positions where it can be found, (d) its maximum kinetic energy, (e) the location where it has maximum kinetic energy, and (f) its *binding energy*—that is, the additional energy that it would have to be given in order for it to move out to $r \to \infty$.

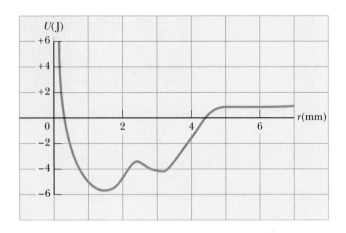

**Figure P7.36**

**37.** A particle of mass 1.18 kg is attached between two identical springs on a horizontal frictionless tabletop. The springs have force constant $k$ and each is initially unstressed. (a) If the particle is pulled a distance $x$ along a direction perpendicular to the initial configuration of the springs, as in Figure P7.37, show that the potential energy of the system is

$$U(x) = kx^2 + 2kL(L - \sqrt{x^2 + L^2})$$

(*Hint:* See Problem 50 in Chapter 6.) (b) Make a plot of $U(x)$ versus $x$, and identify all equilibrium points. Assume that $L = 1.20$ m and $k = 40.0$ N/m. (c) If the particle is pulled 0.500 m to the right and then released, what is its speed when it reaches the equilibrium point $x = 0$?

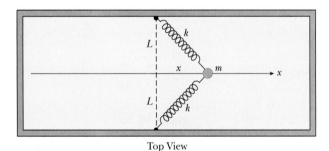

Top View

**Figure P7.37**

### Section 7.8 Context Connection — Escape Speed from the Sun

**38.** *Voyager 1* achieved a maximum speed of 125 000 km/h on its way to photograph Jupiter. At what distance from the Sun is this speed sufficient for a spacecraft to escape the solar system?

**39.** An asteroid is on a collision course with Earth. An astronaut lands on the rock to bury explosive charges that will blow the asteroid apart. Most of the small fragments will miss the Earth, and those that fall into the atmosphere will produce only a beautiful meteor shower. The astronaut finds that the density of the spherical asteroid is equal to the average density of the Earth. To ensure its pulverization, she incorporates into the explosives the rocket fuel and oxidizer intended for her return journey. What maximum radius must the asteroid have for her to be able to leave it entirely simply by jumping straight up? On Earth she can jump to a height of 0.500 m.

### Additional Problems

**40.** In the dangerous "sport" of bungee-jumping, a thrill-seeker jumps from a great height with an elastic cord attached to his ankles. Consider a 70.0-kg person leaping from the top of the world's tallest hotel, the 226-m Westin Stamford in Raffles City, Singapore. Instead of a specially designed elastic cord, suppose that he plans to use 9.00-mm-diameter nylon mountain climber's rope, with effective force constant 4 900 N/m. (a) What length of rope does he need to stop his plunge 10.0 m above the sidewalk? His height is negligible compared with the distance he falls. (b) What maximum force will the rope exert on the daredevil? (c) Express the maximum force as a multiple of the jumper's weight. Has he made a good plan?

**41.** Make an order-of-magnitude estimate of your power output as you climb stairs. In your solution state the physical quantities you take as data and the values you measure or estimate for them. Do you consider your peak power or your sustainable power?

**42.** A 200-g particle is released from rest at point Ⓐ along the horizontal diameter on the inside of a frictionless, hemispherical bowl of radius $R = 30.0$ cm (Fig. P7.42). Calculate (a) the gravitational potential energy of the particle–Earth system when the particle is at point Ⓐ relative to point Ⓑ, (b) the kinetic energy of the particle at point Ⓑ, (c) its speed at point Ⓑ, and (d) its kinetic energy and the potential energy of the system when the particle is at point Ⓒ.

**Figure P7.42** Problems 42 and 43.

**43.** The particle described in Problem 42 (see Fig. P7.42) is
**web** released from rest at Ⓐ, and the surface of the bowl is
rough. The speed of the particle at Ⓑ is 1.50 m/s. (a)
What is its kinetic energy at Ⓑ? (b) How much mechanical energy is transformed due to friction as the particle
moves from Ⓐ to Ⓑ? (c) Is it possible to determine the coefficient of friction from these results in any simple manner? Explain.

**44.** A 50.0-kg block and a 100-kg block are connected by a
string as in Figure P7.44. The pulley is frictionless and of
negligible mass. The coefficient of kinetic friction between
the 50-kg block and incline is 0.250. Determine the
change in the kinetic energy of the 50-kg block as it moves
from Ⓐ to Ⓑ, a distance of 20.0 m.

**Figure P7.44**

**45.** Assume that you attend a state university that started out
as an agricultural college. Close to the center of the campus is a tall silo topped with a hemispherical cap. The cap
is frictionless when wet. Someone has somehow balanced
a pumpkin at the highest point. The line from the center
of curvature of the cap to the pumpkin makes an angle
$\theta_i = 0°$ with the vertical. While you happen to be standing
nearby in the middle of a rainy night, a breath of wind
makes the pumpkin start sliding downward from rest. It
loses contact with the cap when the line from the center
of the hemisphere to the pumpkin makes a certain angle
with the vertical. What is this angle?

**46.** A child's pogo stick (Fig. P7.46) stores energy in a spring
with a force constant of $2.50 \times 10^4$ N/m. At position Ⓐ,
($x_A = -0.100$ m), the spring compression is a maximum

**Figure P7.46**

and the child is momentarily at rest. At position Ⓑ ($x_B =$
0), the spring is relaxed and the child is moving upward.
At position Ⓒ, the child is again momentarily at rest at the
top of the jump. The combined mass of child and pogo
stick is 25.0 kg. (a) Calculate the total energy of the system
if both the gravitational and elastic potential energies are
zero at $x = 0$. (b) Determine $x_C$. (c) Calculate the speed
of the child at Ⓑ. (d) Determine the value of $x$ for which
the kinetic energy of the system is a maximum. (e) Calculate the child's maximum upward speed.

**47.** A 10.0-kg block is released from point Ⓐ in Figure P7.47.
The track is frictionless except for the portion between
points Ⓑ and Ⓒ, which has a length of 6.00 m. The block
travels down the track, hits a spring of force constant
2 250 N/m, and compresses the spring 0.300 m from its
equilibrium position before coming to rest momentarily.
Determine the coefficient of kinetic friction between the
block and the rough surface between Ⓑ and Ⓒ.

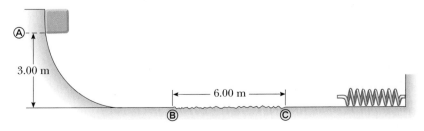

**Figure P7.47**

48. The potential energy function for a system is given by
$U(x) = -x^3 + 2x^2 + 3x$. (a) Determine the force $F_x$ as a
function of $x$. (b) For what values of $x$ is the force equal to
zero? (c) Plot $U(x)$ versus $x$ and $F_x$ versus $x$, and indicate
points of stable and unstable equilibrium.

49. A 20.0-kg block is connected to a 30.0-kg block by a string
that passes over a light frictionless pulley. The 30.0-kg
block is connected to a spring that has negligible mass and
a force constant of 250 N/m, as shown in Figure P7.49.
The spring is unstretched when the system is as shown in
the figure, and the incline is frictionless. The 20.0-kg
block is pulled 20.0 cm down the incline (so that the 30.0-
kg block is 40.0 cm above the floor) and released from
rest. Find the speed of each block when the 30.0-kg block
is 20.0 cm above the floor (i.e., when the spring is unstretched).

**Figure P7.49**

50. A 1.00-kg object slides to the right on a surface having a
coefficient of kinetic friction 0.250 (Fig. P7.50). The object has a speed of $v_i = 3.00$ m/s when it makes contact

**Figure P7.50**

with a light spring that has a force constant of 50.0
N/m. The object comes to rest after the spring has been
compressed a distance $d$. The object is then forced toward the left by the spring and continues to move in
that direction beyond the spring's unstretched position.
Finally the object comes to rest a distance $D$ to the left
of the unstretched spring. Find (a) the distance of compression $d$, (b) the speed $v$ at the unstretched position
when the object is moving to the left, and (c) the
distance $D$.

51. A block of mass 0.500 kg is pushed against a horizontal
web spring of negligible mass until the spring is compressed a
distance $\Delta x$ (Fig. P7.51). The force constant of the spring
is 450 N/m. When it is released, the block travels along a
frictionless, horizontal surface to point $B$, the bottom of a
vertical circular track of radius $R = 1.00$ m, and continues
to move up the track. The speed of the block at the bottom of the track is $v_B = 12.0$ m/s, and the block experiences an average frictional force of 7.00 N while sliding up
the track. (a) What is $\Delta x$? (b) What speed do you predict
for the block at the top of the track? (c) Does the block actually reach the top of the track, or does it fall off before
reaching the top?

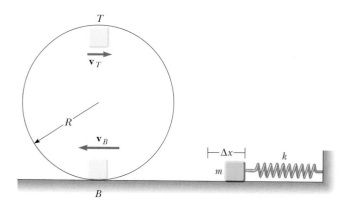

**Figure P7.51**

52. A uniform chain of length 8.00 m initially lies stretched
out on a horizontal table. (a) If the coefficient of static
friction between chain and table is 0.600, show that the
chain will begin to slide off the table if at least 3.00 m of it
hangs over the edge of the table. (b) Determine the speed
of the chain as all of it leaves the table, given that the coefficient of kinetic friction between the chain and the table
is 0.400.

53. A child slides without friction from a height $h$ along a
curved water slide (Fig. P7.53). She is launched from a
height $h/5$ into the pool. Determine her maximum airborne height $y$ in terms of $h$ and $\theta$.

**Figure P7.53**

54. Jane, whose mass is 50.0 kg, needs to swing across a river (having width $D$) filled with crocodiles to save Tarzan from danger. She must swing into a wind exerting constant horizontal force $\mathbf{F}$, on a vine having length $L$, and initially making an angle $\theta$ with the vertical (Fig. P7.54). Taking $D = 50.0$ m, $F = 110$ N, $L = 40.0$ m, and $\theta = 50.0°$, (a) with what minimum speed must Jane begin her swing to just make it to the other side? (*Hint:* First determine the potential energy associated with the wind force.) (b) Once the rescue is complete, Tarzan and Jane must swing back across the river. With what minimum speed must they begin their swing? Assume that Tarzan has a mass of 80.0 kg.

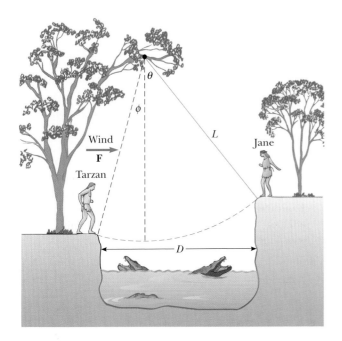

**Figure P7.54**

55. A 5.00-kg block free to move on a horizontal, frictionless surface is attached to one end of a light horizontal spring. The other end of the spring is held fixed. The spring is compressed 0.100 m from equilibrium and released. The speed of the block is 1.20 m/s when it passes the equilib-

rium position of the spring. The same experiment is now repeated with the frictionless surface replaced by a surface for which the coefficient of kinetic friction is 0.300. Determine the speed of the block at the equilibrium position of the spring.

56. A block of mass $M$ rests on a table. It is fastened to the lower end of a light vertical spring. The upper end of the spring is fastened to a block of mass $m$. The upper block is pushed down by an additional force $3mg$, so that the spring compression is $4mg/k$. In this configuration the upper block is released from rest. The spring lifts the lower block off the table. In terms of $m$, what is the greatest possible value for $M$?

57. A roller coaster car is released from rest at the top of the first rise and then moves freely with negligible friction. The roller coaster shown in Figure P7.57 has a circular

**Figure P7.57**

loop of radius $R$ in a vertical plane. (a) Suppose first that the car barely makes it around the loop: At the top of the loop the riders are upside down and feel weightless. Find the required height of the release point above the bottom of the loop, in terms of $R$. (b) Now assume that the release point is at or above the minimum required height. Show that the normal force on the car at the bottom of the loop exceeds the normal force at the top of the loop by six times the weight of the car. The normal force on each rider follows the same rule. Such a large normal force is dangerous and very uncomfortable for the riders. Roller coasters are therefore not built with circular loops in vertical planes. Figure P5.21 and the photograph on page 150 show two actual designs.

58. **Review Problem.** In 1887 in Bridgeport, Connecticut, C. J. Belknap built the water slide shown in Figure P7.58. A rider on a small sled, of total mass 80.0 kg, pushed off to start at the top of the slide (point $A$) with a speed of 2.50 m/s. The chute was 9.76 m high at the top, 54.3 m long,

and 0.51 m wide. Along its length, 725 wheels made friction negligible. Upon leaving the chute horizontally at its bottom end (point *C*), the rider skimmed across the water of Long Island Sound for as much as 50 m, "skipping along like a flat pebble," before at last coming to rest and swimming ashore, pulling his sled after him. According to *Scientific American,* "The facial expression of novices taking their first adventurous slide is quite remarkable, and the sensations felt are correspondingly novel and peculiar." (a) Find the speed of the sled and rider at point *C*. (b) Model the force of water friction as a constant retarding force acting on a particle. Find the work done by water friction in stopping the sled and rider. (c) Find the magnitude of the force the water exerts on the sled. (d) Find the magnitude of the force the chute exerts on the sled at point *B*. (e) At point *C* the chute is horizontal but curving in the vertical plane. Assume its radius of curvature is 20.0 m. Find the force the chute exerts on the sled at point *C*.

**Figure P7.58** *(Photo from* Scientific American, *July 1988)*

## ANSWERS TO QUICK QUIZZES

**7.1** Yes. Unlike kinetic energy, which is always positive, the gravitational potential energy of an object–Earth system can be negative if the object is located below the level at which *U* is defined to be zero.

**7.2** All three balls have the same speed as they reach the ground. Each system of a ball and the Earth has the same potential energy at the top of the building, and all three balls start off from that point with the same kinetic energy. Because energy is conserved, all three balls will have the same kinetic energy and, therefore, the same speed when they strike the ground.

**7.3** (c). The decrease in mechanical energy of the system is $f_k \Delta x$. This is smaller than the value on the horizontal surface for two reasons: (1) the force of kinetic friction $f_k$ is smaller because the normal force is smaller, and (2) the displacement $\Delta x$ is smaller because a component of the gravitational force is pulling on the block in the direction opposite to its velocity.

**7.4** Energy transfer methods include work (you apply forces on objects that move), heat (your body is warmer than the surrounding air), matter transfer (breathing, eating), mechanical waves (you speak and hear), and electromagnetic radiation (you see, as well as absorb and emit radiation from your skin).

During the brief time that a golf club is in contact with the ball, the speed of the ball is changed by a large amount. During that time, Newton's third law tells us that the force on the club has the same magnitude as the force on the ball. But the speed of the club appears to change very little upon hitting the ball in this photograph. In this chapter, we will learn to understand such events.
*(Courtesy of Michael Hans/Photo Researchers, Inc.)*

# Momentum and Collisions

Consider what happens when a golf ball is struck by a club as in the opening photograph for this chapter. The ball is given a very large initial velocity as a result of the collision; consequently, it is able to travel a large distance through the air. Because the ball experiences this change in velocity over a very short time interval, the average force on it during the collision is very large. By Newton's third law, the club experiences a reaction force equal in magnitude to and opposite the force on the ball. This reaction force produces a change in the velocity of the club. Because the club is much more massive than the ball, however, the change in velocity of the club is much less than the change in velocity of the ball.

One of the main objectives of this chapter is to enable you to understand and analyze such events. As a first step, we shall introduce the concept of *momentum,* a term used to describe objects in motion. The concept of momentum leads us to a second conservation law, conservation of momentum, and momentum approaches for treating isolated and nonisolated systems. This conservation law is especially useful for treating problems that involve collisions between objects.

## 8.1 • LINEAR MOMENTUM AND ITS CONSERVATION

In previous chapters, we discussed the velocity of an object as a measure of its motion. We will find in this chapter that this definition does not completely specify the nature of motion. As an illustration, imagine that you have two objects moving toward you, each traveling at 2 m/s. One is a Ping-Pong ball and the other is a large truck. You most likely have an intuitive notion about which one you would

   See the *Core Concepts in Physics CD-ROM,* Screen 6.2

243

rather have strike your body. They both travel with the same speed, but something about the truck represents a larger measure of motion. We can use the notion of momentum to describe the difference between these two moving objects. The **linear momentum** (which we often call simply **momentum**) of an object of mass $m$ moving with a velocity $\mathbf{v}$ is defined to be the product of the mass and velocity:*

- *Definition of linear momentum*

$$\mathbf{p} = m\mathbf{v} \qquad [8.1]$$

Because momentum equals the product of a scalar $m$ and a vector $\mathbf{v}$, it is a vector quantity. Its direction is the same as that for $\mathbf{v}$, and it has dimensions ML/T. In the SI system, momentum has the units kg·m/s.

If an object is moving in an arbitrary direction in three-dimensional space, $\mathbf{p}$ has three components, and Equation 8.1 is equivalent to the component equations

$$p_x = mv_x \qquad p_y = mv_y \qquad p_z = mv_z \qquad [8.2]$$

As you can see from its definition, the concept of momentum provides a quantitative distinction between objects of different masses moving at the same velocity. For example, the momentum of the truck moving at 2 m/s is much greater in magnitude than that of the Ping-Pong ball moving at the same speed. Newton called the product $m\mathbf{v}$ the *quantity of motion*, perhaps a more graphic description than *momentum*, which comes from the Latin word for movement.

### Quick Quiz 8.1

Two objects have equal kinetic energies. How do the magnitudes of their momenta compare? (a) $p_1 < p_2$, (b) $p_1 = p_2$, (c) $p_1 > p_2$, (d) not enough information to determine the answer.

Let us use the particle model for an object in motion. By using Newton's second law of motion, we can relate the linear momentum of a particle to the net force acting on the particle. In Chapter 4 we learned that Newton's second law can be written as $\Sigma \mathbf{F} = m\mathbf{a}$. This form applies only when the mass of the particle remains constant, however. In situations where the mass is changing with time, one must use an alternative statement of Newton's second law: **The time rate of change of momentum of a particle is equal to the net force acting on the particle.**

- *Newton's second law for a particle*

$$\Sigma \mathbf{F} = \frac{d\mathbf{p}}{dt} \qquad [8.3]$$

If the mass of the particle is constant, the preceding equation reduces to our previous expression for Newton's second law:

$$\Sigma \mathbf{F} = \frac{d\mathbf{p}}{dt} = \frac{d(m\mathbf{v})}{dt} = m\frac{d\mathbf{v}}{dt} = m\mathbf{a}$$

It is difficult to imagine a particle whose mass is changing, but if we consider objects, a number of examples emerge. These include a rocket that is ejecting its fuel as it operates, a snowball rolling down a hill and picking up additional snow, and a watertight pickup truck whose bed is collecting water as it moves in the rain.

---

* This expression is nonrelativistic and is valid only when $v \ll c$, where $c$ is the speed of light. In the next chapter, we discuss momentum for high-speed particles.

From Equation 8.3 we see that if the net force on an object is zero, the time derivative of the momentum is zero, and therefore the momentum of the object must be constant. This should sound familiar because it is the case of a particle in equilibrium, expressed in terms of momentum. Of course, if the particle is *isolated* (that is, if it does not interact with its environment), then no forces act on it and **p** remains unchanged—this is Newton's first law.

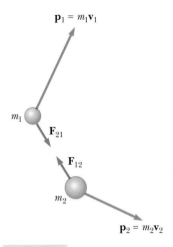

## Momentum and Isolated Systems

To see how to apply this new quantity, momentum, consider a system of two particles that can interact with each other but are isolated from their surroundings (Fig. 8.1). That is, the particles exert forces on each other, but no *external* forces are exerted on the system.

Suppose that at some instant, the momentum of particle 1 is **p**$_1$ and the momentum of particle 2 is **p**$_2$. Applying Newton's second law to each particle gives

$$\mathbf{F}_{21} = \frac{d\mathbf{p}_1}{dt} \quad \text{and} \quad \mathbf{F}_{12} = \frac{d\mathbf{p}_2}{dt}$$

**Figure 8.1**

A system of two interacting particles. At some instant, the momentum of the particle with mass $m_1$ is $\mathbf{p}_1 = m_1\mathbf{v}_1$ and the momentum of the particle with mass $m_2$ is $\mathbf{p}_2 = m_2\mathbf{v}_2$. Note that $\mathbf{F}_{12} = -\mathbf{F}_{21}$. The total momentum $\mathbf{p}_{tot}$ of the system of two particles is $\mathbf{p}_1 + \mathbf{p}_2$.

where **F**$_{21}$ is the force exerted by particle 2 on particle 1, and **F**$_{12}$ is the force exerted by particle 1 on particle 2. (These forces could be gravitational forces, or they could have some other origin. The source of the forces isn't important for the present discussion.) Newton's third law tells us that **F**$_{12}$ and **F**$_{21}$ are equal in magnitude and opposite in direction. That is, they form an action–reaction pair, and $\mathbf{F}_{12} = -\mathbf{F}_{21}$. We can also express this condition as

$$\mathbf{F}_{21} + \mathbf{F}_{12} = 0$$

or, using Equation 8.3,

$$\frac{d\mathbf{p}_1}{dt} + \frac{d\mathbf{p}_2}{dt} = \frac{d}{dt}(\mathbf{p}_1 + \mathbf{p}_2) = 0$$

Because the time derivative of the total system momentum $\mathbf{p}_{tot} = \mathbf{p}_1 + \mathbf{p}_2$ is *zero*, we conclude that the *total* momentum $\mathbf{p}_{tot}$ must remain constant:

$$\mathbf{p}_{tot} = \text{constant} \qquad \qquad \textbf{[8.4]}$$

• *Principle of conservation of momentum for an isolated system*

or, equivalently,

$$\mathbf{p}_{1i} + \mathbf{p}_{2i} = \mathbf{p}_{1f} + \mathbf{p}_{2f} \qquad \qquad \textbf{[8.5]}$$

where $\mathbf{p}_{1i}$ and $\mathbf{p}_{2i}$ are initial values and $\mathbf{p}_{1f}$ and $\mathbf{p}_{2f}$ are final values of the momentum during a period over which the particles interact. Equation 8.5 in component form states that the momentum components of the isolated system in the *x*, *y*, and *z* directions are all *independently constant;* that is,

$$\sum_{\text{system}} p_{ix} = \sum_{\text{system}} p_{fx} \qquad \sum_{\text{system}} p_{iy} = \sum_{\text{system}} p_{fy} \qquad \sum_{\text{system}} p_{iz} = \sum_{\text{system}} p_{fz} \qquad \textbf{[8.6]}$$

This result is known as the law of **conservation of linear momentum.** It is the mathematical representation of the momentum version of the isolated system model. It is considered one of the most important laws of mechanics. We have generated this law for a system of two interacting particles, but it can be shown to be true for a system of any number of particles. We can state it as follows: **The total momentum of an isolated system remains constant.**

Notice that we have made no statement concerning the nature of the forces acting. The only requirement is that the forces must be *internal* to the system. Thus, momentum is conserved for an isolated system *regardless* of the nature of the internal forces, *even if the forces are nonconservative.*

### THINKING PHYSICS 8.1

A baseball is projected into the air at an upward angle to the ground. As it moves through its trajectory, its velocity and therefore its momentum constantly change. Is this a violation of conservation of momentum?

**Reasoning**   The principle of conservation of momentum states that the momentum of an isolated system of particles is conserved. The statement of the question suggests that the ball is to be considered the system. A ball projected through the air is subject to the gravitational force as an external force, so it is *not* an isolated system, and we would not expect its momentum to be conserved. If we consider this in a little more detail, the gravitational force is in the vertical direction, so it is actually only the vertical component of the momentum that changes due to this force. In the horizontal direction, no force is exerted (ignoring air friction), so the horizontal component of the momentum is conserved. In Chapter 3, we used the same idea in stating that the horizontal component of the *velocity* of a projectile remains constant.

If we consider the baseball and the Earth as a system of two particles, the gravitational force is an internal force to this system. The momentum of the ball–Earth system is conserved. The downward force from the thrower's feet as the ball is thrown gives the Earth an initial motion. As the ball rises and falls, the Earth initially sinks and then rises (although imperceptibly!), due to the upward gravitational force from the ball, so that the total momentum of the system remains unchanged.

### Example 8.1   Can We Really Ignore the Kinetic Energy of the Earth?

In Section 7.1, we claimed that we can ignore the kinetic energy of the Earth when considering the energy of a system consisting of the Earth and a dropped ball. Verify this claim.

**Solution**   We will verify this claim by setting up a ratio of the kinetic energy of the Earth to that of the ball:

$$(1) \quad \frac{K_E}{K_b} = \frac{\frac{1}{2}m_E v_E^2}{\frac{1}{2}m_b v_b^2} = \left(\frac{m_E}{m_b}\right)\left(\frac{v_E}{v_b}\right)^2$$

where $v_E$ and $v_b$ are the speeds of the Earth and the ball, respectively, after the ball has fallen through some distance. Now we find a relationship between these two speeds by considering conservation of momentum in the vertical direction for the system of the ball and the Earth. The initial momentum of the system is zero, so the final momentum must also be zero:

$$p_i = p_f \quad \longrightarrow \quad 0 = m_b v_b + m_E v_E$$

$$\longrightarrow \quad \frac{v_E}{v_b} = -\left(\frac{m_b}{m_E}\right)$$

Substituting for $v_E/v_b$ in (1), we have

$$\frac{K_E}{K_b} = \left(\frac{m_E}{m_b}\right)\left(-\frac{m_b}{m_E}\right)^2 = \frac{m_b}{m_E}$$

Substituting order-of-magnitude numbers for the masses, this ratio becomes

$$\frac{K_E}{K_b} = \frac{m_b}{m_E} \sim \frac{1 \text{ kg}}{10^{24} \text{ kg}} \sim 10^{-24}$$

The kinetic energy of the Earth is a very small fraction of the kinetic energy of the ball, so we are justified in ignoring it in the kinetic energy of the system. A similar calculation will verify the statement in Thinking Physics 8.1 that the Earth moves imperceptibly when a ball is thrown.

## Example 8.2 The Recoiling Pitching Machine

A baseball player wishes to maintain his physical condition during the winter. He uses a 50.0-kg pitching machine to help him, placing the machine on the pitcher's mound as in Figure 8.2. The ground is covered with a thin layer of ice, so that friction between the ground and the machine is negligible. The machine fires a 0.15-kg baseball horizontally with a speed of 36 m/s. What is the recoil speed of the machine?

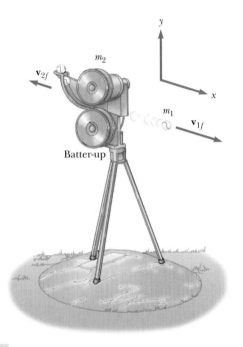

**Figure 8.2**

(Example 8.2) When the baseball is projected to the right, the pitching machine recoils to the left. The momentum of the baseball–machine system is conserved.

**Reasoning**   We identify the isolated system as the baseball and the pitching machine. Because of the gravitational force and the normal force from the ground, the system is not truly isolated. Both of these forces are in the $y$ direction, however, and therefore are directed perpendicularly to the motion of the members of the system. In addition, there is no friction between the machine and the ground. Momentum is therefore conserved in the $x$ direction because there are no external forces in this direction.

We will restrict our attention to the $x$ direction and use the symbol $v$ for *components* of the velocity in the $x$ direction. (We will not use a subscript $x$ because we will already have two subscripts, one to identify the particle and one representing initial or final values!) Thus, for example, a negative value for $v$ means that the velocity component is in the negative direction.

**Solution**   The total momentum of the system before firing is zero. The total momentum after firing must therefore be zero; that is,

$$m_1 v_{1f} + m_2 v_{2f} = 0$$

With $m_1 = 0.15$ kg, $v_{1i} = 36$ m/s, and $m_2 = 50$ kg, solving for $v_{2f}$, we find the recoil speed of the pitching machine to be

$$v_{2f} = -\frac{m_1}{m_2}\, v_{1f} = -\left(\frac{0.15 \text{ kg}}{50 \text{ kg}}\right)(36 \text{ m/s}) = \boxed{-0.11 \text{ m/s}}$$

As alluded to in the Reasoning section, the negative sign for $v_{2f}$ indicates that the pitching machine is moving to the left after firing, in the direction opposite the direction of motion of the baseball. Because the pitching machine is much more massive than the ball, the acceleration and consequent speed of the pitching machine are much smaller than the acceleration and speed of the ball.

## Example 8.3 Decay of the Kaon at Rest

One type of nuclear particle, called the *neutral kaon* ($K^0$), decays into a pair of other particles called *pions* ($\pi^+$ and $\pi^-$), which are oppositely charged but equal in mass, as in Figure 8.3. Assuming the kaon is initially at rest, prove that the two pions must have momenta that are equal in magnitude and opposite in direction.

**Solution**   The isolated system is the kaon before the decay and the two pions afterward. The decay of the kaon, represented in Figure 8.3, can be written

$$K^0 \longrightarrow \pi^+ + \pi^-$$

If we let $\mathbf{p}^+$ be the momentum of the positive pion and $\mathbf{p}^-$ be the momentum of the negative pion after the decay, the

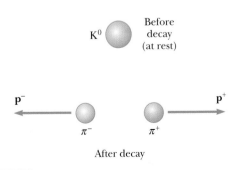

**Figure 8.3**

(Example 8.3) A kaon at rest decays into a pair of oppositely charged pions. The pions move apart with momenta of equal magnitudes but opposite directions.

final momentum of the isolated system of two pions can be written

$$\sum_{\text{system}} \mathbf{p}_f = \mathbf{p}^+ + \mathbf{p}^-$$

Because the kaon is at rest before the decay, we know that $\Sigma_{\text{system}}\, \mathbf{p}_i = 0$. Furthermore, because momentum is conserved, $\Sigma_{\text{system}}\, \mathbf{p}_i = \Sigma_{\text{system}}\, \mathbf{p}_f = 0$, so that $\mathbf{p}^+ + \mathbf{p}^- = 0$, or

$$\mathbf{p}^+ = -\mathbf{p}^-$$

Thus, we see that the two momentum vectors of the pions are equal in magnitude and opposite in direction.

**EXERCISE** A 60-kg boy and a 40-kg girl, both wearing skates, face each other at rest. The girl pushes the boy, sending him eastward with a speed of 4.0 m/s. Describe the subsequent motion of the girl. (Ignore friction.)

*Answer* She moves westward with a speed of 6.0 m/s.

## 8.2 • IMPULSE AND MOMENTUM

*See Screens 6.3 & 6.4*

As described by Equation 8.3, the momentum of a particle changes if a net force acts on the particle. Let us assume that a net force $\Sigma\, \mathbf{F}$ acts on a particle and that this force may vary with time. According to Equation 8.3,

$$d\mathbf{p} = \sum \mathbf{F}\, dt \qquad [8.7]$$

We can integrate this expression to find the change in the momentum of a particle during the time interval $\Delta t = t_f - t_i$. Integrating Equation 8.7 gives

$$\Delta \mathbf{p} = \mathbf{p}_f - \mathbf{p}_i = \int_{t_i}^{t_f} \sum \mathbf{F}\, dt \qquad [8.8]$$

The integral of a force over the time during which it acts is called the **impulse** of the force. The impulse of the net force $\Sigma\, \mathbf{F}$ is a vector defined by

• *Impulse of a net force*

$$\mathbf{I} \equiv \int_{t_i}^{t_f} \sum \mathbf{F}\, dt = \Delta \mathbf{p} \qquad [8.9]$$

Thus, the total **impulse** of the net force $\Sigma\, \mathbf{F}$ on a particle equals the change in the momentum of the particle. This statement, known as the **impulse–momentum theorem,** is equivalent to Newton's second law. It also applies to a system of particles, in which we consider that the net force external to the system causes a change in the total momentum of the system:

• *Impulse–momentum theorem*

$$\mathbf{I} \equiv \int_{t_i}^{t_f} \sum \mathbf{F}_{\text{ext}}\, dt = \Delta \mathbf{p}_{\text{tot}} \qquad [8.10]$$

Impulse is an interaction between the system and its environment. As a result of this interaction, the momentum of the system changes. This is an analog to the continuity equation for energy, in which an interaction with the environment causes the energy of the system to change. Therefore, when we say that an impulse is given to a system, we imply that momentum is transferred from an external agent to that system. In many situations, the system can be modeled as a particle, so Equation 8.9 can be used rather than the more general Equation 8.10.

From the definition, we see that impulse is a vector quantity having a magnitude equal to the area under the curve of the magnitude of the net force versus time, as described in Figure 8.4. In this figure it is assumed that the net force

You experience the same change in momentum in a collision whether your car has air bags or not. The air bag allows you to experience that change in momentum over a longer time interval, however, reducing the peak force on you and increasing your chances of escaping without injury. *(Courtesy of Saab)*

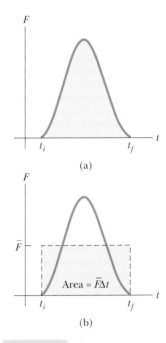

**Figure 8.4**

(a) A net force acting on a particle may vary in time. The impulse is the area under the curve of the magnitude of the net force versus time. (b) The average force (*horizontal dashed line*) gives the same impulse to the particle in the time $\Delta t$ as the time-varying force described in part (a). The area of the rectangle is the same as the area under the curve.

varies in time in the general manner shown and is nonzero in the time interval $\Delta t = t_f - t_i$. The direction of the impulse vector is the same as the direction of the change in momentum. Impulse has the dimensions of momentum, ML / T.

Because the force can generally vary in time as in Figure 8.4a, it is convenient to define a time-averaged net force $\sum \overline{\mathbf{F}}$ given by

$$\sum \overline{\mathbf{F}} \equiv \frac{1}{\Delta t} \int_{t_i}^{t_f} \sum \mathbf{F} \, dt \qquad \textbf{[8.11]}$$

where $\Delta t = t_f - t_i$. Therefore, we can express Equation 8.9 as

$$\mathbf{I} = \Delta \mathbf{p} = \sum \overline{\mathbf{F}} \, \Delta t \qquad \textbf{[8.12]}$$

The magnitude of this average net force, described in Figure 8.4b, can be thought of as the magnitude of the constant net force that would give the same impulse to the particle in the time interval $\Delta t$ as the actual time-varying net force gives over this same interval.

In principle, if $\sum \mathbf{F}$ is known as a function of time, the impulse can be calculated from Equation 8.9. The calculation becomes especially simple if the net force acting on the particle is constant. In this case, $\sum \overline{\mathbf{F}}$ over a time interval is the same as the constant $\sum \mathbf{F}$ at any instant, and Equation 8.12 becomes

$$\mathbf{I} = \Delta \mathbf{p} = \sum \mathbf{F} \, \Delta t \qquad \textbf{[8.13]}$$

In many physical situations, we shall use what is called the **impulse approximation: We assume that one of the forces exerted on a particle acts for a short time but is much greater than any other force present.** This simplification model allows us to ignore the effects of other forces, because these effects are be small during the short time during which the large force acts. This approximation is especially useful in treating collisions, in which the duration of the collision is very short. When this approximation is made, we refer to the force as an *impulsive force*. For example, when a baseball is struck with a bat, the duration of the

• *Impulse approximation*

**WEB**

For information on the use of air bags and seatbelts in automobiles, visit the National Safety Council at **www.nsc.org/airbag.htm**

collision is about 0.01 s, and the average force the bat exerts on the ball during this time interval is typically several thousand newtons. This is much greater than the gravitational force, so we ignore any change in velocity related to the gravitational force during the collision. It is important to remember that $\mathbf{p}_i$ and $\mathbf{p}_f$ represent the momenta *immediately* before and after the collision, respectively. In the impulse approximation, therefore, very little motion of the particle takes place during the collision.

### Quick Quiz 8.2

Two objects are at rest on a frictionless surface. Object 1 has a larger mass than object 2. When a force is applied to object 1, it accelerates through a displacement $\Delta x$. The force is removed from object 1 and is applied to object 2. At the moment when object 2 accelerates through the same displacement $\Delta x$, which statements are true? (a) $p_1 < p_2$, (b) $p_1 = p_2$, (c) $p_1 > p_2$, (d) $K_1 < K_2$, (e) $K_1 = K_2$, (f) $K_1 > K_2$.

**APPLICATION**

Boxing and brain injury

### Quick Quiz 8.3

In boxing matches of the 19th century, bare fists were used. In modern boxing, fighters wear padded gloves. How does this better protect the brain of the boxer from injury?

### THINKING PHYSICS 8.2

A race car travels rapidly around a circular race track at constant speed. For a given portion of the track, what is the direction of the impulse vector? Once the car returns to the starting point, there is no net change in momentum for the entire trip around the track. Thus, according to the impulse— momentum theorem, there must be no impulse. Yet a nonzero force acted over the time interval for one rotation around the track. How can the impulse be zero?

**Reasoning**  For the movement around a short portion of the track, the direction of the impulse vector is the same as that of the change in momentum. The direction of the vector representing the change in momentum is the same as the direction of the vector representing the change in velocity. For a particle in uniform circular motion, we know that the direction of the vector representing the change in velocity is that of the acceleration vector, which is toward the center of the circle. Thus, the impulse vector is directed toward the center of the circle.

If the car makes one rotation around the track, there is indeed no net change in momentum, so there must be zero net impulse. The force on the car causing the circular motion is the friction between the tires and the roadway, and this force is always directed toward the center of the circle. For every location of the car on the track, another point of its motion is diametrically opposed across the circle; at this point the force vector is directed in the opposite direction. Thus, as we add up the vector impulses around the circle to find the total impulse, we discover that they cancel in pairs, for a total impulse of zero.

## Example 8.4  How Good Are the Bumpers?

In a crash test, an automobile of mass 1 500 kg collides with a wall, as in Figure 8.5. The initial and final velocities of the automobile are $\mathbf{v}_i = -15.0\mathbf{i}$ m/s and $\mathbf{v}_f = 2.60\mathbf{i}$ m/s. If the collision lasts for 0.150 s, find the impulse due to the collision and the average force exerted on the automobile.

**Solution**  We identify the automobile as the system. The initial and final momenta of the automobile are

$$\mathbf{p}_i = m\mathbf{v}_i = (1\,500\text{ kg})(-15.0\mathbf{i}\text{ m/s})$$
$$= -2.25 \times 10^4\mathbf{i}\text{ kg} \cdot \text{m/s}$$

$$\mathbf{p}_f = m\mathbf{v}_f = (1\,500\text{ kg})(2.60\mathbf{i}\text{ m/s})$$
$$= 0.390 \times 10^4\mathbf{i}\text{ kg} \cdot \text{m/s}$$

Hence, the impulse is

$$\mathbf{I} = \Delta\mathbf{p} = \mathbf{p}_f - \mathbf{p}_i$$
$$= 0.390 \times 10^4\mathbf{i}\text{ kg} \cdot \text{m/s} - (-2.25 \times 10^4\mathbf{i}\text{ kg} \cdot \text{m/s})$$

$$\mathbf{I} = \boxed{2.64 \times 10^4\mathbf{i}\text{ kg} \cdot \text{m/s}}$$

The average force exerted on the automobile is

$$\overline{\mathbf{F}} = \frac{\Delta\mathbf{p}}{\Delta t} = \frac{2.64 \times 10^4\mathbf{i}\text{ kg} \cdot \text{m/s}}{0.150\text{ s}} = \boxed{1.76 \times 10^5\mathbf{i}\text{ N}}$$

**EXERCISE**  A 0.150-kg baseball is thrown with a speed of 40.0 m/s. It is hit straight back at the pitcher with a speed of 50.0 m/s. (a) What impulse is delivered to the baseball? (b) Find the average force exerted by the bat on the ball if the two are in contact for $2.00 \times 10^{-3}$ s.

*Answer*  (a) 13.5 kg · m/s  (b) 6.75 kN

Before

−15.0 m/s

After

2.60 m/s

(a)

(b)

**Figure 8.5**

(Example 8.4) (a) The car's momentum changes as a result of its collision with the wall. (b) In a crash test, the large force exerted by the wall on the car produces extensive damage to the front end of the car. *(Courtesy of General Motors)*

## 8.3 • COLLISIONS

See Screens 6.5 & 6.6

In this section we use the law of conservation of momentum to describe what happens when two objects collide. The forces due to the collision are assumed to be much larger than any external forces present, so we use the simplification model we call the impulse approximation. The general goal in collision problems is to relate the final conditions of the system to the initial conditions.

A collision may be the result of physical contact between two objects, as described in Figure 8.6a. This is a common observation when two macroscopic objects, such as two billiard balls or a baseball and a bat, collide.

The notion of what we mean by *collision* must be generalized because "contact" on a microscopic scale is ill defined. To understand the distinction between

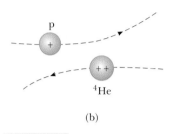

**Figure 8.6**

(a) A collision between two objects as the result of direct contact. (b) A "collision" between two charged particles that do not make contact.

**PITFALL PREVENTION 1.1**

**Momentum and kinetic energy in collisions**

Momentum of an isolated system is conserved in *all* collisions. Kinetic energy of an isolated system is conserved *only* in elastic collisions. This is true because kinetic energy can be transformed into several types of energy or be transferred out of the system (so that the system may *not* be isolated in terms of energy during the collision), but momentum comes in only one type.

macroscopic and microscopic collisions, consider the collision of a proton with an alpha particle (the nucleus of the helium atom), illustrated in Figure 8.6b. Because the two particles are positively charged, they repel each other. A collision has occurred, but the colliding particles were never in "contact."

When two particles of masses $m_1$ and $m_2$ collide, the collision forces may vary in time in a complicated way, as seen in Figure 8.4. As a result, an analysis of the situation with Newton's second law could be very complicated. We find, however, that the momentum concept is similar to the energy concept in Chapters 6 and 7, in that it provides us with a much easier method to solve problems involving isolated systems.

According to Equation 8.5, the momentum of an isolated system is conserved during some interaction event, such as a collision. The kinetic energy of the system, however, is generally *not* conserved in a collision. We define an **inelastic collision** as one in which the kinetic energy of the system is not conserved (even though momentum is conserved). The collision of a rubber ball with a hard surface is inelastic because some of the kinetic energy of the ball is transformed to internal energy when the ball is deformed while in contact with the surface.

When two objects collide and stick together after a collision, the maximum possible fraction of the initial kinetic energy is transformed, and the collision is called **perfectly inelastic.** For example, if two vehicles collide and become entangled, they move with some common velocity after the perfectly inelastic collision. If a meteorite collides with the Earth, it becomes buried in the ground, and the collision is perfectly inelastic.

An **elastic collision** is defined as one in which the kinetic energy of the system is conserved (as well as momentum). Real collisions in the macroscopic world, such as those between billiard balls, are only approximately elastic because some transformation of kinetic energy takes place, and some energy leaves the system by mechanical waves—sound. Imagine a billiard game with truly elastic collisions. The opening break would be completely silent! Truly elastic collisions do occur between atomic and subatomic particles. Elastic and perfectly inelastic collisions are *limiting* cases; a large number of collisions fall in the range between them.

In the remainder of this section we treat collisions in one dimension and consider the two extreme cases: perfectly inelastic and elastic collisions. The important distinction between these two types of collisions is that **momentum is conserved in all cases, but kinetic energy is conserved only in elastic collisions.**

**Quick Quiz 8.4**

A skater is using very low friction rollerblades. A friend throws a Frisbee at her, along the straight line along which she is skating. Describe each of the following events as an elastic, an inelastic, or a perfectly inelastic collision between the skater and the Frisbee: (a) she catches the Frisbee and holds it, (b) she tries to catch the Frisbee but it bounces off her hands and falls to the ground at her feet, (c) she catches the Frisbee and immediately throws it back with the same speed (relative to the ground) to her friend.

**One-Dimensional Perfectly Inelastic Collisions**

Consider two objects of masses $m_1$ and $m_2$ moving with initial velocity components $v_{1i}$ and $v_{2i}$ along a straight line, as in Figure 8.7. If the two objects collide head-on, stick together, and move with some common velocity component $v_f$ after the collision, the collision is perfectly inelastic. Because the total momentum of the two-object isolated system before the collision equals the total momentum of the combined-object system after the collision, we have

$$m_1 v_{1i} + m_2 v_{2i} = (m_1 + m_2) v_f \qquad \text{[8.14]}$$

$$v_f = \frac{m_1 v_{1i} + m_2 v_{2i}}{m_1 + m_2} \qquad \text{[8.15]}$$

Thus, knowing the initial velocity components of the two objects, we can use this single equation to determine the final common velocity component.

## One-Dimensional Elastic Collisions

Now consider two objects that undergo an elastic head-on collision (Fig. 8.8) in one dimension. In this collision, both momentum and kinetic energy are conserved; therefore, we can write*

$$m_1 v_{1i} + m_2 v_{2i} = m_1 v_{1f} + m_2 v_{2f} \qquad \text{[8.16]}$$

$$\tfrac{1}{2} m_1 v_{1i}^2 + \tfrac{1}{2} m_2 v_{2i}^2 = \tfrac{1}{2} m_1 v_{1f}^2 + \tfrac{1}{2} m_2 v_{2f}^2 \qquad \text{[8.17]}$$

In a typical problem involving elastic collisions, two unknown quantities occur, and Equations 8.16 and 8.17 can be solved simultaneously to find them. An alternative approach, employing a little mathematical manipulation of Equation 8.17, often simplifies this process. To see this, let us cancel the factor of $\tfrac{1}{2}$ in Equation 8.17 and rewrite the equation as

$$m_1(v_{1i}^2 - v_{1f}^2) = m_2(v_{2f}^2 - v_{2i}^2)$$

Here we have moved the terms containing $m_1$ to one side of the equation and those containing $m_2$ to the other. Next, let us factor both sides:

$$m_1(v_{1i} - v_{1f})(v_{1i} + v_{1f}) = m_2(v_{2f} - v_{2i})(v_{2f} + v_{2i}) \qquad \text{[8.18]}$$

We now separate the terms containing $m_1$ and $m_2$ in the equation for the conservation of momentum (Eq. 8.16) to obtain

$$m_1(v_{1i} - v_{1f}) = m_2(v_{2f} - v_{2i}) \qquad \text{[8.19]}$$

To obtain our final result, we divide Equation 8.18 by Equation 8.19 and obtain

$$v_{1i} + v_{1f} = v_{2f} + v_{2i}$$

or, gathering initial and final values on opposite sides of the equation,

$$v_{1i} - v_{2i} = -(v_{1f} - v_{2f}) \qquad \text{[8.20]}$$

This equation, in combination with the condition for conservation of momentum, Equation 8.16, can be used to solve problems dealing with one-dimensional

---

* Notice that the kinetic energy of the system is the sum of the kinetic energies of the two particles. In our energy conservation examples in Chapter 7 involving a falling object and the Earth, we ignored the kinetic energy of the Earth because it was so small. Thus, the kinetic energy of the *system* was just the kinetic energy of the falling *object*. This was a special case, in which the mass of one of the objects (the Earth) was so immense that ignoring its kinetic energy introduced no measurable error. In problems such as those described here, however, and for the particle decay problems we will see in Chapters 30 and 31, we need to include the kinetic energies of *all* particles in the system.

Before collision

(a)

After collision

$m_1 + m_2$

(b)

**Figure 8.7**

A perfectly inelastic head-on collision between two particles: (a) before the collision and (b) after the collision.

**PITFALL PREVENTION 8.3**
**Perfectly inelastic collisions**

Keep in mind the distinction between inelastic and perfectly inelastic collisions. If the colliding particles stick together, the collision is perfectly inelastic. If they bounce off each other (and kinetic energy is not conserved), the collision is inelastic. Generally, inelastic collisions are hard to analyze unless additional information is provided. This appears in the mathematical representation as having more unknowns than equations.

Before collision

(a)

After collision

(b)

**Figure 8.8**

An elastic head-on collision between two particles: (a) before the collision and (b) after the collision.

**PITFALL PREVENTION 8.4**

**Not a general equation**

We have spent some effort on deriving Equation 8.20, but remember that it can be used only in a very *specific* situation: a one-dimensional, elastic collision between two objects. The *general* concept is conservation of momentum (and conservation of kinetic energy if the collision is elastic) for an isolated system.

elastic collisions between two objects. According to Equation 8.20, the relative speed* $v_{1i} - v_{2i}$ of the two objects before the collision equals the negative of their relative speed after the collision, $-(v_{1f} - v_{2f})$.

Suppose that the masses and the initial velocity components of both objects are known. Equations 8.16 and 8.20 can be solved for the final velocity components in terms of the initial values, because we have two equations and two unknowns:

$$v_{1f} = \left( \frac{m_1 - m_2}{m_1 + m_2} \right) v_{1i} + \left( \frac{2m_2}{m_1 + m_2} \right) v_{2i} \qquad [8.21]$$

$$v_{2f} = \left( \frac{2m_1}{m_1 + m_2} \right) v_{1i} + \left( \frac{m_2 - m_1}{m_1 + m_2} \right) v_{2i} \qquad [8.22]$$

It is important to remember that the appropriate signs for the velocity components $v_{1i}$ and $v_{2i}$ must be included in Equations 8.21 and 8.22. For example, if $m_2$ is moving to the left initially, as in Figure 8.8, then $v_{2i}$ is negative.

Let us consider some special cases. If $m_1 = m_2$, then Equations 8.21 and 8.22 show us that $v_{1f} = v_{2i}$ and $v_{2f} = v_{1i}$. That is, the objects exchange speeds if they have equal masses. This is what one observes in head-on billiard ball collisions: The initially moving ball stops and the initially stationary ball moves away with approximately the same speed.

If $m_2$ is initially at rest, $v_{2i} = 0$, and Equations 8.21 and 8.22 become

$$v_{1f} = \left( \frac{m_1 - m_2}{m_1 + m_2} \right) v_{1i} \qquad [8.23]$$

$$v_{2f} = \left( \frac{2m_1}{m_1 + m_2} \right) v_{1i} \qquad [8.24]$$

If $m_1$ is very large compared with $m_2$, we see from Equations 8.23 and 8.24 that $v_{1f} \approx v_{1i}$ and $v_{2f} \approx 2v_{1i}$. That is, when a very heavy object collides head-on with a very light one initially at rest, the heavy object continues its motion unaltered after the collision, but the light object rebounds with a speed equal to about twice the initial speed of the heavy object. An example of such a collision is that of a moving heavy atom, such as uranium, with a light atom, such as hydrogen.

If $m_2$ is much larger than $m_1$, and $m_2$ is initially at rest, then we find from Equations 8.23 and 8.24 that $v_{1f} \approx -v_{1i}$ and $v_{2f} \approx 0$. That is, when a very light object collides head-on with a very heavy object initially at rest, the velocity of the light object is reversed, and the heavy object remains approximately at rest. For example, imagine what happens when a marble hits a stationary bowling ball.

**Quick Quiz 8.5**

In a perfectly inelastic collision between two identical objects, under what conditions is all of the initial kinetic energy of the system of two objects transformed to other forms of energy?

---

* See Section 3.6 for a review of relative speed.

## Example 8.5 Kinetic Energy in a Perfectly Inelastic Collision

We claimed that the maximum amount of kinetic energy was transformed to other forms in a perfectly inelastic collision. Prove this statement mathematically for a one-dimensional two-particle collision.

**Solution** We will assume that the maximum kinetic energy is transformed and prove that the collision must be perfectly inelastic. We set up the fraction $f$ of the final kinetic energy after the collision to the initial kinetic energy:

$$f = \frac{K_f}{K_i} = \frac{\frac{1}{2}m_1 v_{1f}^2 + \frac{1}{2}m_2 v_{2f}^2}{\frac{1}{2}m_1 v_{1i}^2 + \frac{1}{2}m_2 v_{2i}^2} = \frac{m_1 v_{1f}^2 + m_2 v_{2f}^2}{m_1 v_{1i}^2 + m_2 v_{2i}^2}$$

The *maximum* amount of energy transformed to other forms corresponds to the *minimum* value of $f$. For fixed initial conditions, we imagine that the final velocity components $v_{1f}$ and $v_{2f}$ are variables. We minimize the fraction $f$ by taking the derivative of $f$ with respect to $v_{1f}$ and setting the result equal to zero:

$$(1) \quad \frac{df}{dv_{1f}} = \frac{d}{dv_{1f}}\left(\frac{m_1 v_{1f}^2 + m_2 v_{2f}^2}{m_1 v_{1i}^2 + m_2 v_{2i}^2}\right)$$

$$= \frac{m_1 2 v_{1f} + m_2 2 v_{2f}\dfrac{dv_{2f}}{dv_{1f}}}{m_1 v_{1i}^2 + m_2 v_{2i}^2} = 0$$

$$\longrightarrow \quad m_1 v_{1f} + m_2 v_{2f}\frac{dv_{2f}}{dv_{1f}} = 0$$

From the conservation of momentum condition, we can evaluate the derivative in (1). We differentiate Equation 8.16 with respect to $v_{1f}$:

$$\frac{d}{dv_{1f}}(m_1 v_{1i} + m_2 v_{2i}) = \frac{d}{dv_{1f}}(m_1 v_{1f} + m_2 v_{2f})$$

$$\longrightarrow \quad 0 = m_1 + m_2\frac{dv_{2f}}{dv_{1f}} \longrightarrow \frac{dv_{2f}}{dv_{1f}} = -\frac{m_1}{m_2}$$

Substituting this expression for the derivative into (1), we find,

$$m_1 v_{1f} - m_2 v_{2f}\frac{m_1}{m_2} = 0 \longrightarrow v_{1f} = v_{2f}$$

If the particles come out of the collision with the same velocity components, they are joined together and this is a perfectly inelastic collision, which is what we set out to prove.

## Example 8.6 Carry Collision Insurance

An 1 800-kg car stopped at a traffic light is struck from the rear by a 900-kg car and the two become entangled. If the smaller car was moving at 20.0 m/s before the collision, what is the speed of the entangled cars after the collision?

**Reasoning** The total momentum of the system (the two cars) before the collision equals the total momentum of the system after the collision because the system is isolated (in the impulse approximation). Notice that we ignore friction with the road in the impulse approximation. Thus, the result we obtain for the final speed will only be approximately true just after the collision. For longer time intervals after the collision, we would use Newton's second law to describe the slowing down of the system due to friction. Because the cars "become entangled," this is a perfectly inelastic collision.

**Solution** The magnitude of the total momentum of the system before the collision is equal to that of only the smaller car because the larger car is initially at rest:

$$p_i = m_1 v_i = (900 \text{ kg})(20.0 \text{ m/s}) = 1.80 \times 10^4 \text{ kg} \cdot \text{m/s}$$

After the collision, the mass that moves is the sum of the masses of the cars. The magnitude of the momentum of the combination is

$$p_f = (m_1 + m_2)v_f = (2\ 700 \text{ kg})v_f$$

Equating the initial momentum to the final momentum and solving for $v_f$, the speed of the entangled cars, we have

$$v_f = \frac{p_i}{m_1 + m_2} = \frac{1.80 \times 10^4 \text{ kg} \cdot \text{m/s}}{2\ 700 \text{ kg}} = \boxed{6.67 \text{ m/s}}$$

## Example 8.7    Slowing Down Neutrons by Collisions

In a nuclear reactor, neutrons are produced when $^{235}_{92}U$ atoms split in a process called *fission*. These neutrons are moving at about $10^7$ m/s and must be slowed down to about $10^3$ m/s before they take part in another fission event. They are slowed down by being passed through a solid or liquid material called a *moderator*. The slowing-down process involves elastic collisions. Let us show that a neutron can lose most of its kinetic energy if it collides elastically with a moderator containing light nuclei, such as deuterium (in "heavy water," $D_2O$).

**Reasoning**    We identify the system as the neutron and a moderator nucleus. Because the momentum and kinetic energy of this system are conserved in an elastic collision, Equations 8.23 and 8.24 can be applied to a one-dimensional collision of these two particles.

**Solution**    Let us assume that the moderator nucleus of mass $m_m$ is at rest initially and that the neutron of mass $m_n$ and initial speed $v_{ni}$ collides head-on with it. The initial kinetic energy of the neutron is

$$K_{ni} = \tfrac{1}{2}m_n v_{ni}^2$$

After the collision, the neutron has kinetic energy $\tfrac{1}{2}m_n v_{nf}^2$, where $v_{nf}$ is given by Equation 8.23.

$$K_{nf} = \tfrac{1}{2}m_n v_{nf}^2 = \tfrac{1}{2}m_n \left(\frac{m_n - m_m}{m_n + m_m}\right)^2 v_{ni}^2$$

Therefore, the fraction of the total kinetic energy possessed by the neutron after the collision is

$$(1)\quad f_n = \frac{K_{nf}}{K_{ni}} = \frac{\tfrac{1}{2}m_n\left(\dfrac{m_n - m_m}{m_n + m_m}\right)^2 v_{ni}^2}{\tfrac{1}{2}m_n v_{ni}^2} = \left(\frac{m_n - m_m}{m_n + m_m}\right)^2$$

From this result, we see that the final kinetic energy of the neutron is small when $m_m$ is close to $m_n$ and is zero when $m_m = m_n$.

We can calculate the kinetic energy of the moderator nucleus after the collision using Equation 8.24:

$$K_{mf} = \tfrac{1}{2}m_m v_{mf}^2 = \frac{2m_n^2 m_m}{(m_n + m_m)^2}v_{ni}^2$$

Hence, the fraction of the total kinetic energy transferred to the moderator nucleus is

$$(2)\quad f_{\text{trans}} = \frac{K_{mf}}{K_{ni}} = \frac{\dfrac{2m_n^2 m_m}{(m_n + m_m)^2}v_{ni}^2}{\tfrac{1}{2}m_n v_{ni}^2} = \frac{4m_n m_m}{(m_n + m_m)^2}$$

If $m_m \approx m_n$, we see that $f_{\text{trans}} \approx 1 = 100\%$. Because the system's kinetic energy is conserved, (2) can also be obtained from (1) with the condition that $f_n + f_m = 1$, so that $f_m = 1 - f_n$.

For collisions of the neutrons with deuterium nuclei in $D_2O$ ($m_m = 2m_n$), $f_n = 1/9$ and $f_{\text{trans}} = 8/9$. That is, 89% of the neutron's kinetic energy is transferred to the deuterium nucleus. In practice, the moderator efficiency is reduced because head-on collisions are very unlikely to occur.

---

## Example 8.8    Two Blocks and a Spring

A block of mass $m_1 = 1.60$ kg, initially moving to the right with a speed of 4.00 m/s on a frictionless horizontal track, collides with a massless spring attached to a second block of mass $m_2 = 2.10$ kg, moving to the left with a speed of 2.50 m/s, as in Figure. 8.9a. The spring has a spring constant of 600 N/m. (a) At the instant when $m_1$ is moving to the right with a speed of 3.00 m/s, as in Figure 8.9b, determine the speed of $m_2$.

**Solution**    We identify the system as the two blocks and the spring. The system satisfies the conditions of the isolated system model, so the total momentum of the system is conserved. We have

$$m_1 v_{1i} + m_2 v_{2i} = m_1 v_{1f} + m_2 v_{2f}$$

$$(1.60\text{ kg})(4.00\text{ m/s}) + (2.10\text{ kg})(-2.50\text{ m/s})$$

$$= (1.60\text{ kg})(3.00\text{ m/s}) + (2.10\text{ kg})v_{2f}$$

$$v_{2f} = -1.74\text{ m/s}$$

Note that the initial velocity component of $m_2$ is $-2.50$ m/s because its direction is to the left. The negative value for $v_{2f}$ means that $m_2$ is still moving to the left at the instant we are considering.

(b) Determine the distance the spring is compressed at that instant.

**Solution**    Because this is an isolated system, we can also approach this problem from the energy version of the isolated system model to determine the compression $x$ in the spring shown in Figure 8.9b. No nonconservative forces are acting within the system, so mechanical energy is conserved:

$$E_i = E_f$$

$$\tfrac{1}{2}m_1 v_{1i}^2 + \tfrac{1}{2}m_2 v_{2i}^2 = \tfrac{1}{2}m_1 v_{1f}^2 + \tfrac{1}{2}m_2 v_{2f}^2 + \tfrac{1}{2}kx^2$$

Substituting the given values and the result to part (a) into this expression gives

$$x = \boxed{0.173 \text{ m}}$$

(c) Determine the maximum distance by which the spring is compressed during the collision.

**Solution**  The maximum compression of the spring occurs when the two blocks are not moving relative to each other. For their relative velocity to be zero, they must be moving with the same velocity in our reference frame as we watch the collision. Thus, we can model the collision *up to this point* as a perfectly inelastic collision:

$$m_1 v_{1i} + m_2 v_{2i} = (m_1 + m_2) v_f$$
$$(1.60 \text{ kg})(4.00 \text{ m/s}) + (2.10 \text{ kg})(-2.50 \text{ m/s})$$
$$= (1.60 \text{ kg} + 2.10 \text{ kg}) v_f$$
$$v_f = 0.311 \text{ m/s}$$

As in part (b), mechanical energy is conserved, so we set up a conservation of mechanical energy expression:

$$E_i = E_f$$
$$\tfrac{1}{2} m_1 v_{1i}^2 + \tfrac{1}{2} m_2 v_{2i}^2 = \tfrac{1}{2}(m_1 + m_2) v_f^2 + \tfrac{1}{2} k x^2$$

Substituting the values into this expression gives

$$x = \boxed{0.358 \text{ m}}$$

**EXERCISE**  Find the velocity of $m_1$ and the compression in the spring at the instant that $m_2$ is at rest.

*Answer*   0.719 m/s to the right; 0.251 m

**EXERCISE**  A 2.50-kg object, moving initially with a speed of 10.0 m/s, makes a perfectly inelastic head-on collision with a 5.00-kg object that is initially at rest. (a) Find the final speed of the composite object. (b) How much of the system kinetic energy is transformed to other forms in the collision?

*Answer*   (a) 3.33 m/s   (b) 83.4 J

$\mathbf{v}_{1i} = (4.00\mathbf{i})$ m/s $\qquad \mathbf{v}_{2i} = (-2.50\mathbf{i})$ m/s $\qquad \mathbf{v}_{1f} = (3.00\mathbf{i})$ m/s $\qquad \mathbf{v}_{2f}$

(a)

(b)

**Figure 8.9**

(Example 8.8) A moving block collides with another moving block with a spring attached: (a) before the collision and (b) at one instant during the collision.

## 8.4 • TWO-DIMENSIONAL COLLISIONS

In Section 8.1 we showed that the total momentum of a system is conserved when the system is isolated (i.e., when no external forces act on the system). For a general collision of two objects in three-dimensional space, the principle of conservation of momentum implies that the total momentum in each direction is conserved. An important subset of collisions takes place in a plane. The game of billiards is a familiar example involving multiple collisions of objects moving on a two-dimensional surface. Let us restrict our attention to a single two-dimensional collision between two objects that takes place in a plane. For such collisions, we obtain two component equations for the conservation of momentum:

$$m_1 v_{1ix} + m_2 v_{2ix} = m_1 v_{1fx} + m_2 v_{2fx}$$
$$m_1 v_{1iy} + m_2 v_{2iy} = m_1 v_{1fy} + m_2 v_{2fy}$$

where we use three subscripts in this general equation to represent, respectively, (1) the identification of the object, (2) initial and final values, and (3) the velocity component.

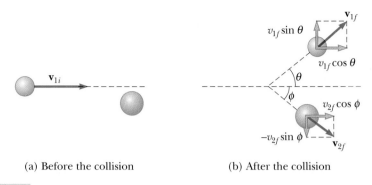

(a) Before the collision          (b) After the collision

**Figure 8.10**

An elastic glancing collision between two particles.

Consider a two-dimensional problem in which an object of mass $m_1$ collides with an object of mass $m_2$ that is initially at rest, as in Figure 8.10. After the collision, $m_1$ moves at an angle $\theta$ with respect to the horizontal, and $m_2$ moves at an angle $\phi$ with respect to the horizontal. This is called a *glancing* collision. Applying the law of conservation of momentum in component form and noting that the initial $y$ component of momentum is zero, we have

$$x \text{ component:} \quad m_1 v_{1i} + 0 = m_1 v_{1f} \cos\theta + m_2 v_{2f} \cos\phi \qquad [8.25]$$

- *Component equations for conservation of momentum in a two-dimensional, two-particle collision*

$$y \text{ component:} \quad 0 + 0 = m_1 v_{1f} \sin\theta - m_2 v_{2f} \sin\phi \qquad [8.26]$$

If the collision is elastic, we can write a third equation for conservation of kinetic energy, in the form

- *Conservation of kinetic energy in a two-dimensional, two-particle collision*

$$\tfrac{1}{2} m_1 v_{1i}^2 = \tfrac{1}{2} m_1 v_{1f}^2 + \tfrac{1}{2} m_2 v_{2f}^2 \qquad [8.27]$$

If we know the initial velocity $v_{1i}$ and the masses, we are left with four unknowns ($v_{1f}$, $v_{2f}$, $\theta$, and $\phi$). Because we have only three equations, one of the four remaining quantities must be given to determine the motion after the collision from conservation principles alone.

If the collision is inelastic, kinetic energy is *not* conserved, and Equation 8.27 does *not* apply.

**PROBLEM-SOLVING STRATEGY**    **Two-Dimensional Collisions**

The following procedure is recommended when dealing with problems involving collisions between two objects.

1. Set up a coordinate system and define your velocities with respect to that system. It is convenient to have the *x* axis coincide with one of the initial velocities.
2. In your sketch of the coordinate system, draw and label all velocity vectors and include all the given information. Write expressions for the *x* and *y* components of the momentum of each object before and after the colli-

sion. Remember to include the appropriate signs for the components of the velocity vectors. It is essential that you pay careful attention to signs.

3. Write expressions for the *total* momentum in the *x* direction *before* and *after* the collision, and equate the two. Repeat this procedure for the total momentum in the *y* direction.

4. Proceed to solve the momentum equations for the unknown quantities. If the collision is inelastic, kinetic energy is *not* conserved, and additional information is probably required. If the collision is perfectly inelastic, the final velocities of the two objects are equal.

5. If the collision is elastic, kinetic energy is conserved, and you can equate the total kinetic energy before the collision to the total kinetic energy after the collision. This provides an additional relationship between the velocity magnitudes.

### Example 8.9   Proton–Proton Collision

A proton collides elastically with another proton that is initially at rest. The incoming proton has an initial speed of $3.5 \times 10^5$ m/s and makes a glancing collision with the second proton, as in Figure 8.10. (At close separations, the protons exert a repulsive electrostatic force on each other.) After the collision, one proton moves off at an angle of $37°$ to the original direction of motion, and the second deflects at an angle of $\phi$ to the same axis. Find the final speeds of the two protons and the angle $\phi$.

**Solution**   The isolated system is the pair of protons. Both momentum and kinetic energy of the system are conserved in this glancing elastic collision. Because $m_1 = m_2$, $\theta = 37°$, and we are given that $v_{1i} = 3.5 \times 10^5$ m/s, Equations 8.25, 8.26, and 8.27 become

(1)   $v_{1f}\cos 37° + v_{2f}\cos \phi = 3.5 \times 10^5$ m/s

(2)   $v_{1f}\sin 37° - v_{2f}\sin \phi = 0$

(3)   $v_{1f}^2 + v_{2f}^2 = (3.5 \times 10^5 \text{ m/s})^2 = 1.2 \times 10^{11} \text{ m}^2/\text{s}^2$

We rewrite (1) and (2) as follows:

$$v_{2f}\cos \phi = 3.5 \times 10^5 \text{ m/s} - v_{1f}\cos 37°$$

$$v_{2f}\sin \phi = v_{1f}\sin 37°$$

Now we square these two equations and add them:

$$v_{2f}^2 \cos^2 \phi + v_{2f}^2 \sin^2 \phi = 1.2 \times 10^{11} \text{ m}^2/\text{s}^2$$
$$- (7.0 \times 10^5 \text{ m/s})v_{1f}\cos 37°$$
$$+ v_{1f}^2 \cos^2 37° + v_{1f}^2 \sin^2 37°$$
$$\longrightarrow \quad v_{2f}^2 = 1.2 \times 10^{11} - (5.6 \times 10^5)v_{1f} + v_{1f}^2$$

Substituting this expression into (3),

$$v_{1f}^2 + [1.2 \times 10^{11} - (5.6 \times 10^5)v_{1f} + v_{1f}^2] = 1.2 \times 10^{11}$$
$$\longrightarrow \quad 2v_{1f}^2 - (5.6 \times 10^5)v_{1f} = (2v_{1f} - 5.6 \times 10^5)v_{1f} = 0$$

One possibility for the solution of this equation is $v_{1f} = 0$, which corresponds to a head-on collision—the first proton stops and the second continues with the same speed in the same direction. This is not what we want. The other possibility is

$$2v_{1f} - 5.6 \times 10^5 = 0 \quad \longrightarrow \quad v_{1f} = \boxed{2.8 \times 10^5 \text{ m/s}}$$

From (3),

$$v_{2f} = \sqrt{1.2 \times 10^{11} - v_{1f}^2} = \sqrt{1.2 \times 10^{11} - (2.8 \times 10^5)^2}$$
$$= \boxed{2.1 \times 10^5 \text{ m/s}}$$

and from (2),

$$\phi = \sin^{-1}\left(\frac{v_{1f}\sin 37°}{v_{2f}}\right) = \sin^{-1}\left(\frac{(2.8 \times 10^5)\sin 37°}{2.1 \times 10^5}\right)$$
$$= \boxed{53°}$$

It is interesting that $\theta + \phi = 90°$. This result is *not* accidental. Whenever two objects of equal mass collide elastically in a glancing collision and one of them is initially at rest, their final velocities are at right angles to each other.

### Example 8.10 Collision at an Intersection

A 1 500-kg car traveling east with a speed of 25.0 m/s collides at an intersection with a 2 500-kg van traveling north at a speed of 20.0 m/s, as shown in Figure 8.11. Find the direction and magnitude of the velocity of the wreckage after the collision, assuming that the vehicles undergo a perfectly inelastic collision (i.e., they stick together).

**Figure 8.11**

(Example 8.10) An eastbound car colliding with a northbound van.

**Solution** Let us choose east to be along the positive $x$ direction and north to be along the positive $y$ direction, as in Figure 8.11. Before the collision, the only object having momentum in the $x$ direction is the car. Thus, the magnitude of the total initial momentum of the system (car plus van) in the $x$ direction is

$$\Sigma\, p_{xi} = (1\,500 \text{ kg})(25.0 \text{ m/s}) = 3.75 \times 10^4 \text{ kg·m/s}$$

The wreckage moves at an angle $\theta$ and speed $v$ after the collision. The magnitude of the total momentum in the $x$ direction after the collision is

$$\Sigma\, p_{xf} = (4\,000 \text{ kg})v \cos\theta$$

Because the total momentum in the $x$ direction is conserved, we can equate these two equations to obtain

(1) $\quad (3.75 \times 10^4) \text{ kg·m/s} = (4000 \text{ kg})\, v \cos\theta$

Similarly, the total initial momentum of the system in the $y$ direction is that of the van, whose magnitude is equal to $(2\,500 \text{ kg})(20.0 \text{ m/s})$. Applying conservation of momentum to the $y$ direction, we have

$$\Sigma\, p_{yi} = \Sigma\, p_{yf}$$

$$(2\,500 \text{ kg})(20.0 \text{ m/s}) = (4\,000 \text{ kg})v \sin\theta$$

(2) $\quad 5.00 \times 10^4 \text{ kg·m/s} = (4\,000 \text{ kg})v \sin\theta$

If we divide (2) by (1), we find

$$\tan\theta = \frac{5.00 \times 10^4}{3.75 \times 10^4} = 1.33$$

$$\theta = \boxed{53.1°}$$

When this angle is substituted into (2), the value of $v$ is

$$v = \frac{5.00 \times 10^4 \text{ kg·m/s}}{(4\,000 \text{ kg})\sin 53.1°} = \boxed{15.6 \text{ m/s}}$$

**EXERCISE** A 3.00-kg object with an initial velocity of $5.00\mathbf{i}$ m/s collides with and sticks to a 2.00-kg object with an initial velocity of $-3.00\mathbf{j}$ m/s. Find the final velocity of the composite object.

*Answer* $(3.00\mathbf{i} - 1.20\mathbf{j})$ m/s

---

## 8.5 • THE CENTER OF MASS

See Screen 6.7

In this section we describe the overall motion of a system of particles in terms of a very special point called the **center of mass** of the system. The notion of the center of mass gives us confidence in the particle model, because we will see that the center of mass accelerates as if all of the system's mass were concentrated at that point and all external forces act there.

Consider a system consisting of a pair of particles connected by a light, rigid rod (Fig. 8.12). The center of mass as indicated in the figure is located on the rod and closer to the larger mass in the figure—we will see why soon. If a single force is applied at some point on the rod that is above the center of mass, the system rotates clockwise (Fig. 8.12a) as it translates through space. If the force is applied at a point on the rod below the center of mass, the system rotates counterclockwise (Fig. 8.12b). If the force is applied exactly at the center of mass, the system moves in the direction of **F** without rotating (Fig. 8.12c), as if the system is behaving as a particle. Thus, in theory, the center of mass can be located with this experiment.

If we were to analyze the motion in Figure 8.12c, we would find that the system moves as if all its mass were concentrated at the center of mass. Furthermore, if the external net force on the system is $\Sigma\,\mathbf{F}$ and the total mass of the system is $M$, the center of mass moves with an acceleration given by $\mathbf{a} = \Sigma\,\mathbf{F}/M$. That is, the system moves as if the resultant external force were applied to a single particle of mass $M$ located at the center of mass. This justifies our particle model for extended objects. We have ignored all rotational effects for extended objects so far, implicitly assuming that forces were provided at just the right position so as to cause no rotation. We will study rotational motion in Chapter 10, where we will apply forces that do not pass through the center of mass.

The position of the center of mass of a system can be described as being the *average position* of the system's mass. For example, the center of mass of the pair of particles described in Figure 8.13 is located on the $x$ axis, somewhere between the particles. The $x$ coordinate of the center of mass in this case is

$$x_{CM} = \frac{m_1 x_1 + m_2 x_2}{m_1 + m_2} \qquad [8.28]$$

For example, if $x_1 = 0$, $x_2 = d$, and $m_2 = 2m_1$, we find $x_{CM} = \frac{2}{3}d$. That is, the center of mass lies closer to the more massive particle. If the two masses are equal, the center of mass lies midway between the particles.

We can extend the concept of center of mass to a system of many particles in three dimensions. The $x$ coordinate of the center of mass of $n$ particles is defined to be

$$x_{CM} \equiv \frac{m_1 x_1 + m_2 x_2 + m_3 x_3 + \cdots + m_n x_n}{m_1 + m_2 + m_3 + \cdots + m_n} = \frac{\sum_i m_i x_i}{\sum_i m_i} = \frac{\sum_i m_i x_i}{M} \qquad [8.29]$$

where $x_i$ is the $x$ coordinate of the $i$th particle and $\Sigma_i m_i$ is the *total mass $M$* of the system. The $y$ and $z$ coordinates of the center of mass are similarly defined by the equations

$$y_{CM} \equiv \frac{\sum_i m_i y_i}{M} \quad \text{and} \quad z_{CM} \equiv \frac{\sum_i m_i z_i}{M} \qquad [8.30]$$

The center of mass can also be located by its position vector, $\mathbf{r}_{CM}$. The rectangular coordinates of this vector are $x_{CM}$, $y_{CM}$, and $z_{CM}$, defined in Equations 8.29 and 8.30. Therefore,

$$\mathbf{r}_{CM} = x_{CM}\mathbf{i} + y_{CM}\mathbf{j} + z_{CM}\mathbf{k} = \frac{\sum_i m_i x_i \mathbf{i} + \sum_i m_i y_i \mathbf{j} + \sum_i m_i z_i \mathbf{k}}{M}$$

$$\mathbf{r}_{CM} = \frac{\sum_i m_i \mathbf{r}_i}{M} \qquad [8.31]$$

where $\mathbf{r}_i$ is the position vector of the $i$th particle, defined by

$$\mathbf{r}_i \equiv x_i \mathbf{i} + y_i \mathbf{j} + z_i \mathbf{k}$$

Although locating the center of mass for an extended object is somewhat more cumbersome than locating the center of mass of a system of particles, this location

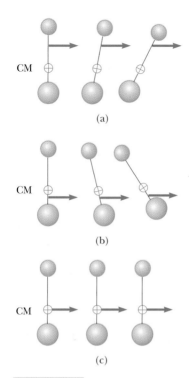

**Figure 8.12**

Two particles of unequal mass are connected by a light, rigid rod. (a) The system rotates clockwise when a force is applied between the less massive particle and the center of mass. (b) The system rotates counterclockwise when a force is applied between the more massive particle and the center of mass. (c) The system moves in the direction of the force without rotating when a force is applied at the center of mass.

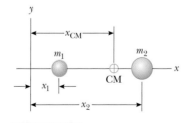

**Figure 8.13**

The center of mass of two particles having unequal mass is located on the $x$ axis at $x_{CM}$, a point between the particles, closer to the one having the larger mass.

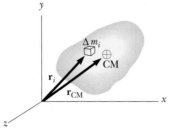

**Figure 8.14**

An extended object can be modeled as a distribution of small elements of mass $\Delta m_i$. The center of mass of the object is located at the vector position $\mathbf{r}_{CM}$, which has coordinates $x_{CM}$, $y_{CM}$, and $z_{CM}$.

is based on the same fundamental ideas. We can model the extended object as a system containing a large number of elements (Fig. 8.14). Each element is modeled as a particle of mass $\Delta m_i$, with coordinates $x_i$, $y_i$, $z_i$. The particle separation is very small, so this model is a good representation of the continuous mass distribution of the object. The $x$ coordinate of the center of mass of the particles representing the object, and therefore of the approximate center of mass of the object, is

$$x_{CM} \approx \frac{\sum_i x_i \Delta m_i}{M}$$

with similar expressions for $y_{CM}$ and $z_{CM}$. If we let the number of elements approach infinity (and, as a consequence, the size and mass of each element approach zero), the model becomes indistinguishable from the continuous mass distribution, and $x_{CM}$ is given precisely. In this limit, we replace the sum by an integral and $\Delta m_i$ by the differential element $dm$:

$$x_{CM} = \lim_{\Delta m_i \to 0} \frac{\sum_i x_i \Delta m_i}{M} = \frac{1}{M} \int x \, dm \qquad \text{[8.32]}$$

where the integration is over the length of the object in the $x$ direction. Likewise, for $y_{CM}$ and $z_{CM}$ we obtain

$$y_{CM} = \frac{1}{M} \int y \, dm \qquad \text{and} \qquad z_{CM} = \frac{1}{M} \int z \, dm \qquad \text{[8.33]}$$

We can express the vector position of the center of mass of an extended object as

$$\mathbf{r}_{CM} \approx \frac{1}{M} \int \mathbf{r} \, dm \qquad \text{[8.34]}$$

which is equivalent to the three expressions in Equations 8.32 and 8.33.

**The center of mass of a homogeneous, symmetric body must lie on an axis of symmetry.** For example, the center of mass of a homogeneous rod must lie midway between the ends of the rod. The center of mass of a homogeneous sphere or a homogeneous cube must lie at the geometric center of the object.

The center of mass of a system is often confused with the **center of gravity** of a system. Each portion of a system is acted on by the gravitational force. The net effect of all of these forces is equivalent to the effect of a single force $M\mathbf{g}$ acting at a special point called the center of gravity. The center of gravity is the average position of the gravitational forces on all parts of the object. If $\mathbf{g}$ is uniform over the system, the center of gravity coincides with the center of mass. If the gravitational field over the system is not uniform, the center of gravity and the center of mass are different. In most cases, for objects or systems of reasonable size, the two points can be considered to be coincident.

One can experimentally determine the center of gravity of an irregularly shaped object, such as a wrench, by suspending the wrench from two different points (Fig. 8.15). An object of this size has virtually no variation in the gravitational field over its dimensions, so this method also locates the center of mass. The wrench is first hung from point $A$, and a vertical line $AB$ is drawn (which can be established with a plumb bob) when the wrench is in equilibrium. The wrench

**Figure 8.15**

An experimental technique for determining the center of gravity of a wrench. The wrench is hung freely from two different pivots, $A$ and $C$. The intersection of the two vertical lines $AB$ and $CD$ locates the center of gravity.

is then hung from point *C*, and a second vertical line *CD* is drawn. The center of mass coincides with the intersection of these two lines. In fact, if the wrench is hung freely from any point, the vertical line through that point will pass through the center of mass.

## Quick Quiz 8.6

Figure 8.16 shows a one-bottle wine holder, which remains stationary in the position shown in the photograph. Describe where the center of mass of the system consisting of the bottle and the wooden holder is located.

**Figure 8.16**

(Quick Quiz 8.6)  *(Charles D. Winters)*

---

### Example 8.11  The Center of Mass of Three Particles

A system consists of three particles located at the corners of a right triangle as in Figure 8.17. Find the center of mass of the system.

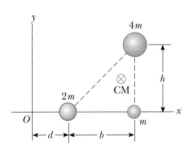

**Figure 8.17**

(Example 8.11) Locating the center of mass for a system of three particles.

**Solution**  Using the basic defining equations for the coordinates of the center of mass, and noting that $z_{CM} = 0$, we have

$$x_{CM} = \frac{\Sigma_i m_i x_i}{M} = \frac{2md + m(d + b) + 4m(d + b)}{7m}$$

$$= d + \frac{5}{7}\,b$$

$$y_{CM} = \frac{\Sigma_i m_i y_i}{M} = \frac{2m(0) + m(0) + 4mh}{7m} = \frac{4}{7}\,h$$

Therefore, we can express the position vector to the center of mass measured from the origin as

$$\mathbf{r}_{CM} = x_{CM}\mathbf{i} + y_{CM}\mathbf{j} + z_{CM}\mathbf{k} = \boxed{\left(d + \frac{5}{7}\,b\right)\mathbf{i} + \frac{4}{7}\,h\mathbf{j}}$$

---

### Example 8.12  The Center of Mass of a Right Triangle

You have been asked to hang a metal sign from a single vertical wire. The sign is of the triangular shape shown in Figure 8.18a. The bottom of the sign is to be parallel to the ground. At what distance from the left end of the sign should you attach the wire?

**Reasoning**  We will need to attach the wire at a point directly above the center of gravity of the sign, which is the same as its center of mass because it is in a uniform gravitational field. We model the sign as a perfect triangle. We assume that the sign has a uniform density and total mass *M*. Because the sign is a continuous distribution of mass, we will need to use the integral expression in Equation 8.32 to find the *x* coordinate of the center of mass.

**Solution**  We divide the triangle into narrow strips of width *dx* and height *y* as shown in Figure 8.18b, where *y* is the

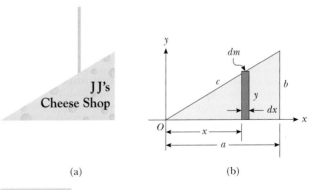

(a)

(b)

**Figure 8.18**

(Example 8.12) (a) A triangular sign to be hung from a single wire. (b) Geometric construction for locating the center of mass.

height to the hypotenuse of the triangle above the $x$ axis for a given value of $x$. The mass of each strip is the product of the volume of the strip and the density $\rho$ of the material from which the sign is made: $dm = \rho y t\, dx$, where $t$ is the thickness of the metal sign. The density of the material is the total mass of the sign divided by its total volume (area of the triangle times thickness), so

$$dm = \rho y t\, dx = \left( \frac{M}{\frac{1}{2}abt} \right) yt\, dx = \frac{2My}{ab}\, dx$$

Using Equation 8.32 to find the $x$ coordinate of the center of mass,

$$x_{CM} = \frac{1}{M} \int x\, dm = \frac{1}{M} \int_0^a x \frac{2My}{ab}\, dx = \frac{2}{ab} \int_0^a xy\, dx$$

To proceed further and evaluate the integral, we must express $y$ in terms of $x$. The line representing the hypotenuse of

the triangle in Figure 8.18b has a slope of $b/a$ and passes through the origin, so the equation of this line is $y = (b/a)x$. With this substitution for $y$ in the integral, we have

$$x_{CM} = \frac{2}{ab} \int_0^a x \left( \frac{b}{a} x \right) dx = \frac{2}{a^2} \int_0^a x^2\, dx = \frac{2}{a^2} \left[ \frac{x^3}{3} \right]_0^a$$

$$= \tfrac{2}{3}a$$

Thus, the wire must be attached to the sign at a distance two thirds of the length of the bottom edge from the left end. We could also find the $y$ coordinate of the center of mass of the sign, but this is not needed to determine where the wire should be attached.

**EXERCISE**   Find the $y$ coordinate of the center of mass of the sign.

*Answer*   $\tfrac{1}{3}b$

---

## 8.6 • MOTION OF A SYSTEM OF PARTICLES

 See Screen 6.8

We can begin to understand the physical significance and utility of the center of mass concept by taking the time derivative of the position vector $\mathbf{r}_{CM}$ of the center of mass, given by Equation 8.31. Assuming that $M$ remains constant—that is, no particles enter or leave the system—we find the following expression for the **velocity of the center of mass:**

- *Velocity of the center of mass*

$$\mathbf{v}_{CM} = \frac{d\mathbf{r}_{CM}}{dt} = \frac{1}{M} \sum_i m_i \frac{d\mathbf{r}_i}{dt} = \frac{1}{M} \sum_i m_i \mathbf{v}_i \qquad [8.35]$$

where $\mathbf{v}_i$ is the velocity of the $i$th particle. Rearranging Equation 8.35 gives

- *Total momentum of a system of particles*

$$M\mathbf{v}_{CM} = \sum_i m_i \mathbf{v}_i = \sum_i \mathbf{p}_i = \mathbf{p}_{tot} \qquad [8.36]$$

This result tells us that **the total momentum of the system equals its total mass multiplied by the velocity of its center of mass**—in other words, the total momentum of a single particle of mass $M$ moving with a velocity $\mathbf{v}_{CM}$—this is the particle model.

If we now differentiate Equation 8.35 with respect to time, we find the **acceleration of the center of mass:**

- *Acceleration of the center of mass for a system of particles*

$$\mathbf{a}_{CM} = \frac{d\mathbf{v}_{CM}}{dt} = \frac{1}{M} \sum_i m_i \frac{d\mathbf{v}_i}{dt} = \frac{1}{M} \sum_i m_i \mathbf{a}_i \qquad [8.37]$$

Rearranging this expression and using Newton's second law, we have

$$M\mathbf{a}_{CM} = \sum_i m_i \mathbf{a}_i = \sum_i \mathbf{F}_i \qquad [8.38]$$

where $\mathbf{F}_i$ is the force on particle $i$.

The forces on any particle in the system may include both external and internal forces. By Newton's third law, however, the force exerted by particle 1 on particle 2, for example, is equal in magnitude and opposite the force exerted by particle 2 on particle 1. When we sum over all internal forces in Equation 8.38, they cancel in pairs. The net force on the system is therefore due *only* to external forces. Thus, we can write Equation 8.38 in the form

$$\sum \mathbf{F}_{ext} = M\mathbf{a}_{CM} = \frac{d\mathbf{p}_{tot}}{dt} \qquad [8.39]$$

- *Newton's second law for a system of particles*

That is, the external net force on the system of particles equals the total mass of the system multiplied by the acceleration of the center of mass, or the time rate of change of the momentum of the system. If we compare this to Newton's second law for a single particle, we see that the center of mass moves like an imaginary particle of mass $M$ under the influence of the external net force on the system. In the absence of external forces, the center of mass moves with uniform velocity, as in the case of the translating and rotating wrench in Figure 8.19. If the net force acts along a line through the center of mass of an extended body such as the wrench, the body is accelerated without rotation. If the net force does not act through the center of mass, the body will undergo rotation in addition to translation. The linear acceleration of the center of mass is the same in either case, as given by Equation 8.39.

Finally, we see that if the external net force is zero, then from Equation 8.39 it follows that

$$\frac{d\mathbf{p}_{tot}}{dt} = M\mathbf{a}_{CM} = 0$$

so that

$$\mathbf{p}_{tot} = M\mathbf{v}_{CM} = \text{constant} \quad \left( \text{when} \sum \mathbf{F}_{ext} = 0 \right) \qquad [8.40]$$

That is, the total linear momentum of a system of particles is constant if no external forces act on the system. It follows that, for an *isolated* system of particles, the total momentum is conserved. The law of conservation of momentum that was derived in Section 8.1 for a two-particle system is thus generalized to a many-particle system.

**Figure 8.19**

Strobe photograph showing an overhead view of a wrench moving on a horizontal surface. The center of mass of the wrench (marked with a white dot) moves in a straight line as the wrench rotates about this point. *(Richard Megna, Fundamental Photographs)*

**Figure 8.20**

(Quick Quiz 8.7)

## Quick Quiz 8.7

A projectile is fired into the air and suddenly explodes into several fragments (Fig. 8.20). What can be said about the motion of the center of mass of the fragments after the explosion?

### THINKING PHYSICS 8.3

A boy stands at one end of a canoe that is stationary relative to the shore (Fig. 8.21). He then walks to the opposite end of the canoe, away from the shore. Does the canoe move?

**Reasoning** Yes, the canoe moves toward the shore. Ignoring friction between the canoe and water, no horizontal force acts on the system consisting of the boy and canoe. The center of mass of the system therefore remains fixed relative to the shore (or any stationary point). As the boy moves away from the shore, the canoe must move toward the shore such that the center of mass of the system remains fixed in position.

**Figure 8.21**

(Thinking Physics 8.3)

**EXERCISE**   A 5.0-kg particle moves along the $x$ axis with a velocity of 3.0 m/s. A 2.0-kg particle moves along the $x$ axis with a velocity of $-2.5$ m/s. Find (a) the velocity of the center of mass and (b) the total momentum of the system.

*Answer*   (a) 1.4 m/s to the right     (b) 10 kg·m/s to the right

**EXERCISE**   A 2.0-kg particle has a velocity $(2.0\mathbf{i} - 3.0\mathbf{j})$ m/s, and a 3.0-kg particle has a velocity $(1.0\mathbf{i} + 6.0\mathbf{j})$ m/s. Find (a) the velocity of the center of mass and (b) the total momentum of the system.

*Answer*   (a) $(1.4\mathbf{i} + 2.4\mathbf{j})$ m/s    (b) $(7.0\mathbf{i} + 12\mathbf{j})$ kg·m/s

### context connection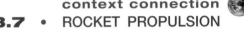
## 8.7 • ROCKET PROPULSION

On our trip to Mars, we will need to control our spacecraft by firing the rocket engines. When ordinary vehicles, such as automobiles and locomotives, are propelled, the driving force for the motion is one of friction. In the case of the automobile, the driving force is the force exerted by the road on the car. A locomotive "pushes" against the tracks; hence, the driving force is the reaction force exerted by the tracks on the locomotive. A rocket moving in space, however, has no road or tracks to "push" against. The source of the propulsion of a rocket must therefore be different. **The operation of a rocket depends on the law of conservation of momentum as applied to a system, where the system is the rocket plus its ejected fuel.**

The propulsion of a rocket can be understood by first considering the pitching machine on ice in Example 8.2. As a baseball is fired from the machine, the baseball receives momentum $m\mathbf{v}$ in one direction, and the machine receives a momentum of equal magnitude in the opposite direction. As each baseball is fired, the machine moves faster, so that a large velocity can be established by firing many baseballs.

In a similar manner, as a rocket moves in free space (a vacuum), its momentum changes when some of its mass is released in the form of ejected gases. Because the ejected gases acquire some momentum, the rocket receives a compensating momentum in the opposite direction. The rocket therefore is accelerated as a result of the "push," or thrust, from the exhaust gases. Note that the rocket represents the *inverse* of an inelastic collision; that is, momentum is conserved, but the kinetic energy of the system is *increased* (at the expense of energy stored in the fuel of the rocket).

Suppose that at some time $t$, the magnitude of the momentum of the rocket plus the fuel is $(M + \Delta m)v$ (Fig. 8.22a). During a short time interval $\Delta t$, the rocket ejects fuel of mass $\Delta m$ and the rocket's speed therefore increases to $v + \Delta v$ (Fig. 8.22b). If the fuel is ejected with velocity $v_e$ *relative to the rocket*, the speed of the fuel relative to a stationary frame of reference is $v - v_e$, according to our discussion of relative velocity in Section 3.6. Thus, if we equate the total initial momentum of the system with the total final momentum, we have

$$(M + \Delta m)v = M(v + \Delta v) + \Delta m(v - v_e)$$

Simplifying this expression gives

$$M\,\Delta v = \Delta m(v_e)$$

$$\mathbf{p}_i = (M + \Delta m)\mathbf{v}$$

(a)

(b)

**Figure 8.22**

Rocket propulsion. (a) The initial mass of the rocket and fuel is $M + \Delta m$ at a time $t$, and its speed is $v$. (b) At a time $t + \Delta t$, the rocket's mass has been reduced to $M$, and an amount of fuel $\Delta m$ has been ejected. The rocket's speed increases by an amount $\Delta v$.

If we now take the limit as $\Delta t$ goes to zero, $\Delta v \rightarrow dv$ and $\Delta m \rightarrow dm$. Furthermore, the increase $dm$ in the exhaust mass corresponds to an equal decrease in the rocket mass, so that $dm = -dM$. Note that the negative sign is introduced into the equation because $dM$ represents a decrease in mass. Using this fact, we have

$$M \, dv = -v_e \, dM \qquad [8.41]$$

Integrating this equation, and taking the initial mass of the rocket plus fuel to be $M_i$ and the final mass of the rocket plus its remaining fuel to be $M_f$, we have

$$\int_{v_i}^{v_f} dv = -v_e \int_{M_i}^{M_f} \frac{dM}{M}$$

$$v_f - v_i = v_e \ln\left(\frac{M_i}{M_f}\right) \qquad [8.42]$$

- *Basic expression for velocity change in rocket propulsion*

This is the basic expression for rocket propulsion. It tells us that the increase in speed is proportional to the exhaust speed $v_e$. The exhaust speed should therefore be very high.

The **thrust** on the rocket is the force exerted on the rocket by the ejected exhaust gases. We can obtain an expression for the instantaneous thrust from Equation 8.41:

- *Rocket thrust*

$$\text{Instantaneous thrust} = Ma = M\frac{dv}{dt} = \left|v_e \frac{dM}{dt}\right| \qquad [8.43]$$

Here we see that the thrust increases as the exhaust speed increases and as the rate of change of mass (burn rate) increases.

We now can determine the amount of fuel needed to set us on our journey to Mars. The fuel requirements are well within the capabilities of current technology, as evidenced by the fact that several missions to Mars have already been accomplished. What if we wanted to visit another *star*, however, rather than another *planet*? This raises many new technological challenges, including the requirement to consider the effects of relativity, which we investigate in the next chapter.

The thrust from a device that expels nitrogen gas allows an astronaut to maneuver in space without using a restrictive tether. *(Courtesy of NASA)*

**WEB**

For an essay on the history of rockets and rocket propulsion by a NASA scientist, visit **www.science.ksc.nasa.gov/ history/rocket-history.txt**

**THINKING PHYSICS 8.4**

When Robert Goddard proposed the possibility of rocket propelled vehicles, the *New York Times* agreed that such vehicles would be useful and successful within the Earth's atmosphere ("Topics of the Times," *New York Times*, January 13, 1920, p. 12). But the *Times* balked at the idea of using such a rocket in the vacuum of space, noting that, ". . . its flight would be neither accelerated nor maintained by the explosion of the charges it then might have left. To claim that it would be is to deny a fundamental law of dynamics, and only Dr. Einstein and his chosen dozen, so few and fit, are licensed to do that . . . That Professor Goddard, with his 'chair' in Clark College and the countenancing of the Smithsonian Institution, does not know the relation of action to reaction, and of the need to have something better than a vacuum against which to react—to say that would be absurd. Of course, he only seems to lack the knowledge ladled out daily in high schools." What did the writer of this passage overlook?

**Reasoning** The writer of this passage was making a common mistake in believing that a rocket works by expelling gases which push on something, propelling the rocket forward. With this belief, it is impossible to see how a rocket fired in empty space would work.

Gases do not need to push on anything—it is the act itself of expelling the gases that pushes the rocket forward. This can be argued from Newton's third law—the rocket pushes the gases backward, resulting in the gases pushing the rocket forward. It can also be argued from conservation of momentum—as the gases gain momentum in one direction, the rocket must gain momentum in the opposite direction to conserve the original momentum of the rocket–gas system.

The *New York Times* did publish a retraction 49 years later ("A Correction," *New York Times,* July 17, 1969, p. 43) while the *Apollo 11* astronauts were on their way to the Moon. It appeared on a page with two other articles entitled, "Fundamentals of Space Travel," and "Spacecraft, Like Squid, Maneuver by 'Squirts'," and contained the following passages: ". . . an editorial feature of the *New York Times* dismissed the notion that a rocket could function in a vacuum and commented on the ideas of Robert H. Goddard . . . Further investigation and experimentation have confirmed the findings of Isaac Newton in the 17th century, and it is now definitely established that a rocket can function in a vacuum as well as in an atmosphere. The *Times* regrets the error."

## Example 8.13  A Rocket in Space

A rocket moving in free space has a speed of $3.0 \times 10^3$ m/s relative to Earth. Its engines are turned on, and fuel is ejected in a direction opposite the rocket's motion at a speed of $5.0 \times 10^3$ m/s relative to the rocket. (a) What is the speed of the rocket relative to Earth once its mass is reduced to one half its mass before ignition?

**Solution**  Applying Equation 8.42, we have

$$v_f = v_i + v_e \ln \left( \frac{M_i}{M_f} \right)$$

$$= 3.0 \times 10^3 \text{ m/s} + (5.0 \times 10^3 \text{ m/s}) \ln \left( \frac{M_i}{0.5 M_i} \right)$$

$$= 6.5 \times 10^3 \text{ m/s}$$

(b) What is the thrust on the rocket if it burns fuel at the rate of 50 kg/s?

**Solution**  Using Equation 8.43, we have

$$\text{Thrust} = \left| v_e \frac{dM}{DT} \right| = (5.0 \times 10^3 \text{ m/s}) (50 \text{kg/s})$$

$$= 2.5 \times 10^5 \text{ N}$$

## SUMMARY

The linear momentum of any object of mass $m$ moving with a velocity **v** is

$$\mathbf{p} = m\mathbf{v} \qquad [8.1]$$

**Conservation of momentum** applied to two interacting objects states that if the two objects form an isolated system, the total momentum of the system at all times equals its initial total momentum:

$$\mathbf{p}_{1i} + \mathbf{p}_{2i} = \mathbf{p}_{1f} + \mathbf{p}_{2f} \qquad [8.5]$$

The **impulse** of a net force $\Sigma \mathbf{F}$ is defined as the integral of the force over the time interval during which it acts. The total impulse on any object is equal to the change in the momentum of the object and is given by

$$\mathbf{I} \equiv \int_{t_i}^{t_f} \Sigma \mathbf{F} \, dt = \Delta \mathbf{p} = \mathbf{p}_f - \mathbf{p}_i \qquad [8.9]$$

This is known as the **impulse–momentum theorem.**

When two objects collide, the total momentum of the isolated system before the collision always equals the total momentum after the collision, regardless of the nature of the collision. An **inelastic collision** is one for which kinetic energy is not conserved. A **perfectly inelastic collision** is one in which the colliding objects stick together after the collision. An **elastic collision** is one in which both momentum and kinetic energy are conserved.

In a two- or three-dimensional collision, the components of momentum in each of the directions are conserved independently.

The **vector position of the center of mass of a system of particles** is defined as

$$\mathbf{r}_{CM} \equiv \frac{\sum_i m_i \mathbf{r}_i}{M} \qquad [8.31]$$

where $M = \sum_i m_i$ is the total mass of the system and $\mathbf{r}_i$ is the vector position of the *i*th particle.

The **velocity of the center of mass for a system of parti-cles** is

$$\mathbf{v}_{CM} = \frac{1}{M} \sum_i m_i \mathbf{v}_i \qquad [8.35]$$

The total momentum of a system of particles equals the total mass multiplied by the velocity of the center of mass, that is, $\mathbf{p}_{tot} = M\mathbf{v}_{CM}$.

**Newton's second law applied to a system of particles** is

$$\sum \mathbf{F}_{ext} = M\mathbf{a}_{CM} = \frac{d\mathbf{p}_{tot}}{dt} \qquad [8.39]$$

where $\mathbf{a}_{CM}$ is the acceleration of the center of mass and the sum is over all external forces. The center of mass therefore moves like an imaginary particle of mass $M$ under the influence of the resultant external force on the system.

## QUESTIONS

1. Does a large force always produce a larger impulse on an object than a smaller force does? Explain.

2. If two objects collide and one is initially at rest, is it possible for both to be at rest after the collision? Is it possible for one to be at rest after the collision? Explain.

3. Explain how linear momentum is conserved when a ball bounces from a floor.

4. Consider a perfectly inelastic collision between a car and a large truck. Which vehicle experiences a larger change in kinetic energy as a result of the collision?

5. Your physical education teacher, who knows something about physics, throws you a tennis ball at a certain velocity, and you catch it. You are now given the following choice: The teacher can throw you a bowling ball, which is much more massive than the tennis ball, with the same velocity as the tennis ball, the same momentum, or the same kinetic energy. Which choice would you make in order to make the easiest catch, and why?

6. You are watching a movie about a superhero and notice that the superhero hovers in the air and throws a piano at some bad guys while remaining stationary in the air. What's wrong with this scenario?

7. Can the center of mass of an object be located at a position at which there is no mass? If so, give examples.

8. Is the center of mass of a mountain higher or lower than the center of gravity?

9. In golf, novice players are often advised to be sure to "follow through" with their swing. Why does this make the ball travel a longer distance?

10. A sharpshooter fires a rifle while standing with the butt of the gun against his shoulder. If the forward momentum of a bullet is the same as the backward momentum of the gun, why isn't it as dangerous to be hit by the gun as by the bullet?

11. A pole-vaulter falls from a height of 6.0 m onto a foam rubber pad. Can you calculate his speed just before he reaches the pad? Could you calculate the force exerted on him due to the collision? Explain.

12. Firefighters must apply large forces to hold a fire hose steady (Fig. Q8.12). What factors related to the projection of the water determine the magnitude of the force needed to keep the end of the fire hose stationary?

**Figure Q8.12** Firefighters attack a burning house with a hose line. (© *Bill Stormont/The Stock Market*)

13. Does the center of mass of a rocket in free space accelerate? Explain. Can the speed of a rocket exceed the exhaust speed of the fuel? Explain.

14. A large bed sheet is held vertically by two students. A third student, who happens to be the star pitcher on the baseball team, throws a raw egg at the sheet. Explain why the egg does not break when it hits the sheet, regardless of its initial speed. (If you try this one, make sure the pitcher hits the sheet near its center, and do not allow the egg to fall on the floor after being caught.)

15. On the subject of the following positions, state your own view and argue to support it: (a) The best theory of motion is that force causes acceleration. (b) The true measure of a force's effectiveness is the work it does, and the best theory of motion is that work on an object changes its energy. (c) The true measure of a force's effect is impulse, and the best theory of motion is that impulse injected into an object changes its momentum.

## PROBLEMS

1, 2, 3 = straightforward, intermediate, challenging     □ = full solution available in the *Student Solutions Manual and Study Guide*

**web** = solution posted at **http://www.harcourtcollege.com/physics/**      = computer useful in solving problem

 = Interactive Physics     ▓ = paired numerical/symbolic problems     ◤ = life science application

### Section 8.1    Linear Momentum and Its Conservation

1. A 3.00-kg particle has a velocity of $(3.00\mathbf{i} - 4.00\mathbf{j})$ m/s. (a) Find its $x$ and $y$ components of momentum. (b) Find the magnitude and direction of its momentum.

2. A 40.0-kg child standing on a frozen pond throws a 0.500-kg stone to the east with a speed of 5.00 m/s. Ignoring friction between the child and the ice, find the recoil velocity of the child.

3. How fast can you set the Earth moving? In particular, when you jump straight up as high as you can, what is the order of magnitude of the maximum recoil speed that you give to the Earth? Model the Earth as a perfectly solid object. In your solution, state the physical quantities you take as data, and the values you measure or estimate for them.

4. Two blocks of masses $M$ and $3M$ are placed on a horizontal, frictionless surface. A light spring is attached to one of them, and the blocks are pushed together with the spring between them (Fig. P8.4). A cord initially holding the blocks together is burned; after this, the block of mass $3M$ moves to the right with a speed of 2.00 m/s. (a) What is the speed of the block of mass $M$? (b) Find the original elastic potential energy in the spring if $M = 0.350$ kg.

5. (a) A particle of mass $m$ moves with momentum $p$. Show that the kinetic energy of the particle is given by $K = p^2/2m$. (b) Express the magnitude of the particle's momentum in terms of its kinetic energy and mass.

6. In research in cardiology and exercise physiology it is often important to know the mass of blood pumped by a person's heart in one stroke. These data can be obtained by means of a *ballistocardiograph*. The instrument works as follows. The subject lies on a horizontal pallet floating on a film of air. Friction on the pallet is negligible. Initially the momentum of the system is zero. When the heart beats, it expels a mass $m$ of blood into the aorta with velocity $v$, and the body and platform move in the opposite direction with velocity $V$. The blood velocity can be determined independently (e.g., by observing the Doppler shift of ultrasound). Assume it is 50.0 cm/s in one typical trial. The mass of the subject plus the pallet is 54.0 kg. The pallet moves $6.00 \times 10^{-5}$ m in 0.160 s after one heartbeat. Calculate the mass of blood that leaves the heart. Assume the mass of blood is negligible compared with the total mass of the person. This simplified example illustrates the principle of ballistocardiography, but in practice a more sophisticated model of heart function is used.

### Section 8.2    Impulse and Momentum

7. An estimated force–time curve for a baseball struck by a bat is shown in Figure P8.7. From this curve, determine (a) the impulse delivered to the ball, (b) the average force exerted on the ball, and (c) the peak force exerted on the ball.

**Figure P8.4**

**Figure P8.7**

**Figure P8.12**

**8.** A car is stopped for a traffic signal. When the light turns green, the car accelerates, increasing its speed from 0 to 5.20 m/s in 0.832 s. What linear impulse and average force does a 70.0-kg passenger in the car experience?

**9.** A 3.00-kg steel ball strikes a wall with a speed of 10.0 m/s
web at an angle of 60.0° with the surface. It bounces off with the same speed and angle (Fig. P8.9). If the ball is in contact with the wall for 0.200 s, what is the average force exerted on the ball by the wall?

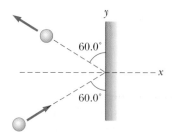

**Figure P8.9**

**10.** A tennis player receives a volley with the ball (0.060 0 kg) traveling horizontally at 50.0 m/s and returns the ball so that it travels horizontally at 40.0 m/s in the opposite direction. (a) What is the impulse delivered to the ball by the racquet? (b) What work does the racquet do on the ball?

**11.** In a slow-pitch softball game, a 0.200-kg softball crosses the plate at 15.0 m/s at an angle of 45.0° below the horizontal. The batter hits the ball toward center field, giving it a velocity of 40.0 m/s at 30.0° above the horizontal. (a) Determine the impulse delivered to the ball. (b) If the force on the ball increases linearly for 4.00 ms, holds constant for 20.0 ms, and then decreases to zero linearly in another 4.00 ms, what is the maximum force on the ball?

**12.** A garden hose is held as shown in Figure P8.12. The hose is originally full of motionless water. What additional force is necessary to hold the nozzle stationary, after the water is turned on, if the discharge rate is 0.600 kg/s with a speed of 25.0 m/s?

**13.** A glider of mass $m$ is free to slide along a horizontal air track. It is pushed against a launcher at one end of the track. Model the launcher as a light spring of force constant $k$, compressed by a distance $x$. The glider is released from rest. (a) Show that the glider attains a speed given by $v = x(k/m)^{1/2}$. (b) Is a greater speed attained by a large or a small mass? (c) Show that the impulse imparted to the glider is given by $I = x(km)^{1/2}$. (d) Is a greater impulse injected into a large or a small mass? (e) Is more work done on a large or a small mass?

### Section 8.3   Collisions

**14.** A 7.00-kg bowling ball collides head-on with a single 2.00-kg bowling pin. The pin flies forward with a speed of 3.00 m/s. If the ball continues forward with a speed of 1.80 m/s, what was the initial speed of the ball? Ignore rotation of the ball.

**15.** A 45.0-kg girl is standing on a plank that has a mass of 150 kg. The plank, originally at rest, is free to slide on a frozen lake, which is a flat, frictionless supporting surface. The girl begins to walk along the plank at a constant speed of 1.50 m/s relative to the plank. (a) What is her speed relative to the ice surface? (b) What is the speed of the plank relative to the ice surface?

**16.** Gayle runs at a speed of 4.00 m/s and dives onto a sled, which is initially at rest on the top of a frictionless snow-covered hill. After she has descended a vertical distance of 5.00 m, her brother, who is initially at rest, hops onto her back and together they continue down the hill. What is their speed at the bottom of the hill if the total vertical drop is 15.0 m? Gayle's mass is 50.0 kg, the sled has a mass of 5.00 kg, and her brother has a mass of 30.0 kg.

**17.** A railroad car of mass $2.50 \times 10^4$ kg is moving with a speed of 4.00 m/s. It collides and couples with three other coupled railroad cars, each of the same mass as the single car and moving in the same direction with an initial speed of 2.00 m/s. (a) What is the speed of the four cars after the collision? (b) How much mechanical energy is lost in the collision?

18. Four railroad cars, each of mass $2.50 \times 10^4$ kg, are coupled together and coasting along horizontal tracks at speed $v_i$ toward the south. A very strong but foolish movie actor, riding on the second car, uncouples the front car and gives it a big push, increasing its speed to 4.00 m/s southward. The remaining three cars continue moving south, now at 2.00 m/s. (a) Find the initial speed of the cars. (b) How much work did the actor do? (c) State the relationship between the process described here and the process in Problem 17.

19. A neutron in a nuclear reactor makes an elastic head-on
**web** collision with the nucleus of a carbon atom initially at rest. (a) What fraction of the neutron's kinetic energy is transferred to the carbon nucleus? (b) If the initial kinetic energy of the neutron is $1.60 \times 10^{-13}$ J, find its final kinetic energy and the kinetic energy of the carbon nucleus after the collision. (The mass of the carbon nucleus is nearly 12.0 times the mass of the neutron.)

20. Consider a frictionless track $ABC$ as shown in Figure P8.20. A block of mass $m_1 = 5.00$ kg is released from $A$. It makes a head-on elastic collision at $B$ with a block having a mass of $m_2 = 10.0$ kg that is initially at rest. Calculate the maximum height to which $m_1$ rises after the collision.

**Figure P8.20**

21. A 12.0-g wad of sticky clay is hurled horizontally at a 100-g wooden block initially at rest on a horizontal surface. The clay sticks to the block. After impact, the block slides 7.50 m before coming to rest. If the coefficient of friction between the block and the surface is 0.650, what was the speed of the clay immediately before impact?

22. Most of us know intuitively that in a head-on collision between a large dump truck and a subcompact car, you are better off being in the truck than in the car. Why is this? Many people imagine that the collision force exerted on the car is much greater than that experienced by the truck. To substantiate this view, they point out that the car is crushed, whereas the truck is only dented. This idea of unequal forces, of course, is false. Newton's third law tells us that both objects experience forces of the same magnitude. The truck suffers less damage because it is made of stronger metal. But what about the two drivers? Do they

experience the same forces? To answer this question, suppose that each vehicle is initially moving at 8.00 m/s and that they undergo a perfectly inelastic head-on collision. Each driver has a mass of 80.0 kg. Including the drivers, the total vehicle masses are 800 kg for the car and 4 000 kg for the truck. If the collision time is 0.120 s, what average force does the seatbelt exert on each driver?

23. As shown in Figure P8.23, a bullet of mass $m$ and speed $v$ passes completely through a pendulum bob of mass $M$. The bullet emerges with a speed of $v/2$. The pendulum bob is suspended by a stiff rod of length $\ell$ and negligible mass. What is the minimum value of $v$ such that the pendulum bob will barely swing through a complete vertical circle?

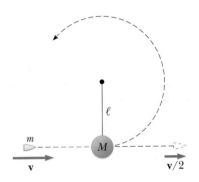

**Figure P8.23**

24. A 7.00-g bullet, when fired from a gun into a 1.00-kg block of wood held in a vise, will penetrate the block to a depth of 8.00 cm. This block of wood is placed on a frictionless horizontal surface, and a 7.00-g bullet is fired from the gun into the block. To what depth will the bullet penetrate the block in this case?

## Section 8.4  Two-Dimensional Collisions

25. A 90.0-kg fullback running east with a speed of 5.00 m/s is tackled by a 95.0-kg opponent running north with a speed of 3.00 m/s. If the collision is perfectly inelastic, (a) calculate the speed and direction of the players just after the tackle and (b) determine the mechanical energy lost as a result of the collision. Account for the missing energy.

26. Two automobiles of equal mass approach an intersection. One vehicle is traveling with velocity 13.0 m/s toward the east, and the other is traveling north with speed $v_{2i}$. Neither driver sees the other. The vehicles collide in the intersection and stick together, leaving parallel skid marks at an angle of 55.0° north of east. The speed limit for both roads is 35 mi/h, and the driver of the northward-moving vehicle claims he was within the speed limit when the collision occurred. Is he telling the truth?

**27.** A billiard ball moving at 5.00 m/s strikes a stationary ball of the same mass. After the collision, the first ball moves at 4.33 m/s, at an angle of 30.0° with respect to the original line of motion. Assuming an elastic collision (and ignoring friction and rotational motion), find the struck ball's velocity.

**28.** A proton, moving with a velocity of $v_i\mathbf{i}$, collides elastically with another proton that is initially at rest. If the two protons have equal speeds after the collision, find (a) the speed of each proton after the collision in terms of $v_i$ and (b) the direction of the velocity vectors after the collision.

**29.** A 3.00-kg object with an initial velocity of $5.00\mathbf{i}$ m/s collides with and sticks to a 2.00-kg object with an initial velocity of $-3.00\mathbf{j}$ m/s. Find the final velocity of the composite object.

**30.** A 0.300-kg puck, initially at rest on a horizontal, frictionless surface, is struck by a 0.200-kg puck moving initially along the *x* axis with a speed of 2.00 m/s. After the collision, the 0.200-kg puck has a speed of 1.00 m/s at an angle of $\theta = 53.0°$ to the positive *x* axis (see Figure 8.10). (a) Determine the velocity of the 0.300-kg puck after the collision. (b) Find the fraction of kinetic energy lost in the collision.

**31.** An unstable atomic nucleus of mass $17.0 \times 10^{-27}$ kg initially at rest disintegrates into three particles. One of the particles, of mass $5.00 \times 10^{-27}$ kg, moves along the *y* axis with a velocity of $6.00 \times 10^6$ m/s. Another particle, of mass $8.40 \times 10^{-27}$ kg, moves along the *x* axis with a speed of $4.00 \times 10^6$ m/s. Find (a) the velocity of the third particle and (b) the total kinetic energy increase in the process.

**32.** Two shuffleboard disks of equal mass, one orange and the other yellow, are involved in an elastic, glancing collision. The yellow disk is initially at rest and is struck by the orange disk moving with a speed $v_i$. After the collision, the orange disk moves along a direction that makes an angle $\theta$ with its initial direction of motion. The velocities of the two disks are perpendicular after the collision. Determine the final speed of each disk.

## Section 8.5   The Center of Mass

**33.** Four objects are situated along the *y* axis as follows: a 2.00-kg object is at $+3.00$ m, a 3.00-kg object is at $+2.50$ m, a 2.50-kg object is at the origin, and a 4.00-kg object is at $-0.500$ m. Where is the center of mass of this system of objects?

**34.** A water molecule consists of an oxygen atom with two hydrogen atoms bound to it (Fig. P8.34). The angle between the two bonds is 106°. If the bonds are 0.100 nm long, where is the center of mass of the molecule?

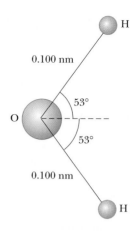

**Figure P8.34**

**35.** A uniform piece of sheet steel is shaped as in Figure P8.35. Compute the *x* and *y* coordinates of the center of mass of the piece.

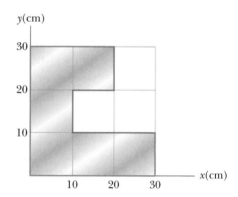

**Figure P8.35**

**36.** A rod of length 30.0 cm has linear density (mass per unit length) given by

$$\lambda = 50.0 \text{ g/m} + (20.0 \text{ g/m}^2)x$$

where *x* is the distance from one end, measured in meters. (a) What is the mass of the rod? (b) How far from the $x = 0$ end is its center of mass?

## Section 8.6   Motion of a System of Particles

**37.** A 2.00-kg particle has a velocity $(2.00\mathbf{i} - 3.00\mathbf{j})$ m/s, and a 3.00-kg particle has a velocity $(1.00\mathbf{i} + 6.00\mathbf{j})$ m/s. Find (a) the velocity of the center of mass and (b) the total momentum of the system.

**38.** Consider a system of two particles in the *xy* plane: $m_1 = 2.00$ kg is at the location $\mathbf{r}_1 = (1.00\mathbf{i} + 2.00\mathbf{j})$ m and has a velocity of $(3.00\mathbf{i} + 0.500\mathbf{j})$ m/s; $m_2 = 3.00$ kg is at $\mathbf{r}_2 = (-4.00\mathbf{i} - 3.00\mathbf{j})$ m and has velocity $(3.00\mathbf{i} - 2.00\mathbf{j})$ m/s.

(a) Plot these particles on a grid or graph paper. Draw their position vectors and show their velocities. (b) Find the position of the center of mass of the system and mark it on the grid. (c) Determine the velocity of the center of mass and also show it on the diagram. (d) What is the total linear momentum of the system?

39. Romeo (77.0 kg) entertains Juliet (55.0 kg) by playing his guitar from the rear of their boat at rest in still water, 2.70 m away from Juliet who is in the front of the boat. After the serenade, Juliet carefully moves to the rear of the boat (away from shore) to plant a kiss on Romeo's cheek. How far does the 80.0-kg boat move toward the shore it is facing?

40. A ball of mass 0.200 kg has a velocity of $1.50\mathbf{i}$ m/s; a ball of mass 0.300 kg has a velocity of $-0.400\mathbf{i}$ m/s. They meet in a head-on elastic collision. (a) Find their velocities after the collision. (b) Find the velocity of their center of mass before and after the collision.

## Section 8.7 Context Connection—Rocket Propulsion

41. The first stage of a Saturn V space vehicle consumes fuel
**web** and oxidizer at the rate of $1.50 \times 10^4$ kg/s, with an exhaust speed of $2.60 \times 10^3$ m/s. (a) Calculate the thrust produced by these engines. (b) Find the acceleration of the vehicle just as it leaves the launch pad if its initial mass is $3.00 \times 10^6$ kg. (*Hint:* You must include the gravitational force to solve part (b).)

42. Model rocket engines are sized by thrust, thrust duration, and total impulse, among other characteristics. A size C5 model rocket engine has an average thrust of 5.26 N, a fuel mass of 12.7 g, and an initial mass of 25.5 g. The duration of its burn is 1.90 s. (a) What is the average exhaust speed of the engine? (b) If this engine is placed in a rocket body of mass 53.5 g, what is the final velocity of the rocket if it is fired in outer space? Assume the fuel burns at a constant rate.

43. A rocket for use in deep space is to have the capability of boosting a total load (payload plus rocket frame and engine) of 3.00 metric tons to a speed of 10 000 m/s. (a) It has an engine and fuel designed to produce an exhaust speed of 2 000 m/s. How much fuel plus oxidizer is required? (b) If a different fuel and engine design could give an exhaust speed of 5 000 m/s, what amount of fuel and oxidizer would be required for the same task?

44. A 3 500-kg spacecraft experiences an acceleration of $2.50 \ \mu g = 2.45 \times 10^{-5}$ m/s$^2$ due to a leak from one of its hydraulic control systems. The fluid is known to escape with a speed of 70.0 m/s into the vacuum of space. How much fluid will be lost in 1.00 h if the leak is not stopped?

45. A rocket has total mass $M_i = 360$ kg, including 330 kg of fuel and oxidizer. In interstellar space it starts from rest.

Its engine is turned on at time $t = 0$, and it puts out exhaust with relative speed $v_e = 1\,500$ m/s at the constant rate 2.50 kg/s. The burn lasts until the fuel runs out, at time 330 kg/(2.50 kg/s) = 132 s. Set up and carry out a computer analysis of the motion according to Euler's method. Find (a) the final velocity of the rocket and (b) the distance it travels during the burn.

## Additional Problems

46. **Review Problem.** A 60.0-kg person running at an initial speed of 4.00 m/s jumps onto a 120-kg cart initially at rest (Figure P8.46). The person slides on the cart's top surface and finally comes to rest relative to the cart. The coefficient of kinetic friction between the person and the cart is 0.400. Friction between the cart and ground can be ignored. (a) Find the final velocity of the person and cart relative to the ground. (b) Find the friction force acting on the person while he is sliding across the top surface of the cart. (c) How long does the friction force act on the person? (d) Find the change in momentum of the person and the change in momentum of the cart. (e) Determine the displacement of the person relative to the ground while he is sliding on the cart. (f) Determine the displacement of the cart relative to the ground while the person is sliding. (g) Find the change in kinetic energy of the person. (h) Find the change in kinetic energy of the cart. (i) Explain why the answers to parts (g) and (h) differ. (What kind of collision is this, and what accounts for the loss of mechanical energy?)

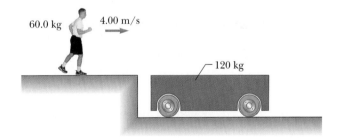

**Figure P8.46**

47. Tarzan, whose mass is 80.0 kg, swings from a 3.00-m vine that is horizontal when he starts. At the bottom of his arc, he picks up 60.0-kg Jane in a perfectly inelastic collision. What is the height of the highest tree limb they can reach on their upward swing?

48. A small block of mass $m_1 = 0.500$ kg is released from rest at the top of a curve-shaped frictionless wedge of mass $m_2 = 3.00$ kg, which sits on a frictionless horizontal surface as in Figure P8.48a. When the block leaves the wedge, its velocity is measured to be 4.00 m/s to the right, as in

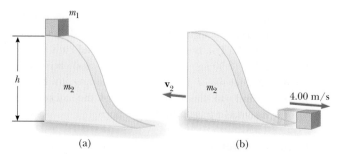

(a)              (b)

**Figure P8.48**

Figure P8.48b. (a) What is the velocity of the wedge after the block reaches the horizontal surface? (b) What is the height $h$ of the wedge?

49. A jet aircraft is traveling at 500 mi/h (223 m/s) in horizontal flight. The engine takes in air at a rate of 80.0 kg/s and burns fuel at a rate of 3.00 kg/s. If the exhaust gases are ejected at 600 m/s relative to the aircraft, find the thrust of the jet engine and the delivered power.

50. An 8.00-g bullet is fired into a 2.50-kg block initially at rest at the edge of a frictionless table of height 1.00 m (Fig. P8.50). The bullet remains in the block, and after impact the block lands 2.00 m from the bottom of the table. Determine the initial speed of the bullet.

**Figure P8.50**

51. A cannon is rigidly attached to a carriage, which can move along horizontal rails but is connected to a post by a large spring, initially unstretched and with force constant $k = 2.00 \times 10^4$ N/m, as in Figure P8.51. The cannon fires a 200-kg projectile at a velocity of 125 m/s directed 45.0° above the horizontal. (a) If the mass of the cannon and its carriage is 5 000 kg, find the recoil speed of the cannon. (b) Determine the maximum extension of the spring. (c) Find the maximum force the spring exerts on the carriage. (d) Consider the system consisting of the cannon, carriage, and shell. Is the momentum of this system conserved during the firing? Why or why not?

**Figure P8.51**

52. Two objects with masses $m$ and $3m$ are moving toward each other along the $x$ axis with the same initial speeds $v_i$. The object with mass $m$ is traveling to the left, and the object with mass $3m$ is traveling to the right. They undergo a head-on elastic collision and each rebounds along the same line as it approached. Find the final speeds of the objects.

53. Two objects with masses $m$ and $3m$ are moving toward each other along the $x$ axis with the same initial speeds $v_i$. The object with mass $m$ is traveling to the left, and the object with mass $3m$ is traveling to the right. They undergo an elastic glancing collision such that the object with mass $m$ is moving downward after the collision at a right angle from its initial direction. (a) Find the final speeds of the two objects. (b) What is the angle $\theta$ at which the object with mass $3m$ is scattered?

54. A 40.0-kg child stands at one end of a 70.0-kg boat that is 4.00 m in length (Fig. P8.54). The boat is initially 3.00 m from the pier. The child notices a turtle on a rock at the far end of the boat and proceeds to walk to that end to catch the turtle. Ignoring friction between boat and water, (a) describe the subsequent motion of the system (child

**Figure P8.54**

plus boat). (b) Where is the child *relative to the pier* when he reaches the far end of the boat? (c) Will he catch the turtle? (Assume he can reach out 1.00 m from the end of the boat.)

55. **Review Problem.** One can say that there are three coequal theories of motion. They are: Newton's second law, stating that the net force on an object causes its acceleration; the work–kinetic energy theorem, stating that the total work on an object causes its change in kinetic energy; and the impulse–momentum theorem, stating that the total impulse on an object causes its change in momentum. In this problem, compare predictions of the three theories in one particular case. A 3.00-kg object has velocity $7.00\mathbf{j}$ m/s. Then, a total force $12.0\mathbf{i}$ N acts on the object for 5.00 s. (a) Calculate the object's final velocity, using the impulse–momentum theorem. (b) Calculate its acceleration from $\mathbf{a} = (\mathbf{v}_f - \mathbf{v}_i)/t$. (c) Calculate its acceleration from $\mathbf{a} = \Sigma \mathbf{F}/m$. (d) Find the object's vector displacement from $\Delta \mathbf{r} = \mathbf{v}_i t + \frac{1}{2}\mathbf{a}t^2$. (e) Find the work done on the object from $W = \mathbf{F} \cdot \Delta \mathbf{r}$. (f) Find the final kinetic energy from $\frac{1}{2}mv_f^2 = \frac{1}{2}m\mathbf{v}_f \cdot \mathbf{v}_f$. (g) Find the final kinetic energy from $\frac{1}{2}mv_i^2 + W$.

56. Two gliders are set in motion on an air track. A spring of force constant $k$ is attached to the near side of one glider. The first glider of mass $m_1$ has velocity $\mathbf{v}_1$ and the second glider of mass $m_2$ moves more slowly, with velocity $\mathbf{v}_2$, as in Figure P8.56. When $m_1$ collides with the spring attached to $m_2$ and compresses the spring to its maximum compres-

sion $x_{max}$, the velocity of the gliders is $\mathbf{v}$. In terms of $\mathbf{v}_1$, $\mathbf{v}_2$, $m_1$, $m_2$, and $k$, find (a) the velocity $\mathbf{v}$ at maximum compression, (b) the maximum compression $x_{max}$, and (c) the velocity of each glider after $m_1$ has lost contact with the spring.

57. A chain of length $L$ and total mass $M$ is released from rest with its lower end just touching the top of a table, as in Figure P8.57a. Find the force exerted by the table on the chain after the chain has fallen through a distance $x$, as in Figure P8.57b. (Assume each link comes to rest the instant it reaches the table.)

**Figure P8.57**

58. Sand from a stationary hopper falls onto a moving conveyor belt at the rate of 5.00 kg/s as in Figure P8.58. The conveyor belt is supported by frictionless rollers and moves at a constant speed of 0.750 m/s under the action of a constant horizontal external force $\mathbf{F}_{ext}$ supplied by the motor that drives the belt. Find (a) the sand's rate of change of momentum in the horizontal direction, (b) the force of friction exerted by the belt on the sand, (c) the external force $\mathbf{F}_{ext}$, (d) the work done by $\mathbf{F}_{ext}$ in 1.00 s, and (e) the kinetic energy acquired by the falling sand each second due to the change in its horizontal motion. (f) Why are the answers to parts (d) and (e) different?

**Figure P8.56**

**Figure P8.58**

## ANSWERS TO QUICK QUIZZES

**8.1** (d). We are given no information about the masses of the objects. If the masses are the same, the speeds must be the same (so that they have equal kinetic energies), and then $p_1 = p_2$. If the masses are not the same, the speeds will be different, as will the momenta, and either $p_1 < p_2$, or $p_1 > p_2$, depending on which particle has more mass. Without information about the masses, we cannot choose among these possibilities.

**8.2** (c) and (e). Because object 1 has larger mass, its acceleration due to the applied force is smaller, and it takes a longer time interval $\Delta t$ to experience the displacement $\Delta x$ than does object 2. Thus, the magnitude of the impulse $F\Delta t$ on object 1 is larger than that on object 2. Consequently, object 1 will experience a larger change in momentum than object 2, which tells us that (c) is true. The same force acts on both objects through the same displacement. Thus, the same work is done on each object, so that each must experience the same change in kinetic energy, which tells us that (e) is true.

**8.3** The brain is immersed in a cushioning fluid inside the skull. If the head is struck suddenly by a bare fist, the skull accelerates rapidly. The brain matches this acceleration only because of the large impulsive force by the inner surface of the skull on the brain. This large, sudden force can severely injure the brain. Padded gloves extend the time over which the force is applied to the head. Thus, for a given impulse, the gloves provide a longer time interval than the bare fist, decreasing the average force. Because the average force is decreased, the acceleration of the skull is decreased, reducing (but not eliminating) the chance for brain injury.

**8.4** (a) Perfectly inelastic—the two "particles," skater and Frisbee, are combined after the collision; (b) Inelastic— because the Frisbee bounced back with almost no speed, kinetic energy has been transformed to other forms of energy; (c) Inelastic—the kinetic energy of the Frisbee is the same before and after the collision. Because momentum of the skater–Frisbee system is conserved, however, the skater must be moving after the catch and the throw, so that the final kinetic energy of the system is larger than the initial kinetic energy. This extra energy comes from the muscles of the skater.

**8.5** If all of the initial kinetic energy is transformed, then nothing is moving after the collision. Consequently, the final momentum of the system is necessarily zero. Because momentum of the system is conserved, the initial momentum of the system must be zero. Finally, because the objects are identical, they must have been initially moving toward each other along the same line with the same speed.

**8.6** The center of mass of the system must be vertically above the support point on the table. If this were not so, because the center of gravity and the center of mass are the same for small objects, the gravitational force acting at the center of gravity would tip the system over.

**8.7** Ignoring air resistance, the only external force on the projectile is the gravitational force. If the projectile did not explode, it would follow the complete parabolic path (solid and dashed) shown in Figure 8.21. In the case of the exploding projectile, all forces involved in the explosion are internal to the system. (The system before the explosion is the projectile; the system after the explosion is all of the fragments.) Thus, the center of mass of the system will follow the same path as if the explosion had not occurred. The center of mass of the system of fragments will land at the same place as the unexploded projectile would.

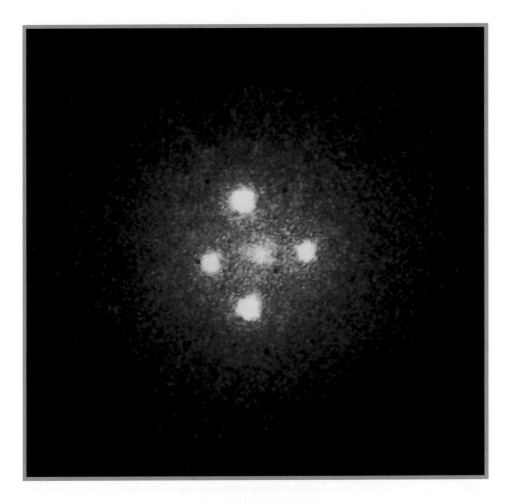

Einstein's cross. The four bright spots are images of the same galaxy formed by light that is bent around a massive object located between the galaxy and the Earth. The massive object acts as a lens, causing the rays of light that were diverging from the distant galaxy to converge at the Earth. If the intervening massive object were perfectly uniform and symmetric, we would see a bright ring of light instead of discrete spots.

*(Courtesy of NASA)*

# Relativity

**M**ost of our everyday experiences and observations are with objects that move at speeds much less than that of light in a vacuum, $c = 3.00 \times 10^8$ m/s. Models based on Newtonian mechanics and early ideas on space and time were formulated to describe the motion of such objects. This formalism is very successful in describing a wide range of phenomena that occur at low speeds, as we have seen in previous chapters. It fails, however, when applied to particles whose speeds approach that of light. Experimentally, the predictions of Newtonian theory can be tested by accelerating electrons or other particles to very high speeds. For example, it is possible to accelerate an electron to a speed of $0.99c$. According to the Newtonian definition of kinetic energy, if the energy transferred to such an electron were increased by a factor of 4, the electron speed should double to $1.98c$. Relativistic calculations, however, show that the speed of the electron—as well as the speeds of all other particles in the Universe—remains less than the speed of light, regardless of the particle energy. Because it places no upper limit on speed, Newtonian mechanics is contrary to modern theoretical predictions and experimental results, and the Newtonian models that we have developed are limited to objects moving much slower than the speed of light. Because Newtonian mechanics does not correctly predict the results of experiments carried out on particles moving at high speeds, we need a new formalism that is valid for these particles.

In 1905, at the age of only 26, Einstein published his *special theory of relativity,* which is the subject of most of this chapter. Regarding the theory, Einstein wrote:

> The relativity theory arose from necessity, from serious and deep contradictions in the old theory from which there seemed no escape. The strength of the new theory lies in the consistency and simplicity with which it solves all these difficulties, using only a few very convincing assumptions. . . .*

Although Einstein made many important contributions to science, special relativity alone represents one of the greatest intellectual achievements of the 20th century. With special relativity, experimental observations can be correctly predicted over the range of speeds from rest to speeds approaching the speed of light. This chapter gives an introduction to special relativity, with emphasis on some of its consequences.

## 9.1 • THE PRINCIPLE OF NEWTONIAN RELATIVITY

We will begin by considering the notion of relativity at low speeds. This discussion was actually begun in Section 3.6, in which we discussed relative velocity. At that time, we discussed the importance of the observer. In a similar way here, we will generate equations that allow us to express one observer's measurements in terms of the other's. This will lead to some rather unexpected and startling results about our understanding of space and time.

As we have mentioned previously, to describe a physical event, it is necessary to establish a frame of reference. You should recall from earlier chapters that Newton's laws are valid in all inertial frames of reference. Because an inertial frame is defined as one in which Newton's first law is valid, it can be said that an **inertial frame is one in which an object experiences no acceleration if no forces act on it.** Furthermore, any frame moving with constant velocity with respect to an inertial frame must also be an inertial frame. The laws predicting the results of an experiment performed in a vehicle moving with uniform velocity will be identical for the driver of the vehicle and a hitchhiker on the side of the road. The formal statement of this result is called the principle of **Newtonian relativity:**

• *Inertial frame of reference*

• *Principle of Newtonian relativity*

> The laws of mechanics must be the same in all inertial frames of reference.

The following observation illustrates the equivalence of the laws of mechanics in different inertial frames. Consider a pickup truck moving with a constant velocity, as in Figure 9.1a. If a passenger in the truck throws a ball straight up in the air, the passenger observes that the ball moves in a vertical path (ignoring air resistance). The motion of the ball appears to be precisely the same as if the ball were thrown by a person at rest on Earth and observed by that person. The kinematic equations of Chapter 3 describe the results correctly whether the truck is at rest or in uniform motion. Now consider the ball thrown in the truck as viewed by an observer at rest on Earth. This observer sees the path of the ball as a parabola, as in Figure 9.1b. Furthermore, according to this observer, the ball has a horizontal component of velocity equal to the speed of the truck. Although the two observers measure different velocities and see different paths of the ball, they see the same forces

---

* A. Einstein and L. Infeld, *The Evolution of Physics,* New York, Simon & Schuster, 1961.

(a)  (b)

**Figure 9.1**

(a) The observer in the truck sees the ball move in a vertical path when thrown upward. (b) The Earth observer sees the path of the ball to be a parabola.

on the ball and agree on the validity of Newton's laws as well as classical principles such as conservation of energy and conservation of momentum. Their measurements differ, but the measurements satisfy the same laws. All differences between the two views stem from the relative motion of one frame with respect to the other.

Suppose that some physical phenomenon, which we call an **event,** occurs. The event's location in space and time of occurrence can be specified by an observer with the coordinates $(x, y, z, t)$. We would like to be able to transform these coordinates from one inertial frame to another moving with uniform relative velocity. This will allow us to express one observer's measurements in terms of the other's.

Consider two inertial frames S and S′ (Fig. 9.2). The frame S′ moves with a constant velocity **v** along the common $x$ and $x'$ axes, where **v** is measured relative to S. We assume that the origins of S and S′ coincide at $t = 0$. Thus, at time $t$, the origin of frame S′ is to the right of the origin of S by a distance $vt$. An event occurs at point $P$. An observer in S describes the event with space–time coordinates $(x, y, z, t)$, and an observer in S′ uses $(x', y', z', t')$ to describe the same event. As we can see from Figure 9.2, a simple geometric model shows that the space coordinates are related by the equations

$$x' = x - vt \qquad y' = y \qquad z' = z \qquad \text{[9.1]}$$

Time is assumed to be the same in both inertial frames. That is, within the framework of classical mechanics, all clocks run at the same rate, regardless of their velocity, so that the time at which an event occurs for an observer in S is the same as the time for the same event in S′:

$$t' = t \qquad \text{[9.2]}$$

Equations 9.1 and 9.2 constitute what is known as the **Galilean transformation of coordinates.**

Now suppose that a particle moves a distance $dx$ in a time interval $dt$ as measured by an observer in S. It follows from the first of Equations 9.1 that the corresponding distance $dx'$ measured by an observer in S′ is $dx' = dx - v\,dt$. Because $dt = dt'$ (Eq. 9.2), we find that

$$\frac{dx'}{dt'} = \frac{dx}{dt} - v$$

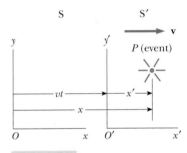

**Figure 9.2**

An event occurs at a point $P$. The event is seen by two observers $O$ and $O'$ in inertial frames S and S′, where S′ moves with a velocity **v** relative to S.

- *Galilean transformation of coordinates*

Keep in mind the relationship between the S and S′ frames. Otherwise, many of the mathematical representations in this chapter are not true. The $x$ and $x'$ axes coincide, except that their origins are different. The $y$ and $y'$ axes (and the $z$ and $z'$ axes), are parallel, but do not coincide due to the displacement of the origin of S′ with respect to that of S. We choose the time $t = 0$ to be the instant at which the origins of the two coordinate systems coincide. If the S′ frame is moving in the positive $x$ direction relative to S, $v$ is positive; otherwise it is negative.

**Albert A. Michelson** (1852–1931)

A German–American physicist, Michelson spent much of his life making accurate measurements of the speed of light, In 1907 he was the first American to be awarded the Nobel prize in physics, which he received for his work in optics. His most famous experiment, conducted with Edward Morley in 1887, indicated that is was impossible to measure the absolute velocity of the Earth with respect to the ether. *(AIP Emilio Segrè Visual Archives, Michelson Collection)*

or

$$u'_x = u_x - v \qquad [9.3]$$

where $u_x$ and $u'_x$ are the instantaneous $x$ components of velocity of the particle* relative to S and S′, respectively. This result, which is called the **Galilean addition law for velocities** (or Galilean velocity transformation), is used in everyday observations and is consistent with our intuitive notion of time and space. It is the same equation we generated in Section 3.6 (Eq. 3.22) when we first discussed relative velocity in one dimension. We find, however, that it leads to serious contradictions when applied to electromagnetic waves.

## 9.2 • THE MICHELSON–MORLEY EXPERIMENT

Many experiments similar to throwing the ball in the pickup truck, described in the preceding section, show us that the laws of classical mechanics are the same in all inertial frames of reference. When similar inquiries are made into the laws of other branches of physics, however, the results are contradictory. In particular, the laws of electricity and magnetism are found to depend on the frame of reference used. It might be argued that these laws are wrong, but that is difficult to accept because the laws are in total agreement with known experimental results. The Michelson–Morley experiment was one of many attempts to investigate this dilemma.

The experiment stemmed from a misconception early physicists had concerning the manner in which light propagates. The properties of mechanical waves, such as water and sound waves, were well known, and all of these waves require a *medium* to support the propagation of the disturbance, as we shall discuss in Chapter 13. For sound from your stereo system, the medium is the air, and for ocean waves, the medium is the water surface. In the 19th century, physicists subscribed to a model for light in which electromagnetic waves also require a medium through which to propagate. They proposed that such a medium exists, and they named it the **luminiferous ether.** The ether would define an **absolute frame of reference** in which the speed of light is $c$.

The most famous experiment designed to show the presence of the ether was performed in 1887 by A. A. Michelson (1852–1931) and E. W. Morley (1838–1923). The objective was to determine the speed of the Earth with respect to the ether, and the experimental tool used was a device called the *interferometer,* shown schematically in Figure 9.3.

Light from the source at the left encounters a beam splitter $M_0$, which is a partially silvered mirror. Part of the light passes through toward mirror $M_2$, and the other part is reflected upward toward mirror $M_1$. Both mirrors are the same distance from the beam splitter. After reflecting from these mirrors, the light returns to the beam splitter, and part of each light beam propagates toward the observer at the bottom.

Suppose one arm of the interferometer ($M_0M_2$, Arm 2, in Fig. 9.3) is aligned along the direction of the velocity **v** of the Earth through space and, therefore, through the ether. The "ether wind" blowing in the direction opposite the Earth's motion should cause the speed of light, as measured in the Earth's frame of refer-

---

* We have used $v$ for the speed of the S′ frame relative to the S frame. To avoid confusion, we will use $u$ for the speed of an object or particle.

ence, to be $c - v$ as the light approaches mirror $M_2$ in Figure 9.3, and $c + v$ after reflection.

The other arm ($M_0M_1$, Arm 1) is perpendicular to the ether wind. For light to travel in this direction, the vector **c** must be aimed "upstream" so that the vector addition of **c** and **v** gives the speed of the light perpendicular to the ether wind as $\sqrt{c^2 - v^2}$. This is similar to Example 3.6, in which a boat crossed a river with a current. The boat is a model for the light beam in the Michelson–Morley experiment, and the river current is a model for the ether wind.

Because they travel in perpendicular directions with different speeds, light beams leaving the beam splitter will arrive back at the beam splitter at different times. The interferometer is designed to detect this time difference. Measurements failed, however, to show any time difference! The Michelson–Morley experiment was repeated by other researchers under varying conditions and at different locations, but the results were always the same: *No time difference of the magnitude required was ever observed.**

The negative result of the Michelson–Morley experiment not only contradicted the ether hypothesis; it also meant that it was impossible to measure the absolute speed of the Earth with respect to the ether frame. From a theoretical viewpoint, this meant that it was impossible to find the absolute frame. As we shall see in the next section, however, Einstein offered a postulate that places a different interpretation on the negative result. In later years, when more was known about the nature of light, the idea of an ether that permeates all space was abandoned. **Light is now understood to be an electromagnetic wave that requires no medium for its propagation.** As a result, an ether through which light travels is an unnecessary construct.

Modern versions of the Michelson–Morley experiment have placed an upper limit of about 5 cm/s on ether wind velocity. These results have shown quite conclusively that the motion of the Earth has no effect on the measured speed of light!

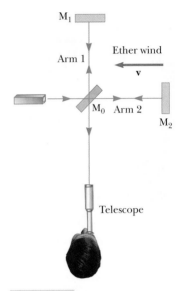

**Figure 9.3**

In the Michelson interferometer, the ether theory claims that the time of travel for a light beam traveling from the beam splitter to mirror $M_1$ will be different from that for a light beam traveling from the beam splitter to mirror $M_2$. The interferometer is sufficiently sensitive to detect this time difference.

## 9.3 • EINSTEIN'S PRINCIPLE OF RELATIVITY

In the preceding section we noted the failure of experiments to measure the speed of the ether with respect to the Earth. Albert Einstein proposed a theory that boldly removed these difficulties and at the same time completely altered our notion of space and time.[†] Einstein based his relativity theory on two postulates:

1. **The principle of relativity:**   All the laws of physics are the same in all inertial reference frames.
2. **The constancy of the speed of light:**   The speed of light in vacuum has the same value in all inertial frames, regardless of the velocity of the observer or the velocity of the source emitting the light.

• *The postulates of special relativity*

* From an Earth observer's point of view, changes in the Earth's speed and direction of motion in the course of a year are viewed as ether wind shifts. Even if the speed of the Earth with respect to the ether were zero at some time, six months later the speed of the Earth would be 60 km/s with respect to the ether and a clear time difference should be detected. None has ever been observed, however.

[†] A. Einstein, "On the Electrodynamics of Moving Bodies," *Ann. Physik* **17**:891, 1905. For an English translation of this article and other publications by Einstein, see the book by H. Lorentz, A. Einstein, H. Minkowski, and H. Weyl, *The Principle of Relativity,* Dover, 1958.

**Albert Einstein** (1879–1955)

Einstein, one of the greatest physicists of all times, was born in Ulm, Germany. In 1905, at the age of 26, he published four scientific papers that revolutionized physics. Two of these papers were concerned with what is now considered his most important contribution of all, the special theory of relativity.

In 1916, Einstein published his work on the general theory of relativity. The most dramatic prediction of this theory is the degree to which light is deflected by a gravitational field. Measurements made by astronomers on bright stars in the vicinity of the eclipsed sun in 1919 confirmed Einstein's prediction, and Einstein suddenly became a world celebrity.

Einstein was deeply disturbed by the development of quantum mechanics in the 1920s despite his own role as a scientific revolutionary. In particular, he could never accept the probabilistic view of events in nature that is a central feature of quantum theory. The last few decades of his life were devoted to an unsuccessful search for a unified theory that would combine gravitation and electromagnetism into one picture. *(AIP Niels Bohr Library)*

These postulates form the basis of **special relativity**, which is the relativity theory applied to observers moving with constant velocity. The first postulate asserts that *all* the laws of physics, those dealing with mechanics, electricity and magnetism, optics, thermodynamics, and so on, are the same in all reference frames moving with constant velocity relative to each other. This postulate is a sweeping generalization of the principle of Newtonian relativity that only refers to the laws of mechanics. From an experimental point of view, Einstein's principle of relativity means that any kind of experiment performed in a laboratory at rest must agree with the same laws of physics as when performed in a laboratory moving at constant velocity relative to the first one. Hence no preferred inertial reference frame exists and it is impossible to detect absolute motion.

Note that postulate 2, the principle of the constancy of the speed of light, is required by postulate 1: If the speed of light were not the same in all inertial frames, it would be possible to experimentally distinguish between inertial frames, and a preferred, absolute frame in which the speed of light is *c* could be identified, in contradiction to postulate 1. Postulate 2 also eliminates the problem of measuring the speed of the ether by denying the existence of the ether and boldly asserting that light always moves with speed *c* relative to all inertial observers.

### Quick Quiz 9.1

You are in a speedboat on a lake. You see ahead of you the crest of a wave, caused by the previous passage of another boat, moving away from you. You accelerate, catch up with, and pass the crest. Is this scenario possible if you are in a spacecraft and you detect the crest of a wave of *light* ahead of you?

## 9.4 • CONSEQUENCES OF SPECIAL RELATIVITY

If we accept the postulates of special relativity, we must conclude that relative motion is unimportant when measuring the speed of light—this is the lesson of the Michelson–Morley experiment. At the same time, we must alter our common-sense notion of space and time and be prepared for some very unexpected consequences, as we shall see now.

### Simultaneity and the Relativity of Time

A basic premise of Newtonian mechanics is that a universal time scale exists that is the same for all observers. In fact, Newton wrote that "Absolute, true, and mathematical time, of itself, and from its own nature, flows equably without relation to anything external." Thus, Newton and his followers simply took simultaneity for granted. In his development of special relativity, Einstein abandoned the notion that two events that appear simultaneous to one observer appear simultaneous to all observers. According to Einstein, a **time measurement depends on the reference frame in which the measurement is made.**

Einstein devised the following thought experiment to illustrate this point. A boxcar moves with uniform velocity, and two lightning bolts strike its ends, as in Figure 9.4a, leaving marks on the boxcar and on the ground. The marks on the boxcar are labeled $A'$ and $B'$, and those on the ground are labeled $A$ and $B$. An observer at $O'$ moving with the boxcar is midway between $A'$ and $B'$, and a ground observer at $O$ is midway between $A$ and $B$. The events recorded by the observers are the light signals from the lightning bolts.

**Figure 9.4**

(a) Two lightning bolts strike the ends of a moving boxcar. (b) The events appear to be simultaneous to the observer at *O*, who is standing on the ground midway between *A* and *B*. The events do not appear to be simultaneous to the observer *O′* riding on the boxcar, who claims that the front of the car is struck before the rear. Note that the leftward-traveling light signal from *B′* has already passed observer *O′*, but the rightward-traveling light signal from *A′* has not yet reached *O′*.

The two light signals reach observer *O* at the same time, as indicated in Figure 9.4b. Because of this, *O* concludes that the events at *A* and *B* occurred simultaneously. Now consider the same events as viewed by the observer on the boxcar at *O′*. From our frame of reference, at rest with respect to the tracks in Figure 9.4, we see the lightning strikes occur as *A′* passes *A*, *O′* passes *O*, and *B′* passes *B*. By the time the light has reached observer *O*, observer *O′* has moved as indicated in Figure 9.4b. Thus, the light signal from *B′* has already swept past *O′* because it had less distance to travel, but the light from *A′* has not yet reached *O′*. According to Einstein, **observer *O′* must find that light travels at the same speed as that measured by observer *O*.** The observer *O′* therefore concludes that the lightning struck the front of the boxcar before it struck the back. This thought experiment clearly demonstrates that the two events, which appear to be simultaneous to observer *O*, do not appear to be simultaneous to observer *O′*. In general, two events separated in space and observed to be simultaneous in one reference frame are not observed to be simultaneous in a second frame moving relative to the first. That is, simultaneity is not an absolute concept but one that depends upon the state of motion of the observer.

### Time Dilation

According to our thought experiment, observers in different inertial frames measure different time intervals between a pair of events. This can be illustrated by considering a vehicle moving to the right with a speed *v* as in the pictorial representation in Figure 9.5a. A mirror is fixed to the ceiling of the vehicle and observer *O′*, at rest in a frame attached to the vehicle, holds a flashlight a distance *d* below the mirror. At some instant, the flashlight is turned on momentarily and emits a pulse of light (event 1) directed toward the mirror. At some later time after reflecting from the mirror, the pulse arrives back at the flashlight (event 2). Observer *O′* carries a clock that she uses to measure the time interval $\Delta t_p$ between these two events. (The subscript "*p*" stands for "proper," as will be discussed shortly.) Because the light pulse has a constant speed *c*, the time it takes the pulse to travel

**Figure 9.5**

A mirror is fixed to a moving vehicle, and two observers measure the time interval between two events: the leaving of a light pulse from a flashlight and the arrival of the reflected light pulse back at the flashlight. (a) Observer $O'$, riding on the vehicle, sees the light pulse travel a total distance of $2d$ and measures a time interval between the events of $\Delta t_p$. (b) Observer $O$ is standing on the Earth and sees the mirror and $O'$ move to the right with speed $v$. Observer $O$ measures the distance that the light pulse travels to be greater than $2d$ and measures a time interval between the events of $\Delta t$. (c) The right triangle for calculating the relationship between $\Delta t$ and $\Delta t_p$.

from $O'$ to the mirror and back to $O'$ can be found by modeling the light pulse as a particle under constant speed from Chapter 2:

$$\Delta t_p = \frac{s}{v} = \frac{2d}{c} \qquad \text{[9.4]}$$

This time interval $\Delta t_p$ is measured by $O'$, for whom the two events occur at the same spatial position.

Now consider the same pair of events as viewed by observer $O$ in a second frame attached to the ground as in Figure 9.5b. According to this observer, the mirror and flashlight are moving to the right with a speed $v$. The geometry appears to be entirely different as viewed by this observer. By the time the light from the flashlight reaches the mirror, the mirror has moved horizontally a distance $v$ $\Delta t/2$, where $\Delta t$ is the time it takes the light to travel from the flashlight to the mirror and back to the flashlight as measured by observer $O$. In other words, the second observer concludes that because of the motion of the vehicle, if the light is to hit the mirror, it must leave the flashlight at an angle with respect to the vertical direction. Comparing Figures 9.5a and 9.5b, we see that the light must travel farther when observed in the second frame than in the first frame.

According to the second postulate of special relativity, both observers must measure $c$ for the speed of light. Because the light travels farther in the second frame, but at the same speed, it follows that the time interval $\Delta t$ measured by the observer in the second frame is longer than the time interval $\Delta t_p$ measured by the observer in the first frame. To obtain a relationship between these two time intervals, it is convenient to use the right triangle geometric model shown in Figure 9.5c. The Pythagorean theorem applied to the triangle gives

$$\left(\frac{c\,\Delta t}{2}\right)^2 = \left(\frac{v\,\Delta t}{2}\right)^2 + d^2$$

Solving for $\Delta t$ gives

$$\Delta t = \frac{2d}{\sqrt{c^2 - v^2}} = \frac{2d}{c\sqrt{1 - \dfrac{v^2}{c^2}}} \qquad [9.5]$$

Because $\Delta t_p = 2d/c$, we can express Equation 9.5 as

$$\Delta t = \frac{\Delta t_p}{\sqrt{1 - \dfrac{v^2}{c^2}}} = \gamma \, \Delta t_p \qquad [9.6]$$

where $\gamma = (1 - v^2/c^2)^{-1/2}$. This result says that the **time interval $\Delta t$ measured by $O$ is longer than the time interval $\Delta t_p$ measured by $O'$** because $\gamma$ is always greater than unity. That is, $\Delta t > \Delta t_p$. This effect is known as **time dilation.**

We can see that time dilation is not observed in our everyday lives by considering the factor $\gamma$. This factor deviates significantly from a value of 1 only for very high speeds, as shown in Table 9.1. For example, for a speed of $0.1c$, the value of $\gamma$ is 1.005. Thus, a time dilation of only 0.5% occurs at one-tenth the speed of light. Speeds we encounter on an everyday basis are far slower than this, so we do not see time dilation in normal situations.

The time interval $\Delta t_p$ in Equation 9.6 is called the **proper time interval.** In general, the proper time interval is defined as the **time interval between two events as measured by an observer for whom the events occur at the same point in space.** In our case, observer $O'$ measures the proper time interval. For us to be able to use Equation 9.6, the events must occur at the same spatial position in *some* inertial frame. Thus, for example, this equation cannot be used for the lightning example described at the beginning of this section.

Time dilation is a verifiable phenomenon—let us look at an example of a situation in which the effects of time dilation can be observed and that served as an important historical confirmation of the predictions of relativity. Muons are unstable elementary particles that have a charge equal to that of the electron and a mass 207 times that of the electron. Muons can be produced as a result of collisions of cosmic radiation with atoms high in the atmosphere. Slow-moving muons in the laboratory have a lifetime measured to be the proper time interval $\Delta t_p = 2.2$ $\mu$s. If we assume that the speed of atmospheric muons is close to the speed of light, we find that these particles can travel a distance of approximately $(3.0 \times 10^8 \text{ m/s})(2.2 \times 10^{-6} \text{ s}) \approx 6 \times 10^2$ m before they decay (Fig. 9.6a). Hence, they are unlikely to reach the surface of the Earth from high in the atmosphere where they are produced; however, experiments show that a large number of muons *do* reach the surface. The phenomenon of time dilation explains this effect. As measured by an observer on Earth, the muons have a dilated lifetime equal to $\gamma \, \Delta t_p$. For example, for $v = 0.99c$, $\gamma \approx 7.1$ and $\gamma \, \Delta t_p \approx 16$ $\mu$s. Hence, the average distance traveled by the muons in this time as measured by an observer on Earth is approximately $(3.0 \times 10^8 \text{ m/s})(16 \times 10^{-6} \text{ s}) \approx 5 \times 10^3$ m, as shown in Figure 9.6b.

The results of an experiment reported by J. C. Hafele and R. E. Keating provided direct evidence of time dilation.* The experiment involved the use of very stable atomic clocks. Time intervals measured with four such clocks in jet flight

| TABLE 9.1 | |
| :--- | :--- |
| **Value for $\gamma$ at Various Speeds** | |
| $v/c$ | $\gamma$ |
| 0.0010 | 1.000 |
| 0.010 | 1.000 |
| 0.10 | 1.005 |
| 0.20 | 1.021 |
| 0.30 | 1.048 |
| 0.40 | 1.091 |
| 0.50 | 1.155 |
| 0.60 | 1.250 |
| 0.70 | 1.400 |
| 0.80 | 1.667 |
| 0.90 | 2.294 |
| 0.92 | 2.552 |
| 0.94 | 2.931 |
| 0.96 | 3.571 |
| 0.98 | 5.025 |
| 0.99 | 7.089 |
| 0.995 | 10.01 |
| 0.999 | 22.37 |

**PITFALL PREVENTION 9.3**

**The proper time interval**

Read this definition carefully. It is *very* important in relativistic calculations to correctly identify the observer who measures the proper time interval. **The proper time interval between two events is always the time interval measured by an observer for whom the two events take place at the same position.**

* J. C. Hafele and R. E. Keating, "Around the World Atomic Clocks: Relativistic Time Gains Observed," *Science*, July 14, 1972, p. 168.

(a)

(b)

(a) Without relativistic considerations, muons created in the atmosphere and traveling downward with a speed of 0.99$c$ would travel only about $6 \times 10^2$ m before decaying with an average lifetime of 2.2 $\mu$s. Thus, very few muons would reach the surface of the Earth. (b) With relativistic considerations, the muon's lifetime is dilated according to an observer on Earth. As a result, according to this observer, the muon can travel about $5 \times 10^3$ m before decaying. This results in many of them arriving at the surface.

were compared with time intervals measured by reference clocks located at the U.S. Naval Observatory. Their results were in good agreement with the predictions of special relativity and can be explained in terms of the relative motion between the Earth's rotation and the jet aircraft. In their paper, Hafele and Keating report the following: "Relative to the atomic time scale of the U.S. Naval Observatory, the flying clocks lost 59 ± 10 ns during the eastward trip and gained 273 ± 7 ns during the westward trip. . . ."

**Quick Quiz 9.2**

A crew watches a movie that is 2 h long in a spacecraft that is moving at high speed through space. Will an Earth-bound observer measure the duration of the movie to be (a) longer than, (b) shorter than, or (c) equal to 2 h?

**The Twins Paradox**

Because of time dilation, the time interval between ticks of a clock moving with respect to an observer is observed to be longer than the time interval between ticks of an identical clock at rest with respect to the observer—it is often said, *"A moving clock runs slower than a clock at rest by a factor γ."* This is true for ordinary mechanical clocks as well as for the light clock on the vehicle in Figure 9.5. In fact, we can generalize these results by stating that **all physical processes, including chemical and biological ones, are measured by an observer to slow down when those processes occur in a frame moving with respect to the observer.** For example, the heartbeat of an astronaut moving through space will be measured by the astronaut to be the same as when he was at rest on Earth. The astronaut will not have any sensation of life slowing down in the spacecraft. The astronaut's heartbeat, however, will appear to slow down when measured by a clock on Earth.

An intriguing biological consequence of time dilation is the so-called *twins paradox.* Consider an experiment involving a set of twins named Speedo and Goslo (Fig. 9.7) who are 20 years old. Speedo, the more adventuresome of the two, sets out on an epic journey to Planet X, located 20 lightyears (ly) from Earth. Furthermore, his spacecraft is capable of reaching a speed of 0.95$c$ relative to the inertial

(a)                                        (b)

(a) As Speedo leaves his twin brother Goslo on the Earth, both are the same age. (b) When Speedo returns from his journey to Planet X, he is younger than Goslo.

frame of his twin brother on Earth. After reaching Planet X, Speedo becomes homesick and immediately returns to Earth at the same high speed as on the outbound journey. On his return, Speedo is shocked to discover that many things have changed on Earth in his absence. To Speedo, the most significant change to have occurred is that his twin brother Goslo has aged 42 years and is now 62 years old. Speedo, on the other hand, has aged by only 13 years.

At this point, it is fair to raise the following questions: Which twin is the traveler, and which twin is really younger as a result of this experiment? From Goslo's frame of reference, he was at rest while his brother traveled at a high speed out and back. From Speedo's perspective, it is he who was at rest while Goslo and the Earth raced away on a 6.5-year journey and then headed back for another 6.5 years. This leads to an apparent contradiction. Each twin observed the other in motion and might claim that the other's clock runs slow. Which twin has *actually* aged more?

In reality, Speedo does return younger than Goslo. How can we resolve this apparent paradox? Despite our discussion in the previous paragraph, the situation is *not* symmetric. Speedo, the space traveler, must experience a series of accelerations during his journey in order to turn around and come back. As a result, his speed is not always uniform and, as a consequence, Speedo is not in a single inertial frame during the entire trip. He cannot be regarded as always being at rest while Goslo is in uniform motion, because he changes his state of motion at least once. Therefore no paradox exists—only Goslo remains in a single inertial frame and can correctly predict results based on special relativity.

The conclusion that Speedo does not remain in a single inertial frame is inescapable. The time required to accelerate and decelerate Speedo's spacecraft may be made very small using large rocket engines, so that Speedo can claim that he spends most of his time traveling to and from Planet X at $0.95c$ in an inertial frame. However, Speedo must slow down, reverse his motion, and return to Earth in an altogether different inertial frame. Only Goslo, who is in a single inertial frame, can apply the simple time dilation equation to Speedo's trip. Thus, Goslo finds that instead of aging 42 years, Speedo ages only $(1 - v^2/c^2)^{1/2}(42 \text{ years}) = 13 \text{ years}$.

## Quick Quiz 9.3

Suppose astronauts were paid according to the time spent traveling in space. After a long voyage traveling at a speed near that of light, a crew of astronauts return to Earth and open their pay envelopes. What will their reaction be?

### THINKING PHYSICS 9.1

Suppose a student explains time dilation with the following argument: If you start running away from a clock at 12:00 at a speed very close to the speed of light, you would not see the time change, because the light from the clock representing 12:01 would never reach you. What is the flaw in this argument?

**Reasoning**   The implication in this argument is that the velocity of light relative to the runner is approximately *zero*—"the light . . . would never reach you." This is a Galilean point of view, in which the relative velocity is a simple subtraction of running velocity from the light velocity. From the point of view of special relativity, one of the fundamental postulates is that the speed of light is the same for all observers, *including one running away from the light source at the speed of light*. Thus, the light from 12:01 will move toward the runner at the speed of light, as measured by all observers, including the runner.

## Example 9.1   What is the Period of the Pendulum?

The period $T$ of a pendulum on the Earth is measured to be 3.0 s in the rest frame of the Earth. What is the period when measured by an observer in a spacecraft moving at a speed of $0.95c$ relative to the pendulum?

**Reasoning**   We identify two events: two sequential arrivals of the pendulum at the same maximum deflection point. The proper time interval between these events is the period $T_p = 3.0$ s. The observer in the spacecraft will measure a dilated time interval.

**Solution**   For the period measured by the observer in the spacecraft, Equation 9.6 gives

$$T = \gamma T_p = \frac{1}{\sqrt{1 - \frac{(0.95c)^2}{c^2}}} T_p = (3.2)(3.0 \text{ s}) = \boxed{9.6 \text{ s}}$$

That is, measurements by this observer show that it takes longer to complete an oscillation compared with an observer on Earth.

**EXERCISE**   Muons move in a circular orbit of radius 500 m at a speed of $0.9994c$. If a muon at rest decays into other particles after 2.2 $\mu$s (the proper time interval), how many trips around the circular path do we expect the muons to make before they decay?

*Answer*   6.1 revolutions

## Length Contraction

•   *Proper length*

**PITFALL PREVENTION 9.4**

**The proper length**

As with the proper time interval, it is *very* important in relativistic calculations to correctly identify the observer who measures the proper length. **The proper length between two points in space is always the length measured by an observer at rest with respect to the length.** It is often the case that the proper time interval and the proper length are *not* measured by the same observer.

The measured distance between two points also depends on the frame of reference. The **proper length** of an object is defined as the **distance in space between the endpoints of the object measured by someone who is at rest relative to the object.** An observer in a reference frame that is moving with respect to the object will measure a length along the direction of the velocity that is always less than the proper length. This effect is known as **length contraction.** Although we have introduced this through the mental representation of an object, the object is not necessary. **The distance between *any* two points in space is measured by an observer to be contracted along the direction of the velocity of the observer relative to the points.**

Consider a spacecraft traveling with a speed $v$ from one star to another. We will consider the time interval between two events: (1) the leaving of the spacecraft from the first star and (2) the arrival of the spacecraft at the second star. There are two observers: one on Earth and the other in the spacecraft. The observer at rest on Earth (and also at rest with respect to the two stars) measures the distance between the stars to be $L_p$, the proper length. According to this observer, the time it takes the spacecraft to complete the voyage is $\Delta t = L_p/v$. What does an observer in the moving spacecraft measure for the distance between the stars? This observer measures the proper time interval, because the passage of each of the two stars by his spacecraft occurs at the same position in his reference frame—at his spacecraft. Therefore, because of time dilation, the time of travel measured by the space traveler will be smaller than that for the Earth-bound observer. Using the time dilation expression, the proper time interval between events is $\Delta t_p = \Delta t/\gamma$. The space traveler claims to be at rest and sees the destination star moving toward the spacecraft with speed $v$. Because the space traveler reaches the star in the time $\Delta t_p < \Delta t$, he concludes that the distance $L$ between the stars is shorter than $L_p$. This distance measured by the space traveler is

$$L = v \Delta t_p = v \frac{\Delta t}{\gamma}$$

Because $L_p = v\,\Delta t$, we see that

$$L = \frac{L_p}{\gamma} = L_p\left(1 - \frac{v^2}{c^2}\right)^{1/2}$$  [9.7]  • *Length contraction*

Because $(1 - v^2/c^2)^{1/2}$ is less than 1, the space traveler measures a length that is shorter than the proper length. Thus, an observer in motion with respect to two points in space measures the length $L$ between the points (along the direction of motion) to be shorter than the length $L_p$ measured by an observer at rest with respect to the points (the proper length).

Note that **length contraction takes place only along the direction of motion.** For example, suppose a stick moves past an Earth observer with speed $v$, as in Figure 9.8. The length of the stick as measured by an observer in a frame attached to the stick is the proper length $L_p$, as in Figure 9.8a. The length $L$ of the stick measured by the Earth observer is shorter than $L_p$ by the factor $(1 - v^2/c^2)^{1/2}$, but the width is the same. Furthermore, length contraction is a symmetric effect: If the stick is at rest on Earth, an observer in the moving frame would also measure its length to be shorter by the same factor $(1 - v^2/c^2)^{1/2}$.

It is important to emphasize that the proper length and proper time interval are defined differently. The proper length is measured by an observer at rest with respect to the endpoints of the length. The proper time interval is measured by someone for whom the events occur at the same position. As an example of this point, let us return to the decaying muons moving at speeds close to the speed of light. An observer in the muon's reference frame would measure the proper lifetime, and an Earth-based observer would measure the proper length (the distance from creation to decay in Figure 9.6). In the muon's reference frame, no time dilation occurs, but the distance of travel is observed to be shorter when measured in this frame. Likewise, in the Earth observer's reference frame, a time dilation does occur, but the distance of travel is measured to be the proper length. Thus, when calculations on the muon are performed in both frames, the outcome of the experiment in one frame is the same as the outcome in the other frame: More muons reach the surface than would be predicted without relativistic calculations.

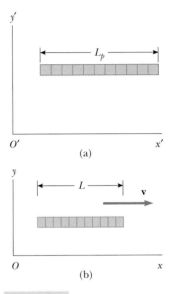

**Figure 9.8**

(a) According to an observer in a frame attached to the stick (that is, both the stick and the frame have the same velocity), the stick is measured to have its proper length $L_p$. (b) According to an observer in a frame in which the stick has a velocity **v** relative to the frame, the stick is measured to be *shorter* than the proper length $L_p$ by a factor $(1 - v^2/c^2)^{1/2}$.

### Quick Quiz 9.4

If you were somehow able to travel at the speed of light, how big would the Universe be according to your measurements?

### Quick Quiz 9.5

You are packing for a trip to another star, to which you will be traveling at $0.99c$. Should you buy smaller sizes of your clothing because you will be skinnier on your trip due to length contraction? Can you sleep in a smaller cabin than usual because you will be shorter when you lie down?

### Quick Quiz 9.6

You are observing a spacecraft moving away from you. You notice that it is measured to be shorter than when it was at rest on the ground next to you, and, through the spacecraft window, you can see a clock. You observe that the passage of time on the clock is measured to be slower than that of the watch on your wrist. What if the spacecraft turns around and comes toward you? Will it appear to be *longer* and will the clock in the spacecraft run *faster*?

## Example 9.2 A Voyage to Sirius

An astronaut takes a trip to Sirius, located 8.00 lightyears from Earth. A lightyear is the distance that light travels in one year: 1 ly = $(3.00 \times 10^8 \text{ m/s})(3.156 \times 10^7 \text{ s}) = 9.46 \times 10^{15}$ m. The astronaut measures the time of the one-way journey to be 6.00 yr. If the spacecraft moves at a constant speed of $0.800c$, how can the 8.00-lightyear distance be reconciled with the 6.00-yr duration measured by the astronaut?

**Reasoning** The 8.00 lightyears represents the proper length (the distance from Earth to Sirius) measured by an observer seeing both Earth and Sirius at rest. The astronaut sees Sirius approaching her at $0.800c$ but also sees the distance contracted to

$$L = \frac{L_p}{\gamma} = \frac{8.00 \text{ ly}}{\gamma} = (8.00 \text{ ly}) \sqrt{1 - \frac{v^2}{c^2}}$$

$$= (8.00 \text{ ly}) \sqrt{1 - \frac{(0.800c)^2}{c^2}} = 4.80 \text{ ly}$$

So the travel time measured on her clock is

$$\Delta t = \frac{L}{v} = \frac{4.80 \text{ ly}}{0.800c} = \frac{4.80 \text{ ly}}{0.800(1.00 \text{ ly/yr})} = 6.00 \text{ yr}$$

Notice that we have used the speed of light as $c = 1.00$ ly/yr to determine this last result.

## Example 9.3 The Triangular Spacecraft

A spacecraft in the form of a triangle flies by an observer on Earth with a speed of $0.95c$ along the $x$ direction. According to an observer on the spacecraft (Fig. 9.9a), the distances $L_p$ and $y$ are measured to be 52 m and 25 m, respectively. What are the dimensions of the spacecraft as measured by the Earth observer when the spacecraft is in motion along the direction shown in Figure 9.9b?

**Solution** In Figure 9.9a, we show the shape of the spacecraft as measured* by the observer on the spacecraft. The

proper length along the direction of motion is $L_p = 52$ m. The Earth observer watching the moving spacecraft measures the horizontal length of the spacecraft to be contracted to

$$L = L_p \sqrt{1 - \frac{v^2}{c^2}} = (52 \text{ m}) \sqrt{1 - \frac{(0.95c)^2}{c^2}} = \boxed{16 \text{ m}}$$

The 25-m vertical height is unchanged because it is perpendicular to the direction of relative motion between observer and spacecraft. Figure 9.9b represents the shape of the spacecraft as measured by the Earth observer.

**EXERCISE** A spacecraft moves at $0.900c$. If its length is $L_0$ when measured from inside the spacecraft, what is its length as measured by a ground observer?

*Answer* $0.436 L_0$

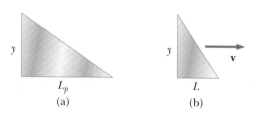

(a)　　　　　　(b)

**Figure 9.9**

(Example 9.3) (a) When the spacecraft is at rest, its shape is measured as shown. (b) The spacecraft is measured to have this shape when it moves to the right with a speed $v$. Note that only its $x$ dimension is contracted in this case.

---

* Notice that we are careful here to say the shape "as measured" by an observer, rather than "as seen" by an observer. What an observer *sees* when he looks at an object is the set of light rays entering the eye at a given instant. These rays left different parts of the object at different times because different parts of the object are at different distances from the eye. Fig. 9.9b is a representation of the light rays leaving different parts of the object *simultaneously*. Viewing an object moving at high speed introduces additional changes to the object besides length contraction, including apparent rotations.

## 9.5 • THE LORENTZ TRANSFORMATION EQUATIONS

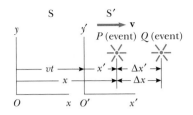

**Figure 9.10**

Events occur at points *P* and *Q* and are observed by an observer at rest in the S frame and another in the S′ frame, which is moving to the right with a speed *v*.

Suppose an event that occurs at some point *P* is reported by two observers: one at rest in a frame S and another in a frame S′ that is moving to the right with speed *v*, as in Figure 9.10. The observer in S reports the event with space–time coordinates $(x, y, z, t)$, and the observer in S′ reports the same event using the coordinates $(x', y', z', t')$. If two events occur at *P* and *Q*, Equation 9.1 predicts that $\Delta x = \Delta x'$; that is, the distance between the two points in space at which the events occur does not depend on the motion of the observer. Because this is contradictory to the notion of length contraction, the Galilean transformation is not valid when *v* approaches the speed of light. In this section, we state the correct transformation equations that apply for all speeds in the range $0 \leq v < c$.

The equations that relate these measurements and enable us to transform coordinates from S to S′ are the **Lorentz transformation equations:**

$$x' = \gamma(x - vt)$$

$$y' = y$$

$$z' = z \qquad \text{[9.8]}$$

$$t' = \gamma\left(t - \frac{v}{c^2}x\right)$$

• *Lorentz transformation for S → S′*

These transformation equations were developed by Hendrik A. Lorentz (1853–1928) in 1890 in connection with electromagnetism. Einstein, however, recognized their physical significance and took the bold step of interpreting them within the framework of special relativity.

We see that the value for $t'$ assigned to an event by observer $O'$ depends both on the time *t* and on the coordinate *x* as measured by observer *O*. Thus, in relativity, space and time are not separate concepts but rather are closely interwoven with each other in what we call **space–time.** This is unlike the case of the Galilean transformation in which $t = t'$.

If we wish to transform coordinates in the S′ frame to coordinates in the S frame, we simply replace *v* by $-v$ and interchange the primed and unprimed coordinates in Equation 9.8:

$$x = \gamma(x' + vt')$$

$$y = y'$$

$$z = z' \qquad \text{[9.9]}$$

$$t = \gamma\left(t' + \frac{v}{c^2}x'\right)$$

• *Inverse Lorentz transformation for S′ → S*

When $v \ll c$, the Lorentz transformation reduces to the Galilean transformation. To check this, note that if $v \ll c$, $v^2/c^2 \ll 1$, so that $\gamma$ approaches 1 and Equation 9.8 reduces in this limit to Equations 9.1 and 9.2:

$$x' = x - vt \qquad y' = y \qquad z' = z \qquad t' = t$$

### Lorentz Velocity Transformation

Let us now derive the Lorentz velocity transformation, which is the relativistic counterpart of the Galilean velocity transformation, Equation 9.3. Once again S is our first frame of reference and S′ is our second frame of reference that moves at a speed $v$ relative to S along the common $x$ and $x'$ axes. Suppose that an object is measured in S′ to have an instantaneous velocity component $u'_x$ given by

$$u'_x = \frac{dx'}{dt'} \qquad [9.10]$$

Using Equations 9.8, we have

$$dx' = \gamma(dx - v\,dt) \qquad \text{and} \qquad dt' = \gamma\left(dt - \frac{v}{c^2}\,dx\right)$$

Substituting these values into Equation 9.10 gives

$$u'_x = \frac{dx'}{dt'} = \frac{dx - v\,dt}{dt - \dfrac{v}{c^2}\,dx} = \frac{\dfrac{dx}{dt} - v}{1 - \dfrac{v}{c^2}\dfrac{dx}{dt}}$$

But $dx/dt$ is the velocity component $u_x$ of the object measured in S, and so this expression becomes

- *Lorentz velocity transformation for S → S′*

$$u'_x = \frac{u_x - v}{1 - \dfrac{u_x v}{c^2}} \qquad [9.11]$$

Similarly, if the object has velocity components along $y$ and $z$, the components in S′ are

$$u'_y = \frac{u_y}{\gamma\left(1 - \dfrac{u_x v}{c^2}\right)} \qquad \text{and} \qquad u'_z = \frac{u_z}{\gamma\left(1 - \dfrac{u_x v}{c^2}\right)} \qquad [9.12]$$

When $u_x$ or $v$ is much smaller than $c$ (the nonrelativistic case), the denominator of Equation 9.11 approaches unity and so $u'_x \approx u_x - v$. This corresponds to the Galilean velocity transformations. In the other extreme, when $u_x = c$, Equation 9.11 becomes

$$u'_x = \frac{c - v}{1 - \dfrac{cv}{c^2}} = \frac{c\left(1 - \dfrac{v}{c}\right)}{1 - \dfrac{v}{c}} = c$$

From this result, we see that an object whose speed approaches $c$ relative to an observer in S also has a speed approaching $c$ relative to an observer in S′—independent of the relative motion of S and S′. Note that this conclusion is consistent with Einstein's second postulate, namely, that the speed of light must be $c$ relative to all inertial frames of reference.

To obtain $u_x$ in terms of $u'_x$, we replace $v$ by $-v$ in Equation 9.11 and interchange the roles of primed and unprimed variables:

- *Inverse Lorentz velocity transformation for S′ → S*

$$u_x = \frac{u'_x + v}{1 + \dfrac{u'_x v}{c^2}} \qquad [9.13]$$

**PITFALL PREVENTION 9.5**

**What can the observers agree on?**

We have seen several measurements on which the two observers $O$ and $O'$ do not agree. These include (1) the time interval between events that take place in the same position in one of the frames, (2) the distance between two points that remain fixed in one of their frames, (3) the velocity components of a moving particle, and (4) whether two events occurring at different locations in both frames are simultaneous. It is worth noting here what the two observers *can* agree on: (1) the relative speed $v$ with which they move with respect to each other, (2) the speed $c$ of any ray of light, and (3) the simultaneity of two events taking place at the same position and time in some frame.

## Example 9.4   Relative Velocity of Spacecraft

Two spacecraft A and B are moving directly toward each other, as in Figure 9.11. An observer on Earth measures the speed of A to be $0.750c$ and the speed of B to be $0.850c$. Find the speed of B with respect to A.

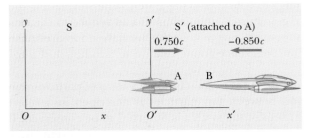

**Solution**   This problem can be solved by taking the S′ frame as being attached to A, so that $v = 0.750c$ relative to the Earth observer (in the S frame). Spacecraft B can be considered as an object moving with a velocity component $u_x = -0.850c$ relative to the Earth observer. Hence, the speed of B with respect to A is the speed of B as measured by the observer O′ on A, and can be obtained using Equation 9.11:

$$u_x' = \frac{u_x - v}{1 - \dfrac{u_x v}{c^2}} = \frac{-0.850c - 0.750c}{1 - \dfrac{(-0.850c)(0.750c)}{c^2}} = \boxed{-0.980c}$$

The negative sign in the result indicates that spacecraft B is moving in the negative $x$ direction as observed by A.

### Figure 9.11

(Example 9.4) Two spacecraft A and B move in opposite directions.

## Example 9.5   Relativistic Leaders of the Pack

Two motorcycle pack leaders named David and Emily are racing at relativistic speeds along perpendicular paths as in Figure 9.12. How fast does Emily recede over David's right shoulder as seen by David?

**Solution**   Figure 9.12 represents the situation as seen by a police officer at rest in frame S, who observes the following:

David:  $u_x = 0.75c$     $u_y = 0$

Emily:  $u_x = 0$     $u_y = -0.90c$

To determine Emily's speed of recession as seen by David, we take S′ to move along with David and we calculate $u_x'$ and $u_y'$ for Emily using Equations 9.11 and 9.12:

$$u_x' = \frac{u_x - v}{1 - \dfrac{u_x v}{c^2}} = \frac{0 - 0.75c}{1 - \dfrac{(0)(0.75c)}{c^2}} = -0.75c$$

$$u_y' = \frac{u_y}{\gamma\left(1 - \dfrac{u_x v}{c^2}\right)} = \frac{\sqrt{1 - \dfrac{(0.75c)^2}{c^2}}\,(-0.90c)}{\left(1 - \dfrac{(0)(0.75c)}{c^2}\right)} = -0.60c$$

Thus, the speed of Emily as observed by David is

$$u' = \sqrt{(u_x')^2 + (u_y')^2} = \sqrt{(-0.75c)^2 + (-0.60c)^2} = \boxed{0.96c}$$

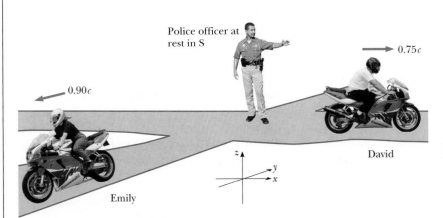

Police officer at rest in S

0.75c

0.90c

David

Emily

### Figure 9.12

(Example 9.5) David moves to the east with a speed $0.75c$ relative to the police officer, and Emily travels south at a speed $0.90c$ relative to the officer.

## 9.6 • RELATIVISTIC MOMENTUM AND THE RELATIVISTIC FORM OF NEWTON'S LAWS

We have seen that to properly describe the motion of particles within the framework of special relativity, the Galilean transformation must be replaced by the Lorentz transformation. Because the laws of physics must remain unchanged under the Lorentz transformation, we must generalize Newton's laws and the definitions of momentum and energy to conform to the Lorentz transformation and the principle of relativity. These generalized definitions should reduce to the classical (nonrelativistic) definitions for $v \ll c$ or $u \ll c$. (As we have done previously, we will use $v$ for the speed of one reference frame relative to another, and $u$ for the speed of a particle.)

First, recall that the total momentum of an isolated system of particles remains constant. Suppose a collision between two particles is described in a reference frame S in which the momentum of the system is observed to be conserved. If the velocities in a second reference frame S′ are calculated using the Lorentz transformation and the Newtonian definition of momentum, $\mathbf{p} = m\mathbf{u}$, it is found that the momentum of the system is *not* observed to be conserved in the second reference frame. This violates one of Einstein's postulates: The laws of physics are the same in all inertial frames. In view of this, and assuming the Lorentz transformation to be correct, we must modify the definition of momentum.

The relativistic equation for the momentum of a particle of mass $m$ that maintains the principle of conservation of momentum is

- *Definition of relativistic momentum*

$$\mathbf{p} \equiv \frac{m\mathbf{u}}{\sqrt{1 - \dfrac{u^2}{c^2}}}$$   [9.14]

where $\mathbf{u}$ is the velocity of the particle. When $u$ is much less than $c$, the denominator of Equation 9.14 approaches unity, so that $\mathbf{p}$ approaches $m\mathbf{u}$. Therefore, the relativistic equation for $\mathbf{p}$ reduces to the classical expression when $u$ is small compared with $c$. Equation 9.14 is often written in simpler form as

$$\mathbf{p} = \gamma m\mathbf{u}$$   [9.15]

using our previously defined expression for $\gamma$.*

The relativistic force $\mathbf{F}$ on a particle whose momentum is $\mathbf{p}$ is defined as

$$\mathbf{F} \equiv \frac{d\mathbf{p}}{dt}$$   [9.16]

where $\mathbf{p}$ is given by Equation 9.14. This expression preserves both classical mechanics in the limit of low velocities and conservation of momentum for an isolated system ($\Sigma\mathbf{F}_{ext} = 0$) both relativistically and classically.

We leave it to Problem 53 at the end of the chapter to show that the acceleration $\mathbf{a}$ of a particle decreases under the action of a constant force, in which case $a \propto (1 - u^2/c^2)^{3/2}$. From this proportionality, note that as the particle's speed approaches $c$, the acceleration caused by any finite force approaches zero. It is therefore impossible to accelerate a particle from rest to a speed $v \geq c$.

---

**PITFALL PREVENTION 9.6**

**Watch out for "relativistic mass"**

Some older treatments of relativity maintained the conservation of momentum principle at high speeds by using a model in which the mass of a particle increases with speed. You might still encounter this notion of "relativistic mass" in your outside reading, especially in older books. Be aware that this notion is no longer widely accepted, and mass is considered as *invariant*, independent of speed. The mass of an object in all frames is considered to be the mass as measured by an observer at rest with respect to the object.

---

\* We defined $\gamma$ previously in terms of the speed $v$ of one frame relative to another frame. The same symbol is also used for $(1 - u^2/c^2)^{-1/2}$ where $u$ is the speed of a particle.

This means that $c$ is an upper limit for the speed of any particle. In fact, it is possible to show that **no matter, energy, or information can travel through space faster than $c$.** Note that the relative speeds of the two spacecraft in Example 9.4 and the two motorcyclists in Example 9.5 were both less than $c$. If we had attempted to solve these examples with Galilean transformations, we would have obtained relative speeds larger than $c$ in both cases.

### Example 9.6  Momentum of an Electron

An electron, which has a mass of $9.11 \times 10^{-31}$ kg, moves with a speed of $0.750c$. Find its relativistic momentum and compare this with the momentum calculated from the classical expression.

**Solution**  Using Equation 9.14 with $u = 0.750c$, we have

$$p = \frac{m_e u}{\sqrt{1 - \dfrac{u^2}{c^2}}}$$

$$p = \frac{(9.11 \times 10^{-31}\ \text{kg})(0.750 \times 3.00 \times 10^8\ \text{m/s})}{\sqrt{1 - \dfrac{(0.750c)^2}{c^2}}}$$

$$= 3.10 \times 10^{-22}\ \text{kg} \cdot \text{m/s}$$

The classical expression, if used (inappropriately) for this high-speed particle, gives

$$p = m_e u = 2.05 \times 10^{-22}\ \text{kg} \cdot \text{m/s}$$

Hence, the correct relativistic result is more than 50% larger than the classical result!

## 9.7 • RELATIVISTIC ENERGY

We have seen that the definition of momentum requires generalization to make it compatible with the principle of relativity. We find that the definition of kinetic energy must also be modified.

To derive the relativistic form of the work–kinetic energy theorem, let us start with the definition of the work done by a force of magnitude $F$ on a particle initially at rest. Recall from Chapter 6 that the work–kinetic energy theorem states that the work done by a net force acting on a particle equals the change in kinetic energy of the particle. Because the initial kinetic energy is zero, we conclude that the work $W$ done in accelerating a particle from rest is equivalent to the relativistic kinetic energy $K$ of the particle:

$$W = \Delta K = K - 0 = K = \int_{x_1}^{x_2} F\, dx = \int_{x_1}^{x_2} \frac{dp}{dt}\, dx \qquad \textbf{[9.17]}$$

where we are considering the special case of force and displacement vectors along the $x$ axis for simplicity. To perform this integration and find the relativistic kinetic energy as a function of $u$, we first evaluate $dp/dt$, using Equation 9.14:

$$\frac{dp}{dt} = \frac{d}{dt} \frac{mu}{\sqrt{1 - \dfrac{u^2}{c^2}}} = \frac{m(du/dt)}{\left(1 - \dfrac{u^2}{c^2}\right)^{3/2}}$$

Substituting this expression for $dp/dt$ and $dx = u\,dt$ into Equation 9.17 gives

$$K = \int_0^t \frac{m(du/dt)\,u\,dt}{\left(1 - \dfrac{u^2}{c^2}\right)^{3/2}} = m \int_0^u \frac{u}{\left(1 - \dfrac{u^2}{c^2}\right)^{3/2}}\,du$$

Evaluating the integral, we find that

● *Relativistic kinetic energy*

$$K = \frac{mc^2}{\sqrt{1 - \dfrac{u^2}{c^2}}} - mc^2 = \gamma mc^2 - mc^2 = (\gamma - 1)\,mc^2 \qquad \textbf{[9.18]}$$

At low speeds, where $u/c \ll 1$, Equation 9.18 should reduce to the classical expression $K = \frac{1}{2}mu^2$. We can check this by using the binomial expansion $(1 - x^2)^{-1/2} \approx 1 + \frac{1}{2}x^2 + \cdots$ for $x \ll 1$, where the higher order powers of $x$ are ignored in the expansion because they are so small. In our case, $x = u/c$, so that

$$\gamma = \frac{1}{\sqrt{1 - \dfrac{u^2}{c^2}}} = \left(1 - \frac{u^2}{c^2}\right)^{-1/2} \approx 1 + \tfrac{1}{2}\frac{u^2}{c^2} + \cdots$$

Substituting this into Equation 9.19 gives

$$K \approx \left(1 + \tfrac{1}{2}\frac{u^2}{c^2} + \cdots\right) mc^2 - mc^2 = \tfrac{1}{2}mu^2$$

which agrees with the classical result. Figure 9.13 shows a comparison of the speed–kinetic energy relationships for a particle using the nonrelativistic expression for $K$ (the blue curve) and the relativistic expression for $K$ (the brown curve). The curves are in good agreement at low speeds, but deviate at higher speeds. The nonrelativistic expression indicates a violation of special relativity because it suggests that sufficient energy can be added to the particle to accelerate it to a speed larger than $c$. In the relativistic case, the particle speed never exceeds $c$, regardless of the kinetic energy, which is consistent with experimental results. When an object's speed is less than one-tenth the speed of light, the classical kinetic energy equation differs by less than 1% from the relativistic equation (which is experi-

**Figure 9.13**

A graph comparing relativistic and nonrelativistic kinetic energy. The energies are plotted as a function of speed. In the relativistic case, $u$ is always less than $c$.

mentally verified at all speeds). Thus, for practical calculations it is valid to use the classical equation when the object's speed is less than $0.1c$.

The constant term $mc^2$ in Equation 9.18, which is independent of the speed, is called the **rest energy** $E_R$ of the particle:

$$E_R = mc^2 \qquad \text{[9.19]}$$

• *Rest energy*

The term $\gamma mc^2$ in Equation 9.18 depends on the particle speed and is the sum of the kinetic and rest energies. We define $\gamma mc^2$ to be the **total energy** $E$, that is, total energy = kinetic energy + rest energy

$$E = \gamma mc^2 = K + mc^2 = K + E_R \qquad \text{[9.20]}$$

or, when $\gamma$ is replaced by its equivalent,

$$E = \frac{mc^2}{\sqrt{1 - \dfrac{u^2}{c^2}}} \qquad \text{[9.21]}$$

• *Definition of total energy*

The relation $E_R = mc^2$ shows that **mass is a manifestation of energy.** The relation shows that a small mass corresponds to an enormous amount of energy. This concept is fundamental to much of the field of nuclear physics.

In many situations, the momentum or energy of a particle is measured rather than its speed. It is therefore useful to have an expression relating the total energy $E$ to the relativistic momentum $p$. This is accomplished by using the expressions $E = \gamma mc^2$ and $p = \gamma mu$. By squaring these equations and subtracting, we can eliminate $u$ (see Problem 35). The result, after some algebra, is

$$E^2 = p^2c^2 + (mc^2)^2 \qquad \text{[9.22]}$$

• *Energy–momentum relationship*

When the particle is at rest, $p = 0$, and so $E = E_R = mc^2$. That is, the total energy equals the rest energy.

For the case of particles that have zero mass, such as photons (massless, chargeless particles of light to be discussed further in Chapter 28), we set $m = 0$ in Equation 9.22, and we see that

$$E = pc \qquad \text{[9.23]}$$

• *Energy–momentum relationship for a photon*

This equation is an exact expression relating energy and momentum for photons, which always travel at the speed of light.

When dealing with subatomic particles, it is convenient to express their energy in a unit called an electron volt (eV). The equality between electron volts and our standard energy unit is

$$1 \text{ eV} = 1.60 \times 10^{-19} \text{ J}$$

For example, the mass of an electron is $9.11 \times 10^{-31}$ kg. Hence, the rest energy of the electron is

$$E_R = m_e c^2 = (9.11 \times 10^{-31} \text{ kg})(3.00 \times 10^8 \text{ m/s})^2 = 8.20 \times 10^{-14} \text{ J}$$

Converting this to eV, we have

$$E_R = m_e c^2 = (8.20 \times 10^{-14} \text{ J}) \left( \frac{1 \text{ eV}}{1.60 \times 10^{-19} \text{ J}} \right) = 0.511 \text{ MeV}$$

**PITFALL PREVENTION 9.7**

**The electron volt**

The electron volt is one of a small number of units in physics whose names are misleading. We will introduce the volt as a unit of *electric potential* in Chapter 20, but the electron volt is *not* a unit of electric potential. It is a unit of *energy*.

### Quick Quiz 9.7

The following *pairs* of energies represent the rest energy and total energy of three different particles: particle 1: $E$, $2E$; particle 2: $E$, $3E$; particle 3: $2E$, $4E$. Rank the particles, from greatest to least, according to their (a) mass; (b) kinetic energy; (c) speed.

### Example 9.7 The Energy of a Speedy Proton

Let us consider the relativistic motion of a proton. (a) Find the proton's rest energy in electron volts. (b) The total energy of a proton is three times its rest energy. With what speed is the proton moving? (c) Determine the kinetic energy of the proton in part (b) in electron volts. (d) What is the magnitude of the proton's momentum in part (b)?

**Solution** (a) To find the rest energy, we use Equation 9.19,

$$E_R = m_p c^2 = (1.67 \times 10^{-27} \text{ kg})(3.00 \times 10^8 \text{ m/s})^2$$

$$= (1.50 \times 10^{-10} \text{ J}) \left( \frac{1.00 \text{ eV}}{1.60 \times 10^{-19} \text{ J}} \right)$$

$$= \boxed{938 \text{ MeV}}$$

(b) Because the total energy $E$ is three times the rest energy, $E = \gamma m c^2$ (Eq. 9.20) gives

$$E = 3 m_p c^2 = \frac{m_p c^2}{\sqrt{1 - \dfrac{u^2}{c^2}}}$$

$$3 = \frac{1}{\sqrt{1 - \dfrac{u^2}{c^2}}}$$

Solving for $u$ gives

$$\left( 1 - \frac{u^2}{c^2} \right) = \frac{1}{9} \quad \text{or} \quad \frac{u^2}{c^2} = \frac{8}{9}$$

$$u = \frac{\sqrt{8}}{3} c = \boxed{2.83 \times 10^8 \text{ m/s}}$$

(c) We use Equation 9.20,

$$K = E - m_p c^2 = 3 m_p c^2 - m_p c^2 = 2 m_p c^2$$

Because $m_p c^2 = 938$ MeV, $K = \boxed{1.88 \text{ GeV}}$

(d) We can use Equation 9.22 to calculate the momentum with $E = 3 m_p c^2$:

$$E^2 = p^2 c^2 + (m_p c^2)^2 = (3 m_p c^2)^2$$

$$p^2 c^2 = 9(m_p c^2)^2 - (m_p c^2)^2 = 8(m_p c^2)^2$$

$$p = \sqrt{8} \frac{m_p c^2}{c} = \sqrt{8} \frac{(938 \text{ MeV})}{c} = \boxed{2.65 \frac{\text{GeV}}{c}}$$

The unit of momentum is written MeV/$c$, which is a momentum unit often used in particle studies.

**EXERCISE** Find the speed of a particle whose total energy is twice its rest energy.

*Answer* 0.866$c$

**EXERCISE** An electron moves with a speed $u = 0.850c$. Find its (a) total energy and (b) kinetic energy in electron volts.

*Answer* (a) 0.970 MeV (b) 0.459 MeV

## 9.8 • MASS AND ENERGY

Equation 9.20, $E = \gamma m c^2$, which represents the total energy of a particle, suggests that even when a particle is at rest ($\gamma = 1$), it still possesses enormous energy through its mass. The clearest experimental proof of the equivalence of mass and energy occurs in nuclear and elementary particle interactions in which the conversion of mass into kinetic energy takes place. Because of this, in relativistic situations, we cannot use the principle of conservation of energy as outlined in Chapter 7. We must include rest energy as another form of energy storage.

This concept is important in atomic and nuclear processes, in which the change in mass during the process is on the order of the initial mass. For example, in a conventional nuclear reactor, the uranium nucleus undergoes *fission,* a reaction that results in several lighter fragments having considerable kinetic energy. In the case of a $^{235}$U atom, which is used as fuel in nuclear power plants, the fragments are two lighter nuclei and a few neutrons. The total mass of the fragments is less than that of the $^{235}$U by an amount $\Delta m$. The corresponding energy $\Delta mc^2$ associated with this mass difference is exactly equal to the total kinetic energy of the fragments. The kinetic energy is absorbed as the fragments move through water, raising the internal energy of the water. This internal energy is used to produce steam for the generation of electric power.

Next, consider a basic *fusion* reaction in which two deuterium atoms combine to form one helium atom. The decrease in mass that results from the creation of one helium atom from two deuterium atoms is $\Delta m = 4.25 \times 10^{-29}$ kg. Hence, the corresponding energy that results from one fusion reaction is $\Delta mc^2 = 3.83 \times 10^{-12}$ J $= 23.9$ MeV. To appreciate the magnitude of this result, if 1 g of deuterium is converted to helium, the energy released is on the order of $10^{12}$ J! At the year 2001 cost of electric energy, this would be worth about \$30 000. We will see more details of these nuclear processes in Chapter 30.

---

### Example 9.8 Mass Change in a Radioactive Decay

The $^{216}$Po nucleus is unstable and exhibits radioactivity, which we will study in detail in Chapter 30. It decays to $^{212}$Pb by emitting an alpha particle, which is a helium nucleus, $^4$He. Find (a) the mass change in this decay and (b) the energy that this represents.

**Solution** Using values in Table A.3, we see that the initial and final masses are

$$m_i = m(^{216}\text{Po}) = 216.001\ 889\ \text{u}$$

$$m_f = m(^{212}\text{Pb}) + m(^4\text{He}) = 211.991\ 872\ \text{u} + 4.002\ 602\ \text{u}$$

$$= 215.994\ 474\ \text{u}$$

Thus, the mass change is,

$$\Delta m = 216.001\ 889\ \text{u} - 215.994\ 474\ \text{u} = 0.007\ 415\ \text{u}$$

$$\boxed{= 1.23 \times 10^{-29}\ \text{kg}}$$

(b) The energy associated with this mass change is,

$$E = \Delta mc^2 = (1.23 \times 10^{-29}\ \text{kg})(3.00 \times 10^8\ \text{m/s})^2$$

$$= 1.11 \times 10^{-12}\ \text{J} = \boxed{6.92\ \text{MeV}}$$

This energy appears as the kinetic energy of the alpha particle and the $^{212}$Pb nucleus after the decay.

---

## 9.9 • GENERAL RELATIVITY

Up to this point, we have sidestepped a curious puzzle. Mass has two seemingly different properties: It determines a force of mutual gravitational attraction between two objects (Newton's law of universal gravitation) and also represents the resistance of a single object to being accelerated (Newton's second law), regardless of the type of force producing the acceleration. How can one quantity have two such different properties? An answer to this question, which puzzled Newton and many other physicists over the years, was provided when Einstein published his theory of gravitation, known as *general relativity,* in 1916. Because it is a mathematically complex theory, we merely offer a hint of its elegance and insight.

In Einstein's view, the dual behavior of mass was evidence for a very intimate and basic connection between the two behaviors. He pointed out that no mechanical experiment (e.g., dropping a mass) could distinguish between the two situations illustrated in Figures 9.14a and 9.14b. In Figure 9.14a, a person is standing in an elevator on the surface of a planet, and feels pressed into the floor, due to the

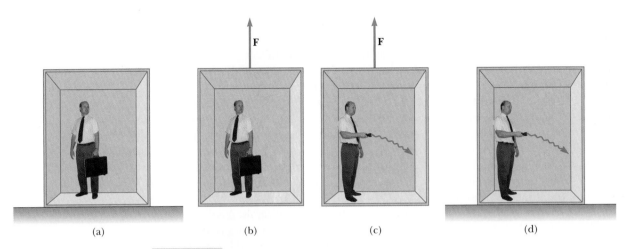

(a)    (b)    (c)    (d)

**Figure 9.14**

(a) The observer is at rest in a uniform gravitational field **g**, directed downward. (b) The observer is in a region where gravity is negligible, but the frame is accelerated by an external force **F** that produces an acceleration **g** directed upward. According to Einstein, the frames of reference in parts (a) and (b) are equivalent in every way. No local experiment can distinguish any difference between the two frames. (c) In the accelerating frame, a ray of light would appear to bend downward due to the acceleration. (d) If parts (a) and (b) are truly equivalent, as Einstein proposed, then part (c) suggests that a ray of light would bend downward in a gravitational field.

gravitational force. In Figure 9.14b, the person is in an elevator in empty space accelerating upward with $a = g$. The person feels pressed into the floor with the same force as in Figure 9.14a. In each case, a mass released by the observer undergoes a downward acceleration of magnitude $g$ relative to the floor. In Figure 9.14a, the person is in an inertial frame in a gravitational field. In Figure 9.14b, the person is in a noninertial frame accelerating in gravity-free space. Einstein's claim is that these two situations are completely equivalent.

Einstein carried this idea further and proposed that *no* experiment, mechanical or otherwise, could distinguish between the two cases. This extension to include all phenomena (not just mechanical ones) has interesting consequences. For example, suppose that a light pulse is sent horizontally across the box, as in Figure 9.14c, in which the elevator is accelerating upward in empty space. From the point of view of an observer in an inertial frame outside the elevator, the light travels in a straight line while the floor of the elevator accelerates upward. According to the observer on the elevator, however, the trajectory of the light pulse bends downward as the floor of the elevator (and the observer) accelerates upward. Therefore, based on the equality of parts (a) and (b) of the figure for all phenomena, Einstein proposed that a **beam of light should also be bent downward by a gravitational field,** as in Figure 9.14d.

The two postulates of Einstein's **general relativity** are as follows:

• *Postulates of general relativity*

- All the laws of nature have the same form for observers in any frame of reference, whether accelerated or not.
- In the vicinity of any given point, a gravitational field is equivalent to an accelerated frame of reference in the absence of gravitational effects. (This is the **principle of equivalence.**)

One interesting effect predicted by general relativity is that the passage of time is altered by gravity. A clock in the presence of gravity runs more slowly than one in which gravity is negligible. Consequently, the frequencies of radiation emitted by atoms in the presence of a strong gravitational field are shifted to lower frequencies compared with the same emissions in a weak field. This gravitational shift has been detected in light emitted by atoms in massive stars. It has also been verified on the Earth by comparing the frequencies of gamma rays (a high-energy form of electromagnetic radiation) emitted from nuclei separated vertically by about 20 m.

The second postulate suggests that a gravitational field may be "transformed away" at any point if we choose an appropriate accelerated frame of reference—a freely falling one. Einstein developed an ingenious method of describing the acceleration necessary to make the gravitational field "disappear." He specified a certain quantity, the *curvature of space–time,* that describes the gravitational effect of a mass. In fact, the curvature of space–time completely replaces Newton's gravitational theory. According to Einstein, there is no such thing as a gravitational force. Rather, the presence of a mass causes a curvature of space–time in the vicinity of the mass, and this curvature dictates the space–time path that all freely moving objects must follow.

One important test of general relativity is the prediction that a light ray passing near the Sun should be deflected by some angle. This prediction was confirmed by astronomers as bending of starlight during a total solar eclipse shortly following World War I (Fig. 9.15).

As an example of the effects of curved space–time, imagine two travelers moving on parallel paths a few meters apart on the surface of the Earth and maintaining an exact northward heading along two longitude lines. As they observe each other near the equator, they will claim that their paths are exactly parallel. As they approach the North Pole, however, they may notice that they are moving closer together, and they will actually meet at the North Pole. Thus, they will claim that they moved along parallel paths, but moved toward each other, *as if there were an attractive force between them.* They will make this conclusion based on their everyday experience of moving on flat surfaces. From our mental representation, however, we realize that they are walking on a curved surface, and the geometry of the curved surface causes them to converge, rather than an attractive force. In a similar way, general relativity replaces the notion of forces with the movement of objects through curved space–time.

This Global Positioning System (GPS) unit incorporates relativistically corrected time calculations in its analysis of signals it receives from orbiting satellites. These corrections allow the unit to determine its position on the Earth's surface to within a few meters. If the corrections were not made, the location error would be about 1 km. *(Courtesy of Trimble Navigation Limited)*

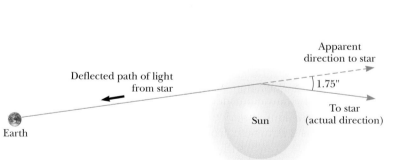

**Figure 9.15**

Deflection of starlight passing near the Sun. Because of this effect, the Sun and other remote objects can act as a *gravitational lens*. In his general theory of relativity, Einstein calculated that starlight just grazing the Sun's surface should be deflected by an angle of 1.75″.

If the concentration of mass becomes very great, as is believed to occur when a large star exhausts its nuclear fuel and collapses to a very small volume, a **black hole** may form. Here the curvature of space−time is so extreme that, within a certain distance from the center of the black hole, all matter and light become trapped. We will say more about black holes in Chapter 11.

### THINKING PHYSICS 9.2

Atomic clocks are extremely accurate; in fact, an error of 1 s in 3 million years is typical. This error can be described as about 1 part in $10^{14}$. On the other hand, the atomic clock in Boulder, Colorado, near Denver, is often 15 ns faster than the one in Washington after only one day. This is an error of about 1 part in $6 \times 10^{12}$, which is about 17 times larger than the previously expressed error. If atomic clocks are so accurate, why does a clock in Boulder not remain in synchronization with one in Washington? (*Hint:* Denver is known as the Mile High City.)

**Reasoning** According to the general theory of relativity, the passage of time depends on gravity—time runs more slowly in strong gravitational fields. Washington is at an elevation very close to sea level, but Boulder is about a mile higher in altitude. This difference results in a weaker gravitational field at Boulder than at Washington. As a result, time runs more rapidly in Boulder than in Washington.

### context connection
### 9.10 • FROM MARS TO THE STARS

In this chapter, we have discussed the strange effects of traveling at high speeds. Do we need to consider these effects in our planned mission to Mars?

To answer this question, let us look back at our Context Connection for Chapter 7. In that discussion, we calculated an upper limit for the speed of our spacecraft to Mars, based on a calculation of the escape speed from the Sun. The result was $u = 4.21 \times 10^4$ m/s. Let us evaluate $\gamma$ for this speed:

$$\gamma = \frac{1}{\sqrt{1 - \dfrac{u^2}{c^2}}} = \frac{1}{\sqrt{1 - \dfrac{(4.21 \times 10^4 \text{ m/s})^2}{(3.00 \times 10^8 \text{ m/s})^2}}} = 1.00000001$$

where we have completely ignored the rules of significant figures, so that we could find the first nonzero digit to the right of the decimal place!

It is clear from this result that relativistic considerations are not important for our trip to Mars. But what about deeper travels into space—suppose we wish to travel to another star? This distance is several orders of magnitude larger. The nearest star is about 4.2 lightyears from Earth. In comparison, Mars is $4.0 \times 10^{-5}$ lightyears at its farthest from Earth. Thus, we are talking about a distance to the nearest star that is five orders of magnitude larger than the distance to Mars. Very long travel times will be needed to reach even the nearest star. At the escape speed from the Sun, for example, assuming that this speed is maintained during the entire trip, the travel time is 30 000 years to the nearest star. This is clearly prohibitive, especially if we would like the people who leave Earth to be the same people who arrive at the star!

We can use the principles of relativity to reduce this travel time by traveling at very high speeds. Suppose our spacecraft travels at a constant speed of $0.99c$. Then, we can use length contraction to calculate the distance from the Earth to the nearest star as measured by the spacecraft occupants:

$$L = \frac{L_p}{\gamma} = L_p \sqrt{1 - \frac{v^2}{c^2}} = (4.2 \text{ ly}) \sqrt{1 - \frac{(0.99c)^2}{c^2}} = 0.59 \text{ ly}$$

The time required to reach the star can now be calculated:

$$\Delta t = \frac{\Delta x}{v} = \frac{0.59 \text{ ly}}{0.99(1.0 \text{ ly/yr})} = 0.60 \text{ yr}$$

This is clearly a reduction in time from the low-speed trip!

There are two major problems with this scenario, however. The first is the technological challenge of designing and building a spacecraft and rocket engine assembly that can attain a speed of $0.99c$. Beyond that is the design of a safety system that will provide early warnings about running into asteroids, meteoroids, or other bits of matter while traveling at almost light speed through space. Even a small piece of rock could be disastrous if struck at $0.99c$. The second problem is related to the twins paradox discussed earlier in this chapter. During the trip to the star, 4.3 years will pass on Earth. If the travelers return to the Earth, another 4.3 years will pass. Thus, the travelers will have aged by only $2(0.6 \text{ y}) = 1.2$ years, but 8.6 years will have passed on Earth. For stars farther away than the nearest star, these effects could result in the personnel assisting with the liftoff from Earth no longer being alive when the travelers return. In conclusion, we see that travel to the stars will be an enormous challenge!

**WEB**

For a discussion of "warp drive" and other inventions of science fiction by a NASA scientist, visit **ssdoo.gsfc.nasa.gov/education/just_for_fun/startrek.html**

## SUMMARY

The two basic postulates of **special relativity** are

- All the laws of physics are the same in all inertial reference frames.
- The speed of light in vacuum has the same value, in all inertial frames, regardless of the velocity of the observer or the velocity of the source emitting the light.

Three consequences of special relativity are

- Events that are simultaneous for one observer may not be simultaneous for another observer who is in motion relative to the first.
- Clocks in motion relative to an observer are measured to be slowed down by a factor $\gamma$. This is known as **time dilation.**
- Lengths of objects in motion are measured to be shorter in the direction of motion. This is known as **length contraction.**

To satisfy the postulates of special relativity, the Galilean transformations must be replaced by the **Lorentz transformations:**

$$x' = \gamma(x - vt)$$

$$y' = y \qquad \text{[9.8]}$$

$$z' = z$$

$$t' = \gamma\left(t - \frac{v}{c^2}x\right)$$

where $\gamma = (1 - v^2/c^2)^{-1/2}$.

The relativistic form of the **velocity transformation** is

$$u_x' = \frac{u_x - v}{1 - \frac{u_x v}{c^2}} \qquad \text{[9.11]}$$

where $u_x$ is the speed of an object as measured in the S frame and $u_x'$ is its speed measured in the S' frame.

The relativistic expression for the **momentum** of a particle moving with a velocity **u** is

$$\mathbf{p} \equiv \frac{m\mathbf{u}}{\sqrt{1 - \frac{u^2}{c^2}}} = \gamma m\mathbf{u} \qquad \text{[9.14]}$$

The relativistic expression for the **kinetic energy** of a particle is

$$K = \gamma mc^2 - mc^2 = (\gamma - 1)mc^2 \qquad \text{[9.18]}$$

where $E_R = mc^2$ is the **rest energy** of the particle.

The **total energy** $E$ of a particle is related to its mass through the expression:

$$E = \gamma mc^2 = \frac{mc^2}{\sqrt{1 - \dfrac{u^2}{c^2}}} \qquad [9.21]$$

The relativistic momentum is related to the total energy through the equation

$$E^2 = p^2c^2 + (mc^2)^2 \qquad [9.22]$$

The **general theory of relativity** claims that no experiment can distinguish between a gravitational field and an accelerating reference frame. It correctly predicts that the path of light is affected by a gravitational field.

## QUESTIONS

1. What two speed measurements do two observers in relative motion *always* agree on?

2. The speed of light in water is $2.3 \times 10^8$ m/s. Suppose an electron is moving through water at $2.5 \times 10^8$ m/s. Does this violate the principles of relativity?

3. Two identical clocks are synchronized. One is put in orbit directed eastward around the Earth, and the other remains on Earth. Which clock runs slower? When the moving clock returns to Earth, do the two clocks still read the same time?

4. A train is approaching you as you stand next to the tracks. Just as an observer on the train passes you, you both begin to play the same Beethoven symphony on portable compact disc players. (a) According to you, whose CD player finishes the symphony first? (b) According to the observer on the train, whose CD player finishes the symphony first? (c) Whose CD player *really* finishes the symphony first?

5. A spacecraft in the shape of a sphere moves past an observer on Earth with a speed $0.5c$. What shape does the observer measure for the spacecraft as it moves past?

6. Explain why it is necessary, when defining length, to specify that the positions of the ends of a rod are to be measured simultaneously.

7. Does saying that a moving clock runs slower than a stationary one imply that something is physically unusual about the moving clock?

8. Give a physical argument that shows that it is impossible to accelerate an object of mass $m$ to the speed of light, even with a continuous force acting on it.

9. List some ways our everyday lives would change if the speed of light were only 50 m/s.

10. It is said that Einstein, in his teenage years, asked the question, "What would I see in a mirror if I carried it in my hands and ran at the speed of light?" How would you answer this question?

11. Some distant astronomical objects, called *quasars*, are receding from us at half the speed of light (or greater). What is the speed of the light we receive from these quasars?

12. How is it possible that photons of light, which have zero mass, have momentum?

13. Relativistic quantities must make a smooth transition to their Newtonian counterparts as the speed of a system becomes small relative to the speed of light. Explain.

14. If the speed of a particle is doubled, what effect does it have on its momentum? (Assume that its initial speed is less than $c/2$.)

15. The upper limit on the speed of an electron is the speed of light $c$. Does that mean that the momentum of the electron has an upper limit?

## PROBLEMS

1, 2, 3 = straightforward, intermediate, challenging   □ = full solution available in the *Student Solutions Manual and Study Guide*

web = solution posted at **http://www.harcourtcollege.com/physics/**   💻 = computer useful in solving problem

🔲 = Interactive Physics   ▪ = paired numerical/symbolic problems   ◨ = life science application

### Section 9.1   The Principle of Newtonian Relativity

1. A 2 000-kg car moving at 20.0 m/s collides and locks together with a 1 500-kg car at rest at a stop sign. Show that momentum is conserved in a reference frame moving at 10.0 m/s in the direction of the moving car.

2. A ball is thrown at 20.0 m/s inside a boxcar moving along the tracks at 40.0 m/s. What is the speed of the ball relative to the ground if the ball is thrown (a) forward? (b) backward? (c) out the side door?

3. In a laboratory frame of reference, an observer notes that Newton's second law is valid. Show that it is also valid for an observer moving at a constant speed, small compared with the speed of light, relative to the laboratory frame.

4. Show that Newton's second law is *not* valid in a reference frame moving past the laboratory frame of Problem 3 with a constant acceleration.

### Section 9.4  Consequences of Special Relativity

5. How fast must a meter stick be moving relative to an observer if the observer measures its length as 0.500 m?

6. At what speed does a clock move if it is measured to run at a rate that is one-half the rate of a clock at rest with respect to an observer?

7. An astronaut is traveling in a space vehicle that has a speed of $0.500c$ relative to the Earth. The astronaut measures his pulse rate at 75.0 beats per minute. Signals generated by the astronaut's pulse are radioed to Earth when the vehicle is moving in a direction perpendicular to the line that connects the vehicle with an observer on the Earth. (a) What pulse rate does the Earth observer measure? (b) What would be the pulse rate if the speed of the space vehicle were increased to $0.990c$?

8. An astronomer on Earth observes a meteoroid in the southern sky approaching the Earth at a speed of $0.800c$. At the time of its discovery the meteoroid is 20.0 lightyears from the Earth. Calculate (a) the time required for the meteoroid to reach the Earth as measured by the Earthbound astronomer, (b) this time as measured by a tourist on the meteoroid, and (c) the distance to the Earth as measured by the tourist.

9. An atomic clock moves at 1 000 km/h for 1.00 h as measured by an identical clock fixed to the Earth. How many nanoseconds slow will the moving atomic clock be compared with the Earth clock at the end of the 1-h interval?

10. A muon formed high in the Earth's atmosphere travels at speed $v = 0.990c$ for a distance of 4.60 km before it decays. (a) How long does the muon live, as measured in its reference frame? (b) How far does the muon travel, as measured in its frame?

11. A spacecraft with a proper length of 300 m takes 0.750 $\mu$s to pass an Earth observer. Determine the speed of the spacecraft as measured by the Earth observer.

12. (a) An object of proper length $L_p$ takes time $\Delta t$ to pass an Earth observer. Determine the speed of the object as measured by the Earth observer. (b) A column of tanks, 300 m long, takes 75.0 s to pass a child waiting at a street corner on her way to school. Determine the speed of the armored vehicles. (c) Show that the answer to part (a) includes the answer to Problem 11 as a special case and includes the answer to part (b) of this problem as another special case.

13. For what value of $v$ does $\gamma = 1.010\,0$? Observe that for speeds less than this value, time dilation and length contraction are effects amounting to less than 1.00%.

14. A friend passes by you in a spacecraft traveling at a high speed. He tells you that his spacecraft is 20.0 m long and that the identically constructed spacecraft you are sitting in is 19.0 m long. According to your observations, (a) how long is your spacecraft, (b) how long is your friend's spacecraft, and (c) what is the speed of your friend's spacecraft?

15. An interstellar space probe is launched from Earth. After a brief period of acceleration, it moves with a constant velocity, with a magnitude of 70.0% of the speed of light. Its nuclear-powered batteries supply the energy to keep its data transmitter active continuously. The batteries have a lifetime of 15.0 yr as measured in a rest frame. (a) How long do the batteries on the space probe last as measured by Mission Control on Earth? (b) How far is the probe from Earth when its batteries fail, as measured by Mission Control? (c) How far is the probe from Earth when its batteries fail, as measured by its built-in trip odometer? (d) For what total time after launch is data received from the probe by Mission Control? Note that radio waves travel at the speed of light and fill the space between the probe and Earth at the time of battery failure.

### Section 9.5  The Lorentz Transformation Equations

16. A moving rod is observed to have a length of 2.00 m and to be oriented at an angle of 30.0° with respect to the direction of motion, as in Figure P9.16. The rod has a speed of $0.995c$. (a) What is the proper length of the rod? (b) What is the orientation angle in the proper frame?

2.00 m

30.0°

Direction of motion

**Figure P9.16**

17. A spacecraft travels at $0.750c$ relative to Earth. If the spacecraft fires a small rocket in the forward direction, how fast (relative to the spacecraft) must it be fired for it to travel at $0.950c$ relative to Earth?

18. A Klingon space ship moves away from the Earth at a speed of $0.800c$ (Fig. P9.18). The starship *Enterprise* pursues at a speed of $0.900c$ relative to the Earth. Observers on Earth see the *Enterprise* overtaking the Klingon ship at a relative speed of $0.100c$. With what speed is the *Enterprise* overtaking the Klingon ship as seen by the crew of the *Enterprise*?

**Figure P9.18**

19. Two jets of material from the center of a radio galaxy are
    web ejected in opposite directions. Both jets move at $0.750c$ relative to the galaxy. Determine the speed of one jet relative to the other.

20. An observer in reference frame S sees two events as simultaneous. Event *A* occurs at the point $(50.0$ m, 0, 0) at the instant 9:00:00 Universal time, 15 January 2001. Event *B* occurs at the point $(150$ m, 0, 0) at the same moment. A second observer, moving past with a velocity of $0.800$ $c$ **i**, also observes the two events. In her reference frame S′, which event occurred first and what time elapsed between the events?

21. A red light flashes at position $x_R = 3.00$ m and time $t_R = 1.00 \times 10^{-9}$ s, and a blue light flashes at $x_B = 5.00$ m and $t_B = 9.00 \times 10^{-9}$ s (all values measured in the S reference frame.) Reference frame S′ has its origin at the same point as S at $t = t' = 0$; frame S′ moves uniformly to the right. Both flashes are observed to occur at the same place in S′. (a) Find the relative speed between S and S′. (b) Find the location of the two flashes in frame S′. (c) At what time does the red flash occur in the S′ frame?

22. A spacecraft is launched from the surface of the Earth with a velocity of magnitude $0.600c$ at an angle of $50.0°$ above the horizontal positive $x$-axis. Another spacecraft is moving past, with a velocity of magnitude $0.700c$ in the negative $x$ direction. Determine the magnitude and direction of the velocity of the first spacecraft as measured by the pilot of the second spacecraft.

### Section 9.6 Relativistic Momentum and the Relativistic Form of Newton's Laws

23. Calculate the momentum of an electron moving with a speed of (a) $0.010\ 0c$, (b) $0.500c$, (c) $0.900c$.

24. The nonrelativistic expression for the magnitude of the momentum of a particle, $p = mu$, can be used if $u \ll c$. For what speed does the use of this equation give an error in the momentum of (a) $1.00\%$ and (b) $10.0\%$?

25. A golf ball travels with a speed of $90.0$ m/s. By what fraction does its relativistic momentum magnitude $p$ differ from its classical value $mu$? That is, find the ratio $(p - mu)/mu$.

26. Show that the speed of an object having momentum of magnitude $p$ and mass $m$ is

$$u = \frac{c}{\sqrt{1 + (mc/p)^2}}$$

27. An unstable particle at rest breaks into two fragments of *un-*
    web *equal* mass. The mass of one fragment is $2.50 \times 10^{-28}$ kg, and that of the other is $1.67 \times 10^{-27}$ kg. If the lighter fragment has a speed of $0.893c$ after the breakup, what is the speed of the heavier fragment?

### Section 9.7 Relativistic Energy

28. Determine the energy required to accelerate an electron from (a) $0.500c$ to $0.900c$ and (b) $0.900c$ to $0.990c$.

29. Find the momentum of a proton in MeV/$c$ units if its total energy is twice its rest energy.

30. Show that, for any object moving at less than one-tenth the speed of light, the relativistic kinetic energy agrees with the result of the classical expression $K = \frac{1}{2}mu^2$ to within less than $1\%$. Thus, for most purposes, the classical equation is good enough to describe these objects, whose motion we call *nonrelativistic*.

31. A proton moves at $0.950c$. Calculate its (a) rest energy,
    web (b) total energy, and (c) kinetic energy.

32. An electron has a kinetic energy five times greater than its rest energy. Find (a) its total energy and (b) its speed.

33. A cube of steel has a volume of $1.00$ cm$^3$ and a mass of $8.00$ g when at rest on the Earth. If this cube is now given a speed $u = 0.900c$, what is its density as measured by a stationary observer? Note that relativistic density is defined as $E_R/c^2V$.

34. An unstable particle with a mass of $3.34 \times 10^{-27}$ kg is initially at rest. The particle decays into two fragments that fly off along the $x$ axis with velocity components of $0.987c$ and $-0.868c$. Find the masses of the fragments. (*Hint:* Conserve both mass–energy and momentum.)

35. Show that the energy–momentum relationship $E^2 = p^2c^2 + (mc^2)^2$ follows from the expressions $E = \gamma mc^2$ and $p = \gamma mu$.

36. Find the kinetic energy of a 78.0-kg spacecraft launched out of the Solar System with speed $106$ km/s by using (a) the classical equation $K = \frac{1}{2}mu^2$ and (b) the relativistic equation.

37. A pion at rest ($m_\pi = 270\ m_e$) decays to a muon ($m_\mu = 207\ m_e$) and an antineutrino ($m_{\bar{\nu}} \approx 0$). The reaction is written $\pi^- \rightarrow \mu^- + \bar{\nu}$. Find the kinetic energy of the muon and the energy of the antineutrino in electron volts. (*Hint:* Relativistic momentum is conserved.)

### Section 9.8 Mass and Energy

38. When $1.00$ g of hydrogen combines with $8.00$ g of oxygen, $9.00$ g of water is formed. During this chemical reaction,

$2.86 \times 10^5$ J of energy is released. How much mass do the constituents of this reaction lose? Is the loss of mass likely to be detectable?

**39.** The power output of the Sun is $3.77 \times 10^{26}$ W. How much mass is converted to energy in the Sun each second?

**40.** In a nuclear power plant, the fuel rods last 3.00 yr before they are replaced. If a plant with rated thermal power 1.00 GW operates at 80.0% capacity for the 3.00 yr, what is the loss of mass of the fuel?

**41.** A gamma ray (a high-energy photon) can produce an electron ($e^-$) and a positron ($e^+$) when it enters the electric field of a heavy nucleus: $\gamma \rightarrow e^+ + e^-$. What minimum gamma-ray energy is required to accomplish this task? (*Hint:* The masses of the electron and the positron are equal.)

**Section 9.10    Context Connection—From Mars to the Stars**

**42. Review Problem.** In 1962, when Mercury astronaut Scott Carpenter orbited the Earth 22 times, the press stated that for each orbit he aged 2 millionths of a second less than he would have had he remained on Earth. (a) Assuming that he was 160 km above the Earth in a circular orbit, determine the time difference between someone on Earth and the orbiting astronaut for the 22 orbits. You will need to use the approximation $\sqrt{1-x} \approx 1 - x/2$, for small $x$. (b) Did the press report accurate information? Explain.

**43.** An astronaut wishes to visit the Andromeda galaxy, making a one-way trip that will take 30.0 yr in the spacecraft's frame of reference. Assume that the galaxy is 2.00 million lightyears away and that his speed is constant. (a) How fast must he travel relative to the Earth? (b) What will be the kinetic energy of his 1 000-metric-ton spacecraft? (c) What is the cost of this energy if it is purchased at a typical consumer price for electric energy: 13.0¢/kWh?

**Additional Problems**

**44.** An electron has a speed of $0.750c$. Find the speed of a proton that has (a) the same kinetic energy as the electron; (b) the same momentum as the electron.

**45.** The cosmic rays of highest energy are protons that have **web** kinetic energy on the order of $10^{13}$ MeV. (a) How long would it take a proton of this energy to travel across the Milky Way galaxy, having a diameter $\sim 10^5$ lightyears, as measured in the proton's frame? (b) From the point of view of the proton, how many kilometers across is the galaxy?

**46.** The net nuclear fusion reaction inside the Sun can be written as $4\,^1H \rightarrow {}^4He + \Delta E$. The rest energy of each hydrogen atom is 938.78 MeV and the rest energy of the

helium-4 atom is 3 728.4 MeV. What is the percentage of the starting mass that is transformed to other forms of energy?

**47.** A rechargeable AA battery with a mass of 25.0 g can supply a power of 1.20 W for 50.0 min. (a) What is the difference in mass between a charged and an uncharged battery? (b) What fraction of the total mass is this mass difference?

**48.** An alien spacecraft traveling at $0.600c$ toward the Earth launches a landing craft with an advance guard of purchasing agents. The lander travels in the same direction with a speed of $0.800c$ relative to the spacecraft. As observed on the Earth, the spacecraft is 0.200 ly from Earth when the lander is launched. (a) With what speed is the lander observed to be approaching by observers on the Earth? (b) What is the distance to the Earth at the time of lander launch, as observed by the aliens? (c) How long does it take the lander to reach the Earth as observed by the aliens on the mother ship? (d) If the lander has a mass of $4.00 \times 10^5$ kg, what is its kinetic energy as observed in the Earth reference frame?

**49.** A physics professor on Earth gives an exam to her students, who are on a spacecraft traveling at speed $v$ relative to Earth. The moment the spacecraft passes the professor, she signals the start of the exam. She wishes her students to have time $T_0$ (spacecraft time) to complete the exam. Show that she should wait a time (Earth time) of

$$T = T_0 \sqrt{\frac{1 - v/c}{1 + v/c}}$$

before sending a light signal telling them to stop. (*Hint:* Remember that it takes some time for the second light signal to travel from the professor to the students.)

**50.** Energy reaches the upper atmosphere of the Earth from the Sun at the rate of $1.79 \times 10^{17}$ W. If all of this energy were absorbed by the Earth and not re-emitted, how much would the mass of the Earth increase in 1.00 yr?

**51.** A supertrain (proper length 100 m) travels at a speed of $0.950c$ as it passes through a tunnel (proper length 50.0 m). As seen by a trackside observer, is the train ever completely within the tunnel? If so, with how much space to spare?

**52.** Imagine that the entire Sun collapses to a sphere of radius $R_g$ such that the work required to remove a small mass $m$ from the surface would be equal to its rest energy $mc^2$. This radius is called the *gravitational radius* for the Sun. Find $R_g$. (It is believed that the ultimate fate of very massive stars is to collapse beyond their gravitational radii into black holes.)

**53.** A particle with electric charge $q$ moves along a straight line in a uniform electric field $\mathbf{E}$ with a speed of $u$. The electric force exerted on the charge is $q\mathbf{E}$. If the motion and the electric field are both in the $x$ direction, (a) show that the acceleration of the charge $q$ in the $x$ direction is given by

$$a = \frac{du}{dt} = \frac{qE}{m}\left(1 - \frac{u^2}{c^2}\right)^{3/2}$$

(b) Discuss the significance of the dependence of the acceleration on the speed. (c) If the particle starts from rest at $x = 0$ at $t = 0$, how would you proceed to find the speed of the particle and its position at time $t$?

54. An observer in a spacecraft moves toward a mirror at speed $v$ relative to the reference frame labeled by S in Figure P9.54. The mirror is stationary with respect to S. A light pulse emitted by the spacecraft travels toward the mirror and is reflected back to the spacecraft. The front of the spacecraft is a distance $d$ from the mirror (as measured by observers in S) at the moment the light pulse leaves the spacecraft. What is the total travel time of the pulse as measured by observers in (a) the S frame and (b) the front of the spacecraft?

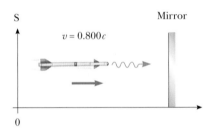

**Figure P9.54**

55. The creation and study of new elementary particles is an important part of contemporary physics. Especially interesting is the discovery of a very massive particle. To create a particle of mass $M$ requires an energy $Mc^2$. With enough energy, an exotic particle can be created by allowing a fast-moving particle of ordinary matter, such as a proton, to collide with a similar target particle. Let us consider a perfectly inelastic collision between two protons: an incident proton with mass $m$, kinetic energy $K$, and momentum magnitude $p$ joins with an originally stationary target proton to form a single product particle of mass $M$. You might think that the creation of a new product particle, nine times more massive than in a previous experiment, would require just nine times more energy for the incident proton. Unfortunately not all of the kinetic energy of the incoming proton is available to create the product particle, because conservation of momentum requires that after the collision the system as a whole still must have some kinetic energy. Only a fraction of the energy of the incident particle is thus available to create a new particle. You will determine how the energy available for particle creation

depends on the energy of the moving proton. Show that the energy available to create a product particle is given by

$$Mc^2 = 2mc^2 \sqrt{1 + \frac{K}{2mc^2}}$$

From this result, when the kinetic energy $K$ of the incident proton is large relative to its rest energy $mc^2$, we see that $M$ approaches $(2mK)^{1/2}/c$. Thus if the energy of the incoming proton is increased by a factor of 9, the mass you can create increases only by a factor of 3. This disappointing result is the main reason that most modern accelerators, such as those at CERN (in Europe), at Fermilab (near Chicago), at SLAC (in California), and at DESY (in Germany), use *colliding beams*. Here the total momentum of a pair of interacting particles can be zero. The center of mass can be at rest after the collision, so in principle all of the initial kinetic energy can be used for particle creation, according to

$$Mc^2 = 2mc^2 + K = 2mc^2\left(1 + \frac{K}{2mc^2}\right)$$

where $K$ is the total kinetic energy of two identical colliding particles. Here if $K \gg mc^2$, we have $M$ directly proportional to $K$, as we would desire. These machines are difficult to build and to operate, but they open new vistas in physics.

56. Ted and Mary are playing a game of catch in frame S′, which is moving at $0.600c$, while Jim in frame S watches the action (Fig. P9.56). Ted throws the ball to Mary at $0.800c$ (according to Ted) and their separation (measured in S′) is $1.80 \times 10^{12}$ m. (a) According to Mary, how fast is the ball moving? (b) According to Mary, how long does it take the ball to reach her? (c) According to Jim, how far apart are Ted and Mary, and how fast is the ball moving? (d) According to Jim, how long does it take the ball to reach Mary?

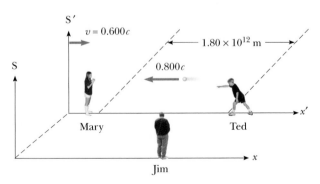

**Figure P9.56**

57. Suppose our Sun is about to explode. In an effort to escape, we depart in a spacecraft at $v = 0.800c$ and head to-

ward the star Tau Ceti, 12.0 ly away. When we reach the midpoint of our journey from the Earth, we see our Sun explode and, unfortunately, at the same instant we see Tau Ceti explode as well. (a) In the spacecraft's frame of reference, should we conclude that the two explosions occurred simultaneously? If not, which occurred first? (b) In a frame of reference in which the Sun and Tau Ceti are at rest, did they explode simultaneously? If not, which exploded first?

58. Prepare a graph of the relativistic kinetic energy and the classical kinetic energy, both as a function of speed, for an object with a mass of your choice. At what speed does the classical kinetic energy underestimate the experimental value by 1%? By 5%? By 50%?

## ANSWERS TO QUICK QUIZZES

**9.1.** This scenario is not possible with light. Water waves move through a medium—the surface of the water—and therefore have an absolute speed relative to the reference frame in which this medium is at rest. Thus, the relative speed of the waves and your boat is determined by Galilean relativity. This makes it possible for you to move faster than the waves. Light waves, which do not require a medium, are described by the principles of special relativity. As you detect the light wave ahead of you and moving away from you (which would be a pretty good trick—think about it!), its speed relative to you is $c$. No matter how fast you accelerate in that direction, trying to catch up to it, its speed relative to you is still $c$. Thus, you will not be able to catch up to the light wave.

**9.2** (a). The two events are the beginning and the end of the movie, both of which take place at rest with respect to the spacecraft crew. Thus, the crew measures the proper time interval of 2 h. Any observer in motion with respect to the spacecraft, which includes the observer on Earth, will measure a longer time interval due to time dilation.

**9.3** Assuming that their on-duty time was kept on Earth, they will be pleasantly surprised with a large paycheck. Less time will have passed for the astronauts in their frame of reference than for their employer back on Earth.

**9.4** According to Equation 9.7, the length of the Universe would contract to zero size. It would maintain the original size in the directions perpendicular to your velocity.

**9.5** The answers to both of these questions is *no*. Both your body and your sleeping cabin are at rest in your reference frame; thus, they will have their proper length according to you. There will be no change in measured lengths of objects within your spacecraft. Another observer, on a spacecraft traveling at a high speed relative to yours, will measure you as thinner (if your body is oriented in a direction *perpendicular* to your velocity vector relative to the other observer) or will claim that you are able to fit into a shorter sleeping cabin (if your body is oriented in a direction *parallel* to your velocity vector relative to the other observer).

**9.6** No. Time dilation and length contraction depend only on the relative speed of one observer relative to another, not on whether the observers are receding or approaching each other.

**9.7** (a) $m_3 > m_2 = m_1$. The rest energy of particle 3 is $2E$, and it is $E$ for particles 1 and 2. (b) $K_3 = K_2 > K_1$; The kinetic energy is the difference between the total energy and the rest energy. The kinetic energy is $4E - 2E = 2E$ for particle 3, $3E - E = 2E$ for particle 2, and $2E - E = E$ for particle 1. (c) $u_2 > u_3 = u_1$; From Equation 9.21, $E = \gamma E_R$. Solving this for the square of the particle speed $u$, we find, $u^2 = c^2(1 - (E_R/E)^2)$. Thus, the particle with the smallest ratio of rest energy to total energy will have the largest speed. Particles 1 and 3 have the same ratio as each other, and the ratio of particle 2 is smaller.

## chapter 10

**Mark Ruiz undergoes a rotation during a dive at the U. S. Olympic trials in June, 2000. He spins at a higher rate when he curls up and grabs his ankles.**

*(Otto Greule/Allsport)*

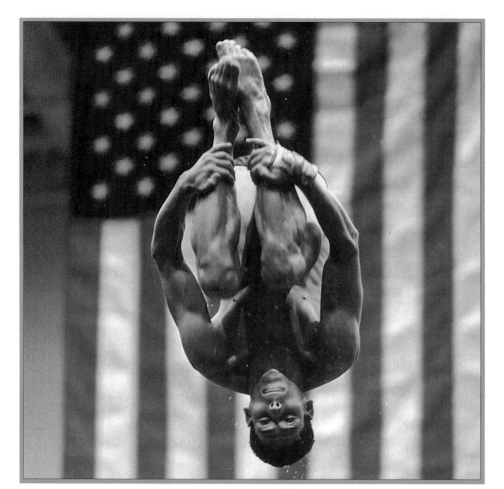

# Rotational Motion

I n previous chapters, we have discussed the motion of particles and systems and have developed a number of analysis models to help us solve problems involving various types of situations. So far, however, we have only considered *translational* motion—motion through space in which the position of the center of mass of the object changes. We do not yet know how to describe the motion of a *rotating* object, such as a compact disc in a stereo system. We will address this type of motion in this chapter.

We will investigate the dynamics of two types of rotational situations. The first type is that of a particle, or an object that can be modeled as a particle, moving along a circular path. We have seen this situation before, beginning in Chapter 3, with the particle in uniform circular motion. The second type of situation involves an extended object (*not* modeled as a particle) rotating about a fixed axis, which we will see for the first time in this chapter. The Earth exhibits both of these types of motion. In its orbit around the Sun, it can be modeled as a particle in a circular path, as we discussed in Chapter 1. In addition, it behaves as an extended object in rotation as it spins about its own axis. Notice that we clearly cannot use the particle model in this second situation. We will develop a new simplification model for objects that we will call the *rigid body model*. The results we derive here will enable us to understand the rotational motions of a diverse range of objects in our environment, from an electron orbiting a nucleus to clusters of galaxies orbiting a common center.

# 10.1 • ANGULAR SPEED AND ANGULAR ACCELERATION

We began our study of translational motion in Chapter 2 by defining the terms *position, velocity*, and *acceleration*. For example, we locate a particle in one-dimensional space with the position variable *x*. In this chapter, we will insert the word *translational* before our previously studied kinematic variables to distinguish them from the analogous rotational variables that we will develop.

Let us think about a rotating object—how would we describe its position in its rotational motion? We do that by describing its *orientation* relative to some fixed reference direction. For example, imagine two soldiers performing a military about-face maneuver. They both begin by facing due north. One soldier, who has been practicing diligently, ends up after the maneuver with his body facing due south. The second, who has not been practicing, ends up facing in a southeast direction. We could describe their respective rotational positions after the maneuver by reporting the angle through which each turned from the original direction. The first soldier turned through 180°, but the second through only 135°. Thus, we can use an angle measured from a reference direction as a measure of **rotational position,** or **angular position,** which is our starting point for our description of rotational motion.

We will now follow up on this mental representation of using an angle to describe rotational position with a more mathematical representation. Let us begin by considering a planar object rotating about a fixed axis that is perpendicular to the object and goes through the point *O* (Fig. 10.1). A particle on the object, indicated by the black dot, is at a fixed distance *r* from the origin and rotates about *O* in a circle of radius *r*. In fact, **every particle on the object undergoes circular motion about *O*.** As a result, there is a strong link between rotational motion of an object and the motion of a particle in a circular path. A rotating object is made up of particles, all undergoing circular motion. The circular motion may not be uniform, however, because the rate of spin of the object may change.

It is convenient to represent the position of the particle with its polar coordinates: $(r, \theta)$. The origin of the polar coordinates is chosen to coincide with the center of the circular paths of the particles. In this representation, the only coordinate for a given particle that changes in time is the angular position $\theta$; *r* remains constant. As a particle on the object moves along the circle of radius *r* from the positive *x* axis ($\theta = 0$) to the point *P*, it moves through an arc of length *s*, which is related to $\theta$ through the relation

$$s = r\theta \tag{10.1a}$$

$$\theta = \frac{s}{r} \tag{10.1b}$$

It is important to note the units of $\theta$ as expressed by Equation 10.1b. The angle $\theta$ is the ratio of an arc length and the radius of the circle and, hence, is a pure (dimensionless) number. We commonly refer to the unit of $\theta$ as a **radian** (rad), however. One radian is the angle subtended by an arc length equal to the radius of the arc. Because the circumference of a circle is $2\pi r$, it follows that 360° corresponds to an angle of $2\pi r/r$ rad or $2\pi$ rad. Hence, 1 rad = $360°/2\pi \approx 57.3°$. To convert an angle in degrees to an angle in radians, we can use the fact that $2\pi$ radians = 360°, or $\pi$ radians = 180°; hence,

$$\theta(\text{rad}) = \left(\frac{2\pi}{360°}\right) \theta(\text{deg}) = \left(\frac{\pi}{180°}\right) \theta(\text{deg})$$

• *Angular position*

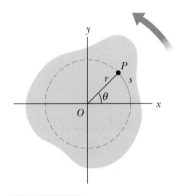

**Figure 10.1**

An object rotating about a fixed axis (the *z* axis) through *O* perpendicular to the plane of the figure. A particle on the object at *P* rotates in a circle of radius *r* centered at *O*.

• *The radian*

**PITFALL PREVENTION 10.1**
**Remember the radian**

Keep in mind that Equation 10.1b defines an angle expressed in *radians*. Don't fall into the trap of using this equation for angles measured in degrees. Also, be sure to set your calculator in radian mode when doing problems in rotation.

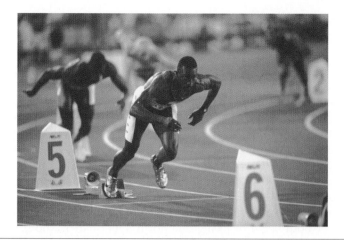

Michael Johnson starts in the 400-m final at the 1996 Olympics (he would go on to win the gold medal). In a short event such as this, in which runners must stay in their lanes, the runners start from staggered positions. If they all started from the same position, Equation 10.1a tells us that the runners farther from the center of a circular portion of the track would have to run farther than those on an inside lane. *(Stephen Dunn/Allsport)*

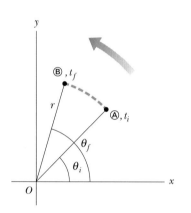

**Figure 10.2**

A particle on a rotating rigid body moves from Ⓐ to Ⓑ along the arc of a circle. In the time interval $\Delta t = t_f - t_i$, the radius vector sweeps out an angle $\Delta \theta = \theta_f - \theta_i$.

For example, 60° equals $\pi/3$ rad, and 45° equals $\pi/4$ rad.

In Figure 10.2, as the particle travels from Ⓐ to Ⓑ in a time $\Delta t$, the radius vector sweeps out an angle of $\Delta \theta = \theta_f - \theta_i$, which equals the **angular displacement** during the time interval $\Delta t$. Note that neither the angular position nor the angular displacement is limited to the range $0 < \theta < 2\pi$. For example, if the particle in Figure 10.2 starts at polar coordinates $(r, 0)$, makes two trips around a circular path, and ends up where it started, its angular displacement is described as $\theta = 4\pi$. Thus, we do not "reset" the angular position at zero each time the particle crosses the $x$ axis. The **number of revolutions** the particles makes in a time interval is the angular displacement during the interval divided by $2\pi$.

We define the **average angular speed** $\overline{\omega}$ (Greek omega) as the ratio of the angular displacement to the time interval $\Delta t$:

- *Average angular speed*

$$\overline{\omega} \equiv \frac{\theta_f - \theta_i}{t_f - t_i} = \frac{\Delta \theta}{\Delta t} \qquad \text{[10.2]}$$

By analogy with translational speed, the **instantaneous angular speed** $\omega$ is defined as the limit of the ratio in Equation 10.2 as $\Delta t$ approaches zero:

- *Instantaneous angular speed*

$$\omega \equiv \lim_{\Delta t \to 0} \frac{\Delta \theta}{\Delta t} = \frac{d\theta}{dt} \qquad \text{[10.3]}$$

Angular speed has units of rad/s (or $s^{-1}$ because radians are not dimensional). Let us adopt the convention that the fixed axis of rotation for an object is the $z$ axis, as in Figure 10.1. We shall take $\omega$ to be positive when $\theta$ is increasing (counterclockwise motion in Figure 10.1) and negative when $\theta$ is decreasing (clockwise motion).

If the instantaneous angular speed of a particle changes from $\omega_i$ to $\omega_f$ in the time interval $\Delta t$, the particle has an angular acceleration. The **average angular acceleration** $\overline{\alpha}$ (Greek alpha) of a particle moving in a circular path is defined as the ratio of the change in the angular speed to the time interval $\Delta t$:

$$\bar{\alpha} \equiv \frac{\omega_f - \omega_i}{t_f - t_i} = \frac{\Delta \omega}{\Delta t}$$

[10.4]     • *Average angular acceleration*

By analogy with translational acceleration, the **instantaneous angular acceleration** is defined as the limit of the ratio $\Delta \omega / \Delta t$ as $\Delta t$ approaches zero:

$$\alpha \equiv \lim_{\Delta t \to 0} \frac{\Delta \omega}{\Delta t} = \frac{d\omega}{dt}$$

[10.5]     • *Instantaneous angular acceleration*

Angular acceleration has units of $\text{rad/s}^2$, or $\text{s}^{-2}$.

Let us now generalize our argument from the rotating object in Figure 10.1 to any rigid body. A **rigid body** is any system of particles in which the particles remain fixed in position with respect to one another. We will call this simplification model, similar to the particle model, the **rigid body model.** If we were to imagine a rotating block of gelatin, the motion is very complicated because of the combination of rotation of the particles and the movement of particles within the deformable block relative to one another. Another example is our own Sun—the region of the Sun near the solar equator is moving with a higher angular speed than the region near the poles—it does not rotate as a rigid body. We shall not analyze such messy problems, however. Instead, our analyses will address only rigid bodies.

• *The rigid body model*

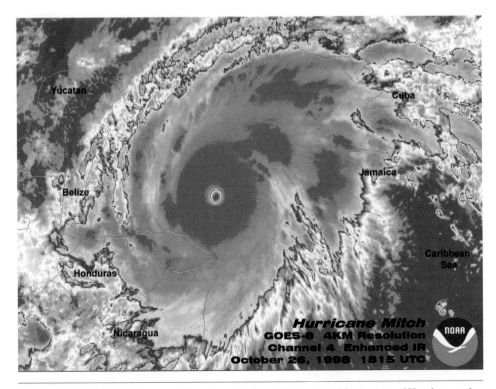

A color-enhanced, infrared image of Hurricane Mitch, which devastated large areas of Honduras and Nicaragua in October 1998. The hurricane does not rotate as a rigid body—regions near the interior rotate with a different angular speed than regions near the edge. *(Courtesy of NOAA)*

**WEB**

For more storm images and information on storms, visit the National Oceanic & Atmospheric Administration (NOAA) at **www.noaa.gov**

As we investigate rotational motion, we shall develop a number of analysis models for rigid bodies that have analogs in our analysis models for particles.

With our simplification model of a rigid body, we can make a statement about the various particles in the rigid body—**for rotation about a fixed axis, every particle on a rigid body has the same angular speed and the same angular acceleration.** That is, the quantities $\omega$ and $\alpha$ that we have discussed for particles characterize the rotational motion of the *entire* rigid body. Using these quantities, we can greatly simplify the analysis of rigid body rotation. Such a statement cannot be made for a nonrigid body, which is why studying the rotating block of gelatin or the Sun would be so difficult.

The angular position $\theta$, angular speed $\omega$, and angular acceleration $\alpha$ are analogous to translational position $x$, translational speed $v$, and translational acceleration $a$, respectively, for the corresponding one-dimensional motion discussed in Chapter 2. The variables $\theta$, $\omega$, and $\alpha$ differ dimensionally from the variables $x$, $v$, and $a$ only by a length factor, as we shall see shortly.

We have not associated any direction with the angular speed and angular acceleration.* Strictly speaking, these variables are the magnitudes of the angular velocity and angular acceleration vectors $\boldsymbol{\omega}$ and $\boldsymbol{\alpha}$. Because we are considering rotation about a fixed axis, we can indicate the directions of these vectors by assigning a positive or negative sign to $\omega$ and $\alpha$, as discussed for $\omega$ after Equation 10.3. For rotation about a fixed axis, the only direction in space that uniquely specifies the rotational motion is the direction along the axis; however, we still must specify one of the two directions along this axis as positive.

The direction of $\boldsymbol{\omega}$ is along the axis of rotation, which is the $z$ axis in Figure 10.1. By convention, we take the direction of $\boldsymbol{\omega}$ to be *out of* the plane of the diagram when the rotation is counterclockwise and *into* the plane of the diagram when the rotation is clockwise. To further illustrate this convention, it is convenient to use the **right-hand rule** illustrated by Figure 10.3. The four fingers of the right hand are wrapped in the direction of the rotation. The extended right thumb points in the direction of $\boldsymbol{\omega}$.

**Figure 10.3**

The right-hand rule for determining the direction of the angular velocity vector.

* Although we do not verify it here, the instantaneous angular velocity and instantaneous angular acceleration are vector quantities, but the corresponding average values are not. This is because angular displacements do not add as vector quantities for finite rotations.

The direction of $\boldsymbol{\alpha}$ follows from its vector definition as $d\boldsymbol{\omega}/dt$. For rotation about a fixed axis, the direction of $\boldsymbol{\alpha}$ is the same as $\boldsymbol{\omega}$ if the angular speed (the magnitude of $\boldsymbol{\omega}$) is increasing in time, and antiparallel to $\boldsymbol{\omega}$ if the angular speed is decreasing in time.

The full vector treatment of rotational motion is beyond the scope of this book and not necessary for our level of understanding, so we will use nonvector notation for most of this chapter.

### Quick Quiz 10.1

A rigid body is rotating in a counterclockwise sense around a fixed axis. Each of the following pairs of quantities represents an initial angular position and a final angular position of the rigid body. Which of the sets can *only* occur if the rigid body rotates through more than 180°? (a) 3 rad, 6 rad; (b) −1 rad, 1 rad; (c) 1 rad, 5 rad.

### Quick Quiz 10.2

Suppose that the change in angular position for each of the pairs of values in Quick Quiz 10.1 occurred in 1 s. Which choice represents the lowest average angular speed?

## 10.2 • ROTATIONAL KINEMATICS: THE RIGID BODY UNDER CONSTANT ANGULAR ACCELERATION

In our study of one-dimensional motion, we found that the simplest accelerated motion to analyze is motion under constant translational acceleration (Chapter 2). Likewise, for rotational motion about a fixed axis, the simplest accelerated motion to analyze is motion under constant angular acceleration.

See the *Core Concepts in Physics CD-ROM*, Screen 7.2

If we write Equation 10.5 in the form $d\omega = \alpha\,dt$ and let $\omega = \omega_i$ at $t_i = 0$, we can integrate this expression directly to find the final angular speed $\omega_f$:

$$\int_{\omega_i}^{\omega_f} d\omega = \int_0^t \alpha\,dt \longrightarrow \omega_f = \omega_i + \alpha t \qquad\qquad \text{[10.6]}$$

• *Rotational kinematic equations* ($\alpha = constant$)

Likewise, rewriting Equation 10.3 and substituting Equation 10.6, we can integrate once more (with $\theta = \theta_i$ at $t_i = 0$):

$$\int_{\theta_i}^{\theta_f} d\theta = \int_0^t \omega\,dt = \int_0^t (\omega_i + \alpha t)\,dt \longrightarrow \theta_f = \theta_i + \omega_i t + \tfrac{1}{2}\alpha t^2 \quad \text{[10.7]}$$

If we eliminate $t$ from Equations 10.6 and 10.7, we obtain

$$\omega_f^2 = \omega_i^2 + 2\alpha(\theta_f - \theta_i) \qquad\qquad \text{[10.8]}$$

If we eliminate $\alpha$, we find

$$\theta_f = \theta_i + \tfrac{1}{2}(\omega_i + \omega_f)t \qquad\qquad \text{[10.9]}$$

**PITFALL PREVENTION 10.3**

**Just like translation?**

Table 10.1 suggests that rotational kinematics is just like translational kinematics. That is almost true, but keep in mind two differences that you must address. (1) In rotational kinematics, as suggested in Pitfall Prevention 10.2, you need to specify a rotation axis. (2) In rotational motion, the object keeps returning to its original orientation—thus, you may be asked for the number of revolutions made by a rigid body, a concept that has no meaning in translational motion.

| **TABLE 10.1** | **A Comparison of Equations for Rotational and Translational Motion: Kinematic Equations** |
|---|---|
| **Rotational Motion About a Fixed Axis with $\alpha$ = Constant (Variables: $\theta_f$ and $\omega_f$)** | **Translational Motion with $a$ = Constant (Variables: $x_f$ and $v_f$)** |
| $\omega_f = \omega_i + \alpha t$ | $v_f = v_i + at$ |
| $\theta_f = \theta_i + \omega_i t + \frac{1}{2}\alpha t^2$ | $x_f = x_i + v_i t + \frac{1}{2}at^2$ |
| $\theta_f = \theta_i + \frac{1}{2}(\omega_i + \omega_f)t$ | $x_f = x_i + \frac{1}{2}(v_i + v_f)t$ |
| $\omega_f^2 = \omega_i^2 + 2\alpha(\theta_f - \theta_i)$ | $v_f^2 = v_i^2 + 2a(x_f - x_i)$ |

Notice that these kinematic expressions for rotational motion under constant angular acceleration are of the *same mathematical form* as those for translational motion under constant translational acceleration, with the substitutions $x \rightarrow \theta$, $v \rightarrow \omega$, and $a \rightarrow \alpha$. The similarities between rotational and translational kinematic equations are shown in Table 10.1.

**Quick Quiz 10.3**

Consider again the pairs of angular positions for the rigid body in Quick Quiz 10.1. If the object starts from rest at the initial angular position, moves counterclockwise with constant angular acceleration, and arrives at the final angular position with the same angular speed in all three cases, for which choice is the angular acceleration the highest?

### Example 10.1 Rotating Wheel

A wheel rotates with a constant angular acceleration of 3.50 rad/s². If the angular speed of the wheel is 2.00 rad/s at $t = 0$, (a) through what angle does the wheel rotate between $t = 0$ and $t = 2.00$ s?

**Solution** We assume that the wheel is perfectly rigid, so that we can use the rigid body model. Because the angular acceleration in the problem is given as constant, we can use the rotational kinematic equations. We use Equation 10.7, setting $\theta_i = 0$,

$$\theta_f = \theta_i + \omega_i t + \tfrac{1}{2}\alpha t^2$$
$$= 0 + (2.00 \text{ rad/s})(2.00 \text{ s}) + \tfrac{1}{2}(3.50 \text{ rad/s}^2)(2.00 \text{ s})^2$$
$$= \boxed{11.0 \text{ rad}}$$

This is equivalent to 11.0 rad/$(2\pi$ rad/rev$)$ = 1.75 rev.

(b) What is the angular speed of the wheel at $t = 2.00$ s?

**Solution** We use Equation 10.6:

$$\omega_f = \omega_i + \alpha t = 2.00 \text{ rad/s} + (3.50 \text{ rad/s}^2)(2.00 \text{ s})$$
$$= \boxed{9.00 \text{ rad/s}}$$

We could also obtain this result using Equation 10.8 and the results of part (a). Try it!

**EXERCISE** Find the angle through which the wheel rotates between $t = 2.00$ s and $t = 3.00$ s.

*Answer* 10.8 rad

**EXERCISE** A wheel starts from rest and rotates with constant angular acceleration to an angular speed of 12.0 rad/s in 3.00 s. Find (a) the angular acceleration of the wheel and (b) the angle in radians through which it rotates in this time.

*Answer* (a) 4.00 rad/s²  (b) 18.0 rad

## 10.3 • RELATIONS BETWEEN ROTATIONAL AND TRANSLATIONAL QUANTITIES

In this section we shall derive some useful relations between the angular speed and angular acceleration of a particle on a rotating rigid body and its translational speed and translational acceleration. Keep in mind that when a rigid body rotates about a fixed axis, *every* particle of the body moves in a circle whose center is the axis of rotation.

Consider a particle on a rotating rigid body, moving in a circle of radius $r$ about the $z$ axis, as in Figure 10.4. Because the particle moves along a circular path, its translational velocity vector **v** is always tangent to the path; hence, we often call this quantity **tangential velocity.** The magnitude of the tangential velocity of the particle is, by definition, the **tangential speed,** given by $v = ds/dt$, where $s$ is the distance traveled by the particle along the circular path. Recalling from Equation 10.1a that $s = r\theta$, and noting that $r$ is a constant, we have

$$v = \frac{ds}{dt} = r\frac{d\theta}{dt}$$

$$v = r\omega \qquad \text{[10.10]}$$

That is, the tangential speed of the particle equals the distance of the particle from the axis of rotation multiplied by the particle's angular speed.

We can relate the angular acceleration of the particle to its tangential acceleration $a_t$ — which is the component of its acceleration tangent to the path of motion — by taking the time derivative of $v$:

$$a_t = \frac{dv}{dt} = r\frac{d\omega}{dt}$$

$$a_t = r\alpha \qquad \text{[10.11]}$$

That is, the tangential component of the translational acceleration of a particle undergoing circular motion equals the distance of the particle from the axis of rotation multiplied by the angular acceleration.

In Chapter 3 we found that a particle rotating in a circular path undergoes a centripetal, or radial, acceleration of magnitude $v^2/r$ directed toward the center of rotation (Fig. 10.5). Because $v = r\omega$, we can express the centripetal acceleration of the particle in terms of the angular speed as

$$a_c = \frac{v^2}{r} = r\omega^2 \qquad \text{[10.12]}$$

The *total translational acceleration* of the particle is $\mathbf{a} = \mathbf{a}_t + \mathbf{a}_r$. The magnitude of the total translational acceleration of the particle is therefore

$$a = \sqrt{a_t^2 + a_r^2} = \sqrt{r^2\alpha^2 + r^2\omega^4} = r\sqrt{\alpha^2 + \omega^4} \qquad \text{[10.13]}$$

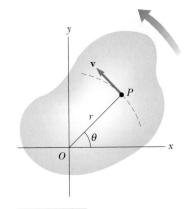

**Figure 10.4**

As a rigid body rotates about the fixed axis through $O$, the point $P$ has a translational velocity **v** that is always tangent to the circular path of radius $r$.

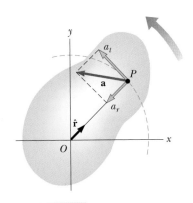

**Figure 10.5**

As a rigid body rotates about a fixed axis through $O$, a particle at point $P$ experiences a tangential component of translational acceleration $a_t$ and a radial component of translational acceleration $a_r$. The total translational acceleration of this particle is $\mathbf{a} = \mathbf{a}_t + \mathbf{a}_r$, where $\mathbf{a}_r = -a_c\hat{\mathbf{r}}$.

### Quick Quiz 10.4

When a wheel of radius $R$ rotates about a fixed axis as in Figure 10.6, (a) do all points on the wheel (including the spokes) have the same angular speed? (b) Do they all have the same tangential speed? (c) If the angular speed is constant and

R

**Figure 10.6**

**Figure 10.6**

(Quick Quiz 10.4) A wheel of radius *R* rotating about its axis.

equal to ω, describe the tangential speeds and total translational accelerations of the points located at $r = 0$, $r = R/2$, and $r = R$, where the points are measured from the center of the wheel.

### THINKING PHYSICS 10.1

A phonograph record (LP, for *long-playing*) is rotated at a constant *angular* speed. A compact disc (CD) is rotated so that the surface sweeps past the laser at a constant *tangential* speed. Consider two circular grooves of information on an LP—one near the outer edge and one near the inner edge. Suppose the outer groove "contains" 1.8 s of music. Does the inner groove also contain 1.8 s of music? How about the CD—do the inner and outer "grooves" contain the same time interval of music?

**Reasoning** On the LP, the inner and outer grooves must both rotate once in the same time period. Thus, each groove, regardless of where it is on the record, contains the same time interval of information. Of course, on the inner grooves, this same information must be compressed into a smaller circumference. On a CD, the constant tangential speed requires that no such compression occur—the digital pits representing the information are spaced uniformly everywhere on the surface. Thus, there is more information in an outer "groove," due to its larger circumference, and, as a result, a longer time interval of music than in the inner "groove."

### THINKING PHYSICS 10.2

The launch area for the European Space Agency is not in Europe—it is in South America. Why?

**Reasoning** Placing a satellite in Earth orbit requires providing a large tangential speed to the satellite, which is the task of the rocket propulsion system. Anything that reduces the requirements on the propulsion system is a welcome contribution. The surface of the Earth is already traveling toward the east at a high speed, due to the rotation of the Earth. Thus, if rockets are launched toward the east, the rotation of the Earth provides some initial tangential speed, reducing somewhat the requirements on the propulsion system. If rockets were launched from Europe, which is at a relatively large latitude, the contribution of the Earth's rotation is relatively small because the distance between Europe and the rotation axis of the Earth is relatively small. The ideal place for launching is at the equator, which is as far as one can be from the rotation axis of the Earth and still be on the surface of the Earth. This results in the largest possible tangential speed due to the Earth's rotation. The European Space Agency exploits this advantage by launching from French Guiana, which is only a few degrees north of the equator.

A second advantage of this launch location is that launching toward the east takes the spacecraft over water. In the event of an accident or a failure, the wreckage will fall into the ocean rather than into populated areas as it would if launched to the east from Europe. This is a primary reason why the United States launches spacecraft from Florida rather than California, despite the more favorable weather conditions in California.

**WEB**

Visit the European Space agency at **www.esrin.esa.it**

## 10.4 • ROTATIONAL KINETIC ENERGY

 See Screen 7.3

Imagine that you begin a workout session on a stationary exercise bicycle. You apply a force with your feet on the pedals, moving them through a displacement—you have done work. The result of this work is the spinning of the wheel. This rotational motion represents kinetic energy because mass is in motion. In this section, we will investigate this kinetic energy. In a later section, we will consider the work done in rotational motion and develop a rotational version of the work–kinetic energy theorem.

Let us model a collection of particles, such as the particles making up the wheel on our exercise bicycle, with the rigid body model, and let us assume that the body rotates about the fixed $z$ axis with an angular speed of $\omega$ (Fig. 10.7). Each particle of the rigid body is in motion and thus has some kinetic energy, determined by its mass and tangential speed. If the mass of the $i$th particle is $m_i$ and its tangential speed is $v_i$, the kinetic energy of this particle is

$$K_i = \tfrac{1}{2} m_i v_i^2$$

We can express the *total* kinetic energy $K_R$ of the rotating rigid body as the sum of the kinetic energies of the individual particles. Thus, incorporating Equation 10.10,

$$K_R = \sum_i K_i = \sum_i \tfrac{1}{2} m_i v_i^2 = \tfrac{1}{2} \sum_i m_i r_i^2 \omega^2$$

$$= \tfrac{1}{2} \left( \sum_i m_i r_i^2 \right) \omega^2$$

where we have factored $\omega^2$ from the sum because it is common to every particle. The quantity in parentheses is called the **moment of inertia** $I$ of the rigid body:

$$I = \sum_i m_i r_i^2 \qquad \textbf{[10.14]}$$

Therefore, we can express the kinetic energy of the rotating rigid body as

$$K_R = \tfrac{1}{2} I \omega^2 \qquad \textbf{[10.15]}$$

From the definition of moment of inertia, we see that it has dimensions of $ML^2$ (kg·m² in SI units). The moment of inertia is a measure of a system's *resistance to change in its angular speed*. Thus, it plays a role in rotational motion identical to that which mass plays in translational motion. Notice that moment of inertia depends not only on the mass of the rigid body but also on *how the mass is distributed around the rotation axis*.

Although we shall commonly refer to the quantity $\tfrac{1}{2} I \omega^2$ as the **rotational kinetic energy,** it is not a new form of energy. It is ordinary kinetic energy because it was derived from a sum over individual kinetic energies of the particles contained in the rigid body. It is a new role for kinetic energy for us, however, because we have only considered kinetic energy associated with translation through space so far. **On the storage side of the continuity equation for energy (see Eq. 6.20), we should now consider that the kinetic energy term should be the sum of the changes in both translational and rotational kinetic energy.** Thus, in energy versions of system models, we should keep in mind the possibility of rotational kinetic energy.

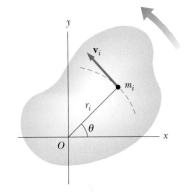

**Figure 10.7**

A rigid body rotating about the $z$ axis with angular speed $\omega$. The kinetic energy of the particle of mass $m_i$ is $\tfrac{1}{2} m_i v_i^2$. The kinetic energy of the rigid body is called its rotational kinetic energy.

• *Moment of inertia for a system of particles*

• *Kinetic energy of a rotating rigid body*

**PITFALL PREVENTION 10.4**
**No single moment of inertia**

 We have pointed out that moment of inertia is analogous to mass, but there is one major difference. Mass is an inherent property of an object. The moment of inertia of an object depends on your choice of rotation axis; therefore an object has no single value of the moment of inertia. An object does have a minimum value of the moment of inertia, which is that calculated around an axis passing through the center of mass of the object.

Equation 10.14 gives the moment of inertia of a collection of particles. For an extended, continuous object, we can calculate the moment of inertia by dividing the object into many small elements with mass $\Delta m_i$. Then, the moment of inertia is approximately $I \approx \Sigma r_i^2 \Delta m_i$, where $r_i$ is the perpendicular distance of the element of mass $\Delta m_i$ from the rotation axis. Now we take the limit as $\Delta m_i \rightarrow 0$, in which case the sum becomes an integral:

$$I = \lim_{\Delta m_i \rightarrow 0} \sum_i r_i^2 \, \Delta m_i = \int r^2 dm \qquad \text{[10.16]}$$

It is usually easier to calculate moments of inertia in terms of the volume of the elements rather than their mass, and we can easily make this change by using Equation 1.1, $\rho = m/V$, where $\rho$ is the density of the object and $V$ is its volume. We can express the mass of an element by writing Equation 1.1 in differential form, $dm = \rho \, dV$. Using this, Equation 10.16 becomes

- *Moment of inertia for an extended, continuous object*

$$I = \int \rho r^2 dV \qquad \text{[10.17]}$$

If the object is homogenous, the density $\rho$ is constant and the integral can be evaluated for a given geometry. If $\rho$ is not uniform over the volume of the object, its variation with position must be known in order to perform the integration.

For symmetric objects, the moment of inertia can be expressed in terms of the total mass of the object and one or more dimensions of the object. Table 10.2 shows the moments of inertia of various common symmetric objects.

**Quick Quiz 10.5**

Two spheres, one hollow and one solid, are rotating with the same angular speed about their centers. Both spheres have the same mass and radius. Which one, if either, has the higher rotational kinetic energy?

## Example 10.2  The Oxygen Molecule

Consider the diatomic oxygen molecule $O_2$, which is rotating in the $xy$ plane about the $z$ axis passing through its center, perpendicular to its length. The mass of each oxygen atom is $2.66 \times 10^{-26}$ kg, and at room temperature, the average separation between the two oxygen atoms is $d = 1.21 \times 10^{-10}$ m. (a) Calculate the moment of inertia of the molecule about the $z$ axis.

**Solution**  We model the molecule as a rigid body, consisting of two particles (the two oxygen atoms), in rotation. Because the distance of each particle from the $z$ axis is $d/2$, the moment of inertia about the $z$ axis is

$$I = \sum_i m_i r_i^2 = m\left(\frac{d}{2}\right)^2 + m\left(\frac{d}{2}\right)^2 = \frac{md^2}{2}$$

$$= \frac{(2.66 \times 10^{-26} \text{ kg})(1.21 \times 10^{-10} \text{ m})^2}{2}$$

$$= 1.95 \times 10^{-46} \text{ kg} \cdot \text{m}^2$$

(b) A typical angular speed of a molecule is $4.60 \times 10^{12}$ rad/s. If the oxygen molecule is rotating with this angular speed about the $z$ axis, what is its rotational kinetic energy?

**Solution**  We use Equation 10.15:

$$K_R = \tfrac{1}{2}I\omega^2$$

$$= \tfrac{1}{2}(1.95 \times 10^{-46} \text{ kg} \cdot \text{m}^2)(4.60 \times 10^{12} \text{ rad/s})^2$$

$$= 2.06 \times 10^{-21} \text{ J}$$

**TABLE 10.2** | **Moments of Inertia of Homogeneous Rigid Bodies with Different Geometries**

Hoop or thin cylindrical shell
$I_{CM} = MR^2$

Hollow cylinder
$I_{CM} = \frac{1}{2}M(R_1^2 + R_2^2)$

Solid cylinder or disk
$I_{CM} = \frac{1}{2}MR^2$

Rectangular plate
$I_{CM} = \frac{1}{12}M(a^2 + b^2)$

Long thin rod with rotation axis through center
$I_{CM} = \frac{1}{12}ML^2$

Long thin rod with rotation axis through end
$I = \frac{1}{3}ML^2$

Solid sphere
$I_{CM} = \frac{2}{5}MR^2$

Thin spherical shell
$I_{CM} = \frac{2}{3}MR^2$

## Example 10.3 Four Rotating Masses

Four small spheres are fastened to the corners of a frame of negligible mass lying in the *xy* plane (Fig. 10.8). (a) If the rotation of the system occurs about the *y* axis with an angular speed $\omega$, find the moment of inertia $I_y$ about the *y* axis and the rotational kinetic energy about this axis.

**Solution** First, note that the two spheres of mass $m$ that lie on the *y* axis do not contribute to $I_y$. Because they are modeled as particles, $r_i = 0$ for these spheres about this axis. Applying Equation 10.14, we have for the two spheres on the *x*

axis,

$$I_y = \sum_i m_i r_i^2 = Ma^2 + Ma^2 = 2Ma^2$$

Therefore, the rotational kinetic energy about the *y* axis is

$$K_R = \frac{1}{2}I_y\omega^2 = \frac{1}{2}(2Ma^2)\omega^2 = Ma^2\omega^2$$

The fact that the spheres of mass $m$ do not enter into this result makes sense, because they have no motion about the chosen axis of rotation; hence, they have no kinetic energy.

(b) Suppose the system rotates in the $xy$ plane about an axis through $O$ (the $z$ axis). Calculate the moment of inertia about the $z$ axis and the rotational energy about this axis.

**Solution** Because $r_i$ in Equation 10.14 is the perpendicular distance to the axis of rotation, we have

$$I_z = \sum_i m_i r_i^2 = Ma^2 + Ma^2 + mb^2 + mb^2 = \boxed{2Ma^2 + 2mb^2}$$

$$K_R = \tfrac{1}{2}I_z\omega^2 = \tfrac{1}{2}(2Ma^2 + 2mb^2)\omega^2 = \boxed{(Ma^2 + mb^2)\omega^2}$$

Comparing the results for (a) and (b), we see explicitly that the moment of inertia and therefore the rotational energy associated with a given angular speed depend on the axis of rotation. In (b), we expect the result to include all masses and distances because all four spheres are in motion for rotation in the $xy$ plane. Furthermore, the fact that the rotational energy in (a) is smaller than in (b) indicates that there is less resistance to changes in rotational motion about the $y$ axis than about the $z$ axis.

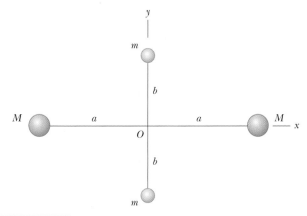

**Figure 10.8**

(Example 10.3) All four spheres are at fixed positions as shown. The moment of inertia of the system depends on the axis about which it is evaluated.

## Example 10.4 Moment of Inertia of a Uniform Solid Cylinder

A uniform solid cylinder has a radius $R$, mass $M$, and length $L$. Calculate its moment of inertia about its central axis (the $z$ axis shown in Figure 10.9).

**Solution** The integral can be evaluated relatively simply by dividing the cylinder into many cylindrical shells of radius $r$, thickness $dr$, and length $L$, as shown in Figure 10.9. The volume $dV$ of a shell is its cross-sectional area multiplied by its length: $dV = (dA)L = (2\pi r dr)L$. Equation 10.17 gives the moment of inertia,

$$I = \int \rho r^2 dV = \int_0^R \rho r^2 (2\pi rL)\,dr = 2\pi\rho L \int_0^R r^3\,dr = \tfrac{1}{2}\pi\rho LR^4$$

The volume of the entire cylinder is $\pi R^2 L$, so the density is $\rho = M/V = M/\pi R^2 L$. Substituting this value of $\rho$ in the above result gives

$$I = \tfrac{1}{2}\pi\left(\frac{M}{\pi R^2 L}\right)Lr^4 = \tfrac{1}{2}MR^2$$

**Figure 10.9**

(Example 10.4) The geometry for calculating the moment of inertia about the central axis of a uniform solid cylinder.

Note that this result does not depend on $L$. Thus, it applies equally well to a long cylinder and a flat disk.

## Example 10.5 Rotating Rod

A uniform rod of length $L$ and mass $M$ is free to rotate on a frictionless pin through one end (Fig. 10.10). The rod is released from rest in the horizontal position. (a) What is the angular speed of the rod at its lowest position?

**Reasoning** We consider the rod and the Earth as an isolated system and use the energy version of the isolated system model. Consider the mechanical energy of the system. When the rod is horizontal, as in Figure 10.10, it has no rotational

**Figure 10.10**

(Example 10.5) A uniform rod rotates freely under the influence of gravity around a pivot at the left end.

kinetic energy. Let us also define this position of the rod as representing the zero of gravitational potential energy of the system. When the center of mass of the rod is at the lowest position, the potential energy of the system is $-MgL/2$ and the rod has rotational kinetic energy $\frac{1}{2}I\omega^2$, where $I$ is the moment of inertia about the pivot.

**Solution**    Because $I = \frac{1}{3}ML^2$ (see Table 10.2) for a geometric model of a long, thin rod, and because mechanical en-

ergy of the isolated system is conserved, we have

$$K_i + U_i = K_f + U_f$$

$$0 + 0 = -\tfrac{1}{2}MgL + \tfrac{1}{2}I\omega^2 = -\tfrac{1}{2}MgL + \tfrac{1}{2}(\tfrac{1}{3}ML^2)\omega^2$$

$$\omega = \sqrt{\frac{3g}{L}}$$

(b) Determine the tangential speed of the center of mass and the tangential speed of the lowest point on the rod in the vertical position.

**Solution**    Using Equation 10.10, we have

$$v_{CM} = r\omega = \frac{L}{2}\,\omega = \tfrac{1}{2}\sqrt{3gL}$$

The lowest point on the rod, because it is twice as far from the pivot as the center of mass, has a tangential speed equal to $2v_{CM} = \sqrt{3gL}$

## 10.5 • TORQUE AND THE VECTOR PRODUCT

Recall our stationary exercise bicycle from the preceding section. We caused the rotational motion of the wheel by applying forces to the pedals. When a force is exerted on a rigid body pivoted about some axis, and the line of action* of the force does not pass through the pivot, the body tends to rotate about that axis. For example, when you push on a door, the door rotates about an axis through the hinges. The tendency of a force to rotate a body about some axis is measured by a vector quantity called **torque.** Torque is the cause of changes in rotational motion, and is analogous to force, which causes changes in translational motion. Consider the wrench pivoted about the axis through $O$ in Figure 10.11. The applied force $\mathbf{F}$ generally can act at an angle $\phi$ with respect to the position vector $\mathbf{r}$ locating the point of application of the force. We define the torque $\tau$ (Greek letter tau) resulting from the force $\mathbf{F}$ with the expression[†]

$$\tau \equiv rF\sin\phi \qquad \text{[10.18]}$$

It is very important to recognize that **torque is defined only when a reference axis is specified,** from which the distance $r$ is determined. We can interpret Equation 10.18 in two different ways. Looking at the force components in Figure 10.11, we see that the component $F\cos\phi$ parallel to $\mathbf{r}$ will not cause a rotation around the pivot point, because its line of action passes right through the pivot

See Screen 7.6

**Figure 10.11**

The force $\mathbf{F}$ has a greater rotating tendency about $O$ as $F$ increases and as the moment arm $d$ increases. The component $F\sin\phi$ tends to rotate the system about $O$.

---

\* The line of action of a force is an imaginary line colinear with the force vector and extending to infinity in both directions.

[†] In general, torque is a vector. For rotation about a fixed axis, however, we will use italic, nonbold notation and specify the direction with a positive or a negative sign, as we did for angular speed and acceleration in Section 10.1. We will treat the vector nature of torque briefly in a short while.

point. You cannot open a door by pushing on the hinges! Thus, only the perpendicular component causes a rotation about the pivot. In this case, we can write Equation 10.18 as

$$\tau = r(F \sin \phi)$$

so that the torque is the product of the distance to the point of application of the force and the perpendicular component of the force. In some problems, this is the easiest way to interpret the calculation of the torque.

The second way to interpret Equation 10.18 is to associate the sine function with the distance $r$, so that we can write

$$\tau = F(r \sin \phi) = Fd$$

The quantity $d = r \sin \phi$, called the **moment arm** (or *lever arm*) of the force $\mathbf{F}$, represents the perpendicular distance from the rotation axis to the line of action of $\mathbf{F}$. In some problems, this approach to the calculation of the torque is easier than that of resolving the force into components.

If two or more forces are acting on a rigid body, as in Figure 10.12, each has a tendency to produce a rotation about the pivot at $O$. For example, if the body is initially at rest, $\mathbf{F}_2$ tends to rotate the body clockwise and $\mathbf{F}_1$ tends to rotate the body counterclockwise. We shall use the convention that the sign of the torque resulting from a force is positive if its turning tendency is counterclockwise around the rotation axis and negative if its turning tendency is clockwise. For example, in Figure 10.12, the torque resulting from $\mathbf{F}_1$, which has a moment arm of $d_1$, is *positive* and equal to $+F_1 d_1$; the torque from $\mathbf{F}_2$ is *negative* and equal to $-F_2 d_2$. Hence, the *net* torque acting on the rigid body about an axis through $O$ is

$$\tau_{\text{net}} = \tau_1 + \tau_2 = F_1 d_1 - F_2 d_2$$

From the definition of torque, we see that the rotating tendency increases as $F$ increases and as $d$ increases. For example, we cause more rotation of a door if (a) we push harder or (b) we push at the doorknob rather than at a point close to the hinges. **Torque should not be confused with force.** Torque *depends* on force, but it also depends on *where the force is applied*. Torque has units of force times length—N · m in SI units.*

So far, we have not discussed the vector nature of torque, aside from assigning a positive or negative value to $\tau$. Consider a force $\mathbf{F}$ acting on a particle located at the vector position $\mathbf{r}$ (Fig. 10.13). The *magnitude* of the torque due to this force relative to an axis through the origin is $|rF \sin \phi|$, where $\phi$ is the angle between $\mathbf{r}$ and $\mathbf{F}$. The axis about which $\mathbf{F}$ would tend to produce rotation is perpendicular to the plane formed by $\mathbf{r}$ and $\mathbf{F}$. If the force lies in the $xy$ plane, as in Figure 10.13, the torque is represented by a vector parallel to the $z$ axis. The force in Figure 10.13 creates a torque that tends to rotate the body counterclockwise when we are looking down the $z$ axis—we define this as the vector $\boldsymbol{\tau}$ being in the positive $z$ direction (i.e., coming toward your eyes). If we reverse the direction of $\mathbf{F}$ in Figure 10.13, $\boldsymbol{\tau}$ is in the negative $z$ direction. With this choice, the torque vector can be defined to be equal to the **vector product,** or **cross product,** of $\mathbf{r}$ and $\mathbf{F}$:

$$\boldsymbol{\tau} \equiv \mathbf{r} \times \mathbf{F} \qquad\qquad [10.19]$$

• *Moment arm*

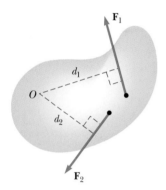

**Figure 10.12**

The force $\mathbf{F}_1$ tends to rotate the object counterclockwise about an axis through $O$, and $\mathbf{F}_2$ tends to rotate the object clockwise.

• *Definition of torque using the cross product*

---

* In Chapter 6, we saw the product of newtons and meters when we defined work, and we called this product a *joule*. We do not use this term here because the joule is only to be used when discussing energy. For torque, the unit is simply the N · m.

We now give a formal definition of the vector product, first introduced in Section 1.9. Given any two vectors **A** and **B**, the vector product **A** × **B** is defined as a third vector **C**, the *magnitude* of which is $AB \sin \theta$, where $\theta$ is the angle between **A** and **B**:

$$\mathbf{C} = \mathbf{A} \times \mathbf{B} \qquad [10.20]$$

$$C = |\mathbf{C}| \equiv AB \sin \theta \qquad [10.21]$$

Note that the quantity $AB \sin \theta$ is equal to the area of the parallelogram formed by **A** and **B**, as shown in Figure 10.14. The *direction* of **A** × **B** is perpendicular to the plane formed by **A** and **B** and is determined by the right-hand rule illustrated in Figure 10.14. The four fingers of the right hand are pointed along **A** and then "wrapped" into **B** through the angle $\theta$. The direction of the upright thumb is the direction of **A** × **B**. Because of the notation, **A** × **B** is often read "**A** cross **B**"; hence the term *cross product*.

Some properties of the vector product follow from its definition:

- Unlike the case of the scalar product, the vector product is not commutative; in fact,

$$\mathbf{A} \times \mathbf{B} = - (\mathbf{B} \times \mathbf{A}) \qquad [10.22]$$

Therefore, if you change the order of the vector product, you must change the sign. One could easily verify this relation with the right-hand rule (see Fig. 10.14).

- If **A** is parallel to **B** ($\theta = 0°$ or $180°$), then **A** × **B** = 0; therefore, it follows that **A** × **A** = 0.
- If **A** is perpendicular to **B**, then $|\mathbf{A} \times \mathbf{B}| = AB$. It is left to Problem 22 to show, from Equations 10.20 and 10.21 and the definition of unit vectors, that the cross products of the unit vectors **i**, **j**, and **k** obey the following expressions:

$$\mathbf{i} \times \mathbf{i} = \mathbf{j} \times \mathbf{j} = \mathbf{k} \times \mathbf{k} = 0$$

$$\mathbf{i} \times \mathbf{j} = -\mathbf{j} \times \mathbf{i} = \mathbf{k}$$

$$\mathbf{j} \times \mathbf{k} = -\mathbf{k} \times \mathbf{j} = \mathbf{i} \qquad [10.23]$$

$$\mathbf{k} \times \mathbf{i} = -\mathbf{i} \times \mathbf{k} = \mathbf{j}$$

Signs are interchangeable. For example, $\mathbf{i} \times (-\mathbf{j}) = -\mathbf{i} \times \mathbf{j} = -\mathbf{k}.$

### Quick Quiz 10.6

Both torque and work are products of force and distance. How are they different?

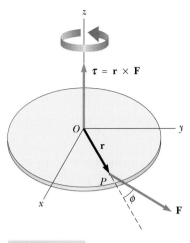

**Figure 10.13**

The torque vector $\boldsymbol{\tau}$ lies in a direction perpendicular to the plane formed by the position vector **r** and the applied force vector **F**.

---

**PITFALL PREVENTION 10.6**

**The cross product is a vector**

Remember that the result of taking a cross product between two vectors is a *third vector*. Equation 10.21 gives only the magnitude of this vector.

---

Right-hand rule    **Figure 10.14**

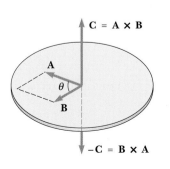

The vector product **A** × **B** is a third vector **C** having a magnitude $AB \sin \theta$ equal to the area of the parallelogram shown. The vector **C** is perpendicular to the plane formed by **A** and **B**, and its direction is determined by the right-hand rule.

## Quick Quiz 10.7

If you are trying to loosen a stubborn screw from a piece of wood with a screwdriver and fail, should you find a screwdriver for which the handle is (a) longer or (b) fatter? Why?

### Example 10.6  The Net Torque on a Cylinder

A one-piece cylinder is shaped as in Figure 10.15, with a core section protruding from the larger drum. The cylinder is free to rotate around the central axis shown in the drawing. A rope wrapped around the drum, of radius $R_1$, exerts a force $\mathbf{T}_1$ to the right on the cylinder. A rope wrapped around

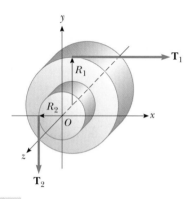

**Figure 10.15**

(Example 10.6) A solid cylinder pivoted about the $z$ axis through $O$. The moment arm of $\mathbf{T}_1$ is $R_1$, and the moment arm of $\mathbf{T}_2$ is $R_2$.

the core, of radius $R_2$, exerts a force $\mathbf{T}_2$ downward on the cylinder. (a) What is the net torque acting on the cylinder about the rotation axis (which is the $z$ axis in Fig. 10.15)?

**Solution**  The torque due to $\mathbf{T}_1$ is $-R_1 T_1$. It is negative because it tends to produce a clockwise rotation from the point of view in Figure 10.15. The torque due to $\mathbf{T}_2$ is $+R_2 T_2$ and is positive because it tends to produce a counterclockwise rotation. Therefore, the net torque about the rotation axis is

$$\tau_{\text{net}} = \tau_1 + \tau_2 = \boxed{R_2 T_2 - R_1 T_1}$$

(b) Suppose $T_1 = 5.0$ N, $R_1 = 1.0$ m, $T_2 = 6.0$ N, and $R_2 = 0.50$ m. What is the net torque about the rotation axis and which way does the cylinder rotate if it starts from rest?

**Solution**  We substitute numerical values in the result from part (a),

$$\tau_{\text{net}} = (6.0\ \text{N})(0.50\ \text{m}) - (5.0\ \text{N})(1.0\ \text{m}) = \boxed{-2.0\ \text{N} \cdot \text{m}}$$

Because the net torque is negative, the cylinder rotates clockwise from rest.

### Example 10.7  The Vector Product

Two vectors lying in the $xy$ plane are given by the equations $\mathbf{A} = 2\mathbf{i} + 3\mathbf{j}$ and $\mathbf{B} = -\mathbf{i} + 2\mathbf{j}$. Find $\mathbf{A} \times \mathbf{B}$, and verify explicitly that $\mathbf{A} \times \mathbf{B} = -\mathbf{B} \times \mathbf{A}$.

**Solution**  Using Equation 10.23 for the vector product of unit vectors gives

$$\mathbf{A} \times \mathbf{B} = (2\mathbf{i} + 3\mathbf{j}) \times (-\mathbf{i} + 2\mathbf{j})$$
$$= 2\mathbf{i} \times (-\mathbf{i}) + 2\mathbf{i} \times 2\mathbf{j} + 3\mathbf{j} \times (-\mathbf{i}) + 3\mathbf{j} \times 2\mathbf{j}$$
$$= 0 + 4\mathbf{k} + 3\mathbf{k} + 0 = \boxed{7\mathbf{k}}$$

$$\mathbf{B} \times \mathbf{A} = (-\mathbf{i} + 2\mathbf{j}) \times (2\mathbf{i} + 3\mathbf{j})$$
$$= -\mathbf{i} \times 2\mathbf{i} + (-\mathbf{i}) \times 3\mathbf{j} + 2\mathbf{j} \times 2\mathbf{i} + 2\mathbf{j} \times 3\mathbf{j}$$
$$= 0 - 3\mathbf{k} - 4\mathbf{k} + 0 = -7\mathbf{k}$$

Therefore, $\mathbf{A} \times \mathbf{B} = -\mathbf{B} \times \mathbf{A}$.

**EXERCISE**  Use the results of this example and Equation 10.21 to find the angle between $\mathbf{A}$ and $\mathbf{B}$.

*Answer*  60.3°

**EXERCISE**  A particle is located at the vector position $\mathbf{r} = (\mathbf{i} + 3\mathbf{j})$ m, and the force acting on it is $\mathbf{F} = (3\mathbf{i} + 2\mathbf{j})$ N. What is the torque on the particle about an axis through (a) the origin and (b) the point having coordinates (0, 6) m?]

*Answer*  (a) $-7\mathbf{k}$ N·m   (b) $11\mathbf{k}$ N·m

**EXERCISE**  If $|\mathbf{A} \times \mathbf{B}| = \mathbf{A} \cdot \mathbf{B}$, what is the angle between $\mathbf{A}$ and $\mathbf{B}$?

*Answer*  45.0°

## 10.6 • THE RIGID BODY IN EQUILIBRIUM

We have defined a rigid body and have discussed torque as the cause of changes in rotational motion of a rigid body. We can now establish models for a rigid body subject to torques that are analogous to those for a particle subject to forces. We begin by imagining a rigid body with balanced torques.

Consider two forces of equal magnitude and opposite direction applied to an object as shown in Figure 10.16a. The force directed to the right tends to rotate the object clockwise about an axis perpendicular to the diagram through *O*, whereas the force directed to the left tends to rotate it counterclockwise about that axis. Because the forces are of equal magnitude and act at the same perpendicular distance from *O*, their torques are equal in magnitude. Thus, the net torque on the rigid body is zero. The situation shown in Figure 10.16b is another case in which the net torque about *O* is zero (although the net *force* on the object is not zero), and we can devise many more cases.

With no net torque, no change occurs in rotational motion, and the rotational motion of the rigid body remains in its original state. This is an equilibrium situation, analogous to translational equilibrium, discussed in Chapter 4.

We now have **two conditions for complete equilibrium of an object,** which can be stated as follows:

• The net external force must equal zero

$$\sum \mathbf{F} = 0 \qquad \text{[10.24]}$$

• The net external torque must be zero about *any* axis

$$\sum \boldsymbol{\tau} = 0 \qquad \text{[10.25]}$$

The first condition is a statement of translational equilibrium. The second condition is a statement of rotational equilibrium. In the special case of **static equilibrium,** the object is at rest, so that it has no translational or angular speed (i.e., $v_{CM} = 0$ and $\omega = 0$).

The two vector expressions given by Equations 10.24 and 10.25 are equivalent, in general, to six scalar equations: three from the first condition of equilibrium, and three from the second (corresponding to *x*, *y*, and *z* components). Hence, in a complex system involving several forces acting in various directions, you would be faced with solving a set of equations with many unknowns. Here, we restrict our discussion to situations in which all the forces lie in the *xy* plane. (Forces whose vector representations are in the same plane are said to be *coplanar.*) With this restriction, we need to deal with only three scalar equations. Two of these come from balancing the forces in the *x* and *y* directions. The third comes from the torque equation, namely, that the net torque about an axis through *any* point in the *xy* plane must be zero. Hence, the two conditions of equilibrium provide the equations

$$\sum F_x = 0 \qquad \sum F_y = 0 \qquad \sum \tau_z = 0 \qquad \text{[10.26]}$$

where the axis of the torque equation is arbitrary.

### Quick Quiz 10.8

(a) Is it possible for a situation to exist in which Equation 10.24 is satisfied, but Equation 10.25 is not? (b) Can Equation 10.25 be satisfied, but Equation 10.24 is not?

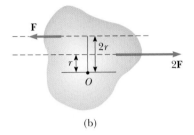

**Figure 10.16**

(a) The two forces acting on the object are equal in magnitude and opposite in direction. Because they also act along the same line of action, the net torque is zero and the object is in equilibrium. (b) Another situation in which two forces act on an object to produce zero net torque (but *not* zero net force).

**PITFALL PREVENTION 10.7**
**Zero torque**

Keep in mind that zero net torque does not mean an absence of rotational motion. An object rotating at a constant angular speed can be under the influence of a net torque of zero. This is analogous to the translational situation—zero net force does not mean an absence of translational motion.

In working static equilibrium problems, it is important to recognize all external forces acting on the object. Failure to do so will result in an incorrect analysis. The following procedure is recommended when analyzing an object in equilibrium under the action of several external forces:

---

**PROBLEM-SOLVING STRATEGY** | **Rigid Body in Equilibrium**

---

1. Make a sketch of the rigid body under consideration.
2. Draw a free-body diagram, and label all external forces acting on the object. Try to guess the correct direction for each force. If you select a direction that leads to a negative sign in your solution for a force, do not be alarmed; this merely means that the direction of the force is the opposite of what you guessed.
3. Resolve all forces into rectangular components, choosing a convenient coordinate system. Then apply the first condition for equilibrium, Equation 10.24. Remember to keep track of the signs of the various force components.
4. Choose a convenient axis for calculating the net torque on the rigid body. Remember that the choice of the axis for the torque equation is arbitrary; therefore, choose an axis that will simplify your calculation as much as possible. Usually, the most convenient axis for calculating torques is one through a point at which several forces act, so that their torques around this axis are zero. If you don't know a force or don't need to know a force, it is often beneficial to choose an axis through the point at which this force acts. Apply the second condition for equilibrium, Equation 10.25.
5. Solve the simultaneous equations for the unknowns in terms of the known quantities.

---

### Example 10.8   Standing on a Horizontal Beam

A uniform horizontal beam of length 8.00 m and weight 200 N is attached to a wall by a pin connection. Its far end is supported by a cable that makes an angle of 53.0° with the horizontal (Fig. 10.17a). If a 600-N man stands 2.00 m from the wall, find the tension in the cable and the force exerted by the wall on the beam.

**Solution**   The beam–man system is at rest and remains at rest, so it is clearly in static equilibrium. First we must identify all the external forces acting on the system, which we do in the free-body diagram in Figure 10.17b. These are the weights of the beam and the man, the force **T** exerted by the cable, the force **R** exerted by the wall at the pivot (the direction of this force is unknown). (The force between the man and the beam is internal to the system, so it is not included in the free-body diagram.) Notice that we have imagined the

gravitational force on the beam as acting at its center of gravity. Because the beam is uniform, the center of gravity is at the geometric center. If we resolve **T** and **R** into horizontal and vertical components (Figure 10.17c) and apply the first condition for equilibrium for the system, we have

$$(1) \quad \sum F_x = R\cos\theta - T\cos 53.0° = 0$$

$$(2) \quad \sum F_y = R\sin\theta + T\sin 53.0° - 600\text{ N} - 200\text{ N} = 0$$

Because we have three unknowns—$R$, $T$, and $\theta$—we cannot obtain a solution from these two expressions alone.

To generate a third expression, let us invoke the condition for rotational equilibrium because the system can be modeled as a rigid body in equilibrium. A convenient axis to choose for our torque equation is the one that passes

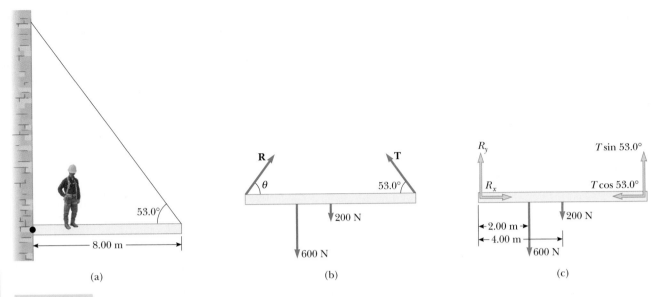

**Figure 10.17**

(Example 10.8) (a) A uniform beam supported by a cable. A man walks out on the beam. (b) The free-body diagram for the beam–man system. (c) The free-body diagram with forces resolved into horizontal and vertical components.

through the pivot at the wall. The feature that makes this point so convenient is that the force **R** and the horizontal component of **T** both have a lever arm of zero, and hence zero torque, about this pivot. Recalling our convention for the sign of the torque about an axis and noting that the lever arms of the 600-N, 200-N, and $T\sin 53°$ forces are 2.00 m, 4.00 m, and 8.00 m, respectively, we have

$$\sum \tau = (T\sin 53.0°)(8.00\ \text{m}) - (600\ \text{N})(2.00\text{m})$$
$$- (200\ \text{N})(4.00\ \text{m}) = 0$$

$$T = \boxed{313\ \text{N}}$$

The torque equation gives us one of the unknowns directly! This is due to our judicious choice of the axis. This value is substituted into (1) and (2) to give

$$R\cos \theta = 188\ \text{N}$$
$$R\sin \theta = 550\ \text{N}$$

We divide these two equations to find

$$\tan \theta = \frac{550\ \text{N}}{188\ \text{N}} = 2.93$$

$$\theta = 71.1°$$

Finally,

$$R = \frac{188\ \text{N}}{\cos \theta} = \frac{188\ \text{N}}{\cos 71.1°} = \boxed{581\ \text{N}}$$

If we had selected some other axis for the torque equation, the results would have been the same, although the details of the solution would be somewhat different. For example, if we had chosen to have the axis pass through the center of gravity of the beam, the torque equation would involve both $T$ and $R$. However, this equation, coupled with (1) and (2), could still be solved for the unknowns $T$, $R$, and $\theta$, yielding the same results. Try it!

## Example 10.9    The Leaning Ladder

A uniform ladder of length $\ell$ and mass $m$ rests against a smooth, vertical wall (Fig. 10.18a). If the coefficient of static friction between ladder and ground is $\mu_s = 0.40$, find the minimum angle $\theta_{min}$ such that the ladder does not slip.

**Solution**   The ladder is at rest and remains at rest, so we model it as a rigid body in equilibrium. The free-body diagram showing all the external forces acting on the ladder is illustrated in Figure 10.18b. The reaction **R** exerted by the ground on the ladder is the vector sum of a normal force **n** and the force of static friction **f**. The reaction force **P**

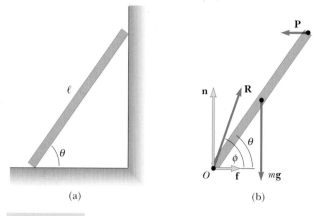

**Figure 10.18**

(Example 10.9) (a) A uniform ladder at rest, leaning against a frictionless wall. (b) The free-body diagram for the ladder.

exerted by the wall on the ladder is horizontal, because the wall is smooth, meaning that it is frictionless. Thus, **P** is simply the normal force on the ladder from the wall. From the first condition of equilibrium applied to the ladder, we have

$$\sum F_x = f - P = 0$$

$$\sum F_y = n - mg = 0$$

We see from the second equation that $n = mg$. Furthermore, when the ladder is on the verge of slipping, the force of static friction must be a maximum, given by $f_{s,\,max} = \mu_s n$.

To find $\theta$, we use the second condition of equilibrium. When the torques are taken about the origin $O$ at the bottom of the ladder, we have

$$\sum \tau_O = P\ell \sin \theta - mg \frac{\ell}{2} \cos \theta = 0$$

This expression gives

$$\tan \theta_{min} = \frac{mg}{2P} = \frac{n}{2f_{s,\,max}} = \frac{n}{2(\mu_s n)} = \frac{1}{2(0.40)} = 1.25$$

$$\theta_{min} = \boxed{51°}$$

It is interesting that the result does not depend on $\ell$ or $m$. The answer depends only on $\mu_s$.

**EXERCISE**   Two children weighing 500 N and 350 N are on a uniform board weighing 40.0 N supported at its center (a "see-saw"). If the 500-N child is 1.50 m from the center, determine (a) the upward force exerted on the board by the support and (b) where the 350-N child must sit to balance the system.

*Answer*   (a) 890 N   (b) 2.14 m from the center

## 10.7 • THE RIGID BODY UNDER A NET TORQUE

See Screen 7.6

In the preceding section, we investigated the equilibrium situation in which the net torque on a rigid body is zero. What if the net torque on a rigid body is not zero? In analogy with Newton's second law for translational motion, we should expect the angular speed of the rigid body to change. The net torque will cause angular acceleration of the rigid body. We investigate this notion in this section.

Let us imagine a rotating rigid body again as a collection of particles. The rigid body will be subject to a number of forces applied at various locations on the rigid body, at which individual particles will be located. Thus, we can imagine that the forces on the rigid body are exerted on individual particles of the rigid body. We will calculate the net torque on the object due to the torques resulting from these forces around the rotation axis of the rotating body. Any applied force can be represented by its radial component and its tangential component. The radial component of an applied force provides no torque because its line of action goes through the rotation axis. Thus, only the tangential component of an applied force contributes to the torque.

On any given particle, described by index variable $i$, within the rigid body, we can use Newton's second law to describe the tangential acceleration of the particle:

$$F_{ti} = m_i a_{ti}$$

where the *t* subscript refers to tangential components. Let us multiply both sides of this expression by $r_i$, the distance of the particle from the rotation axis:

$$r_i F_{ti} = r_i m_i a_{ti}$$

Using Equation 10.11, and recognizing the definition of torque ($\tau = rF \sin \phi = rF_t$ in this case), we can rewrite this as

$$\tau_i = m_i r_i^2 \alpha_i$$

Now, let us add up the torques on all particles of the rigid body:

$$\sum_i \tau_i = \sum_i m_i r_i^2 \alpha_i$$

The left side is the net torque on all particles of the rigid body. The net torque associated with *internal* forces is zero, however. To understand this, recall that Newton's third law tells us that the internal forces occur in equal and opposite pairs that lie along the line of separation of each pair of particles. The torque due to each action–reaction force pair is therefore zero. On summation of all torques, we see that the *net internal torque vanishes*. The term on the left, then, reduces to the net *external* torque.

On the right, we impose the rigid body model by demanding that all particles have the same angular acceleration. Thus, this equation becomes

$$\sum \tau = \left( \sum_i m_i r_i^2 \right) \alpha$$

where the torque and angular acceleration no longer have subscripts because they refer to quantities associated with the rigid body as a whole rather than to individual particles. We recognize the quantity in parentheses as the moment of inertia *I*. Therefore,

$$\sum \tau = I\alpha \qquad \text{[10.27]}$$

• *Relationship between net torque and angular acceleration*

That is, **the net torque acting on the rigid body is proportional to its angular acceleration,** and the proportionality constant is the moment of inertia. It is important to note that $\sum \tau = I\alpha$ is the rotational analog of Newton's second law of motion, $\sum F = ma$.

(a)

### Quick Quiz 10.9

If you turn off your workshop grinding wheel at the same time as your electric drill, the grinding wheel takes much longer to stop rotating. Why?

**F** →

(b)

**Figure 10.19**

(Thinking Physics 10.3) (a) When you step on the gas, the car "noses up." (b) When you step on the brake, the car "noses down."

### THINKING PHYSICS 10.3

When an automobile driver steps on the accelerator, the nose of the car moves upward. When the driver brakes, the nose moves downward. Why do these effects occur?

**Reasoning** When the driver steps on the accelerator, there is an increased force on the tires from the roadway. This force is parallel to the roadway and directed toward the front of the automobile, as suggested in Figure 10.19a. This force provides a torque that tends to cause the car to rotate in the clockwise direction around the center of mass of the car. The result of this rotation is a "nosing up" of the car.

When the driver steps on the brake, there is an increased force on the tires from the roadway, directed toward the rear of the automobile, as suggested in Figure 10.19b. This force results in a torque that causes a counterclockwise rotation and the subsequent "nosing down" of the automobile.

## Example 10.10 An Atwood Machine with a Massive Pulley

In Example 4.4, we analyzed an Atwood machine in which two objects with unequal masses hang from a string that passes over a light, frictionless pulley. Suppose the pulley, which is modeled as a disk, has mass $M$ and radius $R$, and the pulley surface is not frictionless, so that the string does not slide on the pulley (Fig. 10.20a). We will assume that the frictional torque acting at the bearing of the pulley is negligible. Calculate the magnitude of the acceleration of the two objects.

**Reasoning**   Because counterclockwise angular acceleration of the pulley is defined as positive, we define the positive directions for $m_1$ and $m_2$ as shown in Figure 10.20 so that all accelerations, translational and rotational, are positive if $m_1$ accelerates downward. If the pulley has mass and friction, the tensions $T_1$ and $T_2$ in the string on either side of the pulley are not equal in magnitude, as they are in Example 4.4. Indeed, it is the difference in torque due to these different tensions that provides the net torque to cause the angular acceleration of the pulley (Fig. 10.20b). Consequently, some of what we do here will look similar to Example 4.4, with the exception of incorporating $T_1$ and $T_2$ in our mathematical representation instead of just $T$. The forces $M\mathbf{g}$ and $\mathbf{F}$ (the force supporting the pulley) both act through the pulley axle, so these forces do not contribute to the torque on the pulley.

**Solution**   With the help of the free-body diagrams in Figure 10.20b, we apply Newton's second law to $m_1$, so that

$$(1) \ \sum F_y = m_1 g - T_1 = m_1 a$$

and for $m_2$,

$$(2) \ \sum F_y = T_2 - m_2 g = m_2 a$$

We cannot solve these two equations for $a$, as is done in Example 4.4, because we have three unknowns: $a$, $T_1$, and $T_2$. We can find a third equation by applying Equation 10.27 to the pulley (Fig. 10.20b):

$$(3) \ \sum \tau = T_1 R - T_2 R = I\alpha = \left(\tfrac{1}{2}MR^2\right)\left(\frac{a}{R}\right)$$

$$= \tfrac{1}{2}MRa \quad \longrightarrow \quad T_1 - T_2 = \tfrac{1}{2}Ma$$

(a)

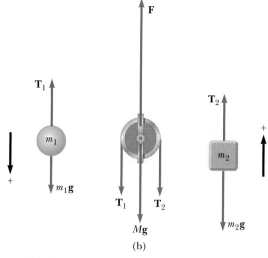

(b)

**Figure 10.20**

(Example 10.10) (a) An Atwood machine with a massive pulley. The pulley is modeled as a disk. (b) Free-body diagrams for the two hanging objects and the pulley.

We substitute into (3) expressions for $T_1$ and $T_2$ from (1) and (2):

$$(m_1 g - m_1 a) - (m_2 a + m_2 g) = \tfrac{1}{2} M a$$

$$a = \left( \frac{m_1 - m_2}{m_1 + m_2 + \tfrac{1}{2}M} \right) g$$

Notice that this differs from the result for Example 4.4 only in the extra term $\tfrac{1}{2}M$ in the denominator. If the pulley mass $M \rightarrow 0$, this expression reduces to that in Example 4.4.

## Work and Energy in Rotational Motion

In translational motion, we found energy concepts, and in particular the reduction of the continuity equation for energy called the work–kinetic energy theorem, to be extremely useful in describing the motion of a system. Energy concepts can be equally useful in simplifying the analysis of rotational motion, as we saw in the isolated system analysis in Example 10.5. From the continuity equation for energy, we expect that for rotation of an object about a fixed axis, the work done by external forces on the object will equal the change in the rotational kinetic energy, as long as energy is not stored by any other means. To show that this is in fact the case, we begin by finding an expression for the work done by a torque.

Consider a rigid body pivoted at the point $O$ in Figure 10.21. Suppose a single external force $\mathbf{F}$ is applied at the point $P$ and $d\mathbf{s}$ is the displacement of the point of application of the force. The small amount of work $dW$ done by $\mathbf{F}$ as the point of application rotates through an infinitesimal distance $ds = r\, d\theta$ in a time $dt$ is

$$dW = \mathbf{F} \cdot d\mathbf{s} = (F \sin \phi)\, r\, d\theta$$

where $F \sin \phi$ is the tangential component of $\mathbf{F}$, or the component of the force along the displacement. Note from Figure 10.21 that the **radial component of F does no work because it is perpendicular to the displacement.**

Because the magnitude of the torque due to $\mathbf{F}$ about the origin is defined as $rF \sin \phi$, we can write the work done for the infinitesimal rotation in the form

$$dW = \tau\, d\theta \qquad \textbf{[10.28]}$$

Notice that this is the product of torque and angular displacement, making it analogous to the work done in translational motion, which is the product of force and translational displacement (Eq. 6.2).

Now, we will combine this result with the rotational form of Newton's second law, $\tau = I\alpha$. Using the chain rule from the calculus, we can express the torque as

$$\tau = I\alpha = I\frac{d\omega}{dt} = I\frac{d\omega}{d\theta}\frac{d\theta}{dt} = I\frac{d\omega}{d\theta}\omega$$

Rearranging this expression and noting that $\tau\, d\theta = dW$ from Equation 10.28, we have

$$\tau\, d\theta = dW = I\omega\, d\omega$$

Integrating this expression, we find the total work done by the torque:

$$W = \int_{\theta_i}^{\theta_f} \tau\, d\theta = \int_{\omega_i}^{\omega_f} I\omega\, d\omega = \tfrac{1}{2}I\omega_f^2 - \tfrac{1}{2}I\omega_i^2 \qquad \textbf{[10.29]}$$

• *Work–kinetic energy theorem for pure rotation*

Notice that this equation, which can be written as $W = \Delta K_R$, has exactly the same mathematical form as the work–kinetic energy theorem for translation.

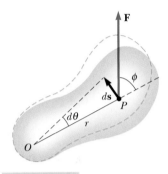

**Figure 10.21**

A rigid body rotates about an axis through $O$ under the action of an external force $\mathbf{F}$ applied at $P$.

We finish this discussion of energy concepts for rotation by investigating the *rate* at which work is being done by **F** on an object rotating about a fixed axis. This rate is obtained by dividing the left and right sides of Equation 10.28 by $dt$:

$$\frac{dW}{dt} = \tau \frac{d\theta}{dt}$$

[**10.30**]

The quantity $dW/dt$ is, by definition, the instantaneous power $\mathcal{P}$ delivered by the force. Furthermore, because $d\theta/dt = \omega$, Equation 10.30 reduces to

$$\mathcal{P} = \frac{dW}{dt} = \tau\omega$$

[**10.31**]

This expression is analogous to $\mathcal{P} = Fv$ in the case of translational motion.

## Example 10.11  Work on an Atwood Machine

Consider the Atwood machine of Example 10.10. Calculate the rate at which work is done on the system of the pulley and the hanging objects by the gravitational force.

**Reasoning**  The only means by which energy enters the system is by work done by the gravitational force. The only types of energy in the system are translational and rotational kinetic energy. Thus, we can apply the reduction of the continuity equation called the work–kinetic energy theorem, $W = \Delta K$. The rate at which work is being done on the system is equal to the rate at which the kinetic energy of the system is changing, so we will evaluate the power as $\mathcal{P} = dK/dt$.

**Solution**  Because the acceleration of the hanging objects in Example 10.10 is constant, we can find the speed of either object at time $t$ by using Equation 2.8. If the objects start from rest, the speed of either at time $t$ is

$$v_f = v_i + at = \left( \frac{m_1 - m_2}{m_1 + m_2 + \frac{1}{2}M} \right) gt$$

Using Equation 10.10, the angular speed of the pulley at time $t$ is

$$\omega_f = \frac{v_f}{R} = \frac{1}{R} \left( \frac{m_1 - m_2}{m_1 + m_2 + \frac{1}{2}M} \right) gt$$

Thus, the kinetic energy of the system is

$$
\begin{aligned}
K_{\text{system}} &= \tfrac{1}{2}m_1 v_{1f}^2 + \tfrac{1}{2}m_2 v_{2f}^2 + \tfrac{1}{2}I\omega_f^2 \\
&= \tfrac{1}{2}(m_1 + m_2)\left( \frac{m_1 - m_2}{m_1 + m_2 + \frac{1}{2}M} \right)^2 g^2 t^2 \\
&\quad + \tfrac{1}{2}(\tfrac{1}{2}MR^2)\frac{1}{R^2}\left( \frac{m_1 - m_2}{m_1 + m_2 + \frac{1}{2}M} \right)^2 g^2 t^2 \\
&= \tfrac{1}{2}(m_1 + m_2 + \tfrac{1}{2}M)\left( \frac{m_1 - m_2}{m_1 + m_2 + \frac{1}{2}M} \right)^2 g^2 t^2
\end{aligned}
$$

**Figure 10.22**

(Example 10.11) The tension in the cord produces a torque about the axle passing through $O$.

$$= \tfrac{1}{2}\frac{(m_1 - m_2)^2}{m_1 + m_2 + \frac{1}{2}M} g^2 t^2$$

Finally, we differentiate this expression with respect to time:

$$
\mathcal{P} = \frac{dK_{\text{system}}}{dt} = \frac{d}{dt}\left[ \tfrac{1}{2}\frac{(m_1 - m_2)^2}{m_1 + m_2 + \frac{1}{2}M} g^2 t^2 \right]
$$

$$= \frac{(m_1 - m_2)^2}{m_1 + m_2 + \frac{1}{2}M} g^2 t$$

Thus, the power increases with time, because the angular speed increases in time.

**EXERCISE**  The wheel in Figure 10.22 is a solid disk with $M = 2.00$ kg, $R = 30.0$ cm, and $I = 0.0900$ kg·m². The suspended object has a mass $m = 0.500$ kg. If the suspended object starts from rest and descends to a point 1.00 m lower, what is its speed at this point?

*Answer*  2.56 m/s

# 10.8 • ANGULAR MOMENTUM

Imagine an object rotating in space with no motion of its center of mass. Each particle in the object is moving in a circular path, so momentum is associated with the motion of each particle. Although the object has no linear momentum (its center of mass is not moving through space), a "quantity of motion" is associated with its rotation. We will investigate **angular momentum** in this section.

Let us first consider a particle of mass $m$, situated at the vector position $\mathbf{r}$ and moving with a momentum $\mathbf{p}$, as shown in Figure 10.23. For now, this is not a particle on a rigid body; it is any particle moving with momentum $\mathbf{p}$. We will apply the result to a rotating rigid body shortly. The **instantaneous angular momentum L** of the particle relative to the origin $O$ is defined by the vector product of its instantaneous vector position $\mathbf{r}$ and the instantaneous linear momentum $\mathbf{p}$:

$$\mathbf{L} = \mathbf{r} \times \mathbf{p} \qquad [10.32]$$

The SI units of angular momentum are $\text{kg} \cdot \text{m}^2/\text{s}$. It is important to note that both the magnitude and the direction of $\mathbf{L}$ depend on the choice of origin. The direction of $\mathbf{L}$ is perpendicular to the plane formed by $\mathbf{r}$ and $\mathbf{p}$, and the sense of $\mathbf{L}$ is governed by the right-hand rule. For example, in Figure 10.23, $\mathbf{r}$ and $\mathbf{p}$ are assumed to be in the $xy$ plane, and $\mathbf{L}$ points in the $z$ direction. Because $\mathbf{p} = m\mathbf{v}$, the magnitude of $\mathbf{L}$ is

$$L = mvr \sin \phi \qquad [10.33]$$

where $\phi$ is the angle between $\mathbf{r}$ and $\mathbf{p}$. It follows that $\mathbf{L}$ is zero when $\mathbf{r}$ is parallel to $\mathbf{p}$ ($\phi = 0°$ or $180°$). In other words, when the particle moves along a line that passes through the origin, it has zero angular momentum with respect to the origin. This is equivalent to stating that the momentum vector is not tangent to *any* circle drawn about the origin. On the other hand, if $\mathbf{r}$ is perpendicular to $\mathbf{p}$ ($\phi = 90°$), $L$ is a maximum and equal to $mvr$. In fact, at that instant the particle moves exactly as though it were on the rim of a wheel rotating about the origin in a plane defined by $\mathbf{r}$ and $\mathbf{p}$. A particle has nonzero angular momentum about some point if the position vector of the particle measured from that point rotates about the point as the particle moves.

For translational motion, we found that the net force on a particle equals the time rate of change of the particle's linear momentum (Eq. 8.3). We shall now show that Newton's second law implies an analogous situation for rotation—that the net torque acting on a particle equals the time rate of change of the particle's angular momentum. Let us start by writing the torque on the particle in the form

$$\boldsymbol{\tau} = \mathbf{r} \times \mathbf{F} = \mathbf{r} \times \frac{d\mathbf{p}}{dt} \qquad [10.34]$$

where we have used the fact that $\mathbf{F} = d\mathbf{p}/dt$ (Eq. 8.3). Now let us differentiate Equation 10.32 with respect to time, using the product rule for differentiation:

$$\frac{d\mathbf{L}}{dt} = \frac{d}{dt}(\mathbf{r} \times \mathbf{p}) = \mathbf{r} \times \frac{d\mathbf{p}}{dt} + \frac{d\mathbf{r}}{dt} \times \mathbf{p}$$

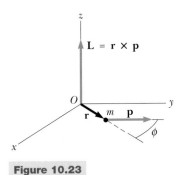

**Figure 10.23**

The angular momentum $\mathbf{L}$ of a particle of mass $m$ and linear momentum $\mathbf{p}$ located at the position $\mathbf{r}$ is a vector given by $\mathbf{L} = \mathbf{r} \times \mathbf{p}$. The value of $\mathbf{L}$ depends on the origin about which it is measured and is a vector perpendicular to both $\mathbf{r}$ and $\mathbf{p}$.

**PITFALL PREVENTION 10.8**

**Is rotation necessary for angular momentum?**

Notice that we can define angular momentum even though there is no motion of the particle in a circular path. Even a particle moving in a straight line has angular momentum about any axis displaced from the path of the particle.

It is important to adhere to the order of factors in the vector product because the vector product is not commutative.

The last term on the right in the preceding equation is zero because $\mathbf{v} = d\mathbf{r}/dt$ is parallel to $\mathbf{p}$. Therefore,

$$\frac{d\mathbf{L}}{dt} = \mathbf{r} \times \frac{d\mathbf{p}}{dt} \tag{10.35}$$

Comparing Equations 10.34 and 10.35, we see that

- *Torque equals time rate of change of angular momentum*

$$\boxed{\boldsymbol{\tau} = \frac{d\mathbf{L}}{dt}} \tag{10.36}$$

This result is the rotational analog of Newton's second law, $\mathbf{F} = d\mathbf{p}/dt$. Equation 10.36 says that the torque acting on a particle is equal to the time rate of change of the particle's angular momentum. It is important to note that Equation 10.36 is valid only if the origins of $\boldsymbol{\tau}$ and $\mathbf{L}$ are the *same*. Equation 10.36 is also valid when several forces are acting on the particle, in which case $\boldsymbol{\tau}$ is the *net* torque on the particle. Of course, the same origin must be used in calculating all torques as well as the angular momentum.

Now, let us apply these ideas to a system of particles. The total angular momentum $\mathbf{L}$ of the system of particles about some point is defined as the vector sum of the angular momenta of the individual particles:

$$\mathbf{L} = \mathbf{L}_1 + \mathbf{L}_2 + \cdots + \mathbf{L}_n = \sum_i \mathbf{L}_i$$

where the vector sum is over all of the $n$ particles in the system.

Because the individual angular momenta of the particles may change in time, the total angular momentum may also vary in time. In fact, from Equations 10.34 and 10.35, we find that the time rate of change of the total angular momentum of the system equals the vector sum of *all* torques, including those associated with internal forces between particles and those associated with external forces.

As we found in our discussion of the rigid body under a net torque, however, the sum of the internal torques is zero. Thus, we conclude that the total angular momentum can vary with time *only* if there is a net *external* torque on the system, so that we have

- *Net external torque on a system equals time rate of change of angular momentum*

$$\sum \boldsymbol{\tau}_{\text{ext}} = \sum_i \frac{d\mathbf{L}_i}{dt} = \frac{d}{dt} \sum_i \mathbf{L}_i = \frac{d\mathbf{L}_{\text{tot}}}{dt} \tag{10.37}$$

That is, the time rate of change of the total angular momentum of the system about some origin in an inertial frame equals the net external torque acting on the system about that origin. Note that Equation 10.37 is the rotational analog of $\sum \mathbf{F}_{\text{ext}} = d\mathbf{p}_{\text{tot}}/dt$ (Eq. 8.39) for a system of particles.

This result is valid for a system of particles that change their positions with respect to one another, that is, a nonrigid body. In this discussion of angular momentum of a system of particles, notice that we never imposed the rigid body condition.

| **TABLE 10.3** | **A Comparison of Equations for Rotational and Translational Motion: Dynamic Equations[a]** | |
|---|---|---|
| | **Rotational Motion About a Fixed Axis** | **Translational Motion** |
| Kinetic energy | $K_R = \frac{1}{2}I\omega^2$ | $K = \frac{1}{2}mv^2$ |
| Equilibrium | $\sum \tau = 0$ | $\sum \mathbf{F} = 0$ |
| Newton's second law | $\sum \tau = I\alpha$ | $\sum \mathbf{F} = m\mathbf{a}$ |
| Newton's second law | $\sum \tau = \dfrac{d\mathbf{L}}{dt}$ | $\sum \mathbf{F} = \dfrac{d\mathbf{p}}{dt}$ |
| Momentum | $L = I\omega$ | $\mathbf{p} = m\mathbf{v}$ |
| Conservation principle | $L_i = L_f$ | $\mathbf{p}_i = \mathbf{p}_f$ |
| Power | $\mathcal{P} = \tau\omega$ | $\mathcal{P} = Fv$ |

[a] Equations in translation motion expressed in terms of vectors have rotational analogs in terms of vectors. Because the full vector treatment of rotation is beyond the scope of this book, however, some rotational equations are given in nonvector form.

Equation 10.37 is the primary equation in the angular momentum version of the nonisolated system model. The system's angular momentum changes in response to an interaction with the environment, described by means of the net torque on the system.

One final result can be obtained for angular momentum, which will serve as an analog to the definition of linear momentum. Let us imagine a rigid body rotating about an axis. Each particle of mass $m_i$ in the rigid body moves in a circular path of radius $r_i$, with a tangential speed $v_i$. Thus, the total angular momentum of the rigid body is

$$L = \sum_i m_i v_i r_i$$

Let us now replace the tangential speed with the product of the radial distance and the angular speed (Eq. 10.10):

$$L = \sum_i m_i v_i r_i = \sum_i m_i(r_i\omega)\, r_i = \left(\sum_i m_i r_i^2\right)\omega$$

We recognize the combination in the parentheses as the moment of inertia, so we can write the angular momentum of the rigid body as

$$L = I\omega$$

• *Angular momentum of an object with moment of inertia I*

This is the rotational analog to $p = mv$. Table 10.3 is a continuation of Table 10.1, with additional translational and rotational analogs that we have developed in the past few sections.

## Example 10.12   The Atwood Machine Once Again

Consider again the Atwood machine with the massive pulley in Examples 10.10 and 10.11. Determine the acceleration of the two objects using an angular momentum approach.

**Reasoning**   This is an example of a nonrigid body experiencing a net torque, so we use the nonisolated system model. We will evaluate the angular momentum of the system at any time and then differentiate the angular momentum, setting it equal to the net external torque. We will solve the resulting expression for the acceleration of the objects.

**Solution**   Let us first calculate the angular momentum of the system, which consists of the two objects plus the pulley. At the instant $m_1$ and $m_2$ have a speed $v$, the angular momentum of $m_1$ around the axle of the pulley is $m_1 vR$, and that of $m_2$ is $m_2 vR$. At the same instant, the angular momentum of the pulley around its center is $L = I\omega = Iv/R$. Therefore, the total angular momentum of the system is

$$(1)\quad L = m_1 vR + m_2 vR + I\frac{v}{R} = m_1 vR + m_2 vR + (\tfrac{1}{2}MR^2)\frac{v}{R}$$

$$= (m_1 + m_2 + \tfrac{1}{2}M)vR$$

Now let us evaluate the total external torque on the system about the axle. The weight of the pulley and the force of the axle upward on the pulley have zero moment arm around the center of the pulley, so they do not contribute to the torque. The external forces on the system that produce torques about the axle are $m_1\mathbf{g}$, with a torque of $m_1 gR$, and $m_2\mathbf{g}$, with a torque of $-m_2 gR$. Combining the net external torque with (1) and Equation 10.37 gives us

$$\tau_{\text{ext}} = \frac{dL}{dt}$$

$$m_1 gR - m_2 gR = \frac{d}{dt}[(m_1 + m_2 + \tfrac{1}{2}M)vR]$$

$$(2)\quad m_1 gR - m_2 gR = (m_1 + m_2 + \tfrac{1}{2}M)R\frac{dv}{dt}$$

Because $dv/dt = a$, we can solve Equation (2) for $a$ to find

$$a = \left(\frac{m_1 - m_2}{m_1 + m_2 + \tfrac{1}{2}M}\right)g$$

which is the same result as that obtained in Example 10.10. You may wonder why we did not include the tension forces that the cord exerts on the objects in evaluating the net torque about the axle. The reason is that these forces are *internal* to the system under consideration. Only the *external* torques contribute to the change in angular momentum.

---

**EXERCISE**   A car of mass 1 500 kg moves on a circular race track of radius 50 m with a speed of 40 m/s. What is the magnitude of its angular momentum relative to the center of the track?

*Answer*   $3.0 \times 10^6 \text{ kg} \cdot \text{m}^2/\text{s}$

**EXERCISE**   What is the angular momentum of the Earth due to its (a) revolution about the Sun, relative to the Sun, and (b) rotation about its axis, relative to the axis? Assume the Earth is a sphere of uniform density.

*Answer*   (a) $2.7 \times 10^{40} \text{ kg} \cdot \text{m}^2/\text{s}$   (b) $7.1 \times 10^{33} \text{ kg} \cdot \text{m}^2/\text{s}$

## 10.9 • CONSERVATION OF ANGULAR MOMENTUM

In Chapter 8 we found that the total linear momentum of a system of particles remains constant when the net external force acting on the system is zero. In rotational motion, we have an analogous conservation law that states that the **total angular momentum of a system remains constant if the net external torque acting on the system is zero.**

Because the net torque acting on the system equals the time rate of change of the system's angular momentum, we see that if

$$\sum \boldsymbol{\tau}_{\text{ext}} = \frac{d\mathbf{L}_{\text{tot}}}{dt} = 0 \qquad\qquad \textbf{[10.38]}$$

then

• *Conservation of angular momentum for an isolated system*

$$\mathbf{L}_{\text{tot}} = \text{constant} \longrightarrow \mathbf{L}_{\text{tot},\, i} = \mathbf{L}_{\text{tot},\, f} \qquad\qquad \textbf{[10.39]}$$

Equation 10.39 represents a third conservation law to add to our list of fundamental conservation principles. We can now state that the **total energy, linear momentum, and angular momentum of an isolated system are all conserved.** We have focused our attention in this chapter on rigid bodies; the conservation of angular momentum principle, however, is a general result of the isolated system model. Thus, **angular momentum of an isolated system is conserved whether the system is a rigid body or not.**

The angular momentum of a rigid body about a fixed axis has a magnitude given by $L = I\omega$, where $I$ is the moment of inertia of the rigid body about the axis. In this case, if the net external torque on the body is zero, we can express the conservation of angular momentum principle as $I\omega$ = constant.

Many examples can be used to demonstrate conservation of angular momentum; some of them should be familiar to you. You may have observed a figure skater spinning (Fig. 10.24). The angular speed of the skater increases as he pulls his hands and feet close to the trunk of his body. Ignoring friction between skater and ice, we see that there are no external torques on the skater. The moment of inertia of his body decreases as his hands and feet are brought in closer to his body.* The resulting change in angular speed is accounted for as follows. Because angular momentum must be conserved, the product $I\omega$ remains constant, and a decrease in $I$ causes a corresponding increase in $\omega$.

An interesting astrophysical example of conservation of angular momentum occurs when, at the end of its lifetime, a massive star uses up all its fuel and collapses under the influence of gravitational forces, causing a gigantic outburst of energy called a supernova explosion. The best studied example of a remnant of a supernova explosion is the Crab Nebula, a chaotic, expanding mass of gas (Fig. 10.25). In a supernova, part of the star's mass is released into space, where it

**Figure 10.24**

Angular momentum is conserved as figure skater Todd Eldridge pulls his arms toward his body. *(© 1998 David Madison)*

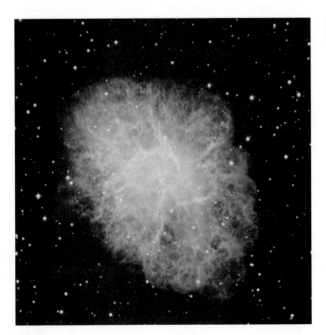

**Figure 10.25**

The Crab Nebula, in the constellation Taurus. This nebula is the remnant of a supernova explosion, which was seen on Earth in the year A.D. 1054. It is located some 6 300 lightyears away and is approximately 6 lightyears in diameter, still expanding outward. *(Max Planck Institute for Astronomy and Calar Alto Observatory; K. Meisenheimer and A. Quetz)*

**WEB**

For detailed Hubble Space Telescope images of the Crab Nebula, visit **www.seds.org/messier/more/m001_hst.html**

---

* We derived $L = I\omega$ for a rigid body. The skater is deformable, so is therefore not a rigid body. On the other hand, we apply $L = I\omega$ only at the instant before he pulls his arms in and the instant after he finishes pulling his arms in. At these two instants, no deformation occurs and he can be modeled as a rigid body.

eventually condenses into new stars and planets. Most of what is left behind typically collapses into a **neutron star**, an extremely dense sphere of matter with a diameter of about 10 km in comparison with the $10^6$-km diameter of the original star, and containing a large fraction of the original mass of the star. As the moment of inertia of the system decreases during the collapse, the star's rotational speed increases, similar to the increase in speed of the skater in Figure 10.24. More than 700 rapidly rotating neutron stars have been identified since the first discovery of such astronomical bodies in 1967, with periods of rotation ranging from a millisecond to several seconds. The neutron star is a most dramatic system—an object with a mass greater than the Sun, rotating about its axis many times each second!

### Quick Quiz 10.10

A skater moves along a straight line on frictionless ice and then grabs hold of a pole in the middle of the ice, so that she rotates around the pole. This represents a change from translational motion to rotational motion. Is the angular momentum of the skater conserved in this process?

### Example 10.13    A Revolving Ball on a Horizontal, Frictionless Surface

A ball of mass $m$ on a horizontal, frictionless table is connected to a string that passes through a small hole in the table. The ball is set into circular motion of radius $R$, at which time its speed is $v_i$ (Fig. 10.26). (a) If the string is pulled from the bottom so that the radius of the circular path is decreased to $r$, what is the final speed $v_f$ of the ball?

**Solution**   We identify the system as the ball. We will calculate torque about the center of rotation $O$. Note that the gravitational force acting on the ball is balanced by the up-

ward normal force, and so these forces cancel, resulting in zero net torque from these forces. The force **F** of the string on the ball acts toward the center of rotation, and the vector position **r** is directed away from $O$. Thus, we see that $\boldsymbol{\tau} = \mathbf{r} \times \mathbf{F} = 0$, so no torque is applied on the ball due to this force. Three forces are acting on the ball, but zero net torque occurs. Thus, **L** is a constant of the motion. The ball can be modeled as a particle moving in a circular path, so $L = mv_iR = mv_fr$, or

$$v_f = \frac{v_iR}{r}$$

From this result, we see that as $r$ decreases, the speed $v$ increases.

(b) Is the kinetic energy of the ball conserved in this process?

**Solution**   We set up the ratio of the final kinetic energy to the initial kinetic energy:

$$\frac{K_f}{K_i} = \frac{\frac{1}{2}mv_f^2}{\frac{1}{2}mv_i^2} = \frac{1}{v_i^2}\left(\frac{v_iR}{r}\right)^2 = \frac{R^2}{r^2}$$

Because this ratio is not equal to 1, kinetic energy is not conserved. Furthermore, because $R > r$, the kinetic energy of the ball has increased. This increase entered the system of the ball by the work done by the person pulling the string.

**Figure 10.26**

(Example 10.13) When the string is pulled downward, the speed of the ball changes.

### Example 10.14 Rotation Period of a Neutron Star

A star undergoes a supernova explosion. The material left behind forms a sphere of radius $8.0 \times 10^6$ m just after the explosion with a rotation period of 15 h. This remaining material collapses into a neutron star of radius 8.0 km. What is the rotation period $T$ of the neutron star?

**Solution** We model the star as an isolated system, so that angular momentum is conserved. As the moment of inertia of the stellar core decreases during its collapse, the angular speed increases. Thus,

$$I_i\omega_i = I_f\omega_f$$

We have no information about the variation in the density of material with radius in either the initial star or the neutron star, so we choose a simplification model in which the density

is uniform. Our result will most likely not be entirely accurate, but we can consider it an estimate. Using the moment of inertia of a sphere of uniform density (see Table 10.2),

$$(\tfrac{2}{5}MR_i^2)\omega_i = (\tfrac{2}{5}MR_f^2)\omega_f \longrightarrow \omega_f = \left(\frac{R_i}{R_f}\right)^2 \omega_i$$

Now, because $\omega = 2\pi/T$, we have

$$\frac{2\pi}{T_f} = \left(\frac{R_i}{R_f}\right)^2 \frac{2\pi}{T_i} \longrightarrow T_f = \left(\frac{R_f}{R_i}\right)^2 T_i$$

Substituting numerical values, we have

$$T_f = \left(\frac{8.0 \times 10^3 \text{ m}}{8.0 \times 10^6 \text{ m}}\right)^2 (15 \text{ h}) = 1.5 \times 10^{-5} \text{ h} = \boxed{0.054 \text{ s}}$$

## 10.10 • PRECESSIONAL MOTION OF GYROSCOPES

Angular momentum is the basis of the operation of a **gyroscope,** which is a spinning object used to control or maintain the orientation in space of the object or a system containing the object. As an example, consider a quarterback passing a football. If he imparts no spin to the ball, there is no angular momentum to be conserved, and forces from the air might cause the ball to tumble as it moves through its trajectory. If a spin is imparted to the ball along the long axis of the football, however, the angular momentum vector stays fixed in direction and the football maintains its orientation throughout the trajectory, resulting in much less air resistance and a longer pass. In this application, the football is acting as a gyroscope to maintain its own orientation in space.

An unusual and fascinating type of motion you probably have observed is that of a top spinning rapidly about its axis of symmetry, as shown in Figure 10.27a. The top is acting as a gyroscope and one might expect the orientation to remain fixed in space. If the top is leaning over, however, it is observed that the symmetry axis rotates about the $z$ axis, sweeping out a cone (see Fig. 10.27b). This phenomenon is called **precessional motion.** The angular speed of the symmetry axis about the vertical is usually slow relative to the angular speed of the top about the symmetry axis.

It is quite natural to wonder why the top does not maintain its direction of spin. Because the center of mass of the top is not directly above the pivot point $O$, a net torque is acting on the top about $O$—a torque resulting from the gravitational force $M\mathbf{g}$. The top would certainly fall over if it were not spinning. Because it is spinning, however, it has an angular momentum $\mathbf{L}$ directed along its symmetry axis. As we shall show, the motion of this symmetry axis about the $z$ axis (the precessional motion) occurs because the torque produces a change in the *direction* of the symmetry axis. This is an excellent example of the importance of the directional nature of angular momentum.

The two forces acting on the top are the downward gravitational force $M\mathbf{g}$ and the normal force $\mathbf{n}$ acting upward at the pivot point $O$. The normal force produces no torque about the pivot because its moment arm through that point is zero. However, the gravitational force produces a torque $\boldsymbol{\tau} = \mathbf{r} \times M\mathbf{g}$ about $O$, where

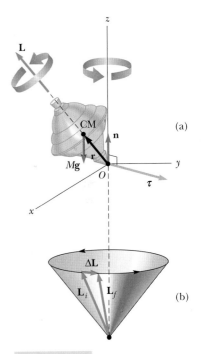

**Figure 10.27**

Precessional motion of a top spinning about its symmetry axis. (a) The only external forces acting on the top are the normal force $\mathbf{n}$ and the gravitational force $M\mathbf{g}$. The direction of the angular momentum $\mathbf{L}$ is along the axis of symmetry. The right-hand rule indicates that $\boldsymbol{\tau} = \mathbf{r} \times \mathbf{F} = \mathbf{r} \times M\mathbf{g}$ is in the $xy$ plane. (b). The direction of $\Delta\mathbf{L}$ is parallel to that of $\boldsymbol{\tau}$ in part (a). The fact that $\mathbf{L}_f = \mathbf{L}_i + \Delta\mathbf{L}$ indicates that the top precesses about the $z$ axis.

the direction of $\boldsymbol{\tau}$ is perpendicular to the plane formed by $\mathbf{r}$ and $M\mathbf{g}$. By necessity, the vector $\boldsymbol{\tau}$ lies in a horizontal plane perpendicular to the angular momentum vector. The net torque and angular momentum of the top are related through Equation 10.37:

$$\sum \boldsymbol{\tau} = \frac{d\mathbf{L}}{dt}$$

From this expression, we see that the nonzero torque produces a change in angular momentum $d\mathbf{L}$—a change that is in the same direction as $\sum \boldsymbol{\tau}$. Therefore, like the torque vector, $d\mathbf{L}$ must also be perpendicular to $\mathbf{L}$. Figure 10.27b illustrates the resulting precessional motion of the symmetry axis of the top. In a time $\Delta t$, the change in angular momentum is $\Delta\mathbf{L} = \mathbf{L}_f - \mathbf{L}_i = \sum \boldsymbol{\tau}\,\Delta t$. Because $\Delta\mathbf{L}$ is perpendicular to $\mathbf{L}$, the magnitude of $\mathbf{L}$ does not change $(|\mathbf{L}_i| = |\mathbf{L}_f|)$. Rather, what is changing is the *direction* of $\mathbf{L}$. Because the change in angular momentum $\Delta\mathbf{L}$ is in the direction of $\sum \boldsymbol{\tau}$, which lies in the $xy$ plane, the top undergoes precessional motion.

With careful manufacturing tolerances, precession due to gravitational torque can be made very small and gyroscopes can be used for guidance systems in vehicles—a change in the direction of the velocity of a vehicle is detected as a change between the direction of the angular momentum of the gyroscope and a reference direction attached to the vehicle. With proper electronic feedback, the deviation from the desired direction of motion can be removed, bringing the angular momentum back in line with the reference direction. Precession rates for highly specialized military gyroscopes are as low as 0.02° per day.

## 10.11 • ROLLING MOTION OF RIGID BODIES

In this section we shall investigate the special case of rotational motion in which a round object rolls on a surface. Many everyday examples exist for such motion, including automobile tires rolling on roads and bowling balls rolling toward the pins.

Suppose a cylinder is rolling on a straight path, as in Figure 10.28. The center of mass moves in a straight line, but a point on the rim moves in a more complex path called a *cycloid*. Let us further assume that the cylinder of radius $R$ is uniform

**Figure 10.28**

Light sources at the center and rim of a rolling cylinder illustrate the different paths these points take. The center moves in a straight line *(green line)*, while a point on the rim moves in the path of a cycloid *(red curve)*. *(Henry Leap and Jim Lehman)*

and rolls on a surface with friction. We make a rather odd, but valid, simplification model here for rolling objects. The surfaces must exert friction forces on each other; otherwise the cylinder would simply slide rather than roll. If the friction force on the cylinder is large enough, the cylinder rolls without slipping. In this situation, the friction force is static rather than kinetic because the contact point of the cylinder with the surface is at rest relative to the surface at any instant. The static friction force acts through no displacement, so that it does no work on the cylinder and causes no decrease in mechanical energy of the cylinder. In real rolling objects, deformations of the surfaces result in some rolling resistance. If both surfaces are hard, however, they will deform very little, and rolling resistance can be negligibly small. Thus, we can model the rolling motion as maintaining constant mechanical energy. The wheel was a great invention!

As the cylinder rotates through an angle $\theta$, its center of mass moves a distance of $s = r\theta$. Therefore, the speed and acceleration of the center of mass for **pure rolling motion** are

$$v_{CM} = \frac{ds}{dt} = R\frac{d\theta}{dt} = R\omega \qquad \text{[10.40]}$$

$$a_{CM} = \frac{dv_{CM}}{dt} = R\frac{d\omega}{dt} = R\alpha \qquad \text{[10.41]}$$

• *Relations between translational and rotational variables for a rolling object*

The translational velocities of various points on the rolling cylinder are illustrated in Figure 10.29. Note that the translational velocity of any point is in a direction perpendicular to the line from that point to the contact point. At any instant, the point $P$ is at rest relative to the surface because sliding does not occur.

We can express the total kinetic energy of a rolling object of mass $M$ and moment of inertia $I$ as the combination of the rotational kinetic energy around the center of mass plus the translational kinetic energy of the center of mass:

$$K = \tfrac{1}{2}I_{CM}\omega^2 + \tfrac{1}{2}Mv_{CM}^2 \qquad \text{[10.42]}$$

• *Total kinetic energy of a rolling object*

A useful theorem called the **parallel axis theorem** enables us to express this energy in terms of the moment of inertia $I_p$ through any axis parallel to the axis through the center of mass of an object. This theorem states that

$$I_p = I_{CM} + MD^2 \qquad \text{[10.43]}$$

where $D$ is the distance from the center-of-mass axis to the parallel axis, and $M$ is the total mass of the object. Let us use this theorem to express the moment of inertia around the contact point $P$ between the rolling object and the surface. The distance from this point to the center of mass of the symmetric object is its radius, so

$$I_p = I_{CM} + MR^2$$

If we write the translational speed of the center of mass of the object in Equation 10.42 in terms of the angular speed, we have

$$K = \tfrac{1}{2}I_{CM}\omega^2 + \tfrac{1}{2}MR^2\omega^2 = \tfrac{1}{2}(I_{CM} + MR^2)\omega^2 = \tfrac{1}{2}I_p\omega^2 \qquad \text{[10.44]}$$

Thus, the kinetic energy of the rolling object can be considered as equivalent to a purely rotational kinetic energy of the object rotating around its contact point.

We can use the energy version of the isolated system model to treat a class of problems concerning the rolling motion of a rigid body down a rough incline. In these types of problems, gravitational potential energy of the body–Earth system decreases as the rotational and translational kinetic energies of the body increase.

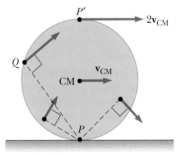

**Figure 10.29**

All points on a rolling object move in a direction perpendicular to an axis through the instantaneous point of contact $P$. The center of the object moves with a velocity $\mathbf{v}_{CM}$, whereas the point $P'$ moves with a velocity $2\mathbf{v}_{CM}$.

As this old-style bicycle rolls along a roadway, the centers of mass of both wheels travel with the same translational speed. Which wheel has the higher angular speed? *(© Steve Lovegrove/Tasmanian Photo Library)*

### Quick Quiz 10.11

Which arrives at the bottom first: a ball rolling without sliding down incline A or a box of the same mass as the ball sliding down a frictionless incline B having the same dimensions as incline A?

### THINKING PHYSICS 10.4

Consider the wheel in Figure 10.29 to be the tire of a truck, moving to the right, with pieces of mud stuck between the treads. As the mud becomes dislodged from the treads and free of the tire, it is thrown in the direction of the tangential velocity vectors shown in Figure 10.29. All of these vectors have a component in the forward direction, that is, in the same direction as the truck is moving. So why do trucks have mud flaps *behind* the tires?

**Reasoning** Consider a piece of mud released from the tire just to the left of point $P$ in Figure 10.29. Although this piece of mud does have a component of velocity in the forward direction, it is smaller than $v_{CM}$. Thus, the tire will move horizontally more rapidly than the mud. Relative to the truck, then, the piece of mud moves *backward*. Without a mud flap, this piece of mud could strike a car driving behind the truck.

### Example 10.15 Sphere Rolling Down an Incline

If the object in Figure 10.30 is a solid sphere, (a) calculate the speed of its center of mass at the bottom and (b) determine the magnitude of the translational acceleration of the center of mass.

**Solution** (a) We shall consider the sphere and the Earth as an isolated system and use the energy version of the isolated system model. The energy of the system when the sphere is at the top of the incline is gravitational potential energy only. We choose the zero of gravitational potential energy to be when the sphere is at the bottom of the incline. Thus, conservation of mechanical energy gives us

$$K_i + U_i = K_f + U_f$$
$$0 + Mgh = \left(\tfrac{1}{2}Mv_{CM,f}^2 + \tfrac{1}{2}I_{CM}\omega_f^2\right) + 0$$

Using Equation 10.40 to relate the translational and angular speeds, and substituting the moment of inertia for a sphere, we have

$$Mgh = \tfrac{1}{2}Mv_{CM,f}^2 + \tfrac{1}{2}\left(\tfrac{2}{5}MR^2\right)\frac{v_{CM,f}^2}{R^2}$$
$$= \tfrac{1}{2}Mv_{CM,f}^2 + \tfrac{1}{5}Mv_{CM,f}^2$$
$$= \tfrac{7}{10}Mv_{CM,f}^2 \longrightarrow v_{CM,f} = \sqrt{\tfrac{10}{7}gh}$$

(b) To find the acceleration, let us recognize that the constant gravitational force should cause a constant acceleration of the center of mass of the sphere. From Equation 2.12,

$$v_{CM,f}^2 = v_{CM,i}^2 + 2a_{CM}(x_{CM,f} - x_{CM,i})$$

we can solve for the acceleration

$$a_{CM} = \frac{v_{CM,f}^2 - v_{CM,i}^2}{2(x_{CM,f} - x_{CM,i})} = \frac{\tfrac{10}{7}gh - 0}{2\left(\dfrac{h}{\sin\theta}\right)} = \tfrac{5}{7}g\sin\theta$$

These results are quite interesting in that both the speed and the acceleration of the center of mass are independent of the mass and radius of the sphere. That is, *all* homogeneous solid spheres experience the same speed and acceleration on a given incline!

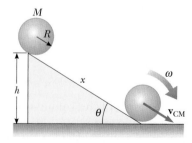

**Figure 10.30**

A round object rolling down an incline. Mechanical energy of the object–Earth system is conserved if no slipping occurs and there is no rolling resistance.

If we repeated the calculations for a hollow sphere, a solid cylinder, or a hoop, we would obtain similar results with different numerical factors appearing in the expressions for $v_{CM,f}$ and $a_{CM}$. These factors depend only on the moment of inertia about the center of mass for the specific object. In all cases, the acceleration of the center of mass is less than $g \sin \theta$, the value it would have if the plane were frictionless and no rolling occurred.

### context connection
## 10.12 • GYROSCOPES IN SPACE

Let us follow up on the gyroscope discussion of Section 10.10. Our technological society offers several examples of gyroscopes. We shall focus on just three such examples that relate to our Context. First, let us imagine that an additional assignment before we set out for Mars is to place a satellite in orbit around the Earth. If you simply hold the stationary satellite outside the orbiting spacecraft and let it go, you will inevitably impart some rotation to it when you let it go, and it will spin according to this influence. If it happens to catch on your spacesuit sleeve just as you release it, it could go into an uncontrolled tumble. This is undesirable. To stabilize the satellite as you let it go, it is more desirable to set it into rotation, so that the satellite has an angular momentum (Fig. 10.31). Then, the direction of the angular momentum vector can be determined before you release the satellite, and this direction will remain fixed according to the principle of conservation of angular momentum. The satellite acts as a gyroscope, and its spin remains stable. The space shuttle has a turntable on which satellites are placed and set into rotation before being released. We would want such a turntable on our spacecraft also.

Our second example is related to the Context Connection discussed in Chapter 4, in which we determined the direction in which to fire our rocket engines so as to return to our desired trajectory. We needed to turn the spacecraft around in

**Figure 10.31**

A spinning satellite is released from the space shuttle. The gyroscopic action due to its spin keeps its rotation axis fixed in direction. *(NASA)*

Spacecraft
rotates
clockwise

Gyroscope rotates
counterclockwise

(a)                                    (b)

**Figure 10.32**

(a) A spacecraft carries a gyroscope that is not spinning. (b) When the gyroscope is set into rotation, the spacecraft turns the other way so that the angular momentum of the system is conserved.

order to fire the engines in the correct direction. But how do you turn a spacecraft around in empty space? One way is to have small rocket engines that fire perpendicularly out the side of the spacecraft, providing a torque around its center of mass. This is desirable, and many spacecraft have such rockets.

Let us consider, however, another possibility related to angular momentum. Suppose the spacecraft carries a gyroscope that is not rotating, as in Figure 10.32a. In this case, the angular momentum of the spacecraft about its center of mass is zero. Suppose the gyroscope is set into rotation. Now, it would appear that the spacecraft system has a nonzero angular momentum, due to the rotation of the gyroscope. But there is no external torque on the system, so the angular momentum of the isolated system must remain zero according to the principle of conservation of angular momentum. This principle can be satisfied by realizing that the spacecraft will turn in the direction opposite to that of the gyroscope, so that the angular momentum vectors of the gyroscope and the spacecraft cancel, resulting in no angular momentum of the system. The result of rotating the gyroscope, as in Figure 10.32b, is that the spacecraft turns around! By including three gyroscopes with mutually perpendicular axles, any desired rotation in space can be achieved. Once the desired gravitation is achieved, the rotation of the gyroscope is halted.

This effect occurred in an undesirable situation with the *Voyager* 2 spacecraft during its flight. The spacecraft carried a tape recorder whose reels rotated at high speeds. Each time the tape recorder was turned on, the reels acted as gyroscopes, and the spacecraft started an undesirable rotation in the opposite direction. This had to be counteracted by Mission Control by using the sideward-firing jets to stop the rotation!

For our third example of gyroscopic motion, let us not look at our spacecraft, but at the Earth. The spinning Earth acts as a gyroscope, and we normally consider

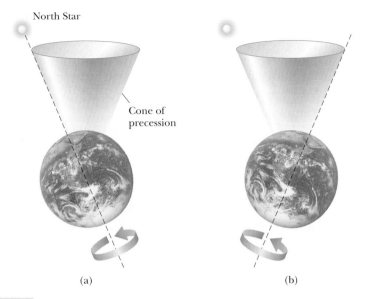

North Star

Cone of precession

(a)                    (b)

**Figure 10.33**

(a) At present, the spin axis of the Earth points toward the North Star. (b) Torque on the spinning Earth will cause it to precess, so that the spin axis will no longer be pointing in this direction in the future.

the rotation axis to be fixed in space, pointing at a location near the North Star. In our first two examples, we applied the principle of conservation of angular momentum because the satellite and the gyroscope–spacecraft system had no external torque on them. We might guess that the Earth rotates torque-free also. Here is a case, however, in which a simplification model of the Earth as a perfect sphere works well over relatively short time periods, but not over a long one. In reality, the Earth is not a perfect sphere. The radius of the Earth is larger at the equator than at the poles, due to the bulging of the Earth at the equator, which is a consequence of its rotation. We could geometrically model the Earth as a perfect sphere with an extra belt of mass around the equator. The gravitational force of the Sun and Moon on this extra mass results in a torque on the Earth.

As discussed in Section 10.10, the torque on a spinning object will cause precession—the angular momentum vector moves along the surface of a cone (Figure 10.33). The time required for the axis to spin around once and trace out a complete cone depends on several factors, including the amount of torque. The torque that the Sun and Moon exert on the Earth is relatively small for an object as massive as the Earth, so that the period for the precession is $2.6 \times 10^4$ years. Thus, 13 000 years from now, the axis of the Earth will not be pointing at the North Star, and 13 000 years later, it will return to the direction it has at present.

## SUMMARY

The **instantaneous angular speed** of a particle rotating in a circle or of a rigid body rotating about a fixed axis is

$$\omega = \frac{d\theta}{dt} \qquad [10.3]$$

where $\omega$ is in rad/s or in $s^{-1}$.

The **instantaneous angular acceleration** of a particle rotating in a circle or of a rigid body rotating about a fixed axis is

$$\alpha = \frac{d\omega}{dt} \qquad [10.5]$$

and has units of rad/s$^2$ or s$^{-2}$.

When a rigid body rotates about a fixed axis, every part of the body has the same angular speed and the same angular acceleration. However, different parts of the body, in general, have different translational speeds and different translational accelerations.

If a particle (or object) undergoes rotational motion about a fixed axis under constant angular acceleration $\alpha$, one can apply equations of kinematics by analogy with kinematic equations for translational motion with constant translational acceleration:

$$\omega_f = \omega_i + \alpha t \qquad [10.6]$$

$$\theta_f = \theta_i + \omega_i t + \tfrac{1}{2}\alpha t^2 \qquad [10.7]$$

$$\omega_f{}^2 = \omega_i{}^2 + 2\alpha(\theta_f - \theta_i) \qquad [10.8]$$

$$\theta_f = \theta_i + \tfrac{1}{2}(\omega_i + \omega_f)t \qquad [10.9]$$

When a particle rotates about a fixed axis, the angular speed and angular acceleration are related to the tangential speed and tangential acceleration through the relationships

$$v = r\omega \qquad [10.10]$$

$$a_t = r\alpha \qquad [10.11]$$

The **moment of inertia** of a system of particles is

$$I = \sum_i m_i r_i^2 \qquad [10.14]$$

If a rigid body rotates about a fixed axis with angular speed $\omega$, its **rotational kinetic energy** can be written

$$K_R = \tfrac{1}{2}I\omega^2 \qquad [10.15]$$

where $I$ is the moment of inertia about the axis of rotation.

The moment of inertia of a continuous object of density $\rho$ is

$$I = \int \rho r^2 dV \qquad [10.17]$$

The **torque** $\tau$ due to a force $\mathbf{F}$ about an origin in an inertial frame is defined to be

$$\boldsymbol{\tau} \equiv \mathbf{r} \times \mathbf{F} \qquad [10.19]$$

where $\mathbf{r}$ is the position vector of the point of application of the force.

Given two vectors $\mathbf{A}$ and $\mathbf{B}$, their **vector product** or **cross product** $\mathbf{A} \times \mathbf{B}$ is a vector $\mathbf{C}$ having the magnitude

$$C \equiv AB \sin \theta \qquad [10.21]$$

where $\theta$ is the angle between $\mathbf{A}$ and $\mathbf{B}$. The direction of $\mathbf{C}$ is perpendicular to the plane formed by $\mathbf{A}$ and $\mathbf{B}$, and is determined by the right-hand rule.

The net torque acting on an object is proportional to the angular acceleration of the object, and the proportionality constant is the moment of inertia $I$:

$$\sum \tau = I\alpha \qquad [10.27]$$

The **angular momentum L** of a particle with linear momentum $\mathbf{p} = m\mathbf{v}$ is

$$\mathbf{L} \equiv \mathbf{r} \times \mathbf{p} \qquad [10.32]$$

where $\mathbf{r}$ is the vector position of the particle relative to the origin. If $\phi$ is the angle between $\mathbf{r}$ and $\mathbf{p}$, the magnitude of $\mathbf{L}$ is

$$L = mvr \sin \phi \qquad [10.33]$$

The net external torque acting on a system is equal to the time rate of change of its angular momentum:

$$\sum \boldsymbol{\tau}_{\text{ext}} = \frac{d\mathbf{L}_{\text{tot}}}{dt} \qquad [10.37]$$

The law of conservation of angular momentum states that the total angular momentum of a system remains constant if the net external torque acting on the system is zero:

$$\mathbf{L}_{\text{tot}, i} = \mathbf{L}_{\text{tot}, f} \qquad [10.39]$$

The **total kinetic energy** of a rigid body, such as a cylinder, that is rolling on a rough surface without slipping equals the rotational kinetic energy $\tfrac{1}{2}I_{\text{CM}}\omega^2$ about the body's center of mass plus the translational kinetic energy $\tfrac{1}{2}Mv_{\text{CM}}^2$ of the center of mass:

$$K = \tfrac{1}{2}I_{\text{CM}}\omega^2 + \tfrac{1}{2}Mv_{\text{CM}}^2 \qquad [10.42]$$

In this expression, $v_{\text{CM}}$ is the speed of the center of mass and $v_{\text{CM}} = R\omega$ for pure rolling motion.

## QUESTIONS

1. In a tape recorder, the tape is pulled past the read and write heads at a constant speed by the drive mechanism. Consider the reel from which the tape is pulled—as the tape is pulled off it, the radius of the roll of remaining tape decreases. How does the torque on the reel change with time? How does the angular speed of the reel change with time? If the tape mechanism is suddenly turned on so that the tape is quickly pulled with a large force, is the tape more likely to break when pulled from a nearly full reel or a nearly empty reel?

2. If a car's wheels are replaced with wheels of a larger diameter, will the reading of the speedometer change? Explain.

3. What is the magnitude of the angular velocity $\boldsymbol{\omega}$ of the second hand of a clock? What is the direction of $\boldsymbol{\omega}$ as you

view a clock hanging vertically? What is the magnitude of the angular acceleration $\alpha$ of the second hand?

4. If you see an object rotating, does a net torque necessarily act on it?

5. Suppose a pencil is balanced on a perfectly frictionless table. If it falls over, what is the path followed by the center of mass of the pencil?

6. For a helicopter to be stable as it flies, it must have two propellers. Why?

7. In some motorcycle races, the riders drive over small hills, and the motorcycle becomes airborne for a short time. If the motorcycle racer keeps the throttle open while leaving the hill and going into the air, the motorcycle tends to nose upward. Why does this happen?

8. The moment of inertia of an object depends on the choice of rotation axis, as suggested by the parallel axis theorem. Argue that an axis passing through the center of mass of an object must be the axis with the smallest moment of inertia.

9. Suppose you remove two eggs from the refrigerator, one hard-boiled and the other uncooked. You wish to determine which is the hard-boiled egg without breaking the eggs. This can be done by spinning the two eggs on the floor and comparing the rotational motions. Which egg spins faster? Which rotates more uniformly? Explain.

10. A ladder rests inclined against a wall. Would you feel safer climbing up the ladder if you were told that the floor is frictionless but the wall is rough, or that the wall is frictionless but the floor is rough? Justify your answer.

11. Often when a high diver wants to flip in midair, she will draw her legs up against her chest. Why does this make her rotate faster? What should she do when she wants to stop rotating?

12. If the net force acting on a system is zero, then is it necessarily true that the net torque on it is also zero?

13. Why do tightrope walkers carry a long pole to help balance themselves?

14. Two uniform solid spheres are rolled down a hill: a large, massive sphere and a small sphere with low mass. Which one reaches the bottom of the hill first? Next, we roll a large, low-density sphere, and a small high-density sphere, and both spheres have the same mass. Which one wins in this case?

15. Suppose you are designing a car for a coasting race. The cars in this race have no engines; they simply coast down a hill. Do you want large wheels or small wheels? Do you want solid, disk-like wheels or hoop-like wheels? Should the wheels be heavy or light?

16. Consider an object in the shape of a hoop lying in the $xy$ plane with all of its mass concentrated on its rim. In two separate experiments, the hoop is rotated by an external agent from rest to an angular speed $\omega$. In one experiment, the rotation occurs about the $z$ axis through the center of the hoop. In the other experiment, the rotation occurs about an axis parallel to $z$ passing through a point $P$ on the rim of the hoop. Which rotation requires more work?

17. Vector $\mathbf{A}$ is in the negative $y$ direction, and vector $\mathbf{B}$ is in the negative $x$ direction. What are the directions of (a) $\mathbf{A} \times \mathbf{B}$ and (b) $\mathbf{B} \times \mathbf{A}$?

18. If global warming occurs over the next century, it is likely that some polar ice will melt and the water will be distributed closer to the equator. How would this change the moment of inertia of the Earth? Would the length of the day (one revolution) increase or decrease?

# PROBLEMS

1, 2, 3 = straightforward, intermediate, challenging    □ = full solution available in the *Student Solutions Manual and Study Guide*

**web** = solution posted at **http://www.harcourtcollege.com/physics/**    💻 = computer useful in solving problem

🔧 = Interactive Physics    ▦ = paired numerical/symbolic problems    🔬 = life science application

## Section 10.1  Angular Speed and Angular Acceleration

## Section 10.2  Rotational Kinematics: The Rigid Body Under Constant Angular Acceleration

1. A motor rotating a grinding wheel at 100 rev/min is
**web** switched off. Assuming constant negative angular acceleration of magnitude 2.00 rad/s², (a) how long does it take the wheel to stop? (b) Through how many radians does it turn while it is slowing down?

2. An airliner arrives at the terminal, and its engines are shut off. The rotor of one of the engines has an initial clockwise angular speed of 2 000 rad/s. The engine's rotation slows with an angular acceleration of magnitude 80.0 rad/s². (a) Determine the angular speed after 10.0 s. (b) How long does it take the rotor to come to rest?

3. A dentist's drill starts from rest. After 3.20 s of constant angular acceleration, the drill turns at a rate of $2.51 \times 10^4$ rev/min. (a) Find the drill's angular acceleration. (b) Determine the angle (in radians) through which the drill rotates during this period.

4. The angular position of a swinging door is described by $\theta = 5.00 + 10.0t + 2.00t^2$ rad. Determine the angular position, angular speed, and angular acceleration of the door (a) at $t = 0$ and (b) at $t = 3.00$ s.

5. The tub of a washer goes into its spin cycle, starting from rest and gaining angular speed uniformly for 8.00 s, at which time it is turning at 5.00 rev/s. At this point the person doing the laundry opens the lid, and a safety switch turns off the washer. The tub smoothly slows to rest in 12.0 s. Through how many revolutions does the tub turn while it is in motion?

6. A rotating wheel requires 3.00 s to rotate 37.0 rev. Its angular speed at the end of the 3.00-s interval is 98.0 rad/s. What is the constant angular acceleration of the wheel?

7. (a) Find the angular speed of the Earth's rotation on its axis. As the Earth turns toward the east, we see the sky turning toward the west at this same rate.
   (b) *The rainy Pleiads wester*
      *And seek beyond the sea*
      *The head that I shall dream of*
      *That shall not dream of me.*
         —A. E. Housman (© Robert E. Symons)
   Cambridge, England, is at longitude 0° and Saskatoon, Saskatchewan, is at longitude 107° west. How much time elapses after the Pleiades set in Cambridge until these stars fall below the western horizon in Saskatoon?

### Section 10.3  Relations Between Rotational and Translational Quantities

8. Make an order-of-magnitude estimate of the number of revolutions through which a typical automobile tire turns in 1 yr. State the quantities you measure or estimate and their values.

9. A disk 8.00 cm in radius rotates about its central axis at a
   **web** constant rate of 1 200 rev/min. Determine (a) its angular speed, (b) the tangential speed at a point 3.00 cm from its center, (c) the radial acceleration of a point on the rim, and (d) the total distance a point on the rim moves in 2.00 s.

10. A digital audio compact disc carries data, each bit of which occupies 0.6 $\mu$m along a continuous spiral track from the inner circumference of the disc to the outside edge. A CD player turns the disc to carry the track counterclockwise above a lens at a constant speed of 1.30 m/s. Find the required angular speed (a) at the beginning of the recording, where the spiral has a radius of 2.30 cm, and (b) at the end of the recording, where the spiral has a radius of 5.80 cm. (c) A full-length recording lasts for 74 min 33 s. Find the average angular acceleration of the disc. (d) Assuming that the acceleration is constant, find the total angular displacement of the disc as it plays. (e) Find the total length of the track.

11. A car accelerates uniformly from rest and reaches a speed of 22.0 m/s in 9.00 s. If the diameter of a tire is 58.0 cm, find (a) the number of revolutions the tire makes during this motion, assuming that no slipping occurs. (b) What is the final rotational speed of a tire in revolutions per second?

12. A wheel 2.00 m in diameter lies in a vertical plane and rotates with a constant angular acceleration of 4.00 rad/s$^2$. The wheel starts at rest at $t = 0$, and the radius vector of a certain point $P$ on the rim makes an angle of 57.3° with the horizontal at this time. At $t = 2.00$ s, find (a) the angular speed of the wheel, (b) the tangential speed and the total acceleration of the point $P$, and (c) the angular position of the point $P$.

### Section 10.4  Rotational Kinetic Energy

13. Rigid rods of negligible mass lying along the y axis connect three small particles (Fig. P10.13). If the system rotates about the x axis with an angular speed of 2.00 rad/s, find (a) the moment of inertia about the $x$ axis and the total rotational kinetic energy evaluated from $\frac{1}{2}I\omega^2$ and (b) the tangential speed of each particle and the total kinetic energy evaluated from $\sum_i \frac{1}{2}m_iv_i^2$.

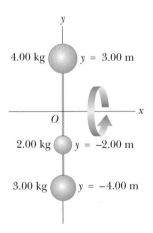

4.00 kg    $y = 3.00$ m

$O$    x

2.00 kg    $y = -2.00$ m

3.00 kg    $y = -4.00$ m

**Figure P10.13**

14. Big Ben, the Parliament tower clock in London, shown in Figure P10.14, has an hour hand 2.70 m long with a mass of 60.0 kg, and a minute hand 4.50 m long with a mass of 100 kg. Calculate the total rotational kinetic energy of the two hands about the axis of rotation. (You may model the hands as long thin uniform rods.)

**Figure P10.14**　Problems 14, 40, and 71. (*John Lawrence/Stone*)

15. This problem describes one experimental method of determining the moment of inertia of an irregularly shaped object such as the payload for a satellite. Figure P10.15 shows a cylinder of mass $m$ suspended by a cord that is wound around a spool of radius $r$, forming part of a turntable supporting the object. When the cylinder is released from rest, it descends through a distance $h$, acquiring a speed $v$. Show that the moment of inertia $I$ of the equipment (including the turntable) is $mr^2(2gh/v^2 - 1)$.

**Figure P10.15**

16. Consider two objects with $m_1 > m_2$ connected by a light string that passes over a pulley having a moment of inertia of $I$ about its axis of rotation, as in Figure P10.16. The string does not slip on the pulley. The pulley turns without friction. The objects are released from rest separated by a vertical distance $2h$. Use the principle of conservation of energy to find the translational speeds of the objects as they pass each other. Find the angular speed of the pulley at this time.

**Figure P10.16**

## Section 10.6　Torque and the Vector Product

17. Find the net torque on the wheel in Figure P10.17 about the axle through $O$ if $a = 10.0$ cm and $b = 25.0$ cm.

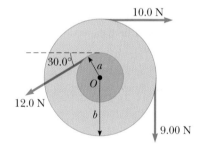

**Figure P10.17**

18. The fishing pole in Figure P10.18 makes an angle of 20.0° with the horizontal. What is the torque exerted by the fish

**Figure P10.18**

about an axis perpendicular to the page and passing through the fisher's hand?

**19.** Given $\mathbf{M} = 6\mathbf{i} + 2\mathbf{j} - \mathbf{k}$ and $\mathbf{N} = 2\mathbf{i} - \mathbf{j} - 3\mathbf{k}$, calculate the vector product $\mathbf{M} \times \mathbf{N}$.

**20.** For the vectors $\mathbf{A} = -3\mathbf{i} + 7\mathbf{j} - 4\mathbf{k}$ and $\mathbf{B} = 6\mathbf{i} - 10\mathbf{j} + 9\mathbf{k}$, evaluate the quantities (a) $\cos^{-1}[\mathbf{A} \cdot \mathbf{B}/AB]$ and (b) $\sin^{-1}[|\mathbf{A} \times \mathbf{B}|/AB]$. (c) Which give(s) the angle between the vectors?

**21.** A force of $\mathbf{F} = (2.00\mathbf{i} + 3.00\mathbf{j})$ N is applied to an object that is pivoted about a fixed axle aligned along the $z$ coordinate axis. If the force is applied at the point $\mathbf{r} = (4.00\mathbf{i} + 5.00\mathbf{j} + 0\mathbf{k})$ m, find (a) the magnitude of the net torque about the $z$ axis and (b) the direction of the torque vector $\boldsymbol{\tau}$.

**22.** Use the definition of the vector product and the definitions of the unit vectors $\mathbf{i}$, $\mathbf{j}$, and $\mathbf{k}$ to prove Equations 10.23. You may assume that the $x$ axis points to the right, the $y$ axis up, and the $z$ axis toward you (not away from you). This choice is said to make the coordinate system *right handed*.

## Section 10.6   The Rigid Body in Equilibrium

**23.** A uniform beam of mass $m_b$ and length $\ell$ supports blocks with masses $m_1$ and $m_2$ at two positions, as in Figure P10.23. The beam rests on two knife edges. For what value of $x$ will the beam be balanced at $P$ such that the normal force at $O$ is zero?

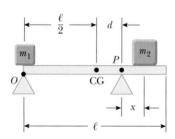

**Figure P10.23**

**24.** A 1 500-kg automobile has a wheel base (the distance between the axles) of 3.00 m. The center of mass of the automobile is on the center line at a point 1.20 m behind the front axle. Find the force exerted by the ground on each wheel.

**25.** A uniform ladder of length $L$ and mass $m_1$ rests against a frictionless wall. The ladder makes an angle $\theta$ with the horizontal. (a) Find the horizontal and vertical forces the ground exerts on the base of the ladder when a firefighter of mass $m_2$ is a distance $x$ from the bottom. (b) If the ladder is just on the verge of slipping when the firefighter is a distance $d$ from the bottom, what is the coefficient of static friction between ladder and ground?

**26.** Figure P10.26 shows a claw hammer as it is being used to pull a nail out of a horizontal board. If a force of magnitude 150 N is exerted horizontally as shown, find (a) the force exerted by the hammer claws on the nail and (b) the force exerted by the surface on the point of contact with the hammer head. Assume that the force the hammer exerts on the nail is parallel to the nail.

**Figure P10.26**

**27.** A uniform sign of weight $F_g$ and width $2L$ hangs from a light, horizontal beam, hinged at the wall and supported by a cable (Fig. P10.27). Determine (a) the tension in the cable and (b) the components of the reaction force exerted by the wall on the beam, in terms of $F_g$, $d$, $L$, and $\theta$.

**Figure P10.27**

**28.** A crane of mass 3 000 kg supports a load of 10 000 kg, as in Figure P10.28. The crane is pivoted with a frictionless pin at $A$ and rests against a smooth support at $B$. Find the reaction forces at $A$ and $B$.

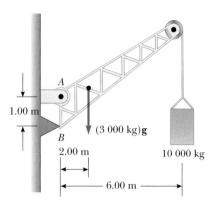

**Figure P10.28**

## Section 10.7 The Rigid Body Under a Net Torque

**29.** A model airplane with mass 0.750 kg is tethered by a wire so that it flies in a circle 30.0 m in radius. The airplane engine provides a net thrust of 0.800 N perpendicular to the tethering wire. (a) Find the torque the net thrust produces about the center of the circle. (b) Find the angular acceleration of the airplane when it is in level flight. (c) Find the tangential acceleration of the airplane.

**30.** The combination of an applied force and a friction force produces a constant total torque of 36.0 N·m on a wheel rotating about a fixed axis. The applied force acts for 6.00 s. During this time the angular speed of the wheel increases from 0 to 10.0 rad/s. The applied force is then removed, and the wheel comes to rest in 60.0 s. Find (a) the moment of inertia of the wheel, (b) the magnitude of the frictional torque, and (c) the total number of revolutions of the wheel.

**31.** An electric motor turns a flywheel through a drive belt that joins a pulley on the motor and a pulley that is rigidly attached to the flywheel, as shown in Figure P10.31. The flywheel is a solid disk with a mass of 80.0 kg and a diameter of 1.25 m. It turns on a frictionless axle. Its pulley has much smaller mass and a radius of 0.230 m. If the tension in the upper (taut) segment of the belt is 135 N and the flywheel has a clockwise angular acceleration of 1.67 rad/s$^2$, find the tension in the lower (slack) segment of the belt.

**32.** A potter's wheel—a thick stone disk with a radius of 0.500 m and a mass of 100 kg—is freely rotating at 50.0 rev/min. The potter can stop the wheel in 6.00 s by pressing a wet rag against the rim and exerting a radially inward force of 70.0 N. Find the effective coefficient of kinetic friction between wheel and rag.

**33.** Two blocks, as shown in Figure P10.33, are connected by a string of negligible mass passing over a pulley of radius 0.250 m and moment of inertia $I$. The block on the frictionless incline is moving up with a constant acceleration of 2.00 m/s$^2$. (a) Determine $T_1$ and $T_2$, the tensions in the two parts of the string. (b) Find the moment of inertia of the pulley.

**Figure P10.33**

**34.** In Figure P10.34 the sliding block has a mass of 0.850 kg, the counterweight has a mass of 0.420 kg, and the pulley is a hollow cylinder with a mass of 0.350 kg, an inner radius of 0.020 0 m, and an outer radius of 0.030 0 m. The coefficient of kinetic friction between the block and the horizontal surface is 0.250. The pulley turns without friction on its axle. The light cord does not stretch and does not slip on the pulley. The block has a velocity of 0.820 m/s toward the pulley when it passes through a photogate. (a) Use energy methods to predict its speed after it has moved

**Figure P10.31**

**Figure P10.34**

to a second photogate, 0.700 m away. (b) Find the angular speed of the pulley at the same moment.

35. An object with weight 50.0 N is attached to the free end of a light string wrapped around a reel with a radius of 0.250 m and a mass of 3.00 kg. The reel is a solid disk, free to rotate in a vertical plane about the horizontal axis passing through its center. The object is released 6.00 m above the floor. (a) Determine the tension in the string, the acceleration of the object, and the speed with which the object hits the floor. (b) Using the principle of conservation of energy, find the speed with which the object hits the floor.

36. A constant torque of 25.0 N·m is applied to a grindstone with a moment of inertia of 0.130 kg·m², which starts from rest and turns without friction. Using energy principles, find the angular speed after the grindstone has made 15.0 rev.

37. A uniform rod of length $L$ and mass $M$ is free to rotate about a frictionless pivot at one end, as in Figure P10.37. The rod is released from rest in the horizontal position. What are the *initial* angular acceleration of the rod and the *initial* translational acceleration of the right end of the rod?

**Figure P10.37**

## Section 10.8    Angular Momentum

38. Heading straight toward the summit of Pike's Peak, an airplane of mass 12 000 kg flies over the plains of Kansas at nearly constant altitude 4.30 km, with constant velocity 175 m/s west. (a) What is the airplane's vector angular momentum relative to a wheat farmer on the ground directly below the airplane? (b) Does this value change as the airplane continues its motion along a straight line? (c) What is its angular momentum relative to the summit of Pike's Peak?

39. The position vector of a particle of mass 2.00 kg is given as a function of time by $\mathbf{r} = (6.00\mathbf{i} + 5.00t\mathbf{j})$ m. Determine the angular momentum of the particle about the origin, as a function of time.

40. The hour and minute hands of Big Ben, the Parliament Building tower clock in London, are 2.70 m and 4.50 m long and have masses of 60.0 kg and 100 kg, respectively (Fig. P10.14). Calculate the total angular momentum of these hands about the center point. Model the hands as long thin uniform rods.

41. A particle of mass 0.400 kg is attached to the 100-cm mark of a meter stick of mass 0.100 kg. The meter stick rotates on a horizontal, frictionless table with an angular speed of 4.00 rad/s. Calculate the angular momentum of the system when the stick is pivoted about an axis (a) perpendicular to the table through the 50.0-cm mark and (b) perpendicular to the table through the 0-cm mark.

42. A 4.00-kg object is attached to a light cord, which is wound around a spool (see Figure 10.22). The spool is a uniform solid cylinder of radius 8.00 cm and mass 2.00 kg. (a) What is the net torque on the system about the point $O$? (b) When the object has a speed $v$, the spool has an angular speed $\omega = v/R$. Determine the total angular momentum of the system about $O$. (c) Using the fact that $\tau = d\mathbf{L}/dt$ and your result from (b), calculate the acceleration of the object.

## Section 10.9    Conservation of Angular Momentum

43. A cylinder with a moment of inertia of $I_1$ rotates about a vertical, frictionless axle with angular speed $\omega_i$. A second cylinder that has a moment of inertia of $I_2$ and initially is not rotating drops onto the first cylinder (Fig. P10.43). Because of friction between the surfaces, the two eventually reach the same angular speed $\omega_f$. (a) Calculate $\omega_f$. (b) Show that the kinetic energy of the system decreases in this interaction and calculate the ratio of the final rotational energy to the initial rotational energy.

**Figure P10.43**

44. A student sits on a freely rotating stool holding two weights, each of which has a mass of 3.00 kg. When his arms are extended horizontally, the weights are 1.00 m from the axis of rotation and he rotates with an angular speed of 0.750 rad/s. The moment of inertia of the student plus stool is 3.00 kg·m² and is assumed to be constant. The student pulls the weights inward horizontally to a position 0.300 m from the rotation axis. (a) Find the new angular speed of the student. (b) Find the kinetic en-

ergy of the rotating system before and after he pulls the weights inward.

**45.** A playground merry-go-round of radius $R = 2.00$ m has a moment of inertia $I = 250$ kg·m² and is rotating at 10.0 rev/min about a frictionless vertical axle. Facing the axle, a 25.0-kg child hops onto the merry-go-round from the ground and manages to sit down on its edge. What is the new angular speed of the merry-go-round?

**46.** A puck with a mass of 80.0 g and a radius of 4.00 cm slides along an air table at a speed of 1.50 m/s, as shown in Figure P10.46a. It makes a glancing collision with a second puck having a radius of 6.00 cm and a mass of 120 g (initially at rest) such that their rims just touch. Because their rims are coated with instant-acting glue, the pucks stick together and spin after the collision (Fig. P10.46b). (a) What is the angular momentum of the system relative to the center of mass? (b) What is the angular speed about the center of mass?

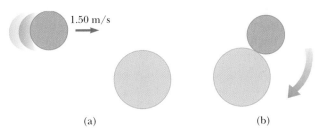

1.50 m/s

(a)                          (b)

**Figure P10.46**

**47.** A space station shaped like a giant wheel has a radius of 100 m and a moment of inertia of $5.00 \times 10^8$ kg·m². A crew of 150 are living on the rim, and the station's rotation causes the crew to experience an acceleration of $1g$ (Fig. P10.47). When 100 people move to the center of the station for a union meeting, the angular speed changes. What acceleration is experienced by the managers remaining at the rim? Assume that the average mass of each inhabitant is 65.0 kg.

**48.** The ball in Figure 10.26 has a mass of 0.120 kg. The distance of the ball from the center of rotation is originally 40.0 cm, and the ball is moving with a speed of 80.0 cm/s. The string is pulled downward 15.0 cm through the hole in the frictionless table. Determine the work done on the ball. (*Hint:* Consider the change of kinetic energy.)

### Section 10.11  Rolling of Rigid Bodies

**49.** A cylinder of mass 10.0 kg rolls without slipping on a horizontal surface. At the instant its center of mass has a speed of 10.0 m/s, determine (a) the translational kinetic energy of its center of mass, (b) the rotational energy about its center of mass, and (c) its total energy.

**50.** A uniform solid disk and a uniform hoop are placed side by side at the top of an incline of height $h$. If they are released from rest and roll without slipping, determine their speeds when they reach the bottom. Which object reaches the bottom first?

**51.** A metal can containing condensed mushroom soup has a mass of 215 g, a height of 10.8 cm, and a diameter of 6.38 cm. It is placed at rest on its side at the top of a 3.00-m-long incline that is at 25.0° to the horizontal, and is then released to roll straight down. Assuming energy conservation, calculate the moment of inertia of the can if it takes 1.50 s to reach the bottom of the incline. Which pieces of data, if any, are unnecessary for calculating the solution?

**52.** A tennis ball is a hollow sphere with a thin wall. It is set rolling without slipping at 4.03 m/s on the horizontal section of a track, as shown in Figure P10.52. It rolls around the inside of a vertical circular loop 90.0 cm in diameter and finally leaves the track at a point 20.0 cm below the horizontal section. (a) Find the speed of the ball at the top of the loop. Demonstrate that it will not fall from the track.

**Figure P10.47**

**Figure P10.52**

(b) Find its speed as it leaves the track. (c) Suppose that static friction between ball and track were negligible, so that the ball slid instead of rolling. Would its speed then be higher, lower, or the same at the top of the loop? Explain.

## Section 10.12  Context Connection — Gyroscopes in Space

53. A spacecraft is in empty space. It carries on board a gyroscope with a moment of inertia $I_g = 20.0$ kg·m² around the axle of the gyroscope. The moment of inertia of the spacecraft around the axle of the gyroscope is $I_s = 5.00 \times 10^5$ kg·m². Neither the spacecraft nor the gyroscope is rotating. The gyroscope can be powered up in a negligible period of time to an angular speed of 100 s⁻¹. If the orientation of the spacecraft is to be changed by 30.0°, for how long should the gyroscope be operated?

54. In a precessing gyroscope, the angular momentum vector sweeps out a cone, as in Figure 10.27, with a *precessional frequency*, given by

$$\omega_p = \frac{\tau}{L}$$

where $\tau$ is the magnitude of the torque on the gyroscope and $L$ is the magnitude of its angular momentum. Given that the precession period of the Earth is $2.58 \times 10^4$ years, calculate the torque on the Earth that is causing this precession. (Model the Earth as a uniform sphere.)

## Additional Problems

55. A 60.0-kg woman stands at the rim of a horizontal turntable having a moment of inertia of 500 kg·m² and a radius of 2.00 m. The turntable is initially at rest and is free to rotate about a frictionless, vertical axle through its center. The woman then starts walking around the rim clockwise (as viewed from above the system) at a constant speed of 1.50 m/s relative to the Earth. (a) In what direction and with what angular speed does the turntable rotate? (b) How much work does the woman do to set herself and the turntable into motion?

56. Two astronauts (Fig. P10.56), each having a mass $M$, are connected by a rope of length $d$ having negligible mass. They are isolated in space, orbiting their center of mass at speeds $v$. Treating the astronauts as particles, calculate (a) the magnitude of the angular momentum of the system and (b) the rotational energy of the system. By pulling on the rope, one of the astronauts shortens the distance between them to $d/2$. (c) What is the new angular momentum of the system? (d) What are the astronauts' new speeds? (e) What is the new rotational energy of the system? (f) How much work does the astronaut do in shortening the rope?

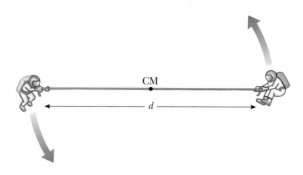

**Figure P10.56**

57. A long, uniform rod of length $L$ and mass $M$ is pivoted about a horizontal, frictionless pin passing through one end. The rod is released from rest in a vertical position, as shown in Figure P10.57. At the instant the rod is horizontal, find (a) its angular speed, (b) the magnitude of its angular acceleration, (c) the $x$ and $y$ components of the acceleration of its center of mass, and (d) the components of the reaction force at the pivot.

**Figure P10.57**

58. **Review Problem.** A mixing beater consists of three thin rods. Each is 10.0 cm long, diverges from a central hub, and is separated from the others by 120°. All turn in the same plane. A ball is attached to the end of each rod. Each ball has cross-sectional area 4.00 cm² and is so shaped that it has a drag coefficient of 0.600. Calculate the power input required to spin the beater at 1 000 rev/min (a) in air and (b) in water.

59. A uniform, hollow, cylindrical spool has inside radius $R/2$, outside radius $R$, and mass $M$ (Fig. P10.59). It is mounted so that it rotates on a stationary horizontal axle. An object of mass $m$ is connected to the end of a string wound around the spool. The object falls from rest through a distance $y$ in time $t$. Show that the torque due to the frictional forces between spool and axle is

$$\tau_f = R\left[m\left(g - \frac{2y}{t^2}\right) - M\frac{5y}{4t^2}\right]$$

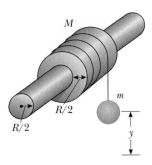

**Figure P10.59**

60. The reel shown in Figure P10.60 has radius $R$ and moment of inertia $I$. The object of mass $m$ is connected between a spring of force constant $k$ and a cord wrapped around the reel. The reel axle and the incline are frictionless. The reel is wound counterclockwise so that the spring stretches a distance $d$ from its unstretched position and is then released from rest. Find (a) the angular speed of the reel when the spring is again unstretched and (b) a numerical value for the angular speed at this point if $I = 1.00 \ \text{kg} \cdot \text{m}^2$, $R = 0.300$ m, $k = 50.0$ N/m, $m = 0.500$ kg, $d = 0.200$ m, and $\theta = 37.0°$.

**Figure P10.60**

61. A block of mass $m_1 = 2.00$ kg and a block of mass $m_2 = 6.00$ kg are connected by a massless string over a pulley in the shape of a solid disk having radius $R = 0.250$ m and mass $M = 10.0$ kg. These blocks are allowed to move on a fixed block wedge of angle $\theta = 30.0°$, as in Figure P10.61. The coefficient of kinetic friction for both blocks is 0.360. Draw free-body diagrams of both blocks and of the pulley. Determine (a) the acceleration of the two blocks, and (b) the tensions in the string on both sides of the pulley.

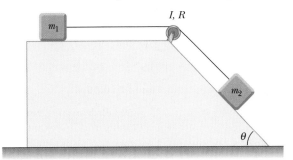

**Figure P10.61**

62. (a) Without the wheels, a bicycle frame has a mass of 8.44 kg. Each of the wheels can be roughly modeled as a uniform solid disk with a mass of 0.820 kg and a radius of 0.343 m. Find the kinetic energy of the whole bicycle when it is moving forward at 3.35 m/s. (b) Before the invention of a wheel turning on an axle, ancient people moved heavy loads by placing rollers under them. (Modern people use rollers too. Any hardware store will sell you a roller bearing for a lazy susan.) A stone block of mass 844 kg moves forward at 0.335 m/s, supported by two uniform cylindrical tree trunks, each of mass 82.0 kg and radius 0.343 m. No slipping occurs between the block and the rollers or between the rollers and the ground. Find the total kinetic energy of the moving objects.

63. A force acts on a uniform rectangular cabinet weighing 400 N, as in Figure P10.63. (a) If the cabinet slides with constant speed when $F = 200$ N and $h = 0.400$ m, find the coefficient of kinetic friction and the position of the resultant normal force. (b) If $F = 300$ N, find the value of $h$ for which the cabinet just begins to tip.

**Figure P10.63**

64. A stepladder of negligible weight is constructed as shown in Figure P10.64. A painter of mass 70.0 kg stands on the

**Figure P10.64**

ladder 3.00 m from the bottom. Assuming the floor is frictionless, find (a) the tension in the horizontal bar connecting the two halves of the ladder, (b) the normal forces at *A* and *B*, and (c) the components of the reaction force at the single hinge *C* that the left half of the ladder exerts on the right half. (*Hint:* Treat each half of the ladder separately.)

65. Figure P10.65 shows a vertical force applied tangentially to a uniform cylinder of weight $F_g$. The coefficient of static friction between the cylinder and both surfaces is 0.500. In terms of $F_g$, find the maximum force **P** that can be applied that does not cause the cylinder to rotate. (*Hint:* When the cylinder is on the verge of slipping, both friction forces are at their maximum values. Why?)

**Figure P10.65**

66. In exercise physiology studies it is sometimes important to determine the location of the center of mass of a person. This can be done with the arrangement shown in Figure P10.66. A light plank rests on two scales, which read $F_{g1}$ = 380 N and $F_{g2}$ = 320 N. The scales are separated by a distance of 2.00 m. How far from the woman's feet is her center of mass?

**Figure P10.66**

67. A person bending forward to lift a load "with his back," rather than "with his knees," can be injured by large forces exerted on the muscles and vertebrae. The spine pivots mainly at the fifth lumbar vertebra, with the principal supporting force provided by the erector spinalis muscle in the back. To see the magnitude of the forces involved and

to understand why back problems are so common among humans, consider the model for a person bending forward to lift a 200-N weight shown in Figure P10.67. The spine and upper body are represented as a uniform horizontal rod of weight 350 N, pivoted at the base of the spine. The erector spinalis muscle, attached at a point two thirds of the way up the spine, maintains the position of the back. The angle between the spine and this muscle is 12.0°. Find the tension in the back muscle and the compressional force in the spine.

(a)                           (b)

**Figure P10.67**

68. A wad of sticky clay of mass *m* and velocity $\mathbf{v}_i$ is fired at a solid cylinder of mass *M* and radius *R* (Fig. P10.68). The cylinder is initially at rest and is mounted on a fixed horizontal axle that runs through the center of mass. The line of motion of the projectile is perpendicular to the axle and at a distance *d*, less than *R*, from the center. (a) Find the angular speed of the system just after the clay strikes and sticks to the surface of the cylinder. (b) Is mechanical energy conserved in this process? Explain your answer.

**Figure P10.68**

69. A string is wound around a uniform disk of radius *R* and mass *M*. The disk is released from rest with the string vertical and its top end tied to a fixed bar (Fig. P10.69). Show

that (a) the tension in the string is one-third the weight of the disk, (b) the magnitude of the acceleration of the center of mass is $2g/3$, and (c) the speed of the center of mass is $(4gh/3)^{1/2}$ as the disk descends. Verify your answer to part (c) using the energy approach.

**Figure P10.69**

70. A common demonstration, illustrated in Figure P10.70, consists of a ball resting at one end of a uniform board of length $\ell$, hinged at the other end and elevated at an angle $\theta$. A light cup is attached to the board at $r_c$ so that it will catch the ball when the support stick is suddenly removed. (a) Show that the ball will lag behind the end of the falling board when $\theta$ is less than $35.3°$. (b) If the board is 1.00 m long and is supported at this limiting angle, show that the cup must be 18.4 cm from the moving end.

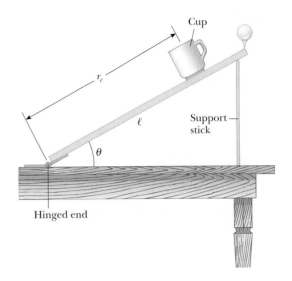

**Figure P10.70**

71. The hour hand and the minute hand of Big Ben, the famous Parliament tower clock in London, are 2.70 m long and 4.50 m long and have masses of 60.0 kg and 100 kg, respectively (see Figure P10.14). (a) Determine the total torque due to the weight of these hands about the axis of rotation when the time reads (i) 3:00, (ii) 5:15, (iii) 6:00, (iv) 8:20, (v) 9:45. (You may model the hands as long thin uniform rods.) (b) Determine all times when the total torque about the axis of rotation is zero. Determine the times to the nearest second, solving a transcendental equation numerically.

## ANSWERS TO QUICK QUIZZES

**10.1** (c). For a rotation of more than $180°$, the angular displacement must be larger than $\pi = 3.14$ rad. The angular displacements in the three choices are (a) 6 rad − 3 rad = 3 rad, (b) 1 rad − (−1) rad = 2 rad, (c) 5 rad − 1 rad = 4 rad.

**10.2** (b). Because all angular displacements occurred in the same time interval, the displacement with the lowest value will be associated with the lowest average angular speed.

**10.3** (b). In Equation 10.8, both the initial and final angular speeds are the same in all three cases. As a result, the angular acceleration is inversely proportional to the angular displacement. Thus, the highest angular acceleration is associated with the lowest angular displacement.

**10.4** (a) Yes. All points on the wheel make one full rotation in the same time interval. (b) No. Points farther from the rotation axis have higher tangential speeds. (c) The point at $r = 0$ has zero tangential speed and zero acceleration; a point at $r = R/2$ has a tangential speed $v = R\omega/2$ and a total translational acceleration equal to the centripetal acceleration $v^2/(R/2) = R\omega^2/2$ (the tangential acceleration is zero at all points because $\omega$ is constant). A point on the rim at $r = R$ has a tangential speed $v = R\omega$ and a total translational acceleration $R\omega^2$.

**10.5** The hollow sphere has the higher rotational kinetic energy. Its mass is located near the spherical surface, and the mass of the solid sphere is located throughout the volume. This results in a higher moment of inertia for the hollow sphere, as can be seen in Table 10.2.

**10.6** Torque and work have two major differences. The primary difference is that the vector displacement in the expression for work is directed *parallel* to the force, but the position vector in the torque expression is *perpendicular* to the force. The second difference depends on whether there is motion or not—in the case of work, work is done only if the force succeeds in causing a displacement of the point of application of the force. On the other hand, a force applied at a perpendicular distance from a rotation axis results in a torque whether there is motion or not.

**10.7**   (b). You are trying to change the rotational motion of the screw, which requires torque. You have failed with the original screwdriver, so you have not been able to supply enough torque to exceed the frictional torque holding the screw in place in the wood. To apply more torque with the same force, you need to apply the force farther from the rotation axis, which you can do by applying the force to a screwdriver with a fatter handle.

**10.8**   (a) Yes. If two forces of equal magnitude but opposite direction are applied to an object along different lines of action, the net force on the object is zero but the net torque is nonzero. (b) Yes. If a number of forces are applied to an object and the lines of action of the forces intersect at a common point, the net torque is zero, but the forces may result in a net force on the object.

**10.9**   Both the grinding wheel and the drill slow down because of frictional torque in the bearings of the motor. Let us estimate that this frictional torque is approximately the same in both tools. The grinding wheel is a large, massive disk, and the spinning bit in an electric drill is a small cylinder of relatively low mass. Thus, the grinding wheel has a larger moment of inertia than the drill bit, because of both its larger mass and its shape. According to Equation 10.27, the larger moment of inertia will result in a smaller angular acceleration for the same torque and a correspondingly longer time to come to rest.

**10.10**  Yes. Even though the skater is not moving in a circular path before grabbing the pole, she has angular momentum about the pole, as pointed out in Pitfall Prevention 10.8. After she grabs the pole, the force on the skater is along her arm, which is parallel to the position vector from the pole to her body. Thus, this force exerts no torque on the skater, so she can be modeled as an isolated system with no external torques. As a result, her angular momentum is conserved.

**10.11**  The box. Both the ball–Earth system and the box–Earth system have the same potential energy before the ball and box are released. As the items move down the inclines, this potential energy is transformed to kinetic energy. For the ball, the transformation is into both rotational and translational kinetic energy. The box has only translational kinetic energy. Because the kinetic energy of the ball is split into two types, its translational kinetic energy is necessarily less than that of the box. Consequently, its translational speed is less than that of the box, so the ball will lag behind.

## chapter 11

Astronauts F. Story Musgrave and Jeffrey A. Hoffman, along with the Hubble Space Telescope and the space shuttle *Endeavor,* are all falling around the Earth.

*(Courtesy NASA)*

**WEB**

For information on the Hubble Space Telescope and links to images, visit
**www.stsci.edu**

# Gravity, Planetary Orbits, and the Hydrogen Atom

At the beginning of our discussion of mechanics in Chapter 1, we introduced the notion of modeling and defined four categories of models: geometric, simplification, analysis, and structural. In this chapter, we apply our analysis models to two very common *structural models*. We shall discuss a structural model for a large system — the Solar System — and a structural model for a small system — the hydrogen atom.

In this chapter we return to Newton's law of universal gravitation — one of the fundamental force laws in nature — and show how it, together with our analysis models, enables us to understand the motions of planets and Earth satellites.

We conclude this chapter with a discussion of Niels Bohr's model of the hydrogen atom, which represents an interesting mixture of classical and nonclassical physics. Despite the hybrid nature of the model, some of its predictions agree with experimental measurements made on hydrogen atoms. This will be our first major venture into the area of *quantum physics,* which we will continue in Chapter 28.

## 11.1 • NEWTON'S LAW OF UNIVERSAL GRAVITATION REVISITED

Prior to 1686, many data had been collected on the motions of the Moon and the planets, but a clear understanding of the forces involved with the motions was not yet attainable. In that year, Isaac Newton provided the key that unlocked the secrets of the heavens. He knew, from the first law of motion, that a net force had to be acting on the Moon. If not, the Moon would move in a straight-line path rather than in its almost circular orbit. Newton reasoned that this force between the Moon and the Earth was an attractive force. He also concluded that there could be nothing special about the Earth–Moon system or the Sun and its planets that would cause gravitational forces to act on them alone.

As you should recall from Chapter 5, every particle in the Universe attracts every other particle with a force that is directly proportional to the product of their masses and inversely proportional to the square of the distance between them. If two particles have masses $m_1$ and $m_2$ and are separated by a distance $r$, the magnitude of the gravitational force between them is

• *Law of universal gravitation*

$$F_g = G\frac{m_1 m_2}{r^2} \qquad [11.1]$$

where $G$ is the *gravitational constant,* whose value in SI units is

$$G = 6.673 \times 10^{-11}\ \mathrm{N \cdot m^2/kg^2} \qquad [11.2]$$

Be sure you understand the difference between *g* and *G*. The symbol *g* represents the magnitude of the free-fall acceleration near a planet. At the surface of the Earth, *g* has the value $9.80\ \mathrm{m/s^2}$. On the other hand, *G* is a universal constant that has the same value everywhere in the Universe.

The force law given by Equation 11.1 is often referred to as an **inverse-square law** because the magnitude of the force varies as the inverse square of the separation of the particles. We can express this attractive force in vector form by defining a unit vector $\hat{\mathbf{r}}_{12}$ directed from $m_1$ toward $m_2$, as shown in Figure 11.1. The force exerted by $m_1$ on $m_2$ is

$$\mathbf{F}_{12} = -G\frac{m_1 m_2}{r^2}\,\hat{\mathbf{r}}_{12} \qquad [11.3]$$

where the negative sign indicates that particle 1 is attracted toward particle 2. Likewise, by Newton's third law, the force exerted by $m_2$ on $m_1$, designated $\mathbf{F}_{21}$, is equal in magnitude to $\mathbf{F}_{12}$ and in the opposite direction. That is, these forces form an action–reaction pair, and

$$\mathbf{F}_{21} = -\mathbf{F}_{12}$$

As Newton demonstrated, the **gravitational force exerted by a finite-sized, spherically symmetric mass distribution on a particle outside the distribution is the same as if the entire mass of the distribution were concentrated at its center.** For example, the force on a particle of mass $m$ at the Earth's surface has the magnitude

$$F_g = G\frac{M_E m}{R_E^{\,2}}$$

where $M_E$ is the Earth's mass and $R_E$ is the Earth's radius. This force is directed toward the center of the Earth.

In formulating his law of universal gravitation, Newton built on an observation that suggests the inverse square nature of the gravitational force between two bodies.

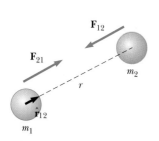

**Figure 11.1**

The gravitational force between two particles is attractive. The unit vector $\hat{\mathbf{r}}_{12}$ is directed from particle 1 toward particle 2. Note that $\mathbf{F}_{21} = -\mathbf{F}_{12}$.

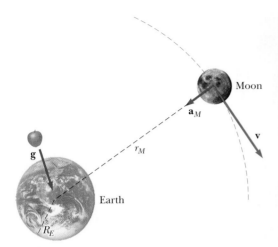

### Figure 11.2

As it revolves about the Earth, the Moon experiences a centripetal acceleration $\mathbf{a}_M$ directed toward the Earth. An object near the Earth's surface, such as the apple shown here, experiences an acceleration $\mathbf{g}$. (Dimensions are not to scale.)

He wrote in his notes, "I deduced that the forces which keep the planets in their orbs must be reciprocally as the squares of their distances from the centers about which they revolve; and thereby compared the force requisite to keep the Moon in her orb with force of gravity at the surface of the Earth; and found them answer pretty nearly."

We will perform some calculations that resemble the steps Newton took in deducing his law of gravitation and that will allow us to understand the phrase "and found them answer pretty nearly." Let us compare the acceleration of the Moon in its orbit with the acceleration of an object falling near the Earth's surface, such as the apple in Figure 11.2. We assume that the two accelerations have the same cause, namely, the gravitational attraction of the Earth. From the inverse-square law, Newton found that the acceleration of the Moon toward the Earth should be proportional to $1/r_M^2$, where $r_M$ is the separation between the centers of the Earth and Moon. Furthermore, the acceleration of the apple toward the Earth should be proportional to $1/R_E^2$, where $R_E$ is the radius of the Earth. When the values $r_M = 3.84 \times 10^8$ m and $R_E = 6.37 \times 10^6$ m are used, the ratio of the Moon's acceleration $a_M$ to the apple's acceleration $g$ is predicted to be

$$\frac{a_M}{g} = \frac{\left(G\dfrac{M_E}{r_M^2}\right)}{\left(G\dfrac{M_E}{R_E^2}\right)} = \left(\frac{R_E}{r_M}\right)^2 = \left(\frac{6.37 \times 10^6 \text{ m}}{3.84 \times 10^8 \text{ m}}\right)^2 = 2.75 \times 10^{-4}$$

Therefore,

$$a_M = (2.75 \times 10^{-4})(9.80 \text{ m/s}^2) = 2.70 \times 10^{-3} \text{ m/s}^2$$

The centripetal acceleration of the Moon can also be calculated. We know the Moon's orbital period $T = 27.32$ days $= 2.36 \times 10^6$ s, and its mean distance $r_M$ from the Earth. Using Equation 10.12,

$$a_M = r_M\omega^2 = r_M\left(\frac{2\pi}{T}\right)^2 = \frac{4\pi^2 r_M}{T^2} = \frac{4\pi^2(3.84 \times 10^8 \text{ m})}{(2.36 \times 10^6 \text{ s})^2} = 2.72 \times 10^{-3} \text{ m/s}^2$$

This agreement between the results of these two *independent* calculations of the acceleration of the Moon — Newton's "answering pretty nearly"—provides strong evidence that the inverse-square law of force is correct.

Artificial Earth satellites behave in the same way that the Moon does. A satellite in orbit around the Earth has a centripetal acceleration equal to the free-fall acceleration at the particular position of the satellite.

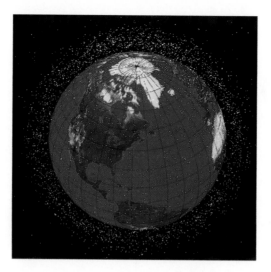

Humans have put many artificial satellites in orbit around the Earth. This diagram shows a plot of all known unclassified satellites and satellite debris larger in size than a baseball as of 1995. *(U.S. Space Command, NORAD)*

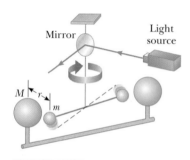

**Figure 11.3**

Schematic diagram of the Cavendish apparatus for measuring *G*. As the small spheres of mass *m* are attracted to the large spheres of mass *M*, the rod rotates through a small angle. A light beam reflected from a mirror on the rotating apparatus measures the angle of rotation. The dashed line represents the original position of the rod. (In reality, the length of wire above the mirror is much larger than that below it.)

## Measurement of the Gravitational Constant

The gravitational constant *G* was first measured in an important experiment by Sir Henry Cavendish in 1798. The apparatus he used consists of two small spheres, each of mass *m*, fixed to the ends of a light horizontal rod suspended by a thin wire, as in Figure 11.3. Two large spheres, each of mass *M*, are then placed near the smaller spheres. The attractive force between the smaller and larger spheres causes the rod to rotate and twist the wire. If the system is oriented as shown in Figure 11.3, the rod rotates clockwise when viewed from the top. The angle through which it rotates is measured by the deflection of a light beam that is reflected from a mirror attached to the wire. The experiment is carefully repeated with different masses at various separations. In addition to providing a value for *G*, the results confirm that the force is attractive, proportional to the product *mM*, and inversely proportional to the square of the distance *r*.

It is interesting that *G* is the least well known of the fundamental constants, with a percentage uncertainty thousands of times larger than those for other constants, such as the speed of light *c* and the fundamental electric charge *e*. Several measurements of *G* made in the 1990s have varied significantly from the previous value and from one another! The search for a more precise value of *G* continues to be an area of active research.

### THINKING PHYSICS 11.1

The novel *Icebound*, by Dean Koontz (Ballantine Books, New York, 1995), is a story of a group of scientists trapped on a floating iceberg near the North Pole. One of the devices that the scientists have with them is a transmitter with which they can fix their position with "the aid of a geosynchronous polar satellite." Can a satellite in a *polar* orbit be *geosynchronous?*

**Reasoning**   A geosynchronous satellite is one that stays over one location on the Earth's surface at all times. Thus, an antenna on the surface that receives signals from the satellite, such as a television dish, can stay pointed in a fixed direction toward the sky. The satellite must be in an orbit with the correct radius such that its orbital period is the same as that of the Earth's rotation. This results in the satellite appearing to have no east–west motion relative to the observer at the chosen location. Another requirement is that a geosynchronous satellite *must be in orbit over the* *equator.* Otherwise it would appear to undergo a north–south oscillation during one orbit. Thus, it would be impossible to have a geosynchronous satellite in a *polar* orbit. Even if such a satellite were at the proper distance from the Earth, it would be moving rapidly in the north–south direction, resulting in the necessity of accurate tracking equipment. What's more, it would be below the horizon for long periods of time, making it useless for determining one's position.

## THINKING PHYSICS 11.2

A satellite is in a circular orbit in space. Is it necessary that a massive object be at the center of the orbit?

**Reasoning**   This is the usual situation—satellites are normally in orbit around the Earth, a massive object. Imagine, however, an analogy—tie a rock somewhere along a length of string, and twirl the rock in a circle, as shown in Figure 11.4a. In this case, there is no entity at the center of the circular path of the rock, nor along its radius, that is exerting a force on the rock. The net force holding the rock in the circular orbit is a combination of the radial components of two forces: one from the string on each side of the rock.

The question posed is not about rocks on strings; it is about satellites. Does this situation have a gravitational counterpart? Yes—imagine moving a satellite from the Earth toward the Sun along the center line between them. At some point along that line, the gravitational force of attraction to the Earth will cancel just enough of the force from the Sun that the net force on the satellite can produce circular motion around the Sun with a period of one year, equal to the period of the Earth. As seen from the Earth, the satellite appears to be stationary at this so-called *Lagrange point*

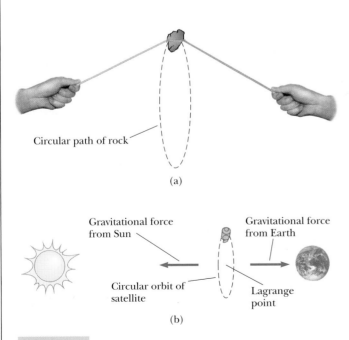

Circular path of rock

(a)

Gravitational force
from Sun

Gravitational force
from Earth

Circular orbit of
satellite

Lagrange
point

(b)

**Figure 11.4**

(Thinking Physics 11.2)

between the Earth and the Sun. This would look similar to the balance along the string of the tension on either side of the rock. Now, imagine pulling the satellite perpendicularly off the Earth–Sun line a small amount. As long as the distance is relatively small, the components of the forces from the Earth and Sun along the Sun–Earth centerline can add as before to cause the satellite to orbit the Sun with a period of one year. The components of the two forces perpendicular to the centerline will add, to provide a net force on the satellite back toward the centerline. Now, suppose we set the satellite into circular motion with just the right velocity. The satellite will be in orbit around the Lagrange point with no mass located there as shown in Figure 11.4b.

The International Sun–Earth Explorer 3 satellite (ISEE-3) was launched in 1978 and placed in just such an orbit in order to study the solar wind. The advantage of this orbit is that the satellite always resides on the side of the Sun toward the Earth, so that it is constantly in communication with Earth. In the absence of a specific mission to Comet Halley from Earth, the ISEE-3 satellite was removed from this orbit in 1982 and placed in orbit around the Moon, in anticipation of a flyby of Comet Halley. On the way to this flyby, it passed through the tail of Comet Giacobini–Zinner and successfully gathered important data on both comets.

## Example 11.1   The Mass of the Earth

Use the gravitational force law to find an approximate value for the mass of the Earth.

**Solution**   Figure 11.5, obviously not to scale, shows a baseball falling toward the Earth at a location where the free-fall acceleration is $g$. We know from Chapter 5 that the magnitude of the gravitational force exerted on the baseball by the Earth is the same as the weight $m_b g$ of the ball. That is,

$$m_b g = G \frac{M_E m_b}{R_E{}^2}$$

We divide each side of this equation by $m_b$ and solve for the mass of the Earth $M_E$:

$$(1) \quad M_E = \frac{g R_E{}^2}{G}$$

The falling baseball is close enough to the Earth that the distance of separation between the center of the ball and the center of the Earth can be taken as the radius of the Earth, $6.37 \times 10^6$ m. Thus, the mass of the Earth is

$$M_E = \frac{(9.80 \text{ m/s}^2)(6.37 \times 10^6 \text{ m})^2}{6.67 \times 10^{-11} \text{ N} \cdot \text{m}^2/\text{kg}^2} = \boxed{5.98 \times 10^{24} \text{ kg}}$$

Notice also that (1) can be written as

$$g = G \frac{M_E}{r^2}$$

which we have seen before as Equation 5.17. This indicates that the free-fall acceleration at a point above the Earth's sur-

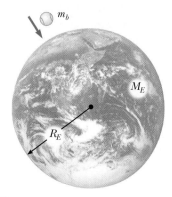

**Figure 11.5**

(Example 11.1) A baseball falling toward the Earth (not drawn to scale).

face decreases as the inverse square of the distance between the point and the center of the Earth.

**EXERCISE**   If an object weighs 270 N at the Earth's surface, what will it weigh at an altitude equal to twice the radius of the Earth?

*Answer*   30 N

**EXERCISE**   Determine the magnitude of the free-fall acceleration at an altitude of 500 km. By what percentage is the weight of a body reduced at this altitude?

*Answer*   8.43 m/s$^2$; 14%

## Example 11.2   An Earth Satellite

A satellite of mass $m$ moves in a circular orbit about the Earth with a constant speed $v$, at a height of $h = 1\,000$ km above the Earth's surface, as in Figure 11.6. (For clarity, this figure is not drawn to scale.) Find the orbital speed of the satellite.

**Solution**   The only external force on the satellite is the gravitational force exerted by the Earth. This force is directed to-ward the center of the satellite's circular path. We apply Newton's second law to the satellite modeled as a particle in uniform circular motion. Because the magnitude of the gravitational force between the Earth and the satellite is $GM_E m/r^2$ we find that

$$F_g = G\frac{M_E m}{r^2} = m\frac{v^2}{r}$$

$$v = \sqrt{\frac{GM_E}{r}}$$

In this expression, the distance $r$ is the Earth's radius plus the height of the satellite; that is, $r = R_E + h = 7.37 \times 10^6$ m, so that

$$v = \sqrt{\frac{(6.67 \times 10^{-11}\ \text{N·m}^2/\text{kg}^2)(5.98 \times 10^{24}\ \text{kg})}{7.37 \times 10^6\ \text{m}}}$$

$$= 7.36 \times 10^3\ \text{m/s} \approx \boxed{16\,400\ \text{mi/h}}$$

Note that $v$ *is independent of the mass of the satellite!*

**EXERCISE**   Calculate the period of revolution $T$ of the satellite.

*Answer*   105 min

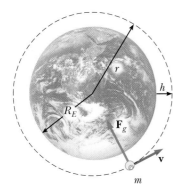

**Figure 11.6**

(Example 11.2) A satellite of mass $m$ moving around the Earth in a circular orbit of radius $r$ with constant speed $v$. The only force acting on the satellite is the gravitational force $\mathbf{F}_g$. (Not drawn to scale.)

## 11.2 • STRUCTURAL MODELS

In Chapter 1, we mentioned that we would discuss four categories of models. The fourth category is **structural models.** In these models, we propose theoretical structures in an attempt to understand the behavior of a system with which we cannot interact directly because it is far different in scale from our macroscopic world—either much smaller or much larger.

One of the earliest structural models to be explored was that of the place of the Earth in the Universe. The movements of the planets, stars, and other celestial bodies have been observed by people for thousands of years. Early in history, scientists regarded the Earth as the center of the Universe. This is a structural model for the Universe called the *geocentric model*. It was elaborated and formalized by the Greek astronomer Claudius Ptolemy in the second century A.D. and was accepted for the next 1400 years. In 1543, the Polish astronomer Nicolaus Copernicus (1473–1543) offered a different structural model in which the Earth is part of a local Solar System, suggesting that the Earth and the other planets revolve in perfectly circular orbits about the Sun (the *heliocentric model*).

In general, a structural model contains the following features:

1. A description of the physical components of the system; in the heliocentric model, the components are the planets and the Sun.
2. A description of where the components are located relative to one another and how they interact; in the heliocentric model, the planets and Sun interact via the gravitational force and the planets are in orbit around the Sun.
3. A description of the time evolution of the system; the heliocentric model assumes a steady-state Solar System, with planets revolving in orbits around the Sun with fixed periods.

• *Features of structural models*

4. A description of the agreement between predictions of the model and actual observations and, possibly, predictions of new effects that have not yet been observed; the heliocentric model predicts Earth-based observations of Mars that are in agreement with historical and present measurements. The geocentric model also was able to find agreement between predictions and observations, but only at the expense of a very complicated structural model in which the planets moved in circles built on other circles. The heliocentric model, along with Newton's law of universal gravitation, predicted that a spacecraft could be sent from the Earth to Mars long before it was actually done in the 1970s.

In Sections 11.3 and 11.4, we explore some of the details of the structural model of the Solar System. In Section 11.5, we investigate a structural model of the hydrogen atom.

## 11.3 • KEPLER'S LAWS

The Danish astronomer Tycho Brahe (1546–1601) made accurate astronomical measurements over a period of 20 years and provided the basis for the currently accepted structural model of the Solar System. It is interesting to note that these precise observations, made on the planets and 777 stars, were carried out with nothing more elaborate than a large sextant and compass; the telescope had not yet been invented.

The German astronomer Johannes Kepler, who was Brahe's assistant, acquired Brahe's astronomical data and spent about 16 years trying to deduce a mathematical model for the motions of the planets. After many laborious calculations, he found that Brahe's precise data on the revolution of Mars about the Sun provided the answer. Kepler's analysis first showed that the concept of circular orbits about the Sun in the heliocentric model had to be abandoned. He discovered that the orbit of Mars could be accurately described by a curve called an *ellipse*. He then generalized this analysis to include the motions of all planets. The complete analysis is summarized in three statements, known as **Kepler's laws,** each of which are discussed in the following sections.

Newton demonstrated that these laws are consequences of the gravitational force that exists between any two masses. Newton's law of universal gravitation, together with his laws of motion, provides the basis for a full mathematical representation of the motion of planets and satellites.

### Kepler's First Law

We are familiar with circular orbits of objects around gravitational force centers from our discussions in this chapter. Kepler's first law indicates that the circular orbit is a very special case, and that elliptical orbits are the general situation:*

**Johannes Kepler** (1571–1630)

The German astronomer Kepler is best known for developing the laws of planetary motion based on the careful observations of Tycho Brahe. *(Art Resource)*

• *Kepler's first law*

Each planet in the Solar System moves in an elliptical orbit with the Sun at one focus.

---

* We choose a simplification model in which a body of mass *m* is in orbit around a body of mass *M*, with $M \gg m$. In this way, we model the body of mass *M* to be stationary. In reality, this is not true—both *M* and *m* move around the center of mass of the system of two objects. This is how we indirectly detect planets around other stars—we see the "wobbling" motion of the star as the planet and the star rotate about the center of mass.

Figure 11.7 shows the geometry of an ellipse, which serves as our geometric model for the elliptical orbit of a planet.* An ellipse is mathematically defined by choosing two points, $F_1$ and $F_2$, each of which is a called a **focus,** and then drawing a curve through points for which the sum of the distances $r_1$ and $r_2$ from $F_1$ and $F_2$ is a constant. The longest distance through the center between points on the ellipse (and passing through both foci) is called the **major axis,** and this distance is $2a$. In Figure 11.7, the major axis is drawn along the $x$ direction. The distance $a$ is called the **semimajor axis.** Similarly, the shortest distance through the center between points on the ellipse is called the **minor axis** of length $2b$, where the distance $b$ is the **semiminor axis.** Either focus of the ellipse is located at a distance $c$ from the center of the ellipse, where $a^2 = b^2 + c^2$. In the elliptical orbit of a planet around the Sun, the Sun is at one focus of the ellipse. Nothing is at the other focus.

The **eccentricity** of an ellipse is defined as $e = c/a$ and describes the general shape of the ellipse. For a circle, $c = 0$, and the eccentricity is therefore zero. The smaller $b$ is than $a$, the shorter the ellipse is along the $y$ direction compared with its extent in the $x$ direction in Figure 11.7. As $b$ decreases, $c$ increases, and the eccentricity $e$ increases. Thus, higher values of eccentricity correspond to longer and thinner ellipses. The range of values of the eccentricity for an ellipse is $e \leq 1$. Eccentricities higher than 1 correspond to hyperbolas.

Eccentricities for planetary orbits vary widely in the Solar System. The eccentricity of the Earth's orbit is 0.017, which makes it nearly circular. On the other hand, the eccentricity of Pluto's orbit is 0.25, the highest of all the nine planets. Figure 11.8a shows an ellipse with the eccentricity of that of Pluto's orbit. Notice

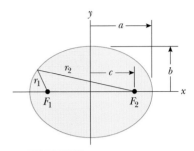

**Figure 11.7**

Plot of an ellipse. The semimajor axis has length $a$, and the semiminor axis has length $b$. A focus is located at a distance $c$ from the center on each side of the center.

**PITFALL PREVENTION 11.2**

**Where is the Sun?**

The Sun is located at one focus of the elliptical orbit of a planet. It is *not* located at the center of the ellipse.

(a)

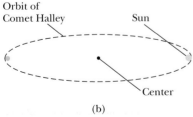

(b)

**Figure 11.8**

(a) The shape of the orbit of Pluto, which has the highest eccentricity ($e = 0.25$) among the planets in the Solar System. The Sun is located at the large yellow dot, which is a focus of the ellipse. Nothing physical is located at the center of the orbit *(the small dot)* or the other focus *(the blue dot)*. (b) The shape of the orbit of Comet Halley.

---

* Actual orbits show perturbations due to moons in orbit around the planet and passages of the planet near other planets. We will ignore these perturbations and adopt a simplification model in which the planet follows a perfectly elliptical orbit.

that even this highest eccentricity orbit is difficult to distinguish from a circle. This is why Kepler's first law is an admirable accomplishment.

The eccentricity of the orbit of Comet Halley is 0.97, describing an orbit whose major axis is much longer than its minor axis, as shown in Figure 11.8b. As a result, Comet Halley spends much of its 76-year period far from the Sun and invisible from the Earth. It is only visible to the naked eye during a small part of its orbit when it is near the Sun.

Let us imagine now a planet in an elliptical orbit such as that shown in Figure 11.7, with the Sun at focus $F_2$. When the planet is at the far left in the diagram, the distance between the planet and the Sun is $a + c$. This point is called the *aphelion*, where the planet is the farthest away from the Sun that it can be in the orbit (for an object in orbit around the Earth, this point is called the *apogee*). Conversely, when the planet is at the right end of the ellipse, the point is called the *perihelion* (for an Earth orbit, the *perigee*), and the distance between the planet and the Sun is $a - c$.

Kepler's first law is a direct result of the inverse-square nature of the gravitational force. We have discussed circular and elliptical orbits. These are the allowed shapes of orbits for objects that are *bound* to the gravitational force center. These objects include planets, asteroids, and comets that move repeatedly around the Sun, as well as moons orbiting a planet. *Unbound* objects also occur such as a meteoroid from deep space that might pass by the Sun once and then never return. The gravitational force between the Sun and these objects also varies as the inverse square of the separation distance, and the allowed paths for these objects are parabolas and hyperbolas.

### Kepler's Second Law

The second of Kepler's laws is

The radius vector drawn from the Sun to any planet sweeps out equal areas in equal time intervals.

This can be shown to be a consequence of angular momentum conservation as follows. Consider a planet of mass $M_p$ moving about the Sun in an elliptical orbit (Fig. 11.9a). Let us consider the planet as a system. We shall assume that the Sun is much more massive than the planet, so that the Sun does not move. The gravitational force acting on the planet is a central force, always along the radius vector, directed toward the Sun. The torque on the planet due to this central force is zero because $\mathbf{F}$ is parallel to $\mathbf{r}$. That is,

$$\boldsymbol{\tau} = \mathbf{r} \times \mathbf{F} = \mathbf{r} \times F(r)\hat{\mathbf{r}} = 0$$

Recall that the external net torque on a system equals the time rate of change of angular momentum of the system; that is, $\boldsymbol{\tau} = d\mathbf{L}/dt$. Therefore, because $\boldsymbol{\tau} = 0$, the angular momentum $\mathbf{L}$ of the planet is a constant of the motion:

$$\mathbf{L} = \mathbf{r} \times \mathbf{p} = M_p \mathbf{r} \times \mathbf{v} = \text{constant}$$

We can relate this result to the following geometric consideration. In a time interval $dt$, the radius vector $\mathbf{r}$ in Figure 11.9b sweeps out the area $dA$, which equals one-half the area $|\mathbf{r} \times d\mathbf{r}|$ of the parallelogram formed by the vectors $\mathbf{r}$ and $d\mathbf{r}$. Be-

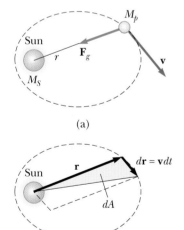

(a)

(b)

**Figure 11.9**

(a) The gravitational force acting on a planet acts toward the Sun, along the radius vector. (b) As a planet orbits the Sun, the area swept out by the radius vector in a time interval $dt$ is equal to one-half the area of the parallelogram formed by the vectors $\mathbf{r}$ and $d\mathbf{r} = \mathbf{v}\, dt$.

cause the displacement of the planet in the time interval $dt$ is given by $d\mathbf{r} = \mathbf{v}dt$, we have

$$dA = \tfrac{1}{2}\left|\mathbf{r} \times d\mathbf{r}\right| = \tfrac{1}{2}\left|\mathbf{r} \times \mathbf{v}dt\right| = \frac{L}{2M_p}\,dt$$

$$\frac{dA}{dt} = \frac{L}{2M_p} = \text{constant} \qquad\qquad \textbf{[11.4]}$$

where $L$ and $M_p$ are both constants. Thus, we conclude that the radius vector from the Sun to any planet sweeps out equal areas in equal times.

It is important to recognize that this result is a consequence of the fact that the gravitational force is a central force, which in turn implies that angular momentum of the planet is constant. Therefore, the law applies to *any* situation that involves a central force, whether inverse-square or not.

**EXERCISE** At its aphelion the planet Mercury is $6.99 \times 10^{10}$ m from the Sun, and at its perihelion it is $4.60 \times 10^{10}$ m from the Sun. If its orbital speed is $3.88 \times 10^4$ m/s at the aphelion, what is its orbital speed at the perihelion?

*Answer* $5.90 \times 10^4$ m/s

### THINKING PHYSICS 11.3

The Earth is closer to the Sun when it is winter in the Northern Hemisphere than in summer. July and January both have 31 days. In which month, if either, does the Earth move through a longer distance in its orbit?

**Reasoning** The Earth is in a slightly elliptical orbit around the Sun. Because of angular momentum conservation, the Earth moves more rapidly when it is close to the Sun and more slowly when it is farther away. Thus, because it is closer to the Sun in January, it is moving faster and will cover more distance in its orbit than it will in July.

## Kepler's Third Law

Kepler's third law is

> The square of the orbital period of any planet is proportional to the cube of the semimajor axis of the elliptical orbit.

This can be shown easily for circular orbits. Consider a planet of mass $M_p$ that is assumed to be moving about the Sun (mass $M_S$) in a circular orbit, as in Figure 11.10. Because the gravitational force provides the centripetal acceleration of the planet as it moves in a circle, we use Newton's second law for a particle in uniform circular motion,

$$\frac{GM_S M_p}{r^2} = \frac{M_p v^2}{r}$$

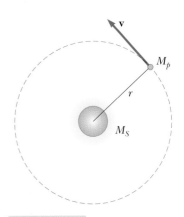

**Figure 11.10**

A planet of mass $M_p$ moving in a circular orbit about the Sun. Kepler's third law relates the period of the orbit to the radius.

The orbital speed of the planet is $2\pi r/T$, where $T$ is the period; therefore, the preceding expression becomes

$$\frac{GM_S}{r^2} = \frac{(2\pi r/T)^2}{r}$$

$$T^2 = \left(\frac{4\pi^2}{GM_S}\right)r^3 = K_S r^3$$

where $K_S$ is a constant given by

$$K_S = \frac{4\pi^2}{GM_S} = 2.97 \times 10^{-19}\ \text{s}^2/\text{m}^3$$

For elliptical orbits Kepler's third law is expressed by starting with $T^2 = K_S r^3$ and replacing $r$ with the length $a$ of the semimajor axis (see Fig. 11.7):

$$T^2 = \left(\frac{4\pi^2}{GM_S}\right)a^3 = K_S a^3 \qquad [11.5]$$

Equation 11.5 is Kepler's third law. Because the semimajor axis of a circular orbit is its radius, Equation 11.5 is valid for both circular and elliptical orbits. Note that the constant of proportionality $K_S$ is independent of the mass of the planet. Equation 11.5 is therefore valid for *any* planet. If we were to consider the orbit of a satellite about the Earth, such as the Moon, then the constant would have a different value, with the Sun's mass replaced by the Earth's mass, that is, $K_E = 4\pi^2/GM_E$.

Table 11.1 is a collection of useful planetary data. The last column verifies that the ratio $T^2/r^3$ is constant. The small variations in the values in this column are due to uncertainties in the data measured for the periods and semimajor axes of the planets.

### Quick Quiz 11.1

Kepler's first law indicates that planets travel in elliptical orbits. This is intimately related to the fact that the gravitational force varies as the inverse square of the

| **TABLE 11.1** | **Useful Planetary Data**[a] | | | |
|---|---|---|---|---|
| Body | Mass (kg) | Mean Radius (m) | Period (s) | Mean Distance from Sun (m) | $\dfrac{T^2}{r^3}\left(\dfrac{\text{s}^2}{\text{m}^3}\right)$ |
| Mercury | $3.18 \times 10^{23}$ | $2.43 \times 10^6$ | $7.60 \times 10^6$ | $5.79 \times 10^{10}$ | $2.97 \times 10^{-19}$ |
| Venus | $4.88 \times 10^{24}$ | $6.06 \times 10^6$ | $1.94 \times 10^7$ | $1.08 \times 10^{11}$ | $2.99 \times 10^{-19}$ |
| Earth | $5.98 \times 10^{24}$ | $6.37 \times 10^6$ | $3.156 \times 10^7$ | $1.496 \times 10^{11}$ | $2.97 \times 10^{-19}$ |
| Mars | $6.42 \times 10^{23}$ | $3.37 \times 10^6$ | $5.94 \times 10^7$ | $2.28 \times 10^{11}$ | $2.98 \times 10^{-19}$ |
| Jupiter | $1.90 \times 10^{27}$ | $6.99 \times 10^7$ | $3.74 \times 10^8$ | $7.78 \times 10^{11}$ | $2.97 \times 10^{-19}$ |
| Saturn | $5.68 \times 10^{26}$ | $5.85 \times 10^7$ | $9.35 \times 10^8$ | $1.43 \times 10^{12}$ | $2.99 \times 10^{-19}$ |
| Uranus | $8.68 \times 10^{25}$ | $2.33 \times 10^7$ | $2.64 \times 10^9$ | $2.87 \times 10^{12}$ | $2.95 \times 10^{-19}$ |
| Neptune | $1.03 \times 10^{26}$ | $2.21 \times 10^7$ | $5.22 \times 10^9$ | $4.50 \times 10^{12}$ | $2.99 \times 10^{-19}$ |
| Pluto | $\approx 1.4 \times 10^{22}$ | $\approx 1.5 \times 10^6$ | $7.82 \times 10^9$ | $5.91 \times 10^{12}$ | $2.96 \times 10^{-19}$ |
| Moon | $7.36 \times 10^{22}$ | $1.74 \times 10^6$ | — | — | — |
| Sun | $1.991 \times 10^{30}$ | $6.96 \times 10^8$ | — | — | — |

[a] For a more complete set of data, see, for example, the *Handbook of Chemistry and Physics*, Boca Raton, FL, The Chemical Rubber Publishing Co.

separation between the Sun and the planet. Suppose the gravitational force varied as the inverse *cube*, rather than the inverse square. This allows for some new types of orbits. What about the second and third laws? Will they change if the force depends on the inverse cube of the separation?

### Quick Quiz 11.2

A comet is in a highly elliptical orbit around the Sun. The period of the comet's orbit is 90 days. Which of the following statements is true about the possibility of a collision between this comet and the Earth? (a) Collision is not possible. (b) Collision is possible. (c) Not enough information is available to determine whether a collision is possible.

### Quick Quiz 11.3

A satellite moves in an elliptical orbit about the Earth such that, at perigee and apogee positions, the distances from the Earth's center are, respectively, $D$ and $4D$. The relationship between the speeds at these two positions is (a) $v_p = v_a$; (b) $v_p = 4v_a$; (c) $v_a = 4v_p$; (d) $v_p = 2v_a$; (e) $v_a = 2v_p$.

## 11.4 • ENERGY CONSIDERATIONS IN PLANETARY AND SATELLITE MOTION

So far we have approached orbital mechanics from the point of view of forces and angular momentum. Let us now investigate the motion of planets in orbit from the *energy* point of view.

Consider an object of mass $m$ moving with a speed $v$ in the vicinity of a massive object of mass $M \gg m$. This two-object system might be a planet moving around the Sun, a satellite orbiting the Earth, or a comet making a one-time flyby past the Sun. We will treat the two objects of mass $m$ and $M$ as an isolated system. If we assume that $M$ is at rest in an inertial reference frame (because $M \gg m$), then the total mechanical energy $E$ of the two-object system is the sum of the kinetic energy of the object of mass $m$ and the gravitational potential energy of the system:

$$E = K + U_g$$

Recall from Chapter 7 that the gravitational potential energy $U_g$ associated with *any pair* of particles of masses $m_1$ and $m_2$ separated by a distance $r$ is given by

$$U_g = -\frac{Gm_1m_2}{r}$$

where we have defined $U_g \to 0$ as $r \to \infty$; therefore, in our case, the mechanical energy of the system of $m$ and $M$ is

$$E = \tfrac{1}{2}mv^2 - \frac{GMm}{r} \qquad \text{[11.6]}$$

Equation 11.6 shows that $E$ may be positive, negative, or zero, depending on the value of $v$ at a particular separation distance $r$. If we consider the energy diagram method of Section 7.7, we can show the potential and total energies of the system as a function of $r$ as in Figure 11.11. A planet moving around the Sun and a satellite in orbit around the Earth are *bound* systems, such as those we discussed in Section 11.3—the Earth will always stay near the Sun and the satellite near the

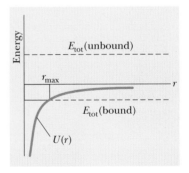

**Figure 11.11**

The lower total energy line represents a bound system. The separation distance $r$ between the two gravitationally bound objects never exceeds $r_{max}$. The upper total energy line represents an unbound system of two objects interacting gravitationally. The separation distance $r$ between the two objects can have any value.

Earth. In Figure 11.11, these are represented by a total energy that is negative. The point at which the total energy line intersects the potential energy curve is a turning point—the maximum separation distance $r_{max}$ between the two bound objects.

A one-time meteoroid flyby represents an unbound system—the meteoroid interacts with the Sun but is not bound to it. Thus, the meteoroid can in theory move infinitely far away from the Sun. This is represented in Figure 11.11 by a total energy line in the positive region of the graph. This line never intersects the potential energy curve, so all values of $r$ are possible.

For a bound system, such as the Earth and Sun, $E$ is necessarily less than zero because we have chosen the convention that $U_g \rightarrow 0$ as $r \rightarrow \infty$. We can easily establish that $E < 0$ for the system consisting of an object of mass $m$ moving in a circular orbit about an object of mass $M \gg m$. Applying Newton's second law to the object of mass $m$ in uniform circular motion,

$$\sum F = ma \quad \longrightarrow \quad \frac{GMm}{r^2} = \frac{mv^2}{r}$$

Multiplying both sides by $r$ and dividing by 2,

$$\tfrac{1}{2} mv^2 = \frac{GMm}{2r} \tag{11.7}$$

Substituting this into Equation 11.6, we obtain

$$E = \frac{GMm}{2r} - \frac{GMm}{r}$$

$$E = -\frac{GMm}{2r} \tag{11.8}$$

This result clearly shows that the **total mechanical energy must be negative in the case of circular orbits.** Furthermore, Equation 11.7 shows that the **kinetic energy of an object in a circular orbit is equal to one-half the magnitude of the potential energy of the system** (when the potential energy is chosen to be zero at infinite separation).

The total mechanical energy is also negative in the case of elliptical orbits. The expression for $E$ for elliptical orbits is the same as Equation 11.8, with $r$ replaced by the semimajor axis length, $a$:

• *Total energy of a planet–star system*

$$E = -\frac{GMm}{2a} \tag{11.9}$$

The total energy, the total angular momentum, and the total linear momentum of a planet–star system are constants of the motion, according to the isolated system model.

### Quick Quiz 11.4

A comet moves in an elliptical orbit around the Sun. Which point in its orbit represents the highest value of (a) the speed of the comet, (b) the potential energy of the comet–Sun system, (c) the kinetic energy of the comet, and (d) the total energy of the comet–Sun system?

## Example 11.3    A Satellite in an Elliptical Orbit

A satellite moves in an elliptical orbit about the Earth, as in Figure 11.12. The minimum and maximum distances from the surface of the Earth are 400 km and 3 000 km, respectively. Find the speeds of the satellite at apogee and perigee.

**Solution**   Because the mass of the satellite is negligible compared with the Earth's mass, we take the center of mass of the Earth to be at rest. Gravity is a central force, and so the angular momentum of the satellite about the Earth's center of mass remains constant with time. With subscripts $a$ and $p$

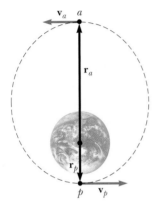

**Figure 11.12**

(Example 11.3) A satellite in an elliptical orbit about the Earth.

for the apogee and perigee positions, conservation of angular momentum for the satellite gives $L_p = L_a$, or

$$m v_p r_p = m v_a r_a$$

$$v_p r_p = v_a r_a$$

Using the Earth's radius of $6.37 \times 10^6$ m and the given data, we find that $r_a = 9.37 \times 10^6$ m and $r_p = 6.77 \times 10^6$ m. Thus,

$$(1) \quad \frac{v_p}{v_a} = \frac{r_a}{r_p} = \frac{9.37 \times 10^6 \text{ m}}{6.77 \times 10^6 \text{ m}} = 1.38$$

Because the satellite and the Earth form an isolated system, we can apply conservation of energy and obtain $E_p = E_a$, or

$$U_p + K_p = U_a + K_a$$

$$- G \frac{M_E m}{r_p} + \tfrac{1}{2} m v_p^2 = - G \frac{M_E m}{r_a} + \tfrac{1}{2} m v_a^2$$

$$(2) \quad 2 G M_E \left( \frac{1}{r_a} - \frac{1}{r_p} \right) = (v_a^2 - v_p^2)$$

Because we know the numerical values of $G$, $M_E$, $r_p$, and $r_a$, we can use Equations (1) and (2) to determine the two unknowns $v_p$ and $v_a$. Solving the equations simultaneously, we obtain

$$v_a = \boxed{8.27 \text{ km/s}} \qquad v_a = \boxed{5.98 \text{ km/s}}$$

## Escape Speed

Suppose an object of mass $m$ is projected vertically from the Earth's surface with an initial speed $v_i$, as in Figure 11.13. We can use energy considerations to find the minimum value of the initial speed such that the object will continue to move away from the Earth forever. Equation 11.6 gives the total energy of the object–Earth system at any point when the speed of the object and its distance from the center of the Earth are known. At the surface of the Earth, $r_i = R_E$. When the object reaches its maximum altitude, $v_f = 0$ and $r_f = r_{max}$. Because the total energy of the system is conserved, substitution of these conditions into Equation 11.6 gives

$$\tfrac{1}{2} m v_i^2 - \frac{G M_E m}{R_E} = - \frac{G M_E m}{r_{max}}$$

Solving for $v_i^2$ gives

$$v_i^2 = 2 G M_E \left( \frac{1}{R_E} - \frac{1}{r_{max}} \right) \qquad \text{[11.10]}$$

If the initial speed is known, this expression can therefore be used to calculate the maximum altitude $h$ because we know that $h = r_{max} - R_E$.

We are now in a position to calculate the minimum speed the object must have at the Earth's surface in order to continue to move away forever. This is

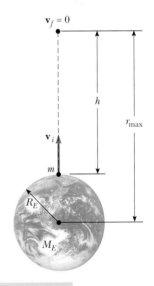

**Figure 11.13**

An object of mass $m$ projected upward from the Earth's surface with an initial speed $v_i$ reaches a maximum altitude $h = r_{max} - R_E$.

- *Escape speed*

Although Equation 11.11 provides the escape speed, remember that this conventional name is misleading. It is impossible to *completely* escape from the Earth's gravitational influence, because the gravitational force is of infinite range. No matter how far away you are, you will always feel some gravitational force due to the Earth. In practice, however, this force will be much smaller than forces due to other astronomical objects closer to you, so the gravitational force from the Earth can be ignored.

called the **escape speed** $v_{esc}$ and results in the speed asymptotically approaching *zero*. Letting $r_{max} \rightarrow \infty$ in Equation 11.10 and setting $v_f = 0$ and $v_i = v_{esc}$, we have

$$v_{esc} = \sqrt{\frac{2GM_E}{R_E}} \qquad [11.11]$$

Note that this expression for $v_{esc}$ is independent of the mass of the object projected from the Earth. For example, a spacecraft has the same escape speed as a molecule. Furthermore, the result is independent of the *direction* of the velocity.

Note also that Equations 11.10 and 11.11 can be applied to objects projected from *any* planet. That is, in general, the escape speed from any planet of mass $M$ and radius $R$ is

$$v_{esc} = \sqrt{\frac{2GM}{R}} \qquad [11.12]$$

A list of escape speeds for the planets, the Moon, and the Sun is given in Table 11.2. Note that the values vary from 1.1 km/s for Pluto to about 618 km/s for the Sun. These results, together with some ideas from the kinetic theory of gases (Chapter 16), explain why our atmosphere does not contain significant amounts of hydrogen, which is the most abundant element in the Universe. As we shall see later, gas molecules have an average kinetic energy that depends on the temperature of the gas. Lighter molecules in an atmosphere have translational speeds that are closer to the escape speed than more massive molecules, so they have a higher probability of escaping from the planet, and the lighter molecules diffuse into space. This mechanism explains why the Earth does not retain hydrogen molecules and helium atoms in its atmosphere but does retain much heavier molecules, such as oxygen and nitrogen. On the other hand, Jupiter has a very large escape speed (60 km/s), which enables it to retain hydrogen, the primary constituent of its atmosphere.

### TABLE 11.2

**Escape Speeds from the Surfaces of the Planets, the Moon, and the Sun**

| Planet | $v_{esc}$ (km/s) |
|---|---|
| Mercury | 4.3 |
| Venus | 10.3 |
| Earth | 11.2 |
| Mars | 5.0 |
| Jupiter | 60 |
| Saturn | 36 |
| Uranus | 22 |
| Neptune | 24 |
| Pluto | 1.1 |
| Moon | 2.3 |
| Sun | 618 |

### Quick Quiz 11.5

If you were a planetary prospector and discovered gold on an asteroid, it probably would not be a good idea to jump up and down in excitement over your find. Why?

### Black Holes

In Chapter 10 we briefly described a rare event called a supernova—the catastrophic explosion of a very massive star. The material that remains in the central core of such an object continues to collapse, and the core's ultimate fate depends on its mass. If the core has a mass less than 1.4 times the mass of our Sun, it gradually cools down and ends its life as a white dwarf star. However, if the core's mass is greater than this, it may collapse further due to gravitational forces. What remains is a neutron star, discussed in Chapter 10, in which the mass of a star is compressed to a radius of about 10 km. (On Earth, a teaspoon of this material would weigh about 5 billion tons!)

An even more unusual star death may occur when the core has a mass greater than about three solar masses. The collapse may continue until the star becomes a very small object in space, commonly referred to as a **black hole.** In effect, black holes are the remains of stars that have collapsed under their own gravitational force. If an object such as a spacecraft comes close to a black hole, it experiences an extremely strong gravitational force and is trapped forever.

The escape speed from any spherical body depends on the mass and radius of the body. The escape speed for a black hole is very high, due to the concentration of the mass of the star into a sphere of very small radius. If the escape speed exceeds the speed of light $c$, radiation from the body (e.g., visible light) cannot escape, and the body appears to be black; hence the origin of the term *black hole.* The critical radius $R_S$ at which the escape speed is $c$ is called the **Schwarzschild radius** (Fig. 11.14). The imaginary surface of a sphere of this radius surrounding the black hole is called the **event horizon.** This is the limit of how close you can approach the black hole and hope to be able to escape.

Although light from a black hole cannot escape, light from events taking place near the black hole should be visible. For example, it is possible for a binary star system to consist of one normal star and one black hole. Material surrounding the ordinary star can be pulled into the black hole, forming an **accretion disk** around the black hole, as suggested in Figure 11.15. Friction among particles in the accretion disk results in transformation of mechanical energy into internal energy. As a result, the orbital height of the material above the event horizon decreases and the temperature rises. This high-temperature material emits a large amount of radiation, extending well into the x-ray region of the electromagnetic spectrum. These x-rays are characteristic of a black hole. Several possible candidates for black holes have been identified by observation of these x-rays.

Evidence also supports the existence of supermassive black holes at the centers of galaxies, with masses very much larger than the Sun. (The evidence is strong for a supermassive black hole of mass 2 to 3 million solar masses at the center of our galaxy.) Theoretical models for these bizarre objects predict that jets of material should be evident along the rotation axis of the black hole. Figure 11.16 shows a Hubble Space Telescope photograph of galaxy M87. The jet of material

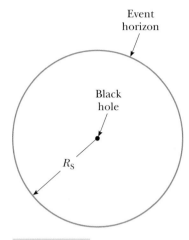

**Figure 11.14**

A black hole. The distance $R_S$ equals the Schwarzschild radius. Any event occurring within the boundary of radius $R_S$, called the event horizon, is invisible to an outside observer.

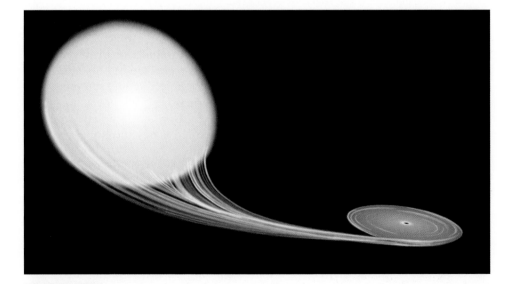

**Figure 11.15**

A binary star system consisting of an ordinary star on the left and a black hole on the right. Matter pulled from the ordinary star forms an accretion disk around the black hole, in which matter is raised to very high temperatures, resulting in the emission of x-rays.

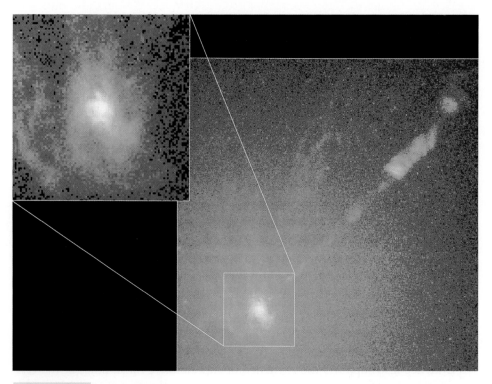

**Figure 11.16**

Hubble Space Telescope images of the galaxy M87. The inset shows the center of the galaxy. The wider view shows a jet of material moving away from the center of the galaxy toward the upper right of the figure at about one-tenth the speed of light. Such jets are believed to be evidence of a supermassive black hole at the galaxy's center. *(H. Ford et al. & NASA)*

**WEB**

For information on the current status of LIGO, visit **www.ligo.caltech.edu/**

coming from this galaxy is believed to be evidence for a supermassive black hole at the center of the galaxy.

Black holes are of considerable interest to those searching for **gravity waves,** which are ripples in space–time caused by changes in a gravitational system. These ripples can be caused by a star collapsing into a black hole, a binary star consisting of a black hole and a visible companion, and supermassive black holes at a galaxy center. A gravity wave detector, the Laser Interferometer Gravitational Wave Observatory (LIGO), is currently being built in the United States and hopes are high for detecting gravitational waves with this instrument.

**EXERCISE**    The escape speed from the surface of the Earth is 11.2 km/s. Estimate the escape speed for a spacecraft from the surface of the Moon. The Moon has a mass 1/81 that of Earth and a radius $\frac{1}{4}$ that of Earth.

*Answer*    2.49 km/s

**EXERCISE**    (a) Calculate the minimum energy required to send a 3 000-kg spacecraft from the Earth to a distant point in space where Earth's gravity is negligible. (b) If the journey is to take three weeks, what *average* power will the engines have to supply over the three-week period?

*Answer*    (a) $1.88 \times 10^{11}$ J    (b) 103 kW

## 11.5 • ATOMIC SPECTRA AND THE BOHR THEORY OF HYDROGEN

In the preceding sections, we described a structural model for a large scale system—the Solar System. Let us now do the same for a very small scale system—the hydrogen atom. We shall find that a Solar System model of the atom, with a few extra features, provides explanations for some of the experimental observations made on the hydrogen atom.

As you may have already learned in a chemistry course, the hydrogen atom is the simplest known atomic system and an especially important one to understand. Much of what is learned about the hydrogen atom (which consists of one proton and one electron) can be extended to single-electron ions such as $He^+$ and $Li^{2+}$. Furthermore, a thorough understanding of the physics underlying the hydrogen atom can then be used to describe more complex atoms and the periodic table of the elements.

Atomic systems can be investigated by observing *electromagnetic waves* emitted from the atom. Our eyes are sensitive to visible light, one type of electromagnetic wave. As you shall learn in more detail in Chapter 13, a mechanical wave is a disturbance that transports energy as it moves through a medium without transporting matter. The wave will be one of our four simplification models around which we will identify analysis models, as we have done for a particle, a system, and a rigid body. A common form of periodic wave is the sinusoidal wave, whose shape is depicted in Figure 11.17. The distance between two consecutive crests of the wave is called the **wavelength** $\lambda$. As the wave travels to the right with a speed $v$, any point on the wave travels a distance of one wavelength in a time interval of one period $T$ (the time for one cycle), so the wave speed is given by $v = \lambda/T$. The inverse of the period, $1/T$, is called the **frequency** $f$ of the wave; it represents the number of cycles per second. Thus, the speed of the wave is often written as $v = \lambda f$. In this section, because we shall deal with electromagnetic waves—which travel at the speed of light $c$—the appropriate relation is

$$c = \lambda f \qquad [11.13]$$

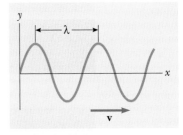

**Figure 11.17**

A sinusoidal wave traveling to the right with wave speed $v$. Any point on the wave moves a distance of one wavelength $\lambda$ in a time interval equal to the period $T$ of the wave.

• *Relation between wavelength, frequency, and wave speed*

Suppose an evacuated glass tube is filled with hydrogen (or some other gas). If a voltage applied between metal electrodes in the tube is great enough to produce an electric current in the gas, the tube emits light with colors that are characteristic of the gas. (This is how a neon sign works.) When the emitted light is analyzed with a device called a spectroscope, a series of discrete **spectral lines** is observed, each line corresponding to a different wavelength, or color, of light. Such a series of spectral lines is commonly referred to as an **emission spectrum.** The wavelengths contained in a given spectrum are characteristic of the element emitting the light. Figure 11.18 is a semigraphical representation of the spectra of various elements. It is semigraphical because the horizontal axis is linear in wavelength, but the vertical axis has no significance. Because no two elements emit the same line spectrum, this phenomenon represents a marvelous and reliable technique for identifying elements in a substance.

The emission spectrum of hydrogen shown in Figure 11.19 includes four prominent lines that occur at wavelengths of 656.3 nm, 486.1 nm, 434.1 nm, and 410.2 nm. In 1885, Johann Balmer (1825–1898) found that the wavelengths of

**PITFALL PREVENTION 11.4**
**Why lines?**

The phrase "spectral lines" is used very often when discussing the radiation from atoms. The shape of a *line* is due to the fact that the light is passed through a long and very narrow slit before being separated by wavelength. You will see many references to "lines" in both physics and chemistry.

**Figure 11.18**

Visible spectra. (a) Line spectra produced by emission in the visible range for the elements hydrogen, helium, and neon. (b) The absorption spectrum for hydrogen. The dark absorption lines occur at the same wavelengths as the emission lines for hydrogen shown in (a).

**Figure 11.19**

A series of spectral lines for atomic hydrogen. The prominent lines labeled are part of the Balmer series.

•  *Rydberg equation*

these and less prominent lines can be described by the following simple empirical equation:

$$\lambda = 364.56 \frac{n^2}{n^2 - 4} \qquad n = 3, 4, 5, \ldots$$

in which $n$ is an integer starting at 3, and the wavelengths given by this expression are in nanometers. These spectral lines are called the **Balmer series.** The first line in the Balmer series, at 656.3 nm, corresponds to $n = 3$; the line at 486.1 nm corresponds to $n = 4$; and so on. At the time this equation was formulated, it had no valid theoretical basis; it simply predicted the wavelengths correctly. Thus, this equation is not based on a model—it is simply a trial-and-error equation that happens to work. A few years later, Johannes Rydberg (1854–1919) recast the equation in the following form:

$$\frac{1}{\lambda} = R_H \left( \frac{1}{2^2} - \frac{1}{n^2} \right) \quad n = 3, 4, 5, \ldots \qquad \text{[11.14]}$$

where $n$ may have integral values of 3, 4, 5, . . . , and $R_H$ is a constant, now called the **Rydberg constant.** If the wavelength is in meters, $R_H$ has the value

$$R_H = 1.0973732 \times 10^7 \text{ m}^{-1}$$

Equation 11.14 is no more based on a model than is Balmer's equation. In this form, however, we can compare it with the predictions of a structural model of the hydrogen atom, which is described as follows.

In addition to emitting light at specific wavelengths, an element can also absorb light at specific wavelengths. The spectral lines corresponding to this process form what is known as an **absorption spectrum.** An absorption spectrum can be obtained by passing a continuous radiation spectrum (one containing all wavelengths) through a vapor of the element being analyzed. The absorption spectrum consists of a series of dark lines superimposed on the otherwise continuous spectrum.

At the beginning of the 20th century, scientists were perplexed by the failure of classical physics to explain the characteristics of atomic spectra. Why did atoms of a given element emit only certain wavelengths of radiation, so that the emission spectrum displayed discrete lines? Furthermore, why did the atoms absorb many of the same wavelengths which they emitted? In 1913, Niels Bohr provided an explanation of atomic spectra that includes some features of the currently accepted theory. Using the simplest atom, hydrogen, Bohr described a structural model for the atom. His model of the hydrogen atom contains some classical features that can be related to our analysis models, as well as some revolutionary postulates that could not be justified within the framework of classical physics. The basic assumptions of the Bohr model as it applies to the hydrogen atom are as follows:

**Niels Bohr** (1885–1962)

Bohr, a Danish physicist, was an active participant in the early development of quantum mechanics and provided much of its philosophical framework. During the 1920s and 1930s, Bohr headed the Institute for Advanced Studies in Copenhagen. The institute was a magnet for many of the world's best physicists and provided a forum for the exchange of ideas. Bohr was awarded the 1922 Nobel prize for his investigation of the structure of atoms and of the radiation emanating from them. *(Photo courtesy of AIP Niels Bohr Library, Margarethe Bohr Collection)*

1. The electron moves in a circular orbit about the proton under the influence of the electric force of attraction, as in Figure 11.20. This is a purely classical notion and is very similar to our previous discussion of planets in orbit around the Sun in our structural model of the Solar System.

2. Only certain electron orbits are stable, and these are the only orbits in which we find the electron. In these orbits the hydrogen atom does not emit energy in the form of radiation. Hence, the total energy of the atom remains constant, and classical mechanics can be used to describe the electron's motion. This restriction to certain orbits is a new idea, which is not consistent with classical physics. As we shall see in Chapter 24, an accelerating electron should emit energy by electromagnetic radiation. Thus, according to the continuity equation for energy, the emission of radiation from the atom should result in a decrease in the energy of the atom. Bohr's postulate boldly claims that this simply does not happen.

3. Radiation is emitted by the hydrogen atom when the atom "jumps" from a more energetic initial state to a lower state. The jump cannot be visualized or treated classically. In particular, the frequency $f$ of the radiation emitted in the jump is related to the change in the atom's energy. The frequency of the emitted radiation is found from

$$E_i - E_f = hf \qquad \text{[11.15]}$$

where $E_i$ is the energy of the initial state, $E_f$ is the energy of the final state, $h$ is **Planck's constant** ($h = 6.63 \times 10^{-34}$ J·s; we will see Planck's constant extensively in our studies of modern physics), and $E_i > E_f$. The notion of energy being emitted only when a jump occurs is nonclassical. Given this notion, however, Equation 11.15 is simply the continuity equation for energy, $\Delta E = \Sigma H \rightarrow E_f - E_i = -hf$. On the left is the change in energy of the system—the atom—and on the right is the energy transferred out of the system by electromagnetic radiation.

4. The size of the allowed electron orbits is determined by a condition imposed on the electron's orbital angular momentum. The allowed orbits are those for which the electron's orbital angular momentum about the nucleus is an integral multiple of $\hbar = h/2\pi$:

$$m_e vr = n\hbar \quad n = 1, 2, 3, \ldots \qquad \text{[11.16]}$$

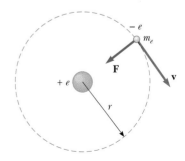

**Figure 11.20**

A pictorial representation of Bohr's model of the hydrogen atom, in which the electron is in a circular orbit about the proton.

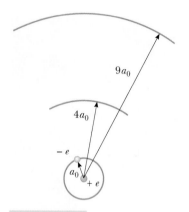

**Figure 11.21**

The first three Bohr orbits for hydrogen.

This new idea cannot be related to any of the models we have developed so far. It can be related, however, to a model that will be developed in later chapters, and we shall return to this idea at that time to see how it is predicted by the model. This is our first introduction to a notion from **quantum mechanics,** which describes the behavior of microscopic particles.

Using these four assumptions, Bohr built a structural model that explains the emission wavelengths of the hydrogen atom. The electrical potential energy of the system shown in Figure 11.20 is given by Equation 7.22, $U_e = -k_e e^2/r$, where $k_e$ is the electric constant, $e$ is the charge on the electron, and $r$ is the electron–proton separation. Thus, the total energy of the atom, which contains both kinetic and potential energy terms, is

$$E = K + U_e = \tfrac{1}{2} m_e v^2 - k_e \frac{e^2}{r} \qquad \textbf{[11.17]}$$

According to assumption 2, the energy of the system remains constant—the system is isolated because the structural model does not allow for electromagnetic radiation for a given orbit. Applying Newton's second law to this system, we see that the magnitude of the electric attractive force on the electron, $k_e e^2/r^2$ (Eq. 5.15), is equal to the product of its mass and its centripetal acceleration ($a_c = v^2/r$):

$$\frac{k_e e^2}{r^2} = \frac{m_e v^2}{r}$$

From this expression, the kinetic energy is found to be

$$K = \tfrac{1}{2} m_e v^2 = \frac{k_e e^2}{2r} \qquad \textbf{[11.18]}$$

Substituting this value of $K$ into Equation 11.17 gives the following expression for the total energy $E$ of the hydrogen atom:

- *Total energy of the hydrogen atom*

$$E = -\frac{k_e e^2}{2r} \qquad \textbf{[11.19]}$$

Note that the total energy is negative,* indicating a bound electron–proton system. This means that energy in the amount of $k_e e^2/2r$ must be added to the atom just to separate the electron and proton by an infinite distance and make the total energy zero.† An expression for $r$, the radius of the allowed orbits, can be obtained by eliminating $v$ by substitution between Equations 11.16 and 11.18:

- *Radii of Bohr orbits in hydrogen*

$$r_n = \frac{n^2 \hbar^2}{m_e k_e e^2} \qquad n = 1, 2, 3, \ldots \qquad \textbf{[11.20]}$$

This result shows that the radii have discrete values, or are *quantized.* The integer $n$ is called a **quantum number** and specifies the particular allowed **quantum state** of the atomic system.

---

\* Compare this expression to Equation 11.8 for a gravitational system.

† This is called *ionizing* the atom. In theory, ionization requires separating the electron and proton by an infinite distance. In reality, however, the electron and proton are in an environment with huge numbers of other particles. Thus, ionization means separating the electron and proton by a distance large enough so that the interaction of these particles with other entities in their environment is larger than the remaining interaction between them.

The orbit for which $n = 1$ has the smallest radius; it is called the **Bohr radius** $a_0$, and has the value

$$a_0 = \frac{\hbar^2}{m_e k_e e^2} = 0.0529 \text{ nm} \qquad [11.21]$$

• *The Bohr radius*

The first three Bohr orbits are shown to scale in Figure 11.21.

The quantization of the orbit radii immediately leads to energy quantization. This can be seen by substituting $r_n = n^2 a_0$ into Equation 11.19. The allowed energy levels are

$$E_n = -\frac{k_e e^2}{2a_0}\left(\frac{1}{n^2}\right) \qquad n = 1, 2, 3, \ldots \qquad [11.22]$$

Insertion of numerical values into Equation 11.22 gives

$$E_n = -\frac{13.606 \text{ eV}}{n^2} \qquad n = 1, 2, 3, \ldots \qquad [11.23]$$

• *Energy levels of the hydrogen atom*

(Recall from Section 9.7 that $1 \text{ eV} = 1.60 \times 10^{-19}$ J.) The lowest quantum state, corresponding to $n = 1$, is called the **ground state,** and has an energy of $E_1 = -13.606$ eV. The next state, the **first excited state,** has $n = 2$ and an energy of $E_2 = E_1/2^2 = -3.401$ eV. Figure 11.22 is an **energy level diagram** showing the energies of these discrete energy states and the corresponding quantum numbers. This is another semigraphical representation. The vertical axis is linear in energy, but the horizontal axis has no significance. The lines correspond to the allowed energies. The atomic system cannot have any energies other than those represented by the lines. The upper limit of the quantized levels, corresponding to $n \to \infty$ (or $r \to \infty$) and $E \to 0$, represents the state for which the electron is removed from the atom.* Above this energy is a continuum of available states for the ionized atom. The minimum energy required to ionize the atom is called the **ionization energy.** As can be seen

The Rosette Nebula is a region of gas and dust surrounding an open cluster of stars. The red color is due to hydrogen atoms, excited by light from the stars in the cluster, making transitions from the $n = 3$ quantum state to the $n = 2$ state. *(© 1987 Royal Observatory Edinburgh/Anglo-Australian Observatory, photography by David F. Malin from U.K. Schmidt plates)*

**Figure 11.22**

An energy level diagram for hydrogen. The discrete allowed energies are plotted on the vertical axis. Nothing is plotted on the horizontal axis, but the horizontal extent of the diagram is made large enough to show allowed transitions. Quantum numbers are given on the left.

---

* The phrase "the electron is removed from the atom" is very commonly used, but, of course, we realize that we mean that the electron and proton are separated *from each other.*

from Figure 11.22, the ionization energy for hydrogen, based on Bohr's calculation, is 13.6 eV. This constituted a major achievement for the Bohr theory because the ionization energy for hydrogen had already been measured to be precisely 13.6 eV.

Figure 11.22 also shows various transitions of the atom from one state to a lower state. These are the jumps referred to in Bohr's assumption 3. As the energy of the atom decreases in a jump, the difference in energy between the states is carried away by electromagnetic radiation. Those transitions ending on $n = 2$ are shown in color, corresponding to the color of the light they represent. The transitions ending on $n = 2$ form the Balmer series of spectral lines, the wavelengths of which are correctly predicted by the Rydberg equation (see Eq. 11.14). Figure 11.22 also shows other spectral series (the Lyman series and the Paschen series) that were found after Balmer's discovery.

Equation 11.22, together with Bohr's third postulate, can be used to calculate the frequency of the radiation that is emitted when the atom makes a transition* from a high-energy state to a low-energy state:

- *Frequency of radiation emitted from hydrogen*

$$f = \frac{E_i - E_f}{h} = \frac{k_e e^2}{2a_0 h}\left(\frac{1}{n_f^2} - \frac{1}{n_i^2}\right) \qquad \text{[11.24]}$$

Because the quantity expressed in the Rydberg equation is wavelength, it is convenient to convert frequency to wavelength, using $c = f\lambda$, to obtain

- *Emission wavelengths of hydrogen*

$$\frac{1}{\lambda} = \frac{f}{c} = \frac{k_e e^2}{2a_0 hc}\left(\frac{1}{n_f^2} - \frac{1}{n_i^2}\right) \qquad \text{[11.25]}$$

Notice the remarkable fact that the *theoretical* expression, Equation 11.25, is identical to the *empirical* Rydberg equation, Equation 11.14, provided that the combination of constants $k_e e^2/2a_0 hc$ is equal to the experimentally determined Rydberg constant, and that $n_f = 2$. After Bohr demonstrated the agreement of the constants in these two equations to a precision of about 1%, it was soon recognized as the crowning achievement of his structural model of the atom.

One question remains—what is the significance of $n_f = 2$? Its importance is due simply to the fact that those transitions ending on $n_f = 2$ result in radiation that happens to lie in the visible; therefore, they were easily observed! As seen in Figure 11.22, other series of lines end on other final states. These lie in regions of the spectrum not visible to the eye—the infrared and ultraviolet. The generalized Rydberg equation for any initial and final states is

$$\frac{1}{\lambda} = R_H\left(\frac{1}{n_f^2} - \frac{1}{n_i^2}\right) \qquad \text{[11.26]}$$

In this equation, different series correspond to different values of $n_f$, and different lines within a series correspond to varying values of $n_i$.

Bohr immediately extended his structural model for hydrogen to other elements in which all but one electron had been removed. Ionized elements such as $He^+$, $Li^{2+}$, and $Be^{3+}$ were suspected to exist in hot stellar atmospheres, where frequent atomic collisions occur with enough energy to completely remove one or more atomic electrons. Bohr showed that many mysterious lines observed in the Sun and several stars could not be due to hydrogen, but were correctly predicted by his theory if attributed to singly ionized helium.

## PITFALL PREVENTION 11.5

### The Bohr model is great, but . . .

The Bohr model is successful in some cases, such as predicting the ionization energy for hydrogen, but keep in mind that it is limited in its predictive power. For example, it cannot account for the spectra of more complex atoms and is unable to predict many subtle spectral details of hydrogen and other simple atoms. As a structural model, it is a simple first step toward understanding the atom. The notion of electrons in well-defined orbits around the nucleus is *not* consistent with current models of the atom, despite the cultural popularity of this model. We will see more sophisticated structural models in later chapters of the book.

* Another commonly used phrase is, "the electron makes a transition," but we will use, "the atom makes a transition" to emphasize the fact that the energy belongs to the system of the atom, not just to the electron. This is similar to our discussion in Chapter 7 of gravitational potential energy belonging to the system of an object and the Earth, not to the object alone.

## Quick Quiz 11.6

A model of the hydrogen atom such as Bohr's is often called a *planetary* model because it is similar in some ways to the structural model of the Solar System. In what ways is the planetary model of the atom similar to the Solar System and in what ways is it different?

### Example 11.4   An Electronic Transition in Hydrogen

A hydrogen atom makes a transition from the $n = 2$ energy state to the ground state (corresponding to $n = 1$). Find the wavelength and frequency of the emitted radiation.

**Solution**   We can use Equation 11.26 directly to obtain $\lambda$, with $n_i = 2$ and $n_f = 1$:

$$\frac{1}{\lambda} = R_H \left( \frac{1}{n_f^2} - \frac{1}{n_i^2} \right) = R_H \left( \frac{1}{1^2} - \frac{1}{2^2} \right) = \frac{3R_H}{4}$$

$$\lambda = \frac{4}{3R_H} = \frac{4}{3(1.097 \times 10^{-7} \, \text{m}^{-1})}$$

$$= 1.215 \times 10^{-7} \, \text{m} = \boxed{121.5 \, \text{nm (ultraviolet)}}$$

Because $c = f\lambda$, the frequency of the radiation is

$$f = \frac{c}{\lambda} = \frac{3.00 \times 10^8 \, \text{m/s}}{1.215 \times 10^{-7} \, \text{m}} = \boxed{2.47 \times 10^{15} \, \text{s}^{-1}}$$

**EXERCISE**   What is the wavelength of the radiation emitted by hydrogen when the atom makes a transition from the $n = 3$ state to the $n = 1$ state?

*Answer*   $\dfrac{9}{8} R_H = 102.6 \, \text{nm}$

### context connection
## 11.6 • CHANGING FROM A CIRCULAR TO AN ELLIPTICAL ORBIT

In the Context Connection of Chapter 3, we discussed our spacecraft in a circular parking orbit around the Earth. From our studies of Kepler's laws in this chapter, we are also aware that an elliptical orbit is possible for our spacecraft. Let us investigate how the motion of our spacecraft can be changed from a circular to an elliptical orbit, which will set us up for the Context Conclusion to our *Mission to Mars* Context.

Let us identify the system as the spacecraft and the Earth, *but not the portion of the fuel in the spacecraft that we use to change the orbit.* In a given orbit, the mechanical energy of the spacecraft–Earth system is given by Equation 11.8,

$$E = -\frac{GMm}{2r}$$

This energy includes the kinetic energy of the spacecraft and the potential energy associated with the gravitational force between the spacecraft and the Earth. If the rocket engines are fired, the exhausted fuel can be seen as doing work on the spacecraft–Earth system because the thrust force moves the spacecraft through a displacement. As a result, the mechanical energy of the spacecraft–Earth system increases.

The spacecraft has a new, higher energy but is constrained to be in an orbit that includes the original starting point. It cannot be in a higher energy circular orbit having a larger radius because this orbit would not contain the starting point. The only possibility is that the orbit is elliptical. Figure 11.23 shows the change from the original circular orbit to the new elliptical orbit for our spacecraft.

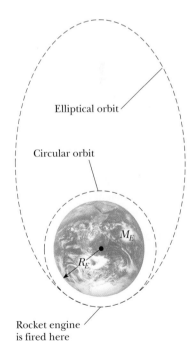

#### Figure 11.23

A spacecraft, originally in a circular orbit about the Earth, fires its engines and enters an elliptical orbit about the Earth.

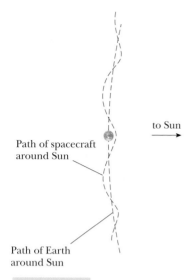

Path of spacecraft around Sun

to Sun →

Path of Earth around Sun

**Figure 11.24**

A spacecraft in orbit about the Earth can be modeled as one in a circular orbit about the Sun, with its orbit about the Earth appearing as small perturbations from the circular orbit.

Equation 11.9 gives the energy of the spacecraft–Earth system for an elliptical orbit. Thus, if we know the new energy of the orbit, we can find the semimajor axis of the elliptical orbit. Conversely, if we know the semimajor axis of an elliptical orbit we would like to achieve, we can calculate how much additional energy is required from the rocket engines. This can then be converted to a required burn time for the rockets.

Larger amounts of energy increase supplied by the rocket engines will move the spacecraft into elliptical orbits with larger semimajor axes. What happens if the burn time of the engines is so long that the total mechanical energy of the space-craft–Earth system becomes *positive?* A positive energy refers to an *unbound* system. Thus, in this case, the spacecraft will *escape* from the Earth, going into a hyperbolic path that would not bring it back to Earth.

This is the essence of what must be done to transfer to Mars. Our rocket engines must be fired to leave the circular parking orbit and escape the Earth. At this point, our thinking must shift to a spacecraft–Sun system rather than a space-craft–Earth system. From this point of view, the spacecraft in orbit around the Earth can also be considered to be in a circular orbit around the Sun, moving along with the Earth, as shown in Figure 11.24. The orbit is not a perfect circle, because there are perturbations corresponding to its extra motion around the Earth, but these perturbations are small compared with the radius of the orbit around the Sun. When our engines are fired to escape from the Earth, our orbit around the Sun changes from a circular orbit (ignoring the perturbations) to an elliptical one with the Sun at one focus. We shall choose the semimajor axis of our elliptical orbit so that it intersects the orbit of Mars! In the Context Conclusion, we shall look at more details of this process.

## Example 11.5　How High Do We Go?

Imagine that you are in a spacecraft in circular orbit around the Earth, at a height of 300 km from the surface. You fire your rocket engines and, as a result, the magnitude of the mechanical energy of the spacecraft–Earth system decreases by 10.0%. (Because the mechanical energy is negative, a decrease in magnitude is an increase in energy.) What is the greatest height of your spacecraft above the surface of the Earth in your new orbit?

**Solution**　We set up a ratio of the energies of the two orbits, using Equations 11.8 and 11.9 for circular and elliptical orbits:

$$\frac{E_{\text{elliptical}}}{E_{\text{circular}}} = \frac{\left(-\dfrac{GMm}{2a}\right)}{\left(-\dfrac{GMm}{2r}\right)} = \frac{r}{a}$$

The ratio on the left is 0.900, due to the 10.0% decrease in magnitude of the mechanical energy. Thus,

$$0.900 = \frac{r}{a} \quad\longrightarrow\quad a = \frac{r}{0.900} = 1.11r$$

From this, we can find the semimajor axis for the elliptical orbit:

$$a = 1.11r = 1.11(6.37 \times 10^3 \text{ km} + 3.00 \times 10^2 \text{ km})$$
$$= 7.40 \times 10^3 \text{ km}$$

The maximum distance from the center of the Earth will occur when the spacecraft is at apogee, and is given by

$$r_{\text{max}} = 2a - r = 2(7.40 \times 10^3 \text{ km}) - (6.67 \times 10^3 \text{ km})$$
$$= 8.14 \times 10^3 \text{ km}$$

Now, if we subtract the radius of the Earth, we have the maximum height above the surface,

$$h_{\text{max}} = r_{\text{max}} - R_E = 8.14 \times 10^3 \text{ km} - 6.37 \times 10^3 \text{ km}$$
$$= \boxed{1.77 \times 10^3 \text{ km}}$$

# SUMMARY

**Newton's law of universal gravitation** states that the gravitational force of attraction between any two particles of masses $m_1$ and $m_2$ separated by a distance $r$ has the magnitude

$$F_g = G\frac{m_1 m_2}{r^2} \qquad \text{[11.1]}$$

where $G$ is the **universal gravitational constant,** whose value is $6.672 \times 10^{-11} \text{ N} \cdot \text{m}^2/\text{kg}^2$.

**Kepler's laws of planetary motion** state that

1. Each planet in the Solar System moves in an elliptical orbit with the Sun at one focus.
2. The radius vector drawn from the Sun to any planet sweeps out equal areas in equal time intervals.
3. The square of the orbital period of any planet is proportional to the cube of the semimajor axis of the elliptical orbit.

**Kepler's first law** is a consequence of the inverse-square nature of the law of universal gravitation. The **semimajor axis** of an ellipse is $a$, where $2a$ is the longest dimension of the ellipse. The **semiminor axis** of the ellipse is $b$, where $2b$ is the shortest dimension of the ellipse. The **eccentricity** of the ellipse is $e = c/a$, where $a^2 = b^2 + c^2$.

**Kepler's second law** is a consequence of the fact that the gravitational force is a central force. For a central force, the angular momentum of the planet is conserved.

**Kepler's third law** is a consequence of the inverse-square nature of the universal law of gravitation. Newton's second law, together with the force law given by Equation 11.1, verifies that the period $T$ and semimajor axis $a$ of the orbit of a planet about the Sun are related by

$$T^2 = \left(\frac{4\pi^2}{GM_S}\right)a^3 \qquad \text{[11.5]}$$

where $M_S$ is the mass of the Sun.

If an isolated system consists of a particle of mass $m$ moving with a speed of $v$ in the vicinity of a massive body of mass $M$, the *total energy* of the system is constant and is

$$E = \tfrac{1}{2}mv^2 - \frac{GMm}{r} \qquad \text{[11.6]}$$

If $m$ moves in an elliptical orbit of major axis $2a$ about $M$, where $M \gg m$, the total energy of the system is

$$E = -\frac{GMm}{2a} \qquad \text{[11.9]}$$

The total energy is negative for any bound system—that is, one in which the orbit is closed, such as a circular or an elliptical orbit.

The Bohr model of the atom successfully describes the spectra of atomic hydrogen and hydrogen-like ions. One of the basic assumptions of this structural model is that the electron can exist only in discrete orbits such that the angular momentum $m_e v r$ is an integral multiple of $h/2\pi = \hbar$. Assuming circular orbits and a simple electrical attraction between the electron and proton, the energies of the quantum states for hydrogen are calculated to be

$$E_n = -\frac{k_e e^2}{2a_0}\left(\frac{1}{n^2}\right) \quad n = 1, 2, 3, \dots \qquad \text{[11.22]}$$

where $k_e$ is the Coulomb constant, $e$ is the fundamental electric charge, $n$ is a positive integer called a **quantum number,** and $a_0 = 0.0529$ nm is the **Bohr radius.**

If the hydrogen atom makes a transition from a state whose quantum number is $n_i$ to one whose quantum number is $n_f$, where $n_f < n_i$, the frequency of the radiation emitted by the atom is

$$f = \frac{k_e e^2}{2a_0 h}\left(\frac{1}{n_f{}^2} - \frac{1}{n_i{}^2}\right) \qquad \text{[11.24]}$$

Using $E_i - E_f = hf = hc/\lambda$, one can calculate the wavelengths of the radiation for various transitions. The calculated wavelengths are in excellent agreement with those in observed atomic spectra.

# QUESTIONS

1. If the gravitational force on an object is directly proportional to its mass, why don't large masses fall with greater acceleration than small ones?
2. At night, you are farther away from the Sun than during the day. What's more, the force from the Sun on you is downward into the Earth at night, and upward into the sky during the day. If you had a sensitive enough bathroom scale, would you appear to weigh more at night than during the day?
3. The gravitational force the Sun exerts on the Moon is about twice as great as the gravitational force that the Earth exerts on the Moon. Why doesn't the Sun pull the Moon away from the Earth during a total eclipse of the Sun?

**4.** Explain why it takes more fuel for a spacecraft to travel from the Earth to the Moon than for the return trip. Estimate the difference.

5. Explain why no work is done on a planet as it moves in a circular orbit around the Sun, even though a gravitational force is acting on the planet. What is the *net* work done on a planet during each revolution as it moves around the Sun in an elliptical orbit?

6. At what position in its elliptical orbit is the speed of a planet a maximum? At what position is the speed a minimum?

**7.** Why don't we put a communications satellite in orbit around the 45th parallel? Wouldn't this be more useful in the United States and Canada than one in orbit around the equator?

8. A satellite in orbit is not truly traveling through a vacuum—it is moving through very, very thin air. Does the resulting air friction cause the satellite to slow down?

**9.** In his 1798 experiment, Cavendish was said to have "weighed the Earth." Explain this statement.

10. If a hole could be dug to the center of the Earth, do you think that the force on an object of mass *m* would still obey Equation 11.1 there? What do you think the force on the object would be at the center of the Earth?

11. The *Voyager* spacecraft was accelerated toward escape speed from the Sun by the gravitational force exerted on the spacecraft by Jupiter. How is this possible?

12. The *Apollo 13* spacecraft developed trouble in its oxygen system about halfway to the Moon. Why did the mission continue on around the Moon, and then return home, rather than immediately turn back to Earth?

13. How would you explain the fact that Saturn and Jupiter have periods much greater than one year?

14. Suppose the energy of a hydrogen atom is − *E*. What is the kinetic energy of the electron? What is the potential energy of the atom?

15. The Bohr theory of the hydrogen atom is based on several assumptions. Discuss those assumptions and their significance. Do any of them contradict classical physics?

16. Explain the significance behind the fact that the total energy of the atom in the Bohr model is negative.

17. Suppose that the electron in the hydrogen atom obeyed classical mechanics rather than quantum mechanics. Why should such a "hypothetical" atom emit a continuous spectrum rather than the observed line spectrum?

## PROBLEMS

$1, 2, 3$ = straightforward, intermediate, challenging    □ = full solution available in the *Student Solutions Manual and Study Guide*

**web** = solution posted at **http://www.harcourtcollege.com/physics/**  = computer useful in solving problem

 = Interactive Physics    ▪ = paired numerical/symbolic problems     = life science application

### Section 11.1   Newton's Law of Universal Gravitation Revisited

*Note:* Problems 33 and 35 in Chapter 5 and Problems 31 through 33 in Chapter 7 can be assigned with this section.

1. Two ocean liners, each with a mass of 40 000 metric tons, are moving on parallel courses, 100 m apart. What is the magnitude of the acceleration of one of the liners toward the other due to their mutual gravitational attraction? Model the ships as particles.

2. A 200-kg object and a 500-kg object are separated by 0.400 m. (a) Find the net gravitational force exerted by these objects on a 50.0-kg object placed midway between them. (b) At what position (other than an infinitely remote one) can the 50.0-kg object be placed so as to experience a net force of zero?

3. Three uniform spheres of mass 2.00 kg, 4.00 kg, and 6.00 kg are placed at the corners of a right triangle, as in Figure P11.3. Calculate the net gravitational force on the

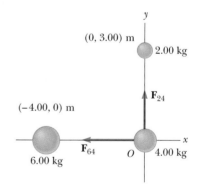

**Figure P11.3**

4.00-kg sphere, assuming the spheres are isolated from the rest of the Universe.

4. During a solar eclipse, the Moon, Earth, and Sun all lie on the same line, with the Moon between the Earth and the Sun. (a) What force is exerted by the Sun on the Moon?

(b) What force is exerted by the Earth on the Moon?
(c) What force is exerted by the Sun on the Earth?

5. The free-fall acceleration on the surface of the Moon is about one-sixth that on the surface of the Earth. If the radius of the Moon is about $0.250R_E$, find the ratio of their average densities, $\rho_{Moon}/\rho_{Earth}$.

6. A student proposes to measure the gravitational constant $G$ by suspending two spherical objects from the ceiling of a tall cathedral and measuring the deflection of the cables from the vertical. Draw a free-body diagram of one of the objects. If two 100.0-kg objects are suspended at the lower ends of cables 45.00 m long, and the cables are attached to the ceiling 1.000 m apart, what is the separation of the objects?

7. On the way to the Moon the *Apollo* astronauts reached a point where the Moon's gravitational pull became stronger than the Earth's. (a) Determine the distance of this point from the center of the Earth. (b) What is the acceleration due to the Earth's gravity at this point?

8. A spacecraft in the shape of a long cylinder has a length of 100 m and its mass with occupants is 1 000 kg. It has strayed too close to a 1.0-m radius black hole having a mass 100 times that of the Sun (Fig. P11.8). The nose of the spacecraft points toward the center of the black hole, and the distance between the nose and the black hole is 10.0 km. (a) Determine the total force on the spacecraft. (b) What is the difference in the gravitational fields acting on the occupants in the nose of the ship and on those in the rear of the ship, farthest from the black hole?

**Figure P11.8**

9. Compute the magnitude and direction of the gravitational field at a point *P* on the perpendicular bisector of two particles with equal masses separated by a distance $2a$ as shown in Figure P11.9.

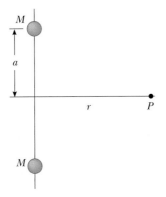

**Figure P11.9**

10. How much work is done by the Moon's gravitational field as a 1 000-kg meteor comes in from outer space and impacts on the Moon's surface?

11. After our Sun exhausts its nuclear fuel, its ultimate fate may be to collapse to a *white dwarf* state, in which it has approximately the same mass as it has now, but a radius equal to the radius of the Earth. Calculate (a) the average density of the white dwarf, (b) the free-fall acceleration at its surface, and (c) the gravitational potential energy associated with a 1.00-kg object at its surface. (Take $U_g = 0$ at infinity.)

## Section 11.3 Kepler's Laws

12. A communication satellite in geosynchronous orbit remains above a single point on the Earth's equator as the planet rotates on its axis. (a) Calculate the radius of its orbit. (b) The satellite relays a radio signal from a transmitter near the North Pole to a receiver, also near the North Pole. Traveling at the speed of light, how long is the radio wave in transit?

13. Plaskett's binary system consists of two stars that revolve in a circular orbit about a center of mass midway between them. This means that the masses of the two stars are equal (Fig. P11.13). If the orbital velocity of each star is 220 km/s and the orbital period of each is 14.4 days, find the mass $M$ of each star. (For comparison, the mass of our Sun is $1.99 \times 10^{30}$ kg.)

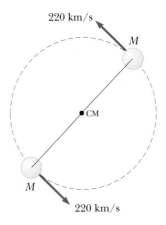

**Figure P11.13**

14. The *Explorer VIII* satellite, placed into orbit November 3, 1960 to investigate the ionosphere, had the following orbit parameters: perigee, 459 km; apogee, 2 289 km (both distances above the Earth's surface); period, 112.7 min. Find the ratio $v_p/v_a$ of the speed at perigee to that at apogee.

15. Io, a satellite of Jupiter, has an orbital period of 1.77 days and an orbital radius of $4.22 \times 10^5$ km. From these data, determine the mass of Jupiter.

16. Comet Halley (Figure P11.16, not to scale) approaches the Sun to within 0.570 AU, and its orbital period is 75.6 years. (AU is the symbol for astronomical unit, where 1 AU = $1.50 \times 10^{11}$ m is the mean Earth–Sun distance.) How far from the Sun will Comet Halley travel before it starts its return journey?

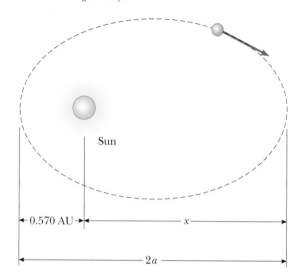

**Figure P11.16**

17. Two planets X and Y travel counterclockwise in circular orbits about a star as in Figure P11.17. The radii of their orbits are in the ratio 3:1. At some time, they are aligned as in Figure P11.17a, making a straight line with the star. During the next five years, the angular displacement of planet X is 90.0°, as in Figure P11.17b. Where is planet Y at this time?

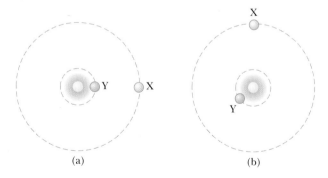

(a)          (b)

**Figure P11.17**

18. Suppose the Sun's gravity were switched off. The planets would leave their nearly circular orbits and fly away in straight lines, as described by Newton's first law. Would Mercury ever be farther from the Sun than Pluto? If so, find how long it would take for Mercury to achieve this passage. If not, give a convincing argument that Pluto is always farther from the Sun.

19. As thermonuclear fusion proceeds in its core, the Sun loses mass at a rate of $3.64 \times 10^{9}$ kg/s. During the 5 000-yr period of recorded history, by how much has the length of the year changed due to the loss of mass from the Sun? (*Hints:* Assume the Earth's orbit is circular. There is no external torque on the Earth–Sun system, so angular momentum is conserved. If $x$ is small compared with 1, then $(1 + x)^{n}$ is nearly equal to $1 + nx$.)

## Section 11.4 Energy Considerations in Planetary and Satellite Motion

*Note:* Problems 41 and 42 in Chapter 6 and 34 and 38 in Chapter 7 can be assigned with this section.

20. A 500-kg satellite is in a circular orbit at an altitude of 500 km above the Earth's surface. Because of air friction, the satellite eventually falls to the Earth's surface, where it hits the ground with a speed of 2.00 km/s. How much energy was transformed into internal energy by means of friction?

21. A "treetop satellite" (Fig. P11.21) moves in a circular orbit just above the surface of a planet, assumed to offer no air resistance. Show that its orbital speed $v$ and the escape speed from the planet are related by the expression $v_{esc} = \sqrt{2}v$.

22. A satellite of mass $m$ is placed into Earth orbit at an altitude $h$. (a) Assuming a circular orbit, how long does the satellite take to complete one orbit? (b) What is the satellite's speed? (c) What is the minimum energy necessary to place this satellite in orbit? Ignore air resistance but include the effect of the planet's daily rotation. At what location on the Earth's surface and in what direction should the satellite be launched to minimize the required energy investment? Represent the mass and radius of the Earth as $M_{E}$ and $R_{E}$.

23. The planet Uranus has a mass about 14 times the Earth's mass, and its radius is equal to about 3.7 Earth radii. (a) By setting up ratios with the corresponding Earth values, find the acceleration due to gravity at the cloud tops of Uranus. (b) Ignoring the rotation of the planet, find the minimum escape speed from Uranus.

24. Derive an expression for the work required to move an Earth satellite of mass $m$ from a circular orbit of radius $2R_{E}$ to one of radius $3R_{E}$.

25. A comet of mass $1.20 \times 10^{10}$ kg moves in an elliptical orbit around the Sun. Its distance from the Sun ranges between 0.500 AU and 50.0 AU. (a) What is the eccentricity of its orbit? (b) What is its period? (c) At aphelion what is the potential energy of the comet–Sun system? (*Note:* 1 AU = one astronomical unit = the average distance from Sun to Earth = $1.496 \times 10^{11}$ m.)

B.C.                                                                                  by John Hart

NO, NO, STUPID... THE OTHER END.

THAT'S **GOTTA** BE THE LOWEST ORBIT ON RECORD.

By permission of John Hart and Field Enterprises, Inc.

**Figure P11.21**

## Section 11.5  Atomic Spectra and the Bohr Theory of Hydrogen

**26.** A hydrogen atom emits light as it undergoes a transition from the $n = 3$ state to the $n = 2$ state. Calculate (a) the energy, (b) the wavelength, and (c) the frequency of the radiation.

**27.** (a) What value of $n_i$ is associated with the 94.96-nm spectral line in the Lyman series of hydrogen? (b) Could this wavelength be associated with the Paschen or Balmer series?

**28.** For a hydrogen atom in its ground state, use the Bohr model to compute (a) the orbital speed of the electron, (b) the kinetic energy of the electron, and (c) the electrical potential energy of the atom.

**29.** A hydrogen atom is in its first excited state ($n = 2$). Using the Bohr theory of the atom, calculate (a) the radius of the orbit, (b) the linear momentum of the electron, (c) the angular momentum of the electron, (d) the kinetic energy, (e) the potential energy, and (f) the total energy.

**30.** Four possible transitions for a hydrogen atom are as follows:

  i. $n_i = 2$; $n_f = 5$      ii. $n_i = 5$; $n_f = 3$

  iii. $n_i = 7$; $n_f = 4$      iv. $n_i = 4$; $n_f = 7$

(a) In which transition is light of the shortest wavelength emitted? (b) In which transition does the atom gain the most energy? (c) In which transition(s) does the atom lose energy?

**31.** How much energy is required to ionize hydrogen (a) when it is in the ground state? (b) When it is in the state for which $n = 3$?

**32.** Show that the speed of the electron in the $n$th Bohr orbit in hydrogen is given by

$$v_n = \frac{k_e e^2}{n\hbar}$$

**33.** Two hydrogen atoms collide head-on and end up with zero kinetic energy. Each atom then emits light with a wavelength of 121.6 nm ($n = 2$ to $n = 1$ transition). At what speed were the atoms moving before the collision?

## Section 11.6  Context Connection — Changing from a Circular Orbit to an Elliptical Orbit

**34.** A spacecraft of mass $1.00 \times 10^4$ kg is in a circular orbit at an altitude of 500 km above the Earth's surface. Mission Control wants to fire the engines so as to put the spacecraft in an elliptical orbit around the Earth with an apogee of $2.00 \times 10^4$ km. How much energy must be used from the fuel to achieve this orbit? (Assume that all of the fuel energy goes into increasing the orbital energy. This will give a lower limit to the required energy because some of the energy from the fuel will appear as internal energy in the hot exhaust gases and engine parts.)

**35.** A spacecraft is approaching Mars after a long trip from Earth. Its velocity is such that it is traveling along a parabolic trajectory under the influence of the gravitational force from Mars. The distance of closest approach will be 300 km above the Martian surface. At this point of closest approach, the engines will be fired to slow down the spacecraft and place it in a circular orbit 300 km above the surface. (a) By what percentage must the speed of the spacecraft be reduced to achieve the desired orbit? (b) How would the answer to part (a) change if the distance of closest approach and the desired circular orbit altitude were 600 km instead of 300 km? (*Hint:* The energy of the spacecraft–Mars system for a parabolic orbit is $E = 0$.)

## Additional Problems

36. The *Solar and Heliospheric Observatory* (SOHO) spacecraft has a special orbit, chosen so that its view of the Sun is never eclipsed and it is always close enough to Earth to transmit data easily. It moves in a near-circle around the Sun in an orbit that is smaller than the Earth's circular orbit and its period equal to one year. It is always located between Earth and Sun along the line joining them. Both objects exert gravitational forces on the observatory. Show that its distance from the Earth must be between $1.47 \times 10^9$ m and $1.48 \times 10^9$ m. In 1772, Joseph Louis Lagrange determined theoretically the special location allowing this orbit. The SOHO spacecraft took this position on February 14, 1996. (*Hint:* Use data that are precise to four digits. The mass of the Earth is $5.983 \times 10^{24}$ kg.)

37. The positron is the antiparticle to the electron. It has the same mass and a positive electric charge of the same magnitude as that of the electron. Positronium is a hydrogen-like atom consisting of a positron and an electron revolving around each other. Using the Bohr model, find the allowed distances between the two particles and the allowed energies of the system.

38. Let $\Delta g_M$ represent the difference in the gravitational fields produced by the Moon at the points on the Earth's surface nearest to and farthest from the Moon. Find the fraction $\Delta g_M/g$, where $g$ is the Earth's gravitational field. (This difference is responsible for the occurrence of the *lunar tides* on the Earth.)

39. In Larry Niven's science fiction novel *Ringworld,* a rigid ring of material rotates about a star (Fig. P11.39). The tangential speed of the ring is $1.25 \times 10^6$ m/s, and its radius is $1.53 \times 10^{11}$ m. (a) Show that the centripetal acceleration of the inhabitants is 10.2 m/s². (b) The inhabitants of this ring world live on the starlit inner surface of the ring. Each person experiences a normal contact force **n**. Acting alone, this normal force would produce an inward acceleration of 9.90 m/s². Additionally, the star at the center of the ring exerts a gravitational force on the ring and its inhabitants. The difference between the total accelera-

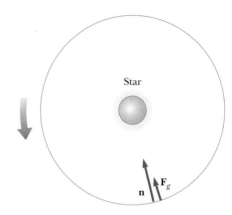

**Figure P11.39**

tion and the acceleration provided by the normal force is due to the gravitational attraction of the central star. Show that the mass of the star is approximately $10^{32}$ kg.

40. (a) Show that the rate of change of the free-fall acceleration with distance above the Earth's surface is

$$\frac{dg}{dr} = -\frac{2GM_E}{R_E{}^3}$$

This rate of change over distance is called a *gradient.* (b) If $h$ is small relative to the radius of the Earth, show that the difference in free-fall acceleration between two points separated by vertical distance $h$ is

$$|\Delta g| = \frac{2GM_E h}{R_E{}^3}$$

(c) Evaluate this difference for $h = 6.00$ m, a typical height for a two-story building.

41. *Voyagers 1* and *2* surveyed the surface of Jupiter's moon Io and photographed active volcanoes spewing liquid sulfur to heights of 70 km above the surface of this moon. Find the speed with which the liquid sulfur left the volcano. Io's mass is $8.9 \times 10^{22}$ kg, and its radius is 1820 km.

42. As an astronaut, you observe a small planet to be spherical. After landing on the planet, you set off, walking always straight ahead, and find yourself returning to your spacecraft from the opposite side after completing a lap of 25.0 km. You hold a hammer and a falcon feather at a height of 1.40 m, release them, and observe that they fall together to the surface in 29.2 s. Determine the mass of the planet.

43. In introductory physics laboratories, a typical Cavendish balance for measuring the gravitational constant $G$ uses lead spheres with masses of 1.50 kg and 15.0 g whose centers are separated by about 4.50 cm. Calculate the gravitational force between these spheres, treating each as a point mass located at the center of the sphere.

44. Many people assume that air resistance acting on a moving object will always make the object slow down. It can ac-

tually be responsible for making the object speed up. Consider a 100-kg Earth satellite in a circular orbit at an altitude of 200 km. A small force of air resistance makes the satellite drop slowly into a circular orbit with an altitude of 100 km. (a) Calculate its original speed. (b) Calculate its final speed in this process. (c) Calculate its original energy. (d) Calculate its final energy. (e) Show that it has lost mechanical energy and find the amount of the loss due to friction. (f) What force makes the object's speed increase? You will find a free-body diagram useful in explaining your answer.

**45.** Two hypothetical planets of masses $m_1$ and $m_2$ and radii $r_1$ and $r_2$, respectively, are nearly at rest when they are an infinite distance apart. Because of their gravitational attraction, they head toward each other on a collision course. (a) When their center-to-center separation is $d$, find expressions for the speed of each planet and for their relative velocity. (b) Find the kinetic energy of each planet just before they collide, assuming $m_1 = 2.00 \times 10^{24}$ kg, $m_2 = 8.00 \times 10^{24}$ kg, $r_1 = 3.00 \times 10^6$ m, and $r_2 = 5.00 \times 10^6$ m. (*Hint:* Both energy and momentum are conserved.)

**46.** The maximum distance from the Earth to the Sun (at aphelion) is $1.521 \times 10^{11}$ m, and the distance of closest approach (at perihelion) is $1.471 \times 10^{11}$ m. If the Earth's orbital speed at perihelion is $3.027 \times 10^4$ m/s, determine (a) the Earth's orbital speed at aphelion, (b) the kinetic and potential energies at perihelion, and (c) the kinetic and potential energies at aphelion. Is the total energy constant? (Ignore the effect of the Moon and other planets.)

**47.** Studies of the relationship of the Sun to its galaxy—the Milky Way—have revealed that the Sun is located near the outer edge of the galactic disc, about 30 000 lightyears from the center. The Sun has an orbital speed of approximately 250 km/s around the galactic center. (a) What is the period of the Sun's galactic motion? (b) What is the order of magnitude of the mass of the Milky Way galaxy? Suppose that the galaxy is made mostly of stars of which the Sun is typical. What is the order of magnitude of the number of stars in the Milky Way?

**48.** The oldest artificial satellite in orbit is *Vanguard I*, launched March 3, 1958. Its mass is 1.60 kg. In its initial orbit its minimum distance from the center of the Earth was 7.02 Mm and its speed at this perigee point was 8.23 km/s. (a) Find its total energy. (b) Find the magnitude of its angular momentum. (c) Find its speed at apogee and its maximum (apogee) distance from the center of the Earth. (d) Find the semimajor axis of its orbit. (e) Determine its period.

**49.** A rocket is given an initial speed vertically upward of $v_i = 2\sqrt{Rg}$ at the surface of the Earth, which has radius $R$

and surface free-fall acceleration $g$. The rocket motors are quickly cut off, and thereafter the rocket coasts under the action of gravitational forces only. (Ignore atmospheric friction and the Earth's rotation.) Derive an expression for the subsequent speed $v$ as a function of the distance $r$ from the center of the Earth in terms of $g$, $R$, and $r$.

**50.** Astronomers detect a distant meteoroid moving along a straight line that, if extended, would pass at a distance $3R_E$ from the center of the Earth, where $R_E$ is the radius of the Earth. What minimum speed must the meteoroid have if the Earth's gravity is not to deflect the meteoroid to make it strike the Earth?

**51.** Two stars of masses $M$ and $m$, separated by a distance $d$, revolve in circular orbits about their center of mass (Fig. P11.51). Show that each star has a period given by

$$T^2 = \frac{4\pi^2 d^3}{G}(M + m)$$

(*Hint:* Apply Newton's second law to each star, and note that the center-of-mass condition requires that $Mr_2 = mr_1$, where $r_1 + r_2 = d$.)

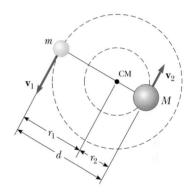

**Figure P11.51**

**52.** A spherical planet has uniform density $\rho$. Show that the minimum period for a satellite in orbit around it is

$$T_{\text{min}} = \sqrt{\frac{3\pi}{G\rho}}$$

independent of the radius of the planet.

**53.** The acceleration of an object moving in the gravitational field of the Earth is

$$\mathbf{a} = -GM_E \frac{\mathbf{r}}{r^3}$$

where $\mathbf{r}$ is the position vector directed from the center of the Earth to the object. Choosing the origin at the center of the Earth and assuming that the small object is moving in the $xy$ plane, we find that the rectangular (cartesian) components of its acceleration are

$$a_x = \frac{-GM_E x}{(x^2 + y^2)^{3/2}} \qquad a_y = \frac{-GM_E y}{(x^2 + y^2)^{3/2}}$$

Use a computer to set up and carry out a numerical prediction of the motion of the object, according to Euler's method. Assume the initial position of the object is $x = 0$ and $y = 2R_E$, where $R_E$ is the radius of the Earth. Give the

object an initial velocity of 5 000 m/s in the $x$ direction. The time increment should be made as small as practical. Try 5 s. Plot the $x$ and $y$ coordinates of the object as time goes on. Does the object hit the Earth? Vary the initial velocity until you find a circular orbit.

## ANSWERS TO QUICK QUIZZES

**11.1** Kepler's second law is a consequence of conservation of angular momentum. Angular momentum is conserved as long as there is no torque, which is true for any central force, regardless of the particular dependence on separation. Thus, Kepler's second law would be unchanged. The particular mathematical form of Kepler's third law depends on the inverse-square character of the force. If the force were an inverse cube, the relationship between a power of the period and a power of the semimajor axis would still be a proportionality, but the powers would be different from that in the existing law.

**11.2** (a). From Kepler's third law and the given period, the major axis of the comet can be calculated. It is found to be $1.2 \times 10^{11}$ m. Because this is smaller than the Earth–Sun distance, the comet cannot possibly collide with the Earth.

**11.3** (b). From conservation of momentum, $mv_p r_p = mv_a r_a$, so that $v_p = (r_a / r_p) v_a = (4D/D) v_a = 4v_a$.

**11.4** (a) Because of conservation of angular momentum, the speed of the comet is highest at its perihelion, its closest position to the Sun. (b) The potential energy of the

comet–Sun system is highest when the comet is at aphelion, its farthest distance from the Sun. (c) The kinetic energy is highest at the point at which the speed is highest, the perihelion. (d) The total energy of the system is the same regardless of where the comet is in its orbit.

**11.5** The mass of the asteroid might be so small that you would be able to exceed escape speed by leg power alone. You would jump up, but you would not come down!

**11.6** The planetary model of the hydrogen atom is similar to a Solar System model in that the electron is moving about the nucleus in a circular orbit, similar to the way we model the planets in orbit about the Sun. It has several differences, however, aside from the sizes of the systems. The atomic model depends on an electric force, but the Solar System depends on a gravitational force. In the Bohr theory, the allowed radii of the electron orbits are quantized, but there is no detectable quantization in the Solar System. Finally, we see the electron in a Bohr atom make transitions between energy levels, but we do not in the case for planets in a Solar System.

# Mission to Mars

## A SUCCESSFUL MISSION PLAN

Now that we have explored the physics of classical mechanics, let us return to our central question for the *Mission to Mars* Context:

> *How can we undertake a successful transfer of a spacecraft from Earth to Mars?*

We make use of the physical principles that we now understand and apply them to our journey from Earth to Mars.

Let us start with a more modest proposition. Suppose that our space shuttle is in a circular orbit around the Earth and you are a passenger on the shuttle. If you toss a wrench in the direction of travel, tangent to the circular path, what orbital path will the wrench follow?

Let us adopt a simplification model in which the space shuttle is much more massive than the wrench. Conservation of momentum for the isolated system of the wrench and the spacecraft tells us that the space shuttle must slow down slightly once the wrench is thrown. Because of the mass difference between the wrench and space shuttle, however, we can ignore the small change in the speed of the shuttle. The wrench now enters a new orbit, from its perigee position, with more energy than that for the circular orbit. Because the energy is related to the major axis, the wrench is injected into an elliptical orbit, as discussed in the Context Connection of Chapter 11 and as shown in Figure 1. Thus, the path of the wrench is changed from a circular to an elliptical orbit by providing the wrench–Earth system with extra energy. The energy is provided by the force you apply to the wrench tangent to the circular orbit—you have done work on the system. The elliptical orbit will take the wrench farther from the Earth than the circular orbit. If another shuttle were in a higher circular orbit than the first shuttle, you could throw the wrench so that it transfers from one shuttle to another, as

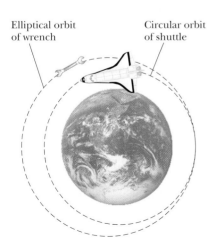

Elliptical orbit of wrench

Circular orbit of shuttle

**Figure 1**

A wrench thrown tangent to the circular orbit of a space shuttle enters an elliptical orbit.

397

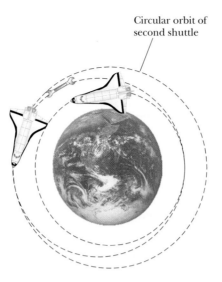

Circular orbit of
second shuttle

**Figure 2**

If a second space shuttle were in a higher circular orbit, the wrench could be carefully thrown so as to be transferred from one space shuttle to the other.

shown in Figure 2. For this to occur, the elliptical orbit of the wrench must intersect with the higher shuttle orbit. Furthermore, the wrench and the second shuttle must arrive at the same point at the same time.

This is the essence of our planned mission from Earth to Mars. Rather than transferring a wrench between two shuttles in orbit around the Earth, we will transfer a spacecraft between two planets in orbit around the Sun. Kinetic energy is added to the wrench–Earth system by throwing the wrench. Kinetic energy is added to the spacecraft–Sun system by firing the engines.

What if you were to throw the wrench harder and harder in the previous example? The wrench would be placed in a larger and larger elliptical orbit around the Earth. As you increased the launch velocity, you could inject the wrench into a *hyperbolic* escape orbit, relative to the Earth, and into an *elliptical* orbit around the *Sun*. This is the approach we will take for the trip from Earth to Mars—we will break free from a circular *parking* orbit around the Earth and move into an elliptical *transfer* orbit around the Sun. The spacecraft will then continue on its journey to Mars, where it will enter a new parking orbit.

Now let us focus our attention on the transfer orbit part of the journey. One simple transfer orbit is called a **Hohmann transfer,** the type of transfer imparted to the wrench in the earlier discussion. The Hohmann transfer involves the least energy expenditure and thus requires the smallest amount of fuel. As might be expected for a lowest energy transfer, the transfer time for a Hohmann transfer is longer than for other types of orbits. We shall investigate the Hohmann transfer due to its simplicity and its general usefulness in planetary transfers.

The rocket is fired from the parking orbit such that the spacecraft enters an elliptical orbit around the Sun at its perihelion and encounters the planet at the spacecraft's aphelion. Thus, the spacecraft makes exactly one half of a revolution about its elliptical path during the transfer, as shown in Figure 3.

This is an energy-efficient process, because fuel is expended only at the beginning and at the end. The movement between parking orbits around Earth and Mars is free—the spacecraft simply follows Kepler's laws while in an elliptical orbit around the Sun.

Let us perform a simple numerical example, to see how to apply the mechanical laws to this process. We assume that the spacecraft is in a parking orbit above

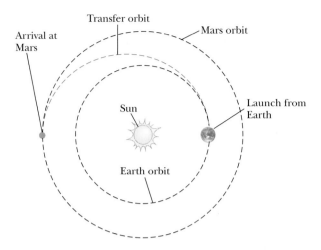

**Figure 3**

The Hohmann transfer orbit from Earth to Mars. This is similar to transferring the wrench from one shuttle to another in Figure 2—we are transferring a spacecraft from one planet to another.

the Earth's surface. Imagine that the spacecraft is in orbit around the Sun, with a perturbation in its orbit caused by the Earth. Thus, the tangential speed of the Earth about the Sun also represents the average speed of the spacecraft around the Sun. This speed is calculated from Newton's second law for a particle in uniform circular motion:

$$F = ma \longrightarrow G\frac{M_{Sun}m_{Earth}}{r^2} = m_{Earth}\frac{v^2}{r}$$

$$\longrightarrow v = \sqrt{\frac{GM_{Sun}}{r}} = \sqrt{\frac{(6.67 \times 10^{-11}\,\text{N}\cdot\text{m}^2/\text{kg}^2)(1.99 \times 10^{30}\,\text{kg})}{1.50 \times 10^{11}\,\text{m}}}$$

$$= 2.97 \times 10^4\,\text{m/s}$$

This is the original speed of the spacecraft, to which we add a change $\Delta v$ in order to inject the spacecraft into the transfer orbit.

The major axis of the elliptical transfer orbit is found by adding together the orbit radii of Earth and Mars (see Fig. 3):

$$\text{Major axis} = 2a = r_{Earth} + r_{Mars}$$
$$= 1.50 \times 10^{11}\,\text{m} + 2.28 \times 10^{11}\,\text{m} = 3.78 \times 10^{11}\,\text{m}$$

Thus,

$$a = 1.89 \times 10^{11}\,\text{m}$$

From this, Kepler's third law is used to find the travel time, which is one half of the period of the orbit:

$$t_{travel} = \tfrac{1}{2}T = \tfrac{1}{2}\sqrt{\frac{4\pi^2}{GM_{Sun}}a^3}$$

$$= \tfrac{1}{2}\sqrt{\frac{4\pi^2}{(6.67 \times 10^{-11}\,\text{N}\cdot\text{m}^2/\text{kg}^2)(1.99 \times 10^{30}\,\text{kg})}(1.89 \times 10^{11}\,\text{m})^3}$$

$$= 2.24 \times 10^7\,\text{s} = 0.711\,\text{yr} = 260\,\text{d}$$

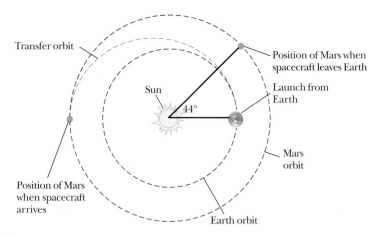

**Figure 4**

The spacecraft must be launched when Mars is 44° ahead of Earth in its orbit.

Thus, the journey to Mars will require 260 Earth days. We can also determine where in their orbits Earth and Mars must be to ensure the planet will be there when the spacecraft arrives. Mars has an orbital period of 687 Earth days. During the transfer time, the angular position *change* of Mars is

$$\Delta \theta_{\text{Mars}} = \frac{260 \text{ d}}{687 \text{ d}} (2\pi) = 2.38 \text{ rad} = 136°$$

Thus, the spacecraft must be launched when Mars is 44° ahead of the Earth in its orbit in order for the spacecraft and Mars to arrive at the same point at the same time. This geometry is shown in Figure 4.

With relatively simple mathematics, this is as far as we can go in describing the details of a trip to Mars—we have found the desired path, the time for the trip, and the position of Mars at launch time. Another important question for the spacecraft captain would be "How much fuel is required for the trip?" This is related to the speed changes necessary to put us into a transfer orbit. These types of calculations involve energy considerations and are explored in Problem 4 below.

Although many considerations necessary for a successful mission to Mars have not been addressed, we have designed a transfer orbit from Earth to Mars that is consistent with the laws of mechanics. We consequently declare success for our endeavor and bring the *Mission to Mars* Context to a close.

**PROBLEMS**

1. Some science fiction stories describe a twin planet to the Earth. It is exactly 180° ahead of us in the same orbit as the Earth, so we will never see it, because it is on the other side of the Sun. If you are in a spacecraft in orbit around the Earth, describe conceptually how you could visit this planet by altering your orbit.

2. You are in an orbiting space shuttle. Another space shuttle is in exactly the same orbit, but 1 km ahead of you, moving in the same direction around the circle. Through an oversight, your food supplies have been exhausted, and there is more than enough food in the other shuttle. The commander of the other shuttle is going to throw you a picnic basket full of sandwiches, from her shuttle to yours. How should she throw it? (There are no numbers—just give a conceptual response.)

3. Consider a Hohmann transfer from Earth to Venus. (a) How long will this transfer take? (b) Should Venus be ahead of the Earth in its orbit or behind the Earth when the space-craft leaves the Earth on its way to the rendezvous? How many degrees is Venus ahead or behind the Earth?

4. Investigate what the engine has to do to make a spacecraft follow the Hohmann transfer orbit from the Earth to Mars described in the text. Short-duration burns of our rocket en-gine are required to change the speed of our spacecraft whenever we alter our orbit. A spacecraft does not have brakes, so fuel is required to both increase and decrease its speed. First ignore the gravitational attraction between the spacecraft and the planets. (a) Calculate the speed change required for switching the craft from a circular orbit around the Sun at the Earth's distance to the transfer orbit to Mars. (b) Calculate the speed change required for switching from the transfer orbit to a circular orbit around the Sun at the distance of Mars. Now consider the effects of the two planets' gravity. (c) Calculate the speed change required for carrying the craft from the Earth's surface to its own independent orbit around the Sun. You may suppose the craft is launched from the Earth's equator toward the east. (d) Model the craft as falling to the surface of Mars from solar orbit. Calculate the magnitude of the speed change required to make a soft landing on Mars at the end of the fall. Mars rotates on its axis with a period of 24.6 h.

# Earthquakes

Earthquakes result in massive movement of the ground, as evidenced by the accompanying photograph of railroad tracks in Mexico, damaged severely by an earthquake in 1985. Anyone who has experienced a serious earthquake can attest to the violent shaking it produces. In this Context, we shall focus on earthquakes as an application of our study of the physics of vibrations and waves.

We might expect that the risk of damage in an earthquake decreases as one moves farther from the epicenter. Over long distances, this is correct. For example, structures in Kansas are not affected by earthquakes in California. In regions close to the earthquake, however, the notion of decrease in risk with distance is not consistent. Consider, for example, the following quotations describing damage in two different earthquakes:

After the Northridge, California, earthquake, January 17, 1994:*

"Although the city [Santa Monica] sits 25 kilometers from the shock's epicenter, it suffered more [damage] than did other areas less than one-third that distance from the jolt."

After the Michoacán earthquake, September 19, 1985:†

" . . . an earthquake rattled the coast of Mexico in the state of Michoacán, about 400 kilometers west of Mexico City. Near the coast, the shaking of the ground was mild and caused little damage. As the seismic waves raced inland, the ground shook even less, and by the time the waves were 100 kilometers from Mexico City, the shaking had nearly subsided. Nevertheless, the seismic waves in-

Severe damage occurred in localized regions of Mexico City in 1985, even though the Michoacán earthquake was hundreds of kilometers away. *(T. Campion/SYGMA)*

The Northridge earthquake in California in 1994 caused billions of dollars in damage. *(Joseph Sohm/ChromoSohm Inc./CORBIS)*

* *Science News,* 1994, p. 287.

† *American Scientist,* November-December, 1992, p. 566.

duced severe shaking in the city, and some areas continued to shake for several minutes after the seismic waves had passed. Some 300 buildings collapsed and more than 20,000 people died."

It is clear from these quotations that the notion of a simple decrease in risk with distance is misleading. We will use these quotations to motivate us to study the physics of vibrations and waves so we can make a more sophisticated analysis of the risk of damage to structures in an earthquake. Our study here will also be important when we investigate electromagnetic waves in Chapters 24 to 27. In this Context, we shall address the central question

*How can we choose locations and build structures to minimize the risk of damage in an earthquake?*

## chapter 12

The Foucault pendulum at the Franklin Institute in Philadelphia. This type of pendulum was first used by French physicist Jean Foucault to verify the Earth's rotation experimentally. As the pendulum swings, the vertical plane in which it oscillates appears to rotate as the bob successively knocks over the indicators arranged in a circle on the floor. In reality, the plane of oscillation is fixed in space, and the Earth rotating beneath the swinging pendulum moves the indicators into position to be knocked down, one after the other.

(*© Bob Emott, Photographer*)

# Oscillatory Motion

You are most likely familiar with several examples of *periodic* motion, such as the oscillations of an object on a spring, the motion of a pendulum, and the vibrations of a stringed musical instrument. Numerous other systems exhibit periodic behavior. For example, the molecules in a solid oscillate about their equilibrium positions; electromagnetic waves, such as light waves, radar, and radio waves, are characterized by oscillating electric and magnetic field vectors; in alternating-current circuits, such as in your household electrical service, voltage and current vary periodically with time. In this chapter, we will investigate mechanical systems that exhibit periodic motion.

We have experienced a number of situations in which the net force on a particle is constant. In these situations, the acceleration of the particle is also constant and we can describe the motion of the particle using the kinematic equations of Chapter 2. If a force acting on a particle varies in time, the acceleration of the particle also changes with time, and the kinematic equations cannot be used.

A special kind of periodic motion occurs when the force that acts on a particle is proportional to the displacement of the particle from the equilibrium position and is always directed toward the equilibrium position. We shall study this special type of varying force in this chapter. When this type of force acts on a particle, the particle exhibits *simple harmonic motion,* which will serve as an analysis model for a large class of oscillation problems.

## 12.1 • MOTION OF A PARTICLE ATTACHED TO A SPRING

At this point in your study of physics, you have probably started to develop a set of mental models associated with the analysis models that we have developed. By mental model, we mean a typical physical situation that comes to your mind each time you identify the analysis model to be used in a problem. For example, a rock falling in the absence of air resistance is a possible mental model for the analysis model of a particle under constant acceleration. Collisions between billiard balls represent a mental model to help understand the momentum version of the isolated system model. A bicycle wheel might offer a mental model for the rigid body under a net torque model.

In the case of oscillatory motion, a useful mental model is an object of mass $m$ attached to a horizontal spring, as in Figure 12.1. If the spring is unstretched, the object is at rest on a frictionless surface at its **equilibrium position,** which is defined as $x = 0$ (Fig. 12.1b). If the object is pulled to the side to position $x$ and released, it will oscillate back and forth as we discussed in Section 6.4. We will use the particle model to remove effects of the size of the object and analyze a particle on a spring as our system.

Recall from Chapter 6 that when the particle attached to an idealized massless spring is displaced to a position $x$, the spring exerts a force on it given by **Hooke's law,**

$$F_s = -kx \qquad [12.1]$$

where $k$ is the **force constant** of the spring and $F_s$ is the component of the spring force $\mathbf{F}_s$. We call the force in Equation 12.1 a **linear restoring force** because it is proportional to the displacement from the equilibrium position and always directed toward the equilibrium position, *opposite* the displacement. That is, when the particle is displaced to the right in Figure 12.1a, $x$ is positive and the spring force has a negative component—to the left. When the particle is displaced to the left of $x = 0$ (Fig. 12.1c), $x$ is negative and the spring force has a positive component—to the right. When a particle is under the effect of a linear restoring force, the motion it follows is a special type of oscillatory motion called **simple harmonic motion.** You can test whether or not a particle will undergo simple harmonic motion by seeing if the force on the particle is linear in $x$. A system undergoing simple harmonic motion is called a **simple harmonic oscillator.**

Let us imagine a particle subject to a linear restoring force such as that given by Equation 12.1. Applying Newton's second law in the $x$ direction to the particle gives us

$$\sum F = F_s = ma \longrightarrow -kx = ma$$

$$a = -\frac{k}{m}x \qquad [12.2]$$

See the *Core Concepts in Physics CD-ROM,* Screens 8.2, 8.3 & 8.10

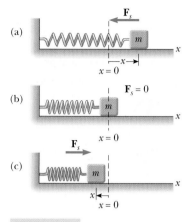

**Figure 12.1**

A block attached to a spring on a frictionless track moves in simple harmonic motion. (a) When the block is displaced to the right of equilibrium, the position is positive and the force and acceleration are negative. (b) At the equilibrium position $x = 0$, the force and acceleration of the block are zero but the speed is a maximum. (c) When the position is negative, the force and acceleration of the block are positive.

**PITFALL PREVENTION 12.1**

**The orientation of the spring**

Figure 12.1 shows the pictorial representation that we will use to study the behavior of systems with springs—a *horizontal* spring, with an attached block sliding on a frictionless surface. Another possible pictorial representation is a block hanging from a *vertical* spring that is attached at the top to a support. All of the results we discuss for the horizontal spring will be the same for the vertical spring, except for one minor difference. When the block is placed on the vertical spring, its weight will cause the spring to extend. If the position of the block at which it hangs at rest on the spring is defined as $x = 0$, the results of this chapter will apply to this system also. We will consider primarily the horizontal spring in this chapter.

• *Acceleration in simple harmonic motion*

See Screens 8.2, 8.3 & 8.10

That is, the **acceleration is proportional to the displacement of the particle from equilibrium and is in the opposite direction.** If the particle is displaced to a position $x = A$ and released from rest, its *initial* acceleration is $-kA/m$. When the particle passes through the equilibrium position $x = 0$, its acceleration is zero. At this instant, its speed is a maximum because the acceleration changes sign. The particle then continues to travel to the left of equilibrium with a positive acceleration and finally reaches $x = -A$, at which time its acceleration is $+kA/m$ and its speed is again zero. The particle completes a full cycle of its motion by returning to the original position, again passing through $x = 0$ with maximum speed. Thus, we see that the particle oscillates between the turning points $x = \pm A$. In the absence of friction, because the force exerted by the spring is conservative, this motion will continue forever. Real systems are generally subject to friction, and so cannot oscillate forever. We explore the details of the situation with friction in Section 12.6.

**Quick Quiz 12.1**

A block on the end of a spring is pulled to position $x = A$ and released. In one full cycle of its motion, through what total distance does it travel? (a) $A/2$ (b) $A$ (c) $2A$ (d) $4A$

## 12.2 • MATHEMATICAL REPRESENTATION OF SIMPLE HARMONIC MOTION

Let us now develop a mathematical representation of the motion we described in the preceding section. Recall that, by definition, $a = dv/dt = d^2x/dt^2$, and so we can express Equation 12.2 as

$$\frac{d^2x}{dt^2} = -\frac{k}{m}x \qquad [12.3]$$

If we denote the ratio $k/m$ with the symbol $\omega^2$ (we choose $\omega^2$ rather than $\omega$ to make the solution simpler in form), then

$$\omega^2 = \frac{k}{m} \qquad [12.4]$$

and Equation 12.3 can be written in the form

$$\frac{d^2x}{dt^2} = -\omega^2 x \qquad [12.5]$$

What we now require is a mathematical solution to Equation 12.5—that is, a function $x(t)$ that satisfies this second-order differential equation. This will be a mathematical representation of the position of the particle as a function of time. We seek a function $x(t)$ such that the second derivative of the function is the same as the original function with a negative sign and multiplied by $\omega^2$. The trigonometric functions sine and cosine exhibit this behavior, so we can build a solution around one or both of these. The following cosine function is a solution to the differential equation:

• *Position as a function of time of a particle undergoing simple harmonic motion*

$$x(t) = A\cos(\omega t + \phi) \qquad [12.6]$$

where $A$, $\omega$, and $\phi$ are constants.* To see this explicitly, note that

$$\frac{dx}{dt} = A\frac{d}{dt}\cos(\omega t + \phi) = -\omega A \sin(\omega t + \phi) \qquad \textbf{[12.7]}$$

$$\frac{d^2x}{dt^2} = -\omega A\frac{d}{dt}\sin(\omega t + \phi) = -\omega^2 A \cos(\omega t + \phi) \qquad \textbf{[12.8]}$$

Comparing Equations 12.6 and 12.8, we see that $d^2x/dt^2 = -\omega^2 x$ and Equation 12.5 is satisfied.

The parameters $A$, $\omega$, and $\phi$ are constants of the motion. To give physical significance to these constants, it is convenient to form a graphical representation of the motion by plotting $x$ as a function of $t$, as in Figure 12.2a. First, we note that $A$, called the **amplitude** of the motion, is simply the **maximum value of the position of the particle in either the positive or negative $x$ direction.** The constant $\omega$ is called the **angular frequency** and has units of rad/s. From Equation 12.4, the angular frequency is

$$\omega = \sqrt{\frac{k}{m}} \qquad \textbf{[12.9]}$$

The constant angle $\phi$ is called the **phase constant** (or phase angle) and, along with the amplitude $A$, is determined uniquely by the position and velocity of the particle at $t = 0$. If the particle is at its maximum position $x = A$ at $t = 0$, the phase constant is $\phi = 0$ and the graphical representation of the motion is shown in Figure 12.2b. The quantity $(\omega t + \phi)$ is called the **phase** of the motion. Note that the function $x(t)$ is periodic and its value is the same each time $\omega t$ increases by $2\pi$ radians.

Equations 12.1, 12.5, and 12.6 form the basis for the analysis model of the simple harmonic motion. If we are analyzing a situation and find that the force on a particle is of the mathematical form of Equation 12.1, we know that the motion will be that of a simple harmonic oscillator and that the position is described by Equation 12.6. If we analyze a system and find that it is described by a differential equation of the form of Equation 12.5, the motion will be that of a simple harmonic oscillator. If we analyze a situation and find that the position of a particle is described by Equation 12.6, we know the particle is undergoing simple harmonic motion.

Let us investigate further the mathematical description of the motion. The **period** $T$ of the motion is the time required for the particle to go through one full cycle of its motion (see Fig. 12.2a). That is, the values of $x$ and $v$ for the particle at time $t$ equal the values of $x$ and $v$ at time $t + T$. We can relate the period to the angular frequency by using the fact that the phase increases by $2\pi$ radians in a time of $T$:

$$[\omega(t + T) + \phi] - (\omega t + \phi) = 2\pi$$

---

* In earlier chapters we saw many examples in which we evaluated a trigonometric function of an angle. The argument of a trigonometric function, such as sine or cosine, *must* be a pure number. The radian is a pure number because it is a ratio of lengths. Degrees are pure simply because the degree is a completely artificial "unit"—it is not related to measurements of lengths. The notion of requiring a pure number for a trigonometric function is important in Equation 12.6, where the angle is expressed in terms of other measurements. Thus, $\omega$ *must* be expressed in rad/s (and not, for example, in revolutions per second) if $t$ is expressed in seconds. Furthermore, the argument of other types of functions must also be pure numbers, including logarithms and exponential functions.

**PITFALL PREVENTION 12.3**
**Where's the triangle?**

Equation 12.6 includes a trigonometric function. We have used trigonometric functions to analyze right triangles as geometric models, but no triangle occurs in this current discussion. This demonstrates that trigonometric functions are just that—mathematical functions that can be used whether they refer to triangles or not. In this case, the cosine function happens to have the correct behavior to provide a mathematical representation of the motion of the particle in simple harmonic motion.

• *Angular frequency*

(a)

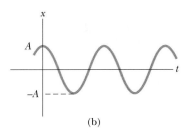

(b)

**Figure 12.2**

(a) A graphical representation (position versus time) for the system in Figure 12.1—a particle undergoing simple harmonic motion. The amplitude of the motion is $A$ and the period is $T$. (b) The $x$-$t$ curve in the special case in which $x = A$ and $v = 0$ at $t = 0$.

Simplifying this expression, we see that $\omega T = 2\pi$, or

- *Period*

$$T = \frac{2\pi}{\omega} \qquad \text{[12.10]}$$

The inverse of the period is called the **frequency** $f$ of the motion. Whereas the period is the time interval per oscillation, the frequency represents the **number of oscillations the particle makes per unit time:**

- *Frequency*

$$f = \frac{1}{T} = \frac{\omega}{2\pi} \qquad \text{[12.11]}$$

The units of $f$ are cycles per second, or **hertz** (Hz). Rearranging Equation 12.11 gives

- *Relation of angular frequency to frequency and period*

$$\omega = 2\pi f = \frac{2\pi}{T} \qquad \text{[12.12]}$$

We can use Equations 12.9, 12.10, and 12.11 to express the period and frequency of the motion for the particle–spring system in terms of the characteristics $m$ and $k$ of the system as

- *Period of a particle–spring system*

$$T = \frac{2\pi}{\omega} = 2\pi\sqrt{\frac{m}{k}} \qquad \text{[12.13]}$$

- *Frequency of a particle–spring system*

$$f = \frac{1}{T} = \frac{1}{2\pi}\sqrt{\frac{k}{m}} \qquad \text{[12.14]}$$

That is, the period and frequency depend *only* on the mass of the particle and the force constant of the spring, and *not* on the parameters of the motion, such as $A$ or

**PITFALL PREVENTION 12.4**

**Two kinds of frequency**

We identify two kinds of frequency for a simple harmonic oscillator: $f$, called simply the *frequency*, is measured in hertz, and $\omega$, the *angular frequency*, is measured in radians per second. Be sure you are clear about which frequency is being discussed or requested in a given problem. Equations 12.11 and 12.12 show the relationship between the two frequencies.

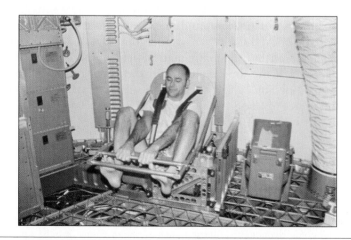

How does an astronaut measure his weight in orbit when a bathroom scale would be useless? He actually measures mass rather than weight by sitting on a seat of known mass that oscillates due to springs of known force constant. In this photo, astronaut Alan L. Bean sits in the oscillating seat. The measured period of oscillation determines the total oscillating mass, according to Equation 12.13. By subtracting the mass of the seat, the mass of the astronaut is calculated. *(NASA)*

$\phi$. As we might expect, the frequency is larger for a stiffer spring (larger value of $k$) and decreases with increasing mass of the particle.

We can obtain the velocity and acceleration* of a particle undergoing simple harmonic motion from Equations 12.7 and 12.8:

$$v = \frac{dx}{dt} = -\omega A \sin(\omega t + \phi)$$  [12.15]

$$a = \frac{d^2x}{dt^2} = -\omega^2 A \cos(\omega t + \phi)$$  [12.16]

- *Velocity as a function of time of a particle undergoing simple harmonic motion*
- *Acceleration as a function of time of a particle undergoing simple harmonic motion*

From Equation 12.15 we see that because the sine and cosine functions oscillate between $\pm 1$, the extreme values of $v$ are $\pm \omega A$. Likewise, Equation 12.16 tells us that the extreme values of the acceleration are $\pm \omega^2 A$. Therefore, the *maximum* values of the magnitudes of the velocity and acceleration are

$$v_{max} = \omega A = \sqrt{\frac{k}{m}}\, A$$  [12.17]

$$a_{max} = \omega^2 A = \frac{k}{m}\, A$$  [12.18]

- *Maximum values of velocity and acceleration of a particle undergoing simple harmonic motion*

Figure 12.3a plots position versus time for an arbitrary value of the phase constant. The associated velocity–time and acceleration–time curves are illustrated in Figures 12.3b and 12.3c. They show that the phase of the velocity differs from the phase of the position by $\pi/2$ rad, or $90°$. That is, when $x$ is a maximum or a minimum, the velocity is zero. Likewise, when $x$ is zero, the speed is a maximum. Furthermore, note that the phase of the acceleration differs from the phase of the position by $\pi$ radians, or $180°$. For example, when $x$ is a maximum, $a$ has a maximum magnitude in the opposite direction.

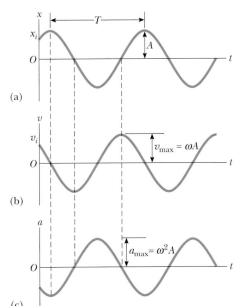

(a)

(b)

(c)

**Figure 12.3**

Graphical representation of three variables in simple harmonic motion: (a) position versus time, (b) velocity versus time, and (c) acceleration versus time. Note that at any specified time the velocity is $90°$ out of phase with the position and the acceleration is $180°$ out of phase with the position.

---

* Because the motion of a simple harmonic oscillator takes place in one dimension, we will denote velocity as $v$ and acceleration as $a$, with the direction indicated by a positive or negative sign, as in Chapter 2.

$t = 0$
$x_i = A$
$v_i = 0$

A block–spring system that is released from rest at $x_i = A$. In this case, $\phi = 0$, and thus $x = A \cos \omega t$.

Equation 12.6 describes simple harmonic motion of a particle in general. Let us now see how to evaluate the constants of the motion. The angular frequency $\omega$ is evaluated using Equation 12.9. The constants $A$ and $\phi$ are evaluated from the initial conditions, that is, the state of the oscillator at $t = 0$.

Suppose we initiate the motion by pulling the particle from equilibrium by a distance $A$ and releasing it from rest at $t = 0$, as in Figure 12.4. We must then require that our solutions for $x(t)$ and $v(t)$ (Eqs. 12.6 and 12.15) obey the initial conditions that $x(0) = A$ and $v(0) = 0$:

$$x(0) = A \cos \phi = A$$

$$v(0) = -\omega A \sin \phi = 0$$

These conditions are met if we choose $\phi = 0$, giving $x = A \cos \omega t$ as our solution. To check this solution, we note that it satisfies the condition that $x(0) = A$, because $\cos 0 = 1$.

Position, velocity, and acceleration are plotted versus time in Figure 12.5a for this special case. The acceleration reaches extreme values of $\pm \omega A$ when the position has extreme values of $\pm A$. Furthermore, the velocity has extreme values of $\pm \omega A$, which both occur at $x = 0$. Hence, the quantitative solution agrees with our qualitative description of this system.

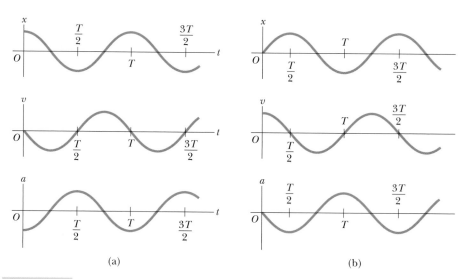

(a)                    (b)

(a) Position, velocity, and acceleration versus time for a block undergoing simple harmonic motion under the initial conditions that at $t = 0$, $x(0) = A$, and $v(0) = 0$. (b) Position, velocity, and acceleration versus time for a block undergoing simple harmonic motion under the initial conditions that at $t = 0$, $x(0) = 0$, and $v(0) = v_i$.

Let us consider another possibility. Suppose that the system is oscillating and we define $t = 0$ as the instant that the particle passes through the unstretched position of the spring while moving to the right (Fig. 12.6). We must then require that our solutions for $x(t)$ and $v(t)$ obey the initial conditions that $x(0) = 0$ and $v(0) = v_i$:

$$x(0) = A \cos \phi = 0$$

$$v(0) = -\omega A \sin \phi = v_i$$

The first of these conditions tells us that $\phi = -\pi/2$. With this value for $\phi$, the second condition tells us that $A = v_i/\omega$. Hence, the solution is given by

$$x = \frac{v_i}{\omega} \cos \left( \omega t - \frac{\pi}{2} \right)$$

Figure 12.5b shows the graphs of position, velocity, and acceleration versus time for this choice of $t = 0$. Note that these curves are the same as those in Figure 12.5a, but shifted to the right by one fourth of a cycle. This is described mathematically by the phase constant $\phi = -\pi/2$, which is one fourth of a full cycle of $2\pi$.

**Figure 12.6**

The block–spring system is undergoing oscillation, and $t = 0$ is defined at an instant when the block passes through the equilibrium position $x = 0$ and is moving to the right with speed $v_i$.

## Quick Quiz 12.2

What would be the solution for $x$ if the particle were initially moving to the left in Figure 12.6?

## Quick Quiz 12.3

Is a bouncing ball an example of simple harmonic motion? Is the daily movement of a student from home to school and back simple harmonic motion?

### THINKING PHYSICS 12.1

We know that the period of oscillation of an object attached to a spring is proportional to the square root of the mass of the object (Eq. 12.13). Thus, if we perform an experiment in which we place objects with a range of masses on the end of a spring and measure the period of oscillation of each mass–spring system, a graph of the square of the period versus the mass will result in a straight line, as suggested in Figure 12.7. But we find that the line does not go through the origin. Why not?

**Reasoning**  The line does not go through the origin because the spring itself has mass. Thus, the resistance to changes in motion of the system is a combination of the mass of the object on the end of the spring and the mass of the oscillating spring coils. The entire mass of the spring is not oscillating in the same way, however. The coil of the spring attached to the object is oscillating over the same amplitude as the object, but the coil at the opposite end of the spring is not oscillating at all. For a cylindrical spring, energy arguments can be used to show that the effective additional mass representing the oscillations of the spring is one third of the mass of the spring. The square of the period is proportional to the total oscillating mass,

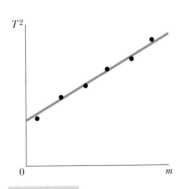

**Figure 12.7**

(Thinking Physics 12.1) A graph of experimental data—the square of the period versus mass of a block in a block–spring system.

but the graph in Figure 12.7 shows the square of the period versus only the mass of the object on the spring. A graph of period squared versus total mass (mass of the object on the spring plus the effective oscillating mass of the spring) would pass through the origin.

## THINKING PHYSICS 12.2

Suppose you attach a stapler to a rubber band and allow the stapler to hang in equilibrium from one end of the loop. You now pull the stapler down a short distance and release it, measuring the frequency $f$ of its oscillations. After making this measurement, you cut the rubber band in one place and unfold it so that you have a single rubber strip that is twice as long as the length of the original loop. You attach the stapler to one end of the rubber strip, let it hang while you hold the other end, and again measure the frequency of oscillation. How does this frequency compare with the original frequency? (Model the rubber band as obeying Hooke's law.)

**Reasoning** First imagine that the cut rubber band is held at the midpoint rather than the end, so the stapler is held from a single rubber strip that has an unstretched length equal to the unstretched length of the original uncut rubber band. Now when the stapler is attached and allowed to hang, the single strip stretches through a distance twice as long as that for the rubber loop, because only one strip is supporting the stapler instead of two in the case of the loop. Now, if the rubber strip is held at the free end and the stapler is hung at the opposite end, each half of the full length of the strip stretches by a distance twice as great as the original stretch of the rubber band. Therefore, the overall stretch will be four times as large as the original rubber band. With four times the stretch for the same hanging weight, the force constant of the rubber strip must be one-fourth as large as that of the original rubber band. Using Equation 12.14, we determine that the frequency of the stapler when oscillating must be $\frac{1}{2} f$.

## Example 12.1 A Block–Spring System

A block with a mass of 200 g is connected to a light horizontal spring of force constant 5.00 N/m and is free to oscillate on a horizontal, frictionless surface. (a) If the block is displaced 5.00 cm from equilibrium and released from rest, as in Figure 12.4, find the period of its motion.

**Solution** The situation (we assume an ideal spring) tells us to use the simple harmonic motion model. Using Equation 12.13,

$$T = 2\pi \sqrt{\frac{m}{k}} = 2\pi \sqrt{\frac{200 \times 10^{-3} \text{ kg}}{5.00 \text{ N/m}}} = 1.26 \text{ s}$$

(b) Determine the maximum speed and maximum acceleration of the block.

**Solution** Using Equations 12.17 and 12.18, with $A = 5.00 \times 10^{-2}$ m, we have

$$v_{\max} = \omega A = \frac{2\pi}{T} A = \left(\frac{2\pi}{1.26 \text{ s}}\right)(5.00 \times 10^{-2} \text{ m})$$
$$= 0.250 \text{ m/s}$$

$$a_{\max} = \omega^2 A = \left(\frac{2\pi}{T}\right)^2 A = \left(\frac{2\pi}{1.26 \text{ s}}\right)^2 (5.00 \times 10^{-2} \text{ m})$$
$$= 1.25 \text{ m/s}^2$$

(c) Express the position, velocity, and acceleration of this object as functions of time, assuming that $\phi = 0$.

**Solution** From Equations 12.6, 12.15, and 12.16,

$$x = A \cos \omega t = (0.0500 \text{ m}) \cos 5.00t$$

$$v = -\omega A \sin \omega t = -(0.250 \text{ m/s}) \sin 5.00t$$

$$a = -\omega^2 A \cos \omega t = -(1.25 \text{ m/s}^2) \cos 5.00t$$

## Example 12.2 An Oscillating Particle

A particle oscillates with simple harmonic motion along the $x$ axis. Its position varies with time according to the equation

$$x = (4.00 \text{ m}) \cos\left(\pi t + \frac{\pi}{4}\right)$$

where $t$ is in seconds.

(a) Determine the amplitude, frequency, and period of the motion.

**Solution** By comparing this equation with the general equation for simple harmonic motion, $x = A \cos(\omega t + \phi)$, we see that $A =$ 4.00 m and $\omega = \pi$ rad/s; therefore we find $f = \omega/2\pi = \pi/2\pi =$ 0.500 Hz and $T = 1/f =$ 2.00 s .

(b) Calculate the velocity and acceleration of the particle at any time $t$.

**Solution** Using Equations 12.15 and 12.16,

$$v = \frac{dx}{dt} = -(4.00\pi \text{ m/s})\sin\left(\pi t + \frac{\pi}{4}\right)$$

$$a = \frac{dv}{dt} = -(4.00\pi^2 \text{ m/s}^2)\cos\left(\pi t + \frac{\pi}{4}\right)$$

(c) What are the position and the velocity of the particle at time $t = 0$?

**Solution** The position function is given in the text of the problem. Evaluating this expression at $t = 0$ gives us

$$x = (4.00 \text{ m})\cos\left(\pi t + \frac{\pi}{4}\right) = (4.00 \text{ m})\cos\left(\frac{\pi}{4}\right) = 2.83 \text{ m}$$

From part (b), we evaluate the velocity function at $t = 0$:

$$v = -(4.00\pi \text{ m/s}) \sin\left(\pi t + \frac{\pi}{4}\right) = -(4.00\pi \text{ m/s}) \sin\left(\frac{\pi}{4}\right)$$

$$= -8.89 \text{ m/s}$$

**EXERCISE** Using the results to part (b), determine the position, velocity, and acceleration of the particle at $t = 1.00$ s.

*Answer* $x = -2.83$ m; $v = 8.89$ m/s; $a = 27.9$ m/s$^2$

**EXERCISE** Determine the maximum velocity and the maximum acceleration of the particle.

*Answer* $v_{\text{max}} = \pm 12.6$ m/s; $a_{\text{max}} = \pm 39.5$ m/s$^2$

## Example 12.3 Initial Conditions

Suppose that the initial position $x_i$ and initial velocity $v_i$ of a harmonic oscillator of known angular frequency are given; that is, $x(0) = x_i$ and $v(0) = v_i$. Find general expressions for the amplitude and the phase constant in terms of these initial parameters.

**Solution** With these initial conditions, Equations 12.6 and 12.15 give us

$$x_i = A \cos \phi \quad \text{and} \quad v_i = -\omega A \sin \phi$$

Dividing these two equations eliminates $A$, giving $v_i/x_i = -\omega \tan \phi$, or

$$\tan \phi = -\frac{v_i}{\omega x_i}$$

Furthermore, if we take the sum $x_i^2 + (v_i/\omega)^2 = A^2 \cos^2 \phi + A^2 \sin^2 \phi = A^2$ (where we have used Eqs. 12.6 and 12.15) and solve for $A$, we find that

$$A = \sqrt{x_i^2 + \left(\frac{v_i}{\omega}\right)^2}$$

**EXERCISE** A 3.0-kg object is attached to a spring and pulled out horizontally to a maximum displacement from equilibrium of 0.50 m. What force constant must the spring have if the object is to achieve an acceleration equal to the free-fall acceleration?

*Answer* 59 N/m

## 12.3 • ENERGY CONSIDERATIONS IN SIMPLE HARMONIC MOTION

If an object attached to a spring slides on a frictionless surface, we can consider the combination of the spring and the attached object to be an isolated system. As a result, we can apply the energy version of the isolated system model to the system. Let us examine the mechanical energy of the system described in Figure 12.1. Because the surface is frictionless, the total mechanical energy of the system is constant. We model the object as a particle. The kinetic energy, in the simplification model in which the spring is massless, is associated only with the motion of the particle of mass *m*. We use Equation 12.15 to express the kinetic energy as

- *Kinetic energy of a simple harmonic oscillator*

$$K = \tfrac{1}{2}mv^2 = \tfrac{1}{2}m\omega^2 A^2 \sin^2(\omega t + \phi)$$ **[12.19]**

Elastic potential energy in this system is associated with the spring, and for any position *x* of the particle is, as we found in Chapter 7, $U = \tfrac{1}{2}kx^2$. Using Equation 12.6, we have

- *Potential energy of a simple harmonic oscillator*

$$U = \tfrac{1}{2}kx^2 = \tfrac{1}{2}kA^2 \cos^2(\omega t + \phi)$$ **[12.20]**

We see that *K* and *U* are always positive quantities, and each varies with time. We can express the *total energy* of the simple harmonic oscillator as

$$E = K + U = \tfrac{1}{2}m\omega^2 A^2 \sin^2(\omega t + \phi) + \tfrac{1}{2}kA^2 \cos^2(\omega t + \phi)$$

Because $\omega^2 = k/m$, we can write this as

$$E = \tfrac{1}{2}kA^2[\sin^2(\omega t + \phi) + \cos^2(\omega t + \phi)]$$

Because $\sin^2\theta + \cos^2\theta = 1$ for any angle $\theta$, this equation reduces to

- *Total energy of a simple harmonic oscillator*

$$E = \tfrac{1}{2}kA^2$$ **[12.21]**

That is, the energy of an isolated simple harmonic oscillator is a constant of the motion and proportional to the square of the amplitude. In fact, the total mechanical energy is just equal to the maximum potential energy stored in the spring when $x = \pm A$. At these points, $v = 0$ and there is no kinetic energy. At the equilibrium position, $x = 0$ and $U = 0$, so that the total energy is all in the form of kinetic energy of the particle,

$$K_{max} = \tfrac{1}{2}mv_{max}^2 = \tfrac{1}{2}kA^2$$

These results are appropriate for the simplification model in which we consider the spring to be massless. In the real situation in which the spring has mass, additional kinetic energy is associated with the motion of the spring. In numerical problems in this book, unless otherwise noted, we will consider only massless springs.

Graphical representations of the kinetic and potential energies versus time for the system of a particle on a massless spring are shown in Figure 12.8a, where $\phi = 0$.

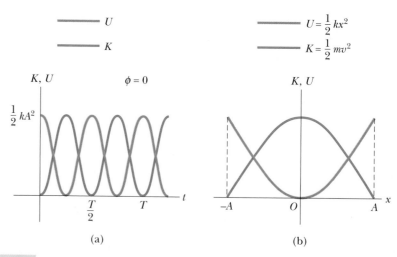

**Figure 12.8**

(a) Kinetic energy and potential energy versus time for a simple harmonic oscillator with $\phi = 0$. (b) Kinetic energy and potential energy versus position for a simple harmonic oscillator. In either plot, note that $K + U = $ constant.

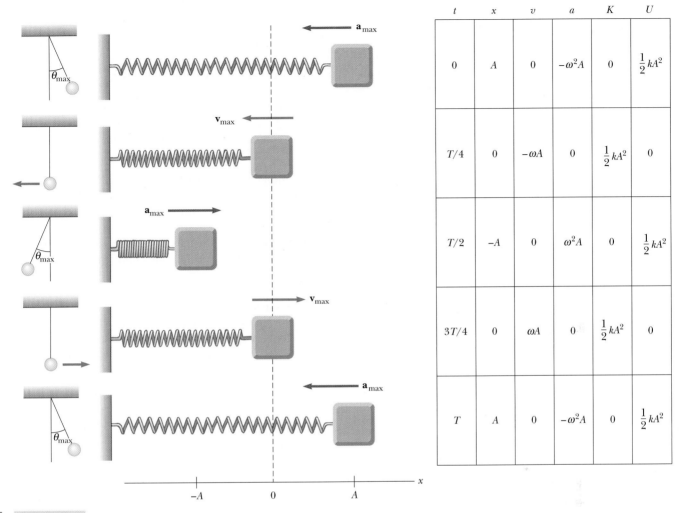

| $t$ | $x$ | $v$ | $a$ | $K$ | $U$ |
|-----|-----|-----|-----|-----|-----|
| $0$ | $A$ | $0$ | $-\omega^2 A$ | $0$ | $\frac{1}{2}kA^2$ |
| $T/4$ | $0$ | $-\omega A$ | $0$ | $\frac{1}{2}kA^2$ | $0$ |
| $T/2$ | $-A$ | $0$ | $\omega^2 A$ | $0$ | $\frac{1}{2}kA^2$ |
| $3T/4$ | $0$ | $\omega A$ | $0$ | $\frac{1}{2}kA^2$ | $0$ |
| $T$ | $A$ | $0$ | $-\omega^2 A$ | $0$ | $\frac{1}{2}kA^2$ |

**Figure 12.9**

Simple harmonic motion for a block–spring system and its analogy to the motion of a simple pendulum (Section 12.4). The parameters in the table at the right refer to the block–spring system, assuming that at $t = 0$, $x = A$ so that $x = A\cos\omega t$.

In this situation, the sum of the kinetic and potential energies at all times is a constant equal to $\frac{1}{2}kA^2$, the total energy of the system. The variations of $K$ and $U$ with position are plotted in Figure 12.8b. Energy in the system is continuously being transformed between potential energy (in the spring) and kinetic energy (of the object attached to the spring). Figure 12.9 illustrates the position, velocity, acceleration, kinetic energy, and potential energy of the particle–spring system for one full period of the motion. Most of the ideas discussed so far for simple harmonic motion are incorporated in this important figure. We suggest that you study it carefully.

Finally, we can use the conservation of mechanical energy expression to obtain the velocity for an arbitrary position $x$ of the particle, expressing the total energy as

$$E = K + U = \tfrac{1}{2}mv^2 + \tfrac{1}{2}kx^2 = \tfrac{1}{2}kA^2$$

$$v = \pm\sqrt{\frac{k}{m}\left(A^2 - x^2\right)} = \pm\omega\sqrt{A^2 - x^2} \qquad \textbf{[12.22]}$$

• *Velocity as a function of position for a particle undergoing simple harmonic motion*

This expression confirms the facts that the speed is a maximum at $x = 0$ and is zero at the turning points $x = \pm A$.

## THINKING PHYSICS 12.3

An object oscillating on the end of a horizontal spring slides back and forth over a frictionless surface. During one oscillation, you set an identical object at the maximum displacement point, with instant-acting glue on its surface. Just as the oscillating object reaches its largest displacement and is momentarily at rest, it adheres to the new object by means of the glue, and the two objects continue the oscillation together. Does the period of the oscillation change? Does the amplitude of oscillation change? Does the energy of the oscillation change?

**Reasoning**  The period of oscillation changes because the period depends on the mass that is oscillating (Eq. 12.13). The amplitude does not change. Because the new object was added while the original object was at rest, the combined objects are at rest at this point also, defining the amplitude as the same as in the original oscillation. The energy does not change, either. At the maximum displacement point, the energy is all potential energy stored in the spring, which depends only on the force constant and the amplitude, not the mass of the object. The object of increased mass will pass through the equilibrium point with lower speed than in the original oscillation, but with the same kinetic energy. Another approach is to think about how energy could be transferred into the oscillating system—no work was done (nor was any other form of energy transferred), so the energy in the system cannot change.

## Example 12.4   Oscillations on a Horizontal Surface

A 0.500-kg object connected to a massless spring of force constant 20.0 N/m oscillates on a horizontal, frictionless track. (a) Calculate the total energy of the system and the maximum velocity of the object if the amplitude of the motion is 3.00 cm.

**Solution**  Because the object slides on a frictionless surface, we can use the isolated system model for the system of the object and the spring. Because only conservative forces are acting within the system, the mechanical energy of the system is conserved. Using Equation 12.21, we have

$$E = \tfrac{1}{2}kA^2 = \tfrac{1}{2}(20.0 \text{ N/m})(3.00 \times 10^{-2} \text{ m})^2 = \boxed{9.00 \times 10^{-3} \text{ J}}$$

When the object is at $x = 0$, $U = 0$ and $E = \tfrac{1}{2}mv_{\text{max}}^2$; therefore

$$\tfrac{1}{2}mv_{\text{max}}^2 = 9.00 \times 10^{-3} \text{ J}$$

$$v_{\text{max}} = \sqrt{\frac{18.0 \times 10^{-3} \text{ J}}{0.500 \text{ kg}}} = \boxed{\pm 0.190 \text{ m/s}}$$

The positive and negative signs indicate that the object could be moving to either the right or the left at this instant.

(b) What is the velocity of the object when the position is equal to 2.00 cm?

**Solution**  We apply Equation 12.22 directly:

$$v = \pm \sqrt{\frac{k}{m}(A^2 - x^2)}$$

$$= \pm \sqrt{\frac{20.0 \text{ N/m}}{0.50 \text{ kg}}[(3.00 \times 10^{-2} \text{ m})^2 - (2.00 \times 10^{-2} \text{ m})^2]}$$

$$= \boxed{\pm 0.141 \text{ m/s}}$$

(c) Compute the kinetic and potential energies of the system when the position equals 2.00 cm.

**Solution**  Using the result to part (b), we find

$$K = \tfrac{1}{2}mv^2 = \tfrac{1}{2}(0.500 \text{ kg})(0.141 \text{ m/s})^2 = \boxed{5.00 \times 10^{-3} \text{ J}}$$

$$U = \tfrac{1}{2}kx^2 = \tfrac{1}{2}(20.0 \text{ N/m})(2.00 \times 10^{-2} \text{ m})^2 = \boxed{4.00 \times 10^{-3} \text{ J}}$$

Note that the sum $K + U$ equals the total mechanical energy $E$ found in part (a).

**EXERCISE**  For what values of $x$ does the speed of the object equal 0.100 m/s?

*Answer*  $\pm 2.55$ cm

## 12.4 • THE SIMPLE PENDULUM

The **simple pendulum** is another mechanical system that exhibits periodic motion. It consists of a point object of mass $m$, suspended by a light string (or rod) of length $L$, where the upper end of the string is fixed, as in Figure 12.10. For a real object, as long as the size of the object is small relative to the length of the string, the pendulum can be modeled as a simple pendulum—this is the particle model. When the object is pulled to the side and released, it oscillates about the lowest point, which is the equilibrium position. The motion occurs in a vertical plane and is driven by the gravitational force.

The forces acting on the object are the force **T** acting along the string and the gravitational force $m\mathbf{g}$. The tangential component of the gravitational force $mg \sin \theta$ always acts toward $\theta = 0$, opposite the displacement. The tangential force is therefore a restoring force, and we can use Newton's second law to write the equation of motion in the tangential direction as

$$F_t = ma_t \longrightarrow -mg \sin \theta = m\frac{d^2 s}{dt^2}$$

where $s$ is the position measured along the circular arc in Figure 12.10, and the negative sign indicates that $F_t$ acts toward the equilibrium position. Because $s = L\theta$ (Eq. 10.1a) and $L$ is constant, this equation reduces to

$$\frac{d^2\theta}{dt^2} = -\frac{g}{L} \sin \theta$$

Compare this equation to Equation 12.5, which is of a similar, but not identical, mathematical form. The right side is proportional to $\sin \theta$ rather than to $\theta$; hence, we conclude that the motion is *not* simple harmonic motion because it is not of the form of Equation 12.5. If we assume that $\theta$ is *small* (less than about $10°$ or $0.2$ rad), however, we can use a simplification model called the **small angle approximation,** in which $\sin \theta \approx \theta$, where $\theta$ is measured in radians. Table 12.1 shows angles in degrees and radians, and the sines of these angles. As long as $\theta$ is less than about $10°$, the angle in radians and its sine are the same, at least to within an accuracy of less than $1.0\%$.

Thus, for small angles, the equation of motion becomes

$$\frac{d^2\theta}{dt^2} = -\frac{g}{L} \theta \qquad\qquad \textbf{[12.23]}$$

 See Screens 8.11 & 8.12

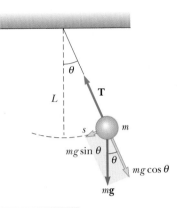

**Figure 12.10**

When $\theta$ is small, the oscillation of the simple pendulum can be modeled as simple harmonic motion about the equilibrium position ($\theta = 0$). The restoring force is $mg \sin \theta$, the component of the gravitational force tangent to the arc.

**PITFALL PREVENTION 12.5**

**Not true simple harmonic motion**

 Remember that the pendulum *does not* exhibit true simple harmonic motion for *any* angle. If the angle is less than about $10°$, the motion can be *modeled* as simple harmonic.

• *Equation of motion for the simple pendulum (small $\theta$)*

| TABLE 12.1 | Angles and Sines of Angles | | |
|---|---|---|---|
| Angle (degrees) | Angle in Radians | Sine of Angle | Percent Difference (%) |
| 0 | 0.0000 | 0.0000 | 0.0 |
| 1 | 0.0175 | 0.0175 | 0.0 |
| 2 | 0.0349 | 0.0349 | 0.0 |
| 3 | 0.0524 | 0.0523 | 0.0 |
| 5 | 0.0873 | 0.0872 | 0.1 |
| 10 | 0.1745 | 0.1736 | 0.5 |
| 15 | 0.2618 | 0.2588 | 1.2 |
| 20 | 0.3491 | 0.3420 | 2.1 |
| 30 | 0.5236 | 0.5000 | 4.7 |

Now we have an expression with exactly the same mathematical form as Equation 12.5, with $\omega^2 = g/L$, and so we conclude that the motion is approximately simple harmonic motion for small amplitudes. Modeling the solution after Equation 12.6, $\theta$ can therefore be written as $\theta = \theta_{max} \cos(\omega t + \phi)$, where $\theta_{max}$ is the **maximum angular position** and the angular frequency $\omega$ is

- *Angular frequency for a simple pendulum*

$$\omega = \sqrt{\frac{g}{L}} \qquad \text{[12.24]}$$

The period of the motion is

- *Period for a simple pendulum*

$$T = \frac{2\pi}{\omega} = 2\pi\sqrt{\frac{L}{g}} \qquad \text{[12.25]}$$

In other words, the **period and frequency of a simple pendulum oscillating at small angles depend only on the length of the string and the free-fall acceleration.** Because the period is *independent* of the mass, we conclude that *all* simple pendula of equal length at the same location oscillate with equal periods. Experiment shows that this is correct. The analogy between the motion of a simple pendulum and the particle–spring system is illustrated in Figure 12.9.

The simple pendulum can be used as a timekeeper. It is also a convenient device for making precise measurements of the free-fall acceleration.

### Quick Quiz 12.4

A simple pendulum is suspended from the ceiling of a stationary elevator, and the period is determined. Describe the changes, if any, in the period when the elevator (a) accelerates upward, (b) accelerates downward, and (c) moves with constant velocity.

### Quick Quiz 12.5

A grandfather clock depends on the period of a pendulum to keep correct time. Suppose a grandfather clock is calibrated correctly and then a mischievous child slides the bob of the pendulum downward on the oscillating rod. Does the grandfather clock run (a) slow, (b) fast, or (c) correctly?

### THINKING PHYSICS 12.4

You set up two oscillating systems: a simple pendulum and a block hanging from a vertical spring. You carefully adjust the length of the pendulum so that both oscillators have the same period. You now take the two oscillators to the Moon. Will they still have the same period as each other? What happens if you observe the two oscillators in an orbiting space shuttle?

**Reasoning** The block hanging from the spring will have the same period on the Moon that it had on the Earth because the period depends on the mass of the block and the force constant of the spring, neither of which have changed. The period of the pendulum on the Moon will be different from its period on the Earth because the period of the pendulum depends on the value of *g*. Because *g* is smaller on the Moon than on the Earth, the pendulum will oscillate with a longer period.

In the orbiting space shuttle, the block–spring system will oscillate with the same period as on the Earth when it is set into motion, because the period does not depend on gravity. The pendulum will not oscillate at all—if you pull it to the side from a direction you define as "vertical" and release it, it stays there. Because the shuttle is in free-fall while in orbit around the Earth, the effective gravity is zero, and there is no restoring force on the pendulum.

### Example 12.5   A Measure of Height

A man enters a tall tower, needing to know its height. He notes that a long pendulum extends from the ceiling almost to the floor and that its period is 12.0 s. How tall is the tower?

**Solution**   We adopt a simplification model in which the height of the tower is equal to the length of the pendulum. If we use $T = 2\pi\sqrt{L/g}$ and solve for $L$, we have

$$L = \frac{gT^2}{4\pi^2} = \frac{(9.80 \text{ m/s}^2)(12.0 \text{ s})^2}{4\pi^2} = \boxed{35.7 \text{ m}}$$

**EXERCISE**   If the pendulum described in this example is taken to the Moon, where the free-fall acceleration is $1.67 \text{ m/s}^2$, what is the period there?

*Answer*   29.1 s

## 12.5 • THE PHYSICAL PENDULUM

If a hanging object oscillates about a fixed axis that does not pass through its center of mass, then it must be treated as a **physical, or compound, pendulum.** For the simple pendulum, we generated Equation 12.23 from the particle under a net force model. For the physical pendulum, we will need to use the rigid body under a net torque model from Chapter 10. Consider a rigid body pivoted at a point $O$ that is a distance $d$ from the center of mass (Fig. 12.11). The torque about $O$ is provided by the gravitational force, and its magnitude is $mgd \sin \theta$. Using Newton's second law for rotation, $\sum \tau = I\alpha$ (Eq. 10.27), where $I$ is the moment of inertia about the axis through $O$, we have

$$-mgd \sin \theta = I\frac{d^2\theta}{dt^2}$$

The negative sign on the left indicates that the torque about $O$ tends to decrease $\theta$. That is, the gravitational force produces a restoring torque.

If we again assume that $\theta$ is small, the small angle approximation $\sin \theta \approx \theta$ is valid and the equation of motion reduces to

$$\frac{d^2\theta}{dt^2} = -\left(\frac{mgd}{I}\right)\theta \qquad \text{[12.26]}$$

Note that this equation has the same mathematical form as Equation 12.5, with $\omega^2 = mgd/I$, and so the motion is approximately simple harmonic motion for small amplitudes. That is, the solution of Equation 12.26 is $\theta = \theta_{max} \cos(\omega t + \phi)$, where $\theta_{max}$ is the maximum angular position and

$$\omega = \sqrt{\frac{mgd}{I}}$$

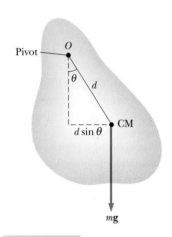

**Figure 12.11**

The physical pendulum consists of a rigid body pivoted at the point $O$, which is not at the center of mass.

• *Angular frequency for a physical pendulum*

The period is

- *Period of motion for a physical pendulum*

$$T = \frac{2\pi}{\omega} = 2\pi\sqrt{\frac{I}{mgd}}$$ [12.27]

One can use this result to measure the moment of inertia of a planar rigid body. If the location of the center of mass and, hence, the distance $d$ are known, the moment of inertia can be obtained through a measurement of the period. Finally, note that Equation 12.27 becomes the equation for the period of a simple pendulum (Eq. 12.25) when $I = md^2$, that is, when all the mass is concentrated at a point, and the physical pendulum reduces to the simple pendulum.

### Quick Quiz 12.6

Two students are watching the swinging of a pendulum with a large bob in a museum, such as that in the opening photograph for this chapter. One student says, "I'm going to sneak past the fence and stick some chewing gum on the top of the pendulum bob, to change its period of oscillation." The other student says, "That won't change the period—the period of a pendulum is independent of mass." Which student is correct?

### Quick Quiz 12.7

In one of the Moon landings, an astronaut walking on the Moon's surface had a belt hanging from his space suit, and the belt oscillated as a physical pendulum. A scientist on Earth observed this motion on television and from it was able to estimate the free-fall acceleration on the Moon. How do you suppose this calculation was done?

### Example 12.6   A Swinging Sign

A circular sign of mass $M$ and radius $R$ is hung on a nail from a small loop located at one edge (Fig. 12.12). After it is placed on the nail, the sign oscillates in a vertical plane. Find the period of oscillation if the amplitude of the motion is small.

**Solution**   The moment of inertia of a disk about an axis through the center is $\frac{1}{2}MR^2$ (see Table 10.2). The pivot point for the sign is through a point on the rim, so we use the parallel axis theorem (see Eq. 10.43) to find the moment of inertia about the pivot:

$$I_p = \frac{1}{2}MR^2 + MR^2 = \frac{3}{2}MR^2$$

The distance $d$ from the pivot to the center of mass is the radius $R$. Substituting these quantities into Equation 12.27 gives

$$T = 2\pi\sqrt{\frac{\frac{3}{2}MR^2}{MgR}} = 2\pi\sqrt{\frac{3R}{2g}}$$

**Figure 12.12**

(Example 12.6) A circular sign oscillating about a pivot as a physical pendulum.

## 12.6 • DAMPED OSCILLATIONS

The oscillatory motions we have considered so far have occurred under the simplification model of an ideal frictionless system, that is, one that oscillates indefinitely under the action of only a linear restoring force. In realistic systems, resistive forces, such as friction, are present and retard the motion of the system. Consequently, the mechanical energy of the system diminishes in time, and the motion is said to be *damped*.

One common type of resistive force, which we discussed in Chapter 5, is proportional to the velocity and acts in the direction opposite the velocity. This type of force is often observed when an object is oscillating slowly in air, for instance. Because the resistive force can be expressed as $\mathbf{R} = -b\mathbf{v}$, where $b$ is a constant related to the strength of the resistive force, and the restoring force exerted on the system is $-kx$, Newton's second law gives us

$$\sum F_x = -kx - bv = ma_x$$

$$-kx - b\frac{dx}{dt} = m\frac{d^2x}{dt^2} \qquad \textbf{[12.28]}$$

The solution of this differential equation requires mathematics that may not yet be familiar to you, and so it will simply be stated without proof. When the parameters of the system are such that $b < \sqrt{4mk}$, so that the resistive force is small, the solution to Equation 12.28 is

$$x = (Ae^{-(b/2m)t})\cos(\omega t + \phi) \qquad \textbf{[12.29]}$$

• *Position as a function of time for a particle undergoing damped oscillations (small damping)*

where the angular frequency of the motion is

$$\omega = \sqrt{\frac{k}{m} - \left(\frac{b}{2m}\right)^2} \qquad \textbf{[12.30]}$$

• *Angular frequency for a particle undergoing damped oscillations (small damping)*

This result can be verified by substituting Equation 12.29 into Equation 12.28. Notice that Equation 12.29 is similar to Equation 12.6, with the new feature that the amplitude (in the parentheses before the cosine function) depends on time. Figure 12.13 shows the position as a function of time in this case. We see that, **when the resistive force is relatively small, the oscillatory character of the motion is preserved but the amplitude of vibration decreases in time** and the motion ultimately ceases. This is known as an **underdamped oscillator.** The dashed blue lines in Figure 12.13, which form the *envelope* of the oscillatory curve, represent the exponential factor that appears in Equation 12.29. The exponential factor shows that the *amplitude decays exponentially with time.*

It is convenient to express the angular frequency of vibration of a damped system (Eq. 12.30) in the form

$$\omega = \sqrt{{\omega_0}^2 - \left(\frac{b}{2m}\right)^2}$$

where $\omega_0 = \sqrt{k/m}$ represents the angular frequency of oscillation in the absence of a resistive force (the undamped oscillator). In other words, when $b = 0$, the resistive force is zero and the system oscillates with angular frequency $\omega_0$, called the **natural frequency.*** As the magnitude of the resistive force increases, the oscillations dampen more rapidly. When $b$ reaches a critical value $b_c$, so that

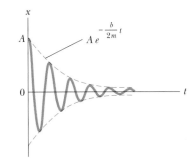

**Figure 12.13**

Graph of the position versus time for a damped oscillator with small damping. Note the decrease in amplitude with time.

---

\* In practice, both $\omega_0$ and $f_0 = \omega_0/2\pi$ are described as the natural frequency. The context of the discussion will help you determine which frequency is being discussed.

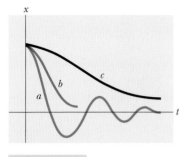

**Figure 12.14**

Plots of position versus time for (a) an underdamped oscillator, (b) a critically damped oscillator and (c) an overdamped oscillator.

$b_c/2m = \omega_0$, the system does not oscillate and is said to be **critically damped.** In this case it returns to equilibrium in an exponential manner with time, as in Figure 12.14.

If the medium is highly viscous, and the parameters meet the condition that $b/2m > \omega_0$, the system is **overdamped.** Again, the displaced system does not oscillate but simply returns to its equilibrium position. As the damping increases, the time it takes the particle to approach the equilibrium position also increases, as indicated in Figure 12.14. In any case, when a resistive force is present, the mechanical energy of the oscillator eventually falls to zero. The mechanical energy is transformed into internal energy in the oscillating system and the resistive medium.

## 12.7 • FORCED OSCILLATIONS

We have seen that the mechanical energy of a damped oscillator decreases in time as a result of the resistive force. It is possible to compensate for this energy decrease by applying an external force that does positive work on the system. At any instant, energy can be transferred into the system by an applied force that acts in the direction of motion of the oscillator. For example, a child on a swing can be kept in motion by appropriately timed "pushes." The amplitude of motion remains constant if the energy input per cycle of motion exactly equals the decrease in mechanical energy each cycle that results from resistive forces. This is an example of the non-isolated system in steady state, discussed in Section 7.5.

A common example of a forced oscillator is a damped oscillator driven by an external force that varies periodically, such as $F(t) = F_0 \sin \omega t$, where $\omega$ is the angular frequency of the driving force and $F_0$ is a constant. In general, the frequency $\omega$ of the driving force is different from the natural frequency $\omega_0$ of the oscillator. Newton's second law in this situation gives

$$\sum F = ma \longrightarrow F_0 \sin \omega t - b\frac{dx}{dt} - kx = m\frac{d^2x}{dt^2} \qquad \text{[12.31]}$$

Again, the solution of this equation is rather lengthy and will not be presented. After the driving force on an initially stationary object begins to act, the amplitude of the oscillation will increase. After a sufficiently long period, when the energy input per cycle from the driving force equals the amount of mechanical energy transformed to internal energy for each cycle, a steady-state condition is reached in which the oscillations proceed with constant amplitude. In this case, Equation 12.31 has the solution

$$x = A \cos(\omega t + \phi) \qquad \text{[12.32]}$$

where

• *Amplitude of oscillation for a forced, damped oscillator*

$$A = \frac{F_0/m}{\sqrt{\left(\omega^2 - \omega_0{}^2\right)^2 + \left(\frac{b\omega}{m}\right)^2}} \qquad \text{[12.33]}$$

and where $\omega_0 = \sqrt{k/m}$ is the natural frequency of the undamped oscillator ($b = 0$).

Equation 12.33 shows that the amplitude of the forced oscillator is constant for a given driving force because it is being driven in steady state by an external force. For small damping, the amplitude becomes large when the frequency of the driving force is near the natural frequency of oscillation, or when $\omega \approx \omega_0$. The

• *Resonance*

dramatic increase in amplitude near the natural frequency is called **resonance,** and the natural frequency $\omega_0$ is also called the **resonance frequency** of the system.

Figure 12.15 is a graph of amplitude as a function of frequency for the forced oscillator, with varying resistive forces. Note that the amplitude increases with decreasing damping ($b \to 0$) and that the resonance curve flattens as the damping increases. In the absence of a damping force ($b = 0$), we see from Equation 12.33 that the steady-state amplitude approaches infinity as $\omega \to \omega_0$. In other words, if there are no resistive forces in the system, and we continue to drive an oscillator with a sinusoidal force at the resonance frequency, the amplitude of motion will build up without limit. This does not occur in practice because some damping is always present in real oscillators.

One experiment that demonstrates a resonance phenomenon is illustrated in Figure 12.16. Several pendula of different lengths are suspended from a stretched string. If one of them, such as $P$, is set swinging, the others will begin to oscillate, because they are coupled by the stretched string. Of those that are forced into oscillation by this coupling, pendulum $Q$, which is the same length as $P$ (hence, the two pendula have the same natural frequency), will oscillate with the greatest amplitude.

Resonance appears in other areas of physics. For example, certain electric circuits have resonance frequencies. This fact is exploited in radio tuners, which allow you to select the station you wish to hear. Vibrating strings and columns of air also have resonance frequencies, which allow them to be used for musical instruments, which we shall discuss in Chapter 14.

### Quick Quiz 12.8

Some parachutes have holes in them to allow air to move smoothly through the chute. Without the holes, the air gathered under the chute as the parachutist falls is sometimes released from under the edges of the chute periodically from one side and then the other. Why might this periodic release of air cause a problem?

### context connection
## 12.8 • RESONANCE IN STRUCTURES

In the preceding section, we investigated the phenomenon of resonance, in which an oscillating system exhibits its maximum response to a periodic driving force when the frequency of the driving force matches the natural frequency of the oscillator. We now apply this understanding to the interaction between the shaking of the ground during an earthquake and structures attached to the ground. The structure is the oscillator—it has a set of natural frequencies, determined by its stiffness, its mass, and the details of its construction. The periodic driving force is supplied by the shaking of the ground.

A disastrous result can occur if a natural frequency of the building matches a frequency contained in the ground shaking. In this case, the resonance vibrations of the building can build to a very large amplitude, large enough to damage or destroy the building. This can be avoided in two ways. The first involves designing the structure so that natural frequencies of the building lie outside the range of earthquake frequencies. (A typical range of earthquake frequencies is 0–15 Hz.) This can be done by varying the size or mass structure of the building. The second method involves incorporating sufficient damping in the building. This may not change the resonance frequency significantly, but it will lower the response to the

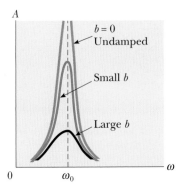

**Figure 12.15**

Graph of amplitude versus frequency for a damped oscillator when a periodic driving force is present. When the frequency of the driving force equals the natural frequency $\omega_0$, resonance occurs. Note that the shape of the resonance curve depends on the size of the damping coefficient $b$.

**Figure 12.16**

If pendulum $P$ is set into oscillation, pendulum $Q$ will eventually oscillate with the greatest amplitude because its length is equal to that of $P$ and so they have the same natural frequency of vibration. The pendula are oscillating in a direction perpendicular to the plane formed by the stationary strings.

(a)                                    (b)

**Figure 12.17**

(a) In 1940, steady winds set up vibrations in the Tacoma Narrows Bridge, causing it to oscillate at a frequency near one of the natural frequencies of the bridge structure. (b) Once established, this resonance condition led to the bridge's collapse. *(Special Collections Division, University of Washington Libraries, photo by Farquharson)*

natural frequency, as in Figure 12.15. It will also flatten the resonance curve, so that the building will respond to a wide range of frequencies, but with relatively small amplitude at any given frequency.

We now describe two examples involving resonance excitations in bridge structures. Soldiers are commanded to break step when marching across a bridge. This command takes into account resonance—if the marching frequency of the soldiers matches that of the bridge, the bridge could be set into resonance oscillation. If the amplitude becomes large enough, the bridge could actually collapse. Just such a situation occurred on April 14, 1831, when the Broughton suspension bridge in England collapsed while troops marched over it. Investigations after the accident showed that the bridge was near failure—the resonance vibration induced by the marching soldiers caused it to fail sooner than it otherwise might have.

The second example of such a structural resonance occurred in 1940, when the Tacoma Narrows Bridge in Washington state was destroyed by resonant vibrations (Fig. 12.17). The winds were not particularly strong on that occasion, but the bridge still collapsed because vortices (turbulences) generated by the wind blowing through the bridge occurred at a frequency that matched a natural frequency of the bridge. The flapping of this wind (similar to the flapping of a flag in a strong breeze) provided the periodic driving force that brought the bridge down into the river.

Resonance gives us our first clue to responding to the central question for this Context. Suppose a building is far from the epicenter of an earthquake, so that the ground shaking is small. If the shaking frequency matches a natural frequency of the building, a very effective energy coupling occurs between the ground and the building. Thus, even for relatively small shaking, the ground, by resonance, can feed energy into the building efficiently enough to cause the failure of the structure. The structure must be carefully designed so as to reduce the resonance response.

In the Context introduction, we indicated that buildings in Mexico City were damaged in 1985 even though they were 400 km from the epicenter of the earthquake. How did a release of energy at the epicenter cause oscillations at such a distant location? In the Loma Prieta earthquake during the World Series of 1989, the Nimitz Freeway in Oakland collapsed because of resonance. But only certain portions of it collapsed while other portions survived the shaking. What was special about the portions that collapsed? The answers to these questions involve *seismic waves*, which we will investigate in the next chapter.

**WEB**

For information on the Loma Prieta earthquake and related links, visit **www.earthquake.org/ lomaprieta.html**

## SUMMARY

The simple harmonic motion model is used for a particle experiencing a linear restoring force, expressed by **Hooke's law:**

$$F_s = -kx \qquad [12.1]$$

where $k$ is the **force constant** of the spring. The motion caused by such a force is called **simple harmonic motion,** and the system is called a **simple harmonic oscillator.** The position of the particle undergoing simple harmonic motion varies periodically in time according to the relation

$$x = A \cos(\omega t + \phi) \qquad [12.6]$$

where $A$ is the **amplitude** of the motion, $\omega$ is the **angular frequency,** and $\phi$ is the **phase constant.** The values of $A$ and $\phi$ depend on the initial position and velocity of the oscillator.

The time for one complete oscillation is called the **period** of the motion, defined by

$$T = \frac{1}{f} = \frac{2\pi}{\omega} \qquad [12.10, 12.11]$$

The inverse of the period is the **frequency** $f$ of the motion, which equals the number of oscillations per second.

The velocity and acceleration of a particle undergoing simple harmonic oscillation are

$$v = \frac{dx}{dt} = -\omega A \sin(\omega t + \phi) \qquad [12.15]$$

$$a = \frac{d^2x}{dt^2} = -\omega^2 A \cos(\omega t + \phi) \qquad [12.16]$$

Thus, the maximum speed is $\omega A$, and the maximum acceleration is of magnitude $\omega^2 A$. The speed is zero when the oscillator is at its turning points, $x = \pm A$, and the speed is a maximum at the equilibrium position, $x = 0$. The magnitude of the acceleration is a maximum at the turning points and is zero at the equilibrium position.

A particle–spring system moving on a frictionless track exhibits simple harmonic motion with the period

$$T = \frac{2\pi}{\omega} = 2\pi \sqrt{\frac{m}{k}} \qquad [12.13]$$

where $m$ is the mass of the particle attached to the spring.

The kinetic energy and potential energy of a simple harmonic oscillator vary with time and are given by

$$K = \tfrac{1}{2}mv^2 = \tfrac{1}{2}m\omega^2 A^2 \sin^2(\omega t + \phi) \qquad [12.19]$$

$$U = \tfrac{1}{2}kx^2 = \tfrac{1}{2}kA^2 \cos^2(\omega t + \phi) \qquad [12.20]$$

The **total energy** of a simple harmonic oscillator is a constant of the motion and is

$$E = \tfrac{1}{2}kA^2 \qquad [12.21]$$

The potential energy of a simple harmonic oscillator is a maximum when the particle is at its turning points (maximum displacement from equilibrium) and is zero at the equilibrium position. The kinetic energy is zero at the turning points and is a maximum at the equilibrium position.

A **simple pendulum** of length $L$ exhibits simple harmonic motion for small angular displacements from the vertical, with a period of

$$T = 2\pi \sqrt{\frac{L}{g}} \qquad [12.25]$$

The period of a simple pendulum is independent of the mass of the suspended object.

A **physical pendulum** exhibits simple harmonic motion for small angular displacements from equilibrium about a pivot that does not go through the center of mass. The period of this motion is

$$T = 2\pi \sqrt{\frac{I}{mgd}} \qquad [12.27]$$

where $I$ is the moment of inertia about an axis through the pivot and $d$ is the distance from the pivot to the center of mass.

**Damped oscillations** occur in a system in which a resistive force opposes the motion of the oscillating object. If such a system is set in motion and then left to itself, its mechanical energy decreases in time because of the presence of the nonconservative resistive force. It is possible to compensate for this transformation of energy by driving the system with an external periodic force. The oscillator in this case is undergoing **forced oscillations.** When the frequency of the driving force matches the natural frequency of the *undamped* oscillator, energy is efficiently transferred to the oscillator and its steady-state amplitude is a maximum. This situation is called **resonance.**

## QUESTIONS

1. Imagine that a pendulum is hanging from the ceiling of a car. As the car coasts freely down a hill, is the equilibrium position of the pendulum vertical? Does the period of oscillation change from that in a stationary car?

2. An object is hung on a spring and the frequency of oscillation $f$ of the system is measured. The object, a second identical object, and the spring are carried in the space shuttle to space. The two objects are attached to the ends of the spring, and the system is taken out into space on a space walk. The spring is extended and the system is released to oscillate while floating in space. What is the frequency of oscillation for this system, in terms of $f$?

3. You stand on the end of a diving board and bounce to set it in oscillation. You find a maximum response, in terms of the amplitude of oscillation of the end of the board, when you bounce at frequency $f$. You now move to the middle of the board, and repeat the experiment. Is the resonance frequency for forced oscillations at this point higher, lower, or the same as $f$? Why?

4. If the position of a particle varies as $x = -A \cos \omega t$, what is the phase constant in Equation 12.6? At what position is the particle at $t = 0$?

5. Does the displacement of an oscillating particle between $t = 0$ and a later time $t$ necessarily equal the position of the particle at time $t$? Explain.

6. Determine whether the following quantities can have the same sign for a simple harmonic oscillator: (a) position and velocity, (b) velocity and acceleration, (c) position and acceleration.

7. Can the amplitude $A$ and phase constant $\phi$ be determined for an oscillator if only the position is specified at $t = 0$? Explain.

8. Explain why the kinetic and potential energies of a block–spring system can never be negative.

9. A block–spring system undergoes simple harmonic motion with an amplitude $A$. Does the total energy change if the mass of the block is doubled but the amplitude is unchanged? Do the kinetic and potential energies depend on the mass? Explain.

10. What happens to the period of a simple pendulum if the pendulum's length is doubled? What happens to the period if the mass of the bob is doubled?

11. A simple pendulum undergoes simple harmonic motion when $\theta$ is small. Is the motion periodic when $\theta$ is large? How does the period of motion change as $\theta$ increases?

12. Will damped oscillations occur for any values of $b$ and $k$? Explain.

13. Is it possible to have damped oscillations when a system is at resonance? Explain.

14. If a grandfather clock were running slow, how could we adjust the length of the pendulum to correct the time?

## PROBLEMS

1, 2, 3 = straightforward, intermediate, challenging      $\square$ = full solution available in the *Student Solutions Manual and Study Guide*

**web** = solution posted at **http://www.harcourtcollege.com/physics/**      = computer useful in solving problem

= Interactive Physics      = paired numerical/symbolic problems      = life science application

---

*Note:* Ignore the mass of every spring, except in Problems 47 and 50.

### Section 12.1 Motion of a Particle Attached to a Spring

1. An archer pulls his bow string back 0.400 m by exerting a force on the string that increases uniformly from zero to 230 N. (a) What is the equivalent force constant of the bow? (b) How much work does he do in pulling the bow?

### Section 12.2 Mathematical Representation of Simple Harmonic Motion

2. A ball dropped from a height of 4.00 m makes a perfectly elastic collision with the ground. Assuming no energy is lost due to air resistance, (a) show that the ensuing motion is periodic and (b) determine the period of the motion. (c) Is the motion simple harmonic? Explain.

3. The position of a particle is given by the expression $x = (4.00 \text{ m}) \cos(3.00\pi t + \pi)$, where $x$ is in meters and $t$ is in seconds. Determine (a) the frequency and period of the motion, (b) the amplitude of the motion, (c) the phase constant, and (d) the position of the particle at $t = 0.250$ s.

4. A particle moves in simple harmonic motion with a frequency of 3.00 Hz and an amplitude of 5.00 cm. (a) Through what total distance does the particle move during one cycle of its motion? (b) What is its maximum speed? Where does this occur? (c) Find the maximum acceleration of the particle. Where in the motion does the maximum acceleration occur?

5. In an engine, a piston oscillates with simple harmonic motion so that its position varies according to the expression

$$x = (5.00 \text{ cm}) \cos(2t + \pi/6)$$

where $x$ is in centimeters and $t$ is in seconds. At $t = 0$, find (a) the position of the particle, (b) its velocity, and (c) its acceleration. (d) Find the period and amplitude of the motion.

6. A spring stretches by 3.90 cm when a 10.0-g object hangs from it at rest. If this object is replaced with a 25.0-g object that is set into simple harmonic motion, calculate the period of the motion.

7. A particle moving along the $x$ axis in simple harmonic mo-
**web** tion starts from its equilibrium position, the origin, at $t = 0$ and moves to the right. The amplitude of its motion is 2.00 cm and the frequency is 1.50 Hz. (a) Show that the position of the particle is given by

$$x = (2.00 \text{ cm}) \sin(3.00 \, \pi t)$$

Determine (b) the maximum speed and the earliest time $(t > 0)$ at which the particle has this speed, (c) the maximum acceleration and the earliest time $(t > 0)$ at which the particle has this acceleration, and (d) the total distance traveled between $t = 0$ and $t = 1.00$ s.

8. A simple harmonic oscillator takes 12.0 s to undergo five complete vibrations. Find (a) the period of its motion, (b) the frequency in hertz, and (c) the angular frequency in radians per second.

9. A 7.00-kg object is hung from the bottom end of a vertical spring fastened to an overhead beam. The object is set into vertical oscillations having a period of 2.60 s. Find the force constant of the spring.

10. The initial position and initial velocity of an object moving in simple harmonic motion are $x_i$, $v_i$, and $a_i$; the angular frequency of oscillation is $\omega$. (a) Show that the position and velocity of the object for all time can be written as

$$x(t) = x_i \cos \omega t + \left(\frac{v_i}{\omega}\right) \sin \omega t$$

$$v(t) = -x_i \omega \sin \omega t + v_i \cos \omega t$$

(b) If the amplitude of the motion is $A$, show that

$$v^2 - ax = v_i^2 - a_i x_i = \omega^2 A^2$$

11. A 0.500-kg object attached to a spring with a force constant of 8.00 N/m vibrates in simple harmonic motion with an amplitude of 10.0 cm. Calculate (a) the maximum value of its speed and acceleration, (b) the speed and acceleration when the object is 6.00 cm from the equilibrium position, and (c) the time it takes the object to move from $x = 0$ to $x = 8.00$ cm.

12. A piston in a gasoline engine is in simple harmonic motion. If the extremes of its displacement away from its center point are $\pm 5.00$ cm, find the maximum velocity and acceleration of the piston when the engine is running at the rate of 3600 rev/min.

13. After a thrilling plunge, bungee-jumpers bounce freely on the bungee cord through many cycles (Fig. P12.13b). Your little brother can make a pest of himself by figuring out the mass of each person, using a proportion which you set up by solving this problem: An object of mass $m$ is oscillating freely on a vertical spring with a period $T$ (Fig. P12.13a). An object of unknown mass $m'$ on the same spring oscillates with a period $T'$. Determine (a) the force constant and (b) the unknown mass.

(a)          (b)

**Figure P12.13**   (a) Mass–spring system for Problems 13 and 50. (b) Bungee-jumping from a bridge. *(Telegraph Colour Library/FPG International)*

14. A 1.00-kg object attached to a spring with a force constant of 25.0 N/m oscillates on a horizontal, frictionless track. At $t = 0$ the object is released from rest at $x = -3.00$ cm (i.e., the spring is compressed by 3.00 cm). Find (a) the period of its motion; (b) the maximum values of its speed and acceleration; and (c) the displacement, velocity, and acceleration as functions of time.

## Section 12.3   Energy Considerations in Simple Harmonic Motion

15. An automobile having a mass of 1 000 kg is driven into a
**web** brick wall in a safety test. The bumper behaves like a spring of constant $5.00 \times 10^6$ N/m and compresses 3.16 cm as the car is brought to rest. What was the speed of the car before impact, assuming that the mechanical energy of the car remains constant during impact with the wall?

16. A block of unknown mass is attached to a spring with a force constant of 6.50 N/m and undergoes simple harmonic motion with an amplitude of 10.0 cm. When the block is halfway between its equilibrium position and the end point, its speed is measured to be 30.0 cm/s. Calculate (a) the mass of the block, (b) the period of the motion, and (c) the maximum acceleration of the block.

17. A 200-g block is attached to a horizontal spring and executes simple harmonic motion on a frictionless surface with a period of 0.250 s. If the total energy of the system is 2.00 J, find (a) the force constant of the spring and (b) the amplitude of the motion.

18. A block–spring system oscillates with an amplitude of 3.50 cm. If the force constant is 250 N/m and the mass is 0.500 kg, determine (a) the mechanical energy of the system, (b) the maximum speed of the block, and (c) the maximum acceleration.

19. A 50.0-g block connected to a spring with a force constant of 35.0 N/m oscillates on a horizontal, frictionless surface with an amplitude of 4.00 cm. Find (a) the total energy of the system and (b) the speed of the block when the displacement is 1.00 cm. Find (c) the kinetic energy and (d) the potential energy when the displacement is 3.00 cm.

20. A 2.00-kg block is attached to a spring and placed on a horizontal, smooth surface. A horizontal force of 20.0 N is required to hold the block at rest when it is pulled 0.200 m from its equilibrium position. The block is now released from rest from this point, and it subsequently undergoes simple harmonic motion. Find (a) the force constant of the spring, (b) the frequency of the oscillations, and (c) the maximum speed of the block. Where does this maximum speed occur? (d) Find the maximum acceleration of the block. Where does it occur? (e) Find the total energy of the oscillating system. Find (f) the speed and (g) the acceleration when the position equals one third of the maximum value.

21. A particle executes simple harmonic motion with an amplitude of 3.00 cm. At what position does its speed equal one half of its maximum speed?

## Section 12.4   The Simple Pendulum

*Note:* Problem 58 in Chapter 1 involves investigation of the $\sin \theta \approx \theta$ approximation.

22. A "seconds pendulum" is one that moves through its equilibrium position once each second. (The period of the pendulum is 2.000 s.) The length of a seconds pendulum is 0.992 7 m at Tokyo and 0.994 2 m at Cambridge, England. What is the ratio of the free-fall accelerations at these two locations?

23. A simple pendulum has a mass of 0.250 kg and a length of 1.00 m. It is displaced through an angle of 15.0° and then released. What are (a) the maximum speed, (b) the maximum angular acceleration, and (c) the maximum restoring force?

24. The angular position of a simple pendulum is represented by the equation $\theta = (0.320 \text{ rad}) \cos \omega t$, where $\theta$ is in radians and $\omega = 4.43$ rad/s. Determine the period and length of the pendulum.

25. A particle of mass $m$ slides without friction inside a hemispherical bowl of radius $R$. Show that, if it starts from rest with a small displacement from equilibrium, the particle moves in simple harmonic motion with an angular frequency equal to that of a simple pendulum of length $R$ (i.e., $\omega = \sqrt{g/R}$).

## Section 12.5   The Physical Pendulum

26. A very light rigid rod with a length of 0.500 m extends straight out from one end of a meter stick. The stick is suspended from a pivot at the far end of the rod and is set into oscillation. (a) Determine the period of oscillation. (*Hint:* Use the parallel-axis theorem from Section 10.11.) (b) By what percentage does the period differ from the period of a simple pendulum 1.00 m long?

27. A physical pendulum in the form of a planar body moves in simple harmonic motion with a frequency of 0.450 Hz. If the pendulum has a mass of 2.20 kg and the pivot is located 0.350 m from the center of mass, determine the moment of inertia of the pendulum about the pivot point.

## Section 12.6   Damped Oscillations

28. Show that the time rate of change of mechanical energy for a damped, undriven oscillator is given by $dE/dt = -bv^2$ and hence is always negative. (*Hint:* Differentiate the expression for the mechanical energy of an oscillator, $E = \frac{1}{2}mv^2 + \frac{1}{2}kx^2$, and use Eq. 12.28.)

29. A pendulum with a length of 1.00 m is released from an initial angle of 15.0°. After 1 000 s, its amplitude has been reduced by friction to 5.50°. What is the value of $b/2m$?

30. Show that Equation 12.29 is a solution of Equation 12.28 provided that $b^2 < 4mk$.

## Section 12.7   Forced Oscillations

31. The front of her sleeper wet from teething, a baby rejoices in the day by crowing and bouncing up and down in her crib. Her mass is 12.5 kg and the crib mattress can be modeled as a light spring with force constant 4.30 kN/m. (a) The baby soon learns to bounce with maximum amplitude and minimum effort by bending her knees at what frequency? (b) She learns to use the mattress as a trampoline—losing contact with it for part of each cycle—when her amplitude exceeds what value?

**32.** A 2.00-kg object attached to a spring is driven by an external force given by $F = (3.00 \text{ N}) \cos(2\pi t)$. If the force constant of the spring is 20.0 N/m, determine (a) the period and (b) the amplitude of the motion. (*Hint:* Assume there is no damping—that is, that $b = 0$—and use Eq. 12.33.)

**33.** Considering an *undamped*, forced oscillator ($b = 0$), show that Equation 12.32 is a solution of Equation 12.31, with an amplitude given by Equation 12.33.

**34.** Damping is negligible for a 0.150-kg object hanging from a light 6.30-N/m spring. The system is driven by a force oscillating with an amplitude of 1.70 N. At what frequency will the force make the mass vibrate with an amplitude of 0.440 m ?

**35.** You are a research biologist. Its batteries are getting low, but you take your emergency pager along to a fine restaurant. You switch the small pager to vibrate instead of beep, and you put it into a side pocket of your suit coat. The arm of your chair presses the light cloth against your body at one spot. Fabric with a length of 8.21 cm hangs freely below that spot, with the pager at the bottom. A co-worker urgently needs instructions and calls you from your laboratory. The motion of the pager makes the hanging part of your coat swing back and forth with remarkably large amplitude. The waiter, maître d', wine steward, and nearby diners notice immediately and fall silent. Your daughter pipes up and says, "Daddy, look! Your cockroaches must have gotten out again!" Find the frequency at which your pager vibrates.

## Section 12.8 Context Connection — Resonance in Structures

**36.** Four people, each with a mass of 72.4 kg, are in a car with a mass of 1 130 kg. An earthquake strikes. The driver manages to pull off the road and stop, as the vertical oscillations of the ground surface make the car bounce up and down on its suspension springs. When the frequency of the shaking is 1.80 Hz, the car exhibits a maximum amplitude of vibration. The earthquake ends and the four people leave the car as fast as they can. By what distance does the car's undamaged suspension lift the car body as the people exit the car?

**37.** People who ride motorcycles and bicycles learn to look out for bumps in the road, and especially for *washboarding*, a condition in which many equally spaced ridges are worn into the road. What is so bad about washboarding? A motorcycle has several springs and shock absorbers in its suspension, but you can model it as a single spring supporting a mass. You can estimate the force constant by thinking about how far the spring compresses when a big biker sits down on the seat. A motorcyclist traveling at highway speed must be particularly careful of washboard

bumps that are a certain distance apart. What is the order of magnitude of their separation distance? State the quantities you take as data and the values you measure or estimate for them.

### Additional Problems

**38.** An object with mass $m_1 = 9.00$ kg is in equilibrium while connected to a light spring of constant $k = 100$ N/m that is fastened to a wall as shown in Figure P12.38a. A second object, of mass $m_2 = 7.00$ kg, is slowly pushed up against $m_1$, compressing the spring by the amount $A = 0.200$ m, (Fig. P12.38b). The system is then released, and both objects start moving to the right on the frictionless surface. (a) When $m_1$ reaches the equilibrium point, $m_2$ loses contact with $m_1$ (Fig. P12.38c) and moves to the right with speed $v$. Determine the value of $v$. (b) How far apart are the objects when the spring is fully stretched for the first time ($D$ in Fig. P12.38d)? (*Hint:* First determine the period of oscillation and the amplitude of the $m_1$–spring system after $m_2$ loses contact with $m_1$.)

**Figure P12.38**

**39.** A large block $P$ executes horizontal simple harmonic motion as it slides across a frictionless surface with a frequency $f = 1.50$ Hz. Block $B$ rests on it, as shown in Figure P12.39, and the coefficient of static friction between the

**Figure P12.39**

**Figure P12.43**

two is $\mu_s = 0.600$. What maximum amplitude of oscillation can the system have if block *B* is not to slip?

40. A particle with a mass of 0.500 kg is attached to a spring with a force constant of 50.0 N/m. At time $t = 0$ the particle has its maximum speed of 20.0 m/s and is moving to the left. (a) Determine the particle's equation of motion, specifying its position as a function of time. (b) Where in the motion is the potential energy three times the kinetic energy? (c) Find the length of a simple pendulum with the same period. (d) Find the minimum time required for the particle to move from $x = 0$ to $x = 1.00$ m.

41. A pendulum of length *L* and mass *M* has a spring of force constant *k* connected to it at a distance *h* below its point of suspension (Fig. P12.41). Find the frequency of vibration of the system for small values of the amplitude (small $\theta$). Assume the vertical suspension of length *L* is rigid, but ignore its mass.

**Figure P12.41**

42. The mass of the deuterium molecule ($D_2$) is twice that of the hydrogen molecule ($H_2$). If the vibrational frequency of $H_2$ is $1.30 \times 10^{14}$ Hz, what is the vibrational frequency of $D_2$? Assume that the "force constant" of attracting forces is the same for the two molecules.

43. A horizontal plank of mass *m* and length *L* is pivoted at one end. The plank's other end is supported by a spring of force constant *k* (Fig P12.43). The moment of inertia of the plank about the pivot is $\frac{1}{3}mL^2$. The plank is displaced by a small angle $\theta$ from its horizontal equilibrium position

and released. (a) Show that it moves with simple harmonic motion with an angular frequency $\omega = \sqrt{3k/m}$. (b) Evaluate the frequency if the mass is 5.00 kg and the spring has a force constant of 100 N/m.

44. **Review Problem.** An object of mass 4.00 kg is attached to a spring with a force constant of 100 N/m. It is oscillating on a horizontal frictionless surface with an amplitude of 2.00 m. A 6.00-kg object is dropped vertically on top of the 4.00-kg object as it passes through its equilibrium point. The two objects stick together. (a) By how much does the amplitude of the vibrating system change as a result of the collision? (b) By how much does the period change? (c) By how much does the energy change? (d) Account for the change in energy.

45. One end of a light spring with force constant 100 N/m is attached to a vertical wall. A light string is tied to the other end of the horizontal spring. The string changes from horizontal to vertical as it passes over a solid pulley of diameter 4.00 cm. The pulley is free to turn on a fixed smooth axle. The vertical section of the string supports a 200-g object. The string does not slip at its contact with the pulley. Find the frequency of oscillation of the object if the mass of the pulley is (a) negligible, (b) 250 g, and (c) 750 g.

46. A 2.00-kg block hangs without vibrating at the bottom end of a spring with a force constant of 500 N/m. The top end of the spring is attached to the ceiling of an elevator car. The car is rising with an upward acceleration of $g/3$ when the acceleration suddenly ceases (at $t = 0$). (a) What is the angular frequency of oscillation of the block after the acceleration ceases? (b) By what amount is the spring stretched during the time that the elevator car is accelerating? (c) What are the amplitude of the oscillation and the initial phase angle observed by a rider in the car? Take the upward direction to be positive.

47. A block of mass *M* is connected to a spring of mass *m* and oscillates in simple harmonic motion on a horizontal, frictionless track (Fig. P12.47). The force constant of the spring is *k* and the equilibrium length is $\ell$. Find (a) the kinetic energy of the system when the block has a speed *v*, and (b) the period of oscillation. [*Hint:* Assume that all portions of the spring oscillate in phase and that the veloc-

**Figure P12.47**

ity $v_x$ of a segment $dx$ is proportional to the position $x$ of the segment from the fixed end; that is, $v_x = (x/\ell)v$. Also, note that the mass of a segment of the spring is $dm = (m/\ell)\,dx$.]

48. A simple pendulum with a length of 2.23 m and a mass of 6.74 kg is given an initial speed of 2.06 m/s at its equilibrium position. Assume it undergoes simple harmonic motion, and determine its (a) period, (b) total energy, and (c) maximum angular position.

49. A ball of mass $m$ is connected to two rubber bands of
web length $L$, each under tension $T$, as in Figure P12.49. The ball is displaced by a small distance $y$ perpendicular to the length of the rubber bands. Assuming that the tension does not change, show that (a) the restoring force is $-(2T/L)y$ and (b) the system exhibits simple harmonic motion with an angular frequency $\omega = \sqrt{2T/mL}$.

**Figure P12.49**

50. When an object of mass $M$, connected to the end of a spring of mass $m_s = 7.40$ g and force constant $k$, is set into simple harmonic motion, the period of its motion is

$$T = 2\pi\sqrt{\frac{M + (m_s/3)}{k}}$$

A two-part experiment is conducted with the use of various masses suspended vertically from the spring, as shown in Figure P12.13a. (a) Static extensions of 17.0, 29.3, 35.3, 41.3, 47.1, and 49.3 cm are measured for $M$ values of 20.0, 40.0, 50.0, 60.0, 70.0, and 80.0 g, respectively. Construct a graph of $Mg$ versus $x$, and perform a linear least-squares fit to the data. From the slope of your graph, determine a value for $k$ for this spring. (b) The system is now set into simple harmonic motion, and periods are measured with a stopwatch. With $M = 80.0$ g, the total time for ten oscillations is measured to be 13.41 s. The experiment is re-

peated with $M$ values of 70.0, 60.0, 50.0, 40.0, and 20.0 g, with corresponding times for ten oscillations of 12.52, 11.67, 10.67, 9.62, and 7.03 s. Compute the experimental value for $T$ from each of these measurements. Plot a graph of $T^2$ versus $M$ and determine a value for $k$ from the slope of the linear least-squares fit through the data points. Compare this value of $k$ with the value obtained in part (a). (c) Obtain a value for $m_s$ from your graph and compare it with the given value of 7.40 g.

51. Imagine that a hole is drilled through the center of the Earth to the other side. It can be shown that an object of mass $m$ at a distance $r$ from the center of the Earth is pulled toward the center of the Earth only by the mass within the sphere of radius $r$ (the reddish region in Fig. P12.51). (a) Write Newton's law of gravitation for an object at the distance $r$ from the center of the Earth, and show that the force on it is of Hooke's law form, $F = -kr$, where the effective force constant is $k = (4/3)\pi\rho Gm$. Here $\rho$ is the density of the Earth, assumed uniform, and $G$ is the gravitational constant. (b) Show that a sack of mail dropped into the hole will execute simple harmonic motion if it moves without friction. How long does it take to arrive at the other side of the Earth?

**Figure P12.51**

52. An object is hung from a spring, set into vertical vibration, and immersed in a beaker of oil. Its motion is graphed in Figure 12.13. Assume the object has mass $m = 375$ g, the spring has force constant $k = 100$ N/m, and $b = 0.100$ N·s/m. (a) How long does it take for the amplitude to drop to half its initial value? (b) How long does it take for the mechanical energy to drop to half its initial value? (c) Show that, in general, the fractional rate at which the amplitude decreases in a damped harmonic oscillator is one-half the fractional rate at which the mechanical energy decreases.

53. A block of mass $m$ is connected to two springs of force constants $k_1$ and $k_2$ as shown in Figures P12.53a and P12.53b. In each case, the block moves on a frictionless table after it is displaced from equilibrium and released.

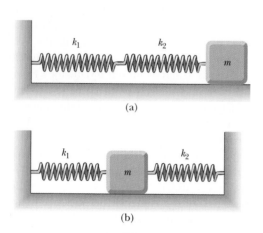

(a)

(b)

**Figure P12.53**

Show that in the two cases the block exhibits simple harmonic motion with periods

(a) $T = 2\pi\sqrt{\dfrac{m(k_1 + k_2)}{k_1 k_2}}$

(b) $T = 2\pi\sqrt{\dfrac{m}{k_1 + k_2}}$

54. Consider a small object on a light rigid rod, forming a simple pendulum of length $L = 1.20$ m. It is displaced from the vertical by an angle $\theta_{max}$ and then released. Predict the subsequent angular positions if $\theta_{max}$ is small and also if it is large. Set up and carry out a numerical method to integrate the equation of motion for the simple pendulum:

$$\frac{d^2\theta}{dt^2} = -\frac{g}{L}\sin\theta$$

Take the initial conditions to be $\theta = \theta_{max}$ and $d\theta/dt = 0$ at $t = 0$. On one trial choose $\theta_{max} = 5.00°$, and on another trial take $\theta_{max} = 100°$. In each case find the angular position $\theta$ as a function of time. Using the same values of $\theta_{max}$, compare your results for $\theta$ with those obtained from the expression $\theta_{max}\cos\omega t$. How does the period for the large value of $\theta_{max}$ compare with that for the small value of $\theta_{max}$? *Note:* Using the Euler method to solve this differential equation, you may find that the amplitude tends to increase with time. If you choose $\Delta t$ small enough, however, the solution that you obtain using Euler's method can still be good.

55. Your thumb squeaks on a plate you have just washed. A hinge creaks. A wheel squeaks. A corkscrew squeals inside the cork. Your sneakers squeak on the gym floor. Car tires squeal when you stop or start abruptly. You can make a goblet "sing" by wiping your moistened finger around its rim. Mortise joints groan in an old table or house. The

**Figure P12.55**

ground shudders in an earthquake. As you slide it across a table, a Styrofoam cup may not make much sound, but it makes the surface of some water inside it dance in a complicated resonance vibration.

As these examples suggest, vibration commonly results when friction acts on a moving elastic object. The oscillation is not simple harmonic motion, but is called *stick-and-slip*. When chalk squeaks on a blackboard, you can see that it makes a chain of regularly spaced dashes. This problem models stick-and-slip motion. A block of mass $m$ is attached to a fixed support by a horizontal spring with force constant $k$ and negligible mass (Fig. P12.55). Hooke's law describes the spring both in extension and in compression. The block sits on a long horizontal board, with which it has a coefficient of static friction $\mu_s$ and a smaller coefficient of static friction $\mu_k$. The board moves to the right at constant speed $v$. Assume that the block spends most of its time sticking to the board and moving to the right, so that the speed $v$ is small in comparison to the average speed of the block as it slips back to the left. (a) Show that the maximum extension of the spring from its unstressed position is very nearly given by $\mu_s mg/k$. (b) Show that the block oscillates around an equilibrium position at which the spring is stretched by $\mu_k mg/k$. (c) Graph the block's position versus time. (d) Show that the amplitude of the block's motion is

$$A = \frac{(\mu_s - \mu_k)mg}{k}$$

(e) Show that the period of the block's motion is

$$T = \frac{2(\mu_s - \mu_k)mg}{vk} + \pi\sqrt{\frac{m}{k}}$$

(f) Evaluate the frequency of the motion if $\mu_s = 0.400$, $\mu_k = 0.250$, $m = 0.300$ kg, $k = 12.0$ N/m, and $v = 2.40$ cm/s. (g) What happens to the frequency if the mass increases? (h) If the spring stiffness increases? (i) If the speed of the board increases? (j) If the coefficient of static

friction increases relative to the coefficient of kinetic friction? Observe that the *difference* between static and kinetic friction is important for the vibration. A squeaky wheel is greased because even a viscous fluid cannot exert a force of static friction.

## ANSWERS TO QUICK QUIZZES

**12.1** (d). In moving from its initial position to the equilibrium position, the block moves through distance $A$. It then continues to the other side of the equilibrium position, stopping after moving through another distance $A$, for a total so far of $2A$. It then returns to the initial position, traveling through another distance of $2A$, for a total distance in the full cycle of $4A$.

**12.2** $x = -\dfrac{v_i}{\omega} \cos\left(\omega t - \dfrac{\pi}{2}\right)$

**12.3** The bouncing ball is not an example of simple harmonic motion. The position of the ball is not described by a sinusoidal function. The daily movement of a student is also not simple harmonic motion because the student stays at a fixed position—school—for a long period. If this motion were sinusoidal, the student would move more and more slowly as she approached her desk, and, as soon as she sat down at the desk, she would start to move back toward home again!

**12.4** According to the principle of equivalence (Section 9.9), an acceleration of the elevator is equivalent to a gravitational field. Thus, if the elevator is (a) accelerating upward, this is equivalent to an increased effective gravitational field magnitude $g$, and, according to Equation 12.25, the period will decrease. Similarly, if the elevator is (b) accelerating downward, the effective value of $g$ is reduced and the period increases. (c) If the elevator moves with constant velocity, the period of the pendulum is the same as that in the stationary elevator.

**12.5** (a). With a longer length, the period of the pendulum increases. Thus, it takes longer to execute each swing, so

that each second according to the clock takes longer than an actual second. Thus, the clock runs *slow*.

**12.6** The first student is correct. Although changing the mass of a simple pendulum does not change the frequency, the fact that the bob is large means that we must model the pendulum as a physical pendulum rather than a simple pendulum. When the gum is placed on top of the bob, the moment of inertia of the physical pendulum is altered slightly. According to Equation 12.27, this alters the period of the pendulum.

**12.7** The period of oscillation $T$ of the swinging belt can be measured by watching the television coverage. Furthermore, the belt can be geometrically modeled as a long thin rod, so that its moment of inertia can be approximated as $\frac{1}{3}mL^2$ from Table 10.2. Finally, because the belt is assumed to be of uniform density, the center of mass is located at $d = L/2$. Thus, Equation 12.27 becomes

$$T = 2\pi \sqrt{\frac{\frac{1}{3}mL^2}{mg(L/2)}} = 2\pi \sqrt{\frac{2L}{3g}}$$

Upon estimating the length of the swinging belt by comparing it to the height of a typical astronaut this equation can be solved for $g$.

**12.8** If the period of the alternating release of air were to match the swinging period of the parachutist, a resonance response could occur. The swinging of the parachutist could become so energetic that the parachute lines could become fouled and the braking action would disappear.

# chapter
# 13

Drops of water fall from a leaf into a pond. The disturbance caused by the falling water moves away from the drop point as circular ripples on the water's surface.

*(Don Bonsey/Stone)*

# Mechanical Waves

**M**ost of us have experienced waves as children, when we dropped pebbles into a pond. The disturbance created by a pebble manifested itself as ripples that moved outward from the point at which the pebble landed in the water, like the ripples due to the falling water drops in the opening photograph. If you were to carefully examine the motion of a leaf floating near the point where the pebble entered the water, you would see that the leaf moves up and down and back and forth about its original position but does not undergo any net displacement away from or toward the source of the disturbance. The disturbance in the water moves from one place to another, *yet the water is not carried with it*—this is the essence of wave motion.

The world is full of other kinds of waves, including sound waves, waves on strings, seismic waves, radio waves, and x-rays. Most waves can be placed in one of two categories. **Mechanical waves** are waves that disturb and propagate through a medium; the ripple in the water due to the pebble and a sound wave, for which air is the medium, are examples of mechanical waves. **Electromagnetic waves** are a special class of waves that do not require a medium in order to propagate, as discussed with regard to the absence of the ether in Section 9.2; light waves and radio waves are two familiar examples. In this chapter we shall confine our attention to the study of mechanical waves, deferring our study of electromagnetic waves to Chapter 24.

The wave concept is abstract. When we observe a water wave, what we see is a rearrangement of the water's surface. Without the water, there would be no wave. In the case of this or any other mechanical wave, what we interpret as a wave corresponds to the disturbance of a medium. Therefore, we can consider a mechanical wave to be the *propagation of a disturbance in a medium.*

## 13.1 • PROPAGATION OF A DISTURBANCE

In the introduction, we alluded to the essence of wave motion—the transfer of a *disturbance* through space without the accompanying transfer of *matter*. The propagation of the disturbance also represents a transfer of energy—thus, we can view waves as a means of energy transfer. In the list of energy transfer mechanisms in Section 6.6, we see two entries that depend on waves: mechanical waves and electromagnetic radiation. These are to be contrasted with another entry—matter transfer—in which the energy transfer is accompanied by a movement of matter through space.

**All mechanical waves require (1) some source of disturbance, (2) a medium that can be disturbed, and (3) some physical mechanism through which particles of the medium can influence one another.** All waves carry energy, but the amount of energy transmitted through a medium and the mechanism responsible for the energy transport differ from case to case. For instance, the power of ocean waves during a storm is much greater than that of sound waves generated by a musical instrument.

One way to demonstrate wave motion is to flip the free end of a long rope that is under tension and has its opposite end fixed, as in Figure 13.1. In this manner, a single **pulse** is formed and travels (to the right in Fig. 13.1) with a definite speed. The rope is the medium through which the pulse travels. Figure 13.1 represents consecutive "snapshots" of the traveling pulse. The shape of the pulse changes very little as it travels along the rope.

 See the *Core Concepts in Physics CD-ROM*, Screens 8.4 & 8.5

**Figure 13.1**

A wave pulse traveling down a stretched rope. The shape of the pulse is approximately unchanged as it travels along the rope.

**Figure 13.2**

**Figure 13.2**

A pulse traveling on a stretched rope is a transverse wave. That is, any element *P* on the rope moves *(blue arrows)* in a direction perpendicular to the propagation of the wave *(red arrows)*.

**As the pulse travels, each rope segment that is disturbed moves in a direction perpendicular to the direction of propagation.** Figure 13.2 illustrates this point for a particular segment, labeled *P*. Note that no motion occurs in any part of the rope in the direction of the wave. A disturbance such as this, in which the particles of the disturbed medium move perpendicularly to the direction of propagation, is called a **transverse wave.**

In another class of waves, called **longitudinal waves,** the particles of the medium undergo displacements *parallel* to the direction of propagation. Sound waves in air, for instance, are longitudinal. Their disturbance corresponds to a series of high- and low-pressure regions that may travel through air or through any material medium with a certain speed. A longitudinal pulse can be easily produced in a stretched spring, as in Figure 13.3. A group of coils at the free end is pushed forward and pulled back. This action produces a pulse in the form of a compressed region of the coil that travels along the spring, parallel to the direction of propagation.

So far we have provided pictorial representations of a traveling pulse and hope you have begun to develop a mental representation of such a pulse. Let us now develop a mathematical representation for the propagation of this pulse. Consider a pulse traveling to the right with constant speed *v* on a long, stretched string, as in Figure 13.4. The pulse moves along the *x* axis (the axis of the string), and the transverse (up-and-down) displacement of the string is measured with the position coordinate *y*.

Compressed          Compressed

Stretched          Stretched

**Figure 13.3**

A longitudinal pulse along a stretched spring. The displacement of the coils is in the direction of the wave motion. For the starting motion described in the text, the compressed region is followed by a stretched region.

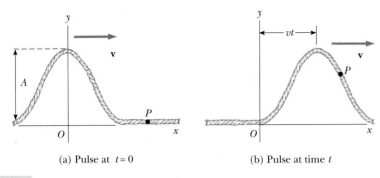

(a) Pulse at $t = 0$    (b) Pulse at time $t$

**Figure 13.4**

A one-dimensional wave pulse traveling to the right with a speed $v$. (a) At $t = 0$, the shape of the pulse is given by $y = f(x)$. (b) At some later time $t$, the shape remains unchanged and the vertical displacement of any point $P$ of the medium is given by $y = f(x - vt)$.

Figure 13.4a represents the shape and position of the pulse at time $t = 0$. At this time, the shape of the pulse, whatever it may be, can be represented by some mathematical function that we will write as $y(x, 0) = f(x)$. This function describes the vertical position $y$ of the element of the string located at each value of $x$ at time $t = 0$. Because the speed of the pulse is $v$, the pulse has traveled to the right a distance $vt$ at the time $t$ (Fig. 13.4b). We adopt a simplification model in which the shape of the pulse does not change with time.* Thus, at time $t$, the shape of the pulse is the same as it was at time $t = 0$, as in Figure 13.4a. Consequently, an element of the string at $x$ at this time has the same $y$ position as an element located at $x - vt$ had at time $t = 0$:

$$y(x, t) = y(x - vt, 0)$$

In general, then, we can represent the displacement $y$ for all positions and times, measured in a stationary frame with the origin at $O$, as

$$y(x, t) = f(x - vt) \qquad [13.1]$$

• *Pulse traveling to the right*

Similarly, if the wave pulse travels to the left, the displacement of the string is

$$y(x, t) = f(x + vt) \qquad [13.2]$$

• *Pulse traveling to the left*

The function $y$, sometimes called the **wave function,** depends on the two variables $x$ and $t$. For this reason, it is often written $y(x, t)$, which is read "$y$ as a function of $x$ and $t$."

It is important to understand the meaning of $y$. Consider a point $P$ on the string, identified by a particular value of its $x$ coordinate. As the pulse passes through $P$, the $y$ coordinate of this point increases, reaches a maximum, and then decreases to zero. **The wave function $y(x, t)$ represents the $y$ coordinate of any point $P$ located at position $x$ at any time $t$.** Furthermore, if $t$ is fixed (e.g., in the case of taking a snapshot of the pulse), then the wave function $y$ as a function

---

* In reality, the pulse changes its shape and gradually spreads out during the motion. This effect is called *dispersion* and is common to many mechanical waves; however we adopt a simplification model that ignores this effect.

of *x*, sometimes called the **waveform,** defines a curve representing the actual geometric shape of the pulse at that time.

### Quick Quiz 13.1

(a) A long line of people is waiting to buy tickets at a movie theater. When the first person leaves, a pulse of motion occurs as people step forward to fill in the gap. As each person steps forward, the gap moves through the line. Is the propagation of this gap transverse or longitudinal? (b) Consider "the wave" at a soccer or American football game: People stand up and shout as the wave arrives at their location, and this pulse moves around the stadium. Is this wave transverse or longitudinal?

### THINKING PHYSICS 13.1

Why is it important to be quiet in avalanche country? In the movie *On Her Majesty's Secret Service* (United Artists, 1969), the bad guys try to stop James Bond, who is escaping on skis, by firing a gun and causing an avalanche. Why did this happen?

**Reasoning**  The essence of a wave is the propagation of a *disturbance* through a medium. An impulsive sound, like the gunshot in the James Bond movie, can cause an acoustic disturbance that propagates through the air and can impact a ledge of unstable snow that is just ready to break free to begin an avalanche. Such a disastrous event occurred in 1916 during World War I, when Austrian soldiers in the Alps were smothered by an avalanche caused by cannon fire.

### Example 13.1    A Pulse Moving to the Right

A wave pulse moving to the right along the *x* axis is represented by the wave function

$$y(x, t) = \frac{2.0}{(x - 3.0t)^2 + 1}$$

where *x* and *y* are measured in centimeters and *t* is in seconds. Let us plot the waveform at *t* = 0, *t* = 1.0 s, and *t* = 2.0 s.

**Solution**  First, note that this function is of the form *y* = *f*(*x* − *vt*). By inspection, we see that the speed of the wave is *v* = 3.0 cm/s. The location of the peak of the pulse occurs at the value of *x* for which the denominator is a minimum, that is, where (*x* − 3.0*t*) = 0. Thus, the peaks occur at *x* = 0.0 cm at *t* = 0, at *x* = 3.0 cm at *t* = 1.0 s, and *x* = 6.0 cm at *t* = 2.0 s.

At times *t* = 0, *t* = 1.0 s, and *t* = 2.0 s, the wave function expressions are

$$y(x, 0) = \frac{2.0}{x^2 + 1} \quad \text{at } t = 0$$

$$y(x, 1.0) = \frac{2.0}{(x - 3.0)^2 + 1} \quad \text{at } t = 1.0 \text{ s}$$

$$y(x, 2.0) = \frac{2.0}{(x - 6.0)^2 + 1} \quad \text{at } t = 2.0 \text{ s}$$

We can now use these expressions to plot the wave function versus *x* at these times. For example, let us evaluate *y*(*x*, 0) at *x* = 0.50 cm:

$$y(0.50, 0) = \frac{2.0}{(0.50)^2 + 1} = 1.6 \text{ cm}$$

Likewise, *y*(1.0, 0) = 1.0 cm, *y*(2.0, 0) = 0.40 cm, and so on. A continuation of this procedure for other values of *x* yields the waveform shown in Figure 13.5a. In a similar manner, one obtains the graphs of *y*(*x*, 1.0) and *y*(*x*, 2.0), shown in Figures 13.5b and 13.5c, respectively. These snapshots show that the wave pulse moves to the right without changing its shape and has a constant speed of 3.0 cm/s.

(a)

(b)

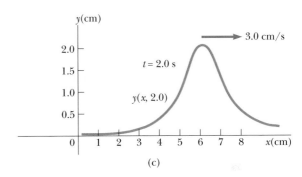

(c)

**Figure 13.5**

(Example 13.1) Graphs of the function $y(x, t) = 2.0/[(x - 3.0t)^2 + 1]$ at (a) $t = 0$, (b) $t = 1.0$ s, and (c) $t = 2.0$ s.

## 13.2 • THE WAVE MODEL

We have discussed creating a disturbance moving through a medium such as a stretched string by a simple up-and-down displacement of the end of the string. This results in a pulse moving along the medium. A continuous wave is created by shaking the end of the string in simple harmonic motion, which we studied in Chapter 12. If we do this, the string will take on the shape shown by the curve in the graph in Figure 13.6a, with this shape remaining the same but moving toward the right. This is what we call a **sinusoidal** wave because the shape in Figure 13.6 is that of a sine wave. The point with the largest positive displacement of the string is called the **crest** of the wave. The lowest point is called the **trough.** The crest and trough move along with the wave, so that a particular point on the string alternates between locations on a crest and a trough. In idealized wave motion in an idealized medium, each particle of the medium undergoes simple harmonic motion around its equilibrium position.

Three physical characteristics are important in describing a sinusoidal wave: **wavelength, frequency,** and **wave speed. One wavelength is the minimum distance between any two identical points on a wave**—for example, adjacent crests or adjacent troughs, as in Figure 13.6a, which is a graph of displacement versus position for a sinusoidal wave at a specific time. The symbol $\lambda$ (Greek lambda) is used to denote wavelength.

**The frequency of sinusoidal waves is the same as the frequency of simple harmonic motion of a particle of the medium.** The **period** $T$ of the wave is the minimum time it takes a particle of the medium to undergo one complete

(a)

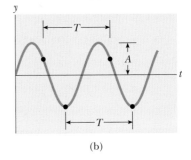

(b)

**Figure 13.6**

(a) The wavelength $\lambda$ of a wave is the distance between adjacent crests or adjacent troughs. (b) The period $T$ of a wave is the time it takes the wave to travel one wavelength.

oscillation, and is equal to the inverse of the frequency:

- *Period of a wave*

$$T = \frac{1}{f}$$

Figure 13.6b shows position versus time for a particle of the medium as a sinusoidal wave is passing through its position. The period is the time between instants when the particle has identical displacements and velocities.

Waves travel through the medium with a specific **wave speed,** which depends on the properties of the medium being disturbed. For instance, sound waves travel through air at 20°C with a speed of about 343 m/s, whereas the speed of sound in most solids is higher than 343 m/s. We will learn more about wavelength, frequency, and wave speed in the next section.

Another important parameter for the wave in Figure 13.6 is the **amplitude** of the wave. This is the maximum displacement of a particle of the medium from the equilibrium position. It is denoted by $A$ and is the same as the amplitude of the simple harmonic motion of the particles of the medium.

One method of producing a traveling sinusoidal wave on a very long string is shown in Figure 13.7. One end of the string is connected to a blade that is set vibrating. As the blade oscillates vertically with simple harmonic motion, a traveling wave moving to the right is set up on the string. Figure 13.7 represents snapshots of the wave at intervals of one quarter of a period. **Each particle of the string, such as $P$, oscillates vertically in the $y$ direction with simple harmonic motion.** Every segment of the string can therefore be treated as a simple harmonic oscillator vibrating with a frequency equal to the frequency of vibration of the blade that drives the string. Although each segment oscillates in the $y$ direction, the wave (or disturbance) travels in the $x$ direction with a speed $v$. Of course, this

**PITFALL PREVENTION 13.1**

**What's the difference between Figure 13.6a and 13.6b?**

Notice the visual similarity between Figures 13.6a and 13.6b. The shapes are the same, but one (a) is a graph of vertical position versus horizontal position and the other (b) is vertical position versus time. Figure 13.6a is a pictorial representation of the wave *for a series of particles of the medium*—this is what you would see at an instant of time. Figure 13.6b is a graphical representation of the position of *one particle of the medium* as a function of time. The fact that both figures have the identical shape represents Equation 13.1—a wave is the *same* function of both $x$ and $t$.

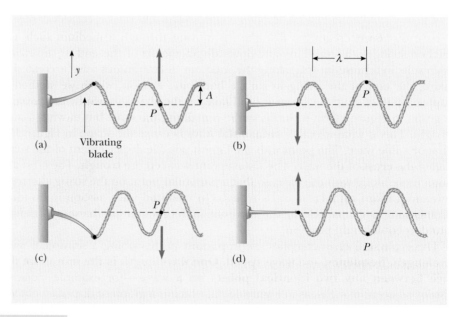

**Figure 13.7**

One method for producing a sinusoidal wave on a continuous string. The left end of the string is connected to a blade that is set into vibration. Every segment of the string, such as the point $P$, oscillates with simple harmonic motion in the vertical direction.

is the definition of a transverse wave. In this case, the energy carried by the traveling wave is supplied by the vibrating blade.

In the early chapters of this book, we developed several analysis models based on the particle model. With our introduction to waves, we can develop a new simplification model, the **wave model,** which will allow us to explore more analysis models for solving problems. An ideal particle has zero size. We can build physical objects with nonzero size as combinations of particles. Thus, the particle can be considered a basic building block. An ideal wave has a single frequency and is infinitely long; that is, the wave exists throughout the Universe. (It is beyond the mathematical scope of this text at this point to prove this, but a wave of finite length must necessarily have a mixture of frequencies.) We will find that we can combine ideal waves, just as we combined particles, and thus the ideal wave can be considered as a basic building block.

The wave model starts with an ideal wave having a single frequency, wavelength, wave speed, and amplitude. From this beginning, we describe waves in a variety of situations that serve as analysis models to help us solve problems. In the next section, we further develop our mathematical representation for the wave model.

## 13.3 • THE TRAVELING WAVE

Let us investigate further the mathematics of a sinusoidal wave (Fig. 13.8). The red curve represents a snapshot of a sinusoidal wave at $t = 0$, and the blue curve represents a snapshot of the wave at some later time $t$. In what follows, we will develop the principal features and mathematical representations of the model of a **traveling wave.** This analysis model is used in situations in which a wave moves through space without interacting with any other waves or particles.

At $t = 0$, the red curve in Figure 13.8 can be described mathematically as

$$y = A \sin\left(\frac{2\pi}{\lambda} x\right) \qquad \text{[13.3]}$$

where the amplitude $A$, as usual, represents the maximum value of the displacement, and $\lambda$ is the wavelength as defined in Figure 13.6a. Thus, we see that the value of $y$ is the same when $x$ is increased by an integral multiple of $\lambda$. If the wave moves to the right with a speed of $v$, the wave function at some later time $t$ is

$$y = A \sin\left[\frac{2\pi}{\lambda}(x - vt)\right] \qquad \text{[13.4]}$$

That is, the sinusoidal wave moves to the right a distance of $vt$ in the time $t$, as in Figure 13.8. Note that the wave function has the form $f(x - vt)$ and represents a wave traveling to the right. If the wave were traveling to the left, the quantity $x - vt$ would be replaced by $x + vt$, just as in the case of the traveling pulse described by Equations 13.1 and 13.2.

Because the period $T$ is the time it takes the wave to travel a distance of one wavelength, the speed, wavelength, and period are related by

$$v = \frac{\lambda}{T} \qquad \text{[13.5]}$$

Substituting Equation 13.5 into Equation 13.4, we find that

$$y = A \sin\left[2\pi\left(\frac{x}{\lambda} - \frac{t}{T}\right)\right] \qquad \text{[13.6]}$$

 See Screens 8.6, 8.8, & 8.9

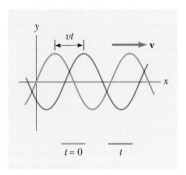

**Figure 13.8**

A one-dimensional sinusoidal wave traveling to the right with a speed $v$. The red curve represents a snapshot of the wave at $t = 0$, and the blue curve represents a snapshot at some later time $t$.

This form of the wave function shows the periodic nature of $y$ in both space and time. That is, at any given time $t$ (a snapshot of the wave), $y$ has the same value at the positions $x$, $x + \lambda$, $x + 2\lambda$, and so on. Furthermore, at any given position $x$ (at which a single particle of the medium is undergoing simple harmonic motion), the values of $y$ at times $t$, $t + T$, $t + 2T$, and so on, are the same.

We can express the sinusoidal wave function in a compact form by defining two other quantities: **angular wave number** $k$ (often called simply the **wave number**) and **angular frequency** $\omega$:

• *Angular wave number*

$$k \equiv \frac{2\pi}{\lambda} \qquad \text{[13.7]}$$

• *Angular frequency*

$$\omega \equiv \frac{2\pi}{T} = 2\pi f \qquad \text{[13.8]}$$

Note that in Equation 13.8, we use the definition of frequency, $f = 1/T$. Using these definitions, Equation 13.6 can be written in the more compact form

• *Wave function for a sinusoidal wave*

$$y = A \sin(kx - \omega t) \qquad \text{[13.9]}$$

We shall use this form most frequently.

Using Equations 13.7 and 13.8, we can express the wave speed $v$ (Eq. 13.5) in the alternative forms

$$v = \frac{\omega}{k} \qquad \text{[13.10]}$$

• *Speed of a traveling sinusoidal wave*

$$v = \lambda f \qquad \text{[13.11]}$$

The wave function given by Equation 13.9 assumes that the displacement $y$ is zero at $x = 0$ and $t = 0$. This need not be the case. If the transverse displacement is not zero at $x = 0$ and $t = 0$, we generally express the wave function in the form

$$y = A \sin(kx - \omega t + \phi) \qquad \text{[13.12]}$$

where $\phi$ is called the **phase constant** and can be determined from the initial conditions.

### Quick Quiz 13.2

Sound waves ($v = 343$ m/s) from a musical performance are encoded onto radio waves for transmission from the radio studio to your home receiver. A radio wave traveling at $c = 3.00 \times 10^8$ m/s (the symbol $c$, which we used in Chapter 9 for the speed of light, is also used for the speed of all electromagnetic waves traveling through a vacuum) with a wavelength of 3.00 m carries a sound wave that has a wavelength of 3.00 m in air. Which has the higher frequency, the radio wave or the sound wave?

## Example 13.2 A Traveling Sinusoidal Wave

A sinusoidal wave traveling in the positive $x$ direction has an amplitude of 15.0 cm, a wavelength of 40.0 cm, and a frequency of 8.00 Hz. The vertical displacement of the medium at $t = 0$ and $x = 0$ is also 15.0 cm, as shown in Figure 13.9. (a) Find the angular wave number, period, angular frequency, and speed of the wave.

**Solution** This is a simple problem in which we apply the traveling wave model. Using Equations 13.7, 13.8, and 13.11, we find the following:

$$k = \frac{2\pi}{\lambda} = \frac{2\pi \text{ rad}}{40.0 \text{ cm}} = \boxed{0.157 \text{ rad/cm}}$$

$$T = \frac{1}{f} = \frac{1}{8.00 \text{ s}^{-1}} = \boxed{0.125 \text{ s}}$$

$$\omega = 2\pi f = 2\pi(8.00 \text{ s}^{-1}) = \boxed{50.3 \text{ rad/s}}$$

$$v = f\lambda = (8.00 \text{ s}^{-1})(40.0 \text{ cm}) = \boxed{320 \text{ cm/s}}$$

(b) Determine the phase constant $\phi$, and write a general expression for the wave function.

**Solution** Because $A = 15.0$ cm and because it is given that $y = 15.0$ cm at $x = 0$ and $t = 0$, substitution into Equation 13.12 gives

$$15.0 = 15.0 \sin \phi \quad \text{or} \quad \sin \phi = 1$$

We see that $\phi = \boxed{\pi/2 \text{ rad}}$ (or 90°). Hence, the wave function is of the form

$$y = A \sin\left(kx - \omega t + \frac{\pi}{2}\right) = A \cos(kx - \omega t)$$

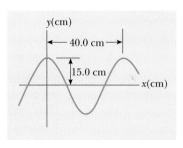

**Figure 13.9**

(Example 13.2) A sinusoidal wave of wavelength $\lambda = 40.0$ cm and amplitude $A = 15.0$ cm. The wave function can be written in the form $y = A \cos(kx - \omega t)$.

As we can see by inspection, the wave function must have this form because the cosine argument is displaced by 90° from the sine function. Substituting the values for $A$, $k$, and $\omega$ into this expression gives

$$y = \boxed{(15.0 \text{ cm}) \cos(0.157x - 50.3t)}$$

**EXERCISE** When a particular wire is vibrating with a frequency of 4.00 Hz, a transverse wave of wavelength 60.0 cm is produced. Determine the speed of wave pulses along the wire.

*Answer* 2.40 m/s

## The Linear Wave Equation

If the waveform at $t = 0$ is as described in Figure 13.7b, the wave function can be written

$$y = A \sin(kx - \omega t)$$

We can use this expression to describe the motion of any point on the string. The point $P$ (or any other point on the string) moves only vertically, and so its $x$ coordinate *remains constant*. The **transverse velocity** $v_y$ of the point $P$ and its **transverse acceleration** $a_y$ are therefore

$$v_y = \frac{dy}{dt}\bigg]_{x=\text{constant}} = \frac{\partial y}{\partial t} = -\omega A \cos(kx - \omega t) \qquad [13.13]$$

$$a_y = \frac{dv_y}{dt}\bigg]_{x=\text{constant}} = \frac{\partial v_y}{\partial t} = \frac{\partial^2 y}{\partial t^2} = -\omega^2 A \sin(kx - \omega t) \qquad [13.14]$$

**PITFALL PREVENTION 13.2**

**Two kinds of speed/velocity**

Be sure to differentiate between $v$, the speed of the wave as it propagates through the medium, and $v_y$, the transverse velocity of a point on the string. The speed $v$ is constant, whereas $v_y$ varies sinusoidally.

The maximum values of these quantities are simply the absolute values of the coefficients of the cosine and sine functions:

$$v_{y, \text{max}} = \omega A \qquad \text{[13.15]}$$

$$a_{y, \text{max}} = \omega^2 A \qquad \text{[13.16]}$$

You should recognize that the transverse velocity and transverse acceleration of any point on the string do not reach their maximum values simultaneously. In fact, the transverse velocity reaches its maximum value ($\omega A$) when the displacement $y = 0$, whereas the transverse acceleration reaches its maximum magnitude ($\omega^2 A$) when $y = \pm A$. Finally, Equations 13.15 and 13.16 are identical to the corresponding equations for simple harmonic motion (Eq. 12.17 and Eq. 12.18).

Let us take derivatives of our wave function with respect to position at a fixed time, similar to the process by which we took derivatives with respect to time in Equations 13.13 and 13.14:

$$\frac{dy}{dx}\bigg]_{t=\text{constant}} = \frac{\partial y}{\partial x} = -kA \cos(kx - \omega t) \qquad \text{[13.17]}$$

$$\frac{d^2 y}{dx^2}\bigg]_{t=\text{constant}} = \frac{\partial^2 y}{\partial x^2} = -k^2 A \sin(kx - \omega t) \qquad \text{[13.18]}$$

Comparing Equations 13.14 and 13.18, we see that

$$A \sin(kx - \omega t) = -\frac{1}{k^2}\frac{\partial^2 y}{\partial x^2} = -\frac{1}{\omega^2}\frac{\partial^2 y}{\partial t^2} \quad \longrightarrow \quad \frac{\partial^2 y}{\partial x^2} = \frac{k^2}{\omega^2}\frac{\partial^2 y}{\partial t^2}$$

Using Equation 13.10, we can rewrite this as

- *Linear wave equation*

$$\frac{\partial^2 y}{\partial x^2} = \frac{1}{v^2}\frac{\partial^2 y}{\partial t^2} \qquad \text{[13.19]}$$

This is known as the **linear wave equation.** If we analyze a situation and find this kind of relationship between derivatives of a function describing the situation, then wave motion is occurring. Equation 13.19 is a differential equation representation of the traveling wave model. The solutions to the equation describe **linear mechanical waves.** We have developed the linear wave equation from a sinusoidal mechanical wave traveling through a medium, but it is much more general. The linear wave equation successfully describes waves on strings, sound waves, and also electromagnetic waves.* What's more, although the sinusoidal wave that we have studied is a solution to Equation 13.19, the general solution to the equation is *any* function of the form $y(x, t) = f(x \pm vt)$, as discussed in Section 13.1.

## Example 13.3   A Solution to the Linear Wave Equation

Verify that the wave function presented in Example 13.1 is a solution to the linear wave equation.

**Solution**   The wave function is

$$y(x, t) = \frac{2.0}{(x - 3.0t)^2 + 1}$$

By taking partial derivatives of this function with respect to $x$ and to $t$, we find

$$\frac{\partial^2 y}{\partial x^2} = \frac{12(x - 3.0t)^2 - 4.0}{[(x - 3.0t)^2 + 1]^3}$$

$$\frac{\partial^2 y}{\partial t^2} = \frac{108(x - 3.0t)^2 - 36}{[(x - 3.0t)^2 + 1]^3}$$

---

* In the case of electromagnetic waves, $y$ is interpreted to represent an electric field, which we will study in Chapter 24.

Comparing these two expressions, we see that

$$\frac{\partial^2 y}{\partial x^2} = \frac{1}{9.0} \frac{\partial^2 y}{\partial t^2}$$

Comparing this with Equation 13.19, we see that the wave function is a solution to the linear wave equation if the speed

at which the pulse moves is 3.0 cm/s. We have already determined in Example 13.1 that this is indeed the speed of the pulse, so we have proven what we set out to do.

## 13.4 • THE SPEED OF TRANSVERSE WAVES ON STRINGS

 See Screen 9.2

An aspect of the behavior of linear mechanical waves is that the **wave speed depends only on the properties of the medium through which the wave travels.** Waves for which the amplitude $A$ is small relative to the wavelength $\lambda$ are well represented as linear waves. In this section we determine the speed of a transverse wave traveling on a stretched string.

Let us use a mechanical analysis to derive the expression for the speed of a pulse traveling on a stretched string under tension $T$. Consider a pulse moving to the right with a uniform speed $v$, measured relative to a stationary (with respect to the Earth) inertial reference frame. Recall from Chapter 9 that Newton's laws are valid in any inertial reference frame. Thus, let us view this pulse from a different inertial reference frame—one that moves along with the pulse at the same speed, so that the pulse appears to be at rest in the frame, as in Figure 13.10a. In this reference frame, the pulse remains fixed and each point on the string moves to the left through the pulse shape.

A short segment of the string, of length $\Delta s$, forms the approximate arc of a circle of radius $R$, as shown in Figure 13.10a and magnified in Figure 13.10b. We use a simplification model in which this is an arc of a perfect circle. In our moving frame of reference, the segment of the string moves to the left with the speed $v$ through the arc. As it travels through the arc, we can model the segment as a particle in uniform circular motion. This segment has a centripetal acceleration of $v^2/R$, which is supplied by components of the force **T** in the string at each end of the segment. The force **T** acts on each side of the segment, tangent to the arc, as

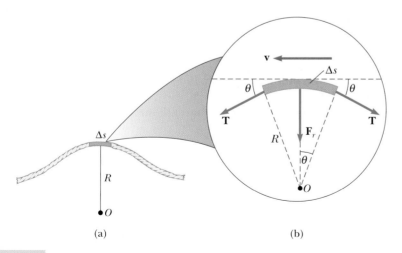

(a)     (b)

**Figure 13.10**

(a) To obtain the speed $v$ of a wave on a stretched string, it is convenient to describe the motion of a small segment of the string in a moving frame of reference. (b) The net force on a small segment of length $\Delta s$ is in the radial direction. The horizontal components of the tension force cancel.

in Figure 13.10b. The horizontal components of **T** cancel, and each vertical component $T \sin \theta$ acts radially inward toward the center of the arc. Hence, the magnitude of the total radial force is $2T \sin \theta$. Because the segment is small, $\theta$ is small and we can use the small-angle approximation $\sin \theta \approx \theta$. Therefore, the magnitude of the total radial force can be expressed as

$$F_r = 2T \sin \theta \approx 2T\theta$$

The segment has mass $m = \mu \, \Delta s$, where $\mu$ is the mass per unit length of the string. Because the segment forms part of a circle and subtends an angle of $2\theta$ at the center, $\Delta s = R(2\theta)$, and hence

$$m = \mu \, \Delta s = 2\mu R\theta$$

The radial component of Newton's second law applied to the segment gives

$$F_r = \frac{mv^2}{R} \quad \longrightarrow \quad 2T\theta = \frac{2\mu R\theta v^2}{R} \quad \longrightarrow \quad T = \mu v^2$$

where $F_r$ is the force that supplies the centripetal acceleration of the segment. Solving for $v$ gives

- *Speed of a wave on a stretched string*

$$v = \sqrt{\frac{T}{\mu}} \qquad \text{[13.20]}$$

Notice that this derivation is based on the linear wave assumption that the pulse height is small relative to the length of the string. Using this assumption, we were able to use the approximation $\sin \theta \approx \theta$. Furthermore, the model assumes that the tension $T$ is not affected by the presence of the pulse, so that $T$ is the same at all points on the string. Finally, this proof does *not* assume any particular shape for the pulse. We therefore conclude that a pulse of *any shape* will travel on the string with speed $v = \sqrt{T/\mu}$, without changing its shape.

### Quick Quiz 13.3

Suppose you create a pulse by moving the free end of a taut string up and down once with your hand. The string is attached at its other end to a distant wall. The pulse reaches the wall in a time interval $\Delta t$. Which of the following actions, taken by itself, decreases the time interval required for the pulse to reach the wall? More than one choice may be correct.

- (a) Moving your hand more quickly, but still only up and down once by the same amount
- (b) Moving your hand more slowly, but still only up and down once by the same amount
- (c) Moving your hand a greater distance up and down in the same amount of time
- (d) Moving your hand a smaller distance up and down in the same amount of time
- (e) Using a heavier string of the same length and under the same tension
- (f) Using a lighter string of the same length and under the same tension
- (g) Using a string of the same linear mass density but under decreased tension
- (h) Using a string of the same linear mass density but under increased tension

## Quick Quiz 13.4

In mechanics, a simplification model of massless strings is often adopted. Why is this not a good model when discussing waves on strings?

### THINKING PHYSICS 13.2

A secret agent is trapped in a building on top of an elevator car at a lower floor. He attempts to signal a fellow agent on the roof by tapping a message in Morse code on the elevator cable, so that transverse pulses move upward on the cable. As the pulses move up the cable toward the accomplice, does the speed with which they move stay the same, increase, or decrease? If the pulses are sent 1 s apart, are they received 1 s apart by his partner?

**Reasoning**  The elevator cable can be modeled as a string. The speed of waves on the cable is a function of the tension in the cable. As the waves move higher on the cable, they encounter increased tension because each higher point on the cable must support the weight of all the cable below it (and the elevator). Thus, the speed of the pulses increases as they move higher on the cable. The frequency of the pulses will not be affected because each pulse takes the same total time to reach the top—they will still arrive at the top of the cable at intervals of 1 s.

### Example 13.4   The Speed of a Pulse on a Cord

A uniform cord has a mass of 0.300 kg and a total length of 6.00 m. Tension is maintained in the cord by suspending an object of mass 2.00 kg from one end (Fig. 13.11). Find the speed of a pulse on this cord. Assume that the tension is not affected by the mass of the cord.

**Solution**  Because the suspended object can be modeled as a particle in equilibrium, the tension $T$ in the cord is equal to the weight of the suspended 2.00-kg object:

$$T = mg = (2.00 \text{ kg})(9.80 \text{ m/s}^2) = 19.6 \text{ N}$$

(This calculation of the tension neglects the small mass of the cord. Strictly speaking, the horizontal portion of the cord can never be exactly straight—it will sag slightly—and therefore the tension is not uniform.)

The mass per unit length $\mu$ is

$$\mu = \frac{m}{\ell} = \frac{0.300 \text{ kg}}{6.00 \text{ m}} = 0.0500 \text{ kg/m}$$

Therefore, the wave speed is

$$v = \sqrt{\frac{T}{\mu}} = \sqrt{\frac{19.6 \text{ N}}{0.0500 \text{ kg/m}}} = \boxed{19.8 \text{ m/s}}$$

**Figure 13.11**

(Example 13.4) The tension $T$ in the cord is maintained by the suspended mass. The wave speed is given by the expression $v = \sqrt{T/\mu}$.

**EXERCISE**  Transverse waves travel with a speed of 20.0 m/s in a string under a tension of 6.00 N. What tension is required for a wave speed of 30.0 m/s in the same string?

*Answer*   13.5 N

See Screen 9.3

Incident
pulse

(a)

(b)

(c)

(d)

(e)    Reflected
pulse

**Figure 13.12**

The reflection of a traveling wave
pulse at the fixed end of a stretched
string. The reflected pulse is inverted,
but its shape remains the same.

Incident
pulse

(a)

(b)

(c)

Reflected
pulse

(d)

**Figure 13.13**

The reflection of a traveling wave
pulse at the free end of a stretched
string. In this case, the reflected pulse
is not inverted.

## 13.5 • REFLECTION AND TRANSMISSION OF WAVES

So far, we have only considered a wave traveling through a medium with no
changes in the medium and no interactions with anything other than the particles
of the medium—this is the traveling wave model. This situation is similar to a particle traveling through empty space and obeying Newton's first law. Although these
situations demonstrate important physics, things become more interesting when
particles and waves interact with something. Let us see what happens when a wave
encounters a boundary between two media.

For simplicity, consider a single pulse once again. When a traveling pulse
reaches a boundary, part or all of the pulse is *reflected*. Any part not reflected is said
to be *transmitted* through the boundary. Suppose a pulse travels on a string that is
fixed at one end (Fig. 13.12). An upward pulse is initiated by grabbing the end of
the string and snapping it upward and then back to the original position. When
the pulse reaches the fixed boundary, it is reflected. In the simplification model in
which the support attaching the string to the wall is rigid, none of the pulse is
transmitted through the wall.

Note that the reflected pulse (Figs. 13.12d and e) has exactly the same amplitude as the incoming pulse but is inverted. The inversion can be explained as follows. The pulse is created initially with an upward and then downward force on
the free end of the string. As the pulse arrives at the fixed end of the string, the
string first produces an upward force on the support. By Newton's third law, the
support exerts an equal and opposite reaction force on the string. Thus, the positive shape of the pulse results in a downward and then upward force on the string
as the entirety of the pulse encounters the rigid end. This is equivalent to a person
replacing the fixed support and applying a downward and then an upward force to
the string. Thus, reflection at a rigid end causes the pulse to invert on reflection.

Now consider a second idealized situation in which reflection is total and
transmission is zero. In this simplification model, the pulse arrives at the end of a
string that is perfectly free to move vertically, as in Figure 13.13. The tension at the
free end is maintained by tying the string to a ring of negligible mass that is free to
slide vertically on a frictionless post. Again, the pulse is reflected, but this time it is
not inverted. As the pulse reaches the post, it exerts a force on the free end, causing the ring to accelerate upward. In the process, the ring reaches the top of its
motion and is then returned to its original position by the downward component
of the tension force. Thus, the ring experiences the same motion as if it were
raised and lowered by hand. This produces a reflected pulse that is not inverted
and whose amplitude is the same as that of the incoming pulse.

Finally, we may have a situation in which the boundary is intermediate between these two extreme cases; that is, it is neither completely rigid nor completely
free. In this case, part of the wave is transmitted and part is reflected. For instance,
suppose a string is attached to a more dense string as in Figure 13.14. When a
pulse traveling on the first string reaches the boundary between the two strings,
part of the pulse is reflected and inverted and part is transmitted to the more
dense string. Both the reflected and transmitted pulses have smaller amplitude
than the incident pulse. The inversion in the reflected wave is similar to the behavior of a pulse meeting a rigid boundary. As the pulse travels from the initial string
to the more dense string, the junction acts more like a rigid end than a free end.
Thus, the reflected pulse is inverted.

When a pulse traveling on a dense string strikes the boundary of a less dense
string, as in Figure 13.15, again part is reflected and part transmitted. This time,
however, the reflected pulse is not inverted. As the pulse travels from the dense

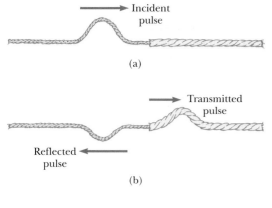

(a)

(b)

**Figure 13.14**

(a) A pulse traveling to the right on a light string attached to a heavier string. (b) Part of the incident pulse is reflected (and inverted), and part is transmitted to the heavier string.

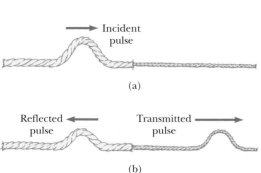

(a)

(b)

**Figure 13.15**

(a) A pulse traveling to the right on a heavy string attached to a lighter string. (b) The incident pulse is partially reflected and partially transmitted. In this case, the reflected pulse is not inverted.

string to the less dense one, the junction acts more like a free end than a rigid end.

The limiting value between these two cases would be that in which both strings have the same linear mass density. In this case, no boundary exists between the two media—both strings are identical. As a result, no reflection occurs and transmission is total.

In the preceding section, we found that the speed of a wave on a string increases as the mass per unit length of the string decreases. In other words, a pulse travels more slowly on a dense string than on a less dense one if both are under the same tension. This is illustrated by the lengths of the vectors in Figures 13.14 and 13.15.

An ultrasound image of a human fetus in the womb after 20 weeks of development, showing the head, body, and arms in profile. Ultrasound refers to sound waves that are higher in frequency than those audible to humans. The sound waves transmit through the body of the mother and reflect from the skin of the fetus. The reflected sound waves are organized by the electronics of the ultrasonic imaging system into a visual image of the fetus. *(U.H.B. Trust/Stone)*

**WEB**

For a number of ultrasound images of a fetus at various stages of development, visit
**www.parenthoodweb.com/
parent_cfmfiles/us.cfm/574**

This discussion has focused on pulses arriving at a boundary. If a sinusoidal wave on a string arrives at a rigid end, the inversion of the waveform is equivalent to shifting the entire wave by half of a wavelength. This equivalence can be seen by looking back at Figure 13.7. The wave in Figure 13.7d is the inversion of the wave in Figure 13.7b. Notice, however, that Figure 13.7b would look like Figure 13.7d if the wave were shifted to the right or left by half of a wavelength. Thus, because a full wavelength can be associated with an angle of 360°, we describe the inversion of a wave at a rigid end as a **180° phase shift.** We will see this effect again in Chapter 27 when we discuss reflection of light waves from materials.

## 13.6 • RATE OF ENERGY TRANSFER BY SINUSOIDAL WAVES ON STRINGS

See Screen 9.5

As waves propagate through a medium, they transport energy. This fact is easily demonstrated by hanging an object on a stretched string and sending a pulse down the string, as in Figure 13.16. When the pulse meets the suspended object, the object is momentarily displaced, as in Figure 13.16b. In the process, energy is transferred to the object because work must be done in moving it upward. This section examines the rate at which energy is transferred along a string. We shall assume a one-dimensional sinusoidal wave in the calculation of the energy transferred.

Consider a sinusoidal wave traveling on a string (Fig. 13.17). The source of the energy is some external agent at the left end of the string, which does work in producing the oscillations. We can consider the string to be a nonisolated system. As the external agent performs work on the end of the string, moving it up and down, energy enters the system of the string and propagates along its length. Let us focus our attention on an element of the string of length $\Delta x$ and mass $\Delta m$. Each such element moves vertically with simple harmonic motion. Thus, we can model each element of the string as a simple harmonic oscillator, with the oscillation in the $y$ direction. All elements have the same angular frequency $\omega$ and the same amplitude $A$. As we found in Chapter 12, the kinetic energy $K$ associated with a particle in simple harmonic motion is $K = \frac{1}{2}mv^2$, where $v$ varies sinusoidally during the oscillation. If we apply this equation to an element of length $\Delta x$, we see that the kinetic energy $\Delta K$ of this element is

$$\Delta K = \tfrac{1}{2}(\Delta m)\,v_y^2$$

If $\mu$ is the mass per unit length of the string, then the element of length $\Delta x$ has a mass $\Delta m$ that is equal to $\mu\,\Delta x$. Hence, we can express the kinetic energy of an ele-

(a)

(b)

**Figure 13.16**

(a) A pulse traveling to the right on a stretched string on which an object has been suspended. (b) Energy is transmitted to the suspended object when the pulse arrives.

**Figure 13.17**

A sinusoidal wave traveling along the $x$ axis on a stretched string. Every segment moves vertically, and each segment has the same total energy. The average power transmitted by the wave equals the energy contained in one wavelength divided by the period of the wave.

ment of the string as

$$\Delta K = \tfrac{1}{2}(\mu\,\Delta x)\,v_y^{\,2} \qquad\qquad\qquad \textbf{[13.21]}$$

As the length of the element of the string shrinks to zero, this becomes a differential relationship:

$$dK = \tfrac{1}{2}(\mu\,dx)\,v_y^{\,2}$$

We substitute for the general velocity of an element of the string using Equation 13.13:

$$dK = \tfrac{1}{2}\mu\,[\omega A\,\cos(kx - \omega t)]^2\,dx$$
$$= \tfrac{1}{2}\mu\omega^2 A^2\,\cos^2(kx - \omega t)\,dx$$

If we take a snapshot of the wave at time $t = 0$, then the kinetic energy in a given element is

$$dK = \tfrac{1}{2}\mu\omega^2 A^2\,\cos^2 kx\,dx$$

Let us integrate this expression over all the string elements in a wavelength of the wave, which will give us the kinetic energy in one wavelength:

$$K_\lambda = \int_0^\lambda \tfrac{1}{2}\mu\omega^2 A^2\,\cos^2 kx\,dx = \tfrac{1}{2}\mu\omega^2 A^2 \int_0^\lambda \cos^2 kx\,dx$$

$$= \tfrac{1}{2}\mu\omega^2 A^2 \left[\tfrac{1}{2}x + \frac{1}{4k}\sin 2kx\right]_0^\lambda = \tfrac{1}{2}\mu\omega^2 A^2 \left[\tfrac{1}{2}\lambda\right] = \tfrac{1}{4}\mu\omega^2 A^2\lambda$$

In addition to this kinetic energy, each element of the string has potential energy, associated with its displacement from the equilibrium position. A similar analysis as that above for the total potential energy in a wavelength gives the same result:

$$U_\lambda = \tfrac{1}{4}\mu\omega^2 A^2\lambda$$

The total energy in one wavelength of the wave is the sum of the kinetic and potential energies:

$$E_\lambda = K_\lambda + U_\lambda = \tfrac{1}{2}\mu\omega^2 A^2\lambda \qquad\qquad \textbf{[13.22]}$$

As the wave moves along the string, this amount of energy passes by a given point on the string during one period of the oscillation. Thus, the power, or rate of energy transfer, associated with the wave is

$$\mathcal{P} = \frac{E_\lambda}{\Delta t} = \frac{\tfrac{1}{2}\mu\omega^2 A^2\lambda}{T} = \tfrac{1}{2}\mu\omega^2 A^2\left(\frac{\lambda}{T}\right) = \tfrac{1}{2}\mu\omega^2 A^2 v \qquad\qquad \textbf{[13.23]}$$

• *Rate of energy transfer for a wave on a string*

This shows that the rate of energy transfer by a sinusoidal wave on a string is proportional to (a) the square of the angular frequency, (b) the square of the amplitude, and (c) the wave speed. In fact, *all* sinusoidal waves have the following general property: **The rate of energy transfer in any sinusoidal wave is proportional to the square of the angular frequency and to the square of the amplitude.**

### Example 13.5    Power Supplied to a Vibrating String

A string with linear mass density $\mu = 5.00 \times 10^{-2}$ kg/m is under a tension of 80.0 N. How much power must be supplied to the string to generate sinusoidal waves at a frequency of 60.0 Hz and an amplitude of 6.00 cm?

**Solution**    The wave speed on the string is

$$v = \sqrt{\frac{T}{\mu}} = \left(\frac{80.0 \text{ N}}{5.00 \times 10^{-2} \text{ kg/m}}\right)^{1/2} = 40.0 \text{ m/s}$$

Because $f = 60.0$ Hz, the angular frequency $\omega$ of the sinusoidal waves on the string has the value

$$\omega = 2\pi f = 2\pi(60.0 \text{ Hz}) = 377 \text{ s}^{-1}$$

Using these values in Equation 13.23 for the power, with $A = 6.00 \times 10^{-2}$ m, gives

$$\begin{aligned}
\mathcal{P} &= \tfrac{1}{2}\mu\omega^2 A^2 v \\
&= \tfrac{1}{2}(5.00 \times 10^{-2} \text{ kg/m})(377 \text{ s}^{-1})^2 \\
&\qquad \times (6.00 \times 10^{-2} \text{ m})^2(40.0 \text{ m/s}) \\
&= \boxed{512 \text{ W}}
\end{aligned}$$

## 13.7 • SOUND WAVES

Let us turn our attention from transverse waves to longitudinal waves. As stated in Section 13.2, in longitudinal waves the particles of the medium undergo displacements parallel to the direction of wave motion. Sound waves in air are the most important examples of longitudinal waves. Sound waves can travel through any material medium, however, and their speed depends on the properties of that medium. Table 13.1 provides examples of the speed of sound in different media.

The displacements accompanying a sound wave in air are longitudinal displacements of small elements of the fluid from their equilibrium positions. Such displacements result if the source of the waves, such as the diaphragm of a loudspeaker, oscillates in air. If the oscillation of the diaphragm is described by simple harmonic motion, a sinusoidal sound wave propagates away from the loudspeaker. For instance, one can produce a one-dimensional sound pulse in a long, narrow tube containing a gas by means of a vibrating piston at one end, as in Figure 13.18.

It is difficult to draw a pictorial representation of longitudinal waves because the displacements of the elements of the medium are in the same direction as that of the propagation of the wave. Figure 13.18 is one way to represent these types of waves. The darker color in the figure represents a region where the gas is compressed; consequently, the density and pressure are *above* their equilibrium values. Such a compressed region of gas, called a **compression,** is formed when the piston is being pushed into the tube. The compression moves down the tube as a pulse, continuously compressing the layers in front of it. When the piston is withdrawn from the tube, the gas in front of it expands, and consequently the pressure and density in this region fall below their equilibrium values. These low-pressure regions, called **rarefactions,** are represented by the lighter areas in Figure 13.18. The rarefactions also propagate along the tube, following the compressions. Both regions move with a speed equal to the speed of sound in that medium.

As the piston oscillates back and forth in a sinusoidal fashion, regions of compression and rarefaction are continuously set up. The distance between two successive compressions (or two successive rarefactions) equals the wavelength $\lambda$. As these regions travel down the tube, any small element of the medium moves with simple harmonic motion parallel to the direction of the wave (in other words, longitudinally). If $s(x, t)$ is the displacement of a small element measured from its equilibrium position, we can express this displacement function as

$$s(x, t) = s_{\text{max}} \sin(kx - \omega t) \qquad [13.24]$$

### TABLE 13.1

**Speeds of Sound in Various Media**

| Medium | $v$ (m/s) |
|---|---|
| *Gases* | |
| Hydrogen (0) | 1 286 |
| Helium (0) | 972 |
| Air (20) | 343 |
| Air (0) | 331 |
| Oxygen (0) | 317 |
| *Liquids (at 25°C)* | |
| Glycerol | 1 904 |
| Sea water | 1 533 |
| Water | 1 493 |
| Mercury | 1 450 |
| Kerosene | 1 324 |
| Methyl alcohol | 1 143 |
| Carbon tetrachloride | 926 |
| *Solids* | |
| Diamond | 12 000 |
| Pyrex glass | 5 640 |
| Iron | 5 130 |
| Aluminum | 5 100 |
| Brass | 4 700 |
| Copper | 3 560 |
| Gold | 3 240 |
| Lucite | 2 680 |
| Lead | 1 322 |
| Rubber | 1 600 |

where $s_{max}$ is the **maximum displacement from equilibrium** (or the **displacement amplitude**), $k$ is the wave number, and $\omega$ is the angular frequency of the piston. The variation $\Delta P$ in the pressure* of the gas measured from its equilibrium value is also sinusoidal; it is given by

$$\Delta P = \Delta P_{max} \cos(kx - \omega t) \qquad [13.25]$$

**The pressure amplitude $\Delta P_{max}$ is the maximum change in pressure from the equilibrium value.** It is proportional to the displacement amplitude $s_{max}$:

$$\Delta P_{max} = \rho v \omega s_{max} \qquad [13.26]$$

where $\rho$ is the density of the medium, $v$ is the wave speed, and $\omega s_{max}$ is the maximum longitudinal speed of an element of the medium in front of the piston. It is these pressure variations in a sound wave that result in an oscillating force on the eardrum, leading to the sensation of hearing.

Thus, we see that a sound wave may be considered as either a displacement wave or a pressure wave. A comparison of Equations 13.24 and 13.25 shows that **the pressure wave is 90° out of phase with the displacement wave.** Graphs of these functions are shown in Figure 13.19. Note that the change in pressure from equilibrium is a maximum when the displacement is zero, whereas the displacement is a maximum when the pressure change is zero.

Note that Figure 13.19 presents two graphical representations of the longitudinal wave: one for position of the elements of the medium and the other for pressure variation. These are *not* pictorial representations, however, for longitudinal waves. For transverse waves, because the particle displacement is perpendicular to the direction of propagation, the pictorial and graphical representations look the same — the perpendicularity of the oscillations and propagation is matched by the perpendicularity of $x$ and $y$ axes. For longitudinal waves, the oscillations and propagation exhibit no perpendicularity, so that pictorial representations look like Figure 13.18.

The speed of a sound wave in air depends only on the temperature of the air. For a small range of temperatures around room temperature, the speed of sound is described by

$$v = 331 \text{ m/s} + (0.6 \text{ m/s} \cdot °\text{C})\, T_C \qquad [13.27]$$

where $T_C$ is the temperature in degrees Celsius and the speed of sound at 0°C is 331 m/s.

(a)

Compressed region

(b)

(c)

(d)

**Figure 13.18**

A sinusoidal longitudinal wave propagating down a tube filled with a compressible gas. The source of the wave is a vibrating piston at the left. The high- and low-pressure regions are dark and light, respectively.

## THINKING PHYSICS 13.3

Garth Brooks had a hit song with "The Thunder Rolls" (Caged Panther Music, Inc., 1990). Why *does* thunder produce an extended "rolling" sound? And how does lightning produce thunder in the first place?

**Reasoning** Let us assume that we are at ground level and ignore ground reflections. When cloud-to-ground lightning strikes, a channel of ionized air carries a very large electric current from the cloud to the ground. (We will study electric current in Chapter 21.) This results in a very rapid temperature increase of this channel of

---

* We will formally introduce pressure in Chapter 15. In the case of longitudinal waves in a gas, each compressed area is a region of higher-than-average pressure and density, and each stretched region is a region of lower-than-average pressure and density.

You can estimate the distance to a lightning strike in kilometers by counting the number of seconds between seeing the strike and hearing the thunder, and then dividing by 3. This works because the speed of sound, about 343 m/s, is approximately one third of a kilometer per second. If you divide the number of seconds by 5, you will have an approximation for the distance in miles. The speed of sound is about 1100 ft/s, which is approximately one fifth of a mile per second. *(Richard Kaylin/Stone)*

**Figure 13.19**

(a) Displacement versus position and (b) pressure versus position for a sinusoidal longitudinal wave. The displacement wave is 90° out of phase with the pressure wave.

air as it carries the current. The temperature increase causes a sudden expansion of the air. This expansion is so sudden and so intense that a tremendous disturbance is produced in the air: thunder. The thunder rolls due to the fact that the lightning channel is a long, extended source—the entire length of the channel produces the sound at essentially the same instant of time. Sound produced at the end of the channel nearest you reaches you first, but sounds from progressively farther portions of the channel reach you shortly thereafter. If the lightning channel were a perfectly straight line, the resulting sound might be a steady roar, but the zigzagged shape of the path results in the rolling variation in loudness.

## 13.8 • THE DOPPLER EFFECT

When a vehicle sounds its horn as it travels along a highway, the frequency of the sound you hear is higher as the vehicle approaches you than it is as the vehicle moves away from you. This is one example of the **Doppler effect,** named after the Austrian physicist Christian Johann Doppler (1803–1853).

The Doppler effect for sound is experienced whenever there is relative motion between the source of sound and the observer. Motion of the source or observer toward the other results in the observer's hearing a frequency that is higher than the true frequency of the source. Motion of the source or observer away from the other results in the observer's hearing a frequency that is lower than the true frequency of the source.

Although we shall restrict our attention to the Doppler effect for sound waves, it is an effect associated with waves of all types. The Doppler effect for electromagnetic waves is used in police radar systems to measure the speeds of motor vehicles. Likewise, astronomers use the effect to determine the relative motions of stars, galaxies, and other celestial objects. In 1842, Doppler first reported the frequency shift in connection with light emitted by two stars revolving about each other in double-star systems. In the early 20th century, the Doppler effect for light from galaxies was used to argue for the expansion of the Universe. This led to the Big Bang theory, which we discuss in Chapter 31.

(a)

(b)

(c)

**Figure 13.20**

(a) Waves moving toward a stationary boat. The waves travel to the left, and their source is far to the right of the boat, out of the frame of the drawing. (b) The boat moving toward the wave source. (c) The boat moving away from the wave source.

To see what causes this apparent frequency change, imagine you are in a boat lying at anchor on a gentle sea where the waves have a period of $T = 2.0$ s. This means that every 2.0 s a crest hits your boat. Figure 13.20a shows this situation with the water waves moving toward the left. If you start a stopwatch at $t = 0$ just as one crest hits, the stopwatch reads 2.0 s when the next crest hits, 4.0 s when the third crest hits, and so on. From these observations you conclude that the wave frequency is $f = 1/T = 0.50$ Hz. Now suppose you start your motor and head directly into the oncoming waves, as shown in Figure 13.20b. Again you set your stopwatch to $t = 0$ as a crest hits the front of your boat. Now, however, because you are moving toward the next wave crest as it moves

**Figure 13.21**

An observer $O$ (the cyclist) moving with a speed $v_O$ toward a stationary point source $S$, the horn of a parked car. The observer hears a frequency $f'$ that is greater than the source frequency.

toward you, it hits you less than 2.0 s after the first hit. In other words, the period you observe is shorter than the 2.0-s period you observed when you were stationary. Because $f = 1/T$, you observe a higher wave frequency than when you were at rest.

If you turn around and move in the same direction as the waves (see Fig. 13.20c), you observe the opposite effect. You set your watch to $t = 0$ as a crest hits the back of the boat. Because you are now moving away from the next crest, more than 2.0 s has elapsed on your watch by the time that crest catches you. Thus, you observe a lower frequency than when you were at rest.

These effects occur because the relative speed between your boat and the water depends on the direction of travel and on the speed of your boat. When you are moving toward the right in Figure 13.20b, this relative speed is higher than that of the wave speed, which leads to the observation of an increased frequency. When you turn around and move to the left, the relative speed is lower, as is the observed frequency of the water waves.

Let us now examine an analogous situation with sound waves, in which we replace the water waves with sound waves, the water surface becomes the air, and the person on the boat becomes an observer listening to the sound. In this case, an observer $O$ is moving with a speed of $v_O$ and a sound source $S$ is stationary. For simplicity, we assume that the air is also stationary and that the observer moves directly toward the source.

The red lines in Figure 13.21 represent circular lines connecting the crests of sound waves moving away from the source. Thus, the radial distance between adjacent red lines is one wavelength. We shall take the frequency of the source to be $f$, the wavelength to be $\lambda$, and the speed of sound to be $v$. A stationary observer would detect a frequency $f$, where $f = v/\lambda$ (i.e., when the source and observer are both at rest, the observed frequency must equal the true frequency of the source). If the observer moves toward the source with the speed $v_O$, however, the relative speed of sound experienced by the observer is higher than the speed of sound in air. Using our relative speed discussion of Section 3.6, if the sound is coming toward the observer at $v$ and the observer is moving toward the sound at $v_O$, the relative speed of sound as measured by the observer is

$$v_{\text{rel}} = v + v_O$$

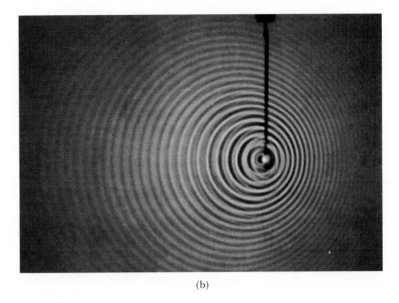

(a)

(b)

**Figure 13.22**

(a) A source $S$ moving with a speed $v_S$ toward a stationary observer A and away from a stationary observer B. Observer A hears an increased frequency, and observer B hears a decreased frequency.
(b) The Doppler effect in water observed in a ripple tank. The vibrating source is moving to the right.
*(Courtesy of the Educational Development Center, Newton, MA)*

Thus, the frequency of sound heard by the observer is based on this apparent speed of sound:

$$f' = \frac{v_{rel}}{\lambda} = \frac{v + v_O}{\lambda} = f\left(\frac{v + v_O}{v}\right) \quad \text{(observer moving toward source)} \qquad \textbf{[13.28]}$$

Now consider the situation in which the source moves with a speed of $v_S$ relative to the medium, and the observer is at rest. Figure 13.22 shows this situation. Because the source is moving, the crest of each new wave is emitted from the source to the right of the position of the emission of the previous crest a distance $v_S T$, where $T$ is the period of the wave being generated by the source. Thus, the center of each colored circle (indicated by the identically colored dot) in Figure 13.22 is shifted to the right by this distance relative to the circle representing the previous crest. If the source moves directly toward observer $O_A$ in Figure 13.22a, the crests detected by the observer along a line between the source and observer are closer to one another than they would be if the source were at rest. As a result, the wavelength $\lambda'$ measured by observer $O_A$ is shorter than the true wavelength $\lambda$ of the source. The wavelength is *shortened* by the distance $v_S T$, and the observed wavelength has the value $\lambda' = \lambda - v_S/f$. Because $\lambda = v/f$, the frequency heard by observer $O_A$ is

$$f' = \frac{v}{\lambda'} = f\left(\frac{v}{v - v_S}\right) \quad \text{(source moving toward observer)} \qquad \textbf{[13.29]}$$

That is, the frequency is *increased* when the source moves toward the observer. In a similar manner, if the source moves away from observer $O_B$ at rest, the sign of $v_S$ is reversed in Equation 13.29, and the frequency is lower.

In Equation 13.29, notice that the denominator approaches zero when the speed of the source approaches the speed of sound, resulting in the frequency $f'$ approaching infinity. Such a situation results in waves that cannot escape from the source in the direction of motion of the source. This concentration of energy in front of the source results in a *shock wave*. Such a disturbance is noted when a jet aircraft flying at a speed equal to or greater than the speed of sound produces a *sonic boom*.

Finally, if both the source and the observer are in motion, one finds the following general equation for the observed frequency:

$$f' = f\left(\frac{v + v_O}{v - v_S}\right)$$

[13.30]

• *Frequency heard with observer and source in motion*

In this expression, the signs for the values substituted for $v_O$ and $v_S$ depend on the direction of the velocity. A positive value is used for motion of the observer or the source *toward* the other, and a negative sign for motion of one *away from* the other.

When working with any Doppler effect problem, remember the following rule concerning signs: The word *toward* is associated with an *increase* in the observed frequency. The words *away from* are associated with a *decrease* in the observed frequency.

A shock wave due to a jet traveling at the speed of sound is made visible as a fog of water vapor. The large pressure variation in the shock wave causes the water in the air to condense into water droplets. *(U. S. Navy. Photo by Ensign John Gay)*

### Quick Quiz 13.5

You are on a cruise and the person next to you at the deck railing falls overboard. During the entire time of fall, the person screams with a constant frequency. How does the apparent frequency of the sound of the scream change with time?

### Quick Quiz 13.6

You are driving toward a cliff and you honk your horn. The frequency of the echo you hear will be Doppler shifted (a) up in frequency, equivalent to a moving source; (b) up in frequency, equivalent to a moving observer; (c) up in frequency, equivalent to a moving source *and* a moving observer; (d) down in frequency, equivalent to a moving source; (e) down in frequency, equivalent to a moving observer; (f) down in frequency, equivalent to a moving source *and* a moving observer.

### THINKING PHYSICS 13.4

Suppose you run past your stereo speakers, from right to left, or left to right. If you run rapidly enough and have excellent pitch discrimination, you may notice that the music that is playing seems to be out of tune when you are between the speakers. Why?

**Reasoning**   When you are between the speakers, you are running away from one of them and toward the other. Thus, there is a Doppler shift downward for the sound from the speaker behind you and a Doppler shift upward for the sound from the speaker ahead of you. As a result, the sound from the two speakers will not be in tune. A calculation shows that a world-class sprint runner could run fast enough to generate about a semitone difference in the sound from the two speakers.

## Example 13.6 The Noisy Siren

An ambulance travels down a highway at a speed of 33.5 m/s (75 mi/h). Its siren emits sound at a frequency of 400 Hz. What is the frequency heard by a passenger in a car traveling at 24.6 m/s (55 mi/h) in the opposite direction (a) as the car approaches the ambulance and (b) as the car moves away from the ambulance?

**Solution** (a) Let us take the speed of sound in air to be $v = 343$ m/s. We can use Equation 13.30 in both cases. As the ambulance and car approach each other, the observed apparent frequency is

$$f' = f\left(\frac{v + v_O}{v - v_S}\right) = (400 \text{ Hz})\left(\frac{343 \text{ m/s} + (+24.6 \text{ m/s})}{343 \text{ m/s} - (+33.5 \text{ m/s})}\right)$$

$$= \boxed{475 \text{ Hz}}$$

(b) Similarly, as they recede from each other, a passenger in the car hears a frequency

$$f' = f\left(\frac{v + v_O}{v - v_S}\right) = (400 \text{ Hz})\left(\frac{343 \text{ m/s} + (-24.6 \text{ m/s})}{343 \text{ m/s} - (-33.5 \text{ m/s})}\right)$$

$$= \boxed{338 \text{ Hz}}$$

The *change* in frequency as detected by the passenger in the car is 475 Hz − 338 Hz = 137 Hz, which is more than 30% of the actual frequency emitted.

**EXERCISE** Suppose that the passenger car is parked on the side of the highway as the ambulance travels down the highway at the speed of 33.5 m/s. What frequency will the passenger in the car hear as the ambulance (a) approaches the parked car and (b) recedes from the parked car?

*Answer* (a) 443 Hz (b) 364 Hz

**EXERCISE** A band is playing on a moving truck. The band strikes the note middle C (262 Hz), but it is heard by spectators ahead of the truck as C# (277 Hz). How fast is the truck moving?

*Answer* 18.6 m/s

 context connection

## 13.9 • SEISMIC WAVES

 See Screen 8.14

When an earthquake occurs, a sudden release of energy takes place at a location called the **focus** or **hypocenter** of the earthquake. (The **epicenter** is the point on the Earth's surface radially above the hypocenter.) This energy will propagate away from the focus of the earthquake by means of **seismic waves.** Seismic waves are like the sound waves that we have studied in the later sections of this chapter—mechanical disturbances moving through a medium.

In discussing mechanical waves in this chapter, we identified two types: transverse and longitudinal. In the case of mechanical waves moving through air, we have only a longitudinal possibility. For mechanical waves moving through a solid, however, both possibilities are available, due to the strong interatomic forces between particles of the solid. Thus, in the case of seismic waves, energy propagates away from the focus both by longitudinal and transverse waves.

In the language used in earthquake studies, these two types of waves are named according to the order of their arrival at a seismograph. The longitudinal wave travels at a higher speed than the transverse wave. As a result, the longitudinal wave arrives at a seismograph first, and is thus called the **P wave,** where *P* stands for *primary.* The slower moving transverse wave arrives next, so it is called the **S wave,** or *secondary* wave.

The wave speed for a seismic wave depends on the medium through which it travels. Typical values are 5 km/s for a *P* wave through granite and 3 km/s for an *S* wave through granite. Figure 13.23 shows a typical seismograph trace of a distant earthquake, with the *S* wave clearly arriving after the *P* wave.

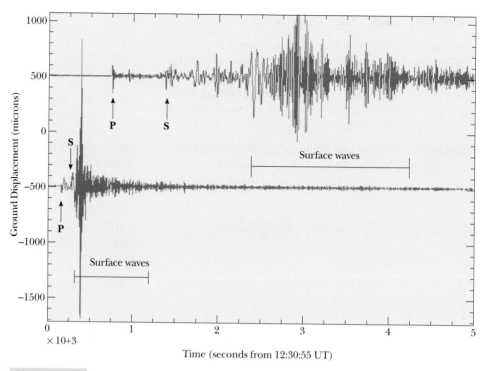

**Figure 13.23**

A seismograph trace, showing the arrival of *P* and *S* waves from the Northridge earthquake at San
Pablo, Spain (*top trace*) and Albuquerque, New Mexico (*bottom trace*). The *P* wave arrives first, because it
travels the fastest, followed by the slower moving *S* wave. The farther the seismograph station is from
the epicenter, the longer is the time interval between the arrivals of the *P* and *S* waves.

The *P* and *S* waves move through the body of the Earth and can be detected by
seismographs at various locations around the Earth. Once these waves reach the
surface, the energy can propagate by additional types of waves along the surface.
In a *Rayleigh* wave, the motion of the particles at the surface is a combination of
longitudinal and transverse displacements, so that the net motion of a point on
the surface is circular or elliptical. This is similar to the path followed by particles
on the ocean surface as a wave passes by, as in Figure 13.24. The *Love* wave is a

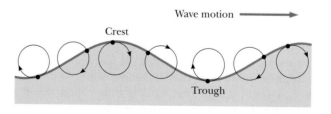

**Figure 13.24**

The motion of particles on the surface of deep water in which a wave is propagating is a combination
of transverse and longitudinal displacements, with the result that the molecules at the surface move in
nearly circular paths. Each molecule is displaced both horizontally and vertically from its equilibrium
position. This is similar to the motion of the Earth surface for a Rayleigh wave.

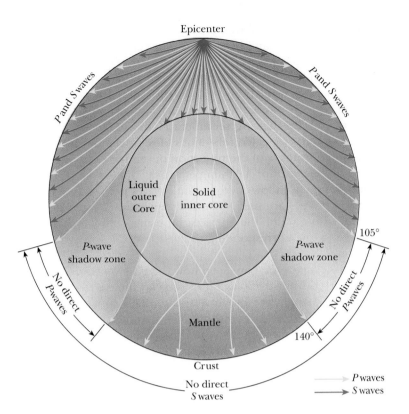

Epicenter

P and S waves

P and S waves

Liquid
outer
Core

Solid
inner core

P-wave
shadow zone

P-wave
shadow zone

105°

No direct
P-waves

No direct
P-waves

Mantle

140°

Crust

No direct
S waves

→ P waves
→ S waves

**Figure 13.25**

Cross-section of the Earth showing
paths of waves produced by an earthquake. Only *P* waves (yellow) can
propagate in the liquid core. The *S*
waves (blue) do not enter the liquid
core. When the *P* waves transmit from
one region to another, such as from
the mantle to the liquid core, they experience *refraction*—a change in the
direction of propagation. We will
study refraction for light in Chapter
25. Because of the refraction for seismic waves, there is a "shadow" zone
between 105° and 140° from the epicenter, in which no waves following a
direct path (i.e., a path with no reflections) arrive.

transverse surface wave in which the transverse oscillations are parallel to the surface. Thus, no vertical displacement of the surface occurs in a Love wave.

It is possible to use the *P* and *S* waves traveling through the body of the Earth
to gain information about the structure of the interior of the Earth. Measurements
of a given earthquake by seismographs at various locations on the surface indicate
that the Earth has an interior region that allows the passage of *P* waves, but not *S*
waves. This fact can be understood if this particular region is modeled as having
liquid characteristics. Similar to a gas, a liquid cannot sustain a transverse force.
For this reason we only have longitudinal sound waves in air. Thus, the transverse
*S* waves cannot pass through this region. This leads to a structural model in which
the Earth has a **liquid core** between radii of approximately $1.2 \times 10^3$ km and
$3.5 \times 10^3$ km.

Other measurements of seismic waves allow additional interpretations of layers
within the interior of the Earth, including a **solid core** at the center, a rocky region called the **mantle,** and a relatively thin outer layer called the **crust.** Figure
13.25 shows this structure. Using x-rays or ultrasound in medicine to provide information about the interior of the human body is somewhat similar to using seismic
waves to provide information about the interior of the Earth.

As *P* and *S* waves propagate in the interior of the Earth, they will encounter
variations in the medium. At each boundary at which the properties of the
medium change, reflection and transmission occur. When the seismic wave arrives
at the surface of the Earth, a small amount of the energy is transmitted into the air

as low-frequency sound waves. Some of the energy spreads out along the surface in the form of Rayleigh and Love waves. The remaining wave energy is reflected back into the interior. As a result, seismic waves can travel over long distances within the Earth and can be detected at seismographs at many locations around the globe. In addition, because a relatively large fraction of the wave energy continues to be reflected at each encounter with the surface, the wave can propagate for a long time. Data are available showing seismograph activity for several hours after an earthquake, due to the repeated reflections of seismic waves from the surface.

Another example of the reflection of seismic waves is available in the technology of oil exploration. A "thumper truck" applies large impulsive forces to the ground, resulting in low-energy seismic waves propagating into the Earth. Specialized microphones are used to detect the waves reflected from various boundaries between layers under the surface. By using computers to map out the underground structure corresponding to these layers, it is possible to detect layers likely to contain oil.

In this chapter, we have seen how the phenomenon of waves allows energy to be transferred from the location of the earthquake to the location of a structure. In the next chapter, we shall investigate what happens when these waves become trapped in a geological depression, such that the amplitude of oscillation becomes very large, representing even greater risk of damage to structures.

## SUMMARY

A **transverse wave** is a wave in which the particles of the medium move in a direction perpendicular to the direction of the wave velocity. An example is a wave moving along a stretched string.

**Longitudinal waves** are waves in which the elements of the medium move back and forth parallel to the direction of the wave speed. Sound waves in air are longitudinal.

Any one-dimensional wave traveling with a speed of $v$ in the positive $x$ direction can be represented by a **wave function** of the form $y = f(x - vt)$. Likewise, the wave function for a wave traveling in the negative $x$ direction has the form $y = f(x + vt)$.

The wave function for a one-dimensional sinusoidal wave traveling to the right can be expressed as

$$y = A \sin\left[\frac{2\pi}{\lambda}(x - vt)\right] = A \sin(kx - \omega t) \qquad \textbf{[13.4, 13.9]}$$

where $A$ is the **amplitude,** $\lambda$ is the **wavelength,** $k$ is the **angular wave number,** and $\omega$ is the **angular frequency.** If $T$ is the **period** and $f$ is the **frequency,** then $v$, $k$, and $\omega$ can be written

$$v = \frac{\lambda}{T} = f\lambda \qquad \textbf{[13.5, 13.11]}$$

$$k \equiv \frac{2\pi}{\lambda} \qquad \textbf{[13.7]}$$

$$\omega \equiv \frac{2\pi}{T} = 2\pi f \qquad \textbf{[13.8]}$$

The speed of a wave traveling on a stretched string of mass per unit length $\mu$ and tension $T$ is

$$v = \sqrt{\frac{T}{\mu}} \qquad \textbf{[13.20]}$$

When a pulse traveling on a string meets a fixed end, the pulse is reflected and inverted. If the pulse reaches a free end, it is reflected but not inverted.

The **power** transmitted by a sinusoidal wave on a stretched string is

$$\mathscr{P} = \tfrac{1}{2}\mu\omega^2 A^2 v \qquad \textbf{[13.23]}$$

The change in frequency heard by an observer whenever there is relative motion between a wave source and the observer is called the **Doppler effect.** When the source and observer are moving toward each other, the observer hears a higher frequency than the true frequency of the source. When the source and observer are moving away from each other, the observer hears a lower frequency than the true frequency of the source. The following general equation provides the observed frequency:

$$f' = f\left(\frac{v + v_O}{v - v_S}\right) \qquad \textbf{[13.30]}$$

A positive value is used for $v_O$ and/or $v_S$ for motion of the observer or source *toward* the other, and a negative sign for motion *away from* the other.

# QUESTIONS

1. Suppose the wind blows. Does this cause a Doppler effect for a sound moving through the air? Is it like a moving source or a moving observer?

2. How would you set up a longitudinal wave in a stretched spring? Would it be possible to set up a transverse wave in a spring?

3. By what factor would you have to increase the tension in a taut string in order to double the wave speed?

4. When traveling on a taut string, does a wave pulse always invert on reflection? Explain.

5. Can two pulses traveling in opposite directions on the same string reflect from each other? Explain.

6. Does the vertical speed of a segment of a horizontal taut string through which a wave is traveling depend on the wave speed?

7. A vibrating source generates a sinusoidal wave on a string under constant tension. If the power delivered to the string is doubled, by what factor does the amplitude change? Does the wave speed change under these circumstances?

8. Consider a wave traveling on a taut rope. What is the difference, if any, between the speed of the wave and the speed of a small section of the rope?

9. If a long rope is hung from a ceiling and waves are sent up the rope from its lower end, they do not ascend with constant speed. Explain.

10. What happens to the wavelength of a wave on a string when the frequency is doubled? Assume the tension in the string remains the same.

11. What happens to the speed of a wave on a taut string when the frequency is doubled? Assume the tension in the string remains the same.

12. When all the strings on a guitar are stretched to the same tension, will the speed of a wave along the more massive bass strings be faster or slower than the speed of a wave on the lighter strings?

13. If you stretch a rubber hose and pluck it, you can observe a pulse traveling up and down the hose. What happens to the speed if you stretch the hose tighter? If you fill the hose with water?

14. In a longitudinal wave in a spring, the coils move back and forth in the direction of wave motion. Does the speed of the wave depend on the maximum speed of each coil?

15. A solid can transport both longitudinal waves and transverse waves, but a fluid can transport only longitudinal waves. Why?

16. In an earthquake both $S$ (transverse) and $P$ (longitudinal) waves propagate from the focus of the earthquake. The $S$ waves travel through the Earth more slowly than the $P$ waves ($\approx 3$ km/s versus $\approx 5$ km/s). By detecting the time of arrival of the waves, how can one determine how far away the epicenter of the quake was? How many detection centers are necessary to pinpoint the location of the epicenter?

# PROBLEMS

1, 2, 3 = straightforward, intermediate, challenging    □ = full solution available in the *Student Solutions Manual and Study Guide*

**web** = solution posted at **http://www.harcourtcollege.com/physics/**    🖳 = computer useful in solving problem

 = Interactive Physics    ▓ = paired numerical/symbolic problems    ▨ = life science application

## Section 13.1  Propagation of a Disturbance

1. At $t = 0$, a transverse wave pulse in a wire is described by the function

$$y = \frac{6}{x^2 + 3}$$

where $x$ and $y$ are in meters. Write the function $y(x, t)$ that describes this wave if it is traveling in the positive $x$ direction with a speed of 4.50 m/s.

2. Ocean waves with a crest-to-crest distance of 10.0 m can be described by the wave function

$$y(x, t) = (0.800 \text{ m}) \sin[0.628(x - vt)]$$

where $v = 1.20$ m/s. (a) Sketch $y(x, t)$ at $t = 0$. (b) Sketch $y(x, t)$ at $t = 2.00$ s. Note how the entire waveform shifts 2.40 m in the positive $x$ direction in this time interval.

## Section 13.2  The Wave Model

## Section 13.3  The Traveling Wave

3. A sinusoidal wave is traveling along a rope. The oscillator that generates the wave completes 40.0 vibrations in 30.0 s. Also, a given maximum travels 425 cm along the rope in 10.0 s. What is the wavelength?

4. For a certain transverse wave, the distance between two successive crests is 1.20 m, and eight crests pass a given point along the direction of travel every 12.0 s. Calculate the wave speed.

5. The wave function for a traveling wave on a taut string is (in SI units)

$$y(x, t) = (0.350 \text{ m}) \sin(10\pi t - 3\pi x + \pi/4)$$

(a) What are the speed and direction of travel of the wave? (b) What is the vertical displacement of the string at $t = 0$, $x = 0.100$ m? (c) What are the wavelength and frequency of the wave? (d) What is the maximum magnitude of the transverse speed of the string?

6. A wave is described by $y = (2.00 \text{ cm}) \sin(kx - \omega t)$, where $k = 2.11$ rad/m, $\omega = 3.62$ rad/s, $x$ is in meters, and $t$ is in seconds. Determine the amplitude, wavelength, frequency, and speed of the wave.

7. The string shown in Figure 13.7 is driven at a frequency of 5.00 Hz. The amplitude of the motion is 12.0 cm, and the wave speed is 20.0 m/s. Furthermore, the wave is such that $y = 0$ at $x = 0$ and $t = 0$. Determine (a) the angular frequency and (b) wave number for this wave. (c) Write an expression for the wave function. Calculate (d) the maximum transverse speed and (e) the maximum transverse acceleration of a point on the string.

8. A transverse wave on a string is described by the wave function

$$y = (0.120 \text{ m}) \sin(\pi x/8 + 4\pi t)$$

(a) Determine the transverse speed and acceleration at $t = 0.200$ s for the point on the string located at $x = 1.60$ m. (b) What are the wavelength, period, and speed of propagation of this wave?

9. (a) Write the expression for $y$ as a function of $x$ and $t$ for a sinusoidal wave traveling along a rope in the *negative x* direction with the following characteristics: $A = 8.00$ cm, $\lambda = 80.0$ cm, $f = 3.00$ Hz, and $y(0,t) = 0$ at $t = 0$. (b) Write the expression for $y$ as a function of $x$ and $t$ for the wave in part (a) assuming that $y(x, 0) = 0$ at the point $x = 10.0$ cm.

10. A transverse sinusoidal wave on a string has a period $T = 25.0$ ms and travels in the negative $x$ direction with a speed of 30.0 m/s. At $t = 0$, a particle on the string at $x = 0$ has a displacement of 2.00 cm and is traveling downward with a speed of 2.00 m/s. (a) What is the amplitude of the wave? (b) What is the initial phase angle? (c) What is the maximum transverse speed of the string? (d) Write the wave function for the wave.

11. Show that the wave function $y = e^{b(x - vt)}$ is a solution of the linear wave equation (see Eq. 13.19), where $b$ is a constant.

## Section 13.4   The Speed of Transverse Waves on Strings

12. A telephone cord is 4.00 m long and has a mass of 0.200 kg. A transverse wave pulse is produced by plucking one end of the taut cord. The pulse makes four trips down and back along the cord in 0.800 s. What is the tension in the cord?

13. A piano string having a mass per unit length equal to $5.00 \times 10^{-3}$ kg/m is under a tension of 1 350 N. Find the speed with which a wave travels on this string.

14. An astronaut on the Moon wishes to measure the local value of $g$ by timing pulses traveling down a wire, clamped at its top end, that has a large mass suspended from it. Assume a wire has a mass of 4.00 g and a length of 1.60 m, and that a 3.00-kg object is suspended from it. A pulse requires 36.1 ms to traverse the length of the wire. Calculate $g_{\text{Moon}}$ from these data. (You may ignore the mass of the wire when calculating the tension in it.)

15. A 30.0-m steel wire and a 20.0-m copper wire, both with 1.00-mm diameters, are connected end to end and stretched to a tension of 150 N. How long does it take a transverse wave to travel the entire length of the two wires?

16. **Review Problem.** A light string with a mass per unit length of 8.00 g/m has its ends tied to two walls separated by a distance equal to three fourths the length of the string (Fig. P13.16). An object of mass $m$ is suspended from the center of the string, creating tension in the string. (a) Find an expression for the transverse wave speed in the string as a function of the hanging mass. (b) How much mass should be suspended from the string to produce a wave speed of 60.0 m/s?

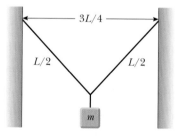

**Figure P13.16**

## Section 13.5   Reflection and Transmission of Waves

17. A series of pulses, each of amplitude 0.150 m, are sent down a string that is attached to a post at one end. The pulses are reflected at the post and travel back along the string without loss of amplitude. When two waves are present on the same string, the net displacement of a given point is the sum of the displacements of the individual waves at that point. What is the net displacement at a

point on the string where two pulses are crossing, (a) if the string is rigidly attached to the post? (b) If the end at which reflection occurs is free to slide up and down?

## Section 13.6 Rate of Energy Transfer by Sinusoidal Waves on Strings

**18.** A taut rope has a mass of 0.180 kg and a length 3.60 m. What power must be supplied to the rope in order to generate sinusoidal waves having an amplitude of 0.100 m and a wavelength of 0.500 m and traveling with a speed of 30.0 m/s?

**19.** Sinusoidal waves 5.00 cm in amplitude are to be transmitted along a string that has a linear mass density equal to $4.00 \times 10^{-2}$ kg/m. If the source can deliver a maximum power of 300 W and the string is under a tension of 100 N, what is the highest vibrational frequency at which the source can operate?

**20.** Transverse waves are being generated on a rope under constant tension. By what factor is the required power increased or decreased if (a) the length of the rope is doubled and the angular frequency remains constant, (b) the amplitude is doubled and the angular frequency is halved, (c) both the wavelength and the amplitude are doubled, and (d) both the length of the rope and the wavelength are halved?

**21.** A 6.00-m segment of a long string has a mass of 180 g. A high-speed photograph shows that the segment contains four complete cycles of a wave. The string is vibrating sinusoidally with a frequency of 50.0 Hz and a peak-to-valley displacement of 15.0 cm. (The "peak-to-valley" displacement is the vertical distance from the farthest positive displacement to the farthest negative displacement.) (a) Write the function that describes this wave traveling in the positive $x$ direction. (b) Determine the power being supplied to the string.

## Section 13.7 Sound Waves

> *Note:* Use the following values as needed unless otherwise specified. The equilibrium density of air is $\rho = 1.20$ kg/m$^3$; the speed of sound in air is $v = 343$ m/s. Problem 52 in Chapter 2 can be assigned with this section.

**22.** A bat (Fig. P13.22) can detect very small objects, such as an insect whose length is approximately equal to one wavelength of the sound the bat makes. If bats emit a chirp at a frequency of 60.0 kHz, and if the speed of sound in air is 340 m/s, what is the smallest insect a bat can detect?

**23.** Suppose that you hear a clap of thunder 16.2 s after seeing the associated lightning stroke. The speed of sound waves in air is 343 m/s and the speed of light in air is $3.00 \times 10^8$ m/s. How far are you from the lightning stroke?

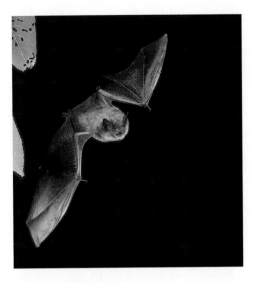

**Figure P13.22** (*Joe McDonald Visuals Unlimited*)

**24.** A flowerpot is knocked off a balcony 20.0 m above the sidewalk and falls toward an unsuspecting 1.75-m-tall man standing below. How close to the sidewalk can the flowerpot fall before it is too late for a warning shouted from the balcony to reach the man in time? Assume that the man below requires 0.300 s to respond to the warning.

**25.** A sinusoidal sound wave is described by the displacement

$$s(x, t) = (2.00 \ \mu\text{m}) \cos[(15.7 \ \text{m}^{-1})x - (858 \ \text{s}^{-1})t]$$

(a) Find the amplitude, wavelength, and speed of this wave. (b) Determine the instantaneous displacement of an element of air at the position $x = 0.050 \ 0$ m at $t = 3.00$ ms. (c) Determine the maximum speed of an element's oscillatory motion.

**26.** Calculate the pressure amplitude of a 2.00-kHz sound wave in air if the displacement amplitude is equal to $2.00 \times 10^{-8}$ m.

**27.** An experimenter wishes to generate in air a sound wave that has a displacement amplitude of $5.50 \times 10^{-6}$ m. The pressure amplitude is to be limited to 0.840 N/m$^2$. What is the minimum wavelength the sound wave can have?

**28.** A sound wave traveling in air has a pressure amplitude of 4.00 N/m$^2$ and a frequency of 5.00 kHz. Take $\Delta P = 0$ at the point $x = 0$ when $t = 0$. (a) What is $\Delta P$ at $x = 0$ when $t = 2.00 \times 10^{-4}$ s? (b) What is $\Delta P$ at $x = 0.020 \ 0$ m when $t = 0$?

**29.** Write an expression that describes the pressure variation as a function of position and time for a sinusoidal sound wave in air, if $\lambda = 0.100$ m and $\Delta P_{\text{max}} = 0.200$ N/m$^2$.

## Section 13.8 The Doppler Effect

**30.** A commuter train passes a passenger platform at a constant speed of 40.0 m/s. The train horn is sounded at its characteristic frequency of 320 Hz. (a) What overall change in frequency is detected by a person on the platform as the train moves from approaching to receding? (b) What wavelength is detected by a person on the platform as the train approaches?

31. Standing at a crosswalk, you hear a frequency of 560 Hz from the siren of an approaching ambulance. After the ambulance passes, the observed frequency of the siren is 480 Hz. Determine the ambulance's speed from these observations.

32. A driver travels northbound on a highway at a speed of 25.0 m/s. A police car, traveling southbound at a speed of 40.0 m/s, approaches with its siren producing sound at a frequency of 2 500 Hz. (a) What frequency does the driver observe as the police car approaches? (b) What frequency does the driver detect after the police car passes him? (c) Repeat parts (a) and (b) for the case when the police car is traveling northbound.

33. A tuning fork vibrating at 512 Hz falls from rest and accelerates at 9.80 m/s². How far below the point of release is the tuning fork when waves of frequency 485 Hz reach the release point? Take the speed of sound in air to be 340 m/s.

34. Expectant parents are thrilled to hear their unborn baby's heartbeat, revealed by an ultrasonic motion detector. Suppose the fetus's ventricular wall moves in simple harmonic motion with an amplitude of 1.80 mm and a frequency of 115/min. (a) Find the maximum linear speed of the heart wall. Suppose the motion detector in contact with the mother's abdomen produces sound at 2 000 000.0 Hz, which travels through tissue at 1.50 km/s. (b) Find the maximum frequency at which sound arrives at the wall of the baby's heart. (c) Find the maximum frequency at which reflected sound is received by the motion detector. By electronically "listening" for echoes at a frequency different from the broadcast frequency, the motion detector can produce beeps of audible sound in synchronization with the fetal heartbeat.

35. A block with a speaker bolted to it is connected to a spring having force constant $k = 20.0$ N/m, as shown in Figure P13.35. The total mass of the block and speaker is 5.00 kg, and the amplitude of this unit's motion is 0.500 m. If the speaker emits sound waves of frequency 440 Hz, determine the highest and lowest frequencies heard by the person to the right of the speaker.

36. At the Winter Olympics, an athlete rides her luge down the track while a bell just above the wall of the chute rings continuously. When her sled passes the bell, she hears the frequency of the bell fall by the musical interval called a minor third. That is, the frequency she hears drops to five sixths of its original value. (a) Find the speed of sound in

**Figure P13.35**

air at the ambient temperature − 10.0°C. (b) Find the speed of the athlete.

## Section 13.9 Context Connection — Seismic Waves

37. Two points, $A$ and $B$, on the surface of the Earth are at the same longitude and 60.0° apart in latitude. Suppose that an earthquake at point $A$ creates a $P$ wave that reaches point $B$ by traveling straight through the body of the Earth at a constant speed of 7.80 km/s. The earthquake also radiates a Rayleigh wave, which travels along the surface of the Earth at 4.50 km/s. (a) Which of these two seismic waves arrives at $B$ first? (b) What is the time difference between the arrivals of the two waves at $B$? Take the radius of the Earth to be 6 370 km.

38. A seismographic station receives $S$ and $P$ waves from an earthquake, 17.3 s apart. Assume the waves have traveled over the same path at speeds of 4.50 km/s and 7.80 km/s. Find the distance from the seismograph to the hypocenter of the quake.

## Additional Problems

39. "The wave" is a particular type of wave pulse that can propagate through a large crowd gathered at a sports arena to watch a soccer or American football match (Figure P13.39). The particles of the medium are the spectators, with zero displacement corresponding to their being seated and maximum displacement corresponding to their standing up and raising their arms. When a large fraction of the spectators participate in the wave motion, a somewhat stable pulse shape can develop. The wave speed depends on people's reaction time, which is typically on the order of 0.1 s. Estimate the order of magnitude, in minutes, of the time required for such a wave pulse to make one circuit around a large sports stadium. State the quantities you measure or estimate and their values.

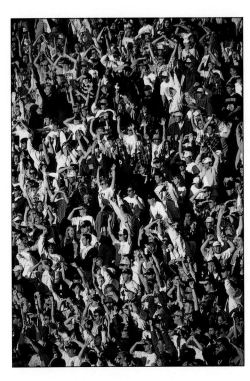

**Figure P13.39** *(Gregg Adams/Stone)*

40. Vocalists sing very high notes most often in ad lib ornaments and cadenzas. The highest note written for a singer in a published score was F-sharp above high C, 1.480 kHz, sung by Zerbinetta in the original version of Richard Strauss's opera *Ariadne auf Naxos*. (a) Find the wavelength of this sound in air. (b) In response to complaints, Strauss later transposed the note down to F above high C, 1.397 kHz. By what increment did the wavelength change?

41. **Review Problem.** A block of mass $M$, supported by a string, rests on an incline making an angle of $\theta$ with the horizontal (Fig. P13.41). The length of the string is $L$ and its mass is $m \ll M$. Derive an expression for the time it takes a transverse wave to travel from one end of the string to the other.

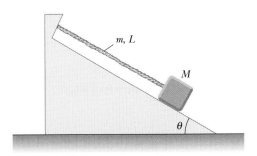

**Figure P13.41**

42. **Review Problem.** A block of mass $M$ hangs from a rubber cord. The block is supported so that the cord is not stretched. The unstretched length of the cord is $L_0$ and its mass is $m$, much less than $M$. The "force constant" for the cord is $k$. The block is released and stops at the lowest point. (a) Determine the tension in the string when the block is at this lowest point. (b) What is the length of the cord in this "stretched" position? (c) Find the speed of a transverse wave in the cord if the block is held in this lowest position.

43. A sound wave in a cylinder is described by Equations 13.24 through 13.26. Show that $\Delta P = \pm \rho v \omega \sqrt{s_{max}^2 - s^2}$.

44. A string on a musical instrument is held under tension $T$ and extends from the point $x = 0$ to the point $x = L$. The string is overwound with wire in such a way that its mass per unit length $\mu(x)$ increases linearly from $\mu_0$ at $x = 0$ to $\mu_L$ at $x = L$. (a) Find an expression for $\mu(x)$ as a function of $x$ over the range $0 \le x \le L$. (b) Show that the time required for a transverse pulse to travel the length of the string is given by

$$t = \frac{2L\left(\mu_L + \mu_0 + \sqrt{\mu_L \mu_0}\right)}{3\sqrt{T}\left(\sqrt{\mu_L} + \sqrt{\mu_0}\right)}$$

45. A rope of total mass $m$ and length $L$ is suspended vertically. Show that a transverse wave pulse travels the length of the rope in a time $t = 2\sqrt{L/g}$. (*Hint:* First find an expression for the wave speed at any point a distance $x$ from the lower end by considering the tension in the rope as resulting from the weight of the segment below that point.)

46. If an object of mass $M$ is suspended from the bottom of the rope in Problem 45, (a) show that the time for a transverse wave to travel the length of the rope is

$$t = 2\sqrt{\frac{L}{mg}}\left(\sqrt{M + m} - \sqrt{M}\right)$$

(b) Show that this reduces to the result of Problem 45 when $M = 0$. (c) Show that for $m \ll M$, the expression in part (a) reduces to

$$t = \sqrt{\frac{mL}{Mg}}$$

47. (a) Show that the speed of longitudinal waves along a spring of force constant $k$ is $v = \sqrt{kL/\mu}$, where $L$ is the unstretched length of the spring and $\mu$ is the mass per unit length. (b) A spring with a mass of 0.400 kg has an unstretched length of 2.00 m and a force constant of 100 N/m. Using the result you obtained in part (a), determine the speed of longitudinal waves along this spring.

48. A wave pulse traveling along a string of linear mass density $\mu$ is described by the wave function

$$y = [A_0 e^{-bx}] \sin(kx - \omega t)$$

where the factor in brackets before the sine function is said to be the amplitude. (a) What is the power $\mathcal{P}(x)$ carried by this wave at a point $x$? (b) What is the power $\mathcal{P}(0)$ carried by this wave at the origin? (c) Compute the ratio $\mathcal{P}(x)/\mathcal{P}(0)$.

49. An earthquake on the ocean floor in the Gulf of Alaska induces a *tsunami* (sometimes called a tidal wave) that reaches Hilo, Hawaii, 4 450 km away, in a time of 9 h 30 min. Tsunamis have enormous wavelengths (100–200 km), and the propagation speed for these waves is $v \approx \sqrt{gd}$, where $\overline{d}$ is the average depth of the water. From the information given, find the average wave speed and the average ocean depth between Alaska and Hawaii. (This method was used in 1856 to estimate the average depth of the Pacific Ocean long before soundings were made to give a direct determination.)

50. On a Saturday morning, pickup trucks and sport utility vehicles carrying garbage to the town dump form a nearly steady procession on a country road, all traveling at 19.7 m/s. From one direction, two trucks arrive at the dump every 3 min. A bicyclist is also traveling this way, at 4.47 m/s. (a) With what frequency do the trucks pass him? (b) A hill does not slow down the trucks, but makes the out-of-shape cyclist's speed drop to 1.56 m/s. How often do the trucks whiz past him now?

51. The ocean floor is underlain by a layer of basalt that constitutes the crust, or uppermost layer, of the Earth in that region. Below this crust is found denser periodotite rock, which forms the Earth's mantle. The boundary between these two layers is called the Mohorovičić discontinuity (Moho for short). If an explosive charge is set off at the surface of the basalt, it generates a seismic wave that is reflected back out at the Moho. If the speed of this wave in basalt is 6.50 km/s and the two-way travel time is 1.85 s, what is the thickness of this oceanic crust?

52. A train whistle ($f = 400$ Hz) appears to be higher or lower in frequency depending on whether it approaches or recedes. (a) Prove that the difference in frequency between the approaching and receding train whistle is

$$\Delta f = \frac{2u/v}{1 - u^2/v^2} f$$

where $u$ is the speed of the train and $v$ is the speed of sound. (b) Calculate this difference for a train moving at a speed of 130 km/h. Take the speed of sound in air to be 340 m/s.

53. A siren mounted on the roof of a firehouse emits sound at a frequency of 900 Hz. A steady wind is blowing with a speed of 15.0 m/s. Taking the speed of sound in calm air to be 343 m/s, find the wavelength of the sound (a) upwind of the siren and (b) downwind of the siren. The volunteer firefighters are approaching the siren from various directions at 15.0 m/s. What frequency does a firefighter hear (c) if he or she is approaching from an upwind position, so that he is moving in the direction in which the wind is blowing? (d) If he or she is approaching from a downwind position and moving against the wind?

54. The Doppler equation presented in the text is valid when the motion between the observer and the source occurs on a straight line, so that the source and observer are moving either directly toward or directly away from each other. If this restriction is relaxed, one must use the more general Doppler equation

$$f' = \left( \frac{v + v_O \cos \theta_O}{v - v_S \cos \theta_S} \right) f$$

where $\theta_O$ and $\theta_S$ are defined in Figure P13.54a. (a) Show that if the observer and source are moving away from each other, the preceding equation reduces to Equation 13.30 with negative values for both $v_O$ and $v_S$. (b) Use the preceding equation to solve the following problem. A train moves at a constant speed of 25.0 m/s toward the intersection shown in Figure P13.54b. A car is stopped near the intersection, 30.0 m from the tracks. If the train's horn emits a frequency of 500 Hz, what is the frequency heard by the passengers in the car when the train is 40.0 m from the intersection? Take the speed of sound to be 343 m/s.

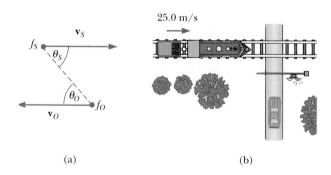

(a)

(b)

**Figure P13.54**

# ANSWERS TO QUICK QUIZZES

**13.1** (a) Longitudinal. The movement of the people is parallel to the direction of propagation of the gap. (b) Transverse. The fans stand up vertically, as the wave sweeps past them horizontally.

**13.2** The radio wave has the higher frequency. Although both waves have the same wavelength, the speed of the radio wave is much higher than that of the sound wave, so Equation 13.11 gives a higher frequency for the radio wave.

**13.3** Only choices (f) and (h) are correct. (a) and (b) affect the transverse speed of a particle of the string but not the wave speed along the string. (c) and (d) change the amplitude. (e) and (g) increase the time by decreasing the wave speed.

**13.4** A wave on a massless string would have an infinite speed of propagation, according to Equation 13.20.

**13.5** The apparent frequency would be Doppler shifted downward in frequency because the screaming person is moving away from you. Because the speed of the source increases due to the gravitational force, the apparent frequency decreases with time.

**13.6** (c). The sound that leaves your horn in the forward direction is Doppler shifted to a higher frequency because it is coming from a moving source. As the sound reflects back and comes toward you, you are a moving observer, so a second Doppler shift occurs to an even higher frequency.

# chapter
# 14

Guitarist Carlos Santana takes advantage of standing waves on strings. He changes to a higher note on the guitar by pushing the strings against the frets on the fingerboard, shortening the lengths of the portions of the strings that vibrate.

*(Corbis)*

# Superposition and Standing Waves

I n the previous chapter, we introduced the wave model. We have seen that waves are very different from particles—a particle is of zero size, but an ideal wave is of infinite length. Another important difference between waves and particles is that we can explore the possibility of two or more waves combining at one point in the same medium. We can combine particles to form extended objects, but the particles must be at different locations. In contrast, two waves can both be present at a given location, and the ramifications of this possibility are explored in this chapter.

One of the ramifications of the combination of waves is that only certain allowed frequencies can exist on systems with boundary conditions—that is, the frequencies are *quantized*. In Chapter 11 we learned about quantized energies of the hydrogen atom. Quantization is at the heart of quantum mechanics, a subject that is introduced formally in Chapter 28. We shall see that waves under boundary conditions explain many of the quantum phenomena. For our present purposes in this chapter, quantization enables us to understand the behavior of the wide array of musical instruments that are based on strings and air columns.

## 14.1 • THE PRINCIPLE OF SUPERPOSITION

Many interesting wave phenomena in nature cannot be described by a single wave. Instead, one must analyze complex waveforms in terms of a combination of many traveling waves. To analyze such wave combinations, we make use of the **principle of superposition,** which states that

> If two or more traveling waves are moving through a medium and combine at a given point, the resultant displacement of the medium at that point is the sum of the displacements of the individual waves.

This rather striking property is exhibited by many waves in nature, including waves on strings, sound waves, surface water waves, and electromagnetic waves. Waves that obey this principle are called *linear waves*. In general, linear waves have an amplitude that is small relative to their wavelength. Waves that violate the superposition principle are called *nonlinear waves* and are often characterized by large amplitudes. In this book, we shall deal only with linear waves.

A simple pictorial representation of the superposition principle is obtained by considering two pulses traveling in opposite directions on a stretched string, as in Figure 14.1. The wave function for the pulse moving to the right is $y_1$, and the

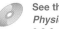
See the *Core Concepts in Physics CD-ROM*, Screens 9.6 & 9.7

• *Principle of superposition*

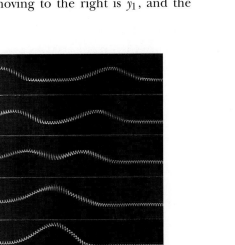

### Figure 14.1

(*Left*) Two wave pulses traveling on a stretched string in opposite directions pass through each other. When the pulses overlap, as in (b) and (c), the net displacement of the string equals the sum of the displacements produced by each pulse. Because each pulse produces positive displacements of the string, we refer to their superposition as *constructive interference*. (*Right*) Photograph of the superposition of two equal and symmetric pulses traveling in opposite directions on a stretched spring. (*Photo, Education Development Center, Newton, MA*) wave function for the pulse moving to the left is $y_2$. The pulses have the same speed but different shapes. Each pulse is assumed to be symmetric (although this is not a necessary condition), and in both cases the vertical displacements of the string are taken to be positive. When the waves begin to overlap (Fig. 14.1b), the resulting waveform is given by $y_1 + y_2$. When the crests of the pulses exactly coincide (Fig. 14.1c), the resulting waveform is symmetric. The two pulses finally separate and continue moving in their original directions (Fig. 14.1d). Note that the final waveforms remain unchanged, as if the two pulses had never met! The combination of separate waves in the same region of space to produce a resultant wave is called **interference.** Notice that the interference exists only while the waves are in the same region of space, and there is no permanent effect on the pulses after they separate.

One consequence of the superposition principle is that **two traveling waves can pass through each other without being destroyed or even altered.** For instance, when two pebbles are thrown into a pond, the expanding circular surface waves do not destroy one another. In fact, the ripples pass through one another. The complex pattern that is observed can be viewed as a combination of two independent sets of expanding circles.

For the two pulses shown in Figure 14.1, the vertical displacements are in the same direction, and so the resultant waveform (when the pulses overlap) ex-

**Figure 14.2**

(*Left*) Two wave pulses traveling in opposite directions with displacements that are inverted relative to each other. When the two overlap as in (c), their displacements subtract from each other. (*Right*) Photograph of the superposition of two symmetric pulses traveling in opposite directions, where one is inverted relative to the other. (*Photo, Education Development Center, Newton, MA*) hibits a displacement greater than those of the individual pulses. Now consider two identical pulses, again traveling in opposite directions on a stretched string, but this time one pulse is inverted relative to the other, as in Figure 14.2. In this case, when the pulses begin to overlap, the resultant waveform is the sum of the two separate displacements again, but one of the displacements is negative. Again, the two pulses pass through each other. When they exactly overlap, they partially *cancel* each other. At this time (Fig. 14.2c), the resultant amplitude is small.

## 14.2 • INTERFERENCE OF WAVES

In this section, we shall investigate the mathematics of waves in interference. Additional applications of this model applied to light waves are presented in Chapter 27.

Let us apply the superposition principle to two sinusoidal waves traveling in the same direction in a medium. If the two waves are traveling to the right and have the same frequency, wavelength, and amplitude but differ in phase, we can express their individual wave functions as

$$y_1 = A \sin(kx - \omega t) \quad \text{and} \quad y_2 = A \sin(kx - \omega t + \phi)$$

where $\phi$ is the phase difference between the two waves. Let us imagine that these two waves coincide in the medium. For example, these expressions might represent two waves traveling along the same string. In this situation, the resultant wave function $y$ is, according to the principle of superposition

$$y = y_1 + y_2 = A[\sin(kx - \omega t) + \sin(kx - \omega t + \phi)]$$

To simplify this expression, it is convenient to use the trigonometric identity

$$\sin a + \sin b = 2 \cos\left(\frac{a - b}{2}\right) \sin\left(\frac{a + b}{2}\right)$$

If we let $a = kx - \omega t$ and $b = kx - \omega t + \phi$, the resultant wave function $y$ reduces to

$$y = \left(2A \cos \frac{\phi}{2}\right) \sin\left(kx - \omega t + \frac{\phi}{2}\right) \qquad \text{[14.1]}$$

This mathematical representation of the resultant wave has several important features. The resultant wave function $y$ is also a sinusoidal wave and has the *same* frequency and wavelength as the individual waves. The amplitude of the resultant wave is $2A \cos(\phi/2)$, and the phase angle is $\phi/2$. If the phase angle $\phi$ equals 0, then $\cos(\phi/2) = \cos 0 = 1$ and the amplitude of the resultant wave is $2A$. In other words, the amplitude of the resultant wave is twice the amplitude of either individual wave. In this case, the waves are said to be everywhere *in phase* ($\phi = 0$) and to **interfere constructively.** That is, the crests of the individual waves occur at the same positions, as is shown by the blue line in Figure 14.3a. In general, constructive interference occurs when $\cos(\phi/2) = \pm 1$, or when $\phi = 0, 2\pi, 4\pi, \ldots$.

On the other hand, if $\phi$ is equal to $\pi$ radians, or to any *odd* multiple of $\pi$, then $\cos(\phi/2) = \cos(\pi/2) = 0$ and the resultant wave has *zero* amplitude everywhere. In this case, the two waves **interfere destructively.** That is, the crest of one wave coincides with the trough of the second (Fig. 14.3b) and their displacements cancel at every point. Finally, when the phase constant has a value between 0 and $\pi$, as in Figure 14.3c, the resultant wave has an amplitude whose value is somewhere between 0 and $2A$.

• *Constructive interference*

• *Destructive interference*

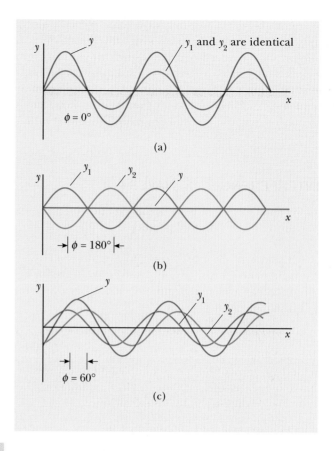

**Figure 14.3**

The superposition of two identical waves $y_1$ and $y_2$. (a) When the two waves are in phase, the result is constructive interference. (b) When the two waves are $\pi$ rad out of phase, the result is destructive interference. (c) When the phase angle has a value other than 0 or $\pi$ rad, the resultant wave $y$ falls somewhere between the extremes shown in (a) and (b).

Although we have used waves having the same amplitude in the preceding discussion, waves of differing amplitudes will interfere in a similar way. If the waves are in phase, the combined amplitude is the sum of the individual amplitudes. If they are 180° out of phase, the combined amplitude is the difference between the individual amplitudes.

It is often useful to express the phase difference between the two waves in terms of a fraction of the wavelength $\lambda$. Because a difference of one wavelength corresponds to a phase difference of $2\pi$ rad, an arbitrary shift $\Delta r$ in position of one wave with respect to another is related to the phase difference as follows:

• *Relationship between wavelength fraction and phase angle*

$$\frac{\Delta r}{\lambda} = \frac{\phi}{2\pi} \quad \longrightarrow \quad \Delta r = \frac{\phi}{2\pi}\lambda \qquad \text{[14.2]}$$

Thus, for example, a phase difference of 180° or $\pi$ rad corresponds to a shift of $\lambda/2$. Conversely, a 1/4-wavelength shift corresponds to a 90° phase difference.

Figure 14.4 shows a simple device for demonstrating interference of sound waves. Sound from speaker S is sent into a tube at $P$, where there is a T-shaped junction. Half the sound energy travels in one direction, and half in the opposite

**Figure 14.4**

An acoustical system for demonstrating interference of sound waves. Sound waves from the speaker propagate into the tube and the energy splits into two parts at point $P$. The waves from the two paths, which superimpose at the opposite side, are detected at the receiver $R$. The upper path length $r_2$ can be varied by sliding the upper section.

direction. Thus, the sound waves that reach receiver R at the other side can travel along either of two paths. The total distance from speaker to receiver is called the **path length** $r$. The length of the lower path is fixed at $r_1$. The upper path length $r_2$ can be varied by sliding the U-shaped tube (similar to that on a slide trombone). When the difference in the path lengths $\Delta r = |r_2 - r_1|$ is either zero or some integral multiple of the wavelength $\lambda$, the two waves reaching the receiver are in phase and interfere constructively, as in Figure 14.3a. In this case, a maximum in the sound intensity is detected at the receiver. If path length $r_2$ is adjusted so that $\Delta r$ is $\lambda/2, 3\lambda/2, \ldots n\lambda/2$ (for $n$ odd), the two waves are exactly 180° out of phase at the receiver and hence cancel each other. In this case of completely destructive interference, no sound is detected at the receiver. This simple experiment is a striking illustration of interference. In addition, it demonstrates the fact that a phase difference may arise between two waves generated by the same source when they travel along paths of unequal lengths.

Nature provides many other examples of interference phenomena. Later in the text we shall describe several interesting interference effects involving light waves.

## THINKING PHYSICS 14.1

If stereo speakers are connected to the amplifier "out of phase," one speaker is moving outward when the other is moving inward. This results in a weakness in the bass notes, which can be corrected by reversing the wires on one of the speaker connections. Why are only the bass notes affected in this case, and not the treble notes?

**Reasoning**  Imagine that you are sitting in front of the speakers, midway between them. Then, the sound from each speaker travels the same distance to you, so the phase difference in the sound is not due to a path difference. Because the speakers are connected out of phase, the sound waves are half a wavelength out of phase on leaving the speaker and, consequently, on arriving at your ear. As a result, the sound for all frequencies cancels, in the simplification model of a zero-size head located exactly on the midpoint between the speakers. If the ideal head were moved off the centerline, an additional phase difference is introduced by the path length difference for the sound from the two speakers. In the case of low-frequency, long-wavelength bass notes, the path length differences are a small fraction of a wavelength, so significant cancellation still occurs. For the high-frequency, short-wavelength treble notes, a small movement of the ideal head results in a much larger fraction of a wavelength in path length difference, or even multiple wavelengths. Thus, the treble notes could be in phase with this head movement. If we now add the fact that the head is not of zero size, and the fact that it has two ears, we can see that complete cancellation is not possible, and, with even small movements of the head, one or both ears will be at or near maxima for the treble notes. The size of the head is much smaller than bass wavelengths, however, so the bass notes are significantly weakened over much of the region in front of the speakers.

### Example 14.1   Two Speakers Driven by the Same Source

Two speakers placed 3.00 m apart are driven in phase by the same oscillator (Fig. 14.5). A listener is originally at point $O$, which is located 8.00 m from the center of the line connecting the two speakers. The listener then moves to point $P$, which is a perpendicular distance 0.350 m from $O$ before reaching the first cancellation of waves, resulting in a minimum in sound intensity. What is the frequency of the oscillator?

**Solution**   The first cancellation occurs when the two waves reaching the listener at $P$ are 180° out of phase—in other words, when their path difference equals $\lambda/2$. To calculate the path difference, we must first find the path lengths $r_1$ and $r_2$. Consider the two geometric model triangles shaded in Figure 14.5. Making use of these triangles, we find the path lengths to be

$$r_1 = \sqrt{(8.00 \text{ m})^2 + (1.15 \text{ m})^2} = 8.08 \text{ m}$$

$$r_2 = \sqrt{(8.00 \text{ m})^2 + (1.85 \text{ m})^2} = 8.21 \text{ m}$$

Hence, the path difference is $r_2 - r_1 = 0.13$ m. Because we require that this path difference be equal to $\lambda/2$ for the first minimum, we find that $\lambda = 0.26$ m.

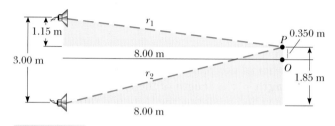

**Figure 14.5**

(Example 14.1).

To obtain the oscillator frequency, we use $v = \lambda f$, where $v$ is the speed of sound in air, 343 m/s:

$$f = \frac{v}{\lambda} = \frac{343 \text{ m/s}}{0.26 \text{ m}} = \boxed{1.3 \text{ kHz}}$$

**EXERCISE**   If the oscillator frequency is adjusted such that the listener hears the first minimum at a distance of 0.75 m from $O$, what is the new frequency?

*Answer*   0.63 kHz

## 14.3 • STANDING WAVES

In the previous section, we considered two waves moving in the same direction and combining according to the principle of superposition. Now imagine that we have two waves that are identical except that they are traveling in opposite directions in the same medium. This happens, for example, when you shake one end of a rope whose opposite end is attached to a rigid wall. The waves reflected from the wall are identical to the waves approaching the wall, except for the direction of travel and an inversion of the wave on reflection. An interesting pattern emerges from this type of combination of waves.

If we ignore any phase angle $\phi$ for simplicity, the wave functions for the two waves we have described can be written

$$y_1 = A \sin(kx - \omega t) \qquad \text{and} \qquad y_2 = A \sin(kx + \omega t)$$

where $y_1$ represents a wave traveling to the right and $y_2$ represents a wave traveling to the left. According to the principle of superposition, adding these two functions

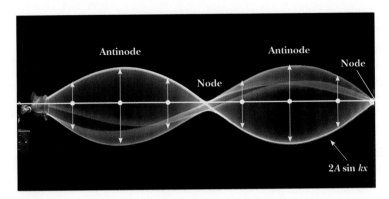

**Figure 14.6**

Multiflash photograph of a standing wave on a string. The vertical displacement from equilibrium of an individual element of the string is given by cos $\omega t$. That is, each element vibrates at an angular frequency $\omega$. The amplitude of the vertical oscillation of any element on the string depends on the horizontal position of the element. Each element vibrates within the confines of the envelope function $2A \sin kx$. (©1991 Richard Megna/Fundamental Photographs)

gives the resultant wave function $y$:

$$y = y_1 + y_2 = A \sin(kx - \omega t) + A \sin(kx + \omega t)$$

Using the trigonometric identity $\sin(a \pm b) = \sin a \cos b \pm \cos a \sin b$, this reduces to

$$y = (2A \sin kx) \cos \omega t \qquad \text{[14.3]}$$

Notice that this function does not look mathematically like a traveling wave—there is no function of $kx - \omega t$. Equation 14.3 represents the wave function of a **standing wave,** such as that shown in Figure 14.6. A standing wave is an oscillation pattern that results from two waves traveling in opposite directions. Mathematically, this equation looks more like simple harmonic motion than wave motion for traveling waves. Every particle of the string vibrates in simple harmonic motion with the same angular frequency $\omega$ (according to the factor cos $\omega t$). The amplitude of motion of a given particle (the factor $2A \sin kx$), however, depends on its position along the string, described by the variable $x$. This is in contrast to the situation involving a traveling sinusoidal wave, in which all particles oscillate with the same amplitude as well as the same frequency. From this result, we see that the simple harmonic motion of every particle has an angular frequency of $\omega$ and a position-dependent amplitude of $2A \sin kx$.

Because the amplitude of the simple harmonic motion at any value of $x$ is equal to $2A \sin kx$, we see that the *maximum* amplitude of the simple harmonic motion has the value $2A$. This is described as the amplitude of the standing wave. This maximum amplitude occurs when the coordinate $x$ satisfies the condition $\sin kx = 1$, or when

$$kx = \frac{\pi}{2}, \frac{3\pi}{2}, \frac{5\pi}{2}, \ldots$$

• *Wave function for a standing wave*

**PITFALL PREVENTION 14.2**
**Three types of amplitude**

We need to distinguish carefully here between the **amplitude of the individual waves,** which is $A$, and the **amplitude of the simple harmonic motion of the particles of the medium,** which is $2A \sin kx$. A given particle in a standing wave vibrates within the constraints of the *envelope* function $2A \sin kx$, where $x$ is that particle's position in the medium. This is in contrast to traveling sinusoidal waves, in which all particles oscillate with the same amplitude and the same frequency, and the amplitude $A$ of the wave is the same as the amplitude $A$ of the simple harmonic motion of the particles. Furthermore, we can identify the **amplitude of the standing wave** as $2A$.

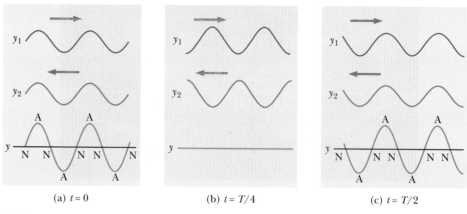

(a) $t = 0$          (b) $t = T/4$          (c) $t = T/2$

**Figure 14.7**

Standing wave patterns at various times produced by two waves of equal amplitude traveling in opposite directions. For the resultant wave $y$, the nodes (N) are points of zero displacement and the antinodes (A) are points of maximum displacement.

Because $k = 2\pi/\lambda$, the positions of maximum amplitude, called **antinodes,** are

- *Position of antinodes*

$$x = \frac{\lambda}{4}, \frac{3\lambda}{4}, \frac{5\lambda}{4}, \ldots = \frac{n\lambda}{4} \qquad \text{[14.4]}$$

where $n = 1, 3, 5, \ldots$ . Note that **adjacent antinodes are separated by a distance of λ/2.**

Similarly, the simple harmonic motion has a *minimum* amplitude of zero when $x$ satisfies the condition $\sin kx = 0$, or when $kx = \pi, 2\pi, 3\pi, \ldots$ giving

- *Position of nodes*

$$x = \frac{\lambda}{2}, \lambda, \frac{3\lambda}{2}, \ldots = \frac{n\lambda}{2} \qquad \text{[14.5]}$$

where $n = 1, 2, 3, \ldots$ **These points of zero amplitude, called nodes, are also spaced by λ/2.** The distance between a node and an adjacent antinode is $\lambda/4$.

The standing wave patterns produced at various times by two waves traveling in opposite directions are represented graphically in Figure 14.7. The upper part of each figure represents the individual traveling waves, and the lower part represents the standing wave patterns. The nodes of the standing wave are labeled N, and the antinodes are labeled A. At $t = 0$ (Fig. 14.7a), the two waves are in phase, giving a wave pattern with amplitude $2A$. One quarter of a period later, at $t = T/4$ (Fig. 14.7b), the individual waves have moved one quarter of a wavelength (one to the right and the other to the left). At this time, the waves are 180° out of phase. The individual displacements are of equal magnitude and opposite direction for all values of $x$, and hence the resultant wave has zero displacement everywhere. At $t = T/2$ (Fig. 14.7c), the individual waves are again in phase, producing a wave pattern that is inverted relative to the $t = 0$ pattern.

**Quick Quiz 14.1**

For the standing wave described by Equation 14.3, what is the maximum transverse speed of a particle of the medium located at a node? What is the maximum transverse speed of a particle of the medium located at an antinode?

### Example 14.2   Formation of a Standing Wave

Two waves traveling in opposite directions produce a standing wave. The individual wave functions are

$$y_1 = (4.0 \text{ cm}) \sin(3.0x - 2.0t)$$

$$y_2 = (4.0 \text{ cm}) \sin(3.0x + 2.0t)$$

where $x$ and $y$ are in centimeters. (a) Find the maximum displacement of a particle of the medium at $x = 2.3$ cm.

**Solution**   When the two waves are summed, the result is a standing wave whose mathematical representation is given by Equation 14.3, with $A = 4.0$ cm and $k = 3.0$ rad/cm:

$$y = (2A \sin kx) \cos \omega t = [(8.0 \text{ cm}) \sin 3.0x] \cos 2.0t$$

Thus, the maximum displacement of a particle at the position $x = 2.3$ cm is

$$y_{max} = [(8.0 \text{ cm}) \sin 3.0x]_{x=2.3 \text{ cm}}$$
$$= (8.0 \text{ cm}) \sin(6.9 \text{ rad}) = \boxed{4.6 \text{ cm}}$$

(b) Find the positions of the nodes and antinodes.

**Solution**   Because $k = 2\pi/\lambda = 3.0$ rad/cm, we see that $\lambda = 2\pi/3$ cm. Therefore, from Equation 14.4 we find that the antinodes are located at

$$x = n\left(\frac{\pi}{6.0}\right) \text{ cm} \quad (n = 1, 3, 5, \ldots)$$

and from Equation 14.5 we find that the nodes are located at

$$x = n\frac{\lambda}{2} = n\left(\frac{\pi}{3.0}\right) \text{ cm} \quad (n = 1, 2, 3, \ldots)$$

## 14.4 • STANDING WAVES IN STRINGS

In the preceding section, we discussed standing waves formed by identical waves moving in opposite directions in the same medium. As you will recall, one way to do this on a string is to combine incoming and reflected waves from a rigid end. If a string is stretched between *two* rigid supports (Fig. 14.8a) and waves are established on the string, standing waves will be set up in the string by the continuous superposition of the waves incident on and reflected from the ends. This physical system is a model for the source of sound in any stringed instrument, such as the guitar, the violin, and the piano. The string has a number of natural patterns of

 See Screen 9.9

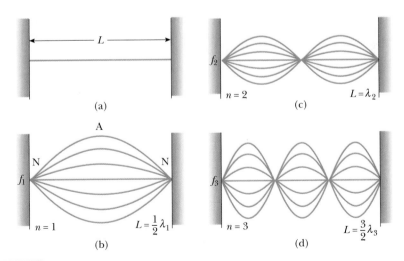

### Figure 14.8

(a) A string of length $L$ fixed at both ends. The normal modes of vibration form a harmonic series:
(b) the fundamental frequency, or first harmonic; (c) the second harmonic; and (d) the third harmonic.

vibration, called **normal modes.** Each of these modes has a characteristic frequency.

This discussion is our first introduction to an important analysis model—the **wave under boundary conditions.** When boundary conditions are applied to a wave, we find very interesting behavior that has no analog in the physics of particles. The most prominent aspect of this behavior is **quantization.** We shall find that only certain waves are allowed—those that satisfy the boundary conditions. The notion of quantization was introduced in Chapter 11 when we discussed the Bohr model of the atom. In that model, angular momentum was quantized. As we shall see in Chapter 29, this is just an application of the wave under boundary conditions model.

In the standing wave pattern on a stretched string, the ends of the string must be nodes, because these points are fixed. This is the boundary condition on the waves. The rest of the pattern can be built from this boundary condition along with the requirement that nodes and antinodes are equally spaced and separated by one fourth of a wavelength. The simplest pattern that satisfies these conditions has the required nodes at the ends of the string and an antinode at the center point (Fig. 14.8b). For this normal mode, the length of the string equals $\lambda/2$ (the distance between adjacent nodes):

$$L = \frac{\lambda_1}{2} \qquad \text{or} \qquad \lambda_1 = 2L$$

The next normal mode, of wavelength $\lambda_2$ (Fig. 14.8c), occurs when the length of the string equals one wavelength, that is, when $\lambda_2 = L$. In this mode, the two halves of the string are moving in opposite directions at a given instant—we sometimes say that two *loops* occur. The third normal mode (Fig. 14.8d) corresponds to the case when the length equals $3\lambda/2$; therefore, $\lambda_3 = 2L/3$. In general, the wavelengths of the various normal modes can be conveniently expressed as

- *Wavelengths of normal modes*

$$\lambda_n = \frac{2L}{n} \quad (n = 1, 2, 3, \ldots) \qquad \textbf{[14.6]}$$

where the index $n$ refers to the $n$th mode of vibration. The natural frequencies associated with these modes are obtained from the relationship $f = v/\lambda$, where the wave speed $v$ is determined by the tension $T$ and linear mass density $\mu$ of the string and, therefore, is the same for all frequencies. Using Equation 14.6, we find that the frequencies of the normal modes are

- *Frequencies of normal modes as functions of wave speed and length of string*

$$f_n = \frac{v}{\lambda_n} = \frac{n}{2L} v \quad (n = 1, 2, 3, \ldots) \qquad \textbf{[14.7]}$$

Because $v = \sqrt{T/\mu}$ (Equation 13.20), we can express the natural frequencies of a stretched string as

- *Frequencies of normal modes as functions of string tension and linear mass density*

$$f_n = \frac{n}{2L} \sqrt{\frac{T}{\mu}} \quad (n = 1, 2, 3, \ldots) \qquad \textbf{[14.8]}$$

Equation 14.8 demonstrates the quantization that we mentioned as a feature of the wave under boundary conditions model. The frequencies are quantized—only certain frequencies of waves satisfy the boundary conditions and can exist on the string. The lowest frequency, corresponding to $n = 1$, is called the **fundamental frequency** $f_1$ and is

$$f_1 = \frac{1}{2L}\sqrt{\frac{T}{\mu}}$$    [14.9]

• *Fundamental frequency of a taut string*

Equation 14.8 shows that the frequencies of the higher modes are integral multiples of the fundamental frequency—that is, $2f_1$, $3f_1$, $4f_1$, and so on. These higher natural frequencies, together with the fundamental frequency, form a **harmonic series** and the various frequencies are called **harmonics.** The fundamental $f_1$ is the first harmonic; the frequency $f_2 = 2f_1$ is the second harmonic; the frequency $f_n$ is the $n$th harmonic.

We can obtain the foregoing results in an alternative manner. Because the string is fixed at $x = 0$ and $x = L$, we require that the wave function $y(x, t)$ given by Equation 14.3 must be *zero* at these points for *all* times—this is a mathematical representation for the boundary conditions on the waves. Thus, the boundary conditions require that $y(0, t) = 0$ and $y(L, t) = 0$ for all values of $t$. Because $y = (2A \sin kx) \cos \omega t$, the first condition, $y(0, t) = 0$, is automatically satisfied because $\sin kx = 0$ at $x = 0$. To meet the second condition, $y(L, t) = 0$, we require that $\sin kL = 0$. This condition is satisfied when the angle $kL$ equals an integral multiple of $\pi$ radians or 180°. Therefore, the allowed values of $k$ are*

$$k_n L = n\pi \quad (n = 1, 2, 3, \dots)$$    [14.10]

where the notation $k_n$ indicates that each value of $n$ has an allowed value of $k$. Because $k_n = 2\pi/\lambda_n$, we find that

$$\left(\frac{2\pi}{\lambda_n}\right) L = n\pi \qquad \text{or} \qquad \lambda_n = \frac{2L}{n}$$

which is identical to Equation 14.6.

If a stretched string is distorted to a shape that corresponds to any one of its harmonics, after being released it will vibrate at the frequency of that harmonic. If the string is plucked, bowed, or struck, however, as occurs when playing a stringed instrument, the resulting vibration will include frequencies of many modes, including the fundamental. In effect, the string "selects" a mixture of normal-mode frequencies when disturbed by a finger or a bow. The frequency of the combination is that of the fundamental because that is the rate at which the waveform repeats—the frequency associated with the string by a listener is that of the fundamental.

Plucking the bottom string on the guitar in the photograph without pressing it to the fingerboard produces the note E. By pressing the string to the fingerboard at the first fret, the length of the string is reduced and the fundamental frequency of vibration is increased. The first fret is located so that the note being played is F. *(Charles D. Winters)*

---

* We exclude $n = 0$ because this corresponds to the trivial case when no wave exists ($k = 0$).

The frequency of a given string on a stringed instrument can be changed either by varying the string's tension $T$ or by changing the length $L$ of the vibrating portion of the string. For example, the tension in the strings of guitars and violins is adjusted by a screw mechanism or by tuning pegs on the neck of the instrument. As the tension increases, the frequencies of the normal modes increase according to Equation 14.8. Once the instrument is "tuned," the player varies the frequency by moving his or her fingers along the neck, thereby changing the length of the vibrating portion of the string. As this length is reduced, the frequency increases because the normal-mode frequencies are inversely proportional to the length of the vibrating portion of the string.

Imagine that we have several strings of the same length under the same tension but varying linear mass density $\mu$. The strings will have different wave speeds and, therefore, different fundamental frequencies. The linear mass density can be changed either by varying the diameter of the string or by wrapping extra mass around the string. Both of these possibilities can be seen on the guitar, on which the higher frequency strings vary in diameter and the lower frequency strings have additional wire wrapped around them.

**Quick Quiz  14.2**

The bass strings on a piano are longer and thicker than the treble strings. Why?

---

### Example 14.3   Give Me a C Note

A middle C string on a piano has a fundamental frequency of 262 Hz, and the A note has a fundamental frequency of 440 Hz. (a) Calculate the frequencies of the next two harmonics of the C string.

**Solution**   Because $f_1 = 262$ Hz for the C string, we can use Equation 14.7 to find the frequencies $f_2$ and $f_3$:

$$f_2 = 2f_1 = \boxed{524 \text{ Hz}}$$

$$f_3 = 3f_1 = \boxed{786 \text{ Hz}}$$

(b) If the strings for the A and C notes are assumed to have the same mass per unit length and the same length, determine the ratio of tensions in the two strings.

**Solution**   Using Equation 14.9 for the two strings vibrating at their fundamental frequencies gives

$$f_{1A} = \frac{1}{2L}\sqrt{\frac{T_A}{\mu}}$$

$$f_{1C} = \frac{1}{2L}\sqrt{\frac{T_C}{\mu}} \longrightarrow \frac{f_{1A}}{f_{1C}} = \sqrt{\frac{T_A}{T_C}}$$

$$\frac{T_A}{T_C} = \left(\frac{f_{1A}}{f_{1C}}\right)^2 = \left(\frac{440 \text{ Hz}}{262 \text{ Hz}}\right)^2 = \boxed{2.82}$$

(c) In a real piano, the assumption we made in part (b) is only partially true. The string densities are equal, but the A string is 64% as long as the C string. What is the ratio of their tensions?

**Solution**   We start from the same point as in part (b), but the string lengths do not cancel in the ratio:

$$f_{1A} = \frac{1}{2L_A}\sqrt{\frac{T_A}{\mu}}$$

$$f_{1C} = \frac{1}{2L_C}\sqrt{\frac{T_C}{\mu}} \longrightarrow \frac{f_{1A}}{f_{1C}} = \frac{L_C}{L_A}\sqrt{\frac{T_A}{T_C}}$$

$$\frac{T_A}{T_C} = \left(\frac{L_A}{L_C}\right)^2\left(\frac{f_{1A}}{f_{1C}}\right)^2 = (0.64)^2\left(\frac{440 \text{ Hz}}{262 \text{ Hz}}\right)^2 = \boxed{1.16}$$

**EXERCISE**   A stretched string is 160 cm long and has a linear mass density of 0.0150 g/cm. What tension in the string will result in a second harmonic of 460 Hz?

*Answer*   813 N

## 14.5 • STANDING WAVES IN AIR COLUMNS

See Screen 9.9

We have discussed musical instruments that use strings, which include guitars, violins, and pianos. What about instruments classified as brasses or woodwinds? These produce music using a column of air. Standing longitudinal waves can be set up in an air column, such as an organ pipe, as the result of interference between longitudinal sound waves traveling in opposite directions. Whether a node or an anti-node occurs at the end of an air column depends on whether that end is open or closed. **The closed end of an air column is a displacement node,** just as the fixed end of a vibrating string is a displacement node. Furthermore, because the pressure wave is 90° out of phase with the displacement wave (Section 13.7), the **closed end of an air column corresponds to a pressure antinode** (i.e., a point of maximum pressure variation). On the other hand, the **open end of an air column is approximately a displacement antinode and a pressure node.**

Strictly speaking, the open end of an air column is not exactly an antinode. A compression in the sound wave does not reach full expansion until it passes somewhat beyond the open end. Thus, to accurately calculate frequencies of the normal modes, an **end correction** must be added to the length of the air column at each open end. For a thin-walled tube of circular cross-section, this end correction is about $0.6R$, where $R$ is the tube's radius. Hence, the effective length of the tube is somewhat greater than the true length $L$.

We can determine the modes of vibration of an air column by applying the appropriate boundary condition at the end of the column, along with the requirement that nodes and antinodes be separated by one fourth of a wavelength. We shall find that the frequency for sound waves in air columns is quantized, similar to the results found for waves on strings under boundary conditions.

The first three modes of vibration of an air column that is open at both ends are shown in Figure 14.9a. Note that the ends are displacement antinodes (approximately). In the fundamental mode, the wavelength is twice the length of the air column, and hence the frequency of the fundamental $f_1$ is $v/2L$. Similarly, the frequencies of the higher harmonics are $2f_1$, $3f_1$, . . . . Thus, in an air column that is open at both ends, the natural frequencies of vibration form a harmonic series; that is, the higher harmonics are integral multiples of the fundamental frequency. Because all harmonics are present, we can express the natural frequencies of vibration as

$$f_n = n\frac{v}{2L} \quad (n = 1, 2, 3, \ldots) \qquad \text{[14.11]}$$

• *Natural frequencies of an air column open at both ends*

where $v$ is the speed of sound in air.

If an air column is closed at one end and open at the other, the closed end is a displacement node (Fig. 14.9b). In this case, the wavelength for the fundamental mode is four times the length of the column. Hence, the fundamental frequency $f_1$ is equal to $v/4L$, and the frequencies of the higher harmonics are equal to $3f_1$, $5f_1$, . . . . That is, **in an air column that is closed at one end, only odd harmonics are present,** and these are

$$f_n = n\frac{v}{4L} \quad (n = 1, 3, 5, \ldots) \qquad \text{[14.12]}$$

• *Natural frequencies of an air column closed at one end and open at the other*

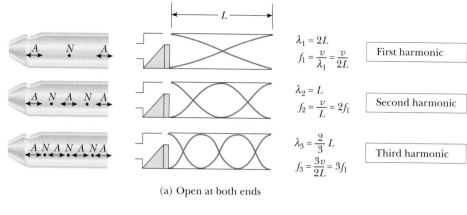

$\lambda_1 = 2L$
$f_1 = \dfrac{v}{\lambda_1} = \dfrac{v}{2L}$ — First harmonic

$\lambda_2 = L$
$f_2 = \dfrac{v}{L} = 2f_1$ — Second harmonic

$\lambda_3 = \dfrac{2}{3}L$
$f_3 = \dfrac{3v}{2L} = 3f_1$ — Third harmonic

(a) Open at both ends

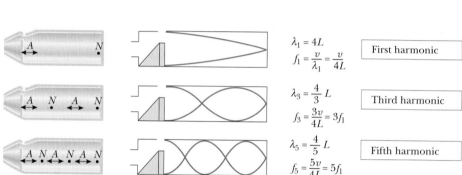

$\lambda_1 = 4L$
$f_1 = \dfrac{v}{\lambda_1} = \dfrac{v}{4L}$ — First harmonic

$\lambda_3 = \dfrac{4}{3}L$
$f_3 = \dfrac{3v}{4L} = 3f_1$ — Third harmonic

$\lambda_5 = \dfrac{4}{5}L$
$f_5 = \dfrac{5v}{4L} = 5f_1$ — Fifth harmonic

(b) Closed at one end, open at the other

**Figure 14.9**

Motion of elements of air in standing longitudinal waves in an air column, along with graphical representations of the displacements of the elements. (a) In an air column open at both ends, the harmonic series created consists of all integer multiples of the fundamental frequency: $f_1, 2f_1, 3f_1, \ldots$ (b) In an air column closed at one end and open at the other, the harmonic series consists of only odd-integer multiples of the fundamental frequency: $f_1, 3f_1, 5f_1, \ldots$

**PITFALL PREVENTION 14.4**

**Sound waves in air are not transverse**

Note that the standing longitudinal waves are drawn as transverse waves in Figure 14.9. This is because it is difficult to draw longitudinal displacements—they are in the same direction as the propagation. Thus, it is best to interpret the curves in Figure 14.9 as a graphical representation of the waves (our diagrams of string waves are pictorial representations), with the vertical axis representing horizontal displacement of the elements of the medium.

Standing waves in air columns are the primary sources of the sounds produced by wind instruments. In a woodwind instrument, a key is pressed, opening a hole in the side of the column. This hole defines the end of the vibrating column of air (because the hole acts as an open end—pressure can be released), so that the column is effectively shortened and the fundamental frequency rises. In a brass instrument, the length of the air column is changed by an adjustable section, as in a trombone, or by adding segments of tubing, as is done in a trumpet when a valve is pressed.

**Quick Quiz 14.3**

A pipe open at both ends resonates at a fundamental frequency $f_{\text{open}}$. When one end is closed and the pipe is again made to resonate, the fundamental frequency is $f_{\text{closed}}$. Which of the following expressions describes how these two resonance frequencies compare? (a) $f_{\text{closed}} = f_{\text{open}}$ (b) $f_{\text{closed}} = \frac{1}{2}f_{\text{open}}$ (c) $f_{\text{closed}} = 2f_{\text{open}}$ (d) $f_{\text{closed}} = \frac{3}{2}f_{\text{open}}$

## Quick Quiz 14.4

Balboa Park in San Diego, California, has an outdoor organ. Does the fundamental frequency of a particular air column on this organ change on hot and cold days? How about on days with high and low atmospheric pressure?

### THINKING PHYSICS 14.2

A bugle has no valves, keys, slides, or finger holes—how can it play a song?

**Reasoning**  Songs for the bugle are limited to harmonics of the fundamental frequency because the bugle has no control over frequencies by means of valves, keys, slides, or finger holes. The player obtains different notes by changing the tension in the lips as the bugle is played, in order to excite different harmonics. The normal playing range of a bugle is among the third, fourth, fifth, and sixth harmonics of the fundamental. For example, "Reveille" is played with just the three notes D (294 Hz), G (392 Hz), and B (494 Hz), and "Taps" is played with these same three notes and the D one octave above the lower D (588 Hz).

### THINKING PHYSICS 14.3

If an orchestra doesn't warm up before a performance, the strings go flat and the wind instruments go sharp during the performance. Why?

**Reasoning**  Without warming up, all the instruments will be at room temperature at the beginning of the concert. As the wind instruments are played, they fill with warm air from the player's exhalation. The increase in temperature of the air in the instrument causes an increase in the speed of sound, which raises the resonance frequencies of the air columns. As a result, the wind instruments go sharp. The strings on the stringed instruments also increase in temperature due to the friction of rubbing with the bow. This results in thermal expansion, which causes a decrease in the tension in the strings. (We will study thermal expansion in Chapter 16.) With a decrease in tension, the wave speed on the strings drops, and the fundamental frequencies decrease. Thus, the stringed instruments go flat.

### Example 14.4   Resonance in a Pipe

A pipe has a length of 1.23 m. (a) Determine the frequencies of the first three harmonics if the pipe is open at each end. Take $v = 343$ m/s as the speed of sound in air.

**Solution**  The first harmonic of a pipe open at both ends is

$$f_1 = \frac{v}{2L} = \frac{343 \text{ m/s}}{2(1.23 \text{ m})} = \boxed{139 \text{ Hz}}$$

Because all harmonics are possible for a pipe open at both ends, the second and third harmonics are $f_2 = 2f_1 = \boxed{278 \text{ Hz}}$ and $f_3 = 3f_1 = \boxed{417 \text{ Hz}}$.

(b) What are the three frequencies determined in part (a) if the pipe is closed at one end?

**Solution**  The fundamental frequency of a pipe closed at one end is

$$f_1 = \frac{v}{4L} = \frac{343 \text{ m/s}}{4(1.23 \text{ m})} = \boxed{69.7 \text{ Hz}}$$

In this case, only odd harmonics are present, and so the next two resonances have frequencies $f_3 = 3f_1 = \boxed{209 \text{ Hz}}$ and $f_5 = 5f_1 = \boxed{349 \text{ Hz}}$.

(c) For the pipe open at both ends, how many harmonics are present in the normal human hearing range (20–20 000 Hz)?

**Solution** Because all harmonics are present, $f_n = nf_1$. For $f_n = 20\ 000$ Hz, we have $n = 20\ 000/139 = 144$, so that 144

harmonics are present in the audible range. Actually, only the first few harmonics have sufficient amplitude to be heard.

---

## Example 14.5 Measuring the Frequency of a Tuning Fork

A simple apparatus for demonstrating resonance in a tube is described in Figure 14.10a. A long, vertical tube open at both ends is partially submerged in a beaker of water, and a vibrating tuning fork of unknown frequency is placed near the top. The length $L$ of the air column is adjusted by moving the tube vertically. The sound waves generated by the fork are reinforced when the length of the air column corresponds to one of the resonant frequencies of the tube.

For a certain tube, the smallest value of $L$ for which a peak occurs in the sound intensity is 9.00 cm. From this measurement, determine the frequency of the tuning fork and the value of $L$ for the next two resonant modes.

**Reasoning** Although the tube is open at the bottom end to allow the water in, the water surface acts like a rigid barrier at that end. This setup can therefore be modeled as an air column closed at one end, and the fundamental has a frequency of $v/4L$, where $L$ is the length of the tube from the open end to the water surface (Fig. 14.10b).

**Solution** Taking $v = 343$ m/s for the speed of sound in air and $L = 0.090\ 0$ m, we have

$$f_1 = \frac{v}{4L} = \frac{343 \text{ m/s}}{4(0.0900 \text{ m})} = \boxed{953 \text{ Hz}}$$

From the information about the fundamental mode, we see that the wavelength is $\lambda = 4L = 0.360$ m. Because the frequency of the source is constant, the next two resonance modes (Fig. 14.10b) correspond to lengths of $3\lambda/4 = \boxed{0.270 \text{ m}}$ and $5\lambda/4 = \boxed{0.450 \text{ m}}$.

**EXERCISE** The longest pipe on a certain organ is 4.88 m.

First resonance

Second resonance (third harmonic)

Third resonance (fifth harmonic)

(b)

**Figure 14.10**

(Example 14.5) (a) Apparatus for demonstrating the resonance of sound waves in a tube closed at one end. The length $L$ of the air column is varied by moving the tube vertically while it is partially submerged in water. (b) The first three normal modes of the system shown in part (a).

What is the fundamental frequency (at 0.0°C) if the non-driven end of the pipe is (a) closed and (b) open?

*Answer* (a) 17.0 Hz (b) 34.0 Hz

**EXERCISE** An air column 2.00 m in length is open at both ends. The frequency of a certain harmonic is 410 Hz, and the frequency of the next higher harmonic is 492 Hz. Determine the speed of sound in the air column.

*Answer* 328 m/s

---

## 14.6 • BEATS: INTERFERENCE IN TIME

The interference phenomena that we have discussed so far involve the superposition of two or more waves with the same frequency. Because the resultant displacement of a particle in the medium in this case depends on the position of the particle, we can refer to the phenomenon as **spatial interference.** Standing waves in strings and air columns are common examples of spatial interference.

We now consider another type of interference effect, one that results from the superposition of two waves with slightly *different* frequencies. In this case, when the two waves of amplitudes $A_1$ and $A_2$ are observed at a given point, they are alternately in and out of phase. We refer to this phenomenon as **interference in time** or **temporal interference.** When the waves are in phase, the combined amplitude is $A_1 + A_2$. When they are out of phase, the combined amplitude is $|A_1 - A_2|$. The combination therefore varies between small and large amplitudes, resulting in what we call **beats.**

Although beats occur for all types of waves, they are particularly noticeable for sound waves. For example, if two tuning forks of slightly different frequencies are struck, you hear a sound of pulsating intensity.

The number of beats you hear per second, the **beat frequency,** equals the difference in frequency between the two sources. The maximum beat frequency that the human ear can detect is about 20 beats/s. When the beat frequency exceeds this value, it blends with the sounds producing the beats.

One can use beats to tune a stringed instrument, such as a piano, by beating a note against a reference tone of known frequency. The frequency of the string can then be adjusted to equal the frequency of the reference by changing the string's tension until the beats disappear—then the two frequencies are the same.

Let us look at the mathematical representation of beats. Consider two waves with equal amplitudes, traveling through a medium with slightly different frequencies, $f_1$ and $f_2$. We can represent the displacement that each wave would produce at a fixed point, which we choose as $x = 0$, as

$$y_1 = A \cos 2\pi f_1 t \qquad \text{and} \qquad y_2 = A \cos 2\pi f_2 t$$

Using the superposition principle, we find that the resultant displacement at that point is given by

$$y = y_1 + y_2 = A(\cos 2\pi f_1 t + \cos 2\pi f_2 t)$$

It is convenient to write this in a form that uses the trigonometric identity

$$\cos a + \cos b = 2 \cos\left(\frac{a-b}{2}\right) \cos\left(\frac{a+b}{2}\right)$$

Letting $a = 2\pi f_1 t$ and $b = 2\pi f_2 t$, we find that

$$y = \left[2A \cos 2\pi\left(\frac{f_1 - f_2}{2}\right)t\right] \cos 2\pi\left(\frac{f_1 + f_2}{2}\right)t \qquad \textbf{[14.13]}$$

Graphs demonstrating the individual waveforms as well as the resultant wave are shown in Figure 14.11. From the factors in Equation 14.13, we see that the resultant simple harmonic motion at a point has an effective frequency equal to the average frequency $(f_1 + f_2)/2$ and an amplitude of

$$A_{x=0} = 2A \cos 2\pi\left(\frac{f_1 - f_2}{2}\right)t \qquad \textbf{[14.14]}$$

That is, the *amplitude varies in time* with a frequency of $(f_1 - f_2)/2$. When $f_1$ is close to $f_2$, this amplitude variation is slow compared with the frequency of the individual waves, as illustrated by the envelope (broken line) of the resultant waveform in Figure 14.11b.

Note that a maximum in amplitude will be detected whenever

$$\cos 2\pi\left(\frac{f_1 - f_2}{2}\right)t = \pm 1$$

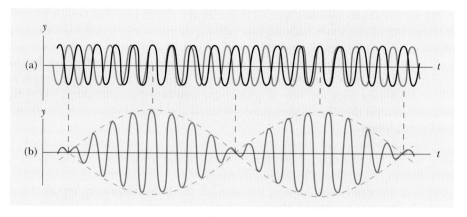

**Figure 14.11**

Beats are formed by the combination of two waves of slightly different frequencies. (a) The individual waves. (b) The combined wave has an amplitude (broken line) that oscillates in time.

That is, the amplitude maximizes twice in each cycle of the function on the left in the preceding expression. Thus, the number of beats per second, or the beat frequency $f_b$ is twice the frequency of this function:

- *Beat frequency*

$$f_b = |f_1 - f_2| \qquad \text{[14.15]}$$

For instance, if two tuning forks vibrate individually at frequencies of 438 Hz and 442 Hz, respectively, the resultant sound wave of the combination has a frequency of $(f_1 + f_2)/2 = 440$ Hz (the musical note A) and a beat frequency of $|f_1 - f_2| = 4$ Hz. That is, the listener hears the 440-Hz sound wave go through an intensity maximum four times every second.

**Quick Quiz 14.5**

You are tuning a guitar by comparing the sound of the string with that of a standard tuning fork. You notice a beat frequency of 5 Hz when both sounds are present. You tighten the guitar string and the beat frequency rises to 8 Hz. Should you continue to tighten the string or should you loosen it to match the frequency of the string to that of the tuning fork?

## 14.7 • NONSINUSOIDAL WAVE PATTERNS

See Screen 9.6

The sound wave patterns produced by most instruments are not sinusoidal. Some characteristic waveforms produced by a tuning fork, a flute, and a clarinet are shown in Figure 14.12. Although each instrument has its own characteristic pattern, Figure 14.12 shows that all three waveforms are periodic. A struck tuning fork produces primarily one harmonic (the fundamental), whereas the flute and clarinet produce many frequencies, which include the fundamental and various harmonics. The nonsinusoidal waveforms produced by a violin or clarinet, and the corresponding richness of musical tones, are the result of the superposition of various harmonics.

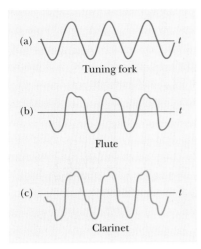

(a) Tuning fork

(b) Flute

(c) Clarinet

**Figure 14.12**

Waveforms of sound produced by (a) a tuning fork, (b) a flute, and (c) a clarinet, each at approximately the same frequency. *(Adapted from C. A. Culver,* Musical Acoustics, *4th ed., New York, McGraw-Hill, 1956, p. 128.)*

**PITFALL PREVENTION 14.5**
**Pitch versus frequency**

A very common mistake made in speech when talking about sound is to use the term *pitch* when one means *frequency*. Frequency is the physical measurement of the number of oscillations per second, as we have defined. Pitch is a psychological reaction of humans to sound that enables a human to place the sound on a scale from high to low, or from treble to bass. Thus, frequency is the stimulus and pitch is the response. Although pitch is related mostly (but not completely) to frequency, they are not the same. A phrase such as "the pitch of the sound" is incorrect because pitch is not a physical property of the sound.

This phenomenon is in contrast to a percussive musical instrument, such as the drum, in which the combination of frequencies do not form a harmonic series. When frequencies that are integer multiples of a fundamental frequency are combined, the result is a *musical* sound. A listener can assign a pitch to the sound, based on the fundamental frequency. Pitch is a psychological reaction to a sound that allows the listener to place the sound on a scale of low to high (bass to treble). Combinations of frequencies that are not integer multiples of a fundamental result in a *noise,* rather than a musical sound. It is much harder for a listener to assign a pitch to a noise than for a musical sound.

Analysis of nonsinusoidal waveforms appears at first sight to be a formidable task. If the waveform is periodic, however, it can be represented with arbitrary precision by the combination of a sufficiently large number of sinusoidal waves that form a harmonic series. In fact, one can represent any periodic function or any function over a finite interval as a series of sine and cosine terms by using a mathematical technique based on *Fourier's theorem.* The corresponding sum of terms that represents the periodic waveform is called a **Fourier series.**

Let $y(t)$ be any function that is periodic in time, with a period of $T$, so that $y(t + T) = y(t)$. **Fourier's theorem** states that this function can be written

$$y(t) = \sum_n (A_n \sin 2\pi f_n t + B_n \cos 2\pi f_n t) \qquad \text{[14.16]}$$

•  *Fourier's theorem*

where the lowest frequency is $f_1 = 1/T$. The higher frequencies are integral multiples of the fundamental, and so $f_n = nf_1$. The coefficients $A_n$ and $B_n$ represent the amplitudes of the various harmonics.

Figure 14.13 represents a harmonic analysis of the waveforms shown in Figure 14.12. Note the variation of relative intensity with harmonic content for the flute and clarinet. In general, any musical sound contains components that are members of a harmonic series with varying relative intensities.

We have discussed the *analysis* of a wave pattern using Fourier's theorem. The analysis involves determining the coefficients of the harmonics in Equation 14.16 from a knowledge of the wave pattern. We can also perform the reverse process, *Fourier synthesis.* In this process, the various harmonics are added together to form a resultant wave pattern. As an example of Fourier synthesis, consider the building

**Figure 14.13**

Harmonics of the waveforms shown in Figure 14.12. Note the variations in intensity of the various harmonics. *(Adapted from C. A. Culver, Musical Acoustics, 4th ed., New York, McGraw-Hill, 1956.)*

of a square wave, as shown in Figure 14.14. The symmetry of the square wave results in only odd multiples of the fundamental combining in the synthesis. In Figure 14.14a, the orange curve shows the combination of $f$ and $3f$. In Figure 14.14b, we have added $5f$ to the combination and obtained the green curve. Notice how the general shape of the square wave is approximated, even though the upper and lower portions are not as flat as they should be.

Figure 14.14c shows the result of adding odd frequencies up to $9f$—the purple curve. This approximation to the square wave (black curve) is better than in parts a and b. To approximate the square wave as closely as possible, we would need to add all odd multiples of the fundamental frequency up to infinite frequency.

Using modern technology, one can produce a variety of musical sounds electronically by mixing harmonics with varying amplitudes. Electronic music synthesizers that do this are now widely used.

The physical mixture of harmonics can be described as the **spectrum** of the sound, with the spectrum displayed in a graphical representation such as Figure 14.13. The psychological reaction to changes in the spectrum of a sound is the detection of a change in the **timbre** or the **quality** of the sound. If a clarinet and a trumpet are both playing the same note, then you will assign the same pitch to the two notes. Yet, if only one of the instruments then plays the note, you will likely be able to tell which instrument is playing. The sounds you hear from the two instruments differ in timbre, due to a different physical mixture of harmonics. For example, the timbre due to the sound of a trumpet is different from that of a clarinet. You have probably developed words to describe timbres of various instruments, such as "brassy," "mellow," "tinny," and the like.

Fourier's theorem allows us to understand the excitation process of musical instruments. In a stringed instrument that is plucked, such as a guitar, the string is pulled aside and released. After release, the string oscillates almost freely; a small damping causes the amplitude to eventually decay to zero. The mixture of harmonic frequencies depends on the length of the string, its linear mass density, and the plucking point.

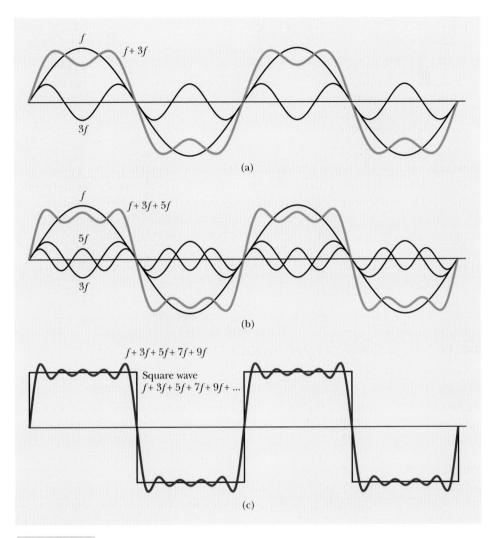

**Figure 14.14**

Fourier synthesis of a square wave, which is represented by the sum of odd multiples of the first harmonic, which has frequency $f$. (a) Waves of frequency $f$ and $3f$ are added. (b) One more odd frequency of $5f$ is added. (c) The synthesis curve approaches the square wave when odd frequencies up to $9f$ are combined.

On the other hand, a bowed string instrument, such as a violin, or a wind instrument, is a forced oscillator. In the case of the violin, the alternate sticking and slipping of the bow on the string provides the periodic driving force. In the case of a wind instrument, the vibration of a reed (in a woodwind), of the lips of the player (in a brass), or the blowing of air across an edge (as in a flute) provides the periodic driving force. According to Fourier's theorem, these periodic driving forces contain a mixture of harmonic frequencies. The violin string or the air column in a wind instrument is therefore driven with a wide variety of frequencies. The frequency actually played is determined by *resonance,* which we studied in Chapter 12. The maximum response of the instrument will be to those frequencies that match or are very close to the harmonic frequencies of the instrument. The spectrum of the instrument, then, depends heavily on the strengths of the various harmonics in the initial periodic driving force.

**THINKING PHYSICS 14.4**

A professor performs a demonstration in which he breathes a small amount of helium and then speaks with a comical voice. One student explains, "the velocity of sound in helium is higher than in air, so the fundamental frequency of the standing waves in the mouth is increased." Another student says, "No, the fundamental frequency is determined by the vocal folds and cannot be changed. Only the *quality* of the voice has changed." Which student is correct?

**Reasoning** The second student is correct. The fundamental frequency of the complex tone from the voice is determined by the vibration of the vocal folds and is not changed by substituting a different gas in the mouth. The introduction of the helium into the mouth results in higher harmonics being excited more than in the normal voice, but the fundamental frequency of the voice is the same—only the spectrum has changed. The unusual inclusion of the higher frequency harmonics results in a common description of this effect as a "high-pitched" voice, but this is incorrect.

**context connection** 🌐
**14.8** • BUILDING ON ANTINODES

As an example of the application of standing waves to earthquakes, we consider the effects of standing waves in *sedimentary basins*. Many of the world's major cities are built on sedimentary basins, which are topographic depressions that over geologic time have filled with sediment. These areas provide large expanses of flat land, often surrounded by attractive mountains, as in the Los Angeles basin. Flat land for building and attractive scenery attracted early settlers and led to today's cities.

Destruction from an earthquake can increase dramatically if the natural frequencies of buildings or other structures coincide with the resonant frequencies of the underlying basin. These resonant frequencies are associated with three-dimensional standing waves, formed from seismic waves reflecting from the boundaries of the basin.

To understand these standing waves, let us assume a simple model of a basin shaped like a half-ellipsoid, similar to an egg sliced in half along its long diameter. Four possible patterns of vertical ground motion in such a basin are shown in the pictorial representation in Figure 14.15. The long axis of the ellipsoid is designated by *x* and the short axis is *y*. Each of the four patterns is shown as a pair, separated in time by half of a period. In Figure 14.15a, the entire surface of the ground moves up and down except at a nodal curve running around the edge of the basin. The drawing on the left shows the ground level raised above its equilibrium position. Half a period later, the drawing on the right shows the ground forming a depression whose surface is below that of the flat surface.

In Figures 14.15b and 14.15c, half the basin lies above and half lies below the equilibrium position, and each half oscillates up and down on either side of a nodal line. The nodal line is along the *y* axis in Figure 14.15b and along the *x* axis in Figure 14.15c. In Figure 14.15d, nodal lines occur along both the *x* and *y* axes and the surface oscillates in four segments, with two above the equilibrium position at any time and the other two below.

The standing wave patterns in a basin arise from seismic waves traveling vertically between the top surface of the ground and the bottom of the basin, and hori-

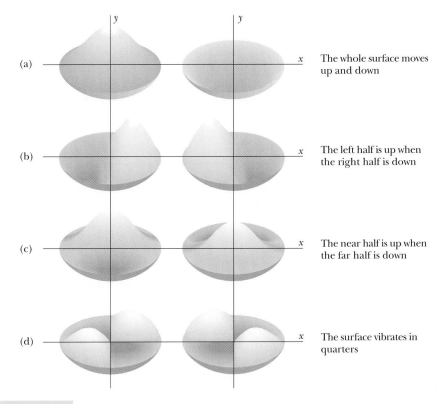

(a) The whole surface moves up and down

(b) The left half is up when the right half is down

(c) The near half is up when the far half is down

(d) The surface vibrates in quarters

**Figure 14.15**

Standing waves in a basin shaped like a half-ellipsoid. Displacements from equilibrium are exaggerated in order to clearly show the nature of the oscillations. Actual displacements are much smaller than those shown. (a) In the fundamental mode of oscillation, the only nodal line is the rim of the basin and the entire surface moves up and down together. (b) The surface oscillates in two halves, with a nodal line along the $y$ axis. (c) The surface oscillates in two halves, with a nodal line along the $x$ axis. (d) The surface oscillates in four quarters, with nodal lines along both the $x$ and $y$ axes.

zontally between the boundaries of the basin. For structures built on sedimentary basins, the degree of seismic risk will depend on the standing wave modes excited by the interference of seismic waves trapped in the basin. It is clear that structures built on regions of maximum ground motion (i.e., the antinodes) will suffer maximum shaking, whereas structures residing near nodes will experience relatively mild ground motion. These considerations appear to have played an important role in the selective destruction that occurred in Mexico City in the Michoacán earthquake in 1985 and, more recently, in the 1989 Loma Prieta earthquake, which caused the collapse of a section of the Nimitz Freeway in Oakland.

A similar effect occurs in bounded bodies of water, such as harbors and bays. A standing wave pattern established in such a body of water is called a **seiche.** This can result in variations in the water level that exhibit a period of several minutes, superposed on the longer period tidal variations. Seiches can be caused by earthquakes, tsunamis, winds, and weather disturbances. You can create a seiche in your bathtub by sliding back and forth at just the right frequency such that the water sloshes back and forth at such a large amplitude that much of it spills out onto the floor.

During the Northridge earthquake of 1994, swimming pools throughout southern California overflowed due to seiches set up by the shaking of the ground.

**WEB**

For an animation of a seiche, with ability to change parameters, visit **www.coastal.udel.edu/faculty/ rad/seiche.html**

**WEB**

For information on the Northridge earthquake and related links, visit **www.earthquake.org/ northridge.html**

In a more dramatic example, seismic waves from the 1964 Alaska earthquake caused severe seiches in the bays and bayous of Louisiana, some causing the water level to shift by 2 m.

We have now considered the role of standing waves in the damage caused by an earthquake. In the Context Conclusion, we will gather together the principles of vibrations and waves that we have learned to respond more fully to the central question of this Context.

## SUMMARY

The **principle of superposition** states that if two or more traveling waves are moving through a medium and combine at a given point, the resultant displacement of the medium at that point is the sum of the displacements of the individual waves.

When two waves with equal amplitudes and frequencies superpose, the resultant wave has an amplitude that depends on the phase angle $\phi$ between the two waves. **Constructive interference** occurs when the two waves are *in phase* everywhere, corresponding to $\phi = 0, 2\pi, 4\pi, \ldots$ . **Destructive interference** occurs when the two waves are 180° out of phase everywhere, corresponding to $\phi = \pi, 3\pi, 5\pi, \ldots$ .

**Standing waves** are formed from the superposition of two sinusoidal waves that have the same frequency, amplitude, and wavelength but are traveling in *opposite* directions. The resultant standing wave is described by the wave function

$$y = (2A \sin kx) \cos \omega t \qquad \textbf{[14.3]}$$

The maximum amplitude points (called **antinodes**) are separated by a distance $\lambda/2$. Halfway between antinodes are points of zero amplitude (called **nodes**).

One can set up standing waves with quantized frequencies in such systems as stretched strings and air columns. The nat-ural frequencies of vibration of a stretched string of length $L$, fixed at both ends, are

$$f_n = \frac{n}{2L} \sqrt{\frac{T}{\mu}} \quad (n = 1, 2, 3, \ldots) \qquad \textbf{[14.8]}$$

where $T$ is the tension in the string and $\mu$ is its mass per unit length. The natural frequencies of vibration form a **harmonic series,** that is, $f_1, 2f_1, 3f_1, \ldots$

The standing wave patterns for longitudinal waves in an air column depend on whether the ends of the column are open or closed. If the column is open at both ends, the natural frequencies of vibration form a harmonic series. If one end is closed, only odd harmonics of the fundamental are present.

The phenomenon of **beats** occurs as a result of the superposition of two traveling waves of slightly different frequencies. For sound waves at a given point, one hears an alternation in sound intensity with time.

Any periodic waveform can be represented by the combination of sinusoidal waves that form a harmonic series. The process is based on **Fourier's theorem.**

## QUESTIONS

1. When two waves interfere constructively or destructively, is there any gain or loss in energy? Explain.

2. Does the phenomenon of wave interference apply only to sinusoidal waves?

3. If you have a series of identical glass bottles, with varying amounts of water in them, you can play musical notes by either striking the bottles with a spoon, or blowing across the open tops of the bottles. When striking the bottles, the frequency of the note decreases as the water level rises. When blowing on the bottles, the frequency of the note increases as the water level rises. Why is the behavior of the frequency different in these two cases?

4. Some singers claim to be able to shatter a wine glass by maintaining a certain vocal pitch over a period of several seconds (Fig. Q14.4). What mechanism causes the glass to break? (The glass must be very clean in order to break.)

5. What limits the amplitude of motion of a real vibrating system that is driven at one of its resonant frequencies?

6. Guitarists sometimes play a "harmonic," by lightly touching a string at the exact center and plucking the string. The result is a clear note one octave higher than the fundamental of the string, even though the string is not pressed to the fingerboard. Why does this happen?

7. An archer shoots an arrow from a bow. Does the string of the bow exhibit standing waves after the arrow leaves? If so, and if the bow is perfectly symmetric, so that the arrow leaves from the center of the string, what harmonics are excited?

8. Explain why all harmonics are present in an organ pipe open at both ends, but only the odd harmonics are present in a pipe closed at one end.

(a)                            (b)

**Figure Q14.4** (a) Standing wave pattern in a vibrating wine glass. *(Courtesy of Professor Thomas D. Rossing, Northern Illinois University)* (b) The glass shatters if the amplitude of the vibrations becomes too large. In this photograph, the vibrations are caused by the amplified sound of a human voice. *(©1992 Ben Rose/The Image Bank)*

**9.** You have a standard tuning fork whose frequency is known and a second tuning fork whose frequency you don't know. When you play them together, you hear a beat frequency of 4 Hz. You know that the frequency of the mystery tuning fork is different from that of the standard fork by 4 Hz, but you can't tell whether it is higher or lower. What could you do to determine this? (*Hint:* You are chewing gum at the time that you perform these measurements.)

**10.** An airplane mechanic notices that the sound from a twin-engine aircraft rapidly varies in loudness when both engines are running. What could be causing this variation from loud to soft?

**11.** Why does a vibrating guitar string sound louder when placed on the instrument than it would if allowed to vibrate in the air while off the instrument?

**12.** When the base of a vibrating tuning fork is placed against a chalk board, the sound becomes louder. How does this affect the length of time for which the fork vibrates? Does this agree with conservation of energy?

**13.** Stereo speakers are supposed to be "phased" when set up. That is, the waves emitted from them should be in phase with each other. What would the sound be like along the center line of the speakers if one speaker were wired up backwards, that is, out of phase?

**14.** To keep animals away from their cars, some people mount short, thin pipes on the fenders. The pipes give out a high-pitched wail when the cars are moving. How do they create the sound?

**15.** If you wet your fingers and lightly run them around the rim of a fine wine glass, a high-pitched sound is heard. Why? How could you produce various musical notes with a set of wine glasses?

**16.** When a bell is rung, standing waves are set up around the bell's circumference. What boundary conditions must be satisfied by the resonant wavelengths? How does a crack in the bell, such as in the Liberty Bell, affect the satisfying of the boundary conditions and the sound emanating from the bell?

**17.** Despite a reasonably steady hand, a certain person often spills his coffee when carrying it to his seat. Discuss resonance as a possible cause of this difficulty, and devise a means for solving the problem.

**18.** Explain why your voice seems to sound better than usual when you sing in the shower.

**19.** What is the purpose of the slide on a trombone or the valves on a trumpet?

## PROBLEMS

1, 2, 3 = straightforward, intermediate, challenging     □ = full solution available in the *Student Solutions Manual and Study Guide*

**web** = solution posted at **http://www.harcourtcollege.com/physics/**      = computer useful in solving problem

= Interactive Physics     = paired numerical/symbolic problems     = life science application

### Section 14.1  The Principle of Superposition

**1.** Two waves on one string are described by the wave functions

$$y_1 = 3.0 \cos(4.0x - 1.6t)$$

and

$$y_2 = 4.0 \sin(5.0x - 2.0t)$$

where $y$ and $x$ are in centimeters and $t$ is in seconds. Find the superposition of the waves $y_1 + y_2$ at the points (a) $x = 1.00$, $t = 1.00$, (b) $x = 1.00$, $t = 0.500$, and (c) $x = 0.500$, $t = 0$. (Remember that the arguments of the trigonometric functions are in radians.)

2. Two wave pulses A and B are moving in opposite directions along a taut string with a speed of 2.00 cm/s. The amplitude of A is twice the amplitude of B. The pulses are shown in Figure P14.2 at $t = 0$. Sketch the shape of the string at $t = 1.0, 1.5, 2.0, 2.5,$ and $3.0$ s.

**Figure P14.2**

3. Two pulses traveling on the same string are described by

$$y_1 = \frac{5}{(3x - 4t)^2 + 2}$$

and

$$y_2 = \frac{-5}{(3x + 4t - 6)^2 + 2}$$

(a) In which direction does each pulse travel? (b) At what time do the two cancel everywhere? (c) At what point do the two waves always cancel?

**Section 14.2**    Interference of Waves

4. Two waves are traveling in the same direction along a stretched string. The waves are 90.0° out of phase. Each wave has an amplitude of 4.00 cm. Find the amplitude of the resultant wave.

5. Two sinusoidal waves are described by the wave functions
web

$$y_1 = (5.00 \text{ m}) \sin[\pi(4.00x - 1\ 200t)]$$

and

$$y_2 = (5.00 \text{ m}) \sin[\pi(4.00x - 1\ 200t - 0.250)]$$

where $x$, $y_1$, and $y_2$ are in meters and $t$ is in seconds. (a) What is the amplitude of the resultant wave? (b) What is the frequency of the resultant wave?

6. Two identical sinusoidal waves with wavelengths of 3.00 m travel in the same direction at a speed of 2.00 m/s. The second wave originates from the same point as the first, but at a later time. Determine the minimum possible time interval between the starting moments of the two waves if the amplitude of the resultant wave is the same as that of each of the two initial waves.

7. A tuning fork generates sound waves with a frequency of 246 Hz. The waves travel in opposite directions along a hallway, are reflected by end walls, and return. The hallway is 47.0 m long, and the tuning fork is located 14.0 m from one end. What is the phase difference between the reflected waves when they meet at the tuning fork? The speed of sound in air is 343 m/s.

8. Two loudspeakers are placed on a wall 2.00 m apart. A listener stands 3.00 m from the wall directly in front of one of the speakers. A single oscillator is driving the speakers in phase at a frequency of 300 Hz. (a) What is the phase difference between the two waves when they reach the observer? (b) What is the frequency closest to 300 Hz to which the oscillator may be adjusted such that the observer hears minimal sound?

9. Two speakers are driven in phase by the same oscillator of frequency $f$. They are located a distance $d$ from each other on a vertical pole. A man walks straight toward the lower speaker in a direction perpendicular to the pole, as shown in Figure P14.9. (a) How many times will he hear a minimum in sound intensity, and (b) how far is he from the pole at these moments? Let $v$ represent the speed of sound and assume that the ground does not reflect sound.

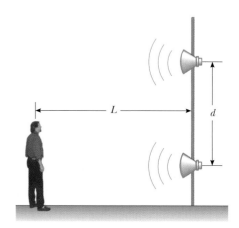

**Figure P14.9**

**Section 14.3**    Standing Waves

10. Two sinusoidal waves traveling in opposite directions interfere to produce a standing wave with the wave function

$$y = (1.50 \text{ m}) \sin(0.400x) \cos(200t)$$

where $x$ is in meters and $t$ is in seconds. Determine the wavelength, frequency, and speed of the interfering waves.

11. Two speakers are driven in phase by a common oscillator
web   at 800 Hz and face each other at a distance of 1.25 m. Locate the points along the line between the speakers where

minima of the sound pressure amplitude would be expected. (Use $v = 343$ m/s.)

12. Two waves that set up a standing wave in a long string are given by the wave functions

$$y_1 = A \sin(kx - \omega t + \phi)$$

and

$$y_2 = A \sin(kx + \omega t)$$

Show (a) that the addition of the arbitrary phase constant changes only the position of the nodes, and in particular (b) that the distance between nodes is still one-half the wavelength.

13. Two sinusoidal waves combining in a medium are described by the wave functions

$$y_1 = (3.0 \text{ cm}) \sin \pi(x + 0.60t)$$

and

$$y_2 = (3.0 \text{ cm}) \sin \pi(x - 0.60t)$$

where $x$ is in centimeters and $t$ is in seconds. Determine the *maximum* displacement of the motion at (a) $x = 0.250$ cm, (b) $x = 0.500$ cm, and (c) $x = 1.50$ cm. (d) Find the three smallest values of $x$ corresponding to antinodes.

14. Verify by direct substitution that the wave function for a standing wave given in Equation 14.3,

$$y = (2A \sin kx) \cos \omega t$$

is a solution of the general linear wave equation, Equation 13.19:

$$\frac{\partial^2 y}{\partial x^2} = \frac{1}{v^2} \frac{\partial^2 y}{\partial t^2}$$

## Section 14.4  Standing Waves in Strings

15. A student wants to establish a standing wave on a wire that is 1.80 m long and clamped at both ends. The wave speed is 540 m/s. What is the minimum frequency the student should apply to set up a standing wave?

16. One end of a string is attached to a wall. The other end is draped over a small pulley 5.00 m away from the wall and is attached to a hanging mass of 4.00 kg. If the string is plucked, what is the fundamental frequency of vibration? The portion of the string that vibrates transversely has a mass of 8.00 g.

17. Find the fundamental frequency and the next three frequencies that could cause standing wave patterns on a 30.0-m string that has a mass per length of $9.00 \times 10^{-3}$ kg/m and is stretched to a tension of 20.0 N.

18. A vibrating string with a uniform linear mass density exhibits a single-loop standing wave pattern with a frequency of 800 Hz. (a) If the tension in the string is changed to reduce the fundamental frequency to 500 Hz, determine the ratio of the new tension to the old. (b) Alternatively, if the original tension in the string is increased by a factor of 4, determine the new fundamental frequency.

19. A standing wave is established in a 120-cm string fixed at both ends. The string vibrates in four segments when driven at 120 Hz. (a) Determine the wavelength of the wave. (b) What is the fundamental frequency of the string?

20. In the arrangement shown in Figure P14.20, an object can be hung from a string (with linear mass density $\mu = 0.002\ 00$ kg/m) that passes over a light pulley. The string is connected to a vibrator (of constant frequency $f$), and the length of the string between point $P$ and the pulley is $L = 2.00$ m. When the mass $m$ is either 16.0 kg or 25.0 kg, standing waves are observed; however, no standing waves are observed with any mass between these values. (a) What is the frequency of the vibrator? (*Hint:* The greater the tension in the string, the smaller the number of nodes in the standing wave.) (b) What is the largest mass for which standing waves could be observed?

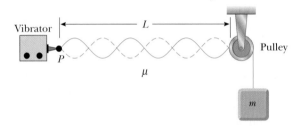

**Figure P14.20**

21. A cello A-string vibrates in its first normal mode with a frequency of 220 vibrations/s. The vibrating segment is 70.0 cm long and has a mass of 1.20 g. (a) Find the tension in the string. (b) Determine the frequency of vibration when the string vibrates in three segments.

22. A violin string has a length of 0.350 m and is tuned to concert G, with $f_G = 392$ Hz. Where must the violinist place her finger to play concert A, with $f_A = 440$ Hz? If this position is to remain correct to one-half the width of a finger (i.e., to within 0.600 cm), what is the maximum allowable percentage change in the string tension?

23. A standing wave pattern is observed in a thin wire with a length of 3.00 m. The equation of the wave is

$$y = (0.002 \text{ m}) \sin(\pi x) \cos(100\pi t)$$

where $x$ is in meters and $t$ is in seconds. (a) How many loops does this pattern exhibit? (b) What is the fundamental frequency of vibration of the wire? (c) If the original frequency is held constant and the tension in the wire is increased by a factor of 9, how many loops are present in the new pattern?

## Section 14.5   Standing Waves in Air Columns

Assume that the speed of sound in air is 343 m/s unless otherwise stated.

**24.** A glass tube (open at both ends) of length $L$ is positioned near an audio speaker of frequency $f = 680$ Hz. For what values of $L$ will the tube resonate with the speaker?

**25.** Calculate the length of a pipe that has a fundamental frequency of 240 Hz if the pipe is (a) closed at one end and (b) open at both ends.

**26.** The overall length of a piccolo is 32.0 cm. The resonating air column vibrates in the same way as a pipe open at both ends. (a) Find the frequency of the lowest note that a piccolo can play, assuming that the speed of sound in air is 340 m/s. (b) Opening holes in the side effectively shortens the length of the resonant column. If the highest note a piccolo can sound is 4 000 Hz, find the distance between adjacent antinodes for this mode of vibration.

**27.** The fundamental frequency of an open organ pipe corresponds to middle C (261.6 Hz on the chromatic musical scale). The third resonance of a closed organ pipe has the same frequency. What are the lengths of the two pipes?

**28.** Do not stick anything into your ear! Estimate the length of your ear canal, from its opening at the external ear to the eardrum. If you regard the canal as a tube that is open at one end and closed at the other, at approximately what fundamental frequency would you expect your hearing to be most sensitive? Explain why you can hear especially soft sounds just around this frequency.

**29.** A pipe that is open at both ends has a fundamental frequency of 300 Hz when the speed of sound in air is 333 m/s. (a) What is the length of the pipe? (b) What is the frequency of the second harmonic when the temperature of the air is increased so that the speed of sound in the pipe is 344 m/s?

**30.** As shown in Figure P14.30, water is pumped into a tall vertical cylinder at a volume flow rate $R$. The radius of the cylinder is $r$, and at the open top of the cylinder a tuning fork vibrates with a frequency $f$. As the water rises, how much time elapses between successive resonances?

**31.** A shower stall measures 86.0 cm × 86.0 cm × 210 cm. If you were singing in this shower, which frequencies would sound the richest (because of resonance)? Assume that the stall acts as a pipe closed at both ends, with nodes at opposite sides. Assume that the voices of various singers range from 130 Hz to 2 000 Hz. Let the speed of sound in the hot shower stall be 355 m/s.

**32.** A tuning fork with a frequency of 512 Hz is placed near the top of the tube shown in Figure 14.10. The water level is lowered so that the length $L$ slowly increases from an initial value of 20.0 cm. Determine the next two values of $L$ that correspond to resonant modes.

**Figure P14.30**

**33.** A glass tube is open at one end and closed at the other by a movable piston. The tube is filled with air warmer than that at room temperature, and a 384-Hz tuning fork is held at the open end. Resonance is heard when the piston is 22.8 cm from the open end and again when it is 68.3 cm from the open end. (a) What speed of sound is implied by these data? (b) How far from the open end will the piston be when the next resonance is heard?

**34.** A student uses an audio oscillator of adjustable frequency to measure the depth of a water well. Two successive resonances are heard at 51.5 Hz and 60.0 Hz. How deep is the well?

## Section 14.6   Beats: Interference in Time

**35.** In certain ranges of a piano keyboard, more than one string is tuned to the same note to provide extra loudness. For example, the note at 110 Hz has two strings at this frequency. If one string slips from its normal tension of 600 N to 540 N, what beat frequency is heard when the hammer strikes the two strings simultaneously?

**36.** While attempting to tune the note C at 523 Hz, a piano tuner hears 2 beats/s between a reference oscillator and the string. (a) What are the possible frequencies of the string? (b) When she tightens the string slightly, she hears 3 beats/s. What is the frequency of the string now? (c) By what percentage should the piano tuner now change the tension in the string to bring it into tune?

**37.** A student holds a tuning fork oscillating at 256 Hz. He walks toward a wall at a constant speed of 1.33 m/s. (a) What beat frequency does he observe between the tuning fork and its echo? (b) How fast must he walk away from the wall to observe a beat frequency of 5.00 Hz?

## Section 14.7 Nonsinusoidal Wave Patterns

38. Suppose that a flutist plays a 523-Hz C note with first harmonic displacement amplitude $A_1 = 100$ nm. From Figure 14.13b read, by proportion, the displacement amplitudes of harmonics 2 through 7. Take these as the values $A_2$ through $A_7$ in the Fourier analysis of the sound, and assume that $B_1 = B_2 = \cdots = B_7 = 0$. Construct a graph of the waveform of the sound. Your waveform will not look exactly like the flute waveform in Figure 14.12b because you simplify by ignoring cosine terms; nevertheless, it produces the same sensation to human hearing.

## Section 14.8 Context Connection — Building on Antinodes

39. An earthquake can produce a seiche in a lake, in which the water sloshes back and forth from end to end with remarkably large amplitude and long period. Consider a seiche produced in a rectangular farm pond, as diagrammed in the cross-sectional view of Figure P14.39. (The figure is not drawn to scale.) Suppose that the pond is 9.15 m long and of uniform width and depth. You measure that a wave pulse produced at one end reaches the other end in 2.50 s. (a) What is the wave speed? (b) To produce the seiche, you have several people stand on the bank at one end and paddle together with snow shovels, moving them in simple harmonic motion. What should be the frequency of this motion?

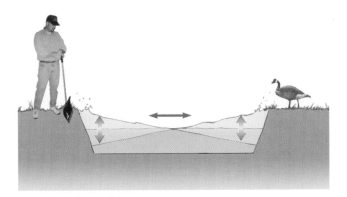

**Figure P14.39**

40. The Bay of Fundy, Nova Scotia, has the highest tides in the world. Assume that in midocean and at the mouth of the bay, the Moon's gravity gradient and the Earth's rotation make the water surface oscillate with an amplitude of a few centimeters and a period of 12 h 24 min. At the head of the bay, the amplitude is several meters. Argue for or against the proposition that the tide is amplified by standing-wave resonance. Assume the bay has a length of 210 km and a uniform depth of 36.1 m. The speed of long-wavelength water waves is given by $\sqrt{gd}$, where $d$ is the water's depth.

### Additional Problems

41. Two identical speakers 10.0 m apart are driven by the same oscillator with a frequency of $f = 21.5$ Hz (Fig. P14.41). (a) Explain why a receiver at point $A$ records a minimum in sound intensity from the two speakers. (b) If the receiver is moved in the plane of the speakers, what path should it take so that the intensity remains at a minimum? That is, determine the relationship between $x$ and $y$ (the coordinates of the receiver) that causes the receiver to record a minimum in sound intensity. Take the speed of sound to be 343 m/s.

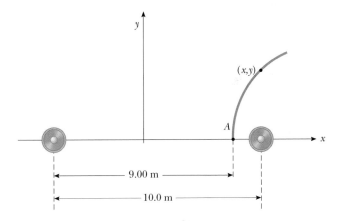

**Figure P14.41**

42. A major chord on the physical pitch musical scale has frequencies in the ratios $4:5:6:8$. A set of pipes, closed at one end, must be cut so that when they are sounded in their first normal mode, they will produce a major chord. (a) What is the ratio of the lengths of the pipes? (b) What are the lengths of the pipes needed if the lowest frequency of the chord is 256 Hz? (c) What are the frequencies of the notes in this chord?

43. Standing-wave vibrations are set up in a crystal goblet with four nodes and four antinodes equally spaced around the 20.0-cm circumference of its rim. If transverse waves move around the glass at 900 m/s, an opera singer would have to produce a high harmonic with what frequency to shatter the glass with a resonant vibration?

44. When a metal pipe is cut into two pieces, the lowest resonance frequency for one piece is 256 Hz and for the other is 440 Hz. (a) What resonant frequency would have been produced by the original length of pipe? (b) How long was the original pipe?

45. On a marimba (Fig. P14.45), the wooden bar that sounds a tone when struck vibrates in a transverse standing wave

**Figure P14.49**

**Figure P14.45**   Marimba players in Mexico City. *(Murray Greenberg)*

having three antinodes and two nodes. The lowest frequency note is 87.0 Hz, produced by a bar 40.0 cm long. (a) Find the speed of transverse waves on the bar. (b) The loudness of the emitted sound is enhanced by a resonant pipe suspended vertically below the center of the bar. If the pipe is open at the top end only and the speed of sound in air is 340 m/s, what is the length of the pipe required to resonate with the bar in part (a)?

46. Two train whistles have identical frequencies of 180 Hz. When one train is at rest in the station and the other is moving nearby, a commuter standing on the station platform hears beats with a frequency of 2.00 beats/s. What are the two possible speeds and directions that the moving train can have?

47. A string fixed at both ends and having a mass of 4.80 g, a length of 2.00 m, and a tension of 48.0 N vibrates in its second ($n = 2$) normal mode. What is the wavelength in air of the sound emitted by this vibrating string?

48. A string 0.400 m long has a mass per unit length of $9.00 \times 10^{-3}$ kg/m. What must be the tension in the string if its second harmonic has the same frequency as the second resonance mode of a 1.75-m-long pipe open at one end?

49. **Review Problem.** For the arrangement shown in Figure P14.49, $\theta = 30.0°$, the inclined plane and the small pulley are frictionless, the string supports the object of mass $M$ at the bottom of the plane, and the string has a mass $m$ that is small relative to $M$. The system is in equilibrium and the vertical part of the string has a length $h$. Standing waves are set up in the vertical section of the string. (a) Find the tension in the string. (b) Model the shape of the string as one leg and the hypotenuse of a right triangle. Find the whole length of the string. (c) Find the mass per unit

length of the string. (d) Find the speed of waves on the string. (e) Find the lowest frequency for a standing wave. (f) Find the period of the standing wave having three nodes. (g) Find the wavelength of the standing wave having three nodes. (h) Find the frequency of the beats resulting from the interference of the sound wave of lowest frequency generated by the string with another sound wave having a frequency that is 2.00% greater.

50. Two wires are welded together end to end. They are made of the same material, but the diameter of one is twice that of the other. They are subjected to a tension of 4.60 N. The thin wire has a length of 40.0 cm and a linear mass density of 2.00 g/m. The combination is fixed at both ends and vibrated in such a way that two antinodes are present, with the node between them being exactly at the location of the weld. (a) What is the frequency of vibration? (b) How long is the thick wire?

51. A standing wave is set up in a string of variable length and tension by a vibrator of variable frequency. Both ends of the string are fixed. When the vibrator has a frequency $f$, in a string of length $L$ and under tension $T$, $n$ antinodes are set up in the string. (a) If the length of the string is doubled, by what factor should the frequency be changed so that the same number of antinodes is produced? (b) If the frequency and length are held constant, what tension will produce $n + 1$ antinodes? (c) If the frequency is tripled and the length of the string is halved, by what factor should the tension be changed so that twice as many antinodes are produced?

52. Two identical strings, each fixed at both ends, are arranged near each other. If string A starts oscillating in its first normal mode, it is observed that string B will begin vibrating in its third ($n = 3$) normal mode. Determine the ratio of the tension of string B to the tension of string A.

53. If two adjacent natural frequencies of an organ pipe are determined to be 550 Hz and 650 Hz, calculate the fundamental frequency and length of this pipe. (Use $v = 340$ m/s.)

**54.** A 0.010 0-kg wire, 2.00 m long, is fixed at both ends and vibrates in its fundamental mode under a tension of 200 N. When a vibrating tuning fork is placed near the wire, a beat frequency of 5.00 Hz is heard. (a) What could be the frequency of the tuning fork? (b) What should the tension in the wire be to make the beats disappear?

**55.** Two waves are described by the wave functions

$$y_1(x, t) = 5.0 \sin(2.0x - 10t)$$

and

$$y_2(x, t) = 10 \cos(2.0x - 10t)$$

where $x$ is in meters and $t$ is in seconds. Show that the wave resulting from their superposition is sinusoidal. Determine the amplitude and phase of this sinusoidal wave.

**56.** The wave function for a standing wave is given in Equation 14.3 as $y = 2A \sin kx \cos \omega t$. (a) Rewrite this wave function in terms of the wavelength $\lambda$ and the wave speed $v$ of the wave. (b) Write the wave function of the simplest standing-wave vibration of a stretched string of length $L$. (c) Write the wave function for the second harmonic. (d) Generalize these results and write the wave function for the $n$th resonance vibration.

**57.** **Review Problem.** A 12.0-kg object hangs in equilibrium from a string with a total length of $L = 5.00$ m and a linear mass density of $\mu = 0.00100$ kg/m. The string is wrapped around two light, frictionless pulleys that are separated by a distance of $d = 2.00$ m (Fig. P14.57a). (a) Determine the tension in the string. (b) At what frequency must the string between the pulleys vibrate in order to form the standing wave pattern shown in Figure P14.57b?

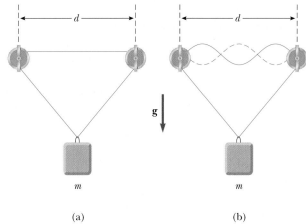

**Figure P14.57**

**58.** A quartz watch contains a crystal oscillator in the form of a block of quartz that vibrates by contracting and expanding. Two opposite faces of the block, 7.05 mm apart, are antinodes, moving alternately toward each other and away from each other. The plane halfway between these two faces is a node of the vibration. The speed of sound in quartz is 3.70 km/s. Find the frequency of the vibration. An oscillating electric voltage accompanies the mechanical oscillation—the quartz is described as *piezoelectric*. An electric circuit feeds in energy to maintain the oscillation and also counts the voltage pulses to keep time.

## ANSWERS TO QUICK QUIZZES

**14.1** At a node, the speed of the particle of the medium at all times is zero. We can find the transverse velocity of a particle at any nonnode point by taking the partial derivative with respect to time of the transverse position:

$$v_y = \frac{\partial y}{\partial t} = \frac{\partial}{\partial t}\left[(2A \sin kx) \cos \omega t\right] = -\omega(2A \sin kx) \sin \omega t$$

At an antinode, the factor in parentheses has its maximum value of $2A$. The maximum speed of the particle is found by allowing the $\sin \omega t$ term to have its maximum value of 1. Thus, the maximum transverse speed is $2\omega A$.

**14.2** The bass strings have low fundamental frequencies. To achieve these low frequencies, we can make the string longer than the treble strings—as indicated in Equation 14.9, if $L$ increases, $f_1$ decreases. If we depend only on string length to lower the frequency, however, pianos would need to be many meters long to accommodate the lowest strings. The frequency of the bass strings is also lowered by increasing the mass per unit length, which is also indicated in Equation 14.9—as $\mu$ increases, $f_1$ decreases. Thus, the bass strings are thicker. Between these two effects, the frequencies of the bass strings can be achieved in a piano of reasonable size.

**14.3** (b). The open tube has a fundamental frequency given by Equation 14.11: $f_{open} = v/2L$. The closed tube's fundamental frequency is given by Equation 14.12:

$$f_{closed} = \frac{v}{4L} = \frac{1}{2}\frac{v}{2L} = \frac{1}{2}f_{open}$$

**14.4** A change in temperature will result in a change in the speed of sound and, therefore, in the fundamental frequency of a particular pipe. The speed of sound for audible frequencies in the open atmosphere is only a function of temperature and does not depend on the pressure. Thus, variations in atmospheric pressure will have no effect on the fundamental frequency.

**14.5** By tightening the string, you have caused an increase in the beat frequency from 5 to 8 Hz. Thus, the difference between the frequency of the guitar string and that of the tuning fork has increased. To reduce the frequency of the guitar string to make the beats vanish and match the frequency of the tuning fork, you must loosen the guitar string.

# Earthquakes

## MINIMIZING THE RISK

We have explored the physics of vibrations and waves. Let us now return to our central question for the *Earthquakes* Context:

> *How can we choose locations and build structures to minimize the risk of damage in an earthquake?*

To answer this question, we shall use the physical principles that we now understand more clearly and apply them to our choices of locations and structural design.

In our discussion of simple harmonic oscillation, we learned about resonance. Designs for structures in earthquake-prone areas need to pay careful attention to resonance vibrations from shaking of the ground. The design features to be considered include ensuring that the resonance frequency of the building does not match typical earthquake frequencies. In addition, the structural details should include sufficient damping to ensure that the amplitude of resonance vibration does not destroy the structure.

In Chapter 13, we discussed the role of the medium in the propagation of a wave. For seismic waves moving across the surface of the Earth, the soil on the surface is the medium. Because soil varies from one location to another, the speed of seismic waves will vary at different locations. A particularly dangerous situation exists for structures built on loose soil or mudfill. In these types of media, the interparticle forces are much weaker than in a more solid foundation, such as granite bedrock. As a result, the wave speed is less in loose soil than in bedrock.

Consider Equation 13.23, which provides an expression for the rate of energy transfer by waves. Although this equation was derived for waves on strings, the general dependence on the square of the angular frequency, the square of the amplitude, and the speed is general. To conserve energy, the rate of energy transfer for a wave must remain constant regardless of the medium. Thus, according to Equation 13.23, **if the wave speed decreases, as it does for seismic waves moving through loose soil, the amplitude must increase.** As a result, the shaking of structures built on loose soil is of larger magnitude than for those built on solid bedrock.

This factor contributed to the collapse of the Nimitz Freeway during the Loma Prieta earthquake, near San Francisco, in 1989. Figure 1 shows the results of the earthquake on the freeway. The portion of the freeway that collapsed was built on mudfill, but the surviving portion was built on bedrock. The amplitude of oscillation in the portion built on mudfill was more than five times as large as the amplitude of other portions.

Another danger for structures on loose soil is the possibility of **liquefaction** of the soil. When soil is shaken, the particles can move with respect to one another, and the soil tends to act like a liquid rather than a solid. It is possible for the structure to sink into the soil during an earthquake. If the liquefaction is not uniform over the foundation of the structure, the structure can tip over, as seen in Figure 2. As a result, even if the earthquake is not sufficient to damage the structure, it will be unusable in its tipped-over orientation.

**503**

**WEB**

For a variety of information and photographs about the Loma Prieta earthquake from the Museum of the City of San Francisco, visit **www.sfmuseum.org/1906/ 89.html**

**Figure 1**

Portions of the double-decked Nimitz Freeway in Oakland collapsed during the Loma Prieta earthquake of 1989. *(Paul X. Scott/SYGMA)*

**WEB**

For a report on the Kobe earthquake, visit **mceer.buffalo.edu/publications/ sp_pubs/kobereport/ KobeReport.asp**

**Figure 2**

The collapse of this crane during the Kobe, Japan, earthquake of 1995 was caused by liquefaction of the underlying soil. *(Photo by M. Hamada; used by permission from the Multidisciplinary Center for Earthquake Engineering Research, University at Buffalo)*

As discussed in Chapter 14, building structures where standing seismic waves can be established is dangerous. This was a factor in the Michoacán Earthquake of 1985. The shape of the bedrock under Mexico City resulted in standing waves, with severe damage to buildings located at antinodes.

In summary, to minimize risk of damage in an earthquake, architects and engineers must design structures to prevent destructive resonances, avoid building on loose soil, and pay attention to the underground rock formations so as to be aware of possible standing wave patterns. Other precautions can also be taken. For example, buildings can be constructed with **seismic isolation** from the ground. This involves mounting the structure on **isolation dampers**, heavy-duty bearings that dampen the oscillations of the building, resulting in reduced amplitude of vibration.

We have not addressed many other considerations for earthquake safety in structures, but we have been able to apply many of our concepts from oscillations and waves so as to understand some aspects of logical choices in locating and designing structures.

### PROBLEMS

1. For seismic waves spreading out from a point (the epicenter) on the surface of the Earth, the intensity of the waves falls off with distance with an inverse dependence on the distance, that is, as $1/r$, where $r$ is the distance from the epicenter to the observation point. This assumes a uniform medium. We have also shown that the energy of vibration of an oscillator is proportional to the square of the amplitude of the vibration of the oscillator. Suppose an earthquake occurs, and, at a distance of 10 km from the epicenter, the *amplitude* of the ground shaking is 5.0 cm. Assuming a uniform medium, what is the *amplitude* of the ground shaking at a point 20 km from the epicenter?

2. As mentioned in the text, the amplitudes of oscillation during the Loma Prieta earthquake of 1989 were five times larger in areas of mudfill as in areas of bedrock. From this information, estimate the percentage by which the seismic wave speed changed as the waves moved from the bedrock to the mudfill.

3. Figure 3 is a graphical representation of the travel time for $P$ and $S$ waves from the epicenter to a seismograph as a function of the distance between the epicenter and the seismograph. The following table shows the times of day for arrival of $P$ waves from an earthquake at three seismograph locations. Fill in the times of day in the last column for the arrival of the $S$ waves at the three seismograph locations.

| Seismograph Station | Distance from Epicenter (km) | $P$ Wave Arrival Time | $S$ Wave Arrival Time |
|---|---|---|---|
| #1 | 170 | 15:46:05 | |
| #2 | 160 | 15:46:03 | |
| #3 | 105 | 15:45:54 | |

**Figure 3**

A graph of travel time versus distance from the epicenter for $P$ and $S$ waves.

# Search for the *Titanic*

The *Titanic* and her sister ships, the *Olympic* and the *Britannic*, were designed to be the largest and most luxurious liners sailing the ocean at the time. The *Titanic* was lost in the North Atlantic on its maiden voyage from Southampton, England, to New York, on April 15, 1912. It was claimed by many (although not by the White Star Line, which operated the *Titanic*) that the ship was unsinkable. This was proven to be untrue when the *Titanic* struck an iceberg at 11:40 PM on April 14 and sank less than 3 h later. This event was one of the worst maritime disasters of all time, with over 1500 lives lost due to a severe shortage of lifeboats. Amazingly, British ships of the time were not required to carry enough lifeboats for all passengers on board. It is fortunate that the *Titanic* was not completely full on its maiden voyage, as there would have been only enough lifeboats for a third of its passengers. As it was, there were enough lifeboats for a little over half of the 2200 passengers on board, but only 705 were saved due to the partial filling of the boats, which occurred for a number of reasons.

The mass of the *Titanic* was over $4.2 \times 10^7$ kg and it was 269 m long. It was designed to be able to travel at 24 to 25 knots (about 12–13 m/s), and the safety of the ship was ensured by lateral bulkheads

During the sinking process in the early morning of April 15, 1912, the bow of the *Titanic* is underwater and the stern is lifted out of the water. The huge forces necessary to hold the stern aloft caused the *Titanic* to split in the middle before sinking. *(Ken Marschall Collection)*

at several places across the ship with electrically operated watertight doors. It is ironic to note that the design of the bulkheads actually worked against the safety of the ship, and the closing of the watertight doors, according to some experts after the disaster, caused the ship to sink more rapidly than if they were left open.

The accidental sinking of the *Titanic* resulted from a remarkable confluence of bad luck, complacency, and poor policy. Several events occurred during the last day of its voyage that would not have led to the foundering of the ship if they had happened in just a slightly different way.

The *Titanic* has captured the interest of the public for many years. Four theatrical movies related to the ship have been produced: *Titanic* (1953), *A Night to Remember* (1958), *Raise the Titanic* (1980), and *Titanic* (1997), along with a Broadway and movie musical, *The Unsinkable Molly Brown* (1964), and a more recent musical play, *Titanic* (1997). A large number of books have been written on the disaster, many of which were reissued when the *Titanic* became wildly popular in response to the 1997 film.

The *Titanic* on its way from Southampton to New York. *(Ken Marschall Collection)*

The bow of the *Titanic* as it rests on the ocean floor 80 years after it sank in the North Atlantic. *(Courtesy of RMS Titanic, Inc.)*

Finding the *Titanic* on the ocean floor was a dream of many individuals ever since she was lost. The wreck of the *Titanic* was discovered in 1985 by a research team from Woods Hole Oceanographic Institute, led by Dr. Robert Ballard. This discovery led to interesting ethical questions as subsequent expeditions to the site salvaged items from the wreck, making them available for sale to the public. We will use the *Titanic*, the search for its wreckage, and the underwater visits to its gravesite in this short Context on fluids, as we address the central question

*How can we safely visit the wreck of the* **Titanic** *?*

# chapter 15

An iceberg floats in the cold waters of the North Atlantic. Although the visible portion of the iceberg may tower over a passing ship, only about 11% of the iceberg is above water.

*(Geraldine Prentice/Stone)*

# Fluid Mechanics

Matter is normally classified as being in one of three states: solid, liquid, or gas. Everyday experience tells us that a solid has a definite volume and shape. A brick maintains its familiar shape and size over a long time. We also know that a liquid has a definite volume but no definite shape—a cup of liquid water has a fixed volume but assumes the shape of its container. Finally, an unconfined gas has neither definite volume nor definite shape. These definitions help us to picture the states of matter, but they are somewhat artificial. For example, asphalt, glass, and plastics are normally considered solids, but over a long time they tend to flow like liquids. Likewise, most substances can be a solid, liquid, or gas (or combinations of these), depending on the temperature and pressure. In general, the time it takes a particular substance to change its shape in response to an external force determines whether we treat the substance as a solid, liquid, or gas.

A **fluid** is a collection of molecules that are randomly arranged and held together by weak cohesive forces between molecules and forces exerted by the walls of a container. Both liquids and gases are fluids. In our treatment of the mechanics of fluids, we shall see that no new physical principles are needed to explain such effects as the buoyant force on a submerged object and vascular flutter in an artery. In this chapter, we shall apply a number of familiar analysis models to the physics of fluids.

## 15.1 • PRESSURE

Our first task in understanding the physics of fluids is to define a new quantity to describe fluids. Imagine applying a force to the surface of an object, with the force having components both parallel to and perpendicular to the surface. If the object is a solid at rest on a table, the force component perpendicular to the surface may cause the object to flatten, depending on how hard the object is. Assuming that the object does not slide on the table, the component of the force parallel to the surface of the object will cause the object to distort. As an example, suppose you place your physics book flat on a table and apply a force with your hand parallel to the front cover and perpendicular to the spine. The book will distort, with the bottom pages staying fixed at their original location and the top pages shifting horizontally by some distance. The cross-section of the book changes from a rectangle to a parallelogram. This kind of force parallel to the surface is called a *shearing force*.

We shall adopt a simplification model in which the fluids we study will be non-viscous; that is, no friction exists between adjacent layers of the fluid. Nonviscous fluids do not sustain shearing forces—if you imagine placing your hand on a water surface and pushing parallel to the surface, your hand simply slides over the water—you cannot distort the water as you did the book. This phenomenon occurs because the interatomic forces in a fluid are not strong enough to lock atoms in place with respect to one another. The fluid cannot be modeled as a rigid body, as for rotating objects in Chapter 10. If we try to apply a shearing force, the molecules of the fluid simply slide past one another.

Thus, the only type of force that can exist in a fluid is one that is perpendicular to a surface. For example, the forces exerted by the fluid on the object in Figure 15.1 are everywhere perpendicular to the surfaces of the object.

The force that a fluid exerts on a surface originates in the collisions of molecules of the fluid with the surface. Each collision results in the reversal of the component of the velocity vector of the molecule perpendicular to the surface. By the impulse–momentum theorem and Newton's third law, each collision results in a force on the surface. A huge number of these impulsive forces occur every second, resulting in a constant macroscopic force on the surface. This force is spread out over the area of the surface and is related to a new quantity called **pressure.**

The pressure at a specific point in a fluid can be measured with the device pictured in Figure 15.2. The device consists of an evacuated cylinder enclosing a light piston connected to a spring. As the device is submerged in a fluid, the fluid presses in on the top of the piston and compresses the spring until the inward force of the fluid is balanced by the outward force of the spring. The force exerted on the piston by the fluid can be measured if the spring is calibrated in advance.

If $F$ is the magnitude of the force exerted by the fluid on the piston and $A$ is the surface area of the piston, then the **pressure $P$ of the fluid at the level to which the device has been submerged is defined as the ratio of force to area:**

$$P \equiv \frac{F}{A}$$

[15.1]

**Figure 15.1**

The force of the fluid on a submerged object at any point is perpendicular to the surface of the object. The force of the fluid on the walls of the container is perpendicular to the walls at all points.

**Figure 15.2**

A simple device for measuring pressure in a fluid.

**PITFALL PREVENTION 15.1**

**Force and pressure**

Equation 15.1 makes a clear distinction between force and pressure. Another important distinction is that *force is a vector* and *pressure is a scalar*. No direction is associated with pressure, but the direction of the force associated with the pressure is perpendicular to the surface of interest.

• *Definition of pressure*

Although pressure is defined in terms of our device in Figure 15.2, the definition is general. Because pressure is force per unit area, it has units of $N/m^2$ in the SI system. Another name for the SI unit of pressure is the **pascal** (Pa):

•  *The pascal*

$$1 \text{ Pa} \equiv 1 \text{ N}/m^2 \qquad\qquad [15.2]$$

**APPLICATION**

Hypodermic needles

Notice that pressure and force are different quantities. We can have a very large pressure from a relatively small force by making the area over which the force is applied small. This is the case with hypodermic needles. The area of the tip of the needle is very small, so a small force pushing on the needle is sufficient to cause a pressure large enough to puncture the skin. We can also create a small pressure from a large force by enlarging the area over which the force acts. This is the principle behind the design of snowshoes. If a person were to walk on deep snow with regular shoes, it is possible for his or her feet to break through the snow and sink. Snowshoes, however, allow the force on the snow due to the weight of the person to spread out over a larger area, reducing the pressure enough so that the snow surface is not broken.

The atmosphere exerts a pressure on the surface of the Earth and all objects at the surface. This pressure is responsible for the action of suction cups, drinking straws, vacuum cleaners, and many other devices. In our calculations and end-of-chapter problems, we usually take atmospheric pressure to be

•  *Atmospheric pressure*

$$P_0 = 1.00 \text{ atm} \approx 1.013 \times 10^5 \text{ Pa} \qquad\qquad [15.3]$$

**Quick Quiz 15.1**

Suppose someone standing directly in front of you steps back and accidentally stomps on your foot with the heel of one shoe. Would you be better off if that person were a professional basketball player wearing sneakers or a petite woman wearing spike-heeled shoes?

Snowshoes prevent a person from sinking into soft snow because the person's weight is spread over a larger area, reducing the pressure on the snow's surface. *(Earl Young/FPG)*

**THINKING PHYSICS 15.1**

Suction cups can be used to hold objects onto surfaces. Why don't astronauts use suction cups to hold onto the outside surface of the space shuttle?

**Reasoning** A suction cup works because air is pushed out from under the cup when it is pressed against a surface. When the cup is released, it tends to spring back a bit, causing the trapped air under the cup to expand. This expansion causes a reduced pressure inside the cup. Thus, the difference between the atmospheric pressure on the outside of the cup and the reduced pressure inside provides a net force pushing the cup against the surface. For astronauts in orbit around the Earth, almost no air exists outside the surface of the spacecraft. Thus, if a suction cup were to be pressed against the outside surface of the spacecraft, the pressure differential needed to press the cup to the surface is not present.

**Figure 15.3**

(Thinking Physics 15.2) The daring physics professor takes a nap on a bed of nails. (*Jim Lehman*)

### THINKING PHYSICS 15.2

The daring physics professor, after a long lecture, stretches out for a nap on a bed of nails as shown in Figure 15.3. How is this possible?

**Reasoning** If you try to support your entire weight on a single nail, the pressure on your body is your weight divided by the very small area of the nail. This pressure is sufficiently large to penetrate the skin, as in the case of the hypodermic needle discussed earlier. If you distribute your weight over several hundred nails, however, as the professor is doing, the pressure is considerably reduced because the area that supports your weight is the total area of all nails in contact with your body. (Note that lying on a bed of nails is much more comfortable than sitting on a bed of nails. Standing on the bed of nails without shoes is not recommended.)

**EXERCISE** Estimate the order of magnitude of the pressure on the floor under the chair legs when you sit in a typical chair.

*Answer* $\approx 10^5 - 10^6$ Pa

## 15.2 • VARIATION OF PRESSURE WITH DEPTH

The study of fluid mechanics involves the density of a substance, defined in Equation 1.1 as the mass per unit volume for the substance. Table 15.1 lists the densities of various substances. These values vary slightly with temperature, because the volume of a substance is temperature-dependent (as we shall see in Chapter 16). Note that under standard conditions (0°C and atmospheric pressure) the densities of gases are on the order of 1/1000 the densities of solids and liquids. This difference implies that the average molecular spacing in a gas under these conditions is about ten times greater in each dimension than in a solid or liquid.

As divers know well, the pressure in the sea or a lake increases as they dive to greater depths. Likewise, atmospheric pressure decreases with increasing altitude. For this reason, aircraft flying at high altitudes must have pressurized cabins in order to provide sufficient oxygen for the passengers.

We now show mathematically how the pressure in a liquid increases with depth. Consider a liquid of density $\rho$ at rest as in Figure 15.4. Let us select a sample of the liquid contained within an imaginary cylinder of cross-sectional area $A$ extending from depth $d$ to depth $d + h$. This sample of liquid is at rest. Thus,

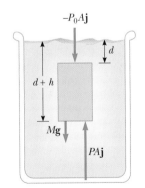

**Figure 15.4**

The net force on the sample of liquid within the darker region must be zero because the sample is in equilibrium.

| TABLE 15.1 | | Densities of Some Common Substances | |
|---|---|---|---|
| Substance | $\rho$ $(\text{kg/m}^3)^a$ | Substance | $\rho$ $(\text{kg/m}^3)^a$ |
| Ice | $0.917 \times 10^3$ | Water | $1.00 \times 10^3$ |
| Aluminum | $2.70 \times 10^3$ | Sea water | $1.03 \times 10^3$ |
| Iron | $7.86 \times 10^3$ | Ethyl alcohol | $0.806 \times 10^3$ |
| Copper | $8.92 \times 10^3$ | Benzene | $0.879 \times 10^3$ |
| Silver | $10.5 \times 10^3$ | Mercury | $13.6 \times 10^3$ |
| Lead | $11.3 \times 10^3$ | Air | $1.29$ |
| Gold | $19.3 \times 10^3$ | Oxygen | $1.43$ |
| Platinum | $21.4 \times 10^3$ | Hydrogen | $8.99 \times 10^{-2}$ |
| Glycerin | $1.26 \times 10^3$ | Helium | $1.79 \times 10^{-1}$ |

a All values are at standard atmospheric pressure and temperature (STP), that is, atmospheric pressure and 0°C. To convert the density in $\text{kg/m}^3$ to units of $\text{g/cm}^3$, multiply by $10^{-3}$.

according to Newton's second law, the net force on the sample must be equal to zero. We will investigate the forces on the sample related to the pressure on it.

The liquid external to our sample exerts forces at all points on the surface of the sample, perpendicular to the surface. The pressure exerted by the liquid on the bottom face of the sample is $P$, and the pressure on the top face is $P_0$. Therefore, from Equation 15.1, the magnitude of the upward force exerted by the liquid on the bottom of the sample is $PA$, and the magnitude of the downward force exerted by the liquid on the top is $P_0A$. In addition, a gravitational force is exerted on the sample. Because the sample is in equilibrium, the net force in the vertical direction must be zero:

$$\sum F_y = 0 \quad \longrightarrow \quad PA - P_0A - mg = 0$$

Because the mass of liquid in the sample is $m = \rho V = \rho Ah$, the gravitational force on the liquid in the sample is $mg = \rho gAh$. Thus,

$$PA = P_0A + \rho gAh$$

or

- *Variation of pressure with depth*

$$P = P_0 + \rho gh \qquad [15.4]$$

If the top surface of our sample is at $d = 0$, so that it is open to the atmosphere, then $P_0$ is atmospheric pressure. Equation 15.4 indicates that the pressure in a liquid depends only on the depth $h$ within the liquid. The pressure is therefore the same at all points having the same depth, independent of the shape of the container.

In view of Equation 15.4, any increase in pressure at the surface must be transmitted to every point in the liquid. This was first recognized by the French scientist Blaise Pascal (1623–1662) and is called **Pascal's law: A change in the pressure applied to an enclosed liquid is transmitted undiminished to every point of the fluid and to the walls of the container.**

- *Pascal's law*

An important application of Pascal's law is the hydraulic press illustrated by Figure 15.5. A force $\mathbf{F}_1$ is applied to a small piston of area $A_1$. The pressure is transmitted through a liquid to a larger piston of area $A_2$, and force $\mathbf{F}_2$ is exerted

(a)                                                              (b)

**Figure 15.5**

(a) Diagram of a hydraulic press. Because the increase in pressure is the same at the left and right sides, a small force $\mathbf{F}_1$ at the left produces a much larger force $\mathbf{F}_2$ at the right. (b) A vehicle under repair is supported by a hydraulic lift in a garage. *(David Frazier)*

by the liquid on this piston. Because the pressure is the same at both pistons, we see that $P = F_1/A_1 = F_2/A_2$. The force magnitude $F_2$ is therefore larger than $F_1$ by the multiplying factor $A_2/A_1$. Hydraulic brakes, car lifts, hydraulic jacks, and forklifts all make use of this principle.

## THINKING PHYSICS 15.3

Blood pressure is normally measured with the cuff of the sphygmomanometer around the arm. Suppose that the blood pressure were measured with the cuff around the calf of the leg of a standing person. Would the reading of the blood pressure be the same here as it is for the arm?

**Reasoning**   The blood pressure measured at the calf would be larger than that measured at the arm. If we imagine the vascular system of the body to be a vessel containing a liquid (blood), the pressure in the liquid will increase with depth. The blood at the calf is deeper in the liquid than that at the arm and is at a higher pressure.

Blood pressures are normally taken at the arm because it is at approximately the same height as the heart. If blood pressures at the calf were used as a standard, adjustments would need to be made for the height of the person, and the blood pressure would be different if the person were lying down.

 **APPLICATION**

Measuring blood pressure

## Example 15.1  The Car Lift

In a car lift used in a service station, compressed air exerts a force on a small piston of circular cross-section having a radius of 5.00 cm. This pressure is transmitted by an incompressible liquid to a second piston of radius 15.0 cm. (a) What force must the compressed air exert in order to lift a car weighing 13 300 N? (b) What air pressure will produce this force? (c) Consider the lift as a nonisolated system, and show that the input energy transfer is equal in magnitude to the output energy transfer.

**Solution**  (a) Because the pressure exerted by the compressed air is transmitted undiminished throughout the liquid, we have

$$F_1 = \left(\frac{A_1}{A_2}\right)F_2 = \frac{\pi(5.00 \times 10^{-2}\,\text{m})^2}{\pi(15.0 \times 10^{-2}\,\text{m})^2}\,(1.33 \times 10^4\,\text{N})$$

$$= \boxed{1.48 \times 10^3\,\text{N}}$$

(b) The air pressure that will produce this force is

$$P = \frac{F_1}{A_1} = \frac{1.48 \times 10^3\,\text{N}}{\pi(5.00 \times 10^{-2}\,\text{m})^2} = \boxed{1.88 \times 10^5\,\text{Pa}}$$

This pressure is approximately twice atmospheric pressure.

(c) The energy input and output are by means of work done by the forces as the pistons move. To determine the work done, we must find the magnitude of the displacement through which each force acts. Because the liquid is modeled to be incompressible, the volume of the cylinder through which the input piston moves must equal that through which the output piston moves. The lengths of these cylinders are the magnitudes $\Delta x_1$ and $\Delta x_2$ of the displacements of the forces (see Fig. 15.5). Setting the volumes equal, we have

$$V_1 = V_2 \quad\longrightarrow\quad A_1\,\Delta x_1 = A_2\,\Delta x_2$$

$$\frac{A_1}{A_2} = \frac{\Delta x_2}{\Delta x_1}$$

Evaluating the ratio of the input work to the output work, we find

$$\frac{W_1}{W_2} = \frac{F_1\,\Delta x_1}{F_2\,\Delta x_2} = \left(\frac{F_1}{F_2}\right)\left(\frac{\Delta x_1}{\Delta x_2}\right) = \left(\frac{A_1}{A_2}\right)\left(\frac{A_2}{A_1}\right) = 1$$

which verifies that the work input and output are the same, as they must be to conserve energy.

## Example 15.2  The Force on a Dam

Water is filled to a height $H$ behind a dam of width $w$ (Fig. 15.6). Determine the resultant force on the dam.

**Reasoning**  We cannot calculate the force on the dam by simply multiplying the area by the pressure, because the pressure varies with depth. The problem can be solved by finding the force $dF$ on a narrow horizontal strip at depth $h$, and then integrating the expression to find the total force on the dam.

**Solution**  The pressure at the depth $h$ beneath the surface at the shaded portion in Figure 15.6 is

$$P = \rho g h = \rho g(H - y)$$

(We have not included atmospheric pressure in our calculation because it acts on both sides of the dam, resulting in a net contribution of zero to the total force.) From Equation 15.1, we find the force on the shaded strip of area $dA$:

$$F = PA \quad\longrightarrow\quad dF = P\,dA$$

Because $dA = w\,dy$, we have

$$dF = P\,dA = \rho g(H - y)\,w\,dy$$

**Figure 15.6**

(Example 15.2) The total force on a dam is obtained from the expression $F = \int P\,dA$, where $dA$ is the area of the dark strip.

Therefore, the total force on the dam is

$$F = \int_0^H \rho g(H - y)\,w\,dy = \boxed{\tfrac{1}{2}\rho g w H^2}$$

Note that because the pressure increases with depth, the dam is designed such that its thickness increases with depth, as in Figure 15.6.

**EXERCISE**   Use the fact that the pressure increases linearly with depth to find the average pressure on the dam.

*Answer* $\frac{1}{2}\rho g H$

**EXERCISE**   What is the hydrostatic force on the back of

Grand Coulee Dam if the water in the reservoir is 150 m deep and the width of the dam is 1 200 m?

*Answer*   $1.32 \times 10^{11}$ N

Exercise   In some places, the Greenland ice sheet is 1.0 km thick. Estimate the pressure on the ground underneath the ice. ($\rho_{ice} = 920$ kg/m$^3$)

*Answer*   $9 \times 10^6$ Pa

## 15.3 • PRESSURE MEASUREMENTS

During the weather report on a television news program, the *barometric pressure* is often provided. This is the current pressure of the atmosphere, which varies over a small range from the standard value provided earlier. How is this pressure measured?

One instrument used to measure atmospheric pressure is the common barometer, invented by Evangelista Torricelli (1608–1647). A long tube closed at one end is filled with mercury and then inverted into a dish of mercury (Fig. 15.7a). The closed end of the tube is nearly a vacuum, so the pressure at the top of the mercury column can be taken as zero. In Figure 15.7a, the pressure at point $A$ due to the column of mercury must equal the pressure at point $B$ due to the atmosphere. If this were not the case, a net force would move mercury from one point to the other until equilibrium was established. It therefore follows that $P_0 = \rho_{Hg} g h$, where $\rho_{Hg}$ is the density of the mercury and $h$ is the height of the mercury column. As atmospheric pressure varies, the height of the mercury column varies, so the height can be calibrated to measure atmospheric pressure. Let us determine the height of a mercury column for one atmosphere of pressure, $P_0 = 1$ atm $= 1.013 \times 10^5$ Pa:

(a)

$$P_0 = \rho_{Hg} g h \quad \longrightarrow \quad h = \frac{P_0}{\rho_{Hg} g} = \frac{1.013 \times 10^5 \text{ Pa}}{(13.6 \times 10^3 \text{ kg/m}^3)(9.80 \text{ m/s}^2)} = 0.760 \text{ m}$$

Based on a calculation such as this, one atmosphere of pressure is defined as the pressure equivalent of a column of mercury that is exactly 0.760 0 m in height at 0°C.

The open-tube manometer illustrated in Figure 15.7b is a device for measuring the pressure of a gas contained in a vessel. One end of a U-shaped tube containing a liquid is open to the atmosphere, and the other end is connected to a system of unknown pressure $P$. The pressures at points $A$ and $B$ must be the same (otherwise, the curved portion of the liquid would experience a net force and would accelerate), and the pressure at $A$ is the unknown pressure of the gas. Therefore, equating the unknown pressure $P$ to the pressure at point $B$, we see that $P = P_0 + \rho g h$. The difference in pressure $P - P_0$ is equal to $\rho g h$. The pressure $P$ is called the **absolute pressure,** and the difference $P - P_0$ is called the **gauge pressure.** For example, the pressure you measure in your bicycle tire is gauge pressure.

(b)

**Figure 15.7**

Two devices for measuring pressure: (a) a mercury barometer and (b) an open-tube manometer.

**Quick Quiz 15.2**

It is possible to build a barometer like that in Figure 15.7a with water rather than mercury. Why is this impractical?

**Archimedes** (287–212 BC)

Archimedes, a Greek mathematician, physicist, and engineer, was perhaps the greatest scientist of antiquity. He was the first to compute accurately the ratio of a circle's circumference to its diameter, and he showed how to calculate the volume and surface area of spheres, cylinders, and other geometric shapes. He is well known for discovering the nature of the buoyant force.

**Figure 15.8**

The external forces on the cube of liquid are the gravitational force $\mathbf{F}_g$ and the buoyant force $\mathbf{B}$. Under equilibrium conditions, $B = F_g$.

•   *Archimedes's principle*

## 15.4 • BUOYANT FORCES AND ARCHIMEDES'S PRINCIPLE

In this section, we investigate the origin of a **buoyant force,** which is **an upward force exerted on an object by the surrounding fluid.** Buoyant forces are evident in many situations—anyone who has ridden in a boat has experienced a buoyant force. Another common example is the relative ease with which you can lift someone in a swimming pool compared with lifting that same individual on dry land. Evidently, water provides partial support to any object placed in it. According to **Archimedes's principle,**

> Any object completely or partially submerged in a fluid experiences an upward buoyant force whose magnitude is equal to the weight of the fluid displaced by the object.

Archimedes's principle can be verified in the following manner. Suppose we focus our attention on the indicated cube of fluid in the container of Figure 15.8. This cube of fluid is in equilibrium under the action of the forces on it. One of these forces in the vertical direction is the gravitational force. Because the cube is in equilibrium, the net force on it in the vertical direction must be zero. What cancels the downward gravitational force so that the cube remains in equilibrium? Apparently, the rest of the fluid inside the container is applying an upward force—the buoyant force. Thus, the magnitude $B$ of the buoyant force must be exactly equal to the weight of the fluid inside the cube:

$$\sum F_y = 0 \quad \longrightarrow \quad B - F_g = 0 \quad \longrightarrow \quad B = mg$$

where $m$ is the mass of the fluid in the cube.

Now, imagine that the cube of fluid is replaced by a cube of steel of the same dimensions. What is the buoyant force on the steel? The fluid surrounding a cube behaves in the same way whether it is exerting pressure on a cube of fluid or a cube of steel; therefore, the **buoyant force acting on the steel is the same as the buoyant force acting on a cube of fluid of the same dimensions.** This result applies for a submerged object of any shape, size, or density.

Let us now show more explicitly that the magnitude of the buoyant force is equal to the weight of the displaced fluid. Although this is true for both liquids and gases, we will perform the derivation for a liquid. On the sides of the cube of liquid in Figure 15.8, forces due to the pressure act horizontally and cancel in pairs on opposite sides of the cube for a net horizontal force of zero. In a liquid, the pressure at the bottom of the cube is greater than the pressure at the top by an amount $\rho_f g h$, where $\rho_f$ is the density of the liquid and $h$ is the height of the cube. Thus, the upward force $F_{bot}$ on the bottom is greater than the downward force $F_{top}$ on the top of the cube. The net vertical force *exerted by the liquid* (we are ignoring the gravitational force for now) is

$$\sum F_{liquid} = B = F_{bot} - F_{top}$$

Expressing this in terms of pressure gives us

$$B = P_{bot}A - P_{top}A = \Delta PA = \rho_f ghA$$

$$B = \rho_f gV \qquad\qquad [15.5]$$

where $V$ is the volume of the cube. Because the mass of the liquid in the cube is

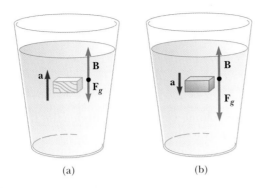

(a)                              (b)

**Figure 15.9**

(a) A totally submerged object that is less dense than the fluid in which it is submerged experiences a net upward force. (b) A totally submerged object that is denser than the fluid sinks.

**PITFALL PREVENTION 15.2**

**Buoyant force is exerted by the fluid**

Notice a very important point in this discussion. The buoyant force is a force **exerted by the fluid.** It is not determined by properties of the object, except for the amount of fluid displaced by the object. Thus, if several objects of different densities but the same volume are immersed in a fluid, they will all experience the same buoyant force. Whether they sink or float is determined by the relationship between the buoyant force and their weight. It is not the natural tendency for helium to rise—a helium balloon rises because its overall density is less than that of air.

$M = \rho_f V$, we see that

$$B = Mg$$

which is the weight of the displaced liquid.

Before proceeding with a few examples, it is instructive to compare two common cases: the buoyant force acting on a totally submerged object and that acting on a floating object.

**Case I: A Totally Submerged Object**  When an object is totally submerged in a liquid of density $\rho_f$, the magnitude of the upward buoyant force is $B = \rho_f g V_O$, where $V_O$ is the volume of the object. Because the object is totally submerged, the volume of the object and the volume of liquid displaced by the object are the same, $V = V_O$. If the object has a density $\rho_O$, its weight is $Mg = \rho_O V_O g$. Thus, the net force on it is $\Sigma F = B - Mg = (\rho_f - \rho_O) V_O g$. We see that if the density of the object is less than the density of the liquid as in Figure 15.9a, the net force is positive and the unsupported object accelerates upward. If the density of the object is greater than the density of the liquid as in Figure 15.9b, the net force is negative and the unsupported object sinks.

The same behavior is exhibited by an object immersed in a gas, such as the air in the atmosphere.* If the object is less dense than air, like a helium-filled balloon, the object floats upward. If it is denser, like a rock, it falls downward.

**Case II: A Floating Object**  Now consider an object in static equilibrium floating on the surface of a fluid, that is, an object that is only partially submerged, such as the ice cube floating in water in Figure 15.10. Because it is only partially submerged, the volume $V$ of fluid displaced by the object is only a fraction of the total volume $V_O$ of the object. The volume of the fluid displaced by the object corresponds to that volume of the object beneath the fluid surface. Because the object is in equilibrium, the upward buoyant force is balanced by the downward gravitational force exerted on the object. The buoyant force has a magnitude $B = \rho_f g V$.

**Figure 15.10**

An object that is less dense than a fluid floats in the fluid, such as this ice cube in water.

* The general behavior is the same, but the buoyant force varies with height in the atmosphere due to the variation in density of the air.

Notice that the sinking or floating behavior of an object is determined by a comparison of the fluid and object *densities*. In the case of aircraft, one often hears descriptive phrases such as "heavier than air" (for an airplane) or "lighter than air" (for a hot-air balloon). These are misguided phrases, because they describe a comparison between the *weight* of an object and that of an unspecified amount of air. Despite the popularity of these phrases, try to train yourself to say "denser than air" and "less dense than air."

Hot-air balloons. Because hot air is less dense than cold air, a net upward force acts on the balloons. *(Richard Megna/Fundamental Photographs)*

Because the weight of the object is $Mg = \rho_O V_O g$, and because Newton's second law tells us that $\Sigma F = 0$ in the vertical direction, $Mg = B$, and we see that $\rho_f g V = \rho_O V_O g$, or

$$\frac{\rho_O}{\rho_f} = \frac{V}{V_O} \qquad \text{[15.6]}$$

Thus, the fraction of the volume of the object under the fluid surface is equal to the ratio of the object density to the fluid density.

Let us consider examples of both cases. Under normal conditions, the average density of a fish is slightly greater than the density of water. This being the case, a fish would sink if it did not have some mechanism for adjusting its density. The fish accomplishes this by internally regulating the size of its swim bladder, a gas-filled cavity within the fish's body. Increasing its size increases the amount of water displaced, which increases the buoyant force. In this manner, fish are able to swim to various depths. Because the fish is totally submerged in the water, this is an example of Case I.

As an example of Case II, imagine a large cargo ship. When the ship is at rest, the upward buoyant force from the water balances the weight so that the ship is in equilibrium. Only part of the volume of the ship is under water. If the ship is loaded with heavy cargo, it sinks deeper into the water. The increased weight of the ship due to the cargo is balanced by the extra buoyant force related to the extra volume of the ship that is now beneath the water surface.

**Quick Quiz 15.3**

Atmospheric pressure varies from day to day. Does a ship float higher in the water on a high-pressure day than on a low-pressure day?

**Quick Quiz 15.4**

A pound of styrofoam and a pound of lead have the same weight. If they are placed on an equal-arm balance surrounded by air, will it balance?

**Quick Quiz 15.5**

What is the origin of the saying, "That's only the tip of the iceberg"?

**THINKING PHYSICS 15.4**

Suppose an office party is taking place on the top floor of a tall building. Carrying an iced soft drink, you step on the elevator, which begins to accelerate downward. What happens to the ice in the drink? Does it rise farther out of the liquid? Sink deeper into the liquid? Or is it unaffected by the motion?

**Reasoning** The acceleration of the elevator is equivalent to a change in the gravitational field, according to the principle of equivalence (Section 9.9). If the elevator accelerates downward, one might be tempted to say that the effect is the same as if gravity decreases—the weight of the ice cube decreases, causing it to float higher in the liquid. Recall, however, that the magnitude of the buoyant force is equal to the weight of the liquid displaced by the ice cube. The weight of the liquid also decreases with the effectively decreased gravity. Because both the weight of the ice cube and the buoyant force decrease by the same factor, the level of the ice cube in the liquid is unaffected.

## THINKING PHYSICS 15.5

A florist delivery person is delivering a flower basket to a home. The basket includes an attached helium-filled balloon, which suddenly comes loose from the basket and begins to accelerate upward toward the sky. Startled by the release of the balloon, the delivery person drops the flower basket. As the basket falls, the basket–Earth system experiences an increase in kinetic energy and a decrease in gravitational potential energy, consistent with conservation of mechanical energy. The balloon–Earth system, however, experiences an increase in *both* gravitational potential energy and kinetic energy. Is this inconsistent with the principle of conservation of mechanical energy? If not, from where is the extra energy coming?

**Reasoning**   In the case of the system of the flower basket and the Earth, a good approximation to the motion of the basket can be made by ignoring the effects of the air. Thus, the basket–Earth system can be analyzed with the isolated system model—mechanical energy is conserved. For the balloon–Earth system, we cannot ignore the effects of the air—it is the buoyant force of the air that causes the balloon to rise. Thus, the balloon–Earth system is analyzed with the nonisolated system model. The buoyant force of the air does work across the boundary of the system, and that work results in an increase in both the kinetic and gravitational potential energies of the system.

## Example 15.3   A Submerged Object

A piece of aluminum is suspended from a string and then completely immersed in a container of water (Fig. 15.11). The mass of the aluminum is 1.0 kg, and its density is $2.7 \times 10^3 \ kg/m^3$. Calculate the tension in the string before and after the aluminum is immersed.

**Solution**   When the aluminum is suspended in air as in Figure 15.11a, Newton's second law tells us that the tension $T_1$ in the string (the reading on the scale) is equal to the weight $Mg$ of the aluminum, assuming that the buoyant force of air can be ignored:

$$T_1 = Mg = (1.0 \ kg)(9.80 \ m/s^2) = \boxed{9.8 \ N}$$

When immersed in water, we use Newton's second law again. The aluminum experiences an upward buoyant force **B** as in Figure 15.11b, which reduces the tension in the string. To calculate $B$, we must first calculate the volume of the aluminum:

$$V_{Al} = \frac{M}{\rho_{Al}} = \frac{1.0 \ kg}{2.7 \times 10^3 \ kg/m^3} = 3.7 \times 10^{-4} \ m^3$$

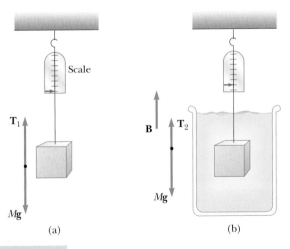

**Figure 15.11**

(Example 15.3) (a) When the aluminum is suspended in air, the scale reads the true weight $Mg$ (ignoring the buoyancy of air).
(b) When the aluminum is immersed in water, the buoyant force **B** reduces the scale reading to $T_2 = Mg - B$.

From Equation 15.5,

$$B = \rho_w V_{Al} g = (1.0 \times 10^3 \text{ kg/m}^3)(3.7 \times 10^{-4} \text{ m}^3)(9.80 \text{ m/s}^2)$$

$$= 3.6 \text{ N}$$

Now we apply Newton's second law to the aluminum,

$$\sum F = T_2 + B - Mg = 0$$

$$T_2 = Mg - B = 9.8 \text{ N} - 3.6 \text{ N} = \boxed{6.2 \text{ N}}$$

## Example 15.4 Changing String Vibration with Water

One end of a horizontal string is attached to a vibrating blade, and the other end passes over a pulley as in Figure 15.12a. A sphere of mass 2.00 kg hangs on the end of the string. The string is vibrating in its second harmonic. A container of water is raised under the sphere so that the sphere is completely submerged. After this is done, the string vibrates in its fifth harmonic, as shown in Figure 15.12b. What is the radius of the sphere?

**Solution** In Figure 15.12a, Newton's second law applied to the sphere tells us that the initial tension $T_i$ in the string is equal to the weight of the sphere:

$$T_i - mg = 0 \longrightarrow T_i = mg$$

$$T_i = (2.00 \text{ kg})(9.80 \text{ m/s}^2) = 19.6 \text{ N}$$

where the subscript $i$ is used to indicate initial variables before we immerse the sphere in water. Once the sphere is immersed in water, the tension in the string will decrease to $T_f$. Applying Newton's second law to the sphere again in this situation, we have

$$T_f + B - mg = 0 \longrightarrow B = mg - T_f \quad (1)$$

The desired quantity, the radius of the sphere, will appear in the expression for the buoyant force $B$. Before we can proceed in this direction, however, we need to evaluate $T_f$. We do this from the standing wave information. We write the equation for the frequency of a standing wave on a string (Equation 14.8) twice: once before we immerse the sphere and once after, and divide the equations:

$$f = \frac{n_i}{2L}\sqrt{\frac{T_i}{\mu}}$$
$$f = \frac{n_f}{2L}\sqrt{\frac{T_f}{\mu}} \longrightarrow 1 = \frac{n_i}{n_f}\sqrt{\frac{T_i}{T_f}} \quad (2)$$

where the frequency $f$ is the same in both cases, because it is determined by the vibrating blade. In addition, the linear mass density $\mu$ and the length $L$ of the vibrating portion of the string are the same in both cases. Solving (2) for $T_f$ gives

$$T_f = \left(\frac{n_i}{n_f}\right)^2 T_i = \left(\frac{2}{5}\right)^2 (19.6 \text{ N}) = 3.14 \text{ N}$$

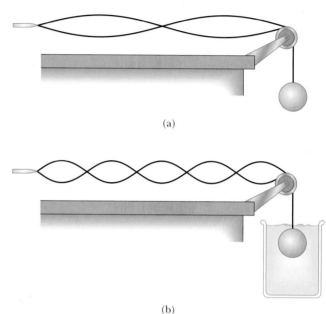

(a)

(b)

**Figure 15.12**

(Example 15.4) When the sphere hangs in air, the string vibrates in its second harmonic. When the sphere is immersed in water, the string vibrates in its fifth harmonic.

Substituting this into equation (1), we can evaluate the buoyant force on the sphere:

$$B = mg - T_f = 19.6 \text{ N} - 3.14 \text{ N} = 16.5 \text{ N}$$

Finally, expressing the buoyant force in terms of the radius of the sphere, we solve for the radius,

$$B = \rho_{water} g V_{sphere} = \rho_{water} g (\tfrac{4}{3}\pi r^3)$$

$$r = \sqrt[3]{\frac{3B}{4\pi\rho_{water}g}} = \sqrt[3]{\frac{3(16.5 \text{ N})}{4\pi(1\,000 \text{ kg/m}^3)(9.80 \text{ m/s}^2)}}$$

$$= 7.38 \times 10^{-2} \text{ m} = \boxed{7.38 \text{ cm}}$$

## 15.5 • FLUID DYNAMICS

Thus far, our study of fluids has been restricted to fluids at rest—**fluid statics.** We now turn our attention to **fluid dynamics,** that is, the study of fluids in motion. Instead of trying to study the motion of each particle of the fluid as a function of time, we describe the properties of the fluid as a whole.

### Flow Characteristics

When fluid is in motion, its flow is of one of two main types. The flow is said to be **steady,** or **laminar,** if each particle of the fluid follows a smooth path, so that the paths of different particles never cross each other, as in Figure 15.13. Thus, in steady flow, the velocity of the fluid at any point remains constant in time.

Above a certain critical speed, fluid flow becomes **turbulent.** Turbulent flow is an irregular flow characterized by small whirlpool-like regions as in Figure 15.14. As an example, the flow of water in a river becomes turbulent in regions where rocks and other obstructions are encountered, often forming "white water" rapids.

The term **viscosity** is commonly used in fluid flow to characterize the degree of internal friction in the fluid. This internal friction, or viscous force, is associated with the resistance of two adjacent layers of the fluid against moving relative to each other. Because viscosity represents a nonconservative force, part of the kinetic energy of a fluid is converted to internal energy when layers of fluid slide past one another. This is similar to the mechanism by which an object sliding on a rough horizontal surface experiences a transformation of kinetic energy to internal energy.

Because the motion of a real fluid is very complex and not yet fully understood, we adopt a simplification model. As we shall see, many features of real fluids in motion can be understood by considering the behavior of an ideal fluid. In our

**Figure 15.14**

Turbulent flow: the tip of a rotating blade (the dark region at the top) forms a vortex in air that is being heated by an alcohol lamp (the wick is at the bottom). Note the air turbulence on both sides of the rotating blade. *(Kim Vandiver and Harold Edgerton, Palm Press, Inc.)*

**Figure 15.13**

An illustration of steady flow around an automobile in a test wind tunnel. The streamlines in the airflow are made visible by smoke particles. *(Andy Sacks/Stone)*

simplification model, we make the following four assumptions:

- **Nonviscous fluid** In a nonviscous fluid, internal friction is ignored. An object moving through the fluid experiences no viscous force.
- **Incompressible fluid** The density of the fluid is assumed to remain constant regardless of the pressure in the fluid.
- **Steady flow** In steady flow, we assume that the velocity of the fluid at each point remains constant in time.
- **Irrotational flow** Fluid flow is irrotational if the fluid about any point has no angular momentum. If a small paddle wheel placed anywhere in the fluid does not rotate about the wheel's center of mass, the flow is irrotational. (If the wheel were to rotate, as it would if turbulence were present, the flow would be rotational.)

The first two assumptions in our simplification model are properties of our ideal fluid. The last two are descriptions of the way that the fluid flows.

- *Properties of an ideal fluid under ideal flow*

**Figure 15.15**

This diagram represents a set of streamlines (*blue lines*). A particle at $P$ follows one of these streamlines, and its velocity is tangent to the streamline at each point along its path.

## 15.6 • STREAMLINES AND THE CONTINUITY EQUATION FOR FLUIDS

If you are watering your garden, and your garden hose is too short, you might do one of two things to help you reach the garden with the water (before you look for a longer hose!). You might attach a nozzle to the end of the hose, or, in the absence of a nozzle, you might place your thumb over the end of the hose, allowing the water to come out of a narrower opening. Why does either of these techniques cause the water to come out faster, so that it can be projected over a longer range? We shall see the answer to this question in this section.

The path taken by a particle of the fluid under steady flow is called a **streamline.** The velocity of the particle is always tangent to the streamline, as shown in Figure 15.15. No two streamlines can cross each other, for if they did, a particle could move either way at the crossover point and then the flow would not be steady. A set of streamlines as shown in Figure 15.15 forms what is called a *tube of flow.* Particles of the fluid cannot flow into or out of the sides of this tube, because if they did, the streamlines would be crossing each other.

Consider an ideal fluid flowing through a pipe of nonuniform size as in Figure 15.16. The particles in the fluid move along the streamlines in steady flow. Let us analyze this situation using the nonisolated system in steady-state model. We have seen this model used for energy in Chapter 7, but we were told at that time that the model can be used for any conserved quantity. The volume of an incompressible fluid is a conserved quantity. Assuming no leaks in our pipe, we can neither create nor destroy fluid, just as we could not create nor destroy energy in Chapters 6 and 7.

We choose as our system the region of space in the pipe from point 1 to point 2 in Figure 15.16. Let us assume that this region is filled with fluid at all times. As the fluid flows in the pipe, fluid enters the system at point 1 and leaves the system at point 2. Imagine that the fluid moves through a distance $\Delta x_1$ at point 1, and moves a distance $\Delta x_2$ at point 2 as it leaves the system. The volume of fluid entering the system at point 1 is $A_1 \Delta x_1$ and the volume leaving at point 2 is $A_2 \Delta x_2$. Because the volume of an incompressible fluid is a conserved quantity, these two volumes must be equal for the system to be in steady state. If this were not true, the volume of fluid in the system would be changing. Thus,

$$A_1 \Delta x_1 = A_2 \Delta x_2$$

**Figure 15.16**

A fluid moving with steady flow through a pipe of varying cross-sectional area. The volume of fluid flowing through $A_1$ in a time interval $\Delta t$ must equal the volume flowing through $A_2$ in the same time interval.

Let us divide this equation by the time interval during which the fluid moves:

$$\frac{A_1 \, \Delta x_1}{\Delta t} = \frac{A_2 \, \Delta x_2}{\Delta t}$$

In the limit as the time interval shrinks to zero, the ratio of distance traveled to the time interval is the instantaneous speed of the fluid, so we can write this as

$$A_1 v_1 = A_2 v_2 \qquad \text{[15.7]}$$

• *Continuity equation for fluids*

This expression, called the **continuity equation for fluids,** says that the **product of the area and the fluid speed at all points along the pipe is a constant.** Therefore, the speed is high where the tube is constricted and low where the tube is wide. The product $Av$, which has the dimensions of volume per time, is called the **volume flow rate.** This is why a nozzle or your thumb over the garden hose allows you to project the water farther. By reducing the area through which the water flows, you increase its speed. Thus, you project the water from the hose with a high initial velocity, resulting in a large value of the range, as discussed for projectiles in Chapter 3.

**PITFALL PREVENTION 15.4**

**A new continuity equation**

This is the first time that we have seen a continuity equation for anything other than energy. A continuity equation exists for *any* conserved quantity. Equation 15.7 states that the rate of flow of fluid into the system equals the rate of flow out of the system because fluid cannot be stored in the system once the pipe is full.

### Quick Quiz 15.6

As water exits a faucet at a low setting so that a streamline flow of water results, the stream becomes narrower as it descends. Explain why this happens.

### Example 15.5    Watering a Garden

A water hose 2.50 cm in diameter is used by a gardener to fill a 30.0-L bucket. The gardener notes that it takes 1.00 min to fill the bucket. A nozzle with an opening of cross-sectional area 0.500 cm² is then attached to the hose. The nozzle is held so that water is projected horizontally from a point 1.00 m above the ground. Over what horizontal distance can the water be projected?

**Solution**   We identify point 1 within the hose and point 2 at the exit of the nozzle. We first find the speed of the water in the hose from the bucket-filling information. The cross-sectional area of the hose is

$$A_1 = \pi r^2 = \pi \frac{d^2}{4} = \pi \left[ \frac{(2.50 \text{ cm})^2}{4} \right] = 4.91 \text{ cm}^2$$

According to the data given, the volume flow rate is equal to 30.0 liters per minute:

$$A_1 v_1 = 30.0 \text{ L/min} = \frac{30.0 \times 10^3 \text{ cm}^3}{60.0 \text{ s}} = 500 \text{ cm}^3/\text{s}$$

$$v_1 = \frac{500 \text{ cm}^3/\text{s}}{4.91 \text{ cm}^2} = 102 \text{ cm/s} = 1.02 \text{ m/s}$$

Now we use the continuity equation for fluids to find the speed $v_2 = v_{xi}$ with which the water exits the nozzle. The subscript $i$ anticipates that this will be the *initial* velocity component of the water projected from the hose, and the sub-

script $x$ indicates that the initial velocity vector of the projected water is in the horizontal direction.

$$A_1 v_1 = A_2 v_2 = A_2 v_{ix}$$

$$v_{ix} = \frac{A_1}{A_2} v_1 = \frac{4.91 \text{ cm}^2}{0.500 \text{ cm}^2} (1.02 \text{ m/s}) = 10.0 \text{ m/s}$$

We now shift our thinking away from fluids and to projectile motion because the water is in free-fall once it exits the nozzle. A particle of the water falls through a vertical distance of 1.00 m starting from rest, which requires a time, which we find from Equation 3.13:

$$y_f = y_i + v_{yi}t - \tfrac{1}{2} gt^2$$

$$-1.00 \text{ m} = 0 + 0 - \tfrac{1}{2} (9.80 \text{ m/s}^2) \, t^2$$

$$t = \sqrt{\frac{2(1.00 \text{ m})}{9.80 \text{ m/s}^2}} = 0.452 \text{ s}$$

In the horizontal direction, we apply Equation 3.12 to a particle of water to find the horizontal distance:

$$x_f = x_i + v_{xi}t = 0 + (10.0 \text{ m/s})(0.452 \text{ s}) = \boxed{4.52 \text{ m}}$$

**EXERCISE**   The Garfield Thomas water tunnel at Pennsylvania State University has a circular cross-section that constricts from a diameter of 3.6 m to the test section, which is 1.2 m in diameter. If the speed of flow is 3.0 m/s in the larger diameter pipe, determine the speed of flow in the test section.

*Answer*   27 m/s

**Daniel Bernoulli** (1700–1782)

Bernoulli, a Swiss physicist and mathematician, made important discoveries in fluid dynamics. His most famous work, *Hydrodynamica,* published in 1738, is both a theoretical and a practical study of equilibrium, pressure, and speed in fluids. In this publication, Bernoulli also attempted the first explanation of the behavior of gases with changing pressure and temperature; this was the beginning of the kinetic theory of gases, which we will study in Chapter 16. *(Corbis-Bettmann)*

**Figure 15.17**

A fluid flowing through a constricted pipe with streamline flow. In a time interval $\Delta t$, the change in the system is the same as if the fluid in the section of length $\Delta x_1$ moves to the position of the section of length $\Delta x_2$. The volumes of fluid in the two sections are equal.

## 15.7 • BERNOULLI'S PRINCIPLE

You have probably had the experience of driving on a highway and having a large truck pass by you at high speed. In that situation, you may have had the frightening feeling that your car was being pulled in toward the truck as it passed. We will see the origin for this effect in this section.

As a fluid moves through a region in which its speed or elevation above the Earth's surface changes, the pressure in the fluid varies with these changes. In 1738, the Swiss physicist Daniel Bernoulli first derived an expression that relates the pressure to fluid speed and elevation.

Let us develop a mathematical representation for **Bernoulli's principle,** which shows explicitly the dependence of pressure on speed and elevation. Consider the flow of an ideal fluid (Section 15.5) through a nonuniform pipe in a time $\Delta t$, as illustrated in Figure 15.17. The system is chosen to be the section of fluid that is initially between points 1 and 2 in Figure 15.17 and the Earth. This choice of system is different from that for the derivation of the continuity equation for fluids, in which we chose a region of space as the system. We analyze the current situation with the nonisolated system model. Work is being done on our system by the external fluid that is in contact with the two ends of the fluid in the system, and the kinetic and gravitational potential energy of the system is changing as a result. Thus, the continuity equation for energy (Eq. 6.20) for the system in this situation is

$$\Delta K + \Delta U = W \qquad [15.8]$$

Let us evaluate each of the terms in Equation 15.8. First notice that the shaded elements of fluid of lengths $\Delta x_1$ and $\Delta x_2$ in Figure 15.17 represent the only change between the initial situation and the final situation. The fluid in between these elements experiences no change in either kinetic or gravitational potential energy. Thus, we can restrict our attention to the shaded elements of fluid. The difference in kinetic energy is that between the element at point 2 and the element at point 1:

$$\Delta K = \tfrac{1}{2}(\Delta m)v_2{}^2 - \tfrac{1}{2}(\Delta m)v_1{}^2 \qquad [15.9]$$

We have used the same increment of mass $\Delta m$ for both elements of fluid—both elements have the same volume and, because the fluid is incompressible, the same mass. Now, the change in potential energy of the fluid–Earth system is that associated with moving the element of fluid located at point 1 up to the location at point 2:

$$\Delta U = (\Delta m)gy_2 - (\Delta m)gy_1 \qquad [15.10]$$

Finally, we evaluate the work done on the section of fluid. The fluid to the left of point 1 does positive work on our section because it applies a force in the same direction as the displacement. The fluid to the right of the element at point 2 does negative work because the force and displacement vectors are in opposite directions. Thus, the net work done on the system is

$$W = F_1 \Delta x_1 - F_2 \Delta x_2$$

We replace the force with the product of pressure and cross-sectional area of the pipe:

$$W = P_1 A_1 \Delta x_1 - P_2 A_2 \Delta x_2$$

Finally, we recognize the product of cross-sectional area and displacement as the volume of the elements of fluid:

$$W = P_1 \, \Delta V - P_2 \, \Delta V \qquad [15.11]$$

where we have not included subscripts on the volume element $\Delta V$ because both elements have the same volume for an incompressible fluid. We now use Equations 15.9, 15.10, and 15.11 to substitute for each term in Equation 15.8:

$$\Delta K + \Delta U = W$$

$$\tfrac{1}{2}(\Delta m)v_2^2 - \tfrac{1}{2}(\Delta m)v_1^2 + (\Delta m)gy_2 - (\Delta m)gy_1 = P_1 \, \Delta V - P_2 \, \Delta V$$

Dividing each term by the volume element $\Delta V$ gives

$$\tfrac{1}{2}\left(\frac{\Delta m}{\Delta V}\right)v_2^2 - \tfrac{1}{2}\left(\frac{\Delta m}{\Delta V}\right)v_1^2 + \left(\frac{\Delta m}{\Delta V}\right)gy_2 - \left(\frac{\Delta m}{\Delta V}\right)gy_1 = P_1 - P_2$$

The ratio of the mass $\Delta m$ for a fluid element to the volume $\Delta V$ of the element is the density $\rho$ of the fluid. Incorporating this fact and rearranging terms, we have

$$P_1 + \tfrac{1}{2}\rho v_1^2 + \rho gy_1 = P_2 + \tfrac{1}{2}\rho v_2^2 + \rho gy_2 \qquad [15.12]$$

• *Bernoulli's equation*

This is **Bernoulli's equation** applied to an ideal fluid. It is often expressed as

$$P + \tfrac{1}{2}\rho v^2 + \rho gy = \text{constant} \qquad [15.13]$$

Bernoulli's equation says that the sum of the pressure $P$, the kinetic energy per unit volume $\tfrac{1}{2}\rho v^2$, and gravitational potential energy per unit volume $\rho gy$ has the same value at all points along a streamline.

When the fluid is at rest, $v_1 = v_2 = 0$ and Equation 15.12 becomes

$$P_1 - P_2 = \rho g(y_2 - y_1) = \rho gh$$

which agrees with Equation 15.4.

**PITFALL PREVENTION 15.5**
**Bernoulli's principle for gases**

 For gases, we cannot use the assumption that the fluid is incompressible. Thus, Equation 15.12 is not true for gases. The qualitative behavior is the same, however—as the speed of the gas increases, its pressure decreases.

## Quick Quiz 15.7

You observe two helium balloons floating next to each other at the ends of strings secured to a table. The facing surfaces of the balloons are separated by 1 to 2 cm. You blow through the opening between the balloons. What happens to the balloons? (a) They move toward each other; (b) they move away from each other; (c) they are unaffected.

## Quick Quiz 15.8

You may have noticed a convertible passing you on a highway with its cloth top bulging upward, as if the interior of the car is pressurized. Why does this happen?

## THINKING PHYSICS 15.6

During a tornado, the windows of a house sometimes explode outward. At other times, the roof of a house is lifted off the house. What causes these events to happen? How does opening the windows in a tornado (which may seem counterintuitive!) help?

**WEB**

For information on tornadoes and
tornado safety, visit
**www.usatoday.com/weather/
wtwisto.htm**

**Reasoning** The wind speed in a tornado is very high. According to Bernoulli's principle, such high wind speeds will result in low pressures. When these winds pass by a closed window, the pressure outside the window is reduced well below that of the pressure on the inside surface, which is due to the still air inside the house. As a result, the net outward force can burst the window. When these high-speed winds pass over the roof of a house, the pressure above the roof is much lower than the pressure below the roof, again due to the still air inside the house. The pressure difference can be enough to lift the roof off the house.

## Example 15.6 Sinking the Cruise Ship

A scuba diver is hunting for fish with a spear gun. He accidentally fires the gun so that a spear punctures the side of a cruise ship. The hole is located at a depth of 10.0 m below the water surface. With what speed does the water enter the cruise ship through the hole?

**Reasoning** We identify point 1 as the water surface outside the ship, which we will assign as $y = 0$. At this point, the water is static, so $v_1 = 0$. We identify point 2 as a point just inside the hole in the interior of the ship because that is the point at which we wish to evaluate the speed of the water.

This point is at a depth $y = -h = -10.0$ m below the water surface. We use Bernoulli's equation to compare these two points. At both points, the water is open to atmospheric pressure, so that $P_1 = P_2 = P_0$.

**Solution** Based on the arguments made in the Reasoning, Bernoulli's equation becomes

$$P_0 + \tfrac{1}{2}\rho(0)^2 + \rho g(0) = P_0 + \tfrac{1}{2}\rho v_2^2 + \rho g(-h)$$

$$\longrightarrow v_2 = \sqrt{2gh} = \sqrt{2(9.80 \text{ m/s}^2)(10.0 \text{ m})} = \boxed{14 \text{ m/s}}$$

## Example 15.7 The Venturi Tube

The horizontal constricted pipe illustrated in Figure 15.18, known as a *Venturi tube*, can be used to measure the flow speed of an incompressible fluid. Let us determine the flow speed at point 2 if the pressure difference $P_1 - P_2$ is known.

**Solution** Because the pipe is horizontal, $y_1 = y_2$ and Equation 15.12 applied to points 1 and 2 gives

$$P_1 + \tfrac{1}{2}\rho v_1^2 = P_2 + \tfrac{1}{2}\rho v_2^2$$

From the continuity equation for fluids (Eq. 15.7), we see that $A_1 v_1 = A_2 v_2$ or

$$v_1 = \frac{A_2}{A_1} v_2$$

Substituting this expression into the previous equation gives

$$P_1 + \tfrac{1}{2}\rho \left(\frac{A_2}{A_1}\right)^2 v_2^2 = P_2 + \tfrac{1}{2}\rho v_2^2$$

$$v_2 = A_1 \sqrt{\frac{2(P_1 - P_2)}{\rho(A_1^2 - A_2^2)}}$$

(a)                     (b)

**Figure 15.18**

(a) (Example 15.7) The pressure $P_1$ is greater that the pressure $P_2$, because $v_1 < v_2$. This device can be used to measure the speed of fluid flow. (b) A Venturi tube. The higher level of fluid in the middle column shows that the pressure at the top of the column, which is in the constricted region of the Venturi tube, is lower. *(Courtesy of Central Scientific Company)*

We can also obtain an expression for $v_1$ using this result and the continuity equation for fluids. Note that because $A_2 < A_1$, it follows that $P_1 > P_2$. In other words, the pressure is reduced in the constricted part of the pipe.

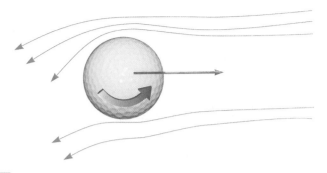

**Figure 15.20**

A spinning golf ball experiences a lifting force that allows it to travel much farther than it would if it were not spinning.

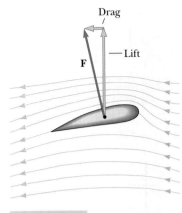

**Figure 15.19**

Streamline flow around a moving airplane wing. The air approaching from the right is deflected downward by the wing. Because the airstream is deflected, it exerts a force **F** on the wing.

## 15.8 • OTHER APPLICATIONS OF FLUID DYNAMICS

Consider the streamlines that flow around an airplane wing as shown in Figure 15.19. Let us assume that the airstream approaches the wing horizontally from the right. The tilt of the wing causes the airstream to be deflected downward. Because the airstream is deflected by the wing, the wing must exert a force on the airstream. According to Newton's third law, the airstream must exert an equal and opposite force **F** on the wing. This force has a vertical component called the **lift** (or aerodynamic lift) and a horizontal component called **drag.** The lift depends on several factors, such as the speed of the airplane, the area of the wing, its curvature, and the angle between the wing and the horizontal. As this angle increases, turbulent flow can set in above the wing to reduce the lift.

In general, an object experiences lift by any effect that causes the fluid to change its direction as it flows past the object. Some factors that influence lift are the shape of the object, its orientation with respect to the fluid flow, spinning motion (for example, a spinning golf ball as in Figure 15.20), and the texture of the object's surface.

A number of devices operate in a manner similar to the *atomizer* in Figure 15.21. A stream of air passing over an open tube reduces the pressure above the tube. This reduction in pressure causes the liquid to rise into the air stream. The liquid is then dispersed into a fine spray of droplets. This type of system is used in perfume bottles and paint sprayers.

Bernoulli's principle explains one symptom of advanced arteriosclerosis called *vascular flutter.* The artery is constricted as a result of an accumulation of plaque on its inner walls (Fig. 15.22). Plaque is a combination of fat, cell debris, connective tissue, and sometimes calcium that forms a flat patch inside a blood vessel. The blood speed through the constriction is higher than elsewhere according to the continuity equation for fluids. According to Bernoulli's principle, the pressure in the constriction is lower than elsewhere. If the blood speed is sufficiently high in the constricted region, the artery may collapse under the larger external pressure, causing a momentary interruption in blood flow. At this point, the speed of the blood goes to zero, its pressure rises again, and the vessel reopens. As the blood rushes through the constricted artery, the internal pressure drops and again the artery closes. Such variations in blood flow can be heard with a stethoscope.

**APPLICATION**

Vascular flutter

**Figure 15.21**

A stream of air passing over a tube dipped into a liquid will cause the liquid to rise in the tube.

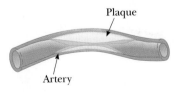

**Figure 15.22**

Blood must travel faster than normal through a constricted region of an artery.

Figure 15.23

(Thinking Physics 15.7) What is the purpose of the vent—the vertical pipe open to the atmosphere?

### THINKING PHYSICS 15.7

Consider the portion of a home plumbing system shown in Figure 15.23. The water trap in the pipe below the sink captures a plug of water that prevents sewer gas from finding its way from the sewer pipe, up the sink drain, and into the home. Suppose the dishwasher is draining, so that water is moving to the left in the sewer pipe. What is the purpose of the vent, which is open to the air above the roof of the house? In which direction is air moving at the opening of the vent, upward or downward?

**Reasoning**   Let us imagine that the vent is not present, so that the drain pipe for the sink is simply connected through the trap to the sewer pipe. As water from the dishwasher moves to the left in the sewer pipe, the pressure in the sewer pipe is reduced below atmospheric pressure, according to Bernoulli's principle. The pressure at the drain in the sink is still at atmospheric pressure. Thus, this pressure differential pushes the plug of water in the water trap of the sink down the drain pipe into the sewer pipe, removing it as a barrier to sewer gas. With the addition of the vent to the roof, the reduced pressure of the dishwasher water causes air to enter the vent pipe at the roof. This keeps the pressure in the vent pipe and the right-hand side of the sink drain pipe close to atmospheric pressure, so that the plug of water in the water trap remains in the trap and prevents sewer gas from entering the home.

### context connection
### 15.9 • A NEAR MISS EVEN BEFORE LEAVING SOUTHAMPTON

As the *Titanic* began its maiden voyage, it experienced a potentially disastrous incident before it left Southampton Harbor. (If this disaster had occurred, however, it is highly likely that it would have been much less disastrous in terms of lives lost than the incident that actually occurred later in the voyage.) The *Titanic* passed closely by the *New York*, which was tied securely next to the *Oceanic* at the dock, with the keels of the two ships parallel. As the *Titanic* passed by, the *New York* was forced toward it, the *New York*'s mooring ropes snapped, and its stern swung out toward the *Titanic*. It was only quick thinking by the harbor pilot on the *Titanic*, who reversed the engines, causing the *Titanic* to slow and allow the *New York* to pass by safely, that saved the two ships from a collision. As it was, a collision was averted by only a few feet, and the *Titanic* was delayed by over an hour in her departure. Figure 15.24 is a photograph taken from the *Titanic*, showing how close the ships came to colliding.

It is ironic that the captain of the *Titanic*, E. J. Smith, who watched the near miss from the bridge, was captain on one of the *Titanic*'s sister ships, *Olympic*, when a similar incident occurred seven months before the *New York* incident. In this case, the cruiser *Hawke* was pulled toward the *Olympic* and a collision was not averted. The *Hawke*'s bow was seriously damaged in the collision, and the hull of the *Olympic* was punctured above and below the waterline. Both ships were able to return to port but needed extensive repairs.

Why did these events occur? The answer lies in Bernoulli's principle. As ships move through the water, they push water out of the way, and the water moves around the sides of the ship. Imagine now that a ship such as the *Titanic* passes near another ship, such as the *New York*, with their keels parallel. The water moving around the side of the *Titanic* toward the *New York* is forced into a

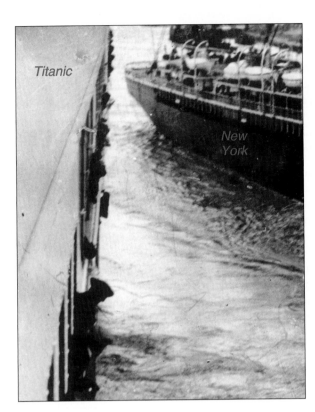
Titanic

New York

**Figure 15.24**

While leaving Southampton harbor, the *Titanic (left)* experienced a near miss with the *New York (right)*, due to Bernoulli's principle. If this accident had actually occurred, it may have changed the timing enough that the *Titanic,* once underway, might not have been sunk by an iceberg.

narrow channel between the ships. Because water is incompressible, its volume remains constant when it is squeezed into a narrow channel. The initial tendency is for the compressed water to rise into the air between the ships because the air above the water offers little resistance to being compressed. As soon as the water level between the ships rises, however, the water will begin to flow in a direction parallel to the keels toward the lower level water near the bow and stern of the ships. Thus, the *water between the ships is moving at a higher speed than the water on the opposite sides of the ships.* According to Bernoulli's principle, this rapidly moving water exerts less pressure on the sides of the ships than the slower moving water on the outer sides. The result is a *net force pushing the two ships toward each other.*

Thus, captains of boats and ships are advised not to pass too close by other boats in a parallel direction. If this occurs, then the boats could be pushed into each other. This effect occurs for air, explaining the effect of the passing truck at the beginning of Section 15.7. As the truck passes, air is squeezed into the channel between the truck and your car. The high-speed air exerts less pressure on your car than the slower moving air on the opposite side, resulting in a net force toward the truck.

In this Context Connection section, we investigated an application of Bernoulli's principle. In the Context Conclusion, we shall explore the difficulties in visiting the *Titanic* due to its great depth under the ocean surface.

## SUMMARY

The **pressure** $P$ in a fluid is the force per unit area that the fluid exerts on a surface:

$$P \equiv \frac{F}{A} \qquad [15.1]$$

In the SI system, pressure has units of $N/m^2$, and $1\ N/m^2 = 1$ pascal (Pa).

The pressure in a liquid varies with depth $h$ according to the expression

$$P = P_0 + \rho g h \qquad [15.4]$$

where $P_0$ is atmospheric pressure $(= 1.013 \times 10^5\ N/m^2)$ and $\rho$ is the density of the liquid, assumed uniform.

**Pascal's law** states that when a change in pressure is applied to a fluid, the change in pressure is transmitted undiminished to every point in the fluid and to every point on the walls of the container.

When an object is partially or fully submerged in a fluid, the fluid exerts an upward force on the object called the **buoyant force.** According to **Archimedes's principle,** the buoyant force is equal to the weight of the fluid displaced by the object.

Various aspects of fluid dynamics can be understood by adopting a simplification model in which the fluid is nonviscous and incompressible and the fluid motion is a steady flow with no turbulence.

Using this model, two important results regarding fluid flow through a pipe of nonuniform size can be obtained:

- The flow rate through the pipe is a constant, which is equivalent to stating that the product of the cross-sectional area $A$ and the speed $v$ at any point is a constant. This gives the **continuity equation for fluids:**

$$A_1 v_1 = A_2 v_2 = \text{constant} \qquad [15.7]$$

- The sum of the pressure, kinetic energy per unit volume, and gravitational potential energy per unit volume has the same value at all points along a streamline. This corresponds to **Bernoulli's equation:**

$$P_1 + \tfrac{1}{2}\rho v_1^2 + \rho g y_1 = P_2 + \tfrac{1}{2}\rho v_2^2 + \rho g y_2 \qquad [15.12]$$

## QUESTIONS

1. Suppose a damaged ship can just barely float in the ocean. It is pulled toward shore and into a river, heading toward a dry dock for repair. As it is pulled up the river, it sinks. Why?

2. Figure Q15.2 shows aerial views from directly above two dams. Both dams are equally long (the vertical dimension in the diagram) and equally deep (into the page in the diagram). The dam on the left holds back a very large lake, while the dam on the right holds back a narrow river. Which dam has to be built more strongly?

Dam          Dam

**Figure Q15.2**

3. Some physics students attach a long tube to the opening of a balloon. Leaving the balloon on the ground, the other end of the tube is hoisted to the roof of a multistory campus building. Students at the top of the building start pouring water into the tube. The students on the ground watch the balloon fill with water. On the roof, the students

are surprised to see that the tube never seems to fill up—more and more water can continue to be poured into the tube. On the ground, the balloon bursts and drenches the students there. Why did the balloon–tube system never fill up, and why did the balloon burst?

4. A barge is carrying a load of gravel along a river. It approaches a low bridge, and the captain realizes that the top of the pile of gravel is not going to make it under the bridge. The captain orders the crew to quickly shovel gravel from the pile into the water. Is this a good decision?

5. A person in a boat floating in a small pond throws an anchor overboard. Does the level of the pond rise, fall, or remain the same?

6. Two drinking glasses having equal weights but different shapes and different cross-sectional areas are filled to the same level with water. According to the expression $P = P_0 + \rho g h$, the pressure is the same at the bottom of both glasses. In view of this, why does one weigh more than the other?

7. A helium-filled balloon rises until its density becomes the same as that of the air. If a sealed submarine begins to sink, will it go all the way to the bottom of the ocean or will it stop when its density becomes the same as that of the surrounding water?

8. A fish rests on the bottom of a bucket of water while the bucket is being weighed. When the fish begins to swim around, does the weight change?

9. Will a ship ride higher in the water of an inland lake or in the ocean? Why?

10. Lead has a greater density than iron, and both are denser than water. Is the buoyant force on a lead object greater than, less than, or equal to the buoyant force on an iron object of the same volume?

11. An ice cube is placed in a glass of water. What happens to the level of the water as the ice melts?

12. The water supply for a city is often provided from reservoirs built on high ground. Water flows from the reservoir, through pipes, and into your home when you turn the tap on your faucet. Why is the water flow faster out of a faucet on the first floor of a building than in an apartment on a higher floor?

13. If air from a hair dryer is blown upward under a Ping-Pong ball, the ball can be suspended in air. Explain.

14. When ski-jumpers are airborne, why do they bend their bodies forward and keep their hands at their sides?

15. Explain why a sealed bottle partially filled with a liquid can float.

16. When is the buoyant force on a swimmer greater—after exhaling or after inhaling?

17. A piece of unpainted wood is partially submerged in a container filled with water. If the container is sealed and pressurized above atmospheric pressure, does the wood rise, fall, or remain at the same level? (*Hint:* Wood is porous.)

18. Because atmospheric pressure is about $10^5$ N/m$^2$ and the area of a person's chest is about 0.13 m$^2$, the force of the atmosphere on one's chest is approximately 13 000 N. In view of this enormous force, why don't our bodies collapse?

19. If you release a ball while inside a freely falling elevator, the ball remains in front of you rather than falling to the floor, because the ball, the elevator, and you all experience the same downward acceleration $g$. What happens if you repeat this experiment with a helium-filled balloon? (This one is tricky.)

20. Two identical ships set out to sea. One is loaded with a cargo of Styrofoam, and the other is empty. Which ship is more submerged?

21. A small piece of steel is tied to a block of wood. When the wood is placed in a tub of water with the steel on top, half of the block is submerged. If the block is inverted so that the steel is under water, does the amount of the block submerged increase, decrease, or remain the same? What happens to the water level in the tub when the block is inverted?

22. Prairie dogs ventilate their burrows by building a mound over one entrance, which is open to a stream of air. A second entrance at ground level is open to almost stagnant air. How does this construction create an air flow through the burrow?

23. An unopened can of diet cola floats when placed in a tank of water, whereas a can of regular cola of the same brand sinks in the tank. What do you suppose could explain this behavior?

# PROBLEMS

1, 2, 3 = straightforward, intermediate, challenging    □ = full solution available in the *Student Solutions Manual and Study Guide*

**web** = solution posted at **http://www.harcourtcollege.com/physics/**    🖥 = computer useful in solving problem

🖥 = Interactive Physics    ■ = paired numerical/symbolic problems    ▨ = life science application

## Section 15.1  Pressure

1. Calculate the mass of a solid iron sphere that has a diameter of 3.00 cm.

2. A king orders a gold crown having a mass of 0.500 kg. When it arrives from the metalsmith, the volume of the crown is found to be 185 cm$^3$. Is the crown made of solid gold?

3. A 50.0-kg woman balances on one heel of a pair of high-heeled shoes. If the heel is circular and has a radius of 0.500 cm, what pressure does she exert on the floor?

4. The four tires of an automobile are inflated to a gauge pressure of 200 kPa. Each tire has an area of 0.024 0 m$^2$ in contact with the ground. Determine the weight of the automobile.

5. What is the total mass of the Earth's atmosphere? (The radius of the Earth is $6.37 \times 10^6$ m, and atmospheric pressure at the surface is $1.013 \times 10^5$ N/m$^2$.)

## Section 15.2  Variation of Pressure with Depth

6. (a) Calculate the absolute pressure at an ocean depth of 1 000 m. Assume the density of sea water is 1 024 kg/m$^3$ and that the air above exerts a pressure of 101.3 kPa. (b) At this depth, what force must the frame around a circular submarine porthole having a diameter of 30.0 cm exert to counterbalance the force exerted by the water?

7. The spring of the pressure gauge shown in Figure 15.2 has a force constant of 1 000 N/m, and the piston has a

(a)

(b)

**Figure P15.10**

diameter of 2.00 cm. As the gauge is lowered into water, what change in depth causes the piston to move in by 0.500 cm?

8. The small piston of a hydraulic lift has a cross-sectional area of 3.00 cm², and its large piston has a cross-sectional area of 200 cm² (see Figure 15.5). What force must be applied to the small piston for the lift to raise a load of 15.0 kN? (In service stations, this force is usually exerted by compressed air.)

9. What must be the contact area between a suction cup (completely exhausted) and a ceiling if the cup is to support the weight of an 80.0-kg student?

web

10. (a) A very powerful vacuum cleaner has a hose 2.86 cm in diameter. With no nozzle on the hose, what is the weight of the heaviest brick that the cleaner can lift? (Fig. P15.10) (b) A very powerful octopus uses one sucker of diameter 2.86 cm on each of the two shells of a clam in an attempt to pull the shells apart. Find the greatest force the octopus can exert in salt water 32.3 m deep. (*Caution:* Experimental verification can be interesting, but do not drop a brick on your foot. Do not overheat the motor of a vacuum cleaner. Do not get an octopus mad at you.)

11. A swimming pool has dimensions 30.0 m × 10.0 m and a flat bottom. When the pool is filled to a depth of 2.00 m with fresh water, what is the force caused by the water on the bottom? On each end? On each side?

## Section 15.3 Pressure Measurements

12. Figure P15.12 shows Superman attempting to drink water through a very long straw. With his great strength he achieves maximum possible suction. The walls of the tubular straw do not collapse. (a) Find the maximum height through which he can lift the water. (b) Still thirsty, the Man of Steel repeats his attempt on the Moon, which has

**Figure P15.12**

no atmosphere. Find the difference between the water levels inside and outside the straw.

13. Blaise Pascal duplicated Torricelli's barometer using a red Bordeaux wine, of density 984 kg/m³, as the working liquid (Fig. P15.13). What was the height *h* of the wine column for normal atmospheric pressure? Would you expect the vacuum above the column to be as good as for mercury?

web

**Figure P15.13**

**Figure P15.16**

14. Mercury is poured into a U-tube as in Figure P15.14a. The left arm of the tube has cross-sectional area $A_1$ of 10.0 cm², and the right arm has a cross-sectional area $A_2$ of 5.00 cm². One hundred grams of water is then poured into the right arm as in Figure P15.14b. (a) Determine the length of the water column in the right arm of the U-tube. (b) The density of mercury is 13.6 g/cm³. Through what distance $h$ does the mercury rise in the left arm?

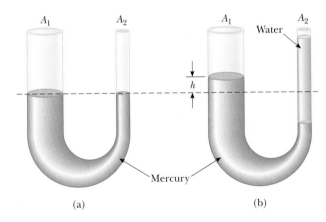

(a)    (b)

**Figure P15.14**

15. Normal atmospheric pressure is $1.013 \times 10^5$ Pa. The approach of a storm causes the height of a mercury barometer to drop by 20.0 mm from the normal height. What is the atmospheric pressure? (The density of mercury is 13.59 g/cm³.)

16. The human brain and spinal cord are immersed in the cerebrospinal fluid. The fluid is normally continuous between the cranial and spinal cavities. It normally exerts a pressure of 100 to 200 mm of $H_2O$ above the prevailing at-

mospheric pressure. In medical work pressures are often measured in units of millimeters of $H_2O$ because body fluids, including the cerebrospinal fluid, typically have the same density as water. The pressure of the cerebrospinal fluid can be measured be means of a *spinal tap*, as illustrated in Figure P15.16. A hollow tube is inserted into the spinal column, and the height to which the fluid rises is observed. If the fluid rises to a height of 160 mm, its gauge pressure is recorded as 160 mm $H_2O$. (a) Express this pressure in pascals, in atmospheres, and in millimeters of mercury. (b) Sometimes it is necessary to determine if an accident victim has suffered a crushed vertebra that is blocking flow of the cerebrospinal fluid in the spinal column. In other cases a physician may suspect a tumor or other growth is blocking the spinal column and inhibiting flow of cerebrospinal fluid. Such conditions can be investigated by means of the *Queckenstedt test*. In this procedure, the veins in the patient's neck are compressed to make the blood pressure rise in the brain. The increase in pressure in the blood vessels is transmitted to the cerebrospinal fluid. What should be the normal effect on the height of the fluid in the spinal tap? (c) Suppose that compressing the veins had no effect on the fluid level. What might account for this?

### Section 15.4 Buoyant Forces and Archimedes's Principle

17. A Ping-Pong ball has a diameter of 3.80 cm and average density of 0.084 0 g/cm³. What force is required to hold it completely submerged under water?

18. A swimmer of mass $m$ rests on top of a Styrofoam slab, which has thickness $h$ and density $\rho_s$. What is the area of the slab if it floats in fresh water with its upper surface just awash?

19. A 10.0-kg block of metal measuring 12.0 cm × 10.0 cm × 10.0 cm is suspended from a scale and immersed in water as in Figure P15.19. The 12.0-cm dimension is vertical, and the top of the block is 5.00 cm below the

**Figure P15.19**

**Figure P15.22**

surface of the water. (a) What are the forces acting on the top and on the bottom of the block? (Take atmospheric pressure to be $1.013\ 0 \times 10^5$ N/m$^2$.) (b) What is the reading of the spring scale? (c) Show that the buoyant force equals the difference between the forces at the top and bottom of the block.

20. To an order of magnitude, how many helium-filled toy balloons would be required to lift you? Because helium is an irreplaceable resource, develop a theoretical answer rather than an experimental one. In your solution state what physical quantities you take as data and the values you measure or estimate for them.

21. A cube of wood having an edge dimension of 20.0 cm and a density of 650 kg/m$^3$ floats on water. (a) What is the distance from the horizontal top surface of the cube to the water level? (b) How much lead weight has to be placed on top of the cube so that its top is just level with the water?

**web**

22. Determination of the density of a fluid has many important applications. A car battery contains sulfuric acid, and the battery will not function properly if the acid density is too low. Similarly, the effectiveness of antifreeze in your car's engine coolant depends on the density of the mixture (usually ethylene glycol and water). When you donate blood to a blood bank, its screening includes determination of the density of the blood, because higher density indicates higher hemoglobin content. A *hydrometer* is an instrument used to determine liquid density. A simple one is sketched in Figure P15.22. The bulb of a syringe is squeezed and released to lift a sample of the liquid of interest into a tube containing a calibrated rod of known density. The rod, of length $L$ and average density $\rho_0$, floats partially immersed in the liquid of density $\rho$. A length $h$ of the rod protrudes above the surface of the liquid. Show

that the density of the liquid is given by

$$\rho = \frac{\rho_0 L}{L - h}$$

23. How many cubic meters of helium are required to lift a balloon with a 400-kg payload to a height of 8 000 m? (Take $\rho_{\text{He}} = 0.180$ kg/m$^3$.) Assume that the balloon maintains a constant volume and that the density of air decreases with the altitude $z$ according to the expression $\rho_{\text{air}} = \rho_0 e^{-z/8000}$, where $z$ is in meters and $\rho_0 = 1.25$ kg/m$^3$ is the density of air at sea level.

24. A bathysphere used for deep-sea exploration has a radius of 1.50 m and a mass of $1.20 \times 10^4$ kg. To dive, this submarine takes on mass in the form of sea water. Determine the amount of mass the submarine must take on if it is to descend at a constant speed of 1.20 m/s, when the resistive force on it is 1 100 N in the upward direction. The density of sea water is $1.03 \times 10^3$ kg/m$^3$.

25. A plastic sphere floats in water with 50.0% of its volume submerged. This same sphere floats in glycerin with 40.0% of its volume submerged. Determine the densities of the glycerin and the sphere.

26. **Review Problem.** A long cylindrical rod of radius $r$ is weighted on one end so that it floats upright in a fluid having a density $\rho$. It is pushed down a distance $x$ from its equilibrium position and released. Show that the rod will execute simple harmonic motion if the resistive effects of the fluid are ignored, and determine the period of the oscillations.

## Section 15.5 Fluid Dynamics

## Section 15.6 Streamlines and the Continuity Equation for Fluids

## Section 15.7 Bernoulli's Principle

27. (a) A water hose 2.00 cm in diameter is used to fill a 20.0-L bucket. If it takes 1.00 min to fill the bucket, what is the speed $v$ at which water moves through the hose? (*Note:* 1 L = 1 000 cm³.) (b) If the hose has a nozzle 1.00 cm in diameter, find the speed of the water at the nozzle.

28. A horizontal pipe 10.0 cm in diameter has a smooth reduction to a pipe 5.00 cm in diameter. If the pressure of the water in the larger pipe is $8.00 \times 10^4$ Pa and the pressure in the smaller pipe is $6.00 \times 10^4$ Pa, at what rate does water flow through the pipes?

29. Water falls over a dam of height $h$ with a mass flow rate of $R$, in units of kg/s. (a) Show that the power available from the water is

$$\mathcal{P} = Rgh$$

where $g$ is the acceleration due to gravity. (b) Each hydroelectric unit at the Grand Coulee Dam takes in water at a rate of $8.50 \times 10^5$ kg/s from a height of 87.0 m. The power developed by the falling water is converted to electric power with an efficiency of 85.0%. How much electric power is produced by each hydroelectric unit?

30. *Torricelli's Law:* A tank with a cover, containing a liquid of density $\rho$, has a hole in its side at a distance $y_1$ from the bottom (Fig. P15.30). The diameter of the hole is small relative to the diameter of the tank. The air above the liquid is maintained at a pressure $P$. Assume steady frictionless flow. Show that the speed at which the fluid leaves the hole when the liquid level is a distance $h$ above the hole is

$$v_1 = \sqrt{\frac{2(P - P_0)}{\rho} + 2gh}$$

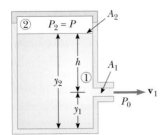

**Figure P15.30**

31. A large storage tank with an open top is filled to a height $h_0$. The tank is punctured at a height $h$ above the bottom of the tank (Fig. P15.31). Find an expression for how far from the tank the exiting stream lands.

**Figure P15.31** Problems 31 and 32.

32. A hole is punched at height $h$ in the side of an open tank of height $h_0$. The container stands on level ground and is full of water, as shown in Figure P15.31. If the water is to shoot as far as possible horizontally, (a) how far from the bottom of the container should the hole be punched? (b) Ignoring friction losses, how far (initially) from the side of the container will the water land?

33. A legendary Dutch boy saved Holland by plugging a hole in a dike with his finger, 1.20 cm in diameter. If the hole was located 2.00 m below the surface of the North Sea (density 1 030 kg/m³), (a) what was the force on his finger? (b) If he pulled his finger out of the hole, how long would it take the released water to fill 1 acre of land to a depth of 1 foot, assuming the hole remained constant in size? (A typical U.S. family of four uses 1 acre-foot of water, 1 234 m³, in 1 year.)

34. Water is pumped through a pipe 15.0 cm in diameter, from the Colorado River up to Grand Canyon Village, located on the rim of the canyon. The river is at an elevation of 564 m and the village is at an elevation of 2 096 m. (a) What is the minimum pressure at which the water must be pumped if it is to arrive at the village? (b) If 4 500 m³ is pumped per day, what is the speed of the water in the pipe? (c) What additional pressure is necessary to deliver this flow? (*Note:* You may assume that the acceleration due to gravity and the density of air are constant over this range of elevations.)

35. Old Faithful Geyser in Yellowstone Park erupts at approximately 1-h intervals, and the height of the water column reaches 40.0 m. (Fig. P15.35). (a) Consider the rising stream as a series of separate drops. Analyze the free-fall motion of one of the drops to determine the speed at which the water leaves the ground. (b) Treat the rising stream as an ideal fluid in streamline flow. Use Bernoulli's equation to determine the speed of the water as it leaves ground level. (c) What is the pressure (above

atmospheric) in the heated underground chamber if its depth is 175 m? You may assume that the chamber is large compared with the geyser's vent.

**Figure P15.35** *(Stan Osolinski/Dembinsky Photo Associates)*

**36.** A Venturi tube may be used as a fluid flow meter (see Fig. 15.18a). The radius of the outlet tube is 1.00 cm, the radius of the inlet tube is 2.00 cm, and the fluid is gasoline ($\rho = 700$ kg/m$^3$). If the difference in pressure is $P_1 - P_2 = 21.0$ kPa, find the fluid flow rate in cubic meters per second.

## Section 15.8 Other Applications of Fluid Dynamics

**37.** An airplane has a mass of $1.60 \times 10^4$ kg, and each wing has an area of 40.0 m$^2$. During level flight, the pressure on the lower wing surface is $7.00 \times 10^4$ Pa. Determine the pressure on the upper wing surface.

**38.** A siphon is used to drain water from a tank, as illustrated in Figure P15.38. The siphon has a uniform diameter. Assume steady flow without friction. (a) If the distance $h = 1.00$ m, find the speed of outflow at the end of the siphon. (b) What is the limitation on the height of the top of the siphon above the water surface? (For the flow of the liquid to be continuous, the pressure must not drop below the vapor pressure of the liquid.)

**Figure P15.38**

**39.** Bernoulli's principle can have important consequences for the design of buildings. For example, wind can move around a skyscraper at remarkably high speed, creating low pressure. The higher atmospheric pressure in the still air inside the buildings can cause windows to pop out. As originally constructed, the John Hancock building in Boston popped window panes, which fell many stories to the sidewalk below. (a) Suppose that a horizontal wind moves with a speed of 11.2 m/s outside a large pane of plate glass with dimensions 4.00 m $\times$ 1.50 m. Assume the density of the air to be 1.30 kg/m$^3$. The air inside the building is at atmospheric pressure. What is the total force exerted by air on the window pane? (b) If a second skyscraper is built nearby, the air speed can be especially high where wind passes through the narrow separation between the buildings. Solve part (a) again if the wind speed is 22.4 m/s, twice as high.

**40.** A hypodermic syringe contains a medicine with the density of water (Figure P15.40). The barrel of the syringe has a cross-sectional area $A = 2.50 \times 10^{-5}$ m$^2$, and the needle has a cross-sectional area $1.00 \times 10^{-8}$ m$^2$. In the absence of a force on the plunger, the pressure everywhere is 1 atm. A force **F** of magnitude 2.00 N acts on the plunger, making medicine squirt horizontally from the needle. Determine the speed of the medicine as it leaves the tip of the needle.

**Figure P15.40**

## Section 15.9 Context Connection—A Near Miss Even Before Leaving Southampton

**41.** According to the caption of the chapter-opening photograph, about 11% of an iceberg is above water. (a) Confirm this value mathematically. (b) Suppose an iceberg were floating in fresh water rather than sea water. Would a larger or smaller percentage be above the water? Calculate this percentage.

**42.** The *Titanic* is docked in Southampton harbor just before boarding. You, as a ticket agent for White Star Lines, notice where the water level is on a scale of numbers marked on the side of the vessel. Because your ticket-collecting job is so boring that you need something to occupy your mind, you make a note of the water level with the intent to check it after everyone boards. During the next 2 h, 2 205 passengers, of average mass 75.0 kg, board the *Titanic*. You notice that the ship has sunk 1.00 cm deeper in the water with the passengers on board. What is the cross-sectional area of the hull of the *Titanic* right at the waterline?

43. **Review Problem.** The *Titanic* is drifting in Southampton harbor before its fateful journey, and the captain wishes to stop the drift by dropping an anchor. The anchor is made of iron and has mass 2 000 kg. It is attached to a massless rope. The rope is wrapped around a reel in the form of a solid disk of radius 0.250 m and mass 300 kg, which rotates on a frictionless axle. (a) Find the angular displacement of the reel when the anchor moves down 15.0 m. (b) Find the acceleration of the anchor as it falls through the air, which offers negligible resistance. (c) While the anchor continues to drop through the water, the water exerts a friction force of 2 500 N on it. With what acceleration does the anchor move through the water? (d) While the anchor drops through the water, what torque is exerted on the reel?

## Additional Problems

44. Water is forced out of a fire extinguisher by air pressure, as shown in Figure P15.44. How much gauge air pressure in the tank (above atmospheric) is required for the water jet to have a speed of 30.0 m/s when the water level is 0.500 m below the nozzle?

**Figure P15.44**

45. The true weight of an object can be measured in a vacuum, where buoyant forces are absent. A body of volume $V$ is weighed in air on a balance with the use of weights of density $\rho$. If the density of air is $\rho_{air}$ and the balance reads $F'_g$, show that the true weight $F_g$ of the object is

$$F_g = F'_g + \left(V - \frac{F'_g}{\rho g}\right)\rho_{air}g$$

46. A light spring with force constant $k = 90.0$ N/m is attached vertically to a table (Fig. P15.46a). A 2.00-g balloon is filled with helium (density $= 0.180$ kg/m$^3$) to a volume of 5.00 m$^3$ and is then connected to the spring, causing it to stretch as in Figure P15.46b. Determine the extension distance $L$ when the balloon is in equilibrium.

47. **Review Problem.** With reference to Figure 15.6, show that the total torque exerted by the water behind the dam about a horizontal axis through $O$ is $\frac{1}{6}\rho gwH^3$. Show that

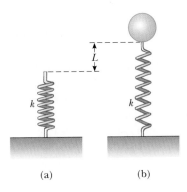

(a)  (b)

**Figure P15.46**

the effective line of action of the total force exerted by the water is at a distance $\frac{1}{3}H$ above $O$.

48. Evangelista Torricelli was the first person to realize that we live at the bottom of an ocean of air. He correctly surmised that the pressure of our atmosphere is attributable to the weight of the air. The density of air at 0°C at the Earth's surface is 1.29 kg/m$^3$. The density decreases with increasing altitude (as the atmosphere thins). On the other hand, if we assume that the density is constant at 1.29 kg/m$^3$ up to some altitude $h$, and zero above that altitude, then $h$ would represent the depth of the ocean of air. Use this model to determine the value of $h$ that gives a pressure of 1.00 atm at the surface of the Earth. Would the peak of Mount Everest rise above the surface of such an atmosphere?

49. A beaker of mass $m_b$ containing oil of mass $m_o$ (density = $\rho_o$) rests on a scale. A block of iron of mass $m_i$ is suspended from a spring scale and completely submerged in the oil as in Figure P15.49. Determine the equilibrium readings of both scales.

**Figure P15.49**

50. **Review Problem.** A copper cylinder hangs at the bottom of a steel wire of negligible mass. The top end of the wire is fixed. When the wire is struck, it emits sound with a fundamental frequency of 300 Hz. If the copper cylinder is then submerged in water so that half its volume is below the water line, determine the new fundamental frequency.

51. In about 1657, Otto von Guericke, inventor of the air pump, evacuated a sphere made of two brass hemispheres. Two teams of eight horses each could pull the hemispheres apart only on some trials, and then "with greatest difficulty," with the resulting sound likened to a cannon firing (Fig. P15.51). (a) Show that the force $F$ required to pull the evacuated hemispheres apart is $\pi R^2(P_0 - P)$, where $R$ is the radius of the hemispheres and $P$ is the pressure inside the hemispheres, which is much less than $P_0$. (b) Determine the force if $P = 0.100P_0$ and $R = 0.300$ m.

**Figure P15.51**   The colored engraving, dated 1672, illustrates Otto von Guericke's demonstration of the force due to air pressure as performed before Emperor Ferdinand III in 1657. (*The Granger Collection*)

52. Show that the variation of atmospheric pressure with altitude is given by $P = P_0 e^{-\alpha h}$, where $\alpha = \rho_0 g/P_0$, $P_0$ is atmospheric pressure at some reference level $y = 0$, and $\rho_0$ is the atmospheric density at this level. Assume that the decrease in atmospheric pressure over an infinitesimal change in altitude (so that the density is approximately uniform) is given by $dP = -\rho g \, dy$, and that the density of air is proportional to the pressure.

53. An incompressible, nonviscous fluid is initially at rest in the vertical portion of the pipe shown in Figure P15.53a, where $L = 2.00$ m. When the valve is opened, the fluid flows into the horizontal section of the pipe. What is the speed of the fluid when all of it is in the horizontal section, as in Figure P15.53b? Assume the cross-sectional area of the entire pipe is constant.

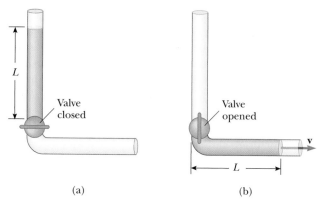

**Figure P15.53**

54. A cube of ice whose edges measure 20.0 mm is floating in a glass of ice-cold water with one of its faces parallel to the water's surface. (a) How far below the water surface is the bottom face of the block? (b) Ice-cold ethyl alcohol is gently poured onto the water surface to form a layer 5.00 mm thick above the water. The alcohol does not mix with the water. When the ice cube again attains hydrostatic equilibrium, what is the distance from the top of the water to the bottom face of the block? (c) Additional cold ethyl alcohol is poured onto the water's surface until the top surface of the alcohol coincides with the top surface of the ice cube (in hydrostatic equilibrium). How thick is the required layer of ethyl alcohol?

55. A U-tube open at both ends is partially filled with water (Fig. P15.55a). Oil having a density of 750 kg/m$^3$ is then poured into the right arm and forms a column of height $L = 5.00$ cm (Fig. P 15.55b). (a) Determine the difference $h$ in the heights of the two liquid surfaces. (b) The right

**Figure P15.55**

arm is then shielded from any air motion while air is blown across the top of the left arm until the surfaces of the two liquids are at the same height (Fig. P15.55c). Determine the speed of the air being blown across the left arm. Take the density of air as 1.29 kg/m³.

56. The water supply of a building is fed through a main pipe 6.00 cm in diameter. A 2.00-cm-diameter faucet tap, located 2.00 m above the main pipe, is observed to fill a 25.0-L container in 30.0 s. (a) What is the speed at which the water leaves the faucet? (b) What is the gauge pressure in the 6.00-cm main pipe? (Assume the faucet is the only "leak" in the building.)

57. The *spirit-in-glass thermometer,* invented in Florence, Italy, around 1654, consists of a tube of liquid (the spirit) containing a number of submerged glass spheres with slightly different masses (Fig. P15.57). At sufficiently low temperatures all the spheres float, but as the temperature rises, the spheres sink one after another. The device is a crude but interesting tool for measuring temperature. Suppose that the tube is filled with ethyl alcohol, whose density is 0.789 45 g/cm³ at 20.0°C and decreases to 0.780 97 g/cm³ at 30.0°C. (a) If one of the spheres has a radius of 1.000 cm and is in equilibrium halfway up the tube at 20.0°C, determine its mass. (b) When the temperature increases to 30.0°C, what mass must a second sphere of the same radius have to be in equilibrium at the halfway point? (c) At 30.0°C the first sphere has fallen to the bottom of the tube. What upward force does the bottom of the tube exert on this sphere?

**Figure P15.57** *(Courtesy of Jeanne Maier)*

## ANSWERS TO QUICK QUIZZES

**15.1** You would be better off with the basketball player. The basketball player has a larger weight, but it is distributed over a large area of the bottom of the sneaker, so that the pressure, which is force per unit area, is relatively small. The woman's smaller weight is distributed over the very small area of the spiked heel, resulting in a large pressure. Some museums require women in high-heeled shoes to wear slippers or special heel attachments so that they do not damage the wood floors.

**15.2** Because water is so much less dense than mercury, the column for a water barometer would have to be $h = P_0/\rho g = 10.3$ m high, making it inconveniently tall.

**15.3** The level of floating of a ship in the water is unaffected by the atmospheric pressure. The buoyant force results from the pressure *differential* in the fluid. On a high-pressure day, the pressure at all points in the water is higher than on a low-pressure day. Because water is almost in-compressible, however, the rate of change of pressure with depth is the same, resulting in no change in the buoyant force.

**15.4** The balance will not be in equilibrium—the lead side will be lower. Despite the fact that the weights on both sides of the balance are the same, the Styrofoam, due to its larger volume, experiences a larger buoyant force from the surrounding air. Thus, the net force of gravity plus the buoyant force is larger, in the downward direction, for the lead than for the Styrofoam.

**15.5** Because the density of ice is about 89% that of sea water, 89% of a floating iceberg is underwater. Thus, only 11% is visible above water—this is the "tip of the iceberg."

**15.6** As the water falls, its speed increases. Because the flow rate $Av$ must remain constant at all points (Eq. 15.7), the stream must become narrower as the speed increases.

**15.7** (a). The air between the balloons is moving at high speed. Therefore, it is at lower pressure than the stagnant area on the outer edges of the balloons, according to Bernoulli's principle. Thus, the balloons move toward each other.

**15.8** The air passing over the roof of the car has a high speed relative to the car. The air inside the car is moving with the car and thus has a speed of zero relative to the car. As a result, the air inside the car has a higher pressure than the air above the roof and the flexible top is pushed upward.

(*CALVIN AND HOBBES* ©1992 Watterson. Reprinted with permission of UNIVERSAL PRESS SYNDICATE. All rights reserved.)

# Search for the *Titanic*

## FINDING AND VISITING THE *TITANIC*

We have now investigated the physics of fluids and can respond to our central question for the *Search for the Titanic* Context:

### *How can we safely visit the wreck of the* Titanic?

Many individuals felt that the *Titanic* was unsinkable. One of the factors in this belief was the series of watertight bulkheads that divided the hull of the ship into several watertight compartments. Even if the hull were breached so that a compartment became flooded, the incoming water could be isolated to that compartment by closing watertight doors in the bulkhead.

According to the design of the ship, the *Titanic* could be kept afloat if her four forwardmost compartments were flooded. Unfortunately, the collision with the iceberg caused a breach in the first five compartments. As these forward compartments filled, the extra weight of the water in the bow of the ship resulted in the bow sinking into the water and the stern lifting out of the water (Figure 1).

**Figure 1**

The *Titanic* struck the iceberg near the bow, so that the forward compartments filled with water and sank, lifting the stern of the ship above the water. *(Painting by Ken Marschall)*

Despite the shipbuilders' pride in their watertight compartments, they were not watertight at the top. The bulkheads only went up to a certain height in the ship and then ended. Thus, as the *Titanic* tilted forward, water from one compartment simply spilled over the top of the bulkhead into the next compartment, and the compartments filled one by one.

Some experts after the disaster claimed that opening the watertight doors in the bulkheads would have kept the *Titanic* afloat longer, with an increased possibility of another ship arriving in time to save those who were not able to leave in the lifeboats. According to this hypothesis, if the water entering the forward compartments had been allowed to distribute evenly along the ship by passing through the doors in the bulkheads, the ship would not have tilted so that water could spill over the tops of the bulkheads. The sinking of a ship is a complicated event, however, and this hypothesis is not universally accepted.

In the region of the sinking, the depth of the ocean is about 4 km. When the *Titanic* was located and visited, the depth was measured to be 3 784 m. Let us use Equation 15.3 to calculate the pressure at this depth of sea water:

$$P = P_0 + \rho g h$$
$$= 1.013 \times 10^5 \text{ Pa} + (1.03 \times 10^3 \text{ kg/m}^3)(9.80 \text{ m/s}^2)(3\ 784 \text{ m})$$
$$= 3.83 \times 10^7 \text{ Pa} = 378 \text{ atm}$$

Thus, the pressure is 378 times that at the surface! A human being could not survive at this pressure.

Plans for finding and possibly salvaging the *Titanic* began immediately after she sank. Families of some of the wealthy victims contacted salvage companies with requests for a salvage operation. One plan suggested filling the *Titanic* with Ping-Pong balls, so that its overall density would be less than that of water, and she would float to the surface! This plan, of course, ignores the obvious problems of the Ping-Pong balls' failure to withstand the tremendous pressure at that depth.

An early expedition to find the *Titanic* occurred in 1980 and met with failure. After unsuccessful searches were carried out by a number of teams, Dr. Robert Ballard, of Woods Hole Oceanographic Institute, discovered the wreck in 1985, in cooperation with a team from IFREMER, the French National Institute of Oceanography. The search began by towing a sonar device, which emitted sound waves through the water and analyzed the reflection of the waves from solid objects such as the hull of the *Titanic*. The search pattern was a tedious back-and-forth sweeping of the area near the reported sinking of the ship, looking for sonar reflections, and checking possible sites with a magnetometer for the presence of an iron hull.

After failing to find the *Titanic* with the sonar system, Ballard switched to a visual search using an underwater video system called *Argo*. After three more grueling weeks with no reward, the searchers saw one of the *Titanic's* boilers in the early morning of September 1, 1985. This first evidence of the wreck having been found, the remainder was quickly located.

The visual evidence indicated clearly that the *Titanic* had split in two, as had been reported by some of the survivors in 1912. As it tilted steeply in the water due to the sinking of the bow, the midsection was subjected to forces that it was not designed to sustain. After the break occurred, but while the two sections were still connected, the stern section settled back into the water, with the bow section hanging from it underwater. As more water entered the bow section, it pulled the stern section into a vertical orientation and then broke free, beginning its trip to the bottom. The stern bobbed for a while as it filled with water and then sank into the ocean.

**WEB**

For information on Woods Hole Oceanographic Institute, visit **www.whoi.edu/**
For information on IFREMER, visit **www.ifremer.fr/anglais/**

**Figure 2**

The bow section of the *Titanic* rests on the ocean floor relatively intact. *(Painting by Ken Marschall)*

The two sections of the *Titanic* lie about 600 m apart on the ocean floor. The bow section (Figure 2) is fairly intact, but the stern section (Figure 3) is tremendously damaged. As the bow section sank, it was already filled with water. As the pressure of the water outside the bow section increased during the plummet to the bottom, the pressure inside the section increased. On the other hand, the stern section spent most of the time in the air before sinking. Thus, as it sank, a significant volume of air was still trapped inside the stern section. As the water pressure increased while the stern section sank, the air pressure inside could not increase along with the external water pressure because many air pockets existed in the relatively sealed sections of the structure. Thus, some areas of the hull of the stern section experienced very large pressure on the outside surface, with relatively low pressure on the inside surface. This extreme imbalance in pressures possibly caused an implosion of the stern section at some depth, causing severe destruction of the structure. Further damage was caused by the sudden impact of hitting the bottom. With little structural integrity left, the decks pancaked downward as the stern hit the ocean floor.

The *Titanic* has been visited by a number of teams, for purposes of research, salvage, and even filmmaking, by James Cameron, the director of the 1997 version of the film *Titanic*. What is necessary to travel to such depths? Our calculation of the pressure at the location of the *Titanic* indicates that special submarines must be used that can withstand such high pressure while maintaining normal atmospheric pressure inside for the human occupants. This was first done by Ballard in the summer of 1986, using a deep-sea submersible called *Alvin*, with a remote-controlled robot named *Jason Junior*. Figure 4 shows the structure of *Alvin*.

*Alvin* has a titanium alloy hull that can withstand the pressure at the depth of the *Titanic*. The submersible has room for three occupants although they are quite cramped. A number of air tanks on the craft can be flooded with water. Blocks of

**Figure 3**

The stern section of the *Titanic* is heavily damaged and lies in pieces on the ocean floor. *(Painting by Ken Marschall)*

**Figure 4**

The submersible *Alvin,* which can carry three scientists to the great depths at which the *Titanic* currently lies. *(E. Paul Oberlander/Woods Hole Oceanographic Institution)* iron can also be jettisoned. When the tanks are filled with air and the iron blocks are attached, *Alvin* floats on water. When the air tanks are flooded with water, the submersible sinks. In visiting the *Titanic*, the first step is to fill the air tanks with water, and then wait for 2.5 h to sink to the bottom. After visiting the wreckage, the iron blocks are jettisoned. After doing this, the buoyant force on *Alvin* is larger than its weight, and it starts upward on another long journey to the surface. The descent and ascent of *Alvin* are effective examples of applications of the physics described in Case I in Section 15.4.

Once *Alvin* reaches the *Titanic*, the only means of viewing it are by visual inspection through the portholes or by video, using *Jason Junior*. It is impossible to don scuba gear and exit the submersible because of the tremendous pressure.

There is much more to the story of the *Titanic*, but we need to return to our investigations into physics.

## PROBLEMS

1. When the *Titanic* is in its normal sailing position, the torque about the midpoint of the ship is zero. Imagine now that the bow of the ship is under water and the stern is in the air above the water during the sinking process, as in Figure 1. The waterline is at the midpoint of the ship, and the keel of the ship makes a 45° angle with the horizontal. The torque is zero here also, because the *Titanic* is in equilibrium. However, the torque counteracting the weight of the stern section must be applied by the structure of the ship at the midpoint. Calculate the torque about the midpoint of the *Titanic* required to hold the stern section in the air. Model the *Titanic* as a uniform rod of length 269 m and mass $4.2 \times 10^7$ kg. This torque caused the *Titanic* to split near the middle of the ship during the sinking process.

2. The *Titanic* had two almost identical sister ships: the *Britannic* and the *Olympic*. The *Britannic* sank in 1916 off the coast of Athens, Greece, possibly due to a mine planted during World War I. It now sits below 119 m of sea water. (a) What is the pressure at the location of the *Britannic*? (b) Can you visit the *Britannic* by scuba diving?

3. The submersible *Alvin* requires 2.5 h to sink to the location of the *Titanic*. (a) What is the average speed during descent? (b) Suppose the speed of *Alvin* remains constant during the entire descent. Is the average density of *Alvin* greater, less than, or equal to that of sea water? (*Hint:* Include the resistive force on *Alvin* as it moves through the water in your analysis.)

4. The *Titanic* is only one of many maritime disasters. In 1956, despite the presence of radar, which was not invented at the time of the *Titanic*, a collision occurred between the Italian luxury liner *Andrea Doria* and the Swedish liner *Stockholm*. Deaths were few because of the long time interval between the collision and the sinking of the *Andrea Doria*, but a remarkable event occurred. A 14-year-old girl was sleeping in her bed on the *Andrea Doria* before the collision and awoke on the bow of the *Stockholm* after the collision. The bow of the *Stockholm* pierced the hull of the *Andrea Doria* at the point where the girl was sleeping and she was miraculously transferred from one ship to the other! Her only injuries were two broken kneecaps.

   Let us imagine that the collision between the *Andrea Doria* and the *Stockholm* is perfectly inelastic. (This was not the case, as the *Stockholm* pulled away after the collision, but it is a reasonable model.) The weight of the *Andrea Doria* is 29 100 tons, and it is traveling at full speed of 23 knots at the time of the collision, 15° south of west. The *Stockholm* has a weight of 12 165 tons and is traveling at 18 knots 30° east of south at the time of the collision. (a) Just after the collision, what is the velocity, in knots, of the combined wreckage? (b) What fraction of the initial kinetic energy was transformed or transferred away in the collision?

# Global Warming

**N**umerous news stories have detailed the increase in temperature of the Earth and the subsequent results, including the melting of ice from the polar ice caps and changes in climate, with corresponding effects on vegetation. Data taken over the past few decades are interpreted by some scientists as showing a measurable global temperature increase. Life on Earth depends on a delicate balance that keeps the global temperature in a narrow range necessary for our survival. How is this temperature determined? What factors need to be in balance to keep the temperature constant? If we can devise an adequate structural model to calculate the temperature of the Earth, we can predict changes in the temperature as we vary the parameters.

You most likely have an intuitive sense for the temperature of an object, and as long as the object is small (and the object is not undergoing combustion or some other rapid process), no significant temperature variation occurs between different points on the object. But what about a huge object like the Earth? It is clear that no single temperature describes the entire planet—we know that it is summer in Australia when it is winter in Canada. The polar ice caps clearly have different temperatures from the tropical regions. Variations also occur in temperature within a single large body of water such as an ocean. Temperature varies greatly with altitude in a relatively local region, such as in and near Palm Springs, California. Thus, when we speak of

Temperature variations with altitude can exist in a local region on the Earth. Here in Palm Springs, California, palm trees grow in the city while snow is present at the top of the local mountains. *(FPG International)*

the temperature of the Earth, we will refer to an *average* surface temperature, taking into account all of the variations across the surface. It is this average temperature that we would like to calculate by building a structural model of the atmosphere.

One of the primary factors in determining the surface temperature of the Earth is the existence of our atmosphere. The atmosphere is a relatively thin (compared with the radius of the Earth) layer of gas above the surface that provides us with life-supporting oxygen. In addition to providing this important element for life, the atmosphere plays a major role in the energy balance that determines the average temperature. As we proceed with this Context, we shall focus on the physics of gases

The concentration of atmospheric carbon dioxide in parts per million (ppm) of dry air as a function of time during the latter part of the 20th century. These data were recorded at the Mauna Loa Observatory in Hawaii. The yearly variations (*red curve*) coincide with growing seasons because vegetation supplies carbon dioxide to the air. The steady increase (*black curve*) is of concern to scientists.

and apply the principles that we learn to the atmosphere.

One of the important components of the global warming problem is the concentration of carbon dioxide in the atmosphere. Carbon dioxide plays an important role in absorbing energy and raising the temperature of the atmosphere. As seen in the accompanying graph, the amount of carbon dioxide in the atmosphere has been steadily increasing since the middle of the 20th century. This graph shows hard data that indicate that the atmosphere is undergoing a distinct change, although not all scientists agree on the interpretation of what that change means in terms of global temperatures.

In addition to its scientific aspects, global warming is a social issue with many facets. These aspects encompass international politics and economics, because this is a global problem. Changing our policies requires real costs to solve the problem. It also has technological aspects; new methods of manufacturing, transportation, and energy supply must be designed to slow down or reverse the increase in temperature. We shall restrict our attention to the physical aspects of global warming as we address the central question

*What factors determine the average temperature at the Earth's surface?*

## chapter 16

In an automobile engine, a gaseous mixture of fuel and air is compressed in a cylinder by a moving piston. This cutaway diagram of an engine shows two of these cylinders with the pistons at different positions. When the gas in the cylinder is compressed, its temperature and pressure increase. In this chapter, we study the behavior of gases in processes such as these.

*(Courtesy of Ford Motor Company)*

# Temperature and the Kinetic Theory of Gases

Our study thus far has focused mainly on Newtonian mechanics, which explains a wide range of phenomena, such as the motion of baseballs, rockets, and planets. We have applied these principles to oscillating systems, the propagation of mechanical waves through a medium, and the properties of fluids at rest and in motion. In Chapter 6, we introduced the notions of temperature and internal energy. We now extend the study of notions such as these as we focus on **thermodynamics,** which is concerned with the concepts of energy transfers between a system and its environment and the resulting variations in temperature or changes of state. As we shall see, thermodynamics explains the bulk properties of matter and the correlation between these properties and the mechanics of atoms and molecules.

Have you ever wondered how a refrigerator cools or what types of transformations occur in an automobile engine or why a bicycle pump becomes warm as you inflate a tire? The laws of thermodynamics enable us to answer such questions. In general, thermodynamics deals with the physical and chemical transformations of matter in all of its states: solid, liquid, gas, and plasma.

This chapter concludes with a study of ideal gases, which we shall approach on two levels. The first examines ideal gases on the macroscopic scale. Here we shall be concerned with the relationships among such quantities as pressure, volume, and temperature of the gas. On the second level, we shall examine gases on a microscopic (molecular) scale, using a structural model that treats the gas as a collection of particles. The latter approach will help us understand how behavior on the atomic level affects such macroscopic properties as pressure and temperature.

## 16.1 • TEMPERATURE AND THE ZEROTH LAW OF THERMODYNAMICS

See the *Core Concepts in Physics CD-ROM*, Screens 10.3 & 10.4

We often associate the concept of temperature with how hot or cold an object feels to the touch. Thus, our sense of touch provides us with a qualitative indication of temperature; however, our senses are unreliable and often misleading. For example, if you remove a metal ice tray and a cardboard package of frozen vegetables from the freezer, the ice tray feels colder to your hand than the vegetables even though the two are at the same temperature. This is because the properties of the metal are such that the transfer of energy (by heat) to the ice tray from your hand is more rapid than to the cardboard package. Your skin is sensitive to the *rate* of energy transfer—power—not the temperature of the object. Of course, the larger the difference in temperature between that of the object and that of your hand, the faster the energy transfer, so temperature and your sense of touch are related in *some* way. What we need is a reliable and reproducible method for establishing the relative "hotness" or "coldness" of objects that is solely related to the temperature of the object. Scientists have developed a variety of thermometers for making such quantitative measurements.

**APPLICATION**

Sense of Warm and Cold.

We are all familiar with experiences in which two objects at different initial temperatures eventually reach some intermediate temperature when placed in contact with each other. For example, if you combine hot and cold water in a bathtub from separate faucets, the combined water rapidly reaches an equilibrium temperature between the temperatures of the hot and cold water. Likewise, if an ice cube is placed in a cup of hot coffee, the ice eventually melts and the temperature of the coffee decreases.

We shall use these familiar examples to develop the scientific notion of temperature. Imagine two objects placed in an insulated container so that they form an isolated system. If the objects are at different temperatures, energy can be exchanged between them by, for example, heat or electromagnetic radiation. Objects that can exchange energy with each other in this way are said to be in **thermal contact.** Eventually, the temperatures of the two objects will become the same, one becoming warmer and the other cooler, as in our preceding examples. **Thermal equilibrium** is the situation in which two objects in thermal contact cease to have any exchange of energy.

Using these ideas, we can develop a formal definition of temperature. Consider two objects A and B that are not in thermal contact, and a third object C that will be our **thermometer**—a device calibrated to measure the temperature of an object. We wish to determine whether A and B would be in thermal equilibrium if they were placed in thermal contact. The thermometer is first placed in thermal contact with A, and its reading is recorded, as shown in Figure 16.1a. The ther-

(a)  (b)  (c)

**Figure 16.1**

The zeroth law of thermodynamics. (a) and (b) If the temperatures of A and B are measured to be the same by placing them in thermal contact with a thermometer (object C), no energy will be exchanged between them when they are placed in thermal contact with each other (c).

mometer is then placed in thermal contact with B, and its reading is recorded (Fig. 16.1b). If the two readings are the same, then A and B are in thermal equilibrium with each other. If they are placed in thermal contact with each other, as in Figure 16.1c, there is no net transfer of energy between them.

We can summarize these results in a statement known as the **zeroth law of thermodynamics** (the law of equilibrium):

• *Zeroth law of thermodynamics*

> If objects A and B are separately in thermal equilibrium with a third object C, then A and B are in thermal equilibrium with each other.

This statement, elementary as it may seem, is easily proved experimentally and is very important because it can be used to define the notion of temperature. We can think of temperature as the property that determines whether an object is in thermal equilibrium with other objects. **Two objects in thermal equilibrium with each other are at the same temperature.**

## 16.2 • THERMOMETERS AND TEMPERATURE SCALES

In our discussion of the zeroth law, we mentioned a thermometer. Thermometers are devices used to measure the temperature of an object or a system with which the thermometer is in thermal equilibrium. All thermometers make use of some physical property that exhibits a change with temperature that can be calibrated to make the temperature measurable. Some of the physical properties used are (1) the volume of a liquid, (2) the length of a solid, (3) the pressure of a gas held at constant volume, (4) the volume of a gas held at constant pressure, (5) the electric resistance of a conductor, and (6) the color of a hot object.

A common thermometer in everyday use consists of a liquid—usually mercury or alcohol—that expands into a glass capillary tube when its temperature rises (Fig. 16.2). In this case the physical property that changes is the volume of a liq-

**Figure 16.2**

As a result of thermal expansion, the level of the mercury in the thermometer rises as the mercury is heated by water in the test tube.
*(Charles D. Winters)*

uid. Because the cross-sectional area of the capillary tube is uniform, the change in volume of the liquid varies linearly with its length along the tube. We can then define a temperature to be related to the length of the liquid column. The thermometer can be calibrated by placing it in thermal contact with some environments that remain at constant temperature and marking the end of the liquid column on the thermometer. One such environment is a mixture of water and ice in thermal equilibrium with each other at atmospheric pressure. Once we have marked the ends of the liquid column for our chosen environments on our thermometer, we need to define a scale of numbers associated with various temperatures. One such scale is the **Celsius temperature scale.** On the Celsius scale the temperature of the ice–water mixture is defined as zero degrees Celsius, written 0°C; this temperature is called the **ice point** or **freezing point** of water. Another commonly used environment is a mixture of water and steam in thermal equilibrium with each other at atmospheric pressure. On the Celsius scale, this temperature is defined as 100°C, the **steam point** or **boiling point** of water. Once the ends of the liquid column in the thermometer have been marked at these two points, the distance between the marks is divided into 100 equal segments, each denoting a change in temperature of one degree Celsius.

Thermometers calibrated in this way present problems when extremely accurate readings are needed. For instance, an alcohol thermometer calibrated at the ice and steam points of water might agree with a mercury thermometer only at the calibration points. Because mercury and alcohol have different thermal expansion properties, when one indicates a given temperature, the other may indicate a slightly different value. The discrepancies between different types of thermometers are especially large when the temperatures to be measured are far from the calibration points.

**PITFALL PREVENTION 16.1**
**Measuring a perturbed system**

You are likely familiar with the process of measuring a temperature with a common thermometer—placing a mercury thermometer in a beaker of hot water, for example. This raises a common notion in experimental science—when you make a measurement, you interact with the system being measured and often perturb the system. Thus, you measure the perturbed system rather than the original system. In the case of the temperature measurement just mentioned, the thermometer has an initial temperature (we will assume it's lower than that of the water) before it is dipped into the hot water. Thus, the thermometer and the water come to thermal equilibrium at a temperature below the initial temperature of the water and you measure the temperature of the perturbed system. In practice, one can perform calculations to determine the unperturbed measurement. In addition, one can take experimental steps to minimize the perturbation, such as by ensuring that the mass of the system is much larger than that of the thermometer or by preheating the thermometer to a temperature close to the expected temperature of the water.

Scale

$h$

0

$P$
Gas

Mercury
reservoir

$A$        $B$

Bath or
environment
to be measured

Flexible
hose

**Figure 16.3**

A constant-volume gas thermometer measures the pressure of the gas contained in the flask immersed in the bath. The volume of gas in the flask is kept constant by raising or lowering reservoir $B$ such that the mercury level in column $A$ remains constant.

$P$

0°C        100°C        $T(°C)$

**Figure 16.4**

A typical graph of pressure versus temperature taken with a constant-volume gas thermometer. The dots represent known reference temperatures (the ice point and the steam point).

## The Constant-Volume Gas Thermometer and the Kelvin Scale

Although practical devices such as the mercury thermometer can measure temperature, they do not define it in a fundamental way. Only one thermometer offers a way to define temperature and relate it to internal energy directly: the **gas thermometer.** In a gas thermometer, the temperature readings are nearly independent of the substance used in the thermometer. One type of gas thermometer is the constant-volume example shown in Figure 16.3. The behavior observed in this device is the pressure variation with temperature of a fixed volume of gas. When the constant-volume gas thermometer was developed, it was calibrated using the ice and steam points of water as follows. (A different calibration procedure, to be discussed shortly, is now used.) The gas flask is inserted into an ice bath, and mercury reservoir $B$ is raised or lowered until the volume of the confined gas is at some value, indicated by the zero point on the scale. The height $h$, the difference between the levels in the reservoir and column $A$, indicates the pressure in the flask at 0°C, according to Equation 15.4. The flask is inserted into water at the steam point, and reservoir $B$ is readjusted until the height in column $A$ is again brought to zero on the scale, ensuring that the gas volume is the same as it had been in the ice bath (hence the designation "constant-volume"). A measure of the new value for $h$ gives a value for the pressure at 100°C. These pressure and temperature values are then plotted on a graph, as in Figure 16.4. Based on experimental observations that the pressure of a gas varies linearly with its temperature, which is discussed in more detail in Section 16.4, we draw a straight line through our two points. The line connecting the two points serves as a calibration curve for measuring unknown temperatures. If we want to measure the temperature of a substance, we place the gas flask in thermal contact with the substance and adjust the column of mercury until the level in column $A$ again returns to zero. The height of the mercury column tells us the pressure of the gas, and we can then find the temperature of the substance from the calibration curve.

Now suppose that temperatures are measured with various gas thermometers containing different gases. Experiments show that the thermometer readings are nearly independent of the type of gas used, as long as the gas pressure is low and the temperature is well above the point at which the gas liquefies.

We can also perform the temperature measurements with the gas in the flask at different starting pressures at 0°C. As long as the pressure is low, we will generate straight-line calibration curves for each different starting pressure, as shown for three experimental trials (solid lines) in Figure 16.5.

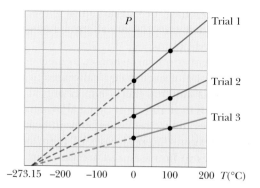

$P$                        Trial 1

Trial 2

Trial 3

−273.15   −200      −100        0        100        200   $T(°C)$

**Figure 16.5**

Pressure versus temperature for dilute gases. Note that, for all three experimental trials, the pressure extrapolates to zero at the unique temperature of −273.15°C.

If the curves in Figure 16.5 are extended back toward negative temperatures, we find a startling result. **In every case, regardless of the type of gas or the value of the low starting pressure, the pressure extrapolates to zero when the temperature is −273.15°C.** This result suggests that this particular temperature is universal in its importance because it does not depend on the substance used in the thermometer. In addition, because the lowest possible pressure is $P = 0$, which would be a perfect vacuum, this temperature must represent a lower bound for physical processes. Thus, we define this temperature as **absolute zero.** Some interesting effects occur at temperatures near absolute zero, such as the phenomenon of *superconductivity,* which we shall study in Chapter 21.

This significant temperature is used as the basis for the **Kelvin temperature scale,** which sets −273.15°C as its zero point (0 K). The size of a degree on the Kelvin scale is chosen to be identical to the size of a degree on the Celsius scale. Thus, the following relationship enables conversion between these temperatures:

$$T_C = T - 273.15 \qquad [16.1]$$

where $T_C$ is the Celsius temperature and $T$ is the Kelvin temperature (sometimes called the **absolute temperature**). The primary difference between these two temperature scales is a shift in the zero of the scale. The zero of the Celsius scale is arbitrary—it depends on a property associated with only one substance—water. The zero on the Kelvin scale is not arbitrary, because it is characteristic of a behavior associated with all substances.

Equation 16.1 shows that the Celsius temperature $T_C$ is shifted from the absolute temperature $T$ by 273.15°. Because the size of a degree is the same on the two scales, a temperature difference of 5°C is equal to a temperature difference of 5 K. The two scales differ only in the choice of the zero point. Thus, the ice point (273.15 K) corresponds to 0.00°C, and the steam point (373.15 K) is equivalent to 100.00°C.

Early gas thermometers made use of ice and steam points according to the procedure just described. These points are experimentally difficult to duplicate, however. For this reason, a new procedure based on two new points was adopted in 1954 by the International Committee on Weights and Measures. The first point is absolute zero. The second point is the **triple point of water,** which corresponds to the **single temperature and pressure at which water, water vapor, and ice can coexist in equilibrium.** This point is a convenient and reproducible reference temperature for the Kelvin scale. It occurs at a temperature of 0.01°C and a very low pressure of 4.58 mm of mercury. The temperature at the triple point of water on the Kelvin scale has a value of 273.16 K. Thus, the SI unit of temperature, the **kelvin,** is defined as **1/273.16 of the temperature of the triple point of water.**

Figure 16.6 shows the Kelvin temperatures for various physical processes and conditions. As the figure reveals, absolute zero has never been achieved, although laboratory experiments have created conditions that are very close to absolute zero.

What would happen to a gas if its temperature could reach 0 K? As Figure 16.5 indicates (if we ignore the liquefaction and solidification of the substance), the pressure it would exert on the walls of its container would be zero. In Section 16.5 we shall show that the pressure of a gas is proportional to the kinetic energy of the molecules of that gas. Thus, according to classical physics, the kinetic energy of the gas would go to zero, and there would be no motion at all of the individual components of the gas; hence, the molecules would settle out on the bottom of

**PITFALL PREVENTION 16.2**

**A matter of degree**

Note that notations for temperatures in the Kelvin scale do not use the degree sign. The unit for a Kelvin temperature is simply "kelvins" and not "degrees Kelvin."

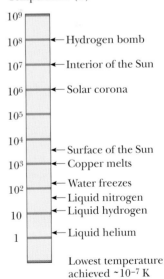

**Figure 16.6**

Absolute temperatures at which various selected physical processes occur. Note that the scale is logarithmic.

the container. Quantum theory, to be discussed in Chapter 28, modifies this statement to indicate that there would be some residual energy, called the *zero-point energy*, at this low temperature.

### The Fahrenheit Scale

The most common temperature scale in everyday use in the United States is the **Fahrenheit scale.** This scale sets the temperature of the ice point at 32°F and the temperature of the steam point at 212°F. The relationship between the Celsius and Fahrenheit temperature scales is

$$T_F = \tfrac{9}{5}T_C + 32°F \qquad \text{[16.2]}$$

**WEB**

For more information about the temperature standard, visit the National Institute of Standards and Technology at **www.nist.gov**

Equation 16.2 can easily be used to find a relationship between changes in temperature on the Celsius and Fahrenheit scales. It is left as a problem for you to show that if the Celsius temperature changes by $\Delta T_C$, the Fahrenheit temperature changes by an amount $\Delta T_F$ given by

$$\Delta T_F = \tfrac{9}{5}\Delta T_C \qquad \text{[16.3]}$$

### Quick Quiz 16.1

In an astronomy class, the temperature at the core of a star is given by the teacher as $1.5 \times 10^7$ degrees. A student asks if this is on the Kelvin or Celsius scale. How would you respond?

### THINKING PHYSICS 16.1

A group of future astronauts lands on an inhabited planet. They strike up a conversation with the aliens about temperature scales. It turns out that the inhabitants of this planet have a temperature scale based on the freezing and boiling points of water, which are separated by 100 of the inhabitants' degrees. Would these two temperatures on this planet be the same as those on Earth? Would the size of the aliens' degrees be the same as ours? Suppose that the aliens have also devised a scale similar to the Kelvin scale. Would their absolute zero be the same as ours?

**Reasoning** The values of 0°C and 100°C for the freezing and boiling points of water are defined at atmospheric pressure. On another planet, it is unlikely that atmospheric pressure would be exactly the same as that on Earth. Thus, water would freeze and boil at different temperatures on the alien planet. The aliens may call these temperatures 0° and 100°, but they would not be the same temperatures as our 0°C and 100°C. If the aliens did assign values of 0° and 100° for these temperatures, then their degrees would not be the same size as our Celsius degrees (unless their atmospheric pressure were the same as ours). For an alien version of the Kelvin scale, the absolute zero would be the same as ours because it is based on a natural, universal definition rather than being associated with a particular substance or a given atmospheric pressure.

### Example 16.1    Converting Temperatures

On a day when the temperature reaches 50°F, what is the temperature in degrees Celsius and in kelvins?

**Solution**    Solving Equation 16.2 for the Celsius temperature and substituting $T_F = 50°F$, we have

$$T_C = \tfrac{5}{9}(T_F - 32°F) = \tfrac{5}{9}(50°F - 32°F) = \boxed{10°C}$$

From Equation 16.1, we find that

$$T = T_C + 273.15 = 10°C + 273.15 = \boxed{283\ K}$$

**EXERCISE**    Gold has a melting point of 1 064°C and a boiling point of 2 660°C. (a) Express these temperatures in kelvins. (b) Compute the difference between these temperatures in Celsius degrees and kelvins.

*Answer*    (a) 1 337 K; 2 933 K    (b) 1 596°C; 1 596 K

## 16.3 • THERMAL EXPANSION OF SOLIDS AND LIQUIDS

Our discussion of the liquid thermometer makes use of one of the best known changes that occur in most substances: as temperature increases, volume increases. This phenomenon, known as **thermal expansion,** plays an important role in numerous applications. For example, thermal expansion joints must be included in buildings, concrete highways, railroad tracks, and bridges to compensate for changes in dimensions with temperature variations.

The overall thermal expansion of an object is a consequence of the change in the average separation between its constituent atoms or molecules. To understand this, consider how the atoms in a solid substance behave. These atoms are located at fixed equilibrium positions; if an atom is pulled away from its position, a restoring force pulls it back. We can build a structural model in which we imagine that the atoms are particles at their equilibrium positions connected by springs to their neighboring atoms (Fig. 16.7). If an atom is pulled away from its equilibrium position, the distortion of the springs provides a restoring force. If the atom is released, it oscillates, and we can apply the simple harmonic motion model to it. A number of macroscopic properties of the substance can be understood with this type of structural model on the atomic level.

**Figure 16.7**

A structural model of the atomic configuration in a solid. The atoms (spheres) are imagined to be connected to one another by springs that reflect the elastic nature of the interatomic forces.

(a)

(b)

(a) Thermal expansion joints are used to separate sections of roadways on bridges. Without these joints, the surfaces would buckle due to thermal expansion on very hot days or crack due to contraction on very cold days. (b) The long, vertical joint is filled with a soft material that allows the wall to expand and contract as the temperature of the bricks changes. *(a, Frank Siteman, Stock/Boston; b, George Semple)*

Thermal expansion: the extremely high temperature of a July day in Asbury Park, New Jersey, caused these railroad tracks to buckle. *(Wide World Photos)*

In Chapter 6, we introduced the notion of internal energy and pointed out that it is related to the temperature of a system. For a solid, the internal energy is associated with the kinetic and potential energy of the vibrations of the atoms around their equilibrium positions. At ordinary temperatures, the atoms vibrate with an amplitude of about $10^{-11}$ m, and the average spacing between the atoms is about $10^{-10}$ m. As the temperature of the solid increases, the average separation between atoms increases. The increase in average separation with increasing temperature (and subsequent thermal expansion) is due to a breakdown in the model of simple harmonic motion. Figure 7.16 in Chapter 7 shows the potential energy curve for an ideal simple harmonic oscillator. The potential energy curve for atoms in a solid is similar to this, but not exactly the same—it is slightly asymmetric around the equilibrium position. It is this asymmetry that leads to thermal expansion.

If the thermal expansion of an object is sufficiently small compared with the object's initial dimensions, then the change in any dimension is, to a good approximation, dependent on the first power of the temperature change. For most situations, we can adopt a simplification model in which this dependence is true. Suppose an object has an initial length $L_i$ along some direction at some temperature. The length increases by $\Delta L$ for a change in temperature $\Delta T$. Experiments show that, when $\Delta T$ is small enough, $\Delta L$ is proportional to $\Delta T$ and to $L_i$:

$$\Delta L = \alpha L_i \Delta T \qquad \text{[16.4]}$$

or

$$L_f - L_i = \alpha L_i (T_f - T_i) \qquad \text{[16.5]}$$

where $L_f$ is the final length, $T_f$ is the final temperature, and the proportionality constant $\alpha$ is called the **average coefficient of linear expansion** for a given material and has units of $(°C)^{-1}$.

Table 16.1 lists the average coefficient of linear expansion for various materials. Note that for these materials $\alpha$ is positive, indicating an increase in length with increasing temperature. This is not always the case. For example, some substances, such as calcite ($CaCO_3$), expand along one dimension (positive $\alpha$) and contract along another (negative $\alpha$) with increasing temperature.

| TABLE 16.1 | Expansion Coefficients for Some Materials Near Room Temperature | | |
|---|---|---|---|
| **Material** | **Linear Expansion Coefficient $\alpha$ $(°C)^{-1}$** | **Material** | **Volume Expansion Coefficient $\beta$ $(°C)^{-1}$** |
| Aluminum | $24 \times 10^{-6}$ | Alcohol, ethyl | $1.12 \times 10^{-4}$ |
| Brass and bronze | $19 \times 10^{-6}$ | Benzene | $1.24 \times 10^{-4}$ |
| Copper | $17 \times 10^{-6}$ | Acetone | $1.5 \times 10^{-4}$ |
| Glass (ordinary) | $9 \times 10^{-6}$ | Glycerin | $4.85 \times 10^{-4}$ |
| Glass (Pyrex) | $3.2 \times 10^{-6}$ | Mercury | $1.82 \times 10^{-4}$ |
| Lead | $29 \times 10^{-6}$ | Turpentine | $9.0 \times 10^{-4}$ |
| Steel | $11 \times 10^{-6}$ | Gasoline | $9.6 \times 10^{-4}$ |
| Invar (Ni–Fe alloy) | $0.9 \times 10^{-6}$ | Air at 0°C | $3.67 \times 10^{-3}$ |
| Concrete | $12 \times 10^{-6}$ | Helium at 0°C | $3.665 \times 10^{-3}$ |

It may be helpful to think of thermal expansion as a magnification or a photographic enlargement. For example, as a metal washer is heated (Fig. 16.8), all dimensions, including the radius of the hole, increase according to Equation 16.4. Because the linear dimensions of an object change with temperature, it follows that volume and surface area also change with temperature. Consider a cube having an initial edge length $L_i$ and, therefore, an initial volume $V_i = L_i^3$. As the temperature is increased, the length of each side increases to

$$L_f = L_i + \alpha L_i \Delta T$$

The new volume, $V_f = L_f^3$, is

$$L_f^3 = (L_i + \alpha L_i \Delta T)^3 = L_i^3 + 3\alpha L_i^3 \Delta T + 3\alpha^2 L_i^3 (\Delta T)^2 + \alpha^3 L_i^3 (\Delta T)^3$$

The last two terms in this expression contain the quantity $\alpha \Delta T$ raised to the second and third powers. Because $\alpha \Delta T$ is much less than 1, raising it to a power makes it even smaller. Therefore, we can ignore these terms to obtain a simpler expression:

$$V_f = L_f^3 = L_i^3 + 3\alpha L_i^3 \Delta T = V_i + 3\alpha V_i \Delta T$$

or

$$\Delta V = V_f - V_i = \beta V_i \Delta T \qquad \text{[16.6]}$$

where $\beta = 3\alpha$. The quantity $\beta$ is called the **average coefficient of volume expansion.** We considered a cubic shape in deriving this equation, but Equation 16.6 describes a sample of any shape, as long as the average coefficient of linear expansion is the same in all directions.

By a similar procedure, we can show that the *increase in area* of an object accompanying an increase in temperature is

$$\Delta A = \gamma A_i \Delta T \qquad \text{[16.7]}$$

where $\gamma$, the **average coefficient of area expansion,** is given by $\gamma = 2\alpha$.

As Table 16.1 indicates, each substance has its own characteristic coefficients of expansion. For example, when the temperatures of a brass rod and a steel rod of equal length are raised by the same amount from some common initial value, the brass rod expands more than the steel rod because brass has a larger coefficient of expansion than steel. A simple device called a bimetallic strip that demonstrates this principle is found in practical devices such as thermostats in home furnace systems. The strip is made by securely bonding two different metals together along their surfaces. As the temperature of the strip increases, the two metals expand by different amounts, and the strip bends as in Figure 16.9.

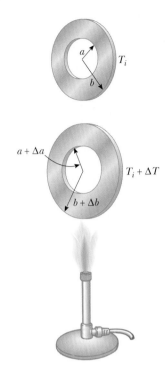

**Figure 16.8**

Thermal expansion of a homogeneous metal washer. Note that as the washer is heated, all dimensions increase. (The expansion is exaggerated in this figure.)

**PITFALL PREVENTION 16.3**

**Do holes become larger or smaller?**

When an object's temperature is raised, every linear dimension increases in size. This includes any holes in the material, which expand in the same way as if the hole were filled with the material, as shown in Figure 16.8. Keep in mind the notion of thermal expansion as being similar to a photographic enlargement.

Steel

Brass

Room temperature

Higher temperature

On        25°C        Off        30°C

Bimetallic strip

(a)        (b)

**Figure 16.9**

(a) A bimetallic strip bends as the temperature changes because the two metals have different expansion coefficients. (b) A bimetallic strip used in a thermostat to make or break electrical contact.

### Quick Quiz 16.2

Common thermometers are made of a mercury column in a glass tube. Based on the operation of these common thermometers, which has the larger coefficient of linear expansion—glass or mercury? (Don't answer this by looking in a table!)

### Quick Quiz 16.3

Two spheres are made of the same metal and have the same radius, but one is hollow and the other is solid. The spheres are taken through the same temperature increase. Which sphere expands more?

**EXERCISE** An automobile fuel tank is filled to the brim with 45 L of gasoline at 10°C. Immediately afterward, the vehicle is parked in the Sun where the temperature is 35°C. How much gasoline overflows from the tank as a result of expansion? (Ignore the expansion of the tank.)

*Answer* 1.1 L

### THINKING PHYSICS 16.2

A bar of aluminum floats in mercury. The level of the mercury is carefully marked on the aluminum bar. The temperature of the aluminum–mercury system is now raised. Is the level of the mercury on the floating aluminum bar now higher than, lower than, or at the same place as before? Can the temperature be changed to a point at which the aluminum sinks?

**Reasoning** The densities in Table 15.1 tell us that aluminum floats in mercury— the density of aluminum, 2 700 kg/m³, is less than the density of mercury, 13 600 kg/m³. The aluminum bar floats in equilibrium with its weight equal in magnitude to the buoyant force on it. The magnitude of the buoyant force, in turn, is the weight of the mercury that would fit into the volume of aluminum below the wet line on the bar. The fraction of the bar submerged is then the ratio of the densities, 2 700/13 600 = 19.9%. In Table 16.1 observe that the volume expansion coefficient for aluminum, $3\,[14 \times 10^{-6}\,(\text{°C})^{-1}] = 42 \times 10^{-6}\,(\text{°C})^{-1}$, is less than that of mercury, $182 \times 10^{-6}(\text{°C})^{-1}$. As the temperature rises, the densities of both metals decrease, but the density of the mercury decreases more. Thus the fraction $\rho_{Al}/\rho_{Hg}$ increases. As a result, a greater fraction of the bar is immersed. The level of the mercury on the bar is higher than before and covers up your mark. In principle, the temperature of the system could be raised to the point at which the density of the expanded mercury would equal that of the expanded aluminum, and the aluminum would then sink. A calculation, however, shows that the temperature at which this would occur is far above the boiling point of the mercury (as well as the melting and boiling points of the aluminum!).

## THINKING PHYSICS 16.3

A homeowner is painting a ceiling, and a drop of paint falls from the brush onto an operating incandescent lightbulb. The bulb breaks. Why?

**Reasoning**   The glass envelope of an incandescent lightbulb receives energy on the inside surface by electromagnetic radiation from the very hot filament. In addition, because the bulb contains gas, the glass envelope receives energy by matter transfer related to the movement of the hot gas near the filament to the colder glass. Thus, the glass can become very hot. If a drop of relatively cold paint falls onto the glass, that portion of the glass envelope suddenly becomes colder than the other portions, and the contraction of this region can cause thermal stresses that might break the glass.

## Example 16.2   Does the Hole Become Bigger or Smaller?

A hole of cross-sectional area 100 cm² is cut in a piece of steel at 20°C. What is the change in the area of the hole if the steel is heated from 20°C to 100°C?

**Solution**   A hole in a substance expands in exactly the same way as does a piece of the substance having the same shape as the hole. The change in the area of the hole can be found by using Equation 16.7:

$$\Delta A = \gamma A_i \Delta T$$
$$= [22 \times 10^{-6}(°C)^{-1}](100 \text{ cm}^2)(80°C) = \boxed{0.18 \text{ cm}^2}$$

## The Unusual Behavior of Water

Liquids generally increase in volume with increasing temperature and have volume expansion coefficients on the order ten times greater than those of solids. Water is an exception to this rule over a small temperature range, as we can see from its density-versus-temperature curve in Figure 16.10. As the temperature increases from 0°C to 4°C, water contracts and thus its density increases. Above 4°C,

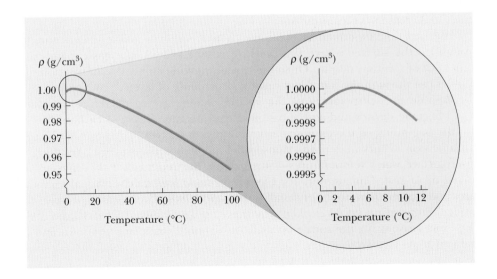

**Figure 16.10**

The variation of density with temperature for water at atmospheric pressure. The inset at the right shows that the maximum density of water occurs at 4°C.

water exhibits the expected expansion with increasing temperature. Thus, the density of water reaches a maximum value of 1 000 kg/m$^3$ at 4°C.

We can use this unusual thermal expansion behavior of water to explain why a pond freezes at the surface. When the atmospheric temperature drops from 7°C to 6°C, for example, the water at the surface of the pond also cools and consequently decreases in volume. This means that the surface water is denser than the water below it, which has not cooled and decreased in volume. As a result, the surface water sinks and warmer water from below moves to the surface to be cooled, a process called upwelling. When the atmospheric temperature is between 4°C and 0°C, however, the surface water expands as it cools, becoming less dense than the water below it. The sinking process stops, and eventually the surface water freezes. As the water freezes, the ice remains on the surface because ice is less dense than water. The ice continues to build up on the surface, while water near the bottom of the pool remains at 4°C. If this did not happen, fish and other forms of marine life would not survive through the winter.

A vivid example of the dangers of the absence of the upwelling and mixing processes is the sudden and deadly release of carbon dioxide gas from Lake Monoun in August 1984 and Lake Nyos in August 1986. Both lakes are located in the rain forest country of Cameroon in Africa. Over 1 700 natives of Cameroon died in these events.

In a lake located in a temperate zone such as the United States, significant temperature variations occur during the day and during the entire year. For example, imagine the Sun going down in the evening. As the temperature of the surface water drops, because of the absence of sunlight, the sinking process tends to mix the upper and lower layers of water.

**APPLICATION**

Suffocation by Explosive Release of Carbon Dioxide.

**WEB**

For information on Lake Nyos from PBS and a list of related links, visit **www.pbs.org/wnet/savage planet/01volcano/01/ indexmid.html**

This mixing process does not normally occur in Lakes Monoun and Nyos because of two characteristics that contributed significantly to the disasters. First, the lakes are very deep, so that mixing the various layers of water over such a large vertical distance is difficult. This factor also results in a very large pressure at the bottom of the lake so that a large amount of carbon dioxide from local rocks and deep springs dissolves into the water. Second, both lakes are located in an equatorial rain forest region where the temperature variation is much smaller than in temperate zones. This results in little driving force to mix the layers of water in the lakes. Water near the bottom of the lake stays there for a long time and collects a large amount of dissolved carbon dioxide. In the absence of a mixing process, this carbon dioxide cannot be brought to the surface and released safely. It simply continues to increase in concentration.

The situation described is explosive. If the carbon dioxide–laden water is brought to the surface where the pressure is much lower, the gas expands and comes out of solution rapidly. Once the carbon dioxide comes out of solution, the bubbles rise through the water and cause more mixing of layers.

Suppose the temperature of the surface water were to decrease; this water would become denser and sink, possibly triggering the release of carbon dioxide and the beginning of the explosive situation just described. The monsoon season in Cameroon occurs in August. Monsoon clouds block the sunlight, resulting in lower surface water temperatures—this may be the reason that the disasters occurred in August. Climate data for Cameroon show lower than normal temperatures and higher than normal rainfall in the mid-1980s. The resulting decrease in surface temperature could explain why these events occurred in 1984 and 1986. The exact reasons for the sudden release of carbon dioxide are unknown and remain an area of active research.

(*Top*) Lake Nyos, in Cameroon, Africa, after an explosive outpouring of carbon dioxide. (*Bottom*) The carbon dioxide caused many deaths, both of humans and animals, such as the cattle shown here. (© *Peter Turnley/Corbis*)

Finally, once the carbon dioxide was released from the lakes, it stayed near the ground because carbon dioxide is denser than air. Thus, a layer of carbon dioxide gas spread out over the land around the lake, representing a deadly suffocating gas for all humans and animals in its path.

See Screen 10.5

**PITFALL PREVENTION 16.4**

**So many *m*'s**

We have previously used and will again use $m$ and $M$ in later chapters to represent the mass of an object or of a system. In this chapter we use $m_{sample}$ for the mass of a specific quantity of a substance and $m$ to stand for the mass of one molecule. These two masses are related by $m = m_{sample}/N$, where $N$ is the number of molecules in the sample. We use $M$ for the molar mass of the substance of which the sample is composed.

**Figure 16.11**

An ideal gas confined to a cylinder whose volume can be varied with a movable piston.

# 16.4 • MACROSCOPIC DESCRIPTION OF AN IDEAL GAS

The properties of gases are very important in a number of thermal processes. Our everyday weather is a perfect example of the types of processes that depend on the behavior of gases.

If we introduce a gas into a container, it expands to fill the container uniformly. Thus, the gas does not have a fixed volume or pressure. Its volume is that of the container, and its pressure depends on the size of the container. In this section we shall be concerned with the properties of a gas with pressure $P$ and temperature $T$, confined to a container of volume $V$. It is useful to know how these quantities are related. In general, the equation that interrelates these quantities, called the **equation of state**, can be complicated. If the gas is maintained at a very low pressure (or low density), however, the equation of state is found experimentally to be relatively simple. Such a low-density gas is commonly referred to as an **ideal gas.** Most gases at room temperature and atmospheric pressure behave approximately as ideal gases. We shall adopt a simplification model, called the **ideal gas model,** for these types of studies. An ideal gas is **a collection of atoms or molecules that move randomly, exert no long-range forces on one another, and occupy a negligible fraction of the volume of their container.**

It is convenient to express the amount of gas in a given volume in terms of the number of moles. One **mole** of any substance is that mass of the substance that contains **Avogadro's number,** $N_A = 6.022 \times 10^{23}$, of molecules. The number of moles $n$ of a substance in a sample is related to its mass $m_{sample}$ through the expression

$$n = \frac{m_{sample}}{M} \qquad [16.8]$$

where $M$ is the **molar mass** of the substance, usually expressed in grams per mole. For example, the molar mass of molecular oxygen $O_2$ is 32.0 g/mol. The mass of one mole of oxygen is therefore 32.0 g. We can calculate the mass of one molecule by dividing the molar mass by the number of molecules, which is Avogadro's number. Thus, for oxygen,

$$m = \frac{M}{N_A} = \frac{32.0 \times 10^{-3}\,\text{kg/mol}}{6.02 \times 10^{23}\,\text{molecule/mol}} = 5.32 \times 10^{-26}\,\text{kg/molecule}$$

Now suppose an ideal gas is confined to a cylindrical container whose volume can be varied by means of a movable piston, as in Figure 16.11. We shall assume that the cylinder does not leak, and so the number of moles of gas remains constant. For such a system, experiments provide the following information. First, when the gas is kept at a constant temperature, its pressure is inversely proportional to the volume. (This is known historically as Boyle's law.) Second, when the pressure of the gas is kept constant, the volume is directly proportional to the temperature. (This is known historically as the law of Charles and Gay-Lussac.) These observations can be summarized by the following equation of state, known as the **ideal gas law:**

• *Ideal gas law*

$$PV = nRT \qquad [16.9]$$

In this expression $R$ is a constant for a specific gas that can be determined from experiments, and $T$ is the absolute temperature in kelvins. Experiments on several gases show that as the pressure approaches zero, the quantity $PV/nT$ approaches the same value of $R$ for all gases. For this reason, $R$ is called the **universal gas constant.** In SI units, where pressure is expressed in pascals and volume in cubic meters, $R$ has the value

$$R = 8.315 \text{ J/mol} \cdot \text{K} \qquad \textbf{[16.10]}$$

• *The universal gas constant*

If the pressure is expressed in atmospheres and the volume in liters ($1 \text{ L} = 10^3 \text{ cm}^3 = 10^{-3} \text{ m}^3$), then $R$ has the value

$$R = 0.0821 \text{ L} \cdot \text{atm/mol} \cdot \text{K}$$

Using this value of $R$ and Equation 16.9, one finds that the volume occupied by 1 mol of any gas at atmospheric pressure and 0°C (273 K) is 22.4 L.

The ideal gas law is often expressed in terms of the total number of molecules $N$. Because the total number of molecules equals the product of the number of moles and Avogadro's number $N_A$, we can write Equation 16.9 as

$$PV = nRT = \frac{N}{N_A} RT$$

$$PV = Nk_B T \qquad \textbf{[16.11]}$$

where $k_B$ is called **Boltzmann's constant** and has the value

$$k_B = \frac{R}{N_A} = 1.38 \times 10^{-23} \text{ J/K} \qquad \textbf{[16.12]}$$

**PITFALL PREVENTION 16.5**
**So many $k$'s**

In a variety of situations in physics the letter $k$ is used—we have seen two previously, the force constant for a spring (Chapter 12) and the wave number for a mechanical wave (Chapter 13). We also saw $k_e$, the Coulomb constant, in Chapter 5. Boltzmann's constant is another $k$, and we will see $k$ used for thermal conductivity in Chapter 17. To make some sense of this confusing state of affairs, we will use a subscript for Boltzmann's constant to help us recognize it. In this book, we will see Boltzmann's constant as $k_B$, but you may see Boltzmann's constant in other resources as simply $k$.

### Quick Quiz 16.4

Some picnickers stop at the convenience store and buy food, including bags of snack chips. They drive up the mountains to their picnic site. When they unload the food, they notice that the bags of chips are puffed up like balloons. Why did this happen?

### Quick Quiz 16.5

A common material for cushioning objects in packages is made by trapping small bubbles of air between sheets of plastic. Many individuals find satisfaction in popping the bubbles after receiving a package (other individuals who are nearby and listening to the sound may not be so satisfied). Is this material more effective at cushioning the package contents on a hot day or a cold day?

## Example 16.3   Squeezing a Tank of Gas

Pure helium gas is admitted into a tank containing a movable piston. The initial volume, pressure, and temperature of the gas are $15.0 \times 10^{-3}$ m³, 200 kPa, and 300 K. If the volume is decreased to $12.0 \times 10^{-3}$ m³ and the pressure increased to 350 kPa, find the final temperature of the gas.

**Solution**   Let us model the helium as an ideal gas. If no gas escapes from the tank, the number of moles of gas remains constant; therefore, using $PV = nRT$ at the initial and final points gives

$$nR = \frac{P_i V_i}{T_i} = \frac{P_f V_f}{T_f}$$

where $i$ and $f$ refer to the initial and final values, respectively. Solving for $T_f$, we find

$$T_f = \frac{P_f V_f}{P_i V_i} T_i = \frac{(350 \text{ kPa})(12.0 \times 10^{-3} \text{ m}^3)}{(200 \text{ kPa})(15.0 \times 10^{-3} \text{ m}^3)} (300 \text{ K})$$

$$= \boxed{420 \text{ K}}$$

## Example 16.4   How Many Moles?

An ideal gas occupies a volume of 100 cm³ at 20.0°C and 100 Pa. Find the number of moles of gas in the container.

**Solution**   The quantities given are volume, temperature, and pressure. Using Equation 16.9, we find that

$$n = \frac{PV}{RT} = \frac{(100 \text{ Pa})(1.00 \times 10^{-4} \text{ m}^3)}{(8.315 \text{ J/mol} \cdot \text{K})(293 \text{ K})} = \boxed{4.10 \times 10^{-6} \text{ mol}}$$

**EXERCISE**   How many molecules of gas are in the container?

*Answer*   $2.47 \times 10^{18}$ molecules

**EXERCISE**   An ideal gas is held in a container at constant volume. Initially, its temperature is 10.0°C and its pressure is 2.50 atm. What is its pressure when its temperature is 80.0°C?

*Answer*   3.12 atm

   See Screen 10.5

**Ludwig Boltzmann (1844–1906)**

Austrian theoretical physicist Boltzmann made many important contributions to the development of the kinetic theory of gases, electromagnetism, and thermodynamics. His pioneering work in the field of kinetic theory led to the branch of physics known as *statistical mechanics*. *(Courtesy of AIP Niels Bohr Library, Lande Collection)*

## 16.5 • THE KINETIC THEORY OF GASES

In the preceding section we discussed the macroscopic properties of an ideal gas using such quantities as pressure, volume, number of moles, and temperature. From a macroscopic point of view, the mathematical representation of the ideal gas model is the ideal gas law. In this section we consider the microscopic point of view of the ideal gas model. We shall show that the macroscopic properties can be understood on the basis of what is happening on the atomic scale. In addition, we shall re-examine the ideal gas law in terms of the behavior of the individual molecules that make up the gas.

Using the ideal gas model, we shall build a structural model of a gas enclosed in a container. The mathematical structure and the predictions made by this model constitute what is known as the **kinetic theory of gases.** With this theory, we shall interpret the pressure and temperature of an ideal gas in terms of microscopic variables. In our structural model, we make the following assumptions:

1. **The number of molecules in the gas is large, and the average separation between them is large compared with their dimensions.** This means that the molecules occupy a negligible volume in the container. This assumption is consistent with the ideal gas model, in which we imagine the molecules to be point-like.

2. **The molecules obey Newton's laws of motion, but as a whole they move randomly.** By "randomly" we mean that any molecule can move in any direction with any speed. At any given moment, a certain percentage of molecules move at high speeds, and a certain percentage move at low speeds.

3. **The molecules interact only by short-range forces during elastic collisions.** This is consistent with the ideal gas model, in which the molecules exert no long-range forces on one another.
4. **The molecules make elastic collisions with the walls.**
5. **The gas under consideration is a pure substance; that is, all molecules are identical.**

• *Assumptions of kinetic theory for an ideal gas*

Although we often picture an ideal gas as consisting of single atoms, molecular gases exhibit equally good approximations to ideal gas behavior at low pressures. Effects associated with molecular structure have no influence on the motions considered here. Thus, we can apply the results of the following development to molecular gases as well as monatomic gases.

## Molecular Interpretation of the Pressure of an Ideal Gas

For our first application of kinetic theory, let us derive an expression for the pressure of $N$ molecules of an ideal gas in a container of volume $V$ in terms of microscopic quantities. The container is a cube with edges of length $d$ (Fig. 16.12). We shall focus our attention on one of these molecules of mass $m$ and assumed to be moving so that its component of velocity in the $x$ direction is $v_{xi}$ as in Figure 16.13. (The subscript $i$ here refers to the $i$th molecule, not to an initial value. We will combine the effects of all of the molecules shortly.) As the molecule collides elastically with any wall (assumption 4), its velocity component perpendicular to the wall is reversed because the mass of the wall is far greater than the mass of the molecule. Because the momentum component $p_{xi}$ of the molecule is $mv_{xi}$ before the collision and $-mv_{xi}$ after the collision, the change in momentum of the molecule in the $x$ direction is

$$\Delta p_{xi} = -mv_{xi} - (mv_{xi}) = -2mv_{xi}$$

Applying the impulse–momentum theorem (Eq. 8.12) to the molecule gives

$$\overline{F}_{i,\text{on molecule}} \, \Delta t_{\text{collision}} = \Delta p_{xi} = -2mv_{xi}$$

where $\overline{F}_{i,\text{on molecule}}$ is the average force component, perpendicular to the wall, for the force that the wall exerts on the molecule during the collision and $\Delta t_{\text{collision}}$ is the duration of the collision. For the molecule to make another collision with the same wall after this first collision, it must travel a distance of $2d$ in the $x$ direction (across the container and back). The time interval between two collisions with the same wall is therefore

$$\Delta t = \frac{2d}{v_{xi}}$$

The force that causes the change in momentum of the molecule in the collision with the wall occurs only during the collision. However, we can average the force over the time interval for the molecule to move across the cube and back. Sometime during this time interval, the collision occurs, so that the change in momentum for this time interval is the same as that for the short duration of the collision. Thus, we can rewrite the impulse–momentum theorem as

$$\overline{F}_i \, \Delta t = -2mv_{xi}$$

where $\overline{F}_i$ is the average force component over the time for the molecule to move across the cube and back. Because exactly one collision occurs for each such time interval, this is also the long-term average force component on the molecule, over long time intervals containing any number of multiples of $\Delta t$.

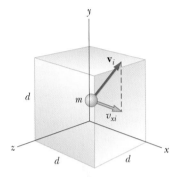

**Figure 16.12**

A cubical box with sides of length $d$ containing an ideal gas. The molecule shown moves with velocity $\mathbf{v}_i$.

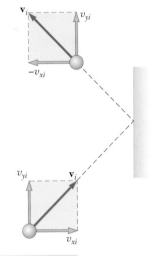

**Figure 16.13**

A molecule makes an elastic collision with the wall of the container. Its $x$ component of momentum is reversed, thereby exerting a force on the wall, while its $y$ component remains unchanged. In this construction, we assume that the molecule moves in the $xy$ plane.

The substitution of $\Delta t$ into the impulse–momentum equation enables us to express the long-term average force component of the wall on the molecule:

$$\overline{F_i} = \frac{-2mv_{xi}}{\Delta t} = \frac{-2mv_{xi}^2}{2d} = \frac{-mv_{xi}^2}{d}$$

Now, by Newton's third law, the force component of the molecule on the wall is equal in magnitude and opposite in direction:

$$\overline{F}_{i,\text{on wall}} = -\overline{F_i} = -\left(\frac{-mv_{xi}^2}{d}\right) = \frac{mv_{xi}^2}{d}$$

The magnitude of the total average force $F$ exerted on the wall by the gas is found by adding the average force components exerted by the individual molecules—we add terms such as those shown in the preceding equations for all molecules:

$$\overline{F} = \sum_{i=1}^{N} \frac{mv_{xi}^2}{d} = \frac{m}{d} \sum_{i=1}^{N} v_{xi}^2$$

where we have factored out the length of the box and the mass $m$ because assumption 5 tells us that all of the molecules are the same. We now impose assumption 1, that the number of molecules is large. For a small number of molecules, the actual force on the wall would vary with time. It would be nonzero during the short interval of a collision of a molecule with the wall and zero when no molecule happens to be hitting the wall. For a very large number of molecules, however, such as Avogadro's number, these variations in force are smoothed out, so that the average force is the same over *any* time interval. Thus, the *constant* force on the wall due to the molecular collisions is of magnitude

$$F = \frac{m}{d} \sum_{i=1}^{N} v_{xi}^2$$

To proceed further, let us consider how we express the average value of the square of the $x$ component of the velocity for the $N$ molecules. The traditional average of a value is the sum of the values over the number of values:

$$\overline{v_x^2} = \frac{\displaystyle\sum_{i=1}^{N} v_{xi}^2}{N}$$

The numerator of this expression is contained in the right-hand side of the previous equation. Thus, combining the two expressions, the total force on the wall can be written

$$F = \frac{m}{d} N\overline{v_x^2}$$

Now let us focus again on one molecule with velocity components $v_{xi}$, $v_{yi}$, and $v_{zi}$. The Pythagorean theorem relates the square of the speed of the molecule to the squares of the velocity components:

$$v_i^2 = v_{xi}^2 + v_{yi}^2 + v_{zi}^2$$

If we take an average of both sides of this equation (sum over all particles and divide by $N$), the average value of $v^2$ for all the molecules in the container is related to the average values of $v_x^2$, $v_y^2$, and $v_z^2$ according to the expression

$$\overline{v^2} = \overline{v_x^2} + \overline{v_y^2} + \overline{v_z^2}$$

Now we assume that the motion is completely random (assumption 2). This implies that no direction is preferred. On the average, the *x*, *y* and *z* directions are equivalent, so

$$\overline{v_x^2} = \overline{v_y^2} = \overline{v_z^2}$$

which allows us to write

$$\overline{v^2} = 3\overline{v_x^2}$$

Thus, the total force on the wall is

$$F = \frac{m}{d} N\left(\tfrac{1}{3}\overline{v^2}\right) = \frac{N}{3}\left(\frac{m\overline{v^2}}{d}\right)$$

From this expression, we can find the pressure exerted on the wall by dividing this force by the area of the wall:

$$P = \frac{F}{A} = \frac{F}{d^2} = \tfrac{1}{3}\frac{N}{d^3}\left(m\overline{v^2}\right) = \tfrac{1}{3}\left(\frac{N}{V}\right)\left(m\overline{v^2}\right)$$

$$P = \tfrac{2}{3}\left(\frac{N}{V}\right)\left(\tfrac{1}{2}m\overline{v^2}\right) \qquad\qquad \textbf{[16.13]}$$

•   *Pressure of an ideal gas*

This result shows that the **pressure is proportional to the number of molecules per unit volume and to the average translational kinetic energy of the molecules, $\tfrac{1}{2}m\overline{v^2}$.** With this structural model of an ideal gas, we have arrived at an important result that relates the macroscopic quantity of pressure to a microscopic quantity, the average value of the molecular translational kinetic energy. Thus, we have a key link between the atomic world and the large-scale world.

   Equation 16.13 verifies some features of pressure that are probably familiar to you. One way to increase the pressure inside a container is to increase the number of molecules per unit volume in the container ($N/V$). You do this when you add air to a tire. The pressure in the tire can also be increased by increasing the average translational kinetic energy of the molecules in the tire. As we shall see shortly, this can be accomplished by increasing the temperature of the gas inside the tire. That is why the pressure inside a tire increases as the tire warms up during long trips. The continuous flexing of the tires as they move along the road surface results in work done as parts of the tire distort and in an increase in internal energy of the rubber. The increased temperature of the rubber results in transfer of energy into the air by heat, increasing the average translational kinetic energy of the molecules, which in turn produces an increase in pressure.

## Molecular Interpretation of the Temperature of an Ideal Gas

   See Screen 10.3

We have related the pressure to the average kinetic energy of molecules; let us now relate temperature to a microscopic description of the gas. We can obtain some insight into the meaning of temperature by first writing Equation 16.13 in the form

$$PV = \tfrac{2}{3}N\left(\tfrac{1}{2}m\overline{v^2}\right)$$

Let us now compare this with the equation of state for an ideal gas:

$$PV = Nk_B T$$

The left-hand sides of these two equations are identical. Equating the right-hand sides of these expressions, we find that

- *Temperature is proportional to average kinetic energy*

$$T = \frac{2}{3k_B}\left(\frac{1}{2}\,m\overline{v^2}\right) \qquad [16.14]$$

This tells us that the **temperature of a gas is a direct measure of average translational molecular kinetic energy.** Thus, as the temperature of a gas increases, the molecules move with higher average kinetic energy.

By rearranging Equation 16.14, we can relate the average translational molecular kinetic energy to the temperature:

- *Average kinetic energy per molecule*

$$\frac{1}{2}m\overline{v^2} = \frac{3}{2}k_B T \qquad [16.15]$$

That is, the average translational kinetic energy per molecule is $\frac{3}{2}\,k_B T$. Because $\overline{v_x^2} = \frac{1}{3}\,\overline{v^2}$, it follows that

$$\frac{1}{2}m\overline{v_x^2} = \frac{1}{2}k_B T \qquad [16.16]$$

In a similar manner, for the $y$ and $z$ motions we find

$$\frac{1}{2}m\overline{v_y^2} = \frac{1}{2}k_B T \qquad \text{and} \qquad \frac{1}{2}m\overline{v_z^2} = \frac{1}{2}k_B T$$

Thus, each translational degree of freedom contributes an equal amount of energy to the gas, namely, $\frac{1}{2}k_B T$ per molecule. (In general, the phrase "degrees of freedom" refers to the number of independent means by which a molecule can possess energy.) A generalization of this result, known as the **theorem of equipartition of energy,** says that **the energy of a system in thermal equilibrium is equally divided among all degrees of freedom.** Furthermore, each degree of freedom contributes the same amount of average energy to the total, $\frac{1}{2}k_B T$ per molecule.

The total translational kinetic energy of $N$ molecules of gas is simply $N$ times the average translational kinetic energy per molecule, which is given by Equation 16.15:

- *Total kinetic energy of N molecules*

$$E_{\text{total}} = N\left(\tfrac{1}{2}m\overline{v^2}\right) = \tfrac{3}{2}Nk_B T = \tfrac{3}{2}nRT \qquad [16.17]$$

where we have used $k_B = R/N_A$ for Boltzmann's constant and $n = N/N_A$ for the number of moles of gas. From this result, we see that the **total translational kinetic energy of a system of molecules is proportional to the absolute temperature of the system.**

For a monatomic gas, translational kinetic energy is the only type of energy the molecules can have. Thus, Equation 16.17 gives the **internal energy for a monatomic gas:**

$$E_{\text{int}} = \tfrac{3}{2}nRT \quad \text{(monatomic gas)} \qquad [16.18]$$

This equation mathematically justifies our claim that internal energy is related to the temperature of a system, which we introduced in Chapter 6. For diatomic and polyatomic molecules, additional possibilities for energy storage are available in the vibration and rotation of the molecule but the proportionality between $E_{\text{int}}$ and $T$ remains.

The square root of $\overline{v^2}$ is called the *root-mean-square* (rms) *speed* of the molecules. From Equation 16.15 we find, for the rms speed

- *Root-mean-square speed*

$$v_{\text{rms}} = \sqrt{\overline{v^2}} = \sqrt{\frac{3k_B T}{m}} = \sqrt{\frac{3RT}{M}} \qquad [16.19]$$

where $M$ is the molar mass in kilograms per mole. This expression shows that, at a given temperature, lighter molecules move faster, on the average, than heavier

| TABLE 16.2 | Some rms Speeds | |
|---|---|---|
| Gas | Molar Mass (g/mol) | $v_{rms}$ at 20°C (m/s) |
| $H_2$ | 2.02 | 1902 |
| He | 4.0 | 1352 |
| $H_2O$ | 18 | 637 |
| Ne | 20.1 | 603 |
| $N_2$ or CO | 28 | 511 |
| NO | 30 | 494 |
| $CO_2$ | 44 | 408 |
| $SO_2$ | 64 | 338 |

molecules. For example, hydrogen, with a molar mass of $2.0 \times 10^{-3}$ kg/mol, moves four times as fast as oxygen, whose molar mass is $32 \times 10^{-3}$ kg/mol. If we calculate the rms speed for hydrogen at room temperature ($\approx 300$ K), we find

$$v_{rms} = \sqrt{\frac{3RT}{M}} = \sqrt{\frac{3(8.31 \text{ J/mol} \cdot \text{K})(300 \text{ K})}{2.0 \times 10^{-3} \text{ kg/mol}}} = 1.9 \times 10^3 \text{ m/s}$$

This value is about 17% of the escape speed for the Earth, which we calculated in Chapter 11. Because this is an average speed, a large number of molecules having speeds much higher than this can escape from the Earth's atmosphere. This is why the Earth's atmosphere does not at present contain hydrogen—it has all bled off into space.

Table 16.2 lists the rms speeds for various molecules at 20°C.

## Quick Quiz 16.6

Small planets tend to have little or no atmosphere. Why is this?

## THINKING PHYSICS 16.4

Imagine a gas in an insulated cylinder with a movable piston. The piston is pushed inward, compressing the gas, and then released. As the molecules of the gas strike the piston, they move it outward. According to the ideal gas law, as the volume of the gas increases, its temperature decreases. From the points of view of (a) energy principles and (b) kinetic theory, explain how the expansion of this gas causes its temperature to drop.

**Reasoning** (a) From the point of view of energy principles, let us identify the gas as a nonisolated system. The gas molecules strike the piston and move it through a distance. Thus, the molecules do work on the piston, which represents a transfer of energy out of the system. As a result, the internal energy of the gas drops. Because temperature is related to internal energy, the temperature of the gas drops.

(b) From the point of view of kinetic theory, a molecule colliding with the piston causes the piston to move upward with some velocity. According to conservation of momentum, then, the molecule must rebound with less speed than it had before the collision. Thus, as these collisions occur, the average speed of the collection of molecules is reduced. Because temperature is related to the average speed of molecules, the temperature of the gas drops.

**Example 16.5 A Tank of Helium**

A tank of volume 0.300 m³ contains 2.00 mol of helium gas at 20.0°C. Assuming the helium behaves like an ideal gas, (a) find the total internal energy of the gas.

**Solution** Helium is a monatomic gas, so Equation 16.18 can be used for the internal energy. With $n = 2.00$ and $T = 293$ K, we have

$$E_{int} = \tfrac{3}{2} nRT = \tfrac{3}{2}(2.00 \text{ mol})(8.31 \text{ J/mol·K})(293 \text{ K})$$
$$= 7.30 \times 10^3 \text{ J}$$

(b) What is the average kinetic energy per molecule?

**Solution** From Equation 16.15, we see that the average kinetic energy per molecule is

$$\tfrac{1}{2} m \overline{v^2} = \tfrac{3}{2} k_B T = \tfrac{3}{2}(1.38 \times 10^{-23} \text{ J/K})(293 \text{ K})$$
$$= 6.07 \times 10^{-21} \text{ J}$$

**EXERCISE** Using the fact that the molar mass of helium is $4.00 \times 10^{-3}$ kg/mol, determine the rms speed of the atoms at 20.0°C.

*Answer* $1.35 \times 10^3$ m/s

## 16.6 • DISTRIBUTION OF MOLECULAR SPEEDS

In the preceding section, we derived an expression for the average speed of a gas molecule but made no mention of the actual distribution of molecular speeds among all possible values. In 1860, James Clerk Maxwell (1831–1879) derived an expression that describes this distribution of molecular speeds. His work and developments by other scientists shortly thereafter were highly controversial because experiments at that time could not directly detect molecules. About 60 years later, however, experiments confirmed Maxwell's predictions.

Consider a container of gas whose molecules have some distribution of speeds. Suppose we want to determine how many gas molecules have a speed in the range from, for example, 400 to 410 m/s. Intuitively, we expect that the speed distribution depends on temperature. Furthermore, we expect that the distribution peaks in the vicinity of $v_{rms}$. That is, few molecules are expected to have speeds much less than or much greater than $v_{rms}$ because these extreme speeds will result only from an unlikely chain of collisions.

The observed speed distribution of gas molecules in thermal equilibrium is shown in Figure 16.14. The quantity $N_v$, called the **Maxwell–Boltzmann distribution function,** is defined as follows: If $N$ is the total number of molecules, then the number of molecules with speeds between $v$ and $v + dv$ is $dN = N_v \, dv$. This number is also equal to the area of the shaded rectangle in Figure 16.14. Furthermore, the fraction of molecules with speeds between $v$ and $v + dv$ is $N_v \, dv/N$. This fraction is also equal to the *probability* that a molecule has a speed in the range $v$ to $v + dv$.

The fundamental expression that describes the distribution of speeds of $N$ gas molecules is

$$N_v = 4\pi N \left( \frac{m}{2\pi k_B T} \right)^{3/2} v^2 e^{-mv^2/2k_B T} \qquad \text{[16.20]}$$

where $m$ is the mass of a gas molecule, $k_B$ is Boltzmann's constant, and $T$ is the absolute temperature.*

**Figure 16.14**

The speed distribution of gas molecules at some temperature. The number of molecules having speeds in the range $dv$ is equal to the area $N_v \, dv$ of the shaded rectangle. The function $N_v$ approaches zero as $v$ approaches infinity.

*Maxwell–Boltzmann distribution function*

---

* For the derivation of this expression, see a text on thermodynamics, such as that by R. P. Bauman, *Modern Thermodynamics and Statistical Mechanics,* New York, Macmillan, 1992.

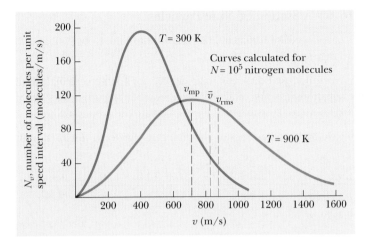

**Figure 16.15**

The speed distribution function for $10^5$ nitrogen molecules at 300 K and 900 K. The total area under either curve is equal to the total number of molecules. Note that $v_{rms} > \bar{v} > v_{mp}$.

As indicated in Figure 16.14, the average speed $\bar{v}$ is somewhat lower than the rms speed. The most probable speed $v_{mp}$ is the speed at which the distribution curve reaches a peak. Using Equation 16.20, we find that

$$v_{rms} = \sqrt{\overline{v^2}} = \sqrt{3k_B T/m} = 1.73\sqrt{k_B T/m} \qquad \text{[16.21]}$$

$$\bar{v} = \sqrt{8k_B T/\pi m} = 1.60\sqrt{k_B T/m} \qquad \text{[16.22]}$$

$$v_{mp} = \sqrt{2k_B T/m} = 1.41\sqrt{k_B T/m} \qquad \text{[16.23]}$$

From these equations we see that $v_{rms} > \bar{v} > v_{mp}$.

Figure 16.15 represents speed distribution curves for nitrogen molecules. The curves were obtained using Equation 16.20 to evaluate the distribution function at various speeds and at two temperatures. Note that the peak in the curve shifts to the right as $T$ increases, indicating that the average speed increases with increasing temperature, as expected. The asymmetric shape of the curves is due to the fact that the lowest speed possible is zero, while the upper classical limit of the speed is infinity.

The speed distribution curve for molecules in a liquid is similar to those shown in Figure 16.15. The phenomenon of evaporation of a liquid can be understood from this distribution in speeds using the fact that some molecules in the liquid are more energetic than others. Some of the faster moving molecules in the liquid penetrate the surface and leave the liquid even at temperatures well below the boiling point. The molecules that escape the liquid by evaporation are those that have sufficient energy to overcome the attractive forces of the molecules in the liquid phase. Consequently, the molecules left behind in the liquid phase have a lower average kinetic energy, causing the temperature of the liquid to decrease. Hence evaporation is a cooling process. For example, an alcohol-soaked cloth is often placed on a feverish head to cool and comfort the patient.

### Example 16.6    A System of Nine Particles

Nine particles have speeds of 5.00, 8.00, 12.0, 12.0, 12.0, 14.0, 14.0, 17.0, and 20.0 m/s. (a) Find the average speed.

**Reasoning**   We cannot use Equations 16.21, 16.22, and 16.23 to calculate average speeds because these equations are only valid for a large number of gas particles. On the other hand, because we have so few particles, we can calculate the averages directly.

**Solution**   The average speed is the sum of the speeds divided by the total number of particles:

$$\bar{v} = \frac{(5.00 + 8.00 + 12.0 + 12.0 + 12.0 + 14.0 + 14.0 + 17.0 + 20.0) \text{ m/s}}{9}$$

$$= \boxed{12.7 \text{ m/s}}$$

(b) What is the rms speed?

**Solution**   The average value of the square of the speed is

$$\overline{v^2} = \frac{(5.00^2 + 8.00^2 + 12.0^2 + 12.0^2 + 12.0^2 + 14.0^2 + 14.0^2 + 17.0^2 + 20.0^2) \text{ m}^2/\text{s}^2}{9}$$

$$= 178 \text{ m}^2/\text{s}^2$$

Hence, the rms speed is

$$v_{\text{rms}} = \sqrt{\overline{v^2}} = \sqrt{178 \text{ m}^2/\text{s}^2} = \boxed{13.3 \text{ m/s}}$$

(c) What is the most probable speed of the particles?

**Solution**   Three of the particles have a speed of 12.0 m/s, two have a speed of 14.0 m/s, and the remaining particles have different speeds. Hence, we see that the most probable speed $v_{\text{mp}}$ is $\boxed{12.0 \text{ m/s}}$.

---

### context connection
### 16.7 • THE ATMOSPHERIC LAPSE RATE

We have discussed the temperature of a gas with the assumption that all parts of the gas are at the same temperature. For small volumes of gas, this is a relatively good assumption. What about a *huge* volume of gas, however, such as the atmosphere? It is clear that the assumption of a uniform temperature throughout the gas is not valid in this case. When it is a hot summer day in Los Angeles, it is a cold winter day in Melbourne—different parts of the atmosphere are clearly at different temperatures.

We can address this, as we discussed in the opening section of this Context, by considering the global average of the air temperature at the surface of the Earth. But variations also occur in temperature at different *heights* in the atmosphere. It is this variation of temperature with height that we explore here.

Figure 16.16 shows graphical representations of average air temperature in January at various heights in four states.* These data are taken at locations on the surface of the Earth, but at varying elevations, such as at sea level and on moun-

---

* Data are taken from *Climates of the States,* National Oceanic and Atmospheric Administration, U. S. Department of Commerce, Water Information Center Inc., Port Washington, NY, 1974

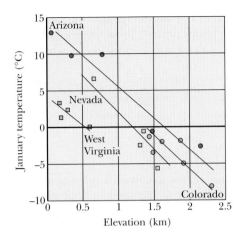

**Figure 16.16**

Variation of average temperature with elevation for four states. Note that the slopes of all four lines are approximately the same.

tains. For all four states, we see a clear indication that the temperature decreases as we move to higher elevations. Of course, one look at snow-capped mountains tells us that this is the case. We also see that the data tend to lie along straight lines, although some of the data are scattered. This suggests that the temperature decreases approximately linearly with height above the surface.

We can argue conceptually why the temperature decreases with height. Imagine a parcel of air moving upward along the slope of a mountain. As this parcel rises into higher elevations, the pressure on it from the rest of the air decreases. The pressure difference between the interior and the exterior of the parcel causes the parcel to expand. In doing so, the parcel is pushing the surrounding air outward, doing work on it. Because the system (the parcel of air) is doing work on the environment, the energy in the parcel decreases. The decreased energy is manifested as a decrease in temperature.

If this process is reversed, so that the parcel moves toward lower elevations, work is done on the parcel, which increases its energy so that it becomes warmer. This occurs during Santa Ana wind conditions in the Los Angeles basin, in which air is pushed from the mountains down into the low elevations of the basin, resulting in hot, dry winds. Similar conditions go by other names in other regions, such as the *chinook* from the Rocky Mountains, and the *foehn* from the Swiss Alps.

Notice in Figure 16.16 that the slopes of all four lines are relatively close together. This suggests that the decrease in temperature with height—called the **atmospheric lapse rate**—is similar at various locations across the surface of the Earth, so that we might define an average lapse rate for the entire surface.

This is indeed the case, and we find that the average global lapse rate is $-6.7°C/km$. If you determine the slopes of the lines in Figure 16.16, you will find that they are close to this value.

The linear decrease with temperature only occurs in the lower part of the atmosphere called the **troposphere.** This is the part of the atmosphere in which weather occurs and airplanes fly. Above the troposphere is the **stratosphere,** with an imaginary boundary called the **tropopause** separating the two layers. In the stratosphere, temperature tends to be relatively constant.

The decrease in temperature with height in the troposphere is one component to a structural model of the atmosphere that will allow us to predict the surface temperature of the Earth. If we can find the temperature of the stratosphere and the height of the tropopause, we can extrapolate to the surface, using the lapse rate to find the temperature at the surface. The lapse rate and the height of the tropopause can be measured. To find the temperature of the stratosphere, we need to know more about energy exchanges in the Earth's atmosphere, which we will investigate in the next chapter.

## SUMMARY

The **zeroth law of thermodynamics** states that if two objects, A and B, are separately in thermal equilibrium with a third object, then A and B are in thermal equilibrium with each other.

The relationship between $T_C$, the **Celsius temperature,** and $T$, the **Kelvin (absolute) temperature,** is

$$T_C = T - 273.15 \qquad \textbf{[16.1]}$$

The relationship between the **Fahrenheit** and **Celsius** temperatures is

$$T_F = \tfrac{9}{5}T_C + 32°\text{F} \qquad \textbf{[16.2]}$$

When the temperature of a substance is raised, it generally expands. If an object has an initial length of $L_i$ at some temperature and undergoes a change in temperature $\Delta T$, its length changes by the amount $\Delta L$, which is proportional to the object's initial length and the temperature change:

$$\Delta L = \alpha L_i \Delta T \qquad \textbf{[16.4]}$$

The parameter $\alpha$ is called the **average coefficient of linear expansion.**

The change in volume of most substances is proportional to the initial volume $V_i$ and the temperature change $\Delta T$:

$$\Delta V = \beta V_i \Delta T \qquad \textbf{[16.6]}$$

where $\beta$ is the **average coefficient of volume expansion** and is equal to $3\alpha$.

The change in area of a substance is given by

$$\Delta A = \gamma A_i \Delta T \qquad \textbf{[16.7]}$$

where $\gamma$ is the **average coefficient of area expansion** and is equal to $2\alpha$.

The **ideal gas model** refers to a collection of gas molecules that move randomly and are of negligible size. An ideal gas obeys the equation

$$PV = nRT \qquad \textbf{[16.9]}$$

where $P$ is the pressure of the gas, $V$ is its volume, $n$ is the number of moles of gas, $R$ is the universal gas constant ($8.315 \text{ J/mol·K}$), and $T$ is the absolute temperature in kelvins. A real gas at very low pressures behaves approximately as an ideal gas.

The **pressure** of $N$ molecules of an ideal gas contained in a volume $V$ is given by

$$P = \tfrac{2}{3}\left(\frac{N}{V}\right)\left(\tfrac{1}{2}m\overline{v^2}\right) \qquad \textbf{[16.13]}$$

where $\tfrac{1}{2}m\overline{v^2}$ is the **average kinetic energy per molecule.**

The average kinetic energy of the molecules of a gas is directly proportional to the absolute temperature of the gas:

$$\tfrac{1}{2}m\overline{v^2} = \tfrac{3}{2}k_B T \qquad \textbf{[16.15]}$$

where $k_B$ is **Boltzmann's constant** ($1.38 \times 10^{-23} \text{ J/K}$).

For a monatomic gas, the internal energy of the gas is the total translational kinetic energy

$$E_{\text{int}} = \tfrac{3}{2}nRT \qquad \text{(monatomic gas)} \qquad \textbf{[16.18]}$$

The **root-mean-square** (rms) speed of the molecules of a gas is

$$v_{\text{rms}} = \sqrt{\overline{v^2}} = \sqrt{\frac{3k_B T}{m}} = \sqrt{\frac{3RT}{M}} \qquad \textbf{[16.19]}$$

The **Maxwell–Boltzmann distribution function** describes the distribution of speeds of $N$ gas molecules:

$$N_v = 4\pi N\left(\frac{m}{2\pi k_B T}\right)^{3/2} v^2 e^{-mv^2/2k_B T} \qquad \textbf{[16.20]}$$

where $m$ is the mass of a gas molecule, $k_B$ is Boltzmann's constant, and $T$ is the absolute temperature.

## QUESTIONS

1. A piece of copper is dropped into a beaker of water. If the water's temperature rises, what happens to the temperature of the copper? Under what conditions are the water and copper in thermal equilibrium?

2. In the movie *Apollo 13* (Universal, 1995), astronaut Jim Lovell says, in describing his upcoming trip to the Moon, "I'll be walking in a place where there's 400° difference between sunlight and shadow." What exactly is it

that is hot in sunlight and cold in shadow? Suppose an astronaut standing on the Moon holds a thermometer in his or her gloved hand. Is it reading the temperature of the vacuum at the Moon's surface? Does it read a temperature? If so, what object or substance has that temperature?

3. When the metal ring and metal sphere in Figure Q16.3 are both at room temperature, the sphere does not fit through the ring. After the ring is heated, the sphere can be passed through the ring. Explain.

**Figure Q16.3** *(Courtesy of Central Scientific Company)*

4. Explain why a column of mercury in a thermometer first descends slightly and then rises when placed in hot water.

5. Markings to indicate length are placed on a steel tape in a room that has a temperature of 22°C. Are measurements made with the tape on a day when the temperature is 27°C too long, too short, or accurate? Defend your answer.

6. If a helium-filled balloon initially at room temperature is placed in a freezer, will its volume increase, decrease, or remain the same?

7. What does the ideal gas law predict about the volume of a gas at absolute zero? Why is this prediction incorrect?

8. What happens to a helium-filled balloon released into the air? Does it expand or contract? Does it stop rising at some height?

9. An inflated rubber balloon filled with air is immersed in a flask of liquid nitrogen that is at 77 K. Describe what happens to the balloon, assuming that it remains flexible while being cooled.

10. Two cylinders at the same temperature contain the same kind and quantity of gas. If the volume of cylinder A is three times greater than the volume of cylinder B, what can you say about the relative pressures in the cylinders?

11. After food is cooked in a pressure cooker, why is it very important to cool the container with cold water before attempting to remove the lid?

12. The shore at a particular point of the ocean is very rocky. At one location, the rock formation is such that a cave is located with an opening that is underwater, as shown in Figure Q16.12a. (a) Inside the cave is a pocket of trapped air. As the level of the ocean rises and falls due to the tides, will the level of the water in the cave rise and fall? If so, will it have the same amplitude as that of the ocean? (b) Suppose the cave is deeper in the water, so that it is completely submerged and filled with water at high tide, as shown in Figure Q16.12b. At low tide, will the level of the water in the cave be the same as that of the ocean?

**Figure Q16.12**

13. An alcohol rub is sometimes used to help lower the temperature of a sick patient. Why does this help?

14. Ideal gas is contained in a vessel at 300 K. If the temperature is increased to 900 K, by what factor does each one of the following change? (a) The average kinetic energy of the molecules. (b) The rms molecular speed. (c) The average momentum change in one collision of a molecule with a wall. (d) The rate of collisions of molecules with the walls. (d) The pressure of the gas.

15. In *Colonization: Second Contact* (Harry Turtledove, Ballantine Publishing Group, 1999), the Earth has been partially settled by aliens (called *Lizards* by the humans) from another planet. Laboratory study by humans of Lizard science requires "shifting back and forth between the metric system and the one the Lizards used, which was also based on powers of ten but used different basic quantities for everything but temperature." Why would temperature be an exception?

# PROBLEMS

$1, 2, 3$ = straightforward, intermediate, challenging    □ = full solution available in the *Student Solutions Manual and Study Guide*

**web** = solution posted at **http://www.harcourtcollege.com/physics/**    💻 = computer useful in solving problem

🖥 = Interactive Physics    ■ = paired numerical/symbolic problems    📰 = life science application

## Section 16.2   Thermometers and Temperature Scales

**1.** A constant-volume gas thermometer is calibrated in dry
**web** ice (i.e., carbon dioxide in the solid state, which has a temperature of $-80.0°C$) and in boiling ethyl alcohol ($78.0°C$). The two pressures are 0.900 atm and 1.635 atm. (a) What Celsius value of absolute zero does the calibration yield? What is the pressure at (b) the freezing point of water and (c) the boiling point of water?

**2.** In a constant-volume gas thermometer, the pressure at $20.0°C$ is 0.980 atm. (a) What is the pressure at $45.0°C$? (b) What is the temperature if the pressure is 0.500 atm?

**3.** Liquid nitrogen has a boiling point of $-195.81°C$ at atmospheric pressure. Express this temperature (a) in degrees Fahrenheit and (b) in kelvins.

**4.** Convert the following to equivalent temperatures on the Celsius and Kelvin scales: (a) the normal human body temperature, $98.6°F$; (b) the air temperature on a cold day, $-5.00°F$.

**5.** The temperature difference between the inside and the outside of an automobile engine is $450°C$. Express this temperature difference on (a) the Fahrenheit scale and (b) the Kelvin scale.

## Section 16.3   Thermal Expansion of Solids and Liquids

*Note:* When solving problems in this section, use Table 16.1.

**6.** The New River Gorge bridge in West Virginia is a steel arch bridge 518 m in length. How much does the total length of the roadway decking change between temperature extremes of $-20.0°C$ and $35.0°C$? The result indicates the size of the expansion joints that must be built into the structure.

**7.** An aluminum tube is 3.000 0 m long at $20.0°C$. What is its length at (a) $100.0°C$ and (b) $0.0°C$?

**8.** A pair of eyeglass frames is made of epoxy plastic. At room temperature ($20.0°C$), the frames have circular lens holes 2.20 cm in radius. To what temperature must the frames be heated if lenses 2.21 cm in radius are to be inserted in them? The average coefficient of linear expansion for epoxy is $1.30 \times 10^{-4} \ (°C)^{-1}$.

**9.** The active element of a certain laser is made of a glass rod
**web** 30.0 cm long by 1.50 cm in diameter. If the temperature of the rod increases by $65.0°C$, what is the increase in (a) its

length, (b) its diameter, and (c) its volume? Assume that the average coefficient of linear expansion of the glass is $9.00 \times 10^{-6}(°C)^{-1}$.

**10.** A square hole 8.00 cm along each side is cut in a sheet of copper. (a) Calculate the change in the area of this hole if the temperature of the sheet is increased by 50.0 K. (b) Does this change represent an increase or a decrease in the area enclosed by the hole?

**11.** The average coefficient of volume expansion for carbon tetrachloride is $5.81 \times 10^{-4} \ (°C)^{-1}$. If a 50.0-gal steel container is filled completely with carbon tetrachloride when the temperature is $10.0°C$, how much will spill over when the temperature rises to $30.0°C$?

**12.** At $20.0°C$, an aluminum ring has an inner diameter of 5.000 0 cm and a brass rod has a diameter of 5.050 0 cm. (a) If only the ring is heated, what temperature must it reach so that it will just slip over the rod? (b) If both are heated together, what temperature must they both reach so that the ring just slips over the rod? Would this latter process work?

**13.** A hollow aluminum cylinder 20.0 cm deep has an internal capacity of 2.000 L at $20.0°C$. It is completely filled with turpentine, and then slowly warmed to $80.0°C$. (a) How much turpentine overflows? (b) If the cylinder is then cooled back to $20.0°C$, how far below the cylinder's rim does the turpentine's surface recede?

**14.** A certain telescope forms an image of part of a cluster of stars on a square silicon charge-coupled detector (CCD) chip 2.00 cm on each side. A star field is focused on the CCD when it is first turned on and its temperature is $20.0°C$. The star field contains 5 342 stars scattered uniformly. To make the detector more sensitive, it is cooled to $-100°C$. How many star images then fit onto the chip? The average coefficient of linear expansion of silicon is $4.68 \times 10^{-6} \ (°C)^{-1}$.

**15. Review Problem.** Inside the wall of a house, an L-shaped section of hot water pipe consists of a straight horizontal piece 28.0 cm long, an elbow, and a straight vertical piece 134 cm long (Figure P16.15). A stud and a second-story floorboard hold stationary the ends of this section of copper pipe. Find the magnitude and direction of the displacement of the pipe elbow when the water flow is turned on, raising the temperature of the pipe from $18.0°C$ to $46.5°C$.

**Figure P16.15**

## Section 16.4 Macroscopic Description of an Ideal Gas

16. An automobile tire is inflated with air originally at 10.0°C and normal atmospheric pressure. During the process, the air is compressed to 28.0% of its original volume and the temperature is increased to 40.0°C. (a) What is the tire pressure? (b) After the car is driven at high speed, the tire air temperature rises to 85.0°C and the interior volume of the tire increases by 2.00%. What is the new tire pressure (absolute) in pascals?

17. A school auditorium has dimensions 10.0 m × 20.0 m × 30.0 m. How many molecules of air fill the auditorium at 20.0°C and a pressure of 101 kPa?

18. A tank having a volume of 0.100 m³ contains helium gas at 150 atm. How many balloons can the tank inflate if each filled balloon is a sphere 0.300 m in diameter at an absolute pressure of 1.20 atm?

19. The mass of a hot-air balloon and its cargo (not including the air inside) is 200 kg. The air outside is at 10.0°C and 101 kPa. The volume of the balloon is 400 m³. To what temperature must the air in the balloon be heated before the balloon will lift off? (Air density at 10.0°C is 1.25 kg/m³.)

20. A cube 10.0 cm on each edge contains air (with equivalent molar mass 28.9 g/mol) at atmospheric pressure and temperature 300 K. Find (a) the mass of the gas, (b) its weight, and (c) the force it exerts on each face of the

cube. (d) Comment on the physical reason why such a small sample can exert such a great force.

21. At 25.0 m below the surface of the sea (density = 1 025 kg/m³), where the temperature is 5.00°C, a diver exhales an air bubble having a volume of 1.00 cm³. If the surface temperature of the sea is 20.0°C, what is the volume of the bubble just before it breaks the surface?

22. Estimate the mass of the air in your bedroom. State the quantities you take as data and the value you measure or estimate for each.

23. The pressure gauge on a tank registers the gauge pressure, which is the difference between the interior and exterior pressure. When the tank is filled with oxygen ($O_2$), it contains 12.0 kg of the gas at a gauge pressure of 40.0 atm. Determine the mass of oxygen that has been withdrawn from the tank when the pressure reading is 25.0 atm. Assume that the temperature of the tank remains constant.

24. A popular brand of cola contains 6.50 g of carbon dioxide dissolved in 1.00 L of soft drink. If the evaporating carbon dioxide is trapped in a cylinder at 1.00 atm and 20.0°C, what volume does the gas occupy?

25. In state-of-the-art vacuum systems, pressures as low as $10^{-9}$ Pa are being attained. Calculate the number of molecules in a 1.00-m³ vessel at a pressure of $1.00 \times 10^{-9}$ Pa if the temperature is 27.0°C.

26. A room of volume $V$ contains air having equivalent molar mass $M$ (in grams per mole). If the temperature of the room is raised from $T_1$ to $T_2$, what mass of air will leave the room? Assume that the air pressure in the room is maintained at $P_0$.

## Section 16.5 The Kinetic Theory of Gases

27. In a time $t$, $N$ hailstones strike a glass window of area $A$ at an angle $\theta$ to the window surface. Each hailstone has a mass $m$ and a speed $v$. If the collisions are elastic, what are the average force and pressure on the window?

28. In a period of 1.00 s, $5.00 \times 10^{23}$ nitrogen molecules strike a wall with an area of 8.00 cm². If the molecules move with a speed of 300 m/s and strike the wall head-on in perfectly elastic collisions, what is the pressure exerted on the wall? (The mass of one $N_2$ molecule is $4.68 \times 10^{-26}$ kg.)

29. (a) How many atoms of helium gas fill a balloon of diameter 30.0 cm at 20.0°C and 1.00 atm? (b) What is the average kinetic energy of the helium atoms? (c) What is the rms speed of the helium atoms?

30. Use the definition of Avogadro's number to find the mass of a helium atom.

31. A cylinder contains a mixture of helium and argon gas in equilibrium at 150°C. (a) What is the average kinetic en-

ergy for each type of gas molecule? (b) What is the rms speed of each type of molecule?

32. A 5.00-L vessel contains nitrogen gas at 27.0°C and 3.00 atm. (a) Find the total translational kinetic energy of the gas molecules and (b) the average kinetic energy per molecule.

## Section 16.6 Distribution of Molecular Speeds

33. Fifteen identical particles have various speeds: one has a speed of 2.00 m/s; two have speeds of 3.00 m/s; three have speeds of 5.00 m/s; four have speeds of 7.00 m/s; three have speeds of 9.00 m/s; and two have speeds of 12.0 m/s. Find (a) the average speed, (b) the rms speed, and (c) the most probable speed of these particles.

34. Gaseous helium is in thermal equilibrium with liquid helium at 4.20 K. Even though it is on the point of condensation, model the gas as ideal and determine the most probable speed of a helium atom (mass = $6.64 \times 10^{-27}$ kg) in it.

35. From the Maxwell–Boltzmann speed distribution, show that the most probable speed of a gas molecule is given by Equation 16.23. Note that the most probable speed corresponds to the point at which the slope of the speed distribution curve $dN_v/dv$ is zero.

36. **Review Problem.** At what temperature would the average speed of helium atoms equal (a) the escape speed from Earth, $1.12 \times 10^4$ m/s, and (b) the escape speed from the Moon, $2.37 \times 10^3$ m/s ? (See Chapter 11 for a discussion of escape speed, and note that the mass of a helium atom is $6.64 \times 10^{-27}$ kg.)

## Section 16.7 Context Connection — The Atmospheric Lapse Rate

37. The summit of Mount Whitney, in California, is 3 660 m higher than a point in the foothills. Assume that the atmospheric lapse rate in the Mount Whitney area is the same as the global average of $-6.7°$C/km. What is the temperature at the summit of Mount Whitney when eager hikers depart from the foothill location at a temperature of 30°C?

38. The theoretical lapse rate for dry air (no water vapor) in an atmosphere is given by

$$\frac{dT}{dy} = -\frac{\gamma - 1}{\gamma}\frac{gM}{R}$$

where $g$ is the acceleration due to gravity, $M$ is the molar mass of the ideal gas in the atmosphere, $R$ is the gas constant, and $\gamma$ is the ratio of molar specific heats, which we will study in Chapter 17. (a) Calculate the theoretical lapse rate on Earth, given that $\gamma = 1.40$, and the effective

molar mass of air is 28.9 g/mol. (b) Why is this value different from the value of $-6.7°$C/km given in the text? (c) The atmosphere of Mars is mostly carbon dioxide, with a molar mass of 44.0 g/mol and a ratio of molar specific heats $\gamma = 1.30$. The mass of Mars is $6.42 \times 10^{23}$ kg, and the radius is $3.37 \times 10^6$ m. If the atmosphere is dry, what is the lapse rate for the Martian troposphere? (d) A typical surface atmospheric temperature on Mars is $-40.0°$C. Using the lapse rate calculated in part (c), find the height in the Martian troposphere at which the temperature is $-60.0°$C. (e) Data from the *Mariner* flights in 1969 indicated a lapse rate in the Martian troposphere of about $-1.5°$C/km. The *Viking* missions in 1976 gave measured lapse rates of about $-2°$C/km. These values deviate from the ideal value calculated in part (c) because of *dust* in the Martian atmosphere. Why would dust alter the lapse rate, and which missions occurred in dustier conditions— *Mariner* or *Viking*?

## Additional Problems

39. A student measures the length of a brass rod with a steel tape at 20.0°C. The reading is 95.00 cm. What will the tape indicate for the length of the rod when the rod and the tape are at (a) $-15.0°$C, (b) 55.0°C?

40. The density of gasoline is 730 kg/m³ at 0°C. Its average coefficient of volume expansion is $9.60 \times 10^{-4}$ (°C)$^{-1}$. If 1.00 gal of gasoline occupies 0.003 80 m³, how many extra kilograms of gasoline would you obtain if you bought 10.0 gal of gasoline at 0°C rather than at 20.0°C from a pump that does not compensate for temperature change?

41. A mercury thermometer is constructed as shown in Figure P16.41. The capillary tube has a diameter of 0.004 00 cm, and the bulb has a diameter of 0.250 cm. Ignoring the expansion of the glass, find the change in height of the mercury column that occurs with a temperature change of 30.0°C.

**Figure P16.41** Problems 41 and 42.

42. A liquid with a coefficient of volume expansion $\beta$ just fills a spherical shell of volume $V_i$ at a temperature of $T_i$ (see Fig. P16.41). The shell is made of a material that has an average coefficient of linear expansion $\alpha$. The liquid is free to expand into an open capillary of area $A$ projecting from the top of the sphere. (a) If the temperature increases by $\Delta T$, show that the liquid rises in the capillary by the amount $\Delta h$ given by the equation $\Delta h = (V_i/A)(\beta - 3\alpha)\Delta T$. (b) For a typical system, such as a mercury thermometer, why is it a good approximation to ignore the expansion of the shell?

43. A liquid has a density $\rho$. (a) Show that the fractional change in density for a change in temperature $\Delta T$ is $\Delta\rho/\rho = -\beta\,\Delta T$. What does the negative sign signify? (b) Fresh water has a maximum density of $1.000\ 0\ \text{g/cm}^3$ at $4.0°C$. At $10.0°C$, its density is $0.999\ 7\ \text{g/cm}^3$. What is $\beta$ for water over this temperature interval?

44. Long-term space missions require reclamation of the oxygen in the carbon dioxide exhaled by the crew. In one method of reclamation, 1.00 mol of carbon dioxide produces 1.00 mol of oxygen and 1.00 mol of methane as a byproduct. The methane is stored in a tank under pressure and is available to control the attitude of the spacecraft by controlled venting. A single astronaut exhales 1.09 kg of carbon dioxide each day. If the methane generated in the respiration recycling of three astronauts during one week of flight is stored in an originally empty 150-L tank at $-45.0°C$, what is the final pressure in the tank?

45. A vertical cylinder of cross-sectional area $A$ is fitted with a **web** tight-fitting, frictionless piston of mass $m$ (Fig. P16.45). (a) If $n$ moles of an ideal gas is in the cylinder at a temperature of $T$, what is the height $h$ at which the piston is in equilibrium under its own weight? (b) What is the value for $h$ if $n = 0.200$ mol, $T = 400$ K, $A = 0.008\ 00\ \text{m}^2$, and $m = 20.0$ kg?

46. A bimetallic strip is made of two ribbons of dissimilar metals bonded together. (a) First assume the strip is originally straight. As the strip is heated, the metal with the greater average coefficient of expansion expands more, forcing the strip into an arc, with the outer radius having a greater circumference (Fig. P16.46a). Derive an expression for the angle of bending $\theta$ as a function of the initial length of the strips, their average coefficients of linear expansion, the change in temperature, and the separation of the centers of the strips ($\Delta r = r_2 - r_1$). (b) Show that the angle of bending decreases to zero when $\Delta T$ decreases to zero and also when the two average coefficients of expansion become equal. (c) What happens if the strip is cooled? (d) Figure P16.46b shows a compact spiral bimetallic strip in a home thermostat. The equation from part (a) applies to it as well if $\theta$ is interpreted as the angle of additional bending caused by a change in temperature. The inner end of the spiral strip is fixed and the outer end is free to move. Assume the metals are bronze and invar, the thickness of the strip is $2\,\Delta r = 0.500$ mm, and the overall length of the spiral strip is 20.0 cm. Find the angle through which the free end of the strip turns for each one-degree change in temperature. The free end of the strip supports a cap-

(a)

(b)

**Figure P16.46** *(b, Charles D. Winters)*

**Figure P16.45**

sule partly filled with mercury, visible above the strip in Figure P16.46b. When the capsule tilts, the mercury shifts from one end to the other, to make or break an electrical contact, switching the furnace on or off.

47. **Review Problem.** A clock with a brass pendulum has a period of 1.000 s at 20.0°C. If the temperature increases to 30.0°C, (a) by how much does the period change, and (b) how much time does the clock gain or lose in one week?

48. The rectangular plate shown in Figure P16.48 has an area $A_i$ equal to $\ell w$. If the temperature increases by $\Delta T$, show that the increase in area is $\Delta A = 2\alpha A_i \Delta T$, where $\alpha$ is the average coefficient of linear expansion. What approximation does this expression assume? (*Hint:* Note that each dimension increases according to the equation $\Delta L = \alpha L_i \Delta T$.)

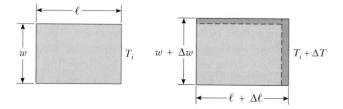

**Figure P16.48**

49. **Review Problem.** Consider an object with any one of the shapes displayed in Table 10.2. What is the percentage increase in the moment of inertia of the object when it is heated from 0°C to 100°C if it is composed of (a) copper or (b) aluminum? (See Table 16.1. Assume that the average linear expansion coefficients do not vary between 0°C and 100°C.)

50. (a) Show that the density of an ideal gas occupying a volume $V$ is given by $\rho = PM/RT$, where $M$ is the molar mass. (b) Determine the density of oxygen gas at atmospheric pressure and 20.0°C.

51. ~Two concrete spans of a 250-m-long bridge are placed end to end so that no room is allowed for expansion (Fig. P16.51a). If a temperature increase of 20.0°C occurs, what is the height $y$ to which the spans rise when they buckle (Fig. P16.51b)?

(a)                    (b)

**Figure P16.51** Problems 51 and 52.

52. Two concrete spans of a bridge of length $L$ are placed end to end so that no room is allowed for expansion (Fig. P16.51a). If a temperature increase of $\Delta T$ occurs, what is the height $y$ to which the spans rise when they buckle (Fig. P16.51b)?

53. (a) Use the equation of state for an ideal gas and the definition of the coefficient of volume expansion, in the form $\beta = (1/V) \, dV/dT$, to show that the coefficient of volume expansion for an ideal gas at constant pressure is given by $\beta = 1/T$, where $T$ is the absolute temperature. (b) What value does this expression predict for $\beta$ at 0°C? Compare this result with the experimental values for helium and air in Table 16.1. Note that it is much larger than the coefficients of volume expansion for most liquids and solids.

54. A copper rod and steel rod are heated. At 0°C the copper rod has length $L_C$, and the steel rod has length $L_S$. When the rods are being heated or cooled, the difference between their lengths stays constant at 5.00 cm. Determine the values of $L_C$ and $L_S$.

55. A cylinder that has a 40.0-cm radius and is 50.0 cm deep is filled with air at 20.0°C and 1.00 atm (Fig. P16.55a). A 20.0-kg piston is now lowered into the cylinder, compressing the air trapped inside (Fig. P16.55b). Finally, a 75.0-kg man stands on the piston, further compressing the air, which remains at 20.0°C (Fig. P16.55c). (a) How far down ($\Delta h$) does the piston move when the man steps onto it? (b) To what temperature should the gas be heated to raise the piston and man back to $h_i$?

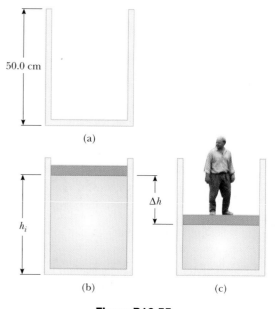

**Figure P16.55**

56. The relationship $L_f = L_i(1 + \alpha \Delta T)$ is an approximation that works when the average coefficient of expansion is small. If $\alpha$ is large, one must integrate the relationship $dL/dT = \alpha L$ to determine the final length. (a) Assuming that the coefficient of linear expansion is constant as $L$ varies, determine a general expression for the final length. (b) Given a rod of length 1.00 m and a temperature change of 100.0°C, determine the error caused by the approximation when $\alpha = 2.00 \times 10^{-5}$ (°C)$^{-1}$ (a typical value for a metal) and when $\alpha = 0.0200$ (°C)$^{-1}$ (an unrealistically large value for comparison).

57. In a chemical processing plant, a reaction chamber of fixed volume $V_0$ is connected to a reservoir chamber of fixed volume $4V_0$ by a passage containing a thermally insulating porous plug. The plug permits the chambers to be at different temperatures. The plug allows gas to pass from either chamber to the other, ensuring that the pressure is the same in both. At one point in the processing, both chambers contain gas at a pressure of 1.00 atm and a temperature of 27.0°C. Intake and exhaust valves to the pair of chambers are closed. The reservoir is maintained at 27.0°C while the reaction chamber is heated to 400°C. What is the pressure in both chambers after this is done?

58. **Review Problem.** Oxygen at pressures much greater than 1 atm is toxic to lung cells. By weight, what ratio of helium gas (He) to oxygen gas (O$_2$) must be used by a diver who is to descend to an ocean depth of 50.0 m?

59. A 1.00-km steel railroad rail is fastened securely at both ends when the temperature is 20.0°C. As the temperature increases, the rail begins to buckle. If its shape is an arc of a vertical circle, find the height $h$ of the center of the rail when the temperature is 25.0°C. You will need to solve a transcendental equation.

60. A vessel contains $1.00 \times 10^4$ oxygen molecules at 500 K. (a) Make an accurate graph of the Maxwell speed distribution function versus speed with points at speed intervals of 100 m/s. (b) Determine the most probable speed from this graph. (c) Calculate the average and rms speeds for the molecules, and label these points on your graph. (d) From the graph, estimate the fraction of molecules with speeds in the range 300 m/s to 600 m/s.

61. For a Maxwellian gas, use a computer or programmable calculator to find the numerical value of the ratio $N_v(v)/N_v(v_{mp})$ for the following values of $v$: $v = (v_{mp}/50)$, $(v_{mp}/10)$, $(v_{mp}/2)$, $v_{mp}$, $2v_{mp}$, $10v_{mp}$, $50v_{mp}$. Give your results to three significant figures.

## ANSWERS TO QUICK QUIZZES

16.1 The correct response would be that it does not matter! Temperatures on the Kelvin and Celsius scales differ by 273 degrees. This difference is insignificant for temperatures on the order of $10^7$. If we imagine that the temperature given is in Kelvins, and ignore any problems with significant figures, the Celsius temperature is $1.4999727 \times 10^7$ °C.

16.2 Mercury must have the larger coefficient of expansion. As the temperature of a thermometer rises, both the mercury and the glass expand. If they both have the same coefficient of linear expansion, the mercury and the cavity in the glass will both expand by the same amount, and there will be no apparent movement of the end of the mercury column relative to the calibration scale on the glass. If the glass expands more than the mercury, the reading would go down as the temperature goes up! Now that we have argued this conceptually, we can look in Table 16.1, and find that the coefficient for mercury is about 20 times as large as that for glass, so that the expansion of the glass can effectively be ignored.

16.3 A cavity in a material expands in exactly the same way as if the cavity were filled with material. Thus, both spheres will expand by the same amount.

16.4 The chip bags contain a sealed sample of air. When the bags are taken up the mountain, the external atmospheric pressure on the bags is reduced. As a result, the difference between the pressure of the air inside the bags and the reduced pressure outside results in a net force pushing the plastic of the bag outward.

16.5 On a cold day, the trapped air in the bubbles is reduced in pressure, according to the ideal gas law. Thus, the volume of the bubbles may be smaller than on a hot day, and the material would not be as effective in cushioning the package contents.

16.6 The existence of an atmosphere on a planet is due to the planet's gravity. On a small planet, the gravitational force is very small, and the escape speed (Section 11.4) is correspondingly small. If a small planet starts its existence with an atmosphere, the molecules of the gas will have a distribution of speeds, according to kinetic theory. Some of these molecules will have speeds higher than the escape speed of the planet and will leave the atmosphere. As the remaining atmosphere is warmed by radiation from the Sun, more molecules will attain speeds high enough to escape. As a result, the atmosphere bleeds off into space.

# chapter
# 17

In this photograph of Moraine Lake, in Banff National Park, Alberta, we can see evidence of water in all three phases. In the lake is liquid water, and solid water in the form of snow appears on the mountains. The clouds in the sky consist of liquid water droplets that have condensed from the gaseous water vapor in the air. Changes of a substance from one phase to another are a result of energy transfer.

*(Toyohiro Yamada/FPG)*

# Energy in Thermal Processes: The First Law of Thermodynamics

In Chapters 6 and 7, we hinted at the relationship between energy in mechanics and energy in thermodynamics. We discussed the transformation of mechanical energy to internal energy in cases in which a nonconservative force such as friction is acting. In Chapter 16, we discussed additional concepts of the relationship between internal energy and temperature. In this chapter, we extend these discussions into a complete treatment of energy in thermal processes.

Until about 1850, the fields of thermodynamics and mechanics were considered to be two distinct branches of science, and the law of conservation of energy seemed to describe only certain kinds of mechanical systems. Mid 19th-century experiments performed by the English physicist James Joule (1818–1889) and others showed that energy may enter or leave a system by heat and by work. Today, as we discussed in Chapter 6, internal energy is treated as a form of energy that can be transformed into mechanical energy and vice versa. Once the concept of energy was broadened to include internal energy, the law of conservation of energy emerged as a universal law of nature.

This chapter focuses on developing the concept of heat, extending our concept of work to thermal processes, introducing the first law of thermodynamics, and investigating some important applications.

**James Prescott Joule**
(1818–1889)

Joule received some formal education in mathematics, philosophy, and chemistry but was in large part self-educated. His research led to the establishment of the principle of conservation of energy. His study of the quantitative relationship among electrical, mechanical, and chemical effects of heat culminated in his announcement in 1843 of the amount of work required to produce a unit of internal energy, called the mechanical equivalent of heat. *(By kind permission of the President and Council of the Royal Society)*

## 17.1 • HEAT AND INTERNAL ENERGY

A major distinction must be made between internal energy and heat because these terms tend to be used interchangeably in everyday communication. You should read the following descriptions carefully and try to use these terms correctly, because they are not interchangeable—they have very different meanings.

We introduced internal energy in Chapter 6, and we formally define it here:

**Internal energy** $E_{\text{int}}$ is the energy associated with the microscopic components of a system—atoms and molecules—when viewed from a reference frame at rest with respect to the system. It includes kinetic and potential energy associated with the random translational, rotational, and vibrational motion of the atoms or molecules that make up the system as well as intermolecular potential energy.

In the preceding chapter we showed that the internal energy of a monatomic ideal gas is associated with the translational motion of its atoms. In this special case, the internal energy is simply the total translational kinetic energy of the atoms; the higher the temperature of the gas, the greater the kinetic energy of the atoms and the greater the internal energy of the gas. For more complicated diatomic and polyatomic gases, internal energy includes other forms of molecular energy, such as rotational kinetic energy and the kinetic and potential energy associated with molecular vibrations. We shall study phase changes in this chapter and find that internal energy is also associated with the intermolecular potential energy ("bond energy") between molecules in a liquid or solid.

Heat was introduced in Chapter 6 as one possible method of energy transfer, and we provide a formal definition here:

**Heat** is a mechanism by which energy is transferred between a system and its environment because of a temperature difference between them. It is also the amount of energy $Q$ transferred by this mechanism.

Figure 17.1 shows a pan of water in contact with a gas flame. Energy enters the water by heat from the hot gases in the flame and the internal energy of the water increases as a result. It is *incorrect* to say that the water has more heat as time goes by.

As further clarification of the use of the word *heat*, consider the distinction between work and energy. The work done on (or by) a system is a measure of the amount of energy transferred between the system and its surroundings, whereas the mechanical energy of the system (kinetic or potential) is a consequence of its motion and coordinates. Thus, when a person does work on a system, energy is

 See *Core Concepts in Physics*, Screen 10.3

**Figure 17.1**

A pan of boiling water is warmed by a gas flame. Energy enters the water through the bottom of the pan by heat. *(Charles D. Winters)*

**PITFALL PREVENTION 17.1**

**Internal energy, thermal energy, and bond energy**

In reading other physics books, you may see terms such as *thermal energy* and *bond energy*. Thermal energy can be interpreted as that part of the internal energy associated with random motion of molecules and, therefore, is related to temperature. Bond energy is the intermolecular potential energy. Thus,

Internal energy

= thermal energy + bond energy

We present this breakdown here to clarify our definition relative to other texts, but we will not use these terms because they serve no other purpose and only introduce confusion. In particular, in some other texts, much confusion is caused by using internal energy and thermal energy interchangeably.

• *Mechanical equivalent of heat*

transferred from the person to the system. It makes no sense to talk about the work *in* a system—one refers only to the work done *on* or *by* a system when some process has occurred in which energy has been transferred to or from the system. Likewise, it makes no sense to use the term *heat* unless energy has been transferred as a result of a temperature difference.

## Units of Heat

Early in the development of thermodynamics, before scientists recognized the connection between thermodynamics and mechanics, heat was defined in terms of the temperature changes it produced in an object, and a separate unit of energy, the calorie, was used for heat. The **calorie** (cal) was defined as **the heat necessary to raise the temperature of 1 g of water from 14.5° to 15.5°C.*** (The "Calorie," with a capital C, used in describing the energy content of foods, is actually a kilocalorie.) Likewise, the unit of heat in the British system, the **British thermal unit** (Btu), was defined as **the heat required to raise the temperature of 1 lb of water from 63°F to 64°F.**

In 1948, scientists agreed that because heat (like work) is a measure of the transfer of energy, its SI unit should be the joule. The calorie is now defined to be exactly 4.186 J:

$$1 \text{ cal} \equiv 4.186 \text{ J} \qquad \text{[17.1]}$$

Note that this definition makes no reference to the heating of water. The calorie is a general energy unit. We could have used it in Chapter 6 for the kinetic energy of an object, for example. It is introduced here for historical reasons, but we shall make little use of it as an energy unit. The definition in Equation 17.1 is known as the **mechanical equivalent of heat.**

---

### Example 17.1 Losing Weight the Hard Way

A student eats a dinner containing 2 000 Calories of energy. He wishes to do an equivalent amount of work in the gymnasium by lifting a 50.0-kg object. How many times must he raise the object to expend this much energy? Assume that he raises it a distance of 2.00 m each time.

**Solution** The student desires to transfer 2 000 Calories of energy from his body by doing work on the object–Earth system. Because 1 Calorie = $1.00 \times 10^3$ cal, the total work required is $2.00 \times 10^6$ cal. Converting this to joules, we have for the total work required

$$W = (2.00 \times 10^6 \text{ cal})(4.186 \text{ J/cal}) = 8.37 \times 10^6 \text{ J}$$

The work done in lifting the object of mass *m* a distance *h* is equal to *mgh*, and the work done in lifting it *n* times is *nmgh*.

We equate this to the total work required:

$$W = nmgh = 8.37 \times 10^6 \text{ J}$$

Solving for *n*,

$$n = \frac{8.37 \times 10^6 \text{ J}}{(50.0 \text{ kg})(9.80 \text{ m/s}^2)(2.00 \text{ m})} = \boxed{8.54 \times 10^3 \text{ times}}$$

If the student is in good shape and lifts the weight once every 5 s, it will take him about 12 h to perform this feat. Clearly, it is much easier to lose weight by dieting than by lifting weights!

---

* Originally, the calorie was defined as the heat necessary to raise the temperature of 1 g of water by 1°C at any initial temperature. However, careful measurements showed that the energy required depends somewhat on temperature; hence, a more precise definition evolved.

See Screen 10.3

## 17.2 • SPECIFIC HEAT

The definition of the calorie indicates the amount of energy necessary to raise the temperature of one gram of a specific substance—water—by one degree, which is 4.186 J. To raise the temperature of one kilogram of water by one degree, we need to transfer 4 186 J of energy. The quantity of energy required to raise the temperature of one kilogram of an arbitrary substance by one degree varies with the substance. For example, the energy required to raise the temperature of one kilogram of copper by 1.0°C is 387 J, which is significantly less than that required for water. Every substance requires a unique amount of energy per unit mass to change the temperature of that substance by 1.0°C.

Suppose that a quantity of energy $Q$ is transferred to a mass $m$ of a substance, thereby changing its temperature by $\Delta T$. The **specific heat** $c$ of the substance is defined as

$$c \equiv \frac{Q}{m\,\Delta T} \qquad\qquad \textbf{[17.2]}$$

The units of specific heat are J/kg·°C. Table 17.1 lists specific heats for several substances. From the definition of the calorie, the specific heat of water is 4 186 J/kg·°C.

From this definition, we can express the energy $Q$ transferred between a system of mass $m$ and its surroundings in terms of the resulting temperature change $\Delta T$ as

$$Q = mc\,\Delta T \qquad\qquad \textbf{[17.3]}$$

For example, the energy required to raise the temperature of 0.500 kg of water by 3.00°C is $Q = (0.500\text{ kg})(4\,186\text{ J/kg}\cdot\text{°C})(3.00\text{°C}) = 6.28 \times 10^3$ J. Note that when the temperature increases, $\Delta T$ and $Q$ are taken to be *positive*, corresponding to energy flowing *into* the system. When the temperature decreases, $\Delta T$ and $Q$ are *negative* and energy flows *out* of the system. These sign conventions are consistent with those in our discussion of the continuity equation for energy, Equation 6.20.

Table 17.1 shows that water has a high specific heat relative to most other common substances. (The specific heats of hydrogen and helium are higher.) The high specific heat of water is responsible for the moderate temperatures found in regions near large bodies of water. As the temperature of a body of water decreases during the winter, the water transfers energy to the air, which carries the energy landward when prevailing winds are toward the land. For example, the prevailing winds off the western coast of the United States are toward the land, and the energy liberated by the Pacific Ocean as it cools keeps coastal areas much warmer than they would be otherwise. This explains why the western coastal states generally have warmer winter weather than the eastern coastal states, where the winds do not transfer energy toward land.

The fact that the specific heat of water is higher than that of sand accounts for the pattern of air flow at a beach. During the day, the Sun adds roughly equal amounts of energy to beach and water, but the lower specific heat of sand causes the beach to reach a higher temperature than the water. As a result, the air above the land reaches a higher temperature than the air above the water. The denser cold air pushes the less dense hot air upward (due to Archimedes's principle). This results in a breeze from ocean to land during the day. Because the hot air gradually cools as it rises, it subsequently sinks, setting up the circulating pattern shown in Figure 17.2a. During the night, the sand cools more quickly than the water, and the circulating pattern reverses because the hotter air is now over the water (Fig. 17.2b). These offshore and onshore breezes are well known to sailors.

**PITFALL PREVENTION 17.2**

**Heat, temperature, and internal energy are different**

Considering the definitions of heat and internal energy in Section 17.1 and our discussion of temperature in Chapter 16, let us look at a few common everyday statements that you may have heard:

1. "As the truck braked to a stop, a large amount of heat was generated by friction." This is a classic misuse of the word *heat*, in which it is substituted for the correct term, internal energy. No transfer of energy takes place due to a temperature difference in the braking process (ignoring the transfer from the hot brakes into the air), but rather mechanical energy is transformed into internal energy due to friction.

2. "The heat of a hot summer day. . . ." This misuses the word *heat* as a substitute for temperature. The intent of this phrase is to describe a very high temperature, but no energy is transferred by heat.

As you read the newspaper or listen to the radio, be alert for these kinds of phrases, and think about the proper word to be used in place of *heat*.

**PITFALL PREVENTION 17.3**

**Energy can be transferred by any method**

We use $Q$ to represent the amount of energy transferred, but keep in mind that the transfer in Equation 17.3 could be by *any* method; it does not have to be heat. See Thinking Physics 17.1 for examples.

The name *specific heat* is an unfortunate holdover from the days when thermodynamics and mechanics developed separately. As noted in Pitfall Prevention 17.4, the energy transfer could be by any method. A better name would be *specific energy transfer*, but the existing term is too entrenched to be replaced.

| TABLE 17.1 | **Specific Heats of Some Substances at 25.0°C and Atmospheric Pressure** | |
|---|---|---|
| | **Specific heat, $c$** | |
| **Substance** | **J/kg·°C** | **cal/g·°C** |
| **Elemental Solids** | | |
| Aluminum | 900 | 0.215 |
| Beryllium | 1830 | 0.436 |
| Cadmium | 230 | 0.055 |
| Copper | 387 | 0.0924 |
| Germanium | 322 | 0.077 |
| Gold | 129 | 0.0308 |
| Iron | 448 | 0.107 |
| Lead | 128 | 0.0305 |
| Silicon | 703 | 0.168 |
| Silver | 234 | 0.056 |
| **Other Solids** | | |
| Brass | 380 | 0.092 |
| Wood | 1700 | 0.41 |
| Glass | 837 | 0.200 |
| Ice (−5.0°C) | 2090 | 0.50 |
| Marble | 860 | 0.21 |
| **Liquids** | | |
| Alcohol (ethyl) | 2400 | 0.58 |
| Mercury | 140 | 0.033 |
| Water (15.0°C) | 4186 | 1.00 |

**Quick Quiz 17.1**

Imagine you have 1.00 kg each of iron, glass, and water and that all three samples are at 10.0°C. (a) Rank them from lowest to highest temperature after 100 J of energy is added to each. (b) Rank them from lowest to greatest amount of energy transfer by heat if each increases in temperature by 20.0°C.

Day

Night

(a)

(b)

**Figure 17.2**

Circulation of air at the beach. (a) On a hot day, the air above the warm sand warms faster than the air above the cooler water. The cooler air pushes the warmer air upward according to Archimedes's Principle. (b) At night, the sand cools more rapidly than the water and hence the air currents reverse their directions.

## Calorimetry

One technique for measuring the specific heat of a solid or liquid is to raise the temperature of the substance to some value, place it into a vessel containing water of known mass and temperature, and measure the temperature of the combination after equilibrium is reached. Let us define the system as the substance and the water. If the vessel is assumed to be a good insulator, so that energy does not leave the system by heat (nor by any other means), then we can use the isolated system model. Vessels having this property are called **calorimeters,** and the analysis performed by using such vessels is called **calorimetry.**

The principle of conservation of energy for this isolated system requires that the energy leaving by heat from the warmer substance (of unknown specific heat) equals the energy entering the water.* Thus, we can write

$$Q_{\text{cold}} = - Q_{\text{hot}} \qquad [17.4]$$

To see how to set up a calorimetry problem, suppose that $m_x$ is the mass of a substance whose specific heat we wish to determine, $c_x$ its specific heat, and $T_x$ its initial temperature. Let $m_w$, $c_w$, and $T_w$ represent corresponding values for the water. If $T$ is the final equilibrium temperature after the substance and the water are combined, from Equation 17.3 we find that the energy gained by the water is $m_w c_w(T - T_w)$, and the energy lost by the substance of unknown specific heat is $m_x c_x(T - T_x)$. Substituting these values into Equation 17.4, we have

$$m_w c_w(T - T_w) = - m_x c_x(T - T_x)$$

Solving for $c_x$ gives

$$c_x = \frac{m_w c_w(T - T_w)}{m_x(T_x - T)}$$

By substituting the known values in the right-hand side, we can calculate the specific heat of the substance.

### THINKING PHYSICS 17.1

The equation $Q = mc\,\Delta T$ indicates the relationship between energy $Q$ transferred to an object of mass $m$ and specific heat $c$ by means of heat and the resulting temperature change $\Delta T$. In reality, the energy transfer on the left-hand side of the equation could be made by any method, not just heat. Give a few examples in which this equation could be used to calculate a temperature change of an object due to an energy transfer process other than heat.

**Reasoning**  The following are a few of several possible examples.

During the first few seconds after turning on a toaster, the temperature of the electrical coils rises. The transfer mechanism here is *electrical transmission* of energy through the power cord.

---

* For precise measurements, the container holding the water should be included in our calculations, because it also changes temperature. This would require a knowledge of its mass and composition. If, however, the mass of the water is large compared with that of the container, we can adopt a simplification model in which we ignore the energy gained by the container.

The temperature of a potato in a microwave oven increases due to the absorption of microwaves. In this case, the energy transfer mechanism is by *electromagnetic radiation*—the microwaves.

A carpenter attempts to use a dull drill bit to bore a hole in a piece of wood. The bit fails to make much headway but becomes very warm. The increase in temperature in this case is due to *work* done on the bit by the wood.

In each of these cases, as well as many other possibilities, the $Q$ on the left of the equation of interest is not a measure of heat, but is replaced with the energy transferred or transformed by other means. Regardless of the fact that heat is not involved, the equation can still be used to calculate the temperature change.

---

### Example 17.2   Cooling a Hot Ingot

The temperature of a 0.0500-kg ingot of metal is raised to 200.0°C and the ingot is then dropped into a light, insulated beaker containing 0.400 kg of water initially at 20.0°C. If the final equilibrium temperature of the mixed system is 22.4°C, find the specific heat of the metal.

**Solution**   Using Equations 17.3 and 17.4, we can write

$$Q_{cold} = -Q_{hot} \longrightarrow m_w c_w (T - T_w) = -m_x c_x (T - T_x)$$

Substituting numerical values, we have

$$(0.400 \text{ kg})(4\,186 \text{ J/kg} \cdot {}^\circ\text{C})(22.4{}^\circ\text{C} - 20.0{}^\circ\text{C})$$
$$= -(0.0500 \text{ kg})(c_x)(22.4{}^\circ\text{ C} - 200.0{}^\circ\text{C})$$

from which we find that

$$c_x = \boxed{453 \text{ J/kg} \cdot {}^\circ\text{C}}$$

The ingot is most likely iron, as can be seen by comparing this result with the data in Table 17.1.

**EXERCISE**   What is the total energy transferred to the water as the ingot is cooled?

*Answer*   $4.02 \times 10^3$ J

**EXERCISE**   A cowboy fires a silver bullet of mass 2.00 g with a muzzle speed of 200 m/s into the pine wall of a saloon, where the bullet stops. Assume that all the internal energy generated by the impact remains within the bullet. What is the temperature change of the bullet?

*Answer*   85.5°C

---

## 17.3 • LATENT HEAT AND PHASE CHANGES

A substance often undergoes a change in temperature when energy is transferred between the substance and its environment. In some situations, however, the transfer of energy does not result in a change in temperature. This can occur when the physical characteristics of the substance change from one form to another, commonly referred to as a **phase change.** Some common phase changes are solid to liquid (melting), liquid to gas (boiling), and a change in crystalline structure of a solid. All such phase changes involve a change in internal energy, but no change in temperature. The energy that enters the substance during melting and boiling appears as increased intermolecular potential energy as bonds are broken, rather than as an increase in random motion of the molecules.

In Chapter 11, we discussed two types of bound systems. An object is gravitationally bound to the Earth, and a certain amount of energy must be added to the object–Earth system to separate the Earth and the object by an infinite distance. Knowing this energy allowed us to calculate the escape speed (Section 11.4). The input of energy breaks the bond between the object and the Earth. In Section 11.5, we discussed the bound system of an electron and a proton in the hydrogen atom. The minimum energy required to separate the electron and the proton of a hydrogen atom was called the ionization energy. An input of energy of at least the ionization energy breaks the bond between the electron and the proton. The energy input during a phase change is similar to these examples. During phase changes, the energy added to the system of all of the molecules of a substance modifies or breaks the bonds between the molecules.

The energy transfer required to change the phase of a given mass $m$ of a pure substance is

$$Q = \pm \, mL \qquad\qquad [17.5]$$

• *Latent heat*

where $L$ is called the **latent heat**\* of the substance and depends on the nature of the phase change as well as on the substance. The proper sign in Equation 17.5 is chosen according to the direction of the energy flow. When an ice cube melts into water, we express Equation 17.5 as $Q = mL$, but in the case of liquid water freezing into ice, because we are removing energy from the water, we express Equation 17.5 with the negative sign: $Q = -\, mL$.

**Heat of fusion** $L_f$ is the term used when the phase change occurs during melting or freezing, and **heat of vaporization** $L_v$ is the term used when the phase change occurs during boiling or condensing. For example, the latent heat of fusion for water at atmospheric pressure is $3.33 \times 10^5$ J/kg, and the latent heat of vaporization of water is $2.26 \times 10^6$ J/kg. The latent heats of different substances vary considerably, as is seen in Table 17.2.

The latent heat of fusion is the energy required to break all of the intermolecular bonds in one kilogram of a substance so as to convert the solid phase to the liquid phase. The latent heat of vaporization is the energy that must be added to one kilogram of the liquid phase of a substance to break all of the liquid bonds so as to form a gas.

As you can see from Table 17.2, the latent heat of vaporization for a given substance is usually larger than the latent heat of fusion. In the change from solid to liquid phase, solid bonds between molecules are transformed into somewhat

**PITFALL PREVENTION 17.7**
**Signs are critical**

Sign errors occur very often when students perform calorimetry equations, so we will make this point once again. For phase changes, use the correct explicit sign in Equation 17.5, depending on whether you are adding or removing energy from the substance. Equation 17.3 has no explicit sign to consider, but be sure that your $\Delta T$ is *always* the final temperature minus the initial temperature. In addition, make sure that you *always* include the negative sign on the right-hand side of Equation 17.4.

| TABLE 17.2 | Latent Heats of Fusion and Vaporization | | | |
|---|---|---|---|---|
| **Substance** | **Melting Point (°C)** | **Latent Heat of Fusion (J/kg)** | **Boiling Point (°C)** | **Latent Heat of Vaporization (J/kg)** |
| Helium | − 269.65 | $5.23 \times 10^3$ | − 268.93 | $2.09 \times 10^4$ |
| Nitrogen | − 209.97 | $2.55 \times 10^4$ | − 195.81 | $2.01 \times 10^5$ |
| Oxygen | − 218.79 | $1.38 \times 10^4$ | − 182.97 | $2.13 \times 10^5$ |
| Ethyl alcohol | − 114 | $1.04 \times 10^5$ | 78 | $8.54 \times 10^5$ |
| Water | 0.00 | $3.33 \times 10^5$ | 100.00 | $2.26 \times 10^6$ |
| Sulfur | 119 | $3.81 \times 10^4$ | 444.60 | $3.26 \times 10^5$ |
| Lead | 327.3 | $2.45 \times 10^4$ | 1750 | $8.70 \times 10^5$ |
| Aluminum | 660 | $3.97 \times 10^5$ | 2450 | $1.14 \times 10^7$ |
| Silver | 960.80 | $8.82 \times 10^4$ | 2193 | $2.33 \times 10^6$ |
| Gold | 1063.00 | $6.44 \times 10^4$ | 2660 | $1.58 \times 10^6$ |
| Copper | 1083 | $1.34 \times 10^5$ | 1187 | $5.06 \times 10^6$ |

\* The word *latent* is from the Latin *latere*, meaning *hidden* or *concealed*. Notice that this phrase uses the term *heat* incorrectly, but we will still use it because it is well ingrained in the terminology of physics.

weaker liquid bonds. In the change from liquid to gas phase, however, liquid bonds are broken, creating a situation in which the molecules of the gas have essentially no bonding to one another. More energy is therefore required to vaporize a given mass of substance than to melt it.

With our knowledge of latent heat, we can understand the full behavior of a substance as energy is added to it. Consider, for example, the addition of energy to a system of a 1-g block of ice at $-30.0°C$ in a container held at constant pressure. Suppose that this energy results in the ice turning to steam (water vapor) at $120.0°C$. Figure 17.3 indicates the experimental measurement of temperature as energy is added to the system. Let us examine each portion of the curve separately.

**Part A**   During this portion of the curve, the temperature of the ice changes from $-30.0°C$ to $0.0°C$. Because the specific heat of ice is $2\,090 \ J/kg \cdot °C$, we can calculate the amount of energy added from Equation 17.3:

$$Q = mc_{ice} \ \Delta T = (1.00 \times 10^{-3} \ kg)(2\,090 \ J/kg \cdot °C)(30.0°C) = 62.7 \ J$$

The transferred energy appears in the system as internal energy associated with random motion of the molecules of the ice, as represented by the increasing temperature of the ice.

**Part B**   When the ice reaches $0.0°C$, the ice–water mixture remains at this temperature—even though energy is being added—until all the ice melts to become water at $0.0°C$. According to Equation 17.5, the energy required to melt $1.00$ g of ice at $0.0°C$ is

$$Q = mL_f = (1.00 \times 10^{-3} \ kg)(3.33 \times 10^5 \ J/kg) = 333 \ J$$

During this process, the transferred energy appears in the system as internal energy associated with the increase in intermolecular potential energy as the bonds between water molecules in the ice break.

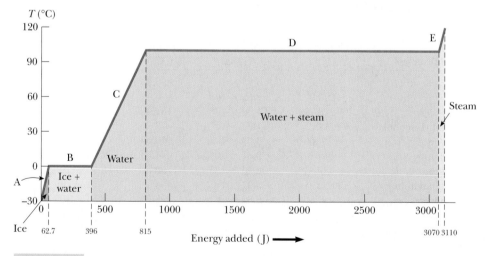

**Figure 17.3**

A plot of temperature versus energy added when 1 g of ice initially at $-30.0°C$ is converted to steam at $120.0°C$.

**Part C**  Between 0.0°C and 100.0°C, no phase change occurs. The energy added to the water is used to increase its temperature, as it was in Part A. The amount of energy necessary to increase the temperature from 0.0°C to 100.0°C is

$$Q = mc_{water}\, \Delta T = (1.00 \times 10^{-3}\, kg)(4.19 \times 10^3\, J/kg \cdot °C)(100.0°C) = 419\, J$$

**Part D**  At 100.0°C, another phase change occurs as the water changes to steam at 100.0°C. Similar to the behavior in Part B, the water–steam mixture remains at a constant temperature, this time at 100.0°C—and the energy goes into breaking bonds so that the gas molecules move far apart—until all the liquid has been converted to steam. The energy required to convert 1.00 g of water to steam at 100.0°C is

$$Q = mL_v = (1.00 \times 10^{-3}\, kg)(2.26 \times 10^6\, J/kg) = 2.26 \times 10^3\, J$$

**Part E**  On this portion of the curve, as in parts A and C, no phase change occurs, so all energy added is used to increase the temperature of the steam. The energy that must be added to raise the temperature of the steam to 120.0°C is

$$Q = mc_{steam}\, \Delta T = (1.00 \times 10^{-3}\, kg)(2.01 \times 10^3\, J/kg \cdot °C)(20.0°C) = 40.2\, J$$

The total amount of energy that must be added to change 1 g of ice at $-30.0°C$ to steam at 120.0°C is the sum of the results from all five parts of the curve, $3.11 \times 10^3$ J. Conversely, to cool 1 g of steam at 120.0°C to ice at $-30.0°C$, we must remove $3.11 \times 10^3$ J of energy.

### Quick Quiz 17.2

Suppose the same process of adding energy to the ice cube is performed, but we graph the internal energy of the system as a function of energy input. What would this graph look like?

### Quick Quiz 17.3

Calculate the slopes for the A, C, and E portions of Figure 17.3. Rank the slopes from least steep to steepest, and explain what this ordering means.

### Example 17.3  Boiling Liquid Helium

Liquid helium has a very low boiling point, 4.2 K, and a very low heat of vaporization, $2.09 \times 10^4$ J/kg (see Table 17.2). Energy is transferred at a constant rate of 10.0 W to a container of liquid helium from an immersed electric heater. At this rate, how long does it take to vaporize 1.00 kg of liquid helium?

**Reasoning and Solution**  We divide Equation 17.5 by our desired time interval and recognize the ratio of the energy transfer to the time interval as the power:

$$Q = mL_v \quad \longrightarrow \quad \mathcal{P} = \frac{Q}{\Delta t} = \frac{mL_v}{\Delta t}$$

Solving for the time interval, we have

$$\Delta t = \frac{mL_v}{\mathcal{P}}$$

Now, substituting the numerical values,

$$\Delta t = \frac{(1.00\, kg)(2.09 \times 10^4\, J/kg)}{10.0\, W} = 2.09 \times 10^3\, s \approx \boxed{35\ min}$$

**EXERCISE**  If 10.0 W of power is supplied to 1.00 kg of water at 100.0°C, how long will it take for the water to completely vaporize?

*Answer*  62.8 h

See Screen 10.6

• *State variables*

• *Transfer variables*

## 17.4 • WORK IN THERMODYNAMIC PROCESSES

In the macroscopic approach to thermodynamics, we describe the *state* of a system with such quantities as pressure, volume, temperature, and internal energy. As a result, these quantities belong to a category called **state variables.** For any given condition of the system, we can identify values of the state variables. It is important to note, however, that a macroscopic state of a system can be specified only if the system is in internal thermal equilibrium. In the case of a gas in a container, internal thermal equilibrium requires that every part of the gas be at the same pressure and temperature. If the temperature varies from one part of the gas to another, for example, we cannot specify a single temperature for the entire gas to be used in the ideal gas law.

A second category of variables in situations involving energy is **transfer variables.** These variables only have a nonzero value if a process occurs in which energy is transferred across the boundary of the system. Because a transfer of energy across the boundary represents a change in the system, transfer variables are not associated with a given state of the system, but with a *change* in the state of the system. In the previous sections, we discussed heat as a transfer variable. For a given set of conditions of a system, the heat has no defined value. We can assign a value to the heat only if energy crosses the boundary by heat, resulting in a change in the system. State variables are characteristic of a system in thermal equilibrium. Transfer variables are characteristic of a process in which energy is transferred between a system and its environment.

We have seen this notion before, but we have not used the language of state variables and transfer variables. In the continuity equation for energy, $\Delta E_{\text{system}} = \Sigma H$, we can identify the terms on the right-hand side as transfer variables: work, heat, mechanical waves, matter transfer, electromagnetic radiation, and electrical transmission. The left side of the continuity equation represents *changes* in state variables, such as kinetic energy, potential energy, and internal energy. For a gas, we have additional state variables, such as pressure, volume, and temperature.

In this section, we study an important transfer variable for thermodynamic systems—work. Work performed on particles was studied extensively in Chapter 6, and here we investigate the work done on a deformable system—a gas. Consider a gas contained in a cylinder fitted with a frictionless, movable piston of face area $A$ (Fig. 17.4) and in thermal equilibrium. The gas occupies a volume $V$ and exerts a uniform pressure $P$ on the cylinder walls and the piston. Now let us adopt a simplification model in which the gas is compressed **quasi-statically,** that is, slowly enough to allow the system to remain in thermal equilibrium at all times. As the piston is pushed inward by an external force through a distance of $d\mathbf{r} = dy\mathbf{j}$ (Fig. 17.4b), the work done on the gas is, according to our definition of work in Chapter 6,

$$dW = \mathbf{F} \cdot d\mathbf{r} = -F\mathbf{j} \cdot dy\mathbf{j} = -F\,dy = -PA\,dy$$

where we have set the magnitude $F$ of the external force equal to $PA$ because the piston is always in equilibrium. Because $A\,dy$ is the change in volume of the gas $dV$, we can express the work done *on* the gas as

$$dW = -P\,dV \qquad \qquad \textbf{[17.6]}$$

If the gas is compressed, $dV$ is negative and the work done on the gas is positive. If the gas expands, $dV$ is positive and the work done on the gas is negative. If the volume remains constant, the work done on the gas is zero. The total work done on

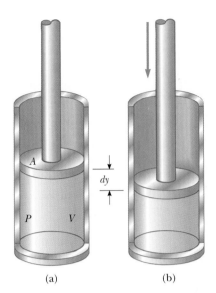

**Figure 17.4**

Work is done on a gas contained in a cylinder at pressure *P* as the piston is pushed downward so that the gas is compressed.

(a)    (b)

the gas as its volume changes from $V_i$ to $V_f$ is given by the integral of Equation 17.6:

$$W = -\int_{V_i}^{V_f} P\, dV$$    [**17.7**]    •    *Work done on a gas*

To evaluate this integral, one must know how the pressure varies with volume during the expansion process.

In general, the pressure is not constant during a process that takes a gas from its initial state to its final state, but depends on the volume and temperature. If the pressure and volume are known at each step of the process, the state of the gas at each step can be plotted on a specialized graphical representation that is very important in thermodynamics—a ***PV* diagram,** as in Figure 17.5. This type of diagram allows us to visualize a process through which a gas is progressing. The curve on such a graphical representation is called the *path* taken between the initial and final states.

Considering the integral in Equation 17.7 and recognizing the significance of the integral as an area under a curve, we can identify an important use for *PV* diagrams:

The work done on a gas in a quasi-static process that takes the gas from an initial state to a final state is the negative of the area under the curve on a *PV* diagram, evaluated between the initial and final states.

As Figure 17.5 suggests, for our process of compressing a gas in the cylinder, the work done depends on the particular path taken between the initial and final states. To illustrate this important point, consider several different paths connecting *i* and *f* (Fig. 17.6). In the process depicted in Figure 17.6a, the volume of the gas is first reduced from $V_i$ to $V_f$ at constant pressure $P_i$ and the pressure of the gas then increases from $P_i$ to $P_f$ by cooling at constant volume $V_f$. The work done on the gas along this path is $-P_i(V_f - V_i)$. In Figure 17.6b, the pressure of the gas is

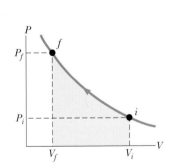

**Figure 17.5**

A gas is compressed quasi-statically from state *i* to state *f*. The work done on the gas equals the negative of the area under the *PV* curve.

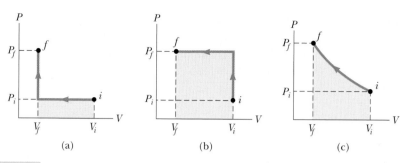

**Figure 17.6**

The work done on a gas as it is taken from an initial state to a final state depends on the path between these states.

increased from $P_i$ to $P_f$ at constant volume $V_i$ and then the volume of the gas is reduced from $V_i$ to $V_f$ at constant pressure $P_f$. The work done on the gas along this path is $-P_f(V_f - V_i)$, which is greater in magnitude than that for the process described in Figure 17.6a because the piston was displaced through the same distance by a larger force than for the situation in Figure 17.6a. Finally, for the process described in Figure 17.6c, where both $P$ and $V$ change continuously, the work done on the gas has some value intermediate between the values obtained in the first two processes. To evaluate the work in this case, the function $P(V)$ must be known, so that we can evaluate the integral in Equation 17.7.

In a similar manner, the energy transferred by heat into or out of the gas is also found to depend on the process. This can be demonstrated by the situations depicted in Figure 17.7. In each case, the gas has the same initial volume, temperature, and pressure and is assumed to be ideal. In Figure 17.7a, the gas is thermally insulated from its surroundings except at the bottom where it is in thermal contact with an energy reservoir. An **energy reservoir** is a source of internal energy that is considered to be so large that a finite transfer of energy from the reservoir does not change its temperature. The piston is held at its initial position by an external agent—your hand, for instance. Now, the force holding the piston is reduced slightly, so that the piston rises very slowly to its final position. Because the piston is moving upward, negative work is done on the gas—the gas is doing work on the

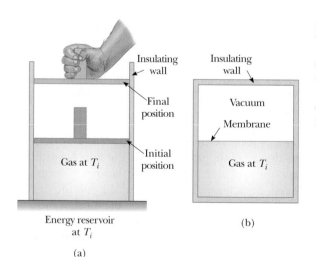

**Figure 17.7**

(a) A gas at temperature $T_i$ expands slowly by absorbing energy from a reservoir at the same temperature. (b) A gas expands rapidly into an evacuated region after a membrane is broken.

piston. During this expansion to the final volume $V_f$, just enough energy to maintain a constant temperature $T_i$ is transferred by heat from the reservoir to the gas.

Now consider the completely thermally insulated system shown in Figure 17.7b. When the membrane is broken, the gas expands rapidly into the vacuum until it occupies a volume $V_f$ and is at a pressure $P_f$. In this case, no work is done on the gas because no force is exerted on the gas. Furthermore, no energy is transferred by heat through the insulating wall. As we shall show in Section 17.6, the initial and final temperatures of the gas are the same in this type of expansion.

The initial and final states of the ideal gas in Figure 17.7a are identical to the initial and final states in Figure 17.7b, but the paths are different. In the first case, the gas does work on the piston, and energy is transferred slowly to the gas. In the second case, no energy is transferred and the work done is zero. We therefore conclude that **energy transfer by heat, like work done, depends on the process followed between the initial and final states of the system.**

## Example 17.4   Comparing Processes

An ideal gas is taken through two processes in which $P_f = 1.00 \times 10^5$ Pa, $V_f = 2.00$ m$^3$, $P_i = 0.200 \times 10^5$ Pa, and $V_i = 10.0$ m$^3$. For process 1 shown in Figure 17.5, the temperature remains constant. For process 2 shown in Figure 17.6a, the pressure remains constant and then the volume remains constant. What is the ratio of the work $W_1$ done on the gas in the first process to the work $W_2$ done in the second process?

**Solution**   For process 1, the pressure as a function of volume can be expressed using the ideal gas law:

$$P = \frac{nRT}{V}$$

For process 2, no work is done during the portion at constant volume, because the piston does not move through a displacement. During the first part of the process, the pressure is constant at $P = P_i$. Let us use these results in the ratio of the work done:

$$\frac{W_1}{W_2} = \frac{-\int_{process\ 1} P\,dV}{-\int_{process\ 2} P\,dV} = \frac{\int_{V_i}^{V_f} \frac{nRT}{V}\,dV}{\int_{V_i}^{V_f} P_i\,dV} = \frac{nRT\int_{V_i}^{V_f} \frac{dV}{V}}{P_i\int_{V_i}^{V_f} dV}$$

$$= \frac{nRT\ln\left(\frac{V_f}{V_i}\right)}{P_i(V_f - V_i)} = \frac{P_iV_i\ln\left(\frac{V_f}{V_i}\right)}{P_i(V_f - V_i)} = \frac{V_i\ln\left(\frac{V_f}{V_i}\right)}{V_f - V_i}$$

Substituting the numerical values for the initial and final volumes gives us

$$\frac{W_1}{W_2} = \frac{(10.0\ m^3)\ln\left(\frac{2.00\ m^3}{10.0 m^3}\right)}{(2.00\ m^3 - 10.0\ m^3)} = 2.01$$

**EXERCISE**   An ideal gas is enclosed in a cylinder that has a movable piston on top. The piston has a mass of $8.00 \times 10^3$ g and an area of 5.00 cm$^2$ and is free to slide up and down, keeping the pressure of the gas constant. How much work is done on the gas as the temperature of 0.200 mol of the gas is raised from 20.0°C to 300°C?

*Answer*   $-466$ J

## 17.5 • THE FIRST LAW OF THERMODYNAMICS

 See Screen 10.6

In Chapter 6, we discussed the continuity equation for energy, Equation 6.20. Let us consider a special case of this general principle, in which the only change in the energy of a system is in its internal energy $E_{int}$, and the only transfer mechanisms are heat $Q$ and work $W$, which we have discussed in this chapter. This leads to an equation, called the **first law of thermodynamics,** which can be used to analyze many problems in thermodynamics.

This special case of the continuity equation can be written as

• *First law of thermodynamics*

$$\Delta E_{int} = Q + W \qquad \text{[17.8]}$$

This indicates that the change in the internal energy of a system is equal to the sum of the energy transferred across the system boundary by heat and the energy transferred by work.

Figure 17.8 shows the energy transfers and change in internal energy for a gas in a cylinder consistent with the first law. Equation 17.8 can be used in a variety of problems in which the only energy considerations are internal energy, heat, and work. We shall consider several examples shortly. Some problems may not fit the conditions of the first law. For example, the internal energy of the coils in your toaster does not increase due to heat or work, but rather due to electrical transmission. Keep in mind that the first law is a special case of the continuity equation for energy, and the latter is the more general equation that covers the widest range of possible situations.

When a system undergoes an infinitesimal change in state, such that a small amount of energy $dQ$ is transferred by heat and a small amount of work $dW$ is done on the system, the internal energy also changes by a small amount $dE_{int}$. Thus, for infinitesimal processes we can express the first law as*

$$dE_{int} = dQ + dW \qquad \text{[17.9]}$$

No practical distinction exists between the results of heat and work on a microscopic scale. Each can produce a change in the internal energy of a system. Although the macroscopic quantities $Q$ and $W$ are *not* properties of a system, they are related to changes of the internal energy of a stationary system through the first law of thermodynamics. Once a process or path is defined, $Q$ and $W$ can be either calculated or measured, and the change in internal energy can be found from the first law.

**Figure 17.8**

The first law of thermodynamics equates the change in internal energy in a system to the net energy transfer to the system by heat $Q$ and work $W$.

### Quick Quiz 17.4

Fill in the boxes in the following table with $-$, $+$, or 0 for each of the three terms in the first law of thermodynamics. For each situation, the system to be considered is identified.

| Situation | System | $Q$ | $W$ | $\Delta E_{int}$ |
|---|---|---|---|---|
| 1. Rapidly pumping up a bicycle tire | Air in the pump | | | |
| 2. Pan of room-temperature water sitting on a hot stove | Water in the pan | | | |
| 3. Air quickly leaking out of a balloon | Air originally in balloon | | | |

* It should be noted that $dQ$ and $dW$ are not true differential quantities, because $Q$ and $W$ are not state variables, although $dE_{int}$ is a true differential. For further details on this point, see R.P. Bauman, *Modern Thermodynamics and Statistical Mechanics*, New York, Macmillan, 1992.

## Quick Quiz 17.5

Consider the human body performing a strenuous exercise, such as lifting weights or riding a bicycle. Work is being done by the body, and energy is leaving by heat from the skin into the surrounding air. According to the first law of thermodynamics, the temperature of the body should be decreasing steadily during the exercise. This is not what happens, however. Is the first law invalid in this situation? Explain.

## THINKING PHYSICS 17.2

In the late 1970s, casino gambling was approved in Atlantic City, New Jersey, which can become quite cold in the winter. Energy projections that were performed for the design of the casinos showed that the air conditioning system would need to operate in the casino even in the middle of a very cold January. Why?

**Reasoning** If we consider the air in the casino to be the gas to which we apply the first law, imagine a simplification model in which there is no air conditioning and no ventilation, so that this gas simply stays in the room. No work is being done on the gas, so we focus on the energy transferred by heat. A casino contains a large number of people, many of whom are active (throwing dice, cheering, etc.), and many of whom are in excited states (celebration, frustration, panic, etc.). As a result, these people have large rates of energy flow by heat from their bodies into the air. This energy results in an increase in internal energy of the air in the casino. With the large number of excited people in a casino (along with the large number of machines and incandescent lights), the temperature of the gas can rise quickly, and to a very high value. To keep the temperature at a comfortable level, energy must be transferred out of the air to compensate for the energy input. Calculations show that energy transfer by heat through the walls even on a 10°F January day is not sufficient to provide the required energy transfer, so the air conditioning system must be in almost continuous use throughout the year.

## 17.6 • SOME APPLICATIONS OF THE FIRST LAW OF THERMODYNAMICS

To apply the first law of thermodynamics to specific systems, it is useful to first define some common thermodynamic processes. We shall identify four special processes used as simplification models to approximate real processes. For each of the following processes, we build a mental representation by imagining that the process occurs for the gas in Figure 17.8.

During an **adiabatic process** no energy enters or leaves the system by heat; that is, $Q = 0$. For the piston in Figure 17.8, imagine that all surfaces of the piston are perfect insulators, so that energy transfer by heat does not exist. (Another way of achieving an adiabatic process is by performing the process very rapidly, because energy transfer by heat tends to be relatively slow.) Applying the first law in this case, we see that

$$\Delta E_{\text{int}} = W \qquad\qquad [17.10]$$

**APPLICATION**

Temperature of the Body During Exercise

**PITFALL PREVENTION 17.8**

**Dual sign conventions**

Equation 17.8 follows directly from the continuity equation for energy. A significant number of physics and engineering textbooks, however, present the first law as $\Delta E_{\text{int}} = Q - W$, with a minus sign between the heat and work. The reason for this is that work is defined in these treatments as the work done *by* the gas rather than *on* the gas, as in our treatment. The equivalent equation to Equation 17.7 in these treatments defines work as $W = \int_{V_i}^{V_f} P\, dV$. Thus, if positive work is done by the gas, energy is leaving the system, leading to the negative sign in the first law.

Chemistry textbooks are almost unanimous in using the plus sign in the first law, and the International Union of Pure and Applied Chemistry lists the first law with the plus sign in its current recommendations. The International Union of Pure and Applied Physics, on the other hand, is silent on the sign convention. Some movement toward a uniform treatment has been made by the Advanced Placement Physics Development Committee, which, in 1999, agreed to list the first law in the form of Equation 17.8 in its list of equations for the Advanced Placement Examination.

In your studies in other chemistry or engineering courses, or in your reading of other physics textbooks, be sure to note which sign convention is being used for the first law.

• *Adiabatic process*

- *Isobaric process (constant pressure)*

- *Isovolumetric process (constant volume)*

From this result, we see that when a gas is compressed adiabatically, both $W$ and $\Delta E_{\text{int}}$ are positive—work is done on the gas, representing a transfer of energy into the system, so the internal energy increases. Conversely, when the gas expands adiabatically, $\Delta E_{\text{int}}$ is negative.

Adiabatic processes are very important in engineering practice. Common applications include the expansion of hot gases in an internal combustion engine, the liquefaction of gases in a cooling system, and the compression stroke in a diesel engine. We study adiabatic processes in more detail in Section 17.8.

The **free expansion** depicted in Figure 17.7b is a unique adiabatic process in which no work is done on the gas. Because $Q = 0$ and $W = 0$, we see from the first law that $\Delta E_{\text{int}} = 0$ for this process. That is, the **initial and final internal energies of a gas are equal in a free expansion.** As we saw in Chapter 16, the internal energy of an ideal gas depends only on its temperature. Thus, we expect no change in temperature during an adiabatic free expansion. This is in accord with experiments performed at low pressures. Experiments with real gases at high pressures show a slight increase or decrease in temperature after the expansion due to interactions between molecules.

A process that occurs at constant pressure is called an **isobaric process.** In Figure 17.8, as long as the piston is perfectly free to move, the pressure of the gas inside the cylinder is due to atmospheric pressure and the weight of the piston. Hence, the piston can be modeled as a particle in equilibrium. When such a process occurs, the work done on the gas is simply the negative of the constant pressure multiplied by the change in volume, or $-P(V_f - V_i)$. On a $PV$ diagram, an isobaric process appears as a horizontal line, such as the first portion of the process in Figure 17.6a or the second portion of the process in Figure 17.6b.

A process that takes place at constant volume is called an **isovolumetric process.** In Figure 17.8, an isovolumetric process is created by clamping the piston so that it cannot move. In such a process, the work done is zero because the volume does not change. Hence, the first law applied to an isovolumetric process gives

$$\Delta E_{\text{int}} = Q \qquad \text{[17.11]}$$

This equation tells us that if energy is added by heat to a system kept at constant volume, all of the energy goes into increasing the internal energy of the system—none enters or leaves the system by work. For example, when an aerosol can is thrown into a fire, energy enters the system (the gas in the can) by heat through the metal walls of the can. Consequently, the temperature and pressure of the gas rise until the can possibly explodes. On a $PV$ diagram, an isovolumetric process appears as a vertical line, such as the second portion of the process in Figure 17.6a or the first portion of the process in Figure 17.6b.

A process that occurs at constant temperature is called an **isothermal process.** Because the internal energy of an ideal gas is a function of temperature only, in an isothermal process for an ideal gas, $\Delta E_{\text{int}} = 0$. Hence, the first law applied to an isothermal process gives

$$Q = -W$$

Whatever energy enters the gas by work leaves the gas by heat in an isothermal process so that the internal energy remains fixed. On a $PV$ diagram, an isothermal process appears as a curved line, such as that in Figure 17.9. The work done on the ideal gas in an isothermal process was calculated in Example 17.4:

$$W = -nRT \ln\left(\frac{V_f}{V_i}\right) \qquad \text{[17.12]}$$

The isothermal process can be analyzed as a model of a nonisolated system in steady-state. There is a transfer of energy across the boundary of the system, but no change occurs in the internal energy of the system. The adiabatic, isobaric, and isovolumetric processes are examples of the nonisolated system model.

Next consider the case in which a nonisolated system is taken through a **cyclic process,** that is, one that originates and ends at the same state. In this case, the change in the internal energy must be zero because internal energy is a state variable and the initial and final states are identical. The energy added by heat to the system must therefore equal the negative of the work done on the system during the cycle. That is, in a cyclic process

$$\Delta E_{\text{int}} = 0 \qquad \text{and} \qquad Q = -W$$

The net work done per cycle equals the area enclosed by the path representing the process on a *PV* diagram. As we shall see in Chapter 18, cyclic processes are very important in describing the thermodynamics of **heat engines**—thermal devices in which a fraction of the energy added by heat to the system is extracted by mechanical work.

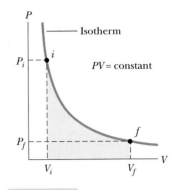

**Figure 17.9**

The *PV* diagram for an isothermal expansion of an ideal gas from an initial state to a final state. The curve is a hyperbola.

**Quick Quiz 17.6**

Characterize the paths in Figure 17.10 as isobaric, isovolumetric, isothermal, or adiabatic. Note that $Q = 0$ in path B.

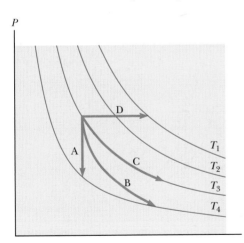

**Figure 17.10**

Identify the nature of paths A, B, C, and D.

### Example 17.5  Boiling Water

Suppose 1.00 g of water vaporizes isobarically at atmospheric pressure ($1.013 \times 10^5$ Pa). Its volume in the liquid state is $V_\ell = 1.00$ cm$^3$, and its volume in the vapor state is $V_v = 1\,671$ cm$^3$. Find the work done on the steam in the expan-

sion and the change in internal energy of the system. Ignore any mixing of the steam and the surrounding air—we shall adopt a simplification model in which the steam simply pushes the surrounding air out of the way.

**Solution**  Because the latent heat of vaporization of water is $2.26 \times 10^6$ J/kg at atmospheric pressure, the energy required to boil 1.00 g is

$$Q = mL_v = (1.00 \times 10^{-3}\,\text{kg})(2.26 \times 10^6\,\text{J/kg}) = 2\,260\,\text{J}$$

The work done on the system is negative and equal to

$$W = -P(V_v - V_\ell)$$
$$= -(1.013 \times 10^5\,\text{N/m}^2)[(1\,671 - 1.00) \times 10^{-6}\,\text{m}^3]$$
$$= -169\,\text{J}$$

Hence, the change in internal energy is

$$\Delta E_{\text{int}} = Q + W = 2\,260\,\text{J} + (-169\,\text{J}) = 2.09\,\text{kJ}$$

The positive value indicates that the internal energy of the system increases. We see that most (93%) of the energy transferred to the liquid stays in the system as an increase in the internal energy. Only 7% leaves the system by work. In this situation, because the temperature does not change, all of the increase in internal energy is due to breaking bonds between the water molecules to form the steam.

## Example 17.6   Cylinder in an Ice Water Bath

The cylinder in Figure 17.11a has thermally conducting walls and is immersed in an ice water bath. The gas in the cylinder undergoes three processes: (1) the piston is rapidly pushed downward, compressing the gas in the cylinder; (2) the piston is held at the final position of the previous process while the gas returns to the temperature of the ice water bath; and (3) the piston is very slowly raised back to the original position. (a) For the system of the gas, what kind of special thermodynamic process does each of these represent?

**Solution**  Because process (1) occurs rapidly, it is very close to an adiabatic compression. In process (2), the piston is held fixed, so this process is isovolumetric. In the very slow process (3), the gas and the ice water bath can be approximated as remaining in thermal equilibrium at all times, so the process is very close to isothermal.

(b) Draw the complete cycle on a *PV* diagram.

**Solution**  The cycle is shown in Figure 17.11b.

(c) The work done on the gas during the cycle is 500 J. What mass of ice in the ice water bath melts during the cycle?

**Solution**  For the entire cycle, the change in internal energy is zero. Thus, from the first law, the energy transfer by heat must equal the negative of the work done on the gas, $Q = -W = -500$ J. This indicates that energy leaves the system of the gas by heat during the cycle, entering the ice water bath (so $Q_{\text{ice}} = +500$ J), where it melts some of the ice. The amount of ice that melts is found using Equation 17.5:

$$Q_{\text{ice}} = mL_f$$

$$m = \frac{Q_{\text{ice}}}{L_f} = \frac{500\,\text{J}}{3.33 \times 10^5\,\text{J/kg}} = 1.5 \times 10^{-3}\,\text{kg}$$

$$= 1.50\,\text{g}$$

(a)

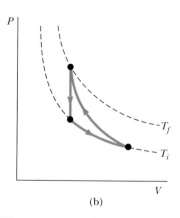

(b)

**Figure 17.11**

(Example 17.6) (a) Cut-away view of a cylinder containing an ideal gas is immersed in an ice water bath. (b) The *PV* diagram for the cycle described.

## 17.7 • MOLAR SPECIFIC HEATS OF IDEAL GASES

 See Screen 10.5

In Section 17.2, we considered the energy necessary to change the temperature of a mass $m$ of a substance by $\Delta T$. In this section, we focus our attention on ideal gases, and the amount of gas is measured by the number of moles $n$ rather than the mass $m$. In doing so, some important new connections are found between thermodynamics and mechanics.

The energy required to raise the temperature of $n$ moles of gas from $T_i$ to $T_f$ depends on the path taken between the initial and final states. To understand this, consider an ideal gas undergoing several processes such that the change in temperature is $\Delta T = T_f - T_i$ for all processes. The temperature change can be achieved by traveling along a variety of paths from one isotherm to another, as in Figure 17.12. Because $\Delta T$ is the same for all paths, the change in internal energy $\Delta E_{int}$ is the same for all paths. From the first law, $Q = \Delta E_{int} - W$, however, we see that the heat $Q$ for each path is different because $W$ (the negative of the area under the curves) is different for each path. Thus the energy required to produce a given change in temperature does not have a unique value.

This difficulty is addressed by defining specific heats for two processes that frequently occur: isovolumetric and isobaric processes. Modifying Equation 17.3 so that the amount of gas is measured in moles, we define the **molar specific heats** associated with these processes with the following equations:

$$Q = nC_V\Delta T \quad \text{(constant volume)} \qquad [17.13]$$

$$Q = nC_P\Delta T \quad \text{(constant pressure)} \qquad [17.14]$$

where $C_V$ is the **molar specific heat at constant volume** and $C_P$ is the **molar specific heat at constant pressure.**

In Chapter 16, we found that the temperature of a monatomic gas is a measure of the average translational kinetic energy of the gas molecules. In view of this, let us first consider the simplest case of an ideal monatomic gas (i.e., a gas containing one atom per molecule), such as helium, neon, or argon. When energy is added to a monatomic gas in a container of fixed volume (e.g., by heating), all of the added energy goes into increasing the translational kinetic energy of the atoms. There is no other way to store the energy in a monatomic gas. The constant-volume process from $i$ to $f$ is described in Figure 17.13, where $\Delta T$ is the temperature difference between the two isotherms. From Equation 16.18 we see that the total internal energy $E_{int}$ of $N$ molecules (or $n$ mol) of an ideal monatomic gas is

$$E_{int} = \tfrac{3}{2}nRT \qquad [17.15]$$

If energy is transferred by heat to the system at constant volume, the work done on the system is zero. That is, $W = -\int P\,dV = 0$ for a constant-volume process. Hence, from the first law of thermodynamics and Equation 17.15 we find

$$Q = \Delta E_{int} = \tfrac{3}{2}nR\,\Delta T \qquad [17.16]$$

Substituting the value for $Q$ given by Equation 17.13 into Equation 17.16, we have

$$nC_V\Delta T = \tfrac{3}{2}nR\,\Delta T$$

$$C_V = \tfrac{3}{2}R = 12.5\ \text{J/mol·K} \qquad [17.17]$$

This expression predicts a value of $C_V = \tfrac{3}{2}R$ for *all* monatomic gases, regardless of the type of gas. This prediction is based on our structural model of kinetic theory, in which the atoms interact with one another only via short-range forces. The third

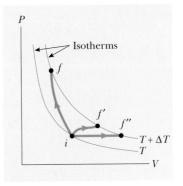

**Figure 17.12**

An ideal gas follows three different paths between an isotherm at temperature $T$ and a second isotherm at temperature $T + \Delta T$.

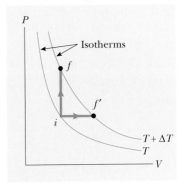

**Figure 17.13**

Energy is transferred by heat to an ideal gas in two ways. For the constant-volume path $i \rightarrow f$, all the energy goes into increasing the internal energy of the gas because no work is done. Along the constant-pressure path $i \rightarrow f'$, part of the energy transferred into the gas by heat is transferred out by work.

• *Molar specific heat at constant volume for an ideal monatomic gas*

| **TABLE 17.3** | **Molar Specific Heats of Various Gases** | | | |
|---|---|---|---|---|
| | **Molar Specific Heat (J/mol · K)$^a$** | | | |
| | $C_P$ | $C_V$ | $C_P - C_V$ | $\gamma = C_P/C_V$ |
| **Monatomic Gases** | | | | |
| He | 20.8 | 12.5 | 8.33 | 1.67 |
| Ar | 20.8 | 12.5 | 8.33 | 1.67 |
| Ne | 20.8 | 12.7 | 8.12 | 1.64 |
| Kr | 20.8 | 12.3 | 8.49 | 1.69 |
| **Diatomic Gases** | | | | |
| $H_2$ | 28.8 | 20.4 | 8.33 | 1.41 |
| $N_2$ | 29.1 | 20.8 | 8.33 | 1.40 |
| $O_2$ | 29.4 | 21.1 | 8.33 | 1.40 |
| CO | 29.3 | 21.0 | 8.33 | 1.40 |
| $Cl_2$ | 34.7 | 25.7 | 8.96 | 1.35 |
| **Polyatomic Gases** | | | | |
| $CO_2$ | 37.0 | 28.5 | 8.50 | 1.30 |
| $SO_2$ | 40.4 | 31.4 | 9.00 | 1.29 |
| $H_2O$ | 35.4 | 27.0 | 8.37 | 1.30 |
| $CH_4$ | 35.5 | 27.1 | 8.41 | 1.31 |

$^a$ All values except that for water were obtained at 300 K.

column of Table 17.3 indicates that this prediction is in excellent agreement with measured values of molar specific heats for monatomic gases. It also indicates that this prediction is not in agreement with values of molar specific heats for diatomic and polyatomic gases. We address these types of gases shortly.

Because no work is done on an ideal gas undergoing an isovolumetric process, the energy transfer by heat is equal to the change in internal energy. Thus, the change in internal energy can be expressed as

$$\Delta E_{int} = nC_V \Delta T \qquad \text{[17.18]}$$

Because internal energy is a state function, the change in internal energy does not depend on the path followed between the initial and final states. Thus, **Equation 17.18 gives the change in internal energy of an ideal gas for *any* process in which the temperature change is $\Delta T$, not just an isovolumetric process. Furthermore, it is true for monatomic, diatomic and polyatomic gases.**

In the case of infinitesimal changes, we can use Equation 17.18 to express the molar specific heat at constant volume as

$$C_V = \frac{1}{n} \frac{dE_{int}}{dT} \qquad \text{[17.19]}$$

Now suppose that the gas is taken along the constant-pressure path $i \rightarrow f'$ in Figure 17.13. Along this path, the temperature again increases by $\Delta T$. The energy transferred to the gas by heat in this process is $Q = nC_P \Delta T$. Because the volume changes in this process, the work done on the gas is $W = -P \Delta V$. Applying the first law to this process gives

$$\Delta E_{int} = Q + W = nC_P \Delta T - P \Delta V \qquad \text{[17.20]}$$

The change in internal energy for the process $i \rightarrow f'$ is equal to that for the process $i \rightarrow f$ because $E_{int}$ depends only on temperature for an ideal gas and $\Delta T$ is the same for both processes. Because $PV = nRT$, for a constant-pressure process, $P \Delta V = nR \Delta T$. Substituting this value for $P \Delta V$ into Equation 17.20 with $\Delta E_{int} = nC_V \Delta T$ (Eq. 17.18) gives

$$nC_V \Delta T = nC_P \Delta T - nR \Delta T \longrightarrow C_P - C_V = R \qquad [17.21]$$

• *Relation between molar specific heats*

This expression applies to *any* ideal gas. It shows that the molar specific heat of an ideal gas at constant pressure is greater than the molar specific heat at constant volume by an amount $R$, the universal gas constant. As shown by the fourth column in Table 17.3, this result is in good agreement with real gases regardless of the number of atoms in the molecule.

Because $C_V = \frac{3}{2}R$ for a monatomic ideal gas, Equation 17.21 predicts a value $C_P = \frac{5}{2}R = 20.8 \text{ J/mol} \cdot \text{K}$ for the molar specific heat of a monatomic gas at constant pressure. The second column of Table 17.3 shows the validity of this prediction for monatomic gases.

The ratio of molar specific heats is a dimensionless quantity $\gamma$:

$$\gamma = \frac{C_P}{C_V} \qquad [17.22]$$

For a monatomic gas, this ratio has the value,

$$\gamma = \frac{C_P}{C_V} = \frac{\frac{5}{2}R}{\frac{3}{2}R} = \frac{5}{3} = 1.67$$

The last column in Table 17.3 shows good agreement between this predicted value for $\gamma$ and experimentally measured values for monatomic gases.

---

## Example 17.7  Heating a Cylinder of Helium

A cylinder contains 3.00 mol of helium gas at a temperature of 300 K. (a) How much energy must be transferred to the gas by heat to increase its temperature to 500 K if it is heated at constant volume?

**Solution**   For the constant-volume process

$$Q = nC_V \, \Delta T = \tfrac{3}{2}nR \, \Delta T$$

$$Q = \tfrac{3}{2}(3.00 \text{ mol})(8.31 \text{ J/mol} \cdot \text{K})(500 \text{ K} - 300 \text{ K})$$

$$= 7.48 \times 10^3 \text{ J}$$

(b) How much energy must be transferred to the gas by heat at constant pressure to raise the temperature to 500 K?

**Solution**   For the constant-pressure process,

$$Q = nC_P \, \Delta T = \tfrac{5}{2}nR \, \Delta T$$

$$Q = \tfrac{5}{2}(3.00 \text{ mol})(8.31 \text{ J/mol} \cdot \text{K})(500 \text{ K} - 300 \text{ K})$$

$$= 12.5 \times 10^3 \text{ J}$$

Notice that the energy transfer for part (b) is larger than that for part (a), because positive work must be done *by* the gas pushing the piston outward to maintain the constant pressure in an isobaric process, but no work is done in an isovolumetric process.

**EXERCISE**   What is the work done *on* the gas in the isobaric process?

*Answer*   $-5.00 \times 10^3 \text{ J}$

---

## 17.8 • ADIABATIC PROCESSES FOR AN IDEAL GAS

In Section 17.6, we identified four special processes of interest for ideal gases. In three of them, a state variable is held constant: $P =$ constant for an isobaric process, $V =$ constant for an isovolumetric process, and $T =$ constant for an isothermal process. What about our fourth special process, the adiabatic process? Is anything constant in this process? As you recall, an adiabatic process is one in which no energy is transferred by heat between a system and its surroundings. In reality, true adiabatic processes on Earth cannot occur because there is no such thing as a perfect thermal insulator. However, some processes are nearly adiabatic. For example, if a gas is compressed (or expanded) very rapidly, very little energy flows into (or out of) the system by heat, and so the process is nearly adiabatic. Another example of an adiabatic process is the very slow expansion of a gas that is thermally insulated from its surroundings.

Suppose an ideal gas undergoes a quasi-static adiabatic expansion. We find that all three variables in the ideal gas law—$P$, $V$, and $T$—change during an adiabatic process. At any time during the process, however, the ideal gas law $PV = nRT$ describes the correct relationship among these variables. Although none of the three variables alone are constant in this process, we find that a *combination* of some of these variables remains constant. This relationship is derived in the following discussion.

Imagine a gas expanding adiabatically in a thermally insulated cylinder, so that $Q = 0$. Let us take the infinitesimal change in volume to be $dV$ and the infinitesimal change in temperature to be $dT$. The work done on the gas is $-P\,dV$. The change in internal energy is given by Equation 17.18, $dE_{\text{int}} = nC_V\,dT$. Hence, the first law of thermodynamics becomes

$$dE_{\text{int}} = dQ + dW \longrightarrow nC_V\,dT = 0 - P\,dV \qquad \textbf{[17.23]}$$

Taking the differential of the equation of state for an ideal gas, $PV = nRT$, gives

$$P\,dV + V\,dP = nR\,dT$$

Eliminating $n\,dT$ from these last two equations, we find that

$$P\,dV + V\,dP = -\frac{R}{C_V}P\,dV$$

From Equation 17.21, we substitute $R = C_P - C_V$ and divide by $PV$ to obtain

$$\frac{dV}{V} + \frac{dP}{P} = -\left(\frac{C_P - C_V}{C_V}\right)\frac{dV}{V} = (1 - \gamma)\frac{dV}{V}$$

$$\frac{dP}{P} + \gamma\frac{dV}{V} = 0$$

Integrating this expression gives

$$\ln P + \gamma \ln V = \text{constant}$$

which we can write as

$$\boxed{PV^\gamma = \text{constant}} \qquad \textbf{[17.24]}$$

- *Relation between pressure and volume in an adiabatic process for an ideal gas*

The $PV$ diagram for an adiabatic expansion is shown in Figure 17.14. Because $\gamma > 1$, the $PV$ curve is steeper than that for an isothermal expansion, in which $PV = $ constant. Equation 17.23 shows that during an adiabatic expansion, $\Delta E_{\text{int}}$ is negative and so $\Delta T$ is also negative. Thus, the gas cools during an adiabatic expansion. Conversely, the temperature increases if the gas is compressed adiabatically.

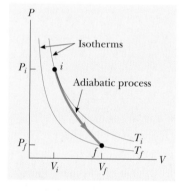

**Figure 17.14**

The $PV$ diagram for an adiabatic expansion of an ideal gas. Note that $T_f < T_i$ in this process, so the gas cools.

Expressing Equation 17.24 in terms of initial and final states,

$$P_i V_i^{\gamma} = P_f V_f^{\gamma} \qquad \text{[17.25]}$$

Using the ideal gas law, Equation 17.25 can also be expressed as

$$T_i V_i^{\gamma-1} = T_f V_f^{\gamma-1} \qquad \text{[17.26]}$$

### Example 17.8 A Diesel Engine Cylinder

The fuel–air mixture in the cylinder of a diesel engine at 20.0°C is compressed from an initial pressure of 1.00 atm and volume of 800 cm$^3$ to a volume of 60.0 cm$^3$. Assuming that the mixture behaves as an ideal gas with $\gamma = 1.40$ and that the compression is adiabatic, find the final pressure and temperature of the mixture.

**Solution**   Using Equation 17.25, we find that

$$P_f = P_i \left( \frac{V_i}{V_f} \right)^{\gamma} = (1.00 \text{ atm}) \left( \frac{800 \text{ cm}^3}{60.0 \text{ cm}^3} \right)^{1.40} = \boxed{37.6 \text{ atm}}$$

Because $PV = nRT$ for an ideal gas, we can compare initial and final conditions in another way as

$$\frac{P_i V_i}{T_i} = \frac{P_f V_f}{T_f}$$

Solving for the final temperature and substituting the given and calculated quantities,

$$T_f = \frac{P_f V_f}{P_i V_i} T_i = \frac{(37.6 \text{ atm})(60.0 \text{ cm}^3)}{(1.00 \text{ atm})(800 \text{ cm}^3)} (293 \text{ K}) = 826 \text{ K}$$
$$= \boxed{553°C}$$

The high compression in a diesel engine raises the temperature of the fuel enough to cause its combustion without the need for spark plugs.

## 17.9 • MOLAR SPECIFIC HEATS AND THE EQUIPARTITION OF ENERGY

We have found that predictions of molar specific heats based on kinetic theory agree quite well with the behavior of monatomic gases but not with the behavior of complex gases (Table 17.3). To explain the variations in $C_V$ and $C_P$ between monatomic gases and more complex gases, let us explore the origin of specific heat by extending our structural model of kinetic theory in Chapter 16. In Section 16.5, we discussed the fact that the sole contribution to the internal energy of a monatomic gas is the translational kinetic energy of the molecules. We also discussed the theorem of equipartition of energy, which states that, at equilibrium, each degree of freedom contributes, on the average, $\frac{1}{2}k_B T$ of energy per molecule. The monatomic gas has three degrees of freedom, one associated with each of the independent directions of translational motion.

For more complex molecules, other types of motion exist in addition to translation. The internal energy of a diatomic or polyatomic gas includes contributions from the vibrational and rotational motion of the molecules in addition to translation. The rotational and vibrational motions of molecules with structure can be activated by collisions and therefore are "coupled" to the translational motion of the molecules. The branch of physics known as *statistical mechanics* suggests that the average energy for each of these additional degrees of freedom is the same as that for translation. This suggests that the determination of the internal energy of a gas is a simple matter of counting the degrees of freedom. We will find that this is true, although the model must be modified with some notions from quantum physics for us to completely explain the experimental data.

Let us consider a diatomic gas, which we can model as consisting of dumbbell-shaped molecules (Fig. 17.15), and apply concepts that we studied in Chapter 10.

(a)

(b)

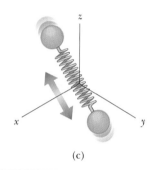

(c)

**Figure 17.15**

Possible motions of a diatomic molecule: (a) translational motion of the center of mass, (b) rotational motion about the various axes, and (c) vibrational motion along the molecular axis.

In this model, the center of mass of the molecule can translate in the $x$, $y$, and $z$ directions (Fig. 17.15a). In addition, the molecule can rotate about three mutually perpendicular axes (Fig. 17.15b). We can ignore the rotation about the $y$ axis because the moment of inertia and the rotational energy $\frac{1}{2}I\omega^2$ about this axis are negligible compared with those associated with the $x$ and $z$ axes. Thus, there are five degrees of freedom: three associated with the translational motion and two associated with the rotational motion. Because each degree of freedom contributes, on average, $\frac{1}{2}k_B T$ of energy per molecule, the total internal energy for a diatomic gas consisting of $N$ molecules and considering both translation and rotation is

$$E_{\text{int}} = 3N(\tfrac{1}{2}k_B T) + 2N(\tfrac{1}{2}k_B T) = \tfrac{5}{2}Nk_B T = \tfrac{5}{2}nRT$$

We can use this result and Equation 17.19 to predict the molar specific heat at constant volume:

$$C_V = \frac{1}{n}\frac{dE_{\text{int}}}{dT} = \frac{1}{n}\frac{d}{dT}(\tfrac{5}{2}nRT) = \tfrac{5}{2}R = 20.8\,\text{J/mol·K} \qquad \textbf{[17.27]}$$

From Equations 17.21 and 17.22 we find that the model predicts

$$C_P = C_V + R = \tfrac{7}{2}R \qquad \textbf{[17.28]}$$

$$\gamma = \frac{C_P}{C_V} = \frac{\tfrac{7}{2}R}{\tfrac{5}{2}R} = \frac{7}{5} = 1.40 \qquad \textbf{[17.29]}$$

Let us now incorporate the vibration of the molecule in the model. We use the structural model for the diatomic molecule in which the two atoms are joined by an imaginary spring (Fig. 17.15c) and apply the concepts of Chapter 12. The vibrational motion has two types of energy associated with the vibrations along the length of the molecule—kinetic energy of the atoms and potential energy in the model spring. This adds two more degrees of freedom for a total of seven for translation, rotation, and vibration. Because each degree of freedom contributes $\frac{1}{2}k_B T$ of energy per molecule, the total internal energy for a diatomic gas consisting of $N$ molecules and considering all types of motion is

$$E_{\text{int}} = 3N(\tfrac{1}{2}k_B T) + 2N(\tfrac{1}{2}k_B T) + 2N(\tfrac{1}{2}k_B T) = \tfrac{7}{2}Nk_B T = \tfrac{7}{2}nRT$$

Thus, the molar specific heat at constant volume is predicted to be

$$C_V = \frac{1}{n}\frac{dE_{\text{int}}}{dT} = \frac{1}{n}\frac{d}{dT}(\tfrac{7}{2}nRT) = \tfrac{7}{2}R = 29.1\,\text{J/mol·K} \qquad \textbf{[17.30]}$$

From Equations 17.21 and 17.22

$$C_P = C_V + R = \tfrac{9}{2}R \qquad \textbf{[17.31]}$$

$$\gamma = \frac{C_P}{C_V} = \frac{\tfrac{9}{2}R}{\tfrac{7}{2}R} = \frac{9}{7} = 1.29 \qquad \textbf{[17.32]}$$

When we compare our predictions to the section of Table 17.3 corresponding to diatomic gases, we find a curious result. For the first four gases, hydrogen, nitrogen, oxygen, and carbon monoxide, the value of $C_V$ is close to that predicted by Equation 17.27, which includes rotation but not vibration. The value for the fifth gas, chlorine, lies between the prediction including rotation and the prediction that includes rotation and vibration. None of the values agree with Equation 17.30, which is based on the most complete model for motion of the diatomic molecule!

It might seem that our model is a failure for predicting molar specific heats for diatomic gases. We can claim success for our model, however, if measurements of molar specific heat are made over a wide temperature range, rather than at the

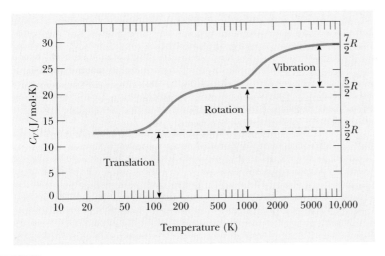

**Figure 17.16**

The molar specific heat of hydrogen as a function of temperature. The horizontal scale is logarithmic. Note that hydrogen liquefies at 20 K.

single temperature that gives us the values in Table 17.3. Figure 17.16 shows the molar specific heat of hydrogen as a function of temperature. The curve has three plateaus and they are at the values of the molar specific heat predicted by Equations 17.17, 17.27, and 17.30! For low temperatures, the diatomic hydrogen gas behaves like a monatomic gas. As the temperature rises to room temperature, its molar specific heat rises to a value for a diatomic gas, consistent with the inclusion of rotation but not vibration. For high temperatures, the molar specific heat is consistent with a model including all types of motion.

Before addressing the reason for this mysterious behavior, let us make a brief remark about polyatomic gases. For molecules with more than two atoms, the number of degrees of freedom is even larger and the vibrations are more complex than for diatomic molecules. This results in an even higher predicted molar specific heat, which is in qualitative agreement with experiment. For the polyatomic gases shown in Table 17.3, we see that the molar specific heats are higher than those for diatomic gases. The more degrees of freedom available to a molecule, the more "ways" of storing energy are available, resulting in a higher molar specific heat.

## A Hint of Energy Quantization

Our model for molar specific heats has been based so far on purely classical notions. It predicts a value of the specific heat for a diatomic gas that, according to Figure 17.16, only agrees with experimental measurements made at high temperatures. To explain why this value is only true at high temperatures and why the plateaus exist in Figure 17.16, we must go beyond classical physics and introduce some quantum physics into the model. In Section 11.5, we discussed energy quantization for the hydrogen atom. Only certain energies were allowed for the system, and an energy level diagram was drawn to illustrate those allowed energies. For a molecule, quantum physics tells us that the rotational and vibrational energies are quantized. Figure 17.17 shows an energy level diagram for the rotational and vibrational quantum states of a diatomic molecule. Notice that vibrational states are separated by larger energy gaps than rotational states.

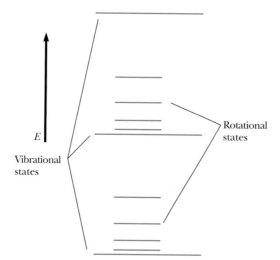

**Figure 17.17**

An energy level diagram for vibrational and rotational states of a diatomic molecule. Note that the rotational states lie closer together in energy than the vibrational states.

At low temperatures, the energy that a molecule gains in collisions with its neighbors is generally not large enough to raise it to the first excited state of either rotation or vibration. Thus, even though rotation and vibration are classically allowed, they do not occur at low temperatures. All molecules are in the ground state for rotation and vibration. Thus, at low temperatures, the only contribution to the molecules' average energy is from translation, and the specific heat is that predicted by Equation 17.17.

As the temperature is raised, the average energy of the molecules increases. In some collisions, a molecule may have enough energy transferred to it from another molecule to excite the first rotational state. As the temperature is raised further, more molecules can be excited to this state. The result is that rotation begins to contribute to the internal energy and the molar specific heat rises. At about room temperature in Figure 17.16, the second plateau is reached and rotation contributes fully to the molar specific heat. The molar specific heat is now equal to the value predicted by Equation 17.27.

Vibration contributes nothing at room temperature, because the molecules are still in the ground vibrational state. The temperature must be raised even further to excite the first vibrational state. This happens in Figure 17.16 between 1 000 K and 10 000 K. At 10 000 K on the right side of the figure, vibration is contributing fully to the internal energy and the molar specific heat has the value predicted by Equation 17.30.

The predictions of this structural model are supportive of the theorem of equipartition of energy. In addition, the inclusion in the model of energy quantization from quantum physics allows a full understanding of Figure 17.16. This is an excellent example of the power of the modeling approach.

## 17.10 • ENERGY TRANSFER MECHANISMS IN THERMAL PROCESSES

In Chapter 6, we introduced the continuity equation for energy $\Delta E_{\text{system}} = \Sigma H$ as a principle allowing a global approach to energy considerations in physical processes. In this chapter, we discussed two of the terms on the right-hand side of the continuity equation: work and heat. In this section, we consider more details of heat and two other energy transfer methods that are often related to temperature changes: convection (a form of matter transfer) and electromagnetic radiation.

## Conduction

The process of energy transfer by heat can also be called **conduction** or **thermal conduction.** In this process, the transfer mechanism can be viewed on an atomic scale as an exchange of kinetic energy between molecules, in which the less energetic molecules gain energy by colliding with the more energetic molecules. For example, if you hold one end of a long metal bar and insert the other end into a flame, the temperature of the metal in your hand soon increases. The energy reaches your hand through conduction. How this happens can be understood by examining what is happening to the atoms in the metal. Initially, before the rod is inserted into the flame, the atoms are vibrating about their equilibrium positions. As the flame provides energy to the rod, those atoms near the flame begin to vibrate with larger and larger amplitudes. These, in turn, collide with their neighbors and transfer some of their energy in the collisions. Slowly, metal atoms farther and farther from the flame increase their amplitude of vibration, until eventually those in the metal near your hand are affected. This increased vibration represents an increase in temperature of the metal (and possibly a burned hand).

Although the transfer of energy through a material can be partially explained by atomic vibrations, the rate of conduction also depends on the properties of the substance. For example, it is possible to hold a piece of asbestos in a flame indefinitely. This implies that very little energy is being conducted through the asbestos. In general, metals are good thermal conductors because they contain large numbers of electrons that are relatively free to move through the metal and can transport energy from one region to another. Thus, in a good thermal conductor, such as copper, conduction takes place via the vibration of atoms and via the motion of free electrons. Materials such as asbestos, cork, paper, and fiber glass are poor conductors. Gases also are poor thermal conductors because of the large distance between the molecules.

Conduction occurs only if the temperatures differ between two parts of the conducting medium. This temperature difference drives the flow of energy. Consider a slab of material of thickness $\Delta x$ and cross-sectional area $A$ with its opposite faces at different temperatures $T_C$ and $T_H$, where $T_H > T_C$ (Fig. 17.18). The slab allows energy to transfer from the region of high temperature to that of low temperature by thermal conduction. The rate of energy transfer by heat, $\mathscr{P} = Q/\Delta t$, is proportional to the cross-sectional area of the slab and the temperature difference and inversely proportional to the thickness of the slab:

$$\mathscr{P} = \frac{Q}{\Delta t} \propto A \frac{\Delta T}{\Delta x}$$

Note that $\mathscr{P}$ has units of watts when $Q$ is in joules and $\Delta t$ is in seconds. This is not surprising, because $\mathscr{P}$ is *power*—the rate of transfer of energy by heat. For a slab of infinitesimal thickness $dx$ and temperature difference $dT$, we can write the **law of conduction** as

$$\mathscr{P} = kA \left| \frac{dT}{dx} \right|$$

[17.33]   •   *Law of conduction*

**Figure 17.18**

Thermal conduction through a slab of cross-sectional area $A$ and thickness $\Delta x$. The opposite faces are at different temperatures $T_C$ and $T_H$.

**Figure 17.19**

Thermal conduction through a uniform, insulated rod of length $L$. The opposite ends are in thermal contact with reservoirs at different temperatures.

where the proportionality constant $k$ is called the **thermal conductivity** of the material, and $dT/dx$ is the **temperature gradient** (the variation of temperature with position). It is the higher thermal conductivity of metal relative to cardboard that makes the metal ice tray feel colder than the cardboard package of vegetables in the discussion at the beginning of Chapter 16.

Suppose a substance is in the shape of a long uniform rod of length $L$, as in Figure 17.19, and is insulated so that energy cannot escape by heat from its surface except at the ends, which are in thermal contact with reservoirs having temperatures $T_C$ and $T_H$. When steady state is reached, the temperature at each point along the rod is constant in time. In this case, the temperature gradient is the same everywhere along the rod and is

$$\left|\frac{dT}{dx}\right| = \frac{T_H - T_C}{L}$$

Thus the rate of energy transfer by heat is

$$\mathcal{P} = kA\frac{(T_H - T_C)}{L} \qquad [17.34]$$

Substances that are good thermal conductors have large thermal conductivity values, whereas good thermal insulators have low thermal conductivity values. Table 17.4 lists thermal conductivities for various substances.

**TABLE 17.4**

**Thermal Conductivities of Some Substances**

| Substance | Thermal Conductivity (W/m · °C) |
|---|---|
| **Metals (at 25°C)** | |
| Aluminum | 238 |
| Copper | 397 |
| Gold | 314 |
| Iron | 79.5 |
| Lead | 34.7 |
| Silver | 427 |
| **Gases (at 20°C)** | |
| Air | 0.0234 |
| Helium | 0.138 |
| Hydrogen | 0.172 |
| Nitrogen | 0.0234 |
| Oxygen | 0.0238 |
| **Nonmetals (approximate values)** | |
| Asbestos | 0.08 |
| Concrete | 0.8 |
| Diamond | 2300 |
| Glass | 0.8 |
| Ice | 2 |
| Water | 0.6 |
| Wood | 0.08 |

**Quick Quiz 17.7**

Rub the palm of your hand on a metal surface for 30 to 45 s. Place the palm of your other hand on an unrubbed portion of the surface and then the rubbed portion. The rubbed portion will feel warmer. Now repeat this process on a wooden surface. You should notice that the difference in temperature between the rubbed and unrubbed portions of the wood surface seems larger than that of the metal surface. Why is this?

The absence of snow on some parts of the roof show that energy is conducted from the inside of the residence to the exterior more rapidly on those parts of the roof. The dormer appears to have been added and insulated. The main roof does not appear to be well insulated. *(Courtesy of Dr. Albert A. Bartlett, University of Colorado, Boulder, CO)*

## Example 17.9 The Leaky Window

A window of area 2.0 m$^3$ is glazed with glass of thickness 4.0 mm. The window is in the wall of a house, and the outside temperature is 10°C. The temperature inside the house is 25°C. (a) How much energy transfers through the window by heat in 1.0 h?

**Solution** We use Equation 17.34 to find the rate of energy transfer by heat:

$$\mathcal{P} = kA\frac{(T_H - T_C)}{L}$$

$$= (0.8 \text{ W/m} \cdot °\text{C})(2.0 \text{ m}^3)\frac{(25°\text{C} - 10°\text{C})}{4.0 \times 10^{-3} \text{ m}}$$

$$= 6 \times 10^3 \text{ W}$$

where $k$ for glass is from Table 17.4. From the definition of power as the rate of energy transfer, we find the energy transferred at this rate in 1.0 h:

$$Q = \mathcal{P}\,\Delta t = (6 \times 10^3 \text{ W})(3.6 \times 10^3 \text{ s}) = \boxed{2 \times 10^7 \text{ J}}$$

(b) If electrical energy costs 12¢/kWh, how much does the transfer of energy in part (a) cost to replace with electrical heating?

**Solution** We cast the answer to part (a) in units of kilowatt-hours:

$$Q = \mathcal{P}\,\Delta t = (6 \times 10^3 \text{ W})(1.0 \text{ h}) = 6 \times 10^3 \text{ Wh} = 6 \text{ kWh}$$

Thus, the cost to replace the energy transferred through the window is (6 kWh)(12¢/kWh) $\approx$ 72¢.

If you imagine paying this much for each hour for each window in your home, your electric bill will be extremely high! For example, for ten such windows, your electric bill would be over $5000 for one month. It seems like something is wrong here, because electric bills are not that high. In reality, a thin layer of air adheres to each of the two surfaces of the window. This air provides additional insulation to that of the glass. As seen in Table 17.4, air is a much poorer thermal conductor than glass, so that most of the insulation is performed by the air, not the glass, in a window!

## Convection

At one time or another you may have warmed your hands by holding them over an open flame. In this situation, the air directly above the flame is heated and expands. As a result, the density of the air decreases and the air rises. This warmed mass of air transfers energy by heat into your hands as it flows by. The transfer of energy from the flame to your hands is performed by matter transfer—the energy travels with the air. Energy transferred by the movement of a fluid is a process called **convection.** When the movement results from differences in density, as in the example of air around a fire, the process is called **natural convection.** When the fluid is forced to move by a fan or pump, as in some hot-air and hot-water heating systems, the process is called **forced convection.**

The circulating pattern of air flow at a beach (Fig 17.2) is an example of convection in nature. The mixing that occurs as water is cooled and eventually freezes at its surface (Section 16.3) is another example.

If it were not for convection currents, it would be very difficult to boil water. As water is heated in a teakettle, the lower layers are warmed first. These regions expand and rise to the top because their density is lower than that of the cooler water. At the same time, the denser cool water falls to the bottom of the kettle so that it can be heated.

The same process occurs when a room is heated by a radiator. The radiator warms the air in the lower regions of the room by heat at the interface between the radiator surface and the air. The warm air expands and floats to the ceiling be-

**Figure 17.20**

Convection currents are set up in a room heated by a radiator.

cause of its lower density, setting up the continuous air current pattern shown in Figure 17.20.

### Radiation

Another method of transferring energy that can be related to a temperature change is **electromagnetic radiation.** All objects radiate energy continuously in the form of electromagnetic waves. As we shall find out in Chapter 24, electromagnetic radiation arises from accelerating electric charges. From our discussion of temperature, we know that temperature corresponds to random motion of molecules which are constantly changing direction, and therefore are accelerating. Because the molecules contain electric charges, the charges also accelerate. Thus, any object emits electromagnetic radiation due to the thermal motion of its molecules. This radiation is called **thermal radiation.**

Through electromagnetic radiation, approximately 1 340 J of energy from the Sun strikes each square meter at the top of the Earth's atmosphere every second. Some of this energy is reflected back into space and some is absorbed by the atmosphere, but enough arrives at the surface of the Earth each day to supply all of our energy needs on this planet hundreds of times over—if it could be captured and used efficiently. The growth in the number of solar houses in this country is one example of an attempt to make use of this abundant energy.

The rate at which an object emits energy by thermal radiation from its surface is proportional to the fourth power of its absolute surface temperature. This principle, known as **Stefan's law,** is expressed in equation form as

- *Stefan's law*

$$\mathcal{P} = \sigma A e T^4 \qquad \text{[17.35]}$$

where $\mathcal{P}$ is the power radiated by the object in watts, $\sigma$ is the **Stefan–Boltzmann constant,** equal to $5.669\ 6 \times 10^{-8}$ W/m$^2 \cdot$K$^4$, $A$ is the surface area of the object in square meters, $e$ is a constant called the **emissivity,** and $T$ is the surface temperature of the body in kelvins. The value of $e$ can vary between zero and unity, depending on the properties of the surface. The emissivity is equal to the fraction of incoming radiation that the surface absorbs.

At the same time as it radiates, the object also absorbs electromagnetic radiation from the environment. If the latter process did not occur, an object would continuously radiate its energy, and its temperature would eventually decrease spontaneously to absolute zero. If an object is at a temperature $T$ and its environment is at a temperature $T_0$, the net rate of energy change for the object as a result of radiation is

$$\mathcal{P}_{\text{net}} = \sigma A e (T^4 - T_0^4) \qquad \textbf{(17.36)}$$

When an object is in equilibrium with its environment, it radiates and absorbs energy at the same rate, and so its temperature remains constant—this is the nonisolated system in steady state model. When an object is hotter than its surroundings, it radiates more energy than it absorbs and so it cools—this is the nonisolated system model.

### Quick Quiz 17.8

If you observe frost on a cold morning, the frost tends to form on the tops of objects such as your car and a metal mailbox more than on the sides. Why is this?

## Example 17.10    The Temperature of a Lightbulb Filament

Estimate the order of magnitude of the temperature of the filament of a 100-W lightbulb when it is operating.

**Solution**    We assume that the filament is so much hotter than the environment that the energy it absorbs by radiation from the environment is negligible. From Equation 17.35, we can solve for the temperature of the filament:

$$T = \sqrt[4]{\frac{\mathscr{P}}{\sigma A e}}$$

Now, we make several simplifying assumptions in our model:

1. The rate of energy transfer $\mathscr{P}$ out of the filament by electromagnetic radiation is equal to the rate of energy transferred to the bulb by electrical transmission, 100 W. For a bulb containing gas, this is not true because another transfer mechanism is available — the gas can carry energy by convection from the hot filament to the glass envelope.

2. The filament has emissivity $e = 1$. This is a simplification model, because a perfect emitter is an ideal that does not exist in nature. A filament that looks dark

when cool has $e > 0.5$, so our value is of the correct order of magnitude.

3. Let us model the filament as a cylinder 10 cm long with a radius of 0.050 mm. This allows us to calculate the surface area:

$$A = (2\pi r)\ell = 2\pi(0.05 \times 10^{-3} \text{ m})(0.10 \text{ m})$$
$$\approx 3 \times 10^{-5} \text{ m}^2$$

With these assumptions, our estimate of the temperature is,

$$T = \sqrt[4]{\frac{\mathscr{P}}{\sigma A e}}$$
$$= \sqrt[4]{\frac{100 \text{ W}}{(5.7 \times 10^{-8} \text{ W/m}^2 \cdot \text{K}^4)(3 \times 10^{-5} \text{ m}^2)(1)}}$$
$$= 2.7 \times 10^3 \text{ K} \sim 10^3 \text{ K}$$

This order of magnitude agrees with typical filament temperatures in lightbulbs, despite our somewhat unrealistic assumptions.

### THINKING PHYSICS 17.3

If you sit in front of a fire with your eyes closed, you can feel significant warmth in your eyelids. If you now put on a pair of eyeglasses and repeat this activity, your eyelids will not feel nearly so warm. Why?

**Reasoning**    Much of the warmth you feel is due to electromagnetic radiation from the fire. A large fraction of this radiation is in the infrared part of the electromagnetic spectrum. (We will study the electromagnetic spectrum in detail in Chapter 24.) Your eyelids are particularly sensitive to infrared radiation. On the other hand, glass is very opaque to infrared radiation. Thus, when you put on the glasses, you block much of the radiation from reaching your eyelids, and they feel cooler.

### THINKING PHYSICS 17.4

If you inspect a lightbulb that has been operating for a long time, a dark region appears on the inner surface of the bulb. This region is located on the highest parts of the bulb's glass envelope. What is the origin of this dark region, and why is it located at the high point?

**Reasoning**    The dark region is tungsten that vaporized from the filament of the lightbulb and collected on the inner surface of the glass. Many lightbulbs contain a gas that allows convection to occur within the bulb. The gas near the filament is at a very high temperature, causing it to expand and float upward, due to Archimedes's principle. As it floats upward, it carries the vaporized tungsten with it, so that the tungsten collects on the surface at the top of the lightbulb.

**THINKING PHYSICS 17.5**

Imagine you are in empty space holding a thermometer. Does the thermometer measure the temperature of empty space? If not, what does it measure?

**Reasoning**  The thermometer does not measure the temperature of empty space because temperature is not defined for a vacuum. We define temperature in terms of internal energy, which is related to the energy of a collection of molecules. There are no molecules in a vacuum.

The thermometer in empty space measures the same thing that all thermometers measure—the temperature of the thermometer. In space, a thermometer's reading depends on its proximity to radiating objects. The thermometer in space can only transfer energy by radiation because all other mechanisms for the stationary thermometer require contact with molecules. Thus, if the thermometer is near a star, its temperature reading may go up as it absorbs energy by electromagnetic radiation until its internal energy stabilizes. If it is carried far away from the star, its reading will go down as its emits more radiation into the environment than it absorbs from the environment.

**context connection**
## 17.11 • ENERGY BALANCE FOR THE EARTH

Let us follow up on our discussion of energy transfer by radiation for the Earth from Section 17.10. This will allow us to perform an initial calculation of the temperature of the Earth.

As mentioned previously, energy arrives at the Earth by electromagnetic radiation from the Sun.* This energy is absorbed by the surface of the Earth and re-radiated out into space according to Stefan's law, Equation 17.35. The only type of energy in the system that can change due to radiation is internal energy. Let us assume that any change in temperature of the Earth is so small over a time interval that we can approximate the change in internal energy as zero. This leads to the following reduction of the continuity equation, Equation 6.20:

$$0 = H_{ER}$$

Two energy transfer mechanisms occur by electromagnetic radiation, so we can write this as

$$0 = H_{ER} \text{ (in)} + H_{ER} \text{ (out)} \quad \longrightarrow \quad H_{ER} \text{ (in)} = -H_{ER} \text{ (out)} \qquad \textbf{[17.37]}$$

where the "in" and "out" refer to energy transfers across the boundary of the system of the Earth. The energy coming into the system is from the Sun, and the energy going out of the system is by thermal radiation emitted from the Earth's surface. Figure 17.21 depicts these energy exchanges. The energy coming in from the Sun comes from only one direction, but the energy radiated out from the Earth's surface leaves in all directions. This distinction will be important in setting up our calculation of the equilibrium temperature.

As mentioned in Section 17.10, the rate of energy transfer per unit area from the Sun is approximately 1 340 W/m$^2$ at the top of the atmosphere. The rate of en-

---

* Some energy arrives at the surface of the Earth from the interior. The source of this energy is radioactive decay (Chapter 30) deep underground. We will ignore this energy because its contribution is much smaller than that due to electromagnetic radiation from the Sun.

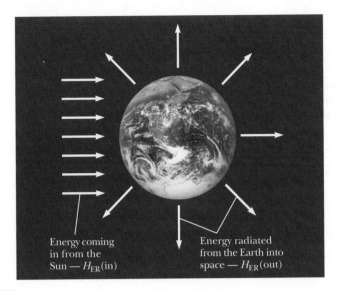

Energy coming in from the Sun — $H_{ER}(\text{in})$

Energy radiated from the Earth into space — $H_{ER}(\text{out})$

**Figure 17.21**

Energy exchanges by electromagnetic radiation for the Earth. The Sun is at the left of the diagram and is not visible. When the rates of energy transfer due to these exchanges are equal, the temperature of the Earth remains constant.

ergy transfer per area is called **intensity,** and the intensity of radiation from the Sun at the top of the atmosphere is called the **Solar constant** $I_S$. A large amount of this energy is in the form of visible radiation, to which the atmosphere is very transparent. The radiation emitted from the Earth's surface, however, is not visible. For a radiating object at the temperature of the Earth's surface, the radiation peaks in the infrared, with greatest intensity at a wavelength of about 10 $\mu$m. In general, objects with typical household temperatures have wavelength distributions in the infrared, so that we do not see them glowing visibly. Only much hotter items emit enough radiation to be seen visibly. An example is a household electric stove burner. When turned off, it emits a small amount of radiation, mostly in the infrared. When turned to its highest setting, its much higher temperature results in significant radiation, with much of it in the visible. As a result, it appears to glow with a reddish color—it is described as *red-hot*.

Let us divide Equation 17.37 by the time interval $\Delta t$ during which the energy transfer occurs, which gives us

$$\mathcal{P}_{ER}\,(\text{in}) = -\mathcal{P}_{ER}\,(\text{out}) \qquad\qquad \textbf{[17.38]}$$

We can express the rate of energy transfer into the top of the atmosphere of the Earth in terms of the Solar constant $I_S$

$$\mathcal{P}_{ER}\,(\text{in}) = I_S\,A_c$$

where $A_c$ is the circular cross-sectional area of the Earth. Not all of the radiation arriving at the top of the atmosphere reaches the ground. A fraction of it is reflected from clouds and the ground and escapes back into space. For the Earth, this fraction is about 30%, so only 70% of the incident radiation reaches the surface. Using this fact, we modify the input power, assuming 70.0% reaches the surface:

$$\mathcal{P}_{ER}\,(\text{in}) = (0.700)I_S\,A_c$$

Stefan's law can be used to express the outgoing power, assuming that the Earth is a perfect emitter:

$$\mathscr{P}_{ER}\,(\text{out}) = -\sigma A T^4$$

In this expression, $A$ is the surface area of the Earth, and the negative sign indicates that energy is leaving the Earth. Substituting the expressions for the input and output power into Equation 17.38, we have

$$(0.700)I_S A_c = -(-\sigma A T^4)$$

Solving for the temperature of the Earth's surface gives

$$T = \sqrt[4]{\frac{(0.700)I_S A_c}{\sigma A}}$$

Substituting the numbers, we find

$$T = \sqrt[4]{\frac{(0.700)(1\,340\text{ W/m}^2)(\pi R_E^2)}{(5.67 \times 10^{-8}\text{ W/m}^2 \cdot \text{K}^4)(4\pi R_E^2)}} = 254\text{ K} \qquad \text{[17.39]}$$

Measurements show that the average global temperature at the surface of the Earth is 288 K, about 34 K warmer than the temperature from our calculation. This difference indicates that a major factor was left out of our analysis. The major factor is the thermodynamic effects of the atmosphere, which result in additional energy from the Sun being "trapped" in the system of the Earth and raising the temperature. This effect is not included in the simple energy balance calculation we performed. To evaluate this effect, we must incorporate into our model the principles of thermodynamics of gases for the air in the atmosphere. The details of this incorporation are explored in the next chapter and in the Context Conclusion.

## SUMMARY

**Heat** is a process by which energy is transferred as a consequence of a temperature difference. It is also the amount of energy $Q$ transferred by this process. The **internal energy** $E_{int}$ of a system is the total of the kinetic and potential energies of the system associated with its microscopic components.

The energy required to change the temperature of a substance by $\Delta T$ is

$$Q = mc\,\Delta T \qquad \text{[17.3]}$$

where $m$ is the mass of the substance and $c$ is its **specific heat.**

The energy required to change the phase of a pure substance of mass $m$ is

$$Q = \pm\,mL \qquad \text{[17.5]}$$

The parameter $L$ is called the **latent heat** of the substance and depends on the nature of the phase change and the properties of the substance.

A **quasi-static process** is one that proceeds slowly enough to allow the system to always be in a state of thermal equilibrium.

The **work** done on a gas as its volume changes from some initial value $V_i$ to some final value $V_f$ is

$$W = -\int_{V_i}^{V_f} P\,dV \qquad \text{[17.7]}$$

where $P$ is the pressure, which may vary during the process.

The **first law of thermodynamics** is a special case of the continuity equation for energy, relating the internal energy of a system to energy transfer by heat and work:

$$\Delta E_{int} = Q + W \qquad \text{[17.8]}$$

where $Q$ is the energy transferred across the boundary of the system by heat and $W$ is the work done on the system. Although $Q$ and $W$ both depend on the path taken from the initial state to the final state, internal energy is a **state variable,** so that the quantity $\Delta E_{int}$ is independent of path.

An **adiabatic process** is one in which no energy is transferred by heat between the system and its surroundings ($Q = 0$). In this case, the first law gives $\Delta E_{int} = W$.

An **isobaric process** is one that occurs at constant pressure. The work done on the gas in such a process is $-P\,\Delta V$.

An **isovolumetric process** is one that occurs at constant volume. No work is done in such a process.

An **isothermal process** is one that occurs at constant temperature. The work done on an ideal gas during an isothermal process is

$$W = -nRT \ln\left(\frac{V_f}{V_i}\right) \qquad \text{[17.12]}$$

In a **cyclic process** (one that originates and terminates at the same state), $\Delta E_{int} = 0$, and therefore $Q = -W$.

We define the **molar specific heats** of an ideal gas with the following equations:

$$Q = nC_V \Delta T \quad \text{(constant volume)} \qquad \text{[17.13]}$$

$$Q = nC_P \Delta T \quad \text{(constant pressure)} \qquad \text{[17.14]}$$

where $C_V$ is the **molar specific heat at constant volume** and $C_P$ is the **molar specific heat at constant pressure.**

The molar specific heat at constant volume is related to internal energy as follows:

$$C_V = \frac{1}{n} \frac{dE_{int}}{dT} \qquad \text{[17.19]}$$

The change in internal energy of an ideal gas for *any* process in which the temperature change is $\Delta T$ is

$$\Delta E_{int} = nC_V \Delta T \qquad \text{[17.18]}$$

The **molar specific heat at constant volume** and **molar specific heat at constant pressure** for all ideal gases are related as follows

$$C_P - C_V = R \qquad \text{[17.21]}$$

For an ideal gas undergoing an adiabatic process, the pressure and volume are related as

$$PV^\gamma = \text{constant} \qquad \text{[17.24]}$$

where

$$\gamma = \frac{C_P}{C_V} \qquad \text{[17.22]}$$

The **theorem of equipartition of energy** can be used to predict the molar specific heat at constant volume for various types of gases. Monatomic gases can only store energy by means of translational motion of the molecules of the gas. Diatomic and polyatomic gases can also store energy by means of rotation and vibration of the molecules. For a given molecule, the rotational and vibrational energies are quantized, so their contribution does not enter into the internal energy until the temperature is raised to a sufficiently high value.

**Thermal conduction** is the transfer of energy by molecular collisions. It is driven by a temperature difference, and the rate of energy transfer is

$$\mathcal{P} = kA \left| \frac{dT}{dx} \right| \qquad \text{[17.33]}$$

where the proportionality constant $k$ is called the **thermal conductivity** of the material, and $dT/dx$ is the **temperature gradient** (the variation of temperature with position).

**Convection** is energy transfer by means of a moving fluid.

All objects emit **electromagnetic radiation** continuously in the form of **thermal radiation,** which depends on temperature according to **Stefan's law:**

$$\mathcal{P} = \sigma A e T^4 \qquad \text{[17.35]}$$

## QUESTIONS

1. Clearly distinguish among temperature, heat, and internal energy.
2. When a sealed Thermos bottle full of hot coffee is shaken, what changes, if any, occur in (a) the temperature of the coffee and (b) the internal energy of the coffee?
3. Using the first law of thermodynamics, explain why the *total* energy of an isolated system is always constant.
4. Ethyl alcohol has about one-half the specific heat of water. If equal masses of alcohol and water in separate beakers are supplied with the same amount of energy, compare the temperature increases of the two liquids.
5. A small crucible is taken from a 200°C oven and immersed in a tub full of water at room temperature (this process is often referred to as *quenching*). What is the approximate final equilibrium temperature?
6. The U.S. penny is now made of copper-coated zinc. Can a calorimetric experiment be devised to test for the metal content in a collection of pennies? If so, describe the procedure.
7. What is wrong with the statement: "Given any two bodies, the one with the higher temperature contains more heat"?
8. What is a major problem that arises in measuring specific heats if a sample with a temperature above 100°C is placed in water?
9. In a daring lecture demonstration, an instructor dips his wetted fingers into molten lead (327°C) and withdraws them quickly, without getting burned. How is this possible? (This is a dangerous experiment, which you should *not* attempt.)
10. The air temperature above coastal areas is profoundly influenced by the large specific heat of water. One reason is that the energy released when 1 m³ of water cools by 1.0°C will raise the temperature of an enormously larger volume of air by 1.0°C. Estimate this volume of air. The specific

heat of air is approximately 1.0 kJ/kg · °C. Take the density of air to be 1.3 kg/m³.

**11.** You need to pick up a very hot cooking pot in your kitchen. You have a pair of hot pads. Should you soak them in cold water or keep them dry to be able to pick up the pot most comfortably?

**12.** Why is it more comfortable to hold a cup of hot tea by the handle rather than by wrapping your hands around the teacup?

**13.** A tile floor in a bathroom may feel uncomfortably cold to your bare feet, but a carpeted floor in an adjoining room at the same temperature will feel warm. Why?

**14.** Why can potatoes be baked more quickly when a metal skewer has been inserted through them?

**15.** Why do heavy draperies over the windows help keep a home warm in the winter and cool in the summer?

**16.** If you wish to cook a piece of meat thoroughly on an open fire, why should you not use a high flame? (*Note:* Carbon is a good thermal insulator.)

**17.** Concrete has a higher specific heat than soil. Use this fact to explain (partially) why cities have a higher average night-time temperature than the surrounding countryside. If a city is hotter than the surrounding countryside, would you expect breezes to blow from city to country or from country to city? Explain.

**18.** Pioneers stored fruits and vegetables in underground cellars. Discuss as fully as possible this choice for a storage site.

**19.** In winter, the pioneers mentioned in Question 18 stored an open barrel of water alongside their produce. Why?

**20.** Suppose you pour hot coffee for your guests, and one of them chooses to drink the coffee after it has been in the cup for several minutes. To have the warmest coffee, should the person add the cream just after the coffee is poured or just before drinking? Explain.

**21.** The following warning sign is often seen on highways just before a bridge: "Caution: Bridge surface freezes before road surface." Which of the three energy transfer processes discussed in Section 17.10 is most important in causing a bridge surface to freeze before a road surface on very cold days?

## PROBLEMS

1, 2, 3 = straightforward, intermediate, challenging      □ = full solution available in the *Student Solutions Manual and Study Guide*

**web** = solution posted at **http://www.harcourtcollege.com/physics/**      💻 = computer useful in solving problem

🔲 = Interactive Physics      ▨ = paired numerical/symbolic problems      ▧ = life science application

### Section 17.1   Heat and Internal Energy

**1.** On his honeymoon James Joule traveled from England to Switzerland. He attempted to verify his idea of the interconvertibility of mechanical energy and internal energy by measuring the increase in temperature of water that fell in a waterfall. For the waterfall near Chamonix in the French Alps, which has a 120-m drop, what maximum temperature rise could Joule expect? He did not succeed in measuring it, partly because evaporation cooled the falling water, and also because his thermometer was not sufficiently sensitive.

**2.** Consider Joule's apparatus shown in Figure P17.2. Each of the two masses is 1.50 kg, and the insulated tank is filled with 200 g of water. What is the increase in the temperature of the water after the masses fall through a distance of 3.00 m?

### Section 17.2   Specific Heat

**3.** A 50.0-g sample of copper is at 25.0°C. If 1 200 J of energy is added to it by heat, what is the final temperature of the copper?

**Figure P17.2**   The falling weights rotate the paddles, causing the temperature of the water to increase.

**4.** Systematic use of solar energy can yield a large saving in the cost of winter space heating for a typical home in the north central United States. If the house has good insulation, you may model it as losing energy by heat steadily at the rate 6 000 W on a day in April when the average exterior temperature is 4°C, and when the conventional heating system is not used at all. The passive solar energy collector can consist simply of very large windows in a room facing south. Sunlight shining in during the daytime is absorbed by the floor, interior walls, and objects in the room, raising their temperature to 38.0°C. As the Sun goes down, insulating draperies or shutters are closed over the windows. During the period between 5:00 PM and 7:00 AM, the temperature of the house drops, and a sufficiently large "thermal mass" is required to keep it from dropping too far. The thermal mass can be a large quantity of stone (with specific heat 850 J/kg·°C) in the floor and the interior walls exposed to sunlight. What mass of stone is required if the temperature is not to drop below 18.0°C overnight?

5. A 1.50-kg iron horseshoe initially at 600°C is dropped into a bucket containing 20.0 kg of water at 25.0°C. What is the final temperature? (Ignore the specific heat of the container, and assume that a negligible amount of water boils away.)

6. An aluminum calorimeter with a mass of 100 g contains 250 g of water. The calorimeter and water are in thermal equilibrium at 10.0°C. Two metallic blocks are placed into the water. One is a 50.0-g piece of copper at 80.0°C. The other block has a mass of 70.0 g and is originally at a temperature of 100°C. The entire system stabilizes at a final temperature of 20.0°C. (a) Determine the specific heat of the unknown sample. (b) Guess the material of the unknown, using the data in Table 17.1.

7. If a mass $m_h$ of water at $T_h$ is poured into an aluminum cup of mass $m_{Al}$ containing mass $m_c$ of water at $T_c$, where $T_h > T_c$, what is the equilibrium temperature of the system?

8. A water heater is operated by solar power. If the solar collector has an area of 6.00 m², and the intensity delivered by sunlight is 550 W/m², how long does it take to increase the temperature of 1.00 m³ of water from 20.0°C to 60.0°C?

## Section 17.3   Latent Heat and Phase Changes

9. How much energy is required to change a 40.0-g ice cube from ice at −10.0°C to steam at 110°C?

10. A 50.0-g copper calorimeter contains 250 g of water at 20.0°C. How much steam must be condensed into the water if the final temperature of the system is to reach 50.0°C?

11. A 3.00-g lead bullet at 30.0°C is fired at a speed of 240 m/s into a large block of ice at 0°C, in which it becomes embedded. What quantity of ice melts?

12. A 1.00-kg block of copper at 20.0°C is dropped into a large vessel of liquid nitrogen at 77.3 K. How many kilograms of nitrogen boil away by the time the copper reaches 77.3 K? (The specific heat of copper is 0.092 0 cal/g ·°C. The latent heat of vaporization of nitrogen is 48.0 cal/g.)

13. In an insulated vessel, 250 g of ice at 0°C is added to 600 g of water at 18.0°C. (a) What is the final temperature of the system? (b) How much ice remains when the system reaches equilibrium?

14. Assume that a hailstone at 0°C falls through air at a uniform temperature of 0°C and lands on a sidewalk also at this temperature. From what initial height must the hailstone fall in order to entirely melt on impact?

15. If 90.0 g of molten lead at 327.3°C is poured into a 300-g casting form made of iron and initially at 20.0°C, what is the final temperature of the system? (Assume no energy loss to the environment occurs.)

16. A resting adult of average size converts chemical energy in food into internal energy at the rate 120 W, called her *basal metabolic rate*. To stay at constant temperature, the body must transfer energy out of the body at the same rate. Several processes exhaust energy from your body. Usually the most important is thermal conduction into the air in contact with your exposed skin. If you are not wearing a hat, a convection current of warm air rises vertically from your head like a plume from a smokestack. Your body also loses energy by electromagnetic radiation, by your exhaling warm air, and by evaporation of perspiration. In this problem consider still another pathway for energy loss: moisture in exhaled breath. Suppose you breathe out 22.0 breaths per minute, each with a volume of 0.600 L. Assume you inhale dry air and exhale air at 37°C containing water vapor with a vapor pressure of 3.20 kPa. The vapor comes from evaporation of liquid water in your body. Model the water vapor as an ideal gas. Assume its latent heat of evaporation at 37°C is the same as its heat of vaporization at 100°C. Calculate the rate at which you lose energy by exhaling humid air.

## Section 17.4   Work in Thermodynamic Processes

17. A sample of ideal gas is expanded to twice its original volume of 1.00 m³ in a quasi-static process for which $P = \alpha V^2$, with $\alpha = 5.00$ atm/m⁶, as shown in Figure P17.17. How much work is done on the expanding gas?

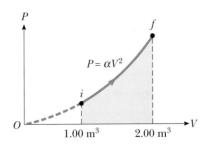

**Figure P17.17**

18. (a) Determine the work done on a fluid that expands from $i$ to $f$ as indicated in Figure P17.18. (b) How much work is performed on the fluid if it is compressed from $f$ to $i$ along the same path?

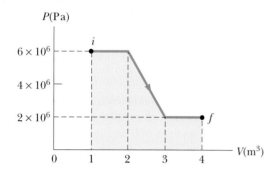

**Figure P17.18**

19. An ideal gas is enclosed in a cylinder with a movable piston on top. The piston has a mass of 8 000 g and an area of 5.00 cm$^2$ and is free to slide up and down, keeping the pressure of the gas constant. How much work is done on the gas as the temperature of 0.200 mol of the gas is raised from 20.0°C to 300°C?

20. An ideal gas is enclosed in a cylinder that has a movable piston on top. The piston has a mass $m$ and an area $A$ and is free to slide up and down, keeping the pressure of the gas constant. How much work is done on the gas as the temperature of $n$ mol of the gas is raised from $T_1$ to $T_2$?

## Section 17.5  The First Law of Thermodynamics

21. A thermodynamic system undergoes a process in which its internal energy decreases by 500 J. At the same time, 220 J of work is done on the system. Find the energy transferred to or from it by heat.

22. A gas is taken through the cyclic process described in Figure P17.22. (a) Find the net energy transferred to the system by heat during one complete cycle. (b) If the cycle is reversed—that is, if the process follows the path $ACBA$—what is the net energy input per cycle by heat?

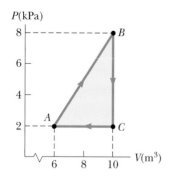

**Figure P17.22**  Problems 22 and 23.

23. Consider the cyclic process depicted in Figure P17.22. If $Q$ is negative for the process $BC$, and if $\Delta E_{int}$ is negative for the process $CA$, what signs for $Q$, $W$, and $\Delta E_{int}$ are associated with each process?

## Section 17.6  Some Applications of the First Law of Thermodynamics

24. One mole of an ideal gas does 3 000 J of work on its surroundings as it expands isothermally to a final pressure of 1.00 atm and volume of 25.0 L. Determine (a) the initial volume and (b) the temperature of the gas.

25. An ideal gas initially at 300 K undergoes an isobaric expansion at 2.50 kPa. If the volume increases from 1.00 m$^3$ to 3.00 m$^3$ and 12.5 kJ is transferred to the gas by heat, what are (a) the change in its internal energy and (b) its final temperature?

26. How much work is done on the steam when 1.00 mol of water at 100°C boils and becomes 1.00 mol of steam at 100°C at 1.00-atm pressure? Assuming the steam behaves as an ideal gas, determine the change in internal energy of the steam as it vaporizes.

27. A 2.00-mol sample of helium gas initially at 300 K and 0.400 atm is compressed isothermally to 1.20 atm. Assuming the helium behaves as an ideal gas, find (a) the final volume of the gas, (b) the work done on the gas, and (c) the energy transferred by heat.

28. A 1.00-kg block of aluminum is heated at atmospheric pressure such that its temperature increases from 22.0°C to 40.0°C. Find (a) the work done on the aluminum, (b) the energy added to it by heat, and (c) the change in its internal energy.

29. In Figure P17.29, the change in internal energy of a gas that is taken from $A$ to $C$ is $+800$ J. The work done on the gas along path $ABC$ is $-500$ J. (a) How much energy must be added to the system by heat as it goes from $A$ through $B$ and on to $C$? (b) If the pressure at point $A$ is five times that at point $C$, what is the work done on the system in going

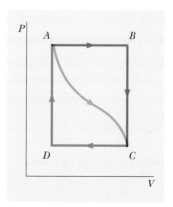

**Figure P17.29**

from *C* to *D*? (c) What is the energy exchanged with the surroundings by heat as the cycle goes from *C* to *A* along the green path? (d) If the change in internal energy in going from point *D* to point *A* is +500 J, how much energy must be added to the system by heat as it goes from point *C* to point *D*?

## Section 17.7   Molar Specific Heats of Ideal Gases

*Note:* You may use data in Table 17.3.

**30.** Calculate the change in internal energy of 3.00 mol of helium gas when its temperature is increased by 2.00 K.

**31.** One mole of hydrogen gas is heated at constant pressure
**web** from 300 K to 420 K. Calculate (a) the energy transferred to the gas by heat, (b) the increase in its internal energy, and (c) the work done on the gas.

**32.** In a constant-volume process, 209 J of energy is transferred by heat to 1.00 mol of an ideal monatomic gas initially at 300 K. Find (a) the increase in internal energy of the gas, (b) the work done on it, and (c) its final temperature.

**33.** A vertical cylinder with a heavy piston contains air at 300 K. The initial pressure is 200 kPa and the initial volume is 0.350 m³. Take the molar mass of air as 28.9 g/mol and assume that $C_V = 5R/2$. (a) Find the specific heat of air at constant volume in units of J/kg·°C. (b) Calculate the mass of the air in the cylinder. (c) Suppose the piston is held fixed. Find the energy input required to raise the temperature of the air to 700 K. (d) Assume again the conditions of the initial state and that the heavy piston is free to move. Find the energy input required to raise the temperature to 700 K.

**34.** A 1-L Thermos bottle is full of tea at 90°C. You pour out one cup and immediately screw the stopper back on. Estimate the order of magnitude of the change in temperature of the tea remaining in the flask that results from the admission of air at room temperature. State the quantities you take as data and the values you measure or estimate for them.

**35.** For a diatomic ideal gas, $C_V = 5R/2$. One mole of this gas has pressure *P* and volume *V*. When the gas is heated, its pressure triples and its volume doubles. If this heating process includes two steps, the first at constant pressure and the second at constant volume, determine the amount of energy transferred to the gas by heat.

## Section 17.8   Adiabatic Processes for an Ideal Gas

**36.** During the compression stroke of a certain gasoline engine, the pressure increases from 1.00 atm to 20.0 atm. Assuming that the process is adiabatic and the gas is ideal with γ = 1.40, (a) by what factor does the volume change and (b) by what factor does the temperature change? (c) If the compression starts with 0.016 0 mol of gas at 27.0°C, what values of *Q*, *W*, and $\Delta E_{int}$ characterize the process?

**37.** A 2.00-mol sample of an ideal gas with γ = 1.40 expands slowly and adiabatically from a pressure of 5.00 atm and a volume of 12.0 L to a final volume of 30.0 L. (a) What is the final pressure of the gas? (b) What are the initial and final temperatures? (c) Find *Q*, *W*, and $\Delta E_{int}$.

**38.** Air in a thundercloud expands as it rises. If its initial temperature is 300 K and no energy is lost by thermal conduction on expansion, what is its temperature when the initial volume has doubled?

**39.** How much work is required to compress 5.00 mol of air at 20.0°C and 1.00 atm to one tenth of the original volume by (a) an isothermal process and (b) an adiabatic process? (c) What is the final pressure in each of these two cases?

**40.** A 4.00-L sample of a diatomic ideal gas with specific heat ratio 1.40, confined to a cylinder, is carried through a closed cycle. The gas is initially at 1.00 atm and at 300 K. First, its pressure is tripled under constant volume. Then, it expands adiabatically to its original pressure. Finally, the gas is compressed isobarically to its original volume. (a) Draw a *PV* diagram of this cycle. (b) Determine the volume of the gas at the end of the adiabatic expansion. (c) Find the temperature of the gas at the start of the adiabatic expansion. (d) Find the temperature at the end of the cycle. (e) What is the net work done on the gas for this cycle?

**41.** During the power stroke in a four-stroke automobile engine, the piston is forced down as the mixture of combustion products and air undergoes an adiabatic expansion. Assume that (1) the engine is running at the rate of 2 500 cycles/min, (2) the gauge pressure right before the expansion is 20.0 atm, (3) the volumes of the mixture right before and after the expansion are 50.0 and 400 cm³, respectively (Fig. P17.41), (4) the time involved

Before

After

**Figure P17.41**

in the expansion is one-fourth that of the total cycle, and (5) the mixture behaves like an ideal gas with specific heat ratio 1.40. Find the average power generated during the expansion.

## Section 17.9  Molar Specific Heats and the Equipartition of Energy

**42.** A certain molecule has *f* degrees of freedom. Show that an ideal gas consisting of such molecules has the following properties: (1) its total internal energy is $fnRT/2$; (2) its molar specific heat at constant volume is $fR/2$; (3) its molar specific heat at constant pressure is $(f + 2)R/2$; (4) its specific heat ratio is $\gamma = C_P/C_V = (f + 2)/f$.

**43.** The *heat capacity* of a sample of a substance is the product
**web** of the mass of the sample and the specific heat of the substance. Consider 2.00 mol of an ideal diatomic gas. Find the total heat capacity at constant volume and at constant pressure (a) if the molecules rotate but do not vibrate and (b) if the molecules both rotate and vibrate.

## Section 17.10  Energy Transfer Mechanisms

**44.** A box with a total surface area of 1.20 m² and a wall thickness of 4.00 cm is made of an insulating material. A 10.0-W electric heater inside the box maintains the inside temperature at 15.0°C above the outside temperature. Find the thermal conductivity *k* of the insulating material.

**45.** A glass window pane has an area of 3.00 m² and a thickness of 0.600 cm. If the temperature difference between its faces is 25.0°C, what is the rate of energy transfer by conduction through the window?

**46.** The human body must maintain its core temperature inside a rather narrow range around 37°C. Metabolic processes, notably muscular exertion, convert chemical energy into internal energy deep in the body's interior. Energy must flow from the interior out to the skin or lungs to be lost by heat to the environment. During moderate exercise, an 80-kg man can metabolize food energy at the rate 300 kcal/h, do 60 kcal/h of mechanical work, and exhaust the remaining 240 kcal/h of energy by heat. Most of the energy is carried from the body interior out to the skin by forced convection: blood is warmed in the interior and then cooled at the skin, which is a few degrees cooler than the body core. Without blood flow, living tissue is a good thermal insulator, with thermal conductivity about 0.210 W/m · °C. Show that blood flow is essential to cool the body by calculating the rate of energy conduction in kcal/h through the tissue layer under the skin. Assume that its area is 1.40 m², its thickness is 2.50 cm, and it is maintained at 37.0°C on one side and at 34.0°C on the other side.

**47.** A bar of gold is in thermal contact with a bar of silver of the same length and area (Fig. P17.47). One end of the compound bar is maintained at 80.0°C, and the opposite end is at 30.0°C. When the energy transfer reaches steady state, what is the temperature at the junction?

**Figure P17.47**

**48.** A power transistor is a solid-state electronic device. Assume it transforms electric energy into internal energy at the rate 1.50 W. The surface area of the transistor is so small that it tends to overheat. To prevent overheating, the transistor is attached to a larger metal heat sink with fins. The surrounding air keeps the heat sink at 35.0°C. The transistor is electrically insulated from the heat sink by a rectangular sheet of mica measuring 8.25 mm by 6.25 mm and 0.085 2 mm thick. The thermal conductivity of mica is 0.075 3 W/m · °C. What is the operating temperature of the transistor?

**49.** The tungsten filament of a certain 100-W lightbulb radiates 2.00 W of light. (The other 98 W is carried away by convection and conduction.) The filament has a surface area of 0.250 mm² and an emissivity of 0.950. Find the filament's temperature. (The melting point of tungsten is 3 683 K.)

**50.** A student is trying to decide what to wear. His bedroom is at 20°C. His skin temperature is 35°C. The area of his exposed skin is 1.50 m². People of all races have skin that is dark in the infrared, with emissivity about 0.900. Find the net energy loss from his body by radiation in 10.0 min.

**51.** The surface of the Sun has a temperature of about 5 800 K. The radius of the Sun is $6.96 \times 10^8$ m. Calculate the total energy radiated by the Sun each second. Assume that the emissivity is 0.965.

## Section 17.11  Context Connection — Energy Balance for the Earth

**52.** At high noon, the Sun delivers 1 000 W to each square meter of a blacktop road. If the hot asphalt loses energy only by radiation, what is its equilibrium temperature?

**53.** At our distance from the Sun, the intensity of solar radiation is 1 340 W/m². The temperature of the Earth is affected by the so-called greenhouse effect of the atmosphere. For comparison, consider a spherical object with no atmosphere, at the same distance from the Sun as the

Earth. Assume that its emissivity is the same for all kinds of electromagnetic waves and that its temperature is uniform over its surface. Identify the projected area over which it absorbs sunlight and the surface area over which it radiates. Compute its equilibrium temperature. Chilly, isn't it? Your calculation applies to (a) the average temperature of the Moon, (b) astronauts in mortal danger aboard the crippled *Apollo 13* spacecraft, and (c) global catastrophe on the Earth if widespread fires caused a layer of soot to accumulate throughout the upper atmosphere so that most of the radiation from the Sun was absorbed there rather than at the surface below the atmosphere.

54. *A theoretical atmospheric lapse rate.* Section 16.7 described experimental data on the decrease in temperature with altitude in the Earth's atmosphere. Model the troposphere as an ideal gas with equivalent molar mass $M$ and ratio of specific heats $\gamma$. Absorption of sunlight at the Earth's surface warms the troposphere from below, so vertical convection currents are continually mixing the air. As a parcel of air rises, its pressure drops and it expands. The parcel does work on its surroundings, so that its internal energy and, therefore, its temperature drop. Assume that the vertical mixing is so rapid as to be adiabatic. (a) Show that the quantity $TP^{(1-\gamma)/\gamma}$ has a uniform value through the layers of the troposphere. (b) By differentiating with respect to altitude $y$, show that the lapse rate is given by

$$\frac{dT}{dy} = \frac{T}{P}\left(1 - \frac{1}{\gamma}\right)\frac{dP}{dy}$$

(c) A lower layer of air must support the weight of the layers above. From Equation 15.4, observe that mechanical equilibrium of the atmosphere requires that the pressure decrease with altitude according to $dP/dy = -\rho g$. The depth of the troposphere is small compared with the radius of the Earth, so you may assume that the free-fall acceleration is uniform. Prove that the lapse rate is

$$\frac{dT}{dy} = -\left(1 - \frac{1}{\gamma}\right)\frac{Mg}{R}$$

Problem 38 in Chapter 16 calls for an evaluation of this theoretical lapse rate on the Earth and on Mars, and for comparison with experimental results.

## Additional Problems

55. A 75.0-kg cross-country skier moves across the snow (Fig. P17.55). The coefficient of friction between the skis and the snow is 0.200. Assume that all the snow beneath his skis is at 0°C and that all the internal energy generated by friction is added to the snow, which sticks to his skis until it melts. How far would he have to ski to melt 1.00 kg of snow?

**Figure P17.55**  A cross-country skier. *(Nathan Bilow/Leo de Wys, Inc.)*

56. An aluminum rod 0.500 m in length and with a cross-sectional area of 2.50 cm$^2$ is inserted into a thermally insulated vessel containing liquid helium at 4.20 K. The rod is initially at 300 K. (a) If one half of the rod is inserted into the helium, how many liters of helium boil off by the time the inserted half cools to 4.20 K? (Assume the upper half does not yet cool.) (b) If the upper end of the rod is maintained at 300 K, what is the approximate boil-off rate of liquid helium after the lower half has reached 4.20 K? (Aluminum has thermal conductivity of 31.0 J/s · cm · K at 4.2 K; ignore its temperature variation. Aluminum has a specific heat of 0.210 cal/g · °C and density of 2.70 g/cm$^3$. The density of liquid helium is 0.125 g/cm$^3$.)

57. On a cold winter day you buy roasted chestnuts from a street vendor. Into the pocket of your down parka you put the change he gives you: coins constituting 9.00 g of copper at − 12.0°C. Your pocket already contains 14.0 g of silver coins at 30.0°C. A short time later the temperature of the copper coins is 4.00°C and is increasing at a rate of 0.500 °C/s. At this time (a) what is the temperature of the silver coins, and (b) at what rate is it changing?

58. An ideal gas initially at $P_i$, $V_i$, and $T_i$ is taken through a cycle as in Figure P17.58. (a) Find the net work done on the gas per cycle. (b) What is the net energy added by heat to the system per cycle? (c) Obtain a numerical value for the net work done per cycle for 1.00 mol of gas initially at 0°C.

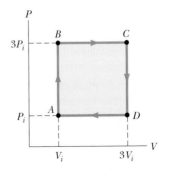

**Figure P17.58**

59. A *flow calorimeter* is an apparatus used to measure the specific heat of a liquid. The technique of flow calorimetry involves measuring the temperature difference between the input and output points of a flowing stream of the liquid while energy is added by heat at a known rate. A liquid of density $\rho$ flows through the calorimeter with volume flow rate $R$. At steady state, a temperature difference $\Delta T$ is established between the input and output points when energy is supplied at the rate $\mathcal{P}$. What is the specific heat of the liquid?

60. One mole of an ideal gas is contained in a cylinder with a movable piston. The initial pressure, volume, and temperature are $P_i$, $V_i$, and $T_i$, respectively. Find the work done on the gas for the following processes, and show each process on a *PV* diagram: (a) An isobaric compression in which the final volume is one-half the initial volume. (b) An isothermal compression in which the final pressure is four times the initial pressure. (c) An isovolumetric process in which the final pressure is triple the initial pressure.

61. A solar cooker consists of a curved reflecting surface that
web focuses sunlight onto the object to be warmed (Fig. P17.61). The solar power per unit area reaching the Earth's surface at the location is 600 W/m², and the cooker has a diameter of 0.600 m. Assume that 40.0% of the incident energy is transferred to 0.500 L of water in an open container, initially at 20.0°C. How long does it take to completely boil away the water? (Ignore the specific heat of the container.)

**Figure P17.61**

62. An iron plate is held against an iron wheel so that a sliding friction force of 50.0 N acts between the two pieces of metal. The relative speed at which the two surfaces slide over each other is 40.0 m/s. (a) Calculate the rate at which mechanical energy is converted to internal energy.

(b) The plate and the wheel each have a mass of 5.00 kg, and each receives 50.0% of the internal energy. If the system is run as described for 10.0 s and each object is then allowed to reach a uniform internal temperature, what is the resultant temperature increase?

63. Water in an electric tea kettle is boiling. The power absorbed by the water is 1.00 kW. Assuming that the pressure of vapor in the kettle equals atmospheric pressure, determine the speed of effusion of vapor from the kettle's spout if the spout has a cross-sectional area of 2.00 cm².

64. *Smokin'!* A pitcher throws a 0.142-kg baseball at 47.2 m/s (Fig. P17.64). As it travels a distance of 19.4 m, the ball slows to 42.5 m/s because of air resistance. Find the change in temperature of the air through which it passes. To find the greatest possible temperature change, you may make the following assumptions: Air has a molar specific heat of $C_P = 7R/2$ and an equivalent molar mass of 28.9 g/mol. The process is so rapid that the cover of the baseball acts as thermal insulation, and the temperature of the ball itself does not change. A change in temperature happens initially only for the air in a cylinder 19.4 m in length and 3.70 cm in radius. This air is initially at 20.0°C.

**Figure P17.64** Orlando Hernandez hurls another strikeout. (*Jed Jacobson/Allsport*)

65. (a) In air at 0°C, a 1.60-kg copper block at 0°C is set sliding at 2.50 m/s over a sheet of ice at 0°C. Friction brings the block to rest. Find the mass of the ice that melts. To describe the process of slowing down, identify the energy input $Q$, the work input $W$, the change in internal energy $\Delta E_{\text{int}}$, and the change in mechanical energy $\Delta K$ for both the block and the ice. (b) A 1.60-kg block of ice at 0°C is set sliding at 2.50 m/s over a sheet of copper at 0°C. Friction brings the block to rest. Find the mass of the ice that melts. Identify $Q$, $W$, $\Delta E_{\text{int}}$, and $\Delta K$ for the block and for

the metal sheet during the process. (c) A thin 1.60-kg slab of copper at 20°C is set sliding at 2.50 m/s over an identical stationary slab at the same temperature. Friction quickly stops the motion. If no energy is lost to the environment by heat, find the change in temperature of both objects. Identify $Q$, $W$, $\Delta E_{int}$, and $\Delta K$ for each object during the process.

66. The function $E_{int} = 3.50nRT$ describes the internal energy of a certain ideal gas. A sample consisting of 2.00 mol of the gas always starts at pressure 100 kPa and temperature 300 K. For each one of the following processes, determine the final pressure, volume, and temperature; the change in internal energy of the gas; the energy added to the gas by heat; and the work done on the gas. (a) The gas is heated at constant pressure to 400 K. (b) The gas is heated at constant volume to 400 K. (c) The gas is compressed at constant temperature to 120 kPa. (d) The gas is compressed adiabatically to 120 kPa.

67. A sample of monatomic ideal gas occupies 5.00 L at atmospheric pressure and 300 K (point $A$ in Figure P17.67). It is heated at constant volume to 3.00 atm (point $B$). Then it is allowed to expand isothermally to 1.00 atm (point $C$) and at last compressed isobarically to its original state. (a) Find the number of moles in the sample. (b) Find the temperature at points $B$ and $C$ and the volume at point $C$. (c) Assuming that the specific heat does not depend on temperature, so that $E_{int} = 3nRT/2$, find the internal energy at points $A$, $B$, and $C$. (d) Tabulate $P$, $V$, $T$, and $E_{int}$ at the states at points $A$, $B$, and $C$. (e) Now consider the processes $A \rightarrow B$, $B \rightarrow C$, and $C \rightarrow A$. Describe just how to carry out each process experimentally. (f) Find $Q$, $W$, and $\Delta E_{int}$ for each of the processes. (g) For the whole cycle $A \rightarrow B \rightarrow C \rightarrow A$ find $Q$, $W$, and $\Delta E_{int}$.

68. A pond of water at 0°C is covered with a layer of ice 4.00 cm thick. If the air temperature stays constant at −10.0°C, how long does it take for the ice thickness to increase to 8.00 cm? *Hint:* Use Equation 17.33 in the form

$$\frac{dQ}{dt} = kA\frac{\Delta T}{x}$$

and note that the incremental energy $dQ$ extracted from the water through the thickness $x$ of ice is the amount required to freeze a thickness $dx$ of ice. That is, $dQ = L\rho A\,dx$, where $\rho$ is the density of the ice, $A$ is the area, and $L$ is the latent heat of fusion.

69. The average thermal conductivity of the walls (including the windows) and roof of the house depicted in Figure P17.69 is 0.480 W/m·°C, and their average thickness is 21.0 cm. The house is heated with natural gas having a heat of combustion (i.e., the energy provided per cubic meter of gas burned) of 9 300 kcal/m³. How many cubic meters of gas must be burned each day to maintain an inside temperature of 25.0°C if the outside temperature is 0.0°C? Disregard radiation and the energy lost by heat through the ground.

**Figure P17.69**

70. The inside of a hollow cylinder is maintained at a temperature $T_a$, and the outside is at a lower temperature $T_b$ (Fig. P17.70). The wall of the cylinder has a thermal conductivity $k$. Ignoring end effects, show that the rate of en-

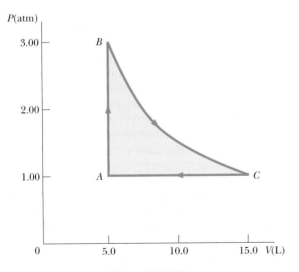

$P$(atm)

3.00 — $B$

2.00 —

1.00 — $A$                 $C$

0      5.0      10.0      15.0   $V$(L)

**Figure P17.67**

**Figure P17.70**

ergy conduction from the inner to the outer surface in the radial direction is

$$\frac{dQ}{dt} = 2\pi Lk \left[ \frac{T_a - T_b}{\ln(b/a)} \right]$$

(*Hint:* The temperature gradient is $dT/dr$. Note that a radial energy current passes through a concentric cylinder of area $2\pi rL$.)

71. The passenger section of a jet airliner is in the shape of a cylindrical tube with a length of 35.0 m and an inner radius of 2.50 m. Its walls are lined with an insulating material 6.00 cm in thickness and having a thermal conductivity of $4.00 \times 10^{-5}$ cal/s·cm·°C. A heater must maintain the interior temperature at 25.0°C when the outside temperature is $-35.0$°C. What power must be supplied to the heater? (Use the result from Problem 70.)

72. A student obtains the following data in a calorimetry experiment designed to measure the specific heat of aluminum:

| | |
|---|---|
| Initial temperature of water and calorimeter | 70°C |
| Mass of water | 0.400 kg |
| Mass of calorimeter | 0.040 kg |
| Specific heat of calorimeter | 0.63 kJ/kg·°C |
| Initial temperature of aluminum | 27°C |
| Mass of aluminum | 0.200 kg |
| Final temperature of mixture | 66.3°C |

Use these data to determine the specific heat of aluminum. Your result should be within 15% of the value listed in Table 17.1.

## ANSWERS TO QUICK QUIZZES

**17.1** (a) Water, glass, iron. Because water has the highest specific heat (4 186 J/kg·°C), it undergoes the smallest change in temperature. Glass is next (837 J/kg·°C), and iron (448 J/kg·°C) is last. (b) Iron, glass, water. For a given temperature increase, the energy transfer by heat is proportional to the specific heat.

**17.2** The following figure shows a graphical representation of the internal energy of the ice in parts A to E as a function of energy added. Notice that this graph looks quite different from Figure 17.3—it doesn't have the flat portions during the phase changes. This graph is just a plot of the continuity equation for energy, $\Delta E_{int} = Q$. Regardless of how the temperature varies in Figure 17.3, the internal energy of the system simply increases linearly with energy input. As we have mentioned, the *form* of the internal energy change is different for the different portions of Fig-

ure 17.3, but the *amount* of internal energy change for a given energy input is constant.

**17.3** C, A, E. The slope is the ratio of the temperature change to the amount of energy input. Thus, the slope is proportional to the reciprocal of the specific heat. Water, which has the highest specific heat, has the lowest slope.

**17.4**

| Situation | System | Q | W | $\Delta E_{int}$ |
|---|---|---|---|---|
| 1. Rapidly pumping up a bicycle tire | Air in the pump | 0 | + | + |
| 2. Pan of room-temperature water sitting on a hot stove | Water in the pan | + | 0 | + |
| 3. Air quickly leaking out of a balloon | Air originally in balloon | 0 | − | − |

(a) Because the pumping is rapid, no energy enters or leaves the system by heat. Because $W > 0$ when work is done *on* the system, it is positive here. Thus, $\Delta E_{int} = Q + W$ must be positive. The air in the pump is warmer. (b) No work is done either by or on the system, but energy flows into the water by heat from the hot burner, making both $Q$ and $\Delta E_{int}$ positive. (c) Again no energy flows into or out of the system by heat, but the air molecules escaping from the balloon do work on the surrounding air molecules as they push them out of the way. Thus $W$ is negative and $\Delta E_{int}$ is negative. The decrease in internal energy is evidenced by the fact that the escaping air becomes cooler.

**Figure QQA17.2**

**17.5** The first law does not adequately describe this situation because a source of internal energy exists within the system—food from previous meals. The body can use this source to maintain body temperature. In fact, strenuous exercise *raises* the body temperature because of the conversion of this energy from stored energy to internal energy, and the body must be cooled, which is the function of perspiration.

**17.6** *A* is isovolumetric, *B* is adiabatic, *C* is isothermal, *D* is isobaric.

**17.7** On a metal surface, energy can move relatively rapidly by thermal conduction from the rubbed region to other regions of the surface. Wood, on the other hand, is not a good thermal conductor. As a result, internal energy at the rubbed region cannot move rapidly to other areas of the surface and the different areas exhibit a distinct temperature difference.

**17.8** The mailbox is both emitting and absorbing radiation. On the sides, it is absorbing radiation from other objects in the environment—cars, trees, houses, and so on. On the top, during the night, it absorbs radiation from the sky. The amount of radiation from the sky is much smaller than that from nearby objects. Thus, the top can become cold enough to form frost during the night, while the absorption of radiation on the sides can keep the temperature high enough to prevent frost from forming.

# chapter 18

This steam-driven locomotive obtains its energy by burning wood or coal. The generated energy vaporizes water into steam, which powers the locomotive. Modern locomotives use diesel fuel instead of wood or coal. Whether old-fashioned or modern, such locomotives are heat engines that extract energy from a burning fuel and convert a fraction of it to mechanical energy.

*(Richard A. Cook III/Stone)*

# Heat Engines, Entropy, and the Second Law of Thermodynamics

The first law of thermodynamics that we studied in the preceding chapter is a statement of conservation of energy. This law places no restrictions on the types of energy conversions that can occur. Contrary to what the first law might suggest, only certain types of energy conversions are observed to take place. Consider the following examples of processes that are consistent with the first law of thermodynamics in either direction, but that proceed only in a particular direction in practice.

1. When two objects at different temperatures are placed in thermal contact with each other, energy always flows by heat from the warmer to the cooler object. We never see energy flow from the cooler object to the warmer object.

2. A rubber ball dropped to the ground bounces several times and eventually comes to rest, the original gravitational potential energy having been transformed to internal energy in the ball and the ground. A ball lying on the ground, however, never gathers internal energy up from the ground and begins bouncing on its own.

3. If oxygen and nitrogen are maintained in separate halves of a container by a membrane, and the membrane is punctured, the oxygen and nitrogen molecules mix together. We never see a mixture of oxygen and nitrogen spontaneously separate into separate sides of the container.

These are all *irreversible processes;* that is, they occur naturally in only one direction.* In this chapter we investigate a new fundamental principle that allows us to understand why these processes occur in one direction only.* The second law of thermodynamics, which is the primary focus of this chapter, establishes which natural processes do and which do not occur.

From an engineering viewpoint, perhaps the most important application of the concepts in this chapter is the limited efficiency of heat engines. We shall explore the theoretical background of the fact that a machine capable of continuously converting internal energy in a cyclic process *completely* to other forms of energy cannot be constructed.

## 18.1 • HEAT ENGINES AND THE SECOND LAW OF THERMODYNAMICS

See *Core Concepts in Physics*, Screen 10.8

One device that is very useful in understanding the second law of thermodynamics is the heat engine. A **heat engine** is a device that converts internal energy to other useful forms, such as kinetic energy. In a typical process for producing electricity in a power plant, for instance, coal or some other fuel is burned, and the resulting internal energy is used to convert water to steam. This steam is directed at the blades of a turbine, setting it into rotation. Finally, the mechanical energy associated with this rotation is used to drive an electric generator. In another heat engine, the internal combustion engine in your automobile, energy enters the engine by matter transfer as the fuel is injected into the cylinder and a fraction of this energy is converted to mechanical energy.

In general, a heat engine carries some working substance through cyclic processes† during which (1) energy is transferred from a source at a high temperature, (2) work is done by the engine, and (3) energy is expelled by the engine to a source at a lower temperature. As an example, consider the operation of a steam engine in which the working substance is water. The water in the engine is carried through a cycle in which it first evaporates into steam in a boiler and then expands against a piston. After the steam is condensed with cooling water, it is returned to the boiler, and the process is repeated.

**Lord Kelvin** (1824–1907)

British physicist and mathematician. Born William Thomson in Belfast, Northern Ireland, Kelvin was the first to propose the use of an absolute scale of temperature. The Kelvin temperature scale is named in his honor. Kelvin's work in thermodynamics led to the idea that energy cannot pass spontaneously from a colder body to a hotter body. (*J. L. Charmet/SPL/Photo Researchers*)

---

\* As we shall see in this chapter, it is more proper to say that the set of events in the time-reversed sense is highly improbable. From this viewpoint, events in one direction are vastly more probable than those in the opposite direction.

† The automobile engine is not strictly a heat engine according to the cyclic process description, because the substance (the air–fuel mixture) undergoes only one cycle and is then expelled through the exhaust system.

It is useful to draw a heat engine schematically as in the pictorial representation in Figure 18.1. The engine absorbs a quantity of energy $Q_h$ from the hot reservoir.* The engine does work $W_{eng}$ (so that *negative* work $W = -W_{eng}$ is done *on* the engine), and then gives up energy $Q_c$ to the cold reservoir. Because the working substance goes through a cycle, its initial and final internal energies are equal, so $\Delta E_{int} = 0$. The engine can be modeled as a nonisolated system in steady state. Hence, from the first law

$$\Delta E_{int} = Q + W \longrightarrow Q_{net} = -W = W_{eng}$$

and we see that the **work $W_{eng}$ done by a heat engine equals the net energy absorbed by the engine.** As we can see from Figure 18.1, $Q_{net} = |Q_h| - |Q_c|$. Therefore,

$$W_{eng} = |Q_h| - |Q_c| \tag{18.1}$$

If the working substance is a gas, the **net work done by the engine for a cyclic process is the area enclosed by the curve representing the process on a *PV* diagram.** This is shown for an arbitrary cyclic process in Figure 18.2.

The **thermal efficiency** $e$ of a heat engine is defined as the ratio of the net work done by the engine to the energy absorbed at the higher temperature during one cycle:

- *Thermal efficiency of a heat engine*

$$e = \frac{W_{eng}}{|Q_h|} = \frac{|Q_h| - |Q_c|}{|Q_h|} = 1 - \frac{|Q_c|}{|Q_h|} \tag{18.2}$$

**Figure 18.1**

Schematic representation of a heat engine. The engine absorbs energy $Q_h$ from the hot reservoir, expels energy $Q_c$ to the cold reservoir, and does work $W_{eng}$.

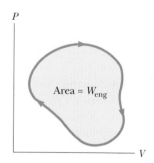

**Figure 18.2**

The *PV* diagram for an arbitrary cyclic process. The net work done by the engine equals the area enclosed by the curve.

* We will adopt a simplification model in which we assume the energy transfer from the reservoir is by heat, but will realize that other transfer mechanisms are possible. For example, as mentioned earlier, energy is brought into the cylinder of an automobile engine by matter transfer.

Hot reservoir at $T_h$

$Q_h$

$W_{eng}$

Engine

Cold reservoir at $T_c$

The impossible engine

**Figure 18.3**

Schematic representation of a heat engine that absorbs energy $Q_h$ from a hot reservoir and does an equivalent amount of work. It is not possible to construct such a perfect engine.

We can think of the efficiency as the ratio of what you gain (work) to what you give (energy transfer at the higher temperature). Equation 18.2 shows that a heat engine has 100% efficiency ($e = 1$) only if $Q_c = 0$ (i.e., if no energy is expelled to the cold reservoir). In other words, a heat engine with perfect efficiency would have to expel all the input energy by mechanical work.

The Kelvin–Planck statement of the **second law of thermodynamics** can be stated as follows: **It is impossible to construct a heat engine that, operating in a cycle, produces no effect other than the absorption of energy from a reservoir and the performance of an equal amount of work.** The essence of this form of the second law is that it is theoretically impossible to construct an engine such as that in Figure 18.3 that works with 100% efficiency.

• *Second law of thermodynamics; Kelvin–Planck statement*

## Example 18.1 The Efficiency of an Engine

An engine transfers $2.00 \times 10^3$ J of energy from a hot reservoir during a cycle and transfers $1.50 \times 10^3$ J to a cold reservoir. (a) Find the efficiency of the engine.

**Solution** The efficiency of the engine is calculated using Equation 18.2:

$$e = 1 - \frac{|Q_c|}{|Q_h|} = 1 - \frac{1.50 \times 10^3 \, \text{J}}{2.00 \times 10^3 \, \text{J}} = \boxed{0.250, \text{ or } 25.0\%}$$

(b) How much work does this engine do in one cycle?

**Solution** The work done is the difference between the input and output energies:

$$W_{eng} = |Q_h| - |Q_c| = 2.00 \times 10^3 \, \text{J} - 1.50 \times 10^3 \, \text{J} = \boxed{500 \, \text{J}}$$

**EXERCISE** If an engine has an efficiency of 20.0% and exhausts $3.00 \times 10^3$ J to water used to cool the engine, how much work is done by the engine?

*Answer* 750 J

**EXERCISE** The energy absorbed by an engine is three times greater than the work it performs. (a) What is its thermal efficiency? (b) What fraction of the energy absorbed is expelled to the cold reservoir?

*Answer* (a) 1/3 (b) 2/3

Sand

Energy reservoir

**Figure 18.4**

A gas in thermal contact with an energy reservoir is compressed slowly as individual grains of sand are dropped onto the piston. The compression is isothermal and reversible.

See Screen 10.9

**Sadi Carnot** (1796–1832)

French physicist Carnot was the first to show the quantitative relationship between work and heat. In 1824 he published his only work — *Reflections on the Motive Power of Heat* —which reviewed the industrial, political, and economic importance of the steam engine. In it, he defined work as "weight lifted through a height." *(FPG)*

## 18.2 • REVERSIBLE AND IRREVERSIBLE PROCESSES

In the next section we shall discuss a theoretical heat engine that is the most efficient engine possible. To understand its nature, we must first examine the meaning of reversible and irreversible processes. A **reversible** process is one for which the system can be returned to its initial conditions along the same path and for which every point along the path is an equilibrium state. A process that does not satisfy these requirements is **irreversible.**

Most natural processes are known to be irreversible—the reversible process is an idealization. The three processes described in the introduction to this chapter are irreversible—we see them proceed in only one direction. The free expansion of a gas discussed in Section 17.6 is irreversible. When the membrane is removed, the gas rushes into the empty half of the vessel and the environment is not changed. No matter how long we watched, we would never see the gas in the full volume spontaneously rush back into only half the volume. The only way we could cause this to happen is to interact with the gas, perhaps by pushing it inward with a piston, but this would result in a change in the environment.

If a real process occurs very slowly so that the system is always very nearly in equilibrium, the process can be modeled as reversible. For example, imagine compressing a gas very slowly by dropping some grains of sand onto a frictionless piston as in Figure 18.4. The pressure, volume, and temperature of the gas are well defined during this isothermal compression. Each added grain of sand represents a small change to a new equilibrium state. The process can be reversed by the slow removal of grains of sand from the piston.

## 18.3 • THE CARNOT ENGINE

In 1824, a French engineer named Sadi Carnot described a theoretical engine, now called a **Carnot engine,** that is of great importance from both practical and theoretical viewpoints. He showed that a heat engine operating in an ideal, reversible cycle—called a **Carnot cycle—**between two energy reservoirs is the most efficient engine possible. Such an ideal engine establishes an upper limit on the efficiencies of all real engines. That is, **the net work done by a working substance taken through the Carnot cycle is the greatest amount of work possible for a given amount of energy supplied to the substance at the upper temperature.**

To describe the Carnot cycle, we shall assume that the working substance in the engine is an ideal gas contained in a cylinder with a movable piston at one end. The cylinder walls and the piston are thermally nonconducting. Four stages of the Carnot cycle are shown in Figure 18.5; Figure 18.6 is the *PV* diagram for the cycle, which consists of two adiabatic and two isothermal processes, all reversible.

- The process $A \rightarrow B$ is an isothermal expansion at temperature $T_h$, in which the gas is placed in thermal contact with an energy reservoir at temperature $T_h$ (Fig. 18.5a). During the process, the gas absorbs energy $Q_h$ by heat from the reservoir and does work $W_{AB}$ in raising the piston.
- In the process $B \rightarrow C$, the base of the cylinder is replaced by a thermally nonconducting wall and the gas expands adiabatically; that is, no energy enters or leaves the system by heat (Fig. 18.5b). During the process, the temperature falls from $T_h$ to $T_c$ and the gas does work $W_{BC}$ in raising the piston.
- In the process $C \rightarrow D$, the gas is placed in thermal contact with an energy reservoir at temperature $T_c$ (Fig. 18.5c) and is compressed isothermally at tempera-

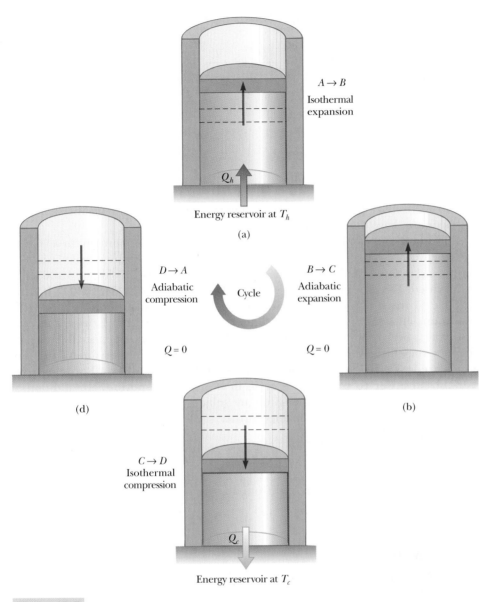

$A \rightarrow B$
Isothermal
expansion

$Q_h$

Energy reservoir at $T_h$

(a)

$D \rightarrow A$
Adiabatic
compression

Cycle

$B \rightarrow C$
Adiabatic
expansion

$Q = 0$

$Q = 0$

(d)

(b)

$C \rightarrow D$
Isothermal
compression

$Q_c$

Energy reservoir at $T_c$

**Figure 18.5**

The Carnot cycle. (a) In process $A \rightarrow B$, the gas expands isothermally while in contact with an energy reservoir at $T_h$. (b) In process $B \rightarrow C$, the gas expands adiabatically ($Q = 0$). (c) In process $C \rightarrow D$, the gas is compressed isothermally while in contact with an energy reservoir at $T_c < T_h$. (d) In process $D \rightarrow A$, the gas is compressed adiabatically. The upward arrows on the piston in (a) and (b) indicate that weights are being removed during the expansions, and the downward arrows in (c) and (d) indicate that weights are being added during the compressions.

ture $T_c$. During this time, the gas expels energy $Q_c$ to the reservoir, and the work done on the gas is $W_{CD}$.

- In the final process $D \rightarrow A$, the base of the cylinder is again replaced by a thermally nonconducting wall (Fig. 18.5d) and the gas is compressed adiabatically. The temperature of the gas increases to $T_h$, and the work done on the gas is $W_{DA}$.

**WEB**

For a java applet showing animations of both pictorial and graphical representations of the Carnot cycle, visit **www.phy.ntnu.edu.tw/java/ carnot/carnot.html**

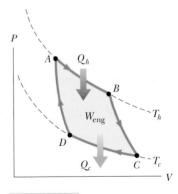

$P$

$A$     $Q_h$

$B$

$W_{eng}$

$T_h$

$D$

$Q_c$     $C$     $T_c$

$V$

**Figure 18.6**

The $PV$ diagram for the Carnot cycle. The net work done $W_{eng}$ equals the net energy received in one cycle, $|Q_h| - |Q_c|$. Note that $\Delta E_{int} = 0$ for the cycle.

Carnot showed that for this cycle

$$\frac{|Q_c|}{|Q_h|} = \frac{T_c}{T_h}$$

[18.3]

Thus, using Equation 18.2, the thermal efficiency of a Carnot engine is

- *Carnot efficiency*

$$e_c = 1 - \frac{T_c}{T_h}$$

[18.4]

**PITFALL PREVENTION 18.1**

**Don't shop for a Carnot engine**

The Carnot engine is an idealization—do not expect a Carnot engine to be developed for commercial use. If a Carnot engine were developed in an effort to maximize the efficiency, it would have zero power output because for all of the processes to be reversible, the engine would have to run infinitesimally slowly.

From this result, we see that all Carnot engines operating between the same two temperatures have the same efficiency.

Equation 18.4 can be applied to any working substance operating in a Carnot cycle between two energy reservoirs. According to this result, the efficiency is zero if $T_c = T_h$, as one would expect. The efficiency increases as $T_c$ is lowered and as $T_h$ is increased. The efficiency can be unity (100%), however, only if $T_c = 0$ K. It is impossible to reach absolute zero,* so such reservoirs are not available. Thus, the maximum efficiency is always less than unity. In most practical cases, the cold reservoir is near room temperature, about 300 K. Therefore, one usually strives to increase the efficiency by raising the temperature of the hot reservoir. **All real engines are less efficient than the Carnot engine because they all operate irreversibly in order to complete a cycle in a brief time period.**[†] In addition to this theoretical limitation, real engines are subject to practical difficulties, including friction, that reduce the efficiency further.

**Quick Quiz 18.1**

If friction could be totally removed from a Carnot engine, would it be 100% efficient?

**Quick Quiz 18.2**

Three engines operate between reservoirs separated in temperature by 300 K. The reservoir temperatures are as follows: engine A: $T_h = 1\,000$ K, $T_c = 700$ K; engine B: $T_h = 800$ K, $T_c = 500$ K; engine C: $T_h = 600$ K, $T_c = 300$ K. Rank the engines in order of theoretically possible efficiency, from highest to lowest.

---

* The inability to reach absolute zero is known as the *third law of thermodynamics*. It would require an infinite amount of energy to lower the temperature of a substance to absolute zero.

[†] In order for the processes in the Carnot cycle to be reversible, they must be carried out infinitesimally slowly. Thus, although the Carnot engine is the most efficient engine possible, it has zero power output because it takes an infinite time interval to complete one cycle! For a real engine, the short time interval for each cycle results in the working substance reaching a high temperature lower than that of the hot reservoir and a low temperature higher than that of the cold reservoir. An engine undergoing a Carnot cycle between this narrower temperature range was analyzed by Curzon and Ahlborn (*Am. J. Phys.*, 43(1), 1975), who found that the efficiency at maximum power output depends only on the reservoir temperatures $T_c$ and $T_h$, and is given by $e_{C-A} = 1 - (T_c/T_h)^{1/2}$. The Curzon–Ahlborn efficiency $e_{C-A}$ provides a closer approximation to the efficiencies of real engines than does the Carnot efficiency.

## Example 18.2  The Steam Engine

A steam engine has a boiler that operates at 500 K. The energy resulting from the burning of the fuel changes water to steam, and this steam then drives the piston. The exhaust temperature is that of the outside air, approximately 300 K. What is the maximum thermal efficiency of this steam engine?

**Solution**  From the expression for the efficiency of a Carnot engine, we find the maximum thermal efficiency for any engine operating between these temperatures:

$$e_c = 1 - \frac{T_c}{T_h} = 1 - \frac{300 \text{ K}}{500 \text{ K}} = \boxed{0.400, \text{ or } 40.0\%}$$

Note that this is the highest theoretical efficiency of the engine. In practice, the efficiency is considerably lower.

**EXERCISE**  Determine the maximum work the engine can perform in each cycle of operation if it absorbs 200 J of energy from the hot reservoir during each cycle.

*Answer*  80.0 J

**EXERCISE**  A heat engine operates between two reservoirs at 20.0°C and 300°C. What is the maximum efficiency possible for this engine?

*Answer*  48.9%

## 18.4 • HEAT PUMPS AND REFRIGERATORS

In a heat engine, the direction of energy flow is from the hot reservoir to the cold reservoir, which is the natural direction. The role of the heat engine is to process the energy from the hot reservoir so as to do useful work. What if we wanted to move energy from the cold reservoir to the hot reservoir? Because this is not the natural direction of flow, we must transfer some energy into a device to cause this to occur. Devices that perform this task are called **heat pumps** or **refrigerators.**

Figure 18.7 is a schematic representation of a heat pump. The cold reservoir temperature is $T_c$, the hot reservoir temperature is $T_h$, and the energy absorbed by the heat pump is $Q_c$. Energy is transferred into the system, which we model as work* $W$, and the energy transferred out of the pump is $Q_h$.

Heat pumps have long been popular for cooling in homes, where they are called *air conditioners,* and are now becoming increasingly popular for heating purposes as well. In the heating mode, a circulating coolant fluid absorbs energy from the outside air (the cold reservoir) and releases energy to the interior of the structure (the hot reservoir). The fluid is usually in the form of a low-pressure vapor when in the coils of the exterior part of the unit, where it absorbs energy from either the air or the ground by heat. This gas is then compressed into a hot, high-pressure vapor and enters the interior part of the unit, where it condenses to a liquid and releases its stored energy. An air conditioner is simply a heat pump installed backward, with "exterior" and "interior" interchanged. The inside of the home is the cold reservoir and the outside air is the hot reservoir.

The effectiveness of a heat pump is described in terms of a number called the **coefficient of performance,** COP. In the heating mode, this is defined as the

**Figure 18.7**

Schematic representation of a heat pump, which absorbs energy $Q_c$ from a cold reservoir and expels energy $Q_h$ to a hot reservoir. The work done on the heat pump is $W$.

---

* The traditional notation is to model the input energy as transferred by work, although most heat pumps operate from electricity, so the more appropriate transfer mechanism *into the device as the system* is *electrical transmission.* If we identify the refrigerant fluid in a heat pump as the system, then the energy transfers into the fluid by work done by a piston attached to a compressor operated electrically. In keeping with tradition, we will schematize the heat pump with input by work regardless of the choice of system.

ratio of the energy transferred by heat into the hot reservoir to the work required to transfer that energy:

$$\text{COP (heat pump)} \equiv \frac{\text{energy transferred to hot reservoir}}{\text{work done on pump}} \qquad \text{[18.5]}$$

$$= \frac{|Q_h|}{W}$$

If the outside temperature is $-4°C$ (25°F) or higher, the COP for a heat pump is about 4. That is, the energy transferred into the house is about four times greater than the work done by the compressor in the heat pump. As the outside temperature decreases, however, it becomes more difficult for the heat pump to extract sufficient energy from the air, and the COP drops. In fact, the COP can fall below unity for temperatures below about $-10°C$ (14°F).

A Carnot cycle heat engine operating in reverse constitutes an ideal heat pump—the heat pump with the highest possible COP for the temperatures between which it operates. The maximum coefficient of performance is

$$\text{COP}_{\text{Carnot}}\text{(heat pump)} = \frac{T_h}{T_h - T_c}$$

Although heat pumps are relatively new products in heating, the refrigerator has been a standard appliance in homes for years. The refrigerator cools its interior by pumping energy from the food storage compartments into the warmer air outside. During its operation, a refrigerator removes energy $Q_c$ from the interior of the refrigerator, and in the process its motor does work $W$. The COP of a refrigerator or of a heat pump used in its cooling cycle is

$$\text{COP(refrigerator)} = \frac{|Q_c|}{W} \qquad \text{[18.6]}$$

An efficient refrigerator is one that removes the greatest amount of energy from the cold reservoir with the least amount of work. Thus, a good refrigerator should have a high coefficient of performance, typically 5 or 6.

The highest possible COP is again that of a refrigerator whose working substance is carried through the Carnot heat engine cycle in reverse:

$$\text{COP}_{\text{Carnot}}\text{(refrigerator)} = \frac{|Q_c|}{|Q_h| - |Q_c|} = \frac{T_c}{T_h - T_c}$$

As the difference between the temperatures of the two reservoirs approaches zero, the theoretical coefficient of performance of a Carnot heat pump approaches infinity. In practice, the low temperature of the cooling coils and the high temperature at the compressor limit the COP to values below 10.

**Quick Quiz 18.3**

Can an electric heat pump deliver more energy than you pay for in electricity costs?

### THINKING PHYSICS 18.1

Suppose the temperature inside a house is held constant at 72°F. Assume that it is 92°F outside in the summer, and the air conditioner (heat pump) keeps the house at 72°F. Later, during the winter, it is 52°F outside and the heat pump is used to keep the house at 72°F again. In both cases, the difference between the inside and outside temperatures is 20°F. The electric bill for the heat pump is higher in the summer than in the winter. Why?

**Reasoning**   In the summer, the interior of the house serves as the low-temperature (cold) reservoir for the heat engine, and the outside air is the high-temperature (hot) reservoir. Energy drawn from the interior air combines with energy entering the heat pump by electrical transmission and the combination is transferred to the external air. Thus, the only advantage of the energy entering by electrical transmission is to pump the energy from the interior to the exterior. In comparison, consider the situation in the winter. In this case, energy drawn from the exterior air combines with energy entering the heat pump by electrical transmission and the combination is transferred to the internal air. Thus, the energy entering by electricity does two things—it operates the heat pump *and* it adds to the internal energy of the interior air. Thus, less energy must be drawn from the external air in winter than from the internal air in the summer, and it costs less to run the heat pump in the winter.

### THINKING PHYSICS 18.2

It is a sweltering summer day and your air conditioning system is not operating. In your kitchen, you have a working refrigerator, and an ice chest full of ice. Which should you open and leave open to cool the room more effectively?

**Reasoning**   The high-temperature reservoir for your kitchen refrigerator is the air in the kitchen. If the refrigerator door were left open, energy would be drawn from the air in the kitchen, passed through the refrigeration system and transferred right back into the air. The result would be that the kitchen would become *warmer,* due to the addition of the energy coming in by electricity to run the refrigeration system. If the ice chest were opened, energy in the air would enter the ice, raising its temperature and causing it to melt. The transfer of energy from the air would cause its temperature to drop. Thus, it would be more effective to open the ice chest.

## 18.5 • AN ALTERNATIVE STATEMENT OF THE SECOND LAW

Suppose you wish to cool off a hot piece of pizza by placing it on a block of ice. You will certainly be successful, because in every similar situation, energy transfer has always taken place from a hot object to a cooler one. Yet, nothing in the first law of thermodynamics says that this energy transfer could not proceed in the opposite direction. (Imagine your astonishment if someday you place a piece of hot pizza on ice and the pizza becomes warmer!) It is the second law that determines the directions of such natural phenomena.

**Figure 18.8**

Schematic representation of a heat pump or refrigerator that absorbs energy from a cold reservoir and expels an equivalent amount of energy to a hot reservoir with no input by work. Such a perfect heat pump is impossible.

 See Screens 10.10 & 10.11

**PITFALL PREVENTION 18.2**

**Entropy is abstract**

 Entropy is one of the most abstract notions in physics, so follow the discussion in this and the subsequent sections very carefully. Be sure that you do not confuse energy and entropy—even though the names sound similar, they are very different.

An analogy can be made with the impossible sequence of events seen in a movie film running backward—such as a person rising out of a swimming pool and landing back on the diving board, an apple rising from the ground and latching onto the branch of a tree, or a pot of hot water becoming colder as it rests over an open flame. Such events occurring backward in time are impossible because they violate the second law of thermodynamics. **Real processes proceed in a preferred direction.**

The second law can be stated in several different ways, but all the statements can be shown to be equivalent. Which form you use depends on the application you have in mind. For example, if you were concerned about the energy transfer between pizza and ice, you might choose to concentrate on the **Clausius statement of the second law: Energy does not flow spontaneously from a cold object to a hot object.** Figure 18.8 shows a heat pump that violates this statement of the second law. Energy is transferring from the cold reservoir to the hot reservoir without an input of work. At first glance, this statement of the second law seems to be radically different from that in Section 18.1. The two are, in fact, equivalent in all respects. Although we shall not prove it here, it can be shown that if either statement is false, so is the other.

## 18.6 • ENTROPY

The zeroth law of thermodynamics involves the concept of temperature and the first law involves the concept of internal energy. Temperature and internal energy are both state variables; that is, they can be used to describe the thermodynamic state of a system. Another state variable, this one related to the second law of thermodynamics, is **entropy** *S*. In this section we define entropy on a macroscopic scale as the German physicist Rudolf Clausius (1822–1888) first expressed it in 1865.

Equation 18.3, which describes the Carnot engine, can be rewritten as

$$\frac{|Q_c|}{T_c} = \frac{|Q_h|}{T_h}$$

Thus, the ratio of the energy transfer by heat in a Carnot cycle to the (constant) temperature at which the transfer takes place has the same magnitude for both isothermal processes. To generalize the current discussion beyond heat engines, let us drop the absolute value notation and revive our original sign convention, in which $Q_c$ represents energy leaving the system of the gas and is therefore a negative number. Thus, we need an explicit negative sign to maintain the equality:

$$-\frac{Q_c}{T_c} = \frac{Q_h}{T_h}$$

We can write this equation as

$$\frac{Q_h}{T_h} + \frac{Q_c}{T_c} = 0 \quad \longrightarrow \quad \sum \frac{Q}{T} = 0 \qquad \text{[18.7]}$$

We have not specified a particular Carnot cycle in generating this equation, so this must be true for all Carnot cycles. Furthermore, by approximating a general reversible cycle with a series of Carnot cycles, we can show that this equation is true for *any* reversible cycle. This suggests that the ratio $Q/T$ may have some special significance. It does indeed, as we note in the following discussion.

Consider any infinitesimal process for a system between two equilibrium states. If $dQ_r$ is the energy transferred by heat as the system follows a reversible path between the states, the change in entropy, regardless of the actual path followed, is equal to this energy transferred by heat along the reversible path divided by the absolute temperature of the system:

$$dS = \frac{dQ_r}{T}$$    [18.8]

• *Change in entropy for an infinitesimal process*

The subscript $r$ on the term $dQ_r$ is a reminder that the heat is to be determined along a reversible path, even though the system may actually follow some irreversible path. Thus, we *must* model a nonreversible process by a reversible process between the same initial and final states in order to calculate the entropy change. In this case, the model might not be close to the actual process at all, but this is not a concern because the entropy change depends only on the initial and final states. The only requirements are that the model process must be reversible and must connect the given initial and final states.

When energy is absorbed by the system, $dQ_r$ is positive and hence the entropy increases. When energy is expelled by the system, $dQ_r$ is negative and the entropy decreases. Note that Equation 18.8 defines not entropy, but rather the *change* in entropy. Hence, the meaningful quantity in a description of a process is the *change* in entropy.

With Equation 18.8, we have a mathematical representation of the change in entropy, but we have developed no mental representation of what entropy means. In this and the next few sections, we explore various aspects of entropy that will allow us to gain a conceptual understanding of it.

Entropy originally found its place in thermodynamics, but its importance grew tremendously as the field of physics called *statistical mechanics* developed, because this method of analysis provided an alternative way of interpreting entropy. In statistical mechanics, a substance's behavior is described in terms of the statistical behavior of its large number of atoms and molecules. Kinetic theory, which we studied in Chapter 16, is an excellent example of the statistical mechanics approach. One of the main outcomes of this treatment is the principle that **isolated systems tend toward disorder, and entropy is a measure of that disorder.**

To understand this notion, we introduce the distinction between **microstates** and **macrostates** for a system. We can do this by looking at an example far removed from thermodynamics, the throwing of dice at a craps table in a casino. For two dice, a *microstate* is the particular combination of numbers on the upturned faces of the dice; for example, 1-3 and 2-4 are two different microstates (Fig. 18.9). The *macrostate* is the sum of the numbers. Thus, the macrostates for the two example microstates given are 4 and 6. Now, here is the central notion that we will need to understand entropy: **the number of microstates associated with a given macrostate is not the same for all macrostates, and the most probable macrostate is that with the largest number of possible microstates.** A macrostate of 7 on our pair of dice has six possible microstates: 1-6, 2-5, 3-4, 4-3, 5-2, and 6-1 (Fig. 18.10a). For a macrostate of 2, there is only one possible microstate: 1-1 (Fig. 18.10b). Thus, a macrostate of 7 has six times as many microstates as a macrostate of 2 and is, therefore, six times as probable. In fact, the macrostate of 7 is the most probable macrostate for two dice. The game of craps is built around these probabilities of various macrostates.

Consider the low probability macrostate 2. The *only* way of achieving this is to have a 1 on each die. We say that this macrostate has a high degree of *order*—we *must* have a 1 on each die for this macrostate to exist. Considering the possible

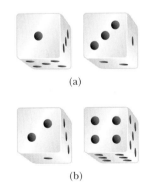

(a)

(b)

**Figure 18.9**

Two different microstates for a throw of two dice. These correspond to two macrostates, having values of (a) 4 and (b) 6.

Possible two-dice microstates for a macrostate of (a) 7 and (b) 2. The macrostate of 7 is more probable because there are more ways of achieving it—more microstates are possible.

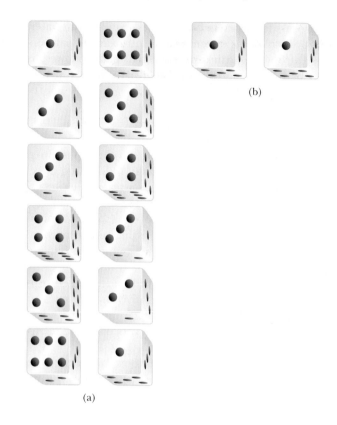

(b)

(a)

microstates for a macrostate of 7, however, we see six possibilities. This macrostate is more *disordered*, because several microstates are possible that will result in the same macrostate. Thus, we conclude that **high-probability macrostates are disordered macrostates, and low-probability macrostates are ordered macrostates.**

As a more physical example, consider the molecules in the air in your room. Let us compare two possible macrostates. Macrostate 1 is the condition in which the oxygen and nitrogen molecules are mixed evenly throughout the room. Macrostate 2 is that in which the oxygen molecules are in the front half of the room and the nitrogen molecules are in the back half. From our everyday experience, it is *extremely unlikely* for macrostate 2 to exist. On the other hand, macrostate 1 is what we would normally expect to see. Let us relate this to the microstates, which correspond to the possible positions of molecules of each type. For macrostate 2 to exist, every molecule of oxygen would have to be in one half of the room and every molecule of nitrogen in the other half—this is a highly ordered and unlikely situation. The probability of this occurring is infinitesimal. For macrostate 1 to exist, both types of molecules are simply distributed evenly around the room—this is a much lower level of order and a highly probable situation. Thus, the mixed state is much more likely than the separated state, and that is what we normally see.

Let us look now at the notion that isolated systems tend toward disorder. The cause of this tendency toward disorder is easily seen. For any given energy of the system, only certain microstates are possible, or accessible. Let us assume that all microstates are equally probable. When the possible macrostates associated with the microstates are examined, however, far more of them are disordered macrostates than ordered macrostates. Because each of the microstates is equally probable, it is highly probable that the actual macrostate will be one of the highly disordered macrostates.

Let us relate this to our dice example. A major difference between dice and physical systems is that once a pair of dice is thrown, the macrostate is determined and *remains* determined. Until the dice are picked up and thrown again, the result of a given throw is fixed in time. On the other hand, physical systems are constantly changing in time. Let us use the dice to model a physical system by repeatedly and violently bumping the table on which the dice sit, so that the results on the dice change every few fractions of a second. As we watch the macrostates of the dice over a long time, a 7 turns up far more often than a 2 or a 12 because the 7 can be formed from more microstates than a 2 or a 12. We could also interpret this by saying that a 2 turns into a 7 more often than a 7 turns into a 2. Thus, for the system of two dice, it is more likely for a less probable macrostate to change to a more probable macrostate than for the reverse to occur.

In physical systems, we are not talking about two molecules; we are talking about a number on the order of Avogadro's number of molecules. If you imagine throwing Avogadro's number of dice, the game of craps would be meaningless—you could make an almost perfect prediction of the result when the numbers on the faces are all added up (if the numbers on the face of the dice were added once per second, more than nineteen thousand trillion years would be needed to tabulate the results for only one throw!) because you are dealing with the statistics of a huge number of dice. It is these kinds of statistics that we are facing with Avogadro's number of molecules. The macrostate can be well predicted—even if a system starts off in a very low probability state (e.g., the nitrogen and oxygen molecules separated in a room by a membrane that is then punctured); it quickly develops into a high-probability state (the molecules rapidly mix evenly throughout the room).

We can now present this as a general principle for physical processes—**all physical processes tend toward more probable states for the system and its surroundings. The more probable state is always one of higher disorder.**

### Quick Quiz 18.4

Suppose that you select four cards at random from a standard deck of playing cards and end up with a macrostate of four deuces. How many microstates are associated with this macrostate? Suppose you pick up two cards and end up with a macrostate of two aces. How many microstates are associated with this macrostate? Which macrostate is more probable?

Now, what does all this talk about dice and states have to do with entropy? To answer this question, we shall show that entropy is a measure of the disorder of a state. Then, we shall use these ideas to generate a new statement of the second law of thermodynamics.

As we have seen, entropy can be defined using the macroscopic concepts of heat and temperature. Entropy can also be treated from a microscopic viewpoint through statistical analysis of molecular motions. We shall use a microscopic model to investigate the free expansion of an ideal gas.

In the kinetic theory of gases, gas molecules are taken to be particles moving in a random fashion. Suppose all the gas is confined initially to one side of a container so that the gas occupies volume $V_i$ (Fig. 18.11a). When the partition separating the two sides of the larger container is removed, the molecules eventually are distributed in some fashion throughout the greater volume $V_f$ (Fig. 18.11b). The exact nature of the distribution is a matter of probability.

Such a probability can be determined by first finding the probabilities for the variety of molecular locations involved in the free expansion process. The instant after the partition is removed (and before the molecules have had a chance to

(a)

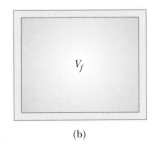

(b)

**Figure 18.11**

In a free expansion, the gas is allowed to expand into a larger volume that was previously a vacuum.

rush into the other half of the container), all the molecules are in the initial volume $V_i$. Let us estimate the probability of the molecules' arriving at a particular configuration through natural random motions within some larger volume $V_f$. Assume each molecule occupies some microscopic volume $V_m$ and that each location of a molecule is equally probable. Then the total number of possible locations of that molecule in a macroscopic initial volume $V_i$ is the ratio $w_i = V_i/V_m$, which is a huge number. In this expression, we use $w$ to represent the number of *ways* of locating the molecule in the volume. Each way is a microstate.

As more molecules are added to the system, the numbers of possible microstates multiply together. Ignoring the small probability of two molecules occupying the same location, each molecule may go into any of the $(V_i/V_m)$ locations. Hence, the number of possible microstates for $N$ molecules is $W_i = w_i^N = (V_i/V_m)^N$. Although a large number of the possible microstates would be considered unusual (e.g., all of the molecules in one corner of the available space), the total number of microstates is so enormous that these unusual microstates are an infinitesimal fraction of all possible microstates. Similarly, if the volume is increased to $V_f$, the number of microstates increases to $W_f = w_f^N = (V_f/V_m)^N$.

The probability of a given macrostate is proportional to the number of microstates corresponding to the macrostate. Hence, the probabilities of given initial and final macrostates occurring are $P_i = cW_i$ and $P_f = cW_f$, where the constant $c$ has been left undetermined. The ratio of these probabilities is

$$\frac{P_f}{P_i} = \frac{cW_f}{cW_i} = \frac{W_f}{W_i} = \frac{\left(\dfrac{V_f}{V_m}\right)^N}{\left(\dfrac{V_i}{V_m}\right)^N} = \left(\frac{V_f}{V_i}\right)^N$$

Taking the natural logarithm of this equation and multiplying by Boltzmann's constant $k_B$ gives

$$k_B \ln\left(\frac{W_f}{W_i}\right) = nN_A k_B \ln\left(\frac{V_f}{V_i}\right)$$

where we write the number of molecules $N$ as $nN_A$, the number of moles times Avogadro's number. From Section 16.4 we know that $N_A k_B$ is the universal gas constant $R$, so this equation may be written as

$$k_B \ln W_f - k_B \ln W_i = nR \ln\left(\frac{V_f}{V_i}\right) \tag{18.9}$$

From macroscopic thermodynamic considerations (we will see this calculation in Section 18.8), we know that when a gas undergoes a free expansion from $V_i$ to $V_f$, the change in entropy $\Delta S$ is

$$S_f - S_i = nR \ln\left(\frac{V_f}{V_i}\right) \tag{18.10}$$

Note that the right-hand sides of Equations 18.9 and 18.10 are identical. We thus make the following connection between *entropy* and the *number of microstates associated with a given macrostate:*

• *Entropy (microscopic definition)*
$$S \equiv k_B \ln W \tag{18.11}$$

Because the more probable macrostates are the ones with larger numbers of microstates, and the larger numbers of microstates are associated with more disorder, Equation 18.11 tells us that **entropy is a measure of microscopic disorder.** Although our discussion used the specific example of the free expansion of an ideal gas, a more rigorous development of the statistical interpretation of entropy leads to the same conclusion.

### Quick Quiz 18.5

Boxes containing many flowers have been hastily unloaded and piled on shelves in the back room of a flower shop. A florist selects 200 flowers to fill an order. She makes ten beautiful arrangements from the 200 flowers and delivers them to a church for a wedding ceremony. After the ceremony, the flowers have wilted from the heat and are thrown into a dumpster. Rank the three states of the 200 flowers (in boxes in the back room, in arrangements, in the dumpster) in order of increasing entropy.

### THINKING PHYSICS 18.3

Consider a reversible adiabatic expansion of a gas. According to the definition of entropy, no entropy change takes place because no energy is transferred by heat in an adiabatic process. Following the expansion, the gas molecules occupy a larger volume, so disorder seems to increase. An increase in disorder means an increase in entropy. How can we reconcile these notions, one that suggests no entropy change and the other that suggests an increase in entropy?

**Reasoning**   From either point of view, entropy does not increase. In an adiabatic expansion, the temperature decreases. Thus, although the molecules occupy a larger volume, they move with lower speeds. The decrease in speed represents more order because the random jostling of the molecules is reduced. A calculation shows that the decrease in entropy due to the reduced speeds exactly compensates for the increase due to the increased volume. Thus, the net entropy change is zero.

## 18.7 • ENTROPY AND THE SECOND LAW OF THERMODYNAMICS

Because entropy is a measure of disorder, and physical systems tend toward disordered macrostates, we can state that **the entropy of the Universe increases in all natural processes.** This statement is yet another way of stating the second law of thermodynamics.

• *Second law of thermodynamics; entropy statement*

To calculate the change in entropy for a finite process, we must recognize that $T$ is generally not constant. If $dQ_r$ is the energy transferred reversibly by heat when the system is at temperature $T$, then the change in entropy in an arbitrary reversible process between an initial state and a final state is

$$\Delta S = \int_i^f dS = \int_i^f \frac{dQ_r}{T} \quad \text{(reversible path)} \qquad [18.12]$$

• *Changes in entropy for a finite process*

The change in entropy of a system depends only on the properties of the initial and final equilibrium states because entropy is a state variable, like internal energy.

This is consistent with the relationship of entropy to disorder. For a given macrostate of a system, a given amount of disorder exists, measured by $W$, the number of microstates corresponding to the macrostate. This number does not depend on the path followed as a system goes from one state to another.

In the case of a reversible, adiabatic process, no energy is transferred by heat between the system and its surroundings, and therefore $\Delta S = 0$. Because no change in entropy occurs, such a process is often referred to as an **isentropic process.**

Consider the changes in entropy that occur in a Carnot heat engine operating between the temperatures $T_c$ and $T_h$. Equation 18.7 tells us that for a Carnot cycle

$$\Delta S = 0$$

• $\Delta S = 0$ *for any reversible cycle*

Now consider a system taken through an arbitrary reversible cycle. Because entropy is a state variable and hence depends only on the properties of a given equilibrium state, we conclude that $\Delta S = 0$ for *any* reversible cycle. In general, we can write this condition in the mathematical form

$$\oint \frac{dQ_r}{T} = 0 \qquad \text{[18.13]}$$

where the symbol $\oint$ indicates that the integration is over a *closed* path.

### Quick Quiz 18.6

Which of the following is true for the entropy change of a system that undergoes a reversible, adiabatic process? (a) $\Delta S < 0$, (b) $\Delta S = 0$, (c) $\Delta S > 0$.

### Quick Quiz 18.7

An ideal gas is taken from an initial temperature $T_i$ to a higher final temperature $T_f$ along two different reversible paths starting from the same point on a $PV$ diagram: Path A is at constant pressure; path B is at constant volume. The relation between the entropy changes of the gas for these paths is (a) $\Delta S_A > \Delta S_B$; (b) $\Delta S_A = \Delta S_B$; (c) $\Delta S_A < \Delta S_B$.

### THINKING PHYSICS 18.4

A box contains five gas molecules, spread throughout the box. At some time, all five are in one half of the box. Does this violate the second law of thermodynamics? Is the second law valid for this system?

**Reasoning**   Strictly speaking, this situation does violate the second law of thermodynamics. In response to the second question, however, the second law is not valid for small numbers of particles. The second law is based on collections of huge numbers of particles for which disordered states have astronomically higher probabilities than ordered states. Because the macroscopic world is built from these huge numbers of particles, the second law is valid as real processes proceed from order to disorder. In the five-molecule system, the general idea of the second law is valid, in that more disordered states occur than ordered ones, but the relatively high probability of the ordered states results in their existence from time to time.

## THINKING PHYSICS 18.5

Touch a rubber band to your forehead and note the sensation of temperature. Remove the rubber band from your forehead, stretch it, and touch it to your forehead again. It should now feel warmer. Remove the rubber band again from your forehead, let it return to its original length, and touch it to your forehead again. It now feels cooler. (a) Why do these temperature changes occur? (b) Someone proposes that the answer to part (a) is *internal friction,* as portions of the rubber band rub against each other during the size changes. How would you respond to this proposal?

**Reasoning**    (a) The answer lies in the second law of thermodynamics. Rubber consists of long, coiled molecules, whose axes lie in random directions relative to the length of the rubber band. The force of the stretch tends to cause the coiled molecules to line up parallel with the sides of the rubber band. This represents an increase in the degree of *order* of these molecules, corresponding to a *decrease* in entropy. If the rubber band is stretched quickly, however, the stretching process is very close to adiabatic, and the net entropy change is close to zero. Thus, some change must occur to compensate for the decrease in entropy due to the coils lining up. We can compensate for the decrease *if the molecules increase their vibrational motion,* so that their positions in the rubber represent a lower degree of order. Increased vibrational motion is represented by more internal energy, which in turn means that the temperature increases—this is the warmth that is felt on the forehead.

When the rubber band returns to its original length, the coiled molecules become more disordered, resulting in an increase in entropy. The increase in entropy of the directions of the molecules is accompanied by a decrease in entropy of the vibrations of the molecules. This decreased vibration corresponds to a lower internal energy and a cooling of the rubber.

(b) The internal friction argument is consistent with the increase in temperature associated with stretching the rubber band. When the rubber band returns to its original length, however, *additional* portions of the rubber band rub together, and the internal friction results in a further temperature *increase.* This is inconsistent with the observation that the temperature decreases when the rubber band returns to its original length.

## Example 18.3   Change in Entropy—Melting Process

A solid substance that has a latent heat of fusion $L_f$ melts at a temperature $T_m$. Calculate the change in entropy when a mass $m$ of this substance is melted.

**Solution**    Let us assume that the melting process occurs so slowly that it can be considered a reversible process—we can reverse the process by extracting energy very slowly to freeze the liquid into the solid form. In this case the temperature can be regarded as constant and equal to $T_m$. Making use of Equation 18.12 and the fact that the latent heat of fusion $Q = mL_f$, (Eq. 17.5), we find that

$$\Delta S = \int \frac{dQ_r}{T} = \frac{1}{T_m} \int dQ = \frac{Q}{T_m} = \boxed{\frac{mL_f}{T_m}}$$

Note that we were able to remove $T_m$ from the integral because the process is isothermal.

**EXERCISE**    Calculate the change in entropy when 0.30 kg of lead melts at 327°C. Lead has a latent heat of fusion equal to 24.5 kJ/kg.

*Answer*    $\Delta S = 12.3$ J/K

## 18.8 • ENTROPY CHANGES IN IRREVERSIBLE PROCESSES

So far, we have calculated changes in entropy using information about a reversible path connecting the initial and final equilibrium states. To calculate changes in entropy for real (irreversible) processes, first recall that entropy is a state variable. Hence, the change in entropy when a system moves between any two equilibrium states depends only on the initial and final states.

We can calculate entropy changes for irreversible processes by devising a reversible process (or a series of reversible processes) between the same two equilibrium states and computing $\int dQ_r/T$ for the reversible process. In irreversible processes, it is critically important to distinguish between $Q$, the actual energy transfer in the process, and $Q_r$, the energy that would have been transferred by heat along a reversible path between the same states. Only the second value gives the correct entropy change. For example, as we shall see, if an ideal gas expands adiabatically into a vacuum, $Q = 0$, but $\Delta S \neq 0$ because $Q_r \neq 0$. The reversible path between the same two states is the reversible, isothermal expansion that gives $\Delta S > 0$.

As we shall see in the following examples, the change in entropy for the system plus its environment is always positive for an irreversible process. In general, the total entropy (and disorder) always increases in irreversible processes. From these considerations, the second law of thermodynamics can be stated as follows: **The total entropy of an isolated system that undergoes a change cannot decrease.** Furthermore, if the process is *irreversible,* the total entropy of an isolated system always *increases.* On the other hand, in a reversible process, the total entropy of an isolated system remains constant.

When dealing with interacting objects that are not isolated from the environment, we must consider the change of entropy for the system *and* its environment. When two objects interact in an irreversible process, the increase in entropy of one part of the Universe is greater than the decrease in entropy of the other part. Hence, we conclude that the change in entropy of the Universe must be greater than zero for an irreversible process and equal to zero for a reversible process. Ultimately, the entropy of the Universe should reach a maximum value. At this point the Universe will be in a state of uniform temperature and density. All physical, chemical, and biological processes will cease because a state of perfect disorder implies that no energy is available for doing work. This gloomy state of affairs is sometimes referred to as the "heat death" of the Universe.

**PITFALL PREVENTION 18.3**

**All real processes are irreversible**

The reversible process is an idealization—all real processes on Earth are irreversible. Thus, the entropy of the Universe always increases in any real process.

### Quick Quiz 18.8

True or false: The entropy change in an adiabatic process must be zero because $Q = 0$.

### THINKING PHYSICS 18.6

According to the entropy statement of the second law, the entropy of the Universe increases in irreversible processes. This sounds very different from the Kelvin–Planck and Clausius forms of the second law. Can these two statements be made consistent with the entropy interpretation of the second law?

**Reasoning** These three forms are consistent. In the Kelvin–Planck statement, the energy in the reservoir is disordered internal energy—the random motion of molecules. Performing work results in ordered energy—pushing a piston through a displacement, for example. In this case, the motion of all molecules of the piston is in the same direction. If a heat engine absorbed energy by heat and performed an equal amount of work, then it would have converted disorder into order, in violation of the entropy statement. In the Clausius statement, we start with an ordered system—higher temperature in the hot object, lower in the cold object. This separation of temperatures is an example of order. Energy flowing from the cold object to the hot object, so that the temperatures spread even farther apart, would be an increase in order, in violation of the entropy statement.

## Entropy Changes in a Free Expansion

An ideal gas in an insulated container initially occupies a volume of $V_i$ (see Fig. 18.11). A partition separating the gas from an evacuated region is broken so that the gas expands (irreversibly) to the volume $V_f$. Let us find the change in entropy of the gas and the Universe.

The process is neither reversible nor quasi-static. The work done on the gas is zero, and because the walls are insulating, no energy is transferred by heat during the expansion. That is, $W = 0$ and $Q = 0$. The first law tells us that the change in internal energy $E_{int}$ is zero; therefore, $E_{int,i} = E_{int,f}$. Because the gas is ideal, $E_{int}$ depends on temperature only, so we conclude that $T_i = T_f$.

To apply Equation 18.12, we must find $Q_r$; that is, we must find an equivalent reversible path that shares the same initial and final states. A simple choice is an isothermal, reversible expansion in which the gas pushes slowly against a piston. Because $T$ is constant in this process, Equation 18.12 gives

$$\Delta S = \int \frac{dQ_r}{T} = \frac{1}{T} \int_i^f dQ_r$$

Because this is an isothermal process, $\Delta E_{int} = 0$, so the first law of thermodynamics tells us that the energy input by heat is equal to the negative of the work done on the gas, $dQ_r = -dW = P\,dV$. Using this result, we find that

$$\Delta S = \frac{1}{T} \int dQ_r = \frac{1}{T} \int P\,dV = \frac{1}{T} \int \frac{nRT}{V} dV = nR \int_{V_i}^{V_f} \frac{dV}{V}$$

$$= nR \ln \frac{V_f}{V_i} \qquad\qquad [18.14]$$

Because $V_f > V_i$, we conclude that $\Delta S$ is positive, and so both the entropy and the disorder of the gas (and the Universe) increase as a result of the irreversible, adiabatic expansion.

### Example 18.4 Free Expansion of an Ideal Gas—Revisited

Again consider the free expansion of an ideal gas. Let us verify that the macroscopic and microscopic approaches lead to the same conclusion. Suppose that 1 mol of an ideal gas undergoes a free expansion to four times its initial volume. The initial and final temperatures are, as we have seen, the same. (a) Using a macroscopic approach, calculate the entropy change. (b) Find the number of ways all molecules can be distributed in the original volume. (c) Using the considerations of part (b), calculate the change in entropy for the free expansion, and show that it agrees with part (a).

**Solution** (a) From Equation 18.14 we obtain

$$\Delta S = nR \ln \frac{V_f}{V_i} = (1)R \ln \left( \frac{4V_i}{V_i} \right) = \boxed{R \ln 4}$$

(b) As discussed in Section 18.6 the number of states available to a single molecule in the initial volume $V_i$ is $w_i = (V_i/V_m)$. For one mole ($N_A$ molecules), the number of available states is

$$W_i = w_i{}^{N_A} = \left(\frac{V_i}{V_m}\right)^{N_A}$$

(c) The number of states for all $N_A$ molecules in the volume $V_f = 4V_i$ is

$$W_f = w_f{}^{N_A} = \left(\frac{V_f}{V_m}\right)^{N_A} = \left(\frac{4V_i}{V_m}\right)^{N_A}$$

From Equation 18.9 we obtain

$$\Delta S = k_B \ln W_f - k_B \ln W_i = k_B \ln \left(\frac{4V_i}{V_i}\right)^{N_A}$$

$$= k_B \ln(4)^{N_A} = N_A k_B \ln 4 = R \ln 4$$

The answer is the same as that in part (a), which dealt with macroscopic parameters.

## context connection
## 18.9 • THE ATMOSPHERE AS A HEAT ENGINE

In Chapter 17, we predicted a global temperature based on the notion of energy balance between incoming visible radiation from the Sun and outgoing infrared radiation from the Earth. This model leads to a global temperature that is well below the measured temperature. This discrepancy results because atmospheric effects are not included in our model. In this chapter, we shall introduce some of these effects and show that the atmosphere can be modeled as a heat engine. In the Context Conclusion, we shall use concepts learned in the thermodynamics chapters to build a model that is more successful at predicting the correct temperature of the Earth.

What happens to the energy that enters the atmosphere by radiation from the Sun? Figure 18.12 helps answer this question by showing how the input energy undergoes various processes. If we identify the incoming energy as 100%, we find that 30% of it is reflected back into space, as we mentioned in Chapter 17. This includes 6% back-scattered from air molecules, 20% reflected from clouds, and 4% reflected from the surface of the Earth. The remaining 70% is absorbed by either the air or the surface. Before reaching the surface, 20% of the original radiation is absorbed in the air; 4% by clouds; and 16% by water, dust particles, and ozone in the atmosphere. Of the original radiation striking the top of the atmosphere, the ground absorbs 50%.

The ground emits radiation upward and transfers energy to the atmosphere by several processes. Of the original 100% of the incoming energy, 6% simply passes back through the atmosphere into space (at the right in Figure 18.12). In addition, 14% of the original incoming energy emitted as radiation from the ground is absorbed by water and carbon dioxide molecules. The air warmed by the surface rises upward by convection, carrying 6% of the original energy into the atmosphere. The hydrological cycle results in 24% of the original energy being carried upward as water vapor and released into the atmosphere when the water vapor condenses into liquid water.

These processes result in a total of 64% of the original energy being absorbed in the atmosphere, with another 6% from the surface passing back through into space. Because the atmosphere is in steady state, this 64% is also emitted from the atmosphere into space. The emission is divided into two types. The first is infrared radiation from molecules in the atmosphere, including water vapor, carbon dioxide, and the nitrogen and oxygen molecules of the air. This accounts for emission

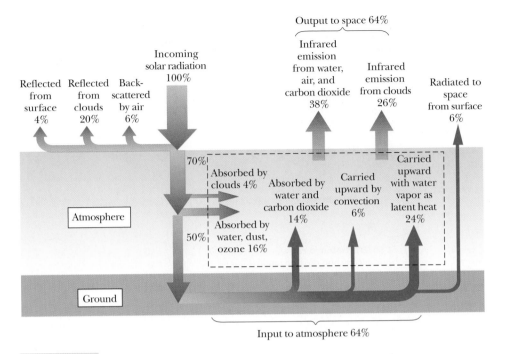

**Figure 18.12**

Energy input to the atmosphere from the Sun is divided into several components. Only 50% of the incident radiation is absorbed by the ground, and it is then re-radiated into the atmosphere. The dashed line represents the atmospheric heat engine that is shown schematically in Figure 18.13.

of 38% of the original energy. The remaining 26% is emitted as infrared radiation from clouds.

Figure 18.12 accounts for all of the energy—the amount of energy input equals the amount of energy output, which is the premise used in the Context Connection of Chapter 17. A major difference from our discussion in that chapter, however, is the notion of absorption of the energy by the atmosphere. It is this absorption that creates thermodynamic processes in the atmosphere to raise the surface temperature above the value we determined in Chapter 17. We shall explore more about these processes and the temperature profile of the atmosphere in the Context Conclusion.

To close the present chapter, let us discuss one more process that is not included in Figure 18.12. The various processes depicted in this figure result in a small amount of work done on the air, which appears as the kinetic energy of the prevailing winds in the atmosphere.

The amount of the original solar energy that is converted to kinetic energy of prevailing winds is about 0.5%. The process of generating the winds does not change the energy balance shown in Figure 18.12. The kinetic energy of the wind is converted to internal energy as masses of air move past one another. This internal energy produces an increased infrared emission of the atmosphere into space, so the 0.5% is only temporarily in the form of kinetic energy before being emitted as radiation.

We can model the atmosphere as a heat engine, which is indicated in Figure 18.12 by the dotted rectangle. A schematic diagram of this heat engine is shown in Figure 18.13. The warm reservoir is the surface and the atmosphere, and the cold

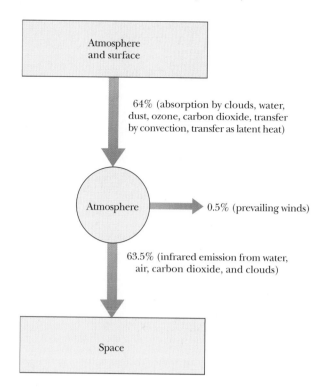

**Figure 18.13**

A schematic representation of the atmosphere as a heat engine.

reservoir is empty space. We can calculate the efficiency of the atmospheric engine using Equation 18.2:

$$e = \frac{W_{eng}}{|Q_h|} = \frac{0.5\%}{64\%} = 0.008 = 0.8\%$$

This is a very low efficiency. But keep in mind that a tremendous amount of energy enters the atmosphere from the Sun, so that even a very small fraction of it can create a very complex and powerful wind system.

Notice that the output energy in Figure 18.13 is less than that in Figure 18.12 by 0.5%. However, the 0.5% transferred to the atmosphere by generating winds is eventually transformed to internal energy in the atmosphere by friction, and then radiated into space as thermal radiation. We cannot separate the heat engine and the winds in the atmosphere in a diagram because the atmosphere *is* the heat engine, and the winds are generated *in* the atmosphere!

We now have all of the pieces we need to put together the puzzle of the temperature of the Earth. We shall discuss this in the Context Conclusion.

## SUMMARY

A **heat engine** is a device that converts internal energy to other useful forms of energy by carrying a substance through a cyclic process. The net work done by a heat engine is

$$W_{eng} = |Q_h| - |Q_c| \qquad [18.1]$$

where $Q_h$ is the energy absorbed from a hot reservoir and $Q_c$ is the energy rejected to a cold reservoir.

The **thermal efficiency** $e$ of a heat engine is defined as the ratio of the net work done to the energy absorbed per cycle from the higher temperature reservoir:

$$e = \frac{W_{eng}}{|Q_h|} = 1 - \frac{|Q_c|}{|Q_h|} \qquad [18.2]$$

The Kelvin–Planck statement of the **second law of thermodynamics** can be stated as follows: **It is impossible to** construct a heat engine that, operating in a cycle, produces no effect other than the absorption of energy from a reservoir and the performance of an equal amount of work.

A **reversible** process is one for which the system can be returned to its initial conditions along the same path, and for which every point along the path is an equilibrium state. A process that does not satisfy these requirements is **irreversible.**

The thermal efficiency of a heat engine operating in a **Carnot cycle** is given by

$$e_c = 1 - \frac{T_c}{T_h} \qquad [18.4]$$

where $T_c$ is the absolute temperature of the cold reservoir and $T_h$ is the absolute temperature of the hot reservoir.

No real heat engine operating between the temperatures $T_c$ and $T_h$ can be more efficient than an engine operating reversibly in a Carnot cycle between the same two temperatures.

The **Clausius statement** of the second law states that **energy will not flow spontaneously from a cold object to a hot object.**

The second law of thermodynamics states that when real (irreversible) processes occur, the degree of disorder in the system plus the surroundings increases. When a process occurs in an isolated system, ordered energy is converted into disordered energy. The measure of disorder in a system is called **entropy** $S$.

The **change in entropy** $dS$ of a system moving through an infinitesimal process between two equilibrium states is

$$dS = \frac{dQ_r}{T} \qquad [18.8]$$

where $dQ_r$ is the energy transferred by heat in a reversible process between the same states.

From a microscopic viewpoint, the **entropy** $S$ associated with a macrostate of a system is defined as

$$S \equiv k_B \ln W \qquad [18.11]$$

where $k_B$ is Boltzmann's constant and $W$ is the number of microstates corresponding to a given macrostate. Thus, **entropy is a measure of microscopic disorder.** Because of the statistical tendency of systems to proceed toward states of greater probability and greater disorder, all natural processes are irreversible and result in an increase in entropy.

The change in entropy of a system moving between two general equilibrium states is

$$\Delta S = \int_i^f \frac{dQ_r}{T} \qquad [18.12]$$

The value of $\Delta S$ is the same for all reversible paths connecting the initial and final states.

The change in entropy for any reversible, cyclic process is zero, and when such a process occurs, the entropy of the Universe remains constant.

In an irreversible process, the total entropy of an isolated system always increases. In general, the total entropy (and disorder) always increases in any irreversible process. Furthermore, the change in entropy of the Universe is greater than zero for an irreversible process.

## QUESTIONS

1. What are some factors that affect the efficiency of automobile engines?

2. A steam-driven turbine is one major component of an electric power plant. Why is it advantageous to have the temperature of the steam as high as possible?

3. Is it possible to construct a heat engine that creates no thermal pollution?

4. Discuss the change in entropy of a gas that expands (a) at constant temperature and (b) adiabatically.

5. In solar ponds constructed in Israel, the Sun's energy is concentrated near the bottom of a salty pond. With the proper layering of salt in the water, convection is prevented, and temperatures of 100°C may be reached. Can you guess the maximum efficiency with which useful energy can be extracted from the pond?

6. Why does your automobile burn more gas in winter than in summer?

7. Can a heat pump have a coefficient of performance less than unity? Explain.

8. The Carnot cycle was illustrated in the chapter on a *PV* diagram. What would the Carnot cycle look like on a *TS* diagram?

9. Give some examples of irreversible processes that occur in nature. Give an example of a process in nature that is nearly reversible.

10. A thermodynamic process occurs in which the entropy of a system changes by $-8.0$ J/K. According to the second law of thermodynamics, what can you conclude about the entropy change of the environment?

11. If a supersaturated sugar solution is allowed to slowly evaporate, sugar crystals form in the container. Hence, sugar molecules go from a disordered form (in solution) to a highly ordered crystalline form. Does this process violate the second law of thermodynamics? Explain.

12. How could you increase the entropy of 1 mol of a metal that is at room temperature? How could you decrease its entropy?

13. A heat pump is to be installed in a region where the average outdoor temperature in the winter months is −20°C. In view of this, why would it be advisable to place the outdoor compressor unit deep in the ground? Why are heat pumps not commonly used for heating in cold climates?

14. Suppose your roommate is "Mr. Clean" and tidies up your messy room after a big party. Is your roommate violating the second law of thermodynamics by cleaning the room?

15. Discuss the entropy changes that occur when you (a) bake a loaf of bread and (b) consume the bread.

16. A diesel engine operates with a compression ratio of about 15. A gasoline engine has a compression ratio of about 6. Which engine runs hotter? Which engine (at least theoretically) can operate more efficiently?

17. If you shake a jar full of jellybeans of different sizes, the larger beans tend to appear near the top, and the smaller ones tend to fall to the bottom. Why? Does this process violate the second law of thermodynamics?

18. The device in Figure Q18.18, called a thermoelectric converter, uses a series of semiconductor cells to convert internal energy to electric energy. In the photograph at the left, both legs of the device are at the same temperature, and no electric energy is produced. When one leg is at a higher temperature than the other, however, as in the photograph on the right, electric energy is produced as the device extracts energy from the hot reservoir and drives a small electric motor. (a) Why does the temperature differential produce electric energy in this demonstration? (b) In what sense does this intriguing experiment demonstrate the second law of thermodynamics?

**Figure Q18.18** *(Courtesy of PASCO Scientific Company)*

# PROBLEMS

1, 2, 3 = straightforward, intermediate, challenging    ☐ = full solution available in the *Student Solutions Manual and Study Guide*

web = solution posted at **http://www.harcourtcollege.com/physics/**    🖥 = computer useful in solving problem

🖲 = Interactive Physics    ▪ = paired numerical/symbolic problems    🖾 = life science application

### Section 18.1   Heat Engines and the Second Law of Thermodynamics

1. A heat engine absorbs 360 J of energy by heat and performs 25.0 J of work in each cycle. Find (a) the efficiency of the engine and (b) the energy expelled to the cold reservoir in each cycle.

2. A heat engine performs 200 J of work in each cycle and has an efficiency of 30.0%. For each cycle, how much energy is (a) absorbed and (b) expelled by heat?

3. A particular heat engine has a useful power output of 5.00 kW and an efficiency of 25.0%. The engine expels 8 000 J of exhaust energy in each cycle. Find (a) the

energy absorbed in each cycle and (b) the time for each cycle.

4. Suppose a heat engine is connected to two energy reservoirs, one a pool of molten aluminum (660°C) and the other a block of solid mercury ($-38.9$°C). The engine runs by freezing 1.00 g of aluminum and melting 15.0 g of mercury during each cycle. The heat of fusion of aluminum is $3.97 \times 10^5$ J/kg; the heat of fusion of mercury is $1.18 \times 10^4$ J/kg. What is the efficiency of this engine?

## Section 18.2  Reversible and Irreversible Processes

## Section 18.3  The Carnot Engine

5. One of the most efficient engines ever built operates between 430°C and 1 870°C. (a) What is its maximum theoretical efficiency? (b) The actual efficiency of the engine is 42.0%. How much useful power does the engine deliver if it absorbs $1.40 \times 10^5$ J of energy each second from its hot reservoir?

6. A heat engine operating between 200°C and 80.0°C achieves 20.0% of the maximum possible efficiency. What energy input will enable the engine to perform 10.0 kJ of work?

7. A Carnot engine has a useful power output of 150 kW. The engine operates between two reservoirs at 20.0°C and 500°C. (a) How much energy does it absorb per hour? (b) How much energy is expelled per hour in its exhaust?

8. The exhaust temperature of a Carnot heat engine is 300°C. What is the intake temperature if the efficiency of the engine is 30.0%?

9. An ideal gas is taken through a Carnot cycle. The isothermal expansion occurs at 250°C, and the isothermal compression takes place at 50.0°C. The gas absorbs 1 200 J of energy from the hot reservoir during the isothermal expansion. Find (a) the energy expelled to the cold reservoir in each cycle and (b) the net work done by the gas in each cycle.

web

10. A Carnot heat engine uses a steam boiler at 100°C as the high-temperature reservoir. The low-temperature reservoir is the outside environment at 20.0°C. Energy is exhausted to the low-temperature reservoir at the rate of 15.4 W. (a) Determine the useful power output of the heat engine. (b) How much steam will it cause to condense in the high-temperature reservoir in one hour?

11. A power plant operates at a 32.0% efficiency during the summer when the sea water used for cooling is at 20.0°C. The plant uses 350°C steam to drive turbines. Assuming that the plant's efficiency changes in the same proportion as the ideal efficiency, what would be the plant's efficiency in the winter, when the sea water is 10.0°C?

12. A power plant that would make use of the temperature gradient in the ocean has been proposed. The system is to operate between 20.0°C (surface water temperature) and 5.00°C (water temperature at a depth of about 1 km). (a) What is the maximum efficiency of such a system? (b) If the useful power output of the plant is 75.0 MW, how much energy is absorbed per hour? (c) In view of your answer to part (a), do you think such a system is worthwhile (considering that the "fuel" is free)?

13. At point *A* in a Carnot cycle, 2.34 mol of a monatomic ideal gas has a pressure of 1 400 kPa, a volume of 10.0 L, and a temperature of 720 K. It expands isothermally to point *B* and then expands adiabatically to point *C*, where its volume is 24.0 L. An isothermal compression brings it to point *D*, where its volume is 15.0 L. An adiabatic process returns the gas to point *A*. (a) Determine all the unknown pressures, volumes, and temperatures as you fill in the following table:

|   | *P* | *V* | *T* |
|---|-----|-----|-----|
| *A* | 1 400 kPa | 10.0 L | 720 K |
| *B* | | | |
| *C* | | 24.0 L | |
| *D* | | 15.0 L | |

(b) Find the energy added by heat, the work done by the engine, and the change in internal energy for each of the steps $A \rightarrow B$, $B \rightarrow C$, $C \rightarrow D$, and $D \rightarrow A$. (c) Calculate the efficiency $W_{eng}/|Q_{in}|$. Show that it is equal to $1 - T_C/T_A$, the Carnot efficiency.

## Section 18.4  Heat Pumps and Refrigerators

14. What is the coefficient of performance (COP) of a refrigerator that operates with Carnot efficiency between temperatures $-3.00$°C and $+27.0$°C?

15. What is the maximum possible COP of a heat pump that brings energy from outdoors at $-3.00$°C into a 22.0°C house? (*Hint:* The heat pump has work *W* done on it, which is also available to warm the house.)

16. A heat pump, shown in Figure P18.16, is essentially a heat engine run backward. It extracts energy from colder air outside and deposits it in a warmer room. Suppose that the ratio of the actual energy entering the room to the work done by the device's motor is 10.0% of the theoretical maximum ratio. Determine the energy entering the room per joule of work done by the motor when the inside temperature is 20.0°C and the outside temperature is $-5.00$°C.

**Figure P18.16**

**17.** An ideal refrigerator or ideal heat pump is equivalent to a Carnot engine running in reverse. That is, energy $Q_c$ is absorbed from a cold reservoir and energy $Q_h$ is rejected to a hot reservoir. (a) Show that the work that must be supplied to run the refrigerator or heat pump is

$$W = \frac{T_h - T_c}{T_c} Q_c$$

(b) Show that the COP of the ideal refrigerator is

$$\text{COP} = \frac{T_c}{T_h - T_c}$$

**18.** How much work does an ideal Carnot refrigerator use to change 0.500 kg of tap water at 10.0°C into ice at −20.0°C? Assume the freezer compartment is held at −20.0°C and that the refrigerator exhausts energy into a room at 20.0°C.

**19.** How much work does an ideal Carnot refrigerator require to remove 1.00 J of energy from helium at 4.00 K and reject this energy to a room-temperature (293-K) environment?

### Section 18.6   Entropy

### Section 18.7   Entropy and the Second Law of Thermodynamics

**20.** An ice tray contains 500 g of liquid water at 0°C. Calculate the change in entropy of the water as it freezes slowly and completely at 0°C.

**21.** Calculate the change in entropy of 250 g of water heated slowly from 20.0°C to 80.0°C. (*Hint:* Note that $dQ = mc\, dT$.)

**22.** An airtight freezer holds 2.50 mol of air at 25.0°C and 1.00 atm. The air is then cooled to −18.0°C. (a) What is the change in entropy of the air if the volume is held con-

stant? (b) What would the change be if the pressure were maintained at 1 atm during the cooling?

**23.** What change in entropy occurs when a 27.0-g ice cube at −12.0°C is transformed at constant pressure into steam at 115°C?

**24.** If you toss two dice, what is the total number of ways in which you can obtain (a) a 12 and (b) a 7?

**25.** A bag contains 50 red marbles and 50 green marbles. (a) You draw a marble at random from the bag, notice its color, return it to the bag, and repeat for a total of three draws. You record the result as the number of red marbles and the number of green marbles in the set of three. Construct a table listing each possible macrostate and the number of microstates within it. For example, RRG, RGR, and GRR are the three microstates constituting the macrostate 1G,2R. (b) Construct a table for the case in which you draw five marbles rather than three.

**26.** You toss four coins into the air simultaneously and record the number of heads and the number of tails that result. Prepare a table listing each macrostate and each of the microstates included in it. For example, the two microstates HHTH and HTHH, together with some others, are included in the macrostate of three heads and one tail. (a) On the basis of your table, what is the most probable result recorded for a toss? In terms of entropy, (b) what is the most ordered macrostate and (c) what is the most disordered?

### Section 18.8   Entropy Changes in Irreversible Processes

**27.** The temperature at the surface of the Sun is approximately 5 700 K, and the temperature at the surface of the Earth is approximately 290 K. What entropy change occurs when 1 000 J of energy is transferred by radiation from the Sun to the Earth?

**28.** A 1.00-kg iron horseshoe is taken from a forge at 900°C and dropped into 4.00 kg of water at 10.0°C. Assuming that no energy is lost by heat to the surroundings, determine the total entropy change of the horseshoe–water system.

**29.** A 1 500-kg car is moving at 20.0 m/s. The driver brakes to a stop. The brakes cool off to the temperature of the surrounding air, which is nearly constant at 20.0°C. What is the total entropy change?

**30.** Every second at Niagara Falls, some 5 000 m³ of water falls a distance of 50.0 m (Fig. P18.30). What is the increase in entropy per second due to the falling water? (Assume that the mass of the surroundings is so great that its temperature and that of the water stay nearly constant at 20.0°C. Suppose that a negligible amount of water evaporates.)

**Figure P18.30**   Niagara Falls. *(Jan Kopec/Tony Stone Images)*

**31.** A 100 000-kg iceberg at $-5.00°C$ breaks away from the polar ice shelf and floats away into the ocean, at $5.00°C$. What is the final change in the entropy of the system, after the iceberg has completely melted? (The specific heat of ice is $2\,010\,J/kg\cdot°C$.)

**32.** One mole of $H_2$ gas is contained in the left-hand side of the container shown in Figure P18.32, which has equal volumes left and right. The right-hand side is evacuated. When the valve is opened, the gas streams into the right-hand side. What is the final entropy change of the gas? Does the temperature of the gas change?

**Figure P18.32**

**33.** One mole of an ideal monatomic gas, initially at a pressure of 1.00 atm and a volume of $0.025\,0\,m^3$, is heated to a final state with a pressure of 2.00 atm and a volume of $0.040\,0\,m^3$. Determine the change in entropy of the gas in this process.

**34.** A 2.00-L container has a center partition that divides it in half, as shown in Figure P18.34. The left-hand side contains $H_2$ gas, and the right-hand side contains $O_2$ gas. Both gases are at room temperature and at atmospheric pressure. The partition is removed, and the gases are allowed to mix. What is the entropy increase of the system?

**Figure P18.34**

### Section 18.9   Context Connection — The Atmosphere as a Heat Engine

**35.** We found the efficiency of the atmospheric heat engine to be 0.8%. Considering the solar constant $I_S = 1\,340\,W/m^2$ and the fact that 64% of this energy is absorbed in the atmosphere, find the "wind power"—that is, the rate at which energy becomes available for driving the winds.

**36.** (a) Find the kinetic energy of the moving air in a hurricane, modeled as a disk 600 km in diameter and 11 km thick, with wind blowing at a uniform speed of 60 km/h. (b) For how long a time must sunlight with intensity of $1\,000\,W/m^2$ fall on the 600-km-diameter circular top of the hurricane to deliver this much energy?

### *Additional Problems*

**37.** A firebox is at 750 K, and the ambient temperature is 300 K. The efficiency of a Carnot engine doing 150 J of work as it transports energy between these constant-temperature baths is 60.0%. The Carnot engine must absorb energy $150\,J/0.600 = 250\,J$ from the hot reservoir and must put out 100 J of exhaust into the environment. To follow Carnot's reasoning, suppose that some other heat engine S could have efficiency 70.0%. (a) Find the energy input and wasted energy output of engine S as it does 150 J of work. (b) Let engine S operate as in part (a) and run the Carnot engine in reverse. Find the total energy the firebox puts out as both engines operate together, and the total energy the environment absorbs. Show that the Clausius statement of the second law of thermodynamics is violated. (c) Find the energy input and work output of engine S as it puts out exhaust energy of 100 J. (d) Let engine S operate as in (c) and contribute 150 J of its work output to running the Carnot engine in reverse. Find the total energy the firebox puts out as both engines operate together, the total work output, and the total energy the environment absorbs. Show that the Kelvin–Planck statement of the second law is violated. Thus our assumption about the efficiency of engine S must be false. (e) Let the engines operate together through one cycle as in part (d). Find the change in entropy of the Universe. Show that the entropy statement of the second law is violated.

**38.** How fast are you personally making the entropy of the Universe increase right now? Make an order-of-magnitude estimate, stating what quantities you take as data and the values you measure or estimate for them.

**39.** A house loses energy by conduction through the exterior
web   walls and roof at a rate of $5\,000\,J/s = 5.00\,kW$ when the interior temperature is $22.0°C$ and the outside temperature is $-5.00°C$. Calculate the electric power required to maintain the interior temperature at $22.0°C$ for the following two cases: (a) The electric power is used in electric re-

sistance heaters (which convert all of the electric energy supplied into internal energy). (b) The electric power is used to drive an electric motor that operates the compressor of a heat pump that has a coefficient of performance equal to 60.0% of the Carnot-cycle value.

40. On the *PV* diagram for an ideal gas, one isothermal curve and one adiabatic curve pass through each point. Prove that the slope of the adiabat is steeper than the slope of the isotherm by the factor $\gamma$.

41. In 1827, Robert Sterling, a Scottish minister, invented the **web** *Sterling engine,* which has found a variety of applications ever since. Fuel is burned externally to warm one of the engine's two cylinders. A fixed quantity of inert gas moves cyclically between the cylinders, expanding in the hot one and contracting in the cold one. Figure P18.41 represents a model for its thermodynamic cycle. Consider *n* mol of an ideal monatomic gas being taken once through the cycle, consisting of two isothermal processes at temperatures $3T_i$ and $T_i$ and two constant-volume processes. Determine, in terms of *n*, *R*, and $T_i$, (a) the net energy transferred by heat to the gas and (b) the efficiency of the engine.

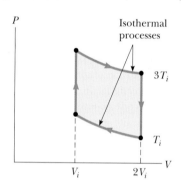

**Figure P18.41**

42. A heat engine operates between two reservoirs at $T_2 = 600$ K and $T_1 = 350$ K. It absorbs 1 000 J of energy from the higher temperature reservoir and performs 250 J of work. Find (a) the entropy change of the Universe $\Delta S_u$ for this process and (b) the work that could have been done by an ideal Carnot engine operating between these two reservoirs. (c) Show that the difference between the amounts of work done in parts (a) and (b) is $T_1 \Delta S_u$.

43. A power plant, having a Carnot efficiency, produces 1 000 MW of electric power from turbines that take in steam at 500 K and reject water at 300 K into a flowing river. The water downstream is 6.00 K warmer due to the output of the power plant. Determine the flow rate of the river.

44. A power plant, having a Carnot efficiency, produces electric power $\mathcal{P}$ from turbines that take in energy from steam at temperature $T_h$ and discharge energy at temperature $T_c$

through a heat exchanger into a flowing river. The water downstream is warmer by $\Delta T$ due to the output of the power plant. Determine the flow rate of the river.

45. An ideal (Carnot) freezer in a kitchen has a constant temperature of 260 K, and the air in the kitchen has a constant temperature of 300 K. Suppose that the insulation for the freezer is not perfect, but conducts energy into the freezer at a rate of 0.150 W. Determine the average power required for the freezer's motor to maintain the constant temperature in the freezer.

46. One mole of an ideal monatomic gas is taken through the cycle shown in Figure P18.46. The process $A \rightarrow B$ is a reversible isothermal expansion. Calculate (a) the net work done by the gas, (b) the energy added to the gas, (c) the energy exhausted from the gas, and (d) the efficiency of the cycle.

**Figure P18.46**

47. Calculate the increase in entropy of the Universe when you add 20.0 g of 5.00°C cream to 200 g of 60.0°C coffee. Assume that the specific heats of cream and coffee are both 4.20 J/g · °C.

48. A biology laboratory is maintained at a constant temperature of 7.00°C by an air conditioner, which is vented to the air outside. On a typical summer day the outside temperature is 27.0°C and the air conditioning unit emits energy to the outside at a rate of 10.0 kW. Model the unit as having a coefficient of performance equal to 40.0% of the coefficient of performance of an ideal Carnot refrigerator. (a) At what rate does the air conditioner remove energy from the laboratory? (b) Calculate the power required for the work input. (c) Find the change in entropy produced by the air conditioner in 1.00 h. (d) The outside temperature increases to 32.0°C. Find the fractional change in the coefficient of performance of the air conditioner.

49. One mole of a monatomic ideal gas is taken through the cycle shown in Figure P18.49. At point *A*, the pressure, volume, and temperature are $P_i$, $V_i$, and $T_i$, respectively. In terms of *R* and $T_i$, find (a) the total energy entering the system by heat per cycle, (b) the total energy leaving the

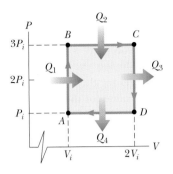

**Figure P18.49**

system by heat per cycle, (c) the efficiency of an engine operating in this cycle, and (d) the efficiency of an engine operating in a Carnot cycle between the same temperature extremes.

50. In 1993, the federal government instituted a requirement that all room air conditioners sold in the United States have an energy efficiency ratio (EER) of 10 or more. The EER is defined as the ratio of the cooling capacity of the air conditioner, measured in Btu/h, to its electric power requirement in watts. (a) Convert the EER of 10.0 to dimensionless form, using the conversion 1 Btu = 1 055 J. (b) What is the appropriate name for this dimensionless quantity? (c) In the 1970s it was common to find room air conditioners with EERs of 5 or lower. Compare the operating costs for 10 000-Btu/h air conditioners with EERs of 5.00 and 10.0 if each air conditioner operates for 1 500 h during the summer in a city where the cost of electricity is 10.0¢/kWh.

51. One mole of an ideal gas ($\gamma = 1.40$) is carried through the Carnot cycle described in Figure 18.5. At point $A$, the pressure is 25.0 atm and the temperature is 600 K. At point $C$, the pressure is 1.00 atm and the temperature is 400 K. (a) Determine the pressures and volumes at points $A$, $B$, $C$, and $D$. (b) Calculate the net work done by the gas per cycle. (c) Determine the efficiency of an engine operating in this cycle.

52. A sample consisting of $n$ mol of an ideal gas undergoes a reversible isobaric expansion from volume $V_i$ to volume

$3V_i$. Find the change in entropy of the gas by calculating $\int_i^f dQ/T$, where $dQ = nC_P \, dT$.

53. A system consisting of $n$ mol of an ideal gas undergoes two reversible processes. It starts with pressure $P_i$ and volume $V_i$, expands isothermally, and then contracts adiabatically to reach a final state with pressure $P_i$ and volume $3V_i$. (a) Find its change in entropy in the isothermal process. Note that no entropy change occurs in the adiabatic process. (b) Explain why the answer to part (a) must be the same as the answer to Problem 52.

54. The *Otto cycle* in Figure P18.54 models the operation of the internal combustion engine in an automobile. A mixture of gasoline vapor and air is drawn into a cylinder as the piston moves down during the intake stroke $O \rightarrow A$. The piston moves up toward the closed end of the cylinder to compress the mixture adiabatically in process $A \rightarrow B$. The ratio $r = V_1/V_2$ is the *compression ratio* of the engine. At $B$ the gasoline is ignited by the spark plug and the pressure rises rapidly as it burns in process $B \rightarrow C$. In the power stroke $C \rightarrow D$ the combustion products expand adiabatically as they drive the piston down. The combustion products cool further in an isovolumetric process $D \rightarrow A$ and in the exhaust stroke $A \rightarrow O$, when the exhaust gases are pushed out of the cylinder. Assume that a single value of the specific heat ratio $\gamma$ characterizes both the fuel–air mixture and the exhaust gases after combustion. Prove that the efficiency of the engine is $1 - r^{1-\gamma}$.

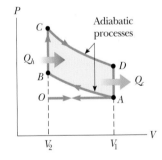

**Figure P18.54**

## ANSWERS TO QUICK QUIZZES

**18.1** No. Friction is one factor that reduces the efficiency of an engine, but even a frictionless engine is restricted to an efficiency less than or equal to that of the Carnot engine, which is less than 100%.

**18.2** C, B, A. Although all three engines operate over a 300-K temperature difference, the efficiency depends on the ratio of temperatures.

**18.3** Yes. A heat pump can deliver significantly more energy than it consumes. In the winter, the energy delivered to the interior of the home is a combination of the energy drawn from the exterior air and the energy coming into the unit by electricity. If the COP of the heat pump is 5, then it is delivering five times as much energy as it consumes. According to Equation 18.5, as long as the COP is greater than 1, the heat pump delivers more energy ($Q_h$) than it consumes ($W$).

**18.4** Four deuces: one microstate—all four deuces. Two aces: six microstates—club-diamond, club-heart, club-spade, diamond-heart, diamond-spade, heart-spade. The macrostate of two aces is more probable because this particular macrostate has six times as many microstates as the macrostate of four deuces. Thus, in a hand of poker, two of a kind is less valuable than four of a kind.

**18.5** In arrangements, in boxes in the back room, in the dumpster. In arrangements, the flowers are highly ordered. In the back room, the flowers are distributed among several boxes. In the dumpster, the flowers are mixed together and not all oriented in the same direction. In this sequence, the flowers are decreasing in order and, therefore, increasing in entropy.

**18.6** (b). Because the process is reversible and adiabatic, $Q_r = 0$; therefore $\Delta S = 0$.

**18.7** (a). From the first law of thermodynamics, for these two reversible processes, $Q_r = \Delta E_{int} - W$. During the constant-volume process, $W = 0$, and the work $W$ is nonzero and negative during the constant-pressure expansion. Thus, $Q_r$ is larger for the constant-pressure process, leading to a larger value for the change in entropy. In terms of entropy as disorder, during the constant-pressure process, the gas must expand. The increase in volume results in more ways of locating the molecules of the gas in a container, resulting in a larger increase in entropy.

**18.8** False. The determining factor for the entropy change is $Q_r$, not $Q$. If the adiabatic process is not reversible, the entropy change is not necessarily zero because a reversible path between the same initial and final states may involve energy transfer by heat.

# Global Warming

## PREDICTING THE CORRECT SURFACE TEMPERATURE

Now that we have investigated the principles of thermodynamics, we respond to our central question for the Context on global warming:

*What factors determine the average temperature at the Earth's surface?*

We discussed some of these factors in Chapter 17—the energy input from the Sun and the energy output by thermal radiation from the surface of the Earth. In Chapter 18, we introduced the role of the atmosphere in absorbing radiation by means of various molecules. In the following discussion, we explore how the atmosphere modifies the temperature calculation performed in Chapter 17. This leads to a structural model that predicts a temperature in agreement with observations.

We first ask the question—is the temperature of 254 K that we found in Chapter 17 valid and, if so, what does this represent? The answer to the first question is *yes*. The energy balance concept is certainly valid and the Earth, as a system, must emit energy at the same rate as it absorbs energy. The temperature of 254 K is representative of the radiation leaving the atmosphere. A space traveler taking a reading of radiation from the Earth would determine that the temperature representing this radiation is indeed 254 K. But this is the temperature associated with radiation leaving the *top* of the atmosphere—it is not the temperature at the surface of the Earth.

As we have mentioned, the atmosphere is almost transparent to the visible radiation from the Sun, but not to the infrared radiation emitted by the surface of the Earth. Let us build a model in which we assume that all radiation with wavelengths less than about 5 $\mu$m is allowed to pass through the atmosphere. Thus, almost *all* incoming radiation from the Sun (except for the 30% reflected) reaches the Earth's surface. In addition, let us assume that all radiation above about 5 $\mu$m (which is *infrared* radiation, including that emitted by the Earth surface) is absorbed by the atmosphere.

On average, energy emitted from the Earth's surface is absorbed twice before it escapes into space. As a result, we identify two layers to the atmosphere in our model (Fig. 1). The lower part of the atmosphere is the *troposphere*—this is the convective region in which the temperature decreases with height according to the lapse rate, as discussed in Section 16.7. This is also the region of the atmosphere in which our familiar weather occurs. Above this is the *stratosphere*, which we assume to be perfectly transmitting for solar radiation and almost so for infrared radiation, so that the temperature remains approximately constant with height. Between these two layers is the *tropopause*, which is about 11 km from the Earth's surface. In reality, the tropopause is a thin region in which the primary transfer mechanism changes continuously from convection to radiation. In our model, we imagine the tropopause to be a sharp boundary.

659

**Figure 1**

In our structural model of the atmosphere, we consider two layers. The upper layer is the stratosphere, in which the temperature is modeled as constant. The lower layer is the troposphere, in which the temperature increases linearly from the tropopause to the ground.

The first task is to find the temperature, assumed constant, of the stratosphere. We appeal again to Stefan's law, and consider the energy transfer into and out of the stratosphere, as indicated in Figure 2. Radiation from the troposphere (to which we assign an effective average temperature of $T_t = 254$ K, so that that is the temperature associated with the radiation coming through the stratosphere to

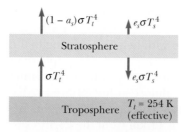

**Figure 2**

The stratosphere is modeled as an object with a temperature, emitting thermal radiation from both upper and lower surfaces. The input of energy to the stratosphere is a portion of the radiation passing through it from the troposphere. The energy transfers indicated are expressed as power per unit area.

our imagined outer space observer) passes through the stratosphere, with a fraction $a_s$ absorbed. The stratosphere, at temperature $T_s$, radiates both upward and downward, according to its emissivity $e_s$. Thus, at equilibrium, the power balance equation for the stratosphere is

$$\mathcal{P}_{ER}\,(\text{in}) = -\mathcal{P}_{ER}\,(\text{out})$$

$$a_s\sigma A T_t^4 = 2e_s\sigma A T_s^4$$

where the factor of 2 arises from the output radiation of the stratosphere from both top and bottom surfaces. We can solve this for the temperature of the stratosphere:

$$T_s = \sqrt[4]{\frac{a_s\sigma T_t^4}{2e_s\sigma}} = \sqrt[4]{\frac{a_s}{2e_s}}\,T_t = \sqrt[4]{\frac{1}{2}}\,(254\ \text{K}) = 214\ \text{K}$$

where we have used the fact that the absorptivity and the emissivity of the stratosphere are the same number.

Now, we have all of the pieces—the temperature of the stratosphere, the height of the tropopause, and the lapse rate. We simply need to extrapolate, using the lapse rate, from the temperature at the tropopause, which is the temperature of the stratosphere, to that at the Earth's surface.

If the tropopause is 11 km from the surface, which is a typical measured value, and the lapse rate is $-6.7°C/km$ (Section 16.7), then the net change in temperature from the surface to the tropopause is

$$\Delta T = T_{\text{tropopause}} - T_{\text{surface}} = \left(\frac{\Delta T}{\Delta y}\right)\Delta y = (-6.7°C/km)(11\ \text{km})$$

$$= -74°C = -74\ \text{K}$$

Because the tropopause temperature is 214 K, we can now find the surface temperature:

$$\Delta T = T_{\text{tropopause}} - T_{\text{surface}}$$

$$-74\ \text{K} = 214\ \text{K} - T_{\text{surface}} \quad\longrightarrow\quad T_{\text{surface}} = 288\ \text{K}$$

This agrees with the measured average temperature as discussed in Chapter 17! Figure 3 shows a graphical representation (height vs. temperature) of the temperature in the troposphere.

The absorption of radiation from the Earth's surface is dependent on molecules in the atmosphere. Our industrialized society is changing the atmospheric concentrations of molecules such as water, carbon dioxide, and methane. As a result, we are altering the energy balance and putting the Earth at risk of a change in temperature. Some data taken since the mid-19th century show a temperature increase of 0.5 to 1.0°C in the last 150 years. Although this increase may seem small, a slightly warmer Earth results in some melting of the polar ice caps and subsequent rises in the level of the oceans. Measured ocean levels show rises by as much as 50 cm in the 20th century. Further rises will cause severe problems for coastal populations. In addition, changes in temperature will have major effects on balanced ecosystems in various parts of the Earth.

The model we have described in this Context is successful in predicting the correct surface temperature. If we extend the model to predict changes in the surface temperature as we add more carbon dioxide to the atmosphere, we find that the predictions are not in agreement with more sophisticated models. The atmosphere is a very complicated entity, and the models used by atmos-

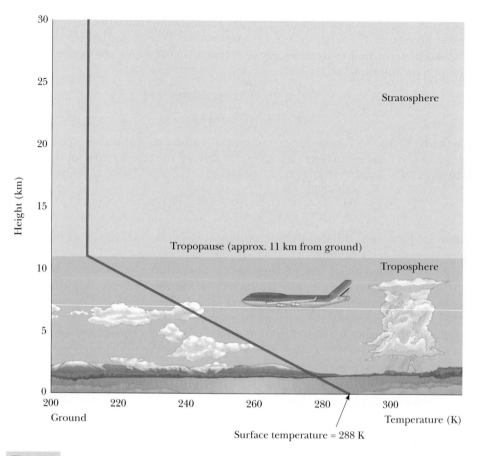

**Figure 3**

A graphical representation of the temperature variation with altitude in our model atmosphere. The predicted surface temperature agrees with measurements.

pheric scientists are far more sophisticated than the one we have studied here. For our purposes, however, our successful prediction of the surface temperature is sufficient.

**PROBLEMS**

1. A simple model of absorption in the atmosphere shows that doubling the amount of carbon dioxide in the next century will raise the altitude of the tropopause from 11 km to about 13 km. If the stratosphere temperature and the lapse rate stay the same, what is the surface temperature in this case? The result you obtain is much larger than the temperature predicted by sophisticated computer models—this displays a weakness in our simple model.

2. The stratosphere of Venus has a temperature of 200 K. The lapse rate in the Venutian troposphere is about $-8.8°C/km$. The measured temperature on the surface of Venus is 732 K. What is the altitude of the Venutian tropopause?

3. Another atmospheric model is based on splitting the atmosphere into $N$ layers of gas. The depth of a given layer is one *radiation thickness*; that is, the probability of absorption of infrared radiation in the layer is 100%. Because the density of the gas, and therefore, the probability of absorption, varies with altitude, the layers have different physical

thicknesses. A given layer has temperature $T_i$, where $i$ runs from 1 (the top layer) to $N$ (the layer just above the surface). Each layer emits thermal radiation from its top and bottom surfaces and absorbs radiation from the layers above and below it. The lowest layer emits radiation from its bottom surface into the surface of the planet, of temperature $T_s$, and also absorbs radiation from the planet. The highest layer emits into space from its top surface, but does not have a higher layer from which to absorb radiation. (a) For the Earth, show that the temperature $T_1$ of the top layer is 254 K. (b) Show that the surface temperature for an arbitrary number $N$ of layers is $T_s = \sqrt[4]{N + 1}\, T_1$. (c) Considering the troposphere and stratosphere of the Earth as a two-layer system, what surface temperature does this model predict? (d) Why is this prediction so bad for the Earth? (e) Consider the atmosphere of Venus, from which 77% of incident radiation is reflected. What is the temperature $T_1$ of the top layer of the Venutian atmosphere? (f) Given that the surface temperature of Venus is 732 K, how many layers are in the Venutian atmosphere? (g) Do you think the multilayer model will be more successful in describing the atmosphere of Venus than that of Earth? Why?

# Lightning

Lightning occurs all over the world, but more often in some places than others. Florida, for example, experiences lightning storms very often, but lightning is rare in Southern California. We begin this Context by looking at the details of a flash of lightning in a *qualitative* way. As we continue deeper into the Context, we will return to this description and attach more quantitative structure to it.

In general, we shall consider a flash of lightning to be an electric discharge occurring between a charged cloud and the ground—in other words, an enormous spark. But lightning can occur in *any* situation in which a large electric charge (which we discuss in Chapter 19) can result in electrical breakdown of the air, including snowstorms, sandstorms, and over erupting volcanoes. If we consider lightning associated with clouds, we observe cloud-to-ground discharge, cloud-to-cloud discharge, intracloud discharge, and cloud-to-air discharge. In this Context we shall consider only the most commonly described discharge, *cloud to ground*. Intracloud discharge actually occurs more often, but is not the type of lightning that we regularly observe.

Because a flash of lightning occurs in a very short time, the structure of the process is hidden from normal human observation. A *flash* of lightning is composed of a number of individual *strokes* of lightning, separated by tens of milliseconds. A typical number of strokes is 3 or 4, although as many as 26 strokes (for a total duration of 2 seconds) have been measured in a flash.

Although a stroke of lightning may appear as a sudden single event, several steps are involved in the process. The process begins with an electrical breakdown in the air near the cloud that results in a column of negative charge, called a *stepped leader,* moving toward the ground at a typical speed of $10^5$ m/s. The term stepped leader refers to the fact that the movement occurs in discrete steps of length about 50 m, with a delay of about 50 $\mu$s before the next step. This step-wise process is due to the random variations in free electron density in the air. The stepped leader is only faintly luminous and is not the bright flash we ordinarily think of as lightning. The radius of the channel of charge carried by the stepped leader is typically several meters.

As the tip of the stepped leader approaches the ground, it can initiate an electrical breakdown in the air near the ground, often at the tip of a pointed object. As a result, a column of positive charge begins to move upward. This is the beginning of the *return stroke*. At 20 to 100 m above the ground, the return stroke meets the stepped leader, and we have an effective short circuit between the cloud and the ground. As a result, electrons pour downward into the ground, with a leading edge of electron drainage moving upward at speeds up to half that of light. This results

During an eruption of the Sakurajima volcano in Japan, lightning is prevalent in the charged atmosphere above the volcano. Although lightning is possible in this as well as many other situations, in this Context we will study the familiar lightning that occurs in a thunderstorm. *(M. Zhilin/M. Newman/Photo Researchers, Inc.)*

in a very large electric current moving through a channel with a radius measured in centimeters. This high current rapidly raises the temperature of the air, ionizing atoms and providing the bright light flash we associate with lightning. Emission spectra of lightning show many spectral lines from oxygen and nitrogen, the major components of air.

This photograph shows a lightning stroke as well as the individual components of the stroke. The bright channel represents a lightning stroke in progress, just after a stepped leader and a return stroke have connected and the channel becomes conducting. Several stepped leaders can be seen at the top of the photograph, branching off from the bright channel. These are less luminous than the bright channel because they have not yet connected with return strokes. A return stroke can be seen, just to the left of the bright channel, moving upward from the tree, in search of a stepped leader. Another very faint return stroke can be seen leaving the top of the power pole at the left of the photograph. *(© Johnny Autery)*

After the return stroke, the conducting channel retains its conductivity for a short time (measured in tens of milliseconds). If more negative charge from the cloud is made available at the top of the conducting channel, it can move downward and result in a new stroke. In this case, because the conducting channel is "open," the leader does not move in a stepped fashion but moves downward smoothly and quickly. For this reason, it is called a *dart leader*. Once again, as the dart leader approaches the ground, a return stroke is initiated and a bright flash of light occurs.

Just after the current has passed through the conducting channel, the air is turned into a plasma at a typical temperature of 30 000 K. As a result, there is a sudden increase of pressure causing a rapid expansion of the plasma and generating a shock wave in the surrounding gas. This is the origin of the *thunder* associated with lightning.

Having taken this first qualitative step into the understanding of lightning, let us now seek more details. After investigating the physics of lightning, we shall respond to our central question

*How can we determine the number of lightning flashes on the Earth in a typical day?*

## chapter 19

In a photocopy machine, an image of the original document is created on a selenium-coated drum in the form of a pattern of positive electrical charge. Particles of a black powder called toner are given a negative charge and are attracted to the drum at the dark locations on the original image. The toner is then transferred to a blank sheet of paper, which carries a larger positive charge, forming a copy of the original document.

*(Charles D. Winters)*

# Electric Forces and Electric Fields

**T**his is the first of three chapters on *electricity*. You are probably familiar with electrical effects, such as the static cling between articles of clothing removed from the dryer. You may also be familiar with the spark that jumps from your finger to a doorknob after you have walked across a carpet. Much of your daily experience involves working with devices that operate on energy transferred to the device by means of electrical transmission and provided by the electric power company.

This chapter begins with a review of some of the basic properties of the electrostatic force that we introduced in Chapter 5, as well as some properties of the electric field associated with stationary charged particles. Our study of electrostatics then continues with the concept of an electric field that is associated with a continuous charge distribution and the effect of this field on other charged particles. In these studies, we shall apply the models of a particle in a field and a particle under a net force that we have seen in earlier chapters.

## 19.1 • HISTORICAL OVERVIEW

The laws of electricity and magnetism play a central role in the operation of devices such as radios, televisions, electric motors, computers, high-energy particle accelerators, and a host of electronic devices used in medicine. More fundamental, however, is the fact that the interatomic and intermolecular forces responsible for the formation of solids and liquids are electric in origin. Furthermore, such forces as the pushes and pulls between objects in contact and the elastic force in a spring arise from electric forces at the atomic level.

Chinese documents suggest that magnetism was recognized as early as about 2000 B.C. The ancient Greeks observed electric and magnetic phenomena possibly as early as 700 B.C. They found that a piece of amber, when rubbed, attracted pieces of straw or feathers. The existence of magnetic forces was known from observations that pieces of a naturally occurring stone called *magnetite* ($Fe_3O_4$) were attracted to iron. (The word *electric* comes from the Greek word for amber, *elektron*. The word *magnetic* comes from *Magnesia*, on the coast of Turkey where magnetite was found.)

In 1600, the Englishman William Gilbert discovered that electrification was not limited to amber but was a general phenomenon. Scientists went on to electrify a variety of objects, including chickens and people! Experiments by Charles Coulomb in 1785 confirmed the inverse-square law for the electrostatic force.

It was not until the early part of the 19th century that scientists established that electricity and magnetism are related phenomena. In 1820, Hans Oersted discovered that a compass needle, which is magnetic, is deflected when placed near an electric current. In 1831, Michael Faraday in England and, almost simultaneously, Joseph Henry in the United States showed that, when a wire loop is moved near a magnet (or, equivalently, when a magnet is moved near a wire loop), an electric current is observed in the wire. In 1873, James Clerk Maxwell used these observations and other experimental facts as a basis for formulating the laws of electromagnetism as we know them today. Shortly thereafter (around 1888), Heinrich Hertz verified Maxwell's predictions by producing electromagnetic waves in the laboratory. This achievement was followed by such practical developments as radio and television.

Maxwell's contributions to the science of electromagnetism were especially significant because the laws he formulated are basic to *all* forms of electromagnetic phenomena. His work is comparable in importance to Newton's discovery of the laws of motion and the theory of gravitation.

**Figure 19.1**

Rubbing a balloon against your hair on a dry day causes the balloon and your hair to become electrically charged. *(Charles D. Winters)*

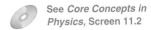

See *Core Concepts in Physics*, Screen 11.2

## 19.2 • PROPERTIES OF ELECTRIC CHARGES

A number of simple experiments demonstrate the existence of electrostatic forces. For example, after running a comb through your hair, you will find that the comb attracts bits of paper. The attractive electrostatic force is often strong enough to suspend the bits. The same effect occurs with other rubbed materials, such as glass or rubber.

Another simple experiment is to rub an inflated balloon with wool, or across your hair (Fig. 19.1). On a dry day, the rubbed balloon will stick to the wall of a room, often for hours. When materials behave this way, they are said to have become **electrically charged.** You can give your body an electric charge by walking across a wool rug or by sliding across a car seat. You can then feel,

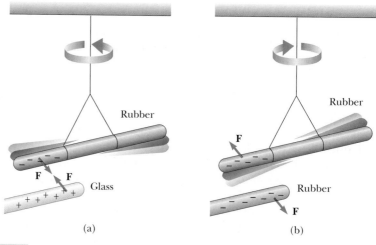

**Figure 19.2**

(a) A negatively charged rubber rod, suspended by a thread, is attracted to a positively charged glass rod. (b) A negatively charged rubber rod is repelled by another negatively charged rubber rod.

**WEB**

For information on Benjamin Franklin, visit **library.thinkquest.org/22254/**

**Figure 19.3**

When a glass rod is rubbed with silk, electrons are transferred from the glass to the silk. Because of conservation of charge, each electron adds negative charge to the silk, and an equal positive charge is left behind on the rod. Also, because the charges are transferred in discrete bundles, the charges on the two objects are $\pm e$ or $\pm 2e$ or $\pm 3e$, and so on.

and remove, the charge on your body by lightly touching another person. Under the right conditions, a visible spark is seen when you touch, and a slight tingle is felt by both parties. (Such an experiment works best on a dry day because excessive moisture in the air can provide a pathway for charge to leak off a charged object.)

Experiments also demonstrate that there are two kinds of electric charge, given the names **positive** and **negative** by Benjamin Franklin (1706–1790). Figure 19.2 illustrates the interactions of the two charges. A hard rubber (or plastic) rod that has been rubbed with fur (or an acrylic material) is suspended by a piece of string. When a glass rod that has been rubbed with silk is brought near the rubber rod, the rubber rod is attracted toward the glass rod (Fig. 19.2a). If two charged rubber rods (or two charged glass rods) are brought near each other, as in Figure 19.2b, the force between them is repulsive. This observation demonstrates that the rubber and glass have different kinds of charge. We use the convention suggested by Franklin; the electric charge on the glass rod is called positive and that on the rubber rod is called negative. On the basis of observations such as these, we conclude that **like charges repel each other, and unlike charges attract each other.**

We know that only two kinds of electric charge exist because any unknown charge that is found experimentally to be attracted to a positive charge is also repelled by a negative charge. No one has ever observed a charged object that is repelled by both a positive and a negative charge or attracted to both.

Another important characteristic of electric charge is that the **net charge in an isolated system is always conserved.** This is the electric charge version of the isolated system model. We first introduced isolated system models in Chapter 7, in which we discussed conservation of energy; we now see a principle of **conservation of electric charge** for an isolated system. When two initially neutral objects are charged by being rubbed together, charge is not created in the process. The objects become charged because *electrons are transferred* from one object to the other. One object gains some amount of negative charge from the electrons transferred to it while the other loses an equal amount of negative charge and hence is left with a positive charge. For the isolated system of the two objects, no transfer of charge occurs across the boundary of the system. The only change is that charge has been transferred between two members of the system. For example, when a glass rod is rubbed with silk, as in Figure 19.3, the silk obtains a negative charge that is equal in magnitude to the positive charge on the glass rod as negatively charged electrons are transferred from the glass to the silk. Likewise, when rubber is rubbed with fur, electrons are transferred from the fur to the rubber. An *uncharged object* contains an enormous number of electrons (on the order of $10^{23}$). However, for every negative electron, a positively charged proton is also present; hence, an uncharged object has no net charge of either sign.

**PITFALL PREVENTION 19.1**

**Like charges**

The phrase "like charges" means two charges with the same sign; we are not implying that they have to be identical in magnitude.

### Quick Quiz 19.1

Two insulated rods are oppositely charged on their ends. They are mounted at their centers so that they are free to rotate, and then held in the position shown in a view from above (Fig. 19.4). The plane of the rotation of the rods is in the plane of the paper. Do the rods return to these positions if slightly displaced from them and released? If not, into what position(s) will they move? Does the final configuration(s) represent stable equilibrium?

**Figure 19.4**

(Quick Quiz 19.1) If the system is disturbed, will it return to this configuration?

## 19.3 • INSULATORS AND CONDUCTORS

We have discussed the transfer of charge from one object to another. It is also possible for electric charges to move from one location to another within an object—such motion of charge is called **electrical conduction.** It is convenient to classify substances in terms of the ability of charges to move within the substance:

See Screen 11.3

> **Conductors** are materials in which electric charges move relatively freely, and **insulators** are materials in which electric charges do not move freely.

Materials such as glass, rubber, and Lucite are insulators. When such materials are charged by rubbing, only the rubbed area becomes charged, and the charge does not tend to move to other regions of the material. In contrast, materials such as copper, aluminum, and silver are good conductors. When such materials are charged in some small region, the charge readily distributes itself over the entire surface of the material. If you hold a copper rod in your hand and rub it with wool or fur, it will not attract a small piece of paper. This might suggest that a metal cannot be charged. If you hold the copper rod by an insulating handle and then rub it, however, the rod remains charged and attracts the piece of paper. In the first case, the electric charges produced by rubbing readily move from the copper through your body, which is a conductor, and finally to Earth. In the second case, the insulating handle prevents the flow of charge to your hand.

*Semiconductors* are a third class of materials, and their electrical properties are somewhere between those of insulators and those of conductors. Charges can move somewhat freely in a semiconductor, but far fewer charges are moving through a semiconductor than in a conductor. Silicon and germanium are well-known examples of semiconductors that are widely used in the fabrication of a

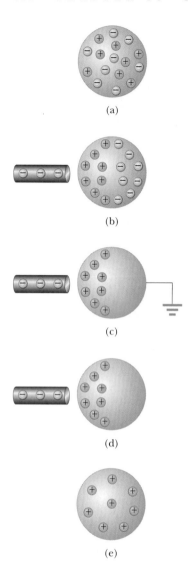

**Figure 19.5**

Charging a metallic object by induction. (a) A neutral metallic sphere, with equal numbers of positive and negative charge. (b) The charge on the neutral sphere is redistributed when a charged rubber rod is placed near the sphere. (c) When the sphere is grounded, some of its electrons leave through the ground wire. (d) When the ground connection is removed, the sphere has excess positive charge that is nonuniformly distributed. (e) When the rubber rod is removed, the excess positive charge becomes uniformly distributed over the surface of the sphere.

variety of electronic devices. The electrical properties of semiconductors can be changed over many orders of magnitude by adding controlled amounts of certain foreign atoms to the materials.

### Charging by Induction

When a conductor is connected to Earth by means of a conducting wire or pipe, it is said to be **grounded.** For present purposes, the Earth can be modeled as an infinite reservoir for electrons, which means that it can accept or supply an unlimited number of electrons. In this context, the Earth serves a purpose similar to our energy reservoirs introduced in Chapter 17. With this in mind, we can understand how to charge a conductor by a process known as **charging by induction.**

Consider a neutral (uncharged) conducting sphere that is insulated so that there is no conducting path to ground (Fig. 19.5a). A negatively charged rubber rod is brought near the sphere. The repulsive force between the electrons in the rod and those in the sphere causes a redistribution of charge on the sphere so that some electrons move to the side of the sphere farthest away from the rod (Fig. 19.5b). The region of the sphere nearest the rod has an excess of positive charge because of the migration of electrons away from this location. If a grounded conducting wire is then connected to the sphere near the accumulation of electrons, as in Figure 19.5c, some of the electrons leave the sphere and travel to the Earth. If the wire to ground is then removed (Fig. 19.5d), the conducting sphere is left with an excess of induced positive charge. Finally, when the rubber rod is removed from the vicinity of the sphere (Fig. 19.5e), the induced positive charge remains on the ungrounded sphere. This excess positive charge becomes uniformly distributed over the surface of the ungrounded sphere because of the repulsive forces among the like charges and the high mobility of charge carriers in a metal.

In the process of inducing a charge on the sphere, the charged rubber rod loses none of its negative charge, because it never comes in contact with the sphere. **Charging an object by induction requires no contact with the object inducing the charge.** This is in contrast to charging an object by rubbing, which does require contact between the two objects.

A process similar to the first step in charging by induction in conductors takes place in insulators. In most neutral atoms and molecules, the average position of the positive charge coincides with the average position of the negative charge. In the presence of a charged object, however, these positions may shift slightly, due to the attractive and repulsive forces from the charged object, resulting in more positive charge on one side of the molecule than on the other. This effect is known as **polarization.** The polarization of individual molecules produces a layer of charge on the surface of the insulator, as shown in Figure 19.6a, in which a charged balloon on the left is placed against a wall on the right. In the figure, the negative charge layer in the wall is closer to the positively charged balloon than the positive charges at the other ends of the molecules. Thus, the attractive force between the positive and negative charges is larger than the repulsive force between the positive charges. The result is a net attractive force between the charged balloon and the neutral insulator. It is this polarization effect that explains why a comb that has been rubbed through hair attracts bits of neutral paper (Fig. 19.6b), or why a balloon that has been rubbed against your hair can stick to a neutral wall.

Wall

Charged
balloon

Induced
charges

(a)

(b)

**Figure 19.6**

(a) A charged balloon induces charges on the surface of a wall. (b) A charged comb attracts bits of paper because charges are displaced in the paper. The paper is neutral but polarized. *(Fundamental Photographs, NYC)*

### THINKING PHYSICS 19.1

A positively charged ball, hanging from a string, is brought near a non-conducting object. The ball is attracted to the object. From this experiment, it is not possible to determine whether the object is negatively charged or neutral. Why not? What additional experiment would help you decide between these two possibilities?

**Reasoning** The attraction between the ball and the object could be an attraction of unlike charges *or* it could be an attraction between a charged object and a neutral object, due to polarization of the molecules of the neutral object. Two possible additional experiments would help to determine if the object is charged. First, a known neutral ball could be brought near the object—if the ball is attracted to the object, the object is negatively charged. Another possibility is to bring a known negatively charged ball near the object—if the ball is repelled by the object, the object is negatively charged. If the ball is attracted, it is neutral.

## 19.4 • COULOMB'S LAW

Electric forces between charged objects were measured quantitatively by Charles Coulomb using the torsion balance, which he invented (Fig. 19.7). Coulomb confirmed that the electric force between two small charged spheres is proportional to the inverse square of their separation distance $r$, that is, $F_e \propto 1/r^2$. The operating principle of the torsion balance is the same as that of the apparatus used by Cavendish to measure the gravitational constant (Section 11.1), with the electrically neutral spheres replaced by charged ones. The electric force between charged spheres A and B in Figure 19.7 causes the spheres to either attract or repel each other, and the resulting motion causes the suspended fiber to twist. Because the restoring torque of the twisted fiber is proportional to the angle through which it rotates, a measurement of this angle provides a quantitative measure of the electric force of attraction or repulsion. Once the spheres are charged by

Suspension
head

Fiber

B

A

**Figure 19.7**

Coulomb's torsion balance, which was used to establish the inverse-square law for the electrostatic force between two charges.

See Screen 11.4

• *Coulomb's law*

• *Permittivity of free space*

rubbing, the electric force between them is very large compared with the gravitational attraction, and so the gravitational force can be ignored.

In Chapter 5 we introduced **Coulomb's law,** which describes the magnitude of the electrostatic force between two charged particles with charges $q_1$ and $q_2$ and separated by a distance $r$:

$$F_e = k_e \frac{|q_1||q_2|}{r^2} \qquad [19.1]$$

where $k_e$ ($= 8.99 \times 10^9 \text{ N} \cdot \text{m}^2/\text{C}^2$) is the **Coulomb constant** and the force is in newtons if the charges are in coulombs and the separation distance in meters. The constant $k_e$ is also written

$$k_e = \frac{1}{4\pi\epsilon_0}$$

where the constant $\epsilon_0$, known as the **permittivity of free space,** has the value

$$\epsilon_0 = 8.8542 \times 10^{-12} \text{ C}^2/\text{N} \cdot \text{m}^2$$

It is important to note that Equation 19.1 gives only the magnitude of the force. The direction of the force on a given particle must be found by considering where the particles are located with respect to each other and the sign of each charge. Thus, a pictorial representation of a problem in electrostatics is very important for an understanding of the problem.

The charge of an electron is $q = -e = -1.60 \times 10^{-19}$ C, and the proton has a charge of $q = +e = 1.60 \times 10^{-19}$ C; therefore, 1 C of charge is equal to the magnitude of the charge of $(1.60 \times 10^{-19})^{-1} = 6.25 \times 10^{18}$ electrons. Note that 1 C is a substantial amount of charge. In typical electrostatic experiments, where a rubber or glass rod is charged by friction, a net charge on the order of $10^{-6}$ C ($= 1 \ \mu$C) is obtained. In other words, only a very small number of the total available electrons (on the order of $10^{23}$ in a 1-cm$^3$ sample) are transferred between the rod and the rubbing material. The experimentally measured values of the charges and masses of the electron, proton, and neutron are given in Table 19.1.

When dealing with Coulomb's law, remember that force is a *vector* quantity and must be treated accordingly. Furthermore, **Coulomb's law applies exactly only to particles.**\* The electrostatic force exerted by $q_1$ on $q_2$, written **F**$_{12}$, can be expressed in vector form as$^\dagger$

$$\mathbf{F}_{12} = k_e \frac{q_1 q_2}{r^2} \hat{\mathbf{r}}_{12} \qquad [19.2]$$

**Charles Coulomb** (1736–1806)

French physicist. Coulomb's major contributions to science were in the areas of electrostatics and magnetism. During his lifetime, he also investigated the strengths of materials and determined the forces that affect objects on beams, thereby contributing to the field of structural mechanics. In the field of ergonomics, his research provided a fundamental understanding of the ways in which people and animals can best do work. *(Photo courtesy of AIP Niels Bohr Library/E. Scott Barr Collection)*

| TABLE 19.1 | Charge and Mass of the Electron, Proton, and Neutron | |
|---|---|---|
| **Particle** | **Charge (C)** | **Mass (kg)** |
| Electron (e) | $-1.602\ 191\ 7 \times 10^{-19}$ | $9.109\ 5 \times 10^{-31}$ |
| Proton (p) | $+1.602\ 191\ 7 \times 10^{-19}$ | $1.672\ 61 \times 10^{-27}$ |
| Neutron (n) | $0$ | $1.674\ 92 \times 10^{-27}$ |

\* Coulomb's law can also be used for larger objects to which the particle model can be applied.

$^\dagger$ Notice that we use "$q_2$" as short-hand notation for "the particle with charge $q_2$." This is common usage when discussing charged particles, similar to the use in mechanics of "$m_2$" for "the particle with mass $m_2$." The context of the sentence will tell you whether the symbol represents an amount of charge or a particle with that charge.

where $\hat{\mathbf{r}}_{12}$ is a unit vector directed from $q_1$ toward $q_2$ as in Figure 19.8a. Equation 19.2 can be used to find the actual direction of the force in space, although a carefully drawn pictorial representation is needed in order to clearly identify the direction of $\hat{\mathbf{r}}_{12}$. From Newton's third law, we see that the electric force exerted by $q_2$ on $q_1$ is equal in magnitude to the force exerted by $q_1$ on $q_2$ and in the opposite direction; that is, $\mathbf{F}_{21} = -\mathbf{F}_{12}$. From Equation 19.2 we see that if $q_1$ and $q_2$ have the same sign, the product $q_1 q_2$ is positive and the force is repulsive, as in Figure 19.8a. The force on $q_2$ is in the same direction as $\hat{\mathbf{r}}_{12}$, and is directed away from $q_1$. If $q_1$ and $q_2$ are of opposite sign as in Figure 19.8b, the product $q_1 q_2$ is negative and the force is attractive. In this case, the force on $q_2$ is in the direction opposite to $\hat{\mathbf{r}}_{12}$, directed toward $q_1$.

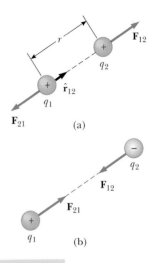

### Quick Quiz 19.2

Object A has a charge of $+2\ \mu\text{C}$, and object B has a charge of $+6\ \mu\text{C}$. Which statement is true? (a) $\mathbf{F}_{AB} = -3\mathbf{F}_{BA}$  (b) $\mathbf{F}_{AB} = -\mathbf{F}_{BA}$  (c) $3\mathbf{F}_{AB} = -\mathbf{F}_{BA}$

When more than two charged particles are present, the force between any pair is given by Equation 19.2. Therefore, the **resultant force on any one particle equals the *vector* sum of the individual forces due to all other particles.** This **principle of superposition** as applied to electrostatic forces is an experimentally observed fact and simply represents the traditional vector sum of forces introduced in Chapter 4. As an example, if four charged particles are present, the resultant force on particle 1 due to particles 2, 3, and 4 is given by the vector sum

$$\mathbf{F}_1 = \mathbf{F}_{21} + \mathbf{F}_{31} + \mathbf{F}_{41}$$

**Figure 19.8**

Two point charges separated by a distance $r$ exert a force on each other given by Coulomb's law. Note that the force $\mathbf{F}_{21}$ exerted by $q_2$ on $q_1$ is equal in magnitude and opposite in direction to the force $\mathbf{F}_{12}$ exerted by $q_1$ on $q_2$. (a) When the charges are of the same sign, the force is repulsive. (b) When the charges are of opposite signs, the force is attractive.

---

### Example 19.1   Where Is the Resultant Force Zero?

Three charged particles lie along the $x$ axis as in Figure 19.9. The particle with charge $q_1 = +15.0\ \mu\text{C}$ is at $x = 2.00$ m, and the particle with charge $q_2 = +6.00\ \mu\text{C}$ is at the origin. Where on the $x$ axis can a particle with negative charge $q_3$ be placed such that the resultant force on it is zero?

**Reasoning**   The requested resultant force of zero indicates that $q_3$ is a particle in equilibrium, so that the two forces on $q_3$ cancel. Because $q_3$ is negative and both $q_1$ and $q_2$ are positive, the forces $\mathbf{F}_{13}$ and $\mathbf{F}_{23}$ are both attractive. In order to cancel, the forces on $q_3$ must be in opposite directions. If $q_3$ is placed to the left of $q_2$ or to the right of $q_1$, the forces on $q_3$ will be in the same direction. Thus, the only possibility of having forces in opposite directions is to place $q_3$ between $q_1$ and $q_2$, as indicated in Figure 19.9.

**Solution**   If we let $x$ be the coordinate of $q_3$, then the forces $\mathbf{F}_{13}$ and $\mathbf{F}_{23}$ can be written

$$\mathbf{F}_{13} = k_e \frac{q_1 q_3}{(2.00 - x)^2} \hat{\mathbf{r}}_{13} = -k_e \frac{q_1 q_3}{(2.00 - x)^2} \mathbf{i}$$

$$\mathbf{F}_{23} = k_e \frac{q_2 q_3}{x^2} \hat{\mathbf{r}}_{23} = k_e \frac{q_2 q_3}{x^2} \mathbf{i}$$

where we have recognized that $\hat{\mathbf{r}}_{13} = -\mathbf{i}$ for the force due to $q_1$, because $q_1$ is to the right of $q_3$. We now add these two forces and set the resultant equal to zero:

$$\sum \mathbf{F} = \mathbf{F}_{13} + \mathbf{F}_{23} = -k_e \frac{q_1 q_3}{(2.00 - x)^2} \mathbf{i} + k_e \frac{q_2 q_3}{x^2} \mathbf{i} = 0$$

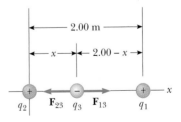

**Figure 19.9**

(Example 19.1) Three point charges are placed along the $x$ axis. If the net force on $q_3$ is zero, the force $\mathbf{F}_{13}$ exerted by $q_1$ on $q_3$ must be equal in magnitude and opposite in direction to the force $\mathbf{F}_{23}$ exerted by $q_2$ on $q_3$.

Because $k_e$ and $q_3$ are common to both terms, they cancel, and we can solve for $x$:

$$\frac{q_1}{(2.00 - x)^2} = \frac{q_2}{x^2}$$

$$(4.00 - 4.00x + x^2)(6.00 \times 10^{-6} \text{ C}) = x^2(15.0 \times 10^{-6} \text{ C})$$

which simplifies to

$$9.00x^2 + 24.0x - 24.0 = 0$$

Solving this quadratic equation for $x$, we find that $x = \boxed{0.775 \text{ m}}$. Why is the negative root not acceptable?

---

### Example 19.2 The Hydrogen Atom

The electron and proton of a hydrogen atom are separated (on average) by a distance of approximately $5.3 \times 10^{-11}$ m. Find the magnitudes of the electrostatic force and the gravitational force that either particle exerts on the other.

**Solution** From Coulomb's law, we find that the magnitude of the attractive electric force is

$$F_e = k_e \frac{e^2}{r^2} = (8.99 \times 10^9 \text{ N} \cdot \text{m}^2/\text{C}^2) \frac{(1.60 \times 10^{-19} \text{ C})^2}{(5.3 \times 10^{-11} \text{ m})^2} = \boxed{8.2 \times 10^{-8} \text{ N}}$$

Using Newton's law of universal gravitation (Section 5.6) and Table 19.1 for the particle masses, we find that the magnitude of the gravitational force is

$$F_g = G \frac{m_e m_p}{r^2} = (6.7 \times 10^{-11} \text{ N} \cdot \text{m}^2/\text{kg}^2) \frac{(9.11 \times 10^{-31} \text{ kg})(1.67 \times 10^{-27} \text{ kg})}{(5.3 \times 10^{-11} \text{ m})^2}$$

$$= \boxed{3.6 \times 10^{-47} \text{ N}}$$

The ratio $F_g/F_e \approx 4 \times 10^{-40}$. Thus the gravitational force between charged atomic particles is negligible compared with the electric force.

**EXERCISE** Two point charges of magnitude 3.0 nC and 6.0 nC are separated by a distance of 0.30 m. Find the magnitude of the electric force each exerts on the other.

*Answer* $1.8 \times 10^{-6}$ N

---

See Screen 11.5

## 19.5 • ELECTRIC FIELDS

The gravitational field **g** at a point in space was defined in Chapter 5 to be equal to the gravitational force $\mathbf{F}_g$ acting on a test particle of mass $m_0$ divided by the mass of the test particle: $\mathbf{g} = \mathbf{F}_g/m_0$. This is the gravitational version of the model of a particle in a field. In similar manner, an electric field at a point in space can be defined in terms of the electric force acting on a test particle with charge $q_0$ placed at that point. Because charge exists in two varieties, we must choose a convention for our test particle. We choose the convention that **a test particle always carries a positive electric charge.** With this convention, we can introduce the electric version of the particle in a field model—the **electric field E** at a point in space is defined as **the electric force $\mathbf{F}_e$ acting on a test particle placed at that point divided by the charge $q_0$ of the test particle:**

• *Definition of electric field*

$$\mathbf{E} \equiv \frac{\mathbf{F}_e}{q_0} \qquad [19.3]$$

Thus, **an electric field exists at a point if a charged test particle placed at rest at that point experiences an electric force.** Because force is a vector, the electric field is also a vector. Note that **E** is the field produced by some charged particle(s) separate from the test particle—it is *not* the field produced by the test particle. We call the particle(s) creating the electric field the **source particle(s).** This is analogous to the gravitational field set up by some body such as the Earth. This gravitational field exists regardless of whether a test particle of mass $m_0$ is present or not. Similarly, the electric field of the source particles is present whether or not we introduce a test particle into the field. The test particle is used only to measure the force and thus detect the existence of the field and evaluate its strength.

When using Equation 19.3, we must assume that the test charge $q_0$ is small enough that it does not disturb the charge distribution responsible for the electric field. If a vanishingly small test charge $q_0$ is placed near a uniformly charged metallic sphere as in Figure 19.10a, the charge on the metallic sphere remains uniformly distributed. If the test charge is large enough ($q_0' \gg q_0$), as in Figure 19.10b, the charge on the metallic sphere is redistributed and the ratio of the force to the test charge is different: ($F_e'/q_0' \neq F_e/q_0$). That is, because of this redistribution of charge on the metallic sphere, the electric field it sets up is different from the field it sets up in the presence of the much smaller $q_0$.

The vector **E** has the SI units of newtons per coulomb (N/C), analogous to the units N/kg for the gravitational field. The direction of **E** is the same as the direction of **F**$_e$ because we have used the convention of a positive charge on the test particle.

Once the electric field is known at some point, the force on *any* particle with charge $q$ placed at that point can be calculated from a rearrangement of Equation 19.3:

$$\mathbf{F}_e = q\mathbf{E} \qquad [19.4]$$

Once the electric force is evaluated, the situation can be described with the particle under a net force model (the electric force may have to be combined with other forces acting on the particle), and the techniques of earlier chapters can be used to find the motion of the particle.

Consider a point charge* $q$ located a distance $r$ from a test particle with charge $q_0$. According to Coulomb's law, the force exerted on the test particle by $q$ is

$$\mathbf{F}_e = k_e \frac{qq_0}{r^2}\,\hat{\mathbf{r}}$$

Using Equation 19.3, we find that the electric field created by $q$ at a point $P$, which is the position of $q_0$, is

$$\mathbf{E} = k_e \frac{q}{r^2}\,\hat{\mathbf{r}} \qquad [19.5]$$

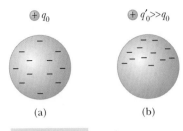

**Figure 19.10**

(a) For a small enough test charge $q_0$, the charge distribution on the sphere is undisturbed. (b) When the test charge $q_0'$ is larger, the charge distribution on the sphere is disturbed as the result of the proximity of $q_0'$.

**PITFALL PREVENTION 19.2**
**Particles only**

Keep in mind that Equation 19.4 is only valid for a charged *particle*—an object of zero size. For a charged object of finite size in an electric field, the field may vary in magnitude and direction over the size of the object, so the corresponding force equation would be more complicated.

•  *Electric field of a point charge*

---

* We have used the phrases "charged particle" or "particle with a charge" so far. The phrase "point charge" is somewhat misleading because charge is a property of a particle, it is not a physical entity. It is similar to misleading phrasing in mechanics such as "a mass $m$ is placed . . ." (which we have avoided) rather than "a particle with mass $m$ is placed. . . ." This phrase is so ingrained in physics usage, however, that we will use it and hope that this footnote suffices to clarify its use.

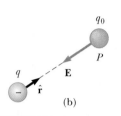

**Figure 19.11**

A test charge $q_0$ at point $P$ is a distance $r$ from a point charge $q$. (a) If $q$ is positive, the electric field at $P$ due to $q$ points radially *outward* from $q$. (b) If $q$ is negative, the electric field at $P$ due to $q$ points radially *inward* toward $q$. The electric field due to $q$ exists even in the absence of $q_0$.

where $\hat{\mathbf{r}}$ is a unit vector directed from $q$ toward $P$ (Fig. 19.11). If $q$ is positive as in Figure 19.11a, the electric field is directed radially outward from it. If $q$ is negative as in Figure 19.11b, the field is directed toward it.

To calculate the electric field at a point $P$ due to a group of point charges, we first calculate the electric field vectors at $P$ individually using Equation 19.5 and then add them vectorially. In other words, **the total electric field at a point in space due to a group of charged particles equals the vector sum of the electric fields at that point due to all the particles.** This superposition principle applied to fields follows directly from the vector addition property of forces. Thus, the electric field at point $P$ of a group of source charges can be expressed as

$$\mathbf{E} = k_e \sum_i \frac{q_i}{r_i^2} \hat{\mathbf{r}}_i \qquad [19.6]$$

where $r_i$ is the distance from the $i$th charge $q_i$ to the point $P$ (the location at which the field is to be evaluated) and $\hat{\mathbf{r}}_i$ is a unit vector directed from $q_i$ toward $P$.

**Quick Quiz 19.3**

A test charge of $+3\ \mu C$ is at a point $P$ where the electric field due to source charges is directed to the right and has a magnitude of $4 \times 10^6$ N/C. If the test charge is replaced with a $-3\ \mu C$ charge, what happens to the electric field at $P$?

**Quick Quiz 19.4**

A very small uncharged metallic-coated Styrofoam ball is suspended in the region between two vertical metal plates in which a uniform electric field occurs. If the two plates are charged, one positive and one negative, describe the motion of the ball after it is brought into contact with one of the plates.

**Example 19.3 Electric Field of a Dipole**

An **electric dipole** consists of a point charge $q$ and a point charge $-q$ separated by a distance of $2a$, as in Figure 19.12. As we shall see in later chapters, neutral atoms and molecules behave as dipoles when placed in an external electric field. Furthermore, many molecules, such as HCl, are permanent dipoles. (HCl can be effectively modeled as an $H^+$ ion combined with a $Cl^-$ ion.) The effect of such dipoles on the behavior of materials subjected to electric fields is discussed in Chapter 20. (a) Find the electric field $\mathbf{E}$ due to the dipole along the $y$ axis at the point $P$, which is a distance $y$ from the origin. (b) Find the electric field for points $y \gg a$ far from the dipole.

**Solution** (a) At $P$, the fields $\mathbf{E}_1$ and $\mathbf{E}_2$ due to the two particles are equal in magnitude because $P$ is equidistant from the two charges. The total field at $P$ is $\mathbf{E} = \mathbf{E}_1 + \mathbf{E}_2$, where the magnitudes of the fields are

$$E_1 = E_2 = k_e \frac{q}{r^2} = k_e \frac{q}{y^2 + a^2}$$

The $y$ components of $\mathbf{E}_1$ and $\mathbf{E}_2$ are equal in magnitude and opposite in sign, so they cancel. The $x$ components are equal and add, because they have the same sign. The total field $\mathbf{E}$ is therefore parallel to the $x$ axis and has a magnitude

$$E = 2k_e \frac{q}{y^2 + a^2} \cos \theta$$

From Figure 19.12 we see that $\cos \theta = a/r = a/(y^2 + a^2)^{1/2}$. Therefore,

$$E = 2k_e \frac{q}{y^2 + a^2} \cos \theta = 2k_e \frac{q}{(y^2 + a^2)} \frac{a}{(y^2 + a^2)^{1/2}}$$

$$= k_e \frac{2qa}{(y^2 + a^2)^{3/2}}$$

(b) The preceding equation gives the value of the electric field at all values of $y$. For points far from the dipole, for

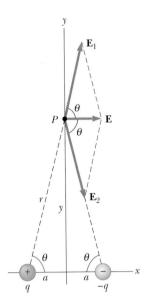

**Figure 19.12**

(Example 19.3) The total electric field **E** at *P* due to two equal and opposite charges (an electric dipole) equals the vector sum **E**$_1$ + **E**$_2$. The field **E**$_1$ is due to the positive charge *q*, and **E**$_2$ is the field due to the negative charge − *q*.

which $y \gg a$, we can ignore $a^2$ in the denominator and write

$$E \approx k_e \frac{2qa}{y^3}$$

Thus, we see that along the *y* axis the field of a dipole at a distant point varies as $1/r^3$, whereas the more slowly varying field of a point charge varies as $1/r^2$. (*Note:* In the geometry of this example, $r = y$.) This is because at distant points, the fields of the two charges in the dipole almost cancel each other. The $1/r^3$ variation in *E* for the dipole is also obtained for a distant point along the *x* axis (Problem 15) and for a general distant point.

**EXERCISE** A piece of aluminum foil of mass $5.0 \times 10^{-2}$ kg is suspended by a string in an electric field directed vertically upward. If the charge on the foil is 3.0 $\mu$C, find the strength of the field that will reduce the tension in the string to zero.

*Answer* $1.6 \times 10^5$ N/C

**EXERCISE** The nucleus of a hydrogen atom, a proton, sets up an electric field. The average distance between the proton and the electron of a hydrogen atom is about $5.3 \times 10^{-11}$ m. What is the magnitude of the electric field at this distance from the proton?

*Answer* $5.1 \times 10^{11}$ N/C

## Electric Field Due to Continuous Charge Distributions

In most practical situations (e.g., an object charged by rubbing), the average separation between source charges is small compared with their distances from the point at which the field is to be evaluated. In such cases, the system of source charges can be modeled as *continuous*. That is, we imagine that the system of closely spaced charges is equivalent to a total charge that is continuously distributed through some volume or over some surface.

To evaluate the electric field of a continuous charge distribution, the following procedure is used. First, we divide the charge distribution into small elements, each of which contains a small amount of charge $\Delta q$ as in Figure 19.13. Next, modeling the element as a point charge, we use Equation 19.5 to calculate the electric field $\Delta$**E** at a point *P* due to one of these elements. Finally, we evaluate the total field at *P* due to the charge distribution by performing a vector sum of the contributions of all the charge elements (i.e., by applying the superposition principle).

**Figure 19.13**

The electric field at *P* due to a continuous charge distribution is the vector sum of the fields due to all the elements $\Delta q$ of the charge distribution.

The electric field at $P$ due to one element of charge $\Delta q_i$ is given by

$$\Delta \mathbf{E}_i = k_e \frac{\Delta q_i}{r_i^2} \hat{\mathbf{r}}_i$$

where the index $i$ refers to the $i$th element in the distribution, $r_i$ is the distance from the element to point $P$, and $\hat{\mathbf{r}}_i$ is a unit vector directed from the element toward $P$. The total electric field $\mathbf{E}$ at $P$ due to all elements in the charge distribution is approximately

$$\mathbf{E} \approx k_e \sum_i \frac{\Delta q_i}{r_i^2} \hat{\mathbf{r}}_i$$

Now, we apply the model in which the charge distribution is continuous—we let the elements of charge become infinitesimally small. With this model, the total field at $P$ in the limit $\Delta q_i \rightarrow 0$ becomes

- *Electric field of a continuous charge distribution*

$$\mathbf{E} = \lim_{\Delta q_i \rightarrow 0} k_e \sum_i \frac{\Delta q_i}{r_i^2} \hat{\mathbf{r}}_i = k_e \int \frac{dq}{r^2} \hat{\mathbf{r}} \qquad [19.7]$$

where $dq$ is an infinitesimal amount of charge and the integration is over all the charge creating the electric field. The integration is a *vector* operation and must be treated with caution. It can be evaluated in terms of individual components, or perhaps symmetry arguments can be used to reduce it to a scalar integral. We shall illustrate this type of calculation with several examples in which we assume that the charge is *uniformly* distributed on a line or a surface or throughout some volume. When performing such calculations, it is convenient to use the concept of a *charge density* along with the following notations:

- If a total charge $Q$ is uniformly distributed throughout a volume $V$, the **volume charge density** $\rho$ is defined by

- *Volume charge density*

$$\rho \equiv \frac{Q}{V} \qquad [19.8]$$

where $\rho$ has units of coulombs per cubic meter.
- If $Q$ is uniformly distributed on a surface of area $A$, the **surface charge density** $\sigma$ is defined by

- *Surface charge density*

$$\sigma \equiv \frac{Q}{A} \qquad [19.9]$$

where $\sigma$ has units of coulombs per square meter.
- If $Q$ is uniformly distributed along a line of length $\ell$, the **linear charge density** $\lambda$ is defined by

- *Linear charge density*

$$\lambda \equiv \frac{Q}{\ell} \qquad [19.10]$$

where $\lambda$ has units of coulombs per meter.

| **PROBLEM-SOLVING STRATEGY** | **Calculating the Electric Field** |

1. *Units:* When performing calculations that involve the Coulomb constant, $k_e$ $(= 1/4\pi\epsilon_0)$, charges must be in coulombs and distances in meters. If they appear in other units, you must convert them.

2. *Applying Coulomb's law to point charges:* It is important to use the superposition principle properly when dealing with a collection of point charges. When several point charges are present, the resultant force on any one is the *vector sum* of the individual forces due to the other point charges. Be very careful in the manipulation of vector quantities. It may be useful to review the material on vector addition in Chapter 1.

3. *Calculating the electric field of point charges:* Remember that the superposition principle can be applied to electric fields, which are also vector quantities. To find the total electric field at a given point, first calculate the electric field at that point due to each individual point charge. The resultant field at the point is the vector sum of the fields due to the individual point charges.

4. *Continuous charge distributions:* When confronted with problems that involve a continuous distribution of charge, at some point you must replace the vector sums for evaluating the total electric field by vector integrals. The charge distribution is divided into infinitesimal pieces, and the vector sum is carried out by integrating over the entire charge distribution. Examples 19.4 and 19.5 demonstrate such procedures.

5. *Symmetry:* Whenever dealing with either a distribution of point charges or a continuous charge distribution, take advantage of any symmetry in the system to simplify your calculations. The cancellation of field components parallel to the $y$ axis in Example 19.3 and perpendicular to the axis in Example 19.5 are examples of the application of symmetry.

**PITFALL PREVENTION 19.3**

**Limiting techniques**

The use of the limiting technique in Example 19.4 [(distance from source particle)/(size of source distribution) $\rightarrow \infty$] is often a good method for checking an algebraic expression. If this limit approaches the field of a point charge, it doesn't prove that the result is correct, but if it does *not* approach the field of a point charge, it tells you that the result is wrong.

## Example 19.4   The Electric Field Due to a Charged Rod

A rod of length $\ell$ has a uniform linear charge density $\lambda$ and a total charge $Q$. Calculate the electric field at a point $P$ along the axis of the rod, a distance $a$ from one end (Fig. 19.14).

**Reasoning**   For clarity, let us describe the steps needed to carry out integrations such as this and then follow up by doing it explicitly. First, we choose an element of the charge distribution whose parts are all equidistant from the point at which the field is being calculated. Next, we express the charge $dq$ of the element in terms of the other variables within the integral (in this example, there is one variable, $x$). If necessary, the integral is expressed in terms of components. Then we reduce the integral to one over a single variable (or multiple integrals, each over a single variable).

**Figure 19.14**

(Example 19.4) The electric field at $P$ due to a uniformly charged rod lying along the $x$ axis. The field at $P$ due to the segment of charge $dq$ is $k_e dq/x^2$. The total field at $P$ is the vector sum over all segments of the rod.

**Solution** For this calculation, the rod is taken to be along the *x* axis. Let us use *dx* to represent the length of one small segment of the rod and let *dq* be the charge on the segment. The charge *dq* on the small segment is $dq = \lambda \, dx$.

The field *d***E** due to this segment at the point *P* is in the negative *x* direction, and its magnitude is

$$dE = k_e \frac{dq}{x^2} = k_e \frac{\lambda \, dx}{x^2}$$

Each element of the charge distribution produces a field in the negative *x* direction, and so the vector sum of their contributions reduces to an algebraic sum. The total field at *P* due to all segments of the rod, which are at different distances from *P*, is given by Equation 19.7, which in this case becomes

$$E = \int_a^{\ell + a} k_e \lambda \frac{dx}{x^2}$$

where the limits on the integral extend from one end of the rod ($x = a$) to the other ($x = \ell + a$). Because $k_e$ and $\lambda$ are constants, they can be removed from the integral. Thus, we find that

$$E = k_e \lambda \int_a^{\ell + a} \frac{dx}{x^2} = k_e \lambda \left[ -\frac{1}{x} \right]_a^{\ell + a}$$

$$= k_e \lambda \left( \frac{1}{a} - \frac{1}{\ell + a} \right) = \boxed{\frac{k_e Q}{a(\ell + a)}}$$

where we have used the fact that the linear charge density is $\lambda = Q/\ell$.

From this result we see that if the point *P* is far from the rod ($a \gg \ell$), then $\ell$ in the denominator can be ignored, and $E \approx k_e Q/a^2$. This is just the form you would expect for a point charge. At large values of *a*, therefore, the charge distribution appears to be a point charge of magnitude *Q* as you should expect.

---

## Example 19.5   The Electric Field of a Uniform Ring of Charge

A ring of radius *a* has a uniform positive charge per unit length, with a total charge *Q*. Calculate the electric field at a point *P* on the axis of the ring at a distance *x* from the center of the ring (Fig. 19.15a).

**Reasoning and Solution** The magnitude of the electric field at *P* due to the segment of charge *dq* is

$$dE = k_e \frac{dq}{r^2}$$

This field has an *x* component $dE_x = dE \cos \theta$ along the axis of the ring and a component $dE_\perp$ perpendicular to the axis.

The perpendicular component of any element is canceled by the perpendicular component of an element on the opposite side of the ring, as for the elements 1 and 2 in Figure 19.15b. Thus, the perpendicular components of the field for the entire ring sum to zero and the resultant field at *P* must lie along the *x* axis. Because $r = (x^2 + a^2)^{1/2}$ and $\cos \theta = x/r$, we find that

$$dE_x = dE \cos \theta = \left( k_e \frac{dq}{r^2} \right) \left( \frac{x}{r} \right) = \frac{k_e x}{(x^2 + a^2)^{3/2}} \, dq$$

We integrate this expression to find the total field at *P*. In this case, all segments of the ring give the same contribution

(a)

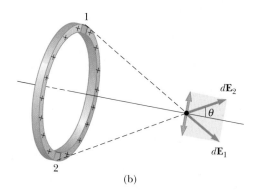

(b)

**Figure 19.15**

(Example 19.5) A uniformly charged ring of radius *a*. (a) The field at *P* on the *x* axis due to an element of charge *dq*. (b) The total electric field at *P* is along the *x* axis. The perpendicular component of the electric field at *P* due to segment 1 is canceled by the perpendicular component due to segment 2.

to the field at *P* because they are all equidistant from this point. Thus,

$$E_x = \int \frac{k_e x}{(x^2 + a^2)^{3/2}} \, dq = \frac{k_e x}{(x^2 + a^2)^{3/2}} \int dq$$

$$= \frac{k_e x}{(x^2 + a^2)^{3/2}} \, Q$$

This result shows that the field is zero at the center point of the ring, $x = 0$. Does this surprise you?

**EXERCISE** Show that at large distances from the ring $(x \gg a)$ the electric field along the axis approaches that of a point charge of magnitude $Q$.

## 19.6 • ELECTRIC FIELD LINES

A convenient specialized pictorial representation for visualizing electric field patterns is created by drawing lines showing the direction of the electric field vector at any point. These lines, called **electric field lines,** are related to the electric field in any region of space in the following manner:

- The electric field vector **E** is *tangent* to the electric field line at each point.
- The number of electric field lines per unit area through a surface that is perpendicular to the lines is proportional to the magnitude of the electric field in that region. Thus, *E* is large where the field lines are close together and small where they are far apart.

These properties are illustrated in Figure 19.16. The density of lines through surface A is greater than the density of lines through surface B. Therefore, the magnitude of the electric field on surface A is larger than on surface B. Furthermore, the field drawn in Figure 19.16 is nonuniform because the lines at different locations point in different directions.

Some representative electric field lines for a single positive point charge are shown in Figure 19.17a. Note that in this two-dimensional drawing we show only

See Screen 11.5

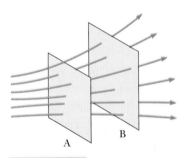

**Figure 19.16**

Electric field lines penetrating two surfaces. The magnitude of the field is greater on surface A than on surface B.

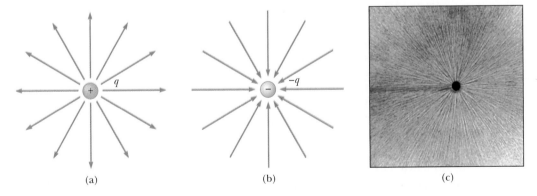

(a)  (b)  (c)

**Figure 19.17**

The electric field lines for a point charge. (a) For a positive point charge, the lines are directed radially outward. (b) For a negative point charge, the lines are directed radially inward. Note that the figures show only those field lines that lie in the plane containing the charge. (c) The dark areas are small pieces of thread suspended in oil, which align with the electric field produced by a small charged conductor at the center. (© *Courtesy of Harold M. Waage, Princeton University*) the field lines that lie in the plane containing the point charge. The lines are actually directed radially outward in *all* directions from the charge, somewhat like the needles of a porcupine. Because a positively charged test particle placed in this field would be repelled by the charge $q$, the lines are directed radially away from $q$. Similarly, the electric field lines for a single negative point charge are directed toward the charge (Fig. 19.17b). In either case, the lines are radial and extend to infinity. Note that the lines are closer together as they come nearer to the charge, indicating that the magnitude of the field is increasing.

### Quick Quiz 19.5

In fair weather, an electric field occurs at the surface of the Earth, pointing down into the ground. What is the sign of the electric charge on the ground in fair weather?

The rules for drawing electric field lines for any charge distribution are as follows:

- The lines for a group of point charges must begin on positive charges and end on negative ones. In the case of an excess of one type of charge, some lines will begin or end infinitely far away.
- The number of lines drawn beginning on a positive charge or ending on a negative one is proportional to the magnitude of the charge.
- No two field lines can cross each other.

Is this visualization of the electric field in terms of field lines consistent with Equation 19.5? To answer this question, consider an imaginary spherical surface of radius $r$, concentric with the charge. From symmetry, we see that the magnitude of the electric field is the same everywhere on the surface of the sphere. The number of lines $N$ emerging from the charge is equal to the number penetrating the spherical surface. Hence, the number of lines per unit area on the sphere is $N/4\pi r^2$ (where the surface area of the sphere is $4\pi r^2$). Because $E$ is proportional to the number of lines per unit area, we see that $E$ varies as $1/r^2$. This is consistent with the result obtained from Equation 19.5, that is, $E = k_e q/r^2$.

Because charge is quantized, the number of lines leaving any positively charged object must be 0, $ae$, $2ae$, . . . , where $a$ is an arbitrary (but fixed) proportionality constant chosen by the person drawing the lines. Once $a$ is chosen, the number of lines is no longer arbitrary. For example, if object 1 has charge $Q_1$ and object 2 has charge $Q_2$, then the ratio of the number of lines connected to object 2 to those connected to object 1 is $N_2/N_1 = Q_2/Q_1$.

The electric field lines for two point charges of equal magnitude but opposite signs (the electric dipole) are shown in Figure 19.18. In this case, the number of lines that begin at the positive charge must equal the number that terminate at the negative charge. At points very near the charges, the lines are nearly radial. The high density of lines between the charges indicates a region of strong electric field. The attractive nature of the force between the particles is also suggested by Figure 19.18, with the lines from one particle ending on the other particle.

Figure 19.19 shows the electric field lines in the vicinity of two equal positive point charges. Again, close to either charge the lines are nearly radial. The same number of lines emerges from each particle because the charges are equal in magnitude. At great distances from the particles, the field is approximately equal to that of a single point charge of magnitude $2q$. The repulsive nature of the electric force between particles of like charge is suggested in the figure by the fact that no

(a)

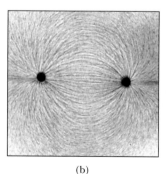

(b)

**Figure 19.18**

(a) The electric field lines for two charges of equal magnitude and opposite sign (an electric dipole). Note that the number of lines leaving the positive charge equals the number terminating at the negative charge. (b) Small pieces of thread suspended in oil align with the electric field. *(b, Courtesy of Harold M. Waage, Princeton University)*

**Figure 19.19**

(a) The electric field lines for two positive point charges. (The locations *A*, *B*, and *C* are discussed in Quick Quiz 19.6.) (b) Small pieces of thread suspended in oil align with the electric field. *(b, Courtesy of Harold M. Waage, Princeton University)*

lines connect the particles and the lines bend away from the region between the charges.

Finally, in Figure 19.20 we sketch the electric field lines associated with a positive point charge $+2q$ and a negative point charge $-q$. In this case, the number of lines leaving $+2q$ is twice the number terminating on $-q$. Hence, only half of the lines that leave the positive charge end at the negative charge. The remaining half terminate on hypothetical negative charges we assume to be located infinitely far away. At large distances from the particles (large compared with the particle separation), the electric field lines are equivalent to those of a single point charge $+q$.

### Quick Quiz 19.6

Rank the magnitudes of the electric field at points *A*, *B*, and *C* in Figure 19.19, largest magnitude first.

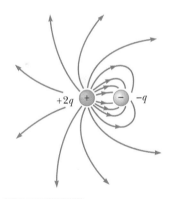

**Figure 19.20**

The electric field lines for a point charge $+2q$ and a second point charge $-q$. Note that two lines leave the charge $+2q$ for every one that terminates on $-q$.

## 19.7 • MOTION OF CHARGED PARTICLES IN A UNIFORM ELECTRIC FIELD

When a particle of charge *q* and mass *m* is placed in an electric field **E**, the electric force exerted on the charge is given by Equation 19.4, $\mathbf{F}_e = q\mathbf{E}$. If this is the only force exerted on the particle, it is the net force. The net force causes the particle to accelerate—this is the particle under a net force from Chapter 4. In this case, Newton's second law applied to the particle gives

$$\mathbf{F}_e = q\mathbf{E} = m\mathbf{a}$$

The acceleration of the particle is therefore

$$\mathbf{a} = \frac{q\mathbf{E}}{m} \qquad \text{[19.11]}$$

If **E** is uniform (i.e., constant in magnitude and direction), the acceleration is constant. If the particle has a positive charge, its acceleration is in the direction of the electric field. If the particle has a negative charge, its acceleration is in the direction opposite the electric field.

### Example 19.6    An Accelerating Positive Charge

A particle with positive charge $q$ and mass $m$ is released from rest in a uniform electric field $\mathbf{E}$ directed along the $x$ axis as in Figure 19.21. Describe its motion.

**Solution**    The acceleration is constant and given by $q\mathbf{E}/m$ (Eq. 19.11). The motion is simple linear motion along the $x$ axis. We can therefore apply the model of a particle under constant acceleration and use the equations of kinematics in one dimension (from Chapter 2):

$$x_f = x_i + v_i t + \tfrac{1}{2} a t^2$$

$$v_f = v_i + at$$

$$v_f{}^2 = v_i{}^2 + 2a(x_f - x_i)$$

Choosing $x_i = 0$ and $v_i = 0$ gives

$$x_f = \tfrac{1}{2} a t^2 = \frac{qE}{2m} \, t^2$$

$$v_f = at = \frac{qE}{m} \, t$$

$$v_f{}^2 = 2ax_f = \left( \frac{2qE}{m} \right) x_f$$

The kinetic energy of the particle after it has moved a distance $x = x_f - x_i$ is

$$K = \tfrac{1}{2} m v^2 = \tfrac{1}{2} m \left( \frac{2qE}{m} \right) x = qEx$$

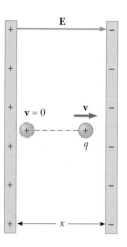

**Figure 19.21**

(Example 19.6) A positive point charge $q$ in a uniform electric field $\mathbf{E}$ undergoes constant acceleration in the direction of the field.

This result can also be obtained by identifying the particle as a nonisolated system and applying the nonisolated system model. Energy is transferred from the environment (the electric field) by work, so the work–kinetic energy theorem gives the same result as the calculation above. Try it!

---

The electric field in the region between two oppositely charged flat metal plates is approximately uniform (Fig. 19.22). Suppose an electron of charge $-e$ is projected horizontally into this field with an initial velocity $v_i \mathbf{i}$. Because the electric field $\mathbf{E}$ in Figure 19.22 is in the positive $y$ direction, the acceleration of the electron is in the negative $y$ direction. That is,

$$\mathbf{a} = -\frac{eE}{m} \, \mathbf{j} \qquad\qquad [19.12]$$

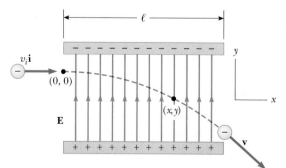

**Figure 19.22**

An electron is projected horizontally into a uniform electric field produced by two charged plates. The electron undergoes a downward acceleration (opposite $\mathbf{E}$), and its motion is parabolic while it is between the plates.

Because the acceleration is constant, we can apply the kinematic equations from Chapter 3 with $v_{xi} = v_i$ and $v_{yi} = 0$. At time $t$, the components of its velocity are

$$v_x = v_i = \text{constant} \qquad\qquad \textbf{[19.13]}$$

$$v_y = a_y t = -\frac{eE}{m} t \qquad\qquad \textbf{[19.14]}$$

Its position coordinates at time $t$ are

$$x_f = v_i t \qquad\qquad \textbf{[19.15]}$$

$$y_f = \tfrac{1}{2} a_y t^2 = -\tfrac{1}{2} \frac{eE}{m} t^2 \qquad\qquad \textbf{[19.16]}$$

Substituting the value $t = x_f/v_i$ from Equation 19.15 into Equation 19.16, we see that $y_f$ is proportional to $x_f^2$. Hence, the trajectory is a parabola. The trajectory of the electron in a uniform electric field **E** under the action of a constant force of magnitude $qE$ has the same shape as that of a particle in a uniform gravitational field **g** under the action of a constant force of magnitude $mg$. After the electron leaves the field, it continues to move in a straight line, obeying Newton's first law, with a speed $v > v_i$.

Note that we have ignored the gravitational force on the electron. This is a good approximation when dealing with atomic particles. For an electric field of $10^4$ N/C, the ratio of the magnitude of the electric force $eE$ to the magnitude of the gravitational force $mg$ is of the order of $10^{14}$ for an electron and of the order of $10^{11}$ for a proton.

**PITFALL PREVENTION 19.6**
**Notice the connection to the past**

Be sure to notice the important link between our new knowledge of electric fields and our understanding of forces and motion. Although $\mathbf{F}_e$ is a new, specific type of force about which we are learning, it causes a particle to move in the same way as if the force were a gravitational force or a contact force such as those we studied in Chapters 4 and 5. The electric force may be abstract, but it *is* simply a force.

---

### Example 19.7   An Accelerated Electron

An electron enters the region of a uniform electric field as in Figure 19.22, with $v_i = 3.00 \times 10^6$ m/s and $E = 200$ N/C. The horizontal length of the plates is $\ell = 0.100$ m. (a) Find the acceleration of the electron while it is in the electric field.

**Solution**   The charge on the electron is $-e$ and its mass is $m_e = 9.11 \times 10^{-31}$ kg. Therefore Equation 19.12 gives

$$\mathbf{a} = -\frac{eE}{m_e}\mathbf{j} = -\frac{(1.60 \times 10^{-19}\ \text{C})(200\ \text{N/C})}{9.11 \times 10^{-31}\ \text{kg}}\mathbf{j}$$

$$= -3.51 \times 10^{13}\,\mathbf{j}\ \text{m/s}^2$$

(b) Find the time it takes the electron to travel through the field.

**Solution**   The horizontal distance through the field is $\ell = 0.100$ m. Modeling the electron as a particle under constant velocity in the horizontal direction, we find that the time spent in the electric field is

$$\Delta t = \frac{\ell}{v_i} = \frac{0.100\ \text{m}}{3.00 \times 10^6\ \text{m/s}} = 3.33 \times 10^{-8}\ \text{s}$$

(c) What is the vertical displacement $\Delta y$ of the electron while it is in the field?

**Solution**   Modeling the electron as a particle under constant acceleration in the vertical direction and using the results from (a) and (b), we find that

$$\Delta y = y_f - y_i = \tfrac{1}{2} a_y t^2$$
$$= -\tfrac{1}{2}(3.51 \times 10^{13}\ \text{m/s}^2)(3.33 \times 10^{-8}\ \text{s})^2$$
$$= -0.0195\ \text{m} = -1.95\ \text{cm}$$

**EXERCISE**   Find the speed of the electron as it emerges from the field.

*Answer*   $3.22 \times 10^6$ m/s

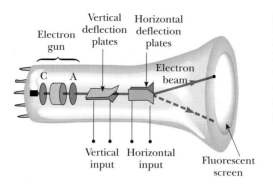

**Figure 19.23**

Schematic diagram of a cathode-ray tube. Electrons leaving the hot cathode C are accelerated to the anode A. In addition to accelerating electrons, the electron gun is also used to focus the beam of electrons, and the plates deflect the beam.

**The Cathode-Ray Tube**

The previous example describes a portion of a cathode-ray tube (CRT). This tube, illustrated in Figure 19.23, is commonly used to obtain a visual display of electronic information in oscilloscopes, radar systems, television receivers, and computer monitors. The CRT is a vacuum tube in which a beam of electrons is accelerated and deflected under the influence of electric or magnetic fields. The electron beam is produced by an *electron gun* located in the neck of the tube. These electrons, if left undisturbed, travel in a straight-line path until they strike the "screen" of the CRT. The screen is coated with a material that emits visible light when bombarded with electrons.

In an oscilloscope, the electrons are deflected in various directions by two sets of plates placed at right angles to each other in the neck of the tube. (A television CRT steers the beam with a magnetic field, which we will discuss in Chapter 22.) An external electric circuit is used to control the amount of charge present on the plates. Placing positive charge on one horizontal deflection plate and negative charge on the other creates an electric field between the plates and allows the beam to be steered from side to side. The vertical deflection plates act in the same way, except that changing the charge on them deflects the beam vertically.

**WEB**

For more information on the structure and operation of a cathode ray tube, visit
**www.montest.com/
mon-apps.html**

**See Screen 11.6**

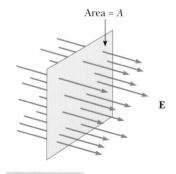

**Figure 19.24**

Field lines of a uniform electric field penetrating a plane of area $A$ perpendicular to the field. The electric flux $\Phi_E$ through this area is equal to $EA$.

## 19.8 • ELECTRIC FLUX

Now that we have described the concept of electric field lines qualitatively, let us use a new concept, *electric flux*, to approach electric field lines on a quantitative basis. Electric flux is a quantity proportional to the number of electric field lines penetrating some surface. (We can define only a proportionality because the number of lines we choose to draw is arbitrary.)

First consider an electric field that is uniform in both magnitude and direction, as in Figure 19.24. The field lines penetrate a plane rectangular surface of area $A$, which is perpendicular to the field. Recall that the number of lines per unit area is proportional to the magnitude of the electric field. The number of lines penetrating the surface of area $A$ is therefore proportional to the product $EA$. The product of the electric field magnitude $E$ and a surface area $A$ perpendicular to the field is called the **electric flux** $\Phi_E$:

$$\Phi_E = EA \qquad \text{[19.17]}$$

From the SI units of $E$ and $A$, we see that electric flux has the units $\mathrm{N \cdot m^2 / C}$.

If the surface under consideration is not perpendicular to the field, the number of lines through it must be less than that given by Equation 19.17. This can be

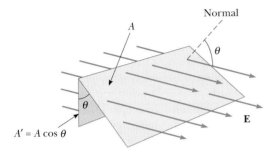

Field lines for a uniform electric field through an area $A$ that is at an angle $\theta$ to the field. Because the number of lines that go through the shaded area $A'$ is the same as the number that go through $A$, we conclude that the total flux through $A'$ is equal to the flux through $A$ and is given by $\Phi_E = EA \cos \theta$.

understood by considering Figure 19.25 where the normal to the surface of area $A$ is at an angle of $\theta$ to the uniform electric field. Note that the number of lines that cross this area is equal to the number that cross the projected area $A'$, which is perpendicular to the field. From Figure 19.25 we see that the two areas are related by $A' = A \cos \theta$. Because the flux through area $A$ equals the flux through $A'$, we conclude that the desired flux is

$$\Phi_E = EA \cos \theta \qquad [19.18]$$

From this result, we see that the flux through a surface of fixed area has the maximum value $EA$ when the surface is perpendicular to the field (in other words, when the *normal* to the surface is parallel to the field, that is, $\theta = 0°$); the flux is zero when the surface is parallel to the field (when the normal to the surface is perpendicular to the field, that is, $\theta = 90°$).

In more general situations, the electric field may vary in both magnitude and direction over the surface in question. Unless the field is uniform, our definition of flux given by Equation 19.18 therefore has meaning only over a small element of area. Consider a general surface divided up into a large number of small elements, each of area $\Delta A$. The variation in the electric field over the element can be ignored if the element is small enough. It is convenient to define a vector $\Delta \mathbf{A}_i$ whose magnitude represents the area of the $i$th element and whose direction is defined to be perpendicular to the surface, as in Figure 19.26. The electric flux $\Delta \Phi_E$ through this small element is

$$\Delta \Phi_E = E_i \Delta A_i \cos \theta_i = \mathbf{E}_i \cdot \Delta \mathbf{A}_i$$

where we have used the definition of the scalar product of two vectors ($\mathbf{A} \cdot \mathbf{B} = AB \cos \theta$). By summing the contributions of all elements, we obtain the total flux through the surface. If we let the area of each element approach zero, then the number of elements approaches infinity and the sum is replaced by an integral. The general definition of electric flux is therefore

$$\Phi_E \equiv \lim_{\Delta A_i \to 0} \sum \mathbf{E}_i \cdot \Delta \mathbf{A}_i = \int_{\text{surface}} \mathbf{E} \cdot d\mathbf{A} \qquad [19.19]$$

• *Electric flux*

Equation 19.19 is a surface integral, which must be evaluated over the surface in question. In general, the value of $\Phi_E$ depends both on the field pattern and on the specified surface.

We shall often be interested in evaluating electric flux through a *closed surface*. A closed surface is defined as one that completely divides space into an inside region and an outside region, so that movement cannot take place from one region to the other without penetrating the surface. This is similar to the system bound-

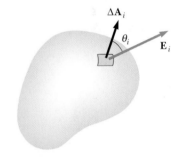

A small element of a surface of area $\Delta A_i$. The electric field makes an angle $\theta_i$ with the normal to the surface (the direction of $\Delta \mathbf{A}_i$), and the flux through the element is equal to $E_i \Delta A_i \cos \theta_i$.

ary in system models, in which the boundary divides space into a region inside the system and the outer region, the environment. The surface of a sphere is an example of a closed surface, whereas a drinking glass is an open surface.

Consider the closed surface in Figure 19.27. Note that the vectors $\Delta \mathbf{A}_i$ point in different directions for the various surface elements. At each point, these vectors are *perpendicular* to the surface and, by convention, always point *outward* from the inside region. At the element labeled ①, $\mathbf{E}$ is outward and $\theta_i < 90°$; hence, the flux $\Delta \Phi_E = \mathbf{E} \cdot \Delta \mathbf{A}$ through this element is positive. For element ②, the field lines graze the surface (perpendicular to the vector $\Delta \mathbf{A}_i$); thus, $\theta_i = 90°$, and the flux is zero. For elements such as ③, where the field lines are crossing the surface from the outside to the inside, $180° > \theta_i > 90°$, and the flux is negative because $\cos \theta_i$ is negative. The net flux through the surface is proportional to the net number of lines penetrating the surface, where the net number means **the number leaving the volume surrounded by the surface minus the number entering the volume.** If more lines are leaving the surface than entering, the net flux is positive. If more lines enter than leave the surface, the net flux is negative. Using the symbol $\oint$ to represent an integral over a closed surface, we can write the net flux $\Phi_E$ through a closed surface as

$$\Phi_E = \oint \mathbf{E} \cdot d\mathbf{A} = \oint E_n \, dA \qquad \qquad [19.20]$$

where $E_n$ represents the component of the electric field normal to the surface.

Evaluating the net flux through a closed surface can be very cumbersome. If the field is perpendicular to the surface at each point and constant in magnitude, however, the calculation is straightforward. The following example illustrates this point.

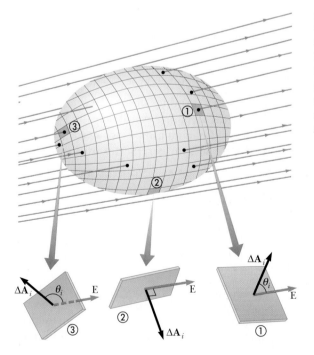

**Figure 19.27**

A closed surface in an electric field. The area vectors $\Delta \mathbf{A}_i$ are, by convention, normal to the surface and point outward. The flux through an area element can be positive (element ①), zero (element ②), or negative (element ③).

## Example 19.8   Flux Through a Cube

Consider a uniform electric field **E** directed along the $+x$ axis. Find the net electric flux through the surface of a cube of edges $\ell$ oriented as shown in Figure 19.28.

**Solution**   The net flux can be evaluated by summing up the fluxes through each face of the cube. First, note that the flux through four of the faces is zero, because **E** is perpendicular to $d\mathbf{A}$ on these faces. In particular, the orientation of $d\mathbf{A}$ is perpendicular to **E** for the faces labeled ③ and ④ in Figure 19.28. Therefore, $\theta = 90°$, so that $\mathbf{E} \cdot d\mathbf{A} = E\, dA \cos 90° = 0$. The flux through each of the planes parallel to the $xy$ plane is also zero for the same reason.

Now consider the faces labeled ① and ②. The net flux through these faces is

$$\Phi_E = \int_1 \mathbf{E} \cdot d\mathbf{A} + \int_2 \mathbf{E} \cdot d\mathbf{A}$$

For face ①, **E** is constant and inward, whereas $d\mathbf{A}$ is outward ($\theta = 180°$), so that the flux through this face is

$$\int_1 \mathbf{E} \cdot d\mathbf{A} = \int_1 E\, dA \cos 180° = -E \int_1 dA = -EA = -E\ell^2$$

because the area of each face is $A = \ell^2$.

Likewise, for ②, **E** is constant and outward and in the same direction as $d\mathbf{A}$ ($\theta = 0°$), so that the flux through this face is

$$\int_2 \mathbf{E} \cdot d\mathbf{A} = \int_2 E\, dA \cos 0° = E \int_2 dA = +EA = E\ell^2$$

Hence, the net flux over all faces is zero because

$$\Phi_E = -E\ell^2 + E\ell^2 = \boxed{0}$$

**Figure 19.28**

(Example 19.8) A hypothetical surface in the shape of a cube in a uniform electric field parallel to the $x$ axis. The net flux through the surface is zero. Side ④ is the bottom of the cube and side ① is opposite side ②.

**EXERCISE**   A flat surface having an area of 3.20 m² is placed in various orientations in a uniform electric field of magnitude $E = 6.20 \times 10^5$ N/C. Calculate the electric flux through this area when the electric field (a) is perpendicular to the surface; (b) is parallel to the surface; (c) makes an angle of 75° with the plane of the surface.

*Answer*   (a) $1.98 \times 10^6$ N·m²/C   (b) zero
(c) $1.92 \times 10^6$ N·m²/C

## 19.9 • GAUSS'S LAW

In this section we describe a general relation between the net electric flux through a closed surface and the charge *enclosed* by the surface. This relation, known as **Gauss's law,** is of fundamental importance in the study of electrostatic fields.

First, let us consider a positive point charge $q$ located at the center of a spherical surface of radius $r$, as in Figure 19.29. The field lines radiate outward, and hence are perpendicular (or normal) to the surface at each point. That is, at each point on the surface, **E** is parallel to the vector $\Delta\mathbf{A}_i$ representing the local element of area $\Delta A_i$. Therefore, at all points on the surface

$$\mathbf{E} \cdot \Delta\mathbf{A}_i = E_n \, \Delta A_i = E \, \Delta A_i$$

and from Equation 19.20 we find that the net flux through the surface is

$$\Phi_E = \oint E_n \, dA = \oint E \, dA = E \oint dA = EA$$

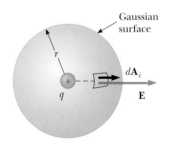

**Figure 19.29**

A spherical gaussian surface of radius $r$ surrounding a point charge $q$. When the charge is at the center of the sphere, the electric field is normal to the surface and constant in magnitude everywhere on the surface.

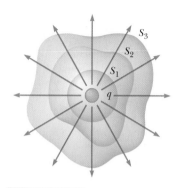

**Figure 19.30**

Closed surfaces of various shapes surrounding a charge $q$. The net electric flux through each surface is the same.

because $E$ is constant over the surface. From Equation 19.5, we know that the magnitude of the electric field everywhere on the surface of the sphere is $E = k_e q / r^2$. Furthermore, for a spherical surface, $A = 4\pi r^2$ (the surface area of a sphere). Hence, the net flux through the surface is

$$\Phi_E = EA = \left(\frac{k_e q}{r^2}\right)(4\pi r^2) = 4\pi k_e q$$

Recalling that $k_e = 1/4\pi\epsilon_0$, we can write this in the form

$$\Phi_E = \frac{q}{\epsilon_0} \qquad \text{[19.21]}$$

This result, which is independent of $r$, says that the net flux through a spherical surface is proportional to the charge $q$ *inside* the surface. This is a mathematical representation of the fact that (1) the net flux is proportional to the number of field lines, (2) the number of field lines is proportional to the charge inside the surface, and (3) every field line from the charge must pass through the surface. The fact that the net flux is independent of the radius is a consequence of the inverse-square dependence of the electric field given by Equation 19.5. That is, $E$ varies as $1/r^2$, but the area of the sphere varies as $r^2$. Their combined effect produces a flux that is independent of $r$.

Now consider several closed surfaces surrounding a charge $q$, as in Figure 19.30. Surface $S_1$ is spherical, whereas surfaces $S_2$ and $S_3$ are nonspherical. The flux that passes through surface $S_1$ has the value $q/\epsilon_0$. As we discussed in the preceding section, the flux is proportional to the number of electric field lines passing through that surface. The construction in Figure 19.30 shows that the number of electric field lines through the spherical surface $S_1$ is equal to the number of electric field lines through the nonspherical surfaces $S_2$ and $S_3$. It is therefore reasonable to conclude that the net flux through any closed surface is independent of the shape of that surface. (One can prove this is the case because $E \propto 1/r^2$.) In fact, **the net flux through any closed surface surrounding the point charge $q$ is given by $q/\epsilon_0$.**

Now consider a point charge located *outside* a closed surface of arbitrary shape, as in Figure 19.31. As you can see from this construction, electric field lines enter the surface, and field lines leave it. However, **the number of electric field lines entering the surface equals the number leaving the surface.** Therefore, we conclude that **the net electric flux through a closed surface that surrounds no net charge is zero.** If we apply this result to Example 19.8, we see that the net flux through the cube is zero, because it was assumed there was no charge inside the cube.

Let us extend these arguments to the generalized case of many point charges. We shall again make use of the superposition principle. That is, we can express the net flux through any closed surface as

$$\oint \mathbf{E} \cdot d\mathbf{A} = \oint (\mathbf{E}_1 + \mathbf{E}_2 + \mathbf{E}_3) \cdot d\mathbf{A}$$

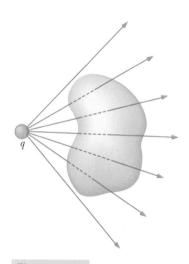

**Figure 19.31**

A point charge located *outside* a closed surface. The number of lines entering the surface equals the number leaving the surface.

where $\mathbf{E}$ is the total electric field at any point on the surface and $\mathbf{E}_1$, $\mathbf{E}_2$, and $\mathbf{E}_3$ are the fields produced by the individual charges at that point. Consider the system of charges shown in Figure 19.32. The surface $S$ surrounds only one charge, $q_1$; hence, the net flux through $S$ is $q_1/\epsilon_0$. The flux through $S$ due to the charges outside it is zero because each electric field line from these charges that enters $S$ at one point leaves it at another. The surface $S'$ surrounds charges $q_2$ and $q_3$; hence, the net flux through $S'$ is $(q_2 + q_3)/\epsilon_0$. Finally, the net flux through surface $S''$ is zero because no charge exists inside this surface. That is, *all* electric field lines that enter $S''$ at one point leave $S''$ at another.

**Gauss's law,** which is a generalization of the foregoing discussion, states that the net flux through *any* closed surface is

$$\Phi_E = \oint \mathbf{E} \cdot d\mathbf{A} = \frac{q_{in}}{\epsilon_0}$$     [19.22]

where $q_{in}$ represents the *net charge inside the surface* and $\mathbf{E}$ represents the electric field at any point on the surface. In words, Gauss's law states that **the net electric flux through any closed surface is equal to the net charge inside the surface divided by $\epsilon_0$.**

In principle, Gauss's law is valid for the electric field of any system of charges or continuous distribution of charge. In practice, however, **the technique is useful for calculating the electric field only in situations where the degree of symmetry is high.** As we shall see in the next section, **Gauss's law can be used to evaluate the electric field for charge distributions that have spherical, cylindrical, or plane symmetry.** We do this by choosing an appropriate gaussian surface that allows $\mathbf{E}$ to be removed from the integral in Gauss's law and performing the integration. Note that a gaussian surface is a mathematical surface and need not coincide with any real physical surface.

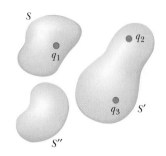

**Figure 19.32**

The net electric flux through any closed surface depends only on the charge *inside* that surface. The net flux through surface $S$ is $q_1/\epsilon_0$, the net flux through surface $S'$ is $(q_2 + q_3)/\epsilon_0$, and the net flux through surface $S''$ is zero.

### Quick Quiz 19.7

For a gaussian surface through which the net flux is zero, the following four statements *could be true*. Which of the statements *must be true?* (a) No charges are inside the surface. (b) The net charge inside the surface is zero. (c) The electric field is zero everywhere on the surface. (d) The number of electric field lines entering the surface equals the number leaving the surface.

### THINKING PHYSICS 19.2

A spherical gaussian surface surrounds a point charge $q$. Describe what happens to the net flux through the surface if (a) the charge is tripled, (b) the volume of the sphere is doubled, (c) the surface is changed to a cube, and (d) the charge is moved to another location *inside* the surface.

**Reasoning**   (a) If the charge is tripled, the flux through the surface is also tripled, because the net flux is proportional to the charge inside the surface. (b) The net flux remains constant when the volume changes because the surface surrounds the same amount of charge, regardless of its volume. (c) The net flux does not change when the shape of the closed surface changes. (d) The net flux through the closed surface remains unchanged as the charge inside the surface is moved to another location, as long as the new location remains inside the surface.

## 19.10 • APPLICATION OF GAUSS'S LAW TO SYMMETRIC CHARGE DISTRIBUTIONS

As mentioned earlier, Gauss's law is useful in determining electric fields when the charge distribution has a high degree of symmetry. The following examples show ways of choosing the gaussian surface over which the surface integral given by

Equation 19.22 can be simplified and the electric field determined. The surface should always be chosen to take advantage of the symmetry of the charge distribution, so that we can remove $E$ from the integral and solve for it. The goal in this type of calculation is to determine a surface that satisfies one or more of the following conditions:

1. The value of the electric field can be argued by symmetry to be constant over the surface.
2. The dot product in Equation 19.22 can be expressed as a simple algebraic product $E\,dA$ because $\mathbf{E}$ and $d\mathbf{A}$ are parallel.
3. The dot product in Equation 19.22 is zero because $\mathbf{E}$ and $d\mathbf{A}$ are perpendicular.
4. The field can be argued to be zero everywhere on the surface.

We will see all four of these conditions used in the examples through the remainder of this chapter.

### Example 19.9    The Electric Field Due to a Point Charge

Starting with Gauss's law, calculate the electric field due to an isolated point charge $q$.

**Solution**    A single charge is the simplest possible charge distribution and we will use this familiar example to show the technique of solving for the electric field with Gauss's law. We choose a spherical gaussian surface of radius $r$ centered on the point charge, as in Figure 19.33. The electric field of a positive point charge is radial outward by symmetry and is therefore normal to the surface at every point. As in condition (2), $\mathbf{E}$ is therefore parallel to $d\mathbf{A}$ at each point on the surface, and so $\mathbf{E} \cdot d\mathbf{A} = E\,dA$ and Gauss's law gives

$$\Phi_E = \oint \mathbf{E} \cdot d\mathbf{A} = \oint E\,dA = \frac{q}{\epsilon_0}$$

By symmetry, $E$ is constant everywhere on the surface, which satisfies condition (1), and so it can be removed from the integral. Therefore,

$$\oint E\,dA = E \oint dA = E(4\pi r^2) = \frac{q}{\epsilon_0}$$

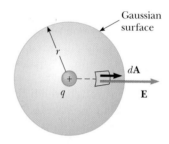

**Figure 19.33**

(Example 19.9) The point charge $q$ is at the center of the spherical gaussian surface, and $\mathbf{E}$ is parallel to $d\mathbf{A}$ at every point on the surface.

where we have used the fact that the surface area of a sphere is $4\pi r^2$. Now, we solve for the electric field:

$$E = \frac{q}{4\pi\epsilon_0 r^2} = k_e \frac{q}{r^2}$$

which is the familiar electric field of a point charge that we developed from Coulomb's law earlier in this chapter.

### Example 19.10    A Spherically Symmetric Charge Distribution

An insulating solid sphere of radius $a$ has a uniform volume charge density $\rho$ and carries a total positive charge $Q$ (Fig. 19.34). (a) Calculate the magnitude of the electric field at a point outside the sphere.

**Solution**    Because the charge distribution is spherically symmetric, we again select a spherical gaussian surface of radius

$r$, concentric with the sphere, as in Figure 19.34a. For this choice, conditions (1) and (2) are satisfied, as they were for the point charge in Example 19.9. Following the line of reasoning given in Example 19.9, we find that

$$E = k_e \frac{Q}{r^2} \quad \text{(for } r > a\text{)}$$

Note that this result is identical to that obtained for a point charge. We therefore conclude that, for a uniformly charged sphere, the field in the region external to the sphere is *equivalent* to that of a point charge located at the center of the sphere.

(b) Find the magnitude of the electric field at a point inside the sphere.

**Solution** In this case we select a spherical gaussian surface having radius $r < a$, concentric with the insulated sphere (Fig. 19.34b). Let us denote the volume of this smaller sphere by $V'$. To apply Gauss's law in this situation, it is important to recognize that the charge $q_{in}$ within the gaussian surface of volume $V'$ is less than $Q$. To calculate $q_{in}$, we use the fact that $q_{in} = \rho V'$:

$$q_{in} = \rho V' = \rho(\tfrac{4}{3}\pi r^3)$$

By symmetry, the magnitude of the electric field is constant everywhere on the spherical gaussian surface, and the field is normal to the surface at each point—both conditions (1) and (2) are satisfied. Gauss's law in the region $r < a$ therefore gives

$$\oint E\, dA = E \oint dA = E(4\pi r^2) = \frac{q_{in}}{\epsilon_0}$$

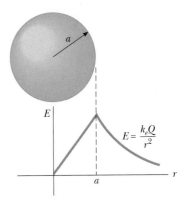

**Figure 19.35**

(Example 19.10) A plot of $E$ versus $r$ for a uniformly charged insulating sphere. The electric field inside the sphere ($r < a$) varies linearly with $r$. The electric field outside the sphere ($r > a$) is the same as that of a point charge $Q$ located at $r = 0$.

Solving for $E$ gives

$$E = \frac{q_{in}}{4\pi\epsilon_0 r^2} = \frac{\rho(\tfrac{4}{3}\pi r^3)}{4\pi\epsilon_0 r^2} = \frac{\rho}{3\epsilon_0}r$$

Because $\rho = Q/\tfrac{4}{3}\pi a^3$ by definition and $k_e = 1/4\pi\epsilon_0$, this expression for $E$ can be written

$$E = \frac{Qr}{4\pi\epsilon_0 a^3} = \frac{k_e Q}{a^3}r \quad (\text{for } r < a)$$

This result for $E$ differs from that obtained in part (a). It shows that $E \rightarrow 0$ as $r \rightarrow 0$. A plot of $E$ versus $r$ is shown in Figure 19.35. Note that the expressions for parts (a) and (b) match when $r = a$.

**EXERCISE** The electric field in the Earth's atmosphere is $E = 1.0 \times 10^2$ N/C, pointing downward. Determine the electric charge on the Earth.

*Answer* $-4.5 \times 10^5$ C

**Figure 19.34**

(Example 19.10) A uniformly charged insulating sphere of radius $a$ and total charge $Q$. (a) The magnitude of the electric field at a point exterior to the sphere is $k_e Q/r^2$. (b) The magnitude of the electric field inside the sphere is due only to the charge *within* the gaussian surface (dashed circle) and is given by $k_e Qr/a^3$.

## Example 19.11  A Cylindrically Symmetric Charge Distribution

Find the electric field a distance $r$ from a line of positive charge of infinite length and constant charge per unit length $\lambda$ (Fig. 19.36a).

**Solution** The symmetry of the charge distribution requires that **E** must be perpendicular to the line charge and directed outward as in Figures 19.36a and 19.36b. To reflect the symmetry of the charge distribution, we select a cylindrical gaussian surface of radius $r$ and length $\ell$ that is coaxial with the line charge. For the curved part of this surface, **E** is constant in magnitude and perpendicular to the surface at each point—conditions (1) and (2). Furthermore, the flux

(a)

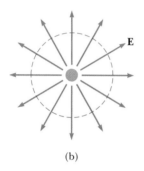

(b)

**Figure 19.36**

(Example 19.11) (a) An infinite line of charge surrounded by a cylindrical gaussian surface concentric with the line charge. (b) An end view shows that the electric field on the cylindrical surface is constant in magnitude and perpendicular to the surface.

through the ends of the gaussian cylinder is zero because **E** is parallel to these surfaces (and, therefore, perpendicular to $d\mathbf{A}$) — the first application we have seen of condition (3).

The surface integral in Gauss's law is taken over the entire gaussian surface. Because of the zero value of $\mathbf{E} \cdot d\mathbf{A}$ for the ends of the cylinder, however, we can restrict our attention to only the curved surface of the cylinder.

The total charge inside our gaussian surface is $q_{in} = \lambda \ell$. Applying Gauss's law and applying conditions (1) and (2), we find for the curved surface,

$$\Phi_E = \oint \mathbf{E} \cdot d\mathbf{A} = E \oint dA = EA = \frac{q_{in}}{\epsilon_0} = \frac{\lambda \ell}{\epsilon_0}$$

The area of the curved surface is $A = 2\pi r \ell$. Therefore

$$E(2\pi r \ell) = \frac{\lambda \ell}{\epsilon_0}$$

$$E = \frac{\lambda}{2\pi\epsilon_0 r} = \boxed{2 k_e \frac{\lambda}{r}} \qquad [19.23]$$

Thus, we see that the electric field of a cylindrically symmetric charge distribution varies as $1/r$, whereas the field external to a spherically symmetric charge distribution varies as $1/r^2$. Equation 19.23 can also be obtained using Equation 19.7; however, the mathematical techniques necessary for this calculation are more cumbersome.

If the line charge in this example were of finite length, the result for $E$ is not that given by Equation 19.23. A finite line charge does not possess sufficient symmetry to use Gauss's law. This is because the magnitude of the electric field is no longer constant over the surface of the gaussian cylinder — the field near the ends of the line would be different from that far from the ends. Thus, condition (1) is not satisfied in this situation. Furthermore, **E** is not perpendicular to the cylindrical surface at all points — the field vectors near the ends would have a component parallel to the line. Condition (2) is not satisfied. When the symmetry in the charge distribution is insufficient, as in this situation, it is necessary to calculate **E** using Equation 19.7.

For points close to a finite line charge and far from the ends, Equation 19.23 gives a good approximation of the value of the field.

It is left as a problem (Problem 37) to show that the electric field inside a uniformly charged rod of finite thickness and infinite length is proportional to $r$.

## Example 19.12　A Nonconducting Plane Sheet of Charge

Find the electric field due to a nonconducting, infinite plane with uniform surface charge density $\sigma$.

**Solution**　Symmetry tells us that **E** must be perpendicular to the plane and that the field will have the same magnitude at points on opposite sides of the plane and equidistant from it.

The fact that the direction of **E** is away from positive charges tells us that the direction of **E** on one side of the plane must be opposite its direction on the other side, as in Figure 19.37. A gaussian surface that reflects the symmetry is a small cylinder whose axis is perpendicular to the plane and whose ends each have an area $A$ and are equidistant from the

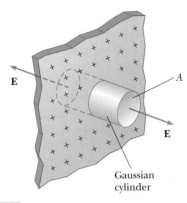

**Figure 19.37**

(Example 19.12) A cylindrical gaussian surface penetrating an infinite sheet of charge. The flux is *EA* through each end of the gaussian surface and zero through its curved surface.

plane. Because **E** is parallel to the curved surface, and therefore perpendicular to *d***A** everywhere on the surface, condition (3) is satisfied and this surface makes no contribution to the surface integral. For the flat ends of the cylinder, condi-

tions (1) and (2) are satisfied. The flux through each end of the cylinder is *EA*; hence, the total flux through the entire gaussian surface is just that through the ends, $\Phi_E = 2EA$.

Noting that the total charge inside the surface is $q_{in} = \sigma A$, we use Gauss's law to obtain

$$\Phi_E = 2EA = \frac{q_{in}}{\epsilon_0} = \frac{\sigma A}{\epsilon_0}$$

$$E = \frac{\sigma}{2\epsilon_0} \qquad \text{[19.24]}$$

Because the distance of the flat end of the cylinder from the plane does not appear in Equation 19.24, we conclude that $E = \sigma/2\epsilon_0$ at *any* distance from the plane. That is, the field is uniform everywhere.

An important charge configuration related to this example is two parallel planes each with a surface charge density $\sigma$, with one plane positively charged and the other negatively charged (Problem 60). In this situation, the electric fields from the two planes add in the region between the planes, resulting in a field with a magnitude of $\sigma/\epsilon_0$, and cancel to give a field of zero elsewhere.

## 19.11 • CONDUCTORS IN ELECTROSTATIC EQUILIBRIUM

A good electrical conductor, such as copper, contains charges (electrons) that are not bound to any atom and are free to move about within the material. When no motion of charge occurs within the conductor, the conductor is in **electrostatic equilibrium.** In this situation, *every* charge in the conductor is a particle in equilibrium, experiencing zero net force. As we shall see, an isolated conductor (one that is insulated from ground) in electrostatic equilibrium has the following properties:

1. The electric field is zero everywhere inside the conductor.
2. If the isolated conductor carries a net charge, the net charge resides entirely on its surface.
3. The electric field just outside the charged conductor is perpendicular to the conductor surface and has a magnitude $\sigma/\epsilon_0$, where $\sigma$ is the surface charge density at that point.
4. On an irregularly shaped conductor, the surface charge density is highest at locations where the radius of curvature of the surface is smallest.

We will verify the first three properties in the following discussion. The fourth property is presented here so that we have a complete list of properties for conductors in electrostatic equilibrium. The verification of it, however, requires concepts from Chapter 20, so we will postpone its verification until then.

The first property can be understood by considering a conducting slab placed in an external field **E** (Fig. 19.38). The electric field inside the conductor *must* be zero under the assumption that we have electrostatic equilibrium. If the field were

**Figure 19.38**

A conducting slab in an external electric field **E**. The charges induced on the surfaces of the slab produce an electric field that opposes the external field, giving a resultant field of zero *inside* the conductor.

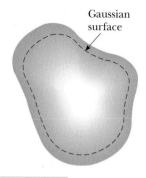

**Figure 19.39**

An insulated conductor of arbitrary shape. The broken line represents a gaussian surface just inside the physical surface of the conductor.

not zero, free charges in the conductor would accelerate under the action of the electric force. This motion of electrons, however, would mean that the conductor is not in electrostatic equilibrium. Thus, the existence of electrostatic equilibrium is consistent only with a zero field in the conductor.

Let us investigate how this zero field is accomplished. Before the external field is applied, free electrons are uniformly distributed throughout the conductor. When the external field is applied, the free electrons accelerate to the left in Figure 19.38, causing a plane of negative charge to be present on the left surface. The movement of electrons to the left results in a plane of positive charge on the right surface. These planes of charge create an additional electric field inside the conductor that opposes the external field. As the electrons move, the surface charge density increases until the magnitude of the internal field equals that of the external field, giving a net field of zero inside the conductor.

We can use Gauss's law to verify the second property of a conductor in electrostatic equilibrium. Figure 19.39 shows an arbitrarily shaped conductor. A gaussian surface is drawn just inside the conductor and can be as close to the surface as we wish. As we have just shown, the electric field everywhere inside a conductor in electrostatic equilibrium is zero. Therefore the electric field must be zero at every point on the gaussian surface—condition (4) in Section 19.10. From this result and Gauss's law, we conclude that the net charge inside the gaussian surface is zero. Because there can be no net charge inside the gaussian surface (which is arbitrarily close to the conductor's surface), any net charge on the conductor must reside on its surface. Gauss's law does not tell us how this excess charge is distributed on the surface, only that it must reside on the surface.

Conceptually, we can understand the location of the charges on the surface by imagining placing many charges at the center of the conductor. The mutual repulsion of the charges causes them to move apart. They will move as far as they can, which is to the surface.

To verify the third property, we can also use Gauss's law. We draw a gaussian surface in the shape of a small cylinder having its end faces parallel to the surface (Fig. 19.40). Part of the cylinder is just outside the conductor, and part is inside. The field is normal to the surface because the conductor is in electrostatic equilibrium: If **E** had a component parallel to the surface, an electric force would be exerted on the charges parallel to the surface, the free charges would move along the surface, and so the conductor would not be in equilibrium. Thus, we satisfy condition (3) in Section 19.10 for the curved part of the cylinder—no flux occurs through this part of the gaussian surface because **E** is parallel to this part of the surface. No flux occurs through the flat face of the cylinder inside the conductor because **E** = 0—this is condition (4). Hence, the net flux through the gaussian surface is the flux through the flat face outside the conductor where the field is perpendicular to the surface. Using conditions (1) and (2) for this face, the flux is $EA$, where $E$ is the electric field just outside the conductor and $A$ is the area of the cylinder's face. Applying Gauss's law to this surface gives

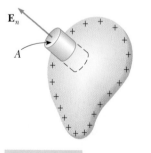

**Figure 19.40**

A gaussian surface in the shape of a small cylinder is used to calculate the electric field just outside a charged conductor. The flux through the gaussian surface is $E_n A$.

$$\Phi_E = \oint E \, dA = EA = \frac{q_{in}}{\epsilon_0} = \frac{\sigma A}{\epsilon_0}$$

where we have used the fact that $q_{in} = \sigma A$. Solving for $E$ gives

$$E = \frac{\sigma}{\epsilon_0}$$

[19.25]

**THINKING PHYSICS 19.3**

Suppose a point charge $+Q$ is in empty space. We surround the charge with a spherical conducting shell so that the charge is at the center of the shell. What effect does this have on the field lines from the charge?

**Reasoning**   When the spherical shell is placed around the charge, the charges in the shell adjust so as to satisfy the rules for a conductor in equilibrium and Gauss's law. A net charge of $-Q$ moves to the interior surface of the conductor, so that the electric field within the conductor is zero (a spherical gaussian surface totally within the shell encloses no *net* charge). A net charge of $+Q$ resides on the outer surface, so that a gaussian surface outside the sphere encloses a net charge of $+Q$, just as if the shell were not there. Thus, the only change in the field lines from the initial situation is the absence of field lines within the conducting shell.

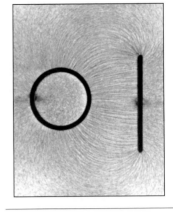

Electric field pattern of a charged conducting plate near an oppositely charged conducting cylinder. Small pieces of thread suspended in oil align with the electric field lines. Note that (1) the electric field lines are perpendicular to the conductors and (2) no lines are inside the cylinder ($E = 0$). *(Courtesy of Harold M. Waage, Princeton University)*

## context connection
## 19.12 • THE ATMOSPHERIC ELECTRIC FIELD

In this chapter we discussed the electric field due to various charge distributions. On the surface of the Earth and in the atmosphere, a number of processes create charge distributions, resulting in an electric field in the atmosphere. These processes include cosmic rays entering the atmosphere, radioactive decay at the Earth's surface, and lightning, the focus of our study in this Context.

The result of these processes is an average negative charge distributed over the surface of the Earth of about $5 \times 10^5$ C, which is a tremendous amount of charge. (The Earth is neutral overall; the corresponding positive charges are spread through the atmosphere, as we shall discuss in Chapter 20.) We can calculate the average surface charge density over the surface of the Earth:

$$\overline{\sigma} = \frac{Q}{A} = \frac{Q}{4\pi r^2} = \frac{5 \times 10^5 \text{ C}}{4\pi(6.37 \times 10^6 \text{ m})^2} \sim 10^{-9} \text{ C/m}^2$$

Throughout this Context, we will be adopting a number of simplification models. Consequently, we will consider our calculations to be order-of-magnitude estimates of the actual values, as suggested by the $\sim$ sign above.

The Earth is a good conductor. Thus, we can use the third property of conductors in Section 19.11 to find the average magnitude of the electric field at the surface of the Earth:

$$\overline{E} = \frac{\overline{\sigma}}{\epsilon_0} = \frac{10^{-9} \text{ C/m}^2}{8.85 \times 10^{-12} \text{ C}^2/\text{N} \cdot \text{m}^2} \sim 10^2 \text{ N/C}$$

This is a typical value of the **fair-weather electric field** that exists in the absence of a thunderstorm. The direction of the field is downward, because the charge on the Earth's surface is negative. During a thunderstorm, the electric field under the thundercloud is significantly higher than the fair-weather electric field, due to the charge distribution in the thundercloud.

Figure 19.41 shows a typical charge distribution in a thundercloud. The charge distribution can be modeled as a *tripole*, although the positive charge at the bottom of the cloud tends to be smaller than the other two charges. The mechanism of charging in thunderclouds is not well understood and continues to be an active area of research.

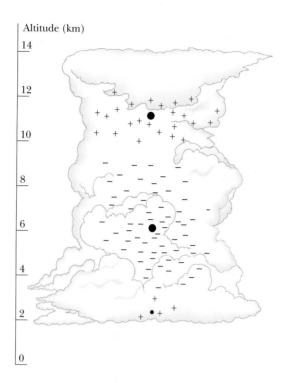

Altitude (km)

**Figure 19.41**

A typical tripolar charge distribution in a thundercloud. The amounts of positive charge at the top and negative charge in the middle are approximately the same, but the amount of positive charge at the bottom is less.

It is this high concentration of charge in the thundercloud that is responsible for the very strong electric fields that cause lightning discharge between the cloud and the ground. Typical electric fields during a thunderstorm are as high as 25 000 N/C.

Atmospheric electric fields can be measured with an instrument called a **field mill.** Figure 19.42 shows the operation of the field mill. In Figure 19.42a, a metallic plate is connected to the ground by a wire. A meter measures the flow of charge through the wire. Because the ground is negatively charged, electrons will flow from the ground into the metal plate. These electrons represent the ends of some of the electric field lines in the atmosphere.

Now, as shown in Figure 19.42b, this plate is covered with a second plate also attached to the ground. The electric field lines that previously ended on the lower plate now end on the upper plate. The charges on the lower plate are repelled by those in the upper plate and pass through the meter into the ground. The meter measures the amount of charge flowing through the wire. This charge is related to how much charge is on the lower plate, which, in turn, is related to the magnitude of the electric field. Thus, the meter can be calibrated to measure the atmospheric electric field. In operation, the plates are similar to the blades of a fan. As one set of blades rotates over a second stationary set, charge pulses back and forth through the meter.

In this chapter, we have analyzed the atmosphere in terms of the electric field. In Chapter 20, we shall learn about electric potential and analyze the atmosphere again in terms of this new parameter.

**WEB**

Recently discovered effects associated with atmospheric electric fields are called *red sprites* and *blue jets*. For information and photos of these fascinating phenomena, visit **umbra.nascom.nasa.gov/spd/sprites.html** and **elf.gi.alaska.edu/**

**Figure 19.42**

A field mill for measuring the atmospheric electric field. When the upper plate is moved over the lower plate, charges move through the meter.

## SUMMARY

**Electric charges** have the following important properties:

- Two kinds of charges exist in nature, **positive** and **negative,** with the property that unlike charges attract each other and like charges repel each other.
- The force between charged particles varies as the inverse square of their separation distance.
- Charge is conserved.
- Charge is quantized.

**Conductors** are materials in which charges move relatively freely. **Insulators** are materials in which charges do not move freely.

**Coulomb's law** states that the electrostatic force between two stationary, charged particles separated by a distance $r$ has the magnitude

$$F_e = k_e \frac{|q_1||q_2|}{r^2} \qquad [19.1]$$

where the Coulomb constant $k_e = 8.99 \times 10^9 \ \mathrm{N \cdot m^2/C^2}$. The vector form of Coulomb's law is

$$\mathbf{F}_{12} = k_e \frac{q_1 q_2}{r^2} \hat{\mathbf{r}}_{12} \qquad [19.2]$$

An **electric field** exists at a point in space if a positive test charge $q_0$ placed at that point experiences an electric force. The electric field is defined as

$$\mathbf{E} \equiv \frac{\mathbf{F}_e}{q_0} \qquad [19.3]$$

The force on a particle with charge $q$ placed in an electric field **E** is

$$\mathbf{F}_e = q\mathbf{E} \qquad [19.4]$$

The electric field due to the point charge $q$ at a distance of $r$ from the charge is

$$\mathbf{E} = k_e \frac{q}{r^2} \hat{\mathbf{r}} \qquad [19.5]$$

where $\hat{\mathbf{r}}$ is a unit vector directed from the charge toward the point in question. The electric field is directed radially outward from a positive charge and is directed toward a negative charge.

The electric field due to a group of charges can be obtained using the superposition principle. That is, the total electric field equals the vector sum of the electric fields of all the charges at some point:

$$\mathbf{E} = k_e \sum_i \frac{q_i}{r_i^2} \hat{\mathbf{r}}_i \qquad [19.6]$$

Similarly, the electric field of a continuous charge distribution at some point is

$$\mathbf{E} = k_e \int \frac{dq}{r^2} \hat{\mathbf{r}} \qquad [19.7]$$

where $dq$ is the charge on one element of the charge distribution and $r$ is the distance from the element to the point in question.

**Electric field lines** are useful for describing the electric field in any region of space. The electric field vector $\mathbf{E}$ is always tangent to the electric field lines at every point. Furthermore, the number of lines per unit area through a surface perpendicular to the lines is proportional to the magnitude of $\mathbf{E}$ in that region.

**Electric flux** is proportional to the number of electric field lines that penetrate a surface. If the electric field is uniform and makes an angle of $\theta$ with the normal to the surface, the electric flux through the surface is

$$\Phi_E = EA \cos \theta \qquad [19.18]$$

In general, the electric flux through a surface is defined by the expression

$$\Phi_E = \int_{\text{surface}} \mathbf{E} \cdot d\mathbf{A} \qquad [19.19]$$

**Gauss's law** says that the net electric flux $\Phi_E$ through any closed gaussian surface is equal to the *net* charge *inside* the surface divided by $\epsilon_0$:

$$\Phi_E = \oint \mathbf{E} \cdot d\mathbf{A} = \frac{q_{\text{in}}}{\epsilon_0} \qquad [19.22]$$

Using Gauss's law, one can calculate the electric field due to various symmetric charge distributions.

A **conductor in electrostatic equilibrium** has the following properties:

1. The electric field is zero everywhere inside the conductor.
2. If the isolated conductor carries a net charge, the net charge resides entirely on its surface.
3. The electric field just outside the charged conductor is perpendicular to the conductor surface and has a magnitude $\sigma/\epsilon_0$, where $\sigma$ is the surface charge density at that point.
4. On an irregularly shaped conductor, the surface charge density is highest at locations where the radius of curvature of the surface is smallest.

## QUESTIONS

1. Explain from an atomic viewpoint why charge is usually transferred by electrons.
2. Sparks are often observed (or heard) on a dry day when clothes are removed from a dryer in dim light. Explain.
3. A balloon is negatively charged by rubbing and then clings to a wall. Does this mean that the wall is positively charged? Why does the balloon eventually fall?
4. A charged comb often attracts small bits of dry paper that then fly away when they touch the comb. Explain.
5. Operating-room personnel must wear special conducting shoes while working around oxygen. Why? Contrast this procedure with what might happen if personnel wore rubber shoes.
6. Would life be different if the electron were positively charged and the proton were negatively charged? Does

the choice of signs have any bearing on physical and chemical interactions? Explain.
7. Explain the similarities and differences between Newton's law of universal gravitation and Coulomb's law.
8. When defining the electric field, why is it necessary to specify that the magnitude of the test charge be very small?
9. Consider two equal point charges separated by some distance $d$. At what point (other than $\infty$) would a third test charge experience no net force?
10. A negative point charge $-q$ is placed at the point $P$ near the positively charged ring shown in Figure 19.15 (Example 19.5). If $x \ll a$, describe the motion of the point charge if it is released from rest.

11. A charge $4q$ is a distance $r$ from a charge $-q$. Compare the number of electric field lines beginning on the charge $4q$ with the number ending on the charge $-q$. Where do the extra lines beginning on $4q$ end?

12. A very large, thin, flat plate of aluminum of area $A$ has a total charge of $Q$ uniformly distributed over its surfaces. The same charge is spread uniformly over only the *upper* surface of an otherwise identical glass plate. Compare the electric fields just above the centers of the plates' upper surfaces.

13. If more electric field lines are leaving a gaussian surface than entering, what can you conclude about the net charge enclosed by that surface?

14. A uniform electric field exists in a region of space with no charges. What can you conclude about the *net* electric flux through a gaussian surface placed in this region of space?

15. If the total charge inside a closed surface is known but the distribution of the charge is unspecified, can you use Gauss's law to find the electric field? Explain.

16. Explain why the electric flux through a closed surface with a given enclosed charge is independent of the size or shape of the surface.

17. A common demonstration involves charging a balloon, which is an insulator, by rubbing it on your head, and touching the balloon to a ceiling or wall, which is also an insulator. The electrical attraction between the charged balloon and the neutral wall results in the balloon sticking to the wall. Imagine now that we have two infinitely large flat sheets of insulating material. One is charged and the other is neutral. If these are brought into contact, will an attractive force exist between them, as there was for the balloon and the wall?

18. Explain why excess charge on an isolated conductor must reside on its surface, using the repulsive nature of the force between like charges and the freedom of motion of charge within the conductor.

19. When is it possible to model a charge distribution as a point charge?

20. A person is placed in a large hollow metallic sphere that is insulated from ground. If a large charge is placed on the sphere, will the person be harmed on touching the inside of the sphere? Explain what will happen if the person also has an initial charge whose sign is opposite that of the charge on the sphere.

21. Imagine two electric dipoles in empty space. Each dipole has zero net charge. Does an electric force exist between the dipoles—that is, can two objects with *zero net charge* exert electric forces on each other? If so, is the force attractive or repulsive?

22. Two solid spheres, both of radius $R$, carry identical total charges, $Q$. One sphere is a good conductor and the other is an insulator. If the charge on the insulating sphere is uniformly distributed throughout its interior volume, how do the electric fields outside these two spheres compare? Are the fields identical inside the two spheres?

# PROBLEMS

1, 2, 3 = straightforward, intermediate, challenging    □ = full solution available in the *Student Solutions Manual and Study Guide*

web = solution posted at **http://www.harcourtcollege.com/physics/**    🖳 = computer useful in solving problem

🌐 = Interactive Physics    ▥ = paired numerical/symbolic problems    ▨ = life science application

## Section 19.2   Properties of Electric Charges

## Section 19.3   Insulators and Conductors

## Section 19.4   Coulomb's Law

1. (a) Calculate the number of electrons in a small, electrically neutral silver pin that has a mass of 10.0 g. Silver has 47 electrons per atom, and its molar mass is 107.87 g/mol. (b) Electrons are added to the pin until the net negative charge is 1.00 mC. How many electrons are added for every $1.00 \times 10^9$ electrons already present?

2. Suppose that 1.00 g of hydrogen is separated into electrons and protons. Suppose also that the protons are placed at the Earth's North Pole and the electrons are placed at the South Pole. What is the resulting compressional force on the Earth?

3. The Nobel laureate in physics Richard Feynman once said that if two persons stood at arm's length from each other and each person had 1% more electrons than protons, the force of repulsion between them would be enough to lift a "weight" equal to that of the entire Earth. Carry out an order-of-magnitude calculation to substantiate this assertion.

4. Two identical conducting small spheres are placed with their centers 0.300 m apart. One is given a charge of 12.0 nC and the other a charge of $-18.0$ nC. (a) Find the

electric force exerted on one sphere by the other. (b) The spheres are connected by a conducting wire. Find the electric force between the two after they have come to equilibrium.

5. Two protons in an atomic nucleus are typically separated by a distance of $2 \times 10^{-15}$ m. The electric repulsion force between the protons is huge, but the attractive nuclear force is even stronger and keeps the nucleus from bursting apart. What is the magnitude of the electric force between two protons separated by $2.00 \times 10^{-15}$ m?

6. **Review Problem.** In the Bohr theory of the hydrogen atom, an electron moves in a circular orbit about a proton, where the radius of the orbit is $5.29 \times 10^{-11}$ m. (a) Find the electric force between the two. (b) If this force causes the centripetal acceleration of the electron, what is the speed of the electron?

7. Three point charges are located at the corners of an equilateral triangle as shown in Figure P19.7. Calculate the net electric force on the 7.00-$\mu$C charge.

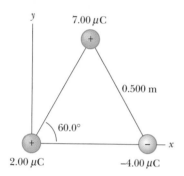

7.00 $\mu$C

0.500 m

60.0°

2.00 $\mu$C

−4.00 $\mu$C

**Figure P19.7**

## Section 19.5   Electric Fields

8. What are the magnitude and direction of the electric field that will balance the weight of (a) an electron and (b) a proton? (Use the data in Table 19.1.)

9. In Figure P19.9, determine the point (other than infinity) at which the electric field is zero.

−2.50 $\mu$C

6.00 $\mu$C

1.00 m

**Figure P19.9**

10. Three point charges are arranged as shown in Figure P19.10. (a) Find the vector electric field that the 6.00-nC and −3.00-nC charges together create at the origin. (b) Find the vector force on the 5.00-nC charge.

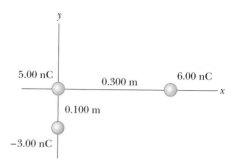

5.00 nC

0.300 m

6.00 nC

0.100 m

−3.00 nC

**Figure P19.10**

11. Three equal positive charges $q$ are at the corners of an equilateral triangle of side $a$ as in Figure P19.11. (a) Assume that the three charges together create an electric field. Find the location of a point (other than ∞) where the electric field is zero. (*Hint:* Sketch the field lines in the plane of the charges.) (b) What are the magnitude and direction of the electric field at $P$ due to the two charges at the base?

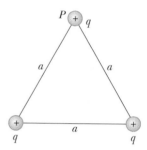

P

q

a

a

q

a

q

**Figure P19.11**

12. Two 2.00-$\mu$C point charges are located on the $x$ axis. One is located at $x = 1.00$ m, and the other is located at $x = -1.00$ m. (a) Determine the electric field on the $y$ axis at $y = 0.500$ m. (b) Calculate the electric force on a $-3.00$-$\mu$C charge placed on the $y$ axis at $y = 0.500$ m.

13. Four point charges are at the corners of a square of side $a$ as shown in Figure P19.13. (a) Determine the magnitude and direction of the electric field at the location of charge $q$. (b) What is the resultant force on $q$?

2q

a

q

a

a

3q

a

4q

**Figure P19.13**

**14.** A rod 14.0 cm long is uniformly charged and has a total charge of $-22.0\ \mu\text{C}$. Determine the magnitude and direction of the electric field along the axis of the rod at a point 36.0 cm from its center.

**15.** Consider the electric dipole shown in Figure P19.15. Show that the electric field at a *distant* point along the $x$ axis is $E_x \approx 4k_eqa/x^3$.

**Figure P19.15**

**16.** A uniformly charged ring of radius 10.0 cm has a total charge of $75.0\ \mu\text{C}$. Find the electric field on the axis of the ring at (a) 1.00 cm, (b) 5.00 cm, (c) 30.0 cm, and (d) 100 cm from the center of the ring.

**17.** A uniformly charged insulating rod of length 14.0 cm is
web bent into the shape of a semicircle, as shown in Figure P19.17. The rod has a total charge of $-7.50\ \mu\text{C}$. Find the magnitude and direction of the electric field at $O$, the center of the semicircle.

**Figure P19.17**

**18.** A continuous line of charge lies along the $x$ axis, extending from $x = +x_0$ to positive infinity. The line carries charge with a uniform linear charge density $\lambda_0$. What are the magnitude and direction of the electric field at the origin?

**19.** A thin rod of length $\ell$ and uniform charge per unit length $\lambda$ lies along the $x$ axis, as shown in Figure P19.19. (a) Show that the electric field at $P$, a distance $y$ from the rod along its perpendicular bisector, has no $x$ component and is given by $E = 2k_e\lambda \sin \theta_0/y$. (b) Using your result to part (a), show that the field of a rod of infinite length is $E = 2k_e\lambda/y$. (*Hint:* First calculate the field at $P$ due to an element of length $dx$, which has a charge $\lambda\ dx$. Then change variables from $x$ to $\theta$, using the relationships $x = y \tan \theta$ and $dx = y \sec^2\theta\ d\theta$, and integrate over $\theta$.)

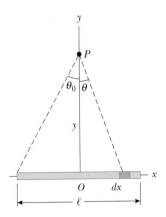

**Figure P19.19**

**20.** Three solid plastic cylinders all have a radius of 2.50 cm and a length of 6.00 cm. One (a) carries charge with uniform density $15.0\ \text{nC/m}^2$ everywhere on its surface. Another (b) carries charge with the same uniform density on its curved lateral surface only. The third (c) carries charge with uniform density $500\ \text{nC/m}^3$ throughout the plastic. Find the charge of each cylinder.

### Section 19.6 Electric Field Lines

**21.** A negatively charged rod of finite length has a uniform charge per unit length. Sketch the electric field lines in a plane containing the rod.

**22.** Figure P19.22 shows the electric field lines for two point charges separated by a small distance. (a) Determine the ratio $q_1/q_2$. (b) What are the signs of $q_1$ and $q_2$?

**Figure P19.22**

### Section 19.7 Motion of Charged Particles in a Uniform Electric Field

**23.** A proton accelerates from rest in a uniform electric field of 640 N/C. At some later time, its speed has reached $1.20 \times 10^6$ m/s (nonrelativistic, because $v$ is much less than the speed of light). (a) Find the acceleration of the proton. (b) How long does it take the proton to reach this speed? (c) How far has it moved in this time? (d) What is its kinetic energy at this time?

24. The electrons in a particle beam each have a kinetic energy $K$. What are the magnitude and direction of the electric field that stops these electrons in a distance $d$?

25. A positively charged bead having a mass of 1.00 g falls from rest in a vacuum from a height of 5.00 m in a uniform vertical electric field with a magnitude of $1.00 \times 10^4$ N/C. The bead hits the ground at a speed of 21.0 m/s. Determine (a) the direction of the electric field (up or down), and (b) the charge on the bead.

26. A proton moves at $4.50 \times 10^5$ m/s in the horizontal direction. It enters a uniform vertical electric field with a magnitude of $9.60 \times 10^3$ N/C. Ignoring any gravitational effects, find (a) the time it takes the proton to travel 5.00 cm horizontally, (b) its vertical displacement after it has traveled 5.00 cm horizontally, and (c) the horizontal and vertical components of its velocity after it has traveled 5.00 cm horizontally.

27. Two horizontal metal plates, each 100 mm square, are aligned 10.0 mm apart, with one above the other. They are given equal and opposite charges so that a uniform downward electric field of 2 000 N/C exists in the region between them. A particle of mass $2.00 \times 10^{-16}$ kg and with a positive charge of $1.00 \times 10^{-6}$ C leaves the center of the bottom negative plate with an initial speed of $1.00 \times 10^5$ m/s at an angle of 37.0° above the horizontal. Describe the trajectory of the particle. Which plate does it strike? Where does it strike, relative to its starting point?

## Section 19.8 Electric Flux

28. A vertical electric field of magnitude $2.00 \times 10^4$ N/C exists above the Earth's surface on a day when a thunderstorm is brewing. A car with a rectangular size of 6.00 m by 3.00 m is traveling along a roadway sloping downward at 10.0°. Determine the electric flux through the bottom of the car.

29. A 40.0-cm-diameter loop is rotated in a uniform electric field until the position of maximum electric flux is found. The flux in this position is measured to be $5.20 \times 10^5$ N·m²/C. What is the magnitude of the electric field?

## Section 19.9 Gauss's Law

30. The electric field everywhere on the surface of a thin spherical shell of radius 0.750 m is measured to be 890 N/C and points radially toward the center of the sphere. (a) What is the net charge within the sphere's surface? (b) What can you conclude about the nature and distribution of the charge inside the spherical shell?

31. The following charges are located inside a submarine: 5.00 μC, −9.00 μC, 27.0 μC, and −84.0 μC. (a) Calculate the net electric flux through the hull of the submarine. (b) Is the number of electric field lines leaving the submarine greater than, equal to, or less than the number entering it?

32. A charge of 170 μC is at the center of a cube of side 80.0 cm. (a) Find the total flux through each face of the cube. (b) Find the flux through the whole surface of the cube. (c) Would your answers to parts (a) or (b) change if the charge were not at the center? Explain.

33. A point charge $Q$ is located just above the center of the flat face of a hemisphere of radius $R$, as shown in Figure P19.33. What is the electric flux (a) through the curved surface and (b) through the flat face?

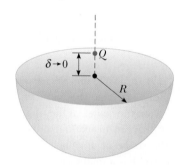

**Figure P19.33**

## Section 19.10 Application of Gauss's Law to Symmetric Charge Distributions

34. A solid sphere of radius 40.0 cm has a total positive charge of 26.0 μC uniformly distributed throughout its volume. Calculate the magnitude of the electric field (a) 0 cm, (b) 10.0 cm, (c) 40.0 cm, and (d) 60.0 cm from the center of the sphere.

35. Consider a thin spherical shell of radius 14.0 cm with a total charge of 32.0 μC distributed uniformly on its surface. Find the electric field (a) 10.0 cm and (b) 20.0 cm from the center of the charge distribution.

36. A cylindrical shell of radius 7.00 cm and length 240 cm has its charge uniformly distributed on its curved surface. The magnitude of the electric field at a point 19.0 cm radially outward from its axis (measured from the midpoint of the shell) is 36.0 kN/C. Find (a) the net charge on the shell and (b) the electric field at a point 4.00 cm from the axis, measured radially outward from the midpoint of the shell.

37. Consider a long cylindrical charge distribution of radius $R$ with a uniform charge density $\rho$. Find the electric field at distance $r$ from the axis where $r < R$.

38. A 10.0-g piece of Styrofoam carries a net charge of −0.700 μC and floats above the center of a large horizontal sheet of plastic that has a uniform charge density on its surface. What is the charge per area on the plastic sheet?

**39.** In a nuclear fission reaction, a nucleus of uranium-235, which contains 92 protons, divides into two smaller spheres, each having 46 protons and each having a radius of $5.90 \times 10^{-15}$ m. What is the magnitude of the repulsive electric force pushing the two spheres apart?

## Section 19.11   Conductors in Electrostatic Equilibrium

**40.** A long, straight metal rod has a radius of 5.00 cm and a charge per unit length of 30.0 nC/m. Find the electric field (a) 3.00 cm, (b) 10.0 cm, and (c) 100 cm from the axis of the rod, where distances are measured perpendicular to the rod.

**41.** A square plate of copper with 50.0-cm sides has no net charge and is placed in a region of uniform electric field of 80.0 kN/C directed perpendicular to the plate. Find (a) the charge density of each face of the plate and (b) the total charge on each face.

**42.** The electric field on the surface of an irregularly shaped conductor varies from 56.0 kN/C to 28.0 kN/C. Calculate the local surface charge density at the point on the surface where the radius of curvature of the surface is (a) greatest and (b) smallest.

**43.** A thin square conducting plate 50.0 cm on a side lies in the *xy* plane. If a total charge of $4.00 \times 10^{-8}$ C is placed on the plate, find (a) the charge density on the plate, (b) the electric field just above the plate, and (c) the electric field just below the plate. You may assume that the charge density is uniform.

**44.** A long, straight wire is surrounded by a hollow metal cylinder whose axis coincides with that of the wire. The wire has a charge per unit length of $\lambda$, and the cylinder has a net charge per unit length of $2\lambda$. From this information, use Gauss's law to find (a) the charge per unit length on the inner and outer surfaces of the cylinder and (b) the electric field outside the cylinder, a distance $r$ from the axis.

**45.** A solid conducting sphere of radius 2.00 cm has a charge 8.00 $\mu$C. A conducting spherical shell of inner radius 4.00 cm and outer radius 5.00 cm is concentric with the solid sphere and has a total charge $-4.00$ $\mu$C. Find the electric field at (a) $r = 1.00$ cm, (b) $r = 3.00$ cm, (c) $r = 4.50$ cm, and (d) $r = 7.00$ cm from the center of this charge configuration.

## Section 19.12   Context Connection — The Atmospheric Electric Field

**46.** In fair weather the electric field in the air at a particular location just above the Earth's surface is 120 N/C directed downward. (a) What is the surface charge density on the ground surface? Is it positive or negative? (b) If the weather were fair everywhere and the surface charge density were uniform, what would be the charge of the whole Earth? How many excess electrons (or protons) would be on the entire surface of the Earth to produce an atmospheric field of 120 N/C down?

**47.** Over a particular region in the air at an altitude of 500 m above the ground, the electric field is 120 N/C directed downward. At 600 m above the ground the electric field is 100 N/C downward. What is the average volume charge density in the layer of air between these two elevations? Is it positive or negative?

**48.** The electric field in the Earth's atmosphere suggests that the solid and liquid surface of the Earth has a charge of about $-5 \times 10^{5}$ C. Imagine that the planet as a whole had a charge of $-5.00 \times 10^{5}$ C, and that the Moon, with 27.3% of the radius of the Earth, had a charge of $-1.37 \times 10^{5}$ C. (a) Find the electric force that the Earth would then exert on the Moon. (b) Compare it with the gravitational force that the Earth exerts on the Moon. As your calculation suggests, for the purpose of accounting for astronomical motions, we may treat the actual forces as purely gravitational. We may say that astronomical objects have negligible total charges.

### Additional Problems

**49.** A small, 2.00-g plastic ball is suspended by a 20.0-cm-long string in a uniform electric field as shown in Figure P19.49. If the ball is in equilibrium when the string makes a 15.0° angle with the vertical, what is the net charge on the ball?

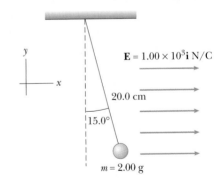

$\mathbf{E} = 1.00 \times 10^{3}\mathbf{i}$ N/C

20.0 cm

15.0°

$m = 2.00$ g

**Figure P19.49**

**50.** Three identical small Styrofoam balls ($m = 2.00$ g) are suspended from a fixed point by three nonconducting threads, each with a length of 50.0 cm and with negligible mass. At equilibrium the three balls form an equilateral triangle with sides of 30.0 cm. What is the common charge $q$ carried by each ball?

51. Two identical metallic blocks resting on a frictionless horizontal surface are connected by a light metallic spring having the spring constant 100 N/m and an unstretched length of 0.300 m, as shown in Figure P19.51a. A total charge of $Q$ is slowly placed on the system, causing the spring to stretch to an equilibrium length of 0.400 m, as shown in Figure P19.51b. Determine the value of $Q$, assuming that all the charge resides on the blocks and that the blocks are like point charges.

(a)

(b)

**Figure P19.51** Problems 51 and 52.

52. Two identical metallic blocks resting on a frictionless horizontal surface are connected by a light metallic spring having a spring constant $k$ and an unstretched length $L_0$ as shown in Figure P19.51a. A total charge of $Q$ is slowly placed on the system, causing the spring to stretch to an equilibrium length $L$, as shown in Figure P19.51b. Determine the value of $Q$, assuming that all the charge resides on the blocks and that the blocks are like point charges.

53. Four identical point charges ($q = +10.0\ \mu C$) are located on the corners of a rectangle, as shown in Figure P19.53. The dimensions of the rectangle are $L = 60.0$ cm and $W = 15.0$ cm. Calculate the magnitude and direction of the net electric force exerted on the charge at the lower left corner by the other three charges.

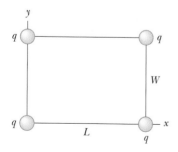

**Figure P19.53**

54. Three charges of equal magnitude $q$ are fixed in position at the vertices of an equilateral triangle (Fig. P19.54). A fourth charge $Q$ is free to move along the positive $x$ axis

under the influence of the forces exerted by the three fixed charges. Find a value for $s$ for which $Q$ is in equilibrium. You will need to solve a transcendental equation.

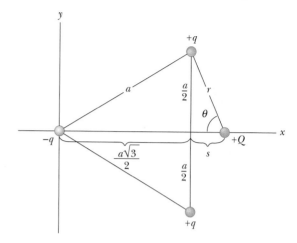

**Figure P19.54**

55. Two small spheres of mass $m$ are suspended from strings of length $\ell$ that are connected at a common point. One sphere has charge $Q$; the other has charge $2Q$. Assume the angles $\theta_1$ and $\theta_2$ that the strings make with the vertical are small. (a) How are $\theta_1$ and $\theta_2$ related? (b) Show that the distance $r$ between the spheres is given by

$$r \approx \left( \frac{4k_e Q^2 \ell}{mg} \right)^{1/3}$$

56. A line of charge with a uniform density of 35.0 nC/m lies along the line $y = -15.0$ cm, between the points with coordinates $x = 0$ and $x = 40.0$ cm. Find the electric field it creates at the origin.

57. Consider the charge distribution shown in Figure P19.57. (a) Show that the magnitude of the electric field at the center of any face of the cube has a value of $2.18 k_e q / s^2$.

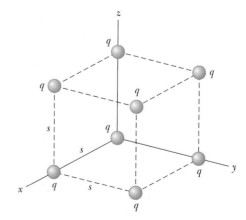

**Figure P19.57**

(b) What is the direction of the electric field at the center of the top face of the cube?

58. **Review Problem.** A negatively charged particle $-q$ is placed at the center of a uniformly charged ring, where the ring has a total positive charge $Q$ as in Example 19.5. The particle, confined to move along the $x$ axis, is displaced a small distance $x$ along the axis (where $x \ll a$) and released. Show that the particle oscillates with simple harmonic motion and that its frequency is given by

$$f = \frac{1}{2\pi} \left( \frac{k_e qQ}{ma^3} \right)^{1/2}$$

59. A solid, insulating sphere of radius $a$ has a uniform charge density $\rho$ and a total charge $Q$. Concentric with this sphere is an uncharged, conducting hollow sphere whose inner and outer radii are $b$ and $c$, as shown in Figure P19.59. (a) Find the magnitude of the electric field in the regions $r < a$, $a < r < b$, $b < r < c$, and $r > c$. (b) Determine the induced charge per unit area on the inner and outer surfaces of the hollow sphere.

**Figure P19.59**

60. Two infinite, nonconducting sheets of charge are parallel to each other, as shown in Figure P19.60. The sheet on the left has a uniform surface charge density $\sigma$, and the one on the right has a uniform charge density $-\sigma$. Calculate the electric field at points (a) to the left of, (b) in between, and (c) to the right of the two sheets. (*Hint:* See Example 19.12.)

61. Repeat the calculations for Problem 60 when both sheets have *positive* uniform surface charge densities of value $\sigma$.

62. A sphere of radius $2a$ is made of a nonconducting material that has a uniform volume charge density $\rho$. (Assume that the material does not affect the electric field.) A spherical cavity of radius $a$ is now removed from the sphere, as shown in Figure P19.62. Show that the electric field within the cavity is uniform and is given by $E_x = 0$ and $E_y = \rho a / 3\epsilon_0$. (*Hint:* The field within the cavity is the superposition of the field due to the original uncut sphere, plus the field due to a sphere the size of the cavity with a uniform negative charge density $-\rho$.)

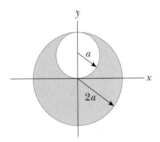

**Figure P19.62**

63. A solid insulating sphere of radius $R$ has a nonuniform charge density that varies with $r$ according to the expression $\rho = Ar^2$, where $A$ is a constant and $r < R$ is measured from the center of the sphere. (a) Show that the electric field outside $(r > R)$ the sphere is $E = AR^5/5\epsilon_0 r^2$. (b) Show that the electric field inside $(r < R)$ the sphere is $E = Ar^3/5\epsilon_0$. (*Hint:* Note that the total charge $Q$ on the sphere is equal to the integral of $\rho \, dV$, where $r$ extends from 0 to $R$; also note that the charge $q$ within a radius $r < R$ is less than $Q$. To evaluate the integrals, note that the volume element $dV$ for a spherical shell of radius $r$ and thickness $dr$ is equal to $4\pi r^2 \, dr$.)

**Figure P19.60**

## ANSWERS TO QUICK QUIZZES

**19.1** The configuration shown is inherently unstable. The negative charges repel each other. If any slight rotation of one of the rods occurs, the repulsion can result in further rotation away from this configuration. Three possible final configurations are shown in the following diagram: configuration (a) is stable—if the positive upper ends are pushed toward each other, they will repel and move the system back to the original configuration. Configuration (b) is an equilibrium configuration but is unstable—if the upper ends are moved toward each other, the attraction of the upper ends will be larger than that of the lower ends and the configuration will shift to (c). Configuration (c) is stable.

(a)                            (b)

(c)

**Figure QQA19.1**

**19.2** (b). From Newton's third law, the electric force exerted by B on A is equal in magnitude to the force exerted by A on B and in the opposite direction, that is, $\mathbf{F}_{AB} = -\mathbf{F}_{BA}$.

**19.3** Nothing, assuming the source charges producing the field are not disturbed by our actions. Remember that the electric field is created not by the $+3$-$\mu$C charge or the $-3$-$\mu$C charge but by the source charges.

**19.4** The two charged plates create a region of uniform electric field between them, directed from the positive toward the negative plate. Once the ball is disturbed so as to touch one plate, say the negative one, some negative charge is transferred to the ball and it experiences an electric force that accelerates it to the positive plate. Once the charge touches the positive plate, it releases its negative charge, acquires a positive charge, and accelerates back to the negative plate. The ball continues to move back and forth between the plates until it has transferred all their net charge thereby making both plates neutral.

**19.5** Negative. Because the electric field lines point downward, the ground must have negative charges.

**19.6** *A, B, C.* The field is strongest at *A* because that is where the field lines are closest together. No lines at *C* indicate that the electric field there is zero.

**19.7** (b) and (d). (a) is not necessarily true because equal number of positive and negative charges could be inside the surface. (c) is not necessarily true as can be seen in Figure 19.31, where a nonzero electric field everywhere is on the surface, but the charge is not enclosed within the surface, so the net flux is zero.

This device is a *variable capacitor,* used to tune radios to a selected station. When one set of metal plates is rotated so as to lie between a fixed set of plates, the *capacitance* of the device changes. Capacitance is a parameter that depends on *electric potential,* the primary topic of this chapter.

*(George Semple)*

# Electric Potential and Capacitance

The concept of potential energy was introduced in Chapter 7 in connection with such conservative forces as gravity and the elastic force of a spring. By using the principle of conservation of energy in an isolated system, we are often able to avoid working directly with forces when solving mechanical problems. In this chapter we shall use the energy concept in our study of electricity. Because the electrostatic force (given by Coulomb's law) is conservative, electrostatic phenomena can conveniently be described in terms of an *electric* potential energy function. This concept enables us to define a quantity called *electric potential,* which is a scalar function of position and thus leads to a simpler means of describing some electrostatic phenomena than the electric field method. As we shall see in subsequent chapters, the concept of electric potential is of great practical value.

This chapter also addresses the properties of capacitors, devices that store charge. The ability of a capacitor to store charge is measured by its *capacitance.* Capacitors are used in such common applications as frequency tuners in radio receivers, as filters in power supplies, as dampers to eliminate unwanted sparking in automobile ignition systems, and as energy-storing devices in electronic flash units.

Both electric potential and capacitance are abstract ideas somewhat different from those we have encountered previously. On the other hand, they both are extensions of our earlier work. Because these new ideas are abstract, we shall provide more than the usual number of Quick Quizzes and Pitfall Preventions in this chapter to help you understand them.

*See Core Concepts in Physics, Screens 11.8 & 11.9*

## 20.1 • POTENTIAL DIFFERENCE AND ELECTRIC POTENTIAL

When a point charge $q_0$ is placed in an electric field $\mathbf{E}$, the electric force on the particle is $q_0\mathbf{E}$. This force is the vector sum of the individual forces exerted on $q_0$ by the various charges producing the field $\mathbf{E}$. It follows that the force $q_0\mathbf{E}$ is conservative because the individual forces governed by Coulomb's law are conservative. (See Section 7.3 for a review of conservative forces.) Let us consider a system consisting of the point charge and all of the source charges creating the electric field. Because the field represents the effect of the source charges, we can also consider the system to be the electric field and the charge $q_0$ that we place in the field, without referring specifically to the source charges. When the point charge moves in response to the electric force within the electric field, work is done on the particle by the field. For an infinitesimal displacement $d\mathbf{s}$ of a point charge $q_0$, the work done by the electric field on the charge is $\mathbf{F}_e \cdot d\mathbf{s} = q_0\mathbf{E} \cdot d\mathbf{s}$.

The work done by the field on a point charge is similar to the work done by a gravitational field on a falling object. We found in Chapter 7 that the gravitational potential energy of an isolated object–field system changes by an amount equal to the negative of the work done by the field on the object (Equation 7.3). Similarly, the work done by the electric field on a charged particle changes the potential energy of the isolated charge–field system by an amount $dU = -dW = -q_0\mathbf{E} \cdot d\mathbf{s}$. For a finite displacement of a test particle of charge $q_0$ between points $A$ and $B$, the **change in potential energy** of the charge–field system is

- *Change in potential energy*

$$\Delta U = U_B - U_A = -q_0 \int_A^B \mathbf{E} \cdot d\mathbf{s} \qquad [20.1]$$

The integral in Equation 20.1 is performed along the path by which the particle moves from $A$ to $B$ and is called either a **path integral** or a **line integral.** Because the force $q_0\mathbf{E}$ is conservative, **this integral does not depend on the path taken between $A$ and $B$.**

### Quick Quiz 20.1

If the path between $A$ and $B$ does not make any difference in Equation 20.1, why don't we just use the expression $\Delta U = -q_0Ed$, where $d$ is the straight-line distance between $A$ and $B$?

The potential energy $U$ of the system per unit charge $q_0$ is independent of the value of $q_0$ and has a unique value at every point in an electric field. The quantity $U/q_0$ is called the **electric potential** $V$ (or simply the **potential**):

- *Electric potential*

$$V \equiv \frac{U}{q_0} \qquad [20.2]$$

Because potential energy is a scalar, electric potential is also a scalar quantity. Note that potential is not a property of the charge–field system because we have divided the potential energy of the system by the charge. It is a property only of the field. Thus, in the physical situation, we can imagine removing the test charge from the field. The potential still exists at the point the test charge occupied and is due to the source charges that establish the electric field.

The **potential difference** $\Delta V = V_B - V_A$ between the points $A$ and $B$ is defined as the change in potential energy of the charge–field system when the test particle is moved between the points divided by the charge $q_0$ on the test particle:

$$\Delta V = \frac{\Delta U}{q_0} = -\int_A^B \mathbf{E} \cdot d\mathbf{s} \qquad \text{[20.3]}$$

● *Potential difference between two points*

**Potential difference should not be confused with potential energy difference.** The potential difference between two points in an electric field is *proportional* to the potential energy difference of the charge–field system when the charge is at the two points, and we see from Equation 20.3 that the two are related by $\Delta U = q_0 \, \Delta V$.

Equation 20.3 defines potential difference only. The potential is often taken to be zero at some convenient point, sometimes called a *ground*. We usually set the potential due to one or more source charges at zero for a point at infinity (i.e., a point infinitely remote from the source charges producing the electric field). With this choice, we can say that **the electric potential at an arbitrary point due to source charges equals the work required to bring a test particle from infinity to that point divided by the charge on the test particle.** Thus, if we take $V_A = 0$ at infinity in Equation 20.3, then the potential at any point $P$ is

$$V_P = -\int_\infty^P \mathbf{E} \cdot d\mathbf{s} \qquad \text{[20.4]}$$

where $\mathbf{E}$ is the electric field established by the source charges. In reality, $V_P$ represents the potential difference between the point $P$ and a point at infinity. (Equation 20.4 is a special case of Equation 20.3.) When discussing potentials in an electric circuit, we shall set $V = 0$ at some selected point in the circuit.

Because potential is a measure of energy per unit charge, the SI units of potential are joules per coulomb, called the **volt** (V):

$$1 \text{ V} \equiv 1 \text{ J/C}$$

That is, if we release a particle with a charge of 1 C in an electric field and it moves from a point of high potential to a point of low potential through a potential difference of $-1$ V, it will have 1 J of work done on it by the field and, therefore, will attain a kinetic energy of 1 J (according to the work–kinetic energy theorem). Alternatively, 1 J of work must be done by an external agent to take a particle with a charge of 1 C through a potential difference of $+1$ V at constant velocity (according to the nonisolated system model). Equation 20.3 shows that the potential difference also has the same units as electric field times distance. From this, it follows that the SI units of electric field, newtons per coulomb, can be expressed as volts per meter:

$$1 \text{ N/C} = 1 \text{ V/m}$$

This suggests that the electric field can be interpreted as the rate of change in space of the electric potential. A strong electric field corresponds to a potential that changes rapidly in space while a weak field represents a slowly changing potential.

**PITFALL PREVENTION 20.1**
**Potential and potential energy**

The distinction between potential and potential energy is often a difficult concept for students. The potential is characteristic of the field only, independent of a charged test particle that may be placed in the field. On the other hand, potential energy refers to a characteristic of the charge–field system due to an interaction between the field and a charged particle placed in the field. Despite the similarities between electric and gravitational fields, we did not define a gravitational potential in our discussions of mechanics (although we *could* have), because it does not happen to be a useful tool for solving gravitational problems. It *is* very useful in electrical problems.

As we learned in Section 9.7, a unit of energy commonly used in physics is the **electron volt** (eV):

$$1 \text{ eV} = (1e)(1 \text{ V}) = (1.60 \times 10^{-19} \text{ C})(1 \text{ J/C}) = 1.60 \times 10^{-19} \text{ J} \quad \textbf{[20.5]}$$

One eV is the kinetic energy gained by a particle with charge $e$ being accelerated through a potential difference of magnitude 1 V. Equation 20.5 can be used to convert any energy in joules to electron volts. For instance, an electron in the beam of a typical TV picture tube may have a speed of $3.5 \times 10^7$ m/s. This corresponds to a kinetic energy of $5.6 \times 10^{-16}$ J, which is equivalent to $3.5 \times 10^3$ eV. Such an electron has to be accelerated from rest through a potential difference of 3.5 kV to reach this speed.

**PITFALL PREVENTION 20.2**

**The electron volt revisited**

Now that we have defined the volt, we provide a reminder of Pitfall Prevention 9.1 in Chapter 9. The electron volt may sound like a unit of potential but remember that it is a unit of *energy*.

**THINKING PHYSICS 20.1**

Suppose scientists had chosen to measure small energies in *proton volts* rather than electron volts. What difference would this make?

**Reasoning**   There would be no difference at all. An electron volt is the kinetic energy gained by an electron that is accelerated through a potential difference of one volt. A proton accelerated through one volt will have the same kinetic energy because it carries a charge of the same magnitude as the electron. The proton will be moving more slowly after accelerating through one volt due to its larger mass, but it will still gain one electron volt, or one proton volt, of kinetic energy.

## 20.2 • POTENTIAL DIFFERENCES IN A UNIFORM ELECTRIC FIELD

In this section we describe the potential difference between any two points in a *uniform* electric field. Consider a uniform electric field directed along the negative $y$ axis, as in Figure 20.1a. Let us calculate the potential difference between two points $A$ and $B$, separated by a distance $d$, where $d$ is measured parallel to the field lines. If we apply Equation 20.3 to this situation, we have

$$V_B - V_A = \Delta V = -\int_A^B \mathbf{E} \cdot d\mathbf{s} = -\int_A^B E \cos 0° \, ds = -\int_A^B E \, ds$$

**Figure 20.1**

(a) When the electric field **E** is directed downward, point $B$ is at a lower electric potential than point $A$. When a positive test charge moves from $A$ to $B$, the charge–field system loses electric potential energy. (b) When an object with mass $m$ moves downward in the direction of the gravitational field **g**, the object–field system loses gravitational potential energy.

(a)

(b)

Because $E$ is constant, it can be removed from the integral, giving

$$\Delta V = -E \int_A^B ds = -Ed \qquad \text{[20.6]}$$

• *Potential difference between two points in a uniform* **E** *field*

The negative sign results from the fact that point $B$ is at a lower potential than point $A$; that is, $V_B < V_A$. In general, **electric field lines always point in the direction of decreasing electric potential.**

Now suppose that a test particle with charge $q_0$ moves from $A$ to $B$. The change in the electric potential energy of the charge–field system can be found from Equations 20.3 and 20.6:

$$\Delta U = q_0 \, \Delta V = -q_0 Ed \qquad \text{[20.7]}$$

From this result, we see that if $q_0$ is positive, $\Delta U$ is negative. This means that **when a positive charge moves in the direction of the electric field, the electric potential energy of the charge–field system decreases.** (This is analogous to the change in gravitational potential energy $-mgd$ of an object–field system when an object with mass $m$ falls through a height $d$ in a uniform gravitational field, as suggested in Figure 20.1b.) If a particle with a positive charge $q_0$ is released from rest in the electric field, it experiences an electric force $q_0\mathbf{E}$ in the direction of $\mathbf{E}$ (downward in Fig. 20.1a). Therefore, it accelerates downward, gaining kinetic energy. **As the charged particle gains kinetic energy, the charge–field system loses an equal amount of potential energy.** This familiar result is similar to what we have seen for gravitational situations (Fig. 20.1b). The statement is simply the principle of conservation of mechanical energy in the isolated system model for electric fields.

If $q_0$ is negative, then $\Delta U$ in Equation 20.7 is positive and the situation is reversed. If a negatively charged particle is released from rest in the field $\mathbf{E}$, it accelerates in a direction opposite the electric field. **The charge–field system loses electric potential energy when a negative charge moves in the direction opposite to the electric field.** We have no analog for this situation in the gravitational case because no negative mass has been observed.

Now consider the more general case of a charged particle moving between any two points in a uniform electric field, as in Figure 20.2. If $\Delta\mathbf{r}$ represents the displacement vector between points $A$ and $B$, Equation 20.3 gives

$$\Delta V = -\int_A^B \mathbf{E} \cdot d\mathbf{s} = -\mathbf{E} \cdot \int_A^B d\mathbf{s} = -\mathbf{E} \cdot \Delta\mathbf{r} \qquad \text{[20.8]}$$

where again we are able to remove $\mathbf{E}$ from the integral because it is constant. Furthermore, the change in electric potential energy of the charge–field system is

$$\Delta U = q_0 \, \Delta V = -q_0 \mathbf{E} \cdot \Delta\mathbf{r} \qquad \text{[20.9]}$$

Finally, our results show that all points in a plane *perpendicular* to a uniform electric field are at the same potential. This can be seen in Figure 20.2, where the potential difference $V_B - V_A = -\mathbf{E} \cdot \Delta\mathbf{r} = -E\,\Delta r \cos\theta = -Ed = V_C - V_A$. Therefore, $V_B = V_C$. The name **equipotential surface** is given to any surface consisting of a continuous distribution of points having the same electric potential. Note that because $\Delta U = q_0 \, \Delta V$, no work is required to move a test particle between any two points on an equipotential surface. The equipotential surfaces of a uniform electric field consist of a family of planes, all perpendicular to the field. Equipotential surfaces for fields with other symmetries will be described in later sections.

**PITFALL PREVENTION 20.3**

**Voltage**

In practice, a variety of phrases are used to describe the potential difference between two points, the most common being **voltage,** arising from the unit for potential. A voltage *applied* to a device, such as a television, or *across* a device has the same meaning as the potential difference across the device. For example, if we say that the voltage applied to a lightbulb is 120 V, we mean that the potential difference between the two electrical contacts on the lightbulb is 120 V.

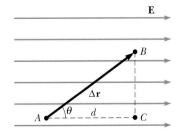

**Figure 20.2**

A particle is moved in a uniform electric field. Point $B$ is at a lower potential than point $A$. Points $B$ and $C$ are at the *same* potential.

If an electron is released from rest in a uniform electric field, does the electric potential energy of the charge–field system increase, decrease, or remain the same?

## Example 20.1 The Electric Field Between Two Parallel Plates of Opposite Charge

A 12-V battery is connected between two parallel plates as in Figure 20.3. The separation between the plates is 0.30 cm, and the electric field is assumed to be uniform. (This simplification model is reasonable if the plate separation is small relative to the plate size and if we do not consider points near the edges of the plates.) Find the magnitude of the electric field between the plates.

**Solution** The electric field is directed from the positive plate toward the negative plate. The positive plate is at a higher potential than the negative plate. Note that the potential difference between plates must equal the potential difference between the battery terminals. This can be understood by recognizing that all points on a conductor in equilibrium are at the same potential.* Hence no potential difference occurs between a terminal of the battery and any portion of the plate to which it is connected. The magnitude of the uniform electric field between the plates is therefore

$$E = \frac{|V_B - V_A|}{d} = \frac{12\ \text{V}}{0.30 \times 10^{-2}\ \text{m}} = \boxed{4.0 \times 10^3\ \text{V/m}}$$

**Figure 20.3**

(Example 20.1) A 12-V battery connected to two parallel plates. The electric field between the plates has a magnitude given by the potential difference divided by the plate separation *d*.

This configuration, which is called a *parallel-plate capacitor*, is examined in more detail later in this chapter.

## Example 20.2 Motion of a Proton in a Uniform Electric Field

A proton is released from rest in a uniform electric field of magnitude $8.0 \times 10^4$ V/m directed along the positive *x* axis (Fig. 20.4). The proton undergoes a displacement of magnitude $d = 0.50$ m in the direction of **E**. (a) Find the change in the electric potential between the points *A* and *B*.

**Solution** The change in electric potential does not depend on the presence of the proton. From Equation 20.6 we have

$$\Delta V = -Ed = -(8.0 \times 10^4\ \text{V/m})(0.50\ \text{m}) = \boxed{-4.0 \times 10^4\ \text{V}}$$

This negative result tells us that the electric potential decreases between points *A* and *B*.

(b) Find the change in potential energy of the charge–field system for this displacement.

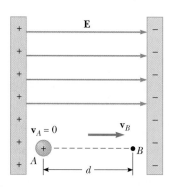

**Figure 20.4**

(Example 20.2) A proton accelerates from *A* to *B* in the direction of the electric field.

---

* The electric field vanishes within a conductor in electrostatic equilibrium, and so the path integral ∫ **E** · *d***s** between any two points within the conductor must be zero. A fuller discussion of this point is given in Section 20.6.

**Solution**    From Equation 20.3, we have

$$\Delta U = q\,\Delta V = e\,\Delta V$$

$$= (1.6 \times 10^{-19}\,\text{C})(-4.0 \times 10^4\,\text{V}) = \boxed{-6.4 \times 10^{-15}\,\text{J}}$$

The negative sign here means that the potential energy of the system decreases as the proton moves in the direction of the electric field. This is consistent with conservation of energy in an isolated system—as the proton accelerates in the direction of the field, it gains kinetic energy and at the same time the system loses electric potential energy. The increase

in kinetic energy of a charged particle in an electric field is exploited in many devices, including electron guns for television picture tubes and particle accelerators for research in particle physics.

**EXERCISE**    Apply the principle of energy conservation to find the speed of the proton after it has moved 0.50 m, starting from rest.

*Answer*    $2.8 \times 10^6\,\text{m/s}$

## 20.3 • ELECTRIC POTENTIAL AND ELECTRIC POTENTIAL ENERGY DUE TO POINT CHARGES

In establishing the concept of electric potential, we imagined placing a test particle in an electric field set up by some undescribed source charges. As a simplification model, the field was assumed to be uniform in Section 20.2 in order to firmly plant the idea of electric potential in our minds. Let us now focus our attention on point charges, which we know set up electric fields that are not uniform.

Consider an isolated positive point charge $q$ (Fig. 20.5). Recall that such a charge is a source of an electric field that is directed radially outward from the charge. To find the electric potential at a distance of $r$ from the charge, we begin with the general expression for potential difference, Equation 20.3:

$$V_B - V_A = -\int_A^B \mathbf{E} \cdot d\mathbf{s}$$

Because the electric field due to the point charge is given by $\mathbf{E} = k_e q\hat{\mathbf{r}}/r^2$ (Equation 19.5), where $\hat{\mathbf{r}}$ is a unit vector directed from the charge toward the field point, the quantity $\mathbf{E} \cdot d\mathbf{s}$ can be expressed as

$$\mathbf{E} \cdot d\mathbf{s} = k_e \frac{q}{r^2} \hat{\mathbf{r}} \cdot d\mathbf{s}$$

The dot product $\hat{\mathbf{r}} \cdot d\mathbf{s} = ds \cos\theta$, where $\theta$ is the angle between $\hat{\mathbf{r}}$ and $d\mathbf{s}$ as in Figure 20.5. Furthermore, note that $ds \cos\theta$ is the projection of $d\mathbf{s}$ onto $\mathbf{r}$, so that $ds \cos\theta = dr$. With these substitutions, we find that $\mathbf{E} \cdot d\mathbf{s} = (k_e q/r^2)\,dr$, so that the expression for the potential difference becomes

$$V_B - V_A = -\int_{r_A}^{r_B} k_e \frac{q}{r^2}\,dr = -k_e q \int_{r_A}^{r_B} \frac{dr}{r^2} = \left. \frac{k_e q}{r} \right]_{r_A}^{r_B}$$

$$= k_e q \left[ \frac{1}{r_B} - \frac{1}{r_A} \right] \qquad\qquad \textbf{[20.10]}$$

The line integral of $\mathbf{E} \cdot d\mathbf{s}$ is *independent* of the path between $A$ and $B$—as it must be, because the electric field of a point charge is conservative.* Furthermore, Equation 20.10 expresses the important result that the potential difference

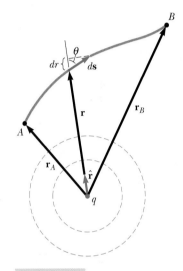

**Figure 20.5**

The potential difference between points $A$ and $B$ due to a point charge $q$ depends *only* on the initial and final radial coordinates $r_A$ and $r_B$. The two dashed circles represent cross-sections of spherical equipotential surfaces.

---

* A conservative field is one that exerts a conservative force on an object placed within it. Both gravitational and electric fields are conservative.

Be sure to avoid confusion between Equation 20.11 for the electric potential of a point charge and Equation 19.5 for the electric field of a point charge. The equations look very similar, but potential is proportional to $1/r$, whereas the field is proportional to $1/r^2$. The effect of a charge on the space surrounding it can be described in two ways. The charge sets up a vector electric field **E**, which is related to the force experienced by a test charge placed in the field. It also sets up a scalar potential $V$, which is related to the potential energy of the two-charge system when a test charge is placed in the field.

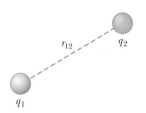

**Figure 20.6**

If two point charges are separated by a distance $r_{12}$, the electric potential energy of the pair of charges is given by $k_e q_1 q_2 / r_{12}$.

between any two points $A$ and $B$ depends *only* on the *radial* coordinates $r_A$ and $r_B$. As we learned in Section 20.1, it is customary to choose the reference of potential to be zero at $r_A = \infty$. With this choice, the electric potential due to a point charge at any distance $r$ from the charge is

$$V = k_e \frac{q}{r} \qquad \text{[20.11]}$$

From this we see that $V$ is constant on a spherical surface of radius $r$ centered on the point charge. Hence, we conclude that **the equipotential surfaces for an isolated point charge consist of a family of spheres concentric with the charge,** as shown in Figure 20.5. Note that the equipotential surfaces are perpendicular to the lines of electric force, as is the case for a uniform electric field.

The electric potential of two or more point charges is obtained by applying the superposition principle. That is, the total potential at some point $P$ due to multiple point charges is the sum of the potentials due to the individual charges. For a group of charges, we can write the total potential at $P$ in the form

$$V = k_e \sum_i \frac{q_i}{r_i} \qquad \text{[20.12]}$$

where the potential is again taken to be zero at infinity and $r_i$ is the distance from point $P$ to the charge $q_i$. Note that the sum in Equation 20.12 is an *algebraic* sum of scalars rather than a *vector* sum (which is used to calculate the electric field of a group of charges, as in Equation 19.6). Thus, it is much easier to evaluate $V$ for multiple charges than to evaluate **E**.

We now consider the electric potential energy of interaction of a system of charged particles. If $V_1$ is the electric potential at point $P$ due to charge $q_1$, the work required to bring a second charge $q_2$ from infinity to point $P$ without acceleration is $q_2 \Delta V = q_2 V_1$. By definition, this work equals the potential energy $U$ of the two-particle system (Fig. 20.6) when the particles are separated by a distance of $r_{12}$—the external agent has changed the energy of the system by doing work on it. We can therefore express the potential energy as

$$U = q_2 V_1 = k_e \frac{q_1 q_2}{r_{12}} \qquad \text{[20.13]}$$

Note that if the charges are of the same sign, $U$ is positive. This is consistent with the fact that like charges repel, and so positive work must be done on the system to bring the two charges near each other, and the potential energy of the system increases due to this work. If the charges are of opposite sign, the force is at-

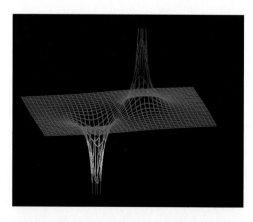

Computer-generated three-dimensional plot of the electric potential associated with an electric dipole. The values of the potential are plotted vertically in the figure. The charges lie in the horizontal plane, at the centers of the potential spikes. The positive charge in the dipole is on the right and the negative charge is on the left. The contour lines help visualize the potential. (*Richard Megna 1990, Fundamental Photographs*)

tractive and *U* is negative. This means that negative work is done when the unlike charges are brought near one another. Because the force is attractive, the particles will move together naturally. In this case, the system does work on the external agent (or the external agent does negative work) as the particles move near one another. This represents energy transferring out of the system, and the potential energy of the system decreases.

If the system consists of more than two charged particles, the total electric potential energy can be obtained by calculating *U* for every pair of charges and summing the terms algebraically. The total electric potential energy of a system of point charges is equal to the work required to bring the charges, one at a time, from an infinite separation to their final positions.

### Quick Quiz 20.3

If the electric potential at some point is zero, can you conclude that no charges exist in the vicinity of that point?

### Quick Quiz 20.4

A spherical balloon contains a positively charged particle at its center. As the balloon is inflated to a larger volume while the charged particle remains at the center, which of the following changes? (a) electric potential at the surface of the balloon, (b) magnitude of the electric field at the surface of the balloon, (c) electric flux through the balloon.

### Example 20.3   The Potential Due to Two Point Charges

A 2.00-$\mu$C point charge is located at the origin, and a second point charge of $-6.00$ $\mu$C is located on the *y* axis at the position (0, 3.00) m, as in Figure 20.7a. (a) Find the total electric potential due to these charges at point *P*, whose coordinates are (4.00, 0) m.

**Solution**   For two point charges, the sum in Equation 20.12 gives

$$V_P = k_e \left( \frac{q_1}{r_1} + \frac{q_2}{r_2} \right)$$

(a)

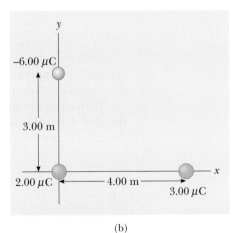

(b)

**Figure 20.7**

(Example 20.3) (a) The electric potential at point *P* due to the two point charges $q_1$ and $q_2$ is the algebraic sum of the potentials due to the individual charges. (b) How much work is done to bring a 3.00-$\mu$C charge from infinity to point *P*?

In this example, $q_1 = 2.00 \ \mu C$, $r_1 = 4.00$ m, $q_2 = -6.00 \ \mu C$, and $r_2 = 5.00$ m. Therefore, $V_P$ reduces to

$$V_P = (8.99 \times 10^9 \text{ N} \cdot \text{m}^2/\text{C}^2)$$

$$\times \left( \frac{2.00 \times 10^{-6} \text{ C}}{4.00 \text{ m}} + \frac{-6.00 \times 10^{-6} \text{ C}}{5.00 \text{ m}} \right) = -6.29 \times 10^3 \text{ V}$$

(b) How much work is required to bring a $3.00$-$\mu C$ point charge from infinity to the point $P$ (Fig. 20.7b)?

**Solution** The work done is equal to the change in potential energy given by Equation 20.3:

$$W = \Delta U = q_3 \, \Delta V = q_3 (V_P - 0)$$

$$= (3.00 \times 10^{-6} \text{ C})(-6.29 \times 10^3 \text{ V})$$

$$= -18.9 \times 10^{-3} \text{ J}$$

The negative sign is due to the fact that the $3.00$-$\mu C$ charge is attracted to the combination of $q_1$ and $q_2$, which has a net negative charge. The $3.00$-$\mu C$ charge would naturally move toward the other charges if released, so the external agent does not have to do anything to cause them to move together. To keep the charge from accelerating, however, the agent must apply a force *away* from point $P$. Thus, the force exerted by the agent is opposite the displacement of the charge, leading to a negative value of the work. Positive work would have to be done by an external agent to remove the charge from $P$ back to infinity.

**EXERCISE** Find the total potential energy of the system of three charges in the configuration shown in Figure 20.7b.

*Answer* $-5.48 \times 10^{-2}$ J

## 20.4 • OBTAINING ELECTRIC FIELD FROM ELECTRIC POTENTIAL

The electric field $\mathbf{E}$ and the electric potential $V$ are related by Equation 20.3. Both quantities are determined by a specific source charge distribution. We now show how to calculate the electric field if the electric potential is known in a certain region.

From Equation 20.3 we can express the potential difference $dV$ between two points a distance $ds$ apart as

$$dV = -\mathbf{E} \cdot d\mathbf{s} \qquad \text{[20.14]}$$

If the electric field has only *one* component, $E_x$, for example, then $\mathbf{E} \cdot d\mathbf{s} = E_x \, dx$. Therefore, Equation 20.14 becomes $dV = -E_x \, dx$, or

$$E_x = -\frac{dV}{dx} \qquad \text{[20.15]}$$

• *Relation between electric field and electric potential*

That is, the electric field is equal to the negative of the derivative of the electric potential with respect to some coordinate. The potential change is zero for any displacement perpendicular to the electric field. This is consistent with the notion that equipotential surfaces are perpendicular to the field, as in Figure 20.8a.

**If the charge distribution has spherical symmetry, such that the charge density depends only on the radial distance $r$, the electric field is radial.** In this case, $\mathbf{E} \cdot d\mathbf{s} = E_r \, dr$, and we can express $dV$ as $dV = -E_r \, dr$. Therefore,

$$E_r = -\frac{dV}{dr} \qquad \text{[20.16]}$$

For example, the potential of a point charge is $V = k_e q / r$. Because $V$ is a function of $r$ only, the potential function has spherical symmetry. Applying Equation 20.16, we find that the magnitude of the electric field due to the point charge is $E_r = k_e q / r^2$, a familiar result. Note that the potential changes only in the radial direction, not in a direction perpendicular to $r$. Thus, $V$ (like $E_r$) is a function only of $r$.

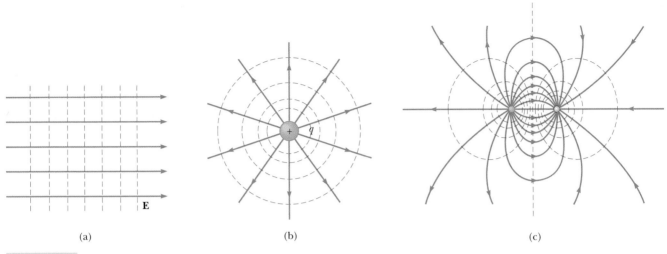

**Figure 20.8**

Equipotential surfaces (dashed blue lines) and electric field lines (red lines) for (a) a uniform electric field produced by an infinite sheet of charge, (b) a point charge, and (c) an electric dipole. In all cases, the equipotential surfaces are *perpendicular* to the electric field lines at every point.

Again, this is consistent with the idea that equipotential surfaces are perpendicular to field lines. In this case the equipotential surfaces are a family of spheres concentric with the spherically symmetric charge distribution (Fig. 20.8b). The equipotential surfaces for the electric dipole are sketched in Figure 20.8c.

In general, the electric potential is a function of all three spatial coordinates. If $V$ is given in terms of rectangular coordinates, the electric field components $E_x$, $E_y$, and $E_z$ can be found from $V(x, y, z)$ as the partial derivatives

$$E_x = -\frac{\partial V}{\partial x} \qquad E_y = -\frac{\partial V}{\partial y} \qquad E_z = -\frac{\partial V}{\partial z}$$

For example, if $V = 3x^2 y + y^2 + yz$, then

$$E_x = -\frac{\partial V}{\partial x} = -\frac{\partial}{\partial x}(3x^2 y + y^2 + yz) = -\frac{\partial}{\partial x}(3x^2 y) = -3y\frac{d}{dx}(x^2) = -6xy$$

**Quick Quiz 20.5**

Suppose you know the value of the electric potential at one point. Can you find the electric field at that point from this information?

**Quick Quiz 20.6**

If the electric potential is constant in some region, what can you conclude about the electric field in that region? If the electric field is zero in some region, what can you say about the electric potential in that region?

## Example 20.4 The Electric Potential of a Dipole

An electric dipole consists of two equal and opposite charges separated by a distance $2a$, as in Figure 20.9. The dipole is along the $x$ axis and is centered at the origin. Calculate
(a) the electric potential at any point $P$ along the $x$ axis and
(b) the electric field at points very far from the dipole.

**Solution** (a) Using Equation 20.12, we have

$$V = k_e \sum_i \frac{q_i}{r_i} = k_e \left( \frac{q}{x - a} + \frac{-q}{x + a} \right) = \boxed{\frac{2k_e qa}{x^2 - a^2}}$$

(b) If $P$ is far from the dipole, so that $x \gg a$, then $a^2$ can be ignored in the term $x^2 - a^2$ and $V$ becomes

$$V \approx \frac{2k_e qa}{x^2} \quad (x \gg a)$$

Using Equation 20.15 and this result, we calculate the electric field at $P$:

$$E_x = -\frac{dV}{dx} = \boxed{\frac{4k_e qa}{x^3} \quad \text{for } x \gg a}$$

Comparing this to the result from Example 19.3, we see a factor of 2 difference between the results for the field far from the dipole. In the previous example, we were looking at the field along a line perpendicular to the line connecting the

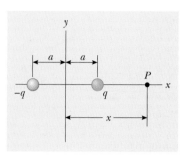

**Figure 20.9**

(Example 20.4) An electric dipole located on the $x$ axis.

charges. As we see in Figure 19.10, the vertical components of the field cancel. Thus, only the very small horizontal components of the individual fields contribute to the total field. In the present example, we are looking at the field along an extension of the line connecting the charges. For points along this line, the field vectors have components only along the line and the full field vectors combine to provide the total field. As a result, the field is larger than that along the perpendicular direction by a factor of 2.

## 20.5 • ELECTRIC POTENTIAL DUE TO CONTINUOUS CHARGE DISTRIBUTIONS

The electric potential due to a continuous charge distribution can be calculated in two ways. If the charge distribution is known, we can start with Equation 20.11 for the potential of a point charge. We then consider the potential due to a small charge element $dq$, treating this element as a point charge (Fig. 20.10). The potential $dV$ at some point $P$ due to the charge element $dq$ is

$$dV = k_e \frac{dq}{r} \qquad \text{[20.17]}$$

where $r$ is the distance from the charge element to $P$. To find the total potential at $P$, we integrate Equation 20.17 to include contributions from all elements of the charge distribution. Because each element is, in general, at a different distance from $P$ and because $k_e$ is a constant, we can express $V$ as

$$V = k_e \int \frac{dq}{r} \qquad \text{[20.18]}$$

In effect, we have replaced the sum in Equation 20.12 with an integral.

The second method for calculating the potential of a continuous charge distribution makes use of Equation 20.3. This procedure is useful when the electric field is already known from other considerations, such as Gauss's law. In this case,

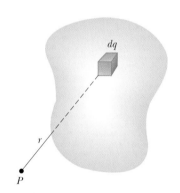

**Figure 20.10**

The electric potential at the point $P$ due to a continuous charge distribution can be calculated by dividing the charge distribution into segments of charge $dq$ and summing the potential contributions over all segments.

we substitute the electric field into Equation 20.3 to determine the potential difference between any two points. We then choose $V$ to be zero at some convenient point. We illustrate both methods with examples below.

---

**PROBLEM-SOLVING STRATEGY** | **Calculating Electric Potential**

1. When working problems involving electric potential, remember that it is a *scalar quantity*, and so there are no components to consider. Therefore, when using the superposition principle to evaluate the electric potential at a point due to a system of point charges, simply take the algebraic sum of the potentials due to each charge. You must keep track of signs, however.

2. As with potential energy in mechanics, only *changes* in electric potential are significant; hence, the point where the potential is set at zero is arbitrary. When dealing with point charges or a finite-sized charge distribution, we usually define $V = 0$ to be at a point infinitely far from the charges. If the charge distribution itself extends to infinity, however, some other nearby point must be selected as the reference point.

3. The electric potential at some point $P$ due to a continuous distribution of charge can be evaluated by dividing the charge distribution into infinitesimal elements of charge $dq$ located at a distance of $r$ from the point $P$. This element is then treated as a point charge, and so the potential at $P$ due to the element is $dV = k_e \, dq/r$. The total potential at $P$ is obtained by integrating $dV$ over the entire charge distribution. For many problems, it is possible in performing the integration to express $dq$ and $r$ in terms of a single variable. To simplify the integration, it is important to give careful consideration to the geometry involved in the problem. Review Example 20.5 as a guide for using this method.

4. Another method that can be used to obtain the potential due to a finite continuous charge distribution is to start with the definition of the potential difference given by Equation 20.3. If $\mathbf{E}$ is known or can be obtained easily (e.g., from Gauss's law), then the line integral of $\mathbf{E} \cdot d\mathbf{s}$ can be evaluated. Example 20.6 uses this method.

5. Once you have a functional description of the electric potential, it is possible to obtain the electric field by remembering that **the electric field is equal to the negative of the derivative of the potential with respect to an appropriate coordinate.** Example 20.5 illustrates how to use this procedure.

---

## Example 20.5   Potential Due to a Uniformly Charged Ring

Find the electric potential and electric field at a point $P$ located on the axis of a uniformly charged ring of radius $a$ and total charge $Q$. The plane of the ring is chosen perpendicular to the $x$ axis (Fig. 20.11).

**Reasoning and Solution**   Let us take $P$ to be at a distance $x$ from the center of the ring, as in Figure 20.11. The charge element $dq$ is at a distance equal to $r = \sqrt{x^2 + a^2}$ from point $P$. Hence, we can express $V$ as

$$V = k_e \int \frac{dq}{r} = k_e \int \frac{dq}{\sqrt{x^2 + a^2}}$$

In this case, each element $dq$ is at the same distance from $P$. The term $\sqrt{x^2 + a^2}$ can therefore be removed from the

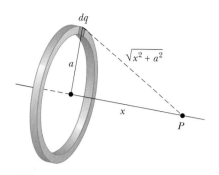

**Figure 20.11**

(Example 20.5) A uniformly charged ring of radius $a$, whose plane is perpendicular to the $x$ axis. All segments $dq$ of the ring are at the same distance from any point $P$ on the $x$ axis.

integral, and $V$ reduces to

$$V = \frac{k_e}{\sqrt{x^2 + a^2}} \int dq = \frac{k_e Q}{\sqrt{x^2 + a^2}}$$

The only variable in this expression for $V$ is $x$. From the symmetry, we see that along the $x$ axis **E** can have only an $x$ component. We can therefore use Equation 20.15 to find the magnitude of the electric field at $P$:

$$E_x = -\frac{dV}{dx} = -k_e Q \frac{d}{dx} (x^2 + a^2)^{-1/2}$$

$$= -k_e Q(-\tfrac{1}{2})(x^2 + a^2)^{-3/2} (2x)$$

$$= \frac{k_e Q x}{(x^2 + a^2)^{3/2}}$$

This result agrees with that obtained by direct integration (see Example 19.5).

**EXERCISE** What is the electric potential at the center of the uniformly charged ring? What does the value of the field at the center imply about this result?

*Answer* $V = k_e Q / a$ at $x = 0$. Because $E_x = 0$ at the center, $V$ must have a maximum or minimum value; it is in fact a maximum at $x = 0$.

## Example 20.6 Potential of a Uniformly Charged Sphere

An insulating solid sphere of radius $R$ has a total charge of $Q$, which is distributed uniformly throughout the volume of the sphere (Fig. 20.12a). (a) Find the electric potential at a point outside the sphere, that is, for $r > R$. Take the potential to be zero at $r = \infty$.

**Solution** In Example 19.8 we found from Gauss's law that the magnitude of the electric field outside a spherically sym-

metric charge distribution is

$$E_r = k_e \frac{Q}{r^2} \quad (\text{for } r > R)$$

where the field is directed radially outward when $Q$ is positive. To obtain the potential at an exterior point, such as $B$ in Figure 20.12a, we substitute this expression for $E$ into Equa-

(a)

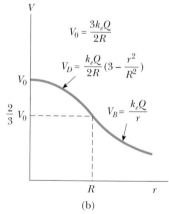

(b)

**Figure 20.12**

(Example 20.6) (a) A uniformly charged insulating sphere of radius $R$ and total charge $Q$. The electric potential at points $B$ and $C$ is equivalent to that of a point charge $Q$ located at the center of the sphere. (b) A plot of the electric potential $V$ versus the distance $r$ from the center of a uniformly charged, insulating sphere of radius $R$. The curve for $V_D$ inside the sphere is parabolic and joins smoothly with the curve for $V_B$ outside the sphere, which is a hyperbola. The potential has a maximum value $V_0$ at the center of the sphere.

tion 20.4. Because $\mathbf{E} \cdot d\mathbf{s} = E_r \, dr$ in this case, we have

$$V_B = -\int_{\infty}^{r} E_r \, dr = -k_e Q \int_{\infty}^{r} \frac{dr}{r^2}$$

$$= k_e \frac{Q}{r} \quad \text{(for } r > R)$$

Note that the result is identical to that for the electric potential due to a point charge. Because the potential must be continuous at $r = R$, we can use this expression to obtain the potential at the surface of the sphere. That is, the potential at a point such as $C$ in Figure 20.12a is

$$V_C = k_e \frac{Q}{R} \quad \text{(for } r = R)$$

   (b) Find the potential at a point inside the charged sphere, that is, for $r < R$.

**Solution**   In Example 19.8 we found that the electric field inside a uniformly charged sphere is

$$E_r = k_e \frac{Q}{R^3} r \quad \text{(for } r < R)$$

We can use this result and Equation 20.3 to evaluate the potential difference $V_D - V_C$, where $D$ is an interior point:

$$V_D - V_C = -\int_{R}^{r} E_r \, dr = -\frac{k_e Q}{R^3} \int_{R}^{r} r \, dr = \frac{k_e Q}{2R^3} (R^2 - r^2)$$

Substituting $V_C = k_e Q / R$ into this expression and solving for $V_D$, we find

$$V_D = \frac{k_e Q}{2R} \left( 3 - \frac{r^2}{R^2} \right) \quad \text{(for } r < R)$$

At $r = R$, this expression gives a result for the potential that agrees with the potential $V_C$ at the surface. A plot of $V$ versus $r$ for this charge distribution is given in Figure 20.12b.

**EXERCISE**   What are the electric field and electric potential at the center of a uniformly charged sphere?

*Answer*   $E = 0$ and $V = 3k_e Q / 2R$.

---

## 20.6 • ELECTRIC POTENTIAL OF A CHARGED CONDUCTOR

In Chapter 19 we found that when a solid conductor in electrostatic equilibrium carries a net charge, the charge resides on the outer surface of the conductor. Furthermore, we showed that the electric field just outside the surface of a conductor in equilibrium is perpendicular to the surface, whereas the field *inside* the conductor is zero.

   We shall now show that **every point on the surface of a charged conductor in electrostatic equilibrium is at the same electric potential.** Consider two points $A$ and $B$ on the surface of a charged conductor, as in Figure 20.13. Along a surface path connecting these points, $\mathbf{E}$ is always perpendicular to the displacement $d\mathbf{s}$; therefore, $\mathbf{E} \cdot d\mathbf{s} = 0$. Using this result and Equation 20.3, we conclude that the potential difference between $A$ and $B$ is necessarily zero. That is,

$$V_B - V_A = -\int_{A}^{B} \mathbf{E} \cdot d\mathbf{s} = 0$$

This result applies to *any* two points on the surface. Therefore, $V$ is constant everywhere on the surface of a charged conductor in equilibrium, and so such a surface is an equipotential surface. Furthermore, because the electric field is zero inside the conductor, we conclude that the potential is constant everywhere inside the conductor and equal to its value at the surface. It follows that no work is required to move a test charge from the interior of a charged conductor to its surface.

   For example, consider a solid metal sphere of radius $R$ and total positive charge $Q$, as in Figure 20.14a. The electric field outside the sphere has magnitude $k_e Q / r^2$ and points radially outward. Following Example 20.6, we see that the potential at the interior and surface of the sphere must be $k_e Q / R$ relative to infinity. The potential outside the sphere is $k_e Q / r$. Figure 20.14b is a plot of the potential as a function of $r$, and Figure 20.14c shows the variations of the electric field with $r$.

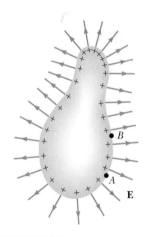

**Figure 20.13**

An arbitrarily shaped conductor with an excess positive charge. When the conductor is in electrostatic equilibrium, all of the charge resides at the surface, $\mathbf{E} = 0$ inside the conductor, and the electric field just outside the conductor is perpendicular to the surface. The potential is constant inside the conductor and is equal to the potential at the surface. The surface charge density is nonuniform.

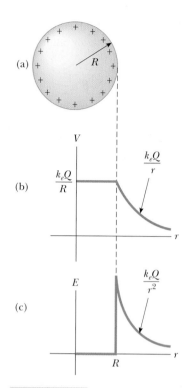

**Figure 20.14**

(a) The excess charge on a conducting sphere of radius $R$ is uniformly distributed on its surface. (b) Electric potential versus distance $r$ from the center of the charged conducting sphere. (c) Electric field intensity versus distance $r$ from the center of the charged conducting sphere.

When a net charge resides on a spherical conductor, the surface charge density is uniform as indicated in Figure 20.14a. If, however, the conductor is non-spherical as in Figure 20.13, the surface charge density is not uniform. To determine how the charge distributes on a nonspherical conductor, imagine a simplification model in which a nonspherical conductor is represented by the system shown in Figure 20.15. The system consists of two charged conducting spheres of radii $r_1$ and $r_2$, where $r_1 > r_2$, connected by a thin conducting wire. Imagine that the spheres are so far apart that the electric field of one does not influence the other (much farther apart than shown in Figure 20.15). As a result, the electric field of each sphere can be modeled as that due to a spherically symmetric distribution of charge, which is the same as that due to a point charge.

Because the spheres are connected by a conducting wire, the entire system is a single conductor and all points must be at the same potential. In particular, the potentials at the surfaces of the two spheres must be equal. Using Equation 20.11 for the potential of a point charge, we set the potentials at the surfaces of the spheres equal:

$$k_e \frac{q_1}{r_1} = k_e \frac{q_2}{r_2} \quad \longrightarrow \quad \frac{q_1}{q_2} = \frac{r_1}{r_2}$$

Thus, the larger sphere has the larger amount of charge. However, let us compare the surface charge densities on the two spheres:

$$\frac{\sigma_2}{\sigma_1} = \frac{\left(\dfrac{q_2}{4\pi r_2^2}\right)}{\left(\dfrac{q_1}{4\pi r_1^2}\right)} = \frac{q_2}{q_1}\frac{r_1^2}{r_2^2} = \frac{r_2}{r_1}\frac{r_1^2}{r_2^2} = \frac{r_1}{r_2}$$

Thus, although the larger sphere has the larger total charge, the smaller sphere has the larger surface charge density. This is the fourth property listed in Section 19.11. Equation 19.18 tells us that the electric field near the surface of a conductor is proportional to the surface charge density. Thus, the field near the smaller sphere is larger than the field close to the larger sphere.

We generalize this result by stating that **the electric field due to a charged conductor is large near convex surfaces of the conductor having small radii of curvature and is small near convex surfaces of the conductor having large radii of curvature.** A sharp point on a conductor is a region with an extremely small radius of curvature, so the field is very high near points on conductors.

**Figure 20.15**

Two charged spherical conductors connected by a conducting wire. The spheres are at the same potential $V$.

### THINKING PHYSICS 20.2

Why is the end of a lightning rod pointed?

**Reasoning** The role of a lightning rod is to serve as a location at which the lightning strikes, so that the charge delivered by the lightning will pass safely to the ground. If the lightning rod is pointed, the electric field is very strong near the point, because the radius of curvature of the conductor is very small. This large electric field will greatly increase the likelihood that the lightning discharge will occur near the tip of the lightning rod rather than elsewhere.

### A Cavity Within a Conductor in Equilibrium

Now consider a conductor of arbitrary shape containing a cavity as in Figure 20.16. Let us assume that no charges are inside the cavity. We shall show that **the electric field inside the cavity must be zero,** regardless of the charge distribution on the outside surface of the conductor. Furthermore, the field in the cavity is zero even if an electric field exists outside the conductor.

To prove this point, we shall use the fact that every point on the conductor is at the same potential, and therefore any two points $A$ and $B$ on the surface of the cavity must be at the same potential. Now imagine that a field $\mathbf{E}$ exists in the cavity, and evaluate the potential difference $V_B - V_A$, defined by the expression

$$V_B - V_A = -\int_A^B \mathbf{E} \cdot d\mathbf{s}$$

If $\mathbf{E}$ is nonzero, we can invariably find a path between $A$ and $B$ for which $\mathbf{E} \cdot d\mathbf{s}$ is always a positive number (a path along the direction of $\mathbf{E}$), and so the integral must be positive. Because $V_B - V_A = 0$, however, the integral must also be zero. This contradiction can be reconciled only if $\mathbf{E} = 0$ inside the cavity. Thus, we conclude that a cavity surrounded by conducting walls is a field-free region as long as no charges are inside the cavity.

This result has some interesting applications. For example, it is possible to shield an electronic device or even an entire laboratory from external fields by surrounding it with conducting walls. Shielding is often necessary during highly sensitive electrical measurements. During a thunderstorm, the safest location is inside an automobile. Even if lightning strikes the car, the metal body guarantees that you will not receive a shock inside, where $\mathbf{E} = 0$.

**Figure 20.16**

A conductor in electrostatic equilibrium containing an empty cavity. The electric field in the cavity is zero, regardless of the charge on the conductor.

## 20.7 • CAPACITANCE

As we continue with our discussion of electricity and, in later chapters, magnetism, we shall build *circuits* consisting of *circuit elements*. A circuit generally consists of a number of electrical components (circuit elements) connected together by conducting wires and forming one or more closed loops. These circuits can be considered as systems that exhibit a particular type of behavior. The first circuit element we shall consider is a **capacitor.**

In general, a capacitor consists of two conductors of any shape. Consider two conductors having a potential difference of $\Delta V$ between them. Let us assume that the conductors have charges of equal magnitude and opposite sign as in Figure 20.17. This can be accomplished by connecting two uncharged conductors to the terminals of a battery. Once this is done and the battery is disconnected, the charges remain on the conductors. We describe this by saying that **the capacitor stores charge.**

The potential difference $\Delta V$ *across* the capacitor is the magnitude of the potential difference between the two conductors. This potential difference is proportional to the charge $Q$ on the capacitor, which is defined as the magnitude of the charge on *either* of the two conductors. The **capacitance** $C$ of a capacitor is defined as the ratio of the charge on the capacitor to the magnitude of the potential difference across the capacitor:

$$C \equiv \frac{Q}{\Delta V} \qquad \text{[20.19]}$$

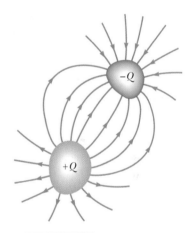

**Figure 20.17**

A capacitor consists of two conductors electrically isolated from each other and their surroundings. Once the capacitor is charged, the two conductors carry charges of equal magnitude but opposite sign.

• *Definition of capacitance*

By definition, **capacitance is always a positive quantity.** Because the potential difference is proportional to the charge, the ratio $Q/\Delta V$ is constant for a given capacitor. Equation 20.19 tells us that the capacitance of a system is a measure of the amount of charge that can be stored on the capacitor for a given potential difference.

• *Definition of the farad*

From Equation 20.19, we see that capacitance has the SI units coulombs per volt, which is called a **farad** (F) in honor of Michael Faraday. The farad is a very large unit of capacitance. In practice, typical devices have capacitances ranging from microfarads to picofarads.

The capacitance of a device depends on the geometric arrangement of the conductors. To illustrate this point, let us calculate the capacitance of an isolated spherical conductor of radius $R$ and charge $Q$. (Based on the shape of the field lines from a single spherical conductor, we can model the second conductor as a concentric spherical shell of infinite radius.) Because the potential of the sphere is simply $k_e Q/R$ (and $V = 0$ for the shell of infinite radius), the capacitance of the sphere is

$$C = \frac{Q}{\Delta V} = \frac{Q}{k_e Q/R} = \frac{R}{k_e} = 4\pi\epsilon_0 R \qquad \text{[20.20]}$$

(Remember from Section 19.4 that the Coulomb constant $k_e = 1/4\pi\epsilon_0$.) Equation 20.20 shows that the capacitance of an isolated charged sphere is proportional to the sphere's radius and is independent of both the charge and the potential difference.

The capacitance of a pair of oppositely charged conductors can be calculated in the following manner. A convenient charge of magnitude $Q$ is assumed, and the potential difference is calculated using the techniques described in Section 20.5. One then uses $C = Q/\Delta V$ to evaluate the capacitance. As you might expect, the calculation is relatively straightforward if the geometry of the capacitor is simple.

Let us illustrate this with two familiar geometries: parallel plates and concentric cylinders. In these examples, we shall assume that the charged conductors are separated by a vacuum. (The effect of a material between the conductors will be treated in Section 20.10.)

**PITFALL PREVENTION 20.6**

**Capacitance is a capacity**

To help you understand the concept of capacitance, think of similar notions that use a similar word. The *capacity* of a milk carton is the volume of milk it can store. The *heat capacity* of an object is the amount of energy an object can store per unit of temperature difference. The *capacitance* of a capacitor is the amount of charge the capacitor can store per unit of potential difference.

### The Parallel-Plate Capacitor

A parallel-plate capacitor consists of two parallel plates of equal area $A$ separated by a distance $d$ as in Figure 20.18. One plate has charge $Q$; the other, charge $-Q$. The magnitude of the charge per unit area on either plate is $\sigma = Q/A$. If the plates are very close together (compared with their length and width), we adopt a simplification model in which the electric field is uniform between the plates and zero elsewhere, as we discussed in Example 19.12. According to Example 19.12, the electric field between the plates is

$$E = \frac{\sigma}{\epsilon_0} = \frac{Q}{\epsilon_0 A}$$

Because the field is uniform, the potential difference across the capacitor can be found from Equation 20.6. Therefore,

$$\Delta V = Ed = \frac{Qd}{\epsilon_0 A}$$

Substituting this result into Equation 20.19, we find that the capacitance is

$$C = \frac{Q}{\Delta V} = \frac{Q}{Qd/\epsilon_0 A}$$

**Figure 20.18**

A parallel-plate capacitor consists of two parallel conducting plates, each of area $A$, separated by a distance $d$. When the capacitor is charged, the plates carry charges of equal magnitude but opposite sign.

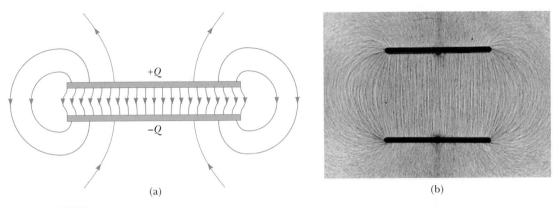

**Figure 20.19**

(a) The electric field between the plates of a parallel-plate capacitor is uniform near the center but nonuniform near the edges. (b) Electric field pattern of two oppositely charged conducting parallel plates. Small pieces of thread on an oil surface align with the electric field. *(b, Courtesy of Harold M. Waage, Princeton University)*

$$C = \frac{\epsilon_0 A}{d}$$    [20.21]

That is, **the capacitance of a parallel-plate capacitor is proportional to the area of its plates and inversely proportional to the plate separation.**

As you can see from the definition of capacitance, $C = Q/\Delta V$, the amount of charge a given capacitor can store for a given potential difference across its plates increases as the capacitance increases. It seems reasonable, therefore, that a capacitor constructed from plates having large areas should be able to store a large charge.

A careful inspection of the electric field lines for a parallel-plate capacitor reveals that the field is uniform in the central region between the plates, but nonuniform at the edges of the plates. Figure 20.19 shows a drawing and a photograph of the electric field pattern of a parallel-plate capacitor, showing the nonuniform field lines at the plates' edges. As long as the separation between the plates is small compared with the dimensions of the plates (unlike Figure 20.19b), the edge effects can be ignored and we can use the simplification model in which the electric field is uniform everywhere between the plates.

Figure 20.20 shows a battery connected to a single parallel-plate capacitor with a switch in the circuit. Let us identify the circuit as a system. When the switch is closed, the battery establishes an electric field in the wires and charges flow between the wires and the capacitor. As this occurs, energy is transformed within the system. Before the switch is closed, energy is stored as chemical energy in the battery. This type of energy is associated with chemical bonds and is transformed during the chemical reaction that occurs within the battery when it is operating in an electric circuit. When the switch is closed, some of the chemical energy in the battery is converted to electric potential energy related to the separation of positive and negative charges on the plates. As a result, we can describe a capacitor as a device that stores energy as well as charge. We will explore this energy storage in more detail in Section 20.9.

**PITFALL PREVENTION 20.7**
**Potential difference is $\Delta V$, not $V$**

In this book, we will use the symbol $\Delta V$ for the potential difference across a circuit element or a device because this is consistent with our definition of potential difference and our traditional use of the delta sign. It is common practice, however, in almost every other book, to use the symbol $V$, without the delta sign, for a potential difference across a circuit element. We feel that this usage leads to confusion, so we will use the more correct $\Delta V$. Keep this Pitfall Prevention in mind, however, if you consult other texts.

**PITFALL PREVENTION 20.8**
**Too many Cs**

Be sure not to confuse italic $C$ for capacitance with regular C for the unit coulomb.

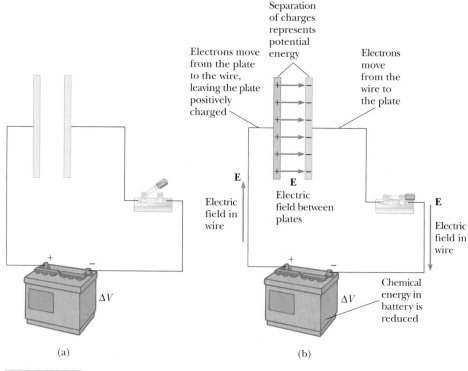

**Figure 20.20**

(a) A circuit consisting of a capacitor, a battery, and a switch. (b) When the switch is closed, the battery establishes an electric field in the wire that causes electrons to move from the left plate into the wire and into the right plate from the wire. As a result, a separation of charge exists on the plates, which represents electric potential energy. This energy in the system of the circuit has been transformed from chemical energy in the battery.

**Quick Quiz 20.7**

Why is it dangerous to touch the terminals of a high-voltage capacitor even after the voltage source that charged the capacitor is disconnected from the capacitor? What can be done to make the capacitor safe to handle after the voltage source has been removed?

### Example 20.7 Parallel-Plate Capacitor

A parallel-plate capacitor has an area $A = 2.00 \times 10^{-4}$ m$^2$ and a plate separation $d = 1.00$ mm. Find its capacitance.

**Solution**   From Equation 20.21, we find

$$C = \frac{\epsilon_0 A}{d} = (8.85 \times 10^{-12} \, \text{C}^2/\text{N} \cdot \text{m}^2) \left( \frac{2.00 \times 10^{-4} \, \text{m}^2}{1.00 \times 10^{-3} \, \text{m}} \right)$$

$$= 1.77 \times 10^{-12} \, \text{F} = \boxed{1.77 \, \text{pF}}$$

**EXERCISE**   If the plate separation is increased to 3.00 mm, find the capacitance.

*Answer*   0.590 pF

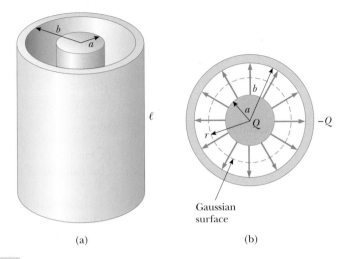

(a)                                          (b)

## Figure 20.21

(a) A cylindrical capacitor consists of a solid cylindrical conductor of radius $a$ and length $\ell$ surrounded by a coaxial cylindrical shell of radius $b$. (b) End view. The dashed line represents the end of the cylindrical gaussian surface of radius $r$ and length $\ell$.

## The Cylindrical Capacitor

A cylindrical capacitor consists of a cylindrical conductor of radius $a$ and charge $Q$ coaxial with a larger cylindrical shell of radius $b$ and charge $-Q$ (Fig. 20.21a). Let us find the capacitance of this device if its length is $\ell$. If we assume that $\ell$ is large compared with $a$ and $b$, we can adopt a simplification model in which we ignore end effects. In this case, the field is perpendicular to the axis of the cylinders and is confined to the region between them (Fig. 20.21b). We first calculate the potential difference between the two cylinders, which is given in general by

$$V_b - V_a = - \int_a^b \mathbf{E} \cdot d\mathbf{s}$$

where $\mathbf{E}$ is the electric field in the region $a < r < b$. In Chapter 19, using Gauss's law, we showed that the electric field of a cylinder with charge per unit length $\lambda$ has the magnitude $E = 2k_e\lambda/r$. The same result applies here, because the outer cylinder does not contribute to the electric field inside it. Using this result and noting that the direction of $\mathbf{E}$ is radially away from the inner cylinder in Figure 20.21b, we find that

$$V_b - V_a = - \int_a^b E_r \, dr = - 2k_e\lambda \int_a^b \frac{dr}{r} = - 2k_e\lambda \ln\!\left(\frac{b}{a}\right)$$

Substituting this into Equation 20.19 and using the fact that $\lambda = Q/\ell$, we find

$$C = \frac{Q}{|\Delta V|} = \frac{Q}{\dfrac{2k_eQ}{\ell} \ln\!\left(\dfrac{b}{a}\right)} = \frac{\ell}{2k_e \ln\!\left(\dfrac{b}{a}\right)} \qquad \textbf{[20.22]}$$

where we have recognized that $\Delta V$ in Equation 20.19 is the magnitude of the potential difference. Our result for $C$ shows that the capacitance is proportional to the length of the cylinders. As you might expect, the capacitance also depends on the radii of the two cylindrical conductors. As an example, a coaxial cable consists of two concentric cylindrical conductors of radii $a$ and $b$ separated by an insulator. The cable carries currents in opposite directions in the inner and outer conductors. Such a geometry is especially useful for shielding an electrical signal from

A collection of capacitors used in a variety of applications. (*Henry Leap and Jim Lehman*)

external influences. From Equation 20.22, we see that the capacitance per unit length of a coaxial cable is

$$\frac{C}{\ell} = \frac{1}{2k_e \ln\left(\dfrac{b}{a}\right)}$$

## 20.8 • COMBINATIONS OF CAPACITORS

Two or more capacitors are often combined in electric circuits in several ways. The equivalent capacitance of certain combinations can be calculated using methods described in this section.

In studying electric circuits, we use a specialized simplified pictorial representation called a **circuit diagram.** Such a diagram uses **circuit symbols** to represent various circuit elements. The circuit symbols are connected by straight lines that represent the wires between the circuit elements. Figure 20.22 shows the circuit symbols for a capacitor, a battery, and an open switch, together with their color codes. Notice that the circuit symbol for a capacitor consists of two parallel lines of equal length, representing the plates in a parallel-plate capacitor, and the lines in the battery symbol are of different lengths. The positive terminal of the battery is at the higher potential and is represented by the longer line in the battery symbol.

### Parallel Combination

Two capacitors connected as shown in the pictorial representation in Figure 20.23a are known as a **parallel combination** of capacitors. Figure 20.23b shows the circuit diagram for this configuration. The left plates of both capacitors are

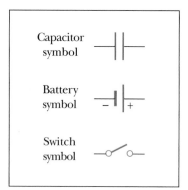

**Figure 20.22**

Circuit symbols for a capacitor, a battery, and an open switch. Note that capacitors are in blue, and batteries and switches are in red.

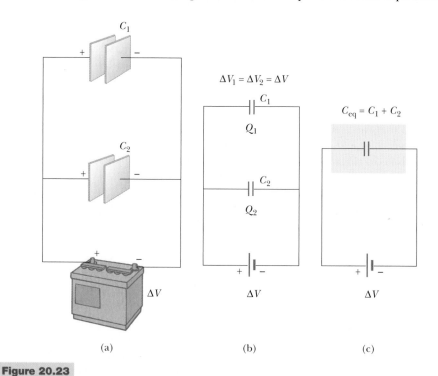

(a)  (b)  (c)

**Figure 20.23**

(a) A parallel combination of two capacitors connected to a battery. (b) The circuit diagram for the parallel combination. The potential difference is the same across each capacitor. (c) The equivalent capacitance is $C_{eq} = C_1 + C_2$.

connected by a conducting wire to the positive terminal of the battery, and therefore both plates are at the same potential as that of the battery terminal. Likewise, the right plates are connected to the negative terminal of the battery and are at the same potential as that terminal. The voltage applied across the combination is therefore the terminal voltage of the battery.* Furthermore, the voltage across *each capacitor* is the same as the terminal voltage of the battery.

When the capacitors are first connected in the circuit, electrons are transferred between the wires and the plates, causing the left plates to become positively charged and the right plates negatively charged. The flow of charge ceases when the voltage across the capacitors is equal to that across the battery terminals. At this point, the capacitors have reached their maximum charge. Let us call the maximum charges on the two capacitors $Q_1$ and $Q_2$. Then the *total charge Q* stored by the two capacitors is

$$Q = Q_1 + Q_2 \qquad \text{[20.23]}$$

Suppose we wish to replace the two capacitors in Figure 20.23b with one equivalent capacitor having the capacitance $C_{eq}$. This equivalent capacitor (Fig. 20.23c) must have exactly the same result in the circuit as the original two. That is, it must store charge $Q$ when connected to the battery. From Figure 20.23c, we see that the voltage across the equivalent capacitor is $\Delta V$. Thus, we have

$$Q = C_{eq}\, \Delta V$$

and, for the individual capacitors

$$Q_1 = C_1\, \Delta V \qquad Q_2 = C_2\, \Delta V$$

Substitution of these relations into Equation 20.23 gives

$$C_{eq}\, \Delta V = C_1\, \Delta V + C_2\, \Delta V$$

or

$$C_{eq} = C_1 + C_2 \quad \text{(parallel combination)} \qquad \text{[20.24]}$$

If we extend this treatment to three or more capacitors connected in parallel, the equivalent capacitance is

$$C_{eq} = C_1 + C_2 + C_3 + \cdots \quad \text{(parallel combination)} \qquad \text{[20.25]}$$

• *Equivalent capacitance of several capacitors in parallel*

Thus, we see that **the equivalent capacitance of a parallel combination of capacitors is the algebraic sum of the individual capacitances and is larger than any of the individual capacitances.**

**PITFALL PREVENTION 20.9**

**Micro- might be "u" in the electrical industry**

To avoid using Greek symbols, the electrical industry sometimes uses the unit "uF" for $10^{-6}$ F rather than "$\mu$F."

**EXERCISE**  Two capacitors, with capacitances $C_1 = 5.0\ \mu\text{F}$ and $C_2 = 12\ \mu\text{F}$, are connected in parallel, and the resulting combination is connected to a 9.0-V battery. What is (a) the value of the equivalent capacitance of the combination? (b) the potential difference across each capacitor and (c) the charge stored on each capacitor?

*Answer*  (a) 17 $\mu$F   (b) 9.0 V   (c) 108 $\mu$C and 45 $\mu$C

---

* In some situations the parallel combination may be in a circuit with other circuit elements so that the potential difference across the combination is not that of a battery in the circuit, but must be determined by analyzing the entire circuit.

### Series Combination

Now consider two capacitors connected in **series,** as illustrated in Figure 20.24a. Figure 20.24b shows the circuit diagram. For this series combination of capacitors, **the magnitude of the charge is the same on all the plates.**

To see why this is true, let us consider the charge transfer process in some detail. We start with uncharged capacitors and follow what happens just after a battery is connected to the circuit. When the connection is made, the right plate of $C_1$ and the left plate of $C_2$ form an isolated conductor. Thus, whatever negative charge enters one plate from the connecting wire must be equal to the positive charge of the other plate, to maintain neutrality of the isolated conductor—this is the electric charge version of the isolated system model. As a result, both capacitors must have the same charge $Q$.

Suppose we wish to determine the capacitance of an equivalent capacitor that has the same effect in the circuit as the series combination. Thus, as the equivalent capacitor is being charged, charge $-Q$ must enter its right plate from the wires and the charge on its left plate must be $+Q$. By applying the definition of capacitance to the circuit shown in Figure 20.24c, we have

$$\Delta V = \frac{Q}{C_{eq}}$$

where $\Delta V$ is the potential difference between the terminals of the battery and $C_{eq}$ is the equivalent capacitance.

Because the right plate of $C_1$ and the left plate of $C_2$ form an isolated conductor, both plates are at the same potential $V_i$, where the $i$ stands for the isolated conductor. The notation $V_{left}$ represents the potential of the left plate of $C_1$ and $V_{right}$ represents the potential of the right plate of $C_2$. Because these latter two

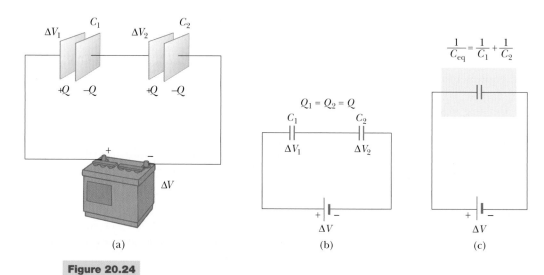

**Figure 20.24**

(a) A series combination of two capacitors connected to a battery. (b) The circuit diagram for the series combination. The charge on each capacitor is the same. (c) The equivalent capacitance can be calculated from the relationship

$$\frac{1}{C_{eq}} = \frac{1}{C_1} + \frac{1}{C_2}$$

plates are connected directly to the battery, the potential difference between them must be

$$\Delta V = V_{\text{left}} - V_{\text{right}}$$

If we add and subtract $V_i$ to this equation, we have

$$\Delta V = (V_{\text{left}} - V_i) + (V_i - V_{\text{right}})$$

which we can write as

$$\Delta V = \Delta V_1 + \Delta V_2 \qquad \text{[20.26]}$$

where $\Delta V_1$ and $\Delta V_2$ are the potential differences across capacitors $C_1$ and $C_2$. In general, the potential difference across any number of capacitors in series is equal to the sum of the potential differences across the individual capacitors. Because $Q = C\Delta V$ can be applied to each capacitor, the potential difference across each is

$$\Delta V_1 = \frac{Q}{C_1} \qquad \Delta V_2 = \frac{Q}{C_2}$$

Substituting these expressions into Equation 20.26, and recalling that $\Delta V = Q/C_{\text{eq}}$, we have

$$\frac{Q}{C_{\text{eq}}} = \frac{Q}{C_1} + \frac{Q}{C_2}$$

Canceling $Q$, we arrive at the relationship

$$\frac{1}{C_{\text{eq}}} = \frac{1}{C_1} + \frac{1}{C_2} \quad \text{(series combination)} \qquad \text{[20.27]}$$

If this analysis is applied to three or more capacitors connected in series, the equivalent capacitance is found to be

$$\frac{1}{C_{\text{eq}}} = \frac{1}{C_1} + \frac{1}{C_2} + \frac{1}{C_3} + \cdots \quad \text{(series combination)} \qquad \text{[20.28]}$$

• *Equivalent capacitance of several capacitors in series*

This shows that **the inverse of the equivalent capacitance is the algebraic sum of the inverses of the individual capacitances and the equivalent capacitance of a series combination is always less than any individual capacitance in the combination.**

## Example 20.8  Equivalent Capacitance

Find the equivalent capacitance between $a$ and $b$ for the combination of capacitors shown in Figure 20.25a. All capacitances are in microfarads.

**Solution**  Using Equations 20.25 and 20.28, we reduce the combination step by step as indicated in the figure. The 1.0-$\mu$F and 3.0-$\mu$F capacitors are in parallel and combine according to $C_{\text{eq}} = C_1 + C_2$. Their equivalent capacitance is

4.0 $\mu$F. Likewise, the 2.0-$\mu$F and 6.0-$\mu$F capacitors are also in parallel and have an equivalent capacitance of 8.0 $\mu$F. The upper branch in Figure 20.25b now consists of two 4.0-$\mu$F capacitors in series, which combine according to

$$\frac{1}{C_{\text{eq}}} = \frac{1}{C_1} + \frac{1}{C_2} = \frac{1}{4.0\ \mu\text{F}} + \frac{1}{4.0\ \mu\text{F}} = \frac{1}{2.0\ \mu\text{F}}$$

$$C_{\text{eq}} = 2.0\ \mu\text{F}$$

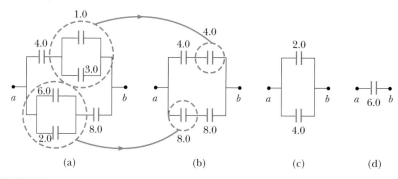

**Figure 20.25**

(Example 20.8) To find the equivalent combination of the capacitors in (a), the various combinations are reduced in steps as indicated in (b), (c), and (d), using the series and parallel rules described in the text.

Likewise, the lower branch in Figure 20.25b consists of two 8.0-$\mu$F capacitors in series, which give an equivalent of 4.0 $\mu$F. Finally, the 2.0-$\mu$F and 4.0 $\mu$F capacitors in Figure 20.25c are in parallel and have an equivalent capacitance of 6.0 $\mu$F. Hence, the equivalent capacitance of the circuit is 6.0 $\mu$F, as shown in Figure 20.25d.

**EXERCISE**   Consider three capacitors having capacitances of 3.0 $\mu$F, 6.0 $\mu$F, and 12 $\mu$F. Find their equivalent capacitance if they are connected (a) in parallel and (b) in series.

*Answer*   (a) 21 $\mu$F,   (b) 1.7 $\mu$F

## 20.9 • ENERGY STORED IN A CHARGED CAPACITOR

Almost everyone who works with electronic equipment has at some time verified that a capacitor can store energy. If the plates of a charged capacitor are connected by a conductor, such as a wire, charge transfers through the wire until the two plates are uncharged. The discharge can often be observed as a visible spark. If you accidentally touch the opposite plates of a charged capacitor, your fingers act as pathways by which the capacitor discharges, resulting in an electric shock. The degree of shock depends on the capacitance and the voltage applied to the capacitor. Where high voltages are present, such as in the power supply of a television set, the shock can be fatal.

Consider a parallel-plate capacitor that is initially uncharged, so that the initial potential difference across the plates is zero. Now imagine that the capacitor is connected to a battery and develops a maximum charge of $Q$. The final potential difference across the capacitor is $\Delta V = Q/C$.

To calculate the energy stored in the capacitor, imagine charging the capacitor in a different way that achieves the same result. An external agent reaches in and grabs small bits of charge and transfers them from one plate to the other. Suppose that $q$ is the charge on the capacitor at some instant during this charging process. At the same instant, the potential difference across the capacitor is $\Delta V = q/C$. Now imagine that the external agent transfers an additional increment of charge $dq$ from the plate of charge $-q$ to the plate of charge $q$ (which is at the higher potential) by applying a force on the charge $dq$ to move it through the electric field between the plates. The work required to transfer an increment of charge $dq$ from one plate to the other is

$$dW = \Delta V \, dq = \frac{q}{C} \, dq$$

Thus, the total work required to charge the capacitor from $q = 0$ to the final charge $q = Q$ is

$$W = \int_0^Q \frac{q}{C} \, dq = \frac{Q^2}{2C}$$

The capacitor can be modeled as a nonisolated system for this discussion. The work done by the external agent on the system in charging the capacitor appears as potential energy $U$ stored in the capacitor. In reality, of course, this energy is not the result of mechanical work done by an external agent moving charge from one plate to the other, but is due to transformation of chemical energy in the battery. We have used a model of work done by an external agent that gives us a result that is also valid for the actual situation. Using $Q = C\Delta V$, the energy stored in a charged capacitor can be expressed in the following alternative forms:

$$U = \frac{Q^2}{2C} = \tfrac{1}{2} Q \Delta V = \tfrac{1}{2} C(\Delta V)^2 \qquad \text{[20.29]}$$

• *Energy stored in a charged capacitor*

This result applies to *any* capacitor, regardless of its geometry. We see that the stored energy increases as $C$ increases and as the potential difference increases. In practice, the maximum energy (or charge) that can be stored is limited. This is because electric discharge ultimately occurs between the plates of the capacitor at a sufficiently large value of $\Delta V$. For this reason, capacitors are usually labeled with a maximum operating voltage.

For a mass on an extended spring, the elastic potential energy can be modeled as being stored *in the spring*. Internal energy of a substance associated with its temperature is located *throughout the substance*. Where is the energy in a capacitor located? The energy stored in a capacitor can be modeled as being stored *in the electric field between the plates of the capacitor.* For a parallel-plate capacitor, the potential difference is related to the electric field through the relationship $\Delta V = Ed$. Furthermore, the capacitance is $C = \epsilon_0 A/d$. Substituting these expressions into Equation 20.29 gives

$$U = \tfrac{1}{2} \left( \frac{\epsilon_0 A}{d} \right) (Ed)^2 = \tfrac{1}{2} (\epsilon_0 Ad) E^2 \qquad \text{[20.30]}$$

Because the volume of a parallel-plate capacitor that is occupied by the electric field is $Ad$, the energy per unit volume $u = U/Ad$, called the **energy density,** is

$$u = \tfrac{1}{2} \epsilon_0 E^2 \qquad \text{[20.31]}$$

• *Energy density in an electric field*

Although Equation 20.31 was derived for a parallel-plate capacitor, the expression is generally valid. That is, **the energy density in any electric field is proportional to the square of the magnitude of the electric field at a given point.**

**PITFALL PREVENTION 20.10**
**Not a new kind of energy**

The energy given by Equation 20.30 is not a new kind of energy. It is familiar electric potential energy associated with a system of separated charges. Equation 20.30 provides a new *interpretation*, or a new way of *modeling* the energy. We can think of the energy in the old way, as electric potential energy of separated charges, or this new way, as energy associated with the electric field. The value of the new way is that it allows us to assign energy to an electric field even if we have no idea what charge distribution establishes the field.

Suppose you have three capacitors and a battery. How should you connect the capacitors and the battery so that the capacitors will store the maximum possible energy?

### THINKING PHYSICS 20.3

You charge a capacitor and then remove it from the battery. The capacitor consists of large movable plates, with air between them. You pull the plates farther apart a small distance. What happens to the charge on the capacitor? To the potential difference? To the energy stored in the capacitor? To the capacitance? To the electric field between the plates? Is work done in pulling the plates apart?

**Reasoning**    Because the capacitor is removed from the battery, charges on the plates have nowhere to go. Thus, the charge on the capacitor remains the same as the plates are pulled apart. Because the electric field of large plates is independent of distance for uniform fields, the electric field remains constant. Because the electric field is a measure of the rate of change of potential with distance, the potential difference between the plates increases as the separation distance increases. Because the same charge is stored at a higher potential difference, the capacitance decreases. Because energy stored is proportional to both charge and potential difference, the energy stored in the capacitor increases. This energy must be transferred into the system from somewhere—the plates attract each other, so work is done by you on the system of two plates when you pull them apart.

### Example 20.9    Rewiring Two Charged Capacitors

Two capacitors with capacitances $C_1$ and $C_2$ (where $C_1 > C_2$) are charged to the same potential difference $\Delta V_i$. The charged capacitors are removed from the battery, and their plates are connected as shown in Figure 20.26a. The switches $S_1$ and $S_2$ are then closed as in Figure 20.26b. (a) Find the final potential difference $\Delta V_f$ between $a$ and $b$ after the switches are closed.

**Solution**    Let us identify the left-hand plates of the capacitors as an isolated system, because they are not connected to the right-hand plates by conductors. The charges on the left-hand plates before the switches are closed are

$$Q_{1i} = C_1 \, \Delta V_i \quad \text{and} \quad Q_{2i} = -C_2 \, \Delta V_i$$

The negative sign for $Q_{2i}$ is necessary because the charge on the left plate of capacitor $C_2$ is negative. The total charge $Q$ in the system is

$$Q = Q_{1i} + Q_{2i} = (C_1 - C_2)\Delta V_i \quad (1)$$

After the switches are closed, the electric charge version of the isolated system model tells us that the total charge on the left-hand plates remains the same:

$$Q = Q_{1f} + Q_{2f} \quad (2)$$

The charges will redistribute on the left-hand plates until the entire conductor in the system is at the same potential $V_{\text{left}}$. Similarly, charges will distribute on the right-hand plate until the entire conductor in the system is at the same potential

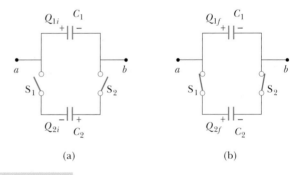

**Figure 20.26**

(Example 20.9)

$V_{\text{right}}$. Thus, the final potential difference $\Delta V_f = |V_{\text{left}} - V_{\text{right}}|$ across both capacitors is the same. To satisfy this requirement, the charges on the capacitors after the switches are closed are

$$Q_{1f} = C_1 \, \Delta V_f$$

$$Q_{2f} = C_2 \, \Delta V_f$$

Dividing these equations, we have,

$$\frac{Q_{1f}}{Q_{2f}} = \frac{C_1 \, \Delta V_f}{C_2 \, \Delta V_f} = \frac{C_1}{C_2} \quad \longrightarrow \quad Q_{1f} = \frac{C_1}{C_2} Q_{2f} \quad (3)$$

Combining (2) and (3),

$$Q = Q_{1f} + Q_{2f} = \frac{C_1}{C_2} Q_{2f} + Q_{2f} = Q_{2f}\left(1 + \frac{C_1}{C_2}\right)$$

$$Q_{2f} = Q\left(\frac{C_2}{C_1 + C_2}\right) \qquad \qquad (4)$$

Using (3) and (4) to find $Q_{1f}$, we have

$$Q_{1f} = \frac{C_1}{C_2} Q_{2f} = \frac{C_1}{C_2} Q\left(\frac{C_2}{C_1 + C_2}\right) = Q\left(\frac{C_1}{C_1 + C_2}\right)$$

Finally, we use Equation 20.19 to find the voltage across each capacitor:

$$\Delta V_{1f} = \frac{Q_{1f}}{C_1} = \frac{Q\left(\dfrac{C_1}{C_1 + C_2}\right)}{C_1} = \frac{Q}{C_1 + C_2}$$

$$\Delta V_{2f} = \frac{Q_{2f}}{C_2} = \frac{Q\left(\dfrac{C_2}{C_1 + C_2}\right)}{C_2} = \frac{Q}{C_1 + C_2}$$

Notice that $\Delta V_{1f} = \Delta V_{2f} = \Delta V_f$, which is the expected result.

(b) Find the total energy stored in the capacitors before and after the switches are closed and the ratio of the final energy to the initial energy.

**Solution** Before the switches are closed, the total energy stored in the capacitors is

$$U_i = \tfrac{1}{2}C_1(\Delta V_i)^2 + \tfrac{1}{2}C_2(\Delta V_i)^2 = \tfrac{1}{2}(C_1 + C_2)(\Delta V_i)^2$$

After the switches are closed, the total energy stored in the capacitors is

$$U_f = \tfrac{1}{2}C_1(\Delta V_f)^2 + \tfrac{1}{2}C_2(\Delta V_f)^2 = \tfrac{1}{2}(C_1 + C_2)(\Delta V_f)^2$$

$$= \tfrac{1}{2}(C_1 + C_2)\left(\frac{Q}{C_1 + C_2}\right)^2 = \tfrac{1}{2}\frac{Q^2}{C_1 + C_2}$$

Using (1), this can be expressed as

$$U_f = \tfrac{1}{2}\frac{Q^2}{C_1 + C_2} = \tfrac{1}{2}\frac{(C_1 - C_2)^2(\Delta V_i)^2}{C_1 + C_2}$$

Therefore, the ratio of the final energy stored to the initial energy stored is

$$\frac{U_f}{U_i} = \frac{\left(\dfrac{1}{2}\dfrac{(C_1 - C_2)^2(\Delta V_i)^2}{C_1 + C_2}\right)}{\left(\dfrac{1}{2}(C_1 + C_2)(\Delta V_i)^2\right)} = \left(\frac{C_1 - C_2}{C_1 + C_2}\right)^2$$

This shows that the final energy is less than the initial energy. Thus, even though we correctly used an isolated system model for electric charge in this problem, we see that this is *not* an isolated system for energy. This begs the question as to how energy is transferred out of the system. The transfer method is electromagnetic radiation, which may not be clear to you at this point, but will become clearer once we study the material in Chapter 24.

**EXERCISE** A 3.00-$\mu$F capacitor is connected to a 12-V battery. How much energy is stored in the capacitor?

*Answer* $2.2 \times 10^{-4}$ J

**EXERCISE** A parallel-plate capacitor is charged and then disconnected from a battery. By what fraction does the stored energy change (increase or decrease) when the plate separation is doubled?

*Answer* The stored energy doubles.

## 20.10 • CAPACITORS WITH DIELECTRICS

A **dielectric** is a nonconducting material such as rubber, glass, or waxed paper. When a dielectric material is inserted between the plates of a capacitor, the capacitance increases. If the dielectric completely fills the space between the plates, the capacitance increases by the dimensionless factor $\kappa$, called the **dielectric constant** of the material.

The following experiment can be performed to illustrate the effect of a dielectric in a capacitor. Consider a parallel-plate capacitor of charge $Q_0$ and capacitance $C_0$ in the absence of a dielectric. The potential difference across the capacitor as measured by a voltmeter is $\Delta V_0 = Q_0/C_0$ (Fig. 20.27a). Notice that the capacitor circuit is *open;* that is, the plates of the capacitor are *not* connected to a battery and charge cannot flow through an ideal voltmeter. Hence, there is *no* path by which charge can flow and alter the charge on the capacitor. If a dielectric

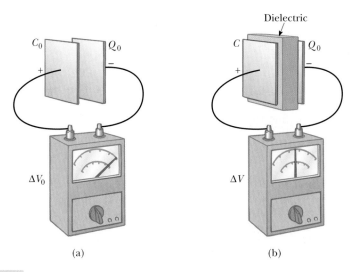

**Figure 20.27**

A charged capacitor (a) before and (b) after insertion of a dielectric between the plates. The charge on the plates remains unchanged, but the potential difference decreases from $\Delta V_0$ to $\Delta V = \Delta V_0/\kappa$. Thus, the capacitance *increases* from $C_0$ to $\kappa C_0$.

**PITFALL PREVENTION 20.11**

**Is the capacitor connected to a battery?**

Be careful with problems in which you are modifying a capacitor. It is important to note whether modifications to the capacitor are being made while the capacitor is connected to the battery or after it is disconnected. If the capacitor remains connected to the battery, the voltage across the capacitor necessarily remains the same (equal to the battery voltage), and the charge is proportional to the capacitance, *regardless of how it may be modified* (e.g., by insertion of a dielectric). On the other hand, if you disconnect the capacitor from the battery before making any modifications to the capacitor, the capacitor is an isolated system and its charge remains the same. In this case, as you vary the capacitance, the voltage across the plates changes in inverse proportion to capacitance, according to $\Delta V = Q/C$.

is now inserted between the plates as in Figure 20.27b, it is found that the voltmeter reading *decreases* by a factor of $\kappa$ to the value $\Delta V$, where

$$\Delta V = \frac{\Delta V_0}{\kappa}$$

Because $\Delta V < \Delta V_0$, we see that $\kappa > 1$.

Because the charge $Q_0$ on the capacitor *does not change*, we conclude that the capacitance must change to the value

$$C = \frac{Q_0}{\Delta V} = \frac{Q_0}{\Delta V_0/\kappa} = \kappa \frac{Q_0}{\Delta V_0}$$

$$= \kappa C_0 \qquad \textbf{[20.32]}$$

where $C_0$ is the capacitance in the absence of the dielectric. That is, the capacitance *increases* by the factor $\kappa$ when the dielectric completely fills the region between the plates.* For a parallel-plate capacitor, where $C_0 = \epsilon_0 A/d$, we can express the capacitance when the capacitor is filled with a dielectric as

$$C = \kappa \frac{\epsilon_0 A}{d} \qquad \textbf{[20.33]}$$

From this result, it would appear that the capacitance could be made very large by decreasing $d$, the distance between the plates. In practice, however, the lowest value of $d$ is limited by the electric discharge that could occur through the dielectric medium separating the plates. For any given separation $d$, the maximum

---

\* If another experiment is performed in which the dielectric is introduced while the potential difference is held constant by means of a battery, the charge increases to the value $Q = \kappa Q_0$. The additional charge is transferred from the connecting wires, and the capacitance still increases by the factor $\kappa$.

| TABLE 20.1 | Dielectric Constants and Dielectric Strengths of Various Materials at Room Temperature | |
|---|---|---|
| **Material** | **Dielectric Constant $\kappa$** | **Dielectric Strength[a] (V/m)** |
| Vacuum | 1.00000 | — |
| Air (dry) | 1.00059 | $3 \times 10^6$ |
| Bakelite | 4.9 | $24 \times 10^6$ |
| Fused quartz | 3.78 | $8 \times 10^6$ |
| Pyrex glass | 5.6 | $14 \times 10^6$ |
| Polystyrene | 2.56 | $24 \times 10^6$ |
| Teflon | 2.1 | $60 \times 10^6$ |
| Neoprene rubber | 6.7 | $12 \times 10^6$ |
| Nylon | 3.4 | $14 \times 10^6$ |
| Paper | 3.7 | $16 \times 10^6$ |
| Strontium titanate | 233 | $8 \times 10^6$ |
| Water | 80 | — |
| Silicone oil | 2.5 | $15 \times 10^6$ |

[a] The dielectric strength equals the maximum electric field that can exist in a dielectric without electrical breakdown.

voltage that can be applied to a capacitor without causing a discharge depends on the **dielectric strength** (maximum electric field) of the dielectric, which for dry air is equal to $3 \times 10^6$ V/m. If the electric field in the medium exceeds the dielectric strength, the insulating properties break down and the medium begins to conduct. Most insulating materials have dielectric strengths and dielectric constants greater than those of air, as Table 20.1 indicates. Thus, we see that a dielectric provides the following advantages:

- It increases the capacitance of a capacitor.
- It increases the maximum operating voltage of a capacitor.
- It may provide mechanical support between the conducting plates.

We can understand the effects of a dielectric by considering the polarization of molecules that we discussed in Section 19.3. Figure 20.28a shows polarized molecules of a dielectric in random orientations in the absence of an electric field. Figure 20.28b shows the polarization of the molecules when the dielectric is placed between the plates of the charged capacitor—the polarized molecules tend to line up parallel to the field lines. The plates set up an electric field $\mathbf{E}_0$ in a direction to the right in Figure 20.28b. In the body of the dielectric, a general homogeneity of charge exists, but look along the edges. There is a layer of negative charge along the left edge of the dielectric and a layer of positive charge along the right edge. These layers of charge can be modeled as additional charged parallel plates, as in Figure 20.28c. Because the polarity is opposite that of the real plates, these charges set up an induced electric field $\mathbf{E}_{ind}$ directed to the left in the diagram. This partially cancels the electric field due to the real plates. Thus, for the charged capacitor removed from a battery, the electric field and, hence, the voltage between the plates, is reduced by the introduction of the dielectric. The charge on the plates is stored at a lower potential difference, so the capacitance increases.

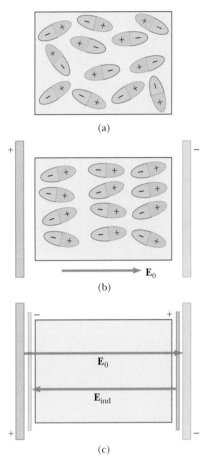

(a)

(b)

(c)

**Figure 20.28**

(a) Polar molecules are randomly oriented in the absence of an external electric field. (b) When an external electric field is applied, the molecules partially align with the field. (c) The charged edges of the dielectric can be modeled as an additional pair of parallel plates establishing an electric field $\mathbf{E}_{ind}$ in the direction opposite to that of $\mathbf{E}_0$.

## Types of Capacitors

Commercial capacitors are often made using metal foil interlaced with a dielectric such as thin sheets of paraffin-impregnated paper. These alternating layers of metal foil and dielectric are then rolled into the shape of a cylinder to form a small package (Fig. 20.29a). High-voltage capacitors commonly consist of interwoven metal plates immersed in silicone oil (Fig. 20.29b). Small capacitors are often constructed from ceramic materials. Variable capacitors (typically 10–500 pF) usually consist of two interwoven sets of metal plates, one fixed and the other movable, with air as the dielectric.

An *electrolytic capacitor* is often used to store large amounts of charge at relatively low voltages. This device, shown in Figure 20.29c, consists of a metal foil in contact with an electrolyte—a solution that conducts electricity by virtue of the motion of ions contained in the solution. When a voltage is applied between the foil and the electrolyte, a thin layer of metal oxide (an insulator) is formed on the foil, and this layer serves as the dielectric. Very large capacitance values can be attained because the dielectric layer is very thin.

When electrolytic capacitors are used in circuits, they must be installed with the proper polarity. If the polarity of the applied voltage is opposite what is intended, the oxide layer will be removed and the capacitor will not be able to store charge.

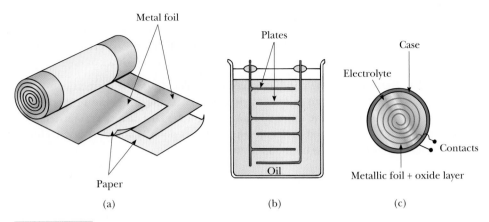

**Figure 20.29**

Three commercial capacitor designs. (a) A tubular capacitor whose plates are separated by paper and then rolled into a cylinder. (b) A high-voltage capacitor consists of many parallel plates separated by insulating oil. (c) An electrolytic capacitor.

## Quick Quiz 20.9

If you have ever tried to hang a picture, you know it can be difficult to locate a wooden stud in which to anchor your nail or screw. A carpenter's electric stud finder will locate the studs in a wall. It consists of a parallel-plate capacitor, with the plates next to each other, as shown in Figure 20.30. Does the capacitance increase or decrease when the device is moved over a stud?

## Quick Quiz 20.10

A fully charged parallel-plate capacitor remains connected to a battery while you slide a dielectric between the plates. Do the following quantities increase, decrease, or stay the same? (a) $C$; (b) $Q$; (c) $E$ between the plates; (d) $\Delta V$; (e) energy stored in the capacitor.

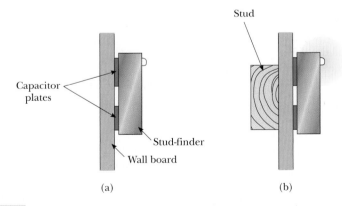

**Figure 20.30**

(Quick Quiz 20.9) An electric stud finder. (a) The materials between the plates of the capacitor are the wall board and air. (b) When the capacitor moves across a stud in the wall, the materials between the plates are wall board and wood. The change in the dielectric constant causes a signal light to illuminate.

**THINKING PHYSICS 20.4**

Consider a parallel-plate capacitor with a dielectric material between the plates. Is the capacitance higher on a cold day or a hot day?

**Reasoning**   The polarization of the molecules in the dielectric increases the capacitance when the dielectric is added. As the temperature increases, the polarized molecules have more vibrational motion. This disturbs the orderly arrangement of the polarized molecules, and the net polarization decreases. Thus, the capacitance decreases as the temperature increases.

---

### Example 20.10   A Paper-Filled Capacitor

A parallel-plate capacitor has plates of dimensions 2.0 cm $\times$ 3.0 cm separated by a 1.0-mm thickness of paper. (a) Find the capacitance of this device.

**Solution**   Because $\kappa = 3.7$ for paper (see Table 20.1), we have

$$C = \kappa \frac{\epsilon_0 A}{d}$$

$$= 3.7 \, (8.85 \times 10^{-12} \, \mathrm{C^2/N \cdot m^2}) \left( \frac{6.0 \times 10^{-4} \, \mathrm{m^2}}{1.0 \times 10^{-3} \, \mathrm{m}} \right)$$

$$= 20 \times 10^{-12} \, \mathrm{F} = \boxed{20 \ \mathrm{pF}}$$

(b) What is the maximum charge that can be placed on the capacitor?

**Solution**   From Table 20.1 we see that the dielectric strength of paper is $16 \times 10^6$ V/m. Because the thickness of the paper is 1.0 mm, the maximum voltage that can be applied before breakdown is

$$\Delta V_{max} = E_{max} d = (16 \times 10^6 \, \mathrm{V/m})(1.0 \times 10^{-3} \, \mathrm{m})$$

$$= 16 \times 10^3 \, \mathrm{V}$$

Hence, the maximum charge is

$$Q_{max} = C \Delta V_{max} = (20 \times 10^{-12} \, \mathrm{F})(16 \times 10^3 \, \mathrm{V}) = \boxed{0.32 \ \mu\mathrm{C}}$$

**EXERCISE**   What is the maximum energy that can be stored in the capacitor?

*Answer*   $2.5 \times 10^{-3}$ J

---

### Example 20.11   Energy Stored Before and After

A parallel-plate capacitor is charged with a battery to a charge $Q_0$, as in Figure 20.31a. The battery is then removed, and a slab of material that has a dielectric constant $\kappa$ is inserted between the plates, as in Figure 20.31b. Find the energy stored in the capacitor before and after the dielectric is inserted.

**Solution**   The energy stored in the capacitor in the absence of the dielectric is

$$U_0 = \tfrac{1}{2} C_0 (\Delta V_0)^2$$

Because $\Delta V_0 = Q_0/C_0$, this can be expressed as

$$U_0 = \frac{Q_0^2}{2C_0}$$

After the battery is removed and the dielectric is inserted between the plates, the charge on the capacitor remains the same, because the unconnected capacitor is an isolated sys-

tem. Hence, the energy stored in the presence of the dielectric is

$$U = \frac{Q_0^2}{2C}$$

But the capacitance in the presence of the dielectric is given by $C = \kappa C_0$, and so $U$ becomes

$$U = \frac{Q_0^2}{2\kappa C_0} = \frac{U_0}{\kappa}$$

Because $\kappa > 1$, we see that the final energy is less than the initial energy by the factor $1/\kappa$.

This missing energy can be accounted for as follows. We identify the system as the capacitor and the dielectric. As the dielectric is brought near the capacitor, so that electric field lines from the plates pass through the dielectric, the molecules of the dielectric become polarized. The edges of the dielectric take on a charge opposite to the plate nearest the

Dielectric

(b)

**Figure 20.31**

(Example 20.11)

Dielectric

**Figure 20.32**

(Example 20.11) When a dielectric approaches an empty capacitor, the charge distributions induced on the edges cause an attraction between the dielectric and the capacitor.

edge as in Figure 20.32. Thus, an attractive force exists between the dielectric and the plates. If the dielectric were released, it would be pulled into the plates and would pass through the plates with a kinetic energy, emerging on the other side, and exhibiting oscillatory motion. If an external agent such as your hand holds the dielectric, allowing it to move into the plates at constant speed, the agent is doing negative work on the system—the displacement of the dielectric is into the plates, but the applied force is away from the plates. This represents a decrease in energy for the system, so that the energy change is $\Delta U = U - U_0$. Thus, the unconnected capacitor and dielectric form an isolated system for electric charge, but when considering energy, it is a nonisolated system.

**EXERCISE** Suppose that the capacitance in the absence of a dielectric is 8.50 pF, and the capacitor is charged to a potential difference of 12.0 V. If the battery is disconnected, and a slab of polystyrene ($\kappa = 2.56$) is inserted between the plates, calculate the energy difference $U_0 - U$.

*Answer* 373 pJ

**EXERCISE** (a) How much charge can be placed on a capacitor with air between the plates before it breaks down, if the area of each of the plates is 5.00 cm$^2$? (b) Find the maximum charge if polystyrene is used between the plates instead of air.

*Answer* (a) 13.3 nC (b) 272 nC

**context connection**

## 20.11 • THE ATMOSPHERE AS A CAPACITOR

In the Context Connection of Chapter 19, we mentioned some processes occurring on the surface of the Earth and in the atmosphere that result in charge distributions. These result in a negative charge on the Earth's surface and positive charges distributed throughout the air.

This separation of charge can be modeled as a capacitor. The surface of the Earth is one plate and the positive charge in the air is the other plate. The positive charge in the atmosphere is not all located at one height but is spread throughout the atmosphere. Thus, the position of the upper plate must be modeled, based on the charge distribution. Models of the atmosphere show that an appropriate effective height of the upper plate is about 5 km from the surface. The model atmospheric capacitor is shown in Figure 20.33.

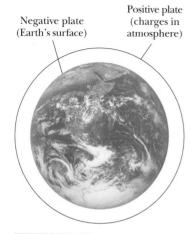

Negative plate (Earth's surface)    Positive plate (charges in atmosphere)

**Figure 20.33**

The atmospheric capacitor. The Earth serves as the negative plate, and the positive plate is modeled at a height in the atmosphere that represents positive charges spread through the atmosphere.

Considering the charge distribution on the surface of the Earth to be spherically symmetric, we can use the result from Example 20.6 to claim that the potential at a point above the Earth's surface is

$$V = k_e \frac{Q}{r}$$

where $Q$ is the charge on the surface. The potential difference between the plates of our atmospheric capacitor is

$$\Delta V = \frac{Q}{4\pi\epsilon_0}\left(\frac{1}{r_{\text{surface}}} - \frac{1}{r_{\text{upper plate}}}\right)$$

$$= \frac{Q}{4\pi\epsilon_0}\left(\frac{1}{R_E} - \frac{1}{R_E + h}\right) = \frac{Q}{4\pi\epsilon_0}\left(\frac{h}{R_E(R_E + h)}\right)$$

where $R_E$ is the radius of the Earth and $h = 5$ km. From this expression, we can calculate the capacitance of the atmospheric capacitor:

$$C = \frac{Q}{\Delta V} = \frac{Q}{\dfrac{Q}{4\pi\epsilon_0}\left[\dfrac{h}{R_E(R_E + h)}\right]} = \frac{4\pi\epsilon_0 R_E(R_E + h)}{h}$$

Substituting the numerical values, we have,

$$C = \frac{4\pi\epsilon_0 R_E(R_E + h)}{h}$$

$$= \frac{4\pi(8.85 \times 10^{-12}\,\text{C}^2/\text{N}\cdot\text{m}^2)(6.4 \times 10^3\,\text{km})(6.4 \times 10^3\,\text{km} + 5\,\text{km})}{5\,\text{km}}\left(\frac{1000\,\text{m}}{1\,\text{km}}\right)$$

$$\approx 0.9\,\text{F}$$

This is extremely large, compared with the *picofarads* and *microfarads* that are typical values for capacitors in electrical circuits, especially for a capacitor having plates that are 5 km apart! We shall use this model of the atmosphere as a capacitor in our Context Conclusion, in which we calculate the number of lightning strikes on the Earth in one day.

**EXERCISE**  Let us model the Earth and a cloud layer 800 m above the Earth as the "plates" of a capacitor. Calculate the capacitance if the cloud layer has an area of 1.0 km$^2$. If an electric field of magnitude $3.0 \times 10^6$ N/C makes the air break down and conduct electricity (i.e., causes lightning), what is the maximum charge the cloud can hold?

*Answer*  $C = 11$ nF; $Q = 27$ C

## SUMMARY

When a positive test charge $q_0$ is moved between points $A$ and $B$ in an electric field $\mathbf{E}$, the **change in potential energy** is

$$\Delta U = -q_0 \int_A^B \mathbf{E}\cdot d\mathbf{s} \qquad \textbf{[20.1]}$$

The **potential difference** $\Delta V$ between points $A$ and $B$ in an electric field $\mathbf{E}$ is defined as the change in potential energy divided by the test charge $q_0$:

$$\Delta V = \frac{\Delta U}{q_0} = -\int_A^B \mathbf{E}\cdot d\mathbf{s} \qquad \textbf{[20.3]}$$

where electric potential $V$ is a scalar and has the units joules per coulomb, defined as 1 **volt** (V).

The potential difference between two points $A$ and $B$ in a uniform electric field $\mathbf{E}$ is

$$\Delta V = -\mathbf{E}\cdot\Delta\mathbf{r} \qquad \textbf{[20.8]}$$

where $\Delta\mathbf{r}$ is the displacement vector between $A$ and $B$.

**Equipotential surfaces** are surfaces on which the electric potential remains constant. Equipotential surfaces are *perpendicular* to electric field lines.

The electric potential due to a point charge $q$ at a distance $r$ from the charge is

$$V = k_e \frac{q}{r} \qquad [20.11]$$

The electric potential due to a group of point charges is obtained by summing the potentials due to the individual charges. Because $V$ is a scalar, the sum is a simple algebraic operation.

The **electric potential energy of a pair of point charges** separated by a distance $r_{12}$ is

$$U = k_e \frac{q_1 q_2}{r_{12}} \qquad [20.13]$$

This represents the work required to bring the charges from an infinite separation to the separation $r_{12}$. The potential energy of a distribution of point charges is obtained by summing terms like Equation 20.13 over *all pairs* of particles.

If the electric potential is known as a function of coordinates $x$, $y$, $z$, the components of the electric field can be obtained by taking the negative derivative of the potential with respect to the coordinates. For example, the $x$ component of an electric field in the $x$ direction is

$$E_x = -\frac{dV}{dx} \qquad [20.15]$$

The **electric potential due to a continuous charge distribution** is

$$V = k_e \int \frac{dq}{r} \qquad [20.18]$$

Every point on the surface of a charged conductor in electrostatic equilibrium is at the same potential. Furthermore, the potential is constant everywhere inside the conductor and equal to its value at the surface.

A capacitor is a device for storing charge. It consists of two equal and oppositely charged conductors with a potential difference $\Delta V$ between them. The **capacitance** $C$ of any capacitor is defined as the ratio of the magnitude of the charge $Q$ on either conductor to the magnitude of the potential difference $\Delta V$:

$$C \equiv \frac{Q}{\Delta V} \qquad [20.19]$$

The SI units of capacitance are coulombs per volt, or the **farad** (F), and 1 F = 1 C/V.

If two or more capacitors are connected in parallel, the potential differences across them must be the same. The equivalent capacitance of a parallel combination of capacitors is

$$C_{eq} = C_1 + C_2 + C_3 + \cdots \qquad [20.25]$$

If two or more capacitors are connected in series, the charges on them are the same, and the equivalent capacitance of the series combination is given by

$$\frac{1}{C_{eq}} = \frac{1}{C_1} + \frac{1}{C_2} + \frac{1}{C_3} + \cdots \qquad [20.28]$$

Energy is required to charge a capacitor, because the charging process is equivalent to transferring charges from one conductor at a lower potential to another conductor at a higher potential. The electric potential energy $U$ stored in the capacitor is

$$U = \frac{Q^2}{2C} = \tfrac{1}{2} Q \Delta V = \tfrac{1}{2} C (\Delta V)^2 \qquad [20.29]$$

When a dielectric material is inserted between the plates of a capacitor, the capacitance generally increases by the dimensionless factor $\kappa$, called the **dielectric constant.** That is,

$$C = \kappa C_0 \qquad [20.32]$$

where $C_0$ is the capacitance in the absence of the dielectric.

# QUESTIONS

1. Distinguish between electric potential and electric potential energy.
2. Give a physical explanation of the fact that the potential energy of a pair of like charges is positive, whereas the potential energy of a pair of unlike charges is negative.
3. Explain why, under static conditions, all points in a conductor must be at the same electric potential.
4. Why is it important to avoid sharp edges or points on conductors used in high-voltage equipment?
5. In what type of weather would a car battery be more likely to discharge and why?
6. How would you shield an electronic circuit or laboratory from stray electric fields? Why does this work?
7. If you are given three capacitors with different capacitances $C_1$, $C_2$, and $C_3$, how many different combinations of capacitance can you produce?
8. The plates of a capacitor are connected to a battery. What happens to the charge on the plates if the connecting wires are removed from the battery? What happens to the charge if the wires are removed from the battery and connected to each other?
9. You want to increase the maximum operating voltage of a parallel-plate capacitor. Describe how you can do this for a fixed plate separation.
10. If the potential difference across a capacitor is doubled, by what factor does the energy stored change?

**11.** Because the charges on the plates of a parallel-plate capacitor are of opposite sign, they attract each other. Hence, it takes positive work to increase the plate separation. What happens to the external work done in this process? (Assume the capacitor plates are removed from the charging battery.)

**12.** You are asked to design a capacitor with a small size and large capacitance. What factors would be important in your design?

**13.** Explain why a dielectric increases the maximum operating voltage of a capacitor although the physical size of the capacitor does not change.

**14.** A pair of capacitors are connected in parallel, and an identical pair are connected in series. Which pair would be more dangerous to handle after being connected to the same voltage source? Explain.

**15.** The energy stored in a particular capacitor is increased fourfold. What is the accompanying change in (a) the charge and (b) the potential difference across the capacitor?

# PROBLEMS

1, 2, 3 = straightforward, intermediate, challenging     □ = full solution available in the *Student Solutions Manual and Study Guide*

**web** = solution posted at **http://www.harcourtcollege.com/physics/**     🖥 = computer useful in solving problem

🌐 = Interactive Physics     ▪ = paired numerical/symbolic problems     �kg = life science application

## Section 20.1   Potential Difference and Electric Potential

**1.** (a) Calculate the speed of a proton that is accelerated from rest through a potential difference of 120 V. (b) Calculate the speed of an electron that is accelerated through the same potential difference.

**2.** How much work is done (by a battery, generator, or some other source of energy) in moving Avogadro's number of electrons from an initial point where the electric potential is 9.00 V to a point where the potential is − 5.00 V? (The potential in each case is measured relative to a common reference point.)

**3. Review Problem.** Through what potential difference would an electron need to be accelerated for it to achieve a speed of 40.0% of the speed of light, starting from rest? The speed of light is $c = 3.00 \times 10^8$ m/s; review Chapter 9.

**4. Review Problem.** An electron starts from rest and is accelerated through a potential difference of 20.0 kV. (a) Find its actual relativistic speed. (b) Find the speed that classical physics would predict for it. By how much is this in error?

## Section 20.2   Potential Difference in a Uniform Electric Field

**5.** A uniform electric field of magnitude 250 V/m is directed in the positive *x* direction. A +12.0-$\mu$C charge moves from the origin to the point $(x, y) = (20.0$ cm, $50.0$ cm$)$. (a) What is the change in the potential energy of this charge–field system? (b) Through what potential difference does the charge move?

**6.** The difference in potential between the accelerating plates in the electron gun of a TV picture tube is about 25 000 V. If the distance between these plates is 1.50 cm, find the magnitude of the uniform electric field in this region.

**7.** An electron moving parallel to the *x* axis has an initial speed of $3.70 \times 10^6$ m/s at the origin. Its speed is reduced to $1.40 \times 10^5$ m/s at the point $x = 2.00$ cm. Calculate the potential difference between the origin and that point. Which point is at the higher potential?

**8.** Suppose an electron is released from rest in a uniform electric field whose magnitude is $5.90 \times 10^3$ V/m. (a) Through what potential difference will it have passed after moving 1.00 cm? (b) How fast will the electron be moving after it has traveled 1.00 cm?

**9. Review Problem.** A block having mass *m* and charge *Q* is connected to a spring having constant *k*. The block lies on a frictionless horizontal track, and the system is immersed in a uniform electric field of magnitude *E*, directed as shown in Figure P20.9. If the block is released from rest when the spring is unstretched (at $x = 0$), (a) by what maximum amount does the spring expand? (b) What is

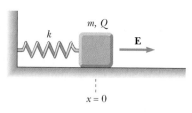

**Figure P20.9**

the equilibrium position of the block? (c) Show that the block's motion is simple harmonic, and determine its period. (d) Repeat part (a) if the coefficient of kinetic friction between block and surface is $\mu_k$.

## Section 20.3 Electric Potential and Potential Energy Due to Point Charges

*Note:* Unless stated otherwise, assume a reference level of potential $V = 0$ at $r = \infty$.

10. (a) Find the potential at a distance of 1.00 cm from a proton. (b) What is the potential difference between two points that are 1.00 cm and 2.00 cm from a proton? (c) Repeat parts (a) and (b) for an electron.

11. Given two 2.00-$\mu$C charges, as in Figure P20.11, and a positive test charge $q = 1.28 \times 10^{-18}$ C at the origin, (a) what is the net force exerted on $q$ by the two 2.00-$\mu$C charges? (b) What is the electric field at the origin due to the two 2.00-$\mu$C charges? (c) What is the electrical potential at the origin due to the two 2.00-$\mu$C charges?

**Figure P20.11**

12. A charge $+q$ is at the origin. A charge $-2q$ is at $x = 2.00$ m on the $x$ axis. For what finite value(s) of $x$ is (a) the electric field zero? (b) the electric potential zero?

13. The three charges in Figure P20.13 are at the vertices of an isosceles triangle. Calculate the electric potential at the midpoint of the base, taking $q = 7.00$ $\mu$C.

**Figure P20.13**

14. The Bohr model of the hydrogen atom states that the single electron can exist only in certain allowed orbits around the proton. The radius of each Bohr orbit is given by $r = n^2(0.052\ 9$ nm) where $n = 1, 2, 3, \ldots$ . Calculate the electric potential energy of a hydrogen atom when the electron is in the (a) first allowed orbit, with $n = 1$; (b) second allowed orbit, $n = 2$; and (c) when the electron has escaped from the atom, with $r = \infty$. Express your answers in electron volts.

15. *Compare this problem with Problem 53 in Chapter 19.* Four identical point charges ($q = +10.0$ $\mu$C) are located on the corners of a rectangle, as shown in Figure P19.53. The dimensions of the rectangle are $L = 60.0$ cm and $W = 15.0$ cm. Calculate the electric potential energy of the charge at the lower left corner due to the other three charges.

16. *Compare this problem with Problem 12 in Chapter 19.* Two point charges each of magnitude 2.00 $\mu$C are located on the $x$ axis. One is at $x = 1.00$ m, and the other is at $x = -1.00$ m. (a) Determine the electric potential on the $y$ axis at $y = 0.500$ m. (b) Calculate the change in electric potential energy of the system of these two particles and a third particle if the third particle, of charge $-3.00$ $\mu$C, is brought from infinitely far away and placed on the $y$ axis at $y = 0.500$ m.

17. Show that the amount of work required to assemble four identical point charges of magnitude $Q$ at the corners of a square of side $s$ is $5.41 k_e Q^2/s$.

18. *Compare this problem with Problem 11 in Chapter 19.* Three equal positive charges $q$ are at the corners of an equilateral triangle of side $a$ as shown in Figure P19.11. (a) At what point, if any, in the plane of the charges is the electric potential zero? (b) What is the electric potential at the point $P$ due to the two charges at the base of the triangle?

19. **Review Problem.** Two insulating spheres have radii 0.300 cm and 0.500 cm, masses 0.100 kg and 0.700 kg, and uniformly distributed charges $-2.00$ $\mu$C and 3.00 $\mu$C. They are released from rest when their centers are separated by 1.00 m. (a) How fast will each be moving when they collide? (*Hint:* Consider conservation of energy and of linear momentum.) (b) If the spheres were conductors, would the speeds be greater or less than those calculated in part (a)? Explain.

20. **Review Problem.** Two insulating spheres have radii $r_1$ and $r_2$, masses $m_1$ and $m_2$, and uniformly distributed charges $-q_1$ and $q_2$. They are released from rest when their centers are separated by a distance $d$. (a) How fast is each moving when they collide? (*Hint:* Consider conservation of energy and conservation of linear momentum.) (b) If the spheres were conductors, would the speeds be greater or less than those calculated in part (a)? Explain.

21. In 1911 Ernest Rutherford and his assistants Geiger and Marsden conducted an experiment in which they scattered alpha particles from thin sheets of gold. An alpha particle, having charge $+2e$ and mass $6.64 \times 10^{-27}$ kg, is a product of certain radioactive decays. The results of the experiment led Rutherford to the idea that most of the mass of an atom is in a very small nucleus, with electrons in orbit around it—his planetary model of the atom. Assume an alpha particle, initially very far from a gold nucleus, is fired with a velocity of $2.00 \times 10^7$ m/s directly toward the nucleus (charge $+79e$). How close does the alpha particle come to the nucleus before turning around? Assume the gold nucleus remains stationary.

22. Two particles, with charges of 20.0 nC and $-20.0$ nC, are placed at the points with coordinates (0, 4.00 cm) and (0, $-4.00$ cm), as shown in Figure P20.22. A particle with charge 10.0 nC is located at the origin. (a) Find the energy of the configuration of the three fixed charges. (b) A fourth particle, with a mass of $2.00 \times 10^{-13}$ kg and a charge of 40.0 nC, is released from rest at the point (3.00 cm, 0). Find its speed after it has moved freely to a very large distance away.

20.0 nC

4.00 cm

10.0 nC    3.00 cm    40.0 nC

4.00 cm

−20.0 nC

**Figure P20.22**

## Section 20.4 Obtaining Electric Field from Electric Potential

23. The potential in a region between $x = 0$ and $x = 6.00$ m is $V = a + bx$, where $a = 10.0$ V and $b = -7.00$ V/m. Determine (a) the potential at $x = 0$, 3.00 m, and 6.00 m, and (b) the magnitude and direction of the electric field at $x = 0$, 3.00 m, and 6.00 m.

24. The electric potential inside a charged spherical conductor of radius $R$ is given by $V = k_e Q/R$ and outside the conductor is given by $V = k_e Q/r$. Using $E_r = -dV/dr$, derive the electric field (a) inside and (b) outside this charge distribution.

25. Over a certain region of space, the electric potential is given by $V = 5x - 3x^2y + 2yz^2$. Find the expressions for the $x$, $y$, and $z$ components of the electric field over this region. What is the magnitude of the field at the point $P$, which has coordinates (1, 0, $-2$) m?

## Section 20.5 Electric Potential Due to Continuous Charge Distributions

26. Consider a ring of radius $R$ with the total charge $Q$ spread uniformly over its perimeter. What is the potential difference between the point at the center of the ring and a point on its axis a distance $2R$ from the center?

27. A rod of length $L$ (Fig. P20.27) lies along the $x$ axis with its left end at the origin. It has a nonuniform charge density $\lambda = \alpha x$, where $\alpha$ is a positive constant. (a) What are the units of $\alpha$? (b) Calculate the electric potential at $A$.

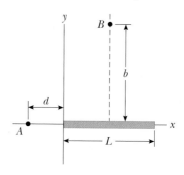

**Figure P20.27** Problems 27 and 28.

28. For the arrangement described in the previous problem, calculate the electric potential at point $B$ that lies on the perpendicular bisector of the rod a distance $b$ above the $x$ axis.

29. *Compare this problem with Problem 17 in Chapter 19.* A uniformly charged insulating rod of length 14.0 cm is bent into the shape of a semicircle as in Figure P19.17. If the rod has a total charge of $-7.50$ μC, find the electric potential at $O$, the center of the semicircle.

## Section 25.6 Electric Potential of a Charged Conductor

30. How many electrons should be removed from an initially uncharged spherical conductor of radius 0.300 m to produce a potential of 7.50 kV at the surface?

31. A spherical conductor has a radius of 14.0 cm and charge of 26.0 μC. Calculate the electric field and the electric potential at (a) $r = 10.0$ cm, (b) $r = 20.0$ cm, and (c) $r = 14.0$ cm from the center.

32. Electric charge can accumulate on an airplane in flight. You may have observed needle-shaped metal extensions on the wing tips and tail of an airplane. Their purpose is

to allow charge to leak off before much of it accumulates. The electric field around the needle is much larger than around the body of the airplane and can become large enough to produce dielectric breakdown of the air, discharging the airplane. To model this process, assume that two charged spherical conductors are connected by a long conducting wire, and a charge of 1.20 $\mu$C is placed on the combination. One sphere, representing the body of the airplane, has a radius of 6.00 cm and the other, representing the tip of the needle, has a radius of 2.00 cm. (a) What is the electric potential of each sphere? (b) What is the electric field at the surface of each sphere?

## Section 20.7  Capacitance

33. (a) How much charge is on each plate of a 4.00-$\mu$F capacitor when it is connected to a 12.0-V battery? (b) If this same capacitor is connected to a 1.50-V battery, what charge is stored?

34. Two conductors having net charges of $+10.0$ $\mu$C and $-10.0$ $\mu$C have a potential difference of 10.0 V. Determine (a) the capacitance of the system and (b) the potential difference between the two conductors if the charges on them are increased to $+100$ $\mu$C and $-100$ $\mu$C.

35. An air-filled capacitor consists of two parallel plates, each with an area of 7.60 cm$^2$, separated by a distance of 1.80 mm. If a 20.0-V potential difference is applied to these plates, calculate (a) the electric field between the plates, (b) the surface charge density, (c) the capacitance, and (d) the charge on each plate.

36. An isolated charged conducting sphere of radius 12.0 cm creates an electric field of $4.90 \times 10^4$ N/C at a distance 21.0 cm from its center. (a) What is its surface charge density? (b) What is its capacitance?

37. A variable air capacitor used in a radio tuning circuit is made of $N$ semicircular plates each of radius $R$ and positioned a distance $d$ from its neighbors, to which it is electrically connected. As shown in Figure P20.37, a second identical set of plates is enmeshed with its plates halfway between those of the first set. The second set can rotate as

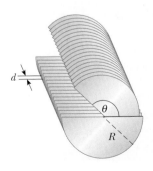

**Figure P20.37**

a unit. Determine the capacitance as a function of the angle of rotation $\theta$, where $\theta = 0$ corresponds to the maximum capacitance.

38. A 50.0-m length of coaxial cable has an inner conductor with a diameter of 2.58 mm and a charge of 8.10 $\mu$C. The surrounding conductor has an inner diameter of 7.27 mm and a charge of $-8.10$ $\mu$C. (a) What is the capacitance of this cable? (b) What is the potential difference between the two conductors? Assume the region between the conductors is air.

39. A small object of mass $m$ carries a charge $q$ and is suspended by a thread between the vertical plates of a parallel-plate capacitor. The plate separation is $d$. If the thread makes an angle $\theta$ with the vertical, what is the potential difference between the plates?

40. A *spherical capacitor* consists of a spherical conducting shell of radius $b$ and charge $-Q$ that is concentric with a smaller conducting sphere of radius $a$ and charge $+Q$ (Fig. P20.40). (a) Show that its capacitance is

$$C = \frac{ab}{k_e(b-a)}$$

(b) Show that as $b$ approaches infinity, the capacitance approaches the value $a/k_e = 4\pi\epsilon_0 a$.

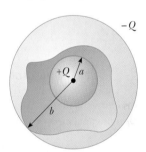

**Figure P20.40**

## Section 20.8  Combinations of Capacitors

41. Two capacitors $C_1 = 5.00$ $\mu$F and $C_2 = 12.0$ $\mu$F are connected in parallel, and the resulting combination is connected to a 9.00-V battery. (a) What is the equivalent capacitance of the combination? What are (b) the potential difference across each capacitor and (c) the charge stored on each capacitor?

42. The two capacitors of Problem 41 are now connected in series and to a 9.00-V battery. Find (a) the equivalent capacitance of the combination, (b) the voltage across each capacitor, and (c) the charge on each capacitor.

43. Four capacitors are connected as shown in Figure P20.43.
web (a) Find the equivalent capacitance between points $a$ and $b$. (b) Calculate the charge on each capacitor if $\Delta V_{ab} = 15.0$ V.

**Figure P20.43**

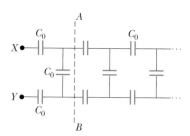

**Figure P20.47**

**44.** Two capacitors when connected in parallel give an equivalent capacitance of $C_p$ and an equivalent capacitance of $C_s$ when connected in series. What is the capacitance of each capacitor?

**45.** Consider the circuit shown in Figure P20.45, where $C_1 = 6.00\ \mu F$, $C_2 = 3.00\ \mu F$, and $\Delta V = 20.0\ V$. Capacitor $C_1$ is first charged by the closing of switch $S_1$. Switch $S_1$ is then opened, and the charged capacitor is connected to the uncharged capacitor by closing $S_2$. Calculate the initial charge on $C_1$ and the final charge on each capacitor.

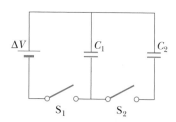

**Figure P20.45**

**46.** Find the equivalent capacitance between points $a$ and $b$ in the combination of capacitors shown in Figure P20.46.

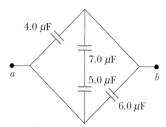

**Figure P20.46**

**47.** Some physical systems possessing capacitance continuously distributed over space can be modeled as an infinite array of discrete circuit elements. Examples are a microwave waveguide and the axon of a nerve cell. To practice analysis of an infinite array, determine the equivalent capacitance $C$ between terminals $X$ and $Y$ of the infinite set of capacitors represented in Figure P20.47. Each capacitor has capacitance $C_0$. (*Hint:* Imagine that the ladder is cut at the line $AB$, and note that the equivalent capacitance of the infinite section to the right of $AB$ is also $C$.)

## Section 20.9 Energy Stored in a Charged Capacitor

**48.** (a) A 3.00-$\mu F$ capacitor is connected to a 12.0-V battery. How much energy is stored in the capacitor? (b) If the capacitor had been connected to a 6.00-V battery, how much energy would have been stored?

**49.** The immediate cause of many deaths is ventricular fibrillation, uncoordinated quivering of the heart as opposed to proper beating. An electric shock to the chest can cause momentary paralysis of the heart muscle, after which the heart will sometimes start organized beating again. A *defibrillator* (Fig. P20.49) is a device that applies a strong electric shock to the chest over a few milliseconds. The device contains a capacitor of several microfarads, charged to several thousand volts. Electrodes called paddles, about 8 cm across and coated with conducting paste, are held against the chest on both sides of the heart. Their handles are insulated to prevent injury to the operator, who calls, "Clear!" (meaning that no one should touch the patient) and pushes a button on one paddle to discharge the ca-

**Figure P20.49** (*Adam Hart-Davis/Science Photo Library/Photo Researchers, Inc.*)

pacitor through the patient's chest. Assume that an energy of 300 J is to be delivered from a 30.0-$\mu$F capacitor. To what potential difference must it be charged?

50. Two capacitors, $C_1 = 25.0$ $\mu$F and $C_2 = 5.00$ $\mu$F, are connected in parallel and charged with a 100-V power supply. (a) Draw a circuit diagram, and calculate the total energy stored in the two capacitors. (b) What potential difference is required across the same two capacitors connected in series so that the combination stores the same energy as in part (a)? Draw a circuit diagram of this circuit.

51. A parallel-plate capacitor has a charge $Q$ and plates of area $A$. What force acts on one plate to attract it toward the other plate? Because the electric field between the plates is $E = Q/A\epsilon_0$, you might think that the force is $F = QE = Q^2/A\epsilon_0$. This is wrong, because the field $E$ includes contributions from both plates, and the field created by the positive plate cannot exert any force on the positive plate. Show that the force exerted on each plate is actually $F = Q^2/2\epsilon_0 A$. (*Hint:* Let $C = \epsilon_0 A/x$ for an arbitrary plate separation $x$; then require that the work done in separating the two charged plates be $W = \int F\,dx$.) The force exerted on one charged plate by another is sometimes used in a machine shop to hold a workpiece stationary.

52. As a person moves about in a dry environment, electric charge accumulates on his body. Once it is at high voltage, either positive or negative, the body can discharge via sometimes noticeable sparks and shocks. Consider a human body well separated from ground, with the typical capacitance 150 pF. (a) What charge on the body will produce a potential of 10.0 kV? (b) Sensitive electronic devices can be destroyed by electrostatic discharge from a person. A particular device can be destroyed by a discharge releasing an energy of 250 $\mu$J. To what voltage on the body does this correspond?

53. A uniform electric field of magnitude $E = 3\,000$ V/m exists within a certain region. What volume of space contains an energy equal to $1.00 \times 10^{-7}$ J? Express your answer in cubic meters and in liters.

## Section 20.10 Capacitors with Dielectrics

54. Determine (a) the capacitance and (b) the maximum potential difference that can be applied to a Teflon-filled parallel-plate capacitor having a plate area of 1.75 cm$^2$ and plate separation of 0.040 0 mm.

55. (a) How much charge can be placed on a capacitor with air between the plates before it breaks down if the area of each of the plates is 5.00 cm$^2$? (b) Find the maximum charge if polystyrene is used between the plates instead of air.

56. A commercial capacitor is constructed as shown in Figure 20.29a. This particular capacitor is rolled from two strips of aluminum separated by two strips of paraffin-coated pa-

per. Each strip of foil and paper is 7.00 cm wide. The foil is 0.004 00 mm thick, and the paper is 0.025 0 mm thick and has a dielectric constant of 3.70. What length should the strips be if a capacitance of $9.50 \times 10^{-8}$ F is desired? (Model the capacitor as parallel plates.)

57. A parallel-plate capacitor is constructed using a dielectric material whose dielectric constant is 3.00 and whose dielectric strength is $2.00 \times 10^8$ V/m. The desired capacitance is 0.250 $\mu$F, and the capacitor must withstand a maximum potential difference of 4 000 V. Find the minimum area of the capacitor plates.

58. A parallel-plate capacitor in air has a plate separation of 1.50 cm and a plate area of 25.0 cm$^2$. The plates are charged to a potential difference of 250 V and disconnected from the source. The capacitor is then immersed in distilled water. Determine (a) the charge on the plates before and after immersion, (b) the capacitance and potential difference after immersion, and (c) the change in energy of the capacitor. Assume the liquid is an insulator.

## Section 20.11 Context Connection — The Atmosphere as a Capacitor

59. Lightning can be studied with a Van de Graaff generator, essentially consisting of a spherical dome on which charge is continuously deposited by a moving belt. Charge can be added until the electric field at the surface of the dome becomes equal to the dielectric strength of air. Any more charge leaks off in sparks. Assume the dome has a diameter of 30.0 cm and is surrounded by dry air. (a) What is the maximum potential of the dome? (b) What is the maximum charge on the dome?

60. **Review Problem.** A certain storm cloud has a potential difference of $1.00 \times 10^8$ V relative to a tree. If, during a lightning storm, 50.0 C of charge is transferred through this potential difference and 1.00% of the energy is absorbed by the tree, how much sap in the tree can be boiled away? Model the sap as water initially at 30.0°C. Water has a specific heat of 4 186 J/kg·°C, a boiling point of 100°C, and a heat of vaporization of $2.26 \times 10^6$ J/kg.

## Additional Problems

61. The $x$ axis is the symmetry axis of a uniformly charged ring of radius $R$ and charge $Q$ (Fig. P20.61). A point charge $Q$ of mass $M$ is located at the center of the ring. When it is displaced slightly, the point charge accelerates along the $x$ axis to infinity. Show that the ultimate speed of the point charge is

$$ v = \left( \frac{2k_eQ^2}{MR} \right)^{1/2} $$

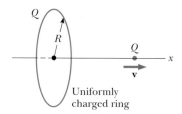

**Figure P20.61**

62. On a dry winter day you scuff your leather-soled shoes across a carpet and feel a shock when you extend the tip of one finger toward a metal doorknob. In a dark room you see a spark perhaps 5 mm long. Make order-of-magnitude estimates of (a) your electric potential and (b) the charge on your body before you touch the doorknob. Explain your reasoning.

63. Calculate the work that must be done to charge a spherical shell of radius $R$ to a total charge $Q$.

64. A 10.0-$\mu$F capacitor has plates with vacuum between them. Each plate carries a charge of magnitude 1 000 $\mu$C. A particle with charge $-3.00$ $\mu$C and mass $2.00 \times 10^{-16}$ kg is fired from the positive plate toward the negative plate with an initial speed of $2.00 \times 10^6$ m/s. Does it reach the negative plate? If so, find its impact speed. If not, what fraction of the way across the capacitor does it travel?

65. A 2.00-nF parallel-plate capacitor is charged to an initial potential difference $\Delta V_i = 100$ V and then isolated. The dielectric material between the plates is mica, with a dielectric constant of 5.00. (a) How much work is required to withdraw the mica sheet? (b) What is the potential difference of the capacitor after the mica is withdrawn?

66. A capacitor is constructed from two square plates of sides $\ell$ and separation $d$. A material of dielectric constant $\kappa$ is inserted a distance $x$ into the capacitor, as shown in Figure P20.66. Assume that $d$ is much smaller than $x$. (a) Find the equivalent capacitance of the device. (b) Calculate the energy stored in the capacitor if the potential difference is $\Delta V$. (c) Find the direction and magnitude of the force exerted on the dielectric, assuming a constant potential difference $\Delta V$. Ignore friction. (d) Obtain a numerical value for the force assuming that $\ell = 5.00$ cm, $\Delta V = 2\,000$ V,

**Figure P20.66**  Problems 66 and 67.

$d = 2.00$ mm, and the dielectric is glass ($\kappa = 4.50$). (*Hint:* The system can be considered as two capacitors connected in parallel.)

67. A capacitor is constructed from two square plates of sides $\ell$ and separation $d$, as suggested in Figure P20.66. You may assume that $d$ is much less than $\ell$. The plates carry charges $+Q_0$ and $-Q_0$. A block of metal has width $\ell$, length $\ell$, and thickness slightly less than $d$. It is inserted a distance $x$ into the capacitor. The charges on the plates are not disturbed as the block slides in. In a static situation, a metal prevents an electric field from penetrating inside it. The metal can be thought of as a perfect dielectric, with $\kappa \to \infty$. (a) Calculate the stored energy as a function of $x$. (b) Find the direction and magnitude of the force that acts on the metallic block. (c) The area of the advancing front face of the block is essentially equal to $\ell d$. Considering the force on the block as acting on this face, find the stress (force per area) on it. (d) For comparison, express the energy density in the electric field between the capacitor plates in terms of $Q_0$, $\ell$, $d$, and $\epsilon_0$.

68. Determine the equivalent capacitance of the combination shown in Figure P20.68. (*Hint:* Consider the symmetry involved.)

**Figure P20.68**

69. When considering the energy supply for an automobile, the energy per unit mass of the energy source is an important parameter. Using the following data, compare the energy per unit mass (J/kg) for gasoline, lead–acid batteries, and capacitors. (The ampere A will be introduced in the next chapter as the SI unit of electric current. Note that 1 A = 1 C/s.)

> *Gasoline:*   126 000 Btu/gal; density = 670 kg/m$^3$.

> *Lead–acid battery:*   12.0 V; 100 A·h; mass = 16.0 kg.

> *Capacitor:*   potential difference at full charge = 12.0 V;
> capacitance = 0.100 F; mass = 0.100 kg.

70. A 10.0-$\mu$F capacitor is charged to 15.0 V. It is next connected in series with an uncharged 5.00-$\mu$F capacitor. The series combination is finally connected across a 50.0-V battery, as diagrammed in Figure P20.70. Find the new potential differences across the 5-$\mu$F and 10-$\mu$F capacitors.

5.00 μF   10.0 μF

$\Delta V_i = 15.0$ V

50.0 V

**Figure P20.70**

71. It is possible to obtain large potential differences by first charging a bank of capacitors connected in parallel and then activating a switch arrangement that in effect disconnects the capacitors from the charging source and from each other and reconnects them all in series. The group of charged capacitors is then discharged in series. What is the maximum potential difference that can be obtained in this manner by using ten capacitors each of 500 $\mu$F and a charging source of 800 V?

72. A Geiger–Mueller tube is a radiation detector that essentially consists of a closed, hollow metal cylinder (the cathode) of inner radius $r_a$ and a coaxial cylindrical wire (the anode) of radius $r_b$ (Fig. P20.72). The charge per unit length on the anode is $\lambda$, and the charge per unit length on the cathode is $-\lambda$. A gas fills the space between the electrodes. When a high-energy elementary particle passes through this space, it can ionize an atom of the gas. The strong electric field makes the resulting ion and electron accelerate in opposite directions. They strike other molecules of the gas to ionize them, producing an avalanche of electric discharge. The pulse of electric current between the wire and the cylinder is counted by an external circuit. (a) Show that the magnitude of the potential difference between the wire and the cylinder is

$$\Delta V = 2k_e\lambda \ln\left(\frac{r_a}{r_b}\right)$$

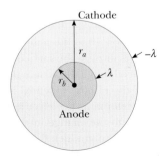

Cathode

$r_a$

$-\lambda$

$r_b$

$\lambda$

Anode

**Figure P20.72**

(b) Show that the magnitude of the electric field in the space between cathode and anode is given by

$$E = \frac{\Delta V}{\ln(r_a/r_b)}\left(\frac{1}{r}\right)$$

where $r$ is the distance from the axis of the anode to the point where the field is to be calculated.

73. Suppose that the internal diameter of the Geiger–Mueller tube described in Problem 72 is 2.50 cm and that the wire along the axis has a diameter of 0.200 mm. The dielectric strength of the gas between the central wire and the cylinder is $1.20 \times 10^6$ V/m. Use the results of Problem 72 to calculate the maximum voltage that can be applied between the wire and the cylinder before breakdown occurs in the gas.

74. The results of Problem 72 also apply to an electrostatic precipitator (Figure P20.74). This pollution-control device consists of a vertical cylindrical duct with a wire along its axis at a high negative voltage. Corona discharge ionizes the air around the wire to produce free electrons and positive and negative molecular ions. The electrons and negative ions accelerate outward. As air passes through the cylinder, the dirt particles become electrically charged by collisions and ion capture. Then they are swept out of the air by the horizontal electric field between the wire and the cylinder. In a particular case, an applied voltage $\Delta V = 50.0$ kV is to produce an electric field of magnitude 5.50 MV/m at the surface of the central wire. The outer cylindrical wall has uniform radius $r_a = 0.850$ m. (a) What should be the radius $r_b$ of the central wire? You will need to solve a transcendental equation. (b) What is the magnitude of the electric field at the outer wall?

Insulator

Clean air out

Dirty air in

Weight

Dirt out

**Figure P20.74**

**75.** The liquid-drop model of the atomic nucleus suggests that high-energy oscillations of certain nuclei can split the nucleus into two unequal fragments plus a few neutrons. The fission products acquire kinetic energy from their mutual Coulomb repulsion. Calculate the electric potential energy (in electron volts) of two spherical fragments from a uranium nucleus having the following charges and radii: $38e$ and $5.50 \times 10^{-15}$ m; $54e$ and $6.20 \times 10^{-15}$ m. Assume that the charge is distributed uniformly throughout the volume of each spherical fragment and that just before separating they are at rest with their surfaces in contact. The electrons surrounding the nucleus can be ignored.

**76.** An electric dipole is located along the $y$ axis as shown in Figure P20.76. The magnitude of its electric dipole moment is defined as $p = 2qa$. (a) At a point $P$, far from the dipole ($r \gg a$), show that the electric potential is

$$V \approx \frac{k_e p \cos \theta}{r^2}$$

(b) Calculate the radial component $E_r$ and the perpendicular component $E_\theta$ of the associated electric field. Note that $E_\theta = -(1/r)(\partial V/\partial \theta)$. Do these results seem reasonable for $\theta = 90°$ and $0°$? For $r = 0$? (c) For the dipole arrangement shown, express $V$ in terms of cartesian coordinates using $r = (x^2 + y^2)^{1/2}$ and the expression

$$\cos \theta = \frac{y}{(x^2 + y^2)^{1/2}}$$

Using these results and again taking $r \gg a$ calculate the field components $E_x$ and $E_y$.

**77.** A parallel-plate capacitor of plate separation $d$ is charged to a potential difference $\Delta V_0$. A dielectric slab of thickness $d$ and dielectric constant $\kappa$ is introduced between the plates *while the battery remains connected to the plates.* (a) Show that the ratio of energy stored after the dielectric is introduced to the energy stored in the empty capacitor is $U/U_0 = \kappa$. Give a physical explanation for this increase in stored energy. (b) What happens to the charge on the capacitor? (*Note:* This situation is not the same as Example 20.11, in which the battery was removed from the circuit before the dielectric was introduced.)

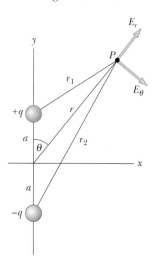

**Figure P20.76**

## ANSWERS TO QUICK QUIZZES

**20.1** In general, an electric field changes from one place to another, so that the proposed expression does not produce the correct result.

**20.2** The electric potential energy decreases if an electron (in fact, if any charged particle) is released in an electric field. The electric force causes the electron to accelerate, and potential energy of the charge–field system decreases as the kinetic energy of the electron increases. This is analogous to the decrease in potential energy and increase in kinetic energy of an object falling due to gravity.

**20.3** No. Suppose several charges are in the vicinity of the point in question. If some charges are positive and some negative, their contributions to the potential at the point may cancel. For example, the electric potential at the midpoint between two charges of equal magnitude but opposite sign is zero.

**20.4** (a), (b). The electric potential decreases in inverse proportion to the radius (Equation 20.11). The electric field magnitude decreases as the reciprocal of the radius squared (Equation 19.5). Because the same number of field lines passes through the surface regardless of its size, the electric flux through the surface remains constant.

**20.5** The value of the electric potential at one point is not sufficient to determine the electric field. The electric field is related to the change in the potential through space, so the nature in which the potential varies in the neighborhood of the point must be known.

**20.6**   If $V$ is constant in a certain region, the electric field in that region must be zero because the electric field is related to the change in the potential through space. (In one dimension, $E_x = -\,dV/dx$, so if $V$ = constant, $E$ = 0.) Likewise, if $E = 0$ in some region, one can only conclude that $V$ is a constant in that region (e.g., the interior of a charged conductor in equilibrium).

**20.7**   The capacitor often remains charged long after the voltage source is disconnected. This residual charge can be lethal. The capacitor can be safely handled after discharging the plates by short circuiting the device with a conductor, such as a screwdriver with an insulating handle.

**20.8**   The energy stored in a capacitor is proportional to the capacitance and the square of the potential difference. Thus, we would like to maximize the equivalent capacitance; the potential difference is that of the battery, over which we have no control. We can maximize the equivalent capacitance by connecting the three capacitors in *parallel* so that the capacitances add.

**20.9**   Increase. The electric field lines penetrate into the wall. The dielectric constant of wood (and all other insulating materials, for that matter) is greater than 1; therefore the capacitance increases (Equation 20.32). This is sensed by special circuitry that causes the light to illuminate.

**20.10**   (a) $C$ increases (Equation 20.32). (b) $Q$ increases. Because the battery maintains a constant $\Delta V$, $Q$ must increase if $C$ ($= Q/\Delta V$) increases. (c) $E$ between the plates remains constant because $\Delta V = Ed$, and neither $\Delta V$ nor $d$ change. The electric field due to the charges on the plates increases because more charge has flowed onto the plates. The induced surface charges on the dielectric create a field that opposes the increase in the field caused by the greater number of charges on the plates. (d) The battery maintains a constant $\Delta V$. (e) The energy stored in the capacitor increases (Equation 20.29). You would have to push the dielectric into the capacitor, just as you would have to do positive work to raise a mass and increase its gravitational potential energy.

These power lines transfer energy from the electrical power company to homes and businesses. The energy is transferred at a very high voltage, possibly hundreds of thousands of volts in some cases. Despite the fact that this makes power lines very dangerous, the high voltage results in less loss of power due to resistance in the wires. We will study both resistance and power in this chapter.

*(Telegraph Colour Library/FPG)*

# Current and Direct Current Circuits

**T**hus far, our discussion of electrical phenomena has focused on charges at rest, or the study of *electrostatics*. We shall now consider situations involving electric charges in motion. The term *electric current*, or simply *current*, is used to describe the flow of charge through some region of space. Most practical applications of electricity involve electric currents. For example, in a flashlight, charges flow through the filament of the lightbulb after the switch is turned on. In most common situations, the flow of charge takes place in a conductor, such as a copper wire. It is also possible, however, for currents to exist outside a conductor. For instance, a beam of electrons in a TV picture tube constitutes a current in which charge flows through a vacuum.

In the preceding chapter, we introduced the notion of a *circuit*. As we continue our investigations into circuits in this chapter, we introduce the *resistor* as a new circuit element.

See the *Core Concepts in Physics CD-ROM*, Screen 13.2

## 21.1 • ELECTRIC CURRENT

Whenever charge is flowing, a **current** is said to exist. To define current mathematically, suppose charged particles are moving perpendicular to a surface of area *A*, as in Figure 21.1. (This area could be the cross-sectional area of a wire, for example.) The current is defined as the **rate at which electric charge flows through this surface.** If $\Delta Q$ is the amount of charge that passes through this area in a time interval $\Delta t$, the average current $I_{av}$ over the time interval is the ratio of the charge to the time interval:

$$I_{av} = \frac{\Delta Q}{\Delta t} \qquad [21.1]$$

It is possible for the rate at which charge flows to vary in time. We define the **instantaneous current** *I* as the limit of the preceding expression as $\Delta t$ goes to zero:

$$I \equiv \lim_{\Delta t \to 0} \frac{\Delta Q}{\Delta t} = \frac{dQ}{dt} \qquad [21.2]$$

The SI unit of current is the **ampere** (A):

$$1 \text{ A} = 1 \text{ C/s} \qquad [21.3]$$

That is, 1 A of current is equivalent to 1 C of charge passing through a surface in 1 s.

The particles flowing through a surface as in Figure 21.1 can be charged positively or negatively, or we can have two or more types of particles moving, with charges of both signs in the flow. **Conventionally, we define the direction of the current as the direction of flow of positive charge,** regardless of the sign of the actual charged particles in motion.* In a common conductor such as copper, the current is physically due to the motion of the negatively charged electrons. Therefore, when we speak of current in such a conductor, **the direction of the current is opposite the direction of flow of electrons.** On the other hand, if one considers a beam of positively charged protons in a particle accelerator, the current is in the direction of motion of the protons. In some cases—gases and electrolytes, for example—the current is the result of the flow of both positive and negative charged particles. It is common to refer to a moving charged particle (whether it is positive or negative) as a mobile **charge carrier.** For example, the charge carriers in a metal are electrons.

We now build a structural model that will allow us to relate the macroscopic current to the motion of the charged particles. Consider identically charged particles moving in a conductor of cross-sectional area *A* (Fig. 21.2). The volume of an element of the conductor of length $\Delta x_e$ is $A \, \Delta x_e$, where the subscript *e* indicates that we are referring to the length of the element of the conductor. If *n* represents the number of mobile charge carriers per unit volume, then the number of carriers in the volume element is $nA \, \Delta x_e$. The mobile charge $\Delta Q$ in this element is therefore

$$\Delta Q = \text{number of carriers} \times \text{charge per carrier} = (nA \, \Delta x_e) q$$

---

* Even though we discuss a direction for current, current is not a vector. As we shall see later in the chapter, currents add algebraically and not vectorially.

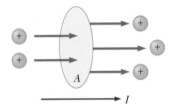

**Figure 21.1**

Charges in motion through an area *A*. The time rate at which charge flows through the area is defined as the current *I*. The direction of the current is the direction in which positive charges flow when free to do so.

**Figure 21.2**

A section of a uniform cylindrical conductor of cross-sectional area *A*. The charge carriers move with a speed $v_d$, and the distance they travel in a time $\Delta t$ is given by $\Delta x = v_d \, \Delta t$. The number of mobile charge carriers in the section of length $\Delta x_e$ is given by $nA \, \Delta x_e$, where *n* is the number of charge carriers per unit volume.

• *Current in terms of microscopic parameters*

where $q$ is the charge on each carrier. If the carriers move along the length of the conductor and across its cross-section with a constant average speed called the **drift speed** $v_d$, the distance they move in a time interval $\Delta t$ is $\Delta x_d = v_d \Delta t$.

Now imagine that $\Delta t$ is chosen such that during this time interval all of the charge carriers in the element of the conductor move to the right by a distance equal to the length of the element. In this case, $\Delta x_d = \Delta x_e$. In Figure 21.2, this process would be equivalent to moving the gray cylinder to the right by a distance equal to its own length. In doing so, every charge contained in the element of the conductor passes through the cross-sectional area marked $A$ in Figure 21.2. The amount of charge passing through this area is

$$\Delta Q = (nA \, \Delta x_e)q = (nA \, \Delta x_d)q = (nAv_d \, \Delta t)q$$

If we divide both sides of this equation by the time interval $\Delta t$ during which the flow of charge occurs, we see that the current in the conductor is*

$$I = \frac{\Delta Q}{\Delta t} = nqv_d A \qquad\qquad [21.4]$$

Equation 21.4 relates a macroscopically measured current $I$ to the microscopic origin of the current—the density of charge carriers $n$, the charge per carrier $q$, and the drift speed $v_d$.

### Quick Quiz 21.1

Consider positive and negative charges moving horizontally through the four regions shown in Figure 21.3. Rank the currents in these four regions, from lowest to highest.

Let us investigate further the notion of drift speed. We have identified this as an average speed along the wire, but the charge carriers are by no means moving in a straight line with speed $v_d$. Consider a conductor in which the charge carriers are free electrons. In the absence of a potential difference across the conductor, these electrons undergo random motion similar to that of gas molecules in the structural model of kinetic theory that we studied in Chapter 16. This random mo-

(a)    (b)    (c)    (d)

**Figure 21.3**

(Quick Quiz 21.1)

---

* Because we have assumed that the average drift speed is constant, the current is constant and, in turn, the average current over any time interval is the same as the instantaneous current at any instant.

tion is related to the temperature of the conductor. The electrons undergo repeated collisions with the metal atoms, and the result is a complicated zigzag motion (Fig. 21.4). When a potential difference is applied across the conductor, an electric field is established in the conductor.* The electric field exerts an electric force on the electrons (Eq. 19.4). This force accelerates the electrons and hence produces a current. The motion of the electrons due to the electric force is superimposed on their random motion to provide an average velocity whose magnitude is the drift speed.

When electrons make collisions with metal atoms during their motion, they transfer energy to the atoms. This energy transfer causes an increase in the vibrational energy of the atoms and a corresponding increase in the temperature of the conductor.† This is a three-step process involving all three types of energy storage in the continuity equation for energy, Equation 6.20. If we consider the system to be the electrons, the metal atoms, and the electric field (which is established by an external source such as a battery), then the energy at the instant when the potential difference is applied across the conductor is electric potential energy associated with the field and the electrons. This energy is transformed by work done by the field on the electrons to kinetic energy of electrons. When the electrons strike the metal atoms, some of the kinetic energy is transferred to the atoms, which adds to the internal energy of the system.

The **current density** $J$ in the conductor is defined as the current per unit area. From Equation 21.4, the current density is

$$J \equiv \frac{I}{A} = nqv_d \qquad \text{[21.5]}$$

where $J$ has the SI units amperes per square meter.

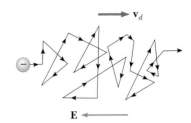

**Figure 21.4**

A schematic representation of the zigzag motion of a charge carrier in a conductor. The changes in direction are due to collisions with atoms in the conductor. Note that the net motion of electrons is opposite the direction of the electric field. Because of the acceleration of the charge carriers due to the electric force, the paths are actually parabolic. The drift speed is much smaller, however, than the average speed, so the parabolic shape is not visible on this scale.

**PITFALL PREVENTION 21.2**

**The drift speed does not contribute to internal energy**

We do not identify the motion associated with the drift speed of the electrons as internal energy because this motion is not *random*. This is a subtle and interesting point, in that the random motion of the electrons as they bounce around inside the conductor *is* identified as internal energy, but the motion associated with the drift speed superimposed on this random motion is not!

### THINKING PHYSICS 21.1

Suppose a current-carrying wire has a cross-sectional area that gradually becomes smaller along the wire, so that the wire has the shape of a very long cone. How does the drift speed of electrons vary along the wire?

**Reasoning**   Every portion of the wire is carrying the same amount of current—otherwise, charge would build up or disappear somewhere along the wire. Thus, for Equation 21.4 to be satisfied, as the cross-sectional area decreases, the drift speed must increase to maintain the constant current. This increased drift speed is a result of the electric field lines in the wire being distributed over a smaller area, thus increasing the magnitude of the field, and, in turn, increasing the electric force on the electrons.

---

* You may be tempted to say that the electric field in a conductor is zero, according to Chapter 19. But the field is zero only in a conductor in *equilibrium*, that is, a conductor in which the charges are at rest after having moved to equilibrium positions. In a current-carrying conductor, the charges are not at rest, so the requirement for a zero field is not imposed. The electric field in a conductor in a circuit is due to a distribution of charge over the surface of the conductor, which can be quite complicated. See Chapter 6 in R. Chabay and B. Sherwood, *Electric and Magnetic Interactions*, Wiley, 1995 for details on this charge distribution.

† This increase in temperature is sometimes called *Joule heating*, but this is a misnomer—no heat is involved. We will not use this term.

## Example 21.1 Drift Speed in a Copper Wire

A copper wire of cross-sectional area $3.00 \times 10^{-6} \text{ m}^2$ carries a current of 10.0 A. Find the drift speed of the electrons in this wire. The density of copper is $8.95 \text{ g/cm}^3$.

**Solution** From the periodic table of the elements in Appendix C, we find that the molar mass of copper is 63.5 g/mol. Knowing the density of copper enables us to calculate the volume occupied by 63.5 g of copper:

$$V = \frac{M}{\rho} = \frac{63.5 \text{ g/mol}}{8.95 \text{ g/cm}^3} = 7.09 \text{ cm}^3/\text{mol}$$

Recall that 1 mole of any substance contains Avogadro's number of atoms, $6.02 \times 10^{23}$ atoms. If we now assume that each copper atom contributes one free electron to the body of the material, the density of free electrons is

$$n = \frac{6.02 \times 10^{23} \text{ electrons/mol}}{7.09 \text{ cm}^3/\text{mol}}$$

$$= (8.48 \times 10^{22} \text{ electrons/cm}^3) \left( \frac{10^6 \text{ cm}^3}{1 \text{ m}^3} \right)$$

$$= 8.48 \times 10^{28} \text{ electrons/m}^3$$

From Equation 21.4, we find that the drift speed is

$$v_d = \frac{I}{nqA}$$

$$= \frac{10.0 \text{ C/s}}{(8.48 \times 10^{28} \text{ m}^{-3})(1.60 \times 10^{-19} \text{ C})(3.00 \times 10^{-6} \text{ m}^2)}$$

$$= 2.46 \times 10^{-4} \text{ m/s}$$

**EXERCISE** An aluminum wire having a cross-sectional area of $4.0 \times 10^{-6} \text{ m}^2$ carries a current of 5.0 A. Find the drift speed of the electrons in the wire. The density of aluminum is $2.7 \text{ g/cm}^3$. (Assume one electron is supplied by each atom.)

*Answer* 0.13 mm/s

---

**PITFALL PREVENTION 21.3**

**Electrons are available everywhere**

Let us emphasize the point being made here—*electrons do not have to travel from the light switch to the light for the light to operate.* Electrons already in the filament of the lightbulb move in response to the electric field set up by the battery. Notice also that a battery does not provide electrons to the circuit. It establishes the electric field that exerts a force on electrons already in the wires and elements of the circuit.

See Screen 13.3

Example 21.1 shows that typical drift speeds in conductors are very small. In fact, the drift speed is much smaller than the average speed between collisions. For instance, electrons traveling with the drift speed calculated in Example 21.1 would take about 68 min to travel 1 m! In view of this low speed, you might wonder why a light turns on almost instantaneously when a switch is thrown. In a conductor, the electric field that drives the free electrons is established in the conductor almost instantaneously. Thus, when you flip a light switch, the electric force that causes the electrons to start moving through the wire begins immediately. Electrons already in the filament of the lightbulb begin to move in response to this force, and the lightbulb begins to emit light.

## 21.2 • RESISTANCE AND OHM'S LAW

The drift speed of electrons in a current-carrying wire is related to the electric field in the wire—if the field is increased, the electric force on the electrons is stronger and the drift speed increases. We shall show in Section 21.4 that this is a linear relationship—the drift speed is directly proportional to the electric field. For a uniform field in a conductor of uniform cross-section, the potential difference across the conductor is proportional to the electric field, as in Equation 20.6. Thus, when a potential difference $\Delta V$ is applied across the ends of a metallic conductor as in Figure 21.5, the current in the conductor is found to be proportional to the applied voltage; that is, $I \propto \Delta V$. We can write this proportionality as $\Delta V = IR$, where the proportionality constant $R$ is called the **resistance** of the conductor. We define this resistance according to the equation we have just written, as the ratio of

**Figure 21.5**

A uniform conductor of length $\ell$ and cross-sectional area $A$. A potential difference $V_b - V_a$ is maintained across the conductor so that an electric field $\mathbf{E}$ exists in the conductor, and this field produces a current $I$ that is proportional to the potential difference.

the voltage across the conductor to the current it carries:

$$R \equiv \frac{\Delta V}{I}$$

**[21.6]**

• *Definition of resistance*

Resistance has the SI units volts per ampere, called **ohms** ($\Omega$). Thus, if a potential difference of 1 V across a conductor produces a current of 1 A, the resistance of the conductor is 1 $\Omega$. As another example, if an electrical appliance connected to a 120-V source carries a current of 6.0 A, its resistance is 20 $\Omega$.

Resistance is the quantity that determines the current due to a given voltage in a simple circuit. If the resistance increases, the current decreases. If the resistance decreases, the current increases.

It might be useful for you to build a mental model for current, voltage, and resistance by comparing these concepts to analogous concepts for the flow of water in a river. As water flows downhill in a river of constant width and depth, the rate of flow of water (analogous to current) depends on the angle the river bottom makes with the horizontal (analogous to voltage) and on the width and depth as well as on the effects of rocks, the river bank, and other obstructions (analogous to resistance). Likewise, electric current in a uniform conductor depends on the applied voltage and the resistance of the conductor caused by collisions of the electrons with atoms in the conductor.

For many materials, including most metals, experiments show that the resistance is constant over a wide range of applied voltages. This behavior is known as **Ohm's law** after Georg Simon Ohm (1787–1854), who was the first to conduct a systematic study of electrical resistance.

Ohm's law is *not* a fundamental law of nature, but an empirical relationship that is valid only for certain materials and devices, and only over a limited range of conditions. Materials or devices that obey Ohm's law, and hence have a constant resistance over a wide range of voltages, are said to be **ohmic.** Materials or devices that do not obey Ohm's law are **nonohmic.** Ohmic materials or devices have a linear current–voltage relationship over a large range of applied voltages as shown in the graphical representation in Figure 21.6a. Nonohmic materials or devices have a nonlinear current–voltage relationship as in Figure 21.6b. One common semiconducting device that is nonohmic is the *diode*, a circuit element that acts like a one-way valve for current. Its resistance is small for currents in one direction (positive $\Delta V$) and large for currents in the reverse direction (negative $\Delta V$). Most modern electronic devices, such as transistors, have nonlinear current–voltage relationships; their operation depends on the particular ways they violate Ohm's law.

**PITFALL PREVENTION 21.4**

**We've seen something like Equation 21.6 before**

In Chapter 4, we introduced Newton's second law, $\Sigma F = ma$, for a net force on an object of mass $m$. This can be written as

$$m = \frac{\Sigma F}{a}$$

In that chapter, we defined mass as *resistance to a change in motion in response to an external force.* Mass as resistance to changes in motion is analogous to electrical resistance to charge flow, and Equation 21.6 is analogous to the form of Newton's second law shown here.

**PITFALL PREVENTION 21.5**

**Equation 21.6 is not Ohm's law**

Many individuals call Equation 21.6 Ohm's law, but this is incorrect. This equation is simply the definition of resistance and provides an important relationship among voltage, current, and resistance. Ohm's law refers to a linear relationship between current and voltage, which, from Equation 21.6, indicates that the resistance is constant, independent of the applied voltage.

## Quick Quiz 21.2

In Figure 21.6b, does the resistance of the diode (a) increase or (b) decrease as the positive voltage $\Delta V$ increases?

(a)

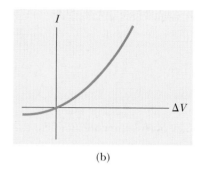

(b)

### Figure 21.6

(a) The current–voltage curve for an ohmic device. The curve is linear, and the slope gives the resistance of the conductor. (b) A nonlinear current–voltage curve for a semiconducting diode. This device does not obey Ohm's law.

A **resistor** is a simple circuit element that provides a specified resistance in an electric circuit. The symbol for a resistor in circuit diagrams is a zigzag line, (—⌁—). We can express Equation 21.6 in the form

$$\Delta V = IR \qquad [21.7]$$

This equation tells us that the voltage across a resistor is the product of the resistance and the current in the resistor.

The resistance of an ohmic conducting wire is found to be proportional to its length and inversely proportional to its cross-sectional area. That is,

- *Relation between resistance and resistivity*

$$R = \rho \frac{\ell}{A} \qquad [21.8]$$

where the constant of proportionality $\rho$ is called the **resistivity** of the material,* which has the unit ohm-meter ($\Omega \cdot m$). To understand this relationship between resistance and resistivity, note that every ohmic material has a characteristic resistiv-

### PITFALL PREVENTION 21.6
#### Resistance and resistivity

Resistivity is a property of a *substance*, but resistance is a property of an *object*. We have seen similar pairs of variables before. For example, density is a property of a substance, whereas mass is a property of an object with chosen points of exit and entry for the current. Equation 21.8 relates resistance to resistivity, and a previous equation (Eq. 1.1) relates mass to density.

An assortment of resistors used in electric circuits. *(Henry Leap and Jim Lehman)*

* The symbol $\rho$ used for resistivity should not be confused with the same symbol used earlier in the text for mass density and volume charge density.

| TABLE 21.1 | Resistivities and Temperature Coefficients of Resistivity for Various Materials | |
|---|---|---|
| Material | Resistivity[a] $(\Omega \cdot m)$ | Temperature Coefficient $\alpha[(°C)^{-1}]$ |
| Silver | $1.59 \times 10^{-8}$ | $3.8 \times 10^{-3}$ |
| Copper | $1.7 \times 10^{-8}$ | $3.9 \times 10^{-3}$ |
| Gold | $2.44 \times 10^{-8}$ | $3.4 \times 10^{-3}$ |
| Aluminum | $2.82 \times 10^{-8}$ | $3.9 \times 10^{-3}$ |
| Tungsten | $5.6 \times 10^{-8}$ | $4.5 \times 10^{-3}$ |
| Iron | $10 \times 10^{-8}$ | $5.0 \times 10^{-3}$ |
| Platinum | $11 \times 10^{-8}$ | $3.92 \times 10^{-3}$ |
| Lead | $22 \times 10^{-8}$ | $3.9 \times 10^{-3}$ |
| Nichrome[b] | $1.50 \times 10^{-6}$ | $0.4 \times 10^{-3}$ |
| Carbon | $3.5 \times 10^{-5}$ | $-0.5 \times 10^{-3}$ |
| Germanium | $0.46$ | $-48 \times 10^{-3}$ |
| Silicon | $640$ | $-75 \times 10^{-3}$ |
| Glass | $10^{10} - 10^{14}$ | |
| Hard rubber | $\sim 10^{13}$ | |
| Sulfur | $10^{15}$ | |
| Quartz (fused) | $75 \times 10^{16}$ | |

[a] All values at 20°C.
[b] A nickel–chromium alloy commonly used in heating elements.

ity, a parameter that depends on the properties of the material and on temperature. On the other hand, as you can see from Equation 21.8, the resistance of a conductor depends on its size and shape as well as on the resistivity of the material. Table 21.1 provides a list of resistivities for various materials measured at 20°C.

The inverse of the resistivity is defined* as the **conductivity $\sigma$.** Hence, the resistance of an ohmic conductor can be expressed in terms of its conductivity as

$$R = \frac{\ell}{\sigma A} \tag{21.9}$$

where $\sigma \ (= 1/\rho)$ has the unit $(\Omega \cdot m)^{-1}$.

Equation 21.9 shows that the resistance of a conductor is proportional to its length and inversely proportional to its cross-sectional area, similar to the flow of liquid through a pipe. As the length of the pipe is increased and the pressure difference between the ends of the pipe is held constant, the pressure difference between any two points separated by a fixed distance decreases, and less force is pushing the fluid between these points. As its cross-sectional area is increased, the pipe can transport more fluid in a given time interval, so its resistance drops.

As another analogy between electric circuits and our previous studies, let us combine Equations 21.6 and 21.9:

$$R = \frac{\ell}{\sigma A} = \frac{\Delta V}{I} \quad \longrightarrow \quad I = \sigma A \frac{\Delta V}{\ell} \quad \longrightarrow \quad \frac{Q}{\Delta t} = \sigma A \frac{\Delta V}{\ell}$$

* Again, do not confuse the symbol $\sigma$ for conductivity with the same symbol used earlier for the Stefan–Boltzmann constant and surface charge density.

**Figure 21.7**

The colored bands on a resistor represent a code for determining resistance. The first two colors give the first two digits in the resistance value. The third color represents the power of ten for the multiplier of the resistance value. The last color is the tolerance of the resistance value. As an example, the four colors on the circled resistors are red (= 2), black (= 0), orange (= $10^3$), and gold (= 5%), and so the resistance value is $20 \times 10^3\ \Omega = 20\ \mathrm{k}\Omega$ with a tolerance value of 5% = 1 kΩ. (The values for the colors come from Table 21.2.) *(SuperStock)*

**APPLICATION**

Shocking Error!

where $Q$ is the amount of charge transferred in a time interval $\Delta t$. Let us compare this to Equation 17.34 for conduction of energy through a slab of material of area $A$, length $\ell$, and thermal conductivity $k$, which we reproduce here:

$$\mathcal{P} = kA\,\frac{(T_\mathrm{H} - T_\mathrm{C})}{L} \longrightarrow \frac{Q}{\Delta t} = kA\,\frac{\Delta T}{L}$$

In this equation, $Q$ is the amount of energy transferred by heat in a time interval $\Delta t$. Both of the preceding equations have exactly the same mathematical form. Each has a time rate of change on the left, and each has the product of a conductivity, an area, and a ratio of a difference in a variable to a length on the right. This type of equation is a *transport equation*—we transport either energy or charge. The difference in a variable in each equation is what drives the transport—a temperature difference drives energy flow by heat, and a potential difference drives a flow of charge.

Most electric circuits use resistors to control the current level in the various parts of the circuit. Two common types of resistors are the *composition* resistor containing carbon, and the *wire-wound* resistor, which consists of a coil of wire. Resistors are normally color coded to give their values in ohms, as shown in Figure 21.7 and Table 21.2.

**Quick Quiz 21.3**

Household electrical devices are often rated with a voltage and a current, for example, 120 V, 5 A. Batteries, however, are only rated with a voltage, such as 1.5 V. Why isn't current labeled on a battery?

**Quick Quiz 21.4**

Newspaper articles often have statements such as "10,000 volts of electricity surged through the victim's body." What is wrong with this statement?

| TABLE 21.2 | Color Code for Resistors | | |
|---|---|---|---|
| **Color** | **Number** | **Multiplier** | **Tolerance (%)** |
| Black | 0 | 1 | |
| Brown | 1 | $10^1$ | |
| Red | 2 | $10^2$ | |
| Orange | 3 | $10^3$ | |
| Yellow | 4 | $10^4$ | |
| Green | 5 | $10^5$ | |
| Blue | 6 | $10^6$ | |
| Violet | 7 | $10^7$ | |
| Gray | 8 | $10^8$ | |
| White | 9 | $10^9$ | |
| Gold | | $10^{-1}$ | 5 |
| Silver | | $10^{-2}$ | 10 |
| Colorless | | | 20 |

## Example 21.2 The Resistance of Nichrome Wire

(a) Calculate the resistance per unit length of a 22-gauge Nichrome wire, which has a radius of 0.321 mm.

**Solution** The cross-sectional area of this wire is

$$A = \pi r^2 = \pi (0.321 \times 10^{-3} \text{ m})^2 = 3.24 \times 10^{-7} \text{ m}^2$$

The resistivity of Nichrome is $1.5 \times 10^{-6} \ \Omega \cdot \text{m}$ (see Table 21.1). We use Equation 21.8 to find the resistance per unit length:

$$\frac{R}{\ell} = \frac{\rho}{A} = \frac{1.5 \times 10^{-6} \ \Omega \cdot \text{m}}{3.24 \times 10^{-7} \text{ m}^2} = \boxed{4.6 \ \Omega/\text{m}}$$

(b) If a potential difference of 10 V is maintained across a 1.0-m length of the Nichrome wire, what is the current in the wire?

**Solution** Because a 1.0-m length of this wire has a resistance of 4.6 $\Omega$, we have

$$I = \frac{\Delta V}{R} = \frac{10 \text{ V}}{4.6 \ \Omega} = \boxed{2.2 \text{ A}}$$

Note from Table 21.1 that the resistivity of Nichrome wire is two orders of magnitude larger than that of copper. A cop-per wire of the same radius would have a resistance per unit length of only 0.052 $\Omega/\text{m}$. A 1.0-m length of copper wire of the same radius would carry the same current (2.2 A) with an applied voltage of only 0.11 V.

Because of its high resistivity and its resistance to oxida-tion, Nichrome is often used for heating elements in toast-ers, irons, and electric heaters.

**EXERCISE** Calculate the current density and electric field in the wire assuming that it carries a current of 2.2 A.

*Answer* $6.7 \times 10^6 \text{ A/m}^2$; 10 N/C

**EXERCISE** What is the resistance of a 6.0-m length of 22-gauge Nichrome wire? How much current does it carry when connected to a 120-V source?

*Answer* 28 $\Omega$; 4.3 A

**EXERCISE** A 0.90-V potential difference is maintained across a 1.5-m length of tungsten wire that has a cross-sec-tional area of 0.60 mm². What is the current in the wire?

*Answer* 6.4 A

## Change in Resistivity with Temperature

Resistivity depends on a number of factors, one of which is temperature. For most metals, resistivity increases approximately linearly with increasing temperature over a limited temperature range, according to the expression

$$\rho = \rho_0[1 + \alpha(T - T_0)] \qquad \text{[21.10]}$$

• *Variation of resistivity with temperature*

where $\rho$ is the resistivity at some temperature $T$ (in degrees Celsius), $\rho_0$ is the resis-tivity at some reference temperature $T_0$ (usually 20°C), and $\alpha$ is called the **temper-ature coefficient of resistivity** (not to be confused with the average coefficient of linear expansion $\alpha$ in Chapter 16). From Equation 21.10, we see that $\alpha$ can be expressed as

$$\alpha = \frac{1}{\rho_0} \frac{\Delta \rho}{\Delta T} \qquad \text{[21.11]}$$

• *Temperature coefficient of resistivity*

where $\Delta \rho = \rho - \rho_0$ is the change in resistivity in the temperature interval $\Delta T = T - T_0$.

The resistivities and temperature coefficients of certain materials are listed in Table 21.1. Note the enormous range in resistivities, from very low values for good conductors, such as copper and silver, to very high values for good insulators, such as glass and rubber. An ideal, or "perfect," conductor would have zero resistivity, and an ideal insulator would have infinite resistivity.

Because resistance is proportional to resistivity according to Equation 21.8, the temperature variation of the resistance can be written

$$R = R_0[1 + \alpha(T - T_0)] \qquad \text{[21.12]}$$

• *Variation of resistance with temperature*

Precise temperature measurements are often made using this property, as shown in Example 21.3.

### Quick Quiz 21.5

Aliens with strange powers visit Earth and double the linear dimensions of every object on the planet's surface. Does the electrical cord from the wall socket to your floor lamp now have (a) more resistance than before, (b) less resistance, or (c) the same resistance? Does the lightbulb filament glow (d) more brightly than before, (e) less brightly, or (f) the same? (Assume the resistivities of materials remain the same before and after the doubling.)

### Example 21.3   A Platinum Resistance Thermometer

A resistance thermometer, which measures temperature by measuring the change in resistance of a conductor, is made from platinum and has a resistance of 50.0 Ω at 20.0°C. When immersed in a vessel containing melting indium, its resistance increases to 76.8 Ω. Assuming that the resistance varies linearly with temperature over the temperature range in question, what is the melting point of indium?

**Solution**   Solving Equation 21.12 for $\Delta T$, and obtaining $\alpha$ from Table 21.1, we have

$$\Delta T = \frac{R - R_0}{\alpha R_0} = \frac{76.8\ \Omega - 50.0\ \Omega}{[3.92 \times 10^{-3}\ (°C)^{-1}](50.0\ \Omega)} = 137°C$$

Because $T_0 = 20.0°C$, we find that $T = \boxed{157°C.}$

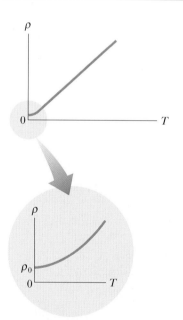

**Figure 21.8**

Resistivity versus temperature for a normal metal, such as copper. The curve is linear over a wide range of temperatures, and $\rho$ increases with increasing temperature. As $T$ approaches absolute zero (inset), the resistivity approaches a finite value $\rho_0$.

For several metals, resistivity is nearly proportional to temperature, as shown in Figure 21.8. In reality, however, a nonlinear region always occurs at very low temperatures, and the resistivity usually approaches some finite value near absolute zero (see the inset in Figure 21.8). This residual resistivity near absolute zero is due primarily to collisions of electrons with impurities and to imperfections in the metal. In contrast, the high-temperature resistivity (the linear region) is dominated by collisions of electrons with the vibrating metal atoms. We shall describe this process in more detail in Section 21.4.

**EXERCISE**   A certain lightbulb has a tungsten filament with a resistance of 19.0 Ω when cold and 140 Ω when hot. Assume that Equation 21.10 can be used over the large temperature range involved here, and find the temperature of the filament when hot. Assume an initial temperature of 20.0°C.

*Answer*   $1.44 \times 10^{3}°C$

## 21.3 • SUPERCONDUCTORS

For a class of metals and compounds known as **superconductors** the resistance goes to zero below a certain **critical temperature** $T_c$. The resistance–temperature graph for a superconductor follows that of a normal metal at temperatures above $T_c$. When the temperature reaches $T_c$, the resistivity drops suddenly to zero (Fig. 21.9). This phenomenon was discovered by the Dutch physicist Heike Kamerlingh Onnes in 1911 as he worked with mercury, which is a superconductor below 4.2 K. Recent measurements have shown that the resistivities of superconductors below $T_c$ are less than $4 \times 10^{-25}$ Ω·m, which is about $10^{17}$ times smaller than the resistivity of copper and considered to be zero in practice.

Today thousands of superconductors are known. Such common metals as aluminum, tin, lead, zinc, and indium are superconductors. Table 21.3 lists the critical temperatures of several superconductors. The value of $T_c$ is sensitive to chemical composition, pressure, and crystalline structure. It is interesting that copper, silver, and gold, which are excellent conductors at room temperature, do not exhibit superconductivity.

One of the truly remarkable features of superconductors is the fact that once a current is set up in them, it persists *without any applied voltage* (because $R = 0$). In fact, steady currents have been observed to persist in superconducting loops for several years with no apparent decay!

An important development in physics that created much excitement in the scientific community in the latter part of the 20th century is the discovery of high-temperature copper-oxide–based superconductors. The excitement began with a 1986 publication by Georg Bednorz and K. Alex Müller, scientists at the IBM Zurich Research Laboratory in Switzerland, in which they reported evidence for superconductivity at a temperature near 30 K in an oxide of barium, lanthanum, and copper. Bednorz and Müller were awarded the Nobel Prize in physics in 1987 for their remarkable discovery. Shortly thereafter, a new family of compounds was open for investigation, and research activity in the field of superconductivity proceeded vigorously. In early 1987, groups at the University of Alabama at Huntsville and the University of Houston announced the discovery of superconductivity at about 92 K in an oxide of yttrium, barium, and copper ($YBa_2Cu_3O_7$). Late in 1987, teams of scientists from Japan and the United States reported superconductivity at 105 K in an oxide of bismuth, strontium, calcium, and copper. More recently, scientists have reported superconductivity at temperatures as high as 134 K in a compound containing mercury. At this point one cannot rule out the possibility of room-temperature superconductivity, and the search for novel superconducting materials continues. It is an important search, both for scientific reasons and because practical applications become more probable and widespread as the critical temperature is raised.

An important and useful application is superconducting magnets in which the magnetic field magnitudes are about ten times greater than those of the best normal electromagnets. (We will study magnetism in Chapter 22.) Such superconducting magnets are being considered as a means of storing energy. The idea of using superconducting power lines for transmitting power efficiently is also receiving some consideration. Modern superconducting electronic devices consisting of

**Figure 21.9**

Resistance versus temperature for a sample of mercury. The graph follows that of a normal metal above the critical temperature $T_c$. The resistance drops to zero at $T_c$, which is 4.2 K for mercury.

A small permanent magnet levitated above a disk of the superconductor $YBa_2Cu_3O_7$, which is at 77 K. *(Courtesy of IBM Research Laboratory)*

**WEB**

For updates on recent developments in superconductivity, visit **www.iitap.iastate.edu/htcu/htcu. html**

**TABLE 21.3**

**Critical Temperatures for Various Superconductors**

| Material | $T_c$ (K) |
|---|---|
| $HgBa_2Ca_2Cu_3O_8$ | 134 |
| Tl-Ba-Ca-Cu-O | 125 |
| Bi-Sr-Ca-Cu-O | 105 |
| $YBa_2Cu_3O_7$ | 92 |
| $Nb_3Ge$ | 23.2 |
| $Nb_3Sn$ | 21.05 |
| Nb | 9.46 |
| Pb | 7.18 |
| Hg | 4.15 |
| Sn | 3.72 |
| Al | 1.19 |
| Zn | 0.88 |

two thin-film superconductors separated by a thin insulator have been constructed. They include magnetometers (magnetic field–measuring devices) and various microwave devices.

## 21.4 • A STRUCTURAL MODEL FOR ELECTRICAL CONDUCTION

In Section 21.1, a structural model of electrical conduction was developed by relating the macroscopic current to the drift speed of microscopic charge carriers in a material. This section expands that model by introducing the microscopic origin of resistance. Once the model is completed, we shall compare its predictions to experimental measurements.

Consider a conductor as a regular array of atoms containing free electrons (sometimes called *conduction* electrons). Such electrons are free to move through the conductor (as we learned in our discussion of drift speed in Section 21.1) and are approximately equal in number to the number of atoms in the conductor. In the absence of an electric field, the free electrons move in random directions with average speeds on the order of $10^6$ m/s. The situation is similar to the motion of gas molecules confined in a vessel, which we studied in kinetic theory in Chapter 16. In fact, the conduction electrons in a metal are often called an *electron gas*.

Conduction electrons are not totally free because they are confined to the interior of the conductor and undergo frequent collisions with the array of atoms. The collisions are the predominant mechanism contributing to the resistivity of a metal at normal temperatures. Note that no current exists in a conductor in the absence of an electric field because the average velocity of the free electrons is zero. On average, just as many electrons move in one direction as in the opposite direction, and so no net flow of charge occurs.

The situation is modified, however, when an electric field is applied to the metal. In addition to random thermal motion, the free electrons drift slowly in a direction opposite that of the electric field, with an average drift speed of $v_d$, which is much less (typically $10^{-4}$ m/s; see Example 21.1) than the average speed between collisions (typically $10^6$ m/s).

In our structural model, we shall assume that the excess kinetic energy acquired by the electrons in the electric field is lost to the conductor in the collision process. The energy given up to the atoms in the collisions increases the total vibrational energy of the atoms, causing the conductor to warm up. The model also assumes that an electron's motion after a collision is independent of its motion before the collision.

Given this basis for our model, we now take the first step toward obtaining an expression for the drift speed. When a mobile, charged particle of mass $m$ and charge $q$ is subjected to an electric field $\mathbf{E}$, it experiences a force $q\mathbf{E}$. For electrons in a metal, this gives us $\mathbf{F}_e = -e\mathbf{E}$. The motion of the electron can be determined from Newton's second law, $\Sigma \mathbf{F} = m_e\mathbf{a}$. The acceleration of the electron is

$$\mathbf{a} = \frac{\Sigma\mathbf{F}}{m_e} = \frac{\mathbf{F}_e}{m_e} = \frac{-e\mathbf{E}}{m_e} \qquad [21.13]$$

The acceleration, which occurs for only a short time interval between collisions, changes the velocity of the electron. Because the force is constant, the acceleration is constant, and we can model the electron as a particle under constant acceleration. If $\mathbf{v}_0$ is the velocity of the electron just after a collision, at which we define the time as $t = 0$, the velocity of the electron at time $t$ is

$$\mathbf{v} = \mathbf{v}_0 + \mathbf{a}t = \mathbf{v}_0 - \frac{e\mathbf{E}}{m_e} t \qquad \text{[21.14]}$$

The motion of the electron through the metal is characterized by a very large number of collisions per second. Consequently, we consider the average value of $\mathbf{v}$ over an interval that is long compared with $t$, which gives us the drift velocity $\mathbf{v}_d$. Because the velocity of the electron after a collision is assumed to be independent of its velocity before the collision, the initial velocities are randomly distributed in direction, so that the average value of $\mathbf{v}_0$ is zero. In the second term on the right of Equation 21.14, the charge, electric field, and mass are all constant. Thus, the only factor affected by the averaging process is the time. The average value of this term is $(- e\mathbf{E}/m_e)\tau$, where $\tau$ is the *average time between collisions*. Thus, Equation 21.14 becomes, after the averaging process,

$$\mathbf{v}_d = \frac{-e\mathbf{E}}{m_e} \tau \qquad \text{[21.15]} \qquad \bullet \quad \textit{Drift velocity}$$

Substituting the magnitude of this drift velocity (the drift speed) into Equation 21.4, we have

$$I = nev_d A = ne\left(\frac{eE}{m_e} \tau\right) A = \frac{ne^2 E}{m_e} \tau A \qquad \text{[21.16]}$$

According to Equation 21.6, the current is related to the macroscopic variables of potential difference and resistance:

$$I = \frac{\Delta V}{R}$$

Incorporating Equation 21.8, we can write this as

$$I = \frac{\Delta V}{\left(\rho \dfrac{\ell}{A}\right)} = \frac{\Delta V}{\rho\ell} A$$

In the conductor, the electric field is uniform, so we use Equation 20.6, $\Delta V = E\ell$, to substitute for the magnitude of the potential difference across the conductor:

$$I = \frac{E\ell}{\rho\ell} A = \frac{E}{\rho} A \qquad \text{[21.17]}$$

Setting the two expressions for the current, Equations 21.16 and 21.17, equal, we solve for the resistivity:

$$I = \frac{ne^2 E}{m_e} \tau A = \frac{E}{\rho} A \quad \longrightarrow \quad \rho = \frac{m_e}{ne^2 \tau} \qquad \text{[21.18]} \qquad \bullet \quad \textit{Resistivity in terms of micro-}$$
$$\textit{scopic parameters}$$

According to this structural model, resistivity does not depend on the electric field or, equivalently, the potential difference, but depends only on fixed parameters associated with the material and the electron. This feature is characteristic of a conductor obeying Ohm's law. The model shows that the resistivity can be calculated from a knowledge of the density of the electrons, their charge and mass, and the average time $\bar{\tau}$ between collisions, which is related to the average distance between collisions $\bar{\ell}$ (the *mean free path*) and the average speed $\bar{v}$ through the expression*

$$\tau = \frac{\bar{\ell}}{\bar{v}}$$    [21.19]

## Example 21.4   Electron Collisions in Copper

(a) Using the data and results from Example 21.1 and the structural model of electron conduction, estimate the average time between collisions for electrons in copper at 20°C.

**Solution**   From Equation 21.18 we see that

$$\tau = \frac{m_e}{ne^2\rho}$$

where $\rho = 1.7 \times 10^{-8}\ \Omega\cdot\text{m}$ for copper and the carrier density $n = 8.48 \times 10^{28}$ electrons/m$^3$ for the wire described in Example 21.1. Substitution of these values into the expression above gives

$$\tau = \frac{9.11 \times 10^{-31}\ \text{kg}}{(8.48 \times 10^{28}\ \text{m}^{-3})(1.6 \times 10^{-19}\ \text{C})^2(1.7 \times 10^{-8}\ \Omega\cdot\text{m})}$$
$$= 2.5 \times 10^{-14}\ \text{s}$$

Note that this is a very short time—the electrons make a very large number of collisions per second.

(b) Assuming the average speed for free electrons in copper to be $1.6 \times 10^6$ m/s and using the result from part (a), calculate the mean free path for electrons in copper.

**Solution**   Using Equation 21.19,

$$\bar{\ell} = \bar{v}\tau = (1.6 \times 10^6\ \text{m/s})(2.5 \times 10^{-14}\ \text{s}) = 4.0 \times 10^{-8}\ \text{m}$$

which is equivalent to 40 nm (compared with atomic spacings of about 0.2 nm). Thus, although the time between collisions is very short, the electrons travel about 200 atomic distances before colliding with an atom.

Although this structural model of conduction is consistent with Ohm's law, it does not correctly predict the values of resistivity or the behavior of the resistivity with temperature. For example, the results of classical calculations for $\bar{v}$ using the ideal gas model for the electrons are about a factor of 10 smaller than the actual values, which results in incorrect predictions of values of resistivity from Equation 21.18. Furthermore, according to Equations 21.18 and 21.19, the temperature variation of the resistivity is predicted to vary as $\bar{v}$, which, according to an ideal gas model (Chapter 16, Eq. 16.22), is proportional to $\sqrt{T}$. This is in disagreement with the linear dependence of resistivity with temperature for pure metals (Fig. 21.8a). Because of these incorrect predictions, we must modify our structural model. We shall call the model we have developed so far the *classical* model for electrical conduction. To account for the incorrect predictions of the classical model, we will develop it further into a *quantum mechanical* model, which we shall describe briefly.

We discussed two important simplification models in earlier chapters, the particle model and the wave model. Although we discussed these two simplification models separately, quantum physics tells us that this separation is not so clear-cut. As we shall discuss in detail in Chapter 28, particles have wave-like properties. The predictions of some models can only be matched to experimental results if the

---

* Recall that the average speed is the speed a particle has as a consequence of the temperature of its surroundings (Chapter 16).

model includes the wave-like behavior of particles. The structural model for electrical conduction in metals is one of these cases.

Let us imagine that the electrons moving through the metal have wave-like properties. If the array of atoms in a conductor is regularly spaced (i.e., periodic), the wave-like character of the electrons makes it possible for them to move freely through the conductor, and a collision with an atom is unlikely. For an idealized conductor, no collisions would occur, the mean free path would be infinite, and the resistivity would be zero. Electrons are scattered only if the atomic arrangement is irregular (not periodic) — for example, as a result of structural defects or impurities. At low temperatures, the resistivity of metals is dominated by scattering caused by collisions between the electrons and impurities. At high temperatures, the resistivity is dominated by scattering caused by collisions between the electrons and the atoms of the conductor, which are continuously displaced as a result of thermal agitation, destroying the perfect periodicity. The thermal motion of the atoms makes the structure irregular (compared with an atomic array at rest), thereby reducing the electron's mean free path.

Although it is beyond the scope of this text to show this in detail, the classical model modified with the wave-like character of the electrons results in predictions of resistivity values that are in agreement with measured values and predicts a linear temperature dependence. When discussing the hydrogen atom in Chapter 11, we had to introduce some quantum notions to understand experimental observations such as atomic spectra. Likewise, we had to introduce quantum notions in Chapter 17 to understand the temperature behavior of molar specific heats of gases. Here is another case where quantum physics is necessary for the model to agree with experiment. Although classical physics can explain a tremendous range of phenomena, we continue to see hints that quantum physics must be incorporated into our models. We shall study quantum physics in detail in Chapters 28–31.

## 21.5 • ELECTRIC ENERGY AND POWER

See Screen 13.3

In Section 21.1, we discussed the energy transformations occurring in a circuit. If a battery is used to establish an electric current in a conductor, there is a continuous transformation of chemical energy in the battery to kinetic energy of the electrons to internal energy in the conductor, resulting in an increase in the temperature of the conductor.

In typical electric circuits, energy is transferred from a source such as a battery, to some device, such as a lightbulb or a radio receiver. Let us determine an expression that will allow us to calculate the rate of this energy transfer. First, consider the simple circuit in Figure 21.10, where we imagine energy is being delivered to a resistor. Because the connecting wires also have resistance, some energy is delivered to the wires and some energy to the resistor. Unless noted otherwise, we will adopt a simplification model in which the resistance of the wires is so small compared with the resistance of the circuit element that we ignore the energy delivered to the wires.

Let us now analyze the energetics of the circuit in which a battery is connected to a resistor of resistance $R$ as in Figure 21.10. Imagine following a positive quantity of charge $Q$ around the circuit from point $a$ through the battery and resistor and back to $a$. Point $a$ is a reference point at which the potential is defined as zero. We identify the entire circuit as our system. As the charge moves from $a$ to $b$ through the battery whose potential difference is $\Delta V$, the electrical potential en-

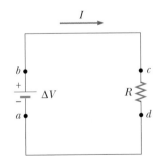

**Figure 21.10**

A circuit consisting of a resistor of resistance $R$ and a battery having a potential difference $\Delta V$ across its terminals. Positive charge flows in the clockwise direction.

**PITFALL PREVENTION 21.7**

**Misconceptions about current**

Several common misconceptions are associated with current in a circuit like that in Figure 21.10. One is that current comes out of one terminal of the battery and is then "used up" as it passes through the resistor. Thus, current exists in only one part of the circuit. The correct understanding, however, is that the current is the same *everywhere* in the circuit. A related misconception has the current coming out of the resistor being smaller than that going in, because some of the current is "used up." Another misconception has current coming out of both terminals of the battery, in opposite directions, and then "clashing" in the resistor, delivering the energy in this manner. We know this is not the case—the charges flow in the same rotational sense at *all* points in the circuit. Be sure your conceptual understanding of current is valid.

ergy of the system increases by the amount $Q\,\Delta V$, and the chemical energy in the battery decreases by the same amount. (Recall from Chapter 20 that $\Delta U = q\,\Delta V$.) As the charge moves from $c$ to $d$ through the resistor, however, the system loses this electrical potential energy during collisions with atoms in the resistor. In this process, the energy is transformed to internal energy corresponding to increased vibrational motion of the atoms in the resistor. Because we have ignored the resistance of the interconnecting wires, no energy transformation occurs for paths $bc$ and $da$. When the charge returns to point $a$, the net result is that some of the chemical energy in the battery has been delivered to the resistor and resides in the resistor as internal energy associated with molecular vibration.

The resistor is normally in contact with air, so its increased temperature results in a transfer of energy by heat into the air. In addition, thermal radiation occurs from the resistor, representing another means of escape for the energy. After some time has passed, the resistor remains at a constant temperature, at which time the input of energy from the battery is balanced by the output of energy by heat and radiation. Some electrical devices include *heat sinks** connected to parts of the circuit to prevent these parts from reaching dangerously high temperatures. These are pieces of metal with many fins. The high thermal conductivity of the metal provides a rapid transfer of energy by heat away from the hot component, and the many fins provide a large surface area in contact with the air, so that energy can transfer by radiation and into the air by heat rapidly.

Let us consider now the rate at which the system loses electric potential energy as the charge $Q$ passes through the resistor:

$$\frac{dU}{dt} = \frac{d}{dt}\left(Q\,\Delta V\right) = \frac{dQ}{dt}\,\Delta V = I\,\Delta V$$

where $I$ is the current in the circuit. Of course, the system regains this potential energy when the charge passes through the battery, at the expense of chemical energy in the battery. The rate at which the system loses potential energy as the charge passes through the resistor is equal to the rate at which the system gains internal energy in the resistor. Thus, the power $\mathcal{P}$, representing the rate at which energy is delivered to the resistor, is

• *Power delivered to a device*

$$\boxed{\mathcal{P} = I\,\Delta V} \qquad\qquad \textbf{[21.20]}$$

We have developed this result by considering a battery delivering energy to a resistor. However, Equation 21.20 can be used to determine the power transferred from a voltage source to *any* device carrying a current $I$ and having a potential difference $\Delta V$ between its terminals.

Using Equation 21.20 and the fact that $\Delta V = IR$ for a resistor, we can express the power delivered to the resistor in the alternative forms

• *Power delivered to a resistor*

$$\boxed{\mathcal{P} = I^2 R = \frac{(\Delta V)^2}{R}} \qquad\qquad \textbf{[21.21]}$$

The SI unit of power is the watt, introduced in Chapter 6. If you analyze the units in Equations 21.20 and 21.21, you will see that the result of the calculation provides a watt as the unit. The power delivered to a conductor of resistance $R$ is often referred to as an $I^2R$ *loss.*

---

* This is another misuse of the word *heat* that is ingrained in our common language.

As we learned in Section 6.8, the unit of energy your electric company uses to calculate energy transfer, the kilowatt hour, is the amount of energy transferred in 1 h at the constant rate of 1 kW. Because 1 W = 1 J/s, we have

$$1 \text{ kWh} = (1.0 \times 10^3 \text{ W})(3\,600 \text{ s}) = 3.6 \times 10^6 \text{ J} \qquad \textbf{[21.22]}$$

**Figure 21.11**

(Quick Quiz 21.6 and Thinking Physics 21.2)

### Quick Quiz 21.6

For the two lightbulbs shown in Figure 21.11, rank the currents at points *a* through *f*, from largest to smallest.

## THINKING PHYSICS 21.2

Two lightbulbs A and B are connected across the same potential difference as in Figure 21.11. The electric input powers to the lightbulbs are shown. Which lightbulb has the higher resistance? Which carries the greater current?

**Reasoning** Because the voltage across each lightbulb is the same, and the rate of energy delivered to a resistor is $\mathcal{P} = (\Delta V)^2/R$, the lightbulb with the lower resistance exhibits the higher rate of energy transfer. In this case, the resistance of A is larger than that for B. Furthermore, because $\mathcal{P} = I\,\Delta V$, we see that the current carried by B is larger than that of A.

## THINKING PHYSICS 21.3

When is the rate at which energy is being delivered to a lightbulb higher—just after it is turned on, and the glow of the filament is increasing, or after it has been on for a few seconds and the glow is steady?

**Reasoning** When the switch is closed, the source voltage is immediately applied across the lightbulb. As the voltage is applied across the cold filament when the lightbulb is first turned on, the resistance of the filament is low. Thus, the current is high, and a relatively large amount of energy is delivered to the bulb per unit time. As the filament warms up, its resistance rises, and the current falls. As a result, the rate of energy delivered to the bulb falls. The large current spike at the beginning of operation is the reason that lightbulbs often fail just as they are turned on.

### PITFALL PREVENTION 21.8

**Energy is not "dissipated"**

 In some books, you may see Equation 21.20 described as the power "dissipated in" a resistor. We will avoid this usage because it suggests that energy disappears. We will talk about energy being "delivered to" a resistor, as in Thinking Physics 21.3. The notion of *dissipation* arises from the fact that a warm resistor will expel energy by radiation and heat, so that the energy delivered by the battery leaves the circuit, but we wish to avoid the connotations of this word.

## Example 21.5 Electrical Rating of a Lightbulb

A lightbulb is rated at 120 V/75 W, which means that at its intended operating voltage of 120 V, it has a power rating of 75.0 W. The bulb is powered by a 120-V direct current power supply. Find the current in the bulb and its resistance.

**Solution** Because the power rating of the bulb is 75.0 W and the operating voltage is 120 V, we can use $\mathcal{P} = I\,\Delta V$ to find the current:

$$I = \frac{\mathcal{P}}{\Delta V} = \frac{75.0 \text{ W}}{120 \text{ V}} = \boxed{0.625 \text{ A}}$$

Using $\Delta V = IR$, the resistance is calculated to be

$$R = \frac{\Delta V}{I} = \frac{120 \text{ V}}{0.625 \text{ A}} = \boxed{192 \ \Omega}$$

**EXERCISE** What is the resistance in a lamp rated at 120 V and 100 W?

*Answer* 144 Ω

## Example 21.6  The Cost of Operating a Lightbulb

How much does it cost to operate a 100-W lightbulb for 24 h if electricity costs 12¢/kWh?

**Solution**  Because the energy delivered to the bulb equals power multiplied by time, the amount of energy you must pay for, expressed in kilowatt hours, is

$$\text{Energy} = (0.100 \text{ kW})(24 \text{ h}) = 2.4 \text{ kWh}$$

If energy is purchased at 12¢/kWh, the cost is

$$\text{Cost} = (2.4 \text{ kWh})(\$0.12/\text{kWh}) = \boxed{\$0.29}$$

That is, it costs 29¢ to operate the lightbulb for one day. This is a small amount, but when larger and more complex devices are used, the costs go up rapidly.

Demands on our energy supplies have made it necessary to be aware of the energy requirements of our electrical devices. This is true not only because they are becoming more expensive to operate but also because, with the dwindling of the coal and oil resources that ultimately supply us with electric energy, increased awareness of conservation becomes necessary. Every electrical appliance has a label that contains the information you need to calculate its power requirements. The power consumption in watts is often stated directly, as on a lightbulb. In other cases, the amount of current in the device and the voltage at which it operates are given. This information and Equation 21.20 are sufficient to calculate the operating cost of any electrical device.

**EXERCISE**  If electricity costs 12.0¢/kWh, what does it cost to operate an electric oven, which operates at 20.0 A and 220 V, for 5.00 h?

*Answer*  $2.64

## Example 21.7  Linking Electricity and Thermodynamics

What is the required resistance of an immersion heater that will increase the temperature of 1.50 kg of water from 10.0°C to 50.0°C in 10.0 min while operating at 110 V?

**Reasoning**  This example allows us to link our new understanding of power in electricity with our experience with specific heat in thermodynamics in Chapter 17. An immersion heater is a resistor that is inserted into a container of water. As energy is delivered to the immersion heater, raising its temperature, energy leaves the surface of the resistor by heat, going into the water. When the immersion heater reaches a constant temperature, the rate of energy delivered to the resistance by electrical transmission is equal to the rate of energy delivered by heat to the water.

**Solution**  As a simplification model, we ignore the initial period during which the temperature of the resistor increases, and also ignore any variation of resistance with temperature. Thus, we imagine a constant rate of energy transfer for the entire 10.0 min. Setting the rate of energy delivered to the resistor equal to the rate of energy entering the water by heat, we have

$$\mathcal{P} = \frac{(\Delta V)^2}{R} = \frac{Q}{\Delta t}$$

where $Q$ represents an amount of energy transfer by heat into the water and we have used Equation 21.21 to express the electric power. The amount of energy transfer by heat necessary to raise the temperature of the water is given by Equation 17.3, $Q = mc\,\Delta T$. Thus,

$$\frac{(\Delta V)^2}{R} = \frac{mc\,\Delta T}{\Delta t} \longrightarrow R = \frac{(\Delta V)^2\,\Delta t}{mc\,\Delta T}$$

Substituting the values given in the statement of the problem, we have,

$$R = \frac{(110 \text{ V})^2(600 \text{ s})}{(1.50 \text{ kg})(4\,186 \text{ J/kg}\cdot°\text{C})(50.0°\text{C} - 10.0°\text{C})} = \boxed{28.9\ \Omega}$$

## 21.6 • SOURCES OF emf

The entity that maintains the constant voltage in Figure 21.12 is called a **source of emf.**\* Sources of emf are any devices (e.g., batteries and generators) that increase the potential energy of a circuit system by maintaining a potential difference be-

---

\* The term *emf* was originally an abbreviation for *electromotive force,* but it is not a force, so the long form is discouraged. The name electromotive force was used early in the study of electricity before the understanding of batteries was as sophisticated as it is today.

Battery

Resistor

**Figure 21.12**

A circuit consisting of a resistor connected to the terminals of a battery.

tween points in the circuit while charges move through the circuit. One can think of a source of emf as a "charge pump" that forces electrons to move in a direction opposite the electric field inside the source. The emf $\mathcal{E}$ of a source describes the work done per unit charge, and hence the SI unit of emf is the volt.

At this point, you may wonder why we need to define a second quantity, emf, with the volt as a unit, when we have already defined the potential difference. To see the need for this new quantity, consider the circuit shown in Figure 21.12, consisting of a battery connected to a resistor. We shall assume that the connecting wires have no resistance. We might be tempted to claim that the potential difference across the battery terminals (the terminal voltage) equals the emf of the battery. A real battery, however, always has some **internal resistance** $r$. As a result, the terminal voltage is not equal to the emf, as we shall show.

The circuit shown in Figure 21.12 can be described by the circuit diagram in Figure 21.13a. The battery within the dashed rectangle is represented by an ideal, zero-resistance source of emf $\mathcal{E}$ in series with the internal resistance $r$. Now imagine moving from $a$ to $b$ in Figure 21.13a. As you pass from the negative to the positive terminal within the battery, the potential increases by $\mathcal{E}$. As you move through the resistance $r$, however, the potential decreases by an amount $Ir$, where $I$ is the current in the circuit. Thus, the terminal voltage $\Delta V = V_b - V_a$ of the battery is*

$$\Delta V = \mathcal{E} - Ir \qquad [21.23]$$

Note from this expression that $\mathcal{E}$ is equivalent to the **open-circuit voltage, that is, the terminal voltage when the current is zero.** Figure 21.13b is a graphical representation of the changes in potential as the circuit is traversed clockwise. By inspecting Figure 21.13a, we see that the terminal voltage $\Delta V$ must also equal the potential difference across the external resistance $R$, often called the **load resistance;** that is, $\Delta V = IR$. Combining this with Equation 21.23, we see that

$$\mathcal{E} = IR + Ir \qquad [21.24]$$

Solving for the current gives

$$I = \frac{\mathcal{E}}{R + r} \qquad [21.25]$$

(a)

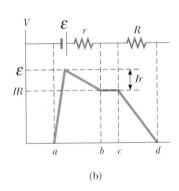

(b)

**Figure 21.13**

(a) Circuit diagram of a source of emf $\mathcal{E}$ (in this case, a battery) with internal resistance $r$, connected to an external resistor of resistance $R$.
(b) Graphical representation showing how the potential changes as the circuit in part (a) is traversed clockwise.

* The terminal voltage in this case is less than the emf by the amount $Ir$. In some situations, the terminal voltage may *exceed* the emf by the amount $Ir$. This happens when the direction of the current is *opposite* that of the emf, as when a battery is being charged with another source of emf.

This shows that the current in this simple circuit depends on both the resistance $R$ external to the battery and the internal resistance $r$. If $R$ is much greater than $r$, we can adopt a simplification model in which we ignore $r$ in our analysis. In many circuits we shall adopt this simplification model.

If we multiply Equation 21.24 by the current $I$, we have

$$I\mathcal{E} = I^2R + I^2r$$

This equation tells us that the total power output $I\mathcal{E}$ of the source of emf is equal to the rate $I^2R$ at which energy is delivered to the load resistance *plus* the rate $I^2r$ at which energy is delivered to the internal resistance. If $r \ll R$, much more of the energy from the battery is delivered to the load resistance than stays in the battery, although the amount of energy is relatively small because the load resistance is large, resulting in a small current. If $r \gg R$, a significant fraction of the energy from the source of emf stays in the battery package because it is delivered to the internal resistance. For example, if a wire is simply connected between the terminals of a flashlight battery, the battery becomes warm. This represents the transfer of energy from the source of emf to the internal resistance, where it appears as internal energy associated with temperature. Problem 54 explores the conditions under which the largest amount of energy is transferred from the battery to the load resistor.

### Quick Quiz 21.7

If the energy transferred to a dead battery during charging is $E$, is the total energy transferred out of the battery to an electrical load during use in which it completely discharges also $E$?

### Quick Quiz 21.8

If you have your headlights on while you start your car, why do they dim while the car is starting?

**EXERCISE**  (a) What is the current in a 5.60-$\Omega$ resistor connected to a battery that has a 0.200-$\Omega$ internal resistance if the terminal voltage of the battery is 10.0 V? (b) What is the emf of the battery?

*Answer*  (a)  1.79 A  (b)  10.4 V

See Screen 13.4

## 21.7 • RESISTORS IN SERIES AND IN PARALLEL

When two or more resistors are connected together end to end, as in Figure 21.14a, they are said to be in *series*. (Compare this with capacitors in series in Figure 20.24.) Note that the current is the *same* in the two resistors because any charge that flows through $R_1$ must also flow through $R_2$. Because the potential difference between $a$ and $b$ in the circuit diagram of Figure 21.14b equals $IR_1$ and the potential difference between $b$ and $c$ equals $IR_2$, the potential difference between $a$ and $c$ is

$$\Delta V = IR_1 + IR_2 = I(R_1 + R_2)$$

The potential difference across the battery is also applied to the **equivalent resistance** in Figure 21.14c:

$$\Delta V = IR_{\text{eq}}$$

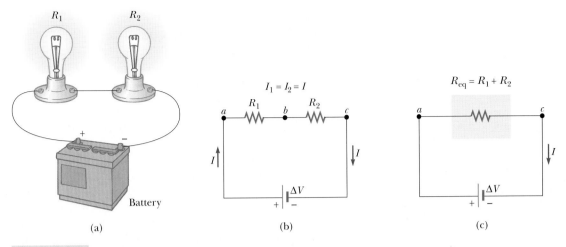

**Figure 21.14**

(a) A series connection of two lightbulbs with resistances $R_1$ and $R_2$. (b) The circuit diagram for the two-resistor circuit. The current in $R_1$ is the same as that in $R_2$. (c) The resistors are replaced with a single resistor having an equivalent resistance $R_{eq} = R_1 + R_2$.

where we have indicated that the equivalent resistance has the same effect on the circuit because it results in the same current in the battery as the combination of resistors. Combining these equations, we see that we can replace the two resistors in series with a single equivalent resistance whose value is the *sum* of the individual resistances:

$$\Delta V = IR_{eq} = I(R_1 + R_2) \quad \longrightarrow \quad R_{eq} = R_1 + R_2 \qquad \text{[21.26]}$$

The equivalent resistance of three or more resistors connected in series is simply

$$R_{eq} = R_1 + R_2 + R_3 \cdots \qquad \text{[21.27]}$$

- *Equivalent resistance of resistors in series*

Therefore, **the equivalent resistance of a series connection of resistors is the algebraic sum of the individual resistances and is always greater than any individual resistance.**

Looking back at Equation 21.25, the denominator is the simple algebraic sum of the external and internal resistances. This is consistent with the fact that internal and external resistances are in series in Figure 21.13a.

Note that if the filament of one lightbulb in Figure 21.14 were to fail, the circuit would no longer be complete (an open-circuit condition would exist), and the second bulb would also go out.

### Quick Quiz 21.9

If a piece of wire were used to connect points $b$ and $c$ in Figure 21.14b, does the brightness of bulb $R_1$ increase, decrease, or stay the same? What happens to the brightness of bulb $R_2$?

**PITFALL PREVENTION 21.11**

**Lightbulbs don't burn**

We will describe the end of the life of a lightbulb by saying that *the filament fails*, rather than by saying that the lightbulb "burns out." The word *burn* suggests a combustion process, which is not what occurs in a lightbulb.

Quick Quiz 21.10, as well as some of the subsequent Quick Quizzes, will test your understanding of the combinations of resistors that we have been developing. The appropriate response to these Quick Quizzes also depends on an important concept that is often overlooked by students—**a local change in one part of a circuit may result in a global change throughout the circuit.** For example, if a single resistance is changed, the current in the battery (-ies), the currents in all resistors, the terminal voltage of the battery (-ies), and the voltages across all resistors may change as a result. In general, therefore, do *not* assume that any currents or voltages that you have determined before the change will remain the same once the change is made. You are likely to need to analyze the circuit again, using the values that do indeed remain constant—the unaltered resistances, the emf(s) of the battery (-ies), and any internal resistance(s) of the battery (-ies).

(a)　　　　　　(b)

**Figure 21.15**

(Quick Quiz 21.10)

## Quick Quiz 21.10

With the switch in the circuit of Figure 21.15a closed, no current exists in $R_2$ because the current has an alternative zero-resistance path through the switch. Current does exist in $R_1$, and this current is measured with the ammeter (a device for measuring current) at the right side of the circuit. If the switch is opened (Fig. 21.15b), current exists in $R_2$. What happens to the reading on the ammeter when the switch is opened? (a) The reading goes up. (b) The reading goes down. (c) The reading does not change.

Now consider two resistors connected in *parallel,* as shown in Figure 21.16a. In this case, the potential differences across the resistors are equal because each resistor is connected directly across the battery terminals. The currents are generally

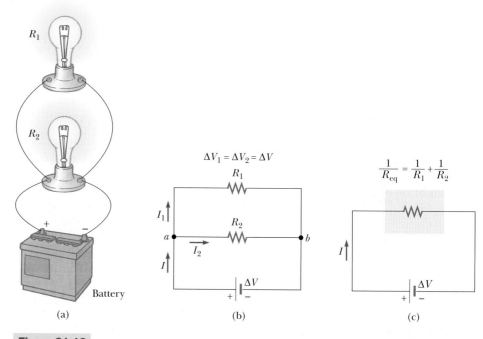

(a)　　　　　　(b)　　　　　　(c)

**Figure 21.16**

(a) A parallel connection of two lightbulbs with resistances $R_1$ and $R_2$. (b) The circuit diagram for the two-resistor circuit. The potential difference across $R_1$ is the same as that across $R_2$. (c) The resistors are replaced with a single resistor having an equivalent resistance $R_{eq} = (1/R_1 + 1/R_2)^{-1}$.

not the same, however. When the charges reach point $a$ (called a *junction*) in the circuit diagram in Figure 21.16b, the current splits into two parts: $I_1$ going through $R_1$ and $I_2$ going through $R_2$. If $R_1$ is greater than $R_2$, then $I_1$ is less than $I_2$. Because charge must be conserved, the current $I$ that enters point $a$ must equal the total current leaving point $a$:

$$I = I_1 + I_2$$

Because the potential differences across the resistors are the same, $I = \Delta V/R$ gives

$$I = I_1 + I_2 = \frac{\Delta V}{R_1} + \frac{\Delta V}{R_2} = \Delta V \left( \frac{1}{R_1} + \frac{1}{R_2} \right) = \frac{\Delta V}{R_{eq}}$$

where $R_{eq}$ is an equivalent single resistance that has the same effect on the circuit; that is, it causes the same current in the battery (Fig. 21.16c). From this result, we see that the equivalent resistance of two resistors in parallel is given by

$$\frac{1}{R_{eq}} = \frac{1}{R_1} + \frac{1}{R_2} \qquad \text{[21.28]}$$

An extension of this analysis to three or more resistors in parallel yields the following general expression:

$$\frac{1}{R_{eq}} = \frac{1}{R_1} + \frac{1}{R_2} + \frac{1}{R_3} \cdots \qquad \text{[21.29]}$$

• *Equivalent resistance of resistors in parallel*

From this expression it can be seen that **the inverse of the equivalent resistance of two or more resistors connected in parallel is the algebraic sum of the inverses of the individual resistances, and the equivalent resistance is always less than the smallest resistance in the group.**

A complicated circuit consisting of resistors can often be reduced to a simple circuit containing only one resistor. To do so, examine the initial circuit and replace any resistors in series or any in parallel with equivalent resistances using Equations 21.27 and 21.29. Draw a sketch of the new circuit after these changes have been made. Examine the new circuit and replace any new series or parallel combinations that now exist. Continue this process until a single equivalent resistance is found for the entire circuit. (This may not be possible; if not, see the techniques of Section 21.8.)

If the current in or the potential difference across a resistor in the complicated circuit is to be found, start with the final circuit and gradually work your way back through the equivalent circuits, finding currents and voltages across resistors using $\Delta V = IR$ and your understanding of series and parallel combinations.

Household circuits are always wired so that the electrical devices are connected in parallel, as in Figure 21.16a. In this manner, each device operates independently of the others, so that if one is switched off, the others remain on. For example, if one of the lightbulbs in Figure 21.16 were removed from its socket, the other would continue to operate. Equally important, each device operates on the same voltage. If devices were connected in series, the voltage applied to the combination would divide among the devices, so that the voltage applied to any one device would depend on how many devices were in the combination.

In many household circuits, circuit breakers are used in series with other circuit elements for safety purposes. A circuit breaker is designed to switch off and open the circuit at some maximum current (typically 15 A or 20 A) whose value depends on the nature of the circuit. If a circuit breaker were not used, excessive

**PITFALL PREVENTION 21.13**

**Current does not take the path of least resistance**

You may have heard a phrase like "current takes the path of least resistance." This is a reference to a parallel combination of current paths, such that the current can take two or more paths. The phrase is incorrect, however. The current takes *all* paths. Those paths with lower resistance will have large currents, but even very high-resistance paths will carry *some* of the current.

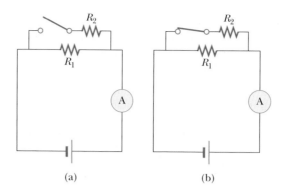

**Figure 21.17**

(Quick Quiz 21.11)

currents caused by turning on many devices could result in excessive temperatures in wires, and perhaps cause a fire. In older home construction, fuses were used in place of circuit breakers. When the current in a circuit exceeds some value, the conductor in a fuse melts and opens the circuit. The disadvantage of fuses is that they are destroyed in the process of opening the circuit, whereas circuit breakers can be reset.

### Quick Quiz 21.11

With the switch in the circuit of Figure 21.17a open, no current exists in $R_2$. Current does exist in $R_1$ and this current is measured with the ammeter at the right side of the circuit. If the switch is closed (Fig. 21.17b), current exists in $R_2$. What happens to the reading on the ammeter when the switch is closed? (a) The reading goes up. (b) The reading goes down. (c) The reading does not change.

### Quick Quiz 21.12

You have a large supply of lightbulbs and a battery. You start with one lightbulb connected to the battery and notice its brightness. You then add one lightbulb at a time, each new bulb being added in series to the previous bulbs. As you add the lightbulbs, what happens (a) to the brightness of the bulbs? (b) to the current in the bulbs? (c) to the power transferred from the battery? (d) to the lifetime of the battery? (e) to the terminal voltage of the battery? Answer the same questions if the lightbulbs are added one by one in parallel with the first.

**Figure 21.18**

(Thinking Physics 21.4)

### THINKING PHYSICS 21.4

Compare the brightnesses of the four identical lightbulbs in Figure 21.18. What happens if bulb A fails, so that it cannot conduct? What if C fails? What if D fails?

**Reasoning**   Bulbs A and B are connected in series across the emf of the battery, and bulb C is connected by itself across this emf. Thus, the emf is split between bulbs A and B. As a result, bulb C will be brighter than bulbs A and B, which should be as bright as each other. Bulb D has a wire connected across it. Thus, no potential difference occurs across D, and it does not glow at all. If bulb A fails, B goes out, but C stays lit. If C fails, there is no effect on the other bulbs. If D fails, the event is undetectable, because D was not glowing initially.

## THINKING PHYSICS 21.5

Figure 21.19 illustrates how a three-way lightbulb is constructed to provide three levels of light intensity. The socket of the lamp is equipped with a three-way switch for selecting different light intensities. The bulb contains two filaments. Why are the filaments connected in parallel? Explain how the two filaments are used to provide three different light intensities.

**Reasoning** If the filaments were connected in series, and one of them were to fail, there would be no current in the bulb, and the bulb would give no illumination, regardless of the switch position. When the filaments are connected in parallel, however, and one of them (say the 75-W filament) fails, the bulb still operates in one of the switch positions because there is current in the other (100-W) filament. The three light intensities are made possible by selecting one of three values of filament resistance, using a single value of 120 V for the applied voltage. The 75-W filament offers one value of resistance, the 100-W filament offers a second value, and the third resistance is obtained by combining the two filaments in parallel. When switch $S_1$ is closed and switch $S_2$ is opened, only the 75-W filament carries current. When switch $S_1$ is open and switch $S_2$ is closed, only the 100-W filament carries current. When both switches are closed, both filaments carry current, and a total illumination corresponding to 175 W is obtained.

**Figure 21.19**

(Thinking Physics 21.5)

## Example 21.8 Find the Equivalent Resistance

Four resistors are connected as shown in Figure 21.20a.
(a) Find the equivalent resistance between *a* and *c*.

**Solution** The circuit can be reduced in steps as shown in Figure 21.20. The 8.0-Ω and 4.0-Ω resistors are in series, and so the equivalent resistance between *a* and *b* is 12.0 Ω (Eq. 21.27). The 6.0-Ω and 3.0-Ω resistors are in parallel, and so from Equation 21.29 we find that the equivalent resistance from *b* to *c* is 2.0 Ω. Hence, the equivalent resistance from *a* to *c* is 14.0 Ω.

(b) What is the current in each resistor if a potential difference of 42 V is maintained between *a* and *c*?

**Solution** The current *I* in the 8.0-Ω and 4.0-Ω resistors is the same because they are in series. Using $\Delta V = IR$ and the results from part (a), we have

$$I = \frac{\Delta V_{ac}}{R_{eq}} = \frac{42 \text{ V}}{14.0 \ \Omega} = 3.0 \text{ A}$$

When this current enters the junction at *b*, it splits. Part of it ($I_1$) passes through the 6.0-Ω resistor and part ($I_2$) goes through the 3.0-Ω resistor. Because the potential differences $\Delta V_{bc}$ across these resistors are the same (they are in parallel), we see that $6.0I_1 = 3.0I_2$, or $I_2 = 2.0I_1$. Using this result and the fact that $I_1 + I_2 = 3.0$ A, we find that $I_1 = 1.0$ A and $I_2 = 2.0$ A. We could have guessed this by noting that the current in the 3.0-Ω resistor has to be twice the current in

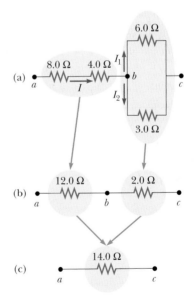

**Figure 21.20**

(Example 21.8) The four resistors shown in (a) can be reduced in steps to an equivalent 14.0-Ω resistor shown in (c).

the 6.0-Ω resistor in view of their relative resistances and the fact that the same voltage is applied to each of them.

As a final check, note that $\Delta V_{bc} = 6.0I_1 = 3.0I_2 = 6.0$ V and $\Delta V_{ab} = 12I = 36$ V; therefore, $\Delta V_{ac} = \Delta V_{ab} + \Delta V_{bc} = 42$ V, as it must.

## Example 21.9   Three Resistors in Parallel

Three resistors are connected in parallel as in Figure 21.21. A potential difference of 18.0 V is maintained between points *a* and *b*. (a) Find the current in each resistor.

**Solution**   The resistors are in parallel, and the potential difference across each is 18.0 V. Applying $\Delta V = IR$ to each resistor gives

$$I_1 = \frac{\Delta V}{R_1} = \frac{18.0\ V}{3.00\ \Omega} = \boxed{6.00\ A}$$

$$I_2 = \frac{\Delta V}{R_2} = \frac{18.0\ V}{6.00\ \Omega} = \boxed{3.00\ A}$$

$$I_3 = \frac{\Delta V}{R_3} = \frac{18.0\ V}{9.00\ \Omega} = \boxed{2.00\ A}$$

(b) Calculate the power delivered to each resistor and the total power delivered to the three resistors.

**Solution**   Applying $\mathcal{P} = I^2 R$ to each resistor gives

3.00-Ω:   $\mathcal{P}_1 = I_1^2 R_1 = (6.00\ A)^2(3.00\ \Omega) = \boxed{108\ W}$

6.00-Ω:   $\mathcal{P}_2 = I_2^2 R_2 = (3.00\ A)^2(6.00\ \Omega) = \boxed{54.0\ W}$

9.00-Ω:   $\mathcal{P}_3 = I_3^2 R_3 = (2.00\ A)^2(9.00\ \Omega) = \boxed{36.0\ W}$

This shows that the smallest resistor receives the most power. (Note that you can also use $\mathcal{P} = (\Delta V)^2/R$ to find the power delivered to each resistor.) Summing the three quantities gives a total power of 198 W.

**Figure 21.21**

(Example 21.9) Three resistors connected in parallel. The voltage across each resistor is 18.0 V.

(c) Calculate the equivalent resistance of the combination of three resistors.

**Solution**   We can use Equation 21.29 to find $R_{eq}$:

$$\frac{1}{R_{eq}} = \frac{1}{3.00\ \Omega} + \frac{1}{6.00\ \Omega} + \frac{1}{9.00\ \Omega}$$

$$R_{eq} = \frac{18.0\ \Omega}{11.0} = \boxed{1.64\ \Omega}$$

**EXERCISE**   Use $R_{eq}$ to calculate the total power delivered to the circuit.

*Answer*   198 W

See Screen 13.4

## 21.8 • KIRCHHOFF'S RULES AND SIMPLE DC CIRCUITS

As indicated in the preceding section, simple circuits can be analyzed using $\Delta V = IR$ and the rules for series and parallel combinations of resistors. However, resistors can be connected so that the circuits formed cannot be reduced to a single equivalent resistor. The procedure for analyzing such complex circuits is greatly simplified by the use of two simple rules called **Kirchhoff's rules:**

- The sum of the currents entering any junction must equal the sum of the currents leaving that junction. (This rule is often referred to as the **junction rule.**)
- The sum of the potential differences across each element around any closed circuit loop must be zero. (This rule is usually called the **loop rule.**)

Kirchhoff's rules are generally used to determine the current in each element in the circuit. In using these rules, we first draw the circuit diagram and assume a direction for the current in each device in the circuit. We draw an arrow representing that direction next to the device and assign a symbol to each independent current, such as $I_1$, $I_2$, and so on. Keep in mind that currents in devices connected in series are the same, so the currents in these devices will have the same assigned symbol.

The junction rule is a statement of **conservation of charge.** Whatever current enters a given point in a circuit must leave that point because charge cannot build up or disappear at a point. If we apply this rule to the junction in Figure 21.22a, we have

$$I_1 = I_2 + I_3$$

Figure 21.22b represents a hydraulic analog to this situation in which water flows through a branched pipe with no leaks. The flow rate into the pipe equals the total flow rate out of the two branches.

The second rule is equivalent to the law of **conservation of energy.** Suppose a charge moves around any closed loop in a circuit (the charge starts and ends at the same point). In this case, the circuit must gain as much energy as it loses. This is the isolated system model for the system of the circuit—no energy is transferred across the boundary of the system, but energy transformations do occur within the system (ignoring energy transfer by radiation and heat into the air from warm circuit elements). The energy of the circuit may decrease due to a potential drop $-IR$ as a charge moves through a resistor or as a result of having the charge move in the reverse direction through an emf. In the latter case, electric potential energy is converted to chemical energy as the battery is charged. The energy increases when the charge moves through a battery in the same direction as the emf.

Another approach to understanding the loop rule is to remember the definition of a conservative force from Chapter 7. One of the mathematical behaviors of a conservative force is that the work done by a conservative force when a member of the system moves around a closed path is zero. A loop in a circuit is a closed path. If we imagine moving a charge around a loop, the total work done by the conservative electric force must be zero. The total work is the sum of positive and negative works as the charge passes through various circuit elements. Because work is related to potential energy changes and because potential energy changes are related to potential differences (Eq. 20.3), the fact that the sum of all the works is zero is equivalent to the sum of all the potential differences being zero, which is Kirchhoff's loop rule.

As an aid in applying the loop rule, the following sign conventions are used. We have already drawn arrows for currents on our diagram and have assigned names to the currents in order to apply the junction rule. To set up the sign

**Gustav Kirchhoff** (1824–1887)

Kirchhoff, a professor at Heidelberg, Germany, and Robert Bunsen invented the spectroscope and founded the science of spectroscopy, which led to atomic spectra such as those seen in Chapter 11. They discovered the elements cesium and rubidium and invented astronomical spectroscopy. Kirchhoff formulated another Kirchhoff's rule, namely, "a cool substance will absorb light of the same wavelengths that it emits when hot." *(AIP ESVA/W. F. Meggers Collection)*

**PITFALL PREVENTION 21.14**

**Charges do not move all the way around a circuit in a short time**

At the beginning of Section 21.5 and in the present discussion of conservation of energy, we have imagined a charge moving completely around a conducting loop. Keep in mind that the movement of a charge all the way around the circuit is not what happens in a circuit, unless you wait for a very long time. Due to the very low value of the drift speed, it might take *hours* for a single electron to make one complete trip around the circuit. In terms of understanding the energy transfer in a circuit, however, it is useful to *imagine* a charge moving all the way around the circuit.

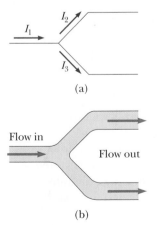

(a)

**Figure 21.22**

(a) A schematic diagram illustrating Kirchhoff's junction rule. Conservation of charge requires that whatever current enters a junction must leave that junction. In this case, therefore, $I_1 = I_2 + I_3$. (b) A mechanical analog of the junction rule: The amount of water flowing out of the branches on the right must equal the amount flowing into the single branch on the left.

Flow in

Flow out

(b)

(a) $\Delta V = -IR$

(b) $\Delta V = +IR$

(c) $\mathcal{E}$ $\Delta V = +\mathcal{E}$

(d) $\mathcal{E}$ $\Delta V = -\mathcal{E}$

**Figure 21.23**

Rules for determining the potential differences across a resistor and a battery. (The battery is assumed to have no internal resistance.) Each circuit element is traversed from *a* to *b*.

conventions, we choose a direction around each loop that we imagine carrying a positive charge—clockwise or counterclockwise. Thus, for any device, two directions need to be considered: one for our chosen current and one for our chosen travel through the device. The sign conventions for potential differences for resistors and batteries based on these two directions are summarized in Figure 21.23, where it is assumed that travel is from point *a* toward point *b*:

- If a resistor is traversed in the direction of the current, the potential difference across the resistor is $-IR$ (Fig. 21.23a).
- If a resistor is traversed in the direction *opposite* the current, the potential difference across the resistor is $+IR$ (Figure 21.23b).
- If a source of emf is traversed in the direction of the emf (from $-$ to $+$ on the terminals), the potential difference is $+\mathcal{E}$ (Fig. 21.23c).
- If a source of emf is traversed in the direction opposite the emf (from $+$ to $-$ on the terminals), the potential difference is $-\mathcal{E}$ (Fig. 21.23d).

The uses of the junction rule and the loop rule have limitations. You may use the junction rule as often as needed so long as each time you write an equation, you include in it a current that has not been used in a previous junction rule equation. In general, the number of times the junction rule can be used is one fewer than the number of junction points in the circuit. The loop rule can be used as often as needed so long as a new circuit element (a resistor or battery) or a new current appears in each new equation. In general, **the number of independent equations you need must equal the number of unknown currents in order to solve a particular circuit problem.**

The following examples illustrate the use of Kirchhoff's rules in analyzing circuits. In all cases, it is assumed that the circuits have reached steady-state conditions; that is, the currents in the various circuit elements are constant. If a capacitor is included as an element in one of the branches, **it acts as an open circuit: the current in the branch containing the capacitor is zero under steady-state conditions.**

**PROBLEM-SOLVING STRATEGY** | Kirchhoff's Rules

1. Draw the circuit diagram, and assign labels to all the known quantities and symbols to all the unknown quantities. You must assign *directions* to the currents in each part of the circuit. Although the assignment of current directions is arbitrary, you must adhere *rigorously* to the directions you assign when you apply Kirchhoff's rules.

2. Apply the junction rule (Kirchhoff's first rule) to all junctions in the circuit except one.

3. Now apply the loop rule (Kirchhoff's second rule) to as many loops in the circuit as are needed to obtain, in combination with step 2, as many equations as unknowns. To apply this rule, you must choose a direction in which to travel around the loop (either clockwise or counterclockwise) and correctly identify the change in potential as you cross each element. Watch out for signs!

4. Solve the equations simultaneously for the unknown quantities. Be careful in your algebraic steps, and check your numerical answers for consistency. Do not be alarmed if any of the resulting currents has a negative value— you have guessed the direction of that current incorrectly, but *its magnitude will be correct.*

## Example 21.10  Applying Kirchhoff's Rules

(a) Find the currents $I_1$, $I_2$, and $I_3$ in the circuit shown in Figure 21.24.

**Reasoning**   We choose the directions of the currents as in Figure 21.24. Applying Kirchhoff's first rule to junction $c$ gives

$$I_1 + I_2 = I_3 \qquad (1)$$

The circuit has three loops: *abcda*, *befcb*, and *aefda* (the outer loop). We need only two loop equations to determine the unknown currents. The third loop equation would give no new information. Applying Kirchhoff's second rule to loops *abcda* and *befcb* and traversing these loops in the clockwise direction, we obtain the expressions

Loop *abcda*:   $10.0\,\text{V} - (6.0\,\Omega)I_1 - (2.0\,\Omega)I_3 = 0 \qquad (2)$

Loop *befcb*:   $-14.0\,\text{V} - 10.0\,\text{V} + (6.0\,\Omega)I_1$
$$- (4.0\,\Omega)I_2 = 0 \quad (3)$$

Note that in loop *befcb*, a positive sign is obtained when traversing the 6.0-$\Omega$ resistor because the direction of the path is opposite the direction of $I_1$. A third loop equation for *aefda* gives $-14.0\,\text{V} - (2.0\,\Omega)I_3 - (4.0\,\Omega)I_2 = 0$, which is just the sum of (2) and (3).

**Solution**   Expressions (1), (2), and (3) represent three independent equations with three unknowns. We can solve the problem as follows: Dropping the units for simplicity and substituting (1) into (2) gives

$$10.0 - 6.0I_1 - 2.0(I_1 + I_2) = 0$$

$$10.0 = 8.0I_1 + 2.0I_2 \qquad (4)$$

Dividing each term in (3) by 2 and rearranging the equation gives

$$-12.0 = -3.0I_1 + 2.0I_2 \qquad (5)$$

Subtracting (5) from (4) eliminates $I_2$, giving

$$22.0 = 11.0I_1$$

$$I_1 = 2.0\,\text{A}$$

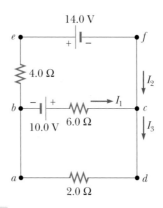

14.0 V

4.0 Ω

$I_1$

10.0 V  6.0 Ω

$I_2$

$I_3$

2.0 Ω

**Figure 21.24**

(Example 21.10) A circuit to be analyzed with Kirchhoff's rules.

Using this value of $I_1$ in (5) gives a value for $I_2$:

$$2.0I_2 = 3.0I_1 - 12.0 = 3.0(2.0) - 12.0 = -6.0$$

$$I_2 = -3.0\,\text{A}$$

Finally, $I_3 = I_1 + I_2 = -1.0\,\text{A}$. Hence, the currents have the values

$$I_1 = \boxed{2.0\,\text{A}} \qquad I_2 = \boxed{-3.0\,\text{A}} \qquad I_3 = \boxed{-1.0\,\text{A}}$$

The fact that $I_2$ and $I_3$ are negative indicates only that we chose the wrong directions for these currents. The numerical values are correct, however.

(b) Find the potential difference between points $b$ and $c$.

**Solution**   In traversing the path from $b$ to $c$ along the central branch, we have

$$V_c - V_b = +10.0\,\text{V} - (6.0\,\Omega)I_1$$

$$= +10.0\,\text{V} - (6.0\,\Omega)(2.0\,\text{A}) = \boxed{-2.0\,\text{V}}$$

**EXERCISE**   Find the potential difference between points $b$ and $c$ by following a path passing through $a$ and $d$.

*Answer*   $V_c - V_b = -2.0\,\text{V}$

## Example 21.11  A Multiloop Circuit

Under steady-state conditions, find the unknown currents in the multiloop circuit shown in Figure 21.25.

**Reasoning**   First note that the capacitor represents an open circuit, and hence no current flows along path *ghab* under

steady-state conditions. Therefore, $I_{fg} = I_{gb} = I_{bc} \equiv I_1$. Labeling the currents as shown in Figure 21.25 and applying Kirchhoff's first rule to junction $c$, we have

$$I_1 + I_2 = I_3 \qquad (1)$$

Kirchhoff's second rule applied to loops *defcd* and *cfgbc*, traveled clockwise, gives

Loop *defcd*:   $4.00 \text{ V} - (3.00 \ \Omega) I_2 - (5.00 \ \Omega) I_3 = 0$   **(2)**

Loop *cfgbc*:   $8.00 \text{ V} - (5.00 \ \Omega) I_1 + (3.00 \ \Omega) I_2 = 0$   **(3)**

**Solution**   From (1) we see that $I_1 = I_3 - I_2$, which when substituted into (3) gives

$$8.00 \text{ V} - (5.00 \ \Omega) I_3 + (8.00 \ \Omega) I_2 = 0 \qquad \textbf{(4)}$$

Subtracting (4) from (2), we eliminate $I_3$ and find

$$I_2 = -\frac{4.00}{11.00} \text{A} = -0.364 \text{ A}$$

Because $I_2$ is negative, we conclude that $I_2$ is from *c* to *f* through the 3.00-$\Omega$ resistor. Using this value of $I_2$ in (3) and (1) gives the following values for $I_1$ and $I_3$:

$$I_1 = 1.38 \text{ A} \qquad I_3 = 1.02 \text{ A}$$

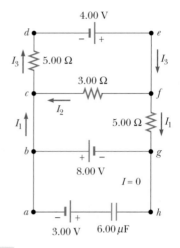

**Figure 21.25**

(Example 21.11) A multiloop circuit.

## 21.9 • *RC* CIRCUITS

So far we have been concerned with circuits with constant currents, or so-called *steady-state circuits*. We now consider circuits containing capacitors in which the currents may vary in time.

### Charging a Capacitor

Consider the series circuit shown in Figure 21.26. Let us assume that the capacitor is initially uncharged. No current exists when switch S is open (Fig. 21.26b). If the switch is closed at $t = 0$, charge begins to flow, setting up a current in the circuit,* and the capacitor begins to charge (Fig. 21.26c). Note that during the charging, charges do not jump across the plates of the capacitor, because the gap between the plates represents an open circuit. Instead, due to the electric field in the wires established by the battery, electrons move into the top plate from the wires and out of the bottom plate into the wires until the capacitor is fully charged. The value of the maximum charge depends on the emf of the battery. Once the maximum charge is reached, the current in the circuit is zero.

To put this discussion on a quantitative basis, let us apply Kirchhoff's second rule to the circuit after the switch is closed. In our sign conventions, we did not specify a convention for the potential difference across a capacitor. From our study of capacitors in Chapter 20, however, it should be clear that carrying a positive charge across a capacitor from − to + would represent an increase in potential energy for the circuit—a positive potential difference. Traversing the capacitor in the opposite direction would correspond to a decrease in potential energy—a negative potential difference.

---

* By "a current in the circuit," we mean current in all parts of the circuit *except* for the region between the plates of the capacitor.

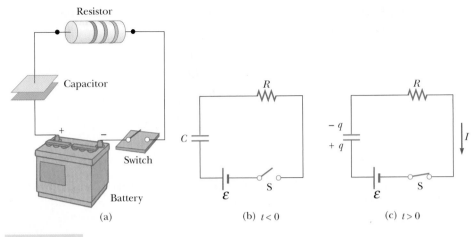

**Figure 21.26**

(a) A capacitor in series with a resistor, switch, and battery. (b) Circuit diagram representing this system at time $t < 0$, before the switch is closed. (c) Circuit diagram at time $t > 0$, after the switch has been closed.

Choosing clockwise as our direction around the circuit in Figure 21.26, and applying the sign convention for capacitors that we have just discussed, we have

$$\mathcal{E} - \frac{q}{C} - IR = 0 \qquad \text{[21.30]}$$

where $-q/C$ is the potential difference across the capacitor and $-IR$ is the potential difference across the resistor consistent with our direction of travel. Note that $q$ and $I$ are *instantaneous* values of the charge and current, respectively, as the capacitor is being charged.

We can use Equation 21.30 to find the initial current in the circuit and the maximum charge on the capacitor. At $t = 0$, when the switch is closed, the charge on the capacitor is zero, and from Equation 21.30 we find that the initial current in the circuit $I_0$ is a maximum and equal to

$$I_0 = \frac{\mathcal{E}}{R} \qquad \text{[21.31]}$$

At this time, **the potential difference is entirely across the resistor.** Later, when the capacitor is charged to its maximum value $Q$, charges cease to flow, the current in the circuit is zero, and **the potential difference is entirely across the capacitor.** Substituting $I = 0$ into Equation 21.30 yields the following expression for $Q$:

$$Q = C\mathcal{E} \qquad \text{(maximum charge)} \qquad \text{[21.32]}$$

To determine analytical expressions for the time dependence of the charge and current, we must solve Equation 21.30. To do this, let us substitute $I = dq/dt$ and rearrange the equation:

$$\frac{dq}{dt} = \frac{\mathcal{E}}{R} - \frac{q}{RC}$$

This is a differential equation whose solution is the time-dependent charge on the capacitor. An expression for $q$ may be found in the following way. We rearrange the equation by placing terms involving $q$ on the left side and those involving $t$ on the right side. Then we integrate both sides from the moment when the switch is closed to an arbitrary later instant:

$$\frac{dq}{(q - C\mathcal{E})} = -\frac{1}{RC} dt$$

$$\int_0^q \frac{dq}{(q - C\mathcal{E})} = -\frac{1}{RC} \int_0^t dt$$

$$\ln\left(\frac{q - C\mathcal{E}}{- C\mathcal{E}}\right) = -\frac{t}{RC}$$

Using the definition of the natural logarithm, we can solve this expression for the charge on the capacitor as a function of time:

- *Charge versus time for a charging capacitor*

$$q(t) = C\mathcal{E}[1 - e^{-t/RC}] = Q[1 - e^{-t/RC}] \qquad \textbf{[21.33]}$$

where $e$ is the base of the natural logarithm (*not* the charge on the electron!) and $Q = C\mathcal{E}$ is the maximum charge on the capacitor.

An expression for the current may be found by differentiating Equation 21.33 with respect to time. Using $I = dq/dt$, we obtain

- *Current versus time for a charging capacitor*

$$I(t) = \frac{\mathcal{E}}{R} e^{-t/RC} \qquad \textbf{[21.34]}$$

where $\mathcal{E}/R$ is the initial current in the circuit.

Plots of charge and current versus time are shown in Figure 21.27. Note that the charge is zero at $t = 0$ and approaches the maximum value of $C\mathcal{E}$ as $t \to \infty$ (Fig. 21.27a). Furthermore, the current has its maximum value of $I_0 = \mathcal{E}/R$ at $t = 0$ and decays exponentially to zero as $t \to \infty$ (Fig. 21.27b). The quantity $RC$ that appears in the exponential of Equations 21.33 and 21.34 is called the **time constant** $\tau$ of the circuit. It represents the time interval during which the current decreases to $1/e$ of its initial value; that is, at the end of the time interval $\tau$,

**WEB**

For a java applet showing the behavior of an *RC* circuit, visit **www.phy.ntnu.edu.tw/java/rc/rc.html**

(a)

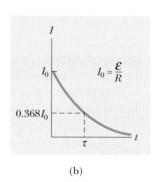

(b)

**Figure 21.27**

(a) Plot of capacitor charge versus time for the circuit shown in Figure 21.26. After a time interval equal to one time constant $\tau$ has passed, the charge is 63.2% of the maximum value $C\mathcal{E}$. The charge approaches its maximum value as $t$ approaches infinity. (b) Plot of current versus time for the *RC* circuit shown in Figure 21.26. The current has its maximum value $I_0 = \mathcal{E}/R$ at $t = 0$ and decays to zero exponentially as $t$ approaches infinity. After a time interval equal to one time constant $\tau$ has passed, the current is 36.8% of its initial value.

$I = e^{-1}I_0 = 0.368I_0$. After the time interval $2\tau$, $I = e^{-2}I_0 = 0.135I_0$, and so forth. Likewise, in a time interval $\tau$ the charge increases from zero to $C\mathcal{E}[1 - e^{-1}] = 0.632C\mathcal{E}$.

The energy decrease of the battery during the charging process is the product of the total charge and the emf, $Q\mathcal{E} = C\mathcal{E}^2$. After the capacitor is fully charged, the energy stored in it is $\frac{1}{2}Q\mathcal{E} = \frac{1}{2}C\mathcal{E}^2$, which is just half the energy decrease of the battery. It is left to an end-of-chapter problem to show that the remaining half of the energy supplied by the battery appears as internal energy in the resistor (Problem 58).

## Discharging a Capacitor

Now consider the circuit in Figure 21.28, consisting of a capacitor with an initial charge $Q$, a resistor, and a switch. When the switch is open (Fig. 21.28a), a potential difference of $Q/C$ exists across the capacitor and zero potential difference exists across the resistor because $I = 0$. If the switch is closed at $t = 0$, the capacitor begins to discharge through the resistor. At some time during the discharge, the current in the circuit is $I$ and the charge on the capacitor is $q$ (Fig. 21.28b).

The circuit of Figure 21.28 is the same as the circuit of Figure 21.26 except for the absence of the battery. Thus, we modify the Kirchhoff's rule expression in Equation 21.30 by dropping the emf:

$$-\frac{q}{C} - IR = 0 \qquad \text{[21.35]}$$

Because $I = dq/dt$, Equation 21.35 becomes

$$-R\frac{dq}{dt} = \frac{q}{C}$$

$$\frac{dq}{q} = -\frac{1}{RC}\,dt$$

Integrating this expression from the moment the switch is closed, at which time $q = Q$, to an arbitrary later instant gives

$$\int_Q^q \frac{dq}{q} = -\frac{1}{RC}\int_0^t dt$$

$$\ln\left(\frac{q}{Q}\right) = -\frac{t}{RC}$$

$$q(t) = Qe^{-t/RC} \qquad \text{[21.36]}$$

Differentiating Equation 21.36 with respect to time gives the current as a function of time:

$$I(t) = \frac{dq}{dt} = -I_0 e^{-t/RC} \qquad \text{[21.37]}$$

where the initial current is $I_0 = Q/RC$. Thus we see that both the charge on the capacitor and the current decay exponentially at a rate characterized by the time constant $\tau = RC$. The negative sign in Equation 21.37 indicates the direction of the current, which is opposite to the direction during the charging process.

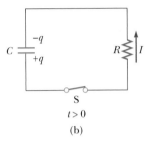

**Figure 21.28**

(a) A charged capacitor connected to a resistor and a switch, which is open at $t < 0$. (b) After the switch is closed at $t = 0$, the current in the circuit and the charge on the capacitor decrease exponentially with time.

- *Charge versus time for a discharging capacitor*

- *Current versus time for a discharging capacitor*

(Thinking Physics 21.6) The *RC* circuit in a roadway construction flasher. When the switch is closed, the charge on the capacitor increases until the voltage across the capacitor (and across the flash lamp) is high enough for the lamp to flash, discharging the capacitor.

### THINKING PHYSICS 21.6

Many roadway construction sites have flashing yellow lights to warn motorists of possible dangers. What causes the lightbulbs to flash?

**Reasoning**   A typical circuit for such a flasher is shown in Figure 21.29. The lamp *L* is a gas-filled lamp that acts as an open circuit until a large potential difference causes an electric discharge in the gas, which gives off a bright light. During this discharge, charge flows through the gas between the electrodes of the lamp. After switch S is closed, the battery charges up the capacitor of capacitance *C*. At the beginning, the current is high and the charge on the capacitor is low, so that most of the potential difference appears across the resistance *R*. As the capacitor charges, more potential difference appears across it, reflecting the lower current and, thus, lower potential difference across the resistor. Eventually, the potential difference across the capacitor reaches a value at which the lamp will conduct, causing a flash. This discharges the capacitor through the lamp and the process of charging begins again. The period between flashes can be adjusted by changing the time constant of the *RC* circuit.

### THINKING PHYSICS 21.7

Many automobiles are equipped with windshield wipers that can be used intermittently during a light rainfall. How does the operation of this feature depend on the charging and discharging of a capacitor?

**Reasoning**   The wipers are part of an *RC* circuit whose time constant can be varied by selecting different values of *R* through a multipositioned switch. The time between sweeps of the wipers is determined by the value of the time constant of the circuit.

### Example 21.12   Charging a Capacitor in an *RC* Circuit

An uncharged capacitor and a resistor are connected in series to a battery as in Figure 21.26. If $\mathcal{E} = 12.0$ V, $C = 5.00$ $\mu$F, and $R = 8.00 \times 10^5$ $\Omega$, find the time constant of the circuit, the maximum charge on the capacitor, the maximum current in the circuit, and the charge and current as a function of time.

**Solution**   Because this is a simple *RC* circuit, the time constant of the circuit is $\tau = RC = (8.00 \times 10^5 \ \Omega)(5.00 \times 10^6 \ \text{F}) = 4.00$ s. The maximum charge on the capacitor is $Q = C\mathcal{E} = (5.00 \times 10^{-6} \ \text{F})(12.0 \ \text{V}) = 60.0 \ \mu$C. The maxi-

mum current in the circuit is $I_0 = \mathcal{E}/R = (12.0 \ \text{V})/(8.00 \times 10^5 \ \Omega) = 15.0 \ \mu$A. Using these values and Equations 21.33 and 21.34, we find that

$$q(t) = (60.0 \ \mu\text{C})[1 - e^{-t/4.00}]$$

$$I(t) = (15.0 \ \mu\text{A})e^{-t/4.00}$$

**EXERCISE**   Calculate the charge on the capacitor and the current in the circuit after one time constant has elapsed.

*Answer*   37.9 $\mu$C, 5.52 $\mu$A

## Example 21.13   Discharging a Capacitor in an *RC* Circuit

Consider a capacitor *C* being discharged through a resistor *R* as in Figure 21.28. (a) After how many time constants is the charge on the capacitor one fourth of its initial value?

**Solution**   The charge on the capacitor varies with time according to Equation 21.36, $q(t) = Qe^{-t/RC}$. To find the time at which the charge $q$ has dropped to one fourth of its initial value, we substitute $q(t) = Q/4$ into this expression and solve for $t$:

$$\tfrac{1}{4}Q = Qe^{-t/RC}$$

$$\tfrac{1}{4} = e^{-t/RC}$$

Taking the natural logarithm of both sides, we find

$$-\ln 4 = -\frac{t}{RC}$$

$$t = RC\ln 4 = \boxed{1.39RC}$$

(b) The energy stored in the capacitor decreases with time as it discharges. After how many time constants is this stored energy one fourth of its initial value?

**Solution**   Using Equations 20.29 and 21.36, we can express the energy stored in the capacitor at any time $t$ as

$$U = \frac{q^2}{2C} = \frac{Q^2}{2C}e^{-2t/RC} = U_0 e^{-2t/RC}$$

where $U_0$ is the initial energy stored in the capacitor. Similar to part (a), we now set $U = U_0/4$ and solve for $t$:

$$\tfrac{1}{4}U_0 = U_0 e^{-2t/RC}$$

$$\tfrac{1}{4} = e^{-2t/RC}$$

Again, taking the natural logarithm of both sides and solving for $t$ gives

$$t = \tfrac{1}{2}RC\ln 4 = \boxed{0.693RC}$$

**EXERCISE**   After how many time constants is the current in the *RC* circuit one half of its initial value?

*Answer*   0.693*RC*

**EXERCISE**   A 10.0-$\mu$F capacitor is charged by a 10.0-V battery through a resistance *R*. The capacitor reaches a potential difference of 4.00 V at a time of 3.00 s after charging begins. Find *R*.

*Answer*   587 k$\Omega$

$$U = U_0 e^{-2t/RC}$$

### context connection
## 21.10 • THE ATMOSPHERE AS A CONDUCTOR

When discussing capacitors with air between the plates in Chapter 20, we adopted the simplification model that air was a perfect insulator. Although this was a good model for typical potential differences encountered in capacitors, we know that it is possible for a current to exist in air. Lightning is a dramatic example of this possibility, but a more mundane example is the common spark that you might receive when touching a doorknob after rubbing your feet across a carpet.

Let us analyze the process that occurs in electric discharge, which is the same for lightning and the doorknob spark, except for the size of the current. At any given time, air contains a number of ionized molecules, due to cosmic ray collisions and other events (Fig. 21.30a). For a relatively weak electric field, such as the fair-weather electric field, these ions and freed electrons accelerate slowly due to the electric force. They collide with other molecules with no effect and eventually neutralize as a freed electron ultimately finds an ion and combines with it. In a strong electric field, however, such as that associated with a thunderstorm, the freed electrons can accelerate to very high speeds (Fig. 21.30b) before making a collision with a molecule (Fig. 21.30c). If the field is strong enough, the electron may have enough energy to ionize the molecule in this collision (Fig. 21.30d). Now two electrons are accelerated by the field, and each of these can strike another molecule at high speed (Fig. 21.30e). The result is a very rapid increase in the number of charge carriers available in the air and a corresponding decrease in resistance of the air. As a result, a large current in the air tends to neutralize the

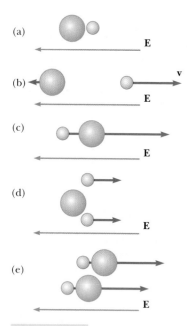

**Figure 21.30**

The anatomy of a spark. (a) An atom is ionized due to a random event. (b) The ion accelerates slowly and the electron accelerates rapidly due to the force from the electric field. (c) The accelerated electron approaches another atom at high speed. (d) The new atom is ionized—the original electron and the new electron accelerate rapidly. (e) These electrons approach other atoms, freeing two more electrons, and an avalanche of ionization proceeds.

charges that established the initial potential difference, such as the charges in the cloud and on the ground. When this happens, we have lightning.

Typical currents during lightning strikes can be very high. While the stepped leader is making its way toward the ground, the current is relatively modest—in the range of 200 to 300 A. This is large compared with typical household currents, but small compared with peak currents in lightning discharges. Once the connection is made between the stepped leader and the return stroke, the current rises rapidly to a typical value of $5 \times 10^4$ A. Considering that typical potential differences between cloud and ground in a thunderstorm can be measured in hundreds of thousands of volts, the power during a lightning stroke is measured in billions of watts. Much of the energy in the stroke is delivered to the air, resulting in a rapid temperature increase and the resultant flash of light and sound of thunder.

Even in the absence of a thundercloud, charge flows through the air. The ions in the air make the air a conductor, although not a very good one. Atmospheric measurements indicate a typical potential difference across our atmospheric capacitor (Section 20.11) of about $3 \times 10^5$ V. As we shall show in the Context Conclusion, the total resistance of the air between the plates in the atmospheric capacitor is about 300 $\Omega$. Thus, the average fair-weather current in the air is

$$I = \frac{\Delta V}{R} = \frac{3 \times 10^5 \text{ V}}{300 \ \Omega} \approx 1 \times 10^3 \text{ A}$$

A number of simplifying assumptions were made in these calculations, but this result is on the right order of magnitude for the global current. Although this might seem surprisingly large, remember that this current is spread out over the entire surface area of the Earth. Thus, the average fair-weather current density is

$$J = \frac{I}{A} = \frac{I}{4\pi r^2} = \frac{1 \times 10^3 \text{ A}}{4\pi (6.4 \times 10^6 \text{ m})^2} \approx 2 \times 10^{-12} \text{ A/m}^2$$

In comparison, the current density in a lightning strike is on the order of $10^5$ A/m$^2$.

The fair-weather current and the lightning current are in opposite directions. The fair-weather current delivers positive charge to the ground, while lightning delivers negative charge. These two effects are in balance,* which is the principle that we shall use to estimate the average number of lightning strikes on the Earth in the Context Conclusion.

## SUMMARY

The **electric current** $I$ in a conductor is defined as

$$I \equiv \frac{dQ}{dt} \qquad \text{[21.2]}$$

where $dQ$ is the charge that passes through a cross section of the conductor in the time interval $dt$. The SI unit of current is the ampere (A); 1 A = 1 C/s.

The current in a conductor is related to the motion of the charge carriers through the relationship

$$I = nqv_d A \qquad \text{[21.4]}$$

where $n$ is the density of charge carriers, $q$ is their charge, $v_d$ is the **drift speed,** and $A$ is the cross-sectional area of the conductor.

---

* There are a number of other effects, too, but we will adopt a simplification model in which these are the only two effects. For more information, see "The Global Electric Circuit" by E. A. Bering, A. A. Few, and J. R. Benbrook in *Physics Today*, October 1998.

The **resistance** $R$ of a conductor is defined as the ratio of the potential difference across the conductor to the current:

$$R \equiv \frac{\Delta V}{I} \qquad \text{[21.6]}$$

The SI units of resistance are volts per ampere, defined as ohms ($\Omega$); $1\ \Omega = 1\ \text{V/A}$.

If the resistance is independent of the applied voltage, the conductor obeys **Ohm's law,** and conductors that have a constant resistance over a wide range of voltages are said to be **ohmic.**

If a conductor has a uniform cross-sectional area $A$ and a length $\ell$, its resistance is

$$R = \rho \frac{\ell}{A} \qquad \text{[21.8]}$$

where $\rho$ is called the **resistivity** of the material from which the conductor is made. The inverse of the resistivity is defined as the **conductivity** $\sigma = 1/\rho$.

The resistivity of a conductor varies with temperature in an approximately linear fashion; that is

$$\rho = \rho_0[1 + \alpha(T - T_0)] \qquad \text{[21.10]}$$

where $\alpha$ is the **temperature coefficient of resistivity** and $\rho_0$ is the resistivity at some reference temperature $T_0$.

In a classical model of electronic conduction in a metal, the electrons are treated as molecules of a gas. In the absence of an electric field, the average velocity of the electrons is zero. When an electric field is applied, the electrons move (on the average) with a drift velocity $\mathbf{v}_d$, which is opposite the electric field:

$$\mathbf{v}_d = \frac{-e\mathbf{E}}{m_e}\tau \qquad \text{[21.15]}$$

where $\tau$ is the average time between collisions with the atoms of the metal. The resistivity of the material according to this model is

$$\rho = \frac{m_e}{ne^2\tau} \qquad \text{[21.18]}$$

where $n$ is the number of free electrons per unit volume.

If a potential difference $\Delta V$ is maintained across a circuit element, the **power,** or the rate at which energy is delivered to the circuit element, is

$$\mathscr{P} = I\,\Delta V \qquad \text{[21.20]}$$

Because the potential difference across a resistor is $\Delta V = IR$, we can express the power delivered to a resistor in the form

$$\mathscr{P} = I^2 R = \frac{(\Delta V)^2}{R} \qquad \text{[21.21]}$$

The **emf** of a battery is the voltage across its terminals when the current is zero. Because of the voltage drop across the **internal resistance** $r$ of a battery, the **terminal voltage** of the battery is less than the emf when a current exists in the battery.

The **equivalent resistance** of a set of resistors connected in **series** is

$$R_{\text{eq}} = R_1 + R_2 + R_3 \cdots \qquad \text{[21.27]}$$

The **equivalent resistance** of a set of resistors connected in **parallel** is given by

$$\frac{1}{R_{\text{eq}}} = \frac{1}{R_1} + \frac{1}{R_2} + \frac{1}{R_3} \cdots \qquad \text{[21.29]}$$

Complex circuits involving more than one loop are analyzed using two simple rules called **Kirchhoff's rules:**

- The sum of the currents entering any junction must equal the sum of the currents leaving that junction.
- The sum of the potential differences across the elements around any closed-circuit loop must be *zero.*

When a resistor is traversed in the direction of the current, the change in potential $\Delta V$ across the resistor is $-IR$. If a resistor is traversed in the direction opposite the current, $\Delta V = +IR$.

If a source of emf is traversed in the direction of the emf (negative to positive) the change in potential is $+\mathcal{E}$. If it is traversed opposite the emf (positive to negative), the change in potential is $-\mathcal{E}$.

If a capacitor is charged with a battery of emf $\mathcal{E}$ through a resistance $R$, the charge on the capacitor and the current in the circuit vary in time according to the expressions

$$q(t) = Q[1 - e^{-t/RC}] \qquad \text{[21.33]}$$

$$I(t) = \frac{\mathcal{E}}{R}\,e^{-t/RC} \qquad \text{[21.34]}$$

where $Q = C\mathcal{E}$ is the *maximum* charge on the capacitor. The product $RC$ is called the **time constant** of the circuit.

If a charged capacitor is discharged through a resistance $R$, the charge and current decrease exponentially in time according to the expressions

$$q(t) = Qe^{-t/RC} \qquad \text{[21.36]}$$

$$I(t) = -I_0 e^{-t/RC} \qquad \text{[21.37]}$$

where $I_0 = Q/RC$ is the initial current in the circuit and $Q$ is the initial charge on the capacitor.

## QUESTIONS

1. In an analogy between automobile traffic flow and electric current, what would correspond to the charge $Q$? What would correspond to the current $I$?

2. What factors affect the resistance of a conductor?

3. We have seen that an electric field must exist inside a conductor that carries a current. How is this possible in view of the fact that in *electrostatics*, we concluded that $E$ must be zero inside a conductor?

4. Two wires A and B of circular cross-section are made of the same metal and have equal lengths, but the resistance of wire A is three times greater than that of wire B. What is the ratio of their cross-sectional areas? How do their radii compare?

5. Use the atomic theory of matter to explain why the resistance of a material should increase as its temperature increases.

6. Explain how a current can persist in a superconductor without any applied voltage.

7. What would happen to the drift velocity of the electrons in a wire and to the current in the wire if the electrons could move freely without resistance through the wire?

8. If charges flow very slowly through a metal, why does it not require several hours for a light to come on when you throw a switch?

9. If you were to design an electric heater using Nichrome wire as the heating element, what parameters of the wire could you vary to meet a specific power output, such as 1 000 W?

10. Car batteries are often rated in ampere-hours. Does this designate the amount of current, power, energy, or charge that can be drawn from the battery?

11. How would you connect resistors so that the equivalent resistance is larger than the individual resistances? Give an example involving two or three resistors.

12. How would you connect resistors so that the equivalent resistance is smaller than the individual resistances? Give an example involving two or three resistors.

13. Why is it possible for a bird to sit on a high-voltage wire without being electrocuted?

14. A "short circuit" is a circuit containing a path of very low resistance in parallel with some other part of the circuit. Discuss the effect of a short circuit on the portion of the circuit it parallels. Use a lamp with a frayed line cord as an example.

15. A series circuit consists of three identical lamps connected to a battery as in Figure Q21.15. When the switch S is closed, what happens (a) to the intensities of lamps A and B; (b) to the intensity of lamp C; (c) to the current in the circuit; and (d) to the voltage drop across the three lamps? (e) Does the power delivered to the circuit increase, decrease, or remain the same?

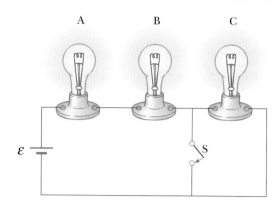

**Figure Q21.15**

16. Two lightbulbs both operate from 110 V, but one has a power rating of 25 W and the other of 100 W. Which bulb has the higher resistance? Which bulb carries the greater current?

17. If electric power is transmitted over long distances, the resistance of the wires becomes significant. Why? Which mode of transmission would result in less energy loss—high current and low voltage or low current and high voltage? Discuss.

18. Two sets of Christmas-tree lights are available. For set A, when one bulb is removed (or fails), the remaining bulbs remain illuminated. For set B, when one bulb is removed, the remaining bulbs do not operate. Explain the difference in wiring for the two sets.

19. Are the two headlights on a car wired in series or in parallel? How can you tell?

20. Embodied in Kirchhoff's rules are two conservation laws. What are they?

21. In Figure Q21.21, describe what happens to the lightbulb after the switch is closed. Assume the capacitor has a large

Switch      Battery

**Figure Q21.21**

capacitance and is initially uncharged, and assume that the light illuminates when connected directly across the battery terminals.

**22.** Figure Q21.22 shows a series connection of three lamps, all rated at 120 V, with power ratings of 60 W, 75 W, and 200 W. Why do the intensities of the lamps differ? Which lamp has the greatest resistance? How would their intensities differ if they were connected in parallel?

**Figure Q21.22**  *(Henry Leap and Jim Lehman)*

**23.** A student claims that a second lightbulb in series is less bright than the first, because the first bulb uses up some of the current. How would you respond to this statement?

**24.** A ski resort consists of a few ski lifts and several interconnecting downhill runs on the side of a mountain. The lifts are analogous to batteries and the runs are analogous to resistors. Sketch two runs in series. Sketch three runs in parallel. Sketch a junction between a lift and two runs. State Kirchhoff's junction rule for ski resorts. One of the skiers happens to be carrying a skydiver's altimeter. She never takes the same set of lifts and runs twice, but nevertheless passes you again and again at the fixed location where you are working. State Kirchhoff's loop rule for ski resorts.

# PROBLEMS

1, 2, 3 = straightforward, intermediate, challenging     ☐ = full solution available in the *Student Solutions Manual and Study Guide*

**web** = solution posted at **http://www.harcourtcollege.com/physics/**     💻 = computer useful in solving problem

🌐 = Interactive Physics     ▨ = paired numerical/symbolic problems     🔖 = life science application

## Section 21.1   Electric Current

**1.** In a particular cathode-ray tube, the measured beam current is 30.0 $\mu$A. How many electrons strike the tube screen every 40.0 s?

**2.** A teapot with a surface area of 700 cm$^2$ is to be silver plated. It is attached to the negative electrode of an electrolytic cell containing silver nitrate ($Ag^+NO_3^-$). If the cell is powered by a 12.0-V battery and has a resistance of 1.80 $\Omega$, how long does it take for a 0.133-mm layer of silver to build up on the teapot? (Note that the density of silver is $10.5 \times 10^3$ kg/m$^3$.)

**3.** A small sphere that carries a charge $q$ is whirled in a circle at the end of an insulating string. The angular frequency of rotation is $\omega$. What average current does this rotating charge represent?

**4.** The quantity of charge $q$ (in coulombs) that passes through a surface of area 2.00 cm$^2$ varies with time according to the equation $q = 4t^3 + 5t + 6$, where $t$ is in seconds. (a) What is the instantaneous current through the surface at $t = 1.00$ s? (b) What is the value of the current density?

**5.** Suppose that the current in a conductor decreases exponentially with time according to the equation $I(t) = I_0 e^{-t/\tau}$, where $I_0$ is the initial current (at $t = 0$) and $\tau$ is a constant having dimensions of time. Consider a fixed observation point within the conductor. (a) How much charge passes this point between $t = 0$ and $t = \tau$? (b) How much charge passes this point between $t = 0$ and $t = 10.0\tau$? (c) How much charge passes this point between $t = 0$ and $t = \infty$?

**6.** An aluminum wire having a cross-sectional area of $4.00 \times 10^{-6}$ m$^2$ carries a current of 5.00 A. Find the drift speed of the electrons in the wire. The density of aluminum is 2.70 g/cm$^3$. Assume that one electron is supplied by each atom.

## Section 21.2    Resistance and Ohm's Law

7. A lightbulb has a resistance of 240 $\Omega$ when operating at a voltage of 120 V. What is the current in the lightbulb?

8. Suppose that you wish to fabricate a uniform wire out of 1.00 g of copper. If the wire is to have a resistance of $R = 0.500\ \Omega$, and if all of the copper is to be used, what will be (a) the length and (b) the diameter of this wire?

9. A 0.900-V potential difference is maintained across a
web 1.50-m length of tungsten wire that has a cross-sectional area of 0.600 mm$^2$. What is the current in the wire?

10. While taking photographs in Death Valley on a day when the temperature is 58.0°C, Bill Hiker finds that a certain voltage applied to a copper wire produces a current of 1.000 A. Bill then travels to Antarctica and applies the same voltage to the same wire. What current does he register there if the temperature is −88.0°C? Assume that no change occurs in the wire's shape and size.

11. An aluminum wire with a diameter of 0.100 mm has a uniform electric field of 0.200 V/m imposed along its entire length. The temperature of the wire is 50.0°C. Assume one free electron per atom. (a) Using the information in Table 21.1, determine the resistivity. (b) What is the current density in the wire? (c) What is the total current in the wire? (d) What is the drift speed of the conduction electrons? (e) What potential difference must exist between the ends of a 2.00-m length of the wire to produce the stated electric field?

12. **Review Problem.** An aluminum rod has a resistance of 1.234 $\Omega$ at 20.0°C. Calculate the resistance of the rod at 120°C by accounting for the changes in both the resistivity and the dimensions of the rod.

## Section 21.4    A Structural Model for Electrical Conduction

13. If the drift velocity of free electrons in a copper wire is $7.84 \times 10^{-4}$ m/s, what is the electric field in the conductor?

14. If the current carried by a conductor is doubled, what happens to the (a) charge carrier density? (b) current density? (c) electron drift velocity? (d) average time between collisions?

## Section 21.5    Electric Energy and Power

15. A toaster is rated at 600 W when connected to a 120-V source. What current does the toaster carry, and what is its resistance?

16. In a hydroelectric installation, a turbine delivers 1 500 hp to a generator, which in turn converts 80.0% of the mechanical energy into electric energy. Under these conditions, what current does the generator deliver at a terminal potential difference of 2 000 V?

17. Suppose that a voltage surge produces 140 V for a moment. By what percentage does the power output of a 120-V, 100-W lightbulb increase? Assume that its resistance does not change.

18. **Review Problem.** The heating element of a coffee maker operates at 120 V and carries a current of 2.00 A. Assuming that the water absorbs all of the energy converted by the resistor, calculate how long it takes to heat 0.500 kg of water from room temperature (23.0°C) to the boiling point.

19. A certain toaster has a heating element made of Nichrome resistance wire. When the toaster is first connected to a 120-V source of potential difference (and the wire is at a temperature of 20.0°C), the initial current is 1.80 A. The current begins to decrease as the resistive element warms up, however. When the toaster has reached its final operating temperature, the current has dropped to 1.53 A. (a) Find the power delivered to the toaster when it is at its operating temperature. (b) What is the final temperature of the heating element?

20. We estimate that there are 270 million plug-in electric clocks in the United States, approximately one clock for each person. The clocks have an average power rating of 2.50 W. To supply the energy to operate these clocks, how many metric tons of coal are burned per hour in coal-fired electric generating plants that are, on average, 25.0% efficient? The heat of combustion for coal is 33.0 MJ/kg.

21. Estimate the cost of one person's routine use of a hair dryer for one year. If you do not use a blow dryer yourself, observe or interview someone who does. State the quantities you estimate and their values.

22. The cost of electricity varies widely through the United States; $0.120/kWh is one typical value. At this unit price, calculate the cost of (a) leaving a 40.0-W porch light on for two weeks while you are on vacation, (b) making a piece of dark toast in 3.00 min with a 970-W toaster, and (c) drying a load of clothes in 40.0 min in a 5 200-W dryer.

## Section 21.6    Sources of emf

23. A battery has an emf of 15.0 V. The terminal voltage of the battery is 11.6 V when it is delivering 20.0 W of power to an external load resistor $R$. (a) What is the value of $R$? (b) What is the internal resistance of the battery?

24. Two 1.50-V batteries—with their positive terminals in the same direction—are inserted in series into the barrel of a flashlight. One battery has an internal resistance of 0.255 $\Omega$, the other an internal resistance of 0.153 $\Omega$. When the switch is closed, a current of 600 mA occurs in the lamp. (a) What is the lamp's resistance? (b) What fraction of the chemical energy converted appears as internal energy in the batteries?

## Section 21.7   Resistors in Series and in Parallel

25. (a) Find the equivalent resistance between points $a$ and $b$ in Figure P21.25. (b) A potential difference of 34.0 V is applied between points $a$ and $b$. Calculate the current in each resistor.

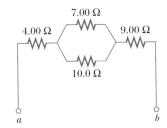

**Figure P21.25**

26. A lightbulb marked "75 W [at] 120 V" is screwed into a socket at one end of a long extension cord in which each of the two conductors has a resistance of 0.800 Ω. The other end of the extension cord is plugged into a 120-V outlet. Draw a circuit diagram, and find the actual power delivered to the bulb in this circuit.

27. Consider the circuit shown in Figure P21.27. Find (a) the current in the 20.0-Ω resistor and (b) the potential difference between points $a$ and $b$.

web

**Figure P21.27**

28. For the purpose of measuring the electrical resistance of shoes through the body of the wearer to a metal ground plate, the American National Standards Institute (ANSI) specifies the circuit shown in Figure P21.28. The potential difference $\Delta V$ across the 1.00-MΩ resistor is measured with a high-resistance voltmeter. (a) Show that the resistance of the footwear is given by

$$R_{\text{shoes}} = 1.00 \text{ M}\Omega \left( \frac{50.0 \text{ V} - \Delta V}{\Delta V} \right)$$

(b) In a medical test, a current in the human body should not exceed 150 μA. Can the current delivered by the ANSI-specified circuit exceed 150 μA? To decide, consider a person standing barefoot on the ground plate.

**Figure P21.28**

29. Calculate the power delivered to each resistor in the circuit shown in Figure P21.29.

**Figure P21.29**

30. Three 100-Ω resistors are connected as shown in Figure P21.30. The maximum power that can safely be delivered to any one resistor is 25.0 W. (a) What is the maximum voltage that can be applied to the terminals $a$ and $b$? (b) For the voltage determined in part (a), what is the power delivered to each resistor? What is the total power delivered?

**Figure P21.30**

**31.** An electric heater is rated at 1 500 W, a toaster at 750 W, and an electric grill at 1 000 W. The three appliances are connected to a common 120-V household circuit. (a) How much current does each draw? (b) Is a 25.0-A circuit breaker sufficient in this situation? Explain your answer.

## Section 21.8   Kirchhoff's Rules and Simple DC Circuits

*Note:* The currents are not necessarily in the direction shown for some circuits.

**32.** The ammeter shown in Figure P21.32 reads 2.00 A. Find $I_1$, $I_2$, and $\mathcal{E}$.

**Figure P21.32**

**33.** Determine the current in each branch of the circuit
**web** shown in Figure P21.33.

**Figure P21.33**    Problems 33, 34, and 35.

**34.** In Figure P21.33, show how to add just enough ammeters to measure every different current that is flowing. Show how to add just enough voltmeters to measure the potential difference across each resistor and across each battery.

**35.** The circuit considered in Problem 33 and shown in Figure P21.33 is connected for 2.00 min. (a) Find the energy delivered by each battery to the circuit. (b) Find the energy delivered to each resistor. (c) Find the total energy converted in the circuit.

**36.** If $R = 1.00$ k$\Omega$ and $\mathcal{E} = 250$ V in Figure P21.36, determine the direction and magnitude of the current in the horizontal wire between $a$ and $e$.

**Figure P21.36**

**37.** A dead battery is charged by connecting it to the live battery of another car with jumper cables (Fig. P21.37). Determine the current in the starter and in the dead battery.

**Figure P21.37**

## Section 21.9   *RC* Circuits

**38.** A 2.00-nF capacitor with an initial charge of 5.10 $\mu$C is discharged through a 1.30-k$\Omega$ resistor. (a) Calculate the current in the resistor 9.00 $\mu$s after the resistor is connected across the terminals of the capacitor. (b) What charge remains on the capacitor after 8.00 $\mu$s? (c) What is the maximum current in the resistor?

**39.** Consider a series *RC* circuit (see Fig. 21.26) for which $R =$
**web** 1.00 M$\Omega$, $C = 5.00$ $\mu$F, and $\mathcal{E} = 30.0$ V. Find (a) the time constant of the circuit and (b) the maximum charge on the capacitor after the switch is closed. (c) If the switch is closed at $t = 0$, find the current in the resistor 10.0 s later.

**40.** In the circuit of Figure P21.40, the switch S has been open for a long time. It is then suddenly closed. Determine the time constant (a) before the switch is closed and (b) after

**Figure P21.40**

the switch is closed. (c) If the switch is closed at $t = 0$, determine the current in the switch as a function of time.

41. The circuit in Figure P21.41 has been connected for a long time. (a) What is the voltage across the capacitor? (b) If the battery is disconnected, how long does it take the capacitor to discharge to one tenth of its initial voltage?

**Figure P21.41**

42. A 10.0-$\mu$F capacitor is charged by a 10.0-V battery through a resistance $R$. The capacitor reaches a potential difference of 4.00 V at a time 3.00 s after charging begins. Find $R$.

## Section 21.10 Context Connection— The Atmosphere as a Conductor

43. A current density of $6.00 \times 10^{-13}$ A/m$^2$ exists in the atmosphere at a location where the electric field is 100 V/m. Calculate the electrical conductivity of the Earth's atmosphere in this region.

44. Assume that global lightning on Earth constitutes a constant current of 1.00 kA between the ground and an atmospheric layer at potential 300 kV. (a) Find the power of terrestrial lightning. (b) For comparison, find the power of sunlight falling on the Earth. Sunlight has an intensity of 1 340 W/m$^2$ above the atmosphere. Sunlight falls perpendicularly on the circular projected area that the Earth presents to the Sun.

## Additional Problems

45. One lightbulb is marked "25 W 120 V," and another "100 W 120 V"; this means that each has its respective power delivered to it when plugged into a constant 120-V potential difference. (a) Find the resistance of each bulb. (b) How long does it take for 1.00 C to pass from the wire into the dim bulb? (c) How long does it take for 1.00 J to be delivered to the dim bulb? (d) Find the cost of running the dim bulb continuously for 30.0 days if the electric company sells its product at $0.070 0/kWh. What product *does* the electric company sell? What is its price for one SI unit of this quantity?

46. Switch S has been closed for a long time, and the electric circuit shown in Figure P21.46 carries a constant current.

Let $C_1 = 3.00$ $\mu$F, $C_2 = 6.00$ $\mu$F, $R_1 = 4.00$ k$\Omega$, and $R_2 = 7.00$ k$\Omega$. The power delivered to $R_2$ is 2.40 W. (a) Find the charge on $C_1$. (b) Now the switch is opened. After many milliseconds, by how much has the charge on $C_2$ changed?

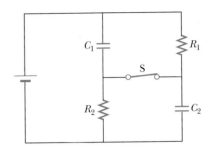

**Figure P21.46**

47. A straight cylindrical wire lying along the $x$ axis has a length of 0.500 m and a diameter of 0.200 mm. It is made of an ohmic material with a resistivity of $\rho = 4.00 \times 10^{-8}$ $\Omega \cdot$m. Assume that a potential of 4.00 V is maintained at $x = 0$, and that $V = 0$ at $x = 0.500$ m. Find (a) the electric field **E** in the wire, (b) the resistance of the wire, (c) the electric current in the wire, and (d) the current density **J** in the wire. Express vectors in vector notation. (e) Show that $\mathbf{E} = \rho \mathbf{J}$.

48. A straight cylindrical wire lying along the $x$ axis has a length $L$ and a diameter $d$. It is made of an ohmic material with a resistivity $\rho$. Assume that potential $V$ is maintained at $x = 0$, and that the potential is zero at $x = L$. In terms of $L$, $d$, $V$, $\rho$, and physical constants, derive expressions for (a) the electric field in the wire, (b) the resistance of the wire, (c) the electric current in the wire, and (d) the current density in the wire. Express vectors in vector notation. (e) Prove that $\mathbf{E} = \rho \mathbf{J}$.

49. An experiment is conducted to measure the electrical resistivity of Nichrome in the form of wires with different lengths and cross-sectional areas. For one set of measurements, a student uses 30-gauge wire, which has a cross-sectional area of $7.30 \times 10^{-8}$ m$^2$. The student measures the potential difference across the wire and the current in the wire with a voltmeter and an ammeter, respectively. For each of the measurements given in the table taken on wires of three different lengths, calculate the resistance of the wires and the corresponding values of the resistivity. What is the average value of the resistivity, and how does this value compare with the value given in Table 21.1?

| $L$, m | $\Delta V$, V | $I$, A | $R$, $\Omega$ | $\rho$, $\Omega \cdot$ m |
|---|---|---|---|---|
| 0.540 | 5.22 | 0.500 | | |
| 1.028 | 5.82 | 0.276 | | |
| 1.543 | 5.94 | 0.187 | | |

**50.** (a) Using symmetry arguments, show that the current in any resistor in the configuration of Figure P21.50 is either $I/3$ or $I/6$. All resistors have the same resistance $r$. (b) Show that the equivalent resistance between points $a$ and $b$ is $(5/6)r$.

**Figure P21.50**

**51.** An electric car is designed to run off a bank of 12.0-V batteries with a total energy storage of $2.00 \times 10^7$ J. (a) If the electric motor draws 8.00 kW, what is the current in the motor? (b) If the electric motor draws 8.00 kW as the car moves at a steady speed of 20.0 m/s, how far will the car travel before it is "out of juice"?

**52.** Four resistors are connected in parallel across a 9.20-V battery. They carry currents of 150 mA, 45.0 mA, 14.00 mA, and 4.00 mA, respectively. (a) If the resistor with the largest resistance is replaced with one having twice the resistance, what is the ratio of the new current in the battery to the original current? (b) If instead the resistor with the smallest resistance is replaced with one having twice the resistance, what is the ratio of the new total current to the original current? (c) On a February night, energy leaves a house by several energy leaks, including the following: 1 500 W by conduction through the ceiling; 450 W by infiltration (air flow) around the windows; 140 W by conduction through the basement wall above the foundation sill; and 40.0 W by conduction through the plywood door to the attic. To produce the biggest saving in heating bills, which one of these energy transfers should be reduced first?

**53.** Figure P21.53 shows a circuit model for the transmission of an electrical signal, such as cable TV, to a large number of subscribers. Each subscriber connects a load resistance $R_L$ between the transmission line and the ground. Assume the ground to be at zero potential and to have negligible resistance. The resistance of the transmission line between the connection points of different subscribers is modeled as the constant resistance $R_T$. Show that the equivalent resistance across the signal source is

$$R_{eq} = \tfrac{1}{2}[(4R_T R_L + R_T^2)^{1/2} + R_T]$$

**Figure P21.53**

(*Hint:* Because the number of subscribers is large, the equivalent resistance would not change noticeably if the first subscriber canceled his service. Consequently, the equivalent resistance of the section of the circuit to the right of the first load resistor is nearly equal to $R_{eq}$.)

**54.** A battery has emf $\mathcal{E}$ and internal resistance $r$. A variable-load resistor $R$ is connected across the terminals of the battery. (a) Determine the value of $R$ such that the potential difference across the terminals is a maximum. (b) Determine the value of $R$ that would maximize the current in the circuit. (c) Determine the value of $R$ that would maximize the power delivered to the load resistor. Choosing the load resistance for maximum power transfer is a case of what is called *impedance matching* in general. Impedance matching is important in shifting gears on a bicycle, in connecting a loudspeaker to an audio amplifier, in connecting a battery charger to a bank of solar photoelectric cells, and in many other applications.

**55.** The circuit shown in Figure P21.55 is set up in the laboratory to measure an unknown capacitance $C$ with the use of a voltmeter of resistance $R = 10.0$ MΩ and a battery whose emf is 6.19 V. The data given in the table are the measured voltages across the capacitor as a function of time, where $t = 0$ represents the instant at which the switch is opened. (a) Construct a graph of $\ln(\mathcal{E}/\Delta V)$ versus $t$, and perform a linear least-squares fit to the data. (b) From the slope of your graph, obtain a value for the time constant of the circuit and a value for the capacitance.

| $\Delta V$, V | $t$, s | $\ln(\mathcal{E}/\Delta V)$ |
|---|---|---|
| 6.19 | 0 | |
| 5.55 | 4.87 | |
| 4.93 | 11.1 | |
| 4.34 | 19.4 | |
| 3.72 | 30.8 | |
| 3.09 | 46.6 | |
| 2.47 | 67.3 | |
| 1.83 | 102.2 | |

**Figure P21.55**

56. The values of the components in a simple series $RC$ circuit containing a switch (Fig. 21.26) are $C = 1.00$ $\mu$F, $R = 2.00 \times 10^6$ $\Omega$, and $\mathcal{E} = 10.0$ V. At the instant 10.0 s after the switch is closed, calculate (a) the charge on the capacitor, (b) the current in the resistor, (c) the rate at which energy is being stored in the capacitor, and (d) the rate at which energy is being delivered by the battery.

57. The switch in Figure P21.57a closes when $\Delta V_c > 2\,\Delta V/3$ and opens when $\Delta V_c < \Delta V/3$. The voltmeter reads a voltage as plotted in Figure P21.57b. What is the period $T$ of the waveform in terms of $R_A$, $R_B$, and $C$?

58. A battery is used to charge a capacitor through a resistor, as shown in Figure 21.26. Show that half the energy supplied by the battery appears as internal energy in the resistor and that half is stored in the capacitor.

59. A battery has an emf of 9.20 V and an internal resistance of 1.20 $\Omega$. (a) What resistance across the battery will extract from it a power of 12.8 W? (b) a power of 21.2 W?

60. An oceanographer is studying how the ion concentration in sea water depends on depth. The probe of her apparatus consists of a pair of concentric metallic cylinders at the end of a cable that is lowered into the sea. The resistance between these cylinders is then measured as a function of depth. The water between the two cylinders is shown in Figure P21.60. The water forms a cylindrical shell of inner radius $r_a$, outer radius $r_b$, and length $L$ much larger than $r_b$. The scientist applies a potential difference $\Delta V$ between the inner and outer surfaces, producing an outward radial current $I$. Let $\rho$ represent the resistivity of the water. (a) Find the resistance of the water between the cylinders in terms of $L$, $\rho$, $r_a$, and $r_b$. (b) Express the resistivity of the water in terms of the measured quantities $L$, $r_a$, $r_b$, $\Delta V$, and $I$.

**Figure P21.60**

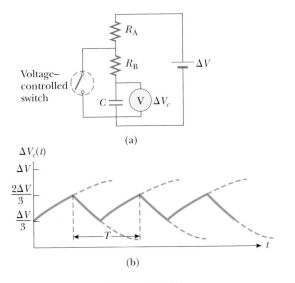

(a)

(b)

**Figure P21.57**

# ANSWERS TO QUICK QUIZZES

**21.1** d, b = c, a. The current in (d) is equivalent to two positive charges moving to the left. Parts (b) and (c) each represent four charges moving in the same direction because negative charges moving to the left are equivalent to positive charges moving to the right. The current in (a) is equivalent to five positive charges moving to the right.

**21.2** (b). According to Equation 21.6, resistance is the ratio of voltage across a device to current in the device. In Figure 21.6b, a line drawn from the origin to a point on the curve will have a slope equal to $I/\Delta V$, which is the inverse of resistance. As $\Delta V$ increases, the slope of this line also increases, so that the resistance decreases.

**21.3**   Household electrical devices are intended to be operated from the 120-V potential difference available at wall sockets. Assuming this voltage, the current in the device can be calculated based on the known resistance of the device. A battery can be used to operate many devices, with varying resistances. As a consequence, the current in the battery will vary depending on the device that is operating.

**21.4**   Voltage is not something that "surges" through a completed circuit. A voltage is a potential difference that is applied *across* a device or a circuit. What goes *through* a device is *electric field,* which causes charges to flow in the device, resulting in a current in the device. Thus, it would be more correct to say, "1 coulomb of electric charge surged into the victim's body per second." Although this current would have disastrous results on the human body, a value of 1 (ampere) doesn't sound as exciting for a newspaper article as 10 000 (volts). Another possibility is to write, "10 000 volts of potential difference were applied across the victim's body," which still doesn't sound quite as exciting!

**21.5**   (b), (d). The length of the line cord will double in this event. This would tend to increase the resistance of the line cord. But the doubling of the radius of the line cord results in the cross-sectional area increasing by a factor of 4. This would reduce the resistance more than the doubling of length increases it. The net result is a decrease in resistance. The same effect would occur for the lightbulb filament. The lowered resistance would result in a larger current in the filament, causing it to glow more brightly.

**21.6**   $I_a = I_b > I_c = I_d > I_e = I_f$. Charges constituting the current $I_a$ leave the positive terminal of the battery and then split to flow through the two bulbs; thus, $I_a = I_c + I_e$. Because the potential difference $\Delta V$ is the same across the two bulbs and because the power delivered to a device is $\mathcal{P} = I\Delta V$, the 60-W bulb with the higher power rating must carry the greater current. Because charge does not accumulate in the bulbs, we know that all the charge flowing into a bulb from the left has to flow out on the right; consequently $I_c = I_d$ and $I_e = I_f$. The two currents leaving the bulbs recombine to form the current back into the battery, $I_f + I_d = I_b$.

**21.7**   The total energy delivered by the battery will be less than $E$. Recall that a battery can be considered to be an ideal, resistanceless battery in series with the internal resistance. When charging, the energy delivered to the battery includes the energy necessary to charge the ideal battery, plus the energy that goes into raising the temperature of the battery in the internal resistance. This latter energy is not available during the discharge

of the battery. During discharge, part of the reduced available energy again transforms into internal energy in the internal resistance, further reducing the available energy below $E$.

**21.8**   The starter in the automobile draws a relatively large current from the battery. This large current causes a significant voltage across the internal resistance of the battery. As a result, the terminal voltage of the battery is reduced, and the headlights dim accordingly.

**21.9**   Bulb $R_1$ becomes brighter. Connecting $b$ to $c$ "shorts out" bulb $R_2$, which changes the total resistance of the circuit from $R_1 + R_2$ to just $R_1$. Because the resistance has decreased (and the potential difference supplied by the battery does not change), the current in the battery increases. This means that the current in bulb $R_1$ increases, making it glow brighter. Bulb $R_2$ goes out because the new piece of wire provides an almost resistance-free path for the current, so that essentially zero current exists in bulb $R_2$.

**21.10**   (b). When the switch is opened, resistors $R_1$ and $R_2$ are in series, so that the total circuit resistance is larger than when the switch was closed. As a result, the current drops.

**21.11**   (a). When the switch is closed, resistors $R_1$ and $R_2$ are in parallel, so that the total circuit resistance is smaller than when the switch was open. As a result, the current increases.

**21.12**   *Series:* (a) decrease; (b) decrease; (c) decrease; (d) increase; (e) increase. As you add more lightbulbs in *series,* the overall resistance of the circuit increases. Thus, the current in the bulbs decreases, resulting in reduced brightness. This decrease in current results in a decrease in power transferred from the battery. As a result, the battery lifetime increases. As the current drops, the terminal voltage across the battery becomes closer and closer to the battery emf.

   *Parallel:* (a) decrease; (b) decrease; (c) increase; (d) decrease; (e) decrease. As you add more lightbulbs in *parallel,* the overall resistance of the circuit decreases. The current leaving the battery increases with the addition of each bulb. This increase in current results in an increase in power transferred from the battery. As a result, the battery lifetime decreases. As the current rises, the terminal voltage across the battery drops further below the battery emf. Because of the drop in terminal voltage, which is applied to each lightbulb, the current in each lightbulb is reduced, leading to a reduction in brightness. The reduction in brightness is much smaller than in the case of adding lightbulbs in series and may not be noticeable until several lightbulbs are added.

# Lightning

## MODELING THE ATMOSPHERE TO DETERMINE THE NUMBER OF LIGHTNING STRIKES

Now that we have investigated the principles of electricity, let us respond to our central question for the *Lightning* Context:

> *How can we determine the number of lightning strikes on the Earth in a typical day?*

We must combine several ideas from our knowledge of electricity to perform this calculation. In Chapter 20, the atmosphere was modeled as a capacitor. This was first done by Lord Kelvin who modeled the ionosphere as the positive plate several tens of kilometers above the Earth's surface. More sophisticated models have shown the effective height of the positive plate to be the 5 km that we used in our earlier calculation.

The plates of the atmospheric capacitor are separated by a layer of air containing a large number of free ions that can carry current. The air is not a perfect conductor, of course; measurements show that the resistivity of air is about $3 \times 10^{13} \, \Omega \cdot m$. Let us calculate the resistance of the air between our capacitor plates. The shape of the resistor is that of a spherical shell between the plates of the atmospheric capacitor. We use Equation 21.8 with the area of our resistor modeled as the surface area of the Earth and the length as the distance between the plates:

$$R = \rho \frac{\ell}{A} = (3 \times 10^{13} \, \Omega \cdot m) \frac{5 \times 10^3 \, m}{4\pi(6.4 \times 10^6 \, m)^2} \approx 3 \times 10^2 \, \Omega$$

The charge on the atmospheric capacitor can pass from the upper plate to the ground by electric current in the air between the plates. Thus, we can model the atmosphere as an *RC* circuit, using the capacitance found in Chapter 20, and the resistance connecting the plates calculated above (Fig. 1). The time constant for this *RC* circuit is

$$\tau = RC = (0.9 \, F)(3 \times 10^2 \, \Omega) \approx 3 \times 10^2 \, s = 5 \, min$$

Thus, the charge on the atmospheric capacitor should fall to $e^{-1} = 37\%$ of its original value after only 5 min! After 30 min, less than 0.3% of the charge would remain! Why doesn't this happen—what keeps the atmospheric capacitor charged? The answer is *lightning*. The processes occurring in cloud charging result in lightning strikes that deliver negative charge to the ground to replace that neutralized by the flow of charge through the air. On the average, a net charge on the atmospheric capacitor results from a balance between these two processes.

Now, let's use this balance to numerically answer our central question. We first address the charge on the atmospheric capacitor. In Chapter 19, we mentioned a charge of $5 \times 10^5 \, C$ that is spread over the surface of the Earth. This is the charge on the atmospheric capacitor.

A typical lightning strike delivers about 25 C of negative charge to the ground—this is the process of charging the capacitor. Dividing the charge on the

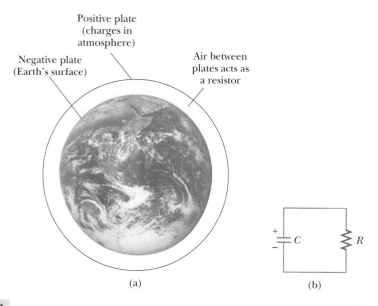

(a)                    (b)

**Figure 1**

(a) The atmosphere can be modeled as a capacitor, with conductive air between the plates. (b) We can imagine an equivalent *RC* circuit for the atmosphere, with the natural discharge of the capacitor in balance with the charging of the capacitor by lightning.

capacitor by the charge per lightning strike tells us the number of lightning strikes required to charge the capacitor:

$$\text{Number of lightning strikes} = \frac{\text{total charge}}{\text{charge per lightning strike}}$$

$$= \frac{5 \times 10^5 \text{ C}}{25 \text{ C per strike}} \approx 2 \times 10^4 \text{ lightning strikes}$$

According to our calculation of the *RC* circuit, the atmospheric capacitor almost completely discharges through the air in about 30 min. Thus, $2 \times 10^4$ lightning strikes must occur every 30 minutes, or $4 \times 10^4$/h, to keep the charging and discharging processes in balance. Multiplying this by the number of hours in a day, we have:

$$\text{Number of lightning strikes per day} = (4 \times 10^4 \text{ strikes/h}) \left( \frac{24 \text{ h}}{1 \text{ d}} \right)$$

$$\approx 1 \times 10^6 \text{ strokes/day}$$

Despite the simplifications that we have adopted in our calculations, this number is on the right order of magnitude for the actual number of lightning strikes on the Earth in a typical day—1 million!

**PROBLEMS**

1. Consider the atmospheric capacitor described in the text, with the ground as one plate and positive charges in the atmosphere as the other. On one particular day, the capacitance of the atmospheric capacitor is 0.800 F. The effective plate separation distance is 4.00 km, and the resistivity of the air is $2.00 \times 10^{13} \ \Omega \cdot \text{m}$. If no lightning events occur, the capacitor will discharge through the air. If there is $4.00 \times 10^4$ C of charge on the

atmospheric capacitor at time $t = 0$, at what later time is the charge reduced (a) to $2.00 \times 10^4$ C? (b) to $5.00 \times 10^3$ C? (c) to zero?

2. Consider this alternative line of reasoning to estimate the number of lightning strikes on the Earth in one day. Using the charge on the Earth of $5 \times 10^5$ C and the atmospheric capacitance of 0.9 F, we find that the potential difference across the capacitor is $\Delta V = Q/C = 5 \times 10^5$ C$/0.9$ F $\approx 6 \times 10^5$ V. The leakage current in the air is $I = \Delta V/R = 6 \times 10^5$ V$/300 \ \Omega \approx 2$ kA. (a) If each lightning strike delivers 25 C of charge to the ground, what is the average time interval between lightning strikes so that the average current due to lightning is 2 kA? (b) Using this average time interval between lightning strikes, calculate the number of lightning strikes in one day.

3. Consider again the atmospheric capacitor discussed in the text. (a) Suppose that atmospheric conditions are such that, for one complete day, the lower 2.50 km of the air between the capacitor plates has resistivity $2.00 \times 10^{13} \ \Omega \cdot$m and the upper 2.50 km has resistivity $0.500 \times 10^{13} \ \Omega \cdot$m. How many lightning strikes occur on this day? (b) Suppose that atmospheric conditions are such that, for one complete day, the resistivity of the air between the plates in the southern hemisphere is $2.00 \times 10^{13} \ \Omega \cdot$m and the resistivity between the plates in the northern hemisphere is $0.200 \times 10^{13} \ \Omega \cdot$m. How many lightning strikes occur on this day?

# Magnetic Levitation Vehicles

**A**ll commercial long-distance ground transportation currently operating in the United States is subject to the force of friction between wheels and a roadway or a track. Recall that friction is a nonconservative force that transforms kinetic energy into internal energy. As discussed in the chapters on thermodynamics, this internal energy is wasted.

Because magnetic levitation (maglev) vehicles are suspended by magnetic forces, they do not make physical contact with a roadway or track. This eliminates mechanical friction, the primary cause of transformation of kinetic energy to internal energy.

Robert Goddard, of rocket fame, published a story in 1907 that describes many features of magnetic levitation. He also published a paper in *Scientific American* in 1909 describing a magnetically levitated vehicle operating in a tunnel between Boston and New York. Emile Bachelet published a paper in 1912 describing a magnetically levitated vehicle for delivering mail. He received a patent for his invention, but it required far too much power to be practical.

The German Transrapid. Although this vehicle differs in technology from the Japanese maglev vehicles, it also can travel at very high speeds. *(Regis Bossu/Corbis SYGMA)*

The Japanese MLX01 test vehicle. Recent tests have shown that this vehicle can travel at speeds over 500 km/h. *(Noboru Hashimoto/Corbis SYGMA)*

After these early ideas, no significant progress in magnetic levitation was made until the 1960s. At that time, advances in superconducting magnets spurred new interest in magnetic levitation, due to the possible savings in energy costs over previous designs such as Bachelet's. In 1963, a physicist at Brookhaven National Laboratory proposed a system using superconducting magnets. Within a few years, projects were underway at Stanford University, MIT, Raytheon, Ford Motor Company, the University of Toronto, and McGill University. Projects were also initiated shortly thereafter in Japan, Germany, and England.

Despite this promising start by several American companies, federal funding for maglev research in the United States ended in 1975. Research in other countries continued, primarily in Germany and Japan. These studies and a variety of full-scale test vehicles have shown that maglev technology is very successful. Research in maglev has seen a recent modest revival in the United States following the National Maglev Initiative signed into law in 1991, but the United States remains far behind Germany and Japan. Several possible maglev

The Shinkansen (bullet train) in Japan. In rail-based transportation systems such as this, much of the input energy is transformed to internal energy due to friction between wheels and the track. *(Stone)*

routes in the United States have been proposed using foreign technology.

In this Context, we shall investigate the physics of magnetic fields and electromagnetism and apply these principles to understanding the processes of lifting, propelling, and braking a maglev vehicle. Two primary mechanisms—the *attractive* and *repulsive* models—are the basis of current research and development efforts. We shall study each of these models so we can respond to our central question

*How can we lift, propel, and brake a vehicle with magnetic forces?*

**WEB**

The development of maglev in the United States is still far behind foreign efforts. For the final report of the National Maglev Initiative (1991), visit **www.bts.gov/NTL/DOCS/TNM.html**
For an updated status of maglev in the U.S. from the Federal Railroad Administration, visit **www.fra.dot.gov/o/hsgt/fedassist/maglev.htm**
For an ABC news story on U.S. maglev, visit **abcnews.go.com/ABC2000/abc2000tech/maglev2000.html**

# chapter

# 22

Aurora Borealis, the Northern Lights. These displays are caused by cosmic ray particles trapped in the magnetic field of the Earth. When the particles collide with atoms in the atmosphere, they cause the emission of light.

*(Johnny Johnson/Stone)*

# Magnetic Forces and Magnetic Fields

The list of technological applications of magnetism is very long. For instance, large electromagnets are used to pick up heavy loads in scrap yards. Magnets are used in such devices as meters, motors, and loudspeakers. Magnetic tapes are routinely used in sound and video recording equipment and for computer data storage. Intense magnetic fields generated by superconducting magnets are currently being used as a means of containing plasmas at temperatures on the order of $10^8$ K used in controlled nuclear fusion research.

As we investigate magnetism in this chapter, we shall find that the subject cannot be divorced from electricity. For example, magnetic fields affect moving electric charges, and moving charges produce magnetic fields. We shall find that the source of all magnetic fields is electric current.

The attraction between magnets and certain metals is used in a junkyard as a magnet is used to move tons of scrap metal. *(Jeffrey Sylvester/FPG International)*

## 22.1 • HISTORICAL OVERVIEW

Many historians of science believe that the compass, which uses a magnetic needle, was used in China as early as the 13th century BC, its invention being of Arab or Indian origin. The phenomenon of magnetism was known to the Greeks as early as about 800 BC. They discovered that certain stones, made of a material now called *magnetite* ($Fe_3O_4$), attracted pieces of iron.

In 1269, Pierre de Maricourt mapped out the directions taken by a needle when it was placed at various points on the surface of a spherical natural magnet. He found that the directions formed lines that encircled the sphere and passed through two points diametrically opposite each other, which he called the **poles** of the magnet. Subsequent experiments have shown that every magnet, regardless of its shape, has two poles, called **north** and **south,** which exhibit forces on each other in a manner analogous to electric charges. That is, like poles repel each other and unlike poles attract each other. The poles received their names because of the behavior of a magnet in the presence of the Earth's magnetic field. If a bar magnet is suspended from its midpoint by a piece of string so that it can swing freely in a horizontal plane, it rotates until its "north" pole points to the north geographic pole of the Earth (which is a south magnetic pole) and its "south" pole points to the Earth's south geographic pole. (The same idea is used to construct a simple compass.)

In 1600, William Gilbert extended these experiments to a variety of materials. Using the fact that a compass needle orients in preferred directions, he suggested that magnets are attracted to land masses. In 1750, John Michell (1724–1793) used a torsion balance to show that magnetic poles exert attractive or repulsive forces on each other and that these forces vary as the inverse square of their separation. Although the force between two magnetic poles is similar to the force

**Hans Christian Oersted**
(1777–1851)

Danish physicist. *(North Wind Picture Archives)*

between two electric charges, an important difference exists. Electric charges can be isolated (witness the electron and proton), whereas magnetic poles cannot be isolated. That is, **magnetic poles are always found in pairs.** No matter how many times a permanent magnet is cut, each piece always has a north pole and a south pole. (Some theories speculate that magnetic monopoles—isolated north or south poles—may exist in nature, and attempts to detect them currently make up an active experimental field of investigation. None of these attempts has yet proven successful, however.)

The relationship between magnetism and electricity was discovered in 1819 when, while preparing for a lecture demonstration, the Danish scientist Hans Oersted found that an electric current in a wire deflected a nearby compass needle. Shortly thereafter, André Ampère (1775–1836) deduced quantitative laws of magnetic force between current-carrying conductors. He also suggested that electric current loops of molecular size are responsible for *all* magnetic phenomena.

In the 1820s, Faraday and independently Joseph Henry (1797–1878) identified further connections between electricity and magnetism. They showed that an electric current could be produced in a circuit either by moving a magnet near the circuit or by changing the current in a nearby circuit. Their observations demonstrated that a changing magnetic field produces an electric field. Years later, theoretical work by James Clerk Maxwell showed that the reverse is also true: a changing electric field gives rise to a magnetic field.

In this chapter, we shall investigate the effects of constant magnetic fields on charges and currents, and study the sources of magnetic fields. In the next chapter, we shall explore the effects of magnetic fields that vary in time.

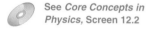

See *Core Concepts in Physics*, Screen 12.2

## 22.2 • THE MAGNETIC FIELD

In earlier chapters we described the interaction between charged objects in terms of electric fields. Recall that an electric field surrounds any stationary electric charge. The region of space surrounding a *moving* charge includes a **magnetic field** in addition to the electric field. A magnetic field also surrounds any material with permanent magnetism. We find that the magnetic field is a vector field, similar to the electric field.

To describe any type of vector field, we must define its magnitude and its direction. The direction of the magnetic field vector **B** at any location is the direction in which the north pole of a compass needle points at that location. Figure 22.1 shows how the magnetic field of a bar magnet can be traced with the aid of a

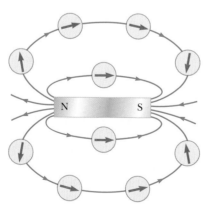

**Figure 22.1**

A small compass can be used to trace the magnetic field lines of a bar magnet.

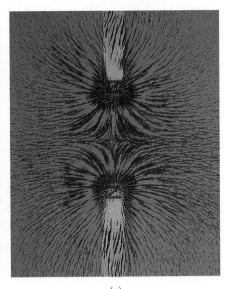

(a)    (b)    (c)

**Figure 22.2**

(a) Magnetic field patterns surrounding a bar magnet as displayed with iron filings. (b) Magnetic field patterns between *unlike* poles of two bar magnets. (c) Magnetic field pattern between *like* poles of two bar magnets. *(Courtesy of Henry Leap and Jim Lehman)*

compass, defining a **magnetic field line,** similar in many ways to the electric field lines we studied in Chapter 19. Several magnetic field lines of a bar magnet traced out in this manner are shown in the two-dimensional pictorial representation in Figure 22.1. Magnetic field patterns can be displayed by small iron filings placed in the vicinity of a magnet, as in Figure 22.2.

We can quantify the magnetic field **B** (sometimes called the *magnetic induction*) by using our model of a particle in a field. The existence of a magnetic field at some point in space can be determined by measuring the force **F**$_B$ exerted on an appropriate test particle placed at that point. This is the same process we followed in defining the electric field in Chapter 19. Our test particle will be an electrically charged particle, such as a proton. If we perform such an experiment, we find the following results:

- The magnetic force **F**$_B$ is proportional to the charge $q$ of the particle as well as to the speed $v$ of the particle.
- The magnitude and direction of the magnetic force on the particle depend on the relative direction between the velocity vector of the particle and the magnetic field vector.
- When a charged particle moves parallel to the magnetic field vector, the magnetic force **F**$_B$ on the charge is zero.
- When the velocity vector makes an angle $\theta$ with the magnetic field, the magnetic force acts in a direction perpendicular to both **v** and **B**; that is, the magnetic force is perpendicular to the plane formed by **v** and **B** (Fig. 22.3a).
- The magnetic force on a negative charge is directed opposite to the force on a positive charge moving in the same direction (Fig. 22.3b).
- If the velocity vector makes an angle $\theta$ with the magnetic field, the magnitude of the magnetic force is proportional to sin $\theta$.

• *Properties of the magnetic force on a charge moving in a **B** field*

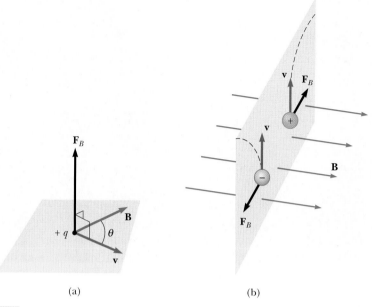

**Figure 22.3**

The direction of the magnetic force on a charged particle moving with a velocity **v** in the presence of a magnetic field **B**. (a) When **v** is at an angle $\theta$ to **B**, the magnetic force is perpendicular to both **v** and **B**. (b) Oppositely directed magnetic forces are exerted on two oppositely charged particles moving with the same velocity in a magnetic field.

These results show that the magnetic force on a particle is more complicated than the electric force. The magnetic force is distinctive because it depends on the velocity of the particle and because its direction is perpendicular to both **v** and **B**. Despite this complicated behavior, these observations can be summarized in a compact way by writing the magnetic force in the form

- *Magnetic force on a charged particle moving in a magnetic field*

$$\mathbf{F}_B = q\mathbf{v} \times \mathbf{B} \tag{22.1}$$

where the direction of the magnetic force is that of **v** × **B**, which, by definition of the cross product, is perpendicular to both **v** and **B**. Equation 22.1 is analogous to Equation 19.4, $\mathbf{F}_e = q\mathbf{E}$, but is clearly more complicated. We can regard Equation 22.1 as an operational definition of the magnetic field at a point in space. The SI unit of magnetic field is the **tesla** (T), where

$$1\ \text{T} = 1\ \text{N} \cdot \text{s}/\text{C} \cdot \text{m}$$

Figure 22.4 reviews the right-hand rule for determining the direction of the cross product **v** × **B**. Point the four fingers of your right hand along the direction of **v**, then curl them until they point along the direction of **B**. The upright thumb then points in the direction of **v** × **B**. Because $\mathbf{F}_B = q\mathbf{v} \times \mathbf{B}$, $\mathbf{F}_B$ is in the direction of **v** × **B** if $q$ is positive (Fig. 22.4a) and *opposite* the direction of **v** × **B** if $q$ is negative (Fig. 22.4b).

The magnitude of the magnetic force is

$$F_B = |q|vB \sin \theta \tag{22.2}$$

where $\theta$ is the angle between **v** and **B**. From this expression, we see that $F_B$ is zero when **v** is either parallel or antiparallel to **B** ($\theta = 0$ or $180°$). Furthermore, the force has its maximum value $F_B = |q|vB$ when **v** is perpendicular to **B** ($\theta = 90°$).

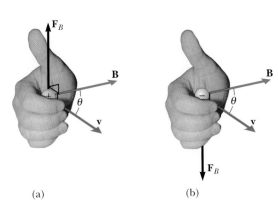

(a)                    (b)

**Figure 22.4**

The right-hand rule for determining the direction of the magnetic force $\mathbf{F}_B = q\mathbf{v} \times \mathbf{B}$ acting on a charge $q$ moving with a velocity $\mathbf{v}$ in a magnetic field $\mathbf{B}$. The direction of the force is that in which the thumb points. If $q$ is positive, $\mathbf{F}_B$ is upward. If $q$ is negative, $\mathbf{F}_B$ is downward.

## Quick Quiz 22.1

If a charged particle moves in a straight line through some region of space, can you say that the magnetic field in that region is zero?

The following list details the important differences between electric and magnetic forces on charged particles:

- The electric force is always parallel or antiparallel to the direction of the electric field, whereas the magnetic force is perpendicular to the magnetic field.
- The electric force acts on a charged particle independent of the particle's velocity, whereas the magnetic force acts on a charged particle only when the particle is in motion.
- The electric force does work in displacing a charged particle, whereas the magnetic force associated with a steady magnetic field does *no* work when a charged particle is displaced.

This last statement is a consequence of the fact that when a charge moves in a constant magnetic field, the magnetic force is always *perpendicular* to the displacement. That is, $\mathbf{F}_B \cdot d\mathbf{s} = (\mathbf{F}_B \cdot \mathbf{v})\,dt = 0$, because the magnetic force is a vector perpendicular to $\mathbf{v}$. From this property and the work−kinetic energy theorem, we conclude that the kinetic energy of a charged particle *cannot* be altered by a constant magnetic field alone. In other words, when a charge moves with a velocity of $\mathbf{v}$, an applied magnetic field can alter the direction of the velocity vector, but it cannot change the speed of the particle.

In Figures 22.3 and 22.4, we used blue vectors to represent magnetic field vectors, which will be the convention in this book. In Figure 22.1, we represented the magnetic field of a bar magnet with blue field lines. Studying magnetic fields presents a complication that we avoided in electric fields. In our study of electric fields, we drew all electric field vectors in the plane of the page or used perspective to represent them directed at an angle to the page. The cross product in Equation 22.1 requires us to think in three dimensions for problems in magnetism. Thus, in addition to drawing vectors pointing left or right and up or down, we will need a method of drawing vectors into or out of the page. These methods of representing the vectors are illustrated in Figure 22.5. A vector coming out of the page is represented by a dot, which we can think of as the tip of the arrowhead representing the vector coming through the paper toward us (Fig. 22.5a). A vector going into the page is represented by a cross, which we can think of as the tail feathers of an arrow going into the page (Fig. 22.5b). This depiction can be used for any type of vector we will encounter: magnetic field, velocity, force, and so on.

• *Differences between electric and magnetic fields*

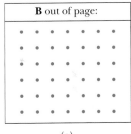

(a)

(b)

**Figure 22.5**

(a) Magnetic field lines coming out of the paper are indicated by dots, representing the tips of arrows coming outward. (b) Magnetic field lines going into the paper are indicated by crosses, representing the feathers of arrows going inward.

**Quick Quiz 22.2**

Electric field lines are sometimes called "lines of force." Is this a proper name for magnetic field lines?

**THINKING PHYSICS 22.1**

On a business trip to Australia, you take along your American-made compass that you used in your Boy Scout days. Does this compass work correctly in Australia?

**Reasoning**   Using the compass in Australia presents no problem. The north pole of the magnet in the compass will be attracted to the south magnetic pole near the north geographic pole, just as it was in the United States. The only difference in the magnetic field lines is that they have an upward component in Australia, whereas they have a downward component in the United States. When you hold it in a horizontal plane, your compass cannot detect the vertical component of the field, however—it only displays the direction of the horizontal component of the magnetic field.

**Example 22.1   An Electron Moving in a Magnetic Field**

An electron in a television picture tube moves toward the front of the tube with a speed of $8.0 \times 10^6$ m/s along the $x$ axis (Fig. 22.6). The neck of the tube is surrounded by a coil of wire that creates a magnetic field of magnitude 0.025 T, directed at an angle of 60° to the $x$ axis and lying in the $xy$ plane. Calculate the magnetic force on and acceleration of the electron.

**Solution**   Using Equation 22.2, we find the magnitude of the magnetic force:

$F_B = |q|vB \sin \theta$

$= (1.60 \times 10^{-19} \text{ C})(8.0 \times 10^6 \text{ m/s})(0.025 \text{ T})(\sin 60°)$

$= 2.8 \times 10^{-14} \text{ N}$

Because $\mathbf{v} \times \mathbf{B}$ is in the positive $z$ direction (from the right-hand rule), and the charge is negative, $\mathbf{F}_B$ is in the negative $z$ direction.

Once we have determined the magnetic force, this becomes a Chapter 4 problem—the acceleration is determined from Newton's second law. The mass of the electron is $m_e = 9.1 \times 10^{-31}$ kg, and so its acceleration is

$$a = \frac{F_B}{m_e} = \frac{2.8 \times 10^{-14} \text{ N}}{9.1 \times 10^{-31} \text{ kg}} = 3.1 \times 10^{16} \text{ m/s}^2$$

in the negative $z$ direction.

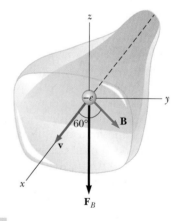

**Figure 22.6**

(Example 22.1) The magnetic force $\mathbf{F}_B$ on the electron is in the negative $z$ direction when $\mathbf{v}$ and $\mathbf{B}$ lie in the $xy$ plane.

**EXERCISE** Calculate the magnitude of the acceleration of a proton that moves through the same magnetic field at the same speed as the electron.

*Answer* $1.7 \times 10^{13} \text{ m/s}^2$

**EXERCISE** A proton moves with a speed of $2.5 \times 10^6$ m/s horizontally, at a right angle to a magnetic field. What magnetic field is required to just balance the weight of the proton and keep it moving horizontally?

*Answer* $4.1 \times 10^{-14}$ T in the horizontal plane

## 22.3 • MOTION OF A CHARGED PARTICLE IN A MAGNETIC FIELD

See Screen 12.3

In Section 22.2 we found that the magnetic force acting on a charged particle moving in a magnetic field is perpendicular to the velocity of the particle and that, consequently, the work done on the particle by the magnetic force is zero. Consider now the special case of a positively charged particle moving in a uniform magnetic field when the initial velocity vector of the particle is perpendicular to the field. Let us assume that the direction of the magnetic field is into the page. Figure 22.7 shows that the particle moves in a circular path whose plane is perpendicular to the magnetic field.

The particle moves in this way because the magnetic force $\mathbf{F}_B$ is perpendicular to $\mathbf{v}$ and $\mathbf{B}$ and has a constant magnitude $qvB$. As the force changes the direction of $\mathbf{v}$, the direction of $\mathbf{F}_B$ changes continuously, as in Figure 22.7. Because $\mathbf{F}_B$ always points toward the center of the circle, the particle can be modeled as being in uniform circular motion. As Figure 22.7 shows, the rotation is counterclockwise for a positive charge in a magnetic field directed into the page. If $q$ were negative, the rotation would be clockwise. We can use Newton's second law to determine the radius of the circular path:

$$\sum F = F_B = ma$$

$$qvB = \frac{mv^2}{r}$$

$$r = \frac{mv}{qB} \qquad [22.3]$$

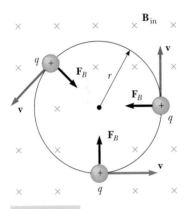

**Figure 22.7**

When the velocity of a charged particle is perpendicular to a uniform magnetic field, the particle moves in a circular path in a plane perpendicular to $\mathbf{B}$. The magnetic force $\mathbf{F}_B$ acting on the charge is always directed toward the center of the circle.

That is, the radius of the path is proportional to the linear momentum $mv$ of the particle and inversely proportional to the magnitude of the charge on the particle and to the magnitude of the magnetic field. The angular speed of the particle is (from Eq. 10.10)

$$\omega = \frac{v}{r} = \frac{qB}{m} \qquad [22.4]$$

• *Cyclotron frequency*

The period of the motion (the time the particle takes to move one revolution) is equal to the circumference of the circular path divided by the speed of the particle:

$$T = \frac{2\pi r}{v} = \frac{2\pi}{\omega} = \frac{2\pi m}{qB} \qquad [22.5]$$

These results show that the angular speed of the particle and the period of the circular motion do not depend on the translational speed of the particle or the radius of the orbit for a given particle in a given uniform magnetic field. The

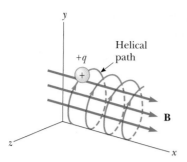

**Figure 22.8**

A charged particle having a velocity vector with a component parallel to a uniform magnetic field moves in a helical path.

angular speed $\omega$ is often referred to as the **cyclotron frequency** because charged particles circulate at this angular speed in one type of accelerator called a *cyclotron*, discussed in Section 22.4.

If a charged particle moves in a uniform magnetic field with its velocity at some arbitrary angle to **B**, its path is a helix. For example, if the field is in the *x* direction as in Figure 22.8, there is no component of force in the *x* direction. As a result, $a_x = 0$, and so the *x* component of velocity remains constant. The magnetic force $q\mathbf{v} \times \mathbf{B}$ causes the components $v_y$ and $v_z$ to change in time, however, and the resulting motion is a helix having its axis parallel to the magnetic field. The projection of the path onto the *yz* plane (viewed along the *x* axis) is a circle. (The projections of the path onto the *xy* and *xz* planes are sinusoids!) Equations 22.3 to 22.5 still apply provided that $v$ is replaced by $v_\perp = \sqrt{v_y{}^2 + v_z{}^2}$.

### Quick Quiz 22.3

A charged particle moves in a circular path in the presence of a constant magnetic field applied perpendicular to the particle's velocity. Does the particle gain energy from the magnetic field?

### Quick Quiz 22.4

How can the motion of a charged particle be used to distinguish between a magnetic field and an electric field in a certain region?

### THINKING PHYSICS 22.2

Suppose a uniform magnetic field exists in a finite region of space as in Figure 22.9. Can you inject a charged particle into this region and have it stay trapped in the region by the magnetic force?

**Reasoning** Consider separately the components of the particle velocity parallel and perpendicular to the field lines in the region. For the component parallel to the field lines, no force is exerted on the particle—it continues to move with the parallel component until it leaves the region of the magnetic field. Now consider the component perpendicular to the field lines. This component results in a magnetic force that is perpendicular to both the field lines and the velocity component. As discussed earlier, if the force acting on a charged particle is always perpendicular to its velocity, the particle moves in a circular path. Thus, the particle follows half of a circular arc and exits the field on the other side of the circle, as shown in Figure 22.9. Thus, a particle injected into a uniform magnetic field cannot stay trapped in the field region.

Magnetic field region (out of page)

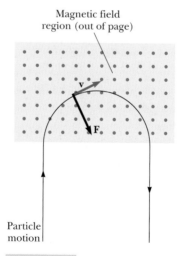

Particle motion

**Figure 22.9**

(Thinking Physics 22.2) A positively charged particle enters a region of magnetic field directed out of the page.

## Example 22.2 A Proton Moving Perpendicular to a Uniform Magnetic Field

A proton is moving in a circular orbit of radius 14.0 cm in a uniform 0.350-T magnetic field directed perpendicular to the velocity of the proton. Find the translational speed of the proton.

**Solution**  From Equation 22.3, we find

$$v = \frac{qBr}{m_p} = \frac{(1.60 \times 10^{-19} \text{ C})(0.350 \text{ T})(14.0 \times 10^{-2} \text{ m})}{1.67 \times 10^{-27} \text{ kg}}$$

$$= 4.69 \times 10^6 \text{ m/s}$$

**EXERCISE**  If an electron moves perpendicular to the same magnetic field with this same translational speed, what is the radius of its circular orbit?

*Answer*  $7.63 \times 10^{-5}$ m

## Example 22.3 Bending an Electron Beam

In an experiment designed to measure the strength of a uniform magnetic field, electrons are accelerated from rest (by means of an electric field) through a potential difference of 350 V. After leaving the region of the electric field, the electrons enter a magnetic field and travel along a curved path because of the magnetic force exerted on them. The radius of the path is measured to be 7.50 cm. Figure 22.10 shows such a curved beam of electrons. Assuming the magnetic field is perpendicular to the beam, (a) what is the magnitude of the field?

**Reasoning**  First we calculate the speed of the electrons from their acceleration in the electric field. We apply the energy version of the isolated system model—the sum of the change in the kinetic energy of an electron and the change in the potential energy $q \Delta V$ of the electron-field system is equal to zero. Then we use Equation 22.3 to find the magnitude of the magnetic field.

**Solution**  Because $K_i = 0$ and $K_f = \frac{1}{2}m_e v^2$, we have

$$\Delta K + \Delta U = 0 \longrightarrow \tfrac{1}{2}m_e v^2 + q \Delta V = 0$$

$$v = \sqrt{\frac{-2q \Delta V}{m_e}} = \sqrt{\frac{-2(-1.60 \times 10^{-19} \text{ C})(350 \text{ V})}{9.11 \times 10^{-31} \text{ kg}}}$$

$$= 1.11 \times 10^7 \text{ m/s}$$

$$B = \frac{m_e v}{|q|r} = \frac{(9.11 \times 10^{-31} \text{ kg})(1.11 \times 10^7 \text{ m/s})}{(1.60 \times 10^{-19} \text{ C})(0.0750 \text{ m})}$$

$$= 8.43 \times 10^{-4} \text{ T}$$

**Figure 22.10**

(Example 22.3) The bending of an electron beam in a magnetic field. *(Henry Leap and Jim Lehman)*

(b) What is the angular speed of the electrons?

**Solution**  Using Equation 10.10, we find

$$\omega = \frac{v}{r} = \frac{1.11 \times 10^7 \text{ m/s}}{0.0750 \text{ m}} = 1.48 \times 10^8 \text{ rad/s}$$

**EXERCISE**  What is the period of revolution of the electrons?

*Answer*  42.5 ns

## 22.4 • APPLICATIONS OF THE MOTION OF CHARGED PARTICLES IN A MAGNETIC FIELD

A charge moving with velocity **v** in the presence of an electric field **E** and a magnetic field **B** experiences both an electric force $q\mathbf{E}$ and a magnetic force $q\mathbf{v} \times \mathbf{B}$. The total force, called the **Lorentz force,** acting on the charge is therefore

$$\mathbf{F} = q\mathbf{E} + q\mathbf{v} \times \mathbf{B} \qquad [22.6]$$

In this section, we look at three applications involving particles experiencing the Lorentz force.

### Velocity Selector

In many experiments involving moving charged particles, it is important to have particles that all move with essentially the same velocity. This can be achieved by applying a combination of an electric field and a magnetic field oriented as shown in Figure 22.11. A uniform electric field is directed vertically downward (in the plane of the page in Fig. 22.11) and a uniform magnetic field is applied perpendicular to the electric field (into the page in Fig. 22.11). Particles moving through this region will experience the Lorentz force, given by Equation 22.6. For a positively charged particle, the magnetic force $q\mathbf{v} \times \mathbf{B}$ is upward and the electric force $q\mathbf{E}$ is downward. When the magnitudes of the two fields are chosen so that $qE = qvB$, the particle is in equilibrium and moves in a straight horizontal line through the region of the fields. From $qE = qvB$ we find that

$$v = \frac{E}{B} \qquad [22.7]$$

Only those particles having this speed are undeflected as they move through the perpendicular electric and magnetic fields and pass through a small opening at the end of the device. The magnetic force exerted on particles moving at speeds greater than this is stronger than the electric force, and these particles are deflected upward. Those moving at speeds less than this are deflected downward.

### The Mass Spectrometer

A **mass spectrometer** separates ions according to their mass-to-charge ratio. In one version, known as the *Bainbridge mass spectrometer,* a beam of ions first passes

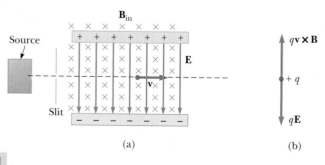

(a)    (b)

**Figure 22.11**

(a) A velocity selector. When a positively charged particle is in the presence of a magnetic field directed into the page and an electric field directed downward, it experiences a downward electric force $q\mathbf{E}$ and an upward magnetic force $q\mathbf{v} \times \mathbf{B}$. (b) When these forces balance, the particle moves in a straight line through the fields.

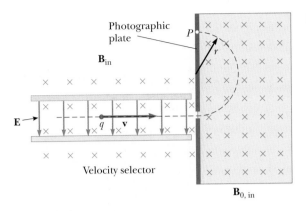

Photographic plate

$\mathbf{B}_{in}$

$E$

$q$ $\mathbf{v}$

Velocity selector

$\mathbf{B}_{0,\, in}$

**Figure 22.12**

A mass spectrometer. Positively charged particles are sent first through a velocity selector and then into a region where the magnetic field $\mathbf{B}_0$ causes the particles to move in a semicircular path and strike a photographic film at $P$.

through a velocity selector and then enters a second region with no electric field and a uniform magnetic field $\mathbf{B}_0$ that has the same direction as the magnetic field in the selector (Fig. 22.12). On entering the second magnetic field, the ions move in a semicircle of radius $r$ before striking a photographic plate at $P$. If the ions are positively charged, the beam deflects upward as in Figure 22.12. If the ions are negatively charged, the beam deflects downward. From Equation 22.3, we can express the ratio $m/q$ as

$$\frac{m}{q} = \frac{rB_0}{v}$$

Using Equation 22.7, we find that

$$\frac{m}{q} = \frac{rB_0 B}{E} \qquad\qquad \textbf{[22.8]}$$

Therefore, $m/q$ can be determined by measuring the radius of curvature and knowing the field magnitudes $B$, $B_0$, and $E$. In practice, one usually measures the masses of various isotopes of a given ion, with the ions all carrying the same charge $q$. In this way, the mass ratios can be determined even if $q$ is unknown.

A variation of this technique was used by J. J. Thomson (1856–1940) in 1897 to measure the ratio $e/m_e$ for electrons. Figure 22.13a shows the basic apparatus he used. Electrons are accelerated from the cathode and pass through two slits. They then drift into a region of perpendicular electric and magnetic fields. The magnitudes of the two fields are first adjusted to produce an undeflected beam. When the magnetic field is turned off, the electric field produces a measurable beam deflection that is recorded on the fluorescent screen. From the size of the deflection and the measured values of $E$ and $B$, the charge-to-mass ratio can be determined. The results of this crucial experiment represent the discovery of the electron as a fundamental particle of nature.

## The Cyclotron

A **cyclotron** can accelerate charged particles to very high speeds. Both electric and magnetic forces play a key role in its operation. The energetic particles produced are used to bombard atomic nuclei and thereby produce nuclear reactions of interest to researchers. A number of hospitals use cyclotron facilities to produce radioactive substances for diagnosis and treatment.

A schematic drawing of a cyclotron is shown in Figure 22.14. The charges move inside two semicircular containers, $D_1$ and $D_2$, referred to as *dees* because

**APPLICATION**

Uses of the cyclotron.

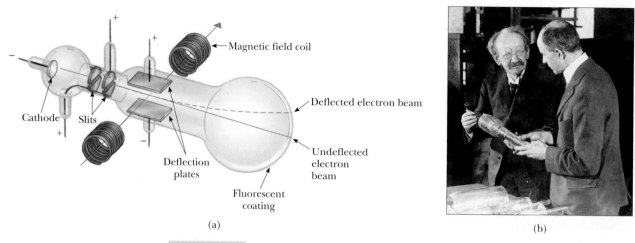

(a)

(b)

**Figure 22.13**

(a) Thomson's apparatus for measuring $e/m_e$. Electrons are accelerated from the cathode, pass through two slits, and are deflected by both an electric field and a magnetic field (directed perpendicular to the electric field). The electrons then strike a fluorescent screen. (b) J. J. Thomson *(left)* in the Cavendish Laboratory, University of Cambridge. It is interesting to note that the man on the right, Frank Baldwin Jewett, is a distant relative of John W. Jewett, Jr., coauthor of this text. *(Bell Telephone Labs/Courtesy of Emilio Segrè Visual Archives)*

they are shaped like the letter "D." A high-frequency alternating potential difference is applied to the dees, and a uniform magnetic field is directed perpendicular to them. A positive ion released at *P* near the center of the magnet moves in a semicircular path in one dee (indicated by the dashed red line in the drawing) and arrives back at the gap in a time *T*/2, where *T* is the time needed to make one

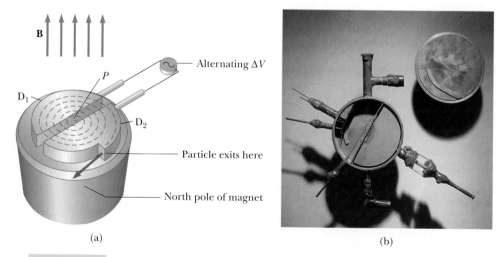

(a)

(b)

**Figure 22.14**

(a) A cyclotron consists of an ion source at *P*, two hollow sections called dees, $D_1$ and $D_2$, across which an alternating potential difference is applied, and a uniform magnetic field. (The south pole of the magnet is not shown.) The red dashed curved lines represent the path of the particles. (b) The first cyclotron, invented by E. O. Lawrence and M. S. Livingston in 1934. *(Courtesy of Lawrence Berkeley Laboratory, University of California)* complete trip around the two dees, given by Equation 22.5. The frequency of the applied potential difference is chosen so that the polarity of the dees is reversed during the time interval in which the ion travels around one dee. If the applied potential difference is adjusted such that $D_2$ is at a lower electric potential than $D_1$ by an amount $\Delta V$, the ion accelerates across the gap to $D_2$ and its kinetic energy increases by an amount $q\,\Delta V$. It then moves around $D_2$ in a semicircular path of larger radius (because its speed has increased). After a time $T/2$, it again arrives at the gap between the dees. By this time, the polarity across the dees has reversed again and the ion is given another "kick" across the gap. The motion continues so that for each half-circle trip around one dee, the ion gains additional kinetic energy equal to $q\,\Delta V$. When the radius of its path is nearly that of the dees, the energetic ion leaves the system through the exit slit. It is important to note that the operation of the cyclotron is based on the fact that $T$ is independent of the speed of the ion and the radius of its circular path (Eq. 22.5).

We can obtain an expression for the kinetic energy of the ion when it exits from the cyclotron in terms of the radius $R$ of the dees. From Equation 22.3 we know that $v = qBR/m$. Hence, the kinetic energy is

$$K = \tfrac{1}{2}mv^2 = \frac{q^2 B^2 R^2}{2m} \qquad \text{[22.9]}$$

When the energy of the ions in a cyclotron exceeds about 20 MeV, relativistic effects come into play. For this reason, the moving ions do not remain in phase with the applied potential difference. Some accelerators solve this problem by modifying the frequency of the applied potential difference so that it remains in phase with the moving ions.

**PITFALL PREVENTION 22.1**

**The cyclotron is not state-of-the-art technology**

The cyclotron is important historically because it was the first particle accelerator to achieve very high particle speeds. Cyclotrons are still in use, for example, to accelerate particles to high speeds in medical applications. The cyclotron is also important as an application of the ideas in our current discussion. Most accelerators built recently and currently in use in research, however, are not cyclotrons. They work on a different principle and are generally called *synchrotrons*.

## 22.5 • MAGNETIC FORCE ON A CURRENT-CARRYING CONDUCTOR

 See Screen 12.3

Because a magnetic force is exerted on a single charged particle when it moves through an external magnetic field, it should not surprise you to find that a current-carrying wire also experiences a magnetic force when placed in an external magnetic field. This follows from the fact that the current represents a collection of many charged particles in motion; hence, the resultant magnetic force on the wire is due to the sum of the individual magnetic forces on the charged particles. The force on the particles is transmitted to the "bulk" of the wire through collisions with the atoms making up the wire.

The magnetic force on a current-carrying conductor can be demonstrated by hanging a wire between the faces of a magnet as in Figure 22.15, where the magnetic field is directed into the page. The wire deflects to the left or right when a current is passed through it.

Let us quantify this discussion by considering a straight segment of wire of length $\ell$ and cross-sectional area $A$, carrying a current $I$ in a uniform external magnetic field **B**, as in Figure 22.16. As a simplification model, we shall ignore the high-speed zigzag motion of the charges in the wire (which is valid because the net velocity associated with this motion is zero) and assume that the charges simply move with the drift velocity $\mathbf{v}_d$. The magnetic force on a charge $q$ moving with drift velocity $\mathbf{v}_d$ is $q\mathbf{v}_d \times \mathbf{B}$. To find the total magnetic force on the wire segment, we multiply the magnetic force on one charge by the number of charges in the segment. Because the volume of the segment is $A\ell$, the number of charges in the seg-

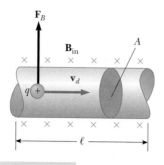

**Figure 22.16**

A section of a wire containing moving charges in a magnetic field **B**. The magnetic force on each charge is $q\mathbf{v}_d \times \mathbf{B}$, and the net force on a segment of length $\ell$ is $I\ell \times \mathbf{B}$.

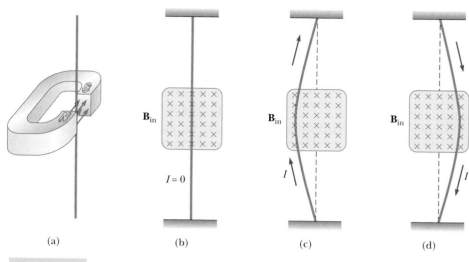

(a)          (b)          (c)          (d)

**Figure 22.15**

(a) A wire suspended vertically between the poles of a magnet. (b) The setup shown in (a) as seen looking at the south pole of the magnet, so that the magnetic field (blue crosses) is directed into the page. When no current is flowing in the wire, it remains vertical. (c) When the current is upward, the wire deflects to the left. (d) When the current is downward, the wire deflects to the right.

ment is $nA\ell$, where $n$ is the number of charges per unit volume. Hence, the total magnetic force on the wire of length $\ell$ is

$$\mathbf{F}_B = (q\mathbf{v}_d \times \mathbf{B})\, nA\ell$$

This can be written in a more convenient form by noting that, from Equation 21.4, the current in the wire is $I = nqv_dA$. Therefore, $\mathbf{F}_B$ can be expressed as

• *Magnetic force on a current-carrying conductor*

$$\mathbf{F}_B = I\ell \times \mathbf{B} \qquad [22.10]$$

where $\ell$ is a vector in the direction of the current $I$; the magnitude of $\ell$ equals the length of the segment. Note that this expression applies only to a straight segment of wire in a uniform external magnetic field.

Now consider an arbitrarily shaped wire of uniform cross-section in an external magnetic field, as in Figure 22.17. It follows from Equation 22.10 that the magnetic force on a very small segment of the wire of length $ds$ in the presence of an external field **B** is

$$d\mathbf{F}_B = I\, d\mathbf{s} \times \mathbf{B} \qquad [22.11]$$

where $d\mathbf{s}$ is a vector representing the length segment, with its direction the same as that of the current, and $d\mathbf{F}_B$ is directed out of the page for the directions assumed in Figure 22.17. We can consider Equation 22.11 as an alternative definition of **B** to Equation 22.1. That is, the field **B** can be defined in terms of a measurable force on a current element, where the force is a maximum when **B** is perpendicular to the element and zero when **B** is parallel to the element.

To obtain the total magnetic force $\mathbf{F}_B$ on a length of the wire between arbitrary points $a$ and $b$, we integrate Equation 22.11 over the length of the wire between these points:

$$\mathbf{F}_B = I \int_a^b d\mathbf{s} \times \mathbf{B} \qquad [22.12]$$

**Figure 22.17**

A wire segment of arbitrary shape carrying a current $I$ in a magnetic field **B** experiences a magnetic force. The force on any length element $d\mathbf{s}$ is $I\, d\mathbf{s} \times \mathbf{B}$ and is directed out of the page.

When this integration is carried out, the magnitude of the magnetic field and the direction the field makes with the vector $d\mathbf{s}$ may vary from point to point.

### Quick Quiz 22.5

What is the net magnetic force on a closed current loop in a uniform magnetic field?

### THINKING PHYSICS 22.3

In a lightning stroke, negative charge rapidly moves from a cloud to the ground. In what direction is a lightning stroke deflected by the Earth's magnetic field?

**Reasoning** The downward flow of negative charge in a lightning stroke is equivalent to an upward-moving current. Thus, the vector $d\mathbf{s}$ is upward, and the magnetic field vector has a northward component. According to the cross product of the length element and magnetic field vectors (Eq. 22.11), the lightning stroke would be deflected to the *west*.

### Example 22.4 Force on a Semicircular Conductor

A wire bent into the shape of a semicircle of radius $R$ forms a closed circuit and carries a current $I$. The circuit lies in the $xy$ plane, and a uniform magnetic field is present along the positive $y$ axis as in Figure 22.18. Find the magnetic force on the straight portion of the wire and on the curved portion.

**Reasoning and Solution** The force on the straight portion of the wire has a magnitude $F_1 = I\ell B = 2IRB$, because $\ell = 2R$ and the wire is perpendicular to $\mathbf{B}$. The direction of $\mathbf{F}_1$ is out of the paper because $\ell \times \mathbf{B}$ is outward. (That is, $\ell$ is to the right, in the direction of the current, and so by the rule of cross products, $\ell \times \mathbf{B}$ is outward.)

To find the magnetic force on the curved part, we first write an expression for the magnetic force $d\mathbf{F}_2$ on the element $d\mathbf{s}$. If $\theta$ is the angle between $\mathbf{B}$ and $d\mathbf{s}$ in Figure 22.18, the magnitude of $d\mathbf{F}_2$ is

$$dF_2 = I\,|\,d\mathbf{s} \times \mathbf{B}\,| = IB \sin\theta \, ds$$

To integrate this expression, we express $ds$ in terms of $\theta$. Because $s = R\theta$, $ds = R\,d\theta$, and the expression for $dF_2$ can be written

$$dF_2 = IRB \sin\theta \, d\theta$$

To obtain the total magnetic force $F_2$ on the curved portion, we integrate this expression to account for contributions from all elements. Note that the direction of the mag-

**Figure 22.18**

(Example 22.4) The net force on a closed current loop in a uniform magnetic field is zero. For the loop shown here, the force on the straight portion is $2IRB$ and out of the page, whereas the force on the curved portion is $2IRB$ and into the page.

netic force on every element is the same: into the paper (because $d\mathbf{s} \times \mathbf{B}$ is inward). Therefore, the resultant magnetic force $\mathbf{F}_2$ on the curved wire must also be into the paper. Integrating $dF_2$ over the limits $\theta = 0$ to $\theta = \pi$ (i.e., the entire semicircle) gives

$$F_2 = IRB \int_0^\pi \sin\theta \, d\theta = IRB[-\cos\theta]_0^\pi$$

$$= -IRB(\cos\pi - \cos 0) = -IRB(-1-1) = 2IRB$$

Because $F_2 = 2IRB$ and the vector $\mathbf{F}_2$ is directed into the paper and the force on the straight wire has magnitude $F_1 = 2IRB$ and is out of the paper, we see that the net magnetic force on the closed loop is zero, as we expect from Quick Quiz 22.5.

**EXERCISE** Calculate the magnitude of the magnetic force per unit length exerted on a conductor carrying a current of 22.0 A in a region where a uniform magnetic field has a magnitude of 0.770 T and is directed perpendicular to the conductor.

*Answer* 16.9 N/m

## 22.6 • TORQUE ON A CURRENT LOOP IN A UNIFORM MAGNETIC FIELD

In the preceding section we showed how a magnetic force is exerted on a current-carrying conductor when the conductor is placed in an external magnetic field. With this as a starting point, we shall show that a torque is exerted on a current loop placed in a magnetic field. The results of this analysis are of great practical value in the design of motors and generators.

Consider a rectangular loop carrying a current $I$ in the presence of a uniform external magnetic field *in the plane of the loop*, as in Figure 22.19a. The magnetic forces on sides ① and ③, of length $b$, are zero because these wires are parallel to the field; hence, $d\mathbf{s} \times \mathbf{B} = 0$ for these sides. However, nonzero magnetic forces act on sides ② and ④ because these sides are oriented perpendicular to the field. The magnitude of these forces is

$$F_2 = F_4 = IaB$$

The direction of $\mathbf{F}_2$, the magnetic force on side ②, is out of the paper, and that of $\mathbf{F}_4$, the magnetic force on side ④, is into the paper. If we view the loop from side ③ as in Figure 22.19b, we see the forces on ② and ④ directed as shown. If we assume that the loop is pivoted so that it can rotate about an axis perpendicular to the page and passing through point $O$, we see that these two magnetic forces produce a torque about this axis that rotates the loop clockwise. The magnitude of the torque, which we will call $\tau_{\max}$, is

$$\tau_{\max} = F_2 \frac{b}{2} + F_4 \frac{b}{2} = (IaB)\frac{b}{2} + (IaB)\frac{b}{2} = IabB$$

where the moment arm about this axis is $b/2$ for each force. Because the area of the loop is $A = ab$, the magnitude of the torque can be expressed as

$$\tau_{\max} = IAB \qquad \text{[22.13]}$$

(a)

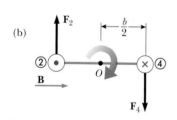

(b)

**Figure 22.19**

(a) Overhead view of a rectangular current loop in a uniform magnetic field. No forces are exerted on sides ① and ③ because these sides are parallel to **B**. Forces are exerted on sides ② and ④, however. (b) Edge view of the loop sighting down ② and ④ shows that the forces $\mathbf{F}_2$ and $\mathbf{F}_4$ exerted on these sides create a torque that tends to rotate the loop clockwise. The purple dot in the left circle represents current in wire ② coming toward you; the purple × in the right circle represents current in wire ④ moving away from you.

Remember that this result is valid only when the field **B** is parallel to the plane of the loop. The sense of the rotation is clockwise when the loop is viewed as in Figure 22.19b. If the current were reversed, the magnetic forces would reverse their directions and the rotational tendency would be counterclockwise.

Now suppose the uniform magnetic field makes an angle of $\theta$ with a line perpendicular to the plane of the loop, as in Figure 22.20. For convenience, we shall assume that **B** is perpendicular to sides ② and ④. (The end view of these sides is shown in Fig. 22.20.) In this case, the magnetic forces on sides ① and ③ cancel each other and produce no torque because they have the same line of action. The magnetic forces $\mathbf{F}_2$ and $\mathbf{F}_4$ acting on sides ② and ④, however, both produce a

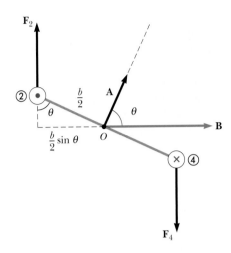

**Figure 22.20**

An end view of the loop in Figure 22.19b rotated through an angle with respect to the magnetic field. If **B** is at an angle $\theta$ with respect to vector **A**, which is perpendicular to the plane of the loop, the torque is $IAB \sin \theta$.

torque about an axis through the center of the loop. Referring to Figure 22.20, we note that the moment arm of $\mathbf{F}_2$ about this axis is $(b/2) \sin \theta$. Likewise, the moment arm of $\mathbf{F}_4$ is also $(b/2) \sin \theta$. Because $F_2 = F_4 = IaB$, the net torque $\tau$ has the magnitude

$$\tau = F_2 \frac{b}{2} \sin \theta + F_4 \frac{b}{2} \sin \theta$$

$$= (IaB)\left(\frac{b}{2} \sin \theta\right) + (IaB)\left(\frac{b}{2} \sin \theta\right) = IabB \sin \theta$$

$$= IAB \sin \theta$$

where $A = ab$ is the area of the loop. This result shows that the torque has its maximum value $IAB$ (Eq. 22.13) when the field is parallel to the plane of the loop ($\theta = 90°$) and is zero when the field is perpendicular to the plane of the loop ($\theta = 0$). As we see in Figure 22.20, the loop tends to rotate in the direction of decreasing values of $\theta$ (i.e., so that the normal to the plane of the loop rotates toward the direction of the magnetic field). A convenient vector expression for the torque is

$$\tau = I\mathbf{A} \times \mathbf{B} \qquad \qquad \textbf{[22.14]}$$

where **A**, a vector perpendicular to the plane of the loop (Fig. 22.20), has a magnitude equal to the area of the loop. The sense of **A** is determined by the right-hand rule illustrated in Figure 22.21. When the four fingers of the right hand are curled in the direction of the current in the loop, the thumb points in the direction of **A**. The product $I\mathbf{A}$ is defined to be the **magnetic dipole moment** $\boldsymbol{\mu}$ of the loop:

$$\boldsymbol{\mu} \equiv I\mathbf{A} \qquad \qquad \textbf{[22.15]}$$

The SI unit of magnetic dipole moment is the ampere-meter$^2$ (A·m$^2$). Using this definition, the torque can be expressed as

$$\tau = \boldsymbol{\mu} \times \mathbf{B} \qquad \qquad \textbf{[22.16]}$$

Although the torque was obtained for a particular orientation of **B** with respect to the loop, Equation 22.16 is valid for any orientation. Furthermore, although the torque expression was derived for a rectangular loop, the result is valid for a loop of any shape. Once the torque is determined, the motion of the coil can be modeled as a rigid body under a net torque.

**Figure 22.21**

Right-hand rule for determining the direction of the vector **A**. The direction of the magnetic moment $\boldsymbol{\mu}$ is the same as the direction of **A**.

- *Magnetic moment of a current loop*

- *Torque on a current loop*

If a coil consists of $N$ turns of wire, each carrying the same current and each having the same area, the total magnetic moment of the coil is the product of the number of turns and the magnetic moment for one turn, $\mu = NIA$. Thus, the torque on an $N$-turn coil is $N$ times greater than that on a one-turn coil.

A common electric motor consists of a coil of wire mounted so that it can rotate in the field of a permanent magnet. The torque on the current-carrying coil is used to rotate a shaft that drives a mechanical device—the power windows in your car, your household fan, or your electric hedge trimmer, for example.

---

### Example 22.5   The Magnetic Moment and Torque on a Coil

A rectangular coil of dimensions 5.40 cm × 8.50 cm consists of 25 turns of wire. The coil carries a current of 15.0 mA. (a) Calculate the magnitude of its magnetic moment.

**Solution**   The magnitude of the magnetic moment of a current loop is $\mu = IA$ (Eq. 22.15). In this case, $A = (0.0540 \text{ m})(0.0850 \text{ m}) = 4.59 \times 10^{-3} \text{ m}^2$. Because the coil has 25 turns, and assuming that each turn has the same area $A$, we have

$$\mu_{\text{coil}} = NIA = (25)(15.0 \times 10^{-3}\,\text{A})(4.59 \times 10^{-3}\,\text{m}^2)$$
$$= 1.72 \times 10^{-3}\,\text{A} \cdot \text{m}^2$$

(b) Suppose a uniform magnetic field of magnitude 0.350 T is applied parallel to the plane of the loop. What is the magnitude of the torque acting on the loop?

**Solution**   The torque is given by Equation 22.16, $\tau = \mu \times B$. In this case, $B$ is perpendicular to $\mu_{\text{coil}}$, so that

$$\tau = \mu_{\text{coil}} B = (1.72 \times 10^{-3}\,\text{A} \cdot \text{m}^2)(0.350\,\text{T})$$
$$= 6.02 \times 10^{-4}\,\text{N} \cdot \text{m}$$

**EXERCISE**   Show that the units of torque, $\text{A} \cdot \text{m}^2 \cdot \text{T}$, reduce to $\text{N} \cdot \text{m}$.

**EXERCISE**   Calculate the magnitude of the torque on the coil when the 0.350-T magnetic field makes angles of (a) 60.0° and (b) 0° with $\mu$.

*Answer*   (a) $5.21 \times 10^{-4}\,\text{N} \cdot \text{m}$   (b) zero

---

## 22.7 • THE BIOT–SAVART LAW

In the previous sections, we investigated the result of placing an object in an existing magnetic field. When a moving charge is placed in the field, it experiences a magnetic force; a current-carrying wire placed in the field also experiences a magnetic force; a current loop in the field experiences a torque.

Now we shift our thinking and investigate the *source* of the magnetic field. Oersted's 1819 discovery (Section 22.1) that an electric current in a wire deflects a nearby compass needle indicates that current acts as a source of a magnetic field. From their early 19th-century investigations on the force between a current-carrying conductor and a magnet, Jean-Baptiste Biot and Félix Savart arrived at an expression for the magnetic field at a point in space in terms of the current that produces the field. No point currents exist comparable to point charges (because we must have a complete circuit for a current). Hence, we must investigate the magnetic field due to an infinitesimally small element of current that is part of a larger current distribution. Suppose that the current distribution is a wire carrying a steady current $I$ as in Figure 22.22. The **Biot–Savart law** says that the magnetic field $d\mathbf{B}$ at point $P$ created by an element of infinitesimal length $ds$ of the wire has the following properties:

**Figure 22.22**

The magnetic field $d\mathbf{B}$ at a point $P$ due to a current $I$ through a length element $d\mathbf{s}$ is given by the Biot–Savart law. The field is out of the page at $P$ and into the page at $P'$.

• The vector $d\mathbf{B}$ is perpendicular both to $d\mathbf{s}$ (which is in the direction of the current) and to the unit vector $\hat{\mathbf{r}}$ directed from the element to $P$.

- The magnitude of $d\mathbf{B}$ is inversely proportional to $r^2$, where $r$ is the distance from the element to $P$.
- The magnitude of $d\mathbf{B}$ is proportional to the current and to the length $ds$ of the element.
- The magnitude of $d\mathbf{B}$ is proportional to $\sin\theta$, where $\theta$ is the angle between $d\mathbf{s}$ and $\hat{\mathbf{r}}$.

• *Properties of the magnetic field due to a current element*

The **Biot–Savart law** can be summarized in the following compact form:

$$d\mathbf{B} = k_m \frac{I\,d\mathbf{s} \times \hat{\mathbf{r}}}{r^2} \qquad [22.17]$$

• *Biot–Savart law*

where $k_m$ is a constant that in SI units is exactly $10^{-7}\ \text{T}\cdot\text{m/A}$. The constant $k_m$ is usually written $\mu_0/4\pi$, where $\mu_0$ is another constant, called the **permeability of free space:**

$$\frac{\mu_0}{4\pi} = k_m = 10^{-7}\ \text{T}\cdot\text{m/A} \qquad [22.18]$$

$$\mu_0 = 4\pi k_m = 4\pi \times 10^{-7}\ \text{T}\cdot\text{m/A} \qquad [22.19]$$

• *Permeability of free space*

Hence, the Biot–Savart law, Equation 22.17, can also be written

$$d\mathbf{B} = \frac{\mu_0}{4\pi}\frac{I\,d\mathbf{s} \times \hat{\mathbf{r}}}{r^2} \qquad [22.20]$$

It is important to note that the Biot–Savart law gives the magnetic field at a point only for a small length element of the conductor. We identify the product $I\,d\mathbf{s}$ as a **current element.** To find the total magnetic field $\mathbf{B}$ at some point due to a conductor of finite size, we must sum up contributions from all current elements making up the conductor. That is, we evaluate $\mathbf{B}$ by integrating Equation 22.20 over the entire conductor.

There are two similarities between the Biot–Savart law of magnetism and Equation 19.7 for the electric field of a charge distribution, and two important differences. The current element $I\,d\mathbf{s}$ produces a magnetic field, whereas the charge element $dq$ produces an electric field. Furthermore, the magnitude of the magnetic field varies as the inverse square of the distance from the current element, as does the electric field due to a charge element. The directions of the two fields are quite different, however. The electric field due to a charge element is radial; in the case of a positive point charge, $\mathbf{E}$ is directed from the charge toward the field point. The magnetic field due to a current element is perpendicular to both the current element and the radius vector. Hence, if the conductor lies in the plane of the page, as in Figure 22.22, $d\mathbf{B}$ points out of the page at the point $P$ and into the page at $P'$. Another important difference is that an electric field can be a result either of a single charge or a distribution of charges, but a magnetic field can only be a result of a current distribution.

Figure 22.23 shows a convenient right-hand rule for determining the direction of the magnetic field due to a current. Note that the field lines generally encircle the current. In the case of current in a long, straight wire, the field lines form circles that are concentric with the wire and are in a plane perpendicular to the wire. If the wire is grasped in the right hand with the thumb in the direction of the current, the fingers will curl in the direction of $\mathbf{B}$.

Although the magnetic field due to an infinitely long, current-carrying wire can be calculated using the Biot–Savart law (Problem 47), in Section 22.9 we use a

**Figure 22.23**

The right-hand rule for determining the direction of the magnetic field surrounding a long, straight wire carrying a current. Note that the magnetic field lines form circles around the wire.

different method to show that the magnitude of this field at a distance $r$ from the wire is

- *Magnetic field due to a long, straight wire*

$$B = \frac{\mu_0 I}{2\pi r}$$ [22.21]

### PITFALL PREVENTION 22.2

**The Biot–Savart law**

When you are applying the Biot–Savart law, it is important to recognize that the magnetic field described in these calculations is the **field *due to* a given current-carrying conductor.** This is not to be confused with any *external* field that may be applied to the conductor from some other source.

### Quick Quiz 22.6

Suppose you move along a line parallel to a metal wire at a speed that is the same as the drift speed of the electrons in the current. Do you now measure a magnetic field of zero?

### THINKING PHYSICS 22.4

In electric circuits, wires carrying currents in opposite directions are often twisted together. What is the advantage of this configuration?

**Reasoning**   If the wires are not twisted together, the combination of the two wires forms a current loop. The magnetic field generated by the loop might affect adjacent circuits or components. (In Example 22.6, we find the specific expression for the magnetic field of a circular current loop.) If the wires are twisted together, they can be modeled as a single wire carrying no current, because the currents in the two wires are in opposite directions. In this model, no magnetic field due to the wires affects adjacent circuits. (The magnetic field is not *precisely* zero because the wires are not in the same location in space, just like the electric field of an electric dipole is not zero because the two charges are at different locations.)

### Example 22.6   Magnetic Field on the Axis of a Circular Current Loop

Consider a circular loop of wire of radius $R$ located in the $yz$ plane and carrying a steady current $I$, as in Figure 22.24. Calculate the magnetic field at an axial point $P$ a distance $x$ from the center of the loop.

**Reasoning**   In this situation, note that any element $d\mathbf{s}$ is perpendicular to $\hat{\mathbf{r}}$. Furthermore, all elements around the loop are at the same distance $r$ from $P$, where $r^2 = x^2 + R^2$. Hence, the magnitude of $d\mathbf{B}$ due to the element $d\mathbf{s}$ is

$$dB = \frac{\mu_0 I}{4\pi} \frac{|d\mathbf{s} \times \hat{\mathbf{r}}|}{r^2} = \frac{\mu_0 I}{4\pi} \frac{ds}{(x^2 + R^2)}$$ [22.22]

The direction of the magnetic field $d\mathbf{B}$ due to the element $d\mathbf{s}$ is perpendicular to the plane formed by $\hat{\mathbf{r}}$ and $d\mathbf{s}$, as in Figure 22.24. The vector $d\mathbf{B}$ can be resolved into a component $dB_x$, along the $x$ axis, and a component $dB_\perp$, which is perpendicular to the $x$ axis. When the components $dB_\perp$ are summed over the whole loop, the result is zero. That is, by

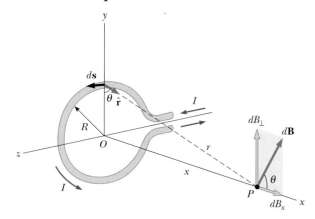

**Figure 22.24**

(Example 22.6) The geometry for calculating the magnetic field at a point $P$ lying on the axis of a current loop. By symmetry, the total field **B** is along this axis.

symmetry any element on one side of the loop sets up a component $dB_\perp$ that cancels the component set up by an element diametrically opposite it.

**Solution** For the reasons given in the Reasoning, the resultant field at $P$ must be along the $x$ axis and can be found by integrating the components $dB_x = dB \cos \theta$, where this expression is obtained from resolving the vector $d\mathbf{B}$ into its components as shown in Figure 22.24. That is, $\mathbf{B} = B_x\mathbf{i}$, where

$$B_x = \oint dB \cos \theta = \frac{\mu_0 I}{4\pi} \oint \frac{ds \cos \theta}{x^2 + R^2}$$

and the integral must be taken over the entire loop. Because $\theta$, $x$, and $R$ are constants for all elements of the loop and because $\cos \theta = R/(x^2 + R^2)^{1/2}$, we obtain

$$B_x = \frac{\mu_0 IR}{4\pi(x^2 + R^2)^{3/2}} \oint ds = \frac{\mu_0 IR^2}{2(x^2 + R^2)^{3/2}} \quad \textbf{[22.23]}$$

where we have used the fact that $\oint ds = 2\pi R$ (the circumference of the loop).

To find the magnetic field at the center of the loop, we set $x = 0$ in Equation 22.23. At this special point, this gives

$$B = \frac{\mu_0 I}{2R} \quad \text{(at } x = 0) \quad \textbf{[22.24]}$$

It is also interesting to determine the behavior of the magnetic field far from the loop, that is, when $x$ is large compared with $R$. In this case, we can ignore the term $R^2$ in the denominator of Equation 22.23 and we find

$$B \approx \frac{\mu_0 IR^2}{2x^3} \quad \text{(for } x \gg R) \quad \textbf{[22.25]}$$

Because the magnitude of the magnetic dipole moment $\mu$ of the loop is defined as the product of the current and the area (Eq. 22.15), $\mu = I(\pi R^2)$, we can express Equation 22.25 in the form

$$B = \frac{\mu_0}{2\pi} \frac{\mu}{x^3} \quad \textbf{[22.26]}$$

This result is similar in form to the expression for the electric field due to an electric dipole, $E = k_e(2qa)/y^3 = k_e p/y^3$ (Example 19.3), where $p$ is the electric dipole moment. The pattern of the magnetic field lines for a circular loop is shown in Figure 22.25. For clarity, the lines are drawn only for one plane that contains the axis of the loop. The field pattern is axially symmetric.

**EXERCISE** A wire with a current of 5.00 A is to be formed into a circular loop of one turn. If the required value of the magnetic field at the center of the loop is 10.0 $\mu$T, what is the required radius?

*Answer* 31.4 cm

(a)

(b)

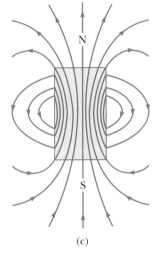

(c)

**Figure 22.25**

(Example 22.6) (a) Magnetic field lines surrounding a current loop. (b) Magnetic field lines surrounding a current loop displayed with iron filings *(Education Development Center, Newton, Mass.)* (c) Magnetic field lines surrounding a bar magnet. Note the similarity between this line pattern and that of a current loop.

**Figure 22.26**

Two parallel wires that each carry a steady current exert a force on each other. The field $\mathbf{B}_2$ due to the current in wire 2 exerts a force of magnitude $F_1 = I_1 \ell B_2$ on wire 1. The force is attractive if the currents are parallel (as shown) and repulsive if the currents are antiparallel.

## 22.8 • THE MAGNETIC FORCE BETWEEN TWO PARALLEL CONDUCTORS

In Section 22.5 we described the magnetic force that acts on a current-carrying conductor when the conductor is placed in an external magnetic field. Because a current in a conductor sets up its own magnetic field, it is easy to understand that two current-carrying conductors exert magnetic forces on each other. As we shall see, such forces can be used as the basis for defining the ampere and the coulomb.

Consider two infinitely long, straight, parallel wires separated by the distance $a$ and carrying currents $I_1$ and $I_2$ in the same direction as in Figure 22.26. We shall adopt a simplification model in which the radii of the wires are much smaller than $a$, so that the radius plays no role in the calculation. We can determine the force on one wire due to the magnetic field set up by the other wire. Wire 2, which carries current $I_2$, sets up a magnetic field $\mathbf{B}_2$ at the position of wire 1. The direction of $\mathbf{B}_2$ is perpendicular to the wire as shown in Figure 22.26. According to Equation 22.10, the magnetic force on a length $\ell$ of wire 1 is $\mathbf{F}_1 = I_1 \boldsymbol{\ell} \times \mathbf{B}_2$. Because $\boldsymbol{\ell}$ is perpendicular to $\mathbf{B}_2$, the magnitude of $\mathbf{F}_1$ is $F_1 = I_1 \ell B_2$. Because the field due to wire 2 is given by Equation 22.21, we see that

$$F_1 = I_1 \ell B_2 = I_1 \ell \left( \frac{\mu_0 I_2}{2\pi a} \right) = \frac{\ell \mu_0 I_1 I_2}{2\pi a}$$

We can rewrite this in terms of the force per unit length as

$$\frac{F_1}{\ell} = \frac{\mu_0 I_1 I_2}{2\pi a}$$

The direction of $\mathbf{F}_1$ is downward, toward wire 2, because $\boldsymbol{\ell} \times \mathbf{B}_2$ is downward. If one considers the field set up at wire 2 due to wire 1, the force $\mathbf{F}_2$ on wire 2 is found to be equal in magnitude and opposite in direction to $\mathbf{F}_1$. That is what one would expect, because Newton's third law must be obeyed. Thus, we can drop the force subscript so that the magnetic force per unit length exerted by each long current-carrying wire on the other is

- *Magnetic force per unit length between parallel current-carrying wires*

$$\frac{F}{\ell} = \frac{\mu_0}{2\pi} \frac{I_1 I_2}{a} \qquad \text{[22.27]}$$

This equation also applies if one of the wires is of finite length.

When the currents are in opposite directions, the magnetic forces are reversed and the wires repel each other. Hence, we find that **parallel conductors carrying currents in the same direction attract each other,** whereas **parallel conductors carrying currents in opposite directions repel each other.**

The magnetic force between two parallel wires, each carrying a current, is used to define the **ampere:** If two long, parallel wires 1 m apart carry the same current and the force per unit length on each wire is $2 \times 10^{-7}$ N/m, then the current is defined to be 1 A. The numerical value of $2 \times 10^{-7}$ N/m is obtained from Equation 22.27, with $I_1 = I_2 = 1$ A and $a = 1$ m.

The SI unit of charge, the **coulomb,** can now be defined in terms of the ampere: If a conductor carries a steady current of 1 A, the quantity of charge that flows through a cross-section of the conductor in 1 s is 1 C.

### Quick Quiz 22.7

If $I_1 = 2$ A and $I_2 = 6$ A in Figure 22.26, which of the following is true: (a) $F_1 = 3F_2$, (b) $F_1 = F_2/3$, (c) $F_1 = F_2$?

## Quick Quiz 22.8

A hanging spring is attached to a powerful battery and a switch. When the switch is closed so that current suddenly exists in the spring, does the spring compress or expand?

## 22.9 • AMPÈRE'S LAW

See Screen 12.4

A simple experiment first carried out by Oersted in 1820 clearly demonstrates that a current-carrying conductor produces a magnetic field. In this experiment, several compass needles are placed in a horizontal plane near a long vertical wire as in Figure 22.27a. When the wire carries no current, all needles point in the same direction (that of the Earth's magnetic field), as one would expect. When the wire carries a strong, steady current, however, the needles all deflect in a direction tangent to the circle as in Figure 22.27b. These observations show that the direction of **B** is consistent with the right-hand rule described in Section 22.7. When the current is reversed, the needles in Figure 22.27b also reverse.

Because the needles point in the direction of **B**, we conclude that the lines of **B** form circles about the wire, as discussed in Section 22.7. By symmetry, the magnitude of **B** is the same everywhere on a circular path that is centered on the wire and lies in a plane perpendicular to the wire. By varying the current and distance from the wire, one finds that **B** is proportional to the current and inversely proportional to the distance from the wire.

In Chapter 19, we investigated Gauss's law, which is a relationship between an electric charge and the electric field it produces. Gauss's law can be used to determine the electric field in highly symmetric situations. We now consider an analogous relationship in magnetism between a current and the magnetic field it produces. This relationship can be used to determine the magnetic field created by a highly symmetric current distribution.

(a) (b) (c)

### Figure 22.27

(a) When no current is present in the vertical wire, all compass needles point in the same direction (toward the Earth's North Pole). (b) When the wire carries a strong current, the compass needles deflect in a direction tangent to the circle, which is the direction of the magnetic field created by the current. (c) Circular magnetic field lines surrounding a current-carrying conductor, displayed with iron filings. *(Henry Leap and Jim Lehman)*

Let us evaluate the product $\mathbf{B} \cdot d\mathbf{s}$ for a small length element $d\mathbf{s}$ on the circular path centered on the wire in Figure 22.27. Along this path, the vectors $d\mathbf{s}$ and $\mathbf{B}$ are parallel at each point (Fig. 22.27b), so that $\mathbf{B} \cdot d\mathbf{s} = B\,ds$. Furthermore, by symmetry, $\mathbf{B}$ is constant in magnitude on this circle and is given by Equation 22.21. Therefore, the sum of the products $B\,ds$ over the closed path, which is equivalent to the line integral of $\mathbf{B} \cdot d\mathbf{s}$, is

• *Ampère's law*

$$\oint \mathbf{B} \cdot d\mathbf{s} = B \oint ds = \frac{\mu_0 I}{2\pi r}(2\pi r) = \mu_0 I \qquad \text{[22.28]}$$

where $\oint ds = 2\pi r$ is the circumference of the circle.

This result, known as **Ampère's law,** was calculated for the special case of a circular path surrounding a wire. It can, however, also be applied in the general case in which a steady current passes through the area surrounded by an arbitrary closed path. That is, Ampère's law says that the line integral of $\mathbf{B} \cdot d\mathbf{s}$ around any closed path equals $\mu_0 I$, where $I$ is the total steady current passing through any surface bounded by the closed path:

$$\oint \mathbf{B} \cdot d\mathbf{s} = \mu_0 I \qquad \text{[22.29]}$$

### Quick Quiz 22.9

Rank the values of $\oint \mathbf{B} \cdot d\mathbf{s}$ for the closed paths in Figure 22.28, from smallest to largest.

### Quick Quiz 22.10

Rank the values of $\oint \mathbf{B} \cdot d\mathbf{s}$ for the closed paths in Figure 22.29, from smallest to largest.

**André-Marie Ampère**
(1775–1836)

Ampère, a Frenchman, is credited with the discovery of electromagnetism — the relationship between electric currents and magnetic fields. Ampère's genius, particularly in mathematics, became evident by the time he was 12 years old; his personal life, however, was filled with tragedy. His father, a wealthy city official, was guillotined during the French Revolution, and his wife died young, in 1803. Ampère died at the age of 61 of pneumonia. His judgment of his life is clear from the epitaph he chose for his gravestone: *Tandem Felix* (Happy at Last). *(AIP Emilio Segrè Visual Archive)*

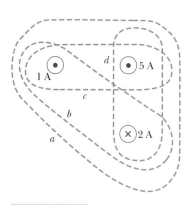

**Figure 22.28**

(Quick Quiz 22.9) Four closed paths around three current-carrying wires.

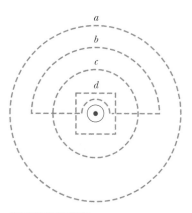

**Figure 22.29**

(Quick Quiz 22.10) Several closed paths near a single current-carrying wire.

**Ampère's law is valid only for steady currents**. Furthermore, even though Ampère's law is *true* for all current configurations, it is only *useful* for calculating the magnetic fields of configurations with high degrees of symmetry.

In Section 19.10, we provided some conditions to be sought when defining a gaussian surface. Similarly, to apply Equation 22.29 to calculate a magnetic field, we must determine a path of integration (sometimes called an *amperian loop*) that satisfies one or more of the following conditions:

1. The value of the magnetic field can be argued by symmetry to be constant over the path.
2. The dot product in Equation 22.29 can be expressed as a simple algebraic product $B\,ds$ because **B** and $d\mathbf{s}$ are parallel.
3. The dot product in Equation 2.29 is zero because **B** and $d\mathbf{s}$ are perpendicular.
4. The magnetic field can be argued to be zero at all points on the path.

The following examples illustrate some symmetric current configurations for which Ampère's law is useful.

---

### Example 22.7   The Magnetic Field Created by a Long Current-Carrying Wire

A long, straight wire of radius $R$ carries a steady current $I_0$ that is uniformly distributed through the cross-section of the wire (Fig. 22.30). Calculate the magnetic field a distance $r$ from the center of the wire in the regions $r \geq R$ and $r < R$.

**Solution**   As mentioned in Section 22.7, we could use the Biot–Savart law to solve this problem, but Ampère's law provides a much simpler solution. For $r \geq R$, let us choose path 1 in Figure 22.30—a circle of radius $r$ centered at the wire. From symmetry, we see that **B** must be constant in magnitude—condition (1)—and parallel to $d\mathbf{s}$—condition (2)—at every point on this circle. Because the total current passing through the plane of the circle is $I_0$, Ampère's law applied to the circular path gives

$$\oint \mathbf{B} \cdot d\mathbf{s} = B \oint ds = B(2\pi r) = \mu_0 I_0$$

$$B = \frac{\mu_0 I_0}{2\pi r} \quad \text{(for } r \geq R)$$

This is the result (Eq. 22.21) referred to in Section 22.7.

Now consider the interior of the wire, where $r < R$. We choose the circular path 2 shown in Figure 22.30. Here the current $I$ passing through the plane of the circle is less than the total current $I_0$. Because the current is uniform over the cross-section of the wire, the fraction of the current enclosed

**Figure 22.30**

(Example 22.7) A long, straight wire of radius $R$ carrying a steady current $I_0$ uniformly distributed across the wire. The magnetic field at any point can be calculated from Ampère's law using a circular path of radius $r$, concentric with the wire.

by the circle of radius $r < R$ must equal the ratio of the area $\pi r^2$ enclosed by circular path 2 and the cross-sectional area $\pi R^2$ of the wire:

$$\frac{I}{I_0} = \frac{\pi r^2}{\pi R^2}$$

$$I = \frac{r^2}{R^2} I_0$$

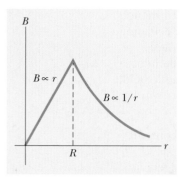

**Figure 22.31**

(Example 22.7) Magnitude of the magnetic field versus $r$ for the wire described in Figure 22.30. The field is proportional to $r$ inside the wire and varies as $1/r$ outside the wire.

Following the same procedure as for circular path 1, we apply Ampère's law to circular path 2:

$$\oint \mathbf{B} \cdot d\mathbf{s} = B(2\pi r) = \mu_0 I = \mu_0 \left( \frac{r^2}{R^2} I_0 \right)$$

$$B = \left( \frac{\mu_0 I_0}{2\pi R^2} \right) r \quad \text{(for } r < R) \qquad \textbf{[22.30]}$$

The magnitude of the magnetic field versus $r$ for this configuration is sketched in Figure 22.31. Note that inside the wire, $B \to 0$ as $r \to 0$. This result is similar in form to that of the electric field inside a uniformly charged rod.

## Example 22.8   The Magnetic Field Created by a Toroid

A device called a *toroid* (Figure 22.32) is often used to create an almost uniform magnetic field in some enclosed area. The device consists of a conducting wire wrapped around a ring (a *torus*) made of a nonconducting material. For a toroid having $N$ closely spaced turns of wire and air in the torus, calculate the magnetic field in the region occupied by the torus, a distance $r$ from the center.

**Reasoning**   To calculate the field inside the toroid, we evaluate the line integral of $\mathbf{B} \cdot d\mathbf{s}$ over a circular path of radius $r$. By symmetry, we see that conditions (1) and (2) apply—the magnetic field is constant in magnitude on this circle and tangent to it, so that $\mathbf{B} \cdot d\mathbf{s} = B \, ds$.

Furthermore, note that the closed path surrounds a circular area through which $N$ loops of wire pass, each of which carries a current $I$. The right side of Equation 22.29 is therefore $\mu_0 NI$ in this case.

**Solution**   Ampère's law applied to the circular path gives

$$\oint \mathbf{B} \cdot d\mathbf{s} = B \oint ds = B(2\pi r) = \mu_0 NI$$

$$B = \frac{\mu_0 NI}{2\pi r} \qquad \textbf{[22.31]}$$

This result shows that $B$ varies as $1/r$ and hence is nonuniform within the coil. If $r$ is large compared with the cross-sectional radius $a$ of the torus, however, the field is approximately uniform inside the coil. Furthermore, for an ideal

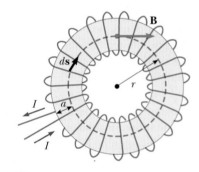

**Figure 22.32**

(Example 22.8) A toroid consisting of many turns of wire wrapped around a doughnut-shaped structure (called a torus). If the coils are closely spaced, the field in the interior of the toroid is tangent to the dashed circle and varies as $1/r$. The magnetic field outside the toroid is zero. The dimension $a$ is the cross-sectional radius of the torus.

toroid, in which the turns are closely spaced, the magnitude of the field external to the toroid is zero. This can be seen by noting that the net current passing through the surface bounded by any circular path lying outside the toroid is zero (including the region of the "hole in the doughnut"). From Ampère's law we therefore find that $B = 0$ in the regions exterior to the coil. In reality, the turns of a toroid form a helix rather than circular loops. As a result, a small field always exists external to the coil.

## 22.10 • THE MAGNETIC FIELD OF A SOLENOID

A solenoid is a long wire wound in the form of a helix. If the turns are closely spaced, this configuration can produce a reasonably uniform magnetic field throughout the volume enclosed by the solenoid, except close to its ends. Each of the turns can be modeled as a circular loop, and the net magnetic field is the vector sum of the fields due to all the turns.

If the turns are closely spaced and the solenoid is of finite length, the field lines are as shown in Figure 22.33a. In this case, the field lines diverge from one end and converge at the opposite end. An inspection of this field distribution exterior to the solenoid shows a similarity to the field of a bar magnet (Fig. 22.33b). Hence, one end of the solenoid behaves like the north pole of a magnet while the opposite end behaves like the south pole. As the length of the solenoid increases, the field within it becomes more and more uniform. When the solenoid's turns are closely spaced and its length is large compared with its radius, it approaches the case of an *ideal solenoid*. For an ideal solenoid, the field outside the solenoid is zero, and the field inside is uniform. We will use the ideal solenoid as a simplification model for a real solenoid.

We can use Ampère's law to obtain an expression for the magnetic field inside an ideal solenoid. A longitudinal cross-section of part of our ideal solenoid (Fig. 22.34) carries current *I*. **B** inside the ideal solenoid is uniform and parallel to the axis, and **B** outside is zero. Consider a rectangular path of length $\ell$ and width *w*, as shown in Figure 22.34. We can apply Ampère's law to this path by evaluating the integral of $\mathbf{B} \cdot d\mathbf{s}$ over each of the four sides of the rectangle. The contribution along side 3 is clearly zero, because **B** = 0 in this region—condition (4) in Section 22.9. The contributions from sides 2 and 4 are both zero because **B** is

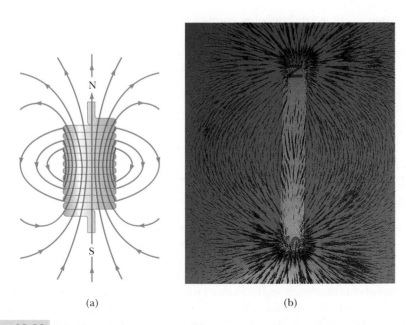

(a)                                    (b)

**Figure 22.33**

(a) Magnetic field lines for a tightly wound solenoid of finite length carrying a steady current. The field in the space enclosed by the solenoid is nearly uniform and strong. Note that the field lines resemble those of a bar magnet, so that the solenoid effectively has north and south poles. (b) The magnetic field pattern of a bar magnet, displayed with iron filings. *(Henry Leap and Jim Lehman)*

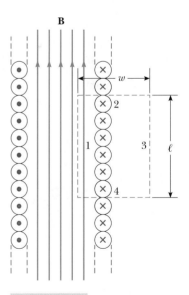

**Figure 22.34**

Cross-sectional view of an ideal solenoid. The magnetic field inside the solenoid is uniform, and the field outside is zero. Ampère's law applied to the red dashed path can be used to calculate the magnitude of the magnetic field inside the solenoid.

perpendicular to $d\mathbf{s}$ along these paths—condition (3). Side 1, whose length is $\ell$, gives a contribution of $B\ell$ to the integral because $\mathbf{B}$ along this portion of the path is constant in magnitude and parallel to $d\mathbf{s}$—conditions (1) and (2). The integral over the closed rectangular path therefore has the value

$$\oint \mathbf{B} \cdot d\mathbf{s} = \int_{\text{side 1}} \mathbf{B} \cdot d\mathbf{s} = B \int_{\text{side 1}} ds = B\ell$$

The right side of Ampère's law involves the *total* current that passes through the surface bound by the path of integration. In our case, the total current through the rectangular path equals the current through each turn of the solenoid multiplied by the number of turns enclosed by the path of integration. If $N$ is the number of turns in the length $\ell$, then the total current through the rectangle equals $NI$. Ampère's law applied to this path therefore gives

$$\oint \mathbf{B} \cdot d\mathbf{s} = B\ell = \mu_0 NI$$

$$B = \mu_0 \frac{N}{\ell} I = \mu_0 nI \qquad \text{[22.32]}$$

- *Magnetic field inside a long solenoid*

where $n = N/\ell$ is the number of turns *per unit length* (not to be confused with $N$, the number of turns).

We also could obtain this result in a simpler manner by reconsidering the magnetic field of a toroidal coil (Example 22.8). If the radius $r$ of the toroidal coil containing $N$ turns is large compared with its cross-sectional radius $a$, then a short section of the toroidal coil approximates a short section of a solenoid, with $n = N/2\pi r$. In this limit, we see that Equation 22.31 derived for the toroidal coil agrees with Equation 22.32.

Equation 22.32 is valid only for points near the center of a very long solenoid. As you might expect, the field near each end is smaller than the value given by Equation 22.32. At the very end of a long solenoid, the magnitude of the field is about one-half that of the field at the center (see Problem 42).

**EXERCISE**   A closely wound, long solenoid of overall length 30.0 cm has a magnetic field of magnitude $5.00 \times 10^{-4}$ T at its center produced by a current of 1.00 A through its windings. How many turns of wire are on the solenoid?

*Answer*   120 turns

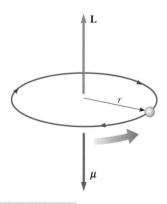

**Figure 22.35**

An electron moving in a circular orbit of radius $r$ has an angular momentum $\mathbf{L}$ in one direction and a magnetic moment $\boldsymbol{\mu}$ in the opposite direction. The motion of the electron in the direction of the gray arrow results in a current in the direction shown by the arrowheads on the circular path.

## 22.11 • MAGNETISM IN MATTER

The magnetic field produced by a current in a coil of wire gives us a hint about what causes certain materials to exhibit strong magnetic properties. To understand why some materials are magnetic, it is instructive to begin our discussion with the Bohr structural model of the atom, in which electrons are assumed to move in circular orbits about the much more massive nucleus. Figure 22.35 shows the angular momentum associated with the electron. In the Bohr model, each electron, with its charge of magnitude $1.6 \times 10^{-19}$ C, circles the atom once in about $10^{-16}$ s. If we divide the electronic charge by this time interval, we find that the orbiting electron is equivalent to a current of $1.6 \times 10^{-3}$ A. Each orbiting electron is therefore viewed as a tiny current loop with a corresponding magnetic moment. Because the charge of the electron is negative, the magnetic moment is directed opposite to the angular momentum, as shown in Figure 22.35.

In most substances, the magnetic moment of one electron in an atom is canceled by that of another electron in the atom, orbiting in the opposite direction. The net result is that **the magnetic effect produced by the orbital motion of the electrons is either zero or very small for most materials.**

In addition to its orbital angular momentum, an electron has an **intrinsic angular momentum,** called **spin,** which also contributes to its magnetic moment. The spin of an electron is an angular momentum separate from its orbital angular momentum, just as the spin of the Earth is separate from its orbital motion about the Sun. Even if the electron is at rest, it still has an angular momentum associated with spin. We shall investigate spin more deeply in Chapter 29.

In atoms or ions containing multiple electrons, many electrons are paired up with their spins in opposite directions—an arrangement that results in a cancellation of the spin magnetic moments. An atom with an odd number of electrons, however, must have at least one "unpaired" electron and a corresponding spin magnetic moment. The net magnetic moment of the atom leads to various types of magnetic behavior. The magnetic moments of several atoms and ions are listed in Table 22.1.

## Ferromagnetic Materials

Iron, cobalt, nickel, gadolinium, and dysprosium are strongly magnetic materials and are said to be **ferromagnetic.** Ferromagnetic substances, used to fabricate permanent magnets, contain atoms with spin magnetic moments that tend to align parallel to each other even in a weak external magnetic field. Once the moments are aligned, the substance remains magnetized after the external field is removed. This permanent alignment is due to strong coupling between neighboring atoms, which can only be understood using quantum physics.

All ferromagnetic materials contain microscopic regions called **domains,** within which all magnetic moments are aligned. The domains range from about $10^{-12}$ to $10^{-8}$ m$^3$ in volume and contain $10^{17}$ to $10^{21}$ atoms. The boundaries between domains having different orientations are called **domain walls.** In an unmagnetized sample, the domains are randomly oriented so that the net magnetic moment is zero, as in Figure 22.36a. When the sample is placed in an external magnetic field, domains with magnetic moment vectors initially oriented along the

**PITFALL PREVENTION 22.3**
**The electron does not spin**

Do not be misled by the word *spin* into believing that the electron is physically spinning. The electron has an intrinsic angular momentum *as if it were spinning,* but the notion of rotation for a point particle is meaningless—remember that we described rotation of a *rigid body,* with an extent in space, in Chapter 10. Spin angular momentum is actually a relativistic effect.

| TABLE 22.1 | Magnetic Moments of Some Atoms and Ions ($\times\ 10^{-24}$ J/T) |
|---|---|
| **Atom (or ion)** | **Magnet Moment per Atom or Ion** |
| H | 9.27 |
| He | 0 |
| Ne | 0 |
| Fe | 2.06 |
| Co | 16.0 |
| Ni | 5.62 |
| Gd | 65.8 |
| Dy | 92.7 |
| Co$^{2+}$ | 44.5 |
| Ni$^{2+}$ | 29.7 |
| Fe$^{2+}$ | 50.1 |
| Ce$^{3+}$ | 19.8 |
| Yb$^{3+}$ | 37.1 |

(a)

(b)

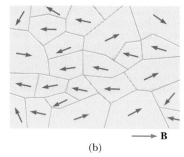

(c)

**Figure 22.36**

(a) Random orientation of atomic magnetic dipoles in the domains of an unmagnetized substance. (b) When an external field **B** is applied, the domains with components of magnetic moment in the same direction as **B** grow larger. (c) As the field is made even stronger, the domains with magnetic moment vectors not aligned with the external field become very small.

external field grow in size at the expense of other domains, which results in a magnetized sample, as in Figures 22.36b and 22.36c. When the external field is removed, the sample may retain most of its magnetism.

The extent to which a ferromagnetic substance retains its magnetism is described by its classification as being magnetically **hard** or **soft.** Soft magnetic materials, such as iron, are easily magnetized but also tend to lose their magnetism easily. When a soft magnetic material is magnetized and the external magnetic field is removed, thermal agitation produces domain motion and the material quickly returns to an unmagnetized state. In contrast, hard magnetic materials, such as cobalt and nickel, are difficult to magnetize but tend to retain their magnetism, and domain alignment persists in them after the external magnetic field is removed. Such hard magnetic materials are referred to as **permanent magnets.** Rare-earth permanent magnets, such as samarium–cobalt, are now regularly used in industry.

### context connection
## 22.12 • THE ATTRACTIVE MODEL FOR MAGNETIC LEVITATION

A number of designs have been developed for magnetic levitation. In this section we shall describe one design model called the *electromagnetic system (EMS)*. This model is conceptually simple because it depends only on the attractive force between magnets and ferromagnetic materials. It has some technological complications, however. The EMS system is used in the German Transrapid design.

In an EMS system, the magnets supporting the vehicle are located below the track, because the attractive force between these magnets and those in the track must lift the vehicle upward. A diagram of the German Transrapid system is shown in Figure 22.37.

Guidance magnet

Steel rail

Levitation magnet

**Figure 22.37**

Schematic diagram of the levitation system for the German Transrapid. The levitation magnets are attached to the vehicle and are located below the rails, so that the attractive force lifts the vehicle upward. The guidance magnets keep the vehicle centered on the track.

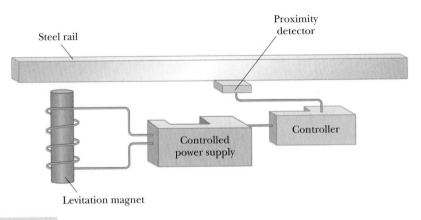

**Figure 22.38**

Control system for maintaining a fixed separation distance between the magnets and the track.

The electromagnets attached to the vehicle are attracted to the steel rail, lifting the car. One disadvantage of this system is the *instability* of the vehicle caused by the variation of the magnetic force with distance. If the vehicle rises slightly, the magnet moves closer to the rail and the strength of the attractive force increases. As a result, the vehicle continues to move upward until the magnet makes contact with the rail. Conversely, if the vehicle drops slightly, the force decreases and the vehicle continues to drop. For these reasons, this system requires a proximity detector and electronic controls that adjust the magnetizing current to keep the vehicle at a constant position relative to the rail.

Figure 22.38 shows a typical method for controlling the separation between the magnets and the rails. The proximity detector is a device that uses magnetic induction (which we shall study in the next chapter) to measure the magnet–rail separation. If the vehicle drops so that the levitation magnet moves farther from the rail, the detector causes the power supply to send more current to the magnet, pulling the vehicle back up. If the magnet rises, the decreased separation distance is detected and the power supply sends less current to the magnet so that the vehicle drops downward.

Another disadvantage of the EMS system is the relatively small separation between the levitating magnets and the track, about 10 mm. This small separation requires careful tolerance in track layout and curvature and steadfast maintenance of the track against problems with snow, ice, and temperature changes.

A major advantage of the EMS system is that the levitation is independent of speed, so that wheels are not required—the vehicle is levitated even when stopped at a station. Wheels are still required, however, for an emergency "landing" system if a loss of power occurs.

The Transrapid system has undergone extensive testing in Germany and has achieved speeds of over 450 km/h. Government approval has been received to begin the construction of a maglev vehicle line between Berlin and Hamburg. The distance between these cities of 292 km will be covered in a maximum time of 1 h. Construction plans estimate that commercial operation of this high-speed line could begin in 2005.

## SUMMARY

The **magnetic force** that acts on a charge $q$ moving with velocity $\mathbf{v}$ in an external **magnetic field B** is

$$\mathbf{F}_B = q\mathbf{v} \times \mathbf{B} \qquad [22.1]$$

This force is in a direction perpendicular both to the velocity of the particle and to the field. The magnitude of the magnetic force is

$$F_B = |q|vB \sin\theta \qquad [22.2]$$

where $\theta$ is the angle between $\mathbf{v}$ and $\mathbf{B}$.

A particle with mass $m$ and charge $q$ moving with velocity $\mathbf{v}$ perpendicular to a uniform magnetic field $\mathbf{B}$ follows a circular path of radius

$$r = \frac{mv}{qB} \qquad [22.3]$$

If a straight conductor of length $\ell$ carries current $I$, the magnetic force on that conductor when placed in a uniform external magnetic field $\mathbf{B}$ is

$$\mathbf{F}_B = I\boldsymbol{\ell} \times \mathbf{B} \qquad [22.10]$$

where $\boldsymbol{\ell}$ is in the direction of the current and $|\boldsymbol{\ell}| = \ell$, the length of the conductor.

If an arbitrarily shaped wire carrying current $I$ is placed in an external magnetic field, the magnetic force on a very small length element $d\mathbf{s}$ is

$$d\mathbf{F}_B = I\,d\mathbf{s} \times \mathbf{B} \qquad [22.11]$$

To determine the total magnetic force on the wire, one must integrate Equation 22.11 over the wire.

The **magnetic moment** $\boldsymbol{\mu}$ of a loop carrying current $I$ is

$$\boldsymbol{\mu} = I\mathbf{A} \qquad [22.15]$$

where $\mathbf{A}$ is perpendicular to the plane of the loop and $|\mathbf{A}|$ is equal to the area of the loop. The SI unit of $\boldsymbol{\mu}$ is the $\text{A}\cdot\text{m}^2$.

The torque $\boldsymbol{\tau}$ on a current loop when the loop is placed in a uniform external magnetic field $\mathbf{B}$ is

$$\boldsymbol{\tau} = \boldsymbol{\mu} \times \mathbf{B} \qquad [22.16]$$

The **Biot–Savart law** says that the magnetic field $d\mathbf{B}$ at the point $P$, due to the wire element $d\mathbf{s}$ carrying a steady current $I$ is

$$d\mathbf{B} = k_m \frac{I\,d\mathbf{s} \times \hat{\mathbf{r}}}{r^2} \qquad [22.17]$$

where $k_m = 10^{-7}\,\text{T}\cdot\text{m/A}$ and $r$ is the distance from the element to the point $P$. To find the total field at $P$ due to a current distribution, one must integrate this vector expression over the entire distribution.

The magnitude of the magnetic field at a distance of $r$ from a long, straight wire carrying current $I$ is

$$B = \frac{\mu_0 I}{2\pi r} \qquad [22.21]$$

where $\mu_0 = 4\pi \times 10^{-7}\,\text{T}\cdot\text{m/A}$ is the **permeability of free space.** The field lines are circles concentric with the wire.

The magnetic force per unit length between two parallel wires (at least one of which is long) separated by a distance $a$ and carrying currents $I_1$ and $I_2$ has the magnitude

$$\frac{F}{\ell} = \frac{\mu_0 I_1 I_2}{2\pi a} \qquad [22.27]$$

The force is attractive if the currents are in the same direction and repulsive if they are in opposite directions.

**Ampère's law** says that the line integral of $\mathbf{B} \cdot d\mathbf{s}$ around any closed path equals $\mu_0 I$, where $I$ is the total steady current passing through any surface bounded by the closed path:

$$\oint \mathbf{B} \cdot d\mathbf{s} = \mu_0 I \qquad [22.29]$$

Using Ampère's law, one finds that the fields inside a toroidal coil and solenoid are

$$B = \frac{\mu_0 NI}{2\pi r} \quad \text{(toroid)} \qquad [22.31]$$

$$B = \mu_0 \frac{N}{\ell} I = \mu_0 nI \quad \text{(solenoid)} \qquad [22.32]$$

where $N$ is the total number of turns and $n$ is the number of turns per unit length.

## QUESTIONS

1. Two charged particles are projected into a region where a magnetic field is perpendicular to their velocities. If the charges are deflected in opposite directions, what can you say about them?

2. A current-carrying conductor experiences no magnetic force when placed in a certain manner in a uniform magnetic field. Explain.

3. Is it possible to orient a current loop in a uniform magnetic field so that the loop does not tend to rotate? Explain.

4. The north-seeking pole of a magnet is attracted toward the geographic North Pole of the Earth. Yet, like poles repel. What is the way out of this dilemma?

5. Which way would a compass point if you were at the north magnetic pole of the Earth?

6. A magnet attracts a piece of iron. The iron can then attract another piece of iron. On the basis of domain alignment, explain what happens in each piece of iron.

7. Will a nail be attracted to either pole of a magnet? Explain what is happening inside the nail when placed near the magnet.

8. Why does hitting a magnet with a hammer cause the magnetism to be reduced?

9. Parallel current-carrying wires exert magnetic forces on each other. What about perpendicular wires? Imagine two such wires oriented perpendicular to each other, and almost touching. Does a magnetic force exist between the wires?

10. Explain why it is not possible to determine the charge and mass of a charged particle separately via electric and magnetic forces.

11. How can a current loop be used to determine the presence of a magnetic field in a given region of space?

12. It is found that charged particles from outer space, called cosmic rays, strike the Earth more frequently at the poles than at the equator. Why?

13. Explain why two parallel wires carrying currents in opposite directions repel each other.

14. A hollow copper tube carries a current along its length. Why is **B** = 0 inside the tube? Is **B** nonzero outside the tube?

15. Describe the change in the magnetic field in the space enclosed by a solenoid carrying a steady current $I$ if (a) the length of the solenoid is doubled but the number of turns remains the same and (b) the number of turns is doubled but the length remains the same.

16. Consider an electron near the magnetic equator of the Earth. In which direction will it tend to be deflected if its velocity is directed (a) downward? (b) northward? (c) westward? (d) southeastward?

17. You are an astronaut stranded on a planet with no test equipment for minerals available. The planet does not even have a magnetic field. You have two bars of iron in your possession: one is magnetized, one is not. How could you determine which is which?

18. A Hindu ruler once suggested that he be entombed in a magnetic coffin with the polarity arranged so that he would be forever suspended between heaven and Earth. Is such magnetic levitation possible? Discuss.

19. Should the surface of a computer disk be made from a hard or a soft ferromagnetic substance?

20. Figure Q22.20 shows two permanent magnets, each having a hole through its center. Note that the upper magnet is levitated above the lower one. (a) How does this occur? (b) What purpose does the pencil serve? (c) What can you say about the poles of the magnets from this observation?

(d) If the upper magnet were inverted, what do you suppose would happen?

**Figure Q22.20** Magnetic levitation using two ceramic magnets. *(Courtesy of Central Scientific Company)*

21. The *bubble chamber* is a device used for observing tracks of particles that pass through the chamber, which is immersed in a magnetic field. If some of the tracks are spirals and others are straight lines, what can you say about the particles?

22. The electron beam in Figure Q22.22 is projected to the right. The beam deflects downward in the presence of a magnetic field produced by a pair of current-carrying coils. (a) What is the direction of the magnetic field? (b) What would happen to the beam if the current in the coils were reversed?

**Figure Q22.22** Bending of a beam of electrons in a magnetic field. *(Courtesy of Central Scientific Company)*

# PROBLEMS

1, 2, 3 = straightforward, intermediate, challenging □ = full solution available in the *Student Solutions Manual and Study Guide*

**web** = solution posted at **http://www.harcourtcollege.com/physics/** 🖥 = computer useful in solving problem

🖱 = Interactive Physics ▨ = paired numerical/symbolic problems ☊ = life science application

## Section 22.2 The Magnetic Field

1. Determine the initial direction of the deflection of **web** charged particles as they enter the magnetic fields as shown in Figure P22.1.

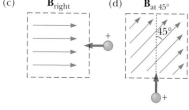

**Figure P22.1**

2. A proton travels with a velocity having a magnitude of $3.00 \times 10^6$ m/s at an angle of $37.0°$ with a magnetic field of magnitude 0.300 T in the $+y$ direction. What are (a) the magnitude of the magnetic force on the proton and (b) its acceleration?

3. An electron is accelerated through 2 400 V from rest and then enters a region with a uniform 1.70-T magnetic field. What are (a) the maximum and (b) the minimum values of the magnetic force this charge can experience?

4. At the equator, near the surface of the Earth, the magnetic field is approximately 50.0 $\mu$T northward, and the electric field is about 100 N/C downward in fair weather. Find the gravitational, electric, and magnetic forces on an electron in this environment if it is moving with an instantaneous velocity of $6.00 \times 10^6$ m/s directed to the east.

5. A proton moves with a velocity of $\mathbf{v} = (2\mathbf{i} - 4\mathbf{j} + \mathbf{k})$ m/s in a region in which the magnetic field is $\mathbf{B} = (\mathbf{i} + 2\mathbf{j} - 3\mathbf{k})$ T. What is the magnitude of the magnetic force this charge experiences?

## Section 22.3 Motion of a Charged Particle in a Magnetic Field

6. **Review Problem.** An electron moves in a circular path perpendicular to a constant magnetic field with a magnitude of 1.00 mT. If the angular momentum of the electron about the center of the circle is $4.00 \times 10^{-25}$ J·s, determine (a) the radius of the circular path and (b) the speed of the electron.

7. A cosmic-ray proton in interstellar space has an energy of 10.0 MeV and executes a circular orbit having a radius equal to that of Mercury's orbit around the Sun ($5.80 \times 10^{10}$ m). What is the magnetic field in that region of space?

## Section 22.4 Applications of the Motion of Charged Particles in a Magnetic Field

8. A velocity selector consists of electric and magnetic fields described by the expressions $\mathbf{E} = E\mathbf{k}$ and $\mathbf{B} = B\mathbf{j}$, with $B = 15.0$ mT. Find the value of $E$ such that a 750-eV electron moving along the positive $x$ axis is undeflected.

9. Consider the mass spectrometer shown schematically in Figure 22.12. The electric field between the plates of the velocity selector is 2 500 V/m, and the magnetic field in both the velocity selector and the deflection chamber has a magnitude of 0.035 0 T. Calculate the radius of the path for a singly charged ion of mass $m = 2.18 \times 10^{-26}$ kg.

10. A cyclotron designed to accelerate protons has an outer radius of 0.350 m. The protons are emitted nearly at rest from a source at the center and are accelerated through 600 V each time they cross the gap between the dees. The dees are between the poles of an electromagnet where the field has a magnitude of 0.800 T. (a) Find the cyclotron frequency. (b) Find the speed at which protons exit the cyclotron and (c) their maximum kinetic energy. (d) How many revolutions does a proton make in the cyclotron? (e) For what time interval is one proton in the cyclotron?

11. The picture tube in a television uses magnetic deflection coils rather than electric deflection plates. Suppose an electron beam is accelerated through a 50.0-kV potential difference and then through a region of uniform magnetic field 1.00 cm wide. The screen is located 10.0 cm from the center of the coils and is 50.0 cm wide. When the field is turned off, the electron beam hits the center of the screen. What field magnitude is necessary to deflect the beam to the side of the screen? Ignore relativistic corrections.

12. The *Hall effect* finds important applications in the electronics industry. It is used to determine the sign and density of

charge carriers in semiconductor chips. The arrangement is shown in Figure P22.12. A semiconductor of thickness $t$ and width $d$ carries a current $I$ in the $x$ direction. A uniform magnetic field $B$ is applied in the $y$ direction. If the charge carriers are positive, the magnetic force deflects them in the $z$ direction. Positive charge accumulates on the top surface of the sample and negative charge on the bottom surface, creating a downward electric field. In equilibrium the downward electric force on the charge carriers balances the upward magnetic force, and the carriers move through the sample without deflection. The *Hall voltage* $\Delta V_H = V_c - V_a$ between the top and bottom surfaces is measured, and the density of the charge carriers can be calculated from it. (a) Show that if the charge carriers are negative, the Hall voltage will be negative. Hence the Hall effect reveals the sign of the charge carriers, so that the sample can be classified as *p*-type (with positive majority charge carriers) or *n*-type (with negative). (b) Determine the number of charge carriers per unit volume $n$ in terms of $I$, $t$, $B$, $\Delta V_H$, and the magnitude $q$ of the charge carrier.

**Figure P22.14**

**15.** *A nonuniform magnetic field exerts a net force on a magnetic dipole.* A strong magnet is placed under a horizontal conducting ring of radius $r$ that carries current $I$, as shown in Figure P22.15. If the magnetic field **B** makes an angle $\theta$ with the vertical at the ring's location, what are the magnitude and direction of the resultant force on the ring?

**Figure P22.15**

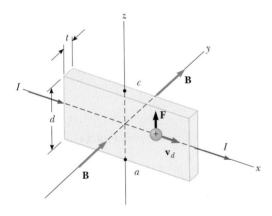

**Figure P22.12**

## Section 22.5 Magnetic Force on a Current-Carrying Conductor

**13.** A wire carries a steady current of 2.40 A. A straight section of the wire is 0.750 m long and lies along the $x$ axis within a uniform magnetic field, **B** $= 1.60\mathbf{k}$ T. If the current is in the $+x$ direction, what is the magnetic force on the section of wire?

**14.** A conductor suspended by two flexible wires as shown in Figure P22.14 has a mass per unit length of 0.040 0 kg/m. What current must exist in the conductor for the tension in the supporting wires to be zero when the magnetic field is 3.60 T into the page? What is the required direction for the current?

**16.** In Figure P22.16, the cube is 40.0 cm on each edge. Four straight segments of wire—*ab*, *bc*, *cd*, and *da*—form a closed loop that carries a current $I = 5.00$ A, in the direction shown. The loop is placed in a uniform magnetic field of magnitude $B = 0.020\ 0$ T in the positive $y$ direction. Determine the magnitude and direction of the magnetic force on each segment.

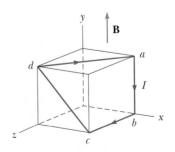

**Figure P22.16**

$$f_B = \frac{I\ell \times B}{\ell} = \frac{mg}{\ell}$$

$$I \times B = mg$$

## Section 22.6   Torque on a Current Loop in a Uniform Magnetic Field

**17.** A rectangular loop consists of $N = 100$ closely wrapped turns and has dimensions $a = 0.400$ m and $b = 0.300$ m. The loop is hinged along the $y$ axis, and its plane makes an angle $\theta = 30.0°$ with the $x$ axis (Fig. P22.17). What is the magnitude of the torque exerted on the loop by a uniform magnetic field $B = 0.800$ T directed along the $x$ axis when the current is $I = 1.20$ A in the direction shown? What is the expected direction of rotation of the loop?

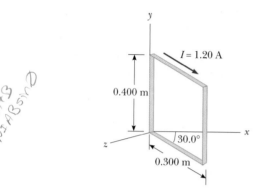

**Figure P22.17**

**18.** A current of 17.0 mA is maintained in a single circular loop with a circumference of 2.00 m. A magnetic field of 0.800 T is directed parallel to the plane of the loop. (a) Calculate the magnetic moment of the loop. (b) What is the magnitude of the torque exerted on the loop by the magnetic field?

**19.** A long piece of wire with a mass of 0.100 kg and a total length of 4.00 m is used to make a square coil with a side of 0.100 m. The coil is hinged along a horizontal side, carries a 3.40-A current, and is placed in a vertical magnetic field with a magnitude of 0.010 0 T. (a) Determine the angle that the plane of the coil makes with the vertical when the coil is in equilibrium. (b) Find the torque acting on the coil due to the magnetic force at equilibrium.

**20.** A current loop with magnetic dipole moment $\boldsymbol{\mu}$ is placed in a uniform magnetic field $\mathbf{B}$, with its moment making angle $\theta$ with the field. With the arbitrary choice of $U = 0$ for $\theta = 90°$, prove that the potential energy of the dipole–field system is $U = -\boldsymbol{\mu} \cdot \mathbf{B}$.

## Section 22.7   The Biot–Savart Law

**21.** In Niels Bohr's 1913 model of the hydrogen atom, an electron circles the proton at a distance of $5.29 \times 10^{-11}$ m with a speed of $2.19 \times 10^6$ m/s. Compute the magnitude of the magnetic field that this motion produces at the location of the proton.

**22.** A current path shaped as shown in Figure P22.22 produces a magnetic field at $P$, the center of the arc. If the arc subtends an angle of $30.0°$ and the radius of the arc is 0.600 m, what are the magnitude and direction of the field produced at $P$ if the current is 3.00 A?

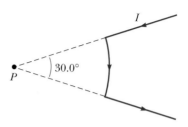

**Figure P22.22**

**23.** Calculate the magnitude of the magnetic field at a point 100 cm from a long, thin conductor carrying a current of 1.00 A.

**24.** A conductor consists of a circular loop of radius $R$ and two straight, long sections, as shown in Figure P22.24. The wire lies in the plane of the paper and carries a current $I$. Find an expression for the magnetic field at the center of the loop.

**Figure P22.24**

**25.** Determine the magnetic field at a point $P$ located a distance $x$ from the corner of an infinitely long wire bent at a right angle, as shown in Figure P22.25. The wire carries a steady current $I$.

**Figure P22.25**

**26.** Consider a flat circular current loop of radius $R$ carrying current $I$. Choose the $x$ axis to be along the axis of the loop, with the origin at the center of the loop. Graph the ratio of the magnitude of the magnetic field at coordinate

$x$ to that at the origin, for $x = 0$ to $x = 5R$. It may be helpful to use a programmable calculator or a computer to solve this problem.

27. Three long, parallel conductors carry currents of $I = 2.00$ A. Figure P22.27 is an end view of the conductors, with each current coming out of the page. If $a = 1.00$ cm, determine the magnitude and direction of the magnetic field at points $A$, $B$, and $C$.

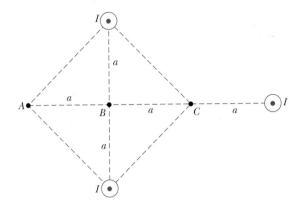

**Figure P22.27**

28. A long, straight wire carries current $I$. A right-angle bend formed in the middle of the wire is in the shape of a circular arc of radius $r$, as in Figure P22.28. Determine the magnetic field at the center of the arc.

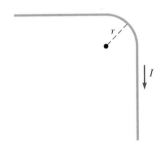

**Figure P22.28**

29. One long wire carries current 30.0 A to the left along the $x$ axis. A second long wire carries current 50.0 A to the right along the line ($y = 0.280$ m, $z = 0$). (a) Where in the plane of the two wires is the total magnetic field equal to zero? (b) A particle having a charge of $-2.00$ $\mu$C moves with a velocity of $150\mathbf{i}$ Mm/s along the line ($y = 0.100$ m, $z = 0$). Calculate the magnetic force acting on the particle. (c) A uniform electric field is applied to allow this particle to pass through this region undeflected. Calculate the required electric field.

## Section 22.8   The Magnetic Force Between Two Parallel Conductors

30. Two long, parallel conductors separated by 10.0 cm carry currents in the same direction. The first wire carries a current $I_1 = 5.00$ A, and the second carries a current $I_2 = 8.00$ A. (a) What is the magnitude of the magnetic field created by $I_1$ and acting on $I_2$? (b) What is the magnetic force per unit length exerted on $I_2$ by $I_1$? (c) What is the magnitude of the magnetic field created by $I_2$ at the location of $I_1$? (d) What is the magnetic force per length exerted by $I_2$ on $I_1$?

31. In Figure P22.31, the current in the long, straight wire is $I_1 = 5.00$ A and the wire lies in the plane of the rectangular loop, which carries 10.0 A. The dimensions are $c = 0.100$ m, $a = 0.150$ m, and $\ell = 0.450$ m. Find the magnitude and direction of the net force exerted on the loop by the magnetic field created by the wire.

**Figure P22.31**

32. Two long, parallel wires are attracted to each other by a magnetic force per unit length of 320 $\mu$N/m when they are separated by 0.500 m. The current in the upper wire is 20.0 A to the right. Determine the line in the plane of the two wires along which the total magnetic field is zero.

## Section 22.9   Ampère's Law

33. Four long, parallel conductors carry equal currents of $I = 5.00$ A. Figure P22.33 is an end view of the conductors. The current direction is into the page at points $A$ and $B$ (indicated by the crosses) and out of the page at $C$ and $D$ (indicated by the dots). Calculate the magnitude and direction of the magnetic field at point $P$, located at the center of the square with an edge length of 0.200 m.

**Figure P22.33**

34. A long, straight wire lies on a horizontal table and carries a current of 1.20 $\mu$A. In a vacuum, a proton moves parallel to the wire (opposite the current) with a constant velocity of 2.30 $\times$ 10$^4$ m/s at a distance $d$ above the wire. Determine the value of $d$. You may ignore the magnetic field due to the Earth.

35. Figure P22.35 is a cross-sectional view of a coaxial cable. The center conductor is surrounded by a rubber layer, which is surrounded by an outer conductor, which is surrounded by another rubber layer. In a particular application, the current in the inner conductor is 1.00 A out of the page and the current in the outer conductor is 3.00 A into the page. Determine the magnitude and direction of the magnetic field at points $a$ and $b$.

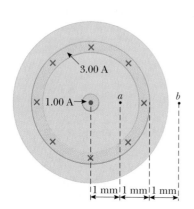

**Figure P22.35**

36. A tokamak fusion reactor has magnetic coils in the shape of a toroid with an inner radius of 0.700 m and an outer radius of 1.30 m. If the toroid has 900 turns of large-diameter wire, each of which carries a current of 14.0 kA, find the magnitude of the magnetic field inside the toroid along (a) the inner radius and (b) the outer radius.

37. A packed bundle of 100 long, straight, insulated wires
**web** forms a cylinder of radius $R$ = 0.500 cm. (a) If each wire carries 2.00 A, what are the magnitude and direction of the magnetic force per unit length acting on a wire lo-

cated 0.200 cm from the center of the bundle? (b) Would a wire on the outer edge of the bundle experience a force greater or less than the value calculated in part (a)?

38. Niobium metal becomes a superconductor when cooled below 9 K. If its superconductivity is destroyed when the surface magnetic field exceeds 0.100 T, determine the maximum current a 2.00-mm-diameter niobium wire can carry and remain superconducting in the absence of any external magnetic field.

39. Consider a column of electric current passing through plasma (ionized gas). Filaments of current within the column are magnetically attracted to one another. They can crowd together to yield a high current density and a strong magnetic field in a small region. Sometimes the current can be cut off momentarily by this so-called *pinch effect.* (In a metallic wire, the pinch effect is not important because the current-carrying electrons repel one another with electric forces.) The pinch effect can be demonstrated by passing a current through a long, hollow metallic pipe, oriented vertically. Let $R$ represent the radius of the cylindrical pipe and $I$ the upward current, uniformly distributed over its curved wall. Determine the magnetic field (a) just inside the wall and (b) just outside. (c) Determine the pressure on the wall.

### Section 22.10  The Magnetic Field of a Solenoid

40. **Review Problem.** A solenoid of radius $R$ = 5.00 cm is made of a long piece of wire of radius $r$ = 2.00 mm, length $\ell$ = 10.0 m and resistivity $\rho$ = 1.70 $\times$ 10$^{-8}$ $\Omega \cdot$ m. (Note that $\ell \gg R$.) Find the magnetic field at the center of the solenoid if the wire is connected to a battery having an emf of 20.0 V.

41. What current is required in the windings of a long solenoid that has 1 000 turns uniformly distributed over a length of 0.400 m, to produce at the center of the solenoid a magnetic field of magnitude 1.00 $\times$ 10$^{-4}$ T?

42. Consider a solenoid of length $\ell$ and radius $R$, containing $N$ closely spaced turns and carrying a steady current $I$. (a) In terms of these parameters, find the magnetic field at a point along the axis as a function of distance $a$ from the end of the solenoid. (b) Show that as $\ell$ becomes very long, $B$ approaches $\mu_0 NI/2\ell$ at each end of the solenoid.

### Section 22.11  Magnetism in Matter

43. In Bohr's 1913 model of the hydrogen atom, the electron is in a circular orbit of radius 5.29 $\times$ 10$^{-11}$ m and its speed is 2.19 $\times$ 10$^6$ m/s. (a) What is the magnitude of the magnetic moment due to the electron's motion? (b) If the electron orbits counterclockwise in a horizontal circle, what is the direction of this magnetic moment vector?

**44.** At *saturation,* when nearly all of the atoms have their magnetic moments aligned, the magnetic field in a sample of iron can be 2.00 T. If each electron contributes a magnetic moment of $9.27 \times 10^{-24}$ A·m$^2$(called a Bohr magneton), how many electrons per atom contribute to the saturated field of iron? (*Hint:* Iron contains approximately $8.50 \times 10^{28}$ atoms/m$^3$.)

## Section 22.12 Context Connection — The Attractive Model for Magnetic Levitation

**45.** The following represents a crude model for levitating a commercial transportation vehicle. Suppose the levitation is achieved by mounting small electrically charged spheres below the vehicle. The spheres pass through a magnetic field established by permanent magnets placed along the track. Let us assume that the permanent magnets produce a uniform magnetic field of 0.1 T at the location of the spheres and that an electronic control system maintains a charge of 1 $\mu$C on each sphere. The vehicle has a mass of $5 \times 10^4$ kg and travels at a speed of 400 km/h. How many charged spheres are required to support the weight of the vehicle at this speed? Your answer should suggest that this crude model would not be successful as a means of magnetic levitation.

**46.** Data from the Transrapid maglev system show that the input electric power required to operate the vehicle is on the order of $10^2$ kW. (a) Imagine that the Transrapid vehicle moves at 400 km/h. Approximately how much energy, in joules, is used for each mile of travel for the vehicle? (b) Calculate the energy per mile used by an automobile that achieves 20 mi/gal. The energy available from gasoline is approximately 40 MJ/kg, a typical automobile engine efficiency is 20%, and the density of gasoline is 754 kg/m$^3$. (c) Considering 1 passenger in the automobile and 100 on the Transrapid vehicle, the energy per mile necessary for each passenger in the Transrapid is what fraction of that for an automobile?

### *Additional Problems*

**47.** Consider a thin, straight wire segment carrying a constant current $I$ and placed along the $x$ axis, as shown in Figure P22.47. (a) Use the Biot–Savart law to show that the total magnetic field at the point $P$, located a distance $a$ from the wire, is

$$B = \frac{\mu_0 I}{4\pi a} (\cos \theta_1 - \cos \theta_2)$$

(b) If the wire is infinitely long, show that the result in part (a) gives a magnetic field that agrees with that obtained by using Ampère's law in Example 22.7.

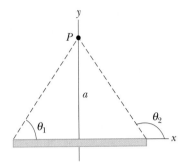

**Figure P22.47**

**48.** Sodium melts at 99°C. Liquid sodium, an excellent thermal conductor, is used in some nuclear reactors to cool the reactor core. The liquid sodium is moved through pipes by pumps that exploit the force on a moving charge in a magnetic field. The principle is as follows: Assume the liquid metal to be in an electrically insulating pipe having a rectangular cross-section of width $w$ and height $h$. A uniform magnetic field perpendicular to the pipe affects a section of length $L$ (Fig. P22.48). An electric current directed perpendicular to the pipe and to the magnetic field produces a current density $J$ in the liquid sodium. (a) Explain why this arrangement produces on the liquid a force that is directed along the length of the pipe. (b) Show that the section of liquid in the magnetic field experiences a pressure increase $JLB$.

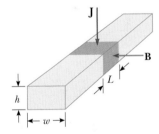

**Figure P22.48**

**49.** An electron enters a region of magnetic field of magnitude 1.00 mT, traveling perpendicular to the linear boundary of the region. The direction of the field is perpendicular to the velocity of the electron. (a) Determine the time it takes for the electron to leave the "field-filled" region, noting that its path is a semicircle. (b) Find the kinetic energy of the electron if the maximum depth of penetration in the field is 2.00 cm.

**50.** Protons having a kinetic energy of 5.00 MeV are moving in the positive $x$ direction and enter a magnetic field **B** = $(0.050\ 0\mathbf{k})$ T directed out of the plane of the page and extending from $x = 0$ to $x = 1.00$ m, as shown in Figure P22.50. (a) Calculate the $y$ component of the protons' momentum as they leave the magnetic field. (b) Find the angle $\alpha$ between the initial velocity vector of the proton beam and the velocity vector after the beam emerges from the field. (*Hint:* Ignore relativistic effects and note that $1 \text{ eV} = 1.60 \times 10^{-19} \text{ J}$.)

**Figure P22.50**

51. A positive charge $q = 3.20 \times 10^{-19}$ C moves with velocity $\mathbf{v} = (2\mathbf{i} + 3\mathbf{j} - \mathbf{k})$ m/s through a region in which both a uniform magnetic field and a uniform electric field exist. (a) What is the total force on the moving charge (in unit-vector notation) if $\mathbf{B} = (2\mathbf{i} + 4\mathbf{j} + \mathbf{k})$ T and $\mathbf{E} = (4\mathbf{i} - \mathbf{j} - 2\mathbf{k})$ V/m? (b) What angle does the force vector make with the positive $x$ axis?

52. A uniform magnetic field of magnitude 0.150 T is directed along the positive $x$ axis. A positron moving at $5.00 \times 10^6$ m/s enters the field along a direction that makes an angle of 85.0° with the $x$ axis (Fig. P22.52). The motion of the particle is expected to be a helix, as described in Section 22.3. Calculate (a) the pitch $p$ and (b) the radius $r$ of the trajectory.

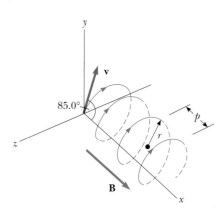

**Figure P22.52**

53. A 0.200-kg metal rod carrying a current of 10.0 A glides on two horizontal rails 0.500 m apart. What vertical magnetic field is required to keep the rod moving at a constant speed if the coefficient of kinetic friction between the rod and rails is 0.100?

54. A heart surgeon monitors the flow rate of blood through an artery using an electromagnetic flowmeter (Fig. P22.54). Electrodes $A$ and $B$ make contact with the outer surface of the blood vessel, which has interior diameter 3.00 mm. (a) For a magnetic field magnitude of 0.040 0 T, an emf of 160 $\mu$V appears between the electrodes. Calculate the speed of the blood. (b) Verify that electrode $A$ is positive, as shown. Does the sign of the emf depend on whether the mobile ions in the blood are predominantly positively or negatively charged? Explain.

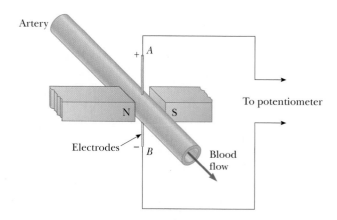

**Figure P22.54**

55. A hand-held electric mixer contains an electric motor. Model the motor as a single flat compact circular coil carrying electric current in a region where a magnetic field is produced by an external permanent magnet. You need consider only one instant in the operation of the motor. (We will consider motors again in Chapter 23.) The coil moves because the magnetic field exerts a torque on the coil, as described in Section 22.6. Make order-of-magnitude estimates of the magnetic field, the torque on the coil, the current in it, its area, and the number of turns in the coil, so that they are related according to Equation 22.16. Note that the input power to the motor is electric, given by $\mathcal{P} = I\,\Delta V$, and the useful output power is mechanical, given by $\mathcal{P} = \tau\omega$.

56. Two circular coils of radius $R$ are each perpendicular to a common axis. The coil centers are a distance $R$ apart and each coil carries a steady current $I$ in the same direction as shown in Figure P22.56. (a) Show that the magnetic field on the axis at a distance $x$ from the center of one coil is

$$B = \frac{\mu_0 I R^2}{2}\left[\frac{1}{(R^2 + x^2)^{3/2}} + \frac{1}{(2R^2 + x^2 - 2Rx)^{3/2}}\right]$$

(b) Show that $dB/dx$ and $d^2B/dx^2$ are both zero at the point midway between the coils. This means the magnetic field in the region midway between the coils is uniform. Coils in this configuration are called *Helmholtz coils*.

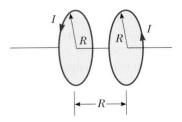

**Figure P22.56**

57. The magnetic moment of the Earth is approximately $8.00 \times 10^{22}$ A·m². (a) If this were caused by the complete magnetization of a huge iron deposit, how many unpaired electrons would this correspond to? (b) At two unpaired electrons per iron atom, how many kilograms of iron would this correspond to? (Iron has a density of 7 900 kg/m³, and approximately $8.50 \times 10^{28}$ iron atoms/m³.)

58. Two circular loops are parallel, coaxial, and almost in contact, 1.00 mm apart (Fig. P22.58). Each loop is 10.0 cm in radius. The top loop carries a clockwise current of 140 A. The bottom loop carries a counterclockwise current of 140 A. (a) Calculate the magnetic force that the bottom loop exerts on the top loop. (b) The upper loop has a mass of 0.021 0 kg. Calculate its acceleration, assuming that the only forces acting on it are the force in part (a) and its weight. (*Hint:* Think about how one loop looks to a bug perched on the other loop.)

**Figure P22.58**

59. A lightning bolt may carry a current of $1.00 \times 10^4$ A for a short time. What is the resulting magnetic field 100 m from the bolt? Assume that the bolt extends far above and below the point of observation.

60. *Rail guns* have been suggested for launching projectiles into space without chemical rockets, and for ground-to-air antimissile weapons of war. A tabletop model rail gun (Fig. P22.60) consists of two long parallel horizontal rails 3.50 cm apart, bridged by a bar *BD* of mass 3.00 g. The bar is originally at rest at the midpoint of the rails and is free to slide without friction. When the switch is closed, electric current is quickly established in the circuit *ABCDEA*. The rails and bar have low electrical resistance, and the current is limited to a constant 24.0 A by the power supply. (a) Find the magnitude of the magnetic field 1.75 cm from a long, straight wire carrying current

24.0 A. (b) Find the magnetic field at point *C* in the diagram, the midpoint of the bar, immediately after the switch is closed. (*Hint:* Consider what conclusions you can draw from the Biot–Savart law.) (c) At other points along the bar *BD*, the field is in the same direction as at point *C*, but greater in magnitude. Assume that the average effective magnetic field along *BD* is five times larger than the field at *C*. With this assumption, find the force on the bar. (d) Find the acceleration with which the bar starts to move. (e) Does the bar move with constant acceleration? (f) Find the speed of the bar after it has traveled 130 cm to the end of the rails.

**Figure P22.60**

61. A nonconducting ring of radius *R* is uniformly charged with a total positive charge *q*. The ring rotates at a constant angular speed $\omega$ about an axis through its center, perpendicular to the plane of the ring. What is the magnitude of the magnetic field on the axis of the ring a distance $R/2$ from its center?

62. An infinite sheet of current lying in the *yz* plane carries a surface current of density $\mathbf{J}_s$. The current is in the *y* direction, and $J_s$ represents the current per unit length measured along the *z* axis. Figure P22.62 is an edge view of the sheet. Find the magnetic field near the sheet. (*Hint:* Use Ampère's law, and evaluate the line integral for a rectangular path around the sheet, represented by the dashed line in Fig. P22.62.)

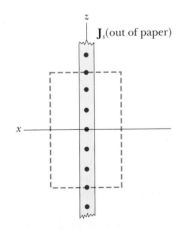

**Figure P22.62**

**63.** The magnitude of the Earth's magnetic field at either pole is approximately $7.00 \times 10^{-5}$ T. Suppose that the field fades away before its next reversal. Scouts, sailors, and politicians around the world join together in a program to replace the field. One plan is to use a current loop around the equator, without relying on magnetization of any materials inside the Earth. Determine the current that would generate such a field if this plan were carried out. (Take the radius of the Earth as $R_E = 6.37 \times 10^6$ m.)

**64.** A long, straight wire—wire 1—oriented along the $y$ axis carries a steady current $I_1$, as shown in Figure P22.64. A rectangular circuit located to the right of the wire carries current $I_2$. Consider wire 2, which is the upper horizontal segment of the circuit that runs from $x = a$ to $x = a + b$. Show that the magnetic force exerted on wire 2 is upward and has a magnitude given by

$$F = \frac{\mu_0 I_1 I_2}{2\pi} \ln\left(1 + \frac{b}{a}\right)$$

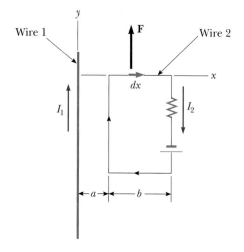

**Figure P22.64**

**65.** A very long, thin strip of metal of width $w$ carries a current $I$ along its length, as shown in Figure P22.65. Find the magnetic field at the point $P$ in the diagram. Point $P$ is in the plane of the strip at a distance $b$ away from it.

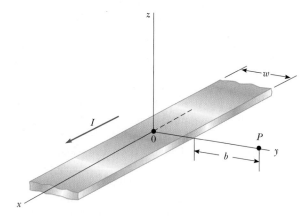

**Figure P22.65**

---

## ANSWERS TO QUICK QUIZZES

**22.1** No. Another field, such as an electric field or gravitational field, may produce another force on the charged particle that has the same magnitude but is opposite the magnetic force, resulting in a net force of zero. Also, a charged particle moving parallel to a magnetic field would experience no magnetic force.

**22.2** No. The magnetic force on a charged particle moving in a magnetic field is perpendicular to the magnetic field lines.

**22.3** No. The magnetic field produces a magnetic force that is directed toward the center of the circular path of the particle, and this causes a change in direction of the particle's velocity vector. Because the magnetic force is always perpendicular to the displacement of the particle, however, no work is done on the particle, and its speed remains constant. Hence, its kinetic energy also remains constant.

**22.4** The magnetic force on a moving charged particle is always perpendicular to the direction of motion. No magnetic force is exerted on the charge when it moves parallel to the direction of the magnetic field. The electric force on a charged particle moving in an electric field is never zero, however, and is always parallel to the direction of the electric field. By projecting the charged particle in different directions, it is therefore possible to determine the nature of the field.

**22.5** Zero. A net torque exists, but the net force is zero.

**22.6** If you are moving along with the electrons, you would measure zero current for the electrons, so the electrons would not produce a magnetic field according to your observations. On the other hand, the positive crystal lattice ions in the metal would be moving backward relative to you and creating a current equivalent to the forward movement of electrons when you were stationary. Thus, you would measure the same magnetic field as when you were stationary, but it would be due to the moving positive crystal lattice ions in your reference frame.

**22.7** (c). The magnetic force each wire exerts on the other depends on the product of the currents. This choice can also be argued by Newton's third law.

**22.8** Each coil of the spring becomes a magnet because a coil acts as a current loop. Because the sense of rotation of the current is the same in all coils, each coil becomes a magnet with the same orientation of poles. Thus, all of the coils attract, and the spring compresses.

**22.9** *b, d, a, c*. Equation 22.29 tells us that the value of the line integral depends only on the net current through each closed path. Path *b* encloses 1 A, path *d* encloses 3 A, path *a* encloses 4 A, and path *c* encloses 6 A.

**22.10** *b*, then *a = c = d*. Paths *a*, *c*, and *d* all have the same nonzero value $\mu_0 I$, because the size and shape of the paths do not matter. Path *b* does not enclose the current, and so its line integral is zero.

An airport metal detector contains a large coil of wire around the frame. This coil of wire has a property called *inductance,* which we study in this chapter. When a passenger carries metal through the detector, the inductance changes, and the change in inductance signals an alarm to sound.

*(Ryan Williams/*
*International Stock Photography)*

 See *Core Concepts in Physics,* Screens 12.6 & 12.7

# Faraday's Law and Inductance

O ur studies in electromagnetism so far have been concerned with the electric fields due to stationary charges and the magnetic fields produced by moving charges. This chapter introduces a new type of electric field—one that is due to a changing magnetic field.

As we learned in Section 19.1, experiments conducted by Michael Faraday in England in the early 1800s and independently by Joseph Henry in the United States showed that an electric current can be induced in a circuit by a changing magnetic field. The results of those experiments led to a very basic and important law of electromagnetism known as *Faraday's law of induction.* Faraday's law explains how a generator works, as well as other practical devices.

## 23.1 • FARADAY'S LAW OF INDUCTION

We begin discussing the concepts in this chapter by considering a simple experiment that builds on material presented in Chapter 22. Imagine that a straight metal wire resides in a uniform magnetic field directed into the page, as in Figure

23.1. Within the wire, there are free electrons. Suppose the wire is now moved with a velocity **v** toward the right. Equation 22.1 tells us that a magnetic force acts on the electrons in the wire. Using the right hand rule, the force on the electrons is downward in Figure 23.1 (remember that the electrons carry a negative charge). Because this direction is along the wire, the electrons move along the wire in response to this force. Thus, a *current* is produced in the wire as it moves through a magnetic field!

Let us consider another simple experiment that demonstrates that an electric current can be produced by a magnetic field. Consider a loop of wire connected to a galvanometer, a device that measures current, as in Figure 23.2. If a magnet is moved toward the loop, the galvanometer needle deflects in one direction as in Figure 23.2a. When the magnet is held stationary as in Figure 23.2b, the needle is not deflected. If the magnet is moved away from the loop as in Figure 23.2c, the galvanometer needle deflects in the opposite direction from that caused by the motion of the magnet toward the galvanometer. Finally, if the magnet is held stationary and the coil is moved either toward or away from it, the needle deflects. From these observations comes the conclusion that **an electric current is set up**

**Figure 23.1**

A straight electric conductor of length $\ell$ moving with a velocity **v** through a uniform magnetic field **B** directed perpendicular to **v**. A current is induced in the conductor due to the magnetic force on charged particles in the conductor.

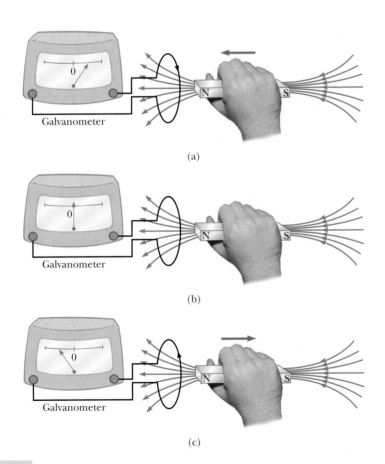

(a)

(b)

(c)

**Figure 23.2**

(a) When a magnet is moved toward a loop of wire connected to a galvanometer, the galvanometer deflects as shown, indicating that a current is induced in the loop. (b) When the magnet is held stationary, no current is induced in the loop, even when the magnet is inside the loop. (c) When the magnet is moved away from the loop, the galvanometer deflects in the opposite direction, indicating that the induced current is opposite that shown in part (a).

**Michael Faraday** (1791–1867)

Faraday, a British physicist and chemist, is often regarded as the greatest experimental scientist of the 1800s. His many contributions to the study of electricity include the invention of the electric motor, electric generator, and transformer, as well as the discovery of electro-magnetic induction and the laws of electrolysis. Greatly influenced by his religious beliefs, he refused to work on the development of poison gas for the British military. *(By kind permission of the President and Council of the Royal Society)*

**in the galvanometer circuit as long as relative motion occurs between the magnet and the coil.**

These results are quite remarkable in view of the fact that a current exists in a wire even though no batteries are connected to the wire. We call such a current an **induced current,** and it is produced by an **induced emf.**

Another experiment, first conducted by Faraday, is illustrated in Figure 23.3. Part of the apparatus consists of a coil of insulated wire connected to a switch and a battery. We shall refer to this coil as the *primary coil* of wire and to the corresponding circuit as the primary circuit. The coil is wrapped around an iron ring to intensify the magnetic field produced by the current through the coil. A second coil of insulated wire at the right is also wrapped around the iron ring and is connected to a galvanometer. We shall refer to this as the *secondary coil* and to the corresponding circuit as the secondary circuit. The secondary circuit has no battery, and the secondary coil is not electrically connected to the primary coil. The purpose of this apparatus is to detect any current that might be generated in the secondary circuit by a change in the magnetic field produced by the primary circuit.

At first sight, you might guess that no current would ever be detected in the secondary circuit. Something quite surprising happens, however, when the switch in the primary circuit is suddenly closed or opened. At the instant the switch is thrown closed, the galvanometer deflects in one direction and then returns to zero. When the switch is opened, the galvanometer deflects in the opposite direction and then again returns to zero. Finally, the galvanometer reads zero when the primary circuit carries a steady current.

As a result of these observations, Faraday concluded that **an electric current can be produced by a time-varying magnetic field.** A current cannot be produced by a steady magnetic field. In the experiment shown in Figure 23.2, the changing magnetic field is a result of the relative motion between the magnet and the loop of wire. As long as the motion persists, the current is maintained. In the experiment shown in Figure 23.3, the current produced in the secondary circuit occurs for only an instant while the magnetic field acting on the secondary coil builds from its zero value to its final value. In effect, the secondary circuit behaves as though a source of emf were connected to it for an instant. It is customary to say that an emf is induced in the secondary circuit by the changing magnetic field produced by the current in the primary circuit.

To quantify such observations, we define a quantity called **magnetic flux.** The flux associated with a magnetic field is defined in a similar manner to the electric flux (Section 19.8) and is proportional to the number of magnetic field lines passing through an area. Consider an element of area $dA$ on an arbitrarily shaped

**Figure 23.3**

Faraday's experiment. When the switch in the primary circuit is closed, the galvanometer in the secondary circuit deflects momentarily. The emf induced in the secondary circuit is caused by the changing magnetic field through the secondary coil.

open surface, as in Figure 23.4. If the magnetic field at the location of this element is **B**, the magnetic flux through the element is $\mathbf{B} \cdot d\mathbf{A}$, where $d\mathbf{A}$ is a vector perpendicular to the surface whose magnitude equals the area $dA$. Hence, the total magnetic flux $\Phi_B$ through the surface is

$$\Phi_B = \int \mathbf{B} \cdot d\mathbf{A} \qquad \text{[23.1]}$$

The SI unit of magnetic flux is the tesla-meter squared, which is named the *weber* (Wb): 1 Wb = 1 T·m².

The two experiments illustrated in Figures 23.2 and 23.3 have one thing in common. In both cases, an **emf is induced in a circuit when the magnetic flux through the surface bounded by the circuit changes with time.** In fact, a general statement summarizes such experiments involving induced emfs:

> The emf induced in a circuit is equal to the time rate of change of magnetic flux through the circuit.

This statement, known as **Faraday's law of induction,** can be written

$$\mathcal{E} = -\frac{d\Phi_B}{dt} \qquad \text{[23.2]}$$

where $\Phi_B$ is the magnetic flux through the surface bounded by the circuit and is given by Equation 23.1. The negative sign in Equation 23.2 shall be discussed in Section 23.3. If the circuit is a coil consisting of $N$ identical and concentric loops and if the field lines pass through all loops, the induced emf is

$$\mathcal{E} = -N\frac{d\Phi_B}{dt} \qquad \text{[23.3]}$$

The emf is increased by the factor $N$ because all of the loops are in series, so that the emfs in the individual loops add to give the total emf.

Suppose the magnetic field is uniform over the area $A$ bounded by a loop lying in a plane as in Figure 23.5. In this case, the magnetic flux through the loop is

$$\Phi_B = \int \mathbf{B} \cdot d\mathbf{A} = \int B \, dA \cos \theta = B \cos \theta \int dA = BA \cos \theta$$

Hence, the induced emf is

$$\mathcal{E} = -\frac{d}{dt} (BA \cos \theta) \qquad \text{[23.4]}$$

This expression shows that an emf can be induced in a circuit by changing the magnetic flux in several ways: (1) the magnitude of **B** can vary with time; (2) the area $A$ of the circuit can change with time; (3) the angle $\theta$ between **B** and the normal to the plane can change with time; and (4) any combination of these can occur.

## Quick Quiz 23.1

Suppose you would like to steal power for your home from the electric company by placing a loop of wire near a transmission cable, so as to induce an emf in the loop (an illegal procedure). Should you orient your loop so that the transmission cable passes through your loop, or simply place your loop near the transmission cable?

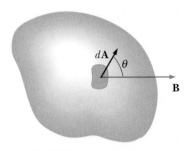

**Figure 23.4**

The magnetic flux through an area element $d\mathbf{A}$ is given by $\mathbf{B} \cdot d\mathbf{A} = B \, dA \cos \theta$. Note that vector $d\mathbf{A}$ is perpendicular to the surface.

• *Faraday's law*

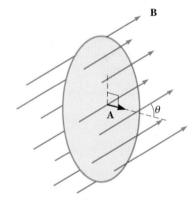

**Figure 23.5**

A conducting loop that encloses an area $A$ in the presence of a uniform magnetic field **B**. The angle between **B** and the normal to the loop is $\theta$.

**PITFALL PREVENTION 23.1**

**Induced current requires a change in flux**

Remember that the *existence* of a magnetic flux through an area is not sufficient to create an induced emf. A *change* in the magnetic flux must occur for an emf to be induced.

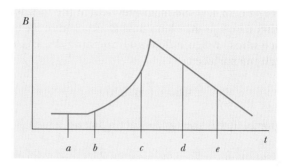

**Figure 23.6**

(Quick Quiz 23.2) The time behavior of a magnetic field through a loop.

## Quick Quiz 23.2

Figure 23.6 is a graph of magnitude versus time for a magnetic field that passes through a fixed loop and is oriented perpendicular to the plane of the loop. The magnitude of the magnetic field at any time is uniform over the area of the loop. Rank the magnitudes of the emf generated in the loop at the five instants indicated, from largest to smallest.

**WEB**

For a history of the electric guitar, visit **www.history-of-rock.com/guitars.htm**

## Quick Quiz 23.3

An electric guitar pickup consists of a permanent magnet surrounded by a coil of wire (Fig. 23.7). How does the pickup detect the movement of a steel guitar string?

**Figure 23.7**

(Quick Quiz 23.3) (a) In an electric guitar, a pickup coil wrapped around a magnet is located near each string. (b) Several pickups allow the vibration to be detected from different portions of the string *(b, Charles D. Winters)*

## THINKING PHYSICS 23.1

Argentina has more land area ($2.8 \times 10^6$ km$^2$) than Greenland ($2.2 \times 10^6$ km$^2$). Yet the magnetic flux of the Earth's magnetic field is larger through Greenland than through Argentina. Why?

**Reasoning**  The larger magnetic flux occurs through Greenland for two reasons. Greenland (latitude 60° north to 80° north) is closer to a magnetic pole than is Argentina (latitude 20° south to 50° south). As a result, the angle that the field lines make with the vertical is smaller in Greenland than in Argentina. The dot product in the definition of the magnetic flux through an area incorporates the cosine of this angle, and smaller angles have larger values of the cosine function. The second reason is also due to Greenland's proximity to a magnetic pole. The magnitude of the magnetic field is larger in the vicinity of a magnetic pole than farther from the pole. Thus, **B** in Equation 23.1 has a larger magnitude in Greenland than in Argentina. These two influences will dominate over the larger area of Argentina.

## THINKING PHYSICS 23.2

The ground fault interrupter (GFI) is a safety device that protects users of electric power against electric shock when they touch appliances. Its essential parts are shown in Figure 23.8. How does its operation make use of Faraday's law?

**Reasoning**  Wire 1 leads from the wall outlet to the appliance being protected, and wire 2 leads from the appliance back to the wall outlet. An iron ring surrounds the two wires. A sensing coil wrapped around part of the iron ring activates a circuit breaker when changes in magnetic flux occur. Because the currents in the two wires are in opposite directions during normal operation of the appliance, the net magnetic field through the sensing coil due to the currents is zero. A change in magnetic flux through the sensing coil can happen, however, if one of the wires on the appliance loses its insulation and accidentally touches the metal case of the appliance, providing a direct path to ground. When such a short to ground occurs, a net magnetic flux occurs through the sensing coil that alternates in time because household current is alternating. This produces an induced voltage in the coil. This induced voltage triggers a circuit breaker, stopping the current before it reaches a level that might be harmful to the person using the appliance.

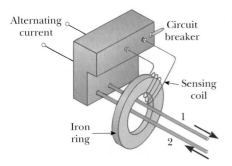

**Figure 23.8**

(Thinking Physics 23.2) Essential components of a ground fault interrupter.

## Example 23.1 One Way to Induce an emf in a Coil

A coil is wrapped with 200 turns of wire on the perimeter of a square frame with sides of 18 cm. Each turn has the same area, equal to that of the frame, and the total resistance of the coil is 2.0 $\Omega$. A magnetic field is perpendicular to the plane of the coil and has the same magnitude at all points within the area of the coil at any time. If the field magnitude changes at a constant rate from 0 to 0.50 T in a time of 0.80 s, find the magnitude of the induced emf in the coil while the field is changing.

**Solution** Because the field is uniform across the area of the coil and perpendicular to the turns of wire, the magnetic flux at any time is simply the product of the magnitude of the field and the area of the turns, and Equation 23.3 becomes

$$\mathcal{E} = -N\frac{d\Phi_B}{dt} = -N\frac{d(BA)}{dt} = -NA\frac{dB}{dt}$$

Because the magnetic field changes at a constant rate, the derivative of the field is equal to the ratio of the change in field to the time interval during which that change occurs:

$$|\mathcal{E}| = NA\frac{dB}{dt} = NA\frac{\Delta B}{\Delta t}$$

$$= (200)(0.18 \text{ m})^2\frac{0.50 \text{ T} - 0}{0.80 \text{ s}} = \boxed{4.1 \text{ V}}$$

**EXERCISE** What is the magnitude of the induced current in the coil while the field is changing?

*Answer* 2.0 A

## Example 23.2 An Exponentially Decaying B Field

A plane loop of wire of area $A$ is placed in a region where the magnetic field is at an angle $\theta$ to the normal to the plane and has the same magnitude at all points within the area of the coil at any time. The magnitude of the magnetic field varies with time according to the expression $B = B_{max}e^{-at}$. That is, at $t = 0$ the field is $B_{max}$, and for $t > 0$, the field decreases exponentially with time (Fig. 23.9). Find the induced emf in the loop as a function of time.

**Solution** Because **B** is uniform across the area of the coil, the magnetic flux through the loop at time $t > 0$ is

$$\Phi_B = BA\cos\theta = AB_{max}\cos\theta\, e^{-at}$$

Because the coefficient $AB_{max}$ and the parameter $a$ are constants, the induced emf from Equation 23.2 is

$$\mathcal{E} = -\frac{d\Phi_B}{dt} = -AB_{max}\cos\theta\frac{d}{dt}e^{-at}$$

$$= \boxed{aAB_{max}\cos\theta\, e^{-at}}$$

That is, the induced emf decays exponentially with time. Note that the maximum emf occurs at $t = 0$, where $\mathcal{E}_{max} = aAB_{max}$. The plot of $\mathcal{E}$ versus $t$ is similar to the $B$ versus $t$ curve shown in Figure 23.9.

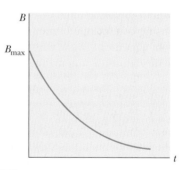

**Figure 23.9**

(Example 23.2) Exponential decrease in the magnitude of the magnetic field with time. The induced emf and induced current vary with time in the same way.

**EXERCISE** A flat loop of wire consisting of a single turn of cross-sectional area 8.00 cm$^2$ is perpendicular to a magnetic field that increases uniformly in magnitude from 0.50 to 2.50 T in 1.00 s. What is the resulting induced current if the loop has a resistance of 2.00 $\Omega$?

*Answer* 800 $\mu$A

## 23.2 • MOTIONAL EMF

Examples 23.1 and 23.2 are cases in which an emf is produced in a circuit when the magnetic field changes with time. In this section we describe **motional emf,** in which an emf is induced in a conductor moving through a magnetic field. This is the situation we described in Figure 23.1 at the beginning of Section 23.1.

Consider a straight conductor of length $\ell$ moving with constant velocity through a uniform magnetic field directed into the page as in Figure 23.10. For simplicity, we shall assume that the conductor is moving perpendicularly to the field. The electrons in the conductor experience a force along the conductor with magnitude $|\mathbf{F}_B| = |q\mathbf{v} \times \mathbf{B}| = qvB$. According to Newton's second law, the electrons accelerate in response to this force and move along the wire. Once the electrons move to the lower end of the wire, they accumulate there, leaving a net positive charge at the upper end. As a result of this charge separation, an electric field $\mathbf{E}$ is produced within the conductor. The charge at the ends of the conductor builds up until the magnetic force $qvB$ on an electron in the conductor is balanced by the electric force $qE$ on the electron as shown in Fig. 23.10. At this point, charge stops flowing. In this situation, the net force on an electron allows us to relate the electric field to the magnetic field:

$$\sum \mathbf{F} = \mathbf{F}_e - \mathbf{F}_B = 0 \longrightarrow qE = qvB \longrightarrow E = vB$$

Because the electric field produced in the conductor is uniform, it is related to the potential difference across the ends of the conductor according to the relation $\Delta V = E\ell$ (Section 20.2). Thus,

$$\Delta V = E\ell = B\ell v$$

where the upper end is at a higher potential than the lower end. Thus, **a potential difference is maintained as long as the conductor is moving through the magnetic field. If the motion is reversed, the polarity of $\Delta V$ is also reversed.**

An interesting situation occurs if we now consider what happens when the moving conductor is part of a closed circuit. Consider a circuit consisting of a conducting bar of length $\ell$ sliding along two fixed parallel conducting rails as in Figure 23.11a. For simplicity, we assume that the moving bar has zero electrical resistance and that the stationary part of the circuit has a resistance $R$. A uniform and constant magnetic field $\mathbf{B}$ is applied perpendicular to the plane of the circuit.

As the bar is pulled to the right with a velocity $\mathbf{v}$ under the influence of an applied force $\mathbf{F}_{\text{app}}$, free charges in the bar experience a magnetic force along the length of the bar. This magnetic force sets up an induced current because the charges on which the force is acting are free to move in the circuit. In this case, the rate of change of magnetic flux through the loop and the accompanying induced emf across the moving bar are proportional to the change in loop area as the bar moves through the magnetic field.

Because the area of the circuit at any instant is $\ell x$, the external magnetic flux through the circuit is

$$\Phi_B = B\ell x$$

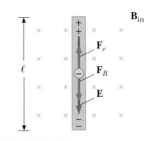

**Figure 23.10**

A conductor is moved through a magnetic field. Due to the magnetic force on electrons, the ends of the wire become oppositely charged. This establishes an electric field in the wire. In steady state, the electric and magnetic forces on an electron in the wire are balanced.

(a)

(b)

**Figure 23.11**

(a) A conducting bar sliding with a velocity $\mathbf{v}$ along two conducting rails under the action of an applied force $\mathbf{F}_{\text{app}}$. The magnetic force $\mathbf{F}_B$ opposes the motion, and a counterclockwise current $I$ is induced in the loop.
(b) The equivalent circuit diagram for the pictorial representation in part (a).

where $x$ is the width of the circuit, a parameter that changes with time. Using Faraday's law, we find that the induced emf is

•  *Motional emf*

$$\mathcal{E} = -\frac{d\Phi_B}{dt} = -\frac{d}{dt}(B\ell x) = -B\ell\frac{dx}{dt} = -B\ell v \qquad \text{[23.5]}$$

Because the resistance of the circuit is $R$, the magnitude of the induced current is

$$I = \frac{|\mathcal{E}|}{R} = \frac{B\ell v}{R} \qquad \text{[23.6]}$$

The equivalent circuit diagram for this example is shown in Figure 23.11b. The moving bar is behaving like a battery in that it is a source of emf as long as the bar continues to move.

Let us examine this situation using energy considerations in the nonisolated system model, with the system being the entire circuit. Because the circuit has no battery, you might wonder about the origin of the induced current and the energy delivered to the resistor. This can be understood by noting that the external force does work on the conductor, thereby moving charges through a magnetic field. This causes the charges to move along the conductor with some average drift velocity, and hence a current is established. From the viewpoint of the continuity equation for energy, the total work done on the system by the applied force while the bar moves with constant speed must equal the increase in internal energy in the resistor during this time interval. (This assumes that the energy stays in the resistor; in reality, energy leaves the resistor by heat and electromagnetic radiation.)

As the conductor of length $\ell$ moves through the uniform magnetic field **B**, it experiences a magnetic force $\mathbf{F}_B$ of magnitude $I\ell B$ (Eq. 22.10), where $I$ is the current induced due to its motion. The direction of this force is opposite the motion of the bar, or to the left in Figure 23.11a.

If the bar is to move with a *constant* velocity, the applied force $\mathbf{F}_{app}$ must be equal in magnitude and opposite in direction to the magnetic force, or to the right in Figure 23.11a. (If the magnetic force acted in the direction of motion, it would cause the bar to accelerate once it was in motion, thereby increasing its speed. This state of affairs would represent a violation of the principle of energy conservation.) Using Equation 23.6 and the fact that $F_{app} = F_B = I\ell B$, we find that the power delivered by the applied force is

$$\mathcal{P} = F_{app}v = (I\ell B)v = \frac{B^2\ell^2 v^2}{R} = \left(\frac{B\ell v}{R}\right)^2 R = I^2 R \qquad \text{[23.7]}$$

This power is equal to the rate at which energy is delivered to the resistor as we expect. It is also equal to the power $I\mathcal{E}$ supplied by the induced emf. This example demonstrates the conversion of mechanical energy in the moving bar to internal energy in the resistor.

### Quick Quiz 23.4

You wish to move a rectangular loop of wire into a region of uniform magnetic field at a given speed so as to induce an emf in the loop. The plane of the loop must remain perpendicular to the magnetic field lines. In which orientation should you hold the loop while you move it into the region of magnetic field so as to generate the largest emf? (a) With the long dimension of the loop parallel to the velocity vector. (b) With the short dimension of the loop parallel to the velocity vector. (c) Either way—the emf is the same regardless of orientation.

## THINKING PHYSICS 23.3

As an airplane flies from Los Angeles to Seattle, it cuts through the lines of the Earth's magnetic field. As a result, an emf is developed between the wingtips of the airplane. Which wingtip is positively charged and which is negatively charged? Would the wingtips be charged in the same way on an airplane traveling from Antarctica to Australia?

**Reasoning**   The magnetic field lines of the Earth have a downward component in the Northern Hemisphere. As the airplane flies northward, the right-hand rule indicates that positive charges experience a force to the left side of the airplane. Thus, the left wingtip becomes positively charged and the right wingtip negatively charged. In the Southern Hemisphere, the field lines have an upward component, so the charging of the wingtips would be in the opposite sense.

## Example 23.3   Motional emf Induced in a Rotating Bar

A conducting bar of length $\ell$ rotates with a constant angular speed $\omega$ about a pivot at one end. A uniform magnetic field **B** is directed perpendicular to the plane of rotation as in Figure 23.12. Find the emf induced between the ends of the bar.

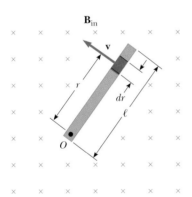

**Figure 23.12**

(Example 23.3) A conducting bar rotating around a pivot at one end in a uniform magnetic field that is perpendicular to the plane of rotation. An emf is induced across the ends of the bar.

**Solution**   Consider a segment of the bar of length $dr$, whose velocity is **v**. According to Equation 23.5, the emf induced in a conductor of length $dr$ moving perpendicular to a field **B** is

$$(1) \quad d\mathcal{E} = Bv \, dr$$

Each segment of the bar is moving perpendicular to **B**, so an emf is generated across each segment, the value of which is given by Equation (1). Summing up the emfs induced across all elements, which are in series, gives the total emf between the ends of the bar. That is,

$$\mathcal{E} = \int Bv \, dr$$

To integrate this expression, note that the linear speed of an element is related to the angular speed $\omega$ through the relationship $v = r\omega$ (Equation 10.10). Because $B$ and $\omega$ are constants, we therefore find

$$\mathcal{E} = B \int_0^\ell v \, dr = B\omega \int_0^\ell r \, dr = \tfrac{1}{2}B\omega \ell^2$$

## Example 23.4   A Sliding Bar in a Magnetic Field

A bar of mass $m$ and length $\ell$ moves on two frictionless parallel rails in the presence of a uniform magnetic field directed into the paper (Fig. 23.13). The bar is given an initial velocity $\mathbf{v}_i$ to the right and is released. Find the velocity of the bar as a function of time.

**Solution**   First note that the induced current is counterclockwise and the magnetic force on the bar is $\mathbf{F}_B = -I\ell B\mathbf{i}$, where the negative sign means that the force is to the left

and retards the motion. This is the only horizontal force acting on the bar. Hence, Newton's second law applied to motion in the $x$ direction gives

$$F_x = ma_x \quad \longrightarrow \quad -I\ell B = m\frac{dv}{dt}$$

Because the force depends on the current, and the current depends on the speed, the force is not constant and the acceleration of the bar is not constant. Because the induced current is given by Equation 23.6, $I = B\ell v/R$, we can write this expression as

$$m \frac{dv}{dt} = -\frac{B^2\ell^2}{R} v$$

Rearranging this expression gives

$$\frac{dv}{v} = -\left(\frac{B^2\ell^2}{mR}\right) dt$$

Integrating this last equation using the initial condition that $v = v_i$ at $t = 0$, we find

$$\int_{v_i}^{v} \frac{dv}{v} = \frac{-B^2\ell^2}{mR} \int_{0}^{t} dt$$

$$\ln\left(\frac{v}{v_i}\right) = -\left(\frac{B^2\ell^2}{mR}\right) t = -\frac{t}{\tau}$$

where the constant $\tau = mR/B^2\ell^2$. From this, we see that the speed of the bar can be expressed in the exponential form

$$v = v_i e^{-t/\tau}$$

The speed of the bar therefore decreases exponentially with time under the action of the magnetic retarding force. Furthermore, if we substitute this result into Equation 23.6, we

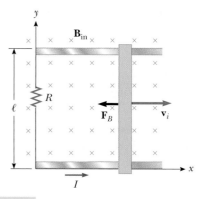

**Figure 23.13**

(Example 23.4) A conducting bar of length $\ell$ sliding on two fixed conducting rails is given an initial velocity $\mathbf{v}_i$ in the positive $x$ direction.

find that the induced emf and induced current also decrease exponentially with time. That is,

$$I = \frac{B\ell v}{R} = \frac{B\ell v_i}{R} e^{-t/\tau}$$

$$\mathcal{E} = IR = B\ell v_i e^{-t/\tau}$$

## The Alternating-Current Generator

The alternating current (ac) generator is a device in which energy is transferred in by work and out by electrical transmission. A simplified pictorial representation of such a generator is shown in Figure 23.14a. It consists of a coil of wire rotated in

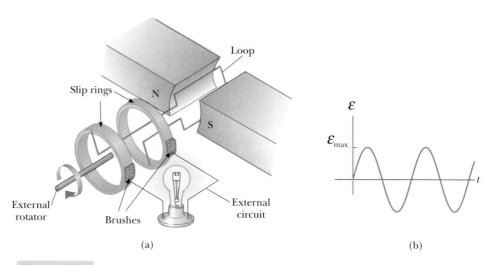

(a)                                        (b)

**Figure 23.14**

(a) Schematic diagram of an ac generator. An emf is induced in a loop that rotates in a magnetic field.
(b) A graphical representation of the alternating emf induced in the loop as a function of time.

an external magnetic field by some external agent—this is the work input. In commercial power plants, the energy required to rotate the loop can be derived from a variety of sources. For example, in a hydroelectric plant, falling water directed against the blades of a turbine produces the rotary motion; in a coal-fired plant, the high temperature produced by burning the coal is used to convert water to steam and this steam is directed against turbine blades. As the loop rotates, the magnetic flux through it changes with time, inducing an emf and a current in a circuit connected to the coil.

Suppose that the coil has $N$ turns, all of the same area $A$, and suppose that the coil rotates with a constant angular speed $\omega$ about an axis perpendicular to the magnetic field. If $\theta$ is the angle between the magnetic field and the direction perpendicular to the plane of the coil, the magnetic flux through the loop at any time $t$ is given by

$$\Phi_B = BA \cos \theta = BA \cos \omega t$$

where we have used the relationship between angular displacement and angular speed, $\theta = \omega t$. Hence, the induced emf in the coil is

$$\mathcal{E} = -N\frac{d\Phi_B}{dt} = -NAB\frac{d}{dt}(\cos \omega t) = NAB\omega \sin \omega t \qquad \text{[23.8]}$$

This result shows that the emf varies sinusoidally with time as shown in Figure 23.14b. From Equation 23.8 we see that the maximum emf has the value $\mathcal{E}_{max} = NAB\omega$, which occurs when $\omega t = 90°$ or $270°$. In other words, $\mathcal{E} = \mathcal{E}_{max}$ when the magnetic field is in the plane of the coil, and the time rate of change of flux is a maximum. In this position, the velocity vector for a wire in the loop is perpendicular to the magnetic field vector. Furthermore, the emf is *zero* when $\omega t = 0$ or $180°$, that is, when **B** is perpendicular to the plane of the coil, and the time rate of change of flux is zero. In this orientation, the velocity vector for a wire in the loop is parallel to the magnetic field vector.

The sinusoidally varying emf in Equation 23.8 is the source of *alternating current* delivered to customers of electrical utility companies. It is called **ac voltage** as opposed to the dc voltage from a source such as a battery.

**EXERCISE**  Coils rotating in a magnetic field are often used to measure unknown magnetic fields. As an example, consider a coil of radius 1.0 cm with 50 turns that is rotated about an axis perpendicular to the field at a rate of 20 Hz. If the maximum induced emf in the coil is 3.0 V, find the magnitude of the magnetic field.

*Answer*  1.5 T

## 23.3 • LENZ'S LAW

 See Screen 12.7

Let us now address the negative sign in Faraday's law. When a change occurs in the magnetic flux, the direction of the induced emf and induced current can be found from **Lenz's law:**

> The polarity of the induced emf in a loop is such that it produces a current whose magnetic field opposes the change in magnetic flux through the loop. That is, the induced current is in a direction such that the induced magnetic field attempts to maintain the original flux through the loop.

**Figure 23.15**

(a) As the conducting bar slides on the two fixed conducting rails, the magnetic field directed inward through the area enclosed by the loop increases in time. By Lenz's law, the induced current must be counterclockwise so as to produce a counteracting magnetic field directed outward from the page. (b) When the bar moves to the left, the induced current must be clockwise. Why?

Notice that no equation is associated with Lenz's law. The law is in words only and provides a means for determining the direction of the current in a circuit when a magnetic change occurs.

To attain a better understanding of Lenz's law, let us return to the example of a bar moving to the right on two parallel rails in the presence of a uniform magnetic field directed into the page (Fig. 23.15a). As the bar moves to the right, the magnetic flux through the circuit increases with time because the area of the loop increases. Lenz's law says that the induced current must be in such a direction that the magnetic field *it* produces opposes the *change* in the magnetic flux of the external magnetic field. Because the flux is due to an external field *into* the page and is increasing, the induced current, if it is to oppose the change, must produce a magnetic field through the circuit *out* of the page. Hence, the induced current must be counterclockwise when the bar moves to the right to give a counteracting field out of the page in the region inside the loop. (Use the right-hand rule to verify this direction.) If the bar is moving to the left, as in Figure 23.15b, the magnetic flux through the loop decreases with time. Because the magnetic field is into the page, the induced current has to be clockwise to produce a magnetic field into the page inside the loop. In either case, the induced current attempts to maintain the original flux through the circuit.

Let us examine this situation from the viewpoint of energy considerations. Suppose that the bar is given a slight push to the right. In the preceding analysis, we found that this motion leads to a counterclockwise current in the loop. What happens if we incorrectly assume that the current is clockwise? For a clockwise current *I*, the direction of the magnetic force *IℓB* on the sliding bar would be to the right. According to Newton's second law, this force would accelerate the rod and increase its speed, which would in turn cause the area of the loop to increase more

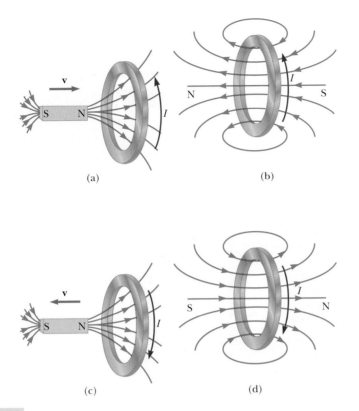

**Figure 23.16**

(a) When the magnet is moved toward the stationary conducting loop, a current is induced in the direction shown. (b) This induced current produces its own magnetic field that is directed to the left within the loop to counteract the increasing external flux. (c) When the magnet is moved away from the stationary conducting loop, a current is induced in the direction shown. (d) This induced current produces its own magnetic field that is directed to the right within the loop to counteract the decreasing external flux.

rapidly. This would increase the induced current, which would increase the force, which would increase the current, which would . . . In effect, the system would acquire energy with no additional energy input. This result is clearly inconsistent with all experience and with the continuity equation for energy. Thus, we are forced to conclude that the current must be counterclockwise.

Consider another situation, one in which a bar magnet is moved to the right toward a stationary loop of wire, as in Figure 23.16a. As the magnet moves toward the loop, the magnetic flux through the loop increases with time. To counteract this increase in flux due to a magnetic field directed toward the right, the induced current produces a magnetic field to the left as in Figure 23.16b; hence, the induced current is in the direction shown. The left face of the current loop is therefore a north pole, and the right face is a south pole.

If the magnet is moved to the left as in Figure 23.16c, the magnetic field through the loop, which is toward the right, decreases with time. Under these circumstances, the induced current in the loop sets up a magnetic field through the loop from left to right, in an effort to maintain a constant flux. Hence, the direction of the induced current in the loop is as shown in Figure 23.16d. In this case, the left face of the loop is a south pole, and the right face is a north pole.

**Figure 23.17**

(Quick Quiz 23.5) In an old-fashioned equal-arm balance, an aluminum sheet hangs between the poles of a magnet. *(Photos by John Jewett)*

**Quick Quiz 23.5**

In equal-arm balances from the early 20th century (Fig. 23.17), it is sometimes observed that an aluminum sheet hangs from one of the arms and passes between the poles of a magnet—why?

### Example 23.5 Application of Lenz's Law

A coil of wire is placed near an electromagnet as in Figure 23.18a. Find the direction of the induced current in the coil (a) at the instant the switch is closed, (b) after the switch has been closed for several seconds, and (c) when the switch is opened.

**Reasoning** (a) When the switch is closed, the situation changes from a condition in which no magnetic flux occurs

through the coil to one in which flux does occur due to a magnetic field in the direction shown in Figure 23.18b. To counteract this increase in magnetic flux, the coil must set up a field from left to right in the figure. This requires a current directed as shown in Figure 23.18b.

(b) After the switch has been closed for several seconds, the magnetic flux through the loop does not change; hence the induced current is zero.

(a)  (b)

(c) Opening the switch causes the magnetic field to change from a condition in which magnetic field lines pass through the coil from right to left to a condition of zero field. The induced current must then be as shown in Figure 23.18c, so as to set up its own magnetic field from right to left.

(c)

**Figure 23.18**

(Example 23.5)

## 23.4 • INDUCED EMFS AND ELECTRIC FIELDS

See Screen 12.8

We have seen that a changing magnetic flux induces an emf and a current in a conducting loop. We can also interpret this from another point of view. Because the normal flow of charges in a circuit is due to an electric field in the wires set up by a source such as a battery, we can interpret the changing magnetic field as creating an induced electric field. This electric field applies a force on the charges to cause them to move. With this approach, then, we see that **an electric field is created in the conductor as a result of changing magnetic flux.** In fact, the law of electromagnetic induction can be interpreted as follows: **An electric field is always generated by a changing magnetic flux,** even in free space where no charges are present. This induced electric field, however, has quite different properties from those of the electrostatic field produced by stationary charges.

Let us illustrate this point by considering a conducting loop of radius $r$, situated in a uniform magnetic field that is perpendicular to the plane of the loop as in Figure 23.19. If the magnetic field changes with time, Faraday's law tells us that an emf $\mathcal{E} = -d\Phi_B/dt$ is induced in the loop. The induced current thus produced implies the presence of an induced electric field $\mathbf{E}$ that must be tangent to the loop in order to provide an electric force on the charges around the loop. The work done by the electric field on the loop in moving a test charge $q$ once around the loop is equal to $q\mathcal{E}$. Because the magnitude of the electric force on the charge is $qE$, the work done by the electric field can also be expressed as $qE(2\pi r)$, where $2\pi r$ is the circumference of the loop. These two expressions for the work must be equal; therefore, we see that

$$q\mathcal{E} = qE(2\pi r)$$

$$E = \frac{\mathcal{E}}{2\pi r}$$

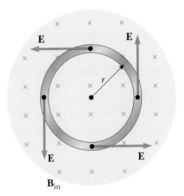

**Figure 23.19**

A conducting loop of radius $r$ in a uniform magnetic field perpendicular to the plane of the loop. If **B** changes in time, an electric field is induced in a direction tangent to the loop.

Using this result along with Faraday's law and the fact that $\Phi_B = BA = B\pi r^2$ for a circular loop, we find that the induced electric field can be expressed as

$$E = -\frac{1}{2\pi r}\frac{d\Phi_B}{dt} = -\frac{1}{2\pi r}\frac{d}{dt}(B\pi r^2) = -\frac{r}{2}\frac{dB}{dt}$$

This expression can be used to calculate the induced electric field if the time variation of the magnetic field is specified. The negative sign indicates that the induced electric field **E** results in a current that opposes the change in the magnetic field. It is important to understand that **this result is also valid in the absence of a conductor or charges.** That is, the same electric field is induced by the changing magnetic field in empty space.

In general, the emf for any closed path can be expressed as the line integral of **E** · *d***s** over that path. Hence, the general form of Faraday's law of induction is

• *Faraday's law in general form*

$$\varepsilon = \oint \mathbf{E}\cdot d\mathbf{s} = -\frac{d\Phi_B}{dt} \tag{23.9}$$

It is important to recognize that **the induced electric field E that appears in Equation 23.9 is a nonconservative field that is generated by a changing magnetic field.** We call it a nonconservative field because the work done in moving a charge around a closed path (the loop in Figure 23.19) is not zero. This type of electric field is very different from an electrostatic field.

### THINKING PHYSICS 23.4

In studying electric fields, we noted that electric field lines begin on positive charges and end on negative charges. Do *all* electric field lines begin and end on charges?

**Reasoning**   The statement that electric field lines begin and end on charges is true only for *electrostatic* fields—electric fields due to stationary charges. Electric field lines due to changing magnetic fields form closed loops, with no beginning and no end, and are independent of the presence of charges.

### Example 23.6   Electric Field Induced by a Changing Magnetic Field in a Solenoid

A long solenoid of radius $R$ has $n$ turns of wire per unit length and carries a time-varying current that varies sinusoidally as $I = I_{max}\cos \omega t$, where $I_{max}$ is the maximum current and $\omega$ is the angular frequency of the ac source (Fig. 23.20). (a) Determine the magnitude of the induced electric field outside the solenoid, a distance $r > R$ from its long central axis.

**Solution**   First consider an external point and take the path for our line integral to be a circle of radius $r$ centered on the solenoid, as illustrated in Figure 23.20. By symmetry, we see that the magnitude of **E** is constant on this path and that **E** is

tangent to it. The magnetic flux through the area enclosed by the path is $\mathbf{B}\cdot\mathbf{A} = B\pi R^2$; hence, Equation 23.9 gives

$$(1) \qquad \oint \mathbf{E}\cdot d\mathbf{s} = -\frac{d}{dt}(B\pi R^2) = -\pi R^2\frac{dB}{dt}$$

Based on the symmetry of the situation,

$$(2) \qquad \oint \mathbf{E}\cdot d\mathbf{s} = \oint E\,ds = E\oint ds = E(2\pi r)$$

Setting these two expressions equal, we find

$$(3) \qquad E = -\frac{R^2}{2r}\frac{dB}{dt}$$

Path of integration

R

$I_{max} \cos \omega t$

(Example 23.6) A long solenoid carrying a time-varying current given by $I = I_{max} \cos \omega t$. An electric field is induced both inside and outside the solenoid.

The magnetic field inside a long solenoid is given by Equation 22.32, $B = \mu_0 nI$. When we substitute $I = I_{max} \cos \omega t$ into this equation, and then substitute the result into Equation (3), we find that

$$E = -\frac{R^2}{2r} \frac{d}{dt} (\mu_0 nI_{max} \cos \omega t)$$

$$= \frac{\mu_0 nI_{max}\omega R^2}{2r} \sin \omega t \quad (\text{for } r > R)$$

Hence, the electric field varies sinusoidally with time, and its amplitude falls off as $1/r$ outside the solenoid.

(b) What is the magnitude of the induced electric field inside the solenoid, a distance $r$ from its axis?

For an interior point $(r < R)$, the flux passing through the area bounded by a path of integration is given by $B\pi r^2$. Thus, the analogs to Equations (1), (2), and (3) become

$$(4) \quad \oint \mathbf{E} \cdot d\mathbf{s} = -\frac{d}{dt}(B\pi r^2) = -\pi r^2 \frac{dB}{dt}$$

$$(5) \quad \oint \mathbf{E} \cdot d\mathbf{s} = \oint E \, ds = E \oint ds = E(2\pi r)$$

$$(6) \quad E = -\frac{r}{2}\frac{dB}{dt}$$

Substituting the expression for the current into Equation (6) gives

$$E = -\frac{r}{2}\frac{d}{dt}(\mu_0 nI_{max} \cos \omega t)$$

$$= \frac{\mu_0 nI_{max}\omega}{2} r \sin \omega t \quad (\text{for } r < R)$$

This shows that the amplitude of the electric field induced inside the solenoid by the changing magnetic field varies linearly with $r$ and varies sinusoidally with time.

## 23.5 • SELF-INDUCTANCE

Consider an isolated circuit consisting of a switch, a resistor, and a source of emf as in Figure 23.21. The circuit diagram is represented in perspective so that we can see the orientations of some of the magnetic field lines due to the current in the circuit. When the switch is closed, the current doesn't immediately jump from zero to its maximum value $\mathcal{E}/R$; the law of electromagnetic induction (Faraday's law) describes the actual behavior. As the current increases with time, the magnetic flux through the loop of the circuit itself due to the current also increases with time. This increasing magnetic flux *from* the circuit induces an emf *in* the circuit that opposes the change in the net magnetic flux through the loop of the circuit. By Lenz's law, the induced electric field in the wires must therefore be opposite the direction of the current, and the opposing emf results in a *gradual* increase in the current. This effect is called *self-induction* because the changing magnetic flux through the circuit arises from the circuit itself. The emf set up in this case is called a **self-induced emf.**

To obtain a quantitative description of self-induction, we recall from Faraday's law that the induced emf is the negative time rate of change of the magnetic flux. The magnetic flux is proportional to the magnetic field, which in turn is proportional to the current in the circuit. Therefore, **the self-induced emf is always proportional to the time rate of change of the current.** For a closely spaced

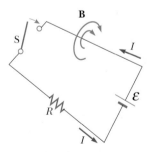

After the switch is closed, the current produces a magnetic flux through the area enclosed by the loop of the circuit. As the current increases toward its final value, this magnetic flux changes with time and induces an emf in the loop.

coil of $N$ turns of fixed geometry (a toroidal coil or the ideal solenoid), we can express this proportionality as follows:

- *Self-induced emf*

$$\mathcal{E}_L = -N\frac{d\Phi_B}{dt} = -L\frac{dI}{dt} \qquad [23.10]$$

where $L$ is a proportionality constant, called the **inductance** of the coil, that depends on the geometric features of the coil and other physical characteristics. From this expression, we see that the inductance of a coil containing $N$ turns is

- *Inductance of an N-turn coil*

$$L = \frac{N\Phi_B}{I} \qquad [23.11]$$

where it is assumed that the same magnetic flux passes through each turn. Later we shall use this equation to calculate the inductance of some special coil geometries.

From Equation 23.10 we can also write the inductance as the ratio

- *Inductance*

$$L = -\frac{\mathcal{E}_L}{dI/dt} \qquad [23.12]$$

This is usually taken to be the defining equation for the inductance of any coil, regardless of its shape, size, or material characteristics. If we compare Equation 23.10 with the equation in Figure 21.23a, $\Delta V = -IR$, we see that resistance is a measure of opposition to current, whereas inductance is a measure of opposition to the *change* in current.

The SI unit of inductance is the **henry (H),** which, from Equation 23.12, is seen to be equal to 1 volt-second per ampere:

$$1\ \text{H} = 1\ \text{V}\cdot\text{s/A}$$

As we shall see, the **inductance of a coil depends on its geometry.** Because inductance calculations can be quite difficult for complicated geometries, the examples we shall explore involve simple situations for which inductances are easily evaluated.

**Joseph Henry** (1797–1878)

Henry, an American physicist, became the first director of the Smithsonian Institution and first president of the Academy of Natural Science. He improved the design of the electromagnet and constructed one of the first motors. He also discovered the phenomenon of self-induction but failed to publish his findings. The unit of inductance, the henry, is named in his honor. *(North Wind Picture Archives)*

### Quick Quiz 23.6

The ends of an inductor having zero resistance are labeled $a$ and $b$. The potential at $a$ is higher than at $b$. Which of the following could be consistent with this situation? (a) The current is constant and is directed from $a$ to $b$. (b) The current is constant and is directed from $b$ to $a$. (c) The current is increasing and is directed from $a$ to $b$. (d) The current is decreasing and is directed from $a$ to $b$. (e) The current is increasing and is directed from $b$ to $a$. (f) The current is decreasing and is directed from $b$ to $a$.

## THINKING PHYSICS 23.5

In some circuits, a spark occurs between the poles of a switch when the switch is opened. Yet, when the switch for this circuit is closed, no spark occurs. Why is there a difference?

**Reasoning**  According to Lenz's law, induced emfs are in a direction such as to attempt to maintain the original magnetic flux when a change occurs. When the switch is opened, the sudden drop in the magnetic field in the circuit induces an emf in a direction that attempts to keep the original current flowing. This can cause a spark as the current bridges the air gap between the poles of the switch. (The spark is more likely in circuits with large inductance.) The spark does not occur when the switch is closed, because the original current is zero, and the induced emf attempts to maintain it at zero.

## THINKING PHYSICS 23.6

A transformer (Fig. 23.22) consists of a pair of coils wrapped around an iron form. When ac voltage is applied to one coil, the *primary*, the magnetic field lines cutting through the other coil, the *secondary*, induce an emf. (This is the arrangement in Faraday's experiment shown in Figure 23.3.) By varying the number of turns of wire on each coil, the ac voltage in the secondary can be made larger or smaller than that in the primary. Clearly, this device cannot work with dc voltage. What's more, if dc voltage is applied, the primary coil sometimes overheats and burns. Why?

**Reasoning**  The primary coil of the transformer is an inductor. When an ac voltage is applied, the opposing emf due to Lenz's law limits the current in the coil. If dc voltage is applied, no opposing emf occurs, and the current can rise to a higher value. This increased current causes the temperature of the coil to rise, to the point where the insulation on the wire sometimes burns.

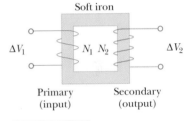

**Figure 23.22**

(Thinking Physics 23.6) An ideal transformer consists of two coils of wire wound on the same iron core. An alternating voltage $\Delta V_1$ is applied to the primary coil, and the output voltage $\Delta V_2$ appears across the secondary coil.

## THINKING PHYSICS 23.7

If inductors are connected in series or in parallel, the inductances combine like resistors—they add in series, and the inverses add in parallel. The previous statement holds true for toroidal inductors in tight quarters in an electric circuit, but not for inductors formed from solenoids. Why?

**Reasoning**  The inductance of a coil is *self*-inductance—an effect due to the magnetic field of a toroid or solenoid passing through its own coils. In an ideal toroidal inductor, the magnetic field is contained within the interior of the toroid—there is zero field outside the toroid. Thus, if toroidal inductors are placed close to each other, little or no magnetic interaction exists between the inductors. Finite solenoids, however, have a nonzero magnetic field outside the solenoid. If solenoids are placed in close proximity, the field of one solenoid can interact with the coils of the other. Thus, *mutual* inductance in addition to self-inductance occurs, and the simple inductance combination rules do not apply.

## Example 23.7    Inductance of a Solenoid

Find the inductance of a uniformly wound solenoid having $N$ turns and length $\ell$. Assume that $\ell$ is long compared with the radius and that the core of the solenoid is air.

**Solution**    Because $\ell$ is long compared with the radius, we can model the solenoid as an ideal solenoid. In this case, the interior magnetic field is uniform and given by Equation 22.32:

$$B = \mu_0 nI = \mu_0 \frac{N}{\ell} I$$

where $n = N/\ell$ is the number of turns per unit length. The magnetic flux through each turn is

$$\Phi_B = BA = \mu_0 \frac{NA}{\ell} I$$

where $A$ is the cross-sectional area of the solenoid. Using this expression and Equation 23.11 we find that

$$L = \frac{N\Phi_B}{I} = \frac{\mu_0 N^2 A}{\ell}$$

This shows that $L$ depends on geometry and is proportional to the square of the number of turns. Because $N = n\ell$, we can also express the result in the form

$$L = \mu_0 \frac{(n\ell)^2}{\ell} A = \mu_0 n^2 A\ell = \mu_0 n^2 V$$

where $V = A\ell$ is the volume of the solenoid.

## Example 23.8    Calculating Inductance and emf

(a) Calculate the inductance of a solenoid containing 300 turns if the length of the solenoid is 25.0 cm and its cross-sectional area is $4.00\ \text{cm}^2 = 4.00 \times 10^{-4}\ \text{m}^2$.

**Solution**    Using the expression for $L$ from Example 23.7 we find

$$L = \frac{\mu_0 N^2 A}{\ell}$$

$$= (4\pi \times 10^{-7}\ \text{T·m/A}) \frac{(300)^2 (4.00 \times 10^{-4}\ \text{m}^2)}{25.0 \times 10^{-2}\ \text{m}}$$

$$= 1.81 \times 10^{-4}\ \text{T·m}^2/\text{A} = 0.181\ \text{mH}$$

(b) Calculate the self-induced emf in the solenoid described in (a) if the current through it is decreasing at the rate of 50.0 A/s.

**Solution**    Using Equation 23.10 and given that $dI/dt = 50.0$ A/s, we have

$$\mathcal{E}_L = -L\frac{dI}{dt} = -(1.81 \times 10^{-4}\ \text{H})(-50.0\ \text{A/s}) = 9.05\ \text{mV}$$

**EXERCISE**    A 0.388-mH inductor in the form of a solenoid has a length four times its diameter. If it is wound with 22 turns per centimeter, what is its length?

*Answer*    0.109 m

See Screen 13.6

## 23.6 • *RL* CIRCUITS

A circuit that contains a coil, such as a solenoid, has a self-inductance that prevents the current from increasing or decreasing instantaneously. A circuit element whose main purpose is to provide inductance in a circuit is called an **inductor.** The circuit symbol for an inductor is ⎓⎓⎓. As a simplification model, we shall always assume that the self-inductance of the remainder of the circuit is negligible compared with that of the inductor.

Consider the circuit shown in Figure 23.23, consisting of a resistor, an inductor, a switch, and a battery. The internal resistance of the battery will be ignored as a further simplification model. Suppose the switch S is closed at $t = 0$. The current begins to increase, and, due to the increasing current, the inductor produces an emf (sometimes referred to as a *back emf*) that opposes the increasing current. The back emf produced by the inductor is

$$\mathcal{E}_L = -L\frac{dI}{dt}$$

Because the current is increasing, $dI/dt$ is positive; therefore $\mathcal{E}_L$ is negative. This corresponds to the fact that a potential drop occurs from $a$ to $b$ across the inductor. For this reason, point $a$ is at a higher potential than point $b$, as illustrated in Figure 23.23.

With this in mind, we can apply Kirchhoff's loop rule to this circuit. If we begin at the battery and travel clockwise, we have

$$\mathcal{E} - IR - L\frac{dI}{dt} = 0 \qquad \text{[23.13]}$$

where $IR$ is the voltage across the resistor. The potential difference across the inductor is given a negative sign because its emf is in the opposite sense to that of the battery. We must now look for a solution to this differential equation, which is a mathematical representation of the behavior of the RL circuit. It is similar to Equation 21.30 for the RC circuit.

To obtain a mathematical solution of Equation 23.13, it is convenient to change variables by letting $x = (\mathcal{E}/R) - I$, so that $dx = -dI$. With these substitutions, Equation 23.13 can be written

$$\frac{\mathcal{E}}{R} - I - \frac{L}{R}\frac{dI}{dt} = x + \frac{L}{R}\frac{dx}{dt} = 0$$

$$\frac{dx}{x} = -\frac{R}{L}dt$$

Integrating this last expression from an initial instant to some later time $t$ gives

$$\int_{x_i}^{x} \frac{dx}{x} = -\frac{R}{L}\int_0^t dt \quad \longrightarrow \quad \ln\frac{x}{x_i} = -\frac{R}{L}t$$

where the value of $x$ at $t = 0$ is expressed as $x_i = \mathcal{E}/R$, because $I = 0$ at $t = 0$. Taking the antilog of this result gives

$$x = x_i e^{-Rt/L}$$

Hence, the preceding expression is equivalent to

$$\frac{\mathcal{E}}{R} - I = \frac{\mathcal{E}}{R}e^{-Rt/L}$$

$$I = \frac{\mathcal{E}}{R}(1 - e^{-Rt/L})$$

This expression represents the solution of Equation 23.13, the current as a function of time. It can also be written

$$I(t) = \frac{\mathcal{E}}{R}(1 - e^{-t/\tau}) \qquad \text{[23.14]}$$

where the constant $\tau$ is the **time constant** of the RL circuit:

$$\tau = L/R \qquad \text{[23.15]}$$

**Figure 23.23**

A series *RL* circuit. As the current increases toward its maximum value, an emf that opposes the increasing current is induced in the inductor.

• *Time constant of the RL circuit*

**Figure 23.24**

Plot of current versus time for the *RL* circuit shown in Figure 23.23. The switch is closed at $t = 0$, and the current increases toward its maximum value $\mathcal{E}/R$. The time constant $\tau$ is the time at which *I* reaches 63.2% of its maximum value.

It is left to an exercise to show that the dimension of $\tau$ is time. Physically, $\tau$ is the time it takes the current to reach $(1 - e^{-1}) = 0.632$ of its final value $\mathcal{E}/R$.

Figure 23.24 plots current versus time, where $I = 0$ at $t = 0$. Note that the final steady-state value of the current, which occurs as $t \rightarrow \infty$, is $\mathcal{E}/R$. This can be seen by setting $dI/dt$ equal to zero in Equation 23.13 (in steady state, the change in the current is zero) and solving for the current. Thus, we see that the current rises very rapidly initially and then gradually approaches the maximum value $\mathcal{E}/R$ as $t \rightarrow \infty$. Notice that the final current does not involve *L* because the inductor has no effect on the circuit (ignoring any resistance associated with it) if the current is not changing.

Taking the first time derivative of Equation 23.14, we obtain

$$\frac{dI}{dt} = \frac{\mathcal{E}}{L} e^{-t/\tau}$$  [23.16]

From this equation we see that the rate of increase of current $dI/dt$ is a *maximum* (equal to $\mathcal{E}/L$) at $t = 0$ and falls off exponentially to zero as $t \rightarrow \infty$ (Fig. 23.25).

Now consider the *RL* circuit arranged as shown in Figure 23.26. The circuit contains two switches that operate so that when one is closed, the other is opened. Suppose that $S_1$ is closed for a long enough time to allow the current to reach its steady-state value $\mathcal{E}/R$. If $S_1$ is now opened and $S_2$ is closed at $t = 0$, we have a circuit with no battery ($\mathcal{E} = 0$). Applying Kirchhoff's loop rule to this circuit, gives

$$IR + L\frac{dI}{dt} = 0$$  [23.17]

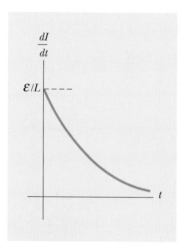

**Figure 23.25**

Plot of $dI/dt$ versus time for the *RL* circuit shown in Figure 23.23. The time rate of change of current is a maximum at $t = 0$, which is the instant at which the switch is closed. The rate decreases exponentially with time as *I* increases toward its maximum value.

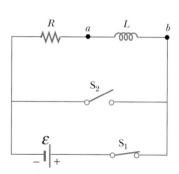

**Figure 23.26**

An *RL* circuit containing two switches. When $S_1$ is closed and $S_2$ is open as shown, the battery is in the circuit. At the instant $S_2$ is closed, $S_1$ is opened and the battery is no longer part of the circuit.

It is left to Problem 32 to show that the solution of this differential equation is

$$I(t) = \frac{\mathcal{E}}{R} e^{-t/\tau} = I_i e^{-t/\tau}$$    [23.18]

where the current at $t = 0$ is $I_i = \mathcal{E}/R$ and $\tau = L/R$.

The graph of current versus time (Fig. 23.27) for the circuit of Figure 23.26 shows that the current is continuously decreasing with time, as one would expect. Furthermore, the slope $dI/dt$ is always negative and has its maximum value at $t = 0$. The negative slope signifies that $\mathcal{E}_L = -L(dI/dt)$ is now *positive;* that is, point $a$ in Figure 23.6 is at a lower potential than point $b$.

### Quick Quiz 23.7

The circuit in Figure 23.28 includes an ac power source so that the magnetic field in the inductor is constantly changing. The inductor is a simple air-core solenoid. The switch in the circuit is closed and the lightbulb glows steadily. An iron rod is inserted into the interior of the solenoid, which increases the magnitude of the magnetic field in the solenoid. What happens to the brightness of the lightbulb?

### Quick Quiz 23.8

Two circuits like the one shown in Figure 23.26 are identical except for the value of $L$. In circuit A the inductance of the inductor is $L_A$, and in circuit B it is $L_B$. Switch $S_1$ is closed at $t = 0$ while switch $S_2$ remains open. At $t = 10$ s, switch $S_1$ is opened and switch $S_2$ is closed. The resulting graphical representation of the current as a function of time is shown in Figure 23.29. Assuming that the time constant of each circuit is much less than 10 s, which of the following is true? (a) $L_A > L_B$; (b) $L_A < L_B$; (c) not enough information to determine the relative values.

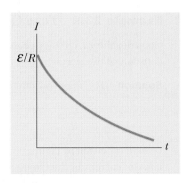

**Figure 23.27**

Current versus time for the upper loop of the circuit shown in Figure 23.26. For $t < 0$, $S_1$ is closed and $S_2$ is open. At $t = 0$, $S_2$ is closed, $S_1$ is opened, and the current has its maximum value $\mathcal{E}/R$. For $t > 0$, the current decreases exponentially.

**Figure 23.28**

(Quick Quiz 23.7) A lightbulb is powered by an ac source with an inductor in the circuit. When the iron bar is inserted into the coil, what happens to the brightness of the lightbulb?

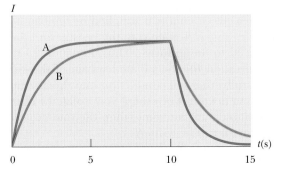

**Figure 23.29**

(Quick Quiz 23.8)

## Example 23.9 Time Constant of an *RL* Circuit

Consider the *RL* circuit in Figure 23.30a. (a) Find the time constant of the circuit.

**Solution** The time constant is given by Equation 23.15:

$$\tau = \frac{L}{R} = \frac{30.0 \times 10^{-3} \text{ H}}{6.00 \ \Omega} = \boxed{5.00 \text{ ms}}$$

(b) The switch in Figure 23.30a is closed at $t = 0$. Calculate the current in the circuit at $t = 2.00$ ms.

**Solution** Using Equation 23.14 for the current as a function of time (with $t$ and $\tau$ in milliseconds), we find that at $t = 2.00$ ms,

$$I = \frac{\mathcal{E}}{R} (1 - e^{-t/\tau}) = \frac{12.0 \text{ V}}{6.00 \ \Omega} (1 - e^{-0.400}) = \boxed{0.659 \text{ A}}$$

A plot of Equation 23.14 for this circuit is given in Figure 23.30b.

**EXERCISE** Calculate the current in the circuit and the voltage across the resistor after one time constant has elapsed.

*Answer* 1.26 A, 7.56 V

**EXERCISE** Calculate the inductance in a series *RL* circuit in which $R = 0.50 \ \Omega$ and the current increases to one-fourth its final value in 1.5 s.

*Answer* 2.6 H

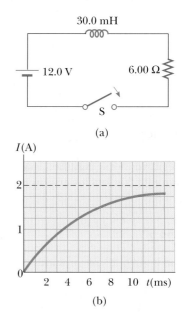

(a)

(b)

**Figure 23.30**

(Example 23.8) (a) The switch in this *RL* circuit is closed at $t = 0$. (b) A graph of the current versus time for the circuit in part (a).

---

See Screen 13.6

## 23.7 • ENERGY STORED IN A MAGNETIC FIELD

In the preceding section we found that the induced emf set up by an inductor prevents a battery from establishing an instantaneous current. Part of the energy supplied by the battery goes into internal energy in the resistor, and the remaining energy is stored in the inductor. If we multiply each term in Equation 23.13 by the current $I$ and rearrange the expression, we have

$$I\mathcal{E} = I^2 R + LI \frac{dI}{dt} \qquad \text{[23.19]}$$

This expression tells us that the rate $I\mathcal{E}$ at which energy is supplied by the battery equals the sum of the rate $I^2 R$ at which energy is delivered to the resistor and the rate $LI \, (dI/dt)$ at which energy is delivered to the inductor. Thus, Equation 23.19 is simply an expression of energy conservation for the isolated system of the circuit. (Actually, energy can leave the circuit by thermal conduction into the air and by electromagnetic radiation, so the system need not be completely isolated.) If we let $U_B$ denote the energy stored in the inductor at any time, then the rate $dU_B/dt$ at which energy is transferred to the inductor can be written

$$\frac{dU_B}{dt} = LI \frac{dI}{dt}$$

To find the total energy stored in the inductor at any instant, we can rewrite this expression as $dU_B = LI\,dI$ and integrate:

$$U_B = \int_0^{U_B} dU_B = \int_0^I LI\,dI$$

$$U_B = \tfrac{1}{2}LI^2 \qquad\qquad \textbf{[23.20]}$$

where $L$ is constant and so has been removed from the integral. Equation 23.20 represents the energy stored in the magnetic field of the inductor when the current is $I$.

Equation 23.20 is similar to the equation for the energy stored in the electric field of a capacitor, $U_E = Q^2/2C$ (Eq. 20.29). In either case, we see that energy from a battery is required to establish a field and that energy is stored in the field. In the case of the capacitor, we can conceptually relate the energy stored in the capacitor to the electric potential energy associated with the separated charge on the plates. We have not discussed a magnetic analogy to electric potential energy, so the storage of energy in an inductor is not as easy to conceptualize.

To argue that energy is stored in an inductor, consider the circuit in Figure 23.31a, which is the same as that in Figure 23.26 with the addition of switch $S_3$ across the resistor. With switches $S_1$ and $S_3$ closed as shown and $S_2$ open, a current is established in the inductor. Now, as in Figure 23.31b, switch $S_1$ is opened and $S_2$ is closed. The current persists in this (ideally) resistance-free circuit, consisting of only the inductor and a conducting path between its ends. There is no current in the resistor (because the path around it through $S_3$ is resistance free), so no energy is being delivered to it. The next step is to open switch $S_3$ as shown in Figure 23.31c. This puts the resistor into the circuit. The current in the circuit now passes through the resistor, and energy is delivered to the resistor. Where is the energy coming from? The only element in the circuit previous to opening switch $S_3$ was

• *Energy stored in an inductor*

**PITFALL PREVENTION 23.2**

**Compare energy in a capacitor, resistor, and inductor**

We have now seen three circuit elements to which we can transfer energy. Keep in mind the difference in energy transfer mechanisms. A capacitor stores a given amount of energy for a fixed charge on its plates. Further energy is delivered to the capacitor as a current in the wires connected to the capacitor delivers more charge to the plates. An inductor stores a given amount of energy if the current remains constant. Further energy is delivered to the inductor by increasing the current. A resistor acts differently because the energy is not stored as potential energy but rather is transformed to internal energy. Energy is delivered to the resistor as long as it carries a current.

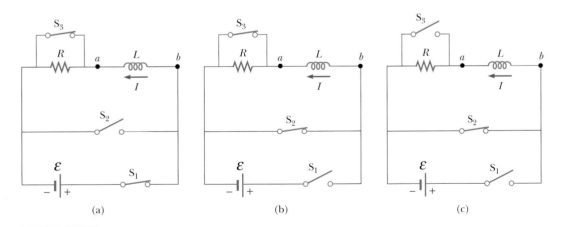

(a)  (b)  (c)

**Figure 23.31**

The $RL$ circuit from Figure 23.26, with an extra switch $S_3$, is used for conceptualizing energy storage in an inductor. (a) With the switches as shown, the battery establishes a current through the inductor. (b) Switch $S_1$ is opened and $S_2$ is closed. Because the ends of the inductor are connected by a resistance-free path, the current continues to flow through the inductor. (c) Switch $S_3$ is opened, adding the resistor to the circuit, and energy is delivered to the resistor. This energy can only have been stored in the inductor because that is the only other element in the circuit.

the inductor. Energy must therefore have been stored in the inductor and is now being delivered to the resistor.

Now let us determine the energy per unit volume, or energy density, stored in a magnetic field. For simplicity, consider a solenoid whose inductance is $L = \mu_0 n^2 A\ell$ (see Example 23.7). The magnetic field of the solenoid is $B = \mu_0 nI$. Substituting the expression for $L$ and $I = B/\mu_0 n$ into Equation 23.20 gives

$$U_B = \tfrac{1}{2} LI^2 = \tfrac{1}{2}\, \mu_0 n^2 A\ell \left( \frac{B}{\mu_0 n} \right)^2 = \frac{B^2}{2\mu_0}\, (A\ell) \qquad [23.21]$$

Because $A\ell$ is the volume of the solenoid, the energy stored per unit volume in a magnetic field—in other words, the *energy density*—is

• *Magnetic energy density*

$$u_B = \frac{U_B}{A\ell} = \frac{B^2}{2\mu_0} \qquad [23.22]$$

Although Equation 23.22 was derived for the special case of a solenoid, **it is valid for any region of space in which a magnetic field exists.** Note that it is similar to the equation for the energy per unit volume stored in an electric field, given by $\tfrac{1}{2}\epsilon_0 E^2$ (Eq. 20.31). In both cases, the energy density is proportional to the square of the magnitude of the field.

### Quick Quiz 23.9

You are performing an experiment in which you need the highest possible energy density in the interior of a very long solenoid. Which of the following would increase the energy density? (a) Increasing only the number of turns per unit length on the solenoid. (b) Increasing only the cross-sectional area of the solenoid. (c) Increasing only the length of the solenoid. (d) Increasing only the current in the wires in the solenoid.

### Example 23.10    What Happens to the Energy in the Inductor?

Consider once again the *RL* circuit shown in Figure 23.26, in which switch $S_2$ is closed at the instant $S_1$ is opened (at $t = 0$). Recall that the current in the upper loop decays exponentially with time according to the expression $I = I_i e^{-t/\tau}$, where $I_i = \mathcal{E}/R$ is the initial current in the circuit and $\tau = L/R$ is the time constant. Let us show explicitly that all the energy stored in the magnetic field of the inductor is transferred to the resistor.

**Solution**    The rate at which energy is transferred to the resistor is $I^2 R$, where $I$ is the instantaneous current. Using $I$ from Equation 23.18,

$$\mathcal{P} = I^2 R = (I_i e^{-Rt/L})^2 R = I_i^2 R e^{-2Rt/L}$$

To find the total energy transferred to the resistor, we integrate this expression over the limits $t = 0$ to $t = \infty$ (because it takes an infinite time for the current to reach zero):

$$(1) \qquad E = \int_0^\infty \mathcal{P}\, dt = \int_0^\infty I_i^2 R e^{-2Rt/L}\, dt = I_i^2 R \int_0^\infty e^{-2Rt/L}\, dt$$

The value of the definite integral is $L/2R$, and so Equation (1) becomes

$$E = I_i^2 R \left( \frac{L}{2R} \right) = \tfrac{1}{2} LI_i^2$$

Note that this is equal to the initial energy stored in the magnetic field of the inductor, given by Equation 23.20, as we set out to prove.

**EXERCISE**    Show that the integral on the right-hand side of Equation (1) has the value $L/2R$.

## Example 23.11 The Coaxial Cable

A long coaxial cable consists of two concentric cylindrical conductors of radii $a$ and $b$ and length $\ell$ as in Figure 23.32. The inner conductor is assumed to be a thin cylindrical shell. The conductors carry current $I$ in opposite directions.
(a) Calculate the self-inductance $L$ of this cable.

**Solution** To obtain $L$, we must know the magnetic flux through any cross-section between the two conductors. From Ampère's law (Equation 22.29) we know that the magnetic field between the conductors is $B = \mu_0 I/2\pi r$. The magnetic field is zero outside the conductors ($r > b$) because the net current through a circular path surrounding both wires is zero, and hence from Ampère's law, $\oint \mathbf{B} \cdot d\mathbf{s} = 0$. The magnetic field is zero inside the inner conductor because it is hollow and no current flows within a radius $r < a$.

The magnetic field is perpendicular to the light blue rectangular strip of length $\ell$ and width $(b - a)$, the cross-section of interest. Dividing this rectangle into strips of width $dr$, such as the dark blue strip in Figure 23.32, we see that the area of each strip is $\ell\, dr$ and the flux through each strip is $B\, dA = B\ell\, dr$. Hence, the total magnetic flux through the entire cross-section is

$$\Phi_B = \int B\, dA = \int_a^b \frac{\mu_0 I}{2\pi r}\, \ell\, dr = \frac{\mu_0 I\ell}{2\pi} \int_a^b \frac{dr}{r}$$

$$= \frac{\mu_0 I\ell}{2\pi} \ln\left(\frac{b}{a}\right)$$

Using this result, we find that the self-inductance of the cable is

$$L = \frac{\Phi_B}{I} = \frac{\mu_0 \ell}{2\pi} \ln\left(\frac{b}{a}\right)$$

(b) Calculate the total energy stored in the magnetic field of the cable.

**Solution** Using Equation 23.20 and the results to part (a) gives

$$U_B = \tfrac{1}{2}LI^2 = \frac{\mu_0 \ell I^2}{4\pi} \ln\left(\frac{b}{a}\right)$$

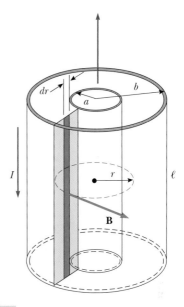

**Figure 23.32**

(Example 23.11) Section of a long coaxial cable. The inner and outer conductors carry equal currents in opposite directions.

**EXERCISE** A 10.0-V battery, a 5.00-$\Omega$ resistor, and a 10.0-H inductor are connected in series. After the current in the circuit has reached its maximum value, calculate (a) the power supplied by the battery, (b) the power transferred to the resistor, (c) the power transferred to the inductor, and (d) the energy stored in the magnetic field of the inductor.

*Answer* (a) 20.0 W (b) 20.0 W (c) 0 (d) 20.0 J

### context connection
## 23.8 • THE REPULSIVE MODEL FOR MAGNETIC LEVITATION

In the preceding chapter, we considered a model for magnetic levitation that is based on the attractive force between a magnet and a rail made of magnetic material. The second major model for magnetically levitated vehicles is the EDS *(electrodynamic system)* model. The EDS model is used in Japan Railways's system and the

Levitation and
guidance coils

Guidance wheels

Take-off and
landing wheels

Magnets

**Figure 23.33**

Schematic diagram of the levitation system for the Japan Railways's vehicles. The magnets induce currents in the coils at the side of the track, so that, by Lenz's law, the repulsive force pushes the vehicle upward.

Magneplane, developed in response to the National Maglev Initiative by Magneplane International in Massachusetts.

In this model, we appeal to Lenz's law. In the simplest form of the model, the magnetic vehicle carries a magnet. As the magnet passes over a metal plate or coils of wire, currents are induced in the plate or coils that tend to oppose the original change. This results in a *repulsive* force, which lifts the vehicle.

Although the idea of inducing a current in a metal plate is a valid concept, it represents a large expense in terms of the amount of metal required for a long track. Another technique is used in Japan Railways's maglev vehicle. In this vehicle, the current is induced by magnets passing by coils located on the side of the railway channel. A schematic illustration of such a vehicle is given in Figure 23.33.

One of the disadvantages of the EMS model discussed in the preceding chapter is the instability of the attractive force, requiring feedback electronics. The EDS model, however, has a stabilizing feature—if the vehicle drops, the repulsion becomes stronger and pushes the vehicle back up. If the vehicle rises, the force decreases and the vehicle drops back down. Another advantage of the EDS system is the larger separation between track and vehicle—about 10 cm, as opposed to 10 mm in the EMS model.

A disadvantage of the EDS system is that *levitation only exists while the vehicle is moving*, because it depends on Faraday's law—a magnetic *change* must occur. Thus, the vehicle must have landing wheels for stopping and starting at stations; these wheels are indicated in Figure 23.33.

Another disadvantage of the EDS system is that the induced currents result in a *drag* force as well as a lift force. This requires more power for propulsion. The drag force is larger than the lift force for small speeds, but the drag force maximizes at some speed and then begins to decrease. The lift force continues to in-

**WEB**

For information on the magnetic levitation vehicles of Japan Railways, visit **www.rtri.or.jp/rd/ maglev/html/english/maglev_ frame_E.html** crease as the speed increases. Thus, it is advantageous to travel at high speeds, but the significant drag force at low speeds must be overcome every time the vehicle starts up.

Japan Railways has performed exhaustive tests on five generations of maglev vehicles, beginning with the four-seater ML100, built in 1972 to celebrate Japan Railways's 100th anniversary. Current tests are being performed on the sixth-generation vehicle, the MLX01, a multicar train that can carry over 100 passengers in its commercial form. A 43-km test line between Sakaigawa and Akiyama in Yamanashi Prefecture was opened in 1997. A five-car MLX01 attained a speed of 552 km/h on this line in 1999, with humans on board. The Yamanashi Test Line is funded by the Japanese government, with the intention of final confirmation of maglev feasibility and commercial operation early in the first decade of the 21st century.

## SUMMARY

**Faraday's law of induction** states that the emf induced in a circuit is directly proportional to the time rate of change of magnetic flux through the circuit:

$$\mathcal{E} = -N\frac{d\Phi_B}{dt} \qquad [23.3]$$

where $N$ is the number of turns and $\Phi_B$ is the magnetic flux through each, given by

$$\Phi_B = \int \mathbf{B} \cdot d\mathbf{A} \qquad [23.1]$$

When a conducting bar of length $\ell$ moves through a magnetic field $\mathbf{B}$ with a velocity $\mathbf{v}$ so that $\mathbf{v}$ is perpendicular to $\mathbf{B}$, the emf induced in the bar (called the **motional emf**) is

$$\mathcal{E} = -B\ell v \qquad [23.5]$$

**Lenz's law** states that the induced current and induced emf in a conductor are in such a direction as to oppose the change that produced them.

A general form of **Faraday's law of induction** is

$$\mathcal{E} = \oint \mathbf{E} \cdot d\mathbf{s} = -\frac{d\Phi_B}{dt} \qquad [23.9]$$

where $\mathbf{E}$ is a nonconservative electric field produced by the changing magnetic flux.

When the current in a coil changes with time, an emf is induced in the coil according to Faraday's law. The **self-induced emf** is described by the expression

$$\mathcal{E}_L = -L\frac{dI}{dt} \qquad [23.10]$$

where $L$ is the *inductance* of the coil. Inductance is a measure of the opposition of a device to a change in current. Inductance has the SI unit the **henry** (H), where 1 H = 1 V·s/A.

The **inductance** of a coil is

$$L = \frac{N\Phi_B}{I} \qquad [23.11]$$

where $\Phi_B$ is the magnetic flux through the coil and $N$ is the total number of turns.

If a resistor and inductor are connected in series to a battery of emf $\mathcal{E}$, as shown in Figure 23.23, and a switch in the circuit is closed at $t = 0$, the current in the circuit varies with time according to the expression

$$I(t) = \frac{\mathcal{E}}{R}(1 - e^{-Rt/L}) = \frac{\mathcal{E}}{R}(1 - e^{-t/\tau}) \qquad [23.14]$$

where $\tau = L/R$ is the **time constant** of the $RL$ circuit.

If the battery is removed from an $RL$ circuit, as in Figure 23.26 with $S_1$ open and $S_2$ closed, the current decays exponentially with time according to the expression

$$I(t) = \frac{\mathcal{E}}{R}e^{-t/\tau} \qquad [23.18]$$

where $\mathcal{E}/R$ is the initial current in the circuit.

The **energy** stored in the magnetic field of an inductor carrying a current $I$ is

$$U_B = \tfrac{1}{2}LI^2 \qquad [23.20]$$

The **energy per unit volume** (or energy density) at a point where the magnetic field is $B$ is

$$u_B = \frac{B^2}{2\mu_0} \qquad [23.22]$$

$$u_E = \frac{\varepsilon_0 E^2}{2}$$

## QUESTIONS

1. A loop of wire is placed in a uniform magnetic field. For what orientation of the loop is the magnetic flux a maximum? For what orientation is the flux zero?

2. As the conducting bar in Figure Q23.2 moves to the right, an electric field is set up directed downward. If the bar were moving to the left, explain why the electric field would be upward.

**Figure Q23.2**

3. As the bar in Figure Q23.2 moves perpendicular to the field, is an external force required to keep it moving with constant speed?

4. Magnetic storms on the Sun can cause difficulties with communications here on the Earth. Why do these sunspots affect us in this way?

5. Wearing a metal bracelet in a region of strong magnetic field could be hazardous. Discuss.

6. How is electric energy produced in dams (i.e., how is the energy of motion of the water converted to ac electricity)?

7. A piece of aluminum is dropped vertically downward between the poles of an electromagnet. Does the magnetic field affect the velocity of the aluminum?

8. The bar in Figure Q23.8 moves on rails to the right with a velocity **v**, and the uniform, constant magnetic field is directed out of the page. Why is the induced current clock-

9. In a beam balance scale, an aluminum plate is sometimes used to slow the oscillations of the beam near equilibrium. The plate is mounted at the end of the beam and moves between the poles of a small horseshoe magnet attached to the frame. Why are the oscillations of the beam strongly damped near equilibrium?

wise? If the bar were moving to the left, what would be the direction of the induced current?

10. When the switch in Figure Q23.10a is closed, a current is set up in the coil and the metal ring springs upward (Fig. Q23.10b). Explain this behavior.

Iron core
Metal ring

(a)                                    (b)

**Figure Q23.10**

11. Assume that the battery in Figure Q23.10a is replaced by an ac source and the switch is held closed. If held down, the metal ring on top of the solenoid becomes hot. Why?

12. What is the inductance of two inductors connected in series?

13. A bar magnet is held above a loop of wire in a horizontal plane, as Figure Q23.13. The south end of the magnet is

**Figure Q23.8**

**Figure Q23.13**

toward the loop of wire. The magnet is dropped toward the loop. Find the direction of the current through the resistor (a) while the magnet is falling toward the loop and (b) after the magnet has passed through the loop and moves away from it.

14. Find the direction of the current in the resistor in Figure Q23.14 (a) at the instant the switch is closed, (b) after the switch has been closed for several minutes, and (c) at the instant the switch is opened.

**Figure Q23.14**

15. Discuss the similarities between the energy stored in the electric field of a charged capacitor and the energy stored in the magnetic field of a current-carrying coil.

16. If the current in an inductor is doubled, by what factor does the stored energy change?

17. Suppose the switch in Figure 23.23 has been closed for a long time and is suddenly opened. Does the current in-stantaneously drop to zero? Why does a spark appear at the switch contacts at the instant the switch is thrown open?

18. Thinking Physics 23.3 describes the emf induced between the wingtips of an airplane by its motion in the Earth's magnetic field. Can this emf be used to power a light in the passenger compartment? Explain your answer.

19. Can an object exert a force on itself? Can a circuit induce an emf in itself? Explain your answers.

20. Consider this thesis: "Joseph Henry, America's first professional physicist, caused the last basic change in the human view of the Universe when he discovered self-induction during a school vacation at the Albany Academy about 1830. Before that time, one could think of the Universe as composed of just one thing: matter. The energy that temporarily maintains the current after a battery is removed from a coil, on the other hand, is not energy that belongs to any chunk of matter. It is energy in the massless magnetic field surrounding the coil. With Henry's discovery, Nature forced us to admit that the Universe consists of fields as well as matter." Argue for or against the thesis. What in your view comprises the Universe?

# PROBLEMS

1, 2, 3 = straightforward, intermediate, challenging     □ = full solution available in the *Student Solutions Manual and Study Guide*

**web** = solution posted at **http://www.harcourtcollege.com/physics/**     🖥 = computer useful in solving problem

🧩 = Interactive Physics     ▥ = paired numerical/symbolic problems     🔬 = life science application

### Section 23.1 Faraday's Law of Induction

### Section 23.3 Lenz's Law

1. A flat loop of wire consisting of a single turn of cross-sectional area 8.00 cm² is perpendicular to a magnetic field that increases uniformly in magnitude from 0.500 T to 2.50 T in 1.00 s. What is the resulting induced current if the loop has a resistance of 2.00 Ω ?

2. A 25-turn circular coil of wire has a diameter of 1.00 m. It is placed with its axis along the direction of the Earth's magnetic field of 50.0 $\mu$T, and then in 0.200 s it is flipped 180°. An average emf of what magnitude is generated in the coil?

3. A strong electromagnet produces a uniform magnetic **web** field having a magnitude of 1.60 T over a cross-sectional area of 0.200 m². A coil having 200 turns and a total resistance of 20.0 Ω is placed around the electromagnet. The current is then smoothly decreased in the electromagnet until it reaches zero in 20.0 ms. What is the current induced in the coil?

4. An aluminum ring of radius $r_1$ and resistance $R$ is placed around the top of a long air-core solenoid with $n$ turns per meter and smaller radius $r_2$ as in Figure P23.4. Assume that the axial component of the field produced by the solenoid over the area of the end of the solenoid is half as strong as at the center of the solenoid. Assume that the solenoid produces negligible field outside its cross-sectional area. The current in the solenoid is increasing at a rate of $\Delta I/\Delta t$. (a) What is the induced current in the ring? (b) At the center of the ring, what is the magnetic field produced by the induced current in the ring? (c) What is the direction of this field ?

**Figure P23.4**

5. A rectangular loop of area $A$ is placed in a region where the magnetic field is perpendicular to the plane of the loop. The magnitude of the field is allowed to vary with time according to $B = B_{max}e^{-t/\tau}$, where $B_{max}$ and $\tau$ are constants. The field has the constant value $B_{max}$ for $t < 0$. (a) Use Faraday's law to show that the emf induced in the loop is given by

$$\mathcal{E} = \frac{AB_{max}}{\tau}e^{-t/\tau}$$

(b) Obtain a numerical value for $\mathcal{E}$ at $t = 4.00$ s when $A = 0.160$ m$^2$, $B_{max} = 0.350$ T, and $\tau = 2.00$ s. (c) For the values of $A$, $B_{max}$, and $\tau$ given in (b), what is the maximum value of $\mathcal{E}$?

6. (a) A loop of wire in the shape of a rectangle of width $w$ and length $L$ and a long, straight wire carrying a current $I$ lie on a tabletop as shown in Figure P23.6. (a) Determine the magnetic flux through the loop due to the current $I$. (b) Suppose the current is changing with time according

**Figure P23.6**  Problems 6 and 55.

to $I = a + bt$, where $a$ and $b$ are constants. Determine the induced emf in the loop if $b = 10.0$ A/s, $h = 1.00$ cm, $w = 10.0$ cm, and $L = 100$ cm. What is the direction of the induced current in the rectangle?

7. A 30-turn circular coil of radius 4.00 cm and resistance 1.00 $\Omega$ is placed in a magnetic field directed perpendicular to the plane of the coil. The magnitude of the magnetic field varies with time according to the expression $B = 0.010\ 0t + 0.040\ 0t^2$, where $t$ is in seconds and $B$ is in teslas. Calculate the induced emf in the coil at $t = 5.00$ s.

8. A *transformer* is used to move power from one ac electric circuit to another, changing the current and voltage as it does so. A particular transformer consists of a coil of 15 turns and radius 10.0 cm surrounding a long solenoid of radius 2.00 cm and $1.00 \times 10^3$ turns/m (Fig. P23.8). If the current in the solenoid changes as $I = (5.00$ A$) \sin(120t)$, find the induced emf in the 15-turn coil as a function of time.

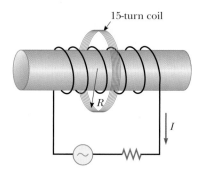

**Figure P23.8**

## Section 23.2  Motional emf

## Section 23.3  Lenz's Law

9. Use Lenz's law to answer the following questions concerning the direction of induced currents. (a) What is the direction of the induced current in resistor $R$ in Figure P23.9a when the bar magnet is moved to the left? (b) What is the direction of the current induced in the resistor $R$ immediately after the switch S in Figure P23.9b is closed? (c) What is the direction of the induced current in $R$ when the current $I$ in Figure P23.9c decreases rapidly to zero? (d) A copper bar is moved to the right while its axis is maintained in a direction perpendicular to a magnetic field, as shown in Figure P23.9d. If the top of the bar becomes positive relative to the bottom, what is the direction of the magnetic field?

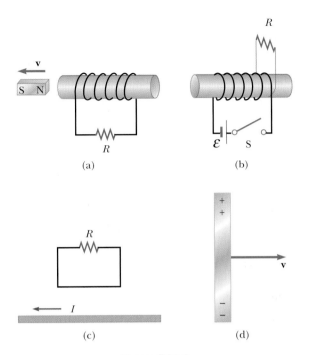

(a)

(b)

(c)

(d)

**Figure P23.9**

**10.** Consider the arrangement shown in Figure P23.10. Assume that $R = 6.00\ \Omega$, $\ell = 1.20$ m, and a uniform 2.50-T magnetic field is directed into the page. At what speed should the bar be moved to produce a current of 0.500 A in the resistor?

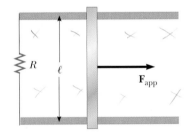

**Figure P23.10**   Problems 10, 11, and 12.

**11.** Figure P23.10 shows a top view of a bar that can slide without friction. The resistor is 6.00 $\Omega$ and a 2.50-T magnetic field is directed perpendicularly downward, into the paper. Let $\ell = 1.20$ m. (a) Calculate the applied force required to move the bar to the right at a constant speed of 2.00 m/s. (b) At what rate is energy delivered to the resistor?

$F_{app} = I\,d\ell \times B$

$I = \frac{\varepsilon}{R} = \frac{B\ell v}{R}$

$P = (F_{app})(v)$

**12.** A conducting rod of length $\ell$ moves on two horizontal, frictionless rails, as shown in Figure P23.10. If a constant force of 1.00 N moves the bar at 2.00 m/s through a magnetic field **B** that is directed into the page, (a) what is the current in the 8.00-$\Omega$ resistor $R$? (b) At what rate is energy

delivered to the resistor? (c) What is the mechanical power delivered by the force $\mathbf{F}_{app}$?

**13.** An automobile has a vertical radio antenna 1.20 m long. The automobile travels at 65.0 km/h on a horizontal road where the Earth's magnetic field is 50.0 $\mu$T directed toward the north and downward at an angle of 65.0° below the horizontal. (a) Specify the direction that the automobile should move in order to generate the maximum motional emf in the antenna, with the top of the antenna positive relative to the bottom. (b) Calculate the magnitude of this induced emf.

**14.** A helicopter has blades with a length of 3.00 m, extending outward from a central hub and rotating at 2.00 rev/s. If the vertical component of the Earth's magnetic field is 50.0 $\mu$T, what is the emf induced between the blade tip and the center hub?

**15.** When a wire carries an ac current with a known frequency, you can use a *Rogowski coil* to determine the amplitude $I_{max}$ of the current without disconnecting the wire to shunt the current though a meter. The Rogowski coil, shown in Figure P23.15, simply clips around the wire. It consists of a toroidal conductor wrapped around a circular return cord. The toroid has $n$ turns per unit distance, its cross-sectional area is $A$, and the current to be measured is given by $I(t) = I_{max} \sin \omega t$. (a) Show that the amplitude of the emf induced in the coil is $\mathcal{E}_{max} = \mu_0 n A \omega I_{max}$. (b) Explain why the wire carrying the current need not be at the center of the Rogowski coil and why the coil will not respond to nearby currents that it does not enclose.

**Figure P23.15**

**16.** A rectangular coil with resistance $R$ has $N$ turns, each of length $\ell$ and width $w$ as shown in Figure P23.16. The coil moves into a uniform magnetic field **B** with constant velocity **v**. What are the magnitude and direction of the total magnetic force on the coil (a) as it enters the magnetic field, (b) as it moves within the field, and (c) as it leaves the field?

**Figure P23.16**

**17.** A conducting rectangular loop of mass $M$, resistance $R$,
web   and dimensions $w$ by $\ell$ falls from rest into a magnetic field
**B** as in Figure P23.17. While only the bottom edge of the
loop is immersed in the field, the loop approaches terminal speed $v_T$. (a) Show that

$$v_T = \frac{MgR}{B^2 w^2}$$

(b) Why is $v_T$ proportional to $R$? (c) Why is it inversely
proportional to $B^2$?

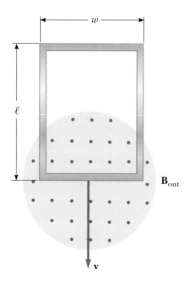

**Figure P23.17**

**18.** The *homopolar generator*, also called the *Faraday disk*, is a
low-voltage, high-current electric generator. It consists of a
rotating conducting disk with one stationary brush (a sliding electric contact) at its axle and another at a point on
its circumference, as in Figure P23.18. A magnetic field is
applied perpendicular to the plane of the disk. Assume
the field is 0.900 T, the angular speed is 3 200 rev/min,
and the radius of the disk is 0.400 m. Find the emf generated between the brushes. When superconducting coils
are used to produce a large magnetic field, a homopolar
generator can have a power output of several megawatts.
Such a generator is useful, for example, in purifying metals by electrolysis. If a voltage is applied to the output ter-

minals of the generator, it runs in reverse as a *homopolar
motor* capable of providing great torque, useful in ship
propulsion.

**Figure P23.18**

**19.** A coil of area 0.100 m$^2$ is rotating at 60.0 rev/s with the
web   axis of rotation perpendicular to a 0.200-T magnetic field.
(a) If the coil has 1 000 turns, what is the maximum voltage generated in it? (b) What is the orientation of the coil
with respect to the magnetic field when the maximum induced voltage occurs?

**20.** A long solenoid, with its axis along the $x$ axis, consists of
200 turns/m of wire and carries a steady current of 15.0
A. A coil is formed by wrapping 30 turns of thin wire
around a circular frame that has a radius of 8.00 cm. The
coil is placed inside the solenoid and mounted on an axis
that is a diameter of the coil and coincides with the $y$ axis.
The coil is then rotated with an angular speed of $4.00\pi$
rad/s. (The plane of the coil is in the $yz$ plane at $t = 0$.)
Determine the emf generated in the coil as a function of
time.

**Section 23.4   Induced emfs and Electric Fields**

**21.** A magnetic field directed into the page changes with time
according to the expression $B = (0.030\ 0t^2 + 1.40)$ T,
where $t$ is in seconds. The field has a circular cross-section
of radius $R = 2.50$ cm (Fig. P23.21). What are the magni-

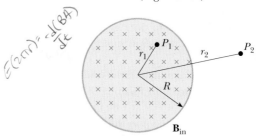

**Figure P23.21**   Problems 21 and 22.

tude and direction of the electric field at point $P_1$ when $t = 3.00$ s and $r_1 = 0.020\ 0$ m?

22. For the situation shown in Figure P23.21, the magnetic field in the circular region changes with time according to the expression $B = (2.00t^3 - 4.00t^2 + 0.800)$ T, and $r_2 = 2R = 5.00$ cm. (a) Calculate the magnitude and direction of the force exerted on an electron located at point $P_2$ when $t = 2.00$ s. (b) At what time is this force equal to zero?

## Section 23.5   Self-Inductance

23. A coil has an inductance of 3.00 mH, and the current through it changes from 0.200 A to 1.50 A in a time of 0.200 s. Find the magnitude of the average induced emf in the coil during this time.

24. A coiled telephone cord forms a spiral with 70 turns, a diameter of 1.30 cm, and an unstretched length of 60.0 cm. Determine the self-inductance of one conductor in the unstretched cord.

25. A 10.0-mH inductor carries a current $I = I_{max} \sin \omega t$, with $I_{max} = 5.00$ A and $\omega/2\pi = 60.0$ Hz. What is the back emf as a function of time?
web
$$\mathcal{E} = L\frac{dI}{dt}$$

26. A 2.00-H inductor carries a steady current of 0.500 A. When the switch in the circuit is opened, the current is effectively zero after 10.0 ms. What is the average induced emf in the inductor during this time?

27. An emf of 24.0 mV is induced in a 500-turn coil at an instant when the current is 4.00 A and is changing at the rate of 10.0 A/s. What is the magnetic flux through each turn of the coil?
$$\mathcal{E} = N\frac{d\Phi}{dt}$$ 
$$\frac{}{I}\frac{d}{dt}$$

28. The current in a 90.0-mH inductor changes with time as $I = t^2 - 6.00t$ (in SI units). Find the magnitude of the induced emf at (a) $t = 1.00$ s and (b) $t = 4.00$ s. (c) At what time is the emf zero?

29. A toroid has a major radius $R$ and a minor radius $r$ and is tightly wound with $N$ turns of wire, as shown in Figure P23.29. If $R \gg r$, the magnetic field within the region of the torus, of cross-sectional area $A = \pi r^2$, is essentially that of a long solenoid that has been bent into a large circle of radius $R$. Modeling the field as the uniform field of

a long solenoid, show that the self-inductance of such a toroid is approximately

$$L \approx \frac{\mu_0 N^2 A}{2\pi R}$$

(An exact expression of the inductance of a toroid with a rectangular cross-section is derived in Problem 56.)

## Section 23.6   *RL* Circuits

30. Calculate the resistance in an *RL* circuit in which the current increases to 90.0% of its final value in 3.00 s and $L = 2.50$ H.

31. A 12.0-V battery is connected into a series circuit containing a 10.0-$\Omega$ resistor and a 2.00-H inductor. How long will it take the current to reach (a) 50.0% and (b) 90.0% of its final value?
$$I = \frac{\mathcal{E}}{R}(1 - e^{-tR/L})$$

32. Show that $I = I_i e^{-t/\tau}$ is a solution of the differential equation

$$IR + L\frac{dI}{dt} = 0$$

where $\tau = L/R$ and $I_i$ is the current at $t = 0$.

33. Consider the circuit in Figure P23.33, taking $\mathcal{E} = 6.00$ V, $L = 8.00$ mH, and $R = 4.00\ \Omega$. (a) What is the inductive time constant of the circuit? (b) Calculate the current in the circuit 250 $\mu$s after the switch is closed. (c) What is the value of the final steady-state current? (d) How long does it take the current to reach 80.0% of its maximum value?

**Figure P23.33**   Problems 33, 34, and 35

34. When the switch in Figure P23.33 is closed, the current takes 3.00 ms to reach 98.0% of its final value. If $R = 10.0\ \Omega$, what is the inductance ?

35. For the *RL* circuit shown in Figure P23.33, let the inductance be 3.00 H, the resistance 8.00 $\Omega$, and the battery emf 36.0 V. (a) Calculate the ratio of the voltage across the resistor to that across the inductor when the current is 2.00 A. (b) Calculate the voltage across the inductor when the current is 4.50 A.

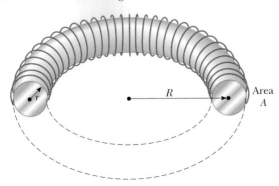

**Figure P23.29**

**36.** The switch in Figure P23.36 is closed at time $t = 0$. Find the current in the inductor and the current through the switch as functions of time thereafter.

For solenoid
$L = \mu_0 N^2 \dfrac{A}{\ell}$
$U_B = \frac{1}{2} L I^2$

**Figure P23.36**

**37.** A 140-mH inductor and a 4.90-$\Omega$ resistor are connected with a switch to a 6.00-V battery as shown in Figure P23.37. (a) If the switch is thrown to the left (connecting the battery), how much time elapses before the current reaches 220 mA? (b) What is the current in the inductor 10.0 s after the switch is closed? (c) Now the switch is quickly thrown from A to B. How much time elapses before the current falls to 160 mA?

web

**Figure P23.37**

**38.** One application of an *RL* circuit is the generation of time-varying high voltage from a low-voltage source, as shown in Figure P23.38. (a) What is the current in the circuit a long time after the switch has been in position A? (b) Now the switch is thrown quickly from A to B. Compute the initial voltage across each resistor and across the inductor.

**Figure P23.38**

**(c)** How much time elapses before the voltage across the inductor drops to 12.0 V?

### Section 23.7   Energy in a Magnetic Field

**39.** An air-core solenoid with 68 turns is 8.00 cm long and has a diameter of 1.20 cm. How much energy is stored in its magnetic field when it carries a current of 0.770 A?

**40.** The magnetic field inside a superconducting solenoid is 4.50 T. The solenoid has an inner diameter of 6.20 cm and a length of 26.0 cm. Determine (a) the magnetic energy density in the field and (b) the energy stored in the magnetic field within the solenoid.

**41.** On a clear day at a certain location, a 100-V/m vertical electric field exists near the Earth's surface. At the same place, the Earth's magnetic field has a magnitude of $0.500 \times 10^{-4}$ T. Compute the energy densities of the two fields.

**42.** An *RL* circuit in which $L = 4.00$ H and $R = 5.00$ $\Omega$ is connected to a 22.0-V battery at $t = 0$. (a) What energy is stored in the inductor when the current is 0.500 A? (b) At what rate is energy being stored in the inductor when $I = 1.00$ A? (c) What power is being delivered to the circuit by the battery when $I = 0.500$ A?

### Section 23.8   Context Connection — The Repulsive Model for Magnetic Levitation

**43.** The following represents a crude model for levitating a commercial transportation vehicle using Faraday's law. Suppose magnets are used to establish regions of magnetic field across the track as in Figure P23.43. Rectangular loops of wire are mounted on the vehicle so that the lower 20 cm of each loop passes into these regions of mag-

**Figure P23.43**

netic field. The upper portion of the loop contains a 25-$\Omega$ resistor. As the leading edge of a loop passes into the magnetic field, a current is induced in the loop as in Figure P23.43, and the magnetic force on this current in the bottom edge of the loop of length 10 cm results in an upward force on the vehicle. (By means of sophisticated electronics, a switch is opened in the loop before the trailing edge of the loop enters the region of magnetic field, so that a current is not induced in the opposite direction to apply a downward force on the vehicle.) The vehicle has a mass of $5 \times 10^4$ kg and travels at a speed of 400 km/h. If the vehicle has 100 loops carrying current at any moment, what is the approximate magnitude of the magnetic field required to levitate the vehicle? Assume that the magnetic force acts on the entire 10-cm length of the horizontal wire. Your answer should suggest that this crude model would not be successful as a means of magnetic levitation.

44. **Review Problem.** *The Meissner effect.* Compare this problem with Problem 66 in Chapter 20, on the force attracting a perfect dielectric into a strong electric field. A fundamental property of a type I superconducting material is *perfect diamagnetism,* or demonstration of the *Meissner effect,* illustrated in the photograph of the levitating magnet on page 767 and described as follows: The superconducting material has **B** = 0 everywhere inside it. If a sample of the material is placed into an externally produced magnetic field, or if it is cooled to become superconducting while it is in a magnetic field, electric currents appear on the surface of the sample. The currents have precisely the strength and orientation required to make the total magnetic field zero throughout the interior of the sample. The following problem will help you understand the magnetic force that can act on the superconducting sample.

A vertical solenoid having a length of 120 cm and a diameter of 2.50 cm consists of 1 400 turns of copper wire carrying a counterclockwise current of 2.00 A as in Figure P23.44a. (a) Find the magnetic field in the vacuum inside the solenoid. (b) Find the energy density of the magnetic field. (Note that the units J/m$^3$ of energy density are the same as the units N/m$^2$ = Pa of pressure.) (c) A superconducting bar 2.20 cm in diameter is inserted partway into the solenoid. Its upper end is far outside the solenoid, where the magnetic field is negligible. The lower end of the bar is deep inside the solenoid. Identify the direction required for the current on the curved surface of the bar, so that the total magnetic field is zero within the bar. The field created by the supercurrents is sketched in Figure P23.44b, and the total field is sketched in Figure P23.44c. (d) The field of the solenoid exerts a force on the current in the superconductor. Identify the direction of the force on the bar. (e) Calculate the magnitude of the

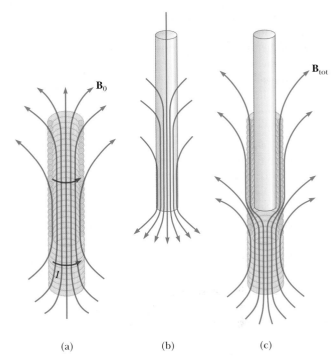

**Figure P23.44**

force by multiplying the energy density of the solenoid field by the area of the bottom end of the superconducting bar.

### Additional Problems

45. A steel guitar string vibrates (see Figure 23.7). The component of the magnetic field perpendicular to the area of a pickup coil nearby is given by

$$B = 50.0 \text{ mT} + (3.20 \text{ mT}) \sin(2\pi 523t).$$

The circular pickup coil has 30 turns and a radius of 2.70 mm. Find the emf induced in the coil as a function of time.

46. Figure P23.46 is a graph of the induced emf versus time for a coil of $N$ turns rotating with angular velocity $\omega$ in a uniform magnetic field directed perpendicular to the axis

**Figure P23.46**

of rotation of the coil. Copy this graph (on a larger scale), and on the same set of axes show the graph of emf versus $t$ (a) if the number of turns in the coil is doubled; (b) if instead the angular velocity is doubled; and (c) if the angular velocity is doubled while the number of turns in the coil is halved.

47. The magnetic flux through a metal ring varies with time $t$ according to $\Phi_B = 3(at^3 - bt^2)$ T·m², with $a = 2.00$ s⁻³ and $b = 6.00$ s⁻². The resistance of the ring is 3.00 Ω. Determine the maximum current induced in the ring during the interval from $t = 0$ to $t = 2.00$ s.

48. **Review Problem.** A particle of mass $2.00 \times 10^{-16}$ kg and charge 30.0 nC starts from rest, is accelerated by a strong electric field, and is fired from a small source inside a region of uniform constant magnetic field 0.600 T. The velocity of the particle is perpendicular to the magnetic field. The circular orbit of the particle encloses a magnetic flux of 15.0 μWb. (a) Calculate the speed of the particle. (b) Calculate the potential difference through which the particle accelerated inside the source.

49. A bar of mass $m$, length $d$, and resistance $R$ slides without friction in a horizontal plane, moving on parallel rails as shown in Figure P23.49. A battery that maintains a constant emf $\mathcal{E}$ is connected between the rails, and a constant magnetic field **B** is directed perpendicularly to the plane of the page. If the bar starts from rest at $t = 0$, show that at time $t$ it moves with a speed

$$v = \frac{\mathcal{E}}{Bd}(1 - e^{-B^2 d^2 t / mR})$$

**Figure P23.49**

50. An *induction furnace* uses electromagnetic induction to set up eddy currents in a conductor, thereby heating the conductor. Commercial units operate at frequencies ranging from 60 Hz to 1 MHz and deliver powers from a few watts to several megawatts. Induction heating can be used for welding in a vacuum chamber, to avoid oxidation or contamination of the metal. At high frequencies, induced currents appear only near the surface of the conductor—this is the "skin effect." By creating an induced current for a short time at an appropriately high frequency, a sample can be heated down to a controlled depth. For example, the surface of a farm tiller can be tempered to make it hard and brittle for effective cutting while keeping metal's interior soft and ductile to resist breakage.

To explore induction heating, consider a flat conducting disk of radius $R$, thickness $b$, and resistivity $\rho$. A magnetic field $B_{max} \cos \omega t$ is applied perpendicular to the disk. Assume that the frequency is so low that the skin effect is not important. Assume the eddy currents flow in circles concentric with the disk. (a) Calculate the average power delivered to the disk. (b) By what factor does the power change when the amplitude of the field doubles? (c) When the frequency doubles? (d) When the radius of the disk doubles?

51. The plane of a square loop of wire with edge length $a = 0.200$ m is perpendicular to the Earth's magnetic field at a point where $B = 15.0$ μT, as in Figure P23.51. The total resistance of the loop and the wires connecting it to the galvanometer is 0.500 Ω. If the loop is suddenly collapsed by horizontal forces as shown, what total charge passes through the galvanometer?

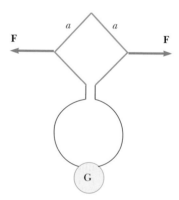

**Figure P23.51**

52. To monitor the breathing of a hospital patient, a thin belt is girded around the patient's chest. The belt is a 200-turn coil. When the patient inhales, the area encircled by the coil increases by 39.0 cm². The magnitude of the Earth's magnetic field is 50.0 μT and makes an angle of 28.0° with the plane of the coil. If a patient takes 1.80 s to inhale, find the average induced emf in the coil during this time.

53. A solenoid wound with 2 000 turns/m is supplied with current that varies in time according to $I = 4 \sin(120\pi t)$, where $I$ is in amperes and $t$ is in seconds. A small coaxial circular coil of 40 turns and radius $r = 5.00$ cm is located inside the solenoid near its center. (a) Derive an expression that describes the manner in which the emf in the small coil varies in time. (b) At what average rate is energy delivered to the small coil if the windings have a total resistance of 8.00 Ω?

54. A *betatron* accelerates electrons to energies in the MeV range by means of electromagnetic induction. Electrons in a vacuum chamber are held in a circular orbit by a magnetic field perpendicular to the orbital plane. The mag-

netic field is gradually increased to induce an electric field around the orbit. (a) Show that the electric field is in the correct direction to make the electrons speed up. (b) Assume that the radius of the orbit remains constant. Show that the average magnetic field over the area enclosed by the orbit must be twice as large as the magnetic field at the circumference of the circle.

55. A long, straight wire carries a current $I = I_{max} \sin(\omega t + \phi)$ and lies in the plane of a rectangular coil of $N$ turns of wire, as shown in Figure P23.6. The quantities $I_{max}$, $\omega$, and $\phi$ are all constants. Determine the emf induced in the coil by the magnetic field created by the current in the straight wire. Assume $I_{max} = 50.0$ A, $\omega = 200\pi \, s^{-1}$, $N = 100$, $h = w = 5.00$ cm, and $L = 20.0$ cm.

56. The toroid in Figure P23.56 consists of $N$ turns and has a rectangular cross-section. Its inner and outer radii are $a$ and $b$, respectively. (a) Show that

$$L = \frac{\mu_0 N^2 h}{2\pi} \ln \frac{b}{a}$$

(b) Using this result, compute the self-inductance of a 500-turn toroid for which $a = 10.0$ cm, $b = 12.0$ cm, and $h = 1.00$ cm. (c) In Problem 29, an approximate equation for the inductance of a toroid with $R \gg r$ was derived. To test the accuracy of that result, use the expression in Problem 29 to compute the approximate inductance of the toroid described in part (b). Compare the result with the answer to part (b).

**Figure P23.56**

57. (a) A flat circular coil does not really produce a uniform magnetic field in the area it encloses, but estimate the self-inductance of a flat circular coil, with radius $R$ and $N$ turns, by supposing the field at its center is uniform over its area. (b) A circuit on a laboratory table consists of a 1.5-V battery, a 270-$\Omega$ resistor, a switch, and three 30-cm-long cords connecting them. Suppose the circuit is arranged to be circular. Think of it as a flat coil with one turn. Compute the order of magnitude of its self-inductance and (c) of the time constant describing how fast the current increases when you close the switch.

58. To prevent damage from arcing in an electric motor, a discharge resistor is sometimes placed in parallel with the armature. If the motor is suddenly unplugged while running, this resistor limits the voltage that appears across the

armature coils. Consider a 12.0-V dc motor with an armature that has a resistance of 7.50 $\Omega$ and an inductance of 450 mH. Assume the back emf in the armature coils is 10.0 V when the motor is running at normal speed. (The equivalent circuit for the armature is shown in Figure P23.58.) Calculate the maximum resistance $R$ that limits the voltage across the armature to 80.0 V when the motor is unplugged.

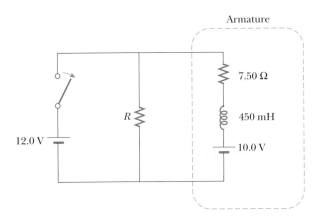

**Figure P23.58**

> **Review Problems.** Problems 59 through 61 apply ideas from this chapter and earlier chapters to some properties of superconductors, which were introduced in Section 21.3.

59. *The resistance of a superconductor.* In an experiment carried out by S. C. Collins between 1955 and 1958, a current was maintained in a superconducting lead ring for 2.50 yr with no observed loss. If the inductance of the ring was $3.14 \times 10^{-8}$ H and the sensitivity of the experiment was 1 part in $10^9$, what was the maximum resistance of the ring? (*Hint:* Treat this as a decaying current in an *RL* circuit, and recall that $e^{-x} \approx 1 - x$ for small $x$.)

60. A novel method of storing electric energy has been proposed. A huge underground superconducting coil, 1.00 km in diameter, would be fabricated. It would carry a maximum current of 50.0 kA through each winding of a 150-turn Nb$_3$Sn solenoid. (a) If the inductance of this huge coil were 50.0 H, what would be the total energy stored? (b) What would be the compressive force per meter length acting between two adjacent windings 0.250 m apart?

61. *Superconducting power transmission.* The use of superconductors has been proposed for power transmission lines. A single coaxial cable (Fig. P23.61) could carry $1.00 \times 10^3$ MW (the output of a large power plant) at 200 kV, dc, over a distance of 1 000 km without loss. An inner wire of radius 2.00 cm, made from the superconductor Nb$_3$Sn, carries the current $I$ in one direction. A

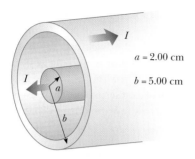

$a = 2.00$ cm

$b = 5.00$ cm

**Figure P23.61**

surrounding superconducting cylinder, of radius 5.00 cm, would carry the return current *I*. In such a system, what is the magnetic field (a) at the surface of the inner conductor and (b) at the inner surface of the outer conductor? (c) How much energy would be stored in the space between the conductors in a 1 000-km superconducting line? (d) What is the pressure exerted on the outer conductor?

## ANSWERS TO QUICK QUIZZES

**23.1** The magnetic field lines around the transmission cable will be circular, centered on the cable. If you place your loop around the cable, no field lines pass through the loop, so no emf is induced. The loop must be placed next to the cable, with the plane of the loop parallel to the cable to maximize the flux through its area.

**23.2** *c, d = e, b, a.* The magnitude of the emf is proportional to the rate of change of the magnetic flux. For the situation described, the rate of change of magnetic flux is proportional to the rate of change of the magnetic field. This rate of change is the slope of the graph in Figure 23.6. The magnitude of the slope is largest at *c.* Points *d* and *e* are on a straight line, so the slope is the same at each point. Point *b* represents a point of relatively small slope, but *a* is at a point of zero slope because the curve is horizontal at that point.

**23.3** The pickup coil is placed near the vibrating guitar string, which is made of a metal that can be magnetized. The permanent magnet inside the coil magnetizes the portion of the string nearest the coil. When the guitar string vibrates at some frequency, its magnetized segment produces a changing magnetic flux through the pickup coil. The changing flux induces a voltage in the coil and this voltage is fed to an amplifier. The output of the amplifier is sent to the loudspeakers, producing the sound waves that we hear.

**23.4** (b). According to Equation 23.5, because *B* and *v* are fixed, the emf depends only on the length of the wire moving in the magnetic field. Thus, you want the long dimension moving through the magnetic field lines so that it is perpendicular to the velocity vector. In this case, the short dimension is parallel to the velocity vector. From a more conceptual point of view, you want the rate of change of area in the magnetic field to be the largest, which you do by thrusting the long dimension into the field.

**23.5** When the aluminum sheet moves between the poles of the magnet, circular currents called *eddy currents* are established in the aluminum. According to Lenz's law,

these currents are in a direction so as to oppose the original change, which is the movement of the aluminum sheet in the magnetic field. Thus, the effect of the eddy currents is to impose *magnetic braking*, so that the oscillations of the equal arm balance settle down and a reading of the mass can take place. In the absence of such magnetic braking, the oscillation might continue for a very long time, so that the experimenter would have to wait to take a reading.

**23.6** (d), (e). For the constant current in (a) and (b), no potential difference exists across the resistanceless inductor. In (c), if the current increases, the emf induced in the inductor is in the opposite direction, from *b* to *a*, making *b* higher in potential than *a.* Similarly, in (f), the decreasing current induces an emf in the same direction as the current, from *b* to *a*, again making the potential higher at *b* than *a.*

**23.7** When the iron rod is inserted into the solenoid, the inductance of the coil increases. As a result, more potential difference appears across the coil than before. Consequently, less potential difference appears across the bulb and its brightness decreases.

**23.8** (b). Figure 23.28 shows that circuit B has the larger time constant because in this circuit it takes longer for the current to reach its maximum value and then longer for this current to drop back down to zero after switch $S_2$ is closed. Equation 23.15 indicates that for equal resistances $R_A$ and $R_B$, the condition $\tau_B > \tau_A$ means that $L_A < L_B$.

**23.9** (a), (d). Because the energy density depends on the magnitude of the magnetic field, to increase the energy density, we must increase the magnetic field. For a solenoid, $B = \mu_0 n I$, where *n* is the number of turns per unit length. In (a), we increase *n* to increase the magnetic field. In (b), the change in cross-sectional area has no effect on the magnetic field. In (c), increasing the length but keeping *n* fixed has no effect on the magnetic field. Increasing the current in (d) increases the magnetic field in the solenoid.

# Magnetic Levitation Vehicles

## PROPELLING AND BRAKING THE VEHICLE

Now that we have investigated the principles of electromagnetism, let us respond to our central question for the *Magnetic Levitation Vehicles* Context:

### How can we lift, propel, and brake a vehicle with magnetic forces?

We have addressed two mechanisms for lifting the vehicle, one in each of the preceding chapters. Once the vehicle is suspended above the track, we must control its speed with propulsion and braking. We shall discuss these processes in this Context Conclusion section.

If you have a toy car made of iron, you can imagine how to propel the car with a magnet. You hold the magnet near the front of the car so that it exerts an attractive force on the car, causing it to accelerate from rest. As the car moves, you move the magnet in the same direction, not allowing the magnet and the car to touch. In essence, the car "chases" the moving magnet due to the attractive force.

You could increase the propelling force on the car by mounting a bar magnet on it, with the north pole near the front of the car and the south pole near the back. Now, you use two other bar magnets, one with its south pole in front of the car and another with its south pole near the back of the car. The magnet on the car is attracted to the front magnet and repelled by the rear magnet. In this manner, the car experiences a strong propelling force.

Although it is difficult to move both magnets along with the car without allowing any magnets to touch, this is the fundamental idea behind magnetic propulsion. Electromagnets are mounted on the magnetic levitation vehicle and a series of propulsion coils are placed along the side of the track, as shown in Figure 1. The polarity of the magnets on the vehicle remains constant. An electrical signal is passed through the coils such that each magnet on the vehicle "sees" a coil with opposite polarity ahead of it and a coil with the same polarity behind it. As a result, the vehicle experiences an attractive force from the coil ahead of it and a repulsive force from the coil behind it, both forces pushing the vehicle in the forward direction. Figure 1 shows this situation at a single instant of time. As the sinusoidal wave in Figure 1 moves along the track, the vehicle "chases" it due to the magnetic forces.

Electromagnetic transportation has the added benefit of a built-in braking mechanism. Lenz's law tells us that a magnetic change induces a current that acts to oppose the original change. This represents a natural braking mechanism. For example, if an aluminum plate is dropped between the poles of a very strong magnet, the plate drops slowly because currents established in the plate feel a magnetic force that opposes the fall.

In the case of magnetic levitation vehicles, imagine that the propulsion system is deactivated so that the vehicle begins to coast. The relative motion of magnets and coils in the train and track induces currents that slow the train down, according to Lenz's law. The propulsion system can be used in combination with this magnetic braking for complete control over the stopping process.

This principle is not new—railroads operating in the Alps at the beginning of the 20th century included mechanisms for connecting the electric drive motor to a resis-

**893**

**Figure 1**

Magnets on the vehicle are attracted and repelled by electromagnetic poles created by a sinusoidal current passed through coils on the side of the track. As the wave of current moves along the track, the vehicle moves with it.

tance when the train moved downhill. As the motion of the train caused the motor to turn, the motor acted as a generator. The generator produced a current in the resistance, resulting in a back emf. Because this back emf opposed the original change (the rotation of the motor), it controlled the motion of the train down the hill.

In Chapters 22 and 23, we discussed models for magnetic levitation that are under study in Germany and Japan, both of which require strong electromagnets. At Lawrence Livermore Laboratory in California, scientists are working on a magnetic levitation scheme that involves permanent magnets. The scheme is called the *Inductrack*. One of the attractive features of this approach is that no electric energy is required to power the magnets, resulting in savings for energy costs to operate the system.

The Inductrack system uses a *Halbach array* of magnets. Figure 2 shows a one-dimensional Halbach array. Underneath the array, the magnetic field lines of adjacent magnets with poles oriented vertically combine with those with poles oriented horizontally to create a very strong magnetic field. Above the array, the magnetic field lines of vertically oriented magnets are in the opposite direction from those created by the horizontally oriented magnets, resulting in a weak field in this region.

The Halbach array, attached to the underside of the vehicle, passes over a series of coils of wire and induces currents in these coils. By Faraday's law, similar to the sit-

**WEB**

For more information on the Inductrack system, visit
**www.llnl.gov/str/Post.html**

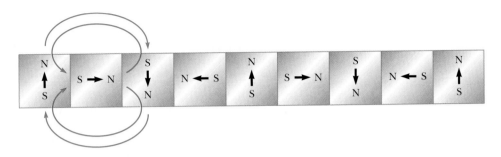

**Figure 2**

A Halbach array of magnets. A field line from the north pole of the leftmost magnet to the south pole of the third magnet is clockwise. The field line above the array from the north pole of the second magnet to the south pole of the same magnet is counterclockwise. The result is a relatively weak field above the array. Below the array, the field line from the north pole of the third magnet to the south pole of the leftmost magnet is clockwise. The field line below the array from the second magnet is also clockwise. The combination of these field lines results in a strong magnetic field below the array.

uation in the electrodynamic model of Section 23.8, the magnetic field resulting from the current in the coils exerts a repulsive force on the magnets, creating a levitation force on the vehicle. A levitation force of 40 metric tons per square meter can be achieved using high-field alloy magnets. A working model of the Inductrack system successfully propelled and levitated a 22-kg vehicle over a 20-m test track. Feasibility studies show that a full-scale Inductrack vehicle would cost about half that of an equivalent Transrapid vehicle, with significant further savings in operating costs.

What does the future of magnetic levitation transportation hold? Will the Inductrack become the system of choice, or will it be the Japanese or German system? Or will all three coexist? At this point, it is impossible to predict, and we invite you to watch the newspapers for further developments!

## PROBLEMS

1. Suppose the vehicle shown in Figure 1 is moving at 400 km/h. The distance between adjacent magnets on the vehicle is 10.0 m. What is the frequency of the alternating current in the coils on the sides of the track required to propel the vehicle?

2. Figure 3 represents an electromagnetic brake that uses eddy currents. An electromagnet hangs from a conventional railroad car near one steel rail. To stop the car, a large current is sent through the coils of the electromagnet. The moving electromagnet induces eddy currents in the rail, whose fields oppose the change in the field of the electromagnet. The magnetic fields of the eddy currents exert forces on the current in the electromagnet, thereby slowing the car. The direction of the car's motion and the direction of the current in the electromagnet are shown correctly in the picture. Determine which of the eddy currents shown on the rails is correct. Explain your answer.

**Figure 3**

Magnets on a moving railroad car induce small current loops in the rail, called eddy currents.

# Lasers

The invention of the laser was popularly credited to Arthur Schawlow and Charles H. Townes for many years after their publication of a proposal for the laser in a 1958 issue of *Physical Review*. Schawlow and Townes received a patent for the device in 1959. In 1960, the first laser was built and operated by Theodore Maiman. This device used a ruby crystal to create the laser light, which was emitted in pulses from the end of a ruby cylinder. A flash lamp was used to excite the laser action.

In 1977, the first victory in a 30-year-long legal battle was completed in which Gordon Gould, who was a graduate student at Columbia University in the late 1950s, received a patent for inventing the laser in 1957, as well as coining the term. Believing erroneously that he had to have a working prototype before he could file for a patent, he did not file until later in 1959 than had Schawlow and Townes. Gould's legal battles ended in 1987. By this time, Gould's technology was being widely used in industry and medicine. His victory finally resulted in his control of patent rights to perhaps 90% of the lasers used and sold in the United States.

Since the development of the first device, laser technology has experienced tremendous growth. Lasers that cover wavelengths in the infrared, visible, and ultraviolet regions are now available. Various types of lasers use solids, liquids, and gases as the active medium. Although the original laser emitted light over a very narrow range around a fixed wavelength, tunable lasers are now available, in which the wavelength can be varied.

The laser is an omnipresent technological tool in our daily life. Applications include surgical "welding" of detached retinas, precision surveying and length measurement, a potential source for inducing nuclear fusion reactions, precision cutting of metals and other materials, and telephone communication along optical fibers.

We also use lasers to read information from compact discs for use in audio entertainment and computer applications. Digital videodiscs use lasers to read video information. Lasers are used in retail stores to read price and inventory information from product labels.

The original ruby laser emitted red light, as did many lasers developed soon afterward. Today, lasers are available in a variety of colors and various regions of the electromagnetic spectrum. In this photograph, a blue laser is used to perform scientific research. *(Hank Morgan/Photo Researchers, Inc.)*

Photograph of an early ruby laser, showing the flash lamp (glass helix) surrounding the ruby rod (red cylinder). *(Courtesy of HRL Laboratories LLC, Malibu, CA)*

A laser is used on a human eye to perform a surgical procedure. The word *laser* is an acronym meaning light amplification by stimulated emission of radiation. *(Will & Deni McIntyre/Photo Researchers, Inc.)*

This robotic device carries laser scissors, which can cut up to 50 layers of fabric at a time. This is one of the many technological uses of lasers in our society. *(Philippe Plailly/SPL/Photo Researchers, Inc.)*

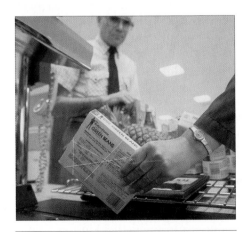

A supermarket scanner uses light from a laser to identify products being purchased. The reflections from the bar code on the package are read and entered into the computer in order to determine the price of the item. *(Paul Shambroom/Photo Researchers, Inc.)*

These and other applications are possible because of the unique characteristics of laser light. In addition to its being highly monochromatic, laser light is also highly directional and can therefore be sharply focused to produce regions of extreme intensity.

In this Context, we shall investigate the physics of electromagnetic radiation and optics and apply the principles to an understanding of the behavior of laser light and its applications. A major focus of our study will be on the technology of optical fibers and how they are used in industry and medicine. We shall study the nature of light as we respond to our central question

*What is special about laser light, and how is it used in technological applications?*

# chapter
# 24

This photograph involves two types of electromagnetic waves. Light waves create the rainbow, as will be discussed in Chapter 25. The radio telescopes form the Very Large Array (VLA) in New Mexico. The 27 telescopes are laid out in a giant "Y" with a diameter of 27 km. The VLA is so large that the weather is sometimes different from one side to the other. In this chapter, we explore electromagnetic waves of all types.

*(Riccardo Giovanelli and Martha Haynes, Cornell University)*

# Electromagnetic Waves

Although we are not always aware of their presence, electromagnetic waves permeate our environment. In the form of visible light, they enable us to view the world around us with our eyes; infrared waves from the surface of the Earth warm our environment; radio-frequency waves carry our favorite radio entertainment; microwaves cook our food and are used in radar communication systems. The list goes on and on. The waves described in Chapter 13 are mechanical waves, which require a medium through which to propagate. Electromagnetic waves, in contrast, can propagate through a vacuum. Despite this difference between mechanical and electromagnetic waves, much of the behavior in the wave models of Chapters 13 and 14 is similar for electromagnetic waves.

The purpose of this chapter is to explore the properties of electromagnetic waves. The fundamental laws of electricity and magnetism—Maxwell's equations—form the basis of all electromagnetic phenomena. One of these equations predicts that a time-varying electric field produces a magnetic field just as a time-varying magnetic field produces an electric field. From this generalization, Maxwell provided the final important link between electric and magnetic fields. The most dramatic prediction of his equations is the existence of electromagnetic waves that propagate through empty space with the speed of light. This discovery led to many practical applications, such as radio and television, and to the realization that light is one form of electromagnetic radiation.

## 24.1 • DISPLACEMENT CURRENT AND THE GENERALIZED AMPÈRE'S LAW

 See the *Core Concepts in Physics CD-ROM,* Screen 12.9

We have seen that charges in motion, or currents, produce magnetic fields. When a current-carrying conductor has high symmetry, we can calculate the magnetic field using Ampère's law, given by Equation 22.29:

$$\oint \mathbf{B} \cdot d\mathbf{s} = \mu_0 I \qquad [22.29]$$

where **the line integral is over any closed path through which the conduction current passes,** and the current is defined by $I = dQ/dt$.

In this section, we shall use the term *conduction current* to refer to the current carried by charged particles in a wire in order to differentiate this from a different type of current we will introduce shortly. **Ampère's law in this form is valid only if the conduction current is continuous in space.** Maxwell recognized this limitation and modified Ampère's law to include all possible situations.

This problem can be understood by considering a capacitor being charged as in Figure 24.1. When conduction current exists in the wires, the charge on the plates changes, but **no conduction current exists between the plates.** Consider the two surfaces $S_1$ (a circle, shown in blue) and $S_2$ (a paraboloid, in orange, passing between the plates) in Figure 24.1 bounded by the same path $P$. Ampère's law says that the line integral of $\mathbf{B} \cdot d\mathbf{s}$ around this path must equal $\mu_0 I$, where $I$ is the total current through *any* surface bounded by the path $P$.

When the path $P$ is considered as bounding $S_1$, the result of the integral is $\mu_0 I$ because the conduction current passes through $S_1$ while the capacitor is charging. When the path bounds $S_2$, however, the result is zero because no conduction current passes through $S_2$. Thus, a contradictory situation arises because of the discontinuity of the current! Maxwell solved this problem by postulating an additional term on the right side of Equation 22.29, called the **displacement current** $I_d$, defined as

$$I_d \equiv \epsilon_0 \frac{d\Phi_E}{dt} \qquad [24.1]$$

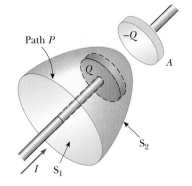

**Figure 24.1**

The surfaces $S_1$ and $S_2$ are bounded by the same path $P$. The conduction current in the wire passes only through the flat surface $S_1$. This leads to a contradiction in Ampère's law that is resolved only if one postulates a displacement current through $S_2$.

• *Displacement current*

Recall that $\Phi_E$ is the flux of the electric field, defined as $\Phi_E = \int \mathbf{E} \cdot d\mathbf{A}$.

As the capacitor is being charged (or discharged), the changing electric field between the plates may be considered as equivalent to a current that acts as a continuation of the conduction current in the wire. When the expression for the displacement current given by Equation 24.1 is added to the conduction current on the right side of Ampère's law, the difficulty represented in Figure 24.1 is resolved. No matter what surface bounded by the path $P$ is chosen, either conduction current or displacement current passes through it. With this new notion of displacement current, we can express the general form of Ampère's law (sometimes called the **Ampère–Maxwell law**) as*

$$\oint \mathbf{B} \cdot d\mathbf{s} = \mu_0 (I + I_d) = \mu_0 I + \mu_0 \epsilon_0 \frac{d\Phi_E}{dt} \qquad [24.2]$$

**PITFALL PREVENTION 24.1**

**Displacement is different here**

The word *displacement* in the phrase *displacement current* does not have the same meaning in electromagnetism as in Chapter 2. Despite this difference, it is historically entrenched in the language of physics, so we continue to use it.

• *Ampère–Maxwell law*

* Strictly speaking, this expression is valid only in a vacuum. If a magnetic material is present, a magnetizing current must also be included on the right side of Equation 24.2 to make Ampère's law fully general.

**James Clerk Maxwell**
(1831–1879)

Scottish theoretical physicist
Maxwell developed the electromagnetic theory of light and the
kinetic theory of gases, and he explained the nature of color vision
and of Saturn's rings. His successful interpretation of electromagnetic fields produced the field
equations that bear his name. Formidable mathematical ability combined with great insight enabled
Maxwell to lead the way in the
study of electromagnetism and
kinetic theory. He died of cancer
before he was 50. (*North Wind
Picture Archives*)

See Screen 12.10

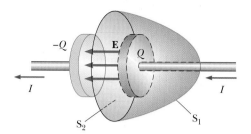

**Figure 24.2**

Because it exists only in the wires, the conduction current $I = dQ/dt$ passes through the curved surface $S_1$ but not the flat surface $S_2$. Only the displacement current $I_d = \epsilon_0 \, d\Phi_E/dt$ passes through $S_2$. The two currents must be equal for continuity.

The meaning of this expression can be understood by referring to Figure 24.2. The electric flux through $S_2$ (a circle, shown in gray, between the plates) is $\Phi_E = \int \mathbf{E} \cdot d\mathbf{A} = EA$, where $A$ is the area of the capacitor plates and $E$ is the uniform electric field magnitude between the plates. If $Q$ is the charge on the plates at any instant, then $E = Q/\epsilon_0 A$ (Section 20.7). Therefore, the electric flux through $S_2$ is simply

$$\Phi_E = EA = \frac{Q}{\epsilon_0}$$

Hence, the displacement current $I_d$ through $S_2$ is

$$I_d = \epsilon_0 \frac{d\Phi_E}{dt} = \frac{dQ}{dt} \qquad [24.3]$$

That is, the displacement current through $S_2$ is precisely equal to the conduction current $I$ through $S_1$! The central point of this formalism is the fact that **magnetic fields are produced both by conduction currents and by changing electric fields.** This result is a remarkable example of theoretical work by Maxwell and of his major contributions in advancing the understanding of electromagnetism.

## 24.2 • MAXWELL'S EQUATIONS

In this section we gather together four equations from our studies in recent chapters that together can be regarded as the basis of all electric and magnetic phenomena. These relationships, known as Maxwell's equations after James Clerk Maxwell, are as fundamental to electromagnetic phenomena as Newton's laws are to mechanical phenomena. In fact, the theory developed by Maxwell was more far reaching than even he imagined, because it was shown by Einstein in 1905 to be in agreement with the special theory of relativity. As we shall see, Maxwell's equations represent laws of electricity and magnetism that have already been discussed. The equations have additional important consequences, however, in that they predict the existence of electromagnetic waves (traveling patterns of electric and magnetic fields) that travel in vacuum with a speed of $c = 1/\sqrt{\epsilon_0 \mu_0} = 3.00 \times 10^8$ m/s, the speed of light. Furthermore, Maxwell's equations show that electromagnetic waves are radiated by accelerating charges, as we discussed in Chapter 17 with thermal radiation.

For simplicity, we present **Maxwell's equations** as applied to free space—that is, in the absence of any dielectric or magnetic material. The four equations are

$$\oint \mathbf{E} \cdot d\mathbf{A} = \frac{Q}{\epsilon_0} \qquad [24.4]$$

$$\oint \mathbf{B} \cdot d\mathbf{A} = 0 \qquad [24.5]$$

$$\oint \mathbf{E} \cdot d\mathbf{s} = -\frac{d\Phi_B}{dt} \qquad [24.6]$$ • *Maxwell's equations*

$$\oint \mathbf{B} \cdot d\mathbf{s} = \mu_0 I + \epsilon_0 \mu_0 \frac{d\Phi_E}{dt} \qquad [24.7]$$

Equation 24.4 is **Gauss's law,** which states that **the total electric flux through any closed surface equals the net charge inside that surface divided by $\epsilon_0$** (Chapter 19). This law describes how charges create electric fields, in which the electric field lines originate on positive charges and terminate on negative charges.

Equation 24.5, which can be considered **Gauss's law for magnetism,** says that **the net magnetic flux through a closed surface is zero.** That is, the number of magnetic field lines entering a closed volume must equal the number leaving that volume. This implies that magnetic field lines cannot begin or end at any point. If they did, this would mean that isolated magnetic monopoles existed at those points. The fact that isolated magnetic monopoles have not been observed in nature can be taken as a basis of Equation 24.5.

Equation 24.6 is **Faraday's law of induction** (Eq. 23.9), which describes how a changing magnetic field creates an electric field. This law states that **the line integral of the electric field around any closed path (which equals the emf) equals the rate of change of magnetic flux through any surface area bounded by that path.**

Equation 24.7 is the generalized form of Ampère's law and describes how both an electric current and a changing electric field create a magnetic field. That is, **the line integral of the magnetic field around any closed path is determined by the net current and the rate of change of electric flux through any surface bounded by that path.**

Once the electric and magnetic fields are known at some point in space, the force those fields exert on a particle of charge $q$ can be calculated from the expression

$$\mathbf{F} = q\mathbf{E} + q\mathbf{v} \times \mathbf{B} \qquad [24.8]$$ • *Lorentz force*

This is called the **Lorentz force** (Section 22.4). Maxwell's equations and this force law give a complete description of all classical electromagnetic interactions.

Note the interesting symmetry of Maxwell's equations. Equations 24.4 and 24.5 are symmetric, apart from the absence of a magnetic monopole term in Equation 24.5. Furthermore, Equations 24.6 and 24.7 are symmetric in that the line integrals of **E** and **B** around a closed path are related to the rate of change of magnetic flux and electric flux, respectively.

## 24.3 • ELECTROMAGNETIC WAVES

The fundamental laws governing the behavior of electric and magnetic fields are Maxwell's equations, discussed in Section 24.2. In his unified theory of electromagnetism, Maxwell showed that time-dependent electric and magnetic fields satisfy a

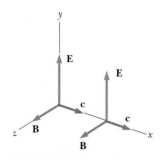

**Figure 24.3**

An electromagnetic wave traveling at velocity **c** in the positive *x* direction. The wave is shown at two instants at which the electric field is along the *y* direction and has its maximum magnitude, and the magnetic field is along the *z* direction, also with its maximum magnitude. These fields depend only on *x* and *t*.

wave equation (see Section 13.3). The most significant outcome of this theory is the prediction of the existence of electromagnetic waves.

Maxwell's equations predict that an electromagnetic wave consists of oscillating electric and magnetic fields. The changing fields create each other to maintain the propagation of the wave—a changing electric field induces a magnetic field, and a changing magnetic field induces an electric field. The **E** and **B** vectors are perpendicular to each other, and to the direction of propagation, as shown in Figure 24.3. The direction of propagation is the direction of the vector product **E × B**, which we explore more fully in Section 24.5.

To understand the prediction of electromagnetic waves, let us focus our attention on an electromagnetic wave traveling in the *x* direction. For this wave, the electric field **E** is in the *y* direction, and the magnetic field **B** is in the *z* direction as in Figure 24.3. Waves in which the electric and magnetic fields are restricted to being parallel to certain directions are said to be **linearly polarized waves.**\* Furthermore, in Figure 24.3 let us assume that at any point in space, the magnitudes *E* and *B* of the fields depend on *x* and *t* only, and not on the *y* or *z* coordinates.

Let us also imagine that the source of the electromagnetic waves is such that a wave radiated from *any* position in the *yz* plane (not just from the origin as might be suggested by Figure 24.3) propagates in the *x* direction, and all such waves are emitted in phase. If we define a **ray** as the line along which the wave travels, then all rays for these waves are parallel. This whole collection of waves is often called a **plane wave.** A surface connecting points of equal phase on all waves, which we call a **wave front,** is a geometric plane. In comparison, a point source of radiation sends waves out in all directions. A surface connecting points of equal phase for this situation is a sphere, so we call this a **spherical wave.**

The properties of electromagnetic waves can be deduced from Maxwell's equations. We start with Faraday's law, Equation 24.6:

$$\oint \mathbf{E} \cdot d\mathbf{s} = -\frac{d\Phi_B}{dt}$$

Let us assume that a plane electromagnetic wave is traveling in the *x* direction, with the electric field **E** in the positive *y* direction and the magnetic field **B** in the positive *z* direction.

Consider a rectangle of width *dx* and height $\ell$ lying in the *xy* plane as in Figure 24.4. To apply Equation 24.6, we first evaluate the line integral of **E** · *d***s** around this rectangle. The contributions from the top and bottom of the rectangle are zero because **E** is perpendicular to *d***s** for these paths. We can express the electric field on the right side of the rectangle as[†]

$$E(x + dx, t) \approx E(x, t) + \left.\frac{dE}{dx}\right]_{t\,\text{constant}} dx = E(x, t) + \frac{\partial E}{\partial x}\, dx$$

whereas the field on the left side of the rectangle is simply $E(x, t)$. The line integral over this rectangle is therefore approximately

$$\oint \mathbf{E} \cdot d\mathbf{s} = [E(x + dx, t)]\ell - [E(x, t)]\ell \approx \ell \left(\frac{\partial E}{\partial x}\right) dx \qquad \text{[24.9]}$$

**Figure 24.4**

As a plane wave moving in the +*x* direction passes through a rectangular path of width *dx* lying in the *xy* plane, the electric field in the *y* direction varies from **E** to **E** + *d***E**. This construction allows us to evaluate the line integral of **E** over the perimeter of the rectangle.

\* Waves with other particular patterns of vibrations of **E** and **B** include **circularly polarized waves.** The most general polarization pattern is **elliptical.**

[†] Because *dE/dx* in this equation is expressed as the change in *E* with *x* at a given instant *t*, *dE/dx* is equivalent to the partial derivative $\partial E/\partial x$. Likewise, we will shortly require *dB/dt*, which means the change in *B* with time at a particular position *x*, and so we can replace *dB/dt* by $\partial B/\partial t$.

Because the magnetic field is in the $z$ direction, the magnetic flux through the rectangle of area $\ell \, dx$ is approximately $\Phi_B = B\ell \, dx$. (This assumes that $dx$ is very small compared with the wavelength of the wave.) Taking the time derivative of the magnetic flux at the location of the rectangle on the $x$ axis,

$$\frac{d\Phi_B}{dt} = \ell \, dx \frac{dB}{dt}\bigg]_{x\,constant} = \ell \, \frac{\partial B}{\partial t} \, dx \qquad \text{[24.10]}$$

Substituting Equations 24.9 and 24.10 into Equation 24.6 gives

$$\ell \left( \frac{\partial E}{\partial x} \right) dx = -\ell \, \frac{\partial B}{\partial t} \, dx$$

$$\frac{\partial E}{\partial x} = -\frac{\partial B}{\partial t} \qquad \text{[24.11]}$$

We can derive a second equation by starting with Maxwell's fourth equation in empty space (Eq. 24.7). In this case, we evaluate the line integral of $\mathbf{B} \cdot d\mathbf{s}$ around a rectangle lying in the $xz$ plane and having width $dx$ and length $\ell$, as in Figure 24.5. Using the sense of the integration shown and noting that the magnetic field changes from $B(x, t)$ to $B(x + dx, t)$ over the width $dx$, we find

$$\oint \mathbf{B} \cdot d\mathbf{s} = [B(x, t)]\ell - [B(x + dx, t)]\ell = -\ell \left( \frac{\partial B}{\partial x} \right) dx \qquad \text{[24.12]}$$

The electric flux through the rectangle is $\Phi_E = E\ell \, dx$, which when differentiated with respect to time gives

$$\frac{d\Phi_E}{dt} = \ell \left( \frac{\partial E}{\partial t} \right) dx \qquad \text{[24.13]}$$

Substituting Equations 24.12 and 24.13 into Equation 24.7 gives

$$-\ell \left( \frac{\partial B}{\partial x} \right) dx = \epsilon_0 \mu_0 \ell \left( \frac{\partial E}{\partial t} \right) dx$$

$$\frac{\partial B}{\partial x} = -\epsilon_0 \mu_0 \frac{\partial E}{\partial t} \qquad \text{[24.14]}$$

Taking the derivative of Equation 24.11 with respect to $x$ and combining this with Equation 24.14 we have

$$\frac{\partial^2 E}{\partial x^2} = -\frac{\partial}{\partial x} \left( \frac{\partial B}{\partial t} \right) = -\frac{\partial}{\partial t} \left( \frac{\partial B}{\partial x} \right) = -\frac{\partial}{\partial t} \left( -\epsilon_0 \mu_0 \frac{\partial E}{\partial t} \right)$$

$$\frac{\partial^2 E}{\partial x^2} = \epsilon_0 \mu_0 \frac{\partial^2 E}{\partial t^2} \qquad \text{[24.15]}$$

In the same manner, taking a derivative of Equation 24.14 with respect to $x$ and combining it with Equation 24.11, we find

$$\frac{\partial^2 B}{\partial x^2} = \epsilon_0 \mu_0 \frac{\partial^2 B}{\partial t^2} \qquad \text{[24.16]}$$

Equations 24.15 and 24.16 both have the form of a linear wave equation (Eq. 13.20). As indicated in Chapter 13, such an equation is a mathematical representation

**Figure 24.5**

As a plane wave moving in the $+x$ direction passes through a rectangular path of width $dx$ lying in the $xz$ plane, the magnetic field in the $z$ direction varies from $\mathbf{B}$ to $\mathbf{B} + d\mathbf{B}$. This construction allows us to evaluate the line integral of $\mathbf{B}$ over the perimeter of the rectangle.

• *Electric field wave equation for electromagnetic waves in free space*

• *Magnetic field wave equation for electromagnetic waves in free space* of the traveling wave model. In the present discussion, this equation represents traveling electromagnetic waves. These waves travel with a speed $c$ of

• *The speed of electromagnetic waves*

$$c = \frac{1}{\sqrt{\epsilon_0 \mu_0}} \qquad \text{[24.17]}$$

Substituting $\epsilon_0 = 8.854\ 18 \times 10^{-12}\ \text{C}^2/\text{N} \cdot \text{m}^2$ and $\mu_0 = 4\pi \times 10^{-7}\ \text{T} \cdot \text{m}/\text{A}$ in Equation 24.17, we find that $c = 2.997\ 92 \times 10^8$ m/s. Because this speed is precisely the same as the speed of light in empty space,* one is led to believe (correctly) that **light is an electromagnetic wave.**

Because electromagnetic waves are described by the traveling wave model, we can adopt another mathematical representation from the model, first seen in Equation 13.11 for mechanical waves. This is the relationship between wave speed, wavelength, and frequency, $v = \lambda f$, which we can write as $c = \lambda f$ for continuous electromagnetic waves.

The simplest wave solutions to Equations 24.15 and 24.16 are those for which the field amplitudes $E$ and $B$ vary with $x$ and $t$ according to the expressions

$$E = E_{\text{max}} \cos(kx - \omega t) \qquad \text{[24.18]}$$

$$B = B_{\text{max}} \cos(kx - \omega t) \qquad \text{[24.19]}$$

In these expressions, $E_{\text{max}}$ and $B_{\text{max}}$ are the maximum values of the fields, the wave number $k = 2\pi/\lambda$, where $\lambda$ is the wavelength, and the angular frequency $\omega = 2\pi f$, where $f$ is the frequency. Figure 24.6 represents one instant of a sinusoidal, linearly polarized electromagnetic wave moving in the positive $x$ direction.

The electric and magnetic fields of a plane electromagnetic wave are perpendicular to each other and to the direction of propagation. Thus, **electromagnetic waves are transverse waves.** The transverse mechanical waves studied in Chapter 13 exhibited physical displacements of the particles of the medium that were per-

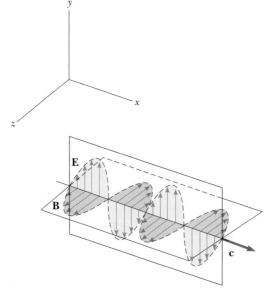

**Figure 24.6**

Representation of a sinusoidal, plane-polarized electromagnetic wave moving in the positive $x$ direction with a speed $c$.

---

* Because of the redefinition of the meter in 1983, the speed of light is now a *defined* quantity with an *exact* value of $c = 2.997\ 924\ 58 \times 10^8$ m/s.

pendicular to the direction of propagation of the wave. Electromagnetic waves do not require a medium for propagation, so there are no particles to be displaced. The transverse nature of an electromagnetic wave reflects the direction of the field vectors with respect to the direction of propagation.

Taking partial derivatives of Equations 24.18 (with respect to $x$) and 24.19 (with respect to $t$), and substituting into Equation 24.11 we find that

$$\frac{E_{max}}{B_{max}} = c$$

Substituting from Equations 24.18 and 24.19 gives

$$\frac{E}{B} = c \qquad\qquad \text{[24.20]}$$

That is, **at every instant the ratio of the electric field to the magnetic field of an electromagnetic wave equals the speed of light.**

Finally, electromagnetic waves obey the **superposition principle** because the differential equations involving $E$ and $B$ are linear equations. For example, the resultant electric field magnitude of two waves coinciding in space with their **E** vectors parallel can be found by simply adding the individual expressions for $E$ given by Equation 24.18.

### Quick Quiz 24.1

What is the phase difference between the two sine waves representing **E** and **B** in Figure 24.6?

### THINKING PHYSICS 24.1

Light exhibits a Doppler effect, which is demonstrated in astronomical observations by the shift of spectral lines from receding galaxies toward the red end of the visible spectrum. The equation for the Doppler effect for light is not the same equation as that for sound. Why would the equation be different?

**Reasoning**   In general, for waves requiring a medium, the speeds of the source and observer can be separately measured with respect to a third entity, the medium. In the Doppler effect for sound, these two speeds are that of the source and that of the observer relative to the air. Because light does not require a medium, no third entity is required. Thus, we cannot identify separate speeds for the source and observer— only their relative speed can be identified. As a result, a different equation must be used,* one that contains only this single speed. This equation can be generated from the laws of relativity.

---

* The appropriate equation for the Doppler effect for light is

$$f' = f\sqrt{\frac{c + v}{c - v}}$$

where $v$ is the relative speed between the source and the observer, $c$ is the speed of light, $f'$ is the frequency of light detected by the observer, and $f$ is the frequency emitted by the source.

## Example 24.1 An Electromagnetic Wave

An electromagnetic sinusoidal wave of frequency 40.0 MHz travels in free space in the *x* direction, as in Figure 24.7. At some point and at some instant, the electric field has its maximum value of 750 N/C and is along the *z* axis. (a) Determine the wavelength and period of the wave.

**Solution** Because $c = \lambda f$ and we know that $f = 40.0$ MHz $= 4.00 \times 10^7$ Hz, we have

$$\lambda = \frac{c}{f} = \frac{3.00 \times 10^8 \text{ m/s}}{4.00 \times 10^7 \text{ Hz}} = \boxed{7.50 \text{ m}}$$

The period of the wave *T* equals the inverse of the frequency, and so

$$T = \frac{1}{f} = \frac{1}{4.00 \times 10^7 \text{ Hz}} = \boxed{2.50 \times 10^{-8} \text{ s}}$$

(b) Calculate the magnitude and direction of the magnetic field when $\mathbf{E} = 750\mathbf{k}$ N/C.

**Solution** From Equation 24.20 we see that

$$B_{\text{max}} = \frac{E_{\text{max}}}{c} = \frac{750 \text{ N/C}}{3.00 \times 10^8 \text{ m/s}} = \boxed{2.50 \times 10^{-6} \text{ T}}$$

Because **E** and **B** must be perpendicular to each other and $\mathbf{E} \times \mathbf{B}$ must be in the direction of wave propagation (*x* in this case), we conclude that **B** is in the $-y$ direction.

(c) Write expressions for the space–time variation of the electric and magnetic field components for this wave.

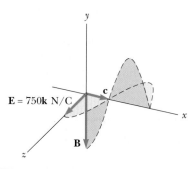

**Figure 24.7**

(Example 24.1) At some instant, a plane electromagnetic wave moving in the *x* direction has a maximum electric field of 750 N/C in the positive *z* direction. The corresponding magnetic field at that point has a magnitude $E/c$ and is in the $-y$ direction.

**Solution** We can apply Equations 24.18 and 24.19 directly:

$$E = E_{\text{max}} \cos(kx - \omega t) = (750 \text{ N/C}) \cos(kx - \omega t)$$

$$B = B_{\text{max}} \cos(kx - \omega t) = (2.50 \times 10^{-6} \text{ T}) \cos(kx - \omega t)$$

where

$$\omega = 2\pi f = 2\pi(4.00 \times 10^7 \text{ Hz}) = 2.51 \times 10^8 \text{ rad/s}$$

$$k = \frac{2\pi}{\lambda} = \frac{2\pi}{7.50 \text{ m}} = 0.838 \text{ rad/m}$$

## 24.4 • HERTZ'S DISCOVERIES

In 1888, Heinrich Hertz (1857–1894) was the first to generate and detect electromagnetic waves in a laboratory setting. To appreciate the details of his experiment, let us first examine the properties of an *LC* circuit. In such a circuit, a charged capacitor is connected to an inductor, as in Figure 24.8. When the switch is closed, both the current in the circuit and the charge on the capacitor oscillate in a manner closely related to our simple harmonic motion model in Chapter 12. If resistance is ignored, no energy is transformed to internal energy and the oscillations continue.

Let us investigate these oscillations in a way similar to our energy analysis of the simple harmonic motion model in Chapter 12. We assume that the capacitor has an initial charge of $Q_{\text{max}}$ and that the switch is closed at $t = 0$. When the capacitor is fully charged, the total energy in the circuit is stored in the electric field of the capacitor and is equal to $Q_{\text{max}}^2/2C$. At this time, the current is zero and so no energy is stored in the inductor. As the capacitor begins to discharge, the energy stored in its electric field decreases. At the same time, the current increases and an amount of energy equal to $\frac{1}{2}LI^2$ is now stored in the magnetic field of the inductor. Thus, energy is transferred from the electric field of the capacitor to the magnetic field of the inductor. When the capacitor is fully discharged, it stores no

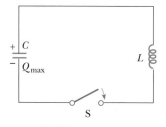

**Figure 24.8**

A simple *LC* circuit. The capacitor has an initial charge $Q_{\text{max}}$ and the switch is closed at $t = 0$.

energy. At this time, the current reaches its maximum value and all of the energy is stored in the inductor. The process then repeats in the reverse direction. The energy continues to transfer between the inductor and the capacitor, corresponding to oscillations of both current and charge.

A representation of this energy transfer is shown in Figure 24.9. As mentioned, the behavior of the circuit is analogous to that of the oscillating particle–spring system studied in Chapter 12. The potential energy $\frac{1}{2}kx^2$ stored in a stretched spring is analogous to the potential energy $Q_{max}^2/2C$ stored in the capacitor. The kinetic energy $\frac{1}{2}mv^2$ of the moving particle is analogous to the magnetic energy $\frac{1}{2}LI^2$ stored in the inductor, which requires the presence of moving charges. In Figure 24.9a, all of the energy is stored as electric potential energy in the capacitor at $t = 0$ (because $I = 0$), just like all the energy in a particle–spring system is initially stored as potential energy in the spring if it is stretched and released at $t = 0$. In Figure 24.9b, all of the energy is stored as magnetic energy $\frac{1}{2}LI_{max}^2$ in the inductor, where $I_{max}$ is the maximum current. Figures 24.9c through 24.9e show subsequent quarter-cycle situations in which the energy is all electric or all magnetic. At intermediate points, part of the energy is electric and part is magnetic.

We now describe an alternative approach to the analogy between the $LC$ circuit and the particle–spring system of Chapter 12. Recall Equation 12.3, which is the differential equation describing the position of the particle in the simple harmonic motion model:

$$\frac{d^2x}{dt^2} = -\frac{k}{m}x \qquad [12.3]$$

Applying Kirchhoff's loop rule to the circuit in Figure 24.8 gives

$$\frac{Q}{C} + L\frac{dI}{dt} = 0$$

Because $I = dQ/dt$, we can rewrite this equation as

$$\frac{Q}{C} = -L\frac{d}{dt}\left(\frac{dQ}{dt}\right) \longrightarrow \frac{d^2Q}{dt^2} = -\frac{1}{LC}Q$$

This equation has exactly the same mathematical form as Equation 12.3 for the particle–spring system. Thus, we conclude that the charge in the circuit will oscillate in a way analogous to the particle on the spring.

In Chapter 12, we recognized the coefficient of $x$ in Equation 12.3 as the square of the angular frequency (Eq. 12.4):

$$\omega^2 = \frac{k}{m}$$

Because of the identical mathematical form of the equation describing the $LC$ circuit, we can identify the coefficient of $Q$ as the square of the angular frequency:

$$\omega^2 = \frac{1}{LC}$$

Thus, the frequency of oscillation of an $LC$ circuit, called the *resonance frequency*, is

$$f_0 = \frac{1}{2\pi\sqrt{LC}} \qquad [24.21]$$

**Heinrich Rudolf Hertz**
(1857–1894)

German physicist Hertz made his greatest discovery—radio waves—in 1887. After finding that the speed of a radio wave is the same as that of light, he showed that radio waves, like light waves, could be reflected, refracted, and diffracted. Hertz died of blood poisoning at age 36. During his short life, he made many contributions to science. The hertz, equal to one complete vibration or cycle per second, is named after him. *(The Bettman Archive)*

● *Resonance frequency of an LC circuit*

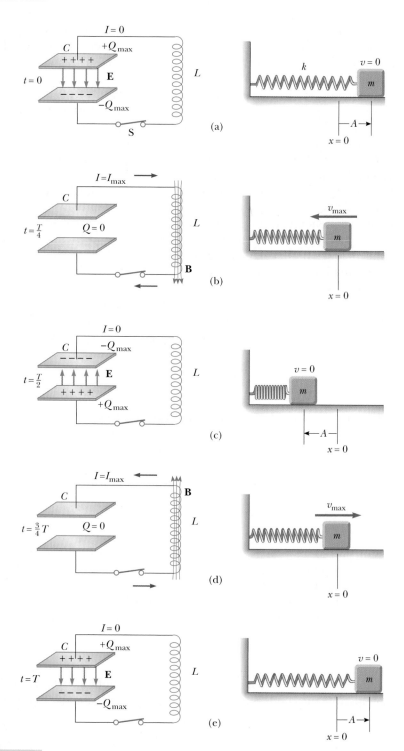

**Figure 24.9**

Energy transfer in a resistanceless *LC* circuit. The capacitor has a charge $Q_{max}$ at $t = 0$ when the switch is closed. The mechanical analog of this circuit, the mass–spring system, is shown at the right.

The circuit Hertz used in his investigations of electromagnetic waves is shown schematically in Figure 24.10. An induction coil (a large coil of wire) is connected to two metal spheres with a narrow gap between them to form a capacitor. Oscillations are initiated in the circuit by sending short voltage pulses via the coil to the spheres, charging one positive, the other negative. Because $L$ and $C$ are quite small in this circuit, the frequency of oscillation is quite high, $f \approx 100$ MHz. This circuit is called a transmitter because it produces electromagnetic waves.

Hertz placed a second circuit, the receiver, several meters from the transmitter circuit. This receiver circuit, which consisted of a single loop of wire connected to two spheres, had its own effective inductance, capacitance, and natural frequency of oscillation. Hertz found that energy was being sent from the transmitter to the receiver when the resonance frequency of the receiver was adjusted to match that of the transmitter.* The energy transfer was detected when the voltage across the spheres in the receiver circuit became high enough to ionize air molecules, causing sparks to appear in the air gap separating the spheres. Hertz's experiment is analogous to the mechanical phenomenon in which one tuning fork responds to acoustic vibrations from an identical vibrating fork. In the case of the tuning fork, the energy transfer from one fork to another is by means of sound, whereas the transfer mechanism is electromagnetic radiation for Hertz's apparatus.

Hertz assumed that the energy transferred from the transmitter to the receiver was carried in the form of waves, which are now known to have been electromagnetic waves. In a series of experiments, he also showed that the radiation generated by the transmitter exhibited the wave properties of interference, diffraction, reflection, refraction, and polarization. As we shall see shortly, all of these properties are exhibited by light. Thus, it became evident that the waves observed by Hertz had properties similar to those of light waves and differed only in frequency and wavelength.

Perhaps Hertz's most convincing experiment was his measurement of the speed of the waves from the transmitter. Waves of known frequency from the transmitter were reflected from a metal sheet so that a pattern of nodes and antinodes was set up, much like the standing wave pattern on a stretched string. As we saw in our discussion of standing waves (Chapter 14), the distance between nodes is $\lambda/2$, and so Hertz was able to determine the wavelength $\lambda$. Using the relationship $v = f\lambda$, Hertz found that $v$ was close to $3.00 \times 10^8$ m/s, the known speed of visible light. Thus, Hertz's experiments provided the first evidence in support of Maxwell's theory.

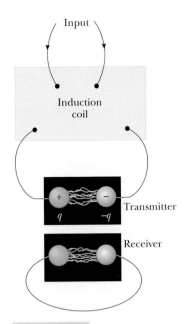

**Figure 24.10**

Schematic diagram of Hertz's apparatus for generating and detecting electromagnetic waves. The transmitter consists of two spherical electrodes connected to an induction coil, which provides short voltage surges to the spheres, setting up oscillations in the discharge. The receiver is a nearby loop containing a second spark gap.

### THINKING PHYSICS 24.2

In radio transmission, a radio wave serves as a carrier wave, and the sound wave is superimposed on the carrier wave. In amplitude modulation (AM radio), the amplitude of the carrier wave varies according to the sound wave. (The word *modulate* means "to change.") In frequency modulation (FM radio), the frequency of the carrier wave varies according to the sound wave. The navy sometimes uses flashing lights to send Morse code to neighboring ships, a process that is similar to radio broadcasting. Is this AM or FM? What is the carrier

---

* Following Hertz's discoveries, Guglielmo Marconi succeeded in developing a practical, long-range radio communication system. This system is still in common use.

frequency? What is the signal frequency? What is the broadcasting antenna? What is the receiving antenna?

**Reasoning**   The flashing of the light according to Morse code is a drastic amplitude modulation—the amplitude is changing from a maximum to zero. In this sense, it is similar to the on-and-off binary code used in computers and compact discs. The carrier frequency is that of the visible light, on the order of $10^{14}$ Hz. The signal frequency depends on the skill of the signal operator, but is on the order of a few hertz, as the light is flashed on and off. The broadcasting antenna for this modulated signal is the filament of the lightbulb in the signal source. The receiving antenna is the eye.

## 24.5 • ENERGY CARRIED BY ELECTROMAGNETIC WAVES

In Section 13.6, we found that mechanical waves carry energy. Electromagnetic waves also carry energy, and as they propagate through space they can transfer energy to objects placed in their path. This notion was introduced in Chapter 6 when we discussed the transfer mechanisms in the continuity equation for energy, and it was noted again in Chapter 17 in the discussion of thermal radiation. The rate of flow of energy in an electromagnetic wave is described by a vector **S**, called the **Poynting vector,** defined by the expression

$$S \equiv \frac{1}{\mu_0} \mathbf{E} \times \mathbf{B} \qquad [24.22]$$

The magnitude of the Poynting vector represents the rate at which energy flows through a unit surface area perpendicular to the flow and its direction is along the direction of wave propagation (Fig. 24.11). Thus, the Poynting vector represents *power per unit area.* The SI units of the Poynting vector are $J/s \cdot m^2 = W/m^2$.

As an example, let us evaluate the magnitude of **S** for a plane electromagnetic wave. We have $|\mathbf{E} \times \mathbf{B}| = EB$, because **E** and **B** are perpendicular to each other. In this case,

$$S = \frac{EB}{\mu_0} \qquad [24.23]$$

Because $B = E/c$, we can also express this as

$$S = \frac{E^2}{\mu_0 c} = \frac{cB^2}{\mu_0}$$

These equations for $S$ apply at any instant of time.

What is of more interest for a sinusoidal electromagnetic wave (Eqs. 24.18 and 24.19) is the time average of $S$ over one or more cycles, which is the **intensity** $I$. When this average is taken, we obtain an expression involving the time average of $\cos^2(kx - \omega t)$, which equals $\frac{1}{2}$. Hence, the average value of $S$ (or the intensity of the wave) is

$$I = S_{av} = \frac{E_{max} B_{max}}{2\mu_0} = \frac{E_{max}^2}{2\mu_0 c} = \frac{cB_{max}^2}{2\mu_0} \qquad [24.24]$$

**Figure 24.11**

The Poynting vector **S** for an electromagnetic wave is along the direction of wave propagation.

• *Intensity*

Recall that the energy per unit volume $u_E$, which is the instantaneous energy density associated with an electric field (Section 20.9), is given by Equation 20.31:

$$u_E = \tfrac{1}{2}\epsilon_0 E^2 \qquad \text{[24.25]}$$

and that the instantaneous energy density $u_B$ associated with a magnetic field (Section 23.7) is given by Equation 23.22:

$$u_B = \frac{B^2}{2\mu_0} \qquad \text{[24.26]}$$

Because $E$ and $B$ vary with time for an electromagnetic wave, the energy densities also vary with time. Using the relationships $B = E/c$ and $c = 1/\sqrt{\epsilon_0\mu_0}$, Equation 24.26 becomes

$$u_B = \frac{(E/c)^2}{2\mu_0} = \frac{\epsilon_0\mu_0}{2\mu_0}E^2 = \tfrac{1}{2}\epsilon_0 E^2$$

Comparing this result with the expression for $u_E$, we see that

$$u_B = u_E$$

That is, **for an electromagnetic wave, the instantaneous energy density associated with the magnetic field equals the instantaneous energy density associated with the electric field.** Hence, in a given volume the energy is equally shared by the two fields.

The **total instantaneous energy density** $u$ is equal to the sum of the energy densities associated with the electric and magnetic fields:

$$u = u_E + u_B = \epsilon_0 E^2 = \frac{B^2}{\mu_0}$$

- *Total instantaneous energy density of an electromagnetic wave*

When this is averaged over one or more cycles of an electromagnetic wave, we again obtain a factor of $\tfrac{1}{2}$. Hence, the total average energy per unit volume of an electromagnetic wave is

$$u_{av} = \epsilon_0(E^2)_{av} = \tfrac{1}{2}\epsilon_0 E_{max}^2 = \frac{B_{max}^2}{2\mu_0} \qquad \text{[24.27]}$$

- *Average energy density of an electromagnetic wave*

Comparing this result with Equation 24.24 for the average value of $S$, we see that

$$I = S_{av} = cu_{av} \qquad \text{[24.28]}$$

In other words, **the intensity of an electromagnetic wave equals the average energy density multiplied by the speed of light.**

## Quick Quiz 24.2

An electromagnetic wave transfers energy along the positive $z$ axis. At an instant of time, the electric field vector is directed in the negative $x$ direction. In what direction is the magnetic field vector pointing at this instant?

## Example 24.2   Fields Due to a Point Source

A point source of electromagnetic radiation has an average power output of 800 W. Calculate the maximum values of the electric and magnetic fields at a point 3.50 m from the source.

**Solution**   For waves propagating uniformly from a point source, the energy of the wave at a distance $r$ from the source is distributed over the surface area of an imaginary sphere of radius $r$. Thus, the intensity of the radiation at a point on the sphere is

$$I = \frac{\mathcal{P}_{av}}{4\pi r^2}$$

where $\mathcal{P}_{av}$ is the average power output of the source and $4\pi r^2$ is the area of the sphere centered on the source. Because the intensity of an electromagnetic wave is also given by Equation 24.24, we have

$$I = \frac{\mathcal{P}_{av}}{4\pi r^2} = \frac{E_{max}^2}{2\mu_0 c}$$

Solving for $E_{max}$ gives

$$E_{max} = \sqrt{\frac{\mu_0 c \mathcal{P}_{av}}{2\pi r^2}}$$

$$= \sqrt{\frac{(4\pi \times 10^{-7}\ \text{T·m/A})(3.00 \times 10^8\ \text{m/s})(800\ \text{W})}{2\pi(3.50\ \text{m})^2}}$$

$$= 62.6\ \text{V/m}$$

We calculate the maximum value of the magnetic field using this result and Equation 24.20:

$$B_{max} = \frac{E_{max}}{c} = \frac{62.6\ \text{V/m}}{3.00 \times 10^8\ \text{m/s}} = 2.09 \times 10^{-7}\ \text{T}$$

**EXERCISE**   Calculate the energy density 3.50 m from the point source.

*Answer*   $1.73 \times 10^{-8}\ \text{J/m}^3$

## 24.6 • MOMENTUM AND RADIATION PRESSURE

Electromagnetic waves transport linear momentum as well as energy. Hence, it follows that pressure is exerted on a surface when an electromagnetic wave impinges on it. In what follows, let us assume that the electromagnetic wave transports a total energy $U$ to a surface in a time interval $\Delta t$. If the surface absorbs all the incident energy $U$ in this time, Maxwell showed that the total momentum **p** delivered to this surface has a magnitude

• *Momentum delivered to a perfectly absorbing surface*

$$p = \frac{U}{c} \quad \text{(complete absorption)} \qquad \qquad \textbf{[24.29]}$$

Furthermore, if the intensity of the wave is $I$, the radiation pressure $P$ (force per unit area) exerted on the perfectly absorbing surface is

• *Radiation pressure exerted on a perfect absorbing surface*

$$P = \frac{I}{c} \qquad \qquad \textbf{[24.30]}$$

An absorbing surface for which all of the incident energy is absorbed (none is reflected) is called a **black body.** A more detailed discussion of a black body will be presented in Chapter 28.

If the surface is a perfect reflector, then the momentum delivered in a time interval $\Delta t$ for normal incidence is twice that given by Equation 24.29, or $p = 2U/c$. That is, a momentum $U/c$ is delivered first by the incident wave and then again by the reflected wave, in analogy with a ball colliding elastically with a wall.* Finally,

---

\* For *oblique* incidence, the momentum transferred is $2U \cos \theta/c$ and the pressure is given by $P = 2S \cos^2 \theta/c$, where $\theta$ is the angle between the normal to the surface and the direction of propagation.

the radiation pressure exerted on a perfect reflecting surface for normal incidence of the wave is twice that given by Equation 24.30, or $P = 2I/c$.

Although radiation pressures are very small (about $5 \times 10^{-6}$ N/m² for direct sunlight), they have been measured using torsion balances such as the one shown in Figure 24.12. Light is allowed to strike either a mirror or a black disk, both of which are suspended from a fine fiber. Light striking the black disk is completely absorbed, and so all of its momentum is transferred to the disk. Light striking the mirror (normal incidence) is totally reflected, hence the momentum transfer is twice as great as that transferred to the disk. The radiation pressure is determined by measuring the angle through which the horizontal connecting rod rotates. The apparatus must be placed in a high vacuum to eliminate the effects of air currents.

**Figure 24.12**

An apparatus for measuring the pressure exerted by light. In practice, the system is contained in a high vacuum.

## Quick Quiz 24.3

In an apparatus such as that in Figure 24.12, suppose the black disk is replaced by one with half the radius. Which of the following are different after the disk is replaced? (a) radiation pressure on the disk; (b) radiation force on the disk; (c) radiation momentum delivered to the disk in a given time interval.

### THINKING PHYSICS 24.3

A large amount of dust occurs in the interplanetary space in the Solar System. Although this dust can theoretically have a variety of sizes, from molecular size upward, very little of it is smaller than about 0.2 μm in our Solar System. Why? (*Hint:* The Solar System originally contained dust particles of all sizes.)

**Reasoning** Dust particles in the Solar System are subject to two forces: the gravitational force toward the Sun, and the force from radiation pressure, which is away from the Sun. The gravitational force is proportional to the cube of the radius of a spherical dust particle because it is proportional to the mass of the particle. The radiation pressure is proportional to the square of the radius because it depends on the planar cross section of the particle. For large particles, the gravitational force is larger than the force from radiation pressure. For small particles, less than about 0.2 μm, the larger force from radiation pressure sweeps these particles out of the Solar System.

### Example 24.3 Solar Energy

The Sun delivers about 1 000 W/m² of energy to the Earth's surface. (a) Calculate the total power incident on a roof of dimensions 8.00 m × 20.0 m.

**Solution** The Poynting vector has an average magnitude $I = S_{av} = 1\,000$ W/m², which represents the power per unit area. Assuming the radiation is incident normal to the roof, we can find the power for the whole roof:

$$\mathcal{P} = IA = (1\,000 \text{ W/m}^2)(8.00 \times 20.0 \text{ m}^2) = \boxed{1.60 \times 10^5 \text{ W}}$$

If solar energy could all be converted to electric energy, it would provide more than enough power for the average home. Solar energy is not easily harnessed, however, and the prospects for large-scale conversion are not as bright as they may appear from this simple calculation. For example, the conversion efficiency from solar to electric energy is typically 10% for photovoltaic cells. Roof systems for converting solar energy to internal energy are approximately 50% efficient; however, solar energy has other practical problems that must be considered, such as overcast days, geographic location, and energy storage.

(b) Determine the radiation pressure and radiation force on the roof assuming the roof covering is a perfect absorber.

**Solution**   Using Equation 24.30 with $I = 1\,000$ W/m$^2$, we find that the radiation pressure is

$$P = \frac{I}{c} = \frac{1\,000 \text{ W/m}^2}{3.00 \times 10^8 \text{ m/s}} = \boxed{3.33 \times 10^{-6} \text{ N/m}^2}$$

Because pressure equals force per unit area, this corresponds to a radiation force of

$$F = PA = (3.33 \times 10^{-6} \text{ N/m}^2)(160 \text{ m}^2) = \boxed{5.33 \times 10^{-4} \text{ N}}$$

**EXERCISE**   How much solar energy (in joules) is incident on the roof in 1.00 h?

*Answer*   $5.76 \times 10^8$ J.

### Example 24.4   Pressure from a Laser Pointer

Many people giving presentations use a laser pointer to direct the attention of their audience. If a 3.0-mW pointer creates a spot that is 2.0 mm in diameter, determine the radiation pressure on a screen that reflects 70% of the light striking it. The power of 3.0 mW is a time-averaged power.

**Solution**   We certainly don't expect the pressure to be very large. Before we can calculate it, we must first determine the intensity of the beam by dividing the time-averaged power delivered via the electromagnetic wave by the cross-sectional area of the beam.

$$I = \frac{\mathcal{P}}{A} = \frac{\mathcal{P}}{\pi r^2} = \frac{3.0 \times 10^{-3} \text{ W}}{\pi \left(\dfrac{2.0 \times 10^{-3} \text{ m}}{2}\right)^2} = 9.6 \times 10^2 \text{ W/m}^2$$

This is about the same as the intensity of sunlight at the Earth's surface. (Thus, it is not safe to shine a laser pointer's beam into a person's eyes, just as it is dangerous to look directly at the Sun.)

Now we must account for the radiation pressure from the laser beam. A completely reflected beam would apply a pressure of $P = 2I/c$. We can model the actual reflection process for a perfect reflector as follows. Imagine that the surface absorbs the beam, resulting in pressure $P = I/c$. Then, the surface emits the beam, resulting in additional pressure $P = I/c$. If the surface emits only a fraction $f$ of the beam (so that $f$ is the amount of the incident beam reflected), then the pressure due to the emitted beam is $P = fI/c$. Thus, the total pressure on the surface due to absorption and re-emission (reflection) is

$$P = \frac{I}{c} + f\frac{I}{c} = (1 + f)\frac{I}{c}$$

For a beam that is 70% reflected, the pressure is

$$P = (1 + 0.70)\frac{9.6 \times 10^2 \text{ W/m}^2}{3.0 \times 10^8 \text{ m/s}} = \boxed{5.4 \times 10^{-6} \text{ N/m}^2}$$

This is an extremely small value, as expected. (Recall from Section 15.1 that atmospheric pressure is approximately $10^5$ N/m$^2$.)

### Space Sailing

When imagining a trip to another planet, we normally think of traditional rocket engines that convert chemical energy in fuel carried on the spacecraft to kinetic energy of the spacecraft. An interesting alternative to this approach is that of **space sailing.** A space sailing craft includes a very large sail that reflects light. The motion of the spacecraft depends on pressure from light—that is, the force exerted on the sail by the reflection of light from the Sun. Calculations performed (before U.S. government budget cutbacks shelved early space sailing projects) showed that sailing craft could travel to and from the planets in times similar to those for traditional rockets, but for less cost.

Let us set up the physics of operation for a sailcraft and perform some calculations on expected forces and accelerations from the photon model for light (Section 9.7). We shall assume a simplification model—the sail is perfectly reflecting and oriented perpendicularly to the Sun's rays. The sailcraft is in orbit around the Earth. A photon striking the sail reflects straight back, experiencing a momentum change of twice its original momentum $p$. Thus, according to Newton's second law, the force on the sail due to this photon reflection is

$$F = \frac{\Delta p}{\Delta t} = \frac{2p}{\Delta t}$$

For a photon, the relationship between momentum and energy is (Equation 9.23)

$$p = \frac{E}{c}$$

So, for a single photon, the force imparted to the sail is

$$F = \frac{2E}{c \, \Delta t}$$

If we consider all of the photons coming from the Sun, we can calculate a total force on the sail based on $\Delta E$, the radiant energy flowing into the sail in the time $\Delta t$:

$$F_{\text{total}} = \frac{2}{c} \frac{\Delta E}{\Delta t} = \frac{2}{c} \mathcal{P}$$

where $\mathcal{P}$ is the power associated with the sunlight. The power is related to the solar constant $I_S$, which is the intensity of sunlight at the position of the Earth. We use this relationship to find the solar pressure $P$ on the sail:

$$P = \frac{F_{\text{total}}}{A} = \frac{2\mathcal{P}}{cA} = \frac{2I_S}{c}$$

This is the same result that Maxwell obtained using the wave model for light.

We now evaluate the pressure on a sail at one Earth orbit radius from the Sun, using the solar constant $I_S = 1\,340\ \text{W/m}^2$:

$$P = \frac{2I_S}{c} = \frac{2(1\,340\ \text{W/m}^2)}{3.0 \times 10^8\ \text{m/s}} = 8.9 \times 10^{-6}\ \text{N/m}^2$$

A larger sail would have more total force exerted on it, so we might opt for as large a sail as possible. A larger sail also has more mass, however. In space sailing, a parameter called the *sail loading* $\sigma$ is defined as the mass (of the sail and the payload) per unit area (of the sail):

$$\sigma = \frac{m_{\text{sail+payload}}}{A_{\text{sail}}}$$

Using the sail loading parameter and Newton's second law, we can calculate the acceleration of the sailcraft:

$$a_{\text{ideal}} = \frac{F}{m_{\text{sail+payload}}} = \frac{PA_{\text{sail}}}{m_{\text{sail+payload}}} = \frac{P}{\sigma} = \frac{8.9 \times 10^{-6}\ \text{N/m}^2}{\sigma}$$

This is an ideal acceleration because we have assumed that every photon reflects from the sail. In reality, some photons are absorbed and some may transmit. Current estimates are that materials can be developed in sail-type size for which the reflection coefficient is 90%. Thus, using the technique of Example 24.4, the pressure on the sail is $1.90 I_S/c$ rather than $2I_S/c$ and the expected acceleration is

$$a_{\text{expected}} = \frac{1.90}{2} a_{\text{ideal}} = \frac{8.5 \times 10^{-6}\ \text{N/m}^2}{\sigma}$$

**PITFALL PREVENTION 24.4**

**So many $p$'s**

Be careful in this discussion of space sailing. We have $p$ for momentum, $P$ for pressure and $\mathcal{P}$ for power. Be sure you keep these all straight!

**WEB**

For more information on solar sailing, visit **www.ugcs.caltech.edu/~diedrich/solarsails/**
For a NASA article on solar sailing, visit **science.nasa.gov/headlines/y2000/ast28jun_1m.htm**

For very high performance sails, an expected value of the sail loading is $\sigma = 1.4 \times 10^{-3}$ kg/m$^2$. For this value, we expect an acceleration of

$$a_{\text{expected}} = \frac{8.5 \times 10^{-6} \text{ N/m}^2}{1.4 \times 10^{-3} \text{ kg/m}^2} = 6.1 \times 10^{-3} \text{ m/s}^2 = 6.1 \text{ mm/s}^2$$

This calculation assumes that the only force on the sailcraft is that due to the reflected light. Let us consider also the gravitational force on the sailcraft exerted by the Sun. In the absence of any force from the light from the Sun, the acceleration of the spacecraft toward the Sun due to the gravitational force would be equal to the value of the gravitational field due to the Sun at the location of the spacecraft:

$$g = G \frac{M_{\text{Sun}}}{r^2} = (6.67 \times 10^{-11} \text{ N} \cdot \text{m}^2/\text{kg}^2) \frac{2.0 \times 10^{30} \text{ kg}}{(1.5 \times 10^{11} \text{ m})^2}$$

$$= 5.9 \times 10^{-3} \text{ m/s}^2 = 5.9 \text{ mm/s}^2$$

This is approximately the same value as the acceleration of our sailcraft due to the force exerted by the sunlight! If these two values are comparable, the sailcraft can be modeled as a particle in equilibrium—the inward gravitational force of the Sun balances the outward force exerted by the light from the Sun. If the sailcraft has an initial velocity in some direction away from the Sun, it would move in a straight line under the action of these two forces, with no necessity for fuel. A traditional spacecraft with its rocket engines turned off, on the other hand, would slow down due to the gravitational force on it due to the Sun. Both the force on the sail and the gravitational force from the Sun fall off as the inverse square of the Sun–sailcraft separation. Thus, in theory, the straight-line motion of the sailcraft would continue forever with no fuel requirement.

Before leaving the topic of space sailing, we mention one hypothesized method of sending a sailcraft to another star. By using just the motion imparted to a sailcraft by the Sun, the craft could reach Alpha Centauri in about 10 000 years. This time interval can be reduced to 30 to 100 years using a *beamed power system*. In this concept, light from the Sun is gathered by a transformation device in orbit around the Earth and is converted to a laser beam or microwave beam aimed at the sailcraft. The force from this intense beam of radiation increases the acceleration of the craft, so that the transit time is significantly reduced. Calculations indicate that the sailcraft could achieve design speeds of up to 20% of the speed of light using this technique.

## 24.7 • THE SPECTRUM OF ELECTROMAGNETIC WAVES

We have seen that all electromagnetic (EM) waves travel in a vacuum with the speed of light $c$. These waves transport energy and momentum away from a source. In 1888, Hertz successfully generated and detected the radio-frequency electromagnetic waves predicted by Maxwell. Maxwell himself had recognized as EM waves both visible light and the near-infrared radiation discovered in 1800 by William Herschel. It is now known that other forms of electromagnetic waves exist; they are distinguished by their frequencies and wavelengths.

Electromagnetic waves travel through vacuum with speed $c$, frequency $f$, and wavelength $\lambda$. The various types of electromagnetic waves, all produced by accelerating charges, are shown in Figure 24.13. Note the wide range of frequencies and

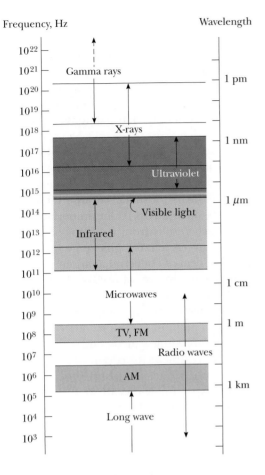

Frequency, Hz

Wavelength

**Figure 24.13**

The electromagnetic spectrum. Note the overlap between adjacent wave types.

wavelengths. For instance, a radio wave of frequency 94.7 MHz (a typical value) has the wavelength

$$\lambda = \frac{c}{f} = \frac{3.00 \times 10^8 \text{ m/s}}{94.7 \times 10^6 \text{ Hz}} = 3.17 \text{ m}$$

Let us briefly describe the wave types shown in Figure 24.13.

**Radio waves** are the result of charges accelerating, for example, through conducting wires in a radio antenna. They are generated by such electronic devices as $LC$ oscillators and are used in radio and television communication systems.

**Microwaves** (short-wavelength radio waves) have wavelengths ranging between about 1 mm and 30 cm and are also generated by electronic devices. Because of their short wavelengths, they are well suited for radar systems used in aircraft navigation and for studying the atomic and molecular properties of matter. Microwave ovens are a domestic application of these waves.

**Infrared waves** have wavelengths ranging from about 1 mm to the longest wavelength of visible light, $7 \times 10^{-7}$ m. These waves, produced by objects at room temperature and by molecules, are readily absorbed by most materials. Infrared radiation has many practical and scientific applications, including physical therapy, infrared photography, and vibrational spectroscopy. Your remote control for your television, VCR, or DVD player likely uses an infrared beam to communicate with the video device.

**PITFALL PREVENTION 24.5**

**Heat rays**

Infrared rays are often called "heat rays." This is a misnomer. Although infrared radiation is used to raise or maintain temperature, as in the case of keeping food warm with "heat lamps" at a fast-food restaurant, all wavelengths of electromagnetic radiation carry energy that can cause the temperature of a system to increase. As an example, consider using your microwave oven to bake a potato, whose temperature increases because of microwaves.

**Visible light,** the most familiar form of electromagnetic waves, is that part of the spectrum the human eye can detect. Light is produced by hot objects like lightbulb filaments and by the rearrangement of electrons in atoms and molecules. The wavelengths of visible light are classified by color, ranging from violet ($\lambda \approx 4 \times 10^{-7}$ m) to red ($\lambda \approx 7 \times 10^{-7}$ m). The eye's sensitivity is a function of wavelength and is a maximum at a wavelength of about $5.5 \times 10^{-7}$ m (yellow-green). Light is the basis of the science of optics and optical instruments, to be discussed in Chapters 25 through 27.

**Ultraviolet light** covers wavelengths ranging from about $4 \times 10^{-7}$ m (400 nm) down to $6 \times 10^{-10}$ m (0.6 nm). The Sun is an important source of ultraviolet waves, which is the main cause of suntans and sunburns. Atoms in the stratosphere absorb most of the ultraviolet waves from the Sun. (This is fortunate because ultraviolet waves in large quantities have harmful effects on humans.) One important constituent of the stratosphere is ozone ($O_3$), which results from reactions of oxygen with ultraviolet radiation. This ozone shield converts lethal high-energy ultraviolet radiation to harmless infrared radiation. A great deal of concern has arisen concerning the depletion of the protective ozone layer by the use of a class of chemicals called chlorofluorocarbons (e.g., Freon) in aerosol spray cans and as refrigerants.

**X-rays** are electromagnetic waves with wavelengths in the range of about $10^{-8}$ m (10 nm) down to $10^{-13}$ m ($10^{-4}$ nm). The most common source of x-rays is the acceleration of high-energy electrons bombarding a metal target. X-rays are used as a diagnostic tool in medicine and as a treatment for certain forms of cancer. Because x-rays damage or destroy living tissues and organisms, care must be taken to avoid unnecessary exposure and overexposure. X-rays are also used in the study of crystal structure; x-ray wavelengths are comparable to the atomic separation distances ($\approx 0.1$ nm) in solids.

**Gamma rays** are electromagnetic waves emitted by radioactive nuclei and during certain nuclear reactions. They have wavelengths ranging from about $10^{-10}$ m to less than $10^{-14}$ m. Gamma rays are highly penetrating and produce serious damage when absorbed by living tissues. Consequently, those working near such dangerous radiation must be protected with heavily absorbing materials, such as layers of lead.

**Quick Quiz 24.4**

As mentioned in Thinking Physics 24.2, the *AM* in AM radio stands for *amplitude modulation,* and *FM* stands for *frequency modulation.* If our eyes could see the electromagnetic waves from a radio antenna, what visual aspects associated with the waves (e.g., brightness, color) would you observe changing in an AM wave and an FM wave?

**APPLICATION**

Center of eyesight sensitivity

**THINKING PHYSICS 24.4**

The center of sensitivity of our eyes coincides with the center of the wavelength distribution of the Sun. Is this an amazing coincidence?

**Reasoning** This is not a coincidence—it is the result of biological evolution. Humans have evolved so as to be most visually sensitive to the wavelengths that are strongest from the Sun. It is an interesting conjecture to imagine aliens from another planet, with a Sun with a different temperature, arriving at Earth. Their eyes would have the center of sensitivity at different wavelengths than ours. How would their vision of the Earth compare to ours?

## 24.8 • POLARIZATION

As we learned in Section 24.3, the electric and magnetic vectors associated with an electromagnetic wave are at right angles to each other and also to the direction of wave propagation as shown in Figure 24.6. The phenomenon of polarization described in this section is a property that specifies the directions of the electric and magnetic fields associated with an electromagnetic wave.

An ordinary beam of light consists of a large number of waves emitted by the atoms or molecules of the light source. Each atom produces a wave with its own orientation of the electric field **E**, corresponding to the direction of atomic vibration. The direction of polarization of the electromagnetic wave is defined to be the direction in which **E** is vibrating. Because all directions of vibration are possible, however, the resultant beam of electromagnetic radiation is a superposition of waves produced by the individual atomic sources. The result is an **unpolarized** light wave, represented schematically in Figure 24.14a. The direction of wave propagation in this figure is perpendicular to the page. Note that *all* directions of the electric field vector lying in a plane perpendicular to the direction of propagation are equally probable.

A wave is said to be **linearly polarized** if the orientation of **E** is the same for all individual waves *at all times* at a particular point, as in Figure 24.14b. (Sometimes such a wave is described as **plane polarized.**) The wave described in Figure 24.6 is an example of a wave linearly polarized along the *y* axis. As the field propagates in the *x* direction, **E** is always along the *y* axis. The plane formed by **E** and the direction of propagation is called the **plane of polarization** of the wave. In Figure 24.6, the plane of polarization is the *xy* plane. It is possible to obtain a linearly polarized wave from an unpolarized wave by removing from the unpolarized wave all components of electric field vectors except those that lie in a single plane.

The most common technique for polarizing light is to send it through a material that passes only components of electric field vectors that are parallel to a characteristic direction of the material called the **polarizing direction.**

In 1938, E. H. Land discovered a material, which he called **Polaroid,** that polarizes light through selective absorption by oriented molecules. This material is fabricated in thin sheets of long-chain hydrocarbons, which are stretched during manufacture so that the molecules align. After a sheet is dipped into a solution containing iodine, the molecules become good electric conductors. However, the conduction takes place primarily along the hydrocarbon chains because the valence electrons of the molecules can move easily only along the chains (valence electrons are "free" electrons that can readily move through the conductor). As a result, the molecules readily *absorb* light whose electric field vector is parallel to their length and *transmit* light whose electric field vector is perpendicular to their length. It is common to refer to the direction perpendicular to the molecular chains as the **transmission axis.** An ideal polarizer passes the components of electric vectors that are parallel to the transmission axis. Components perpendicular to the transmission axis are absorbed. If light passes through several polarizers, whatever is transmitted has the plane of polarization parallel to the polarizing direction of the last polarizer through which it passed.

Let us now obtain an expression for the intensity of light that passes through a polarizing material. In Figure 24.15, an unpolarized light beam is incident on the first polarizing sheet, called the **polarizer,** where the transmission axis is as indicated. The light that passes through this sheet is polarized vertically, and the transmitted electric field vector is $\mathbf{E}_0$. A second polarizing sheet, called the **analyzer,**

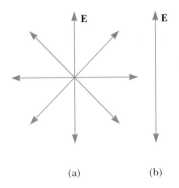

**Figure 24.14**

(a) An unpolarized light beam viewed along the direction of propagation (perpendicular to the page). The time-varying electric field vector can be in any direction in the plane of the page with equal probability.
(b) A linearly polarized light beam with the time-varying electric field vector in the vertical direction.

**PITFALL PREVENTION 24.6**
**A new type of polarization**

The polarization of light discussed here is different from the polarization of molecules discussed in Chapters 19 and 20. Be sure you understand the difference between the two meanings of polarization.

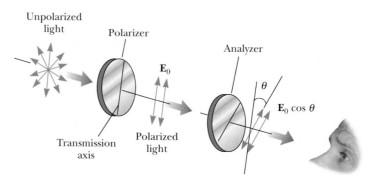

**Figure 24.15**

Two polarizing sheets whose transmission axes make an angle $\theta$ with each other. Only a fraction of the polarized light incident on the analyzer is transmitted through it.

intercepts this beam with its transmission axis at an angle of $\theta$ to the axis of the polarizer. The component of $\mathbf{E}_0$ that is perpendicular to the axis of the analyzer is completely absorbed, and the component parallel to that axis is $E_0 \cos \theta$. We know from Equation 24.24 that the transmitted intensity varies as the *square* of the transmitted amplitude, and so we conclude that the intensity of the transmitted (polarized) light varies as

• *Malus's law*

$$I = I_0 \cos^2 \theta \qquad \text{[24.31]}$$

where $I_0$ is the intensity of the polarized wave incident on the analyzer. This expression, known as **Malus's law,** applies to any two polarizing materials whose transmission axes are at an angle of $\theta$ to each other. From this expression, note that the transmitted intensity is a maximum when the transmission axes are parallel ($\theta = 0$ or $180°$), and zero (complete absorption by the analyzer) when the transmission axes are perpendicular to each other. This variation in transmitted intensity through a pair of polarizing sheets is illustrated in Figure 24.16. Because the average value of $\cos^2\theta$ is $\frac{1}{2}$, the intensity of initially unpolarized light is reduced by a factor of one half as the light passes through a single ideal polarizer.

 **THINKING   PHYSICS   24.5**

A polarizer for microwaves can be made as a grid of parallel metal wires about a centimeter apart. Is the electric field vector for microwaves transmitted through this polarizer parallel to, or perpendicular to, the metal wires?

**Reasoning**   Electric field vectors parallel to the metal wires cause electrons in the metal to oscillate parallel to the wires. Thus, the energy from the waves with these electric field vectors are transferred to the metal by accelerating these electrons and are eventually transformed to internal energy through the resistance of the metal. Waves with electric field vectors perpendicular to the metal wires are not able to accelerate electrons and pass through. Thus, the electric field polarization is perpendicular to the metal wires.

(a)

(b)

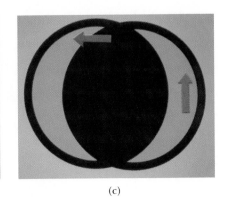
(c)

**Figure 24.16**

The intensity of light transmitted through two polarizers depends on the relative orientation of their transmission axes. (a) The transmitted light has maximum intensity when the transmission axes are aligned with each other. (b) The transmitted light intensity diminishes when the transmission axes are at an angle of 45° with each other. (c) The transmitted light intensity is a minimum when the transmission axes are at right angles to each other. *(Henry Leap and Jim Lehman)*

### context connection
## 24.9 • THE SPECIAL PROPERTIES OF LASER LIGHT

In this chapter and the next three, we shall explore the nature of laser light and a variety of applications of lasers in our technological society. The primary properties of laser light that make it useful in these applications are

- The light is coherent—The individual rays of light in a laser beam maintain a fixed phase relationship with each other, resulting in no destructive interference.
- The light is monochromatic—Laser light has a very small range of wavelengths.
- The light has a small angle of divergence—The beam spreads out very little, even over large distances.

To understand the origin of these properties, let us combine our knowledge of atomic energy levels from Chapter 11 with some special requirements for the atoms that emit laser light.

As we found in Chapter 11, the energies of an atom are quantized—only certain energies are allowed. We used a semigraphical representation called an *energy level diagram* in that chapter to help us understand the quantized energies in an atom. The production of laser light depends heavily on the properties of these energy levels in the atoms, the source of the laser light.

The word *laser* is an acronym for **l**ight **a**mplification by **s**timulated **e**mission of **r**adiation. The full name indicates one of the requirements for laser light—the process of **stimulated emission** must occur to achieve laser action.

Suppose an atom is in the excited state $E_2$, as in Figure 24.17, and a photon with energy $hf = E_2 - E_1$ is incident on it. The incoming photon can stimulate the excited atom to return to the ground state and thereby emit a second photon having the same energy $hf$ and traveling in the same direction. Note that the incident photon is not absorbed, so after the stimulated emission, two identical photons

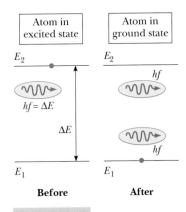

**Figure 24.17**

Stimulated emission of a photon by an incoming photon of energy $hf$. Initially, the atom is in the excited state. The incoming photon stimulates the atom to emit a second photon of energy $hf = E_2 - E_1$.

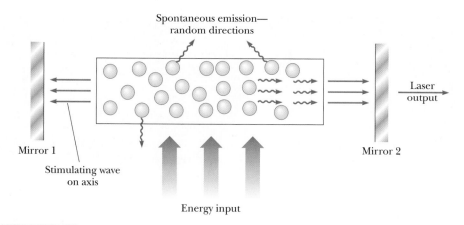

**Figure 24.18**

A schematic of a laser design. The tube contains atoms, which represent the active medium. An external source of energy (optical, electric, etc.) is needed to "pump" the atoms to excited energy states. The parallel end mirrors confine the photons to the tube. Mirror 2 is slightly transparent so that laser light leaves the tube through this mirror.

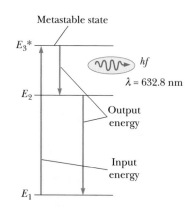

**Figure 24.19**

Energy level diagram for the neon atom, which emits photons at a wavelength of 632.8 nm through stimulated emission. The photon at this wavelength arises from the transition $E_3^* \rightarrow E_2$. This is the source of coherent light in the helium–neon gas laser.

exist: the incident photon and the emitted photon. The emitted photon is in phase with the incident photon. These photons can stimulate other atoms to emit photons in a chain of similar processes. The many photons produced in this fashion are the source of the intense, coherent light in a laser.

In order that the stimulated emission result in laser light, we must have a buildup of photons in the system. The following three conditions must be satisfied to achieve this buildup:

- The system must be in a state of **population inversion**—more atoms must be in an excited state than in the ground state. This must be true because we must have more emission of photons than absorption.
- The excited state of the system must be a *metastable state,* which means its lifetime must be long compared with the usually short lifetime of excited states, which is typically $10^{-8}$ s. In this case, stimulated emission is likely to occur before spontaneous emission. The energy of a metastable state is indicated with an asterisk $E^*$.
- The emitted photons must be confined in the system long enough to enable them to stimulate further emission from other excited atoms. This is achieved by using reflecting mirrors at the ends of the system. One end is made totally reflecting, and the other is slightly transparent to allow the laser beam to escape (Fig. 24.18).

One device that exhibits stimulated emission of radiation is the helium–neon gas laser. Figure 24.19 is an energy level diagram for the neon atom in this system. The mixture of helium and neon is confined to a glass tube that is sealed at the ends by mirrors. A voltage applied across the tube causes electrons to sweep through the tube, colliding with the atoms of the gases and raising them into excited states. Neon atoms are excited to state $E_3^*$ through this process and also as a result of collisions with excited helium atoms. Stimulated emission occurs as the neon atoms make a transition to state $E_2$ and neighboring excited atoms are stimulated. This results in the production of coherent light at a wavelength of 632.8 nm.

### Figure 24.21

An optical trap for atoms is formed at the intersection point of six counterpropagating laser beams along mutually perpendicular axes. The frequency of the laser light is tuned to be just below that for absorption by the trapped atoms. If an atom moves away from the trap, it absorbs the Doppler-shifted laser light, and the momentum of the light pushes the atom back into the trap.

### Figure 24.20

A staff member of the National Institute of Standards and Technology views a sample of trapped sodium atoms (the small yellow dot in the center of the vacuum chamber) cooled to a temperature of less than 1 mK. *(Courtesy of Mark Helfer/ National Institute of Standards and Technology)*

**WEB**

For a *Scientific American* article on laser trapping written by Steven Chu, visit **www.sciam.com/ 0292issue/0292chu.html**

An exciting area of research and technological applications began in the 1990s with the development of *laser trapping* of atoms (Fig. 24.20). One scheme, called *optical molasses* and developed by Steven Chu of Stanford University and his colleagues, involves focusing six laser beams onto a small region in which atoms are to be trapped. Each pair of lasers is along one of the $x$, $y$ and $z$ axes and emits light in opposite directions (Fig. 24.21). The frequency of the laser light is tuned to be just below the absorption frequency of the subject atom. Imagine that an atom has been placed into the trap region and moves along the positive $x$ axis toward the laser that is emitting light toward it (the rightmost laser in Figure 24.21). Because the atom is moving, the light from the laser appears Doppler shifted upward in frequency in the reference frame of the atom. This creates a match between the Doppler-shifted laser frequency and the absorption frequency of the atom, and the atom absorbs photons.* The momentum carried by these photons results in the atom being pushed back to the center of the trap. By incorporating six lasers, the atoms are pushed back into the trap regardless of which way they move along any axis.

In 1986, Chu developed *optical tweezers,* in which a single tightly focused laser beam can be used to trap and manipulate small particles. In combination with microscopes, optical tweezers have opened up many new possibilities for biologists. Optical tweezers have been used to manipulate live bacteria without damage, move chromosomes within a cell nucleus, and measure the elastic properties of a single DNA molecule. Steven Chu shared the 1997 Nobel Prize in physics with Claude Cohen-Tannoudji (Collège de France) and William Phillips (National Institute of Standards and Technology), for the development of the techniques of optical trapping.

An extension of laser trapping, *laser cooling*, is due to the fact that the normal high speeds of the atoms are reduced because they are restricted to the region of

**APPLICATION**

Optical Tweezers

**WEB**

For information on laser cooling at the National Institute of Standards and Technology, visit **physlab.nist.gov/Divisions/ Div842/Gp4/group4.html**

---

* The laser light traveling in the same direction as the atom (from the leftmost laser in Figure 24.21) is Doppler shifted further downward in frequency, so that no absorption occurs. Thus, the atom is not pushed out of the trap by the diametrically opposed laser.

the trap. As a result, the temperature of the collection of atoms can be reduced to a few microkelvins. This laser cooling allows scientists to study the behavior of atoms at extremely low temperatures.

We have explored general properties of laser light in this chapter. In the Context Connection of the next chapter, we shall explore the technology of optical fibers, in which lasers are used in a variety of applications.

# SUMMARY

**Displacement current** $I_d$ is defined as

$$I_d \equiv \epsilon_0 \frac{d\Phi_E}{dt} \qquad \textbf{[24.1]}$$

and represents an effective current through a region of space in which an electric field is changing in time.

When used with the Lorentz force law ($\mathbf{F} = q\mathbf{E} + q\mathbf{v} \times \mathbf{B}$) **Maxwell's equations** describe *all* electromagnetic phenomena:

$$\oint \mathbf{E} \cdot d\mathbf{A} = \frac{Q}{\epsilon_0} \qquad \textbf{[24.4]}$$

$$\oint \mathbf{B} \cdot d\mathbf{A} = 0 \qquad \textbf{[24.5]}$$

$$\oint \mathbf{E} \cdot d\mathbf{s} = -\frac{d\Phi_B}{dt} \qquad \textbf{[24.6]}$$

$$\oint \mathbf{B} \cdot d\mathbf{s} = \mu_0 I + \epsilon_0 \mu_0 \frac{d\Phi_E}{dt} \qquad \textbf{[24.7]}$$

**Electromagnetic waves,** which are predicted by Maxwell's equations, have the following properties:

- The electric and magnetic fields satisfy the following wave equations, which can be obtained from Maxwell's third and fourth equations:

$$\frac{\partial^2 E}{\partial x^2} = \epsilon_0 \mu_0 \frac{\partial^2 E}{\partial t^2} \qquad \textbf{[24.15]}$$

$$\frac{\partial^2 B}{\partial x^2} = \epsilon_0 \mu_0 \frac{\partial^2 B}{\partial t^2} \qquad \textbf{[24.16]}$$

- Electromagnetic waves travel through a vacuum with the speed of light $c$, where

$$c = \frac{1}{\sqrt{\epsilon_0 \mu_0}} = 3.00 \times 10^8 \text{ m/s} \qquad \textbf{[24.17]}$$

- The electric and magnetic fields of an electromagnetic wave are perpendicular to each other and perpendicular to the direction of wave propagation; hence, electromagnetic waves are transverse waves.
- The instantaneous magnitudes of $\mathbf{E}$ and $\mathbf{B}$ in an electromagnetic wave are related by the expression

$$\frac{E}{B} = c \qquad \textbf{[24.20]}$$

- Electromagnetic waves carry energy. The rate of flow of energy crossing a unit area is described by the **Poynting vector S**, where

$$\mathbf{S} \equiv \frac{1}{\mu_0} \mathbf{E} \times \mathbf{B} \qquad \textbf{[24.22]}$$

- Electromagnetic waves carry momentum and hence can exert pressure on surfaces. If an electromagnetic wave whose intensity is $I$ is completely absorbed by a surface on which it is normally incident, the radiation pressure on that surface is

$$P = \frac{I}{c} \quad \text{(complete absorption)} \qquad \textbf{[24.30]}$$

If the surface totally reflects a normally incident wave, the pressure is doubled.

The electric and magnetic fields of a sinusoidal plane electromagnetic wave propagating in the positive $x$ direction can be written

$$E = E_{\max} \cos(kx - \omega t) \qquad \textbf{[24.18]}$$

$$B = B_{\max} \cos(kx - \omega t) \qquad \textbf{[24.19]}$$

where $\omega$ is the angular frequency of the wave and $k$ is the angular wave number. These equations represent special solutions to the wave equations for $\mathbf{E}$ and $\mathbf{B}$.

The average value of the Poynting vector for a plane electromagnetic wave has the magnitude

$$I = S_{\text{av}} = \frac{E_{\max} B_{\max}}{2\mu_0} = \frac{E_{\max}^2}{2\mu_0 c} = \frac{c B_{\max}^2}{2\mu_0} \qquad \textbf{[24.24]}$$

The average power per unit area (intensity) of a sinusoidal plane electromagnetic wave equals the average value of the Poynting vector taken over one or more cycles.

The **electromagnetic spectrum** includes waves covering a broad range of frequencies and wavelengths.

When polarized light of intensity $I_0$ is incident on a polarizing film, the light transmitted through the film has an intensity equal to $I_0 \cos^2 \theta$, where $\theta$ is the angle between the transmission axis of the polarizing film and the electric field vector of the incident light.

## QUESTIONS

1. What is the fundamental source of electromagnetic radiation?
2. Does a wire connected to the terminals of a battery emit electromagnetic waves? Explain.
3. Describe the physical significance of the Poynting vector.
4. If a high-frequency current is passed through a solenoid containing a metallic core, the core becomes warm due to induction. Explain why the core becomes warm in this situation.
5. Certain orientations of the receiving antenna on a television set give better reception than other orientations. Furthermore, the best orientation varies from station to station. Explain.
6. If you charge a comb by running it through your hair and then hold the comb next to a bar magnet, do the electric and magnetic fields produced constitute an electromagnetic wave?
7. An empty plastic or glass dish removed from a microwave oven is cool to the touch right after it is removed from the oven. How is this possible? (You can assume that your electric bill has been paid.)
8. List as many similarities and differences between sound waves and light waves as you can.
9. When light (or another form of electromagnetic radiation) travels across a given region, what is it that moves?
10. Suppose an alien from another planet had eyes that were sensitive to infrared radiation. Describe what the alien would see if he looked around the room you are now in. That is, what would be bright and what would be dim?
11. Why should an infrared photograph of a person look different from a photograph taken with visible light?
12. A welder must wear protective glasses and clothing to prevent eye damage and sunburn. What does this imply about the light produced by the welding?
13. Radio stations often advertise "instant news." If what they mean is that you hear the news at the instant they speak it, is their claim true? About how long would it take for a message to travel across this country by radio waves, assuming that these waves could travel this great distance and still be detected?
14. Light from the Sun takes approximately $8\frac{1}{3}$ min to reach the Earth. During this time the Earth has continued to move in its orbit around the Sun. How far is the actual location of the Sun from its image in the sky?

## PROBLEMS

1, 2, 3 = straightforward, intermediate, challenging     □ = full solution available in the *Student Solutions Manual and Study Guide*

**web** = solution posted at **http://www.harcourtcollege.com/physics/**      = computer useful in solving problem

 = Interactive Physics     ▩ = paired numerical/symbolic problems     ◩ = life science application

### Section 24.1 Displacement Current and the Generalized Ampère's Law

1. Consider the situation shown in Figure P24.1. An electric field of 300 V/m is confined to a circular area 10.0 cm in diameter and directed outward perpendicular to the plane of the figure. If the field is increasing at a rate of 20.0 V/m·s, what are the direction and magnitude of the magnetic field at the point $P$, 15.0 cm from the center of the circle?

### Section 24.2 Maxwell's Equations

2. A very long, thin rod lying along the $x$ axis has a linear charge density of 35.0 nC/m. The rod moves in the $x$ direction at a speed of 15.0 Mm/s. Find (a) the electric field the rod creates at the point (0, 20.0 cm, 0), (b) the magnetic field it creates at the same point, and (c) the force exerted on an electron at this point when its velocity is 240**i** Mm/s.

### Section 24.3 Electromagnetic Waves

*Note:* Assume that the medium is vacuum unless specified otherwise.

**Figure P24.1**

3. (a) The distance to the North Star, Polaris, is approximately $6.44 \times 10^{18}$ m. If Polaris were to burn out today, in what year would we see it disappear? (b) How long does it take for sunlight to reach the Earth? (c) How long does it take for a microwave radar signal to travel from the Earth to the Moon and back? (d) How long does it take for a radio wave to travel once around the Earth in a great circle, close to the planet's surface? (e) How long does it take for light to reach you from a lightning stroke 10.0 km away?

4. The speed of an electromagnetic wave traveling in a transparent nonmagnetic substance is $v = 1/\sqrt{\kappa \epsilon_0 \mu_0}$, where $\kappa$ is the dielectric constant of the substance. Determine the speed of light in water, which has a dielectric constant at optical frequencies of 1.78.

5. Figure 24.6 shows a plane electromagnetic sinusoidal wave propagating in the $x$ direction. Suppose that the wavelength is 50.0 m and the electric field vibrates in the $xy$ plane with an amplitude of 22.0 V/m. Calculate (a) the frequency of the wave and (b) the magnitude and direction of **B** when the electric field has its maximum value in the negative $y$ direction. (c) Write an expression for $B$ in the form

$$B = B_{max} \cos(kx - \omega t)$$

with numerical values for $B_{max}$, $k$, and $\omega$.

6. An electromagnetic wave in vacuum has an electric field amplitude of 220 V/m. Calculate the amplitude of the corresponding magnetic field.

7. In SI units, the electric field in an electromagnetic wave is described by

$$E_y = 100 \sin(1.00 \times 10^7 x - \omega t)$$

Find (a) the amplitude of the corresponding magnetic field oscillations, (b) the wavelength $\lambda$, and (c) the frequency $f$.

8. Verify by substitution that the following equations are solutions to Equations 24.15 and 24.16, respectively:

$$E = E_{max} \cos(kx - \omega t)$$
$$B = B_{max} \cos(kx - \omega t)$$

9. **Review Problem.** A standing wave interference pattern is set up by radio waves between two metal sheets 2.00 m apart. This is the shortest distance between the plates that will produce a standing wave pattern. What is the fundamental frequency?

10. A microwave oven is powered by an electron tube, called a magnetron, that generates electromagnetic waves of frequency 2.45 GHz. The microwaves enter the oven and are reflected by the walls. The standing wave pattern produced in the oven can cook food unevenly, with hot spots in the food at antinodes and cool spots at nodes, so a turntable is often used to rotate the food and distribute the energy. If a microwave oven intended for use with a turntable is instead used with a cooking dish in a fixed position, the antinodes can appear as burn marks on foods such as carrot strips or cheese. The separation distance between the burns is measured to be 6 cm ± 5%. From these data, calculate the speed of the microwaves.

11. *The red shift.* A light source recedes from an observer with a speed $v_{source}$, which is small compared with $c$. (a) Show that the fractional shift in the measured wavelength is given by the approximate expression

$$\frac{\Delta\lambda}{\lambda} \approx \frac{v_{source}}{c}$$

This phenomenon is known as the red shift, because the visible light is shifted toward the red. (b) Spectroscopic measurements of light at $\lambda = 397$ nm coming from a galaxy in the constellation Ursa Major reveal a red shift of 20.0 nm. What is the recessional speed of the galaxy?

12. (a) Show that the Doppler shift $\Delta\lambda$ in the wavelength of light is described by the expression

$$\frac{\Delta\lambda}{\lambda} + 1 = \sqrt{\frac{c-v}{c+v}}$$

where $\lambda$ is the source wavelength and $v$ is the speed of relative approach between source and observer. (b) How fast would a motorist have to be going to make a red light appear green? Consider 650 nm to be a typical wavelength for red light and 550 nm as typical for green.

## Section 24.4 Hertz's Discoveries

13. A fixed inductance of 1.05 $\mu$H is used in series with a variable capacitor in the tuning section of a radio. What capacitance tunes the circuit to the signal from a station broadcasting at 6.30 MHz?

14. Calculate the inductance of an $LC$ circuit that oscillates at 120 Hz when the capacitance is 8.00 $\mu$F.

15. The switch in Figure P24.15 is connected to point $a$ for a long time. After the switch is thrown to point $b$, what are (a) the frequency of oscillation of the $LC$ circuit, (b) the maximum charge that appears on the capacitor, (c) the maximum current in the inductor, and (d) the total energy the circuit possesses at $t = 3.00$ s?

**Figure P24.15**

## Section 24.5 Energy Carried by Electromagnetic Waves

16. How much electromagnetic energy per cubic meter is contained in sunlight, if the intensity of sunlight at the Earth's surface under a fairly clear sky is 1 000 W/m²?

17. An AM radio station broadcasts isotropically (equally in all directions) with an average power of 4.00 kW. A dipole receiving antenna 65.0 cm long is at a location 4.00 miles from the transmitter. Compute the amplitude of the emf induced by this signal between the ends of the receiving antenna.

18. What is the average magnitude of the Poynting vector 5.00 miles from a radio transmitter broadcasting isotropically with an average power of 250 kW?

19. A community plans to build a facility to convert solar radiation to electric power. They require 1.00 MW of power, and the system to be installed has an efficiency of 30.0% (i.e., 30.0% of the solar energy incident on the surface is converted to electric energy). What must be the effective area of a perfectly absorbing surface used in such an installation, assuming a constant intensity of 1 000 W/m²?

20. In a region of free space the electric field at an instant of time is $\mathbf{E} = (80.0\mathbf{i} + 32.0\mathbf{j} - 64.0\mathbf{k})$ N/C and the magnetic field is $\mathbf{B} = (0.200\mathbf{i} + 0.080\ 0\mathbf{j} + 0.290\mathbf{k})$ $\mu$T. (a) Show that the two fields are perpendicular to each other. (b) Determine the Poynting vector for these fields.

21. The filament of an incandescent lamp has a 150-$\Omega$ resistance and carries a direct current of 1.00 A. The filament is 8.00 cm long and 0.900 mm in radius. (a) Calculate the Poynting vector at the surface of the filament. (b) Find the magnitude of the electric and magnetic fields at the surface of the filament.

22. At what distance from a 100-W electromagnetic wave point source does $E_{max} = 15.0$ V/m?

23. One of the weapons being considered for an antimissile system is a laser that could be used to destroy ballistic missiles. When a high-power laser is used in the Earth's atmosphere, the electric field can ionize the air, turning it into a conducting plasma that reflects back the laser light. In dry air at 0°C and 1 atm, electric breakdown occurs for fields with amplitudes above about 3.00 MV/m. What intensity for the laser beam would produce such a field?

## Section 24.6 Momentum and Radiation Pressure

24. A plane electromagnetic wave of intensity 6.00 W/m² strikes a small pocket mirror, of area 40.0 cm², held perpendicular to the approaching wave. (a) What momentum does the wave transfer to the mirror each second? (b) Find the force that the wave exerts on the mirror.

25. A radio wave transmits 25.0 W/m² of power per unit area. A flat surface of area $A$ is perpendicular to the direction of propagation of the wave. Calculate the radiation pressure on it if the surface is a perfect absorber.

26. A possible means of space flight is to place a perfectly reflecting aluminized sheet into orbit around the Earth and then use the light from the Sun to push this "solar sail." Suppose a sail of area $6.00 \times 10^5$ m² and mass 6 000 kg is placed in orbit facing the Sun. (a) What force is exerted on the sail? (b) What is the sail's acceleration? (c) How long does it take the sail to reach the Moon, $3.84 \times 10^8$ m away? Ignore all gravitational effects, assume that the acceleration calculated in part (b) remains constant, and assume a solar intensity of 1 340 W/m².

27. A 15.0-mW helium–neon laser ($\lambda = 632.8$ nm) emits a beam of circular cross-section with a diameter of 2.00 mm. (a) Find the maximum electric field in the beam. (b) What total energy is contained in a 1.00-m length of the beam? (c) Find the momentum carried by a 1.00-m length of the beam.

28. Given that the intensity of solar radiation incident on the upper atmosphere of the Earth is 1 340 W/m², determine (a) the intensity of solar radiation incident on Mars, (b) the total power incident on Mars, and (c) the radiation force acting on the planet, if it absorbs nearly all of the light. (d) Compare this force to the gravitational attraction between Mars and the Sun. (See Table 11.1.)

29. **Review Problem.** Accelerating charges radiate electromagnetic waves. Calculate the wavelength of radiation produced by a proton in a cyclotron with a radius of 0.500 m and magnetic field of 0.350 T.

## Section 24.7 The Spectrum of Electromagnetic Waves

30. Classify waves with frequencies of 2 Hz, 2 kHz, 2 MHz, 2 GHz, 2 THz, 2 PHz, 2 EHz, 2 ZHz, and 2 YHz on the electromagnetic spectrum. Classify waves with wavelengths of 2 km, 2 m, 2 mm, 2 $\mu$m, 2 nm, 2 pm, 2 fm, and 2 am.

31. Compute an order-of-magnitude estimate for the frequency of an electromagnetic wave with a wavelength equal to (a) your height; (b) the thickness of this sheet of paper. How is each wave classified on the electromagnetic spectrum?

32. The human eye is most sensitive to light having a wavelength of $5.50 \times 10^{-7}$ m, which is in the green-yellow region of the visible electromagnetic spectrum. What is the frequency of this light?

33. What wavelengths of electromagnetic waves in free space have frequencies of (a) $5.00 \times 10^{19}$ Hz and (b) $4.00 \times 10^9$ Hz?

34. Suppose you are located 180 m from a radio transmitter. (a) How many wavelengths are you from the transmitter if the station calls itself 1 150 AM? (The AM band frequen-

cies are in kilohertz.) (b) What if this station were 98.1 FM? (The FM band frequencies are in megahertz.)

35. *This just in!* An important news announcement is transmitted by radio waves to people sitting next to their radios 100 km from the station, and by sound waves to people sitting across the newsroom, 3.00 m from the newscaster. Who receives the news first? Explain. Take the speed of sound in air to be 343 m/s.

36. Twelve VHF television channels (Channels 2–13) lie in the range of frequencies between 54.0 MHz and 216 MHz. Each channel is assigned a width of 6.0 MHz, with the two ranges 72.0–76.0 MHz and 88.0–174 MHz reserved for non-TV purposes. (Channel 2, for example, lies between 54.0 and 60.0 MHz.) Calculate the wavelength range for (a) Channel 4, (b) Channel 6, and (c) Channel 8.

## Section 24.8    Polarization

37. Plane-polarized light is incident on a single polarizing disk with the direction of $E_0$ parallel to the direction of the transmission axis. Through what angle should the disk be rotated so that the intensity in the transmitted beam is reduced by a factor of (a) 3.00, (b) 5.00, (c) 10.0?

38. Unpolarized light passes through two polaroid sheets. The axis of the first is vertical, and that of the second is at 30.0° to the vertical. What fraction of the incident light is transmitted?

39. Two polarizing sheets are placed together with their transmission axes crossed so that no light is transmitted. A third sheet is inserted between them with its transmission axis at an angle of 45.0° with respect to each of the other axes. The combination is shown in Figure P24.39 if we take $\theta_1 = 0$, $\theta_2 = 45.0°$, and $\theta_3 = 90.0°$. Assume each polarizing sheet is ideal. Find the fraction of the incident unpolarized light intensity transmitted by the three-sheet combination.

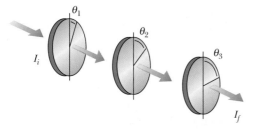

**Figure P24.39**    Problems 39 and 41.

40. You want to rotate the plane of polarization of a polarized light beam by 45.0° with a maximum intensity reduction of 10.0%. (a) How many sheets of perfect polarizers do you need to achieve your goal? (b) What is the angle between adjacent polarizers?

41. In Figure P24.39, suppose that the transmission axes of the left and right polarizing disks are perpendicular to each other. Also, let the center disk be rotated on the common axis with an angular speed $\omega$. Show that if unpolarized light is incident on the left disk with an intensity $I_{max}$, the intensity of the beam emerging from the right disk is

$$I = \tfrac{1}{16} I_{max}(1 - \cos 4\omega t)$$

This means that the intensity of the emerging beam is modulated at a rate four times the rate of rotation of the center disk. [*Hint:* Use the trigonometric identities $\cos^2 \theta = (1 + \cos 2\theta)/2$ and $\sin^2\theta = (1 - \cos 2\theta)/2$, and recall that $\theta = \omega t$.]

## Section 24.9    Context Connection—The Special Properties of Laser Light

42. High-power lasers in factories are used to cut through cloth and metal (Fig. P24.42). One such laser has a beam diameter of 1.00 mm and generates an electric field having an amplitude of 0.700 MV/m at the target. Find (a) the amplitude of the magnetic field produced, (b) the intensity of the laser, and (c) the power delivered by the laser.

**Figure P24.42**    A laser cutting device mounted on a robot arm is being used to cut through a metallic plate. *(Philippe Plailly/SPL/Photo Researchers)*

43. The carbon dioxide laser is one of the most powerful developed. The energy difference between the two laser levels is 0.117 eV. Determine the frequency and wavelength of the radiation emitted by this laser. In what portion of the electromagnetic spectrum does this radiation occur?

44. The number $N$ of atoms in a particular state is called the population of that state. This number depends on the energy of that state and the temperature. In thermal equilibrium the population of atoms in a state of energy $E_n$ is given by a Boltzmann distribution expression

$$N = N_g e^{-(E_n - E_g)/k_B T}$$

where $T$ is the absolute temperature and $N_g$ is the population of the ground state, of energy $E_g$. For simplicity, we assume that each level has only one quantum state associated with it. (a) Before the power is switched on, the neon atoms in a laser are in thermal equilibrium at 27.0°C. Find the equilibrium ratio of the populations of the states $E_3^*$ and $E_2$ shown in Figure 24.19. Lasers operate by a clever artificial production of a "population inversion" between the upper and lower atomic energy states involved in the lasing transition. This means that more atoms are in the upper excited state then in the lower one. Consider the helium–neon laser transition at 632.8 nm. Assume that 2.00% more atoms are in the upper state than in the lower. (b) To demonstrate how unnatural such a situation is, find the temperature for which the Boltzmann distribution describes a 2.00% population inversion. (c) Why does such a situation not occur naturally?

45. A Nd:YAG laser used in eye surgery emits a 3.00-mJ pulse in 1.00 ns, focused to a spot 30.0 $\mu$m in diameter on the retina. (a) Find (in SI units) the power per unit area at the retina. (This quantity is called the irradiance.) (b) What energy is delivered to an area of molecular size, taken as a circular area 0.600 nm in diameter?

46. A pulsed ruby laser emits light at 694.3 nm. For a 14.0-ps pulse containing 3.00 J of energy, find (a) the physical length of the pulse as it travels through space and (b) the number of photons in it. (c) If the beam has a circular cross-section of 0.600 cm diameter, find the number of photons per cubic millimeter.

47. Figure P24.47 shows portions of the energy level diagrams of the helium and neon atoms. An electric discharge ex-

cites the He atom from its ground state to its excited state of 20.61 eV. The excited He atom collides with a Ne atom in its ground state and excites this atom to the state at 20.66 eV. Lasing action takes place for electron transitions from $E_3^*$ to $E_2$ in the Ne atoms. From the data in the figure, show that the wavelength of the red He-Ne laser light is approximately 633 nm.

### Additional Problems

48. Assume that the intensity of solar radiation incident on the cloudtops of Earth is 1 340 W/m². (a) Calculate the total power radiated by the Sun, taking the average Earth–Sun separation to be $1.496 \times 10^{11}$ m. (b) Determine the maximum values of the electric and magnetic fields in the sunlight at the Earth's location.

49. The intensity of solar radiation at the top of the Earth's atmosphere is 1 340 W/m². Assuming that 60% of the incoming solar energy reaches the Earth's surface and assuming that you absorb 50% of the incident energy, make an order-of-magnitude estimate of the amount of solar energy you absorb in a 60-min sunbath.

50. Consider a small, spherical particle of radius $r$ located in space a distance $R$ from the Sun. (a) Show that the ratio $F_{rad}/F_{grav}$ is proportional to $1/r$, where $F_{rad}$ is the force exerted by solar radiation and $F_{grav}$ is the force of gravitational attraction. (b) The result of part (a) means that, for a sufficiently small value of $r$, the force exerted on the particle by solar radiation exceeds the force of gravitational attraction. Calculate the value of $r$ for which the particle is in equilibrium under the two forces. (Assume that the particle has a perfectly absorbing surface and a mass density of 1.50 g/cm³. Let the particle be located $3.75 \times 10^{11}$ m from the Sun, and use 214 W/m² as the value of the solar intensity at that point.)

51. Lasers have been used to suspend spherical glass beads in the Earth's gravitational field. (a) If a bead has a mass $m$ and a density $\rho$, determine the radiation intensity needed to support the bead. (b) If the beam has a radius $r$, what is the power required for this laser?

52. A dish antenna having a diameter of 20.0 m receives (at normal incidence) a radio signal from a distant source, as shown in Figure P24.52. The radio signal is a continuous sinusoidal wave with amplitude $E_{max} = 0.200$ $\mu$V/m. Assume the antenna absorbs all the radiation that falls on the dish. (a) What is the amplitude of the magnetic field in this wave? (b) What is the intensity of the radiation received by this antenna? (c) What is the power received by the antenna? (d) What force is exerted on the antenna by the radio waves?

**Figure P24.47**

**Figure P24.52**

**53.** In 1965, Arno Penzias and Robert Wilson discovered that cosmic microwave radiation was left over from the Big Bang expansion of the Universe. Suppose the energy density of this background radiation is $4.00 \times 10^{-14}$ J/m$^3$. Determine the corresponding electric field amplitude.

**54.** A hand-held cellular telephone operates in the 860- to 900-MHz band and has a power output of 0.600 W from an antenna 10.0 cm long (Fig. P24.54). (a) Find the average magnitude of the Poynting vector 4.00 cm from the antenna, at the location of a typical person's head. Assume that the antenna emits energy with cylindrical wave fronts. (The actual radiation from antennas follows a more complicated pattern.) (b) The ANSI/IEEE C95.1-1991 maximum exposure standard is 0.57 mW/cm$^2$ for persons living near cellular telephone base stations, who would be continuously exposed to the radiation. Compare the answer to part (a) with this standard.

**Figure P24.54** (© *1998 Adam Smith/FPG International*)

**55.** A linearly polarized microwave of wavelength 1.50 cm is directed along the positive *x* axis. The electric field vector has a maximum value of 175 V/m and vibrates in the *xy* plane. (a) Assume that the magnetic field component of the wave can be written in the form $B = B_{max} \sin(kx - \omega t)$

and give values for $B_{max}$, *k*, and $\omega$. Also, determine in which plane the magnetic field vector vibrates. (b) Calculate the average value of the Poynting vector for this wave. (c) What radiation pressure would this wave exert if it were directed at normal incidence onto a perfectly reflecting sheet? (d) What acceleration would be imparted to a 500-g sheet (perfectly reflecting and at normal incidence) with dimensions of 1.00 m × 0.750 m?

**56.** **Review Problem.** (a) A couple installs a solar water heater on the roof of their house (Fig. P24.56). The solar energy collector consists of a flat closed box with extraordinarily good thermal insulation. Its interior is painted black, and its front face is made of insulating glass. Assume that its emissivity for visible light is 0.900 and its emissivity for infrared light is 0.700. Assume that the noon Sun shines in perpendicular to the glass with intensity 1 000 W/m$^2$, and that no water enters or leaves the box. Find the steady-state temperature of the interior of the box. (b) The couple now builds an identical box with no water tubes. It lies flat on the ground in front of the house. They use it as a cold frame, where they plant seeds in early spring. If the same noon Sun is at an elevation angle of 50.0°, find the steady-state temperature of the interior of this box, assuming that the ventilation slots are tightly closed.

**Figure P24.56** (© *Bill Banaszewski/Visuals Unlimited*)

**57.** An astronaut, stranded in space 10.0 m from his spacecraft and at rest relative to it, has a mass (including equipment) of 110 kg. Because he has a 100-W light source that forms a directed beam, he considers using the beam as a photon rocket to propel himself continuously toward the spacecraft. (a) Calculate how long it takes him to reach the spacecraft by this method. (b) Suppose, instead, that he decides to throw the light source away in a direction opposite the spacecraft. If the mass of the light source is 3.00 kg and, after being thrown, moves at 12.0 m/s *relative to the recoiling astronaut*, how long does it take for the astronaut to reach the spacecraft?

**58.** The Earth reflects approximately 38.0% of the incident sunlight from its clouds and surface. (a) Given that the intensity of solar radiation is 1 340 W/m², what is the radiation pressure on the Earth, in pascals, when the Sun is straight overhead? (b) Compare this with normal atmospheric pressure at the Earth's surface, which is 101 kPa.

**59.** **Review Problem.** In the absence of cable input or a satellite dish, a television set can use a dipole-receiving antenna for VHF channels and a loop antenna for UHF channels (Fig. P24.59). The UHF antenna produces an emf from the changing magnetic flux through the loop.

**Figure P24.59** *(George Semple)*

The TV station broadcasts a signal with a frequency $f$, and the signal has an electric field amplitude $E_{max}$ and a magnetic field amplitude $B_{max}$ at the location of the receiving antenna. (a) Using Faraday's law, derive an expression for the amplitude of the emf that appears in a single-turn, circular loop antenna with a radius $r$, which is small compared with the wavelength of the wave. (b) If the electric field in the signal points vertically, what orientation of the loop produces the best reception?

**60.** **Review Problem.** A 1.00-m-diameter mirror focuses the Sun's rays onto an absorbing plate 2.00 cm in radius, which holds a can containing 1.00 L of water at 20.0°C. (a) If the solar intensity is 1.00 kW/m², what is the intensity on the absorbing plate? (b) What are the maximum magnitudes of the fields **E** and **B**? (c) If 40.0% of the energy is absorbed, how long does it take to bring the water to its boiling point?

**61.** The electromagnetic power radiated by a nonrelativistic moving point charge $q$ having an acceleration $a$ is

$$\mathcal{P} = \frac{q^2 a^2}{6\pi\epsilon_0 c^3}$$

where $\epsilon_0$ is the permittivity of free space and $c$ is the speed of light in vacuum. (a) Show that the right side of this equation is in watts. (b) If an electron is placed in a constant electric field of 100 N/C, determine the acceleration of the electron and the electromagnetic power radiated by this electron. (c) If a proton is placed in a cyclotron with a radius of 0.500 m and a magnetic field of magnitude 0.350 T, what electromagnetic power does this proton radiate?

## ANSWERS TO QUICK QUIZZES

**24.1** Zero. Figure 24.6 shows that the **B** and **E** vectors reach their maximum and minimum values at the same time.

**24.2** $-y$. This can be determined by using the right-hand rule for the cross product $\mathbf{E} \times \mathbf{B}$.

**24.3** (b), (c). The radiation pressure (a) does not change because pressure is force per unit area. In (b), the smaller disk absorbs less radiation, resulting in a smaller force. For the same reason, the momentum in (c) is reduced.

**24.4** The AM wave, because its amplitude is changing, would appear to vary in brightness. The FM wave would have changing colors because the color we perceive is related to the frequency of the light.

# chapter
# 25

This photograph of a rainbow shows a distinct secondary rainbow with the colors reversed. The appearance of the rainbow depends on three optical phenomena discussed in this chapter: reflection, refraction, and dispersion.

*(Mark D. Phillips/Photo Researchers, Inc.)*

# Reflection and Refraction of Light

The preceding chapter serves as a bridge between electromagnetism and the area of physics called *optics*. Now that we have established the wave nature of electromagnetic radiation, we shall study the behavior of visible light and apply what we learn to all electromagnetic radiation. Our emphasis in this chapter will be on the behavior of light as it encounters an interface between two media.

So far, we have focused on the wave nature of light and discussed it in terms of our wave simplification model. As we learn more about the behavior of light, however, we shall return to our particle simplification model, especially as we incorporate the notions of quantum physics, beginning in Chapter 28. As we discuss in the opening section of the current chapter, a long historical debate took place between proponents of wave and particle models for light.

## 25.1 • THE NATURE OF LIGHT

Although we encounter light every day, its behavior is complicated. We have discussed both the particle model and the wave model in previous chapters as simplification models to help us gain physical understanding. Both of these models have been applied to the behavior of light. Until the beginning of the 19th century, most scientists thought that light was a stream of particles emitted by a light source. According to this model, the light particles stimulated the sense of sight on entering the eye. The chief architect of this particle model of light was Isaac Newton. The model provided a simple explanation of some known experimental facts concerning the nature of light—namely, the laws of reflection and refraction—to be discussed in this chapter.

Most scientists accepted the particle model of light. During Newton's lifetime, however, another model was proposed—a model that views light as having wave-like properties. In 1678, a Dutch physicist and astronomer, Christiaan Huygens (1629–1695), showed that a wave model of light can also explain the laws of reflection and refraction. The wave model did not receive immediate acceptance for several reasons. All the waves known at the time (sound, water, and so on) traveled through a medium, but light from the Sun could travel to Earth through empty space. Furthermore, it was argued that if light were based on wave motion, the waves would be able to bend around obstacles; hence, we should be able to see around corners. It is now known that light does indeed bend around the edges of objects. This phenomenon, known as *diffraction,* is not easy to observe because light waves have such short wavelengths. (We shall discuss diffraction in Chapter 27.) Even though experimental evidence for the diffraction of light was discovered by Francesco Grimaldi (1618–1663) around 1660, most scientists rejected the wave model for more than a century and adhered to Newton's particle model. This was, for the most part, due to Newton's great reputation as a scientist.

The first clear demonstration of the wave nature of light was provided in 1801 by Thomas Young (1773–1829), who showed that under appropriate conditions, light exhibits interference behavior. That is, light waves emitted by a single source and traveling along two different paths can arrive at some point, combine, and cancel each other by destructive interference. Such behavior could not be explained at that time by a particle model, because scientists could not imagine how two or more particles could come together and cancel one another. Several years later, a French physicist, Augustin Fresnel (1788–1827), performed a number of experiments on interference, building a stronger foundation for the wave model. In 1850, Jean Foucault (1819–1868) provided further evidence of the inadequacy of the particle model by showing that the speed of light in liquids is less than that in air; according to the particle model, the speed of light would be *higher* in glasses and liquids than in air. Additional developments during the 19th century led to the general acceptance of the wave model of light.

A critical development concerning the understanding of light was the work of James Clerk Maxwell, who in 1865 mathematically predicted that light is a form of high-frequency electromagnetic wave. As discussed in Chapter 24, Hertz in 1887 provided experimental confirmation of Maxwell's theory by producing and detecting other electromagnetic waves. Furthermore, Hertz and other investigators showed that these waves exhibited reflection, refraction, and all the other characteristic properties of waves.

Although the electromagnetic wave model seemed to be well established and could explain most known properties of light, some experiments could not be explained by the assumption that light was a wave. The most striking of these was the *photoelectric effect,* discovered by Hertz, in which electrons are ejected from a metal when its surface is exposed to light. We shall explore this experiment in detail in Chapter 28.

In view of these developments, light must be regarded as having a dual nature: **in some cases light acts like a wave and in others it acts like a particle.** The classical electromagnetic wave model provides an adequate explanation of light propagation and interference, whereas the photoelectric effect and other experiments involving the interaction of light with matter are best explained by assuming that light is a particle. Light is light, to be sure. The question "Is light a wave or a particle?" is inappropriate; in some experiments, we measure its wave properties; in other experiments we measure its particle properties. This curious dual nature of light may be unsettling at this point, but will be clarified when we introduce the notion of a *quantum particle.* The photon, a particle of light, is our first example of a quantum particle, which we shall explore more fully in Chapter 28. Until then, let us focus our attention on the properties of light that can be satisfactorily explained with the wave model.

See the *Core Concepts in Physics CD-ROM,* Screen 14.2

## 25.2 • THE RAY MODEL IN GEOMETRIC OPTICS

In the beginning of our study of optics, we shall use a simplification model called the **ray model** or the **ray approximation.** A **ray** is a straight line drawn along the direction of propagation of a single wave, showing the path of the wave as it travels through space. The ray approximation involves geometric models based on these straight lines. Phenomena explained with the ray approximation do not depend explicitly on the wave nature of light, other than its propagation along a straight line.

A set of light waves can be represented by wave fronts, as illustrated in the pictorial representation in Figure 25.1 for a plane wave, which was introduced in Section 24.3. The definition of a wave front requires that the rays are perpendicular to the wave front at every location in space.

If a plane wave meets a barrier containing a circular opening whose diameter is large relative to the wavelength as in Figure 25.2a, the individual waves emerging from the opening continue to move in a straight line (apart from some small edge

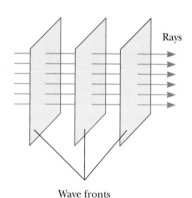

Rays

Wave fronts

**Figure 25.1**

A plane wave propagating to the right. Note that the rays, which always point in the direction of wave motion, are straight lines perpendicular to the wave fronts.

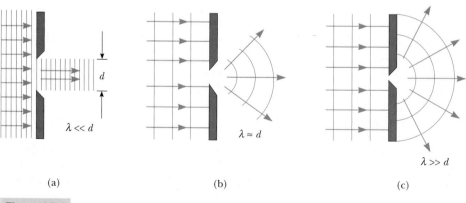

$\lambda \ll d$

(a)

$\lambda \approx d$

(b)

$\lambda \gg d$

(c)

**Figure 25.2**

A plane wave of wavelength $\lambda$ is incident on a barrier in which an opening of diameter $d$ occurs.
(a) When $\lambda \ll d$, almost no observable diffraction takes place and the ray approximation remains valid.
(b) When $\lambda \approx d$, diffraction becomes significant. (c) When $\lambda \gg d$, the opening behaves as a point source emitting spherical waves.

effects); hence, the ray approximation continues to be valid. If the diameter of the opening is on the order of the wavelength as in Figure 25.2b, the waves (and, consequently, the rays we draw) spread out from the opening in all directions. We say that the incoming plane wave undergoes *diffraction* as it passes through the opening. If the opening is small relative to the wavelength, the diffraction is so strong that the opening can be approximated as a point source of waves (Fig. 25.2c). Thus, diffraction is more pronounced as the ratio $d/\lambda$ approaches zero.

The ray approximation assumes that $\lambda \ll d$ so that we do not concern ourselves with diffraction effects, which depend on the full wave nature of light. We shall delay studying diffraction until Chapter 27. The ray approximation is used in the current chapter and in Chapter 26. The material in these chapters is often called *geometric optics*. The ray approximation is very good for the study of mirrors, lenses, prisms, and associated optical instruments, such as telescopes, cameras, and eyeglasses.

## 25.3 • THE WAVE UNDER REFLECTION

 See Screen 14.3

In Chapter 13, we introduced a one-dimensional version of the wave under reflection by considering waves on strings. When such a wave meets a discontinuity between strings representing different wave speeds, some of the energy is reflected and some of the energy is transmitted. In that discussion, the waves are constrained to move along the one-dimensional string. In our present discussion of optics, we are not subject to that restriction. Light waves can move in three dimensions.

Figure 25.3 shows several rays of light incident on a surface. Unless the surface is perfectly absorbing, some portion of the light is reflected from the surface. (The

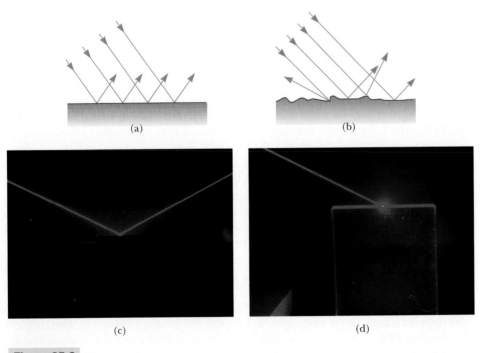

**Figure 25.3**

Schematic representation of (a) specular reflection, in which the reflected rays are all parallel, and (b) diffuse reflection, in which the reflected rays travel in scattered directions. (c) and (d) Photographs of specular and diffuse reflection using laser light. *(Photographs courtesy of Henry Leap and Jim Lehman)*

**Figure 25.4**

This photograph, taken in Salamanca, Spain, shows the reflection of the New Cathedral in the Tormes River. Because the water is so calm, the reflection is specular. *(David Parker/ Photo Researchers, Inc.)*

transmitted portion will be discussed in Section 25.4.) If the surface is very smooth, the reflected rays are parallel as indicated in Figure 25.3a. Reflection of light from such a smooth surface is called **specular reflection.** If the reflecting surface is rough as in Figure 25.3b, it reflects the rays in various directions. Reflection from a rough surface is known as **diffuse reflection.** A surface behaves as a smooth surface as long as the surface variations are small compared with the wavelength of the incident light. For example, many radio telescopes have reflecting dishes that are made of metal latticework rather than solid metal. As long as the spaces between pieces of metal are small relative to the radio wavelengths, the dish "appears" to be a solid metal dish and radio waves will exhibit specular reflection. Similarly, light passes through the small holes in a microwave oven door, allowing you to see the interior, but the large-wavelength microwaves reflect from the door as if it were a solid piece of metal.

Figures 25.3c and 25.3d are photographs of specular reflection and diffuse reflection using laser light, made visible by dust in the air, which scatters the light toward the camera. The reflected laser beam is clearly visible in Figure 25.3c. In Figure 25.3d, the diffuse reflection has caused the incident beam to be reflected in many directions, so that no clear outgoing beam is visible.

Specular reflection is necessary for the formation of clear images from reflecting surfaces, a topic we shall investigate in Chapter 26. Figure 25.4 shows an image resulting from specular reflection from a smooth water surface. If the water surface were rough, diffuse reflection would occur and the image would not be visible.

Both types of reflection can occur from a road surface that you observe when you drive at night. On a dry night, light from oncoming vehicles is scattered off the road in different directions (diffuse reflection) and the road is quite visible. On a rainy night, the small irregularities in the road surface are filled with water. Because the water surface is smooth, the light undergoes specular reflection and the glare from reflected light makes the road less visible.

Let us now develop the mathematical representation for the wave under reflection. Consider a light ray that travels in air and is incident at an angle on a flat, smooth surface as in Figure 25.5. The incident and reflected rays make angles of $\theta_1$ and $\theta_1'$, respectively, with a line drawn normal to the surface at the point where the incident ray strikes the surface. Experiments show that the **angle of reflection equals the angle of incidence:**

$$\theta_1' = \theta_1 \qquad [25.1]$$

- *Law of reflection*

By convention, the angles of incidence and reflection are measured from the normal to the surface rather than from the surface itself.

In diffuse reflection, the angle of incidence and angle of reflection obey Equation 25.1, *with respect to the local normal.* Because of the roughness of the surface, the local normal varies significantly from one location to another. In this book we shall concern ourselves only with specular reflection and shall use the term *reflection* to mean specular reflection.

As you might guess from Equation 25.1 and the figures we have seen so far, geometric models are used extensively in the study of optics. As we represent physical situations with geometric constructions, the mathematics of triangles and the principles of trigonometry will find many applications.

**The path of a light ray is reversible.** For example, the ray in Figure 25.5 travels from the upper left, reflects from the mirror, and then moves toward a point at the upper right. If the ray originated at the same point at the upper right, it would follow the same path to reach the same point at the upper left. This reversible property will be useful when we set up geometric constructions for finding the paths of light rays.

## Quick Quiz 25.1

Which part of Figure 25.6, (a) or (b), shows specular reflection of light from the roadway?

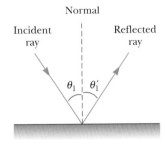

**Figure 25.5**

According to the law of reflection, $\theta_1 = \theta_1'$. The incident ray, the reflected ray, and the normal all lie in the same plane.

**PITFALL PREVENTION 25.1**
**Subscript notation**

We will use the subscript 1 to refer to parameters for the light in the initial medium. When light travels from one medium to another, we will use the subscript 2 for the parameters associated with the light in the new medium. In the current discussion, the light stays in the same medium, so we will have subscripts 1 only. These may look unnecessary in Equation 25.1, but we are simply preparing ourselves for a future discussion when the light enters a second medium.

(a)

(b)

**Figure 25.6**

(Quick Quiz 25.1) Light reflecting from a roadway. Which reflection is specular and which is diffuse? *(Charles D. Winters)*

### THINKING PHYSICS 25.1

An observer on the west-facing beach of a large lake is watching the beginning of a sunset. The water is very smooth except for some areas with small ripples. The observer notices that some areas of the water appear blue and some appear pink. Why does the water appear to be different colors in different areas?

**Reasoning**   The different colors arise from specular and diffuse reflection. The smooth areas of the water specularly reflect the light from the west, which is the pink light from the sunset. The areas with small ripples reflect the light diffusely. Thus, light from all parts of the sky is reflected into the observers' eyes. Because most of the sky is still blue at the beginning of the sunset, these areas appear to be blue.

### THINKING PHYSICS 25.2

When looking through a glass window to the outdoors at night, you sometimes see a *double* image of yourself. Why?

**Reasoning**   Reflection occurs whenever light encounters an interface between two optical media. For the glass in the window, two such interfaces exist. The first is the inner surface of the glass and the second is the outer surface. Each of these interfaces results in an image.

---

### Example 25.1   The Double-Reflected Light Ray

Two mirrors make an angle of 120° with each other as in Figure 25.7. A ray is incident on mirror $M_1$ at an angle of 65° to the normal. Find the direction of the ray after it is reflected from mirror $M_2$.

**Reasoning and Solution**   From the law of reflection, we know that the first reflected ray also makes an angle of 65° with the normal. Thus, this ray makes an angle of 90° − 65°, or 25°, with the horizontal. We build the geometric model triangle shown in Figure 25.7 from the triangle made by the first reflected ray and the two mirrors. The first reflected ray makes an angle of 35° with $M_2$ (because the sum of the interior angles of any triangle is 180°). This means that this ray makes an angle of 55° with the normal to $M_2$. Hence, from the law of reflection, the second reflected ray makes an angle of 55° with the normal to $M_2$.

**Figure 25.7**

(Example 25.1) Mirrors $M_1$ and $M_2$ make an angle of 120° with each other.

---

*See Screen 14.4*

## 25.4 • THE WAVE UNDER REFRACTION

Referring again to our discussion of string waves in Chapter 13, we discussed the fact that some of the energy of a wave incident on a discontinuity in the string is transmitted through the discontinuity. As a light wave moves through three dimensions, understanding the transmitted light wave involves new principles that we now discuss.

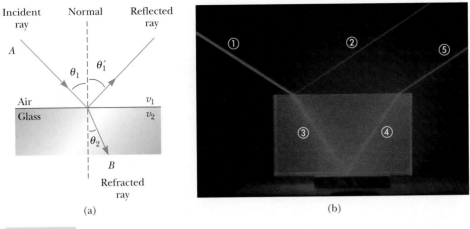

**Figure 25.8**

(a) A light ray obliquely incident on an air–glass interface. The refracted ray is deviated toward the normal because $v_2 < v_1$. All rays and the normal lie in the same plane. (b) (Quick Quiz 25.2) Of light rays 2 through 5, which are reflected and which are refracted? *(Henry Leap and Jim Lehman)*

When a ray of light traveling through a transparent medium is obliquely incident on a boundary leading into another transparent medium as in Figure 25.8a, part of the ray is reflected but part is transmitted into the second medium. The ray that enters the second medium experiences a change in direction at the boundary and is said to undergo **refraction. The incident ray, the reflected ray, and the refracted ray all lie in the same plane.** The **angle of refraction** $\theta_2$ in Figure 25.8a depends on the properties of the two media and on the angle of incidence, through the relationship

$$\frac{\sin \theta_2}{\sin \theta_1} = \frac{v_2}{v_1} = \text{constant} \qquad [25.2]$$

where $v_1$ is the speed of light in medium 1 and $v_2$ is the speed of light in medium 2. The experimental discovery of this relationship is usually credited to Willebrord Snell (1591–1626) and is therefore known as **Snell's law.**\* Equation 25.2 is a mathematical representation of the wave under refraction, although we find a more commonly used form in Equation 25.7.

The path of a light ray through a refracting surface is reversible, as was the case for reflection. For example, the ray in Figure 25.8a travels from point $A$ to point $B$. If the ray originated at $B$, it would follow the same path to reach point $A$. In the latter case, however, the reflected ray would be in the glass.

## Quick Quiz 25.2

If beam ① is the incoming beam in Figure 25.8b, which of the other four beams are reflected and which are refracted?

When light moves from a material in which its speed is high to a material in which its speed is lower, Snell's law shows that the angle of refraction $\theta_2$ is less than

---

\* The same law was deduced from the particle theory of light in 1637 by René Descartes (1596–1650) and hence is known as *Descartes's law* in France.

**Figure 25.9**

(a) When the light ray moves from air into glass, its path deviates toward the normal. (b) When the ray moves from glass into air, its path deviates away from the normal.

The pencil partially immersed in water appears bent because light from the lower part of the pencil is refracted as it travels across the boundary between water and air. (*Jim Lehman*)

**Figure 25.10**

Light passing from one atom to another in a medium. The dots are atoms, and the vertical arrows represent their oscillations.

the angle of incidence. The refracted ray therefore deviates toward the normal as shown in Figure 25.9a. If the ray moves from a material in which it travels slowly to a material in which it travels more rapidly, $\theta_2$ is greater than $\theta_1$, and so the ray deviates away from the normal as shown in Figure 25.9b. In both cases, the change in direction of the ray is the angle of deviation $\delta$, given by $\delta = |\theta_1 - \theta_2|$.

The behavior of light as it passes from air into another substance and then re-emerges into air is often a source of confusion to students. Why is this behavior so different from other occurrences in our daily lives? When light travels in air, its speed is $3.0 \times 10^8$ m/s; on entry into a block of glass, its speed is reduced to approximately $2.0 \times 10^8$ m/s. When the light re-emerges into air, its speed increases to its original value $3.0 \times 10^8$ m/s. This is very different from what happens, for example, when a bullet is fired through a block of wood. In this case, the speed of the bullet is reduced as it moves through the wood because some of its original energy is used to tear apart the fibers of the wood. When the bullet enters the air again, it emerges at the speed it had just before leaving the block of wood.

To see why light behaves as it does, consider Figure 25.10, which represents a beam of light entering a piece of glass from the left. Once inside the glass, the light may encounter an atom, represented by point $A$ in the figure. Let us assume that light is absorbed by the atom, causing it to oscillate (a detail represented by the double-headed arrow in the drawing). The oscillating atom then radiates (emits) the beam of light toward an atom at point $B$, where the light is again absorbed. The details of these absorptions and emissions are best explained in terms of quantum physics, a subject we shall study in Chapter 28. For now, think of the process as one in which the light passes from one atom to another through the glass. (The situation is somewhat analogous to a relay race in which a baton is passed between runners on the same team.) Although light travels from one atom to another through the empty space between the atoms with a speed of $3.0 \times 10^8$ m/s, the absorptions and emissions of light by the atoms, which require time to occur, cause the *average* speed of light through the glass to be lowered. Once the light emerges into the air, the absorptions and emissions cease and the light's average speed returns to its original value.* Thus, whether the light is inside the material or outside, it always travels through vacuum with the same speed.

Light passing from one medium to another is refracted because the average speed of light is different in the two media. In fact, **light travels at its maximum speed in vacuum.** It is convenient to define the **index of refraction** $n$ of a medium to be the ratio

$$n \equiv \frac{\text{speed of light in vacuum}}{\text{average speed of light in the medium}} = \frac{c}{v} \qquad \text{[25.3]}$$

• *Index of refraction*

From this definition we see that the index of refraction is a dimensionless number greater than or equal to unity because $v$ in a medium is less than $c$. Furthermore, $n$ is equal to unity for vacuum. The indices of refraction for various substances are listed in Table 25.1.

**As a wave travels from one medium to another, its frequency does not change.** Let us first consider this notion for waves passing from a light string to a heavier string. If the frequencies of the incident and transmitted waves on the two strings at the junction point were different, the strings could not remain tied together—the joined ends of the two pieces of string would not move up and down in unison!

For a light wave passing from one medium to another, the frequency also remains constant. To see why this is so, consider Figure 25.11. Wave fronts pass an observer at point $A$ in medium 1 with a certain frequency and are incident on the boundary between medium 1 and medium 2. The frequency at which the wave

**PITFALL PREVENTION 25.2**

**n is not an integer here**

We have seen $n$ used in Chapter 11 to indicate the quantum number of a Bohr orbit, and in Chapter 14 to indicate the standing wave mode on a string or in an air column. In those cases, $n$ was an integer. The index of refraction $n$ is *not* an integer.

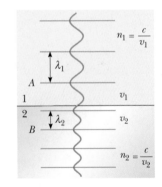

**Figure 25.11**

As a wave front moves from medium 1 to medium 2, its wavelength changes but its frequency remains constant.

| TABLE 25.1 | Index of Refraction for Various Substances Measured with Light of Vacuum Wavelength $\lambda_0 = 589$ nm | | | |
|---|---|---|---|---|
| **Substance** | **Index of Refraction** | **Substance** | **Index of Refraction** |
| **Solids at 20°C** | | **Liquids at 20°C** | |
| Cubic zirconia | 2.21 | Benzene | 1.501 |
| Diamond (C) | 2.419 | Carbon disulfide | 1.628 |
| Fluorite ($CaF_2$) | 1.434 | Carbon tetrachloride | 1.461 |
| Fused quartz ($SiO_2$) | 1.458 | Ethyl alcohol | 1.361 |
| Glass, crown | 1.52 | Glycerin | 1.473 |
| Glass, flint | 1.66 | Corn syrup | 2.21 |
| Ice ($H_2O$) | 1.309 | Water | 1.333 |
| Polystyrene | 1.49 | **Gases at 0°C, 1 atm** | |
| Sodium chloride (NaCl) | 1.544 | Air | 1.000293 |
| Zircon | 1.923 | Carbon dioxide | 1.0045 |

---

\* As an analogy, consider a subway entering a city at a constant speed $v$ and then stopping at several stations in the downtown region of the city. Although the subway may achieve the instantaneous speed $v$ between stations, the *average* speed across the city is less than $v$. Once the subway leaves the city and makes no stops, it moves again at a constant speed $v$. The analogy, as is the case with many analogies, is not perfect because the subway requires time to accelerate to the speed $v$ between stations, whereas light achieves speed $c$ immediately as it travels between atoms.

fronts pass an observer at point *B* in medium 2 must equal the frequency at which they arrive at point *A*. If this were not the case, the wave fronts would either pile up at the boundary or be destroyed or created at the boundary. Because this does not occur, the frequency must be a constant as a light ray passes from one medium into another.

Therefore, because the relation $v = f\lambda$ (Eq. 13.11) must be valid in both media and because $f_1 = f_2 = f$, we see that

$$v_1 = f\lambda_1 \qquad \text{and} \qquad v_2 = f\lambda_2$$

Because $v_1 \neq v_2$, it follows that $\lambda_1 \neq \lambda_2$. A relationship between index of refraction and wavelength can be obtained by dividing these two equations and making use of the definition of the index of refraction given by Equation 25.3:

$$\frac{\lambda_1}{\lambda_2} = \frac{v_1}{v_2} = \frac{c/n_1}{c/n_2} = \frac{n_2}{n_1} \qquad [25.4]$$

which gives

$$\lambda_1 n_1 = \lambda_2 n_2 \qquad [25.5]$$

It follows from Equation 25.5 that the index of refraction of any medium can be expressed as the ratio

$$n = \frac{\lambda_0}{\lambda_n} \qquad [25.6]$$

where $\lambda_0$ is the wavelength of light in vacuum and $\lambda_n$ is the wavelength in the medium whose index of refraction is *n*.

We are now in a position to express Snell's law (see Eq. 25.2) in an alternative form. If we combine Equation 25.3 and Equation 25.2, we find

• *Snell's law of refraction*

$$n_1 \sin \theta_1 = n_2 \sin \theta_2 \qquad [25.7]$$

This is the conventional form of Snell's law used in optics, expressed in terms of *n* values rather than the speeds as in Equation 25.2.

### Quick Quiz 25.3

Why do stars appear to twinkle in the night sky? (*Hint:* You are looking through turbulent air in the atmosphere.)

### Quick Quiz 25.4

As light from the Sun enters the atmosphere, it refracts due to the small difference between the speeds of light in air and in vacuum. The *optical* length of the day is defined as the time interval between the instant when the top of the Sun is just visibly observed above the horizon to the instant at which the top of the Sun just disappears below the horizon. The *geometric* length of the day is defined as the time interval between the instant when a geometric straight line drawn from the observer to the top of the Sun just clears the horizon to the instant at which this line just dips below the horizon. Which is longer, the optical day, or the geometric day?

---

### THINKING PHYSICS 25.3

APPLICATION

Underwater vision

Why do face masks make vision clearer under water? A face mask includes a flat piece of glass—the mask does not have lenses like those in eyeglasses.

**Reasoning**   The refraction necessary for focused viewing in the eye occurs at the air–cornea interface. The lens of the eye only performs some fine tuning of this image, allowing for accommodation for objects at various distances. When the eye is opened underwater, the interface is water–cornea, rather than air–cornea. Thus, the light from the scene is not focused on the retina, and the scene is blurry. The face mask simply provides a layer of air in front of the eyes, so that the air–cornea interface is re-established, and the refraction is correct to focus the light on the retina.

---

### Example 25.2   An Index of Refraction Measurement

A beam of light of wavelength 550 nm traveling in air is incident on a slab of transparent material. The incident beam makes an angle of 40.0° with the normal, and the refracted beam makes an angle of 26.0° with the normal. Find the index of refraction of the material.

**Solution**   Because the index of refraction of air is so close to that of a vacuum, we set $n_1$ equal to 1.00 exactly. Snell's law of refraction (see Eq. 25.7) with these data gives

$$n_1 \sin \theta_1 = n_2 \sin \theta_2$$

$$n_2 = n_1 \frac{\sin \theta_1}{\sin \theta_2} = (1.00) \frac{\sin 40.0°}{\sin 26.0°} = \frac{0.643}{0.438} = \boxed{1.47}$$

**EXERCISE**   What is the wavelength of light in the material?

*Answer*   374 nm

---

### Example 25.3   The Speed of Light in Silica

Light of wavelength 589 nm in vacuum passes through a piece of silica ($n = 1.458$). (a) Find the speed of light in silica.

**Solution**   The speed of light in silica can be easily obtained from Equation 25.3:

$$v = \frac{c}{n} = \frac{3.00 \times 10^8 \text{ m/s}}{1.458} = \boxed{2.06 \times 10^8 \text{ m/s}}$$

(b) What is the wavelength of this light in silica?

**Solution**   We use Equation 25.6 to calculate the wavelength in silica, noting that we are given the wavelength in vacuum to be $\lambda_0 = 589$ nm:

$$\lambda_n = \frac{\lambda_0}{n} = \frac{589 \text{ nm}}{1.458} = \boxed{404 \text{ nm}}$$

**EXERCISE**   Find the frequency of the light.

*Answer*   $5.09 \times 10^{14}$ Hz

## Example 25.4 Light Passing Through a Slab

A light beam passes from medium 1 to medium 2, with the latter being a thick slab of material whose index of refraction is $n_2$ (Fig. 25.12). Show that the emerging beam is parallel to the incident beam.

**Solution** First, let us apply Snell's law to the upper surface:

$$\sin \theta_2 = \frac{n_1}{n_2} \sin \theta_1 \tag{1}$$

Applying Snell's law to the lower surface gives

$$\sin \theta_3 = \frac{n_2}{n_1} \sin \theta_2 \tag{2}$$

Substituting Equation (1) into (2) gives

$$\sin \theta_3 = \frac{n_2}{n_1} \left( \frac{n_1}{n_2} \sin \theta_1 \right) = \sin \theta_1$$

Thus, $\theta_3 = \theta_1$, and so the layer does not alter the direction of the beam as shown by the dashed line in Figure 25.12. It does, however, produce a lateral displacement $d$ of the beam.

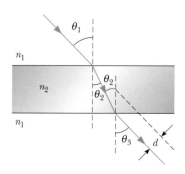

**Figure 25.12**

(Example 25.4) When light passes through a flat slab of material, the emerging beam is parallel to the incident beam, and therefore $\theta_1 = \theta_3$. The dashed line represents the path the light would take if the slab were not there.

---

See Screen 14.4

## 25.5 • DISPERSION AND PRISMS

In the preceding section, we developed Snell's law, which incorporates the index of refraction of a material. In Table 25.1, we presented index of refraction values for a number of materials. If we make careful measurements, however, we find that the value of the index of refraction in anything but vacuum depends on the wavelength of light. The dependence of the index of refraction on wavelength, which results from the dependence of the wave speed on wavelength, is called **dispersion.** Figure 25.13 is a graphical representation of this variation in index of refraction with wavelength. Because $n$ is a function of wavelength, Snell's law indicates that **the angle of refraction when light enters a material depends on the wavelength of the light.** As we see from Figure 25.13, the index of refraction for a material generally decreases with increasing wavelength in the visible range. This means that violet light ($\lambda \approx 400$ nm) refracts more than red light ($\lambda \approx 650$ nm) when passing from air into a material.

To understand the effects of dispersion on light, consider what happens when light strikes a prism as in Figure 25.14. A ray of light of a single wavelength that is incident on the prism from the left emerges in a direction deviated from its original direction of travel by an angle of deviation $\delta$. Now suppose a beam of white light (a combination of all visible wavelengths) is incident on a prism. Because of dispersion, the different colors refract through different angles of deviation, and the rays that emerge from the second face of the prism spread out in a series of colors known as a **visible spectrum** as shown in Figure 25.15. These colors, in order of decreasing wavelength, are red, orange, yellow, green, blue, and violet.* Vi-

**Figure 25.13**

Variation of index of refraction with vacuum wavelength for three materials.

---

* In Newton's time, the colors we now call teal and blue were called blue and indigo. Your "blue jeans" are dyed with indigo. A mnemonic device for remembering the colors of the spectrum is the acronym ROYGBIV, from the first letters of the colors—red, orange, yellow, green, blue, indigo, violet. Some individuals think of this as the name of a person, Roy G. Biv!

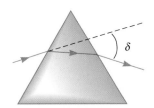

**Figure 25.14**

A prism refracts single-wavelength light and deviates the light through an angle δ.

**Figure 25.15**

Different colors are refracted at different angles because the index of refraction of the glass depends on wavelength. Violet light deviates the most; red light deviates the least. *(Photograph courtesy of Bausch and Lomb)*

olet light deviates the most, red light the least, and the remaining colors in the visible spectrum fall between these extremes.

The dispersion of light into a spectrum is demonstrated most vividly in nature through the formation of a rainbow, often seen by an observer positioned between the Sun and a rain shower. To understand how a rainbow is formed, consider Figure 25.16. A ray of light passing overhead strikes a spherical drop of water in the atmosphere and is refracted and reflected as follows. It is first refracted at the front surface of the drop, with the violet light deviating the most and the red light the least. At the back surface of the drop, the light is reflected and returns to the front surface, where it again undergoes refraction as it moves from water into air.

Because light enters the front surface of the raindrop at all locations, there is a range of exit angles for the light leaving the raindrop after reflecting from the back surface. A careful analysis of the spherical shape of the water drop, however, shows that the exit angle of highest light intensity is 42° for the red light and 40° for the violet light. Thus, the light from the raindrop seen by the observer is brightest for these angles, and the observer sees a rainbow. Figure 25.17 shows the geometry for the observer. The colors of the rainbow are seen in a range of 40° to 42° from the antisolar point, which is the point exactly 180° from the Sun. If red light is seen coming from a raindrop high in the sky, the violet light from this drop passes over the observer's head and is not seen. Thus, the portion of the rainbow in the vicinity of this drop is red. The violet portion of the rainbow seen by an observer is supplied by drops lower in the sky, which send violet light to the observer's eyes and red light below the eyes.

The opening photograph for this chapter shows a *double rainbow*. The secondary rainbow is fainter than the primary rainbow, and the colors are reversed. The secondary rainbow arises from light that makes two reflections from the interior surface before exiting the raindrop. In the laboratory, rainbows have been observed in which the light makes over 30 reflections before exiting the water drop. Because each reflection involves some loss of light due to refraction out of the water drop, the intensity of these higher order rainbows is very small.

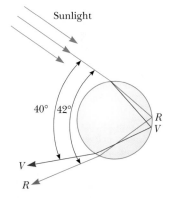

**Figure 25.16**

Refraction of sunlight by a spherical raindrop.

**WEB**

For a java applet which allows you to vary the entrance location of a ray of light to a raindrop and see the effect on the outgoing ray, visit **www.phy.ntnu.edu.tw/java/ Rainbow/rainbow.html**

**Quick Quiz 25.5**

In dispersive materials, the angle of refraction for a light ray depends on the wavelength of the light. Does the angle of reflection from the surface of the material depend on the wavelength?

Pictorial representations such as Figure 25.16 are subject to misinterpretation. The figure shows one ray of light entering the raindrop and undergoing reflection and refraction, exiting the raindrop in a range of 40° to 42° from the entering ray. This might be interpreted as meaning that *all* light entering the raindrop exits in this small range of angles. In reality, light exits the raindrop over a much larger range of angles, from 0° to 42°. A careful analysis of the reflection and refraction from the spherical raindrop shows that the range of 40° to 42° is where the *largest* amount of light exits the raindrop. The light coming from other angles provides much of the bright region below the rainbow in Figure 25.17b. In the region to the right of the boat, we see a combination of small-angle light from the rainbow and scattering of light from the mist due to the falls. To the left of the boat, however, we see a clear contrast between the small-angle light below the rainbow and the absence of this light above the rainbow.

(a)

(b)

**Figure 25.17**

(a) The formation of a rainbow. (b) A rainbow over Niagara Falls, in Ontario, Canada. *(b, John Edwards/Tony Stone Images)*

### THINKING PHYSICS 25.4

When a beam of light enters a glass prism with nonparallel sides, the spectrum of colors exiting the prism is a testimonial to the dispersion occurring in the glass. Suppose a beam of light enters a slab of material with *parallel* sides at an angle to the normal. When the beam exits the other side, traveling in the same direction as the original beam, is there any evidence of dispersion?

**Reasoning** Due to dispersion, light at the violet end of the spectrum exhibits a larger angle of refraction on entering the glass than light at the red end. All colors of light return to the original *direction* of propagation on refracting back out into

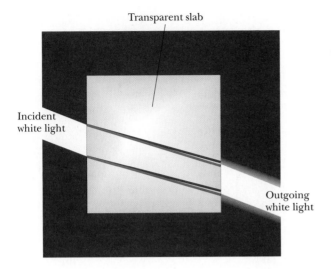

Transparent slab

Incident
white light

Outgoing
white light

**Figure 25.18**

(Thinking Physics 25.4)

the air. Thus, the outgoing beam is white. But the net lateral shift in the *position* of the violet light along the edge of the beam is larger than that for the red light. Thus, one edge of the outgoing beam has a bluish tinge to it (it appears blue rather than violet, because the eye is not very sensitive to violet light), whereas the other edge has a reddish tinge. This effect is indicated in Figure 25.18. The colored edges of the outgoing beam of white light are the evidence for dispersion.

## 25.6 • HUYGENS'S PRINCIPLE

In this section, we introduce a geometric construction proposed by Huygens in 1678. Huygens assumed that light consists of waves rather than a stream of particles. He had no knowledge of the nature of light or of its electromagnetic character. Nevertheless, his geometric model is adequate for understanding many practical aspects of the propagation of light.

**Huygens's principle** is a geometric model that allows us to determine the position of a wave front from a knowledge of an earlier wave front. In Huygens's construction, all points on a given wave front are taken as point sources for the production of spherical secondary waves, called *wavelets,* that propagate outward with speeds characteristic of waves in that medium. After some time interval has elapsed, the new position of the wave front is the surface tangent to the wavelets.

Figure 25.19 illustrates two simple examples of a Huygens's principle construction. First, consider a plane wave moving through free space as in Figure 25.19a. At $t = 0$, the wave front is indicated by the plane labeled $AA'$. Each point on this wave front is a point source for a wavelet. Showing three of these points, we draw circles, each of radius $c\,\Delta t$, where $c$ is the speed of light in free space and $\Delta t$ is the time interval during which the wave propagates. The surface drawn tangent to the wavelets is the plane $BB'$, which is parallel to $AA'$. This is the wave front at the end of the time interval $\Delta t$. In a similar manner, Figure 25.19b shows Huygens's construction for an outgoing spherical wave.

**Christiaan Huygens** (1629–1695)

Huygens, a Dutch physicist and astronomer, is best known for his contributions to the fields of optics and dynamics. To Huygens, light was a type of vibratory motion, spreading out and producing the sensation of sight when impinging on the eye. On the basis of this theory, he deduced the laws of reflection and refraction. *(Courtesy of Rijksmuseum voor de Geschiedenis der Natuurwetenschappen. Courtesy AIP Niels Bohr Library)*

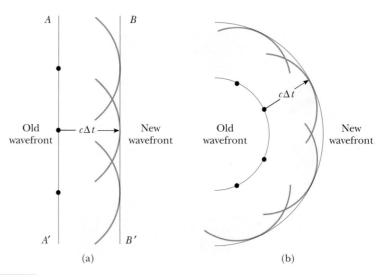

**Figure 25.19**

Huygens' construction for (a) a plane wave propagating to the right and (b) a spherical wave.

A convincing demonstration of the existence of Huygens wavelets is obtained with water waves in a shallow tank (called a ripple tank) as in Figure 25.20. Plane waves produced below the slits emerge above the slits as two-dimensional circular waves propagating outward. In the plane wave, each point on the wave front acts as a source of circular waves on the two-dimensional water surface. At a later time, the tangent of the circular wave fronts remains a straight line. As the wave front encounters the barrier, however, waves at all points on the wave front are reflected except those that encounter the openings. For very small openings, we can model this as if only one source of Huygens wavelets exists at each of the two openings. As a result, the Huygens wavelets from those single sources are seen as the outgoing circular waves in the upper portion of Figure 25.20. This is a dramatic example of diffraction that was mentioned in the opening section of this chapter, a phenomenon we shall study in more detail in Chapter 27.

**Figure 25.20**

Water waves in a ripple tank—a demonstration of Huygens wavelets. A plane wave is incident on a barrier with two small openings. The openings act as sources of circular wavelets. *(Erich Schrempp/Photo Researchers)*

## Example 25.5  Deriving the Laws of Reflection and Refraction

Use Huygens's principle to derive the (a) law of reflection, and (b) the law of refraction.

**Solution**  (a) To derive the law of reflection, consider the rays shown in Figure 25.21a. The line $AA'$ represents a wave front of the incident light. As ray 2 travels from $A'$ to $C$, ray 1 reflects from $A$ and produces a spherical wavelet of radius $AD$. Recall that the radius of a Huygens wavelet is $c\,\Delta t$—because the two wavelets having radii $A'C$ and $AD$ are in the same medium, they have the same speed $c$; therefore, $A'C = AD$.

From Huygens's principle, we find that the reflected wave front is $CD$, a line tangent to the outgoing spherical wavelets. Consider now the geometric model triangles in Figure 25.21b. The right triangles $ADC$ and $AA'C$ are congruent because they have the same hypotenuse $AC$ and because $AD = A'C$. Thus, from Figure 25.21b,

$$\sin \theta_1 = \frac{A'C}{AC} \quad \text{and} \quad \sin \theta_1' = \frac{AD}{AC}$$

Thus,

$$\sin \theta_1 = \sin \theta_1'$$

$$\theta_1 = \theta_1'$$

which is the law of reflection.

(b) For the law of refraction, consider the geometric construction shown in Figure 25.22. In a time interval $\Delta t$, ray 1

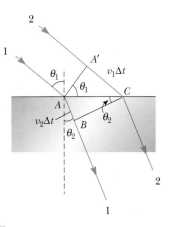

**Figure 25.22**

(Example 25.5) Huygens's construction for proving Snell's law of refraction.

moves from $A$ to $B$ and ray 2 moves from $A'$ to $C$. The radius of the outgoing spherical wavelet centered at $A$ is $AB = v_2\,\Delta t$. Similarly, the distance $A'C$ is equal to $v_1\,\Delta t$. Geometry shows that angle $A'AC$ equals $\theta_1$ and that angle $ACB$ equals $\theta_2$. From the geometric model right triangles $AA'C$ and $ABC$, we find that

$$\sin \theta_1 = \frac{v_1\,\Delta t}{AC} \quad \text{and} \quad \sin \theta_2 = \frac{v_2\,\Delta t}{AC}$$

Dividing the first equation by the second, we obtain,

$$\frac{\sin \theta_1}{\sin \theta_2} = \frac{v_1}{v_2}$$

From Equation 25.3, we know that $v_1 = c/n_1$ and $v_2 = c/n_2$. Therefore,

$$\frac{\sin \theta_1}{\sin \theta_2} = \frac{c/n_1}{c/n_2} = \frac{n_2}{n_1}$$

$$n_1 \sin \theta_1 = n_2 \sin \theta_2$$

which is Snell's law of refraction.

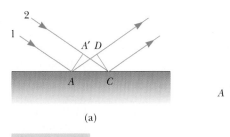

(a)

(b)

**Figure 25.21**

(Example 25.5) (a) Huygens's construction for proving the law of reflection. (b) Triangle $ADC$ is congruent with triangle $AA'C$.

## 25.7 • TOTAL INTERNAL REFLECTION

 See Screen 14.5

An interesting effect called **total internal reflection** can occur when light travels from a medium with a high index of refraction to one with a lower index of refraction. Consider a light ray traveling in medium 1 and meeting the boundary between media 1 and 2, where $n_1 > n_2$ (Fig. 25.23). Various possible directions of the ray are indicated by rays 1 through 5. The refracted rays are bent away from

(a)

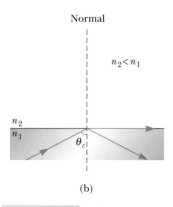

(b)

**Figure 25.23**

(a) Rays travel from a medium of index of refraction $n_1$ into a medium of index of refraction $n_2$, where $n_2 < n_1$. As the angle of incidence increases, the angle of refraction $\theta_2$ increases until $\theta_2$ is 90° (ray 4). For even larger angles of incidence, total internal reflection occurs (ray 5). (b) The angle of incidence producing an angle of refraction equal to 90° is the critical angle $\theta_c$.

the normal because $n_1 > n_2$. (Remember that when light refracts at the interface between the two media, it is also partially reflected. For simplicity, we ignore these reflected rays here, except for ray 5.) At some particular angle of incidence $\theta_c$, called the **critical angle,** the refracted light ray moves parallel to the boundary so that $\theta_2 = 90°$ (Fig. 25.23b). For angles of incidence greater than $\theta_c$, no ray is refracted and the incident ray is entirely reflected at the boundary, as is ray 5 in Figure 25.23a. This ray is reflected at the boundary as though it had struck a perfectly reflecting surface. It obeys the law of reflection; that is, the angle of incidence equals the angle of reflection.

We can use Snell's law to find the critical angle. When $\theta_1 = \theta_c$, $\theta_2 = 90°$, and Snell's law (Eq. 25.7) gives

$$n_1 \sin \theta_c = n_2 \sin 90° = n_2$$

$$\sin \theta_c = \frac{n_2}{n_1} \quad \text{(for } n_1 > n_2) \qquad \textbf{[25.8]}$$

This equation can be used only when $n_1$ is greater than $n_2$. That is, **total internal reflection occurs only when light travels from a medium of high index of refraction to a medium of lower index of refraction.** This is why the word *internal* is in the name—the light must initially be *inside* a material of higher index of refraction than the medium outside the material. If $n_1$ were less than $n_2$, Equation 25.8 would give $\sin \theta_c > 1$, which is meaningless because the sine of an angle can never be greater than unity.

The critical angle for total internal reflection is small when $n_1$ is considerably larger than $n_2$. Examples of this situation are diamond ($n = 2.42$ and $\theta_c = 24°$) and crown glass ($n = 1.52$ and $\theta_c = 41°$), where the angles given correspond to light refracting from the material into air. Total internal reflection combined with proper faceting causes diamonds and crystal glass to sparkle when observed in light.

### Quick Quiz 25.6

Look closely at Figure 25.24. You will see five beams of light exiting the bottom of the prism. Why are the two beams at the right of the set of five much brighter than the other three?

A prism and the phenomenon of total internal reflection can be used to alter the direction of travel of a light beam. Two such possibilities are illustrated in Figure 25.25. In one case the light beam is deflected by 90° (Fig. 25.25a), and in the

**Figure 25.24**

(Quick Quiz 25.6)

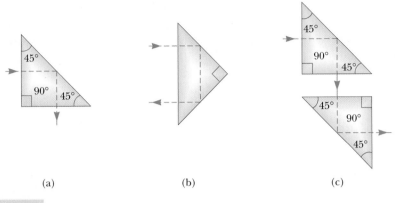

(a)                              (b)                              (c)

**Figure 25.25**

Internal reflection in a prism. (a) The ray is deviated by 90°. (b) The direction of the ray is reversed. (c) Two prisms used as a periscope.

second case the path of the beam is reversed (Fig. 25.25b). A common application of total internal reflection is in a submarine periscope. In this device, two prisms are arranged as in Figure 25.25c so that an incident beam of light follows the path shown and the periscope user is able to "see around corners."

### THINKING PHYSICS 25.5

A beam of white light is incident on the curved edge of a semicircular piece of glass as shown in Figure 25.26. The light enters the curved surface along the normal, so it experiences no refraction. It encounters the straight side at the center of curvature of the curved side and refracts into the air. The incoming beam is moved clockwise (so that the angle $\theta$ increases), such that the beam always enters along the normal to the curved side and encounters the straight side at the center of curvature of the curved side. As the refracted beam approaches a direction parallel to the straight side, it becomes redder. Why?

**Reasoning**   When the outgoing beam approaches the direction parallel to the straight side, the incident angle is approaching the critical angle for total internal reflection. The index of refraction for light at the violet end of the visible spectrum is larger than that at the red end. Thus, as the outgoing beam approaches the

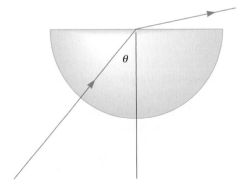

**Figure 25.26**

(Thinking Physics 25.5)

straight side, the violet light experiences total internal reflection first, followed by the other colors. The red light is the last to experience total internal reflection. Hence, just before the outgoing beam disappears, it is composed of light from the red end of the visible spectrum.

### Example 25.6   A View from the Fish's Eye

(a) Find the critical angle for a water–air boundary if the index of refraction of water is 1.33.

**Solution**   Applying Equation 25.8, we find the critical angle to be

$$\sin \theta_c = \frac{n_2}{n_1} = \frac{1}{1.33} = 0.752$$

$$\theta_c = \boxed{48.8°}$$

(b) Use the results of part (a) to predict what a fish sees when it looks upward toward the water surface at an angle of 40°, 48.8°, and 60° (Fig. 25.27).

**Reasoning**   Examine Figure 25.23a. Note that because the path of a light ray is reversible, light traveling from medium 2 into medium 1 follows the paths shown in Figure 25.23a, but in the *opposite* direction. This situation is illustrated in Figure 25.27 for the case of a fish looking upward toward the water's surface. A fish can see out of the water if it looks toward the surface at an angle less than the critical angle. Thus, at 40°, the fish can see into the air above the water. At an angle of 48.8°, the critical angle for water, the light that reaches the fish has to skim along the water's surface before

**Figure 25.27**

(Example 25.6)

being refracted to the fish's eye. Within this angle, the fish can see in principle the whole shore of the pond. At angles greater than the critical angle, the light reaching the fish comes via internal reflection at the surface. Thus, at 60°, the fish sees a reflection of some object on the bottom of the pool.

### context connection
### 25.8 • OPTICAL FIBERS

An interesting application of total internal reflection is the use of glass or transparent plastic rods to "pipe" light from one place to another. In the communication industry, digital pulses of laser light move along these light pipes, carrying information at an extremely high rate. In this Context Connection, we investigate the physics of this technological advance.

Figure 25.28 shows light traveling within a narrow transparent rod. The light is limited to traveling within the rod, even around gentle curves, as the result of successive internal reflections. Such a light pipe can be flexible if thin fibers are used rather than thick rods—these are called **optical fibers.** If a bundle of parallel optical fibers is used to construct an optical transmission line, images can be transferred from one point to another, as we shall discuss in the Context Connection of Chapter 26. Typical diameters for optical fibers are measured in tens of micrometers.

An optical fiber consists of a transparent core surrounded by a *cladding,* a material that has a lower index of refraction than the core. The combination may be surrounded by a plastic *jacket* to prevent mechanical damage. Figure 25.29 shows a

**Figure 25.28**

Light travels in a curved transparent rod by multiple internal reflections.

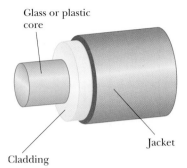

Glass or plastic core

Jacket

Cladding

**Figure 25.29**

The construction of an optical fiber. Light travels in the core, which is surrounded by a cladding and a protective jacket.

Strands of glass or plastic optical fibers are used to carry voice, video, and data signals in telecommunication networks. Typical fibers have diameters of 60 $\mu$m. *(Dennis O'Clair/Tony Stone Images)*

cut-away view of this construction. Because the index of refraction of the cladding is less than that of the core, light traveling in the core experiences total internal reflection if it arrives at the interface between the core and the cladding at an angle of incidence that exceeds the critical angle. In this case, light "bounces" along the core of the optical fiber, losing very little of its intensity as it travels.

Figure 25.30 shows a side view of a simple type of optical fiber known as a *multimode, stepped index fiber.* The term *stepped index* refers to the discontinuity in index of refraction between the core and the cladding, and *multimode* refers to the fact that light entering the fiber at many angles is transmitted. This type of fiber is acceptable for transmitting signals over a short distance, but not long distances because a digital pulse spreads with distance. Let us imagine that we input a perfectly rectangular pulse of laser light to the core of the optical fiber. Figure 25.31a shows the idealized time behavior of the laser light intensity for the input pulse. The laser light intensity rises instantaneously to its highest value, stays constant for the duration of the pulse and then instantaneously drops to zero. The light from the pulse entering along the axis in Figure 25.30 travels the shortest distance and arrives at the other end first. The other light paths represent longer distance of travel due to the angled bounces. As a result, the light from the pulse arrives at the other end over a longer period—the pulse is smeared out as in Figure 25.31b. If

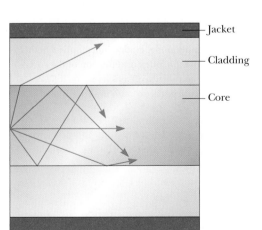

Jacket

Cladding

Core

**Figure 25.30**

A multimode, stepped index optical fiber. Rays of light entering over a wide range of angles pass through the core. Those making large angles with the axis take longer to travel the length of the fiber than those making small angles.

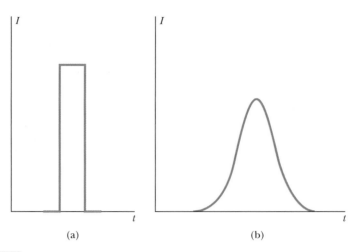

**Figure 25.31**

(a) A rectangular pulse of laser light to be sent into an optical fiber. (b) The output pulse of light, which has been broadened due to light taking different paths through the fiber.

the pulses represent zeroes and ones for a binary signal, this could cause overlap of the pulses or a reduction of the peak intensity below the detection threshold; either situation would result in obliteration of the information.

One way to improve optical transmission in such a situation is to use a *multi-mode, graded index fiber.* This fiber has a core whose index of refraction is smaller at larger radii from the center, as suggested by the shading in Figure 25.32. With a graded index core, off-axis rays of light experience continuous refraction and curve gradually away from the edges and back toward the center as shown by the light path in Figure 25.32. This reduces the transit time through the fiber for off-axis rays and also reduces the smearing out of the pulse. The transit time is reduced for two reasons: (1) The path length is reduced; and (2) much of the time the wave travels in the lower index of refraction region, where the speed of light is higher than at the center.

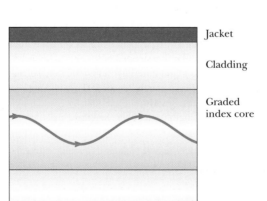

Jacket

Cladding

Graded
index core

**Figure 25.32**

A multimode, graded index optical fiber. Because the index of refraction of the core varies radially, off-axis light rays follow curved paths through the core.

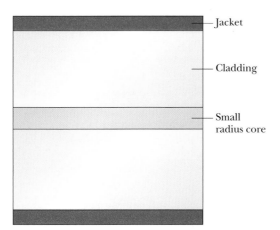

Jacket

Cladding

Small radius core

**Figure 25.33**

A single-mode, stepped index optical fiber. The small radius of the core and the small difference between the indices of refraction of the core and cladding reduce the broadening of light pulses.

The spreading effect in Figure 25.31 can be further reduced and almost eliminated by designing the fiber with two changes: (1) The core is made very small so that all paths within it are more nearly the same length; and (2) the difference in index of refraction between core and cladding is made relatively small, so that off-axis rays enter the cladding and are absorbed. This is suggested in Figure 25.33. This kind of fiber is called a *single-mode, stepped index fiber.* It can carry information at high bit rates because the pulses are minimally smeared out.

In reality, the material of the core is not perfectly transparent. Some absorption and scattering occurs as the light travels down the fiber. Absorption transforms light energy into increased internal energy in the fiber. Scattering causes light to strike the core–cladding interface at angles less than the critical angle for total internal reflection, resulting in some loss in the cladding or jacket. Even with these problems, optical fibers can transmit about 95% of the input energy over a kilometer. These problems are minimized by using as long a wavelength as possible for which the core material is transparent. The scattering and absorption centers then appear as small as possible to the waves and minimize the probability of interaction. Much of optical fiber communication occurs with light from infrared lasers, having wavelengths of about 1300 nm.

The field of developing applications for optical fibers is called **fiber optics.** One common application is the use of optical fibers in telecommunications because the fibers can carry a much higher volume of telephone calls, or other forms of communication, than electric wires. Optical fibers are also used in "smart buildings." In this application, sensors are located at various points within a building, and an optical fiber carries laser light to the sensor, which reflects it back to a control system. If any distortion occurs in the building due to earthquake or other causes, the intensity of the reflected light from the sensor changes, and the control system locates the point of distortion by identifying the particular sensor involved.

**WEB**

For information on optical fibers and other aspects of optical communication, explore *Telecom Glossary 2000,* which is an updated version of Federal Standard 1037C, *Glossary of Telecommunication Terms,* by visiting **ntia.its.bldrdoc.gov/projects/ t1glossary2000/**

**WEB**

An online course on optical fibers and lasers used in fiber communications is available at **www.fotec.com/fiberu-online/ fuonline.htm**

### THINKING PHYSICS 25.6

A cone of angles, called the *acceptance cone,* exists at the entrance to the core of the optical fiber. Incoming light at angles within this cone is transmitted through the core, and light entering the core from angles outside the cone strikes the core–cladding interface at an angle less than the critical angle and enters the cladding. Figure 25.34 shows a light ray entering the fiber

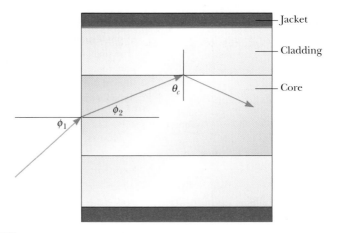

**Figure 25.34**

(Thinking Physics 25.6)

just within the acceptance cone and experiencing total internal reflection at the interface between the core and the cladding. If it is technologically difficult to produce light entering the fiber from a small range of angles, how could you adjust the indices of refraction of the core and cladding to increase the size of the acceptance cone—would you design them to be farther apart, or closer together?

**Reasoning** The acceptance cone becomes larger if the critical angle ($\theta_c$ in Figure 25.34) is made smaller. This can be accomplished by reducing the index of refraction of the cladding, so that the indices of refraction of the core and cladding become farther apart.

A single optical fiber can carry a digital signal, as we already described. If it is desired for optical fibers to carry an image of a scene, it is necessary to use a bundle of optical fibers. A popular use of such bundles is in the use of *fiberscopes* in medicine. In the Context Connection of Chapter 26, we shall investigate these devices.

## SUMMARY

In geometric optics, we use the **ray approximation,** in which we assume that a wave travels through a medium in straight lines in the direction of the rays of that wave. We ignore diffraction effects, which is a good approximation as long as the wavelength is short compared with any aperture dimensions.

The **index of refraction** $n$ of a material is defined as

$$n \equiv \frac{c}{v} \qquad \text{[25.3]}$$

where $c$ is the speed of light in a vacuum and $v$ is the speed of light in the material. In general, $n$ varies with wavelength, which is called **dispersion.**

The **law of reflection** states that part of a wave incident on a surface reflects from the surface so that the angle of reflection $\theta_1'$ equals the angle of incidence $\theta_1$.

Part of a light wave striking an interface between two media is transmitted into the second medium and undergoes a change in the direction of propagation. The **law of refraction,** or **Snell's law,** states that

$$n_1 \sin \theta_1 = n_2 \sin \theta_2 \qquad \text{[25.7]}$$

**Huygens's principle** states that all points on a wave front can be taken as point sources for the production of secondary wavelets. At some later time, the new position of the wave front is the surface tangent to these secondary wavelets.

**Total internal reflection** can occur when light travels from a medium of high index of refraction to one of lower index of refraction. The **critical angle** of incidence $\theta_c$ for which total internal reflection occurs at an interface is

$$\sin \theta_c = \frac{n_2}{n_1} \quad \text{(for } n_1 > n_2) \qquad \text{[25.8]}$$

## QUESTIONS

1. You can make a corner reflector by placing three flat mirrors in the corner of a room where the ceiling meets the walls. Show that no matter where you are in the room, you can see yourself reflected in the mirrors—upside down.

2. Several corner reflectors were left on the Moon's Sea of Tranquillity by the astronauts of *Apollo 11*. How can scientists use a laser beam sent from Earth even today to determine the precise distance from the Earth to the Moon?

3. The rectangular aquarium sketched in Figure Q25.3 contains only one goldfish. When the fish is near a corner of the aquarium and is viewed along a direction that makes an equal angle with two adjacent faces, the observer sees two fish mirroring each other, as shown. Explain this observation.

**Figure Q25.3**

4. As light travels from a vacuum ($n = 1$) to a medium such as glass ($n > 1$), does the wavelength of the light change? Does the frequency change? Does the speed change? Explain.

5. Explain why an oar in the water appears bent.

6. Explain why a diamond loses most of its sparkle when submerged in carbon disulfide, and why an imitation diamond of cubic zirconia loses all of its sparkle in corn syrup.

7. A laser beam passing through a nonhomogeneous sugar solution follows a curved path. Explain.

8. Light of wavelength $\lambda$ is incident on a slit of width $d$. Under what conditions is the ray approximation valid? Under what circumstances does the slit produce enough diffraction to make the ray approximation invalid?

9. The level of water in a clear, colorless glass is easily observed with the naked eye. The level of liquid helium in a clear glass vessel is extremely difficult to see with the naked eye. Explain.

10. Describe an experiment in which internal reflection is used to determine the index of refraction of a medium.

11. Why does a diamond show flashes of color when observed under white light?

12. Explain why a diamond sparkles more than a glass crystal of the same shape and size.

13. Redesign the periscope of Figure 25.25c so that it shows you an image of a scene behind you. Will the image be right side up or upside down?

14. The display windows of some department stores are slanted slightly inward at the bottom. This is to decrease the glare from streetlights or the Sun, which would make it difficult for shoppers to see the display. Sketch a light ray reflecting off such a window to show how this technique works.

15. Figure 25.16 represents sunlight striking a drop of water in the atmosphere. Use the laws of refraction and reflection and the fact that sunlight consists of a wide range of wavelengths to discuss the formation of rainbows.

16. Why does the arc of a rainbow appear with red on top and violet on the bottom?

17. Under what conditions is a mirage formed? On a hot day, what are we seeing when we observe "water on the road"?

18. Suppose you are told that only two colors of light (X and Y) are sent through a glass prism and that X is bent more than Y. Which color travels more slowly in the prism?

## PROBLEMS

1, 2, 3 = straightforward, intermediate, challenging  □ = full solution available in the *Student Solutions Manual and Study Guide*

**web** = solution posted at **http://www.harcourtcollege.com/physics/**   = computer useful in solving problem

= Interactive Physics  ▧ = paired numerical/symbolic problems  = life science application

### Section 25.2  The Ray Model in Geometric Optics

### Section 25.3  The Wave Under Reflection

### Section 25.4  The Wave Under Refraction

*Note:* If the value of an index of refraction is not given, refer to Table 25.1.

1. A narrow beam of sodium yellow light, with wavelength 589 nm in vacuum, is incident from air onto a smooth water surface at an angle $\theta_1 = 35.0°$. Determine the angle of refraction $\theta_2$ and the wavelength of the light in water.

2. The wavelength of red helium–neon laser light in air is 632.8 nm. (a) What is its frequency? (b) What is its wavelength in glass that has an index of refraction of 1.50? (c) What is its speed in the glass?

**3.** An underwater scuba diver sees the Sun at an apparent angle of 45.0° from the vertical. What is the actual direction of the Sun?

**4.** A laser beam is incident at an angle of 30.0° from the vertical onto a solution of corn syrup in water. If the beam is refracted to 19.24° from the vertical, (a) what is the index of refraction of the syrup solution? Suppose the light is red, with a vacuum wavelength of 632.8 nm. Find its (b) wavelength, (c) frequency, and (d) speed in the solution.

**5.** Find the speed of light in (a) flint glass, (b) water, and (c) cubic zirconia.

**6.** A light ray initially in water enters a transparent substance at an angle of incidence of 37.0°, and the transmitted ray is refracted at an angle of 25.0°. Calculate the speed of light in the transparent substance.

**7.** A ray of light strikes a flat block of glass ($n = 1.50$) of
web thickness 2.00 cm at an angle of 30.0° with the normal. Trace the light beam through the glass, and find the angles of incidence and refraction at each surface.

**8.** An opaque cylindrical tank with an open top has a diameter of 3.00 m and is completely filled with water. When the setting Sun reaches an angle of 28.0° above the horizon, sunlight ceases to illuminate any part of the bottom of the tank. How deep is the tank?

**9.** The angle between the two mirrors illustrated in Figure P25.9 is a right angle. The beam of light in the vertical plane $P$ strikes mirror 1 as shown. (a) Determine the distance that the reflected light beam travels before striking mirror 2. (b) In what direction does the light beam travel after being reflected from mirror 2?

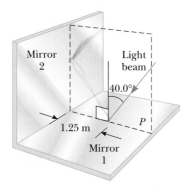

**Figure P25.9**

**10.** Two flat rectangular mirrors, both perpendicular to a horizontal sheet of paper, are set edge to edge with their reflecting surfaces perpendicular to each other. (a) A light ray in the plane of the paper strikes one of the mirrors at an arbitrary angle of incidence $\theta_1$. Prove that the final direction of the ray, after reflection from both mirrors, is opposite to its initial direction. In a clothing store, such a pair of mirrors shows you an image of yourself as others see you, with no apparent right–left reversal. (b) Now as-

sume that the paper is replaced with a third flat mirror, touching edges with the other two and perpendicular to both. The set of three mirrors is called a *corner cube reflector*. A ray of light is incident from any direction within the octant of space bounded by the reflecting surfaces. Argue that the ray reflects once from each mirror and that its final direction is opposite to its original direction. [The *Apollo 11* astronauts placed a panel of corner cube retroreflectors on the Moon. A pulse of laser light from Earth can be reflected back on itself and its transit time measured. This information is used to determine the distance to the Moon with an uncertainty of 15 cm. The radius of the Moon's orbit is increasing at the rate of 3.8 cm/yr as it loses mechanical energy because of tidal friction.]

**11.** How many times will the incident beam shown in Figure P25.11 be reflected by each of the parallel mirrors?

**Figure P25.11**

**12.** Unpolarized light in vacuum is incident onto a sheet of glass with index of refraction $n$. The reflected and refracted rays are perpendicular to each other. Find the angle of incidence. This angle is called *Brewster's angle* or the *polarizing angle*. In this situation the reflected light is linearly polarized, with its electric field restricted to the direction perpendicular to the plane containing the rays and the normal.

**13.** When the light illustrated in Figure P25.13 passes through the glass block, it is shifted laterally by the distance $d$. If $n = 1.50$, what is the value of $d$?

**Figure P25.13**   Problems 13 and 14.

14. Find the time required for the light to pass through the glass block described in Problem 13.

15. The light beam shown in Figure P25.15 makes an angle of 20.0° with the normal line $NN'$ in the linseed oil. Determine the angles $\theta$ and $\theta'$. (The index of refraction for linseed oil is 1.48.)

**Figure P25.15**

16. When you look through a window, by how much time is the light you see delayed by having to go through glass instead of air? Make an order-of-magnitude estimate on the basis of data you specify. By how many wavelengths is it delayed?

17. The reflecting surfaces of two intersecting flat mirrors are at an angle of $\theta$ ($0° < \theta < 90°$), as shown in Figure P25.17. If a light ray strikes the horizontal mirror, show that the emerging ray will intersect the incident ray at an angle of $\beta = 180° - 2\theta$.

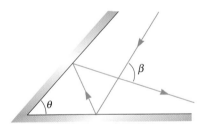

**Figure P25.17**

18. A light ray enters the atmosphere of a planet and descends vertically a distance $h$ to the surface. The index of refraction where the light enters the atmosphere is 1.000, and it increases linearly to the surface where it has the value $n$. (a) How long does it take the ray to traverse this path? (b) Compare this with the time it takes in the absence of an atmosphere.

## Section 25.5 Dispersion and Prisms

19. A narrow white light beam is incident on a block of fused quartz at an angle of 30.0°. Find the angular spread of the light beam inside the quartz.

20. A ray of light strikes the midpoint of one face of an equiangular glass prism ($n = 1.50$) at an angle of incidence of 30.0°. Trace the path of the light ray through the glass, and find the angles of incidence and refraction at each surface.

21. The refractive index for violet light in silica flint glass is 1.66, and that for red light is 1.62. A prism has an apex angle of 60.0°, measured between the surface at which light enters the prism and the surface at which the light leaves the prism. What is the angular dispersion of visible light passing through the prism if the angle of incidence is 50.0°? (See Fig. P25.21.)

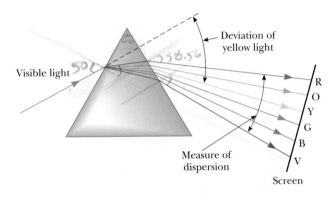

**Figure P25.21**

22. A triangular glass prism with an apex angle of $\Phi = 60.0°$ has an index of refraction $n = 1.50$ (Fig. P25.22). What is the smallest angle of incidence $\theta_1$ for which a light ray can emerge from the other side?

**Figure P25.22** Problems 22 and 23.

23. A triangular glass prism with apex angle $\Phi$ has index of refraction $n$ (see Fig. P25.22). What is the smallest angle of incidence $\theta_1$ for which a light ray can emerge from the other side?

## Section 25.6 Huygens's Principle

24. The speed of a water wave is described by $v = \sqrt{gd}$, where $d$ is the water depth, assumed to be small compared with the wavelength. Because their speed depends on the depth, water waves refract when moving into a region of different depth. Sketch a map of an ocean beach on the

eastern side of a landmass. Show contour lines of constant depth under water, assuming a reasonably uniform slope. (a) Suppose that waves approach the coast from a storm far away to the north-northeast. Show that the waves move nearly perpendicular to the shoreline when they reach the beach. (b) Sketch a map of a coastline with alternating bays and headlands as suggested in Figure P25.24. Again make a reasonable guess about the shape of contour lines of constant depth. Suppose that waves approach the coast, carrying energy with uniform density along originally straight wave fronts. Show that the energy reaching the coast is concentrated at the headlands and has lower intensity in the bays.

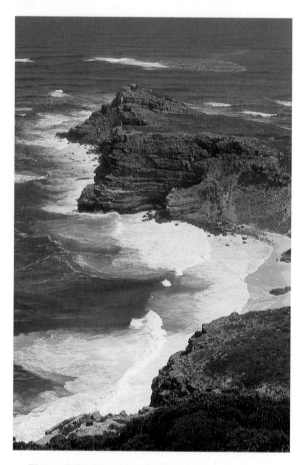

**Figure P25.24** *(Mark D. Phillips/Photo Researchers, Inc.)*

## Section 25.7 Total Internal Reflection

25. For 589-nm light, calculate the critical angle for the following materials surrounded by air: (a) diamond, (b) flint glass, and (c) ice.

26. Repeat Problem 25 for the situation in which the materials are surrounded by water.

27. Consider a common mirage formed by super-heated air just above a roadway. A truck driver whose eyes are 2.00 m above the road, where $n = 1.000\ 3$, looks forward. She perceives the illusion of a patch of water ahead on the road, where her line of sight makes an angle of 1.20° below the horizontal. Find the index of refraction of the air just above the road surface. (*Hint:* Treat this as a problem in total internal reflection.)

28. A room contains air in which the speed of sound is 343 m/s. The walls of the room are made of concrete, in which the speed of sound is 1 850 m/s. (a) Find the critical angle for total internal reflection of sound at the concrete–air boundary. (b) In which medium must the sound be traveling in order to undergo total internal reflection? (c) "A bare concrete wall is a highly efficient mirror for sound." Give evidence for or against this statement.

29. In about 1965, engineers at the Toro Company invented a gasoline gauge for small engines, diagrammed in Figure P25.29. The gauge has no moving parts. It consists of a flat slab of transparent plastic fitting vertically into a slot in the cap on the gas tank. None of the plastic has a reflective coating. The plastic projects from the horizontal top down nearly to the bottom of the opaque tank. Its lower edge is cut with facets making angles of 45° with the horizontal. A lawnmower operator looks down from above and sees a boundary between bright and dark on the gauge. The location of the boundary, across the width of the plastic, indicates the quantity of gasoline in the tank. Explain how the gauge works. Explain the design requirements, if any, for the index of refraction of the plastic.

**Figure P25.29**

## Section 25.8 Context Connection — Optical Fibers

30. Determine the maximum angle $\theta$ for which the light rays incident on the end of the pipe in Figure P25.30 are subject to total internal reflection along the walls of the pipe. Assume that the pipe has an index of refraction of 1.36 and that the outside medium is air.

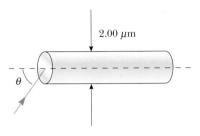

**Figure P25.30**

**31.** A laser beam strikes one end of a slab of material, as
web shown in Figure P25.31. The index of refraction of the
slab is 1.48. Determine the number of internal reflections
of the beam before it emerges from the opposite end of
the slab.

**Figure P25.31**

**32.** A glass fiber ($n = 1.50$) is submerged in water ($n = 1.33$).
What is the critical angle for light to stay inside the optical
fiber?

**33.** An optical fiber of index of refraction $n$ and diameter $d$ is
surrounded by air. Light is sent into the fiber along its axis
as in Figure P25.33. (a) Find the smallest outside radius $R$
permitted for a bend in the fiber if no light is to escape.
(b) Does the result for part (a) predict reasonable behavior as $d$ approaches zero? As $n$ increases? As $n$ approaches
1? (c) Evaluate $R$ if the fiber diameter is 100 $\mu$m and its index of refraction is 1.40.

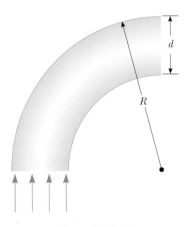

**Figure P25.33**

## Additional Problems

**34.** (a) Consider a horizontal interface between air above and
glass with an index of 1.55 below. Draw a light ray incident
from the air at an angle of incidence of 30.0°. Determine
the angles of the reflected and refracted rays and show
them on the diagram. (b) Suppose instead that the light
ray is incident from the glass at an angle of incidence of
30.0°. Determine the angles of the reflected and refracted
rays and show all three rays on a new diagram. (c) For rays
incident from the air onto the air–glass surface, determine and tabulate the angles of reflection and refraction
for all the angles of incidence at 10.0° intervals from 0° to
90.0°. (d) Do the same for light rays traveling up to the interface through the glass.

**35.** A small underwater pool light is 1.00 m below the surface.
web The light emerging from the water forms a circle on the
water's surface. What is the diameter of this circle?

**36.** One technique for measuring the angle of a prism is
shown in Figure P25.36. A parallel beam of light is directed on the angle so that the beam reflects from opposite sides. Show that the angular separation of the two
beams is given by $B = 2A$.

**Figure P25.36**

**37.** The walls of a prison cell are perpendicular to the four
cardinal compass directions. On the first day of spring,
light from the rising Sun enters a rectangular window in
the eastern wall. The light traverses 2.37 m horizontally to
shine perpendicularly on the wall opposite the window. A
young prisoner observes the patch of light moving across
this western wall and for the first time forms his own understanding of the rotation of the Earth. (a) With what
speed does the illuminated rectangle move? (b) The prisoner holds a small square mirror flat against the wall at
one corner of the rectangle of light. The mirror reflects
light back to a spot on the eastern wall close beside the
window. How fast does the smaller square of light move
across that wall? (c) Seen from a latitude of 40.0° north,
the rising Sun moves through the sky along a line making
a 50.0° angle with the southeastern horizon. In what direction does the rectangular patch of light on the western

wall of the prisoner's cell move? (d) In what direction does the smaller square of light on the eastern wall move?

38. Figure P25.38 shows a top view of a square enclosure. The inner surfaces are plane mirrors. A ray of light enters a small hole in the center of one mirror. (a) At what angle $\theta$ must the ray enter to exit through the hole after being reflected once by each of the other three mirrors? (b) Can the ray exit after multiple reflections for other values of $\theta$? If so, make a sketch of one of the ray's paths.

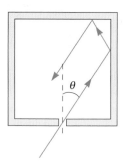

**Figure P25.38**

39. A hiker stands on an isolated mountain peak near sunset and observes a rainbow caused by water droplets in the air 8.00 km away. The valley is 2.00 km below the mountain peak and entirely flat. What fraction of the complete circular arc of the rainbow is visible to the hiker? (See Figure 25.17.)

40. The laws of refraction and reflection are the same for sound as for light. The speed of sound in air is 340 m/s, and the speed of sound in water is 1 510 m/s. If a sound wave approaches a plane water surface at an angle of incidence of 12.0°, what is the angle of refraction?

41. A glass block having an index of refraction of 1.52 and surrounded by air measures 10.0 cm × 10.0 cm. Light is incident on one side at an angle of 45.0°, as shown in Figure P25.41. What is the maximum distance $x$ so that the ray will emerge from the opposite side?

**Figure P25.41**

42. When light is incident normally on the interface between two transparent optical media, the intensity of the reflected light is given by the expression

$$S_1' = \left(\frac{n_2 - n_1}{n_2 + n_1}\right)^2 S_1$$

In this equation $S_1$ represents the average magnitude of the Poynting vector in the incident light (the incident intensity), $S_1'$ is the reflected intensity, and $n_1$ and $n_2$ are the refractive indices of the two media. (a) What fraction of the incident intensity is reflected for 589-nm light normally incident on an interface between air and crown glass? (b) In part (a), does it matter whether the light is in the air or in the glass as it strikes the interface?

43. Refer to Problem 42 for its description of the reflected intensity of light normally incident on an interface between two transparent media. (a) When light is normally incident on an interface between vacuum and a transparent medium of index $n$, show that the intensity $S_2$ of the transmitted light is given by the expression $S_2/S_1 = 4n/(n + 1)^2$. (b) Light travels perpendicularly through a diamond slab, surrounded by air, with parallel surfaces of entry and exit. Apply the transmission fraction in part (a) to find the approximate overall transmission through the slab of diamond as a percentage. Ignore light reflected back and forth within the slab.

44. This problem builds on the results of Problems 42 and 43. Light travels perpendicularly through a diamond slab, surrounded by air, with parallel surfaces of entry and exit. What fraction of the incident intensity is the intensity of the transmitted light? Include the effects of light reflected back and forth inside the slab.

45. The light beam shown in Figure P25.45 strikes surface 2 at the critical angle. Determine the angle of incidence $\theta_1$.

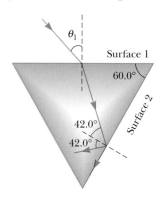

**Figure P25.45**

46. A 4.00-m-long pole stands vertically in a lake having a depth of 2.00 m. When the Sun is 40.0° above the horizontal, determine the length of the pole's shadow on the bottom of the lake. Take the index of refraction for water to be 1.33.

47. A light ray of wavelength 589 nm is incident at an angle $\theta$ on the top surface of a block of polystyrene, as shown in Figure P25.47. (a) Find the maximum value of $\theta$ for which the refracted ray undergoes total internal reflection at the left vertical face of the block. Repeat the calculation for the case in which the polystyrene block is immersed in (b) water and (c) carbon disulfide.

**Figure P25.47**

48. Builders use a leveling instrument with the beam from a fixed helium–neon laser reflecting in a horizontal plane from a small flat mirror mounted on a vertical rotating shaft. The light is sufficiently bright and the rotation rate is sufficiently high that the reflected light appears as a horizontal line, wherever it falls on a wall. (a) Assume the mirror is at the center of a circular grain elevator of radius $R$. The mirror spins with constant angular velocity $\omega_m$. Find the speed of the spot of laser light on the wall. (b) Assume the spinning mirror is at a perpendicular distance $d$ from point $O$ on a flat vertical wall. When the spot of laser light on the wall is at distance $x$ from point $O$, what is its speed?

49. A shallow glass dish is 4.00 cm wide at the bottom, as shown in Figure P25.49. When an observer's eye is posi-

tioned as shown, the observer sees the edge of the bottom of the empty dish. When this dish is filled with water, the observer sees the center of the bottom of the glass. Find the height of the glass.

50. A ray of light passes from air into water. For its deviation angle $\delta = |\theta_1 - \theta_2|$ to be 10.0°, what must be its angle of incidence?

51. A material having an index of refraction $n$ is surrounded by a vacuum and is in the shape of a quarter circle of radius $R$ (Fig. P25.51). A light ray parallel to the base of the material is incident from the left at a distance of $L$ above the base and emerges out of the material at the angle $\theta$. Determine an expression for $\theta$.

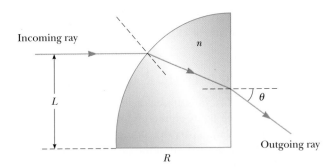

**Figure P25.51**

52. A transparent cylinder of radius $R = 2.00$ m has a mirrored surface on its right half, as shown in Figure P25.52. A light ray traveling in air is incident on the left side of the cylinder. The incident light ray and the exiting light ray are parallel and $d = 2.00$ m. Determine the index of refraction of the material.

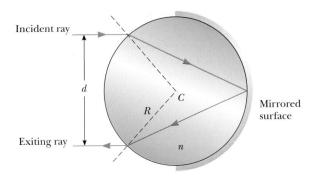

**Figure P25.52**

53. A light ray enters a rectangular block of plastic at an angle of $\theta_1 = 45.0°$ and emerges at an angle of $\theta_2 = 76.0°$, as in Figure P25.53. (a) Determine the index of refraction for the plastic. (b) If the light ray enters the plastic at a point $L = 50.0$ cm from the bottom edge, how long does it take the light ray to travel through the plastic?

**Figure P25.49**

**Figure P25.53**

in the accompanying table. Use the data to verify Snell's law of refraction by plotting the sine of the angle of incidence versus the sine of the angle of refraction. Use the resulting plot to deduce the index of refraction of water.

| Angle of Incidence (degrees) | Angle of Refraction (degrees) |
|---|---|
| 10.0 | 7.5 |
| 20.0 | 15.1 |
| 30.0 | 22.3 |
| 40.0 | 28.7 |
| 50.0 | 35.2 |
| 60.0 | 40.3 |
| 70.0 | 45.3 |
| 80.0 | 47.7 |

54. Students allow a narrow beam of laser light to strike a water surface. They arrange to measure the angle of refraction for selected angles of incidence and record the data shown

# ANSWERS TO QUICK QUIZZES

**25.1** (a). In part (a), you can see clear reflections of the headlights and the lights on the top of the truck. The reflection is specular. In part (b), although bright areas appear on the roadway in front of the headlights, the reflection is not as clear and no separate reflection of the lights from the top of the truck is visible. The reflection in part (b) is mostly diffuse.

**25.2** Beams ② and ④ are reflected; beams ③ and ⑤ are refracted.

**25.3** The index of refraction of air depends on its density. As the light from a star passes through the turbulent air, where the density varies from point to point, continuous refractions result in the path of the light being slightly zigzagged. Thus, the light entering your eye comes from directions that vary rapidly over a small angular range, resulting in the twinkling appearance.

**25.4** The optical day is longer than the geometric day. Due to the refraction of light by air, light rays from the Sun deviate slightly downward toward the surface of the Earth as

the light enters the atmosphere. Thus, in the morning, light rays from the upper edge of the Sun arrive at your eyes before the geometric line from your eyes to the top of the Sun clears the horizon. In the evening, light rays from the top of the Sun continue to arrive at your eyes even after the geometric line from your eyes to the top of the Sun dips below the horizon.

**25.5** The angle of reflection does not depend on wavelength because the light does not enter deeply into the material during reflection—it reflects from the surface. Thus, the properties of the material do not affect the angle of reflection.

**25.6** The two beams on the right result from total internal reflection at the right face of the prism. Because all of the light in these beams is reflected (rather than partly refracted), these two beams are the brightest. The light from the other three beams is divided into reflected and refracted parts.

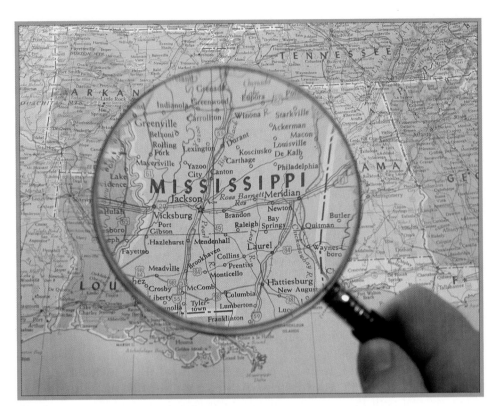

A magnifying glass is used to view an enlarged image of a portion of a map. A magnifying glass is a simple lens. We will explore the images formed by both mirrors and lenses in this chapter.

*(George Semple)*

# Image Formation by Mirrors and Lenses

**T**his chapter is concerned with the images formed when light interacts with flat and curved surfaces. We find that images of an object can be formed by reflection or by refraction and that mirrors and lenses work because of these phenomena. In this chapter, we continue to use the ray approximation. We shall find that some images are upside down and that they can vary in size relative to the object.

Images formed by reflection and refraction are used in a variety of everyday devices. These include the rearview mirror in your car, a shaving or make-up mirror, a camera, your eyeglasses, and a magnifying glass. In addition, more scientific devices, such as telescopes and microscopes, take advantage of the image formation principles discussed in this chapter.

We shall make extensive use of geometric models developed from the principles of reflection and refraction. These constructions allow us to develop mathematical representations for the image locations of various types of mirrors and lenses.

See the *Core Concepts in Physics CD-ROM,* Screen 14.6

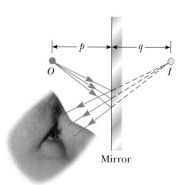

**Figure 26.1**

An image formed by reflection from a flat mirror. The image point *I* is located behind the mirror at a distance *q*, which is equal to the object distance *p*.

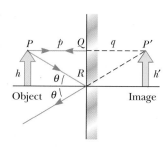

**Figure 26.2**

Geometric construction used to locate the image of an object placed in front of a flat mirror. Because the triangles *PQR* and *P'QR* are congruent, $p = |q|$ and $h = h'$.

**PITFALL PREVENTION 26.1**

**We are *choosing* a small number of rays**

Keep in mind that a *huge* number of light rays leaves each point on an object (and passes through each point on an image). We choose only a few major ones to draw in ray diagrams that allow us to locate the image.

## 26.1 • IMAGES FORMED BY FLAT MIRRORS

We begin our investigation by considering the simplest possible optical element, the flat mirror. Consider a point source of light* placed at *O* in Figure 26.1, a distance *p* in front of a flat mirror. The distance *p* is called the **object distance.** Light rays diverge from the source and are reflected from the mirror. After reflection, the rays continue to diverge. If the rays moving toward the observer's eyes in Figure 26.1 are extended backward until they intersect, the rays appear to diverge from a point *I* located behind the mirror. The apparent source of light rays at point *I* is called the **image** of the object at *O*. Point *I* is called the **image point.** Because the rays in Figure 26.1 appear to originate at *I*, which is located at a distance *q* behind the mirror, this is the location of the image.† The distance *q* is called the **image distance.**

Figure 26.2 is an example of a specialized pictorial representation, called a **ray diagram,** which is very useful in studies of mirrors and lenses. In a ray diagram, a small number of the myriad rays leaving a point source are drawn, and the location of the image is found by applying the laws of reflection (and refraction, in the case of refracting surfaces and lenses) to these rays. A carefully drawn ray diagram allows us to build a geometric model so that geometry and trigonometry can be used to solve a problem mathematically.

We examine the properties of the images formed by flat mirrors using the simple ray diagram shown in Figure 26.2. Let us locate the image of the tip of the blue arrow. To find out where the image is formed, it is necessary to follow at least two rays of light as they reflect from the mirror. One of those rays starts at *P*, follows the horizontal path *PQ* to the mirror, and reflects back on itself. The second ray follows the oblique path *PR* and reflects at the same angle according to the law of reflection. We can extend the two reflected rays back to the point from which they appear to have diverged, point *I*, the image point. A continuation of this process for points other than *P* on the object would result in an image (drawn as a yellow arrow) to the right of the mirror. These rays and the extensions of the rays allow us to build a geometric model for the image formation based on triangles *PQR* and *P'QR*. Because these two triangles are identical, $PQ = P'Q$, or $p = |q|$. (We use the absolute value notation because, as we shall see shortly, a sign convention is associated with the values of *p* and *q*.) Hence, we conclude that **the image formed by an object placed in front of a flat mirror is as far behind the mirror as the object is in front of the mirror.**

Our geometric model also shows that the object height *h* equals the image height $h'$. We define the **lateral magnification** (or simply the **magnification**) *M* of an image as follows:

$$M \equiv \frac{\text{image height}}{\text{object height}} = \frac{h'}{h} \qquad [26.1]$$

---

* We imagine the object to be a point source of light. This could actually *be* a point source, such as a very small lightbulb but, more often, it is a single point on some extended object that is illuminated from the exterior by a light source. Thus, the reflected light leaves the point on the object as if the point were a source of light.

† Your eyes and brain interpret diverging light rays as originating at the point from which the rays diverge. Your eye–brain system can detect the rays only *as they enter your eye,* and has no access to information about what experiences the rays underwent before reaching your eyes. Thus, despite the fact that the light rays did not *actually originate* at point *I*, they enter the eye *as if they had,* so that *I* is the point at which your brain locates the object.

This is a general definition of the magnification for any type of image formed by a mirror or lens. $M = 1$ for a flat mirror because $h' = h$ in this case. We also note that the image is **upright**—the image arrow points in the same direction as the object arrow. An upright image is indicated mathematically by a positive value of the magnification. (Later we discuss situations in which *inverted* images are formed.)

Finally, note that a flat mirror produces an image having an *apparent* left–right reversal. This reversal can be seen by standing in front of a mirror and raising your right hand. The image you see raises its left hand. Likewise, your hair appears to be parted on the opposite side, and a mole on your right cheek appears to be on your left cheek.

This reversal is not *actually* a left–right reversal. Imagine, for example, lying on your left side on the floor, with your body parallel to the mirror surface. Now, your head is on the left and your feet are on the right as you face the mirror. If you shake your feet, the image does not shake its head! If you raise your right hand, however, the image raises its left hand. Thus, it again appears like a left–right reversal, but in an up–down direction!

The apparent left–right reversal is actually a *front–back* reversal caused by the light rays going forward toward the mirror and then reflecting back from it. Figure 26.3 shows a person's right hand and its image in a flat mirror. Notice that no left–right reversal takes place—the thumbs on both the real hand and the image are on the left side. It is the front–back reversal that makes the image of the right hand appear similar to the real left hand at the left side of the photograph.

An interesting experience with front–back reversal is to stand in front of a mirror while holding an overhead transparency in front of you so that you can read the writing on the transparency. You are also able to read the writing on the image of the transparency. You might have had a similar experience if you have a transparent decal with words on it on the rear window of your car. If the decal is placed so that it can be read from outside the car, you can also read it when looking into your rearview mirror from the front seat.

### Quick Quiz 26.1

Tape a photograph on a bathroom mirror. Stand about one meter away from the mirror. Can you focus your eyes on *both* the photograph *and* your image in the mirror *at the same time?*

### Quick Quiz 26.2

In the movies, you sometimes see an actor looking in a mirror and you can see the face of the actor in the mirror. During the filming of this scene, what does the actor see in the mirror?

## THINKING PHYSICS 26.1

The professor in the box shown in Figure 26.4 appears to be balancing himself on a few fingers with both of his feet elevated from the floor. The professor can maintain this position for a long time, and he appears to defy gravity. How do you suppose this illusion was created?

**Reasoning** This is one example of an optical illusion, used by magicians, that makes use of a mirror. The box the professor is standing in is a cubical frame that contains a flat vertical mirror through a diagonal plane of the box. The professor

**PITFALL PREVENTION 26.2**

**Magnification does not necessarily imply enlargement**

For optical elements other than flat mirrors, the magnification defined in Equation 26.1 can result in a number with magnitude larger *or* smaller than 1. Thus, despite the cultural usage of the word *magnification* to mean *enlargement*, the image could be smaller than the object. We shall see several examples of such a situation in this chapter.

**Figure 26.3**

The image in the mirror of a person's right hand is reversed front to back. This makes the image in the mirror appear to be a left hand. Notice that the thumb is on the left side of both real hands and on the left side of the image. That the thumb is not on the right side of the image indicates that the reversal is not left–right.

*(George Semple)*

straddles the mirror so that the foot you see is in front of the mirror, and the other foot (which remains flat on the floor) is behind the mirror so you cannot see it. When he raises the foot in front of the mirror, the reflection of this foot also rises, so he appears to float in air.

## THINKING PHYSICS 26.2

Most rearview mirrors on cars have a day setting and a night setting. The night setting greatly diminishes the intensity of the image so that lights from trailing vehicles do not blind the driver. How does such a mirror work?

**Reasoning**  Figure 26.5 represents a cross-sectional view of the mirror for the two settings. The mirror is a wedge of glass with a reflecting surface on the back side. When the mirror is in the day setting, as in Figure 26.5a, the light from an object behind the car strikes the mirror at point 1. Most of the light enters the wedge, is refracted, and reflects from the back surface to return to the front surface, where it is refracted again as it re-enters the air as ray *B* (for *bright*). In addition, a small portion of the light is reflected at the front surface, as indicated by ray *D* (for *dim*). This dim reflected light is responsible for the image observed when the mirror is in the night setting, as in Figure 26.5b. In this case, the wedge is rotated so that the path followed by the bright light (ray *B*) does not lead to the eye. Instead, the dim light reflected from the front surface travels to the eye, and the brightness of trailing headlights does not become a hazard.

**Figure 26.4**

(Thinking Physics 26.1) *(Courtesy of Henry Leap and Jim Lehman)*

**Figure 26.5**

(Thinking Physics 26.2)

## Example 26.1 Multiple Images Formed by Two Mirrors

Two flat mirrors are at right angles, as in Figure 26.6, and an object is placed at point *O*. In this situation, multiple images are formed. Locate the positions of these images.

**Reasoning**  The image of the object is at $I_1$ in mirror 1 and at $I_2$ in mirror 2. In addition, a third image is formed at $I_3$, which is the image of $I_1$ in mirror 2 or, equivalently, the image of $I_2$ in mirror 1. That is, the image at $I_1$ (or $I_2$) serves as the object for $I_3$. Note that to form this image at $I_3$, the rays reflect twice after leaving the object at $O$.

**EXERCISE** Sketch the rays corresponding to viewing the images at $I_1$ and $I_2$, and show that the light is reflected only once in these cases.

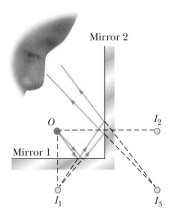

**Figure 26.6**

(Example 26.1) When an object is placed in front of two mutually perpendicular mirrors as shown, three images are formed.

## 26.2 • IMAGES FORMED BY SPHERICAL MIRRORS

See Screen 14.7

### Concave Mirrors

A **spherical mirror,** as its name implies, has the shape of a segment of a sphere. Figure 26.7a shows the cross-section of a spherical mirror with light reflecting from its surface, represented by the solid curved line. Such a mirror in which light is reflected from the inner, concave surface, is called a **concave mirror.** The mirror's radius of curvature is $R$, and its center of curvature is at point $C$. Point $V$ is the center of the spherical segment, and a line drawn from $C$ to $V$ is called the **principal axis** of the mirror.

Now consider a point source of light placed at point $O$ in Figure 26.7b, on the principal axis and outside point $C$. Two diverging rays that originate at $O$ are

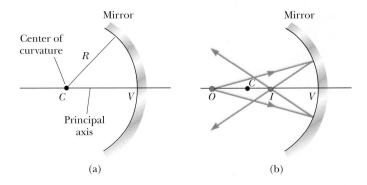

(a)            (b)

**Figure 26.7**

(a) A concave mirror of radius $R$. The center of curvature $C$ is located on the principal axis. (b) A point object placed at $O$ in front of a concave spherical mirror of radius $R$, where $O$ is any point on the principal axis farther than $R$ from the mirror surface, forms a real image at $I$. If the rays diverge from $O$ at small angles, they all reflect through the same image point.

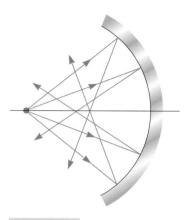

**Figure 26.8**

Rays diverging from an object at large angles from the principal axis reflect from a spherical concave mirror to intersect the principal axis at different points, resulting in a blurred image.

shown. After reflecting from the mirror, these rays converge and meet at *I*, the image point. They then continue to diverge from *I* as if a source of light existed there. Thus, if your eyes detect the rays diverging from point *I*, you would claim that a light source is located at that point.

This is the second example we have seen of rays diverging from an image point. But there is a distinct difference between the image point in Section 26.1 and the one we have just described. For the flat mirror, the light rays *seemed* to diverge from point *I*, but did not actually pass through the point. For the curved mirror, we see in Figure 26.7 that light rays pass through point *I*.

This distinction allows images to be classified into two categories. A **real image** is one in which light passes through and diverges from the image point as in Figure 26.7b; a **virtual image** is one in which the light only appears to diverge from that point as in Figure 26.2. Real images can be displayed on a screen (as at a movie), but virtual images cannot be displayed on a screen. The image of a real object seen in a flat mirror is *always virtual*. (We will discuss a *virtual object* in Section 26.4.)

In what follows, we shall adopt a simplification model which assumes that all rays diverging from an object make small angles with the principal axis. Such rays, called **paraxial rays,** always reflect through the image point as in Figure 26.7. Rays that make large angles with the principal axis as in Figure 26.8 converge at other points on the principal axis, producing a blurred image.

We can use a geometric model based on the ray diagram in Figure 26.9 to calculate the image distance *q* if we know the object distance *p* and radius of curvature *R*. By convention, these distances are measured from point *V*. Figure 26.9 shows two of the many rays of light leaving the tip of the object. One passes through the center of curvature *C* of the mirror, hitting the mirror perpendicular to the mirror surface and reflecting back on itself. The second ray strikes the mirror at the center point *V* and reflects as shown, obeying the law of reflection. The image of the tip of the arrow is at the point where these two reflected rays inter-

Red, blue, and green light rays are reflected by a parabolic mirror. Note that the focal point where the three colors meet is white light. *(Ken Kay/Fundamental Photographs)*

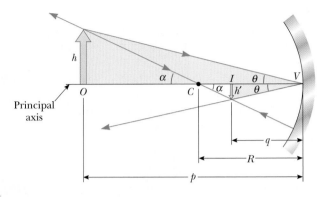

**Figure 26.9**

The image formed by a spherical concave mirror when the object $O$ lies outside the center of curvature $C$.

sect. Using these rays, we identify the gold and blue model right triangles in Figure 26.9. From the gold triangle, we see that $\tan\theta = h/p$, whereas the blue triangle gives $\tan\theta = -h'/q$. The negative sign signifies that the image is inverted, so $h'$ is a negative number. Thus, from Equation 26.1 and these results, we find that the magnification of the image is

$$M = \frac{h'}{h} = -\frac{q}{p} \qquad \text{[26.2]}$$

We can identify two additional right triangles in the figure, with a common point at $C$, and with angle $\alpha$. These triangles tell us that

$$\tan\alpha = \frac{h}{p-R} \qquad \text{and} \qquad \tan\alpha = -\frac{h'}{R-q}$$

from which we find that

$$\frac{h'}{h} = -\frac{R-q}{p-R} \qquad \text{[26.3]}$$

If we compare Equations 26.2 and 26.3, we see that

$$\frac{R-q}{p-R} = \frac{q}{p}$$

Algebra reduces this to

$$\frac{1}{p} + \frac{1}{q} = \frac{2}{R} \qquad \text{[26.4]}$$

• *Mirror equation in terms of the radius of curvature*

This expression is called the **mirror equation.** It is applicable only to the paraxial ray simplification model.

If the object is very far from the mirror—that is, if the object distance $p$ is large compared with $R$, so that $p$ can be said to approach infinity, then $1/p \to 0$, and we see from Equation 26.4 that $q \approx R/2$. In other words, **when the object is very far from the mirror, the image point is halfway between the center of curvature and the center of the mirror,** as in Figure 26.10a. The rays are essentially parallel in this figure because only those few rays traveling parallel to the axis from the distant object encounter the mirror. Rays not parallel to the axis miss the

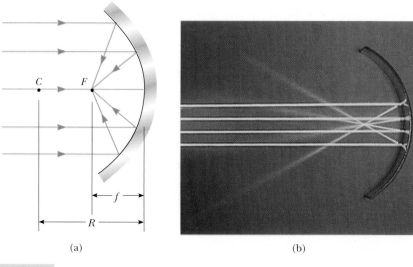

(a)                                              (b)

**Figure 26.10**

(a) Light rays from a distant object ( $p \approx \infty$ ) reflect from a concave mirror through the focal point *F*. In this case, the image distance $q \approx R/2 = f$, where *f* is the focal length of the mirror. (b) Reflection of parallel rays from a concave mirror. *(Henry Leap and Jim Lehman)*

**PITFALL PREVENTION 26.3**

**The *focal* point is not the *focus* point**

Do not misinterpret the focal point as the point at which the light rays focus to form an image. The focal point is determined by the curvature of the mirror and is determined solely by the mirror—it does not depend on the location of any object. In general, an image forms at a point different from the focal point of a mirror (or a lens, as we shall see in a later section). The *only* exception occurs when the object is located infinitely far away from the mirror. In this situation, the light rays do focus at the focal point.

mirror. Figure 26.10b shows an experimental setup of this situation, demonstrating the crossing of the light rays at a single point. The point at which the parallel rays intersect after reflecting from the mirror is called the **focal point** of the mirror. The focal point is a distance *f* from the mirror, called the **focal length.** The focal length is a parameter associated with the mirror and is given by

- *Focal length of a mirror*

$$f = \frac{R}{2} \qquad [26.5]$$

The mirror equation can therefore be expressed in terms of the focal length:

- *Mirror equation in terms of focal length*

$$\frac{1}{p} + \frac{1}{q} = \frac{1}{f} \qquad [26.6]$$

This equation is the commonly used mirror equation, in terms of the focal length of the mirror rather than its radius of curvature, as in Equation 26.4. We shall see how to use this equation in examples that follow shortly.

### Convex Mirrors

Figure 26.11 shows the formation of an image by a **convex mirror**—a mirror that is silvered so that light is reflected from the outer, convex surface. Convex mirrors are sometimes called **diverging mirrors** because the rays from any point on an object diverge after reflection as though they were coming from some point behind the mirror. The image in Figure 26.11 is virtual rather than real because it lies behind the mirror at the point from which the reflected rays appear to diverge. In general, as shown in the figure, the image formed by a convex mirror is always upright, virtual, and smaller than the object.

We can set up a geometric model for a convex mirror, using the ray diagram in Figure 26.11. The equations developed for concave mirrors can also be used

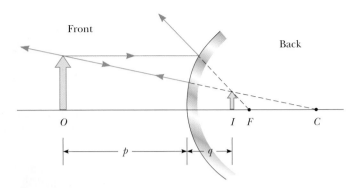

**Figure 26.11**

Formation of an image by a spherical convex mirror. The image formed by the real object is virtual and upright.

with convex mirrors if we adhere to a particular sign convention. Let us refer to the region in which light rays move as the *front side* of the mirror, and the other side, where virtual images are formed, as the *back side*. For example, in Figures 26.9 and 26.11, the side to the left of the mirror is the front side, and that to the right of the mirror is the back side. Table 26.1 summarizes the sign conventions for all the necessary quantities. One entry that may appear strange is a "virtual object." This will only be used when combining two or more optical elements as we shall see in Section 26.4.

## Constructing Ray Diagrams for Mirrors

We have been using the specialized pictorial representations called ray diagrams to help us locate images for flat and curved mirrors. Let us now formalize the procedure for drawing accurate ray diagrams. To construct such a diagram, we must know the position of the object and the locations of the focal point and center of curvature of the mirror. Three rays are constructed as shown by the examples in Figure 26.12. In each part of the figure, the right-hand portion shows a photograph of the situation described by the ray diagram in the left-hand portion. All three rays start from the same object point; in these examples, the top of the arrow is chosen as the starting point. For the concave mirrors in Figure 26.12a and b, the rays are drawn as follows:

• Ray 1 is drawn parallel to the principal axis and is reflected back through the focal point *F*. (This would be a light path followed by light from an object infinitely far from the mirror.)

| **TABLE 26.1** | **Sign Conventions for Mirrors** |
|---|---|

*p* is **positive** if the object is in **front** of the mirror (real object).
*p* is **negative** if the object is in **back** of the mirror (virtual object).
*q* is **positive** if the image is in **front** of the mirror (real image).
*q* is **negative** if the image is in **back** of the mirror (virtual image).
Both *f* and *R* are **positive** if the center of curvature is in **front** of the mirror (concave mirror).
Both *f* and *R* are **negative** if the center of curvature is in **back** of the mirror (convex mirror).
If *M* is **positive,** the image is **upright.**
If *M* is **negative,** the image is **inverted.**

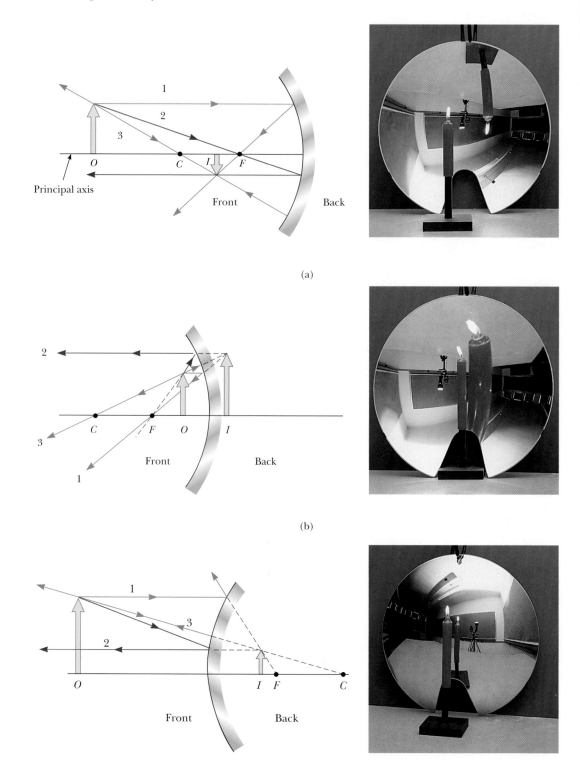

**Figure 26.12**

Ray diagrams for spherical mirrors, along with corresponding photographs of the images of a candle as the object. (a) When the object is located so that the center of curvature lies between the object and a concave mirror surface, the image is real and inverted. (b) When the object is located between the focal point and a concave mirror surface, the image is virtual, upright, and enlarged. (c) When the object is in front of a spherical convex mirror, the image is virtual, upright, and reduced in size.

- Ray 2 is drawn through the focal point (or as if coming from the focal point if $p < f$, as in Figure 26.12b). Thus, it is reflected parallel to the principal axis. (This would be a light path followed by light from an object at the focal point and is the reverse of a ray approaching the mirror from an object infinitely far away.)
- Ray 3 is drawn through the center of curvature $C$ and is reflected back on itself. (This ray follows the law of reflection for light incident along the normal to the surface.)

Note that only two rays are needed to locate the image. The third ray serves as a check on your construction—all three rays must pass through the image point. The image point obtained in this fashion must always agree with the value of $q$ calculated from the mirror equation.

With concave mirrors, note what happens as the object is moved closer to the mirror from infinity. The real, inverted image in Figure 26.12a moves to the left as the object approaches the mirror. When the object is at the center of curvature, the object and image are at the same distance from the mirror and are the same size. (This is the case in the photograph of Figure 26.12a.) When the object is at the focal point, the image is infinitely far to the left. (Check these last three sentences with the mirror equation!)

When the object lies between the focal point and the mirror surface as in Figure 26.12b, the image is virtual, upright, and located to the right of the mirror. The image is also larger than the object in this case. This is the principle behind a shaving mirror or a makeup mirror. Your face is located closer to the concave mirror than the focal point, so you see an enlarged, upright image of your face, to assist you with shaving or applying makeup. If you have such a mirror, look into it and move your face farther from the mirror. Your head will pass through a point where the image is indistinct and then the image will reappear with your face upside down as you continue to move farther away. The region where the image is indistinct is where your head passes through the focal point and the image is infinitely far away.

For a convex mirror as shown in Figure 26.12c, the rays are drawn as follows:

- Ray 1 is drawn parallel to the principal axis and is reflected as if coming from the focal point $F$.
- Ray 2 is drawn heading toward the focal point on the back side of the mirror. It is reflected parallel to the principal axis.
- Ray 3 is drawn heading toward the center of curvature $C$ on the back side of the mirror and is reflected back on itself.

The image of a real object in a convex mirror is always virtual and upright. As the object distance increases, the virtual image becomes smaller and approaches the focal point as $p$ approaches infinity. You should construct other diagrams to verify how the image position varies with object position.

Convex mirrors are often used as security devices in large stores, where they are hung at a high position on the wall. The large field of view of the store is made smaller by the convex mirror, so that the store personnel can observe possible shoplifting activity in several aisles at once. Right-hand rearview mirrors on automobiles are also often made with a convex surface. This allows a wider field of view behind the automobile to be available to the driver than is the case with a flat mirror. These mirrors introduce a perceptual distortion, however, in that they cause cars behind the driver to appear smaller, and, therefore, farther away. This is why these mirrors carry the inscription, "Objects in this mirror are closer than they appear."

### THINKING PHYSICS 26.3

For a concave mirror, a virtual image can be anywhere behind the mirror. For a convex mirror, however, there is a maximum distance at which the image can exist behind the mirror. Why?

**Reasoning** First consider the concave mirror, and imagine two different light rays leaving a tiny object and striking the mirror. If the object is inside the focal point, but infinitesimally close to it, the light rays reflecting from the mirror are parallel to the mirror axis. They can be interpreted as forming a virtual image infinitely far away behind the mirror. As the object is brought closer to the mirror, the reflected rays diverge through larger and larger angles, resulting in their extensions converging closer and closer to the back of the mirror. When the object is much closer to the mirror compared with the focal length, the mirror looks like a plane mirror, so the image is just as far behind the mirror as the object is in front of it. Thus, the image can be anywhere from infinitely far away to right at the surface of the mirror.

For the convex mirror, an object at infinity produces a virtual image at the focal point. As the object is brought closer to the mirror, the reflected rays diverge more sharply and the image moves closer to the mirror. Thus, the virtual image is restricted to the region between the mirror and the focal point.

### Example 26.2 The Image for a Concave Mirror

Assume that a certain concave spherical mirror has a focal length of 10.0 cm. Find the location of the image for object distances of (a) 25.0 cm, (b) 10.0 cm, and (c) 5.00 cm. Describe the image in each case.

**Solution** (a) For an object distance of 25.0 cm, we find the image distance using the mirror equation:

$$\frac{1}{p} + \frac{1}{q} = \frac{1}{f}$$

$$\frac{1}{25.0 \text{ cm}} + \frac{1}{q} = \frac{1}{10.0 \text{ cm}}$$

$$q = \boxed{16.7 \text{ cm}}$$

The magnification is given by Equation 26.2:

$$M = -\frac{q}{p} = -\frac{16.7 \text{ cm}}{25.0 \text{ cm}} = -0.668$$

The magnitude of $M$ less than unity tells us that the image is smaller than the object. The negative sign for $M$ tells us that the image is inverted. Finally, because $q$ is positive, the image is located on the front side of the mirror and is real. This situation is pictured in Figure 26.12a.

(b) When the object distance is 10.0 cm, the object is located at the focal point. Substituting the values $p = 10.0$ cm and $f = 10.0$ cm into the mirror equation, we find

$$\frac{1}{10.0 \text{ cm}} + \frac{1}{q} = \frac{1}{10.0 \text{ cm}}$$

$$q = \infty$$

Thus, we see that rays of light originating from an object located at the focal point of a mirror are reflected so that the image is formed at an infinite distance from the mirror; that is, the rays travel parallel to one another after reflection.

(c) When the object is at the position $p = 5.00$ cm, it lies between the focal point and the mirror surface. In this case, the mirror equation gives

$$\frac{1}{5.00 \text{ cm}} + \frac{1}{q} = \frac{1}{10.0 \text{ cm}}$$

$$q = \boxed{-10.0 \text{ cm}}$$

The negative value for $q$ tells us the image is virtual and located behind the mirror. The magnification is

$$M = -\frac{q}{p} = -\left(\frac{-10.0 \text{ cm}}{5.00 \text{ cm}}\right) = 2.00$$

From this, we see that the image is larger than the object by a factor of 2.00, and the positive sign for $M$ indicates that the image is upright (see Fig. 26.12b).

Note the characteristics of the images formed by a concave spherical mirror. When the object is farther from the mirror than the focal point, the image is inverted and real; with the object at the focal point, the image is formed at infinity; with the object between the focal point and mirror surface, the image is upright and virtual.

**EXERCISE** If the object distance is 20.0 cm, find the image distance and the magnification of the image.

*Answer* $q = 20.0$ cm, $M = -1.00$

---

### Example 26.3   The Image for a Convex Mirror

An object 3.00 cm high is placed 20.0 cm from a convex mirror having a focal length of 8.00 cm. Find (a) the position of the final image and (b) the height of the image.

**Solution** (a) Because the mirror is convex, its focal length is negative. To find the image position, we use the mirror equation:

$$\frac{1}{p} + \frac{1}{q} = \frac{1}{f} = \frac{1}{-8.00 \text{ cm}}$$

$$\frac{1}{q} = -\frac{1}{8.00 \text{ cm}} - \frac{1}{20.0 \text{ cm}}$$

$$q = \boxed{-5.71 \text{ cm}}$$

The negative value of $q$ indicates that the image is virtual, or behind the mirror, as in Figure 26.12c.

(b) The magnification is

$$M = -\frac{q}{p} = -\left(\frac{-5.71 \text{ cm}}{20.0 \text{ cm}}\right) = 0.286$$

The image is upright because $M$ is positive. Its height is

$$h' = Mh = (0.286)(3.00 \text{ cm}) = \boxed{0.858 \text{ cm}}$$

---

## 26.3 • IMAGES FORMED BY REFRACTION

In this section we describe how images are formed by the refraction of rays at the surface of a transparent material. We shall apply the law of refraction and use the simplification model in which we consider only paraxial rays.

Consider two transparent media with indices of refraction $n_1$ and $n_2$, where the boundary between the two media is a spherical surface with radius of curvature $R$ (Fig. 26.13). We shall assume that the object at point $O$ is in the medium with index of refraction $n_1$. As we shall see, all paraxial rays are refracted at the spherical surface and converge to a single point $I$, the image point.

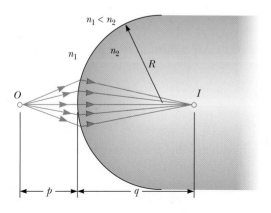

**Figure 26.13**

An image formed by refraction at a spherical surface. Rays making small angles with the principal axis diverge from a point object at $O$ and are refracted through the image point $I$.

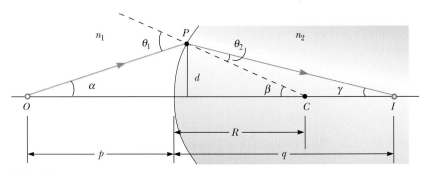

**Figure 26.14**

Geometry used to derive Equation 26.8.

Let us proceed by considering the geometric construction in Figure 26.14, which shows a single ray leaving point $O$ and passing through point $I$. Snell's law applied to this refracted ray gives

$$n_1 \sin \theta_1 = n_2 \sin \theta_2$$

Because the angles $\theta_1$ and $\theta_2$ are small for paraxial rays, we can use the approximation $\sin \theta \approx \theta$ (angles in radians). Therefore, Snell's law becomes

$$n_1 \theta_1 = n_2 \theta_2$$

Now we make use of geometric model triangles and recall that an exterior angle of any triangle equals the sum of the two opposite interior angles. Applying this to the triangles $OPC$ and $PIC$ in Figure 26.14 gives

$$\theta_1 = \alpha + \beta$$

$$\beta = \theta_2 + \gamma$$

If we combine the last three equations and eliminate $\theta_1$ and $\theta_2$, we find

$$n_1 \alpha + n_2 \gamma = (n_2 - n_1)\beta \qquad \text{[26.7]}$$

In the small angle approximation, $\tan \theta \approx \theta$, and so we can write the approximate relations

$$\tan \alpha \approx \alpha \approx \frac{d}{p} \qquad \tan \beta \approx \beta \approx \frac{d}{R} \qquad \tan \gamma \approx \gamma \approx \frac{d}{q}$$

where $d$ is the distance shown in Figure 26.14. We substitute these into Equation 26.7 and divide through by $d$ to give

$$\frac{n_1}{p} + \frac{n_2}{q} = \frac{n_2 - n_1}{R} \qquad \text{[26.8]}$$

By setting up a geometric construction with an object and a refracting surface, we can show that the magnification of a refracting surface is

$$M = -\frac{n_1 q}{n_2 p} \qquad \text{[26.9]}$$

| TABLE 26.2 | Sign Conventions for Refracting Surfaces |
|---|---|

$p$ is **positive** if the object is in **front** of the surface (real object).
$p$ is **negative** if the object is in **back** of the surface (virtual object).
$q$ is **positive** if the image is in **back** of the surface (real image).
$q$ is **negative** if the image is in **front** of the surface (virtual image).
$R$ is **positive** if the center of curvature is in **back** of the surface.
$R$ is **negative** if the center of curvature is in **front** of the surface.

As with mirrors, we must use a sign convention if we are to apply Equations 26.8 and 26.9 to a variety of circumstances. First note that real images are formed on the side of the surface that is *opposite* the side from which the light comes. This is in contrast to mirrors, for which real images are formed on the side where the light originates. Therefore, **the sign conventions for spherical refracting surfaces are similar to the conventions for mirrors, recognizing the change in sides of the surface for real and virtual images.** For example, in Figure 26.14, $p$, $q$, and $R$ are all positive.

The sign conventions for spherical refracting surfaces are summarized in Table 26.2. The same conventions will be used for thin lenses discussed in the next section. As with mirrors, we assume that the front of the refracting surface is the side from which the light approaches the surface.

## Flat Refracting Surfaces

If the refracting surface is flat, then $R$ approaches infinity and Equation 26.8 reduces to

$$\frac{n_1}{p} = -\frac{n_2}{q}$$

or

$$q = -\frac{n_2}{n_1} p \qquad [26.10]$$

From Equation 26.10 we see that the sign of $q$ is opposite that of $p$. Thus, **the image formed by a flat refracting surface is on the same side of the surface as the object.** This is illustrated in Figure 26.15 for the situation in which $n_1$ is greater than $n_2$, where a virtual image is formed between the object and the surface. Note that the refracted ray bends *away* from the normal in this case, because $n_1 > n_2$.

The value of $q$ given by Equation 26.10 is always smaller in magnitude than $p$ when $n_1 > n_2$. This indicates that the image of an object located within a material with higher index of refraction than that of the material from which it is viewed is always closer to the flat refracting surface than the object. For this reason, transparent bodies of water such as streams and swimming pools always appear shallower than they are—the image of the bottom of the body of water is closer to the surface than the bottom is in reality.

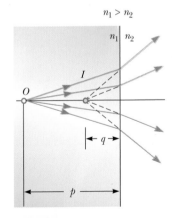

**Figure 26.15**

The image formed by a flat refracting surface is virtual; that is, it forms on the same side of the refracting surface as the object. All rays are assumed to be paraxial.

## Example 26.4    Gaze into the Crystal Ball

A coin 2.00 cm in diameter is embedded in a solid glass ball of radius 30.0 cm (Fig. 26.16). The index of refraction of the ball is $n_1 = 1.50$, and the coin is 20.0 cm from the surface. (a) Find the position of the image.

**Solution**    Because $n_1 > n_2$, where $n_2 = 1.00$ is the index of refraction for air, the rays originating from the object are refracted away from the normal at the surface and diverge outward. Hence, the image is formed inside the glass and is virtual. Applying Equation 26.8, we have

$$\frac{n_1}{p} + \frac{n_2}{q} = \frac{n_2 - n_1}{R}$$

$$\frac{1.50}{20.0 \text{ cm}} + \frac{1.00}{q} = \frac{1.00 - 1.50}{-30.0 \text{ cm}}$$

where the radius of curvature is indicated as negative because the center of curvature is in front of the concave surface (see Table 26.2). Solving for $q$ gives

$$q = -17.1 \text{ cm}$$

The negative sign indicates that the image is in the same medium as the object (the side of the incident light), in agreement with our ray diagram. Because the light rays do not pass through the image point, the image is virtual.

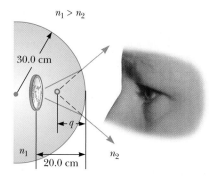

**Figure 26.16**

(Example 26.4) A coin embedded in a glass ball forms a virtual image between the coin and the glass surface. All rays are assumed to be paraxial.

(b) What is the magnification of the image?

**Solution**    Using Equation 26.9, we have

$$M = -\frac{n_1 q}{n_2 p} = -\frac{(1.50)(-17.1 \text{ cm})}{(1.00)(20.0 \text{ cm})} = 1.28$$

Thus, the image appears 28% larger than the actual object.

## Example 26.5    The One That Got Away

A small fish is swimming at a depth $d$ below the surface of a pond (Fig. 26.17). What is the apparent depth of the fish as viewed from directly overhead?

**Figure 26.17**

(Example 26.5) The apparent depth $q$ of the fish is less than the true depth $d$. All rays are assumed to be paraxial.

**Solution**    In this example, the refracting surface is flat, and so $R$ is infinite. Hence, we can use Equation 26.10 to determine the location of the image. Using the facts that $n_1 = 1.33$ for water and $p = d$ gives

$$q = -\frac{n_2}{n_1} p = -\frac{1.00}{1.33} d = -0.752 \, d$$

Again, because $q$ is negative, the image is virtual as indicated in Figure 26.17. The apparent depth is approximately three-fourths the actual depth.

# 26.4 • THIN LENSES

See Screen 14.8

A typical **thin lens** consists of a piece of glass or plastic, ground so that its two surfaces are either segments of spheres or planes. Lenses are commonly used in optical instruments such as cameras, telescopes, and microscopes, to form images by refraction.

Figure 26.18 shows some representative shapes of lenses. These lenses have been placed in two groups. Those in Figure 26.18a are thicker at the center than at the rim, and those in Figure 26.18b are thinner at the center than at the rim. The lenses in the first group are examples of **converging lenses,** and those in the second group are called **diverging lenses.** The reason for these names will become apparent shortly.

As with mirrors, it is convenient to define a point called the **focal point** for a lens. For example, on the left side of Figure 26.19a, a group of rays parallel to the principal axis passes through the focal point $F_2$ after being converged by the lens. The distance from the focal point to the lens is again called the **focal length** $f$. **The focal length is the image distance that corresponds to an infinite object distance.**

To avoid the complications arising from the thickness of the lens, we adopt a simplification model called the **thin lens approximation: the thickness of the lens is assumed to be negligible.** As a result, it makes no difference whether we take the focal length to be the distance from the focal point to the surface of the lens or from the focal point to the center of the lens, because the difference in these two lengths is assumed to be negligible. A thin lens has one focal length and *two* focal points as illustrated in Figure 26.19, corresponding to parallel light rays traveling from the left or right.

Rays parallel to the axis diverge after passing through a lens of the shape shown in Figure 26.19b. In this case, the focal point is defined as the point from which the diverging rays appear to originate, labeled $F_1$ on the left side of Figure 26.19b. Figures 26.19a and 26.19b indicate why the names *converging* and *diverging* are applied to these lenses in Figure 26.18.

**Figure 26.18**

Various lens shapes: (a) Converging lenses have a positive focal length and are thickest at the middle. (b) Diverging lenses have a negative focal length and are thickest at the edges.

Biconvex       Convex–         Plano–
               concave         convex

(a)

Biconcave      Convex–         Plano–
               concave         concave

(b)

**Figure 26.19**

(*Left*) Effects of a converging lens (*top*) and a diverging lens (*bottom*) on parallel rays. (*Henry Leap and Jim Lehman*) (*Right*) The focal points of (a) a converging lens and (b) a diverging lens.

**Figure 26.20**

A ray passing through the center of the lens is deviated from its path by a distance of $\delta$. In the thin lens approximation, this deviation is ignored.

Suppose a ray of light passes through the center of a lens as in Figure 26.20. Applying Snell's law at both surfaces, we find that this ray is shifted from its original direction of travel by a distance $\delta$. In the thin-lens simplification model, however, $\delta$ becomes vanishingly small and the ray passes through the lens unshifted.

Consider now the ray diagram in Figure 26.21. Ray 1 passes through the center of the lens. Ray 2 is parallel to the principal axis of the lens (the horizontal axis passing through $Q$), and as a result it passes through the focal point $F$ after refraction. The point at which these two rays intersect is the image point.

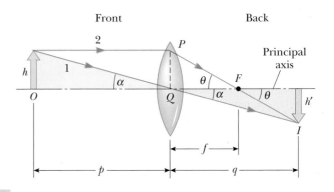

**Figure 26.21**

A geometric construction for developing the thin lens equation.

The tangent of the angle $\alpha$ can be found by using the blue and gold geometric model triangles in Figure 26.21:

$$\tan \alpha = \frac{h}{p} \quad \text{and} \quad \tan \alpha = -\frac{h'}{q}$$

from which

$$M = \frac{h'}{h} = -\frac{q}{p} \qquad \text{[26.11]}$$

Thus, the equation for magnification by a lens is the same as the equation for magnification by a mirror (Eq. 26.2). We also note from Figure 26.21 that

$$\tan \theta = \frac{PQ}{f} \quad \text{and} \quad \tan \theta = -\frac{h'}{q - f}$$

However, the height $PQ$ is the same as $h$. Therefore,

$$\frac{h}{f} = -\frac{h'}{q - f}$$

$$\frac{h'}{h} = -\frac{q - f}{f}$$

Using this expression in combination with Equation 26.11 gives

$$\frac{q}{p} = \frac{q - f}{f}$$

which reduces to

$$\frac{1}{p} + \frac{1}{q} = \frac{1}{f} \qquad \text{[26.12]}$$

• *Thin lens equation*

This equation, called the **thin lens equation** (which is identical to the mirror equation, Equation 26.6), can be used with either converging or diverging lenses if we adhere to a set of sign conventions. Figure 26.22 is useful for obtaining the signs of $p$ and $q$. (As with mirrors, we call the side from which the light approaches the *front* of the lens.) The complete sign conventions for lenses are provided in Table 26.3. Note that **a converging lens has a positive focal length** under this convention and **a diverging lens has a negative focal length.** Hence, the names *positive* and *negative* are often given to these lenses.

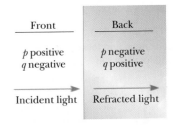

**Figure 26.22**

A diagram for obtaining the signs of $p$ and $q$ for a thin lens or a refracting surface.

| **TABLE 26.3** | **Sign Conventions for Thin Lenses** |
|---|---|

$p$ is **positive** if the object is in **front** of the lens.
$p$ is **negative** if the object is in **back** of the lens.
$q$ is **positive** if the image is in **back** of the lens.
$q$ is **negative** if the image is in **front** of the lens.
$R_1$ and $R_2$ are **positive** if the center of curvature is in **back** of the lens.
$R_1$ and $R_2$ are **negative** if the center of curvature is in **front** of the lens.
$f$ is **positive** if the lens is **converging.**
$f$ is **negative** if the lens is **diverging.**

### Quick Quiz 26.3

Suppose you take a photograph with a camera that uses an ultrasonic range finder to measure the distance to the subject. The range finder sends out an ultrasonic sound wave and measures the time interval between when the sound wave is emitted and when it arrives back at the camera after reflection. When you take a picture of yourself in a mirror with this camera, your image is likely to be out of focus. Why?

### Quick Quiz 26.4

A virtual image is often described as one through which light rays do not travel, as they do for a real image. Can a virtual image be photographed?

### Quick Quiz 26.5

Suppose you want to use a converging lens to project the image of two trees onto a screen as in Figure 26.23. One tree is at a distance $x$ from the lens, the other is at $2x$. You adjust the screen so that the near tree is in focus. If you now want to move the screen so that the far tree is in focus, should you move the screen toward or away from the lens?

The focal length for a lens in air is related to the curvatures of its surfaces and to the index of refraction $n$ of the lens material by

• *Lens makers' equation*

$$\frac{1}{f} = (n - 1)\left(\frac{1}{R_1} - \frac{1}{R_2}\right)$$    [26.13]

where $R_1$ is the radius of curvature of the front surface and $R_2$ is the radius of curvature of the back surface. Equation 26.13 enables us to calculate the focal length from the known properties of the lens. It is called the **lens makers' equation.** Table 26.3 includes the sign conventions for determining the signs of the radii $R_1$ and $R_2$.

**PITFALL PREVENTION 26.5**

**Lens have two focal points but one focal length**

A lens has a focal point on each side, front and back. As seen in Figure 26.19, however, only one focal length exists—each of the two focal points is located the same distance from the lens. This can be seen mathematically by interchanging $R_1$ and $R_2$ in Equation 26.13 (as well as changing the signs of the radii because back and front have been swapped). This results in no change in the focal length. As a result, the lens forms an image of an object at the same point if it is turned around. In practice, this claim might not be verified experimentally because real lenses are not infinitesimally thin.

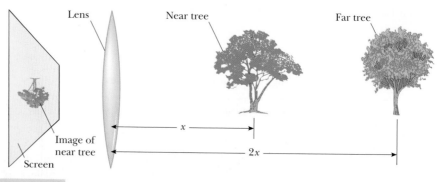

**Figure 26.23**

(Quick Quiz 26.5)

## Quick Quiz 26.6

For each of the lens shapes in Figure 26.18, provide the sign of the radii of curvature (or indicate that the radius is infinite), assuming that the left side of each lens is the front side, $R_1$ is the radius of curvature of the left surface of a lens, and $R_2$ is that of the right surface.

## Quick Quiz 26.7

Why does the focal length of a lens depend on the lens material, but the focal length for a mirror is independent of the material from which the mirror is made?

## Ray Diagrams for Thin Lenses

Our specialized pictorial representations called ray diagrams are very convenient for locating the image of a thin lens or system of lenses. They should also help clarify the sign conventions we have already discussed. Figure 26.24 illustrates this method for three single-lens situations. To locate the image of a converging lens (Figs. 26.24a and b), the following three rays are drawn from the top of the object:

- Ray 1 is drawn parallel to the principal axis. After being refracted by the lens, this ray passes through the focal point on the back side of the lens.
- Ray 2 is drawn through the center of the lens and continues in a straight line.
- Ray 3 is drawn through the focal point on the front side of the lens (or as if coming from the focal point if $p < f$, as in Figure 26.24b) and emerges from the lens parallel to the principal axis.

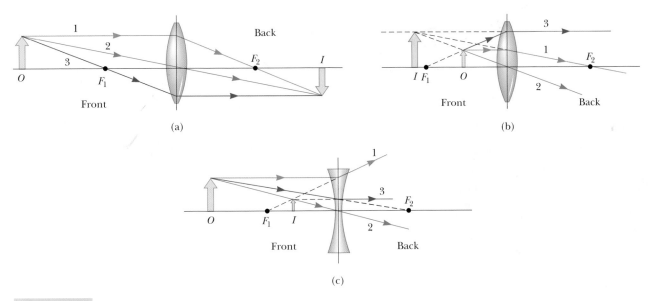

(a)

(b)

(c)

### Figure 26.24

Ray diagrams for locating the image formed by a thin lens. (a) When the object is in front of and outside the object focal point $F_1$ of a converging lens, the image is real, inverted, and on the back side of the lens. (b) When the object is between $F_1$ and a converging lens, the image is virtual, upright, larger than the object and on the front side of the lens. (c) When an object is anywhere in front of a diverging lens, the image is virtual, upright, smaller than the object, and on the front side of the lens.

To locate the image of a diverging lens (Figure 26.24c), the following three rays are drawn from the top of the object:

- Ray 1 is drawn parallel to the principal axis. After being refracted by the lens, this ray emerges directed away from the focal point on the front side of the lens.
- Ray 2 is drawn through the center of the lens and continues in a straight line.
- Ray 3 is drawn in the direction toward the focal point on the back side of the lens and emerges from the lens parallel to the principal axis.

In these ray diagrams, the point of intersection of *any two* of the rays can be used to locate the image. The third ray serves as a check of construction.

For the converging lens in Figure 26.24a where the object is *outside* the front focal point ($p > f$), the image is real and inverted and is located on the back side of the lens. This diagram would be representative of a movie projector, for which the film is the object, the lens is in the projector, and the image is projected on a large screen for the audience to watch. The film is placed in the projector with the scene upside down so that the inverted image is right side up for the audience.

When the real object is *inside* the front focal point ($p < f$) as in Figure 26.24b, the image is virtual and upright. When used in this way, the lens is acting as a magnifying glass, providing an enlarged upright image for closer study of the object as in the case of the opening photograph for this chapter. The object might be a stamp, a fingerprint, or a printed page for someone with failing eyesight.

For the diverging lens of Figure 26.24c, the image is virtual and upright for all object locations. A diverging lens is used in a security peephole in a door, to give a wide-angle view. Nearsighted individuals use diverging eyeglass lenses or contact lenses. Another use is for a panoramic lens for a camera (although a sophisticated camera "lens" is actually a combination of several lenses). A diverging lens in this application creates a small image of a wide field of view.

### Quick Quiz 26.8

In Figure 26.24a, the blue object arrow is replaced by one that is much taller than the lens. How many rays from the object strike the lens?

### Quick Quiz 26.9

If you cover the top half of the lens in Figure 26.24a with a piece of paper, which of the following happens to the appearance of the image of the object? (a) the bottom half disappears, (b) the top half disappears, (c) the entire image is visible but dimmer, (d) no change, (e) the entire image disappears.

APPLICATION

Corrective lenses on diving masks

**THINKING PHYSICS 26.4**

Diving masks often have lenses built into the glass for divers who do not have perfect vision. This allows the individual to dive without the necessity for glasses, because the lenses in the faceplate perform the necessary refraction to provide clear vision. Normal glasses have lenses that are curved on both the front and rear surfaces. The lenses in a diving mask faceplate often only have curved surfaces on the *inside* of the glass. Why is this design desirable?

**Reasoning**   The main reason for curving only the inner surface of the lenses in the diving mask faceplate is so that the diver can see clearly when looking at objects straight ahead while underwater *and* in the air. Consider light rays approaching the mask along a normal to the plane of the faceplate. If curved surfaces were on both the front and the back of the diving lens on the faceplate, refraction would occur at each surface. The lens could be designed so that these two refractions would give clear vision while the diver is in air. When the diver is underwater, however, the refraction between the water and the glass at the first interface is now different because the index of refraction of water is different from that of air. Thus, the vision would not be clear underwater.

By making the outer surface of the lens flat, light is not refracted at normal incidence to the faceplate at the outer surface *in either air or water*—all of the refraction occurs at the inner glass–air surface. Thus, the same refractive correction exists in water and in air, and the diver can see clearly in both environments.

## THINKING PHYSICS 26.5

Consider a glass planoconvex lens—flat on one side and convex on the other. You project three parallel laser beams through it, as in Figure 26.25a, and measure the focal length *f*, that is, the distance from the lens to the point at which the three beams cross. Now, you hold the flat side of the lens against the glass of an aquarium filled with water as in Figure 26.25b. When you shine the laser beams through the lens from the outside of the aquarium, do the beams cross at a point closer to the lens than before, farther away, or at the same distance? What happens if you direct the laser beams through the lens from the other side of the aquarium as in Figure 26.25c?

**Reasoning**   The laser beam going through the center of the lens is unaffected by the lens in all three cases. We need to look at what happens to the outer beams. In Figure 26.25a, these two laser beams refract as they pass from the air into the glass of the lens so as to be closer to the normal than when in the air. Thus, they deviate *toward* the central beam. As they reach the other side and pass from the glass back into the air, they deviate away from the normal, causing additional deviation toward the central beam. As a result, all three beams cross at the focal point. When the flat side of the lens is held against the glass side of the aquarium as in Figure 26.25b, the outgoing laser beams pass through the glass of the aquarium without additional refraction, but then enter *water*. The change in index of refraction in going from glass to water is much smaller than that for glass to air. Thus, the outgoing refraction is smaller than in the previous case, and the deviation toward the central beam is less. As a result, the crossing point for the three laser beams is farther from the lens. When the laser beams are directed through the water and then through the lens, as in Figure 26.25c, we must consider two factors. First, the focal length of a lens is independent of which way the light passes through it. Secondly, in the situation shown, the laser beams experience no refraction on entering the flat side of the lens because they strike the glass at normal incidence. Thus, all of the refraction occurs at the interface of the air with the curved side of the lens. This is the same situation as if the laser beams had struck the flat side of the lens while it was in air. Thus, the water in the aquarium has no effect. The laser beams cross at the same distance from the lens as they did in the first diagram, although on the opposite side of the lens. This third situation is similar to the curvature of a lens on the inside of a diving mask discussed in Thinking Physics 26.4.

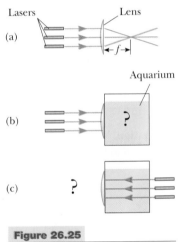

**Figure 26.25**

(Thinking Physics 26.5)

## Example 26.6   The Lens Makers' Equation

The biconvex lens of Figure 26.26 has an index of refraction of 1.50. The radius of curvature of the front surface is 10 cm, and that of the back surface is 15 cm. Find the focal length of the lens.

**Solution**   From the sign conventions in Table 26.3 we find that $R_1 = +10$ cm and $R_2 = -15$ cm. Thus, using the lens makers' equation, we have

$$\frac{1}{f} = (n-1)\left(\frac{1}{R_1} - \frac{1}{R_2}\right)$$

$$= (1.50 - 1)\left(\frac{1}{10\ \text{cm}} - \frac{1}{-15\ \text{cm}}\right)$$

$$f = \boxed{12\ \text{cm}}$$

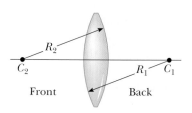

**Figure 26.26**

(Example 26.6) This lens has two curved surfaces with radii of curvature $R_1$ and $R_2$.

## Example 26.7   An Image Formed by a Converging Lens

A converging lens of focal length 10.0 cm forms an image of an object placed (a) 30.0 cm, (b) 10.0 cm, and (c) 5.00 cm from the lens. Find the image distance and describe the image in each case.

**Solution**   (a) The thin lens equation, Equation 26.12, can be used to find the image distance:

$$\frac{1}{p} + \frac{1}{q} = \frac{1}{f}$$

$$\frac{1}{30.0\ \text{cm}} + \frac{1}{q} = \frac{1}{10.0\ \text{cm}}$$

$$q = \boxed{15.0\ \text{cm}}$$

The positive sign tells us that the image is real. The magnification is

$$M = -\frac{q}{p} = -\frac{15.0\ \text{cm}}{30.0\ \text{cm}} = \boxed{-0.500}$$

Thus, the image is reduced in size by one half, and the negative sign for $M$ tells us that the image is inverted. The situation is qualitatively like that pictured in Figure 26.24a.

(b) No calculation is necessary for this case because we know that when the object is placed at the focal point, the image is formed at infinity. This is verified by substituting $p = 10.0$ cm into the lens equation.

(c) We now move inside the focal point, to an object distance of 5.00 cm. In this case, the thin lens equation gives

$$\frac{1}{5.00\ \text{cm}} + \frac{1}{q} = \frac{1}{10.0\ \text{cm}}$$

$$q = \boxed{-10.0\ \text{cm}}$$

$$M = -\frac{q}{p} = -\left(\frac{-10.0\ \text{cm}}{5.00\ \text{cm}}\right) = \boxed{2.00}$$

The negative image distance tells us that the image is virtual. The image is enlarged, and the positive sign for $M$ tells us that the image is upright, as in Figure 26.24b.

### Combinations of Thin Lenses

If two thin lenses are used to form an image, the system can be treated in the following manner. First, the image of the first lens is calculated as though the second lens were not present. The light then approaches the second lens *as if* it had originally come from the image formed by the first lens. Hence, the image of the first lens is treated as the object of the second lens. The image of the second lens is

Light from a distant object brought into focus by two converging lenses. *(Henry Leap and Jim Lehman)*

the final image of the system. If the image of the first lens lies on the back side of the second lens, then the image is treated as a *virtual object* for the second lens (i.e., $p$ is negative). The same procedure can be extended to a system of three or more lenses. The overall magnification of a system of thin lenses equals the *product* of the magnifications of the separate lenses.

## Example 26.8 Where Is the Final Image?

Two thin converging lenses of focal lengths 10.0 cm and 20.0 cm are separated by 20.0 cm, as in Figure 26.27. An object is placed 15.0 cm to the left of the first lens. Find the position and magnification of the final image.

**Solution** First, we find the image position for the first lens while ignoring the presence of the second lens:

$$\frac{1}{p_1} + \frac{1}{q_1} = \frac{1}{f} \longrightarrow \frac{1}{15.0\ \text{cm}} + \frac{1}{q_1} = \frac{1}{10.0\ \text{cm}}$$

$$q_1 = 30.0\ \text{cm}$$

where $q_1$ is measured from the first lens.

Because $q_1$ is greater than the 20.0 cm separation between the two lenses, the image of the first lens lies 10.0 cm to the right of the second lens. We recognize that this identifies the image of the first lens as a virtual object for the second lens. Thus, we apply the thin lens equation to the second lens with $p_2 = -10.0$ cm, where distances are now measured from the second lens, whose focal length is 20.0 cm:

$$\frac{1}{p_2} + \frac{1}{q_2} = \frac{1}{f_2} \longrightarrow \frac{1}{-10.0\ \text{cm}} + \frac{1}{q_2} = \frac{1}{20.0\ \text{cm}}$$

$$q_2 = \boxed{6.67\ \text{cm}}$$

Thus, the final image lies 6.67 cm to the right of the second lens.

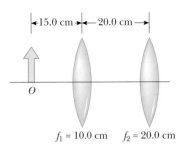

**Figure 26.27**

(Example 26.8) A combination of two converging lenses.

The magnification of each lens separately is given by

$$M_1 = -\frac{q_1}{p_1} = -\frac{30.0\ \text{cm}}{15.0\ \text{cm}} = -2.00$$

$$M_2 = -\frac{q_2}{p_2} = -\frac{6.67\ \text{cm}}{-10.0\ \text{cm}} = 0.667$$

The total magnification $M$ of the two lenses is the product

$$M = M_1 M_2 = (-2.00)(0.667) = -1.33$$

Hence, the final image is real, inverted, and larger than the object.

## 26.5 • LENS ABERRATIONS

One of the basic problems of lenses and lens systems is the imperfect quality of the images. The paraxial ray model for images formed by mirrors and lenses assumes that rays make small angles with the principal axis and that all rays reaching the mirror or lens from a point source are focused at a single point, producing a sharp image. Clearly, this is not always true in the real world. Where the approximations used in this model do not hold, imperfect images are formed.

If you want to precisely analyze image formation, you must trace each ray and apply the law of reflection at each reflecting surface and Snell's law at each refracting surface. This procedure shows that there is no single point at which an image occurs; instead, the image is *blurred*. The departures of real (imperfect) images from the ideal predicted by the simple model are called **aberrations.** Two types of aberrations shall now be described.

### Spherical Aberration

**Spherical aberration** results from the fact that the focal points of light rays far from the principal axis of a spherical lens (or mirror) are different from those for rays of the same wavelength passing near the axis. Figure 26.28 illustrates spherical aberration for parallel rays passing through a converging lens. Rays near the middle of the lens are focused farther from the lens than rays at the edges. Hence, the lens has no single focal length.

Most cameras are equipped with an adjustable aperture to control the light intensity and reduce spherical aberration when possible. (An aperture is an opening that controls the amount of light transmitted through the lens.) Sharper images are produced as the aperture size is reduced, because only the central portion of the lens is exposed to the incident light when the aperture is very small. At the same time, however, less light passes through the lens. To compensate for this loss in light intensity, a longer exposure time is used.

In the case of mirrors used for very distant objects, one can eliminate or minimize spherical aberration by employing a surface with a parabolic cross section rather than a spherical surface. Parallel light rays incident on such a surface focus at a common point, regardless of their distance from the principal axis. These surfaces are not used in many applications, however, because they are very expensive to make. Parabolic reflecting surfaces are used in many astronomical telescopes to enhance the image quality. They are also used in searchlights, for which a nearly parallel light beam is produced from a small lamp placed at the focal point of the reflecting surface.

### Chromatic Aberration

From our discussion of lenses, we find that the focal length of a lens is determined by the refraction of light as it passes through the lens. As we saw in Chapter 25, however, the refraction of light depends on the wavelength of the light—this is the phenomenon of dispersion. Thus, the focal length of a given lens depends on the color of light that passes through the lens. The focal length is larger for red light than for violet light as shown in Figure 26.29. Other wavelengths (not shown in Fig. 26.29) would have intermediate focal points. If white light from an object passes through the lens, each color focuses to form a sharp image at a different location. This shortcoming of a lens is called **chromatic aberration.** The chromatic aberration for a diverging lens is opposite that for a converging lens. Chromatic aberration can be greatly reduced by using a combination of a converging and diverging lens made from two different types of glass.

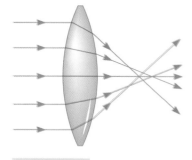

**Figure 26.28**

Spherical aberration caused by a converging lens. Does a diverging lens cause spherical aberration?

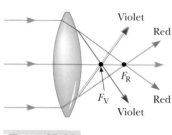

**Figure 26.29**

Chromatic aberration caused by a converging lens. Rays of different wavelengths focus at different points.

context connection
## 26.6 • MEDICAL FIBERSCOPES

Electromagnetic radiation has played a role in the transfer of information in medicine for decades. Of particular interest is the ability to gain information about the relatively inaccessible regions inside the body without using invasive procedures such as surgery. An early advance in this area was the use of x-rays to create shadowy images of bones and other internal structures. In this section, we consider advances that have been made in image formation using optical fibers in medical instruments. These advances have in turn opened up new uses for lasers in medicine.

The first use of optical fibers in medicine appeared with the invention of the *fiberscope* in 1957. Figure 26.30 indicates the construction of a fiberscope. The fiberscope consists of two bundles of optical fibers. The *illuminating bundle* is an *incoherent* bundle, meaning that no effort is made to match the relative positions of the fibers at the two ends. This matching is not necessary because the sole purpose of this bundle is to deliver light to illuminate the scene. A lens (called the *objective lens*) is used at the internal end of the fiberscope to create a real image of the illuminated scene on the ends of the *viewing bundle* of fibers. The light from the image is transmitted along the fibers to the viewing end. An eyepiece lens is used at this end to magnify the image appearing on the ends of the fibers in the viewing bundle.

The viewing bundle is coherent, so that the fibers have the same relative relationships at both ends of the bundle. If one end of an individual fiber is at the very top of the eyepiece end of the bundle, the other end of the fiber must be at the very top of the interior end of the bundle. This is necessary because each fiber in the viewing bundle collects light from a particular part of the real image of the scene formed by the objective lens on the ends of the fibers. That part of the scene's image must appear in the correct place with all of the parts at the other end in order for the image to make sense!

The diameter of such a fiberscope can be as small as 1 mm and still provide excellent optical imaging of the scene to be viewed. Fiber densities are currently

APPLICATION

Electromagnetic radiation in medicine

APPLICATION

Medical uses of the fiberscope

**Figure 26.30**

The construction of a fiberscope for viewing the interior of the body. The objective lens forms a real image of the scene on the end of a bundle of optical fibers. This image is carried to the other end of the bundle, where an eyepiece lens is used to magnify the image for the physician.

about 10 000 fibers for a 1-mm-diameter scope. Resolution is as high as 70 $\mu$m. Thus, the fiberscope can be inserted through very small surgical openings in the skin and threaded through narrow areas such as arteries.

As another example, a fiberscope can be passed through the esophagus and into the stomach to enable a physician to look for ulcers. The resulting image can be viewed directly by the physician through the eyepiece lens, but most often is displayed on a television monitor or captured on film.

*Endoscopes* are fiberscopes with additional channels besides those for the illuminating and viewing fibers. These channels may be used for

withdrawing fluids

introducing fluids

vacuum suction

wire manipulators

scalpels for cutting tissue

needles for injections

lasers for surgical applications

These additional channels require more room, so that endoscopes range from 2 to 15 mm in diameter. Despite this larger size, however, endoscopes can be used to perform surgery within the body using incisions that are much smaller than those in traditional surgery.

Lasers are used with endoscopes in a variety of medical diagnostic and treatment procedures. As a diagnostic example, the dependence on wavelength of the amount of reflection from a surface allows a fiberscope to be used to make a direct measurement of the oxygen content of the blood. Using two laser sources, red light and infrared light are both sent into the blood through optical fibers. Hemoglobin reflects a known fraction of infrared light, regardless of the oxygen carried. Thus, the measurement of the infrared reflection gives a total hemoglobin count. Red light is reflected much more by hemoglobin carrying oxygen than by hemoglobin that does not. Thus, the amount of red laser light reflected allows a measurement of the ability of the patient's blood to carry oxygen.

Lasers are used to treat medical conditions such as *hydrocephalus,* which occurs in about 0.1% of births. This condition involves an increase in intracranial pressure due to an overproduction of cerebrospinal fluid (CSF), an obstruction of the flow of CSF, or insufficient absorption of CSF. In addition to congenital hydrocephalus, the condition can be acquired later in life due to trauma to the head, brain tumors, or other factors.

The older treatment method for obstructive hydrocephalus involved placing a shunt (tube) between ventricular chambers in the brain to allow passage of CSF. A new alternative is *laser-assisted ventriculostomy,* in which a new pathway for CSF is made with an infrared laser beam and an endoscope having a spherical end as shown in Figure 26.31. As the laser beam strikes the spherical end, refraction at the spherical surface causes light waves to spread out in all directions as if the end of the endoscope were a point source of radiation. This results in a rapid decrease in intensity with distance from the sphere, avoiding damage to vital structures in the brain that are close to the area in which a new passageway is to be made. The surface of the spherical end is coated with an infrared radiation-absorbing material, so that the absorbed laser energy raises the temperature of the sphere. As the sphere is placed in contact with the location of the desired passageway, the combination of the high temperature and laser radiation leaving the sphere burn a new

Endoscope    Optical    Spherical
covering    fiber bundle    end

**Figure 26.31**

An endoscope probe used to open new passageways for cerebrospinal fluid in the treatment of hydrocephalus. Laser light raises the temperature of the sphere and radiates from the sphere to provide energy to tissues for cutting the new passageway.

passageway for the CSF. This treatment requires much less recovery time than that associated with the placement of shunts as well as significantly less postoperative care.

In the next chapter, we shall investigate another application of lasers that has grown tremendously in recent years. This is the technology of *holography,* in which three-dimensional images of objects are recorded on film.

**WEB**

For information on another use of lasers and fiberscopes in the heart (transmyocardial revascularization), visit **archive.abcnews.go.com/ sections/living/hightech1113/ index.html**

## SUMMARY

An **image** of an object is a point from which light either diverges or seems to diverge after interacting with a mirror or lens. If light passes through the image point, the image is a **real image.** If light only appears to diverge from the image point, the image is a **virtual image.**

In the **paraxial ray** simplification model, the object distance $p$ and image distance $q$ for a spherical mirror of radius $R$ are related by the **mirror equation**

$$\frac{1}{p} + \frac{1}{q} = \frac{2}{R} = \frac{1}{f} \qquad [26.4, 26.6]$$

where $f = R/2$ is the **focal length** of the mirror.

The **magnification** $M$ of a mirror or lens is defined as the ratio of the image height $h'$ to the object height $h$:

$$M = \frac{h'}{h} = -\frac{q}{p} \qquad [26.2, 26.11]$$

An image can be formed by refraction from a spherical surface of radius $R$. The object and image distances for refraction from such a surface are related by

$$\frac{n_1}{p} + \frac{n_2}{q} = \frac{n_2 - n_1}{R} \qquad [26.8]$$

where the light is incident from the medium of index of refraction $n_1$ and is refracted in the medium whose index of refraction is $n_2$.

For a thin lens, and in the paraxial ray approximation, the object and image distances are related by the **thin lens equation:**

$$\frac{1}{p} + \frac{1}{q} = \frac{1}{f} \qquad [26.12]$$

The **focal length** $f$ of a thin lens in air is related to the curvature of its surfaces and to the index of refraction $n$ of the lens material by

$$\frac{1}{f} = (n - 1)\left(\frac{1}{R_1} - \frac{1}{R_2}\right) \qquad [26.13]$$

**Converging lenses** have positive focal lengths, and **diverging lenses** have negative focal lengths.

## QUESTIONS

1. Why do some emergency vehicles have the symbol ƎƆИA⅃UᗷMA written on the front?
2. The sideview mirror on late-model cars warns the user that objects may be closer than they appear. What kind of mirror is being used and why was this type selected?
3. Consider a concave spherical mirror with a real object. Is the image always inverted? Is the image always real? Give conditions for your answers.
4. Does the mirror equation apply to a plane mirror? What relationship between $p$ and $q$ does it imply? Does the magnification equation apply to a plane mirror?
5. Why does a clear stream always appear to be shallower than it actually is?
6. A person spearfishing from a boat sees a fish that is 3 m from the boat at a depth of 1 m. In order to spear the fish, should the person aim at, above, or below the image of the fish?
7. Explain why a fish in a spherical goldfish bowl appears larger than it really is.

8. If a cylinder of solid glass or clear plastic is placed above the words LEAD OXIDE and viewed from the side as shown in Figure Q26.8, the LEAD appears inverted but the OXIDE does not. Explain.

**Figure Q26.8** (*Courtesy of Henry Leap and Jim Lehman*)

9. A mirage is formed when the air becomes gradually cooler as the height above the ground increases. What might happen if the air grows gradually warmer as the height is increased? This often happens over bodies of water or snow-covered ground; the effect is called *looming.*

**10.** Lenses used in eyeglasses, whether converging or diverging, are always designed such that the middle of the lens curves away from the eye, like the center lenses of Figure 26.18a and 26.18b. Why?

**11.** Describe lenses that can be used to start a fire.

**12.** Explain this statement: "The focal point of a lens is the location of the image of a point object at infinity." Discuss the notion of infinity in real terms as it applies to object distances. Based on this statement, can you think of a "quick and dirty" method for determining the focal length of a positive lens?

**13.** Consider the image formed by a thin converging lens. Under what conditions is the image (a) inverted, (b) upright, (c) real, (d) virtual, (e) larger than the object, and (f) smaller than the object?

**14.** Discuss the proper position of a slide relative to the lens in a slide projector. What type of lens must the slide projector have?

**15.** Discuss why each of the following statements is true or false. (a) A virtual image can be a virtual object. (b) A virtual image can be a real object. (c) A real image can be a virtual object. (d) A real image can be a real object.

**16.** Discuss the type of aberration involved in each of the following situations. (a) The edges of the image appear reddish. (b) A clear focus of the image's central portion cannot be obtained. (c) A clear focus of the image's outer portion cannot be obtained. (d) The central portion of the image is enlarged relative to the outer portions.

**17.** In a Jules Verne novel, a piece of ice is shaped to form a magnifying lens to focus sunlight to start a fire. Is this possible?

**18.** A solar furnace can be constructed by using a concave mirror to reflect and focus sunlight into a furnace enclosure. What factors in the design of the reflecting mirror would guarantee very high temperatures?

**19.** A pinhole camera can be constructed by punching a small hole in one side of a cardboard box. If the opposite side is cut out and replaced with tissue paper, the image of a distant object is formed by light passing through the pinhole onto the tissue paper (the screen). No lens is involved here. In effect, the pinhole replaces the lens of the camera. Explain why an image is formed, and describe the nature of the image.

**20.** If you want to examine the fine detail of an object with a magnifying lens of focal length 15 cm, where should the object be placed to observe a magnified image of the object?

**21.** Large telescopes are usually reflecting rather than refracting. List some reasons for this choice.

**22.** A classic science-fiction story, "The Invisible Man," tells of a person who becomes invisible by changing the index of refraction of his body to that of air. This story has been criticized by students who know how the eye works; they claim the invisible man would be unable to see. On the basis of your knowledge of the eye, could he see or not?

---

## PROBLEMS

1, 2, 3 = straightforward, intermediate, challenging        ☐ = full solution available in the *Student Solutions Manual and Study Guide*

web = solution posted at **http://www.harcourtcollege.com/physics/**        💻 = computer useful in solving problem

🖥️ = Interactive Physics        ◼ = paired numerical/symbolic problems         = life science application

---

### Section 26.1 Images Formed by Flat Mirrors

**1.** Does your bathroom mirror show you older or younger than you actually are? Compute an order-of-magnitude estimate for the age difference, based on data that you specify.

**2.** In a church choir loft, two parallel walls are 5.30 m apart. The singers stand against the north wall. The organist faces the south wall, sitting 0.800 m away from it. To enable her to see the choir, a flat mirror 0.600 m wide is mounted on the south wall, straight in front of her. What width of the north wall can she see? (*Hint:* Draw a top-view diagram to justify your answer.)

**3.** Determine the minimum height of a vertical flat mirror in which a person 5′10″ in height can see his or her full image. (A ray diagram would be helpful.)

**4.** In a physics laboratory experiment, a torque is applied to a small-diameter wire suspended vertically under tension. The small angle through which the wire turns as a consequence of the torque is measured by attaching a small mirror to the wire and reflecting a beam of light off the mirror and onto a circular scale. Such an arrangement is known as an optical lever and is shown as a top view in Figure P26.4. Show that when the mirror turns through an angle $\theta$, the reflected beam turns through angle $2\theta$.

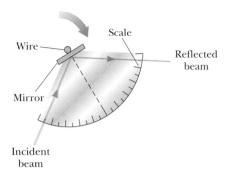

**Figure P26.4**

5. Two flat mirrors have their reflecting surfaces facing each other, with the edge of one mirror in contact with an edge of the other, so that the angle between the mirrors is $\alpha$. When an object is placed between the mirrors, a number of images are formed. In general, if the angle $\alpha$ is such that $n\alpha = 360°$, where $n$ is an integer, the number of images formed is $n - 1$. Graphically, find all the image positions for the case $n = 6$ when an object is between the mirrors (but not on the angle bisector).

6. A person walks into a room with two flat mirrors on opposite walls, which produce multiple images. When the person is 5.00 ft from the mirror on the left wall and 10.0 ft from the mirror on the right wall, find the distances from the person to the first three images seen in the mirror on the left.

7. A periscope (Figure P26.7) is useful for viewing objects that cannot be seen directly. It finds use in submarines and in watching golf matches and other athletic events from behind a crowd of people. Suppose the object is a distance $p_1$ from the upper mirror and the two flat mirrors are separated by a distance $h$. (a) What is the distance of the final

**Figure P26.7**

image from the lower mirror? (b) Is the final image real or virtual? (c) Is it upright or inverted? (d) What is its magnification? (e) Does it appear to be reversed from right to left?

### Section 26.2 Images Formed by Spherical Mirrors

8. A concave spherical mirror has a radius of curvature of 20.0 cm. Find the location of the image for object distances of (a) 40.0 cm, (b) 20.0 cm, and (c) 10.0 cm. For each case, state whether the image is real or virtual and upright or inverted. Find the magnification in each case.

9. A concave mirror has a radius of curvature of 60.0 cm. Calculate the image position and magnification of an object placed in front of the mirror (a) at a distance of 90.0 cm and (b) at a distance of 20.0 cm. (c) In each case, draw ray diagrams to obtain the image characteristics.

10. A large church has a niche in one wall. On the floor plan it appears as a semicircular indentation of radius 2.50 m. A worshiper stands on the center line of the niche, 2.00 m out from its deepest point, and whispers a prayer. Where is the sound concentrated after reflection from the back wall of the niche?

11. A spherical convex mirror (Fig. P26.11) has a radius of curvature with a magnitude of 40.0 cm. Determine the position of the virtual image and the magnification (a) for an object distance of 30.0 cm and (b) for an object distance of 60.0 cm. (c) Are the images upright or inverted?

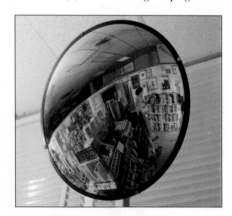

**Figure P26.11** Convex mirrors, often used for security in department stores, provide wide-angle viewing. (© *Paul Silverman 1990, Fundamental Photographs*)

12. The height of the real image formed by a concave mirror is four times larger than the object height when the object is 30.0 cm in front of the mirror. (a) What is the radius of curvature of the mirror? (b) Use a ray diagram to locate this image.

13. (a) A concave mirror forms an inverted image four times larger than the object. Find the focal length of the mirror if the distance between object and image is 0.600 m. (b) A

convex mirror forms a virtual image half the size of the object. If the distance between image and object is 20.0 cm, determine the radius of curvature of the mirror.

14. To fit a contact lens to a patient's eye, a *keratometer* can be used to measure the curvature of the front surface of the eye, the cornea. This instrument places an illuminated object of known size at a known distance $p$ from the cornea. The cornea reflects some light from the object, forming an image of the object. The magnification $M$ of the image is measured using a small viewing telescope that allows comparison of the image formed by the cornea with a second calibrated image projected into the field of view by a prism arrangement. Determine the radius of curvature of the cornea for the case $p = 30.0$ cm and $M = 0.013\ 0$.

15. A spherical mirror is to be used to form, on a screen located 5.00 m from the object, an image five times the size of the object. (a) Describe the type of mirror required. (b) Where should the mirror be positioned relative to the object?

16. A dedicated sports-car enthusiast polishes the inside and outside surfaces of a hubcap that is a section of a sphere. When she looks into one side of the hubcap, she sees an image of her face 30.0 cm in back of the hubcap. She then flips the hubcap over and sees another image of her face 10.0 cm in back of the hubcap. (a) How far is her face from the hubcap? (b) What is the radius of curvature of the hubcap?

17. You unconsciously estimate the distance to an object from the angle it subtends in your field of view. This angle $\theta$ in radians is related to the linear height of the object $h$ and to the distance $d$ by $\theta = h/d$. Assume that you are driving a car and another car, 1.50 m high, is 24.0 m behind you. (a) Suppose your car has a flat passenger-side rearview mirror, 1.55 m from your eyes. How far from your eyes is the image of the car following you? (b) What angle does the image subtend in your field of view? (c) Suppose instead that your car has a convex rearview mirror with a radius of curvature of magnitude 2.00 m (Fig. P26.17). How far from your eyes is the image of the car following you? (d) What angle does the image subtend at your eyes? (e) Based on its angular size, how far away does the following car appear to be?

## Section 26.3   Images Formed by Refraction

18. A flint glass plate ($n = 1.66$) rests on the bottom of an aquarium tank. The plate is 8.00 cm thick (vertical dimension) and is covered with a layer of water ($n = 1.33$) 12.0 cm deep. Calculate the apparent thickness of the plate as viewed from above the water. (Assume nearly normal incidence.)

19. A cubical block of ice 50.0 cm on a side is placed on a level floor over a speck of dust. Find the location of the image of the speck if the index of refraction of ice is 1.309.

20. A simple model of the human eye ignores its lens entirely. Most of what the eye does to light happens at the transparent cornea. Assume that this outer surface has a 6.00 mm radius of curvature, and assume that the eyeball contains just one fluid with a refractive index of 1.40. Prove that a very distant object will be imaged on the retina, 21.0 mm behind the cornea. Describe the image.

21. A glass sphere ($n = 1.50$) with a radius of 15.0 cm has a tiny air bubble 5.00 cm above its center. The sphere is viewed looking down along the extended radius containing the bubble. What is the apparent depth of the bubble below the surface of the sphere?

22. One end of a long glass rod ($n = 1.50$) is formed into a convex surface of radius 6.00 cm. An object is positioned in air along the axis of the rod. Find the image positions corresponding to object distances of (a) 20.0 cm, (b) 10.0 cm, and (c) 3.00 cm from the end of the rod.

## Section 26.4   Thin Lenses

23. The left face of a biconvex lens has a radius of curvature of magnitude 12.0 cm, and the right face has a radius of curvature of magnitude 18.0 cm. The index of refraction of the glass is 1.44. (a) Calculate the focal length of the lens. (b) Calculate the focal length if the radii of curvature of the two faces are interchanged.

24. A contact lens is made of plastic with an index of refraction of 1.50. The lens has an outer radius of curvature of

**THE FAR SIDE**        By GARY LARSON

OBJECTS IN MIRROR ARE CLOSER THAN THEY APPEAR

10-3

© 1985 FarWorks, Inc.

**Figure P26.17**    *(THE FAR SIDE ©1985 FARWORKS, INC. Used by permission. All rights reserved.)*

+ 2.00 cm and an inner radius of curvature of + 2.50 cm. What is the focal length of the lens?

25. A converging lens has a focal length of 20.0 cm. Locate the image for object distances of (a) 40.0 cm, (b) 20.0 cm, and (c) 10.0 cm. For each case, state whether the image is real or virtual and upright or inverted. Find the magnification in each case.

26. A thin lens has a focal length of 25.0 cm. Locate and describe the image when the object is placed (a) 26.0 cm and (b) 24.0 cm in front of the lens.

27. The nickel's image in Figure P26.27 has twice the diameter of the nickel and is 2.84 cm from the lens. Determine the focal length of the lens.
web

**Figure P26.27**

28. An object positioned 32.0 cm in front of a lens forms an image on a screen 8.00 cm behind the lens. (a) Find the focal length of the lens. (b) Determine the magnification. (c) Is the lens converging or diverging?

29. A magnifying glass is a converging lens of focal length 15.0 cm. At what distance from a postage stamp should you hold this lens to obtain a magnification of + 2.00?

30. A transparent photographic slide is placed in front of a converging lens with a focal length of 2.44 cm. The lens forms an image of the slide 12.9 cm from the slide. How far is the lens from the slide if the image is (a) real? (b) virtual?

31. Suppose an object has thickness $dp$ so that it extends from object distance $p$ to $p + dp$. Prove that the thickness $dq$ of its image is given by $(- q^2/p^2)\,dp$, so the longitudinal magnification $dq/dp = - M^2$, where $M$ is the lateral magnification.

32. The projection lens in a certain slide projector is a single thin lens. A slide 24.0 mm high is to be projected so that its image fills a screen 1.80 m high. The slide-to-screen distance is 3.00 m. (a) Determine the focal length of the projection lens. (b) How far from the slide should the lens of the projector be placed to form the image on the screen?

33. An object is positioned 20.0 cm to the left of a diverging lens with focal length $f = - 32.0$ cm. Determine (a) the location and (b) the magnification of the image. (c) Construct a ray diagram for this arrangement.

34. Figure P26.34 shows a thin glass ($n = 1.50$) converging lens for which the radii of curvature are $R_1 = 15.0$ cm and $R_2 = - 12.0$ cm. To the left of the lens is a cube having a face area of 100 cm². The base of the cube is on the axis of the lens, and the right face is 20.0 cm to the left of the lens. (a) Determine the focal length of the lens. (b) Draw the image of the square face formed by the lens. What type of geometric figure is this? (c) Determine the area of the image.

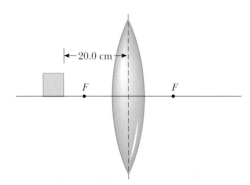

**Figure P26.34**

35. An object is at a distance $d$ to the left of a flat screen. A converging lens with focal length $f < d/4$ is placed between object and screen. (a) Show that two lens positions exist that form an image on the screen, and determine how far these positions are from the object. (b) How do the two images differ from each other?

36. Figure P26.36 diagrams a cross-section of a camera. It has a single lens with a focal length of 65.0 mm, which is to form an image on the film at the back of the camera. Suppose the position of the lens has been adjusted to focus the image of a distant object. How far and in what direction must the lens be moved to form a sharp image of an object that is 2.00 m away?

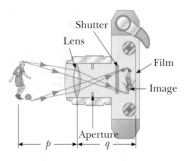

**Figure P26.36**

## Section 26.5 Lens Aberrations

37. The magnitudes of the radii of curvature are 32.5 cm and 42.5 cm for the two faces of a biconcave lens. The glass has index of refraction 1.53 for violet light and 1.51 for red light. For a very distant object, locate and describe

(a) the image formed by violet light, and (b) the image formed by red light.

38. Two rays traveling parallel to the principal axis strike a large planoconvex lens having a refractive index of 1.60 (Fig. P26.38). If the convex face is spherical, a ray near the edge does not pass through the focal point (spherical aberration occurs). If this face has a radius of curvature of 20.0 cm and the two rays are $h_1 = 0.500$ cm and $h_2 = 12.0$ cm from the principal axis, find the difference in the positions where they cross the principal axis.

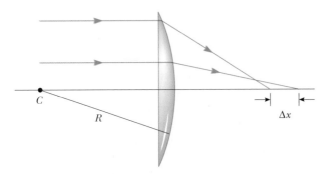

**Figure P26.38**

## Section 26.6   Context Connection — Medical Fiberscopes

39. You are designing an endoscope for use inside an air-filled body cavity. A lens at the end of the endoscope will form an image covering the end of a bundle of optical fibers. This image will then be carried by the optical fibers to an eyepiece lens at the outside end of the fiberscope. The radius of the bundle is 1.00 mm. The scene within the body that is to appear within the image fills a circle of radius 6.00 cm. The lens will be located 5.00 cm from the tissues you wish to observe. (a) How far should the lens be located from the end of the optical fiber bundle? (b) What is the focal length of the lens required?

40. Consider the endoscope probe used for treating hydrocephalus and shown in Figure 26.31. The spherical end, of index of refraction $n = 1.50$, is attached to an optical fiber bundle of radius 1.00 mm, which is smaller than the radius of the sphere. The center of the spherical end is on the central axis of the bundle. Consider laser light that travels precisely parallel to the central axis of the bundle and then refracts out from the surface of the sphere into air. (a) In Figure 26.31, does light that refracts out of the sphere and travels toward the upper right come from the top half or the bottom half of the sphere? (b) If laser light that travels along the edge of the optical fiber refracts out of the sphere tangent to the surface of the sphere, what is the radius of the sphere? (c) Find the angle of deviation of the ray considered in part (b)—that is, the angle

by which its direction changes as it leaves the sphere. (d) Show that the ray considered in part (b) has a greater angle of deviation than any other ray. Show that the light from all parts of the optical fiber bundle does not refract out of the sphere with spherical symmetry, but rather fills a cone around the forward direction. Find the angular diameter of the cone. (e) In reality, however, laser light can diverge from the sphere with approximate spherical symmetry. What considerations that we have not addressed will lead to this approximate spherical symmetry in practice?

## *Additional Problems*

41. An astronomical telescope used for photography can consist of a single mirror with a photographic plate at its focus. A certain telescope is a mirror with an aperture diameter of 200 mm and a focal length of 2 000 mm. It captures the image of a nebula on photographic film at its prime focus with an exposure time of 1.50 min. To produce the same light energy per unit area on the film, what is the required exposure time to photograph the same nebula with a smaller telescope, which has a lens with an aperture diameter of 60.0 mm and a focal length of 900 mm?

42. The distance between an object and its upright image is $d$. If the magnification is $M$, what is the focal length of the lens that is being used to form the image?

43. The lens and mirror in Figure P26.43 have focal lengths of $+80.0$ cm and $-50.0$ cm, respectively. An object is placed 1.00 m to the left of the lens as shown. Locate the final image, formed by light that has gone through the lens twice. State whether the image is upright or inverted, and determine the overall magnification.

**Figure P26.43**

44. A real object is located at the zero end of a meter stick. A large concave mirror at the 100-cm end of the meter stick forms an image of the object at the 70.0-cm position. A small convex mirror placed at the 20.0-cm position forms a final image at the 10.0-cm point. What is the radius of curvature of the convex mirror?

**45.** Derive the lens makers' equation as follows: Consider an object in vacuum at $p_1 = \infty$ from a first refracting surface of radius of curvature $R_1$. Locate its image. Use this image as the object for the second refracting surface, which has nearly the same location as the first, because the lens is thin. Locate the final image, proving it is at the image distance $q_2$ given by

$$\frac{1}{q_2} = (n - 1) \left( \frac{1}{R_1} - \frac{1}{R_2} \right)$$

**46.** The object in Figure P26.46 is midway between the lens and the mirror. The mirror's radius of curvature is 20.0 cm, and the lens has a focal length of $-16.7$ cm. Considering only the light that leaves the object and travels first toward the mirror, locate the final image formed by this system. Is this image real or virtual? Is it upright or inverted? What is the overall magnification?

**Figure P26.46**

**47.** In many applications it is necessary to expand or to decrease the diameter of a beam of parallel rays of light. This change can be made using a converging lens and a diverging lens in combination. Suppose you have a converging lens of focal length 21.0 cm and a diverging lens of focal length $-12.0$ cm. How can you arrange these lenses to increase the diameter of a beam of parallel rays? By what factor will the diameter increase?

**48.** **Review Problem.** A spherical lightbulb of diameter 3.20 cm radiates light equally in all directions, with a power of 4.50 W. (a) Find the light intensity at the surface of the bulb. (b) Find the light intensity 7.20 m from the center of the bulb. (c) At this 7.20-m distance a lens is set up with its axis pointing toward the bulb. The lens has a circular face with a diameter of 15.0 cm and a focal length of 35.0 cm. Find the diameter of the image of the bulb. (d) Find the light intensity at the image.

**49.** A parallel beam of light enters a glass hemisphere perpendicular to the flat face, as shown in Figure P26.49. The magnitude of the radius is 6.00 cm, and the index of refraction is $n = 1.560$. Determine the point at which the beam is focused. (Assume paraxial rays.)

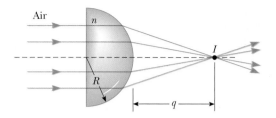

**Figure P26.49**

**50.** Assume that the intensity of sunlight is 1.00 kW/m² at a particular location. A highly reflecting concave mirror is to be pointed toward the Sun to produce a power of at least 350 W at the image. (a) Find the required radius $R_a$ of the circular face area of the mirror. (b) Now suppose the light intensity is to be at least 120 kW/m² at the image. Find the required relationship between $R_a$ and the radius of curvature $R$ of the mirror. The disk of the Sun subtends an angle of 0.533° at the Earth.

**51.** The disk of the Sun subtends an angle of 0.533° at the Earth. What are the position and diameter of the solar image formed by a concave spherical mirror with a radius of curvature of 3.00 m?

**52.** An observer to the right of the mirror–lens combination shown in Figure P26.52 sees two real images that are the same size and in the same location. One image is upright and the other is inverted. Both images are 1.50 times larger than the object. The lens has a focal length of 10.0 cm. The lens and mirror are separated by 40.0 cm. Determine the focal length of the mirror. Do not assume that the figure is drawn to scale.

**Figure P26.52**

**53.** An object is placed 12.0 cm to the left of a diverging lens with a focal length of $-6.00$ cm. A converging lens with a focal length of 12.0 cm is placed a distance $d$ to the right of the diverging lens. Find the distance $d$ that corresponds to a final image at infinity. Draw a ray diagram for this case.

**54.** Figure P26.54 shows a thin converging lens for which the radii of curvature are $R_1 = 9.00$ cm and $R_2 = -11.0$ cm. The lens is in front of a concave spherical mirror with the radius of curvature $R = 8.00$ cm. (a) If its focal points $F_1$ and $F_2$ are 5.00 cm from the center of the lens, determine its index of refraction. (b) If the lens and mirror are 20.0 cm apart, and an object is placed 8.00 cm to the left of the lens, determine the position of the final image and its magnification as seen by the eye in the figure. (c) Is the final image inverted or upright? Explain.

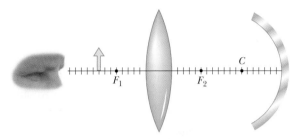

**Figure P26.54**

55. In a darkened room, a burning candle is placed 1.50 m from a white wall. A lens is placed between candle and wall at a location that causes a larger, inverted image to form on the wall. When the lens is moved 90.0 cm toward the wall, another image of the candle is formed. Find (a) the two object distances that produce the specified images and (b) the focal length of the lens. (c) Characterize the second image.

56. A floating strawberry illusion consists of two parabolic mirrors, each with a focal length of 7.50 cm, facing each other so that their centers are 7.50 cm apart (Fig. P26.56). If a strawberry is placed on the lower mirror, an image of the strawberry is formed at the small opening at the center of the top mirror. Show that the final image is formed at that location, and describe its characteristics. (*Note:* A very startling effect is to shine a flashlight beam on the image. Even at a glancing angle, the incoming light beam is seemingly reflected off the image! Do you understand why?)

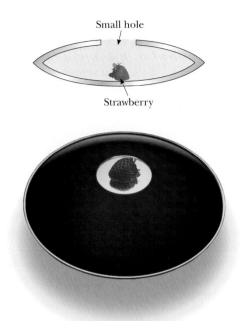

**Figure P26.56** (© *Michael Levin/Opti-Gone Associates*)

57. An object 2.00 cm high is placed 40.0 cm to the left of a converging lens with a focal length of 30.0 cm. A diverging lens with a focal length of −20.0 cm is placed 110 cm to the right of the converging lens. (a) Determine the position and magnification of the final image. (b) Is the image upright or inverted? (c) Repeat parts (a) and (b) for the case in which the second lens is a converging lens with a focal length of +20.0 cm.

## ANSWERS TO QUICK QUIZZES

**26.1** No. To focus on the photograph, you must adjust your eyes so that an object about 1 m away (the photograph) is in focus. Thus, you are focusing *on the mirror surface.* But your image in the mirror is behind the mirror, as far behind the mirror as you are in front of it. Thus, you must focus your eyes beyond the mirror, twice as far away as the photograph, to bring the image into focus.

**26.2** The actor will see the *movie camera.* Light from the actor's face must reflect from the mirror in the direction of the camera lens, so that his face is recorded on film. If we reverse that light path, light from the camera reflects from the mirror into the eyes of the actor.

**26.3** Using the time interval between sending the ultrasonic sound and receiving it, the camera calculates the distance to the subject and sets the camera lens for the appropriate object distance. When facing a mirror, the ultrasonic signal reflects from the mirror surface and the camera adjusts its focus so that the mirror surface is at the correct object distance from the camera. But your image in the mirror is twice this distance from the camera, so your image is blurry on the film. What is in focus on the film is the surface of the mirror!

**26.4** Yes. Although light rays don't pass through a virtual image point, they diverge as if coming from that image point. These light rays can be focused by a camera lens to form an image on photographic film just like light rays diverging from an object. For example, it is possible to photograph a virtual image of yourself in your bathroom mirror.

**26.5** We consider the two trees to be two separate objects. The far tree is an object that is farther from the lens than the near tree. Thus, the image of the far tree is closer to the lens than the image of the near tree. The screen must be moved closer to the lens to put the far tree in focus.

**26.6** (a, *left*): $R_1$ is positive, $R_2$ is negative; (a, *middle*): $R_1$ is positive, $R_2$ is positive; (a, *right*): $R_1$ is positive, $R_2$ is infinite; (b, *left*): $R_1$ is negative, $R_2$ is positive; (b, *middle*): $R_1$ is positive, $R_2$ is positive; (b, *right*): $R_1$ is negative, $R_2$ is infinite.

**26.7** The focal length for a mirror is determined by the law of reflection from the mirror surface. The law of reflection is independent of the material of which the mirror is made and of the surrounding medium. Thus, the focal length depends only on the radius of curvature and not the material. The focal length of a lens depends on the law of refraction, and refraction is dependent on the indices of refraction of the lens material and surrounding medium. Thus, the focal length of a lens depends on the lens material.

**26.8** An infinite number. In general, an infinite number of rays leave each point of any object and travel outward in all directions. The three principal rays we use to locate an image are a selected subset of rays used for construction of the ray diagram. When an object is taller than a lens, you merely extend the plane containing the lens, as shown in the next column.

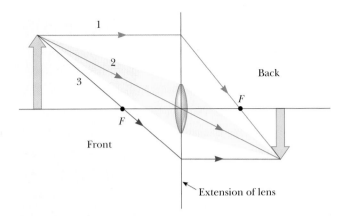

**Figure QQAns 26.8**

**26.9** (c). The entire image is visible but has half the intensity. Each point on the object is a source of rays that travel in all directions. Thus, light from all parts of the object goes through all parts of the lens and forms an image. If you block part of the lens, you are blocking some of the rays, but the remaining ones still come from all parts of the object.

# chapter 27

In reflected light, a compact disc exhibits bright colors. These colors are due to interference of light reflecting from parallel regions on the surface of the disc created by the tracks of recorded information. In this chapter, we study various effects of interference of light.

*(© Kristen Brochmann 1991, Fundamental Photographs)*

# Wave Optics

In the preceding two chapters, we used the ray approximation to examine what happens when light reflects from a surface or refracts into a new medium. We used the general term *geometric optics* for these discussions. This chapter is concerned with **wave optics,** a subject that addresses the optical phenomena of interference and diffraction. These phenomena cannot be adequately explained with the ray approximation. We must address the wave nature of light to be able to understand these phenomena.

We introduced the concept of wave interference in Chapter 14 for one-dimensional waves. This phenomenon depends on the principle of superposition, which tells us that when two or more traveling mechanical waves combine at a given point, the resultant displacement of the particles of the medium at that point is the sum of the displacements due to the individual waves.

We shall see the full richness of the waves in interference model in this chapter as we apply it to light. We used one-dimensional waves on strings to introduce interference, as in Figures 14.1 and 14.2. As we discuss the interference of light waves, two major changes from this previous discussion must be noted. First, we shall no longer focus on one-dimensional waves, so we must build geometric models to analyze the situation in two or three dimensions. Second, we shall study electromagnetic waves rather than mechanical waves. Thus, the principle of superposition needs to be cast in terms of addition of fields rather than particle displacements.

## 27.1 • CONDITIONS FOR INTERFERENCE

In our discussion of wave interference for mechanical waves in Chapter 14, we found that two waves can add together either constructively or destructively. In constructive interference between waves, the amplitude of the resultant wave is greater than that of either individual wave, whereas in destructive interference, the resultant amplitude is less than that of either individual wave. Electromagnetic waves also undergo interference. Fundamentally, all interference associated with electromagnetic waves arises as a result of combining the electric and magnetic fields that constitute the individual waves.

In Figure 14.4 of Chapter 14, we described a device that allows interference to be observed for sound waves. Interference effects in visible electromagnetic waves are not easy to observe because of their short wavelengths (from about $4 \times 10^{-7}$ to $7 \times 10^{-7}$ m). Two sources producing two waves of identical wavelengths are needed to create interference. To produce a stable interference pattern, however, the individual waves must maintain a constant phase relationship with one another—they must be **coherent.** As an example, the sound waves emitted by two side-by-side loudspeakers driven by a single amplifier can produce interference because the two speakers respond to the amplifier in the same way at the same time.

Now, if two separate light sources are placed side by side, no interference effects are observed because the light waves from one source are emitted independently of the other source; hence, the emissions from the two sources do not maintain a constant phase relationship with each other over the time of observation. An ordinary light source undergoes random changes about once every $10^{-8}$ s. Therefore, the conditions for constructive interference or destructive interference last for times on the order of $10^{-8}$ s. The result is that no interference effects are observed because the eye cannot follow such rapid changes. Such light sources are said to be **incoherent.**

A common method for producing two coherent light sources of identical wavelength is to use a single light source emitting a single wavelength to illuminate a screen containing two narrow slits. The light emerging from the two slits is coherent because a single source produces the original light beam and the two slits serve only to separate the original beam into two parts (which is exactly what was done to the sound signal discussed earlier—the signal from the amplifier is split and goes to separate speakers). A random change in the light emitted by the source occurs in the two separate beams at the same time, and interference effects can be observed.

## 27.2 • YOUNG'S DOUBLE-SLIT EXPERIMENT

Interference in light waves from two slits was first demonstrated by Thomas Young in 1801. A schematic diagram of the apparatus used in this experiment is shown in Figure 27.1a. (Young used pinholes in his original experiments, rather than slits.) Light is incident on a screen with a narrow slit $S_0$. The light waves emerging from this slit arrive at a second screen that contains two narrow, parallel slits $S_1$ and $S_2$. These two slits serve as a pair of coherent light sources because waves emerging

**WEB**

For an animated version of Figure 27.1a, with adjustable wavelength and slit separation, visit **www.phy.ntnu.edu.tw/java/ doubleSlit/doubleSlit.html**

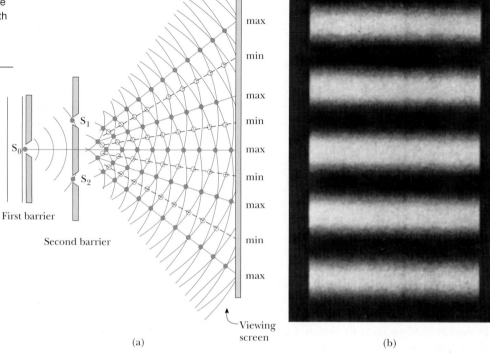

(a)

(b)

**Figure 27.1**

(a) Schematic diagram of Young's double-slit experiment. Slits $S_1$ and $S_2$ behave as coherent sources of light waves that produce an interference pattern on the viewing screen (drawing not to scale). (b) An enlargement of the center of a fringe pattern formed on the viewing screen could look like this.

**PITFALL PREVENTION 27.1**

**Interference patterns are not standing waves**

The interference pattern in Figure 27.1b shows bright and dark regions that you might be tempted to identify with the antinodes and nodes of a standing wave pattern on a string, which we studied in Section 14.4. Although both patterns depend on the principle of superposition, major differences exist between them. For example, the string pattern exists on a one-dimensional system, but the interference pattern exists in three dimensions. Another difference is that the standing wave pattern represents no net energy flow, but a net energy flow occurs from the slits to the screen in an interference pattern. Quantitatively, in practial experiments the distance $\Delta y$ between interference maxima is much larger than the distance $\lambda/2$ between antinodes in a standing wave.

(and spreading out because of *diffraction*) from them originate from the same source $S_0$ and therefore maintain a constant phase relationship. The light from the two slits produces a visible pattern on the viewing screen; the pattern consists of a series of bright and dark parallel bands called **fringes** (Fig. 27.1b). When the light waves from $S_1$ and $S_2$ arrive at a point on the screen at which constructive interference is occurring, a bright fringe appears. When the light waves from the two slits arrive at a point of destructive interference, a dark fringe results. *Partial interference* describes the ways that waves combine at regions between the midpoint of a bright fringe and that of a dark fringe.

Figure 27.2 is a schematic diagram that allows us to generate a mathematical representation by modeling the interference as if waves combine at the viewing screen.* In Figure 27.2a, two waves leave the two slits in phase and strike the screen at the central point $P$. Because these waves travel equal distances, they arrive in phase at $P$. As a result, constructive interference occurs at this location and a bright fringe is observed. In Figure 27.2b, the two light waves again start in

---

* The interference occurs everywhere between the slits and the screen, not only at the screen. See Thinking Physics 27.1. The model we have proposed will give us a valid result.

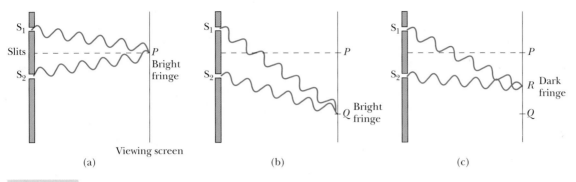

**Figure 27.2**

(a) Constructive interference occurs at point *P* when the waves combine. (b) Constructive interference also occurs at point *Q*. (c) Destructive interference occurs at *R* because the wave from the upper slit falls half a wavelength behind the wave from the lower slit. (All figures not to scale.)

phase, but the upper wave has to travel one wavelength farther to reach point *Q* on the screen. Because the upper wave falls behind the lower one by exactly one wavelength, they still arrive in phase at *Q*. Hence, a second bright fringe appears at this location. Now consider point *R* located between *P* and *Q* in Figure 27.2c. At this location, the upper wave has fallen half a wavelength behind the lower wave when they arrive at the screen. This means that the trough from the lower wave overlaps the crest from the upper wave, giving rise to destructive interference at *R*. For this reason, one observes a dark fringe at this location.

Young's double-slit experiment is the prototype for many interference effects. In the next section, we develop the mathematical representation for interference of light.

## 27.3 • LIGHT WAVES IN INTERFERENCE

We can obtain a quantitative description of Young's experiment with the help of a geometric model constructed from Figure 27.3a. The viewing screen is located a perpendicular distance *L* from the slits $S_1$ and $S_2$, which are separated by a distance *d*. Consider point *P* on the screen. Angle $\theta$ is measured from a line perpendicular to the screen from the midpoint between the slits and a line from the midpoint to point *P*. We identify $r_1$ and $r_2$ as the distances the waves travel from slit to screen. Let us assume that the source is monochromatic. Under these conditions, the waves emerging from $S_1$ and $S_2$ have the same wavelength and amplitude and are in phase. The light intensity on the screen at *P* is the result of the superposition of the light coming from both slits. Note from the geometric model triangle in gold in Figure 27.3a that a wave from the lower slit travels farther than a wave from the upper slit by an amount $\delta$. This distance is called the **path difference.**

If *L* is much greater than *d*, the two paths are very close to being parallel. We shall adopt a simplification model in which the two paths are exactly parallel. In this case, from Figure 27.3b, we see that

$$\delta = r_2 - r_1 = d \sin \theta \qquad\qquad \text{[27.1]} \qquad \bullet \quad \textit{Path difference}$$

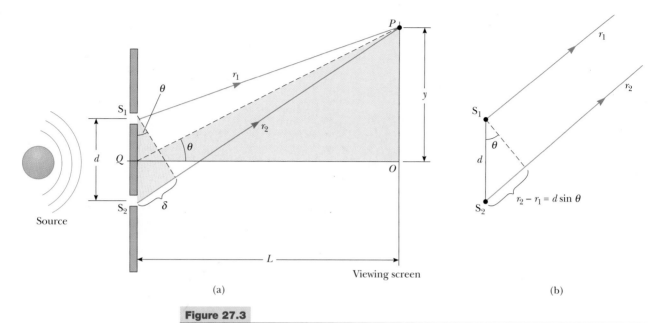

**Figure 27.3**

(a) Geometric construction for describing Young's double-slit experiment (not to scale). (b) When we assume that path $r_1$ is parallel to $r_2$, the path difference between the two rays is $\delta = r_2 - r_1 = d \sin \theta$. For this approximation to be valid, it is essential that $L \gg d$.

In Figure 27.3a, the condition $L \gg d$ is not satisfied—the figure is not to scale; in Figure 27.3b, the rays leave the slits as if the condition is satisfied. As noted earlier, the value of this path difference determines whether the two waves are in phase or out of phase when they arrive at $P$. If the path difference is either zero or some integral multiple of the wavelength, the two waves are in phase at $P$ and **constructive interference** results. The condition for bright fringes at $P$ is therefore

• *Conditions for constructive interference for two slits*

$$\delta = d \sin \theta_{\text{bright}} = m\lambda \quad (m = 0, \pm 1, \pm 2, \ldots) \qquad \textbf{[27.2]}$$

The number $m$ is an integer called the **order number.** The central bright fringe at $\theta_{\text{bright}} = 0$ ($m = 0$) is associated with the order number $m = 0$ and is called the **zeroth-order maximum.** The first maximum on either side, for which $m = \pm 1$, is called the **first-order maximum,** and so forth.

Similarly, when the path difference is an odd multiple of $\lambda/2$, the two waves arriving at $P$ are 180° out of phase and give rise to **destructive interference.** Therefore, the condition for dark fringes at $P$ is

• *Conditions for destructive interference for two slits*

$$\delta = d \sin \theta_{\text{dark}} = (m + \tfrac{1}{2})\lambda \quad (m = 0, \pm 1, \pm 2, \ldots) \qquad \textbf{[27.3]}$$

These equations provide the *angular* positions of the fringes; it is also useful to obtain expressions for the *linear* positions measured along the screen from $O$ to $P$. In addition to our assumption that $L \gg d$, we shall assume that $d \gg \lambda$; that is, the distance between the two slits is much larger than the wavelength. This situation prevails in practice for visible light because $L$ is often on the order of 1 m whereas $d$ is a fraction of a millimeter and $\lambda$ is a fraction of a micrometer. Under these conditions, $\theta$ is small for the first several fringes, and so we can use the small angle approximation $\sin\theta \approx \tan\theta$. From the geometric model triangle $OPQ$ in Figure 27.3a, we see that

$$\sin\theta \approx \tan\theta = \frac{y}{L} \qquad [27.4]$$

Using this result and making the substitution $\sin\theta = m\lambda/d$ from Equation 27.2, we see that the bright fringes measured from $O$ are located at

$$y_{\text{bright}} = m\frac{\lambda L}{d} \qquad [27.5]$$

Similarly, using Equations 27.3 and 27.4, we find that the dark fringes are located at

$$y_{\text{dark}} = \left(m + \tfrac{1}{2}\right)\frac{\lambda L}{d} \qquad [27.6]$$

Equations 27.1 through 27.6 are various mathematical representations for two-slit interference patterns. These equations describe the results when two waves interfere according to the principle of superposition. Interference of waves occurs relatively commonly in technological applications, so this is an important analysis model to understand.

As we shall demonstrate in Example 27.1, Young's double-slit experiment provides a method for measuring the wavelength of light. In fact, Young used this technique to make the first measurement of the wavelength of light. Additionally, the experiment gave the wave model of light a great deal of credibility. Today we still use the phenomenon of interference to describe many observations of wave-like behavior.

**PITFALL PREVENTION 27.2**
**It may not be true that $\theta$ is small**

The small-angle approximation $\sin\theta \approx \tan\theta$ is true to three-digit precision only for angles less than about 4°. If you are considering fringes far removed from the central fringe, the identification $\tan\theta = y/L$ is still true, but the small-angle approximation may not be valid. In this case, Equations 27.5, 27.6, and 27.9 cannot be used. These problems can be solved, but the geometry is not as simple.

**PITFALL PREVENTION 27.3**
**It may not be true that $L \gg d$**

Equations 27.2, 27.3, and 27.5 through 27.9 are developed under the assumption that $L \gg d$. This is a very common situation when investigating interference with visible light. You are likely, however, to encounter some situations in which this assumption is not valid, especially if you are analyzing interference in longer wavelength radiation, such as microwaves, for which $d$ might be large. In those cases, the geometric construction will be more complicated, but the general approach outlined here will be similar.

## THINKING PHYSICS 27.1

Consider a double-slit experiment in which a laser beam is passed through a pair of very closely spaced slits and a clear interference pattern is displayed on a distant screen. Now suppose you place smoke particles between the double slit and the screen. With the presence of the smoke particles, will you see the effects of the interference in the space between the slits and the screen, or will you only see the effects on the screen?

**Reasoning** You see the effects in the area filled with smoke. Bright beams of light are directed toward the bright areas on the screen and dark regions are directed toward the dark areas on the screen. The geometrical construction shown in Figure 27.3 is important for developing the mathematical description of interference. It is subject to misinterpretation, however, because it might suggest that the interference can only occur at the position of the screen. A better diagram for this situation is Figure 27.1a, which shows *paths* of destructive and constructive interference all the way from the slits to the screen. These paths are made visible by the smoke.

## Example 27.1 Measuring the Wavelength of Laser Light

A laser is used to illuminate a double slit. The distance between the two slits is 0.030 mm. A viewing screen is separated from the double slit by 1.2 m. The second-order bright fringe ($m = 2$) is 5.1 cm from the center line. (a) Determine the wavelength of the laser light.

**Solution** Because the distance between the screen and the slits is much larger than the slit separation, Equation 27.5 is a valid mathematical representation of this situation. With $m = 2$, $y_2 = 5.1 \times 10^{-2}$ m, $L = 1.2$ m, and $d = 3.0 \times 10^{-5}$ m, we have

$$\lambda = \frac{dy_2}{mL} = \frac{(3.0 \times 10^{-5} \text{ m})(5.1 \times 10^{-2} \text{ m})}{2(1.2 \text{ m})}$$

$$= 6.4 \times 10^{-7} \text{ m} = \boxed{6.4 \times 10^2 \text{ nm}}$$

(b) Calculate the distance between adjacent bright fringes.

**Solution** From Equation 27.5 and the results from part (a), we have

$$y_{m+1} - y_m = \frac{\lambda L(m + 1)}{d} - \frac{\lambda Lm}{d}$$

$$= \frac{\lambda L}{d} = \frac{(6.4 \times 10^{-7} \text{ m})(1.2 \text{ m})}{3.0 \times 10^{-5} \text{ m}}$$

$$= 2.6 \times 10^{-2} \text{ m} = \boxed{2.6 \text{ cm}}$$

### Intensity Distribution of the Double-Slit Interference Pattern

We shall now discuss briefly the distribution of light intensity $I$ (the energy delivered by the light per unit area per unit time) associated with the double-slit interference pattern. Again, suppose that the two slits represent coherent sources of sinusoidal waves. In this case, the two waves have the same angular frequency $\omega$ and a constant phase difference $\phi$. Although the waves have equal phase at the slits, their phase difference $\phi$ at $P$ depends on the path difference $\delta = r_2 - r_1 = d \sin \theta$. Because a path difference of $\lambda$ corresponds to a phase difference of $2\pi$ rad, we can establish the equality of the ratios,

$$\frac{\delta}{\phi} = \frac{\lambda}{2\pi}$$

- *Phase difference*

$$\phi = \frac{2\pi}{\lambda} \delta = \frac{2\pi}{\lambda} d \sin \theta \qquad [27.7]$$

This equation tells us how the phase difference $\phi$ depends on the angle $\theta$.

Although we shall not prove it here, a careful analysis of the electric fields arriving at the screen from the two very narrow slits shows that the time-averaged light intensity at a given angle $\theta$ is

- *Average light intensity for the double-slit interference pattern*

$$I_{av} = I_{max} \cos^2\left(\frac{\pi d \sin \theta}{\lambda}\right) \qquad [27.8]$$

where $I_{max}$ is the intensity at point $O$ in Figure 27.3, directly behind the midpoint between the slits. Alternatively, because $\sin \theta \approx y/L$ for small values of $\theta$, we can write Equation 27.8 in the form

$$I_{av} = I_{max} \cos^2\left(\frac{\pi d}{\lambda L} y\right) \qquad [27.9]$$

Constructive interference, which produces intensity maxima, occurs when the quantity $(\pi d/\lambda L)y$ is an integral multiple of $\pi$, corresponding to $y = m(\lambda L/d)$. This is consistent with Equation 27.5. Intensity versus $d \sin \theta$ is plotted in Figure 27.4.

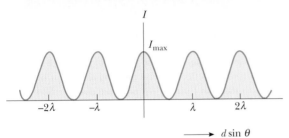

## Figure 27.4

Light intensity versus $d \sin \theta$ for the double-slit interference pattern when the screen is far from the two slits ($L \gg d$). *(Photograph from M. Cagnet, M. Francon, and J.C. Thierr, Atlas of Optical Phenomena, Berlin, Springer-Verlag, 1962)*

## Quick Quiz 27.1

Consider a dark fringe in a two-slit interference pattern, at which almost no light energy is arriving. Waves (and, therefore, energy) from both slits travel to this point, but the waves cancel. Where does the energy go?

## 27.4 • CHANGE OF PHASE DUE TO REFLECTION

Young's method of producing two coherent light sources involves illuminating a pair of slits with a single source. Another simple arrangement for producing an interference pattern with a single light source is known as *Lloyd's mirror*. A light source is placed at point $S$ close to a mirror as illustrated in Figure 27.5. Waves can reach the point $P$ either by the direct path $SP$ or by the indirect path involving reflection from the mirror. The reflected ray strikes the screen as if it originated from a source at $S'$ located below the mirror.

At points far from the source, one would expect an interference pattern due to waves from $S$ and $S'$, just as is observed for two real coherent sources at these points. An interference pattern is indeed observed. The positions of the dark and bright fringes, however, are *reversed* relative to the pattern of two real coherent sources (Young's experiment). This is because the coherent sources at $S$ and $S'$ differ in phase by 180°. This 180° phase change is produced on reflection.

To illustrate this further, consider the point $P'$, where the mirror meets the screen. This point is equidistant from $S$ and $S'$. If path difference alone were responsible for the phase difference, one would expect to see a bright fringe at $P'$ (because the path difference is zero for this point), corresponding to the central fringe of the two-slit interference pattern. Instead, one observes a *dark* fringe at $P'$ because of the 180° phase change produced by reflection. In general, **an electromagnetic wave undergoes a phase change of 180° on reflection from a medium of higher index of refraction than the one in which it is traveling.**

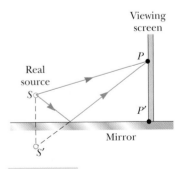

## Figure 27.5

Lloyd's mirror. An interference pattern is produced on a screen at point $P$ as a result of the combination of the direct ray (blue) and the reflected ray (red). The reflected ray undergoes a phase change of 180°.

180° phase change                                   No phase change

$n_1$  $n_2$                                         $n_1$  $n_2$
$n_1 < n_2$                Rigid support             $n_1 > n_2$              Free support

(a)                                                  (b)

(a) For $n_1 < n_2$, a light ray traveling in medium 1 and reflected from the surface of medium 2 undergoes a 180° phase change. The same thing happens with a reflected pulse traveling along a string fixed at one end. (b) For $n_1 > n_2$, a light ray traveling in medium 1 undergoes no phase change when reflected from the surface of medium 2. The same is true of a reflected wave pulse on a string whose supported end is free to move.

It is useful to draw an analogy between reflected light waves and the reflections of a transverse wave on a stretched string when the wave meets a boundary (Section 13.5), as in Figure 27.6. The reflected pulse on a string undergoes a phase change of 180° when it is reflected from a rigid end, and no phase change when it is reflected from a free end. If the boundary is between two strings, the transmitted wave exhibits no phase change. Similarly, an electromagnetic wave undergoes a 180° phase change when reflected from the boundary of a medium of higher index of refraction than the one in which it is traveling. There is no phase change for the reflected ray when the wave is incident on a boundary leading to a medium of lower index of refraction. In either case, the transmitted wave exhibits no phase change.

## 27.5 • INTERFERENCE IN THIN FILMS

Interference effects can be observed in many situations in which one beam of light is split and then recombined. A common occurrence is the appearance of colored bands in a film of oil on water or in a soap bubble illuminated with white light. The colors in these situations result from the interference of waves reflected from the opposite surfaces of the film.

Consider a film of uniform thickness $t$ and index of refraction $n$ as in Figure 27.7. We adopt a simplification model in which the light ray is incident on the film from above and nearly normal to the surface of the film. Two rays are reflected from the film, one from the upper surface and one from the lower surface after the refracted ray has traveled through the film. Because the film is thin and has parallel sides, the reflected rays are parallel. Hence, rays reflected from the top surface can interfere with rays reflected from the bottom surface. To determine

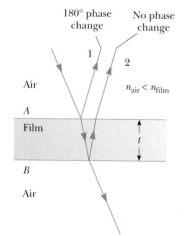

180° phase   No phase
change        change
      1
           2
Air             $n_{air} < n_{film}$

A
Film                        $t$

B
Air

Interference in light reflected from a thin film is due to a combination of rays reflected from the upper and lower surfaces of the film.

whether the reflected rays interfere constructively or destructively, we first note the following facts:

- An electromagnetic wave traveling from a medium of index of refraction $n_1$ toward a medium of index of refraction $n_2$ undergoes a 180° phase change on reflection when $n_2 > n_1$. No phase change occurs in the reflected wave if $n_2 < n_1$.
- The wavelength $\lambda_n$ of light in a medium whose index of refraction $n$ is

$$\lambda_n = \frac{\lambda}{n} \qquad \text{[27.10]}$$

where $\lambda$ is the wavelength of light in free space.

Interference in soap bubbles. The colors are due to interference between light rays reflected from the front and back surfaces of the thin film of soap making up the bubble. The color depends on the thickness of the film, ranging from black where the film is thinnest to red where it is thickest. (*Dr. Jeremy Burgess/Science Photo Library*)

Let us apply these rules to the film of Figure 27.7. According to the first rule, ray 1, which is reflected from the upper surface (A), undergoes a phase change of 180° with respect to the incident wave. Ray 2, which is reflected from the lower surface (B), undergoes no phase change with respect to the incident wave. Therefore, ignoring the path difference for now, outgoing ray 1 is 180° out of phase with respect to ray 2, a phase difference that is equivalent to a path difference of $\lambda_n/2$. However, we must also consider that ray 2 travels an extra distance approximately equal to $2t$ before the waves recombine. The *total* phase difference arises from a combination of the path difference and the 180° phase change on reflection. For example, if $2t = \lambda_n/2$, rays 1 and 2 will recombine in phase, and constructive interference will result. In general, the condition for constructive interference is

$$2t = (m + \tfrac{1}{2})\lambda_n \quad (m = 0, 1, 2, \ldots) \qquad \text{[27.11]}$$

This condition takes into account two factors: (a) the difference in optical path length for the two rays (the term $m\lambda_n$) and (b) the 180° phase change on reflection (the term $\lambda_n/2$). Because $\lambda_n = \lambda/n$, we can write Equation 27.11 in the form

$$2nt = (m + \tfrac{1}{2})\lambda \quad (m = 0, 1, 2, \ldots) \qquad \text{[27.12]}$$

- *Constructive interference in thin films*

If the extra distance $2t$ traveled by ray 2 corresponds to a multiple of $\lambda_n$, the two waves will combine out of phase and destructive interference results. The general equation for destructive interference is

$$2nt = m\lambda \quad (m = 0, 1, 2, \ldots) \qquad \text{[27.13]}$$

- *Destructive interference in thin films*

The preceding conditions for constructive and destructive interference are valid when the medium above the top surface of the film is the same as the medium below the bottom surface. The surrounding medium may have a refractive index less than or greater than that of the film. In either case, the rays reflected from the two surfaces will be out of phase by 180°. The conditions are also valid if different media are above and below the film and both have $n$ less than or larger than that of the film.

If the film is placed between two different media, one with $n < n_{\text{film}}$ and the other with $n > n_{\text{film}}$, the conditions for constructive and destructive interference are reversed. In this case, either a phase change of 180° takes place for both ray 1 reflecting from surface A and ray 2 reflecting from surface B, or no phase change occurs for either ray; hence, the net change in relative phase due to the reflections is zero.

**PITFALL PREVENTION 27.4**

**Be careful with thin films**

Be sure to include *both* effects—path length and phase change—when analyzing an interference pattern resulting from a thin film. The necessity of including the phase change is a new twist that we did not need to consider for double-slit interference.

## Quick Quiz 27.2

In a laboratory accident, you spill two liquids onto water, neither of which mixes with the water. They both form thin films on the water surface. Although the films become thinner as they spread, you notice that one film becomes bright and the other dark in reflected light. Explain this observation.

## Quick Quiz 27.3

In our discussion of thin-film interference, we looked at light *reflecting* from a thin film. Now consider the light *transmitting* through a thin film, with air on both sides of the film. First, consider the direct ray that transmits through the film without reflecting. Next, consider the second ray that transmits through the first surface, reflects back from the second, reflects again from the first and then transmits out into the air at the second surface, parallel to the direct ray. For normal incidence, how thick must the film be, in terms of the wavelength of the light, for the outgoing rays to interfere destructively? Is it the same thickness as for reflected destructive interference?

---

**PROBLEM-SOLVING STRATEGY** | **Thin-Film Interference**

The following features should be kept in mind while working thin-film interference problems:

- Identify the thin film causing the interference.
- The type of interference that occurs is determined by the phase relationship between the portion of the wave reflected at the upper surface of the film and the portion reflected at the lower surface.
- Phase differences between the two portions of the wave have two causes: (a) differences in the distances traveled by the two portions and (b) phase changes occurring on reflection. *Both* causes must be considered when determining which type of interference occurs.

---

## Example 27.2   Interference in a Soap Film

Calculate the minimum thickness of a soap film ($n = 1.33$) that results in constructive interference in reflected light if the film is illuminated with light whose wavelength in free space is 600 nm.

**Solution**   The minimum film thickness for constructive interference in the reflected light corresponds to $m = 0$ in Equation 27.13. This gives $2nt = \lambda/2$, or

$$t = \frac{\lambda}{4n} = \frac{600 \text{ nm}}{4(1.33)} = \boxed{113 \text{ nm}}$$

**EXERCISE**   What other film thickness produces constructive interference?

*Answer*   338 nm, 564 nm, 789 nm, . . . $(2m + 1)(113 \text{ nm})$

## Example 27.3 Nonreflecting Coatings for Solar Cells

Semiconductors such as silicon are used to fabricate solar cells—devices that absorb energy by electromagnetic radiation (e.g., sunlight), resulting in a potential difference, so that the cell can transfer energy to a device by electrical transmission. Solar cells are often coated with a transparent thin film, such as silicon monoxide (SiO, $n = 1.45$), in order to minimize reflective losses from the surface. Suppose a silicon solar cell ($n = 3.5$) is coated with a thin film of silicon monoxide for this purpose (Fig. 27.8). Determine the minimum film thickness that produces the least reflection at a wavelength of 550 nm, which is the center of the visible spectrum.

**Reasoning** The reflected light is a minimum when rays 1 and 2 in Figure 27.8 meet the condition of destructive interference. Note that both rays undergo a 180° phase change on reflection in this case, one from the upper surface and one from the lower surface. Hence, the net change in phase is zero due to reflection, and the condition for a reflection minimum requires a path difference of $\lambda_n/2$; therefore, $2t = \lambda/2n$.

**Solution** Because $2t = \lambda/2n$, the required thickness is

$$t = \frac{\lambda}{4n} = \frac{550 \text{ nm}}{4(1.45)} = \boxed{94.8 \text{ nm}}$$

Typically, such antireflecting coatings reduce the reflective loss from 30% (with no coating) to 10% (with coating),

**Figure 27.9**

(Example 27.3) This camera lens has several coatings (of different thicknesses) that minimize reflection of light waves having wavelengths near the center of the visible spectrum. As a result, the little light that is reflected by the lens has a greater proportion of the far ends of the spectrum and appears purple. *(Kristen Brochmann/Fundamental Photographs)*

thereby increasing the cell's efficiency because more light is available to provide energy to the cell. In reality, the coating is never perfectly nonreflecting for all light because the required thickness is wavelength dependent and the incident light covers a wide range of wavelengths.

Glass lenses used in cameras and other optical instruments are usually coated with a transparent thin film, such as magnesium fluoride ($MgF_2$), to reduce or eliminate unwanted reflection. This results in the enhancement of the transmission of light through the lenses. Figure 27.9 shows such a camera lens. Notice that the light reflecting from the lens is tinged purple. The coating on the lens is designed with a thickness such that light near the center of the visible spectrum experiences little reflection. Light near the ends of the spectrum is reflected from the coating. The combination of red and violet light from the ends of the spectrum provides the purple tinge.

**Figure 27.8**

(Example 27.3) Reflective losses from a silicon solar cell are minimized by coating it with a thin film of silicon monoxide.

## Example 27.4   Interference in a Wedge-Shaped Film

A thin, wedge-shaped film of refractive index $n$ is illuminated with monochromatic light of wavelength $\lambda$ as illustrated in Figure 27.10a. Describe the interference pattern observed for this case.

**Reasoning and Solution**   The interference pattern is that of a thin film of variable thickness surrounded by air. Hence, the pattern is a series of alternating bright and dark parallel bands. A dark band corresponding to destructive interference appears at point $O$ (where the path length difference is zero) because the upper reflected ray undergoes a 180° phase change but the lower one does not. According to Equation 27.13, other dark bands appear when $2nt = m\lambda$, so that $t_1 = \lambda/2n$, $t_2 = \lambda/n$, $t_3 = 3\lambda/2n$, and so on. Similarly, bright bands are observed when the thickness satisfies the condition $2nt = (m + \frac{1}{2})\lambda_n$, corresponding to thicknesses of $\lambda/4n$, $3\lambda/4n$, $5\lambda/4n$, and so on. If white light is used, bands of different colors are observed at different points, corresponding to the different wavelengths of light. This is shown in the soap film in Figure 27.10b.

(a)                               (b)

**Figure 27.10**

(Example 27.4) (a) Interference bands in reflected light can be observed by illuminating a wedge-shaped film with monochromatic light. The darker areas correspond to regions at which rays cancel due to destructive interference. (b) Interference in a vertical film of variable thickness. The top of the film appears darkest where the film is thinnest. *(Richard Megna/Fundamental Photographs)*

## 27.6 • DIFFRACTION PATTERNS

In Figure 25.20, we showed plane water waves striking a pair of small openings in a barrier. According to Huygens's principle (Section 25.6), this resulted in the slits acting as sources of circular waves. We identified this behavior as **diffraction**, and now we shall investigate this phenomenon more fully for light waves.

Suppose a light beam is incident on two slits as in Young's double-slit experiment. If the light traveled in straight-line paths after passing through the slits as in Figure 27.11a, the waves would not overlap and no interference pattern would be seen on the screen. Instead, Huygens's principle requires that the waves spread

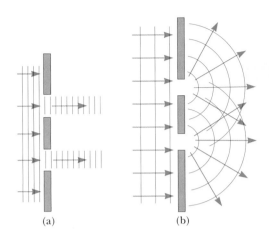

(a)    (b)

**Figure 27.11**

(a) If light waves did not spread out on passing through the slits, no interference would occur. (b) The light waves from the two slits overlap as they spread out, filling the expected shadowed regions with light and producing interference fringes.

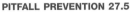

**PITFALL PREVENTION 27.5**
**Diffraction versus diffraction pattern**

The word *diffraction* refers to the general behavior of waves spreading out as they pass through a slit. We used diffraction in explaining the existence of an interference pattern. A *diffraction pattern* is actually a misnomer, but is deeply entrenched in the language of physics. We describe here the diffraction pattern seen on a screen when a single slit is illuminated. In reality, this is another interference pattern. The interference is between parts of the incident light illuminating different regions of the slit.

out from the slits as in Figure 27.11b, so that an interference pattern appears on the screen.

In general, diffraction occurs when waves pass through small openings, around obstacles, or by sharp edges. We might expect that the light passing through one such small opening would simply result in a broad region of light on a screen, due to the spreading of the light as it passes through the opening. We find something more interesting, however. A **diffraction pattern** consisting of light and dark areas is observed, somewhat similar to the interference patterns discussed earlier. For example, when a narrow slit is placed between a distant light source (or a laser beam) and a screen, the light produces a diffraction pattern like that in Figure 27.12. The pattern consists of a broad, intense central band (called the **central maximum**), flanked by a series of narrower, less intense additional bands (called **side maxima**) and a series of dark bands (or **minima**).

Figure 27.13 shows the shadow of a penny, which displays bright and dark rings of a diffraction pattern. The bright spot at the center (called the *Arago bright spot* after its discoverer, Dominique Arago) can be explained using the wave theory

**Figure 27.12**

The diffraction pattern that appears on a screen when light passes through a narrow vertical slit. The pattern consists of a broad central band and a series of less intense and narrower side bands.

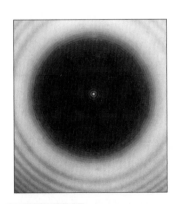

**Figure 27.13**

Diffraction pattern of a penny, taken with the penny midway between screen and source. *(Courtesy of P.M. Rinard, from Am. J. Phys. 44:70, 1976)*

## Figure 27.14

(a) Fraunhofer diffraction pattern of a single slit. The pattern consists of a central bright region flanked by much weaker maxima alternating with dark bands. (Drawing not to scale.) (b) Photograph of a single-slit Fraunhofer diffraction pattern. *(From M. Cagnet, M. Francon, and J.C. Thierr, Atlas of Optical Phenomena, Berlin, Springer-Verlag, 1962, plate 18)*

Slit

Incoming wave

Viewing screen

(b)

(a)

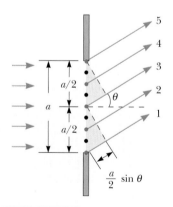

## Figure 27.15

Diffraction of light by a narrow slit of width *a*. Each portion of the slit acts as a point source of waves. The path difference between rays 1 and 3 or between rays 3 and 5 is $(a/2) \sin \theta$ (drawing not to scale).

of light. Waves that diffract from all points on the edge of the penny travel the same distance to the midpoint on the screen. Thus, the midpoint is a region of constructive interference and a bright spot appears. In contrast, from the viewpoint of geometric optics, the center of the pattern would be completely screened by the penny, and so an approach that does not include the wave nature of light would not predict a central bright spot.

Let us consider a common situation, that of light passing through a narrow opening modeled as a slit, and projected onto a screen. As a simplification model, we assume that the observing screen is far from the slit, so that the rays reaching the screen are approximately parallel. This can also be achieved experimentally by using a converging lens to focus the parallel rays on a nearby screen. In this model, the pattern on the screen is called a **Fraunhofer diffraction pattern.**[*]

Figure 27.14a shows light entering a single slit from the left and diffracting as it propagates toward a screen. Figure 27.14b is a photograph of a single-slit Fraunhofer diffraction pattern. A bright fringe is observed along the axis at $\theta = 0$, with alternating dark and bright fringes on each side of the central bright fringe.

Until now we assumed that slits act as point sources of light. In this section we shall determine how their finite widths are the basis for understanding the nature of the Fraunhofer diffraction pattern produced by a single slit. We can deduce some important features of this problem by examining waves coming from various portions of the slit, as shown in the geometric model of Figure 27.15. According to Huygens's principle, **each portion of the slit acts as a source of waves. Hence, light from one portion of the slit can interfere with light from another portion,** and the resultant intensity on the screen depends on the direction $\theta$.

To analyze the diffraction pattern, it is convenient to divide the slit into two halves as in Figure 27.15. All the waves that originate at the slit are in phase. Consider waves 1 and 3, which originate at the bottom and center of the slit, respectively. To reach the same point on the viewing screen, wave 1 travels farther than

---

[*] If the screen were brought close to the slit (and no lens is used), the pattern is a *Fresnel* diffraction pattern. The Fresnel pattern is more difficult to analyze, so we shall restrict our discussion to Fraunhofer diffraction.

wave 3 by an amount equal to the path difference $(a/2) \sin \theta$, where $a$ is the width of the slit. Similarly, the path difference between waves 3 and 5 is also $(a/2) \sin \theta$. If the path difference between two waves is exactly one half of a wavelength (corresponding to a phase difference of $180°$), the two waves cancel each other and destructive interference results. This is true, in fact, for any two waves that originate at points separated by half the slit width because the phase difference between two such points is $180°$. Therefore, waves from the upper half of the slit interfere *destructively* with waves from the lower half of the slit when

$$\frac{a}{2} \sin \theta = \frac{\lambda}{2}$$

or when

$$\sin \theta = \frac{\lambda}{a}$$

If we divide the slit into four parts rather than two and use similar reasoning, we find that the screen is also dark when

$$\sin \theta = \frac{2\lambda}{a}$$

Likewise, we can divide the slit into six parts and show that darkness occurs on the screen when

$$\sin \theta = \frac{3\lambda}{a}$$

Therefore, the general condition for destructive interference is

$$\sin \theta_{\text{dark}} = m \frac{\lambda}{a} \quad (m = \pm 1, \pm 2, \pm 3, \ldots) \qquad \text{[27.14]}$$

Equation 27.14 gives the values of $\theta$ for which the diffraction pattern has zero intensity—that is, a dark fringe is formed. However, Equation 27.14 tells us nothing about the variation in intensity along the screen. The general features of the intensity distribution are shown in Figure 27.16: a broad central bright fringe flanked by much weaker, alternating bright fringes. The various dark fringes (points of zero intensity) occur at the values of $\theta$ that satisfy Equation 27.14. The

**PITFALL PREVENTION 27.6**

**Similar equations**

Equation 27.14 has exactly the same form as Equation 27.2, with $d$, the slit separation, used in Equation 27.2 and $a$, the slit width, in Equation 27.14. Keep in mind, however, that Equation 27.2 describes the *bright* regions in a two-slit interference pattern, but Equation 27.14 describes the *dark* regions in a single-slit diffraction pattern. Furthermore, $m = 0$ does not represent a dark fringe in the diffraction pattern.

- *Condition for destructive interference in a diffraction pattern*

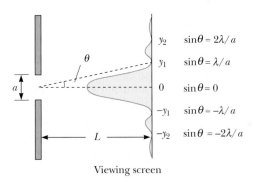

**Figure 27.16**

Light intensity distribution for the Fraunhofer diffraction pattern of a single slit of width $a$. The positions of two minima on each side of the central maximum are labeled (drawing not to scale).

position of the points of constructive interference lie approximately halfway between the dark fringes. Note that the central bright fringe is twice as wide as the weaker maxima.

### THINKING PHYSICS 27.2

If a classroom door is open slightly, you can hear sounds coming from the hallway. Yet you cannot see what is happening in the hallway. What accounts for the difference?

**Reasoning**   The space between the slightly open door and the wall is acting as a single slit for waves. Sound waves have wavelengths larger than the slit width, so sound is effectively diffracted by the opening and spread throughout the room. Light wavelengths are much smaller than the slit width, so virtually no diffraction for the light occurs. You must have a direct line of sight to detect the light waves.

### Example 27.5   Where Are the Dark Fringes?

Light of wavelength 580 nm is incident on a slit of width 0.300 mm. The observing screen is 2.00 m from the slit. Find the positions of the first dark fringes and the width of the central bright fringe.

**Solution**   The first dark fringes that flank the central bright fringe correspond to $m = \pm 1$ in Equation 27.14. Hence, we find that

$$\sin \theta_{\text{dark}} = \pm \frac{\lambda}{a} = \pm \frac{5.80 \times 10^{-7}\ \text{m}}{0.300 \times 10^{-3}\ \text{m}} = \pm 1.93 \times 10^{-3}$$

From the triangle in Figure 27.16, note that $\tan \theta = y_1/L$. Because $\theta$ is very small, we can use the approximation $\sin \theta \approx \tan \theta$, so that $\sin \theta \approx y_1/L$. The positions of the first minima measured from the central axis are therefore given by

$$y_1 \approx L \sin \theta_{\text{dark}} = (2.00\ \text{m})(\pm 1.93 \times 10^{-3})$$

$$= \boxed{\pm 3.86 \times 10^{-3}\ \text{m}}$$

The positive and negative signs correspond to the dark fringes on either side of the central bright fringe. Hence, the width of the central bright fringe is equal to $2|y_1| = 7.72 \times 10^{-3}\ \text{m} = 7.72\ \text{mm}$. Note that this value is much larger than the width of the slit. As the width of the slit is increased, the diffraction pattern narrows, corresponding to smaller values of $\theta$. In fact, for large values of $a$, the various maxima and minima are so closely spaced that the only pattern observed is a large bright area resembling the geometric image of the slit. This matter is of great importance in the design of lenses used in telescopes, microscopes, and other optical instruments.

**EXERCISE**   Determine the width of the first-order bright fringe.

*Answer*   3.86 mm

## 27.7 • RESOLUTION OF SINGLE-SLIT AND CIRCULAR APERTURES

Imagine you are driving in the middle of a dark desert at night, along a road that is perfectly straight and flat for many kilometers. You see another vehicle coming toward you from a distance. When the vehicle is far away, you might be unable to determine whether it is an automobile with two headlights or a motorcycle with one. As it approaches you, at some point you will be able to distinguish the two headlights and determine that it is an automobile. Once you are able to see two separate headlights, you describe the light sources as being **resolved.**

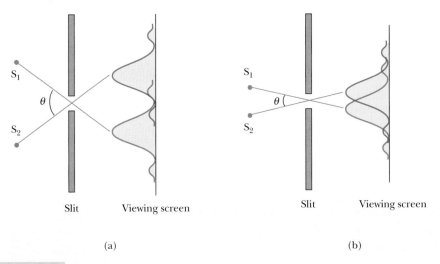

**Figure 27.17**

Two point sources far from a small aperture each produce a diffraction pattern. (a) The angle sub-tended by the sources at the aperture is large enough for the diffraction patterns to be distinguishable. (b) The angle subtended by the sources is so small that their diffraction patterns overlap and the sources are not well resolved. (Note that the angles are greatly exaggerated. The drawings are not to scale.)

The ability of optical systems to distinguish between closely spaced objects is limited because of the wave nature of light. To understand this limitation, consider Figure 27.17, which shows two light sources far from a narrow slit. The sources can be considered as two point sources $S_1$ and $S_2$ that are incoherent. For example, they could be two distant stars observed through the aperture of a telescope tube. If no diffraction occurred, one would observe two distinct bright spots (or images) on the screen at the right in the figure. Because of diffraction, however, each source is imaged as a bright central region flanked by weaker bright and dark bands. What is observed on the screen is the sum of two diffraction patterns: one from $S_1$ and the other from $S_2$.

If the two sources are far enough apart to ensure that their central maxima do not overlap as in Figure 27.17a, their images can be distinguished and are said to be resolved. If the sources are close together, however, as in Figure 27.17b, the two central maxima may overlap and the sources are not resolved. To decide when two sources are resolved, the following condition is often used:

When the central maximum of the diffraction pattern of one source falls on the first minimum of the diffraction pattern of another source, the sources are said to be just resolved. This limiting condition of resolution is known as **Rayleigh's criterion.**

• *Rayleigh's criterion*

Figure 27.18 shows the diffraction patterns from circular apertures for three situations. When the objects are far apart, they are well resolved (Fig. 27.18a). They are just resolved when their angular separation satisfies Rayleigh's criterion (Fig. 27.18b). Finally, the sources are not resolved in Figure 27.18c.

From Rayleigh's criterion, we can determine the minimum angular separation $\theta_{min}$ subtended by the sources at a slit such that the sources are just resolved. In

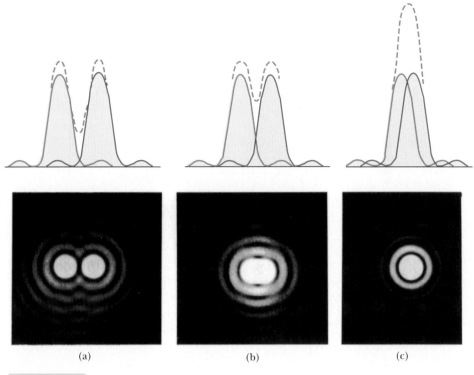

**Figure 27.18**

Individual diffraction patterns of two point sources (solid curves) and the resultant pattern (dashed-curves), for various angular separations of the sources. In each case, the dashed curve is the sum of the two solid curves. (a) The sources are far apart, and are well resolved. (b) The sources are closer together such that the patterns satisfy Rayleigh's criterion, and the sources are just resolved. (c) The sources are so close together that they are not resolved. *(From M. Cagnet, M. Francon, and J.C. Thierr, Atlas of Optical Phenomena, Berlin, Springer-Verlag, 1962, plate 16)*

Section 27.4, we found that the first minimum in a single-slit diffraction pattern occurs at the angle that satisfies the relationship

$$\sin \theta = \frac{\lambda}{a}$$

where $a$ is the width of the slit. According to Rayleigh's criterion, this expression gives the smallest angular separation for which the two sources are resolved. Because $\lambda \ll a$ in most situations, $\sin \theta$ is small and we can use the approximation $\sin \theta \approx \theta$. Therefore, the limiting angle of resolution for a slit of width $a$ is

- *Limiting angle of resolution for a slit*

$$\theta_{\min} = \frac{\lambda}{a} \qquad [27.15]$$

where $\theta_{\min}$ is expressed in radians. Hence, the angle subtended by the two sources at the slit must be *greater* than $\lambda/a$ if the sources are to be resolved.

Many optical systems use circular apertures rather than slits. The diffraction pattern of a circular aperture, as seen in Figure 27.18, consists of a central circular bright disk surrounded by progressively fainter rings. Analysis shows that the limiting angle of resolution of the circular aperture is

$$\theta_{min} = 1.22 \frac{\lambda}{D}$$    [27.16]    • *Limiting angle of resolution for a circular aperture*

where $D$ is the diameter of the aperture. Note that Equation 27.16 is similar to Equation 27.15 except for the factor of 1.22, which arises from a mathematical analysis of diffraction from a circular aperture. This equation is related to the difficulty we had seeing the two headlights at the beginning of this section. When observing with the eye, $D$ in Equation 27.16 is the diameter of the pupil. The diffraction pattern formed when light passes through the pupil causes the difficulty in resolving the headlights.

Another example of the effect of diffraction on resolution for circular apertures is the astronomical telescope. The end of the tube through which the light passes is circular, so the ability of the telescope to resolve light from closely spaced stars is limited by the diameter of this opening.

**Quick Quiz 27.4**

Suppose you are observing a binary star with a telescope and are having difficulty resolving the two stars. You decide to use a colored filter to help you. Should you choose a blue filter or a red filter?

### THINKING PHYSICS 27.3

Cat's eyes have pupils that can be modeled as vertical slits. At night, are cats more successful in resolving headlights on a distant car, or vertically separated lights on the mast of a distant boat?

**Reasoning**   The effective slit width in the vertical direction of the cat's eye is larger than that in the horizontal direction. Thus, the eye has more resolving power for lights separated in the vertical direction and would be more effective at resolving the mast lights on the boat.

### Example 27.6   Resolution of a Telescope

The Hale Telescope at Mount Palomar has a diameter of 200 in. What is its limiting angle of resolution for 600-nm light?

**Solution**   Because $D = 200$ in. $= 5.08$ m and $\lambda = 6.00 \times 10^{-7}$ m, Equation 27.16 gives

$$\theta_{min} = 1.22 \frac{\lambda}{D} = 1.22 \left( \frac{6.00 \times 10^{-7} \text{ m}}{5.08 \text{ m}} \right)$$

$$= 1.44 \times 10^{-7} \text{ rad} \approx 0.03 \text{ s of arc}$$

Therefore, assuming ideal atmospheric conditions, any two stars that subtend an angle greater than or equal to this value could be resolved. The Hale Telescope can never reach its diffraction limit. Instead, its minimum angle of resolution is

always increased by atmospheric blurring. This seeing limit is usually approximately 1 s of arc and is never smaller than approximately 0.1 s of arc. The Hubble Space Telescope can take higher quality photographs than Earth-based telescopes because it does not suffer from atmospheric limitations.

**EXERCISE**   The large radio telescope at Arecibo, Puerto Rico, has a diameter of 305 m and is designed to detect 0.75-m radio waves. Calculate the minimum angle of resolution for this telescope and compare your answer with that of the Hale Telescope.

*Answer*   $3.0 \times 10^{-3}$ rad (10 min of arc), more than 10 000 times larger than the Hale minimum.

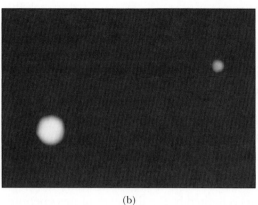

(a)                 (b)

**Figure 27.19**

(a) The photograph on which Charon, the moon of Pluto, was discovered in 1978. From an Earth-based telescope, atmospheric turbulence results in Charon appearing only as a subtle bump on the edge of Pluto. (b) A Hubble Space Telescope photo of Pluto and Charon, clearly resolving the two objects. *(a, U.S. Naval Observatory/James W. Christy, U.S. Navy photograph; b, Photo courtesy of Gemini Observatory)*

As an example of the effects of atmospheric turbulence discussed in Example 27.6, consider telescopic images of Pluto and its moon Charon. Figure 27.19a shows the image taken in 1978 which represents the discovery of Charon. In this photograph from an Earth-based telescope, atmospheric turbulence results in Charon appearing only as a bump on the edge of Pluto. In comparison, Figure 27.19b shows a photograph taken with the Hubble Space Telescope. Without the problems of atmospheric turbulence, Pluto and its moon are clearly resolved.

**WEB**

For more information on Pluto and Charon, visit **www.seds.org/billa/tnp/pluto.html**

## 27.8 • THE DIFFRACTION GRATING

**PITFALL PREVENTION 27.7**

**A diffraction grating is an interference grating**

As with the term *diffraction pattern*, *diffraction grating* is a misnomer but is deeply entrenched. The diffraction grating depends on diffraction in the same way as the double slit— spreading the light so that light from different slits can interfere. It would be more correct to call it an *interference grating*. It is unlikely, however, that you will hear anything other than *diffraction grating* for this device.

The **diffraction grating,** a useful device for analyzing light sources, consists of a large number of equally spaced parallel slits. A grating can be made by cutting parallel, equally spaced grooves on a glass or metal plate with a precision ruling machine. In a *transmission grating*, the spaces between lines are transparent to the light and hence act as separate slits. In a *reflection grating*, the spaces between lines are highly reflective. Gratings with many lines very close to each other can have very small slit spacings. For example, a grating ruled with 5 000 lines/cm has a slit spacing of $d = (1/5\ 000)$ cm $= 2 \times 10^{-4}$ cm.

Figure 27.20 shows a pictorial representation of a section of a flat diffraction grating. A plane wave is incident from the left, normal to the plane of the grating. The pattern observed on the screen (far to the right of Figure 27.20) is the result of the combined effects of interference and diffraction. Each slit produces diffraction, and the diffracted beams interfere with one another to produce the final pattern. Each slit acts as a source of waves, and all waves start at the slits in phase. For some arbitrary direction $\theta$ measured from the horizontal, however, the waves must travel different path lengths before reaching a particular point on the screen. From Figure 27.20, note that the path difference between waves from any two adja-

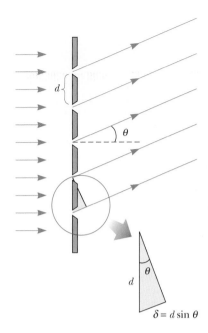

**Figure 27.20**

Side view of a diffraction grating. The slit separation is $d$ and the path difference between adjacent slits is $d \sin \theta$.

cent slits is equal to $d \sin \theta$. (We assume once again that the distance $L$ to the screen is much larger than $d$.) If this path difference equals one wavelength or some integral multiple of a wavelength, waves from all slits will be in phase at the screen and a bright line will be observed. When the light is incident normally on the plane of the grating, the condition for *maxima* in the interference pattern at the angle $\theta$ is therefore*

$$d \sin \theta_{\text{bright}} = m\lambda \quad (m = 0, 1, 2, 3, \ldots) \qquad \text{[27.17]}$$

• *Conditions for constructive interference for a diffraction grating*

This expression can be used to calculate the wavelength from a knowledge of the grating spacing $d$ and the angle of deviation $\theta$. If the incident radiation contains several wavelengths, the $m$th-order maximum for each wavelength occurs at an angle determined from Equation 27.17. All wavelengths are mixed together at $\theta = 0$, corresponding to $m = 0$.

The intensity distribution for a diffraction grating is shown in Figure 27.21. If the source contains various wavelengths, a spectrum of lines at different positions for different order numbers will be observed. Note the sharpness of the principal maxima and the broad range of dark areas. This is in contrast to the broad bright fringes characteristic of the two-slit interference pattern (see Fig. 27.4).

A simple arrangement for measuring the wavelength of light is shown in Figure 27.22. This is a *diffraction grating spectrometer*. The light to be analyzed passes through a slit,[†] and a parallel beam of light exits from the collimator perpendicular to the grating. The diffracted light leaves the grating and exhibits constructive interference at angles that satisfy Equation 27.17. A telescope is used to view the

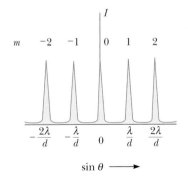

**Figure 27.21**

Intensity versus $\sin \theta$ for a diffraction grating. The zeroth-, first-, and second-order maxima are shown.

---

\* Notice that this equation is identical to Equation 27.2. This equation can be used for a number of slits from two to any number $N$. The intensity distribution will change with the number of slits, but the locations of the maxima are the same.

† A long, narrow slit enables us to observe *line* spectra in the light coming from atomic and molecular systems, as discussed in Chapter 11.

**Figure 27.22**

Diagram of a diffraction grating spectrometer. The collimated beam incident on the grating is diffracted into the various orders at the angles $\theta$ that satisfy the equation $d \sin \theta = m\lambda$, where $m = 0, 1, 2, \ldots$.

image of the slit. The wavelength can be determined by measuring the precise angles at which the images of the slit appear for the various orders.

**Quick Quiz 27.5**

If laser light is reflected from a phonograph record or a compact disc, a diffraction pattern appears. This occurs because both devices contain parallel tracks of information that act as a reflection diffraction grating. Which device, record or compact disc, results in diffraction maxima that are farther apart?

**THINKING PHYSICS 27.4**

White light reflected from the surface of a compact disc has a multicolored appearance as shown in the opening photograph of this chapter. Furthermore, the observation depends on the orientation of the disc relative to the eye and the position of the light source. Explain how this works.

**Reasoning** The surface of a compact disc has a spiral track (with a spacing of approximately 1 $\mu$m) that acts as a reflection grating. The light scattered by these closely spaced tracks interferes constructively in directions that depend on the wavelength and on the direction of the incident light. Any one section of the disc serves as a diffraction grating for white light, sending beams of constructive interference for different colors in different directions. The different colors you see when viewing one section of the disc change as the light source, the disc, or you move to change the angle of incidence or the viewing angle.

## THINKING PHYSICS 27.5

A diffraction grating is a series of transparent openings between opaque areas. Venetian blinds satisfy this definition. Will venetian blinds result in an observable diffraction pattern for light? For sound? For radio waves?

**Reasoning**  For radiation passing through a diffraction grating to exhibit a diffraction pattern, the wavelength of the radiation must be smaller than the slit separation distance. Radio waves and audible sound waves are longer in wavelength than the separation between typical venetian blind slats. Thus, no diffraction effects can be detected for radio or sound waves. Light waves satisfy the requirement that the wavelength is smaller than the slit separation, but light waves are *so* much smaller in wavelength that the diffraction peaks would be indistinguishable. An order-of-magnitude calculation for visible light shows that hundreds of interference peaks exist within 1° from the center line. If microwaves were transmitted through the venetian blinds, we would expect to detect a diffraction pattern because the microwave wavelengths are on the same order of magnitude as the slit width.

## THINKING PHYSICS 27.6

White light enters through an opening in an opaque box, exits through an opening on the other side of the box, and a spectrum of colors appears on the wall. How would you determine whether the box contains a prism or a diffraction grating?

**Reasoning**  The determination could be made by noticing the order of the colors in the spectrum relative to the direction of the original beam of white light. For a prism, in which the separation of light is a result of dispersion, the violet light will be refracted more than the red light, so the order of the spectrum will be from red, closest to the original direction, to violet. For a diffraction grating, the angle of diffraction increases with wavelength. Thus, the spectrum from the diffraction grating will have colors in the order of violet, closest to the original direction, to red.

## Example 27.7  The Orders of a Diffraction Grating

Monochromatic light from a helium-neon laser ($\lambda = 632.8$ nm) is incident normally on a diffraction grating containing 6 000 lines/cm. Find the angles at which the first-order, second-order, and third-order maxima can be observed.

**Solution**  First, we calculate the slit separation, which is equal to the inverse of the number of lines per cm:

$$d = (1/6\ 000) \text{ cm} = 1.667 \times 10^{-4} \text{ cm} = 1\ 667 \text{ nm}$$

For the first-order maximum ($m = 1$), we find

$$\sin \theta_1 = \frac{\lambda}{d} = \frac{632.8 \text{ nm}}{1\ 667 \text{ nm}} = 0.379\ 7$$

$$\theta_1 = \boxed{22.31°}$$

For $m = 2$, we find

$$\sin \theta_2 = \frac{2\lambda}{d} = \frac{2(632.8 \text{ nm})}{1\ 667 \text{ nm}} = 0.759\ 2$$

$$\theta_2 = \boxed{49.41°}$$

For $m = 3$, we find $\sin \theta_3 = 1.139$. Because $\sin \theta$ cannot exceed unity, this does not represent a realistic solution. Hence, only zeroth-, first-, and second-order maxima are observed for this situation.

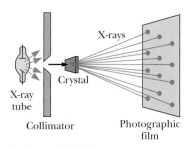

**Figure 27.23**

Schematic diagram of the technique used to observe the diffraction of x-rays by a crystal. The array of spots formed on the film is called a Laue pattern.

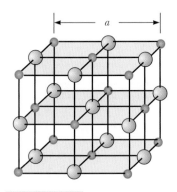

**Figure 27.24**

Crystalline structure of sodium chloride (NaCl). The blue spheres represent Cl⁻ ions, and the red spheres represent Na⁺ ions.

• *Bragg's law*

## 27.9 • DIFFRACTION OF X-RAYS BY CRYSTALS

In principle, the wavelength of any electromagnetic wave can be determined if a grating of the proper spacing (on the order of λ) is available. **X-rays,** discovered in 1895 by Wilhelm Roentgen (1845–1923), are electromagnetic waves with very short wavelengths (on the order of $10^{-10}$ m = 0.1 nm). In 1913, Max von Laue (1879–1960) suggested that the regular array of atoms in a crystal, whose spacing is known to be about $10^{-10}$ m, could act as a three-dimensional diffraction grating for x-rays. Subsequent experiments confirmed his prediction. The observed diffraction patterns are complicated because of the three-dimensional nature of the crystal. Nevertheless, x-ray diffraction is an invaluable technique for elucidating crystalline structures and for understanding the structure of matter.

Figure 27.23 is one experimental arrangement for observing x-ray diffraction from a crystal. A collimated beam of x-rays with a continuous range of wavelengths is incident on a crystal. The diffracted beams are very intense in certain directions, corresponding to constructive interference from waves reflected from layers of atoms in the crystal. The diffracted beams can be detected by a photographic film and form an array of spots known as a *Laue pattern*. The crystalline structure is deduced by analyzing the positions and intensities of the various spots in the pattern.

The arrangement of atoms in a crystal of NaCl is shown in Figure 27.24. The red spheres represent Na⁺ ions, and the blue spheres represent Cl⁻ ions. Each unit cell (the geometric shape that repeats through the crystal) contains four Na⁺ and four Cl⁻ ions. The unit cell is a cube whose edge length is *a*.

The ions in a crystal lie in various planes, as shown in Figure 27.25. Suppose an incident x-ray beam makes an angle θ with one of the planes as in Figure 27.25. The beam can be reflected from both the upper plane and the lower one; however, the geometric construction in Figure 27.25 shows that the beam reflected from the lower surface travels farther than the beam reflected from the upper surface. The path difference between the two beams is $2d \sin \theta$ where *d* is the distance between the planes. The two beams reinforce each other (constructive interference) when this path difference equals some integral multiple of the wavelength λ. The same is true of reflection from the entire family of parallel planes. Hence, the condition for constructive interference (maxima in the reflected wave) is

$$2d \sin \theta = m\lambda \quad (m = 1, 2, 3, \ldots) \qquad \textbf{[27.18]}$$

This condition is known as **Bragg's law** after W. Lawrence Bragg (1890–1971), who first derived the relationship. If the wavelength and diffraction angle are measured, Equation 27.18 can be used to calculate the spacing between atomic planes.

**PITFALL PREVENTION 27.8**

**Different angles**

Notice in Figure 27.25 that the angle θ is measured from the reflecting surface, rather than from the normal, as in the case of the law of reflection in Chapter 25. With slits and diffraction gratings, we also measured the angle θ from the normal to the array of slits. Because of historical tradition, the angle is measured differently in Bragg diffraction, so keep this difference in mind.

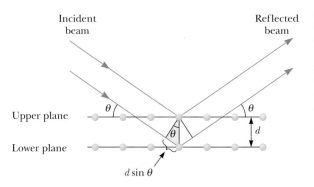

**Figure 27.25**

A two-dimensional description of the reflection of an x-ray beam from two parallel crystalline planes separated by a distance *d*. The beam reflected from the lower plane travels farther than the one reflected from the upper plane by a distance equal to $2d \sin \theta$.

## context connection
# 27.10 • HOLOGRAPHY

One interesting application of the laser is **holography,** the production of three-dimensional images of objects. The physics of holography was developed by Dennis Gabor in 1948, for which he was awarded the 1971 Nobel Prize for physics. The requirement of coherent light for holography, however, delayed the realization of holographic images from Gabor's work until the development of lasers in the 1960s. Figure 27.26 shows a hologram and the three-dimensional character of its image.

Figure 27.27 shows how a hologram is made. Light from the laser is split into two parts by a half-silvered mirror at *B.* One part of the beam reflects off the object to be photographed and strikes an ordinary photographic film. The other half of the beam is diverged by lens $L_2$, reflects from mirrors $M_1$ and $M_2$, and finally strikes the film. The two beams overlap to form an extremely complicated interference pattern on the film. Such an interference pattern can be produced only if the phase relationship of the two waves is constant throughout the exposure of the film. This condition is met by illuminating the scene with light coming through a pinhole or with coherent laser radiation. The hologram records not only the intensity of the light scattered from the object (as in a conventional photograph), but also the phase difference between the reference beam and the beam scattered from the object. This phase difference results in an interference pattern that produces an image with full three-dimensional perspective.

In a normal photographic image, a lens is used to focus the image so that each point on the object corresponds to a single point on the film. Notice that no lens is used in Figure 27.27 to focus the light onto the film. Thus, light from each point on the object reaches *all* points on the film. As a result, each region of the photographic film on which the hologram is recorded contains information about all illuminated points on the object. This leads to a remarkable result: If a small section

**Figure 27.26**

In this hologram, a circuit board is shown from two different views. Notice the difference in the appearance of the measuring tape and the view through the magnifying lens. *(Photo by Ronald R. Erickson; hologram by Nicklaus Phillips)*

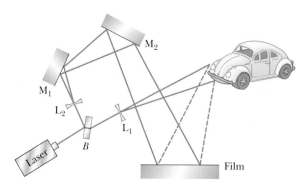

**Figure 27.27**

Experimental arrangement for producing a hologram.

of the hologram is cut from the film, the complete image can be formed from the small piece!

A hologram is best viewed by allowing coherent light to pass through the developed film as one looks back along the direction from which the beam comes. The interference pattern on the film acts as a diffraction grating. Figure 27.28 shows two rays of light striking the film and passing through. For each ray, the $m = 0$ and $m = \pm 1$ rays in the diffraction pattern are shown emerging from the right side of the film. Notice that the $m = +1$ rays converge to form a real image of the scene, which is not the image that is normally viewed. By extending the light rays corresponding to $m = -1$ back behind the film, we see that there is a virtual image located there, with light coming from it in exactly the same way that light came from the actual object when the film was exposed. This is the image that we see by looking through the holographic film.

Holograms are finding a number of applications in displays and in precision measurements. You may have a hologram on your credit card. This special type of hologram is called a *rainbow hologram,* designed to be viewed in reflected white light.

Holograms represent a means of storing visual information using lasers. In the Context Conclusion, we will investigate means of using lasers to store digital information that can be converted into sound waves or video displays.

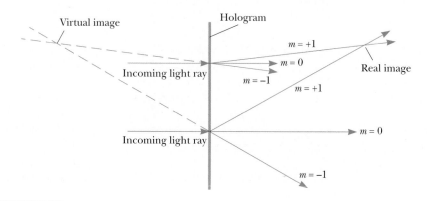

**Figure 27.28**

Two light rays strike a hologram at normal incidence. For each ray, outgoing rays corresponding to $m = 0$ and $m = \pm 1$ are shown. If the $m = -1$ rays are extended backward, a virtual image of the object photographed in the hologram exists on the front side of the hologram.

# SUMMARY

Interference of light waves is the result of the linear superposition of two or more waves at a given point. A sustained interference pattern is observed if (1) the sources have identical wavelengths and (2) the sources are coherent.

In Young's double-slit experiment, two slits separated by a distance $d$ are illuminated by a monochromatic light source. An interference pattern consisting of bright and dark fringes is observed on a screen that is a distance of $L \gg d$ from the slits. The condition for **constructive interference** is

$$\delta = d \sin \theta_{bright} = m\lambda \quad (m = 0, \pm 1, \pm 2, \ldots) \quad [27.2]$$

The condition for **destructive interference** is

$$\delta = d \sin \theta_{dark} = (m + \tfrac{1}{2}) \lambda \quad (m = 0, \pm 1, \pm 2, \ldots) \quad [27.3]$$

The number $m$ is called the **order number** of the fringe.

The **average intensity** of the double-slit interference pattern is

$$I_{av} = I_{max} \cos^2 \left( \frac{\pi d \sin \theta}{\lambda} \right) \quad [27.8]$$

where $I_{max}$ is the maximum intensity on the screen.

An electromagnetic wave traveling from a medium with an index of refraction $n_1$ toward a medium with index of refraction $n_2$ undergoes a 180° phase change on reflection when $n_2 > n_1$. No phase change occurs in the reflected wave if $n_2 < n_1$.

The condition for constructive interference in a film of thickness $t$ and refractive index $n$ with the same medium on both sides of the film is given by

$$2nt = (m + \tfrac{1}{2}) \lambda \quad (m = 0, 1, 2, \ldots) \quad [27.12]$$

Similarly, the condition for destructive interference is

$$2nt = m\lambda \quad (m = 0, 1, 2, \ldots) \quad [27.13]$$

**Diffraction** is the spreading of light from a straight-line path when the light passes through an aperture or around obstacles. A **diffraction pattern** can be analyzed as the interference of a large number of coherent Huygens sources spread across the aperture.

The diffraction pattern produced by a single slit of width $a$ on a distant screen consists of a central, bright maximum and alternating bright and dark regions of much lower intensities. The angles $\theta$ at which the diffraction pattern has *zero* intensity are given by

$$\sin \theta_{dark} = m \frac{\lambda}{a} \quad (m = \pm 1, \pm 2, \pm 3, \ldots) \quad [27.14]$$

**Rayleigh's criterion,** which is a limiting condition of resolution, says that two images formed by an aperture are just distinguishable if the central maximum of the diffraction pattern for one image falls on the first minimum of the other image. The limiting angle of resolution for a slit of width $a$ is given by $\theta_{min} = \lambda / a$, and the limiting angle of resolution for a circular aperture of diameter $D$ is given by $\theta_{min} = 1.22\lambda / D$.

A **diffraction grating** consists of a large number of equally spaced, identical slits. The condition for intensity maxima in the interference pattern of a diffraction grating for normal incidence is

$$d \sin \theta_{bright} = m\lambda \quad (m = 0, 1, 2, 3, \ldots) \quad [27.17]$$

where $d$ is the spacing between adjacent slits and $m$ is the order number of the diffraction maximum.

# QUESTIONS

1. What is the necessary condition on the path length difference between two waves that interfere (a) constructively and (b) destructively?

2. Explain why two flashlights held close together do not produce an interference pattern on a distant screen.

3. If Young's double-slit experiment were performed under water, how would the observed interference pattern be affected?

4. In Young's double-slit experiment, why do we use monochromatic light? If white light were used, how would the pattern change?

5. As a soap bubble evaporates, it appears black just before it breaks. Explain this phenomenon in terms of the phase changes that occur on reflection from the two surfaces of the soap film.

6. An oil film on water appears brightest at the outer regions, where it is thinnest. From this information, what

can you say about the index of refraction of oil relative to that of water?

7. A soap film on a wire loop and held in air appears dark in the thinnest regions when observed by reflected light and shows a variety of colors in thicker regions, as in Figure 27.10b. Explain.

8. A simple way of observing an interference pattern is to look at a distant light source through a stretched handkerchief or an opened umbrella. Explain how this works.

9. Why must a thin film not be very thick (on the order of a few wavelengths) for interference to be observed?

10. A lens with outer radius of curvature $R$ and index of refraction $n$ rests on a flat glass plate, and the combination is illuminated with white light from above. Does a dark spot or a light spot occur at the center of the lens? What does it mean if the observed rings are noncircular?

**11.** Why is the lens on a good-quality camera coated with a thin film?

**12.** Would it be possible to place a nonreflective coating on an airplane to cancel radar waves of wavelength 3 cm?

**13.** Why is it so much easier to perform interference experiments with a laser than with an ordinary light source?

**14.** Observe the shadow of your book when it is held a few inches above a table with a lamp several feet above the book. Why is the shadow somewhat fuzzy at the edges?

**15.** Although we can hear around corners, we cannot see around corners. How can you explain this in view of the fact that sound and light are both waves?

**16.** Assuming that the headlights of a car are point sources, estimate the maximum distance from an observer to the car at which the headlights are distinguishable from each other.

**17.** A laser beam is incident at a shallow angle on a machinist's ruler that has a finely calibrated scale. The rulings on the scale give rise to a diffraction pattern on a screen. Discuss how you can use this technique to obtain a measure of the wavelength of the laser light.

**18.** If a fine wire is stretched across the path of a laser beam, is it possible to produce a diffraction pattern?

**19.** A typical AM radio station broadcasts at 1 000 kHz. An FM radio station broadcasts at a frequency on the order of 100 MHz. Explain why FM radio stations often fade out when you drive your car through a short tunnel or underpass, but AM radio stations do not.

**20.** Describe the change in width of the central maximum of the single-slit diffraction pattern as the width of the slit is made smaller.

**21.** Suppose that reflected white light is used to observe a thin, transparent coating on glass as the coating material is gradually deposited by evaporation in a vacuum. De-scribe possible color changes that might occur during the process of building up the thickness of the coating.

**22.** The brilliant colors of peacock feathers (Fig. Q27.22) are due to a phenomenon known as *iridescence*. The melanin fibers in the feathers act as a natural transmission grating. How do you explain the different colors? Why do the colors often change as the bird moves?

**23.** John William Strutt, Lord Rayleigh (1842–1919), has been called "the last person to understand all of physics and all of mathematics." He invented an improved foghorn. To warn ships of a coastline, a foghorn should radiate sound in a wide horizontal sheet over the ocean's surface. It should not waste sound energy by broadcasting it upward. It should not emit sound downward, because the water in front of the foghorn would reflect that sound upward. Rayleigh's foghorn trumpet is shown in Figure Q27.23. Is it installed in the correct orientation? Decide whether the long dimension of the rectangular opening should be horizontal or vertical, and argue for your decision.

**Figure Q27.23**

**Figure Q27.22** Iridescence in peacock feathers. (© *Diane Schiumo 1988, Fundamental Photographs*)

# PROBLEMS

1, 2, 3 = straightforward, intermediate, challenging    □ = full solution available in the *Student Solutions Manual and Study Guide*

**web** = solution posted at **http://www.harcourtcollege.com/physics/**    ▢ = computer useful in solving problem

▧ = Interactive Physics    ■ = paired numerical/symbolic problems    ▨ = life science application

**Section 27.1  Conditions for Interference**

**Section 27.2  Young's Double-Slit Experiment**

**Section 27.3  Light Waves in Interference**

> Problems 8, 9, or 41 in Chapter 14 can be assigned with this section.

1. A Young's interference experiment is performed with monochromatic light. The separation between the slits is 0.500 mm, and the interference pattern on a screen 3.30 m away shows the first side maximum at 3.40 mm from the center of the pattern. What is the wavelength?

2. In a location where the speed of sound is 354 m/s, a 2 000-Hz sound wave impinges on two slits 30.0 cm apart. (a) At what angle is the first maximum located? (b) If the sound wave is replaced by 3.00-cm microwaves, what slit separation gives the same angle for the first maximum? (c) If the slit separation is 1.00 $\mu$m, what frequency of light gives the same first maximum angle?

3. Two radio antennas separated by 300 m as in Figure P27.3 simultaneously broadcast identical signals at the same wavelength. A radio in a car traveling due north receives the signals. (a) If the car is at the position of the second maximum, what is the wavelength of the signals? (b) How much farther must the car travel to encounter the next minimum in reception? (*Note:* Do not use the small-angle approximation in this problem.)

4. The two speakers of a boom box are 35.0 cm apart. A single oscillator makes the speakers vibrate in phase at a frequency of 2.00 kHz. At what angles, measured from the perpendicular bisector of the line joining the speakers, would a distant observer hear maximum sound intensity? minimum sound intensity? (Take the speed of sound as 340 m/s.)

5. Young's double-slit experiment is performed with 589-nm light and a distance of 2.00 m between the slits and the screen. The tenth interference minimum is observed 7.26 mm from the central maximum. Determine the spacing of the slits.

6. Two slits are separated by 0.320 mm. A beam of 500-nm light strikes the slits, producing an interference pattern. Determine the number of maxima observed in the angular range $-30.0° < \theta < 30.0°$.

7. Young's double-slit experiment underlies the *instrument landing system* used to guide aircraft to safe landings when the visibility is poor. Although real systems are more complicated than the example described here, they operate on the same principles. A pilot is trying to align her plane with a runway as suggested in Figure P27.7a. Two radio antennas $A_1$ and $A_2$ separated by 40.0 m are positioned adjacent to the runway. The antennas broadcast unmodulated

(a)

(b)

**Figure P27.7**

**Figure P27.3**

coherent radio waves at 30.0 MHz. (a) Find the wavelength of the waves. The pilot "locks onto" the strong signal radiated along an interference maximum and steers the plane to keep the received signal strong. If she detects the central maximum, the plane will have the right heading to land when it reaches the runway. (b) Suppose instead that the plane is flying along the first side maximum (Fig. P27.7b). How far to the side of the runway centerline is the plane when it is 2.00 km from the antennas? (c) It is possible to tell the pilot she is on the wrong maximum by sending out two signals from each antenna and equipping the aircraft with a two-channel receiver. The ratio of the two frequencies must not be the ratio of small integers (such as 3/4). Explain how this two-frequency system would work, and why it would not necessarily work if the frequencies were related by an integer ratio.

8. In Figure 27.3 let $L = 1.20$ m and $d = 0.120$ mm, and assume that the slit system is illuminated with monochromatic 500-nm light. Calculate the phase difference between the two wavefronts arriving at point $P$ when (a) $\theta = 0.500°$ and (b) $y = 5.00$ mm. (c) What is the value of $\theta$ for which the phase difference is 0.333 rad? (d) What is the value of $\theta$ for which the path difference is $\lambda/4$?

9. In Figure 27.3, let $L = 120$ cm and $d = 0.250$ cm. The slits are illuminated with coherent 600-nm light. Calculate the distance $y$ above the central maximum for which the intensity on the screen is 75.0% of the maximum.

10. The intensity on the screen at a certain point in a double-slit interference pattern is 64.0% of the maximum value. (a) What minimum phase difference (in radians) between sources produces this result? (b) Express this phase difference as a path difference for 486.1-nm light.

11. Two slits are separated by 0.180 mm. An interference pattern is formed on a screen 80.0 cm away by 656.3-nm light. Calculate the fraction of the maximum intensity 0.600 cm above the central maximum.

## Section 27.4   Change in Phase Due to Reflection

## Section 27.5   Interference in Thin Films

12. A soap bubble ($n = 1.33$) is floating in air. If the thickness of the bubble wall is 115 nm, what is the wavelength of the light that is most strongly reflected?

13. An oil film ($n = 1.45$) floating on water is illuminated by white light at normal incidence. The film is 280 nm thick. Find (a) the dominant observed color in the reflected light and (b) the dominant color in the transmitted light. Explain your reasoning.

14. A possible means for making an airplane invisible to radar is to coat the plane with an antireflective polymer. If radar waves have a wavelength of 3.00 cm and the index of refraction of the polymer is $n = 1.50$, how thick would you make the coating?

15. An air wedge is formed between two glass plates separated at one edge by a very fine wire as in Figure P27.15. When the wedge is illuminated from above by 600-nm light, 30 dark fringes are observed. Calculate the radius of the wire.

**Figure P27.15**

16. A material having an index of refraction of 1.30 is used to coat a piece of glass ($n = 1.50$). What should the minimum thickness of this film be to minimize reflection of 500-nm light?

## Section 27.6   Diffraction Patterns

17. Helium–neon laser light ($\lambda = 632.8$ nm) is sent through a 0.300-mm-wide single slit. What is the width of the central maximum on a screen 1.00 m from the slit?

18. A beam of green light is diffracted by a slit of width 0.550 mm. The diffraction pattern forms on a wall 2.06 m beyond the slit. The distance between the positions of zero intensity on both sides of the central bright fringe is 4.10 mm. Calculate the wavelength of the laser light.

19. A screen is placed 50.0 cm from a single slit, which is illuminated with 690-nm light. If the distance between the first and third minima in the diffraction pattern is 3.00 mm, what is the width of the slit?

20. Coherent microwaves of wavelength 5.00 cm enter a long, narrow window in a building otherwise essentially opaque to the microwaves. If the window is 36.0 cm wide, what is the distance from the central maximum to the first-order minimum along a wall 6.50 m from the window?

21. Sound with a frequency of 650 Hz from a distant source passes through a doorway 1.10 m wide in a sound-absorbing wall. Find the number and approximate directions of the diffraction-maximum beams radiated into the space beyond.

## Section 27.7   Resolution of Single-Slit and Circular Apertures

22. The pupil of a cat's eye narrows to a vertical slit of width 0.500 mm in daylight. What is the angular resolution for horizontally separated mice? Assume that the average wavelength of the light is 500 nm.

23. A helium–neon laser emits light that has a wavelength of 632.8 nm. The circular aperture through which the beam

emerges has a diameter of 0.500 cm. Estimate the diameter of the beam 10.0 km from the laser.

24. Suppose you are standing on a straight highway and a car passes you at time $t = 0$. You watch the car move away from you at speed $v$. The air is perfectly clear and steady, and at a time $t$ the taillights appear to merge into one. Assuming the diameter of your pupil is $d$, estimate the width of the car.

25. The Impressionist painter Georges Seurat created paintings with an enormous number of dots of pure pigment, each of which was approximately 2.00 mm in diameter. The idea was to have colors such as red and green next to each other to form a scintillating canvas (Fig. P27.25). Outside what distance would one be unable to discern individual dots on the canvas? (Assume that $\lambda = 500$ nm and that the pupil diameter is 4.00 mm.)

**Figure P27.25** *Sunday Afternoon on the Isle of La Grande Jatte,* by Georges Seurat (*SuperStock*)

26. A circular radar antenna on a Coast Guard ship has a diameter of 2.10 m and radiates at a frequency of 15.0 GHz. Two small boats are located 9.00 km away from the ship. How close together could the boats be and still be detected as two objects?

27. Assume a beam of light from a ruby laser (with a wavelength of 694.3 nm) is sent outward from the barrel of a telescope 2.70 m in diameter. What will be the diameter of the big red spot when the beam hits the Moon, 384 000 km away? (Ignore atmospheric dispersion.)

28. A spy satellite consists of a large-diameter concave mirror that forms an image on a television-camera detector and sends the picture to a ground receiver by radio waves. In effect, it is an astronomical telescope in orbit, looking down instead of up. Can a spy satellite read a license plate? Can it read the date on a dime? Argue for your answers by making an order-of-magnitude calculation, specifying the data you estimate.

## Section 27.8  The Diffraction Grating

In the following problems, assume that the light is incident normally on the gratings.

29. Light from an argon laser strikes a diffraction grating that has 5 310 lines/cm. The central and first-order principal maxima are separated by 0.488 m on a wall 1.72 m from the grating. Determine the wavelength of the laser light.

30. A helium–neon laser ($\lambda = 632.8$ nm) is used to calibrate a diffraction grating. If the first-order maximum occurs at $20.5°$, what is the spacing between adjacent grooves in the grating?

31. web The hydrogen spectrum has a red line at 656 nm and a blue line at 434 nm. What are the angular separations between two spectral lines obtained with a diffraction grating that has 4 500 lines/cm?

32. Light of wavelength 500 nm is incident normally on a diffraction grating. If the third-order maximum of the diffraction pattern is observed at $32.0°$, (a) what is the number of rulings per centimeter for the grating? (b) Determine the total number of primary maxima that can be observed in this situation.

33. A diffraction grating has 800 rulings/mm. A beam of light containing wavelengths from 500 to 700 nm is incident on the grating. Do the spectra of different orders overlap? Explain.

34. Two wavelengths $\lambda$ and $\lambda + \Delta\lambda$ (with $\Delta\lambda \ll \lambda$) are incident on a diffraction grating. Show that the angular separation between the spectral lines in the $m$th-order spectrum is

$$\Delta\theta = \frac{\Delta\lambda}{\sqrt{(d/m)^2 - \lambda^2}}$$

where $d$ is the slit spacing and $m$ is the order number.

35. A grating with 250 lines/mm is used with an incandescent light source. Assume the visible spectrum to range in wavelength from 400 to 700 nm. In how many orders can one see (a) the entire visible spectrum and (b) the short-wavelength region?

## Section 27.9  Diffraction of X-Rays by Crystals

36. Potassium iodide (KI) has the same crystalline structure as NaCl, with atomic planes separated by 0.353 nm. A monochromatic x-ray beam shows a first-order diffraction maximum when the grazing angle is $7.60°$. Calculate the x-ray wavelength.

37. If the interplanar spacing of NaCl is 0.281 nm, what is the predicted angle at which 0.140-nm x-rays are diffracted in a first-order maximum?

38. In water of uniform depth, a wide pier is supported on pilings in several parallel rows 2.80 m apart. Ocean waves of uniform wavelength roll in, moving in a direction that

makes an angle of 80.0° with the rows of posts. Find the three longest wavelengths of waves that will be strongly reflected by the pilings.

## Section 27.10   Context Connection — Holography

39. A wide beam of laser light of wavelength 632.8 nm passes through several narrow parallel slits separated by 1.20 mm and falls on a sheet of photographic film 1.40 m away. The exposure time is chosen so that the film stays unexposed everywhere except at the central region of each bright fringe. (a) Find the distance between these interference maxima. The film is printed as a transparency—it is opaque everywhere except at the exposed lines. Next, the same beam of laser light is directed through the transparency and allowed to fall on a screen 1.40 m beyond. (b) Argue that several narrow parallel bright regions separated by 1.20 mm will appear on the screen as real images of the original slits. If the screen is removed, light diverges from the images of the original slits with the same reconstructed wavefronts as the original slits produced. (*Hint:* You may find it useful to draw diagrams similar to Figure 27.2. A train of thought like this, at a soccer game, led Dennis Gabor to the invention of holography.)

40. A helium–neon laser can produce a green laser beam instead of red. Refer to Figure 24.19, which omits some energy levels between $E_2$ and $E_1$. After a population inversion is established, neon atoms will make a variety of downward transitions in falling from the state labeled $E_3^*$ down eventually to level $E_1$. The atoms emit both red light with a wavelength of 632.8 nm and green light with a wavelength of 543 nm in a subsequent transition. Assume the atoms are in a cavity between mirrors designed to reflect the green light with high efficiency but to allow the red light to leave the cavity immediately. Then stimulated emission can lead to the buildup of a collimated beam of green light between the mirrors, having a greater intensity than does the red light. A small fraction of the green light is permitted to escape by transmission through one mirror, to constitute the radiated laser beam. The mirrors forming the resonant cavity are made of layers of silicon dioxide and titanium dioxide. (a) How thick a layer of silicon dioxide, between layers of titanium dioxide, would minimize reflection of the red light? (b) What should be the thickness of a similar but separate layer of silicon dioxide to maximize reflection of the green light?

## Additional Problems

41. Raise your hand and hold it flat. Think of the space between your index finger and your middle finger as one slit, and think of the space between middle finger and ring finger as a second slit. (a) Consider the interference resulting from sending coherent visible light perpendicularly through this pair of openings. Compute an order-of-magnitude estimate for the angle between adjacent zones of constructive interference. (b) To make the angles in the interference pattern easy to measure with a plastic protractor, you should use an electromagnetic wave with frequency of what order of magnitude? How is this wave classified on the electromagnetic spectrum?

42. **Review Problem.**   A flat piece of glass is held stationary and horizontal above the flat top end of a 10.0-cm-long vertical metal rod that has a rigidly fixed lower end. The thin film of air between the rod and the glass is observed to be bright by reflected light when it is illuminated by light of wavelength 500 nm. As the temperature is slowly increased by 25.0°C, the film changes from bright to dark and back to bright 200 times. What is the coefficient of linear expansion of the metal?

43. Interference effects are produced at point $P$ on a screen as a result of direct rays from a 500-nm source and reflected rays off the mirror, as in Figure P27.43. Assume the source is 100 m to the left of the screen and 1.00 cm above the plane of the mirror. Find the distance $y$ to the first dark band above the mirror.

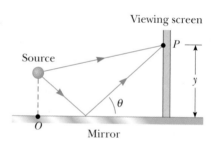

**Figure P27.43**

44. The waves from a radio station can reach a home receiver by two paths. One is a straight-line path from transmitter to home, a distance of 30.0 km. The second path is by reflection from the ionosphere (a layer of ionized air molecules high in the atmosphere). Assume that this reflection takes place at a point midway between receiver and transmitter. The wavelength broadcast by the radio station is 350 m. Find the minimum height of the ionospheric layer that produces destructive interference between the direct and reflected beams. (Assume that no phase change occurs on reflection.)

45. Many cells are transparent and colorless. Structures of great interest in biology and medicine can be practically invisible to ordinary microscopy. An *interference microscope* reveals a difference in refractive index as a shift in interference fringes, to indicate the size and shape of cell structures. The idea is exemplified in the following problem: An air wedge is formed between two glass plates in contact

along one edge and slightly separated at the opposite edge. When the plates are illuminated with monochromatic light from above, the reflected light has 85 dark fringes. Calculate the number of dark fringes that appear if water ($n = 1.33$) replaces the air between the plates.

46. (a) Both sides of a uniform film with an index of refraction $n$ and thickness $d$ are in contact with air. For normal incidence of light, an intensity minimum is observed in the reflected light at $\lambda_2$ and an intensity maximum is observed at $\lambda_1$, where $\lambda_1 > \lambda_2$. If no intensity minima are observed between $\lambda_1$ and $\lambda_2$, show that the integer $m$ in Equations 27.12 and 27.13 is given by $m = \lambda_1/2(\lambda_1 - \lambda_2)$. (b) Determine the thickness of the film if $n = 1.40$, $\lambda_1 = 500$ nm, and $\lambda_2 = 370$ nm.

47. The condition for constructive interference by reflection from a thin film in air as developed in Section 27.5 assumes nearly normal incidence. Show that if the light is incident on the film at a nonzero angle $\phi_1$ (relative to the normal), then the condition for constructive interference is $2nt \cos \theta_2 = (m + \frac{1}{2}) \lambda$, where $\theta_2$ is the angle of refraction.

48. A soap film ($n = 1.33$) is contained within a rectangular wire frame. The frame is held vertically so that the film drains downward and forms a wedge with flat faces. The thickness of the film at the top is essentially zero. The film is viewed by reflected white light with near-normal incidence, and the first violet ($\lambda = 420$ nm) interference band is observed 3.00 cm from the top edge of the film. (a) Locate the first red ($\lambda = 680$ nm) interference band. (b) Determine the film thickness at the positions of the violet and red bands. (c) What is the wedge angle of the film?

49. Light from a helium–neon laser ($\lambda = 632.8$ nm) is incident on a single slit. What is the maximum width for which no diffraction minima are observed?

50. Figure P27.50 shows a megaphone in use. Construct a theoretical description of how a megaphone works. You may assume that the sound of your voice radiates just through the opening of your mouth. Most of the information in speech is carried not in a signal at the fundamental frequency, but rather in noises and in harmonics, with frequencies of a few thousand hertz. Does your theory allow any prediction that is simple to test?

51. **Review Problem.** A beam of 541-nm light is incident on a diffraction grating that has 400 lines/mm. (a) Determine the angle of the second-order ray. (b) If the entire apparatus is immersed in water, what is the new second-order angle of diffraction? (c) Show that the two diffracted rays of parts (a) and (b) are related through the law of refraction.

52. The *Very Large Array* (VLA) is a set of 27 radio telescope dishes in Caton and Socorro Counties, New Mexico (Fig P27.52). The antennas can be moved apart on railroad tracks, and their combined signals give the resolving power of a synthetic aperture 36.0 km in diameter. (a) If the detectors are tuned to a frequency of 1.40 GHz, what is the angular resolution of the VLA? (b) Clouds of hydrogen radiate at this frequency. What must be the separation distance of two clouds at the center of the galaxy, 26 000 lightyears away, if they are to be resolved? (c) As the telescope looks up, a circling hawk looks down. For comparison, find the angular resolution of the hawk's eye. Assume that it is most sensitive to green light having a wavelength of 500 nm and that it has a pupil of diameter 12.0 mm. (d) A mouse is on the ground 30.0 m below. By what distance must the mouse's whiskers be separated if the hawk can resolve them?

**Figure P27.52** A rancher in New Mexico rides past one of the 27 radio telescopes that make up the Very Large Array (VLA). *(© Danny Lehman)*

**Figure P27.50** *(Susan Allen Sigmon/Allsport USA)*

**53.** A *pinhole camera* has a small circular aperture of diameter $D$. Light from distant objects passes through the aperture into an otherwise dark box, falling on a screen located a distance $L$ away. If $D$ is too large, the display on the screen will be fuzzy because a bright point in the field of view will send light onto a circle of diameter slightly larger than $D$. On the other hand, if $D$ is too small, diffraction will blur the display on the screen. The screen shows a reasonably sharp image if the diameter of the central disk of the diffraction pattern, specified by Equation 27.16, is equal to $D$ at the screen. (a) Show that for monochromatic light with plane wavefronts and $L \gg D$, the condition for a sharp view is fulfilled if $D^2 = 2.44 \lambda L$. (b) Find the optimum pinhole diameter if 500-nm light is projected onto a screen 15.0 cm away.

## ANSWERS TO QUICK QUIZZES

**27.1** The result of the double slit is to *redistribute* the energy arriving at the screen. Although no energy arrives at the location of a dark fringe, more energy arrives at the location of a bright fringe than would be the case without the double slit. The total amount of energy arriving at the screen is the same as that without the slit, as it must be according to the principle of conservation of energy.

**27.2** One of the materials has a higher index of refraction than water, the other lower. The material with a higher index of refraction than water appears dark as it approaches zero thickness. The light reflected from the upper surface has a 180° phase shift, but no such phase change occurs from the lower surface because the index of refraction for water on the other side is lower than that of the film. Thus, the two reflections are out of phase and interfere destructively. The material with index of refraction lower than water will have a phase change for the light reflected from both upper and lower surfaces, so that the reflections from the zero-thickness film will be in phase, and the film will appear bright.

**27.3** For normal incidence, the extra path length followed by the reflected ray is twice the thickness of the film. For destructive interference, this must be a distance of half a wavelength of the light in the material of the film. Thus, the thickness of the film must be one-quarter wavelength, which is the same as that for reflected destructive interference.

**27.4** We would like to reduce the minimum angular separation for two objects below the angle subtended by the two stars in the binary system. We can do that by reducing the wavelength of the light—this in essence makes the aperture larger, relative to the light wavelength, increasing the resolving power. Thus, we should choose a blue filter.

**27.5** The tracks of information on a compact disc are much closer together than on a phonograph record. As a result, the diffraction maxima from the compact disc will be farther apart than those from the record.

# Lasers

## USING LASERS TO STORE INFORMATION

We have now investigated the principles of optics and can respond to our central question for the *Lasers* Context:

> *What is special about laser light and how is it used in technological applications?*

In the Context Connections in Chapters 24 to 27 we discussed two primary technological applications of lasers: optical fibers and holography. In this Context Conclusion, we will choose two more from the vast number of possibilities, both involving storage and retrieval of information using lasers: compact discs (as well as CD-ROM and DVD) and magnetooptical storage.

### Compact Discs

The storage of information in a small volume of space is a goal toward which humans have worked for several decades. In the early days of computing, information was stored on punched cards. This seems humorous in today's world, especially because the area taken up by laying the cards representing a page of text out on a table was larger than the original page of text.

The magnetic disc recording and storage technique was introduced in the 1950s and allowed a reduction in space over that taken up by the original data.

The beginning of optical storage occurred in the 1970s, with the introduction of videodiscs. These plastic discs included encoded pits representing the analog information associated with a video signal. A laser, focused by lenses to a spot of diameter about a micron, is used to read the data. When the laser light reflects off the flat area of the disc, the light is reflected back into the system and detected. When the light encounters a pit, some of it is scattered and the light reflected from the bottom of the pit interferes destructively with that reflected from the surface, so that very little of the incident light finds its way back to the detection system.

The next step in the optical recording story involves the digital revolution, exemplified by the introduction of the compact disc, or CD. The reading of the disc is similar to that of the videodisc, but the information is stored in a *digital* format. Musical CDs were rapidly accepted by the public with much more enthusiasm than videodiscs. Shortly after the introduction of CDs, plans were announced to market an optical disc for storage of information for computers—the CD-ROM.

Figure 1 shows the detection system of a CD player. The optical components are mounted on a track (not shown in the figure) that rolls radially so that the system can access all regions of the disc. The laser is located near the bottom of the figure, directing its light upward. The light is collimated by a lens into a parallel beam and passes through a beam splitter. This serves no purpose on the way up, but is important for the return light. The laser beam is then focused to a very small spot on the disc by the objective lens.

If the light encounters a pit in the disc, the light is scattered and very little light returns along the original path. If the light encounters a flat region of the disc at which a pit has not been recorded, the light reflects back along its original path. The reflected light moves downward in the diagram, arriving at the beam

1037

**Figure 1**

The detection system of a compact disc player. The laser (bottom) sends a beam of light upward. Laser light reflected back from the disc and then reflected to the right by the beam splitter enters a photocell. The digital information entering the photocell as pulses of light is converted to audio information.

splitter, so that it is partially reflected to the right. Lenses focus the beam, which is then detected by the photocell.

Digital information is recorded on the disc by means of the pits and the flats. A change from a pit to a flat or a flat to a pit signals a one, whereas no change indicates a zero. The electronic circuitry in the compact disc player converts the series of zeroes and ones back into an audible signal. This technology can also be used to store video information on a disc, leading to the rapid growth of DVD (digital video disc, or digital versatile disc) in the later years of the 20th century.

## Magnetooptical Storage

Magnetooptical storage is a nice mixture of our Contexts on thermodynamics, magnetism, and optics. A magnetooptical disc contains ferromagnetic material (Section 22.11), with its magnetization directed perpendicularly to the surface (Fig. 2a). Ferromagnetism is a phase of matter, similar in some ways to the more familiar phases of solid, liquid, and gas. As with these phases, the phase can be changed by raising the temperature. In the case of a solid, the temperature at which the phase changes from solid to liquid is called the melting point. In magnetism, the temperature at which the ferromagnetic phase changes to the nonferromagnetic phase is called the *Curie temperature*.

A "pit" is recorded on a magnetooptical disc by focusing a high-power laser on a spot while an electromagnet nearby applies a magnetic field *opposite* to the original direction, as shown in Figure 2b. The laser warms the spot to a temperature above the Curie temperature, at which the magnetization of the spot can be reversed. When the laser is removed, the spot cools and maintains the reverse magnetization.

Reading the disc depends on the *Kerr effect*, which is the conversion of linearly polarized light into elliptically polarized light on reflection from ferromagnetic surfaces. A weak laser beam is reflected from the disc, as in Figure 2c. This involves an extra beam splitter in the optics, which splits the polarizations of the returned light and allows the detection of a change when the light reflects from a region with a different magnetization.

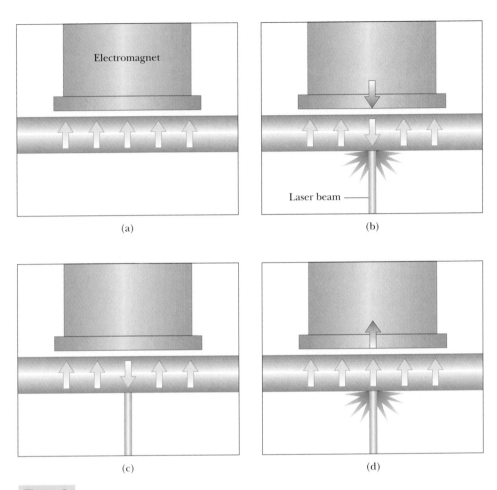

**Figure 2**

(a) A magnetooptical disc. The arrows indicate the direction of the original magnetization. (b) A high-powered laser is used to raise the ferromagnetic material above its Curie temperature, so that a magnetization opposite the original direction can be established. (c) The recorded disc is read by reflecting a low-power laser from the surface. The reflection from the different directions of magnetization varies according to the Kerr effect. (d) The disc is erased by applying a magnetic field in the original direction and using a laser to raise the material above the Curie temperature once again.

To erase the disc, the electromagnet applies a magnetic field in the same direction as the original field, and the laser warms up all spots on the disc, restoring the original condition, as in Figure 2d.

With this discussion, we close the *Lasers* Context. We have not been able to explore all technological applications of this versatile light source, but you have been introduced to enough to convince you, we hope, of its value in our technical society.

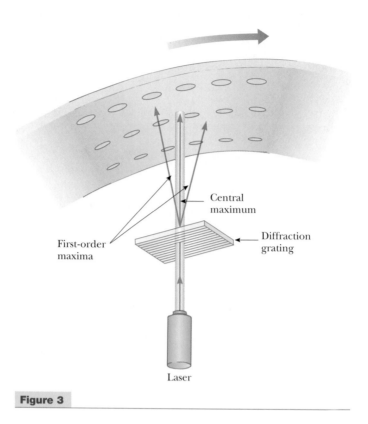

**Figure 3**

**PROBLEMS**

1. Compact disc and digital video disc players use interference to obtain a strong signal from a tiny bump. The depth of a pit is chosen to be one quarter of the wavelength of the laser light used to read the disc. Then light reflected from the pit and light reflected from the adjoining flat differ in path length traveled by one-half wavelength, to interfere destructively at the detector. As the disc rotates, the light intensity drops significantly every time light is reflected from near a pit edge. The time intervals between variations in light intensity are decoded into a series of zeros and ones that carries the stored information. Assume that infrared light with a wavelength of 780 nm in vacuum is used in a CD player. The disc is coated with plastic having a refractive index of 1.50. What should be the depth of each pit? A DVD player uses light of a shorter wavelength, and the pit dimensions are correspondingly smaller. This is one factor resulting in greater storage capacity on a DVD than on a CD.

2. The laser in a CD player must precisely follow the spiral track, along which the distance between one loop of the spiral and the next is only about 1.25 μm. A feedback mechanism lets the player know if the laser drifts off the track, so that the player can steer it back again. Figure 3 shows how a diffraction grating is used to provide information to keep the beam on track. The laser light passes through a diffraction grating just before it reaches the disc. The strong central maximum of the diffraction pattern is used to read the information in the track of pits. The two first-order side maxima are used for steering. The grating is designed so that the first-order maxima fall on the flat surfaces on both sides of the information track. Both side beams are reflected into their own detectors. As long as both beams are reflecting from smooth nonpitted surfaces they are detected with constant high intensity. If the main beam wanders off the track, however, one of the side beams will begin to strike pits on the information track and the reflected light will diminish. This change is used with an electronic circuit to guide the beam back to the desired location. Assume that the laser light has a wavelength of 780 nm and that the

diffraction grating is positioned 6.90 $\mu$m from the disc. Assume that the first-order beams are to fall on the disc 0.400 $\mu$m on either side of the information track. What should be the number of tracks per millimeter in the grating?

3. Consider the recording process for a magnetooptical disc. To record a pit, its location on the ferromagnetic layer must be raised above the Curie temperature. Imagine that the surface moves past the laser at about 1 m/s and that the pit is modeled as a cylinder 1 $\mu$m deep with a radius of about 1 $\mu$m. The Curie temperature of the ferromagnetic material is 600 K. The specific heat of the ferromagnetic material is 300 J/kg°·C. The density of the ferromagnetic material is $2 \times 10^3$ kg/m$^3$. What is the order of magnitude of the intensity of the laser beam necessary to raise the pit above the Curie temperature?

# The Cosmic Connection

In this Context, we investigate the principles included in the area of physics commonly called *modern physics*. This encompasses the revolution in physics that commenced at the beginning of the 20th century. We began our discussion of modern physics in Chapter 9 when we studied relativity. Other aspects of modern physics appeared at various locations throughout the book. These included atomic spectra and the Bohr model in Chapter 11, quantization of angular momentum and energy in Chapter 11, black holes in Chapter 11, black bodies in Chapter 24, and the discussion of the photon in Chapter 24.

In this book, we stress the importance of models in understanding physical phenomena. At the turn of the 20th century, classical physics was well established and provided many principles on which models for phenomena could be built. Many experimental observations, however, could not be brought into agreement with theory using classical models. Attempts to apply the laws of classical physics to atomic systems were consistently unsuccessful in making accurate predictions of the behavior of matter on the atomic scale. Various phenomena such as blackbody radiation, the photoelectric effect, and the emission of sharp spectral lines by atoms in a gas discharge could not be understood within the framework of classical physics. Between 1900 and 1930, however, new models collectively called *quantum physics* or *quantum mechanics* were highly successful in explaining the behavior of atoms, molecules, and nuclei. Like relativity, quantum physics requires a modification of our ideas concerning the physical world. Quantum mechanics does not, however, directly contradict or invalidate classical mechanics. As with relativity, the equations of quantum physics reduce to classical equations in the appropriate realm, that is, when the quantum equations are used to describe macroscopic systems.

An extensive study of quantum physics is certainly beyond the scope of this book, and therefore this Context is simply an introduction to its underlying ideas. One of the true successes of quantum physics is the connection it makes between microscopic phenomena and the structure and evolution of the Universe. Ironically, recent developments in physics that probe smaller and smaller scales allow us to advance our understanding of the larger and larger systems that are familiar to us.

For example, consider your experiences with a hand-held calculator or a computer. The events you observe are macroscopic—the appearance of numbers on an LCD display or of a spreadsheet on a computer screen. But what controls these macroscopic events? They are controlled by a microprocessor within the calculator or the computer. The operation of the microprocessor depends on the behavior of electrons within the solid-state material in an integrated circuit chip. The operation of the macroscopic calculator or computer is not possible without an understanding of this behavior.

A computer monitor. The display on the monitor is due to the behavior of microscopic electrons in the circuitry of the computer. *(George Semple)*

(a)

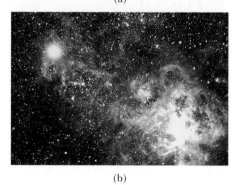

(b)

Supernova 1987A. (a) The region of the Taran-
tula Nebula (lower right) of the Large Magellanic Cloud before the supernova. (b) The supernova appears at the upper left on February
24, 1987. An understanding of this cosmic explosion is found in the interactions between
the microscopic particles within the nucleus.
*(© 1987 Anglo-Australian Observatory, photo-
graph by David Malin)*

As a second example, the supernova
explosion in the photograph shown here is
clearly a macroscopic event—a star with a
radius on the order of millions of meters is
undergoing a violent event. We have been
able to advance our understanding of such
events by studying the atomic nucleus,
which is on the order of $10^{-15}$ m in size.

If we imagine an even larger system
than a star—the entire Universe—we can
advance our understanding of its origin by
thinking about particles even smaller than
the nucleus. Consider the constituents of
protons and neutrons, called *quarks*. Models based on quarks lead to a theory of the
origin of the Universe called the *Big Bang*.
In this Context, we shall study both quarks
and the Big Bang.

It seems that the larger the system we
wish to investigate, the smaller are the particles whose behavior we must understand!
We shall explore this relationship and study
the principles of quantum physics as we respond to our central question

*How can we connect the physics
of microscopic particles to the
physics of the Universe?*

# chapter
# 28

This lightbulb filament is raised to a high temperature as it carries current. As a result, it glows visibly. The distribution of wavelengths of radiation emitted from the filament cannot be described by classical physics. Quantum physics is needed to describe it theoretically.

*(George Semple)*

# Quantum Physics

In the earlier chapters of this book, we focused on the physics of particles. We also combined particles into systems and rigid bodies. The particle model was a simplification model that allowed us to ignore the unnecessary details of an actual object when studying its behavior. In Chapter 13, we introduced the wave as another simplification model, and found that we could understand the motion of vibrating strings and the intricacies of sound by studying simple waves. In Chapters 24 to 27, we found that the wave model for light helped us understand many phenomena associated with optics.

This, we hope, has led you to have confidence in your abilities to analyze problems in the very different worlds of particles and waves. Your confidence may have been shaken somewhat by the discussion at the beginning of Chapter 25, in which we indicated that light has both wave-like and particle-like behaviors.

In this chapter, we return to this dual nature of light and study it in more detail. This study leads to two final analysis models: the quantum particle model and the model of the quantum particle under boundary conditions. Investigations into these two models should restore your confidence in your ability to understand particles and waves and show that they are not as unrelated as you might expect.

# 28.1 • BLACKBODY RADIATION AND PLANCK'S THEORY

As we discussed in Chapter 17, an object at any temperature emits energy referred to as **thermal radiation.** The characteristics of this radiation depend on the temperature and properties of the object. At room temperature, the wavelengths of the thermal radiation are primarily in the infrared region and hence are not observed by the eye. As the temperature of the object increases, the object eventually begins to glow red. At sufficiently high temperatures, the object appears to be white, as in the glow of the hot tungsten filament of a lightbulb. A careful study of thermal radiation shows that it consists of a continuous distribution of wavelengths from all portions of the spectrum.

From a classical viewpoint, thermal radiation originates from accelerated charged particles near the surface of the object. The thermally agitated charges can have a distribution of accelerations, which accounts for the continuous spectrum of radiation emitted by the object. By the end of the 19th century, it had become apparent that this classical explanation of thermal radiation was inadequate. The basic problem was in understanding the observed distribution of wavelengths in the radiation emitted by an ideal object called a black body. As mentioned in Chapter 24, a **black body** is an ideal system that absorbs all radiation incident on it. A good approximation of a black body is a small hole leading to the inside of a hollow object, as shown in Figure 28.1. The nature of the radiation emitted from the hole depends only on the temperature of the cavity walls.

The wavelength distribution of radiation from cavities was studied extensively in the late 19th century. Experimental data for the distribution of energy in blackbody radiation at three temperatures are shown in Figure 28.2. The distribution of radiated energy varies with wavelength and temperature. The following two regular features of the distribution were noted in these experiments:

**PITFALL PREVENTION 28.1**
**Expect to be confused**

The discussions that we will undertake in the next few chapters will involve many aspects of quantum physics, and it is likely that these discussions will be strange and confusing for you. That is because your whole life experience has taken place in the macroscopic world, where quantum effects are not evident. As you struggle through this material, keep in mind the words of the Nobel Prize–winning physicist Richard Feynman—"Nobody understands quantum mechanics."

The glow emanating from the spaces between these hot charcoal briquettes is, to a close approximation, blackbody radiation. The color of the light depends on the temperature of the briquettes. *(Corbis)*

**Figure 28.1**

A small hole that leads to a cavity is a good approximation of a black body. Light entering the small opening strikes the far wall, where some of it is absorbed but some is reflected at a random angle. The light continues to be reflected, and each time it is, part of it is absorbed by the cavity walls. After many reflections, essentially all of the incident energy is absorbed.

**Figure 28.2**

Intensity of blackbody radiation versus wavelength at three temperatures. With increasing temperature, note that the amount of radiation emitted (the area under a curve) increases (Stefan's law) and that the peak in the distribution shifts toward shorter wavelengths (Wien's law).

### Feature 1   *The total power of emitted radiation increases with temperature.*

We discussed this briefly in Chapter 17, where we introduced **Stefan's law**

- *Stefan's law*

$$\mathcal{P} = \sigma A e T^4 \tag{17.35}$$

For a black body, the emissivity is $e = 1$ exactly.

### Feature 2   *The peak of the wavelength distribution shifts to shorter wavelengths as the temperature increases.*

This shift was found to obey the following relationship, called **Wien's displacement law:**

- *Wien's displacement law*

$$\lambda_{\max} T = 2.898 \times 10^{-3} \text{ m} \cdot \text{K} \tag{28.1}$$

where $\lambda_{\max}$ is the wavelength at which the curve peaks and $T$ is the absolute temperature of the object emitting the radiation.

A successful model for blackbody radiation must predict the shape of the curve in Figure 28.2, the temperature dependence expressed in Stefan's law, and the shift of the peak with temperature described by Wien's displacement law. Early attempts to use classical ideas to explain the shapes of the curves in Figure 28.2 failed. Figure 28.3 shows an experimental plot of the blackbody radiation spectrum (red) together with the curve predicted by classical theory (blue). At long wavelengths, classical theory is in good agreement with the experimental data. At short wavelengths, however, major disagreement exists between classical theory and experiment. This disagreement is often called the **ultraviolet catastrophe.** (This name is somewhat of a misnomer, because the "catastrophe"—infinite energy—occurs as the wavelength approaches zero, not ultraviolet wavelengths.)

In 1900, Max Planck developed a structural model for blackbody radiation that leads to a theoretical equation for the wavelength distribution that is in complete agreement with experimental results at all wavelengths. In his model, Planck imagined that oscillators exist at the surface of the black body, related to the charges within the molecules. He made two bold and controversial assumptions concerning the nature of these oscillators:

- The energy of the oscillator is quantized—that is, it can have only certain *discrete* amounts of energy $E_n$ given by

$$E_n = nhf \tag{28.2}$$

where $n$ is a positive integer called a **quantum number,**\* $f$ is the frequency of oscillation of the oscillator, and $h$ is **Planck's constant,** first introduced in Chapter 11. Because the energy of each oscillator can have only discrete values given by Equation 28.2, we say the energy is **quantized.** Each discrete energy value corresponds to a different **quantum state,** represented by the quantum number $n$. When the oscillator is in the $n = 1$ quantum state, its energy is $hf$; when it is in the $n = 2$ quantum state, its energy is $2hf$; and so on.

- The oscillators emit or absorb energy in discrete units. They emit or absorb these energy units by making a transition from one quantum state to another, similar to the transitions discussed in the Bohr model in Chapter 11. The entire

**Max Planck** (1858–1947)

Planck introduced the concept of a "quantum of action" (Planck's constant, *h*) in an attempt to explain the spectral distribution of blackbody radiation, which laid the foundations for quantum theory. In 1918, he was awarded the Nobel Prize in physics for this discovery of the quantized nature of energy. *(Photo courtesy of AIP Niels Bohr Library, W. F. Meggers Collection)*

---

\* We first introduced the notion of a quantum number for microscopic systems in Section 11.5, in which we incorporated it into the Bohr model of the hydrogen atom. We put it in bold again here because it is an important notion for the remainder of the chapters in this book.

**Figure 28.3**

Comparison of the experimental results with the curve predicted by classical theory for the distribution of blackbody radiation.

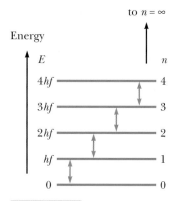

**Figure 28.4**

Allowed energy levels for an oscillator with a natural frequency $f$. Allowed transitions are indicated by vertical arrows.

energy difference between the initial and final states in the transition is emitted as a single quantum of radiation. If the transition is from one state to an adjacent state—say, from the $n = 3$ state to the $n = 2$ state—Equation 28.2 shows that the amount of energy radiated by the oscillator is

$$E = hf \qquad \text{[28.3]}$$

An oscillator radiates or absorbs energy only when it changes quantum states. If it remains in one quantum state, no energy is absorbed or emitted. Figure 28.4 shows the quantized energy levels and allowed transitions proposed by Planck.

These assumptions may not sound bold to you because we have seen them in the Bohr model of the hydrogen atom in Chapter 11. It is important to keep in mind, however, that the Bohr model was not introduced until 1913, whereas Planck made his assumptions in 1900. The key point in Planck's theory is the radical assumption of quantized energy states. This development marked the birth of the quantum theory. Using these assumptions, Planck was able to generate a theoretical expression for the wavelength distribution that agreed remarkably well with the experimental curves in Figure 28.2. When Planck presented his theory, most scientists (including Planck!) did not consider the quantum concept realistic. It was felt to be a mathematical trick that happened to predict the correct results. Hence, Planck and others continued to search for what they felt to be a more rational explanation of blackbody radiation. Subsequent developments, however, showed that a theory based on the quantum concept (rather than on classical concepts) was required to explain a number of other phenomena at the atomic level.

### Quick **Quiz** 28.1

Is a black body black?

### Quick **Quiz** 28.2

All objects radiate energy. Why, then, are we not able to see all objects in a dark room?

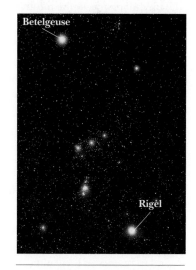

The bright stars in this photograph make up the constellation Orion. Because Rigel is hotter than Betelgeuse, it appears blue, and Betelgeuse appears red. (*John Chumack/Photo Researchers, Inc.*)

## THINKING PHYSICS 28.1

You are observing a yellow candle flame, and your laboratory partner claims that the light from the flame is atomic in origin. You disagree, claiming that the candle flame is hot, so the radiation must be thermal in origin. Before this disagreement leads to fisticuffs, how could you determine who is correct?

**Reasoning** A simple determination could be made by observing the light from the candle flame through a diffraction grating spectrometer, which was discussed in Section 27.8. If the spectrum of the light is continuous, it is thermal in origin. If the spectrum shows discrete lines, it is atomic in origin. The results of the experiment show that the light is indeed thermal in origin and originates in the hot particles of soot in the candle flame.

---

### Example 28.1 Thermal Radiation from the Human Body

The temperature of your skin is approximately 35°C.
(a) What is the peak wavelength of the radiation it emits?

**Solution** From Wien's displacement law (Eq. 28.1), we have

$$\lambda_{max} T = 2.898 \times 10^{-3} \text{ m} \cdot \text{K}$$

Solving for $\lambda_{max}$, noting that 35°C corresponds to an absolute temperature of 308 K, we have

$$\lambda_{max} = \frac{2.898 \times 10^{-3} \text{ m} \cdot \text{K}}{308 \text{ K}} = \boxed{9.41 \ \mu m}$$

This radiation is in the infrared region of the spectrum.
(b) What total power is emitted by your skin, assuming it emits like a black body?

**Solution** We need to make an estimate of the surface area of your skin. If we model your body as a rectangular box of height 2 m, width 0.3 m, and depth 0.2 m, the total surface area is

$$A = 2(2 \text{ m})(0.3 \text{ m}) + 2(2 \text{ m})(0.2 \text{ m}) + 2(0.2 \text{ m})(0.3 \text{ m})$$
$$\approx 2 \text{ m}^2$$

Thus, from Stefan's law, we have

$$\mathcal{P} = \sigma A e T^4 \approx (5.7 \times 10^{-8} \text{ W/m}^2 \cdot \text{K}^4)(2 \text{ m}^2)(1)(308 \text{ K})^4$$
$$\sim \boxed{10^3 \text{ W}}$$

(c) Based on your answer to part (b), why don't you glow as brightly as several lightbulbs?

**Solution** The answer to part (b) indicates that your skin is radiating approximately the same amount of energy as that which enters ten 100-W lightbulbs by electrical transmission. You are not visibly glowing, however, because most of this radiation is in the infrared range, as we found in part (a), and our eyes are not sensitive to infrared radiation.

---

### Example 28.2 The Quantized Oscillator

A 2.0-kg block is attached to a massless spring of force constant $k = 25$ N/m. The spring is stretched 0.40 m from its equilibrium position and released. (a) Find the total energy and frequency of oscillation according to classical calculations.

**Solution** From the simple harmonic motion model in Chapter 12, we can use Equation 12.21 for the total energy of a simple harmonic oscillator with amplitude $A$ to find

$$E = \tfrac{1}{2}kA^2 = \tfrac{1}{2}(25 \text{ N/m})(0.40 \text{ m})^2 = \boxed{2.0 \text{ J}}$$

The frequency of oscillation is, from Equation 12.14,

$$f = \frac{1}{2\pi}\sqrt{\frac{k}{m}} = \frac{1}{2\pi}\sqrt{\frac{25 \text{ N/m}}{2.0 \text{ kg}}} = \boxed{0.56 \text{ Hz}}$$

(b) Assume that the energy is quantized, and find the quantum number $n$ for the system.

**Solution** If the energy is quantized, we have $E_n = nhf$, and from the result of (a) we have

$$E_n = nhf = n(6.63 \times 10^{-34} \text{ J} \cdot \text{s})(0.56 \text{ Hz}) = 2.0 \text{ J}$$

Therefore,

$$n = \frac{2.0 \text{ J}}{(6.63 \times 10^{-34} \text{ J} \cdot \text{s})(0.56 \text{ Hz})} = \boxed{5.4 \times 10^{33}}$$

(c) How much energy is carried away when the oscillator makes a transition to the next lowest quantum state?

**Solution** The energy difference between adjacent quantum states is $\Delta E = hf$. Thus, the energy carried away is

$$E = hf = (6.63 \times 10^{-34} \text{ J} \cdot \text{s})(0.56 \text{ Hz})$$
$$= \boxed{3.7 \times 10^{-34} \text{ J}}$$

The energy carried away due to a transition between adjacent states is such a small fraction of the total energy of the oscillator (about one part in ten million billion billion billion, or $1:10^{34}$!) that we could never expect to see such a small change in the system. Thus, even though the decrease in energy of a macroscopic block–spring system is indeed quantized and proceeds by small quantum jumps, our senses perceive the decrease as continuous. **Quantum effects become important and measurable only on the submicroscopic level of atoms and molecules.** Furthermore, quantum results must blend smoothly with classical results when the quantum number becomes large. This is known as the **correspondence principle.**

## 28.2 • THE PHOTOELECTRIC EFFECT

Blackbody radiation was historically the first phenomenon to be explained with a quantum model. In the latter part of the 19th century, at the same time as data were being taken on thermal radiation, experiments showed that light incident on certain metallic surfaces causes electrons to be emitted from the surfaces. As mentioned in Section 25.1, this phenomenon, first discovered by Hertz, is known as the **photoelectric effect.** The emitted electrons are called **photoelectrons.***

Figure 28.5 is a schematic diagram of a photoelectric effect apparatus. An evacuated glass tube contains a metal plate E connected to the negative terminal of a battery. Another metal plate C is maintained at a positive potential by the battery. When the tube is kept in the dark, the ammeter reads zero, indicating that no current is flowing in the circuit. When light of the appropriate wavelength shines on plate E, however, a current is detected by the ammeter, indicating a flow of charges across the gap between E and C. This current arises from electrons emitted from the negative plate E (the emitter) and collected at the positive plate C (the collector).

Figure 28.6 is a graphical representation of the results of a photoelectric experiment—a plot of the photoelectric current versus the potential difference $\Delta V$ between E and C for two light intensities. For large positive values of $\Delta V$, the current reaches a maximum value. In addition, the current increases as the incident light intensity increases, as you might expect. Finally, when $\Delta V$ is negative—that is, when the battery polarity is reversed to make E positive and C negative—the current drops because many of the photoelectrons emitted from E are repelled by the negative collecting plate C. Only those electrons ejected from the metal with a kinetic energy greater than $e|\Delta V|$ will reach C, where $e$ is the magnitude of the charge on the electron. When the magnitude of $\Delta V$ is equal to $\Delta V_s$, the **stopping potential,** no electrons reach C and the current is zero.

Let us consider the combination of the electric field between the plates and an electron ejected from plate E with the maximum kinetic energy to be an iso-

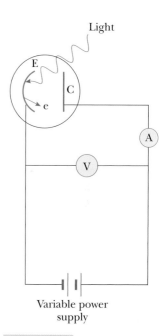

Light

**Figure 28.5**

A circuit diagram for observing the photoelectric effect. When light strikes the plate E (the emitter), photoelectrons are ejected from the plate. Electrons collected at C (the collector) constitute a current in the circuit.

Variable power supply

---

* Photoelectrons are not different from other electrons. They are given this name solely because of their ejection from the metal by photons in the photoelectric effect.

Current

High intensity

Low intensity

$-\Delta V_s$ | Applied potential difference

**Figure 28.6**

Photoelectric current versus applied potential difference for two light intensities. The current increases with intensity but reaches a saturation level for large values of $\Delta V$. At voltages equal to or less than $-\Delta V_s$, the stopping potential, the current is zero.

lated system. Suppose this electron stops just as it reaches plate C. Applying the isolated system model, the total energy of the system must be conserved:

$$E_1 = E_2 \longrightarrow K_1 + U_1 = K_2 + U_2$$

where condition 1 refers to the instant that the electron leaves the metal with the maximum possible kinetic energy $K_{max}$ and condition 2 is when the electron stops just before touching plate C. If we define the electric potential energy of the system in condition 1 to be zero, we have

$$K_{max} + 0 = 0 + (-e)(-\Delta V_s)$$

$$K_{max} = e\,\Delta V_s \qquad\qquad [28.4]$$

This equation allows us to experimentally measure $K_{max}$ by measuring the voltage at which the current drops to zero.

The following are several features of the photoelectric effect in which the predictions made by a classical approach are compared, using the wave model for light, with the experimental results. Notice the strong contrast between the predictions and the results:

### Feature 1 *Dependence of photoelectron kinetic energy on light intensity*

**Classical Prediction:** Electrons should absorb energy continuously from the electromagnetic waves. A more intense light should transfer energy into the metal faster, and the electrons should be ejected with more kinetic energy.

**Experimental Result:** The maximum kinetic energy of the photoelectrons is *independent* of light intensity. This is shown in Figure 28.6 by the fact that both curves fall to zero at the *same* negative voltage.

### Feature 2 *Time between incidence of light and ejection of photoelectrons*

**Classical Prediction:** For very weak light, a measurable time interval should pass between the incidence of the light and the ejection of an electron. This time is required for the electron to absorb the incident radiation before it acquires enough energy to escape from the metal.

**Experimental Result:** Electrons are emitted from the surface almost *instantaneously* (less than $10^{-9}$ s after the surface is illuminated), even at very low light intensities.

### Feature 3 *Dependence of ejection of electrons on light frequency*

**Classical Prediction:** Electrons should be ejected at any frequency of the incident light, as long as the intensity is high enough, because energy is being transferred to the metal regardless of the frequency.

**Experimental Result:** No electrons are emitted if the incident light frequency falls below some **cutoff frequency** $f_c$, which is characteristic of the material being illuminated. No electrons are ejected below this cutoff frequency *regardless* of how intense the light is.

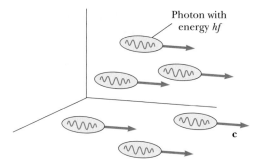

Photon with energy *hf*

**Figure 28.7**

A representation of light quanta. Each photon has a discrete energy *hf*.

## Feature 4    *Dependence of photoelectron kinetic energy on light frequency*

**Classical Prediction:**   *No* relationship should exist between the frequency of the light and the electron kinetic energy. The kinetic energy should be related to the intensity of the light.

**Experimental Result:**   The maximum kinetic energy of the photoelectrons increases with increasing light frequency.

   Notice that *all four* predictions of the classical model are incorrect. A successful explanation of the photoelectric effect was given by Einstein in 1905, the same year he published his special theory of relativity. As part of a general paper on electromagnetic radiation, for which he received the Nobel Prize in physics in 1921, Einstein extended Planck's concept of quantization to electromagnetic waves. He assumed that light (or any other electromagnetic wave) of frequency *f* can be considered to be a stream of quanta, regardless of the source of the radiation. Today we call these quanta **photons.** Each photon has an energy *E* given by Equation 28.3, $E = hf$. A suggestive image of several photons, not to be taken too literally, is shown in Figure 28.7. Each photon moves with a velocity **c** directed to the right in Figure 28.7. The magnitude of **c** is $c = 3.00 \times 10^8$ m/s.

### Quick Quiz 28.3

While standing outdoors on a dark night, you are subjected to the following four types of electromagnetic radiation: yellow light from a sodium street lamp, radio waves from a nearby AM radio station, radio waves from a nearby FM radio station, and microwaves from a nearby antenna of a communications system. Rank these types of waves in terms of increasing photon energy, lowest first.

   In Einstein's model, a photon of the incident light gives *all* its energy *hf* to a *single* electron in the metal. Thus, the absorption of energy by the electrons is not a continuous process as envisioned in the wave model, but rather a discontinuous process in which energy is delivered to the electrons in bundles. The energy transfer is accomplished via a one-photon/one-electron event.
   Electrons ejected from the surface of the metal possess the maximum kinetic energy $K_{max}$. According to Einstein, the maximum kinetic energy for these liberated electrons is

$$K_{max} = hf - \phi \qquad \text{[28.5]} \qquad \bullet \quad \textit{Photoelectric effect equation}$$

**TABLE 28.1**

**Work Functions of Selected Metals**

| Metal | $\phi$ (eV) |
|-------|-------------|
| Na | 2.28 |
| Al | 4.08 |
| Cu | 4.70 |
| Zn | 4.31 |
| Ag | 4.73 |
| Pt | 6.35 |
| Pb | 4.14 |
| Fe | 4.50 |

where $\phi$ is called the **work function** of the metal. **The work function represents the minimum energy with which an electron is bound in the metal** and is on the order of a few electron volts. Table 28.1 lists selected values.

Equation 28.5 is a statement of the continuity equation for energy from Chapter 6:

$$\Delta E_{\text{system}} = \Sigma H \qquad [6.20]$$

We imagine the system to consist of an electron that is to be ejected and the remainder of the metal, and then apply the nonisolated system model for energy. Energy is transferred into the system by electromagnetic radiation—the photon. The system has two types of energy: the potential energy of the metal–electron system and the kinetic energy of the electron. Thus, we can write the continuity equation as

$$\Delta K + \Delta U = H_{\text{ER}} \qquad [28.6]$$

The energy transfer is that of the photon, $H_{\text{ER}} = hf$. During the process, the kinetic energy of the electron increases from zero to its final value, which we assume to be the maximum possible value $K_{\text{max}}$. The potential energy of the system increases because the electron is pulled away from the metal to which it is attracted. We define the potential energy of the system when the electron is outside the metal as zero. The potential energy of the system when the electron is in the metal is $U = -\phi$, where $\phi$ is the work function. Thus, the increase in potential energy when the electron is removed from the metal is the work function $\phi$. Substituting these energies into Equation 28.6, we have

$$K_{\text{max}} + \phi = hf$$

which is the same as Equation 28.5. If the electron makes collisions with other electrons or metal ions as it is being ejected, some of the incoming energy is transferred to the metal and the electron is ejected with less kinetic energy than $K_{\text{max}}$.

With the photon model of light, one can explain the observed features of the photoelectric effect that cannot be understood using classical concepts:

### Feature 1   *Dependence of photoelectron kinetic energy on light intensity*

That $K_{\text{max}}$ is independent of the light intensity can be understood with the following argument. The maximum kinetic energy of any one electron, which equals $hf - \phi$, depends only on the light frequency and the work function, not on the light intensity. If the light intensity is doubled, the number of photons arriving per unit time is doubled, which doubles the rate at which photoelectrons are emitted. The maximum kinetic energy of all emitted electrons, however, is unchanged.

### Feature 2   *Time between incidence of light and ejection of photoelectrons*

That the electrons are emitted almost instantaneously is consistent with the photon model of light, in which the incident energy appears in small packets and the interaction between photons and electrons is one to one. Thus, for very weak incident light, very few photons may arrive per unit time, but each one has sufficient energy to eject an electron immediately.

## Feature 3   *Dependence of ejection of electrons on light frequency*

That the effect is not observed below a certain cutoff frequency follows from the fact that the photon must have energy greater than the work function $\phi$ in order to eject an electron. If the energy of an incoming photon does not satisfy this requirement, an electron cannot be ejected from the surface, regardless of light intensity.

## Feature 4   *Dependence of photoelectron kinetic energy on light frequency*

That $K_{max}$ increases with increasing frequency is easily understood with Equation 28.5.

Einstein's result (Eq. 28.5) predicts a linear relationship between the maximum electron kinetic energy $K_{max}$ and the light frequency $f$. Experimental observation of such a linear relationship would be a final confirmation of Einstein's theory. Indeed, such a linear relationship is observed, as sketched in Figure 28.8. The slope of the curves for all metals is $h$. The absolute value of the intercept on the vertical axis is the work function $\phi$, which varies from one metal to another. The intercept on the horizontal axis is the cutoff frequency, which is related to the work function through the relation $f_c = \phi/h$. This corresponds to a **cutoff wavelength** of

$$\lambda_c = \frac{c}{f_c} = \frac{c}{\phi/h} = \frac{hc}{\phi} \qquad [28.7] \qquad \bullet \quad \textit{Cutoff wavelength}$$

where $c$ is the speed of light. Light with wavelength *greater* than $\lambda_c$ incident on a material with a work function of $\phi$ does not result in the emission of photoelectrons.

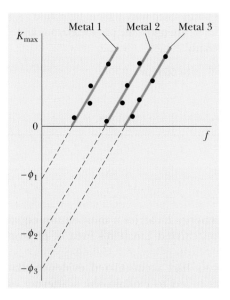

**Figure 28.8**

A graphical representation of results for $K_{max}$ versus frequency of incident light for photoelectrons in a typical photoelectric effect experiment. Photons with frequency less than the critical frequency for a given metal do not have sufficient energy to eject an electron from the metal.

The explanation of the photoelectric effect with a quantum model, combined with Planck's quantum model for blackbody radiation, laid a strong foundation for further investigation into quantum physics. In the next section, we present a third experimental result that provides further strong evidence of the quantum nature of light.

### Quick Quiz 28.4

A student claims that he is going to eject electrons from a piece of metal by placing a radio transmitter antenna adjacent to the metal and sending a strong AM radio signal into the antenna. Will this work?

### Quick Quiz 28.5

Suppose classical physicists had come up with the idea of predicting the appearance of a plot of $K_{max}$ versus $f$, as in Figure 28.8. What would their expected plot look like, based on the wave model for light?

---

### Example 28.3 The Photoelectric Effect for Sodium

A sodium surface is illuminated with light of wavelength 300 nm. The work function for sodium metal is 2.46 eV. Find (a) the maximum kinetic energy of the ejected photoelectrons and (b) the cutoff wavelength for sodium.

**Solution** (a) The energy of photons in the illuminating light beam is

$$E = hf = \frac{hc}{\lambda} = \frac{(6.626 \times 10^{-34} \, \text{J} \cdot \text{s})(3.00 \times 10^8 \, \text{m/s})}{300 \times 10^{-9} \, \text{m}}$$

$$= 6.63 \times 10^{-19} \, \text{J} = 4.14 \, \text{eV}$$

Using Equation 28.5 gives

$$K_{max} = hf - \phi = 4.14 \, \text{eV} - 2.46 \, \text{eV} = \boxed{1.68 \, \text{eV}}$$

(b) The cutoff wavelength can be calculated from Equation 28.7 after we convert $\phi$ from electron volts to joules:

$$\phi = 2.46 \, \text{eV} \left( \frac{1.60 \times 10^{-19} \, \text{J}}{1 \, \text{eV}} \right) = 3.94 \times 10^{-19} \, \text{J}$$

$$\lambda_c = \frac{hc}{\phi} = \frac{(6.626 \times 10^{-34} \, \text{J} \cdot \text{s})(3.00 \times 10^8 \, \text{m/s})}{3.94 \times 10^{-19} \, \text{J}}$$

$$= 5.05 \times 10^{-7} \, \text{m} = \boxed{505 \, \text{nm}}$$

This wavelength is in the green region of the visible spectrum.

**EXERCISE** Calculate the maximum speed of the photoelectrons under the conditions described in this example.

*Answer* $7.68 \times 10^5 \, \text{m/s}$

**EXERCISE** The work function for potassium is 2.24 eV. If potassium metal is illuminated with 480-nm light, find (a) the maximum kinetic energy of the photoelectrons and (b) the cutoff wavelength.

*Answer* (a) 0.350 eV (b) 554 nm

---

## 28.3 • THE COMPTON EFFECT

In 1919, Einstein proposed that a photon of energy $E$ carries a momentum equal to $E/c = hf/c$. In 1923, Arthur Holly Compton carried Einstein's idea of photon momentum further.

Prior to 1922, Compton and his coworkers had accumulated evidence that showed that the classical wave theory of light failed to explain the scattering of x-rays from electrons. According to classical theory, incident electromagnetic waves of frequency $f_0$ should have two effects: (1) the electrons should accelerate in the direction of propagation of the x-ray by radiation pressure (see Section 24.6), and (2) the oscillating electric field should set the electrons into oscillation at the apparent frequency of the radiation as detected by the moving electron. Both of these effects are included in Figure 28.9a, which shows the classical expectation. The apparent frequency detected by the electron differs from $f_0$ due to the Doppler effect (see Section 13.8 and Thinking Physics 24.1) because the electron absorbs as a moving particle. The electron then re-radiates as a moving particle, exhibiting another Doppler shift in the frequency of emitted radiation.

Because different electrons move at different speeds, depending on the amount of energy absorbed from the electromagnetic waves, the scattered wave frequency at a given angle should show a distribution of Doppler-shifted values. Contrary to this prediction, Compton's experiment showed that, at a given angle, only *one* frequency of radiation was observed that was different from that of the incident radiation. Compton and his coworkers realized that the scattering of x-ray

**Arthur Holly Compton**
(1892–1962)

Compton measured and explained the effect named for him at the University of Chicago in 1923 and shared the 1927 Nobel Prize. He went on to demonstrate that cosmic rays are charged particles and to direct research on producing plutonium for nuclear weapons. *(Courtesy of AIP Niels Bohr Library)*

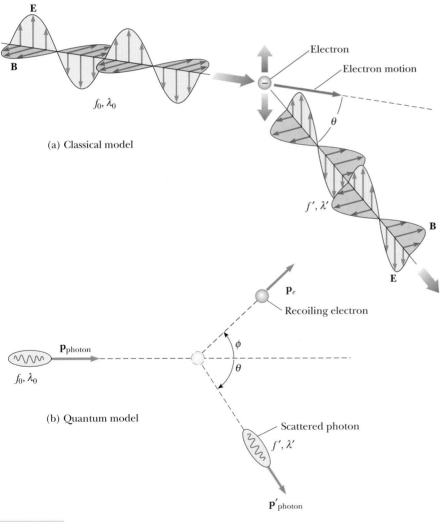

(a) Classical model

(b) Quantum model

**Figure 28.9**

X-ray scattering from an electron: (a) the classical model and (b) the quantum model.

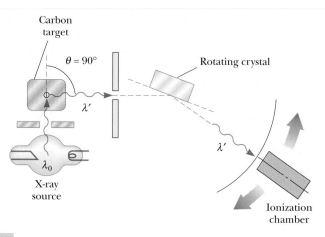

Carbon target

$\theta = 90°$

Rotating crystal

$\lambda'$

X-ray source

$\lambda_0$

$\lambda'$

Ionization chamber

**Figure 28.10**

Schematic diagram of Compton's apparatus. Photons are scattered through 90° from a carbon target. The wavelength is measured with a rotating crystal spectrometer using Bragg's law (Section 27.9).

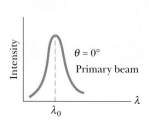

$\theta = 0°$
Primary beam

$\lambda_0$

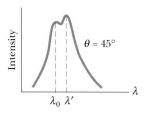

$\theta = 45°$

$\lambda_0 \ \lambda'$

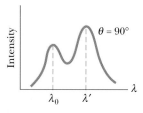

$\theta = 90°$

$\lambda_0 \quad \lambda'$

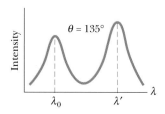

$\theta = 135°$

$\lambda_0 \qquad \lambda'$

**Figure 28.11**

Scattered intensity versus wavelength of Compton scattering at $\theta = 0°$, 45°, 90°, and 135°.

photons from electrons could be explained by treating photons as point-like particles with energy $hf$ and momentum $hf/c$ and by assuming that the energy and momentum of the isolated system of the photon and the electron are conserved in a collision. By doing this, Compton was adopting a particle model for something that was well-known as a wave, as had Einstein in his explanation of the photoelectric effect. Figure 28.9b shows the quantum picture of the exchange of momentum and energy between an individual x-ray photon and an electron. A major distinction between the classical and quantum models is shown in Figure 28.9. In the classical model, the electron is pushed along the direction of propagation of the incident x-ray by radiation pressure. In the quantum model, the electron is scattered through an angle $\phi$ with respect to this direction, as if this were a billiard-ball type collision.

Figure 28.10 is a schematic diagram of the apparatus used by Compton. In his original experiment, Compton measured how scattered x-ray intensity depends on wavelength at various scattering angles. The incident beam consisted of monochromatic x-rays of wavelength $\lambda_0 = 0.071$ nm. The experimental plots of intensity versus wavelength obtained by Compton for four scattering angles are shown in Figure 28.11. They show two peaks: one at $\lambda_0$ and the other at a longer wavelength $\lambda'$. The peak at $\lambda_0$ is caused by x-rays scattered from electrons that are tightly bound to the target atoms, and the shifted peak at $\lambda'$ is caused by x-rays scattered from free electrons in the target. In his analysis, Compton predicted that the shifted peak should depend on scattering angle $\theta$ as

• *Compton shift equation*

$$\lambda' - \lambda_0 = \frac{h}{m_e c}(1 - \cos\theta) \qquad \textbf{[28.8]}$$

In this expression, known as the **Compton shift equation,** $m_e$ is the mass of the electron; $h/m_e c$ is called the **Compton wavelength** $\lambda_C$ of the electron and has a currently accepted value of

• *Compton wavelength*

$$\lambda_C = \frac{h}{m_e c} = 0.00243 \text{ nm} \qquad \textbf{[28.9]}$$

Compton's measurements were in excellent agreement with the predictions of Equation 28.8. These were the first experimental results to convince most physicists of the fundamental validity of the quantum theory!

The Compton effect should be kept in mind by x-ray technicians working in hospitals and radiology laboratories. X-rays directed into the patient's body are Compton scattered by electrons in the body in all directions. Equation 28.8 shows that the scattered wavelength is still well within the x-ray region so that these scattered x-rays can damage human tissue. In general, technicians operate the x-ray machine from behind an absorbing wall to avoid exposure to the scattered x-rays. Furthermore, when dental x-rays are taken, a lead apron is placed over the patient to reduce the absorption of scattered x-rays by other parts of the patient's body.

**APPLICATION**

The Compton effect and x-ray technicians

## THINKING PHYSICS 28.2

The Compton effect involves a change in wavelength as photons are scattered through different angles. Suppose we illuminate a piece of material with a beam of light and then view the material from different angles relative to the beam of light. Will we see a *color change* corresponding to the change in wavelength of the scattered light?

**Reasoning**   Visible light scattered by the material undergoes a change in wavelength, but the change is far too small to detect as a color change. The largest possible change in wavelength, at 180° scattering, is twice the Compton wavelength, about 0.005 nm. This represents a change of less than 0.001% of the wavelength of red light. The Compton effect is only detectable for wavelengths that are very short to begin with, so that the Compton wavelength is an appreciable fraction of the incident wavelength. As a result, the usual radiation for observing the Compton effect is in the x-ray range of the electromagnetic spectrum.

## Example 28.4   Compton Scattering at 45°

X-rays of wavelength $\lambda_0 = 0.200\,000$ nm are scattered from a block of material. The scattered x-rays are observed at an angle of 45.0° to the incident beam. Calculate the wavelength of the x-rays scattered at this angle.

**Solution**   The shift in wavelength of the scattered x-rays is given by Equation 28.8:

$$\Delta \lambda = \frac{h}{m_e c} (1 - \cos \theta)$$

$$= \frac{6.626 \times 10^{-34}\,\text{J}\cdot\text{s}}{(9.11 \times 10^{-31}\,\text{kg})(3.00 \times 10^8\,\text{m/s})} (1 - \cos 45°)$$

$$= 7.10 \times 10^{-13}\,\text{m} = 0.000\,710\,\text{nm}$$

Hence, the wavelength of the scattered x-rays at this angle is

$$\lambda' = \Delta \lambda + \lambda_0 = \boxed{0.200\,710\,\text{nm}}$$

**EXERCISE**   Find the fraction of energy lost by the photon in this collision.

*Answer*   $|\Delta E|/E = 0.003\,54$

**EXERCISE**   X-rays of wavelength 0.200 00 nm are scattered from a carbon target. If the scattered radiation is detected at 60.0° to the incident beam, find (a) the Compton shift $\Delta \lambda$ and (b) the kinetic energy imparted to the recoiling electron.

*Answer*   (a) 0.001 21 nm   (b) 37.4 eV

## 28.4 • PHOTONS AND ELECTROMAGNETIC WAVES

The agreement between experimental measurements and theoretical predictions based on quantum models for phenomena such as the photoelectric effect and the Compton effect offer clear evidence that when light and matter interact, the light behaves as if it were composed of particles with energy $hf$ and momentum $h/\lambda$. An obvious question at this point is "How can light be considered a photon when it exhibits wave-like properties?" We describe light in terms of photons having energy and momentum—this is the particle model. We recognize, however, that light and other electromagnetic waves exhibit interference and diffraction effects, which are consistent only with the wave model.

• *The dual nature of light*

Which model is correct? Is light a wave or a particle? The answer depends on the phenomenon being observed. Some experiments can be explained better, or solely, with the photon model, whereas others are best described, or can only be described, with a wave model. The end result is that **we must accept both models and admit that the true nature of light is not describable in terms of any single classical picture.** Hence, **light has a dual nature: It exhibits both wave and particle characteristics.** You should recognize, however, that the same beam of light that can eject photoelectrons from a metal can also be diffracted by a grating. In other words, the **particle model and the wave model of light complement each other.**

The success of the particle model of light in explaining the photoelectric effect and the Compton effect raises many other questions. Because a photon is a particle, what is the meaning of its "frequency" and "wavelength," and which of these determines its energy and momentum? Is light in some sense simultaneously a wave and a particle? Although photons have no rest energy, can some simple expression describe the effective mass of a "moving" photon? If a "moving" photon has mass, do photons experience gravitational attraction? What is the spatial extent of a photon, and how does an electron absorb or scatter one photon? Some of these questions can be answered, but others demand a view of atomic processes that is too pictorial and literal. Furthermore, many of these questions stem from classical analogies such as colliding billiard balls and water waves breaking on a shore. Quantum mechanics gives light a more fluid and flexible nature by treating the particle model and wave model of light as both necessary and complementary. Neither model can be used exclusively to describe all properties of light. A complete understanding of the observed behavior of light can be attained only if the two models are combined in a complementary manner. Before discussing this combination in more detail, we now turn our attention from electromagnetic waves to the behavior of entities that we have called particles.

## 28.5 • THE WAVE PROPERTIES OF PARTICLES

Students introduced to the dual nature of light often find the concept very difficult to accept. Let us now reconsider our study of matter. We feel quite comfortable in adopting a particle model for matter, because we have studied such concepts as conservation of energy and momentum for particles as well as extended objects. It might therefore be even more difficult to accept that *matter* also has a dual nature!

In 1923, in his doctoral dissertation, Louis Victor de Broglie postulated that, **because photons have wave and particle characteristics, perhaps all forms of matter have wave as well as particle properties.** This was a highly revolu-

**Louis de Broglie** (1892–1987)

A French physicist, de Broglie was awarded the Nobel Prize for physics in 1929 for his prediction of the wave nature of electrons. *(AIP Niels Bohr Library)* tionary idea with no experimental confirmation at that time. According to de Broglie, an electron in motion exhibits both wave and particle characteristics. De Broglie explained the source of this assertion in his 1929 Nobel Prize acceptance speech:

> On the one hand the quantum theory of light cannot be considered satisfactory since it defines the energy of a light corpuscle by the equation $E = hf$ containing the frequency $f$. Now a purely corpuscular theory contains nothing that enables us to define a frequency; for this reason alone, therefore, we are compelled, in the case of light, to introduce the idea of a corpuscle and that of periodicity simultaneously. On the other hand, determination of the stable motion of electrons in the atom introduces integers, and up to this point the only phenomena involving integers in physics were those of interference and of normal modes of vibration. This fact suggested to me the idea that electrons too could not be considered simply as corpuscles, but that periodicity must be assigned to them also.

In Chapter 9 we found that the relationship between energy and momentum for a photon is $p = E/c$. We also know from Equation 28.3 that the energy of a photon is $E = hf = hc/\lambda$. Thus, the momentum of a photon can be expressed as

$$p = \frac{E}{c} = \frac{hf}{c} = \frac{hc}{c\lambda} = \frac{h}{\lambda}$$

• *Momentum of a photon*

From this equation we see that the photon wavelength can be specified by its momentum: $\lambda = h/p$. De Broglie suggested that material particles of momentum $p$ should also have wave properties and a corresponding wavelength. Because the magnitude of the momentum of a nonrelativistic particle of mass $m$ and speed $v$ is $p = mv$, the **de Broglie wavelength** of that particle is*

$$\lambda = \frac{h}{p} = \frac{h}{mv}$$  **[28.10]**

• *De Broglie wavelength of a particle*

Furthermore, in analogy with photons, de Broglie postulated that particles obey the Einstein relation $E = hf$, so that the frequency of a particle is

$$f = \frac{E}{h}$$  **[28.11]**

• *Frequency of a particle*

The dual nature of matter is apparent in these last two equations because each contains both particle concepts ($p$ and $E$) and wave concepts ($\lambda$ and $f$). The fact that these relationships are established experimentally for photons makes the de Broglie hypothesis that much easier to accept.

## The Davisson–Germer Experiment

De Broglie's proposal that any kind of particle exhibits both wave and particle properties was first regarded as pure speculation. If particles such as electrons had wave-like properties, then under the correct conditions they should exhibit diffraction effects. In 1927, three years after de Broglie published his work, C. J. Davisson and L. H. Germer of the United States succeeded in measuring the wavelength of

---

* The de Broglie wavelength for a particle moving at any speed $v$ is $\lambda = h/\gamma mv$, where $\gamma = (1 - v^2/c^2)^{-1/2}$.

electrons. Their important discovery provided the first experimental confirmation of the wave nature of particles proposed by de Broglie.

Interestingly, the intent of the initial Davisson–Germer experiment was not to confirm the de Broglie hypothesis. In fact, their discovery was made by accident (as is often the case). The experiment involved the scattering of low-energy electrons ($\approx 54$ eV) projected toward a nickel target in a vacuum. During one experiment, the nickel surface was badly oxidized because of an accidental break in the vacuum system. After the nickel target was heated in a flowing stream of hydrogen to remove the oxide coating, electrons scattered by it exhibited intensity maxima and minima at specific angles. The experimenters finally realized that the nickel had formed large crystal regions on heating and that the regularly spaced planes of atoms in the crystalline regions served as a diffraction grating (Section 27.8) for electrons.

Shortly thereafter, Davisson and Germer performed more extensive diffraction measurements on electrons scattered from single-crystal targets. Their results showed conclusively the wave nature of electrons and confirmed the de Broglie relation $p = h/\lambda$. A year later in 1928, G. P. Thomson of Scotland observed electron diffraction patterns by passing electrons through very thin gold foils. Diffraction patterns have since been observed for helium atoms, hydrogen atoms, and neutrons. Hence, the universal nature of the wave nature of particles has been established in a variety of ways.

### Quick Quiz 28.6

A nonrelativistic electron and a nonrelativistic proton are moving and have the same de Broglie wavelength. Which of the following are also the same for the two particles? (a) speed; (b) kinetic energy; (c) momentum; (d) frequency.

### Quick Quiz 28.7

We have seen two wavelengths assigned to the electron: the Compton wavelength and the de Broglie wavelength. Which is an actual *physical* wavelength associated with the electron? (a) the Compton wavelength; (b) the de Broglie wavelength; (c) both wavelengths, (d) neither wavelength.

### THINKING PHYSICS 28.3

**WEB**

For information on the electron microscope and several images taken with such a microscope, visit **www.lifelong.com/programs/ k12/science/sem/default.html**

Electron microscopes (Fig. 28.12) take advantage of the wave nature of particles. Electrons are accelerated to high speeds, giving them a short de Broglie wavelength. Imagine an electron microscope using electrons with a de Broglie wavelength of 0.2 nm. Why don't we design a microscope using 0.2-nm *photons* to do the same thing?

**Reasoning** Because electrons are charged particles, they interact electrically with the sample in the microscope and scatter according to the shape and density of various portions of the sample, providing a means of viewing the sample. Photons of wavelength 0.2 nm are uncharged and in the x-ray region of the spectrum. They tend to simply pass through the sample without interacting.

**Figure 28.12**

(Thinking Physics 28.3) (a) Diagram of a transmission electron microscope for viewing a thinly sectioned sample. The "lenses" that control the electron beam are magnetic deflection coils. (b) An electron microscope in use. *(b, W. Ormerod/Visuals Unlimited)*

### Example 28.5   The Wavelength of an Electron

Calculate the de Broglie wavelength for an electron ($m_e = 9.11 \times 10^{-31}$ kg) moving with a speed of $1.00 \times 10^7$ m/s.

**Solution**   Equation 28.10 gives

$$\lambda = \frac{h}{m_e v} = \frac{6.626 \times 10^{-34} \,\text{J} \cdot \text{s}}{(9.11 \times 10^{-31} \,\text{kg})(1.00 \times 10^7 \,\text{m/s})}$$

$$= 7.28 \times 10^{-11} \,\text{m}$$

This wavelength corresponds to that of typical x-rays.

**EXERCISE**   Find the de Broglie wavelength of a proton moving with a speed of $1.00 \times 10^7$ m/s.

*Answer*   $3.97 \times 10^{-14}$ m.

## Example 28.6   The Wavelength of a Rock

A rock of mass 50.0 g is thrown with a speed of 40.0 m/s. What is its de Broglie wavelength?

**Solution**   From Equation 28.10 we have

$$\lambda = \frac{h}{mv} = \frac{6.63 \times 10^{-34}\,\text{J}\cdot\text{s}}{(50 \times 10^{-3}\,\text{kg})(40\,\text{m/s})}$$

$$= \boxed{3.32 \times 10^{-34}\,\text{m}}$$

This wavelength is much smaller than any aperture through which the rock could possibly pass. This means that we could not observe diffraction effects, and as a result the wave properties of large-scale objects cannot be observed.

## Example 28.7   An Accelerated Charge

A particle of charge $q$ and mass $m$ is accelerated from rest through a potential difference $\Delta V$. Assuming the particle moves with a nonrelativistic speed, find its de Broglie wavelength.

**Solution**   We use the isolated system model for the particle and the electric field associated with the potential difference. When a charge is accelerated from rest through a potential difference $\Delta V$, its gain in kinetic energy $\frac{1}{2} mv^2$ must equal the loss in potential energy $q\,\Delta V$ of the particle–field system:

$$\tfrac{1}{2} mv^2 = q\,\Delta V$$

Because $p = mv$, we can express this in the form

$$\frac{p^2}{2m} = q\,\Delta V \qquad \text{or} \qquad p = \sqrt{2mq\,\Delta V}$$

Substituting this expression for $p$ in the de Broglie relation $\lambda = h/p$ gives

$$\lambda = \frac{h}{\sqrt{2mq\,\Delta V}}$$

**EXERCISE**   Calculate the de Broglie wavelength of an electron accelerated through a potential difference of 50 V.

*Answer:*   0.17 nm

## 28.6 • THE QUANTUM PARTICLE

The discussions presented in previous sections may be quite disturbing, because we considered the particle and wave models to be distinct in the past. The notion that both light and material particles have both particle and wave properties does not fit with this distinction. We have experimental evidence, however, that this is just what we must accept. The recognition of this dual nature leads to a new simplification model, the **quantum particle.** In this model, entities have both particle and wave characteristics, and we must choose one appropriate behavior—particle or wave—to understand a particular phenomenon.

In this section, we shall investigate a model that might bring you more comfort with this idea. We shall do this by demonstrating that we can construct from waves an entity that exhibits properties of a particle.

Let us first remind ourselves of the characteristics of ideal particles and waves. An ideal particle has zero size. As mentioned in Section 13.2, an ideal wave has a single frequency and is infinitely long. Thus, an essential feature of a particle is that it is *localized* in space. Let us show that we can build a localized entity from infinitely long waves. Imagine drawing one wave along the $x$ axis, with one of its crests located at $x = 0$, as in Figure 28.13a. Now, draw a second wave, of the same amplitude but a different frequency, with one of its crests also at $x = 0$. The result

(a)

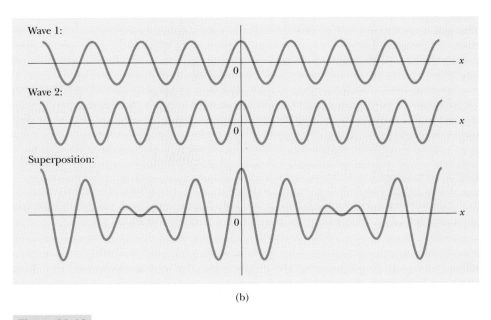

(b)

**Figure 28.13**

(a) An idealized wave of an exact single frequency is the same throughout space and time. (b) If two ideal waves with slightly different frequencies are combined, beats result (Section 14.6). The regions of space at which constructive interference occurs are different from those at which destructive interference happens.

of the superposition of these two waves is a *beat,* as the waves are alternately in phase and out of phase. (Beats were discussed in Section 14.6.) Figure 28.13b shows the results of superposing these two waves.

Notice that we have already introduced some localization by doing this. A single wave has the same amplitude everywhere in space—no point in space is any different from any other point. By adding a second wave, however, something is different between the in-phase and the out-of-phase points.

Now imagine that more and more waves are added to our original two, each new wave having a new frequency. Each new wave is added so that one of its crests is at $x = 0$. The result at $x = 0$ is that all of the waves add constructively. When we consider a large number of waves, the probability of a positive value of a wave function at any point $x$ is equal to the probability of a negative value, and destructive interference occurs *everywhere* except near $x = 0$, where we superposed all of the crests. The result of this is shown in Figure 28.14. The small region of constructive interference is called a **wave packet.** This is a localized region of space that is different from all other regions. We can identify the wave packet as a particle—it has the localized nature of what we have come to recognize as a particle!

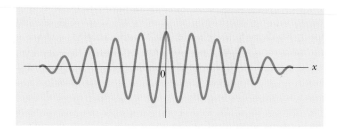

**Figure 28.14**

If many waves are combined, the result is a wave packet, which represents a particle.

The localized nature of this entity is the *only* characteristic of a particle that was generated with this process. We have not addressed how the wave packet might achieve such particle characteristics as mass, electric charge, spin, and so on. Thus, you may not be completely convinced that we have built a particle. As further evidence that the wave packet can represent the particle, let us show that the wave packet has another characteristic of a particle.

We return to our combination of two waves so as to make the mathematical representation simple. Consider two waves with equal amplitudes but different frequencies $f_1$ and $f_2$. We can represent the waves mathematically as

$$y_1 = A \cos(k_1 x - \omega_1 t) \qquad \text{and} \qquad y_2 = A \cos(k_2 x - \omega_2 t)$$

where, as in Chapter 13, $\omega = 2\pi f$ and $k = 2\pi/\lambda$. Using the superposition principle, we add the waves:

$$y = y_1 + y_2 = A \cos(k_1 x - \omega_1 t) + A \cos(k_2 x - \omega_2 t)$$

It is convenient to write this in a form that uses the trigonometric identity

$$\cos a + \cos b = 2 \cos\left(\frac{a - b}{2}\right) \cos\left(\frac{a + b}{2}\right)$$

Letting $a = k_1 x - \omega_1 t$ and $b = k_2 x - \omega_2 t$, we find that

$$y = 2A \cos\left[\frac{(k_1 x - \omega_1 t) - (k_2 x - \omega_2 t)}{2}\right] \cos\left[\frac{(k_1 x - \omega_1 t) + (k_2 x - \omega_2 t)}{2}\right]$$

$$= \left[2A \cos\left(\frac{\Delta k}{2} x - \frac{\Delta\omega}{2} t\right)\right] \cos\left(\frac{k_1 + k_2}{2} x - \frac{\omega_1 + \omega_2}{2} t\right) \qquad \text{[28.12]}$$

The second cosine factor represents a wave with a wave number and frequency equal to the averages of the values for the individual waves.

The factor in brackets represents the envelope of the wave, as shown in Figure 28.15. Notice that this factor also has the mathematical form of a wave. **This envelope of the combination can travel through space with a different speed than the individual waves.** As an extreme example of this possibility, imagine combining two identical waves moving in opposite directions. The two waves move with the same speed, but the envelope has a speed of *zero*, because we have built a standing wave, which we studied in Section 14.3.

For an individual wave, the speed is given by Equation 13.10

• *Phase speed for a wave*

$$v_{\text{phase}} = \frac{\omega}{k}$$

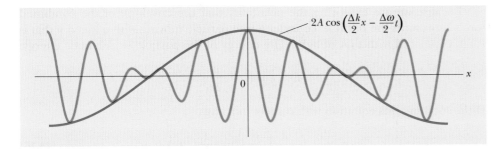

**Figure 28.15**

The beat pattern of Figure 28.13b, with an envelope function superimposed.

It is called the **phase speed** because it is the rate of advance of a crest on a single wave, which is a point of fixed phase. This equation can be interpreted as the following: the phase speed of a wave is the ratio of the coefficient of the time variable $t$ to the coefficient of the space variable $x$ in the equation for the wave, $y = A \cos(kx - \omega t)$.

The factor in brackets in Equation 28.12 is of the form of a wave, so it moves with a speed given by this same ratio:

$$v_g = \frac{\text{coefficient of time variable } t}{\text{coefficient of space variable } x} = \frac{(\Delta \omega / 2)}{(\Delta k / 2)} = \frac{\Delta \omega}{\Delta k}$$

The subscript $g$ on the speed indicates that this is commonly called the **group speed,** or the speed of the wave packet (the *group* of waves) that we have built. We have generated this expression for a simple addition of two waves. For a superposition of a very large number of waves to form a wave packet, this ratio becomes a derivative:

$$v_g = \frac{d\omega}{dk} \qquad \text{[28.13]}$$

• *Group speed for a wave packet*

Let us multiply the numerator and the denominator by $\hbar$, where $\hbar = h/2\pi$:

$$v_g = \frac{\hbar \, d\omega}{\hbar \, dk} = \frac{d(\hbar \omega)}{d(\hbar k)} \qquad \text{[28.14]}$$

We look at the terms in the parentheses in the numerator and denominator in this equation separately. For the numerator

$$\hbar \omega = \frac{h}{2\pi} (2\pi f) = hf = E$$

For the denominator

$$\hbar k = \frac{h}{2\pi} \left( \frac{2\pi}{\lambda} \right) = \frac{h}{\lambda} = p$$

Thus, Equation 28.14 can be written as

$$v_g = \frac{d(\hbar \omega)}{d(\hbar k)} = \frac{dE}{dp} \qquad \text{[28.15]}$$

Because we are exploring the possibility that the envelope of the combined waves represents the particle, consider a free particle moving with a speed $u$ that is small compared with that of light. The energy of the particle is its kinetic energy:

$$E = \tfrac{1}{2} mu^2 = \frac{p^2}{2m}$$

Differentiating this equation with respect to $p$ gives

$$v_g = \frac{dE}{dp} = \frac{d}{dp}\left(\frac{p^2}{2m}\right) = \frac{1}{2m}(2p) = u \qquad \textbf{[28.16]}$$

Thus, the group speed of the wave packet is identical to the speed of the particle that it is modeled to represent! This gives us further confidence that the wave packet is a reasonable way to build a particle.

## 28.7 • THE DOUBLE-SLIT EXPERIMENT REVISITED

One way to crystallize our ideas about the electron's wave–particle duality is to consider a hypothetical experiment in which electrons are fired at a double slit. Consider a parallel beam of monoenergetic electrons that is incident on a double slit as in Figure 28.16. We shall assume that the slit widths are small compared with

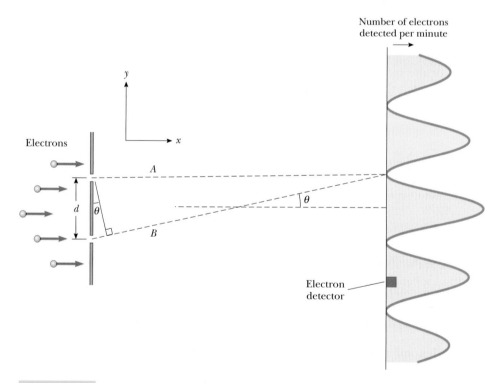

**Figure 28.16**

Electron diffraction. The slit separation $d$ is much greater than the individual slit widths and much less than the distance between the slit and the detector. The electron detector is movable along the $y$ direction in the drawing and so can detect electrons diffracted at different values of $\theta$. The detector acts like the "viewing screen" of Young's double-slit experiment with light discussed in Chapter 27.

the electron wavelength, so that we need not worry about diffraction maxima and minima, as we discussed for light in Section 27.6. An electron detector is positioned far from the slits at a distance much greater than $d$, the separation distance of the slits. If the detector collects electrons for a long enough time, one finds a typical wave interference pattern for the counts per minute, or probability of arrival of electrons. Such an interference pattern would not be expected if the electrons behaved as classical particles. This is clear evidence that electrons are interfering—a distinct wave-like behavior.

If the experiment is carried out at lower electron beam intensities, the interference pattern is still observed if the time of the measurement is sufficiently long. This is illustrated by the computer-simulated patterns in Figure 28.17. Note that the interference pattern becomes clearer as the number of electrons reaching the screen increases.

If one imagines a single electron in the beam producing in-phase "wavelets" as it reaches one of the slits, standard wave theory can be used to find the angular separation $\theta$ between the central probability maximum and its neighboring minimum. The minimum occurs when the path length difference between $A$ and $B$ is half a wavelength, or when

$$d \sin \theta = \frac{\lambda}{2}$$

Because an electron's wavelength is given by $\lambda = h/p_x$, we see that for small $\theta$

$$\sin \theta \approx \theta = \frac{h}{2p_x d}$$

Thus, the dual nature of the electron is clearly shown in this experiment: **The electrons are detected as particles at a localized spot at some instant of time, but the probability of arrival at that spot is determined by finding the intensity of two interfering waves.**

Let us now look at this experiment from another point of view. If one slit is covered during the experiment, a symmetric curve peaked around the center of the open slit is observed—the central maximum of the single-slit diffraction pattern. Plots of the counts per minute (probability of arrival of electrons) with the lower or upper slit closed are shown as blue curves in the central part of Figure 28.18.

If another experiment is now performed with slit 2 blocked half of the time and then slit 1 blocked during the remaining time, the accumulated pattern of counts per minute shown by the blue curve on the right side of Figure 28.18 is completely different from the case with both slits open (red curve). A maximum probability of arrival of an electron at $\theta = 0$ no longer exists. In fact, **the interference pattern has been lost, and the accumulated result is simply the sum of the individual results.** When only one slit is open at a time, we know the electron has the same localizability and indivisibility at the slits as we measure at the detector, because the electron clearly goes through slit 1 or slit 2. Thus, the total must be analyzed as the sum of those electrons that come through slit 1, and those that come through slit 2.

When both slits are open, it is tempting to assume that the electron goes through either slit 1 or slit 2, and that the counts per minute are again given by the combination of the single-slit distributions. We know, however, that the experimental results indicated by the red interference pattern in Figure 28.18 contradict

(a) After 28 electrons

(b) After 1 000 electrons

(c) After 10 000 electrons

(d) Two-slit electron pattern

**Figure 28.17**

(a, b, c) Computer-simulated interference patterns for a beam of electrons incident on a double slit. *(From E. R. Huggins, Physics I, New York, W. A. Benjamin, 1968)* (d) Photograph of a double-slit interference pattern produced by electrons. *(From C. Jönsson, Zeitschrift fur Physik 161:454, 1961; used with permission)*

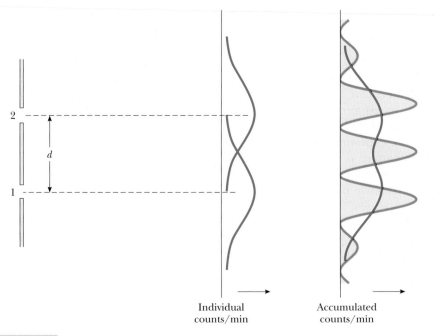

Individual
counts/min

Accumulated
counts/min

**Figure 28.18**

Accumulated results of the two-slit electron diffraction experiment with each slit closed half the time. The result with both slits open is shown in red.

this. Hence, our assumption that the electron is localized and goes through only one slit when both slits are open must be wrong (a painful conclusion!).

To interpret these results, we are forced to conclude that **an electron interacts with both slits simultaneously.** If we attempt to determine experimentally which slit the electron goes through, the act of measuring destroys the interference pattern. It is impossible to determine which slit the electron goes through. In effect, **we can say only that the electron passes through both slits!** The same arguments apply to photons.

If we restrict ourselves to a pure particle model, it is an uncomfortable notion that the electron can be present at both slits at once. From the quantum particle model, however, the particle can be considered to be built from waves that exist throughout space. Thus, the wave components of the electron are present at both slits at the same time, and this model leads to a more comfortable interpretation of this experiment.

## 28.8 • THE UNCERTAINTY PRINCIPLE

Whenever one measures the position or velocity of a particle at any instant, experimental uncertainties are built into the measurements. According to classical mechanics, there is no fundamental barrier to an ultimate refinement of the apparatus or experimental procedures. In other words, it is possible, in principle, to make such measurements with arbitrarily small uncertainty. Quantum theory predicts, however, that **it is fundamentally impossible to make simultaneous measurements of a particle's position and momentum with infinite accuracy.**

In 1927, Werner Heisenberg introduced this notion, which is now known as the **Heisenberg uncertainty principle:**

If a measurement of the position of a particle is made with uncertainty $\Delta x$ and a simultaneous measurement of its momentum is made with uncertainty $\Delta p_x$, the product of the two uncertainties can never be smaller than $\hbar/2$:

$$\Delta x \, \Delta p_x \geq \frac{\hbar}{2}$$

[28.17]

- *Uncertainty principle for momentum and position*

That is, **it is physically impossible to simultaneously measure the exact position and exact momentum of a particle.** Heisenberg was careful to point out that the inescapable uncertainties $\Delta x$ and $\Delta p_x$ do not arise from imperfections in practical measuring instruments. Rather, **the uncertainties arise from the quantum structure of matter.**

To understand the uncertainty principle, consider a particle for which we know the wavelength *exactly*. According to the de Broglie relations $\lambda = h/p$, we would know the momentum to infinite accuracy.

In reality, as we have mentioned, a single-wavelength wave would exist throughout space. Any region along this wave is the same as any other region (see Fig. 28.13a). If we were to ask, "Where is the particle that this wave represents?", no special location in space along the wave could be identified with the particle—all points along the wave are the same. Thus, we have *infinite* uncertainty in the position of the particle—we know nothing about where it is. Perfect knowledge of the momentum has cost us all information about the position.

In comparison, now consider a particle with some uncertainty in momentum, so that a range of values of momentum are possible. According to the de Broglie relation, this results in a range of wavelengths. Thus, the particle is not represented by a single wavelength, but a combination of wavelengths within this range. This combination forms a wave packet, as we discussed in Section 28.6 and illustrated in Figure 28.14. Now, if we are asked to determine the location of the particle, we can only say that it is somewhere in the region defined by the wave packet—because a distinct difference exists between this region and the rest of space. Thus, by losing some information about the momentum of the particle, we have gained information about its position.

If we were to lose all information about the momentum, then we would be adding together waves of all possible wavelengths. This would result in a wave packet of zero length. Thus, if we know nothing about the momentum, we know exactly where the particle is.

The mathematical form of the uncertainty principle argues that the product of the uncertainties in position and momentum will always be larger than some minimum value. This value can be calculated from the types of arguments discussed earlier, which result in the value of $\hbar/2$ in Equation 28.17.

Another form of the uncertainty principle can be generated by reconsidering Figure 28.14. Imagine that the horizontal axis is time rather than spatial position *x*. We can then make the same arguments that we made about knowledge of wavelength and position in the time domain. The corresponding variables would be

**Werner Heisenberg** (1901–1976)

A German theoretical physicist, Heisenberg made many significant contributions to physics, including his famous uncertainty principle, for which he received the Nobel Prize for physics in 1932; the development of an abstract model of quantum mechanics called matrix mechanics; the prediction of two forms of molecular hydrogen; and theoretical models of the nucleus. *(Courtesy of the University of Hamburg)* frequency and time. Because frequency is related to the energy of the particle by $E = hf$, the uncertainty principle in this form is

- *Uncertainty principle for energy and time*

$$\Delta E\, \Delta t \geq \frac{\hbar}{2}$$

[28.18]

This form of the uncertainty principle suggests that we can violate energy conservation by an amount $\Delta E$ as long as we do it only for a short time $\Delta t$ consistent with Equation 28.18. We shall use this notion to estimate the rest energies of particles in Chapter 31.

**Quick Quiz 28.8**

The location of a particle is measured and specified as being exactly at $x = 0$, with *zero* uncertainty. How does this affect our ability to measure its velocity component in the $y$ direction?

**THINKING PHYSICS 28.4**

A common, but erroneous, description of the absolute zero of temperature is *that temperature at which all molecular motion ceases.* How can the uncertainty principle be used to argue against this description?

**Reasoning**   Let us imagine molecules in a finite piece of material. The molecules are constrained to be within the piece of material, so a fixed uncertainty $\Delta x$ exists in their position along one axis, which is the size of the piece of material along this axis. If the molecular motion were to cease at absolute zero, we would be claiming that the molecular velocity is *zero,* with no uncertainty: $\Delta v = 0$. The product of zero uncertainty in velocity and a non-zero uncertainty in position will be zero, violating the uncertainty principle. Thus, even at absolute zero, some molecular motion must occur.

---

**Example 28.8   Locating an Electron**

The speed of an electron is measured to be $5.00 \times 10^3$ m/s $\pm$ 0.003%. Within what limits could one determine the position of this electron along the direction of its velocity vector?

**Solution**   The momentum of the electron is

$$p = m_e v = (9.11 \times 10^{-31}\ \text{kg})(5.00 \times 10^3\ \text{m/s})$$

$$= 4.56 \times 10^{-27}\ \text{kg·m/s}$$

Because the uncertainty is 0.003% of this value, we have

$$\Delta p = 0.00003 p = 1.37 \times 10^{-31}\ \text{kg·m/s}$$

The minimum uncertainty in position can now be calculated by using this value of $\Delta p$ and Equation 28.17:

$$\Delta x \geq \frac{\hbar}{2\,\Delta p} = \frac{1.05 \times 10^{-34}\ \text{J·s}}{2(1.37 \times 10^{-31}\ \text{kg·m/s})}$$

$$= 0.384 \times 10^{-3}\ \text{m} = \boxed{0.384\ \text{mm}}$$

## 28.9 • AN INTERPRETATION OF QUANTUM MECHANICS

We have been introduced to some new and strange ideas so far in this chapter. Let us continue to strive toward further understanding of quantum physics by making another bridge between particles and waves. We first think about electromagnetic radiation from the particle point of view. The probability per unit volume of finding a photon in a given region of space at an instant of time is proportional to the number of photons per unit volume at that time:

$$\frac{probability}{V} \propto \frac{N}{V}$$

The number of photons per unit volume is proportional to the intensity of the radiation:

$$\frac{N}{V} \propto I$$

Now, we form the bridge to the wave model by recalling that the intensity of electromagnetic radiation is proportional to the square of the electric field amplitude for the electromagnetic wave (Section 24.5):

$$I \propto E^2$$

Equating the beginning and the end of this string of proportionalities, we have

$$\frac{probability}{V} \propto E^2 \qquad\qquad\text{[28.19]}$$

Thus, for electromagnetic radiation, the probability per unit volume of finding a particle associated with this radiation (the photon) is proportional to the square of the amplitude of the wave. Recognizing the wave–particle duality of both electromagnetic radiation and matter, we can argue that a parallel proportionality exists for a material particle. The probability per unit volume of finding the particle is proportional to the square of the amplitude of the wave representing the particle. So far, we have discussed the fact that a wave is associated with a particle but have not described it specifically. We find that the amplitude of the wave associated with a particle is not a measurable quantity like the electric field is for an electromagnetic wave. Thus, we cannot relate this amplitude to other physical behavior like we can the electric field—the only connection we have is the matter analog to Equation 28.19, relating the square of the amplitude of the wave to the probability per unit volume. As a result, we simply call the amplitude of the wave associated with the particle the **probability amplitude,** or the **wave function,** and give it the symbol $\psi$. In general, the wave function $\psi$ is a complex-valued function. The absolute square $|\psi|^2 = \psi^*\psi$, where $\psi^*$ is the complex conjugate of $\psi$, will always be real and positive and is proportional to the probability per unit volume of finding a particle at a given point at some instant. The wave function contains within it all the information that can be known about the particle.

This probability interpretation of the wave function was first suggested by Max Born (1882–1970) in 1928. In 1926, Erwin Schrödinger (1887–1961) proposed a wave equation that describes the manner in which the wave function changes in space and time. The *Schrödinger wave equation,* which we shall examine in Section 28.12, represents a key element in the theory of quantum mechanics.

The concepts of quantum mechanics, strange as they sometimes may seem, developed from classical ideas. In fact, when the techniques of quantum mechanics are applied to macroscopic systems, the results are essentially identical to those of classical physics. This blending of the two approaches occurs when the de Broglie wavelength is small compared with the dimensions of the system. The situation is similar to the agreement between relativistic mechanics and classical mechanics when $v \ll c$.

The wave function $\psi$ that we have introduced contains only spatial information about the particle. In general, a more complete wave function $\Psi$ depends on the positions of all the particles in a system and on time, and therefore is often written $\Psi(x, y, z, t)$. For most systems of interest for us, the full wave function $\Psi$ can be written as a product of $\psi$ and a simple time-dependent function, so the understanding of $\psi$ is the critical part of a given problem. If $\psi$ is known for a system, the particular properties of that system can be described.

In Section 28.5, we found that the de Broglie equation relates the momentum of a particle to its wavelength through the relation $p = h/\lambda$. If an ideal free particle has a precisely known momentum $p_x$, its wave function is a sinusoidal wave of wavelength $\lambda = h/p_x$, and the particle has equal probability of being at any point along the $x$ axis. The wave function for such a free particle moving along the $x$ axis can be written as

$$\psi(x) = A \sin\left(\frac{2\pi x}{\lambda}\right) = A \sin(kx) \qquad \text{[28.20]}$$

where $k = 2\pi/\lambda$ is the angular wave number and $A$ is a constant amplitude.

Although we cannot measure $\psi$, we can measure the quantity $|\psi|^2$, which can be interpreted as follows. If $\psi$ represents a single particle, then $|\psi|^2$—called the **probability density**—is the relative probability per unit volume that the particle will be found at any given point in the volume. This interpretation can also be stated in the following manner. If $dV$ is a small volume element surrounding some point, then the probability of finding the particle in that volume element is $|\psi|^2 \, dV$. In this section we deal only with one-dimensional systems, where the particle must be located along the $x$ axis, and thus we replace $dV$ with $dx$. In this case, the probability $P(x) \, dx$ that the particle will be found in the infinitesimal interval $dx$ around the point $x$ is

- $|\psi|^2$ *equals the probability density*

$$P(x) \, dx = |\psi|^2 \, dx \qquad \text{[28.21]}$$

Because the particle must be somewhere along the $x$ axis, the sum of the probabilities over all values of $x$ must be 1:

- *Normalization condition on $\psi$*

$$\int_{-\infty}^{\infty} |\psi|^2 \, dx = 1 \qquad \text{[28.22]}$$

Any wave function satisfying Equation 28.22 is said to be **normalized.** Normalization is simply a statement that the particle exists at some point at all times.

Although it is not possible to specify the position of a particle with complete certainty, it is possible through $|\psi|^2$ to specify the probability of observing it in a small region surrounding a given point. **The probability of finding the particle in the arbitrarily sized interval $a \le x \le b$ is**

$$P_{ab} = \int_{a}^{b} |\psi|^2 \, dx \qquad \text{[28.23]}$$

The probability $P_{ab}$ is the area under the curve of $|\psi|^2$ versus $x$ between the points $x = a$ and $x = b$, as in Figure 28.19.

Experimentally, the probability is finite of finding a particle in an interval near some point at some instant. The value of that probability must lie between the limits 0 and 1. For example, if the probability is 0.3, there is a 30% chance of finding the particle in the interval.

The wave function $\psi$ satisfies a wave equation, just as the electric field associated with an electromagnetic wave satisfies a wave equation that follows from Maxwell's equations. The wave equation satisfied by $\psi$ is the Schrödinger equation (Section 28.12), and $\psi$ can be computed from it. Although $\psi$ is not a measurable quantity, all the measurable quantities of a particle, such as its energy and momentum, can be derived from a knowledge of $\psi$. For example, once the wave function for a particle is known, it is possible to calculate the average position at which you would find the particle after many measurements. This average position is called the **expectation value** of $x$ and is defined by the equation

$$\langle x \rangle \equiv \int_{-\infty}^{\infty} \psi^* x \psi \, dx \qquad \textbf{[28.24]}$$

(Brackets, $\langle \quad \rangle$ are used to denote expectation values.) Furthermore, one can find the expectation value of any function $f(x)$ associated with the particle by using the following equation:

$$\langle f(x) \rangle \equiv \int_{-\infty}^{\infty} \psi^* f(x) \psi \, dx \qquad \textbf{[28.25]}$$

# 28.10 • A PARTICLE IN A BOX

In this section, we shall apply some of the ideas we have developed to a sample problem. Let us choose a simple problem—a particle confined to a one-dimensional region of space, called the *particle in a box* (even though the "box" is one-dimensional!). From a classical viewpoint, if a particle is confined to bouncing back and forth along the $x$ axis between two impenetrable walls, as in the pictorial representation in Figure 28.20a, its motion is easy to describe. If the speed of the particle is $v$, the magnitude of its momentum $mv$ remains constant, as does its kinetic energy. Classical physics places no restrictions on the values of a particle's momentum and energy. The quantum mechanics approach to this problem is quite different and requires that we find the appropriate wave function consistent with the conditions of the situation.*

Because the walls are impenetrable, the probability of finding the particle outside the box is zero, so the wave function $\psi(x)$ must be zero for $x < 0$ and for $x > L$, where $L$ is the distance between the two walls. A mathematical condition for any wave function is that it must be continuous in space.† Thus, if $\psi$ is zero outside the walls, it must also be zero *at* the walls; that is, $\psi(0) = 0$ and $\psi(L) = 0$. Only those wave functions that satisfy this condition are allowed.

Figure 28.20b shows a graphical representation of the particle in a box problem, which graphs the potential energy of the particle–environment system as a function of the position of the particle. When the particle is inside the box, the potential energy of the system does not depend on the location of the particle and

---

* Before continuing, you might want to review Sections 14.3 and 14.4 on standing mechanical waves.

† If the wave function is not continuous at a point, the derivative of the wave function at that point is infinite. This leads to problems in the Schrödinger equation, for which the wave function is a solution and which is discussed in Section 28.12.

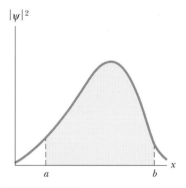

$|\psi|^2$

**Figure 28.19**

The probability that a particle is in the interval $a \le x \le b$ is represented by the area from $a$ to $b$ under the curve of the probability density function $|\psi(x)|^2$.

(a)

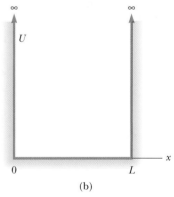

(b)

**Figure 28.20**

(a) A particle of mass $m$ and velocity **v**, confined to bouncing between two impenetrable walls. (b) The potential energy function for the system.

we can choose its value to be zero. Outside the box, we have to ensure that the wave function is zero. We can do this by defining the potential energy of the system as infinitely large if the particle were outside the box. Because kinetic energy is necessarily positive, the only way a particle could be outside the box is if the system has an infinite amount of energy.

Except for limitations on the particle defined by the walls, the particle is a free particle, so we identify the wave functions for the particle in the box as sinusoidal (see Eq. 28.20):

$$\psi(x) = A \sin\left(\frac{2\pi x}{\lambda}\right) \tag{28.26}$$

This wave function must satisfy the boundary conditions at the walls. The boundary condition at $x = 0$ is satisfied already because the sine function is zero when $x = 0$. For the boundary condition at $x = L$, we have

$$\psi(L) = 0 = A \sin\left(\frac{2\pi L}{\lambda}\right)$$

which can only be true if

$$\frac{2\pi L}{\lambda} = n\pi \longrightarrow \lambda = \frac{2L}{n} \tag{28.27}$$

where $n = 1, 2, 3 \ldots$. Thus, only certain wavelengths for the particle are allowed! Each of the allowed wavelengths corresponds to a quantum state for the system, and $n$ is the quantum number. Expressing the wave function in terms of the quantum number $n$, we have

- *Allowed wave functions for a particle in a box*

$$\psi(x) = A \sin\left(\frac{2\pi x}{\lambda}\right) = A \sin\left(\frac{2\pi x}{2L/n}\right) = A \sin\left(\frac{n\pi x}{L}\right) \tag{28.28}$$

Figures 28.21a and b are graphical representations of $\psi$ versus $x$ and $|\psi|^2$ versus $x$ for $n = 1, 2,$ and 3 for the particle in a box. Note that although $\psi$ can be posi-

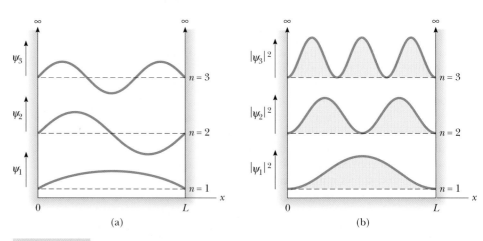

**Figure 28.21**

The first three allowed stationary states for a particle confined to a one-dimensional box. Each function is graphed on a higher horizontal axis so that the functions can be seen separately. (a) The wave functions for $n = 1, 2,$ and 3. (b) The probability density graphs for $n = 1, 2,$ and 3.

tive or negative, $|\psi|^2$ is always positive. Because $|\psi|^2$ represents a probability density, a negative value for $|\psi|^2$ is meaningless.

Further inspection of Figure 28.21b shows that $|\psi|^2$ is zero at the boundaries, satisfying our boundary condition. In addition, $|\psi|^2$ is zero at other points, depending on the value of $n$. For $n = 2$, $|\psi|^2 = 0$ at $x = L/2$; for $n = 3$, $|\psi|^2 = 0$ at $x = L/3$ and $x = 2L/3$. The number of zero points increases by one each time the quantum number increases by one.

Because the wavelengths of the particle are restricted by the condition $\lambda = 2L/n$, the magnitude of the momentum of the particle is also restricted to specific values that we can find from the expression for the de Broglie wavelength, Equation 28.10:

$$p = \frac{h}{\lambda} = \frac{h}{2L/n} = \frac{nh}{2L}$$

From this expression, we find that the allowed values of the energy, which is simply the kinetic energy of the particle, are

$$E_n = \tfrac{1}{2}mv^2 = \frac{p^2}{2m} = \frac{(nh/2L)^2}{2m}$$

$$= \left(\frac{h^2}{8mL^2}\right)n^2 \quad n = 1, 2, 3, \ldots \qquad \textbf{[28.29]}$$

As we see from this expression, the **energy of the particle is quantized,** similar to our quantization of energy in the hydrogen atom in Chapter 11. The lowest allowed energy corresponds to $n = 1$, for which $E_1 = h^2/8mL^2$. Because $E_n = n^2 E_1$, the excited states corresponding to $n = 2, 3, 4, \ldots$ have energies given by $4E_1$, $9E_1$, $16E_1$, $\ldots$ .

Figure 28.22 is an energy level diagram* describing the energy values of the allowed states. Note that the state $n = 0$, for which $E$ would be equal to zero, is not allowed. This means that according to quantum mechanics, the particle can never be at rest. The least energy it can have, corresponding to $n = 1$, is called the **zero-point energy.** This result is clearly contradictory to the classical viewpoint, in which $E = 0$ is an acceptable state, as are all positive values of $E$.

● *Allowed energies for a particle in a box*

### Quick Quiz 28.9

Redraw Figure 28.21b, the probability of finding a particle at a particular location in a box, on the basis of expectations from classical physics rather than quantum physics.

### Quick Quiz 28.10

A particle is in a box of length $L$. Suddenly, the length of the box is increased to $2L$. What happens to the energy levels shown in Figure 28.22? (a) Nothing—they are unaffected. (b) They move farther apart. (c) They move closer together.

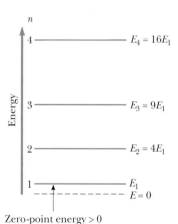

Zero-point energy > 0

**Figure 28.22**

An energy level diagram for a particle confined to a one-dimensional box of width $L$. The lowest allowed energy is $E_1$ and has the value $E_1 = h^2/8mL^2$.

* We introduced the energy level diagram as a specialized semigraphical representation in Chapter 11.

## Example 28.9    A Bound Electron

An electron is confined between two impenetrable walls 0.200 nm apart. Determine the allowed energies of the particle for the quantum states described by $n = 1, 2,$ and $3$.

**Solution**    We apply Equation 28.29, using $m_e = 9.11 \times 10^{-31}$ kg for the electron. For $n = 1$, we have

$$E_1 = \frac{h^2}{8 m_e L^2} = \frac{(6.63 \times 10^{-34}\,\mathrm{J \cdot s})^2}{8(9.11 \times 10^{-31}\,\mathrm{kg})(2.00 \times 10^{-10}\,\mathrm{m})^2}$$

$$= 1.51 \times 10^{-18}\,\mathrm{J} = \boxed{9.42\ \mathrm{eV}}$$

For $n = 2$ and $n = 3$, we find that $E_2 = 4E_1 = \boxed{37.7\ \mathrm{eV}}$ and $E_3 = 9E_1 = \boxed{84.8\ \mathrm{eV}.}$

## Example 28.10    Energy Quantization for a Macroscopic Object

A 1.00-mg object is confined between two rigid walls separated by 1.00 cm. (a) Calculate its minimum speed.

**Solution**    The minimum speed corresponds to the state for which $n = 1$. Using Equation 28.29 with $n = 1$ gives the zero-point energy:

$$E_1 = \frac{h^2}{8 m L^2} = \frac{(6.63 \times 10^{-34}\,\mathrm{J \cdot s})^2}{8(1.00 \times 10^{-6}\,\mathrm{kg})(1.00 \times 10^{-2}\,\mathrm{m})^2}$$

$$= 5.49 \times 10^{-58}\,\mathrm{J}$$

Because $E = K = \frac{1}{2} m v^2$, we can find $v$ as follows:

$$\tfrac{1}{2} m v^2 = 5.49 \times 10^{-58}\,\mathrm{J}$$

$$v = \left[ \frac{2(5.49 \times 10^{-58}\,\mathrm{J})}{1.00 \times 10^{-6}\,\mathrm{kg}} \right]^{1/2} = \boxed{3.31 \times 10^{-26}\ \mathrm{m/s}}$$

This result is so small that the object appears to be at rest, which is what one would expect for the zero-point speed of a macroscopic object.

  (b) If the speed of the object is $3.00 \times 10^{-2}$ m/s, find the corresponding value of $n$.

**Solution**    The kinetic energy of the object is

$$K = \tfrac{1}{2} m v^2 = \tfrac{1}{2} (1.00 \times 10^{-6}\,\mathrm{kg})(3.00 \times 10^{-2}\,\mathrm{m/s})$$

$$= 4.50 \times 10^{-10}\,\mathrm{J}$$

Because $E_n = n^2 E_1$ and $E_1 = 5.49 \times 10^{-58}$ J, we find that

$$K = E_n = n^2 E_1 = 4.50 \times 10^{-10}\,\mathrm{J}$$

$$n = \left( \frac{4.50 \times 10^{-10}\,\mathrm{J}}{E_1} \right)^{1/2} = \left( \frac{4.50 \times 10^{-10}\,\mathrm{J}}{5.49 \times 10^{-58}\,\mathrm{J}} \right)^{1/2}$$

$$= \boxed{9.05 \times 10^{23}}$$

This value of $n$ is so large that we would never be able to distinguish the quantized nature of the energy levels. That is, the difference in energy between the two states $n_1 = 9.05 \times 10^{23}$ and $n_2 = (9.05 \times 10^{23}) + 1$ is too small to be detected experimentally. Like example 28.2, this example illustrates the working of the correspondence principle; that is, as $m$ or $L$ become large, the quantum description must agree with the classical result.

**EXERCISE**    An alpha particle in a nucleus can be modeled as a particle moving in a box of width $1.0 \times 10^{-14}$ m (the approximate diameter of a nucleus). Using this model, estimate (a) the energy and (b) the momentum of an alpha particle in its lowest energy state. ($m_\alpha = 6.6 \times 10^{-27}$ kg)

*Answer*    (a) 0.52 MeV    (b) $3.3 \times 10^{-20}$ kg $\cdot$ m/s

## 28.11  •  THE QUANTUM PARTICLE UNDER BOUNDARY CONDITIONS

The particle in a box discussed in the preceding section is an example of how all quantum problems can be addressed. To begin to understand this, consider Equation 28.27 and compare it with Equation 14.6. The allowed wavelengths for the particle in a box are identical to the allowed wavelengths for mechanical waves on a string fixed at both ends. In both the string wave and the particle wave, we applied **boundary conditions** to determine the allowed states of the system. For the string fixed at both ends, the boundary condition was that the displacement of the string at the boundaries was zero. For the particle in a box, the probability amplitude at the boundaries was zero. In both cases, this resulted in quantized wavelengths. In the case of the vibrating string, wavelength is related to the frequency, so we have a set of harmonics, or quantized frequencies, given by Equation 14.8. In the case of the particle in a box, we also have quantized frequencies. We can go further in this case, however, because the frequency is related to the energy through $E = hf$, and we generate a set of quantized *energies*.

The quantization of energy, therefore, is no more surprising than the quantization of frequencies for the vibrating guitar string. The essential feature of the analysis model of the quantum particle under boundary conditions is the recognition that **any constraint on the motion of particles in a system represents one or more boundary conditions, resulting in quantization of the energy of the system.** Thus, because particles have wave-like characteristics, the allowed quantum states of a system are those in which the boundary conditions are satisfied.

The only quantization of energy we have seen before this chapter is that of the hydrogen atom in Chapter 11. In that case, the electric force between the proton and the electron creates a constraint that requires the electron and the proton to stay near each other (assuming we have not supplied enough energy to ionize the atom). This constraint results in boundary conditions that limit the energies of the atom to those corresponding to specific allowed wave functions.

In general, boundary conditions are related to the coordinates describing the problem. For the particle in a box, we required a zero value of the wave function at two values of *x*. In the case of the hydrogen atom, the problem is best presented in *spherical coordinates*. These are an extension of the polar coordinates introduced in Chapter 1 and consist of a radial coordinate *r* and two angular coordinates. Boundary conditions on the wave function related to *r* are that the radial part of the wave function must approach zero as $r \rightarrow \infty$ (so that the wave function can be normalized) and remain finite as $r \rightarrow 0$. A boundary condition on the wave function related to an angular coordinate is that adding $2\pi$ to the angle must return the wave function to the same value because an addition of $2\pi$ results in the same angular position. The generation of the wave function and application of the boundary conditions for the hydrogen atom are beyond the scope of this book. We shall, however, examine the behavior of some of the wave functions in Section 29.4.

## 28.12 • THE SCHRÖDINGER EQUATION

In Section 24.3, we discussed a wave equation for electromagnetic radiation. The waves associated with particles also satisfy a wave equation. We might guess that the wave equation for material particles is different from that associated with photons due to the fact that material particles have a nonzero rest energy. This wave equation was developed by Schrödinger in 1926. In analyzing the behavior of a quantum system, the approach is to determine a solution to this equation, and then apply the appropriate boundary conditions to the solution. The solution yields the allowed wave functions and energy levels of the system under consideration. Proper manipulation of the wave function then enables one to calculate all measurable features of the system.

**Erwin Schrödinger** (1887–1961)

An Austrian theoretical physicist, Schrödinger is best known as the creator of quantum mechanics. He also produced important papers in the fields of statistical mechanics, color vision, and general relativity. Schrödinger did much to hasten the universal acceptance of quantum theory by demonstrating the mathematical equivalence between his wave mechanics and the more abstract matrix mechanics developed by Heisenberg. *(AIP Emilio Segré Visual Archives)*

The Schrödinger equation as it applies to a particle of mass $m$ confined to moving along the $x$ axis and interacting with its environment through a potential energy function $U(x)$ is

- *Time-independent Schrödinger equation*

$$-\frac{\hbar^2}{2m}\frac{d^2\psi}{dx^2} + U\psi = E\psi \qquad\qquad [28.30]$$

where $E$ is the total energy of the system (particle and environment). Because this equation is independent of time, it is commonly referred to as the **time-independent Schrödinger equation.** (We shall not discuss the time-dependent Schrödinger equation in this text.)

This equation is consistent with the energy version of the isolated system model. The system is the particle and its environment. Problem 42 shows, both for a free particle and a particle in a box, that the first term in the Schrödinger equation reduces to the kinetic energy of the particle multiplied by the wave function. Note that Equation 28.30 tells us that the total energy is the sum of the kinetic energy and the potential energy and that the total energy is a constant: $K + U = E = $ constant.

In principle, if the potential energy $U(x)$ for the system is known, one can solve Equation 28.30 and obtain the wave functions and energies for the allowed states of the system. Because $U$ may vary with position, it may be necessary to solve the equation in pieces. In the process, the wave functions for the different regions must join smoothly at the boundaries—we require that $\psi(x)$ be *continuous*. Furthermore, so that $\psi(x)$ obeys the normalization condition, we require that $\psi(x)$ approach zero as $x$ approaches $\pm\infty$. Finally, $\psi(x)$ must be *single-valued* and $d\psi/dx$ must also be continuous for finite values of $U(x)$.*

The task of solving the Schrödinger equation may be very difficult, depending on the form of the potential energy function. As it turns out, the Schrödinger equation has been extremely successful in explaining the behavior of atomic and nuclear systems, whereas classical physics has failed to do so. Furthermore, when quantum mechanics is applied to macroscopic objects, the results agree with classical physics, as required by the correspondence principle.

### The Particle in a Box via the Schrödinger Equation

To see how the Schrödinger equation is applied to a problem, let us return to our particle in a one-dimensional box of width $L$ (see Fig. 28.20) and analyze it with the Schrödinger equation. In association with Figure 28.20b, we discussed the potential energy diagram that describes the problem. A potential energy diagram such as this is a useful representation for understanding and solving problems with the Schrödinger equation.

Because of the shape of the curve in Figure 28.20b, the particle in a box is sometimes said to be in a **square well,**† where a **well** is an upward-facing region of the curve in a potential energy diagram. (A downward-facing region is called a *barrier*, which we shall investigate in Section 28.13.)

**PITFALL PREVENTION 28.5**

**Potential wells**

A potential well, such as that in Figure 28.20, is a graphical representation of energy, not a pictorial representation, so you would not see this shape if you were able to observe the situation. A particle moves *only horizontally* in a potential energy diagram, representing the conserved energy of the system of the particle and its environment.

---

\* If $d\psi/dx$ were not continuous, we would not be able to evaluate $d^2\psi/dx^2$ in Equation 28.30 at the point of discontinuity.

† It is called a square well even if it has a rectangular shape in a potential energy diagram.

In the region $0 < x < L$, where $U = 0$, we can express the Schrödinger equation in the form

$$\frac{d^2\psi}{dx^2} = -\frac{2mE}{\hbar^2}\psi = -k^2\psi \qquad \text{[28.31]}$$

where

$$k = \frac{\sqrt{2mE}}{\hbar} \qquad \text{[28.32]}$$

The solution to Equation 28.31 is a function whose second derivative is the negative of the same function multiplied by a constant $k^2$. We recognize both the sine and cosine functions as satisfying this requirement. Thus, the most general solution to the equation is a linear combination of both solutions:

$$\psi(x) = A\sin kx + B\cos kx$$

where $A$ and $B$ are constants determined by the boundary conditions.

Our first boundary condition is that $\psi(0) = 0$:

$$\psi(0) = A\sin 0 + B\cos 0 = 0 + B = 0$$

Thus, our solution reduces to

$$\psi(x) = A\sin kx$$

The second boundary condition, $\psi(L) = 0$, when applied to the reduced solution, gives

$$\psi(L) = A\sin kL = 0$$

This is satisfied only if $kL$ is an integral multiple of $\pi$, that is, if $kL = n\pi$, where $n$ is an integer. Because $k = \sqrt{2mE}/\hbar$, we have

$$kL = \frac{\sqrt{2mE}}{\hbar}L = n\pi$$

For each integer choice for $n$, this equation determines a quantized energy $E_n$. Solving for the allowed energies $E_n$ gives

$$E_n = \left(\frac{h^2}{8mL^2}\right)n^2 \qquad \text{[28.33]}$$

which are identical to the allowed energies in Equation 28.29.

Substituting the values of $k$ in the wave function, the allowed wave functions $\psi_n(x)$ are given by

$$\psi_n(x) = A\sin\left(\frac{n\pi x}{L}\right) \qquad \text{[28.34]}$$

This wave function agrees with Equation 28.28.

Normalizing this relationship shows that $A = \sqrt{(2/L)}$. (See Problem 38.) Thus, the normalized wave function is

$$\psi(x) = \sqrt{\frac{2}{L}}\sin\left(\frac{n\pi x}{L}\right) \qquad \text{[28.35]}$$

### Quick Quiz 28.11

Suppose a particle is confined to a box discussed earlier is of length $L$, but the box located with its left end at $x = -L/2$ and its right end at $x = +L/2$. What is the wave function for the particle now; that is, what is the equivalent to Equation 28.35? (*Hint:* You will find different mathematical functions for odd and even values of $n$.)

---

### Example 28.11    The Expectation Values for the Particle in a Box

A particle of mass $m$ is confined to a one-dimensional box between $x = 0$ and $x = L$. Find the expectation value of the position $x$ of the particle for a state with quantum number $n$.

**Solution** Using Equation 28.24 and the wave function in Equation 28.35, we can set up the expectation value for $x$:

$$\langle x \rangle = \int_{-\infty}^{\infty} \psi^* x \psi \, dx = \int_0^L x \left( \sqrt{\frac{2}{L}} \sin \left( \frac{n \pi x}{L} \right) \right)^2 dx$$

$$= \frac{2}{L} \int_0^L x \sin^2 \left( \frac{n \pi x}{L} \right) dx$$

where we have reduced the limits on the integral to 0 to $L$ because the value of the wave function is zero elsewhere. Evaluating the integral gives us,

$$\langle x \rangle = \frac{2}{L} \left[ \frac{x^2}{4} - \frac{x \sin \left( 2\frac{n \pi x}{L} \right)}{4 \frac{n \pi}{L}} - \frac{\cos \left( 2\frac{n \pi x}{L} \right)}{8 \left( \frac{n \pi}{L} \right)^2} \right]_0^L$$

$$= \frac{2}{L} \left( \frac{L^2}{4} \right) = \boxed{\frac{L}{2}}$$

Notice that the expectation value is right at the center of the box, which we would expect from the symmetry of the square of the wave function about the center (see Fig. 28.21b). Because the squares of all wave functions are symmetric about the midpoint, the expectation value does not depend on $n$.

---

## 28.13 • TUNNELING THROUGH A POTENTIAL ENERGY BARRIER

Consider the potential energy function shown in Figure 28.23, in which the potential energy of the system is zero everywhere except for a region of width $L$ in which the potential energy has a constant value of $U$. This is called a **square barrier,** and $U$ is called the **barrier height.** A very interesting and peculiar phenom-

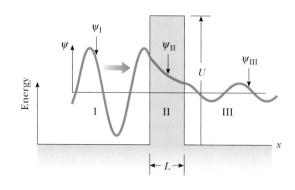

**Figure 28.23**

Wave function for a particle of energy $E$ incident from the left on a barrier of height $U > E$. Note that the wave function is sinusoidal in regions I and III but is exponentially decaying in region II. Energy is plotted vertically from the $x$ axis. The wave function is plotted vertically from an axis at the energy of the particle.

enon occurs when a moving particle encounters such a barrier of finite height and width. Consider a particle of energy $E < U$ that is incident on the barrier from the left (see Fig. 28.23). Classically, the particle is reflected by the barrier. If the particle were to exist in region II, its kinetic energy would be negative, which is not allowed classically. Thus, region II, and therefore region III, are both classically *forbidden* to the particle incident from the left. According to quantum mechanics, however, **all regions are accessible to the particle, regardless of its energy.** (Although all regions are accessible, the probability of the particle being in a region that is classically forbidden is very low.) From the point of view of the uncertainty principle, we can accept a violation of conservation of energy while the particle is within the barrier, as long as its time in the barrier is short and consistent with Equation 28.18. If the barrier is relatively thin, this short time can allow the particle to move across the width of the barrier. Thus, it is possible for us to understand the passing of the particle through the barrier with the help of the uncertainty principle.

Let us approach this situation using a mathematical representation. The Schrödinger equation has valid solutions in all three regions I, II, and III. The solutions in regions I and III are sinusoidal like Equation 28.20. In region II, the solution is exponential. Applying the boundary conditions that the wave functions in the three regions must join smoothly at the boundaries, we find that a full solution can be found, such as that represented by the curve in Figure 28.23. Thus, Schrödinger's equation and the boundary conditions are satisfied, which tells us mathematically that such a process can theoretically occur according to the quantum theory.

Because the probability of locating the particle is proportional to $|\psi|^2$, we conclude that the chance of finding the particle beyond the barrier in region III is nonzero. This result is in complete disagreement with classical physics. The movement of the particle to the far side of the barrier is called **tunneling** or **barrier penetration.**

The probability of tunneling can be described with a **transmission coefficient** $T$ and a **reflection coefficient** $R$. **The transmission coefficient represents the probability that the particle penetrates to the other side of the barrier, and the reflection coefficient is the probability that the particle is reflected by the barrier.** Because the incident particle is either reflected or transmitted, we require that $T + R = 1$. An approximate expression for the transmission coefficient, obtained when $T \ll 1$ (a very wide barrier or a very high barrier, that is, $U \gg E$), is

$$T \approx e^{-2CL} \qquad\qquad \textbf{[28.36]}$$

where

$$C = \frac{\sqrt{2m(U-E)}}{\hbar} \qquad\qquad \textbf{[28.37]}$$

Note that $T = 0$ from a classical point of view, but that Equation 28.36 tells us that $T$ can be nonzero from a quantum approach. The fact that we experimentally observe the phenomenon of tunneling provides further confidence in the principles of quantum physics.

## Example 28.12    Transmission Coefficient for an Electron

A 30-eV electron is incident on a square barrier of height 40 eV. What is the probability that the electron will tunnel through if the barrier width is (a) 1.0 nm? (b) 0.10 nm?

**Solution**    Let us assume that the probability of transmission is low, so that we can use the approximation in Equation 28.36. For the given barrier height and electron energy, the quantity $U - E$ has the value

$$U - E = (40 \text{ eV} - 30 \text{ eV}) = 10 \text{ eV} = 1.6 \times 10^{-18} \text{ J}$$

(a) Using Equation 28.37, the quantity $2CL$ is

$$2CL = 2 \frac{\sqrt{2(9.11 \times 10^{-31} \text{ kg})(1.6 \times 10^{-18} \text{ J})}}{1.054 \times 10^{-34} \text{ J} \cdot \text{s}} (1.0 \times 10^{-9} \text{ m})$$

$$= 32$$

Thus, the probability of tunneling through the barrier is

$$T \approx e^{-2CL} = e^{-32} = \boxed{1.3 \times 10^{-14}}$$

That is, the electron has only about 1 chance in $10^{14}$ to tunnel through the 1.0-nm-wide barrier.

(b) For $L = 0.10$ nm, we find $2CL = 3.2$, and

$$T \approx e^{-2CL} = e^{-3.2} = \boxed{0.041}$$

This result shows that the electron has a relatively high probability, about 4%, compared with $10^{-12}$% in part (a), of penetrating the 0.10-nm barrier. Notice an important behavior that leads to effective practical applications for tunneling—reducing the width of the barrier by only one order of magnitude increases the probability of tunneling by about 12 orders of magnitude!

## Applications of Tunneling

As we have seen, tunneling is a quantum phenomenon, a result of the wave nature of matter. Many applications may be understood only on the basis of tunneling.

- **Tunnel diode.**    The tunnel diode is a semiconductor device consisting of two oppositely charged regions separated by a very narrow neutral region representing the barrier. The current in this device is largely due to tunneling of electrons through the neutral region. The current, or rate of tunneling, can be controlled over a wide range by varying the applied voltage, which changes the height of the barrier.

- **Josephson junction.**    A Josephson junction consists of two superconductors separated by a thin insulating oxide layer, 1 to 2 nm thick. Under appropriate conditions, electrons in the superconductors bond as pairs and tunnel from one superconductor to the other through the oxide layer. Several effects have been observed in this type of junction. For example, a **dc current is observed across the junction in the absence of electric and magnetic fields.** When a voltage $\Delta V$ is applied across the junction, one observes the current oscillating with a frequency $f = 2e \Delta V / h$, where $e$ is the charge on the electron.

- **Alpha decay.**    One form of radioactive decay is the emission of alpha particles (the nuclei of helium atoms) by unstable, heavy nuclei (Chapter 30). For an alpha particle to escape from the nucleus, it must penetrate a barrier whose height is several times larger than the energy of the nucleus–alpha particle system. The barrier is due to a combination of the attractive nuclear force (introduced in Chapter 5 and discussed further in Chapter 31) and the Coulomb repulsion (discussed in detail in Chapter 19) between the alpha particle and the rest of the nucleus. Occasionally an alpha particle tunnels through the barrier, which explains the basic mechanism for this type of decay and the large variations in the mean lifetimes of various radioactive nuclei.

- **Scanning tunneling microscope.**    The scanning tunneling microscope, or STM, is a remarkable device that uses tunneling to create images of surfaces with resolution comparable to the size of a single atom. A small probe with a very fine tip is made to scan very close to the surface of a specimen. A tunneling current

**WEB**

For more images taken with a scanning tunneling microscope, visit **www.physics.purdue.edu/ nanophys/stm.html**

**Figure 28.24**

This photograph is a demonstration of a "quantum-corral" consisting of a ring of 48 iron atoms located on a copper surface. The diameter of the ring is 143 nm, and the photograph was obtained using a low-temperature scanning tunneling microscope (STM). Corrals and other structures are able to confine surface electron waves. The study of such structures will play an important role in determining the future of small electronic devices. *(IBM Corporation Research Division)*

is maintained between the probe and specimen; the current (which is related to the probability of tunneling) is very sensitive to the barrier height (which is related to the separation between the tip and specimen), as seen in Example 28.12. Maintaining a constant tunneling current produces a feedback signal that is used to raise and lower the probe as the surface is scanned. Because the vertical motion of the probe follows the contour of the specimen's surface, an image of the surface is obtained.

Figure 28.24 shows an image made with a scanning tunneling microscope.

### context connection
### 28.14 • THE COSMIC TEMPERATURE

Now that we have introduced the concepts of quantum physics for microscopic particles and systems, let us see how we can connect these concepts to processes occurring on a cosmic scale. For our first such connection, consider the Universe as a system. It is widely believed that the Universe began with a cataclysmic explosion called the **Big Bang,** first mentioned in Chapter 5. Because of this explosion, all of the material in the Universe is moving apart. This expansion causes a Doppler shift in radiation left over from the Big Bang, such that the wavelength of the radiation lengthens (as discussed in Section 13.8 for sound waves). In the 1940s, Ralph Alpher, George Gamow, and George Hermann developed a structural model of the universe in which they predicted that the thermal radiation from the Big Bang should still be present and that it should now have a wavelength distribution consistent with a black body with a temperature of a few kelvins.

In 1965, Arno Penzias and Robert Wilson of Bell Telephone Laboratories, were measuring radiation from the Milky Way galaxy using a special 20-ft antenna as a radio telescope. They noticed a consistent background "noise" of radiation in the signals from the antenna. Despite their great efforts to test alternative hypotheses for the origin of the noise in terms of interference from the Sun, an unknown source in the Milky Way, structural problems in the antenna, and even the presence of pigeon droppings in the antenna, none of the hypotheses were sufficient to explain the noise.

What Penzias and Wilson were detecting was the thermal radiation from the Big Bang. The fact that it was detected by their system regardless of the direction of the antenna was consistent with the fact that the radiation is spread throughout the Universe, as the Big Bang model predicts. A measurement of the intensity of this radiation suggested that the temperature associated with the radiation was about 3 K, consistent with Alpher, Gamow, and Hermann's prediction from the 1940s. Although the measurement was consistent with their prediction, the measurement was taken at only a single wavelength. Full agreement with the model of the Universe as a black body would come only if measurements at many wavelengths demonstrated a distribution in wavelengths consistent with Figure 28.2.

In the years following Penzias and Wilson's discovery, other researchers made measurements at different wavelengths. In 1989, the COBE (*COsmic Background Explorer*) satellite was launched by NASA and added critical measurements at wavelengths below 0.1 cm. The results of these measurements are shown in Figure 28.25. The series of measurements taken since 1965 are consistent with thermal radiation associated with a temperature of 2.7 K. The whole story of the cosmic temperature is a remarkable example of science at work—building a model, making a prediction, taking measurements, and testing the measurements against the predictions.

In this first chapter of our Context on *The Cosmic Connection*, we have seen our first example of this connection. By looking at the thermal radiation from microscopic vibrating objects, we learn something about the origin of our Universe. In the next chapter, we shall see more examples of this fascinating connection.

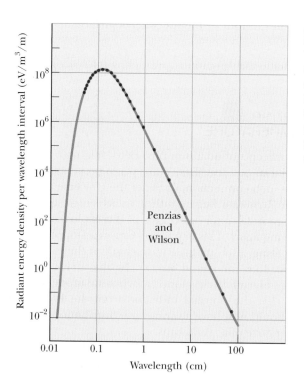

**Figure 28.25**

Theoretical blackbody wavelength distribution (red curve) and measured data points for radiation from the Big Bang. Most of the data were collected from the Cosmic Background Explorer (COBE) satellite. The data point of Penzias and Wilson is indicated.

# SUMMARY

The characteristics of **blackbody radiation** cannot be explained by classical concepts. Planck introduced the first model of **quantum physics** when he argued that the atomic oscillators responsible for this radiation exist only in discrete **quantum states.**

In the **photoelectric effect,** electrons are ejected from a metallic surface when light is incident on that surface. Einstein provided a successful explanation of this effect by extending Planck's quantum theory to the electromagnetic field. In this model, light is viewed as a stream of particles called **photons,** each with energy $E = hf$, where $f$ is the frequency and $h$ is **Planck's constant.** The maximum kinetic energy of the ejected photoelectron is given by

$$K_{max} = hf - \phi \qquad [28.5]$$

where $\phi$ is the **work function** of the metal.

X-rays striking a target are scattered at various angles by electrons in the target. A shift in wavelength is observed for the scattered x-rays, and the phenomenon is known as the **Compton effect.** Classical physics does not correctly explain the experimental results of this effect. If the x-ray is treated as a photon, conservation of energy and momentum applied to the isolated system of the photon and the electron yields for the Compton shift the expression

$$\lambda' - \lambda_0 = \frac{h}{m_e c} (1 - \cos \theta) \qquad [28.8]$$

where $m_e$ is the mass of the electron, $c$ is the speed of light, and $\theta$ is the scattering angle.

Every object of mass $m$ and momentum $p$ has wave-like properties, with a **de Broglie wavelength** given by the relation

$$\lambda = \frac{h}{p} = \frac{h}{mv} \qquad [28.10]$$

The wave–particle duality is the basis of the **quantum particle model.** It can be interpreted by imagining a particle to be made up of a combination of a large number of waves. These waves interfere constructively in a small region of space called a **wave packet.**

The **uncertainty principle** states that if a measurement of position is made with uncertainty $\Delta x$ and a *simultaneous* measurement of momentum is made with uncertainty $\Delta p_x$, then the product of the two uncertainties can never be less than a number on the order of $\hbar$.

$$\Delta x \, \Delta p_x \geq \frac{\hbar}{2} \qquad [28.17]$$

The uncertainty principle is a natural outgrowth of the wave packet model.

Matter waves are represented by the **wave function** $\psi(x, y, z)$. The **probability density** that a particle will be found at a point is $|\psi|^2$. If the particle is confined to moving along the $x$ axis, then the probability that it will be located in an interval $dx$ is given by $|\psi|^2 \, dx$. Furthermore, the wave function must be **normalized:**

$$\int_{-\infty}^{\infty} |\psi|^2 \, dx = 1 \qquad [28.22]$$

The measured position $x$ of the particle, averaged over many trials, is called the **expectation value** of $x$ and is defined by

$$\langle x \rangle \equiv \int_{-\infty}^{\infty} \psi^* x \psi \, dx \qquad [28.24]$$

If a particle of mass $m$ is confined to moving in a one-dimensional box of width $L$ whose walls are perfectly rigid, the allowed wave functions for the particle are

$$\psi(x) = A \sin \left( \frac{n \pi x}{L} \right) \quad n = 1, 2, 3, \dots \qquad [28.28]$$

The particle has a well-defined wavelength $\lambda$ whose values are such that the width of the box $L$ is equal to an integral number of half wavelengths, that is, $L = n\lambda/2$. The energies of a particle in a box are quantized and are given by

$$E_n = \left( \frac{h^2}{8mL^2} \right) n^2 \quad n = 1, 2, 3, \dots \qquad [28.29]$$

Quantum systems generally involve a particle under some constraints imposed by the environment. This is the basis of the **quantum particle under boundary conditions model.** In this model, the wave function for a system is found and application of boundary conditions on the system allow a determination of the allowed energies and unknown constants in the wave function.

The wave function must satisfy the **Schrödinger equation.** The time-independent Schrödinger equation for a particle confined to moving along the $x$ axis is

$$-\frac{\hbar^2}{2m} \frac{d^2 \psi}{dx^2} + U\psi = E\psi \qquad [28.30]$$

where $E$ is the total energy of the system and $U$ is the potential energy of the system.

When a particle of energy $E$ meets a barrier of height $U$, where $E < U$, the particle has a finite probability of penetrating the barrier. This process, called **tunneling,** is the basic mechanism that explains the operation of the Josephson junction and the phenomenon of alpha decay in some radioactive nuclei.

## QUESTIONS

1. Why is it impossible to simultaneously measure with infinite accuracy the position and speed of a particle along a given direction?

2. If matter has a wave nature, why is this wave-like characteristic not observable in our daily experiences?

3. An electron and a proton are accelerated from rest through the same potential difference. Which particle has the longer wavelength?

4. In describing the passage of electrons through a slit and arriving at a screen, the physicist Richard Feynman said that "electrons arrive in lumps, like particles, but the probability of arrival of these lumps is determined as the intensity of the waves would be. It is in this sense that the electron behaves sometimes like a particle and sometimes like a wave." Elaborate on this point in your own words. (For a further discussion of this point, see R. Feynman, *The Character of Physical Law,* Cambridge, Mass., MIT Press, 1980, Chapter 6.)

5. In the photoelectric effect, explain why the photocurrent depends on the intensity of the light source but not on the frequency.

6. Why does the existence of a cutoff frequency in the photoelectric effect favor a particle theory of light over a wave theory?

7. Is light a wave or a particle? Support your answer by citing specific experimental evidence.

8. Is an electron a wave or a particle? Support your answer by citing specific experimental evidence.

9. Suppose a photograph were made of a person's face using only a few photons. Would the result be simply a very faint image of the face? Discuss.

10. Which has more energy, a photon of ultraviolet radiation or a photon of yellow light?

11. The brightest star in the constellation Lyra is the bluish star Vega, and the brightest star in Bootes is the reddish star Arcturus. How do you account for the difference in color of the two stars?

12. An x-ray photon is scattered by an electron. What happens to the frequency of the scattered photon relative to that of the incident photon?

13. Using Wien's displacement law, estimate the wavelength of highest intensity given off by a human body. Using this information, explain why an infrared detector would be a useful alarm for security work.

14. Why was the Davisson–Germer diffraction of electrons an important experiment?

15. What is the significance of the wave function $\psi$?

16. For a particle in a box, the probability density at certain points is zero (see Fig. 28.21). Does this imply that the particle cannot move across these points? Explain.

17. What is the Schrödinger equation? How is it useful in describing atomic phenomena?

18. If the photoelectric effect is observed for one metal, can you conclude that the effect will also be observed for another metal under the same conditions? Explain.

## PROBLEMS

1, 2, 3 = straightforward, intermediate, challenging    □ = full solution available in the *Student Solutions Manual and Study Guide*

web = solution posted at **http://www.harcourtcollege.com/physics/**    ▯ = computer useful in solving problem

▦ = Interactive Physics    ▩ = paired numerical/symbolic problems    ▨ = life science application

### Section 28.1 Blackbody Radiation and Planck's Theory

1. (a) Assuming that the tungsten filament of a lightbulb is a black body with a temperature of 2 900 K, determine the wavelength at which it radiates with highest intensity. (b) Why does your answer to part (a) suggest that more energy from a lightbulb goes into infrared radiation than into visible light?

2. The radius of our Sun is $6.96 \times 10^8$ m, and its total power output is $3.77 \times 10^{26}$ W. (a) Assuming that the Sun's surface emits as a black body, calculate its surface temperature. (b) Using the result of part (a), find $\lambda_{max}$ for the Sun.

3. Calculate the energy, in electron volts, of a photon whose frequency is (a) 620 THz, (b) 3.10 GHz, (c) 46.0 MHz. (d) Determine the corresponding wavelengths for these photons, and state the classification of each on the electromagnetic spectrum.

4. The average threshold of dark-adapted (scotopic) vision is $4.00 \times 10^{-11}$ W/m$^2$ at a central wavelength of 500 nm. If light having this intensity and wavelength enters the eye and the pupil is open to its maximum diameter of 8.50 mm, how many photons per second enter the eye?

5. An FM radio transmitter has a power output of 150 kW and operates at a frequency of 99.7 MHz. How many photons per second does the transmitter emit?

6. A simple pendulum has a length of 1.00 m and a mass of 1.00 kg. The amplitude of oscillations of the pendulum is 3.00 cm. Estimate the quantum number for the pendulum.

### Section 28.2 The Photoelectric Effect

7. Molybdenum has a work function of 4.20 eV. (a) Find the cutoff wavelength and cutoff frequency for the photoelectric effect. (b) What is the stopping potential if the incident light has a wavelength of 180 nm?

8. Electrons are ejected from a metallic surface with speeds ranging up to $4.60 \times 10^5$ m/s when light with a wavelength of 625 nm is used. (a) What is the work function of the surface? (b) What is the cutoff frequency for this surface?

9. Two light sources are used in a photoelectric experiment to determine the work function for a particular metal surface. When green light from a mercury lamp ($\lambda = 546.1$ nm) is used, a stopping potential of 0.376 V reduces the photocurrent to zero. (a) Based on this measurement, what is the work function for this metal? (b) What stopping potential is observed when using the yellow light from a helium discharge tube ($\lambda = 587.5$ nm)?

10. From the scattering of sunlight, Thomson calculated the classical radius of the electron as having a value of $2.82 \times 10^{-15}$ m. Sunlight with an intensity of 500 W/m² falls on a disk with this radius. Calculate the time required to accumulate 1.00 eV of energy. Assume that light is a classical wave and that the light striking the disk is completely absorbed. How does your result compare with the observation that photoelectrons are emitted promptly (within $10^{-9}$ s)?

11. **Review Problem.** An isolated copper sphere of radius 5.00 cm, initially uncharged, is illuminated by ultraviolet light of wavelength 200 nm. What charge will the photoelectric effect induce on the sphere? The work function for copper is 4.70 eV.

### Section 28.3 The Compton Effect

12. Calculate the energy and momentum of a photon of wavelength 700 nm.

13. X-rays having an energy of 300 keV undergo Compton scattering from a target. The scattered rays are detected at 37.0° relative to the incident rays. Find (a) the Compton shift at this angle, (b) the energy of the scattered x-ray, and (c) the energy of the recoiling electron.

14. A 0.110-nm photon collides with a stationary electron. After the collision, the electron moves forward and the photon recoils backwards. Find the momentum and the kinetic energy of the electron.

15. A 0.001 60-nm photon scatters from a free electron. For what (photon) scattering angle does the recoiling electron have kinetic energy equal to the energy of the scattered photon?

16. After a 0.800-nm x-ray photon scatters from a free electron, the electron recoils at $1.40 \times 10^6$ m/s. (a) What was the Compton shift in the photon's wavelength? (b) Through what angle was the photon scattered?

### Section 28.4 Photons and Electromagnetic Waves

### Section 28.5 The Wave Properties of Particles

17. Calculate the de Broglie wavelength for a proton moving with a speed of $1.00 \times 10^6$ m/s.

18. (a) An electron has kinetic energy 3.00 eV. Find its wavelength. (b) A photon has energy 3.00 eV. Find its wavelength.

19. In the Davisson–Germer experiment, 54.0-eV electrons were diffracted from a nickel lattice. If the first maximum in the diffraction pattern was observed at $\phi = 50.0°$ (Fig. P28.19), what was the lattice spacing $a$?

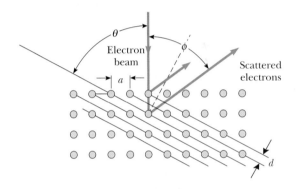

**Figure P28.19**

20. After learning about de Broglie's hypothesis that particles of momentum $p$ have wave characteristics with wavelength $\lambda = h/p$, an 80.0-kg student has grown concerned about being diffracted when passing through a doorway that is 75.0 cm wide. Assume that significant diffraction occurs when the width of the diffraction aperture is less that 10.0 times the wavelength of the wave being diffracted. (a) Determine the maximum speed at which the student can pass through the doorway so as to be significantly diffracted. (b) With that speed, how long will it take the student to pass through the doorway if it is 15.0 cm thick? Compare your result to the currently accepted age of the Universe, which is $4 \times 10^{17}$ s. (c) Should this student worry about being diffracted?

21. The nucleus of an atom is on the order of $10^{-14}$ m in diameter. For an electron to be confined to a nucleus, its de Broglie wavelength would have to be of this order of magnitude or smaller. (a) What would be the kinetic energy of an electron confined to this region? (b) On the basis of this result, would you expect to find an electron in a nucleus? Explain.

**22.** Robert Hofstadter won the 1961 Nobel Prize in physics for his pioneering work in scattering 20-GeV electrons from nuclei. (a) What is the gamma factor for a 20.0-GeV electron, where $\gamma = (1 - v^2/c^2)^{-1/2}$? (b) What is the momentum of the electron in kg·m/s? (c) What is the wavelength of a 20.0-GeV electron, and how does it compare with the size of a nucleus?

## Section 28.6 The Quantum Particle

**23.** For a free relativistic quantum particle moving with speed $v$, the total energy is $E = hf = \hbar\omega = \sqrt{p^2c^2 + m^2c^4}$ and the momentum is $p = h/\lambda = \hbar k = \gamma mv$. For the wave function representing the particle, the group speed is $v_g = d\omega/dk$. Prove that the group speed of the wave function is the same as the speed of the particle.

## Section 28.7 The Double-Slit Experiment Revisited

**24.** A modified oscilloscope is used to perform an electron interference experiment. Electrons are incident on a pair of narrow slits 0.060 0 $\mu$m apart. The bright bands in the interference pattern are separated by 0.400 mm on a screen 20.0 cm from the slits. Determine the potential difference through which the electrons were accelerated to give this pattern.

**25.** Neutrons traveling at 0.400 m/s are directed through a double slit having a 1.00-mm separation. An array of detectors is placed 10.0 m from the slit. (a) What is the de Broglie wavelength of the neutrons? (b) How far off axis is the first zero-intensity point on the detector array? (c) When a neutron reaches a detector, can we say which slit the neutron passed through? Explain.

**26.** The resolving power of a microscope depends on the wavelength used. If one wished to "see" an atom, a resolution of approximately $1.00 \times 10^{-11}$ m would be required. (a) If electrons are used (in an electron microscope), what minimum kinetic energy is required for the electrons? (b) If photons are used, what minimum photon energy is needed to obtain the required resolution?

## Section 28.8 The Uncertainty Principle

**27.** An electron ($m_e = 9.11 \times 10^{-31}$ kg) and a bullet ($m = 0.020\ 0$ kg) each have a velocity of magnitude 500 m/s, accurate to within 0.010 0 %. Within what limits could we determine the position of the objects along the direction of the velocity?

**28.** Suppose Fuzzy, a quantum-mechanical duck, lives in a world in which $h = 2\pi$ J·s. Fuzzy has a mass of 2.00 kg and is initially known to be swimming within a pond 1.00 m wide. (a) What is the minimum uncertainty in his velocity component parallel to the width? (b) Assuming that this

uncertainty in speed prevails for 5.00 s, determine the uncertainty in position along the width after this time.

**29.** An air rifle is used to shoot 1.00-g particles at 100 m/s through a hole of diameter 2.00 mm. How far from the rifle must an observer be to see the beam spread by 1.00 cm because of the uncertainty principle? Compare this answer with the diameter of the visible Universe ($2 \times 10^{26}$ m).

**30.** A $\pi^0$ meson is an unstable particle produced in high-energy particle collisions. Its rest energy is about 135 MeV, and it exists for an average lifetime of only $8.70 \times 10^{-17}$ s before decaying into two gamma rays. Using the uncertainty principle, estimate the fractional uncertainty $\Delta m/m$ in its mass determination.

**31.** A woman on a ladder drops small pellets toward a point target on the floor. (a) Show that, according to the uncertainty principle, the average miss distance must be at least

$$\Delta x_f = (2\hbar/m)^{1/2}(2H/g)^{1/4}$$

where $H$ is the initial height of each pellet above the floor and $m$ is the mass of each pellet. Assume that the spread in impact points is given by $\Delta x_f = \Delta x_i + (\Delta v_x)t$. (b) If $H = 2.00$ m and $m = 0.500$ g, what is $\Delta x_f$?

## Section 28.9 An Interpretation of Quantum Mechanics

**32.** The wave function for a particle is

$$\psi(x) = \sqrt{\frac{a}{\pi(x^2 + a^2)}}$$

for $a > 0$ and $-\infty < x < +\infty$. Determine the probability that the particle is located somewhere between $x = -a$ and $x = +a$.

**33.** A free electron has a wave function

$$\psi(x) = A\sin(5.00 \times 10^{10}\ x)$$

where $x$ is in meters. Find (a) the de Broglie wavelength, (b) the linear momentum, and (c) the kinetic energy in electron volts.

## Section 28.10 A Particle in a Box

## Section 28.11 The Quantum Particle Under Boundary Conditions

**34.** An electron that has an energy of approximately 6 eV moves between rigid walls 1.00 nm apart. Find (a) the quantum number $n$ for the energy state that the electron occupies and (b) the precise energy of the electron.

**35.** An electron is contained in a one-dimensional box of width 0.100 nm. (a) Draw an energy level diagram for the electron for levels up to $n = 4$. (b) Find the wavelengths of all photons that can be emitted by the electron in making spontaneous transitions that could eventually carry it from the $n = 4$ state to the $n = 1$ state.

**36.** The nuclear potential energy that binds protons and neutrons in a nucleus is often approximated by a square well. Imagine a proton confined in an infinitely high square well of width 10.0 fm, a typical nuclear diameter. Calculate the wavelength and energy associated with the photon emitted when the proton moves from the $n = 2$ state to the ground state. In what region of the electromagnetic spectrum does this wavelength belong?

**37.** A particle in an infinitely deep square well has a wave function that is given by

$$\psi_2(x) = \sqrt{\frac{2}{L}} \sin\left(\frac{2\pi x}{L}\right)$$

for $0 \leq x \leq L$ and is zero otherwise. (a) Determine the expectation value of $x$. (b) Determine the probability of finding the particle near $L/2$, by calculating the probability that the particle lies in the range $0.490\,L \leq x \leq 0.510\,L$. (c) Determine the probability of finding the particle near $L/4$, by calculating the probability that the particle lies in the range $0.240\,L \leq x \leq 0.260\,L$. (d) Argue that no contradiction exists between the result of part (a) and the results of parts (b) and (c).

**38.** The wave function for a particle confined to moving in a one-dimensional box is

$$\psi(x) = A \sin\left(\frac{n\pi x}{L}\right)$$

Use the normalization condition on $\psi$ to show that

$$A = \sqrt{\frac{2}{L}}$$

*Hint:* Because the box width is $L$, the wave function is zero for $x < 0$ and for $x > L$, so the normalization condition (see Eq. 28.22) reduces to

$$\int_0^L |\psi|^2 \, dx = 1$$

**39.** The wave function of an electron is

$$\psi_2(x) = \sqrt{\frac{2}{L}} \sin\left(\frac{2\pi x}{L}\right)$$

Calculate the probability of finding the electron between $x = 0$ and $x = L/4$.

## Section 28.12  The Schrödinger Equation

**40.** The wave function of a particle is given by

$$\psi(x) = A \cos(kx) + B \sin(kx)$$

where $A$, $B$, and $k$ are constants. Show that $\psi$ is a solution of the Schrödinger equation (see Eq. 28.30), assuming the particle is free ($U = 0$), and find the corresponding energy $E$ of the particle.

**41.** Show that the wave function $\psi = Ae^{i(kx-\omega t)}$ is a solution to the Schrödinger equation (see Eq. 28.30), where $k = 2\pi/\lambda$ and $U = 0$.

**42.** Prove that the first term in the Schrödinger equation, $-(\hbar^2/2m)(d^2\psi/dx^2)$, reduces to the kinetic energy of the particle multiplied by the wave function (a) for a freely moving particle, with the wave function given by Equation 28.20, and (b) for a particle in a box, with the wave function given by Equation 28.28.

**43.** A particle of mass $m$ moves in a potential well of width $2L$. Its potential energy is infinite for $x < -L$ and for $x > +L$. Inside the region $-L < x < L$, its potential energy is given by

$$U(x) = \frac{-\hbar^2 x^2}{mL^2(L^2 - x^2)}$$

In addition, the particle is in a stationary state that is described by the wave function $\psi(x) = A(1 - x^2/L^2)$ for $-L < x < +L$, and by $\psi(x) = 0$ elsewhere. (a) Determine the energy of the particle in terms of $\hbar$, $m$, and $L$. (*Hint:* Use the Schrödinger equation, Eq. 28.30.) (b) Show that $A = (15/16L)^{1/2}$. (c) Determine the probability that the particle is located between $x = -L/3$ and $x = +L/3$.

## Section 28.13  Tunneling Through a Potential Energy Barrier

**44.** An electron having total energy $E = 4.50$ eV approaches a rectangular energy barrier with $U = 5.00$ eV and $L = 950$ pm as in Figure P28.44. Classically, the electron cannot pass through the barrier because $E < U$. However, quantum-mechanically the probability of tunneling is not zero. Calculate this probability, which is the transmission coefficient.

**Figure P28.44**  Problems 44 and 45.

**45.** An electron with kinetic energy $E = 5.00$ eV is incident on a barrier with thickness $L = 0.200$ nm and height $U = 10.0$ eV (see Fig. P28.44). What is the probability that the electron (a) will tunnel through the barrier? (b) will be reflected?

### Section 28.14 Context Connection — The Cosmic Temperature

> Problems 11, 12, and 53 in Chapter 24 can be assigned with this section.

46. **Review Problem.** A star moving away from the Earth at $0.280c$ emits radiation that we measure to be most intense at the wavelength 500 nm. Determine the surface temperature of this star. (*Hint:* See Section 24.3 on the Doppler effect for light.)

47. The cosmic background radiation is blackbody radiation at a temperature of 2.73 K. (a) Determine the wavelength at which this radiation has its maximum intensity. (b) In what part of the electromagnetic spectrum is the peak of the distribution?

48. Find the intensity of the cosmic background radiation, emitted by the fireball of the Big Bang at a temperature of 2.73 K.

### Additional Problems

49. Figure P28.49 shows the stopping potential versus the incident photon frequency for the photoelectric effect for sodium. Use the graph to find (a) the work function, (b) the ratio $h/e$, and (c) the cutoff wavelength. The data are taken from R. A. Millikan, *Phys. Rev.* **7**:362, 1916.

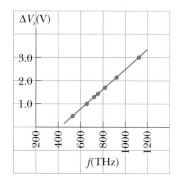

**Figure P28.49**

50. **Review Problem.** Photons of wavelength $\lambda$ are incident on a metal. The most energetic electrons ejected from the metal are bent into a circular arc of radius $R$ by a magnetic field having a magnitude $B$. What is the work function of the metal?

51. The following table shows data obtained in a photoelectric experiment. (a) Using these data, make a graph similar to Figure 28.8 that plots as a straight line. From the graph, determine (b) an experimental value for Planck's constant (in joule-seconds) and (c) the work function (in electron volts) for the surface. (Two significant figures for each answer are sufficient.)

| Wavelength (nm) | Maximum Kinetic Energy of Photoelectrons (eV) |
|---|---|
| 588 | 0.67 |
| 505 | 0.98 |
| 445 | 1.35 |
| 399 | 1.63 |

52. Show that the speed of a particle having de Broglie wavelength $\lambda$ and Compton wavelength $\lambda_C = h/(mc)$ is

$$v = \frac{c}{\sqrt{1 + (\lambda/\lambda_C)^2}}$$

53. The neutron has a mass of $1.67 \times 10^{-27}$ kg. Neutrons emitted in nuclear reactions can be slowed down via collisions with matter. They are referred to as thermal neutrons once they come into thermal equilibrium with their surroundings. The average kinetic energy $(3k_B T/2)$ of a thermal neutron is approximately 0.04 eV. Calculate the de Broglie wavelength of a neutron with a kinetic energy of 0.040 0 eV. How does it compare with the characteristic atomic spacing in a crystal? Would you expect thermal neutrons to exhibit diffraction effects when scattered by a crystal?

54. A particle of mass $2.00 \times 10^{-28}$ kg is confined to a one-dimensional box of width $1.00 \times 10^{-10}$ m. For $n = 1$, what are (a) the particle's wavelength, (b) its momentum, and (c) its ground-state energy?

55. An electron is represented by the time-independent wave function

$$\psi(x) = \begin{cases} Ae^{-\alpha x} & \text{for } x > 0 \\ Ae^{+\alpha x} & \text{for } x < 0 \end{cases}$$

(a) Sketch the wave function as a function of $x$. (b) Sketch the probability that the electron is found between $x$ and $x + dx$. (c) Argue that this can be a physically reasonable wave function. (d) Normalize the wave function. (e) Determine the probability of finding the electron somewhere in the range

$$x_1 = -\frac{1}{2\alpha} \quad \text{to} \quad x_2 = \frac{1}{2\alpha}$$

56. Particles incident from the left are confronted with a step in potential energy as shown in Figure P28.56. The step has a height $U$, and the particles have energy $E > U$. Classically, we would expect all of the particles to continue on, although with reduced speed. According to quantum mechanics, a fraction of the particles are reflected at the barrier. (a) Prove that the reflection coefficient $R$ for this case is

$$R = \frac{(k_1 - k_2)^2}{(k_1 + k_2)^2}$$

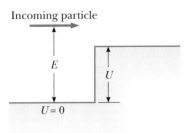

Incoming particle

$E$

$U$

$U = 0$

**Figure P28.56**

where $k_1 = 2\pi/\lambda_1$ and $k_2 = 2\pi/\lambda_2$ are the angular wave numbers for the incident and transmitted particles. Proceed as follows: Show that the wave function $\psi_1 = A \cos k_1 x + B \cos(-k_1 x)$ satisfies the Schrödinger equation in region 1, where $x < 0$. Here $A \cos k_1 x$ represents the incident beam and $B \cos(-k_1 x)$ represents the reflected particles. Show that $\psi_{\text{II}} = C \cos k_2 x$ satisfies the Schrödinger equation in region 2, for $x > 0$. Impose the boundary conditions $\psi_1 = \psi_2$ and $d\psi_1/dx = d\psi_2/dx$ at $x = 0$, to find the relationship between $B$ and $A$. Then evaluate $R = B^2/A^2$. (b) A particle that has kinetic energy $E = 7.00$ eV is incident from a region where the potential energy is zero onto one in which $U = 5.00$ eV. Find its probability of being reflected and its probability of being transmitted.

57. Johnny Jumper's favorite trick is to step out of his 16th-story window and fall 50.0 m into a pool. A news reporter takes a picture of 75.0-kg Johnny just before he makes a splash, using an exposure time of 5.00 ms. Find (a) Johnny's de Broglie wavelength at this moment, (b) the uncertainty of his kinetic energy measurement during such a period of time, and (c) the percent error caused by such an uncertainty.

58. For a particle described by a wave function $\psi(x)$, the expectation value of a physical quantity $f(x)$ associated with the particle is defined by

$$\langle f(x) \rangle = \int_{-\infty}^{\infty} f(x) \, |\psi|^2 \, dx$$

For a particle in a one-dimensional box extending from $x = 0$ to $x = L$, show that

$$\langle x^2 \rangle = \frac{L^2}{3} - \frac{L^2}{2n^2\pi^2}$$

59. A particle has a wave function

$$\psi(x) = \begin{cases} \sqrt{\dfrac{2}{a}} \, e^{-x/a} & \text{for } x > 0 \\ 0 & \text{for } x < 0 \end{cases}$$

(a) Find and sketch the probability density. (b) Find the probability that the particle will be found at any point where $x < 0$. (c) Show that $\psi$ is normalized, and then find the probability that the particle will be found between $x = 0$ and $x = a$.

(*ROSE IS ROSE reprinted by permission of United Feature Syndicate, Inc.*)

## ANSWERS TO QUICK QUIZZES

**28.1** No. The "blackness" of a black body refers to its ideal property of absorbing all radiation incident on it. If a room-temperature object absorbs all radiation, we describe it as (visibly) black. The black appearance, however, is due to the fact that our eyes are sensitive only to visible light. If we could detect infrared radiation with our eyes, we would see the object emitting radiation. If the temperature of the black body is raised, Wien's law and Stefan's law tell us that sufficient radiation will be emitted in the visible range for the object to glow visibly.

Thus, the black body could appear possibly as red, white, or blue, depending on its temperature.

**28.2** All objects indeed radiate energy, but at room temperature, this energy is primarily in the infrared region of the electromagnetic spectrum, which our eyes cannot detect. Some snakes have a membrane located above their eyes that is sensitive to infrared radiation; thus they can seek out their prey in what we would consider to be complete darkness.

**28.3** AM radio, FM radio, microwaves, sodium light. The order of photon energy will be the same as the order of frequency. See Figure 24.13 for a pictorial representation of electromagnetic radiation in order of frequency.

**28.4** No. Most metals have cutoff frequencies corresponding to photons in or near the visible range of the electromagnetic spectrum. AM radio wave photons will have far too little energy to eject electrons from the metal.

**28.5** Classical physics predicts that light of sufficient intensity causes electrons to be emitted, independent of frequency and certainly without a cutoff frequency. Also, the greater the intensity, the larger the maximum kinetic energy, with some time delay in emission at low intensities. Thus the classical expectation (which did not match experiment) yields a graph like the following:

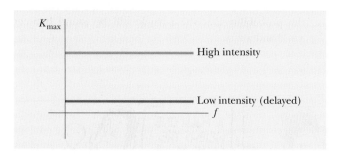

**Figure QQA28.5**

**28.6** (c). According to Equation 28.10, two particles with the same de Broglie wavelength will have the same momentum $p = mv$. If the electron and proton have the same momentum, they cannot have the same speed (a) because of the difference in their masses. For the same reason, remembering that $K = p^2/2m$, they cannot have the same kinetic energy (b). Because kinetic energy is the only type of energy a particle can have, and we have argued that the particles have different energies, Equation 28.11 tells us that the particles do not have the same frequency (d).

**28.7** (b). The Compton wavelength (Eq. 28.9) is a combination of constants and has no relation to the motion of the electron. The de Broglie wavelength (Eq. 28.10) is associated with the motion of the electron through its momentum.

**28.8** This does not affect our ability to measure its velocity component in the $y$ direction at all. The uncertainty principle relates uncertainty in position and velocity along the same axis. The zero uncertainty in position along the $x$ axis results in infinite uncertainty in its velocity component in the $x$ direction, but it is unrelated to the $y$ direction.

**28.9** Classically, we expect the particle to bounce back and forth between the two walls at constant speed. Thus, we are as likely to find the particle at any one point as any other point. Our graph of probability density versus $x$ would therefore be a straight line, with a total area under the line of unity:

**Figure QQA28.9**

**28.10** (c). According to Equation 28.29, if the length $L$ is increased, all quantized energies become smaller. Thus, the energy levels move closer together. As $L$ becomes macroscopic, the energy levels are so close together that we do not observe the quantized behavior.

**28.11** The movement of the origin will not affect the shapes of the wavefunctions. The general solution to Equation 28.31 is still $\psi(x) = A \sin kx + B \cos kx$. In Figure 28.21a, consider the wave functions for $n = 1, 2,$ and 3 with $x = 0$ at the midpoint. For odd $n$, the shape of the curve is a cosine function—it is a maximum at $x = 0$. For even $n$, the shape is a sine function—it has zero value at $x = 0$. Thus, the equivalent to Equation 28.35 is

$$\psi(x) = \begin{cases} \sqrt{\dfrac{2}{L}} \cos\left(\dfrac{n\pi x}{L}\right) & \text{(odd } n) \\ \sqrt{\dfrac{2}{L}} \sin\left(\dfrac{n\pi x}{L}\right) & \text{(even } n) \end{cases}$$

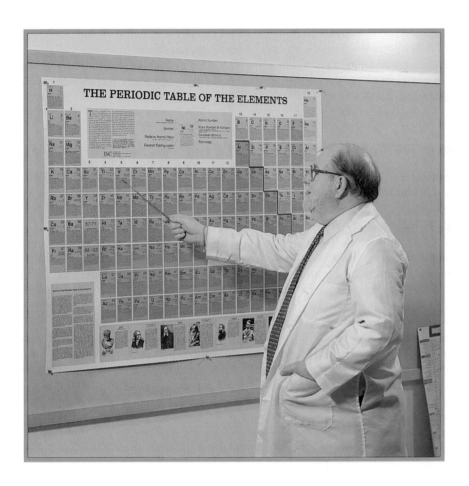

**A professor, teaching with the assistance of the periodic table, points to one of the elements. The periodic table is one of the triumphs of quantum physics applied to atomic systems.**

*(Charles D. Winters)*

# Atomic Physics

I n Chapter 28 we introduced some of the basic concepts and techniques used in quantum physics, along with their applications to various simple systems. This chapter describes the application of quantum physics to more sophisticated structural models of atoms than we have seen previously.

A large portion of this chapter is concerned with the study of the hydrogen atom from the viewpoint of quantum physics. We studied the hydrogen atom in Chapter 11 using the semiclassical approach of Bohr. In this chapter, we shall analyze the hydrogen atom with a full quantum model. Although the hydrogen atom is the simplest atomic system, it is an especially important system to understand, for several reasons:

- Much of what we learn about the hydrogen atom, with its single electron, can be extended to such single-electron ions as $He^+$ and $Li^{2+}$.
- The hydrogen atom is an ideal system for performing precise tests of theory against experiment and for improving our overall understanding of atomic structure.
- The quantum numbers used to characterize the allowed states of hydrogen can be used to qualitatively describe the allowed states of more complex atoms. This

Electron

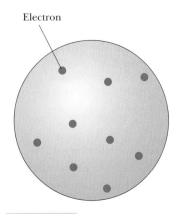

**Figure 29.1**

Thomson's model of the atom with the electrons embedded inside the positive charge like seeds in a water-melon.

characterization enables us to understand the periodic table of the elements, which is one of the greatest triumphs of quantum physics.

• The basic ideas about atomic structure must be well understood before we attempt to deal with the complexities of molecular structures and the electronic structures of solids.

## 29.1 • EARLY STRUCTURAL MODELS OF THE ATOM

The structural model of the atom in Newton's day described the atom only as a tiny, hard, indestructible sphere—a particle model that ignored any internal structure of the atom. Although this model was a good basis for the kinetic theory of gases (Chapter 16), new structural models had to be devised when later experiments revealed the electrical nature of atoms. J. J. Thomson suggested a structural model that describes the atom as a continuous volume of positive charge with electrons embedded throughout it, much like the seeds in a watermelon (Fig. 29.1).

In 1911, Ernest Rutherford and his students Hans Geiger and Ernst Marsden performed a critical experiment that showed that Thomson's model could not be correct. In this experiment, a beam of positively charged alpha particles was projected into a thin metal foil, as in Figure 29.2a. Most of the particles passed through the foil as if it were empty space—this is consistent with the Thomson model. But some of the results of the experiment were astounding: Many alpha particles were deflected from their original direction of travel through large angles. Some particles were even deflected backward, reversing their direction of travel. When Geiger informed Rutherford of these results, Rutherford wrote, "It was quite the most incredible event that has ever happened to me in my life. It was almost as incredible as if you fired a 15-inch [artillery] shell at a piece of tissue paper and it came back and hit you."

Such large deflections were not expected on the basis of Thomson's model. According to this model, a positively charged alpha particle would never come

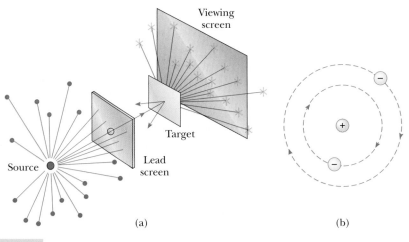

**Figure 29.2**

(a) Rutherford's technique for observing the scattering of alpha particles from a thin foil target. The source is a naturally occurring radioactive substance, such as radium. (b) Rutherford's planetary model of the atom.

close enough to a sufficiently large concentration of positive charge to cause any large-angle deflections. Rutherford explained his astounding results with a new structural model—he assumed that the positive charge was concentrated in a region that was small relative to the size of the atom. He called this concentration of positive charge the **nucleus** of the atom. Any electrons belonging to the atom were assumed to be outside the nucleus. To explain why these electrons were not pulled into the nucleus by the attractive electric force, Rutherford imagined that the electrons move in orbits about the nucleus in the same manner as the planets orbit the Sun, as in Figure 29.2b.

There are two basic difficulties with Rutherford's planetary structural model. As we saw in Chapter 11, an atom emits discrete characteristic frequencies of electromagnetic radiation and no others; the Rutherford model is unable to explain this phenomenon. A second difficulty is that Rutherford's electrons experience a centripetal acceleration. According to Maxwell's equations in electromagnetism, charges orbiting with frequency $f$ experience centripetal acceleration and therefore should radiate electromagnetic waves of frequency $f$. Unfortunately, this classical model leads to disaster when applied to the atom. As the electron radiates energy from the electron–proton system, the radius of the orbit of the electron steadily decreases and its frequency of revolution increases. Energy is continuously transferred out of the system by electromagnetic radiation. As a result, the energy of the system decreases, resulting in the decay of the orbit of the electron. This decrease in total energy leads to an increase in the kinetic energy of the electron,* an ever-increasing frequency of emitted radiation, and a rapid collapse of the atom as the electron plunges into the nucleus (Fig. 29.3).

The stage was set for Bohr! To circumvent the erroneous predictions of the Rutherford model—electrons falling into the nucleus and a continuous emission spectrum from elements—Bohr postulated that classical radiation theory does not hold for atomic-sized systems. He overcame the problem of an atom that continuously loses energy by applying Planck's ideas of quantized energy levels to orbiting atomic electrons. Thus, as described in Section 11.5, Bohr postulated that electrons in atoms are generally confined to stable, nonradiating orbits called stationary states. Furthermore, he applied Einstein's concept of the photon to arrive at an expression for the frequency of radiation emitted when the atom makes a transition from one stationary state to another.

One of the first indications that the Bohr theory needed modification arose when improved spectroscopic techniques were used to examine the spectral lines of hydrogen. It was found that many of the lines in the Balmer and other series were not single lines at all. Instead, each was a group of closely spaced lines. An additional difficulty arose when it was observed that, in some situations, some single spectral lines were split into three closely spaced lines when the atoms were placed in a strong magnetic field.

Efforts to explain these deviations from the Bohr model led to improvements in the structural model of the atom. One of the changes introduced was the concept that the electron has an intrinsic angular momentum called *spin*, which we discussed in Chapter 22 in terms of the contribution of spin to the magnetic properties of materials. We shall discuss spin in more detail in this chapter.

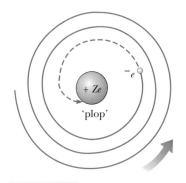

**Figure 29.3**

The classical expectation for the nuclear atom. Because the accelerating electron radiates energy, the orbit decays until the electron falls into the nucleus.

---

* As an orbital system that interacts via an inverse square force law loses energy, the kinetic energy of the orbiting object increases, but the potential energy of the system decreases by a larger amount, so that the change in the total energy of the system is negative.

## 29.2 • THE HYDROGEN ATOM REVISITED

A quantum treatment of the hydrogen atom requires a solution to the Schrödinger equation (Equation 28.30), with $U$ being the electric potential energy of the electron–proton system. The full mathematical solution of the Schrödinger equation as applied to the hydrogen atom gives a complete and beautiful description of the atom's properties. The mathematical procedures that make up the solution are beyond the scope of this text, however, and so the details shall be omitted. The solutions for some states of hydrogen will be discussed, together with the quantum numbers used to characterize allowed stationary states. We also discuss the physical significance of the quantum numbers.

Let us outline the steps that we take in developing a quantum structural model for the hydrogen atom. We apply the quantum particle under boundary conditions model by solving the Schrödinger equation and then applying boundary conditions to the solution to determine the allowed wave functions and energies of the atom. For the particle in a one-dimensional box example from Chapter 28, we found that the imposition of boundary conditions generated a single quantum number. For the three-dimensional system of the hydrogen atom, each dimension introduces a quantum number, so the model will generate three quantum numbers. We also find the need for a fourth quantum number, representing the spin, which cannot be extracted from the Schrödinger equation.

To set up the Schrödinger equation, we first need to specify the potential energy function for the system. For the hydrogen atom, this function is

$$U(r) = -k_e \frac{e^2}{r} \qquad [29.1]$$

where $k_e$ is the Coulomb constant and $r$ is the radial distance between the proton (situated at $r = 0$) and the electron.

The formal procedure for solving the problem of the hydrogen atom is to substitute $U(r)$ into the Schrödinger equation and find appropriate solutions to the equation. We did this for the particle in a box in Section 28.12. The current problem is more complicated, however, because it is three-dimensional and because $U$ is not constant. In addition, $U$ depends on the radial coordinate $r$ rather than a Cartesian coordinate $x$, $y$, or $z$. This requires the use of spherical coordinates. We shall not attempt to carry out these solutions because they are quite complicated. Rather, we shall simply describe their properties and some of their implications with regard to atomic structure.

When the boundary conditions are applied to the solutions of the Schrödinger equation, we find that the energies of the allowed states for the hydrogen atom are

$$E_n = -\left(\frac{k_e e^2}{2a_0}\right)\frac{1}{n^2} = -\frac{13.606 \text{ eV}}{n^2} \qquad n = 1, 2, 3, \ldots \qquad [29.2]$$

This result is in precise agreement with the Bohr model and with observed spectral lines—a triumph for both the Bohr approach and the quantum approach! Note that the allowed energies in our model depend only on the quantum number $n$, which is called the **principal quantum number.**

The imposition of boundary conditions also leads to two new quantum numbers that do not appear in the Bohr model. The quantum number $\ell$ is called the **orbital quantum number,** and $m_\ell$ is called the **orbital magnetic quantum number.** Although $n$ is related to the energy of the atom, $\ell$ and $m_\ell$ are related to the

**PITFALL PREVENTION 29.1**

**Energy depends on *n* only for hydrogen**

The statement after Equation 29.2 that the energy depends only on the quantum number $n$ is only true for the hydrogen atom. For more complicated atoms, we will use the same quantum numbers that we develop here for hydrogen. The energy levels for these atoms will depend primarily on $n$, but will also depend to a lesser degree on other quantum numbers, as we shall discuss later.

• *Allowed energies for the hydrogen atom*

| TABLE 29.1 | Three Quantum Numbers for the Hydrogen Atom | | |
|---|---|---|---|
| Quantum Number | Name | Allowed Values | Number of Allowed Values |
| $n$ | Principal quantum number | $1, 2, 3, \ldots$ | Any number |
| $\ell$ | Orbital quantum number | $0, 1, 2, \ldots, n-1$ | $n$ |
| $m_\ell$ | Orbital magnetic quantum number | $-\ell, -\ell+1, \ldots,$ $0, \ldots, \ell-1, \ell$ | $2\ell + 1$ |

angular momentum of the atom described in a subsequent section. From the solution to the Schrödinger equation, we find the following allowed values for these three quantum numbers:

- $n$ is an integer that can range from 1 to $\infty$

For a particular value of $n$,

- $\ell$ is an integer that can range from 0 to $n-1$

For a particular value of $\ell$,

- $m_\ell$ is an integer that can range from $-\ell$ to $\ell$

Table 29.1 summarizes the rules for determining the allowed values of $\ell$ and $m_\ell$ for a given value of $n$.

For historical reasons, all states with the same principal quantum number are said to form a **shell.** Shells are identified by the letters K, L, M, . . . , which designate the states for which $n = 1, 2, 3, \ldots$ . Likewise, all states with given values of $n$ and $\ell$ are said to form a **subshell.** Based on early practices in spectroscopy, the letters $s, p, d, f, g, h, \ldots$ are used to designate the subshells for which $\ell = 0, 1, 2, 3, 4, 5, \ldots$.* For example, the subshell designated by $3p$ has the quantum numbers $n = 3$ and $\ell = 1$; the $2s$ subshell has the quantum numbers $n = 2$ and $\ell = 0$. These notations are summarized in Table 29.2.

States with quantum numbers that violate the rules given in Table 29.1 cannot exist—they do not satisfy the boundary conditions on the wave function of the system. For instance, a $2d$ state, which would have $n = 2$ and $\ell = 2$, cannot exist—the highest allowed value of $\ell$ is $n - 1$, or 1 in this case. Thus, for $n = 2$, $2s$ and $2p$ are allowed states but $2d, 2f, \ldots$ are not. For $n = 3$, the allowed subshells are $3s$, $3p$, and $3d$.

| TABLE 29.2 | | | |
|---|---|---|---|
| **Atomic Shell and Subshell Notations** | | | |
| | Shell | | Subshell |
| $n$ | Symbol | $\ell$ | Symbol |
| 1 | K | 0 | $s$ |
| 2 | L | 1 | $p$ |
| 3 | M | 2 | $d$ |
| 4 | N | 3 | $f$ |
| 5 | O | 4 | $g$ |
| 6 | P | 5 | $h$ |
| . . . | | . . . | |

**Quick Quiz 29.1**

How many possible orbital states are there for (a) the $n = 3$ level of hydrogen? (b) the $n = 4$ level?

**Quick Quiz 29.2**

When the principal quantum number is $n = 5$, how many different values of (a) $\ell$ and (b) $m_\ell$ are possible?

---

* These seemingly strange letter designations come from descriptions of spectral lines in the early history of spectroscopy: $s$—sharp; $p$—principal; $d$—diffuse; $f$—fine. After $s, p, d,$ and $f$, the subsequent letters follow alphabetically from $f$.

In the hydrogen atom, the quantum number *n* can increase without limit. Because of this, does the frequency of possible spectrum lines from hydrogen also increase without limit?

### Example 29.1 The *n* = 2 Level of Hydrogen

For a hydrogen atom, determine the number of orbital states corresponding to the principal quantum number *n* = 2 and calculate the energies of these states.

**Solution** When $n = 2$, $\ell$ can be 0 or 1. For $\ell = 0$, $m_\ell$ can only be 0; for $\ell = 1$, $m_\ell$ can be $-1$, 0, or 1. Hence, we have one orbital state designated as the 2*s* state associated with the quantum numbers $n = 2$, $\ell = 0$, and $m_\ell = 0$, and three orbital states designated as 2*p* states for which the quantum

numbers are $n = 2$, $\ell = 1$, $m_\ell = -1$; $n = 2$, $\ell = 1$, $m_\ell = 0$; and $n = 2$, $\ell = 1$, $m_\ell = 1$, for a total of four orbital states.

Because all these states have the same principal quantum number, they also have the same energy, which can be calculated with Equation 29.2, with $n = 2$:

$$E_2 = -\frac{13.606 \text{ eV}}{2^2} = -3.401 \text{ eV}$$

## 29.3 • THE SPIN MAGNETIC QUANTUM NUMBER

**PITFALL PREVENTION 29.2**

**The electron is not spinning**

Keep in mind Pitfall Prevention 22.3—the electron is *not* physically spinning. Although the analogy between the spin of the Earth and the electron spin is conceptually useful, it should not be taken literally. The spin of the Earth is a physical rotation. Electron spin is a purely quantum effect that gives the electron an angular momentum as if it were physically spinning.

Example 29.1 was presented to give you practice in manipulating quantum numbers, but, as we shall see in this section, there are *eight* electron states for *n* = 2 rather than the four we found. These extra states can be explained by requiring a fourth quantum number for each state, the **spin magnetic quantum number** $m_s$.

Evidence of the need for this new quantum number came about because of an unusual feature in the spectra of certain gases such as sodium vapor. Close examination of one of the prominent lines of sodium shows that it is, in fact, two very closely spaced lines called a doublet. The wavelengths of these lines occur in the yellow region at 589.0 nm and 589.6 nm. In 1925, when this doublet was first noticed, atomic models could not explain it. To resolve this dilemma, Samuel Goudsmidt and George Uhlenbeck, following a suggestion by the Austrian physicist Wolfgang Pauli, proposed a new quantum number, called the spin quantum number.

To describe the spin quantum number, it is convenient (but incorrect!) to think of the electron as spinning on its axis as it orbits the nucleus in a planetary model, just as the Earth spins on its axis as it orbits the Sun. The direction in which the spin angular momentum vector can point is quantized—it can have only two directions, as shown in Figure 29.4. If the direction of spin is as shown in Figure 29.4a, the electron is said to have "spin up." If the direction of spin is as shown in Figure 29.4b, the electron is said to have "spin down." In the presence of a magnetic field, the energy of the system of the electron and the magnetic field is slightly different for the two spin directions, and this energy difference accounts for the sodium doublet. The quantum numbers associated with electron spin are $m_s = \frac{1}{2}$ for the spin-up state and $m_s = -\frac{1}{2}$ for the spin-down state. As we shall see in Example 29.2, this added quantum number doubles the number of allowed states specified by the quantum numbers $n$, $\ell$, and $m_\ell$.

The electron cannot be considered to be physically spinning. In spite of this conceptual difficulty, all experimental evidence supports the fact that the electron has some intrinsic property that is consistent with an angular momentum and can be described by the spin quantum number. The origin of this fourth quantum number was shown by Arnold Sommerfeld and Paul Dirac to lie in the relativistic properties of the electron, which requires four quantum numbers to describe it in four-dimensional space-time.

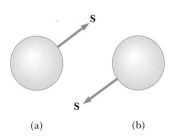

**Figure 29.4**

The spin of an electron can be described as either (a) spin up or (b) spin down.

## Example 29.2   Putting a Spin on Hydrogen

For a hydrogen atom, determine the quantum numbers associated with the possible states that correspond to the principal quantum number $n = 2$.

**Solution**   With the results from Example 29.1 and the addition of the spin quantum number, we have the possibilities given in the following table. Thus, there are eight possible states.

**EXERCISE**   Show that for $n = 3$, 18 states are possible.

| $n$ | $\ell$ | $m_\ell$ | $m_s$ | Subshell | Shell | Number of States in Subshell |
|---|---|---|---|---|---|---|
| 2 | 0 | 0 | $\frac{1}{2}$ | | | |
| 2 | 0 | 0 | $-\frac{1}{2}$ | $2s$ | L | 2 |
| 2 | 1 | 1 | $\frac{1}{2}$ | | | |
| 2 | 1 | 1 | $-\frac{1}{2}$ | | | |
| 2 | 1 | 0 | $\frac{1}{2}$ | | | |
| 2 | 1 | 0 | $-\frac{1}{2}$ | $2p$ | L | 6 |
| 2 | 1 | $-1$ | $\frac{1}{2}$ | | | |
| 2 | 1 | $-1$ | $-\frac{1}{2}$ | | | |

## 29.4 • THE WAVE FUNCTIONS FOR HYDROGEN

The potential energy of the hydrogen atom depends only on the radial distance $r$ between nucleus and electron. We therefore expect that some of the allowed states for this atom can be represented by wave functions that depend only on $r$. This indeed is the case. (Other wave functions depend on $r$ and on the angular coordinates.) The simplest wave function for the hydrogen atom describes the $1s$ state and is designated $\psi_{1s}(r)$:

$$\psi_{1s}(r) = \frac{1}{\sqrt{\pi a_0{}^3}} e^{-r/a_0} \qquad \text{[29.3]}$$

• *Wave function for hydrogen in its ground state*

where $a_0$ is the Bohr radius. This wave function satisfies the boundary conditions mentioned in Section 28.11; that is, $\psi_{1s}$ approaches zero as $r \rightarrow \infty$ and remains finite as $r \rightarrow 0$. The function in Equation 29.3 is normalized as presented. Because $\psi_{1s}$ depends only on $r$, it is spherically symmetric. In fact, **all $s$ states have spherical symmetry.**

Recall that the probability of finding the electron in any region is equal to an integral of $|\psi|^2$ over the region, if $\psi$ is normalized. The probability density for the $1s$ state is

$$|\psi_{1s}|^2 = \left(\frac{1}{\pi a_0{}^3}\right) e^{-2r/a_0} \qquad \text{[29.4]}$$

The probability of finding the electron in a volume element $dV$ is $|\psi|^2\, dV$. It is convenient to define the **radial probability density function** $P(r)$ as the probability per unit radial distance of finding the electron in a spherical shell of radius $r$ and

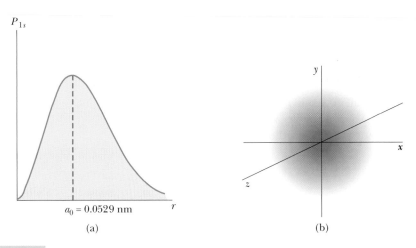

$P_{1s}$

$a_0 = 0.0529$ nm

(a)

(b)

### Figure 29.6

(a) The probability of finding the electron as a function of distance from the nucleus for the hydrogen atom in the 1*s* (ground) state. Note that the probability has its maximum value when *r* equals the first Bohr radius, $a_0$ (see Example 29.3). (b) The spherical electron charge distribution for the hydrogen atom in its 1*s* state.

### Figure 29.5

A spherical shell of radius *r* and thickness *dr* has a volume equal to $4\pi r^2\, dr$.

• *Radial probability density for the 1s state of hydrogen*

thickness *dr*. The volume of such a shell equals its surface area $4\pi r^2$ multiplied by the shell thickness *dr* (Fig. 29.5), so that

$$P(r)\, dr = |\psi|^2\, dV = |\psi|^2\, 4\pi r^2\, dr \qquad [29.5]$$

$$P(r) = 4\pi r^2 |\psi|^2 \qquad [29.6]$$

Substituting Equation 29.4 into Equation 29.6 gives the radial probability density function for the hydrogen atom in its ground state:

$$P_{1s}(r) = \left(\frac{4r^2}{a_0^3}\right) e^{-2r/a_0} \qquad [29.7]$$

A graphical representation of the function $P_{1s}(r)$ versus *r* is presented in Figure 29.6a. The peak of the curve corresponds to the most probable value of *r* for this particular state. The spherical symmetry of the distribution function is shown in Figure 29.6b.

### Example 29.3   The Ground State of Hydrogen

Calculate the most probable value of *r* for a hydrogen atom in its ground state.

**Solution**   The most probable value of *r* corresponds to the peak of the plot of $P(r)$ versus *r*. The slope of the curve at this point is zero, and so we can evaluate the most probable value of *r* by setting $dP/dr = 0$ and solving for *r*. Using Equation 29.7, we find

$$\frac{dP_{1s}(r)}{dr} = \frac{d}{dr}\left[\left(\frac{4r^2}{a_0^3}\right) e^{-2r/a_0}\right] = 0$$

Carrying out the derivative operation and simplifying the expression, we have

$$e^{-2r/a_0}\frac{d}{dr}(r^2) + r^2\frac{d}{dr}(e^{-2r/a_0}) = 0$$

$$2re^{-2r/a_0} + r^2\left(-\frac{2}{a_0}\right)e^{-2r/a_0} = 0$$

$$2r\left[1 - \left(\frac{r}{a_0}\right)\right]e^{-2r/a_0} = 0$$

This expression is satisfied if

$$1 - \left(\frac{r}{a_0}\right) = 0 \quad \longrightarrow \quad r = \boxed{a_0}$$

Notice that although the quantum model differs from the Bohr model in that the electron has a finite probability of being at *any* distance from the nucleus, the most probable distance is the same as the orbital radius in the Bohr model!

## Example 29.4 Probabilities for the Electron in Hydrogen

Calculate the probability that the electron in the ground state of hydrogen will be found outside the Bohr radius.

**Solution** The probability is found by integrating the radial probability density $P_{1s}(r)$ for this state from the Bohr radius $a_0$ to $\infty$. Using Equation 29.7, we have

$$P = \int_{a_0}^{\infty} P_{1s}(r) \; dr = \frac{4}{a_0{}^3} \int_{a_0}^{\infty} r^2 e^{-2r/a_0} \; dr$$

We can put the integral in dimensionless form by changing variables from $r$ to $z = 2r/a_0$. Noting that $z = 2$ when $r = a_0$, and that $dr = (a_0/2) \; dz$, we find

$$P = \frac{1}{2} \int_2^{\infty} z^2 e^{-z} \; dz = -\frac{1}{2} \left. (z^2 + 2z + 2) e^{-z} \right|_2^{\infty}$$

$$P = 5e^{-2} = 0.677 \qquad \text{or} \qquad \boxed{67.7\%}$$

## Example 29.5 The Quantized Solar System

Consider the Schrödinger equation for the Earth and the Sun as a system of two particles interacting via the gravitational force. What is the quantum number of the system with the Earth in its present orbit?

**Reasoning** The potential energy function for the system is

$$U(r) = -G\frac{m_E m_S}{r}$$

where $m_E$ is the mass of the Earth and $m_S$ is the mass of the Sun. Comparing this to Equation 29.1 for the hydrogen atom, $U(r) = -k_e e^2/r$, we see that it has the same mathematical form and that the constant $Gm_E m_S$ plays the role of $k_e e^2$. Thus, the solutions to the Schrödinger equation for the Earth–Sun system will be the same as those of the hydrogen atom with the appropriate change in the constants.

**Solution** If we make the substitution for the constants in Equation 29.2, we find the allowed energies of the quantized states of the Earth–Sun system:

$$E_n = -\left(\frac{Gm_E m_S}{2a_0}\right)\frac{1}{n^2} \quad n = 1, 2, 3, \ldots$$

From Equation 11.21, we can find the Bohr radius for the Earth–Sun system:

$$a_0 = \frac{\hbar^2}{m_E(Gm_E m_S)} = \frac{\hbar^2}{Gm_E{}^2 m_S}$$

$$= \frac{1.055 \times 10^{-34} \, \text{J} \cdot \text{s}}{(6.67 \times 10^{-11} \, \text{N} \cdot \text{m}^2/\text{kg}^2)(5.98 \times 10^{24} \, \text{kg})^2(1.99 \times 10^{30} \, \text{kg})}$$

$$= 2.22 \times 10^{-104} \, \text{m}$$

Thus, evaluating the allowed energies of the system, we have

$$E_n = -\left(\frac{Gm_E m_S}{2a_0}\right)\frac{1}{n^2}$$

$$= -\frac{(6.67 \times 10^{-11} \text{ N} \cdot \text{m}^2/\text{kg}^2)(5.98 \times 10^{24} \text{ kg})(1.99 \times 10^{30} \text{ kg})}{2(2.22 \times 10^{-104} \text{ m})}$$

$$= -\frac{1.79 \times 10^{148} \text{J}}{n^2} \quad n = 1, 2, 3, \ldots$$

We now evaluate the energy of the Earth–Sun system from Equation 11.8, assuming a circular orbit:

$$E = -\frac{Gm_E m_S}{2r} = -\frac{(6.67 \times 10^{-11} \text{ N} \cdot \text{m}^2/\text{kg}^2)(5.98 \times 10^{24} \text{ kg})(1.99 \times 10^{30} \text{ kg})}{2(1.50 \times 10^{11} \text{ m})}$$

$$= -2.65 \times 10^{33} \text{ J}$$

Finally, we find the quantum number associated with this state:

$$E_n = -\frac{1.79 \times 10^{148} \text{J}}{n^2}$$

$$n = \sqrt{\frac{-1.79 \times 10^{148} \text{ J}}{E}} = \sqrt{\frac{-1.79 \times 10^{148} \text{ J}}{-2.65 \times 10^{33} \text{ J}}} = 2.60 \times 10^{57}$$

This is a tremendously large quantum number. According to the correspondence principle, classical mechanics describes the Earth's motion as well as quantum mechanics does. The energies of quantum states for adjacent values of $n$ are so close together that we do not see the quantized nature of the energy. For example, if the Earth were to move into the next higher quantum state, calculations show that it would be farther from the Sun by a distance on the order of $10^{-80}$ m. Even on a nuclear scale of $10^{-15}$ m, this is undetectable.

---

Example 29.3 shows that the most probable value of $r$ for the ground state of hydrogen equals the Bohr radius $a_0$. It turns out that the average value of $r$ for the ground state of hydrogen is $\frac{3}{2} a_0$, which is 50% larger than the most probable value of $r$ (see Problem 41). The reason for this is the large asymmetry in the radial distribution function shown in Figure 29.6a. According to quantum mechanics, the atom has no sharply defined boundary. The probability distribution for the electron can be viewed as being a diffuse region of space, commonly referred to as an **electron cloud.** This is quite different from the Bohr model, which places the electron at a fixed distance from the nucleus.

For an atom in a quantum state that is a solution to the Schrödinger equation, the electron cloud structure remains the same, on the average, over time. Thus, *the atom does not radiate when it is in one particular quantum state*. This removes the problem that plagued the Rutherford model, in which the atom continuously radiates until the electron spirals into the nucleus. Because no change occurs in the charge structure in the electron cloud, the atom does not radiate. Radiation occurs only when a transition is made, so that the structure of the electron cloud changes in time.

The next simplest wave function for the hydrogen atom is the one corresponding to the $2s$ state ($n = 2$, $\ell = 0$). The normalized wave function for this state is

• *Wave function for hydrogen in the 2s state*

$$\psi_{2s}(r) = \frac{1}{4\sqrt{2\pi}} \left(\frac{1}{a_0}\right)^{3/2} \left[2 - \frac{r}{a_0}\right] e^{-r/2a_0} \tag{29.8}$$

Like the $\psi_{1s}$ function, $\psi_{2s}$ depends only on $r$ and is spherically symmetric. The energy corresponding to this state is $E_2 = -(13.6 \text{ eV}/4) = -3.4$ eV. This energy level represents the first excited state of hydrogen. Plots of the radial probability

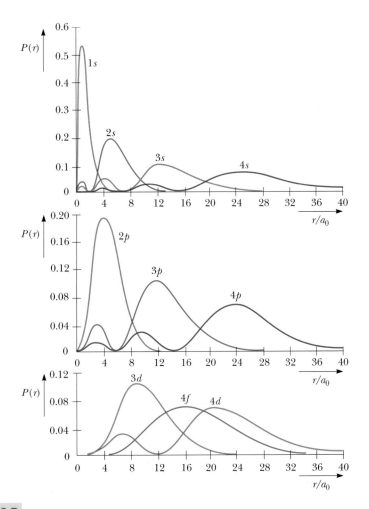

**Figure 29.7**

The radial probability density function versus $r/a_0$ for several states of the hydrogen atom. *(From E. U. Condon and G. H. Shortley, The Theory of Atomic Spectra, Cambridge, Cambridge University Press, 1953; used with permission)*

density function for this state and several other states of hydrogen are shown in Figure 29.7. The plot for the $2s$ state has two peaks. In this case, the most probable value corresponds to that value of $r$ with the higher value of $P(\approx 5a_0)$. An electron in an atom in the $2s$ state would be much farther from the nucleus (on the average) than an electron in an atom in the $1s$ state. The average value of $r$ is even greater for the $3d$, $3p$, and $4d$ states.

# 29.5 • PHYSICAL INTERPRETATION OF THE QUANTUM NUMBERS

As discussed in Section 29.2, the energy of a particular state in our model depends on the principal quantum number. Now let us see what the other three quantum numbers contribute to the physical nature of our quantum structural model of the atom.

## The Orbital Quantum Number

If a particle moves in a circle of radius $r$, the magnitude of its angular momentum relative to the center of the circle is $L = mvr$. The direction of **L** is perpendicular to the plane of the circle, and the sense of **L** is given by a right-hand rule.* According to classical physics, $L$ can have any value. The Bohr model of hydrogen, however, postulates that the angular momentum is restricted to integer multiples of $\hbar$; that is, $mvr = n\hbar$. This model must be modified because it predicts (incorrectly) that the ground state of hydrogen ($n = 1$) has one unit of angular momentum. Our quantum model shows that the lowest value of the orbital quantum number, which is related to the orbital momentum, is $\ell = 0$, which corresponds to zero angular momentum.

According to the quantum model, an atom in a state whose principal quantum number is $n$ can take on the following *discrete* values for the magnitude of the orbital angular momentum vector:[†]

• *Allowed values of L*

$$|\mathbf{L}| = L = \sqrt{\ell(\ell + 1)}\,\hbar \quad \ell = 0, 1, 2, \ldots, n - 1 \qquad \text{[29.9]}$$

The fact that $L$ can be zero in this model points out the difficulties inherent in any attempt to describe results based on quantum mechanics in terms of a purely particle-like model. We cannot think in terms of electrons traveling in well-defined orbits of circular shape or any other shape, for that matter. It is more consistent with the probability notions of quantum physics to imagine the electron smeared out in space in an electron cloud, with the "density" of the cloud highest where the probability is highest. In the quantum mechanical interpretation, the electron cloud for the $L = 0$ state is spherically symmetric and has no fundamental axis of rotation.

---

## Example 29.6 Calculating $L$ for a $p$ State

Calculate the orbital angular momentum for a $p$ state of hydrogen.

**Solution** Because we know that $\hbar = 1.055 \times 10^{-34}\,\text{J}\cdot\text{s}$, we can use Equation 29.9 to calculate $L$. With $\ell = 1$ for a $p$ state, we have

$$L = \sqrt{\ell(\ell + 1)}\,\hbar = \sqrt{2}\,\hbar = \boxed{1.49 \times 10^{-34}\,\text{J}\cdot\text{s}}$$

This value is extremely small relative to that of the orbital angular momentum of a macroscopic system, such as the Earth orbiting the Sun, which is about $2.7 \times 10^{40}\,\text{J}\cdot\text{s}$. The quantum number that describes $L$ for macroscopic systems, such as the Earth and the Sun, is so large that the separation between adjacent states cannot be measured. We do not see quantized angular momentum for macroscopic systems. Once again, the correspondence principle is upheld.

---

## The Magnetic Orbital Quantum Number

We have seen in the previous discussion that the magnitude of the orbital angular momentum is quantized. Because angular momentum is a vector, its direction must also be specified. An orbiting electron can be considered an effective current loop with a corresponding magnetic moment. Such a moment placed in a magnetic field **B** will interact with the field.

---

* See Sections 10.8 and 10.9 for a review of this material on angular momentum.

[†] Equation 29.9 is a direct result of the mathematical solution of the Schrödinger equation and the application of angular boundary conditions. This development, however, is beyond the scope of this text and will not be presented.

Suppose a weak magnetic field is applied to an atom and we define the direction of the field as the $z$ axis. According to quantum mechanics, we find a startling result—the *direction* of the angular momentum vector relative to the $z$ axis is quantized! Once a $z$ axis is specified, the angular momentum vector can only point in certain directions with respect to this axis. The fact that the direction of **L** is quantized is often referred to as **space quantization,** because we are quantizing a *direction* rather than a *magnitude*.

The quantization of the direction of **L** is described by giving the allowed $z$ components of the vector. The magnetic orbital quantum number $m_\ell$ specifies the allowed values of $L_z$ according to the expression

$$L_z = m_\ell \hbar \qquad \text{[29.10]}$$

• *Allowed values of $L_z$*

Let us look at the possible orientations of **L** for a given value of $\ell$. Recall that $m_\ell$ can have values ranging from $-\ell$ to $\ell$. If $\ell = 0$, then $L = 0$ and there is no vector for which to consider a direction. If $\ell = 1$, then the possible values of $m_\ell$ are $-1$, $0$, and $1$, so that $L_z$ may be $-\hbar$, $0$, or $\hbar$. If $\ell = 2$, $m_\ell$ can be $-2$, $-1$, $0$, $1$, or $2$, corresponding to $L_z$ values of $-2\hbar$, $-\hbar$, $0$, $\hbar$, or $2\hbar$, and so on.

A useful specialized pictorial representation for understanding space quantization is commonly called a **vector model.** A vector model for $\ell = 2$ is shown in Figure 29.8a. Note that **L can never be aligned parallel or antiparallel to the $z$ axis** because $L_z$ must be smaller than the magnitude of the angular momentum **L**. The vector **L** can be *perpendicular* to the $z$ axis, which is the case if $m_\ell = 0$. From a three-dimensional viewpoint, **L** lies on the surface of a cone that makes an angle of $\theta$ with the $z$ axis, as shown in Figure 29.8b. From the figure, we see that $\theta$ is also quantized and that its values are specified through a relation based on a geometric model triangle with the **L** vector as the hypotenuse and the $z$ component as one leg:

$$\cos \theta = \frac{L_z}{|\mathbf{L}|} = \frac{m_\ell}{\sqrt{\ell(\ell + 1)}} \qquad \text{[29.11]}$$

Note that $m_\ell$ is never greater than $\ell$, and therefore $\theta$ can never be zero, consistent with our restriction on **L** not being parallel to the $z$ axis.

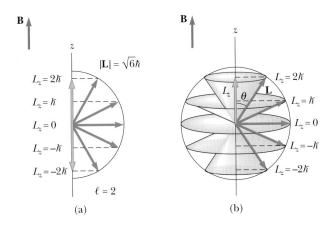

(a)

(b)

**Figure 29.8**

(a) The allowed projections of the orbital angular momentum **L** for the case $\ell = 2$. (b) The orbital angular momentum vector **L** lies on the surface of a cone and precesses about the $z$ axis when a magnetic field **B** is applied in this direction.

Because of the uncertainty principle, **L** does not point in a specific direction but rather lies somewhere on a cone in space. If **L** had a definite direction, all three components $L_x$, $L_y$, and $L_z$ would be exactly specified. For the moment, let us assume that this is the case, and let us suppose that the electron moves in the $xy$ plane, so that the uncertainty $\Delta z = 0$. Because the electron moves in the $xy$ plane, $p_z = 0$. This means that $p_z$ is precisely known, so that $\Delta p_z = 0$. The product of these two uncertainties is $\Delta z \, \Delta p_z = 0$, which is in violation of the uncertainty principle, which requires that $\Delta z \, \Delta p_z \geq \hbar/2$. In reality, only the magnitude of **L** and one component (which is traditionally chosen as $L_z$) can have definite values at the same time. In other words, quantum mechanics allows us to specify $L$ and $L_z$ but not $L_x$ and $L_y$. Because the direction of **L** is constantly changing as it precesses about the $z$ axis, the average values of $L_x$ and $L_y$ are zero and $L_z$ maintains a fixed value $m_\ell \hbar$.

### Quick Quiz 29.4

Sketch a vector model (shown in Figure 29.8 for $\ell = 2$) for $\ell = 1$.

### Example 29.7   Space Quantization for Hydrogen

For the hydrogen atom in the $\ell = 3$ state, calculate the magnitude of **L** and the allowed values of $L_z$ and $\theta$.

**Solution**   We use Equation 29.9 with $\ell = 3$:

$$L = \sqrt{\ell(\ell + 1)}\, \hbar = \sqrt{3(3 + 1)}\, \hbar = 2\sqrt{3}\, \hbar$$

The allowed values of $L_z$ are $L_z = m_\ell \hbar$ with $m_\ell = -3, -2, -1, 0, 1, 2,$ and 3:

$$L_z = \boxed{-3\hbar, -2\hbar, -\hbar, 0, \hbar, 2\hbar, 3\hbar}$$

Finally, we use Equation 29.11 to calculate the allowed values of $\theta$. Because $L = 2\sqrt{3}\, \hbar$, we have

$$\cos \theta = \frac{m_\ell}{2\sqrt{3}}$$

Substitution of the allowed values of $m_\ell$ gives

$$\cos \theta = \pm 0.866, \pm 0.577, \pm 0.289, 0$$

$$\theta = \boxed{30.0°, 54.8°, 73.2°, 90.0°, 107°, 125°, 150°}$$

Wolfgang Pauli and Niels Bohr watch a spinning top. *(Courtesy of AIP Niels Bohr Library, Margarethe Bohr Collection)*

### Electron Spin

The three quantum numbers that we have discussed so far are generated by the solution to the Schrödinger equation, and we can assign a physical interpretation to each of the quantum numbers. Let us now consider electron spin, which does *not* come from the Schrödinger equation.

Electron spin is a second type of angular momentum in the atom. Spin, which we also call intrinsic angular momentum, has a magnetic moment associated with it. Recall that a magnetic moment placed in a magnetic field results in a torque on the magnetic moment (Eq. 22.16). If we were to repeat the development of Section 22.6 for a nonuniform magnetic field, we would find that such a field also exerts a net *force* on a magnetic moment in addition to the torque. The direction and magnitude of this force depend on the direction of a given magnetic moment with respect to the magnetic field gradient.

In 1921 Otto Stern and Walther Gerlach performed an experiment (Fig. 29.9) that detected the effects of the force on a magnetic moment in a nonuniform magnetic field. The experiment demonstrated that the angular momentum of an atom is quantized. In their experiment, a beam of neutral silver atoms was sent through a nonuniform magnetic field. In such a situation, the atoms experience a

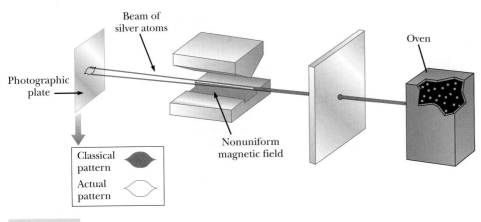

**Figure 29.9**

The apparatus used by Stern and Gerlach to verify space quantization. A beam of neutral silver atoms is split into two components by a nonuniform magnetic field, as shown by the actual pattern in the box.

force (in the vertical direction in Figure 29.9) due to their magnetic moments in this field. Classically, we would expect the beam to be spread out into a continuous distribution on the photographic plate in Figure 29.9 because all possible directions of the atomic magnetic moments are allowed. Stern and Gerlach found, however, that the beam split into two *discrete* components. The experiment was repeated using other atoms, and in each case the beam split into two or more discrete components.

These results are clearly inconsistent with the prediction of a classical model. According to a quantum model, however, the direction of the total angular momentum of the atom, and hence the direction of its magnetic moment, is quantized. Thus, the deflected beam has an integral number of discrete components, and the number of components determines the number of possible values of $\mu_z$. Because the Stern–Gerlach experiment showed discrete beams, space quantization was at least qualitatively verified.

For the moment, let us assume that $\mu_z$ is due to the orbital angular momentum.* Because $\mu_z$ is proportional to $m_\ell$, the number of possible values of $\mu_z$ is $2\ell + 1$. Furthermore, because $\ell$ is an integer, the number of values of $\mu_z$ is always odd. This prediction was not consistent with the observations of Stern and Gerlach, who observed two components in the deflected beam of silver atoms. Thus, although the Stern–Gerlach experiment demonstrated space quantization, the number of components was not consistent with the quantum model developed at that time.

In 1927, T. E. Phipps and J. B. Taylor repeated the Stern–Gerlach experiment using a beam of hydrogen atoms. This experiment is important because it deals with an atom with a single electron in its ground state, for which the quantum model makes reliable predictions. At room temperature, almost all hydrogen atoms are in the ground state. Recall that $\ell = 0$ for hydrogen in its ground state, and so $m_\ell = 0$. Hence, from the orbital angular momentum approach, one would not expect the beam to be deflected by the field at all, because $\mu_z$ would be zero. The beam in the Phipps–Taylor experiment, however, was again split into two

---

* The Stern–Gerlach experiment was performed in 1921, before spin was hypothesized, so orbital angular momentum was the only type of angular momentum in the quantum model at the time.

components. On the basis of this result, one can conclude only one thing: there is some contribution to the angular momentum of the atom and its magnetic moment other than the orbital angular momentum.

In 1925, Goudsmit and Uhlenbeck proposed that the electron has an intrinsic angular momentum apart from its orbital angular momentum. This intrinsic angular momentum is **electron spin,** which we have discussed previously, but for which we now recognize the experimental verification. The total angular momentum of the atom in a particular state contains both an orbital contribution **L** and a spin contribution **S**. A quantum number $s$ exists for spin that is analogous to $\ell$ for orbital angular momentum. The value of $s$ for an electron, however, is *always* $s = \frac{1}{2}$, unlike $\ell$, which varies for different states of the atom. Like **L**, however, the spin angular momentum vector **S** must obey the rules of the quantum model.

In analogy with Equation 29.9, the magnitude of the **spin angular momentum S** for the electron is

- *Spin angular momentum of an electron*

$$S = \sqrt{s(s + 1)}\hbar = \frac{\sqrt{3}}{2}\hbar \qquad \text{[29.12]}$$

This is the only allowed value for the magnitude of the spin angular momentum vector for an electron, so we usually do not include $s$ in a list of quantum numbers describing states of the atom. Like orbital angular momentum, spin angular momentum is quantized in space, as described in Figure 29.10. It can have two orientations, specified by the spin magnetic quantum number $m_s$, where $m_s$ has two possible values, $\pm\frac{1}{2}$. In analogy with Equation 29.10, the $z$ component of spin angular momentum is

$$S_z = m_s\hbar = \pm\frac{1}{2}\hbar \qquad \text{[29.13]}$$

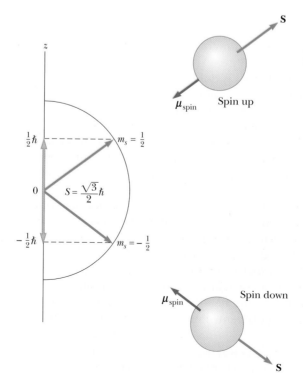

**Figure 29.10**

Spin angular momentum **S** also exhibits space quantization. This figure shows the two allowed orientations of the spin angular momentum vector **S** and the spin magnetic moment vector $\boldsymbol{\mu}_{\text{spin}}$ for a spin $\frac{1}{2}$ particle, such as the electron.

The two values $\pm \hbar/2$ for $S_z$ correspond to the two possible orientations for $\mathbf{S}$ shown in Figure 29.10. Because $m_s$ has more than one value, this quantum number is listed as the fourth quantum number describing a particular state of the atom.

The spin magnetic moment $\boldsymbol{\mu}_s$ of the electron is related to its spin angular momentum $\mathbf{S}$ by the expression

$$\boldsymbol{\mu}_s = -\frac{e}{m_e}\mathbf{S} \tag{29.14}$$

Because $S_z = \pm\frac{1}{2}\hbar$, the $z$ component of the spin magnetic moment can have the values

$$\mu_{sz} = \pm\frac{e\hbar}{2m_e} \tag{29.15}$$

The quantity $e\hbar/2m_e$ is called the **Bohr magneton** $\mu_B$ and has the numerical value $9.274 \times 10^{-24}$ J/T.

Today physicists explain the outcome of the Stern–Gerlach experiment as follows. The observed moments for both silver and hydrogen are due to spin angular momentum alone and not to orbital angular momentum. (The hydrogen atom in its ground state has $\ell = 0$; for silver, used in the Stern–Gerlach experiment, the net orbital angular momentum is $|\mathbf{L}| = 0$.) A single-electron atom such as hydrogen has its electron spin quantized in the magnetic field in such a way that its $z$ component of spin angular momentum is either $\frac{1}{2}\hbar$ or $-\frac{1}{2}\hbar$, corresponding to $m_s = \pm\frac{1}{2}$. Electrons with spin $+\frac{1}{2}$ are deflected in one direction by the nonuniform magnetic field, and those with spin $-\frac{1}{2}$ are deflected in the opposite direction.

The Stern–Gerlach experiment provided two important results. First, it verified the concept of space quantization. Second, it showed that spin angular momentum exists even though this property was not recognized until long after the experiments were performed.

## THINKING PHYSICS 29.1

Does the Stern–Gerlach experiment differentiate between orbital angular momentum and spin angular momentum?

**Reasoning**  A magnetic force on the magnetic moment arises from both orbital angular momentum and spin angular momentum. In this sense, the experiment does not differentiate between the two. The number of lines on the screen tells us something, however, due to the fact that orbital angular momenta are described by an integral quantum number, whereas spin angular momentum depends on a half-integral quantum number. If an odd number of lines occur on the screen, three possibilities arise: The atom has orbital angular momentum only, an even number of electrons with spin angular momentum, or a combination of orbital angular momentum and an even number of electrons with spin angular momentum. If an even number of lines occur on the screen, then at least one unpaired spin angular momentum exists, possibly in combination with orbital angular momentum. The only numbers of lines for which we can specify the type of angular momentum is one line (no orbital, no spin) and two lines (spin). Once we see more than two lines, multiple possibilities arise. For example, three lines could result from two electron spins or one electron with orbital angular momentum quantum number $\ell = 1$.

**PITFALL PREVENTION 29.3**

**Quantum numbers describe a system**

The common usage is to assign the quantum numbers to an electron. Remember, however, that these quantum numbers arise from the Schrödinger equation, which involves a potential energy function for the *system* of the electron and the nucleus. Thus, it is more *proper* to assign the quantum numbers to the atom, but it is more *popular* to assign them to an electron. We will follow this latter usage because it is so common, but keep the notion of the system in the back of your mind.

**Wolfgang Pauli** (1900–1958)

An extremely talented Austrian theoretical physicist, Pauli made important contributions in many areas of modern physics. Pauli gained public recognition at the age of 21 with a masterful review article on relativity, which is still considered one of the finest and most comprehensive introductions to the subject. Other major contributions were the discovery of the exclusion principle, the explanation of the connection between particle spin and statistics, and theories of relativistic quantum electrodynamics, the neutrino hypothesis, and the hypothesis of nuclear spin.

*(Photo taken by S. A. Goudsmit, AIP Niels Bohr Library)*

## 29.6 • THE EXCLUSION PRINCIPLE AND THE PERIODIC TABLE

The quantum model generated from the Schrödinger equation is based on the hydrogen atom—a system of one electron and one proton. As soon as we consider the next atom, helium, we introduce complicating factors. The two electrons in helium both interact with the nucleus, so that we can define a potential energy function for those interactions. But they also interact with each other. The line of action of the electron–nucleus interaction is along a line between the electron and the nucleus. The line of action of the electron–electron interaction is along the line between the two electrons, which is constantly changing as the electrons move. This makes the Schrödinger equation extremely difficult to solve. As we consider atoms with more and more electrons, the possibility of an algebraic solution of the Schrödinger equation becomes hopeless.

We find, however, that despite our inability to solve the Schrödinger equation, **we can use the same four quantum numbers developed for hydrogen for the electrons in heavier atoms.**

We are not able to easily calculate the quantized energy levels, but we can gain information about the levels from theoretical models and experimental measurements.

Because a quantum state in any atom is specified by four quantum numbers, $n$, $\ell$, $m_\ell$, and $m_s$, an obvious and important question is "How many electrons in an atom can have a particular set of quantum numbers?" Pauli provided an answer in 1925 in a powerful statement known as the **exclusion principle:**

> No two electrons in an atom can ever be in the same quantum state; that is, no two electrons in the same atom can have the same set of quantum numbers.

It is interesting that if this principle were not valid, every atom would radiate energy by means of photons and end up with all electrons in the lowest energy state. The chemical behavior of the elements would be grossly modified, because this behavior depends on the electronic structure of atoms. Nature as we know it would not exist! In reality, we can view the electronic structure of complex atoms as a succession of filled levels increasing in energy, where the outermost electrons are primarily responsible for the chemical properties of the element.

Imagine building an atom by forming the nucleus and then filling in the available quantum states with electrons until the atom is neutral. We shall use the common language here that "electrons go into available states." Keep in mind, however, that the states are those of the *system* of the atom. As a general rule, the order of filling of an atom's subshells with electrons is as follows. Once one subshell is filled, the next electron goes into the vacant subshell that is lowest in energy. One can understand this principle by recognizing that if the atom were not in the lowest energy state available to it, the atom would radiate energy until it reached this state.

Before we discuss the electronic configurations of some elements, it is convenient to define an **orbital** as the state of an electron characterized by the quantum numbers $n$, $\ell$, and $m_\ell$. From the exclusion principle, we see that **at most two electrons can be in any orbital.** One of these electrons has $m_s = +\frac{1}{2}$, and the other has $m_s = -\frac{1}{2}$. Because each orbital is limited to two electrons, the numbers of electrons that can occupy the shells are also limited.

Table 29.3 shows the allowed quantum states for an atom up to $n = 3$. Each square in the bottom row of the table represents one orbital, with the ↑ arrows

| TABLE 29.3 | Allowed Quantum Numbers for an Atom up to $n = 3$ | | | | | | | | | | | | | |
|---|---|---|---|---|---|---|---|---|---|---|---|---|---|---|
| $n$ | 1 | 2 | | | | 3 | | | | | | | | |
| $\ell$ | 0 | 0 | 1 | | | 0 | 1 | | | 2 | | | | |
| $m_\ell$ | 0 | 0 | 1 | 0 | −1 | 0 | 1 | 0 | −1 | 2 | 1 | 0 | −1 | −2 |
| $m_s$ | ↑↓ | ↑↓ | ↑↓ | ↑↓ | ↑↓ | ↑↓ | ↑↓ | ↑↓ | ↑↓ | ↑↓ | ↑↓ | ↑↓ | ↑↓ | ↑↓ |

representing $m_s = +\frac{1}{2}$ and the ↓ arrows representing $m_s = -\frac{1}{2}$. The $n = 1$ shell can accommodate only two electrons, because only one orbital is allowed with $m_\ell = 0$. The $n = 2$ shell has two subshells, with $\ell = 0$ and $\ell = 1$. The $\ell = 0$ subshell is limited to only two electrons because $m_\ell = 0$. The $\ell = 1$ subshell has three allowed orbitals, corresponding to $m_\ell = 1$, 0, and −1. Because each orbital can accommodate two electrons, the $\ell = 1$ subshell can hold six electrons (and the $n = 2$ shell can hold eight). The $n = 3$ shell has three subshells and nine orbitals

| Atom | 1s | 2s | 2p | | | Electronic configuration |
|---|---|---|---|---|---|---|
| Li | ↑↓ | ↑ | | | | $1s^2 2s^1$ |
| Be | ↑↓ | ↑↓ | | | | $1s^2 2s^2$ |
| B | ↑↓ | ↑↓ | ↑ | | | $1s^2 2s^2 2p^1$ |
| C | ↑↓ | ↑↓ | ↑ | ↑ | | $1s^2 2s^2 2p^2$ |
| N | ↑↓ | ↑↓ | ↑ | ↑ | ↑ | $1s^2 2s^2 2p^3$ |
| O | ↑↓ | ↑↓ | ↑↓ | ↑ | ↑ | $1s^2 2s^2 2p^4$ |
| F | ↑↓ | ↑↓ | ↑↓ | ↑↓ | ↑ | $1s^2 2s^2 2p^5$ |
| Ne | ↑↓ | ↑↓ | ↑↓ | ↑↓ | ↑↓ | $1s^2 2s^2 2p^6$ |

**Figure 29.11**

The filling of electronic states obeys both the Pauli exclusion principle and Hund's rules.

| TABLE 29.4 | Electronic Configuration of the Elements | | | | | | |

| Atomic Number Z | Symbol | Ground-state Configuration | Ionization Energy (eV) | Atomic Number Z | Symbol | Ground-state Configuration | Ionization Energy (eV) |
|---|---|---|---|---|---|---|---|
| 1 | H | $1s^1$ | 13.595 | 27 | Co | $3d^74s^2$ | 7.86 |
| 2 | He | $1s^2$ | 24.581 | 28 | Ni | $3d^84s^2$ | 7.633 |
| 3 | Li | [He] $2s^1$ | 5.39 | 29 | Cu | $3d^{10}4s^1$ | 7.724 |
| 4 | Be | $2s^2$ | 9.320 | 30 | Zn | $3d^{10}4s^2$ | 9.391 |
| 5 | B | $2s^22p^1$ | 8.296 | 31 | Ga | $3d^{10}4s^24p^1$ | 6.00 |
| 6 | C | $2s^22p^2$ | 11.256 | 32 | Ge | $3d^{10}4s^24p^2$ | 7.88 |
| 7 | N | $2s^22p^3$ | 14.545 | 33 | As | $3d^{10}4s^24p^3$ | 9.81 |
| 8 | O | $2s^22p^4$ | 13.614 | 34 | Se | $3d^{10}4s^24p^4$ | 9.75 |
| 9 | F | $2s^22p^5$ | 17.418 | 35 | Br | $3d^{10}4s^24p^5$ | 11.84 |
| 10 | Ne | $2s^22p^6$ | 21.559 | 36 | Kr | $3d^{10}4s^24p^6$ | 13.996 |
| 11 | Na | [Ne] $3s^1$ | 5.138 | 37 | Rb | [Kr] $5s^1$ | 4.176 |
| 12 | Mg | $3s^2$ | 7.644 | 38 | Sr | $5s^2$ | 5.692 |
| 13 | Al | $3s^23p^1$ | 5.984 | 39 | Y | $4d^15s^2$ | 6.377 |
| 14 | Si | $3s^23p^2$ | 8.149 | 40 | Zr | $4d^25s^2$ | |
| 15 | P | $3s^23p^3$ | 10.484 | 41 | Nb | $4d^45s^1$ | 6.881 |
| 16 | S | $3s^23p^4$ | 10.357 | 42 | Mo | $4d^55s^1$ | 7.10 |
| 17 | Cl | $3s^23p^5$ | 13.01 | 43 | Tc | $4d^55s^2$ | 7.228 |
| 18 | Ar | $3s^23p^6$ | 15.755 | 44 | Ru | $4d^75s^1$ | 7.365 |
| 19 | K | [Ar] $4s^1$ | 4.339 | 45 | Rh | $4d^85s^1$ | 7.461 |
| 20 | Ca | $4s^2$ | 6.111 | 46 | Pd | $4d^{10}$ | 8.33 |
| 21 | Sc | $3d^14s^2$ | 6.54 | 47 | Ag | $4d^{10}5s^1$ | 7.574 |
| 22 | Ti | $3d^24s^2$ | 6.83 | 48 | Cd | $4d^{10}5s^2$ | 8.991 |
| 23 | V | $3d^34s^2$ | 6.74 | 49 | In | [Kr] $4d^{10}5s^25p^1$ | |
| 24 | Cr | $3d^54s^1$ | 6.76 | 50 | Sn | $5p^2$ | 7.342 |
| 25 | Mn | $3d^54s^2$ | 7.432 | 51 | Sb | $4d^{10}5s^25p^3$ | 8.639 |
| 26 | Fe | $3d^64s^2$ | 7.87 | 52 | Te | $4d^{10}5s^25p^4$ | 9.01 |

and can accommodate up to 18 electrons. In general, each shell can accommodate up to $2n^2$ electrons.

The results of the exclusion principle can be illustrated by an examination of the electronic arrangement in a few of the lighter atoms.

**Hydrogen** has only one electron, which, in its ground state, can be described by either of two sets of quantum numbers: 1, 0, 0, $+\frac{1}{2}$ or 1, 0, 0, $-\frac{1}{2}$. The electronic configuration of this atom is often designated as $1s^1$. The notation $1s$ refers to a state for which $n = 1$ and $\ell = 0$, and the superscript indicates that one electron is present in the $s$ subshell.

Neutral **helium** has two electrons. In the ground state, the quantum numbers for these two electrons are 1, 0, 0, $+\frac{1}{2}$ and 1, 0, 0, $-\frac{1}{2}$. No other combinations of quantum numbers are possible for this level, and we say that the K shell is filled. The electronic configuration of helium is designated as $1s^2$.

The electronic configurations of some successive elements are given in Figure 29.11. Neutral **lithium** has three electrons. In the ground state, two of these are in the $1s$ subshell and the third is in the $2s$ subshell, because this subshell is lower in energy than the $2p$ subshell. (In addition to the simple dependence of $E$ on $n$ in Equation 29.2, there is an additional dependence on $\ell$ which will be addressed in Section 29.7.) Hence, the electronic configuration for lithium is $1s^22s^1$.

Note that the electronic configuration of **beryllium,** with its four electrons, is $1s^22s^2$, and **boron** has a configuration of $1s^22s^22p^1$. The $2p$ electron in boron may

| TABLE 29.4 | | Electronic Configuration of the Elements *(Continued)* | | | | | |

| Atomic Number Z | Symbol | Ground-state Configuration | Ionization Energy (eV) | Atomic Number Z | Symbol | Ground-state Configuration | Ionization Energy (eV) |
|---|---|---|---|---|---|---|---|
| 53 | I | $4d^{10}5s^25p^5$ | 10.454 | 79 | Au | $[\text{Xe}, 4f^{14}5d^{10}]\,6s^1$ | 9.22 |
| 54 | Xe | $4d^{10}5s^25p^6$ | 12.127 | 80 | Hg | $6s^2$ | 10.434 |
| 55 | Cs | $[\text{Xe}]\,6s^1$ | 3.893 | 81 | Tl | $6s^26p^1$ | 6.106 |
| 56 | Ba | $6s^2$ | 5.210 | 82 | Pb | $6s^26p^2$ | 7.415 |
| 57 | La | $5d^16s^2$ | 5.61 | 83 | Bi | $6s^26p^3$ | 7.287 |
| 58 | Ce | $4f^15d^16s^2$ | 6.54 | 84 | Po | $6s^26p^4$ | 8.43 |
| 59 | Pr | $4f^36s^2$ | 5.48 | 85 | At | $6s^26p^5$ | |
| 60 | Nd | $4f^46s^2$ | 5.51 | 86 | Rn | $6s^26p^6$ | 10.745 |
| 61 | Pm | $4f^56s^2$ | | 87 | Fr | $[\text{Rn}]\,7s^1$ | |
| 62 | Fm | $4f^66s^2$ | 5.6 | 88 | Ra | $7s^2$ | 5.277 |
| 63 | Eu | $4f^76s^2$ | 5.67 | 89 | Ac | $6d^17s^2$ | 6.9 |
| 64 | Gd | $4f^75d^16s^2$ | 6.16 | 90 | Th | $6d^27s^2$ | |
| 65 | Tb | $4f^96s^2$ | 6.74 | 91 | Pa | $5f^26d^17s^2$ | |
| 66 | Dy | $4f^{10}6s^2$ | | 92 | U | $5f^36d^17s^2$ | 4.0 |
| 67 | Ho | $4f^{11}6s^2$ | | 93 | Np | $5f^46d^17s^2$ | |
| 68 | Er | $4f^{12}6s^2$ | | 94 | Pu | $5f^67s^2$ | |
| 69 | Tm | $4f^{13}6s^2$ | | 95 | Am | $5f^77s^2$ | |
| 70 | Yb | $4f^{14}6s^2$ | 6.22 | 96 | Cm | $5f^76d^17s^2$ | |
| 71 | Lu | $4f^{14}5d^16s^2$ | 6.15 | 97 | Bk | $5f^86d^17s^2$ | |
| 72 | Hf | $4f^{14}5d^26s^2$ | 7.0 | 98 | Cf | $5f^{10}7s^2$ | |
| 73 | Ta | $4f^{14}5d^36s^2$ | 7.88 | 99 | Es | $5f^{11}7s^2$ | |
| 74 | W | $4f^{14}5d^46s^2$ | 7.98 | 100 | Fm | $5f^{12}7s^2$ | |
| 75 | Re | $4f^{14}5d^56s^2$ | 7.87 | 101 | Mv | $5f^{13}7s^2$ | |
| 76 | Os | $4f^{14}5d^66s^2$ | 8.7 | 102 | No | $5f^{14}7s^2$ | |
| 77 | Ir | $4f^{14}5d^76s^2$ | 9.2 | 103 | Lw | $5f^{14}6d^17s^2$ | |
| 78 | Pt | $4f^{14}5d^86s^2$ | 8.88 | 104 | Ku | $5f^{14}6d^27s^2$ | |

be described by one of six sets of quantum numbers, corresponding to six states of equal energy.

**Carbon** has six electrons, and a question arises concerning how to assign the two $2p$ electrons. Do they go into the same orbital with paired spins ($\uparrow\downarrow$), or do they occupy different orbitals with unpaired spins ($\uparrow\uparrow$ or $\downarrow\downarrow$)? Experimental data show that the lowest energy configuration is the latter, where the spins are unpaired. Hence, the two $2p$ electrons in carbon and the three $2p$ electrons in nitrogen have unpaired spins in the ground state (see Fig. 29.11). The general rules that govern such situations throughout the periodic table are called **Hund's rules.** The rule appropriate for elements like carbon is that **when an atom has orbitals of equal energy, the order in which they are filled by electrons is such that a maximum number of electrons will have unpaired spins.** Some exceptions to this rule occur in elements having subshells close to being filled or half-filled.

A complete list of electronic configurations is provided in the tabular representation in Table 29.4. An early attempt to find some order among the elements was made by a Russian chemist, Dmitri Mendeleev, in 1871. He developed a tabular representation of the elements, which has become one of the most important, as well as well-recognized, tools of science. He arranged the atoms in a table (similar to that in Appendix C) according to their atomic masses and chemical similarities. Thus was born the first **periodic table of the elements.** The first table Mendeleev proposed contained many blank spaces, and he boldly stated that the

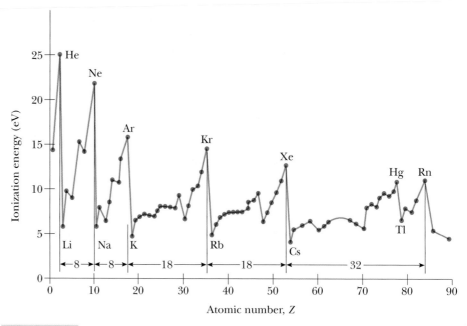

**Figure 29.12**

Ionization energy of the elements versus atomic number Z. *(Adapted from J. Orear, Physics, New York, Macmillan, 1979)*

gaps were there only because the elements had not yet been discovered. By noting the columns in which these missing elements should be located, he was able to make rough predictions about their chemical properties. Within 20 years of Mendeleev's announcement, the missing elements were indeed discovered. This is an excellent example of the power of presenting information in an alternative representation.

The elements in the periodic table are arranged so that all those in a vertical column have similar chemical properties. For example, consider the elements in the last column: He (helium), Ne (neon), Ar (argon), Kr (krypton), Xe (xenon), and Rn (radon). The outstanding characteristic of these elements is that they do not normally take part in chemical reactions—that is, they do not join with other atoms to form molecules—and are therefore classified as being inert. Because of this aloofness, they are referred to as the **inert gases,** or **noble gases.**

We can partially understand this behavior by looking at the electronic configurations in Table 29.4. This table also lists the ionization energies for certain elements. The element helium is one in which the electronic configuration is $1s^2$—in other words, one shell is filled. Additionally, it is found that the energy of the electrons in this filled shell is considerably lower than the energy of the next available level, the $2s$ level. Next, look at the electronic configuration for neon, $1s^2 2s^2 2p^6$. Again, the outermost shell is filled, and a gap in energy occurs between the $2p$ level and the $3s$ level. Argon has the configuration $1s^2 2s^2 2p^6 3s^2 3p^6$. Here, the $3p$ subshell is filled, and a gap in energy arises between the $3p$ subshell and the $3d$ subshell. We could continue this procedure through all the inert gases; the pattern remains the same. An inert gas is formed when either a shell or a subshell is filled, and a gap in energy occurs before the next possible level is encountered.

It is interesting to plot ionization energy versus the atomic number Z, as in Figure 29.12. Note the pattern of differences in atomic numbers between the peaks in the graph: 8, 8, 18, 18, 32. This pattern follows from the Pauli exclusion principle and helps explain why the elements repeat their chemical properties in groups. For example, the peaks at $Z = 2$, 10, 18, and 36 correspond to the elements He, Ne, Ar, and Kr, which have filled shells. These elements have similar chemical behavior.

### THINKING PHYSICS 29.2

As one moves from left to right across one row of the periodic table, the effective size of the atoms first decreases and then increases. What would cause this behavior?

**Reasoning**   Many results determine the effective size of an atom. In a simple model, as one begins at the left side of the periodic table and moves toward the middle, the nuclear charge increases. As a result, the Coulomb attraction increases between the nucleus and the electrons, and the electrons are pulled into an average position that is closer to the nucleus. From the middle of a row to the right side, the increasing number of electrons being placed in proximity to one another results in an accumulated stronger repulsion that increases their average distance from the nucleus and causes the atomic size to grow.

## 29.7 • ATOMIC SPECTRA: VISIBLE AND X-RAY

In Chapter 11 we briefly discussed the origin of the spectral lines for hydrogen and hydrogen-like ions. Recall that an atom in an excited state will emit electromagnetic radiation if it makes a transition to a lower energy state.

The energy level diagram for hydrogen is shown in Figure 29.13. This semigraphical representation is different from Figure 11.22 in that the individual states corresponding to different values of $\ell$ within a given value of $n$ are spread out horizontally. Figure 29.13 shows only those states up to $\ell = 2$; the shells from $n = 4$ upward would have more sets of states to the right, which are not shown.

The diagonal lines in Figure 29.13 represent allowed transitions between stationary states. Whenever an atom makes a transition from a higher energy state to a lower one, a photon of light is emitted. The frequency of this photon is

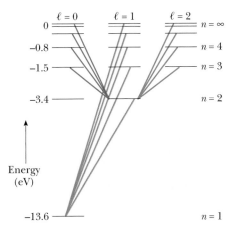

**Figure 29.13**

Some allowed electronic transitions for hydrogen, represented by the colored lines. These transitions must obey the selection rule $\Delta \ell = \pm 1$.

$f = \Delta E/h$, where $\Delta E$ is the energy difference between the two levels and $h$ is Planck's constant. The **selection rules** for the allowed transitions are

$$\Delta \ell = \pm 1 \qquad \text{and} \qquad \Delta m_\ell = 0 \text{ or } \pm 1 \qquad \textbf{[29.16]}$$

Transitions that do not obey the above selection rules are said to be **forbidden.** (Such transitions can occur, but their probability is negligible relative to the probability of the allowed transitions.) For example, any transition represented by a vertical line in Figure 29.13 is forbidden because the quantum number $\ell$ does not change.

Because the orbital angular momentum of an atom changes when a photon is emitted or absorbed (i.e., as a result of a transition) and because angular momentum of the isolated system of the atom and the photon must be conserved, we conclude that **the photon involved in the process must carry angular momentum.** In fact, the photon has an intrinsic angular momentum equivalent to that of a particle with a spin of $s = 1$, compared with the electron with $s = \frac{1}{2}$. Hence, **a photon possesses energy, linear momentum, and angular momentum.** This is the first example we have seen of a single particle with *integral* spin.

Equation 29.2 gives the energies of the allowed quantum states for hydrogen. We can also apply the Schrödinger equation to other one-electron systems, such as the $He^+$ and $Li^{++}$ ions. The primary difference between these ions and the hydrogen atom is the different number of protons $Z$ in the nucleus. This results in a generalization of Equation 29.2 for these other one-electron systems:

$$E_n = -\frac{(13.606 \text{ eV}) Z^2}{n^2} \qquad \textbf{[29.17]}$$

For outer electrons in multielectron atoms, the nuclear charge $Ze$ is largely canceled or shielded by the negative charge of the inner-core electrons. Hence, the outer electrons interact with a net charge that is reduced below the actual charge of the nucleus. (According to Gauss's law, the electric field at the position of an outer electron depends on the net charge of the nucleus and the electrons closer to the nucleus.) The expression for the allowed energies for multielectron atoms has the same form as Equation 29.17, with $Z$ replaced by an effective atomic number $Z_{\text{eff}}$. That is,

$$E_n = -\frac{(13.606 \text{ eV}) Z_{\text{eff}}^2}{n^2} \qquad \textbf{[29.18]}$$

where $Z_{\text{eff}}$ depends on $n$ and $\ell$.

## THINKING PHYSICS 29.3

A physics student is watching a meteor shower in the early morning hours. She notices that the streaks of light from the meteoroids entering the very high regions of the atmosphere last for up to 2 or 3 s before fading.

She also notices a lightning storm off in the distance. The streaks of light from the lightning fade away almost immediately after the flash, certainly in much less than 1 s. Both lightning and meteors cause the air to turn into a plasma because of the very high temperatures generated. The light is emitted when the stripped electrons in the plasma recombine with the ionized atoms. Why would this light last longer for meteors than for lightning?

**Reasoning**  The answer lies in the subtle phrase in the description of the meteoroids ". . . entering the very high regions of the atmosphere. . . ." In the very high regions of the atmosphere, the pressure is very low. The *density* is therefore very low, so that atoms of the gas are relatively far apart. Thus, after the air is ionized by the passing meteoroid, the probability per unit time of freed electrons encountering an ionized atom with which to recombine is relatively low. As a result, the recombination process for all freed electrons occurs over a relatively long time, measured in seconds.

On the other hand, lightning occurs in the lower regions of the atmosphere (the troposphere) where the pressure and density are relatively high. After the ionization by the lightning flash, the freed electrons and ionized atoms are much closer together than in the upper atmosphere. The probability per unit time of a recombination is much higher, and the time for the recombination of all the electrons and ions to occur is much shorter.

## X-Ray Spectra

X-rays are emitted from a metal target that is being bombarded by high-energy electrons. The continuity equation for energy (see Eq. 6.20) for this process can be written as

$$\Delta E_{\text{system}} = \sum H \longrightarrow \Delta E_{\text{int}} = H_{\text{MT}} + H_{\text{ER}}$$

The matter-transfer term on the right-hand side represents the process by which energy enters the system of the target—it travels with the electron. The second term has a negative value and represents the transfer of energy out of the system by x-rays. On the left, we have an increase in internal energy of the target, which recognizes the fact that only a fraction of the incoming energy actually leaves as x-rays. A large fraction of the incoming energy results in an increase in temperature of the target.

The x-ray spectrum typically consists of a broad continuous band and a series of sharp lines that depend on the type of material used for the target, as shown in Figure 29.14. In Chapter 24, we mentioned the fact that an accelerated electric charge emits electromagnetic radiation. The x-rays we see in Figure 29.14 are the result of the slowing down of high-energy electrons as they strike the target. It may take several interactions with the atoms of the target before the electron loses all of its kinetic energy. The amount of kinetic energy lost in any given interaction can vary from zero up to the entire kinetic energy of the electron. Thus, the wavelength of radiation from these interactions lies in a continuous range from some minimum value up to infinity. It is this general slowing down of the electrons that provides the continuous curve in Figure 29.14, which shows the cutoff of x-rays below a minimum wavelength value which depends on the kinetic energy of the incoming electrons. X-radiation with its origin in the slowing down of electrons is called **bremsstrahlung,** the German word for "braking radiation."

The discrete lines in Figure 29.14, called **characteristic x-rays** and discovered in 1908, have a different origin. Their origin remained unexplained until the details of atomic structure were understood. The first step in the production of characteristic x-rays occurs when a bombarding electron collides with a target atom. The electron must have sufficient energy to remove an inner-shell electron from the atom. The vacancy created in the shell is filled when an electron in a higher

**Figure 29.14**

The x-ray spectrum of a metal target consists of a broad continuous spectrum *(bremsstrahlung)* plus a number of sharp lines, which are due to *characteristic x-rays*. The data shown were obtained when 37-keV electrons bombarded a molybdenum target.

level drops down into the level containing the vacancy. The time it takes for this to happen is very short, less than $10^{-9}$ s. As usual, this transition is accompanied by the emission of a photon whose energy equals the difference in energy between the two levels. Typically, the energy of such transitions is greater than 1 000 eV, and the emitted x-ray photons have wavelengths in the range of 0.01 to 1 nm.

Let us assume that the incoming electron has dislodged an atomic electron from the innermost shell, the K shell. If the vacancy is filled by an electron dropping from the next higher shell, the L shell, the photon emitted in the process has an energy corresponding to the $K_\alpha$ line on the curve of Figure 29.14. If the vacancy is filled by an electron dropping from the M shell, the line produced is called the $K_\beta$ line. In this notation, the letter K represents the final shell into which the electron drops, and the subscript provides a Greek letter corresponding to the number of the shell above the final shell in which the electron originates. Thus, $K_\alpha$ indicates that the final shell is the K shell, whereas the initial shell is the first shell above K (because $\alpha$ is the first letter in the Greek alphabet), which is the L shell.

Other characteristic x-ray lines are formed when electrons drop from upper levels to vacancies in shells other than the K shell. For example, L lines are produced when vacancies in the L shell are filled by electrons dropping from higher shells. An $L_\alpha$ line is produced as an electron drops from the M shell to the L shell, and an $L_\beta$ line is produced by a transition from the N shell to the L shell.

**WEB**

For detailed x-ray spectra of many elements, visit
**ie.lbl.gov/xray/mainpage.htm**

### Quick Quiz 29.5

What are the initial and final shells for an $M_\beta$ line in an x-ray spectrum?

Although multielectron atoms cannot be analyzed exactly using either the Bohr model or the Schrödinger equation, we can apply our knowledge of Gauss's law from Chapter 19 to make some surprisingly accurate estimates of expected x-ray energies and wavelengths. Consider an atom of atomic number $Z$ in which one of the two electrons in the K shell has been ejected. Imagine that we draw a gaussian sphere just inside the most probable radius of the L electrons. The electric field at the position of the L electrons is a combination of that due to the nucleus, the single K electron, the other L electrons, and the outer electrons. The wave functions of the outer electrons are such that they have a very high probability of being farther from the nucleus than the L electrons are. Thus, they are much more likely to be outside the gaussian surface than inside, and, on the average, do not contribute significantly to the electric field at the position of the L electrons. The effective charge inside the gaussian surface is the positive nuclear charge and one negative charge due to the single K electron. If we ignore the interactions between L electrons, a single L electron behaves as if it experiences an electric field due to a charge enclosed by the gaussian surface of $(Z - 1)e$. The nuclear charge is shielded by the electron in the K shell such that $Z_{\text{eff}}$ in Equation 29.18 is $Z - 1$. For higher level shells, the nuclear charge is shielded by electrons in all of the inner shells.

We can now use Equation 29.18 to estimate the energy associated with an electron in the L shell:

$$E_{\text{L}} = -(Z - 1)^2 \frac{13.6 \text{ eV}}{2^2}$$

The final state of the atom after it makes the transition is such that there are two electrons in the K shell. We can use a similar argument by drawing a gaussian surface just inside the most probable radius for one K electron. The energy associated with one of these electrons is approximately that of a one-electron atom with the nuclear charge reduced by the negative charge of the other electron. Thus,

$$E_K \approx -(Z-1)^2(13.6 \text{ eV}) \qquad \text{[29.19]}$$

As we show in Example 29.8, the energy of the atom with an electron in an M shell can be estimated in a similar fashion. Taking the energy difference between the initial and final levels, the energy and wavelength of the emitted photon can then be calculated.

In 1914, Henry G. J. Moseley plotted the $Z$ values for a number of elements versus $\sqrt{1/\lambda}$, where $\lambda$ is the wavelength of the $K_\alpha$ line of each element. He found that the curve is a straight line, as in Figure 29.15. This is consistent with rough calculations of the energy levels given by Equation 29.19. From this plot, Moseley was able to determine the $Z$ values of some missing elements, which provided a periodic chart in excellent agreement with the known chemical properties of the elements.

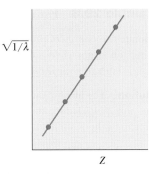

**Figure 29.15**

A Moseley plot of $\sqrt{1/\lambda}$ versus $Z$, where $\lambda$ is the wavelength of the $K_\alpha$ x-ray line of the element with atomic number $Z$.

### Quick Quiz 29.6

In an x-ray tube, if the energy with which the electrons strike the metal target is increased, the wavelengths of the characteristic x-rays do not change. Why not?

### Quick Quiz 29.7

Is it possible for an x-ray spectrum to show the continuous spectrum of x-rays without the presence of the characteristic x-rays?

### THINKING PHYSICS 29.4

Bohr theory gives good theoretical agreement with experimentally measured frequencies of spectral lines for the hydrogen atom, but fails completely to predict the correct optical frequencies for heavier atoms. Moseley, however, used Bohr theory to predict the frequencies of characteristic x-rays from heavy atoms. Why would Bohr theory give good predictions for x-rays, but not for visible photons?

**Reasoning** Let us consider an electron ejected from the K shell and the resulting electrical interactions of an L shell electron with the other electrons. If we look first at the electrons in shells higher than L, these electrons spend most of their time farther from the nucleus than the L electrons. Furthermore, when averaged over time, the charge distribution of these outer electrons is isotropic—no direction is preferred. Thus, the L electrons find themselves *inside* a set of almost perfectly symmetric shells of charge. According to Gauss's Law, then, **the L electron is largely unaffected by the outer electrons**—the time-averaged electric field from these electrons is zero. What remains for the L electrons is a nuclear charge shielded by the one remaining K electron. As a result, Bohr theory works relatively well in describing the energy of the L electron. The model is not perfect, however, because we have ignored the other L electrons.

Visible photons, on the other hand, arise from transitions associated with the outer electrons. For these electrons, the interactions with the remaining electrons is a significant effect, so that Bohr theory does not describe the visible frequencies accurately.

## Example 29.8 Estimating the Energy of an X-Ray

Estimate the energy of the characteristic x-ray emitted from a tungsten target when an electron drops from an M shell ($n = 3$ state) to a vacancy in the K shell ($n = 1$ state).

**Solution** The atomic number for tungsten is $Z = 74$. Using Equation 29.19, we see that the energy associated with the electron in the K shell is approximately

$$E_K \approx -(74 - 1)^2 (13.6 \text{ eV}) = -7.2 \times 10^4 \text{ eV}$$

An electron in the M shell is subject to an effective nuclear charge that depends on the number of electrons in the $n = 1$ and $n = 2$ states, which shield the nucleus. Because eight electrons are in the $n = 2$ state and one electron is in the $n = 1$ state, nine electrons shield the nucleus, and so

$Z_{eff} = Z - 9$. Hence, the energy of the M shell, following Equation 29.18, is approximately

$$E_3 \approx -\frac{(13.6 \text{ eV})(74 - 9)^2}{(3)^2} \approx -6.4 \times 10^3 \text{ eV}$$

The emitted x-ray therefore has an energy equal to $E_3 - E_1 \approx -6.4 \times 10^3 \text{ eV} - (-7.2 \times 10^4 \text{ eV}) \approx 6.6 \times 10^4 \text{ eV} = $ 66 keV . Consultation of x-ray tables shows that the M–K transition energies in tungsten vary from 66.9 to 67.7 keV, where the range of energies is due to slightly different energy values for states of different $\ell$. Thus, our estimate differs from the midpoint of this experimentally measured range by about 2%.

context connection

## 29.8 • ATOMS IN SPACE

We have spent quite a bit of time on the hydrogen atom in this chapter. Let us now consider hydrogen atoms located in space. Because hydrogen is the most abundant element in the Universe, its role in astronomy and cosmology is very important.

Let us begin by considering pictures of some nebulae you might have seen in an astronomy text, such as Figure 29.16. Time exposure photographs of these objects show a variety of colors. What causes the colors in these clouds of gas and grains of dust? Let us imagine a cloud of hydrogen atoms in space near a very hot star. The high-energy photons from the star can interact with the hydrogen atoms, either raising them to a high-energy state or ionizing them. As the atoms fall back

(a)

(b)

(c)

**Figure 29.16**

Types of astronomical nebulae. (a) The central part of the Orion Nebula represents an emission nebula, from which colored light is emitted from atoms. (b) The Pleiades. The clouds of light surrounding the stars represent a reflection nebula, from which starlight is reflected by dust particles. (c) The Lagoon Nebula shows the effects of a dark nebula, in which clouds of dust block starlight and appear as a dark silhouette against the light from stars farther away. *(a, C.R. O'Dell (Rice University) and NASA; b, © 1985 Royal Observatory Edinburgh/Anglo-Australian Observatory photograph from original U.K. Schmidt Plates by David Malin; c, A. Caulet (ST-ECF, ESA) and NASA)* to the lower states, many atoms emit the Balmer series of wavelengths. Thus, these atoms provide red, green, blue, and violet colors to the nebula, corresponding to the colors seen in the hydrogen spectrum in Chapter 11.

In practice, nebulae are classified into three groups, depending on the transitions occurring in the hydrogen atoms. **Emission nebulae** (Fig. 29.16a) are near a hot star, so that hydrogen atoms are excited by light from the star. Thus, the light from an emission nebula is dominated by discrete emission spectral lines and contains colors. **Reflection nebulae** (Fig. 29.16b) are near a cool star. In these cases, most of the light from the nebula is the starlight reflected from larger grains of material in the nebula rather than emitted by excited atoms. Thus, the spectrum of the light from the nebula is the same as that from the star—an absorption spectrum with dark lines corresponding to atoms and ions in the outer regions of the star. The light from these nebulae tends to be white. Finally, **dark nebulae** (Fig. 29.16c) are not close to a star. Thus, little radiation is available to excite atoms or reflect from grains of dust. As a result, the material in these nebulae screen out light from stars beyond them, so that they appear as black patches against the brightness of the more distant stars.

In addition to hydrogen, some other atoms and ions in space are raised to higher energy states by radiation from stars and proceed to emit various colors. Some of the more prominent colors are violet (373 nm) from the $O^+$ ion and green (496 nm and 501 nm) from the $O^{++}$ ion. Helium and nitrogen also provide strong colors.

In our discussion of the quantum numbers for the hydrogen atom, we claimed that two states are possible in the $1s$ shell, corresponding to up or down spin, and that these two states are equivalent in energy. If we modify our structural model to include the spin of the proton, however, we find that the two atomic states corresponding to the electron spin are not the same in energy. The state in which the electron and proton spins are parallel is slightly higher in energy than the state in which they are antiparallel. The energy difference is only $5.9 \times 10^{-6}$ eV. Because these two states differ in energy, it is possible for the atom to make a transition between the states. If the transition is from the parallel state to the antiparallel state, a photon is emitted, with energy equal to the difference in energy between the states. The wavelength of this photon is

$$\lambda = \frac{c}{f} = \frac{hc}{hf} = \frac{hc}{E} = \frac{(6.63 \times 10^{-34}\,\text{J} \cdot \text{s})(3.00 \times 10^8\,\text{m/s})}{5.9 \times 10^{-6}\,\text{eV}} \left( \frac{1\,\text{eV}}{1.6 \times 10^{-19}\,\text{J}} \right)$$

$$= 0.21\,\text{m} = 21\,\text{cm}$$

This radiation is called, for obvious reasons, **21-cm radiation.** It is radiation with a wavelength that is identifiable with the hydrogen atom. Thus, by looking for this radiation in space, we can detect hydrogen atoms. Furthermore, if the wavelength of the observed radiation is not equal to 21 cm, we can infer that it has been Doppler shifted due to relative motion between the Earth and the source. This Doppler shift can then be used to measure the relative speed of the source toward or away from the Earth. This technique has been extensively used to study the hydrogen distribution in the Milky Way galaxy and to detect the presence of spiral arms in our galaxy, similar to the spiral arms in other galaxies.

Our study of atomic physics allows us to understand an important connection between the microscopic world of quantum physics and the macroscopic Universe. Atoms throughout the Universe act as transmitters of information to us about the local conditions. In the next chapter, which deals with nuclear physics, we shall see how our understanding of microscopic processes helps us to understand the local conditions at the center of a star.

**WEB**

For galleries of photographs of nebulae, visit **www.seds.org/billa/twn/** and **www.ne.jp/asahi/stellar/ scenes/english/**

**WEB**

For an image of the distribution of hydrogen in our galaxy made with 21-cm radiation, visit **www.astronomica.org/Gallery/ multiwave/multiwave4.html**

## SUMMARY

The methods of quantum mechanics can be applied to the hydrogen atom using the appropriate potential energy function $U(r) = -k_e e^2/r$ in the Schrödinger equation. The solution to this equation yields the wave functions for the allowed states and the allowed energies, given by

$$E_n = -\left(\frac{k_e e^2}{2a_0}\right)\frac{1}{n^2} = -\frac{13.606 \text{ eV}}{n^2} \quad n = 1, 2, 3, \ldots \quad \textbf{[29.2]}$$

This is precisely the result obtained in the Bohr theory. The allowed energy depends only on the **principal quantum number** $n$. The allowed wave functions depend on three quantum numbers, $n$, $\ell$, and $m_\ell$, where $\ell$ is the **orbital quantum number** and $m_\ell$ is the **orbital magnetic quantum number**. The restrictions on the quantum numbers are as follows:

$$n = 1, 2, 3, \ldots$$
$$\ell = 0, 1, 2, \ldots, (n-1)$$
$$m_\ell = -\ell, -\ell + 1, \ldots, \ell - 1, \ell$$

All states with the same principal quantum number $n$ form a **shell**, identified by the letters K, L, M, . . . (corresponding to $n = 1, 2, 3, \ldots$). All states with the same values of both $n$ and $\ell$ form a **subshell**, designated by the letters, $s$, $p$, $d$, $f$, . . . (corresponding to $\ell = 0, 1, 2, 3, \ldots$).

To completely describe a quantum state of the hydrogen atom, it is necessary to include a fourth quantum number $m_s$, called the **spin magnetic quantum number.** This quantum number can have only two values, $\pm\frac{1}{2}$. In effect, this doubles the number of allowed states specified by the quantum numbers $n$, $\ell$, and $m_\ell$.

An atom in a state characterized by a specific $n$ can have the following values of **orbital angular momentum** $L$:

$$|\textbf{L}| = L = \sqrt{\ell(\ell + 1)}\hbar \quad \ell = 0, 1, 2, \ldots, n - 1 \quad \textbf{[29.9]}$$

The allowed values of the projection of the angular momentum vector **L** along the $z$ axis are given by

$$L_z = m_\ell \hbar \quad \textbf{[29.10]}$$

where $m_\ell$ is restricted to integer values lying between $-\ell$ and $\ell$. Only discrete values of $L_z$ are allowed, and these are determined by the restrictions on $m_\ell$. This quantization of $L_z$ is referred to as **space quantization.**

The electron has an intrinsic angular momentum called **spin angular momentum.** That is, the total angular momentum of an atom can have two contributions: one arising from the spin of the electron (**S**) and one arising from the orbital motion of the electron (**L**).

Electronic spin can be described by a quantum number $s = \frac{1}{2}$. The **magnitude of the spin angular momentum** is

$$S = \frac{\sqrt{3}}{2}\hbar \quad \textbf{[29.12]}$$

and the $z$ component of **S** is

$$S_z = m_s\hbar = \pm\frac{1}{2}\hbar \quad \textbf{[29.13]}$$

The magnetic moment $\boldsymbol{\mu}_s$ associated with the spin angular momentum of an electron is

$$\boldsymbol{\mu}_s = -\frac{e}{m_e}\textbf{S} \quad \textbf{[29.14]}$$

The $z$ component of $\boldsymbol{\mu}_s$ can have the values

$$\mu_{sz} = \pm\frac{e\hbar}{2m_e} \quad \textbf{[29.15]}$$

The **exclusion principle** states that **no two electrons in an atom can have the same set of quantum numbers** $n$, $\ell$, $m_\ell$, and $m_s$. Using this principle, one can determine the electronic configuration of the elements. This serves as a basis for understanding atomic structure and the chemical properties of the elements.

The allowed electronic transitions between any two levels in an atom are governed by the selection rules

$$\Delta\ell = \pm 1 \quad \text{and} \quad \Delta m_\ell = 0 \text{ or } \pm 1 \quad \textbf{[29.16]}$$

The **x-ray spectrum** of a metal target consists of a set of sharp characteristic lines superimposed on a broad, continuous spectrum. **Bremsstrahlung** is x-radiation with its origin in the slowing down of high-energy electrons as they encounter the target. **Characteristic x-rays** are emitted by atoms when an electron undergoes a transition from an outer shell into an electron vacancy in one of the inner shells.

## QUESTIONS

1. Does the light emitted by a neon sign constitute a continuous spectrum or only a few colors? Defend your answer.
2. Must an atom first be ionized before it can emit light?
3. Discuss why the term *electron clouds* is used to describe the electronic arrangement in the quantum mechanical view of the atom.
4. It was stated in the text that, if the exclusion principle were not valid, every electron would end up in the lowest energy state of the atom and the chemical behavior of the elements would be grossly modified. Explain this statement.
5. Why are three quantum numbers (ignoring spin) needed to describe a state of a one-electron atom?

**6.** Why is a nonuniform magnetic field used in the Stern–Gerlach experiment?

**7.** Could the Stern–Gerlach experiment be performed with ions rather than neutral atoms? Explain.

**8.** Describe some experiments that support the conclusion that the spin quantum number for electrons can only have the values $\pm\frac{1}{2}$.

**9.** Discuss some of the consequences of the exclusion principle.

**10.** Why do lithium, potassium, and sodium exhibit similar chemical properties?

**11.** From Table 29.4, we find that the ionization energies for Li, Na, K, Rb, and Cs are 5.390, 5.138, 4.339, 4.176, and 3.893 eV, respectively. Explain why these values are to be expected in terms of the atomic structures.

**12.** The emission spectrum of a gas consists of lines that broaden in energy as the density of the gas increases. Why does this occur?

**13.** How is it possible that electrons, which have a probability distribution around a nucleus, can exist in states of definite energy (e.g., $1s$, $2p$, $3d$, . . .)?

**14.** It is easy to understand how two electrons (one spin up, one spin down) can fill the $1s$ shell for a helium atom. How is it possible that eight more electrons can fit into the $2s$, $2p$ level to complete the $1s^2 2s^2 2p^6$ shell for a neon atom?

**15.** In 1914, Henry Moseley was able to define the atomic number of an element from its characteristic x-ray spectrum. How was this possible? (*Hint:* See Figs. 29.14 and 29.15.)

# PROBLEMS

1, 2, 3 = straightforward, intermediate, challenging    □ = full solution available in the *Student Solutions Manual and Study Guide*

**web** = solution posted at **http://www.harcourtcollege.com/physics/**    💻 = computer useful in solving problem

🌐 = Interactive Physics    ■ = paired numerical/symbolic problems    📰 = life science application

### Section 28.1  Early Structural Models of the Atom

**1. Review Problem.** In the Rutherford scattering experiment, 4.00-MeV alpha particles ($^4$He nuclei containing 2 protons and 2 neutrons) scatter off gold nuclei (containing 79 protons and 118 neutrons). Assume that a particular alpha particle makes a direct head-on collision with the gold nucleus and scatters backward at 180°. Determine (a) the distance of closest approach of the alpha particle to the gold nucleus and (b) the maximum force exerted on the alpha particle. Assume that the gold nucleus remains fixed throughout the entire process.

### Section 29.2  The Hydrogen Atom Revisited

**2.** The Balmer series for the hydrogen atom corresponds to electronic transitions that terminate in the state with quantum number $n = 2$, as shown in Figure P29.2. (a) Consider the photon of longest wavelength; determine its energy and wavelength. (b) Consider the spectral line of shortest wavelength; find its photon energy and wavelength.

**3.** A general expression for the energy levels of one-electron atoms and ions is

$$E_n = -\frac{\mu k_e^2 q_1^2 q_2^2}{2\hbar^2 n^2}$$

where $k_e$ is the Coulomb constant, $q_1$ and $q_2$ are the charges of the electron and the nucleus, and $\mu$ is the reduced mass, given by $\mu = m_1 m_2/(m_1 + m_2)$. In Problem 2 we found that the wavelength for the $n = 3$ to $n = 2$ transition of the hydrogen atom is 656.3 nm (visible red light). What are the wavelengths for this same transition in (a) positronium, which consists of an electron and a positron, and (b) singly ionized helium? (*Note:* A positron is a positively charged electron.)

**4.** Ordinary hydrogen gas is a mixture of two kinds of atoms (isotopes) containing either one- or two-particle nuclei. These isotopes are hydrogen-1 with a proton nucleus, and hydrogen-2, called deuterium, with a deuteron nucleus. A deuteron is one proton and one neutron bound together. Hydrogen-1 and deuterium have identical chemical properties, but can be separated via an ultracentrifuge, or other methods. Their emission spectra show lines of the same colors at very slightly different wavelengths. (a) Use the equation given in Problem 3 to show that the difference in wavelength, between the hydrogen-1 and deu-

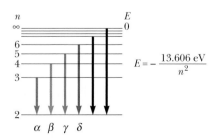

**Figure P29.2**  An energy level diagram for hydrogen showing the Balmer series (not drawn to scale).

terium spectral lines associated with a particular electron transition, is given by

$$\lambda_H - \lambda_D = \left(1 - \frac{\mu_H}{\mu_D}\right)\lambda_H$$

(b) Evaluate the wavelength difference for the Balmer alpha line of hydrogen, with wavelength 656.3 nm, emitted by an atom making a transition from an $n = 3$ state to an $n = 2$ state. This wavelength difference was observed in 1931 by Harold Urey, to confirm his discovery of deuterium.

5. A photon with energy 2.28 eV is barely capable of causing a photoelectric effect when it strikes a sodium plate. Suppose that the photon is instead absorbed by hydrogen. Find (a) the minimum $n$ for a hydrogen atom that can be ionized by such a photon and (b) the speed of the released electron far from the nucleus.

### Section 29.3   The Spin Magnetic Quantum Number

6. List the possible sets of quantum numbers for electrons in (a) the $3d$ subshell and (b) the $3p$ subshell.

### Section 29.4   The Wave Functions for Hydrogen

7. The ground-state wave function for the electron in a hydrogen atom is

$$\psi_{1s}(r) = \frac{1}{\sqrt{\pi a_0^3}}\, e^{-r/a_0}$$

where $r$ is the radial coordinate of the electron and $a_0$ is the Bohr radius. (a) Show that the wave function as given is normalized. (b) Find the probability of locating the electron between $r_1 = a_0/2$ and $r_2 = 3a_0/2$.

8. Plot the wave function $\psi_{1s}(r)$ (see Eq. 29.3) and the radial probability density function $P_{1s}(r)$ (see Eq. 29.7) for hydrogen. Let $r$ range from 0 to $1.5a_0$, where $a_0$ is the Bohr radius.

9. For a spherically symmetric state of a hydrogen atom, the Schrödinger equation in spherical coordinates is

$$-\frac{\hbar^2}{2m}\left(\frac{d^2\psi}{dr^2} + \frac{2}{r}\frac{d\psi}{dr}\right) - \frac{k_e e^2}{r}\psi = E\psi$$

Show that the $1s$ wave function for an electron in hydrogen,

$$\psi_{1s}(r) = \frac{1}{\sqrt{\pi a_0^3}}\, e^{-r/a_0}$$

satisfies the Schrödinger equation.

10. The wave function for an electron in the $2p$ state of hydrogen is described by the expression

$$\psi_{2p} = \frac{1}{\sqrt{3}(2a_0)^{3/2}}\frac{r}{a_0}\, e^{-r/2a_0}$$

What is the most likely distance from the nucleus to find an electron in the $2p$ state? (See Fig. 29.7.)

11. During a particular period of time, an electron in the ground state of a hydrogen atom is "observed" 1 000 times at a distance $a_0/2$ from the nucleus. How many times is this electron observed at a distance $2a_0$ from the nucleus during this period of observation?

### Section 29.5   Physical Interpretation of the Quantum Numbers

12. Calculate the angular momentum for an electron in (a) the $4d$ state and (b) the $6f$ state.

13. If an electron has orbital angular momentum equal to $4.714 \times 10^{-34}$ J·s, what is the orbital quantum number for the state of the electron?

14. An electron is in the N shell. Determine the maximum value of the $z$ component of its angular momentum.

15. A hydrogen atom is in its fifth excited state. The atom emits a photon with a wavelength of 1 090 nm. Determine the maximum possible orbital angular momentum of the electron after emission.

16. Find all possible values of $L$, $L_z$, and $\theta$ for an electron in a $3d$ state of hydrogen.

17. How many sets of quantum numbers are possible for an electron for which (a) $n = 1$, (b) $n = 2$, (c) $n = 3$, (d) $n = 4$, and (e) $n = 5$? Check your results to show that they agree with the general rule that the number of sets of quantum numbers for a shell is equal to $2n^2$.

18. (a) Find the mass density of a proton, picturing it as a solid sphere of radius $1.00 \times 10^{-15}$ m. (b) Consider a classical model of an electron as a solid sphere with the same density as the proton. Find its radius. (c) Imagine that this electron possesses spin angular momentum $I\omega = \hbar/2$ because of classical rotation about the $z$ axis. Determine the speed of a point on the equator of the electron and (d) compare this speed to the speed of light.

19. The $\rho$-meson has a charge of $-e$, a spin quantum number of 1, and a mass 1 507 times that of the electron. Imagine that the electrons in atoms were replaced by $\rho$-mesons. List the possible sets of quantum numbers for $\rho$-mesons in the $3d$ subshell.

20. All objects, large and small, behave quantum mechanically. (a) Estimate the quantum number $\ell$ for the Earth in its orbit about the Sun. (b) What energy change (in joules) would occur if the Earth made a transition to an adjacent allowed state?

### Section 29.6   The Exclusion Principle and The Periodic Table

21. (a) Write out the electronic configuration for the ground state of oxygen ($Z = 8$). (b) Write out values for the set of quantum numbers $n$, $\ell$, $m_\ell$, and $m_s$ for each electron in oxygen.

22. Devise a table similar to that shown in Figure 29.11 for atoms containing 11 through 19 electrons. Use Hund's rule and educated guesswork.

23. Going down the periodic table, which subshell is filled first, the $3d$ or the $4s$ subshell? Which electronic configuration has a lower energy: $[Ar]3d^4 4s^2$ or $[Ar]3d^5 4s^1$? Which has the greater number of unpaired spins? Identify this element and discuss Hund's rule in this case. (*Note:* The notation [Ar] represents the filled configuration for argon.)

24. Two electrons in the same atom both have $n = 3$ and $\ell = 1$. List the quantum numbers for the possible states of the atom. (b) How many states would be possible if the exclusion principle were inoperative?

25. Consider an atom in its ground state, with its outer electrons completely filling the M shell. (a) Identify the atom. (b) List the number of electrons in each subshell.

26. A certain element has its outermost electron in a $3p$ state. It has valence $+3$, because it has three more electrons than a certain noble gas. What element is it?

27. (a) Scanning through Table 29.4 in order of increasing atomic number, note that the electrons fill the subshells in such a way that the subshells with the lowest values of $n + \ell$ are filled first. If two subshells have the same value of $n + \ell$, the one with the lower value of $n$ is filled first. Using these two rules, write the order in which the subshells are filled through $n + \ell = 7$. (b) Predict the chemical valence for the elements that have atomic numbers 15, 47, and 86, and compare your predictions with the actual valences.

28. For a neutral atom of element 110, what would be the probable ground-state electronic configuration?

29. For an electron with magnetic moment $\boldsymbol{\mu}_s$ in a magnetic field **B**, the result of Problem 20 in Chapter 22 implies the following: The electron can be in a higher energy state with the $z$ component of its magnetic moment opposite to the field, or in a lower energy state with the $z$ component of its magnetic moment in the direction of the field. The difference in energy between the two states is $2\mu_B B$.

Under high resolution, many spectral lines are observed to be doublets. The most famous of these are the two yellow lines in the spectrum of sodium (the D lines), with wavelengths of 588.995 nm and 589.592 nm. Their existence was explained in 1925 by Goudsmit and Uhlenbeck, who postulated that an electron has intrinsic spin angular momentum. When the sodium atom is excited with its outermost electron in a $3p$ state, the orbital motion of the outermost electron creates a magnetic field. The atom's energy is somewhat different depending on whether the electron is itself spin up or spin down in this field. Then the photon energy the atom radiates as it falls back into its ground state depends on the energy of the excited state. Calculate the magnitude of the internal magnetic field mediating this so-called *spin–orbit coupling*.

## Section 29.7 Atomic Spectra: Visible and X-Ray

30. (a) Determine the possible values of the quantum numbers $\ell$ and $m_\ell$ for the $He^+$ ion in the state corresponding to $n = 3$. (b) What is the energy of this state?

31. If you wish to produce 10.0-nm x-rays in the laboratory, what is the minimum voltage you must use in accelerating the electrons?

32. In x-ray production, electrons are accelerated through a high voltage $\Delta V$ and then decelerated by striking a target. Show that the shortest wavelength of an x-ray that can be produced is

$$\lambda_{min} = \frac{1\,240 \text{ nm} \cdot \text{V}}{\Delta V}$$

33. Use the method illustrated in Example 29.8 to calculate the wavelength of the x-ray emitted from a molybdenum target ($Z = 42$) when an electron moves from the L shell ($n = 2$) to the K shell ($n = 1$).

34. The wavelength of characteristic x-rays corresponding to the $K_\beta$ line is 0.152 nm. Determine the material in the target.

35. Electrons are shot into a bismuth target, and x-rays are emitted. Determine (a) the M- to L-shell transitional energy for Bi and (b) the wavelength of the x-ray emitted when an electron falls from the M shell into the L shell.

36. The K series of the discrete spectrum of tungsten contains wavelengths of 0.018 5 nm, 0.020 9 nm, and 0.021 5 nm. The K-shell ionization energy is 69.5 keV. Determine the ionization energies of the L, M, and N shells. Sketch the transitions.

## Section 29.8 Context Connection — Atoms in Space

37. In interstellar space, atomic hydrogen produces the sharp spectral line called the 21-cm radiation, which astronomers find most helpful in detecting clouds of hydrogen between stars. This radiation is useful because it is the only signal cold hydrogen emits and because interstellar dust that obscures visible light is transparent to these radio waves. The radiation is not generated by an electron transition between energy states characterized by different values of $n$. Instead, in the ground state ($n = 1$), the electron and proton spins may be parallel or antiparallel, with a resultant slight difference in these energy states. (a) Which condition has the higher energy? (b) More precisely, the line has wavelength 21.11 cm. What is the energy difference between the states? (c) The average lifetime in the excited state is about $10^7$ years. Calculate the associated uncertainty in energy of the excited energy level.

**38. Review Problem.**    Refer to Section 24.3. Prove that the Doppler shift in wavelength of electromagnetic waves is described by

$$\lambda' = \lambda\sqrt{\frac{1 + v/c}{1 - v/c}}$$

where $\lambda'$ is the wavelength measured by an observer moving at speed $v$ away from a source radiating waves of wavelength $\lambda$.

**39.** A distant quasar is moving away from Earth at such high
**web** speed that the blue 434-nm $H_\gamma$ line of hydrogen is observed at 510 nm, in the green portion of the spectrum (Fig.

P29.39). (a) How fast is the quasar receding? You may use the result of Problem 38. (b) Edwin Hubble discovered that all objects outside the local group of galaxies are moving away from us, with speeds proportional to their distances. Hubble's law is expressed as $v = HR$, where Hubble's constant has the approximate value $H = 17 \times 10^{-3}$ m/s·ly. Determine the distance from Earth to this quasar.

### Additional Problems

**40. Review Problem.**    (a) How much energy is required to cause an electron in hydrogen to move from the $n = 1$ state to the $n = 2$ state? (b) Suppose the electrons gain

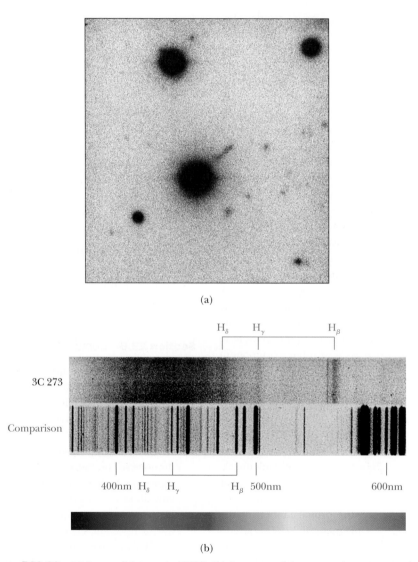

(a)

(b)

**Figure P29.39**    (a) Image of the quasar 3C273. (b) Spectrum of the quasar above a comparison spectrum emitted by stationary hydrogen and helium atoms. Both parts of the figure are printed as black-and-white photographic negatives to reveal detail. *(Maarten Schmidt/Palomar Observatory/California Institute of Technology photograph)* this energy through collisions among hydrogen atoms at a high temperature. At what temperature would the average atomic kinetic energy $3k_B T/2$, where $k_B$ is the Boltzmann constant, be great enough to excite the electrons?

41. Show that the average value of $r$ for the $1s$ state of hydrogen has the value $3a_0/2$. (*Hint:* Use Eq. 29.7.)

42. Assume that three identical uncharged particles of mass $m$ and spin $\frac{1}{2}$ are contained in a one-dimensional box of length $L$. What is the ground-state energy of this system?

43. Suppose a hydrogen atom is in the $2s$ state. Taking $r = a_0$, calculate values for (a) $\psi_{2s}(a_0)$, (b) $|\psi_{2s}(a_0)|^2$, and (c) $P_{2s}(a_0)$. (*Hint:* Use Eq. 29.8.)

44. The states of matter are solid, liquid, gas, and plasma. Plasma can be described as a gas of charged particles, or a gas of ionized atoms. Most of the matter in the Solar System is plasma (throughout the interior of the Sun). In fact, most of the matter in the Universe is plasma; so is a candle flame. Use the information in Figure 29.12 to make an order-of-magnitude estimate for the temperature to which a typical chemical element must be raised to turn into plasma by ionizing most of the atoms in a sample. Explain your reasoning.

45. (a) Show that the most probable radial position for an electron in the $2s$ state of hydrogen is $r = 5.236a_0$. (b) Show that the wave function given by Equation 29.8 is normalized.

46. The force on a magnetic moment $\mu_z$ in a nonuniform magnetic field $B_z$ is given by $F_z = \mu_z(dB_z/dz)$. If a beam of silver atoms travels a horizontal distance of 1.00 m through such a field and each atom has a speed of 100 m/s, how strong must be the field gradient $dB_z/dz$ to deflect the beam 1.00 mm?

47. An electron in chromium moves from the $n = 2$ state to the $n = 1$ state without emitting a photon. Instead, the excess energy is transferred to an outer electron (one in the $n = 4$ state), which is then ejected by the atom. (This is called an Auger [pronounced 'ohjay'] process, and the ejected electron is referred to as an Auger electron.) Use the Bohr theory to find the kinetic energy of the Auger electron.

48. Suppose the ionization energy of an atom is 4.10 eV. In the spectrum of this same atom, we observe emission lines with wavelengths 310 nm, 400 nm, and 1 377.8 nm. Use this information to construct the energy-level diagram with the fewest levels. Assume that the higher levels are closer together.

49. All atoms have the same size, to an order of magnitude. (a) To show this, estimate the diameters for aluminum (with molar mass 27.0 g/mol and density 2.70 g/cm³) and uranium (molar mass 238 g/mol and density 18.9 g/cm³). (b) What do the results imply about the wave functions for inner-shell electrons as we progress to higher and higher atomic mass atoms? (*Hint:* The molar volume is approxi-

mately $D^3 N_A$, where $D$ is the atomic diameter and $N_A$ is Avogadro's number.)

50. For hydrogen in the $1s$ state, what is the probability of finding the electron farther than $2.50a_0$ from the nucleus?

51. According to classical physics, a charge $e$ moving with an acceleration $a$ radiates at a rate

$$\frac{dE}{dt} = -\frac{1}{6\pi\epsilon_0} \frac{e^2 a^2}{c^3}$$

(a) Show that an electron in a classical hydrogen atom (see Fig. 29.3) spirals into the nucleus at a rate

$$\frac{dr}{dt} = -\frac{e^4}{12\pi^2 \epsilon_0^2 r^2 m_3^2 c^3}$$

(b) Find the time it takes the electron to reach $r = 0$, starting from $r_0 = 2.00 \times 10^{-10}$ m.

52. In the technique known as electron spin resonance (ESR), a sample containing unpaired electrons is placed in a magnetic field. Consider the simplest situation, in which only one electron is present and therefore only two energy states are possible, corresponding to $m_s = \pm\frac{1}{2}$. In ESR, the absorption of a photon causes the electron's spin magnetic moment to flip from the lower to the higher energy state. According to the result of Problem 20 in Chapter 22, the change in energy is $2\mu_B B$. (The lower energy state corresponds to the case when the magnetic moment $\mu_{spin}$ is aligned with the magnetic field, and the higher energy state to when $\mu_{spin}$ is aligned opposite to the field.) What is the photon frequency required to excite an ESR transition in a 0.350-T magnetic field?

53. Show that the wave function for an electron in the $2s$ state in hydrogen

$$\psi_{2s}(r) = \frac{1}{4\sqrt{2\pi}} \left(\frac{1}{a_0}\right)^{3/2} \left(2 - \frac{r}{a_0}\right) e^{-r/2a_0}$$

satisfies the spherically symmetric Schrödinger equation given in Problem 9.

54. **Review Problem.** Steven Chu, Claude Cohen-Tannoudji, and William Phillips received the 1997 Nobel Prize in physics for "the development of methods to cool and trap atoms with laser light." One part of their work was with a beam of atoms (mass $\sim 10^{-25}$ kg) that move at a speed on the order of 1 km/s, similar to the speed of molecules in air at room temperature. An intense laser light beam tuned to a visible atomic transition (assume 500 nm) is directed straight into the atomic beam. That is, the atomic beam and light beam are traveling in opposite directions. An atom in the ground state immediately absorbs a photon. Total momentum is conserved in the absorption process. After a lifetime on the order of $10^{-8}$ s, the excited atom radiates by spontaneous emission. It has an

equal probability of emitting a photon in any direction. Thus, the average "recoil" of the atom is zero over many absorption and emission cycles. (a) Estimate the average deceleration of the atomic beam. (b) What is the order of magnitude of the distance over which the atoms in the beam will be brought to a halt?

55. Use the energy levels for hydrogen to show that when the electron falls from state $n$ to state $n - 1$, the frequency of the emitted light is

$$f = \left( \frac{2\pi^2 m_e k_e^2 e^4}{h^3 n^2} \right) \frac{2n - 1}{(n - 1)^2}$$

56. Calculate the frequency of the light that a hydrogen atom should emit according to classical theory. To do so, note that the frequency of revolution of the electron is $v/2\pi r$, where $r$ is the radius of its Bohr orbit. Show that as $n$ approaches infinity in the equation in Problem 55, the expression varies as $1/n^3$ and reduces to the classical frequency the atom is expected to emit. This is one example of the *correspondence principle*, which requires that the classical and quantum models agree for large values of $n$.

57. **LENINGRAD, 1930**—Lev Davidovich Landau solved the Schrödinger equation for a charged particle moving in a uniform magnetic field. A single electron moving perpendicular to a field **B** can be considered as an atom without a nucleus. Landau proved that its energy is quantized in uniform steps of $e\hbar B/m_e$.

**HARVARD, 1999**—Gerald Gabrielse traps a single electron in an evacuated centimeter-sized metal can cooled to a temperature of 80 mK. In a magnetic field of magnitude 5.26 T, the electron circulates for hours in its lowest energy level, generating a measurable signal as it moves. (a) Evaluate the size of a quantum jump in the electron's energy. (b) For comparison, evaluate $k_B T$ as a measure of the energy available to the electron in blackbody radiation from the walls of its container. (c) Microwave radiation can be introduced to excite the electron. Calculate the frequency and wavelength of the photon that the electron absorbs as it jumps to its second energy level. Measurement of the resonant absorption frequency verifies the theory and permits precise determination of properties of the electron.

## ANSWERS TO QUICK QUIZZES

**29.1** (a) 9. For $n = 3$, the allowed values of $\ell$ are 0, 1, and 2. The $\ell = 0$ state has only one possible value of $m_\ell$. The $m_\ell$ for $\ell = 1$ has three values $(-1, 0, 1)$ and five values $(-2, -1, 0, 1, 2)$ for $\ell = 2$. (b) 16. In addition to the nine states up to $\ell = 2$ in part (a), we add the seven values $(-3, -2, -1, 0, 1, 2, 3)$ of $m_\ell$ for $\ell = 3$.

**29.2** Five values $(0, 1, 2, 3, 4)$ of $\ell$ and nine different values $(-4, -3, -2, -1, 0, 1, 2, 3, 4)$ of $m_\ell$ as follows:

| $\ell$ | $m_\ell$ |
|---|---|
| 0 | 0 |
| 1 | $-1, 0, 1$ |
| 2 | $-2, -1, 0, 1, 2$ |
| 3 | $-3, -2, -1, 0, 1, 2, 3$ |
| 4 | $-4, -3, -2, -1, 0, 1, 2, 3, 4$ |

**29.3** No. If the energy of the hydrogen atom were proportional to $n$ (or any power of $n$), the energy would become infinite as $n$ grows to infinity. But the energy of the atom is *inversely* proportional to $n^2$. Thus, as $n$ grows to infinity, the energy of the atom approaches zero from the negative side. As a result, the maximum frequency of emitted radiation approaches a value determined by the difference in energy between zero and the (negative) energy of the ground state.

**29.4** The vector model for $\ell = 1$ is shown in the figure in the next column.

**29.5** Final: M. Initial: O (because the subscript $\beta$ indicates that the initial shell is the second shell higher than M).

**Figure QQAns29.4**

**29.6** The wavelengths of the characteristic x-rays are determined by the separation between energy levels in the atoms of the target, which is unrelated to the energy with which electrons are fired at the target. The only dependence is that the incoming electrons must have enough energy to eject an atomic electron from an inner shell.

**29.7** Yes. If the electrons arrive at the target with very low energy, atomic electrons cannot be ejected and characteristic x-rays do not appear. Because the incoming electrons experience accelerations, the continuous spectrum appears.

A whole-body bone scan performed by injecting a radioactive element into the body and detecting gamma rays given off by the concentration of the element in the bones. Gamma rays are emitted from nuclei that are undergoing radioactive decay.

*(Alexander Tsiaris/Science Source/Photo Researchers, Inc.)*

# chapter

# 30

# Nuclear Physics

I n 1896, the year that marked the birth of nuclear physics, Antoine Henri Becquerel (1852–1908) introduced the world of science to radioactivity in uranium compounds by accidentally discovering that uranyl potassium sulfate crystals emitted an invisible radiation that could darken a photographic plate when the plate was covered to exclude light. After a series of experiments, he concluded that the radiation emitted by the crystals was of a new type, one that required no external stimulation and was so penetrating that it could darken protected photographic plates and ionize gases.

A great deal of research followed as scientists attempted to understand the radiation emitted by radioactive nuclei. Pioneering work by Rutherford showed that the radiation was of three types, which he called alpha, beta, and gamma rays. Later experiments showed that alpha rays are helium nuclei, beta rays are electrons or related particles called positrons, and gamma rays are high-energy photons.

As we saw in Section 29.1, the 1911 experiments of Rutherford established that the nucleus of an atom has a very small volume and that most of the atomic mass is contained in the nucleus. Furthermore, such studies demonstrated a new type of force, the nuclear force, first introduced in Section 5.6, that is predominant at distances on the order of $10^{-15}$ m and essentially zero at distances larger than this.

In this chapter we discuss the properties and structure of the atomic nucleus. We shall describe the basic properties of nuclei, nuclear forces, nuclear binding energy, the phenomenon of radioactivity, and nuclear reactions.

## 30.1 • SOME PROPERTIES OF NUCLEI

In the well-accepted structural model of the nucleus, all nuclei are composed of two types of particles: protons and neutrons. The only exception is the ordinary hydrogen nucleus, which is a single proton with no neutrons. In describing the atomic nucleus, we identify the following integer quantities:

- The **atomic number** $Z$ (introduced in Chapter 29) equals the number of protons in the nucleus; (the atomic number is sometimes called the charge number)
- The **neutron number** $N$ equals the number of neutrons in the nucleus
- The **mass number** $A$ equals the number of nucleons (neutrons plus protons) in the nucleus. That is, $A = N + Z$.

In representing nuclei, it is convenient to have a symbolic representation that shows how many protons and neutrons are present. The symbol used is ${}^{A}_{Z}\text{X}$, where X represents the chemical symbol for the element. For example, ${}^{56}_{26}\text{Fe}$ (iron) has a mass number of 56 and an atomic number of 26; therefore it contains 26 protons and 30 neutrons. When no confusion is likely to arise, we omit the presubscript $Z$ because the chemical symbol can be used to determine $Z$.

The nuclei of all atoms of a particular element contain the same number of protons but often contain different numbers of neutrons. Nuclei that are related in this way are called **isotopes. The isotopes of an element have the same $Z$ value but different $N$ and $A$ values.** The natural abundances of isotopes can differ substantially. For example, ${}^{11}_{6}\text{C}$, ${}^{12}_{6}\text{C}$, ${}^{13}_{6}\text{C}$, and ${}^{14}_{6}\text{C}$ are four isotopes of carbon. The natural abundance of the ${}^{12}_{6}\text{C}$ isotope is about 98.9%, whereas that of the ${}^{13}_{6}\text{C}$, isotope is only about 1.1%. (${}^{11}_{6}\text{C}$ and ${}^{14}_{6}\text{C}$ exist in trace amounts.) Even the simplest element, hydrogen, has isotopes: ${}^{1}_{1}\text{H}$, the ordinary hydrogen nucleus; ${}^{2}_{1}\text{H}$, deuterium; and ${}^{3}_{1}\text{H}$, tritium. Some isotopes do not occur naturally but can be produced in the laboratory through nuclear reactions.

### Charge and Mass

The proton carries a single positive charge $+e$, the electron carries a single negative charge $-e$, where $e = 1.60 \times 10^{-19}$ C, and the neutron is electrically neutral. Because the neutron has no charge, it does not interact electrically with other particles and is difficult to detect.

In Chapter 1, we defined the atomic mass unit u in such a way that the atomic mass of the isotope ${}^{12}_{6}\text{C}$ is exactly 12 u, where 1 u = $1.660\ 559 \times 10^{-27}$ kg. The proton and neutron each have a mass of approximately 1 u, and the electron has a mass that is only a small fraction of an atomic mass unit:

$$\text{Mass of proton} = 1.007\ 276\ \text{u}$$

$$\text{Mass of neutron} = 1.008\ 665\ \text{u}$$

$$\text{Mass of electron} = 0.000\ 549\ \text{u}$$

| TABLE 30.1 | Masses of the Proton, Neutron, and Electron in Various Units | | |
|---|---|---|---|
| | | Mass | |
| Particle | kg | u | MeV/$c^2$ |
| Proton | $1.6726 \times 10^{-27}$ | 1.007 276 | 938.28 |
| Neutron | $1.6750 \times 10^{-27}$ | 1.008 665 | 939.57 |
| Electron | $9.109 \times 10^{-31}$ | $5.486 \times 10^{-4}$ | 0.511 |

Because the rest energy of a particle is given by $E_R = mc^2$ (Section 9.7), it is often convenient to express the atomic mass unit in terms of its rest energy equivalent. For one atomic mass unit, we have

$$E_R = mc^2 = (1.660\ 559 \times 10^{-27}\ \text{kg})(2.997\ 92 \times 10^8\ \text{m/s})^2$$

$$= 931.494\ \text{MeV}$$

Using this equivalence, nuclear physicists often express mass in terms of the unit MeV/$c^2$, where

$$1\ \text{u} = 931.494\ \text{MeV}/c^2$$

The masses of several simple particles are given in Table 30.1. The masses and some other properties of selected isotopes are provided in Table 30.3 and Table A.3 in Appendix A.

## The Size of Nuclei

The size and structure of nuclei were first investigated in the scattering experiments of Rutherford, discussed in Section 29.1. Using the principle of conservation of energy, Rutherford found an expression for how close an alpha particle moving directly toward the nucleus can approach the nucleus before being turned around by Coulomb repulsion.

Let us consider the system of the incoming alpha particle and the nucleus, and apply the energy version of the isolated system model. In a head-on collision, the kinetic energy of the incoming alpha particle (which, because the nucleus is assumed to be much more massive than the alpha particle, we take as the kinetic energy of the system) is converted completely to electrical potential energy of the system when the particle stops (Fig. 30.1). Equating the initial kinetic energy of the alpha particle ($Z = 2$) to the electrical potential energy of the system when the alpha particle is stopped, we have

$$\tfrac{1}{2}mv^2 = k_e \frac{q_1 q_2}{r} = k_e \frac{(2e)(Ze)}{d}$$

where $d$ is the distance of closest approach, $Z$ is the atomic number of the target nucleus, and we have used the nonrelativistic expression for kinetic energy because speeds of alpha particles from radioactive decay are small relative to $c$. Solving for $d$, we find

$$d = \frac{4k_e Ze^2}{mv^2}$$

**Figure 30.1**

An alpha particle on a head-on collision course with a nucleus of charge $Ze$. Because of the Coulomb repulsion between the like charges, the alpha particle stops instantaneously at a distance $d$ from the target nucleus, called the distance of closest approach.

**Figure 30.2**

A nucleus can be modeled as a cluster of tightly packed spheres, each of which is a nucleon.

From this expression, Rutherford found that alpha particles approached to within $3.2 \times 10^{-14}$ m of a nucleus when the foil was made of gold. Based on this calculation and his analysis of results for collisions that were not head-on, Rutherford argued that the radius of the gold nucleus must be less than this value. For silver atoms, the distance of closest approach was found to be $2 \times 10^{-14}$ m. From these results, Rutherford reached his conclusion that the positive charge in an atom is concentrated in a small sphere called the nucleus, whose radius is no greater than about $10^{-14}$ m. Note that this radius is on the order of $10^{-4}$ of the Bohr radius, corresponding to a nuclear volume which is on the order of $10^{-12}$ of the volume of a hydrogen atom. The nucleus is an incredibly small part of the atom! Because such small lengths are common in nuclear physics, a convenient unit of length is the *femtometer* (fm), sometimes called the **fermi,** defined as

$$1 \text{ fm} \equiv 10^{-15} \text{ m}$$

Since the time of Rutherford's scattering experiments, a multitude of other experiments have shown that most nuclei can be geometrically modeled as being approximately spherical with an average radius of

- *Radius of a nucleus*

$$r = r_0 A^{1/3} \qquad \text{[30.1]}$$

where $A$ is the mass number and $r_0$ is a constant equal to $1.2 \times 10^{-15}$ m. Because the volume of a sphere is proportional to the cube of the radius, it follows from Equation 30.1 that the volume of a nucleus (assumed to be spherical) is directly proportional to $A$, the total number of nucleons. This suggests that **all nuclei have nearly the same density.** Nucleons combine to form a nucleus as though they were tightly packed spheres (Fig. 30.2).

---

**Example 30.1    Nuclear Volume and Density**

Find (a) an approximate expression for the mass of a nucleus of mass number $A$, (b) an expression for the volume of this nucleus in terms of the mass number, and (c) a numerical value for its density.

**Solution**    (a) The mass of the proton is approximately equal to that of the neutron. Thus, if the mass of one of these particles is $m$, the mass of the nucleus is approximately $Am$.

(b) Assuming the nucleus is spherical and using Equation 30.1, we find that the volume is

$$V = \tfrac{4}{3} \pi r^3 = \tfrac{4}{3} \pi r_0^3 A$$

(c) The nuclear density is

$$\rho_n = \frac{m_{\text{nucleus}}}{V} = \frac{Am}{\tfrac{4}{3} \pi r_0^3 A} = \frac{3m}{4 \pi r_0^3}$$

$$= \frac{3(1.67 \times 10^{-27} \text{ kg})}{4 \pi (1.2 \times 10^{-15} \text{ m})^3} = 2.3 \times 10^{17} \text{ kg/m}^3$$

Note that the nuclear density is about $2.3 \times 10^{14}$ times greater than the density of water ($10^3$ kg/m³)!

---

**Nuclear Stability**

Because the nucleus consists of a closely packed collection of protons and neutrons, you might be surprised that it can exist at all. The very large repulsive electrostatic forces between protons in close proximity should cause the nucleus to fly apart. Nuclei are stable, however, because of the presence of another force, the **nuclear force** (see Section 5.6). This force, which is short-range (it exists for particle separations less than about 2 fm), is an attractive force that acts between all nuclear particles. The protons attract one another via the nuclear force, and at the

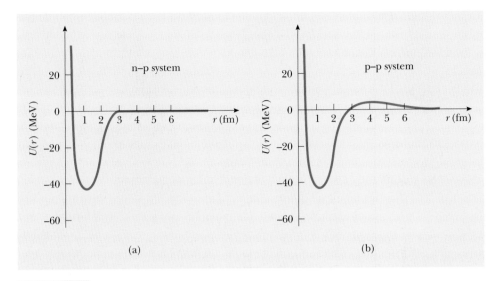

**Figure 30.3**

(a) Potential energy versus separation distance for the neutron–proton system. (b) Potential energy versus separation distance for the proton–proton system. The difference in the two curves is due to the Coulomb repulsion in the case of the proton–proton interaction.

same time they repel one another through the Coulomb force. The nuclear force also acts between pairs of neutrons and between neutrons and protons.

The nuclear force dominates the Coulomb repulsive force within the nucleus (at short ranges). If this were not the case, stable nuclei would not exist. Moreover, the nuclear force is independent of charge. In other words, the forces associated with the proton–proton, proton–neutron, and neutron–neutron interactions are the same, apart from the additional repulsive Coulomb force for the proton–proton interaction.

Evidence for the limited range of nuclear forces comes from scattering experiments and from studies of nuclear binding energies, which we shall discuss shortly. The short range of the nuclear force is shown in the neutron–proton (n–p) potential energy plot of Figure 30.3a obtained by scattering neutrons from a target containing hydrogen. The depth of the n–p potential energy well is 40 to 50 MeV, and a strong repulsive component prevents the nucleons from approaching much closer than 0.4 fm. Another interesting feature of the nuclear force is that its size depends on the relative spin orientations of the nucleons, as shown by scattering experiments using spin–polarized beams and targets.

The nuclear force does not affect electrons, enabling energetic electrons to serve as point-like probes of the charge density of nuclei. The charge independence of the nuclear force also means that the main difference between the n–p and p–p interactions is that the p–p potential energy consists of a *superposition* of nuclear and Coulomb interactions as shown in Figure 30.3b. At distances less than 2 fm, both p–p and n–p potential energies are nearly identical, but for distances greater than this, the p–p potential has a positive energy barrier with a maximum at 4 fm.

About 260 stable nuclei exist; hundreds of other nuclei have been observed but are unstable. A useful graphical representation in nuclear physics is a plot of $N$ versus $Z$ for stable nuclei, as shown in Figure 30.4. Note that light nuclei are stable

**WEB**

For a very detailed, "clickable" version of Figure 30.4, visit Brookhaven National Lab at **www.dne.bnl.gov/CoN/**

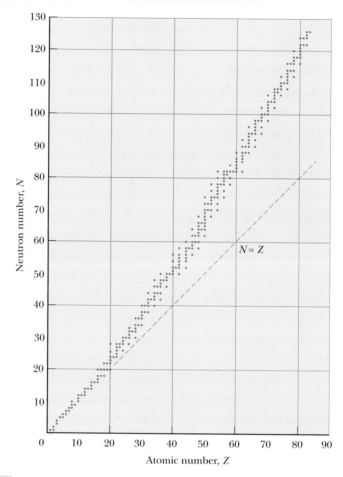

**Figure 30.4**

Neutron number $N$ versus atomic number $Z$ for the stable nuclei (blue dots). The dashed line corresponds to the condition $N = Z$.

if they contain equal numbers of protons and neutrons—that is, if $N = Z$—but heavy nuclei are stable if $N > Z$. This can be partially understood by recognizing that as the number of protons increases, the strength of the Coulomb force increases, which tends to break the nucleus apart. As a result, more neutrons are needed to keep the nucleus stable because neutrons experience only the attractive nuclear forces. Eventually, when $Z = 83$, the repulsive forces between protons cannot be compensated by the addition of more neutrons. Elements that contain more than 83 protons do not have stable nuclei.

Interestingly, most stable nuclei have even values of $A$. In fact, certain values of $Z$ and $N$ correspond to nuclei with unusually high stability. These values of $N$ and $Z$, called **magic numbers,** are

$$Z \text{ or } N = 2, 8, 20, 28, 50, 82, 126 \qquad \text{[30.2]}$$

For example, the helium nucleus (two protons and two neutrons), which has $Z = 2$ and $N = 2$, is very stable. This is reminiscent of the chemical stability of inert gases and suggests quantized nuclear energy levels, which we indeed find to be the case. Some structural models of the nucleus predict a shell structure similar to that for the atom.

### Quick Quiz 30.1

Figure 30.4 shows many instances of vertical groups of several blue dots. What is the relationship among the nuclei in any one of these groups?

### Nuclear Spin and Magnetic Moment

In Chapter 29 we discussed the fact that an electron has an intrinsic angular momentum that is called spin. Nuclei, like electrons, also have a net intrinsic angular momentum that arises from the individual spins of the protons and neutrons. This angular momentum must obey the same quantum rules as orbital angular momentum and spin (Section 29.5). Thus, the magnitude of the **nuclear angular momentum** is due to the combination of all nucleons and is equal to $\sqrt{I(I+1)}\hbar$, where $I$ is a quantum number called the **nuclear spin** and may be an integer or a half-integer. The maximum component of the nuclear angular momentum projected along any direction is $I\hbar$. Figure 30.5 illustrates the possible orientations of the nuclear spin and its projections along the $z$ axis for the case where $I = \frac{3}{2}$.

The nuclear angular momentum has a nuclear magnetic moment associated with it. The magnetic moment of a nucleus is measured in terms of the **nuclear magneton** $\mu_n$, a unit of magnetic moment defined as

$$\mu_n \equiv \frac{e\hbar}{2m_p} = 5.05 \times 10^{-27}\,\text{J/T} \qquad \text{[30.3]}$$

• *Nuclear magneton*

This definition is analogous to Equation 29.15 for the $z$ component of the spin magnetic moment for an electron, which is the Bohr magneton $\mu_B$. Note that $\mu_n$ is smaller than $\mu_B$ by a factor of about 2 000, due to the large difference in masses of the proton and electron.

The magnetic moment of a free proton is $2.792\ 8\mu_n$. Unfortunately, no general theory of nuclear magnetism explains this value. Another surprising point is the fact that a neutron, despite having no electric charge, also has a magnetic moment, which has a value of $-1.913\ 5\mu_n$. The negative sign indicates that the neutron's magnetic moment is opposite its spin angular momentum. Such a magnetic moment for a neutral particle suggests that we need to design a structural model for the neutron that explains such an observation. This structural model, the *quark model,* will be discussed in Chapter 31.

### Quick Quiz 30.2

Which do you expect to show very little variation among different isotopes of an element? (a) atomic mass; (b) nuclear spin magnetic moment; (c) chemical behavior.

### Nuclear Magnetic Resonance and Magnetic Resonance Imaging

The potential energy associated with a magnetic dipole moment in an external magnetic field is $-\boldsymbol{\mu} \cdot \mathbf{B}$. When the projection of $\boldsymbol{\mu}$ is along the field, the potential energy associated with the dipole moment is $-\mu B$; this is its minimum value. When the projection of $\boldsymbol{\mu}$ is opposite the field, the potential energy has its maximum value $\mu B$. Because the direction of the magnetic moment for a particle is quantized, the energies of the magnetic moment in the field are also quantized. In addition, because the spin vector cannot align exactly with the direction of the magnetic field, the extreme values of the energy are $\pm \mu_z B$, where $\mu_z$ is the $z$ component of the magnetic moment. These two energy states for a nucleus with a spin of $\frac{1}{2}$ are shown in Figure 30.6. These states are often called **spin states** because they differ in energy as a result of the direction of the spin.

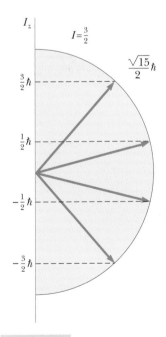

### Figure 30.5

Possible orientations of the nuclear spin angular momentum vector and its projections along the $z$ axis for the case $I = \frac{3}{2}$.

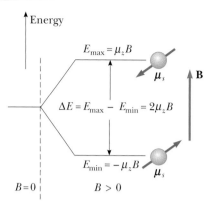

**Figure 30.6**

A nucleus, whose spin is $\frac{1}{2}$, can occupy one of two energy states when placed in an external magnetic field. The lower energy state $E_{min}$ corresponds to the case in which the spin magnetic moment vector $\boldsymbol{\mu}_s$ has a component in the same direction as the magnetic field, and the higher energy state $E_{max}$ corresponds to the case in which the component is opposite the field.

It is possible to observe transitions between these two spin states in a sample using a technique known as **nuclear magnetic resonance** (NMR). A dc magnetic field changes the energy associated with the spin states, splitting them apart in energy (Fig. 30.6). In addition, the sample is irradiated with electromagnetic waves in the radio range of the electromagnetic spectrum. When the frequency of the radio waves is adjusted such that the photon energy matches the separation energy between spin states, the photon is absorbed by a nucleus in the ground state, raising the nucleus–magnetic field system to the higher energy spin state. This results in a net absorption of energy by the system, which is detected by the experimental control and measurement system. A diagram of the apparatus used to detect an NMR signal is illustrated in Figure 30.7. The absorbed energy is supplied by the generator producing the radio waves. Nuclear magnetic resonance and a related technique called electron spin resonance are extremely important methods for studying nuclear and atomic systems and how these systems interact with their surroundings. A typical NMR spectrum is shown in Figure 30.8.

A widely used medical diagnostic technique called **MRI,** for **magnetic resonance imaging,** is based on nuclear magnetic resonance. In MRI, the patient is

**APPLICATION**

Magnetic resonance imaging

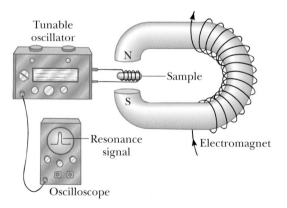

**Figure 30.7**

Experimental arrangement for nuclear magnetic resonance. The radio-frequency magnetic field created by the coil surrounding the sample and provided by the variable-frequency oscillator, must be perpendicular to the constant magnetic field created by the electromagnet. When the nuclei in the sample meet the resonance condition, the nuclei absorb energy from the radio-frequency field of the coil, which changes the characteristics of the circuit in which the coil is included.

Frequency

**Figure 30.8**

An NMR spectrum of $^{31}$P in a bridged metallic complex containing platinum. The lines that flank the central strong peak result from the interaction between $^{31}$P and other distant $^{31}$P nuclei. The outermost set of lines is due to the interaction between $^{31}$P and neighboring platinum nuclei. The spectrum was recorded at a fixed field of about 4 T, and the mean frequency was 200 MHz.

placed inside a large solenoid that supplies a spatially varying magnetic field. Because of the gradient in the magnetic field, protons in hydrogen atoms in water molecules in different parts of the body have different splittings in energy between spin states, so that the resonance signal can be used to provide information on the positions of the protons. A computer is used to analyze the position information to provide data for constructing a final image. An MRI scan showing incredible detail in internal body structure is shown in Figure 30.9. The main advantage of MRI over other imaging techniques in medical diagnostics is that it does not cause damage to cellular structures as x-rays do. Photons associated with the radio-frequency signals used in MRI have energies of only about $10^{-7}$ eV. Because molecular bond strengths are much larger (on the order of 1 eV), the radio-frequency radiation cannot cause cellular damage. In comparison, x-rays or $\gamma$-rays have energies ranging from $10^4$ to $10^6$ eV and can cause considerable cellular damage. Thus, despite some individuals' fears of the word *nuclear* associated with magnetic resonance imaging, the radio-frequency radiation involved is overwhelmingly safer than x-rays!

**WEB**

For several MRI images of a human embryo, visit **embryo.soad.umich.edu/ animal/home.html**

**Figure 30.9**

An MRI scan of a knee. *(Susie Leavines/Photo Researchers, Inc.)*

## 30.2 • BINDING ENERGY

The mass of a nucleus is always less than the sum of the masses of its nucleons. Because mass is a manifestation of energy, the **total rest energy of the bound system (the nucleus) is less than the combined rest energy of the separated nucleons.** This difference in energy is called the **binding energy** $E_b$ of the nucleus and represents the energy that must be added to a nucleus to break it apart into its components:

• *Binding energy of a nucleus*

$$E_b(\text{MeV}) = [ZM(\text{H}) + Nm_n - M(^A_Z\text{X})] \times 931.494\ \text{MeV/u} \qquad \text{[30.4]}$$

where $M(\text{H})$ is the atomic mass of the neutral hydrogen atom, $M(^A_Z\text{X})$ represents the atomic mass of the element $^A_Z\text{X}$, $m_n$ is the mass of the neutron, and the masses are all in atomic mass units. Note that the mass of the $Z$ electrons included in the first term in Equation 30.4 cancels with the mass of the $Z$ electrons included in the term $M(^A_Z\text{X})$ within a small difference associated with the atomic binding energy of the electrons. Because atomic binding energies are typically several electron volts and nuclear binding energies are several megaelectron volts, this difference is negligible, and we adopt a simplification model in which we ignore this difference.

### Example 30.2   The Binding Energy of the Deuteron

Calculate the binding energy of the deuteron (the nucleus of a deuterium atom), which consists of a proton and a neutron, given that the atomic mass of deuterium is 2.014 102 u.

**Solution**   From Table 30.1, we see that the mass of the hydrogen atom, representing the proton, is $M(\text{H}) = 1.007\ 825$ u, and the neutron mass $m_n = 1.008\ 665$ u. Therefore,

$$M(\text{H}) + m_n = 2.016\ 490\ \text{u}$$

To calculate the mass difference, we subtract the deuterium mass from this value:

$$\Delta m = (M(\text{H}) + m_n) - M(^2_1\text{H})$$
$$= 2.016\ 490\ \text{u} - 2.014\ 102\ \text{u}$$
$$= 0.002\ 388\ \text{u}$$

The binding energy is $\Delta mc^2$:

$$E_b = \Delta mc^2 = (0.002\ 388\ \text{u})(931.494\ \text{MeV/u}) = \boxed{2.224\ \text{MeV}}$$

This result tells us that, to separate a deuteron into its constituent proton and neutron, it is necessary to add 2.224 MeV of energy to the deuteron. One way of supplying the deuteron with this energy is by bombarding it with energetic particles.

A plot of binding energy per nucleon $E_b/A$ as a function of mass number for various stable nuclei is shown in Figure 30.10. Note that the curve has a maximum in the vicinity of $A = 60$, corresponding to isotopes of iron, cobalt, and nickel. That is, nuclei with mass numbers greater or less than 60 are not as strongly bound as are those with $A \approx 60$.

The binding energy per nucleon in Figure 30.10 is approximately constant at 8 MeV for $A > 20$. In this case the nuclear forces between a particular nucleon and all the other nucleons in the nucleus are said to be *saturated;* that is, a particu-

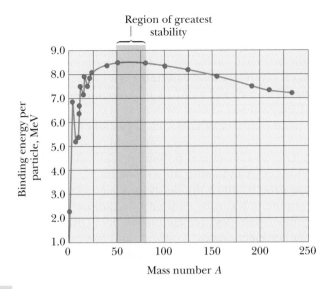

Region of greatest stability

**Figure 30.10**

Binding energy per nucleon versus mass number for nuclei that lie along the line of stability in Figure 30.4.

lar nucleon interacts with only a limited number of other nucleons because of the short range character of the nuclear force. These other nucleons can be viewed as being the nearest neighbors in the close packed structure shown in Figure 30.2.

## THINKING PHYSICS 30.1

Figure 30.10 shows a graph of the average amount of energy necessary to remove a nucleon from the nucleus. Figure 29.12 in the preceding chapter shows the energy necessary to remove an electron from an atom. Why does Figure 30.10 show an *approximately constant* amount of energy necessary to remove a nucleon (above about $A = 20$), but Figure 29.12 shows *widely varying* amounts of energy necessary to remove an electron from the atom?

**Reasoning** In the case of Figure 30.10, the approximately constant value of the nuclear binding energy is a result of the short-range nature of the nuclear strong force. A given nucleon interacts only with its few nearest neighbors, rather than with all of the nucleons in the nucleus. Thus, no matter how many nucleons are present in the nucleus, removing one nucleon involves separating it only from its nearest neighbors. The energy to do this, therefore, is approximately independent of how many nucleons are present.

On the other hand, the electric force holding the electrons to the nucleus in an atom is a long range force. An electron in the atom interacts with all of the protons in the nucleus. When the nuclear charge increases, a stronger attraction occurs between the nucleus and the electrons. As a result, as the nuclear charge increases, more energy is necessary to remove an electron. This is demonstrated by the upward tendency of the ionization energy in Figure 29.12 for each period.

**Marie Curie** (1867–1934)

A Polish scientist, Marie Curie shared the Nobel Prize in physics in 1903 with her husband, Pierre, and with Henri Becquerel for their work on spontaneous radioactivity and the radiation emitted by radioactive substances. "I persist in believing that the ideas that then guided us are the only ones which can lead to the true social progress. We cannot hope to build a better world without improving the individual. Toward this end, each of us must work toward his own highest development, accepting at the same time his share of responsibility in the general life of humanity." *(FPG International)*

**PITFALL PREVENTION 30.2**
**Rays or particles?**

Early in the history of nuclear physics, the term *radiation* was used to describe the emanations from radioactive nuclei. We now know that two out of the three types, namely alpha radiation and beta radiation, involve the emission of particles. Even though these are not examples of electromagnetic radiation, the use of the term *radiation* for all three types is deeply entrenched in our language. We will use this term because of its wide usage in the physics community.

## 30.3 • RADIOACTIVITY

At the beginning of this chapter, we discussed the discovery of radioactivity by Becquerel, which indicated that nuclei emitted particles and radiation. This spontaneous emission was soon to be called **radioactivity.**

The most significant investigations of this phenomenon were conducted by Marie Curie and Pierre Curie (1859–1906). After several years of careful and laborious chemical separation processes on tons of pitchblende, a radioactive ore, the Curies reported the discovery of two previously unknown elements, both of which were radioactive. These were named polonium and radium. Subsequent experiments, including Rutherford's famous work on alpha particle scattering, suggested that radioactivity was the result of the decay, or disintegration, of unstable nuclei.

Three types of radiation can be emitted by a radioactive substance: alpha ($\alpha$) rays, where the emitted particles are $^4$He nuclei; beta ($\beta$) rays, in which the emitted particles are either electrons or positrons; and gamma ($\gamma$) rays, in which the emitted rays are high-energy photons. A **positron** is a particle similar to the electron in all respects except that it has a charge of $+e$ (the positron is said to be the **antiparticle** of the electron; we shall discuss antiparticles further in Chapter 31). The symbol $e^-$ is used to designate an electron and $e^+$ designates a positron.

It is possible to distinguish these three forms of radiation using the scheme illustrated in Figure 30.11. The radiation from a variety of radioactive samples is directed into a region with a magnetic field. The radiation is separated into three components by the magnetic field, two bending in opposite directions and the third experiencing no change in direction. From this simple observation, one can conclude that the radiation of the undeflected beam carries no charge (the gamma ray), the component deflected upward corresponds to positively charged particles (alpha particles), and the component deflected downward corresponds to negatively charged particles ($e^-$). If the beam includes positrons ($e^+$), these particles are deflected upward.

The three types of radiation have quite different penetrating powers. Alpha particles barely penetrate a sheet of paper, beta particles can penetrate a few millimeters of aluminum, and gamma rays can penetrate several centimeters of lead.

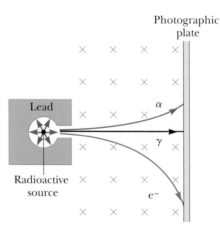

**Figure 30.11**

The radiation from radioactive sources can be separated into three components by using a magnetic field to deflect the charged particles. The photographic plate at the right records the events. The gamma ray is not deflected by the magnetic field.

The rate at which a decay process occurs in a radioactive sample is proportional to the number of radioactive nuclei present in the sample (i.e., those nuclei that have not yet decayed). This is similar to the behavior of population growth—the rate at which babies are born is proportional to the number of people currently alive. If $N$ is the number of radioactive nuclei present at some instant, the rate of change of $N$ is

$$\frac{dN}{dt} = -\lambda N \qquad \text{[30.5]}$$

where $\lambda$ is called either the **decay constant** or the **disintegration constant** and has a different value for different nuclei. The negative sign indicates that $dN/dt$ is a negative number; that is, $N$ decreases in time.

If we write Equation 30.5 in the form

$$\frac{dN}{N} = -\lambda \, dt$$

we can integrate from an arbitrary initial instant $t = 0$ to a later time $t$:

$$\int_{N_0}^{N} \frac{dN}{N} = -\lambda \int_{0}^{t} dt$$

$$\ln\left(\frac{N}{N_0}\right) = -\lambda t$$

$$N = N_0 e^{-\lambda t} \qquad \text{[30.6]}$$

The constant $N_0$ represents the number of undecayed radioactive nuclei at $t = 0$. We have seen exponential behaviors before, for example, with the discharging of a capacitor in Section 21.9. Based on these experiences, we can identify the inverse of the decay constant $1/\lambda$ as the time required for the number of undecayed nuclei to fall to $1/e$ of its original value. Thus, $1/\lambda$ is the **time constant** for this decay, similar to the time constants we investigated for the decay of the current in an *RC* circuit in Section 21.9 and an *RL* circuit in Section 23.6.

The **decay rate** $R = |dN/dt|$ is obtained by differentiating Equation 30.6 with respect to time:

$$R = \left|\frac{dN}{dt}\right| = N_0 \lambda e^{-\lambda t} = R_0 e^{-\lambda t} \qquad \text{[30.7]}$$

• *Decay rate in a radioactive sample*

where $R = N\lambda$ and $R_0 = N_0\lambda$ is the decay rate at $t = 0$. The decay rate of a sample is often referred to as its **activity.** Note that both $N$ and $R$ decrease exponentially with time. The plot of $N$ versus $t$ in Figure 30.12 illustrates the exponential decay law.

A common unit of activity for a radioactive sample is the **curie** (Ci), defined as

$$1 \text{ Ci} \equiv 3.7 \times 10^{10} \text{ decays/s}$$

• *The curie*

This unit was selected as the original unit of activity because it is the approximate activity of 1 g of radium. The SI unit of activity is called the **becquerel** (Bq):

$$1 \text{ Bq} \equiv 1 \text{ decay/s}$$

• *The becquerel*

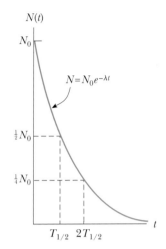

**Figure 30.12**

Plot of the exponential decay law for radioactive nuclei. The vertical axis represents the number of radioactive nuclei present at any time $t$, and the horizontal axis is time. The time $T_{1/2}$ is the half-life of the sample.

Therefore, $1 \text{ Ci} = 3.7 \times 10^{10}$ Bq. The most commonly used units of activity are millicuries (mCi) and microcuries ($\mu$Ci).

A useful parameter for characterizing radioactive decay is the **half-life** $T_{1/2}$. **The half-life of a radioactive substance is the time it takes half of a given number of radioactive nuclei to decay.** Setting $N = N_0/2$ and $t = T_{1/2}$ in Equation 30.6 gives

$$\frac{N_0}{2} = N_0 e^{-\lambda T_{1/2}}$$

Writing this in the form $e^{\lambda T_{1/2}} = 2$ and taking the natural logarithm of both sides, we have

- *Relationship between half-life and decay constant*

$$T_{1/2} = \frac{\ln 2}{\lambda} = \frac{0.693}{\lambda} \qquad [30.8]$$

This is a convenient expression relating the half-life to the decay constant. Note that after an elapsed time of one half-life, $N_0/2$ radioactive nuclei remain (by definition); after two half-lives, half of these have decayed and $N_0/4$ radioactive nuclei remain; after three half-lives, $N_0/8$ remain; and so on. In general, after $n$ half-lives, the number of radioactive nuclei remaining is $N_0/2^n$.

**PITFALL PREVENTION 30.3**

**Half-life**

Be sure that you are clear on the concept of the half-life. In particular, be sure you do not believe that all of the nuclei have decayed after two half-lives.

**Quick Quiz 30.3**

On your birthday, you measure the activity of a sample of $^{210}$Bi, which has a half-life of 5.01 days. The activity you measure is 1.000 $\mu$Ci. What is the activity of this sample on your next birthday? (a) 1.000 $\mu$Ci (b) 0 (c) $\approx 0.2$ $\mu$Ci (d) $\approx 0.01$ $\mu$Ci (e) $\approx 10^{-22}$ $\mu$Ci

**THINKING PHYSICS 30.2**

The isotope $^{14}_{6}$C is radioactive and has a half-life of 5 730 years. If you start with a sample of 1 000 carbon-14 nuclei, how many remain (have not decayed) after 17 190 years?

**Reasoning** In 5 730 years, half the sample will have decayed, leaving 500 radioactive $^{14}_{6}$C nuclei. In another 5 730 years (for a total elapsed time of 11 460 years), the number remaining is 250 nuclei. After another 5 730 years (total time 17 190 years), 125 remain.

These numbers represent ideal circumstances. Radioactive decay is an averaging process over a very large number of atoms, and the actual outcome depends on statistics. Our original sample in this example contained only 1 000 nuclei, certainly not a very large number when we are dealing with atoms, for which we measure the numbers in macroscopic samples in terms of Avogadro's number. Thus, if we were actually to count the number remaining after one half-life for this small sample, it probably would not be exactly 500.

## Example 30.3 The Activity of Radium

The half-life of the radioactive nucleus $^{226}_{86}$Ra is $1.6 \times 10^3$ years. If a sample contains $3.0 \times 10^{16}$ such nuclei at time $t$, determine the activity at this time.

**Solution** First, let us convert the half-life to seconds:

$$T_{1/2} = (1.6 \times 10^3 \text{ yr})(3.16 \times 10^7 \text{ s/yr}) = 5.0 \times 10^{10} \text{ s}$$

Now we can use this value in Equation 30.8 to find the decay constant:

$$\lambda = \frac{0.693}{T_{1/2}} = \frac{0.693}{5.0 \times 10^{10} \text{ s}} = 1.4 \times 10^{-11} \text{ s}^{-1}$$

We can calculate the activity $R$ of the sample at time $t$ using $R = \lambda N$, where $N$ is the number of radioactive nuclei present at this time:

$$R = \lambda N = (1.4 \times 10^{-11} \text{ s}^{-1})(3.0 \times 10^{16}) = 4.1 \times 10^5 \text{ Bq}$$

Because 1 Ci = $3.7 \times 10^{10}$ Bq, the activity, or decay rate, is

$$R = (4.1 \times 10^5 \text{ Bq})\left(\frac{1 \text{ Ci}}{3.7 \times 10^{10} \text{ Bq}}\right) = \boxed{11 \ \mu\text{Ci}}$$

## Example 30.4 The Activity of Carbon

A radioactive sample contains 3.50 $\mu$g of pure $^{11}_{6}$C, which has a half-life of 20.4 min. (a) Determine the number of nuclei present initially.

**Solution** The atomic mass of $^{11}_{6}$C is approximately 11.0 u, and therefore 11.0 g contains Avogadro's number of nuclei. Hence, 3.50 $\mu$g contains $N_0$ nuclei, where

$$\frac{N_0}{6.02 \times 10^{23} \text{ nuclei/mol}} = \frac{3.50 \times 10^{-6} \text{ g}}{11.0 \text{ g/mol}}$$

$$N_0 = \boxed{1.92 \times 10^{17} \text{ nuclei}}$$

(b) What is the activity of the sample initially and after 8.00 h?

**Solution** Because $T_{1/2} = 20.4$ min $= 1.22 \times 10^3$ s, the decay constant is

$$\lambda = \frac{0.693}{T_{1/2}} = \frac{0.693}{1.22 \times 10^3 \text{ s}} = 5.66 \times 10^{-4} \text{ s}^{-1}$$

Therefore, the initial activity of the sample is

$$R_0 = \lambda N_0 = (5.66 \times 10^{-4} \text{ s}^{-1})(1.92 \times 10^{17})$$

$$= \boxed{1.08 \times 10^{14} \text{ Bq}}$$

We can use Equation 30.7 to find the activity at any time $t$. For $t = 8.00$ h $= 2.88 \times 10^4$ s, we see that $\lambda t = 16.3$, and so

$$R = R_0 e^{-\lambda t} = (1.08 \times 10^{14} \text{ Bq}) e^{-16.3}$$

$$= \boxed{8.96 \times 10^6 \text{ Bq}}$$

**EXERCISE** Calculate the number of radioactive nuclei remaining after 8.00 h.

*Answer* $N = 1.58 \times 10^{10}$ nuclei

## 30.4 • THE RADIOACTIVE DECAY PROCESSES

When one nucleus changes into another without external influence, the process is called **spontaneous decay.** As we stated in the preceding section, a radioactive nucleus spontaneously decays via alpha, beta, or gamma decay. Let us discuss these three processes in more detail.

## Alpha Decay

If a nucleus emits an alpha particle ($^4_2$He) in a spontaneous decay, it loses two protons and two neutrons. Therefore, $N$ decreases by 2, $Z$ decreases by 2, and $A$ decreases by 4. The decay can be written with a symbolic representation as

$$^A_Z\text{X} \longrightarrow \ ^{A-4}_{Z-2}\text{Y} + \ ^4_2\text{He} \qquad\qquad \textbf{[30.9]} \quad \bullet \quad \textit{Alpha decay}$$

Before decay

$K_{Rn}$

$^{222}_{86}Rn$

$\mathbf{p}_{Rn}$

$K_\alpha$

$\alpha$

$\mathbf{p}_\alpha$

After decay

### Figure 30.13

The alpha decay of radium. The radium nucleus is initially at rest. After the decay, the radon nucleus has kinetic energy $K_{Rn}$ and momentum $\mathbf{p}_{Rn}$, and the alpha particle has kinetic energy $K_\alpha$ and momentum $\mathbf{p}_\alpha$.

### PITFALL PREVENTION 30.4

#### Another *Q*

We have seen the symbol *Q* before. Be sure that you understand that this is a brand new meaning for this symbol—the disintegration energy. It is not heat nor charge, for which we have used *Q* before.

where X is called the **parent nucleus** and Y the **daughter nucleus.** As a general rule, (1) the sum of the mass numbers must be the same on both sides of the symbolic representation, and (2) the sum of the atomic numbers must be the same on both sides. As examples, $^{238}U$ and $^{226}Ra$ are both alpha emitters and decay according to the schemes

$$^{238}_{92}U \longrightarrow ^{234}_{90}Th + ^4_2He \qquad [30.10]$$

$$^{226}_{88}Ra \longrightarrow ^{222}_{86}Rn + ^4_2He \qquad [30.11]$$

The half-life for $^{238}U$ decay is $4.47 \times 10^9$ years, and the half-life for $^{226}Ra$ decay is $1.60 \times 10^3$ years. In both cases, note that the *A* of the daughter nucleus is 4 less than that of the parent nucleus. Likewise, *Z* is reduced by 2.

The decay of $^{226}Ra$ is shown in Figure 30.13. In addition to the rules for the mass number and the atomic number, the total energy of the system must be conserved in the decay. If we call $M_X$ the mass of the parent nucleus, $M_Y$ the mass of the daughter nucleus, and $M_\alpha$ the mass of the alpha particle, we can define the **disintegration energy** *Q*:

$$Q \equiv (M_X - M_Y - M_\alpha)c^2 \qquad [30.12]$$

Note that the value of *Q* will be in joules if the masses are in kilograms, and $c = 3.00 \times 10^8$ m/s. When the nuclear masses are expressed in the more convenient atomic mass unit u, however, the value of *Q* can be calculated in MeV units using the expression

$$Q = (M_X - M_Y - M_\alpha) \times 931.494 \text{ MeV/u} \qquad [30.13]$$

The disintegration energy *Q* represents the decrease in binding energy of the system and appears in the form of kinetic energy of the daughter nucleus and the alpha particle. This is a nuclear example of the energy version of the isolated system model. No energy is entering or leaving the system. The energy in the system simply transforms from mass energy to kinetic energy, and Equation 30.13 gives the amount of energy transformed in the process. This quantity is sometimes referred to as the **Q value** of the nuclear reaction.

In addition to energy conservation, we can also apply the momentum version of the isolated system model to the decay. Because momentum of the isolated system must be conserved, the lighter alpha particle moves with a much higher speed than the daughter nucleus after the decay occurs. As a result, most of the available kinetic energy is associated with the alpha particle. Generally, light particles carry off most of the energy in nuclear decays.

Equation 30.13 suggests that the alpha particles are emitted with a discrete energy. Such an energy is calculated in Example 30.5. In practice, we find that alpha particles are emitted with a *set* of discrete energies, with the *maximum* value calculated as in Example 30.5. This occurs because the energy of the nucleus is quantized, similar to the quantized energies in an atom. In Equation 30.13, we assume that the daughter nucleus is left in the ground state. If the daughter nucleus is left in an excited state, however, less energy is available for the decay and the alpha particle is emitted with less than the maximum energy. The fact that the alpha particles have a discrete set of energies is direct evidence for the quantization of energy in the nucleus. This is consistent with the model of a quantum particle under boundary conditions, because the nucleons are quantum particles and they are subject to the constraints imposed by their mutual forces.

Finally, it is interesting to note that if one assumes that $^{238}U$ (or other alpha emitters) decays by emitting protons and neutrons, the mass of the decay products exceeds that of the parent nucleus, corresponding to negative $Q$ values. This cannot occur for an isolated system so such spontaneous decays do not occur.

## Quick Quiz 30.4

A student claims that a heavy form of hydrogen decays by alpha emission. How do you respond?

---

### Example 30.5    The Energy Liberated When Radium Decays

The $^{226}$Ra nucleus undergoes alpha decay according to Equation 30.11. Calculate the $Q$ value for this process.

**Solution**    Using Equation 30.13 and values in Table A.3, we see that

$$Q = [M(^{226}\text{Ra}) - M(^{222}\text{Rn}) - M(^4\text{He})] \times 931.494 \text{ MeV/u}$$

$$= (226.025\,406 \text{ u} - 222.017\,574 \text{ u} - 4.002\,603 \text{ u})$$

$$\times 931.494 \text{ MeV/u}$$

$$= (0.005\,229 \text{ u}) \times (931.494 \text{ MeV/u}) = \boxed{4.87 \text{ MeV}}$$

It is left to end-of-chapter Problem 49 to show that the kinetic energy of the alpha particle is about 4.8 MeV, whereas that of the recoiling daughter nucleus is only about 0.1 MeV.

---

We now turn to a structural model for the mechanism of alpha decay that allows some understanding of the decay process. Imagine that the alpha particle forms within the parent nucleus so that the parent nucleus is a combination of the alpha particle and the remaining daughter nucleus. Figure 30.14 is a graphical representation of the potential energy of this system of the alpha particle and the daughter nucleus. The horizontal axis is the separation distance $r$ between the center of the daughter nucleus and the center of the alpha particle. The distance $R$ is the range of the nuclear force. The curve represents the combined effects of (1) the repulsive Coulomb force, which gives the positive peak for $r > R$, and (2) the attractive nuclear force, which causes the energy curve to be negative for $r < R$. As we saw in Example 30.5, a typical disintegration energy is a few MeV, which is the approximate kinetic energy of the emitted alpha particle, represented by the lower dotted line in Figure 30.14. According to classical physics, the alpha particle is trapped in the potential well. How, then, does it ever escape from the nucleus?

The answer to this question was provided by Gamow and, independently, Ronald Gurney and Edward Condon in 1928, using quantum mechanics. The view of quantum mechanics is that there is always some probability that the particle can *tunnel* through the barrier, as we discussed in Section 28.13. Our model of the potential energy curve, combined with the possibility of tunneling, predicts that the probability of tunneling should increase as the particle energy increases, due to the narrowing of the barrier for higher energies. This increased probability should be reflected as an increased activity and consequently a shorter half-life. Experimental data show just this relationship—nuclei with higher alpha particle energies have shorter half-lives. If the potential energy curve in Figure 30.14 is modeled as a series of square barriers whose heights vary with particle separation according to the curve, we can generate a theoretical relationship between particle energy and half-life that is in excellent agreement with the experimental results.

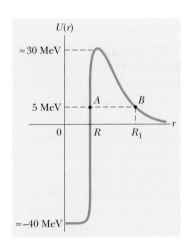

**Figure 30.14**

Potential energy versus separation distance for a system consisting of an alpha particle and a daughter nucleus. Classically, the energy associated with the alpha particle is not sufficiently large to overcome the energy barrier, and so the particle should not be able to escape the nucleus. In reality, the alpha particle does escape by tunneling through the barrier.

This particular application of modeling and quantum physics is a very effective demonstration of the power of these approaches.

Figure 28.23 shows the wave function of a particle tunneling through a barrier in one dimension. A similar wave function having spherical symmetry describes the barrier penetration of an alpha particle leaving a nucleus. The wave function exists both inside and outside the nucleus and its amplitude is constant in time. In this way, the wave function correctly describes the small but constant probability that the nucleus will decay. The moment of decay cannot be predicted. In general, quantum mechanics implies that the future is indeterminate. (This feature is in contrast to classical mechanics, from which the trajectory of an object can be calculated to arbitrarily high precision from precise knowledge of its initial position and velocity and of the forces exerted on it.) We must conclude that the fundamental laws of nature are probabilistic.

A radiation detector can be used to show that a radioactive nucleus decays by radiating a particle at a particular moment and in a particular direction. To point out the contrast between this experimental result and the wave function describing it, Schrödinger imagined a box containing a cat, a radioactive sample, a radiation counter, and a vial of poison. When a nucleus in the sample decays, the counter triggers the administration of lethal poison to the cat. Quantum mechanics correctly predicts the probability of finding the cat dead when the box is opened. Before the box is opened, does the animal have a wave function describing it as a fractionally dead cat, with some chance of being alive?

This question is currently under investigation, never with actual cats, but sometimes with interference experiments building upon the experiment described in Section 28.7. Does the act of measurement change the system from a probabilistic to a definite state? When a particle emitted by a radioactive nucleus is detected at one particular location, does the wave function describing the particle drop instantaneously to zero everywhere else in the Universe? (Einstein called such a state change a "spooky action at a distance.") Is there a fundamental difference between a quantum system and a macroscopic system? The answers to these questions are basically unknown.

### Beta Decay

When a radioactive nucleus undergoes beta decay, the daughter nucleus has the same number of nucleons as the parent nucleus, but the atomic number is changed by 1:

$$^{A}_{Z}\text{X} \longrightarrow \,^{A}_{Z+1}\text{Y} + \text{e}^{-} \qquad [30.14]$$

$$^{A}_{Z}\text{X} \longrightarrow \,^{A}_{Z-1}\text{Y} + \text{e}^{+} \qquad [30.15]$$

Again, note that nucleon number and total charge are both conserved in these decays. As we shall see later, however, these processes are not described completely by these expressions. We shall give reasons for this shortly.

It is important to note that the electron or positron involved in these decays is created within the nucleus as an initial step in the decay process. For example, during beta-minus decay, a neutron in the nucleus is transformed into a proton and an electron:

$$\text{n} \longrightarrow \text{p} + \text{e}^{-}$$

For beta-plus decay, we have a proton transformed into a neutron and a positron:

$$\text{p} \longrightarrow \text{n} + \text{e}^{+}$$

**Figure 30.15**

(a) Distribution of beta particle energies in a typical beta decay. All energies are observed up to a maximum value. (b) In contrast, the energies of alpha particles from an alpha decay are discrete.

Outside the nucleus, this process will not occur because the neutron and electron have more total mass than the proton. This process can occur within the nucleus, however, because we consider the mass–energy changes of the entire nuclear system, not just the individual particles. In beta-plus decay, the process indicated earlier does result in a decrease in the mass of the nucleus, so the process does indeed occur spontaneously. After either of the previous processes occurs within the nucleus, the electron or positron is emitted.

As with alpha decay, the energy of the isolated system of the nucleus and the emitted particle must be conserved in beta decay. Experimentally, one finds that the beta particles are emitted over a continuous range of energies (Fig. 30.15a), unlike alpha particles, which are emitted with discrete energies (Fig. 30.15b). The kinetic energy increase of the system must be balanced by the decrease in mass of the system; either of these is the $Q$ value. Because all decaying nuclei have the same initial mass, however, **the $Q$ value must be the same for each decay.** In view of this, why do the emitted electrons have a range of kinetic energies? The energy version of the isolated system model seems to make an incorrect prediction! Further experimentation shows that, according to the decay processes given by Equations 30.14 and 30.15, the angular momentum (spin) and linear momentum versions of the isolated system model fail too—neither angular momentum nor linear momentum of the system is conserved!

Clearly, the structural model for beta decay must differ from that for alpha decay. In 1930, after a great deal of experimental and theoretical study, Pauli proposed that a third particle must be involved in the decay to account for the "missing" energy, linear momentum, and angular momentum. Enrico Fermi later named this particle the **neutrino** (little neutral one) because it has to be electrically neutral and have little or no rest energy. Although it eluded detection for many years, the neutrino (symbol $\nu$, Greek *nu*) was finally detected experimentally in 1956 by Frederick Reines and Clyde Cowan. It has the following properties:

- Zero electric charge
- A mass smaller than that of the electron, and possibly zero (Recent experiments suggest that mass of the neutrino is less than $7 \text{ eV}/c^2$.)
- A spin of $\frac{1}{2}$
- Very weak interaction with matter, making it therefore quite difficult to detect

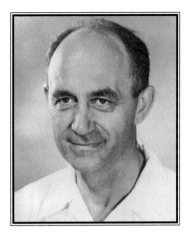

**Enrico Fermi** (1901–1954)

An Italian physicist who immigrated to the United States to escape the Fascists, Fermi was awarded the Nobel Prize in physics in 1938 for producing the transuranic elements by neutron irradiation and for his discovery of nuclear reactions brought about by slow neutrons. He made many other outstanding contributions to physics including his theory of beta decay, the free electron theory of metals, and the development of the world's first fission reactor in 1942. Fermi was truly a gifted theoretical and experimental physicist. He was also well known for his ability to present physics in a clear and exciting manner. "Whatever Nature has in store for mankind, unpleasant as it may be, men must accept, for ignorance is never better than knowledge." (*National Accelerator Laboratory*)

• *Properties of the neutrino*

We can now write the beta decay processes (Equations 30.14 and 30.15) in their correct form:

$$\,^{A}_{Z}X \longrightarrow \,^{A}_{Z+1}Y + e^{-} + \bar{\nu} \qquad [30.16]$$

$$\,^{A}_{Z}X \longrightarrow \,^{A}_{Z-1}Y + e^{+} + \nu \qquad [30.17]$$

- *Beta decay*

where $\bar{\nu}$ represents the **antineutrino,** the antiparticle to the neutrino. We shall discuss antiparticles further in Chapter 31. For now, it suffices to say that **a neutrino is emitted in positron decay, and an antineutrino is emitted in electron decay.** The spin of the neutrino allows angular momentum to be conserved in the decay processes. Despite its small mass, the neutrino does carry momentum, which allows linear momentum to be conserved. The decays of the neutron and proton within the nucleus are more properly written as

$$n \longrightarrow p + e^{-} + \bar{\nu}$$

$$p \longrightarrow n + e^{+} + \nu$$

As examples of beta decay, we can write the decay schemes for carbon-14 and nitrogen-12:

$$\,^{14}_{6}C \longrightarrow \,^{14}_{7}N + e^{-} + \bar{\nu} \qquad [30.18]$$

$$\,^{12}_{7}N \longrightarrow \,^{12}_{6}C + e^{+} + \nu$$

In beta-plus decay, the final system consists of the daughter nucleus, the ejected positron and neutrino, and an electron shed from the atom to neutralize the daughter atom. In some cases, this represents an overall increase in mass-energy, so this process does not occur. There is an alternate process that allows some proton−rich nuclei to decay and become more stable. This process, called **electron capture,** occurs when a parent nucleus captures one of its own orbital electrons and emits a neutrino. The final product after decay is a nucleus whose charge is $Z - 1$:

- *Electron capture process*

$$\,^{A}_{Z}X + e^{-} \longrightarrow \,^{A}_{Z-1}Y + \nu \qquad [30.19]$$

In most cases, an inner K-shell electron is captured, a process referred to as **K capture.** In this process, the only outgoing particles are the neutrino and x-ray photons, originating in higher shell electrons falling into the vacancy left by the captured K electron.

## Carbon Dating

**APPLICATION**

Carbon dating

The beta decay of $^{14}C$ given by Equation 30.18 is used to date organic samples. Cosmic rays (high-energy particles from outer space) in the upper atmosphere cause nuclear reactions that create $^{14}C$. The ratio of $^{14}C$ to $^{12}C$ in the carbon dioxide molecules of our atmosphere has a constant value of about $1.3 \times 10^{-12}$. All living organisms have the same ratio of $^{14}C$ to $^{12}C$ because they continuously exchange carbon dioxide with their surroundings. When an organism dies, however, it no longer absorbs $^{14}C$ from the atmosphere, and so the ratio of $^{14}C$ to $^{12}C$ decreases as the result of the beta decay of $^{14}C$, which has a half-life of 5 730 years. It is therefore possible to determine the age of a material by measuring its activity per unit mass due to the decay of $^{14}C$. Using carbon dating, samples of wood, charcoal, bone, and shell have been identified as having lived 1 000 to 25 000 years ago.

A particularly interesting example is the dating of the Dead Sea Scrolls, a group of manuscripts discovered by a shepherd in 1947 (Fig. 30.16). Translation showed them to be religious documents, including most of the books of the Old Testament. Carbon dating applied to the material in which they were wrapped established their age at approximately 1 950 years.

(a)

### Quick Quiz 30.5

Can carbon-14 dating be used to measure the age of a stone?

## THINKING PHYSICS 30.3

In 1991, a German tourist discovered the well-preserved remains of the Ice Man trapped in a glacier in the Italian Alps (Fig. 30.17). Radioactive dating of a sample of Ice Man revealed an age of 5 300 years. Why did scientists date the sample using the isotope $^{14}$C, rather than $^{11}$C, a beta emitter with a half-life of 20.4 min?

**Reasoning**  $^{14}$C has a long half-life of 5 730 years, so the fraction of $^{14}$C nuclei remaining after one half-life is high enough to measure accurate changes in the sample's activity. The $^{11}$C isotope, which has a very short half-life, is not useful because its activity decreases to a vanishingly small value over 5 300 years, making it impossible to detect.

(b)

**Figure 30.16**

(a) A fragment of the Dead Sea Scrolls, which were discovered in the caves in the photograph (b). The packing material of the scrolls was analyzed by carbon dating to determine their age. *(M. Milner/Corbis Sygma)*

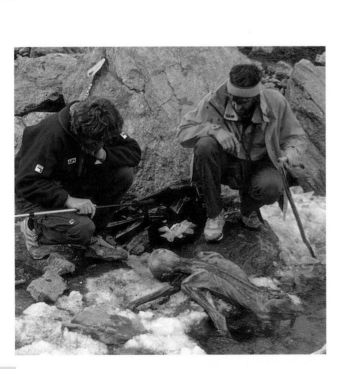

**Figure 30.17**

The Ice Man, discovered in 1991 when an Italian glacier melted enough to expose his remains. His possessions, particularly his tools, have shed light on the way people lived in the Bronze Age. Carbon-14 dating was used to determine how long ago this person lived. *(Paul Hanny/Gamma Liaison)*

  **APPLICATION**

Carbon dating and the Ice Man

If a sample to be dated is not very old, say about 50 years, then you should select the isotope of some other element whose half-life is comparable with the age of the sample. For example, if the sample contained hydrogen, you could measure the activity of $^3$H (tritium), a beta emitter of half-life 12.3 years. As a general rule, the expected age of the sample should be long enough to measure a change in activity, but not so long that its activity cannot be detected.

### THINKING PHYSICS 30.4

A wooden coffin is found and contains a skeleton holding a gold statue. Which of the three objects (coffin, skeleton, and gold statue) can be carbon dated to find out how old it is?

**Reasoning**   Only the coffin and the skeleton can be carbon dated. These two items exchanged air with the environment during their lifetimes, resulting in a fixed ratio of carbon-14 to carbon-12. Once the human and the tree from which the coffin was made died, the amount of carbon-14 began to decrease due to radioactive decay. The gold in the statue was never alive and did not breathe air, so it has no carbon-14 to detect.

### Example 30.6   Radioactive Dating

A piece of charcoal of mass 25.0 g is found in the ruins of an ancient city. The sample shows a $^{14}$C ($T_{1/2}$ = 5 730 years) activity of 250 decays/min. How long has the tree from which this charcoal came been dead?

**Solution**   First, let us calculate the decay constant for $^{14}$C:

$$\lambda = \frac{0.693}{T_{1/2}} = \frac{0.693}{(5\,730 \text{ yr})(3.16 \times 10^7 \text{ s/yr})}$$

$$= 3.83 \times 10^{-12} \text{ s}^{-1}$$

The number of $^{14}$C nuclei can be calculated in two steps. First, the number of $^{12}$C nuclei in 25 g of carbon is

$$N(^{12}\text{C}) = \frac{6.02 \times 10^{23} \text{ nuclei/mol}}{12.0 \text{ g/mol}} (25.0 \text{ g})$$

$$= 1.25 \times 10^{24} \text{ nuclei}$$

Assuming that the ratio of $^{14}$C to $^{12}$C is $1.30 \times 10^{-12}$, we see that the number of $^{14}$C nuclei in 25.0 g *before* decay is

$$N_0(^{14}\text{C}) = (1.30 \times 10^{-12})(1.25 \times 10^{24})$$

$$= 1.63 \times 10^{12} \text{ nuclei}$$

Hence, the initial activity of the sample is

$$R_0 = \lambda N_0 = (3.83 \times 10^{-12} \text{ s}^{-1})(1.6 \times 10^{12} \text{ nuclei})$$

$$= 6.24 \text{ decays/s} = 374 \text{ decays/min}$$

We can now calculate the age of the charcoal using Equation 30.7, which relates the activity $R$ at any time $t$ to the initial activity $R_0$:

$$R = R_0 e^{-\lambda t} \qquad \text{or} \qquad e^{-\lambda t} = \frac{R}{R_0}$$

Because it is given that $R$ = 250 decays/min, and because we found that $R_0$ = 374 decays/min, we can calculate $t$ by taking the natural logarithm of both sides of the last equation:

$$-\lambda t = \ln\left(\frac{R}{R_0}\right) = \ln\left(\frac{250}{374}\right) = -0.403$$

$$t = \frac{0.403}{\lambda} = \frac{0.403}{3.83 \times 10^{-12} \text{ s}^{-1}}$$

$$= 1.05 \times 10^{11} \text{ s} = 3.33 \times 10^3 \text{ yr}$$

This technique of radioactive carbon dating has been successfully used to measure the ages of many organic relics up to 25 000 years old.

| TABLE 30.2 | Decay Pathways |
|------------|----------------|
| Alpha decay | $^{A}_{Z}X \longrightarrow ^{A-4}_{Z-2}Y + ^{4}_{2}He$ |
| Beta decay ($\beta^-$) | $^{A}_{Z}X \longrightarrow ^{A}_{Z+1}Y + e^- + \bar{\nu}$ |
| Beta decay ($\beta^+$) | $^{A}_{Z}X \longrightarrow ^{A}_{Z-1}Y + e^+ + \nu$ |
| Electron capture | $^{A}_{Z}X + e^- \longrightarrow ^{A}_{Z-1}Y + \nu$ |
| Gamma decay | $^{A}_{Z}X^* \longrightarrow ^{A}_{Z}X + \gamma$ |

## Gamma Decay

As we mentioned in the discussion of alpha decay, the energy of a nucleus is quantized in a way similar to that of the atom, although the allowed energy states are separated by larger differences, measured in megaelectron volts. Very often, a nucleus that undergoes radioactive decay is left in an excited energy state. It can then undergo a second decay to a lower energy state, perhaps to the ground state, by emitting a photon. Photons emitted in such a de-excitation process are called **gamma rays.** Such photons have very high energy ($\sim 1$ MeV) relative to the energy of visible light (on the order of a few electron volts). Recall from Chapter 29 that the energy of photons emitted (or absorbed) by an atom equals the difference in energy between the two quantum states involved in the transition. Similarly, a gamma ray photon has an energy $hf$ that equals the energy difference $\Delta E$ between two nuclear quantum states. When a nucleus decays by emitting a gamma ray, the nucleus does not change, apart from the fact that it ends up in a lower energy state. We can represent gamma decay as

$$^{A}_{Z}X^* \longrightarrow ^{A}_{Z}X + \gamma \qquad [30.20]$$

where X* indicates a nucleus in an excited state.

A nucleus may reach an excited state as the result of a violent collision with another particle. It is more common, however, for a nucleus to be in an excited state after it has undergone an alpha or beta decay. The following sequence of events represents a typical situation in which gamma decay occurs:

$$^{12}_{5}B \longrightarrow ^{12}_{6}C^* + e^- + \bar{\nu} \qquad [30.21]$$

$$^{12}_{6}C^* \longrightarrow ^{12}C + \gamma \qquad [30.22]$$

Figure 30.18 shows the decay scheme for $^{12}B$, which undergoes beta decay to either of two levels of $^{12}C$. It can either (1) decay directly to the ground state of $^{12}C$ by emitting a 13.4-MeV electron or (2) undergo beta-minus decay to an excited state of $^{12}C^*$, followed by gamma decay to the ground state. The latter process results in the emission of a 9.0-MeV electron and a 4.4-MeV photon.

The pathways by which a radioactive nucleus can undergo decay are summarized in Table 30.2.

• *Gamma decay*

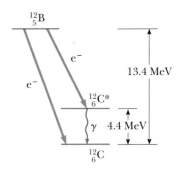

**Figure 30.18**

The $^{12}B$ nucleus undergoes beta decay to either of two levels of $^{12}C$: directly to the ground state or to an excited state. Decay to the excited level, $^{12}C^*$, is followed by gamma decay to the ground state.

## 30.5 • NUCLEAR REACTIONS

In Section 30.4, we discussed the processes by which nuclei can *spontaneously* change to another nucleus by undergoing a radioactive decay process. It is also possible to change the structures and properties of nuclei by bombarding them with energetic particles. Such changes are called **nuclear reactions.** In 1919, Rutherford was the first to observe nuclear reactions, using naturally occurring radioactive sources for the bombarding particles. Since then, thousands of nuclear reactions have been observed following the development of charged-particle accelerators in the 1930s. With today's advanced technology in particle accelerators and particle detectors, it is possible to achieve particle energies of more than 1 000 GeV = 1 TeV. These high-energy particles are used to create new particles whose properties are helping to solve the mysteries of the nucleus.

Consider a reaction (Fig. 30.19) in which a target nucleus X is bombarded by an incoming particle a, resulting in a different nucleus Y and an outgoing particle b:

• *Nuclear reaction*

$$a + X \longrightarrow Y + b \qquad \text{[30.23]}$$

Sometimes this reaction is written in the equivalent symbolic representation

$$X(a, b)Y$$

In the preceding section, the *Q* value, or disintegration energy, associated with radioactive decay was defined as the change in the rest energy, which is the amount of the rest energy transformed to kinetic energy during the decay process. In a similar way, we define the **reaction energy** *Q* associated with a nuclear reaction as the **total change in rest energy which results from the reaction**

• *Reaction energy* Q

$$Q = (M_a + M_X - M_Y - M_b)c^2 \qquad \text{[30.24]}$$

A reaction for which *Q* is positive is called **exothermic.** After the reaction, the transformed rest energy appears as an increase in kinetic energy of Y and b over that of a and X.

A reaction for which *Q* is negative is called **endothermic** and represents an increase in rest energy. An endothermic reaction will not occur unless the bombarding particle has a kinetic energy greater than $|Q|$. The minimum kinetic energy of the incoming particle necessary for such a reaction to occur is called the **threshold energy.** The threshold energy is larger than $|Q|$ because we must also conserve linear momentum in the isolated system of the initial and final particles. If an incoming particle has just energy $|Q|$, enough energy is present to increase the rest energy of the system, but none is left over for kinetic energy of the final particles—nothing is moving after the reaction. Thus, the incoming particle has momentum before the reaction but there is no momentum of the system afterward—a violation of the law of conservation of momentum.

If a nuclear reaction occurs in which particles a and b are identical, so that X and Y are also necessarily identical, the reaction is called a **scattering event.** If kinetic energy is conserved during the scattering (i.e., if $Q = 0$), the event is classified as **elastic scattering.** On the other hand, if $Q \neq 0$, kinetic energy is not conserved and the reaction is called **inelastic scattering.** For example, if the target nucleus is left in an excited state by the event, its total energy is higher than before the scattering, and the kinetic energy of the final particles is less than in the elastic case. This terminology is identical to that used in dealing with collisions between macroscopic objects (Section 8.3).

In addition to energy and momentum, the total charge and total number of nucleons must be conserved in the system of particles for a nuclear reaction. For example, consider the reaction $^{19}\text{F}(p, \alpha)^{16}\text{O}$, which has a *Q* value of 8.124 MeV. We can show this reaction more completely as

$$^{1}_{1}\text{H} + {}^{19}_{9}\text{F} \longrightarrow {}^{16}_{8}\text{O} + {}^{4}_{2}\text{He}$$

We see that the total number of nucleons before the reaction (1 + 19 = 20) is equal to the total number after the reaction (16 + 4 = 20). Furthermore, the total charge ($Z = 10$) is the same before and after the reaction.

Before reaction

After reaction

**Figure 30.19**

A nuclear reaction. Before the reaction, an incoming particle a moves toward a target nucleus X. After the reaction, the target nucleus has changed to nucleus Y and an outgoing particle b moves away from the reaction site.

**context connection**

## 30.6 • THE ENGINE OF THE STARS

One of the important features of nuclear reactions is that much more energy is released (i.e., converted from mass–energy) than in normal chemical reactions, such as the burning of fossil fuels. Let us look back at our binding energy curve (see Fig. 30.10) and consider two important nuclear reactions that relate to that curve. If a heavy nucleus at the right of the graph splits into two lighter nuclei, the total binding energy within the system increases, representing energy released from the nuclei. This type of reaction was observed and reported in 1939 by Otto Hahn and Fritz Strassman. This reaction is known as **fission** and was of great scientific and political interest at the time of World War II because of the development of the first nuclear weapon.

In the fission reaction, a fissionable nucleus (the target nucleus X), which is often $^{235}$U, absorbs a slowly moving neutron (the incoming particle a) and the nucleus splits into two smaller nuclei (two nuclei $Y_1$ and $Y_2$), releasing energy and more neutrons (several particles b). These neutrons can then go on to be absorbed within other nuclei, causing other fissions. With no means of control, the result is a chain reaction explosion, as suggested by Figure 30.20. With proper control, the fission process is used in nuclear power generating stations.

Examining the other end of the binding energy curve, we see that we could also increase the binding energy of the system and release energy by combining two light nuclei. This is the process of **fusion** and is made difficult by the fact that the nuclei must overcome a very strong Coulomb repulsion before they become

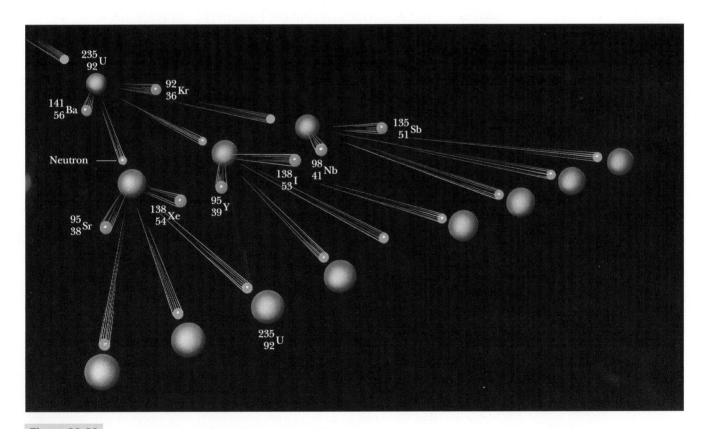

**Figure 30.20**

A nuclear chain reaction initiated by the capture of a neutron by a $^{235}$U nucleus.

close enough together to fuse. One way to assist the nuclei in overcoming this repulsion is to cause them to move with very high kinetic energy by raising the system of nuclei to a very high temperature. If the density of nuclei is high also, the probability of nuclei colliding is high, and fusion can occur. The technological problem of creating very high temperatures and densities is holding up progress in the area of Earth-based controlled fusion research.

At some natural locations (e.g., the cores of stars), the necessary high temperatures and densities exist. Consider a collection of gas and dust somewhere in the Universe to be an isolated system. What happens as this system collapses under its own gravitational attraction? Energy of the system is conserved, and the gravitational potential energy associated with the separated particles decreases while the kinetic energy of the particles increases. This is just like a falling ball; cosmic particles are "falling" into a gravitational center. As the falling particles collide with the particles that have already fallen into the central region of collapse, their kinetic energy is distributed to the other particles by collisions—it becomes internal energy, which is related to the temperature of the collection of particles.

If the system has enough material so that the temperature and density of the core of the system rise to the point where fusion can occur, the system becomes a star. The primary constituent of the Universe is hydrogen, so the fusion reaction at the center of a star combines hydrogen nuclei—protons—into helium nuclei. A common reaction process for stars with relatively cool cores ($T < 15 \times 10^6$ K) is the **proton–proton cycle.** In the first step of the process, two protons combine to form deuterium:

$$^1_1\text{H} + {}^1_1\text{H} \longrightarrow {}^2_1\text{H} + e^+ + \nu$$

Notice the implicit $^2_2\text{He}$ nucleus that is formed, but which does not appear in the reaction equation. This nucleus is highly unstable and decays very rapidly by beta-plus decay to the deuterium nucleus, a positron and a neutrino.

In the next step, the deuterium nucleus undergoes fusion with another proton to form a helium-3 nucleus:

$$^1_1\text{H} + {}^2_1\text{H} \longrightarrow {}^3_2\text{He} + \gamma$$

Finally, two helium-3 nuclei formed in such reactions can fuse to form helium-4 and two protons:

$$^3_2\text{He} + {}^3_2\text{He} \longrightarrow {}^4_2\text{He} + {}^1_1\text{H} + {}^1_1\text{H}$$

The net result of this cycle has been the joining of four protons to form a helium-4 nucleus, with the release of energy that eventually leaves the star as electromagnetic radiation from its surface. In addition, notice that the reaction releases neutrinos, which serve as a signal for beta decay occurring within the star. The observation of increased neutrino flow from a supernova is an important tool in analyzing the event.

For stars with hotter cores ($T > 15 \times 10^6$ K), another process, called the **carbon cycle,** dominates. In temperatures this high, hydrogen nuclei can fuse into nuclei heavier than helium, such as carbon. In the first of six steps in the cycle, a carbon nucleus fuses with a proton to form nitrogen:

$$^1_1\text{H} + {}^{12}_6\text{C} \longrightarrow {}^{13}_7\text{N}$$

The nitrogen nucleus is proton-rich and undergoes beta-plus decay:

$$^{13}_7\text{N} \longrightarrow {}^{13}_6\text{C} + e^+ + \nu$$

The resulting carbon-13 nucleus fuses with another proton, with the emission of a gamma ray:

$$\frac{1}{1}H + \frac{13}{6}C \longrightarrow \frac{14}{7}N + \gamma$$

The nitrogen-14 fuses with another proton, with more gamma emission:

$$\frac{1}{1}H + \frac{14}{7}N \longrightarrow \frac{15}{8}O + \gamma$$

The oxygen nucleus undergoes beta-plus decay:

$$\frac{15}{8}O \longrightarrow \frac{15}{7}N + e^+ + \nu$$

And, finally, the nitrogen-15 fuses with another proton:

$$\frac{1}{1}H + \frac{15}{7}N \longrightarrow \frac{12}{6}C + \frac{4}{2}He$$

Notice that the net effect of this process is to combine four protons into a helium nucleus, just like the proton–proton cycle. The carbon-12 with which we began the process is returned at the end, so it acts only as a catalyst to the process and is not consumed.

Depending on its mass, a star converts energy in its core at a rate between $10^{23}$ and $10^{33}$ W. The energy released from the nuclei in the core is transferred outward through the surrounding envelope by radiation and convection. The energy is then radiated from the surface of the star as infrared, visible, and ultraviolet light. The weight of the envelope keeps the core from exploding. The whole system is stable as long as the supply of hydrogen in the core lasts.

In the previous chapters, we presented examples of the applications of quantum physics and atomic physics to processes in space. In this chapter, we have seen that nuclear processes also have an important role in the cosmos. The formation of stars is a critical process in the development of the Universe. The energy provided by stars is crucial to life on planets such as the Earth. In our next, and final, chapter, we shall discuss the processes that occur on an even smaller scale, the scale of *elementary particles*. We shall find again that looking at a smaller scale allows us to advance our understanding of the largest scale system, the Universe.

## SUMMARY

A nuclear species can be represented by $\frac{A}{Z}X$, where $A$ is the **mass number,** the total number of nucleons, and $Z$ is the **atomic number,** the total number of protons. The total number of neutrons in a nucleus is the **neutron number** $N$, where $A = N + Z$. Elements with the same $Z$ but different $A$ and $N$ values are called **isotopes.**

Assuming that a nucleus is spherical, its radius is

$$r = r_0 A^{1/3} \qquad \textbf{[30.1]}$$

where $r_0 = 1.2$ fm.

Nuclei are stable because of the **nuclear force** between nucleons. This short-range force dominates the Coulomb repulsive force at distances of less than about 2 fm and is independent of charge.

Light nuclei are most stable when the number of protons equals the number of neutrons. Heavy nuclei are most stable when the number of neutrons exceeds the number of protons.

In addition, many stable nuclei have $Z$ and $N$ values that are both even. Nuclei with unusually high stability have $Z$ or $N$ values of 2, 8, 20, 28, 50, 82, and 126, called **magic numbers.**

Nuclei have an intrinsic spin angular momentum of magnitude $\sqrt{I(I + 1)}\hbar$, where $I$ is the **nuclear spin quantum number.** The magnetic moment of a nucleus is measured in terms of the **nuclear magneton** $\mu_n$, where

$$\mu_n \equiv \frac{e\hbar}{2m_p} = 5.05 \times 10^{-27} \text{ J/T} \qquad \textbf{[30.3]}$$

The difference in mass between the separate nucleons and the nucleus containing these nucleons, when multiplied by $c^2$, gives the **binding energy** $E_b$ of the nucleus: $E_b = \Delta mc^2$. We can calculate the binding energy of any nucleus $\frac{A}{Z}X$ using the expression

$$E_b(\text{MeV}) = [ZM(\text{H}) + Nm_n - M(\frac{A}{Z}X)]$$
$$\times \ 931.494 \text{ MeV/u} \qquad \textbf{[30.4]}$$

Radioactive processes include alpha decay, beta decay, and gamma decay. An alpha particle is a $^4$He nucleus; a beta particle is either an electron ($e^-$) or a positron ($e^+$); a gamma particle is a high-energy photon.

If a radioactive material contains $N_0$ radioactive nuclei at $t = 0$, the number $N$ of nuclei remaining at time $t$ is

$$N = N_0 e^{-\lambda t} \qquad \text{[30.6]}$$

where $\lambda$ is the **decay constant,** or **disintegration constant.** The **decay rate,** or **activity,** of a radioactive substance is given by

$$R = \left| \frac{dN}{dt} \right| = N_0 \lambda e^{-\lambda t} = R_0 e^{-\lambda t} \qquad \text{[30.7]}$$

where $R_0 = N_0 \lambda$ is the activity at $t = 0$. The **half-life** $T_{1/2}$ is defined as the time it takes half of a given number of radioactive nuclei to decay, where

$$T_{1/2} = \frac{\ln 2}{\lambda} = \frac{0.693}{\lambda} \qquad \text{[30.8]}$$

**Alpha decay** can occur because, according to quantum mechanics, some nuclei have barriers that can be penetrated by the alpha particles (the tunneling process). This process is energetically more favorable for those nuclei having large excesses of neutrons. A nucleus can undergo **beta decay** in two ways. It can emit either an electron ($e^-$) and an antineutrino ($\bar{\nu}$) or a positron ($e^+$) and a neutrino ($\nu$). In the **electron capture** process, the nucleus of an atom absorbs one of its own electrons (usually from the K shell) and emits a neutrino. In **gamma decay,** a nucleus in an excited state decays to its ground state and emits a gamma ray.

**Nuclear reactions** can occur when a target nucleus X is bombarded by a particle a, resulting in a nucleus Y and an outgoing particle b:

$$a + X \longrightarrow Y + b \qquad \text{or} \qquad X(a, b)Y \qquad \text{[30.23]}$$

The rest energy transformed to kinetic energy in such a reaction, called the **reaction energy** $Q$, is

$$Q = (M_a + M_X - M_Y - M_b)c^2 \qquad \text{[30.24]}$$

A reaction for which $Q$ is positive is called **exothermic.** A reaction for which $Q$ is negative is called **endothermic.** The minimum kinetic energy of the incoming particle necessary for such a reaction to occur is called the **threshold energy.**

## QUESTIONS

1. In Rutherford's experiment, assume that an alpha particle is headed directly toward the nucleus of an atom. Why doesn't the alpha particle make physical contact with the nucleus?

2. Explain the main differences between alpha, beta, and gamma rays.

3. Why do heavier elements require more neutrons to maintain stability?

4. The magnetic moment of a proton precesses with a frequency $\omega_p$ in the presence of a magnetic field. If the magnetic field intensity is doubled, what happens to the precessional frequency?

5. Why do nearly all the naturally occurring isotopes lie above the $N = Z$ line in Figure 30.4?

6. If a nucleus has a half-life of 1 year, does this mean it will be completely decayed after 2 years? Explain.

7. What fraction of a radioactive sample has decayed after two half-lives have elapsed?

8. Two samples of the same radioactive nuclide are prepared. Sample A has twice the initial activity of sample B. How does the half-life of A compare with the half-life of B? After each has passed through five half-lives, what is the ratio of their activities?

9. If no more people were to be born, the law of population growth would strongly resemble the radioactive decay law. Discuss this statement.

10. If a nucleus captures a slow-moving neutron, the product is left in a highly excited state with an energy approximately 8.0 MeV above the ground state. Explain the source of the excitation energy.

11. Use the analogy of a bullet and a rifle to explain why the recoiling nucleus carries off only a very small fraction of the disintegration energy in the $\alpha$ decay of a nucleus.

12. If a nucleus such as $^{226}$Ra initially at rest undergoes alpha decay, which has more kinetic energy after the decay, the alpha particle or the daughter nucleus?

13. Explain why many heavy nuclei undergo alpha decay but do not spontaneously emit neutrons or protons.

14. If an alpha particle and an electron have the same kinetic energy, which undergoes the greater deflection when passed through a magnetic field?

15. If film is kept in a wooden box, alpha particles from a radioactive source outside the box cannot expose the film but beta particles can. Explain.

16. Suppose it could be shown that the cosmic ray intensity at the Earth's surface was much greater 10 000 years ago. How would this difference affect what we accept as valid carbon-dated values of the age of ancient samples of once-living matter?

17. The radioactive nucleus $^{226}_{88}$Ra has a half-life of approximately $1.6 \times 10^3$ years. Being that the Solar System is about 5 billion years old, why do we still find this nucleus in nature?

**18.** From Table A.3, identify the four stable nuclei that have magic numbers in both $Z$ and $N$.

**19.** *And swift, and swift past comprehension*
*Turn round Earth's beauty and her might.*
*The heavens blaze in alternation*
*With deep and chill and rainy night.*
*In mighty currents foams the ocean*
*Up from the rocks' abyssal base,*
*With rock and sea torn into motion*
*In ever-swift celestial race.*
*And tempests bluster in a contest*
*From sea to land, from land to sea.*
*In rage they forge a chain around us*

*Of deepest meaning, energy.*
*There flames a lightning disaster*
*Before the thunder, in its way.*
*But all Your servants honor, Master,*
*The gentle order of Your day.*

Johann Wolfgang von Goethe wrote this song of the archangels in *Faust* a half-century before the law of conservation of energy was recognized. Students often find it convenient to think of a list of several "forms of energy," from kinetic to nuclear. Argue for or against the view that these lines of poetry make an obvious or an oblique reference to every form of energy and energy transfer.

## PROBLEMS

1, 2, 3 = straightforward, intermediate, challenging      □ = full solution available in the *Student Solutions Manual and Study Guide*

**web** = solution posted at **http://www.harcourtcollege.com/physics/**      🖥 = computer useful in solving problem

🌐 = Interactive Physics      ▪ = paired numerical/symbolic problems      📠 = life science application

---

*Note:* Table 30.3 will be useful for many of these problems. A more complete list of atomic masses is given in Table A.3 in Appendix A.

**TABLE 30.3**    **Some Atomic Masses**

| Element (or Particle) | Atomic Mass (u) | Element | Atomic Mass (u) |
|---|---|---|---|
| (electron) | 0.000 549 | $^{23}_{11}$Na | 22.989 770 |
| (neutron) | 1.008 665 | $^{23}_{12}$Mg | 22.994 124 |
| $^{1}_{1}$H | 1.007 825 | $^{27}_{13}$Al | 26.981 538 |
| $^{2}_{1}$H | 2.014 102 | $^{30}_{15}$P | 29.978 307 |
| $^{4}_{2}$He | 4.002 602 | $^{40}_{20}$Ca | 39.962 591 |
| $^{7}_{3}$Li | 7.016 003 | $^{42}_{20}$Ca | 41.958 618 |
| $^{9}_{4}$Be | 9.012 174 | $^{43}_{20}$Ca | 42.958 767 |
| $^{10}_{5}$B | 10.012 936 | $^{56}_{26}$Fe | 55.934 940 |
| $^{12}_{6}$C | 12.000 000 | $^{64}_{30}$Zn | 63.929 144 |
| $^{13}_{6}$C | 13.003 355 | $^{64}_{29}$Cu | 63.929 599 |
| $^{14}_{7}$N | 14.003 074 | $^{93}_{41}$Nb | 92.906 376 |
| $^{15}_{7}$N | 15.000 108 | $^{197}_{79}$Au | 196.966 543 |
| $^{15}_{8}$O | 15.003 065 | $^{202}_{80}$Hg | 201.970 617 |
| $^{17}_{8}$O | 16.999 132 | $^{216}_{84}$Po | 216.001 889 |
| $^{18}_{8}$O | 17.999 160 | $^{220}_{86}$Rn | 220.011 369 |
| $^{18}_{9}$F | 18.000 937 | $^{234}_{90}$Th | 234.043 593 |
| $^{20}_{10}$Ne | 19.992 435 | $^{238}_{92}$U | 238.050 784 |

### Section 30.1    Some Properties of Nuclei

**1.** What order of magnitude of protons are in your body? Of the number of neutrons? Of the number of electrons?

**2.** Find the radius of (a) a nucleus of $^{4}_{2}$He and (b) a nucleus of $^{238}_{92}$U.

**3.** How much energy (in megaelectron volts) must an $\alpha$ particle have to reach the surface of a gold nucleus ($Z = 79$, $A = 197$)? Assume the gold nucleus remains stationary.

**4.** **Review Problem.** What would be the gravitational force between two golf balls (each with a 4.30-cm diameter), 1.00 m apart, if they were made of nuclear matter?

**5.** A star ending its life with a mass of two times the mass of the Sun is expected to collapse, combining its protons and electrons to form a neutron star. Such a star could be thought of as a gigantic atomic nucleus. If a star of mass $2 \times 1.99 \times 10^{30}$ kg collapsed into neutrons ($m_n = 1.67 \times 10^{-27}$ kg), what would its radius be? (Assume that $r = r_0 A^{1/3}$.)

**6.** From Table A.3, identify the stable nuclei that correspond to the magic numbers given by Equation 30.2.

**7.** Construct a diagram like that of Figure 30.5 for the cases when $I$ equals (a) $\frac{5}{2}$ and (b) 4.

**8.** The radio frequency at which a nucleus displays resonance absorption between spin states is called the Larmor precessional frequency and is given by

$$f = \frac{\Delta E}{h} = \frac{2\mu B}{h}$$

where $\mu$ is the magnetic moment of the nucleus. Calculate the Larmor frequency for (a) free neutrons in a magnetic field of 1.00 T, (b) free protons in a magnetic field of

1.00 T, and (c) free protons in the Earth's magnetic field at a location where the magnitude of the field is 50.0 $\mu$T.

## Section 30.2 Binding Energy

**9.** Using the graph in Figure 30.10, estimate how much energy is released when a nucleus of mass number 200 fissions into two nuclei each of mass number 100.

**10.** Calculate the binding energy per nucleon for (a) $^2$H, (b) $^4$He, (c) $^{56}$Fe, and (d) $^{238}$U.

**11.** A pair of nuclei for which $Z_1 = N_2$ and $Z_2 = N_1$ are called
web *mirror isobars* (the atomic and neutron numbers are interchanged). Binding energy measurements on these nuclei can be used to obtain evidence of the charge independence of nuclear forces (i.e., proton–proton, proton–neutron, and neutron–neutron nuclear forces are equal). Calculate the difference in binding energy for the two mirror isobars $^{15}_{8}$O and $^{15}_{7}$N. The electric repulsion among eight protons rather than seven accounts for the difference.

**12.** Nuclei having the same mass numbers are called *isobars*. The isotope $^{139}_{57}$La is stable. A radioactive isobar $^{139}_{59}$Pr is located below the line of stable nuclei in Figure 30.4 and decays by e$^+$ emission. Another radioactive isobar of $^{139}$La, $^{139}_{55}$Cs, decays by e$^-$ emission and is located above the line of stable nuclei in Figure 30.4. (a) Which of these three isobars has the highest neutron-to-proton ratio? (b) Which has the greatest binding energy per nucleon? (c) Which do you expect to be heavier, $^{139}$Pr or $^{139}$Cs?

## Section 30.3 Radioactivity

**13.** A sample of radioactive material contains $1.00 \times 10^{15}$ atoms and has an activity of $6.00 \times 10^{11}$ Bq. What is its half-life?

**14.** The half-life of $^{131}$I is 8.04 days. On a certain day, the activity of an iodine-131 sample is 6.40 mCi. What is its activity 40.2 days later?

**15.** A freshly prepared sample of a certain radioactive isotope
web has an activity of 10.0 mCi. After 4.00 h, its activity is 8.00 mCi. (a) Find the decay constant and half-life. (b) How many atoms of the isotope were contained in the freshly prepared sample? (c) What is the sample's activity 30.0 h after it is prepared?

**16.** How much time elapses before 90.0% of the radioactivity of a sample of $^{72}_{33}$As disappears, as measured by its activity? The half-life of $^{72}_{33}$As is 26 h.

**17.** A radioactive nucleus has half-life $T_{1/2}$. A sample containing these nuclei has initial activity $R_0$. Calculate the number of nuclei that decay during the interval between the times $t_1$ and $t_2$.

**18.** *Indoor air pollution.* Uranium is naturally present in rock and soil. At one step in its series of radioactive decays,

$^{238}$U produces the chemically inert gas radon-222, with a half-life of 3.82 days. The radon seeps out of the ground to mix into the atmosphere, typically making open air radioactive with activity 0.3 pCi/L. In homes $^{222}$Rn can be a serious pollutant, accumulating to reach much higher activities in enclosed spaces. If the radon radioactivity exceeds 4 pCi/L, the Environmental Protection Agency suggests taking action to reduce it, by reducing infiltration of air from the ground. (a) Convert the activity 4 pCi/L to units of becquerel per cubic meter. (b) How many $^{222}$Rn atoms are in one cubic meter of air displaying this activity? (c) What fraction of the mass of the air does the radon constitute?

**19.** A building has become accidentally contaminated with radioactivity. The longest lived material in the building is strontium-90. ($^{90}_{38}$Sr has an atomic mass 89.907 7 u, and its half-life is 29.1 years. It is particularly dangerous because it substitutes for calcium in bones.) Assume that the building initially contained 5.00 kg of this substance uniformly distributed throughout the building (a very unlikely situation) and that the safe level is defined as less than 10.0 decays/min (to be small in comparison with background radiation). How long will the building be unsafe?

## Section 30.4 The Radioactive Decay Processes

**20.** Identify the missing nuclide or particle (X):
   (a) X $\rightarrow$ $^{65}_{28}$Ni $+ \gamma$
   (b) $^{215}_{84}$Po $\rightarrow$ X $+ \alpha$
   (c) X $\rightarrow$ $^{55}_{26}$Fe $+$ e$^+$ $+ \nu$
   (d) $^{109}_{48}$Cd $+$ X $\rightarrow$ $^{109}_{47}$Ag $+ \nu$
   (e) $^{14}_{7}$N $+$ $^4_2$He $\rightarrow$ X $+$ $^{17}_{8}$O

**21.** Find the energy released in the alpha decay

$$^{238}_{92}\text{U} \longrightarrow {}^{234}_{90}\text{Th} + {}^4_2\text{He}$$

You will find Table 30.3 useful.

**22.** A living specimen in equilibrium with the atmosphere contains one atom of $^{14}$C (half-life $= 5730$ years) for every $7.7 \times 10^{11}$ stable carbon atoms. An archeological sample of wood (cellulose, $C_{12}H_{22}O_{11}$) contains 21.0 mg of carbon. When the sample is placed inside a shielded beta counter with 88.0% counting efficiency, 837 counts are accumulated in one week. Assuming that the cosmic-ray flux and the Earth's atmosphere have not changed appreciably since the sample was formed, find the age of the sample.

**23.** **Review Problem.** Singly ionized carbon is accelerated through 1 000 V and passed into a mass spectrometer to determine the isotopes present (see Chapter 22). The magnitude of the magnetic field in the spectrometer is 0.200 T. (a) Determine the orbit radii for $^{12}$C and $^{13}$C isotopes as they pass through the field. (b) Show that the ratio of radii may be written in the form

$$\frac{r_1}{r_2} = \sqrt{\frac{m_1}{m_2}}$$

and verify that your radii in part (a) agree with this.

24. A $^3$H nucleus undergoes beta decay into $^3$He by creating an electron and an antineutrino according to the reaction

$$^3_1\text{H} \longrightarrow ^3_2\text{He} + e^- + \bar{\nu}$$

The symbols in this reaction refer to nuclei. Write the reaction referring to neutral atoms by adding one electron to both sides. Then use Table A.3 to determine the total energy released in this reaction.

25. The nucleus $^{15}_8$O decays by electron capture. The nuclear reaction is written

$$^{15}_8\text{O} + e^- \longrightarrow ^{15}_7\text{N} + \nu$$

(a) Write the process going on for a single particle within the nucleus. (b) Write the decay process referring to neutral atoms. (c) Determine the energy of the neutrino. Disregard the daughter's recoil.

26. (a) Why is the inverse beta decay $p \rightarrow n + e^+ + \nu$ forbidden for a free proton? (b) Why is the same reaction possible if the proton is bound in a nucleus? For example, the following reaction occurs:

$$^{13}_7\text{N} \longrightarrow ^{13}_6\text{C} + e^+ + \nu$$

(c) How much energy is released in the reaction given in part (b)? (*Hint:* Add seven electrons to both sides of the reaction to write it for neutral atoms. You may use the masses $m(e^+) = 0.000\ 549$ u, $M(^{13}\text{C}) = 13.003\ 355$ u, $M(^{13}\text{N}) = 13.005\ 739$ u.)

27. Enter the correct isotope symbol in each open square in Figure P30.27, which shows the sequences of decays in the natural radioactive series starting with the long-lived isotope uranium-235 and ending with the stable nucleus lead-207.

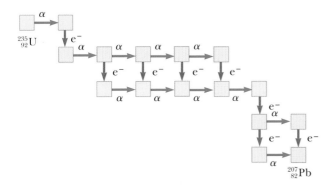

**Figure P30.27**

## Section 30.5 Nuclear Reactions

28. Identify the unknown nuclei and particles X and X′ in the following nuclear reactions:
    (a) $X + ^4_2\text{He} \rightarrow ^{24}_{12}\text{Mg} + ^1_0\text{n}$
    (b) $^{235}_{92}\text{U} + ^1_0\text{n} \rightarrow ^{90}_{38}\text{Sr} + X + 2^1_0\text{n}$
    (c) $2^1_1\text{H} \rightarrow ^2_1\text{H} + X + X′$

29. Natural gold has only one isotope, $^{197}_{79}$Au. If natural gold is irradiated by a flux of slow neutrons, electrons are emitted. (a) Write the reaction equation. (b) Calculate the maximum energy of the emitted electrons. The mass of $^{198}_{80}$Hg is 197.966 743 u.

30. A beam of 6.61-MeV protons is incident on a target of $^{27}_{13}$Al. Those that collide produce the reaction

$$p + ^{27}_{13}\text{Al} \longrightarrow ^{27}_{14}\text{Si} + n$$

($^{27}_{14}$Si has mass 26.986 721 u.) Ignoring any recoil of the product nucleus, determine the kinetic energy of the emerging neutrons.

31. (a) The following fission reaction is typical of those occurring in a nuclear electric generating station:

$$^1_0\text{n} + ^{235}_{92}\text{U} \longrightarrow ^{141}_{56}\text{Ba} + ^{92}_{36}\text{Kr} + 3(^1_0\text{n})$$

Find the energy released. The required masses are

$$M(^1_0\text{n}) = 1.008\ 665\ \text{u}$$

$$M(^{235}_{92}\text{U}) = 235.043\ 915\ \text{u}$$

$$M(^{141}_{56}\text{Ba}) = 140.913\ 9\ \text{u}$$

$$M(^{92}_{36}\text{Kr}) = 91.897\ 3\ \text{u}$$

(b) What fraction of the initial mass of the system is transformed?

32. Burning one metric ton (1 000 kg) of coal can yield $3.30 \times 10^{10}$ J of energy. Fission of one nucleus of uranium-235 yields an average energy of about 208 MeV. What mass of uranium produces the same energy as a ton of coal?

33. **Review Problem.** Suppose enriched uranium containing 3.40% of the fissionable isotope $^{235}_{92}$U is used to fuel a ship. The water exerts an average friction force of magnitude $1.00 \times 10^5$ N on the ship. How far can the ship travel per kilogram of fuel? Assume that the energy released per fission event is 208 MeV and that the ship's engine has an efficiency of 20.0%.

34. Of all the hydrogen in the oceans, 0.030 0% of the mass is deuterium. The oceans have a volume of $3.17 \times 10^8$ mi$^3$. (a) If nuclear fusion were controlled and all the deuterium in the oceans were fused to $^4_2$He, how many joules of energy would be released? (b) World power consumption is about $7.00 \times 10^{12}$ W. If consumption were 100 times greater, how many years would the energy calculated in part (a) last?

**35.** It has been estimated that about $10^9$ tons of natural uranium is available at concentrations exceeding 100 parts per million, of which 0.7% is the fissionable isotope $^{235}U$. Assume that all the world's energy use ($7 \times 10^{12}$ J/s) were supplied by $^{235}U$ fission in conventional nuclear reactors, releasing 208 MeV for each reaction. How long would the supply last? The estimate of uranium supply is taken from K. S. Deffeyes and I. D. MacGregor, "World Uranium Resources." *Sci. Am.* **242(1)**:66, 1980.

## Section 30.6 Context Connection—
## The Engine of the Stars

**36.** The sun radiates energy at the rate of $3.77 \times 10^{26}$ W. Suppose that the net reaction

$$4(^1_1H) + 2(e^-) \longrightarrow {}^4_2He + 2\nu + \gamma$$

accounts for all the energy released. Calculate the number of protons fused per second.

**37.** Consider the two nuclear reactions

$$\text{(I)} \quad A + B \rightarrow C + E$$
$$\text{(II)} \quad C + D \rightarrow F + G$$

(a) Show that the net disintegration energy for these two reactions ($Q_{net} = Q_I + Q_{II}$) is identical to the disintegration energy for the net reaction

$$A + B + D \longrightarrow E + F + G$$

(b) One possible chain of reactions in the proton–proton cycle in the Sun's core is

$$^1_1H + {}^1_1H \longrightarrow {}^2_1H + e^+ + \nu$$
$$e^+ + e^- \longrightarrow 2\gamma$$
$$^1_1H + {}^2_1H \longrightarrow {}^3_2H + \gamma$$
$$^1_1H + {}^3_2H \longrightarrow {}^4_2H + e^+ + \nu$$
$$e^+ + e^- \longrightarrow 2\gamma$$

Based on part (a), what is $Q_{net}$ for this sequence?

**38.** In addition to the proton–proton cycle described in the previous problem, the carbon cycle, first proposed by Hans Bethe in 1939, is another cycle by which energy is released in stars as hydrogen is converted to helium. The carbon cycle requires higher temperatures than the proton–proton cycle. The series of reactions is

$$^{12}C + {}^1H \longrightarrow {}^{13}N + \gamma$$
$$^{13}N \longrightarrow {}^{13}C + e^+ + \nu$$
$$e^+ + e^- \longrightarrow 2\gamma$$
$$^{13}C + {}^1H \longrightarrow {}^{14}N + \gamma$$
$$^{14}N + {}^1H \longrightarrow {}^{15}O + \gamma$$

$$^{15}O \longrightarrow {}^{15}N + e^+ + \nu$$
$$e^+ + e^- \longrightarrow 2\gamma$$
$$^{15}N + {}^1H \longrightarrow {}^{12}C + {}^4He$$

(a) If the proton–proton cycle requires a temperature of $1.5 \times 10^7$ K, estimate by proportion the temperature required for the carbon cycle. (b) Calculate the $Q$ value for each step in the carbon cycle and the overall energy released. (c) Do you think the energy carried off by the neutrinos is deposited in the star? Explain.

**39.** After determining that the Sun has existed for hundreds of millions of years, but before the discovery of nuclear physics, scientists could not explain why the Sun has continued to burn for such a long time. For example, if it were a coal fire, it would have burned up in about 3 000 yr. Assume that the Sun, whose mass is $1.99 \times 10^{30}$ kg, originally consisted entirely of hydrogen and that its total power output is $3.77 \times 10^{26}$ W. (a) If the energy-generating mechanism of the Sun is the fusion of hydrogen into helium via the net reaction

$$4(^1_1H) + 2(e^-) \longrightarrow {}^4_2He + 2\nu + \gamma$$

calculate the energy (in joules) given off by this reaction. (b) Determine how many hydrogen atoms constitute the Sun. Take the mass of one hydrogen atom to be $1.67 \times 10^{-27}$ kg. (c) Assuming that the total power output remains constant, after what time will all the hydrogen be converted into helium, making the Sun die? The actual projected lifetime of the Sun is about 10 billion years, because only the hydrogen in the relatively small core is available as a fuel. Only in the core are temperatures and densities high enough for the fusion reaction to be self-sustaining.

**40.** Carbon detonations are powerful nuclear reactions that temporarily tear apart the cores inside massive stars late in their lives. These blasts are produced by carbon fusion, which requires a temperature of about $6 \times 10^8$ K to overcome the strong Coulomb repulsion between carbon nuclei. (a) Estimate the repulsive energy barrier to fusion, using the required ignition temperature for carbon fusion. (In other words, what is the average kinetic energy of a carbon nucleus at $6 \times 10^8$ K?) (b) Calculate the energy (in megaelectron volts) released in each of these "carbon-burning" reactions:

$$^{12}C + {}^{12}C \longrightarrow {}^{20}Ne + {}^4He$$
$$^{12}C + {}^{12}C \longrightarrow {}^{24}Mg + \gamma$$

(c) Calculate the energy (in kilowatt hours) given off when 2.00 kg of carbon completely fuses according to the first reaction.

41. A theory of nuclear astrophysics proposes that all the elements heavier than iron are formed in supernova explosions ending the lives of massive stars. If we assume that at the time of the explosion the amounts of $^{235}$U and $^{238}$U were equal, how long ago did the star(s) explode that released the elements that formed our Earth? The present $^{235}$U/$^{238}$U ratio is 0.007 25. The half-lives of $^{235}$U and $^{238}$U are $0.704 \times 10^9$ years and $4.47 \times 10^9$ years.

## Additional Problems

42. The alpha emitter polonium-210 ($^{210}_{84}$Po) is used as the nuclear energy source on a spacecraft (Fig. P30.42). Determine the initial power output of the source. Assume that it contains 0.155 kg of $^{210}$Po and that the efficiency for conversion of radioactive decay energy to electrical energy is 1.00%.

**Figure P30.42** The *Pioneer 10* spacecraft leaves the Solar System. It carries radioactive power supplies at the ends of two booms. Solar panels would not work far from the Sun. *(Courtesy of NASA Ames Home Page)*

43. (a) Find the radius of the $^{12}_{6}$C nucleus. (b) Find the force of repulsion between a proton at the surface of a $^{12}_{6}$C nucleus and the remaining five protons. (c) How much work (in megaelectron volts) has to be done to overcome this electrostatic repulsion to put the last proton into the nucleus? (d) Repeat parts (a), (b), and (c) for $^{238}_{92}$U.

44. The activity of a radioactive sample was measured over 12.0 h, with the net count rates shown in the table.

| Time (h) | Counting Rate (counts/min) |
|---|---|
| 1.00 | 3 100 |
| 2.00 | 2 450 |
| 4.00 | 1 480 |
| 6.00 | 910 |
| 8.00 | 545 |
| 10.0 | 330 |
| 12.0 | 200 |

(a) Plot the logarithm of counting rate as a function of time. (b) Determine the disintegration constant and half-life of the radioactive nuclei in the sample. (c) What counting rate would you expect for the sample at $t = 0$? (d) Assuming the efficiency of the counting instrument to be 10.0%, calculate the number of radioactive atoms in the sample at $t = 0$.

45. In a piece of rock from the Moon, the $^{87}$Rb content is assayed to be $1.82 \times 10^{10}$ atoms/g of material, and the $^{87}$Sr content is found to be $1.07 \times 10^9$ atoms/g. (a) Calculate the age of the rock. (b) Could the material in the rock actually be much older? What assumption is implicit in using the radioactive dating method? (The relevant decay is $^{87}$Rb $\rightarrow$ $^{87}$Sr $+$ e$^-$. The half-life of the decay is $4.75 \times 10^{10}$ yr.)

46. When, after a reaction or disturbance of any kind, a nucleus is left in an excited state, it can return to its normal (ground) state by emission of a gamma-ray photon (or several photons). This process is illustrated by Equation 30.20. The emitting nucleus must recoil to conserve both energy and momentum. (a) Show that the recoil energy of the nucleus is

$$E_r = \frac{(\Delta E)^2}{2Mc^2}$$

where $\Delta E$ is the difference in energy between the excited and ground states of a nucleus of mass $M$. (b) Calculate the recoil energy of the $^{57}$Fe nucleus when it decays by gamma emission from the 14.4-keV excited state. For this calculation, take the mass to be 57 u. (*Hint:* When writing the equation for conservation of energy, use $(Mv)^2/2M$ for the kinetic energy of the recoiling nucleus. Also, assume that $hf \ll Mc^2$ and use the binomial expansion.)

47. After the sudden release of radioactivity from the Chernobyl nuclear reactor accident in 1986, the radioactivity of milk in Poland rose to 2 000 Bq/L due to the intake of iodine-131 present in the grass eaten by dairy cattle. Radioactive iodine, with half-life 8.04 days, is particularly hazardous because the thyroid gland concentrates iodine.

The Chernobyl accident caused a measurable increase in thyroid cancers among children in Belarus. (a) For comparison, find the activity of milk due to potassium. Assume that 1 L of milk contains 2.00 g of potassium, of which 0.011 7% is the isotope $^{40}$K, which has a half-life $1.28 \times 10^{9}$ years. (b) After what time would the activity due to iodine fall below that due to potassium?

48. Europeans named a certain direction in the sky as between the horns of Taurus the Bull. On the day they named as July 4, 1054 A.D., a brilliant light appeared there. Europeans left no surviving record of the supernova, which could be seen in daylight for some days. It faded but remained visible for years, dimming for a time with the 77.1-day half-life of the radioactive cobalt-56 which had been created in the explosion. (a) The remains of the star now forms the Crab nebula. In it, the cobalt-56 has now decreased to what fraction of its original activity? (b) Suppose that a Native American, of the people called the Anasazi, made a charcoal drawing of the supernova. The carbon-14 in the charcoal has now decayed to what fraction of its original activity?

49. The decay of an unstable nucleus by alpha emission is represented by Equation 30.9. The disintegration energy $Q$ given by Equation 30.12 must be shared by the alpha particle and the daughter nucleus to conserve both energy and momentum in the decay process. (a) Show that $Q$ and $K_{\alpha}$, the kinetic energy of the alpha particle, are related by the expression

$$Q = K_{\alpha}\left(1 + \frac{M_{\alpha}}{M}\right)$$

where $M$ is the mass of the daughter nucleus. (b) Use the result of part (a) to find the energy of the alpha particle emitted in the decay of $^{226}$Ra. (See Example 30.5 for the calculation of $Q$.)

50. When the nuclear reaction represented by Equation 30.23 is endothermic, the disintegration energy $Q$ is negative. For the reaction to proceed, the incoming particle must have a minimum energy called the threshold energy $E_{th}$. Some fraction of the energy of the incident particle is transferred to the compound nucleus to conserve momentum. Therefore, $E_{th}$ must be greater than $Q$. (a) Show that

$$E_{th} = -Q\left(1 + \frac{M_{a}}{M_{X}}\right)$$

(b) Calculate the threshold energy of the incident alpha particle in the reaction

$$^{4}_{2}\text{He} + ^{14}_{7}\text{N} \longrightarrow ^{17}_{8}\text{O} + ^{1}_{1}\text{H}$$

51. *Student determination of the half-life of* $^{137}$Ba. The radioactive barium isotope $^{137}$Ba has a relatively short half-life and can be easily extracted from a solution containing its parent cesium ($^{137}$Cs). This barium isotope is commonly used in an undergraduate laboratory exercise demonstrating the radioactive decay law. Undergraduate students using modest experimental equipment took the data presented in Figure P30.51. Determine the half-life for the decay of $^{137}$Ba using their data.

**Figure P30.51**

52. During the manufacture of a steel engine component, radioactive iron ($^{59}$Fe) is included in the total mass of 0.200 kg. The component is placed in a test engine when the activity due to this isotope is 20.0 $\mu$Ci. After a 1 000-h test period, some of the lubricating oil is removed from the engine and found to contain enough $^{59}$Fe to produce 800 disintegrations/min/L of oil. The total volume of oil in the engine is 6.50 L. Calculate the mass worn from the engine component per hour of operation. (The half-life of $^{59}$Fe is 45.1 days.)

53. *Neutron activation analysis* is a method for chemical analysis at the level of elements. When a sample is irradiated by neutrons, radioactive atoms are produced continuously and then decay according to their characteristic half-lives. (a) Assume that one species of radioactive nuclei is produced at a constant rate $R$ and that its decay is described by the conventional radioactive decay law. Show that the number of radioactive atoms accumulated after an irradiation time $t$ is

$$N = \frac{R}{\lambda}(1 - e^{-\lambda t})$$

(b) What is the maximum number of radioactive atoms that can be produced?

**54.** The atomic bomb dropped on Hiroshima on August 6, 1945, released $5 \times 10^{13}$ J of energy, equivalent to that from 12 000 tons of trinitrotoluene (TNT). The fission of one $^{235}_{92}$U nucleus releases an average of 208 MeV. Estimate (a) the number of nuclei fissioned and (b) the mass of this $^{235}_{92}$U.

**55. Review Problem.** A nuclear power plant operates using the energy released in nuclear fission to convert 20°C water into 400°C steam. How much water could theoretically be converted to steam by the complete fissioning of 1.00 g of $^{235}$U at 200 MeV/fission?

**56.** About 1 of every 3 300 water molecules contains one deuterium atom. (a) If all the deuterium nuclei in 1 L of water are fused in pairs according to reaction $^2$H + $^2$H → $^3$He + n + 3.27 MeV, how much energy in joules is liberated? (b) Burning gasoline produces about $3.40 \times 10^7$ J/L. Compare the energy obtainable from the fusion of the deuterium in a liter of water with the energy liberated from the burning of a liter of gasoline.

**57.** Consider a radioactive sample. Determine the ratio of the number of atoms decaying during the first half of its half-life to the number of atoms decaying during the second half of its half-life.

**58. Review Problem.** The first nuclear bomb was a fissioning mass of plutonium-239, exploded in the Trinity test before the dawn of July 16, 1945, at Alamogordo, New Mexico. Enrico Fermi was 14 km away, lying on the ground facing away from the bomb. After the whole sky had flashed with unbelievable brightness, Fermi stood up and began dropping bits of paper to the ground. They first fell at his feet in the calm and silent air. As the shock wave passed, about 40 s after the explosion, the paper then in flight jumped about 5 cm away from ground zero. (a) Assume that the shock wave in air propagated equally in all directions without absorption. Find the change in volume of a sphere of radius 14 km as it expands by 5 cm. (b) Find the work $P \Delta V$ done by the air in this sphere on the next layer of air farther from the center. (c) Assume the shock wave carried on the order of one tenth of the energy of the explosion. Estimate the order of magnitude of the bomb yield. (d) One ton of exploding TNT releases energy of 4.2 GJ. What was the order of magnitude of the energy of the Trinity test in equivalent tons of TNT? The dawn revealed the mushroom cloud. Fermi's immediate knowledge of the bomb yield agreed with that determined days later by analysis of elaborate measurements.

**59.** (a) Calculate the energy (in kilowatt hours) released if 1.00 kg of $^{239}$Pu undergoes complete fission and the energy released per fission event is 200 MeV. (b) Calculate the energy (in electron volts) released in the deuterium-tritium fusion reaction

$$^2_1\text{H} + ^3_1\text{H} \longrightarrow ^4_2\text{He} + ^1_0\text{n}$$

(c) Calculate the energy (in kilowatt hours) released if 1.00 kg of deuterium undergoes fusion according to this reaction. (d) Calculate the energy (in kilowatt hours) released by the combustion of 1.00 kg of coal if each C + $O_2$ → $CO_2$ reaction yields 4.20 eV. (e) List advantages and disadvantages of each of these methods of energy generation.

## ANSWERS TO QUICK QUIZZES

**30.1** Within a vertical grouping, the dots represent nuclei all having the same proton number, so they must all represent the same element—they are isotopes. Because the number of neutrons varies within a group, each dot must signify a different isotope of the element.

**30.2** Isotopes of a given element correspond to nuclei with different numbers of neutrons. This results in different masses of the atom, and different magnetic moments, because the neutron, despite being uncharged, has a magnetic moment. The chemical behavior, however, is governed by the electrons. All isotopes of a given element have the same number of electrons and, therefore, the same chemical behavior.

**30.3** (e). A year of 365 days is equivalent to 365 d/5.01 d ≈ 73 half-lives. Thus, the activity will be reduced after one year to approximately $(1/2)^{73}$ (1.000 $\mu$Ci) ∼ $10^{-22}$ $\mu$Ci.

**30.4** An alpha particle contains two protons and two neutrons. Because a hydrogen nucleus contains only one proton, it cannot emit an alpha particle.

**30.5** No. The stone was not alive and breathing air containing carbon-14. Only the ages of artifacts that were once alive can be estimated with carbon dating.

# chapter

# 31

A particle detector at the Fermi National Accelerator Laboratory (Fermilab) near Chicago. Such detectors are enormous, especially relative to the subatomic particles they are designed to detect. The person leaning against the fence at the lower right provides a sense of scale.

*(Courtesy of Fermi National Accelerator Laboratory)*

# Particle Physics

n the early chapters of this book, we discussed the particle model, which treats an object as a particle of zero size with no structure. We could not use the particle model for the atom to explain such phenomena as atomic spectra, however. Furthermore, in Chapter 30, we could not ignore the structure of the nucleus to understand behavior such as nuclear stability and radioactive decay. What about the constituents of the nucleus—the protons and neutrons? Can we apply the particle model to these entities? As we shall see, even these entities have structure. This leads to a puzzling question: As we continue to investigate the structure of smaller and smaller "particles," will we ever reach a level at which the building blocks are truly described by the particle model?

In this concluding chapter, we explore this question by examining the properties and classifications of the various known subatomic particles and the fundamental interactions that govern their behavior. We also discuss the current model of elementary particles, in which all matter is believed to be constructed from only two families of particles: quarks and leptons.

The word *atom* is from the Greek *atomos*, which means "indivisible." At one time, atoms were thought to be the indivisible constituents of matter; that is, they were regarded as elementary particles. After 1932, physicists viewed all matter as consisting of only three constituent particles: electrons, protons, and neutrons. With the exception of the free neutron, these particles are very stable. Beginning in 1945, many new particles were discovered in experiments involving high-energy collisions between known particles. These new particles are characteristically very unstable and have very short half-lives, ranging between $10^{-6}$ s and $10^{-23}$ s. So far, more than 300 of these unstable, temporary particles have been catalogued.

During the past 70 years, many powerful particle accelerators have been constructed throughout the world, making it possible to observe collisions of highly energetic particles under controlled laboratory conditions, so as to reveal the subatomic world in finer detail. Until the 1960s, physicists were bewildered by the large number and variety of subatomic particles being discovered. They wondered if the particles were like animals in a zoo, having no systematic relationship connecting them, or whether a pattern was emerging that would provide a better understanding of the elaborate structure in the subnuclear world. Since that time, physicists have advanced our knowledge of the structure of matter tremendously by developing a structural model in which most of the particles in the ever-growing particle zoo are made of smaller particles called quarks. Thus, protons and neutrons, for example, are not truly elementary but are systems of tightly bound quarks.

## 31.1 • THE FUNDAMENTAL FORCES IN NATURE

As we learned in Chapter 5, all natural phenomena can be described by four fundamental forces between particles. In order of decreasing strength, these are the nuclear force, the electromagnetic force, the weak force, and the gravitational force.

The nuclear force, as we mentioned in Chapter 30, holds nucleons together. It is very short range and is negligible for separations greater than about 2 fm (about the size of the nucleus). The electromagnetic force, which binds atoms and molecules together to form ordinary matter, has about $10^{-2}$ times the strength of the nuclear force. It is a long-range force that decreases in strength as the inverse square of the separation between interacting particles. The weak force is a short-range force that tends to produce instability in certain nuclei. It is responsible for radioactive decay processes such as beta decay, and its strength is only about $10^{-5}$ times that of the nuclear force. Finally, the gravitational force is a long-range force that has a strength of only about $10^{-39}$ times that of the nuclear force. Although this familiar interaction is the force that holds the planets, stars, and galaxies together, its effect on elementary particles is negligible.

In modern physics, interactions between particles are often described in terms of a structural model that involves the exchange of **field particles,** or **quanta.** In the case of the familiar electromagnetic interaction, for instance, the field particles are photons. In the language of modern physics, we say that the electromagnetic force is *mediated* by photons, and that photons are the quanta of the electromagnetic field. Likewise, the nuclear force is mediated by field particles called **gluons,** the weak force is mediated by particles called the W and Z **bosons** (in general, all particles with integral spin are called *bosons*), and the gravitational force is mediated by quanta of the gravitational field called **gravitons.** These interactions, their ranges, and their relative strengths are summarized in Table 31.1.

• *Field particles*

| TABLE 31.1 | Particle Interactions | | |
|---|---|---|---|
| **Interaction** | **Relative Strength** | **Range of Force** | **Mediating Field Particle** |
| Nuclear | 1 | Short ($\sim 2$ fm) | Gluon |
| Electromagnetic | $10^{-2}$ | $\infty$ | Photon |
| Weak | $10^{-5}$ | Short ($\sim 10^{-3}$ fm) | $W^{\pm}$, $Z^0$ bosons |
| Gravitational | $10^{-39}$ | $\infty$ | Graviton |

**Figure 31.1**

Dirac's model for the existence of antielectrons (positrons). The states lower in energy than $-m_e c^2$ are filled with electrons (the Dirac sea). One of these electrons can make a transition out of its state only if it is provided with energy equal to or larger than $2m_e c^2$. This leaves a vacancy in the Dirac sea, which can behave as a particle identical to the electron except for its positive charge.

**PITFALL  PREVENTION  31.1**

**Antiparticles**

An antiparticle is not identified solely on the basis of opposite charge; even neutral particles have antiparticles, which are defined in terms of other properties, such as spin.

## 31.2 • POSITRONS AND OTHER ANTIPARTICLES

In the 1920s, the theoretical physicist Paul Adrien Maurice Dirac (1902–1984) developed a version of quantum mechanics that incorporated special relativity. Dirac's theory explained the origin of electron spin and its magnetic moment. It also presented a major difficulty, however. Dirac's relativistic wave equation required solutions corresponding to negative energy states even for free particles. But if negative energy states existed, one would expect an electron in a state of positive energy to make a rapid transition to one of these states, emitting a photon in the process. Dirac avoided this difficulty by postulating a structural model in which all negative energy states are filled. The electrons occupying these negative energy states are collectively called the *Dirac sea*. Electrons in the Dirac sea are not directly observable because the Pauli exclusion principle does not allow them to react to external forces—there are no states available to which an electron can make a transition in response to an external force. Thus, an electron in such a state acts as an isolated system, unless an interaction with the environment is strong enough to excite the electron to a positive energy state. Such an excitation causes one of the negative energy states to be vacant, as in Figure 31.1, leaving a hole in the sea of filled states. *The hole can react to external forces and is observable.* The hole reacts in a way similar to that of the electron, except that it has a positive charge—it is the *antiparticle* to the electron.

The profound implication of this model is that *every particle has a corresponding antiparticle.* The antiparticle has the same mass as the particle, but the opposite charge. For example, the electron's antiparticle, called a **positron,** has a mass of 0.511 MeV/$c^2$ and a positive charge of $1.60 \times 10^{-19}$ C.

The positron was discovered by Carl Anderson in 1932, and in 1936 he was awarded the Nobel Prize in physics for that achievement. Anderson discovered the positron while examining tracks created by electron-like particles of positive charge in a cloud chamber. (These early experiments used cosmic rays—mostly energetic protons passing through interstellar space—to initiate high-energy reactions in the upper atmosphere, which resulted in the production of positrons at ground level.) To discriminate between positive and negative charges, Anderson placed the cloud chamber in a magnetic field, causing moving charged particles to follow curved paths, as discussed in Section 22.3. Some of the electron-like tracks deflected in a direction corresponding to a positively charged particle.

Since Anderson's initial discovery, the positron has been observed in a number of experiments. Perhaps the most common process for producing positrons is **pair production.** In this process, a gamma ray photon with sufficiently high energy interacts with a nucleus and an electron–positron pair is created. In the Dirac sea model, an electron in a negative energy state is excited to a positive energy

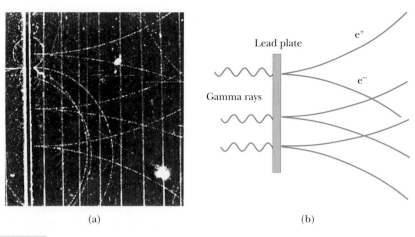

(a)

(b)

**Figure 31.2**

(a) Bubble-chamber tracks of electron–positron pairs produced by 300-MeV gamma rays striking a lead sheet. *(Courtesy Lawrence Berkeley Laboratory, University of California)* (b) Sketch of the pertinent pair-production events. Note that the positrons deflect upward and the electrons deflect downward in an applied magnetic field.

state, resulting in a new observable electron and a hole, which is the positron. Because the total rest energy of the electron–positron pair is $2m_ec^2 = 1.02$ MeV (where $m_e$ is the mass of the electron), the photon must have at least this much energy to create an electron–positron pair. Thus, electromagnetic energy in the form of a gamma ray is converted to mass in accordance with Einstein's relationship $E_R = mc^2$. We can use the isolated system model to describe this process. The energy of the system of the photon and the nucleus is conserved and transformed to mass–energy of the electron and positron, kinetic energy of these particles, and some small amount of kinetic energy associated with the nucleus. Figure 31.2 shows tracks of electron–positron pairs created by 300-MeV gamma rays striking a lead sheet.

### Quick Quiz 31.1

What is the direction of the external magnetic field in Figure 31.2—into or out of the plane of the page?

The reverse process can also occur. Under the proper conditions, an electron and positron can annihilate each other to produce two gamma-ray photons (see Thinking Physics 31.1) that have a combined energy of 1.02 MeV plus the initial kinetic energy of the electron and positron:

$$e^- + e^+ \longrightarrow 2\gamma$$

On the basis of the Dirac theory, it was expected that every particle has a corresponding antiparticle, but antiparticles such as the antiproton and antineutron had not been detected experimentally. In 1955 a team led by Emilio Segrè and Owen Chamberlain used the Bevatron particle accelerator at the University of California, Berkeley, to produce antiprotons and antineutrons, and thus established

**Paul Adrien Maurice Dirac**
(1902–1984)

British physicist Dirac predicted the existence of the positron and other antiparticles. In 1933 he won the Nobel Prize for physics. *(Courtesy of AIP Emilio Segrè Visual Archives)*

with certainty the existence of antiparticles. For this work, Segrè and Chamberlain received the Nobel Prize for Physics in 1959.

It is now established that every particle has a corresponding antiparticle with equal mass and spin, and with charge, magnetic moment, and strangeness of equal magnitude but opposite sign. (The property of strangeness is explained in Section 31.6.) The only special cases are the photon, neutral pion, and eta, each of which is its own antiparticle.

An intriguing aspect of antiparticle theory is that if we replace every proton, neutron, and electron in an atom with its antiparticle, we can create a stable antiatom; combinations of antiatoms should form antimolecules and eventually antiworlds. As far as we know, everything would behave in the same way in an antiworld as in our world. In principle, it is possible that some distant antimatter galaxies exist, separated from normal-matter galaxies by millions of lightyears. Unfortunately, because the photon is its own antiparticle, the light emitted from a normal-matter galaxy is no different from that from an antimatter galaxy, and so astronomical observations cannot determine if the galaxy is composed of matter or antimatter. Although no evidence of antimatter galaxies exists at present, it is awe-inspiring to imagine the cosmic spectacle that would result if matter and antimatter galaxies were to collide: a gigantic eruption of jets of annihilation radiation, transforming the entire galactic mass into neutrinos, antineutrinos, and gamma rays, all fleeing the collision point.

Electron–positron annihilation is used in the medical diagnostic technique called *positron emission tomography* (PET). The patient is injected with a glucose solution containing a radioactive substance that decays by positron emission, and the material is carried by the blood throughout the body. A positron emitted during a decay event in one of the radioactive nuclei in the glucose solution annihilates with an electron in the immediately surrounding tissue, resulting in two gamma-ray photons emitted in opposite directions. A gamma detector surrounding the patient pinpoints the source of the photons and, with the assistance of a computer, displays an image of the sites at which the glucose accumulates. (Glucose is metabolized rapidly in cancerous tumors and accumulates in these sites, providing a strong signal for a PET detector system.) The images from a PET scan can indicate a wide variety of disorders in the brain, including Alzheimer's disease (Fig. 31.3). In addition, because glucose metabolizes more rapidly in active areas of the brain, a PET scan can indicate which areas of the brain are involved when the patient is engaging in such activities as language use, music, or vision.

**APPLICATION**

Positron emission tomography

**WEB**

The Society of Nuclear Medicine has a searchable database for information on positron emission tomography. Visit the society at **www.snm.org**

**Figure 31.3**

PET scans of the brain of a healthy older person and those of patients suffering from Alzheimer's disease. Lighter regions contain higher concentrations of radioactive glucose, indicating higher metabolism rates and therefore increased brain activity. *(Dr. Monty de Leon/New York University Medical Center and National Institute on Aging)*

## THINKING PHYSICS 31.1

When an electron and a positron meet at low speed in free space, why are two 0.511-MeV gamma rays produced, rather than one gamma ray with an energy of 1.022 MeV?

**Reasoning** Gamma rays are photons, and photons carry momentum. We apply the momentum version of the isolated system model to the system, which consists initially of the electron and positron. If the system, assumed to be at rest, transformed to only one photon, momentum would not be conserved because the total momentum of the electron–positron system is zero, whereas a single photon of energy 1.022 MeV would have a nonzero momentum. On the other hand, the two gamma-ray photons travel in *opposite* directions, so the total momentum of the final system—two photons—is zero, and momentum is conserved.

## 31.3 • MESONS AND THE BEGINNING OF PARTICLE PHYSICS

In the mid-1930s, physicists had a fairly simple view of the structure of matter. The building blocks were the proton, the electron, and the neutron. Three other particles were known or had been postulated at the time: the photon, the neutrino, and the positron. These six particles were considered the fundamental constituents of matter. With this marvelously simple picture of the world, however, no one was able to answer the following important question: Because many protons in proximity in any nucleus should strongly repel one another due to their like charges, what is the nature of the force that holds the nucleus together? Scientists recognized that this mysterious force, which we now call the nuclear force, must be much stronger than anything encountered in nature up to that time.

In 1935, the Japanese physicist Hideki Yukawa proposed the first theory to successfully explain the nature of the nuclear force, an effort that later earned him the Nobel Prize. To understand Yukawa's theory, it is useful to first recall that in the modern structural model of electromagnetic interactions, charged particles interact by exchanging photons. Yukawa used this idea to explain the nuclear force by proposing a new particle whose exchange between nucleons in the nucleus produces the nuclear force. Furthermore, he established that the range of the force is inversely proportional to the mass of this particle and predicted that the mass would be about 200 times the mass of the electron. Because the new particle would have a mass between that of the electron and that of the proton, it was called a **meson** (from the Greek *meso*, meaning "middle").

In an effort to substantiate Yukawa's predictions, physicists began an experimental search for the meson by studying cosmic rays entering the Earth's atmosphere. In 1937, Carl Anderson and his collaborators discovered a particle of mass 106 MeV/$c^2$, about 207 times the mass of the electron. Subsequent experiments showed that the particle interacted very weakly with matter, however, and hence could not be the carrier of the nuclear force. The puzzling situation inspired several theoreticians to propose that two mesons existed with slightly different masses. This idea was confirmed by the discovery in 1947 of the **pi ($\pi$) meson,** or simply **pion,** by Cecil Frank Powell and Guiseppe P. S. Occhialini. The particle discovered by Anderson in 1937, the one thought to be Yukawa's meson, is not really a meson. (We shall discuss the requirements for a particle to be a meson in

**Hideki Yukawa** (1907–1981)

Japanese physicist Yukawa was awarded the Nobel Prize in physics in 1949 for predicting the existence of mesons. This photograph of him at work was taken in 1950 in his office at Columbia University.
*(UPI/Corbis-Bettman)*

**PITFALL PREVENTION 31.2**
**The nuclear force and the strong force**

The nuclear force that we discussed in Chapter 30 and continue to discuss here was originally called the *strong* force. Once the quark theory (Section 31.9) was established, however, the phrase *strong force* was reserved for the force between quarks. We shall follow this convention—the strong force is between quarks and the nuclear force is between nucleons. The nuclear force is a secondary result of the strong force, as we discuss in Section 31.10. Because of this historical development of the names for these forces, other books may refer to the nuclear force as the strong force.

Section 31.4). Instead, it takes part in the weak and electromagnetic interactions only and is now called the **muon** ($\mu$). We first discussed the muon in Section 9.4, with regard to time dilation.

The pion, Yukawa's carrier of the nuclear force, comes in three varieties corresponding to three charge states: $\pi^+$, $\pi^-$, and $\pi^0$. The $\pi^+$ and $\pi^-$ particles have masses of 139.6 MeV/$c^2$, and the $\pi^0$ has a mass of 135.0 MeV/$c^2$. Pions and muons are very unstable particles. For example, the $\pi^-$, which has a mean lifetime of $2.6 \times 10^{-8}$ s, first decays to a muon and an antineutrino. The muon, which has a mean lifetime of 2.2 $\mu$s, then decays into an electron, a neutrino, and an antineutrino:

$$\pi^- \longrightarrow \mu^- + \overline{\nu} \qquad\qquad [31.1]$$

$$\mu^- \longrightarrow e^- + \nu + \overline{\nu}$$

Note that for chargeless particles (as well as some charged particles such as the proton), a bar over the symbol indicates an antiparticle.

The interaction between two particles can be represented in a simple qualitative graphical/pictorial representation called a **Feynman diagram,** developed by the American physicist Richard P. Feynman. Figure 31.4 is such a diagram for the electromagnetic interaction between two electrons approaching each other. A Feynman diagram is a qualitative graph of time on the vertical axis versus space on the horizontal axis. It is qualitative in the sense that the actual values of time and space are not important, but the overall appearance of the graph provides a representation of the process. The time evolution of the process can be approximated by starting at the bottom of the diagram and moving your eyes upward.

In the simple case of the electron–electron interaction in Figure 31.4, a photon is the field particle that mediates the electromagnetic force between the electrons. Notice that the entire interaction is represented in these diagrams as if it occurs at a single point in time. Thus, the paths of the electrons appear to undergo a discontinuous change in direction at the moment of interaction. This representation is correct on a microscopic level over a time interval that includes the exchange of one photon. This is different from the paths produced over the much longer interval during which we watch the interaction from a macroscopic point of view. In this case, the paths would be curved (as in Fig. 31.1) due to the continuous exchange of large numbers of field particles. This is another aspect of the qualitative nature of Feynman diagrams.

In the electron–electron interaction, the photon, which transfers energy and momentum from one electron to the other, is called a *virtual photon* because it vanishes during the interaction without being detected. Virtual photons do not violate the law of conservation of energy because, from the uncertainty principle, they have a very short lifetime $\Delta t$ that makes the uncertainty in the energy $\Delta E \approx \hbar/2\Delta t$ of the system of two electrons and the photon greater than the photon energy.

Now consider a pion mediating the nuclear force between a proton and a neutron as in Yukawa's model (Figure 31.5a). The rest energy $\Delta E_R$ needed to create a pion of mass $m_\pi$ is given by Einstein's equation $\Delta E_R = m_\pi c^2$. As with the photon in Figure 31.4, the very existence of the pion would violate the law of conservation of energy if the particle existed for a time greater than $\Delta t \approx \hbar/2\,\Delta E_R$ (from the uncertainty principle), where $\Delta t$ is the time it takes the pion to transfer from one nucleon to the other. Therefore,

$$\Delta t \approx \frac{\hbar}{2\,\Delta E_R} = \frac{\hbar}{2\,m_\pi c^2} \qquad\qquad [31.2]$$

**Figure 31.4**

Feynman diagram representing a photon mediating the electromagnetic force between two electrons.

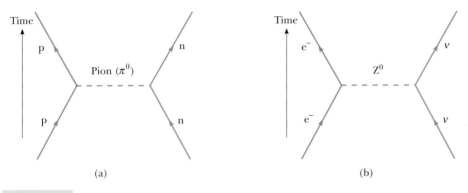

(a)          (b)

**Figure 31.5**

(a) Feynman diagram representing a proton and a neutron interacting via the nuclear force with a pion mediating the force. (b) Feynman diagram for an electron and a neutrino interacting via the weak force with a $Z^0$ boson mediating the force.

Because the pion cannot travel faster than the speed of light, the maximum distance $d$ it can travel in a time $\Delta t$ is $c \, \Delta t$. Thus,

$$d = c \, \Delta t \approx \frac{\hbar}{2 m_\pi c} \qquad [31.3]$$

From Chapter 30, we know that the range of the nuclear force is on the order of $10^{-15}$ fm. Using this value for $d$ in Equation 31.3, we estimate the rest energy of the pion to be

$$m_\pi c^2 \approx \frac{\hbar c}{2d} = \frac{(1.055 \times 10^{-34} \, \text{J} \cdot \text{s})(3.00 \times 10^8 \, \text{m/s})}{2(1 \times 10^{-15} \, \text{m})}$$

$$= 1.6 \times 10^{-11} \, \text{J} \approx 100 \, \text{MeV}$$

This is the same order of magnitude as the observed masses of pions, which provides confidence in the field particle model.

The concept we have just described is quite revolutionary. In effect, it says that a system of two nucleons can change into two nucleons plus a pion, as long as it returns to its original state in a very short time. (Remember that this is the older, historical model, which assumes that the pion is the field particle for the nuclear force.) Physicists often say that a nucleon undergoes *fluctuations* as it emits and absorbs pions. As we have seen, these fluctuations are a consequence of a combination of quantum mechanics (through the uncertainty principle) and special relativity (through Einstein's mass–energy relationship $E_R = mc^2$).

This section has dealt with the particles that mediate the nuclear force, pions, and the mediators of the electromagnetic force, photons. Current ideas indicate that the nuclear force is more accurately described as an average or residual effect of the force between quarks, as will be explained in Section 31.10. The graviton, which is the mediator of the gravitational force, has yet to be observed. The $W^\pm$ and $Z^0$ particles that mediate the weak force were discovered in 1983 by the Italian physicist Carlo Rubbia and his associates using a proton–antiproton collider. Rubbia and Simon van der Meer, both at CERN (European Organization for Nuclear Research), shared the 1984 Nobel Prize for physics for the discovery of the $W^\pm$

**Richard Feynman** (1918–1988)

American physicist Feynman with his son, Carl, after winning the Nobel Prize in physics in 1965. Feynman shared the prize with Julian Schwinger and Sin-itiro Tomonaga. *(UPI Telephotos)*

**WEB**

Visit the European Organization for Nuclear Research at **cern.web.cern.ch/CERN/**

and $Z^0$ particles and the development of the proton–antiproton collider. Figure 31.5b shows a Feynman diagram for a weak interaction mediated by a $Z^0$ boson.

**Quick Quiz 31.2**

What does the infinite range of the electromagnetic and gravitational interactions tell you about the masses of the photon and the graviton?

## 31.4 • CLASSIFICATION OF PARTICLES

As of the 1960s, physicists classified all particles other than field particles into two broad categories—hadrons and leptons—according to their interactions, as described in the following sections. Leptons were perceived as elementary, but the hadrons were not. Thus, the model at that time was a confusing and unsettling patchwork. Table 31.2 provides a summary of the properties of some of these particles.

**TABLE 31.2 | Some Particles and Their Properties**

| Category | Particle Name | Symbol | Antiparticle | Mass (MeV/$c^2$) | $B$ | $L_e$ | $L_\mu$ | $L_\tau$ | $S$ | Lifetime(s) | Principal Decay Modes[a] |
|---|---|---|---|---|---|---|---|---|---|---|---|
| **Leptons** | Electron | $e^-$ | $e^+$ | 0.511 | 0 | +1 | 0 | 0 | 0 | Stable | |
| | Electron neutrino | $\nu_e$ | $\bar{\nu}_e$ | $< 7 \times 10^{-6}$ | 0 | +1 | 0 | 0 | 0 | Stable | |
| | Muon | $\mu^-$ | $\mu^+$ | 105.7 | 0 | 0 | +1 | 0 | 0 | $2.20 \times 10^{-6}$ | $e^- \bar{\nu}_e \nu_\mu$ |
| | Muon neutrino | $\nu_\mu$ | $\bar{\nu}_\mu$ | $< 0.3$ | 0 | 0 | +1 | 0 | 0 | Stable | |
| | Tau | $\tau^-$ | $\tau^+$ | 1 784 | 0 | 0 | 0 | +1 | 0 | $< 4 \times 10^{-13}$ | $\mu^- \bar{\nu}_\mu \nu_\tau, e^- \bar{\nu}_e \nu_\tau$ |
| | Tau neutrino | $\nu_\tau$ | $\bar{\nu}_\tau$ | $< 30$ | 0 | 0 | 0 | +1 | 0 | Stable | |
| **Hadrons** | | | | | | | | | | | |
| **Mesons** | Pion | $\pi^+$ | $\pi^-$ | 139.6 | 0 | 0 | 0 | 0 | 0 | $2.60 \times 10^{-8}$ | $\mu^+ \nu_\mu$ |
| | | $\pi^0$ | Self | 135.0 | 0 | 0 | 0 | 0 | 0 | $0.83 \times 10^{-16}$ | $2\gamma$ |
| | Kaon | $K^+$ | $K^-$ | 493.7 | 0 | 0 | 0 | 0 | +1 | $1.24 \times 10^{-8}$ | $\mu^+ \nu_\mu, \pi^+ \pi^0$ |
| | | $K_S^0$ | $\bar{K}_S^0$ | 497.7 | 0 | 0 | 0 | 0 | +1 | $0.89 \times 10^{-10}$ | $\pi^+ \pi^-, 2\pi^0$ |
| | | $K_L^0$ | $\bar{K}_L^0$ | 497.7 | 0 | 0 | 0 | 0 | +1 | $5.2 \times 10^{-8}$ | $\pi^\pm e^\mp \bar{\nu}_e, 3\pi^0$ |
| | | | | | | | | | | | $\pi^\pm \mu^\mp \bar{\nu}_\mu$ |
| | Eta | $\eta$ | Self | 548.8 | 0 | 0 | 0 | 0 | 0 | $< 10^{-18}$ | $2\gamma, 3\pi$ |
| | | $\eta'$ | Self | 958 | 0 | 0 | 0 | 0 | 0 | $2.2 \times 10^{-21}$ | $\eta \pi^+ \pi^-$ |
| **Baryons** | Proton | p | $\bar{p}$ | 938.3 | +1 | 0 | 0 | 0 | 0 | Stable | |
| | Neutron | n | $\bar{n}$ | 939.6 | +1 | 0 | 0 | 0 | 0 | 920 | $pe^- \bar{\nu}_e$ |
| | Lambda | $\Lambda^0$ | $\bar{\Lambda}^0$ | 1 115.6 | +1 | 0 | 0 | 0 | $-1$ | $2.6 \times 10^{-10}$ | $p\pi^-, n\pi^0$ |
| | Sigma | $\Sigma^+$ | $\bar{\Sigma}^-$ | 1 189.4 | +1 | 0 | 0 | 0 | $-1$ | $0.80 \times 10^{-10}$ | $p\pi^0, n\pi^+$ |
| | | $\Sigma^0$ | $\bar{\Sigma}^0$ | 1 192.5 | +1 | 0 | 0 | 0 | $-1$ | $6 \times 10^{-20}$ | $\Lambda^0 \gamma$ |
| | | $\Sigma^-$ | $\bar{\Sigma}^+$ | 1 197.3 | +1 | 0 | 0 | 0 | $-1$ | $1.5 \times 10^{-10}$ | $n\pi^-$ |
| | Xi | $\Xi^0$ | $\bar{\Xi}^0$ | 1 315 | +1 | 0 | 0 | 0 | $-2$ | $2.9 \times 10^{-10}$ | $\Lambda^0 \pi^0$ |
| | | $\Xi^-$ | $\bar{\Xi}^+$ | 1 321 | +1 | 0 | 0 | 0 | $-2$ | $1.64 \times 10^{-10}$ | $\Lambda^0 \pi^-$ |
| | Omega | $\Omega^-$ | $\Omega^+$ | 1 672 | +1 | 0 | 0 | 0 | $-3$ | $0.82 \times 10^{-10}$ | $\Xi^0 \pi^0, \Lambda^0 K^-$ |

[a] Notations in this column such as $p\pi^-$, $n\pi^0$ mean two possible decay modes. In this case, the two possible decays are $\Lambda^0 \rightarrow p + \pi^-$ and $\Lambda^0 \rightarrow n + \pi^0$.

# Hadrons

Particles that interact through the nuclear force are called **hadrons,** which come in two classes—*mesons* and *baryons*—distinguished by their masses and spins.

**Mesons** all have zero or integer spin (0 or 1).* As indicated in Section 31.3, the name comes from the expectation that Yukawa's proposed meson mass would lie between the mass of the electron and the mass of the proton. Several meson masses do lie in this range, although there are heavier mesons that have masses larger than that of the proton.

All mesons are known to decay into final products including electrons, positrons, neutrinos, and photons. The pions are the lightest known mesons; they have masses of about 140 $\text{MeV}/c^2$ and all three have a spin of 0.

**Baryons,** the second class of hadrons, have masses equal to or greater than the proton mass (the name *baryon* means "heavy" in Greek), and their spin is always a noninteger value ($\frac{1}{2}$ or $\frac{3}{2}$). Protons and neutrons are baryons, as are many other particles. With the exception of the proton, all baryons decay in such a way that the end products include a proton. For example, the baryon called the $\Xi$ hyperon (Greek capital xi) decays to the $\Lambda^0$ baryon (Greek capital lambda) in about $10^{-10}$ s. The $\Lambda^0$ baryon then decays to a proton and a $\pi^-$ in approximately $3 \times 10^{-10}$ s.

Today it is believed that hadrons are not elementary particles, but rather are composed of more elementary units called quarks. We shall discuss quarks in Section 31.9.

# Leptons

**Leptons** (from the Greek *leptos,* meaning "small" or "light") are a group of particles that participate in the electromagnetic (if charged) and weak interactions. All leptons have spins of $\frac{1}{2}$. Although hadrons have size and structure, leptons appear to be truly elementary particles with no structure.

Quite unlike hadrons, the number of known leptons is small. Currently, scientists believe that only six leptons exist: the electron, the muon, and the tau, and a neutrino associated with each:

$$\begin{pmatrix} e^- \\ \nu_e \end{pmatrix} \quad \begin{pmatrix} \mu^- \\ \nu_\mu \end{pmatrix} \quad \begin{pmatrix} \tau^- \\ \nu_\tau \end{pmatrix}$$

The tau lepton, discovered in 1975, has a mass equal to about twice that of the proton. Direct experimental evidence for the neutrino associated with the tau was announced by the Fermi National Accelerator Laboratory (Fermilab) in July, 2000. Each of these six leptons has an antiparticle.

Current studies indicate that neutrinos may have a small but nonzero mass. If they do have mass, they cannot travel at the speed of light. Also, so many neutrinos exist that their combined mass may be sufficient to cause all the matter in the Universe to eventually collapse to infinite density, and then explode and create a completely new Universe! We shall discuss this in more detail in Section 31.12.

---

* This is why the particle discovered by Anderson in 1937, the muon, is not a meson; the muon has spin $\frac{1}{2}$. It belongs in the *lepton* classification described shortly.

## 31.5 • CONSERVATION LAWS

We have seen the importance of conservation laws for isolated systems many times in earlier chapters and have solved problems using conservation of energy, linear momentum, angular momentum, and electric charge. Conservation laws are important in understanding why certain decays and reactions occur but others do not. In general, our familiar conservation laws provide us with a set of rules that all processes must follow.

Certain new conservation laws are important in the study of elementary particles. The isolated system changes identity during the decay or reaction—we define it as the initial particle(s) before the decay or reaction and the final particles afterward. These new conservation laws described here are supported by abundant empirical evidence.

### Baryon Number

• *Conservation of baryon number*

Conservation of baryon number tells us that whenever a baryon is created in a reaction or decay, an antibaryon is also created. This scheme can be quantified by assigning a baryon number $B = +1$ for all baryons, $B = -1$ for all antibaryons, and $B = 0$ for all other particles. Thus, the **law of conservation of baryon number** states that **whenever a reaction or decay occurs, the sum of the baryon numbers of the system before the process must equal the sum of the baryon numbers after the process.** An equivalent statement is that the net number of baryons remains constant in any process.

If baryon number is absolutely conserved, the proton must be absolutely stable, because it is the baryon with the lowest mass. If it were not for the law of conservation of baryon number, the proton could decay to a positron and a neutral pion. Such a decay has never been observed, however. At the present, we can say only that the proton has a half-life of at least $10^{33}$ years (the estimated age of the Universe is only $10^{10}$ years).* Thus, it is extremely unlikely that one would see a given proton undergo a decay process. If we collect a huge number of protons, however, perhaps we might see *some* proton in the collection undergo a decay, as addressed in Example 31.2.

---

### Example 31.1  Checking Baryon Numbers

Use the law of conservation of baryon number to determine whether the following reactions can occur:
(a)  $p + n \rightarrow p + p + n + \bar{p}$
(b)  $p + n \rightarrow p + p + \bar{p}$

**Solution**  (a) The left side of the equation gives a total baryon number of $1 + 1 = 2$. The right side gives a total baryon number of $1 + 1 + 1 + (-1) = 2$. Thus baryon number is conserved and the reaction can occur (provided the incoming proton has sufficient energy so that energy conservation is satisfied).

(b) The left side of the equation gives a total baryon number of $1 + 1 = 2$; however, the right side gives $1 + 1 + (-1) = 1$. Because baryon number is not conserved, the reaction cannot occur.

---

\* In one recent theory, physicists predict that the proton is unstable. According to this theory, baryon number is not absolutely conserved.

## Example 31.2    Detecting Proton Decay

Measurements taken at the Super Kamiokande neutrino detection facility in Japan (Fig. 31.6) indicate that the half-life of protons is at least $10^{33}$ years. (a) Estimate how long we would have to watch, on average, to see a proton in a glass of water decay.

**Solution**    Let us estimate that a glass contains about one fourth of a liter, or 250 g, of water. The number of molecules in a glass of water is about

$$\frac{(250\ \text{g})(6.02 \times 10^{23}\ \text{molecules/mol})}{18\ \text{g/mol}} = 8.4 \times 10^{24}\ \text{molecules}$$

Each water molecule contains one proton in each of its two hydrogen atoms plus eight protons in its oxygen atom, for a total of ten. Thus $8.4 \times 10^{25}$ protons are in the glass of water. The decay constant is given by Equation 30.8:

$$\lambda = \frac{0.693}{T_{1/2}} = \frac{0.693}{10^{33}\ \text{yr}} = 6.9 \times 10^{-34}\ \text{yr}^{-1}$$

This is the probability that any one proton will decay in a year. The probability that some proton in our glass of water will decay in the 1-year interval is (Eqs. 30.5 and 30.7)

$$R = (8.4 \times 10^{25})(6.9 \times 10^{-34}\ \text{yr}^{-1}) = 5.8 \times 10^{-8}\ \text{yr}^{-1}$$

So we have to watch our glass of water for $1/R \approx$ 17 million years!

(b) The Super Kamiokande neutrino facility contains 50 000 tons of water. Estimate the average time between detected proton decays if the half-life is $10^{33}$ yr.

**Solution**    We first find the mass of the water in the neutrino facility:

$$m = 50\ 000\ \text{ton} \left(\frac{2000\ \text{lb}}{1\ \text{ton}}\right)\left(\frac{2.2\ \text{kg}}{1\ \text{lb}}\right)\left(\frac{1\ 000\ \text{g}}{1\ \text{kg}}\right)$$

$$= 2.2 \times 10^{11}\ \text{g}$$

The number of molecules of water is

$$\frac{(2.2 \times 10^{11}\ \text{g})(6.02 \times 10^{23}\ \text{molecules/mol})}{18\ \text{g/mol}}$$

$$= 7.4 \times 10^{33}\ \text{molecules}$$

Because each water molecule contains ten protons, there are $7.4 \times 10^{34}$ protons in the facility. The probability that some

**Figure 31.6**

(Example 31.2) This detector at the Super Kamiokande neutrino facility in Japan is used to study photons and neutrinos. It holds 50 000 tons of highly purified water and 13 000 photomultipliers. The photograph was taken while the detector was being filled. Technicians use a raft to clean the photodetectors before they are submerged. *(Courtesy of KRR [Institute for Cosmic Ray Research], University of Tokyo)*

proton in the facility will decay in the 1-yr interval is

$$R = (7.4 \times 10^{34})(6.9 \times 10^{-34}\ \text{yr}^{-1}) = 51\ \text{yr}^{-1}$$

So we have to observe at the Super Kamiokande neutrino facility for $1/R \approx$ 0.02 yr, which is about one week.

## Lepton Number

• *Conservation of electron-lepton number*

Three conservation laws involve lepton numbers, one for each variety of lepton. The **law of conservation of electron-lepton number** states that **the sum of the electron-lepton numbers of the system before a reaction or decay must equal the sum of the electron-lepton numbers after the reaction or decay.**

The electron and the electron neutrino are assigned a positive electron-lepton number $L_e = +1$, the antileptons $e^+$ and $\bar{\nu}_e$ are assigned a negative electron-lepton number $L_e = -1$; and all others have $L_e = 0$. For example, consider the decay of the neutron

$$n \longrightarrow p + e^- + \bar{\nu}_e$$

Before the decay, the electron-lepton number is $L_e = 0$; after the decay it is $0 + 1 + (-1) = 0$. Thus, the electron-lepton number is conserved. It is important to recognize that the baryon number must also be conserved. This can easily be checked by noting that before the decay $B = +1$, and after the decay $B$ is $+1 + 0 + 0 = +1$.

Similarly, when a decay involves muons, the muon-lepton number $L_\mu$ is conserved. The $\mu^-$ and the $\nu_\mu$ are assigned positive numbers, $L_\mu = +1$, the antimuons $\mu^+$ and $\bar{\nu}_\mu$ are assigned negative numbers, $L_\mu = -1$; and all others have $L_\mu = 0$. Finally, the tau-lepton number $L_\tau$ is conserved, and similar assignments can be made for the tau lepton and its neutrino.

### Quick Quiz 31.3

Suppose a claim is made that the decay of the neutron is given by $n \rightarrow p + e^-$. Which of the following conservation laws are violated by this proposed decay representation? (a) energy, (b) linear momentum, (c) angular momentum, (d) electric charge, (e) lepton number, (f) baryon number.

### THINKING PHYSICS 31.2

A student claims to have observed a decay of an electron into two neutrinos, traveling in opposite directions. What conservation laws would be violated by this decay?

**Reasoning** Several conservation laws are violated. Conservation of electric charge is violated because the negative charge of the electron has disappeared. Conservation of angular momentum is violated because the original angular momentum is that of the electron, with spin $\frac{1}{2}$, but two particles with spin $\frac{1}{2}$ exist after the decay. Conservation of electron-lepton number is also violated because one lepton exists before the decay and two afterward.

Other conservation laws are obeyed by this decay. Energy can be conserved—the rest energy of the electron appears as the kinetic energy (and possibly some small rest energy) of the neutrinos. The opposite directions of the velocities of the two neutrinos allows for conservation of momentum. Conservation of baryon number and conservation of other lepton numbers are also upheld in the decay.

## Example 31.3    Checking Lepton Numbers

Use the law of conservation of electron-lepton number to determine which of the following decay schemes can occur:

(a) $\mu^- \rightarrow e^- + \bar{\nu}_e + \nu_\mu$    (b) $\pi^+ \rightarrow \mu^+ + \nu_\mu + \nu_e$

**Solution** (a) Because this decay involves a muon and an electron, $L_\mu$ and $L_e$ must both be conserved. Before the decay, $L_\mu = +1$ and $L_e = 0$. After the decay, $L_\mu = 0 + 0 + 1 = +1$ and $L = +1 + (-1) + 0 = 0$. Thus, both numbers are conserved, and on this basis the decay is possible.

(b) Before the decay, $L_\mu = 0$ and $L_e = 0$. After the decay, $L_\mu = -1 + 1 + 0 = 0$, but $L_e = 0 + 0 + 1 = 1$. Thus, the decay is not possible because electron-lepton number is not conserved.

**EXERCISE** Determine whether the decay $\mu^- \rightarrow e^- + \bar{\nu}_e$ can occur.

*Answer* No, because $L_\mu$ is $+1$ before the decay and 0 after.

## 31.6 • STRANGE PARTICLES AND STRANGENESS

Many particles discovered in the 1950s were produced by the nuclear interaction of pions with protons and neutrons in the atmosphere. A group of these particles—the kaon (K) lambda ($\Lambda$), and sigma ($\Sigma$) particles—exhibited unusual properties in production and decay and hence were called *strange particles.*

One unusual property is that these particles are always produced in pairs. For example, when a pion collides with a proton, two neutral strange particles are produced with high probability:

$$\pi^- + p^+ \longrightarrow \Lambda^0 + K^0$$

The reaction $\pi^- + p \rightarrow n^0 + K^0$ in which only one of the final particles is strange, never occurs, however, even though no known conservation laws are violated and the energy of the pion is sufficient to initiate the reaction.

The second peculiar feature of strange particles is that, although they are produced by the nuclear force at a high rate, they do not decay at a very high rate into particles that interact via the nuclear force. Instead, they decay very slowly, which is characteristic of the weak interaction as shown in Table 31.1. Their half-lives are in the range $10^{-10}$ s to $10^{-8}$ s; most other particles that interact via the nuclear force have very short lifetimes, on the order of $10^{-20}$ s or less.

Observations such as these indicate the necessity to make modifications in our model. To explain these unusual properties of strange particles, a new quantum number $S$, called **strangeness,** was introduced into our model of elementary particles, along with a new conservation law. The strangeness numbers for some particles are given in Table 31.2. The production of strange particles in pairs is explained by assigning $S = +1$ to one of the particles and $S = -1$ to the other. All nonstrange particles are assigned strangeness $S = 0$. The **law of conservation of strangeness** states that **whenever a reaction or decay occurs, the sum of the strangeness numbers of the system before the process must equal the sum of the strangeness numbers after the process.**

• *Conservation of strangeness*

The slow decay of strange particles can be explained by assuming that the nuclear and electromagnetic interactions obey the law of conservation of strangeness, but the weak interaction does not. Because the decay reaction involves the loss of one strange particle, it violates strangeness conservation and hence proceeds slowly via the weak interaction.

**Example 31.4** **Is Strangeness Conserved?**

(a) Determine whether the following reaction occurs on the basis of conservation of strangeness.

$$\pi^0 + n \longrightarrow K^+ + \Sigma^-$$

**Solution** The initial state has strangeness $S = 0 + 0 = 0$. Because the strangeness of the $K^+$ is $S = +1$ and the strangeness of the $\Sigma^-$ is $S = -1$, the strangeness of the final state is $+1 - 1 = 0$. Thus, strangeness is conserved and the reaction is allowed.

(b) Show that the following reaction does not conserve strangeness.

$$\pi^- + p \longrightarrow \pi^- + \Sigma^+$$

**Solution** The initial state has strangeness $S = 0 + 0 = 0$, and the final state has strangeness $S = 0 + (-1) = -1$. Thus, strangeness is not conserved.

**EXERCISE** Show that the reaction $\pi^- + p \rightarrow K^0 + \Lambda^0$ obeys the law of conservation of strangeness.

## 31.7 • MAKING PARTICLES AND MEASURING THEIR PROPERTIES

Examination of the bewildering array of entries in Table 31.2 leaves one yearning for firm ground. In fact it is natural to wonder about a particle that exists for $10^{-20}$ s and has a mass of 1 192.5 MeV/$c^2$. How is it possible to detect a particle that exists for only $10^{-20}$ s? Furthermore, how can its mass be measured? If a standard attribute of a particle is some type of permanence or stability, in what sense is a fleeting entity that exists for only $10^{-20}$ s a particle? In this section we attempt to answer such questions and explain how particles are produced and how their properties are measured.

Most particles are unstable and are created in nature only rarely, in cosmic ray showers. In the laboratory, however, large numbers of these particles are created in controlled collisions between high-energy particles and a suitable target. The incident particles must have very high energy, and it takes a considerable time for electromagnetic fields to accelerate particles to high energies. Thus, stable charged particles such as electrons or protons generally make up the incident beam. In addition, targets must be simple and stable, and the simplest target, hydrogen, serves nicely as both target (the proton) and detector.

Figure 31.7 shows a typical event in which hydrogen in a bubble chamber served as both target source and detector. Many parallel tracks of negative pions are visible entering the photograph from the bottom. As the labels in the inset drawing show, one of the pions has hit a stationary proton in the hydrogen, producing two strange particles, $\Lambda^0$ and $K^0$ according to the reaction

$$\pi^- + p \longrightarrow \Lambda^0 + K^0$$

Neither neutral strange particle leaves a track, but their subsequent decays into charged particles can be clearly seen as indicated in Figure 31.7. A magnetic field directed into the plane of the photograph causes the track of each charged particle to curve, and from the measured curvature one can determine the particle's charge and linear momentum. If the mass and momentum of the incident particle are known, we can then usually calculate the product particle mass, kinetic energy, and speed from conservation of momentum and energy. Finally, combining a

**Figure 31.7**

This bubble-chamber photograph shows many events, and the inset is a drawing of identified tracks. The strange particles $\Lambda^0$ and $K^0$ are formed at the bottom as the $\pi^-$ interacts with a proton according to $\pi^- + p \rightarrow \Lambda^0 + K^0$. (Note that the neutral particles leave no tracks, as indicated by the dashed lines.) The $\Lambda^0$ and $K^0$ then decay according to $\Lambda^0 \rightarrow \pi^- + p$ and $K^0 \rightarrow \pi^0 + \mu^- + \overline{\nu}_{\mu}$. *(Courtesy Lawrence Berkeley Laboratory, University of California, Photographic Services.)*

product particle's speed with a measurable decay track length, we can calculate the product particle's lifetime. Figure 31.7 shows that sometimes one can use this lifetime technique even for a neutral particle, which leaves no track. As long as the beginning and end points of the missing track are known as well as the particle speed, one can infer the missing track length and find the lifetime of the neutral particle.

## Resonance Particles

With clever experimental technique and much effort, decay track lengths as short as $10^{-6}$ m can be measured. This means that lifetimes as short as $10^{-16}$ s can be measured for high-energy particles traveling at about the speed of light. We arrive at this result by assuming that a decaying particle travels 1 $\mu$m at a speed of $0.99c$ in the reference frame of the laboratory, yielding a lifetime of $\Delta t_{\text{lab}} = 1 \times 10^{-6}$ m$/0.99c \approx 3.4 \times 10^{-15}$ s. This is not our final result, however, because we must account for the relativistic effects of time dilation. Because the proper lifetime $\Delta t_p$ as measured in the decaying particle's reference frame is shorter than the laboratory frame value $\Delta t_{\text{lab}}$ by a factor of $\sqrt{1 - (v^2/c^2)}$ (see Eq. 9.6), we can calculate the proper lifetime:

$$\Delta t_p = \Delta t_{\text{lab}} \sqrt{1 - \frac{v^2}{c^2}} = (3.4 \times 10^{-15}\,\text{s}) \sqrt{1 - \frac{(0.99c)^2}{c^2}} = 4.8 \times 10^{-16}\,\text{s}$$

Unfortunately, even with Einstein's help, the best answer we can obtain with the track length method is several orders of magnitude away from $10^{-20}$ s lifetimes. How then can we detect the presence of particles that exist for time intervals like $10^{-20}$ s? As we shall shortly see, for such short-lived particles, known as **resonance particles,** all we can do is infer their masses, their lifetimes, and, indeed, their very existence from data on their decay products.

Let us consider this in detail by examining the resonance particle called the delta plus $(\Delta^+)$ which has a mass of 1 232 MeV/$c^2$ and a lifetime of about $10^{-23}$ s, even shorter than the most short-lived particle listed in Table 31.2. This particle is produced in the reaction

$$e^- + p \longrightarrow e^- + \Delta^+ \qquad\qquad [31.4]$$

followed in $10^{-23}$ s by the decay

$$\Delta^+ \longrightarrow \pi^+ + n \qquad\qquad [31.5]$$

Because the $\Delta^+$ lifetime is so short, the particle leaves no measurable track in a bubble chamber. It might therefore seem impossible to distinguish the reactions given in Equations 31.4 and 31.5 from the reaction:

$$e^- + p \longrightarrow e^- + \pi^+ + n \qquad\qquad [31.6]$$

in which the reactants of Equation 31.4 decay directly to $e^-$, $\pi^+$, and n with no intermediate step in which a $\Delta^+$ is produced. Distinguishing between these two possibilities is not impossible, however. If a $\Delta^+$ particle exists, it has a distinct rest energy that must come from the kinetic energy of the incoming particles. If we imagine firing electrons with increasing kinetic energy at protons, eventually we will provide enough energy to the system to create the $\Delta^+$ particle. This is very similar to firing photons of increasing energy at an atom until you fire them with enough energy to excite the atom to a higher quantum state. In fact, the $\Delta^+$ particle is an excited state of the proton, which we can understand via the quark theory discussed in Section 31.9. After the $\Delta^+$ particle is formed and decays, its rest energy becomes the energies of the outgoing pion and neutron. Equation 9.22 can be solved for the rest energy of the $\Delta^+$ particle:

$$(m_{\Delta^+}c^2)^2 = E_{\Delta^+}{}^2 - p_{\Delta^+}{}^2 c^2 = E_{\Delta^+}{}^2 - (\mathbf{p}_{\Delta^+} \cdot \mathbf{p}_{\Delta^+})c^2$$

When the $\Delta^+$ particle decays into a pion and a neutron, conservation of energy and momentum require that

$$E_{\Delta^+} = E_{\pi^+} + E_n \qquad\qquad \mathbf{p}_{\Delta^+} = \mathbf{p}_{\pi^+} + \mathbf{p}_n$$

Thus, the rest energy of the $\Delta^+$ particle can be expressed in terms of the energies and momenta of the outgoing particles, which can all be measured in the bubble chamber photograph:

$$(m_{\Delta^+}c^2)^2 = (E_{\pi^+} + E_n)^2 - [(\mathbf{p}_{\pi^+} + \mathbf{p}_n) \cdot (\mathbf{p}_{\pi^+} + \mathbf{p}_n)]c^2$$

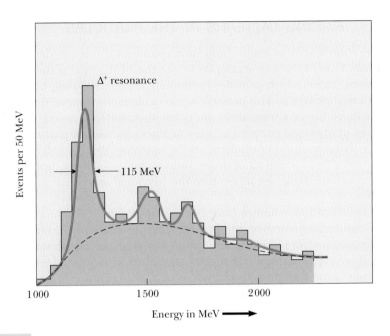

**Figure 31.8**

Experimental evidence for the existence of the $\Delta^+$ particle. The sharp peak at 1 232 MeV was produced by the events in which a $\Delta^+$ formed and promptly decayed to a $\pi^+$ and a neutron.

Any pions and neutrons that come from the decay of a $\Delta^+$ particle must have energies and momenta that combine in this equation to give the rest energy of the $\Delta^+$ particle. Pions and neutrons coming from the reaction in Equation 31.6 will have a variety of energies and momenta with no particular pattern because the energy can divide up in many ways among the three outgoing particles in this reaction. At the energy at which the rest energy of the $\Delta^+$ particle can be created, many reactions can occur, evidenced by the proper combinations of energy and momentum described earlier.

To show the existence of the $\Delta^+$ particle, one analyzes a large number of events in which a $\pi^+$ and a neutron are produced. Then the number of events in a given energy range is plotted versus energy on a graphical representation. Following this procedure, one obtains a slowly varying curve having a sharp peak superimposed on it. The peak represents that incident electron energy at which the rest energy of the $\Delta^+$ particle is created, betraying the existence of the particle.

Figure 31.8 is an experimental plot for the $\Delta^+$ particle. The broad background (dashed) curve is produced by direct events in which no $\Delta^+$ is created (Eq. 31.6). The sharp peak at 1 232 MeV is produced by all those events in which a $\Delta^+$ is formed and decays to a pion and neutron. Thus, the rest energy of the $\Delta^+$ particle is 1 232 MeV. Peaks corresponding to two resonance particles having masses larger than that of the $\Delta^+$ particle can also be seen in Figure 31.8.

Graphs like Figure 31.8 can tell us not only the mass of a short-lived particle but also its lifetime. The width of the resonance peak and the uncertainty relation $\Delta E \, \Delta t \approx \hbar/2$ are used to infer the lifetime $\Delta t$ of the particle. The measured width of 115 MeV in Figure 31.8 leads to a lifetime of $5.7 \times 10^{-24}$ s for the $\Delta^+$ particle. In this incredibly short lifetime, a $\Delta^+$ particle moving at the highest limiting speed of $c$ travels only $10^{-15}$ m, which is about one nuclear diameter.

## 31.8 • FINDING PATTERNS IN THE PARTICLES

One of the tools scientists use to help understand nature is the detection of patterns in data. One of the best examples of the use of this tool is the development of the periodic table, which provides fundamental understanding of the chemical behavior of the elements. The periodic table explains how more than a hundred elements can be formed from three particles, the electron, proton and neutron. The number of observed particles and resonances observed by particle physicists is even larger than the number of elements. Is it possible that a small number of entities could exist from which all of these could be built? Taking a hint from the success of the periodic table, let us explore the historical search for patterns among the particles.

Many classification schemes have been proposed for grouping particles into families. Consider, for instance, the first eight baryons listed in Table 31.2. We can consider these particles to belong to a group based on the fact that they all have a spin of $\frac{1}{2}$. If we create a graphical representation of strangeness versus charge for these baryons using a sloping coordinate system, as in Figure 31.9a, we observe a fascinating pattern: Six of the baryons form a regular hexagon, and the remaining two are at its center.*

As a second example, consider the following nine spin-zero mesons listed in Table 31.2: $\pi^+$, $\pi^0$, $\pi^-$, $K^+$, $K^0$, $K^-$, $\eta$, $\eta'$, and the antiparticle $\overline{K}^0$. Figure 31.9b is a plot of strangeness versus charge for this family. Again, a hexagonal pattern emerges. In this case, each particle on the perimeter of the hexagon lies opposite its antiparticle, and the remaining three (which form their own antiparticles) are at its center. These and related symmetric patterns were developed independently

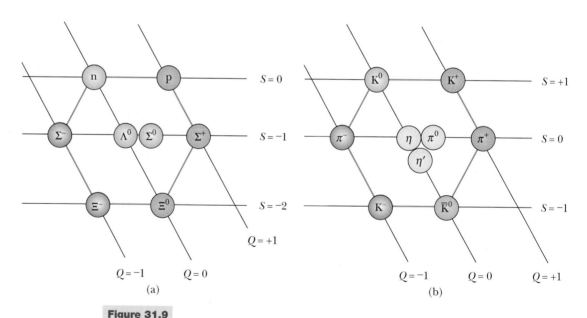

**Figure 31.9**

(a) The hexagonal eightfold-way pattern for the eight spin-$\frac{1}{2}$ baryons. This strangeness-versus-charge plot uses a horizontal axis for the strangeness $S$ values and a sloping axis for the charge number $Q$.
(b) The eightfold-way pattern for the nine spin zero mesons.

* The reason for the sloping coordinate system is so that a *regular* hexagon is formed, one with equal sides. If a normal orthogonal coordinate system is used, the pattern still appears, but the hexagonal shape does not have equal sides. Try it!

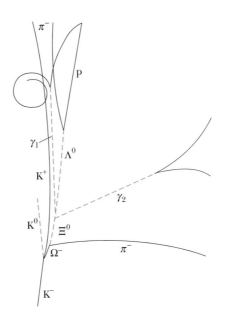

**Figure 31.10**

Discovery of the $\Omega^-$ particle. The $K^-$ particle at the bottom of the photograph collides with a proton to produce the first detected $\Omega^-$ particle plus two other particles. *(Courtesy of Brookhaven National Laboratory)*

in 1961 by Murray Gell-Mann and Yuval Ne'eman. Gell-Mann called the patterns the **eightfold way,** after the eightfold path to nirvana in Buddhism.

Groups of baryons and mesons can be displayed in many other symmetrical patterns within the framework of the eightfold way. For example, the family of spin-$\frac{3}{2}$ baryons contains ten particles arranged in a pattern like the pins in a bowling alley. After this pattern was proposed, an empty spot emerged in the pattern corresponding to a particle that had never been observed. Gell-Mann predicted that the missing particle, which he called the omega minus ($\Omega^-$), should have spin $\frac{3}{2}$, charge $-1$, strangeness $-3$, and rest energy $\approx 1\ 680$ MeV. Shortly thereafter, in 1964, scientists at the Brookhaven National Laboratory found the missing particle through careful analyses of bubble-chamber photographs (Fig. 31.10) and confirmed all its predicted properties.

The prediction of the missing particle from the eightfold way has much in common with the prediction of missing elements in the periodic table. Whenever a vacancy occurs in an organized pattern of information, experimentalists have a guide for their investigations.

**Murray Gell-Mann** (b. 1929)

American physicist Gell-Mann was awarded the Nobel Prize in physics in 1969 for his theoretical studies dealing with subatomic particles. *(Courtesy of Michael R. Dressler)*

## 31.9 • QUARKS

As we have noted, leptons appear to be truly elementary particles because they occur in a small number of types, have no measurable size or internal structure, and do not seem to break down to smaller units. Hadrons, on the other hand, are complex particles having size and structure. The existence of the strangeness–charge patterns of the eightfold way suggests that hadrons have substructure. Furthermore, we know that hundreds of types of hadrons exist and that many of them decay into other hadrons.

Mesons   Baryons

$\pi^+$   p

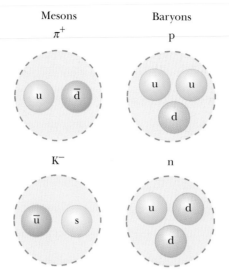

K$^-$   n

**Figure 31.11**

Quark compositions of two mesons and two baryons.

## The Original Quark Model: A Structural Model for Hadrons

In 1963, Gell-Mann and George Zweig independently proposed that hadrons have a more elemental substructure. According to their structural model, all hadrons are composite systems of two or three fundamental constituents called **quarks.** (Gell-Mann borrowed the word *quark* from the passage "Three quarks for Muster Mark" in James Joyce's *Finnegan's Wake.*) The model proposes that three types of quarks exist, designated by the symbols u, d, and s. These are given the arbitrary names **up, down,** and **strange.** The various types of quarks are called **flavors.** Baryons consist of three quarks, and mesons consist of a quark and an antiquark. Figure 31.11 is a pictorial representation of the quark composition of several hadrons.

An unusual property of quarks is that they have fractional multiples of the electronic charge $e$. The u, d, and s quarks have charges of $+\frac{2}{3}e$, $-\frac{1}{3}e$, and $-\frac{1}{3}e$, respectively. These and other properties of quarks and antiquarks are given in Table 31.3. Notice that quarks have spin $\frac{1}{2}$, which means that all quarks are *fermions*, defined as any particle having half-integral spin. As Table 31.3 shows, associated with each quark is an antiquark of opposite charge, baryon number, and strangeness.

The composition of all hadrons known when Gell-Mann and Zweig presented their models can be completely specified by three simple rules:

- A meson consists of one quark and one antiquark, giving it a baryon number of 0, as required.
- A baryon consists of three quarks.
- An antibaryon consists of three antiquarks.

Table 31.4 lists the quark compositions of several mesons and baryons.

### Quick Quiz  31.4

If high-energy electrons, with deBroglie wavelengths smaller than the size of the nucleus, are scattered from nuclei, the behavior of the electrons is consistent with scattering from very massive structures much smaller in size than the nucleus— quarks. Is this similar to another classic experiment that detected small structures in an atom?

| TABLE 31.3 | Properties of Quarks and Antiquarks |
|---|---|

**Quarks**

| Name | Symbol | Spin | Charge | Baryon Number | Strangeness | Charm | Bottomness | Topness |
|---|---|---|---|---|---|---|---|---|
| Up | u | $\frac{1}{2}$ | $+\frac{2}{3}e$ | $\frac{1}{3}$ | 0 | 0 | 0 | 0 |
| Down | d | $\frac{1}{2}$ | $-\frac{1}{3}e$ | $\frac{1}{3}$ | 0 | 0 | 0 | 0 |
| Strange | s | $\frac{1}{2}$ | $-\frac{1}{3}e$ | $\frac{1}{3}$ | $-1$ | 0 | 0 | 0 |
| Charmed | c | $\frac{1}{2}$ | $+\frac{2}{3}e$ | $\frac{1}{3}$ | 0 | $+1$ | 0 | 0 |
| Bottom | b | $\frac{1}{2}$ | $-\frac{1}{3}e$ | $\frac{1}{3}$ | 0 | 0 | $+1$ | 0 |
| Top | t | $\frac{1}{2}$ | $+\frac{2}{3}e$ | $\frac{1}{3}$ | 0 | 0 | 0 | $+1$ |

**Antiquarks**

| Name | Symbol | Spin | Charge | Baryon Number | Strangeness | Charm | Bottomness | Topness |
|---|---|---|---|---|---|---|---|---|
| Antiup | $\bar{u}$ | $\frac{1}{2}$ | $-\frac{2}{3}e$ | $-\frac{1}{3}$ | 0 | 0 | 0 | 0 |
| Antidown | $\bar{d}$ | $\frac{1}{2}$ | $+\frac{1}{3}e$ | $-\frac{1}{3}$ | 0 | 0 | 0 | 0 |
| Antistrange | $\bar{s}$ | $\frac{1}{2}$ | $+\frac{1}{3}e$ | $-\frac{1}{3}$ | $+1$ | 0 | 0 | 0 |
| Anticharmed | $\bar{c}$ | $\frac{1}{2}$ | $-\frac{2}{3}e$ | $-\frac{1}{3}$ | 0 | $-1$ | 0 | 0 |
| Antibottom | $\bar{b}$ | $\frac{1}{2}$ | $+\frac{1}{3}e$ | $-\frac{1}{3}$ | 0 | 0 | $-1$ | 0 |
| Antitop | $\bar{t}$ | $\frac{1}{2}$ | $-\frac{2}{3}e$ | $-\frac{1}{3}$ | 0 | 0 | 0 | $-1$ |

## Charm and Other Developments

Although the original quark model was highly successful in classifying particles into families, some discrepancies were evident between predictions of the model and certain experimental decay rates. It became clear that the structural model needed to be modified to remove these discrepancies. Consequently, a fourth quark was proposed by several physicists in 1967. They argued that if four leptons exist (as was thought at the time: the electron, the muon, and a neutrino associated with each), then four quarks should also exist because of an underlying symmetry in nature. The fourth quark, designated by c, was given a property called **charm.** A *charmed* quark has charge $+2e/3$, but its charm distinguishes it from the other three quarks. This introduces a new quantum number $C$, representing charm. The new quark has charm $C = +1$, its antiquark has charm $C = -1$, and all other quarks have $C = 0$, as indicated in Table 31.3. Charm, like strangeness, is conserved in strong and electromagnetic interactions, but not in weak interactions.

In 1974, a heavy meson called the $J/\Psi$ particle (or simply $\Psi$, upper-case Greek psi) was discovered independently by two groups, one led by Burton Richter at the Stanford Linear Accelerator (SLAC), and the other led by Samuel Ting at the Brookhaven National Laboratory. Richter and Ting were awarded the Nobel Prize in phyics in 1976 for this work. The $J/\Psi$ particle does not fit into the three-quark structural model; instead it has properties of a combination of the proposed charmed quark and its antiquark ($c\bar{c}$). It is much more massive than the other known mesons ($\sim 3\,100$ MeV/$c^2$), and its lifetime is much longer than the lifetimes of particles that decay via the nuclear force. Soon, related mesons were discovered, corresponding to such quark combinations as $\bar{c}d$ and $c\bar{d}$, all of which

| TABLE 31.4 |
|---|
| **Quark Composition of Several Hadrons** |

| Particle | Quark Composition |
|---|---|
| *Mesons* | |
| $\pi^+$ | $u\bar{d}$ |
| $\pi^-$ | $\bar{u}d$ |
| $\pi^0$ | $d\bar{d}$ |
| $K^+$ | $u\bar{s}$ |
| $K^-$ | $\bar{u}s$ |
| $K^0$ | $d\bar{s}$ |
| $J/\Psi$ | $c\bar{c}$ |
| *Baryons* | |
| p | uud |
| n | udd |
| $\Lambda^0$ | uds |
| $\Sigma^+$ | uus |
| $\Sigma^0$ | uds |
| $\Sigma^-$ | dds |
| $\Xi^0$ | uss |
| $\Xi^-$ | dss |
| $\Omega^-$ | sss |

| TABLE 31.5 | The Fundamental Particles and Some of Their Properties | |
|---|---|---|
| **Particle** | **Rest Energy** | **Charge** |
| *Quarks* | | |
| u | 360 MeV | $+\frac{2}{3}e$ |
| d | 360 MeV | $-\frac{1}{3}e$ |
| c | 1 500 MeV | $+\frac{2}{3}e$ |
| s | 540 MeV | $-\frac{1}{3}e$ |
| t | 173 GeV | $+\frac{2}{3}e$ |
| b | 5 GeV | $-\frac{1}{3}e$ |
| *Leptons* | | |
| $e^-$ | 511 keV | $-e$ |
| $\mu^-$ | 105.7 MeV | $-e$ |
| $\tau^-$ | 1 784 MeV | $-e$ |
| $\nu_e$ | $<7$ eV | 0 |
| $\nu_\mu$ | $<0.3$ MeV | 0 |
| $\nu_\tau$ | $<30$ MeV | 0 |

**WEB**

Visit the Stanford Linear Accelerator Center at **www.slac.stanford.edu/**

**WEB**

Visit Brookhaven National Laboratory at **www.bnl.gov/**

**WEB**

Visit the Fermi National Accelerator Laboratory at **www.fnal.gov/**

have large masses and long lifetimes. The existence of these new mesons provided firm evidence for the fourth quark flavor.

In 1975, researchers at Stanford University reported evidence for the tau ($\tau$) lepton, mass 1 784 MeV/$c^2$. This is the fifth type of lepton, which led physicists to propose that more flavors of quarks may exist, based on symmetry arguments similar to those leading to the proposal of the charmed quark. These proposals led to more elaborate quark models, and the prediction of two new quarks: **top** (t) and **bottom** (b). To distinguish these quarks from the original four, quantum numbers called *topness* and *bottomness* (with allowed values $+1$, $0$, $-1$) are assigned to all quarks and antiquarks (Table 31.3). In 1977, researchers at the Fermi National Laboratory, under the direction of Leon Lederman, reported the discovery of a very massive new meson $\Upsilon^-$ (Greek capital upsilon) whose composition is considered to be $b\bar{b}$, providing evidence for the bottom quark. In March 1995, researchers at Fermilab announced the discovery of the top quark (supposedly the last of the quarks to be found), which has a mass of 173 GeV/$c^2$.

You are probably wondering if such discoveries will ever end. How many "building blocks" of matter really exist? At the present, physicists believe that the fundamental particles in nature are six quarks and six leptons (together with their antiparticles). Table 31.5 lists some of the properties of these particles.

Despite extensive experimental effort, no isolated quark has ever been observed. Physicists now believe that quarks are permanently confined inside ordinary particles because of an exceptionally strong force that prevents them from escaping. This force, called the **strong force**\* (discussed in Section 31.10), increases with separation distance (similar to the force of a spring).

---

\* As a reminder, the original meaning of the term *strong force* was the short-range attractive force between nucleons, which we have called the *nuclear force*. As we will discuss in Section 31.10, the nuclear force between nucleons is a secondary effect of the strong force between quarks.

**Quick Quiz 31.5**

Doubly charged baryons, such as the $\Delta^{++}$, are known to exist. Why do no doubly charged mesons occur?

### THINKING PHYSICS 31.3

We have seen a law of conservation of *lepton number* and a law of conservation of *baryon number.* Why isn't there a law of conservation of *meson number?*

**Reasoning**   We can argue this from the point of view of creating particle–antiparticle pairs from available energy. If energy is converted to rest energy of a lepton–antilepton pair, then no net change occurs in lepton number because the lepton has a lepton number of $+1$ and the antilepton $-1$. Energy can also be transformed into rest energy of a baryon–antibaryon pair. The baryon has baryon number $+1$, the antibaryon $-1$, and no net change in baryon number occurs.

But now, suppose energy is transformed into rest energy of a quark–antiquark pair. By definition in quark theory, a quark–antiquark pair *is a meson.* Thus, we have created a meson from energy—no meson existed before, now one does. Thus, meson number is not conserved. With more energy, we can create more mesons, with no restriction from a conservation law other than that of energy.

## 31.10 • COLORED QUARKS

Shortly after the concept of quarks was proposed, scientists recognized that certain particles had quark compositions that were in violation of the Pauli exclusion principle. As noted in Pitfall Prevention 29.4 in Chapter 29, all fermions obey the exclusion principle. Because all quarks have spin $\frac{1}{2}$, and hence are fermions, they are expected to follow the exclusion principle. One example of a particle that appears to violate the exclusion principle is the $\Omega^-$ (sss) baryon that contains three s quarks having parallel spins, giving it a total spin of $\frac{3}{2}$. Other examples of baryons that have identical quarks with parallel spins are the $\Delta^{++}$ (uuu) and the $\Delta^-$ (ddd). To resolve this problem, in 1965 Moo-Young Han and Yoichiro Nambu suggested a modification of the structural model of quarks in which quarks possess a new property called **color.** This property is similar in many respects to electric charge except that it occurs in three varieties called **red, green,** and **blue.** The antiquarks have the colors **antired, antigreen,** and **antiblue.** To satisfy the exclusion principle, all three quarks in a baryon must have different colors. Just as a combination of actual colors of light can produce the neutral color white, a combination of three quarks with different colors is also described as white, or colorless. A meson consists of a quark of one color and an antiquark of the corresponding anticolor. The result is that baryons and mesons are always colorless.

Although the concept of color in the quark model was originally conceived to satisfy the exclusion principle, it also provided a better theory for explaining certain experimental results. For example, the modified theory correctly predicts the lifetime of the $\pi^0$ meson. The theory of how quarks interact with one another is called **quantum chromodynamics,** or QCD, to parallel quantum electrodynamics (the theory of interaction between electric charges). In QCD, the quark is said to carry a **color charge,** in analogy to electric charge. The strong force between quarks is often called the **color force.**

**PITFALL PREVENTION 31.3**

**Color is not really color**

The description of color for a quark has nothing to do with visual sensation from light. It is simply a convenient name for a property analogous to electric charge, except that we need to combine three types of this property to achieve neutrality.

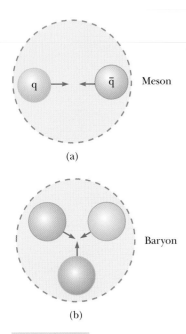

(a)

(b)

**Figure 31.12**

(a) A red quark is attracted to an antired quark to form a meson whose quark structure is (q$\overline{\text{q}}$). (b) Three quarks of different colors attract one another to form a baryon.

As we shall discuss in more detail later, the nuclear interaction between hadrons is actually a secondary effect of the strong force between quarks. **Gluons** are the mediators of the strong force; let us investigate this as the primary effect and then investigate how the gluons also participate in the nuclear force. When a quark emits or absorbs a gluon, its color may change. For example, a blue quark that emits a gluon may become a red quark, and a red quark that absorbs this gluon becomes a blue quark.

The color force between quarks is analogous to the electric force between charges; like colors repel and opposite colors attract. Therefore, two red quarks repel each other, but a red quark is attracted to an antired quark. The attraction between quarks of opposite color to form a meson (q$\overline{\text{q}}$) is indicated in Figure 31.12a. Differently colored quarks also attract one another, but with less strength than opposite colors of quark and antiquark. For example, a cluster of red, blue, and green quarks all attract one another to form a baryon, as indicated in Figure 31.12b. Thus, every baryon contains three quarks of three different colors.

Although the nuclear force between two colorless hadrons is negligible at large separations, the strong force between their constituent quarks does not exactly cancel at small separations. This residual strong force is the nuclear force that binds protons and neutrons to form nuclei. It is similar to the force between two electric dipoles. Each dipole is electrically neutral; however, an electric field surrounds the dipoles because of the separation of the positive and negative charges (see Section 19.5). As a result, an electric force exists between dipoles, albeit weaker than the force between single charges.

According to QCD, a more basic explanation of the nuclear force can be given in terms of quarks and gluons. Figure 31.13a shows the nuclear force interaction between a neutron and a proton by means of Yukawa's pion, in this case a $\pi^-$. In Figure 31.13a, the charged pion carries charge from one nucleon to the other, so the nucleons change identities—the proton becomes a neutron and the neutron becomes a proton. (This differs from Figure 31.5, in which the field particle is a $\pi^0$, resulting in no transfer of charge from one nucleon to the other.)

Let us look at the same interaction from the quark model shown in Figure 31.13b. The proton and neutron are represented in this Feynman diagram by

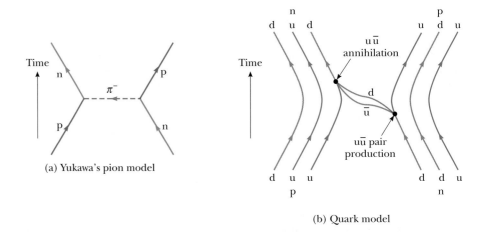

**Figure 31.13**

(a) A nuclear interaction between a proton and a neutron explained in terms of Yukawa's pion exchange model. (b) The same interaction explained in terms of quarks. Note that the exchanged $\overline{\text{u}}$d quark pair makes up a $\pi^-$ meson.

their quark constituents. Each quark in the neutron and proton is continuously emitting and absorbing gluons. The energy of a gluon can result in the creation of quark–antiquark pairs. This is similar to the creation of electron–positron pairs in pair production, which we investigated in Section 31.2. When the neutron and proton approach to within 1–2 fm of each other, these gluons and quarks can be exchanged between the two nucleons, and such exchanges produce the nuclear force. Figure 31.13b depicts one possibility for the process shown in Figure 31.13a. A down quark in the neutron on the right emits a gluon. The energy of the gluon is then transformed to create a u$\bar{\text{u}}$ pair. The u quark stays within the nucleon (which has now changed to a proton) and the recoiling d quark and the $\bar{\text{u}}$ anti-quark are transmitted to the proton on the left side of the diagram. Here the $\bar{\text{u}}$ an-nihilates a u quark within the proton and the d is captured. Thus, the net effect is to change a u quark to a d quark and the proton has changed to a neutron.

As the d quark and $\bar{\text{u}}$ antiquark in Figure 31.13 transfer between the nucleons, they exchange gluons with each other and can be considered to be bound to each other by means of the strong force. If we look back at Table 31.4, we see that this combination is a $\pi^-$—Yukawa's field particle! Thus, the gluon exchange model of interactions between nucleons reduces to the pion exchange model when we con-sider observing the interaction at the level of hadrons rather than at the level of quarks.

## 31.11 • THE STANDARD MODEL

Scientists now believe that elementary particles truly fall into three classifications: leptons, quarks, and field particles. Quarks and leptons have spin $\frac{1}{2}$ and hence are fermions, whereas the field particles have spin 1 or higher and are bosons.

Recall from Section 31.1 that the weak force is believed to be mediated by the $W^+$, $W^-$, and $Z^0$ bosons. These particles are said to have *weak charge* just as quarks have color charge. Thus, each elementary particle can have mass, electric charge, color charge, and weak charge. Of course, one or more of these could be zero.

In 1979, Sheldon Glashow, Abdus Salam, and Steven Weinberg won the Nobel Prize in physics for developing a theory that unified the electromagnetic and weak interactions. This **electroweak theory** postulates that the weak and electromag-netic interactions have the same strength at very high particle energies. Thus, the two interactions are viewed as two different manifestations of a single unifying electroweak interaction. The photon and the three massive bosons ($W^\pm$ and $Z^0$) play a key role in the electroweak theory. The theory makes many concrete predic-tions, but perhaps the most spectacular is the prediction of the masses of the W and Z particles at about 82 GeV/$c^2$ and 93 GeV/$c^2$, respectively. The 1984 Nobel Prize in physics was awarded to Carlo Rubbia and Simon van der Meer for their work leading to the discovery of these particles at just these energies at the CERN Laboratory in Geneva, Switzerland.

The combination of the electroweak theory and QCD for the strong interac-tion form what is referred to in high-energy physics as the **Standard Model,** which is currently the most advanced structural model of elementary particles and their interactions. Although the details of the Standard Model are complex, its es-sential ingredients can be summarized with the help of Figure 31.14. (The Stan-dard Model does not include the gravitational force at present; however, we in-clude gravity in Figure 31.14 because physicists hope to eventually incorporate this force into a unified theory.) This diagram shows that quarks participate in all of the fundamental forces, and leptons participate in all except the strong force.

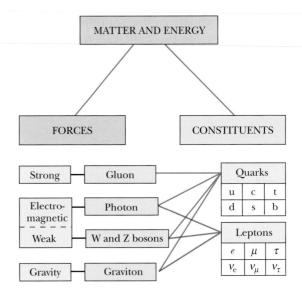

**Figure 31.14**

The Standard Model of particle physics.

The Standard Model does not answer all questions. A major question is why the photon has no mass but the W and Z bosons do. Because of this mass difference, the electromagnetic and weak forces are quite distinct at low energies but become similar in nature at very high energies, at which the rest energies of the W and Z bosons are insignificant fractions of their total energies. Photons and W and Z bosons are indistinguishable at high energies, but quite different at low energies, a process called **symmetry breaking,** which leaves open the question of the origin of particle masses. To resolve this problem, a hypothetical particle called the **Higgs boson,** which provides a mechanism for breaking the electroweak symmetry, has been proposed. The Standard Model, including the Higgs mechanism, provides a logically consistent explanation of the massive nature of the W and Z bosons. Unfortunately, the Higgs boson has not yet been found, although in September 2000 tantalizing evidence delayed the scheduled final shutdown of the Large Electron–Positron (LEP) Collider at CERN. Physicists predict that its mass should be less than 1 TeV ($10^{12}$ eV). To determine whether the Higgs boson exists, two quarks of at least 1 TeV of energy must collide, but calculations show that this requires injecting 40 TeV of energy within the volume of a proton.

Scientists are convinced that because the energy available in conventional accelerators using fixed targets is too limited, it is necessary to build colliding-beam accelerators called **colliders.** The concept of colliders is straightforward. Particles with equal masses and kinetic energies, traveling in opposite directions in an accelerator ring, collide head on to produce the required reaction and the formation of new particles. Because the total momentum of the isolated system of interacting particles is zero, all of their kinetic energy is available for the reaction. The LEP at CERN (Fig. 31.15), near Geneva, Switzerland, and the Stanford Linear Collider in

**Figure 31.15**

A view from inside the Large Electron–Positron (LEP) Collider tunnel, which is 27 km in circumference. *(Courtesy of CERN)*

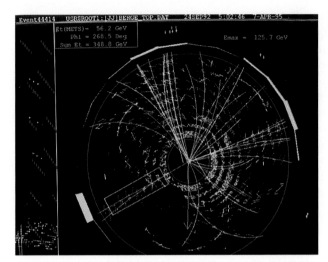

**Figure 31.16**

Computers at Fermilab create a pictorial representation such as this of the paths of particles after a collision. *(Courtesy of Fermi National Accelerator Laboratory)*

California collide electrons and positrons. The Super Proton Synchrotron at CERN accelerates protons and antiprotons to energies of 270 GeV, and the world's highest energy proton accelerator, the Tevatron located at Fermilab in Illinois, produces protons at almost 1 000 GeV (or 1 TeV). The Superconducting Super Collider (SSC), which was being built in Texas, was an accelerator designed to produce 20-TeV protons in a ring 52 miles in circumference. After much debate in Congress, and an investment of almost 2 billion dollars, the SSC project was canceled by the U. S. Department of Energy in October, 1993. CERN expects a 2005 completion date for the Large Hadron Collider (LHC), a proton–proton collider that will provide a center of mass energy of 14 TeV and allow an exploration of Higgs boson physics. The accelerator will be constructed in the same 27-km circumference tunnel as CERN's LEP collider, and many countries are expected to participate in the project.

In addition to increasing energies in modern accelerators, detection techniques have become increasingly sophisticated. The opening photograph for this chapter shows a particle detector at the proton–antiproton collider at Fermilab. Figure 31.16 shows the computer-generated pictorial representation of the tracks of particles after a collision from a modern particle detector.

Following the success of the electroweak theory, scientists attempted to combine it with QCD in a **grand unification theory** (GUT). In this model, the electroweak force is merged with the strong force to form a grand unified force. One version of the theory considers leptons and quarks as members of the same family, which are able to change into each other by exchanging an appropriate field particle.

### THINKING PHYSICS 31.4

Consider a car making a head-on collision with an identical car moving in the opposite direction at the same speed. Compare that collision with one of the cars making a collision with the second car at rest. In which collision is the transformation of kinetic energy to other forms larger? How does this relate to particle accelerators?

**Reasoning**  In the head-on collision with both cars moving, conservation of momentum for the system of two cars requires that the cars come to rest during the collision. Thus, *all* of the original kinetic energy is transformed to other forms. In the collision between a moving car and a stationary car, the cars are still moving with reduced speed after the collision, in the direction of the initially moving car. Thus, *only part* of the kinetic energy is transformed to other forms.

This suggests the importance of colliding beams in a particle accelerator, as opposed to firing a beam into a stationary target. When particles moving in opposite directions collide, all of the kinetic energy is available for transformation into other forms—in this case, the creation of new particles. When a beam is fired into a stationary target, only part of the energy is available for transformation, so higher mass particles cannot be created.

### context connection 🌎
## 31.12 • INVESTIGATING THE SMALLEST SYSTEM TO UNDERSTAND THE LARGEST

In this section we shall describe further one of the most fascinating theories in all of science—the Big Bang theory of the creation of the Universe, introduced in the Context Connection of Chapter 28—and the experimental evidence that supports it. This theory of cosmology states that the Universe had a beginning and, further, that the beginning was so cataclysmic that it is impossible to look back beyond it. According to this theory, the Universe erupted from a singularity with infinite density about 15 to 20 billion years ago. The first few fractions of a second after the Big Bang saw such extremes of energy that all four fundamental forces of physics were believed to be unified.

The evolution of the four fundamental forces from the Big Bang to the present is shown in Figure 31.17. During the first $10^{-43}$ s (the ultrahot epoch, $T \approx$

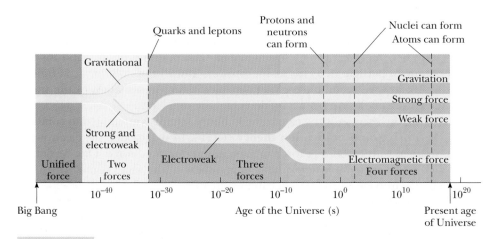

**Figure 31.17**

A brief history of the Universe from the Big Bang to the present. The four forces became distinguishable during the first nanosecond. Following this, all the quarks combined to form particles that interact via the nuclear force. The leptons remained separate, however, and exist as individually observable particles to this day.

$10^{32}$ K), it is presumed that the strong, electroweak, and gravitational forces were joined to form a completely unified force. In the first $10^{-35}$ s following the Big Bang (the hot epoch, $T \approx 10^{29}$ K), gravity broke free of this unification while the strong and electroweak forces remained as one, described by a grand unification theory. This was a period when particle energies were so great ($> 10^{16}$ GeV) that very massive particles as well as quarks, leptons, and their antiparticles existed. Then, after $10^{-35}$ s, the Universe rapidly expanded and cooled (the warm epoch, $T \approx 10^{29}$–$10^{15}$ K), the strong and electroweak forces parted company, and the grand unification scheme was broken. As the Universe continued to cool, the electroweak force split into the weak force and the electromagnetic force about $10^{-10}$ s after the Big Bang.

After a few minutes, protons condensed out of the hot soup. For half an hour the Universe underwent thermonuclear detonation, exploding like a hydrogen bomb and producing most of the helium nuclei now present. The Universe continued to expand and its temperature dropped. Until about 700 000 years after the Big Bang, the Universe was dominated by radiation. Energetic radiation prevented matter from forming clumps or even single hydrogen atoms because collisions would instantly ionize any atoms that might form. Photons experienced continuous Compton scattering from the vast numbers of free electrons, resulting in a Universe that was opaque to radiation. By the time the Universe was about 700 000 years old, it had expanded and cooled to about 3 000 K, and protons could bind to electrons to form neutral hydrogen atoms. Because of the quantized energies of the atoms, far more wavelengths of radiation were not absorbed by atoms than were, and the Universe suddenly became transparent to photons. Radiation no longer dominated the Universe, and clumps of neutral matter steadily grew—first atoms, followed by molecules, gas clouds, stars, and finally galaxies.

## Evidence for the Expanding Universe

In Chapter 28, we discussed the observation of blackbody radiation by Penzias and Wilson that represents the leftover glow from the Big Bang. We discuss here additional relevant astronomical observations. Vesto Melvin Slipher, an American astronomer, reported that most nebulae are receding from the Earth at speeds up to several million miles per hour. Slipher was one of the first to use the methods of Doppler shifts in spectral lines to measure galactic speeds.

In the late 1920s, Edwin P. Hubble made the bold assertion that the whole Universe is expanding. From 1928 to 1936, he and Milton Humason toiled at the Mount Wilson Observatory in California to prove this assertion until they reached the limits of that 100-in. telescope. The results of this work and its continuation on a 200-in. telescope in the 1940s showed that the speeds of galaxies increase in direct proportion to their distance $R$ from us (Fig. 31.18). This linear relationship, known as **Hubble's law,** may be written

$$v = HR \qquad\qquad [\textbf{31.7}]$$

where $H$, called the **Hubble parameter,** has the approximate value

$$H = 17 \times 10^{-3} \, \text{m}/(\text{s} \cdot \text{ly})$$

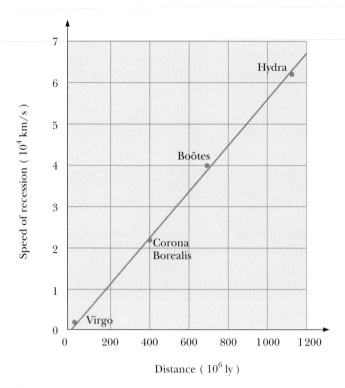

**Figure 31.18**

Hubble's law: The speed of recession is directly proportional to distance. Data points for four galaxies are shown here.

## Example 31.5 Recession of a Quasar

A quasar is a highly luminous object that is very distant from the Earth. Its speed can be measured from Doppler shift measurements in the light it emits. A certain quasar recedes from the Earth at a speed of $0.55c$. How far away is it?

**Solution**  We can find the distance from Hubble's law:

$$R = \frac{v}{H} = \frac{(0.55)(3.00 \times 10^8 \text{ m/s})}{17 \times 10^{-3} \text{ m/(s·ly)}} = \boxed{9.7 \times 10^9 \text{ ly}}$$

**EXERCISE**  Assuming that the quasar has moved with the speed $0.55c$ ever since the Big Bang, estimate the age of the Universe.

*Answer*  $t = R/v = 1/H \approx 18$ billion years, which is in good agreement with other calculations.

### Will the Universe Expand Forever?

In the 1950s and 1960s Allan R. Sandage used the 200-in. telescope at the Mount Palomar Observatory in California to measure the speeds of galaxies at distances of up to 6 billion lightyears. These measurements showed that these very distant galaxies were moving about 10 000 km/s faster than Hubble's law predicted. According to this result, the Universe must have been expanding more rapidly 1 billion years ago and, consequently, the expansion is slowing (Fig. 31.19); today, astronomers and physicists are trying to determine the rate of slowing. (It should be

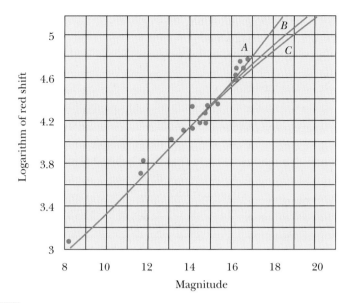

**Figure 31.19**

Red shift, or speed of recession, versus magnitude (which is related to brightness) of 18 faint galaxy clusters. Curve *A* is the trend suggested by the six faintest clusters. Curve *C* corresponds to a Universe having a constant rate of expansion. If more data are taken and the complete set of data shows a curve that falls between *B* and *C*, the expansion will slow but never stop. If the data fall to the left of *B*, expansion will eventually stop and the Universe will begin to contract.

pointed out with regard to Figure 31.19 that the data at large distances have large observational uncertainties and may be systematically in error from effects such as abnormal brightness in the most distant visible clusters.)

If the average density of atoms in the Universe is less than some critical density (about 3 atoms/m$^3$), the galaxies will slow in their outward rush but still escape to infinity. If the average density exceeds the critical value, the expansion will eventually stop and contraction will begin, possibly leading to a new superdense state and another expansion—an **oscillating Universe.**

### Example 31.6 The Critical Density of the Universe

Estimate the critical mass density $\rho_c$ of the Universe, using energy considerations.

**Reasoning and Solution** Figure 31.20 shows a large section of the Universe with radius $R$, containing galaxies with a total mass $M$. Let us apply the isolated system model to an escaping galaxy and the section of the Universe—a galaxy of mass $m$ and speed $v$ at $R$ will just escape to infinity with zero speed if the sum of its kinetic energy and the gravitational potential energy of the system is zero. The Universe may be infinite in extent, but a theorem such as the gravitational

form of Gauss's law implies that only the mass inside the sphere contributes to the gravitational potential energy of the system of the sphere and the galaxy. Thus,

$$E_{\text{total}} = 0 = K + U = \tfrac{1}{2}mv^2 - \frac{GmM}{R}$$

$$\tfrac{1}{2}mv^2 = \frac{Gm\tfrac{4}{3}\pi R^3 \rho_c}{R}$$

$$v^2 = \frac{8\pi G}{3}\, R^2 \rho_c \qquad\qquad (1)$$

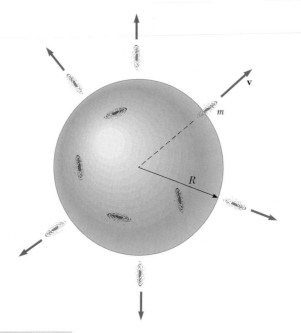

Because the galaxy of mass $m$ obeys the Hubble law, $v = HR$, (1) becomes

$$H^2 = \frac{8\pi G}{3}\rho_c$$

$$\rho_c = \frac{3H^2}{8\pi G} \qquad (2)$$

Using $H = 17 \times 10^{-3}$ m/(s·ly), where 1 ly $= 9.46 \times 10^{12}$ km, and $G = 6.67 \times 10^{-11}$ N·m$^2$/kg$^2$ yields critical density $\rho_c = 6 \times 10^{-30}$ g/cm$^3$. Because the mass of a hydrogen atom is $1.67 \times 10^{-24}$ g, the value calculated for $\rho_c$ corresponds to $3 \times 10^{-6}$ hydrogen atoms per cubic centimeter or 3 atoms per cubic meter.

**Figure 31.20**

(Example 31.6) The galaxy labeled with mass $m$ is escaping from a large cluster of galaxies contained within a spherical volume of radius $R$. Only the mass within the sphere slows the escaping galaxy.

### Missing Mass in the Universe(?)

The luminous matter in galaxies averages out to a Universe density of about $5 \times 10^{-33}$ g/cm$^3$. The radiation in the Universe has a mass equivalent of approximately 2% of the visible matter. The total mass of all nonluminous matter (such as interstellar gas and black holes) may be estimated from the speeds of galaxies orbiting each other in a cluster. The higher the galaxy speeds, the more mass in the cluster. Measurements on the Coma cluster of galaxies indicate that the amount of nonluminous matter is 20 to 30 times the amount of luminous matter present in stars and luminous gas clouds. Yet even this large invisible component of dark matter, if extrapolated to the Universe as a whole, leaves the observed mass density a factor of 10 less than $\rho_c$. The deficit, called *missing mass*, has been the subject of intense theoretical and experimental work. Exotic particles such as axions, photinos, and superstring particles have been suggested as candidates for the missing mass. More mundane proposals argue that the missing mass is present in certain galaxies as neutrinos. In fact, neutrinos are so abundant that a tiny neutrino rest energy on the order of only 20 eV would furnish the missing mass and "close" the Universe. Thus, current experiments designed to measure the rest energy of the neutrino will affect predictions for the future of the Universe—a clear connection between one of the smallest pieces of the Universe and the Universe as a whole!

Although we have some degree of certainty about the beginning of the Universe, we are uncertain about its end. Will the Universe expand forever or will it collapse and repeat its expansion in an endless series of oscillations? Results and answers to these questions remain inconclusive, and the exciting controversy continues.

## Quick Quiz 31.6

Atoms did not exist until hundreds of thousands of years after the Big Bang. Why?

## Quick Quiz 31.7

Observations of other galaxies show that they are moving away from us. Is it therefore correct to propose that we are at the center of the Universe?

## SUMMARY

There are four fundamental forces in nature: **strong, electromagnetic, weak,** and **gravitational.** The strong force is the force between quarks. A residual effect of the strong force is the **nuclear force** between nucleons that keeps the nucleus together. The weak force is responsible for beta decay. The electromagnetic and weak forces are now considered to be manifestations of a single force called the **electroweak force.** Every fundamental interaction is mediated by the exchange of **field particles.** The electromagnetic interaction is mediated by the photon; the weak interaction is mediated by the $W^{\pm}$ **and** $Z^0$ **bosons;** the gravitational interaction is mediated by **gravitons;** the strong interaction is mediated by **gluons.**

An **antiparticle** and a particle have the same mass, but opposite charge, and other properties may have opposite values such as lepton number and baryon number. It is possible to produce particle–antiparticle pairs in nuclear reactions if the available energy is greater than $2mc^2$, where $m$ is the mass of the particle (or antiparticle).

Particles other than field particles are classified as hadrons or leptons. **Hadrons** interact through the strong force. They have size and structure and are not elementary particles. Hadrons are of two types: *baryons* and *mesons.* **Mesons** have baryon number zero and have either zero or integral spin.

**Baryons,** which generally are the most massive particles, have nonzero baryon number and a spin of $\frac{1}{2}$ or $\frac{3}{2}$. The neutron and proton are examples of baryons.

**Leptons** have no structure or size and are considered truly elementary. They interact only through the weak and electromagnetic forces. The six leptons are the electron $e^-$, the muon $\mu^-$, the tau $\tau^-$; and their neutrinos $\nu_e$, $\nu_\mu$, and $\nu_\tau$.

In all reactions and decays, quantities such as energy, linear momentum, angular momentum, electric charge, baryon number, and lepton number are strictly conserved. Certain particles have properties called **strangeness** and **charm.** These unusual properties are conserved only in those reactions and decays that occur via the strong force.

Theories in elementary particle physics have postulated that all hadrons are composed of smaller units known as **quarks.** Quarks have fractional electric charge and come in six "flavors": **up** (u), **down** (d), **strange** (s), **charmed** (c), **top** (t), and **bottom** (b). Each baryon contains three quarks, and each meson contains one quark and one antiquark.

According to the theory of **quantum chromodynamics,** quarks have a property called **color,** and the strong force between quarks is referred to as the **color force.**

## QUESTIONS

1. Name the four fundamental interactions and the field particle that mediates each.
2. Describe the quark model of hadrons, including the properties of quarks.
3. What are the differences between hadrons and leptons?
4. Describe the properties of baryons and mesons and the important differences between them.
5. Particles known as resonances have very short lifetimes, of the order of $10^{-23}$ s. From this information, would you guess they are hadrons or leptons? Explain.
6. Kaons all decay into final states that contain no protons or neutrons. What is the baryon number of kaons?
7. The $\Xi^0$ particle decays by the weak interaction according to the decay mode $\Xi^0 \rightarrow \Lambda^0 + \pi^0$. Would you expect this decay to be fast or slow? Explain.
8. Identify the particle decays listed in Table 31.2 that occur by the weak interaction. Justify your answers.
9. Identify the particle decays listed in Table 31.2 that occur by the electromagnetic interaction. Justify your answers.
10. Two protons in a nucleus interact via the strong interaction. Are they also subject to the weak interaction?
11. Discuss the following conservation laws: energy, linear momentum, angular momentum, electric charge,

baryon number, lepton number, and strangeness. Are all of these laws based on fundamental properties of nature? Explain.

12. An antibaryon interacts with a meson. Can a baryon be produced in such an interaction? Explain.

13. Describe the essential features of the Standard Model of particle physics.

14. How many quarks are in (a) a baryon, (b) an antibaryon, (c) a meson, (d) an antimeson? How do you account for the fact that baryons have half-integral spins but mesons have spins of 0 or 1? (*Hint:* Quarks have spin $\frac{1}{2}$.)

15. In the theory of quantum chromodynamics, quarks come in three colors. How would you justify the statement that "all baryons and mesons are colorless"?

16. Which baryon did Murray Gell-Mann predict in 1961? What is the quark composition of this particle?

17. What is the quark composition of the $\Xi^-$ particle? (See Table 31.4.)

18. The W and Z bosons were first produced at CERN in 1983 (by having a beam of protons and a beam of antiprotons meet at high energy). Why was this an important discovery?

## PROBLEMS

1, 2, 3 = straightforward, intermediate, challenging       □ = full solution available in the *Student Solutions Manual and Study Guide*

**web** = solution posted at **http://www.harcourtcollege.com/physics/**       🖳 = computer useful in solving problem

🌐 = Interactive Physics       ■ = paired numerical/symbolic problems       ▨ = life science application

### Section 31.1   The Fundamental Forces in Nature

### Section 31.2   Positrons and Other Antiparticles

1. A photon produces a proton–antiproton pair according to the reaction $\gamma \rightarrow p + \overline{p}$. What is the minimum possible frequency of the photon? What is its wavelength?

2. You may have undergone or know someone who has experienced a PET scan. The acronym stands for *positron-emission tomography*. In the procedure, a radioactive element that undergoes $e^+$ decay is introduced into your body. The equipment detects the gamma rays that result from pair annihilation when the emitted positron encounters an electron in your body's tissue. Suppose that you receive an injection of glucose containing on the order of $10^{10}$ atoms of $^{14}O$. Assume that the oxygen is uniformly distributed through 2 L of blood after 5 min. What will be the order of magnitude of the activity of the oxygen atoms in 1 cm$^3$ of the blood?

3. A photon with an energy $E_\gamma = 2.09$ GeV creates a proton-antiproton pair in which the proton has a kinetic energy of 95.0 MeV. What is the kinetic energy of the antiproton? ($m_p c^2 = 938.3$ MeV.)

4. Two photons are produced when a proton and antiproton annihilate each other. What are the minimum frequency and corresponding wavelength of each photon?

### Section 31.3   Mesons and the Beginning of Particle Physics

5. One of the mediators of the weak interaction is the $Z^0$
**web** boson, with mass 93 GeV/$c^2$. Use this information to find the order of magnitude of the range of the weak interaction.

6. Occasionally, high-energy muons collide with electrons and produce two neutrinos according to the reaction $\mu^+ + e^- \rightarrow 2\nu$. What kind of neutrinos are these?

7. Calculate the range of the force that might be produced by the virtual exchange of a proton.

8. When a high-energy proton or pion traveling near the speed of light collides with a nucleus, it travels an average distance of $3 \times 10^{-15}$ m before interacting. From this information, find the order of magnitude of the time for the strong interaction to occur.

9. A neutral pion at rest decays into two photons according
**web** to

$$\pi^0 \longrightarrow \gamma + \gamma$$

Find the energy, momentum, and frequency of each photon.

10. A free neutron beta decays by creating a proton, an electron and an antineutrino according to the reaction $n \rightarrow p + e^- + \overline{\nu}$. Imagine that a free neutron decays by creating a proton and electron according to the reaction

$$n \longrightarrow p + e^-$$

and assume that the neutron is initially at rest in the laboratory. (a) Determine the energy released in this reaction. (b) Determine the speeds of the proton and electron after the reaction. (Energy and momentum are conserved in the reaction.) (c) Is either of these particles moving at a relativistic speed? Explain.

### Section 31.4   Classification of Particles

11. Name one possible decay mode (see Table 31.2) for $\Omega^+$, $\overline{K}_S^0$, $\overline{\Lambda}^0$, and $\overline{n}$.

**12.** Identify the unknown particle on the left side of the following reaction:

$$? + p \longrightarrow n + \mu^+$$

## Section 31.5 Conservation Laws

**13.** Each of the following reactions is forbidden. Determine a conservation law that is violated for each reaction.

(a) $p + \bar{p} \rightarrow \mu^+ + e^-$
(b) $\pi^- + p \rightarrow p + \pi^+$
(c) $p + p \rightarrow p + \pi^+$
(d) $p + p \rightarrow p + p + n$
(e) $\gamma + p \rightarrow n + \pi^0$

**14.** (a) Show that baryon number and charge are conserved in the following reactions of a pion with a proton.

$$\pi^+ + p \longrightarrow K^+ + \Sigma^+ \qquad (1)$$

$$\pi^+ + p \longrightarrow \pi^+ + \Sigma^+ \qquad (2)$$

(b) The first reaction is observed, but the second never occurs. Explain.

**15.** The following reactions or decays involve one or more
**web** neutrinos. In each case, supply the missing neutrino ($\nu_e$, $\nu_\mu$, or $\nu_\tau$).

(a) $\pi^- \rightarrow \mu^- + ?$    (d) $? + n \rightarrow p + e^-$
(b) $K^+ \rightarrow \mu^+ + ?$    (e) $? + n \rightarrow p + \mu^-$
(c) $? + p \rightarrow n + e^+$    (f) $\mu^- \rightarrow e^- + ? + ?$

**16.** The first of the following two reactions can occur, but the second cannot. Explain.

$$K_S^0 \longrightarrow \pi^+ + \pi^- \quad \text{(can occur)}$$

$$\Lambda^0 \longrightarrow \pi^+ + \pi^- \quad \text{(cannot occur)}$$

**17.** Determine which of the following reactions can occur. For
**web** those that cannot, determine the conservation law (or laws) violated:

(a) $p \rightarrow \pi^+ + \pi^0$    (d) $\pi^+ \rightarrow \mu^+ + \nu_\mu$
(b) $p + p \rightarrow p + p + \pi^0$    (e) $n \rightarrow p + e^- + \bar{\nu}_e$
(c) $p + p \rightarrow p + \pi^+$    (f) $\pi^+ \rightarrow \mu^+ + n$

**18.** A $K_S^0$ particle at rest decays into a $\pi^+$ and a $\pi^-$. What will be the speed of each of the pions? The mass of the $K_S^0$ is 497.7 MeV/$c^2$, and the mass of each $\pi$ is 139.6 MeV/$c^2$.

**19.** (a) Show that the proton-decay reaction

$$p \longrightarrow e^+ + \gamma$$

cannot occur because it violates conservation of baryon number. (b) Imagine that this reaction does occur, and that the proton is initially at rest. Determine the energy and momentum of the positron and photon after the reaction. (*Hint:* Recall that energy and momentum must be conserved in the reaction.) (c) Determine the speed of the positron after the reaction.

**20.** Determine the type of neutrino or antineutrino involved in each of the following processes:

(a) $\pi^+ \rightarrow \pi^0 + e^+ + ?$
(b) $? + p \rightarrow \mu^- + p + \pi^+$
(c) $\Lambda^0 \rightarrow p + \mu^- + ?$
(d) $\tau^+ \rightarrow \mu^+ + ? + ?$

## Section 31.6 Strange Particles and Strangeness

**21.** Determine whether or not strangeness is conserved in the following decays and reactions.

(a) $\Lambda^0 \rightarrow p + \pi^-$    (d) $\pi^- + p \rightarrow \pi^- + \Sigma^+$
(b) $\pi^- + p \rightarrow \Lambda^0 + K^0$    (e) $\Xi^- \rightarrow \Lambda^0 + \pi^-$
(c) $\bar{p} + p \rightarrow \overline{\Lambda^0} + \Lambda^0$    (f) $\Xi^0 \rightarrow p + \pi^-$

**22.** The neutral meson $\rho^0$ decays by the strong interaction into two pions: $\rho^0 \rightarrow \pi^+ + \pi^-$, half-life $10^{-23}$ s. The neutral kaon also decays into two pions: $K_S^0 \rightarrow \pi^+ + \pi^-$, half-life $10^{-10}$ s. How do you explain the difference in half-lives?

**23.** For each of the following forbidden decays, determine which conservation law is violated:

(a) $\mu^- \rightarrow e^- + \gamma$    (d) $p \rightarrow e^+ + \pi^0$
(b) $n \rightarrow p + e^- + \nu_e$    (e) $\Xi^0 \rightarrow n + \pi^0$
(c) $\Lambda^0 \rightarrow p + \pi^0$

**24.** Which of the following processes are allowed by the strong interaction, the electromagnetic interaction, the weak interaction, or no interaction at all?

(a) $\pi^- + p \rightarrow 2\eta$
(b) $K^- + n \rightarrow \Lambda^0 + \pi^-$
(c) $K^- \rightarrow \pi^- + \pi^0$
(d) $\Omega^- \rightarrow \Xi^- + \pi^0$
(e) $\eta \rightarrow 2\gamma$

**25.** Identify the conserved quantities in the following processes:

(a) $\Xi^- \rightarrow \Lambda^0 + \mu^- + \nu_\mu$
(b) $K_S^0 \rightarrow 2\pi^0$
(c) $K^- + p \rightarrow \Sigma^0 + n$
(d) $\Sigma^0 \rightarrow \Lambda^0 + \gamma$
(e) $e^+ + e^- \rightarrow \mu^+ + \mu^-$
(f) $\bar{p} + n \rightarrow \overline{\Lambda^0} + \Sigma^-$

**26.** Fill in the missing particle. Assume that (a) occurs via the strong interaction and (b) and (c) involve the weak interaction.

(a) $K^+ + p \rightarrow ? + p$
(b) $\Omega^- \rightarrow ? + \pi^-$
(c) $K^+ \rightarrow ? + \mu^+ + \nu_\mu$

## Section 31.7 Making Elementary Particles and Measuring Their Properties

**27.** If a $K_S^0$ meson at rest decays in $0.900 \times 10^{-10}$ s, how far will a $K_S^0$ meson travel if it is moving at $0.960c$ through a bubble chamber?

**Section 31.8**    Finding Patterns in the Particles

**Section 31.9**    Quarks

**Section 31.10**    Colored Quarks

**Section 31.11**    The Standard Model

*Note:* Problem 55 in Chapter 9 can be assigned with this section.

**28.** The quark composition of the proton is uud, and that of the neutron is udd. Show that in each case the charge, baryon number, and strangeness of the particle equal, respectively, the sums of these numbers for the quark constituents.

**29.** (a) Find the number of electrons and the number of each species of quarks in 1 L of water. (b) Make an order-of-magnitude estimate of the number of each kind of fundamental matter particle in your body. State your assumptions and the quantities you take as data.

**30.** The quark compositions of the $K^0$ and $\Lambda^0$ particles are $\bar{s}d$ and uds, respectively. Show that the charge, baryon number, and strangeness of these particles equal, respectively, the sums of these numbers for the quark constituents.

**31.** Imagine that binding energies could be ignored. Find the masses of the u and d quarks from the masses of the proton and neutron.

**32.** The text states that the reaction $\pi^- + p \rightarrow K^0 + \Lambda^0$ occurs with high probability, whereas the reaction $\pi^- + p \rightarrow K^0 + n$ never occurs. Analyze these reactions at the quark level. Show that the first reaction conserves the total number of each type of quark, and the second reaction does not.

**33.** Analyze each reaction in terms of constituent quarks:
   (a) $\pi^- + p \rightarrow K^0 + \Lambda^0$
   (b) $\pi^+ + p \rightarrow K^+ + \Sigma^+$
   (c) $K^- + p \rightarrow K^+ + K^0 + \Omega^-$
   (d) $p + p \rightarrow K^0 + p + \pi^+ + ?$
   In the last reaction, identify the missing particle.

**34.** A $\Sigma^0$ particle traveling through matter strikes a proton; then a $\Sigma^+$ and a gamma ray emerge, as well as a third particle. Use the quark model of each to determine the identity of the third particle.

**35.** Identify the particles corresponding to the quark combinations (a) suu, (b) $\bar{u}d$, (c) $\bar{s}d$, and (d) ssd.

**36.** What is the electric charge of the baryons with the quark compositions (a) $\overline{uud}$ and (b) $\overline{udd}$? What are these baryons called?

**Section 31.12**    Context Connection – Investigating the Smallest System to Understand the Largest

*Note:* Problems 38 and 39 in Chapter 29 and Problem 47 in Chapter 28 can be assigned with this section.

**37.** Using Hubble's law, find the wavelength of the 590-nm sodium line emitted from galaxies (a) $2.00 \times 10^6$ ly away from Earth, (b) $2.00 \times 10^8$ ly away, and (c) $2.00 \times 10^9$ ly away. You may use the result of Problem 38 in Chapter 29.

**38.** The various spectral lines observed in the light from a distant quasar have longer wavelengths $\lambda_n'$ than the wavelengths $\lambda_n$ measured in light from a stationary source. The fractional change in wavelength toward the red is the same for all spectral lines. That is, the redshift parameter $Z$ defined by

$$Z = (\lambda_n' - \lambda_n)/\lambda_n$$

is common to all spectral lines for one object. In terms of $Z$, determine (a) the speed of recession of the quasar and (b) the distance from Earth to this quasar. Use the result of Problem 38 in Chapter 29 and Hubble's law.

**39.** It is mostly your roommate's fault. Nosy astronomers have discovered enough junk and clutter in your dorm room to constitute the missing mass required to close the Universe. After observing your floor, closet, bed, and computer files, they calculate the average density of the observable Universe as $1.20\rho_c$. How many times larger will the Universe become before it begins to collapse? That is, by what factor will the distance between remote galaxies increase in the future?

**40.** The early Universe was dense with gamma-ray photons of energy $\sim k_B T$ and at such a high temperature that protons and antiprotons were created by the process $\gamma \rightarrow p + \bar{p}$ as rapidly as they annihilated each other. As the Universe cooled in adiabatic expansion, its temperature fell below a certain value, and proton pair production became rare. At that time slightly more protons than antiprotons existed, and essentially all of the protons in the Universe today date from that time. (a) Estimate the order of magnitude of the temperature of the Universe when protons condensed out. (b) Estimate the order of magnitude of the temperature of the Universe when electrons condensed out.

**41.** If the average density of the Universe is small compared with the critical density, the expansion of the Universe described by Hubble's law proceeds with speeds that are nearly constant over time. (a) Prove that in this case the age of the Universe is given by the inverse of Hubble's constant. (b) Calculate $1/H$ and express your answer in years.

**42.** Assume that the average density of the Universe is equal to the critical density. (a) Prove that the age of the Universe is given by $2/3H$. (b) Calculate $2/3H$ and express your answer in years.

**43.** Hubble's law can be stated in vector form as $\mathbf{v} = H\mathbf{R}$, where $\mathbf{R}$ is the position vector of a galaxy measured with respect to the Earth. Outside the local group of galaxies, all objects are moving away from us with velocities proportional to their displacements from us. In this form, it sounds as if our location in the Universe is specially privileged. Prove that Hubble's law would be equally true for an observer elsewhere in the Universe. Proceed as follows: Assume that we are at the origin of coordinates, that one galaxy cluster is at location $\mathbf{R}_1$ and has velocity $\mathbf{v}_1 = H\mathbf{R}_1$ relative to us, and that another galaxy cluster has radius vector $\mathbf{R}_2$ and velocity $\mathbf{v}_2 = H\mathbf{R}_2$. Suppose the speeds are nonrelativistic. Consider the frame of reference of an observer in the first of these galaxy clusters. Show that our velocity relative to her, together with the displacement vector of our galaxy cluster from hers, satisfies Hubble's law. Show that the displacement and velocity of cluster 2 relative to cluster 1 satisfy Hubble's law.

### Additional Problems

**44. Review Problem.** Supernova 1987A, located about 170 000 ly from the Earth, is estimated to have emitted a burst of neutrinos carrying energy $\sim 10^{46}$ J (Fig. P31.44). Suppose

Before

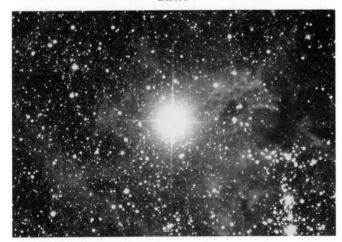

After

**Figure P31.44**  (Problems 44 and 45) Before and after pictures of supernova 1987A in the Large Magellanic Cloud *(Anglo-Australian Telescope Board)*

the average neutrino energy was 6 MeV and your body presented cross-sectional area 5 000 cm². To an order of magnitude, how many of these neutrinos passed through you?

**45.** The most recent supernova observable by the naked eye was Supernova 1987A (Fig. P31.44). It was 170 000 ly away in the next galaxy to ours, the Large Magellanic Cloud. About 3 h before its optical brightening was noticed, two neutrino detection experiments simultaneously registered the first neutrinos from an identified source other than the Sun. The Irvine-Michigan-Brookhaven experiment in a salt mine in Ohio registered eight neutrinos over a 6-s period, and the Kamiokande II experiment in a zinc mine in Japan counted eleven neutrinos in 13 s. (Because the supernova is far south in the sky, these neutrinos entered the detectors from below. They passed through the Earth before they were absorbed by nuclei in the detectors.) The neutrino energies were between about 8 MeV and 40 MeV. If neutrinos have no mass, then neutrinos of all energies should travel together at the speed of light—the data are consistent with this possibility. (The arrival times may show scatter simply because neutrinos were created at different moments as the core of the star collapsed into a neutron star.) If neutrinos have nonzero mass, then lower energy neutrinos should move comparatively slowly. The data are consistent with a 10-MeV neutrino requiring at most about 10 s more than a photon would require to travel from the supernova to us. Find the upper limit this observation sets on the mass of a neutrino. (Other evidence sets an even tighter limit.)

**46.** Name at least one conservation law that prevents each of the following reactions: (a) $\pi^- + p \rightarrow \Sigma^+ + \pi^0$ (b) $\mu^- \rightarrow \pi^- + \nu_e$ (c) $p \rightarrow \pi^+ + \pi^+ + \pi^-$

**47.** The energy flux carried by neutrinos from the Sun is estimated to be on the order of 0.4 W/m² at Earth's surface. Estimate the fractional mass loss of the Sun over $10^9$ years due to the emission of neutrinos. (The mass of the Sun is $2 \times 10^{30}$ kg. The Earth–Sun distance is $1.5 \times 10^{11}$ m.)

**48.** Two protons approach each other head-on, each with 70.4 MeV of kinetic energy, and engage in a reaction in which a proton and positive pion emerge at rest. What third particle, obviously uncharged and therefore difficult to detect, must have been created?

**49.** Free neutrons have a characteristic half-life of 10.4 min. What fraction of a group of free thermal neutrons with kinetic energy 0.040 0 eV will decay before traveling a distance of 10.0 km?

**50.** The nuclear force can be attributed to the exchange of an elementary particle between protons and neutrons if they are sufficiently close. Take the range of the nuclear force as approximately $1.4 \times 10^{-15}$ m. (a) Use the uncertainty principle $\Delta E \, \Delta t \geq \hbar/2$ to estimate the mass of the elementary particle. Assume that it moves at nearly the speed of light. (b) Using Table 31.2, identify the particle.

51. Determine the kinetic energies of the proton and pion resulting from the decay of a $\Lambda^0$ at rest:

$$\Lambda^0 \longrightarrow p + \pi^-$$

52. A gamma-ray photon strikes a stationary electron. Determine the minimum gamma-ray energy to make this reaction go:

$$\gamma + e^- \longrightarrow e^- + e^- + e^+$$

53. An unstable particle, initially at rest, decays into a proton (rest energy 938.3 MeV) and a negative pion (rest energy 139.5 MeV). A uniform magnetic field of 0.250 T exists perpendicular to the velocities of the created particles. The radius of curvature of each track is found to be 1.33 m. What is the mass of the original unstable particle?

54. A $\Sigma^0$ particle at rest decays according to

$$\Sigma^0 \longrightarrow \Lambda^0 + \gamma$$

Find the gamma-ray energy.

55. Two protons approach each other with equal and opposite velocities. What is the minimum kinetic energy of each of the protons if they are to produce a $\pi^+$ meson at rest in the following reaction?

$$p + p \longrightarrow p + n + \pi^+$$

56. A $\pi$-meson at rest decays according to $\pi^- \rightarrow \mu^- + \bar{\nu}_\mu$. What is the energy carried off by the neutrino? (Assume that the neutrino has no mass and moves off with the speed of light. $m_\pi c^2 = 139.5$ MeV, $m_\mu c^2 = 105.7$ MeV. )

57. **Review Problem.** Use the Boltzmann distribution function $e^{-E/k_B T}$ to calculate the temperature at which 1.00% of a population of photons will have energy greater than 1.00 eV. The energy required to excite an atom is on the order of 1 eV. Thus as the temperature of the Universe fell below the value you calculate, neutral atoms could form from plasma, and the Universe became transparent. The cosmic background radiation represents our vastly red-shifted view of the opaque fireball of the Big Bang as it was at this time and temperature. The fireball surrounds us; we are embers.

58. What processes are described by the Feynman diagrams in Figure P31.58? What is the exchanged particle in each process?

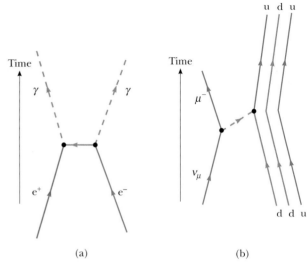

(a)                     (b)

**Figure P31.58**

59. Identify the mediators for the two interactions described in the Feynman diagrams shown in Figure P31.59.

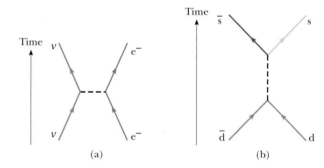

(a)                     (b)

**Figure P31.59**

## ANSWERS TO QUICK QUIZZES

**31.1** Into the page. According to the right-hand rule, this is the direction that will result in a force toward the top of the page on a positive charge moving to the right on the page.

**31.2** Equation 31.3 indicates that the masses of both of these field particles must be zero; otherwise $d$ would be less than infinity.

**31.3** (c), (e). From the representation of the decay only, we have no information about (a) energy or (b) momentum. Angular momentum (c) is not conserved because we begin with one spin-$\frac{1}{2}$ particle and end up with two. Electric charge (d) is conserved because we have zero net charge on both sides of the reaction. Electron lepton number (e) is not conserved because we begin with no leptons but end up with a single electron. Finally, conservation of baryon number is obeyed because we have $B = +1$ on both sides of the reaction.

**31.4** The experiment described is an analogy to the Rutherford scattering experiment described in Section 29.1. In the Rutherford experiment, alpha particles were scattered from atoms, and the scattering was consistent with a small structure in the atom containing the positive charge.

**31.5** The $\Delta^{++}$ particle has three up quarks, each with charge $\frac{2}{3}e$, for a total charge of $+2e$. The largest quark charge is $\frac{2}{3}e$, so a combination of only two particles, a quark and an antiquark, forming a meson, could not possibly have electric charge adding up to $+2e$.

**31.6** Until about 700 000 years after the Big Bang, the temperature of the Universe was high enough for any atoms that formed to be ionized immediately by ambient radiation. Once the average radiation energy dropped below the hydrogen ionization energy of 13.6 eV, hydrogen atoms could form and remain as neutral atoms for a relatively long period of time.

**31.7** No. At any point in the Universe, it will appear that other galaxies are moving away from that point. Imagine that you are cooking raisin bread and the raisins represent galaxies. As the bread expands as it warms (simulating the expansion of the Universe), an observer on *any* raisin will see all of the other raisins moving away.

# The Cosmic Connection

## PROBLEMS AND PERSPECTIVES

We have now investigated the principles of quantum physics and have seen many connections to our central question for the *Cosmic Connection* Context:

> *How can we connect the physics of microscopic particles to the physics of the Universe?*

While particle physicists have been exploring the realm of the very small, cosmologists have been exploring cosmic history back to the first second of the Big Bang. Observation of events that occur when two particles collide in an accelerator is essential in reconstructing the early moments in cosmic history. The key to understanding the early Universe is first to understand the world of elementary particles. Cosmologists and physicists now find they have many common goals and are joining hands to attempt to understand the physical world at its most fundamental level.

### Problems

We have made great progress in understanding the Universe and its underlying structure, but a multitude of questions remain unanswered: Why does so little antimatter exist in the Universe? Do neutrinos have a small rest energy, and if so, how do they contribute to the "dark matter" of the Universe? Is it possible to unify the strong and electroweak forces in a logical and consistent manner? Can gravity be unified with the other forces? Why do quarks and leptons form three similar but distinct families? Are muons the same as electrons (apart from their difference in mass), or do they have other subtle differences that have not been detected? Why are some particles charged and others neutral? Why do quarks carry a fractional charge? What determines the masses of the fundamental constituents? Can isolated quarks exist? Do leptons and quarks have a substructure?

### String Theory: A New Perspective

Let us briefly discuss one current effort at answering some of these questions by proposing a new perspective on particles. As you read this book, you may recall starting off with the particle model and doing quite a bit of physics with it. In the *Earthquakes* Context, we introduced the wave model, and more physics was used to investigate the properties of waves. We used a wave model for light in the *Lasers* Context. Early in the current Context, however, we saw the need to return to the particle model for light. Furthermore, we found that material particles had wave-like characteristics. The quantum particle model of Chapter 28 allowed us to build particles out of waves, suggesting that a wave is the fundamental entity. In Chapter 31, however, we discussed the elementary particles as the fundamental entities. It seems as if we cannot make up our mind! In some sense, this is true—the wave–particle duality is still an area of active research. In this final Context Conclusion, we shall discuss a current research effort to build particles out of waves and vibrations.

1204

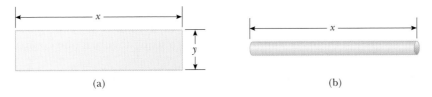

*x*

*x*

*y*

(a)

(b)

**Figure 1**

(a) A piece of paper is cut into a rectangular shape. As a rectangle, the shape has two dimensions.
(b) The paper is rolled up into a soda straw. From far away, it appears to be one-dimensional. The
curled-up second dimension is not visible when viewed from a distance that is large compared with the
diameter of the straw.

**String theory** is an effort to unify the four fundamental forces by modeling
all particles as various vibrational modes of a single entity—an incredibly small
string. The typical length of such a string is on the order of $10^{-35}$ m, called the
**Planck length.** We have seen quantized modes before—the frequencies of vibrating guitar strings in Chapter 14 and the quantized energy levels of atoms in Chapter 29. In string theory, each quantized mode of vibration of the string corresponds to a different elementary particle in the Standard Model.

One of the complicating factors in string theory is that it requires space–time
to have ten dimensions. Despite the theoretical and conceptual difficulties in dealing with ten dimensions, string theory holds promise in incorporating gravity with
the other forces. Four of the ten dimensions are visible to us—three space dimensions and one time dimension—the other six are *compactified*. This is described by
saying that the six dimensions are curled up so tightly that they are not visible in
the macroscopic world.

As an analogy, consider a soda straw. We can build a soda straw by cutting a
rectangular piece of paper (Fig. 1a), which clearly has two dimensions, and rolling
it up into a small tube (Fig. 1b). From far away, the soda straw looks like a one-
dimensional straight line. The second dimension has been curled up and is not
visible. String theory claims that six space–time dimensions are curled up in an
analogous way, with the curling on the size of the Planck length and impossible to
see from our viewpoint.

Another complicating factor with string theory is that it is difficult for string
theorists to guide experimentalists in how and what to look for in an experiment.
The Planck length is so incredibly small that direct experimentation on strings is
impossible. Until the theory has been further developed, string theorists are restricted to applying the theory to known results and testing for consistency.

One of the predictions of string theory is called **supersymmetry** (SUSY). It
suggests that every elementary particle has a superpartner that has not yet been
observed. It is believed that supersymmetry is a broken symmetry (like the broken
electroweak symmetry at low energies) and the masses of the superpartners are
above our current capabilities of detection by accelerators. Some theorists claim
that the mass of superpartners is the missing mass discussed in the Context Conclusion of Chapter 31. Keeping with the whimsical trend in naming particles and
their properties that we saw in Chapter 31, superpartners are given names such as
the *squark* (the superpartner to a quark), the *selectron* (electron), and the *gluinos*
(gluon).

Other theorists are working on **M-theory,** which is an 11-dimensional theory based on membranes rather than strings. In a way reminiscent of the correspondence principle, M-theory is claimed to reduce to string theory if one compactifies from 11 dimensions to 10.

The questions that we listed at the beginning of this Context Conclusion go on and on. Because of the rapid advances and new discoveries in the field of particle physics, by the time you read this book some of these questions will likely be resolved and other new questions may emerge.

### QUESTION

1. A girl and her grandmother grind corn while the woman tells the girl stories that illuminate what is most important. A boy keeps crows away from ripening corn while his grandfather sits in the shade and explains to him the Universe and his place in it. What the children do not understand this summer they will better understand next year. Now you must take the part of the adults. State the most general, most universal, most fundamental truths that you know. If you must repeat someone else's ideas, obtain the best version of those ideas that you can, and state your source. If there is something you do not understand, make a plan to understand it better within the next year.

### PROBLEM

1. Classical general relativity views the structure of space–time as deterministic and well defined down to arbitrarily small distances. On the other hand, quantum general relativity forbids distances smaller than the Planck length given by $L = (\hbar G/c^3)^{1/2}$. (a) Calculate the value of the Planck length. The quantum limitation suggests that after the Big Bang, when all the presently observable section of the Universe was a singularity, nothing could be observed until that singularity grew larger than the Planck length. Because the size of the singularity grew at the speed of light, we can infer that no observations were possible during the time it took for light to travel the Planck length. (b) Calculate this time, known as the Planck time $T$, and compare it with the ultrahot epoch mentioned in the text. (c) Does this suggest we may never know what happened between the time $t = 0$ and the time $t = T$?

### The Meaning of Success

To earn the respect of intelligent people and to win the affection of children;
To appreciate the beauty in nature and all that surrounds us;
To seek out and nurture the best in others;
To give the gift of yourself to others without the slightest thought of return, for it is in giving that we receive;
To have accomplished a task, whether it be saving a lost soul, healing a sick child, writing a book, or risking your life for a friend;
To have celebrated and laughed with great joy and enthusiasm and sung with exaltation;
To have hope even in times of despair, for as long as you have hope, you have life;
To love and be loved;
To be understood and to understand;
To know that even one life has breathed easier because you have lived;
This is the meaning of success.

*Ralph Waldo Emerson and modified by Ray Serway*

# Tables

| TABLE A.1 | Conversion Factors |
| --- | --- |

### Length

| | m | cm | km | in. | ft | mi |
| --- | --- | --- | --- | --- | --- | --- |
| 1 meter | 1 | $10^2$ | $10^{-3}$ | 39.37 | 3.281 | $6.214 \times 10^{-4}$ |
| 1 centimeter | $10^{-2}$ | 1 | $10^{-5}$ | 0.393 7 | $3.281 \times 10^{-2}$ | $6.214 \times 10^{-6}$ |
| 1 kilometer | $10^3$ | $10^5$ | 1 | $3.937 \times 10^4$ | $3.281 \times 10^3$ | 0.621 4 |
| 1 inch | $2.540 \times 10^{-2}$ | 2.540 | $2.540 \times 10^{-5}$ | 1 | $8.333 \times 10^{-2}$ | $1.578 \times 10^{-5}$ |
| 1 foot | 0.304 8 | 30.48 | $3.048 \times 10^{-4}$ | 12 | 1 | $1.894 \times 10^{-4}$ |
| 1 mile | 1 609 | $1.609 \times 10^5$ | 1.609 | $6.336 \times 10^4$ | 5 280 | 1 |

### Mass

| | kg | g | slug | u |
| --- | --- | --- | --- | --- |
| 1 kilogram | 1 | $10^3$ | $6.852 \times 10^{-2}$ | $6.024 \times 10^{26}$ |
| 1 gram | $10^{-3}$ | 1 | $6.852 \times 10^{-5}$ | $6.024 \times 10^{23}$ |
| 1 slug | 14.59 | $1.459 \times 10^4$ | 1 | $8.789 \times 10^{27}$ |
| 1 atomic mass unit | $1.660 \times 10^{-27}$ | $1.660 \times 10^{-24}$ | $1.137 \times 10^{-28}$ | 1 |

*Note:* 1 metric ton = 1 000 kg.

### Time

| | s | min | h | day | yr |
| --- | --- | --- | --- | --- | --- |
| 1 second | 1 | $1.667 \times 10^{-2}$ | $2.778 \times 10^{-4}$ | $1.157 \times 10^{-5}$ | $3.169 \times 10^{-8}$ |
| 1 minute | 60 | 1 | $1.667 \times 10^{-2}$ | $6.994 \times 10^{-4}$ | $1.901 \times 10^{-6}$ |
| 1 hour | 3 600 | 60 | 1 | $4.167 \times 10^{-2}$ | $1.141 \times 10^{-4}$ |
| 1 day | $8.640 \times 10^4$ | 1 440 | 24 | 1 | $2.738 \times 10^{-5}$ |
| 1 year | $3.156 \times 10^7$ | $5.259 \times 10^5$ | $8.766 \times 10^3$ | 365.2 | 1 |

### Speed

| | m/s | cm/s | ft/s | mi/h |
| --- | --- | --- | --- | --- |
| 1 meter per second | 1 | $10^2$ | 3.281 | 2.237 |
| 1 centimeter per second | $10^{-2}$ | 1 | $3.281 \times 10^{-2}$ | $2.237 \times 10^{-2}$ |
| 1 foot per second | 0.304 8 | 30.48 | 1 | 0.681 8 |
| 1 mile per hour | 0.447 0 | 44.70 | 1.467 | 1 |

*Note:* 1 mi/min = 60 mi/h = 88 ft/s.

*continued*

| TABLE A.1 | *Continued* |

### Force

|           | N     | lb      |
|-----------|-------|---------|
| 1 newton  | 1     | 0.224 8 |
| 1 pound   | 4.448 | 1       |

### Work, Energy, Heat

|          | J                       | ft·lb                    | eV                     |
|----------|-------------------------|--------------------------|------------------------|
| 1 joule  | 1                       | 0.737 6                  | $6.242 \times 10^{18}$ |
| 1 ft·lb  | 1.356                   | 1                        | $8.464 \times 10^{18}$ |
| 1 eV     | $1.602 \times 10^{-19}$ | $1.182 \times 10^{-19}$  | 1                      |
| 1 cal    | 4.186                   | 3.087                    | $2.613 \times 10^{19}$ |
| 1 Btu    | $1.055 \times 10^3$     | $7.779 \times 10^2$      | $6.585 \times 10^{21}$ |
| 1 kWh    | $3.600 \times 10^6$     | $2.655 \times 10^6$      | $2.247 \times 10^{25}$ |

|          | cal                     | Btu                     | kWh                     |
|----------|-------------------------|-------------------------|-------------------------|
| 1 joule  | 0.238 9                 | $9.481 \times 10^{-4}$  | $2.778 \times 10^{-7}$  |
| 1 ft·lb  | 0.323 9                 | $1.285 \times 10^{-3}$  | $3.766 \times 10^{-7}$  |
| 1 eV     | $3.827 \times 10^{-20}$ | $1.519 \times 10^{-22}$ | $4.450 \times 10^{-26}$ |
| 1 cal    | 1                       | $3.968 \times 10^{-3}$  | $1.163 \times 10^{-6}$  |
| 1 Btu    | $2.520 \times 10^2$     | 1                       | $2.930 \times 10^{-4}$  |
| 1 kWh    | $8.601 \times 10^5$     | $3.413 \times 10^2$     | 1                       |

### Pressure

|                                 | Pa                  | atm                    |
|---------------------------------|---------------------|------------------------|
| 1 pascal                        | 1                   | $9.869 \times 10^{-6}$ |
| 1 atmosphere                    | $1.013 \times 10^5$ | 1                      |
| 1 centimeter mercury[a]         | $1.333 \times 10^3$ | $1.316 \times 10^{-2}$ |
| 1 pound per inch$^2$            | $6.895 \times 10^3$ | $6.805 \times 10^{-2}$ |
| 1 pound per foot$^2$            | 47.88               | $4.725 \times 10^{-4}$ |

|                                 | cm Hg                  | lb/in.$^2$             | lb/ft$^2$              |
|---------------------------------|------------------------|------------------------|------------------------|
| 1 pascal                        | $7.501 \times 10^{-4}$ | $1.450 \times 10^{-4}$ | $2.089 \times 10^{-2}$ |
| 1 atmosphere                    | 76                     | 14.70                  | $2.116 \times 10^3$    |
| 1 centimeter mercury[a]         | 1                      | 0.194 3                | 27.85                  |
| 1 pound per inch$^2$            | 5.171                  | 1                      | 144                    |
| 1 pound per foot$^2$            | $3.591 \times 10^{-2}$ | $6.944 \times 10^{-3}$ | 1                      |

[a]At 0°C and at a location where the acceleration due to gravity has its "standard" value, 9.806 65 m/s$^2$.

## TABLE A.2 — Symbols, Dimensions, and Units of Physical Quantities

| Quantity | Common Symbol | Unit[a] | Dimensions[b] | Unit in Terms of Base SI Units |
|---|---|---|---|---|
| Acceleration | **a** | $m/s^2$ | $L/T^2$ | $m/s^2$ |
| Amount of substance | $n$ | mole | | mol |
| Angle | $\theta, \phi$ | radian (rad) | | |
| Angular acceleration | $\alpha$ | $rad/s^2$ | $T^{-2}$ | $s^{-2}$ |
| Angular frequency | $\omega$ | $rad/s$ | $T^{-1}$ | $s^{-1}$ |
| Angular momentum | **L** | $kg \cdot m^2/s$ | $ML^2/T$ | $kg \cdot m^2/s$ |
| Angular speed | $\omega$ | $rad/s$ | $T^{-1}$ | $s^{-1}$ |
| Area | $A$ | $m^2$ | $L^2$ | $m^2$ |
| Atomic number | $Z$ | | | |
| Capacitance | $C$ | farad (F) | $Q^2T^2/ML^2$ | $A^2 \cdot s^4/kg \cdot m^2$ |
| Charge | $q, Q, e$ | coulomb (C) | $Q$ | $A \cdot s$ |
| Charge density | | | | |
|   Line | $\lambda$ | $C/m$ | $Q/L$ | $A \cdot s/m$ |
|   Surface | $\sigma$ | $C/m^2$ | $Q/L^2$ | $A \cdot s/m^2$ |
|   Volume | $\rho$ | $C/m^3$ | $Q/L^3$ | $A \cdot s/m^3$ |
| Conductivity | $\sigma$ | $1/\Omega \cdot m$ | $Q^2T/ML^3$ | $A^2 \cdot s^3/kg \cdot m^3$ |
| Current | $I$ | AMPERE | $Q/T$ | A |
| Current density | **J** | $A/m^2$ | $Q/T^2$ | $A/m^2$ |
| Density | $\rho$ | $kg/m^3$ | $M/L^3$ | $kg/m^3$ |
| Dielectric constant | $\kappa$ | | | |
| Displacement | **r, s** | METER | $L$ | m |
|   Distance | $d, h$ | | | |
|   Length | $\ell, L$ | | | |
| Electric dipole moment | **p** | $C \cdot m$ | $QL$ | $A \cdot s \cdot m$ |
| Electric field | **E** | $V/m$ | $ML/QT^2$ | $kg \cdot m/A \cdot s^3$ |
| Electric flux | $\Phi_E$ | $V \cdot m$ | $ML^3/QT^3$ | $kg \cdot m^3/A \cdot s^3$ |
| Electromotive force | $\varepsilon$ | volt (V) | $ML^2/QT^2$ | $kg \cdot m^2/A \cdot s^3$ |
| Energy | $E, U, K$ | joule (J) | $ML^2/T^2$ | $kg \cdot m^2/s^2$ |
| Entropy | $S$ | $J/K$ | $ML^2/T^2 \cdot K$ | $kg \cdot m^2/s^2 \cdot K$ |
| Force | **F** | newton (N) | $ML/T^2$ | $kg \cdot m/s^2$ |
| Frequency | $f$ | hertz (Hz) | $T^{-1}$ | $s^{-1}$ |
| Heat | $Q$ | joule (J) | $ML^2/T^2$ | $kg \cdot m^2/s^2$ |
| Inductance | $L$ | henry (H) | $ML^2/Q^2$ | $kg \cdot m^2/A^2 \cdot s^2$ |
| Magnetic dipole moment | $\boldsymbol{\mu}$ | $N \cdot m/T$ | $QL^2/T$ | $A \cdot m^2$ |
| Magnetic field | **B** | tesla (T) $(= Wb/m^2)$ | $M/QT$ | $kg/A \cdot s^2$ |
| Magnetic flux | $\Phi_B$ | weber (Wb) | $ML^2/QT$ | $kg \cdot m^2/A \cdot s^2$ |
| Mass | $m, M$ | KILOGRAM | $M$ | kg |
| Molar specific heat | $C$ | $J/mol \cdot K$ | | $kg \cdot m^2/s^2 \cdot mol \cdot K$ |
| Moment of inertia | $I$ | $kg \cdot m^2$ | $ML^2$ | $kg \cdot m^2$ |
| Momentum | **p** | $kg \cdot m/s$ | $ML/T$ | $kg \cdot m/s$ |
| Period | $T$ | s | $T$ | s |
| Permeability of space | $\mu_0$ | $N/A^2 (= H/m)$ | $ML/Q^2T$ | $kg \cdot m/A^2 \cdot s^2$ |
| Permittivity of space | $\epsilon_0$ | $C^2/N \cdot m^2 (= F/m)$ | $Q^2T^2/ML^3$ | $A^2 \cdot s^4/kg \cdot m^3$ |
| Potential | $V$ | volt (V) $(= J/C)$ | $ML^2/QT^2$ | $kg \cdot m^2/A \cdot s^3$ |
| Power | $\mathscr{P}$ | watt (W) $(= J/s)$ | $ML^2/T^3$ | $kg \cdot m^2/s^3$ |

*continued*

| TABLE A.2 | *Continued* |

| Quantity | Common Symbol | Unit[a] | Dimensions[b] | Unit in Terms of Base SI Units |
|---|---|---|---|---|
| Pressure | $P$ | pascal (Pa) $= (N/m^2)$ | $M/LT^2$ | $kg/m \cdot s^2$ |
| Resistance | $R$ | ohm ($\Omega$) ($= V/A$) | $ML^2/Q^2T$ | $kg \cdot m^2/A^2 \cdot s^3$ |
| Specific heat | $c$ | $J/kg \cdot K$ | $L^2/T^2 \cdot K$ | $m^2/s^2 \cdot K$ |
| Speed | $v$ | m/s | $L/T$ | m/s |
| Temperature | $T$ | KELVIN | K | K |
| Time | $t$ | SECOND | T | s |
| Torque | $\tau$ | $N \cdot m$ | $ML^2/T^2$ | $kg \cdot m^2/s^2$ |
| Volume | $V$ | $m^3$ | $L^3$ | $m^3$ |
| Wavelength | $\lambda$ | m | L | m |
| Work | $W$ | joule (J) ($= N \cdot m$) | $ML^2/T^2$ | $kg \cdot m^2/s^2$ |

[a] The base SI units are given in uppercase letters.

[b] The symbols M, L, T, Q, and K denote mass, length, time, charge, and temperature, respectively.

| TABLE A.3 | Table of Atomic Masses[a] |

| Atomic Number $Z$ | Element | Symbol | Chemical Atomic Mass (u) | Mass Number (*Indicates Radioactive) $A$ | Atomic Mass (u) | Percent Abundance | Half-Life (If Radioactive) $T_{1/2}$ |
|---|---|---|---|---|---|---|---|
| 0 | (Neutron) | n | | 1* | 1.008 665 | | 10.4 min |
| 1 | Hydrogen | H | 1.007 9 | 1 | 1.007 825 | 99.985 | |
| | Deuterium | D | | 2 | 2.014 102 | 0.015 | |
| | Tritium | T | | 3* | 3.016 049 | | 12.33 yr |
| 2 | Helium | He | 4.002 60 | 3 | 3.016 029 | 0.000 14 | |
| | | | | 4 | 4.002 602 | 99.999 86 | |
| | | | | 6* | 6.018 886 | | 0.81 s |
| 3 | Lithium | Li | 6.941 | 6 | 6.015 121 | 7.5 | |
| | | | | 7 | 7.016 003 | 92.5 | |
| | | | | 8* | 8.022 486 | | 0.84 s |
| 4 | Beryllium | Be | 9.012 2 | 7* | 7.016 928 | | 53.3 days |
| | | | | 9 | 9.012 174 | 100 | |
| | | | | 10* | 10.013 534 | | $1.5 \times 10^6$ yr |
| 5 | Boron | B | 10.81 | 10 | 10.012 936 | 19.9 | |
| | | | | 11 | 11.009 305 | 80.1 | |
| | | | | 12* | 12.014 352 | | 0.020 2 s |
| 6 | Carbon | C | 12.011 | 10* | 10.016 854 | | 19.3 s |
| | | | | 11* | 11.011 433 | | 20.4 min |
| | | | | 12 | 12.000 000 | 98.90 | |
| | | | | 13 | 13.003 355 | 1.10 | |
| | | | | 14* | 14.003 242 | | 5 730 yr |
| | | | | 15* | 15.010 599 | | 2.45 s |
| 7 | Nitrogen | N | 14.006 7 | 12* | 12.018 613 | | 0.011 0 s |
| | | | | 13* | 13.005 738 | | 9.96 min |
| | | | | 14 | 14.003 074 | 99.63 | |
| | | | | 15 | 15.000 108 | 0.37 | |
| | | | | 16* | 16.006 100 | | 7.13 s |
| | | | | 17* | 17.008 450 | | 4.17 s |

*continued*

**TABLE A.3** | *Continued*

| Atomic Number Z | Element | Symbol | Chemical Atomic Mass (u) | Mass Number (*Indicates Radioactive) A | Atomic Mass (u) | Percent Abundance | Half-Life (If Radioactive) $T_{1/2}$ |
|---|---|---|---|---|---|---|---|
| 8 | Oxygen | O | 15.999 4 | 14* | 14.008 595 | | 70.6 s |
| | | | | 15* | 15.003 065 | | 122 s |
| | | | | 16 | 15.994 915 | 99.761 | |
| | | | | 17 | 16.999 132 | 0.039 | |
| | | | | 18 | 17.999 160 | 0.20 | |
| | | | | 19* | 19.003 577 | | 26.9 s |
| 9 | Fluorine | F | 18.998 40 | 17* | 17.002 094 | | 64.5 s |
| | | | | 18* | 18.000 937 | | 109.8 min |
| | | | | 19 | 18.998 404 | 100 | |
| | | | | 20* | 19.999 982 | | 11.0 s |
| | | | | 21* | 20.999 950 | | 4.2 s |
| 10 | Neon | Ne | 20.180 | 18* | 18.005 710 | | 1.67 s |
| | | | | 19* | 19.001 880 | | 17.2 s |
| | | | | 20 | 19.992 435 | 90.48 | |
| | | | | 21 | 20.993 841 | 0.27 | |
| | | | | 22 | 21.991 383 | 9.25 | |
| | | | | 23* | 22.994 465 | | 37.2 s |
| 11 | Sodium | Na | 22.989 87 | 21* | 20.997 650 | | 22.5 s |
| | | | | 22* | 21.994 434 | | 2.61 yr |
| | | | | 23 | 22.989 770 | 100 | |
| | | | | 24* | 23.990 961 | | 14.96 h |
| 12 | Magnesium | Mg | 24.305 | 23* | 22.994 124 | | 11.3 s |
| | | | | 24 | 23.985 042 | 78.99 | |
| | | | | 25 | 24.985 838 | 10.00 | |
| | | | | 26 | 25.982 594 | 11.01 | |
| | | | | 27* | 26.984 341 | | 9.46 min |
| 13 | Aluminum | Al | 26.981 54 | 26* | 25.986 892 | | $7.4 \times 10^5$ yr |
| | | | | 27 | 26.981 538 | 100 | |
| | | | | 28* | 27.981 910 | | 2.24 min |
| 14 | Silicon | Si | 28.086 | 28 | 27.976 927 | 92.23 | |
| | | | | 29 | 28.976 495 | 4.67 | |
| | | | | 30 | 29.973 770 | 3.10 | |
| | | | | 31* | 30.975 362 | | 2.62 h |
| | | | | 32* | 31.974 148 | | 172 yr |
| 15 | Phosphorus | P | 30.973 76 | 30* | 29.978 307 | | 2.50 min |
| | | | | 31 | 30.973 762 | 100 | |
| | | | | 32* | 31.973 908 | | 14.26 days |
| | | | | 33* | 32.971 725 | | 25.3 days |
| 16 | Sulfur | S | 32.066 | 32 | 31.972 071 | 95.02 | |
| | | | | 33 | 32.971 459 | 0.75 | |
| | | | | 34 | 33.967 867 | 4.21 | |
| | | | | 35* | 34.969 033 | | 87.5 days |
| | | | | 36 | 35.967 081 | 0.02 | |
| 17 | Chlorine | Cl | 35.453 | 35 | 34.968 853 | 75.77 | |
| | | | | 36* | 35.968 307 | | $3.0 \times 10^5$ yr |
| | | | | 37 | 36.965 903 | 24.23 | |

*continued*

**TABLE A.3** *Continued*

| Atomic Number Z | Element | Symbol | Chemical Atomic Mass (u) | Mass Number (*Indicates Radioactive) A | Atomic Mass (u) | Percent Abundance | Half-Life (If Radioactive) $T_{1/2}$ |
|---|---|---|---|---|---|---|---|
| 18 | Argon | Ar | 39.948 | 36 | 35.967 547 | 0.337 | |
| | | | | 37* | 36.966 776 | | 35.04 days |
| | | | | 38 | 37.962 732 | 0.063 | |
| | | | | 39* | 38.964 314 | | 269 yr |
| | | | | 40 | 39.962 384 | 99.600 | |
| | | | | 42* | 41.963 049 | | 33 yr |
| 19 | Potassium | K | 39.098 3 | 39 | 38.963 708 | 93.258 1 | |
| | | | | 40* | 39.964 000 | 0.011 7 | $1.28 \times 10^9$ yr |
| | | | | 41 | 40.961 827 | 6.730 2 | |
| 20 | Calcium | Ca | 40.08 | 40 | 39.962 591 | 96.941 | |
| | | | | 41* | 40.962 279 | | $1.0 \times 10^5$ yr |
| | | | | 42 | 41.958 618 | 0.647 | |
| | | | | 43 | 42.958 767 | 0.135 | |
| | | | | 44 | 43.955 481 | 2.086 | |
| | | | | 46 | 45.953 687 | 0.004 | |
| | | | | 48 | 47.952 534 | 0.187 | |
| 21 | Scandium | Sc | 44.955 9 | 41* | 40.969 250 | | 0.596 s |
| | | | | 45 | 44.955 911 | 100 | |
| 22 | Titanium | Ti | 47.88 | 44* | 43.959 691 | | 49 yr |
| | | | | 46 | 45.952 630 | 8.0 | |
| | | | | 47 | 46.951 765 | 7.3 | |
| | | | | 48 | 47.947 947 | 73.8 | |
| | | | | 49 | 48.947 871 | 5.5 | |
| | | | | 50 | 49.944 792 | 5.4 | |
| 23 | Vanadium | V | 50.941 5 | 48* | 47.952 255 | | 15.97 days |
| | | | | 50* | 49.947 161 | 0.25 | $1.5 \times 10^{17}$ yr |
| | | | | 51 | 50.943 962 | 99.75 | |
| 24 | Chromium | Cr | 51.996 | 48* | 47.954 033 | | 21.6 h |
| | | | | 50 | 49.946 047 | 4.345 | |
| | | | | 52 | 51.940 511 | 83.79 | |
| | | | | 53 | 52.940 652 | 9.50 | |
| | | | | 54 | 53.938 883 | 2.365 | |
| 25 | Manganese | Mn | 54.938 05 | 54* | 53.940 361 | | 312.1 days |
| | | | | 55 | 54.938 048 | 100 | |
| 26 | Iron | Fe | 55.847 | 54 | 53.939 613 | 5.9 | |
| | | | | 55* | 54.938 297 | | 2.7 yr |
| | | | | 56 | 55.934 940 | 91.72 | |
| | | | | 57 | 56.935 396 | 2.1 | |
| | | | | 58 | 57.933 278 | 0.28 | |
| | | | | 60* | 59.934 078 | | $1.5 \times 10^6$ yr |
| 27 | Cobalt | Co | 58.933 20 | 59 | 58.933 198 | 100 | |
| | | | | 60* | 59.933 820 | | 5.27 yr |
| 28 | Nickel | Ni | 58.693 | 58 | 57.935 346 | 68.077 | |
| | | | | 59* | 58.934 350 | | $7.5 \times 10^4$ yr |
| | | | | 60 | 59.930 789 | 26.223 | |
| | | | | 61 | 60.931 058 | 1.140 | |
| | | | | 62 | 61.928 346 | 3.634 | |
| | | | | 63* | 62.929 670 | | 100 yr |
| | | | | 64 | 63.927 967 | 0.926 | |

*continued*

## TABLE A.3 *Continued*

| Atomic Number Z | Element | Symbol | Chemical Atomic Mass (u) | Mass Number (*Indicates Radioactive) A | Atomic Mass (u) | Percent Abundance | Half-Life (If Radioactive) $T_{1/2}$ |
|---|---|---|---|---|---|---|---|
| 29 | Copper | Cu | 63.54 | 63 | 62.929 599 | 69.17 | |
| | | | | 65 | 64.927 791 | 30.83 | |
| 30 | Zinc | Zn | 65.39 | 64 | 63.929 144 | 48.6 | |
| | | | | 66 | 65.926 035 | 27.9 | |
| | | | | 67 | 66.927 129 | 4.1 | |
| | | | | 68 | 67.924 845 | 18.8 | |
| | | | | 70 | 69.925 323 | 0.6 | |
| 31 | Gallium | Ga | 69.723 | 69 | 68.925 580 | 60.108 | |
| | | | | 71 | 70.924 703 | 39.892 | |
| 32 | Germanium | Ge | 72.61 | 70 | 69.924 250 | 21.23 | |
| | | | | 72 | 71.922 079 | 27.66 | |
| | | | | 73 | 72.923 462 | 7.73 | |
| | | | | 74 | 73.921 177 | 35.94 | |
| | | | | 76 | 75.921 402 | 7.44 | |
| 33 | Arsenic | As | 74.921 6 | 75 | 74.921 594 | 100 | |
| 34 | Selenium | Se | 78.96 | 74 | 73.922 474 | 0.89 | |
| | | | | 76 | 75.919 212 | 9.36 | |
| | | | | 77 | 76.919 913 | 7.63 | |
| | | | | 78 | 77.917 307 | 23.78 | |
| | | | | 79* | 78.918 497 | | $\leq 6.5 \times 10^4$ yr |
| | | | | 80 | 79.916 519 | 49.61 | |
| | | | | 82* | 81.916 697 | 8.73 | $1.4 \times 10^{20}$ yr |
| 35 | Bromine | Br | 79.904 | 79 | 78.918 336 | 50.69 | |
| | | | | 81 | 80.916 287 | 49.31 | |
| 36 | Krypton | Kr | 83.80 | 78 | 77.920 400 | 0.35 | |
| | | | | 80 | 79.916 377 | 2.25 | |
| | | | | 81* | 80.916 589 | | $2.1 \times 10^5$ yr |
| | | | | 82 | 81.913 481 | 11.6 | |
| | | | | 83 | 82.914 136 | 11.5 | |
| | | | | 84 | 83.911 508 | 57.0 | |
| | | | | 85* | 84.912 531 | | 10.76 yr |
| | | | | 86 | 85.910 615 | 17.3 | |
| 37 | Rubidium | Rb | 85.468 | 85 | 84.911 793 | 72.17 | |
| | | | | 87* | 86.909 186 | 27.83 | $4.75 \times 10^{10}$ yr |
| 38 | Strontium | Sr | 87.62 | 84 | 83.913 428 | 0.56 | |
| | | | | 86 | 85.909 266 | 9.86 | |
| | | | | 87 | 86.908 883 | 7.00 | |
| | | | | 88 | 87.905 618 | 82.58 | |
| | | | | 90* | 89.907 737 | | 29.1 yr |
| 39 | Yttrium | Y | 88.905 8 | 89 | 88.905 847 | 100 | |
| 40 | Zirconium | Zr | 91.224 | 90 | 89.904 702 | 51.45 | |
| | | | | 91 | 90.905 643 | 11.22 | |
| | | | | 92 | 91.905 038 | 17.15 | |
| | | | | 93* | 92.906 473 | | $1.5 \times 10^6$ yr |
| | | | | 94 | 93.906 314 | 17.38 | |
| | | | | 96 | 95.908 274 | 2.80 | |

*continued*

**TABLE A.3**    *Continued*

| Atomic Number Z | Element | Symbol | Chemical Atomic Mass (u) | Mass Number (*Indicates Radioactive) A | Atomic Mass (u) | Percent Abundance | Half-Life (If Radioactive) $T_{1/2}$ |
|---|---|---|---|---|---|---|---|
| 41 | Niobium | Nb | 92.906 4 | 91* | 90.906 988 | | $6.8 \times 10^2$ yr |
| | | | | 92* | 91.907 191 | | $3.5 \times 10^7$ yr |
| | | | | 93 | 92.906 376 | 100 | |
| | | | | 94* | 93.907 280 | | $2 \times 10^4$ yr |
| 42 | Molybdenum | Mo | 95.94 | 92 | 91.906 807 | 14.84 | |
| | | | | 93* | 92.906 811 | | $3.5 \times 10^3$ yr |
| | | | | 94 | 93.905 085 | 9.25 | |
| | | | | 95 | 94.905 841 | 15.92 | |
| | | | | 96 | 95.904 678 | 16.68 | |
| | | | | 97 | 96.906 020 | 9.55 | |
| | | | | 98 | 97.905 407 | 24.13 | |
| | | | | 100 | 99.907 476 | 9.63 | |
| 43 | Technetium | Tc | | 97* | 96.906 363 | | $2.6 \times 10^6$ yr |
| | | | | 98* | 97.907 215 | | $4.2 \times 10^6$ yr |
| | | | | 99* | 98.906 254 | | $2.1 \times 10^5$ yr |
| 44 | Ruthenium | Ru | 101.07 | 96 | 95.907 597 | 5.54 | |
| | | | | 98 | 97.905 287 | 1.86 | |
| | | | | 99 | 98.905 939 | 12.7 | |
| | | | | 100 | 99.904 219 | 12.6 | |
| | | | | 101 | 100.905 558 | 17.1 | |
| | | | | 102 | 101.904 348 | 31.6 | |
| | | | | 104 | 103.905 428 | 18.6 | |
| 45 | Rhodium | Rh | 102.905 5 | 103 | 102.905 502 | 100 | |
| 46 | Palladium | Pd | 106.42 | 102 | 101.905 616 | 1.02 | |
| | | | | 104 | 103.904 033 | 11.14 | |
| | | | | 105 | 104.905 082 | 22.33 | |
| | | | | 106 | 105.903 481 | 27.33 | |
| | | | | 107* | 106.905 126 | | $6.5 \times 10^6$ yr |
| | | | | 108 | 107.903 893 | 26.46 | |
| | | | | 110 | 109.905 158 | 11.72 | |
| 47 | Silver | Ag | 107.868 | 107 | 106.905 091 | 51.84 | |
| | | | | 109 | 108.904 754 | 48.16 | |
| 48 | Cadmium | Cd | 112.41 | 106 | 105.906 457 | 1.25 | |
| | | | | 108 | 107.904 183 | 0.89 | |
| | | | | 109* | 108.904 984 | | 462 days |
| | | | | 110 | 109.903 004 | 12.49 | |
| | | | | 111 | 110.904 182 | 12.80 | |
| | | | | 112 | 111.902 760 | 24.13 | |
| | | | | 113* | 112.904 401 | 12.22 | $9.3 \times 10^{15}$ yr |
| | | | | 114 | 113.903 359 | 28.73 | |
| | | | | 116 | 115.904 755 | 7.49 | |
| 49 | Indium | In | 114.82 | 113 | 112.904 060 | 4.3 | |
| | | | | 115* | 114.903 876 | 95.7 | $4.4 \times 10^{14}$ yr |
| 50 | Tin | Sn | 118.71 | 112 | 111.904 822 | 0.97 | |
| | | | | 114 | 113.902 780 | 0.65 | |
| | | | | 115 | 114.903 345 | 0.36 | |
| | | | | 116 | 115.901 743 | 14.53 | |
| | | | | 117 | 116.902 953 | 7.68 | |

*continued*

**TABLE A.3** *Continued*

| Atomic Number Z | Element | Symbol | Chemical Atomic Mass (u) | Mass Number (*Indicates Radioactive) A | Atomic Mass (u) | Percent Abundance | Half-Life (If Radioactive) $T_{1/2}$ |
|---|---|---|---|---|---|---|---|
| (50) | (Tin) | | | 118 | 117.901 605 | 24.22 | |
| | | | | 119 | 118.903 308 | 8.58 | |
| | | | | 120 | 119.902 197 | 32.59 | |
| | | | | 121* | 120.904 237 | | 55 yr |
| | | | | 122 | 121.903 439 | 4.63 | |
| | | | | 124 | 123.905 274 | 5.79 | |
| 51 | Antimony | Sb | 121.76 | 121 | 120.903 820 | 57.36 | |
| | | | | 123 | 122.904 215 | 42.64 | |
| | | | | 125* | 124.905 251 | | 2.7 yr |
| 52 | Tellurium | Te | 127.60 | 120 | 119.904 040 | 0.095 | |
| | | | | 122 | 121.903 052 | 2.59 | |
| | | | | 123* | 122.904 271 | 0.905 | $1.3 \times 10^{13}$ yr |
| | | | | 124 | 123.902 817 | 4.79 | |
| | | | | 125 | 124.904 429 | 7.12 | |
| | | | | 126 | 125.903 309 | 18.93 | |
| | | | | 128* | 127.904 463 | 31.70 | $> 8 \times 10^{24}$ yr |
| | | | | 130* | 129.906 228 | 33.87 | $\leqslant 1.25 \times 10^{21}$ yr |
| 53 | Iodine | I | 126.904 5 | 127 | 126.904 474 | 100 | |
| | | | | 129* | 128.904 984 | | $1.6 \times 10^{7}$ yr |
| 54 | Xenon | Xe | 131.29 | 124 | 123.905 894 | 0.10 | |
| | | | | 126 | 125.904 268 | 0.09 | |
| | | | | 128 | 127.903 531 | 1.91 | |
| | | | | 129 | 128.904 779 | 26.4 | |
| | | | | 130 | 129.903 509 | 4.1 | |
| | | | | 131 | 130.905 069 | 21.2 | |
| | | | | 132 | 131.904 141 | 26.9 | |
| | | | | 134 | 133.905 394 | 10.4 | |
| | | | | 136* | 135.907 215 | 8.9 | $\geqslant 2.36 \times 10^{21}$ yr |
| 55 | Cesium | Cs | 132.905 4 | 133 | 132.905 436 | 100 | |
| | | | | 134* | 133.906 703 | | 2.1 yr |
| | | | | 135* | 134.905 891 | | $2 \times 10^{6}$ yr |
| | | | | 137* | 136.907 078 | | 30 yr |
| 56 | Barium | Ba | 137.33 | 130 | 129.906 289 | 0.106 | |
| | | | | 132 | 131.905 048 | 0.101 | |
| | | | | 133* | 132.905 990 | | 10.5 yr |
| | | | | 134 | 133.904 492 | 2.42 | |
| | | | | 135 | 134.905 671 | 6.593 | |
| | | | | 136 | 135.904 559 | 7.85 | |
| | | | | 137 | 136.905 816 | 11.23 | |
| | | | | 138 | 137.905 236 | 21.70 | |
| 57 | Lanthanum | La | 138.905 | 137* | 136.906 462 | | $6 \times 10^{4}$ yr |
| | | | | 138* | 137.907 105 | 0.090 2 | $1.05 \times 10^{11}$ yr |
| | | | | 139 | 138.906 346 | 99.909 8 | |
| 58 | Cerium | Ce | 140.12 | 136 | 135.907 139 | 0.19 | |
| | | | | 138 | 137.905 986 | 0.25 | |
| | | | | 140 | 139.905 434 | 88.43 | |
| | | | | 142* | 141.909 241 | 11.13 | $> 5 \times 10^{16}$ yr |
| 59 | Praseodymium | Pr | 140.907 6 | 141 | 140.907 647 | 100 | |

*continued*

**TABLE A.3** *Continued*

| Atomic Number Z | Element | Symbol | Chemical Atomic Mass (u) | Mass Number (*Indicates Radioactive) A | Atomic Mass (u) | Percent Abundance | Half-Life (If Radioactive) $T_{1/2}$ |
|---|---|---|---|---|---|---|---|
| 60 | Neodymium | Nd | 144.24 | 142 | 141.907 718 | 27.13 | |
| | | | | 143 | 142.909 809 | 12.18 | |
| | | | | 144* | 143.910 082 | 23.80 | $2.3 \times 10^{15}$ yr |
| | | | | 145 | 144.912 568 | 8.30 | |
| | | | | 146 | 145.913 113 | 17.19 | |
| | | | | 148 | 147.916 888 | 5.76 | |
| | | | | 150* | 149.920 887 | 5.64 | $>1 \times 10^{18}$ yr |
| 61 | Promethium | Pm | | 143* | 142.910 928 | | 265 days |
| | | | | 145* | 144.912 745 | | 17.7 yr |
| | | | | 146* | 145.914 698 | | 5.5 yr |
| | | | | 147* | 146.915 134 | | 2.623 yr |
| 62 | Samarium | Sm | 150.36 | 144 | 143.911 996 | 3.1 | |
| | | | | 146* | 145.913 043 | | $1.0 \times 10^{8}$ yr |
| | | | | 147* | 146.914 894 | 15.0 | $1.06 \times 10^{11}$ yr |
| | | | | 148* | 147.914 819 | 11.3 | $7 \times 10^{15}$ yr |
| | | | | 149* | 148.917 180 | 13.8 | $>2 \times 10^{15}$ yr |
| | | | | 150 | 149.917 273 | 7.4 | |
| | | | | 151* | 150.919 928 | | 90 yr |
| | | | | 152 | 151.919 728 | 26.7 | |
| | | | | 154 | 153.922 206 | 22.7 | |
| 63 | Europium | Eu | 151.96 | 151 | 150.919 846 | 47.8 | |
| | | | | 152* | 151.921 740 | | 13.5 yr |
| | | | | 153 | 152.921 226 | 52.2 | |
| | | | | 154* | 153.922 975 | | 8.59 yr |
| | | | | 155* | 154.922 888 | | 4.7 yr |
| 64 | Gadolinium | Gd | 157.25 | 148* | 147.918 112 | | 75 yr |
| | | | | 150* | 149.918 657 | | $1.8 \times 10^{6}$ yr |
| | | | | 152* | 151.919 787 | 0.20 | $1.1 \times 10^{14}$ yr |
| | | | | 154 | 153.920 862 | 2.18 | |
| | | | | 155 | 154.922 618 | 14.80 | |
| | | | | 156 | 155.922 119 | 20.47 | |
| | | | | 157 | 156.923 957 | 15.65 | |
| | | | | 158 | 157.924 099 | 24.84 | |
| | | | | 160 | 159.927 050 | 21.86 | |
| 65 | Terbium | Tb | 158.925 3 | 159 | 158.925 345 | 100 | |
| 66 | Dysprosium | Dy | 162.50 | 156 | 155.924 277 | 0.06 | |
| | | | | 158 | 157.924 403 | 0.10 | |
| | | | | 160 | 159.925 193 | 2.34 | |
| | | | | 161 | 160.926 930 | 18.9 | |
| | | | | 162 | 161.926 796 | 25.5 | |
| | | | | 163 | 162.928 729 | 24.9 | |
| | | | | 164 | 163.929 172 | 28.2 | |
| 67 | Holmium | Ho | 164.930 3 | 165 | 164.930 316 | 100 | |
| | | | | 166* | 165.932 282 | | $1.2 \times 10^{3}$ yr |
| 68 | Erbium | Er | 167.26 | 162 | 161.928 775 | 0.14 | |
| | | | | 164 | 163.929 198 | 1.61 | |
| | | | | 166 | 165.930 292 | 33.6 | |

*continued*

| TABLE A.3 | *Continued* | | | | | | |

| Atomic Number Z | Element | Symbol | Chemical Atomic Mass (u) | Mass Number (*Indicates Radioactive) A | Atomic Mass (u) | Percent Abundance | Half-Life (If Radioactive) $T_{1/2}$ |
|---|---|---|---|---|---|---|---|
| (68) | (Erbium) | | | 167 | 166.932 047 | 22.95 | |
| | | | | 168 | 167.932 369 | 27.8 | |
| | | | | 170 | 169.935 462 | 14.9 | |
| 69 | Thulium | Tm | 168.934 2 | 169 | 168.934 213 | 100 | |
| | | | | 171* | 170.936 428 | | 1.92 yr |
| 70 | Ytterbium | Yb | 173.04 | 168 | 167.933 897 | 0.13 | |
| | | | | 170 | 169.934 761 | 3.05 | |
| | | | | 171 | 170.936 324 | 14.3 | |
| | | | | 172 | 171.936 380 | 21.9 | |
| | | | | 173 | 172.938 209 | 16.12 | |
| | | | | 174 | 173.938 861 | 31.8 | |
| | | | | 176 | 175.942 564 | 12.7 | |
| 71 | Lutecium | Lu | 174.967 | 173* | 172.938 930 | | 1.37 yr |
| | | | | 175 | 174.940 772 | 97.41 | |
| | | | | 176* | 175.942 679 | 2.59 | $3.78 \times 10^{10}$ yr |
| 72 | Hafnium | Hf | 178.49 | 174* | 173.940 042 | 0.162 | $2.0 \times 10^{15}$ yr |
| | | | | 176 | 175.941 404 | 5.206 | |
| | | | | 177 | 176.943 218 | 18.606 | |
| | | | | 178 | 177.943 697 | 27.297 | |
| | | | | 179 | 178.945 813 | 13.629 | |
| | | | | 180 | 179.946 547 | 35.100 | |
| 73 | Tantalum | Ta | 180.947 9 | 180 | 179.947 542 | 0.012 | |
| | | | | 181 | 180.947 993 | 99.988 | |
| 74 | Tungsten (Wolfram) | W | 183.85 | 180 | 179.946 702 | 0.12 | |
| | | | | 182 | 181.948 202 | 26.3 | |
| | | | | 183 | 182.950 221 | 14.28 | |
| | | | | 184 | 183.950 929 | 30.7 | |
| | | | | 186 | 185.954 358 | 28.6 | |
| 75 | Rhenium | Re | 186.207 | 185 | 184.952 951 | 37.40 | |
| | | | | 187* | 186.955 746 | 62.60 | $4.4 \times 10^{10}$ yr |
| 76 | Osmium | Os | 190.2 | 184 | 183.952 486 | 0.02 | |
| | | | | 186* | 185.953 834 | 1.58 | $2.0 \times 10^{15}$ yr |
| | | | | 187 | 186.955 744 | 1.6 | |
| | | | | 188 | 187.955 832 | 13.3 | |
| | | | | 189 | 188.958 139 | 16.1 | |
| | | | | 190 | 189.958 439 | 26.4 | |
| | | | | 192 | 191.961 468 | 41.0 | |
| | | | | 194* | 193.965 172 | | 6.0 yr |
| 77 | Iridium | Ir | 192.2 | 191 | 190.960 585 | 37.3 | |
| | | | | 193 | 192.962 916 | 62.7 | |
| 78 | Platinum | Pt | 195.08 | 190* | 189.959 926 | 0.01 | $6.5 \times 10^{11}$ yr |
| | | | | 192 | 191.961 027 | 0.79 | |
| | | | | 194 | 193.962 655 | 32.9 | |
| | | | | 195 | 194.964 765 | 33.8 | |
| | | | | 196 | 195.964 926 | 25.3 | |
| | | | | 198 | 197.967 867 | 7.2 | |
| 79 | Gold | Au | 196.966 5 | 197 | 196.966 543 | 100 | |

*continued*

**TABLE A.3** | *Continued*

| Atomic Number Z | Element | Symbol | Chemical Atomic Mass (u) | Mass Number (*Indicates Radioactive) A | Atomic Mass (u) | Percent Abundance | Half-Life (If Radioactive) $T_{1/2}$ |
|---|---|---|---|---|---|---|---|
| 80 | Mercury | Hg | 200.59 | 196 | 195.965 806 | 0.15 | |
| | | | | 198 | 197.966 743 | 9.97 | |
| | | | | 199 | 198.968 253 | 16.87 | |
| | | | | 200 | 199.968 299 | 23.10 | |
| | | | | 201 | 200.970 276 | 13.10 | |
| | | | | 202 | 201.970 617 | 29.86 | |
| | | | | 204 | 203.973 466 | 6.87 | |
| 81 | Thallium | Tl | 204.383 | 203 | 202.972 320 | 29.524 | |
| | | | | 204* | 203.973 839 | | 3.78 yr |
| | | | | 205 | 204.974 400 | 70.476 | |
| | | (Ra E″) | | 206* | 205.976 084 | | 4.2 min |
| | | (Ac C″) | | 207* | 206.977 403 | | 4.77 min |
| | | (Th C″) | | 208* | 207.981 992 | | 3.053 min |
| | | (Ra C″) | | 210* | 209.990 057 | | 1.30 min |
| 82 | Lead | Pb | 207.2 | 202* | 201.972 134 | | $5 \times 10^4$ yr |
| | | | | 204* | 203.973 020 | 1.4 | $\geq 1.4 \times 10^{17}$ yr |
| | | | | 205* | 204.974 457 | | $1.5 \times 10^7$ yr |
| | | | | 206 | 205.974 440 | 24.1 | |
| | | | | 207 | 206.975 871 | 22.1 | |
| | | | | 208 | 207.976 627 | 52.4 | |
| | | (Ra D) | | 210* | 209.984 163 | | 22.3 yr |
| | | (Ac B) | | 211* | 210.988 734 | | 36.1 min |
| | | (Th B) | | 212* | 211.991 872 | | 10.64 h |
| | | (Ra B) | | 214* | 213.999 798 | | 26.8 min |
| 83 | Bismuth | Bi | 208.980 3 | 207* | 206.978 444 | | 32.2 yr |
| | | | | 208* | 207.979 717 | | $3.7 \times 10^5$ yr |
| | | | | 209 | 208.980 374 | 100 | |
| | | (Ra E) | | 210* | 209.984 096 | | 5.01 days |
| | | (Th C) | | 211* | 210.987 254 | | 2.14 min |
| | | | | 212* | 211.991 259 | | 60.6 min |
| | | (Ra C) | | 214* | 213.998 692 | | 19.9 min |
| | | | | 215* | 215.001 836 | | 7.4 min |
| 84 | Polonium | Po | | 209* | 208.982 405 | | 102 yr |
| | | (Ra F) | | 210* | 209.982 848 | | 138.38 days |
| | | (Ac C′) | | 211* | 210.986 627 | | 0.52 s |
| | | (Th C′) | | 212* | 211.988 842 | | 0.30 $\mu$s |
| | | (Ra C′) | | 214* | 213.995 177 | | 164 $\mu$s |
| | | (Ac A) | | 215* | 214.999 418 | | 0.001 8 s |
| | | (Th A) | | 216* | 216.001 889 | | 0.145 s |
| | | (Ra A) | | 218* | 218.008 965 | | 3.10 min |
| 85 | Astatine | At | | 215* | 214.998 638 | | $\approx$ 100 $\mu$s |
| | | | | 218* | 218.008 685 | | 1.6 s |
| | | | | 219* | 219.011 294 | | 0.9 min |
| 86 | Radon | Rn | | | | | |
| | | (An) | | 219* | 219.009 477 | | 3.96 s |
| | | (Tn) | | 220* | 220.011 369 | | 55.6 s |
| | | (Rn) | | 222* | 222.017 571 | | 3.823 days |
| 87 | Francium | Fr | | | | | |
| | | (Ac K) | | 223* | 223.019 733 | | 22 min |

*continued*

**TABLE A.3** *Continued*

| Atomic Number Z | Element | Symbol | Chemical Atomic Mass (u) | Mass Number (*Indicates Radioactive) A | Atomic Mass (u) | Percent Abundance | Half-Life (If Radioactive) $T_{1/2}$ |
|---|---|---|---|---|---|---|---|
| 88 | Radium | Ra | | | | | |
| | | (Ac X) | | 223* | 223.018 499 | | 11.43 days |
| | | (Th X) | | 224* | 224.020 187 | | 3.66 days |
| | | (Ra) | | 226* | 226.025 402 | | 1 600 yr |
| | | (Ms Th$_1$) | | 228* | 228.031 064 | | 5.75 yr |
| 89 | Actinium | Ac | | 227* | 227.027 749 | | 21.77 yr |
| | | (Ms Th$_2$) | | 228* | 228.031 015 | | 6.15 h |
| 90 | Thorium | Th | 232.038 1 | | | | |
| | | (Rd Ac) | | 227* | 227.027 701 | | 18.72 days |
| | | (Rd Th) | | 228* | 228.028 716 | | 1.913 yr |
| | | | | 229* | 229.031 757 | | 7 300 yr |
| | | (Io) | | 230* | 230.033 127 | | 75.000 yr |
| | | (UY) | | 231* | 231.036 299 | | 25.52 h |
| | | (Th) | | 232* | 232.038 051 | 100 | $1.40 \times 10^{10}$ yr |
| | | (UX1) | | 234* | 234.043 593 | | 24.1 days |
| 91 | Protactinium | Pa | | 231* | 231.035 880 | | 32.760 yr |
| | | (Uz) | | 234* | 234.043 300 | | 6.7 h |
| 92 | Uranium | U | 238.028 9 | 232* | 232.037 131 | | 69 yr |
| | | | | 233* | 233.039 630 | | $1.59 \times 10^5$ yr |
| | | | | 234* | 234.040 946 | 0.005 5 | $2.45 \times 10^5$ yr |
| | | (Ac U) | | 235* | 235.043 924 | 0.720 | $7.04 \times 10^8$ yr |
| | | | | 236* | 236.045 562 | | $2.34 \times 10^7$ yr |
| | | (UI) | | 238* | 238.050 784 | 99.274 5 | $4.47 \times 10^9$ yr |
| 93 | Neptunium | Np | | 235* | 235.044 057 | | 396 days |
| | | | | 236* | 236.046 560 | | $1.15 \times 10^5$ yr |
| | | | | 237* | 237.048 168 | | $2.14 \times 10^6$ yr |
| 94 | Plutonium | Pu | | 236* | 236.046 033 | | 2.87 yr |
| | | | | 238* | 238.049 555 | | 87.7 yr |
| | | | | 239* | 239.052 157 | | $2.412 \times 10^4$ yr |
| | | | | 240* | 240.053 808 | | 6 560 yr |
| | | | | 241* | 241.056 846 | | 14.4 yr |
| | | | | 242* | 242.058 737 | | $3.73 \times 10^6$ yr |
| | | | | 244* | 244.064 200 | | $8.1 \times 10^7$ yr |

[a] The masses in the sixth column are atomic masses, which include the mass of *Z* electrons. Data are from the National Nuclear Data Center, Brookhaven National Laboratory, prepared by Jagdish K. Tuli, July 1990. The data are based on experimental results reported in *Nuclear Data Sheets* and *Nuclear Physics* and also from *Chart of the Nuclides,* 14th ed. Atomic masses are based on those by A. H. Wapstra, G. Audi, and R. Hoekstra. Isotopic abundances are based on those by N. E. Holden.

# Mathematics Review

**T**his appendix in mathematics is intended as a brief review of operations and methods. Early in this course, you should be totally familiar with basic algebraic techniques, analytic geometry, and trigonometry. The sections on differential and integral calculus are more detailed and are intended for those students who have difficulty applying calculus concepts to physical situations.

## B.1 • SCIENTIFIC NOTATION

Many quantities that scientists deal with often have very large or very small values. For example, the speed of light is about 300 000 000 m/s, and the ink required to make the dot over an $i$ in this textbook has a mass of about 0.000 000 001 kg. Obviously, it is very cumbersome to read, write, and keep track of numbers such as these. We avoid this problem by using a method dealing with powers of the number 10:

$$10^0 = 1$$

$$10^1 = 10$$

$$10^2 = 10 \times 10 = 100$$

$$10^3 = 10 \times 10 \times 10 = 1000$$

$$10^4 = 10 \times 10 \times 10 \times 10 = 10\ 000$$

$$10^5 = 10 \times 10 \times 10 \times 10 \times 10 = 100\ 000$$

and so on. The number of zeros corresponds to the power to which 10 is raised, called the **exponent** of 10. For example, the speed of light, 300 000 000 m/s, can be expressed as $3 \times 10^8$ m/s.

In this method, some representative numbers smaller than unity are

$$10^{-1} = \frac{1}{10} = 0.1$$

$$10^{-2} = \frac{1}{10 \times 10} = 0.01$$

$$10^{-3} = \frac{1}{10 \times 10 \times 10} = 0.001$$

$$10^{-4} = \frac{1}{10 \times 10 \times 10 \times 10} = 0.000\ 1$$

$$10^{-5} = \frac{1}{10 \times 10 \times 10 \times 10 \times 10} = 0.000\ 01$$

In these cases, the number of places the decimal point is to the left of the digit 1 equals the value of the (negative) exponent. Numbers expressed as some power of 10 multiplied by another number between 1 and 10 are said to be in **scientific notation.** For example, the scientific notation for 5 943 000 000 is $5.943 \times 10^9$ and that for 0.000 083 2 is $8.32 \times 10^{-5}$.

When numbers expressed in scientific notation are being multiplied, the following general rule is very useful:

$$10^n \times 10^m = 10^{n+m} \tag{B.1}$$

where $n$ and $m$ can be *any* numbers (not necessarily integers). For example, $10^2 \times 10^5 = 10^7$. The rule also applies if one of the exponents is negative: $10^3 \times 10^{-8} = 10^{-5}$.

When dividing numbers expressed in scientific notation, note that

$$\frac{10^n}{10^m} = 10^n \times 10^{-m} = 10^{n-m} \tag{B.2}$$

## EXERCISES

With help from the preceding rules, verify the answers to the following:

1. $86\ 400 = 8.64 \times 10^4$
2. $9\ 816\ 762.5 = 9.816\ 762\ 5 \times 10^6$
3. $0.000\ 000\ 039\ 8 = 3.98 \times 10^{-8}$
4. $(4.0 \times 10^8)(9.0 \times 10^9) = 3.6 \times 10^{18}$
5. $(3.0 \times 10^7)(6.0 \times 10^{-12}) = 1.8 \times 10^{-4}$
6. $\dfrac{75 \times 10^{-11}}{5.0 \times 10^{-3}} = 1.5 \times 10^{-7}$
7. $\dfrac{(3 \times 10^6)(8 \times 10^{-2})}{(2 \times 10^{17})(6 \times 10^5)} = 2 \times 10^{-18}$

# B.2 • ALGEBRA

## Some Basic Rules

When algebraic operations are performed, the laws of arithmetic apply. Symbols such as $x$, $y$, and $z$ are usually used to represent unspecified quantities, called the **unknowns.**

First, consider the equation

$$8x = 32$$

If we wish to solve for $x$, we can divide (or multiply) each side of the equation by the same factor without destroying the equality. In this case, if we divide both sides by 8, we have

$$\frac{8x}{8} = \frac{32}{8}$$

$$x = 4$$

Next consider the equation

$$x + 2 = 8$$

In this type of expression, we can add or subtract the same quantity from each side. If we subtract 2 from each side, we have

$$x + 2 - 2 = 8 - 2$$

$$x = 6$$

In general, if $x + a = b$, then $x = b - a$.

Now consider the equation

$$\frac{x}{5} = 9$$

If we multiply each side by 5, we are left with $x$ on the left by itself and 45 on the right:

$$\left(\frac{x}{5}\right)(5) = 9 \times 5$$

$$x = 45$$

In all cases, *whatever operation is performed on the left side of the equality must also be performed on the right side.*

The following rules for multiplying, dividing, adding, and subtracting fractions should be recalled, where $a$, $b$, and $c$ are three numbers:

| | Rule | Example |
|---|---|---|
| **Multiplying** | $\left(\dfrac{a}{b}\right)\left(\dfrac{c}{d}\right) = \dfrac{ac}{bd}$ | $\left(\dfrac{2}{3}\right)\left(\dfrac{4}{5}\right) = \dfrac{8}{15}$ |
| **Dividing** | $\dfrac{(a/b)}{(c/d)} = \dfrac{ad}{bc}$ | $\dfrac{2/3}{4/5} = \dfrac{(2)(5)}{(4)(3)} = \dfrac{10}{12}$ |
| **Adding** | $\dfrac{a}{b} \pm \dfrac{c}{d} = \dfrac{ad \pm bc}{bd}$ | $\dfrac{2}{3} - \dfrac{4}{5} = \dfrac{(2)(5) - (4)(3)}{(3)(5)} = -\dfrac{2}{15}$ |

## EXERCISES

In the following exercises, solve for $x$:

**Answers**

1. $a = \dfrac{1}{1 + x}$      $x = \dfrac{1 - a}{a}$

2. $3x - 5 = 13$      $x = 6$

3. $ax - 5 = bx + 2$      $x = \dfrac{7}{a - b}$

4. $\dfrac{5}{2x + 6} = \dfrac{3}{4x + 8}$      $x = -\dfrac{11}{7}$

### Powers

When powers of a given quantity $x$ are multiplied, the following rule applies:

$$x^n x^m = x^{n+m} \tag{B.3}$$

For example, $x^2 x^4 = x^{2+4} = x^6$.

When dividing the powers of a given quantity, the rule is

$$\frac{x^n}{x^m} = x^{n-m} \tag{B.4}$$

For example, $x^8/x^2 = x^{8-2} = x^6$.

A power that is a fraction, such as $\frac{1}{3}$, corresponds to a root as follows:

$$x^{1/n} = \sqrt[n]{x} \tag{B.5}$$

For example, $4^{1/3} = \sqrt[3]{4} = 1.5874$. (A scientific calculator is useful for such calculations.)

Finally, any quantity $x^n$ raised to the $m$th power is

$$(x^n)^m = x^{nm} \tag{B.6}$$

Table B.1 summarizes the rules of exponents.

| TABLE B.1 |
| --- |
| **Rules of Exponents** |
| $x^0 = 1$ |
| $x^1 = x$ |
| $x^n x^m = x^{n+m}$ |
| $x^n/x^m = x^{n-m}$ |
| $x^{1/n} = \sqrt[n]{x}$ |
| $(x^n)^m = x^{nm}$ |

## EXERCISES

Verify the following:

1. $3^2 \times 3^3 = 243$
2. $x^5 x^{-8} = x^{-3}$
3. $x^{10}/x^{-5} = x^{15}$
4. $5^{1/3} = 1.709\,975$     (Use your calculator.)
5. $60^{1/4} = 2.783\,158$     (Use your calculator.)
6. $(x^4)^3 = x^{12}$

### Factoring

Some useful formulas for factoring an equation are

$$ax + ay + az = a(x + y + x) \qquad \text{common factor}$$

$$a^2 + 2ab + b^2 = (a + b)^2 \qquad \text{perfect square}$$

$$a^2 - b^2 = (a + b)(a - b) \qquad \text{differences of squares}$$

## Quadratic Equations

The general form of a quadratic equation is

$$ax^2 + bx + c = 0 \tag{B.7}$$

where $x$ is the unknown quantity and $a$, $b$, and $c$ are numerical factors referred to as **coefficients** of the equation. This equation has two roots, given by

$$x = \frac{-b \pm \sqrt{b^2 - 4ac}}{2a} \tag{B.8}$$

If $b^2 \geq 4ac$, the roots are real.

---

**Example 1** The equation $x^2 + 5x + 4 = 0$ has the following roots corresponding to the two signs of the square-root term:

$$x = \frac{-5 \pm \sqrt{5^2 - (4)(1)(4)}}{2(1)} = \frac{-5 \pm \sqrt{9}}{2} = \frac{-5 \pm 3}{2}$$

$$x_+ = \frac{-5 + 3}{2} = -1 \qquad x_- = \frac{-5 - 3}{2} = -4$$

where $x_+$ refers to the root corresponding to the positive sign and $x_-$ refers to the root corresponding to the negative sign.

---

## EXERCISES

Solve the following quadratic equations:

**Answers**

1. $x^2 + 2x - 3 = 0$     $x_+ = 1$             $x_- = -3$
2. $2x^2 - 5x + 2 = 0$     $x_+ = 2$             $x_- = \frac{1}{2}$
3. $2x^2 - 4x - 9 = 0$     $x_+ = 1 + \sqrt{22}/2$     $x_- = 1 - \sqrt{22}/2$

## Linear Equations

A linear equation has the general form

$$y = mx + b \tag{B.9}$$

where $m$ and $b$ are constants. This equation is referred to as being linear because the graph of $y$ versus $x$ is a straight line, as shown in Figure B.1. The constant $b$, called the **y-intercept**, represents the value of $y$ at which the straight line intersects the $y$ axis. The constant $m$ is equal to the **slope** of the straight line and is also equal to the tangent of the angle that the line makes with the $x$ axis. If any two points on

**Figure B.1**

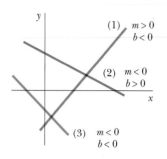

(1)  $m > 0$
     $b < 0$

(2)  $m < 0$
     $b > 0$

(3)  $m < 0$
     $b < 0$

**Figure B.2**

the straight line are specified by the coordinates $(x_1, y_1)$ and $(x_2, y_2)$, as in Figure B.1, then the slope of the straight line can be expressed as

$$\text{Slope} = \frac{y_2 - y_1}{x_2 - x_1} = \frac{\Delta y}{\Delta x} \tag{B.10}$$

Note that $m$ and $b$ can have either positive or negative values. If $m > 0$, the straight line has a *positive* slope, as in Figure B.1. If $m < 0$, the straight line has a *negative* slope. In Figure B.1, both $m$ and $b$ are positive. Three other possible situations are shown in Figure B.2.

## EXERCISES

1. Draw graphs of the following straight lines:
   (a) $y = 5x + 3$   (b) $y = -2x + 4$   (c) $y = -3x - 6$
2. Find the slopes of the straight lines described in Exercise 1.

**Answers**   (a) 5   (b) $-2$   (c) $-3$

3. Find the slopes of the straight lines that pass through the following sets of points:
   (a) $(0, -4)$ and $(4, 2)$   (b) $(0, 0)$ and $(2, -5)$   (c) $(-5, 2)$ and $(4, -2)$

**Answers**   (a) $3/2$   (b) $-5/2$   (c) $-4/9$

### Solving Simultaneous Linear Equations

Consider the equation $3x + 5y = 15$, which has two unknowns, $x$ and $y$. Such an equation does not have a unique solution. For example, note that $(x = 0, y = 3)$, $(x = 5, y = 0)$, and $(x = 2, y = 9/5)$ are all solutions to this equation.

If a problem has two unknowns, a unique solution is possible only if we have *two* equations. In general, if a problem has $n$ unknowns, its solution requires $n$ equations. To solve two simultaneous equations involving two unknowns, $x$ and $y$, we solve one of the equations for $x$ in terms of $y$ and substitute this expression into the other equation.

**Example 2**  Solve the following two simultaneous equations:

(1)   $5x + y = -8$

(2)   $2x - 2y = 4$

**Solution**  From (2), $x = y + 2$. Substitution of this into (1) gives

$$5(y + 2) + y = -8$$

$$6y = -18$$

$$y = -3$$

$$x = y + 2 = \boxed{-1}$$

**Alternative Solution**  Multiply each term in (1) by the factor 2 and add the result to (2):

$$10x + 2y = -16$$

$$\underline{2x - 2y = 4}$$

$$12x = -12$$

$$x = -1$$

$$y = x - 2 = \boxed{-3}$$

Two linear equations containing two unknowns can also be solved by a graphical method. If the straight lines corresponding to the two equations are plotted in a conventional coordinate system, the intersection of the two lines represents the solution. For example, consider the two equations

$$x - y = 2$$

$$x - 2y = -1$$

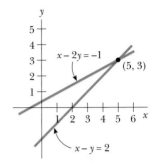

**Figure B.3**

These are plotted in Figure B.3. The intersection of the two lines has the coordinates $x = 5$, $y = 3$. This represents the solution to the equations. You should check this solution by the analytical technique discussed earlier.

## EXERCISES

Solve the following pairs of simultaneous equations involving two unknowns:

**Answers**

1. $x + y = 8$      $x = 5$, $y = 3$
   $x - y = 2$
2. $98 - T = 10a$      $T = 65$, $a = 3.27$
   $T - 49 = 5a$
3. $6x + 2y = 6$      $x = 2$, $y = -3$
   $8x - 4y = 28$

## Logarithms

Suppose that a quantity $x$ is expressed as a power of some quantity $a$:

$$x = a^y \tag{B.11}$$

The number $a$ is called the **base** number. The **logarithm** of $x$ with respect to the base $a$ is equal to the exponent to which the base must be raised to satisfy the expression $x = a^y$:

$$y = \log_a x \tag{B.12}$$

Conversely, the **antilogarithm** of $y$ is the number $x$:

$$x = \text{antilog}_a y \tag{B.13}$$

In practice, the two bases most often used are base 10, called the *common* logarithm base, and base $e = 2.718 \ldots$, called Euler's constant or the *natural* logarithm base. When common logarithms are used,

$$y = \log_{10} x \quad (\text{or } x = 10^y) \tag{B.14}$$

When natural logarithms are used,

$$y = \ln x \quad (\text{or } x = e^y) \tag{B.15}$$

For example, $\log_{10} 52 = 1.716$, so that $\text{antilog}_{10} 1.716 = 10^{1.716} = 52$. Likewise, $\ln 52 = 3.951$, so $\text{antiln } 3.951 = e^{3.951} = 52$.

In general, note that you can convert between base 10 and base $e$ with the equality

$$\ln x = (2.302\ 585) \log_{10} x \tag{B.16}$$

Finally, some useful properties of logarithms are

$$\log(ab) = \log a + \log b$$
$$\log(a/b) = \log a - \log b$$
$$\log(a^n) = n \log a$$
$$\ln e = 1$$
$$\ln e^a = a$$
$$\ln\left(\frac{1}{a}\right) = -\ln a$$

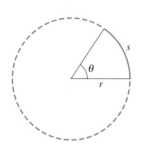

**Figure B.4**

## B.3 • GEOMETRY

The **distance** $d$ between two points having coordinates $(x_1, y_1)$ and $(x_2, y_2)$ is

$$d = \sqrt{(x_2 - x_1)^2 + (y_2 - y_1)^2} \tag{B.17}$$

**Radian measure:** The arc length $s$ of a circular arc (Fig. B.4) is proportional to the radius $r$ for a fixed value of $\theta$ (in radians):

$$s = r\theta$$
$$\theta = \frac{s}{r} \tag{B.18}$$

**Figure B.5**

The equation of a **straight line** (Fig. B.5) is

$$y = mx + b \tag{B.19}$$

where $b$ is the $y$-intercept and $m$ is the slope of the line.

The equation of a **circle** of radius $R$ centered at the origin is

$$x^2 + y^2 = R^2 \tag{B.20}$$

The equation of an **ellipse** having the origin at its center (Fig. B.6) is

$$\frac{x^2}{a^2} + \frac{y^2}{b^2} = 1 \tag{B.21}$$

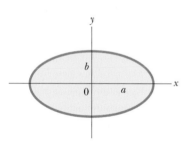

**Figure B.6**

where $a$ is the length of the semimajor axis (the longer one) and $b$ is the length of the semiminor axis (the shorter one).

The equation of a **parabola** the vertex of which is at $y = b$ (Fig. B.7) is

$$y = ax^2 + b \tag{B.22}$$

The equation of a **rectangular hyperbola** (Fig. B.8) is

**Figure B.7**

$$xy = \text{constant} \tag{B.23}$$

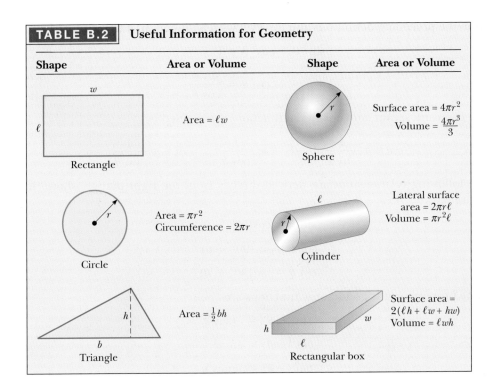

| **TABLE B.2** | **Useful Information for Geometry** |
|---|---|

Table B.2 gives the areas and volumes for several geometric shapes used throughout this text.

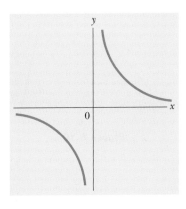

**Figure B.8**

## B.4 • TRIGONOMETRY

That portion of mathematics based on the special properties of the right triangle is called trigonometry. By definition, a right triangle is one containing a 90° angle. Consider the right triangle shown in Figure B.9, where side $a$ is opposite the angle $\theta$, side $b$ is adjacent to the angle $\theta$, and side $c$ is the hypotenuse of the triangle. The three basic trigonometric functions defined by such a triangle are the sine (sin), cosine (cos), and tangent (tan) functions. In terms of the angle $\theta$, these functions are defined by

$$\sin \theta \equiv \frac{\text{side opposite } \theta}{\text{hypotenuse}} = \frac{a}{c} \qquad \text{(B.24)}$$

$$\cos \theta \equiv \frac{\text{side adjacent to } \theta}{\text{hypotenuse}} = \frac{b}{c} \qquad \text{(B.25)}$$

$$\tan \theta \equiv \frac{\text{side opposite } \theta}{\text{side adjacent to } \theta} = \frac{a}{b} \qquad \text{(B.26)}$$

The Pythagorean theorem provides the following relationship between the sides of a right triangle:

$$c^2 = a^2 + b^2 \qquad \text{(B.27)}$$

$a$ = opposite side
$b$ = adjacent side
$c$ = hypotenuse

**Figure B.9**

From the preceding definitions and the Pythagorean theorem, it follows that

$$\sin^2 \theta + \cos^2 \theta = 1$$

$$\tan \theta = \frac{\sin \theta}{\cos \theta}$$

The cosecant, secant, and cotangent functions are defined by

$$\csc \theta \equiv \frac{1}{\sin \theta} \qquad \sec \theta \equiv \frac{1}{\cos \theta} \qquad \cot \theta \equiv \frac{1}{\tan \theta}$$

The following relationships are derived directly from the right triangle shown in Figure B.9:

$$\sin \theta = \cos(90° - \theta)$$

$$\cos \theta = \sin(90° - \theta)$$

$$\cot \theta = \tan(90° - \theta)$$

Some properties of trigonometric functions are

$$\sin(-\theta) = -\sin \theta$$

$$\cos(-\theta) = \cos \theta$$

$$\tan(-\theta) = -\tan \theta$$

The following relationships apply to *any* triangle, as shown in Figure B.10:

$$\alpha + \beta + \gamma = 180°$$

$$\begin{matrix} & a^2 = b^2 + c^2 - 2bc \cos \alpha \\ \text{Law of cosines} & b^2 = a^2 + c^2 - 2ac \cos \beta \\ & c^2 = a^2 + b^2 - 2ab \cos \gamma \end{matrix}$$

$$\text{Law of sines} \qquad \frac{a}{\sin \alpha} = \frac{b}{\sin \beta} = \frac{c}{\sin \gamma}$$

Table B.3 lists a number of useful trigonometric identities.

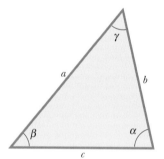

**Figure B.10**

| **TABLE B.3** | **Some Trigonometric Identities** |
|---|---|
| $\sin^2 \theta + \cos^2 \theta = 1$ | $\csc^2 \theta = 1 + \cot^2 \theta$ |
| $\sec^2 \theta = 1 + \tan^2 \theta$ | $\sin^2 \dfrac{\theta}{2} = \frac{1}{2}(1 - \cos \theta)$ |
| $\sin 2\theta = 2 \sin \theta \cos \theta$ | $\cos^2 \dfrac{\theta}{2} = \frac{1}{2}(1 + \cos \theta)$ |
| $\cos 2\theta = \cos^2 \theta - \sin^2 \theta$ | $1 - \cos \theta = 2 \sin^2 \dfrac{\theta}{2}$ |
| $\tan 2\theta = \dfrac{2 \tan \theta}{1 - \tan^2 \theta}$ | $\tan \dfrac{\theta}{2} = \sqrt{\dfrac{1 - \cos \theta}{1 + \cos \theta}}$ |

$$\sin(A \pm B) = \sin A \cos B \pm \cos A \sin B$$
$$\cos(A \pm B) = \cos A \cos B \mp \sin A \sin B$$
$$\sin A \pm \sin B = 2 \sin[\tfrac{1}{2}(A \pm B)]\cos[\tfrac{1}{2}(A \mp B)]$$
$$\cos A + \cos B = 2 \cos[\tfrac{1}{2}(A + B)]\cos[\tfrac{1}{2}(A - B)]$$
$$\cos A - \cos B = 2 \sin[\tfrac{1}{2}(A + B)]\sin[\tfrac{1}{2}(B - A)]$$

• **Example 3** Consider the right triangle in Figure B.11, in which $a = 2$, $b = 5$, and $c$ is unknown. From the Pythagorean theorem, we have

$$c^2 = a^2 + b^2 = 2^2 + 5^2 = 4 + 25 = 29$$

$$c = \sqrt{29} = \boxed{5.39}$$

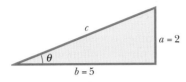

**Figure B.11**

To find the angle $\theta$, note that

$$\tan \theta = \frac{a}{b} = \frac{2}{5} = 0.400$$

From a table of functions or from a calculator, we have

$$\theta = \tan^{-1}(0.400) = \boxed{21.8°}$$

where $\tan^{-1}(0.400)$ is the notation for "angle whose tangent is 0.400," sometimes written as arctan (0.400).

## EXERCISES

1. In Figure B.12, identify (a) the side opposite $\theta$ and (b) the side adjacent to $\phi$ and then find (c) cos $\theta$, (d) sin $\phi$, and (e) tan $\phi$.

**Answers** (a) 3 (b) 3 (c) $\frac{4}{5}$ (d) $\frac{4}{5}$ (e) $\frac{4}{3}$

2. In a certain right triangle, the two sides that are perpendicular to each other are 5m and 7m long. What is the length of the third side?

**Answer** 8.60 m

3. A right triangle has a hypotenuse of length 3 m, and one of its angles is 30°. What is the length of (a) the side opposite the 30° angle and (b) the side adjacent to the 30° angle?

**Answers** (a) 1.5 m (b) 2.60 m

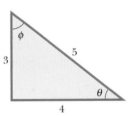

**Figure B.12**

## B.5 • SERIES EXPANSIONS

$$(a + b)^n = a^n + \frac{n}{1!} a^{n-1}b + \frac{n(n-1)}{2!} a^{n-2}b^2 + \cdots$$

$$(1 + x)^n = 1 + nx + \frac{n(n-1)}{2!} x^2 + \cdots$$

$$e^x = 1 + x + \frac{x^2}{2!} + \frac{x^3}{3!} + \cdots$$

$$\ln(1 \pm x) = \pm x - \tfrac{1}{2}x^2 \pm \tfrac{1}{3}x^3 - \cdots$$

$$\left. \begin{array}{l} \sin x = x - \dfrac{x^3}{3!} + \dfrac{x^5}{5!} - \cdots \\[2ex] \cos x = 1 - \dfrac{x^2}{2!} + \dfrac{x^4}{4!} - \cdots \\[2ex] \tan x = x + \dfrac{x^3}{3} + \dfrac{2x^5}{15} + \cdots \quad |x| < \pi/2 \end{array} \right\} \begin{array}{l} x \text{ in radians} \end{array}$$

For $x \ll 1$, the following approximations can be used*:

$$(1 + x)^n \approx 1 + nx \qquad \sin x \approx x$$

$$e^x \approx 1 + x \qquad \cos x \approx 1$$

$$\ln(1 \pm x) \approx \pm x \qquad \tan x \approx x$$

## B.6 • DIFFERENTIAL CALCULUS

In various branches of science, it is sometimes necessary to use the basic tools of calculus, invented by Newton, to describe physical phenomena. The use of calculus is fundamental in the treatment of various problems in Newtonian mechanics, electricity, and magnetism. In this section, we simply state some basic properties and "rules of thumb" that should be a useful review to the student.

First, a **function** must be specified that relates one variable to another (e.g., a coordinate as a function of time). Suppose one of the variables is called $y$ (the dependent variable), the other $x$ (the independent variable). We might have a function relationship such as

$$y(x) = ax^3 + bx^2 + cx + d$$

If $a$, $b$, $c$, and $d$ are specified constants, then $y$ can be calculated for any value of $x$. We usually deal with continuous functions, that is, those for which $y$ varies "smoothly" with $x$.

The **derivative** of $y$ with respect to $x$ is defined as the limit, as $\Delta x$ approaches zero, of the slopes of chords drawn between two points on the $y$ versus $x$ curve. Mathematically, we write this definition as

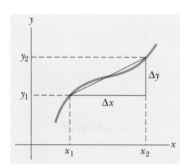

**Figure B.13**

$$\frac{dy}{dx} = \lim_{\Delta x \to 0} \frac{\Delta y}{\Delta x} = \lim_{\Delta x \to 0} \frac{y(x + \Delta x) - y(x)}{\Delta x} \qquad \textbf{(B.28)}$$

where $\Delta y$ and $\Delta x$ are defined as $\Delta x = x_2 - x_1$ and $\Delta y = y_2 - y_1$ (Fig. B.13). It is important to note that $dy/dx$ *does not* mean $dy$ divided by $dx$, but is simply a notation of the limiting process of the derivative as defined by Equation B.28.

A useful expression to remember when $y(x) = ax^n$, where $a$ is a *constant* and $n$ is *any* positive or negative number (integer or fraction), is

$$\frac{dy}{dx} = nax^{n-1} \qquad \textbf{(B.29)}$$

If $y(x)$ is a polynomial or algebraic function of $x$, we apply Equation B.29 to *each* term in the polynomial and take $d[\text{constant}]/dx = 0$. In Examples 4 through 7, we evaluate the derivatives of several functions.

---

**Example 4**  Suppose $y(x)$ (that is, $y$ as a function of $x$) is given by

$$y(x) = ax^3 + bx + c$$

where $a$ and $b$ are constants. Then it follows that

$$y(x + \Delta x) = a(x + \Delta x)^3 + b(x + \Delta x) + c$$

$$y(x + \Delta x) = a(x^3 + 3x^2\Delta x + 3x\Delta x^2 + \Delta x^3) + b(x + \Delta x) + c$$

so

$$\Delta y = y(x + \Delta x) - y(x) = a(3x^2\Delta x + 3x\Delta x^2 + \Delta x^3) + b\Delta x$$

Substituting this into Equation B.28 gives

$$\frac{dy}{dx} = \lim_{\Delta x \to 0} \frac{\Delta y}{\Delta x} = \lim_{\Delta x \to 0} a[3x^2 + 3x\Delta x + \Delta x^2] + b$$

$$\frac{dy}{dx} = 3ax^2 + b$$

---

* The approximations for the functions $\sin x$, $\cos x$, and $\tan x$ are for $x \leq 0.1$ rad.

**Example 5**  Find the derivative of
$$y(x) = 8x^5 + 4x^3 + 2x + 7$$

$$\frac{dy}{dx} = 8(5)x^4 + 4(3)x^2 + 2(1)x^0 + 0$$

**Solution**  Applying Equation B.29 to each term independently, and remembering that $d/dx$ (constant) $= 0$, we have

$$\frac{dy}{dx} = 40x^4 + 12x^2 + 2$$

## Special Properties of the Derivative

**A. Derivative of the product of two functions**  If a function $f(x)$ is given by the product of two functions, say, $g(x)$ and $h(x)$, then the derivative of $f(x)$ is defined as

$$\frac{d}{dx}f(x) = \frac{d}{dx}[g(x)h(x)] = g\frac{dh}{dx} + h\frac{dg}{dx} \qquad \textbf{(B.30)}$$

**B. Derivative of the sum of two functions**  If a function $f(x)$ is equal to the sum of two functions, then the derivative of the sum is equal to the sum of the derivatives:

$$\frac{d}{dx}f(x) = \frac{d}{dx}[g(x) + h(x)] = \frac{dg}{dx} + \frac{dh}{dx} \qquad \textbf{(B.31)}$$

**C. Chain rule of differential calculus**  If $y = f(x)$ and $x = g(z)$, then $dy/dz$ can be written as the product of two derivatives:

$$\frac{dy}{dz} = \frac{dy}{dx}\frac{dx}{dz} \qquad \textbf{(B.32)}$$

**D. The second derivative**  The second derivative of $y$ with respect to $x$ is defined as the derivative of the function $dy/dx$ (the derivative of the derivative). It is usually written

$$\frac{d^2y}{dx^2} = \frac{d}{dx}\left(\frac{dy}{dx}\right) \qquad \textbf{(B.33)}$$

**Example 6**  Find the derivative of $y(x) = x^3/(x + 1)^2$ with respect to $x$.

**Solution**  We can rewrite this function as $y(x) = x^3(x + 1)^{-2}$ and apply Equation B.30:

$$\frac{dy}{dx} = (x + 1)^{-2}\frac{d}{dx}(x^3) + x^3\frac{d}{dx}(x + 1)^{-2}$$

$$= (x + 1)^{-2}3x^2 + x^3(-2)(x + 1)^{-3}$$

$$\frac{dy}{dx} = \frac{3x^2}{(x + 1)^2} - \frac{2x^3}{(x + 1)^3}$$

**Example 7**  A useful formula that follows from Equation B.30 is the derivative of the quotient of two functions. Show that

$$\frac{d}{dx}\left[\frac{g(x)}{h(x)}\right] = \frac{h\dfrac{dg}{dx} - g\dfrac{dh}{dx}}{h^2}$$

**Solution**  We can write the quotient as $gh^{-1}$ and then apply Equations B.29 and B.30:

$$\frac{d}{dx}\left(\frac{g}{h}\right) = \frac{d}{dx}(gh^{-1}) = g\frac{d}{dx}(h^{-1}) + h^{-1}\frac{d}{dx}(g)$$

$$= -gh^{-2}\frac{dh}{dx} + h^{-1}\frac{dg}{dx}$$

$$= \frac{h\dfrac{dg}{dx} - g\dfrac{dh}{dx}}{h^2}$$

### TABLE B.4

**Derivatives for Several Functions**

$$\frac{d}{dx}(a) = 0$$

$$\frac{d}{dx}(ax^n) = nax^{n-1}$$

$$\frac{d}{dx}(e^{ax}) = ae^{ax}$$

$$\frac{d}{dx}(\sin ax) = a\cos ax$$

$$\frac{d}{dx}(\cos ax) = -a\sin ax$$

$$\frac{d}{dx}(\tan ax) = a\sec^2 ax$$

$$\frac{d}{dx}(\cot ax) = -a\csc^2 ax$$

$$\frac{d}{dx}(\sec x) = \tan x \sec x$$

$$\frac{d}{dx}(\csc x) = -\cot x \csc x$$

$$\frac{d}{dx}(\ln ax) = \frac{1}{x}$$

*Note:* The letters *a* and *n* are constants.

Some of the more commonly used derivatives of functions are listed in Table B.4.

## B.7 • INTEGRAL CALCULUS

We think of integration as the inverse of differentiation. As an example, consider the expression

$$f(x) = \frac{dy}{dx} = 3ax^2 + b \tag{B.34}$$

which was the result of differentiating the function

$$y(x) = ax^3 + bx + c$$

in Example 4. We can write Equation B.34 as $dy = f(x)\,dx = (3ax^2 + b)\,dx$ and obtain $y(x)$ by "summing" over all values of $x$. Mathematically, we write this inverse operation

$$y(x) = \int f(x)\,dx$$

For the function $f(x)$ given by Equation B.34, we have

$$y(x) = \int (3ax^2 + b)\,dx = ax^3 + bx + c$$

where $c$ is a constant of the integration. This type of integral is called an *indefinite integral* because its value depends on the choice of $c$.

A general **indefinite integral** $I(x)$ is defined as

$$I(x) = \int f(x)\,dx \tag{B.35}$$

where $f(x)$ is called the *integrand* and $f(x) = dI(x)/dx$.

For a *general continuous* function $f(x)$, the integral can be described as the area under the curve bounded by $f(x)$ and the $x$ axis, between two specified values of $x$, say, $x_1$ and $x_2$, as in Figure B.14.

The area of the blue element is approximately $f(x_i)\Delta x_i$. If we sum all these area elements from $x_1$ and $x_2$ and take the limit of this sum as $\Delta x_i \to 0$, we obtain

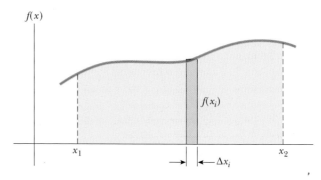

**Figure B.14**

the *true* area under the curve bounded by $f(x)$ and $x$, between the limits $x_1$ and $x_2$:

$$\text{Area} = \lim_{\Delta x_i \to 0} \sum_i f(x_i)\,\Delta x_i = \int_{x_1}^{x_2} f(x)\,dx \qquad \textbf{(B.36)}$$

Integrals of the type defined by Equation B.36 are called **definite integrals.**

One common integral that arises in practical situations has the form

$$\int x^n\,dx = \frac{x^{n+1}}{n+1} + c \quad (n \neq -1) \qquad \textbf{(B.37)}$$

This result is obvious, being that differentiation of the right-hand side with respect to $x$ gives $f(x) = x^n$ directly. If the limits of the integration are known, this integral becomes a *definite integral* and is written

$$\int_{x_1}^{x_2} x^n\,dx = \frac{x_2{}^{n+1} - x_1{}^{n+1}}{n+1} \quad (n \neq -1) \qquad \textbf{(B.38)}$$

### EXAMPLES

1. $\displaystyle\int_0^a x^2\,dx = \frac{x^3}{3}\bigg]_0^a = \frac{a^3}{3}$

2. $\displaystyle\int_0^b x^{3/2}\,dx = \frac{x^{5/2}}{5/2}\bigg]_0^b = \tfrac{2}{5}b^{5/2}$

3. $\displaystyle\int_3^5 x\,dx = \frac{x^2}{2}\bigg]_3^5 = \frac{5^2 - 3^2}{2} = 8$

### Partial Integration

Sometimes it is useful to apply the method of *partial integration* (also called "integrating by parts") to evaluate certain integrals. The method uses the property that

$$\int u\,dv = uv - \int v\,du \qquad \textbf{(B.39)}$$

where $u$ and $v$ are *carefully* chosen so as to reduce a complex integral to a simpler one. In many cases, several reductions have to be made. Consider the function

$$I(x) = \int x^2 e^x\,dx$$

This can be evaluated by integrating by parts twice. First, if we choose $u = x^2$, $v = e^x$, we get

$$\int x^2 e^x\,dx = \int x^2\,d(e^x) = x^2 e^x - 2\int e^x x\,dx + c_1$$

Now, in the second term, choose $u = x$, $v = e^x$, which gives

$$\int x^2 e^x\,dx = x^2 e^x - 2x e^x + 2\int e^x\,dx + c_1$$

or

$$\int x^2 e^x\,dx = x^2 e^x - 2x e^x + 2e^x + c_2$$

## The Perfect Differential

Another useful method to remember is the use of the *perfect differential*, in which we look for a change of variable such that the differential of the function is the differential of the independent variable appearing in the integrand. For example, consider the integral

$$I(x) = \int \cos^2 x \sin x \, dx$$

---

| **TABLE B.5** | **Some Indefinite Integrals (An arbitrary constant should be added to each of these integrals.)** |
|---|---|

$$\int x^n \, dx = \frac{x^{n+1}}{n+1} \quad \text{(provided } n \neq -1\text{)}$$

$$\int \frac{dx}{x} = \int x^{-1} \, dx = \ln x$$

$$\int \frac{dx}{a+bx} = \frac{1}{b} \ln(a+bx)$$

$$\int \frac{x \, dx}{a+bx} = \frac{x}{b} - \frac{a}{b^2} \ln(a+bx)$$

$$\int \frac{dx}{x(x+a)} = -\frac{1}{a} \ln \frac{x+a}{x}$$

$$\int \frac{dx}{(a+bx)^2} = -\frac{1}{b(a+bx)}$$

$$\int \frac{dx}{a^2+x^2} = \frac{1}{a} \tan^{-1} \frac{x}{a}$$

$$\int \frac{dx}{a^2-x^2} = \frac{1}{2a} \ln \frac{a+x}{a-x} \quad (a^2-x^2 > 0)$$

$$\int \frac{dx}{x^2-a^2} = \frac{1}{2a} \ln \frac{x-a}{x+a} \quad (x^2-a^2 > 0)$$

$$\int \frac{x \, dx}{a^2 \pm x^2} = \pm \frac{1}{2} \ln(a^2 \pm x^2)$$

$$\int \frac{dx}{\sqrt{a^2-x^2}} = \sin^{-1} \frac{x}{a} = -\cos^{-1} \frac{x}{a} \quad (a^2-x^2 > 0)$$

$$\int \frac{dx}{\sqrt{x^2 \pm a^2}} = \ln(x + \sqrt{x^2 \pm a^2})$$

$$\int \frac{x \, dx}{\sqrt{a^2-x^2}} = -\sqrt{a^2-x^2}$$

$$\int \frac{x \, dx}{\sqrt{x^2 \pm a^2}} = \sqrt{x^2 \pm a^2}$$

$$\int \sqrt{a^2-x^2} \, dx = \frac{1}{2}\left(x\sqrt{a^2-x^2} + a^2 \sin^{-1} \frac{x}{a}\right)$$

$$\int x\sqrt{a^2-x^2} \, dx = -\frac{1}{3}(a^2-x^2)^{3/2}$$

$$\int \sqrt{x^2 \pm a^2} \, dx = \frac{1}{2}[x\sqrt{x^2 \pm a^2} \pm a^2 \ln(x+\sqrt{x^2 \pm a^2})]$$

$$\int x(\sqrt{x^2 \pm a^2}) \, dx = \frac{1}{3}(x^2 \pm a^2)^{3/2}$$

$$\int e^{ax} \, dx = \frac{1}{a} e^{ax}$$

$$\int \ln ax \, dx = (x \ln ax) - x$$

$$\int xe^{ax} \, dx = \frac{e^{ax}}{a^2}(ax-1)$$

$$\int \frac{dx}{a+be^{cx}} = \frac{x}{a} - \frac{1}{ac} \ln(a+be^{cx})$$

$$\int \sin ax \, dx = -\frac{1}{a} \cos ax$$

$$\int \cos ax \, dx = \frac{1}{a} \sin ax$$

$$\int \tan ax \, dx = -\frac{1}{a} \ln(\cos ax) = \frac{1}{a} \ln(\sec ax)$$

$$\int \cot ax \, dx = \frac{1}{a} \ln(\sin ax)$$

$$\int \sec ax \, dx = \frac{1}{a} \ln(\sec ax + \tan ax) = \frac{1}{a} \ln\left[\tan\left(\frac{ax}{2} + \frac{\pi}{4}\right)\right]$$

$$\int \csc ax \, dx = \frac{1}{a} \ln(\csc ax - \cot ax) = \frac{1}{a} \ln\left(\tan \frac{ax}{2}\right)$$

$$\int \sin^2 ax \, dx = \frac{x}{2} - \frac{\sin 2ax}{4a}$$

$$\int \cos^2 ax \, dx = \frac{x}{2} + \frac{\sin 2ax}{4a}$$

$$\int \frac{dx}{\sin^2 ax} = -\frac{1}{a} \cot ax$$

$$\int \frac{dx}{\cos^2 ax} = \frac{1}{a} \tan ax$$

$$\int \tan^2 ax \, dx = \frac{1}{a}(\tan ax) - x$$

$$\int \cot^2 ax \, dx = -\frac{1}{a}(\cot ax) - x$$

$$\int \sin^{-1} ax \, dx = x(\sin^{-1} ax) + \frac{\sqrt{1-a^2x^2}}{a}$$

$$\int \cos^{-1} ax \, dx = x(\cos^{-1} ax) - \frac{\sqrt{1-a^2x^2}}{a}$$

$$\int \frac{dx}{(x^2+a^2)^{3/2}} = \frac{x}{a^2\sqrt{x^2+a^2}}$$

$$\int \frac{x \, dx}{(x^2+a^2)^{3/2}} = -\frac{1}{\sqrt{x^2+a^2}}$$

| TABLE B.6 | Gauss's Probability Integral and Other Definite Integrals |
|---|---|

$$\int_0^\infty x^n e^{-ax}\, dx = \frac{n!}{a^{n+1}}$$

$$I_0 = \int_0^\infty e^{-ax^2}\, dx = \frac{1}{2}\sqrt{\frac{\pi}{a}} \qquad \text{(Gauss's probability integral)}$$

$$I_1 = \int_0^\infty x e^{-ax^2}\, dx = \frac{1}{2a}$$

$$I_2 = \int_0^\infty x^2 e^{-ax^2}\, dx = -\frac{dI_0}{da} = \frac{1}{4}\sqrt{\frac{\pi}{a^3}}$$

$$I_3 = \int_0^\infty x^3 e^{-ax^2}\, dx = -\frac{dI_1}{da} = \frac{1}{2a^2}$$

$$I_4 = \int_0^\infty x^4 e^{-ax^2}\, dx = \frac{d^2 I_0}{da^2} = \frac{3}{8}\sqrt{\frac{\pi}{a^5}}$$

$$I_5 = \int_0^\infty x^5 e^{-ax^2}\, dx = \frac{d^2 I_1}{da^2} = \frac{1}{a^3}$$

$$\vdots$$

$$I_{2n} = (-1)^n \frac{d^n}{da^n} I_0$$

$$I_{2n+1} = (-1)^n \frac{d^n}{da^n} I_1$$

This becomes easy to evaluate if we rewrite the differential as $d(\cos x) = -\sin x\, dx$. The integral then becomes

$$\int \cos^2 x \sin x\, dx = -\int \cos^2 x\, d(\cos x)$$

If we now change variables, letting $y = \cos x$, we obtain

$$\int \cos^2 x \sin x\, dx = -\int y^2 dy = -\frac{y^3}{3} + c = -\frac{\cos^3 x}{3} + c$$

Table B.5 lists some useful indefinite integrals. Table B.6 gives Gauss's probability integral and other definite integrals. A more complete list can be found in various handbooks, such as *The Handbook of Chemistry and Physics*, CRC Press.

# Periodic Table of the Elements

| Group I | Group II | Transition elements | | | | | | |
|---|---|---|---|---|---|---|---|---|
| **H** 1 <br> 1.008 0 <br> $1s^1$ | | | | | | | | |
| **Li** 3 <br> 6.94 <br> $2s^1$ | **Be** 4 <br> 9.012 <br> $2s^2$ | | | | | | | |
| **Na** 11 <br> 22.99 <br> $3s^1$ | **Mg** 12 <br> 24.31 <br> $3s^2$ | | | | | | | |
| **K** 19 <br> 39.102 <br> $4s^1$ | **Ca** 20 <br> 40.08 <br> $4s^2$ | **Sc** 21 <br> 44.96 <br> $3d^14s^2$ | **Ti** 22 <br> 47.90 <br> $3d^24s^2$ | **V** 23 <br> 50.94 <br> $3d^34s^2$ | **Cr** 24 <br> 51.996 <br> $3d^54s^1$ | **Mn** 25 <br> 54.94 <br> $3d^54s^2$ | **Fe** 26 <br> 55.85 <br> $3d^64s^2$ | **Co** 27 <br> 58.93 <br> $3d^74s^2$ |
| **Rb** 37 <br> 85.47 <br> $5s^1$ | **Sr** 38 <br> 87.62 <br> $5s^2$ | **Y** 39 <br> 88.906 <br> $4d^15s^2$ | **Zr** 40 <br> 91.22 <br> $4d^25s^2$ | **Nb** 41 <br> 92.91 <br> $4d^45s^1$ | **Mo** 42 <br> 95.94 <br> $4d^55s^1$ | **Tc** 43 <br> (99) <br> $4d^55s^2$ | **Ru** 44 <br> 101.1 <br> $4d^75s^1$ | **Rh** 45 <br> 102.91 <br> $4d^85s^1$ |
| **Cs** 55 <br> 132.91 <br> $6s^1$ | **Ba** 56 <br> 137.34 <br> $6s^2$ | 57-71* | **Hf** 72 <br> 178.49 <br> $5d^26s^2$ | **Ta** 73 <br> 180.95 <br> $5d^36s^2$ | **W** 74 <br> 183.85 <br> $5d^46s^2$ | **Re** 75 <br> 186.2 <br> $5d^56s^2$ | **Os** 76 <br> 190.2 <br> $5d^66s^2$ | **Ir** 77 <br> 192.2 <br> $5d^76s^2$ |
| **Fr** 87 <br> (223) <br> $7s^1$ | **Ra** 88 <br> (226) <br> $7s^2$ | 89-103** | **Rf** 104 <br> (261) <br> $6d^27s^2$ | **Db** 105 <br> (262) <br> $6d^37s^2$ | **Sg** 106 <br> (263) | **Bh** 107 <br> (262) | **Hs** 108 <br> (265) | **Mt** 109 <br> (266) |

Symbol — **Ca** 20 — Atomic number  
Atomic mass † — 40.08  
$4s^2$ — Electron configuration

*Lanthanide series

| **La** 57 <br> 138.91 <br> $5d^16s^2$ | **Ce** 58 <br> 140.12 <br> $5d^14f^16s^2$ | **Pr** 59 <br> 140.91 <br> $4f^36s^2$ | **Nd** 60 <br> 144.24 <br> $4f^46s^2$ | **Pm** 61 <br> (147) <br> $4f^56s^2$ | **Sm** 62 <br> 150.4 <br> $4f^66s^2$ |
|---|---|---|---|---|---|
| **Ac** 89 <br> (227) <br> $6d^17s^2$ | **Th** 90 <br> (232) <br> $6d^27s^2$ | **Pa** 91 <br> (231) <br> $5f^26d^17s^2$ | **U** 92 <br> (238) <br> $5f^36d^17s^2$ | **Np** 93 <br> (239) <br> $5f^46d^17s^2$ | **Pu** 94 <br> (239) <br> $5f^66d^07s^2$ |

**Actinide series

**WEB**

For a description of the atomic data, visit **physics.nist.gov/atomic**

† Atomic mass values given are averaged over isotopes in the percentages in which they exist in nature; for an unstable element, mass number of the most stable known isotope is given in parentheses.
†† Elements 110, 111, 112, and 114 have not yet been named.

|  |  | Group III | Group IV | Group V | Group VI | Group VII | Group 0 |
|---|---|---|---|---|---|---|---|
|  |  |  |  |  |  | **H** 1<br>1.008 0<br>$1s^1$ | **He** 2<br>4.002 6<br>$1s^2$ |
|  |  | **B** 5<br>10.81<br>$2p^1$ | **C** 6<br>12.011<br>$2p^2$ | **N** 7<br>14.007<br>$2p^3$ | **O** 8<br>15.999<br>$2p^4$ | **F** 9<br>18.998<br>$2p^5$ | **Ne** 10<br>20.18<br>$2p^6$ |
|  |  | **Al** 13<br>26.98<br>$3p^1$ | **Si** 14<br>28.09<br>$3p^2$ | **P** 15<br>30.97<br>$3p^3$ | **S** 16<br>32.06<br>$3p^4$ | **Cl** 17<br>35.453<br>$3p^5$ | **Ar** 18<br>39.948<br>$3p^6$ |
| **Ni** 28<br>58.71<br>$3d^84s^2$ | **Cu** 29<br>63.54<br>$3d^{10}4s^1$ | **Zn** 30<br>65.37<br>$3d^{10}4s^2$ | **Ga** 31<br>69.72<br>$4p^1$ | **Ge** 32<br>72.59<br>$4p^2$ | **As** 33<br>74.92<br>$4p^3$ | **Se** 34<br>78.96<br>$4p^4$ | **Br** 35<br>79.91<br>$4p^5$ | **Kr** 36<br>83.80<br>$4p^6$ |
| **Pd** 46<br>106.4<br>$4d^{10}$ | **Ag** 47<br>107.87<br>$4d^{10}5s^1$ | **Cd** 48<br>112.40<br>$4d^{10}5s^2$ | **In** 49<br>114.82<br>$5p^1$ | **Sn** 50<br>118.69<br>$5p^2$ | **Sb** 51<br>121.75<br>$5p^3$ | **Te** 52<br>127.60<br>$5p^4$ | **I** 53<br>126.90<br>$5p^5$ | **Xe** 54<br>131.30<br>$5p^6$ |
| **Pt** 78<br>195.09<br>$5d^96s^1$ | **Au** 79<br>196.97<br>$5d^{10}6s^1$ | **Hg** 80<br>200.59<br>$5d^{10}6s^2$ | **Tl** 81<br>204.37<br>$6p^1$ | **Pb** 82<br>207.2<br>$6p^2$ | **Bi** 83<br>208.98<br>$6p^3$ | **Po** 84<br>(210)<br>$6p^4$ | **At** 85<br>(218)<br>$6p^5$ | **Rn** 86<br>(222)<br>$6p^6$ |
| 110††<br>(269) | 111††<br>(272) | 112††<br>(277) |  | 114††<br>(289) |  |  |  |  |

| **Eu** 63<br>152.0<br>$4f^76s^2$ | **Gd** 64<br>157.25<br>$5d^14f^76s^2$ | **Tb** 65<br>158.92<br>$5d^14f^86s^2$ | **Dy** 66<br>162.50<br>$4f^{10}6s^2$ | **Ho** 67<br>164.93<br>$4f^{11}6s^2$ | **Er** 68<br>167.26<br>$4f^{12}6s^2$ | **Tm** 69<br>168.93<br>$4f^{13}6s^2$ | **Yb** 70<br>173.04<br>$4f^{14}6s^2$ | **Lu** 71<br>174.97<br>$5d^14f^{14}6s^2$ |
|---|---|---|---|---|---|---|---|---|
| **Am** 95<br>(243)<br>$5f^76d^07s^2$ | **Cm** 96<br>(245)<br>$5f^76d^17s^2$ | **Bk** 97<br>(247)<br>$5f^86d^17s^2$ | **Cf** 98<br>(249)<br>$5f^{10}6d^07s^2$ | **Es** 99<br>(254)<br>$5f^{11}6d^07s^2$ | **Fm** 100<br>(253)<br>$5f^{12}6d^07s^2$ | **Md** 101<br>(255)<br>$5f^{13}6d^07s^2$ | **No** 102<br>(255)<br>$6d^07s^2$ | **Lr** 103<br>(257)<br>$6d^17s^2$ |

# Appendix
# D

# SI Units

| TABLE D.1 | SI Units | |
| --- | --- | --- |
| | **SI Base Unit** | |
| **Base Quantity** | **Name** | **Symbol** |
| Length | Meter | m |
| Mass | Kilogram | kg |
| Time | Second | s |
| Electric current | Ampere | A |
| Temperature | Kelvin | K |
| Amount of substance | Mole | mol |
| Luminous intensity | Candela | cd |

| TABLE D.2 | Some Derived SI Units | | | |
| --- | --- | --- | --- | --- |
| **Quantity** | **Name** | **Symbol** | **Expression in Terms of Base Units** | **Expression in Terms of Other SI Units** |
| Plane angle | radian | rad | $m/m$ | |
| Frequency | hertz | Hz | $s^{-1}$ | |
| Force | newton | N | $kg \cdot m/s^2$ | $J/m$ |
| Pressure | pascal | Pa | $kg/m \cdot s^2$ | $N/m^2$ |
| Energy; work | joule | J | $kg \cdot m^2/s^2$ | $N \cdot m$ |
| Power | watt | W | $kg \cdot m^2/s^3$ | $J/s$ |
| Electric charge | coulomb | C | $A \cdot s$ | |
| Electric potential | volt | V | $kg \cdot m^2/A \cdot s^3$ | $W/A$ |
| Capacitance | farad | F | $A^2 \cdot s^4/kg \cdot m^2$ | $C/V$ |
| Electric resistance | ohm | $\Omega$ | $kg \cdot m^2/A^2 \cdot s^3$ | $V/A$ |
| Magnetic flux | weber | Wb | $kg \cdot m^2/A \cdot s^2$ | $V \cdot s$ |
| Magnetic field | tesla | T | $kg/A \cdot s^2$ | |
| Inductance | henry | H | $kg \cdot m^2/A^2 \cdot s^2$ | $T \cdot m^2/A$ |

# Nobel Prizes

All Nobel Prizes in physics are listed (and marked with a P), as well as relevant Nobel Prizes in Chemistry (C). The key dates for some of the scientific work are supplied; they often antedate the prize considerably.

**1901**  (P) *Wilhelm Roentgen* for discovering x-rays (1895).

**1902**  (P) *Hendrik A. Lorentz* for predicting the Zeeman effect and *Pieter Zeeman* for discovering the Zeeman effect, the splitting of spectral lines in magnetic fields.

**1903**  (P) *Antoine-Henri Becquerel* for discovering radioactivity (1896) and *Pierre* and *Marie Curie* for studying radioactivity.

**1904**  (P) *John William Strutt, Lord Rayleigh* for studying the density of gases and discovering argon.

(C) *William Ramsay* for discovering the inert gas elements helium, neon, xenon, and krypton, and placing them in the periodic table.

**1905**  (P) *Philipp Lenard* for studying cathode rays, electrons (1898–1899).

**1906**  (P) *J. J. Thomson* for studying electrical discharge through gases and discovering the electron (1897).

**1907**  (P) *Albert A. Michelson* for inventing optical instruments and measuring the speed of light (1880s).

**1908**  (P) *Gabriel Lippmann* for making the first color photographic plate, using interference methods (1891).

(C) *Ernest Rutherford* for discovering that atoms can be broken apart by alpha rays and for studying radioactivity.

**1909**  (P) *Guglielmo Marconi* and *Carl Ferdinand Braun* for developing wireless telegraphy.

**1910**  (P) *Johannes D. van der Waals* for studying the equation of state for gases and liquids (1881).

**1911**  (P) *Wilhelm Wien* for discovering Wien's law giving the peak of a blackbody radiation spectrum (1893).

(C) *Marie Curie* for discovering radium and polonium (1898) and isolating radium.

**1912**  (P) *Nils Dalén* for inventing automatic gas regulators for lighthouses.

**1913**  (P) *Heike Kamerlingh-Onnes* for the discovery of superconductivity and liquefying helium (1908).

**1914**  (P) *Max T. F. von Laue* for studying x-rays from their diffraction by crystals, showing that x-rays are electromagnetic waves (1912).

(C) *Theodore W. Richards* for determining the atomic weights of 60 elements, indicating the existence of isotopes.

**1915** (P) *William Henry Bragg* and *William Lawrence Bragg*, his son, for studying the diffraction of x-rays in crystals.

**1917** (P) *Charles Barkla* for studying atoms by x-ray scattering (1906).

**1918** (P) *Max Planck* for discovering energy quanta (1900).

**1919** (P) *Johannes Stark*, for discovering the Stark effect, the splitting of spectral lines in electric fields (1913).

**1920** (P) *Charles-Édouard Guillaume* for discovering invar, a nickel–steel alloy with low coefficient of expansion.

(C) *Walther Nernst* for studying energy changes in chemical reactions and formulating the third law of thermodynamics (1918).

**1921** (P) *Albert Einstein* for explaining the photoelectric effect and for his services to theoretical physics (1905).

(C) *Frederick Soddy* for studying the chemistry of radioactive substances and discovering isotopes (1912).

**1922** (P) *Niels Bohr* for his model of the atom and its radiation (1913).

(C) *Francis W. Aston* for using the mass spectrograph to study atomic weights, thus discovering 212 of the 287 naturally occurring isotopes.

**1923** (P) *Robert A. Millikan* for measuring the charge on an electron (1911) and for studying the photoelectric effect experimentally (1914).

**1924** (P) *Karl M. G. Siegbahn* for his work in x-ray spectroscopy.

**1925** (P) *James Franck* and *Gustav Hertz* for discovering the Franck–Hertz effect in electron–atom collisions.

**1926** (P) *Jean-Baptiste Perrin* for studying Brownian motion to validate the discontinuous structure of matter and measure the size of atoms.

**1927** (P) *Arthur Holly Compton* for discovering the Compton effect on x-rays, their change in wavelength when they collide with matter (1922), and *Charles T. R. Wilson* for inventing the cloud chamber, used to study charged particles (1906).

**1928** (P) *Owen W. Richardson* for studying the thermionic effect and electrons emitted by hot metals (1911).

**1929** (P) *Louis-Victor de Broglie* for discovering the wave nature of electrons (1923).

**1930** (P) *Chandrasekhara Venkata Raman* for studying Raman scattering, the scattering of light by atoms and molecules with a change in wavelength (1928).

**1932** (P) *Werner Heisenberg* for creating quantum mechanics (1925).

**1933** (P) *Erwin Schrödinger* and *Paul A. M. Dirac* for developing wave mechanics (1925) and relativistic quantum mechanics (1927).

**1934** (C) *Harold Urey* for discovering heavy hydrogen, deuterium (1931).

**1935** (P) *James Chadwick* for discovering the neutron (1932).

(C) *Irène* and *Frédéric Joliot-Curie* for synthesizing new radioactive elements.

**1936** (P) *Carl D. Anderson* for discovering the positron in particular and antimatter in general (1932) and *Victor F. Hess* for discovering cosmic rays.

(C) *Peter J. W. Debye* for studying dipole moments and diffraction of x-rays and electrons in gases.

**1937** (P) *Clinton Davisson* and *George Thomson* for discovering the diffraction of electrons by crystals, confirming de Broglie's hypothesis (1927).

**1938** (P) *Enrico Fermi* for producing the transuranic radioactive elements by neutron irradiation (1934–1937).

**1939** (P) *Ernest O. Lawrence* for inventing the cyclotron.

**1943** (P) *Otto Stern* for developing molecular-beam studies (1923), and using them to discover the magnetic moment of the proton (1933).

**1944** (P) *Isidor I. Rabi* for discovering nuclear magnetic resonance in atomic and molecular beams.

(C) *Otto Hahn* for discovering nuclear fission (1938).

**1945** (P) *Wolfgang Pauli* for discovering the exclusion principle (1924).

**1946** (P) *Percy W. Bridgman* for studying physics at high pressures.

**1947** (P) *Edward V. Appleton* for studying the ionosphere.

**1948** (P) *Patrick M. S. Blackett* for studying nuclear physics with cloud-chamber photographs of cosmic-ray interactions.

**1949** (P) *Hideki Yukawa* for predicting the existence of mesons (1935).

**1950** (P) *Cecil F. Powell* for developing the method of studying cosmic rays with photographic emulsions and discovering new mesons.

**1951** (P) *John D. Cockcroft* and *Ernest T. S. Walton* for transmuting nuclei in an accelerator (1932).

(C) *Edwin M. McMillan* for producing neptunium (1940) and *Glenn T. Seaborg* for producing plutonium (1941) and further transuranic elements.

**1952** (P) *Felix Bloch* and *Edward Mills Purcell* for discovering nuclear magnetic resonance in liquids and gases (1946).

**1953** (P) *Frits Zernike* for inventing the phase-contrast microscope, which uses interference to provide high contrast.

**1954** (P) *Max Born* for interpreting the wave function as a probability (1926) and other quantum-mechanical discoveries and *Walther Bothe* for developing the coincidence method to study subatomic particles (1930–1931), producing, in particular, the particle interpreted by Chadwick as the neutron.

**1955** (P) *Willis E. Lamb, Jr.,* for discovering the Lamb shift in the hydrogen spectrum (1947) and *Polykarp Kusch* for determining the magnetic moment of the electron (1947).

**1956** (P) *John Bardeen, Walter H. Brattain,* and *William Shockley* for inventing the transistor (1956).

**1957** (P) *T.-D. Lee* and *C.-N. Yang* for predicting that parity is not conserved in beta decay (1956).

**1958** (P) *Pavel A. Cerenkov* for discovering Cerenkov radiation (1935) and *Ilya M. Frank* and *Igor Tamm* for interpreting it (1937).

**1959** (P) *Emilio G. Segrè* and *Owen Chamberlain* for discovering the antiproton (1955).

**1960** (P) *Donald A. Glaser* for inventing the bubble chamber to study elementary particles (1952).

(C) *Willard Libby* for developing radiocarbon dating (1947).

**1961** (P) *Robert Hofstadter* for discovering internal structure in protons and neutrons and *Rudolf L. Mössbauer* for discovering the Mössbauer effect of recoilless gamma-ray emission (1957).

**1962** (P) *Lev Davidovich Landau* for studying liquid helium and other condensed matter theoretically.

**1963** (P) *Eugene P. Wigner* for applying symmetry principles to elementary-particle theory and *Maria Goeppert-Mayer* and *J. Hans D. Jensen* for studying the shell model of nuclei (1947).

**1964** (P) *Charles H. Townes, Nikolai G. Basov,* and *Alexandr M. Prokhorov* for developing masers (1951–1952) and lasers.

**1965** (P) *Sin-Itiro Tomonaga, Julian S. Schwinger,* and *Richard P. Feynman* for developing quantum electrodynamics (1948).

**1966** (P) *Alfred Kastler* for his optical methods of studying atomic energy levels.

**1967** (P) *Hans Albrecht Bethe* for discovering the routes of energy production in stars (1939).

**1968** (P) *Luis W. Alvarez* for discovering resonance states of elementary particles.

**1969** (P) *Murray Gell-Mann* for classifying elementary particles (1963).

**1970** (P) *Hannes Alfvén* for developing magnetohydrodynamic theory and *Louis Eugène Félix Néel* for discovering antiferromagnetism and ferrimagnetism (1930s).

**1971** (P) *Dennis Gabor* for developing holography (1947).

(C) *Gerhard Herzberg* for studying the structure of molecules spectroscopically.

**1972** (P) *John Bardeen, Leon N. Cooper,* and *John Robert Schrieffer* for explaining superconductivity (1957).

**1973** (P) *Leo Esaki* for discovering tunneling in semiconductors, *Ivar Giaever* for discovering tunneling in superconductors, and *Brian D. Josephson* for predicting the Josephson effect, which involves tunneling of paired electrons (1958–1962).

**1974** (P) *Anthony Hewish* for discovering pulsars and *Martin Ryle* for developing radio interferometry.

**1975** (P) *Aage N. Bohr, Ben R. Mottelson,* and *James Rainwater* for discovering why some nuclei take asymmetric shapes.

**1976** (P) *Burton Richter* and *Samuel C. C. Ting* for discovering the J/psi particle, the first charmed particle (1974).

**1977** (P) *John H. Van Vleck, Nevill F. Mott,* and *Philip W. Anderson* for studying solids quantum-mechanically.

(C) *Ilya Prigogine* for extending thermodynamics to show how life could arise in the face of the second law.

**1978** (P) *Arno A. Penzias* and *Robert W. Wilson* for discovering the cosmic background radiation (1965) and *Pyotr Kapitsa* for his studies of liquid helium.

**1979** (P) *Sheldon L. Glashow, Abdus Salam,* and *Steven Weinberg* for developing the theory that unified the weak and electromagnetic forces (1958–1971).

**1980** (P) *Val Fitch* and *James W. Cronin* for discovering CP (charge-parity) violation (1964), which possibly explains the cosmological dominance of matter over antimatter.

**1981** (P) *Nicolaas Bloembergen* and *Arthur L. Schawlow* for developing laser spectroscopy and *Kai M. Siegbahn* for developing high-resolution electron spectroscopy (1958).

**1982** (P) *Kenneth G. Wilson* for developing a method of constructing theories of phase transitions to analyze critical phenomena.

**1983** (P) *William A. Fowler* for theoretical studies of astrophysical nucleosynthesis and *Subramanyan Chandrasekhar* for studying physical processes of importance to stellar structure and evolution, including the prediction of white dwarf stars (1930).

**1984** (P) *Carlo Rubbia* for discovering the W and Z particles, verifying the electroweak unification, and *Simon van der Meer,* for developing the method of stochastic cooling of the CERN beam that allowed the discovery (1982–1983).

**1985** (P) *Klaus von Klitzing* for the quantized Hall effect, relating to conductivity in the presence of a magnetic field (1980).

**1986** (P) *Ernst Ruska* for inventing the electron microscope (1931), and *Gerd Binnig* and *Heinrich Rohrer* for inventing the scanning-tunneling microscope (1981).

**1987** (P) *J. Georg Bednorz* and *Karl Alex Müller* for the discovery of high-temperature superconductivity (1986).

**1988** (P) *Leon M. Lederman, Melvin Schwartz,* and *Jack Steinberger* for a collaborative experiment that led to the development of a new tool for studying the weak nuclear force, which affects the radioactive decay of atoms.

**1989** (P) *Norman Ramsey* for various techniques in atomic physics; and *Hans Dehmelt* and *Wolfgang Paul* for the development of techniques for trapping single charge particles.

**1990** (P) *Jerome Friedman, Henry Kendall,* and *Richard Taylor* for experiments important to the development of the quark model.

**1991** (P) *Pierre-Gilles de Gennes* for discovering that methods developed for studying order phenomena in simple systems can be generalized to more complex forms of matter, in particular to liquid crystals and polymers.

**1992** (P) *Georges Charpak* for developing detectors that trace the paths of evanescent subatomic particles produced in particle accelerators.

**1993** (P) *Russell Hulse* and *Joseph Taylor* for discovering evidence of gravitational waves.

**1994** (P) *Bertram N. Brockhouse* and *Clifford G. Shull* for pioneering work in neutron scattering.

**1995** (P) *Martin L. Perl* and *Frederick Reines* for discovering the tau particle and the neutrino, respectively.

**1996** (P) *David M. Lee, Douglas C. Osheroff,* and *Robert C. Richardson* for developing a superfluid using helium-3.

**1997** (P) *Steven Chu, Claude Cohen-Tannoudji,* and *William D. Phillips* for developing methods to cool and trap atoms with laser light.

**1998** (P) *Robert B. Laughlin, Horst L. Störmer,* and *Daniel C. Tsui* for discovering a new form of quantum fluid with fractionally charged excitations.

**1999** (P) *Gerardus 'T Hooft* and *Martinus J. G. Veltman* for elucidating the quantum structure of electroweak interactions in physics.

**2000** (P) *Zhores I. Alferov* and *Herbert Kroemer* for developing semiconductor heterostructures used in high-speed electronics and opto-electronics, and *Jack S. Kilby* for his part in the invention of the integrated circuit.

# Answers to Odd-Numbered Problems

## Chapter 1

1. $2.14 \times 10^4 \, \text{kg/m}^3$
3. $23.0 \, \text{kg}$
5. $8.72 \times 10^{11} \, \text{atom/s}$
7. $2.57 \times 10^{-10} \, \text{m}$
9. No.
11. $0.579 \, t \, \text{ft}^3/\text{s} + 1.19 \times 10^{-9} \, t^2 \, \text{ft}^3/\text{s}^2$
13. (a) $0.071 \, 4 \, \text{gal/s}$  (b) $2.70 \times 10^{-4} \, \text{m}^3/\text{s}$  (c) $1.03 \, \text{h}$
15. $4.05 \times 10^3 \, \text{m}^2$
17. $151 \, \mu\text{m}$
19. $2.86 \, \text{cm}$
21. $\sim 10^6$ balls
23. $\sim 10^2 \, \text{kg}; \sim 10^3 \, \text{kg}$
25. (a) $797$  (b) $1.1$  (c) $17.66$
27. (a) $3$  (b) $4$  (c) $3$  (d) $2$
29. $5.2 \, \text{m}^3, 3\%$
31. (a) $2.24 \, \text{m}$  (b) $2.24 \, \text{m}$ at $26.6°$
33. (a) $r, 180° - \theta$  (b) $2r, 180° + \theta$  (c) $3r, -\theta$
35. (a) $10.0 \, \text{m}$  (b) $15.7 \, \text{m}$  (c) $0$
37. Approximately $420 \, \text{ft}$ at $-3°$
39. $47.2$ units at $122°$
41. $240 \, \text{m}$ at $237°$
43. (a) $49.5\mathbf{i} + 27.1\mathbf{j}$  (b) $56.4$ units at $28.7°$
45. (a) $2.00\mathbf{i} - 6.00\mathbf{j}$  (b) $4.00\mathbf{i} + 2.00\mathbf{j}$  (c) $6.32$
    (d) $4.47$  (e) $288°; 26.6°$
47. $\mathbf{A} + \mathbf{B} = (2.60\mathbf{i} + 4.50\mathbf{j}) \, \text{m}$
49. (a) $8.00\mathbf{i} + 12.0\mathbf{j} - 4.00\mathbf{k}$  (b) $2.00\mathbf{i} + 3.00\mathbf{j} - 1.00\mathbf{k}$
    (c) $-24.0\mathbf{i} - 36.0\mathbf{j} + 12.0\mathbf{k}$
51. $|\mathbf{B}| = 7.81, \alpha = 59.2°, \beta = 39.8°, \gamma = 67.4°$
53. $106°$
55. $0.141 \, \text{nm}$
57. $4.50 \, \text{m}^2$
59. $0.449\%$
61. $\sim 10^{11}$ stars
63. $2 \tan^{-1}(1/n)$
65. (a) zero  (b) zero

## Chapter 2

1. (a) $2.30 \, \text{m/s}$  (b) $16.1 \, \text{m/s}$  (c) $11.5 \, \text{m/s}$
3. (a) $5 \, \text{m/s}$  (b) $1.2 \, \text{m/s}$  (c) $-2.5 \, \text{m/s}$
   (d) $-3.3 \, \text{m/s}$  (e) $0$
5. (a) $2v_1v_2/(v_1 + v_2)$  (b) $0$
7. (a) $-2.4 \, \text{m/s}$  (b) $-3.8 \, \text{m/s}$  (c) $4.0 \, \text{s}$
9. (a) $5.0 \, \text{m/s}$  (b) $-2.5 \, \text{m/s}$  (c) $0$  (d) $5.0 \, \text{m/s}$
11. $5.00 \, \text{m}$
13. (a) $1.3 \, \text{m/s}^2$  (b) $2 \, \text{m/s}^2$ at $3 \, \text{s}$
    (c) at $t = 6 \, \text{s}$ and for $t > 10 \, \text{s}$  (d) $-1.5 \, \text{m/s}^2$ at $8 \, \text{s}$

15. (a) $2.00 \, \text{m}$  (b) $-3.00 \, \text{m/s}$  (c) $-2.00 \, \text{m/s}^2$
17. (a) $20.0 \, \text{m/s}, 5.00 \, \text{m/s}$  (b) $262 \, \text{m}$
19. (a) $4.53 \, \text{s}$  (b) $14.1 \, \text{m/s}$
21. $-16.0 \, \text{cm/s}^2$
23. (a) $12.7 \, \text{m/s}$  (b) $-2.30 \, \text{m/s}$
25. (a) $20.0 \, \text{s}$  (b) no
27. $3.10 \, \text{m/s}$
29. (a) $-202 \, \text{m/s}^2$  (b) $198 \, \text{m}$
31. (a) $35.0 \, \text{s}$  (b) $15.7 \, \text{m/s}$
33. They collide $212 \, \text{m}$ into the tunnel after $11.4 \, \text{s}$.
35. (a) $10.0 \, \text{m/s}$ up  (b) $4.68 \, \text{m/s}$ down
37. (a) $29.4 \, \text{m/s}$  (b) $44.1 \, \text{m}$
39. (a) $7.82 \, \text{m}$  (b) $0.782 \, \text{s}$
41. (a) $1.53 \, \text{s}$  (b) $11.5 \, \text{m}$  (c) $-4.60 \, \text{m/s}, -9.80 \, \text{m/s}^2$
43. $2.74 \times 10^5 \, \text{m/s}^2$, which is $2.79 \times 10^4 \, g$
45. (a) $41.0 \, \text{s}$  (b) $1.73 \, \text{km}$  (c) $-184 \, \text{m/s}$
47. (a) $5.43 \, \text{m/s}^2$ and $3.83 \, \text{m/s}^2$
    (b) $10.9 \, \text{m/s}$ and $11.5 \, \text{m/s}$  (c) Maggie by $2.62 \, \text{m}$
49. $\sim 10^3 \, \text{m/s}^2$
51. (a) $2.99 \, \text{s}$  (b) $-15.4 \, \text{m/s}$
    (c) $31.3 \, \text{m/s}$ down and $34.9 \, \text{m/s}$ down
53. (c) $v_{\text{boy}}^2/h, 0$  (d) $v_{\text{boy}}, 0$
55. $1.60 \, \text{m/s}^2$
57. $0.577 \, v$

## Chapter 3

1. (a) $4.87 \, \text{km}$ at $209°$ from east  (b) $23.3 \, \text{m/s}$
   (c) $13.5 \, \text{m/s}$ at $209°$
3. (a) $18.0 \, t\mathbf{i} + (4.00 \, t - 4.90 \, t^2)\mathbf{j}$
   (b) $18.0 \, \mathbf{i} + (4.00 - 9.80 \, t)\mathbf{j}$  (c) $-9.80 \, \mathbf{j} \, \text{m/s}^2$
   (d) $(54.0 \, \mathbf{i} - 32.1 \, \mathbf{j}) \, \text{m}$  (e) $(18.0 \, \mathbf{i} - 25.4 \, \mathbf{j}) \, \text{m/s}$
   (f) $-9.80 \, \mathbf{j} \, \text{m/s}^2$
5. (a) $(0.800 \, \mathbf{i} - 0.300 \, \mathbf{j}) \, \text{m/s}^2$  (b) $339°$
   (c) $(360 \, \mathbf{i} - 72.7 \, \mathbf{j}) \, \text{m}, -15.2°$
7. (a) $x = 0.010 \, 0 \, \text{m}, y = 2.41 \times 10^{-4} \, \text{m}$
   (b) $\mathbf{v} = (1.84 \times 10^7 \, \mathbf{i} + 8.78 \times 10^5 \, \mathbf{j}) \, \text{m/s}$
   (c) $v = 1.85 \times 10^7 \, \text{m/s}$  (d) $\theta = 2.73°$
9. (a) $3.34 \, \mathbf{i} \, \text{m/s}$  (b) $-50.9°$
11. $48.6 \, \text{m/s}$
13. (a) $22.6 \, \text{m}$  (b) $52.3 \, \text{m}$  (c) $1.18 \, \text{s}$
15. (a) The ball clears by $0.889 \, \text{m}$ while (b) descending
17. $67.8°$
19. $d \tan \theta_i - gd^2/(2v_i^2 \cos^2\theta_i)$
21. (a) $18.1 \, \text{m/s}$  (b) $1.13 \, \text{m}$  (c) $2.79 \, \text{m}$
23. $\tan^{-1}[(2gh)^{1/2}/v]$
25. $377 \, \text{m/s}^2$
27. $10.5 \, \text{m/s}, 219 \, \text{m/s}^2$ inward

**29.** 1.48 m/s$^2$ inward and 29.9° backward

**31.** (a) 13.0 m/s$^2$   (b) 5.70 m/s   (c) 7.50 m/s$^2$

**33.** 2.02 × 10$^3$ s; 21.0% longer

**35.** 153 km/h at 11.3° north of west

**37.** 15.3 m

**39.** 7.58 × 10$^3$ m/s, 5.80 × 10$^3$ s

**41.** (a) 26.6°   (b) 0.949

**43.** (b) 45° + $\phi/2$; $v_i^2 (1 - \sin \phi)/g \cos^2 \phi$

**45.** (a) 41.7 m/s   (b) 3.81 s
   (c) $(34.1 \mathbf{i} - 13.4 \mathbf{j})$ m/s; 36.6 m/s

**47.** 10.7 m/s

**49.** (a) $v_i > (gR)^{1/2}$   (b) $(\sqrt{2} - 1)R$

**51.** 7.50 m/s in the direction the ball was thrown

**53.** (a) 20.0 m/s, 5.00 s   (b) $(16.0 \mathbf{i} - 27.1 \mathbf{j})$ m/s
   (c) 6.54 s   (d) 24.6 $\mathbf{i}$ m

**55.** (a) 43.2 m   (b) $(9.66 \mathbf{i} - 25.6 \mathbf{j})$ m/s

**57.** Imagine you have a sick child and are shaking down the
   mercury in a fever thermometer. Starting with your hand
   at the level of your shoulder, move your hand down as fast
   as you can and snap it around an arc at the bottom.
   ~ 100 m/s$^2$ ~ 10 $g$

## Chapter 4

**1.** (a) 1/3   (b) 0.750 m/s$^2$

**3.** $(6.00 \mathbf{i} + 15.0 \mathbf{j})$ N; 16.2 N

**5.** (a) $(-45.0 \mathbf{i} + 15.0 \mathbf{j})$ m/s   (b) 162° from the $+x$ axis
   (c) $(-225 \mathbf{i} + 75.0 \mathbf{j})$ m   (d) $(-227, 79.0)$ m

**7.** (a) 5.00 m/s$^2$ at 36.9°   (b) 6.08 m/s$^2$ at 25.3°

**9.** 112 N

**11.** 2.38 kN

**13.** (a) 3.64 × 10$^{-18}$ N
   (b) 8.93 × 10$^{-30}$ N is 408 billion times smaller

**15.** 2.55 N for an 88.7-kg person

**17.** (a) ~ 10$^{-22}$ m/s$^2$   (b) ~ 10$^{-23}$ m

**19.** (a) 15.0 lb up   (b) 5.00 lb up   (c) 0

**23.** 8.66 N east

**25.** (a) 49.0 N   (b) 98.0 N   (c) 24.5 N

**27.** $a = F/(m_1 + m_2)$; $T = F m_1/(m_1 + m_2)$

**29.** 3.73 m

**31.** (a) 36.8 N   (b) 2.45 m/s$^2$   (c) 1.23 m

**33.** (a) $F_x > 19.6$ N   (b) $F_x \leq -78.4$ N
   (c) Figure at top of next column.

| $F_x$ (N) | $a_x$ (m/s$^2$) |
|---|---|
| − 100 | − 12.5 |
| − 78.4 | − 9.80 |
| − 50 | − 6.96 |
| 0 | − 1.96 |
| 50 | 3.04 |
| 100 | 8.04 |

**35.** (a) 706 N   (b) 814 N   (c) 706 N   (d) 648 N

**37.** (a) $(2.50 \mathbf{i} + 5.00 \mathbf{j})$ N   (b) 5.59 N

**39.** (a) Figure in middle of next column.   (b) 0.408 m/s$^2$
   (c) 83.3 N

**41.** 1.18 kN

**43.** (a) 2.20 m/s$^2$   (b) 27.4 N

**Chapter 4, Problem 33(c)**

**Chapter 4, Problem 39(a)**

**45.** 173 lb

**47.** (a) $Mg/2$, $Mg/2$, $Mg/2$, $3Mg/2$, $Mg$   (b) $Mg/2$

**49.** (a) 4.90 m/s$^2$   (b) 3.13 m/s   (c) 1.35 m
   (d) 1.14 s   (e) no

**51.** $(M + m_1 + m_2)(m_2 g/m_1)$

**53.** (a) 30.7°   (b) 0.843 N

**55.** (a) 0.515 s, 5.05 m/s   (b) 182 s, 0.014 3 m/s

## Chapter 5

**1.** $\mu_s = 0.306$; $\mu_k = 0.245$

**3.** (a) 1.11 s   (b) 0.875 s

**5.** (a) 14.7 m   (b) Neither mass is necessary.

**7.** (a) 1.78 m/s$^2$   (b) 0.368   (c) 9.37 N
   (d) 2.67 m/s

**9.** (a) Figure at top of next page.   (b) 27.2 N, 1.29 m/s$^2$

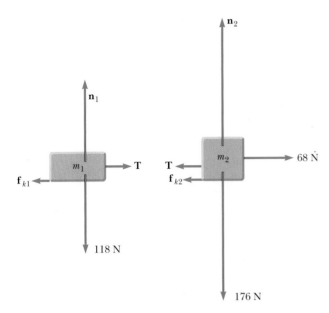

**Chapter 5, Problem 9(a)**

11. Any value between 31.7 N and 48.6 N
13. Any speed up to 8.08 m/s
15. $v \leq 14.3$ m/s
17. No. The jungle-lord needs a vine of tensile strength 1.38 kN.
19. 3.13 m/s
21. (a) 8.62 m    (b) $Mg$ downward    (c) 8.45 m/s²
    Unless they have belts, the riders will fall from the cars.
23. (a) 1.47 N·s/m    (b) $2.04 \times 10^{-3}$ s
    (c) $2.94 \times 10^{-2}$ N
25. (a) 0.034 7 s⁻¹    (b) 2.50 m/s    (c) $a = -cv$
27. (a) 13.7 m/s down    (b)

| $t$ (s) | $x$ (m) | $v$ (m/s) |
|---------|---------|-----------|
| 0       | 0       | 0         |
| 0.2     | 0       | − 1.96    |
| 0.4     | − 0.392 | − 3.88    |
| . . .   | . . .   | . . .     |
| 1.0     | − 3.77  | − 8.71    |
| . . . 2.0 | − 14.4 | − 12.56  |
| . . . 4.0 | − 41.0 | − 13.67  |

29. (a) $7.70 \times 10^{-4}$ kg/m    (b) 0.998 N
    (c) approximately 49 m, 6.3 s, 27 m/s
31. 2.97 nN
33. $\sim 10^{-7}$ N toward you
35. 0.613 m/s² toward the Earth
37. (a) 1.52 m/s²    (b) 1.66 km/s    (c) 6.82 ks
39. (b)

| $\theta$ | $P$ (N) |
|----------|---------|
| 0        | 40.0    |
| 15°      | 46.4    |
| 30°      | 60.1    |
| 45°      | 94.3    |
| 60°      | 260     |

41. They do not; 29.4 N
43. (a) 2.13 s    (b) 1.67 m
45. 6.84 m
47. (a) $m_2 g$    (b) $m_2 g$    (c) $(m_2 g R / m_1)^{1/2}$
49. (b) 732 N down at the equator and 735 N down at the poles
51. 2.14 rev/min
53. (b) 2.54 s; 23.6 rev/min
55. 12.8 N
57. $\Sigma \mathbf{F} = -km\mathbf{v}$

**Chapter 6**

1. (a) 31.9 J    (b) 0    (c) 0    (d) 31.9 J
3. − 4.70 kJ
5. 28.9
7. (a) 16.0 J    (b) 36.9°
9. (a) 11.3°    (b) 156°    (c) 82.3°
11. (a) 7.50 J    (b) 15.0 J    (c) 7.50 J    (d) 30.0 J
13. (a) 0.938 cm    (b) 1.25 J
15. 0.299 m/s
17. 50.0 J
19. (a) 1.20 J    (b) 5.00 m/s    (c) 6.30 J
21. (a) 60.0 J    (b) 60.0 J
23. 878 kN up
25. 0.116 m
27. 1.25 m/s
29. (a) 650 J    (b) $\Delta E_{int} = 588$ J    (c) 0    (d) 0
    (e) 62.0 J    (f) 1.76 m/s
31. 2.04 m
33. (a) − 168 J    (b) $\Delta E_{int} = 184$ J    (c) 500 J
    (d) 148 J    (e) 5.65 m/s
35. 875 W
37. (a) 20.6 kJ    (b) 686 W
39. (a) $2.91 \times 10^3$ times; impractical
    (b) 90.5 W = 0.121 hp
41. $1.66 \times 10^4$ m/s
43. 90.0 J
45. (a) $(2 + 24t^2 + 72t^4)$ J    (b) $12t$ m/s²; $48t$ N
    (c) $(48t + 288t^3)$ W    (d) 1 250 J
47. − 0.047 5 J
49. (b) 240 W
51. (a) 4.12 m    (b) 3.35 m
53. (a) $\mathbf{F}_1 = (20.5\,\mathbf{i} + 14.3\,\mathbf{j})$ N    $\mathbf{F}_2 = (-36.4\,\mathbf{i} + 21.0\,\mathbf{j})$ N    (b) $(-15.9\,\mathbf{i} + 35.3\,\mathbf{j})$ N    (c) $(-3.18\,\mathbf{i} + 7.07\,\mathbf{j})$ m/s²
    (d) $(-5.54\,\mathbf{i} + 23.7\,\mathbf{j})$ m/s    (e) $(-2.30\,\mathbf{i} + 39.3\,\mathbf{j})$ m
    (f) 1.48 kJ (g) 1.48 kJ
55. 1.68 m/s
57. (a) 14.5 m/s    (b) 1.75 kg    (c) 0.350 kg
59. 0.799 J

**Chapter 7**

1. (a) 259 kJ, 0, − 259 kJ    (b) 0, − 259 kJ, − 259 kJ
3. 22.0 kW
5. (a) $v = (3\,g\,R)^{1/2}$    (b) 0.098 0 N down
7. (a) 2.29 m/s    (b) 1.98 m/s
9. (a) 4.43 m/s    (b) 5.00 m

**11.** (a) 22.0 J, 40.0 J     (b) Yes, $\Delta K + \Delta U \neq 0$

**13.** (a) 125 J    (b) 50.0 J    (c) 66.7 J
    (d) Nonconservative. The results differ.

**15.** 7.04 m/s

**17.** $d = kx^2/(2mg \sin \theta) - x$

**19.** 10.2 m

**21.** 3.74 m/s

**23.** (a) $-160$ J    (b) 73.5 J    (c) 28.8 N    (d) 0.679

**25.** (a) 0.381 m    (b) 0.143 m    (c) 0.371 m

**27.** (a) 40.0 J    (b) $-40.0$ J    (c) 62.5 J

**29.** $(A/r^2)$ away from the other particle

**31.** (a) $-4.77 \times 10^9$ J    (b) 569 N    **(c)** 569 N up

**33.** $2.52 \times 10^7$ m

**35.** (a) $+$ at Ⓑ, $-$ at Ⓓ, 0 at Ⓐ, Ⓒ, and Ⓔ
    (b) Ⓒ stable; Ⓐ and Ⓔ unstable
    (c)

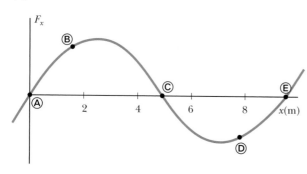

**37.** (b) Equilibrium at $x = 0$

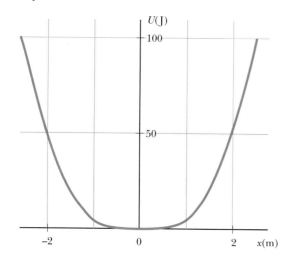

    (c) 0.823 m/s

**39.** 1.78 km

**41.** $\sim 10^3$ W peak or $\sim 10^2$ W sustainable

**43.** (a) 0.225 J    (b) $\Delta E_f = -0.363$ J
    (c) No; the normal force changes in a complicated way.

**45.** 48.2°

**47.** 0.328

**49.** 1.24 m/s

**51.** (a) 0.400 m    (b) 4.10 m/s
    (c) The block stays on the track.

**53.** $(h/5)(4 \sin^2 \theta + 1)$

**55.** 0.923 m/s

**57.** (a) $2.5R$

## Chapter 8

**1.** (a) $(9.00\mathbf{i} - 12.0\mathbf{j})$ kg·m/s    (b) 15.0 kg·m/s at 307°

**3.** $\sim 10^{-23}$ m/s

**5.** (b) $p = \sqrt{2mK}$

**7.** (a) 13.5 N·s    (b) 9.00 kN    (c) 18.0 kN

**9.** 260 N normal to the wall

**11.** (a) $(9.05\mathbf{i} + 6.12\mathbf{j})$ N·s    (b) $(377\mathbf{i} + 255\mathbf{j})$ N

**13.** (b) Small    (d) large    (e) no difference

**15.** (a) $v_{gx} = 1.15$ m/s    (b) $v_{px} = -0.346$ m/s

**17.** (a) 2.50 m/s    (b) $\Delta K = -37.5$ kJ

**19.** (a) 0.284    (b) 115 fJ and 45.4 fJ

**21.** 91.2 m/s

**23.** $(4M/m)\sqrt{g\ell}$

**25.** (a) 2.88 m/s at 32.3° north of east
    (b) 783 J into internal energy

**27.** 2.50 m/s at $-60.0°$

**29.** $(3.00\mathbf{i} - 1.20\mathbf{j})$ m/s

**31.** (a) $(-9.33\mathbf{i} - 8.33\mathbf{j})$ Mm/s    (b) 439 fJ

**33.** $\mathbf{r}_{CM} = (0\mathbf{i} + 1.00\mathbf{j})$ m

**35.** $\mathbf{r}_{CM} = (11.7\mathbf{i} + 13.3\mathbf{j})$ cm

**37.** (a) $(1.40\mathbf{i} + 2.40\mathbf{j})$ m/s    (b) $(7.00\mathbf{i} + 12.0\mathbf{j})$ kg·m/s

**39.** 0.700 m

**41.** (a) 39.0 MN    (b) 3.20 m/s² up

**43.** (a) 442 metric tons    (b) 19.2 metric tons

**45.** (a) 3.7 km/s    (b) 153 km

**47.** 0.980 m

**49.** $3.20 \times 10^4$ N, 7.13 MW

**51.** (a) 3.54 m/s    (b) 1.77 m
    (c) $3.54 \times 10^4$ N to the right
    (d) No; the normal force of the rails contributes upward momentum to the system.

**53.** (a) $\sqrt{2}\, v_i$ and $\sqrt{2/3}\, v_i$    (b) 35.3°

**55.** (a) $(20.0\mathbf{i} + 7.00\mathbf{j})$ m/s    (b) 4.00$\mathbf{i}$ m/s²
    (c) 4.00$\mathbf{i}$ m/s²    (d) $(50.0\mathbf{i} + 35.0\mathbf{j})$ m    (e) 600 J
    (f) 674 J    (g) 674 J

**57.** $(3Mgx/L)\,\mathbf{j}$

## Chapter 9

**5.** $0.866c$

**7.** (a) 64.9/min    (b) 10.6/min

**9.** 1.54 ns

**11.** $0.800c$

**13.** $0.140c$

**15.** (a) 21.0 yr    (b) 14.7 ly    (c) 10.5 ly    (d) 35.7 yr

**17.** $0.696c$

**19.** $0.960c$

**21.** (a) $2.50 \times 10^8$ m/s    (b) 4.97 m    (c) $-1.33 \times 10^{-8}$ s

**23.** (a) $2.73 \times 10^{-24}$ kg·m/s    (b) $1.58 \times 10^{-22}$ kg·m/s
    (c) $5.64 \times 10^{-22}$ kg·m/s

**25.** $4.50 \times 10^{-14}$

**27.** $0.285c$

**29.** $1.63 \times 10^3$ MeV/c

31. (a) 938 MeV   (b) 3.00 GeV   (c) 2.07 GeV
33. 18.4 g/cm$^3$
37. 3.88 MeV and 28.8 MeV
39. 4.19 × 10$^9$ kg/s
41. 1.02 MeV
43. (a) $v/c = 1 - 1.12 \times 10^{-10}$   (b) 6.00 × 10$^{27}$ J
    (c) $2.17 \times 10^{20}
45. (a) A few hundred seconds   (b) ~ 10$^8$ km
47. (a) The charged battery has a mass greater by 4.00 × 10$^{-14}$ kg   (b) 1.60 × 10$^{-12}$
51. Yes, with 18.8 m to spare
53. (b) For $u$ small compared with $c$, the relativistic expression agrees with the classical expression. As $u$ approaches $c$, the acceleration approaches zero, so that the object can never reach or surpass the speed of light.
    (c) Perform $\int (1 - u^2/c^2)^{-3/2} du = (qE/m) \int dt$ to obtain $u = qEct(m^2c^2 + q^2E^2t^2)^{-1/2}$ and then $\int dx = \int qEct(m^2c^2 + q^2E^2t^2)^{-1/2} dt$ to obtain $x = (c/qE)[(m^2c^2 + q^2E^2t^2)^{1/2} - mc]$
57. (a) The refugees conclude that Tau Ceti exploded 16.0 yr before the Sun.   (b) An observer at the midpoint and at rest with respect to the Sun and Tau Ceti concludes that they exploded simultaneously.

## Chapter 10

1. (a) 5.24 s   (b) 27.4 rad
3. (a) 822 rad/s$^2$   (b) 4.21 × 10$^3$ rad
5. 50.0 rev
7. (a) 7.27 × 10$^{-5}$ rad/s   (b) 428 min
9. (a) 126 rad/s   (b) 3.77 m/s   (c) 1.26 km/s$^2$
   (d) 20.1 m
11. (a) 54.3 rev   (b) 12.1 rev/s
13. (a) 92.0 kg·m$^2$, 184 J
    (b) 6.00 m/s, 4.00 m/s, 8.00 m/s, 184 J
17. − 3.55 N·m
19. − 7.00**i** + 16.0**j** − 10.0**k**
21. (a) 2.00 N·m   (b) **k**
23. $\dfrac{[(m_1 + m_b)d + m_1\ell/2]}{m_2}$
25. (a) $\left[\dfrac{m_1g}{2} + \dfrac{xm_2g}{L}\right] \cot\theta, (m_1 + m_2)g$
    (b) $\dfrac{m_1/2 + dm_2/L}{(m_1 + m_2)\tan\theta}$
27. (a) $T = \dfrac{F_g(L + d)}{\sin\theta\,(2L + d)}$
    (b) $R_x = \dfrac{F_g(L + d)\cot\theta}{(2L + d)}; R_y = \dfrac{F_gL}{(2L + d)}$
29. (a) 24.0 N·m   (b) 0.035 6 rad/s$^2$   (c) 1.07 m/s$^2$
31. 21.5 N
33. (a) 118 N and 156 N   (b) 1.17 kg·m$^2$
35. (a) 11.4 N, 7.57 m/s$^2$, 9.53 m/s down   (b) 9.53 m/s
37. 3g/2L; − (3g/2)**j**
39. (60.0**k**) kg·m$^2$/s
41. (a) 0.433 kg·m$^2$/s   (b) 1.73 kg·m$^2$/s
43. (a) $\omega_f = \omega_i I_1/(I_1 + I_2)$   (b) $I_1/(I_1 + I_2)$

45. 7.14 rev/min
47. 12.3 m/s$^2$
49. (a) 500 J   (b) 250 J   (c) 750 J
51. 1.21 × 10$^{-4}$ kg·m$^2$; height is unnecessary.
53. 131 s
55. (a) 0.360 rad/s counterclockwise   (b) 99.9 J
57. (a) $(3g/L)^{1/2}$   (b) 3g/2L   (c) − (3/2)g**i** − (3/4)g**j**
    (d) − (3/2)Mg**i** + (1/4)Mg**j**
61. (a) 0.309 m/s$^2$   (b) 7.67 N and 9.22 N
63. (a) $\mu_k = 0.571$; the normal force acts 20.1 cm to the left of the front edge of the sliding cabinet.   (b) 0.501 m
65. $(3/8)F_g$
67. $T = 2.71$ kN; $R_x = 2.65$ kN
71. (a) − 794 N·m; − 2 510 N·m; 0; − 1 160 N·m; − 2 940 N·m
    (b) At the following times:

| | | | | |
|---|---|---|---|---|
| 12:00:00 | 12:30:55 | 12:58:19 | 1:32:31 | 1:57:01 |
| 2:33:25 | 2:56:29 | 3:33:22 | 3:56:55 | 4:32:24 |
| 4:58:14 | 5:30:52 | 6:00:00 | 6:29:08 | 7:01:46 |
| 7:27:36 | 8:03:05 | 8:26:38 | 9:03:31 | 9:26:35 |
| 10:02:59 | 10:27:29 | 11:01:41 | 11:29:05 | |

## Chapter 11

1. 2.67 × 10$^{-7}$ m/s$^2$
3. (− 100**i** + 59.3**j**) pN
5. $\rho_M/\rho_E = 2/3$
7. (a) 3.46 × 10$^8$ m   (b) 3.34 × 10$^{-3}$ m/s$^2$ toward the Earth
9. **g** = $2MGr(r^2 + a^2)^{-3/2}$ toward the center of mass
11. (a) 1.84 × 10$^9$ kg/m$^3$   (b) 3.27 × 10$^6$ m/s$^2$
    (c) − 2.08 × 10$^{13}$ J
13. 12.6 × 10$^{31}$ kg
15. 1.90 × 10$^{27}$ kg
17. Planet Y has gone through 1.30 revolutions.
19. 18.2 ms
23. (a) 10.0 m/s$^2$   (b) 21.8 km/s
25. (a) 0.980   (b) 127 yr   (c) − 2.13 × 10$^{17}$ J
27. (a) 5   (b) No; no
29. (a) 0.212 nm   (b) 9.95 × 10$^{-25}$ kg·m/s
    (c) 2.11 × 10$^{-34}$ kg·m$^2$/s   (d) 3.40 eV   (e) − 6.80 eV
    (f) − 3.40 eV
31. (a) 13.6 eV   (b) 1.51 eV
33. 4.42 × 10$^4$ m/s
35. (a) 29.3%   (b) no change
37. $r_n = (0.106\text{ nm})n^2$,    $E_n = − 6.80\text{ eV}/n^2$, for $n = 1, 2, 3, \ldots$
39. (b) 1.10 × 10$^{32}$ kg
41. $v = 492$ m/s
43. 7.41 × 10$^{-10}$ N
45. (a) $m_2(2G/d)^{1/2}(m_1 + m_2)^{-1/2}$ and $m_1(2G/d)^{1/2}(m_1 + m_2)^{-1/2}$ and $(2G/d)^{1/2}(m_1 + m_2)^{1/2}$
    (b) 1.07 × 10$^{32}$ J and 2.67 × 10$^{31}$ J
47. (a) 200 Myr   (b) ~ 10$^{41}$ kg; ~ 10$^{11}$ stars

**49.** $v = \sqrt{2R^2 g\left(\dfrac{1}{R} + \dfrac{1}{r}\right)}$

**53.**

| t (s) | x (m) | y (m) | $v_x$ (m/s) | $v_y$ (m/s) |
|---|---|---|---|---|
| 0 | 0 | 12 740 000 | 5 000 | 0 |
| 10 | 50 000 | 12 740 000 | 4 999.9 | − 24.6 |
| 20 | 99 999 | 12 739 754 | 4 999.7 | − 49.1 |
| 30 | 149 996 | 12 739 263 | 4 999.4 | − 73.7 . . . |

The object does not hit the Earth; its minimum radius is $1.33R_E$. Its period is $1.09 \times 10^4$ s. A circular orbit would require speed 5.60 km/s.

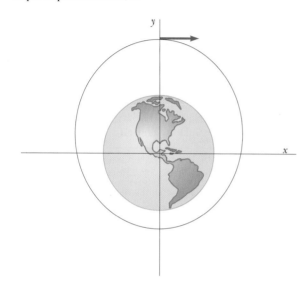

### Context 1 Conclusion

**3.** (a) 146 d    (b) Venus 53.9° behind the Earth
**4.** (a) 2.95 km/s    (b) 2.65 km/s    (c) 10.7 km/s
    (d) 4.80 km/s

## Chapter 12

**1.** (a) 575 N/m    (b) 46.0 J
**3.** (a) 1.50 Hz, 0.667 s    (b) 4.00 m    (c) $\pi$ rad    (d) 2.83 m
**5.** (a) 4.33 cm    (b) − 5.00 cm/s    (c) − 17.3 cm/s²
    (d) 3.14 s, 5.00 cm
**7.** (b) 18.8 cm/s, 0.333 s    (c) 178 cm/s², 0.500 s
    (d) 12.0 cm
**9.** 40.9 N/m
**11.** (a) 40.0 cm/s, 160 cm/s²    (b) 32.0 cm/s, − 96.0 cm/s²
    (c) 0.232 s
**13.** (a) $4\pi^2 m/T^2$    (b) $m\,(T'/T)^2$
**15.** 2.23 m/s
**17.** (a) 126 N/m    (b) 0.178 m
**19.** (a) 28.0 mJ    (b) 1.02 m/s    (c) 12.2 mJ    (d) 15.8 mJ
**21.** 2.60 cm and − 2.60 cm
**23.** (a) 0.817 m/s    (b) 2.54 rad/s²    (c) 0.634 N
**27.** 0.944 kg·m²
**29.** $1.00 \times 10^{-3}$ s$^{-1}$

**31.** (a) 2.95 Hz    (b) 2.85 cm
**35.** 1.74 Hz
**37.** $\sim 10^1$ m
**39.** 6.62 cm
**41.** $f = (2\pi L)^{-1}\,(gL + kh^2/M)^{1/2}$
**43.** (b) 1.23 Hz
**45.** (a) 3.56 Hz    (b) 2.79 Hz    (c) 2.10 Hz
**47.** (a) $\frac{1}{2}\,(M + m/3)v^2$    (b) $2\pi(M + m/3)^{1/2}k^{-1/2}$
**51.** (b) After 42.1 min
**55.** (c)

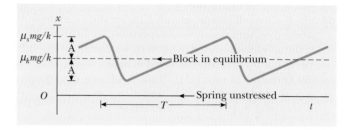

    (f) 0.281 Hz    (g) frequency decreases
    (h) frequency increases    (i) frequency increases
    (j) frequency decreases

## Chapter 13

**1.** $y = 6\,[(x - 4.5t)^2 + 3]^{-1}$
**3.** 0.319 m
**5.** (a) $3.33\mathbf{i}$ m/s    (b) − 5.48 cm    (c) 0.667m, 5.00 Hz
    (d) 11.0 m/s
**7.** (a) 31.4 rad/s    (b) 1.57 rad/m
    (c) $y = (0.120$ m$)\sin(1.57x - 31.4t)$    (d) 3.77 m/s
    (e) 118 m/s²
**9.** (a) $y = (8.00$ cm$)\sin(7.85x + 6\pi t)$
    (b) $y = (8.00$ cm$)\sin(7.85x + 6\pi t - 0.785)$
**13.** 520 m/s
**15.** 0.329 s
**17.** (a) Zero    (b) 0.300 m
**19.** 55.1 Hz
**21.** (a) $y = (7.50$ cm$)\sin(4.19x - 314t)$    (b) 625 W
**23.** 5.56 km
**25.** (a) 2.00 μm, 40.0 cm, 54.6 m/s
    (b) − 0.433 μm (c) 1.72 mm/s
**27.** 5.81 m
**29.** $\Delta P = (0.200$ N/m²$)\sin(62.8x/$m$ - 2.16 \times 10^4 t/$s$)$
**31.** 26.4 m/s
**33.** 19.3 m
**35.** 439 Hz and 441 Hz
**37.** (a) The *P* wave    (b) 665 s
**39.** $\sim 1$ min
**41.** $\left(\dfrac{Lm}{Mg\sin\theta}\right)^{1/2}$
**47.** (b) 31.6 m/s
**49.** 130 m/s, 1.73 km
**51.** 6.01 km
**53.** (a) 0.364 m    (b) 0.398 m    (c) 941 Hz    (d) 938 Hz

## Chapter 14

1. (a) $-1.65$ cm   (b) $-6.02$ cm   (c) 1.15 cm
3. (a) $y_1$ in the $+x$ direction, $y_2$ in the $-x$ direction
   (b) 0.750 s   (c) 1.00 m
5. (a) 9.24 m   (b) 600 Hz
7. 91.3°
9. (a) The number of minima is the greatest integer less than or equal to $df/v + \frac{1}{2}$
   (b) $L_n = \dfrac{d^2 - (n - 1/2)^2 (v/f)^2}{2(n - 1/2)(v/f)}$,
   where $n = 1, 2, 3, \ldots$, number of minima
11. At 0.089 1 m, 0.303 m, 0.518 m, 0.732 m, 0.947 m, 1.16 m from one speaker
13. (a) 4.24 cm   (b) 6.00 cm   (c) 6.00 cm
    (d) 0.500 cm, 1.50 cm, 2.50 cm
15. 150 Hz
17. 0.786 Hz, 1.57 Hz, 2.36 Hz, 3.14 Hz
19. (a) 0.600 m   (b) 30.0 Hz
21. (a) 163 N   (b) 660 Hz
23. (a) Three loops   (b) 16.7 Hz   (c) One loop
25. (a) 0.357 m   (b) 0.715 m
27. 0.656 m and 1.64 m
29. (a) 0.555 m   (b) 620 Hz
31. $n(206\text{ Hz})$ for $n = 1$ to 9 and $n(84.5\text{ Hz})$ for $n = 2$ to 23
33. (a) 350 m/s   (b) 1.14 m
35. 5.64 beats/s
37. (a) 1.99 beats/s   (b) 3.38 m/s
39. (a) 3.66 m/s   (b) 0.200 Hz
41. (a) $\Delta r = \lambda/2$   (b) $9.00\,x^2 - 16.0\,y^2 = 144$
43. 9.00 kHz
45. (a) 34.8 m/s   (b) 0.977 m
47. 4.85 m
49. (a) $\frac{1}{2}Mg$   (b) $3h$   (c) $m/3h$   (d) $\sqrt{\dfrac{3Mgh}{2m}}$   (e) $\sqrt{\dfrac{3Mg}{8mh}}$
    (f) $\sqrt{\dfrac{2mh}{3Mg}}$   (g) $h$   (h) $(2.00 \times 10^{-2})\sqrt{\dfrac{3Mg}{8mh}}$
51. (a) $\frac{1}{2}$   (b) $\left[\dfrac{n}{n+1}\right]^2 T$   (c) $\frac{9}{16}$
53. 50.0 Hz, 1.70 m
55. $y_1 + y_2 = 11.2\sin(2.00x - 10.0t + 63.4°)$
57. (a) 78.9 Hz   (b) 211 Hz

## Context 2 Conclusion

1. 3.5 cm
2. $-96\%$
3. Station 1: 15:46:26
   Station 2: 15:46:23
   Station 3: 15:46:08

## Chapter 15

1. 0.111 kg
3. 6.24 MPa
5. $5.27 \times 10^{18}$ kg
7. 1.62 m
9. $7.76 \times 10^{-3}$ m$^2$

11. 5.88 MN, 196 kN, 588 kN
13. 10.5 m; no; some alcohol and water evaporate.
15. 98.6 kPa
17. 0.258 N
19. (a) $1.017\,9 \times 10^3$ N down, $1.029\,7 \times 10^3$ N up
    (b) 86.2 N   (c) 11.8 N
21. (a) 7.00 cm   (b) 2.80 kg
23. 1 430 m$^3$
25. $\rho_{\text{glycerin}} = 1\,250$ kg/m$^3$ and $\rho_{\text{sphere}} = 500$ kg/m$^3$
27. (a) 1.06 m/s   (b) 4.24 m/s
29. 616 MW
31. $2\sqrt{h(h_0 - h)}$
33. (a) 2.28 N   (b) $1.74 \times 10^6$ s
35. (a) 28.0 m/s   (b) 28.0 m/s   (c) 2.11 MPa
37. 68.0 kPa
39. (a) 489 N outward   (b) 1.96 kN outward
41. (b) Smaller; 8.30%
43. (a) 60.0 rad   (b) 9.12 m/s$^2$   (c) 6.79 m/s$^2$
    (d) 255 N·m
49. The top scale reads $(1 - \rho_0/\rho_{\text{Fe}})m_{\text{Fe}}g$. The bottom scale reads $[m_b + m_0 + \rho_0 m_{\text{Fe}}/\rho_{\text{Fe}}]g$.
51. (b) $2.58 \times 10^4$ N
53. 4.43 m/s
55. (a) 1.25 cm   (b) 13.8 m/s
57. (a) 3.307 g   (b) 3.271 g   (c) $3.48 \times 10^{-4}$ N

## Context 3 Conclusion

1. $2.0 \times 10^{10}$ N·m
2. (a) $1.30 \times 10^6$ Pa   (b) Yes, but only with specialized equipment and techniques
3. (a) 0.42 m/s   (b) greater
4. (a) 16 knots, 56° west of south   (b) 47%

## Chapter 16

1. (a) $-274$°C   (b) 1.27 atm   (c) 1.74 atm
3. (a) $-320$°F   (b) 77.3 K
5. (a) 810°F   (b) 450 K
7. (a) 3.005 8 m   (b) 2.998 6 m
9. (a) 0.176 mm   (b) 8.78 $\mu$m   (c) 0.093 0 cm$^3$
11. 0.548 gal
13. (a) 99.4 cm$^3$   (b) 0.943 cm
15. 0.663 mm at 78.2° below the horizontal
17. $1.50 \times 10^{29}$ molecules
19. 472 K
21. 3.67 cm$^3$
23. 4.39 kg
25. $2.41 \times 10^{11}$ molecules
27. $2Nmv\sin\theta/t$ perpendicular to the window; $2Nmv\sin\theta/At$
29. (a) $3.53 \times 10^{23}$ atoms
    (b) $6.07 \times 10^{-21}$ J
    (c) 1.35 km/s
31. (a) $8.76 \times 10^{-21}$ J for both
    (b) 1.62 km/s for helium and 514 m/s for argon
33. (a) 6.80 m/s   (b) 7.41 m/s   (c) 7.00 m/s
37. 5.5°C
39. (a) 94.97 cm   (b) 95.03 cm

**41.** 3.55 cm

**43.** (a) Expansion makes density drop.
(b) $5 \times 10^{-5}$ $(°C)^{-1}$

**45.** (a) $h = nRT/(mg + P_0A)$ (b) 0.661 m

**47.** (a) $9.49 \times 10^{-5}$ s (b) 57.4 s lost

**49.** (a) 0.340%
(b) 0.480%

**51.** 2.74 m

**53.** (b) $3.66 \times 10^{-3}$ $(°C)^{-1}$; it agrees within 0.06% for helium and within 0.2% for air.

**55.** (a) 7.06 mm (b) 297 K

**57.** 1.12 atm

**59.** 4.54 m

**61.** $1.09 \times 10^{-3}$; $2.69 \times 10^{-2}$; 0.529; 1.00; 0.199; $1.01 \times 10^{-41}$; $1.25 \times 10^{-1082}$

## Chapter 17

**1.** 0.281°C

**3.** 87.0°C

**5.** 29.6°C

**7.** $\dfrac{(m_{Al}c_{Al} + m_c c_w) T_c + m_h c_w T_h}{m_{Al}c_{Al} + m_c c_w + m_h c_w}$

**9.** $1.22 \times 10^5$ J

**11.** 0.294 g

**13.** (a) 0°C (b) 114 g

**15.** 59.4°C

**17.** $-1.18$ MJ

**19.** $-466$ J

**21.** $Q = -720$ J

**23.**

| | Q | W | $\Delta E_{int}$ |
|---|---|---|---|
| BC | $-$ | 0 | $-$ |
| CA | $-$ | $+$ | $-$ |
| AB | $+$ | $-$ | $+$ |

**25.** (a) 7.50 kJ (b) 900 K

**27.** (a) 0.041 0 m³ (b) 5.48 kJ (c) $-5.48$ kJ

**29.** (a) 1 300 J (b) 100 J (c) $-900$ J (d) $-1\,400$ J

**31.** (a) 3.46 kJ (b) 2.45 kJ (c) $-1.01$ kJ

**33.** (a) 719 J/kg·°C (b) 0.811 kg (c) 233 kJ (d) 327 kJ

**35.** 13.5$PV$

**37.** (a) 1.39 atm (b) 365 K, 253 K (c) 0, $-4.66$ kJ, $-4.66$ kJ

**39.** (a) 28.0 kJ (b) 46.1 kJ
(c) isothermal process, $P_f = 10.0$ atm; adiabatic process, $P_f = 25.1$ atm

**41.** 25.0 kW

**43.** (a) 41.6 J/K, 58.2 J/K (b) 58.2 J/K, 74.8 J/K

**45.** 10.0 kW

**47.** 51.2°C

**49.** $3.49 \times 10^3$ K

**51.** $3.77 \times 10^{26}$ J/s

**53.** 277 K = 4°C

**55.** 2.27 km

**57.** (a) 13.0°C (b) $-0.532$ °C/s

**59.** $c = \dfrac{\mathcal{P}}{\rho R\,\Delta T}$

**61.** 5.31 h

**63.** 3.76 m/s

**65.** (a) 15.0 mg; block: $Q = 0$, $W = -5.00$ J, $\Delta E_{int} = 0$, $\Delta K = -5.00$ J. ice: $Q = 0$, $W = +5.00$ J, $\Delta E_{int} = 5.00$ J, $\Delta K = 0$ (b) 15.0 mg; slab: $Q = 0$, $W = 0$, $\Delta E_{int} = 5.00$ J, $\Delta K = -5.00$ J. metal: $Q = 0$, $W = 0$, $\Delta E_{int} = 0$, $\Delta K = 0$ (c) 0.004 04°C; moving slab: $Q = 0$, $W = -2.50$ J, $\Delta E_{int} = 2.50$ J, $\Delta K = -5.00$ J. stationary slab: $Q = 0$, $W = +2.50$ J, $\Delta E_{int} = 2.50$ J, $\Delta K = 0$

**67.** (a) 0.203 mol (b) $T_B = T_C = 900$ K $V_C = 15.0$ L

| (c, d) | P, atm | V, L | T, K | $E_{int}$, kJ |
|---|---|---|---|---|
| A | 1.00 | 5.00 | 300 | 0.760 |
| B | 3.00 | 5.00 | 900 | 2.28 |
| C | 1.00 | 15.0 | 900 | 2.28 |
| A | 1.00 | 5.00 | 300 | 0.760 |

(e) For $A \to B$, lock the piston in place and put the cylinder into an oven at 900 K. For $B \to C$, keep the gas in the oven while gradually letting the gas expand to lift a load on the piston as far as it can. For $C \to A$, move the cylinder from the oven back to the 300-K room and let the gas cool and contract.

| (f, g) | Q, kJ | W, kJ | $\Delta E_{int}$, kJ |
|---|---|---|---|
| $A \to B$ | 1.52 | 0 | 1.52 |
| $B \to C$ | 1.67 | $-1.67$ | 0 |
| $C \to D$ | $-2.53$ | $+1.01$ | $-1.52$ |
| ABCA | 0.656 | $-0.656$ | 0 |

**69.** 38.6 m³/d

**71.** 9.32 kW

## Chapter 18

**1.** (a) 6.94% (b) 335 J

**3.** (a) 10.7 kJ (b) 0.533 s

**5.** (a) 67.2% (b) 58.8 kW

**7.** (a) 869 MJ (b) 330 MJ

**9.** (a) 741 J (b) 459 J

**11.** 0.330, or 33.0%

**13.** (a)

| State | P, kPa | V, L | T, K |
|---|---|---|---|
| A | 1 400 | 10.0 | 720 |
| B | 875 | 16.0 | 720 |
| C | 445 | 24.0 | 549 |
| D | 712 | 15.0 | 549 |

(b)

| Process | Q, kJ | $W_{eng}$, kJ | $\Delta E_{int}$, kJ |
|---|---|---|---|
| $A \to B$ | 6.58 | 6.58 | 0 |
| $B \to C$ | 0 | 4.99 | $-4.99$ |
| $C \to D$ | $-5.02$ | $-5.02$ | 0 |
| $D \to A$ | 0 | $-4.99$ | 4.99 |
| ABCDA | 1.56 | 1.56 | 0 |

(c) $e = e_C = 0.237$, or 23.7%

15. 11.8
19. 72.2 J
21. 195 J/K
23. 236 J/K
25. (a)

| Result | Number of ways to draw |
|---|---|
| All red | 1 |
| 2 red, 1 green | 3 |
| 1 red, 2 green | 3 |
| All green | 1 |

(b)

| Result | Number of ways to draw |
|---|---|
| 5 red | 1 |
| 4 red, 1 green | 5 |
| 3 red, 2 green | 10 |
| 2 red, 3 green | 10 |
| 1 red, 4 green | 5 |
| 5 green | 1 |

27. 3.27 J/K
29. 1.02 kJ/K
31. $1.26 \times 10^8$ J/K
33. 18.4 J/K
35. $9 \times 10^{14}$ W
37. (a) 214 J, 64.3 J
(b) $-35.7$ J, $-35.7$ J. The net effect is the transport of energy by heat from the cold to the hot reservoir without expenditure of external work.
(c) 333 J, 233 J
(d) 83.3 J, 83.3 J, 0. The net effect is the conversion of energy, taken in as heat, entirely into work in a cyclic process.
(e) $-0.111$ J/K. The entropy of the Universe has decreased.
39. (a) 5.00 kW   (b) 763 W
41. (a) $2nRT_i \ln 2$   (b) 0.273
43. $5.97 \times 10^4$ kg/s
45. 23.1 mW
47. 1.18 J/K
49. (a) $10.5nRT_i$   (b) $8.50nRT_i$   (c) 0.190   (d) 0.833
51. (a) $V_A = 1.97$ L, $V_B = 11.9$ L, $V_C = 32.8$ L, $V_D = 5.44$ L, $P_B = 4.14$ atm, $P_D = 6.03$ atm
(b) 2.99 kJ   (c) 0.333
53. (a) $nC_P \ln 3$
(b) Both ask for the change in entropy between the same two states of the same system. Entropy is a function of state. The change in entropy does not depend on path, but only on original and final states.

**Context 4 Conclusion**

1. 301 K
2. 60 km
3. (c) 334 K
(d) Layers are too thick to be modeled as having a single temperature.   (e) 226 K   (f) 109
(g) Yes. There are many layers, so the model of a single temperature for each layer is reasonable.

**Chapter 19**

1. (a) $2.62 \times 10^{24}$ electrons   (b) 2.38 electrons
3. The force is $\sim 10^{26}$ N
5. 57.5 N
7. 0.872 N at 330°
9. 1.82 m to the left of the negative charge
11. (a) The field is zero at the center of the triangle.
(b) $1.73k_e q\mathbf{j}/a^2$
13. (a) $5.91k_e q/a^2$ at 58.8°   (b) $5.91k_e q^2/a^2$ at 58.8°
17. $-21.6\mathbf{i}$ MN/C
21.

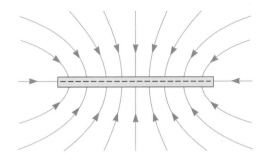

23. (a) $61.4$ Gm/s$^2$   (b) 19.5 $\mu$s   (c) 11.7 m   (d) 1.20 fJ
25. (a) Downward   (b) 3.43 $\mu$C
27. The particle strikes the negative plate after moving in a parabola with a height of 0.181 mm and a width of 0.961 mm.
29. 4.14 MN/C
31. (a) $-6.89$ MN·m$^2$/C
(b) The number of lines entering exceeds the number leaving by 2.91 times or more.
33. (a) $+Q/2\epsilon_0$   (b) $-Q/2\epsilon_0$
35. (a) 0   (b) 7.19 MN/C away from the center
37. $\mathbf{E} = \rho r/2\epsilon_0$ away from the axis
39. 3.50 kN
41. (a) $+708$ nC/m$^2$ and $-708$ nC/m$^2$
(b) $+177$ nC and $-177$ nC
43. (a) 80.0 nC/m$^2$ on each face   (b) $9.04\mathbf{k}$ kN/C
(c) $-9.04\mathbf{k}$ kN/C
45. (a) 0   (b) 79.9 MN/C radially outward   (c) 0
(d) 7.34 MN/C radially outward
47. positive, $1.77 \times 10^{-12}$ C/m$^3$
49. 5.25 $\mu$C
51. 26.7 $\mu$C
53. 40.9 N at 263°
55. (a) $\theta_1 = \theta_2$
57. (b) in the $+z$ direction
59. (a) $\rho r/3\epsilon_0$; $Q/4\pi\epsilon_0 r^2$; 0; $Q/4\pi\epsilon_0 r^2$, all radially outward
(b) $-Q/4\pi b^2$ and $+Q/4\pi c^2$
61. (a) $\sigma/\epsilon_0$ away from both plates   (b) 0
(c) $\sigma/\epsilon_0$ away from both plates

**Chapter 20**

1. (a) 152 km/s   (b) 6.49 Mm/s
3. 46.7 kV
5. (a) $-600$ $\mu$J   (b) $-50.0$ V

7. $-38.9$ V; the origin
9. (a) $2QE/k$    (b) $QE/k$    (c) $2\pi\sqrt{m/k}$
   (d) $2(QE - \mu_k mg)/k$
11. (a) $\mathbf{F} = 0$    (b) $\mathbf{E} = 0$    (c) 45.0 kV
13. $-11.0$ MV
15. 8.95 J
19. (a) 10.8 m/s and 1.55 m/s    (b) greater
21. 27.4 fm
23. (a) 10.0 V, $-11.0$ V, $-32.0$ V
   (b) 7.00 N/C in the $+x$ direction
25. $\mathbf{E} = (-5 + 6xy)\,\mathbf{i} + (3x^2 - 2z^2)\mathbf{j} - 4yz\mathbf{k}$; 7.07 N/C
27. (a) $C/m^2$    (b) $k_e\alpha[L - d\ln(1 + L/d)]$
29. $-1.51$ MV
31. (a) 0, 1.67 MV    (b) 5.84 MN/C away, 1.17 MV
   (c) 11.9 MN/C away, 1.67 MV
33. (a) 48.0 $\mu$C    (b) 6.00 $\mu$C
35. (a) 11.1 kV/m toward the negative plate
   (b) 98.3 nC/m$^2$    (c) 3.74 pF    (d) 74.7 pC
37. $(2N - 1)(\pi - \theta)R^2\epsilon_0/d$
39. $mgd\tan\theta/q$
41. (a) 17.0 $\mu$F    (b) 9.00 V    (c) 45.0 $\mu$C and 108 $\mu$C
43. (a) 5.96 $\mu$F
   (b) 89.5 $\mu$C on the 20-$\mu$F capacitor, 63.2 $\mu$C on the 6-$\mu$F capacitor, 26.3 $\mu$C on the 15-$\mu$F and 3-$\mu$F capacitors
45. 120 $\mu$C; 80.0 $\mu$C and 40.0 $\mu$C
47. $(\sqrt{3} - 1)C_0/2$
49. 4.47 kV
53. $2.51 \times 10^{-3}$ m$^3$ = 2.51 L
55. (a) 13.3 nC    (b) 272 nC
57. 0.188 m$^2$
59. (a) 450 kV    (b) 7.51 $\mu$C
63. $k_eQ^2/2R$
65. (a) 40.0 $\mu$J    (b) 500 V
67. (a) $Q_0^2\,d(\ell - x)/(2\ell^3\epsilon_0)$
   (b) $Q_0^2\,d/(2\ell^3\epsilon_0)$ to the right    (c) $Q_0^2/(2\ell^4\epsilon_0)$
   (d) $Q_0^2/(2\ell^4\epsilon_0)$
69. Gasoline has 194 times the specific energy content of the battery and 727 000 times that of the capacitor.
71. 8.00 kV
73. 579 V
75. 253 MeV
77. (a) The additional energy comes from electrical work done by the battery in separating the additional charge.
   (b) $Q/Q_0 = \kappa$

## Chapter 21

1. $7.50 \times 10^{15}$ electrons
3. $q\omega/2\pi$
5. (a) $0.632I_0\tau$    (b) $0.999\,95I_0\tau$    (c) $I_0\tau$
7. 500 mA
9. 6.43 A
11. (a) 31.5 n$\Omega\cdot$m    (b) 6.35 MA/m$^2$    (c) 49.9 mA
   (d) 659 $\mu$m/s    (e) 0.400 V
13. 0.181 V/m
15. 5.00 A; 24.0 $\Omega$
17. 36.1%

19. (a) 184 W    (b) 461°C
21. ~$1
23. (a) 6.73 $\Omega$    (b) 1.97 $\Omega$
25. (a) 17.1 $\Omega$
   (b) 1.99 A for 4.00 $\Omega$ and 9.00 $\Omega$, 1.17 A for 7.00 $\Omega$, 0.818 A for 10.0 $\Omega$
27. (a) 227 mA    (b) 5.68 V
29. 14.2 W, 28.4 W, 1.33 W, 4.00 W
31. (a) Heater, 12.5 A; toaster, 6.25 A; grill, 8.33 A
   (b) No; together they would require 27.1 A
33. 846 mA down in the 8-$\Omega$ resistor; 462 mA down in the middle branch; 1.31 A up in the right-hand branch
35. (a) $-222$ J and 1.88 kJ
   (b) 687 J, 128 J, 25.6 J, 616 J, 205 J    (c) 1.66 kJ
37. Starter 171 A; battery 0.283 A
39. (a) 5.00 s    (b) 150 $\mu$C    (c) 4.06 $\mu$A
41. (a) 6.00 V    (b) 8.29 $\mu$s
43. $6.00 \times 10^{-15}$ $(\Omega\cdot$m$)^{-1}$
45. (a) 576 $\Omega$, 144 $\Omega$    (b) 4.80 s    (c) 0.040 0 s
   (d) $1.26, energy at $1.94 \times 10^{-8}$ $/J
47. (a) $8.00\mathbf{i}$ V/m    (b) 0.637 $\Omega$    (c) 6.28 A
   (d) $200\mathbf{i}$ MA/m$^2$
49. Experimental resistivity = 1.47 $\mu\Omega\cdot$m $\pm 4\%$, in agreement with 1.50 $\mu\Omega\cdot$m
51. (a) 667 A    (b) 50.0 km
55. (a) $\ln(\mathcal{E}/\Delta V) = 0.011\,8t + 0.088\,2$    (b) 84.7 s, 8.47 $\mu$F
57. $(R_A + 2R_B)C\ln 2$
59. (a) Either 3.84 $\Omega$ or 0.375 $\Omega$
   (b) Impossible; no load resistor can extract 21.2 W from this battery

## Context 5 Conclusion

1. (a) 87.0 s    (b) 261 s    (c) $t \to \infty$
2. (a) 0.01 s    (b) $7 \times 10^6$
3. (a) $3 \times 10^6$    (b) $9 \times 10^6$

## Chapter 22

1. (a) Up    (b) toward you, out of the plane of the paper
   (c) no deflection    (d) into the plane of the paper
3. (a) 7.90 pN    (b) zero
5. 2.34 aN
7. 7.88 pT
9. 0.278 m
11. 70.1 mT
13. $(-2.88\mathbf{j})$ N
15. $2\pi rIB\sin\theta$ up
17. 9.98 N$\cdot$m clockwise as seen looking down from above
19. (a) 3.97°    (b) 3.39 mN$\cdot$m
21. 12.5 T
23. 200 nT
25. $\mu_0 I/4\pi x$ into the paper
27. (a) 53.3 $\mu$T    (b) 20.0 $\mu$T    (c) zero
29. (a) along the line ($y = -0.420$ m, $z = 0$)
   (b) $-34.7\mathbf{j}$ mN    (c) $-17.3\mathbf{j}$ kN/C
31. $-27.0\mathbf{i}$ $\mu$N
33. 20.0 $\mu$T toward the bottom of the page

**35.** 200 $\mu$T toward the top of the page; 133 $\mu$T toward the bottom of the page

**37.** (a) 6.34 mN/m inward    (b) greater

**39.** (a) 0    (b) $\mu_0 I/2\pi R$ tangent to the wall in a counterclockwise sense    (c) $\mu_0 I^2/(2\pi R)^2$ inward

**41.** 31.8 mA

**43.** (a) $9.27 \times 10^{-24}$ A$\cdot$m$^2$    (b) down

**45.** $4 \times 10^{10}$

**49.** (a) 17.9 ns    (b) 35.1 eV

**51.** (a) $(3.52\mathbf{i} - 1.60\mathbf{j})$ aN    (b) 24.4°

**53.** 39.2 mT

**55.** $B \sim 10^{-1}$ T, $\tau \sim 10^{-1}$ N$\cdot$m, $I \sim 10^0$ A, $A \sim 10^{-3}$ m$^2$, $N \sim 10^3$

**57.** (a) $8.63 \times 10^{45}$ electrons    (b) $4.01 \times 10^{20}$ kg

**59.** 20.0 $\mu$T

**61.** $\dfrac{\mu_0 q\omega}{2.5\pi R\sqrt{5}}$

**63.** 2.00 GA west

**65.** $\mu_0 I\mathbf{k} \ln(1 + w/b)/2\pi w$

## Chapter 23

**1.** 0.800 mA

**3.** 160 A

**5.** (b) 3.79 mV    (c) 28.0 mV

**7.** 61.8 mV

**9.** (a) To the right    (b) To the right    (c) To the right    (d) Into the paper

**11.** (a) 3.00 N to the right    (b) 6.00 W

**13.** (a) Eastward    (b) 458 $\mu$V

**15.** (b) The emf induced in the coil is proportional to the line integral of the magnetic field around the circular axis of the toroid. By Ampère's law, this line integral depends only on the current the circle encloses.

**17.** (b) Larger $R$ makes current smaller, so the loop must travel faster to maintain equilibrium between magnetic force and weight.    (c) The magnetic force is proportional to the product of field and current, whereas the current is proportional to the magnetic field. If $B$ becomes two times smaller, the speed must become four times larger to compensate.

**19.** (a) 7.54 kV    (b) The plane of the coil is parallel to **B**

**21.** 1.80 mN/C upward and to the left, perpendicular to $r_1$

**23.** 19.5 mV

**25.** 18.8 V cos(377 $t$)

**27.** 19.2 $\mu$T$\cdot$m$^2$

**31.** (a) 0.139 s    (b) 0.461 s

**33.** (a) 2.00 ms    (b) 0.176 A    (c) 1.50 A    (d) 3.22 ms

**35.** (a) 0.800    (b) 0

**37.** (a) 5.66 ms    (b) 1.22 A    (c) 58.1 ms

**39.** 2.44 $\mu$J

**41.** 44.2 nJ/m$^3$ for the **E** field and 995 $\mu$J/m$^3$ for the **B** field

**43.** $2 \times 10^2$ T

**45.** $(-7.22$ mV$)$ cos$(2\pi\,523t/s)$

**47.** 6.00 A

**51.** 1.20 $\mu$C

**53.** (a) $(1.19$ V$)$ cos$(120\pi t)$    (b) 88.5 mW

**55.** $(-87.1$ mV$)$ cos$(200\pi t + \phi)$

**57.** (a) $L \approx (\pi/2)N^2\mu_0 R$    (b) $\sim$100 nH    (c) $\sim$1 ns

**59.** $3.97 \times 10^{-25}$ $\Omega$

**61.** (a) 50.0 mT    (b) 20.0 mT    (c) 2.29 MJ    (d) 318 Pa

## Context 6 Conclusion

**1.** 5.56 Hz

## Chapter 24

**1.** 1.85 aT up

**3.** (a) About the year 2680    (b) 499 s    (c) 2.56 s    (d) 0.133 s    (e) 33.3 $\mu$s

**5.** (a) 6.00 MHz    (b) 73.3 nT $(-\mathbf{k})$    (c) $\mathbf{B} = (-73.3\mathbf{k}$ nT$)$ cos$(0.126x - 3.77 \times 10^7 t)$

**7.** (a) 0.333 $\mu$T    (b) 0.628 $\mu$m    (c) 477 THz

**9.** 75.0 MHz

**11.** (b) 0.050 4$c$

**13.** 608 pF

**15.** (a) 503 Hz    (b) 12.0 $\mu$C    (c) 37.9 mA    (d) 72.0 $\mu$J

**17.** 49.5 mV

**19.** $3.33 \times 10^3$ m$^2$

**21.** (a) 332 kW/m$^2$ radially inward    (b) 1.88 kV/m and 222 $\mu$T

**23.** 11.9 GW/m$^2$

**25.** 83.3 nPa

**27.** (a) 1.90 kN/C    (b) 50.0 pJ    (c) $1.67 \times 10^{-19}$ kg$\cdot$m/s

**29.** 56.2 m

**31.** (a) $\sim 10^8$ Hz radio wave    (b) $\sim 10^{13}$ Hz infrared light

**33.** (a) 6.00 pm    (b) 7.50 cm

**35.** The radio audience hears it 8.41 ms sooner.

**37.** (a) 54.7°    (b) 63.4°    (c) 71.6°

**39.** $\frac{1}{8}$

**43.** 28.2 THz, 10.6 $\mu$m, infrared

**45.** (a) 4.24 PW/m$^2$    (b) 1.20 pJ = 7.50 MeV

**49.** $\sim 10^6$ J

**51.** (a) $(4\rho gc/3)(3m/4\pi\rho)^{1/3}$    (b) $(4\pi\rho gcr^2/3)(3m/4\pi\rho)^{1/3}$

**53.** 95.1 mV/m

**55.** (a) $B_{max} = 583$ nT, $k = 419$ rad/m, $\omega = 12.6$ Trad/s    (b) $S_{av} = 40.6$ W/m$^2$    (c) 271 nPa    (d) 406 nm/s$^2$

**57.** (a) 22.6 h    (b) 30.6 s

**59.** (a) $2\pi^2 r^2 fB_{max}$ cos $\theta$, where $\theta$ is the angle between the magnetic field and the normal to the loop    (b) The loop should be in the vertical plane containing the line of sight to the transmitter.

**61.** (b) 17.6 Tm/s$^2$, $1.75 \times 10^{-27}$ W    (c) $1.80 \times 10^{-24}$ W

## Chapter 25

**1.** 25.5°, 442 nm

**3.** 19.5° above the horizon

**5.** (a) 181 Mm/s    (b) 225 Mm/s    (c) 136 Mm/s

**7.** 30.0° and 19.5° at entry; 19.5° and 30.0° at exit

9. (a) 1.94 m    (b) 50.0° above the horizontal
11. Six times from mirror one and five times from mirror two
13. 3.88 mm
15. 30.4° and 22.3°
19. 0.171°
21. 4.61°
23. $\sin^{-1}(\sqrt{n^2 - 1}\ \sin\Phi - \cos\Phi)$
25. (a) 24.4°    (b) 37.0°    (c) 49.8°
27. 1.000 08
29. Skylight incident from above travels down the plastic. If
the index of refraction of the plastic is larger than 1.41,
the rays close in direction to the vertical are totally reflected from the side walls of the slab and from both
facets at the bottom of the plastic, where it is not immersed in gasoline. This light returns up inside the plastic
and makes it look bright. Where the plastic is immersed
in gasoline, total internal reflection is frustrated and the
downward-propagating light passes from the plastic out
into the gasoline. Little light is reflected up, and the
gauge looks dark.
31. 82 reflections
33. (a) $dn/(n - 1)$    (b) yes; yes; yes    (c) 350 $\mu$m
35. 2.27 m
37. (a) 0.172 mm/s    (b) 0.345 mm/s    (c) northward
at 50.0° below the horizontal    (d) northward at 50.0°
below the horizontal
39. 62.2%
41. 4.74 cm
43. (b) 68.5%
45. 27.5°
47. (a) It always happens    (b) 30.3°    (c) It cannot happen
49. 2.37 cm
51. $\sin^{-1}\left[\dfrac{L}{R^2}\left(\sqrt{n^2 R^2 - L^2} - \sqrt{R^2 - L^2}\right)\right]$
53. (a) 1.20    (b) 3.40 ns

# Chapter 26

1. $\sim 10^{-9}$ s younger
3. 35.0 in.
5.

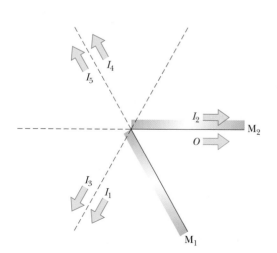

7. (a) $-(p_1 + h)$    (b) virtual    (c) upright    (d) 1.00
(e) no
9. (a) $q = 45.0$ cm $M = -0.500$ real, inverted, diminished
(b) $q = -60.0$ cm $M = 3.00$ virtual, upright, enlarged
(c)

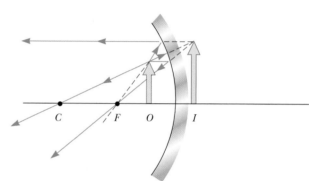

11. (a) $-12.0$ cm; 0.400    (b) $-15.0$ cm; 0.250    (c) upright
13. (a) 160 mm    (b) $-267$ mm
15. (a) A concave mirror with radius of curvature 2.08 m
(b) 1.25 m from the object
17. (a) 25.6 m    (b) 0.058 7 rad    (c) 2.51 m    (d) 0.023 9 rad
(e) 62.8 m from your eyes
19. 38.2 cm below the top surface of the ice
21. 8.57 cm
23. (a) 16.4 cm    (b) 16.4 cm
25. (a) $q = 40.0$ cm, $M = -1.00$, real and inverted
(b) $q \to \infty$, no image is formed    (c) $q = -20.0$ cm,
$M = +2.00$, virtual and upright
27. 2.84 cm
29. 7.50 cm
33. (a) $-12.3$ cm, to the left of the lens    (b) 0.615
(c) Figure at top of next page.
35. (a) $p = \dfrac{d}{2} \pm \sqrt{(d^2/4) - fd}$    (b) Both images are real
and inverted. One is enlarged, the other diminished.
37. (a) $q = -34.7$ cm, virtual, upright, diminished
(b) $q = -36.1$ cm, virtual, upright, diminished
39. (a) 833 $\mu$m    (b) 820 $\mu$m
41. 3.38 min
43. 160 cm to the left of the lens; inverted; $M = -0.800$
47. Align the lenses on the same axis and 9.00 cm apart. Let

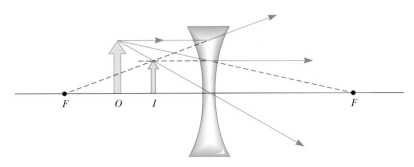

**Chapter 26, Problem 33(c)**

the light pass first through the diverging lens and then through the converging lens. The diameter increases by a factor of 1.75.

**49.** 0.107 m to the right of the vertex of the hemispherical face

**51.** 1.50 m in front of the mirror; 1.40 cm (inverted)

**53.** 8.00 cm

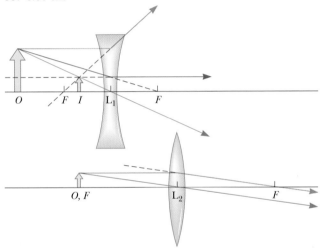

**55.** (a) 30.0 cm and 120 cm   (b) 24.0 cm   (c) real, inverted, diminished with $M = -0.250$

**57.** (a) 20.0 cm to the right of the second lens; $M = -6.00$   (b) inverted   (c) 6.67 cm to the right of the second lens; $M = -2.00$, inverted

## Chapter 27

**1.** 515 nm

**3.** (a) 55.7 m   (b) 124 m

**5.** 1.54 mm

**7.** (a) 10.0 m   (b) 516 m   (c) Only the runway centerline is a maximum for the interference patterns for both frequencies. If the frequencies were related by a ratio of small integers $k/\ell$, the plane could by mistake fly along the $k$th side maximum of one signal where it coincides with the $\ell$th side maximum of the other.

**9.** 48.0 $\mu$m

**11.** 0.987

**13.** (a) Green   (b) violet

**15.** 4.35 $\mu$m

**17.** 4.22 mm

**19.** 0.230 mm

**21.** Three maxima, at $0°$ and near $46°$ to the left and right

**23.** 3.09 m

**25.** 13.1 m

**27.** 241 m

**29.** 514 nm

**31.** $5.91°$ in first order, $13.2°$ in second order, $26.5°$ in third order

**33.** The first-order spectrum is between $23.6°$ and $34.1°$. The second-order spectrum begins at $53.1°$. There is no third-order spectrum. No overlap occurs.

**35.** (a) Five orders   (b) ten orders in the short-wavelength region

**37.** $14.4°$

**39.** (a) 0.738 mm   (b) Individual waves from all of the transparent zones will add crest-on-crest to interfere constructively at the slit images. The grating equation $d \sin \theta = m\lambda$ is satisfied at the slit images. Elsewhere on the screen destructive interference will prevent light from reaching the screen.

**41.** (a) $\sim 10^{-3}$ degree   (b) $\sim 10^{11}$ Hz, microwave

**43.** 2.50 mm

**45.** 113 dark fringes

**49.** 632.8 nm

**51.** (a) $25.6°$   (b) $19.0°$

**53.** (b) 0.428 mm

## Context 7 Conclusion

**1.** 130 nm

**2.** 74.2 tracks/mm

**3.** $\sim 10^8$ W/m$^2$

## Chapter 28

**1.** (a) 999 nm   (b) The infrared region of the spectrum is much wider than the visible region, and the spectral distribution is highest in the infrared.

**3.** (a) 2.57 eV  (b) 12.8 $\mu$eV  (c) 191 neV
(d) 484 nm (visible), 9.68 cm and 6.52 m (radio waves)

**5.** $2.27 \times 10^{30}$ photons/s

**7.** (a) 296 nm, 1.01 PHz  (b) 2.71 V

**9.** (a) 1.90 eV  (b) 0.216 V

**11.** 8.41 pC

**13.** (a) 488 fm  (b) 268 keV  (c) 31.5 keV

**15.** 70.0°

**17.** 397 fm

**19.** 0.218 nm

**21.** (a) $\sim$ 100 MeV or more  (b) No. With kinetic energy much larger than the magnitude of the negative electric potential energy, the electron would immediately escape.

**25.** (a) 992 nm  (b) 4.96 mm  (c) If its detection forms part of an interference pattern, the neutron must have passed through both slits. If we test to see which slit a particular neutron passes through, it will not form part of the interference pattern.

**27.** Within 1.16 mm for the electron, $5.28 \times 10^{-32}$ m for the bullet

**29.** $3.79 \times 10^{28}$ m, 190 times the diameter of the Universe

**31.** (b) 519 am

**33.** (a) 126 pm  (b) $5.27 \times 10^{-24}$ kg·m/s  (c) 95.5 eV

**35.** (a)

(b) 2.20 nm, 2.75 nm, 4.12 nm, 4.71 nm, 6.60 nm, 11.0 nm

**37.** (a) $L/2$  (b) $5.26 \times 10^{-5}$  (c) $3.99 \times 10^{-2}$  (d) The probability density has peaks around $L/4$ and $3L/4$, and a zero at $L/2$. Because the probability density is symmetric about $L/2$, the average experimental value has to be $L/2$.

**39.** 0.250

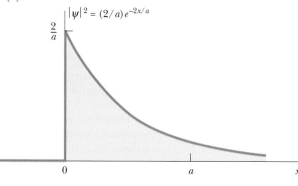

**Chapter 28, Problem 51(a)**

**43.** (a) $E = \hbar^2/mL^2$  (b) Requiring $\int_{-L}^{L} A^2 (1 - x^2/L^2)^2 \, dx = 1$ gives $A = (15/16L)^{1/2}$  (c) $47/81 = 0.580$

**45.** (a) 0.010 3  (b) 0.990

**47.** (a) 1.06 mm  (b) microwave

**49.** (a) 1.7 eV  (b) 4.2 fV·s  (c) 730 nm

**51.** (a) Figure at bottom of preceding column.
(b) $6.4 \times 10^{-34}$ J·s $\pm$ 8%  (c) 1.4 eV

**53.** 0.143 nm. Diffraction effects should appear.

**55.**
(a)

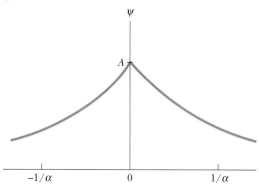

(b)

(c) The wave function is continuous. It shows localization by approaching zero as $x \rightarrow \pm \infty$. It is everywhere finite and can be normalized.  (d) $A = \sqrt{\alpha}$  (e) 0.632

**57.** (a) $2.82 \times 10^{-37}$ m  (b) $1.06 \times 10^{-32}$ J
(c) $2.87 \times 10^{-35}$ % or more

**59.** (a)

(b) 0  (c) 0.865

## Chapter 29

1. (a) 56.8 fm  (b) 11.3 N
3. (a) 1.31 $\mu$m  (b) 164 nm
5. (a) 3  (b) 520 km/s
7. (b) 0.497
9. It does, with $E = -k_e e^2/2a_0$
11. 797 times
13. $\ell = 4$
15. $\sqrt{6}\hbar$
17. (a) 2  (b) 8  (c) 18  (d) 32  (e) 50
19. $n = 3$; $\ell = 2$; $m_\ell = -2, -1, 0, 1,$ or 2; $s = 1$; $m_s = -1, 0,$ or 1, for a total of 15 states
21. (a) $1s^2\, 2s^2\, 2p^4$
    (b)

| $n$ | $\ell$ | $m_\ell$ | $m_s$ |
|---|---|---|---|
| 1 | 0 | 0 | $\frac{1}{2}$ |
| 1 | 0 | 0 | $-\frac{1}{2}$ |
| 2 | 0 | 0 | $\frac{1}{2}$ |
| 2 | 0 | 0 | $-\frac{1}{2}$ |
| 2 | 1 | 1 | $\frac{1}{2}$ |
| 2 | 1 | 0 | $\frac{1}{2}$ |
| 2 | 1 | $-1$ | $\frac{1}{2}$ |
| 2 | 1 | $-1$ | $-\frac{1}{2}$ |

23. The $4s$ subshell is filled first. We would expect $[\text{Ar}]3d^4 4s^2$ to have lower energy, but $[\text{Ar}]3d^5 4s^1$ has more unpaired spins and lower energy, as suggested by Hund's rule. It is the ground-state configuration of chromium.
25. (a) Zn or Cu  (b) $1s^2 2s^2 2p^6 3s^2 3p^6 4s^2 3d^{10}$  or $1s^2 2s^2 2p^6 3s^2 3p^6 4s^1 3d^{10}$
27. (a) $1s, 2s, 2p, 3s, 3p, 4s, 3d, 4p, 5s, 4d, 5p, 6s, 4f, 5d, 6p, 7s$
    (b) Element 15 should have valence $+3$ or $-5$, and it does. Element 47 should have valence $-1$, but it has valence $+1$. Element 86 should be inert, and it is.
29. 18.4 T
31. 124 V
33. 0.072 5 nm
35. (a) 14 keV  (b) 89 pm
37. (a) Antiparallel spins, with parallel magnetic moments
    (b) 5.89 $\mu$eV  (c) $1.04 \times 10^{-30}$ eV
39. (a) $0.160c$  (b) $2.82 \times 10^9$ ly
43. (a) $1.57 \times 10^{14}$ m$^{-3/2}$
    (b) $2.47 \times 10^{28}$ m$^{-3}$
    (c) $8.69 \times 10^8$ m$^{-1}$
47. 5.39 keV
49. (a) For Al, about 0.255 nm $\sim$ 0.1 nm. For U, about 0.276 nm $\sim$ 0.1 nm. For an outer electron, the nuclear charge is screened by the interior electrons.  (b) For an inner-shell electron, the nuclear charge is unscreened. The distance scale of the wave function, representing the orbital size, is proportional to $a_0/Z$.
51. (b) 0.846 ns
57. (a) 609 $\mu$eV  (b) 6.9 $\mu$eV
    (c) 147 GHz, 2.04 mm

## Chapter 30

1. $\sim 10^{28}$; $\sim 10^{28}$; $\sim 10^{28}$
3. 25.6 MeV
5. 16.0 km
7.

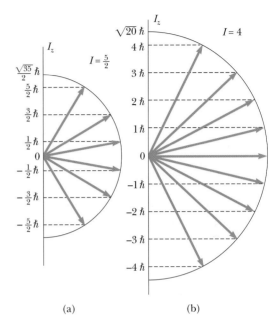

(a)      (b)

9. 200 MeV
11. Greater for N by 3.54 MeV
13. 1.16 ks
15. (a) $1.55 \times 10^{-5}$/s, 12.4 h  (b) $2.39 \times 10^{13}$ atoms
    (c) 1.87 mCi
17. $\dfrac{R_0 T_{1/2}}{\ln 2} \left(2^{-t_1/T_{1/2}} - 2^{-t_2/T_{1/2}}\right)$
19. $1.66 \times 10^3$ yr
21. 4.27 MeV
23. (a) 7.89 cm for $^{12}$C, 8.21 cm for $^{13}$C
25. (a) $e^- + p \rightarrow n + \nu$  (b) $^{15}_{8}$O atom $\rightarrow$ $^{15}_{7}$N atom $+ \nu$
    (c) 2.75 MeV
27.

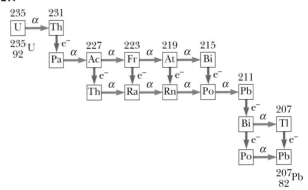

29. (a) $^{197}_{79}$Au $+ ^1_0$n $\rightarrow$ $^{198}_{80}$Hg $+ e^- + \bar{\nu}$  (b) 7.89 MeV
31. (a) 201 MeV  (b) 0.091 3%
33. 5.80 Mm

**35.** About 3 000 yr

**37.** (b) 26.7 MeV

**39.** (a) 4.28 pJ    (b) $1.19 \times 10^{57}$ atoms    (c) 107 Gyr

**41.** 5.94 Gyr

**43.** (a) 2.75 fm    (b) 152 N    (c) 2.62 MeV
(d) 7.44 fm, 379 N, 17.6 MeV

**45.** (a) 3.91 Gyr
(b) No. We must make some assumption about the original quantity of radioactive material. In part (a) we assumed that the rock originally contained no strontium.

**47.** (a) 61.8 Bq    (b) 40.3 d

**49.** (b) 4.78 MeV

**51.** 2.64 min

**53.** (b) $R/\lambda$

**55.** $2.56 \times 10^4$ kg

**57.** $\sqrt{2}$

**59.** (a) $2.24 \times 10^7$ kWh    (b) 17.6 MeV
(c) $2.34 \times 10^8$ kWh    (d) 9.36 kWh
(e) Coal is cheap at this moment in human history. We hope that safety and waste disposal problems can be solved so that nuclear energy can be affordable before scarcity drives up the price of fossil fuels.

## Chapter 31

**1.** 453 ZHz; 662 am

**3.** 118 MeV

**5.** $\sim 10^{-18}$ m

**7.** $\sim 10^{-16}$ m

**9.** 67.5 MeV, 67.5 MeV/$c$, 16.3 ZHz

**11.** $\Omega^+ \rightarrow \overline{\Lambda}{}^0 + K^+$    $\overline{K}{}_S^0 \rightarrow \pi^+ + \pi^-$    $\overline{\Lambda}{}^0 \rightarrow \overline{p} + \pi^+$
$\overline{n} \rightarrow \overline{p} + e^+ + \nu_e$

**13.** (a) $L_\mu$ and $L_e$    (b) charge    (c) baryon number
(d) baryon number    (e) charge

**15.** (a) $\overline{\nu}_\mu$    (b) $\nu_\mu$    (c) $\overline{\nu}_e$    (d) $\nu_e$    (e) $\nu_\mu$    (f) $\overline{\nu}_e + \nu_\mu$

**17.** (a), (c), and (f) violate baryon number conservation.
(b), (d), and (e) can occur.
(f) violates muon-lepton number conservation

**19.** (b) $E_e = E_\gamma = 469$ MeV,      $p_e = p_\gamma = 469$ MeV/$c$
(c) 0.999 999 4$c$

**21.** (b) and (c) conserve strangeness.    (a), (d), (e), and (f) violate strangeness conservation.

**23.** (a) Electron and muon lepton numbers
(b) electron lepton number    (c) charge and strangeness
(d) baryon number    (e) strangeness

**25.** (a) Baryon number, charge, $L_e$ and $L_\tau$
(b) baryon number, charge, $L_e$, $L_\mu$ and $L_\tau$
(c) strangeness, charge, $L_e$, $L_\mu$ and $L_\tau$
(d) baryon number, strangeness, charge, $L_e$, $L_\mu$ and $L_\tau$
(e) baryon number, strangeness, charge, $L_e$, $L_\mu$ and $L_\tau$
(f) baryon number, strangeness, charge, $L_e$, $L_\mu$ and $L_\tau$

**27.** 9.26 cm

**29.** (a) $3.34 \times 10^{26}$ e$^-$, $9.36 \times 10^{26}$ u, $8.70 \times 10^{26}$ d
(b) $\sim 10^{28}$ e$^-$, $\sim 10^{29}$ u, $\sim 10^{29}$ d . My strangeness, charm, topness, and bottomness are zero.

**31.** $m_u = 312$ MeV/$c^2$      $m_d = 314$ MeV/$c^2$

**33.** (a) The reaction $\overline{u}d + uud \rightarrow \overline{s}d + uds$ has a total of 1 u, 2 d and 0 s quarks originally and finally.
(b) The reaction $\overline{d}u + uud \rightarrow \overline{s}u + uus$ has a net of 3 u, 0 d, and 0 s before and after.
(c) $\overline{u}s + uud \rightarrow \overline{s}u + \overline{s}d + sss$ shows conservation at 1 u, 1 d, and 1 s quark.
(d) The process $uud + uud \rightarrow \overline{s}d + uud + \overline{d}u + uds$ nets 4 u, 2 d, 0 s initially and finally; the mystery particle is a $\Lambda^0$ or a $\Sigma^0$.

**35.** (a) $\Sigma^+$    (b) $\pi^-$    (c) $K^0$    (d) $\Xi^-$

**37.** (a) 590.07 nm    (b) 597 nm    (c) 661 nm

**39.** 6.00

**41.** (b) 17.6 Gyr

**45.** 19 eV

**47.** One part in 50 000 000

**49.** 0.400%

**51.** 5.35 MeV and 32.3 MeV

**53.** 1 116 MeV/$c^2$

**55.** 70.4 MeV

**57.** $2.52 \times 10^3$ K

**59.** (a) $Z^0$ boson    (b) gluon or photon

## Context 8 Conclusion

**1.** (a) $1.61 \times 10^{-35}$ m    (b) $5.38 \times 10^{-44}$ s    (c) yes

# Index

Page numbers in *italics* indicate figures; page numbers followed by "n" indicate footnotes; page numbers followed by "t" indicate tables.

## Standard Abbreviations and Symbols for Units

| Symbol | Unit | Symbol | Unit |
|--------|------|--------|------|
| A | ampere | K | kelvin |
| u | atomic mass unit | kg | kilogram |
| atm | atmosphere | kmol | kilomole |
| Btu | British thermal unit | L | liter |
| C | coulomb | lb | pound |
| °C | degree Celsius | ly | lightyear |
| cal | calorie | m | meter |
| d | day | min | minute |
| eV | electron volt | mol | mole |
| °F | degree Fahrenheit | N | newton |
| F | farad | Pa | pascal |
| ft | foot | rad | radian |
| G | gauss | rev | revolution |
| g | gram | s | second |
| H | henry | T | tesla |
| h | hour | V | volt |
| hp | horsepower | W | watt |
| Hz | hertz | Wb | weber |
| in. | inch | yr | year |
| J | joule | $\Omega$ | ohm |

## Mathematical Symbols Used in the Text and Their Meaning

| Symbol | Meaning |
|--------|---------|
| $=$ | is equal to |
| $\equiv$ | is defined as |
| $\neq$ | is not equal to |
| $\propto$ | is proportional to |
| $\sim$ | is on the order of |
| $>$ | is greater than |
| $<$ | is less than |
| $>>(<<)$ | is much greater (less) than |
| $\approx$ | is approximately equal to |
| $\Delta x$ | the change in $x$ |
| $\displaystyle\sum_{i=1}^{N} x_i$ | the sum of all quantities $x_i$ from $i = 1$ to $i = N$ |
| $\lvert x \rvert$ | the magnitude of $x$ (always a nonnegative quantity) |
| $\Delta x \to 0$ | $\Delta x$ approaches zero |
| $\dfrac{dx}{dt}$ | the derivative of $x$ with respect to $t$ |
| $\dfrac{\partial x}{\partial t}$ | the partial derivative of $x$ with respect to $t$ |
| $\displaystyle\int$ | integral |

## Conversions[a]

### Length
1 in. = 2.54 cm (exact)
1 m = 39.37 in. = 3.281 ft
1 ft = 0.304 8 m
12 in. = 1 ft
3 ft = 1 yd
1 yd = 0.914 4 m
1 km = 0.621 mi
1 mi = 1.609 km
1 mi = 5 280 ft
$1 \mu m = 10^{-6}$ m = $10^3$ nm
1 lightyear = $9.461 \times 10^{15}$ m

### Area
$1 m^2 = 10^4 cm^2 = 10.76 ft^2$
$1 ft^2 = 0.092 9 m^2 = 144 in.^2$
$1 in.^2 = 6.452 cm^2$

### Volume
$1 m^3 = 10^6 cm^3 = 6.102 \times 10^4 in.^3$
$1 ft^3 = 1 728 in.^3 = 2.83 \times 10^{-2} m^3$
$1 L = 1 000 cm^3 = 1.057 6$ qt $= 0.035 3 ft^3$
$1 ft^3 = 7.481$ gal $= 28.32 L = 2.832 \times 10^{-2} m^3$
$1$ gal $= 3.786 L = 231 in.^3$

### Mass
1 000 kg = 1 t (metric ton)
1 slug = 14.59 kg
$1 u = 1.66 \times 10^{-27}$ kg $= 931.5$ MeV$/c^2$

### Force
1 N = 0.224 8 lb
1 lb = 4.448 N

### Velocity
1 mi/h = 1.47 ft/s = 0.447 m/s = 1.61 km/h
1 m/s = 100 cm/s = 3.281 ft/s
1 mi/min = 60 mi/h = 88 ft/s

### Acceleration
$1 m/s^2 = 3.28 ft/s^2 = 100 cm/s^2$
$1 ft/s^2 = 0.304 8 m/s^2 = 30.48 cm/s^2$

### Pressure
$1$ bar $= 10^5 N/m^2 = 14.50 lb/in.^2$
1 atm = 760 mm Hg = 76.0 cm Hg
$1$ atm $= 14.7 lb/in.^2 = 1.013 \times 10^5 N/m^2$
$1$ Pa $= 1 N/m^2 = 1.45 \times 10^{-4} lb/in.^2$

### Time
1 yr = 365 days = $3.16 \times 10^7$ s
1 day = 24 h = $1.44 \times 10^3$ min = $8.64 \times 10^4$ s

### Energy
1 J = 0.738 ft·lb
1 cal = 4.186 J
1 Btu = 252 cal = $1.054 \times 10^3$ J
1 eV = $1.602 \times 10^{-19}$ J
1 kWh = $3.60 \times 10^6$ J

### Power
1 hp = 550 ft·lb/s = 0.746 kW
1 W = 1 J/s = 0.738 ft·lb/s
1 Btu/h = 0.293 W

### Some Approximations Useful for Estimation Problems
1 m ≈ 1 yd
1 kg ≈ 2 lb
$1 N \approx \frac{1}{4}$ lb
$1 L \approx \frac{1}{4}$ gal

1 m/s ≈ 2 mi/h
$1$ yr $\approx \pi \times 10^7$ s
60 mi/h ≈ 100 ft/s
$1$ km $\approx \frac{1}{2}$ mi

[a] See Table A.1 of Appendix A for a more complete list.

## The Greek Alphabet

| | | | | | | | | |
|---|---|---|---|---|---|---|---|---|
| Alpha | A | $\alpha$ | Iota | I | $\iota$ | Rho | P | $\rho$ |
| Beta | B | $\beta$ | Kappa | K | $\kappa$ | Sigma | $\Sigma$ | $\sigma$ |
| Gamma | $\Gamma$ | $\gamma$ | Lambda | $\Lambda$ | $\lambda$ | Tau | T | $\tau$ |
| Delta | $\Delta$ | $\delta$ | Mu | M | $\mu$ | Upsilon | $\Upsilon$ | $\upsilon$ |
| Epsilon | E | $\epsilon$ | Nu | N | $\nu$ | Phi | $\Phi$ | $\phi$ |
| Zeta | Z | $\zeta$ | Xi | $\Xi$ | $\xi$ | Chi | X | $\chi$ |
| Eta | H | $\eta$ | Omicron | O | $o$ | Psi | $\Psi$ | $\psi$ |
| Theta | $\Theta$ | $\theta$ | Pi | $\Pi$ | $\pi$ | Omega | $\Omega$ | $\omega$ |